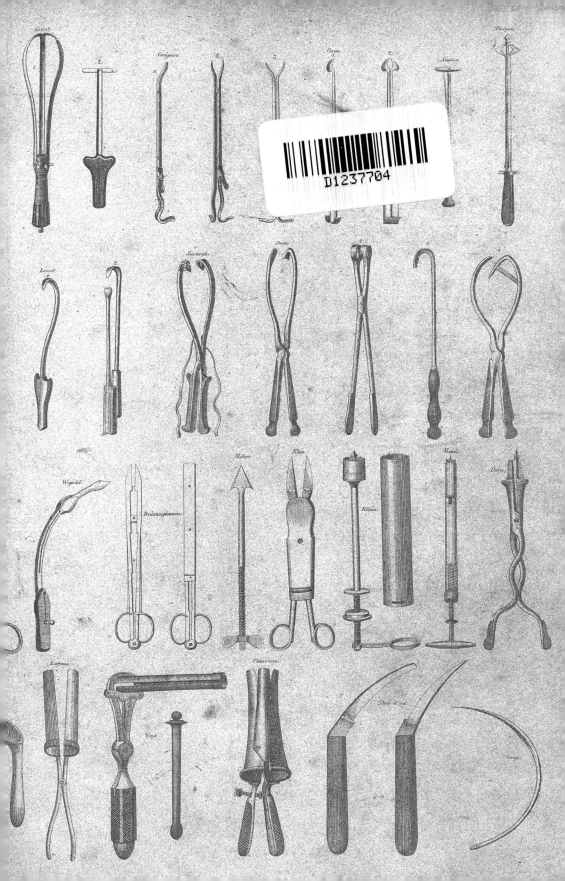

ANATOMY

DESCRIPTIVE AND SURGICAL

BY

HENRY GRAY, F.R.S.

FELLOW OF THE ROYAL COLLEGE OF SURGEONS
LECTURER ON ANATOMY AT ST. GEORGE'S HOSPITAL MEDICAL SCHOOL

THE DRAWINGS BY H. V. CARTER, M.D.
LATE DEMONSTRATOR OF ANATOMY AT ST. GEORGE'S HOSPITAL
WITH ADDITIONAL DRAWINGS IN LATER EDITIONS

FIFTEENTH EDITION

EDITED BY

T. PICKERING PICK, F.R.C.S.

CONSULTING SURGEON TO ST. GEORGE'S HOSPITAL, AND TO THE VICTORIA HOSPITAL FOR CHILDREN
H.M. INSPECTOR OF ANATOMY IN ENGLAND AND WALES

AND BY

ROBERT HOWDEN, M.A., M.B., C.M.

PROFESSOR OF ANATOMY IN THE UNIVERSITY OF DURHAM
EXAMINER IN ANATOMY IN THE UNIVERSITIES OF DURHAM AND EDINBURGH, AND TO
THE BOARD OF EDUCATION, SOUTH KENSINGTON

BARNES & NOBLE

NEW YORK

THIS WORK IS DEDICATED TO

SIR BENJAMIN COLLINS BRODIE, BART.

F.R.S., D.C.L.

SERJEANT-SURGEON TO QUEEN VICTORIA

CORRESPONDING MEMBER OF THE INSTITUTE OF FRANCE

———

PREFACE

TO

THE FIFTEENTH EDITION

IN this edition the entire work has undergone a careful revision. The section on Embryology has been somewhat amplified, and its text rendered more intelligible by the introduction of some sixty additional illustrations after His, Kollmann, Duval, and others. Throughout the rest of the work a considerable number of the diagrams have been re-drawn and new illustrations here and there added.

The preparation of the drawing to illustrate the section on Embryology retarded the printing of the first two sections until after a large portion of the Descriptive and Surgical part of the work was typed and paged, and, as a consequence, the numbers of the pages and illustrations in the sections on Histology and Embryology are contained within brackets [ ] to distinguish them from those in the subsequent part of the book.

———

This 2010 edition published by Barnes & Noble, Inc.

Cover design by Jo Obarowski and Mada Design
Endpapers: Wellcome Library, London/Wellcome Images

Barnes & Noble
122 Fifth Avenue
New York, NY 10011

ISBN: 978-1-4351-1493-7

Printed and bound in China

6 8 10 9 7

CONTENTS

GENERAL ANATOMY OR HISTOLOGY

EMBRYOLOGY

OSTEOLOGY

CONTENTS

CONTENTS

MUSCLES AND FASCIÆ

CONTENTS

CONTENTS

CONTENTS

CONTENTS

THE BLOOD-VASCULAR SYSTEM

CONTENTS

CONTENTS

CONTENTS

CONTENTS

THE LYMPHATIC SYSTEM

THE NERVOUS SYSTEM

CONTENTS

CONTENTS

CONTENTS

CONTENTS

CONTENTS

CONTENTS

THE URINARY ORGANS

CONTENTS

MALE GENERATIVE ORGANS

FEMALE ORGANS OF GENERATION

SURGICAL ANATOMY OF INGUINAL HERNIA

CONTENTS

LIST OF ILLUSTRATIONS

The illustrations, when copied from any other work, have the author's name affixed. When no
such acknowledgment is made the drawing is to be considered original.

GENERAL ANATOMY OR HISTOLOGY

LIST OF ILLUSTRATIONS

EMBRYOLOGY

LIST OF ILLUSTRATIONS

OSTEOLOGY

LIST OF ILLUSTRATIONS

ARTICULATIONS

MUSCLES AND FASCIÆ

BLOOD-VASCULAR SYSTEM

ARTERIES

LIST OF ILLUSTRATIONS

NERVOUS SYSTEM

CRANIAL NERVES

ORGANS OF DIGESTION AND THEIR APPENDAGES

LIST OF ILLUSTRATIONS

ORGANS OF VOICE AND RESPIRATION

THE URINARY AND GENERATIVE ORGANS

SURGICAL ANATOMY OF HERNIA

SURGICAL ANATOMY OF PERINÆUM AND
ISCHIO-RECTAL REGION

GENERAL ANATOMY OR HISTOLOGY

———✳———

THE ANIMAL CELL (fig. [1])

A LL the tissues and organs of which the body is composed were originally
developed from a microscopic body (the *ovum*), consisting of a soft gelatinous granular material enclosed in a membrane, and containing a vesicle, or small
spherical body, inside which are one or more solid spots. This may be regarded
as a perfect cell. Moreover, all the solid tissues can be shown to consist largely
of similar bodies or cells, differing, it is true, in external form, but essentially
similar to an ovum.

In the higher organisms all such cells may be defined as 'nucleated masses of
protoplasm of microscopic size.' The two essentials, therefore, of an animal cell

Fig. [1].—Diagram of a cell. (Modified from Wilson.)

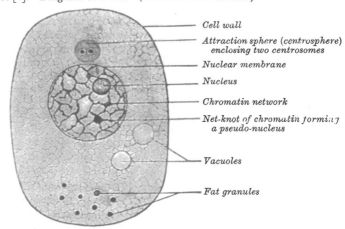

Cell wall

Attraction sphere (centrosphere)
enclosing two centrosomes

Nuclear membrane

Nucleus

Chromatin network

Net-knot of chromatin forming
a pseudo-nucleus

Vacuoles

Fat granules

in the higher organisms are : the presence of a soft gelatinous granular material,
similar to that found in the ovum, and which is usually styled *protoplasm* ; and a
small spherical body embedded in it, and termed a *nucleus* ;* the remaining
constituents of the ovum, viz. its limiting membrane and the solid spot contained
in the nucleus, called the *nucleolus*, are not considered essential to the cell, and
in fact many cells exist without them.

Protoplasm (*cytoplasm*) is a material probably of variable constitution, but
yielding to the chemist on its disintegration bodies chiefly of proteid nature.
Lecithin and cholesterin are constantly found in it, as well as inorganic salts,
chief among which are the phosphates and chlorides of the alkali metals and

* In certain lower forms of life, masses of protoplasm without any nuclei have been
described by Huxley and others, as cells.

calcium. It is of a semi-fluid, viscid consistence, and appears either as a hyaline substance, homogeneous and clear, or else it exhibits a granular appearance. This granular appearance, under a high power of the microscope, is seen to be due to the fact that protoplasm consists of a network or honeycombed reticulum, containing in its meshes a homogeneous substance. The former is known as *spongioplasm*, the latter as *hyaloplasm*. The granular appearance is often caused by the knots of the network being mistaken for granules; but, in addition to this, protoplasm often contains true granules, some of which are proteid in nature and probably essential constituents; others are fat or pigment granules, and are regarded as adventitious material taken in from without. The size and shape of the meshes of the spongioplasm vary in different cells and in different parts of the same cell. In many fixed cells, e.g. epithelial cells, the external layer becomes denser than the rest, and often altered by the deposition in it of some chemical substance, so as to constitute a membrane which encloses the rest of the proto-plasm and forms the *cell wall*. The relative amount of spongioplasm and hyaloplasm varies in different cells; the latter preponderating in the young cell and the former increasing in amount, at the expense of the hyaloplasm, as the cell grows.

The most striking characteristics of protoplasm are its vital properties of *motion* and *nutrition*. By motion is meant the property which protoplasm has of changing its shape and position by some intrinsic power, which enables it to thrust out from its main body an irregular process, into which the whole of the protoplasmic substance is gradually drawn, so that the mass comes to occupy a new position. This, on account of its resemblance to the movements observed in the Amœba or Proteus animalcule, has been termed 'amœboid movement.' Ciliary movement, or the vibration of hair-like processes from the surface of any structure, may also be regarded as a variety of the motion with which protoplasm is endowed.

Nutrition is the power which protoplasm has of attracting to itself the materials necessary for its growth and maintenance from surrounding matter. When any foreign particle comes in contact with the protoplasmic substance, it becomes incorporated in it, being enwrapped by one or more processes projected from the parent mass which enclose it. When thus taken up, it may remain in the substance of the protoplasm for some time without change, or may be again extruded.

The **Nucleus** is a minute body, embedded in the protoplasm, and usually of a spherical or oval form, its size having little relation to the size of the cell. It is surrounded by a well-defined wall, the *nuclear membrane*, which encloses the nuclear contents. These are known as the *nuclear substance* (nuclear matrix), which is composed of a· homogeneous material and a stroma or network. The former is probably of the same nature as the hyaloplasm of the cell; but the latter, which forms also the wall of the nucleus, differs from the spongioplasm of the cell substance. It is sometimes known as the *chromoplasm* or *intranuclear network*, and consists of a network of fibres or filaments arranged in a reticular manner. These filaments stain very readily with certain dyes; they are therefore named *chromatin*; while the interstitial substance does not stain readily, and is hence called *achromatin*. In some resting nuclei, i.e. nuclei which are not undergoing subdivision, the nuclear filaments do not form a network, but present the appear-ance of a convoluted skein, similar to that found in a nucleus about to undergo division, and which will be immediately described.

Within the nuclear matrix are one or more highly refracting bodies, termed *nucleoli*, connected with the nuclear membrane by the nuclear filaments. They are regarded as being of two kinds. Some are mere local condensations of the chromoplasm; these are irregular in shape and are termed *pseudo-nucleoli*; others are distinct bodies differing from the pseudo-nucleoli both in nature and chemical composition; they may be termed *true nucleoli*, and are usually found in resting cells.

The nuclear substance differs chemically from ordinary protoplasm in containing *nuclein*, in its power of resisting the action of acids and alkalies, in its imbibing more intensely the stain of carmine, hæmatoxylin, &c., and in its remaining unstained by some reagents which colour ordinary protoplasm.

Recent investigations tend to show that most living cells contain, in addition to their protoplasm and nucleus, a minute particle which, on account of the power it appears to possess of attracting the surrounding protoplasmic granules, is termed the *attraction particle* or *centrosome* ; it usually lies near the nucleus. The spherical arrangement of fibrillar rows of granules surrounding the central particle is termed the *attraction-sphere* or *centrosphere*. These spheres are usually double, and are connected by a spindle-shaped system of delicate fibrils (*achromatic spindle*). They are best seen in young cells which are about to undergo the process of division, a process believed to commence in these bodies.

The process of reproduction of cells is usually described as being brought about by *indirect* or by *direct division*. *Indirect division* or *karyokinesis* (*karyomitosis*) has been observed in all the tissues—generative cells, epithelial tissue, connective tissue, muscular tissue, and nerve tissue, and probably it will ultimately be shown that the division of cells always takes place in this way, and that the process of reproduction of cells by direct division is, as some observers believe, merely a sort of imperfect or abnormal karyokinesis.

The process of indirect cell division is characterised by a series of complex changes in the nucleus, leading to its subdivision ; this being followed by cleavage of the cell protoplasm. Starting with the nucleus in the quiescent or *resting* stage, these changes may be briefly grouped under the four following phases :

1. *Prophase.*—The nuclear network of chromatin filaments assumes the form of a twisted *skein* or *spirem*, while the nuclear membrane and nucleolus disappear. The convoluted skein of chromatin divides into a definite number of V-shaped loops or *chromosomes*. Coincident with or preceding these changes the centrosome, or attraction particle, which usually lies by the side of the nucleus, undergoes subdivision, and the two resulting centrosomes, each surrounded by a centrosphere, are seen to be connected by a spindle of delicate achromatic fibres, *the achromatic spindle*. These centrosomes move away from each other—one towards each extremity of the nucleus—and the fibrils of the achromatic spindle are correspondingly lengthened. The centrosomes are now situated one at either extremity or pole of the elongated spindle, and each is surrounded by a centrosphere, from which fibrils radiate into the investing protoplasm. A line encircling the spindle midway between its poles is named the *equator*, and around this the V-shaped chromosomes arrange themselves in the form of a star, thus constituting the *mother star* or *monaster*.

2. *Metaphase.*—Each V-shaped chromosome now undergoes longitudinal cleavage into two equal halves or *daughter chromosomes*, the cleavage commencing at the apex of the V and extending along its divergent limbs. The daughter chromosomes, thus separated, travel in opposite directions along the fibrils of the achromatic spindle towards the centrosomes, around which they group themselves, and thus two star-like figures are formed, one at either pole of the achromatic spindle. This is termed the *diaster*.

3. *Anaphase.*—The V-shaped daughter chromosomes now assume the form of a *skein* or *spirem*, and eventually form the network of chromatin which is characteristic of the resting nucleus. The nuclear membrane and nucleolus are also differentiated during this phase. The cell protoplasm begins to appear constricted around the equator of the achromatic spindle, where double rows of granules are also sometimes seen. The constriction deepens and the original cell gradually becomes divided.

4. *Telophase.*—In this stage the cell is completely divided into two new cells, each with its own nucleus, centrosome and centrosphere, which assume the ordinary positions occupied by such structures in the resting stage.

In the case of prickle-cells the subdivision of the cell is incomplete ; here the

FIG. [2].—Karyokinesis: or indirect cell-division

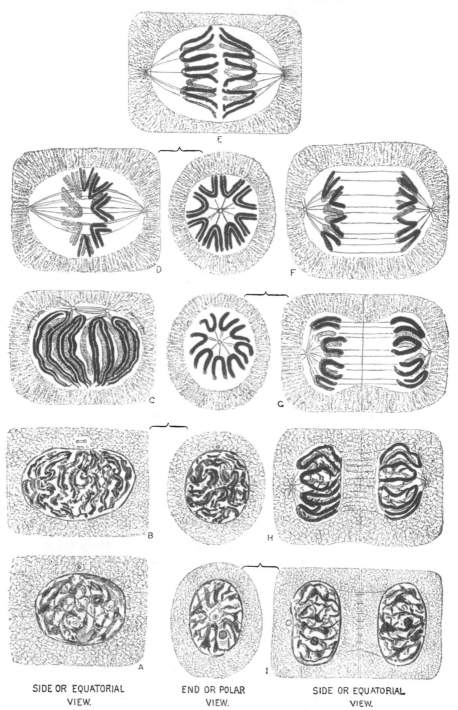

SIDE OR EQUATORIAL   END OR POLAR   SIDE OR EQUATORIAL
       VIEW.   VIEW.   VIEW.

A. Resting nucleus. B. Skein or spirem, close. C. Skein or spirem, open. D. Mother star, monaster. E. Metaphase. F. Daughter stars or diaster. G. Daughter skeins or dispirem, beginning to form. H. Daughter skeins or
dispirem, formed. I. Resting daughter nuclei.

achromatic spindle threads appear to persist and bridge across the intercellular spaces, constituting the prickles.

The series of diagrams (fig. [2]), by Professor S. Delépine, is intended to explain the formation of some of the most important changes observed in nuclei of cells during karyokinesis (mitosis) ; it is based chiefly on the work of Flemming, Strasburger, E. van Beneden, Rabl, O. Hertwig, Henneguy, &c. A. *Resting nucleus.* Nucleolus and nuclear membrane visible. A centrosome is represented near the nucleus. B and C. *Skein* or *spirem.* Chromatic filaments much convoluted. Evidence of longitudinal splitting begins to be distinct in several parts. The centrosome has divided ; the nuclear membrane is becoming indistinct. The two centrosomes are widely separated, and the space between them is occupied by the achromatic spindle. (Two arrows point to the positions which the centrosomes will ultimately occupy ; during their passage to these points the achromatic spindle seems to be within the nucleus.) The nuclear membrane has disappeared. D. *Mother star, monaster.* The nuclear segments (chromosomes) resulting from the breaking-up of the chromatic filament into fragments of nearly equal length have moved towards the equator of the spindle, where they now form an equatorial plate. These segments are all split longitudinally. E. *Metaphase.* One half of each chromosome moves towards one pole and the other half towards the other pole, being guided towards the centrosomes by the achromatic filaments. F. *Daughter stars* or *diaster.* G. *Daughter skeins* or *dispirem,* beginning to form. Segments in the form of thick loops not closely packed. H. *Daughter skeins* or *dispirem,* formed. Segments more closely packed and less distinct, owing to the formation of anastomoses. I. *Resting daughter nuclei.* Cell completely divided into two, but bridges remain between them in the region previously occupied by the achromatic filaments, these being specially distinct in certain cells (e.g. prickle cells). The nucleus has a distinct nuclear membrane and a nucleolus.

In the reproduction of cells by *direct* division the process is brought about either by segmentation or by gemmation. In reproduction by *segmentation* or *fission,* the nucleus becomes constricted in its centre, assuming an hour-glass shape, and then divides into two. This leads to a cleavage or division of the whole protoplasmic mass of the cell ; and thus two daughter cells are formed, each containing a nucleus. These daughter cells are at first smaller than the original mother cell ; but they grow, and the process may be repeated in them, so that multiplication may take place rapidly. In reproduction by *gemmation,* a budding-off or separation of a portion of the nucleus and parent cell takes place, and, becoming separated, forms a new organism.

The **cell-wall,** which is not an essential constituent, and in fact is often absent, is merely the external layer of the protoplasm, firmer than the rest of the cell, and often thickened by the deposit in it of certain chemical substances. It forms a flexible, transparent, finely striated membrane, sometimes furnished with minute pores, so as to be permeable to fluids.

THE NUTRITIVE FLUIDS

The **circulating fluids** of the body, which subserve its nutrition, are the blood, the lymph, and the chyle.

THE BLOOD

The blood is an opaque, rather viscid fluid, of a bright red or scarlet colour when it flows from the arteries, of a dark red or purple colour when it flows from the veins. It is salt to the taste, and has a peculiar faint odour and an alkaline reaction. Its specific gravity is about 1·060, and its temperature is generally about 100° F., though varying slightly in different parts of the body.

General Composition of the Blood.—Blood consists of a faintly yellow fluid, the *plasma* or *liquor sanguinis,* in which are suspended numerous minute particles, the *blood corpuscles*, the majority of which are coloured and give to the blood its red tint. If a drop of blood is placed in a thin layer on a glass slide and examined under the microscope, a number of these corpuscles will be seen immersed in the clear fluid plasma.

The **Blood Corpuscles** are chiefly of two kinds: (1) Coloured corpuscles or Erythrocytes, (2) Colourless corpuscles or Leucocytes. A third variety, the Blood-platelets, are of subsidiary importance.

1. **Coloured or Red Corpuscles** (*erythrocytes*), when examined under the micro-scope, are seen to be circular discs, biconcave in profile. They have no nuclei, but, in consequence of their biconcave shape, present, according to the alteration of focus under an ordinary high power, a central part, sometimes bright, some-times dark, which has the appearance of a nucleus (fig. [3] *a*). It is to their aggregation that the blood owes its red hue, although when examined by transmitted light their colour appears to be only a faint reddish-yellow. Their size varies slightly even in the same drop of blood, but it may be stated that their ordinary diameter is about $\frac{1}{3200}$ of an inch, while their thickness is about $\frac{1}{12000}$ of an inch or nearly one quarter of their diameter. Besides these there are found, especially in disease (e.g. anæmic conditions), certain smaller corpuscles of about one-half or one-third of the size just indicated; these are termed *microcytes*, and are very scarce in human blood. The number of red corpuscles in the blood is enormous; between 4,000,000 and 5,000,000 are contained in a cubic millimetre. Power states that the red corpuscles of an adult would present an aggregate surface of about 3,000 square yards. Each corpuscle consists of a colourless elastic spongework or stroma, condensed at the periphery to form an investing membrane, and uniformly diffused throughout this are the coloured fluid contents. The stroma is composed mainly of *nucleo-proteid* and of the fatty substances, *lecithin* and *cholesterin*, while the coloured material consists chiefly of the respiratory proteid, *hæmoglobin*, which contains a proportion of iron in addition to the ordinary proteid elements. This proteid has a great affinity for oxygen, and, when removed from the body, crystallises readily under certain circum-stances. It is very soluble in water, the addition of which to a drop of blood speedily dissolves out the hæmoglobin from the corpuscles.

If the web of a frog's foot be spread out and examined under the microscope, the blood is seen to flow in a continuous stream through the vessels, and the corpuscles show no tendency to adhere to each other or to the wall of the vessel. Doubt-less the same is the case in the human body; but when the blood is drawn and exa-mined on a slide without reagents, the corpuscles often collect into heaps like rou-leaux of coins (fig. [3] *b*). It has been suggested that this phenomenon may be ex-plained by alteration in sur-face tension. During life the red corpuscles may be seen to change their shape under pressure so as to adapt themselves, to some extent, to the size of the vessel.

FIG. [3].—Human red blood-corpuscles. Highly magnified

a. Seen from the surface. *b*. Seen in profile and forming rouleaux. *c*. Rendered spherical by water. *d*. Rendered crenate by salt solution.

They are, however, highly elastic, and speedily recover their shape when the pressure is removed. They are soon influenced by the medium in which they are placed and by the specific gravity of the medium. In water they swell up, lose their shape, and become globular (fig. [3] *c*). Subsequently the hæmoglobin becomes dissolved out, and the envelope can barely be distinguished as a faint circular outline. Solutions of salt or sugar, denser than the plasma, give them a stellate or crenated appearance (fig. [3] *d*), but the usual shape may be restored by diluting the solution to the same specific gravity as the plasma. The crenated outline may be produced as the first effect of the passage of an electric shock : subsequently, if sufficiently strong, the shock ruptures the envelope. A solution of salt or sugar, of the same specific gravity as the plasma, merely separates the blood corpuscles mechanically, without changing their shape.

The Colourless Corpuscles or *leucocytes* are of various sizes, some no larger, others even smaller, than the red corpuscles. In human blood, however, the majority are rather larger than the red corpuscles, and measure about $\frac{1}{2000}$ to $\frac{1}{2500}$ of an inch in diameter. On the average from 10,000 to 12,000 leucocytes are found in each cubic millimetre of blood.

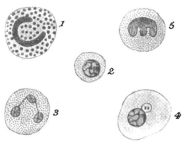

FIG. [4].—Varieties of leucocytes found in human blood.

They consist of minute masses of nucleated protoplasm, and exhibit several varieties, which are differentiated from each other chiefly by the occurrence or non-occurrence of granules in their protoplasm and by the staining reactions of these granules when present (fig. [4]). (1) The most numerous and important are spherical in shape, and are characterised by a nucleus, which often consists of two or three parts (multipartite) connected together by fine threads of chromatin. The protoplasm is clear, and contains a number of very fine granules, which stain with acid dyes, as

1. Eosinophile cell with coarse granules and horse-shoe shaped nucleus. 2. Lymphocyte. 3. Poly-nuclear or finely granular cell. 4. Hyaline cell, showing nucleus with chromatin threads and two centrosomes in clear protoplasm. 5. Finely granular leucocyte ; the nucleus is lobed, the granules stain with basic dyes, such as methylene blue.

eosin (fig. [4], 3). (2) A second variety comprises about 2·4 per cent. of the leucocytes ; they are larger than the previous kind, and are made up of a coarsely granular protoplasm, the granules being highly refractile and grouped round a single nucleus of horse-shoe shape (fig. [4], 1). These granules stain deeply with eosin, and the cells are therefore often termed *eosinophile corpuscles*. (3) A leucocyte, characterised by the presence of a trilobed nucleus, and having in its protoplasm fine granules which stain with basic dyes, such as methylene blue, is found in small numbers (fig. [4], 5). (4) The fourth variety is called the *hyaline* cell (fig. [4], 4). This is usually about the same size as the eosinophile cell, and, when at rest, is spherical in shape and contains a single round or oval nucleus. The protoplasm is free from granules, but is not quite transparent having the appearance of ground glass. (5) The fifth kind of colourless corpuscle is designated the *lymphocyte* (fig. [4], 2), because it is identical with the lymphoid cell derived from the lymphatic glands, the spleen, tonsil, and thymus. It is the smallest of the leucocytes, and consists chiefly of a spheroidal nucleus with very little surrounding protoplasm of a homogeneous nature ; it is regarded as the immature form of the hyaline cell. The fourth and fifth varieties together constitute from 20 to 30 per cent. of the colourless cells, but of these two varieties the lymphocytes are by far the more numerous.

FIG. [5].—Human colourless blood-corpuscle, showing its successive changes of outline within ten minutes when kept moist on a warm stage. (Schofield.)

The white corpuscles are very various in shape in living blood (fig. [5]), because many of them have the power of constantly changing their form by protruding finger shaped or filamentous processes of their own substance, by which they move, and take up granules from the surrounding medium. In locomotion the corpuscle pushes out a process of its substance—a *pseudopodium*, as it is called—and then shifts the rest of the body into it. In the same way when any granule or particle comes in its way it wraps a pseudopodium round it, and then withdrawing it, lodges the particle in its own substance. By means of these amœboid properties the cells have the power of wandering or emigrating from the blood-vessels by penetrating their walls and thus finding their way into the extra-vascular spaces. A chemical investigation of the protoplasm of the leucocytes shows the presence of nucleo-proteid and of a globulin. The occurrence of small amounts of fat and glycogen may also be demonstrated.

The **Blood platelets** are discoid or irregularly shaped, colourless, refractile bodies, much smaller than the red cells. Considerable discussion has arisen as to their significance. In spite of the fact that they have been observed in the blood vessels during life, there is, at present, a tendency to regard them as products of disintegration of the white cells, or as precipitates, possibly of nucleo-proteid, and not as living elements of the blood.

Origin of the Blood Corpuscles.—In the embryo the red corpuscles are developed from mesoblastic cells in the vascular area of the blastoderm. These cells unite with one another to form a network, their nuclei multiply in number, and around some of the nuclei an aggregation of coloured protoplasm takes place. After a time the network becomes hollowed out by an accumulation of fluid, and forms capillary blood-vessels, and in the fluid those nuclei which are surrounded by coloured protoplasm float as the first red blood cells.* The embryonic corpuscles are thus nucleated, and, further, they have the power of amœboid movement. These cells disappear in later embryonic life, to be replaced by smaller non-nucleated corpuscles, having all the characters of the adult erythrocytes, which, according to Schäfer, are formed within certain cells of the connective tissue. Small globules of reddish colouring matter appear in the protoplasm of these cells, and these eventually becoming larger, more uniform in size and disc-shaped, float in a cavity which results from the coalescence of numerous vacuoles. The cells, becoming more hollowed, join with neighbouring cells to form new blood-vessels, and these become connected with previously existing vessels. In post-embryonic life the important source of the red corpuscles is the red marrow in the ends of the long bones and especially in the ribs and sternum. Here are found special, nucleated, coloured cells, termed *erythroblasts*, which are probably direct descendants of the nucleated, embryonic red cells. These erythroblasts by atrophy and disappearance of their nuclei (or, as some observers maintain, by their extrusion) and by assumption of the biconcave form are transformed into the adult red corpuscles. Of the white corpuscles of the blood, the lymphocytes are derived from lymphatic tissue generally, and from the lymphatic glands especially, and enter the blood by way of the lymph stream; the hyaline cells probably develop from the lymphocytes, while the eosinophile cells are believed to originate mainly in the bone marrow and possibly also in the connective tissues.

The Plasma or Liquor Sanguinis, the fluid portion of the blood, has a yellowish tint, is alkaline in reaction, and of a specific gravity of 1·028. It contains in solution about 10 per cent. of solids, of which four-fifths are proteid in nature; the remainder being salts, chiefly chlorides, phosphates, and sulphates of the alkali metals; carbohydrates, chiefly sugar; fats and soaps; cholesterin, urea, and other nitrogenous extractives. The proteids are three in number, *serum albumin, serum globulin,* and *fibrinogen.* Fibrinogen is a body of the globulin class, but differs from serum globulin in several respects. It is the substance from which the *fibrin,* which plays so important a part in the clotting of the blood, is derived.

Coagulation of the Blood.—When blood is drawn from the body and allowed to stand, it solidifies in the course of a very few minutes into a jelly-like mass or

---

* Recent observations tend to show that the endothelial lining of the vessels and the blood corpuscles are of hypoblastic origin.

clot, which has the same appearance and volume as the fluid blood, and, like it, looks quite uniform. Soon, however, drops of a transparent yellowish fluid, the *serum*, begin to ooze from the surface of the mass and to collect around it. Coincidently the clot begins to contract, so that in the course of about twenty-four hours, having become considerably smaller and firmer than the first formed jelly-like mass, it floats in a quantity of yellowish serum. The clotting of the blood is due to the formation of a fine meshwork of the insoluble material, *fibrin*, which entangles and encloses the blood corpuscles. It is supposed that when blood is drawn a nucleo-proteid, termed *prothrombin*, appears in the plasma, probably as the result of disintegration of some of the white cells and perhaps also the blood platelets. This substance interacts with soluble lime salts in the blood, and a fresh body, *thrombin* or *fibrin ferment*, is the result. The thrombin then acts on the fibrinogen in solution in the plasma, converting it into insoluble fibrin, while at the same time a very small amount of a new proteid of the globulin type passes into solution.

Fibrin may be obtained, practically free from corpuscles, by whipping the blood, after it has been withdrawn from the body, with a bundle of twigs, to which the fibrin adheres as it is formed. By various means the clotting of the blood may be retarded so that the plasma may be obtained free from corpuscles ; from this plasma there may be derived fibrin and serum, without the cellular elements. Fibrin thus obtained is a white or buff-coloured stringy substance,

FIG. [6].—Blood-crystals.

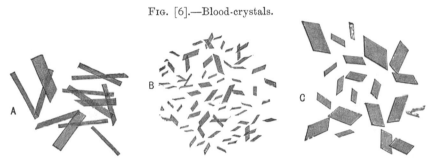

A. Hæmoglobin crystals from human blood. B. Hæmin crystals from blood treated with acetic acid.
C. Hæmatoidin crystals from an old apoplectic clot.

and when observed, in the course of formation, under the microscope, shows a meshwork of fine fibrils. After exposure to the air for some time it becomes hard, dry, brown and brittle. It is one of the class of coagulated proteids, insoluble in hot or cold water, saline solution, alcohol, or ether. Under the action of dilute hydrochloric acid it swells up but does not dissolve, but, when thus swollen, is readily dissolved by a solution of pepsin.

Serum, with the exception of its proteids, has a composition identical with that of plasma. The fibrinogen, characteristic of plasma, has disappeared, and the fibrin ferment or thrombin is found instead, together with the serum albumin and serum globulin which are not involved in the process of coagulation.

The relation of the various constituents of the blood to each other may be easily understood by a reference to the subjoined plan.

$$\text{Blood} \begin{cases} \text{Corpuscles} \\ \text{Plasma} \begin{cases} \text{Fibrin} \\ \text{Serum} \end{cases} \end{cases} \Bigg\} \text{Clot}$$

Gases of the Blood.—When blood is exposed to the vacuum of an air-pump, 100 volumes are found to yield about 60 volumes of gas. The gases present are carbon dioxide, oxygen and nitrogen, and they occur in the following proportions in arterial and venous blood :

	Carbon dioxide	Oxygen	Nitrogen
Arterial blood . .	40 vols.	20 vols.	1 to 2 vols.
Venous blood . .	46 to 50 vols.	10 to 12 vols.	1 to 2 vols.

The greater quantity of the oxygen is in loose chemical combination with the hæmoglobin of the red corpuscles. The carbon dioxide exists in combination for the most part as sodium bicarbonate and carbonate. The nitrogen is in simple solution in the plasma.

Blood Crystals.—Hæmoglobin, as already stated, readily crystallises when separated from the blood corpuscles. In human blood the crystals are elongated prisms (fig. [6], A), and in the majority of animals belong to the rhombic system, though in the squirrel hexagonal plates are met with. Small brown prismatic crystals of *hæmin* (fig. [6], B) may be obtained by mixing dried blood with common salt and boiling with a few drops of glacial acetic acid. A drop of the mixture on a slide will show the characteristic crystals on cooling. *Hæmatoidin* crystals (fig. [6], C) occur sometimes in old blood clots.

LYMPH AND CHYLE

Lymph is a transparent, colourless or slightly yellow fluid, which is conveyed by a set of vessels, named *lymphatics*, into the blood. These vessels arise in nearly all parts of the body as *lymph capillaries*. They take up the blood plasma which has exuded from the blood capillaries into the tissue spaces where it has nourished the tissue elements, and return it into the veins close to the heart, there to be mixed with the mass of blood. The greater number of these lymphatics empty themselves into one main duct, the *thoracic duct*, which passes upwards along the front of the spine and opens into the large veins on the left side of the root of the neck. The remainder empty themselves into a smaller duct which terminates in the corresponding veins on the right side of the neck.

Lymph, as its name implies, is a watery fluid of sp. gr. about 1·015, closely resembling the blood plasma, but more dilute, containing only about 5 per cent. of proteids and 1 per cent. of salt and extractives. When examined under the microscope, leucocytes of the lymphocyte class are found floating in the transparent fluid. They are always increased in number after the passage of the lymph through lymphoid tissue, as in lymphatic glands. They are constantly furnishing a fresh supply of colourless corpuscles to the blood.

Chyle is an opaque, milky-white fluid, absorbed by the villi of the small intestine from the food, and carried by a set of vessels similar to the lymphatics, named *lacteals*, to the commencement of the thoracic duct, where it is intermingled with the lymph and poured into the circulation through the same channels. It must be borne in mind that these two sets of vessels, lymphatics and lacteals, though differing in name, are identical in structure, and that the character of the fluid they convey is different only while digestion is going on. At other times the lacteals convey a transparent, nearly colourless lymph.

Chyle exactly resembles lymph in its physical and chemical properties, except that it has, in addition to the other constituents of lymph, a quantity of finely divided fatty particles, the so-called ' molecular basis of chyle' to which the milky appearance is due. It contains a little more proteid than lymph, but the chief difference lies in the large quantity of fats, soaps, lecithin and cholesterin present in the former. Lymph and chyle, containing, as they do, fibrinogen in solution and leucocytes, clot on removal from the body, the coagulum being free from red cells, and presenting a clear or whitish jelly-like appearance.

EPITHELIUM

All the surfaces of the body—the external surface of the skin, the internal surface of the digestive, respiratory, and genito-urinary tracts, the closed serous cavities, the inner coat of the vessels, the acini and ducts of all secreting and excreting glands, the ventricles of the brain and the central canal of the spinal cord—are covered by one or more layers of simple cells, called *epithelium* or *epithelial cells*. These cells are also present in the terminal parts of the organs of special sense, and in some other structures, as the pituitary and thyroid bodies. They serve various purposes, forming in some cases a protective layer, in others acting as agents in secretion and excretion, and again in others being concerned in the elaboration of the organs of special sense. Thus, in the skin, the main purpose served by the epithelium (here called the *epidermis*) is that of protection. As the surface is worn away by the agency of friction or change of temperature new cells are supplied, and thus the surface of the true skin and the vessels and nerves which it contains, are defended from damage. In the gastro-intestinal mucous membrane and in the glands, the epithelial cells appear to be the principal agents in separating the secretion from the blood or from the alimentary fluids. In other situations (as the nose, fauces, and respiratory passages) the chief office of the epithelial cells appears to be to maintain an equable temperature by the moisture with which they keep the surface always slightly lubricated. In the serous cavities they also keep the opposed layers moist, and thus facilitate their movements on each other. Finally, in all internal parts they insure a perfectly smooth surface.

Of late years there has been a tendency on the part of many histologists to divide these several epithelial structures into two classes: (1) *epithelium*, consisting of nucleated protoplasmic cells, which form continuous masses on the skin and mucous surfaces and the linings of the ducts and alveoli of secreting and excreting glands; and (2) *endothelium*, composed of a single layer of flattened transparent squamous cells, joined edge to edge in such a manner as to form a membrane of cells. This is found on the free surfaces of the serous membranes; as the lining membrane of the heart, blood-vessels, and lymphatics; on the surface of the brain and spinal cord, and in the anterior chamber of the eye. Endothelium originates from the embryonic mesoblast, while epithelium arises, as a rule, from the epiblast or hypoblast.

Epithelium consists of one or more layers of cells, united together by an interstitial cement substance, supported on a basement membrane, and is naturally grouped into two classes according as to whether there is a single layer of cells (*simple epithelium*) or more than one (*stratified epithelium*). A third variety (*transitional epithelium*) is that in which cells, in three or four layers, are so fitted together that the appearance is not one of distinct stratification. The different varieties of simple epithelium are usually spoken of as squamous or pavement, columnar, glandular or spheroidal, and ciliated.

Fig. [7].—Simple pavement epithelium.

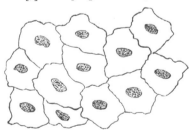

The *pavement* epithelium (fig. [7]) is composed of flat, nucleated scales of different shapes, usually polygonal, and varying in size. These cells fit together by their edges, like the tiles of a mosaic pavement. The nucleus is generally flattened, but may be spheroidal. The flattening depends upon the thinness of the cell. The protoplasm of the cell presents a fine reticulum or honey-combed network, which gives to the cell the appearance of granulation. This kind of epithelium forms the lining of the air-cells of the lungs. The endothelium, which covers the serous membranes, and which lines the heart, blood-vessels, lymphatics, and the anterior chamber of the eye, is also of the pavement type.

The *columnar* or *cylindrical* epithelium (fig. [8]) is formed of cylindrical or rod-shaped cells set together so as to form a complete layer, resembling, when viewed in profile, a palisade. The cells have a prismatic figure, more or less flattened from mutual pressure, and are set upright on the surface on which they are supported. Their protoplasm is always more or less reticulated, and fine longitudinal striæ may be seen in it. They possess a nucleus which is oval in shape and contains an intranuclear network.

FIG. [8].—Columnar epithelium from an intestinal villus.

FIG. [9].—Goblet cells. (From Kirke's 'Physiology.')

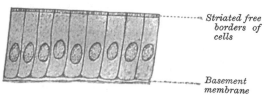

Striated free borders of cells

Basement membrane

This form of epithelium covers the mucous membrane of nearly the whole gastro-intestinal tract and the glands of that part, the greater part of the urethra, the vas deferens, the prostate, Cowper's glands, Bartholini's glands, and a portion of the uterine mucous membrane. In a modified form it also covers the ovary.

Goblet or *chalice* cells are a modification of the columnar cell. They appear to be formed by an alteration in shape of the columnar epithelium (ciliated or otherwise) consequent on the formation of granules which consist of a substance called *mucigen* in the interior of the cell. This distends the upper part of the cell, while the nucleus is pressed down towards its deep part, until the cell bursts and the mucus is discharged on to the surface of the mucous membrane (as shown in fig. [9]), the cell then assuming the shape of an open cup or chalice.

FIG. [10].—Spheroidal epithelium. Magnified 250 times.

FIG. [11].—Ciliated epithelium from the human trachea. Magnified 350 times.

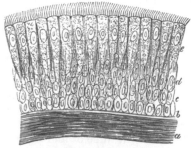

a. Innermost layers of the elastic longitudinal fibres. b. Homogeneous innermost layer of the mucous membrane. c. Deepest round cells. d. Middle elongated cells. e. Superficial cells, bearing cilia.

The *glandular* or *spheroidal* epithelium (fig. [10]) is composed of spheroidal or polyhedral cells, but the cells may be columnar or cubical in shape in some situations. Like other forms of epithelial cells, the protoplasm is a fine reticulum, which gives to the cells the appearance of granulation. They are found in the terminal recesses of secreting glands, and the protoplasm of the cells usually contains the materials which the cells secrete.

Ciliated epithelium (fig. [11]) may be of any of the preceding forms, but usually inclines to the columnar shape. It is distinguished by the presence of minute processes, which are direct prolongations of the cell-protoplasm, like hairs or eyelashes (cilia) standing up from the free surface. If the cells are examined during life or immediately on removal from the living body (for which in the human subject the removal of a nasal polypus offers a convenient opportunity)

Fig. [12].—Epithelial cells from the oral cavity of man. Magnified 350 times.

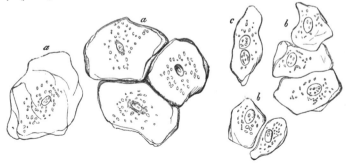

a. Large. *b.* Middle sized. *c.* The same with two nuclei.

in a weak solution of salt, the cilia will be seen in lashing motion ; and if the cells are separate, they will often be seen to be moved about in the.field by this motion.

The situations in which ciliated epithelium is found in the human body are : the respiratory tract from the nose downwards to the smallest ramifications of the bronchial tubes, except a part of the pharynx and the surface of the vocal cords ; the tympanum and Eustachian tube ; the Fallopian tube and upper portion of the uterus ; the vasa efferentia, coni vasculosi and the first part of the excretory duct of the testicle ; and the ventricles of the brain and central canal of the spinal cord.

Stratified epithelium (fig. [13]) consists of several layers of cells superimposed one on the top of the other and varying greatly in shape. The cells of the deepest layer are for the most part columnar in form, and as a rule form a single layer, placed vertically on the supporting membrane ; above these are several layers of spheroidal cells, which as they approach the surface become more and more compressed, until the superficial layers are found to consist of flattened scales (fig. [12]), the margins of which overlap one another so as to present an imbricated appearance. They here undergo a chemical change from the conversion of their protoplasm into a horny substance (keratin).

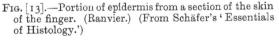

Fig. [13].—Portion of epidermis from a section of the skin of the finger. (Ranvier.) (From Schäfer's 'Essentials of Histology.')

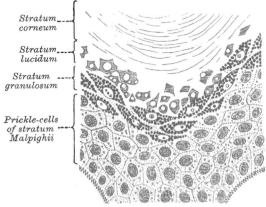

Stratum corneum

Stratum lucidum

Stratum granulosum

Prickle-cells of stratum Malpighii

Certain cells found in the deeper layers of stratified epithelium, and termed *prickle cells* (fig. [13]), constitute a variety of squamous epithelium. These cells possess short, fine fibrils, which pass from their margins to those of neighbouring cells, serving to connect them together. They are not closely connected together by cement-substance, but are separated from each other by intercellular channels, across which these fine fibrils may be seen bridging, and this gives to the cell, when isolated, the appearance of being covered over with a number of short spines, in consequence of the fibrils being broken through.   They were first described by Max Schultze and Virchow, and it was believed by them that the cells were dovetailed together.   Subsequently this was shown not to be so by Martyn, who pointed out that the

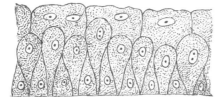

Fig. [14].—Transitional epithelium.

prickles were attached to each other by their apices ; and recently Delépine has stated that he believes the prickles of prickle cells are parts of fibrils forming internuclear bundles between the nuclei of the cells of an epithelium in a state of active growth (see page [3], and fig. [2]).

Transitional epithelium occurs in the ureters and urinary bladder.   Here the cells of the most superficial layer are cubical, with depressions on their under surfaces, which fit on to the rounded ends of the cells of the second layer, which are pear-shaped, the apices touching the basement-membrane.   Between their tapering points is a third variety of cells, filling in the intervals between them, and of smaller size than those of the other two layers (fig. [14]).

CONNECTIVE TISSUES

The term **connective tissue** includes a number of tissues which possess this feature in common, viz. that they serve the general purpose in the animal economy of supporting and connecting the tissues of the body.   These tissues may differ considerably from each other in appearance, but they present nevertheless many points of relationship, and are moreover developed from the same layer of the embryo, the mesoblast.   They are divided into three great groups: (1) the connective tissues proper, (2) cartilage, and (3) bone.   Blood, which has already been described, is, strictly speaking, a form of connective tissue, and is so dealt with by many histologists.

THE CONNECTIVE TISSUES PROPER

Several forms or varieties of connective tissue are recognised : (1) Areolar tissue.   (2) White fibrous tissue.   (3) Yellow elastic tissue.   (4) Mucous tissue. (5) Retiform tissue.   They are all composed of a homogeneous matrix, in which are embedded cells and fibres—the latter of two kinds, white and yellow or elastic.   The distinction between the different forms of tissue depends upon the relative preponderance of one or other kind of fibre, of cells, or of matrix.

Areolar tissue (fig. [15]) is so called because its meshes are easily distended, and thus separated into areolæ or spaces, which open freely into each other, and are consequently easily blown up with air, or permeated by fluid, when injected into any part of the tissue.   Such spaces, however, do not exist in the natural condition of the body, but the whole tissue forms one unbroken membrane composed of a number of interlacing fibres, variously superimposed.   Hence the term ' the cellular membrane ' is in many parts of the body more appropriate than its more modern equivalent.   The chief use of the areolar tissue is to bind parts together ; while by the laxity of its fibres, and the permeability of its areolæ, it allows them to move on each other, and affords a ready exit for inflammatory and other effused fluids.   It is one of the most extensively distributed of all the tissues. It is found beneath the skin, in a continuous layer all over the body, connecting it to the subjacent parts.   In the same way it is situated beneath the mucous and serous membranes.   It is also found between muscles, vessels, and nerves, forming investing sheaths for them, and connecting them with surrounding structures. In addition to this, it is found in the interior of organs, binding together the various lobes and lobules of the compound glands, the various coats of the hollow viscera, and the fibres of muscles &c., and thus forms one of the most important connecting media of the various structures or organs of which the body is made up. In many parts the areolæ or interspaces of areolar tissue are occupied by fat-cells, constituting *adipose tissue*, which will presently be described.

Areolar tissue presents to the naked eye a flocculent appearance, somewhat like spun-silk. When stretched out, it is seen to consist of delicate soft elastic threads interlacing with each other in every direction, and forming a network of extreme delicacy. When examined under the microscope (fig. [15]) it is found to be composed of white fibres and yellow elastic fibres intercrossing in all directions, and united together by a homogeneous cement or ground substance, the *matrix*, showing cell-spaces wherein lie many cellular elements, the *connective tissue corpuscles*; these contain the protoplasm out of which the whole is developed and regenerated.

FIG. [15].—Subcutaneous tissue from a young rabbit. Highly magnified. (Schäfer.)

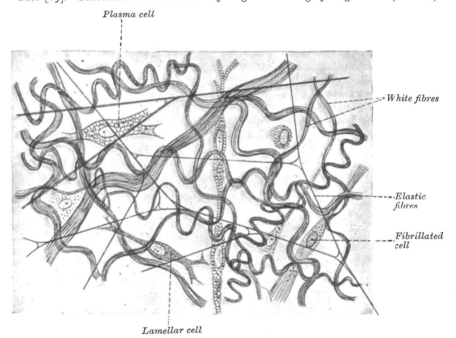

Plasma cell

White fibres

Elastic fibres

Fibrillated cell

Lamellar cell

The *white fibres* are arranged in waving bands or bundles of minute transparent homogeneous filaments or fibrillæ. The bundles have a tendency to split up longitudinally or send off slips to join neighbouring bundles and receive others in return, but the individual fibres are unbranched and never join other fibres ; the *yellow elastic fibres* have a well-defined outline and are considerably larger in size than the white fibrillæ. They vary much, being from the $\frac{1}{24000}$ to the $\frac{1}{4000}$ of an inch in diameter. The fibres form bold and wide curves, branch, and freely anastomose with each other. They are homogeneous in appearance, and tend to curl up, especially at their broken ends.

Connective tissue corpuscles.—The cells of areolar tissue are of three principal kinds : (1) Flattened lamellar cells, which may be either branched or unbranched. The branched lamellar cells are composed of clear cell substance, in which is contained an oval nucleus. The processes of these cells unite so as to form an open network as in the cornea. The unbranched cells are joined edge to edge like the cells of an epithelium. The 'tendon cells,' presently to be described, are an example of this variety. (2) Granule cells, which are ovoid or spheroidal in shape, and formed of a soft protoplasm, containing granules which are albuminous in character and stain deeply with eosin. (3) Plasma cells of Waldeyer, varying greatly in size and form, but always to be distinguished from the other two varieties by containing a largely vacuolated protoplasm. The vacuoles are filled with fluid, and the protoplasm between the spaces is clear, with occasionally a few scattered granules.

FIG. [16].—White fibrous tissue. High power.

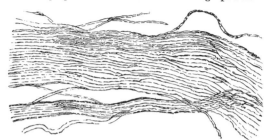

In addition to these three typical forms of connective tissue corpuscles, areolar tissue may be seen to possess *wandering cells*, i.e. leucocytes which have emigrated from the neighbouring vessels, and in some instances, as in the choroid coat of the eye, cells filled with granules of pigment (*pigment-cells*).

The connective tissue corpuscles lie in spaces in the ground substance between the bundles of fibres, and these spaces may be brought into view by treating the tissue with nitrate of silver and exposing it to the light. This will colour the ground substance and leave the cell-spaces unstained.

White fibrous tissue (fig. [16]) is a true connecting structure, and serves three purposes in the animal economy. In the form of ligaments it serves to bind bones together; in the form of tendons it serves to connect muscles to bones or other structures; and it forms an investing or protecting structure to various organs in the

FIG. [17].—Connective tissue. (Klein and Noble Smith.)

FIG. [18].—Tendon of mouse's tail, stained with logwood, showing chains of cells between the tendon-bundles. (From Quain's 'Anatomy.' E. A. Schäfer.)

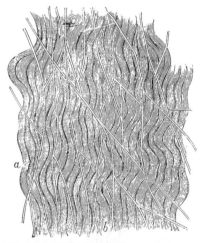

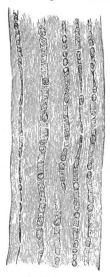

a. The white fibrous element—a layer of more or less sharply outlined, parallel, wavy bundles of connective-tissue fibrils. On the surface of this layer is *b*, a network of fine elastic fibres.

form of membranes. Examples of where it serves this latter office are to be found in the muscular fasciæ or sheaths, the periosteum, and perichondrium; the investments of the various glands (such as the tunica albuginea testis, the capsule of the kidney, &c.), the investing sheath of the nerves (epineurium), and of various organs, as the penis and the eye (sheath of the corpora cavernosa and corpus spongiosum and of the sclerotic). In white fibrous tissue, as its name implies, the white fibres predominate, the matrix being apparent only as a cement substance, the yellow elastic fibres comparatively few, while the tissue-cells are arranged in a special manner. It presents to the naked eye the appearance of silvery white glistening fibres, covered over with a quantity of loose flocculent tissue which binds the fibres together and carries the blood-vessels (fig. [17]). It is not possessed of any elasticity, and only the very slightest extensibility; it is exceedingly strong, so that upon the application of any external violence the bone with which it is connected will fracture before the fibrous tissue will give way. In ligaments and tendons the bundles of fibres run parallel with each other; in membranes they intersect one another in different places. The cells occurring in white fibrous tissue are often called ' tendon-cells.' They are situated on the surface of groups of bundles and are quadrangular in shape, arranged in rows, in single file, each cell being separated from its neighbours by a narrow line of cement-substance. The nucleus is generally situated at one end of the cell, the nucleus of the adjoining cell being in close proximity to it (fig. [18]). Upon the addition of acetic acid to white fibrous tissue it swells up into a glassy-looking indistinguishable mass. When boiled in water it is converted almost completely into gelatin, the white fibres being composed of the albuminoid *collagen*, which is often regarded as the anhydride of gelatin.

Yellow elastic tissue.—In certain parts of the body a tissue is found which when viewed in mass is of a yellowish colour, and is possessed of great elasticity; so that it is capable of considerable extension, and when the extending force is withdrawn returns at once to its original condition. This is *yellow elastic tissue*, which may be regarded as a connective tissue in which the yellow elastic fibres have developed to the practical exclusion of the other elements. It is found in the ligamenta subflava, in the vocal cords, in the longitudinal coat of the trachea and bronchi, in the inner coats of the blood-vessels,

FIG. [19].—Yellow elastic tissue. High power.

especially the larger arteries, and to a very considerable extent in the thyro-hyoid, crico-thyroid, and stylo-hyoid ligaments. It is also found in the ligamentum nuchæ of the lower animals (fig. [19]). In some parts where the fibres are broad and large and the network close, the tissue presents the appearance of a membrane, with gaps or perforations corresponding to the intervening spaces. This is to be found in the inner coat of the arteries, and to it the name of *fenestrated membrane* has been given by Henle. The yellow elastic fibres remain unaltered by acetic acid. Chemically they are composed of the albuminoid body, *elastin*.

Vessels and Nerves of Connective Tissue.—The *blood-vessels* of connective tissue are very few—that is to say, there are few actually destined for the tissue itself, although many vessels may permeate one of its forms, the areolar tissue, carrying blood to other structures. In white fibrous tissue the blood-vessels usually run parallel to the longitudinal bundles and between them, sending transverse communicating branches across; in some forms, as in the periosteum and dura mater, they are fairly numerous. In yellow elastic tissue, the blood-vessels also run between the fibres, and do not penetrate them. *Lymphatic* vessels are very numerous in most forms of connective tissue, especially in the areolar tissue beneath the skin and the mucous and serous surfaces. They are also found in abundance in the sheaths of tendons, as well as in the tendons themselves. *Nerves* are to be found in the white fibrous tissue, where they terminate in a special manner; but it is doubtful whether any nerves terminate in areolar tissue; at all events, they have not yet been demonstrated, and the tissue is possessed of very little sensibility.

Development of Connective Tissue.—Connective tissue is developed from embryonic connective-tissue cells derived from the mesoblast. These cells, at first rounded, become fusiform and branched, and ultimately become the connective tissue corpuscles. A mucinous intercellular substance or matrix, partly formed from the cells themselves and partly from the lymph exuded by the neighbouring blood-vessels, gradually separates the cells. In the matrix the fibres are deposited, probably under the influence of the cells, but not by any transformation of the

FIG. [20].—Mucous tissue.

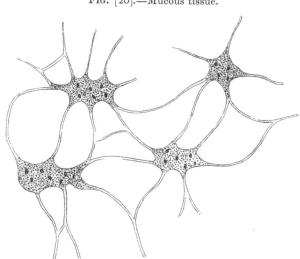

cell protoplasm. In the case of yellow elastic fibres, rows of granules of elastin are first laid down; these eventually fuse to form the fully developed fibre.

Mucous tissue exists chiefly in the 'jelly of Wharton,' which forms the bulk of the umbilical cord, but is also found in other situations in the foetus, chiefly as a stage in the development of connective tissue. It consists of a matrix, largely made up of mucin, in which are nucleated cells with branching and anastomosing processes (fig. [20]). Few fibres are seen in typical mucous tissue, though, at birth, the umbilical cord shows considerable development of fibres. In the adult the vitreous humour of the eye is a persistent form of mucous tissue, in which there are no fibres, and from which the cells have disappeared, leaving only the mucinous ground substance.

Retiform connective tissue is found extensively in many parts of the body, forming the framework of some organs and entering into the construction of many mucous membranes. It is a form of connective tissue, in which the intercellular or ground substance has, in a great measure, disappeared, and has been replaced by fluid. It is apparently composed almost entirely of extremely fine bundles of white fibrous tissue, forming an intricate network, yet chemically it yields, besides gelatin, a fresh substance, *reticulin*. The fibres are covered and concealed by flattened branched connective-tissue cells, and these must be removed or brushed away before the fibres become visible. In many situations the interstices of the network are filled with rounded lymph-corpuscles, and the tissue is then termed **lymphoid** or **adenoid tissue** (fig. [21]).

Basement membranes, formerly described as homogeneous membranes, are really a form of connective tissue. They constitute the supporting membrane, or *membrana propria*, on which is placed the epithelium of mucous membranes or secreting glands, and in other situations. By means of staining with nitrate of silver they may be shown to consist of flattened cells in close apposition, and joined together by their edges, thus forming an example of an epithelioid arrangement of connective-tissue cells. In some situations the cells, instead of adhering by their edges, give off branching processes, which join with similar pro-

Fig. [21].—Retiform connective tissue, from a lymphatic gland; most of the lymph-corpuscles are removed. (From Klein's 'Elements of Histology.')

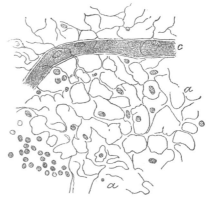

a. The reticulum.   *c.* A capillary blood-vessel.

cesses of other cells and so form a network, rather than a continuous membrane. In other instances basement membranes are composed of elastic tissue, as in the cornea, or, again, in other cases, of condensed ground substance.

ADIPOSE TISSUE

In almost all parts of the body the ordinary areolar tissue contains a variable quantity of fat. The principal situations where it is not found are the subcutaneous tissue of the eyelids, the penis and scrotum, the nymphæ, within the cavity of the cranium, and in the lungs, except near their roots. Nevertheless its distribution is not uniform; in some parts it is collected in great abundance, as

Fig. [22].—Adipose tissue. High power.

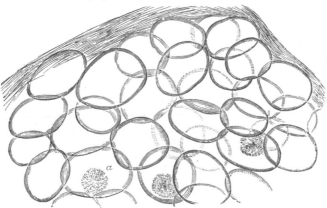

a. Starlike appearance, from crystallisation of fatty acids.

in the subcutaneous tissue, especially of the abdomen; around the kidneys; on the surface of the heart between the furrows, and in some other situations. Lastly, fat enters largely into the formation of the marrow of bones. A distinction must be made between fat and adipose tissue; the latter being a distinct tissue, the former an oily matter, which in addition to forming adipose tissue is also widely present in the body, as in the fat of the brain and liver, and in the blood and chyle, &c.

Adipose tissue consists of small vesicles, *fat-cells*, lodged in the meshes of areolar tissue. The fat-cells (fig. [22]) vary in size, but are of about the average diameter of $\frac{1}{500}$ of an inch. They are formed of an exceedingly delicate protoplasmic membrane, filled with fatty matter, which is liquid during life, but becomes solidified after death. They are round or spherical where they have not been subjected to pressure; otherwise they assume a more or less angular outline. A nucleus is always present and can be easily demonstrated by staining with hæmatoxylin; in the natural condition it is so compressed by the contained oily matter as to be scarcely recognisable. These fat-cells are contained in clusters in the areolæ of fine connective tissue, and are held together mainly by a network of capillary blood-vessels, which are distributed to them.

FIG. [23].—Development of fat. (Klein and Noble Smith.)

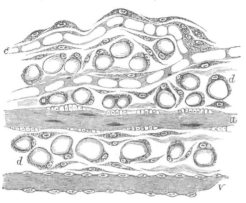

a. Minute artery. v. Minute vein. c. Capillary blood-vessels in the course of formation; they are not yet completely hollowed out, there being still left in them protoplasmic septa. d. The ground-substance, containing numerous nucleated cells, some of which are more distinctly branched and flattened than others, and appear therefore more spindle-shaped.

Chemically the oily material in the cells is composed of the fats, olein, palmitin, and stearin, which are glycerin compounds with fatty acids. Sometimes fat crystals form in the cells after death (fig. [22], *a*). By boiling the tissue in ether or strong alcohol, the fat may be extracted from the vesicle, which is then seen empty and shrunken.

Fat is said to be first detected in the human embryo about the fourteenth week. The fat-cells are formed by the transformation of connective-tissue corpuscles, in which small droplets of oil are formed; these coalesce to produce a larger drop, and this increases until it distends the corpuscle, the remaining protoplasm and the nucleus being crowded to the periphery of the cell (fig. [23]).

PIGMENT

In various parts of the body **pigment** is found; most frequently in epithelial cells and in the cells of connective tissue. Pigmented *epithelial cells* are found in the external layer of the retina and on the posterior surface of the iris. Pigment is also found in the epithelial cells of the deeper layers of the cuticle in some parts of the body—such as the areola of the nipple and in coloured patches of skin and especially in the skin of the coloured races, and also in hair. It is also found in the epithelial cells of the olfactory region, and of the membranous labyrinth of the ear.

FIG. [24].—Pigment-cells from the choroid coat of the eyeball.

In the *connective-tissue cells* pigment is frequently met with in the lower vertebrates. In man it is found in the choroid coat of the eye (fig. [24]), and in the iris of all but the light blue eyes and the albino. It is also occasionally met with in the cells of retiform tissue and in the pia mater of the upper part of the spinal cord. These cells are characterised by their large size and branched processes, which, as well as the body of the cells, are filled with granules. The pigment consists of dark brown or black granules of very small size closely packed together within the cells, but not invading the nucleus. Occasionally the pigment is yellow, and when occurring in the cells of the cuticle constitutes 'freckles.'

Cartilage

Cartilage is a non-vascular structure which is found in various parts of the body—in adult life chiefly in the joints, in the parietes of the thorax, and in various tubes, such as the air-passages, nostrils, and ears, which are to be kept permanently open. In the fœtus, at an early period, the greater part of the skeleton is cartilaginous. As this cartilage is afterwards replaced by bone, it is called *temporary*, in contra-distinction to that which remains unossified during the whole of life, and which is called *permanent*.

Cartilage is divided, according to its minute structure, into hyaline cartilage, white fibro-cartilage, and yellow or elastic fibro-cartilage. Besides these varieties met with in the adult human subject, there is a variety called *cellular cartilage*, which consists entirely, or almost entirely, of cells, united in some cases by a network of very fine fibres, in other cases apparently destitute of any intercellular substance. This is found in the external ear of rats, mice, and some other animals, and is present in the chorda dorsalis of the human embryo, but is not found in any other human structure.

The various cartilages in the body are also classified, according to their function and position, into articular, interarticular, costal, and membraniform.

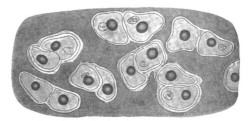

Fig. [25].—Human cartilage-cells from the cricoid cartilage.   × 350.

Hyaline cartilage, which may be taken as the type of this tissue, consists of a gristly mass of a firm consistence, but of considerable elasticity and of a pearly-bluish colour. Except where it coats the articular ends of bones, it is covered externally by a fibrous membrane, the *perichondrium*, from the vessels of which it imbibes its nutritive fluids, being destitute of blood-vessels. It contains no nerves. Its intimate structure is very simple. If a thin slice is examined under the microscope, it will be found to consist of cells of a rounded or bluntly angular form, lying in groups of two or more in a granular or almost homogeneous matrix (fig. [25]). The cells, when arranged in groups of two or more, have generally a straight outline where they are in contact with each other, and in the rest of their circumference are rounded. The cell-contents consist of clear translucent protoplasm, in which fine interlacing filaments and minute granules may sometimes be seen ; embedded in this are one or two round nuclei, having the usual intranuclear network. The cells are embedded in cavities in the matrix, called *cartilage lacunæ*; around these the matrix is arranged in concentric lines, as if it had been formed in successive portions around the cartilage cells. This constitutes the so-called *capsule* of the space. Each lacuna is generally occupied by a single cell, but during the division of the cells it may contain two, four, or eight cartilage-cells. By exposure to the action of an electric shock the cell assumes a jagged outline and shrinks away from the interior of the capsule.

The matrix is transparent and apparently without structure, or else presents a dimly granular appearance, like ground glass. Some observers have shown that the matrix of hyaline cartilage, and especially the articular variety, after prolonged maceration, can be broken up into fine fibrils. These fibrils are probably of the same nature, chemically, as the white fibres of connective tissue.

It is believed by some histologists that the matrix is permeated by a number of fine channels, which connect the lacunæ with each other, and that these canals communicate with the lymphatics of the perichondrium, and thus the structure is permeated with a current of nutrient fluid. This, however, is somewhat doubtful.

Articular cartilage, costal cartilage, and temporary cartilage are all of the hyaline variety. They present minute differences in the size and shape of their cells and in the arrangement of their matrix. In **articular cartilage,** which shows no tendency to ossification, the matrix is finely granular under a high power; the cells and nuclei are small, and are disposed parallel to the surface in the superficial part, while nearer to the bone they become vertical. Articular cartilages have a tendency to split in a vertical direction; in disease this tendency becomes very manifest. Articular cartilage is not covered by perichondrium, on its free surface, where it is exposed to friction, though a layer of connective tissue can be traced in the adult over a small part of its circumference continuous with that of the synovial membrane, and here the cartilage-cells are more or less branched and pass insensibly into the branched connective-tissue corpuscles of the synovial membrane.

FIG. [26].—Costal cartilage from a man seventy-six years of age, showing the development of fibrous structure in the matrix. In several portions of the specimen two or three generations of cells are seen enclosed in a parent cell-wall. High power.

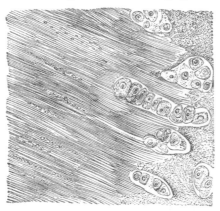

Articular cartilage forms a thin incrustation upon the joint-surfaces of the bones, and its elasticity enables it to break the force of any concussion, while its smoothness affords ease and freedom of movement. It varies in thickness according to the shape of the articular surface on which it lies; where this is convex the cartilage is thickest at the centre, where the greatest pressure is received; and the reverse is the case on the concave articular surfaces. Articular cartilage appears to derive its nutriment partly from the vessels of the neighbouring synovial membrane, partly from those of the bone upon which it is implanted. Toynbee has shown that the minute vessels of the cancellous tissue as they approach the articular lamella dilate and form arches, and then return into the substance of the bone.

In **costal cartilage** the cells and nuclei are large, and the matrix has a tendency to fibrous striation, especially in old age (fig. [26]). In the thickest parts of the costal cartilages a few large vascular channels may be detected. This appears, at first sight, to be an exception to the statement that cartilage is a non-vascular tissue, but is not so really, for the vessels give no branches to the cartilage substance itself, and the channels may rather be looked upon as involutions of the perichondrium. The ensiform cartilage may be regarded as one of the costal cartilages, and the cartilages of the nose and of the larynx and trachea (except the epiglottis and cornicula laryngis, which are composed of elastic fibro-cartilage) resemble them in microscopical characters.

Temporary cartilage and the process of its ossification will be described with bone.

The hyaline cartilages, especially in adult and advanced life, are prone to calcify—that is to say, to have their matrix permeated by the salts of lime without any appearance of true bone. The process of calcification occurs also and still more frequently, according to Rollett, in such cartilages as those of the trachea and in the costal cartilages, which are prone afterwards to conversion into true bone.

White fibro-cartilage consists of a mixture of white fibrous tissue and cartilaginous tissue in various proportions; it is to the first of these two constituents that its flexibility and toughness are chiefly owing, and to the latter its elasticity. When examined under the microscope it is found to be made up of fibrous connective tissue arranged in bundles, with cartilage-cells between the bundles; these to a certain extent resemble tendon-cells, but may be distinguished from them by being surrounded by a concentrically striated area of cartilage matrix and by their being less flattened (fig. [27]). The fibro-cartilages admit of arrangement into four groups —interarticular, connecting, circumferential, and stratiform.

1. The **interarticular fibro-cartilages** (*menisci*) are flattened fibro-cartilaginous plates, of a round, oval, triangular, or sickle-like form, interposed between the articular cartilages of certain joints. They are free on both surfaces, thinner towards their centre than at their circumference, and held in position by the attachment of their margins and extremities to the surrounding ligaments. The synovial membrane of the joint is prolonged over them a short distance from their attached margins. They are found in the temporo-mandibular, sterno-clavicular, acromio-clavicular, wrist and knee joints. These cartilages are usually found in those joints which are most exposed to violent concussion and subject to frequent movement. Their use is—to maintain the apposition of the opposed surfaces in their various motions; to increase the depth of the articular surfaces and give ease to the gliding movement; to moderate the effects of great pressure and deaden the intensity of the shocks to which the parts may be subjected. Humphry has pointed out that these interarticular fibro-cartilages serve an important purpose in increasing the variety of movements in a joint. Thus, in

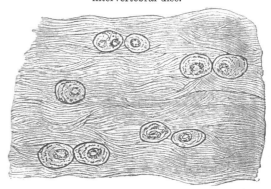

FIG. [27].—White fibro-cartilage from an intervertebral disc.

the knee-joint, there are two kinds of motion, viz. angular movement and rotation, although it is a hinge joint, in which, as a rule, only one variety of motion is permitted; the former movement takes place between the condyles of the femur and the interarticular cartilage, the latter between the cartilage and the head of the tibia. So, also, in the temporo-mandibular joint, the upward and downward movement of opening and shutting the mouth takes place between the fibro-cartilage and the jaw-bone, the grinding movement between the glenoid cavity and the fibro-cartilage, the latter moving with the jaw-bone.

2. The **connecting fibro-cartilages** are interposed between the bony surfaces of those joints which admit of only slight mobility, as between the bodies of the vertebræ and between the pubic bones. They form discs, which adhere closely to both of the opposed surfaces, and are composed of concentric rings of fibrous tissue, with cartilaginous laminæ interposed, the former tissue predominating towards the circumference, the latter towards the centre.

3. The **circumferential fibro-cartilages** consist of a rim of fibro-cartilage, which surrounds the margin of some of the articular cavities, as the cotyloid cavity of the hip, and the glenoid cavity of the shoulder; they serve to deepen the articular surface, and to protect its edges.

4. The **stratiform fibro-cartilages** are those which form a thin coating to osseous grooves through which the tendons of certain muscles glide. Small

masses of fibro-cartilages are also developed in the tendons of some muscles, where they glide over bones, as in the tendons of the Peroneus longus and the Tibialis posticus.

Yellow or **elastic fibro-cartilage** is found in the human body in the auricle of the external ear, the Eustachian tubes, the cornicula laryngis, and the epiglottis. It consists of cartilage-cells and a matrix, the latter being pervaded in every direction, except immediately around each cell, where there is a variable amount of non-fibrillated hyaline, intercellular substance, by a network of yellow elastic fibres, branching and anastomosing in all directions (fig. [28]). The fibres resemble those of yellow elastic tissue, both in appearance and in being unaffected by acetic acid ; and according to Rollett their continuity with the elastic fibres of the neighbouring tissue admits of being demonstrated.

FIG. [28].—Yellow cartilage, ear of horse. High power.

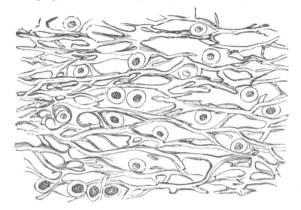

The distinguishing feature of cartilage as to its chemical composition, is that it yields on boiling a substance called *chondrin*, very similar to gelatin, but differing from it in several of its reactions. It is now believed that chondrin is not a simple body, but a mixture of gelatin with mucinoid substances, chief among which, perhaps, is a compound termed *chondro-mucoid*.

BONE

Structure and Physical Properties of Bone.—Bone is one of the hardest structures of the animal body ; it possesses also a certain degree of toughness and elasticity. Its colour, in a fresh state, is pinkish white externally, and deep red within. On examining a section of any bone, it is seen to be composed of two kinds of tissue, one of which is dense in texture, like ivory, and is termed *compact tissue*; the other consists of slender fibres and lamellæ, which join to form a reticular structure; this, from its resemblance to lattice-work, is called *cancellous tissue*. The compact tissue is always placed on the exterior of the bone ; the cancellous is always internal. The relative quantity of these two kinds of tissue varies in different bones, and in different parts of the same bone, as strength or lightness is requisite. Close examination of the compact tissue shows it to be extremely porous, so that the difference in structure between it and the cancellous tissue depends merely upon the different amount of solid matter, and the size and number of spaces in each ; the cavities being small in the compact tissue and the solid matter between them abundant, while in the cancellous tissue the spaces are large and the solid matter is in smaller quantity.

Bone during life is permeated by vessels and is enclosed, except where it is coated with articular cartilage, in a fibrous membrane, the *periosteum*, by means of which many of these vessels reach the hard tissue. If the periosteum is stripped from the surface of the living bone small bleeding points are seen, which mark the entrance of the periosteal vessels ; and on section during life, every part of the bone will be seen to exude blood from the minute vessels which ramify in it. The interior of the bones of the limbs presents a cylindrical cavity filled with marrow and lined by a highly vascular areolar structure, called the *medullary membrane* or *internal periosteum*, which, however, is rather the areolar envelope of the cells of the marrow, than a definite membrane.

The **periosteum** adheres to the surface of the bones in nearly every part, excepting at their cartilaginous extremities. When strong tendons or ligaments are attached to the bone, the periosteum is incorporated with them. It consists of two layers closely united together, the outer one formed chiefly of connective tissue, containing occasionally a few fat-cells; the inner one, of elastic fibres of the finer kind, forming dense membranous networks, which can be again separated into several layers. In young bones the periosteum is thick and very vascular, and is intimately connected at either end of the bone with the epiphysial cartilage, but less closely with the shaft, from which it is separated by a layer of soft tissue, containing a number of granular corpuscles or 'osteoblasts,' in which ossification proceeds on the exterior of the young bone. Later in life the periosteum is thinner, less vascular, and the osteoblasts have become converted into an epithelioid layer, separated from the rest of the periosteum in many places by cleft-like spaces, which are supposed to serve for the transmission of lymph. The periosteum serves as a nidus for the ramification of the vessels previous to their distribution in the bone; hence the liability of bone to exfoliation or necrosis, when denuded of this membrane by injury or disease. Fine nerves and lymphatics, which generally accompany the arteries, may also be demonstrated in the periosteum.

The **marrow** not only fills up the cylindrical cavity in the shafts of the long bones, but also occupies the spaces of the cancellous tissue and extends into the larger bony canals (Haversian canals) which contain the blood-vessels. It differs in composition in different bones. In the shafts of adult long bones the marrow is of a *yellow* colour, and contains, in 100 parts, 96 of fat, 1 of areolar tissue and vessels, and 3 of fluid, with extractive matter, and consists of a basis of connective

FIG. [29].—Cells of red marrow of the guinea-pig. (Schäfer.)

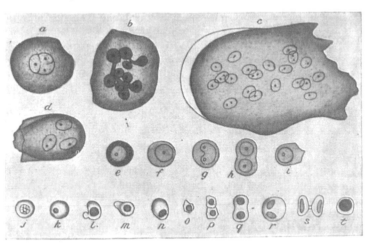

a–d. Myeloplaques. *e–i.* Marrow cells proper. *j–t.* Erythroblasts—some in process of division.

tissue, supporting numerous blood-vessels and cells, most of which are fat-cells, but some are 'marrow-cells,' such as occur in the red marrow, to be immediately described. In the flat and short bones, in the articular ends of the long bones, in the bodies of the vertebræ, in the cranial diploë, and in the sternum and ribs, it is of a *red* colour, and contains, in 100 parts, 75 of water and 25 of solid matter, consisting of cell-globulin, nucleo-proteid, extractives, salts, and only a small proportion of fat. The red marrow consists of a small quantity of connective tissue, blood-vessels, and numerous cells (fig. [29]), some few of which are fat-cells, but the great majority roundish nucleated cells, the true 'marrow-cells' of Kölliker. These marrow-cells proper resemble in appearance lymphoid corpuscles, and like them are amœboid. Among them may be seen smaller cells, which possess a slightly pinkish hue; these are the *erythroblasts*, from which, as we have seen, the red corpuscles of the adult are derived, and which may be regarded as descendants of the nucleated coloured corpuscles of the embryo. *Giant-cells* (*myeloplaques, osteoclasts*), large, multi-nucleated, protoplasmic masses, are also to be found in both sorts of adult marrow, but more particularly in red marrow. They were believed by Kölliker to be concerned in the absorption of bone-matrix, and hence the name which he gave to them—*osteoclasts*. They excavate small shallow pits or cavities, which are named *Howship's lacunæ*, in which they are found lying.

Vessels of Bone.—The blood-vessels of bone are very numerous. Those of the compact tissue are derived from a close and dense network of vessels ramifying in the periosteum. From this membrane, vessels pass into the minute orifices in the compact tissue, and run through the canals which traverse its substance. The cancellous tissue is supplied in a similar way, but by less numerous and larger vessels, which, perforating the outer compact tissue, are distributed to the cavities of the spongy portion of the bone. In the long bones, numerous apertures may be seen at the ends near the articular surfaces, some of which give passage to the arteries of the larger set of vessels referred to; but the most numerous and largest apertures are for the veins of the cancellous tissue, which run separately from the arteries. The medullary canal in the shafts of the long bones is supplied by one large artery (or sometimes more), which enters the bone at the nutrient foramen (situated in most cases near the centre of the shaft), and perforates obliquely the compact structure. The *medullary* or *nutrient* artery, usually accompanied by one or two veins, sends branches upwards and downwards, to supply the medullary membrane, which lines the central cavity and the adjoining canals. The ramifications of this vessel anastomose with the arteries both of the cancellous and compact tissues. In most of the flat, and in many of the short spongy bones, one or more large apertures are observed, which transmit, to the central parts of the bone, vessels corresponding to the medullary arteries and veins. The veins emerge from the long bones in three places (Kölliker)—(1) by one or two large veins, which accompany the artery; (2) by numerous large and small veins at the articular extremities; (3) by many small veins which arise in the compact substance. In the flat cranial bones the veins are large, very numerous, and run in tortuous canals in the diploic tissue, the sides of the canals being formed by a thin lamella of bone, perforated here and there for the passage of branches from the adjacent cancelli. The same condition is also found in all cancellous tissue, the veins being enclosed and supported by osseous structure, and having exceedingly thin coats. When the bony structure is divided, the vessels remain patulous, and do not contract in the canals in which they are contained. Hence the occurrence of purulent absorption after amputation, in those cases where the stump becomes inflamed, and the cancellous tissue is infiltrated and bathed in pus.

Lymphatic vessels, in addition to those found in the periosteum, have been traced by Cruikshank into the substance of bone, and Klein describes them as running in the Haversian canals.

Nerves are distributed freely to the periosteum, and accompany the nutrient arteries into the interior of the bone. They are said by Kölliker to be most numerous in the articular extremities of the long bones, in the vertebræ, and the larger flat bones.

Minute Anatomy.—The intimate structure of bone, which in all essential particulars is identical in the compact and in the cancellous tissue, is most easily studied in a transverse section from the compact wall of one of the long bones after maceration, such as is shown in fig. [30].

If this is examined with a rather low power the bone will be seen to be mapped out into a number of circular districts: each one of which consists of a central hole surrounded by a number of concentric rings. These districts are termed *Haversian systems*; the central hole is an *Haversian canal*, and the rings are layers of bone-tissue arranged concentrically around the central canal, and termed *lamellæ*. Moreover, on closer examination, it will be found that between these lamellæ, and therefore also arranged concentrically around the central canal, are a number of little dark specks, the *lacunæ*, and that these lacunæ are connected with each other and with the central Haversian canal by a number of fine dark lines, which radiate like the spokes of a wheel and are called *canaliculi*. All these structures—the concentric lamellæ, the lacunæ, and the canaliculi—may be seen in any single Haversian system, forming a circular district round a central, Haversian, canal. Between these circular systems, filling in the irregular intervals which are left between them, are other lamellæ, with their lacunæ and canaliculi, running in various directions, but more or less curved (fig. [31]). These are termed *interstitial* lamellæ.

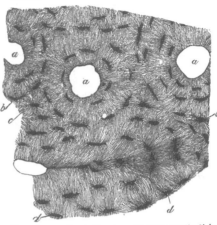

FIG. [30].—From a transverse section of the diaphysis of the humerus. Magnified 350 times.

a. Haversian canals. *b.* Lacunæ, with their canaliculi in the lamellæ of these canals. *c.* Lacunæ, of the interstitial lamellæ. *d.* Others at the surface of the Haversian systems, with canaliculi given off from one side.

Again, other lamellæ, for the most part found on the surface of the bone, are arranged concentrically to the circumference of bone, constituting, as it were, a single Haversian system of the whole bone, of which the medullary cavity would represent the Haversian canal. These latter lamellæ are termed *circumferential*,

FIG. [31].—Transverve section of compact tissue of bone. Magnified about 150 diameters. (Sharpey.)

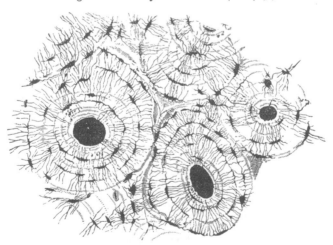

or by some authors *primary* or *fundamental* lamellæ, to distinguish them from those laid down around the axis of the Haversian canals, which are then termed *secondary* or *special* lamellæ.

The *Haversian canals*, seen as round holes in a transverse section of bone at or about the centre of each Haversian system, may be demonstrated to be true canals, if a longitudinal section is made, as in fig. [33]. It will then be seen that these round holes are tubes cut across, which run parallel with the longitudinal axis of the bone for a short distance, and then branch and communicate. They vary considerably in size, some being as large as $\frac{1}{200}$ of an inch in diameter; the average size being, however, about $\frac{1}{500}$ of an inch. Near the medullary cavity the canals are larger than those near the surface of the bone. Each canal contains one or two blood-vessels, with a small quantity of delicate connective tissue and some nerve filaments. In the larger ones there are also lymphatic spaces, and branched cells, the processes of which communicate, through the canaliculi, with the branched processes of certain bone-cells in the substance of the bone. Those canals near the surface of the bone open upon it by minute orifices, and those near the medullary cavity open in the same way into this space, so that the whole of the bone is permeated by a system of blood-vessels running through the bony canals in the centre of the Haversian systems.

The *lamellæ* are thin plates of bone-tissue encircling the central canal, and may be compared, for the sake of illustration, to a number of sheets of paper

FIG. [33].—Section parallel to the surface from the shaft of the femur. Magnified 100 times.

FIG. [32].—Nucleated bone-cells and their processes, contained in the bone-lacunæ and their canaliculi respectively. From a section through the vertebra of an adult mouse. (Klein and Noble Smith.)

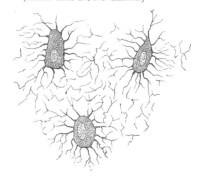

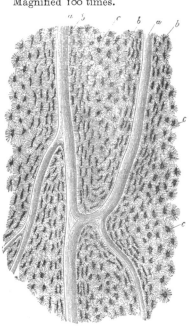

pasted one over another around a central hollow cylinder. After macerating a piece of bone in dilute mineral acid, these lamellæ may be stripped off in a longitudinal direction as thin films. If one of these is examined with a high power under the microscope it will be

a. Haversian canals. *b.* Lacunæ seen from the side. *c.* Others seen from the surface in lamellæ which are cut horizontally.

found to be composed of a finely reticular structure, presenting the appearance of lattice-work made up of very slender, transparent fibres, decussating obliquely, and coalescing at the points of intersection so as to form an exceedingly delicate network. These fibres are composed of fine fibrils, identical with those of white connective tissue. The intercellular matrix between the fibres has been replaced by calcareous deposit which the acid dissolves. In many places the various lamellæ may be seen to be held together by tapering fibres, which run obliquely through them, pinning or bolting them together. These fibres were first described by Sharpey, and were named by him *perforating fibres* (fig. [35]).

The *lacunæ* are situated between the lamellæ, and consist of a number of oblong spaces. In an ordinary microscopic section, viewed by transmitted light, they appear as dark, oblong, opaque spots, and were formerly believed to be solid cells. Subsequently, when it was seen that the Haversian canals were channels which lodge the vessels of the part, and the canaliculi minute tubes by which the plasma of the blood circulates through the tissue, the theory was formulated that the lacunæ were hollow spaces filled during life with the same fluid, and only lined (if lined at all) by a delicate membrane. But this view was subsequently proved to be erroneous, for examination of the structure of bone, when recent, led Virchow to believe that the lacunæ are occupied during life by a branched cell, termed a bone-cell or bone-corpuscle, the processes from which pass down the canaliculi—a view which is now universally received (fig. [32]). It is by means of these cells that the fluids necessary for nutrition are brought into contact with the ultimate tissue of bone.

The *canaliculi* are exceedingly minute channels, which pass *across* the lamellæ and connect the lacunæ with neighbouring lacunæ and also with the Haversian canal. From this central canal a number of the canaliculi are given off, which radiate from it, and open into the first set of lacunæ arranged around the Haversian canal, between the first and second lamellæ. From these lacunæ a second set of canaliculi are given off, which pass outwards to the next series of lacunæ, and so on until they reach the periphery of the Haversian system ; here the canaliculi given off from the last series of lacunæ do not communicate with the lacunæ of neighbouring Haversian systems, but after passing outwards for a short distance form loops and return to their own lacuna. Thus every part of an Haversian system is supplied with nutrient fluids derived from the vessels in the Haversian canal and traversing the canaliculi and lacunæ.

FIG. [34].—Lamellæ torn from a decalcified human parietal bone to show the perforating fibres of Sharpey. (Copied from a drawing by Allen Thomson.)

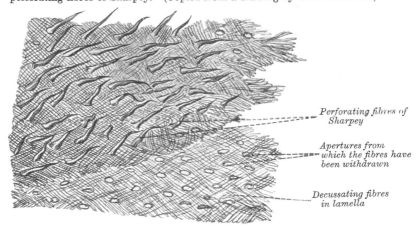

Perforating fibres of Sharpey

Apertures from which the fibres have been withdrawn

Decussating fibres in lamella

The *bone-cells* are contained in the lacunæ, which, however, they do not completely fill. They are flattened nucleated cells, which Virchow has shown are homologous with those of connective tissue. The cells are branched, and the branches, especially in young bones, pass into the canaliculi from the lacunæ.

If a longitudinal section is examined, as in fig. [33], the structure is seen to be the same. The appearance of concentric rings is replaced by that of lamellæ or rows of lacunæ, parallel to the course of the Haversian canals, and these canals appear like half-tubes instead of circular spaces. The tubes are seen to branch and communicate, so that each separate Haversian canal runs only a short distance. In other respects the structure has much the same appearance as in transverse sections.

In sections of thin plates of bone (as in the walls of the cells which form the cancellous tissue) the Haversian canals are absent, and the canaliculi open into the spaces of the cancellous tissue (medullary spaces), which thus have the same function as the Haversian canals in the more compact bone.

Chemical Composition.—Bone consists of an animal and an earthy part intimately combined together.

The animal part may be obtained by immersing the bone for a considerable time in dilute mineral acid, after which process the bone comes out exactly the same shape as before, but perfectly flexible, so that a long bone (one of the ribs, for example) can easily be tied in a knot.

FIG. [35].—Section of bone after the removal of the earthy matter by the action of acids.

If now a transverse section is made (fig. [35]), the same general arrangement of the Haversian canals, lamellæ, lacunæ, and canaliculi is seen, though not so plainly as in the ordinary section.

The earthy part may be separately obtained by calcination, by which the animal matter is completely burnt out. The bone will still retain its original form, but it will be white and brittle, will have lost about one-third of its original weight, and will crumble down with the slightest force. The earthy matter confers on bone its hardness and rigidity, and the animal matter its tenacity.

The animal basis is largely composed of *ossein*, which is identical with the collagen of white fibrous tissue, so that when boiled with water, especially under pressure, it is almost entirely resolved into gelatin.

The organic matter of bone forms about *one-third*, or 33·3 per cent.; the inorganic matter, *two-thirds*, or 66·7 per cent. Of the earthy matter, five-sixths is calcium phosphate, the remainder consisting of calcium carbonate, calcium fluoride, calcium chloride, and magnesium phosphate, with small amounts of sodium chloride and sulphate. Even after the removal of all the marrow, a small percentage of fat is still found in bone.

Some of the diseases to which bones are liable mainly depend on the disproportion between the two constituents of bone. Thus in the disease called rickets, so common in the children of the poor, the bones become bent and curved, either from the superincumbent weight of the body, or under the action of certain muscles. This depends upon some defect of nutrition by which bone becomes deprived of its normal proportion of earthy matter, while the animal matter is of unhealthy quality. In the vertebræ of a rickety subject, Bostock found in 100 parts 79·75 animal and 20·25 earthy matter.

Development of Bone.—In the fœtal skeleton some bones are preceded by membrane, such as those forming the roof and sides of the skull; others, such as the bones of the limbs, are preceded by rods of cartilage. Hence two kinds of ossification are described: the *intramembranous* and the *intracartilaginous*.

Intramembranous Ossification.—In the case of bones which are developed in membrane no cartilaginous mould precedes the appearance of the bone tissue. The membrane, which occupies the place of the future bone, is of the nature of connective tissue, and ultimately forms the periosteum. At this stage it is seen to be composed of fibres and granular cells in a matrix. The outer portion is more fibrous, while, internally, the cells or *osteoblasts* predominate; the whole tissue is richly supplied with blood-vessels. At the outset of the process of bone formation a little network of bony spiculæ is first noticed radiating from the point or centre of ossification. When these rays of growing bone are examined by the microscope they are found to consist at their growing point of a network of fine clear fibres and granular corpuscles with an intervening ground substance (fig. [36]). The fibres are termed *osteogenetic* fibres, and are made up of fine fibrils differing little from those of white fibrous tissue. Like them they are probably deposited in the matrix through the influence of the cells—in this case the osteoblasts. The osteogenetic fibres soon assume a dark and granular appearance from the deposition of calcareous granules in the fibres and in the intervening matrix, and as they calcify they are found to enclose some of the granular corpuscles or osteoblasts. By the fusion of the calcareous granules the bony tissue again assumes a more transparent appearance, but the fibres are no longer so distinctly seen. The involved osteoblasts form the corpuscles of the future bone, the spaces in which they are enclosed constituting the lacunæ. As the osteogenetic fibres grow out to the periphery they continue to calcify, and give rise to fresh bone spicules. Thus a network of bone is formed, the meshes of which contain the blood-vessels and a delicate connective tissue crowded with osteoblasts. The bony trabeculæ thicken by the addition of fresh layers of bone formed by the osteoblasts on their surface, and the meshes are correspondingly encroached upon. Subsequently successive layers of bony tissue are deposited under the periosteum and round the larger vascular channels which become the Haversian canals, so that the bone increases much in thickness.

Intracartilaginous Ossification.—Just before ossification begins the bone is entirely cartilaginous, and in a long bone, which may be taken as an example, the process commences in the centre and proceeds towards the extremities, which for some time remain cartilaginous. Subsequently a similar process

Fig. [36].—Part of the growing edge of the developing parietal bone of a fœtal cat. (After J. Lawrence.)

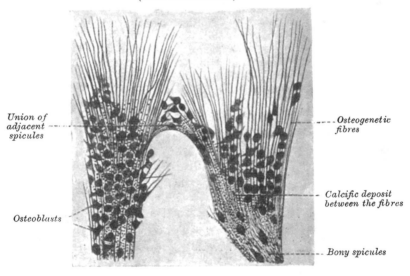

Union of adjacent spicules

Osteoblasts

Osteogenetic fibres

Calcific deposit between the fibres

Bony spicules

commences in one or more places in those extremities and gradually extends through them. The extremities do not, however, become joined to the shaft by bony tissue until growth has ceased, but are attached to it by a layer of cartilaginous tissue termed the *epiphysial cartilage*.

The first step in the ossification of the cartilage is that the cartilage-cells, at the point where ossification is commencing and which is termed a *centre of ossification*, enlarge and arrange themselves in rows (fig. [37]). The matrix in which they are embedded increases in quantity, so that the cells become further separated from each other. A deposit of calcareous material now takes place in this matrix, between the rows of cells, so that they become separated from each other by longitudinal columns of calcified matrix, presenting a granular and opaque appearance. Here and there the matrix between two cells of the same row also becomes calcified, and transverse bars of calcified substance stretch across from one calcareous column to another. Thus there are longitudinal groups of the cartilage-cells enclosed in oblong cavities, the walls of which are formed of calcified matrix, which cuts off all nutrition from the cells, and they, in consequence, waste, leaving spaces called the *primary areolæ* (Sharpey).

At the same time that this process is going on in the centre of the solid bar of cartilage of which the fœtal bone consists, certain changes are taking place on its surface.   This is covered by a very vascular membrane, the *perichondrium*, entirely similar to the embryonic connective tissue already described as constituting the basis of membrane bone, on the inner surface of which, that is to say, on the surface in contact with the cartilage, are gathered the formative cells, the *osteoblasts*.   By the agency of these cells a thin layer of bony tissue is being formed between the perichondrium and the cartilage, by the *intramembranous* mode of ossification just described.   There are then, in this first stage of ossification, two processes going on simultaneously : in the centre of the cartilage the formation

FIG. [37].—Section of fœtal bone of cat.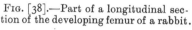

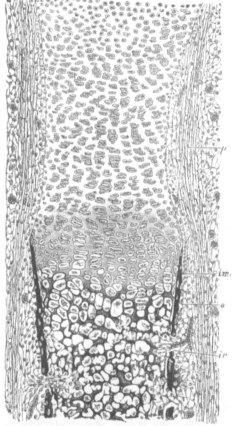
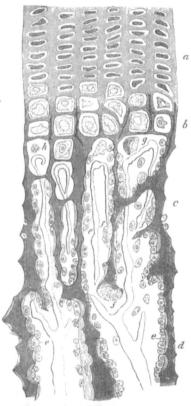

ir. Irruption of the subperiosteal tissue. *p.* Fibrous layer of the periosteum. *o.* Layer of osteoblasts. *im.* Subperiosteal bony deposit. (From Quain's 'Anatomy,' E. A. Schäfer.)

a. Flattened cartilage-cells. *b.* Enlarged cartilage-cells. *c, d.* Newly formed bone. *e.* Osteoblasts. *f.* Giant-cells or osteoclasts. *g, h.* Shrunken cartilage-cells. (From 'Atlas of Histology,' Klein and Noble Smith.)

of a number of oblong spaces, formed of calcified matrix and containing the withered cartilage-cells, and on the surface of the cartilage the formation of a layer of true membrane-bone.   The second stage consists in the prolongation into the cartilage of processes of the deeper or osteogenetic layer of the perichondrium, which has now become periosteum (fig. [37], *ir*).   The processes consist of blood-vessels and cells—*osteoblasts*, or bone-formers, and *osteoclasts*, or bone-destroyers. The latter are similar to the giant-cells (myeloplaques) found in marrow, and they excavate passages through the new-formed bony layer by absorption, and pass through it into the calcified matrix (fig. [37]).   Wherever these processes come in contact with the calcified walls of the primary areolæ they absorb it, and thus cause a fusion of the original cavities and the formation of larger spaces, which are termed the *secondary areolæ* (Sharpey) or *medullary spaces* (Müller). These secondary spaces, the original cartilage-cells having disappeared, become filled with embryonic marrow, consisting of osteoblasts and vessels, and derived, in the manner described above, from the osteogenetic layer of the periosteum (fig. [38]).

Thus far there has been traced the formation of enlarged spaces (secondary areolæ), the perforated walls of which are still formed by calcified cartilage-matrix, containing an embryonic marrow, derived from the processes sent in from the osteogenetic layer of the periosteum, and consisting of blood-vessels and round cells, osteoblasts (fig. [38]). The walls of these secondary areolæ are at this time of only inconsiderable thickness, but they become thickened by the deposition of layers of new bone on their interior. This process takes place in the following manner. Some of the osteoblasts of the embryonic marrow, after undergoing rapid division, arrange themselves as an epithelioid layer on the surface of the wall of the space (fig. [39]). This layer of osteoblasts forms a bony stratum, and thus the wall of the space becomes gradually covered with a layer of true osseous substance. On this a second layer of osteoblasts arrange themselves, and in their turn form an osseous layer. By the repetition of this process the original cavity becomes very much reduced in size, and at last only remains as a small circular hole in the centre, containing the remains of the embryonic marrow—that is, a blood-vessel and a few osteoblasts. This small cavity constitutes the Haversian canal of the perfectly ossified bone. The successive layers of osseous matter which have been laid down, and which encircle this central canal, constitute the lamellæ, of which, as we have seen, each Haversian system is made up. As the successive layers of osteoblasts form osseous tissue, certain of the osteoblastic cells remain included between the various bony layers. These persist as the corpuscles of the future bone, the spaces enclosing them forming the lacunæ

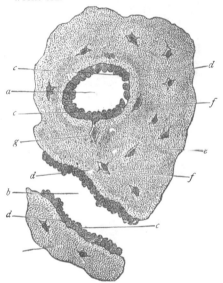

FIG. [39].—Transverse section from the femur of a human embryo about eleven weeks old.

a. A medullary sinus cut transversely ; and *b,* another longitudinally. *c.* Osteoblasts. *d.* Newly formed osseous substance of a lighter colour. *e.* That of greater age. *f.* Lacunæ with their cells. *g.* A cell still united to an osteoblast.

(figs. [39] and [41]). The canaliculi, at first extremely short, are supposed to be extended by absorption, so as to meet those of neighbouring lacunæ.

Such are the changes which may be observed at one particular point, the centre of ossification. While they have been going on a similar process has been set up in the surrounding parts and has been gradually proceeding towards the end of the shaft, so that in the ossifying bone all the changes described above may be seen in different parts, from the true bone in the centre of the shaft to the hyaline cartilage at the extremities. The bone thus formed differs from the bone of the adult in being more spongy and less regularly lamellated.

In this way the steps of a process have been described by which a solid bony mass is produced, having vessels running into it from the periosteum ; Haversian canals in which these vessels run ; medullary spaces filled with fœtal marrow ; lacunæ with their contained bone-cells ; and canaliculi growing out of these lacunæ.

This process of ossification, however, is not the origin of the whole of the skeleton, for even in those bones in which the ossification proceeds in a great measure from a single centre, situated in the cartilaginous shaft of a long bone, a considerable part of the original bone is formed by intramembranous ossification beneath the perichondrium or periosteum ; so that the girth of the bone is increased by bony deposit from the deeper layer of this membrane. The shaft of the bone is at first solid, but a tube is hollowed out in it by absorption around the vessels passing into it, which becomes the medullary canal. This absorption is supposed to be brought about by large giant-cells, the so-called osteoclasts of Kölliker (fig. [38], *f*). They vary in shape and size, and are known by containing a large number of clear nuclei, sometimes as many as twenty. The occurrence of similar cells in some tumours of bones has led to such tumours being denominated 'myeloid.'

As more and more bone is removed by this process of absorption from the interior of the

FIG. [40].—Vertical section from the edge of the ossifying portion of the diaphysis of a metatarsal bone from a fœtal calf. (After Müller.)

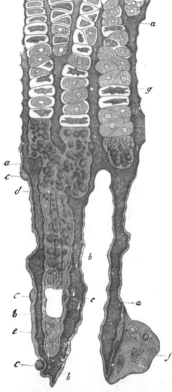

a. Ground-mass of the cartilage. *b.* Of the bone. *c.* Newly formed bone-cells in profile, more or less embedded in intercellular substance. *d.* Medullary canal in process of formation, with vessels and medullary cells. *e, f.* Bone-cells on their broad aspect. *g.* Cartilage-capsules arranged in rows, and partly with shrunken cell-bodies.

FIG. [41].—Osteoblasts from the parietal bone of a human embryo thirteen weeks old. (After Gegenbauer.)

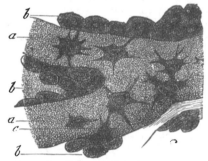

a. Bony septa with the cells of the lacunæ. *b.* Layers of osteoblasts. *c.* The latter in transition to bone-corpuscles.

bone to form the medullary canal, so more and more bone is deposited on the exterior from the periosteum, until at length it has attained the shape and size which it is destined to retain during adult life. As the ossification of the cartilaginous shaft extends towards the articular ends it carries with it, as it were, a layer of cartilage, or the cartilage grows as it ossifies, and thus the bone is increased in length. During this period of growth the articular end, or epiphysis, remains for some time entirely cartilaginous, then a bony centre appears in it, and it commences the same process of intracartilaginous ossification ; but this process never extends to any great distance. The epiphyses remain separated from the shaft by a narrow cartilaginous layer for a definite time. This layer ultimately ossifies, the distinction between shaft and epiphysis is obliterated, and the bone assumes its completed form and shape. The same remarks also apply to the processes of bone which are separately ossified, such as the trochanters of the femur. The bones, having been formed, continue to grow until the body has acquired its full stature. They increase in length by ossification continuing to extend in the epiphysial cartilage, which goes on growing in advance of the ossifying process. They increase in circumference by deposition of new bone, from the deeper layer of the periosteum, on their external surface, and at the same time an absorption takes place from within, by which the medullary cavity is increased.

The medullary spaces which characterise the cancellous tissue are produced by the absorption of the original fœtal bone in the same way as that by which the original medullary canal is formed. The distinction between the cancellous and the compact tissue appears to depend essentially upon the extent to which this process of absorption has been carried ; and we may perhaps remind the reader that in morbid states of the bone inflammatory absorption produces exactly the same change, and converts portions of bone, naturally compact, into cancellous tissue.

The number of ossific centres is different in different bones. In most of the short bones ossification commences by a single point in the centre, and proceeds towards the circumference. In the long bones there is a central point of ossification for the shaft or diaphysis : and one or more for each extremity, the epiphysis. That for the shaft is the first to appear. The union of the epiphyses with the shaft takes place in the reverse order to that in which their ossification began, with the exception of the fibula, and appears to be regulated by the direction of the nutrient artery of the bone. Thus, the nutrient arteries of the bones of the arm and forearm are directed towards the elbow, and the epiphyses of the bones forming this joint become united to the shaft before those at the opposite extremity. In the lower limb, on the other hand, the nutrient arteries pass in a direction from the knee : that is, upwards in the femur, downwards in the tibia and fibula ; and in them it is observed that the upper epiphysis of the femur, and the lower epiphysis of the tibia and fibula, become first united to the shaft.

Where there is only one epiphysis, the medullary artery is directed towards that end of the bone where there is no additional centre ; as towards the acromial end of the clavicle, towards the distal end of the metacarpal bone of the thumb and great toe, and towards the proximal end of the other metacarpal and metatarsal bones.

Besides these epiphyses for the articular ends, there are others for projecting parts or processes, which are formed separately from the bulk of the bone. For an account of these, the reader must be referred to the description of the individual bones in the sequel.

A knowledge of the exact periods when the epiphyses become joined to the shaft is often of great importance in medico-legal inquiries. It also aids the surgeon in the diagnosis of many of the injuries to which the joints are liable ; for it not infrequently happens that, on the application of severe force to a joint, the epiphysis becomes separated from the shaft, and such injuries may be mistaken for fracture or dislocation.

MUSCULAR TISSUE

The **muscles** are formed of bundles of reddish fibres, endowed with the property of contractility. The two principal kinds of muscular tissue found in the body are the voluntary and involuntary. The former of these, from the characteristic appearances which its fibres exhibit under the microscope, is known as ' striped ' muscle, and from the fact that it is capable of being put into action and controlled by the will, as ' voluntary ' muscle. The fibres of the latter do not present any cross-striped appearance, and for the most part are not under the control of the will ; hence they are known as the ' unstriped ' or ' involuntary ' muscles. The muscular fibres of the heart differ in certain particulars from both these groups, and they are therefore separately described as ' cardiac ' muscular fibres.

Thus it will be seen that there are three varieties of muscular fibres : (1) transversely striated muscular fibres, which are for the most part voluntary and under the control of the will, but some of which are not so, such as the muscles of the pharynx and upper part of the œsophagus. This variety of muscle is sometimes called *skeletal*; (2) transversely striated muscular fibres, which are not under the control of the will, i.e. the cardiac muscle; (3) plain or unstriped muscular fibres, which are involuntary and controlled by a different part of the nervous system from that which controls the activity of the voluntary muscles. Such are the muscular walls of the stomach and intestine, of the uterus and bladder, of the blood-vessels, &c.

The **striped** or **voluntary muscles** are composed of bundles of fibres enclosed in a delicate web called the 'perimysium,' in contradistinction to the sheath of areolar tissue which invests the entire muscle, the 'epimysium' (fig. [42]). The bundles are termed 'fasciculi;' they are prismatic in shape, of different sizes in different muscles, and for the most part placed parallel to one another, though they have a tendency to converge towards their tendinous attachments. Each fasciculus is made up of a bundle of *fibres*, which also run parallel with each other, and which are separated from one another by a delicate connective tissue derived from the perimysium, and termed *endomysium*. This does not form the sheath of the fibres, but serves to support the blood-vessels and nerves ramifying between them. The fibres are enclosed in a separate and distinct sheath of their own, but it is not areolar tissue and is therefore not derived from the perimysium.

FIG. [42].—Transverse section from the sterno-mastoid in man. Magnified 50 times.

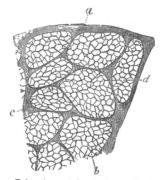

a. Epimysium. *b.* Fasciculus. *c.* Perimysium. *d.* Fibre.

A *muscular fibre* may be said to consist of a soft contractile substance enclosed in a tubular sheath, named by Bowman the *sarcolemma*. The fibres are cylindrical or prismatic in shape, and are of no great length, not exceeding, it is said, an inch and a half. They end either by blending with the tendon or aponeurosis, or else by rounded or tapering extremities which are connected to the neighbouring fibres by means of the sarcolemma. Their breadth varies in man from $\frac{1}{200}$ to $\frac{1}{500}$ of an inch. As a rule, the fibres do not divide or anastomose; but occasionally, especially in the tongue and facial muscles, the fibres may be seen to divide into several branches. The precise mode in which the muscular fibre joins the tendon has been variously described by different observers. It may, perhaps, be sufficient to say that the sarcolemma, or membranous investment of the muscular fibre, appears to become blended with a small bundle of fibres, into which the tendon becomes subdivided, while the muscular substance terminates abruptly and can be readily made to retract from the point of junction. The areolar tissue between the fibres appears to be prolonged more or less into the tendon, so as to form a kind of sheath around the tendon bundles for a longer or shorter distance. When muscular fibres are attached to the skin or mucous membranes, their fibres are described by Hyde Salter as becoming continuous with those of the areolar tissue.

The *sarcolemma*, or tubular sheath of the fibre, is a transparent, elastic, and apparently homogeneous membrane of considerable toughness, so that it will sometimes remain entire when the included substance is ruptured (see fig. [43]). On the internal surface of the sarcolemma in mammalia, and also in the substance of the fibre in the lower animals, elongated nuclei are seen, and in connection with these a row of granules, apparently fatty, is sometimes observed.

Upon examination of a voluntary muscular fibre by transmitted light, it is found to be apparently marked by alternate light and dark bands or striæ, which pass transversely, or somewhat obliquely, round the fibre (fig. [43]). The dark and light bands are of nearly equal breadth, and alternate with great regularity. They vary in breadth from about $\frac{1}{1200}$ to $\frac{1}{1700}$ of an inch. If the surface is carefully focussed, rows of granules will be detected at the point of junction of the dark and light bands, and very fine longitudinal lines may be seen running through the dark bands and joining these granules together. By treating the specimen with certain reagents (e.g. chloride of gold) fine lines may be seen running transversely between the granules uniting them together. This appearance is believed to be due to a reticulum or network of interstitial substance lying between the contractile portions of the muscle. The longitudinal striation gives the fibre the appearance of being made up of a bundle of fibrillæ, which have been termed *sarcostyles* or *muscle columns*, and if the fibre is hardened in alcohol, it can be broken up longitudinally and the sarcostyles separated from each other (fig. [44], A). The reticulum, with its longitudinal and transverse meshes, is called *sarcoplasm*.

If now a transverse section of a muscular fibre is made, it is seen to be divided into a number of areas, called the *areas of Cohnheim*, more or less polyhedral in

Fig. [43].—Two human muscular fibres. Magnified 350 times.

Fig. [44]. — Fragments of striped muscular fibres, showing a cleavage in opposite directions. Magnified 300 diameters.

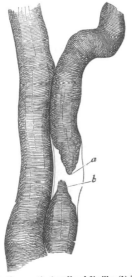

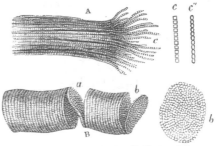

A. Longitudinal cleavage. The longitudinal and transverse lines are both seen. Some longitudinal lines are darker and wider than the rest, and are not continuous from end to end. This results from partial separation of the fibrillæ. *c.* Fibrillæ separated from one another by violence at the broken end of the fibre, and marked by transverse lines equal in width to those on the fibre. *c′ c″* represent two appearances commonly presented by the separated single fibrillæ (more highly magnified). At *c′* the borders and transverse lines are all perfectly rectilinear, and the included spaces perfectly rectangular. At *c″* the borders are scalloped and the spaces bead-like. When most distinct and definite the fibrilla presents the former of these appearances.
B. Transverse cleavage. The longitudinal lines are scarcely visible. *a.* Incomplete fracture following the opposite surfaces of a disc, which stretches across the interval, and retains the two fragments in connection. The edge and surfaces of this disc are seen to be minutely granular, the granules corresponding in size to the thickness of the disc, and to the distance between the faint longitudinal lines. *b.* Another disc nearly detached. *b′.* Detached disc, more highly magnified, showing the sarcous elements.

In the one, the bundle of fibrillæ (*b*) is torn, and the sarcolemma (*a*) is seen as an empty tube.

shape, and consisting of the transversely divided sarcostyles, surrounded by transparent series of sarcoplasm (fig. [44], B, *b*).

Upon closer examination, and by somewhat altering the focus, the appearances become more complicated, and are susceptible of various interpretations. The transverse striation, which in figs. [43] and [44] appears as a mere alternation of dark and light bands, is resolved into the appearance seen in fig. [45], which shows a series of broad dark bands, separated by light bands, which are divided into two by a dark dotted line. This line is termed *Krause's membrane* (fig. [47], K), because it was believed by Krause to be an actual membrane, continuous with the sarcolemma, and dividing the light band into two compartments. It is now more usually regarded as being due to an optical phenomenon, from the light being reflected between discs of different refrangibility. In addition to the membrane of Krause, fine clear lines may be made out, with a sufficiently high power, crossing the centre of the dark band ; these are known as the *lines of Hensen* (fig. [47], H).

Formerly it was supposed by Bowman that a muscular fibre was made up of a number of quadrangular particles, which he named sarcous elements, joined together like so many bricks forming a column, and he came to this conclusion because he found that under the influence of certain reagents the fibre could be broken up transversely into discs, as well as longitudinally into fibrillæ (fig. [44], B). But it is now believed that this cross cleavage is purely artificial, and that a muscular fibre is built up of fibrillæ and not of small quadrangular particles.

Assuming that this is so, we have now to consider a little more in detail the minute structure of these longitudinal fibrillæ, or sarcostyles, as they are termed. Perhaps there are few subjects in histology which have received more attention, and in which the appearances seen under the microscope have been more differently interpreted, than the minute anatomy of muscular fibre. Schäfer has recently worked out this subject, particularly in the wing muscles of insects, which are peculiarly adapted for this purpose on account of the large amount of interstitial sarcoplasm which separates the sarcostyles. In the following description that given by Schäfer will be closely followed (fig. [47]).

FIG. [45].—A. Portion of a medium-sized human muscular fibre. Magnified nearly 800 diameters. B. Separated bundles of fibrils, equally magnified.

FIG. [46].—Part of a striped muscular fibre of the water-beetle, prepared with absolute alcohol. Magnified 300 diameters. (Klein and Noble Smith.)

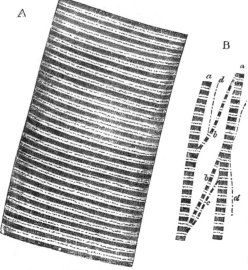

a. Sarcolemma. *b.* Membrane of Krause: owing to contraction during hardening, the sarcolemma shows regular bulgings. At the side of Krause's membrane is the transparent lateral disc.
Several nuclei of muscle-corpuscles are shown, and in them a minute network.

a, a. Larger, and *b, b,* smaller collections. *c.* Still smaller. *d, d.* The smallest which could be detached.

Each sarcostyle may be said to be made up of successive portions, each of which Schäfer terms a *sarcomere*. This is the portion situated between two membranes of Krause, which transversely divides the light band. Each sarcomere consists of a central dark part, which forms a portion of the dark band of the whole fibre, and is named by Schäfer a *sarcous element.** This sarcous element really consists of two parts, superimposed one on the top of the other, and when the fibre is stretched these two parts become separated from each other at the line of Hensen (fig. [47], A). On either side of this central dark portion is a clear layer, most visible when the fibre is extended ; this is situated between the dark centre and the membrane of Krause, and when the sarcomeres are joined together to form the sarcostyle, constitutes the light band of the striated muscular fibre.

When the sarcostyle is extended, the clear intervals are well marked and plainly to be seen ; when, on the other hand, the sarcostyle is contracted, that is to say, the muscle is in a state of contraction, these clear portions are

* This must not be confused with the 'sarcous element of Bowman' (see above).

very small or they may have disappeared altogether (fig. [47], B). When the sarcostyle is stretched to its full extent, not only is the clear portion very well marked, but the dark portion—the sarcous element—will be seen to be separated into its two constituents along the line of Hensen.

The sarcous element does not lie free in the sarcomere, for when the sarcostyle is stretched, so as to render the clear portion visible, very fine lines, which are probably septa, may be seen running through it from the sarcous element to the membrane of Krause.

Schäfer explains these phenomena in the following way. He considers that each sarcous element is made up of a number of longitudinal channels, which open into the clear part towards the membrane of Krause but are closed at the line of Hensen. When the muscular fibre is contracted the *clear* part of the muscular substance finds its way into these channels or tubes and is therefore hidden from sight, but at the same time it swells up the sarcous element and widens and shortens the sarcomere. When, on the other hand, the fibre is extended, this clear substance finds its way out of the tubes and collects between the sarcous element and the membrane of Krause, and gives the appearance of the light part between these two structures; by this means it elongates and narrows the sarcomere.

FIG. [47].—Diagram of a sarcomere. (After Schäfer.)
A. In moderately extended condition. B. In a contracted condition.

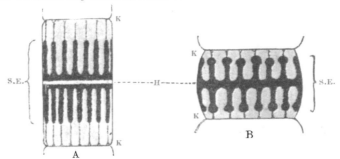

K.K. Membranes of Krause; H. Line or plane of Hensen; s.n. Poriferous sarcous element.

If this view is true it is a matter of great interest, and, as Schäfer has shown, harmonises the contraction of muscle with the amœboid action of protoplasm. In an amœboid cell there is a framework of spongioplasm, which stains with hæmatoxylin and similar reagents, enclosing in its meshes a clear substance, hyaloplasm, which will not stain with these reagents. Under stimulation the hyaloplasm passes into the pores of the spongioplasm; without stimulation it tends to pass out as in the formation of pseudopodia. In muscle there is the same thing, viz. a framework of spongioplasm staining with hæmatoxylin—the substance of the sarcous element—and this encloses a clear hyaloplasm, the clear substance of the sarcomere, which resists staining with this reagent. During contraction of the muscle—i.e. stimulation—this clear substance passes into the pores of the spongioplasm; while during extension of the muscle—i.e. when there is no stimulation—it tends to pass out of the spongioplasm.

In this way the contraction is brought about: under stimulation the protoplasmic material (the clear substance of the sarcomere) recedes into the sarcous element, causing the sarcomere to widen out and shorten. The contraction of the muscle is merely the sum total of this widening out and shortening of these bodies.

The *capillaries* of striped muscle are very abundant, and form a sort of rectangular network, the branches of which run longitudinally in the endomysium between the muscular fibres, and are joined at short intervals by transverse anastomosing branches. The larger vascular channels, arteries and veins, are found only in the perimysium, between the muscular fasciculi.

Nerves are profusely distributed to striped muscle. Their mode of termination will be described on a subsequent page.

The existence of *lymphatic* vessels in striped muscle has not been ascertained, though they have been found in tendons and in the sheath of the muscle.

The **unstriped, plain, or involuntary muscle** is found in the walls of the hollow viscera—viz. the lower half of the œsophagus and the whole of the remainder of the gastro-intestinal tube; in the trachea and bronchi, and the alveoli and infundibula of the lungs; in the gall-bladder and ductus communis choledochus; in the large ducts of the salivary and pancreatic glands; in the pelvis and calyces of the kidney, the ureter, bladder, and urethra; in the female sexual organs—viz. the ovary, the Fallopian tubes, the uterus (enormously developed in pregnancy), the vagina, the broad ligaments, and the erectile tissue of the clitoris; in the male sexual organs—viz. the dartos of the scrotum, the vas deferens and epididymis, the vesiculæ seminales, the prostate gland, and the corpora cavernosa and corpus spongiosum; in the ducts of certain glands, as in Wharton's duct; in the capsule and trabeculæ of the spleen; in the mucous

FIG. [48].—Non-striated muscular fibre. (From Kirke's ' Physiology.')

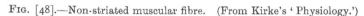
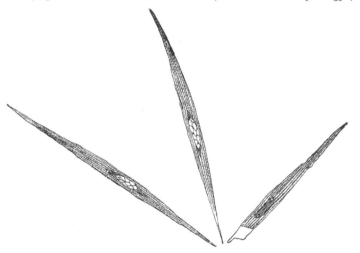

membranes, forming the muscularis mucosæ; in the skin, forming the arrectores pilorum, and also in the sweat-glands; in the arteries, veins, and lymphatics; in the iris and the ciliary muscle.

Plain or unstriped muscle is made up of spindle-shaped cells, called *contractile fibre-cells*, collected into bundles and held together by a cement-substance (fig. [48]). These bundles are further aggregated into larger bundles, or flattened bands, and bound together by ordinary connective tissue.

The *contractile fibre-cells* are elongated, spindle-shaped, nucleated cells of various lengths, averaging from $\frac{1}{600}$ to $\frac{1}{300}$ of an inch in length, and $\frac{1}{4500}$ to $\frac{1}{3500}$ of an inch in breadth. On transverse section they are more or less polyhedral in shape, from mutual pressure. They present a faintly longitudinal striated appearance, and consist of an elastic cell-wall containing a central bundle of fibrillæ, representing the contractile substance, and an oval or rod-like nucleus, which includes, within a membrane, a fine network communicating at the poles of the nucleus with the contractile fibres (Klein). The adhesive interstitial cement substance, which connects the fibre-cells together, represents the endomysium, or delicate connective tissue which binds the fibres of striped muscular tissue into fasciculi; while the tissue connecting the individual bundles together represents the perimysium. The unstriped muscle, as a rule, is not under the control of the will, nor is the contraction rapid and involving the whole muscle, as is the case with the voluntary muscles. The membranes which are composed of the unstriped muscle slowly contract in a part of their extent, generally under the influence of a mechanical stimulus, as that of distension or of cold ; and then the contracted part slowly relaxes while another portion of the membrane takes up the contraction. This peculiarity of action is most strongly marked in the intestines, constituting their *vermicular motion.*

Cardiac Muscular Tissue.—The fibres of the heart differ very remarkably from those of other striped muscles. They are smaller by one-third, and their transverse striæ are by no means so distinct. The fibres are made up of distinct quadrangular cells joined end to end (fig. [49]). Each cell contains a clear oval nucleus, situated near the centre of the cell. The extremities of the cells have a tendency to branch or divide, the subdivisions uniting with offsets from other cells, and thus producing an anastomosis of the fibres. The connective tissue between the bundles of fibres is much less than in ordinary striped muscle, and no sarcolemma has been proved to exist.

Fig. [49].—Anastomosing muscular fibres of the heart seen in a longitudinal section. On the right the limits of the separate cells with their nuclei are exhibited somewhat diagrammatically.

Development of Muscle Fibres. — Voluntary muscular fibres are developed from the meso-blast, the embryonic cells of which elongate, show multiplication of nuclei, and eventually become striated ; the striation is first obvious at the side of the fibre, spreads around the circumference, and ultimately extends to the centre. The nuclei, at first situated centrally, gradually pass out to assume their final position immediately beneath the sarcolemma. In the case of plain muscle the mesoblastic cells assume a pointed shape at the extremities and · become flattened, the nucleus also lengthening out to its permanent rod-like form.

Chemical Composition of Muscle.—In chemical composition the muscular fibres may be said, in round numbers, to consist of 75 per cent. of water, about 20 per cent. of proteids, 2 per cent. of fat, 1 per cent. of nitrogenous extractives and carbohydrates, and 2 per cent. of salts, which are mainly potassium phosphate and carbonate.

NERVOUS TISSUE

The **nervous tissues** of the body are comprised in two great systems—the *cerebro-spinal* and the *sympathetic* ; and each of these systems consists of a *central organ*, or series of central organs, and of *nerves*.

The *cerebro-spinal* system comprises the brain (including the medulla oblongata), the spinal cord, the cranial nerves, the spinal nerves, and the ganglia connected with both these classes of nerves. The *sympathetic* system consists of a double chain of ganglia, with the nerves which go to and come from them. It is not directly connected with the brain or spinal cord, though it is so indirectly by means of its numerous communications with the cranial and spinal nerves.

All these nervous tissues are composed chiefly of two different structures—the *grey* or *cineritious*, and the *white* or *fibrous*. It is in the former, as is generally supposed, that nervous impressions and impulses originate, and by the latter that they are conducted. Hence the grey matter forms the essential constituent of all the ganglionic centres, both those in the isolated ganglia and those aggregated in the cerebro-spinal axis; while the white matter is found in all the commissural portions of the nerve-centres and in all the cerebro-spinal nerves. The nerves of the sympathetic system are chiefly composed of a material of a somewhat different structure, which is named *grey* or *gelatinous* nerve-fibre. This form of nerve-fibre is also found in some of the cerebro-spinal nerves.

The **grey nervous substance** is distinguished by its dark reddish-grey colour and soft consistence. It is found in the brain, spinal cord, and various ganglia intermingled with the fibrous nervous substance, and also in some of the nerves of special sense, and in gangliform enlargements which are found here and there in the course of certain cerebro-spinal nerves. It is composed of cells, commonly called *nerve-cells* or *ganglion-corpuscles*, containing nuclei and nucleoli. The cells, together with the blood-vessels in the grey nerve-substance, and the nerve-fibres and vessels in the white nerve-substance, are embedded in a peculiar ground substance, named by Virchow *neuroglia*. It consists of fibres and cells. Some of the cells are stellate in shape, and their fine processes become neuroglia fibres, which extend radially and unbranched (fig. [50], B) among the nerve-cells and fibres which they

Fig. [50].—Neuroglia cells of brain shown by Golgi's method. (After Andriezen.)
(Copied from Schäfer's ' Essentials of Histology.')

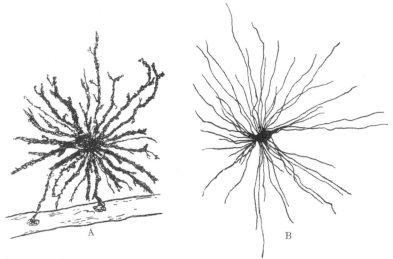

A. Cell with branched processes.   B. Spider cell with unbranched processes.

aid in supporting. Other cells give off fibres which branch repeatedly (fig. [50], A). In addition to these fibres there are others which do not appear to be connected with the neuroglia cells. They start from the epithelial cells lining the ventricles of the brain and central canal of the spinal cord, and pass through the nervous tissue, branching repeatedly to terminate in slight enlargements on the pia mater. Thus, neuroglia is evidently a connective tissue in function but is not so in development; it is epiblastic in origin, whereas all connective tissues are mesoblastic.

Each nerve-cell consists of a finely fibrillated protoplasmic material, of a reddish or yellowish-brown colour, which occasionally presents patches of a deeper tint, caused by the aggregation of pigment-granules at one side of the nucleus, as in the substantia nigra and locus cœruleus. The protoplasm also sometimes contains peculiar angular granules, which stain deeply with basic dyes, such as methylene blue; these are known as *Nissl's granules* (fig. [54]). The nucleus is, as a rule, a large, well-defined, round, vesicular body, often presenting an intra-nuclear network, and containing a nucleolus which is peculiarly clear and brilliant. The nerve-cells vary in shape and size, and have one or more processes. They may be divided for purposes of description into three groups, according to the number of processes which they possess : (1) Unipolar cells, which are found in the spinal ganglia; their single process, after a short course, divides in a T-shaped manner. (2) Bipolar cells, also found in the spinal ganglia (fig. [53]), when the cells are in an embryonic condition. They are best demonstrated in the sympathetic ganglion-cells of a frog. Sometimes the processes come off from opposite poles of the cell, and the cell then assumes a

FIG. [51].—Nerve-cells from the Gasserian ganglion of the human subject.

FIG. [52].—Nerve-cells from the inner part of the grey matter of the convolutions of the human brain. Magnified 350 times.

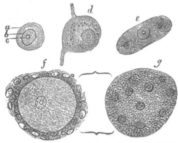

a. A globular one with defined border. b. Its nucleus. c. Its nucleolus. d. Caudate cell. e. Elongated cell with two groups of pigment-particles. f. Cell surrounded by its sheath or capsule of nucleated particles. g. The same, the sheath only being in focus. Magnified 300 diameters.

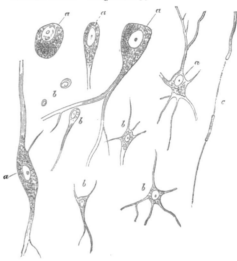

FIG. [53].—Bipolar nerve-cell from the spinal ganglion of the pike. (After Kölliker.)

Nerve-cells : a. Larger. b. Smaller. c. Nerve-fibre with axis-cylinder.

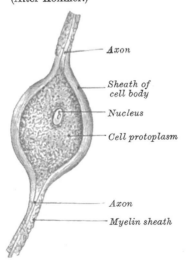

Axon

Sheath of cell body

Nucleus

Cell protoplasm

Axon

Myelin sheath

FIG. [54].—Motor nerve-cell from ventral horn of spinal cord of rabbit. (After Nissl.) The angular and spindle-shaped Nissl bodies are well shown.

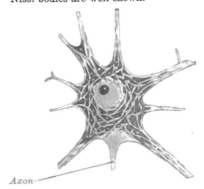

Axon

spindle shape ; at others they both emerge at the same point. In some cases where two fibres are apparently connected with a cell, one of the fibres is really derived from an adjoining nerve-cell and is passing to end in a ramification around the ganglion-cell, or, again, it may be coiled spirally round the nerve process which is issuing from the cell. (3) Multipolar cells, which are caudate or stellate in shape, and characterised by their large size and by the tail-like processes which issue from them. The processes are of two kinds : one of them is termed the *axis-cylinder process* or *axon* because it becomes the axis cylinder of a nerve-fibre (figs. [54], [55], [56]). The others are termed the *protoplasmic processes* or *dendrons*; they begin to divide and subdivide as soon as they emerge from the cell, and finally end in minute twigs and become lost among the other elements of the nervous tissue.

The **white** or **fibrous** nerve-substance or *nerve-fibre* is found universally in the nervous cords, and also constitutes a great part of the brain and spinal cord. The fibres of which it consists are of two kinds, the *medullated* or *white* fibres, and the *non-medullated* or *grey* fibres.

Fig. [55].—Pyramidal cell from the cerebral cortex of a mouse. (After Ramón y Cajal.)

The **medullated** fibres form the white part of the brain and spinal cord, and also the greater part of the cerebro-spinal nerves, and give to these structures their

Fig. [56].—Cell of Purkinje from the cerebellum of a cat. (After Ramón y Cajal.)

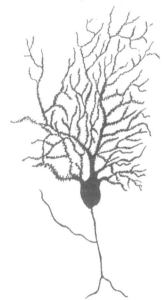

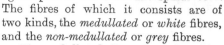

Axon

opaque, white aspect. When perfectly fresh they appear to be homogeneous; but soon after removal from the body they present, when examined by transmitted light, a double outline or contour, as if consisting of two parts (fig. [57]).

The central portion is named the *axis cylinder of Purkinje*; around this is a sort of sheath of fatty material, staining black with osmic acid, named the *white substance of Schwann*, which gives to the fibre its double contour, and the whole is enclosed in a delicate membrane, the *neurilemma, primitive sheath,* or *nucleated sheath of Schwann* (fig. [57]).

The *axis cylinder* is the essential part of the nerve-fibre, and is always present; the other parts, the medullary sheath and the neurilemma, being occasionally absent, especially at the origin and termination of the nerve-fibre. It undergoes no interruption from its origin in the nerve-centre to its peripheral termination, and must be regarded as a direct prolongation of a nerve-cell. It constitutes about one-half or one-third of the nerve-tube, the white substance being greater in proportion in the nerves than in the central organs. It is perfectly transparent, and is therefore indistinguishable in a perfectly fresh and natural state of the nerve. It is made up of exceedingly fine fibrils, which stain darkly with gold chloride (fig. [58]). At its termination the axis cylinder of a nerve-fibre may be seen to break up into fibrillæ, confirming the view of its structure. These fibrillæ have been termed the *primitive fibrillæ* of Schultze. The axis cylinder is said by some to be enveloped in a special, reticular sheath, which separates it

FIG. [57].—White or medullated nerve-fibres showing the sinuous outline and double contours. (After Schäfer.)

FIG. [58].—Longitudinal section through a nerve-fibre from the sciatic nerve of a frog. × 830. (After Böhm and Davidoff.)

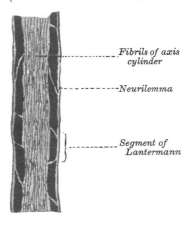

Fibrils of axis cylinder

Neurilemma

Segment of Lantermann

from the white matter of Schwann, and is composed of a substance called *neurokeratin*. The more common opinion is that this network or reticulum is contained in the white matter of Schwann, and by some it is believed to be produced by the action of the reagents employed to show it. The *medullary sheath* or *white matter of Schwann* (fig. [58]) is regarded as being a fatty matter in a fluid state, which insulates and protects the essential part of the nerve—the axis cylinder. The white matter varies in thickness to a very considerable extent, in some forming a layer of extreme thinness, so as to be scarcely distinguishable, in others forming about one-half the nerve-tube. The size of the nerve-fibres, which varies from $\frac{1}{1200}$ to $\frac{1}{3000}$ of an inch, depends mainly upon the amount of the white substance, though the axis cylinder also varies in size within certain limits. The

FIG. [59].—A node of Ranvier of a medullated nerve-fibre, viewed from above, magnified about 750 diameters. The medullary sheath is discontinuous at the node, whereas the axis cylinder passes from one segment into the other. At the node the sheath of Schwann appears thickened. (Klein and Noble Smith.)

white matter of Schwann does not always form a continuous sheath to the axis cylinder, but undergoes interruptions in its continuity at regular intervals, giving to the fibre the appearance of constriction at these points. These were first described by Ranvier, and are known as the *nodes of Ranvier* (fig. [59]). The portion of nerve-fibre between two nodes is called an *internodal segment*. The neurilemma or primitive sheath is not interrupted at the nodes, but passes over them as a continuous membrane. In addition to these interruptions oblique clefts may be seen in the medullary sheath, subdividing it into irregular portions, which are termed *medullary segments*, or *segments of Lantermann* (fig. [58]). There is reason to believe that these clefts are artificially produced in the preparation of the specimens. Medullated nerve-fibres, when examined, frequently present a beaded or varicose appearance : this is due to manipulation and pressure causing the oily matter to collect into drops, and in consequence of the extreme delicacy of the primitive sheath, even slight pressure will cause the transudation of the fatty matter, which collects as drops of oil outside the membrane. This is, of course, promoted by the action of certain reagents.

The *neurilemma* or *primitive sheath* (sometimes called the *tubular membrane* or *sheath of Schwann*) presents the appearance of a delicate, structureless membrane. Here and there beneath it, and situated in depressions in the white matter of Schwann, are nuclei surrounded by a small amount of protoplasm. The nuclei are oval and somewhat flattened, and bear a definite relation to the nodes of Ranvier ; one nucleus generally lying in the centre of each internode. The primitive sheath is not present in all medullated nerve-fibres, being absent in those fibres which are found in the brain and spinal cord.

Non-medullated Fibres.—Most of the nerves of the sympathetic system, and some of the cerebro-spinal, consist of another variety of nervous fibres, which are called the *grey* or *gelatinous* nerve-fibres—*fibres of Remak* (fig. [60]). These consist of a central core or axis cylinder enclosed in a nucleated sheath, which tends to split into fibrillæ, and is probably of the nature of neurokeratin. In external appearance the gelatinous nerves are semi-transparent and grey or yellowish-grey. The individual fibres vary in size, generally averaging about half the size of the medullated fibres.

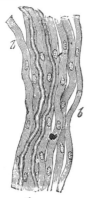

FIG. [60].—A small nervous branch from the sympathetic of a mammal.

a. Two medullated nerve-fibres among a number of grey nerve-fibres, *b.*

Development of Nerve-cells and Fibres.—The nerve-cells are developed from certain of the cells which line the neural canal or form the neural crest of the embryo (see section on development). Some of these cells assume a rounded form and are termed *neuroblasts*, and from each neuroblast there grows out a process, the axis-cylinder process or axon, and subsequently the branching processes or dendrons. The axis cylinders, at first naked, acquire their medullary sheath, possibly by some metamorphosis of their outer layer. The neurilemma is thought to be derived from mesoblastic cells which become flattened and wrapped round the fibre, the cement-substance at their apposed ends forming the material which stains with silver nitrate at the nodes of Ranvier. Nerve-cells in the sympathetic and peripheral ganglia take their origin from small collections of neuroblasts, which are split off from the rudimentary spinal ganglia. Cells which are, originally, similar to neuroblasts seem to give rise to neuroglia cells, numerous processes sprouting from the cell to form the neuroglial fibres.

Chemical Composition.—The amount of water in nervous tissue varies with the situation. Thus in the grey matter of the cerebrum it constitutes about 83 per cent., in the white matter from the same region about 70 per cent., while in the peripheral nerves, such as the sciatic, it may fall to 60 per cent. The solids consist of proteids (in the grey matter they form half the total solids), neurokeratin, nuclein, protagon, lecithin, cerebrosides, cholesterin, nitrogenous extractives, and salts with some gelatin and fat from the adherent connective tissue.

The nervous structures are divided, as before mentioned, into two great systems—viz. the *cerebro-spinal*, comprising the brain and spinal cord, the nerves connected with these structures, and the ganglia situated on them ; and the *sympathetic*, consisting of a double chain of ganglia and the nerves connected with them. All these structures require separate consideration; they are composed of the two kinds of nervous tissue above described, intermingled in various proportions, and having, in some parts, a very intricate arrangement.

The **brain** or **encephalon** is that part of the cerebro-spinal system which is contained in the cavity of the skull. It is divided into several parts, which will be described in the sequel. In these parts the grey or vesicular nervous matter is found partly on the surface of the brain, forming the convolutions of the cerebrum, and the laminæ of the cerebellum. Again, grey matter is found in the interior of the brain, collected into large and distinct masses or ganglionic bodies, such as the corpus striatum, optic thalamus, and corpora quadrigemina. Finally, grey matter is found intermingled intimately with the white, but without definite arrangement, as in the grey matter in the pons Varolii and the floor of the fourth ventricle.

FIG. [61].—Transverse section through a microscopic nerve, representing a compound nerve-bundle, surrounded by perineurium. Magnified 120 diameters.

The medullated fibres are seen as circles with a central dot, viz. medullary sheath and axis cylinder, in transverse section. They are embedded in endoneurium, containing numerous nuclei, which belong to the connective-tissue cells of the latter. (Klein and Noble Smith.)

The white matter of the brain is divisible into three distinct classes of fibres: (1) Diverging or peduncular fibres, which connect the hemispheres with the medulla oblongata and the spinal cord. (2) Commissural fibres, which connect together the two hemispheres. (3) Association fibres, which connect different parts of the same hemisphere.

The manner in which these fibres are intermingled with each other and with the grey matter in the brain and spinal cord is very intricate, and can only be fully understood by a careful study of the details of its descriptive anatomy in the sequel. The further consideration of this subject will therefore be deferred until after the description of the various divisions of which the cerebro-spinal system is made up.

The **nerves** are round or flattened cords, formed of the nerve-fibres already described. They are connected at one end with the

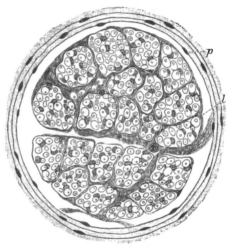

_p._ Perineurium, consisting of laminæ of fibrous connective tissue, alternating with flattened nucleated connective-tissue cells. _l._ Lymph-space between perineurium and surface of nerve-bundle.

cerebro-spinal centre or with the ganglia, and are distributed at the other end to the various textures of the body; they are subdivided into two great classes—the _cerebro-spinal_, which proceed from the cerebro-spinal axis, and the _sympathetic_ or _ganglionic_ nerves, which proceed from the ganglia of the sympathetic.

The **cerebro-spinal nerves** consist of numerous nerve-fibres collected together and enclosed in a membranous sheath (fig. [61]). A small bundle of primitive fibres, enclosed in a tubular sheath, is called a _funiculus_; if the nerve is of small size, it may consist only of a single funiculus; but if large, the funiculi are collected together into larger bundles or fasciculi, which are bound together in a common membranous investment, and constitute the nerve.

In structure, the common membranous investment, or sheath of the whole nerve, which is called the _epineurium_, as well as the septa given off from it, and which separate the fasciculi, consists of connective tissue, composed of white and yellow elastic fibres, the latter existing in great abundance. The tubular sheath of the funiculi, called the _perineurium_, consists of a fine, smooth, transparent membrane, which may be easily separated, in the form of a tube, from the fibres it encloses ; in structure it consists of connective tissue, which has a distinctly lamellar arrangement, consisting of several lamellæ, separated from each other by spaces containing lymph. The nerve-fibres are held together and supported within the funiculus by delicate connective tissue, called the *endoneurium*. It is continuous with septa which pass inwards from the innermost layer of the perineurium, and consists of a ground-substance in which are embedded fine bundles of fibrous connective tissue which run for the most part longitudinally. It serves to support capillary vessels, which are arranged so as to form a net-work with elongated meshes.   The cerebro-spinal nerves consist almost exclusively of the medullated nerve-fibres, the non-medullated existing in very small proportions.

The blood-vessels supplying a nerve terminate in a minute capillary plexus, the vessels composing which pierce the perineurium, and run, for the most part, parallel with the fibres ; they are connected together by short, transverse vessels, forming narrow, oblong meshes, similar to the capillary system of muscle.   Fine non-medullated nerve-fibres accompany these capillary vessels, *vaso-motor fibres*, and break up into elementary fibrils, which form a network around the vessel. Horsley has also demonstrated certain medullated fibres as running in the epineurium and terminating in small spheroidal tactile corpuscles or end-bulbs of Krause.   These nerve-fibres, which Marshall believes to be sensory, and which he has termed *nervi nervorum*, are considered by him to have an important bearing upon certain neuralgic pains.

The nerve-fibres, as far as is at present known, do not coalesce, but pursue an uninterrupted course from the centre to the periphery.   In separating a nerve, however, into its component funiculi, it may be seen that these do not pursue a perfectly insulated course, but occasionally join at a very acute angle with other funiculi proceeding in the same direction ; from this, branches are given off, to join again in like manner with other funiculi.   It must be distinctly understood, however, that in these communications the nerve-fibres do not coalesce, but merely pass into the sheath of the adjacent nerve, become inter-mixed with its nerve-fibres, and again pass on, to become blended with the nerve-fibres in some adjoining funiculus.

Nerves, in their course, subdivide into branches, and these frequently communicate with branches of a neighbouring nerve.   The communications which thus take place form what is called a *plexus*.   Sometimes a plexus is formed by the primary branches of the trunks of the nerves—as the cervical, brachial, lumbar, and sacral plexuses—and occasionally by the terminal funiculi, as in the plexuses formed at the periphery of the body.   In the formation of a plexus, the component nerves divide, then join, and again subdivide in such a complex manner that the individual funiculi become interlaced most intricately ; so that each branch leaving a plexus may contain filaments from each of the primary nervous trunks which form it.   In the formation also of smaller plexuses at the periphery of the body there is a free interchange of the funiculi and primitive fibres.   In each case, however, the individual filaments remain separate and distinct, and do not inosculate with one another.

It is probable that through this interchange of fibres the different branches passing off from a plexus have a more extensive connection with the spinal cord than if they each had proceeded to be distributed without such connection with other nerves.   Consequently the parts supplied by these nerves have more extended relations with the nervous centres ; by this means, also, groups of muscles may be associated for combined action.

The **sympathetic** nerves are constructed in the same manner as the cerebro-spinal nerves, but consist mainly of non-medullated fibres, collected into funiculi, and enclosed in a sheath of connective tissue.   There is, however, in these nerves, a certain admixture of medullated fibres, and the amount varies in different nerves, and may be known by their colour.   Those branches of the sympathetic which present a well-marked grey colour are composed more especially of gelatinous nerve-fibres, intermixed with a few medullated fibres; while those of a white colour contain more of the latter fibres, and a few of the former. Occasionally, the grey and white cords run together in a single nerve, without any intermixture, as in the branches of communication between the sympathetic ganglia and the spinal nerves, or in the communicating cords between the ganglia.

The nerve-fibres, both of the cerebro-spinal and sympathetic system, convey impressions of a twofold kind. The *sensory* nerves, called also *centripetal* or *afferent* nerves, transmit to the nervous centres impressions made upon the peripheral extremities of the nerves, and in this way the mind, through the medium of the brain, becomes conscious of external objects. The *motor* nerves, called also *centrifugal* or *efferent* nerves, transmit impressions from the nervous centres to the parts to which the nerves are distributed, these impressions either exciting muscular contraction, or influencing the processes of nutrition, growth, and secretion.

Origin and Termination of Nerves.—By the expression 'the termination of nerve-fibres' is signified their connections with the nerve-centres, and with the parts they supply. The former are sometimes called their *origin*, or *central* termination; the latter their *peripheral* termination. The origin in some cases is single—that is to say, the whole nerve emerges from the nervous centre by a single root; in other instances the nerve arises by two or more roots, which come off from different parts of the nerve-centre, sometimes widely apart from each other, and it often happens, when a nerve arises in this way by two roots, that the functions of these two roots are different; as, for example, in the spinal nerves, each of which arises by two roots, the anterior of which is motor, and the posterior sensory. The point where the nerve root or roots emerge from the nervous centre is named the *superficial* or *apparent* origin, but the fibres of which the nerve consists can be traced for a certain distance into the nervous centre to some portion of the grey substance, which constitutes the *deep* or *real* origin of the nerve. The manner in which these fibres arise at their deep origin varies with their functions. The centrifugal or efferent nerve-fibres originate in the nerve-cells of the grey substance, the axis-cylinder processes of these cells being prolonged to form the fibres. In the case of the centripetal or afferent nerves the fibres grow inwards either from nerve-cells in the organs of special sense (e.g. the retina) or from nerve-cells in the ganglia. Having entered the nerve-centre they branch and send their ultimate twigs among the cells, without, however, uniting with them.

Peripheral Terminations of Nerves.—Nerve-fibres terminate peripherally in various ways, and these may be conveniently studied in the sensory and motor nerves respectively. **Sensory** nerves would appear to terminate either in *minute primitive fibrillæ* or networks of these; or else in special terminal organs, which have been termed *peripheral end-organs*, and of which there are several principal varieties, viz. the end-bulbs of Krause, the tactile corpuscles of Wagner, the Pacinian corpuscles, and the neuro-tendinous and neuro-muscular spindles.

Termination in Fibrillæ.—When a medullated nerve-fibre approaches its termination, the white matter of Schwann suddenly disappears, leaving only the axis cylinder, surrounded by the neurilemma, and forming a non-medullated fibre. This, after a time, loses its neurilemma, and consists only of an axis cylinder, which can be seen, in preparations stained with chloride of gold, to be made up of fine varicose fibrils. Finally, the axis cylinder breaks up into its constituent primitive nerve-fibrillæ, which often present regular varicosities and anastomose with one another, thus forming a network. This network passes between the elements of the tissue to which the nerves are distributed, which is always epithelial, the nerve-fibrils lying in the interstitial substance between the epithelial cells, and there terminating, though some observers maintain that the actual terminations are within the cells. In this way nerve-fibres have been found to terminate in the epithelium of the skin and mucous membranes, and in the anterior epithelium of the cornea.

The **end-bulbs of Krause** (fig. [62]) are minute cylindrical or oval bodies, consisting of a capsule formed by the expansion of the connective-tissue sheath of a medullated fibre, and containing a soft semi-fluid core in which the termination of the axis cylinder is situated, ending either as a bulbous extremity, or in a coiled-up plexiform mass. End bulbs are found in the conjunctiva of the eye, where they are spheroidal in shape in man but cylindrical in most other animals, in the mucous membrane of the lips and tongue, and in the epineurium of nerve-trunks. They are also found in the genital organs of both sexes, the penis in the male and the clitoris in the female. In this situation they have a mulberry-like appearance, from being constricted by connective-tissue septa into from two to six knob-like masses, and have received the name of *genital corpuscles*. Very similar corpuscles are found in the epineurium of nerve-trunks. In the synovial membrane of certain joints (e.g. those of the fingers), rounded or oval end-bulbs have been found ; these are designated *articular end-bulbs*.

Tactile corpuscles have been described by Grandry as occurring in the papillæ of the beak and tongue of birds, and by Merkel as occurring in the papillæ and

FIG. [62].—End-bulb of Krause.

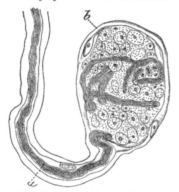

a. Medullated nerve-fibre.   *b.* Capsule of corpuscle.   (From Klein's 'Elements of Histology.')

FIG. [63].—Papilla of the hand, treated with acetic acid.   Magnified 350 times.

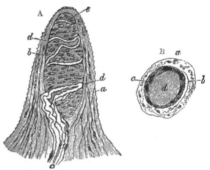

A. Side view of a papilla of the hand. *a.* Cortical layer. *b.* Tactile corpuscle, with transverse nuclei. *c.* Small nerve of the papilla, with neurilemma. *d.* Its two nervous fibres running with spiral coils around the tactile corpuscle. *e.* Apparent termination of one of these fibres. B. A tactile papilla seen from above so as to show its transverse section. *a.* Cortical layer. *b.* Nerve-fibre. *c.* Outer layer of the tactile body, with nuclei. *d.* Clear interior substance.

epithelium of the skin of man and animals, especially in those parts of the skin devoid of hair. They consist of a capsule composed of a very delicate, nucleated membrane, and contain two or more granular, somewhat flattened cells, between which the medullated nerve-fibre, which enters the capsule by piercing its investing membrane, is supposed to terminate.

The **tactile corpuscles** (fig. [63]), described by Wagner and Meissner, are oval-shaped bodies, made up of connective tissue, and consisting of a capsule, and imperfect membranous septa, derived from it, which penetrate its interior. The axis cylinder of the medullated fibres passes through the capsule, and having entered the corpuscle terminates in a small globular or pyriform enlargement, near the inner surface of the capsule. These tactile corpuscles have been described as occurring in the papillæ of the corium of the hand and foot, the front of the forearm, the skin of the lips, and the mucous membrane of the tip of the tongue, the palpebral conjunctiva, and the skin of the nipple. They are not found in all the papillæ ; but from their existence in those parts in which the skin is highly sensitive, it is probable that they are specially concerned in the sense of touch, though their absence from the papillæ of other tactile parts shows that they are not essential to this sense.

Ruffini has described a special variety of nerve-ending in the subcutaneous tissue of the human finger (fig. [64]). These are usually known as *Ruffini's endings*. They are principally situated at the junction of the corium with the subcutaneous

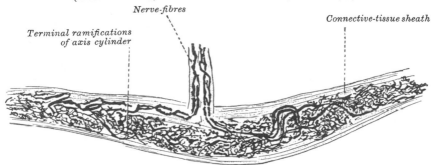

FIG. [64].—Nerve-ending of Ruffini.
(After A. Ruffini, 'Arch. ital. de Biol. Turin,' t. xxi. 1894.)

Nerve-fibres

Terminal ramifications of axis cylinder

Connective-tissue sheath

tissue; they are oval in shape, and consist of a strong connective-tissue sheath, inside which the nerve-fibre divides into numerous branches, which show varicosities and end in small free knobs. They resemble the corpuscles of Golgi.

The **Pacinian corpuscles** * (fig. [65]) are found in the human subject chiefly on the nerves of the palm of the hand and sole of the foot and in the genital organs of both sexes lying in the subcutaneous tissue; but they have also been described as connected with the nerves of the joints, and in some other situations, as the mesentery of the cat and along the tibia of the rabbit. Each of these corpuscles is attached to and encloses the termination of a single nerve-fibre. The corpuscle, which is perfectly visible to the naked eye (and which can be most easily demonstrated in the mesentery of a cat), consists of a number of lamellæ or capsules, arranged more or less concentrically around a central clear space, in which the nerve-fibre is contained. Each lamella is composed of bundles of fine connective-tissue fibres, and is lined on its inner surface by a single layer of flattened epithelioid cells. The central clear space, which is elongated or cylindrical in shape, is filled with a transparent material, in the middle of which is the single medullated fibre, which traverses the space to near its distal extremity. Here it terminates in a rounded knob or end, sometimes bifurcating previously, in which case each branch has a similar arrangement. Todd and Bowman have described minute arteries as entering by the sides of the nerves and forming capillary loops in the intercapsular spaces, and even penetrating into the central space. Other authors describe the artery as entering the corpuscle at the pole opposite to the nerve-fibre.

FIG. [65].—Pacinian corpuscle, with its system of capsules and central cavity.

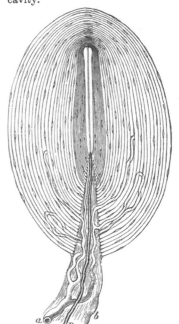

a. Arterial twig, ending in capillaries, which form loops in some of the intercapsular spaces, and one penetrates to the central capsule. *b.* The fibrous tissue of the stalk prolonged from the neurilemma. *n.* Nerve-tube advancing to the central capsule, there losing its white matter, and stretching along the axis to the opposite end, where it is fixed by a tubercular enlargement.

* Often called in German anatomical works 'corpuscles of Vater.'

Herbst has described a somewhat similar 'nerve-ending' to the Pacinian corpuscle, as being found in the mucous membrane of the tongue of the duck, and in some other situations. It differs, however, from the Pacinian corpuscles, in being smaller, its capsule thinner and more closely approximated, and especially in the fact that the axis-cylinder in the central clear space is coated with a continuous row of nuclei. These bodies are known as the *corpuscles of Herbst*.

Neuro-tendinous spindles.—The nerves supplying tendons have a special modification of the terminal fibres, especially numerous at the point where the tendon is becoming muscular. The tendon bundles become enlarged, and the nerve-fibres—one, two, or more in number—penetrate between the fasciculi of the tendon and spread out between the fibres to end in irregular discs or varicosities. A spindle-shaped body is thus formed, composed of tendon bundles and nerve-fibres, which is known as the *organ of Golgi* (fig. [66]).

FIG. [66].—Organ of Golgi (neuro-tendinous spindle) from the human tendo Achillis. (After Ciaccio.)

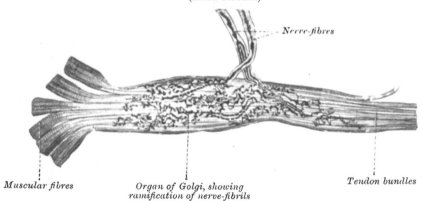

Nerve-fibres

Muscular fibres   *Organ of Golgi, showing ramification of nerve-fibrils*   *Tendon bundles*

Neuro-muscular spindles.—In the majority of voluntary muscles there have been found special end-organs consisting of a small bundle of peculiar muscular fibres (intrafusal fibres), embryonic in type, invested by a capsule within which nerve-fibres, experimentally shown to be sensory in origin, terminate. These neuro-muscular spindles vary in length from $\frac{1}{30}$ to $\frac{1}{5}$ of an inch and have a distinctly fusiform appearance. The large medullated nerve-fibres passing to the end-organ are from one to three or four in number; entering the fibrous capsule they divide several times, and, losing their medulla, ultimately end in naked axis cylinders encircling the intrafusal fibres by flattened expansions, or irregular ovoid or rounded discs (fig. [67]). Neuro-muscular spindles have not yet been demonstrated in the tongue or eye muscles.

In the organs of **special sense** the nerves appear to terminate in cells, which belong to the epithelial class, and have received the name of *sensory* or *nerve-epithelium* cells. This is not, however, the real state of the case; the nerve-fibre is in reality a process from the epithelial cell, and terminates by branching around a ganglion-cell. The stimulus carried by it is continued onwards by an axis cylinder, derived from the ganglion, to the brain. These nerve-epithelium cells must therefore be regarded as modified forms of nerve-cells. They will be more particularly described in the sequel, in connection with the description of the organs of special sense.

Motor nerves are to be traced either into unstriped or striped muscular fibres. In the **unstriped** or **involuntary** muscles the nerves are derived from the sympathetic, and are composed mainly of the non-medullated fibres. Near their termination they divide into a number of branches, which communicate and form an intimate plexus. At the junction of the branches small triangular nuclear bodies (ganglion-cells) are situated. From these plexuses minute branches are given off, which divide and break up into the ultimate fibrillæ of which the nerve is composed. These fibrillæ course between the involuntary muscle-cells, and, according to Elischer, terminate on the surface of the cell, opposite the nucleus, in a minute swelling. Arnold and Frankenhäuser believed that these ultimate

FIG. [67].—Middle third of a terminal plaque in the muscle spindle of an adult cat. (After Ruffini.)

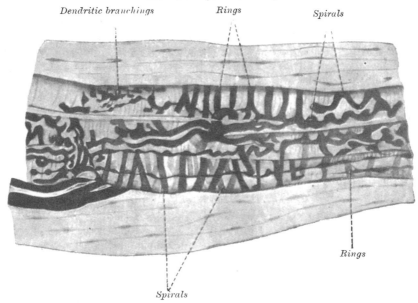

fibrillæ penetrated the muscular cell, and ended in the nucleus. More recent observation has, however, tended to disprove this.

In the **striped** or **voluntary** muscle, the nerves supplying the muscular fibres are derived from the cerebro-spinal nerves, and are composed mainly of medul-

FIG. [68].—Muscular fibres of *Lacerta viridis* with the terminations of nerves.

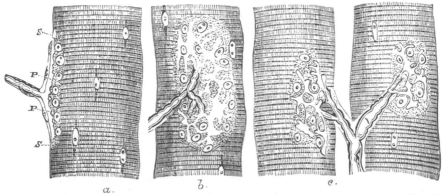

a. Seen in profile. ᴘ ᴘ. The nerve end-plates. ѕ ѕ. The base of the plate, consisting of a granular mass with nuclei. b. The same as seen in looking at a perfectly fresh fibre, the nervous ends being probably still excitable. (The forms of the variously divided plate can hardly be represented in a woodcut by sufficiently delicate and pale contours to reproduce correctly what is seen in nature.) c. The same as seen two hours after death from poisoning by curare.

lated fibres. The nerve, after entering the sheath of the muscle, breaks up into fibres, or bundles of fibres, which form plexuses, and gradually divide until, as a rule, a single nerve-fibre enters a single muscular fibre. Sometimes, however, if the muscular fibre is long, more than one nerve-fibre enters it. Within the muscular fibre the nerve terminates in a special expansion, called by Kühne, who first accurately described them, *motorial end-plates* (fig. [68]).* The nerve-fibre, on approaching the muscular fibre, suddenly loses its white matter of Schwann, which abruptly terminates; the neurilemma becomes continuous with the sarcolemma of the muscle, and only the axis cylinder enters the muscular fibre, where it immediately spreads out, ramifying like the roots of a tree, immediately beneath the sarcolemma, and is embedded in a layer of granular matter, containing a number of clear, oblong nuclei, the whole constituting an end-plate from which the contractile wave of the muscular fibre is said to start.

The **Ganglia** may be regarded as separate small aggregations of nerve-cells, connected with each other, with the cerebro-spinal axis, and with the nerves in various situations. They are found on the posterior root of each of the spinal nerves; on the posterior or sensory root of the fifth cranial nerve; on the facial and auditory nerves; on the glosso-pharyngeal and pneumogastric nerves. They are also found in a connected series along each side of the vertebral column, forming the trunk of the sympathetic; and on the branches of that nerve, generally in the plexuses or at the point of junction of two or more nerves with each other, or with branches of the cerebro-spinal system. On section they are seen to consist of a reddish-grey substance, traversed by numerous white nerve-fibres; they vary considerably in form and size; the largest are found in the cavity of the abdomen; the smallest, not visible to the naked eye, exist in considerable numbers upon the nerves distributed to the different viscera. The ganglia are invested by a smooth and firm, closely adhering, membranous envelope, consisting of dense areolar tissue; this sheath is continuous with the perineurium of the nerves, and sends numerous processes into the interior of the ganglion, which support the blood-vessels supplying its substance.

FIG. [69].—Section through a microscopic ganglion. Magnified 300 diameters. (Klein and Noble Smith.)

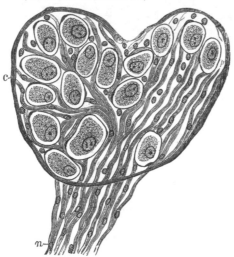

c. Capsule of the ganglion. *n.* Nerve-fibres passing out of the ganglion. The nerve-fibres which entered the ganglion are not represented. The nerve-fibres are ordinary medullated fibres, but the details of their structure are not shown, owing to the low magnifying power. The ganglion-cells are invested by a special capsule, lined by a few nuclei, which are here represented as if contained *in* the capsule.

In structure all ganglia are essentially similar (fig. [69]), consisting of the same structural elements as the other nervous centres—viz. a collection of nerve-cells and nerve-fibres. Each nerve-cell has a nucleated sheath, which is continuous with the sheath of the nerve-fibre with which the cell is connected. The nerve-cells in the ganglia of the spinal nerves are pyriform in shape, and have only one process, the axis cylinder or axon. A short distance from the cell and while still within the ganglion this process divides in a **T**-shaped manner, one limb of the crossbar passing centrally and forming the central portion of a sensory nerve-fibre; the other limb passing peripherally to form the axis-cylinder process of the peripheral nerve-fibre. In the sympathetic ganglia the nerve-cells are multipolar and have one axis-cylinder process or axon and several protoplasmic processes or dendrons. The former of these emerges from the ganglion as a non-medullated nerve-fibre. Similar cells are found in the ganglia connected with the fifth cranial nerve,

* They had, however, previously been noticed, though not accurately described, by Doyère, who named them ' nerve-hillocks.'

and these ganglia are therefore regarded by some as the cranial portions of the sympathetic system. The nerve-cells are disposed in the ganglia in groups of varying size, and these groups are separated from each other by bundles of nerve-fibres, some of which traverse the ganglia without being connected with the cells.

THE VASCULAR SYSTEM

The **Vascular System,** exclusive of its central organ, the heart, is divided into four classes of vessels : the arteries, capillaries, veins, and lymphatics. The minute structure of these vessels will be briefly described here, the reader being referred to the body of the work for the details of their ordinary anatomy.

Structure of Arteries (fig. [70]).—The arteries are composed of three coats : internal or endothelial coat (*tunica intima* of Kölliker) ; middle muscular coat (*tunica media*) ; and external connective-tissue coat (*tunica adventitia*).

The two inner coats together are very easily separated from the external, as by the ordinary operation of tying a ligature on an artery. If a fine string be tied forcibly upon an artery and then taken off, the external coat will be found undivided, but the internal coats are divided in the track of the ligature and can easily be further dissected from the outer coat. The *inner coat* can be separated from the middle by a little maceration, or it may be stripped off in small pieces ; but, on account of its friability, it cannot be separated as a complete membrane. It is a fine, transparent, colourless structure which is highly elastic, and is commonly corrugated into longitudinal wrinkles. The inner coat consists of : (1) A layer of pavement-endothelium, the, cells of which are polygonal, oval, or fusiform, and have very distinct round or oval nuclei. This endothelium is brought into view most distinctly by staining with nitrate of silver. (2) A subendothelial layer, consisting of delicate connective tissue with branched cells lying in the interspaces of the tissue. In arteries of less than a line in diameter the subendothelial layer consists of a single stratum of stellate cells, and the connective tissue is only largely developed in vessels of a considerable size. (3) An elastic or fenestrated layer, which consists of a membrane containing a network of elastic fibres, having principally a longitudinal direction, and in which, under the microscope, small elongated apertures or perforations may be seen, giving it a fenestrated appearance.

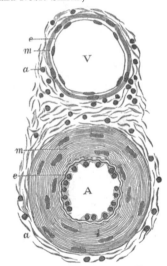

Fig. [70].—Transverse section through a small artery and vein of the mucous membrane of the epiglottis of a child. Magnified about 350 diameters. (Klein and Noble Smith.)

A. Artery, showing the nucleated endothelium, e, which lines it : the vessel being contracted, the endothelial cells appear very thick. Underneath the endothelium is the wavy elastic intima. The chief part of the wall of the vessel is occupied by the circular muscle-coat m : the staff-shaped nuclei of the muscle-cells are well seen. Outside this is a, part of the adventitia. This is composed of bundles of connective-tissue fibres, shown in section, with the nuclei of the connective-tissue corpuscles. The adventitia gradually merges into the surrounding connective tissue. v. Vein showing a thin endothelial membrane, e, raised accidentally from the intima, which on account of its delicacy is seen as a mere line on the media m. This latter is composed of a few circular unstriped muscle-cells. a. The adventitia, similar in structure to that of an artery.

It was therefore called by Henle the *fenestrated membrane.* This membrane forms the chief thickness of the inner coat, and can be separated into several layers, some of which present the appearance of a network of longitudinal elastic fibres, and others present a more membranous character, marked by pale lines having a longitudinal direction

The fenestrated membrane in microscopic arteries is a very thin layer; but in the larger arteries, and especially in the aorta, it has a very considerable thickness.

The *middle coat (tunica media)* is distinguished from the inner by its colour and by the transverse arrangement of its fibres, in contradistinction to the longitudinal direction of those of the inner coat. In the smaller arteries it consists principally of muscular tissue, being made up of plain muscle-fibres in fine bundles, arranged in lamellæ and disposed circularly around the vessel. These lamellæ vary in number according to the size of the vessel; the very small arteries having only a single layer, and those not larger than one-tenth of a line in diameter three or four layers. It is to this coat that the great thickness of the walls of the artery is mainly due (fig. [70], A *m*). In the larger vessels, as the iliac, femoral, and carotid, elastic fibres unite to form lamellæ, which alternate with the layers of muscular fibre and are united by elastic fibres which pass between the muscular bundles, and are connected with the fenestrated membrane of the inner coat (fig. [72]). In the largest arteries, as the aorta and innominate, the amount of elastic tissue is very considerable. In these vessels also bundles of white connective tissue have been found in small quantities in the middle coat. The muscle-fibre cells of which the middle coat is made up are about $\frac{1}{500}$ of an inch in length and contain well-marked, rod-shaped nuclei, which are often slightly curved.

FIG. [71].—Longitudinal section of artery and vein.

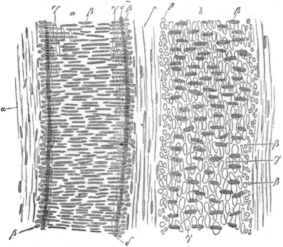

a. An artery from the mesentery of a child, ·062‴; and *b*, vein ·067‴ in diameter, treated with acetic acid and magnified 350 times. *a.* Tunica adventitia, with elongated nuclei. *β.* Nuclei of the contractile fibre-cells of the tunica media, seen partly from the surface, partly apparent in transverse section. *γ.* Nuclei of the endothelial cells. *δ.* Elastic longitudinal fibrous coat.

The *external coat (tunica adventitia)* consists mainly of fine and closely felted bundles of white connective tissue, but also contains elastic fibres in all but the smallest arteries. The elastic tissue is much more abundant next the tunica media, and it is sometimes described as forming here, between the adventitia and media, a special layer, the *tunica elastica externa* of Henle. This layer is most marked in arteries of medium size. In the largest vessels the external coat is relatively thin; but in small arteries it is of greater proportionate thickness. In the smaller arteries it consists of a single layer of white connective tissue and elastic fibres; while in the smallest arteries, just above the capillaries, the elastic fibres are wanting, and the connective tissue, of which the coat is composed, becomes more homogeneous the nearer it approaches the capillaries, and is gradually reduced to a thin membranous envelope, which finally disappears.

Some arteries have extremely thin coats in proportion to their size; this is especially the case in those situated in the cavity of the cranium and spinal canal, the difference depending on the greater thinness of the external and middle coats.

The arteries, in their distribution throughout the body, are included in a thin fibro-areolar investment, which forms what is called their *sheath*. In the limbs this is usually formed by a prolongation of the deep fascia; in the upper part of the thigh it consists of a continuation downwards of the transversalis and iliac fasciæ of the abdomen; in the neck, of a prolongation of the deep cervical fascia

The included vessel is loosely connected with its sheath by delicate areolar tissue; and the sheath usually encloses the accompanying veins, and sometimes a nerve. Some arteries, as those in the cranium, are not included in sheaths.

All the larger arteries are supplied with blood-vessels like the other organs of the body; they are called the *vasa vasorum*. These nutrient vessels arise from a branch of the artery or from a neighbouring vessel, at some considerable distance from the point at which they are distributed; they ramify in the loose areolar tissue connecting the artery with its sheath, and are distributed to the external coat, but do not, in man, penetrate the other coats; though in some of the larger mammals some few vessels have been traced into the middle coat. Minute veins serve to return the blood from these vessels; they empty themselves into the vein or veins accompanying the artery. Lymphatic vessels and lymphatic spaces are also present in the outer coat.

Arteries are also supplied with nerves, which are derived chiefly from the sympathetic, but partly from the cerebro-spinal system. They form intricate plexuses upon the surfaces of the larger trunks, and run along the smaller arteries as single filaments or bundles of filaments, which twist around the vessel and unite with each other in a plexiform manner. The branches derived from these plexuses penetrate the external coat and are principally distributed to the muscular tissue of the middle coat, and thus regulate, by causing the contraction and relaxation of this tissue, the amount of blood sent to any part.

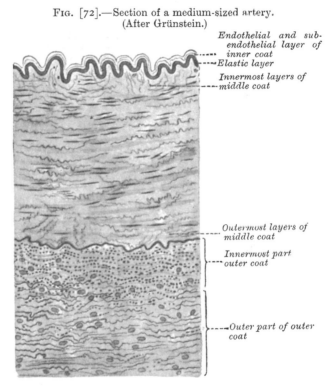

FIG. [72].—Section of a medium-sized artery. (After Grünstein.)

Endothelial and sub-endothelial layer of inner coat
Elastic layer
Innermost layers of middle coat

Outermost layers of middle coat
Innermost part outer coat

Outer part of outer coat

The Capillaries.—The smaller arterial branches (excepting those of the cavernous structure of the sexual organs, of the spleen, and in the uterine placenta) terminate in a network of vessels which pervade nearly every tissue of the body. These vessels, from their minute size, are termed capillaries (*capillus*, a 'hair'). They are interposed between the smallest branches of the arteries and the commencing veins, constituting a network, the branches of which maintain the same diameter throughout; the meshes of the network being more uniform in shape and size than those formed by the anastomoses of the small arteries and veins.

The *diameter* of the capillaries varies in the different tissues o fthe body, their usual size being about $\frac{1}{3000}$ of an inch. The smallest are those of the brain, and the mucous membrane of the intestines; and the largest those of the skin and the marrow of bone, where they are stated to be as large as $\frac{1}{1200}$ of an inch. The *form* of the capillary net varies in the different tissues, the meshes being generally rounded or elongated.

The *rounded form of mesh* is most common, and prevails where there is a dense network, as in the lungs, in most glands and mucous membranes, and in the cutis; here the meshes are more or less angular, sometimes nearly quadrangular, or polygonal, or more often irregular and not of an absolutely circular outline.

Elongated meshes are observed in the muscles and nerves, the meshes being usually of a parallelogram form, the long axis of the mesh running parallel with the long axis of the nerve and fibre. Sometimes the capillaries have a *looped* arrangement; a single vessel projecting from the common network and returning after forming one or more loops, as in the papillæ of the tongue and skin.

The number of the capillaries, and the size of the meshes, determine the degree of vascularity of a part. The closest network and the smallest

Fig. [74].—Finest vessels on the arterial side. From the human brain. Magnified 300 times.

Fig. [73].—Capillaries from the mesentery of a guinea-pig after treatment with solution of nitrate of silver.

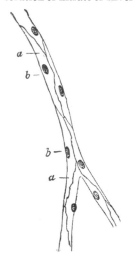

a. Cells. *b.* Their nuclei.

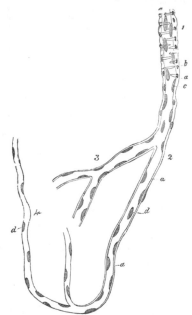

1. Smallest artery. 2. Transition vessel. 3. Coarser capillaries. Finer capillaries. *a.* Structureless membrane still with some nuclei, representative of the tunica adventitia. *b.* Nuclei of the muscular fibre-cells. *c.* Nuclei within the small artery, perhaps appertaining to an endothelium. *d.* Nuclei in the transition vessels.

interspaces are found in the lungs and in the choroid coat of the eye. In these situations the interspaces are smaller than the capillary vessels themselves. In the kidney, in the conjunctiva, and in the cutis the interspaces are from three to four times as large as the capillaries which form them; and in the brain from eight to ten times as large as the capillaries in their long diameter, and from four to six times as large in their transverse diameter. In the adventitia of arteries the width of the meshes is ten times that of the capillary vessels. As a general rule, the more active the function of the organ, the closer is its capillary net and the larger its supply of blood; the meshes of the network are very narrow in all growing parts, in the glands and in the mucous membranes; wider in bones and ligaments, which are comparatively inactive; and nearly altogether absent in tendons, in which very little organic change occurs after their formation.

Structure.—The walls of the capillaries consist of a fine transparent endothelial layer, composed of cells joined edge to edge by an interstitial cement-substance, and continuous with the endothelial cells which line the arteries and veins. When stained with nitrate of silver the edges which bound the epithelial cells are brought into view (fig. [73]). These cells are of large size and of an irregular polygonal or lanceolate shape, each containing an oval nucleus which may be brought into view by carmine or hæmatoxylin. Between their edges, at various points of their meeting, roundish dark spots are sometimes seen, which have been described as stomata, though they are closed by intercellular substance. They have been believed to be the situation through which the white corpuscles of the blood, when migrating from the blood-vessels, emerge; but this view, though probable, is not universally accepted.

Kolossow, a Russian observer, describes these cells as having a rather more complex structure. He states that they consist of two parts : of hyaline ground-plates, and of a protoplasmic granular part, in which is embedded the nucleus, on the outside of the ground-plates. The hyaline internal coat of the capillaries does not form a complete membrane, but consists of ' plates ' which are inelastic, and though in contact with each other, are not continuous; when therefore the capillaries are subjected to intra-vascular pressure, the plates become separated from each other ; the protoplasmic portions of the cells, on the other hand, are united together.

In many situations a delicate sheath or envelope of branched nucleated connective-tissue cells is found around the simple capillary tube, particularly in the larger ones ; and in other places, especially in the glands, the capillaries are invested with retiform connective tissue.

In the largest capillaries (which ought, perhaps, to be described rather as the smallest arteries or pre-capillaries) there is, outside the epithelial layer, a muscular layer, consisting of contractile fibre-cells, arranged transversely, as in the tunica media of the larger arteries (fig. [74]).

The **veins**, like the arteries, are composed of three coats—internal, middle, and external ; and these coats are, with the necessary modifications, analogous to the coats of the arteries ;

the internal being the endothelial, the middle the muscular, and the external the connective or areolar (fig. [75]). The main difference between the veins and the arteries is in the comparative weakness of the middle coat of the former, and to this is due the fact that

FIG. [75].—Transverse section of part of the wall of one of the posterior tibial veins. (After Schäfer.)

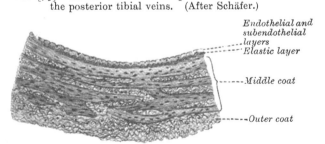

Endothelial and subendothelial layers
Elastic layer

Middle coat

Outer coat

the veins do not stand open when divided, as the arteries do, and that they are passive rather than active organs of the circulation.

In the veins immediately above the capillaries the three coats are hardly to be distinguished. The endothelium is supported on an outer membrane separable into two layers, the outer of which is the thicker, and consists of a delicate, nucleated membrane (adventitia), while the inner is composed of a network of longitudinal elastic fibres (media). In the veins next above these in size (one-fifth of a line according to Kölliker) a muscular layer and a layer of circular fibres can be traced, forming the middle coat, while the elastic and connective elements of the outer coat become more distinctly perceptible. In the middle-sized veins the typical structure of these vessels becomes clear. The endothelium is of the same character as in the arteries, but its cells are more oval and less fusiform. It is supported by a connective-tissue layer, consisting of a delicate network of branched cells, and external to this is a layer of longitudinal elastic fibres, but seldom any appearance of a fenestrated membrane. This constitutes the *internal* coat. The *middle* coat is composed of a thick layer of connective tissue with elastic fibres, intermixed, in some veins, with a transverse layer of muscular fibres. The white fibrous element is in considerable excess, and the elastic fibres are in much smaller proportion in the veins than in the arteries. The *outer* coat consists of areolar tissue, as in the arteries, with longitudinal elastic fibres. In the largest veins the outer coat is from two to five times thicker than the middle coat, and contains a large number of longitudinal muscular fibres This is most distinct in the inferior vena cava, especially at the termination of this vein in the heart, in the trunks of the hepatic veins, in all the large trunks of the vena portæ, in the splenic, superior mesenteric, external iliac, renal and azygos veins. In the renal and portal veins it extends through the whole thickness of the outer coat, but in the other veins mentioned a layer of connective and elastic tissue is found external to the muscular fibres. All the large veins which open into the heart are covered for a short distance with a layer of striped muscular tissue continued on to them from the heart. Muscular tissue is wanting—(1) in the veins of the maternal part of the placenta ; (2) in the venous sinuses of the dura mater and the veins of the pia mater of the brain and spinal cord ; (3) in the veins of the retina ; (4) in the veins. of the cancellous tissue of bones ; (5) in the venous spaces of the corpora cavernosa. The veins of the above-mentioned parts consist of an internal endothelial lining supported on one or more layers of areolar tissue.

Most veins are provided with valves, which serve to prevent the reflux of the blood. They are formed by a reduplication of the inner coat, strengthened by connective tissue and elastic fibres, and are covered on both surfaces with endo- thelium, the arrangement of which differs on the two surfaces. On the surface of the valve next the wall of the vein, the cells are arranged transversely ; while on the other surface, over which the current of blood flows, the cells are arranged vertically in the direction of the current. The valves are semilunar. They are attached by their convex edge to the wall of the vein ; the concave margin is free, directed in the course of the venous current, and lies in close apposition with the wall of the vein as long as the current of blood takes its natural course ; if, how- ever, any regurgitation takes place, the valves become distended, their opposed edges are brought into contact, and the current is interrupted. Most commonly two such valves are found placed opposite one another, more especially in the smaller veins or in the larger trunks at the point where they are joined by smaller branches ; occasionally there are three and sometimes only one. The wall of the vein on the cardiac side of the point of attachment of each segment of the valve is expanded into a pouch or sinus, which gives to the vessel, when injected or distended with blood, a knotted appearance. The valves are very numerous in the veins of the extremities, especially of the lower extremities ; these vessels having to conduct the blood against the force of gravity. They are absent in the very small veins, i.e. those less than $\frac{1}{12}$ of an inch in diameter, also in the venæ cavæ, the hepatic veins, portal vein and most of its branches, the renal, uterine, and ovarian veins. A few valves are found in the spermatic veins, and one also at their point of junction with the renal vein and inferior vena cava respectively. The cerebral and spinal veins, the veins of the cancellated tissue of bone, the pulmonary veins, and the umbilical vein and its branches, are also destitute of valves. They are occasionally found, few in number, in the venæ azygos and intercostal veins.

The veins are supplied with nutrient vessels, *vasa vasorum*, like the arteries. Nerves also are distributed to them in the same manner as to the arteries, but in much less abundance.

The **lymphatic** vessels, including in this term the lacteal vessels, which are identical in structure with them, are composed of three coats. The *internal* is an endothelial and elastic coat. It is thin, transparent, slightly elastic, and ruptures sooner than the other coats. It is composed of a layer of elongated endothelial cells with serrated margins, by which the adjacent cells are dovetailed into one another. These are supported on a single layer of longitudinal elastic fibres. The *middle* coat is composed of smooth muscular and fine elastic fibres, disposed in a transverse direction. The *external*, or fibro-areolar, coat consists of filaments of connective tissue, intermixed with smooth muscular fibres, longitudinally or obliquely disposed. It forms a protective covering to the other coats, and serves to connect the vessel with the neighbouring structures. The above description applies only to the larger lymphatics; in the smaller vessels there is no muscular or elastic coat, and their structure consists only of a connective-tissue coat, lined by endothelium. The thoracic duct (fig. [76]) is a somewhat more complex structure than the other lymphatics; it presents a distinct subendothelial layer of branched corpuscles, similar to that found

FIG. [76].—Transverse section through the coats of the thoracic duct of man. Magnified 30 times.

a. Endothelium, striated lamellæ, and inner elastic coat. *b.* Longitudinal connective tissue of the middle coat. *c.* Transverse muscles of the same. *d.* Tunica adventitia, with *e*, the longitudinal muscular fibres.

in the arteries, and in the middle coat is a layer of connective tissue with its fibres arranged longitudinally. The lymphatics are supplied by nutrient vessels, which are distributed to their outer and middle coats; and here also have been traced many non-medullated nerve-fibres in the form of a fine plexus of fibrils.

The lymphatics are very generally provided with valves, which assist materially in effecting the circulation of the fluid they contain. These valves are formed of a thin layer of fibrous tissue, lined on both surfaces by endothelium, which presents the same arrangement upon the two surfaces as was described in connection with the valves of veins. Their form is semilunar; they are attached by their convex edge to the sides of the vessel, the concave edge being free and directed along the course of the contained current. Usually two such valves, of equal size, are found opposite one another; but occasionally exceptions occur, especially at or near the anastomoses of lymphatic vessels. Thus, one valve may be of very rudimentary size and the other increased in proportion.

The valves in the lymphatic vessels are placed at much shorter intervals than in the veins. They are most numerous near the lymphatic glands, and they are found more frequently in the lymphatics of the neck and upper extremity than in the lower. The wall of the lymphatics immediately above the point of attachment of each segment of a valve is expanded into a pouch or sinus, which gives to these vessels, when distended, the knotted or beaded appearance which they present. Valves are wanting in the vessels composing the plexiform network in which the lymphatics usually originate on the surface of the body.

Origin of Lymphatics.—The finest visible lymphatic vessels (lymphatic capillaries) form a plexiform network in the tissues and organs, and they consist of a single layer of endothelial plates, with more or less sinuous margins. These vessels commence in an intercommunicating system of clefts or spaces, which have no complete endothelial lining, in the connective tissue of the different organs. They have been named the *rootlets* of the lymphatics, and are identical with the spaces in which the connective-tissue corpuscles are contained. This then is properly regarded as one method of their commencement, where the lymphatic vessels are apparently continuous with spaces in the connective tissue, and Klein has described and figured a direct communication between these spaces and the lymphatic vessel.* But the lymphatics have also other modes of origin, for the intestinal lacteals commence by closed extremities, though some observers believe that the closed extremity is continuous with a minute network contained in the substance of the villus, through which the lacteal is connected with the endothelial cells covering it. Again, it seems now to be conclusively proved that

* *Atlas of Histology*, pl. viii. fig. xiv.

the serous membranes present stomata or openings between the endothelial cells (fig. [77]), by which there is an open communication with the lymphatic system and through which the lymph is thought to be pumped by the alternate dilatation and contraction of the serous surface, due to the movements of respiration and circulation, so that the serous and synovial sacs may be regarded, in a certain sense, as large lymph cavities or sinuses.*   Von Recklinghausen was the first to observe the passage of milk and other coloured fluids through these stomata on the peritoneal surface of the central tendon of the diaphragm.   Again, in most glandular structures the lymphatic capillaries have a lacunar origin.   Here they begin in irregular clefts or spaces in the tissue of the part; occupying the penetrating connective tissue and surrounding the lacunæ or tubules of the gland, and in many places separating the capillary network from the alveolus or tubule, so that the interchange between the blood and the secreting cells of the part must be carried on through these lymph-spaces or lacunæ.   Closely allied to this is the mode of origin of lymphatics in perivascular and perineural spaces.   Sometimes a minute artery may be seen to be ensheathed for a certain

FIG. [77].—Stomata of serous membranes.

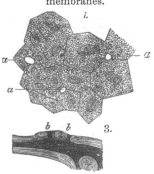

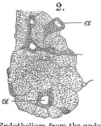

FIG. [78].—Diagrammatic section of lymphatic gland, showing the course of the lymph.

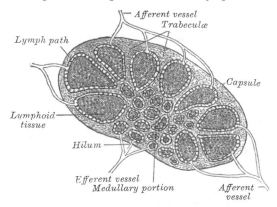

1. Endothelium from the under surface of the *centrum tendineum* of the rabbit. *a.* Stomata. 2. Endothelium of the mediastinum of the dog. *a.* Stomata. 3. Section through the pleura of the same animal. *b.* Free orifices of short lateral passages of the lymph-canals. (Copied from Ludwig, Schweigger-Seidel and Dybkowsky.)

distance by a lymphatic capillary vessel, which is often many times wider than a blood capillary. These are known as perivascular lymphatics.

Terminations of Lymphatics.—The lymphatics, including the lacteals, discharge their contents into the veins at two points : namely, at the angles of junction of the subclavian and internal jugular veins : on the left side by means of the thoracic duct, and on the right side by the right lymphatic duct.   (See description of lymphatics, page 602.)

Lymphatic glands (*conglobate glands*) are small oval or bean-shaped bodies, situated in the course of lymphatic and lacteal vessels so that the lymph and chyle pass through them on their way to the blood.   They generally present on one side a slight depression—the *hilum*—through which the blood-vessels enter and leave the interior.   The efferent lymphatic vessel also emerges from the gland at this spot, while the afferent vessels enter the organ at different parts of the periphery.   On section (fig. [78]), a lymphatic gland displays two different structures : an external, of lighter colour—the *cortical*; and an internal, darker —the *medullary*.   The cortical structure does not form a complete investment,

* The resemblance between lymph and serum led Hewson long ago to regard the serous cavities as sacs into which the lymphatics open.   Recent microscopic discoveries confirm this opinion in a very interesting manner.

but is deficient at the hilum, where the medullary portion reaches the surface of the gland; so that the efferent vessel is derived directly from the medullary structure, while the afferent vessels empty themselves into the cortical substance.

Lymphatic glands consist of (1) a fibrous envelope, or *capsule*, from which a framework of processes (*trabeculæ*) proceeds inwards, dividing the gland into open spaces (*alveoli*) freely communicating with each other; (2) a quantity of lymphoid tissue occupying these spaces without completely filling them; (3) a free supply of blood-vessels, which are supported on the trabeculæ; and (4) the *afferent* and *efferent* vessels. Little is known of the nerves, though Kölliker describes some fine nervous filaments passing into the hilum.

The *capsule* is composed of a layer of connective tissue, and from its internal surface are given off a number of membranous septa or lamellæ, consisting, in man, of connective tissue, with a small admixture of plain muscle-fibres; but in many of the lower animals composed almost entirely of involuntary muscle. They pass inwards, radiating towards the centre of the gland, for a certain distance, that is to say, for about one-third or one-fourth of the space between the circum-

FIG. [79].—Follicle from a lymphatic gland of the dog, in vertical section.

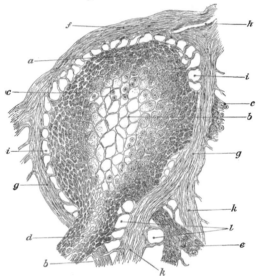

FIG. [80].—From the medullary substance of an inguinal gland of the ox. (After His.)

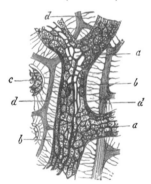

a. Reticular sustentacular substance of the more external portion, *b,* of the more internal, and *c,* of the most external and most finely webbed part on the surface of the follicle. *d.* Origin of a large lymph-tube. *e.* Of a smaller one. *f.* Capsule. *g.* Septa. *h.* Vas afferens. *i.* Investing space of the follicle, with its retinacula. *k.* One of the divisions of the septa. *l l.* Attachment of the lymph-tubes to the septa.

a. Lymph-tube, with its complicated system of vessels. *b.* Retinacula stretched between the tube and the septa. *c.* Portion of another lymph-tube. *d.* Septa.

ference and the centre of the gland. They thus divide the outer part of its interior into a number of oval compartments or alveoli (fig. [78]). This is the cortical portion of the gland. After having penetrated into the gland for some distance, these septa break up into a number of smaller trabeculæ, which form flattened bands or cords, interlacing with each other in all directions, forming in the central part of the organ a number of intercommunicating spaces, also called alveoli. This is the medullary portion of the gland, and the spaces or alveoli in it not only freely communicate with each other, but also with the alveoli of the cortical portion. In these alveoli or spaces (fig. [79]) is contained the proper gland-substance or lymphoid tissue. The gland-pulp does not, however, completely fill the alveolar spaces, but leaves, between its outer margin and the trabeculæ forming the alveoli, a channel or space of uniform width throughout. This is termed the *lymph-path* or *lymph-sinus* (fig. [81]). Running across it are a number of trabeculæ of retiform connective tissue, the fibres of which are, for the most part, covered by ramified cells. This tissue appears to serve the purpose of maintaining the gland-pulp in the centre of the space in its proper position.

On account of the peculiar arrangement of the framework of the organ, the gland-pulp in the cortical portion is disposed in the form of nodules, and in the medullary part in the form of rounded cords. It consists of ordinary lymphoid tissue, being made up of a delicate reticulum of retiform tissue, which is continuous with that in the lymph-paths, but marked off from it by a closer reticulation ; in its meshes are closely packed lymph-corpuscles, traversed by a dense plexus of capillary blood-vessels.

The *afferent vessels*, as above stated, enter at all parts of the periphery of the gland, and after branching and forming a dense plexus in the substance of the capsule open into the lymph-sinuses of the cortical part. In doing this, they lose all their coats except their endothelial lining, which is continuous with a layer of similar cells lining the lymph-paths. In like manner the *efferent* vessel commences from the lymph-sinuses of the medullary portion. The stream of lymph carried to the gland by the afferent vessel thus passes through the plexus in the capsule to the lymph-paths of the cortical portion, where it is

FIG. [81].—Section of lymphatic-gland tissue.

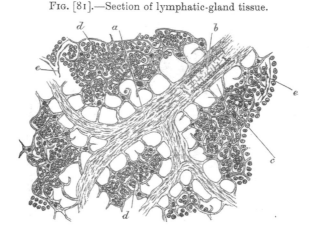

a. Trabeculæ. *b.* Small artery in substance of same. *c.* Lymph-paths.
d. Lymph-corpuscles. *e.* Capillary plexus.

exposed to the action of the gland-pulp ; flowing through these it enters the paths or sinuses of the medullary portion, and finally emerges from the hilum by means of the efferent vessel. The stream of lymph in its passage through the lymph-sinuses is much retarded by the presence of the reticulum. Hence morphological elements, either normal or morbid, are easily arrested and deposited in the sinuses. This is a matter of considerable importance in connection with the subject of poisoned wounds and the absorption of the poison by the lymphatic system, since by this means septic organisms carried along the lymphatic vessels may be arrested in the lymph-sinuses of the gland-tissue, and thus be prevented from entering the general circulation. Many lymph-corpuscles pass with the efferent lymph-stream to join the general blood-stream. The arteries of the gland enter at the hilum, and either pass at once to the gland-pulp, to break up into a capillary plexus, or else run along the trabeculæ, partly to supply them and partly running across the lymph-paths to assist in forming the capillary plexus of the gland-pulp. This plexus traverses the lymphoid tissue, but does not pass into the lymph-sinuses. From it the veins commence and emerge from the organ at the same place as that at which the artery enters.

THE SKIN AND ITS APPENDAGES

The **skin** (fig. [82]) is the principal seat of the sense of touch, and may be regarded as a covering for the protection of the deeper tissues ; it plays an important part in the regulation of the body temperature, and is also an excretory and absorbing organ. It consists principally of a layer of vascular tissue, named the *dermis, corium,* or *cutis vera,* and an external covering of epithelium, termed the *epidermis* or *cuticle.* On the surface of the former layer are the sensitive *papillæ* ; and within, or embedded beneath it, are certain organs with special functions, namely, the *sweat-glands, hair-follicles,* and *sebaceous glands.*

FIG. [82]. —A sectional view of the skin (magnified).

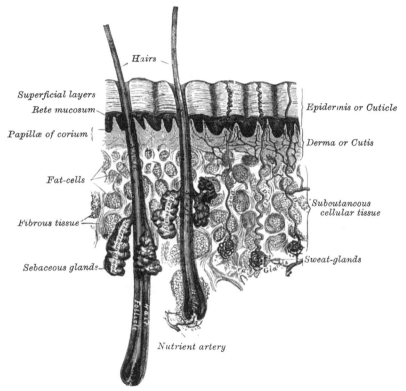

The **epidermis** or **cuticle** (*scarf-skin*) is non-vascular, and consists of stratified epithelium (fig. [83]). It is accurately moulded on the papillary layer of the dermis. It forms a defensive covering to the surface of the true skin, and limits the evaporation of watery vapour from its free surface. It varies in thickness in different parts. In some situations, as in the palms of the hands and soles of the feet, it is thick, hard, and horny in texture. This may be partly due to the fact that these parts are exposed to intermittent pressure, but that this is not the only cause is proved by the fact that the condition exists to a very considerable extent at birth. The more superficial layers of cells, called the *horny* layer (*stratum corneum*), may be separated by maceration from the deeper layers, which are called the *rete mucosum* or *stratum Malpighii,* and which consist of several layers of differently shaped cells. The free surface of the epidermis is marked by a network of linear furrows of variable size, marking out the surface into a number of spaces of polygonal or lozenge-shaped form. Some of these furrows are large, as

opposite the flexures of the joints, and correspond to the folds in the dermis produced by their movements. In other situations, as upon the back of the hand,
they are exceedingly fine, and intersect one another at various angles. Upon the
palmar surface of the hand and fingers, and upon the sole of the foot, these lines
are very distinct, and are disposed in curves; they depend upon the large size
and peculiar arrangement of the papillæ upon which the epidermis is placed.
The deep surface of the epidermis is accurately moulded upon the papillary layer
of the dermis, each papilla being invested by its epidermic sheath; so that when
this layer is removed by maceration, it presents on its under surface a number of
pits or depressions corresponding to the elevations in the papillæ, as well as the
ridges left in the intervals between them. Fine tubular prolongations are
continued from this layer into the ducts of the sudoriferous and sebaceous glands.

In structure, the epidermis consists of several layers of epithelial cells,
agglutinated together and having a laminated arrangement. These several layers
may be described as composed of four different strata from within outwards:

Fig. [83].—Section of epidermis.   (Ranvier.)

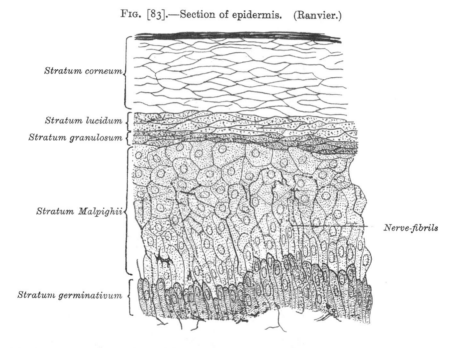

(1) the *stratum Malpighii*, composed of several layers of epithelial cells, of which
the deepest layer is columnar in shape and placed perpendicularly on the surface
of the corium, their lower ends being denticulate, to fit into corresponding den-
ticulations of the true skin; this deepest layer is sometimes termed the basilar
layer or *stratum germinativum*; the succeeding laminæ consist of cells of a more
rounded or polyhedral form, the contents of which are soft, opaque, granular, and
soluble in acetic acid. They are often marked on their surfaces with ridges and
furrows, and are covered with numerous fibrils, which connect the surfaces of the
cells: these are known as *prickle* cells (see page [3]). They contain numerous
epidermic fibrils, which are stained violet with hæmatoxylin and red by carmine,
and form threads of union connecting adjacent cells. Between the cells are fine
intercellular clefts which serve for the passage of lymph and in which lymph-
corpuscles or pigment-granules may be found. (2) Immediately superficial to these
are two or three layers of flattened, spindle-shaped cells, the *stratum granulosum*,
which contain granules that become deeply stained in hæmatoxylin; the granules
consist of a material named *eleidin*, an intermediate substance in the formation of

horn. They are supposed to be cells in a transitional stage between the protoplasmic cells of the stratum Malpighii and the horny cells of the superficial layers.
(3) Above this layer, the cells become indistinct, and appear, in sections, to form
a homogeneous or dimly striated membrane, composed of closely packed scales,
in which traces of a flattened nucleus may be found. It is called the *stratum
lucidum*. (4) As these cells successively approach the surface by the development
of fresh layers from beneath, they assume a flattened form, from the evaporation
of their fluid contents, and consist of many layers of horny epithelial scales in
which no nucleus is discernible, forming the *stratum corneum*. These cells are
unaffected by acetic acid, the protoplasm having become changed into horny
material or *keratin*. According to Ranvier they contain granules of a material
which has the characters of beeswax. The deepest layer of the stratum Malpighii
is separated from the papillæ by an apparently homogeneous basement-membrane,
which is most distinctly brought into view in specimens prepared with chloride
of gold. This, according to Klein, is merely the deepest portion of the epithelium,
and is 'made up of the basis of the individual cells, which have undergone a
chemical and morphological alteration.' The black colour of the skin in the negro,
and the tawny colour among some of the white races, is due to the presence of

FIG. [84].—Microscopic section of skin, showing the epidermis and dermis : a hair in its
follicle : the arrector pili muscle : sebaceous and sudoriferous glands.

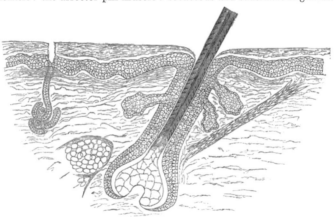

pigment in the cells of the cuticle. This pigment is more especially distinct in
the cells of the deeper layer, or stratum Malpighii, and is similar to that found in
the cells of the pigmentary layer of the retina. As the cells approach the surface
and desiccate, the colour becomes partially lost ; the disappearance of the pigment
from the superficial layers of the epidermis is, however, difficult to explain.

The **dermis, corium,** or **cutis vera** is tough, flexible, and highly elastic, in
order to defend the parts beneath from violence.

It varies in thickness, from a quarter of a line to a line and a half, in different
parts of the body. Thus it is very thick in the palms of the hands and soles of
the feet ; thicker on the posterior aspect of the body than the front, and on the
outer than the inner side of the limbs. In the eyelids, scrotum, and penis it is
exceedingly thin and delicate. The skin generally is thicker in the male than in
the female, and in the adult than in the child.

The corium consists of felted connective tissue, with a varying amount of
elastic fibres and numerous blood-vessels, lymphatics, and nerves. The fibro-
areolar tissue forms the framework of the cutis, and is differently arranged in
different parts, so that it is usual to describe it as consisting of two layers : the
deeper or *reticular* layer, and the superficial or *papillary* layer. Unstriped
muscular fibres are found in the superficial layers of the corium, wherever hairs
are present ; and in the subcutaneous areolar tissue of the scrotum, penis,

labia majora of the female, and the nipples. In the last situation the fibres are arranged in bands, closely reticulated and disposed in superimposed laminæ.

The *reticular* layer consists of strong interlacing fibrous bands, composed chiefly of the white variety of fibrous tissue, but containing, also, some fibres of the yellow elastic tissue, which vary in amount in different parts ; and connective-tissue corpuscles, which are often to be found flattened against the white fibrous tissue bundles. Towards the attached surface the fasciculi are large and coarse, and the areolæ which are left by their interlacement are large, and occupied by adipose tissue and sweat-glands. Below this the elements of the skin become gradually blended with the subcutaneous areolar tissue, which, except in a few situations, contains fat. Towards the free surface the fasciculi are much finer, and their mode of interlacing close and intricate.

The *papillary layer* is situated upon the free surface of the *reticular layer* ; it consists of numerous small, highly sensitive, and vascular eminences, the *papillæ*, which rise perpendicularly from its surface. The papillæ are conical-shaped eminences, having a round or blunted extremity, occasionally divided into two or more parts, and are received into corresponding pits on the under surface of the cuticle. Their average length is about $\frac{1}{100}$ of an inch, and they measure at their base $\frac{1}{250}$ of an inch in diameter. On the general surface of the body, more especially in those parts which are endowed with slight sensibility, they are few in number, short, exceedingly minute, and irregularly scattered over the surface ;

Fig. [85].—Longitudinal section through human nail and its nail groove (sulcus).   × 34.
(From Böhm and Davidoff's ' Histology.')

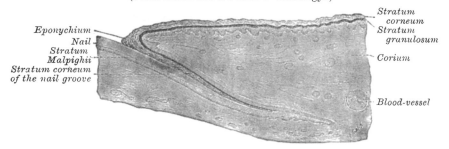

Eponychium
Nail
Stratum
Malpighii
Stratum corneum
of the nail groove

Stratum
corneum
Stratum
granulosum

Corium

Blood-vessel

but in some situations, as upon the palmar surface of the hands and fingers, upon the plantar surface of the feet and toes, and around the nipple, they are long, of large size, closely aggregated together, and arranged in parallel curved lines, forming the elevated ridges seen on the free surface of the epidermis. Each ridge contains two rows of papillæ, and between the two rows the ducts of the sweat-glands pass outwards to open on the summit of the ridges. In structure the papillæ consist of very small and closely interlacing bundles of finely fibrillated tissue, with a few elastic fibres ; within this tissue is a capillary loop, and in some papillæ, especially in the palms of the hands and the fingers, there are tactile corpuscles.

The *arteries* supplying the skin form a network in the subcutaneous tissue, from which branches are given off to supply the sweat-glands, the hair-follicles, and the fat. Other branches are given off which form a plexus immediately beneath the corium ; from this, fine capillary vessels pass into the papillæ, forming, in the smaller ones, a single capillary loop, but in the larger, a more or less convoluted vessel. There are numerous *lymphatics* supplied to the skin, which form two networks, superficial and deep, communicating with each other and with those of the subcutaneous tissue by oblique branches. They originate in the cell-spaces of the tissue.

The *nerves* of the skin terminate partly in the epidermis and partly in the cutis vera. The former are prolonged into the epidermis from a dense plexus in the superficial layer of the corium and terminate between the cells in bulbous extremities ; or, according to some observers, in the deep epithelial cells themselves. The latter terminate in end-bulbs, touch-corpuscles, or Pacinian bodies, in the manner already described ; and, in addition to these, a considerable number of fibrils are distributed to the hair-follicles, which are said to entwine the follicle in a circular manner. Other nerve-fibres are supplied to the plain muscular fibres of the hair-follicles (arrectores pili) and to the muscular coat of the blood-vessels. These are probably non-medullated fibres.

The **appendages of the skin** are the nails, the hairs, the sudoriferous and sebaceous glands, and their ducts.

The nails and hairs are peculiar modifications of the epidermis, consisting essentially of the same cellular structure as that tissue.

The **nails** (figs. [85] and [86]) are flattened, elastic structures of a horny texture, placed upon the dorsal surface of the terminal phalanges of the fingers and toes. Each nail is convex on its outer surface, concave within, and is implanted by a portion, called the *root*, into a groove in the skin ; the exposed portion is called the *body*, and the anterior extremity the *free edge*. The nail has a very firm adhesion to the cutis, being accurately moulded upon its surface, as the epidermis is in other parts. The part of the cutis beneath the body and root of the nail is called the *matrix*, because it is the part from which the nail is produced. Corresponding to the body of the nail, the matrix is thick, and raised into a series of longitudinal

FIG. [86].—Transverse section through human nail and its sulcus. × 34.
(From Böhm and Davidoff's ' Histology.')

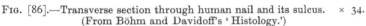

ridges which are very vascular, and the colour is seen through the transparent tissue. Behind this, near the root of the nail, there are papillæ which are small, less vascular, and have no regular arrangement, and here the tissue of the nail is somewhat more opaque ; hence this portion is of a whiter colour, and is called the *lunula* on account of its shape.

The cuticle, as it passes forwards on the dorsal surface of the finger or toe, is attached to the surface of the nail, a little in advance of its root ; at the extremity of the finger it is connected with the under surface of the nail a little behind its free edge. The cuticle and horny substance of the nail (both epidermic structures) are thus directly continuous with each other. The nails consist of a greatly thickened stratum lucidum, the stratum corneum forming merely the thin cuticular fold (*eponychium*) which overlaps the lunula. The cells have a laminated arrangement, and are essentially similar to those composing the epidermis. The deepest layer of cells, which lie in contact with the papillæ of the matrix, are columnar in form and arranged perpendicularly to the surface ; those which succeed them are of a rounded or polygonal form, the more superficial ones becoming broad, thin, and flattened, and so closely compacted as to make the limits of each cell very indistinct. It is by the successive growth of new cells at the root and under surface of the body of the nail that it advances forwards and maintains a due thickness, while, at the same time, the growth of the nail in the proper direction is secured. As these cells in their turn become displaced by the growth of new ones, they assume a flattened form, and finally become closely compacted together into a firm, dense, horny texture. In *chemical composition* the nails resemble the upper layers of the epidermis. According to Mulder, they contain a somewhat larger proportion of carbon and sulphur.

The **hairs** are peculiar modifications of the epidermis, and consist essentially of the same structure as that membrane. They are found on nearly every part of the surface of the body, excepting the palms of the hands, soles of the feet, and the glans penis. They vary much in length, thickness, and colour in different parts of the body and in different races of mankind. In some parts, as in the skin of the eyelids, they are so short as not to project beyond the follicles containing them ; in others, as upon the scalp, they are of considerable length ; again, in other parts, as the eyelashes, the hairs of the pubic region, and the whiskers and beard, they are remarkable for their thickness. Straight hairs are stronger than curly hairs, and present on transverse section a cylindrical or oval outline ; curly hairs, on the other hand, are flattened.

A hair consists of a *root*, the part implanted in the skin ; the *shaft* or *stem*, the portion projecting from its surface ; and the *point*.

The *root of the hair* presents at its extremity a bulbous enlargement, which is whiter in colour and softer in texture than the shaft, and is lodged in a follicular involution of the epidermis called the *hair-follicle* (fig. [84]). When the hair is of considerable length the follicle extends into the subcutaneous cellular tissue. The hair-follicle commences on the surface of the skin with a funnel-shaped opening, and passes inwards in an oblique or curved direction—the latter in curly hairs—to become dilated at its deep extremity, where it corresponds with the bulbous condition of the hair which it contains. It has opening into it, near its free extremity, the orifices of the ducts of one or more sebaceous glands. At the bottom of each hair-follicle is a small conical, vascular eminence or papilla, similar in every respect to those found upon the surface of the skin ; it is continuous with the dermic layer of the follicle, is highly vascular, and probably supplied with nervous fibrils. In structure the hair-follicle consists of two coats—an outer or *dermic*, and an inner or *epidermic*.

FIG. [87].—Transverse section of hair-follicle.

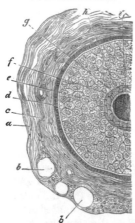

a. Outer layer of dermic coat, with blood-vessels. *b*, *b*. Vessels cut across. *c*. Middle layer. *d*. Inner or hyaline layer. *e*. Outer root-sheath. *f*, *g*. Inner root-sheath. *h*. Cuticle of root-sheath. *i*. Hair. (From Quain's 'Anatomy,' Biesiadecki.)

The *outer* or *dermic* coat is formed mainly of fibrous tissue ; it is continuous with the corium, is highly vascular, and supplied by numerous minute nervous filaments. It consists of three layers (fig. [87]). The most internal, next the cuticular lining of the follicle, consists of a hyaline basement-membrane having a glassy, transparent appearance, which is well marked in the larger hair-follicles, but is not very distinct in the follicles of minute hairs. It is continuous with the basement-membrane of the surface of the corium. External to this is a compact layer of fibres and spindle-shaped cells arranged circularly around the follicle. This layer extends from the bottom of the follicle as high as the entrance of the ducts of the sebaceous glands. Externally is a thick layer of connective tissue, arranged in longitudinal bundles, forming a more open texture and corresponding to the reticular part of the corium. In this are contained blood-vessels and nerves.

The *inner* or *epidermic* layer is closely adherent to the root of the hair, so that when the hair is plucked from its follicle this layer most commonly adheres to it and forms what is called the *root-sheath*. It consists of two strata named respectively the *outer* and *inner root-sheaths* ; the former of these corresponds with the Malpighian layer of the epidermis, and resembles it in the rounded form and soft character of its cells ; at the bottom of the hair-follicle these cells become continuous with those of the root of the hair. The *inner* root-sheath consists of a delicate cuticle next the hair, composed of a thin layer of imbricated scales having a downward direction, so that they fit accurately over the upwardly directed imbricated scales of the hair itself; then of one or two layers of horny, flattened, nucleated cells, known as *Huxley's layer*; and finally of a single layer of horny oblong cells without visible nuclei, called *Henle's layer*.

The hair-follicle contains the root of the hair, which terminates in a bulbous extremity, and is excavated so as to exactly fit the papilla from which it grows. The bulb is composed of polyhedral epithelial cells, which as they pass upwards into the root of the hair become elongated and spindle-shaped, except some in the centre which remain polyhedral. Some of these latter cells contain pigment-granules, which give rise to the colour of the hair. It occasionally happens that these pigment-granules completely fill the cells in the centre of the bulb; this gives rise to the dark tract of pigment often found, of greater or less length, in the axis of the hair.

The *shaft of the hair* consists of a central pith or medulla, the fibrous part of the hair, and the cortex externally. The *medulla* occupies the centre of the shaft and ceases towards the point of the hair. It is usually wanting in the fine hairs covering the surface of the body, and commonly in those of the head. It is more opaque and deeper coloured when viewed by transmitted light than the fibrous part; but when viewed by reflected light it is white. It is composed of rows of polyhedral cells, which contain granules of eleidin and frequently air-bubbles. The *fibrous* portion of the hair constitutes the chief part of the shaft; its cells are elongated and unite to form flattened fusiform fibres. Between the fibres are found minute spaces which contain either pigment-granules in dark hair, or minute air-bubbles in white hair. In addition to this there is also a diffused pigment contained in the fibres. The cells which form the *cortex* of the hair consist of a single layer which surrounds those of the fibrous part; they are converted into thin, flat scales, having an imbricated arrangement.

Connected with the hair-follicles are minute bundles of involuntary muscular fibres, termed the *arrectores pili*. They arise from the superficial layer of the corium, and are inserted into the outer surface of the hair-follicle, below the entrance of the duct of the sebaceous gland. They are placed on the side towards which the hair slopes, and by their action elevate the hair (fig. [84]).*

The **sebaceous glands** are small, sacculated, glandular organs, lodged in the substance of the corium. They are found in most parts of the skin, but are especially abundant in the scalp and face: they are also very numerous around the apertures of the anus, nose, mouth, and external ear, but are wanting in the palms of the hands and soles of the feet. Each gland consists of a single duct, more or less capacious, which terminates in a cluster of small secreting pouches or saccules. The sacculi connected with each duct vary, as a rule, in number from two to five, but, in some instances, may be as many as twenty. They are composed of a transparent, colourless membrane, enclosing a number of epithelial cells. Those of the outer or marginal layer are small and polyhedral, and are continuous with the lining cells of the duct. The remainder of the sac is filled with larger cells, containing fat, except in the centre, where the cells have become broken up, leaving a cavity containing their débris and a mass of fatty matter, which constitutes the sebaceous secretion. The orifices of the ducts open most frequently into the hair-follicles, but occasionally upon the general surface, as in the labia minora and the free margin of the lips. On the nose and face the glands are of large size, distinctly lobulated, and often become much enlarged from the accumulation of pent-up secretion. The largest sebaceous glands are those found in the eyelids—the Meibomian glands.

---

* Arthur Thomson suggests that the contraction of these muscles on follicles which contain weak, flat hairs will tend to produce a permanent curve in the follicle, and this curve will be impressed on the hair which is moulded within it, so that the hair, on emerging through the skin, will be curled. Curved hair-follicles are characteristic of the scalp of the Bushman.

The **sudoriferous** or **sweat glands** are the organs by which a large portion of the aqueous and gaseous materials are excreted by the skin. They are found in almost every part of this structure, and are situated in small pits on the under surface of the corium, or, more frequently, in the subcutaneous areolar tissue, surrounded by a quantity of adipose tissue. They are small, lobular, reddish bodies, consisting of a single convoluted tube, from which the efferent duct proceeds upwards through the corium and cuticle, becomes somewhat dilated at its extremity, and opens on the surface of the cuticle by an oblique valve-like aperture. The efferent duct, as it passes through the epidermis, presents a spiral arrangement, being twisted like a corkscrew, in those parts where the epidermis is thick ; where, however, it is thin, the spiral arrangement does not exist. In the superficial layers of the corium the duct is straight, but in the deeper layers it is convoluted or even twisted. The spiral course of these ducts is especially distinct in the thick cuticle of the palm of the hand and sole of the foot. The size of the glands varies. They are especially large in those regions where the amount of perspiration is great, as in the axillæ, where they form a thin, mammillated layer of a reddish colour, which corresponds exactly to the situation of the hair in this region ; they are large also in the groin. Their number varies. They are most numerous on the palm of the hand, presenting, according to Krause, 2,800 orifices on a square inch of the integument, and are rather less numerous on the sole of the foot. In both of these situations the orifices of the ducts are exceedingly regular, and open on the curved ridges. In other situations they are more irregularly scattered, but the number in a given extent of surface presents a fairly uniform average. In the neck and back they are least numerous, their number amounting to 417 on the square inch (Krause). Their total number is estimated by the same writer at 2,381,248, and, supposing the aperture of each gland to represent a surface of $\frac{1}{56}$ of a line in diameter, he calculates that the whole of these glands would present an evaporating surface of about eight square inches. Each gland consists of a single tube intricately convoluted, terminating at one end by a blind extremity, and opening at the other end upon the surface of the skin. In the larger glands this single duct usually divides and subdivides dichotomously ; the smaller ducts ultimately terminating in short cæcal pouches, rarely anastomosing. The wall of the duct is thick, the width of the canal rarely exceeding one-third of its diameter. The tube, both in the gland and where it forms the excretory duct, consists of two layers—an outer, formed by fine areolar tissue, and an inner layer of epithelium (fig. [88]). The external or fibro-cellular coat is thin, continuous with the superficial layer of the corium, and extends only as high as the surface of the true skin. The epithelial lining in the distal part of the coiled tube of the gland proper consists of a single layer of cubical epithelium, supported on a basement-membrane, and beneath it, between the epithelium and the fibro-cellular coat, a layer of longitudinally or obliquely arranged fibres, which are usually regarded as muscular, though the evidence that this is so is not conclusive. In the duct and the proximal part of the coiled tube of the gland proper there are two or more layers of polyhedral cells, lined on their internal surface, i.e. next the lumen of the tube, by a delicate membrane or

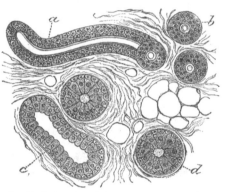

Fig. [88].—Coiled tube of a sweat-gland cut in various directions.

a. Longitudinal section of the proximal part of the coiled tube. *b.* Transverse section of the same. *c.* Longitudinal section of the distal part of the coiled tube. *d.* Transverse section of the same. (From Klein and Noble Smith's ' Atlas of Histology.')

cuticle, and on their outer surface by a limiting membrana propria, but there are no muscular fibres. The epithelium is continuous with the epidermis and with the delicate internal cuticle of the epidermic portion of the tube. When the cuticle is carefully removed from the surface of the cutis, these convoluted tubes of epithelium may be drawn out and form short, thread-like processes on its under surface.

The contents of the smaller sweat-glands are quite fluid; but in the larger glands the contents are semi-fluid and opaque, and contain a number of coloured granules and cells which appear analogous to epithelial cells.

SEROUS MEMBRANES

The **serous membranes** form shut sacs and may be regarded as lymph-sacs, from which lymphatic vessels arise by stomata or openings between the endothelial cells (see page [62]). The sac consists of one portion which is applied to the walls of the cavity which it lines—the *parietal* portion; and another reflected over the surface of the organ or organs contained in the cavity—the *visceral* portion. Sometimes the sac is arranged quite simply, as the tunica vaginalis testis; at others with numerous involutions or recesses, as the peritoneum, in which, nevertheless, the membrane can always be traced continuously around the whole circumference. The sac is completely closed, so that no communication exists between the serous cavity and the parts in its neighbourhood. An apparent exception exists in the peritoneum of the female; for the Fallopian tube opens freely into the peritoneal cavity in the dead subject so that a bristle can be passed from the one into the other. But this communication is closed during life, except at the moment of the passage of the ovum out of the ovary into the tube, as is proved by the fact that no interchange of fluids ever takes place between the two cavities in dropsy of the peritoneum, or in accumulation of fluid in the Fallopian tubes.* The serous membrane is often supported by a firm, fibrous layer, as is the case with the pericardium, and such membranes are sometimes spoken of as 'fibro-serous.'

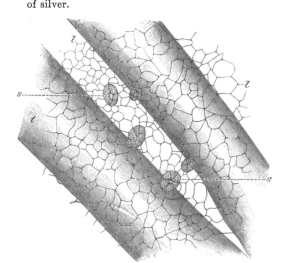

FIG. [89].—Part of peritoneal surface of the central tendon of diaphragm of rabbit, prepared with nitrate of silver.

s. Stomata. l. Lymph-channels. t. Tendon-bundles. The stomata are surrounded by germinating epithelial cells. (From ' Handbook for the Physiological Laboratory.' Klein.)

The various serous membranes are the peritoneum, lining the cavity of the abdomen; the two pleuræ and the pericardium, covering the lungs and heart respectively; and the tunicæ vaginales, surrounding each testicle in the scrotum.†

* The communication between the uterine cavity and the peritoneal sac is not only apparent in the dead subject, but is an anatomical fact, which is established by the continuity of its epithelium with that covering the uterus, Fallopian tubes, and fimbriæ.

† The arachnoid membrane covering the brain and spinal cord was formerly regarded as a serous membrane, but is now no longer classed with them, as it differs from them in structure, and does not form a shut sac as do the other serous membranes.

Serous membranes are thin, transparent, glistening structures, lined on their inner surface by a single layer of polygonal or pavement endothelial cells, supported on a matrix of fibrous connective tissue, with networks of fine elastic fibres, in which are contained numerous capillaries and lymphatics. On the surface of the endothelium between the cells numerous apertures or interruptions are to be seen. Some of these are stomata, surrounded by a ring of cubical endothelium (see fig. [89]), and communicate with a lymphatic capillary; others (*pseudostomata*) are mere interruptions in the endothelial layer, and are occupied by processes of the branched connective-tissue corpuscle of the subjacent tissue or by accumulations of the intercellular cement-substance.

The amount of fluid contained in these closed sacs is, in most cases, only sufficient to moisten the surface, but not to furnish any appreciable quantity of fluid. When a small quantity can be collected, it is found to resemble lymph, and like that fluid coagulates spontaneously; but when secreted in large quantities, as in dropsy, it is a more watery fluid, but still contains a considerable amount of proteid which is coagulated on boiling.

SYNOVIAL MEMBRANES

Synovial membranes, like serous membranes, are connective-tissue membranes placed between two movable tissues, so as to diminish friction, as in movable joints; or between a tendon and a bone, where the former glides over the latter; and between the skin and various subcutaneous bony prominences.

FIG. [90].—Villus of synovial membrane. (After Hammar.)

The synovial membranes are composed essentially of connective tissue, with the cells and fibres of that structure, containing numerous vessels and nerves. It was formerly supposed that these membranes were analogous in structure to the serous membranes, and consisted of a layer of flattened cells on a basement-membrane. No such continuous layer, however, exists, although here and there are patches of cells probably epithelial in nature. They are surrounded and held together by an albuminous ground-substance. Long villus-like processes (fig. [90]) are often found projecting from the surface of synovial membranes; they are covered by small rounded cells, and are supposed to extend the surface for the secretion of the fluid which moistens the membranes and which is named *synovia*. It is a rich lymph, plus a mucin-like substance, and to the latter constituent it owes its viscidity. A further description of the synovial membranes will be found in the descriptive anatomy of the joints.

MUCOUS MEMBRANES

Mucous membranes line all those passages by which the internal parts communicate with the exterior, and are continuous with the skin at the various orifices of the surface of the body. They are soft and velvety, and very vascular, and their surface is coated over by their secretion, *mucus*, which is of a tenacious consistence, and serves to protect them from the foreign substances introduced into the body with which they are brought in contact.

They are described as lining the two tracts—the gastro-pulmonary and the genito-urinary; and all, or almost all, mucous membranes may be classed as belonging to and continuous with the one or the other of these tracts.

The deep surfaces of these membranes are attached to the parts which they line by means of connective tissue, which is sometimes very abundant, forming a loose and lax bed, so as to allow considerable movement of the opposed surfaces on each other. It is then termed the *submucous tissue*. At other times it is exceedingly scanty, and the membrane is closely connected to the tissue beneath; sometimes, for example, to muscle, as in the tongue; sometimes to cartilage, as in the larynx; and sometimes to bone, as in the nasal fossæ and sinuses of the skull.

In structure a mucous membrane is composed of *corium* and *epithelium*. The epithelium is of various forms, including the squamous, columnar, and ciliated, and is often arranged in several layers. This epithelial layer is supported by the corium, which is analogous to the dermis of the skin, and consists of connective tissue, either simply areolar, or containing a greater or less quantity of lymphoid tissue. This tissue is usually covered on its external surface by a transparent basement-membrane generally composed of clear flattened cells, placed edge to edge; on this the epithelium rests. It is only in some situations that the basement-membrane can be demonstrated. The corium is an exceedingly vascular membrane, containing a dense network of capillaries, which lie immediately beneath the epithelium and are derived from small arteries in the submucous tissue.

The fibro-vascular layer of the corium contains, besides the areolar tissue and vessels, unstriped muscle-cells, which form in many situations a definite layer, called the *muscularis mucosæ*. These are situated in the deepest part of the membrane, and are plentifully supplied with nerves. Other nerves pass to the epithelium and terminate between the cells. Lymphatic vessels are found in great abundance, commencing either by cæcal extremities or in networks, and communicating with plexuses in the submucous tissue.

Embedded in the mucous membrane are found numerous glands, and projecting from it are processes (villi and papillæ) analogous to the papillæ of the skin. These glands and processes, however, exist only at certain parts, and it will be more convenient to defer their description to the sequel, where they will be described as they occur.

SECRETING GLANDS

The **secreting glands** are organs whose cells produce, by the metabolism of their protoplasm, certain substances, called 'secretions,' of a more or less definite composition; the material for the secretion being primarily selected from the blood. The essential parts therefore of a secreting gland are *cells*, which have the power of extracting from the blood certain matters, and in some cases converting them into new chemical compounds; and *blood-vessels*, by which the blood is brought into close relationship with these cells. The general arrangement in all secreting structures—that is to say, not only in secreting glands, but also in secreting membranes—is that the cells are arranged on one surface of an extravascular basement-membrane, which supports them, and a minute plexus of capillary vessels ramifies on the other surface of the membrane. The cells then extract from the blood certain constituents which pass through the membrane into the cells, where they are prepared and elaborated. The basement-membrane does not, however, always exist, and any free surface would appear to answer the same purpose in some cases.

By the various modifications of this secreting surface the different glands are formed. This is generally effected by an invagination of the membrane in different ways, the object being to increase the extent of secreting surface within a given bulk.

In the simplest form a single invagination takes place, constituting a *simple* gland; this may be either in the form of an open tube (fig. [91], A), or the walls of the tube may be dilated so as to form a saccule (fig. [91], B). These are named the *simple tubular* or *saccular* glands. Or, instead of a short tube, the invagination may be lengthened to a considerable extent, and then coiled up to occupy less space. This constitutes the *simple convoluted tubular* gland, an example of which may be seen in the sweat-glands of the skin (fig. [91], C).

FIG. [91].—Diagrammatic plan of varieties of secreting glands.

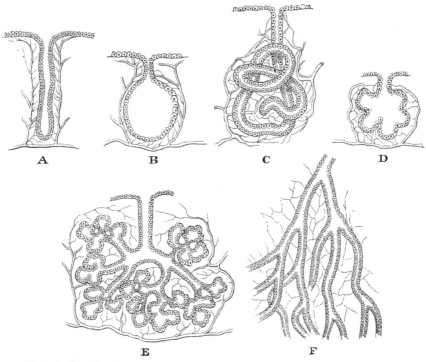

A. Simple gland. B. Sacculated simple gland. C. Simple convoluted tubular gland. D, E. Racemose gland.
F. Compound tubular gland.

If, instead of a single invagination, secondary invaginations take place from the primary one, as in fig. [91], D and E, the gland is then termed a compound one. These secondary invaginations may assume either a saccular or tubular form, and so constitute the two subdivisions—the *compound saccular* or *racemose* gland, and the *compound tubular*. The racemose gland in its simplest form consists of a primary invagination which forms a sort of duct, upon the extremity of which are found a number of secondary invaginations, called saccules or alveoli, as in Brunner's glands (fig. [91], D). But, again, in other instances, the duct, instead of being simple, may divide into branches, and these again into other branches, and so on; each ultimate ramification terminating in a dilated cluster of saccules, and thus we may have the secreting surface almost indefinitely extended, as in the salivary glands (fig. [91], E). In the *compound tubular* glands the division of the primary duct takes place in the same way as in the racemose glands, but the branches retain their tubular form, and do not terminate in saccular recesses, but become greatly lengthened out (fig. [91], F). The best example of this form of gland is to be found in the kidney.   All these varieties of glands are produced by a more or less complicated invagination of a secreting membrane, and they are all identical in structure ; that is to say, the saccules or tubes, as the case may be, are lined with cells, generally spheroidal or columnar in figure, and on their outer surface is an intimate plexus of capillary vessels.   The secretion, whatever it may be, is eliminated by the cells from the blood, and is poured into the saccule or tube, and so finds its way out through the primary invagination on to the free surface of the secreting membrane.   In addition, however, to these glands, which are formed by an *invagination* of the secreting membrane, there are some few others which are formed by a *protrusion* of the same structure, as in the vascular fringes of synovial membranes.   This form of secreting structure is not nearly so frequently met with.

EMBRYOLOGY

THE OVUM

THE whole body is developed out of the **ovum** or female element (figs. [92] and [93]) after it has been fertilised by the spermatozoon or male element. The ovum is a simple nucleated cell, and all the complicated changes by which the various intricate organs of the body are formed from it may be reduced to two general processes, viz. the *segmentation* or *cleavage* of cells, and their *differentiation*. The former process consists in the division of the nucleus and the surrounding cell-substance, whereby the original cell is represented by two. The differentiation of cells is a term used to describe that unknown power or tendency impressed on cells, apparently identical in structure, whereby they grow into different forms; so that (to take one of the first phenomena which occurs in the growth of the embryo) the indifferent cells of the vascular area are differentiated, some of them into blood-globules, others into the solid tissue which forms the blood-vessels. The extreme complexity of the process of development renders it at all times difficult to describe intelligibly, and still more so in a work like this, where adequate space and illustration can hardly be afforded, having respect to the main purpose of the work, and therefore an outline of the principal facts only will be given. Many of the statements which are accepted in human embryology are made on the strength of what has been observed to occur in the lower animals, and their existence in the human subject is merely a matter of inference.

Fig. [92].—Human ovum from a middle-sized follicle. Magnified 350 times.

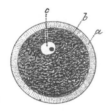

a. Zona pellucida or zona radiata. *b.* External border of the yolk and internal border of the vitelline membrane. *c.* Germinal vesicle and germinal spot.

Within recent years, however, much has been added to our knowledge of the development of the human embryo, and this more especially by the important researches of Professor His and others.

The **ovum** is a small spheroidal body situated in the immature Graafian follicle near its centre, but in the mature one in contact with the membrana granulosa,* at that part of the follicle which projects from the surface of the ovary. The cells of the membrana granulosa are accumulated round the ovum in greater number than at any other part of the follicle, forming a kind of granular zone, the *discus proligerus.*

The **human ovum** (fig. [92]) is extremely minute, measuring from $\frac{1}{150}$ to $\frac{1}{125}$ of an inch in diameter. It is a cell consisting externally of a transparent, striated envelope, the *zona pellucida, zona radiata,* or *vitelline membrane.* The extra-nuclear protoplasm contained within the zona pellucida is known as *the cytoplasm*; it is a sponge-like material, containing in its meshes numerous large fatty and albuminous granules, which constitute the *yolk* or *vitellus*; in the neighbourhood of the nucleus, however, these granules are comparatively few in

* See the description of the ovary at a future page.

number. The nucleus is a large spherical body, which is known by the name of the *germinal vesicle*, and resembles in structure the nucleus of an ordinary cell. Within it there is generally one nucleolus, which is large and well marked, and is known as the *germinal spot*. The zona pellucida is believed to be pierced by numerous pores which are probably channels of nutrition and which give it the appearance of being radially striated, while in some animals (e.g. insects) it presents a small perforation or hole, which is known by the name of the *micropyle*, and is believed to be the means by which the spermatozoa enter the ovum.

The phenomena attending the discharge of the ova from the Graafian follicles, since they belong as much or more to the ordinary function of the ovary than to the general subject of the development of the body, are described with the anatomy of the ovaries on a subsequent page.

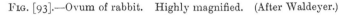

Fɪɢ. [93].—Ovum of rabbit. Highly magnified. (After Waldeyer.)

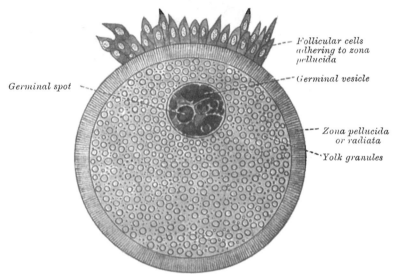

Germinal spot

Follicular cells adhering to zona pellucida

Germinal vesicle

Zona pellucida or radiata

Yolk granules

Maturation of the Ovum previous to Fertilisation.—Either before or immediately after its escape from the Graafian follicle, important changes take place in the nucleus of the ovum, which result in its partial disappearance and in the formation and extrusion from the yolk of two peculiar bodies, the *polar bodies* or *polar globules* of Robin. These changes constitute what is termed the *maturation of the ovum*, and are preparatory to its being fertilised by the male element or spermatozoon. The nucleus approaches the periphery of the ovum and undergoes the changes associated with karyokinesis; it then divides into two, and the upper daughter nucleus, with a thin investment of protoplasm, becomes extruded as the first polar body into a space between the yolk and the vitelline membrane, which has been formed in consequence of a contraction or shrinking of the yolk. The lower daughter nucleus undergoes the same process of division, and forms a second polar body, which is in like manner extruded (fig. [94]). The greater part (three-fourths) of the original nucleus is therefore expelled from the yolk in the form of the two polar bodies, and the remaining fourth, which is now called the *female pro-nucleus*, recedes towards the centre of the ovum. The shrinking of the vitellus still continues, and a fluid—the *peri-vitelline* fluid—collects in the space between it and the zona pellucida; in it, spermatozoa, which have passed through the zona pellucida, may sometimes be seen.

Although the process of maturation has been closely followed in many of the lower animals, it has not yet been successfully demonstrated in mammals.

It is interesting to note that a similar nuclear reduction occurs in connection with the development of spermatozoa. In the germinal ridge, which is to become the future testicle, certain cells, identical with primitive ova, are found. These

Fig. [94].—Formation of polar bodies in asterias glacialis. (Hertwig.)

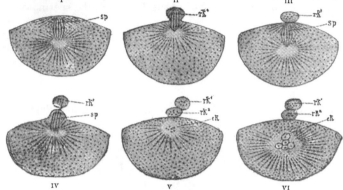

In fig. I. the polar spindle (*sp*) has advanced to the surface of the egg. In fig. II. a small elevation (*rk*) is formed which receives a half of the spindle. In fig. III. the elevation is constricted off, forming a polar body (*rk′*). Out of the remaining half of the previous spindle a second complete spindle (*sp*) has arisen. In fig. IV. is seen a second elevation which in fig. V. has become constricted off as the second polar body. Out of the remainder of the spindle (fig. VI.) is developed the female pronucleus.

are termed *spermatoblasts*, and they become enlarged to form what are called *spermatocytes*, while each spermatocyte ultimately divides into four *spermatids*. The spermatids become changed, without further subdivision, into spermatozoa, and hence the fully developed spermatozoon contains only one-fourth of the nucleus of the original spermatocyte. The matured ovum and the spermatozoon may therefore be looked upon as of the same morphological value.

FERTILISATION AND SEGMENTATION OF THE OVUM

The first changes in the ovum which take place at the time of conception are as follows : 1. **Impregnation.**—One, or perhaps more, spermatozoa penetrate the zona pellucida and are contained in the perivitelline fluid. A single

Fig. [95].—Fertilisation of the ovum of an echinoderm.

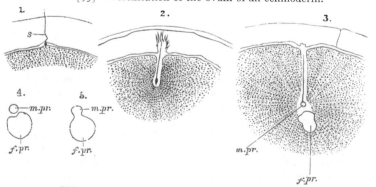

s. Spermatozoon. *m. pr.* Male pronucleus. *. pr.* Female pronucleus. 1. Accession of a spermatozoon to the periphery of the vitellus. 2. Its penetration. 3. Transformation of the head of the spermatozoon into the male pronucleus. 4, 5. Blending of the male and female pronuclei. (From Quain's 'Anatomy,' Selenka.)

spermatozoon, more advanced than the rest, becomes buried in the yolk, the tail disappears, and the head constitutes the *male pronucleus*. This gradually approaches the female pronucleus, and ultimately the two pronuclei come into contact and fuse to form a new nucleus, containing both male and female elements, and named the *segmentation* or *cleavage* nucleus, and the whole cell thus modified is called the blastosphere (fig. [95]). It seems as if this normally occurs in the Fallopian tube,* but it is possible that it sometimes takes place

FIG. [96].—First stages of segmentation of a mammalian ovum : semi-diagrammatic.
(From a drawing by Allen Thomson.)

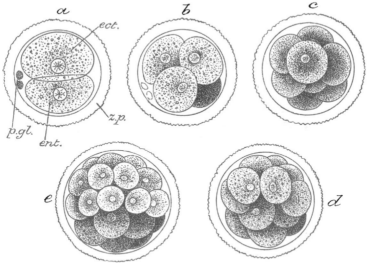

z.p. Zona pellucida. *p. gl.* Polar globules. *ect.* Epiblastic cell. *ent.* Hypoblastic cell. *a.* Division into two spheres. *b.* Stage of four spheres. *c.* Eight spheres, the epiblastic cells partially enclosing the hypoblastic cells. *d, e.* Succeeding stages of segmentation, showing the more rapid division of the epiblastic cells and the enclosure of the hypoblastic cells by them.

before the ovum has entered the tube, or even after it has passed through the tube and reached the cavity of the uterus; abnormally it may take place in the peritoneal cavity.

2. **Segmentation**.—The first result of fertilisation is a cleavage or subdivision of the ovum, which is first cleft into two masses, the segmentation nucleus having

FIG. [97].—Ovum of the rabbit at the end of the process of segmentation.

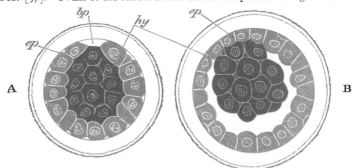

ep. Primitive epiblast. *hy.* Primitive hypoblast. *bp.* Place where the epiblast has not yet grown over the hypoblast.
(From Balfour, after Ed. van Beneden.)

previously split up into two; so that it now consists of two separate masses of protoplasm, each containing a nucleus, and situated within the original zona

* Many physiologists, as Bischoff and Dr. M. Barry, taught that the ovum is fecundated in the ovary, but the reasoning of Dr. Allen Thomson appears very cogent in proving that the usual spot at which the spermatozoa meet the ovum is in the tube, down which it slowly travels to the uterus, in its course becoming surrounded by an albuminous envelope derived from the walls of the tube.

pellucida, which takes no part in this process of division. Then, each of these two divides in like manner, and thus four are formed, and so on, until at length a mulberry-like agglomeration of nucleated masses of protoplasm results (fig. [96]). These masses are sometimes termed *vitelline spheres*.

The manner in which segmentation occurs is somewhat peculiar. The two cells resulting from the first cleavage are of unequal size. One, which for the sake of distinction may be called the *upper*, is slightly larger and paler than the other,

FIG. [98].—Blastodermic vesicle of Vespertilio murinus. (After Van Beneden.)
(Reduced from a drawing in the 'Anatomischer Anzeiger,' XVI Band, Sept. 8, 1899.)

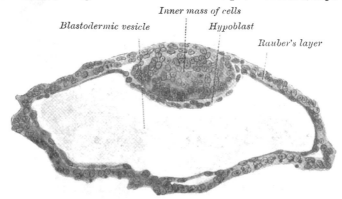

or *lower*. After they have subdivided three or four times the rate of cleavage in the cells derived from the upper becomes more rapid than that in the cells derived from the lower. In addition, the upper cells have a tendency to spread over and enclose the lower cells, so that by the ninth or tenth division there is an external layer of pale cells enclosing a mass of slightly smaller, more opaque cells, which, in consequence of their diminished rate of cleavage, are fewer in number (fig. [97]). Fluid collects between the two sets of cells, except at one part, termed the *embryonal pole*, so that a vesicle, the *blastodermic vesicle*, is formed. This vesicle consists of an outer layer of cells, termed Rauber's layer, derived from the sub-

FIG. [99].—Section through embryonic area of Vespertilio murinus. (After Van Beneden.)
(Reduced from a drawing in the 'Anatomischer Anzeiger,' XVI Band, Sept. 8, 1899.)

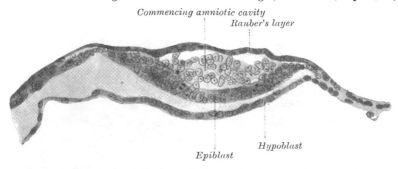

division of the primary upper cell, enclosing at the embryonal area an inner mass of cells (fig. [98]) resulting from the cleavage of the primary lower cell. Rauber's layer takes no share in the formation of the embryo proper, which is entirely developed from the inner mass of cells. The deepest cells of this mass become differentiated as a layer of flattened cells, termed the *hypoblast*, which spreads outwards beneath Rauber's layer. The latter, by subdivision of the cells of its upper hemisphere, is differentiated into two strata, the outer of which becomes rapidly thickened and forms a *plasmodioblast* (i.e. a mass of protoplasm containing numerous nuclei, but not subdivided into individual cells by means of cell-walls) ; the inner layer assumes the form of a prismatic epithelium and is named the *cytoblast* (fig. [100]). These two layers form the *ectoplacenta* or *chorion*, and entirely replace the lining epithelium of the uterus, where the blasto-dermic vesicle comes into contact with it. According to Van Beneden the cells of the inner mass partly undergo atrophy (fig, [99]), giving rise to a cavity, limited

FIG. [100].—Section through embryonic area of Vespertilio murinus (after Van Beneden), to show the formation of the amniotic cavity. (Reduced from a drawing in the 'Anatomischer Anzeiger,' XVI Band, Sept. 8, 1899.)

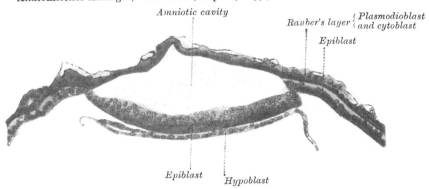

Amniotic cavity

Rauber's layer { *Plasmodioblast and cytoblast*

Epiblast

Epiblast *Hypoblast*

above by the cytoblast and below by a layer of cells, which constitutes the primitive upper layer of the embryo, the *epiblast* or *ectoderm*, and which is con-tinuous peripherally with the cytoblast. The cavity thus formed is the *primitive amniotic cavity*, and becomes the permanent amniotic cavity in man and monkeys, and in some of the bats (fig. [100]). It will thus be seen that from the inner mass of cells two layers are formed an outer of prismatic cells, the epiblast or ecto-derm, and an inner of flattened cells, the hypoblast or entoderm—and this double layer constitutes the *blastodermic membrane*, which at this stage is *bilaminar.**

Formation of the Mesoblast.—At first the area of the blastodermic membrane assumes the form of a small disc, the *germinal disc* or *germinal area*. This disc becomes oval in shape, with its more pointed end situated posteriorly. In it the first traces of the embryo are seen as a faint streak, the *primitive streak* (fig. [101]), which makes its appearance at the posterior or narrow end of the oval disc and from there gradually extends for-wards. The epiblast covering the primitive streak becomes indented by a groove, the *primitive groove*, the anterior end of which communicates through a canal with the yolk sac, forming what is termed the *blastopore*. The primitive streak results from a multiplication of the cells of the epiblast, so that it becomes thickened and grows downwards towards the hypoblast, which also undergoes proliferation. Together they form a thick cellular column, in which it is no longer possible to distinguish the epiblastic from the hypoblastic cells. From the sides of this column a layer of cells grows out between the epiblast and hypoblast, having been derived partly from both ; this layer consti-tutes the *mesoblast* or *mesoderm*.

FIG. [101].—Embryo of a rabbit of eight days. (After Kölliker.)

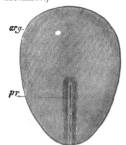

arg

pr

arg. Embryonic area. *pr*. Primitive streak.

In this way the blastodermic membrane comes to consist of three layers, and is now known as the *trilaminar blastoderm*. Each layer has distinctive characters,

* Consult, in this connection, articles by Van Beneden and Kollmann, *Anatomischer Anzeiger*, 1899 and 1900.

the outer and inner presenting the appearance of epithelial cells, while the middle consists of a mass of branched cells without any definite arrangement. The external is termed the *epiblast*, or *ectoderm*; the internal the *hypoblast*, or *entoderm*; and the middle, the *mesoblast*, or *mesoderm* (fig. [103]).

FIG. [102].—Embryonic area of the ovum of a rabbit at the seventh day.

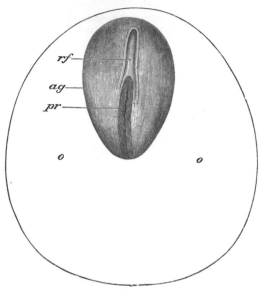

ag. Embryonic area. o, o. Region of the blastodermic vesicle immediately surrounding the embryonic area. pr. Primitive streak. rf. Medullary groove. (From Kölliker.)

The *epiblast* consists of a layer of columnar epithelial cells, which, however, are somewhat flattened towards the circumference of the germinal disc. It forms the whole of the nervous system (central and peripheral), the epidermis of the skin, the hairs and nails, the lining cells of the sebaceous, sweat, and mammary glands, the enamel of the teeth, and the epithelial lining of the nasal passages and of portions of the mouth and pharynx.

The *hypoblast* consists, at first, of flattened epithelial cells, which subsequently become columnar and even larger than those of the epiblast. It forms the epithelial lining of the whole of the alimentary canal except the anus and part of the mouth (which are developed from involutions of the epiblast), the epithelial lining of all the glands opening into the alimentary canal, the epithelium of the Eustachian tube and tympanic cavity, and of the trachea, bronchial tubes, and air-sacs of the lungs, the epithelium of the bladder and urethra, and also that which lines the follicles of the thyroid and thymus glands. The endothelial lining of the heart, blood-vessels, and serous cavities is also of hypoblastic origin, while recent observations tend to show that the primitive red blood-cells are derived from the same source.

FIG. [103].—Section across the anterior part of the medullary groove of an early embryo of the guinea-pig. (By Schäfer. From Quain's 'Anatomy,' 1890.)

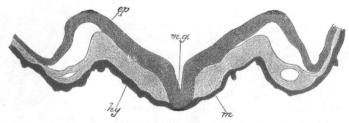

ep. Folds of epiblast rising up on either side of the middle line, and thus bounding the medullary groove. mg. Middle of medullary groove. hy. Hypoblast, which is in contact with the medullary epiblast at the middle of the groove, but is elsewhere separated from it by mesoblast, m, which has burrowed forwards between the two primary layers. A cleft is seen in the mesoblast on either side; this is the commencement of the anterior part of the body-cavity.

The *mesoblast* consists of loosely arranged branched cells, which are surrounded by a considerable amount of intercellular fluid, and which therefore may be considered as resembling embryonic connective tissue. All the other tissues of the embryo are developed from it, including the extra-endothelial portion of the walls of the blood-vessels, the skeleton and voluntary muscles, the connective tissues, the spleen, the generative and urinary organs (except the epithelium of the bladder and urethra), and the involuntary muscles.

FIRST RUDIMENTS OF THE EMBRYO

The *primitive streak* alluded to above is a very transitory structure, which merely marks the direction of the embryonic axis, the embryo proper being developed immediately in front of it in the following manner (figs. [102] and [106]).

First, two longitudinal ridges, caused by a looping or folding up of the epiblast, appear, one on either side of the middle line. These commence in the anterior part of the area germinativa, where they are united, and extend backwards, one on either side of the primitive streak, gradually enclosing it, and thus converting the blastopore into the *neurenteric canal*. This folding up of the epiblast gives rise to a longitudinal groove, the *medullary* or *neural groove* (figs. [102] and [103]), in consequence of the manner in which the cells of the epiblast are heaped up into two longitudinal ridges, with a furrow between them, so that the sides and floor of the groove are formed of epiblastic cells (fig. [103]). The mesoblast fills up the space between the epiblast and hypoblast, so that the sides of the groove are occupied by a longitudinal thickening of mesoblast; the two longitudinal thickenings of mesoblast being at first separated at the bottom of the groove by the junction of the epiblast and hypoblast (fig. [103]). The groove becomes deeper in consequence of the further growing up of the cells to form the ridge on either side. In this way the ridges eventually become two plates, the *laminæ dorsales* or *medullary plates*, which finally coalesce and form a closed tube, the *neural canal*, which is lined and covered by epiblast (figs. [104] and [105]). These lining and covering layers of epiblast are at first in contact with one another but eventually become separated by mesoblast which grows up between them. The coalescence of the medullary plates first takes place in the region of the future hind brain of the embryo, and then extends towards the cephalic and caudal ends. The posterior extremity presents a rhomboidal appearance before the laminæ close; this has been termed the *sinus rhomboidalis* (fig. [106]). The epiblast which lines the neural canal is developed into the nervous centres, that which covers the canal into the epidermis of the back and head. The cephalic extremity of the neural canal is soon seen to be more dilated than the rest, and to present constrictions dividing it imperfectly into three chambers : the brain is developed from this dilated portion ; the spinal cord takes its origin from the remainder of the tube. Below the neural canal, in front of the internal opening of the blastopore, a longitudinal groove forms in the hypoblast; this groove becomes closed off from the roof of the future enteron and forms a rod of cells which lies between the hypoblast and the neural canal. This rod of cells is known as the *notochord* or *chorda dorsalis*, and when fully developed is composed of clear epithelium-like cells enclosed in a homogeneous sheath (fig. [105]). It is essentially an embryonic structure, though traces

FIG. [104].—Transverse sections through the embryo chick before and some time after the closure of the medullary canal, to show the upward and downward inflections of the blastoderm. (After Remak.)

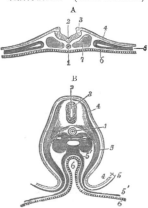

A. At the end of the first day. 1. Notochord. 2. Primitive groove in the medullary canal. 3. Edge of the dorsal lamina. 4. Corneous layer or epiblast. 5. Mesoblast divided in its inner part. 6. Hypoblast or epithelial layer. 7. Section of protovertebral plate. B. On the third day in the lumbar region. 1. Notochord in its sheath. 2. Medullary canal now closed in. 3. Section of the medullary substance of the spinal cord. 4. Corneous layer. 5. Somatopleure of the mesoblast. 5'. Splanchnopleure (one figure is placed in the pleuro-peritoneal cavity). 6. Layers of hypoblast in the intestines spreading also over the yolk. 4×5. Part of the fold of the amnion formed by epiblast and somatopleure.

of it remain in the centre of the intervertebral discs throughout life. The collection of mesoblastic cells, which forms a thick longitudinal column on either side of the neural canal, is termed the *paraxial* mesoblast, as distinguished from the outer or *lateral* part of the mesoblastic layer. The paraxial mesoblast undergoes a series of transverse segmentations and becomes converted into a row of well-defined, dark, square segments or masses, the *protovertebræ* or *mesoblastic somites*, separated by clear, transverse intervals (figs. [105] and [106]). They first make their appearance in the region which afterwards becomes the neck, and from there extend backwards along the entire length of the trunk. These bodies, as will be explained hereafter, are not the representatives of the permanent vertebræ, but are differentiated, partly into the vertebræ and partly into the muscles and true skin. On either side of the protovertebræ the lateral mesoblast splits into two layers; the upper becomes applied to the epiblast, forming with it the *somatopleure* or body wall, while the lower becomes attached to the hypoblast and with it forms the *splanchnopleure* or wall of the alimentary tube (fig. [104]). The space between them is the *cœlum* or *pleuro-peritoneal cavity* (fig. [105]). While the parietes of the body are still unclosed, this cavity is continuous with the space between the amnion and chorion, as seen in fig. [110]. The embryo, which at first seems to be a mere streak, extends longitudinally and laterally. As it grows forward the cephalic end becomes remarkably curved on itself (cephalic flexure), and a

FIG. [105].—Section across the dorsal part of a chick embryo of 45 hours' incubation. (Balfour.)

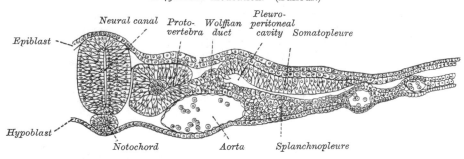

smaller but similar folding-over takes place at its hinder end (caudal flexure). At the same time the sides of the embryo, formed by the somatopleure, grow and curve ventrally towards each other; so that the embryo at this stage is aptly compared to a canoe turned over, and becomes marked off from the general blastoderm by a *limiting sulcus*. In consequence of this incurving of the embryo, both in an antero-posterior and a lateral direction, the blastodermic vesicle becomes nipped by the somatopleure and resembles an hour-glass with two unequal parts. The smaller portion is enclosed within the body of the embryo, and constitutes the *enteron* or primitive alimentary canal, while the larger portion, left outside the embryo, is termed the *yolk-sac* or *umbilical vesicle*. These two parts of the original blastodermic vesicle communicate through the constricted portion, which is the site of the future umbilicus, and, when the body cavity is ultimately closed at the umbilicus, the constriction is narrowed to form a small duct, the *omphalo-mesenteric* or *vitelline duct* (figs. [107], [108], [110]). The cephalic part of the primitive alimentary canal is named the *fore-gut*, the caudal portion the *hind-gut*, while the intermediate portion, which communicates directly with the yolk-sac, is termed the *mid-gut*. The yolk-sac is of small importance and very temporary duration in the human subject. It is for the purpose of supplying nutrition to the embryo during the very earliest period of its existence. In the oviparous animals, however, where no supply of nourishment can be obtained from the mother, since the egg is entirely separated from her, the yolk-sac is large and of great importance, as it supplies nutrition to the chick during the whole of fœtation. Vessels developed in the mesoblast soon cover the yolk-sac, forming the *vascular* area; these are named the *omphalo-mesenteric* vessels, and are two in number (fig. [109]). They appear to absorb the fluid of the yolk-sac which, when the fluid has disappeared, dries up and has no further function. The activity of the yolk-sac ceases about the fifth or sixth week, at the same time that the allantois, which is the great bond of vascular connection

FIG. [106].—Chick embryo of thirty-three hours' incubation, viewed from the dorsal aspect, × 30. (From Duval's ' Atlas d'Embryologie.')

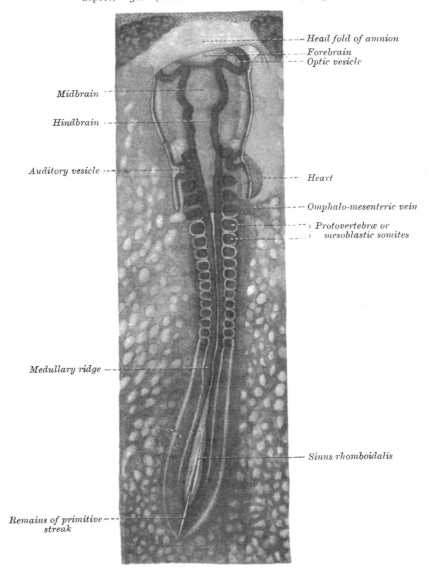

Head fold of amnion
Forebrain
Optic vesicle
Midbrain
Hindbrain
Auditory vesicle
Heart
Omphalo-mesenteric vein
Protovertebræ or mesoblastic somites
Medullary ridge
Sinus rhomboidalis
Remains of primitive streak

between the embryo and the uterine tissues, is formed. The yolk-sac remains visible, however, up to the fourth or fifth month, with its pedicle and the omphalo-mesenteric vessels. The latter vessels become atrophied as the functional activity of the body with which they are connected ceases.

So far we have traced : (1) the segmentation or cleavage of the ovum and the formation of a blastodermic vesicle, consisting of (a) an external envelope, and

FIG. [107].—Diagrams to illustrate the development of the embryo with its yolk-sac, amnion, and allantois.   (From Hertwig's ' Embryology.')

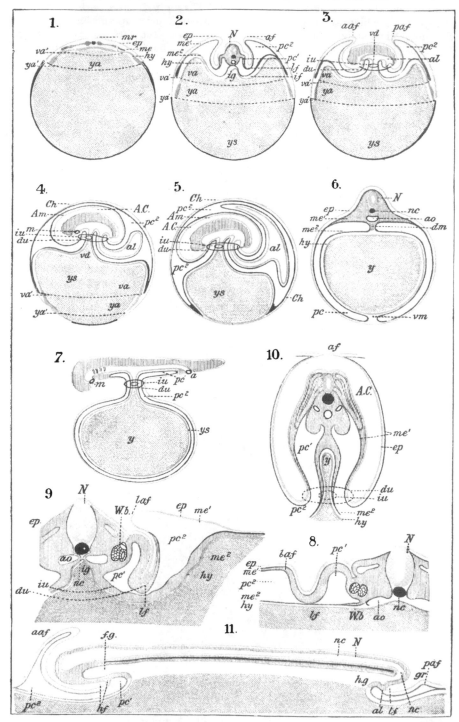

Figs. 1–5 are diagrammatic representations of cross and longitudinal sections through the hen's egg at different stages of incubation. Fig. 6 represents a cross-section through an embryo fish. Fig. 7. A longitudinal section through a Selachian embryo. Figs. 8 and 9. Half of a cross-section through an embryo chick of two and three days respectively (after Kölliker). Fig. 10. Cross-section through a five days' chick embryo (after Remak) ; and fig. 11. Longitudinal

(b) an internal mass of cells applied to it at the embryonal pole, but separated elsewhere by an albuminous fluid. (2) The separation of the hypoblast from the inner surface of this internal mass and its extension as a lining to the external envelope. (3) The development of the epiblast, also from the internal mass of cells, absorption taking place between it and the external layer to form a cavity, the primitive amniotic cavity. (4) The formation of an oval-shaped disc, the germinal disc, and the appearance of the primitive streak at its posterior end.

(5) The development of the mesoblast from the primitive streak and its extension between the epiblast and mesoblast. (6) The formation of the 'neural groove' in front of this primitive streak, caused by the growing-up of the epiblast on either side of it, so as to form two longitudinal ridges, called the 'laminæ dorsales.' (7) The increase and incurvation of these laminæ dorsales, until they meet dorsally and enclose the 'neural canal,' from the epiblastic lining of which the nervous centres are developed. (8) The formation, from the hypoblast immediately under this canal, of a continuous rod of cells, the 'chorda dorsalis' or 'notochord.' (9) The formation from the paraxial mesoblast, on either side of

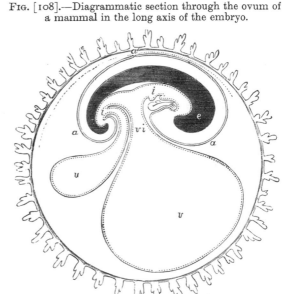

FIG. [108].—Diagrammatic section through the ovum of a mammal in the long axis of the embryo.

e. The cranio-vertebral axis. i, i. The cephalic and caudal portions of the primitive alimentary canal. a. The amnion. a'. The point of reflection into the false amnion. v. Yolk-sac, communicating with the middle part of the intestine by v i, the vitello-intestinal duct. u. The allantois. The ovum is surrounded externally by the villous chorion.

the notochord, of a number of square segments, the 'protovertebræ' or 'mesoblastic somites.' (10) The splitting of the lateral mesoblast into two layers to form the 'somatopleure' and 'splanchnopleure,' the space between the two constituting the 'cœlum' or 'pleuro-peritoneal cavity.' (11) The curving of the embryo on itself, both longitudinally and laterally, so as to be comparable to a canoe, part of the blastodermic vesicle being enclosed within the embryo to form the 'primitive alimentary tube,' part being left outside as the 'yolk-sac,' the two communicating by a duct, the 'omphalo-mesenteric' duct. The yolk-sac provides nutrition to the embryo through the omphalo-mesenteric vessels until such time as the placenta is formed.

FORMATION OF MEMBRANES

In order to have a clear understanding of the manner in which the embryo is developed, it is necessary at this stage to describe the development of the fœtal membranes.

section through a chick embryo. The epiblast is coloured blue; the mesoblast red, the hypoblast green, and the yolk-sac yellow. The reference letters apply to all the figures. (From Hertwig's 'Entwickelungsgeschichte.') ep. Epiblast. me. Mesoblast. hy. Hypoblast. mr. Medullary ridges. N. Neural canal. af. Amniotic fold. aaf, paf, laf. Anterior, posterior, and lateral amniotic folds. AC. Amniotic cavity. Am. Amnion. Ch. Chorion. du. Dermal umbilicus. lf. Lateral fold. hf. Head fold. vd. Vitello-intestinal duct. ig. Intestinal groove. al. Allantois. ys. yolk-sac. iu. Intestinal umbilicus. me¹. Somatopleure mesoblast. me². Splanchnopleure mesoblast. dm. Dorsal mesentery. vm. Ventral mesentery. pc¹, pc². Pleuro-peritoneal cavity. va. Vascular area. va'. Limit of vascular area. ya. Yolk area. ya'. Limit of yolk area. m. Mouth. nc. Notochord. ao. Aorta. a. Anus. y. Yolk mass. In fig. 10, y is placed in the primitive alimentary canal. Wb. Wolffian body. fg. Fore-gut. hg. Hind-gut. tf. Tail fold. nc. Neurenteric canal.

The **membranes** investing the fœtus are the amnion, the chorion, and the decidua. The first two are developed from fœtal structures, and are proper to the fœtus; the last is formed in the uterus, and is derived from the maternal structures.

The Amnion.—The amnion is the innermost of the membranes which surround the embryo. It is at first of small size, but increases considerably towards the middle of pregnancy, as the fœtus acquires the power of independent movement. It exists only in reptiles, birds, and mammals, which are hence called 'Amniota,' but is absent in amphibia and fishes. In man, monkeys, and some of the bats, the primitive amniotic cavity, already described on page [83], persists. In reptiles, birds and certain mammals the amnion is formed in the following manner. At or

FIG. [109].—Magnified view of the human embryo of four weeks, with the membranes opened.   (From Leishmann, after Coste.)

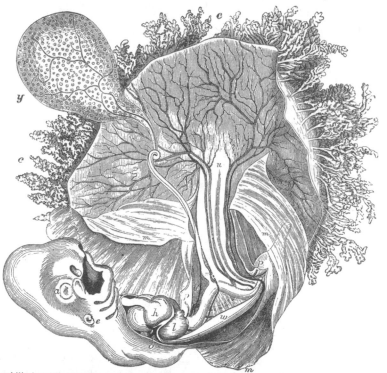

y. The umbilical vesicle with the omphalo-mesenteric vessels, *v*, and its long tubular attachment to the intestine. *c.* The villi of the chorion. *m.* The amnion opened. *u.* Cul-de-sac of the allantois, and on each side of this the umbilical vessels passing out to the chorion. In the embryo: *a.* The eye. *e.* The ear-vesicle. *h.* The heart. *l.* The liver. *o.* The upper; *p*, the lower limb. *w.* Wolffian body, in front of which are the mesentery and fold of intestine. The Wolffian duct and tubes are not represented.

near the extremities of the incurved fœtus—that is to say, at the point of constriction of the blastodermic vesicle where the primitive alimentary canal of the embryo joins the yolk-sac—a reflection or folding backwards of the somatopleure, which has become separated from the splanchnopleure by the formation of the pleuroperitoneal cavity, takes place (fig. [107], 2, 3). This fold commences first at the cephalic extremity, and subsequently at the caudal end and sides, and deepens more and more, in consequence of the sinking of the embryo into the blastodermic vesicle, until, gradually approaching, the different parts meet on the dorsal aspect of the embryo (figs. [107], 10, and [110]). After they come in contact they fuse together, and the septum between them disappears; so that the inner layer of the cephalic fold becomes continuous with the inner layer of the caudal and lateral folds, and the outer with the outer. Thus we have two membranes, one formed by the inner layer of the fold—*the true amnion*—which encloses a space over the back of the embryo—the *amniotic cavity* (fig. [107], 4, 5) —containing a clear fluid, the *liquor amnii*. The other, the outer layer of the fold—the *false amnion*—lines the internal surface of the original zona pellucida. Between the two is an interval, which of course communicates with the pleuro-peritoneal cavity until the body-walls of the embryo have coalesced at the umbilicus. Then the amniotic fold is carried downwards, and encloses the umbilical cord, by which the fœtus is attached to the placenta. The true amnion—or, as it is usually called, the amnion—is formed of two layers, inner and outer, derived respectively from the epiblast and from the parietal layer of the mesoblast.

The amnion is at first in close contact with the surface of the body of the embryo, but about the fourth or fifth week fluid begins to accumulate, and thus separates the two. The quantity of fluid steadily increases up to about the sixth month of pregnancy, after which it diminishes somewhat. The use of the liquor amnii is believed to be chiefly to allow of the movements of the fœtus in the later

Fig. [110].—Diagram of a transverse section of a mammalian embryo, showing the mode of formation of the amnion. The amniotic folds have nearly united in the middle line. (From Quain's 'Anatomy,' vol. i. pt. 1, 1890.)

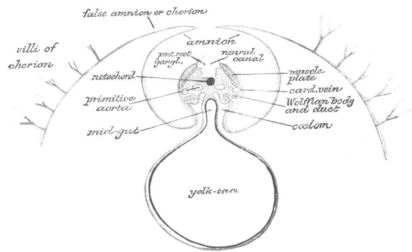

Epiblast, blue ; mesoblast, red ; hypoblast and notochord, black.

stages of pregnancy, though it no doubt serves other purposes. It contains about one per cent. of solid matter, chiefly albumen, with traces of urea, the latter probably derived from the urinary secretion of the fœtus.

The Chorion.—The chorion takes its origin, as has already been seen (page [83]), from the external covering of the blastodermic vesicle ; the cells of the decidua or uterine mucous membrane ·contributing no elements to it. From its outer surface numerous finger-like processes, termed the *villi of the chorion*, project. These increase rapidly in size and at the same time undergo great ramification ; hence they were likened by Dalton to tufts of seaweed (figs. [110] and [113]). They invade the decidua of the uterus and probably absorb from it nutritive materials for the growth of the embryo : they can be forcibly withdrawn from the decidua until the third month of pregnancy. Until about the end of the second month the villi cover the whole surface of the chorion and are of an almost uniform size, but after this they develop unequally. On that part which invades the decidua serotina they increase greatly in size and complexity, and constitute the *chorion frondosum*, which becomes the fœtal part of the placenta (fig. [113]). Over the remainder of the chorion they undergo atrophy, so that by the fourth month hardly any trace of them is left, and hence this part becomes smooth, and is therefore named the *chorion læve*. The chorionic villi are at first non-vascular, but subsequently they become vascularised by the growth into them of the allantoic mesoblast, which carries to them the branches of the allantoic arteries.

The Allantois.—The allantois grows outwards as a hollow bud from the hind gut and is therefore lined by hypoblast and covered by mesoblast (figs. [107], 4, 5, and [108]). It is projected into the space between the amnion and the chorion, and in its mesoblast are carried a pair of arteries, the *allantoic* or *umbilical arteries*, which are continued from the two primary aortæ. The allantoic mesoblast gradually spreads out on the inner surface of the chorion, and, invading the chorionic villi, supplies them with blood-vessels. In this way the allantois becomes the chief agent of the fœtal circulation, since it carries the vessels which convey the blood of the embryo to the chorion, where it is exposed to the influence of the maternal blood circulating in the decidua; from the maternal blood it imbibes the materials of nutrition and to it it gives up effete materials, the removal of which is necessary for the purification of the fœtal blood. In some animals the allantois is a hollow projection, and is usually styled the *allantoic vesicle*; but in most mammals, and especially in man, the external or mesoblastic element undergoes great development, while the internal or hypoblastic element undergoes little increase beyond the body of the embryo, so that it is very doubtful whether any cavity exists in the allantois beyond the limits of the umbilicus, or whether it does not rather consist of a solid mass of material derived from the mesoblastic tissue.* The proximal part of the allantoic vesicle within the body-cavity is eventually destined to form the bladder, while the remainder forms an impervious cord, the *urachus*, stretching from the summit of the bladder to the umbilicus. The part of the allantois external to the fœtus forms the umbilical cord, by which the fœtus is connected with the placenta.

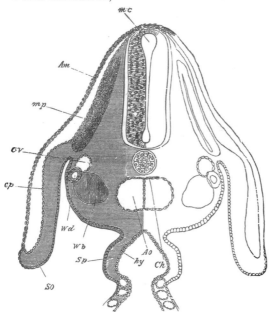

FIG. [111].—Transverse section through the dorsal region of an embryo-chick, end of third day. (From Foster and Balfour.)

Am. Amnion. *mp.* Muscle-plate. *cv.* Cardinal vein. *Ao.* Dorsal aorta, at the point where its two roots begin to join. *Ch.* Notochord. *Wd.* Wolffian duct. *Wb.* Commencement of formation of Wolffian body. *ep.* Epiblast. *So.* Somatopleure. *Hy.* Hypoblast. The section passes through the place where the alimentary canal (*hy*) communicates with the yolk-sac. *Sp.* Splanchnopleure.

The Decidua.—The growth of the chorion and placenta can only be understood by tracing the formation of the decidua.

The decidua is formed from the uterine mucous membrane before the fertilised ovum reaches the cavity of the uterus. The mucous membrane becomes vascular and tumid, its glands are greatly elongated, and their deeper portions are dilated and tortuous, while the interglandular tissue becomes crowded with epithelial-

* Indeed, it would appear, from the researches of His, that in the human embryo the allantois is formed unusually early, being present from a very early period as a stalk of mesoblast connecting the posterior extremity of the embryo with the chorion. This stalk is termed the *abdominal stalk* (Bauchstiel).

like cells (*decidual cells*). The mucous membrane, thus altered, is named the *decidua vera*; it lines the cavity of the uterus as far as the os internum, without, however, occluding the orifices of the Fallopian tubes. When the fertilised ovum reaches the uterus, which is thus prepared for its reception, it becomes attached to the decidua, in most cases in the neighbourhood of the fundus uteri. The decidua then grows up around the ovum and ultimately covers it in. The part of the decidua which grows up to envelop the ovum is named the *decidua reflexa*; that portion to which the ovum originally became attached is termed the *decidua serotina*, and from it the maternal part of the placenta is derived. After conception the cervix uteri is closed by a plug of mucus (fig. [113]).

By the fourth month the decidua vera has acquired a thickness of about half an inch, and consists of the following strata: (1) *Stratum compactum*, next the free surface, in which the uterine glands are little altered and where they preserve a comparatively narrow lumen lined by columnar epithelium; between the glands are large numbers of decidual cells. (2) *Stratum spongiosum*, in which the gland tubes are very tortuous and greatly dilated, while their lining cells are flattened or cubical. (3) *Basal layer*, next the uterine muscular wall, in which the glands are not dilated and where they retain their columnar epithelium. It is through this basal layer that the placenta is separated after the birth of the child (fig. [112]).

The decidua reflexa is gradually expanded by the growing ovum, and ultimately comes into contact and blends with the decidua vera so as to completely obliterate the uterine cavity. This obliteration is followed by the degeneration of the deciduæ; the glands of the decidua reflexa become atrophied and the entire decidua practically disappears, while the decidua vera is much thinned and its glands also disappear, except their deepest portions, which persist in the basal layer.

FIG. [112].—Diagrammatic sections of the uterine mucous membrane: (A) of the non-pregnant uterus; (B) of the pregnant uterus, showing the thickened mucous membrane and the altered condition of the uterine glands.

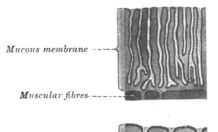

Mucous membrane -----

A

Muscular fibres -----

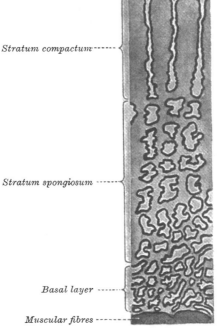

Stratum compactum -----

B

Stratum spongiosum -----

Basal layer -----

Muscular fibres -----

In this manner the embryo becomes surrounded by three membranes: (1) the *amnion*, derived, in the case of reptiles, birds, and some mammals, from the outer layer of the mesoblast and the epiblast; (2) the *chorion*, formed by the allantois (which is derived from the hypoblast and inner layer of the mesoblast) and the false amnion; and (3) the *decidua*, derived from the mucous membrane of the uterus.

Much additional interest has been given to the physiology of the decidua by the fact, which seems to be now established by the researches of Sir John

Williams, that every discharge of an ovum, whether impregnated or not, is, as a rule, accompanied by the formation of a decidua, and that the essence of menstruation consists in the separation of a decidual layer of the mucous membrane from the uterus; while in the case of pregnancy there is no exfoliation of the membrane, but, on the contrary, it undergoes further development in the manner described above.

Formation of the placenta.—The placenta is developed partly from maternal and partly from fœtal tissues—the maternal portion being derived from the decidua serotina, the fœtal, from the villi of the chorion frondosum. These villi penetrate the decidua serotina, which then undergoes a series of complicated and, as yet, imperfectly understood changes. Decidual cells accumulate between the uterine glands, while the glands with their epithelial lining undergo degeneration—a degeneration which does not, however, extend as deep as the basal layer, where the glands persist, and retain their epithelial lining throughout the

Fig. [113].—Sectional plan of the gravid uterus in the third and fourth month.
(Modified from Wagner.)

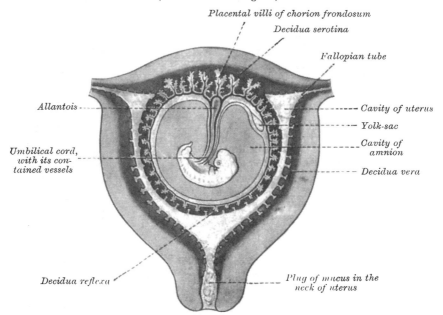

Placental villi of chorion frondosum

Decidua serotina

Fallopian tube

Allantois

Cavity of uterus

Yolk-sac

Cavity of amnion

Umbilical cord, with its contained vessels

Decidua vera

Decidua reflexa

Plug of mucus in the neck of uterus

entire period of gestation. Ultimately the superficial portion of the decidual tissue disappears, and the uterine vessels become expanded to form a labyrinth of freely communicating blood-channels or sinuses, which are filled with maternal blood, and in which are suspended the now greatly ramified tufts of the chorionic villi. These uterine sinuses anastomose freely with one another, and form, at the edge of the placenta, a venous channel with an irregular calibre, which runs round the whole circumference of the placenta, and is termed the *marginal sinus*. Some of the chorionic villi are attached by fibrous bands to the basal layer of the decidua and to the imperfect septa between the sinuses, but the majority of them hang free.

Circulation through the placenta.—The maternal blood is brought to the uterine sinuses by the 'curling arteries' of the uterus and drained away by the uterine veins, while, as already stated, within the chorionic villi are found the ramifications of the fœtal vessels derived from the allantoic or umbilical arteries. Since the villi are suspended in the sinuses they are necessarily bathed in the maternal blood, and hence it follows that the maternal and fœtal blood-currents are brought into close relationship. There is, however, no intermingling of the two currents, or, in other words, no direct communication between the vascular system of the mother and that of the fœtus—the interchange of materials necessary for the growth of the fœtus and for the purification of the fœtal blood taking place through the walls of the villi. The purified blood is carried back to the fœtus by the umbilical vein. From what has been said it will be understood that the placenta is the organ by which the connection between the fœtus and the mother is established, and which subserves the purposes of nutrition, respiration, and excretion.

Placenta.—At the end of the gestation period the placenta presents the form of a disc which weighs about a pound and has a diameter of from six to eight inches. Its average thickness is about an inch and a quarter, but diminishes rapidly near the circumference of the disc. Its outer or *decidual* surface blends with the uterine wall, but if examined after the separation of the placenta, it presents a comparatively smooth surface, which on inspection is seen to be incompletely divided into a number of masses named *cotyledons*. Its inner or *chorionic* surface is smooth, being closely invested by the amnion. The umbilical cord is attached near the centre of this surface, and from this attachment the larger branches of the umbilical vessels are seen radiating under the amnion. On section the placenta presents a soft, spongy appearance, caused by its freely communicating blood-sinuses with their contained villi. Owing to the rapid thinning of the placenta at the periphery of the disc, the decidual and chorionic surfaces come into contact.

Separation of the placenta.—After the birth of the child the placenta and the membranes (i.e. the amnion and chorion) are expelled as the *after-birth* —the separation of the placenta from the uterine wall taking place through the basal layer of the decidua and necessarily causing rupture of the uterine vessels. The orifices of the torn vessels are, however, closed by the firm contraction of the uterine muscular fibres, and this, together with the formation of a blood clot over the placental site, prevents *post-partum hæmorrhage*. The epithelial lining of the uterus is regenerated from the epithelium which lines the persistent portions of the uterine glands in the basal layer of the decidua.

The **umbilical cord** appears about the end of the fifth week after conception. It consists of the coils of two arteries and a single vein, the *umbilical arteries and vein*, united together by a gelatinous tissue (*jelly of Wharton*). There are originally two umbilical veins, but one of these vessels becomes obliterated, as do also the two omphalo-mesenteric arteries and veins and the duct of the umbilical vesicle, all of which are originally contained in the rudimentary cord. The umbilical cord also contains the remains of the allantois, and is covered externally by a layer of the amnion, reflected over it from the umbilicus.

DEVELOPMENT OF THE EMBRYO

The further development of the embryo will, perhaps, be better understood if we follow briefly the principal facts relating to the development of the chief parts of which the body consists—viz. the spine, the cranium, the pharyngeal cavity, mouth, &c., the nervous centres, the organs of the senses, the circulatory system, the alimentary canal and its appendages, the organs of respiration, and the genito-urinary organs. The reader is also referred to the chronological table of the development of the fœtus at the end of this section.

Development of the Spine.—The first steps in the formation of the spine have already been traced, viz.: (1) The looping up of two longitudinal folds from the cells of the epiblast in front of the primitive streak, to form the *neural groove*, and the gradual growing together of the *laminæ dorsales* so as to convert the groove into the *neural canal*. (2) The formation on the ventral aspect of this groove of a continuous cellular cord, the *notochord* or *chorda dorsalis* (fig. [105]), which extends from the cephalic to the caudal extremity of the embryo, and lies in the situation which is afterwards occupied by the bodies of the vertebræ. (3) The segmentation of the paraxial mesoblast on either side of the neural canal into a number of quadrilateral masses, the *protovertebræ* or *mesoblastic somites* (fig. [106]). The process of segmentation commences in the cervical region and proceeds successively through the other regions of the spine until a number of segments are formed, which correspond very closely to the number of the permanent vertebræ. Subsequently the protovertebral somites divide into two parts—a ventral and a dorsal. From the ventral division the vertebræ are formed; the dorsal is termed the *muscle-plate*, and from it the muscles of the back are developed. From the ventral division of the protovertebral somites masses of cells are budded off, which grow inwards towards the middle line, those from opposite sides meeting and enclosing the notochord and extending dorsally around the neural canal which they also envelop. Fusion of the ventral divisions also occurs in the antero-posterior direction so that all trace of their originally segmented condition is lost and the notochord and spinal cord

Fig. [114].—Cervical part of the primitive vertebral column and adjacent parts of an embryo of the sixth day, showing the division of the primitive vertebral segments. (From Kölliker, after Remak.)

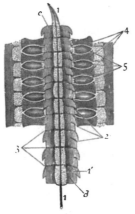

, 1. Chorda dorsalis in its sheath, pointed at its upper end. 2, points by three lines to the original intervals of the primitive vertebræ. 3, in a similar manner, indicates the places of new division into permanent bodies of vertebræ. c indicates the body of the first cervical vertebra; in this and the next the primitive division has disappeared, as also in the two lowest represented, viz. d and the one above; in those intermediate the line of division is shown. 4, points in three places to the vertebral arches; and 5, similarly, to three commencing ganglia of the spinal nerves: the dotted segments outside these parts are the muscular plates.

Fig. [115].—Longitudinal section of vertebral column of an eight weeks' human fœtus. (Kölliker.)

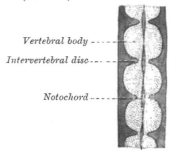

Vertebral body ----

Intervertebral disc ----

Notochord ----

are surrounded by a continuous investment of mesoblast, the *membranous vertebral column*. This investment also extends forwards and envelops the primitive brain, forming the *membranous* or *primordial cranium*. From this investment the base of the skull, the vertebræ and their ligaments, and the membranes of the cerebro-spinal nervous system are developed. The future vertebræ make their first appearance about the beginning of the second month in the form of two small masses of cartilage which are seen in the membranous vertebral column, one on each side of the notochord. These small masses lie opposite to the intervals between the muscle-plates and so alternate with these structures. They are soon joined across the middle line on the ventral aspect of the notochord by a hypochordal cartilaginous bar which ultimately disappears, except in the case of the atlas vertebra, where it forms the anterior arch of that bone. The vertebral bodies are formed immediately to the dorsal aspect of these hypochordal bars, alternating with the muscle-plates which represent the original mesoblastic somites. The notochord contained in the centre of this chondrifying mass does not continue to grow, but becomes in the human subject relatively smaller, so as, at last, to form a mere slender thread, except opposite the intervals between the bodies of the permanent vertebræ. Here it presents thickenings, and forms an irregular network, the remains of which are to be found at all periods of life in the central pulp of the intervertebral discs (figs. [114], [115], and [116]).

Development of the Ribs and Sternum.—The ribs are formed from the muscle-plates of the protovertebral somites, from which also the muscles of the back and the true skin of the body-wall are formed. The ribs consist of extensions of this mesoblastic material, which speedily undergo chondrification, and appear as cartilaginous bars, which become separated from the vertebræ at their posterior extremities. At their anterior ends the nine upper costal bars turn upwards and fuse together so as to form a cartilaginous strip bounding a central median fissure. The strips on either side then join in the middle line from before backwards, and so give rise to a longitudinal piece of cartilage, which represents the manubrium and gladiolus of the sternum. In the process of development the sternal attachment of the eighth rib disappears, while that of the ninth subdivides, one portion remaining attached to the inferior extremity of the cartilaginous sternum and becoming developed into the ensiform cartilage, the other portion receding from the sternum and becoming attached to the cartilage of the eighth rib.

FIG. [116].—Sagittal section through the intervertebral disc and adjacent parts of two vertebræ of an advanced sheep's embryo. (Kölliker.)

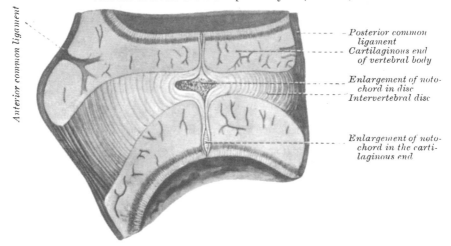

Anterior common ligament

Posterior common ligament
Cartilaginous end of vertebral body

Enlargement of noto-chord in disc
Intervertebral disc

Enlargement of noto-chord in the carti-laginous end

The further development of the vertebræ, ribs, and sternum, and the ossification of their cartilaginous framework, are described in the body of the work.

Development of the Cranium and Face.—It has been seen that the first trace of the embryo consists in the formation of a longitudinal fold of the epiblast on either side of the neural groove, and that these folds or ridges grow backwards and meet in the median line, thus forming the neural canal. This canal, at the cephalic extremity of the embryo, is dilated and forms a bulbous enlargement. The bulbous enlargement soon expands into three vesicular dilatations, the three *primary cerebral vesicles*, from which all the different parts of the encephalon are developed. The primary cerebral vesicles at this time freely communicate with each other at the points of constriction.

The three cavities are lined by epiblast and covered by the same structure. Between these two layers of epiblast, a layer of mesoblast spreads over the whole surface of the cerebral vesicles and forms the membranous cranium. From these structures the cranium and its contents are developed. The external layer of the epiblast forms the epidermis and hairs of the scalp. The mesoblastic layer forms the true skin, the blood-vessels (all but their endothelial lining), muscles,

connective tissue, bones of the skull and membranes of the brain.   The layer of
epiblast lining the vesicles forms the nervous substance of the encephalon, while
the vesicles themselves constitute the ventricles.

The cephalic end of the notochord terminates in a pointed extremity which
extends as far forwards as the situation of the future basi-sphenoid, and is
embedded in a mass of mesoblast, the 'investing mass of Rathke.'   The posterior
part of this mass, which corresponds to the future basi-occipital, shows a subdivision
into four segments, the three roots of the hypoglossal nerve indicating their lines
of separation.   Two cartilaginous bars, the *parachordal cartilages*, then become
developed in this investing mass, and these surround the notochord, meeting first on
its ventral and next on its dorsal aspect to form the *basilar plate*, the anterior margin
of which forms the future dorsum sellæ.   From this plate are developed the basi-
occipital and basi-sphenoid, and by lateral expansions from it the ex-occipitals
and the greater wings of the sphenoid.   On either side of the parachordal cartilage
a cartilaginous capsule, the *labyrinthine* or *periotic cartilage*, surrounds the otic
vesicle, and from it the petrous and mastoid portions of the temporal bone are
developed.   In front of the investing mass of Rathke two lateral bars are directed
forwards, enclosing between them a space, which forms the pituitary fossa, in

Fig. [117].—Diagrams of the cartilaginous cranium.   (Wiedersheim.)

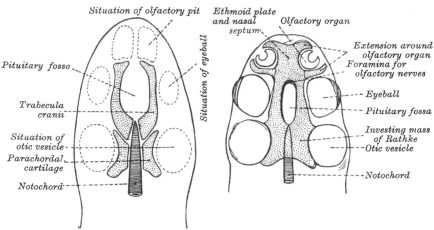

which the pituitary body is eventually developed.   These bars are named the
prechordal cartilages or *trabeculæ cranii*, and extend as far forwards as the
anterior extremity of the head, where they coalesce with each other to form the
ethmoid plate (fig. [117]).   This encloses the olfactory pits forming the cartilaginous nasal capsule, from which the ethmoid and turbinated bones are developed.
A portion of the ethmoid plate remains unossified and constitutes the cartilaginous
part of the nasal septum and the cartilages of the outer nose.   From the
trabeculæ cranii the pre-sphenoid is developed, and from this two lateral expansions extend to form its lesser wings ; each of these arises by two roots, one above
and one below the optic nerve, and, uniting outside the nerve, enclose the optic
foramen.   The base of the primitive cranium therefore consists of two parts,
prechordal and *parachordal* : the former receives the organ of smell and is indented
by the eyeball ; the latter surrounds the auditory vesicle.   Thus it will be seen
that the bones which form the base of the skull are preceded by masses of cartilage, which together form the *chondrocranium*.   Those of the vault of the skull,
on the other hand, are of membranous formation, and are termed *dermal* or
covering bones.   They are developed in the mesoblast which lies superficial to the
primordial cranium, or in that which lies subjacent to the epithelial lining of the
foregut.   They comprise the upper portion of the tabular part of the occipital

(interparietal), the squamous-temporals and tympanic rings, the two parietals, the frontal, the vomer, the internal pterygoid plates, and the bones of the face. Some of them remain distinct throughout life (e.g. parietal and frontal), while others join with the bones of the chondrocranium (e.g. interparietal, squamous-temporal, and internal pterygoid plates).

The head at first consists simply of a cranial cavity, the face and neck being subsequently developed in the manner now to be described.

In all vertebrate animals there is at one period of their development a series of grooves in the upper neck region of the embryo. These are named the *branchial* or *visceral clefts*, and in man are four in number from before backwards. They take origin as paired grooves or pouches from the side of the pharynx, and over each groove a corresponding indentation of the epiblast occurs, so that the latter comes into contact with the hypoblast lining the pharynx, and these two layers unite to form thin septa, along the bottom of the grooves, between the

Fig. [118].—Profile view of the head of a human embryo, estimated as twenty-seven days old. (After His.)

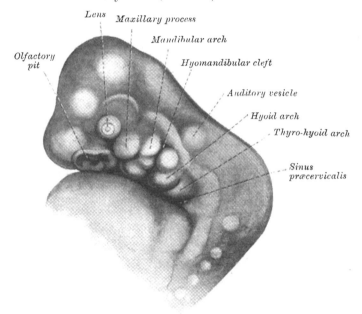

pharyngeal cavity and the exterior. In gill-bearing animals these septa disappear and the grooves become complete clefts, the gill clefts, opening from the pharynx on to the exterior; perforation does not, however, occur in birds and mammals. In front and behind each cleft the mesoblast becomes thickened in the form of arches, the *branchial arches* (figs. [118] and [155]). In the human embryo there are five pairs of these arches, one in front of the first cleft, one behind the last, and the three remaining ones between the first and second, the second and third, and the third and fourth clefts respectively. The first arch is named the *mandibular*; the second the *hyoid*; the third the *thyro-hyoid*, while the fourth and fifth have no distinctive names. In each arch there is developed a carti-laginous bar which gives it firmness and stability, and in each there is also found one of the primitive aortic arches. Continuous with the dorsal end of the first arch and growing forwards from it is a triangular process, the *maxillary process* (figs. [119], [121], and [143]). Ventrally it is separated from the mandibular arch by a > -shaped notch; the first branchial arch may therefore be said to divide into two, viz. the mandibular arch and the maxillary process. In front of the mandibular arch is a pentagonal depression, termed the *oral sinus* or *stomodæum*, since it forms the future mouth. It is bounded anteriorly by a median process, the *fronto-nasal* process, and laterally by the maxillary processes (fig. [119]), and will be again referred to.

These parts must now be considered with a little more detail.

The *fronto-nasal process* covers the forebrain and contains the coalesced portion of the trabeculæ cranii; it consists of a central or *mid-frontal* process and two lateral parts. By the invagination of the *olfactory pits*, which communicate below with the cavity of the mouth, each lateral portion is subdivided into an *outer* and an *inner nasal process*—the latter having been termed by His the *processus globularis*. The lateral nasal process is separated from the maxillary process by a groove which extends from the eye to the olfactory pit; this is the rudiment of the lachrymal duct (figs. [119], [120], and

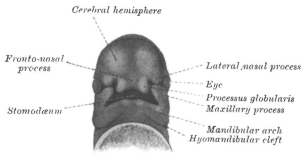

FIG. [119].—Under surface of the head ot a human embryo, about twenty-nine days old. (After His.)

Cerebral hemisphere

Fronto-nasal process

Stomodæum

Lateral nasal process

Eye

Processus globularis

Maxillary process

Mandibular arch

Hyomandibular cleft

[121]). The globular processes are prolonged backwards as plates, termed the *nasal laminæ*; these laminæ are at first some distance apart, but, gradually approaching, they ultimately fuse and form the nasal septum, while the globular processes themselves meet in the middle line and form the præmaxillæ and central part of the upper lip (fig. [122]). The depressed part of the mid-frontal process between the globular processes forms the lower part of the nasal septum, while above this is seen a prominent angle which becomes the future point, and still higher, a flat area, the future bridge of the nose (figs. [122] and [123]). The alæ of the nose are developed from the lateral nasal processes.

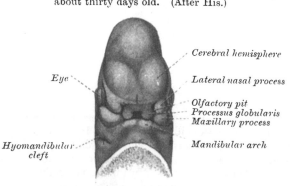

FIG. [120].—Under surface of the head of a human embryo, about thirty days old. (After His.)

Eye

Cerebral hemisphere

Lateral nasal process

Olfactory pit

Processus globularis

Maxillary process

Hyomandibular cleft

Mandibular arch

The *maxillary* processes descend for a short distance, forming the outer wall of the orbit, in which the malar bone is developed; they then incline inwards, and, meeting the lateral nasal process, form the floor of the orbit, and shut it off from the rest of the face; continuing their course downwards and inwards, they join the globular processes, and with them complete the alveolar arch and upper lip. Finally, a pair of palatal processes are formed by inward extensions of the maxillary processes; these coalesce with each other in the median line, thus separating the cavity of the mouth from the nasal fossæ, and completing the palate (fig. [122]). In front the palatal processes join with the premaxillæ, except in the middle line, where a cleft remains which constitutes the naso-palatine canal.

The *mandibular* arch, by its junction with the corresponding process on the other side, forms the lower jaw or mandible. The cartilaginous rod which it contains has long been known as the 'cartilage of Meckel.' The proximal end of this cartilage is in contact with the periotic capsule, and from it are developed

FIG. [121].—The head and neck of a human embryo thirty-two days old, seen from the ventral surface. The floor of the mouth and pharynx have been removed. (His.) (From Marshall's 'Vertebrate Embryology.')

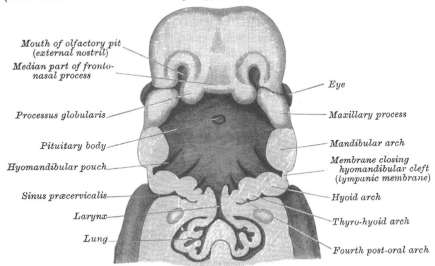

Mouth of olfactory pit (external nostril)

Median part of fronto-nasal process

Processus globularis

Pituitary body

Hyomandibular pouch

Sinus præcervicalis

Larynx

Lung

Eye

Maxillary process

Mandibular arch

Membrane closing hyomandibular cleft (tympanic membrane)

Hyoid arch

Thyro-hyoid arch

Fourth post-oral arch

two of the ossicles of the middle ear, the malleus and incus * (fig. [124]). The remainder of the rod is associated with the formation of the lower jaw, though the greater part of that bone is developed from membrane. The *second visceral arch* is named the *hyoid* arch: from it are formed the styloid

FIG. [122].—The roof of the mouth of a human embryo of about two and a half months old, showing the mode of formation of the palate. (His.) (From Marshall's 'Vertebrate Embryology.')

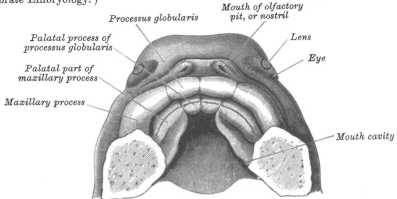

Processus globularis

Mouth of olfactory pit, or nostril

Palatal process of processus globularis

Lens

Eye

Palatal part of maxillary process

Maxillary process

Mouth cavity

process, the stylo-hyoid ligament, and the lesser cornu of the hyoid bone. The third, or *thyro-hyoid* arch, gives origin to the great cornu of the hyoid bone, while the body of this bone is formed between the second and third arches. The *fourth* and *fifth* arches are rudimentary.

* The incus is by some regarded as arising from the proximal end of the hyoid bar.

Between the maxillary processes and the mandibular arch the buccal cavity or mouth is formed. As has been already stated (page [86]) the cephalic end of the embryo becomes remarkably curved on itself, the fore-brain and mid-brain bending downwards over the anterior portion of the original blastodermic vesicle, which is thus enclosed within the body of the embryo and constitutes the fore-gut; the fore-gut terminates in a blind extremity beneath the head (figs. [125] and [164]). Another prominence, the rudimentary heart, appears on the ventral surface of the fore-gut. Between these two prominences, caused by the projection of the fore-brain and the heart, an involution of the epiblast takes place, gradually deepening until it comes in contact with the blind end of the fore-gut. This is the *oral pit* or *stomodæum*, already referred to; it presents the form of a pentangular opening, bounded in front by the fronto-nasal process, behind by the mandibular arch, and laterally by the maxillary processes. From the beginning the mesoblast is absent in the region of the oral pit, and hence its epiblastic lining meets the hypoblastic covering of the blind anterior end of the fore-gut and forms a thin septum, the *pharyngeal septum* (fig. [164]); this soon breaks down, and a communication is established between the mouth and the future pharynx. The oral pit or stomodæum is not equivalent in extent to the adult mouth, since the latter includes the tongue, which is developed from the floor of the pharynx; in fact, as His has pointed out, the anterior pillars of the fauces are developed from the second branchial or hyoid arch.

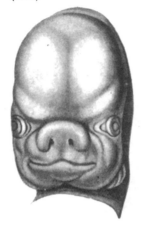

Fig. [123].—Head of a human embryo of about eight weeks, in which the nose and mouth are formed. (His.)

From the upper part of the stomodæum a pocket-like involution of the epiblast, the *pouch of Rathke*, extends upwards between the trabeculæ cranii

Fig. [124].—Head and neck of a human embryo eighteen weeks old, with Meckel's cartilage and hyoid bar exposed. (After Kölliker.)

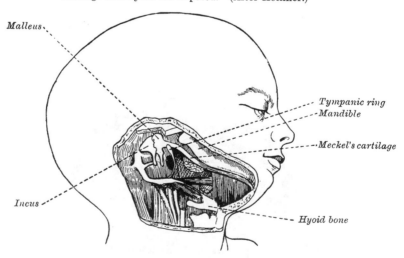

Malleus
Tympanic ring
Mandible
Meckel's cartilage
Incus
Hyoid bone

towards the thalamencephalon. This involution ultimately loses its connection with the stomodæum, and, becoming applied to the infundibulum, forms the anterior lobe of the pituitary body (fig. [125]).

The anterior visceral arches grow more rapidly than the posterior, with the result that the latter become telescoped·within the former, and a deep depression, the *sinus præcervicalis*, is produced. This sinus is bounded in front by the hyoid arch, and ultimately becomes obliterated by the fusion of its anterior and posterior walls.

Before leaving the subject of the visceral arches and clefts, it is necessary to mention that the clefts disappear early in embryonic life, with the exception of portions of the first, which remain permanent—the inner portion, as the Eustachian tube and tympanum; the outer, as the external auditory meatus, while the septum between the two portions becomes invaded by mesoblast and forms the membrana tympani.

Development of the Nervous Centres and the Nerves.

—The medullary or neural groove already described (page [85]) is the rudiment of the cerebro-spinal axis. As has been seen, this groove is converted into a canal (the neural canal): its cephalic end becomes dilated into a sac, from which the brain is developed; the remainder forms the spinal cord. The cavity of the canal becomes the central canal of the spinal cord, and that of the upper dilated portion the ventricles of the brain. The wall of the canal, formed of epiblastic cells, undergoes great changes, and from it the nervous matter and neuroglia are developed. It consists at first of a layer of columnar epithelium, covered on its exterior by a basement-membrane. The wall becomes thickened,

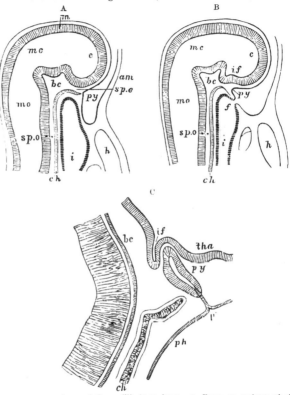

FIG. [125].—Vertical section of the head in early embryos of the rabbit. Magnified. (From Mihalkovics.)

A. From an embryo of five millimètres long. B. From an embryo of six millimètres long. c. Vertical section of the anterior end of the notochord and pituitary body, &c., from an embryo sixteen millimètres long. In A, the faucial opening is still closed. In B, it is formed. c. Anterior cerebral vesicle. *mc.* Meso-cerebrum. *mo.* Medulla oblongata. *co.* Corneous layer. *m.* Medullary layer. *if.* Infundibulum. *am.* Amnion. *spe.* Spheno-ethmoidal. *bc.* Central (dorsum sellæ), and *spo*, spheno-occipital parts of the basis cranii. *h.* Heart. *f.* Anterior extremity of primitive alimentary canal and opening (later) of the fauces. *i.* Cephalic portion of primitive intestine. *tha.* Thalamus. *p'.* Closed opening or the involuted part of the pituitary body (*py*) *ch.* Notochord. *ph.* Pharynx.

partly by the elongation of the columnar cells and partly by the formation of new cells. The elongation of the columnar cells, now called *spongioblasts*, is followed by the breaking up of their outer ends into a reticulum, which is termed the *myelo-spongium*, and eventually forms the neuroglia. The new cells which are formed appear between the inner ends of the columnar cells as rounded masses, which speedily divide, and are termed *neuroblasts*; they become pear-shaped, and projecting from each of them is a tapering process which perforates the basement-membrane. These neuroblasts are the primitive nerve-cells, and their tapering processes the rudimentary axis cylinders of the cells (figs. [126] and [127]).

It will be convenient, in the first place, to trace the changes which take place in the cavity of the cerebro-spinal axis, ignoring for a time those which go on in the enclosing wall. But before doing so, it is necessary to mention that, in consequence of the curve which the cephalic portion of the embryo undergoes, a marked bend forwards of the canal takes place, so that the plane of the ventricles is almost at right angles with the long axis of the central canal of the cord.

The early stage thus consists of a hollow sac, which is the rudimentary brain, and a hollow canal, which is the rudimentary cord; the sac and the canal freely communicating with each other. The sac first of all becomes elongated; then two constrictions appear in it, which partially divide it into three; these are named *anterior, middle,* and *posterior cerebral vesicles,* or the *fore-brain, mid-brain,* and *hind-brain* (fig. [106]). Subsequently the anterior and posterior vesicles each become constricted into two, while the middle one remains undivided. It will thus be seen that at the anterior extremity of the medullary canal there are five dilatations, separated from each other by constrictions, through which, however, they freely communicate with each other. These five vesicles are the five fundamental divisions of the adult brain, and are named from before backwards: prosencephalon, thalamencephalon, mesencephalon, epencephalon, and metencephalon (figs. [128] and [130]). They are at first fairly uniform in size and shape, but soon begin to grow at different rates and assume different forms. The changes are most marked in the first vesicle.

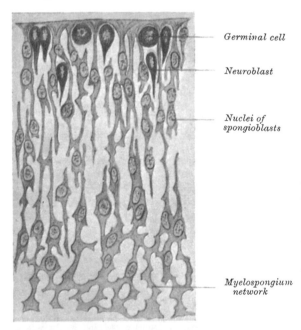

FIG. [126].—Transverse section of the spinal cord of a human embryo at the beginning of the fourth week. (After His.) The top of the figure corresponds to the lining of the central canal.

Germinal cell

Neuroblast

Nuclei of spongioblasts

Myelospongium network

The first secondary vesicle (*prosencephalon*) sends out two hollow protrusions, one on either side, from the fore-part of its lateral surface; these grow rapidly and spread out and extend forwards, laterally, and backwards over the sides of the first and second vesicles, forming large cavities, which become the lateral ventricles (fig. [130], G). From each, three prolongations take place: one, forwards and outwards; a second, backwards and inwards; and a third, at first backwards, outwards, and downwards, and then forwards and inwards; these form the horns of the lateral ventricles. These prolongations far exceed in size the original vesicle from which they sprung, which does not increase to any great extent. It remains as the anterior part of the third ventricle (fig. [130], A), and the communication between it and the future lateral ventricle persists as the *foramen of Monro* (fig. [130], H).

The second vesicle (*thalamencephalon*) becomes elongated from before backwards and compressed laterally so as to form the greater part of the third ventricle (fig. [130], B). From each side of that part of the forebrain which ultimately becomes the second vesicle is budded off a hollow projection, the primary optic vesicle, which is developed eventually into optic nerve and retina : it will be considered later on. The constriction between the first and second vesicle disappears, so as to throw the whole of the cavity (the future third ventricle), formed by the remains of the first vesicle and the whole of the second vesicle, into one.

The third vesicle (*mesencephalon*) is converted into a narrow channel, the *iter a tertio ad quartum ventriculum* (fig. [130], c).

The fourth vesicle (*epencephalon*) becomes widened out, and assumes a triangular form, with its apex directed forwards, and situated at the original point of constriction where the third vesicle joins the fourth. It is at the same time flattened from above downwards, and constitutes the anterior half of the fourth ventricle (fig. [130], D).

The fifth vesicle (*metencephalon*) undergoes the same changes in form as the fourth, becoming triangular in shape and flattened from above downwards, but with this difference, that the apex of the triangle is directed backwards, and is continuous with the portion of the medullary canal which goes to form the central canal of the spinal cord (fig. [130], E). The base is directed forwards,

FIG. [127].—Section of spinal cord of a four weeks' embryo. (His.)

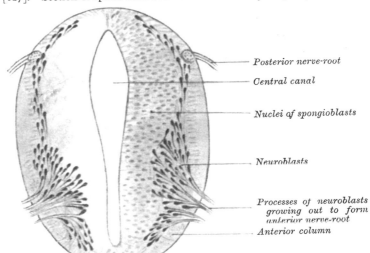

Posterior nerve-root

Central canal

Nuclei of spongioblasts

Neuroblasts

Processes of neuroblasts growing out to form anterior nerve-root
Anterior column

and is continuous with the base of the triangular space formed by the fourth vesicle ; the constriction between the two vesicles having disappeared, the two spaces freely communicate, and together form a rhomboidal cavity which is the fourth ventricle.

These vesicles do not remain in the same plane, but certain definite flexures take place, which result in an alteration of the position of the vesicles to one another. The first of these flexures (*cephalic*) is opposite the base of the middle vesicle, which becomes sharply bent on itself over the end of the notochord. This has the effect of causing the mid-brain to become the most prominent part of the encephalon on the convexity of the curve (fig. [128]). A second flexure (*pontal*), with its curve in the opposite direction, takes place in the epencephalon, and is very abrupt. A third but less marked flexure (*nuchal*) takes place in the metencephalon at its junction with the cord. The first of these curves or flexures remains permanent, but the second and third almost entirely disappear in the further development of the brain.

The manner in which the different parts of the encephalon and cord are formed from the walls of this greatly altered medullary canal must now be considered, and it will be convenient first of all to study the development of the spinal cord.

Fig. [128].—Profile views of the brain of human embryos at three several stages, reconstructed from sections. (His.) (Copied from Quain's 'Anatomy.')

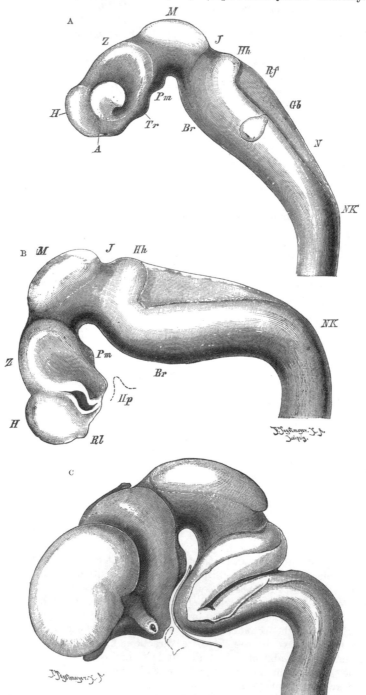

A. Brain of an embryo of about fifteen days, magnified 35 diameters. B. Brain of an embryo about three and a half weeks old. The optic vesicle has been cut away. C. Brain of an embryo about seven and a half weeks old. The optic stalk is cut through. *A.* Optic vesicle. *H.* Vesicle of cerebral hemisphere, first secondary vesicle. *Z.* Thalamencephalon, second secondary vesicle. *M.* Mid-brain. *J.* Isthmus between mid- and hind-brain. *Hh.* Fourth secondary vesicle. *N.* Fifth secondary vesicle. *Gb.* Otic vesicle. *Rf.* Fourth ventricle. *NK.* Neck curvature. *Br.* Pons curvature. *Pm.* Mammillary process. *Tr.* Infundibulum. *Hp.* (in B). Outline of hypophysis-fold of buccal epiblast. *Rl.* Olfactory fold. In C the basilar artery is represented along its whole course.

The lateral walls of the medullary canal become thickened and marked off into two laminæ: a dorsal, or *alar lamina*, and a ventral, or *basal lamina*; the portions of the canal forming in the mid-line both on its dorsal and ventral surfaces remaining thin, and forming the *roof* and *floor plates* respectively (fig. [129]). In the thickened lateral portion the neuroblasts begin to collect into groups; one especially being noticeable in the basal lamina at the situation of the future anterior horn. The processes of this group of cells pass out of the cord and form the anterior nerve-roots: outside this group of cells is the reticulated tissue of the myelospongium, which represents the white matter at this stage, and through which these processes pass obliquely before they leave the cord (fig. [127]). The anterior and posterior columns make their appearance soon after, and as the cornua of grey matter grow out from the central mass the fissures begin to appear. The anterior fissure is a cleft left between the lateral halves of the cord. The mode of formation of the posterior fissure is uncertain; many believe that it is a portion of the neural canal, which, dividing

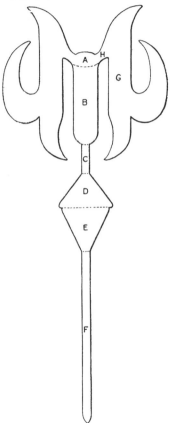

FIG. [130].—Plan showing the mode of formation of the ventricles of the brain and the central canal of the spinal cord. (After Gerrish.)

FIG. [129].—Section of the medulla in the cervical region, at six weeks. Magnified 50 diameters.

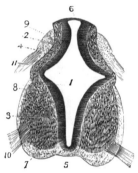

1. Central canal. 2. Its epithelium. 3. Anterior grey matter. 4. Posterior grey matter. 5. Anterior commissure. 6. Posterior portion of the canal, closed by the epithelium only. 7. Anterior column. 8. Lateral column. 9. Posterior column. 10. Anterior roots. 11. Posterior roots.

A. Prosencephalon. B. Thalamencephalon. C. Mesencephalon. D. Epencephalon. E. Metencephalon. F. Central canal of cord. G. Lateral ventricle. H. Foramen of Monro.

into two, forms an anterior part, the *permanent* canal, and a posterior portion, which becomes filled with a septum of connective tissue from the pia mater, and forms the posterior fissure of the cord. Others are of opinion that it is developed independently of the central canal, as a cleft, formed by the enlargement of the lateral halves of the cord, into which an ingrowth of connective tissue from the pia mater takes place.

At first the fœtal spinal cord occupies the whole length of the spinal canal, but after the fourth month the spinal column begins to grow in length more rapidly than the cord, so that the latter no longer occupies the lower part of the canal.

The ventricles of the encephalon are developed in the manner above described from the five secondary vesicles into which the primary expansion of the anterior extremity of the medullary tube is differentiated.

The first vesicle or prosencephalon sends out two hollow protrusions, which spread rapidly, and in the walls of these nervous matter is developed, which constitutes the cerebral hemispheres (fig. [128], H), the cavities remaining as the lateral ventricles.   As these hemispheres extend they grow forwards in front of the anterior extremity of the primitive brain, and lie side by side, separated by the longitudinal fissure ; they also grow upwards, and again lying side by side are separated by another portion of the same fissure, containing a thin layer of mesoblast which forms the falx cerebri ; behind and laterally they overlap the roof and sides of the other cerebral vesicles, so that by the seventh month they project behind them.   In the floor of each of these hemispheres there occurs a local thickening, which forms the *corpus striatum*, which is continuous behind with the optic thalamus, presently to be described.   The surface of the hemisphere is at first smooth, but about the fifth month a sulcus or groove appears in either

Fig. [131].—Median section of brain of human fœtus during the third month.
(After His.)

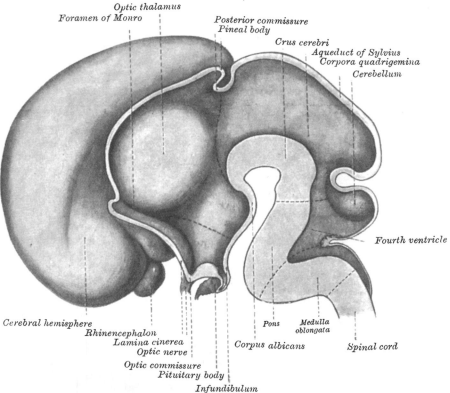

hemisphere just external to the corpus striatum ; this is the fissure of Sylvius : subsequently other fissures appear on the surface, three of which are of sufficient depth to cause a projection into the lateral ventricle.   These are the hippocampal fissure corresponding with the hippocampus major of the lateral ventricle ; the parieto-occipital fissure corresponding with the bend of the posterior horn of the ventricle ; and the calcarine fissure corresponding with the projection of the calcar avis.

The remainder of the first vesicle and the second, as we have seen, form the third ventricle ; in its lateral walls a thickening takes place, which forms the *optic thalamus*.   From the floor of this ventricle a hollow protrusion passes downwards, and is intimately connected with a diverticulum from the stomodæum, to form the *pituitary body* or *hypophysis cerebri* (figs. [125], [128], and [131]).   The greater part of the roof of the third ventricle is very thin, and with the pia mater forms the *velum interpositum*; from its posterior part an outgrowth of cells forms the *pineal body* or *epiphysis cerebri*. Where the cerebral hemispheres are not separated in the middle line by the falx, in front and for some distance backwards over the roof of the third ventricle their mesial surfaces come in contact, and to a certain extent fuse together, leaving however a small portion where no union takes place, and thus a slit-like cavity is left; this is termed the *fifth ventricle*, though it will be at once seen that its development is quite different from that of the other ventricles. Its lateral walls form the *septum lucidum*. The roof of this cavity becomes thickened, and nerve-fibres pass across from the one hemisphere to the other to form the *corpus callosum*, while in its floor longitudinal fibres are developed to form the *fornix*.

The third vesicle, the cavity of which forms the iter a tertio ad quartum ventriculum, develops in its roof four well-marked thickenings, which together form the *corpora quadrigemina*, while its lateral regions become thickened to form the *crura cerebri* (fig. [131]).

The dorsal surface of the fourth vesicle, or epencephalon, forms the covering of the fourth ventricle, and in it a thickening occurs, which is developed into the *cerebellum*; its ventral and lateral regions form the *pons* (fig. [131]).

Fig. [132].—Transverse section of medulla oblongata of human embryo.
(After His.)

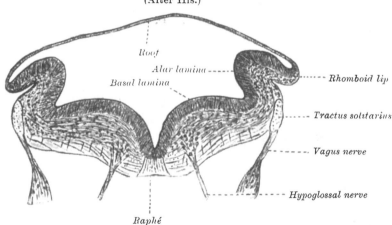

Roof

Alar lamina

Basal lamina

Rhomboid lip

Tractus solitarius

Vagus nerve

Hypoglossal nerve

Raphé

In the fifth vesicle or metencephalon the lateral parts increase and grow downwards on each side towards the middle line forming the *medulla*, while the dorsal surface assists in forming the roof of the fourth ventricle.

On making a transverse section of the lower part of the fourth ventricle, the alar and basal laminæ, already referred to as being present in the cord, are readily recognised, while the thin roof-plate is seen to be greatly expanded laterally. The dorsal part of the alar lamina becomes folded outwards and downwards, forming what is termed the *rhomboid lip* (fig. [132]). This is at first separated by a groove from the lateral aspect of the alar lamina, but ultimately fuses with it. As the central canal of the cord opens out to form the fourth ventricle, the alar and basal laminæ come to occupy the floor of the ventricle— the basal lamina lying nearest the mesial plane.

The Nerves.—The nerves are developed, like the rest of the nervous system, from epiblast. The spinal nerves are developed as follows : close to the point of involution of the epiblast in the median line, that is to say, in the angle of junction of the neural and general epiblast, a cellular swelling, the *neural crest*, appears and forms a continuous ridge of epiblast on the dorsal aspect of the neural canal (fig. [133]). On this crest enlargements occur corresponding with the middle of each protovertebral segment. These enlargements grow downwards between the neural canal and the protovertebræ, and occupy a position on the lateral wall of the canal. They are the rudiments of the ganglia of the posterior roots and are at first attached to the neural crest from which they spring, but subsequently this attachment becomes lost, and they then form isolated masses on either side of the neural canal. They consist of oval cells, from either end of which a process eventually springs: one, growing centrally, passes into the embryonic cord and constitutes the posterior root of the nerve; the other, growing peripherally, joins the fibres of the anterior root to form the spinal nerve.

The anterior root is, according to the researches of His, a direct outgrowth of the neuroblasts which are found in the rudimentary cord (fig. [127]). These cells, at first rounded or oval, become pear-shaped, with their tapering prolongations directed outwards towards the surface of the cord. These prolongations are the future axis cylinders of the anterior nerve-roots; they pass out of the cord in bundles and penetrate the mesoblast to join with the fibres of the posterior root, and from the point of union the nerve grows towards its peripheral termination.

Cranial nerves.—With the exception of the olfactory and optic nerves, which will be specially referred to, the cranial nerves may be developmentally considered as consisting of two sets: (1) those which arise as outgrowths from neuroblasts situated in the brain, similar to the mode of origin of the anterior spinal nerve-

FIG. [133].—Transverse section of a portion of a chick embryo of twenty-nine hours' incubation. (From Duval's ' Atlas d'Embryologie.')

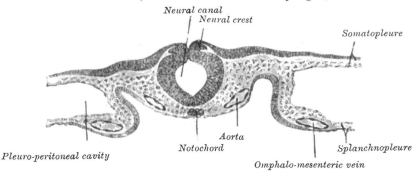

roots; (2) those which arise from ganglionic rudiments situated outside the brain and derived from the neural crest; from the neuroblasts of these ganglionic rudiments one process grows into the brain and the other outwards towards the periphery, similar to the arrangement which exists regarding the posterior spinal nerve-roots. To the first group belong the third, fourth, sixth, seventh, eleventh, and twelfth nerves, together with the motor-roots of the fifth, ninth, and tenth. To the second group belong the eighth, and the sensory roots of the fifth, ninth, and tenth. While, however, the anterior spinal nerve-roots arise in one series from the ventral part of the cord, the cranial motor-fibres arise by two sets of roots, *ventral* and *lateral*; the former include the roots of the sixth and twelfth and probably those of the third and fourth, the latter embrace the spinal accessory and the motor-roots of the fifth, seventh, ninth, and tenth.

The olfactory lobe, or rhinencephalon, arises towards the end of the fourth week as a protrusion of the antero-ventral part of each cerebral hemisphere (fig. [131]), and extends forwards towards the thickened epiblast of the olfactory area (see page [100]). It is subsequently divided by a transverse constriction into two parts: an anterior, which gives rise to the olfactory bulb and tract together with the trigonum olfactorium, and a posterior, which becomes the peduncle of the corpus callosum and the greater part of the anterior perforated space. Neuroblastic cells, formed within the olfactory area, pass out and form a ganglion between the area and the olfactory bulb. From this ganglion cell-processes grow centripetally to form the nerve-roots, and centrifugally to form the olfactory nerves which ramify in the olfactory mucous membrane, while the ganglion itself fuses with the olfactory bulb.

The optic nerve arises as a hollow outgrowth of the brain, which subsequently becomes solid. It will be considered in connection with the development of the eye.

The sympathetic nerves are developed as outgrowths from the ganglia on the roots of the spinal and cranial nerves.

FIG. [134].—Transverse section of head of chick embryo of forty-eight hours' incubation. × 55. (From Duval's 'Atlas d'Embryologie.')

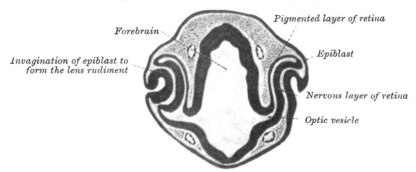

Forebrain

Invagination of epiblast to form the lens rudiment

Pigmented layer of retina

Epiblast

Nervous layer of retina

Optic vesicle

Development of the Eye.—The optic nerve and retina are developed as an outgrowth from the rudimentary brain, which extends towards the side of the head, and is there met by an ingrowth from the epiblast, out of which the lens and the epithelium of the conjunctiva and cornea are developed.

The first appearance of the eye consists in a hollow protrusion of the forebrain; this is called the *primitive optic vesicle*. It is at first an open cavity communicating by a hollow stalk with that of the cerebral vesicle. As it is prolonged forwards, the epiblast lying immediately over it becomes thickened, and then forms a depression which gradually encroaches on the most prominent part of the primitive ocular vesicle; this in its turn appears to recede before it, so as to become at first depressed and then inverted in the manner indicated in figs. [134] and [135], so that the cavity of the vesicle is almost obliterated by the folding back of its anterior half, and the original vesicle converted into a cup, the *optic cup*, in which the involuted epiblastic layer, the rudiment of the lens, is received (fig. [135]); at the same time the proximal part

FIG. [135].—Transverse section of head of chick embryo of fifty-two hours' incubation. (From Duval's 'Atlas d'Embryologie.')

Forebrain

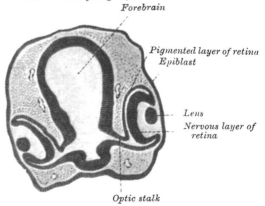

Pigmented layer of retina
Epiblast

Lens
Nervous layer of retina

Optic stalk

of the vesicle becomes elongated and narrowed into a hollow stalk, the *optic stalk*. This cup-shaped cavity consists therefore of two layers : one, the outer, originally the posterior half of the primitive ocular vesicle, is thin, and eventually forms the pigmented layer of the retina ; * the other layer, the inner, originally the

---

* This layer was formerly described as belonging to the choroid, but developmentally it is seen to be a part of the retina.

anterior or more prominent half, which has become folded back, is much thicker, and is converted into the nervous layers of the retina (figs. [135] and [137]). Between the two is the remnant of the cavity of the original primary optic vesicle, which finally becomes obliterated by the union of its two layers. When the retina is established, the optic nerve-fibres originate from its cells and grow backwards towards the brain, along the optic stalk, and thus convert it into a solid optic nerve. The nerve-fibres become ultimately connected with the mesencephalon, a relationship which is permanently maintained. The mouth of the optic cup overlaps the equator of the lens as far as the future aperture of the pupil. In this region the inner or retinal layer of the cup does not become differentiated into nervous elements, but remains as a single layer of columnar cells, which becomes applied to the cells of the pigmented layer, and the conjoined strata form the *pars ciliaris* and *pars iridica retinæ* of the adult (fig. [139]). As development proceeds the optic cup increases in size, and thus a space is formed between it and the rudimentary lens ; this is the *secondary optic vesicle*, and in it the vitreous humour is developed (figs. [136], c and [137]). The folding in of the primary optic vesicle to produce the optic cup not only takes place in front, at its most prominent part, opposite the lens, but also along its postero-inferior aspect,

Fig. [137].—Diagrammatic sketch of a vertical longitudinal section through the eyeball of a human fœtus of four weeks. (After Kölliker.) Magnified 100 diameters. The section is a little to the side, so as to avoid passing through the ocular cleft.

Fig. [136].—Diagram of development of the lens.

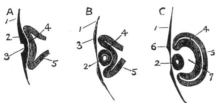

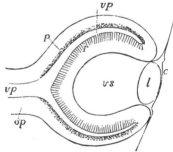

A B C. Different stages of development. 1. Epidermic layer. 2. Thickening of this layer. 3. Crystalline depression. 4. Primitive ocular vesicle, its anterior part pushed back by the crystalline depression. 5. Posterior part of the primitive ocular vesicle, forming the external layer of the secondary ocular vesicle. 6. Point of separation between the lens and the epidermic layer. 7. Cavity of the secondary ocular vesicle, occupied by the vitreous.

c. The cuticle, where it becomes later the epithelium of the cornea. l. The lens. op. Optic nerve formed by the pedicle of the primary optic vesicle. vp. Primary medullary cavity of the optic vesicle. p. The pigment layer of the retina. r. The inner wall forming the nervous layers of the retina. vs.Secondary optic vesicle containing rudiment of the vitreous humour.

where a cleft or fissure is formed, the *choroidal fissure*, through which the mesoblast extends to form the vitreous humour. This gap or cleft is continued for some distance into the stalk of the optic vesicle, and thus allows a process of the mesoblast to extend down the stalk to form the arteria centralis retinæ and its accompanying vein (fig. [138]). After a time the gap or fissure becomes closed, by a coalescence of its margins, but the line of union remains apparent for a considerable period.

The lens is at first a thickening of the epiblast, then a depression or involution takes place, thus forming an open follicle, the margins of which gradually approach each other and coalesce, forming a cavity, the *lens vesicle*, enclosed by epiblastic cells (fig. [136], B C). At the point of involution the external layer of epiblast separates from the lens and passes freely over the surface, so that the lens becomes disconnected from the general epiblast, and recedes into the ocular cup, while the cuticular layer covering it is developed into the corneal epithelium. The cells forming the posterior or inner wall of the lens vesicle rapidly increase in size, becoming elongated and developed into the lens fibres, and, filling up the cavity, convert it into a solid body. The cells on the anterior wall retain their cellular character, and form the anterior lens epithelium of the adult. The secondary optic vesicle, or space between the lens and the hollow of the optic cup (figs. [136], 7, and [139]), contains a quantity of mesoblastic tissue continuous with the general mesoblast through the choroidal fissure. This tissue becomes converted into the vitreous humour, and surrounds the lens with a vascular membrane—the *vascular capsule of the lens*. From the central artery of the retina several branches are prolonged forwards through the vitreous body

FIG. [138].—Optic cup and choroidal fissure seen from below, from a human embryo of about four weeks. (Kollmann.)

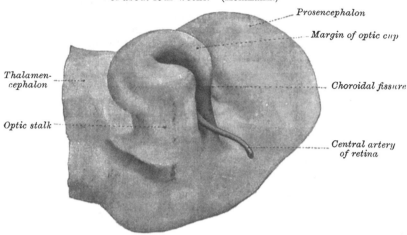

Prosencephalon

Margin of optic cup

Thalamen-cephalon

Choroidal fissure

Optic stalk

Central artery of retina

FIG. [139].—Horizontal section through the eye of an eighteen days' embryo rabbit. × 30. (Kölliker.)

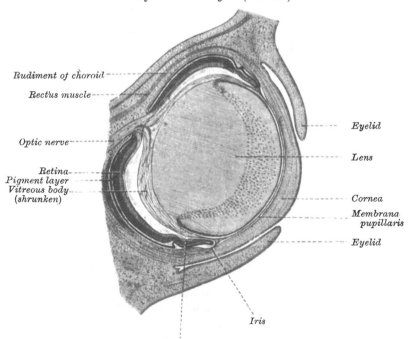

Rudiment of choroid

Rectus muscle

Eyelid

Optic nerve

Lens

Retina
Pigment layer
Vitreous body
(shrunken)

Cornea

Membrana pupillaris

Eyelid

Iris

Pars ciliaris and pars iridica retinæ

to the capsule of the lens, but by the sixth month these have all undergone atrophy except one, which persists till the ninth month as the *arteria hyaloidea*. It disappears, however, before birth, and its position is indicated in the adult by the *canalis hyaloideus of Stilling*. The front part of the vascular capsule of the crystalline lens forms the *membrana pupillaris*, and also attaches the iris to the capsule of the lens. It disappears about the seventh month. The sclera, cornea, and choroid are developed from the mesoblast surrounding the optic vesicle.

The eyelids are formed at the end of the third month, as small cutaneous folds (fig. [139]), which come together and unite in front of the globe and cornea. This union is broken up and the eyelids separate before the end of fœtal life.

The lachrymal sac and nasal duct appear to result from a thickening of the epiblast in the groove between the lateral nasal and maxillary processes. This thickening becomes hollowed out into a channel, and the lips of the groove meet over it, enclose it, and convert it into a duct, which eventually opens into the nasal fossa.

Development of the Ear.—The first rudiment of the ear appears shortly after that of the eye, in the form of a thickening of the epiblast, on the outside of that part of the third primary cerebral vesicle which eventually forms the medulla oblongata. The thickening is then followed by an involution of the epiblast (fig. [140]), which becomes deeper and deeper, and sinking towards the base of the skull, forms a flask-shaped cavity; by the narrowing of the external aperture the neck of the flask constitutes the *recessus labyrinthi*. The mouth of

FIG. [140.]—Section through the head of a human embryo, about twelve days old, in the region of the hindbrain. (Kollmann.)

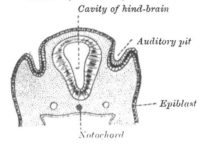

Cavity of hind-brain

Auditory pit

Epiblast

Notochord

FIG. [141].—Section through hindbrain and otic vesicle of an embryo more advanced than that of fig. [140]. (After His.)

Hind-brain

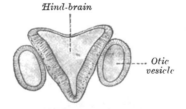

Otic vesicle

FIG. [142].—Left auditory vesicle of a human embryo of four weeks, seen from the outer surface. (W. His, jun.)

Auditory vesicle

Recessus labyrinthi (Aquæductus vestibuli)

the flask then becomes closed, and thus a shut sac is formed, the *primitive auditory* or *otic vesicle* (fig. [141]), which by its sinking inwards comes to be placed between the ali-sphenoid and basi-occipital matrices. From it the epithelial lining of the labyrinth is formed. The primary otic vesicle becomes embedded in a mass of mesoblastic tissue, which rapidly undergoes chondrification and ossification. The vesicle is at first flask- or pear-shaped; the neck of the flask, or *recessus labyrinthi*, prolonged backwards, forms the aquæductus vestibuli. From it are given off certain prolongations or diverticula, from which the various parts of the labyrinth are formed. One from the anterior end gradually elongates, and, forming a tube, bends on itself and becomes the cochlea. Three others, which appear on the surface of the vesicle, form the semicircular canals, of which the external canal is the last to be developed (figs. [143] and [144]). Subsequently, a constriction takes place in the original vesicle, which nearly divides it into two, and from these are formed the utricle and saccule (fig. [144]). Finally, the auditory nerve, which has been developed from the 'neural crest' in the manner above described (page [110]), pierces the auditory capsule in two main divisions—one for the vestibule, the other for the cochlea. The middle ear and Eustachian tube are the remains of the inner part of the first branchial cleft (hyomandibular), and are closed externally by the membrana tympani, which originally consists of a layer of epiblast externally, and a layer of hypoblast

FIG. [143].—Left auditory vesicle of a human embryo of five weeks, seen from the outer surface. (W. His, jun.)

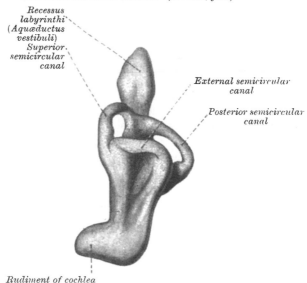

Recessus labyrinthi (Aquæductus vestibuli)

Superior semicircular canal

External semicircular canal

Posterior semicircular canal

Rudiment of cochlea

internally; between these two layers the mesoblast extends to form the substantia propria of the membrane. With regard to the exact mode of development of the ossicles of the middle ear there is considerable difference of opinion. The most

FIG. [144].—Transverse section through head of fœtal sheep, in the region of the labyrinth. × 30. (After Boettcher.)

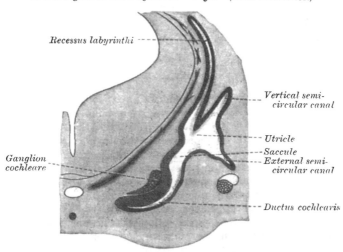

Recessus labyrinthi

Vertical semicircular canal

Utricle

Saccule

External semicircular canal

Ganglion cochleare

Ductus cochlearis

probable view is that the *incus* and *malleus* are developed from the proximal end of the mandibular (Meckel's) cartilage (fig. [124]): that the base of the *stapes* is formed by the ossification of the cartilage which fills in the foramen ovale and its arch from the ossified proximal end of the hyoidean arch.

The external auditory meatus is formed from the outer part of the hyo-mandibular cleft, while the pinna is developed by the gradual differentiation of a series of processes which appear around the outer margin of the cleft (fig. [146]).

Development of the Nose.—The olfactory fossæ, like the primary auditory vesicles, are formed in the first instance by a thickening and involution of the epiblast, which takes place about the fourth week, at a point below and in front of the ocular vesicle (fig. [118]). The borders of the involuted portion very soon become prominent, in consequence of the development of the mesial and lateral nasal processes already referred to (page [100]), and which are formed on either side of the rudimentary fossa (figs. [119] and [120]). As these processes increase, the fossa deepens and becomes converted into a channel, which eventually forms the olfactory region of the nose ; this comprises the portion to which the olfactory nerves are distributed. At this time the nasal cavity is continuous with the buccal cavity, but as the palatal septum is formed, the buccal cavity is divided into two parts, the upper of which forms the lower part of the nasal fossæ, while the remainder forms the permanent mouth. On the mesial wall of the nasal

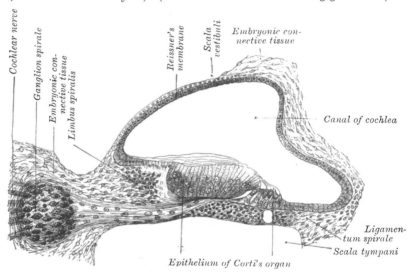

FIG. [145].—Transverse section of the canal of the cochlea of a fœtal cat. (After Boettcher and Ayres.) (From Kollmann's ' Entwickelungsgeschichte.')

fossa a small blind pit of epiblast becomes invaginated and extends backwards into the nasal septum. This forms the rudiment of *Jacobson's organ*, which ultimately becomes partly enclosed in a curved cartilaginous plate derived from the cartilage of the nasal septum.

The development of the external nose has already been described. It is perceptible about the end of the second month. The nostrils are at first closed by epithelium, but this disappears about the fifth month.

The olfactory lobe (rhinencephalon) is formed, as already explained, by an evagination of the anterior cerebral vesicle.

Development of the Skin, Glands, and Soft Parts.—The epidermis and its appendages, consisting of the hairs, nails, sebaceous and sweat glands, are developed from the epiblast, while the corium or true skin is of mesoblastic origin. About the fifth week the epidermis consists of two layers of cells, the deeper one corresponding to the rete mucosum. The subcutaneous fat forms about the fourth month, and the papillæ of the true skin about the sixth. A considerable desquamation of epidermis takes place during fœtal life, and this desquamated epidermis, mixed with a sebaceous secretion, constitutes the *vernix caseosa*, with which the skin is smeared during the last three months of fœtal life. The nails are formed at the third month, and begin to project from the epidermis about the sixth. The hairs appear between the third and fourth months in the form of solid downgrowths of the deeper layer of the epidermis, which then become inverted by papillary projections from the corium. About the fifth month, the

FIG. [146].—Left ears of human embryos, estimated at thirty-five and thirty-eight days respectively. (After His.)

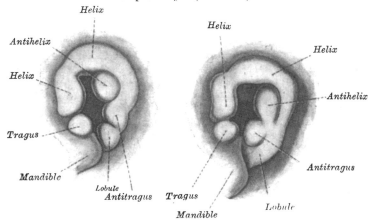

fœtal hairs (*lanugo*) appear, first on the head and then on the other parts they ; drop off after birth, and give place to the permanent hairs. The cellular structure of the sudoriferous and sebaceous glands is formed from the epiblast, while the connective tissue and blood-vessels are derived from the mesoblast. The mammary gland is also formed partly from mesoblast and partly from epiblast —its blood-vessels and connective tissue being derived from the former, its cellular

FIG. [147].—Head of chick embryo of about thirty-eight hours' incubation, viewed from the ventral surface. × 26. (From Duval's ' Atlas d'Embryologie.')

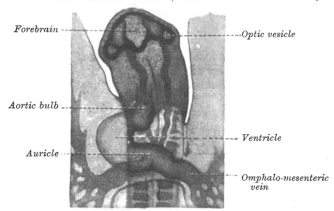

elements from the latter. Its first rudiment is seen about the third month, in the form of a small projection inwards of epithelial elements, which invade the mesoblast ; from this, similar tracts of cellular elements radiate ; these subsequently give rise to the glandular follicles and ducts. The development of the former, however, remains imperfect, except in the adult female.

Development of the Limbs.—The upper and lower limbs begin to project, as buds, from the anterior and posterior part of the embryo about the fourth week.

These buds are formed by a projection of the somatopleure from the point where the mesoblast splits into its parietal and visceral layers, just external to the vertebral somites, of which they may be regarded as lateral extensions. The division of the terminal portion of the bud into fingers and toes is early indicated, and soon a notch or constriction marks the future separation of the hand or foot from the forearm or leg. Next, a similar groove appears at the site of the elbow or knee. The indifferent tissue, of which the whole projection is at first composed, is differentiated into muscle and cartilage, before the appearance of any internal clefts for the joints between the chief bones.

The **muscles** become visible about the seventh or eighth week. The voluntary muscles are developed from the muscle-plates of the protovertebral somites which are at first segmentally arranged on either side of the rudimentary spine. Each muscle-plate becomes differentiated into two parts, superficial and deep. The former is termed the *cutis plate*, and from it the corium or true skin is developed, while the latter becomes developed into longitudinal groups of muscle-fibres, extending forwards into. the neck and head region of the embryo and laterally to enclose the cavities of the thorax and abdomen. The muscles of the limbs are also formed from the same source, being produced by

FIG. [148].—Heart of human embryo of about fifteen days.   (Reconstruction by His.)

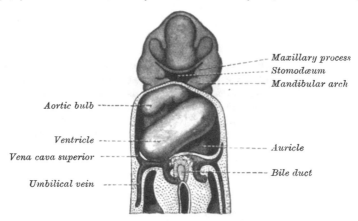

Aortic bulb - - - - - - -

Ventricle - - - - - -

Vena cava superior - - - -

Umbilical vein - - - - -

- - - *Maxillary process*
- - - - *Stomodæum*
- - - - - *Mandibular arch*

- - - - - *Auricle*

- - - - *Bile duct*

outgrowths from the protovertebral somites in those situations where the limb buds appear. The involuntary muscles are derived from the splanchnopleure mesoblast, and are therefore not connected in any way with the protovertebral somites.

Development of the Blood-vascular System.—There are three distinct stages in the development of the circulatory system, each in accordance with the manner in which nourishment is provided for at different periods of the existence of the individual. In the first stage there is the *vitelline circulation*, during which nutriment is extracted from the *vitellus* or contents of the yolk-sac. In the second stage there is the *placental circulation*, during which nutrition is obtained by means of the placenta from the blood of the mother. In the third stage there is the *complete circulation of the adult*, commencing at birth, during which nutrition is provided for by the organs of the individual itself.

1. The **vitelline circulation** is carried on partly within the body of the embryo, and partly external to it in the vascular area of the yolk-sac. It consists of a median tubular heart, from which two vessels (arteries) project anteriorly. These carry the blood to a plexus of capillaries spread over the vascular area, from which the blood is returned by two vessels (veins) which enter the heart posteriorly, and thus a complete circulation is formed (fig. [149]).

In these vessels and the heart a fluid (*blood*) is contained, in which rudimentary corpuscles are found. The mode of formation of these elementary parts must first be considered.

In mammalia the heart is formed by a longitudinal fold of the splanchnopleure with its underlying hypoblast on either side of the median line in front of the anterior extremity of the rudimentary pharynx, at about the level of the posterior primary cerebral vesicle. The folds become tubes, their walls thicken, and present two distinct strata of cells : the inner and thinner layer, derived from the hypoblast, forms the endocardium ; the outer and thicker, derived from the visceral mesoblast, forms the muscular wall of the heart. In its primitive condition, the heart consists therefore of a pair of tubes, one on either side of the body. These, however, soon coalesce in the median line, and, fusing together, form a single central tube.* Each of the two primary tubes receives posteriorly a large vein (the omphalo-mesenteric vein), and is prolonged anteriorly into an artery (the primitive aorta). So that after fusion of the heart-tubes has taken place, there is, in the primitive vitelline circulation, as above mentioned, a single tubular heart,

FIG. [149].—Human embryo of about fourteen days old with yolk sac. (After His.)
(From Kollmann's ' Entwickelungsgeschichte.')

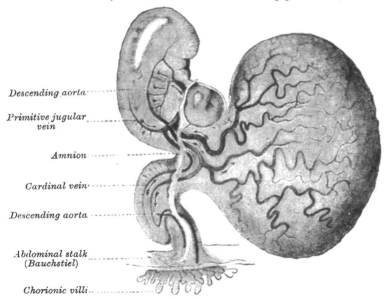

Descending aorta
Primitive jugular vein
Amnion
Cardinal vein
Descending aorta
Abdominal stalk (Bauchstiel)
Chorionic villi

with two arteries proceeding from it and two veins emptying themselves into it. The first blood-vessels are developed as follows : The nucleated embryonic cells of the mesoblast send out processes in various directions. These processes fuse together, and an irregular network is formed. The nuclei of the cells multiply, and, accumulating around themselves a small quantity of the protoplasm of the cell, they acquire a tinge of colour and form the first red blood-corpuscles. The protoplasm of the cells and their branched network becomes hollowed out into a system of canals containing fluid, in which the newly formed corpuscles float (fig. [150]).†

The earliest blood-corpuscles are all nucleated, and in this and other respects—that is, in their possession of amœboid movements and in their capability of undergoing multiplication by division—resemble the white corpuscles. Soon, however, true white corpuscles make their appearance, and it seems that they are derived

* In most fishes and in amphibia the heart originates as a single median tube.
† Recent observers incline to the view that the blood-corpuscles are of hypoblastic origin, being developed from the endothelium of the vessels, the sequence of the development of the different structures being : first the heart, then the blood-vessels, and lastly the blood-corpuscles. (Consult Dr. Ernest Mehnert's *Biomechanik*, Jena, 1898.)

from the rudiments of the thymus gland.* The nucleated condition of the red globules ceases before birth. The vitelline circulation commences about the fifteenth day and lasts till the fifth week. When fully established, it is carried on as follows · Proceeding from the anterior end of the tubular heart are two arteries, the primitive aortæ; these run down in front of the primitive vertebræ and behind the walls of the intestinal cavity into the two *omphalo-mesenteric* arteries, which ramify in the *vascular area* of the yolk-sac. Here they terminate peripherally in a circular vessel — the *terminal sinus*, which surrounds the vascular area. The blood is collected from the capillaries of the vascular area into the two *omphalo-mesenteric* veins, which open into the posterior extremity of the heart.

FIG. [150].—Various forms of mother-cells undergoing development into blood-vessels, from the middle layer of the chick's blastoderm. (Klein.)

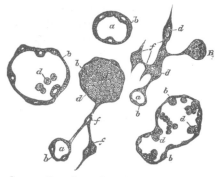

a. Large mother-cell vacuolated, forming the rudimentary vessel. *b.* The wall of this cell formed of protoplasm, with nuclei embedded, and in some cases more or less detached and projecting. *c.* Processes connected with neighbouring cells, formed of the common cellular substance of the germinal area. *d.* Blood-corpuscles. *f.* Small mother-cells — vacuolation commencing. *B.* Mother-cell in which only obscure granular matter is found.

2. **The Placental Circulation.**—As the umbilical vesicle diminishes, the allantois and the placenta are developed in the manner above described (page [92]). When the umbilical vesicle atrophies the placenta becomes the only source of nutrition for the embryo. The allantois carries with it two arteries (*umbilical* or *allantoic*), derived from the primitive aortæ, and two veins; these vessels become much enlarged as the placental circulation is established, but subsequently one of the veins disappears, and in the later stages of uterine life the circulation is carried on between the fœtus and the placenta by two umbilical arteries and one umbilical vein.

During the occurrence of these changes great alterations take place in the primitive heart and blood-vessels, and now require description.

Further Development of the Heart.—The following is an outline of the changes which take place during the further development of the heart.

FIG. [151].—Heart of a human embryo of 5 mm. in length, seen from the front. × 30. (His.)

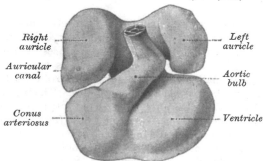

Right auricle
Auricular canal
Conus arteriosus
Left auricle
Aortic bulb
Ventricle

The simple tubular heart, already described, becomes elongated and bent on itself so as to form an S-shaped loop, the anterior part bending to the right and the posterior part to the left. The intermediate portion arches transversely from right to left, and then turns sharply forwards into the anterior part of the loop. Slight constrictions make their appearance in the tube and divide it into four parts, viz.: (1) the *sinus venosus* (*sinus reuniens* of His); (2) the common auricle; (3) the common ventricle; (4) the aortic bulb. The common auricle and ventricle communicate by a short canal, the *auricular* canal (figs. [147], [148], and [151]).

* Consult an article by J. Beard, *Anatomischer Anzeiger*, December 1900.

The sinus venosus is situated in the septum transversum (a layer of mesoblast from which the ventral part of the Diaphragm is developed) behind the common auricle, and is formed by the union of three pairs of veins, viz. : (1) the veins or ducts of Cuvier from the body of the embryo ; (2) the omphalo-mesenteric veins from the yolk-sac ; (3) the umbilical veins from the placenta (fig. [152]). The sinus is at first placed transversely, and opens by a median aperture into the common auricle. Soon, however, it assumes an oblique position, and its right half or horn becomes larger than the left, while the opening into the auricle is now found to be into the right portion of the auricular cavity. The right horn ultimately becomes incorporated with and forms a part of the right auricle, the line of union between it and the auricle proper being indicated in the interior of the adult auricle by a vertical crest, the *crista terminalis* of His. The left horn, which ultimately receives only the left duct of Cuvier, persists as the coronary sinus (fig. [158]). The omphalo-mesenteric and umbilical veins are soon replaced by a single vessel, the inferior vena cava, and the three veins (inferior vena cava and right and left Cuvierian ducts) open into the dorsal aspect of the auricle by a common slit-like aperture. The upper part of this aperture represents the opening of the

FIG. [152].—Heart of a human embryo, 4·2 mm. long, seen from behind. (His.)

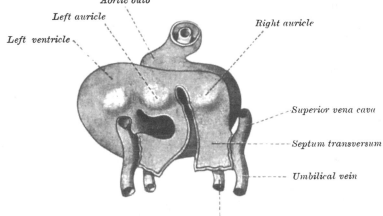

Aortic bulb

Left auricle

Left ventricle

Right auricle

Superior vena cava

Septum transversum

Umbilical vein

Vitelline or omphalo-mesenteric vein

permanent superior vena cava, the lower part that of the inferior vena cava, and the intermediate part the orifice of the coronary sinus. The slit-like aperture lies obliquely, and is bordered on its mesial and lateral aspects by a fold of endocardium. The mesial part of the fold disappears, while from the lateral part the Eustachian and Thebesian valves are developed. At the lower extremity of the slit is a triangular thickening, the *spina vestibuli* of His, which partly closes the aperture between the two auricles, and which, according to His, takes a part in the formation of both the interauricular and interventricular septa.

The common auricle becomes gradually subdivided into right and left auricles by a septum, the *septum superius*, which grows from its dorsal and upper wall so that the two auricles communicate with each other only below the margin of this septum. This communication (*ostium primum* of Born) does not, however, represent the future foramen ovale, for the septum grows downwards and blends with the partition which comes to subdivide the auricular canal. The foramen ovale (*ostium secundum* of Born) results from a perforation of the upper part of the septum superius.

The auricular canal is at first a short straight tube connecting the auricular with the ventricular portion of the heart, but it becomes overlapped by the growing auricles and ventricles so that its position on the surface of the heart is only indicated by an annular constriction (fig. [151]). Its lumen is reduced to a transverse slit, and a thickening appears on its dorsal and ventral walls. These thickenings, or *endocardial cushions* as they are termed, project into the canal, and, meeting in the middle line, divide the canal into two channels, the future right and left auriculo-ventricular orifices.

The common ventricle becomes divided by a septum, the *septum inferius*, which grows upwards from the lower part of the ventricle, its position being indicated on the surface of the heart by a furrow. It extends upwards almost as far as the

FIG. [153].—Diagrams to show the development of the septum of the aortic bulb and of the ventricles. (Born.).

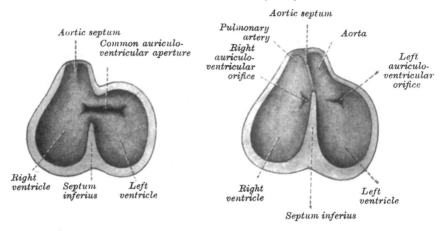

auricular canal, but for some time an interventricular foramen exists between it and the septum of the auricular canal (fig. [153]).

The aortic bulb is divided by the *aortic septum*. This makes its appearance at the distal end of the bulb as two ridge-like thickenings of its endothelial lining ; these increase in size, and, projecting into the lumen, ultimately fuse to form the septum, and thus the aortic bulb is divided into the pulmonary artery and the aorta. The aortic septum takes a spiral course towards the proximal end of the bulb, so that the two vessels lie side by side above ; but near the heart the pulmonary artery is in front of the aorta (fig. [154]). The septum grows down into the ventricle as an oblique partition, which ultimately blends with the septum

FIG. [154].—Transverse sections through the aortic bulb to show the growth of the aortic septum. The lowest section is on the left, the highest on the right of the figure. (After His.)

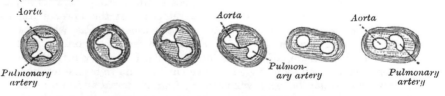

inferius of the ventricles in such a way as to bring the left ventricle into communication with the aorta and the right with the pulmonary artery.

Peculiarities of the fœtal heart.—In early fœtal life the heart is placed directly under the head and is relatively of large size. Later it assumes its position in the thorax but lies at first in the middle line ; towards the end of pregnancy it gradually becomes oblique. Its auricular portion is at first larger than the ventricular part, and the two auricles communicate freely through the foramen ovale. In consequence of the communication, through the ductus arteriosus, between the pulmonary artery and the aorta, the contents of the right ventricle are mainly carried into the latter vessel instead of to the lungs, and hence the wall of the right ventricle is as thick as that of the left. At the end of fœtal life, however, the left ventricle is thicker than the right, a difference which becomes more and more emphasised after birth.

Further Development of the Arteries.—In the vitelline circulation, two arteries were described as coming off from the primitive heart, and running down in front of the developing vertebræ. The first change consists in the fusion of these arteries into one vessel at some distance from the heart ; this vessel is the descending thoracic and abdominal aorta. In consequence of the lengthening of the neck

FIG. [155].—Profile view of a human embryo, estimated at twenty or twenty-one days old. (After His.)

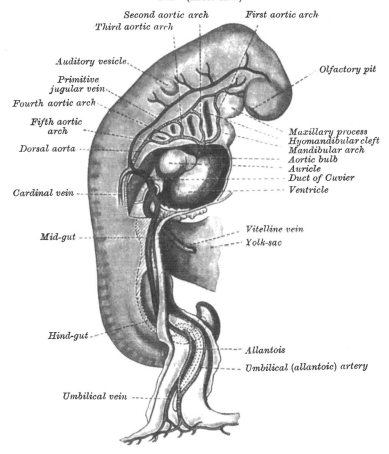

the heart falls backwards to its lower part and then into the thorax, and the two original arteries, proceeding from the heart to their point of fusion in the common descending aorta, become elongated, and assume an arched form, curving backwards on each side, from the front of the body towards the vertebral column (fig. [156], A). These are the *first* or *primitive aortic arches*. As the heart recedes into the thorax, and these arches, which correspond in position to the mandibular arch, become elongated, four additional pairs of arches are formed behind them around the pharynx, one in each branchial arch (fig. [155]). The arches, five in number, remain permanent in fishes, giving off from their convex borders the branchial arteries to supply the gills. In many animals the five pairs do not exist together, for the first two have disappeared before the others are formed ; but this is not so in man, where all five arches are present and pervious during a certain period of embryonic existence (fig. [155]).   Only some of the arches in mammalia remain as permanent structures ; the others, or portions of them, become obliterated or disappear. The first two arches entirely disappear. The third remains as a part of the internal carotid artery, the remainder being formed by the upper part of the posterior aortic root, i.e. the descending part of the original vessel which proceeded from the rudimentary tubular heart.   The common and external carotid arteries are formed from the anterior aortic root, that is, the ascending portion of the same primitive vessel.   The fourth arch on the left side becomes developed into the permanent arch of the aorta in mammals ; but in birds it is the fourth arch on the right side which forms the aortic arch ; in reptiles the fourth arch on both sides persists, so that these animals possess a permanent double aortic arch.   The fourth arch on the right side forms the subclavian artery, and by the junction of its commencement with the anterior aortic root, from which the common carotid is developed, it forms the innominate artery.*   The fifth arch on the left side forms the pulmonary artery and the ductus arteriosus ; that on the right side becomes atrophied and disappears.   The first part of the fifth left arch remains connected with that part of the aortic bulb which is separated as the pulmonary stem, and with it forms the common pulmonary artery.   From about the middle of this arch two branches are given off, which form the right and left pulmonary arteries

FIG. [156].—Diagram of the formation of the aortic arches and the large arteries.

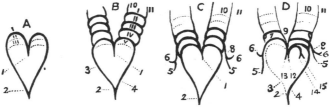

I. II. III. IV. V. First, second, third, fourth, and fifth aortic arches. A. Common trunk from which the first pair spring ; the place where the succeeding pairs are formed is indicated by dotted lines. B. Common trunk, with four arches and a trace of the fifth. C. Common trunk, with the three last pairs, the first two having been obliterated. D. The persistent arteries, those which have disappeared being indicated by dotted lines. 1. Common arterial trunk. 2. Thoracic aorta. 3. Right branch of the common trunk which is only temporary. 4. Left branch, permanent. 5. Axillary artery. 6. Vertebral. 7, 8. Subclavian. 9. Common carotid. 10. External ; and 11, Internal carotid. 12. Aorta. 13. Pulmonary artery. 14. 15. Right and left pulmonary arteries.

respectively, and the remaining portion—that is, the part beyond the origin of the branches—communicates with the left fourth arch, and constitutes the ductus arteriosus.   This duct remains pervious during the whole of fœtal life, but after birth becomes obliterated (fig. [157]). A series of intersegmental or intervertebral arteries arise from the primitive dorsal aortæ, those in the neck alternating with the cervical segments of the spine. The intersegmental artery which lies between the sixth and seventh segments forms the lower part of the vertebral artery ; its upper part is formed by an antero-posterior anastomosis between the higher intersegmental vessels. The subclavian artery is originally a branch of the vertebral, but, owing to the subsequent growth of the upper limb, it comes to exceed in size the parent trunk.

The development of the arteries in the lower part of the body is going on during the same time. It has been seen that originally there were two primitive aortæ coming off from the simple tubular heart. These two vessels course downwards, one on either side of the notochord, and supply the omphalo-mesenteric arteries to the yolk-sac. At the hinder end of the embryo the primitive aortæ give off the two umbilical or allantoic arteries which run in the walls of the allantois to the umbilicus, beyond which they are carried in the umbilical cord to

* This is interesting in connection with the position of the recurrent laryngeal nerve, which is thus seen to hook round the *same* primitive fœtal structure, which becomes on the right side the subclavian artery, on the left the arch of the aorta.

the placenta. The two primitive aortæ soon fuse to form a single vessel, the future descending aorta ; the fusion begins in the thoracic region, and from there proceeds backwards and forwards, and the umbilical arteries now appear as if resulting from the bifurcation of the single vessel ; the part of the fused vessels, beyond their origin, is indicated, however, by the middle sacral artery. The common and internal iliac arteries represent the proximal end of the umbilical artery ; the remainder of the vessel, with the exception of the part which gives off the superior vesical artery, becomes obliterated after birth ; and the obliterated portions of the two umbilical arteries, together with the urachus, carry off the peritoneum from the bladder as its superior false ligament. The external iliac and femoral arteries are developed from a minute branch given off from the umbilical artery near its origin, and are at first of comparatively small size.

FIG. [157].—Diagram to show the destination of the arterial arches in man and mammals. (Modified from Rathke). (From Quain's ' Anatomy,' vol. i. pt. 1, 1890.)

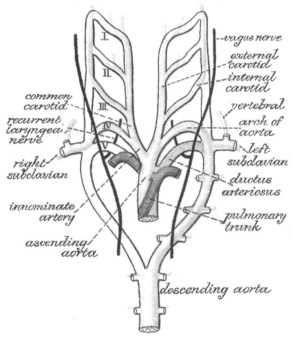

The truncus arteriosus and the five arterial arches springing from it are represented in outline only ; the permanent vessels in colours—those belonging to the aortic system red, to the pulmonary system blue.

Development of the Veins.—The formation of the great veins of the embryo may be best considered under two groups, visceral and parietal.

The *visceral veins* are the two vitelline or omphalo-mesenteric veins bringing the blood from the yolk-sac, and the two umbilical or allantoic veins returning the blood from the placenta ; these four veins open close together into the sinus venosus (fig. [152]).

The vitelline veins run upwards at first in front, and subsequently on either side of the intestinal canal. They unite on the ventral aspect of the canal, before they reach the liver, and then encircle the intestinal tube by forming around it two venous rings, the first on its dorsal, the second on its ventral aspect. The portions of the veins above the upper ring become invaded by the developing liver and broken up by it into a network of smaller vessels, the central part of the network consisting of a capillary plexus. The branches which convey the blood to this plexus are named the *venæ advehentes,* and become the branches of the portal vein ; while the vessels which drain the plexus into the sinus venosus are termed the *venæ revehentes*, and form the future hepatic veins (figs. [158] and [159]).

The lower part of the *portal vein* is formed from the fused vitelline veins which receive the veins from the alimentary canal; its upper part is derived from the venous rings by the persistence of the left half of the lower and the right half of the upper ring, so that the vessel forms a spiral turn round the duodenum (fig. [159]).

The two umbilical veins fuse early to form a single trunk in the allantois, but remain double for some time within the embryo and pass forwards to the sinus venosus in the side walls of the body. Like the vitelline veins, their direct connection with the sinus venosus becomes interrupted by the invasion of the

FIG. [158].—Human embryo with heart and anterior body wall removed to show the sinus venosus and its tributaries. (From Kollmann's ' Entwickelungsgeschichte.') (After His.)

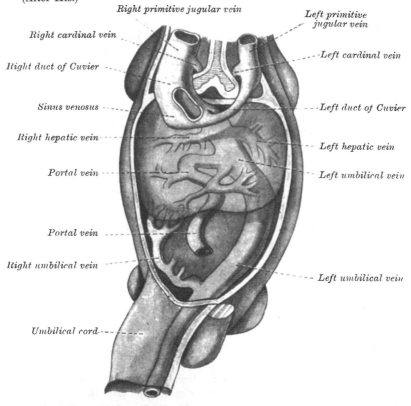

liver, and thus at this stage the whole of the blood from the yolk-sac and placenta passes through the substance of the liver before it reaches the heart. The right umbilical vein shrivels up and almost entirely disappears; the left, on the other hand, becomes much enlarged after the establishment of the placental circulation, and opens into the upper venous ring. Finally a direct branch is established between this ring and the right hepatic vein; this branch is the *ductus venosus* or *vena ascendens*, and, enlarging rapidly, it forms a wide channel through which most of the blood, returned from the placenta, is carried direct to the heart (fig. [159]).

The Parietal Veins.—The first indication of a parietal system consists in the appearance of two short transverse veins (the *ducts of Cuvier*), which open, one on either side, into the auricular portion of the heart. Each of these ducts is formed by an ascending and descending vein. The ascending veins return the blood from the parietes of the trunk and from the Wolffian bodies, and are called *cardinal* veins. The descending veins return the blood from the head, and are called *primitive jugular* veins (fig. [155]). The blood from the lower limbs is collected by the iliac veins, which empty themselves into the cardinal veins. In the earlier stages of development the right and left iliac veins open into the corresponding right and left cardinal veins (fig. [161]), but later on a transverse branch connects the lower ends of the two cardinal veins, and through this the blood from the left iliac vein is carried into the right cardinal vein. By the development of a similar transverse branch higher up the blood from the left kidney is also carried into the right cardinal vein (fig. [160], 2). The portion of the left cardinal vein above the origin of the lower transverse branch becomes atrophied as high as the level of the renal vein, above which it persists as the vena azygos minor. The right cardinal vein, which now receives the blood from both lower extremities, forms a large venous trunk along the posterior

FIG. [159].—The liver, and the veins in connection with it, of a human embryo, twenty-four or twenty-five days old, as seen from the ventral surface. (After His.) (Copied from Milnes Marshall's ' Embryology.')

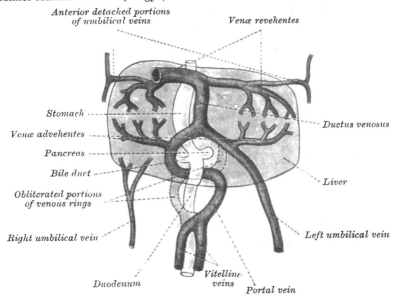

abdominal wall; it receives the renal veins from the kidneys, and forms, up to this level, the inferior vena cava. Above the level of the renal veins the inferior vena cava first makes its appearance as a small vein lying in the tissue between the two kidneys. Superiorly it opens into the sinus venosus, while below it communicates with the right cardinal vein near the level of the renal veins (fig. [160], 1, 2). This small vein ultimately becomes enlarged, and carries the blood upwards from the right cardinal vein, and so forms the upper part of the inferior vena cava. The portion of the right cardinal vein above the renal veins persists as the vena azygos major, and receives the right intercostal veins, while the vena azygos minor is brought into communication with it by the development of transverse branches in front of the spinal column (fig. [160], 2, 3).

In consequence of the atrophy of the Wolffian bodies the cardinal veins diminish in size; the primitive jugular veins, on the other hand, become enlarged, owing to the rapid development of the head and brain. They are further augmented by receiving the vein (*subclavian*) from the upper extremity, and so come to form the chief veins of the Cuvierian ducts; these ducts gradually assume an almost vertical position in consequence of the descent of the heart into the thorax. The right and left Cuvierian ducts are originally of the same diameter, and are frequently termed the *right* and *left superior venæ cavæ*. By the development of a transverse branch (the future *left innominate vein*) between the two ducts the blood is carried from the left duct into the right, which thus

FIG. [160].—Diagram to illustrate the development of the principal systemic veins. (Hertwig.)

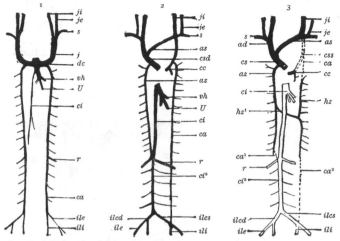

dc. Duct of Cuvier. *je, ji.* External and internal jugular veins. *s.* Subclavian. *vh.* Hepatic veins. *U.* Umbilical. *ci, ci².* Vena cava inferior. *ca* (*ca¹, ca², ca³*). Cardinal veins. *ilcd, ilcs.* Right and left common iliac veins. *ad, as.* Right and left innominate veins. *cs.* Vena cava superior. *css.* Rudimentary portion of left superior vena cava. *cc.* Coronary sinus. *az.* Azygos major. *hz, hz¹.* Azygos minor. *ile.* External iliac. *ili.* Internal iliac. *r.* Renal vein.

becomes much enlarged and forms the permanent *superior vena cava*, and into which the vena azygos major opens. The left duct atrophies: its upper part remains pervious as a small vein, which receives the *left superior intercostal vein*; its intermediate portion is represented by the *vestigial fold* of Marshall; its lower part persists as a small vein, the oblique vein of Marshall, which runs downwards across the back of the left auricle to join the coronary sinus; this

FIG. [161].—Diagrammatic outline of a longitudinal vertical section of the chick on the fourth day.

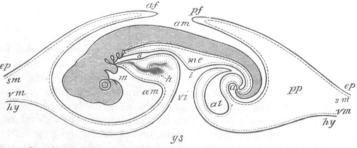

ep. Epiblast. *sm.* Somatic mesoblast. *hy.* Hypoblast. *vm.* Visceral mesoblast. *af.* Cephalic fold. *pf.* Caudal fold. *am.* Cavity of true amnion. *ys.* Yolk-sac. *i.* Intestine. *s.* Foregut. *a.* Future anus, still closed. *m.* The mouth. *me.* The mesentery. *al.* The allantoic vesicle. *pp.* Space between inner and outer folds of amnion. (From Quain's 'Anatomy,' Allen Thomson.)

sinus, as has already been indicated, represents the persistent left horn of the sinus venosus. The primitive jugular veins become the internal jugular veins of the adult; the lower part of the right primitive jugular vein forms also the right innominate vein (figs. [160], 1, 2, 3).

The fœtal circulation is described at a future page.

Development of the Alimentary Canal.—As already indicated (page [86]), the primitive alimentary canal is formed, at an early stage, by the enclosure within

FIG. [162].—Diagram of a longitudinal section of a mammalian embryo. Very early. (After Quain.)

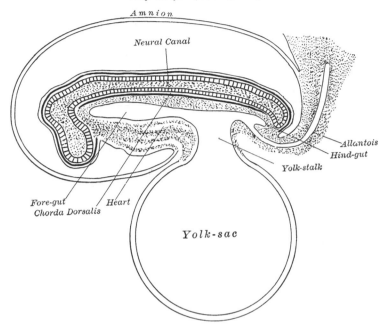

Amnion

Neural Canal

Allantois
Hind-gut

Yolk-stalk

Fore-gut
Chorda Dorsalis   *Heart*

Yolk-sac

the embryo of a portion of the blastodermic vesicle, and is seen to consist of three parts, viz.: (1) the *fore-gut*, within the cephalic flexure and dorsal to the heart; (2) the *mid-gut*, opening freely into the yolk-sac; and (3) the *hind-gut*, within the caudal flexure. The fore-gut and hind-gut end blindly, there being at first neither mouth nor anus (figs. [161] and [162]). The formation of the mouth or stomodæum, and the subsequent communication between it and the cephalic end of the fore-gut, have already been considered; the manner in which the anus is formed will presently be discussed.

From the fore-gut are developed the pharynx, œsophagus, stomach, and duodenum, and further, as diverticula from the duodenum, the liver and pancreas; from the hind-gut, the greater part of the rectum, and as a tubular outgrowth from it the hollow stalk of the allantois; the mid-gut gives origin to the remainder, or

FIG. [163].—Early form of the alimentary canal. (From Kölliker, after Bischoff.)

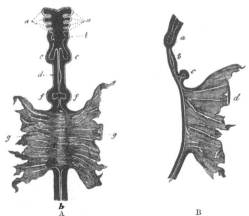

A   B

In A a front view, and in B an antero-posterior section are represented. *a.* Four pharyngeal or visceral plates. *b.* The pharynx. *c, c.* The commencing lungs. *d.* The stomach. *f, f.* The diverticula connected with the formation of the liver. *g.* The yolk-sac into which the middle intestinal groove opens. *h.* The posterior part of the intestine.

longest section, of the alimentary tube, i.e. the portion which reaches from the duodenum to the rectum.

The upper part of the fore-gut becomes dilated to form the pharynx, in relation to which the branchial arches are developed (figs. [121] and [164]) ; the succeeding part remains tubular, and with the descent of the stomach is elongated to form the œsophagus.   Soon a fusiform dilatation, the future stomach, makes its appearance, and beyond this the mid-gut opens freely into the yolk-sac (figs. [164] and [165]). This opening is at first wide, but, as the body-walls close in around the umbilicus, it is gradually narrowed into a tubular stalk, the *yolk-stalk* or *vitello-intestinal duct.*   At this stage, therefore, the alimentary canal forms a nearly straight tube in front of the notochord and primitive aortæ (fig. [162]).   From the stomach to the rectum it is attached to the notochord by a band of mesoblast, from which the common mesentery of the gut is subsequently developed.   The stomach undergoes a further dilatation, and its two curvatures can be recognised (figs. [165] and [169]), the greater directed towards the vertebral column and the

FIG. [164].—Human embryo, about fifteen days old.   Brain and heart represented from right side ; alimentary canal and yolk-sac in mesial section.   (After His.)

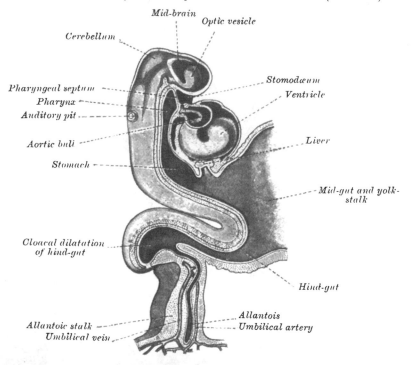

lesser towards the anterior wall of the abdomen, while of its two surfaces one looks to the right and the other to the left.   The mid-gut also undergoes great elongation, and forms a V-shaped loop which projects downwards and forwards ; from the bend or angle of the loop the vitello-intestinal duct passes to the umbilicus (fig. [169]).   For a time a part of the loop extends beyond the abdominal cavity into the umbilical cord, but by the end of the third month this is withdrawn. With the lengthening of the tube, the mesoblast, which attaches it to the future vertebral column and which carries the blood-vessels for the supply of the gut, is thinned and drawn out to form the *primitive* or *common mesentery.*   The portion of this mesentery which is attached to the greater curvature of the stomach is named the *mesogastrium,* and the parts which suspend the colon and rectum are respectively termed the *mesocolon* and *mesorectum* (fig. [169]).   About the sixth week a lateral diverticulum makes its appearance a short distance beyond the vitello-intestinal duct, and indicates the future cæcum or boundary between the small and the large intestine. This cæcal diverticulum has at first a uniform calibre, but its blind extremity remains rudimentary and forms the vermiform

FIG. [165].—Sketches in profile of two stages in the development of the human alimentary canal. (His.) Fig. A × 30. Fig. B × 20.

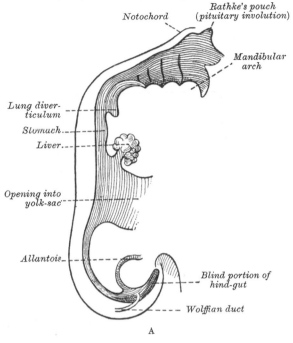

Notochord

Rathke's pouch (pituitary involution)

Mandibular arch

Lung diverticulum

Stomach

Liver

Opening into yolk-sac

Allantois

Blind portion of hind-gut

Wolffian duct

A

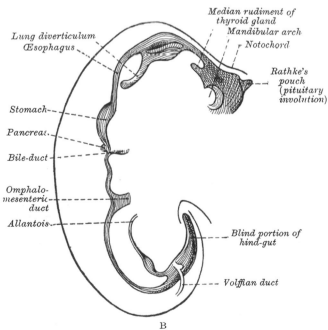

Median rudiment of thyroid gland

Mandibular arch

Notochord

Lung diverticulum
Œsophagus

Rathke's pouch (pituitary involution)

Stomach

Pancreas

Bile-duct

Omphalo-mesenteric duct

Allantois

Blind portion of hind-gut

Wolffian duct

B

appendix (figs. [169] and [170]). Changes also take place in the position and direction of the stomach. It falls over on to its right surface, which henceforth is directed backwards, while its original left surface looks forwards ; further, its greater curvature is drawn downwards and to the left away from the vertebral column, while its lesser curvature is directed upwards, and the commencement

FIG. [166].—Front view of two successive stages in the development of the alimentary canal. (His.)

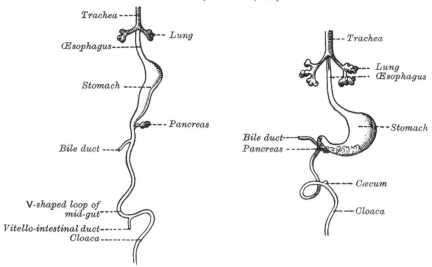

of the duodenum is pushed over to the right side of the middle line.   The meso-gastrium being attached to the greater curvature must necessarily follow its movements, and hence it becomes greatly elongated and drawn outwards from the

FIG. [167].—Schematic and enlarged cross section through the body of a human embryo in the region of the mesogastrium. Beginning of third month. (Toldt.)

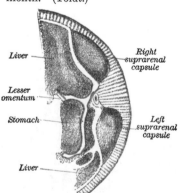

FIG. [168].—Same section as in fig. [167], at end of third month. (Toldt.)

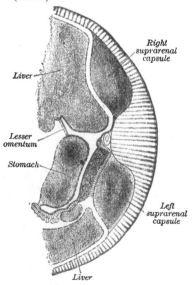

vertebral column, and, like the stomach, what was originally its right surface is now directed backwards and its left forwards.   In this way a pouch, the *bursa omentalis*, is formed behind the stomach ; this pouch is the future lesser sac of the peritoneum, and it increases in size as the alimentary tube undergoes further development ; the entrance to the pouch constitutes the future *foramen of Winslow* (figs. [166], [170], and [173]). The remainder of the canal becomes greatly increased in length, so that the tube is coiled on itself, and this increase in length demands a corresponding increase in the width of the intestinal attachment of the mesentery, so that it becomes plaited or folded.

FIG. [169].—Abdominal part of alimentary canal and its attachment to the primitive or common mesentery. Human embryo of six weeks. (After Toldt.) (From Kollmann's 'Entwickelungsgeschichte.')

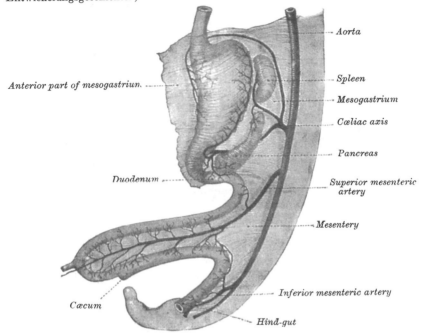

Aorta

Anterior part of mesogastriun.

Spleen

Mesogastrium

Cœliac axis

Pancreas

Duodenum

Superior mesenteric artery

Mesentery

Inferior mesenteric artery

Cæcum

Hind-gut

FIG. [170].—Diagrams to illustrate two stages in the development of the human alimentary canal and its mesentery. The arrow indicates the entrance to the bursa omentalis.

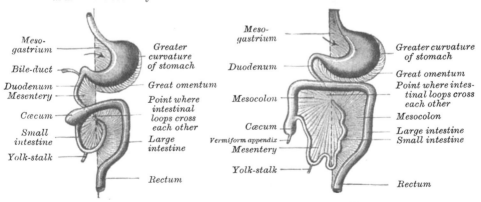

Meso-gastrium

Greater curvature of stomach

Meso-gastrium

Greater curvature of stomach

Bile-duct

Great omentum

Duodenum

Great omentum
Point where intes-tinal loops cross each other

Duodenum
Mesentery

Great omentum

Mesocolon

Point where intestinal loops cross each other

Mesocolon

Cæcum

Large intestine

Cæcum

Small intestine

Large intestine
Small intestine

Vermiform appendix
Mesentery

Yolk-stalk

Large intestine

Yolk-stalk

Rectum

Rectum

At this stage the small and large intestine are attached to the vertebral column by a common mesentery, the coils of the small intestine falling to the right of the middle line, while the large intestine lies on the left side.*

    * Sometimes this condition persists throughout life, and it is then found that the duodenum does not cross from the right to the left side of the vertebral column, but lies entirely on the right side of the mesial plane, where it is continued into the jejunum; the arteries to the small intestine (rami intestini tenuis) also arise from the right instead of the left side of the superior mesenteric artery.

The gut now becomes rotated upon itself, so that the large intestine is carried over in front of the small intestine, and the cæcum is placed immediately below the liver; about the sixth month the cæcum descends into the right iliac fossa, and the large intestine now forms an arch consisting of the ascending, transverse, and descending portions of the colon—the transverse portion crossing in front of the duodenum and lying just below the greater curvature of the stomach; within this arch the coils of the small intestine are disposed (fig. [170]). Sometimes the downward progress of the cæcum is arrested, so that in the adult it may be found lying immediately below the liver instead of in the right iliac region.

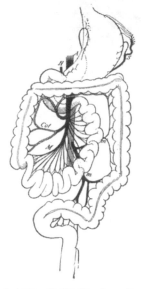

Fig. [171].—Final disposition of the intestines and their vascular relations. (Jonnesco.)

A. Aorta. *H.* Hepatic artery. *S.* Splenic artery. *M, Col.* Branches of superior mesenteric artery. *m, m'.* Branches of inferior mesenteric artery.

Further changes take place in the bursa omentalis and in the common mesentery, and give rise to the peritoneal relations seen in the adult. The bursa omentalis, which at first reaches only as far as the greater curvature of the stomach, grows downwards to form the great omentum, and this downward extension lies in front of the transverse colon and the coils of the small intestine. The anterior layer of the transverse mesocolon is at first quite distinct from the posterior wall of the bursa omentalis, but ultimately the two blend, and hence the great omentum appears as if attached to the transverse colon (figs. [173] and [174]). The mesentery of the duodenum, in which the rudiment of the pancreas is enclosed, disappears, and so this part of the gut becomes fixed to the posterior abdominal wall, and the pancreas lies entirely behind the peritoneal membrane. The mesenteries of the ascending and descending parts of the colon disappear in the majority of cases, while that of the small intestine assumes the oblique attachment characteristic of its adult condition.

The small omentum is formed by a thinning of the mesoblast or *anterior primitive mesentery*, which attaches the lesser curvature of the stomach to the

Fig. [172].—The primitive mesentery of a six weeks' human embryo, half schematic. (Kollmann.)

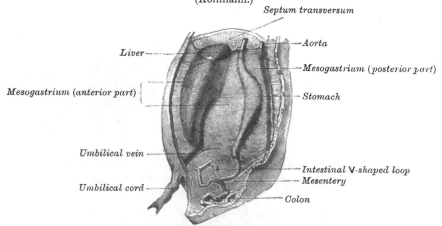

Septum transversum

Liver

Mesogastrium (anterior part)

Umbilical vein

Umbilical cord

Aorta

Mesogastrium (posterior part)

Stomach

Intestinal V*-shaped loop*

Mesentery

Colon

anterior abdominal wall. By the subsequent growth of the liver this leaf of mesoblast is divided into two parts, viz.: the small omentum between the stomach and liver, and the falciform ligament between the liver and the abdominal wall and Diaphragm (fig. [172]).

The anus is developed as a slight invagination of the epiblast a short distance in front of the posterior end of the hind-gut. This invagination is termed the *proctodæum*; the mesoblast between it and the hypoblastic lining of the hind-gut

Fig. [173].—Schematic figure of the bursa ornamentalis, &c. Human embryo of eight weeks. (Kollmann.)

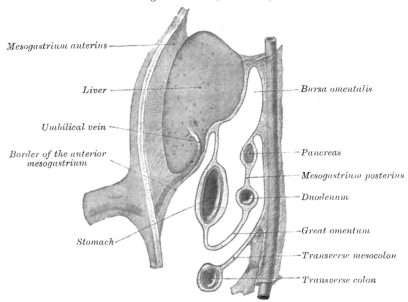

Mesogastrium anterius

Liver

Umbilical vein

Border of the anterior mesogastrium

Stomach

Bursa omentalis

Pancreas

Mesogastrium posterius

Duodenum

Great omentum

Transverse mesocolon

Transverse colon

Fig. [174].—Diagrams to illustrate the development of the great omentum and transverse mesocolon. (O. Hertwig.) In the left figure the early stage is represented : in the right figure the adult condition.

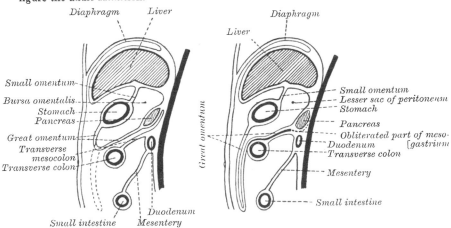

Diaphragm   Liver   Diaphragm

Liver

Small omentum

Bursa omentalis

Stomach

Pancreas

Great omentum

Transverse mesocolon

Transverse colon

Great omentum

Small intestine   Duodenum   Mesentery

Small omentum

Lesser sac of peritoneum

Stomach

Pancreas

Obliterated part of meso-

Duodenum   [gastrium

Transverse colon

Mesentery

Small intestine

is thinned, and ultimately the septum breaks down and disappears, and the hind-gut opens on the surface ; into this part of the hind-gut the urinary and generative organs open for a time, and so it constitutes a *common cloaca*. The small portion of the hind-gut behind the orifice of the anus is named the *caudal* or *post-anal* gut ; it communicates with the neural tube by means of a canal, the *neurenteric canal*, already referred to. Ultimately the post-anal gut becomes obliterated, and it, together with the neurenteric canal, finally disappears.

The peritoneal cavity is the space left between the visceral and parietal layers of the mesoblast, and the serous membrane is developed from these layers.

The tongue originates from the floor of the pharynx. The anterior or papillary portion first appears as a rounded elevation, the *tuberculum impar*, between the ventral ends of the mandibular and hyoid arches. Between the third and fourth arches a second larger elevation arises, in the centre of which is a median groove or furrow. This second elevation is termed the *furcula*, and from it the epiglottis is developed, while the median furrow becomes the entrance to the larynx (fig. [175]). The tuberculum impar and the furcula are at first in apposition, but are soon separated by a ridge produced by the forward growth of the second and third arches. This ridge gives rise to the posterior part of the

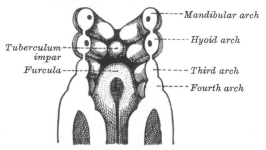

Fig. [175].—The floor of the pharynx of a human embryo about fifteen days old. × 50. (From His.)

tongue and extends forwards in the form of a V, so as to embrace between its two limbs the tuberculum impar. At the apex of the V there is a pit-like invagination to form the middle thyroid rudiment, and this depression persists as the foramen cæcum of the adult. The union of the two parts of the tongue is indicated even in the adult by a V-shaped depression, the apex of which is at the foramen cæcum, while the two limbs run outwards and forwards parallel to but a little behind the circumvallate papillæ which are therefore developed from the tuberculum impar (figs. [176] and [177]). The tonsils are developed from the second branchial cleft, and make their appearance between the fourth and fifth months.

Fig. [176].—The floor of the pharynx of a human embryo about twenty-three days old. × 30. (From His.)

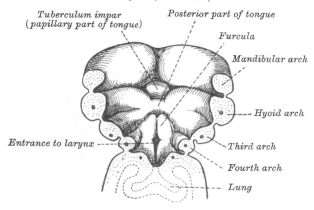

The *liver* arises in the form of two diverticula or hollow outgrowths from the ventral surface of that portion of the fore-gut which afterwards becomes the duodenum (figs. [164] and [165]). The outgrowths, which represent the right and the left lobes respectively of the adult liver, give off solid buds of cells, which grow into columns or cylinders: these unite with one another in every direction to form a close network, in the meshes of which are contained the capillary blood-vessels. Some of these columns become hollowed out and form the bile-ducts, while the remainder constitute the secreting structure. The minute ducts thus produced unite to form the right and left hepatic ducts; while the common bile-duct is developed as a protrusion from the duodenal wall, and as it grows the liver becomes shifted away from the duodenum. The gall-bladder and cystic duct are formed by a hollow evagination from the wall of the common bile-duct. About the third month the liver almost fills the abdominal cavity. From this period the relative development of the liver is less active, more especially that of the left lobe, which now becomes smaller than the right; but up to the end of fœtal life the liver remains relatively larger than in the adult.

The *pancreas* is also an early formation, being far advanced in the second month. It originates as a hollow projection from the hypoblast of the dorsal wall of the duodenum (figs. [165] and [166]), opposite the hepatic diverticula, which, as we have already seen, spring from its ventral wall. This hollow process grows between the two layers of the dorsal mesentery and sends out offshoots, which branch abundantly and form a complicated tubular gland. As torsion of the stomach takes place, the pancreas assumes a transverse position and becomes fixed across the dorsal wall of the abdomen, the posterior layer of its mesentery undergoing absorption. Its duct ultimately opens into the duodenum together with the common bile-duct.

Fig. [177].—Floor of mouth of an embryo slightly older than that shown in fig. 176. × 16. (From His.)

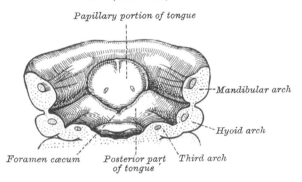

Papillary portion of tongue

Mandibular arch

Hyoid arch

Foramen cæcum *Posterior part of tongue* *Third arch*

The *spleen*, on the other hand, is of mesoblastic origin, for there is never any connection between the intestinal cavity and the substance of this organ. It originates in the mesenteric fold which connects the stomach to the vertebral column (mesogastrium) (fig. [169]).

The thyroid body is developed as a median and two lateral diverticula from the ventral wall of the pharynx. The median diverticulum appears first; it commences at the foramen cæcum, between the anterior and posterior rudiments of the tongue and extends backwards as a tubular duct, the *ductus thyro-glossus*. The lateral diverticula arise from the fourth visceral cleft and fuse with the median part to form the thyroid body. The connection of the lateral diverticula with the pharynx disappears early, but the remains of the ductus thyro-glossus may persist as a tube leading from the foramen cæcum towards the hyoid bone, the *pyramid* of the thyroid probably representing its lower part.*

The thymus is developed from bilateral diverticula, which are principally derived from the third visceral cleft. It increases in size until the second year of life, after which it undergoes atrophy.

* Kanthack (*Journal of Anat. and Physiol.* vol. xxv. p. 155) disputed this view. He examined 100 subjects, 60 of which were fœtuses or infants, and found that in many cases there was no trace of a foramen cæcum and that when it was present it formed a short canal near the surface and was lined with stratified squamous, not columnar, epithelium. Further, after careful microscopical examination, he found no trace of a tubular lumen in the pyramid of the thyroid body.

Development of the Respiratory Organs.—The lungs appear somewhat later than the liver. They are developed from a small median *cul-de-sac* or diverticulum from the upper part of the fore-gut, immediately behind the fourth visceral cleft. During the fourth week a pouch is formed on either side of the central diverticulum, and opens freely through it into the fore-gut (pharynx). These lateral pouches soon become subdivided—the right into three and the left into two parts, these subdivisions being the early indications of the lobes of the lungs (figs. [121] and [166]). The two primary pouches have thus a common tube of communication with the pharynx. This common tube becomes the larynx and trachea, the latter rapidly elongating as development proceeds. The larynx first becomes evident as a dilatation of the upper part of the trachea about the end of the fifth week. The epiglottis is developed from the anterior or median portion of the furcula, and the aryteno-epiglottidean folds from its lateral ridges (fig. [176]). The vocal cords and ventricles of the larynx are formed about the fourth month.

FIG. [178].—Section of the urogenital area of a chick embryo of the fourth day. (Waldeyer.)

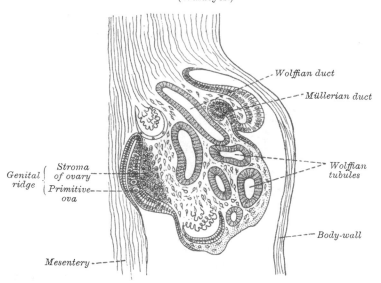

As the lungs grow backwards they project into the anterior part of the cœlum, which becomes shut off from the rest of the body-cavity by the pericardium and Diaphragm to form the pleural cavities.

The Diaphragm is formed in two parts : (*a*) ventral, (*b*) dorsal. The ventral part appears first, and consists of a thick septum of mesoblast, the *septum transversum*, which projects from the anterior and lateral walls of the embryo, and which ends behind in a free edge. The sinus venosus, which receives the vitelline, umbilical, and Cuvierian veins, is placed originally in this septum, and into the posterior part of it also the liver diverticula grow from the duodenum. The sinus separates itself above from the septum, and the greater part of it is incorporated with the right auricle. The liver also becomes separated from it below, except where the veins pass through into the heart. The septum transversum shuts off the greater part of the thoracic from the abdominal cavity, but posteriorly there remain two channels of communication, one on each side of the alimentary tube ; these channels subsequently become the pleural cavities, and are shut off from the abdomen by folds which grow from the lateral and posterior parts of the trunk and which fuse with the posterior edge of the septum transversum. Sometimes the fusion is incomplete, thus leaving a permanent communication between the abdominal and one or other of the pleural cavities, and through which some of the abdominal contents may pass, forming what is termed a *diaphragmatic hernia.*

Development of the Urinary and Generative Organs.—The urinary organs are developed from a ridge of mesoblast at the point where this layer separates into somatopleure and splanchnopleure. As this ridge is situated close to the epiblast, between the paraxial mesoblast and the common pleuro-peritoneal cavity, it has been named the 'intermediate cell-mass.' It is at first solid, and in it is formed a cord-like arrangement of some of the cells, which extends longitudinally from just below the heart to the posterior extremity of the body-cavity. In this cord-like structure a tube is hollowed out; it gradually becomes separated from the rest of the intermediate cell-mass, and is then named the *Wolffian duct* (fig. [105]).* Its posterior end becomes connected with and eventually opens into the hind-gut. Its anterior end becomes connected with pit-like involutions of the peritoneal epithelium, and in the mesoblastic tissue between these invaginations a vascular glomerulus is formed which projects into the peritoneal cavity. It is known as the *head-kidney* or *pronephros* (Lankester), and is a very rudimentary organ which speedily disappears. Behind this body and to the inner side of the Wolffian duct, between it and the body-cavity, a number of tubes are formed, which communicate by one extremity with the Wolffian duct, and, passing

Fig. [179].—Enlarged view from the front of the left Wolffian body before the establishment of the distinction of sex. (From Farre, after Kobelt.)

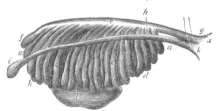

a, a, b, d. Tubular structure of the Wolffian body. *e.* Wolffian duct. *f.* Its upper extremity. *g.* Its termination in *x,* the urogenital sinus. *h.* The duct of Müller. *i.* Its upper, funnel-shaped extremity. *k.* Its lower end, terminating in the urogenital sinus. *l.* The mass of blastema for the reproductive organ, ovary or testicle.

transversely towards the body-cavity, terminate in cæcal extremities. These tubes are called segmental tubes, and the whole mass is known as the *mid-kidney, Wolffian body,* or *mesonephros* (Lankester) (fig. [179]). After a time the cæcal extremities become dilated and enclose a tuft or glomerulus of capillary blood-vessels. As soon as the permanent kidneys are formed, the Wolffian body for the most part disappears. In the male, however, the vasa efferentia and rete testis of the testicle are formed as outgrowths from it. In the female traces of it are left as the *parovarium* and *epoöphoron.* In the male the Wolffian duct becomes the epididymis and vas deferens; in the female it undergoes atrophy, and is represented only by the functionless duct of Gärtner.

Finally, in that portion of the intermediate cell-mass which lies behind the Wolffian body, a differentiation of cells takes place which results in the formation of a number of convoluted tubes; into this a hollow protrusion of the lower end of the Wolffian duct grows up, and thus is formed the *hind-kidney* or *metanephros* (Lankester). This is the permanent kidney. The uriniferous convoluted tubes and Malpighian corpuscles are formed from the intermediate cell-mass, and the collecting tubules and ureter from the protrusion from the posterior end of the Wolffian duct.

Shortly after the formation of the Wolffian body, a second duct becomes developed. It arises on the outer side of this body as a slight thickening of the

---

* By some embryologists the Wolffian duct is regarded as being of epiblastic origin and formed by a longitudinal invagination of the epiblast.

cells lining the pleuro-peritoneal cavity. This thickening then becomes invaginated into the mesoblast and extends as a cord along the outer side of the Wolffian body, to the posterior extremity of the embryo. It speedily acquires a lumen, and is then known as the *Müllerian duct* (fig. [178]). In its passage to the posterior extremity of the embryo it comes into close relation with the Wolffian duct, and the two ducts on either side become connected with their fellows on the opposite side by cellular substance into a single cord, the *genital cord* (fig. [180], G, C), in which the Wolffian ducts lie side by side in front, and the ducts of Müller behind. These latter tubes in the substance of the genital cord become fused together, and open by a single orifice into the hind-gut (cloaca). At their anterior extremities the ducts of Müller open by a somewhat funnel-shaped orifice into the pleuro-peritoneal cavity. In the female the greater part of the Müllerian duct is developed into the Fallopian tube, but the posterior fused portion of the two ducts is converted into the uterus and vagina (fig. [181]). In

FIG. [180].—Diagram of the primitive urogenital organs in the embryo previous to sexual distinction. The parts are shown chiefly in profile, but the Müllerian and Wolffian ducts are seen from the front.

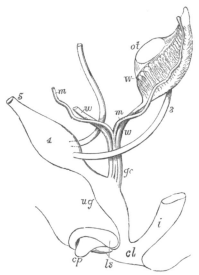

3. Ureter. 4. Urinary bladder. 5. Urachus. *ot.* The mass of blastema from which ovary or testicle is afterwards formed. *W.* Left Wolffian body. *w, w.* Right and left Wolffian ducts. *m, m.* Right and left Müllerian ducts uniting together and with the Wolffian ducts in *gc,* the genital cord. *ug.* Sinus urogenitalis. *i.* Lower part of the intestine. *cl.* Common opening of the intestine and urogenital sinus. *cp.* Elevation which becomes clitoris or penis. *ls.* Ridge from which the labia majora or scrotum are formed.

the male the greater part of the ducts disappear; the posterior fused portion is believed to be represented by the *sinus pocularis (uterus masculinus)* of the urethra.

It has been seen that the Wolffian and Müllerian ducts open into the common cloaca, which is the termination of the intestinal cavity, and into which the allantois also opens in front (fig. [180]). As the allantois expands into the urinary bladder this common cavity is divided into two by a septum, to form the urogenital sinus in front and the rectum behind, and the Wolffian and Müllerian ducts now open into the urogenital sinus.

The *urinary bladder*, as before stated, is formed by a dilatation of a part of the intra-embryonic portion of the allantois. At the end of the second month the middle part of this portion of the allantois becomes dilated into a spindle-shaped cavity, which persists as the *urinary bladder*. Between the lower extremity of the spindle-shaped dilatation and the intestine is the *urogenital sinus*, into which the Müllerian and Wolffian ducts now open, and which becomes the first part of the urethra.   The upper part of the intra-embryonic portion of the allantois, which is not dilated, forms the *urachus* (fig. [180]); this extends into the umbilical cord, and at an early period of embryonic existence forms a tube of communication with the allantois.   It is obliterated before the termination of fœtal life, but the cord formed by its obliteration is perceptible throughout life, passing from the upper part of the bladder to the umbilicus.   It occasionally remains patent after birth, constituting a well-known malformation.

The *suprarenal* bodies are developed from two different sources.   The medullary part of the organ is of epiblastic origin, and is derived from the tissues forming the sympathetic ganglia of the abdomen, while the cortical portion is of mesoblastic origin, and originates as an outgrowth from the upper part of the Wolffian body.   The two parts are at first quite distinct, but become combined in the process of development.   The suprarenal capsules are at first larger than the kidneys; about the tenth week they equal them in size, and from that time decrease relatively to the kidney, though they remain, throughout fœtal life, proportionately much larger than in the adult.

FIG. [181].—Female genital organs of the embryo, with the remains of the Wolffian bodies.   (After J. Müller.)

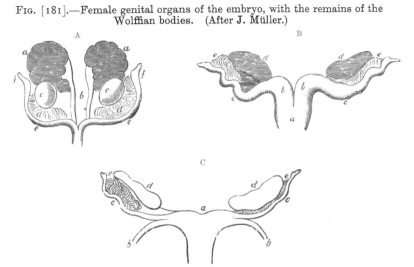

A. From a fœtal sheep : *a.* The kidneys. *b.* The ureters. *c.* The ovaries. *d.* Remains of the Wolffian bodies. *e.* Fallopian tubes. *f.* Their abdominal openings. B. More advanced, from a fœtal deer : *a.* Body of the uterus. *b.* Cornua. *c.* Tubes. *d.* Ovaries. *e.* Remains of the Wolffian bodies. c. Still more advanced, from the human fœtus of three months : *a.* The body of the uterus. *b.* The round ligament. *c.* The Fallopian tubes. *d.* The ovaries. *e.* Remains of the Wolffian bodies.

Ovaries and Testicles.—The first appearance of the reproductive organs is essentially the same in the two sexes, and consists in a thickening of the epithelial layer which lines the peritoneal or body-cavity close to the inner side of the Wolffian body.   Beneath this thickened epithelium an increase in the mesoblast takes place, forming a distinct projection or ridge.   This is termed the *genital ridge* (fig. [178]), and from it the testicle in the one sex, and the ovary in the other, are developed.   As the embryo grows the genital ridge gradually becomes pinched off from the Wolffian body, with which it is at first continuous, though it still remains connected to the remnant of this body by a fold of peritoneum, the *mesorchium* or *mesovarium*.   About the seventh week the distinction of sex begins to be perceptible.

The ovary, thus formed from the genital ridge, consists of a central part of connective tissue covered by a layer or layers of epithelium, the *germinal epithelium*.   Columns of this epithelium, termed *egg-tubes*, grow down into the stroma, and simultaneously with this an upward growth of the connective tissue takes place between the columns of epithelial cells.   It results from this that the columns of cells, become enclosed in meshes of connective tissue (fig. [188]). Each egg-tube or nest represents a primitive Graafian follicle, one cell of which becomes enlarged to form the ovum; the remainder form the epithelium of the follicle. The remains of the germinal epithelium on the surface of the ovary form the permanent epithelial covering of this organ. According to Beard the primitive ova are early set apart during the segmentation of the ovum and migrate into the germinal ridge.

The testicle is developed in a very similar way to the ovary, but the processes are not so well marked. Like the ovary, in its earliest stages it consists of a central mass of connective tissue covered by germinal epithelium. A downward growth of columns of this epithelium into the central connective tissue takes place. From these the seminiferous tubules are developed and become connected with outgrowths from the Wolffian body, which, as before mentioned, form the rete testis and vasa efferentia.

With regard to the other parts of the internal female organs, the Fallopian tube, as has already been mentioned, is developed from the upper part of the duct of Müller, while the lower parts of the two ducts approach each other, and, lying

FIG. [182].—Adult ovary, parovarium, and Fallopian tube.
(From Farre, after Kobelt.)

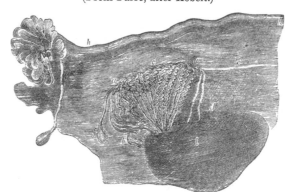

a, a. Epoöphoron formed from the upper part of the Wolffian body. b. Remains of the uppermost tubes sometimes forming hydatids. c. Middle set of tubes. d. Some lower atrophied tubes. e. Atrophied remains of the Wolffian duct. f. The terminal bulb or hydatid. h. The Fallopian tube, originally the duct of Müller. i. Hydatid attached to the extremity. l. The ovary.

side by side, finally coalesce to form the cavity of the uterus and vagina. This coalescence commences in the middle of the genital cord, and corresponds to the body of the uterus. With regard to the further changes in the female organs the only remains of the Wolffian body in the complete condition are two rudimentary or vestigial structures, which can be found, on careful search, in the broad ligament near the ovary: the *parovarium* or *organ of Rosenmüller* and the *epoöphoron* (fig. [182]). The organ of Rosenmüller consists of a number of tubes which converge to a transverse portion, the epoöphoron, and this is sometimes prolonged into a distinct duct, running transversely, the *duct of Gärtner*, which is much more conspicuous and extends further in some of the lower animals. This, as has been pointed out, is the remains of the Wolffian duct. About the fifth month an annular constriction marks the position of the neck of the uterus, and after the sixth month the walls of the uterus begin to thicken. The round ligament is derived from a band containing involuntary muscular fibres, which runs downwards from the lower part of the Wolffian body to the groin, and which in the male forms the gubernaculum testis; the peritoneum constitutes the broad ligament; the superior ligament of the Wolffian body, which serves to connect it with the Diaphragm, disappears with that body.

With regard to the other parts of the male organs, the Müllerian ducts disappear, with the exception of their lower ends. These unite in the middle line, and open by a common orifice into the urogenital sinus. This constitutes the *utriculus hominis* or *sinus prostaticus*. Frequently, however, the upper end of the duct of Müller remains visible in the male as a little pedunculated body, called the hydatid of Morgagni, in the neighbourhood of the epididymis,* between the testis and globus major.

The epididymis, the vas deferens, and ejaculatory duct, are formed from the Wolffian duct. One or more of the tubes of the Wolffian body form the vas aberrans and a structure described by Giraldès, and called, after him, 'the organ of Giraldès,' which bears some resemblance to the organ of Rosenmüller in the other sex. It consists of a number of convoluted tubules, lying in the cellular tissue in front of the cord, and close to the head of the epididymis.

Descent of the Testes.—The testes, at an early period of fœtal life, are placed at the back part of the abdominal cavity, behind the peritoneum and a little below the kidneys. The anterior surface and sides are invested by peritoneum. At about the third month of intra-uterine life a peculiar structure, the *gubernaculum testis*, makes its appearance. This structure is at first

FIG. [183].—Section of the ovary of a newly born child. (Waldeyer.)

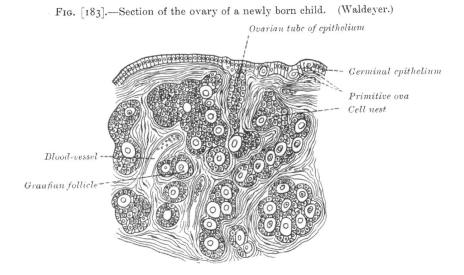

Ovarian tube of epithelium

Germinal epithelium

Primitive ova
Cell nest

Blood-vessel

Graafian follicle

a slender band, extending from that part of the skin of the groin which afterwards forms the scrotum through the inguinal canal to the body and epididymis of the testicle, and is then continued upwards in front of the kidney towards the Diaphragm. As development advances, the peritoneum covering the testicle encloses it and forms a mesentery, the *mesorchium*, which also encloses the gubernaculum and forms two folds, one above the testicle and the other below it. The one above the testicle is the *plica vascularis*, and contains ultimately the spermatic vessels ; the one below, the *plica gubernatrix*, contains the lower part of the gubernaculum, which has now grown into a thick cord ; it terminates below at the internal ring in a tube of peritoneum, the processus vaginalis, which protrudes itself down the inguinal canal. The lower part of the gubernaculum by the fifth month has become a thick cord, while the upper part has disappeared. The lower part can now be seen to consist of a central core of unstriped muscle-fibre, and outside this of a firm layer of striped elements, connected, behind the peritoneum, with the abdominal wall. As the scrotum

* Mr. Osborn, in the *St. Thomas's Hospital Reports*, 1875, has written an interesting paper pointing out the probable connection between this fœtal structure and one form of hydrocele.

develops, the lower end of the gubernaculum is carried with the skin to which it is attached to the bottom of this pouch. The fold of peritoneum, constituting the processus vaginalis, projects itself downwards into the inguinal canal, and emerges at the external abdominal ring, pushing before it a part of the internal oblique muscle and the aponeurosis of the external oblique, which form respectively the cremaster muscle and the external spermatic fascia. It forms a gradually elongating depression or *cul-de-sac*, which eventually reaches the bottom of the scrotum, and into this the testicle is drawn by the growth of the body of the fœtus, for the gubernaculum does not grow commensurately with the growth of other parts, and therefore the testicle, being attached by the gubernaculum to the bottom of the scrotum, is prevented from rising as the body grows, and is drawn first into the inguinal canal and eventually into the scrotum. It seems certain also that the gubernacular cord becomes shortened as development proceeds, and this assists in causing the testicle to reach the bottom of the scrotum. By the eighth month the testicle has reached the scrotum, preceded by the lengthened pouch of peritoneum, the processus vaginalis, which communicates by its upper extremity with the peritoneal cavity. Just before birth the upper part of the pouch usually becomes closed, and this obliteration extends gradually downwards to within a short distance of the testis. The process of peritoneum surrounding the testis, which is now entirely cut off from the general peritoneal cavity, constitutes the *tunica vaginalis.**

In the female there is also a gubernaculum, which effects a considerable change in the position of the ovary, though not so extensive a change as that of the testicle in the male. The gubernaculum in the female, as it lies on either side in contact with the fundus of the uterus formed by the union of the Müllerian ducts, contracts adhesions to this organ, and thus the ovary is prevented from descending below this level. The remains of the gubernaculum—that is to say, the part below the attachment of the cord to the uterus to its termination in the labia majora—ultimately forms the round ligament of the uterus. A pouch of peritoneum accompanies it along the inguinal canal, analogous to the processus vaginalis in the male: it is called the *canal of Nuck*. In rare cases the gubernaculum may fail to contract adhesions to the uterus, and then the ovary descends through the inguinal canal into the labia majora, extending down the canal of Nuck, and under these circumstances resembles in position the testicles in the male.

Surgical Anatomy.—Abnormalities in the formation and in the descent of the testicle may occur. The testicle may fail to be developed : or the testicle may be fully developed, and the vas deferens may be undeveloped in whole or part ; or again, both testicle and vas deferens may be fully developed, but the duct may not become connected to the gland. The testicle may fail in its descent, or it may descend into some abnormal position. Thus it may be retained in the position where it was primarily developed, below the kidney ; or it may descend to the internal abdominal ring, but fail to pass through this opening ; it may be retained in the inguinal canal, which is perhaps the most common position ; or it may pass through the external abdominal ring and remain just outside it, failing to pass to the bottom of the scrotum. On the other hand, it may get into some abnormal position : it may pass the scrotum and reach the perinæum, or it may fail to enter the inguinal canal, and may find its way through the femoral ring into the crural canal, and present itself on the thigh at the saphenous opening. There is still a third class of cases of abnormality of the testicle, where the organ has descended in due course into the scrotum, but is malplaced. The most common form of this is where the testicle is *inverted*: that is to say, the organ is rotated so that the epididymis is connected to the front of the scrotum, and the body, surrounded by the tunica vaginalis, is directed backwards. In these cases the vas deferens is to be felt in the front of the cord. The condition is of importance in connection with hydrocele and hæmatocele, and the position of the testicle should always be carefully ascertained before performing any operation for these affections. Again, more rarely, the testicle may be *reversed*. This is a condition in which the top of the testicle, indicated by the globus major of the epididymis, is at the bottom of the scrotum, and the vas deferens comes off from the summit

* The obliteration of the process of peritoneum which accompanies the cord, and is hence called the *funicular process*, is often incomplete. See section on Inguinal Hernia.

of the organ. Cases sometimes occur, generally in the young adult, in which the spermatic cord becomes twisted. In consequence of this the circulation through it is partially or completely arrested; if the latter, the testicle becomes gangrenous; if the former, it may undergo atrophy.

The **external organs of generation** (fig. [184]), like the internal, pass through a stage in which there is no distinction of sex. It is therefore necessary to describe this stage, and then follow the development of the female and male organs respectively.

As stated above, the anal depression, or proctodæum, at an early period is formed by an involution of the epiblast, and the intestine is still closed at its lower end. When the septum between the two opens, which is about the fourth week, the urachus in front and the intestine behind both communicate with the

Fig. [184].—Stages in the development of the external sexual organs in the male and female. (After the Ecker-Ziegler wax models.) (From Hertwig's 'Entwickelungs-geschichte.')

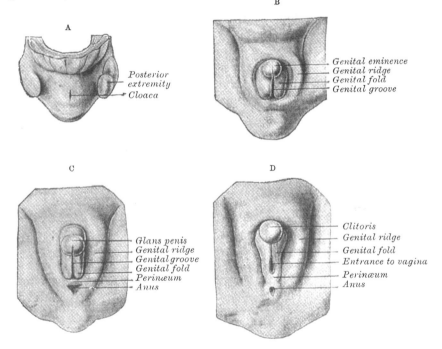

anal depression. This, which is now called the *cloaca*, is afterwards divided by a transverse septum, the *perinæum*, which appears about the second month. Two tubes are thus formed : the posterior becomes the lower part of the rectum, the anterior is the urogenital sinus. In the sixth week a tubercle, the *genital eminence*, is formed in front of the cloaca, and this is soon surrounded by two folds of skin, the *genital ridges*. Towards the end of the second month the genital tubercle presents, on its lower aspect, a groove, the *genital groove*, which extends downwards towards the cloaca. This groove becomes deeper, and is bounded laterally by projecting folds of skin, the *genital folds*. All these parts are well developed by the second month, yet no distinction of sex is possible.

Female Organs.—The female organs are developed by an easy transition from the above. The urogenital sinus persists as the vestibule of the vagina and the urethra. The genital eminence forms the clitoris, the genital ridges the labia majora, and the lips of the genital groove the labia minora, which remain open.   An involution of the epithelium takes place on either side close to the root of the genital tubercle, which becomes the glands of Bartholin.

Male Organs.—In the male the changes are greater.   The genital eminence is developed into the penis, the glans appearing in the third month, the prepuce and corpora cavernosa in the fourth.   The genital groove closes and thus forms a canal, the spongy portion of the urethra.   The urogenital sinus becomes elongated and forms the membranous urethra.   The genital ridges unite in the middle line to form the scrotum, at about the same time as the genital groove closes—viz. between the third and fourth month.   A similar involution of epithelium to that which in the female forms the glands of Bartholin takes place in the male and becomes the glands of Cowper.

The following table is translated from the work of Beaunis and Bouchard, with some alterations, especially in the earlier weeks.   It will serve to present a *résumé* of the above facts in an easily accessible form.*

* It will be noticed that the time assigned in this table for the appearance of the first rudiment of some of the bones varies in some cases from that assigned in the description of the various bones in the sequel.   This is a point on which anatomists differ, and which probably varies in different cases.

CHRONOLOGICAL TABLE

OF

THE DEVELOPMENT OF THE FŒTUS

(FROM BEAUNIS AND BOUCHARD)

—————

First Week.—During this period the ovum is in the Fallopian tube. Having been fertilised in the upper part of the tube, it slowly passes down, undergoing segmentation, and reaches the uterus probably about the end of the first week. During this time it does not undergo much increase in size.

Second Week.—The ovum rapidly increases in size and becomes embedded in the decidua, so that it is completely enclosed in the decidua reflexa by the end of this period. An ovum believed to be of the thirteenth day after conception is described by Reichert. There was no appearance of any embryonic structure. The equatorial margins of the ovum were beset with villi, but the surface in contact with the uterine wall and the one opposite to it were bare. In another ovum, described by His, believed to be of about the fourteenth day, there was a distinct indication of an embryo. There was a medullary groove bounded by folds. In front of this a slightly prominent ridge, the rudimentary heart. The amnion was formed and the embryo was attached by a stalk, the allantois, to the inner surface of the chorion. It may be said, therefore, that these parts, the amnion and the allantois, and the first rudiments of the embryo, the medullary groove and the heart, are formed at the end of the second week.*

Third Week.—By the end of the third week the flexures of the embryo have taken place, so that it is strongly curved. The protovertebral discs, which begin to be formed early in the third week, present their full complement. In the nervous system the primary divisions of the brain are visible, and the primitive ocular and auditory vesicles are already formed. The primary circulation is established. The alimentary canal presents a straight tube communicating with the yolk-sac. The branchial arches are formed. The limbs have appeared as short buds. The Wolffian bodies are visible.

Fourth Week.—The umbilical vesicle has attained its full development. The caudal extremity projects. The upper and the lower limbs and the cloacal aperture appear. The heart separates into a right and left heart. The special ganglia and anterior roots of the spinal nerves, the olfactory fossæ, the lungs and the pancreas can be made out.

Fifth Week.—The allantois is vascular in its whole extent. The first traces of the hands and feet can be seen. The primitive aorta divides into aorta and pulmonary artery. The duct of Müller and genital gland are visible. The ossification of the clavicle and the lower jaw commences. The cartilage of Meckel occupies the first post-oral arch.

---

* [Eternod (*Anat. Anzeiger*, Band XV, 1898) described an ovum which he reconstructed. It had a precise history, from which he concluded that it must have belonged to the end of the second or the beginning of the third week. Including the villi it measured 10 × 8·2 × 6 mm. It was flattened on its embryonal side and the embryo measured 1·3 mm. The amnion was completely formed and the allantois existed as a long canal. The vitelline circulation was established and the villi of the chorion were beginning to be vascularised. The blastopore still opened into the amniotic cavity, with the primitive groove behind it and the rudimentary medullary groove in front. The notochord was closing in and all three layers of the blastoderm were distinct, except around the blastopore, where they formed an undivided mass.—EDS.]

Sixth Week.—The activity of the umbilical vesicle ceases. The pharyngeal clefts disappear. The vertebral column, primitive cranium, and ribs assume the cartilaginous condition. The posterior roots of the nerves, the membranes of the nervous centres, the bladder, kidney, tongue, larynx, thyroid body, the germs of teeth, and the genital tubercle and folds are apparent.

Seventh Week.—The muscles begin to be perceptible. The points of ossification of the ribs, scapula, shaft of humerus, femur, tibia, palate, and upper jaw appear.

Eighth Week.—The distinction of arm and forearm, and of thigh and leg, is apparent, as well as the interdigital clefts. The capsule of the lens and pupillary membrane, the interventricular and commencement of the interauricular septum, the salivary glands, the spleen and suprarenal capsules are distinguishable. The larynx begins to become cartilaginous. All the vertebral bodies are cartilaginous. The points of ossification for the ulna, radius, fibula, and ilium make their appearance. The two halves of the hard palate unite. The sympathetic nerves are now for the first time to be discerned.

Ninth Week.—The corpus striatum and the pericardium are first apparent. The ovary and testicle can be distinguished from each other. The genital furrow appears. The osseous nuclei of the bodies and arches of the vertebræ, of the frontal, vomer, and malar bones, of the shafts of the metacarpal and metatarsal bones, and of the phalanges appear. The union of the hard palate is completed. The gall-bladder is seen.

Third Month.—The formation of the fœtal placenta advances rapidly. The projection of the caudal extremity disappears. It is possible to distinguish the male and female organs from each other. The cloacal aperture is divided into two parts. The cartilaginous arches on the dorsal region of the spine close. The points of ossification for the occipital, sphenoid, lachrymal, nasal, squamous portion of temporal and ischium appear, as well as the orbital centre of the superior maxillary. The pons Varolii and fissure of Sylvius can be made out. The eyelids, the hair, and the nails begin to form. The mammary gland, the epiglottis, and prostate are beginning to develop. The union of the testicle with the canals of the Wolffian body takes place.

Fourth Month.—The closure of the cartilaginous arches of the spine is complete. Osseous points for the first sacral vertebra and os pubis appear. The ossification of the malleus and incus takes place. The corpus callosum, the membrana lamina spiralis, the cartilage of the Eustachian tube, and the tympanic ring are seen. Fat is first developed in the subcutaneous cellular tissue. The tonsils are seen, and the closure of the genital furrow and the formation of the scrotum and prepuce take place.

Fifth Month.—The two layers of the decidua begin to coalesce. Osseous nuclei of the axis and odontoid process, of the lateral points of the first sacral vertebra, of the median points of the second, and of the lateral masses of the ethmoid make their appearance. Ossification of the stapes and the petrous bone, and ossification of the germs of the teeth, take place. The germs of the permanent teeth and the organ of Corti appear. The eruption of hair on the head commences. The sudoriferous glands, Brunner's glands, the follicles of the tonsil and base of the tongue, and the lymphatic glands appear at this period. The differentiation between the uterus and vagina becomes apparent.

Sixth Month.—The points of ossification for the anterior root of the transverse process of the seventh cervical vertebra, the lateral points of the second cervical vertebra, the median points of the third, the manubrium sterni and the os calcis appear. The sacro-vertebral angle forms. The cerebral hemispheres cover the cerebellum. The papillæ of the skin, the sebaceous glands, and Peyer's patches make their appearance. The free border of the nail projects from the corium of the dermis. The walls of the uterus thicken.

Seventh Month.—The additional points of the first sacral vertebra, the lateral points of the third, the median point of the fourth, the first osseous point of the body of the sternum, and the osseous point for the astragalus appear. Meckel's cartilage disappears. The cerebral convolutions, the island of Reil, and the tubercula quadrigemina are apparent. The pupillary membrane atrophies. The testicle passes into the vaginal process of the peritoneum.

Eighth Month.—Additional points for the second sacral vertebra, lateral points for the fourth, and median points for the fifth sacral vertebræ, can be seen.

Ninth Month.—Additional points for the third sacral vertebra, lateral points for the fifth, osseous points for the middle turbinated bone, for the body and great cornu of the hyoid, for the second and third pieces of the body of the sternum, and for the lower end of the femur appear. Ossification of the bony lamina spiralis and axis of the cochlea takes place. The eyelids open, and the testicles are in the scrotum.

DESCRIPTIVE AND SURGICAL
ANATOMY

---✳---

OSTEOLOGY

THE entire skeleton in the adult consists of 200 distinct bones. These are—

The Spine or vertebral column (sacrum and coccyx included) . 26
Cranium 8
Face 14
Hyoid bone, sternum, and ribs 26
Upper extremities 64
Lower extremities 62
                                                   ———
                                                   200

In this enumeration the patellæ are included as separate bones, but the smaller sesamoid bones, and the ossicula auditûs, are not reckoned. The teeth belong to the tegumentary system.

These bones are divisible into four classes : *Long, Short, Flat,* and *Irregular.*

The **Long Bones** are found in the limbs, where they form a system of levers which have to sustain the weight of the trunk, and to confer the power of locomotion. A long bone consists of a shaft and two extremities. The *shaft* is a hollow cylinder, contracted and narrowed to afford greater space for the bellies of the muscles ; the walls consist of dense, compact tissue of great thickness in the middle, but becoming thinner towards the extremities ; the spongy tissue is scanty, and the bone is hollowed out in its interior to form the *medullary canal.* The *extremities* are generally somewhat expanded, for greater convenience of mutual connection, for the purposes of articulation, and to afford a broad surface for muscular attachment. Here the bone is made up of spongy tissue with only a thin coating of compact substance. The long bones are not straight, but curved ; the curve generally taking place in two directions, thus affording greater strength to the bone. The bones belonging to this class are : the *clavicle, humerus, radius, ulna, femur, tibia, fibula, metacarpal* and *metatarsal* bones, and the *phalanges.*

Short Bones.—Where a part of the skeleton is intended for strength and compactness, and its motion is at the same time slight and limited, it is divided into a number of small pieces united together by ligaments, and the separate bones are short and compressed, such as those of the *carpus* and *tarsus.* These bones, in their structure, are spongy throughout, excepting at their surface, where there is a thin crust of compact substance. The *patellæ* also, together with the other sesamoid bones, are by some regarded as short bones.

Flat Bones.—Where the principal requirement is either extensive protection or the provision of broad surfaces for muscular attachment, the osseous structure is expanded into broad, flat plates, as in the bones of the skull and the shoulder-blade. These bones are composed of two thin layers of compact tissue enclosing between them a variable quantity of cancellous tissue. In the cranial bones, these layers of compact tissue are familiarly known as the *tables* of the skull ; the outer one is thick and tough ; the inner one thinner, denser, and more brittle, and hence termed the *vitreous table.* The intervening cancellous tissue is called the *diploë.* The flat bones are : the *occipital, parietal, frontal, nasal, lachrymal, vomer, scapula, os innominatum, sternum, ribs,* and, according to some, the *patella.*

The **Irregular** or **Mixed Bones** are such as, from their peculiar form, cannot be grouped under either of the preceding heads. Their structure is similar to that of other bones, consisting of a layer of compact tissue externally, and of spongy cancellous tissue within. The irregular bones are : the *vertebræ, sacrum, coccyx, temporal, sphenoid, ethmoid, malar, superior maxillary, inferior maxillary, palate, inferior turbinated,* and *hyoid.*

Surfaces of Bones.—If the surface of any bone is examined, certain eminences and depressions are seen, to which descriptive anatomists have given the following names.

These eminences and depressions are of two kinds : *articular* and *non-articular.* Well-marked examples of articular eminences are found in the heads of the humerus and femur ; and of articular depressions, in the glenoid cavity of the scapula, and the acetabulum. Non-articular eminences are designated according to their form. Thus, a broad, rough, uneven elevation is called a *tuberosity ;* a small, rough prominence, a *tubercle ;* a sharp, slender, pointed eminence, a *spine ;* a narrow, rough elevation, running some way along the surface, a *ridge* or *line.*

The non-articular depressions are also of very variable form, and are described as *fossæ, grooves, furrows, fissures, notches,* etc. These non-articular eminences and depressions serve to increase the extent of surface for the attachment of ligaments and muscles, and are usually well marked in proportion to the muscularity of the subject.

A prominent process projecting from the surface of a bone, which it has never been separate from or movable upon, is termed an *apophysis* (from ἀπόφυσις, *an excrescence*) ; but if such process is developed as a separate piece from the rest of the bone, to which it is afterwards joined, it is termed an *epiphysis* (from ἐπίφυσις, *an accretion*). The main part of the bone, or shaft. which is formed from the primary centre of ossification, is termed the *diaphysis,* and is separated, during growth, from the epiphysis by a layer of cartilage, at which growth in length of the bone takes place.

THE SPINE

The **Spine** is a flexuous and flexible column, formed of a series of bones called *vertebræ* (from *vertere,* to turn).

The **Vertebræ** are thirty-three in number, and have received the names *cervical, dorsal, lumbar, sacral,* and *coccygeal,* according to the position which they occupy ; seven being found in the cervical region, twelve in the dorsal, five in the lumbar, five in the sacral, and four in the coccygeal.

This number is sometimes increased by an additional vertebra in one region, or the number may be diminished in one region, the deficiency being supplied by an additional vertebra in another. These observations do not apply to the cervical portion of the spine, the number of bones forming which is seldom increased or diminished.

The Vertebræ in the upper three regions of the spine are separate throughout the whole of life ; but those found in the sacral and coccygeal regions are in the adult firmly united, so as to form two bones—five entering into the formation of the upper bone or *sacrum,* and four into the terminal bone of the spine or *coccyx.*

GENERAL CHARACTERS OF A VERTEBRA

Each **vertebra** consists of two essential parts—an anterior solid segment or *body,* and a posterior segment or *arch.* The arch (neural) is formed of two *pedicles,* and two *laminæ,* supporting seven *processes*—viz. four *articular,* two *transverse,* and one *spinous.*

The bodies of the vertebræ are piled one upon the other, forming a strong pillar, for the support of the cranium and trunk; the arches form a hollow cylinder behind the bodies for the protection of the spinal cord. The different vertebræ are connected together by means of the articular processes and the intervertebral fibro-cartilages; while the transverse and spinous processes serve as levers for the attachment of muscles which move the different parts of the spine. Lastly, between each pair of vertebræ apertures exist through which the spinal nerves pass from the cord. Each of these constituent parts must now be separately examined.

The **Body** or **centrum** is the largest part of a vertebra. Above and below, it is flattened; its upper and lower surfaces are rough, for the attachment of the intervertebral fibro-cartilages, and present a rim around their circumference. In front, it is convex from side to side, concave from above downwards. Behind, it is flat from above downwards and slightly concave from side to side. Its anterior surface is perforated by a few small apertures, for the passage of nutrient vessels; while on the posterior surface is a single large, irregular aperture, or occasionally more than one, for the exit of veins from the body of the vertebra—the *venæ basis vertebræ.*

The **Pedicles** are two short, thick pieces of bone, which project backwards, one on each side, from the upper part of the body of the vertebra, at the line of junction of its posterior and lateral surfaces. The concavities above and below the pedicles are the *intervertebral notches*; they are four in number, two on each side, the inferior ones being generally the deeper. When the vertebræ are articulated, the notches of each contiguous pair of bones form the intervertebral foramina, which communicate with the spinal canal and transmit the spinal nerves and blood-vessels.

The **Laminæ** are two broad plates of bone which complete the neural arch by fusing together in the middle line behind. They enclose a foramen, the *spinal foramen,* which serves for the protection of the spinal cord, and are connected to the body by means of the pedicles. Their upper and lower borders are rough for the attachment of the ligamenta subflava.

The **Spinous Process** projects backwards from the junction of the two laminæ, and serves for the attachment of muscles and ligaments.

The **Articular Processes,** four in number, two on each side, spring from the junction of the pedicles with the laminæ. The two superior project upwards, their articular surfaces being directed more or less backwards; the two inferior project downwards, their articular surfaces looking more or less forwards.*

The **Transverse Processes,** two in number, project one at each side from the point where the lamina joins the pedicle, between the superior and inferior articular processes. They serve for the attachment of muscles and ligaments.

CHARACTER OF THE CERVICAL VERTEBRÆ (fig. 1)

The **Cervical Vertebræ** are smaller than those in any other region of the spine, and may readily be distinguished by a foramen in the transverse process, which does not exist in the transverse process of either the dorsal or lumbar vertebræ.

The **Body** is small, comparatively dense, and broader from side to side than from before backwards. The anterior and posterior surfaces are flattened and of equal depth; the former is placed on a lower level than the latter, and its inferior border is prolonged downwards, so as to overlap the upper and fore part of the vertebra below. Its upper surface is concave transversely, and presents a projecting lip on each side; its lower surface is convex from side to side, concave from before backwards, and presents laterally a shallow concavity, which receives

* It may, perhaps, be as well to remind the reader that the direction of a surface is determined by that of a line drawn at right angles to it.

the corresponding projecting lip of the adjacent vertebra. The *pedicles* are directed outwards and backwards, and are attached to the body midway between the upper and lower borders, so that the superior intervertebral notch is as deep as the inferior, but it is, at the same time, narrower. The *laminæ* are narrow, long, thinner above than below, and overlap each other; they enclose the spinal foramen, which is large, and of a triangular form. The *spinous process* is short, and bifid at the extremity to afford greater extent of surface for the attachment of muscles, the two divisions being often of unequal size. It increases in length from the fourth to the seventh. The *articular processes* are flat, oblique, and of an oval form: the superior are directed backwards and upwards; the inferior forwards and downwards. The *transverse processes* are short, directed downwards, outwards, and forwards, bifid at their extremity, and marked by a groove along their upper surface, which runs downwards and outwards from the superior intervertebral notch, and serves for the transmission of one of the cervical nerves. They are situated in front of the articular processes and on the outer side of the pedicles, and are pierced at their base by a foramen, for the transmission of the vertebral artery, vein, and plexus of nerves. Each process is formed by two roots: the anterior root, sometimes called the *costal process*, arises from the side of the body, and is the homologue of the rib in the dorsal region of the spine; the

FIG. 1.—Cervical vertebra.

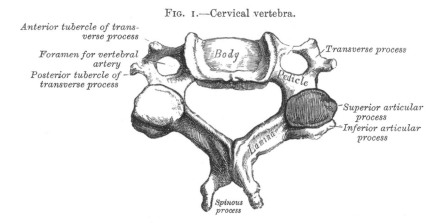

posterior root springs from the junction of the pedicle with the lamina, and corresponds with the transverse process in the dorsal region. It is by the junction of the two that the foramen for the vertebral vessels is formed. The extremity of each of these roots forms the *anterior* and *posterior tubercles* of the transverse processes.*

The peculiar vertebræ in the cervical region are the first, or *Atlas*; the second, or *Axis*; and the seventh, or *Vertebra prominens*. The great modifications in the form of the atlas and axis are designed to admit of the nodding and rotatory movements of the head.

The **Atlas** (fig. 2) is so named from supporting the globe of the head. The chief peculiarities of this bone are that it has neither body nor spinous process. The body is detached from the rest of the bone, and forms the odontoid process of the second vertebra; while the parts corresponding to the pedicles join in front to form the anterior arch. The atlas is ring-like, and consists of an anterior arch, a posterior arch, and two lateral masses. The *anterior* arch forms about one-fifth of the ring: its anterior surface is convex, and presents about its centre

---

* The anterior tubercle of the transverse process of the sixth cervical vertebra is of large size, and is sometimes known as 'Chassaignac's' or the 'carotid tubercle.' It is in close relation with the carotid artery, which lies in front and a little external to it; so that, as was first pointed out by Chassaignac, the vessel can with ease be compressed against it.

a *tubercle*, for the attachment of the Longus colli muscle ; posteriorly it is concave, and marked by a smooth, oval or circular facet, for articulation with the odontoid process of the axis. The upper and lower borders give attachment to the anterior occipito-atlantal and the anterior atlanto-axial ligaments which connect it with the occipital bone above and the axis below. The *posterior* arch forms about two-fifths of the circumference of the bone ; it terminates behind in a *tubercle*, which is the rudiment of a spinous process, and gives origin to the Rectus capitis posticus minor. The diminutive size of this process prevents any interference in the movements between the atlas and the cranium. The posterior part of the arch presents above and behind a rounded edge for the attachment of the posterior occipito-atlantal ligament, while in front, immediately behind each superior articular process, is a groove, sometimes converted into a foramen by a delicate bony spiculum which arches backwards from the posterior extremity of the superior articular process. These grooves represent the superior intervertebral notches, and are peculiar from being situated behind the articular processes, instead of in front of them, as in the other vertebræ. They serve for the transmission of the vertebral artery, which, ascending through the foramen in the transverse process, winds round the lateral mass in a direction backwards and inwards. They also transmit the suboccipital (first spinal) nerve. On the under

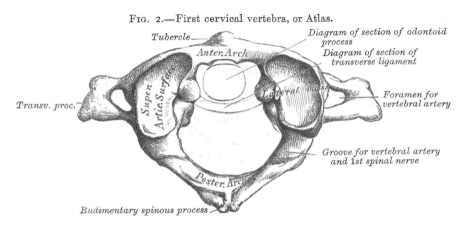

Fig. 2.—First cervical vertebra, or Atlas.

Tubercle

Anter. Arch

Diagram of section of odontoid process

Diagram of section of transverse ligament

Super. Artic. Surface

Lateral mass

Foramen for vertebral artery

Transv. proc.

Groove for vertebral artery and 1st spinal nerve

Poster. Arch

Rudimentary spinous process

surface of the posterior arch, in the same situation, are two other grooves, placed behind the lateral masses, and representing the inferior intervertebral notches of other vertebræ. They are much less marked than the superior. The lower border gives attachment to the posterior atlanto-axial ligament, which connects it with the axis. The *lateral masses* are the most bulky and solid parts of the atlas, in order to support the weight of the head : they present two articulating processes above, and two below. The two superior are of large size, oval, concave, and approach each other in front, but diverge behind : they are directed upwards, inwards, and a little backwards, each forming a kind of cup for the corresponding condyle of the occipital bone, and are admirably adapted to the nodding movements of the head. Not infrequently they are partially subdivided by a more or less deep indentation which encroaches upon each lateral margin. The inferior articular processes are circular in form, flattened or slightly concave and directed downwards and inwards, articulating with the axis, and permitting the rotatory movements. Just below the inner margin of each superior articular surface is a small tubercle, for the attachment of the transverse ligament which, stretching across the ring of the atlas, divides it into two unequal parts—the anterior or smaller segment receiving the odontoid process of the axis, the posterior allowing the transmission of the spinal cord and its membranes. This part of the spinal canal is of considerable size, to afford space for the spinal cord ; and hence lateral displacement of the atlas may occur without compression of this structure. The *transverse processes* are of large size, project directly outwards and downwards from the lateral masses, and serve for the attachment of special muscles which assist in rotating the head. They are long, but not bifid, and are perforated at their base by a canal for the vertebral artery, which is directed from below, upwards and backwards.

The **Axis** (fig. 3) is so named from forming the pivot upon which the first vertebra, carrying the head, rotates. The most distinctive character of this bone is the strong, prominent process, tooth-like in form (hence the name *odontoid*), which rises perpendicularly from the upper surface of the body. The *body* is deeper in front than behind, and prolonged downwards anteriorly so as to overlap the upper and fore part of the next vertebra. It presents in front a median longitudinal ridge, separating two lateral depressions for the attachment of the Longus colli muscle of either side. The *odontoid process* presents two articulating surfaces : one in front, of an oval form, for articulation with the atlas ; another behind for the transverse ligament—the latter frequently encroaching on the sides of the process. The apex is pointed, and gives attachment to the middle fasciculus of the odontoid or check ligaments (*ligamentum suspensorium*). Below the apex, the process is somewhat enlarged and presents on either side a rough impression for the attachment of the lateral fasciculi of the odontoid or check

FIG. 3.—Second cervical vertebra, or Axis.

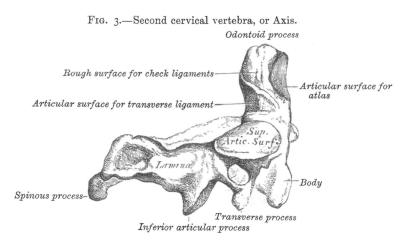

Odontoid process

Rough surface for check ligaments

Articular surface for transverse ligament

Articular surface for atlas

Sup. Artic. Surf.

Lamina

Spinous process

Body

Transverse process

Inferior articular process

ligaments, which connect it to the occipital bone ; the base of the process, where it is attached to the body, is constricted, so as to prevent displacement from the transverse ligament, which binds it in this situation to the anterior arch of the atlas. Sometimes, however, this process does become displaced, especially in children, in whom the ligaments are more relaxed : instant death is the result of this accident. The internal structure of the odontoid process is more compact than that of the body. The *pedicles* are broad and strong, especially their anterior extremities, which coalesce with the sides of the body and the root of the odontoid process. The *laminæ* are thick and strong, and the spinal foramen large, but smaller than that of the atlas. The *transverse processes* are very small, not bifid, but perforated by the foramen for the vertebral artery, which is directed obliquely upwards and outwards. The *superior articular surfaces* are round, slightly convex, directed upwards and outwards, and are peculiar in being supported on the body, pedicles, and transverse processes. The *inferior articular surfaces* have the same direction as those of the other cervical vertebræ. The *superior intervertebral notches* are very shallow, and lie behind the articular processes ; the *inferior* in front of them, as in the other cervical vertebræ. The *spinous process* is of large size, very strong, deeply channelled on its under surface, and presents a bifid, tubercular extremity for the attachment of muscles which serve to rotate the head upon the spine.

Seventh Cervical (fig. 4).—The most distinctive character of this vertebra is the existence of a long and prominent spinous process ; hence the name ' vertebra prominens.' This process is thick, nearly horizontal in direction, not bifurcated, and has attached to it the lower end of the ligamentum nuchæ. The *transverse process* is usually of large size, its posterior tubercles are large and prominent, while the anterior are small and faintly marked ; its upper surface has usually a shallow groove, and it seldom presents more than a trace of bifurcation at its extremity. The foramen in the transverse process is sometimes as large as in the other cervical vertebræ, but is usually smaller on one or both sides, and sometimes wanting. On the left side it occasionally gives passage to the vertebral artery; more frequently the vertebral vein

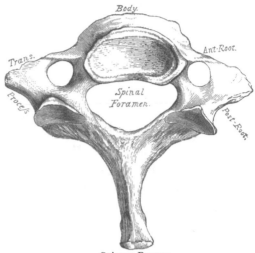

FIG. 4.—Seventh cervical vertebra, or vertebra prominens.

traverses it on both sides; but the usual arrangement is for both artery and vein to pass in front of the transverse process, and not through the foramen. Occasionally the anterior root of the transverse process exists as a separate bone, and attains a large size. It is then known as a ' cervical rib.'

CHARACTERS OF THE DORSAL VERTEBRÆ

The **Dorsal Vertebræ** are intermediate in size between those in the cervical and those in the lumbar region, and increase in size from above downwards, the

FIG. 5.—A dorsal vertebra.

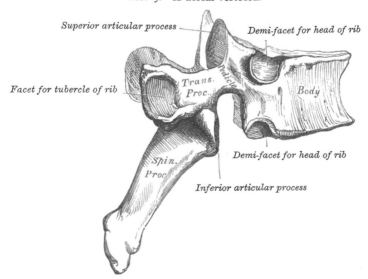

upper vertebræ in this segment of the spine being much smaller than those in the lower part of the region. The dorsal vertebræ may be at once recognised by the presence on the sides of the body of one or more facets or half-facets for the heads of the ribs.

The **bodies** of the dorsal vertebræ resemble those in the cervical and lumbar regions at the respective ends of this portion of the spine; but in the middle of the dorsal region their form is very characteristic, being heart-shaped, and as broad in the antero-posterior as in the lateral direction. They are thicker behind than

FIG. 6.—Peculiar dorsal vertebræ.

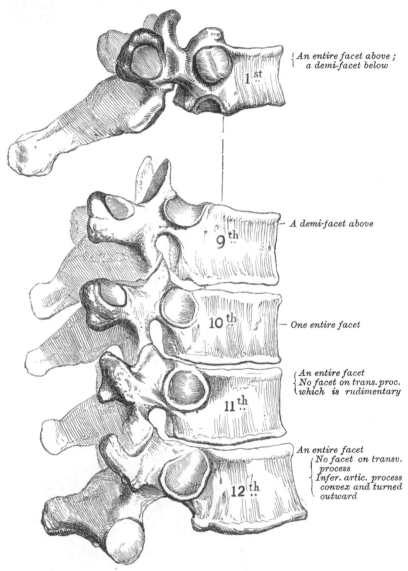

1st { An entire facet above ;
    a demi-facet below

9th — A demi-facet above

10th — One entire facet

11th { An entire facet
      No facet on trans. proc.
      which is rudimentary

12th { An entire facet
      No facet on transv. process
      Infer. artic. process convex and turned outward

in front, flat above and below, convex and prominent in front, deeply concave behind, slightly constricted in front and at the sides, and marked on each side, near the root of the pedicle, by two demi-facets, one above, the other below. These are covered with cartilage in the recent state, and, when the vertebræ are articulated with one another, form, with the intervening fibro-cartilages, oval surfaces for the reception of the heads of the corresponding ribs. The *pedicles* are directed backwards, and the inferior intervertebral notches are of large size, and deeper than in any other region of the spine. The *laminæ* are broad, thick, and imbricated— that is to say, overlapping one another like tiles on a roof. The spinal foramen is small, and of a circular form. The *spinous processes* are long, triangular on transverse section, directed obliquely downwards, and terminate in a tubercular extremity. They overlap one another from the fifth to the eighth, but are less oblique in direction above and below. The *articular processes* are flat, nearly vertical in direction, and project from the upper and lower parts of the pedicles ; the superior being directed backwards and a little outwards and upwards, the inferior forwards and a little inwards and downwards. The *transverse processes* arise from the same parts of the arch as the posterior roots of the transverse processes in the neck, and are situated behind the articular processes and pedicles ; they are thick, strong, and of considerable length, directed obliquely backwards and outwards, presenting a clubbed extremity, which is tipped on its anterior part by a small, concave surface, for articulation with the tubercle of a rib. Besides the articular facet for the rib, three indistinct tubercles may be seen rising from the transverse processes, one at the upper border, one at the lower border, and one externally. In man, they are comparatively of small size, and serve only for the attachment of muscles. But in some animals they attain considerable magnitude, either for the purpose of more closely connecting the segments of this portion of the spine, or for muscular and ligamentous attachment.

The peculiar dorsal vertebræ are the *first, ninth, tenth, eleventh,* and *twelfth* (fig. 6).

The **First Dorsal Vertebra** presents, on each side of the *body*, a single entire articular facet for the head of the first rib, and a half-facet for the upper half of the second. The body is like that of a cervical vertebra, being broad transversely ; its upper surface is concave, and lipped on each side. The *articular surfaces* are oblique, and the *spinous process* thick, long, and almost horizontal.

The **Ninth Dorsal** has no demi-facet below. In some subjects, however, the ninth has two demi-facets on each side ; when this occurs the tenth has only a demi-facet at the upper part.

The **Tenth Dorsal** has (except in the cases just mentioned) an entire articular facet on each side above, which is partly placed on the outer surface of the pedicle. It has no demi-facet below.

In the **Eleventh Dorsal,** the body approaches in its form and size to the lumbar. The articular facets for the heads of the ribs, one on each side, are of large size, and placed chiefly on the pedicles, which are thicker and stronger in this and the next vertebra than in any other part of the dorsal region. The *spinous* process is short, and nearly horizontal in direction. The *transverse* processes are very short, tubercular at their extremities, and have no articular facets for the tubercles of the ribs.

The **Twelfth Dorsal** has the same general characters as the eleventh, but may be distinguished from it by the inferior articular processes being convex and turned outwards, like those of the lumbar vertebræ ; by the general form of the body, laminæ, and spinous process, approaching to that of the lumbar vertebræ ; and by the transverse processes being shorter, and marked by three elevations, the superior, inferior, and external tubercles, which correspond to the mammillary, accessory, and transverse processes of the lumbar vertebræ. Traces of similar elevations are usually to be found upon the other dorsal vertebræ (*vide ut supra*).

CHARACTERS OF THE LUMBAR VERTEBRÆ

The **Lumbar Vertebræ** (fig. 7) are the largest segments of the movable part of the vertebral column, and can at once be distinguished by the absence of the foramen in the transverse process, the characteristic feature of the cervical vertebræ, and by the absence of any articulating facet on the side of the body, the distinguishing mark of the dorsal vertebræ.

The **body** is large, and has a greater diameter from side to side than from before backwards, slightly thicker in front than behind, flattened or slightly concave above and below, concave behind, and deeply constricted in front and at the sides, presenting prominent margins, which afford a broad basis for the support of the superincumbent weight. The *pedicles* are very strong, directed backwards

FIG. 7.—Lumbar vertebra.

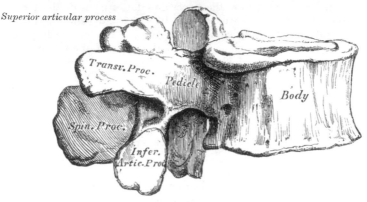

from the upper part of the bodies ; consequently, the inferior intervertebral notches are of considerable depth. The *laminæ* are broad, short, and strong ; and the spinal foramen triangular, larger than in the dorsal, smaller than in the cervical, region. The *spinous processes* are thick and broad, somewhat quadrilateral, horizontal in direction, thicker below than above, and terminating by a rough,

FIG. 8.—Lumbar vertebra.

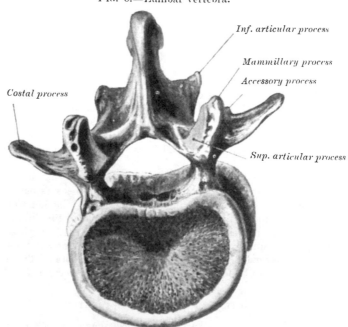

uneven border. The *superior articular* processes are concave, and look backwards and inwards ; the *inferior*, convex, look forwards and outwards ; the former are separated by a much wider interval than the latter, and embrace the lower articulating processes of the vertebra above. The *transverse processes* are long, slender, directed transversely outwards in the upper three lumbar vertebræ, slanting a little upwards in the lower two. They are situated in front of the articular processes instead of behind them as in the dorsal vertebræ, and are homologous with the ribs. Of the three tubercles noticed in connection with the transverse processes of the twelfth dorsal vertebra, the *superior* ones become connected in this region with the back part of the superior articular processes, and have received the name of *mammillary* processes ; the *inferior* are represented by a small process pointing downwards, situated at the back part of the base of the transverse process, and called the *accessory* processes—these are the true transverse processes, which are rudimental in this region of the spine ; the *external* ones are the so-called transverse processes, the homologue of the rib, and hence sometimes called *costal* processes (fig. 8). Although in man these are comparatively small, in some animals they attain considerable size, and serve to lock the vertebræ more closely together.

The **Fifth Lumbar** vertebra is characterised by having the body much thicker in front than behind, which accords with the prominence of the sacro-vertebral articulation ; by the smaller size of its spinous process ; by the wide interval between the inferior articulating processes ; and by the greater size and thickness of its transverse processes, which spring from the body as well as from the pedicles.

Structure of the Vertebræ.—The body is composed of light, spongy, cancellous tissue, having a thin coating of compact tissue on its external surface perforated by numerous orifices, some of large size, for the passage of vessels ; its interior is traversed by one or two large canals, for the reception of veins, which converge towards a single large, irregular aperture, or several small apertures, at the posterior part of the body of each bone. The arch and processes projecting from it have, on the contrary, a thick covering of compact tissue.

Development.—Each vertebra is formed of four primary centres of ossification (fig. 9), one for each lamina and its processes, and two for the body.* Ossification commences in the laminæ about the sixth week of fœtal life, in the situation where the transverse processes afterwards project, the ossific granules shooting backwards to the spine, forwards into the pedicles, and outwards into the transverse and articular processes. Ossification in the body commences in the middle of the cartilage about the eighth week by two closely approximated centres, which speedily coalesce to form one ossific point. According to some authors, ossification commences in the laminæ of the upper vertebræ—i.e. in the cervical and upper dorsal ; the first ossific points in the lower vertebræ being those which are to form the body, the osseous centres for the laminæ appearing at a subsequent period. At birth these three pieces are perfectly separate. During the first year the laminæ become united behind, the union taking place first in the lumbar region and then extending upwards through the dorsal and lower cervical regions. About the third year the body is joined to the arch on each side, in such a manner that the body is formed from the three original centres of ossification, the amount contributed by the pedicles increasing in extent from below upwards. Thus the bodies of the sacral vertebræ are formed almost entirely from the central nuclei ; the bodies of the lumbar are formed laterally and behind by the pedicles ; in the dorsal region, the pedicles advance as far forward as the articular depressions for the head of the ribs, forming these cavities of reception ; and in the neck the lateral portions of the bodies are formed entirely by the advance of the pedicles. The line along which union takes place between the body and the neural arch is named the *neurocentral suture*. Before puberty, no other changes occur, excepting a gradual increase in the growth of these primary centres ; the upper and under surfaces

---

* By many observers it is asserted that the bodies of the vertebræ are developed from a single centre which speedily becomes bilobed, so as to give the appearance of two nuclei ; but that there are two centres, at all events sometimes, is evidenced by the fact that the two halves of the body of the vertebra may remain distinct throughout life, and be separated by a fissure through which a protrusion of the spinal membranes may take place, constituting an anterior spina bifida.

of the bodies, and the ends of the transverse and spinous processes, being tipped with cartilage, in which ossific granules are not as yet deposited.

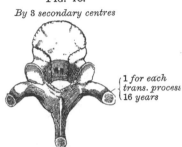

Fig. 9.—Development of a vertebra.

By 4 primary centres

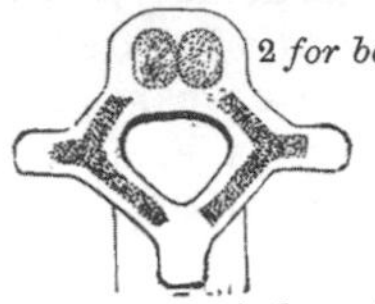

2 *for body* (8th week)

1 *for each lamina* (6th week)

Fig. 10.

By 3 secondary centres

{ 1 *for each trans. process* 16 *years*

1 *for spinous process* (16 years)

Fig. 11.

By 2 additional plates

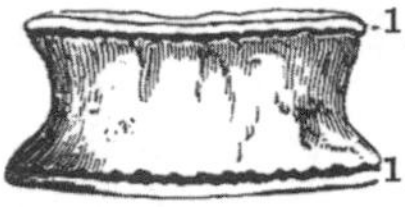

-1 *for upper surface of body* } 21 *years*
1 *for under surface of body*

Fig. 12.—Atlas.

By 3 centres

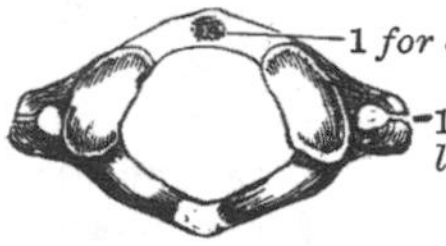

1 *for anter. arch.* (1st yr.) not constant
-1 *for each lateral mass* } 7th week

Fig. 13.—Axis.

By 7 centres

2nd year

6th month
1 *for each lateral mass*
1 *for body* (4th month)
1 *for under surface of body*

Fig. 14.—Lumbar vertebra.

2 *additional centres*

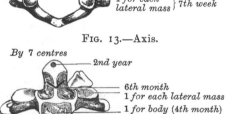

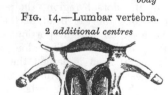

for tubercles on superior articular process

At sixteen years (fig. 10), three secondary centres appear, one for the tip of each transverse process, and one for the extremity of the spinous process. In some of the lumbar vertebræ, especially the first, second and third, a second ossifying centre appears at the base of the spinous process. At twenty-one years (fig. 11), a thin circular epiphysial plate of bone is formed in the layer of cartilage situated on the upper and under surfaces of the body, the former being the thicker of the two. All these become joined, and the bone is completely formed between the twenty-fifth and thirtieth years of life.

Exceptions to this mode of development occur in the first, second, and seventh cervical, and in the vertebræ of the lumbar region.

The Atlas (fig. 12).—The number of centres of ossification of the atlas is very variable. It may be developed from two, three, four, or five centres. The most frequent arrangement is by three centres. Two of these are destined for the two lateral or neural masses, the ossification of which commences about the seventh week near the articular processes, and extends backwards; at birth, these portions of bone are separated from one another behind by a narrow interval filled in with cartilage. Between the third and fourth years they unite either directly or through the medium of a separate centre, developed in the cartilage in the middle line. The anterior arch, at birth, is altogether cartilaginous, and in this a separate nucleus appears about the end of the first year after birth, and extending laterally joins the neural processes in front of the pedicles. Sometimes there are two nuclei developed in the cartilage, one on either side of the median line, which join to form a single mass. And occasionally there is no separate centre, but the

Exceptional cases.

anterior arch is formed by the gradual extension forwards and ultimate junction of the two neural processes.

The **Axis** (fig. 13) is developed by *seven* centres. The body and arch of this bone are formed in the same manner as the corresponding parts in the other vertebræ ; one centre (or two, which speedily coalesce) for the lower part of the body, and one for each lamina. The centres for the laminæ appear about the seventh or eighth week, that for the body about the fourth month. The odontoid process consists originally of an extension upwards of the cartilaginous mass, in which the lower part of the body is formed. At about the sixth month of fœtal life, two osseous nuclei make their appearance in the base of this process : they are placed laterally, and join before birth to form a conical bilobed mass deeply cleft above ; the interval between the cleft and the summit of the process is formed by a wedge-shaped piece of cartilage, the base of the process being separated from the body by a cartilaginous interval, which gradually becomes ossified at its circumference, but remains cartilaginous in its centre until advanced age.* Finally, as Humphry has demonstrated, the apex of the odontoid process has a separate nucleus, which appears in the second year and joins about the twelfth year. In addition to these there is a secondary centre for a thin epiphysial plate on the under surface of the body of the bone.

The Seventh Cervical.—The anterior or costal part of the transverse process of the seventh cervical is developed from a separate osseous centre at about the sixth month of fœtal life, and joins the body and posterior division of the transverse process between the fifth and sixth years. Sometimes this process continues as a separate piece, and, becoming lengthened outwards and forwards, constitutes what is known as a cervical rib. This separate ossific centre for the costal process has also been found in the fourth, fifth, and sixth cervical vertebræ.

The **Lumbar Vertebræ** (fig. 14) have *two additional centres* (besides those common to the vertebræ generally) for the mammillary tubercles, which project from the back part of the superior articular processes. The transverse process of the first lumbar is sometimes developed as a separate piece, which may remain permanently unconnected with the rest of the bone, thus forming a lumbar rib —a peculiarity which is rarely met with.

Progress of Ossification in the Spine generally.—Ossification of the laminæ of the vertebræ commences in the cervical region of the spine, and proceeds gradually downwards. Ossification of the bodies, on the other hand, commences a little below the centre of the spinal column (about the ninth or tenth dorsal vertebra), and extends both upwards and downwards. Although, however, the ossific nuclei make their first appearance in the lower dorsal vertebræ, the lumbar and first sacral are those in which these nuclei are largest at birth.

Attachment of Muscles.—To the *Atlas* are attached nine pairs : the Longus colli, Rectus capitis anticus minor, Rectus lateralis, Obliquus capitis superior and inferior, Splenius colli, Levator anguli scapulæ, First Intertransverse, and Rectus capitis posticus minor.

To the *Axis* are attached eleven pairs : the Longus colli, Levator anguli scapulæ, Splenius colli, Scalenus medius, Transversalis colli, Intertransversales, Obliquus capitis inferior, Rectus capitis posticus major, Semi-spinalis colli, Multifidus spinæ, Interspinales.

To the remaining vertebræ, generally, are attached thirty-five pairs and a single muscle : *anteriorly*, the Rectus capitis anticus major, Longus colli, Scalenus anticus, medius, and posticus, Psoas magnus and parvus, Quadratus lumborum, Diaphragm, Obliquus abdominis internus, and Transversalis abdominis ; *posteriorly*, the Trapezius, Latissimus dorsi, Levator anguli scapulæ Rhomboideus major and minor, Serratus posticus superior and inferior, Splenius Erector spinæ, Ilio-costalis, Longissimus dorsi, Spinalis dorsi, Cervicalis ascendens, Transversalis colli, Trachelo-mastoid, Complexus, Biventer cervicis, Semi-spinalis dorsi et colli, Multifidus spinæ, Rotatores spinæ, Interspinales, Supraspinales, Intertransversales, Levatores costarum.

* See Cunningham, *Journ. Anat.* vol. xx. p. 238.

SACRAL AND COCCYGEAL VERTEBRÆ

The **Sacral** and **Coccygeal Vertebræ** consist at an early period of life of nine separate pieces which are united in the adult, so as to form two bones, five entering into the formation of the sacrum, four into that of the coccyx. Occasionally, the coccyx consists of five bones.*

THE SACRUM

The **Sacrum** (sacer, *sacred*) is a large, triangular bone (fig. 15), situated at the lower part of the vertebral column, and at the upper and back part of the pelvic cavity, where it is inserted like a wedge between the two innominate bones; its upper part or base, articulating with the last lumbar vertebra, its

FIG. 15.—Sacrum, anterior surface.

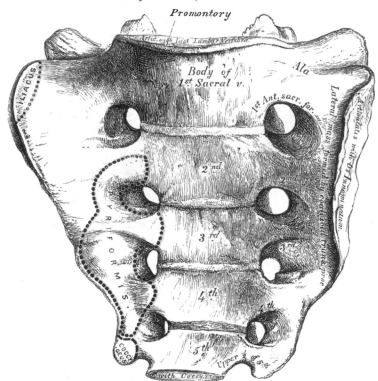

apex with the coccyx. The sacrum is curved upon itself, and placed very obliquely, its upper extremity projecting forwards, and forming, with the last lumbar vertebra, a very prominent angle, called the *promontory* or *sacro-vertebral angle*; while its central part is projected backwards, so as to give increased capacity to the pelvic cavity. It presents for examination an anterior and posterior surface, two lateral surfaces, a base, an apex, and a central canal.

The **Anterior Surface** is concave from above downwards, and slightly so from side to side. In the middle are seen four transverse ridges, indicating the original division of the bone into five separate pieces. The portions of bone intervening between the ridges correspond to the bodies of the vertebræ. The body of the first segment is of large size, and in form resembles that of a lumbar vertebra; the succeeding ones diminish in size from above downwards, are

* The late Sir George Humphry described this as the usual composition of the coccyx.— *On the Skeleton*, p. 456.

flattened from before backwards,. and curved so as to accommodate themselves to the form of the sacrum, being concave in front, convex behind. At each end of the ridges above mentioned are seen the *anterior sacral foramina*, analogous to the intervertebral foramina, four in number on each side, somewhat rounded in form, diminishing in size from above downwards, and directed outwards and forwards : they transmit the anterior branches of the sacral nerves and the lateral sacral arteries. External to these foramina is the *lateral mass*, consisting, at an early period of life, of separate segments ; these become blended, in the adult, with the bodies, with each other, and with the posterior transverse processes.

Each lateral mass is traversed by four broad, shallow grooves, which lodge the anterior sacral nerves as they pass outwards, the grooves being separated by prominent ridges of bone which give attachment to the slips of the Pyriformis muscle.

If a vertical section is made through the centre of the sacrum (fig. 16), the bodies are seen to be united at their circumference by bone, a wide interval being left centrally, which, in the recent state, is filled by intervertebral substance. In some bones, this union is more complete between the lower segments than between the upper ones.

The **Posterior Surface** (fig. 17) is convex and much narrower than the anterior. In the middle line are three or four tubercles, which represent the rudimentary spinous processes of the sacral vertebræ. Of these tubercles, the first is usually prominent, and perfectly distinct from the rest ; the second and third are either separate or united into a tubercular ridge, which diminishes in size from above downwards ; the fourth usually, and the fifth always remaining undeveloped.

FIG. 16.—Vertical section of the sacrum.

External to the spinous processes on each side are the *laminæ*, broad and well marked in the first three pieces ; sometimes the fourth, and generally the fifth, are only partially developed and fail to meet in the middle line. These partially developed laminæ are prolonged downwards, as rounded processes, the *sacral cornua*, and are connected to the cornua of the coccyx. Between them the bony wall of the lower end of the sacral canal is imperfect, and is liable to be opened in the sloughing of bed-sores. External to the laminæ is a linear series of indistinct tubercles representing the *articular processes* ; the upper pair are large, well developed, and correspond in shape and direction to the superior articulating processes of a lumbar vertebra ; the second and third are small ; the fourth and fifth (usually blended together) are situated on each side of the sacral canal and assist in forming the sacral cornua. External to the articular processes are the four *posterior sacral foramina* ; they are smaller in size and less regular in form than the anterior, and transmit the posterior branches of the sacral nerves. On the outer side of the posterior sacral foramina is a series of tubercles, the rudimentary *transverse processes* of the sacral vertebræ. The first pair of transverse tubercles are of large size, very distinct, and correspond with each superior angle of the bone; they, together with the second pair, which are of small size, give attachment to the horizontal part of the sacro-iliac ligament; the third give attachment to the oblique fasciculi of the posterior sacro-iliac ligaments; and the fourth and fifth to the great sacro-sciatic ligaments. The interspace between the spinous and transverse processes on the back of the sacrum presents a wide, shallow concavity, called the *sacral groove*; it is continuous above with the vertebral groove, and lodges the origin of the Multifidus spinæ.

The **Lateral Surface,** broad above, becomes narrowed into a thin edge below. Its upper half presents in front a broad, ear-shaped surface for articulation with the ilium. This is called the *auricular* surface, and in the fresh state is coated with fibro-cartilage. It is bounded posteriorly by deep and uneven impressions, for the attachment of the posterior sacro-iliac ligaments. The lower half is thin

Fig. 17.—Sacrum, posterior surface.

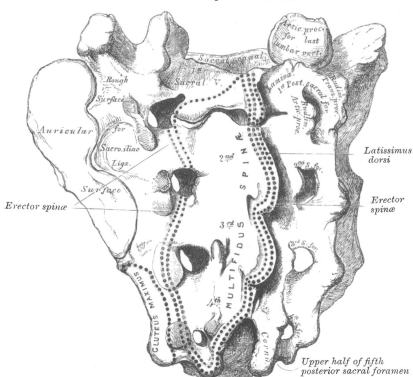

and sharp, and terminates in a projection called the *inferior lateral angle*; below this angle is a notch, which is converted into a foramen by articulation with the transverse process of the upper piece of the coccyx, and transmits the anterior division of the fifth sacral nerve. This lower, sharp border gives attachment to the greater and lesser sacro-sciatic ligaments, and to some fibres of the Gluteus maximus posteriorly, and to the Coccygeus in front.

The **Base** of the sacrum, which is broad and expanded, is directed upwards and forwards. In the middle is seen a large oval articular surface, which is connected with the under surface of the body of the last lumbar vertebra by a fibro-cartilaginous disc. It is bounded behind by the large, triangular orifice of the sacral canal. The orifice is formed behind by the laminæ and spinous process of the first sacral vertebra: the superior articular processes project from it on each side; they are oval, concave, directed backwards and inwards, like the superior articular processes of a lumbar vertebra; and in front of each articular process is an intervertebral notch, which forms the lower part of the foramen between the last lumbar and first sacral vertebra. Lastly, on each side of the large oval articular plate is a broad and flat triangular surface of bone, which extends outwards, supports the Psoas magnus muscle and lumbo-sacral cord, and is continuous on each side with the iliac fossa. This is called the *ala* of the sacrum, and gives attachment to a few of the fibres of the Iliacus muscle. The posterior part of the ala represents the transverse process of the first sacral segment.

The **Apex,** directed downwards and slightly forwards, presents a small, oval, concave surface for articulation with the coccyx.

The **Spinal Canal** runs throughout the greater part of the bone ; it is large and triangular in form above ; small and flattened, from before backwards, below. In this situation its posterior wall is incomplete, from the non-development of the laminæ and spinous processes. It lodges the sacral nerves, and is perforated by the anterior and posterior sacral foramina, through which these pass out.

Structure.—It consists of much loose, spongy tissue within, invested externally by a thin layer of compact tissue.

Differences in the Sacrum of the Male and Female.—The sacrum in the female is shorter and wider than in the male ; the lower half forms a greater angle with the upper ; the upper half of the bone being nearly straight, the lower half presenting the greatest amount of curvature. The bone is also directed more obliquely backwards, which increases the size of the pelvic cavity ; but the sacro-vertebral angle projects less. In the male the curvature is more evenly distributed over the whole length of the bone, and is altogether greater than in the female.

Peculiarities of the Sacrum.—This bone, in some cases, consists of six pieces ; occasionally, the number is reduced to four. Sometimes the bodies of the first and second segments are not joined, or the laminæ and spinous processes have not coalesced. Occasionally, the upper pair of transverse tubercles are not joined to the rest of the bone on one or both sides ; and, lastly, the sacral canal may be open for nearly the lower half of the bone, in consequence of the imperfect development of the laminæ and spinous processes. The sacrum, also, varies considerably with respect to its degree of curvature. From the examination of a large number of skeletons, it appeared that, in one set of cases, the anterior surface of this bone was nearly straight, the curvature, which was very slight, affecting only its lower end. In another set of cases, the bone was curved throughout its whole length, but especially towards its middle. In a third set, the degree of curvature was less marked, and affected especially the lower third of the bone.

Development (fig. 18).—The sacrum, formed by the union of five vertebræ, has *thirty-five* centres of ossification.

The *bodies* of the sacral vertebræ have each three ossific centres : one for the central part, and one for the epiphysial plate on its upper and under surface. Occasionally the primary centres for the bodies of the first and second piece of the sacrum are double.

The arch of each sacral vertebra is developed by two centres, one for each lamina. These unite with each other behind, and subsequently join the body.

The *lateral masses* have six additional centres, two for each of the first three vertebræ. These centres make their appearance above and to the outer side of the anterior sacral foramina (fig. 18), and are developed into separate segments (fig. 19) ; they are subsequently blended with each other, and with the bodies and transverse processes, to form the lateral mass.

Lastly, each *lateral surface* of the sacrum is developed by two epiphysial plates (fig. 20) : one for the auricular surface, and one for the remaining part of the thin lateral edge of the bone.

Period of Development.—At about the eighth or ninth week of fœtal life, ossification of the central part of the bodies of the first three vertebræ commences ;

and, at a somewhat later period, that of the last two. Between the sixth and eighth months ossification of the laminæ takes place; and at about the same period, the centres for the lateral masses of the first three sacral vertebræ make their appearance. The period at which the arch becomes completed by the junction of the laminæ with the bodies in front, and with each other behind, varies in different segments. The junction between the laminæ and the bodies takes place first in the lower vertebræ as early as the second year, but is not effected in the uppermost until the fifth or sixth year. About the sixteenth year the epiphyses for the upper and under surfaces of the bodies are formed; and between the eighteenth and twentieth years, those for each lateral surface of the sacrum make their appearance. The bodies of the sacral vertebræ are, during early life, separated from each other by intervertebral discs. But about the eighteenth year the two lowest segments become joined together by ossification extending through the disc. This process gradually extends upwards until all the segments become united; and the bone is completely formed from the twenty-fifth to the thirtieth year of life.

FIG. 18.—Development of the sacrum.

Additional centres for the first three pieces *

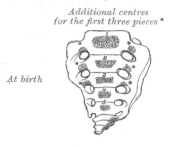

At birth

FIG. 19.

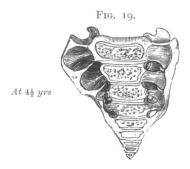

At 4½ yrs

FIG. 20.

*Two epiphysial laminæ for each lateral surface**

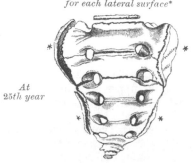

At 25th year

Articulations.—With four bones: the last lumbar vertebra, coccyx, and the two innominate bones.

Attachment of Muscles.—To eight pairs: in front, the Pyriformis and Coccygeus, and a portion of the Iliacus to the base of the bone; behind, the Gluteus maximus, Latissimus dorsi, Multifidus spinæ, and Erector spinæ, and sometimes the Extensor coccygis.

THE COCCYX

The **Coccyx** (κόκκυξ, *cuckoo*), so called from having been compared to a cuckoo's beak (fig. 21), is usually formed of four small segments of bone, and is the most rudimentary part of the vertebral column. In each of the first three segments may be traced a rudimentary body, articular and transverse processes; the last piece (sometimes the third) is a mere nodule of bone, without distinct processes. All the segments are destitute of pedicles, laminæ, and spinous processes; and, consequently, of intervertebral foramina and spinal canal. The first segment is the largest; it resembles the lowermost sacral vertebra, and often exists as a separate piece; the last three, diminishing in size from above downwards, are usually blended together so as to form a single bone. The gradual diminution in the size of the pieces gives this bone a triangular form, the base of the triangle joining the end of the sacrum. It presents for examination an anterior and posterior surface, two borders, a base, and an apex. The *anterior surface* is slightly concave, and marked with three transverse grooves, indicating the points of junction of the different pieces. It has attached to it the anterior sacro-coccygeal ligament and Levator ani muscle, and supports the lower end of the rectum. The *posterior surface* is convex, marked by transverse grooves similar to those on the anterior surface; and presents on each side a lineal row of tubercles, the rudimentary articular processes of the coccygeal vertebræ. Of these, the superior pair are large, and are called the *cornua of the coccyx* : they project upwards, and articulate with the cornua of the sacrum, the junction between these two bones completing the fifth posterior sacral foramen for the transmission of the posterior division of the fifth sacral nerve. The *lateral borders* are thin, and present a series of small eminences, which represent the transverse processes of the coccygeal vertebræ. Of these, the first on each side is the largest, flattened from before backwards, and often ascends to join the lower part of the thin lateral edge of the sacrum, thus completing the fifth anterior sacral foramen for the transmission of the anterior division of the fifth sacral nerve; the others diminish in size from above downwards, and are often wanting. The *borders* of the coccyx are narrow, and give attachment on each side to the sacro-sciatic ligaments, to the Coccygeus muscle in front of the ligaments, and to the Gluteus maximus behind them. The *base* presents an oval surface for articulation with the sacrum. The *apex* is rounded, and has attached to it the tendon of the external Sphincter muscle of the anus. It is occasionally bifid, and sometimes deflected to one or other side.

Fig. 21.—Coccyx.

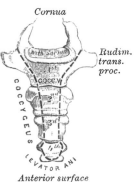

Anterior surface

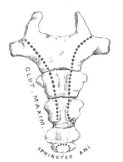

Posterior surface

Development.—The coccyx is developed by *four* centres, one for each piece. Occasionally one of the first three pieces of this bone is developed by two centres, placed side by side. The ossific nuclei make their appearance in the following order : in the first segment, shortly after birth ; in the second piece, at from five to ten years ; in the third, from ten to fifteen years ; in the fourth, from fifteen to twenty years. As age advances, these various segments become united with each other from below upwards, the union between the first and second segments being frequently delayed until after the age of twenty-five or thirty. At a late period of life, especially in females, the coccyx often becomes joined to the end of the sacrum.

Articulation.—With the sacrum.

Attachment of Muscles.—To four pairs and one single muscle: on either side, the Coccygeus ; behind, the Gluteus maximus and Extensor coccygis, when present ; at the apex, the Sphincter ani ; and in front, the Levator ani.

Of the Spine in general

The **Spinal Column,** formed by the junction of the vertebræ, is situated in the median line, at the posterior part of the trunk ; its average length is about two feet two or three inches, measuring along the curved anterior surface of the column. Of this length the cervical part measures about five, the dorsal about eleven, the lumbar about seven inches, and the sacrum and coccyx the remainder. The female spine is about one inch less than the male.

Viewed in front, it presents two pyramids joined together at their bases, the upper one being formed by all the vertebræ from the second cervical to the last lumbar, the lower one by the sacrum and coccyx. When examined more closely, the upper pyramid is seen to be formed of three smaller pyramids   The

FIG. 22.—Lateral view of the spine.

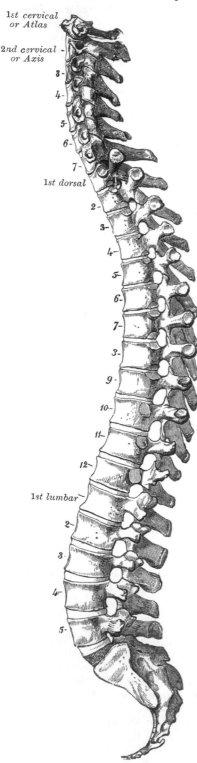

1st cervical
or Atlas

2nd cervical
or Axis

3-

4-

5-

6-

7-

1st dorsal

2-

3-

4-

5-

6-

7-

8-

9-

10-

11-

12-

1st lumbar

2-

3-

4-

5-

uppermost of these consists of the six lower cervical vertebræ; its apex being formed by the axis or second cervical, its base by the first dorsal. The second pyramid, which is inverted, is formed by the four upper dorsal vertebræ, the base being at the first dorsal, the smaller end at the fourth. The third pyramid commences at the fourth dorsal, and gradually increases in size to the fifth lumbar.

Viewed laterally (fig. 22), the spinal column presents several curves, which correspond to the different regions of the column, and are called *cervical, dorsal, lumbar,* and *pelvic.* The *cervical* curve commences at the apex of the odontoid process, and terminates at the middle of the second dorsal vertebra; it is convex in front, and is the least marked of all the curves. The *dorsal* curve, which is concave forwards, commences at the middle of the second and terminates at the middle of the twelfth dorsal vertebra. Its most prominent point behind corresponds to the spine of the seventh dorsal. The *lumbar* curve commences at the middle of the last dorsal vertebra, and terminates at the sacro-vertebral angle. It is convex anteriorly; the convexity of the lower three vertebræ being much greater than that of the upper two. The *pelvic* curve commences at the sacro-vertebral articulation, and terminates at the point of the coccyx. It is concave anteriorly. The dorsal and pelvic curves are the primary curves, and begin to be formed at an early period of fœtal life, and are due to the shape of the bodies of the vertebræ. The cervical and lumbar curves are compensatory or secondary, and are developed after birth in order to maintain the erect position. They are due mainly to the shape of the intervertebral discs.

The spine has also a slight lateral curvature, the convexity of which is directed towards the right side. This is most probably produced, as Bichat first explained, chiefly by muscular action; most persons using the right arm in preference to the left, especially in making long-continued efforts, when the body is curved to the right side. In support of this explanation, it has been found, by Béclard, that in one or two individuals who were left-handed, the lateral curvature was directed to the left side.

The movable part of the spinal column presents for examination an anterior, a posterior, and two lateral surfaces : a base, summit, and spinal canal.

The **anterior surface** presents the bodies of the vertebræ separated in the recent state by the intervertebral discs. The bodies are broad in the cervical region, narrow in the upper part of the dorsal, and broadest in the lumbar region. The whole of this surface is convex transversely, concave from above downwards in the dorsal region, and convex in the same direction in the cervical and lumbar regions.

The **posterior surface** presents in the median line the spinous processes. These are short, horizontal, with bifid extremities, in the cervical region. In the dorsal region, they are directed obliquely above, assume almost a vertical direction in the middle, and are horizontal below, as are also the spines of the lumbar vertebræ. They are separated by considerable intervals in the loins, by narrower intervals in the neck, and are closely approximated in the middle of the dorsal region. Occasionally one of these processes deviates a little from the median line—a fact to be remembered in practice, as irregularities of this sort are attendant also on fractures or displacements of the spine. On either side of the spinous processes, extending the whole length of the column, is the vertebral groove formed by the laminæ in the cervical and lumbar regions, where it is shallow, and by the laminæ and transverse processes in the dorsal region, where it is deep and broad. In the recent state, these grooves lodge the deep muscles of the back. External to the vertebral grooves are the articular processes, and still more externally the transverse processes. In the dorsal region, the latter processes stand backwards, on a plane considerably posterior to the same processes in the cervical and lumbar regions. In the cervical region, the transverse processes are placed in front of the articular processes, and on the outer side of the pedicles, between the intervertebral foramina. In the dorsal region they are posterior to the pedicles, intervertebral foramina, and articular processes. In the lumbar, they are placed in front of the articular processes, but behind the intervertebral foramina.

The **lateral surfaces** are separated from the posterior by the articular processes in the cervical and lumbar regions, and by the transverse processes in the dorsal. These surfaces present in front the sides of the bodies of the vertebræ, marked in the dorsal region by the facets for articulation with the heads of the ribs. More posteriorly are the intervertebral foramina, formed by the juxtaposition of the intervertebral notches, oval in shape, smallest in the cervical and upper part of the dorsal regions, and gradually increasing in size to the last lumbar. They are situated between the transverse processes in the neck, and in front of them in the back and loins, and transmit the spinal nerves.

The **base** of that portion of the vertebral column formed by the twenty-four movable vertebræ consists of the under surface of the body of the fifth lumbar vertebra ; and the **summit** of the upper surface of the atlas.

The **vertebral** or **spinal canal** follows the different curves of the spine ; it is largest in those regions in which the spine enjoys the greatest freedom of movement, as in the neck and loins, where it is wide and triangular ; and narrow and rounded in the back, where motion is more limited.

Surface Form.—The only parts of the vertebral column which lie closely under the skin, and so directly influence surface form, are the apices of the spinous processes. These are always distinguishable at the bottom of a median furrow, which, more or less evident, runs down the mesial line of the back from the external occipital protuberance above to the middle of the sacrum below. In the neck the furrow is broad, and terminates in a conspicuous projection, which is caused by the spinous process of the seventh cervical vertebra (vertebra prominens). Above this the spinous process of the sixth cervical vertebra may sometimes be seen to form a projection ; the other cervical spines are sunken, and are not visible, though the spine of the axis can be felt, and generally also the spines of the third, fourth, and fifth cervical vertebræ. In the dorsal region, the furrow is shallow, and during stooping disappears, and then the spinous processes become more or less visible. The markings produced by these spines are small and close together. In the lumbar region the furrow is deep, and the situation of the lumbar spines is frequently indicated by little pits or depressions, especially if the muscles in the loins are well developed and the spine incurved. They are much larger and farther apart than in the dorsal region. In the sacral region the furrow is shallower, presenting a flattened area which terminates below at the most prominent part of the posterior surface of the sacrum, formed by the spinous process of the third sacral vertebra. At the bottom of the furrow may be felt the irregular posterior surface of the bone. Below this, in the deep groove leading to the anus, the coccyx may be felt. The only other portions of the vertebral column which can be felt from the surface are the transverse processes of three of the cervical vertebræ—viz. the first, the sixth, and the seventh. The transverse process of the atlas can be felt as a rounded nodule of bone just below and in front of the apex of the mastoid process, along the anterior border of the Sterno-mastoid. The transverse process of the sixth cervical vertebra is of surgical importance. If deep pressure be made in the neck, in the course of the carotid artery, opposite the cricoid cartilage, the prominent anterior tubercle of the transverse process of the sixth cervical vertebra can be felt. This has been named *Chassaignac's tubercle*, and against it the carotid artery may be most conveniently compressed by the finger. The transverse process of the seventh cervical vertebra can also be often felt. Occasionally the anterior root, or costal process, is large and segmented off, forming a cervical rib.

Surgical Anatomy.—Occasionally the coalescence of the laminæ is not completed, and consequently a cleft is left in the arches of the vertebræ, through which a protrusion of the spinal membranes (dura mater and arachnoid), and generally of the spinal cord itself, takes place, constituting the malformation known as *spina bifida*. This condition is most common in the lumbo-sacral region, but it may occur in the dorsal or cervical region, or the arches throughout the whole length of the canal may remain unapproximated. In some rare cases, in consequence of the non-coalescence of the two primary centres from which the body is formed, a similar condition may occur in front of the canal, the bodies of the vertebræ being found cleft and the tumour projecting into the thorax, abdomen, or pelvis, between the lateral halves of the bodies affected.

The construction of the spinal column of a number of pieces, securely connected together and enjoying only a slight degree of movement between any two individual pieces, though permitting of a very considerable range of movement as a whole, allows a sufficient degree of mobility without any material diminution of strength. The many joints of which the spine is composed, together with the very varied movements to which it is subjected, render it liable to sprains; but so closely are the individual vertebræ articulated that these sprains are rarely or ever severe, and any amount of violence sufficiently great to produce tearing of the ligaments would tend rather to cause a dislocation or fracture. The further safety of the column and its slight liability to injury is provided for by its disposition in curves, instead of in one straight line. For it is an elastic column, and must first bend before it breaks; under these circumstances, being made up of three curves, it represents three columns, and greater force is required to produce bending of a short column, than of a longer one that is equal to it in breadth and material. Again, the safety of the column is provided for by the interposition of the intervertebral discs between the bodies of the vertebræ, which act as admirable buffers in counteracting the effects of violent jars or shocks. Fracture-dislocation of the spine may be caused by direct or indirect violence or by a combination of the two, as when a person, falling from a height, strikes against some prominence and is doubled over it. The fractures from indirect violence are the more common, and here the bodies of the vertebræ are compressed, while the arches are torn asunder; in fracture from direct violence, on the other hand, the arches are compressed and the bodies of the vertebræ separated from each other. It will therefore be seen that in both classes of injury the spinal marrow is the part least likely to be injured, and may escape damage even where there has been considerable lesion of the bony framework. For, as Mr. Jacobson states, ' being lodged in the centre of the column, it occupies neutral ground in respect to forces which might cause fracture. For it is a law in mechanics that when a beam, as of timber, is exposed to breakage and the force does not exceed the limits of the strength of the material, one division resists compression, another laceration of the particles, while the third, between the two, is in a negative condition.'* Applying this principle to the spine, it will be seen that, whether the fracture-dislocation be produced by direct or indirect violence, one segment, either the anterior or posterior, will be exposed to compression, the other to laceration, and the intermediate part, where the cord is situated, will be in a neutral state. When a fracture-dislocation is produced by indirect violence, the displacement is almost always the same; the upper segment being driven forwards on the lower, so that the cord is compressed between the body of the vertebra below and the arch of the vertebra above.

The parts of the spine most liable to be injured are (1) the dorsi-lumbar region, for this part is near the middle of the column and there is therefore a greater amount of leverage, and moreover the portion above is comparatively fixed, and the vertebræ which form it, though much smaller, have nevertheless to bear almost as great a weight as those below; (2) the cervico-dorsal region, because here the flexible cervical portion of the spine joins the more fixed dorsal region; and (3) the atlanto-axial region, because it enjoys an

* Holmes's *System of Surgery*, vol. i. p. 529. 1883.

extensive range of movement, and, being near the skull, is influenced by violence applied to the head.   In fracture-dislocation it has been proposed to trephine the spine and remove portions of the laminæ and spinous processes.   The operation can be of use only when the paralysis is due to the pressure of bone or the effusion of blood, and not to cases, which are by far the more common, where the cord is crushed to a pulp.   And even in those cases where the cord is compressed by bone, the portion of displaced bone which presses on the cord is generally the body of the vertebra below, and is therefore inaccessible to operation.   The operative proceeding is one of great severity, involving an extensive and deep wound and great risk of septic meningitis, and as the advantages to be derived from it are exceedingly problematical and confined to a very few cases, it is not often resorted to.   Trephining has also been employed in some cases of paraplegia, due to Potts's disease of the spine.   Here the paralysis is due to the pressure of inflammatory products, and where this is new scar-tissue, formed by the organisation of granulation-tissue, its removal has been attended with a very considerable amount of success.

THE SKULL

The **Skull** is supported on the summit of the vertebral column, and is of an oval shape, wider behind than in front.   It is composed of a series of flattened or irregular bones which, with one exception (the lower jaw), are immovably jointed together.   It is divided into two parts, the Cranium and the Face, the former of which constitutes a case for the accommodation and protection of the brain, while opening on the face are the orifices of the nose and mouth ; between the cranium above and the face below the orbital cavities are situated.   The Cranium (κράνος, *a helmet*) is composed of *eight* bones—viz. the *occipital*, two *parietal, frontal*, two *temporal, sphenoid*, and *ethmoid*.   The face is composed of *fourteen* bones—viz. the two *nasal*, two *superior maxillary*, two *lachrymal*, two *malar*, two *palate*, two *inferior turbinated, vomer,* and *inferior maxillary*.   The *ossicula auditûs*, the *teeth*, and *Wormian bones* are not included in this enumeration.

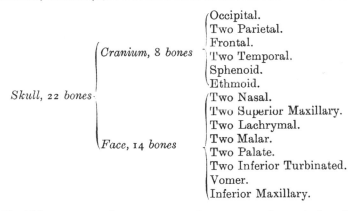

Skull, 22 bones

Cranium, 8 bones
Occipital.
Two Parietal.
Frontal.
Two Temporal.
Sphenoid.
Ethmoid.

Face, 14 bones
Two Nasal.
Two Superior Maxillary.
Two Lachrymal.
Two Malar.
Two Palate.
Two Inferior Turbinated.
Vomer.
Inferior Maxillary.

The Hyoid bone, situated at the root of the tongue and attached to the base of the skull by ligaments, has also to be considered in this section.

The Occipital Bone

The **Occipital Bone** (ob, caput, *against the head*) is situated at the back part and base of the cranium.   It is trapezoid in shape and is much curved on itself (fig. 23) It presents at its front and lower part a large oval aperture, the *foramen magnum*, by which the cranial cavity communicates with the spinal canal.   The portion of bone behind this opening is flat and expanded and forms the *tabula* ;   the portion in front is a thick, elongated mass of bone, the *basilar process* ;   while on either side of the foramen are situated processes bearing the condyles, by which the bone articulates with the atlas.   These processes are known as the *condylar portions*.   It presents for examination two surfaces, four borders, and four angles.

The **external surface** is convex. Midway between the summit of the bone and the posterior margin of the foramen magnum is a prominent tubercle, the *external occipital protuberance*, and descending from it as far as the foramen, a vertical ridge, the *external occipital crest*. This protuberance and crest give attachment to the Ligamentum nuchæ, and vary in prominence in different skulls. Passing outwards from the occipital protuberance is a semicircular ridge on each side, the *superior curved line*. Above this line there is often a second less distinctly marked ridge, called the *highest curved line* (*linea suprema*), to which the epicranial aponeurosis is attached. The bone between these two lines is smoother and denser than the rest of the surface. Running parallel with these from the middle of the crest is another semicircular ridge on each side, the *inferior curved line*. The surface of the bone above the linea suprema is rough and porous, and, in the

FIG. 23.—Occipital bone. Outer surface.

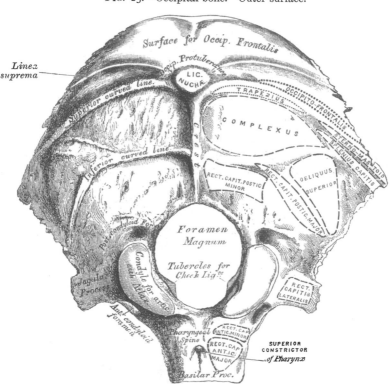

recent state, is covered by the Occipito-frontalis muscle, while the superior and inferior curved lines, together with the surfaces of bone between and below them, serve for the attachment of several muscles. The superior curved line gives attachment internally to the Trapezius, externally to the muscular origin of the Occipito-frontalis, and to the Sterno-cleido-mastoid, to the extent shown in fig. 23 ; the depressions between the curved lines to the Complexus internally, the Splenius capitis and Obliquus capitis superior externally. The inferior curved line, and the depressions below it, afford insertion to the Rectus capitis posticus, major and minor.

The **foramen magnum** is a large, oval aperture, its long diameter extending from before backwards. It transmits the medulla oblongata and its membranes, the spinal accessory nerves, the vertebral arteries, the anterior and posterior spinal arteries, and the occipito-axial ligaments. Its back part is wide for the transmission of the medulla, and the corresponding margin rough for the attachment of the dura mater enclosing it ; the fore part is narrower, being encroached upon by the condyles ; it has projecting towards it, from below, the odontoid process, and its margins are smooth and bevelled internally to support the medulla oblongata. On each side of the foramen magnum are the *condyles*, for articulation with the atlas ; they are convex, oval or reniform in shape, and directed downwards and outwards ; they converge in front, and encroach slightly upon the anterior segment of the foramen. On the inner border of each condyle is a rough tubercle for the attachment of the check ligaments which connect this bone with the odontoid process of the axis ; while external to them is a rough tubercular prominence, the *transverse* or *jugular process*, channelled in front by a deep notch, which forms part of the jugular foramen, or foramen lacerum posterius. The under surface of this process presents an eminence, which represents the *paramastoid process* of some mammals. This eminence is occasionally large, and extends as low as the transverse process of the atlas. This surface affords attachment to the Rectus capitis lateralis muscle and to the lateral occipito-atlantal ligament ; its upper or cerebral surface presents a deep groove which lodges part of the lateral sinus, while its external surface is marked by a quadrilateral rough facet, covered with cartilage in the fresh state, and articulating with a similar surface on the petrous portion of the temporal bone. On the outer side of each condyle, near its fore part, is a foramen, the *anterior condyloid* ; it is directed downwards, outwards, and forwards, and transmits the hypoglossal nerve, and occasionally a meningeal branch of the ascending pharyngeal artery. This foramen is sometimes double. Behind each condyle is a fossa,* sometimes perforated at the bottom by a foramen, the *posterior condyloid*, for the transmission of a vein to the lateral sinus. In front of the foramen magnum is a strong quadrilateral plate of bone, the *basilar process*, wider behind than in front ; its under surface, which is rough, presenting in the median line a tubercular ridge, the *pharyngeal spine*, for the attachment of the tendinous raphé and Superior constrictor of the pharynx ; and on each side of it, rough depressions for the attachment of the Rectus capitis anticus, major and minor.

The **Internal** or **Cerebral Surface** (fig. 24) is deeply concave. The posterior part or tabula is divided by a crucial ridge into four fossæ. The two superior fossæ receive the occipital lobes of the cerebrum, and present slight depressions corresponding to their convolutions. The two inferior, which receive the hemispheres of the cerebellum, are larger than the former, and comparatively smooth ; both are marked by slight grooves for the lodgment of arteries. At the point of meeting of the four divisions of the crucial ridge is an eminence, the *internal occipital protuberance*. It nearly corresponds to that on the outer surface, and is perforated by one or more large vascular foramina. From this eminence, the superior division of the crucial ridge runs upwards to the superior angle of the bone ; it presents a deep groove for the superior longitudinal sinus, the margins of which give attachment to the falx cerebri. The inferior division, the *internal occipital crest*, runs to the posterior margin of the foramen magnum, on the edge of which it becomes gradually lost ; this ridge, which is bifurcated below, serves for the attachment of the falx cerebelli. It is usually marked by a single groove, which commences at the back part of the foramen magnum and lodges the occipital sinus. Occasionally the groove is double where two sinuses exist. The transverse grooves pass outwards to the lateral angles ; they are deeply channelled, for the lodgment of the lateral sinuses, their prominent margins affording attachment to the tentorium cerebelli.† At the point of meeting

* This fossa presents many variations in size. It is usually shallow, and the foramen small ; occasionally wanting on one or both sides. Sometimes both fossa and foramen are large, but confined to one side only ; more rarely, the fossa and foramen are very large on both sides.

† Usually one of the transverse grooves is deeper and broader than the other : occasionally, both grooves are of equal depth and breadth, or both equally indistinct. The broader of the two transverse grooves is nearly always continuous with the vertical groove for the superior longitudinal sinus.

of these grooves is a depression, the *torcular Herophili*,* placed a little to one or the other side of the internal occipital protuberance. More anteriorly is the foramen magnum, and on each side of it, but nearer its anterior than its posterior part, the internal opening of the anterior condyloid foramen; the internal openings of the posterior condyloid foramina are a little external and posterior, protected by a small arch of bone. At this part of the internal surface there is a very deep groove, in which the posterior condyloid foramen, when it exists, has its termination. This groove is continuous, in the complete skull, with the transverse groove on the posterior part of the bone, and lodges the end of the same sinus, the lateral. In front of the foramen magnum is the basilar

FIG. 24.—Occipital bone. Inner surface.

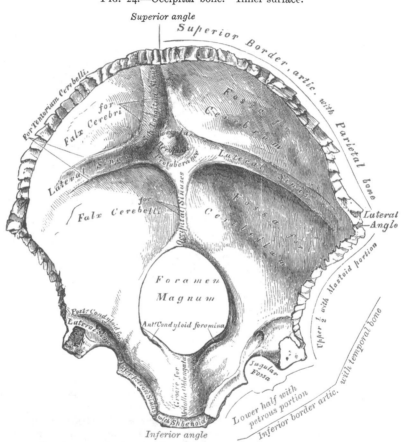

process, presenting a shallow depression, the *basilar groove*, which slopes from behind, upwards and forwards, and supports the medulla oblongata and part of the pons Varolii, and on each side of the basilar process is a narrow channel, which, when united with a similar channel on the petrous portion of the temporal bone, forms a groove, which lodges the inferior petrosal sinus.

Angles.—The *superior* angle is received into the interval between the posterior superior angles of the two parietal bones: it corresponds with that part of the skull in the foetus which is called the *posterior fontanelle*. The *inferior* angle is represented by the square-shaped surface of the basilar process. At an early

* The columns of blood coming in different directions were supposed to be *pressed* together at this point (*torcular*, a wine-press).

period of life, a layer of cartilage separates this part of the bone from the sphenoid but in the adult, the union between them is osseous. The *lateral* angles correspond to the outer ends of the transverse grooves, and are received into the interval between the posterior inferior angles of the parietal and the mastoid portion of the temporal.

Borders.—The *superior* border extends on each side from the superior to the lateral angle, is deeply serrated for articulation with the parietal bone, and forms, by this union, the lambdoid suture ; the *inferior* border extends from the lateral to the inferior angle ; its upper half is rough, and articulates with the mastoid portion of the temporal, forming the masto-occipital suture ; the lower half articulates with the petrous portion of the temporal, forming the petro-occipital suture ; these two portions are separated from one another by the jugular process. In front of this process is a deep notch, which, with a similar one on the petrous portion of the temporal, forms the *foramen lacerum posterius* or *jugular foramen.* This notch is occasionally subdivided into two parts by a small process of bone, and it generally presents an aperture at its upper part, the internal opening of the posterior condyloid foramen.

Structure.—The occipital bone consists of two compact laminæ, called the *outer* and *inner tables,* having between them the diploic tissue ; this bone is especially thick at the ridges, protuberances, condyles, and anterior part of the basilar process ; while at the bottom of the fossæ, especially the inferior, it is thin, semi-transparent, and destitute of diploë.

Development (fig. 25).—At birth the bone consists of four distinct parts : a *tabular* or expanded portion, which lies behind the foramen magnum ; two *condylar* parts, which form the sides of the foramen ; and a *basilar* part, which lies in front of the foramen. The number of nuclei for the tabular part vary. As a rule there are four, but there may be only one (Blandin), or as many as eight (Meckel). They appear about the eighth week of fœtal life, and soon unite to form a single piece ; which is, however, fissured in the direction indicated in the figure. The basilar and two condyloid portions are each

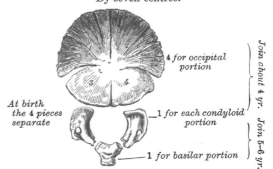

FIG. 25.—Development of occipital bone.
By seven centres.

At birth the 4 pieces separate

4 *for occipital portion*

1 *for each condyloid portion*

1 *for basilar portion*

Join about 4 yr. Join 5–6 yr.

developed from a single nucleus, which appears a little later. The upper portion of the tabular surface, that is to say, the portion above the transverse fissures, is developed from membrane, and may remain separated from the rest of the bone throughout life when it constitutes the *interparietal* bone ; the rest of the bone is developed from cartilage. At about the fourth year, the tabular and the two condyloid pieces join ; and about the sixth year, the bone consists of a single piece. At a later period, between the eighteenth and twenty-fifth years, the occipital and sphenoid become united, forming a single bone.

Articulations.—With six bones : two parietal, two temporal, sphenoid, and atlas.

Attachment of Muscles.—To twelve pairs; to the superior curved line are attached the Occipito-frontalis, Trapezius, and Sterno-cleido-mastoid. To the space between the curved lines, the Complexus,* Splenius capitis, and Obliquus capitis superior ; to the inferior curved line, and the space between it and the

---

* To these the Biventer cervicis should be added, if it is regarded as a separate muscle.

foramen magnum, the Rectus capitis posticus, major and minor ; to the transverse
process, the Rectus capitis lateralis ; and to the basilar process, the Rectus capitis
anticus, major and minor, and the Superior constrictor of the pharynx.

The Parietal Bones

The **Parietal Bones** (paries, *a wall*) form, by their union, the sides and roof of
the skull. Each bone is of an irregular quadrilateral form, and presents for
examination two surfaces, four borders, and four angles.

Surfaces.—The *external surface* (fig. 26) is convex, smooth, and marked
about its centre by an eminence, called the *parietal eminence*, which indicates the

Fig. 26.—Left parietal bone. External surface.

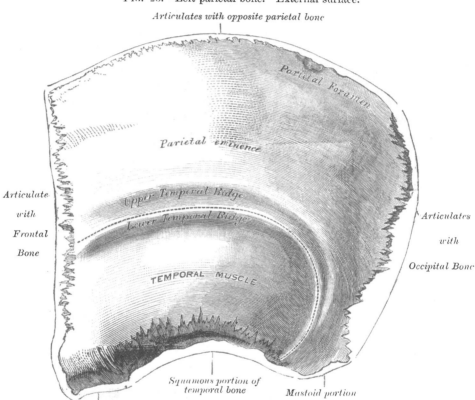

Articulates with opposite parietal bone

Parietal Foramen

Parietal eminence

Upper Temporal Ridge

Lower Temporal Ridge

Articulate with Frontal Bone

Articulates with Occipital Bone

TEMPORAL MUSCLE

Squamous portion of temporal bone

Mastoid portion

Sphenoid

point where ossification commenced. Crossing the middle of the bone in an
arched direction are two well-marked curved lines or ridges, the *upper* and *lower
temporal ridges* ; the former gives attachment to the temporal fascia, while the
latter indicates the upper limit of the origin of the temporal muscle. Above these
ridges, the surface of the bone is rough and porous, and covered by the aponeurosis
of the Occipito-frontalis ; between them the bone is smoother and more polished
than the rest ; below them the bone forms part of the temporal fossa, and affords
attachment to the Temporal muscle. At the back part of the superior border,
close to the sagittal suture, is a small foramen, the *parietal foramen*, which
transmits a vein to the superior longitudinal sinus, and sometimes a small branch
of the occipital artery. Its existence is not constant, and its size varies
considerably.

The **internal surface** (fig. 27), concave, presents depressions for lodging the
convolutions of the cerebrum, and numerous furrows for the ramifications of

the middle meningeal artery; the latter run upwards and backwards from the anterior inferior angle, and from the central and posterior part of the lower border of the bone. Along the upper margin is part of a shallow groove, which, when joined to the opposite parietal, forms a channel for the superior longitudinal sinus, the elevated edges of which afford attachment to the falx cerebri. Near the groove are seen several depressions, especially in the skulls of old persons; they lodge the Pacchionian bodies. The internal opening of the parietal foramen is also seen when that aperture exists.

Borders.—The *superior*, the longest and thickest, is dentated to articulate with its fellow of the opposite side, forming the sagittal suture. The *inferior* is divided into three parts: of these, the anterior is thin and pointed, bevelled at the expense of the outer surface, and overlapped by the tip of the great wing of the sphenoid; the middle portion is arched, bevelled at the expense of the outer

FIG. 27.—Left parietal bone. Internal surface.

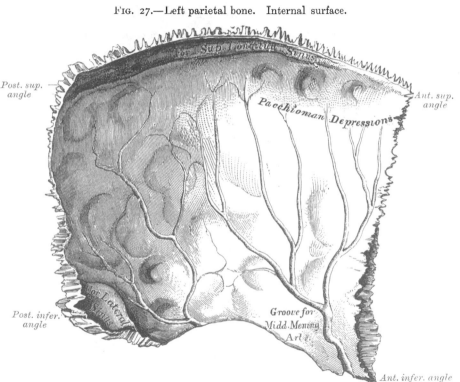

surface, and overlapped by the squamous portion of the temporal; the posterior portion is thick and serrated for articulation with the mastoid portion of the temporal. The *anterior* border, deeply serrated, is bevelled at the expense of the outer surface above and of the inner below; it articulates with the frontal bone, forming the *coronal suture*. The *posterior* border, deeply denticulated, articulates with the occipital, forming the *lambdoid suture*.

Angles.—The *anterior superior* angle, thin and pointed, corresponds with that portion of the skull which in the foetus is membranous, and is called the *anterior fontanelle*. The *anterior inferior* angle is thin and lengthened, being received in the interval between the great wing of the sphenoid and the frontal. Its inner surface is marked by a deep groove, sometimes a canal, for the anterior branch of the middle meningeal artery. The *posterior superior* angle corresponds with the junction of the sagittal and lambdoid sutures. In the foetus this part of the skull is membranous, and is called the *posterior fontanelle*. The *posterior inferior* angle articulates with the mastoid portion of the temporal bone, and generally presents on its inner surface a broad, shallow groove for lodging part of the lateral sinus.

Development.—The parietal bone is formed in membrane, being developed by one centre, which corresponds with the parietal eminence, and makes its first appearance about the seventh or eighth week of fœtal life. Ossification gradually extends from the centre to the circumference of the bone : the angles are consequently the parts last formed, and it is here that the fontanelles exist previous to the completion of the growth of the bone. Occasionally the parietal bone is divided into two parts, upper and lower, by an antero-posterior suture.

Articulations.—With five bones : the opposite parietal, the occipital, frontal, temporal, and sphenoid.

Attachment of Muscles.—One only, the Temporal.

THE FRONTAL BONE

The **Frontal Bone** (frons, *the forehead*) resembles a cockle-shell in form, and consists of two portions—a *vertical* or *frontal* portion situated at the anterior part of the cranium, forming the forehead ; and a *horizontal* or *orbito-nasal* portion which enters into the formation of the roof of the orbits and nasal fossæ.

Vertical Portion.—*External Surface* (fig. 28).—In the median line, traversing the bone from the upper to the lower part, is occasionally seen a slightly elevated

FIG. 28.—Frontal bone. Outer surface.

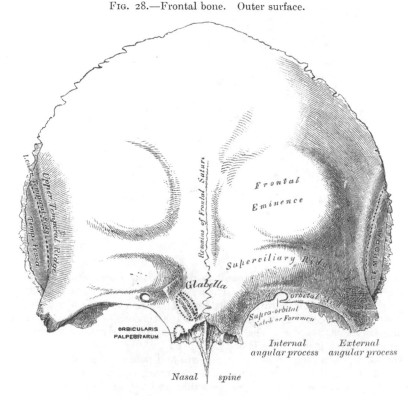

ridge, and in young subjects a suture, which represents the line of union of the two lateral halves of which the bone consists at an early period of life: in the adult this suture is usually obliterated, and the bone forms one piece ; traces of the obliterated suture are, however, generally perceptible at the lower part. On either side of this ridge, a little below the centre of the bone, is a rounded eminence, the *frontal eminence*. These eminences vary in size in different individuals, and are occasionally unsymmetrical in the same subject. They are especially prominent in cases of well-marked cerebral development. The whole surface of the bone above this part is smooth, and covered by the aponeurosis of the Occipito-frontalis muscle. Below the frontal eminence, and separated from it by a shallow groove, is the *superciliary ridge*, broad internally, where it is continuous with the nasal eminence, but less distinct as it arches outwards. These ridges are caused by the projection outwards of the frontal air sinuses,* and give attachment to the Orbicularis palpebrarum and Corrugator supercilii. Between the two superciliary ridges is a smooth surface, the *glabella* or *nasal eminence*. Beneath the superciliary ridge is the *supra-orbital arch*, a curved and prominent margin, which forms the upper boundary of the orbit, and separates the vertical from the horizontal portion of the bone. The outer part of the arch is sharp and prominent, affording to the eye, in that situation, considerable protection from injury ; the inner part is less prominent. At the junction of the internal and middle third of this arch is a notch, sometimes converted into a foramen, and called the *supra-orbital notch* or *foramen*. It transmits the supra-orbital artery, vein, and nerve. A small aperture is seen in the upper part of the notch, which transmits a vein from the diploë to join the supra-orbital vein. The supra-orbital arch terminates externally in the *external angular process*, and internally in the *internal angular process*. The external angular process is strong, prominent, and articulates with the malar bone ; running upwards and backwards from it are two well-marked lines, which, commencing together from the external angular process, soon diverge from each other and run in a curved direction across the bone. These are the *upper* and *lower temporal ridges* ; the upper gives attachment to the temporal fascia, the lower to the temporal muscle. Beneath them is a slight concavity, that forms the anterior part of the temporal fossa, and gives origin to the Temporal muscle. The internal angular processes are less marked than the external, and articulate with the lachrymal bones. Between the internal angular processes is a rough, uneven interval, the *nasal notch*, which articulates in the middle line with the nasal bones, and on either side with the nasal process of the superior maxillary bone. From the concavity of this notch projects a process, the *nasal process*, which extends beneath the nasal bones and nasal processes of the superior maxillary bones, and supports the bridge of the nose. On the under surface of this is a long pointed process, the *nasal spine*, and on either side a small grooved surface enters into the formation of the roof of the nasal fossa. The nasal spine forms part of the septum of the nose, articulating in front with the nasal bones and behind with the perpendicular plate of the ethmoid.

Internal Surface (fig. 29).—Along the middle line is a vertical groove, the edges of which unite below to form a ridge, the *frontal crest* ; the groove lodges the superior longitudinal sinus, while its margins afford attachment to the falx cerebri. The crest terminates below at a small notch which is converted into a foramen by articulation with the ethmoid. It is called the *foramen cæcum*, and varies in size in different subjects ; it is sometimes partially or completely impervious, lodges a process of the falx cerebri, and, when open, transmits a vein from the lining membrane of the nose to the superior longitudinal sinus. On either side of the groove the bone is deeply concave, presenting depressions for the convolutions of the brain, and numerous small furrows for lodging the ramifications of the anterior meningeal arteries. Several small, irregular fossæ

---

* Some confusion is occasioned to students commencing the study of anatomy by the name 'sinus' having been given to two perfectly different kinds of space connected with the skull. It may be as well, therefore, to state here, at the outset, that the 'sinuses' in the interior of the cranium which produce the grooves on the inner surface of the bones are venous channels along which the blood runs in its passage back from the brain, while the 'sinuses' external to the cranial cavity (the frontal, sphenoidal, ethmoidal, and maxillary) are hollow spaces in the bones themselves which communicate with the nostrils, and contain air.

are also seen on either side of the groove, for the reception of the Pacchionian bodies.

Horizontal Portion.—This portion of the bone consists of two thin plates, the *orbital plates*, which form the vault of the orbit, separated from one another by a median gap, the *ethmoidal notch*.

The external surface of each orbital plate consists of a smooth, concave, triangular lamina of bone, marked at its anterior and external part (immediately beneath the external angular process) by a shallow depression, the *lachrymal fossa*, for lodging the lachrymal gland; and at its anterior and internal part, by a depression (sometimes a small tubercle), the *trochlear fossa*, for the attachment of the cartilaginous pulley of the Superior oblique muscle of the eye. The ethmoidal notch separates the two orbital plates; it is quadrilateral, and filled up, when the

FIG. 29.—Frontal bone. Inner surface.

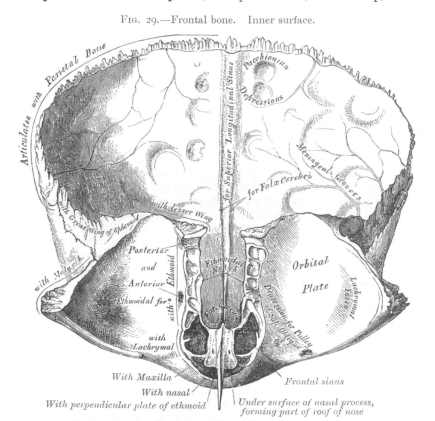

bones are united, by the cribriform plate of the ethmoid. The margins of this notch present several half-cells, which, when united with corresponding half-cells on the upper surface of the ethmoid, complete the ethmoidal cells; two grooves are also seen crossing these edges transversely; they are converted into canals by articulation with the ethmoid, and are called the *anterior* and *posterior ethmoidal* canals: they open on the inner wall of the orbit. The anterior one transmits the nasal nerve and anterior ethmoidal vessels, the posterior one the posterior ethmoidal vessels. In front of the ethmoidal notch, on either side of the nasal spine, are the openings of the frontal sinuses. These are two irregular cavities, which extend upwards and outwards, a variable distance, between the two tables of the skull, and are separated from one another by a thin, bony septum, which is often displaced to one side. They give rise to the prominences above the supra-orbital arches called the *superciliary ridges*. In the child they are generally absent, and they become gradually developed as age advances. These cavities vary in size in different persons, are larger in men than in women, and are frequently of unequal size on the two sides, the right being commonly the larger. They are lined by mucous membrane, and communicate with the nose by the infundibulum, and occasionally with each other by apertures in their septum.

The *internal surface* presents the convex upper surfaces of the orbital plates, separated from each other in the middle line by the ethmoidal notch, and marked by depressions for the convolutions of the frontal lobes of the brain.

Borders.—The border of the vertical portion is thick, strongly serrated, bevelled at the expense of the internal table above, where it rests upon the parietal bones, and at the expense of the external table at each side, where it receives the lateral pressure of those bones; this border is continued below into a tri-angular, rough surface, which articulates with the great wing of the sphenoid. The border of the horizontal portion is thin, serrated, and articulates with the lesser wing of the sphenoid.

Structure.—The vertical portion and external angular processes are very thick, consisting of diploic tissue and the frontal air sinuses contained between two compact laminæ. The horizontal portion is thin, translucent, and composed entirely of compact tissue; hence the facility with which instruments can pene-trate the cranium through this part of the orbit.

Development (fig. 30).—The frontal bone is formed in membrane, being developed by *two* centres, one for each lateral half, which make their appearance about the seventh or eighth week, above the orbital arches. From this point ossi-fication extends, in a radiating manner, upwards into the forehead, and backwards over the orbit. At birth the bone consists of two pieces, which afterwards become united, along the middle line, by a suture which runs from the vertex of the bone to the root of the nose. This suture usually becomes obliterated within a few years after birth; but it occasionally remains throughout life, constituting the *metopic* suture. Secondary centres of ossification may appear for the nasal spine; one on either side at the internal angular process where it articulates with the lachrymal

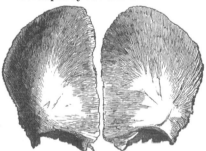

FIG. 30.—Frontal bone at birth. Developed by two lateral halves.

bone; and sometimes there is one on either side at the lower end of the coronal suture. This latter centre sometimes remains ununited, and is known as the *pterion ossicle*; or it may join with the parietal, sphenoid, or temporal bone.

Articulations.—With twelve bones: two parietal, the sphenoid, the ethmoid, two nasal, two superior maxillary, two lachrymal, and two malar.

Attachment of Muscles.—To three pairs: the Corrugator supercilii, Orbicularis palpebrarum, and Temporal on each side.

THE TEMPORAL BONES

The **Temporal Bones** (tempus, *time*) are situated at the sides and base of the skull, and present for examination a *squamous, mastoid,* and *petrous* portion.

The **squamous portion** (squama, *a scale*), the anterior and upper part of the bone, is scale-like in form, and thin and translucent in texture (fig. 31). Its *outer surface* is smooth, convex, and grooved at its back part for the deep tem-poral arteries; it affords attachment to the Temporal muscle, and forms part of the temporal fossa. At its back part may be seen a curved ridge—part of the *temporal ridge*; it serves for the attachment of the temporal fascia, and limits the origin of the Temporal muscle. The boundary between the squamous and mastoid portions of the bone, as indicated by traces of the original suture, lies fully half

an inch below this ridge. Projecting from the lower part of the squamous
portion is a long, arched process of bone, the *zygoma* or zygomatic process. This
process is at first directed outwards, its two surfaces looking upwards and downwards ; it then appears as if twisted upon itself, and runs forwards, its surfaces
now looking inwards and outwards. The superior border of the process is long,
thin, and sharp, and serves for the attachment of the temporal fascia. The
inferior, short, thick, and arched, has attached to it some fibres of the Masseter
muscle. Its outer surface is convex and subcutaneous ; its inner is concave,
and affords attachment to the Masseter muscle. The extremity, broad and deeply
serrated, articulates with the malar bone. The zygomatic process is connected to
the temporal bone by three divisions, called its *roots*—an anterior, middle, and
posterior. The anterior, which is short but broad and strong, is directed inwards,
to terminate in a rounded eminence, the *eminentia articularis*. This eminence
forms the front boundary of the glenoid fossa, and in the recent state is covered

FIG. 31.—Left temporal bone.   Outer surface.

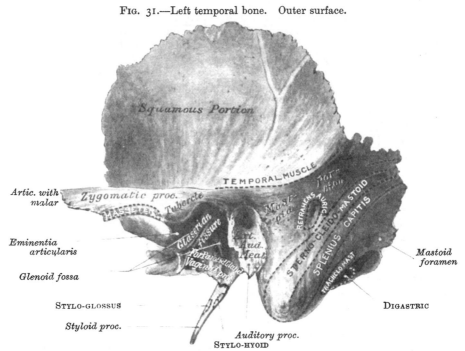

with cartilage. The middle root is known as the *post-glenoid process*, and is very
prominent in young bones. It separates the mandibular portion of the glenoid
fossa from the external auditory meatus, and terminates at the commencement of
a well-marked fissure, the *Glaserian fissure*. The posterior root, which is strongly
marked, runs from the upper border of the zygoma, backwards above the external
auditory meatus. It is termed the *supra-mastoid crest*, and forms the posterior
part of the lower temporal ridge. At the junction of the anterior root with the
zygoma is a projection, called the *tubercle*, for the attachment of the external
lateral ligament of the lower jaw ; and between the anterior and middle roots is
an oval depression, forming part of the glenoid fossa (γλήνη, *a socket*), for the
reception of the condyle of the lower jaw. The glenoid fossa is bounded, in front,
by the eminentia articularis ; behind by the *tympanic plate*, which separates it
from the external auditory meatus ; it is divided into two parts by a narrow
slit, the *Glaserian fissure*. The anterior or mandibular part, formed by the
squamous portion of the bone, is smooth, covered in the recent state with cartilage, and articulates with the condyle of the lower jaw. This part of the fossa
presents posteriorly a small conical eminence, the *post-glenoid process*, already

referred to.   This process is the representative of a prominent tubercle which, in some of the mammalia, descends behind the condyle of the jaw, and prevents it being displaced backwards during mastication (Humphry).   The posterior part of the glenoid fossa, which lodges a portion of the parotid gland, is formed chiefly by the *tympanic plate*, which constitutes the anterior wall of the tympanum and external auditory meatus.   This plate of bone terminates above in the Glaserian fissure, and below forms a sharp edge, the *vaginal* process, which gives origin to some of the fibres of the Tensor palati muscle.   The Glaserian fissure, which leads into the tympanum, lodges the processus gracilis of the malleus, and transmits the tympanic branch of the internal maxillary artery.   The chorda tympani nerve passes through a separate canal (*canal of Huguier*), parallel to the Glaserian fissure, on the outer side of the Eustachian tube, in the retiring angle between the squamous and petrous portions of the temporal bone.*

The **internal surface** of the squamous portion (fig. 32) is concave, presents

Fig. 32.—Left temporal bone.   Inner surface.

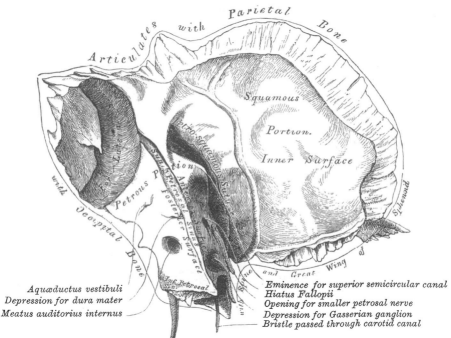

Aquæductus vestibuli
Depression for dura mater
Meatus auditorius internus

Eminence for superior semicircular canal
Hiatus Fallopii
Opening for smaller petrosal nerve
Depression for Gasserian ganglion
Bristle passed through carotid canal

numerous depressions for the convolutions of the cerebrum, and two well-marked grooves for the branches of the middle meningeal artery.

Borders.—The superior border is thin, bevelled at the expense of the internal surface, so as to overlap the lower border of the parietal bone, forming the squamous suture.   The anterior inferior border is thick, serrated, and bevelled, alternately at the expense of the inner and outer surfaces, for articulation with the great wing of the sphenoid.

The **Mastoid Portion** (μαστός, *a nipple* or *teat*) is situated at the posterior part of the bone ; its outer surface is rough, and gives attachment to the Occipito-frontalis and Retrahens auriculam muscles.   It is perforated by numerous foramina ; one of these, of large size, situated near the posterior border of the bone, is termed the *mastoid foramen* ; it transmits a vein to the lateral sinus and a small artery from the occipital, to supply the dura mater.   The position and size of this foramen are very variable.   It is not always present ; sometimes it is situated in

* This small fissure must not be confounded with the large canal which lies above the Eustachian tube and transmits the Tensor tympani muscle.

the occipital bone, or in the suture between the temporal and the occipital. The mastoid portion is continued below into a conical projection, the *mastoid process*, the size and form of which vary somewhat. This process serves for the attachment of the Sterno-mastoid, Splenius capitis, and Trachelo-mastoid muscles. On the inner side of the mastoid process is a deep groove, the *digastric fossa*, for the attachment of the Digastric muscle; and, running parallel with it, but more internal, the *occipital groove*, which lodges the occipital artery. The internal surface of the mastoid portion presents a deep, curved groove, the *fossa sigmoidea*, which lodges part of the lateral sinus; and into it may be seen opening the mastoid foramen. The groove for the lateral sinus is separated from the innermost of the mastoid air-cells by only a thin lamina of bone, and even this may be partly deficient. A section of the mastoid process shows it to be hollowed out into a number of cellular spaces, communicating with each other, called the *mastoid cells*, which exhibit the greatest possible variety as to their size and number. At the upper and front part of the bone these cells are large and irregular

FIG. 33.—Section through the petrous and mastoid portions of the temporal bone, showing the communication of the cavity of the tympanum with the mastoid antrum.

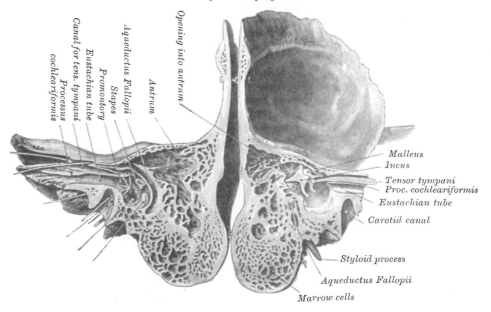

and contain air, but towards the lower part of the bone they diminish in size, while those at the apex of the mastoid process are quite small and usually contain marrow. Occasionally they are entirely absent, and the mastoid is solid throughout. In addition to these a large irregular cavity (fig. 33) is situated at the upper and front part of the bone. It is called the *mastoid antrum*, and must be distinguished from the mastoid cells, though it communicates with them. Like the mastoid cells it is filled with air and lined by a prolongation of the mucous membrane of the tympanic cavity, with which it communicates. The mastoid antrum is bounded above by a thin plate of bone, which separates it from the middle fossa of the base of the skull on the anterior surface of the petrous portion of the temporal bone; below by the mastoid process; externally by the squamous portion of the bone just below the supra-mastoid crest, and internally by the external semicircular canal of the internal ear which projects into its cavity. The opening by which it communicates with the tympanum is situated at the superior internal angle of the posterior wall of that cavity, and opens into that portion of the tympanic cavity which is known as the *attic* or *epitympanic recess*; that is to say, that portion of the tympanum which is above the level of the membrana tympani.

The mastoid cells, like the other sinuses of the cranium, are not developed until towards puberty; hence the prominence of the mastoid process in the adult: the mastoid antrum, on the other hand, is of large size at birth.

In consequence of the communication which exists between the tympanum and mastoid antrum, inflammation of the lining membrane of the former cavity may easily travel backwards to that of the antrum and mastoid air-cells, leading to caries and necrosis of their walls and the risk of transference of the inflammation to the lateral sinus or encephalon.

Borders.—The superior border of the mastoid portion is broad and rough, its serrated edge sloping outwards, for articulation with the posterior inferior angle of the parietal bone. The posterior border, also uneven and serrated, articulates with the inferior border of the occipital bone between its lateral angle and jugular process.

The **Petrous Portion** (πέτρος, *a stone*), so named from its extreme density and hardness, is a pyramidal process of bone, wedged in at the base of the skull between the sphenoid and occipital bones. Its direction from without is inwards, forwards, and a little downwards. It presents for examination a base, an apex, three surfaces, and three borders; and contains, in its interior, the essential parts of the organ of hearing. The *base* is applied against the internal surface of the squamous and mastoid portions, its upper half being concealed; but its lower half is exposed by the divergence of these two portions of the bone, which brings into view the oval expanded orifice of a canal leading into the tympanum, the *meatus auditorius externus*. The curved tympanic plate forms the anterior wall, the floor, and a part of the posterior wall of this meatus, while the squamous-mastoid completes it above and behind. The entrance to the meatus is bounded throughout the greater part of its circumference by the *auditory process*, which is the name applied to the free rough margin of the tympanic plate, and which gives attachment to the cartilaginous portion of the meatus. Superiorly the entrance to the meatus is limited by the posterior root of the zygoma.

The **apex** of the petrous portion, rough and uneven, is received into the angular interval between the posterior border of the greater wing of the sphenoid and the basilar process of the occipital; it presents the anterior or internal orifice of the carotid canal, and forms the posterior and external boundary of the foramen lacerum medium.

The **anterior surface** of the petrous portion (fig. 32) forms the posterior part of the middle fossa of the base of the skull. This surface is continuous with the squamous portion, to which it is united by a suture, the *petro-squamous suture*, the remains of which are distinct even at a late period of life; it presents six points for examination: 1. an eminence near the centre, which indicates the situation of the superior semicircular canal: 2. in front and a little to the outer side of this eminence a depression, indicating the position of the tympanum; here the layer of bone which separates the tympanum from the cranial cavity is extremely thin, and is known as the *tegmen tympani*: 3. a shallow groove, sometimes double, leading outwards and backwards to an oblique opening, the *hiatus Fallopii*, for the passage of the greater petrosal nerve and the petrosal branch of the middle meningeal artery: 4. a smaller opening, occasionally seen external to the latter, for the passage of the smaller petrosal nerve: 5. near the apex of the bone, the termination of the carotid canal, the wall of which in this situation is deficient in front: 6. above this canal a shallow depression for the reception of the Gasserian ganglion.

The **posterior surface** forms the front part of the posterior fossa of the base of the skull, and is continuous with the inner surface of the mastoid portion of the bone. It presents three points for examination: 1. about its centre, a large orifice, the *meatus auditorius internus*, the size of which varies considerably; its margins are smooth and rounded; and it leads into a short canal, about four

lines in length, which runs directly outwards. This canal is closed internally by
a vertical plate, the *lamina cribrosa*, which is divided by a horizontal crest, the
crista falciformis, into two unequal portions (fig. 34). Each portion is subdivided
by a little vertical ridge into two parts, named respectively anterior and poste-
rior. The *lower* portion presents three sets of foramina ; one group, just below the posterior part of the crest, the *area cribrosa media*, consisting of a number of small openings for the nerves to the saccule ; below and posterior to this, the *foramen singulare*, or opening for the nerve to the posterior semi-circular canal ; in front and below the first, the *tractus spiralis foraminosus*, consisting of a number of small spirally arranged open-ings, which terminate in the *canalis cen-tralis cochleæ* and transmit the nerves to the cochlea. The *upper* portion, that above the crista, presents behind a series of small openings, the *area cribrosa superior*, for the passage of filaments to the utricle and superior and external semicircular canals, and, in front, the *area facialis*, with one large opening, the commence-ment of the aquæductus Fallopii, for the passage of the facial nerve : 2. behind the meatus auditorius, a small slit almost hidden by a thin plate of bone, leading to a canal, the *aquæductus*

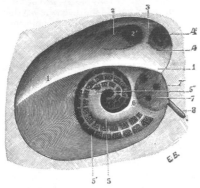

FIG. 34. — Diagrammatic view of the
fundus of the internal auditory meatus
(Testut).

1. Crista falciformis. 2. Area facialis, with (2′) Internal opening of the Aqueductus Fallopii. 3. Ridge separating the area facialis from the area cribrosa superior. 4. Area cribrosa superior, with (4′) Openings for nerve filaments. 5. Anterior inferior cribriform area, with (5′) the tractus spiralis foraminosus, and (5″) the canalis centralis of the cochlea. 6. Ridge separating the tractus spiralis foraminosus from the area cribrosa media. 7. Area cribrosa media, with (7′) orifices for nerves to saccule. 8. Foramen singulare.

vestibuli, which transmits the *ductus endolymphaticus* together with a small artery
and vein : 3. in the interval between these two openings, but above them, an
angular depression which lodges a process of the dura mater, and transmits a
small vein into the cancellous tissue of the bone. In the child this depression is
represented by a large fossa, the *floccular fossa*, which extends backwards as a
blind tunnel under the superior semicircular canal.

The **inferior** or **basilar** surface (fig. 35) is rough and irregular, and forms part
of the base of the skull. Passing from the apex to the base, this surface presents
eleven points for examination : 1. a rough surface, quadrilateral in form, which
serves partly for the attachment of the Levator palati and Tensor tympani
muscles : 2. the large, circular aperture of the carotid canal, which ascends at
first vertically, and then, making a bend, runs horizontally forwards and inwards ;
it transmits the internal carotid artery, and the carotid plexus : 3. the *aquæductus
cochleæ*, a small, triangular opening, lying on the inner side of the latter, close to
the posterior border of the petrous portion ; it transmits a vein from the cochlea,
which joins the internal jugular : 4. behind these openings a deep depression, the
jugular fossa, which varies in depth and size in different skulls ; it lodges the
lateral sinus, and, with a similar depression on the margin of the jugular process
of the occipital bone, forms the foramen lacerum posterius, or jugular foramen ;
5. a small foramen for the passage of Jacobson's nerve (the tympanic branch of
the glosso-pharyngeal) ; this foramen is seen in the bony ridge dividing the carotid
canal from the jugular fossa : 6. a small foramen on the wall of the jugular fossa,
for the entrance of the auricular branch of the pneumogastric (Arnold's nerve) ;
7. behind the jugular fossa, a smooth, square-shaped facet, the *jugular surface* ;
it is covered with cartilage in the recent state, and articulates with the jugular
process of the occipital bone : 8. the *vaginal process*, a very broad, sheath-like plate
of bone, which extends backwards from the carotid canal and gives attachment
to part of the Tensor palati muscle ; this plate divides behind into two laminæ,

the outer of which is continuous with the tympanic plate, the inner with the jugular process : between these laminæ is the 9th point for examination, the *styloid process*, a sharp spine, about an inch in length ; it is directed downwards, forwards, and inwards, varies in size and shape, and sometimes consists of several pieces united by cartilage ; it affords attachment to three muscles, the Stylo-pharyngeus, Stylo-hyoideus, and Stylo-glossus ; and two ligaments, the stylo-hyoid and stylo-maxillary : 10. the *stylo-mastoid foramen*, a rather large orifice, placed between the styloid and mastoid processes ; it is the termination of the aquæductus Fallopii, and transmits the facial nerve and stylomastoid artery : 11. the *auricular fissure*, situated between the tympanic plate and mastoid processes, for the exit of the auricular branch of the pneumogastric nerve.

Borders.—The *superior*, the longest, is grooved for the superior petrosal sinus, and has attached to it the tentorium cerebelli ; at its inner extremity is a

FIG. 35.—Petrous portion.   Inferior surface.

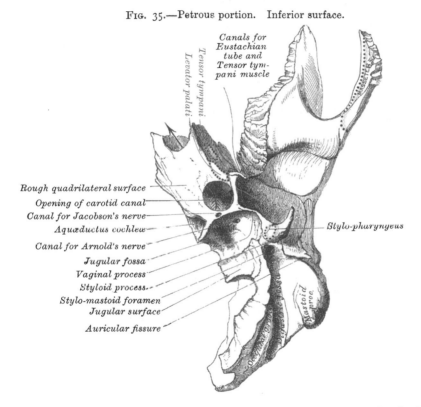

semilunar notch, upon which the fifth nerve lies.   The *posterior* border is intermediate in length between the superior and the anterior.   Its inner half is marked by a groove, which, when completed by its articulation with the occipital, forms the channel for the inferior petrosal sinus.   Its outer half presents a deep excavation—the *jugular fossa*—which, with a similar notch on the occipital, forms the foramen lacerum posterius.   A projecting eminence of bone occasionally stands out from the centre of the notch, and divides the foramen into two parts. The *anterior* border is divided into two parts—an outer joined to the squamous portion by a suture (*petro-squamous*), the remains of which are distinct ; an inner, free, articulating with the spinous process of the sphenoid.   At the angle of junction of the petrous and squamous portions are seen two canals, separated from one another by a thin plate of bone, the *processus cochleariformis* ; they both lead into the tympanum, the upper one transmitting the Tensor tympani muscle, the lower one forming the bony part of the Eustachian tube.

Structure.—The squamous portion is like that of the other cranial bones : the mastoid portion cellular, and the petrous portion dense and hard.

Development (fig. 36).—The temporal bone is developed by *ten* centres, exclusive of those for the internal ear and the ossicula—viz. one for the squamous portion including the zygoma, one for the tympanic plate, six for the petrous and mastoid parts, and two for the styloid process. Just before the close of fœtal life the temporal bone consists of four parts : 1. The *squamo-zygomatic*, ossified in membrane from a single nucleus, which appears at its lower part about the second month. 2. The *tympanic plate*, an imperfect ring in the concavity of which is a groove, the *sulcus tympanicus*, for the attachment of the circumference of the tympanic membrane. This is also ossified from a single centre which appears about the third month. 3. The *petro-mastoid*, developed from six centres, which appear about the fifth or sixth month. Four of these are for the petrous portion, and are placed around the labyrinth, and two for the mastoid (Vrolik). According to Huxley, the centres are more numerous, and are disposed so as to form three portions : (1) including most of the labyrinth with a part of the petrous and mastoid, he has named *prootic*; (2) the rest of the petrous, the *opisthotic*; and (3) the remainder of the mastoid, the *epiotic*. The petro-mastoid is ossified in cartilage. 4. The *styloid process* is also ossified in cartilage from two centres : one for the base, which appears before birth, and is termed the *tympano-hyal*; the other, comprising the rest of the process, is named the *stylo-hyal*, and does not appear until after birth. Shortly before birth the tympanic plate joins with the squamous. The petrous and mastoid join with the squamous during the first year, and the tympano-hyal portion of the styloid process about the same time. The stylo-hyal does not

Fig. 36.—Development of the temporal bone. By ten centres.

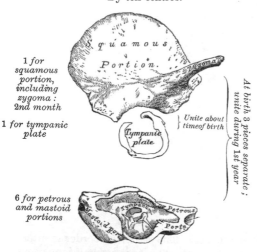

Fig. 36.—Development of the temporal bone. By ten centres.

1 *for squamous portion, including zygoma : 2nd month*

1 *for tympanic plate*

At birth 3 pieces separate ; unite during 1st year

Unite about time of birth

6 *for petrous and mastoid portions*

2 *for styloid process*

join the rest of the bone until after puberty, and in some skulls never becomes united. The chief subsequent changes in this bone are : (1) the tympanic plate extends outwards and backwards so as to form the meatus auditorius externus. The extension of this does not, however, take place at an equal rate all round the circumference of the ring, but occurs most rapidly on its anterior and posterior portions, and these outgrowths meet and blend, and thus, for a time, there exists in the floor of the meatus a foramen, the *foramen of Huschke*: this foramen may persist throughout life. (2) The glenoid cavity is at first extremely shallow, and looks outwards as well as downwards; it becomes deeper and is ultimately directed downwards. Its change in direction is accounted for as follows :

The part of the squamous temporal which supports it lies at first *below* the level of the zygoma. As, however, the base of the skull increases in width, this lower part of the squama is directed horizontally inwards to contribute to the middle fossa of the skull, and its surfaces therefore come to look upwards and downwards. (3) The mastoid portion is at first quite flat, and the stylo-mastoid foramen and rudimentary styloid process lie immediately behind the entrance to the auditory meatus. With the development of the air-cells the outer part of the mastoid portion grows downwards and forwards to form the mastoid process, and the styloid process and stylo-mastoid foramen now come to lie on the under surface. The descent of the foramen is necessarily accompanied by a corresponding lengthening of the aqueduct of Fallopius. (4) The downward and forward growth of the mastoid process also pushes forward the tympanic plate, so that the portion of it which formed the original floor of the meatus and contained the foramen of Huschke is ultimately found in the anterior wall. (5) With the gradual increase in size of the petrous portion, the *floccular fossa* or tunnel under the superior semicircular canal becomes filled up and almost obliterated.

Articulations.—With five bones : occipital, parietal, sphenoid, inferior maxillary, and malar.

Attachment of Muscles.—To fifteen : to the squamous portion, the Temporal ; to the zygoma, the Masseter ; to the mastoid portion, the Occipito-frontalis, Sterno-mastoid, Splenius capitis, Trachelo-mastoid, Digastricus, and Retrahens auriculam ; to the styloid process, the Stylo-pharyngeus, Stylo-hyoid, and Styloglossus ; and to the petrous portion, the Levator palati, Tensor tympani, Tensor palati, and Stapedius.

THE SPHENOID BONE

The **Sphenoid Bone** (σφήν, *a wedge*) is situated at the anterior part of the base of the skull, articulating with all the other cranial bones, and binding them firmly

FIG. 37.—Sphenoid bone.   Superior surface.

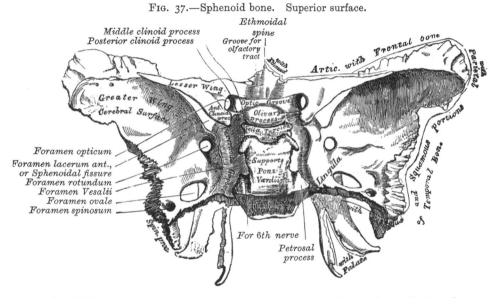

and solidly together.   In its form it somewhat resembles a bat with its wings extended ; and is divided into a central portion or body, two greater and two lesser wings extending outwards on each side of the body, and two processes— the pterygoid processes—which project from it below.

The **body** is of large size, and hollowed out in its interior so as to form a mere shell of bone.   It presents for examination *four* surfaces—a superior, an inferior, an anterior, and a posterior.

The Superior Surface (fig. 37).—In front is seen a prominent spine, the *ethmoidal spine*, for articulation with the cribriform plate of the ethmoid ; behind this a smooth surface presenting, in the median line, a slight longitudinal eminence, with a depression on each side, for lodging the olfactory lobes. This surface is bounded behind by a ridge, which forms the anterior border of a narrow, transverse groove, the *optic groove*, behind which lies the optic commissure ; the groove terminates on either side in the *optic foramen*, for the passage of the optic nerve and ophthalmic artery.   Behind the optic groove is a small eminence, olive-like in shape, the *olivary process* ; and still more posteriorly, a deep depression, the *pituitary fossa*, or *sella turcica*, which lodges the pituitary body. This fossa is perforated by numerous foramina, for the transmission of nutrient vessels into the substance of the bone. It is bounded in front by two small eminences, one on either side, called the *middle clinoid processes* (κλίνη, *a bed*), which are sometimes connected by a spiculum of bone to the anterior clinoid processes, and behind by a square-shaped plate of bone, the *dorsum ephippii* or *dorsum sellæ*, terminating at each superior angle in a tubercle, the *posterior clinoid processes*, the size and form of which vary considerably in different individuals. These processes deepen the pituitary fossa, and serve for the attachment of prolongations from the tentorium cerebelli. The sides of the dorsum ephippii are notched for the passage of the sixth pair of nerves, and below present a sharp process, the *petrosal process*, which is joined to the apex of the petrous portion of the temporal bone, forming the inner boundary of the middle lacerated foramen. Behind this plate, the bone presents a shallow depression, which slopes obliquely backwards, and is continuous with the basilar groove of the occipital bone; it is called the *clivus*, and supports the upper part of the pons Varolii. On either side of the body is a broad groove, curved something like the italic letter *f*; it lodges

FIG. 38.—Sphenoid bone. Anterior surface.*

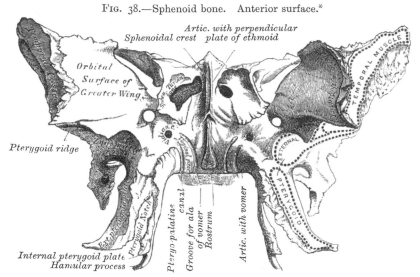

the internal carotid artery and the cavernous sinus, and is called the *carotid* or *cavernous groove*. Along the outer margin of this groove, at its posterior part, is a ridge of bone, in the angle between the body and greater wing, called the **lingula**. The **posterior surface**, quadrilateral in form, is joined to the basilar process of the occipital bone. During childhood these bones are separated by a layer of cartilage; but in after-life (between the eighteenth and twenty-fifth years) this becomes ossified, ossification commencing above and extending downwards; and the two bones then form one piece. The **anterior surface** (fig. 38) presents, in the middle line, a vertical ridge of bone, the *sphenoidal crest*, which articulates with the perpendicular plate of the ethmoid, forming part of the septum of the nose. On either side of it are irregular openings leading into the *sphenoidal air-cells* or *sinuses*. These are two large, irregular cavities hollowed out of the interior of the body of the sphenoid bone, and separated from one another by a more or less complete perpendicular bony septum. Their form and size vary considerably; they are seldom symmetrical, and are often partially subdivided by irregular osseous laminæ. Occasionally, they extend into the basilar process of the occipital nearly as far as the foramen magnum. The septum is seldom quite vertical, being commonly bent to one or the other side.

* In this figure, both the anterior and inferior surfaces of the body of the sphenoid bone are shown, the bone being held with the pterygoid processes almost horizontal.

These sinuses do not exist in children, but they increase in size as age advances. They are partially closed, in front and below, by two thin, curved plates of bone, the *sphenoidal turbinated* bones, leaving a round opening at their upper parts, by which they communicate with the upper and back part of the nose and occasionally with the posterior ethmoidal cells or sinuses. The lateral margins of this surface present a serrated edge, which articulates with the os planum of the ethmoid, completing the posterior ethmoidal cells; the lower margin, also rough and serrated, articulates with the orbital process of the palate bone; and the upper margin with the orbital plate of the frontal bone. The **inferior surface** presents, in the middle line, a triangular spine, the *rostrum*, which is continuous with the sphenoidal crest on the anterior surface, and is received into a deep fissure between the alæ of the vomer. On each side may be seen a projecting lamina of bone, which runs horizontally inwards from near the base of the pterygoid process : these plates, termed the *vaginal processes*, articulate with the edges of the vomer. Close to the root of the pterygoid process is a groove, formed into a complete canal when articulated with the sphenoidal process of the palate bone; it is called the *pterygo-palatine· canal*, and transmits the pterygo-palatine vessels and pharyngeal nerve.

The **Greater Wings** are two strong processes of bone, which arise from the sides of the body, and are curved in a direction upwards, outwards and backwards, being prolonged behind into a sharp-pointed extremity, the *spinous process* of the sphenoid. Each wing presents three surfaces and a circumference. The *superior* or *cerebral* surface (fig. 37) forms part of the middle fossa of the skull; it is deeply concave, and presents depressions for the convolutions of the brain. At its anterior and internal part is seen a circular aperture, the *foramen rotundum*, for the transmission of the second division of the fifth nerve. Behind and external to this is a large, oval foramen, the *foramen ovale*, for the transmission of the third division of the fifth nerve, the small meningeal artery, and sometimes the small petrosal nerve.* At the inner side of the foramen ovale, a small aperture may occasionally be seen opposite the root of the pterygoid process; it is the *foramen Vesalii*, transmitting a small vein. Lastly, in the posterior angle, near to the spine of the sphenoid, is a short canal, sometimes double, the *foramen spinosum*; it transmits the middle meningeal artery. The *external* surface (fig. 38) is convex, and divided by a transverse ridge, the *pterygoid ridge* or *infra-temporal crest*, into two portions. The superior or larger, convex from above downwards, concave from before backwards, enters into the formation of the temporal fossa, and gives attachment to part of the Temporal muscle. The inferior portion, smaller in size and concave, enters into the formation of the zygomatic fossa, and affords attachment to the External pterygoid muscle. It presents, at its posterior part, a sharp-pointed eminence of bone, the *spinous process*, which is frequently grooved on its inner aspect for the corda tympani nerve, and to which are connected the internal lateral ligament of the lower jaw and the Tensor palati muscle. The pterygoid ridge, dividing the temporal and zygomatic portions, gives attachment to part of the External pterygoid muscle. At its inner and anterior extremity is a triangular spine of bone, which serves to increase the extent of origin of this muscle. The *anterior* or *orbital* surface, smooth, and quadrilateral in form, assists in forming the outer wall of the orbit. It is bounded above by a serrated edge, for articulation with the frontal bone; below, by a rounded border, which enters into the formation of the spheno-maxillary fissure. Internally, it presents a sharp border, which forms the lower boundary of the sphenoidal fissure and has projecting from about its centre a little tubercle of bone, which gives origin to one head of the External rectus muscle of the eye; and at its upper part is a notch for the transmission of a recurrent branch of the lachrymal artery; externally, it presents a serrated margin for articulation with the malar bone. One or two small foramina may occasionally be seen for the passage of branches of

---

* The small petrosal nerve sometimes passes through a special foramen between the foramen ovale and foramen spinosum.

the deep temporal arteries : they are called the *external orbital foramina.* *Circumference of the great wing* (fig. 37) : commencing from behind ; that portion of the circumference which extends from the body of the sphenoid to the spine is serrated and articulates by its outer half with the petrous portion of the temporal bone ; while the inner half forms the anterior boundary of the foramen lacerum medium, and presents the posterior aperture of the Vidian canal for the passage of the Vidian nerve and artery. In front of the spine the circumference of the great wing presents a serrated edge, bevelled at the expense of the inner table below, and of the outer table above, which articulates with the squamous portion of the temporal bone. At the tip of the great wing a triangular portion is seen bevelled at the expense of the internal surface, for articulation with the anterior inferior angle of the parietal bone. Internal to this is a triangular, serrated surface, for articulation with the frontal bone : this surface is continuous internally with the sharp inner edge of the orbital plate, which assists in the formation of the sphenoidal fissure, and externally with the serrated margin for articulation with the malar bone.

The **Lesser Wings** (*processes of Ingrassias*) are two thin, triangular plates of bone, which arise from the upper and lateral parts of the body of the sphenoid, and, projecting transversely outwards, terminate in a sharp point (fig. 37). The superior surface of each is smooth, flat, broader internally than externally, and supports part of the frontal lobe of the brain. The inferior surface forms the back part of the roof of the orbit, and the upper boundary of the sphenoidal fissure or foramen lacerum anterius. This fissure is of a triangular form, and leads from the cavity of the cranium into the orbit ; it is bounded internally by the body of the sphenoid ; above, by the lesser wing ; below, by the internal margin of the orbital surface of the great wing ; and is converted into a foramen by the articulation of this bone with the frontal. It transmits the third, the fourth, the three branches of the ophthalmic division of the fifth and the sixth nerve, some filaments from the cavernous plexus of the sympathetic, the orbital branch of the middle meningeal artery, a recurrent branch from the lachrymal artery to the dura mater, and the ophthalmic vein. The anterior border of the lesser wing is serrated for articulation with the frontal bone ; the posterior, smooth and rounded, is received into the fissure of Sylvius of the brain. The inner extremity of this border forms the *anterior clinoid process.*

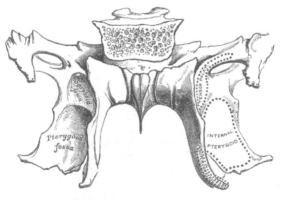

FIG. 39.—Sphenoid bone. Posterior surface.

The lesser wing is connected to the side of the body by two roots, the upper thin and flat, the lower thicker, obliquely directed, and presenting on its outer side, near its junction with the body, a small tubercle, for the attachment of the common tendon of origin of three of the muscles of the eye. Between the two roots is the *optic foramen,* for the transmission of the optic nerve and ophthalmic artery.

The **Pterygoid Processes** (πτέρυξ, *a wing* ; εἶδος, *likeness*), one on each side, descend perpendicularly from the point where the body and greater wing unite (fig. 39). Each process consists of an external and an internal plate, which are joined together by their anterior borders above, but are separated below, leaving an angular cleft, the *pterygoid notch,* in which the pterygoid process or tuberosity of the palate bone is received. The two plates diverge from each other from their line of connection in front, so as to form a V-shaped fossa, the *pterygoid* *fossa.* The *external pterygoid plate* is broad and thin, turned a little outwards, and, by its outer surface, forms part of the inner wall of the zygomatic fossa, giving attachment to the External pterygoid ; its inner surface forms part of the pterygoid fossa, and gives attachment to the Internal pterygoid. The *internal pterygoid plate* is much narrower and longer, curving outwards, at its extremity, into a hook-like process of bone, the *hamular process*, around which the tendon of the Tensor palati muscle turns. The outer surface of this plate forms part of the pterygoid fossa, the inner surface forming the outer boundary of the posterior aperture of the nares. On the posterior surface of the base of the process, above the pterygoid fossa, is a small, oval, shallow depression, the *scaphoid fossa*, from which arises the Tensor palati, and above which is seen the posterior orifice of the Vidian canal. Below and to the inner side of the Vidian canal, on the posterior surface of the base of the internal plate, is a little prominence, which is known by the name of the *pterygoid tubercle*. The Superior constrictor of the pharynx is attached to the posterior edge of the internal plate. The anterior surface of the pterygoid process is very broad at its base, and forms the posterior wall of the spheno-maxillary fossa. It supports Meckel's ganglion. It presents, above, the anterior orifice of the Vidian canal ; and below, a rough margin, which articulates with the perpendicular plate of the palate bone.

The **Sphenoidal Spongy Bones** are two thin, curved plates of bones, which exist as separate pieces until puberty, and occasionally are not joined to the sphenoid in the adult. They are situated at the anterior and inferior part of the body of the sphenoid, an aperture of variable size being left in their anterior wall, through which the sphenoidal sinuses open into the nasal fossæ. They are irregular in form, and taper to a point behind, being broader and thinner in front. Their upper surface, which looks towards the cavity of the sinus, is concave ; their under surface convex. Each bone articulates in front with the ethmoid, externally with the palate ; its pointed posterior extremity is placed above the vomer, and is received between the root of the pterygoid process on the outer side and the rostrum of the sphenoid on the inner.*

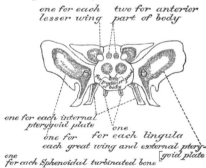

Fig. 40.—Plan of the development of sphenoid. By fourteen centres.

Development.—Up to about the eighth month of fœtal life the sphenoid bone consists of two distinct parts : a posterior or *post-sphenoid* part, which comprises the pituitary fossa, the greater wings, and the pterygoid processes ; and an anterior or *pre-sphenoid* part, to which the anterior part of the body and lesser wings belong. It is developed by fourteen centres : eight for the posterior sphenoid division, and six for the anterior sphenoid. The eight centres for the posterior sphenoid are— one for each greater wing and external pterygoid plate, one for each internal pterygoid plate, two for the posterior part of the body, and one on each side for the lingula. The six for the anterior sphenoid are one for each lesser wing, two for the anterior part of the body, and one for each sphenoidal turbinated bone.

Post-sphenoid Division.—The first nuclei to appear are those for the greater wings (*ali-sphenoids*). They make their appearance between the foramen rotundum and foramen ovale about the eighth week, and from them the external pterygoid plates are also formed. Soon after, the nuclei for the posterior part of the body appear, one on either side of the sella turcica, and become blended together about the middle of fœtal life. About the fourth month the remaining four centres appear, those for the internal pterygoid plates being ossified in membrane and

* A small portion of sphenoidal turbinated bone sometimes enters into the formation of the inner wall of the orbit, between the os planum of the ethmoid in front, the orbital plate of the palate below, and the frontal above. Cleland, *Roy. Soc. Trans.* 1862.

becoming joined to the external pterygoid plate about the sixth month. The centres for the lingulæ speedily become joined to the rest of the bone.

Pre-sphenoid Division.—The first nuclei to appear are those for the lesser wings (*orbito-sphenoids*). They make their appearance about the ninth week, at the outer borders of the optic foramina. A second pair of nuclei appear on the inner side of the foramina shortly after, and becoming united, form the front part of the body of the bone. The remaining two centres for the sphenoidal turbinated bones make their appearance about the fifth month. At birth they consist of small triangular laminæ, and it is not till the third year that they become hollowed out and cone-shaped. About the fourth year they become fused with the lateral masses of the ethmoid, i.e. several years before they unite with the sphenoid, and hence, from an embryological point of view, may be regarded as belonging to the ethmoid.

The pre-sphenoid is united to the body of the post-sphenoid about the eighth month, so that at birth the bone consists of three pieces—viz. the body in the centre, and on each side the great wings with the pterygoid processes. The lesser wings become joined to the body at about the time of birth. At the first year after birth the greater wings and body are united. From the tenth to the twelfth year the spongy bones are partially united to the sphenoid, their junction being complete by the twentieth year. Lastly, the sphenoid joins the occipital from the eighteenth to the twenty-fifth year.

Articulations.—The sphenoid articulates with *all* the bones of the cranium, and five of the face—the two malar, two palate, and vomer: the exact extent of articulation with each bone is shown in the accompanying figures.*

Attachment of Muscles.—To eleven pairs: the Temporal, External pterygoid, Internal pterygoid, Superior constrictor, Tensor palati, Levator palpebræ, and Superior oblique, Superior rectus, Internal rectus, Inferior rectus, External rectus of the eye.

The Ethmoid Bone

The **Ethmoid** (ἠθμός, *a sieve*) is an exceedingly light, spongy bone, of a cubical form, situated at the anterior part of the base of the cranium, between the two orbits, at the root of the nose, and contributing to form each of these cavities. It consists of three parts: a horizontal plate, which forms part of the base of the cranium; a perpendicular plate, which forms part of the septum nasi; and two lateral masses of cells.

Fig. 41.—Ethmoid bone.
Outer surface of right lateral mass (enlarged).

The **Horizontal** or **Cribriform Plate** (fig. 41) forms part of the anterior fossa of the base of the skull, and is received into the ethmoid notch of the frontal bone between the two orbital plates. Projecting upwards from the middle line of this plate is a thick, smooth, triangular process of bone, the *crista galli*, so called from its resemblance to a cock's comb. Its base joins the cribriform plate. Its posterior border, long, thin, and slightly curved, serves for the attachment of the falx cerebri. Its anterior border, short and thick, articulates with the frontal bone, and presents two small projecting alæ, which are received into corresponding depressions in the frontal,

* It also sometimes articulates with the tuberosity of the maxilla (see p. 52).

completing the foramen cæcum behind. Its sides are smooth and sometimes bulging; in which case it is found to enclose a small sinus.* On each side of the crista galli, the cribriform plate is narrow, and deeply grooved, to support the bulb of the olfactory tract, and perforated by foramina for the passage of the olfactory nerves. These foramina are arranged in three rows: the innermost, which are the largest and least numerous, are lost in grooves on the upper part of the septum; the foramina of the outer row are continued on to the surface of the upper spongy bone. The foramina of the middle row are the smallest; they perforate the bone, and transmit nerves to the roof of the nose. At the front part of the cribriform plate, on each side of the crista galli, is a small fissure, which transmits the nasal branch of the ophthalmic nerve; and at its posterior part a triangular notch, which receives the ethmoidal spine of the sphenoid.

The **Perpendicular Plate** (fig. 42) is a thin, flattened lamella of bone, which descends from the under surface of the cribriform plate, and assists in forming the septum of the nose. It is much thinner in the middle than at the circumference, and is generally deflected a little to one side. Its anterior border articulates with the nasal spine of the frontal bone and crest of the nasal bones. Its posterior border, divided into two parts, articulates by its upper half with the sphenoidal crest of the sphenoid, by its lower half with the vomer. The inferior border serves for the attachment of the triangular cartilage of the nose.

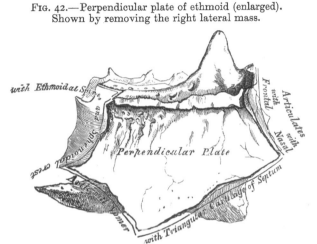

FIG. 42.—Perpendicular plate of ethmoid (enlarged). Shown by removing the right lateral mass.

On each side of the perpendicular plate numerous grooves and canals are seen, leading from foramina on the cribriform plate; they lodge filaments of the olfactory nerves.

The **Lateral Masses** of the ethmoid consist of a number of thin-walled cellular cavities, the *ethmoidal cells*, interposed between two vertical plates of bone, the outer one of which forms part of the orbit, and the inner one part of the nasal fossa of the corresponding side. In the disarticulated bone many of these cells appear to be broken; but when the bones are articulated, they are closed in at every part, except where they open into the nasal fossæ. The upper surface of each lateral mass presents a number of apparently half-broken cellular spaces; these are closed in, when articulated, by the edges of the ethmoidal notch of the frontal bone. Crossing this surface are two grooves on each side, converted into canals by articulation with the frontal; they are the *anterior* and *posterior ethmoidal* canals, and open on the inner wall of the orbit. The posterior surface also presents large, irregular cellular cavities, which are closed in by articulation with the sphenoidal turbinated bones and orbital process of the palate. The cells at the anterior surface are completed by the lachrymal bone and nasal process of the superior maxillary, and those below by the superior maxillary. The outer surface of each lateral mass is formed of a thin, smooth, oblong plate of bone, called the *os planum*; it forms part of the inner wall of the orbit, and articulates, above,

* Sir George Humphry states that the crista galli is commonly inclined to one side, usually the opposite to that towards which the lower part of the perpendicular plate is bent.—*The Human Skeleton*, 1858, p. 277.

with the orbital plate of the frontal ; below, with the superior maxillary ; in front, with the lachrymal ; and behind, with the sphenoid and orbital process of the palate.

From the inferior part of each lateral mass, immediately behind the os planum, there projects downwards and backwards an irregular lamina of bone, called the *unciform process*, from its hook-like form ; it serves to close in the upper part of the orifice of the antrum, and articulates with the ethmoidal process of the inferior turbinated bone. It is often broken in disarticulating the bones.

The inner surface of each lateral mass forms part of the outer wall of the nasal fossa of the corresponding side. It is formed of a thin lamella of bone, which descends from the under surface of the cribriform plate, and terminates below in a free, convoluted margin, the *middle turbinated* bone. The whole of this surface is rough, and marked above by numerous grooves, which run nearly vertically downwards from the cribriform plate ; they lodge branches of the olfactory nerve, which are distributed on the mucous membrane covering the superior turbinated bone. The back part of this surface is subdivided by a narrow oblique fissure, the *superior meatus* of the nose, bounded above by a thin, curved plate of bone, the *superior turbinated* bone. By means of an orifice at the upper part of this fissure, the posterior ethmoidal cells open into the nose. Below, and in front of the superior meatus, is seen the convex surface of the middle turbinated bone. It extends along the whole length of the inner surface of each lateral mass ; its lower margin is free and thick, and its concavity, directed outwards, assists in forming the middle meatus of the nose. It is by a large orifice at the upper and front part of the middle

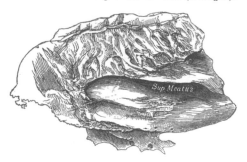

FIG. 43.—Ethmoid bone.
Inner surface of right lateral mass (enlarged).

meatus that the anterior ethmoidal cells, and through them the frontal sinuses, communicate with the nose, by means of a funnel-shaped canal, the *infundibulum*. The cellular cavities of each lateral mass, thus walled in by the os planum on the outer side, and by the other bones already mentioned, are divided by transverse bony partitions into three sets, which do not communicate with each other ; they are termed the *anterior, middle*, and *posterior ethmoidal cells*, or *sinuses*. The anterior cells communicate with the frontal sinuses above, and the middle meatus below, by means of a long, flexuous canal, the *infundibulum* ; the middle also open into the middle meatus ; the posterior open into the superior meatus, and communicate (occasionally) with the sphenoidal sinuses.

Development.—By *three* centres : one for the perpendicular lamella, and one for each lateral mass.

The lateral masses are first developed, ossific granules making their appearance in the os planum between the fourth and fifth months of fœtal life, and extending into the spongy bones. At birth, the bone consists of the two lateral masses, which are small and ill developed. During the first year after birth, the perpendicular plate and crista galli begin to ossify, from a single nucleus, and become joined to the lateral masses about the beginning of the second year. The cribriform plate is ossified partly from the perpendicular plate and partly from the lateral masses. The formation of the ethmoidal cells, which completes the bone, does not commence until about the fourth or fifth year.

Articulations.—With fifteen bones : the sphenoid, two sphenoidal turbinated, the frontal, and eleven of the face—the two nasal, two superior maxillary, two lachrymal, two palate, two inferior turbinated, and the vomer. No muscles are attached to this bone.

DEVELOPMENT OF THE CRANIUM

The early stages of the development of the cranium have already been described in the section on development. We have seen that it is formed from a layer of mesoblast, which is spread over the whole surface of the rudimentary brain, forming a thin, membranous capsule.

Ossification commences in the roof, and is preceded by the deposition of a membranous blastema upon the surface of the cerebral capsule, in which the ossifying process extends: the primitive membranous capsule becoming the internal periosteum, and

FIG. 44.—Skull at birth, showing the anterior and posterior fontanelles.

FIG. 45.—The lateral fontanelles.

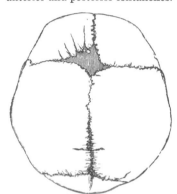
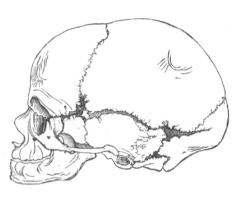

being ultimately blended with the dura mater. The ossification of the bones of the base takes place for the most part in cartilage, and although the bones of the vertex of the skull appear before those at the base, and make considerable progress in their growth, at birth ossification is more advanced in the base, this portion of the skull forming a solid, immovable groundwork.

THE FONTANELLES

Before birth, the bones at the vertex and sides of the cranium are separated from each other by membranous intervals, in which bone is deficient. These intervals are principally found at the four angles of the parietal bones; hence there are six of them. Their formation is due to the wave of ossification being circular and the bones quadrilateral; the ossific matter first meets at the margins of the bones, at the points nearest to their centres of ossification, and vacuities or spaces are left at the angles, which are called *fontanelles*, so named from the pulsations of the brain, which are perceptible at the anterior fontanelle, and were likened to the rising of water in a fountain. The anterior fontanelle is the largest; it is lozenge-shaped, and corresponds to the junction of the sagittal and coronal sutures; the posterior fontanelle, of smaller size, is triangular, and is situated at the junction of the sagittal and lambdoid sutures; the remaining ones are situated at the inferior angles of each parietal bone. The latter are closed soon after birth; the two at the two superior angles remain open longer: the posterior being closed in a few months after birth; the anterior remaining open until the first or second year. These spaces are gradually filled in by an extension of the ossifying process, or by the development of a Wormian bone. Sometimes the anterior fontanelle remains open beyond two years, and is occasionally persistent throughout life.

SUPERNUMERARY OR WORMIAN * BONES

In addition to the constant centres of ossification of the cranium, additional ones are occasionally found in the course of the sutures. These form irregular, isolated bones, interposed between the cranial bones, and have been termed *Wormian bones* or *ossa triquetra*. They are most frequently found in the course of the lambdoid suture, but occasionally also occupy the situation of the fontanelles, especially the posterior and, more rarely, the anterior. Frequently one is found between the anterior inferior angle of the parietal bone and the greater wing of the sphenoid, the *pterion ossicle* (fig. 45). They

---

* Wormius, a physician in Copenhagen, is said to have given the first detailed description of these bones.

have a great tendency to be symmetrical on the two sides of the skull, and they vary much in size, being in some cases not larger than a pin's head, and confined to the outer table; in other cases so large that one pair of these bones may form the whole of the occipital bone above the superior curved lines, as described by Beclard and Ward. Their number is generally limited to two or three; but more than a hundred have been found in the skull of an adult hydrocephalic skeleton. In their development, structure, and mode of articulation, they resemble the other cranial bones.

CONGENITAL FISSURES AND GAPS

An arrest in the ossifying process may give rise to deficiencies, or gaps; or to fissures which are of importance from a medico-legal point of view, as they are liable to be mistaken for fractures. The fissures generally extend from the margins towards the centre of the bone, but the gaps may be found in the middle as well as at the edges. In course of time they may become covered with a thin lamina of bone.

BONES OF THE FACE

The Facial Bones are fourteen in number—viz. the

Two Nasal.	Two Palate.
Two Superior Maxillary.	Two Inferior Turbinated.
Two Lachrymal.	Vomer.
Two Malar.	Inferior Maxillary.

'Of these, the upper and lower jaws are the fundamental bones for mastication, and the others are accessories; for the chief function of the facial bones is to provide an apparatus for mastication, while subsidiary functions are to provide for the sense-organs (eye, nose, tongue) and a vestibule to the respiratory and vocal organs. Hence the variations in the shape of the face in man and the lower animals depend chiefly on the question of the character of their food and their mode of obtaining it.' *

NASAL BONES

The **Nasal** (nasus, *the nose*) are two small oblong bones, varying in size and form in different individuals; they are placed side by side at the middle and

FIG. 46.—Nasal bone and Lachrymal bone *in situ*.

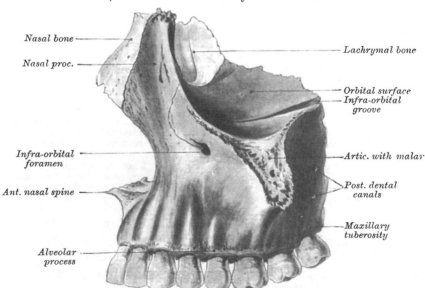

upper part of the face, forming, by their junction, 'the bridge' of the nose (fig. 46). Each bone presents for examination two surfaces and four borders.

* W. W. Keen. American edition, p. 185.

The *outer* surface (fig. 47) is concave from above downwards, convex from side to side ; it is covered by the Pyramidalis and Compressor nasi muscles, and gives attachment at its upper part to a few fibres of the Occipito-frontalis muscle (Theile). It is marked by numerous small arterial furrows, and perforated about its centre by a foramen, sometimes double, for the transmission of a small vein. The *inner* surface (fig. 48) is concave from side to side, convex from above downwards ; in which direction it is traversed by a longitudinal groove (sometimes a canal), for the passage of a branch of the nasal nerve. The *superior* border is narrow, thick, and serrated for articulation with the nasal notch of the frontal bone. The *inferior* border is broad, thin, sharp, inclined obliquely downwards, outwards, and backwards, and serves for the attachment of the lateral cartilage of the nose. It presents, about its middle, a notch, through which passes the branch of the nasal

FIG. 47.—Right nasal bone. Outer surface. FIG. 48.—Left nasal bone. Inner surface.

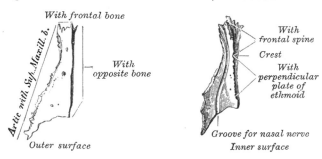

nerve above referred to ; and is prolonged at its inner extremity into a sharp spine, which, when articulated with the opposite bone, forms the *nasal angle*. The *external* border is serrated, bevelled at the expense of the internal surface above, and of the external below, to articulate with the nasal process of the superior maxillary. The *internal* border, thicker above than below, articulates with its fellow of the opposite side, and is prolonged behind into a vertical crest, which forms part of the septum of the nose :. this crest articulates, from above downwards, with the nasal spine of the frontal, the perpendicular plate of the ethmoid, and the triangular septal cartilage of the nose.

Development.—By *one* centre for each bone, which appears about the eighth week.

Articulations.—With four bones : two of the cranium, the frontal and ethmoid, and two of the face, the opposite nasal and the superior maxillary.

Attachment of Muscles.—A few fibres of the Occipito-frontalis muscle.

SUPERIOR MAXILLARY BONES OR MAXILLÆ

The **Superior Maxillary Bones** (maxilla, *the jaw-bone*) are the most important bones of the face from a surgical point of view, on account of the number of diseases to which some of their parts are liable. Their careful examination becomes, therefore, a matter of considerable interest. They are the largest bones of the face, excepting the lower jaw, and form, by their union, the whole of the upper jaw. Each bone assists in the formation of three cavities, the roof of the mouth, the floor and outer wall of the nasal fossæ, and the floor of the orbit ; and also enters into the formation of two fossæ, the zygomatic and spheno-maxillary, and two fissures, the spheno-maxillary and pterygo-maxillary.

The bone presents for examination a body and four processes—malar, nasal, alveolar, and palate.

The **body** is somewhat cuboid, and is hollowed out in its interior to form a large cavity, the *antrum of Highmore*. Its surfaces are four—an external or facial, a posterior or zygomatic, a superior or orbital, and an internal or nasal.

The **external** or **facial surface** (fig. 49) is directed forwards and outwards.   It presents at its lower part a series of eminences corresponding to the position of the teeth.   Just above those for the incisor teeth is a depression, the *incisive* or *myrtiform fossa*, which gives origin to the Depressor alæ nasi ; and, below it, to the alveolar border is attached a slip of the Orbicularis oris.   Above and a little external to it, the Compressor nasi arises.   More external is another depression, the *canine fossa*, larger and deeper than the incisive fossa, from which it is separated by a vertical ridge, the *canine eminence*, corresponding to the socket of the canine tooth.   The canine fossa gives origin to the Levator anguli oris. Above the canine fossa is the *infra-orbital foramen*, the termination of the infra-orbital canal ; it transmits the infra-orbital vessels and nerve.   Sometimes the infra-orbital canal opens by two, very rarely by three, orifices on the face. Above the infra-orbital foramen is the margin of the orbit, which affords partial attachment to the Levator labii superioris proprius.   To the sharp margin of bone

FIG. 49.—Left superior maxillary bone.   Outer surface.

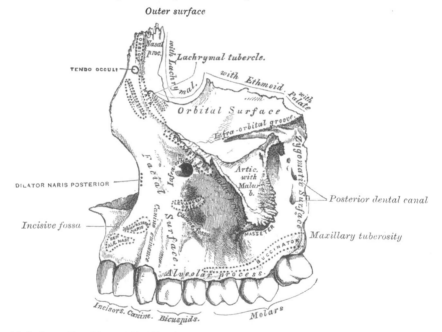

which bounds this surface in front and separates it from the internal surface is attached the Dilator naris posterior.

The **posterior** or **zygomatic surface** is convex, directed backwards and outwards, and forms part of the zygomatic fossa.   It is separated from the facial surface by a strong ridge of bone, which extends upwards from the socket of the second molar tooth.   It presents about its centre several apertures leading to canals in the substance of the bone ; they are termed the *posterior dental canals*, and transmit the posterior dental vessels and nerves.   At the lower part of this surface is a rounded eminence, the *maxillary tuberosity*, especially prominent after the growth of the wisdom-tooth, rough on its inner side for articulation with the tuberosity of the palate-bone and sometimes with the external pterygoid plate.   It gives attachment to a few fibres of origin of the Internal pterygoid muscle.   Immediately above this is a smooth surface, which forms the anterior boundary of the spheno-maxillary fossa ; it presents a groove, which, running obliquely downwards, is converted into a canal by articulation with the palate-bone, forming the *posterior palatine* canal.

The **superior** or **orbital surface** is thin, smooth, triangular, and forms part of the floor of the orbit. It is bounded internally by an irregular margin which in front presents a notch, the *lachrymal notch*, which receives the lachrymal bone ; in the middle it articulates with the os planum of the ethmoid, and behind with the orbital process of the palate-bone ; it is bounded externally by a smooth, rounded edge which enters into the formation of the spheno-maxillary fissure, and which sometimes articulates at its anterior extremity with the orbital plate of the sphenoid ; and it is bounded in front by part of the circumference of the orbit, which is continuous, on the inner side with the nasal, on the outer side with the malar process of the bone. Along the middle line of the orbital surface is a deep groove, the *infra-orbital*, for the passage of the infra-orbital vessels and nerve. The groove commences at the middle of the outer border of this surface, and, passing forwards, terminates in a canal, which subdivides into two branches. One of the canals, the *infra-orbital*, opens just below the margin of the orbit ; the other, which is smaller, runs downwards in the substance of the anterior wall of the antrum ; it is called the *anterior dental* canal, and transmits the anterior dental vessels and

FIG. 50.—Left superior maxillary bone. Inner surface.

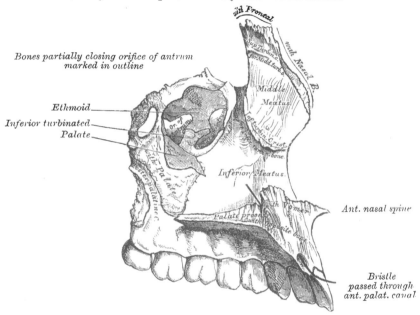

nerve to the front teeth of the upper jaw. From the back part of the infra-orbital canal, a second small canal is sometimes given off, which runs downwards in the outer wall of the antrum, and conveys the middle dental nerve to the bicuspid teeth. Occasionally, this canal is derived from the anterior dental. At the inner and fore part of the orbital surface, just external to the lachrymal groove for the nasal duct, is a depression, which gives origin to the Inferior oblique muscle of the eye.

The **internal surface** (fig. 50) is unequally divided into two parts by a horizontal projection of bone, the *palate process* : the portion above the palate process forms part of the outer wall of the nasal fossæ ; that below it forms part of the cavity of the mouth. The superior division of this surface presents a large, irregular opening leading into the *antrum of Highmore*. At the upper border of this aperture are numerous broken cellular cavities, which, in the articulated skull, are closed in by the ethmoid and lachrymal bones. Below the aperture is a smooth concavity which forms part of the inferior meatus of the nasal fossæ, and behind it is a rough surface which articulates with the perpendicular plate of the palate bone, traversed by a groove which, commencing near the middle of the posterior border, runs obliquely downwards and forwards, and forms, when completed by its articulation with the palate-bone, the *posterior palatine canal*. In front of the opening of the antrum is a deep groove, converted into a canal by the lachrymal and inferior turbinated bones. It is called the *lachrymal groove*, and lodges the nasal duct. More anteriorly is a well-marked rough ridge, the *inferior turbinated crest*, for articulation with the inferior turbinated bone. The shallow concavity above this ridge forms part of the middle meatus of the nose; while that below it forms part of the inferior meatus. The portion of this surface below the palate process is concave, rough and uneven, and perforated by numerous small foramina for the passage of nutrient vessels. It enters into the formation of the roof of the mouth.

The **Antrum of Highmore,** or **Maxillary Sinus,** is a large pyramidal cavity, hollowed out of the body of the maxillary bone : its apex, directed outwards, is formed by the malar process ; its base, by the outer wall of the nose. Its walls are everywhere exceedingly thin, and correspond to the orbital, facial, and zygomatic surfaces of the body of the bone. Its inner wall, or base, presents, in the disarticulated bone, a large, irregular aperture, which communicates with the nasal fossæ. The margins of this aperture are thin and ragged, and the aperture itself is much contracted by its articulation with the ethmoid above, the inferior turbinated below, and the palate-bone behind.* In the articulated skull, this cavity communicates with the middle meatus of the nasal fossæ, generally by two small apertures left between the above-mentioned bones. In the recent state, usually only one small opening exists, near the upper part of the cavity, sufficiently large to admit the end of a probe ; the other being closed by the lining membrane of the sinus.

Crossing the cavity of the antrum are often seen several projecting laminæ of bone, similar to those in the sinuses of the cranium ; and on its posterior wall are the *posterior dental canals*, transmitting the posterior dental vessels and nerves to the teeth. Projecting into the floor are several conical processes, corresponding to the roots of the first and second molar teeth ; † in some cases the floor is perforated by the teeth in this situation.

It is from the extreme thinness of the walls of this cavity that we are enabled to explain how a tumour growing from the antrum encroaches upon the adjacent parts, pushing up the floor of the orbit, and displacing the eyeball, projecting inwards into the nose, protruding forwards on to the cheek, and making its way backwards into the zygomatic fossa, and downwards into the mouth.

The **Malar Process** is a rough triangular eminence, situated at the angle of separation of the facial from the zygomatic surface. In front it is concave, forming part of the facial surface ; behind, it is also concave, and forms part of the zygomatic fossa ; above, it is rough and serrated for articulation with the malar bone ; while below, a prominent ridge marks the division between the facial and zygomatic surfaces. A small part of the Masseter muscle arises from this process.

The **Nasal Process** is a strong, triangular plate of bone, which projects upwards, inwards, and backwards, by the side of the nose, forming part of its lateral boundary. Its *external* surface is concave, smooth, perforated by numerous foramina, and gives attachment to the Levator labii superioris alæque nasi, the Orbicularis palpebrarum, and Tendo oculi. Its *internal* surface forms part of the outer wall of the nasal fossæ : at its upper part it presents a rough, uneven surface, which articulates with the ethmoid bone, closing in the anterior ethmoidal cells ; below this is a transverse ridge, the *superior turbinated crest*, for articulation with the middle turbinated bone of the ethmoid, bounded below by a shallow smooth

---

* In some cases, at any rate, the lachrymal bone also encroaches slightly on the anterior superior portion of the opening, and assists in forming the inner wall of the antrum.

† The number of teeth whose fangs are in relation with the floor of the antrum is variable. The antrum ' may extend so as to be in relation to all the teeth of the true maxilla, from the canine to the *dens sapientiæ*.'—See Mr. Salter on Abscess of the Antrum, in *A System of Surgery*, edited by T. Holmes, 2nd edit. vol iv p. 356.

concavity which forms part of the middle meatus ; below this again is the inferior turbinated crest (already described), where the process joins the body of the bone. Its upper border articulates with the frontal bone. The *anterior* border of the nasal process is thin, directed obliquely upwards and forwards, and presents a serrated edge for articulation with the nasal bone ; its *posterior* border is thick, and hollowed into a groove, the *lachrymal groove*, for the nasal duct : of the two margins of this groove, the inner one articulates with the lachrymal bone, the outer one forms part of the circumference of the orbit. Just where the latter joins the orbital surface is a small tubercle, the *lachrymal tubercle* ; this serves as a guide to the position of the lachrymal sac in the operation for fistula lachrymalis. The lachrymal groove in the articulated skull is converted into a canal by the lachrymal bone and lachrymal process of the inferior turbinated ; it is directed downwards, and a little backwards and outwards, is about the diameter of a goose-quill, slightly narrower in the middle than at either extremity, and terminates below in the inferior meatus. It lodges the nasal duct.

FIG. 51.—The palate and alveolar arch.

Anterior palatine canal

Foramina of Stenson

Foramina of Scarpa

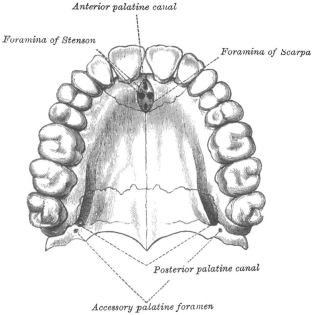

Posterior palatine canal

Accessory palatine foramen

The **Alveolar Process** is the thickest and most spongy part of the bone, broader behind than in front, and excavated into deep cavities for the reception of the teeth. These cavities are eight in number, and vary in size and depth according to the teeth they contain. That for the canine tooth is the deepest ; those for the molars are the widest, and subdivided into minor cavities ; those for the incisors are single, but deep and narrow. The Buccinator muscle arises from the outer surface of this process, as far forward as the first molar tooth.

The **Palate Process,** thick and strong, projects horizontally inwards from the inner surface of the bone. It is much thicker in front than behind, and forms a considerable part of the floor of the nostril and the roof of the mouth.

Its *inferior* surface (fig. 51) is concave, rough and uneven, and forms part of the roof of the mouth. This surface is perforated by numerous foramina for the passage of the nutrient vessels ; channelled at the back part of its alveolar border by a longitudinal groove, sometimes a canal, for the transmission of the posterior palatine vessels, and the anterior and external palatine nerves from Meckel's ganglion ; and presents little depressions for the lodgment of the palatine glands

When the two superior maxillary bones are articulated together, a large orifice may be seen in the middle line, immediately behind the incisor teeth. This is the *anterior palatine canal* or *fossa*. On examining the bottom of this fossa four canals are seen: two branch off laterally to the right and left nasal fossæ, and two, one in front and one behind, lie in the middle line. The lateral canals are named the *foramina of Stenson*, and through each of them passes the anterior or terminal branch of the descending or posterior palatine arteries, which ascends from the mouth to the nasal fossæ. The canals in the middle line are termed the *foramina of Scarpa*, and transmit the naso-palatine nerves, the left passing through the anterior, and the right through the posterior canal. On the palatal surface of the process, a delicate linear suture may sometimes be seen extending from the anterior palatine fossa to the interval between the lateral incisor and the canine tooth. This marks out the intermaxillary, or incisive bone, which in some animals exists permanently as a separate piece. It includes the whole thickness of the alveolus, the corresponding part of the floor of the nose and the anterior nasal spine, and contains the sockets of the incisor teeth. The *upper* surface is concave from side to side, smooth, and forms part of the floor of the nose. It presents the upper orifices of the foramina of Stenson and Scarpa, the former being on each side of the middle line, the latter being situated in the intermaxillary suture, and therefore not visible unless the two bones are placed in apposition. The *outer* border of the palate process is incorporated with the rest of the bone. The *inner* border is thicker in front than behind, and is raised above into a ridge, the *nasal crest*, which, with the corresponding ridge in the opposite bone, forms a groove for the reception of the vomer. In front, this crest rises to a considerable height, and this portion is named the *incisor crest*. The *anterior* margin is bounded by the thin, concave border of the opening of the nose prolonged forwards internally into a sharp process, forming, with a similar process of the opposite bone, the *anterior nasal spine*. The *posterior* border is serrated for articulation with the horizontal plate of the palate-bone.

Development.—This bone commences to ossify at a very early period, and ossification proceeds in it with such rapidity, that it is difficult to ascertain with certainty its precise number of centres. It appears, however, probable that it is ossified from four centres, which are deposited in membrane. 1. One forms that portion of the body of the bone which lies internal to the infra-orbital canal, including the floor of the orbit, the outer wall of the nasal fossa,

FIG. 52.—Development of superior maxillary bone. At birth.

ANTERIOR SURFACE.   INFERIOR SURFACE.

and the nasal process; 2. a second gives origin to that portion of the bone which lies external to the infra-orbital canal and the malar process; 3. a third from which is developed the palatine process posterior to Stenson's canal and the adjoining part of the nasal wall; 4. and a fourth for the front part of the alveolus which carries the incisor teeth and corresponds to the pre-maxillary bone of the lower animals. These centres appear about the eighth week, and by the tenth week the three first-named centres have become fused together and the bone consists of two portions, one the maxilla proper, and the other the pre-maxillary portion. The suture between these two portions persists on the palate till middle life, but is not to be seen on the facial surface. This is believed by

Callender to be due to the fact that the front wall of the sockets of the incisive teeth is not formed by the pre-maxillary bone, but by an outgrowth from the facial part of the superior maxilla. The antrum appears as a shallow groove on the inner surface of the bone at an earlier period than any of the other nasal sinuses, its development commencing about the fourth month of foetal life. The sockets for the teeth are formed by the growing downwards of two plates from the dental groove, which subsequently becomes divided by partitions jutting across from the one to the other.

Articulations.—With *nine* bones : two of the cranium, the frontal and ethmoid, and seven of the face—viz. the nasal, malar, lachrymal, inferior turbinated, palate, vomer, and its fellow of the opposite side. Sometimes it articulates with the orbital plate of the sphenoid, and sometimes with its external pterygoid plate.

Attachment of Muscles.—To twelve : the Orbicularis palpebrarum, Obliquus oculi inferior, Levator labii superioris alæque nasi, Levator labii superioris proprius, Levator anguli oris, Compressor nasi, Depressor alæ nasi, Dilatator naris posterior, Masseter, Buccinator, Internal pterygoid, and Orbicularis oris.

CHANGES PRODUCED IN THE UPPER JAW BY AGE

At birth and during infancy the diameter of the bone is greater in an antero-posterior than in a vertical direction. Its nasal process is long, its orbital surface large, and its tuberosity well marked. In the adult the vertical diameter is the greater, owing to the development of the alveolar process and the increase in size of the antrum. In old age the bone approaches again in character to the infantile condition : its height is diminished, and after the loss of the teeth the alveolar process is absorbed, and the lower part of the bone contracted and diminished in thickness.

THE LACHRYMAL BONES

The **Lachrymal** (lachryma, *a tear*) are the smallest and most fragile bones of the face. They are situated at the front part of the inner wall of the orbit (fig. 46), and resemble somewhat in form, thinness, and size, a finger-nail ; hence they are termed the *ossa unguis*. Each bone presents for examination two surfaces and four borders. The *external* or *orbital* surface (fig. 53) is divided by a vertical ridge, the *lachrymal crest*, into two parts. The portion of bone in front of this ridge presents a smooth, concave, longitudinal groove, the free margin of which unites with the nasal process of the superior maxillary bone, completing the lachrymal groove. The upper part of this groove lodges the lachrymal sac ; the lower part the nasal duct. The portion of bone behind the ridge is smooth, slightly concave, and forms part of the inner wall of the orbit. The ridge, with a part of the orbital surface immediately behind it, affords attachment to the Tensor tarsi muscle : it terminates below in a small, hook-like projection, the *hamular process*, which articulates with the lachrymal tubercle of the superior maxillary bone, and completes the upper orifice of the lachrymal groove. It sometimes exists as a separate piece, which is then called the *lesser lachrymal bone*.

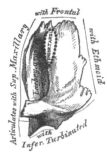

FIG. 53.—Left lachrymal bone. External surface. (Slightly enlarged.)

The *internal* or *nasal* surface presents a depressed furrow, corresponding to the ridge on its outer surface. The surface of bone in front of this forms part of the middle meatus of the nose ; and that behind it articulates with the ethmoid bone, filling in the anterior ethmoidal cells. Of the *four borders* the *anterior* is the longest, and articulates with the nasal process of the superior maxillary bone. The *posterior*, thin and uneven, articulates with the os planum of the ethmoid. The *superior*, the shortest and thickest, articulates with the internal angular process of the frontal bone. The *inferior* is divided by the lower edge of the vertical crest into two parts : the posterior part articulates with the orbital plate of the superior maxillary bone ; the anterior portion is prolonged downwards into a pointed process, which articulates with the lachrymal process of the inferior turbinated bone, and assists in the formation of the lachrymal groove.

Development.—By a single centre, which makes its appearance soon after ossification of the vertebræ has commenced.

Articulations.—With four bones: two of the cranium, the frontal and ethmoid, and two of the face, the superior maxillary and the inferior turbinated.

Attachment of Muscles.—To one muscle, the Tensor tarsi.

THE MALAR BONES

The **Malar** (mala, *the cheek*) are two small, quadrangular bones, situated at the upper and outer part of the face : they form the prominence of the cheek, part of the outer wall and floor of the orbit, and part of the temporal and zygomatic fossæ (fig. 54). Each bone presents for examination an external and an internal surface ; four processes, the frontal, orbital, maxillary, and zygomatic ; and four borders. The **external surface** (fig. 55) is smooth, convex, perforated near its centre by one or two small apertures, the *malar foramina*, for the passage of nerves and vessels, covered by the Orbicularis palpebrarum muscle, and affords attachment to the Zygomaticus major and minor muscles.

The **internal surface** (fig. 56), directed backwards and inwards, is concave, presenting internally a rough, triangular surface, for articulation with the superior maxillary bone, and externally a smooth, concave surface, which above forms the anterior boundary of the temporal fossa, and below, where it is wider, a part of the zygomatic fossa. This surface presents, a little above its centre, the aperture of one or two malar canals, and affords attachment to a portion of the Masseter muscle at its lower part. Of the four processes, the **frontal** is thick and serrated, and articulates with the external angular process of the frontal bone. To its orbital margin is attached the external

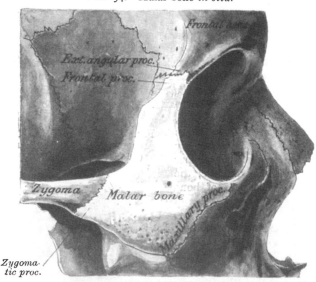

FIG. 54.—Malar bone *in situ*.

Frontal bone
Ext. angular proc.
Frontal proc.
Zygoma *Malar bone*
Maxillary proc.
Zygomatic proc.

tarsal ligament. The **orbital** process is a thick and strong plate, which projects backwards from the orbital margin of the bone. Its *supero-internal* surface, smooth and concave, forms, by its junction with the orbital surface of the superior maxillary bone and with the great wing of the sphenoid, part of the floor and outer wall of the orbit. Its *infero-external* surface, smooth and convex, forms part of the zygomatic and temporal fossæ. Its *anterior* margin is smooth and rounded, forming part of the circumference of the orbit. Its *superior* margin, rough, and directed horizontally, articulates with the frontal bone behind the external angular process. Its *posterior* margin is rough, and serrated for articulation with the sphenoid ; *internally* it is also serrated for articulation with the orbital surface of the superior maxillary. At the angle of junction of the sphenoidal and maxillary portions, a short, rounded, non-articular margin is generally seen ; this forms the anterior boundary of the spheno-maxillary fissure : occasionally, such non-articular margin does not exist, the fissure being completed by the direct junction of the maxillary and sphenoid bones, or by the interposition of a small Wormian bone in the angular interval between them. On the *upper* surface of the orbital process are seen the orifices of one or two temporo-malar canals ; one of these usually opens on the posterior surface, the other (occasionally there are two) on the facial surface: they transmit filaments (temporo-malar) of the orbital branch of the superior maxillary nerve. The **maxillary** process is a rough, triangular surface which articulates with the superior maxillary bone. The **zygomatic** process, long, narrow, and serrated, articulates with the zygomatic process of the temporal bone. Of the **four borders,** the *antero-superior* or *orbital* is smooth, arched, and forms a considerable part of the circumference of the orbit. The *antero-inferior* or *maxillary* border is rough, and bevelled at the expense of its inner table, to articulate with the superior maxillary bone; affording attachment by its margin to the Levator labii superioris proprius, just at its point of junction with the superior maxillary. The *postero-superior* or *temporal* border, curved like an italic letter *f*, is continuous above with the commencement of the temporal ridge ; below, with the upper

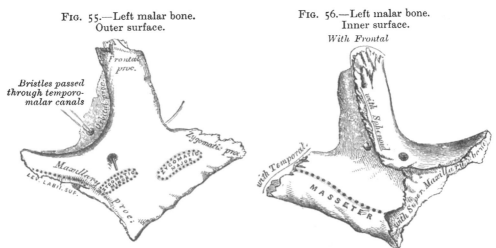

Fig. 55.—Left malar bone.
Outer surface.

Fig. 56.—Left malar bone.
Inner surface.

With Frontal

Bristles passed through temporo-malar canals

border of the zygomatic arch : it affords attachment to the temporal fascia. The *postero-inferior* or *zygomatic* border is continuous with the lower border of the zygomatic arch, affording attachment by its rough edge to the Masseter muscle.

Development.—The malar bone ossifies generally from three centres, which appear about the eighth week—one for the zygomatic and two for the orbital portion—and fuse about the fifth month of foetal life. The bone is sometimes, after birth, seen to be divided, by a horizontal suture, into an upper and larger and a lower and smaller division. In some quadrumana the malar bone consists of two parts, an orbital and a malar, which are ossified by separate centres.

Articulations.—With four bones : three of the cranium, frontal, sphenoid, and temporal ; and one of the face, the superior maxillary.

Attachment of Muscles.—To four : the Levator labii superioris proprius, Zygomaticus major and minor, and Masseter.

The Palate Bones

The **Palate Bones** (palatum, *the palate*) are situated at the back part of the nasal fossæ : they are wedged in between the superior maxillary bones and the pterygoid processes of the sphenoid (fig. 57). Each bone assists in the formation of three cavities : the floor and outer wall of the nose, the roof of the mouth, and the floor of the orbit ; and enters into the formation of two fossæ : the spheno-maxillary and pterygoid ; and one fissure, the spheno-maxillary. In form the palate-bone somewhat resembles the letter L, and may be divided into an inferior or horizontal plate and a superior or vertical plate.

The **Horizontal Plate** is of a quadrilateral form, and presents two surfaces and four borders. The *superior* surface, concave from side to side, forms the back part of the floor of the nose. The *inferior* surface, slightly concave and rough, forms the back part of the hard palate. At its posterior part may be seen a transverse ridge, more or less marked, for the attachment of part of the aponeurosis of the Tensor palati muscle. At the outer extremity of this ridge is a deep groove converted into a canal by its articulation with the tuberosity of the superior maxillary bone, and forming the *posterior palatine canal*. Near this groove, the orifices of one or two small canals, *accessory posterior palatine*, may be seen. The *anterior* border is serrated, bevelled at the expense of its inferior surface, and articulates with the palate process of the superior maxillary bone. The *posterior* border is concave, free, and serves for the attachment of the soft palate. Its inner extremity

FIG. 57.—Palate bone *in situ*.

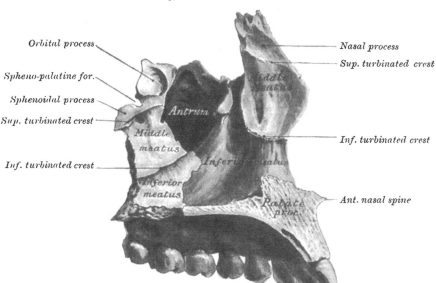

is sharp and pointed, and, when united with the opposite bone, forms a projecting process, the *posterior nasal spine*, for the attachment of the Azygos uvulæ muscle. The *external* border is united with the lower part of the perpendicular plate almost at right angles. The *internal* border, the thickest, is serrated for articulation with its fellow of the opposite side ; its superior edge is raised into a ridge, which, united with the opposite bone, forms a crest in which the vomer is received.

The **Vertical Plate** (fig. 58) is thin, of an oblong form, and directed upwards and a little inwards. It presents two surfaces, an external and an internal, and four borders.

The **internal surface** presents at its lower part a broad, shallow depression, which forms part of the inferior meatus of the nose. Immediately above this is a well-marked, horizontal ridge, the *inferior turbinated crest*, for articulation with the inferior turbinated bone ; above this, a second broad, shallow depression, which forms part of the middle meatus, surmounted above by a horizontal ridge less prominent than the inferior, the *superior turbinated crest*, for articulation with the middle turbinated bone. Above the superior turbinated crest is a narrow, horizontal groove, which forms part of the superior meatus.

The **external surface** is rough and irregular throughout the greater part of its extent, for articulation with the inner surface of the superior maxillary bone, its upper and back part being smooth where it enters into the formation of the spheno-maxillary fossa; it is also smooth in front, where it covers the orifice of the antrum. Towards the back part of this surface is a deep groove, converted into a canal, the *posterior palatine*, by its articulation with the superior maxillary bone. It transmits the posterior or descending palatine vessels, and one of the descending palatine branches from Meckel's ganglion.

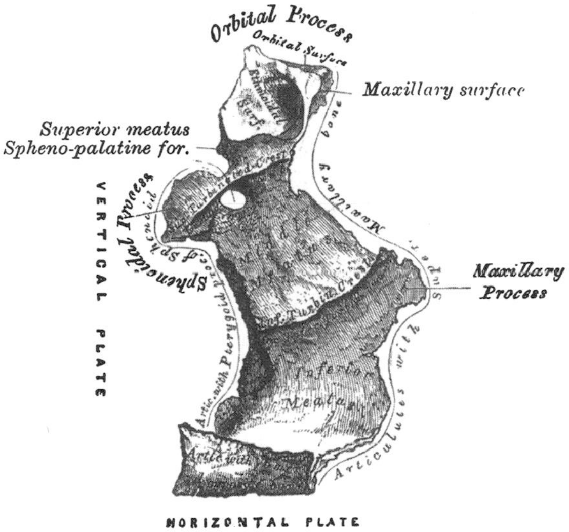

The *anterior* border is thin, irregular, and presents, opposite the inferior turbinated crest, a pointed, projecting lamina, the *maxillary process*, which is directed forwards, and

FIG. 58.—Left palate bone. Internal view. (Enlarged.)

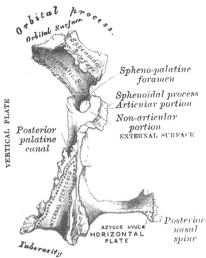

closes in the lower and back part of the opening of the antrum. The *posterior* border (fig. 59) presents a deep groove, the edges of which are serrated for articulation with the pterygoid process of the sphenoid. At the lower part of this border is seen a pyramidal process of bone, the *pterygoid process* or tuberosity of the palate, which is received into the angular interval between the two pterygoid plates of the sphenoid at their inferior extremity. This process presents at its back part a median groove and two lateral surfaces. The groove is smooth, and forms part of the pterygoid fossa affording attachment to the Internal pterygoid muscle; while the lateral surfaces are rough and uneven, for articulation with the anterior border of each pterygoid plate. A few fibres of the Superior constrictor arise from the tuberosity of the palate-bone. The base of this process, continuous with the horizontal portion of the bone, presents the apertures of the *accessory descending palatine canals*, through which pass the two smaller descending branches of Meckel's ganglion; while its outer surface is rough for articulation with the inner surface of the body of the superior maxillary bone.

FIG. 59.—Left palate bone. Posterior view. (Enlarged.)

The *superior* border of the vertical plate presents two well-marked processes separated by an intervening notch or foramen. The anterior, or larger, is called the *orbital process*; the posterior, the *sphenoidal*.

The **Orbital Process**, directed upwards and outwards, is placed on a higher level than the sphenoidal. It presents five surfaces, which enclose a hollow cellular cavity, and is connected to the perpendicular plate by a narrow, constricted neck. Of these five surfaces, three are articular, two non-articular or free surfaces. The three articular are the *anterior* or *maxillary* surface, directed forwards, outwards, and downwards, of an oblong form, and rough for articulation with the superior maxillary bone. The *posterior* or *sphenoidal* surface is directed backwards, upwards, and inwards. It ordinarily presents a small, open cell, which communicates with the sphenoidal cells, and the margins of which are serrated for articulation with the vertical part of the sphenoidal turbinated bone. The *internal* or *ethmoidal* surface is directed inwards, upwards, and forwards, and articulates with the lateral mass of the ethmoid bone. In some cases, the cellular cavity above mentioned opens on this surface of the bone; it then communicates with the posterior ethmoidal cells. More rarely it opens on both surfaces, and then communicates both with the posterior ethmoidal and the sphenoidal cells. The non-articular or free surfaces are the *superior* or *orbital*, directed upwards and outwards, of a triangular shape, concave, smooth, and forming the back part of the floor of the orbit; and the *external* or *zygomatic* surface, directed outwards, backwards, and downwards, of an oblong form, smooth, lying in the sphenomaxillary fossa, and looking into the zygomatic fossa. The latter surface is separated from the orbital by a smooth, rounded border, which enters into the formation of the spheno-maxillary fissure.

The **Sphenoidal Process** of the palate-bone is a thin, compressed plate, much smaller than the orbital, and directed upwards and inwards. It presents three surfaces and two borders. The *superior* surface, the smallest of the three, articulates with the under surface of the sphenoidal turbinated bone; it presents a groove, which contributes to the formation of the pterygo-palatine canal. The *internal* surface is concave, and forms part of the outer wall of the nasal fossa. The *external* surface is divided into an articular and non-articular portion: the former is rough, for articulation with the inner surface of the internal pterygoid plate of the sphenoid; the latter is smooth, and forms part of the spheno-maxillary fossa. The *anterior* border forms the posterior boundary of the spheno-palatine foramen. The *posterior* border, serrated at the expense of the outer table, articulates with the inner surface of the internal pterygoid plate.

The orbital and sphenoidal processes are separated from one another by a deep notch, which is converted into a foramen, the *spheno-palatine*, by articulation with the sphenoidal turbinated bone. Sometimes the two processes are united above, and form between them a complete foramen, or the notch is crossed by one or more spiculæ of bone, so as to form two or more foramina. In the articulated skull, this foramen opens into the back part of the outer wall of the superior meatus, and transmits the spheno-palatine vessels and the superior nasal and naso-palatine nerves.

Development.—From a single centre, which makes its appearance about the second month at the angle of junction of the two plates of the bone. From this point ossification spreads inwards to the horizontal plate, downwards into the tuberosity, and upwards into the vertical plate. In the fœtus, the horizontal plate is much longer than the vertical; and even after it is fully ossified, the whole bone is at first remarkable for its shortness.

Articulations.—With six bones: the sphenoid, ethmoid, superior maxillary, inferior turbinated, vomer, and opposite palate.

Attachment of Muscles.—To four: the Tensor palati, Azygos uvulæ, Internal pterygoid, and Superior constrictor of the pharynx.

THE INFERIOR TURBINATED BONES

The **Inferior Turbinated Bones** (turbo, *a whirl*) are situated one on each side of the outer wall of the nasal fossæ. Each consists of a layer of thin, spongy bone, curled upon itself like a scroll—hence its name 'turbinated'—and extends horizontally along the outer wall of the nasal fossa, immediately below the orifice of the antrum (fig. 60). Each bone presents two surfaces, two borders, and two extremities.

The **internal surface** (fig. 61) is convex, perforated by numerous apertures, and traversed by longitudinal grooves and canals for the lodgment of arteries and veins. In the recent state it is covered by the lining membrane of the nose. The **external surface** is concave (fig. 62), and forms part of the inferior meatus. Its *upper border* is thin, irregular, and connected to various bones along the outer wall of the nose. It may be divided into three portions; of these, the anterior articulates with the inferior turbinated crest of the superior maxillary bone; the posterior with the inferior turbinated crest of the palate-bone; the middle portion of the superior border presents three well-marked processes, which vary

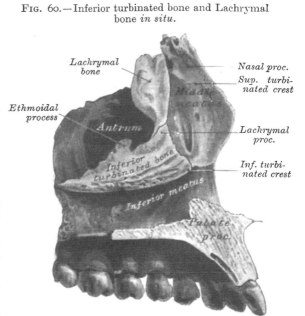

FIG. 60.—Inferior turbinated bone and Lachrymal bone *in situ.*

much in their size and form. Of these, the anterior and smallest is situated at the junction of the anterior fourth with the posterior three-fourths of the bone : it is small and pointed, and is called the *lachrymal process* ; it articulates, by its apex, with the anterior inferior angle of the lachrymal bone, and, by its margins, with the groove on the back of the nasal process of the superior maxillary, and thus assists in forming the canal for the nasal duct. At the junction of the two middle fourths

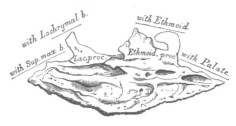

FIG. 61.—Right inferior turbinated bone. Internal surface.

FIG. 62.—Right inferior turbinated bone. External surface.

of the bone, but encroaching on its posterior fourth, a broad, thin plate, the *ethmoidal process*, ascends to join the unciform process of the ethmoid ; from the lower border of this process a thin lamina of bone curves downwards and outwards, hooking over the lower edge of the orifice of the antrum, which it narrows below : it is called the *maxillary process*, and fixes the bone firmly to the outer wall of the nasal fossa. The *inferior border* is free, thick, and cellular in structure, more especially in the middle of the bone. Both *extremities* are more or less narrow and pointed, the posterior being the more tapering. If the bone is held so that its outer concave surface is directed backwards (i.e. towards the holder), and its superior border, from which the lachrymal and ethmoidal processes project, upwards, the lachrymal process will be directed to the side to which the bone belongs.*

Development.—By a single centre, which makes its appearance about the middle of fœtal life.

Articulations.—With four bones : one of the cranium, the ethmoid, and three of the face, the superior maxillary, lachrymal, and palate.

No muscles are attached to this bone.

The Vomer

The **Vomer** (vomer, *a ploughshare*) is a single bone, situated vertically at the back part of the nasal fossæ, forming part of the septum of the nose (fig. 63). It is thin, somewhat like a ploughshare in form; but it varies in different individuals, being frequently bent to one or the other side; it presents for examination two surfaces and four borders. The *lateral surfaces* are smooth, marked by

FIG. 63.—Vomer *in situ.*

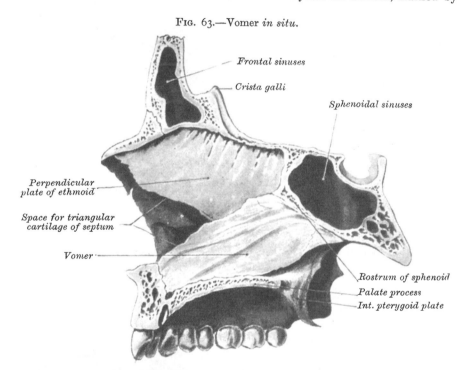

Frontal sinuses

Crista galli

Sphenoidal sinuses

Perpendicular plate of ethmoid

Space for triangular cartilage of septum

Vomer

Rostrum of sphenoid

Palate process

Int. pterygoid plate

small furrows for the lodgment of blood-vessels, and by a groove on each side, sometimes a canal, the *naso-palatine*, which runs obliquely downwards and forwards to the intermaxillary suture; it transmits the naso-palatine nerve. The *superior border*, the thickest, presents a deep groove, bounded on each side by a horizontal projecting ala of bone; the groove receives the rostrum of the sphenoid, while the alæ are overlapped and retained by the *vaginal processes*, which project from the under surface of the body of the sphenoid at the base of the pterygoid processes. At the front of the groove a fissure is left for the transmission of blood-vessels to the substance of the bone. The *inferior border*, the longest, is broad and uneven in front, where it articulates with the two superior maxillary bones; thin and sharp behind, where it joins with the palate-bones. The upper half of the *anterior border* usually consists of two

---

* If the lachrymal process is broken off, as is often the case, the side to which the bone belongs may be known by recollecting that the maxillary process is nearer the back than the front of the bone.

laminæ of bone, between which is received the perpendicular plate of the ethmoid; the lower half, also separated into two lamellæ, receives between them the lower margin of the septal cartilage of the nose. The *posterior border* is free, concave, and separates the nasal fossæ behind. It is thick and bifid above, thin below.

The surfaces of the vomer are covered by mucous membrane, which is intimately connected with the periosteum, with the intervention of very little, if any, submucous connective tissue. Hence polypi are rarely found growing from this surface, though they fre-

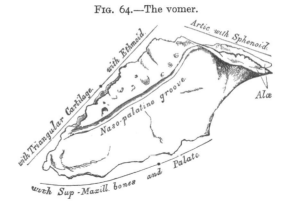

FIG. 64.—The vomer.

quently grow from the outer wall of the nasal fossæ, where the submucous tissue is abundant.

Development.—The vomer at an early period consists of two laminæ, separated by a very considerable interval, and enclosing between them a plate of cartilage, the *vomerine cartilage*, which is prolonged forwards to form the remainder of the septum. Ossification commences in the membrane at the postero-inferior part of this cartilage by two centres, one on each side of the middle line, which extend to form the two laminæ. They begin to coalesce at the lower part, but their union is not complete until after puberty.

Articulations.—With six bones: two of the cranium, the sphenoid and ethmoid; and four of the face, the two superior maxillary and the two palate-bones; and with the cartilage of the septum.

The vomer has no muscles attached to it.

THE INFERIOR MAXILLARY BONE OR MANDIBLE

The **Inferior Maxillary Bone** (the *Mandible*), the largest and strongest bone of the face, serves for the reception of the lower teeth. It consists of a curved, horizontal portion, the *body*, and two perpendicular portions, the *rami*, which join the back part of the body nearly at right angles.

The **Horizontal Portion**, or **Body** (fig. 65), is convex in its general outline, and curved somewhat like a horse-shoe. It presents for examination two surfaces and two borders. The **external surface** is convex from side to side, concave from above downwards. In the median line is a vertical ridge, the *symphysis*, which extends from the upper to the lower border of the bone, and indicates the point of junction of the two pieces of which the bone is composed at an early period of life. The lower part of the ridge terminates in a prominent triangular eminence, the *mental process*. This eminence is rounded below, and often presents a median depression separating two processes, the *mental tubercles*. It forms the chin, a feature peculiar to the human skull. On either side of the symphysis, just below the cavities for the incisor teeth, is a depression, the *incisive fossa*, for the attachment of the Levator menti (or Levator labii inferioris); more externally is attached a portion of the Orbicularis oris (*Accessorii Orbicularis inferioris*), and, still more externally, a foramen, the *mental foramen*, for the passage of the mental vessels and nerve. This foramen is placed just below the interval between the two bicuspid teeth. Running outwards from the base of the mental process on each side is a ridge, the *external oblique* line. The ridge is at first nearly horizontal, but afterwards inclines upwards and backwards, and is continuous with the anterior border of the ramus: it affords attachment to the Depressor

labii inferioris and Depressor anguli oris ; below it the Platysma myoides is
attached.

FIG. 65.—Inferior maxillary bone.   Outer surface.   Side view.

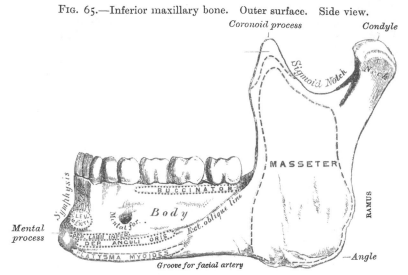

The **internal surface** (fig. 66) is concave from side to side, convex from above
downwards.   In the middle line is an indistinct linear depression, corresponding
to the symphysis externally ; on either side of this depression, just below its
centre, are four prominent tubercles, placed in pairs, two above and two below ;
they are called the *genial tubercles*, and afford attachment, the upper pair to the
Genio-hyo-glossi muscles, the lower pair to the Genio-hyoidei muscles.   Sometimes the tubercles on each side are blended into one, at others they all unite into
an irregular eminence, or again, nothing but an irregularity may be seen on the

FIG. 66.—Inferior maxillary bone.   Inner surface.   Side view.

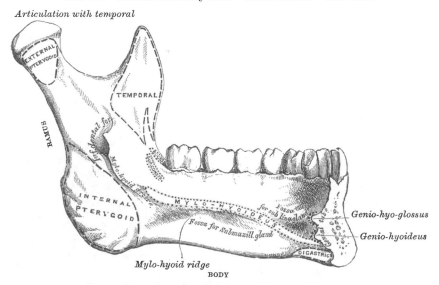

surface of the bone at this part.   On either side of the genial tubercles is an oval
depression, the *sublingual fossa*, for lodging the sublingual gland ; and beneath
the fossa, a rough depression on each side, which gives attachment to the anterior
belly of the Digastric muscle.   At the back part of the sublingual fossa, the

internal oblique line (*mylo-hyoidean*) commences ; it is at first faintly marked, but becomes more distinct as it passes upwards and outwards, and is especially prominent opposite the last two molar teeth ; it affords attachment throughout its whole extent to the Mylo-hyoid muscle : the Superior constrictor of the pharynx with the pterygo-maxillary ligament, being attached above its posterior extremity, near the alveolar margin. The portion of bone above this ridge is smooth, and covered by the mucous membrane of the mouth ; the portion below presents an oblong depression, the *submaxillary fossa*, wider behind than in front, for the lodgment of the submaxillary gland. The external oblique line and the internal or mylo-hyoidean line divide the body of the bone into a superior or alveolar and an inferior or basilar portion.

The **superior** or **alveolar** border is wider, and its margins thicker, behind than in front. It is hollowed into cavities, for the reception of the teeth ; these cavities are sixteen in number, and vary in depth and size according to the teeth which they contain. To its outer side, the Buccinator muscle is attached as far forward as the first molar tooth. The **inferior** border is rounded, longer than the superior and thicker in front than behind ; it presents a shallow groove, just where the body joins the ramus, over which the facial artery turns.

The **Perpendicular Portions**, or **Rami**, are of a quadrilateral form. Each presents for examination two surfaces, four borders, and two processes. The *external surface* is flat, marked with ridges, and gives attachment throughout nearly the whole of its extent to the Masseter muscle. The *internal surface* presents about its centre the oblique aperture of the inferior dental canal, for the passage of the inferior dental vessels and nerve. The margin of this opening is irregular ; it presents in front a prominent ridge, surmounted by a sharp spine, the *lingula*, which gives attachment to the internal lateral ligament of the lower jaw, and at its lower and back part a notch leading to a groove, the *mylo-hyoidean*, which runs oblqiuely downwards to the back part of the submaxillary fossa, and lodges the mylo-hyoid vessels and nerve. Behind the groove is a rough surface, for the insertion of the Internal pterygoid muscle. The inferior dental canal runs obliquely downwards and forwards in the substance of the ramus, and then horizontally forwards in the body ; it is here placed under the alveoli, with which it communicates by small openings. On arriving at the incisor teeth, it turns back to communicate with the mental foramen, giving off two small canals, which run forward, to be lost in the cancellous tissue of the bone beneath the incisor teeth. This canal, in the posterior two-thirds of the bone, is situated nearer the internal surface of the jaw ; and in the anterior third, nearer its external surface. Its walls are composed of compact tissue at either extremity, and of cancellous in the centre. It contains the inferior dental vessels and nerve, from which branches are distributed to the teeth through small apertures at the bases of the alveoli. The *lower border* of the ramus is thick, straight, and continuous with the body of the bone. At its junction with the posterior border is the *angle of the jaw*, which is either inverted or everted, and marked by rough, oblique ridges on each side, for the attachment of the Masseter externally, and the Internal pterygoid internally ; the stylo-maxillary ligament is attached to the angle between these muscles. The *anterior border* is thin above, thicker below, and continuous with the external oblique line. The *posterior border* is thick, smooth, rounded, and covered by the parotid gland. The *upper border* of the ramus is thin, and presents two processes, separated by a deep concavity, the *sigmoid notch*. Of these processes, the anterior is the *coronoid*, the posterior the *condyloid*.

The **Coronoid Process** is a thin, flattened triangular eminence of bone, which varies in shape and size in different subjects, and serves chiefly for the attachment of the Temporal muscle. Its *external surface* is smooth, and affords attachment to the Temporal and Masseter muscles. Its *internal surface* gives attachment to the Temporal muscle, and presents the commencement of a longitudinal ridge, which is continued to the posterior part of the alveolar process. On the outer side of this ridge is a deep groove, continued below on the outer side of the alveolar process; this ridge and part of the groove afford attachment, above, to the Temporal; below, to the Buccinator muscle.

The **Condyloid Process,** shorter but thicker than the coronoid, consists of two portions : the *condyle,* and the constricted portion which supports the condyle, the *neck.* The *condyle* is of an oblong form, its long axis being transverse, and set obliquely on the neck in such a manner that its outer end is a little more forward and a little higher than its inner. It is convex from before backwards, and from side to side, the articular surface extending farther on the posterior than on the anterior aspect. At its outer extremity is a small tubercle for the attachment of the external lateral ligament of the temporo-mandibular joint. The *neck* of the condyle is flattened from before backwards, and strengthened by ridges which descend from the fore part and sides of the condyle. Its lateral margins are narrow, the external one giving attachment to part of the external lateral ligament. Its posterior surface is convex; its anterior is hollowed out on its inner side by a depression (the *pterygoid fossa*) for the attachment of the External pterygoid muscle.

The **Sigmoid Notch,** separating the two processes, is a deep semilunar depression, crossed by the masseteric vessels and nerve.

Development.—The lower jaw is developed principally from membrane, but partly from cartilage. The process of ossification commences early—earlier than any other bone except the clavicle. The greater part of the bone is formed from a centre of ossification (*dentary*), which appears between the fifth and sixth weeks in the membrane on the outer surface of Meckel's cartilage. A second centre (*splenial*) appears in the membrane on the inner surface of the cartilage, and from this centre the inner wall of the sockets of the teeth is formed ; this terminates above in the lingula. The anterior extremity of Meckel's cartilage becomes ossified, forming the body of the bone on each side of the symphysis. Two supplemental patches of cartilage appear at the condyle and at the angle, in each of which a centre of ossification for these parts appears ; the coronoid process is also ossified from a separate centre. At birth the bone consists of two halves, united by a fibrous symphysis, in which ossification takes place during the first year.

Articulation.—With the glenoid fossæ of the two temporal bones.

Attachment of Muscles.—To fifteen pairs ; to its external surface, commencing at the symphysis, and proceeding backwards : Levator menti, Depressor labii inferioris, Depressor anguli oris, Platysma myoides, Buccinator, Masseter; a portion of the Orbicularis oris (Accessorii Orbicularis inferioris) is also attached to this surface. To its internal surface, commencing at the same point : Genio-hyoglossus, Genio-hyoideus, Mylo-hyoideus, Digastric, Superior constrictor, Temporal, Internal pterygoid, External pterygoid.

Changes produced in the Lower Jaw by Age

The changes which the Lower Jaw undergoes after birth, relate (1) to the alterations effected in the body of the bone by the first and second dentitions, the loss of the teeth in the aged, and the subsequent absorption of the alveoli ; (2) to the size and situation of the dental canal; and (3) to the angle at which the ramus joins with the body.

At birth (fig. 67), the bone consists of lateral halves, united by fibrous tissue. The body is a mere shell of bone, containing the sockets of the two incisor, the canine, and the two temporary molar teeth, imperfectly partitioned from one another. The dental canal is of large size, and runs near the lower border of the bone, the mental foramen opening beneath the socket of the first molar. The angle is obtuse (175°), and the condyloid portion nearly in the same horizontal line with the body ; the neck of the condyle is short, and bent backwards. The coronoid process is of comparatively large size, and situated at right angles with the rest of the bone.

After birth (fig. 68), the two segments of the bone become joined at the symphysis, from below upwards, in the first year ; but a trace of separation may be visible in the beginning of the second year, near the alveolar margin. The body becomes elongated in its whole length, but more especially behind the mental foramen, to provide space for the

SIDE VIEW OF THE LOWER JAW AT DIFFERENT PERIODS OF LIFE

FIG. 67.—At birth.

FIG. 68.—At puberty.

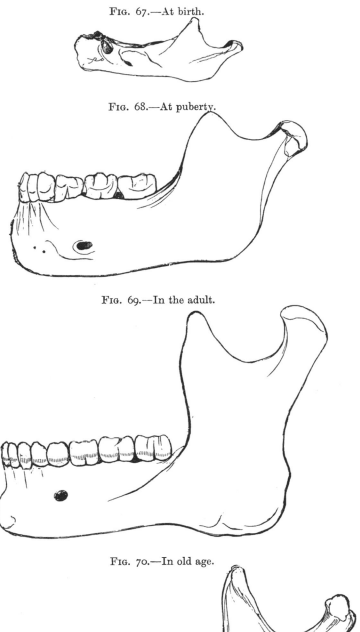

FIG. 69.—In the adult.

FIG. 70.—In old age.

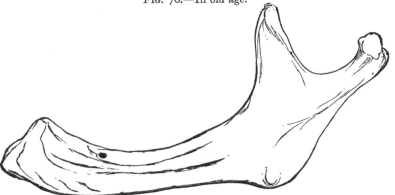

three additional teeth developed in this part. The depth of the body becomes greater, owing to increased growth of the alveolar part, to afford room for the fangs of the teeth, and by thickening of the subdental portion which enables the jaw to withstand the power- ful action of the masticatory muscles ; but the alveolar portion is the deeper of the two, and, consequently, the chief part of the body lies above the oblique line. The dental canal, after the second dentition, is situated just above the level of the mylo-hyoid ridge; and the mental foramen occupies the position usual to it in the adult. The angle becomes less obtuse, owing to the separation of the jaws by the teeth. (About the fourth year it is 140°.)

In the adult (fig. 69), the alveolar and basilar portions of the body are usually of equal depth. The mental foramen opens midway between the upper and lower border of the bone, and the dental canal runs nearly parallel with the mylo-hyoid line. The ramus is almost vertical in direction, and joins the body nearly at right angles.

In old age (fig. 70), the bone becomes greatly reduced in size : for with the loss of the teeth the alveolar process is absorbed, and the basilar part of the bone alone remains ; consequently, the chief part of the bone is *below* the oblique line. The dental canal, with the mental foramen opening from it, is close to the alveolar border. The rami are oblique in direction, the angle obtuse, and the neck of the condyle more or less bent backwards.

THE SUTURES

The bones of the cranium and face are connected to each other by means of *Sutures*. That is, the articulating surfaces or edges of the bones are more or less roughened or uneven, and are closely adapted to each other, a small amount of intervening fibrous tissue, the *sutural ligament*, fastening them together. The *Cranial Sutures* may be divided into three sets : 1. Those at the vertex of the skull. 2. Those at the side of the skull. 3. Those at the base.

The sutures at the vertex of the skull are three : the *sagittal, coronal*, and *lambdoid*.

The **Sagittal Suture** (*interparietal*) is formed by the junction of the two parietal bones, and extends from the middle of the frontal bone, backwards to the superior angle of the occipital. In childhood, and occasionally in the adult, when the two halves of the frontal bone are not united, it is continued forwards to the root of the nose. This suture is sometimes perforated, near its posterior extremity, by the parietal foramen ; and in front, where it joins the coronal suture, a space is occasionally left, which encloses a large Wormian bone.

The **Coronal Suture** (*fronto-parietal*) extends transversely across the vertex of the skull, and connects the frontal with the parietal bones. It commences at the extremity of the greater wing of the sphenoid on one side, and terminates at the same point on the opposite side. The dentations of the suture are more marked at the sides than at the summit, and are so constructed that the frontal rests on the parietal above, while laterally the frontal supports the parietal.

The **Lambdoid Suture** (*occipito-parietal*), so called from its resemblance to the Greek letter Λ, connects the occipital with the parietal bones. It commences on each side at the mastoid portion of the temporal bone, and inclines upwards and backwards to the end of the sagittal suture. The dentations of this suture are very deep and distinct, and are often interrupted by several small Wormian bones.

The sutures at the side of the skull extend from the external angular process of the frontal bone to the lower end of the lambdoid suture behind. The anterior portion is formed between the lateral part of the frontal bone above and the malar and great wing of the sphenoid below, forming the *fronto-malar* and *fronto- sphenoidal* sutures. These sutures can also be seen in the orbit, and form part of the so-called *transverse facial* suture. The posterior portion is formed between the parietal bone above and the great wing of the sphenoid, the squamous and mastoid portions of the temporal bone below, forming the *spheno-parietal, squamo- parietal*, and *masto-parietal* sutures.

The **Spheno-parietal** is very short ; it is formed by the tip of the great wing of the sphenoid, which overlaps the anterior inferior angle of the parietal bone.

The **Squamo-parietal**, or **Squamous Suture**, is arched. It is formed by the squamous portion of the temporal bone overlapping the middle division of the lower border of the parietal.

The **Masto-parietal** is a short suture, deeply dentated, formed by the posterior inferior angle of the parietal, and the superior border of the mastoid portion of the temporal.

The sutures at the base of the skull are, the *basilar* in the centre, and on each side, the *petro-occipital*, the *masto-occipital*, the *petro-sphenoidal*, and the *squamo-sphenoidal*.

The **Basilar Suture** is formed by the junction of the basilar surface of the occipital bone with the posterior surface of the body of the sphenoid. At an early period of life a thin plate of cartilage exists between these bones; but in the adult they become fused into one. Between the outer extremity of the basilar suture and the termination of the lambdoid, an irregular suture exists, which is subdivided into two portions. The inner portion, formed by the union of the petrous part of the temporal with the occipital bone, is termed the **petro-occipital**. The outer portion, formed by the junction of the mastoid part of the temporal with the occipital, is called the **masto-occipital**. Between the bones forming the petro occipital suture, a thin plate of cartilage exists; in the masto-occipital is occasionally found the opening of the mastoid foramen. Between the outer extremity of the basilar suture and the spheno-parietal, an irregular suture may be seen, formed by the union of the sphenoid with the temporal bone. The inner and smaller portion of this suture is termed the **petro-sphenoidal**; it is formed between the petrous portion of the temporal and the great wing of the sphenoid: the outer portion, of greater length, and arched, is formed between the squamous portion of the temporal and the great wing of the sphenoid; it is called the **squamo-sphenoidal**.

The cranial bones are connected with those of the face, and the facial bones with each other, by numerous sutures, which, though distinctly marked, have received no special names. The only remaining suture deserving especial consideration is the **transverse**. This extends across the upper part of the face, and is formed by the junction of the frontal with the facial bones; it extends from the external angular process of one side to the same point on the opposite side, and connects the frontal with the malar, the sphenoid, the ethmoid, the lachrymal, the superior maxillary, and the nasal bones on each side.

The sutures remain separate for a considerable period after the complete formation of the skull. It is probable that they serve the purpose of permitting the growth of the bones at their margins; while their peculiar formation, together with the interposition of the sutural ligament between the bones forming them, prevents the dispersion of blows or jars received upon the skull. Humphry remarks 'that, as a general rule, the sutures are first obliterated at the parts in which the ossification of the skull was last completed—viz. in the neighbourhood of the fontanelles; and the cranial bones seem in this respect to observe a similar law to that which regulates the union of the epiphyses to the shafts of the long bones.' The same author remarks that the time of their disappearance is extremely variable: they are sometimes found well marked in skulls edentulous with age, while in others which have only just reached maturity they can hardly be traced. The obliteration of the sutures takes place sooner on the inner than on the outer surface of the skull. The sagittal and coronal sutures are as a rule the first to become ossified—the process starting near the posterior extremity of the former and the lower ends of the latter.

THE SKULL AS A WHOLE

The Skull, formed by the union of the several cranial and facial bones already described, when considered as a whole, is divisible into five regions: a superior region or vertex, an inferior region or base, two lateral regions, and an anterior region, the face.

Vertex of the Skull

The **Superior Region,** or **Vertex,** presents two surfaces, an external and an internal.

The **external surface** is bounded, in front, by the glabella and supra-orbital ridges; behind, by the occipital protuberance and superior curved lines of the occipital bone; laterally, by an imaginary line extending from the outer end of the superior curved line, along the temporal ridge, to the external angular process of the frontal. This surface includes the greater part of the vertical portion of the frontal, the greater part of the parietal, and the superior third of the occipital bone; it is smooth, convex, of an elongated oval form, crossed transversely by the coronal suture, and from before backwards by the sagittal, which terminates behind in the lambdoid. The point of junction of the coronal and sagittal sutures is named the *bregma,* and is represented by a line drawn vertically upwards from the external auditory meatus, the head being in its normal position. The point of junction of the sagittal and lambdoid sutures is called the *lambda,* and is about two inches and three-quarters above the external occipital protuberance. From before backwards may be seen the frontal eminences and remains of the suture connecting the two lateral halves of the frontal bone; on each side of the sagittal suture are the parietal foramen and parietal eminence, and still more posteriorly the convex surface of the occipital bone. In the neighbourhood of the parietal foramen the skull is often flattened, and the name of *obelion* is sometimes given to that point of the sagittal suture which lies exactly opposite to the parietal foramen.

The **internal surface** is concave, presents depressions for the convolutions of the cerebrum, and numerous furrows for the lodgment of branches of the meningeal arteries. Along the middle line of this surface is a longitudinal groove, narrow in front, where it commences at the frontal crest, but broader behind; it lodges the superior longitudinal sinus, and by its margins affords attachment to the falx cerebri. On either side of it are several depressions for the Pacchionian bodies, and at its back part, the internal openings of the parietal foramina. This surface is crossed, in front, by the coronal suture; from before backwards, by the sagittal; behind by the lambdoid.

Base of the Skull

The **Inferior Region,** or **Base of the Skull,** presents two surfaces—an internal or cerebral, and an external or basilar.

The **internal** or **cerebral surface** (fig. 71) presents three fossæ, called the *anterior, middle,* and *posterior* fossæ of the cranium.

The **Anterior Fossa** is formed by the orbital plates of the frontal, the cribriform plate of the ethmoid, the anterior third of the superior surface of the body, and the upper surface of the lesser wings of the sphenoid. It is the most elevated of the three fossæ, convex externally where it corresponds to the roof of the orbit, concave in the median line in the situation of the cribriform plate of the ethmoid. It is traversed by three sutures, the *ethmo-frontal, ethmo-sphenoidal,* and *fronto-sphenoidal;* and lodges the frontal lobes of the cerebrum. It presents, in the median line, from before backwards, the commencement of the *groove* for the superior longitudinal sinus, and the *frontal crest* for the attachment of the falx cerebri; the *foramen cæcum,* an aperture formed between the frontal bone and the crista galli of the ethmoid, which, if pervious, transmits a small vein from the nose to the superior longitudinal sinus; behind the foramen cæcum, the *crista galli,* the posterior margin of which affords attachment to the falx cerebri; on either side of the crista galli, the cribriform plate, which supports the olfactory bulb, and presents three rows of foramina for the transmission of its nervous filaments, and in front a slit-like opening for the nasal branch of the ophthalmic division of the fifth nerve. On the outer side of each olfactory groove are the internal openings of the *anterior* and *posterior ethmoidal foramina;* the former, situated about the middle of the outer margin of the olfactory groove, transmits the anterior ethmoidal vessels and the nasal nerve, which latter runs in a depression along the surface of the ethmoid, to the slit-like opening above mentioned; while

FIG. 71.—Base of the skull.   Inner or cerebral surface.

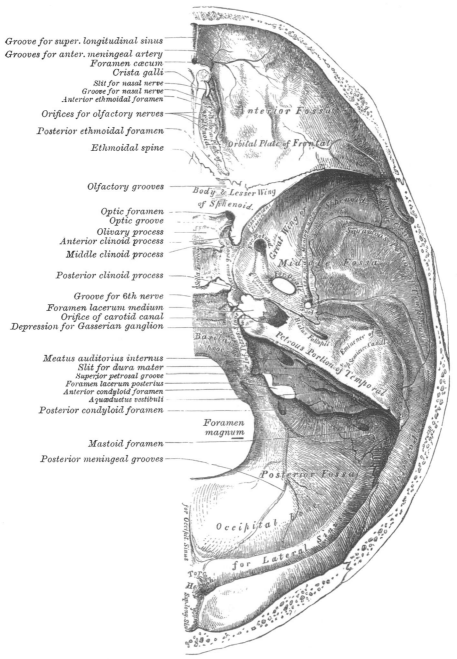

Groove for super. longitudinal sinus
Grooves for anter. meningeal artery
Foramen cæcum
Crista galli
Slit for nasal nerve
Groove for nasal nerve
Anterior ethmoidal foramen
Orifices for olfactory nerves
Posterior ethmoidal foramen
Ethmoidal spine

Olfactory grooves

Optic foramen
Optic groove
Olivary process
Anterior clinoid process
Middle clinoid process

Posterior clinoid process

Groove for 6th nerve
Foramen lacerum medium
Orifice of carotid canal
Depression for Gasserian ganglion

Meatus auditorius internus
Slit for dura mater
Superior petrosal groove
Foramen lacerum posterius
Anterior condyloid foramen
Aquæductus vestibuli
Posterior condyloid foramen

Mastoid foramen
Posterior meningeal grooves

Anterior Fossa
Orbital Plate of Frontal
Body & Lesser Wing of Sphenoid
Great Wing of Sphenoid
Middle Fossa
Petrous Portion of Temporal
Foramen magnum
Posterior Fossa
Occipital
for Lateral Sinus
for Occipit. Sinus
Torc. Heroph.
Sig. long. Sinus

the posterior ethmoidal foramen opens at the back part of this margin under cover of the projecting lamina of the sphenoid, and transmits the posterior ethmoidal vessels.   Farther back in the middle line is the *ethmoidal spine*, bounded behind by a slight elevation, separating two shallow longitudinal grooves which support the olfactory lobes. Behind this is a transverse sharp ridge, running outwards on either side to the anterior margin of the optic foramen, and separating the anterior from the middle fossa of the base of the skull. The anterior fossa presents, laterally, depressions for the convolutions of the brain, and grooves for the lodgment of the anterior meningeal arteries.

The **Middle Fossa,** deeper than the preceding, is narrow in the middle line, but becomes wider at the side of the skull. It is bounded in front by the posterior margin of the lesser wing of the sphenoid, the anterior clinoid process, and the ridge forming the anterior margin of the optic groove; behind, by the superior border of the petrous portion of the temporal, and the dorsum ephippii; externally by the squamous portion of the temporal, anterior inferior angle of the parietal bone, and greater wing of the sphenoid. It is traversed by four sutures, the *squamo-parietal, spheno-parietal, squamo-sphenoidal,* and *petro-sphenoidal.*

In the middle line, from before backwards, is the *optic groove*, behind which lies the optic commissure; the groove terminates on each side in the optic foramen, for the passage of the optic nerve and ophthalmic artery; behind the optic groove is the *olivary process*, and laterally the *anterior clinoid processes*, to which are attached processes of the tentorium cerebelli. Farther back is the *sella turcica*, a deep depression, which lodges the pituitary gland, bounded in front by a small eminence on either side, the *middle clinoid process*, and behind by a broad square plate of bone, the *dorsum ephippii*, surmounted at each superior angle by a tubercle, the *posterior clinoid process*; beneath the latter process is a notch, for the sixth nerve. On each side of the sella turcica is the *cavernous groove*: it is broad, shallow, and curved somewhat like the italic letter *f*; it commences behind at the foramen lacerum medium, and terminates on the inner side of the anterior clinoid process, and presents along its outer margin a ridge of bone. This groove lodges the cavernous sinus, the internal carotid artery, and the nerves of the orbit. The sides of the middle fossa are of considerable depth: they present depressions for the convolutions of the brain, and grooves for the branches of the middle meningeal artery; the latter commence on the outer side of the foramen spinosum, and consist of two large branches, an anterior and a posterior; the former passing upwards and forwards to the anterior inferior angle of the parietal bone, the latter passing upwards and backwards. The following foramina may also be seen from before backwards. Most anteriorly is the *foramen lacerum anterius*, or *sphenoidal fissure*, formed above by the lesser wing of the sphenoid; below, by the greater wing; internally, by the body of the sphenoid; and sometimes completed externally by the orbital plate of the frontal bone. It transmits the third, the fourth, the three branches of the ophthalmic division of the fifth, the sixth nerve, some filaments from the cavernous plexus of the sympathetic, the orbital branch of the middle meningeal artery, a recurrent branch from the lachrymal artery to the dura mater, and the ophthalmic vein. Behind the inner extremity of the sphenoidal fissure is the *foramen rotundum*, for the passage of the second division of the fifth or superior maxillary nerve; still more posteriorly is seen a small orifice, the *foramen Vesalii*, an opening situated between the foramen rotundum and ovale, a little internal to both: it varies in size in different individuals, and is often absent; when present, it transmits a small vein. It opens below into the pterygoid fossa, just at the outer side of the scaphoid depression. Behind and external to the latter opening is the *foramen ovale*, which transmits the third division of the fifth or inferior maxillary nerve, the small meningeal artery, and the small petrosal nerve.* On the outer side of the foramen ovale is the *foramen spinosum*, for the passage of the middle meningeal artery; and on the inner side of the foramen ovale, the *foramen lacerum medium*. The lower part of this aperture is filled up with cartilage in the recent state. The Vidian nerve and a meningeal branch from the

* See footnote, p. 43.

ascending pharyngeal artery pierce this cartilage. On the anterior surface of
the petrous portion of the temporal bone is seen, from without inwards, the
eminence caused by the projection of the superior semicircular canal; in front
of and a little outside this a depression corresponding to the roof of the tympanum;
the groove leading to the hiatus Fallopii, for the transmission of the petrosal
branch of the Vidian nerve and the petrosal branch of the middle meningeal
artery; beneath it, the smaller groove, for the passage of the lesser petrosal
nerve; and, near the apex of the bone, the depression for the Gasserian ganglion
and the orifice of the carotid canal, for the passage of the internal carotid artery
and carotid plexus of nerves.

The **Posterior Fossa,** deeply concave, is the largest of the three, and situated
on a lower level than either of the preceding. It is formed by the posterior third
of the superior surface of the body of the sphenoid, by the occipital, the petrous
and mastoid portions of the temporal, and the posterior inferior angle of the
parietal bone; it is crossed by four sutures, the *petro-occipital*, the *masto-
occipital*, the *masto-parietal*, and the *basilar*; and lodges the cerebellum, pons
Varolii, and medulla oblongata. It is separated from the middle fossa in the
median line by the dorsum ephippii, and on each side by the superior border of
the petrous portion of the temporal bone. This border serves for the attachment
of the tentorium cerebelli, is grooved for the superior petrosal sinus, and at its
inner extremity presents a notch, upon which rests the fifth nerve. The circumference of the fossa is bounded posteriorly by the grooves for the lateral sinuses.
In the centre of this fossa is the *foramen magnum*, bounded on either side by a
rough tubercle, which gives attachment to the odontoid or check ligaments; and
a little above these are seen the internal openings of the *anterior condyloid
foramina*, through which pass the hypoglossal nerve and a meningeal branch from
the ascending pharyngeal artery. In front of the foramen magnum is a grooved
surface, formed by the basilar process of the occipital bone and by the posterior
third of the superior surface of the body of the sphenoid, which supports the
medulla oblongata and pons Varolii, and articulates on each side with the petrous
portion of the temporal bone, forming the *petro-occipital suture*, the anterior half
of which is grooved for the inferior petrosal sinus, the posterior half being encroached upon by the *foramen lacerum posterius*, or *jugular foramen*. This foramen
presents three compartments: through the anterior, passes the inferior petrosal
sinus; through the posterior, the lateral sinus and some meningeal branches
from the occipital and ascending pharyngeal arteries; and through the middle,
the glosso-pharyngeal, pneumo-gastric, and spinal accessory nerves. Above the
jugular foramen is the *internal auditory meatus*, for the facial and auditory nerves
and auditory artery; behind and external to this is the slit-like opening leading
into the aquæductus vestibuli, which lodges the ductus endolymphaticus; while
between the two latter, and near the superior border of the petrous portion, is a
small, triangular depression, the remains of the floccular fossa, which lodges a
process of the dura mater and occasionally transmits a small vein into the substance of the bone. Behind the foramen magnum are the *inferior occipital fossæ*,
which lodge the hemispheres of the cerebellum, separated from one another by
the *internal occipital crest*, which serves for the attachment of the falx cerebelli,
and lodges the occipital sinus. The posterior fossæ are surmounted, above, by
the deep transverse grooves for the lodgment of the *lateral sinuses*. These
channels, in their passage outwards, groove the occipital bone, the posterior
inferior angle of the parietal, the mastoid portion of the temporal, and the jugular
process of the occipital, and terminate at the back part of the jugular foramen.
Where this sinus grooves the mastoid portion of the temporal bone, the orifice
of the *mastoid foramen* may be seen; and, just previous to its termination,
it has opening into it the *posterior condyloid foramen*. Neither foramen is
constant.

The **External Surface** of the Base of the Skull (fig. 72) is extremely irregular.
It is bounded in front by the incisor teeth in the upper jaw; behind, by the

superior curved lines of the occipital bone ; and laterally by the alveolar arch, the lower border of the malar bone, the zygoma, and an imaginary line, extending from the zygoma to the mastoid process and extremity of the superior curved line

FIG. 72.—Base of the skull.   External surface.

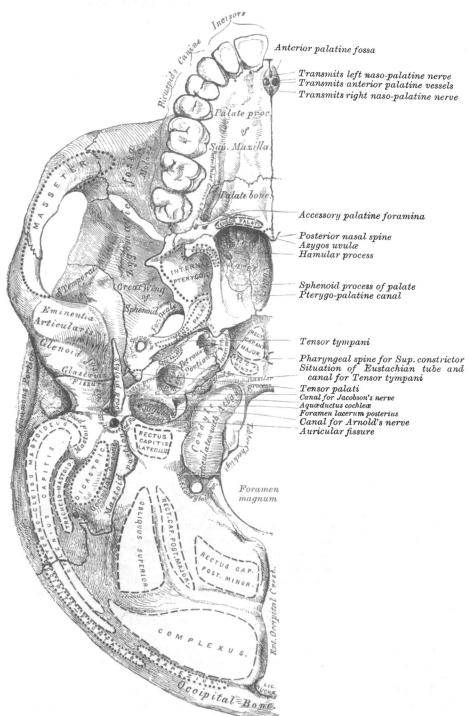

of the occiput. It is formed by the palate processes of the superior maxillary and palate bones, the vomer, the pterygoid processes, under surface of the great wing, spinous processes and part of the body of the sphenoid, the under surface of the squamous, mastoid, and petrous portions of the temporal, and the under surface of the occipital bone. The anterior part of the base of the skull is raised above the level of the rest of this surface (when the skull is turned over for the purpose of examination), and is surrounded by the alveolar process, which is thicker behind than in front, and excavated by sixteen depressions for lodging the teeth of the upper jaw, the cavities varying in depth and size according to the teeth they contain. Immediately behind the incisor teeth is the *anterior palatine fossa*. At the bottom of this fossa may usually be seen four apertures : two placed laterally, the *foramina of Stenson*, which open above, one in the floor of each nostril, and transmit the anterior branch of the posterior palatine vessels, and two in the median line in the intermaxillary suture, the *foramina of Scarpa*, one in front of the other, the anterior transmitting the left, and the posterior (the larger) the right, naso-palatine nerve. These two latter canals are sometimes wanting, or they may join to form a single one, or one of them may open into one of the lateral canals above referred to. The palatine vault is concave, uneven, perforated by numerous foramina, marked by depressions for the palatine glands, and crossed by a crucial suture, formed by the junction of the four bones of which it is composed. At the front part of this surface, a delicate linear suture may frequently be seen, passing outwards and forwards from the anterior palatine fossa to the interval between the lateral incisor and canine teeth, and marking off the pre-maxillary portion of the bone. At each posterior angle of the hard palate is the *posterior palatine foramen*, for the transmission of the posterior palatine vessels and large descending palatine nerve ; and running forwards and inwards from it a groove, for the same vessels and nerve. Behind the posterior palatine foramen is the *tuberosity of the palate-bone*, perforated by one or more accessory posterior palatine canals, and marked by the commencement of a ridge, which runs transversely inwards, and serves for the attachment of the tendinous expansion of the Tensor palati muscle. Projecting backwards from the centre of the posterior border of the hard palate is the *posterior nasal spine*, for the attachment of the Azygos uvulæ muscle. Behind and above the hard palate is the posterior aperture of the nares, divided into two parts by the vomer, bounded above by the body of the sphenoid, below by the horizontal plate of the palate-bone, and laterally by the internal pterygoid plate of the sphenoid. Each aperture measures about an inch in the vertical, and about half an inch in the transverse direction. At the base of the vomer may be seen the expanded alæ of this bone, receiving between them, on each side, the rostrum of the sphenoid. Near the lateral margins of the vomer, at the root of the pterygoid processes, are the *pterygo-palatine canals*. The pterygoid process, which bounds the posterior nares on each side, presents near its base the *pterygoid* or *Vidian canal*, for the Vidian nerve and artery. Each process consists of two plates, which bifurcate at the extremity to receive the tuberosity of the palate-bone, and are separated behind by the pterygoid fossa, which lodges the Internal pterygoid muscle. The internal plate is long and narrow, presenting on the outer side of its base the *scaphoid fossa*, for the origin of the Tensor palati muscle, and at its extremity the *hamular process*, around which the tendon of this muscle turns. The external pterygoid plate is broad, forms the inner boundary of the zygomatic fossa, and affords attachment, by its outer surface, to the External pterygoid muscle.

Behind the nasal fossæ in the middle line is the basilar surface of the occipital bone, presenting in its centre the *pharyngeal spine* for the attachment of the Superior constrictor muscle of the pharynx, with depressions on each side for the insertion of the Rectus capitis anticus major and minor. At the base of the external pterygoid plate is the *foramen ovale*, for the transmission of the third division of the fifth nerve, the small meningeal artery, and sometimes the small petrosal nerve ; behind this, the *foramen spinosum*, which transmits the middle meningeal artery, and the prominent spinous process of the sphenoid, which gives attachment to the internal lateral ligament of the lower jaw and the Tensor palati muscle. External to the spinous process is the *glenoid fossa*, divided into two parts by the Glaserian fissure (page 34), the anterior portion concave, smooth, bounded in front by the eminentia articularis, and serving for the articulation of the condyle of the lower jaw ; the posterior portion rough, bounded behind by the tympanic plate, and serving for the reception of part of the parotid gland. Emerging from between the laminæ of the vaginal process of the tympanic plate is the *styloid process* ; and at the base of this process is the *stylo-mastoid foramen,* for the exit of the facial nerve, and entrance of the stylo-mastoid artery. External to the stylo-mastoid foramen is the *auricular fissure* for the auricular branch of the pneumogastric, bounded behind by the mastoid process. Upon the inner side of the mastoid process is a deep groove, the *digastric fossa* ; and a little more internally, the *occipital groove,* for the occipital artery. At the base of the internal pterygoid

FIG. 73.—Side view of the skull.

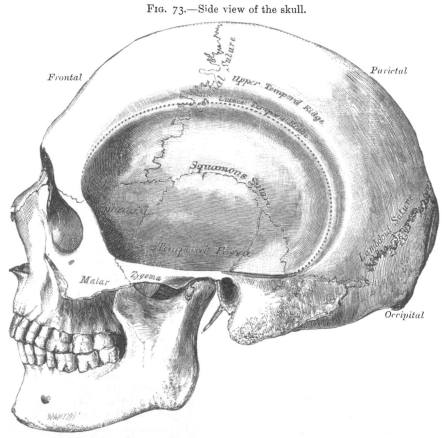

plate is a large and somewhat triangular aperture, the *foramen lacerum medium,* bounded in front by the great wing of the sphenoid, behind by the apex of the petrous portion of the temporal bone, and internally by the body of the sphenoid and basilar process of the occipital bone ; it presents in front the posterior orifice of the Vidian canal ; behind, the aperture of the carotid canal. The basilar surface of this opening is filled up in the recent state by a fibro-cartilaginous substance ; across its upper or cerebral aspect passes the internal carotid artery. External to this aperture, the *petro-sphenoidal suture* is observed, at the outer termination of which is seen the orifice of the canal for the Eustachian tube, and that for the Tensor tympani muscle. Behind this suture is seen the under surface of the petrous portion of the temporal bone, presenting, from within outwards, the quadrilateral, rough surface, part of which affords attachment to the Levator palati and Tensor tympani muscles ; external to this surface the orifices of the carotid canal and the aquæductus cochleæ, the former transmitting the internal carotid artery and the ascending branches of the superior cervical ganglion of the sympathetic, the latter serving for the passage of a small artery and vein to the cochlea. Behind the carotid canal is a large aperture, the *jugular foramen*, formed in front by the petrous portion of the temporal, and behind by the occipital; it is generally larger on the right than on the left side, and is divided into three compartments by processes of dura mater. The anterior is for the passage of the inferior petrosal sinus ; the posterior for the lateral sinus and some meningeal branches from the occipital and ascending pharyngeal arteries ; the central one for the glosso-pharyngeal, pneumogastric, and spinal accessory nerves. On the ridge of bone dividing the carotid canal from the jugular foramen is the small foramen for the transmission of Jacobson's nerve ; and on the wall of the jugular foramen, near the root of the styloid process, is the small aperture for the transmission of Arnold's nerve. Behind the basilar surface of the occipital bone is the *foramen magnum*, bounded on each side by the condyles, rough internally for the attachment of the check or odontoid ligaments, and presenting externally a rough surface, the *jugular process*, which serves for the attachment of the Rectus capitis lateralis muscle and the lateral occipito-atlantal ligament. On either side of each condyle anteriorly is the *anterior condyloid fossa*, perforated by the anterior condyloid foramen, for the passage of the hypoglossal nerve and a meningeal artery. Behind each condyle is the *posterior condyloid fossa*, perforated on one or both sides by the posterior condyloid foramen, for the transmission of a vein to the lateral sinus. Behind the foramen magnum is the *external occipital crest*, terminating above at the *external occipital protuberance*, while on each side are seen the *superior* and *inferior curved lines* ; these, as well as the surfaces of bone between them, are rough for the attachment of the muscles, which are enumerated on page 24.

LATERAL REGION OF THE SKULL

The **Lateral Region** of the Skull is of a somewhat triangular form, the base of the triangle being formed by a line extending from the external angular process of the frontal bone along the temporal ridge backwards to the outer extremity of the superior curved line of the occiput : and the sides by two lines, the one drawn downwards and backwards from the external angular process of the frontal bone to the angle of the lower jaw, the other from the angle of the jaw upwards and backwards to the outer extremity of the superior curved line. This region is divisible into three portions—temporal fossa, mastoid portion, and zygomatic fossa.

THE TEMPORAL FOSSA

The **Temporal Fossa** is bounded above and behind by the temporal ridges, which extend from the external angular process of the frontal upwards and backwards across the frontal and parietal bones, curving downwards behind to terminate in the posterior root of the zygomatic process (*supra-mastoid crest*). In front, it is bounded by the frontal, malar, and great wing of the sphenoid ; externally, by the zygomatic arch, formed conjointly by the malar and temporal bones ; below, it is separated from the zygomatic fossa by the pterygoid ridge, seen on the outer surface of the great wing of the sphenoid. This fossa is formed by five bones, part of the frontal, great wing of the sphenoid, parietal, squamous portion of the temporal, and malar bones, and is traversed by six sutures, part of the *transverse facial, spheno-malar, coronal, spheno-parietal, squamo-parietal*, and *squamo-sphenoidal*. The point where the coronal suture crosses the superior temporal ridge is sometimes named the *stephanion* ; and the region where the four bones, the parietal, the frontal, the squamous, and the greater wing of the sphenoid, meet at the anterior inferior angle of the parietal bone, is named the *pterion*. This point is about on a level with the external angular process of the frontal bone and about an inch and a half behind it. This fossa is deeply concave in front, convex behind, traversed by grooves which lodge branches of the deep temporal arteries, and filled by the Temporal muscles.

THE MASTOID PORTION

The **Mastoid Portion** of the side of the skull is bounded in front by the tubercle of the zygoma ; above, by a line which runs from the posterior root of the zygoma to the end of the mastoid-parietal suture ; behind and below by the masto-occipital suture. It is formed by the mastoid and part of the squamous and petrous portions of the temporal bone ; its surface is convex and rough for the attachment of muscles, and presents, from behind forwards, the mastoid foramen, the mastoid process, the external auditory meatus, surrounded by the tympanic plate, and, most anteriorly, the temporo-maxillary articulation.

The point where the posterior inferior angle of the parietal meets the occipital bone and mastoid portion of the temporal is named the *asterion*.

THE ZYGOMATIC FOSSA

The **Zygomatic Fossa** is an irregularly shaped cavity, situated below and on the inner side of the zygoma ; bounded, in front, by the zygomatic surface of the superior maxillary bone and the ridge which descends from its malar process ; behind, by the posterior border of the external pterygoid plate and the eminentia articularis ; above, by the pterygoid ridge on the outer surface of the great wing

FIG. 74.—Zygomatic fossa.

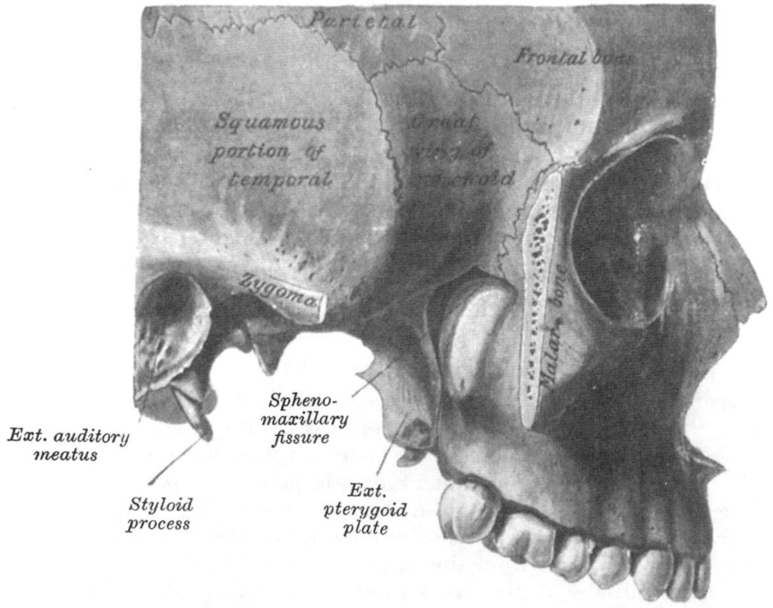

of the sphenoid, and the under part of the squamous portion of the temporal ; below, by the alveolar border of the superior maxilla ; internally, by the external pterygoid plate ; and externally, by the zygomatic arch and ramus of the lower jaw (fig. 74). It contains the lower part of the Temporal, the External and Internal pterygoid muscles, the internal maxillary artery and vein, and inferior maxillary nerve, and their branches. At its upper and inner part may be observed two fissures, the spheno-maxillary and pterygo-maxillary.

The **Spheno-maxillary Fissure,** horizontal in direction, opens into the outer and back part of the orbit. It is formed above by the lower border of the orbital surface of the great wing of the sphenoid ; below, by the external border of the orbital surface of the superior maxilla and a small part of the palate-bone ; externally, by a small part of the malar bone : * internally, it joins at right angles with the pterygo-maxillary fissure. This fissure opens a communication from the orbit into three fossæ—the temporal, zygomatic, and spheno-maxillary ; it transmits the superior maxillary nerve and its orbital branch, the infra-orbital vessels, and ascending branches from the spheno-palatine or Meckel's ganglion.

The **Pterygo-maxillary Fissure** is vertical, and descends at right angles from the inner extremity of the preceding ; it is a V-shaped interval, formed by the divergence of the superior maxillary bone from the pterygoid process of the sphenoid. It serves to connect the spheno-maxillary fossa with the zygomatic fossa, and transmits branches of the internal maxillary artery.

THE SPHENO-MAXILLARY FOSSA

The **Spheno-maxillary Fossa** is a small, triangular space situated at the angle of junction of the spheno-maxillary and pterygo-maxillary fissures, and placed beneath the apex of the orbit. It is formed above by the under surface of the body of the sphenoid and by the orbital process of the palate-bone ; in front, by the superior maxillary bone ; behind, by the anterior surface of the base of the pterygoid process and lower part of the anterior surface of the great wing of the sphenoid ; internally by the vertical plate of the palate. This fossa has three fissures terminating in it, the *sphenoidal, spheno-maxillary,* and *pterygo-maxillary* ; it communicates with the orbit by the spheno-maxillary fissure ; with the nasal fossæ by the spheno-palatine foramen, and with the zygomatic fossa by the pterygo-maxillary fissure. It also communicates with the cavity of the cranium, and has opening into it five foramina. Of these there are three on the posterior wall : the *foramen rotundum* above ; below, and internal to this, the *Vidian canal* ; and still more inferiorly and internally, the *pterygo-palatine canal.* On the inner wall is the *spheno-palatine foramen* by which the spheno-maxillary communicates with the nasal fossa ; and below is the superior orifice of the *posterior palatine canal,* besides occasionally the orifices of the *accessory posterior palatine* canals. The fossa contains the superior maxillary nerve and Meckel's ganglion, and the termination of the internal maxillary artery.

ANTERIOR REGION OF THE SKULL

The **Anterior Region** of the Skull, which forms the face, is of an oval form, presents an irregular surface, and is excavated for the reception of two of the organs of sense, the eye and the nose. It is bounded above by the glabella and margins of the orbit ; below, by the prominence of the chin ; on each side, by the malar bone, and anterior margin of the ramus of the jaw. In the median line are seen from above downwards the glabella, and diverging from it are the *superciliary ridges,* which indicate the situation of the frontal sinuses and support the eyebrows. Beneath the glabella is the fronto-nasal suture, the mid-point of which is termed the *nasion,* and below this is the arch of the nose, formed by the nasal bones, and the nasal processes of the superior maxillary. The nasal arch is convex from side to side, concave from above downwards, presenting in the median line the inter-nasal suture formed between the nasal bones, and laterally the naso-maxillary suture formed between the nasal bone and the nasal process of the superior maxillary bone. Below the nose is seen the opening of the anterior nares, which is heart-shaped, with the narrow end upwards, and presents

* Occasionally the superior maxillary bone and the sphenoid articulate with each other at the anterior extremity of this fissure ; the malar is then excluded from entering into its formation.

laterally the thin, sharp margins serving for the attachment of the lateral cartilages of the nose, and in the middle line below, a prominent process, the *anterior
nasal spine*, bounded by two deep notches. Below this is the *intermaxillary suture*,
and on each side of it the *incisive fossa*. Beneath this fossa are the alveolar
processes of the upper and lower jaws, containing the incisor teeth, and at the
lower part of the median line, the *symphysis of the chin*, the *mental process*, with
its two *mental tubercles*, separated by a median groove, and the *incisive fossa* of
the lower jaw.

On each side, proceeding from above downwards, is the *supra-orbital ridge*,
terminating externally in the external angular process at its junction with the
malar, and internally in the internal angular process ; towards the inner third of
this ridge is the *supra-orbital notch* or *foramen*, for the passage of the supra-
orbital vessels and nerve. Beneath the supra-orbital ridge is the opening of the
orbit, bounded externally by the orbital ridge of the malar bone ; below, by the
orbital ridge formed by the malar and superior maxillary bones ; internally, by
the nasal process of the superior maxillary, and the internal angular process of
the frontal bone. On the outer side of the orbit is the quadrilateral anterior
surface of the malar bone perforated by one or two small malar foramina. Below
the inferior margin of the orbit is the *infra-orbital foramen*, the termination of
the infra-orbital canal, and beneath this the *canine fossa*, which gives attachment
to the Levator anguli oris ; still lower are the alveolar processes, containing
the teeth of the upper and lower jaws. Beneath the alveolar arch of the lower
jaw is the *mental foramen* for the passage of the mental vessels and nerve, the
external oblique line, and at the lower border of the bone, at the point of junction
of the body with the ramus, a shallow groove for the passage of the facial artery.

THE ORBITS

The **Orbits** (fig. 75) are two quadrilateral pyramidal cavities, situated at the
upper and anterior part of the face, their bases being directed forwards and
outwards, and their apices backwards and inwards, so that the axes of the two,
if continued backwards, would meet over the body of the sphenoid bone. Each
orbit is formed of *seven* bones, the frontal, sphenoid, ethmoid, superior maxillary,
malar, lachrymal, and palate ; but three of these, the frontal, ethmoid, and
sphenoid, enter into the formation of *both* orbits, so that the two cavities are
formed of *eleven* bones only. Each cavity presents for examination a roof, a
floor, an inner and an outer wall, four angles, a circumference or base, and an apex.
The **roof** is concave, directed downwards and slightly forwards, and formed in
front by the orbital plate of the frontal ; behind by the lesser wing of the sphenoid.
This surface presents internally the depression for the cartilaginous pulley of the
Superior oblique muscle ; externally, the depression for the lachrymal gland ;
and posteriorly, the suture connecting the frontal and lesser wing of the
sphenoid.

The **floor** is directed upwards and outwards, and is of less extent than the roof ;
it is formed chiefly by the orbital process of the superior maxillary bone ; in front,
to a small extent, by the orbital process of the malar, and behind, by the superior
surface of the orbital process of the palate. This surface presents at its anterior
and internal part, just external to the lachrymal groove, a depression for the
attachment of the Inferior oblique muscle ; externally, the suture between the
malar and superior maxillary bones ; near its middle, the infra-orbital groove ;
and posteriorly, the suture between the maxillary and palate bones.

The **inner wall** is flattened, nearly vertical, and formed from before backwards
by the nasal process of the superior maxilla, the lachrymal, os planum of the
ethmoid, and a small part of the body of the sphenoid. This surface presents the
lachrymal groove, the crest of the lachrymal bone, and the sutures connecting the
lachrymal with the superior maxillary, the ethmoid with the lachrymal in front,
and the ethmoid with the sphenoid behind.

The **outer wall** is directed forwards and inwards, and is formed in front by the orbital process of the malar bone; behind, by the orbital surface of the greater wing of the sphenoid. On it are seen the orifices of one or two malar canals, and the suture connecting the sphenoid and malar bones.

Angles.—The *superior external angle* is formed by the junction of the upper and outer walls; it presents, from before backwards, the suture connecting the frontal with the malar in front, and with the great wing of the sphenoid behind; quite posteriorly is the foramen lacerum anterius, or sphenoidal fissure, which transmits the third, the fourth, the three branches of the ophthalmic division of the fifth, the sixth nerve, some filaments from the cavernous plexus of the sympathetic, the orbital branch of the middle meningeal artery, a recurrent branch

FIG. 75.—Anterior region of the skull.

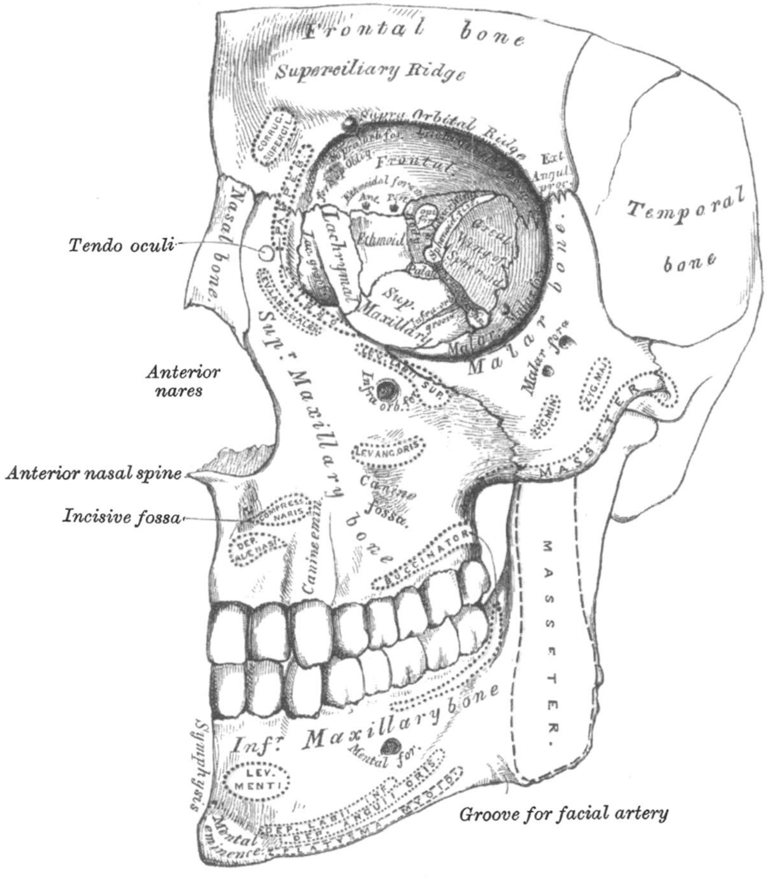

from the lachrymal artery to the dura mater, and the ophthalmic vein. The *superior internal angle* is formed by the junction of the upper and inner wall, and presents the suture connecting the frontal bone with the lachrymal in front, and with the ethmoid behind. The point of junction of the *anterior* border of the lachrymal with the frontal has been named the *dacryon*. This angle presents two foramina, the anterior and posterior ethmoidal, the former transmitting the anterior ethmoidal vessels and nasal nerve, the latter the posterior ethmoidal vessels. The *inferior external angle*, formed by the junction of the outer wall and floor, presents the spheno-maxillary fissure, which transmits the superior maxillary nerve and its orbital branches, the infra-orbital vessels, and the ascending branches from the spheno-palatine or Meckel's ganglion. The *inferior internal* *angle* is formed by the union of the lachrymal and os planum of the ethmoid, with the superior maxillary and palate bones. The *circumference*, or base, of the orbit, quadrilateral in form, is bounded above by the supra-orbital ridge; below, by the anterior border of the orbital plate of the malar and superior maxillary bones; externally, by the external angular process of the frontal and the malar bones; internally, by the internal angular process of the frontal, and the nasal process of the superior maxillary. The circumference is marked by three sutures, the fronto-maxillary internally, the fronto-malar externally, and the malo-maxillary below; it contributes to the formation of the lachrymal groove, and presents, above, the supra-orbital notch (or foramen), for the passage of the supra-orbital vessels and nerve. The *apex*, situated at the back of the orbit, corresponds to the optic foramen,* a short, circular canal, which transmits the optic nerve and ophthalmic artery. It will thus be seen that there are *nine* openings communicating with each orbit—viz. the optic foramen, sphenoidal fissure, spheno-maxillary fissure, supra-orbital foramen, infra-orbital canal, anterior and posterior ethmoidal foramina, malar foramina, and canal for the nasal duct.

THE NASAL FOSSÆ

The **Nasal Fossæ** are two large, irregular cavities, situated on either side of the middle line of the face, extending from the base of the cranium to the roof of the mouth, and separated from each other by a thin vertical septum. They communicate by two large apertures, the *anterior nares*, with the front of the face; and by the two *posterior nares* with the naso-pharynx behind. These fossæ are much narrower above than below, and in the middle than at the anterior or posterior openings: their depth, which is considerable, is much greater in the middle than at either extremity. Each nasal fossa communicates with four sinuses, the frontal above, the sphenoidal behind, and the maxillary and ethmoidal on the outer wall. Each fossa also communicates with four cavities: with the orbit by the lachrymal groove, with the mouth by the anterior palatine canal, with the cranium by the olfactory foramina, and with the spheno-maxillary fossa by the spheno-palatine foramen; and they occasionally communicate with each other by an aperture in the septum. The bones entering into their formation are fourteen in number: three of the cranium, the frontal, sphenoid, and ethmoid, and all the bones of the face, excepting the malar and lower jaw. Each cavity is bounded by a roof, a floor, an inner and an outer wall.

The **upper wall**, or **roof** (fig. 76), is long, narrow, and horizontal in its centre, but sloped downwards at its anterior and posterior extremities; it is formed in front by the nasal bones and nasal spine of the frontal, which are directed downwards and forwards; in the middle, by the cribriform plate of the ethmoid, which is horizontal; and behind, by the under surface of the body of the sphenoid, and sphenoidal turbinated bones, the ala of the vomer and the sphenoidal process of the palate-bone, which are directed downwards and backwards. This surface presents, from before backwards, the internal aspect of the nasal bones; on their outer side, the suture formed between the nasal bone and the nasal process of the superior maxillary; on their inner side, the elevated crest which receives the nasal spine of the frontal and the perpendicular plate of the ethmoid, and articulates with its fellow of the opposite side; while the surface of the bones is perforated by a few small vascular apertures, and presents the longitudinal groove for the nasal nerve: farther back is the transverse suture, connecting the frontal with the nasal in front and the ethmoid behind, the olfactory foramina and nasal slit on the under surface of the cribriform plate, and the suture between it and the sphenoid behind: quite posteriorly are seen the sphenoidal turbinated

---

* Quain, Testut, and others give the apex of the orbit as corresponding with the inner end of the sphenoidal fissure. It seems better, however, to adopt the statement in the text, since the muscles of the eyeball take origin around the optic foramen, and diverge from it to the globe of the eye.

bones, the orifices of the sphenoidal sinuses, and the articulation of the alæ of the vomer with the under surface of the body of the sphenoid.

The **floor** is flattened from before backwards, concave from side to side, and wider in the middle than at either extremity. It is formed in front by the palate process of the superior maxillary; behind, by the palate process of the palatebone. This surface presents, from before backwards, the anterior nasal spine; behind this, the upper orifices of the anterior palatine canal; internally, the elevated crest which articulates with the vomer; and behind, the suture between the palate and superior maxillary bones, and the posterior nasal spine.

The **inner wall**, or septum (fig. 77), is a thin vertical partition, which separates the nasal fossæ from one another; it is occasionally perforated, so that the fossæ communicate, and it is frequently deflected considerably to one side.* It is formed, in front, by the crest of the nasal bones and nasal spine of the frontal; in the middle, by the perpendicular plate of the ethmoid; behind, by the vomer

FIG. 76.—Roof, floor, and outer wall of left nasal fossa.

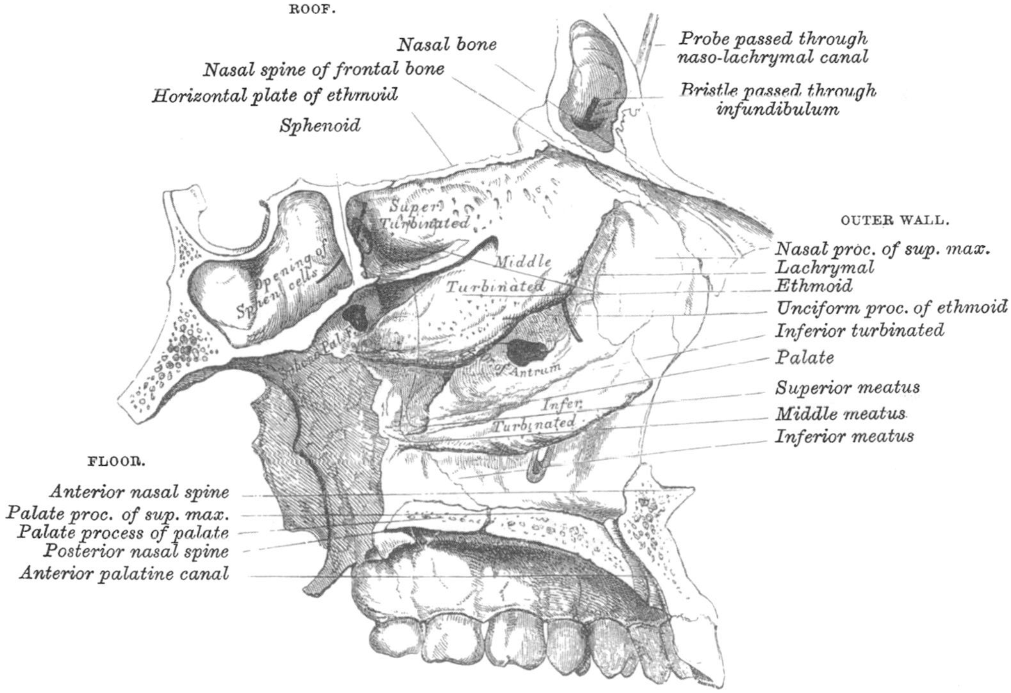

ROOF.

Nasal bone
Nasal spine of frontal bone
Horizontal plate of ethmoid
Sphenoid

Probe passed through naso-lachrymal canal
Bristle passed through infundibulum

OUTER WALL.

Nasal proc. of sup. max.
Lachrymal
Ethmoid
Unciform proc. of ethmoid
Inferior turbinated
Palate
Superior meatus
Middle meatus
Inferior meatus

FLOOR.

Anterior nasal spine
Palate proc. of sup. max.
Palate process of palate
Posterior nasal spine
Anterior palatine canal

and rostrum of the sphenoid; below, by the crest of the superior maxillary and palate bones. It presents, in front, a large, triangular notch, which receives the septal cartilage of the nose; and behind, the grooved edge of the vomer. Its surface is marked by numerous vascular and nervous canals and the groove for the naso-palatine nerve, and is traversed by sutures connecting the bones of which it is formed.

The **outer wall** (fig. 76) is formed, in front, by the nasal process of the superior maxillary and lachrymal bones; in the middle, by the ethmoid and inner surface of the superior maxillary and inferior turbinated bones; behind, by the vertical plate of the palate-bone, and the internal pterygoid plate of the sphenoid. This surface presents three irregular longitudinal passages, or *meatuses*, termed the superior, middle, and inferior meatuses of the nose. The *superior meatus*, the smallest of the three, is situated at the upper and back part of each nasal fossa,

* See footnote, p. 47.

occupying the posterior third of the outer wall. It is situated between the superior and middle turbinated bones, and has opening into it two foramina, the *spheno-palatine* at the back of its outer wall, and the *posterior ethmoidal cells* at the front part of the outer wall. The sphenoidal sinus opens into a recess, the *spheno-ethmoidal recess*, which is situated above and behind the superior turbinated bone. The *middle meatus* is situated between the middle and inferior turbinated bones, and extends from the anterior end of the inferior turbinated bone to the spheno-palatine foramen of the outer wall of the nasal fossa. It presents in front the orifice of the *infundibulum*, by which the middle meatus communicates with the anterior ethmoidal cells, and through these with the frontal sinuses. The middle ethmoidal cells also open into this meatus, while at the centre of the outer wall is the *orifice of the antrum*, which varies somewhat as to its exact position in different skulls. The *inferior meatus*, the largest of the three, is the space between the inferior turbinated bone and the floor of the nasal fossa. It extends along the entire length of the outer wall of the nose, is broader in front than behind, and presents anteriorly the lower *orifice of the canal for the nasal duct.*

FIG. 77.—Inner wall of nasal fossæ, or septum of nose.

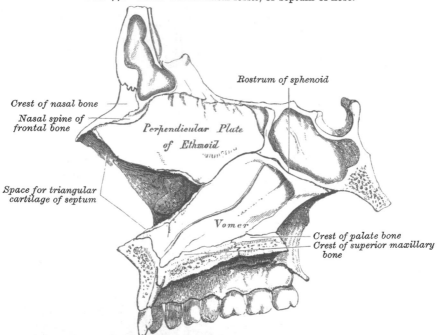

The **anterior nares** present a heart-shaped or pyriform opening, whose long axis is vertical, and narrow extremity upwards. This opening in the recent state is much contracted by the cartilages of the nose. It is bounded above by the inferior border of the nasal bone ; laterally by the thin, sharp margin which separates the facial from the nasal surface of the superior maxillary bone ; and below by the same border, where it slopes inwards to join its fellow of the opposite side at the anterior nasal spine.

The **posterior nares** or **choanæ** are the two posterior oval openings of the nasal fossæ, by which they communicate with the upper part of the pharynx. They are situated immediately in front of the basilar process, and are bounded above by the under surface of the body of the sphenoid and alæ of the vomer ; below, by the posterior border of the horizontal plate of the palate-bone ; externally, by the inner surface of the internal pterygoid plate ; and internally, in the middle line, they are separated from each other by the posterior border of the Vomer.

Surface Form.—The various bony prominences or landmarks which are to be easily felt and recognised in the head and face, and which afford the means of mapping out the important structures comprised in this region, are as follows :—

1. Supra-orbital arch.
2. Internal angular process.
3. External angular process.
4. Zygomatic arch.
5. Mastoid process.
6. External occipital protuberance.
7. Superior curved line of occipital bone.

8. Parietal eminence.
9. Temporal ridge.
10. Frontal eminence.
11. Superciliary ridge.
12. Nasal bone.
13. Lower margin of orbit.
14. Lower jaw.

1. The *supra-orbital arch* is to be felt throughout its entire extent, covered by the eyebrow. It forms the upper boundary of the circumference or base of the orbit and separates the face from the forehead. It is strong and arched, and terminates internally at the root of the nose, in the *internal angular process*, which articulates with the lachrymal bone. Externally it terminates in the *external angular process* which articulates with the malar bone. This arched ridge is sharper and more defined in its outer than in its inner half, and forms an overhanging process which protects and shields the lachrymal gland. It also protects the eye in its most exposed situation and in the direction from which blows are more likely to descend. The supra-orbital arch varies in prominence in different individuals. It is more marked in the male than in the female, and in some races of mankind than others. In the less civilised races, as the forehead recedes backwards, the supra-orbital arch becomes more prominent and approaches more to the characters of the monkey tribe, in which the supra-orbital arches are very largely developed and acquire additional prominence from the oblique direction of the frontal bone. 2. The *internal angular process* is scarcely to be felt. Its position is indicated by the angle formed by the supra-orbital arch with the nasal process of the superior maxillary bone and the lachrymal bone at the inner side of the orbit. Between the internal angular processes of the two sides is a broad surface which assists in forming the root of the nose, and immediately above this a broad, smooth, somewhat triangular surface, the *glabella*, situated between the superciliary ridges. 3. The *external angular process* is much more strongly marked than the internal, and is plainly to be felt. It is formed by the junction or confluence of the supra-orbital and temporal ridges, and, articulating with the malar bone, it serves to a very considerable extent to support the bones of the face. In carnivorous animals the external angular process does not articulate with the malar, and therefore this lateral support to the bones of the face is not present. 4. The *zygomatic arch* can be plainly felt throughout its entire length, being situated almost immediately under the skin. It is formed by the malar bone and the zygomatic process of the temporal bone. At its anterior extremity, where it is formed by the malar bone, it is broad and forms the prominence of the cheek ; the posterior part is narrower, and terminates just in front and a little above the tragus of the external ear. The lower border is more plainly to be felt than the upper, in consequence of the dense temporal fascia being attached to the latter, which somewhat obscures its outline. Its shape differs very much in individuals and in different races of mankind. In the most degraded type of skull—as, for instance, in the skull of the negro of the Guinea Coast—the malar bones project forwards and not outwards, and the zygoma at its posterior extremity extends farther outwards before it is twisted on itself to be prolonged forwards. This makes the zygomatic arch stand out in bold relief, and affords greater space for the Temporal muscle. In skulls which have a more pyramidal shape, as in the Esquimaux or Greenlander, the malar bones do not project forwards and downwards under the eyes, as in the preceding form, but take a direction outwards, forming with the zygoma a large, rounded sweep or segment of a circle. Thus it happens that if two lines are drawn from the zygomatic arches, touching the temporal ridges, they meet above the top of the head, instead of being parallel, or nearly so, as in the European skull, in which the zygomatic arches are not nearly so prominent. This gives to the face a more or less oval type. 5. Behind the ear is the *mastoid portion of the temporal bone*, plainly to be felt, and terminating below in a nipple-shaped process. Its anterior border can be traced immediately behind the concha, and its apex is on about a level with the lobule of the ear. It is rudimentary in infancy, but gradually develops in childhood, and is more marked in the negro than in the European. 6. The *external occipital protuberance* is always plainly to be felt just at the level where the skin of the neck joins that of the head. At this point the skull is thick for the purposes of safety, while radiating from it are numerous curved arches or buttresses of bone which give to this portion of the skull further security. 7. Running outwards on either side from the external occipital protuberance is an arched ridge of bone, which can be more or less plainly perceived. This is the *superior curved line* of the occipital bone, and gives attachment to some of the muscles which keep the head erect on the spine ; accordingly, we find it more developed in the negro tribes, in whom the jaws are much more massive, and therefore require stronger muscles to prevent their extra weight carrying the head forwards. Below this line the surface of bone at the back of the head is obscured by the overlying muscles. Above it, the vault of the cranium is thinly covered with soft structures, so that the form of this part of the head is almost exactly that of the upper portion of the occipital, the parietal,

and the frontal bones themselves ; and, in bald persons, even the lines of junction of the bones, especially the junction of the occipital and parietal at the lambdoid suture, may be defined as a slight depression, caused by the thickening of the borders of the bones in this situation.   8. In the line of the greatest transverse diameter of the head are generally to be found the *parietal eminences*, one on each side of the middle line ; though sometimes these eminences are not situated at the point of the greatest transverse diameter, which is at some other prominent part of the parietal region.   They denote the point where ossification of the parietal bone began.   They are much more prominent and better marked in early life, in consequence of the sharper curve of the bone at this period, so that it describes the segment of a smaller circle.   Later in life, as the bone grows, the curve spreads out and forms the segment of a larger circle, so that the eminence becomes less distinguishable.   In consequence of this sharp curve of the bone in early life, the whole of the vault of the skull has a squarer shape than it has in later life, and this appearance may persist in some rickety skulls.   The eminence is more apparent in the negro's skull than in that of the European.   This is due to greater flattening of the temporal fossa in the former skull to accommodate the larger Temporal muscle which exists in these races.   The parietal eminence is particularly exposed to injury from blows or falls on the head, but fracture is to a certain extent prevented by the shape of the bone, which forms an arch, so that the force of the blow is diffused over the bone in every direction.   9. At the side of the head may be felt the *temporal ridge*.   Commencing at the external angular process, it may be felt as a curved ridge, passing upwards and then curving backwards, on the frontal bone, separating the forehead from the temporal fossa.   It may then be traced, passing backwards in a curved direction, over the parietal bone, and, though less marked, still generally to be recognised.   Finally, the ridge curves downwards, and terminates in the posterior root of the zygoma, which separates the squamous from the subcutaneous mastoid portion of the temporal bone.   Mr. Victor Horsley has shown, in an article on the ' Topography of the Cerebral Cortex,' that the second temporal ridge (see page 28) can be made out on the living body.   10. The *frontal eminences* vary a good deal in different individuals, being considerably more prominent in some than in others, and they are often not symmetrical on the two sides of the body, the one being much more pronounced than the other.   This is often especially noticeable in the skull of the young child or infant, and becomes less marked as age advances.   The prominence of the frontal eminences depends more upon the general shape of the whole bone than upon the size of the protuberances themselves.   As the skull is more highly developed in consequence of increased intellectual capacity, so the frontal bone becomes more upright, and the frontal eminences stand out in bolder relief.   Thus they may be considered as affording, to a certain extent, an indication of the development of the hemispheres of the brain beneath, and of the mental powers of the individual.   They are not so much exposed to injury as the parietal eminences.   In falls forward the upper extremities are involuntarily thrown out, and break the force of the fall, and thus shield the frontal bone from injury.   11. Below the frontal eminences on the forehead are the *superciliary ridges*, which denote the position of the frontal sinuses, and vary according to the size of the sinuses in different individuals, being, as a rule, small in the female, absent in children, and sometimes unusually prominent in the male, when the frontal sinuses are largely developed.   They commence on either side of the glabella, and at first present a rounded form, which gradually fades away at their outer ends.   12. The *nasal bones* form the prominence of the nose.   They vary much in size and shape, and to them is due the varieties in the contour of this organ and much of the character of the face.   Thus, in the Mongolian or Ethiopian they are flat, broad, and thick at their base, giving to these tribes the flattened nose by which they are characterised, and differing very decidedly from the Caucasian, in whom the nose, owing to the shape of the nasal bones, is narrow, elevated at the bridge, and elongated downwards.   Below, the nasal bones are thin and connected with the cartilages of the nose, and the angle or arch formed by their union serves to throw out the bridge of the nose, and is much more marked in some individuals than others.   13. The *lower margin of the orbit*, formed by the superior maxillary bone and the malar bone, is plainly to be felt throughout its entire length.   It is continuous internally with the nasal process of the superior maxillary bone, which forms the inner boundary of the orbit.   At the point of junction of the lower margin of the orbit with the nasal process is to be felt a little tubercle, which can be plainly perceived by running the finger along the bone in this situation.   This tubercle serves as a guide to the position of the lachrymal sac, which is situated above and behind it.   14. The outline of the *lower jaw* is to be felt throughout its entire length.   Just in front of the tragus of the external ear, and below the zygomatic arch, the condyle can be made out.   When the mouth is opened, this prominence of bone can be perceived advancing out of the glenoid fossa on to the eminentia articularis, and receding again when the mouth is closed.   From the condyle the posterior border of the ramus can be felt extending down to the angle.   A line drawn from the condyle to the angle would indicate the exact position of this border.   From the angle to the symphysis of the chin, the lower, rounded border of the body of the bone is plainly to be felt.   At the point of junction of the two halves of the bone is a well-marked, triangular eminence, the *mental process*, which forms the prominence of the chin.

Surgical Anatomy.—An arrest in the ossifying process may give rise to deficiencies or gaps ; or to fissures, which are of importance from a medico-legal point of view, as they are liable to be mistaken for fractures. The fissures generally extend from the margin towards the centre of the bone, but gaps may be found in the middle, as well as at the edges. In course of time they may become covered with a thin lamina of bone.

Occasionally a protrusion of the brain or its membranes may take place through one of these gaps in an imperfectly developed skull. When the protrusion consists of membranes only, and is filled with cerebro-spinal fluid, it is called a *meningocele* ; when the protrusion consists of brain as well as membranes, it is termed an *encephalocele* ; and when the protruded brain is a prolongation from one of the ventricles, and is distended by a collection of fluid from an accumulation in the ventricle, it is termed an *hydrencephalocele*. This latter condition is frequently found at the root of the nose, where a protrusion of the anterior horn of the lateral ventricle takes place through a deficiency of the fronto-nasal suture. These malformations are usually found in the middle line, and most frequently at the back of the head, the protrusion taking place through the fissures which separate the four centres of ossification from which the tabular portion of the occipital bone is originally developed (see page 27). They most frequently occur through the upper part of the vertical fissure, which is the last to ossify, but not uncommonly through the lower part, when the foramen magnum may be incomplete. More rarely these protrusions have been met with in other situations than those two above mentioned, both through normal fissures, as the sagittal, lambdoid, and other sutures, and also through abnormal gaps and deficiencies at the sides, and even at the base of the skull.

Fractures of the skull may be divided into those of the vault and those of the base. Fractures of the vault are usually produced by direct violence. This portion of the skull varies in thickness and strength in different individuals, but, as a rule, is sufficiently strong to resist a very considerable amount of violence without being fractured. This is due to several causes : the rounded shape of the head and its construction of a number of secondary elastic arches, each made up of a single bone ; the fact that it consists of a number of bones, united, at all events in early life, by a sutural ligament, which acts as a sort of buffer and interrupts the continuity of any violence applied to the skull ; the presence of arches or ridges, both on the inside and outside of the skull, which materially strengthen it ; and the mobility of the head upon the spine, which further enables it to withstand violence. The elasticity of the bones of the head is especially marked in the skull of the child, and this fact, together with the wide separation of the individual bones from each other, and the interposition between them of other softer structures, renders fracture of the bones of the head a very uncommon event in infants and quite young children ; as age advances, and the bones become joined, fracture is more common, though still less liable to occur than in the adult. Fractures of the vault may, and generally do, involve the whole thickness of the bone ; but sometimes one table may be fractured without any corresponding injury to the other. Thus, the outer table of the skull may be splintered and driven into the diploë, or in the frontal or mastoid regions into the frontal or mastoid cells, without any injury to the internal table. And, on the other hand, the internal table has been fractured, and portions of it depressed and driven inwards, without any fracture of the outer table. As a rule, in fractures of the skull, the inner table is more splintered and comminuted than the outer, and this is due to several causes. It is thinner and more brittle ; the force of the violence as it passes inwards becomes broken up, and is more diffused by the time it reaches the inner table ; the bone being in the form of an arch bends as a whole and spreads out, and thus presses the particles together on the convex surface of the arch, i.e. the outer table, and forces them asunder on the concave surface or inner table ; and, lastly, there is nothing firm under the inner table to support it and oppose the force. Fractures of the vault may be simple fissures, or starred and comminuted fractures, and these may be depressed or elevated. These latter cases of fracture with elevation of the fractured portion are uncommon, and can only be produced by direct wound. In comminuted fracture, a portion of the skull is broken into several pieces, the lines of fracture radiating from a centre where the chief impact of the blow was felt ; if depressed, a fissure circumscribes the radiating lines, enclosing a portion of skull. If this area is circular it is termed a ' pond ' fracture, and would in all probability have been caused by a round instrument, as a life preserver or hammer ; if elliptical in shape it is termed a ' gutter fracture,' and would owe its shape to the instrument which had produced it, as a poker.

Fractures of the base are most frequently produced by the extension of a fissure from the vault, as in falls on the head, where the fissure starts from the part of the vault which first struck the ground. Sometimes, however, they are caused by direct violence, when foreign bodies have been forced through the thin roof of the orbit, through the cribriform plate of the ethmoid from being thrust up the nose, or through the roof of the pharynx. Other cases of fracture of the base occur from indirect violence, as in fracture of the occipital bone from impaction of the spinal column against its condyles in falls on the buttocks, knees, or feet, or in cases where the glenoid cavity has been fractured by the violent impact of the condyle of the lower jaw against it from blows on the chin.

The most common place for fracture of the base to occur is through the middle fossa, and here the fissure usually takes a fairly definite course. Starting from the point struck, which is generally somewhere in the neighbourhood of the parietal eminence, it runs downwards through the parietal and squamous portion of the temporal bone and across the petrous portion of this bone, frequently traversing and implicating the internal auditory meatus, to the middle lacerated foramen. From this it may pass across the body of the sphenoid, through the pituitary fossa, to the middle lacerated foramen of the other side, and may indeed travel round the whole cranium, so as to completely separate the anterior from the posterior part. The course of the fracture should be borne in mind, as it explains the symptoms to which fracture in this region may give rise : thus, if the fissure pass across the internal auditory meatus, injury to the facial and auditory nerves may result, with consequent facial paralysis and deafness; or the tubular prolongation of the arachnoid around these nerves in the meatus may be torn and thus permit of the escape of the cerebro-spinal fluid should there be a communication between the internal ear and the tympanum together with rupture of the membrana tympani, as is frequently the case : again, if the fissure passes across the pituitary fossa and the muco-periosteum covering the under surface of the body of the sphenoid is torn, blood will find its way into the pharynx and be swallowed, and after a time vomiting of blood will result. Fractures of the anterior fossa, involving the bones forming the roof of the orbit and nasal fossa, are generally the result of blows on the forehead; but fracture of the cribriform plate of the ethmoid may be a complication of fracture of the nasal bone. When the fracture implicates the roof of the orbit, the blood finds its way into this cavity, and, travelling forwards, appears as a subconjunctival ecchymosis. If the roof of the nasal fossa be fractured, the blood escapes from the nose. In rare cases there may be also escape of cerebro-spinal fluid from the nose, where the dura mater and arachnoid have been torn. In fractures of the posterior fossa, extravasation of blood may appear at the nape of the neck.

The bones of the skull are frequently the seat of nodes, and not uncommonly necrosis results from this cause, and also from injury. Necrosis may involve the entire thickness of the skull, but is usually confined to the external table. Necrosis of the internal table alone is rarely met with. The bones of the skull are also occasionally the seat of sarcomatous tumours.

The skull in rickets is peculiar : the forehead is high, square, and projecting ; and the antero-posterior diameter of the skull is long, in relation to the transverse diameter. The bones of the face are small and ill developed, and this gives the appearance of a larger head than actually exists. The bones of the head are often thick, especially in the neighbourhood of the sutures, and the anterior fontanelle is late in closing, sometimes remaining unclosed till the fourth year. The condition of *cranio-tabes* has by some been also believed to be the result of rickets ; by others is believed to be due to inherited syphilis. In all probability it is due to both. In these cases the bone undergoes atrophic changes in patches, so that it becomes greatly thinned in places, generally where there is pressure, as from the pillow or nurse's arm. It is, therefore, usually met with in the parietal bone and vertical plate of the occipital bone.

In congenital syphilis deposits of porous bone are often found at the angles of the parietal bones, and two halves of the frontal bone which bound the anterior fontanelle. These deposits are separated by the coronal and sagittal sutures, and give to the skull an appearance like a ' hot cross bun.' They are known as Parrot's nodes, and such a skull has received the name of *natiform*, from its fancied resemblance to the buttocks.

In connection with the bones of the face a common malformation is *cleft palate*, owing to the non-union of the palatal processes of the maxillary or pre-oral arch. This cleft may involve the whole or only a portion of the hard palate, and usually involves the soft palate also. The cleft is in the middle line, except it involves the alveolus in front, when it follows the suture between the main portion of the bone and the pre-maxillary bone. Sometimes the cleft runs on either side of the pre-maxillary bone, so that this bone is quite isolated from the maxillary bones and hangs from the end of the vomer. The malformation is usually associated with hare-lip, which, when single, is almost always on one side, corresponding to the position of the suture between the lateral incisor and canine tooth. Some few cases of median hare-lip have been described. In double hare-lip there is a cleft on each side of the middle line.

The bones of the face are sometimes fractured as the result of direct violence. The two most commonly broken are the nasal bone and the inferior maxilla, and of these, the latter is by far the most frequently fractured of all the bones of the face. Fracture of the *nasal* bone is for the most part transverse, and takes place about half an inch from the free margin. The broken portion may be displaced backwards or more generally to one side by the force which produced the lesion, as there are no muscles here which can cause displacement. The *malar* bone is probably never broken alone, that is to say, unconnected with a fracture of the other bones of the face. The *zygomatic arch* is occasionally fractured, and when this occurs from direct violence, as is usually the case, the fragments may be displaced inwards. This lesion is often attended with great difficulty or even inability to open and shut the mouth, and this has been stated to be due to the depressed fragments perforating the temporal muscle, but would appear rather to be caused by the injury done to the bony origin of the masseter muscle. Fractures of the *superior maxilla* may vary much in degree, from the chipping off of a portion of the alveolar arch, a frequent accident when the old 'key' instrument was used for the extraction of teeth, to an extensive comminution of the whole bone from severe violence, as the kick of a horse. The most common situation for a fracture of the *inferior maxillary bone* is in the neighbourhood of the canine tooth, as at this spot the jaw is weakened by the deep socket for the fang of this tooth ; it is next most frequently fractured at the angle ; then at the symphysis, and finally the neck of the condyle or the coronoid process may be broken. Occasionally a double fracture may occur, one in either half of the bone. The fractures are usually compound, from laceration of the mucous membrane covering the gums. The displacement is mainly the result of the same violence as produced the injury, but may be further increased by the action of the muscles passing from the neighbourhood of the symphysis to the hyoid bone.

The superior and inferior maxillary bones are both of them frequently the seat of necrosis ; though the disease affects the lower much more frequently than the upper jaw, probably on account of the greater supply of blood to the latter. It may be the result of periostitis, from tooth irritation, injury, or the action of some specific poison, as syphilis, or from salivation by mercury ; it not infrequently occurs in children after attacks of the exanthematous fevers, and a special form occurs from the action of the fumes of phosphorus in persons engaged in the manufacture of matches.

Tumours attack the jaw-bones not infrequently, and these may be either innocent or malignant : in the upper jaw cysts may occur in the antrum, constituting the so-called dropsy of the antrum ; or, again, cysts may form in either jaw in connection with the teeth : either cysts connected with the roots of fully developed teeth, the ' dental cyst ; ' or cysts connected with imperfectly developed teeth, the ' dentigerous cyst.' Solid innocent tumours include the fibroma, the chondroma, and the osteoma. Of malignant tumours there are two classes, the sarcoma and the epithelioma. The sarcomata are of various kinds, the spindle-celled and round-celled of a very malignant character, and the myeloid sarcoma principally affecting the alveolar margin of the bone. Of the epitheliomata we find the squamous variety spreading to the bone from the palate or gum, and the cylindrical epithelioma originating in the antrum or nasal fossæ.

Both superior and inferior maxillary bones occasionally require removal for tumours and for some other conditions. The upper jaw is removed by an incision from the inner canthus of the eye, along the side of the nose, round the ala, and down the middle line of the upper lip. A second incision is carried outwards from the inner canthus of the eye along the lower margin of the orbit as far as the prominence of the malar bone. The flap thus formed is reflected outwards and the surface of the bone exposed. The connections of the bone to the other bones of the face are then divided with a narrow saw. They are (1) the junction with the malar bone, passing into the spheno-maxillary fissure : (2) the nasal process ; a small portion of its upper extremity, connected with the nasal bone in front, the lachrymal bone behind, and the frontal bone above, being left : (3) the connection with the bone on the opposite side and the palate in the roof of the mouth. The bone is now firmly grasped with lion-forceps, and by means of a rocking movement upwards and downwards the remaining attachments of the orbital plate with the ethmoid, and of the back of the bone with the palate, broken through. The soft palate is first separated from the hard with a scalpel and is not removed. Occasionally in removing the upper jaw, it will be found that the orbital plate can be spared, and this should always be done if possible. A horizontal saw-cut is to be made just below the infra-orbital foramen, and the bone cut through with a chisel and mallet.

Removal of one half of the lower jaw is sometimes required. If possible, the section of the bone should be made to one side of the symphysis so as to save the genial tubercles and the origin of the Genio-hyo-glossus muscle, as otherwise the tongue tends to fall backwards and may produce suffocation. Having extracted the central or preferably the lateral incisor tooth, a vertical incision is made down to the bone, commencing at the free margin of the lip, and carried to the lower border of the bone ; it is then carried along its lower border to the angle and up the posterior margin of the ramus to a level with the lobule of the ear. The flap thus formed is raised by separating all the structures attached to the outer surface of the bone. The jaw is now sawn through at the point where the tooth has been extracted, and the knife passed along the inner side of the jaw, separating the structures attached to this surface. The jaw is then grasped by the surgeon and strongly depressed, so as to bring down the coronoid process and enable the operator to sever the tendon of the Temporal muscle. The jaw can be now further depressed, care being taken not to evert it or rotate it outwards, which would endanger the internal maxillary artery, and the External pterygoid torn through or divided. The capsular ligament is now opened in front and the lateral ligaments divided, and the jaw removed with a few final touches of the knife.

The antrum of Highmore occasionally requires tapping for suppuration. This may be done through the socket of a tooth, preferably the first molar, the fangs of which are most intimately connected with the antrum, or through the facial aspect of the bone above the alveolar process. This latter method does not perhaps afford such efficient drainage, but there is less chance of food finding its way into the cavity. The operation may be performed by incising the mucous membrane above the second molar tooth, and driving a trocar or any sharp-pointed instrument into the cavity.

HYOID BONE

The **Hyoid bone** is named from its resemblance to the Greek upsilon ; it is also called the *lingual bone*, because it supports the tongue, and gives attachment to its numerous muscles.   It is a bony arch, shaped like a horse-shoe, and consisting of five segments, a body, two greater cornua, and two lesser cornua.   It is suspended from the tip of the styloid processes of the temporal bones by ligamentous bands, the *stylo-hyoid* ligaments.

The **Body** (*basi-hyal*) forms the central part of the bone, and is of a quadrilateral form ; its *anterior surface* (fig. 78), convex, directed forwards and upwards, is divided into two parts by a vertical ridge which descends along the median line, and is crossed at right angles by a horizontal ridge, so that this surface is divided into four spaces or depressions.   At the point of meeting of these two lines is a prominent elevation, the *tubercle*. The portion above the horizontal ridge is directed upwards, and is sometimes described as the superior border.   The anterior surface gives attachment to the Genio-hyoid in the greater part of its extent ; above, to the Genio-hyo-glossus ; below, to the Mylo-hyoid, Stylo-hyoid,

FIG. 78.—Hyoid bone. Anterior surface (enlarged).

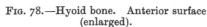

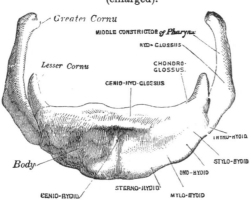

and aponeurosis of the Digastric (suprahyoid aponeurosis) ; and between these to part of the Hyo-glossus.   The *posterior surface* is smooth, concave, directed backwards and downwards, and separated from the epiglottis by the thyro-hyoid membrane, and by a quantity of loose areolar tissue.   The *superior border* is rounded, and gives attachment to the thyro-hyoid membrane, part of the Genio-hyo-glossi, and Chondro-glossi muscles.   The *inferior border* gives attachment, in front, to the Sterno-hyoid ; behind, to the Omo-hyoid and to part of the Thyro-hyoid, at its junction with the great cornu.   It also gives attachment to the Levator glandulæ thyroideæ, when this muscle is present.   The *lateral surfaces*, after middle life, are joined to the greater cornua.   In early life they are connected to the cornua by cartilaginous surfaces, and held together by ligaments, and occasionally a synovial membrane is found between them.

The **Greater Cornua** (*thyro-hyal*) project backwards from the lateral surfaces of the body ; they are flattened from above downwards, diminish in size from before backwards, and terminate posteriorly in a tubercle for the attachment of the lateral thyro-hyoid ligament.   The outer surface gives attachment to the Hyo-glossus, their upper border to the Middle constrictor of the pharynx, their lower border to part of the Thyro-hyoid muscle.

The **Lesser Cornua** (*cerato-hyal*) are two small, conical-shaped eminences, attached by their bases to the angles of junction between the body and greater cornua, and giving attachment by their apices to the stylo-hyoid ligaments.[*] The smaller cornua are connected to the body of the bone by a distinct diarthrodial joint, which usually persists throughout life, but occasionally becomes ankylosed.

Development.—By *five* centres : one for the body, and one for each cornu. Ossification commences in the body about the eighth month, and in the greater

---

[*] These ligaments in many animals are distinct bones, and in man are occasionally ossified to a certain extent.

cornua towards the end of fœtal life. Ossification of the lesser cornua commences some years after birth. Sometimes there are two centres for the body.

Attachment of Muscles.—Sterno-hyoid, Thyro-hyoid, Omo-hyoid, aponeurosis of the Digastric, Stylo-hyoid, Mylo-hyoid, Genio-hyoid, Genio-hyo-glossus, Chondro-glossus, Hyo-glossus, Middle constrictor of the pharynx, and occasionally a few fibres of the Inferior lingualis. It also gives attachment to the thyro-hyoid membrane, and the stylo-hyoid, thyro-hyoid, and hyo-epiglottic ligaments.

Surface Form.—The hyoid bone can be felt in the receding angle below the chin, and the finger can be carried along the whole length of the bone to the greater cornu, which is situated just below the angle of the jaw. This process of bone is best perceived by making pressure on one cornu and so pushing the bone over to the opposite side, when the cornu of this side will be distinctly felt immediately beneath the skin. This process of bone is an important landmark in ligature of the lingual artery.

Surgical Anatomy.—The hyoid bone is occasionally fractured, generally from direct violence, as in the act of garotting or throttling. The great cornu is the part of the bone most frequently broken, but sometimes the fracture takes place through the body of the bone. In consequence of the muscles of the tongue having important connections with this bone, there is great pain upon any attempt being made to move the tongue, as in speaking or swallowing.

THE THORAX

The **Thorax,** or **Chest,** is an osseo-cartilaginous cage, containing and protecting the principal organs of respiration and circulation. It is conical in shape, being narrow above and broad below, flattened from before backwards, and longer behind than in front. It is somewhat reniform on transverse section.

Boundaries.—The *posterior* surface is formed by the twelve dorsal vertebræ and the posterior part of the ribs. It is concave from above downwards, and presents on each side of the middle line a deep groove, in consequence of the direction backwards and outwards which the ribs take from their vertebral extremities to their angles. The *anterior* surface is flattened or slightly convex, and inclined forwards from above downwards. It is formed by the sternum and costal cartilages. The *lateral* surfaces are convex ; they are formed by the ribs, separated from each other by spaces, the *intercostal spaces.* These spaces are eleven in number, and are occupied by the intercostal muscles.

The *upper opening* of the thorax is reniform in shape, being broader from side to side than from before backwards. It is formed by the first dorsal vertebra behind, the upper margin of the sternum in front, and the first rib on each side. It slopes downwards and forwards, so that the anterior part of the ring is on a lower level than the posterior. The antero-posterior diameter is about two inches and the transverse about four. The *lower* opening is formed by the twelfth dorsal vertebra behind, by the twelfth rib at the sides, and in front by the cartilages of the eleventh, tenth, ninth, eighth, and seventh ribs, which ascend on either side and form an angle, the *subcostal angle,* from the apex of which the ensiform cartilage projects. It is wider transversely than from before backwards. It slopes obliquely downwards and backwards ; so that the cavity of the thorax is much deeper behind than in front. The Diaphragm closes in the opening, forming the floor of the thorax.

In the female, the thorax differs as follows from the male : 1. Its general capacity is less. 2. The sternum is shorter. 3. The upper margin of the sternum is on a level with the lower part of the body of the third dorsal vertebra, whereas in the male it is on a level with the lower part of the body of the second dorsal vertebra. 4. The upper ribs are more movable, and so allow a greater enlargement of the upper part of the thorax than in the male.

The Sternum

The **Sternum** (στέρνον, the chest) (figs. 79, 80) is a flat, narrow bone, situated in the median line of the front of the chest, and consisting, in the adult, of three portions. It has been likened to an ancient sword : the upper piece, representing the handle, is termed the *manubrium*; the middle and largest piece, which represents the chief part of the blade, is termed the *gladiolus*; and the inferior piece, which is likened to the point of the sword, is termed the *ensiform* or *xiphoid appendix*. In its natural position its inclination is oblique from above, downwards and forwards. It is slightly convex in front and concave behind,

FIG. 79.—Anterior surface of sternum and costal cartilages.

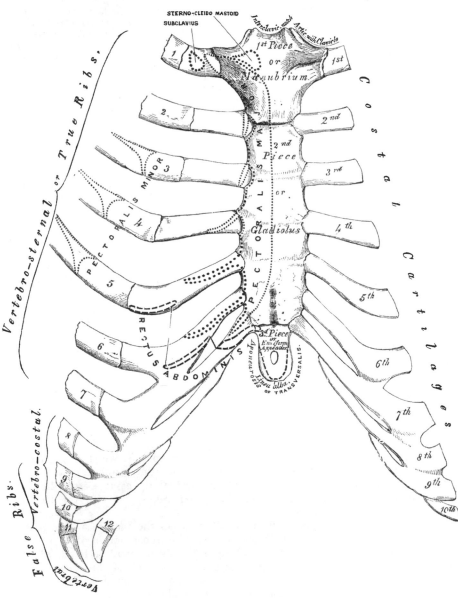

broad above, becoming narrowed at the point where the first and second pieces are connected; after which it again widens a little, and is pointed at its extremity. Its average length in the adult is about seven inches, being rather longer in the male than in the female.

The **First Piece** of the sternum, or **Manubrium** (*pre-sternum*), is of a somewhat triangular form, broad and thick above, narrow below at its junction with the middle piece. Its *anterior surface*, convex from side to side, concave from above downwards, is smooth, and affords attachment on each side to the Pectoralis major and sternal origin of the Sterno-cleido-mastoid muscles. In well-marked bones the ridges limiting the attachment of these muscles are very distinct. Its *posterior surface*, concave and smooth, affords attachment on each side to the Sterno-hyoid and Sterno-thyroid muscles. The *superior border*, the thickest, presents at its centre the *pre-sternal notch*; and on each side an oval articular surface, directed upwards, backwards, and outwards, for articulation with the sternal end of the clavicle. The *inferior border* presents an oval, rough surface, covered in the recent state with a thin layer of cartilage, for articulation with the second portion of the bone. The *lateral borders* are marked above by a depression for the first costal cartilage, and below by a small facet, which, with a similar facet on the upper angle of the middle portion of the bone, forms a notch for the reception of the costal cartilage of the second rib. These articular surfaces are separated by a narrow, curved edge, which slopes from above downwards and inwards.

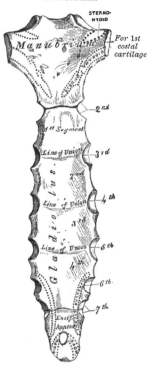

FIG. 80.—Posterior surface of sternum.

The **Second Piece** of the sternum, or **Gladiolus** (*meso-sternum*), considerably longer, narrower, and thinner than the first piece, is broader below than above. Its *anterior surface* is nearly flat, directed upwards and forwards, and marked by three transverse lines which cross the bone opposite the third, fourth, and fifth articular depressions. These lines are produced by the union of the four separate pieces of which this part of the bone consists at an early period of life. At the junction of the third and fourth pieces is occasionally seen an orifice, the *sternal foramen*; it varies in size and form in different individuals, and pierces the bone from before backwards. This surface affords attachment on each side to the sternal origin of the Pectoralis major. The *posterior surface*, slightly concave, is also marked by three transverse lines; but they are less distinct than those in front: this surface affords attachment below, on each side, to the Triangularis sterni muscle, and occasionally presents the posterior opening of the sternal foramen. The *superior border* presents an oval surface for articulation with the manubrium. The *inferior border* is narrow, and articulates with the ensiform appendix. Each *lateral border* presents, at each superior angle, a small facet, which, with a similar facet on the manubrium, forms a cavity for the cartilage of the second rib; the four succeeding angular depressions receive the cartilages of the third, fourth, fifth, and sixth ribs, while each inferior angle presents a small facet, which, with a corresponding one on the ensiform appendix, forms a notch for the cartilage of the seventh rib. These articular depressions are separated by a series of curved interarticular intervals, which diminish in length from above downwards, and correspond to the intercostal spaces. Most of the cartilages belonging to the true ribs, as will be seen from the foregoing description, articulate with the sternum at the line of junction of two of its primitive component segments. This is well seen in many of the lower animals, where the separate parts of the bone remain ununited longer than in man. In this respect a striking analogy exists between the mode of connection of the ribs with the vertebral column, and the connection of the costal cartilages with the sternum.

FIG. 81.—Development of the sternum, by six centres.

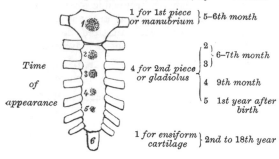

Time of appearance

1 *for 1st piece or manubrium* } 5–6th month

4 *for 2nd piece or gladiolus* {
2 }
3 } 6–7th month
4 9th month
5 1st year after birth

1 *for ensiform cartilage* } 2nd to 18th year

FIG. 82.

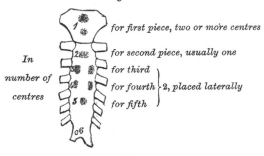

Time of union

Rarely unite, except in old age

Between puberty and the 25th year

Soon after puberty

Partly cartilaginous to advanced life

FIG. 83.—Peculiarities

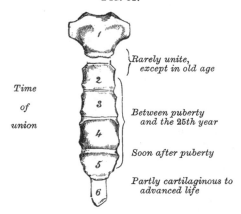

In number of centres

for first piece, two or more centres

for second piece, usually one

for third
for fourth } *2, placed laterally*
for fifth

FIG. 84.

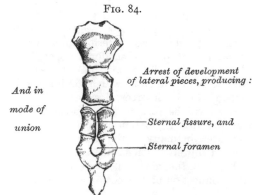

And in mode of union

Arrest of development of lateral pieces, producing :

—— *Sternal fissure, and*

—— *Sternal foramen*

The **Third Piece** of the sternum, the **Ensiform** or **Xiphoid Appendix** (*meta-sternum*), is the smallest of the three ; it is thin and elongated in form, cartilaginous in structure in youth, but more or less ossified at its upper part in the adult. Its *anterior surface* affords attachment to the chondro-xiphoid ligament ; its *posterior surface*, to some of the fibres of the Diaphragm and Triangularis sterni muscles ; its *lateral borders*, to the aponeurosis of the abdominal muscles. Above, it articulates with the lower end of the gladiolus, and at each superior angle presents a facet for the lower half of the cartilage of the seventh rib ; below, by its pointed extremity, it gives attachment to the linea alba. This portion of the sternum is very various in appearance, being sometimes pointed, broad and thin, sometimes bifid, or perforated by a round hole, occasionally curved, or deflected considerably to one or the other side.

Structure.—The bone is composed of delicate cancellous structure, covered by a thin layer of compact tissue, which is thickest in the manubrium, between the articular facets for the clavicles.

Development.—The cartilaginous sternum originally consists of two bars, situated one on either side of the mesial plane and connected with the rib cartilages of its own side. These two bars fuse with each other along the middle line, and the bone, including the ensiform appendix, is developed by *six* centres : one for the first piece or manubrium, four for the second piece or gladiolus, and one for the ensiform appendix. Up to the middle of fœtal life the sternum is entirely cartilaginous, and when ossification takes place the ossific granules are deposited in the middle of the intervals between the articular depressions for the costal cartilages, in the following order: in the first piece, between the fifth and sixth months; in the second and third, between the sixth and seventh months; in the fourth piece, at the ninth month; in the fifth, within the first year, or between the first and second years after birth; and in the ensiform appendix, between the second and the seventeenth or eighteenth years, by a single centre which makes its appearance at the upper part, and proceeds gradually downwards (fig. 81). To these may be added the occasional existence, as described by Breschet, of two small episternal centres, which make their appearance one on each side of the pre-sternal notch. They are probably vestiges of the episternal bone of the mono-tremata and lizards. It occasionally happens that some of the segments are formed from more than one centre, the number and position of which vary (fig. 83). Thus, the first piece may have two, three, or even six centres. When two are present, they are generally situated one above the other, the upper one being the larger; * the second piece has seldom more than one; the third, fourth, and fifth pieces are often formed from two centres placed laterally, the irregular union of which will serve to explain the occasional occurrence of the sternal foramen (fig. 84), or of the vertical fissure which occasionally intersects this part of the bone, and which is further explained by the manner in which the cartilaginous matrix, in which ossification takes place, is formed. Union of the various centres of the gladiolus commences about puberty, from below, and pro-ceeds upwards, so that by the age of twenty-five they are all united, and this portion of bone consists of one piece (fig. 82). The ensiform cartilage becomes joined to the gladiolus about forty. The manubrium is occasionally, but not invariably, joined to the gladiolus in advanced life by bone. When this union takes place, however, it is generally only superficial, a portion of the centre of the sutural cartilage remaining unossified.

Articulations.—With the clavicles and seven costal cartilages on each side.

Attachment of Muscles.—To nine pairs and one single muscle: the Pectoralis major, Sterno-cleido-mastoid, Sterno-hyoid, Sterno-thyroid, Triangularis sterni, the aponeuroses of the Obliquus externus, Obliquus internus, Transversalis, and Rectus muscles, and Diaphragm.

THE RIBS

The **Ribs** are elastic arches of bone, which form the chief part of the thoracic walls. They are twelve in number on each side; but this number may be increased by the development of a cervical or lumbar rib, or may be diminished to eleven. The first seven are connected behind with the spine, and in front with the sternum, through the intervention of the costal cartilages; they are called *true* ribs.† The remaining five are *false* ribs; of these, the first three have their cartilages attached to the cartilage of the rib above: the last two are free at their anterior extremities and are termed *floating* ribs. The ribs vary in their direction, the upper ones being less oblique than the lower. The extent of obliquity reaches its maximum at the ninth rib, and gradually decreases from that rib to the twelfth. The ribs are situated one below the other in such a manner that spaces are left between them, which are called *intercostal spaces*. The length of these spaces corresponds to the length of the ribs and their cartilages; their breadth is greater in front than behind, and between the upper than between the lower ribs. The ribs increase in length from the first to the seventh, when they again diminish to the twelfth. In breadth they

---

* Sir George Humphry states that this is ' probably the more complete condition.'
† Sometimes the eighth rib cartilage articulates with the sternum; this condition occurs more frequently on the right than on the left side.

FIG. 85.—A central rib of left side.

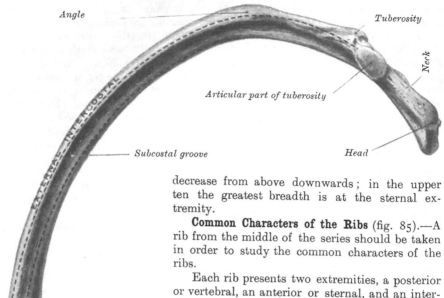

Angle   *Tuberosity*

Neck

Articular part of tuberosity

Subcostal groove   *Head*

Body or shaft

decrease from above downwards ; in the upper ten the greatest breadth is at the sternal extremity.

Common Characters of the Ribs (fig. 85).—A rib from the middle of the series should be taken in order to study the common characters of the ribs.

Each rib presents two extremities, a posterior or vertebral, an anterior or sternal, and an intervening portion—the body or shaft.

The **posterior** or **vertebral extremity** presents for examination a head, neck, and tuberosity. The **head** (fig. 86) is marked by a kidney-shaped articular surface, divided by a horizontal ridge into two facets for articulation with the articular surface formed by the junction of the bodies of two contiguous dorsal vertebræ ; the upper facet is small, the inferior one of larger size ; the ridge separating them serves for the attachment of the interarticular ligament. The **neck** is that flattened portion of the rib which extends outwards from the head ; it is about an inch long, and is placed in front of the transverse process of the lower of the two vertebræ with which the head articulates. Its *anterior surface* is flat and smooth, its *posterior* rough, for the attachment of the middle costo-transverse ligament, and perforated by numerous foramina, the direction of which is less constant than those found on the inner surface of the shaft. Of its two borders the *superior* presents a rough crest for the attachment of the anterior costo-transverse ligament ; its *inferior border* is rounded. On the posterior surface of the neck, just where it joins the shaft, and nearer the lower than the upper border, is an eminence—the **tuberosity**, or **tubercle** ; it consists of an articular and a non-articular portion. The *articular portion*, the more internal and inferior of the two, presents a small, oval surface for articulation with the extremity of the transverse process of the lower of the two vertebræ to which the head is connected. The *non-articular portion* is a rough elevation, which affords attachment to the posterior costo-transverse ligament. The tubercle is much more prominent in the upper than in the lower ribs.

The **shaft** is thin and flat, so as to present two surfaces, an external and an internal ; and two borders, a superior and an inferior. The *external surface* is convex, smooth, and marked, at its back part, a little in front of the tuberosity, by a prominent line, directed obliquely from above downwards and outwards ; this gives attachment to a tendon of the Ilio-costalis muscle, or of one of its accessory portions, and is called the *angle*. At this point the rib is bent in two directions. If the rib is laid upon its lower border, it will be seen that the portion of the shaft in front of the angle rests upon this border, while the portion of the shaft behind the angle is bent inwards and at the same time tilted upwards. The interval between the angle and the tuberosity increases gradually from the second to the tenth rib. The portion of bone between these two parts is rounded, rough, and irregular, and serves for the attachment of the Longissimus dorsi muscle. The portion of bone between the tubercle and sternal extremity is also slightly twisted upon its own axis, the external surface looking downwards behind the angle, a little upwards in front of it. This surface presents, towards its sternal extremity, an oblique line, the *anterior angle*. The *internal surface* is concave, smooth, directed a little upwards behind the angle, a little downwards in front of it, and

Fig. 86.—Vertebral extremity of a rib. External surface.

Facet for body of upper dorsal vertebra
Ridge for interarticular ligament
Facet for body of lower dorsal vertebra

Articular part of tuberosity

Non-articular part of tuberosity

is marked by a ridge which commences at the lower extremity of the head ; it is strongly marked as far as the inner side of the angle, and gradually becomes lost at the junction of the anterior with the middle third of the bone. The interval between it and the inferior border presents a groove, *subcostal*, for the intercostal vessels and nerve. At the back part of the bone, this groove belongs to the inferior border, but just in front of the angle, where it is deepest and broadest, it corresponds to the internal surface. The superior edge of the groove is rounded ; it serves for the attachment of the Internal intercostal muscle. The inferior edge corresponds to the lower margin of the rib, and gives attachment to the External intercostal. Within the groove are seen the orifices of numerous small foramina, which traverse the wall of the shaft obliquely from before backwards. The *superior border*, thick and rounded, is marked by an external and an internal lip, more distinct behind than in front ; they serve for the attachment of the External and Internal inter-costal muscles. The *inferior border*, thin and sharp, has attached to it the External intercostal muscle. The **anterior** or **sternal extremity** is flattened, and presents a porous, oval, concave depression, into which the costal cartilage is received.

PECULIAR RIBS

The ribs which require especial consideration are five in number—viz. the first, second, tenth, eleventh, and twelfth.

The **first rib** (fig. 87) is one of the shortest and the most curved of all the ribs ; it is broad and flat, its surfaces looking upwards and downwards, and its borders inwards and outwards. The *head* is of small size, rounded, and presents only a single articular facet for articulation with the body of the first dorsal vertebra. The *neck* is narrow and rounded. The *tuberosity*, thick and prominent, rests on the outer border. There is no angle, but in this situation the rib is slightly bent, with the convexity of the bend upwards, so that the head of the bone is directed downwards. The *upper surface* of the shaft is marked by two shallow depressions, separated by a small rough surface for the attachment of the Scalenus anticus muscle—the groove in front of it transmitting the subclavian vein, that behind it the subclavian artery. Between the groove for the subclavian artery and the tuberosity is a rough surface for the attachment of the Scalenus medius muscle. The *under surface* is smooth, and destitute of the groove observed

Peculiar ribs.

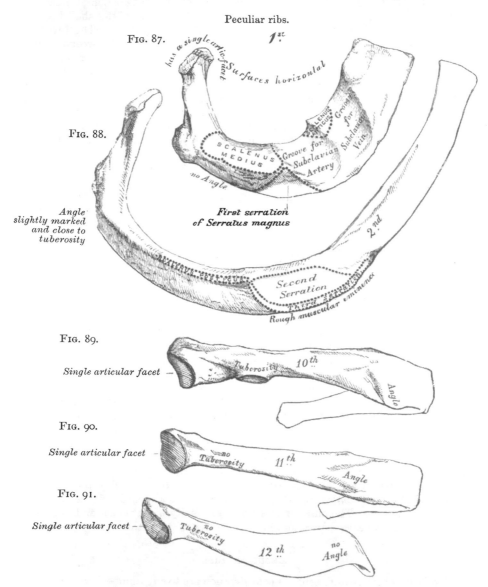

Fig. 87.   1st

Fig. 88.

Fig. 89.

Fig. 90.

Fig. 91.

on the other ribs. The *outer border* is convex, thick, and rounded, and at its posterior part gives attachment to the first serration of the Serratus magnus ; the *inner* is concave, thin, and sharp, and marked about its centre by the commencement of the rough surface for the Scalenus anticus. The *anterior extremity* is larger and thicker than any of the other ribs.

The **second rib** (fig. 88) is much longer than the first, but bears a very considerable resemblance to it in the direction of its curvature. The non-articular portion of the tuberosity is occasionally only slightly marked. The *angle* is slight, and situated close to the tuberosity, and the shaft is not twisted, so that both ends touch any plane surface upon which it may be laid; but there is a similar though slighter bend, with its convexity upwards, to that found in the first rib. The shaft is not horizontal, like that of the first rib; its *outer surface*, which is convex, looking upwards and a little outwards. It presents, near the middle, a rough eminence for the attachment of the second and third digitations of the Serratus magnus; behind and above which is attached the Scalenus posticus. The *inner surface*, smooth and concave, is directed downwards and a little inwards: it presents a short groove towards its posterior part.

The **tenth rib** (fig. 89) has only a single articular facet on its head.

The **eleventh** and **twelfth ribs** (figs. 90 and 91) have each a single articular facet on the head, which is of rather large size; they have no neck or tuberosity, and are pointed at the extremity. The eleventh has a slight angle and a shallow groove on the lower border. The twelfth has neither, and is much shorter than the eleventh, and the head has a slight inclination downwards. Sometimes the twelfth rib is even shorter than the first.

Structure.—The ribs consist of cancellous tissue, enclosed in a thin, compact layer.

Development.—Each rib, with the exception of the last two, is developed by *three* centres: one for the shaft, one for the head, and one for the tubercle. The last two have only *two* centres, that for the tubercle being wanting. Ossification commences in the shaft of the ribs at a very early period, before its appearance in the vertebræ. The epiphysis of the head, which is of a slightly angular shape, and that for the tubercle, of a lenticular form, make their appearance between the sixteenth and twentieth years, and are not united to the rest of the bone until about the twenty-fifth year.

Attachment of Muscles.—To nineteen: the Internal and External intercostals, Scalenus anticus, Scalenus medius, Scalenus posticus, Pectoralis minor, Serratus magnus, Obliquus externus, Quadratus lumborum, Diaphragm, Latissimus dorsi, Serratus posticus superior, Serratus posticus inferior, Ilio-costalis, Musculus accessorius ad ilio-costalem, Longissimus dorsi, Cervicalis ascendens, Levatores costarum, and Infra-costales.

The Costal Cartilages

The **Costal Cartilages** (fig. 79, p. 94) are bars of white, hyaline cartilage, which serve to prolong the ribs forward to the front of the chest, and contribute very materially to the elasticity of its walls. The first seven are connected with the sternum, the next three with the lower border of the cartilage of the preceding rib. The cartilages of the last two ribs have pointed extremities, which terminate in free ends in the walls of the abdomen. Like the ribs, the costal cartilages vary in their length, breadth, and direction. They increase in length from the first to the seventh, then gradually diminish to the last. They diminish in breadth, as well as the intervals between them, from the first to the last. They are broad at their attachment to the ribs, and taper towards their sternal extremities, excepting the first two, which are of the same breadth throughout, and the sixth, seventh, and eighth, which are enlarged where their margins are in contact. In direction they also vary: the first descends a little, the second is horizontal, the third ascends slightly, while all the rest follow the course of the ribs for a short extent, and then ascend to the sternum or preceding cartilage. Each costal cartilage presents two surfaces, two borders, and two extremities. The *anterior surface* is convex, and looks forwards and upwards: that of the first gives attachment to the costo-clavicular ligament and the Subclavius muscle; that of the second, third, fourth, fifth, and sixth, at their sternal ends, to the Pectoralis major.* The others are covered by, and give partial attachment to,

* The first and seventh also, occasionally, give origin to the same muscle.

some of the great flat muscles of the abdomen.   The *posterior surface* is concave, and directed backwards and downwards, the first giving attachment to the Sterno-thyroid, the third to the sixth inclusive to the Triangularis sterni, and the six or seven inferior ones to the Transversalis muscle and the Diaphragm.   Of the two borders, the *superior* is concave, the *inferior* convex : they afford attachment to the internal Intercostal muscles, the upper border of the sixth giving attachment to the Pectoralis major muscle.   The contiguous borders of the sixth, seventh, and eighth, and sometimes the ninth and tenth, costal cartilages present small, smooth, oblong-shaped facets at the points where they articulate.   Of the two extremities, the *outer* one is continuous with the osseous tissue of the rib to which it belongs.   The *inner* extremity of the first is continuous with the sternum ; the six succeeding ones have rounded extremities, which are received into shallow concavities on the lateral margins of the sternum.   The inner extremities of the eighth, ninth, and tenth costal cartilages are pointed, and are connected with the cartilage above.   Those of the eleventh and twelfth are free and pointed.

The costal cartilages are most elastic in youth, those of the false ribs being more so than the true.   In old age they become of a deep yellow colour, and are prone to calcify.

Attachment of Muscles.—To nine : the Subclavius, Sterno-thyroid, Pectoralis major, Internal oblique, Transversalis, Rectus, Diaphragm, Triangularis sterni, and Internal intercostals.

Surface Form.—The bones of the chest are to a very considerable extent covered by muscles, so that in the strongly developed muscular subject they are for the most part concealed.   In the emaciated subject, on the other hand, the ribs, especially in the lower and lateral region, stand out as prominent ridges with the sunken, intercostal spaces between them.

In the middle line, in front, the superficial surface of the sternum is to be felt throughout its entire length, at the bottom of a deep median furrow situated between the two great pectoral muscles and called the *sternal furrow*.   These muscles overlap the anterior surface somewhat, so that the whole of the sternum in its entire width is not subcutaneous, and this overlapping is greater opposite the centre of the bone than above and below, so that the furrow is wide at its upper and lower part, but narrower in the middle. The centre of the upper border of the sternum is visible, constituting the pre-sternal notch, but the lateral parts of this border are obscured by the tendinous origins of the Sterno-mastoid muscles, which present themselves as oblique tendinous cords, which narrow and deepen the notch.   Lower down on the subcutaneous surface a well-defined transverse ridge, the *angle of Ludovic*, is always to be felt.   This denotes the line of junction of the manubrium and body of the bone, and is a useful guide to the second costal cartilage, and thus to the identity of any given rib.   The second rib being found, through its costal cartilage, it is easy to count downwards and find any other.   From the middle of the sternum the furrow spreads out, and, exposing more of the surface of the body of the bone, terminates below in a sudden depression, the *infra-sternal depression* or *pit of the stomach* (*scrobiculus cordis*), which corresponds to the ensiform cartilage. This depression lies between the cartilages of the seventh rib, and in it the ensiform cartilage may be felt.   The sternum in its vertical diameter presents a general convexity forwards, the most prominent point of which is at the joint between the manubrium and gladiolus.

On each side of the sternum the costal cartilages and ribs on the front of the chest are partially obscured by the great pectoral muscle ; through which, however, they are to be felt as ridges, with yielding intervals between them, corresponding to the intercostal spaces.   Of these spaces, the one between the second and third ribs is the widest, the next two somewhat narrower, and the remainder, with the exception of the last two, comparatively narrow.

The lower border of the Pectoralis major muscle corresponds to the fifth rib, and below this, on the front of the chest, the broad, flat outline of the ribs, as they begin to ascend, and the more rounded outline of the costal cartilages, are often visible.   The lower boundary of the front of the thorax, the *abdomino-thoracic arch*, which is most plainly seen by arching the body backwards, is formed by the ensiform cartilage and the cartilages of the seventh, eighth, ninth, and tenth ribs, and the extremities of the eleventh and twelfth ribs or their cartilages.

On each side of the chest, from the axilla downwards, the flattened external surfaces of the ribs may be defined in the form of oblique ridges, separated by depressions corresponding to the intercostal spaces.   They are, however, covered by muscles, which, when strongly developed, obscure their outline to a certain extent.   Nevertheless, the ribs, with the exception of the first, can generally be followed over the front and sides of the chest without difficulty. The first rib, being almost completely covered by the clavicle and scapula, can only be distinguished in a small portion of its extent. At the back, the angles of the ribs form a slightly marked oblique line, on each side of and some distance from the vertebral spines. This line diverges somewhat as it descends, and external to it is a broad, convex surface, caused by the projection of the ribs beyond their angles. Over this surface, except where covered by the scapula, the individual ribs can be distinguished.

Surgical Anatomy.—Malformations of the sternum present nothing of surgical importance beyond the fact that abscesses of the mediastinum may sometimes escape through the sternal foramen. Fracture of the sternum is by no means common, owing, no doubt, to the elasticity of the ribs and their cartilages which support it like so many springs. When broken, it is frequently associated with fracture of the spine, and may be caused by forcibly bending the body either backwards; or forwards until the chin becomes impacted against the top of the sternum. It may also be fractured by direct violence or by muscular action. The fracture usually occurs in the upper half of the body of the bone. Dislocation of the gladiolus from the manubrium also takes place, and is sometimes described as a fracture.

The bone being subcutaneous is frequently the seat of gummatous tumours and not uncommonly is affected with caries. Occasionally the bone, and especially its ensiform appendix, becomes altered in shape and driven inwards by the pressure, in workmen, of tools against their chest.

The ribs are frequently broken, though from their connections and shape they are able to withstand great force, yielding under the injury and recovering themselves like a spring. The middle ones of the series are the most liable to fracture. The first and to a less extent the second, being protected by the clavicle, are rarely fractured; and the eleventh and twelfth on account of their loose and floating condition enjoy a like immunity. The fracture generally occurs from indirect violence, from forcible compression of the chest wall, and the bone then gives way at its weakest part, i.e. just in front of the angle. But the ribs may also be broken by direct violence, when the bone gives way and is driven inwards at the point struck, or they may be broken by muscular action. It seems probable, however, that in these latter cases the bone has undergone some atrophic changes. Fracture of the ribs is frequently complicated with some injury to the viscera contained within the thorax or upper part of the abdominal cavity, and this is most likely to occur in fractures from direct violence.

Fracture of the costal cartilages may also take place, though it is a comparatively rare injury.

The thorax is frequently found to be altered in shape in certain diseases.

The *rickety* thorax is caused chiefly by atmospheric pressure. The balance between the air inside and outside the chest during some stage of respiration is not equal, the preponderance being in favour of the air outside, and this, acting on the softened ribs, causes them to be forced in at the junction of the cartilages with the bones, which is the weakest part. In consequence of this the sternum projects forwards, with a deep depression on either side caused by the sinking in of the softened ribs. The depression is less on the left side, on account of the ribs being supported by the heart. The condition is known as 'pigeon-breast.' The lower ribs, however, are not involved in this deformity, as they are prevented from falling in by the presence of the stomach, liver, and spleen. And when the liver and spleen are enlarged, as they sometimes are in rickets, the lower ribs may be pushed outwards: this causes a transverse constriction just above the costal arch. The anterior extremities of the ribs are usually enlarged in rickets, giving rise to what has been termed the 'rickety rosary.' The *phthisical chest* is often long and narrow, flattened from before backwards, and with great obliquity of the ribs and projection of the scapulæ. In *pulmonary emphysema* the chest is enlarged in all its diameters, and presents on section an almost circular outline. It has received the name of the 'barrel-shaped chest.' In severe cases of *lateral curvature of the spine*, the thorax becomes much distorted. In consequence of the rotation of the bodies of the vertebræ, which takes place in this disease, the ribs opposite the convexity of the dorsal curve become extremely convex behind, being thrown out and bulging, and at the same time flattened in front, so that the two ends of the same rib are almost parallel. Coincident with this the ribs on the opposite side, on the concavity of the curve, are sunk and depressed behind, and bulging and convex in front. In addition to this the ribs become occasionally welded together by bony material.

The ribs are frequently the seat of necrosis leading to abscesses and sinuses, which may burrow to a considerable extent over the wall of the chest. The only special anatomical point in connection with these is that care must be taken in dealing with them, that the intercostal space is not punctured and the pleural cavity opened or the intercostal vessels wounded, as the necrosed portion of bone is generally situated on the internal surface of the rib.

In cases of empyema the chest requires opening to evacuate the pus. There is considerable difference of opinion as to the best position to do this. Probably the best place in most cases will be found to be between the fifth and sixth ribs, in or a little in front of the mid-axillary line. This is the last part of the cavity to be closed by the expansion of the lung; it is not thickly covered by soft parts; the space between the two ribs is sufficiently great to allow of the introduction of a fair-sized drainage tube, and the opening is in a dependent position, when the patient is confined to bed, as he usually inclines towards the affected side, so as to allow the sound lung the freest possible play, and this position permits of efficient drainage.

THE EXTREMITIES

The extremities, or limbs, are those long, jointed appendages of the body which are connected to the trunk by one end, and free in the rest of their extent. They are *four* in number: an *upper* or *thoracic pair*, connected with the thorax through the intervention of the shoulder, and subservient mainly to prehension; and a *lower pair*, connected with the pelvis, intended for support and locomotion. Both pairs of limbs are constructed after one common type, so that they present numerous analogies; while, at the same time, certain differences are observed between the upper and lower pair, dependent on the peculiar offices they have to perform.

The bones by which the upper and lower limbs are attached to the trunk are named respectively the *shoulder* and *pelvic girdles*, and they are constructed on the same general type, though presenting certain modifications relating to the different uses to which the upper and lower limbs are respectively applied. The *shoulder girdle* is formed by the scapulæ and clavicles, and is imperfect in front and behind. In front, however, the girdle is completed by the upper end of the sternum, with which the inner extremities of the clavicles articulate. Behind, the girdle is widely imperfect, and the scapulæ are connected to the trunk by muscles only. The *pelvic girdle* is formed by the innominate bones, and is completed in front, through the symphysis pubis, at which the two innominate bones articulate with each other. It is imperfect behind, but the intervening gap is filled in by the upper part of the sacrum. The pelvic girdle therefore presents, with the sacrum, a complete ring, comparatively fixed, and presenting an arched form which confers upon it a solidity manifestly intended for the support of the trunk, and in marked contrast to the lightness and mobility of the shoulder girdle.

With regard to the morphology of these girdles, the blade of the scapula is generally believed to correspond to the ilium: but with regard to the clavicles there is some difference of opinion; formerly it was believed that they corresponded to the ossa pubis meeting at the symphysis, but it is now generally taught that the clavicle has no homologue in the pelvic girdle, and that the os pubis and ischium are represented by the small coracoid process in man and most mammals.

THE UPPER EXTREMITY

The bones of the upper extremity consist of those of the shoulder girdle, of the arm, the forearm, and the hand.

THE SHOULDER GIRDLE

The Shoulder Girdle consists of two bones, the clavicle and the scapula.

THE CLAVICLE

The **Clavicle** (clavis, *a key*), or collar-bone, forms the anterior portion of the shoulder girdle. It is a long bone, curved somewhat like the italic letter *f*, and placed nearly horizontally at the upper and anterior part of the thorax, immediately above the first rib. It articulates by its inner extremity with the upper border of the sternum, and by its outer extremity with the acromion process of the scapula; serving to sustain the upper extremity in the various positions which it assumes, while, at the same time, it allows of great latitude of motion in the arm.* It presents a double curvature, when looked at in front ; the convexity being forwards at the sternal end, and the concavity at the scapular end. Its outer third is flattened from above downwards, and extends, in the natural position of the bone, from a point opposite the coracoid process to the acromion. Its inner two-thirds are of a prismatic form, and extend from the sternum to a point opposite the coracoid process of the scapula.

External or **Flattened Portion.**—The *outer third* is flattened from above downwards, so as to present two surfaces, an upper and a lower ; and two borders, an anterior and a posterior. The *upper surface* is flattened, rough, marked by impressions for the attachment of the Deltoid in front, and the Trapezius behind ; between these two impressions, externally, a small portion of the bone is subcutaneous. The *under surface* is flattened. At its posterior border, a little external to the point where the prismatic joins with the flattened portion, is a rough eminence, the *conoid tubercle* ; this, in the natural position of the bone, surmounts the coracoid process of the scapula, and gives attachment to the conoid ligament. From this tubercle, an oblique line, occasionally a depression, passes forwards and outwards to near the outer end of the anterior border ; it is called the *oblique line*, or *trapezoid ridge*, and affords attachment to the trapezoid ligament. The *anterior border* is concave, thin, and rough, and gives attachment to the Deltoid ; it occasionally presents, at its inner end, at the commencement of the deltoid impression, a tubercle, the *deltoid tubercle*, which is sometimes to be felt in the living subject. The *posterior border* is convex, rough, broader than the anterior, and gives attachment to the Trapezius.

Internal or **Prismatic Portion.**—The prismatic portion forms the *inner two-thirds* of the bone. It is curved so as to be convex in front, concave behind, and is marked by three borders, separating three surfaces. The *anterior border* is continuous with the anterior margin of the flat portion. At its commencement it is smooth, and corresponds to the interval between the attachment of the Pectoralis major and Deltoid muscles ; at the inner half of the clavicle it forms the lower boundary of an elliptical space for the attachment of the clavicular portion of the Pectoralis major, and approaches the posterior border of the bone. The *superior border* is continuous with the posterior margin of the flat portion, and separates the anterior from the posterior surface. Smooth and rounded at its commencement, it becomes rough towards the inner third for the attachment of the Sterno-mastoid muscle, and terminates at the upper angle of the sternal extremity. The *posterior* or *subclavian border* separates the posterior from the inferior surface, and extends from the conoid tubercle to the rhomboid impression. It forms the posterior boundary of the groove for the Subclavius muscle, and gives attachment to a layer of cervical fascia, enveloping the Omohyoid muscle. The *anterior surface* is included between the superior and anterior borders. It is directed forwards and a little upwards at the sternal end, outwards and still more upwards at the acromial extremity, where it becomes continuous with the upper surface of the flat portion. Externally, it is smooth, convex, nearly subcutaneous, being covered only by the Platysma ; but, corresponding to the inner half of the bone, it is divided by a more or less prominent line into two parts : a lower portion, elliptical in form, rough, and slightly convex, for the attachment of the Pectoralis major ; and an upper part, which is rough, for the attachment of the Sterno-cleido-mastoid. Between the two muscular impressions is a small, subcutaneous interval. The *posterior* or *cervical surface* is smooth, flat, and looks backwards towards the root of the neck. It is limited, above, by the superior border ; below, by the subclavian border ; internally, by the

---

* The clavicle acts especially as a fulcrum to enable the muscles to give lateral motion to the arm. It is accordingly absent in those animals whose fore limbs are used only for progression, but is present for the most part in those animals whose anterior extremities are clawed and used for prehension, though in some of them—as, for instance, in a large number of the carnivora—it is merely a rudimentary bone suspended among the muscles, and not articulating either with the scapula or sternum.

margin of the sternal extremity; externally, it is continuous with the posterior border of the flat portion. It is concave from within outwards, and is in relation, by its lower part, with the suprascapular vessels. This surface, at about the junction of the inner and outer curves, is also in close relation with the brachial plexus and subclavian vessels. It gives attachment, near the sternal extremity, to part of the Sterno-hyoid muscle; and presents, at or near the middle, a foramen, directed obliquely outwards, which transmits the chief nutrient artery of the bone. Sometimes there are two foramina on the posterior surface, or one on the posterior, and another on the inferior surface. The *inferior* or *subclavian surface* is bounded, in front, by the anterior border; behind, by the subclavian border. It is narrow internally, but gradually increases in width externally, and is continuous with the under surface of the flat portion. Commencing at the sternal extremity may be seen a small facet for articulation with the cartilage of the first rib. This is continuous with the articular surface at the sternal end of the bone. External to this is a broad, rough surface, the *rhomboid impression*, rather

FIG. 92.—Left clavicle. Superior surface.

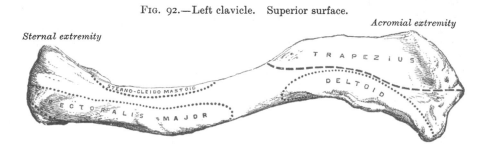

FIG. 93.—Left clavicle. Inferior surface.

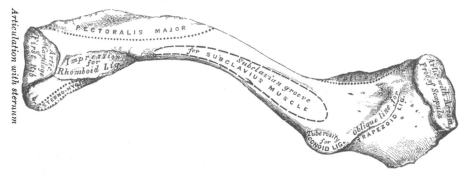

more than an inch in length, for the attachment of the costo-clavicular (rhomboid) ligament. The remaining part of this surface is occupied by a longitudinal groove, the *subclavian groove*, broad and smooth externally, narrow and more uneven internally; it gives attachment to the Subclavius muscle, and, by its margins, to the costo-coracoid membrane, which splits to enclose the muscle. Not infrequently this groove is subdivided into two parts by a longitudinal line, which gives attachment to the intermuscular septum of the Subclavius muscle.

The **internal** or **sternal extremity** of the clavicle is triangular in form, directed inwards, and a little downwards and forwards; and presents an articular facet, concave from before backwards, convex from above downwards, which articulates with the sternum through the intervention of an interarticular fibro-cartilage; the circumference of the articular surface is rough, for the attachment of numerous ligaments. The posterior border of this surface is prolonged backwards so as to increase the size of the articular facet; the upper border gives attachment to the interarticular fibro-cartilage, and the lower border is continuous with the costal facet on the inner end of the inferior or subclavian surface, which articulates with the cartilage of the first rib.

The **outer** or **acromial extremity,** directed outwards and forwards, presents a small, flattened, oval facet, which looks obliquely downwards, for articulation with the acromion process of the scapula. The circumference of the articular facet is rough, especially above, for the attachment of the acromio-clavicular ligaments.

Peculiarities of the Bone in the Sexes and in Individuals.—In the female, the clavicle is generally shorter, thinner, less curved, and smoother than in the male. In those persons who perform considerable manual labour, which brings into constant action the muscles connected with this bone, it becomes thicker and more curved, and its ridges for muscular attachment become prominently marked. The right clavicle is generally longer, thicker, and rougher than the left.

Structure.—The shaft, as well as the extremities, consists of cancellous tissue, invested in a compact layer much thicker in the middle than at either end. The clavicle is highly elastic, by reason of its curves. From the experiments of Mr. Ward, it has been shown that it possesses sufficient longitudinal elastic force to project its own weight nearly two feet on a level surface, when a smart blow is struck on it ; and sufficient transverse elastic force, opposite the centre of its anterior convexity, to throw its own weight about a foot. This extent of elastic power must serve to moderate very considerably the effect of concussions received upon the point of the shoulder.

Development.—By *two* centres : one for the shaft, and one for the sternal extremity. The centre for the shaft appears very early, before any other bone ; according to Beclard, as early as the thirtieth day. The centre for the sternal end makes its appearance about the eighteenth or twentieth year, and unites with the rest of the bone about the twenty-fifth year.

Articulations.—With the sternum, scapula, and cartilage of the first rib.

Attachment of Muscles.—To six : the Sterno-cleido-mastoid, Trapezius, Pectoralis major, Deltoid, Subclavius, and Sterno-hyoid.

Surface Form.—The clavicle can be felt throughout its entire length, even in persons who are very fat. Commencing at the inner end, the enlarged sternal extremity, where the bone projects above the upper margin of the sternum, can be felt, forming with the sternum and the rounded tendon of the Sterno-mastoid a V-shaped notch, the *pre-sternal notch.* Passing outwards, the shaft of the bone can be felt immediately under the skin, with its convexity forwards in the inner two-thirds, the surface partially obscured above and below by the attachments of the Sterno-mastoid and Pectoralis major muscles. In the outer third it forms a gentle curve backwards, and terminates at the outer end in a somewhat enlarged extremity which articulates with the acromial process of the scapula. The direction of the clavicle is almost, if not quite, horizontal when the arm is lying quietly by the side, though in well-developed subjects it may incline a little upwards at its outer end. Its direction is, however, very changeable with the varying movements of the shoulder-joint.

Surgical Anatomy.—The clavicle is the most frequently broken of any single bone in the body. This is due to the fact that it is much exposed to violence, and is the only bony connection between the upper limb and the trunk. The bone, moreover, is slender, and is very superficial. The bone may be broken by direct or indirect violence, or by muscular action. The most common cause is, however, from indirect violence, and the bone then gives way at the junction of its outer with its inner two-thirds : that is to say, at the junction of the two curves, for this is the weakest part of the bone. The fracture is generally oblique, and the displacement of the outer fragment is inwards, away from the surface of the body, hence compound fracture of the clavicle is of rare occurrence. The inner fragment, as a rule, is little displaced. Beneath the bone the main vessels of the upper limb and the great nerve-cords of the brachial plexus lie on the first rib and are liable to be wounded in fracture ; especially in fracture from direct violence, when the force of the blow drives the broken ends inwards. Fortunately the subclavius muscle is interposed between these structures and the clavicle, and this often protects them from injury.

The clavicle is not uncommonly the seat of sarcomatous tumours, rendering the operation of excision of the entire bone necessary. This is an operation of considerable difficulty and danger. It is best performed by exposing the bone freely, disarticulating at the acromial end, and turning it inwards. The removal of the outer part is comparatively easy, but resection of the inner part is fraught with difficulty, the main danger being the risk of wounding the great veins which are in relation with its under surface.

THE SCAPULA

The **Scapula** (σκαπάνη, *a spade*) forms the back part of the shoulder girdle. It is a large, flat bone, triangular in shape, situated at the posterior aspect and side of the thorax, between the second and seventh or sometimes the eighth ribs, its internal border or base being about an inch from, and nearly but not quite parallel with, the spinous processes of the vertebræ, so that it is rather closer to them above than below. It presents for examination two surfaces, three borders, and three angles.

FIG. 94.—Left scapula. Anterior surface, or venter.

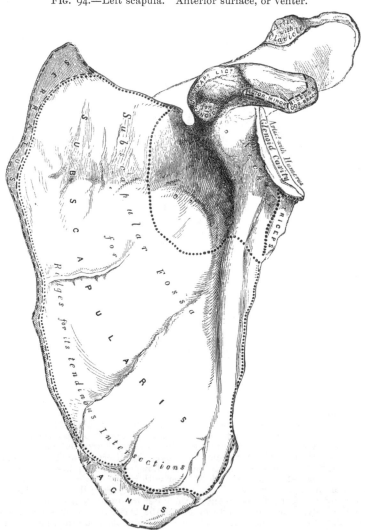

The **anterior surface**, or **venter** (fig. 94), presents a broad concavity, the *subscapular fossa*. It is marked, in the inner two-thirds, by several oblique ridges, which pass from behind outwards and upwards; the outer third is smooth. The oblique ridges give attachment to the tendinous intersections, and the surfaces between them to the fleshy fibres, of the Subscapularis muscle. The outer third of the fossa, which is smooth, is covered by, but does not afford attachment to, the fibres of this muscle. The venter is separated from the internal border by a smooth, triangular margin at the superior and inferior angles, and in the interval between these by a narrow edge which is often deficient. This marginal surface affords attachment throughout its entire extent to the Serratus magnus muscle. The subscapular fossa presents a transverse depression at its upper part, where the bone appears to be bent on itself, forming a considerable angle, called the *subscapular angle*, thus giving greater strength to the body of the bone from its

FIG. 95.—Left scapula. Posterior surface, or dorsum.

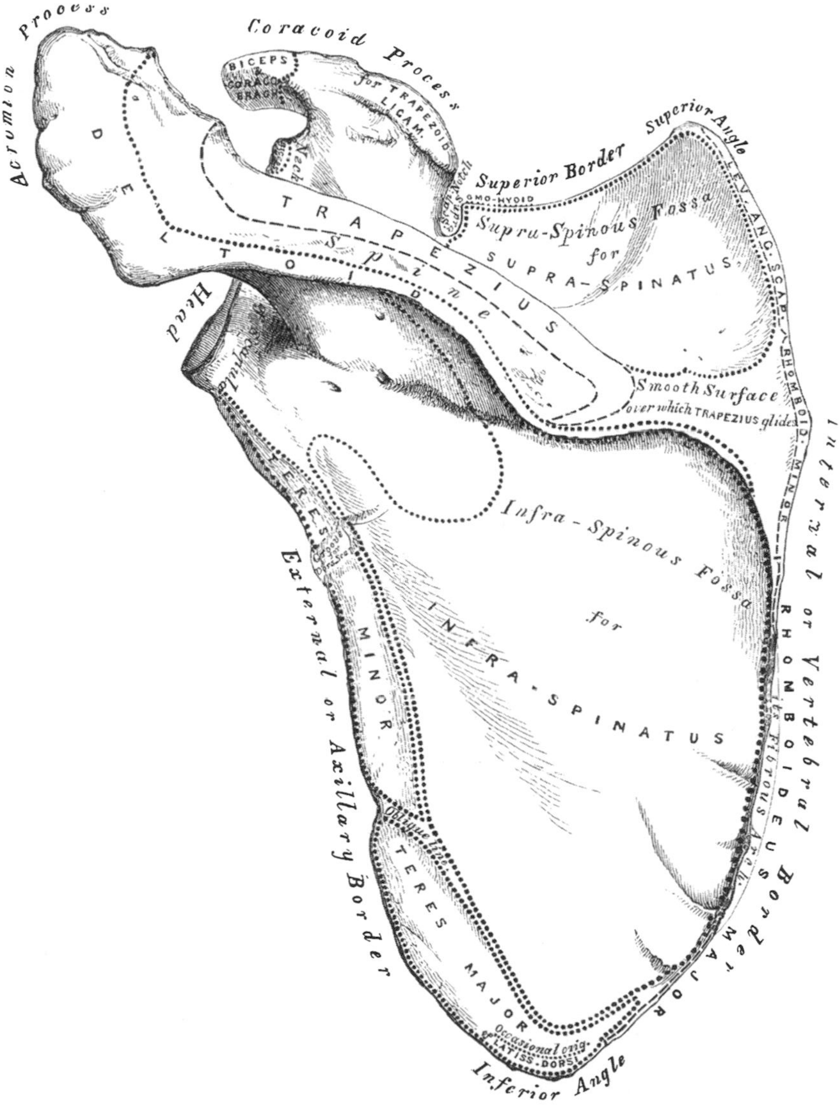

arched form ; while the summit of the arch serves to support the spine and acromion process. It is in this situation that the fossa is deepest ; so that the thickest part of the Subscapularis muscle lies in a line perpendicular to the plane of the glenoid cavity, and must consequently operate most effectively on the head of the humerus, which is contained in that cavity.

The **posterior surface**, or **dorsum** (fig. 95), is arched from above downwards, alternately concave and convex from side to side. It is subdivided unequally into two parts by the *spine*; the portion above the spine is called the *supraspinous fossa*, and that below it the *infraspinous fossa*.

The *supraspinous fossa*, the smaller of the two, is concave, smooth, and broader at the vertebral than at the humeral extremity. It affords attachment by its inner two-thirds to the Supraspinatus muscle.

The *infraspinous fossa* is much larger than the preceding; towards its vertebral margin a shallow concavity is seen at its upper part; its centre presents a prominent convexity, while towards the axillary border is a deep groove which runs from the upper towards the lower part. The inner two-thirds of this surface affords attachment to the Infraspinatus muscle; the outer third is only covered by it, without giving origin to its fibres. This surface is separated from the axillary border by an elevated ridge, which runs from the lower part of the glenoid cavity, downwards and backwards to the posterior border, about an inch above the inferior angle. The ridge serves for the attachment of a strong aponeurosis, which separates the Infraspinatus from the two Teres muscles. The surface of bone between this line and the axillary border is narrow in the upper two-thirds of its extent, and traversed near its centre by a groove for the passage of the dorsalis scapulæ vessels; it affords attachment to the Teres minor. Its lower third presents a broader, somewhat triangular surface, which gives origin to the Teres major, and over which the Latissimus dorsi glides; frequently the latter muscle takes origin by a few fibres from this part. The broad and narrow portions of bone above alluded to are separated by an oblique line, which runs from the axillary border, downwards and backwards, to meet the elevated ridge : to it is attached the aponeurosis separating the two Teres muscles from each other.

The **Spine** is a prominent plate of bone, which crosses obliquely the inner four-fifths of the dorsum of the scapula at its upper part, and separates the supra- from the infra-spinous fossa : it commences at the vertebral border by a smooth, triangular surface over which the Trapezius glides, separated from the bone by a bursa ; and, gradually becoming more elevated as it passes outwards, terminates in the acromion process, which overhangs the shoulder-joint. The spine is triangular, and flattened from above downwards, its apex corresponding to the vertebral border, its base (which is directed outwards) to the neck of the scapula. It presents two surfaces and three borders. Its *superior surface* is concave, assists in forming the supraspinous fossa, and affords attachment to part of the Supraspinatus muscle. Its *inferior surface* forms part of the infraspinous fossa, gives origin to part of the Infraspinatus muscle, and presents near its centre the orifice of a nutrient canal. Of the three borders, the *anterior* is attached to the dorsum of the bone ; the *posterior*, or *crest of the spine*, is broad, and presents two lips and an intervening rough interval. To the superior lip is attached the Trapezius, to the extent shown in the figure. A rough tubercle is generally seen occupying that portion of the spine which receives the insertion of the middle and inferior fibres of this muscle. To the inferior lip, throughout its whole length, is attached the Deltoid. The interval between the lips is also partly covered by the tendinous fibres of these muscles. The *external border*, or *base*, the shortest of the three, is slightly concave, its edge thick and round, continuous above with the under surface of the acromion process, below with the neck of the scapula. The narrow portion of bone external to this border, and separating it from the glenoid cavity, is called the *great scapular notch*, and serves to connect the supra- and infra-spinous fossæ.

The **Acromion Process**, so called from forming the summit of the shoulder (ἄκρον, *a summit* ; ὦμος, *the shoulder*), is a large and somewhat triangular or oblong process, flattened from behind forwards, directed at first a little outwards, and then curving forwards and upwards, so as to overhang the glenoid cavity. Its *upper surface*, directed upwards, backwards, and outwards, is convex, rough, and gives attachment to some fibres of the Deltoid, and in the rest of its extent it is subcutaneous. Its *under surface* is smooth and concave. Its *outer border* is thick and irregular, and presents three or four tubercles for the tendinous origins of the Deltoid muscle. Its *inner margin*, shorter than the outer, is concave, gives attachment to a portion of the Trapezius muscle, and presents about its centre a small, oval surface for articulation with the acromial end of the clavicle. Its *apex*, which corresponds to the point of meeting of these two borders in front, is thin, and has attached to it the coraco-acromial ligament.

Borders.—Of the three borders of the scapula, the *superior* is the shortest and thinnest; it is concave, and extends from the superior angle to the coracoid process. At its outer part is a deep, semicircular notch, the *suprascapular*, formed partly by the base of the coracoid process. This notch is converted into a foramen by the transverse ligament, and serves for the passage of the suprascapular nerve; sometimes this foramen is entirely surrounded by bone. The adjacent margin of the superior border affords attachment to the Omo-hyoid muscle. The *external, or axillary, border* is the thickest of the three. It commences above at the lower margin of the glenoid cavity, and inclines obliquely downwards and backwards to the inferior angle. Immediately below the glenoid cavity is a rough impression (the *infra-glenoid tubercle*), about an inch in length, which affords attachment to the long head of the Triceps muscle; in front of this is a longitudinal groove, which extends as far as its lower third, and affords origin to part of the Subscapularis muscle. The inferior third of this border, which is thin and sharp, serves for the attachment of a few fibres of the Teres major behind and of the Subscapularis in front. The *internal, or vertebral, border*, also named the *base*, is the longest of the three, and extends from the superior to the inferior angle of the bone. It is arched, intermediate in thickness between the superior and the external borders, and the portion of it above the spine is bent considerably outwards, so as to form an obtuse angle with the lower part. The vertebral border presents an anterior lip, a posterior lip, and an intermediate space. The *anterior lip* affords attachment to the Serratus magnus; the *posterior lip*, to the Supraspinatus above the spine, the Infraspinatus below; the interval between the two lips, to the Levator anguli scapulæ above the triangular surface at the commencement of the spine, the Rhomboideus minor to the edge of that surface; the Rhomboideus major being attached by means of a fibrous arch, connected above to the lower part of the triangular surface at the base of the spine, and below to the lower part of the posterior border.

Angles.—Of the three angles, the *superior*, formed by the junction of the superior and internal borders, is thin, smooth, rounded, somewhat inclined outwards, and gives attachment to a few fibres of the Levator anguli scapulæ muscle. The *inferior* angle, thick and rough, is formed by the union of the vertebral and axillary borders, its outer surface affording attachment to the Teres major and frequently to a few fibres of the Latissimus dorsi. The *anterior* angle is the thickest part of the bone, and forms what is called the *head* of the scapula The head presents a shallow, pyriform, articular surface, the *glenoid cavity* (γλήνη, *a socket*), whose longest diameter is from above downwards, and its direction outwards and forwards. It is broader below than above; at its apex is a slight impression (the *supra-glenoid tubercle*), to which is attached the long tendon of the Biceps muscle. It is covered with cartilage in the recent state; and its margins, slightly raised, give attachment to a fibro-cartilaginous structure, the *glenoid ligament*, by which its cavity is deepened. The *neck* of the scapula is the slightly depressed surface which surrounds the head; it is more distinct on the posterior than on the anterior surface, and below than above. In the latter situation it has, arising from it, a thick prominence, the coracoid process.

The **Coracoid Process,** so called from its fancied resemblance to a crow's beak (κόραξ, *a crow*), is a thick, curved process of bone, which arises by a broad base from the upper part of the neck of the scapula; it is directed at first upwards and inwards; then, becoming smaller, it changes its direction, and passes forwards and outwards. The ascending portion, flattened from before backwards, presents in front a smooth, concave surface, over which passes the Subscapularis muscle. The horizontal portion is flattened from above downwards; its upper surface is convex and irregular, and gives attachment to the Pectoralis minor; its under surface is smooth; its inner border is rough, and gives attachment to the Pectoralis minor; its outer border is also rough for the coraco-acromial ligament, while the apex is embraced by the conjoined tendon of origin of the short head of the Biceps and of the Coraco-brachialis and gives attachment to the costo-coracoid ligament. At the inner side of the root of the coracoid process is a rough impression for the attachment of the conoid ligament; and running from it obliquely forwards and outwards, on to the upper surface of the horizontal portion, an elevated ridge for the attachment of the trapezoid ligament.

Structure.—In the head, processes, and all the thickened parts of the bone, the scapula is composed of cancellous tissue, while in the rest of its extent it is

Fig. 96.—Plan of the development of the scapula. By seven centres.

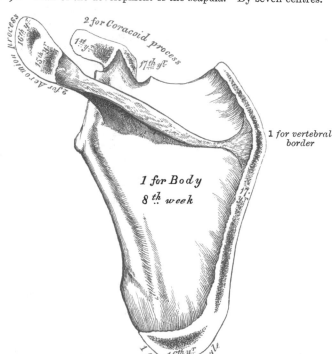

The epiphyses (except one for the coracoid process) appear from fifteen to seventeen years, and unite between twenty-two and twenty-five years of age.

composed of a thin layer of dense, compact tissue. The centre part of the supra-spinous fossa and the upper part of the infra-spinous fossa, but especially the former, are usually so thin as to be semi-transparent; occasionally the bone is found wanting in this situation, and the adjacent muscles come into contact.

Development (fig. 96).—By *seven* or more centres: one for the body, two for the coracoid process, two for the acromion, one for the vertebral border, and one for the inferior angle.

Ossification of the body of the scapula commences about the second month of foetal life, by the formation of an irregular quadrilateral plate of bone, immediately behind the glenoid cavity. This plate extends itself so as to form the chief part of the bone, the spine growing up from its posterior surface about the third month. At birth, a large part of the scapula is osseous, but the glenoid cavity, coracoid and acromion processes, the posterior border, and inferior angle are cartilaginous. From the fifteenth to the eighteenth month after birth, ossification takes place in the middle of the coracoid process, which usually becomes joined with the rest of the bone about the time when the other centres make their appearance. Between the fourteenth and twentieth years, ossification of the remaining centres takes place in quick succession, and in the following order : first, in the root of the coracoid process, in the form of a broad scale ; secondly, near the base of the acromion process ; thirdly, in the inferior angle and contiguous part of the posterior border ; fourthly, near the extremity of the acromion ; fifthly, in the posterior border. The acromion process, besides being formed of two separate nuclei, has its base formed by an extension into it of the centre of ossification which belongs to the spine, the extent of which varies in different cases. The two separate nuclei unite, and then join with the extension from the spine. These various epiphyses become joined to the bone between the ages of twenty-two and twenty-five years. Sometimes failure of union between the acromion process and spine occurs, the junction being effected by fibrous tissue, or by an imperfect articulation ; in some cases of supposed fracture of the acromion with ligamentous union, it is probable that the detached segment was never united to the rest of the bone. The upper third of the glenoid cavity is usually ossified from a separate centre (subcoracoid), which makes its appearance between the tenth and eleventh years. Very often, in addition, an epiphysis appears for the lower part of the glenoid cavity.

Articulations.—With the humerus and clavicle.

Attachment of Muscles.—To seventeen : to the anterior surface, the Subscapularis ; posterior surface, Supraspinatus, Infraspinatus ; spine, Trapezius, Deltoid ; superior border, Omo-hyoid ; vertebral border, Serratus magnus, Levator anguli scapulæ, Rhomboideus minor and major ; axillary border, Triceps, Teres minor, Teres major ; apex of glenoid cavity, long head of the Biceps ; coracoid process, short head of the Biceps, Coraco-brachialis, Pectoralis minor ; and to the inferior angle occasionally a few fibres of the Latissimus dorsi.

Surface Form.—The only parts of the scapula which are truly subcutaneous are the spine and acromion process, but, in addition to these, the coracoid process, the internal or vertebral border and inferior angle, and, to a less extent, the axillary border, may be defined. The acromion process and spine of the scapula are easily felt throughout their entire length, forming, with the clavicle, the arch of the shoulder. The acromion can be ascertained to be connected to the clavicle at the acromio-clavicular joint by running the finger along it, its position being often indicated by an irregularity or bony outgrowth from the clavicle close to the joint. The acromion can be felt forming the point of the shoulder, and from this can be traced backwards to join the spine of the scapula. The place of junction is usually denoted by a prominence, which is sometimes called the angle. From here the spine can be felt as a prominent ridge of bone, marked on the surface as an oblique depression, which becomes less and less distinct and terminates a little external to the spinous processes of the vertebræ. Its termination is usually indicated by a slight dimple in the skin, on a level with the interval between the third and fourth dorsal spines. Below this point the vertebral border of the scapula may be traced, running downwards and outwards, and thus diverging from the vertebral spines, to the inferior angle of the bone, which can be recognised, although covered by the Latissimus dorsi muscle. From this angle the axillary border can usually be traced through its thick muscular covering, forming, with the muscles, the posterior fold of the axilla. The coracoid process may be felt about an inch below the junction of the middle and outer third of the clavicle. Here it is covered by the anterior border of the Deltoid, and lies a little to the outer side of a slight depression, which corresponds to the interval between the Pectoralis major and Deltoid muscles. When the arms are hanging by the side, the upper angle of the scapula corresponds to the upper border of the second rib or the interval between the first and second dorsal spines, the inferior angle to the upper border of the eighth rib or the interval between the seventh and eighth dorsal spines.

Surgical Anatomy.– Fractures of the body of the scapula are rare, owing to the mobility of the bone, the thick layer of muscles by which it is encased on both surfaces, and the elasticity of the ribs on which it rests. Fracture of the neck of the bone is also uncommon. The most frequent course of the fracture is from the suprascapular notch to the infraglenoid tubercle, and it derives its principal interest from its simulation to a subglenoid dislocation of the humerus. The diagnosis can be made by noting the alteration in the position of the coracoid process. A fracture of the neck external to and not including the coracoid process is said to occur, but it is exceedingly doubtful whether such an accident ever takes place. The acromion process is more frequently broken than any other part of the bone, and there is sometimes, in young subjects, a separation of the epiphysis. It is believed that many of the cases of supposed fracture of the acromion, with fibrous union, which have been found on post-mortem examination are really cases of imperfectly united epiphysis. Sir Astley Cooper believed that most fractures of this bone united by fibrous tissue, the cause of this mode of union being the difficulty there is in keeping the fractured ends in constant apposition. The coracoid process is occasionally broken off, either from direct violence or, perhaps rarely, from muscular action.

Tumours of various kinds grow from the scapula. Of the innocent form of tumours probably the osteomata are the most common. When this tumour grows from the venter of the scapula, as it sometimes does, it is of the compact variety, such as usually grows from membrane-formed bones, as the bones of the skull. This would appear to afford evidence that this portion of the bone is formed from membrane, and not like the rest of the bone from cartilage. Sarcomatous tumours sometimes grow from the scapula and may necessitate removal of the bone, with or without amputation of the upper limb. The bone may be excised by a T-shaped incision, and the flaps being reflected, the removal is commenced from the posterior or vertebral border, so that the subscapular vessels which lie along the axillary border are among the last structures divided, and can be at once secured.

THE ARM

The **Arm** is that portion of the upper extremity which is situated between the shoulder and the elbow. Its skeleton consists of a single bone, the humerus.

THE HUMERUS

The **Humerus** is the longest and largest bone of the upper extremity; it presents for examination a shaft and two extremities.

The **Upper Extremity** presents a large, rounded *head*, joined to the shaft by a constricted portion, called the *neck*, and two other eminences, the *greater* and *lesser tuberosities* (fig. 97).

The **head,** nearly hemispherical in form,* is directed upwards, inwards, and a little backwards, and articulates with the glenoid cavity of the scapula; its surface is smooth, and coated with cartilage in the recent state. The circumference of its articular surface is slightly constricted, and is termed the *anatomical neck*, in contradistinction to the constriction which exists below the tuberosities. The latter is called the *surgical neck*, from its being most frequently the seat of fracture. It should be remembered, however, that fracture of the *anatomical neck* does sometimes, though rarely, occur.

The **anatomical neck** is obliquely directed, forming an obtuse angle with the shaft. It is more distinctly marked in the lower half of its circumference than in the upper half, where it presents a narrow groove, separating the head from the tuberosities. Its circumference affords attachment to the capsular ligament, and is perforated by numerous vascular foramina.

The **greater tuberosity** is situated on the outer side of the head and lesser tuberosity. Its upper surface is rounded and marked by three flat facets, separated by two slight ridges: the superior facet gives attachment to the tendon of the Supraspinatus; the middle one to the Infraspinatus; the inferior facet, and the shaft of the bone below it, to the Teres minor. The outer surface of the great tuberosity is convex, rough, and continuous with the outer side of the shaft.

The **lesser tuberosity** is more prominent, although smaller, than the greater : it is situated in front of the head, and is directed inwards and forwards. Its summit presents a prominent facet for the insertion of the tendon of the Subscapularis muscle. The tuberosities are separated from one another by a deep

---

* Though the head is nearly hemispherical in form, its margin, as Sir G. Humphry has shown, is by no means a true circle. Its greatest measurement is, from the top of the bicipital groove in a direction downwards, inwards, and backwards. Hence it follows that the greatest elevation of the arm can be obtained by rolling the articular surface in this direction—that is to say, obliquely upwards, outwards, and forwards.

Fig. 97.—Left humerus.   Anterior view.

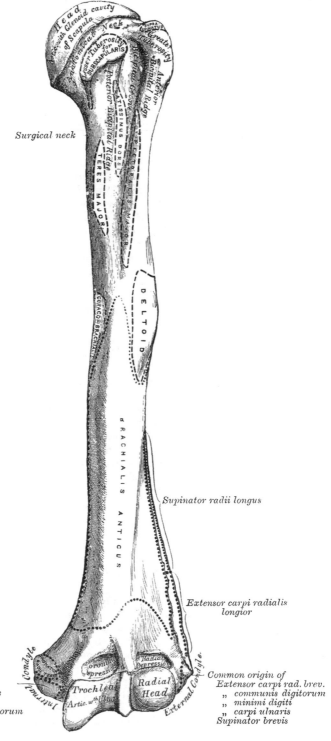

Surgical neck

Supinator radii longus

Extensor carpi radialis
longior

Common origin of
Flexor carpi radialis
Palmaris longus
Flexor sublimis digitorum
Flexor carpi ulnaris

Common origin of
Extensor carpi rad. brev.
      „   communis digitorum
      „   minimi digiti
      „   carpi ulnaris
Supinator brevis

groove, the *bicipital groove*, so called from its lodging the long tendon of the Biceps muscle, with which runs a branch of the anterior circumflex artery. It commences above between the two tuberosities, passes obliquely downwards and a little inwards, and terminates at the junction of the upper with the middle third of the bone. It is deep and narrow at the commencement, and becomes shallow and a little broader as it descends. Its borders are called, respectively, the anterior and posterior *bicipital ridges*, and form the upper part of the anterior and internal borders of the shaft of the bone. In the recent state it is covered with a thin layer of cartilage, lined by a prolongation of the synovial membrane of the shoulder-joint, and receives the tendon of insertion of the Latissimus dorsi muscle.

The **Shaft** of the humerus is almost cylindrical in the upper half of its extent, prismatic and flattened below, and presents three borders and three surfaces for examination.

The **anterior border** runs from the front of the great tuberosity above to the coronoid depression below, separating the internal from the external surface. Its upper part is very prominent and rough, and forms the outer lip of the bicipital groove. It is sometimes called the *anterior bicipital* or *pectoral ridge*, and serves for the attachment of the tendon of the Pectoralis major. About its centre it forms the anterior boundary of the rough deltoid impression ; below, it is smooth and rounded, affording attachment to the Brachialis anticus.

The **external border** runs from the back part of the greater tuberosity to the external condyle, and separates the external from the posterior surface. It is rounded and indistinctly marked in its upper half, serving for the attachment of the lower part of the insertion of the Teres minor, and below this of the external head of the Triceps muscle ; its centre is traversed by a broad but shallow, oblique depression, the *musculo-spiral groove* ; its lower part is marked by a prominent, rough margin, a little curved from behind forwards, the *external supra-condylar ridge*, which presents an anterior lip for the attachment of the Supinator longus above, and Extensor carpi radialis longior below, a posterior lip for the Triceps, and an intermediate space for the attachment of the external intermuscular septum.

The **internal border** extends from the lesser tuberosity to the internal condyle. Its upper third is marked by a prominent ridge, forming the posterior lip of the bicipital groove, and gives attachment to the tendon of the Teres major. About its centre is an impression for the attachment of the Coraco-brachialis, and just below this is seen the entrance of the nutrient canal, directed downwards. Sometimes there is a second canal situated at the commencement of the musculo-spiral groove for a nutrient artery derived from the superior profunda branch of the brachial artery. The inferior third of this border is raised into a slight ridge, the *internal supra-condylar ridge*, which becomes very prominent below ; it presents an anterior lip for the attachment of the Brachialis anticus, a posterior lip for the internal head of the Triceps, and an intermediate space for the attachment of the internal intermuscular septum.

The **external surface** is directed outwards above, where it is smooth, rounded, and covered by the Deltoid muscle ; forwards and outwards below, where it is slightly concave from above downwards, and gives origin to part of the Brachialis anticus muscle. About the middle of this surface is seen a rough, triangular impression for the insertion of the Deltoid muscle ; and below it the musculo-spiral groove, directed obliquely from behind, forwards, and downwards, and transmitting the musculo-spiral nerve and superior profunda artery.

The **internal surface,** less extensive than the external, is directed inwards above, forwards and inwards below ; at its upper part it is narrow, and forms the floor of the bicipital groove : to it is attached the Latissimus dorsi. The middle part of this surface is slightly rough for the attachment of some of the fibres of the tendon of insertion of the Coraco-brachialis ; its lower part is smooth, concave from above downwards, and gives attachment to the Brachialis anticus muscle.*

The **posterior surface** (fig. 98) appears somewhat twisted, so that its upper part is directed a little inwards, its lower part backwards and a little outwards. Nearly the whole of this surface is covered by the external and internal heads of the Triceps, the former of which is attached to its upper and outer part, the latter to its inner and back part, the two being separated by the musculo-spiral groove.

The **Lower Extremity** is flattened from before backwards, and curved slightly forwards; it terminates below in a broad, articular surface, which is divided into two parts by a slight ridge. Projecting on either side are the external and internal condyles. The articular surface extends a little lower than the condyles, and is curved slightly forwards, so as to occupy the more anterior part of the bone; its greatest breadth is in the transverse diameter, and it is obliquely directed, so that its inner extremity occupies a lower level than the outer. The outer portion of the articular surface presents a smooth, rounded eminence, which has received the name of the *capitellum*, or *radial head* of the humerus; it articulates

---

* A small, hook-shaped process of bone, the *supra-condylar process*, varying from $\frac{1}{10}$ to $\frac{3}{4}$ of an inch in length, is not infrequently found projecting from the inner surface of the shaft of the humerus two inches above the internal condyle. It is curved downwards, forwards, and inwards, and its pointed extremity is connected to the internal border, just above the inner condyle, by a ligament or fibrous band, which gives origin to a portion of the Pronator radii teres; through the arch completed by this fibrous band the median nerve and brachial artery pass, when these structures deviate from their usual course. Sometimes the nerve alone is transmitted through it, or the nerve may be accompanied by the ulnar artery, in cases of high division of the brachial. A well-marked groove is usually found behind the process, in which the nerve and artery are lodged. This space is analogous to the supra-condyloid foramen in many animals, and probably serves in them to protect the nerve and artery from compression during the contraction of the muscles in this region. A detailed account of this process is given by Dr. Struthers, in his *Anatomical and Physiological Observations*, p. 202. According to Mr. J. Wood, an accessory portion of the Coracobrachialis muscle is frequently connected with this process (*Journal of Anat. and Phys.* No. 1, Nov. 1866, p. 47).

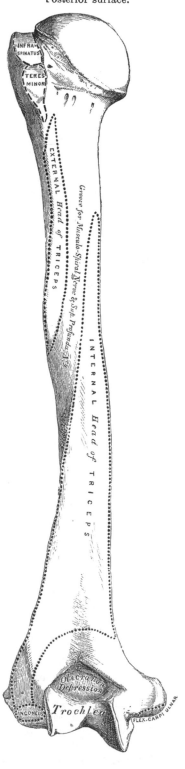

FIG. 98.—Left humerus.
Posterior surface.

with the cup-shaped depression on the head of the radius, and is limited to the front and lower part of the bone, not extending as far back as the other portion of the articular surface. On the inner side of this eminence is a shallow groove, in which is received the inner margin of the head of the radius. Above the front part of the capitellum is a slight depression, the *radial fossa*, which receives the anterior border of the head of the radius, when the fore-arm is flexed. The inner portion of the articular surface, the *trochlea*, presents a deep depression between two well-marked borders. This surface is convex from before backwards, concave from side to side, and occupies the anterior, lower, and posterior parts of the bone. The external border, less prominent than the internal, corresponds to the interval between the radius and the ulna. The internal border is thicker, more prominent, and consequently of greater length, than the external. The grooved portion of the articular surface fits accurately within the greater sigmoid cavity of the ulna; it is broader and deeper on the posterior than on the anterior aspect of the bone, and is inclined obliquely from behind forwards, and from without inwards. Above the front part of the trochlear surface is a smaller depression, the *coronoid fossa*, which receives the coronoid process of the ulna during flexion of the fore-arm. Above the back part of the trochlear surface is a deep, triangular depression, the *olecranon fossa*, in which the summit of the olecranon process is received in extension of the fore-arm. These fossæ are separated from one another by a thin, transparent lamina of bone, which is sometimes perforated, forming the *supratrochlear foramen*; their upper margins afford attachment to the anterior and posterior ligaments of the elbow-joint, and they are lined, in the recent state, by the synovial membrane of this articulation. The articular surfaces, in the recent state, are covered with a thin layer of cartilage. The external condyle (*epicondyle*) is a small, tubercular eminence, less prominent than the internal, curved a little forwards, and giving attachment to the external lateral ligament of the elbow-joint, and to a tendon common to the origin of some of the extensor and supinator muscles. The internal condyle (*epitrochlea*), larger and more prominent, and therefore more liable to fracture, than the external, is directed a little backwards : it gives attachment to the internal lateral ligament, to the Pronator radii teres, and to a tendon common to the origin of some of the flexor muscles of the forearm. The ulnar nerve runs in a groove at the back of the internal condyle, or between it and the olecranon process. These condyles are directly continuous above with the external and internal supra-condylar ridges.

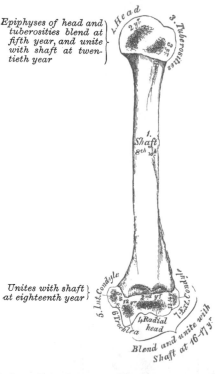

FIG. 99.—Plan of the development of the humerus. By seven centres.

Epiphyses of head and tuberosities blend at fifth year, and unite with shaft at twentieth year

Unites with shaft at eighteenth year

Structure.—The extremities consist of cancellous tissue, covered with a thin, compact layer ; the shaft is composed of a cylinder of compact tissue, thicker at the centre than at the extremities, and hollowed out by a large meuullary canal, which extends along its whole length.

Development.—By *seven*, or occasionally *eight* centres (fig. 99) : one for the shaft, one for the head, one for the tuberosities, one for the radial head, one for the trochlear portion of the articular surface, and one for each condyle. The nucleus for the shaft appears near the centre of the bone in the eighth week, and soon extends towards the extremities. At birth the humerus is ossified nearly in its whole length, the extremities only remaining cartilaginous. During the first year, sometimes before birth, ossification commences in the head of the bone, and during the third year the centre for the tuberosities makes its appearance, usually by a single ossific point, but sometimes, according to Béclard, by one for each tuberosity, that for the lesser being small, and not appearing until the fifth year. By the sixth year the centres for the head and tuberosities have increased in size and become joined, so as to form a single large epiphysis.

The lower end of the humerus is developed in the following manner : At the end of the second year ossification commences in the capitellum, and from this point extends inwards, so as to form the chief part of the articular end of the bone, the centre for the inner part of the trochlea not appearing until about the age of twelve. Ossification commences in the internal condyle about the fifth year, and in the external one not until about the thirteenth or fourteenth year. About sixteen or seventeen years, the outer condyle and both portions of the articulating surface (having already joined) unite with the shaft; at eighteen years the inner condyle becomes joined, while the upper epiphysis, although the first formed, is not united until about the twentieth year.

Articulations.—With the glenoid cavity of the scapula, and with the ulna and radius.

Attachment of Muscles.—To twenty-four : to the greater tuberosity, the Supraspinatus, Infraspinatus, and Teres minor ; to the lesser tuberosity, the Subscapularis ; to the anterior bicipital ridge, the Pectoralis major ; to the posterior bicipital ridge, the Teres major ; to the bicipital groove, the Latissimus dorsi ; to the shaft, the Deltoid, Coraco-brachialis, Brachialis anticus, external and internal heads of the Triceps ; to the internal condyle, the Pronator radii teres, and common tendon of the Flexor carpi radialis, Palmaris longus, Flexor sublimis digitorum, and Flexor carpi ulnaris ; to the external condyloid ridge, the Supinator longus, and Extensor carpi radialis longior ; to the external condyle, the common tendon of the Extensor carpi radialis brevior, Extensor communis digitorum, Extensor minimi digiti, Extensor carpi ulnaris, and Supinator brevis ; to the back of the external condyle, the Anconeus.

Surface Form.—The humerus is almost entirely clothed by the muscles which surround it, and the only parts of this bone which are strictly subcutaneous are small portions of the internal and external condyles. In addition to these, the tuberosities and a part of the head of the bone can be felt under the skin and muscles by which they are covered. Of these the greater tuberosity forms the most prominent bony point of the shoulder, extending beyond the acromion process and covered by the Deltoid muscle. It influences materially the surface form of the shoulder. It is best felt while the arm is lying loosely by the side ; if the arm be raised, it recedes from under the finger. The lesser tuberosity, directed forwards and inwards, is to be felt to the inner side of the greater tuberosity, just below the acromio-clavicular joint. Between the two tuberosities lies the bicipital groove. This can be defined by placing the finger, and making firm pressure, just internal to the greater tuberosity ; then, by rotating the humerus, the groove will be felt to pass under the finger as the bone is rotated. With the arm abducted from the side the lower part of the head of the bone is to be felt by pressing deeply in the axilla. On each side of the elbow-joint, and just above it, the internal and external condyles of the bone are to be felt. Of these the internal is the more prominent, but the internal supra-condylar ridge, passing upwards from it, is much less marked than the external, and, as a rule, is not to be felt. Occasionally, however, we find along this border the hook-shaped process mentioned above. The external condyle is to be seen most plainly during semiflexion of the forearm, and its position is indicated by a depression between the attachment of the adjacent muscles. From it is to be felt a strong bony ridge running up the outer border of the shaft of the bone. This is the external supra-condylar ridge ; it is concave forwards, and corresponds with the curved direction of the lower extremity of the humerus.

Surgical Anatomy.—There are several points of surgical interest connected with the humerus. First, as regards its development. The upper end, though the first to ossify, is the last to join the shaft, and the length of the bone is mainly due to growth from this upper epiphysis. Hence, in cases of amputation of the arm in young subjects, the humerus continues to grow considerably, and the end of the bone which immediately after the operation was covered with a thick cushion of soft tissue begins to project, thinning the soft parts and rendering the stump conical. This may necessitate the removal of a couple of inches or so of the bone, and even after this operation a recurrence of the conical stump may take place.

There are several points of surgical interest in connection with fractures. First, as regards their causation; the bone may be broken by direct or indirect violence like the other long bones, but, in addition to this, it is probably more frequently fractured by muscular action than any other of this class of bone in the body. It is usually the shaft, just below the insertion of the deltoid, which is thus broken. I have seen the accident happen from throwing a stone, and again in an apparently healthy adult, from cutting a piece of hard 'cake tobacco' on a table. In this latter case there was no disease of the bone that could be discovered. Fractures of the upper end may take place through the anatomical neck, through the surgical neck, or separation of the greater tuberosity may occur. Fracture of the anatomical neck is a very rare accident; in fact, it is doubted by some whether it ever occurs. These fractures are usually considered to be intracapsular, but they are probably partly within and partly without the capsule, as the lower part of the capsule is inserted some little distance below the anatomical neck, while the upper part is attached to it. They may be impacted or non-impacted. In most cases there is little or no displacement on account of the capsule, in whole or in part, remaining attached to the lower fragment. But occasionally a very remarkable alteration in position takes place; the upper fragment turns on its own axis, so that the cartilaginous surface of the head rests against the upper end of the lower fragment. When the fractured end is entirely separated from all its surroundings, its vascular supply must be entirely cut off, and one would expect it, theoretically, to necrose. But this must be exceedingly rare, for Gurlt was unable to find a single authenticated case recorded. Separation of the upper epiphysis of the humerus sometimes occurs in the young subject, and is marked by a characteristic deformity, by which the lesion may be at once recognised. This consists in the presence of an abrupt projection at the front of the joint some short distance below the coracoid process, caused by the upper end of the lower fragment. In fractures of the shaft of the humerus the lesion may take place at any point, but appears to be more common in the lower than the upper part of the bone. The points of interest in connection with these fractures are (1) that the musculo-spiral nerve may be injured as it lies in the groove on the bone, or may become involved in the callus which is subsequently thrown out; and (2) the frequency of non-union. This is believed to be more common in the humerus than in any other bone, and various causes have been assigned for it. It would seem most probably to be due to the difficulty that there is in fixing the shoulder-joint and the upper fragment, and possibly the elbow-joint and lower fragment also. Other causes which have been assigned for the non-union are: (1) that in attempting passive motion of the elbow-joint to overcome any rigidity which may exist, the movement does not take place at the articulation but at the seat of fracture; or that the patient, in consequence of the rigidity of the elbow, in attempting to flex or extend the forearm, moves the fragment and not the joint. (2) The presence of small portions of muscular tissue between the broken ends. (3) Want of support to the elbow, so that the weight of the arm tends to drag the lower fragment away from the upper. An important distinction to make in fractures of the lower end of the humerus, is between those that involve the joint and those which do not; the former are always serious, as they may lead to impairment of the utility of the limb. They include the T-shaped fracture and oblique fractures which involve the articular surface. The fractures which do not involve the joint are the transverse above the condyles, and the so-called epitroclear fracture, where the tip of the internal condyle is broken off, generally from direct violence.

Under the head of separation of the epiphysis two separate injuries have been described. One where the whole of the four ossific centres which form the lower extremity of the bone are separated from the shaft; and secondly, where the articular portion is alone separated, the two condyles remaining attached to the shaft of the bone. The epiphysial line between the shaft and lower end runs across the bone just above the tips of the condyles, a point to be borne in mind in performing the operation of excision.

Tumours originating from the humerus are of frequent occurrence. A not uncommon place for a chondroma to grow from is the shaft of the bone somewhere in the neighbourhood of the insertion of the deltoid. Sarcomata frequently grow from this bone.

THE FOREARM

The **Forearm** is that portion of the upper extremity which is situated between the elbow and wrist. Its skeleton is composed of two bones, the ulna and the radius.

THE ULNA

The **Ulna** (figs. 100, 101), so called from its forming the elbow (ὠλένη), is a long bone, prismatic in form, placed at the inner side of the forearm, parallel with the

FIG. 100.—Bones of the left forearm.   Anterior surface.

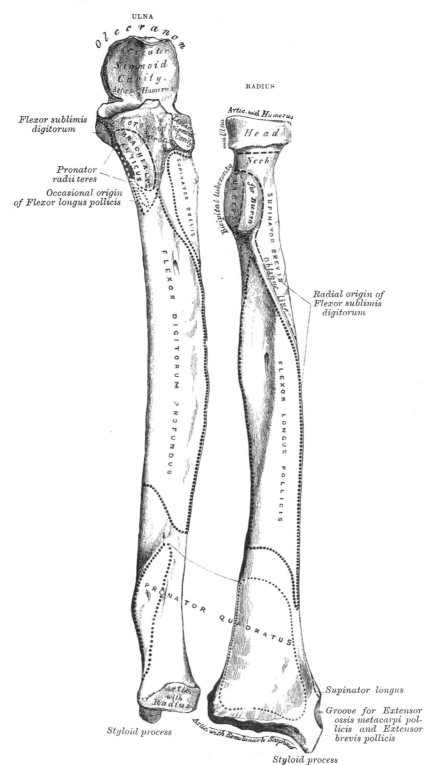

ULNA

Olecranon

Greater
Sigmoid
Cavity.
Artic. with Humerus

*Flexor sublimis
digitorum*

*Pronator
radii teres*

*Occasional origin
of Flexor longus pollicis*

RADIUS

Artic. with Humerus

with Ulna

Head

Neck

Bicipital tuberosity

for Bursa

SUPINATOR BREVIS

Oblique line

FLEXOR DIGITORUM PROFUNDUS

SUPINATOR BREVIS

*Radial origin of
Flexor sublimis
digitorum*

FLEXOR LONGUS POLLICIS

PRONATOR QUADRATUS

*Artic.
with
Radius*

Styloid process

Artic. with Semilunar & Scaphoid

Supinator longus

*Groove for Extensor
ossis metacarpi pol-
licis and Extensor
brevis pollicis*

Styloid process

radius. It is the larger and longer of the two bones. Its upper extremity, of great thickness and strength, forms a large part of the articulation of the elbow-joint; it diminishes in size from above downwards, its lower extremity being very small, and excluded from the wrist-joint by the interposition of an interarticular fibro-cartilage. It is divisible into a shaft and two extremities.

The **Upper Extremity,** the strongest part of the bone, presents for examination two large, curved processes, the Olecranon process and the Coronoid process; and two concave, articular cavities, the greater and lesser sigmoid cavities.

The **Olecranon Process** (ὠλένη, elbow; κρανίον, head) is a large, thick, curved eminence, situated at the upper and back part of the ulna. It is curved forwards at the summit so as to present a prominent tip which is received into the olecranon fossa in extension of the forearm; its base being contracted where it joins the shaft. This is the narrowest part of the upper end of the ulna, and, consequently, the most usual seat of fracture. The posterior surface of the olecranon, directed backwards, is triangular, smooth, subcutaneous, and covered by a bursa. Its upper surface is of quadrilateral form, marked behind by a rough impression for the attachment of the Triceps muscle; and in front, near the margin, by a slight transverse groove for the attachment of part of the posterior ligament of the elbow-joint. Its anterior surface is smooth, concave, covered with cartilage in the recent state, and forms the upper and back part of the great sigmoid cavity. The lateral borders present a continuation of the same groove that was seen on the margin of the superior surface; they serve for the attachment of ligaments —viz. the back part of the internal lateral ligament internally, the posterior ligament externally. To the inner border is also attached a part of the Flexor carpi ulnaris; while to the outer border is attached the Anconeus.

The **Coronoid Process** (κορώνη, anything hooked like a crow's beak) is a triangular eminence of bone which projects horizontally forwards from the upper and front part of the ulna. Its base is continuous with the shaft, and of considerable strength; so much so that fracture of it is an accident of rare occurrence. Its apex is pointed, slightly curved upwards, and received into the coronoid depression of the humerus in flexion of the forearm. Its upper surface is smooth, concave, and forms the lower part of the greater sigmoid cavity. The under surface is concave, and marked internally by a rough impression for the insertion of the Brachialis anticus. At the junction of this surface with the shaft is a rough eminence, the *tubercle of the ulna,* for the attachment of the oblique ligament. Its outer surface presents a narrow, oblong, articular depression, the *lesser sigmoid* cavity. The inner surface, by its prominent, free margin, serves for the attachment of part of the internal lateral ligament. At the front part of this surface is a small, rounded eminence for the attachment of one head of the Flexor sublimis digitorum; behind the eminence, a depression for part of the origin of the Flexor profundus digitorum; and, descending from the eminence, a ridge, which gives attachment to one head of the Pronator radii teres. Generally, the Flexor longus pollicis arises from the lower part of the coronoid process by a rounded bundle of muscular fibres.

The **Greater Sigmoid Cavity,** so called from its resemblance to the old shape of the Greek letter Σ, is a semilunar depression of large size, formed by the olecranon and coronoid processes, and serving for articulation with the trochlear surface of the humerus. About the middle of either lateral border of this cavity is a notch, which contracts it somewhat, and serves to indicate the junction of the two processes of which it is formed. The cavity is concave from above downwards, and divided into two lateral parts by a smooth, elevated ridge which runs from the summit of the olecranon to the tip of the coronoid process. Of these two portions, the internal is the larger, and is slightly concave transversely; the external portion is convex above, slightly concave below. The articular surface, in the recent state, is covered with a thin layer of cartilage.

The **Lesser Sigmoid Cavity** is a narrow, oblong, articular depression, placed on the outer side of the coronoid process, and receives the lateral articular surface

FIG. 101.—Bones of the left forearm. Posterior surface.

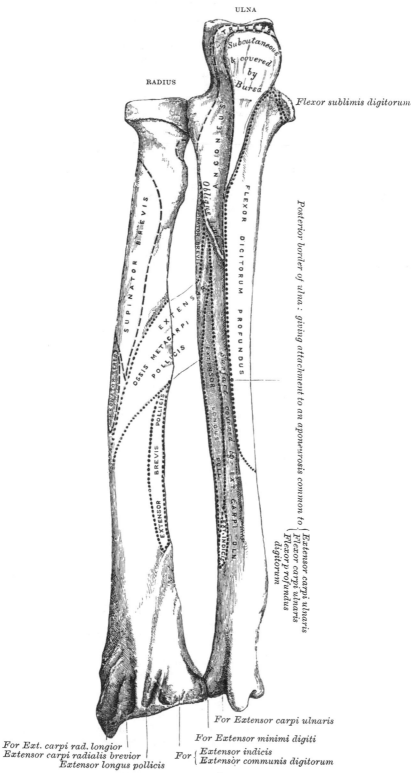

ULNA

RADIUS

Suboutaneous & covered by Bursa

Flexor sublimis digitorum

TRICEPS

ANCONEUS

Oblique Line

SUPINATOR BREVIS

SUPINATOR BREVIS

FLEXOR DIGITORUM PROFUNDUS

EXTENSOR OSSIS METACARPI POLLICIS

EXTENSOR BREVIS POLLICIS

Posterior border of ulna : giving attachment to an aponeurosis common to { Extensor carpi ulnaris / Flexor carpi ulnaris / Flexor profundus digitorum }

For Ext. carpi rad. longior
Extensor carpi radialis brevior
Extensor longus pollicis

For Extensor minimi digiti

For { Extensor indicis / Extensor communis digitorum }

For Extensor carpi ulnaris

of the head of the radius. It is concave from before backwards; and its extremities, which are prominent, serve for the attachment of the orbicular ligament. In the recent state, it is covered with a thin layer of cartilage.

The **Shaft,** at its upper part, is prismatic in form, and curved from behind forwards, and from without inwards, so as to be convex behind and externally; its central part is quite straight; its lower part is rounded, smooth, and bent a little outwards; it tapers gradually from above downwards, and presents for examination three borders, and three surfaces.

The **anterior border** commences above at the prominent inner angle of the coronoid process, and terminates below in front of the styloid process. It is well marked above, smooth and rounded in the middle of its extent, and affords attachment to the Flexor profundus digitorum : its lower fourth, marked off from the rest of the border by the commencement of an oblique ridge on the anterior surface, serves for the attachment of the Pronator quadratus. It separates the anterior from the internal surface.

The **posterior border** commences above at the apex of the triangular subcutaneous surface at the back part of the olecranon, and terminates below at the back part of the styloid process; it is well marked in the upper three-fourths, and gives attachment to the aponeurosis common to the Flexor carpi ulnaris, the Extensor carpi ulnaris, and the Flexor profundus digitorum muscles ; its lower fourth is smooth and rounded. This border separates the internal from the posterior surface.

The **external** or **interosseous border** commences above by the union of two lines, which converge one from each extremity of the lesser sigmoid cavity, enclosing between them a triangular space for the attachment of part of the Supinator brevis ; it terminates below at the middle of the head of the ulna. Its two middle fourths are very prominent, its lower fourth is smooth and rounded. This border gives attachment to the interosseous membrane, and separates the anterior from the posterior surface.

The **anterior surface,** much broader above than below, is concave in the upper three-fourths of its extent, and affords attachment to the Flexor profundus digitorum ; its lower fourth, also concave, is covered by the Pronator quadratus. The lower fourth is separated from the remaining portion of the bone by a prominent ridge, directed obliquely from above downwards and inwards; this ridge (the *oblique* or *Pronator ridge*) marks the extent of attachment of the Pronator quadratus. At the junction of the upper with the middle third of the bone is the nutrient canal, directed obliquely upwards and inwards.

The **posterior surface,** directed backwards and outwards, is broad and concave above, somewhat narrower and convex in the middle of its course, narrow, smooth, and rounded below. It presents, above, an oblique ridge, which runs from the posterior extremity of the lesser sigmoid cavity, downwards to the posterior border ; the triangular surface above this ridge receives the insertion of the Anconeus muscle, while the upper part of the ridge itself affords attachment to the Supinator brevis. The surface of bone below this is subdivided by a longitudinal ridge, sometimes called the *perpendicular line*, into two parts : the internal part is smooth, and covered by the Extensor carpi ulnaris ; the external portion, wider and rougher, gives attachment from above downwards to part of the Supinator brevis, the Extensor ossis metacarpi pollicis, the Extensor longus pollicis, and the Extensor indicis muscles.

The **internal surface** is broad and concave above, narrow and convex below. It gives attachment by its upper three-fourths to the Flexor profundus digitorum muscle : its lower fourth is subcutaneous.

The **Lower Extremity** of the ulna is of small size, and excluded from the articulation of the wrist-joint. It presents for examination two eminences, the outer and larger of which is a rounded, articular eminence, termed the *head* of the ulna ; the inner, narrower and more projecting, is a non-articular eminence, the *styloid process*. The *head* presents an articular facet, part of which, of an oval or semilunar form, is directed downwards, and articulates with the upper surface of the interarticular fibro-cartilage which separates it from the wrist-joint; the remaining portion, directed outwards, is narrow, convex, and received into the sigmoid cavity of the radius. The *styloid process* projects from the inner and back part of the bone, and descends a little lower than the head, terminating in a rounded summit, which affords attachment to the internal lateral ligament of the wrist. The head is separated from the styloid process by a depression for the attachment of the triangular interarticular fibro-cartilage; and behind, by a shallow groove for the passage of the tendon of the Extensor carpi ulnaris.

Structure.—Similar to that of the other long bones.

Development.—By *three* centres: one for the shaft, one for the inferior extremity, and one for the olecranon (fig. 102). Ossification commences near the middle of the shaft about the eighth week, and soon extends through the greater part of the bone. At birth the ends are cartilaginous. About the fourth year, a separate osseous nucleus appears in the middle of the head, which soon extends into the styloid process. About the tenth year, ossific matter appears in the olecranon near its extremity, the chief part of this process being formed from an extension of the shaft of the bone into it. The upper epiphysis is joined to the shaft about the sixteenth year, the lower about the twentieth year.

Articulations.—With the humerus and radius.

Attachment of Muscles.—To fourteen: to the olecranon, the Triceps, Anconeus, and one head of the Flexor carpi ulnaris. To the coronoid process, the Brachialis anticus, Pronator radii teres, Flexor sublimis digitorum, and Flexor profundus digitorum; generally also the Flexor longus pollicis. To the shaft, the Flexor profundus digitorum, Pronator quadratus, Flexor carpi ulnaris, Extensor carpi ulnaris, Anconeus, Supinator brevis, Extensor ossis metacarpi pollicis, Extensor longus pollicis, and Extensor indicis.

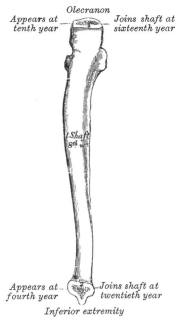

FIG. 102.—Plan of the development of the ulna. By three centres.

Olecranon

Appears at tenth year — *Joins shaft at sixteenth year*

l. Shaft 8th w.

Appears at fourth year — *Joins shaft at twentieth year*

Inferior extremity

Surface Form.—The most prominent part of the ulna on the surface of the body is the olecranon process, which can always be felt at the back of the elbow-joint. When the forearm is flexed, the upper quadrilateral surface can be felt, directed backwards; during extension it recedes into the olecranon fossa, and the contracting fibres of the Triceps prevent its being perceived. At the back of the olecranon is the smooth, triangular subcutaneous surface, continuous below with the posterior border of the shaft of the bone, and felt in every position of the forearm. During extension, the upper border of the olecranon is slightly above the level of the internal condyle, and the process itself is nearer to this condyle than the outer one. Running down the back of the forearm, from the apex of the triangular surface which forms the posterior surface of the olecranon, is a prominent ridge of bone, the posterior border of the ulna. This is to be felt throughout the entire length of the shaft of the bone, from the olecranon above to the styloid process below. As it passes down the forearm it pursues a sinuous course and inclines to the inner side, so that, though it is situated in the middle of the back of the limb above, it is on the inner side of the wrist at its termination. It becomes rounded off in its lower third, and may be traced below to the small, subcutaneous surface of the styloid process. Internal to this border the lower fourth of the inner surface is to be felt. The styloid process is to be felt as a prominent tubercle of bone, continuous above with the posterior subcutaneous border of the ulna, and terminating below in a blunt apex, which lies a little internal. and behind, but on a level with, the wrist-joint. The styloid process is best felt when the hand is in the same line as the bones of the forearm, and in a position midway between supination and pronation. If the forearm is pronated while the finger is placed on the process, it will be felt to recede, and another prominence of bone will appear just behind and above it. This is the head of the ulna, which articulates with the lower end of the radius and the triangular interarticular fibro-cartilage, and now projects between the tendons of the Extensor carpi ulnaris and the Extensor minimi digiti muscles.

The Radius

The **Radius** (radius, *a ray*, or *spoke of a wheel*) is situated on the outer side of the forearm, lying side by side with the ulna, which exceeds it in length and size (figs. 100, 101). Its upper end is small, and forms only a small part of the elbow-joint; but its lower end is large, and forms the chief part of the wrist. It is a long bone, prismatic in form, slightly curved longitudinally, and, like other long bones, has a shaft and two extremities.

The **Upper Extremity** presents a head, neck, and tuberosity. The *head* is of a cylindrical form, depressed on its upper surface into a shallow cup which articulates with the capitellum or radial head of the humerus. In the recent state it is covered with a layer of cartilage, which is thinnest at its centre. Around the circumference of the head is a smooth, articular surface, broad internally where it articulates with the lesser sigmoid cavity of the ulna ; narrow in the rest of its circumference, where it rotates within the orbicular ligament. It is coated with cartilage in the recent state. The head is supported on a round, smooth, and constricted portion of bone, called the *neck*, which presents, behind, a slight ridge, for the attachment of part of the Supinator brevis. Beneath the neck, at the inner and front aspect of the bone, is a rough eminence, the *bicipital tuberosity*. Its surface is divided into two parts by a vertical line— a posterior, rough portion, for the insertion of the tendon of the Biceps muscle ; and an anterior, smooth portion, on which a bursa is interposed between the tendon and the bone.

The **Shaft** of the bone is prismoid in form, narrower above than below, and slightly curved, so as to be convex outwards. It presents three surfaces, separated by three borders.

The **anterior border** extends from the lower part of the tuberosity above, to the anterior part of the base of the styloid process below. It separates the anterior from the external surface. Its upper third is very prominent ; and from its oblique direction, downwards and outwards, has received the name of the *oblique line* of the radius. It gives attachment, externally, to the Supinator brevis ; internally, to the Flexor longus pollicis, and between these to the Flexor sublimis digitorum. The middle third of the anterior border is indistinct and rounded. Its lower fourth is sharp, prominent, affords attachment to the Pronator quadratus and to the posterior annular ligament of the wrist, and terminates in a small tubercle, into which is inserted the tendon of the Supinator longus.

The **posterior border** commences above, at the back part of the neck of the radius, and terminates below, at the posterior part of the base of the styloid process ; it separates the posterior from the external surface. It is indistinct above and below, but well marked in the middle third of the bone.

The **internal** or **interosseous border** commences above, at the back part of the tuberosity, where it is rounded and indistinct, becomes sharp and prominent as it descends, and at its lower part divides into two ridges, which descend to the anterior and posterior margins of the sigmoid cavity. This border separates the anterior from the posterior surface, and has the interosseous membrane attached to it throughout the greater part of its extent.

The **anterior surface** is concave for its upper three-fourths, and gives attachment to the Flexor longus pollicis muscle ; it is broad and flat for its lower fourth, and gives attachment to the Pronator quadratus. A prominent ridge limits the attachment of the Pronator quadratus below, and between this and the inferior border is a triangular rough surface for the attachment of the anterior ligament of the wrist-joint. At the junction of the upper and middle third of this surface is the nutrient foramen, which is directed obliquely upwards.

The **posterior surface** is rounded, convex, and smooth in the upper third of its extent, and covered by the Supinator brevis muscle. Its middle third is broad, slightly concave, and gives attachment to the Extensor ossis metacarpi pollicis above, the Extensor brevis pollicis below. Its lower third is broad, convex, and covered by the tendons of the muscles, which subsequently run in the grooves on the lower end of the bone.

The **external surface** is rounded and convex throughout its entire extent. Its upper third gives attachment to the Supinator brevis muscle. About its centre is seen a rough ridge, for the insertion of the Pronator radii teres muscle. Its lower part is narrow, and covered by the tendons of the Extensor ossis metacarpi pollicis and Extensor brevis pollicis muscles.

The **Lower Extremity** of the radius is large, of quadrilateral form, and provided with two articular surfaces—one at the extremity, for articulation with the carpus, and one at the inner side of the bone, for articulation with the ulna. The carpal articular surface is of triangular form, concave, smooth, and divided by a slight antero-posterior ridge into two parts. Of these, the external is of a triangular form, and articulates with the scaphoid bone ; the inner, quadrilateral, articulates with the semilunar. The articular surface for the ulna is called the *sigmoid cavity* of the radius ; it is narrow, concave, smooth, and articulates with the head of the ulna. The circumference of this end of the bone presents three surfaces—an anterior, external, and posterior. The *anterior surface*, rough and irregular, affords attachment to the anterior ligament of the wrist-joint. The *external surface* is prolonged obliquely downwards into a strong, conical projection, the *styloid process*, which gives attachment by its base to the tendon of the Supinator longus, and by its apex to the external lateral ligament of the wrist-joint. The outer surface of this process is marked by a flat groove, which runs obliquely downwards and forwards, and gives passage to the tendons of the Extensor ossis metacarpi pollicis, and the Extensor brevis pollicis. The *posterior surface* is convex, affords attachment to the posterior ligament of the wrist, and is marked by three grooves. Proceeding from without inwards, the first groove is broad, but shallow, and subdivided into two by a slightly elevated ridge : the outer of these two transmits the tendon of the Extensor carpi radialis longior, the inner the tendon of the Extensor carpi radialis brevior. The second, which is near the centre of the bone, is a deep but narrow groove, bounded on its outer side by a sharply defined ridge ; it is directed obliquely from above, downwards and outwards, and transmits the tendon of the Extensor longus pollicis. The third, lying most internally, is a broad groove, for the passage of the tendons of the Extensor indicis and Extensor communis digitorum.

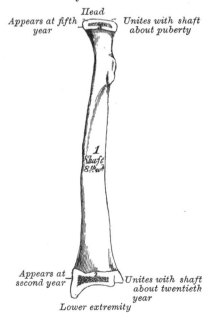

Fig. 103.—Plan of the development of the radius. By three centres.

Head

Appears at fifth year

Unites with shaft about puberty

1 Shaft 8th w

Appears at second year

Unites with shaft about twentieth year

Lower extremity

Structure.—Similar to that of the other long bones.

Development (fig. 103).—By *three* centres : one for the shaft, and one for each extremity. That for the shaft makes its appearance near the centre of the bone, about the eighth week of fœtal life. About the end of the second year, ossification commences in the lower end ; and about the fifth year, in the upper end. At the age of seventeen or eighteen the upper epiphysis becomes joined to the shaft ; the lower epiphysis becoming united about the twentieth year.

Articulations.—With four bones : the humerus, ulna, scaphoid, and semi-lunar.

Attachment of Muscles.—To nine : to the tuberosity, the Biceps ; to the oblique ridge, the Supinator brevis, Flexor sublimis digitorum, and Flexor longus pollicis ; to the shaft (its anterior surface), the Flexor longus pollicis and Pronator quadratus ; (its posterior surface), the Extensor ossis metacarpi pollicis and Extensor brevis pollicis ; (its outer surface), the Pronator radii teres ; and to the styloid process, the Supinator longus.

Surface Form.—Just below, and a little in front of, the posterior surface of the external condyle a part of the head of the radius may be felt, covered by the orbicular and external lateral ligaments. There is in this situation a little dimple in the skin, which is most visible when the arm is extended, and which marks the position of the head of the bone. If the finger is placed on this dimple, and the forearm pronated and supinated, the head of the bone will be distinctly perceived rotating in the lesser sigmoid cavity. The upper half of the shaft of the radius cannot be felt, as it is surrounded by the fleshy bellies of the muscles arising from the external condyle. The lower half of the shaft can be readily examined, though covered by tendons and muscles and not strictly subcutaneous. If traced downwards, the shaft will be felt to terminate in a lozenge-shaped, convex surface on the outer side of the base of the styloid process. This is the only subcutaneous part of the bone, and from its lower extremity the apex of the styloid process will be felt bending inwards towards the wrist. About the middle of the posterior aspect of the lower extremity of the bone is a well-marked ridge, best perceived when the hand is slightly flexed on the wrist. It forms the outer boundary of the oblique groove on the posterior surface of the bone, through which the tendon of the Extensor longus pollicis runs, and serves to keep that tendon in its place.

Surgical Anatomy.—The two bones of the forearm are more often broken together, than is either the radius or ulna separately. It is therefore convenient to consider the fractures of both bones in the first instance and subsequently to mention the principal fractures which take place in each bone. These fractures may be produced by either direct or indirect violence, though more commonly by direct violence. When indirect force is applied to the forearm the radius generally alone gives way, though both bones may suffer. The fracture from indirect force generally takes place somewhere about the middle of the bones, fracture from direct violence may occur at any part, more often, however, in the lower half of the bones. The fracture is usually transverse, but may be more or less oblique. A point of interest in connection with these fractures is the tendency that there is for the two bones to unite across the interosseous membrane ; the limb should therefore be put up in a position midway between supination and pronation, which is not only the most comfortable position, but also separates the bones most widely from each other and therefore diminishes the risk of the bones becoming united across the interosseous membrane. The splints, anterior and posterior, which are applied in these cases should be rather wider than the limb, so as to prevent any lateral pressure on the bones. In these cases there is a greater liability to gangrene from the pressure of the splints than in other parts of the body. This is no doubt due principally to two causes : (1) the flexion of the forearm compressing to a certain extent the brachial artery and retarding the flow of blood to the limb ; and (2) the superficial position of the two main arteries of the forearm in a part of their course, and their liability to be compressed by the splints. The special fractures of the ulna are : (1) fracture of the olecranon. This may be caused by direct violence, falls on the elbow with the forearm flexed, or by muscular action by the sudden contraction of the Triceps. The most common place for the fracture to occur is at the constricted portion where the olecranon joins the shaft of the bone, and the fracture may be either transverse or oblique ; but any part may be broken, even a thin shell may be torn off. Fractures from direct violence are occasionally comminuted. The displacement is sometimes very slight, owing to the fibrous structures around the process not being torn. (2) Fracture of the coronoid process sometimes occurs as a complication of dislocation backwards of the bones of the forearm, but it is doubtful if it ever occurs as an uncomplicated injury. (3) Fractures of the shaft of the ulna may occur at any part, but usually take place at the middle of the bone or a little below it. They are almost always the result of direct violence. (4) The styloid process may be knocked off by direct violence. Fractures of the radius consist of (1) fracture of the head of the bone ; this generally occurs in conjunction with some other lesion, but may occur as an uncomplicated injury. (2) Fracture of the neck may also take place, but is generally complicated with other injury. (3) Fractures of the shaft of the radius are very common, and may take place at any part of the bone. They may take place from either direct or indirect violence. In fractures

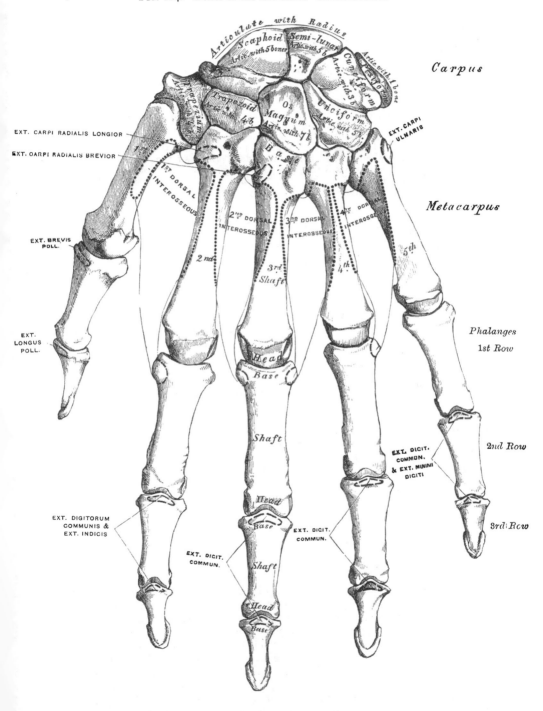

Fig. 104.—Bones of the left hand. Dorsal surface.

of the upper third of the shaft of the bone, that is to say, above the insertion of the Pronator radii teres, the displacement is very great. The upper fragment is strongly supinated by the Biceps and Supinator brevis, and flexed by the Biceps; while the lower fragment is pronated and drawn towards the ulna by the two pronators. If such a fracture is put up in the ordinary position, midway between supination and pronation, the fracture will unite with the upper fragment in a position of supination, and the lower one in the mid-position, and thus considerable impairment of the movements of the hand will result. The limb should be put up with the forearm supinated. (4) The most important fracture of the radius is that of the lower end (Colles's fracture). The fracture is transverse, and generally takes place about an inch from the lower extremity. It is caused by falls on the palm of the hand, and is an injury of advanced life, occurring more frequently in the female than the male. In consequence of the manner in which the fracture is caused, the upper fragment becomes driven into the lower, and impaction is the result; or else the lower fragment becomes split up into two or more pieces, so that no fixation occurs. Separation of the lower epiphysis of the radius may take place in the young. This injury and Colles's fracture may be distinguished from other injuries in this neighbourhood—especially dislocation, with which it is liable to be confounded—by observing the relative positions of the styloid processes of the ulna and radius. In the natural condition of parts, with the arm hanging by the side, the styloid process of the radius is on a lower level than that of the ulna, that is to say, nearer the ground. After fracture or separation of the epiphysis this process is on the same or on a higher level than that of the ulna, whereas it would be unaltered in position in dislocation.

THE HAND

The skeleton of the Hand is subdivided into three segments—the Carpus or wrist bones ; the Metacarpus or bones of the palm ; and the Phalanges or bones of the digits.

THE CARPUS

The bones of the **Carpus** (καρπός, *the wrist*), eight in number, are arranged in two rows. Those of the upper row, enumerated from the radial to the ulnar side, are the scaphoid, semilunar, cuneiform, and pisiform ; those of the lower row, enumerated in the same order, are the trapezium, trapezoid, os magnum, and unciform.

COMMON CHARACTERS OF THE CARPAL BONES

Each bone (excepting the pisiform) presents six surfaces. Of these the *anterior* or *palmar* and the *posterior* or *dorsal* are rough, for ligamentous attachment ; the dorsal surface being the broader, except in the scaphoid and semilunar. The *superior* or *proximal* and *inferior* or *distal* are articular, the superior generally convex, the inferior concave ; and the *internal* and *external* are also articular when in contact with contiguous bones, otherwise rough and tubercular. The structure in all is similar, consisting of cancellous tissue enclosed in a layer of compact bone. Each bone is also developed from a single centre of ossification.

BONES OF THE UPPER ROW

SCAPHOID (fig. 106)

The **Scaphoid** is the largest bone of the first row. It has received its name from its fancied resemblance to a boat, being broad at one end, and narrowed, like a prow, at the opposite (σκάφη, *a boat* ; εἶδος, *like*). It is situated at the upper and outer part of the carpus, its long axis being from above downwards, outwards, and forwards. The *superior surface* is convex, smooth, of triangular shape, and articulates with the lower end of the radius. The *inferior surface*, directed downwards, outwards, and backwards, is smooth, convex, also triangular, and divided by a slight ridge into two parts, the external of which articulates with the trapezium, the inner with the trapezoid. The *posterior* or *dorsal surface* presents a narrow, rough groove, which runs the entire length of the bone, and serves for the attachment of ligaments. The *anterior* or *palmar*

FIG. 105.—Bones of the left hand.   Palmar surface.

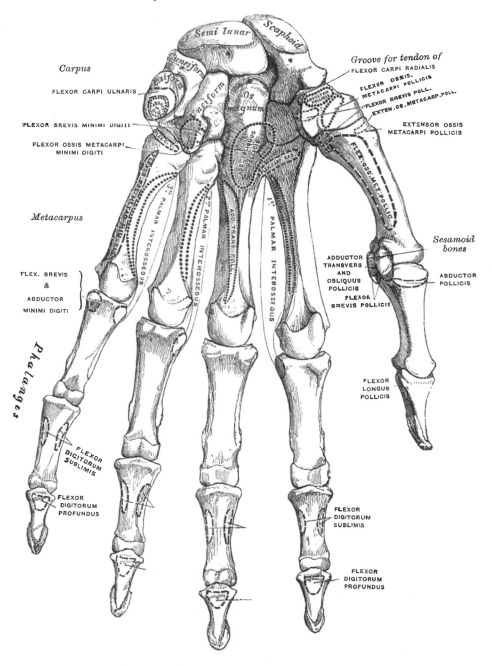

surface is concave above, and elevated at its lower and outer part into a prominent, rounded tuberosity, which projects forwards from the front of the carpus and gives attachment to the anterior annular ligament of the wrist and sometimes a few fibres of the Abductor pollicis. The *external surface* is rough and narrow, and gives attachment to the external lateral ligament of the wrist. The *internal surface* presents two articular facets; of these, the superior or smaller one is

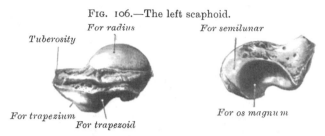

FIG. 106.—The left scaphoid.

For radius

Tuberosity

For semilunar

For trapezium

For trapezoid

For os magnum

flattened, of semilunar form, and articulates with the semilunar; the inferior or larger is concave, forming, with the semilunar bone, a concavity for the head of the os magnum.

To ascertain to which hand the bone belongs, hold it with the superior or radial, convex, articular surface upwards, and the posterior surface—i.e. the narrow, non-articular, grooved surface—towards you. The tubercle on the outer surface points to the side to which the bone belongs.*

Articulations.—With five bones: the radius above, trapezium and trapezoid below, os magnum and semilunar internally.

Attachment of Muscles.—Occasionally a few fibres of the Abductor pollicis.

SEMILUNAR (fig. 107)

The **Semilunar** bone may be distinguished by its deep concavity and crescentic outline (*semi*, half; *luna*, moon). It is situated in the centre of the upper row of the cárpus, between the scaphoid and cuneiform. The *superior surface*, convex, smooth, and bounded by four edges, articulates with the radius. The *inferior*

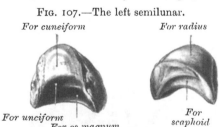

FIG. 107.—The left semilunar.

For cuneiform

For radius

For unciform

For os magnum

For scaphoid

surface is deeply concave, and of greater extent from before backwards than transversely: it articulates with the head of the os magnum, and, by a long, narrow facet (separated by a ridge from the general surface), with the unciform bone. The *anterior* or *palmar* and *posterior* or *dorsal surfaces* are rough, for the attachment of ligaments, the former being the broader, and of a somewhat rounded form. The *external surface* presents a narrow, flattened, semilunar facet for articulation with the scaphoid. The *internal surface* is marked by a smooth, quadrilateral facet, for articulation with the cuneiform.

Hold it with the convex articular surface for the radius upwards, and the narrowest non-articular surface towards you. The semilunar facet for the scaphoid will be on the side to which the bone belongs.

Articulations.—With five bones: the radius above, os magnum and unciform below, scaphoid and cuneiform on either side.

---

* In these directions each bone is supposed to be placed in its natural position—that is, such a position as it would occupy when the arm is hanging by the side, the forearm in a position of supination, the thumb being directed outwards and the palm of the hand looking forwards.

CUNEIFORM (fig. 108)

The **Cuneiform** (cuneus, *a wedge* ; forma, *likeness*) may be distinguished by its pyramidal shape (*os pyramidal*), and by its having an oval, isolated facet for articulation with the pisiform bone. It is situated at the upper and inner side of the carpus. The *superior surface* presents an internal, rough, non-articular portion, and an external or articular portion, which is convex, smooth, and articulates with the tri- angular interarticular fibro-cartilage of the wrist. The *inferior surface*, directed outwards, is concave, sinuously curved, and smooth for articulation with the unciform. The *posterior* or *dorsal surface* is rough, for the attachment of ligaments. The *anterior* or *palmar surface* presents, at its inner side, an oval facet, for articulation with the pisiform ; and is rough externally, for ligamentous attachment. The *external surface*, the base of the pyramid, is marked by a flat, quadrilateral, smooth facet, for articulation with the semilunar. The *internal surface*, the summit of the pyramid, is pointed and roughened, for the attachment of the internal lateral ligament of the wrist.

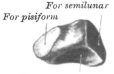

FIG. 108.—The left cuneiform.

For semilunar

For pisiform

For unciform

Hold the bone with the surface supporting the pisiform facet away from you, and the concavo-convex surface for the unciform downwards. The base of the wedge (i.e. the broad end of the bone) will be on the side to which it belongs.

Articulations.—With three bones : the semilunar externally, the pisiform in front, the unciform below ; and with the triangular, interarticular fibro-cartilage which separates it from the lower end of the ulna.

PISIFORM (fig. 109)

The **Pisiform** (pisum, *a pea* ; forma, *likeness*) may be known by its small size, and by its presenting a single articular facet. It is situated on a plane anterior to the other bones of the carpus ; it is spheroidal in form, with its long diameter directed vertically. Its *posterior* surface presents a smooth, oval facet, for articulation with the cuneiform : this facet approaches the superior, but not the inferior, border of the bone. The *anterior* or *palmar surface* is rounded and rough, and gives attachment to the anterior annular ligament and to the Flexor carpi ulnaris and Abductor minimi digiti muscles. The *outer* and *inner surfaces* are also rough, the former being concave, the latter usually convex.

FIG. 109.
The left pisiform.

For cuneiform

Hold the bone with the posterior surface—that which presents the articular facet—towards you, in such a manner that the faceted portion of the surface is uppermost. The outer, concave surface will point to the side to which it belongs.

Articulation.—With one bone, the cuneiform.

Attachment of Muscles.—To two : the Flexor carpi ulnaris and Abductor minimi digiti ; and to the anterior annular ligament.

BONES OF THE LOWER ROW

TRAPEZIUM (fig. 110)

The **Trapezium** (τράπεζα, *a table*) is of very irregular form. It may be distin- guished by a deep groove, for the tendon of the Flexor carpi radialis muscle. It is situated at the external and inferior part of the carpus, between the scaphoid and first metacarpal bone. The *superior surface*, concave and smooth, is directed upwards and inwards, and articulates with the scaphoid. The *inferior surface*, directed downwards and inwards, is oval, concave from side to side, convex from before backwards, so as to form a saddle-shaped surface, for articulation with the base of the first metacarpal bone. The *anterior* or *palmar surface* is narrow and rough. At its upper part is a deep groove, running from above obliquely downwards and inwards : it transmits the tendon of the Flexor carpi radialis, and is bounded externally by a prominent ridge, the *oblique ridge* of the trape-

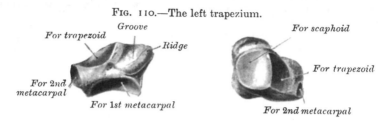

FIG. 110.—The left trapezium.

For trapezoid   Groove   Ridge   For scaphoid

For 2nd metacarpal   For trapezoid

For 1st metacarpal   For 2nd metacarpal

zium. This surface gives attachment to the Abductor pollicis, Flexor ossis metacarpi pollicis, and Flexor brevis pollicis muscles, and the anterior annular ligament. The *posterior* or *dorsal surface* is rough. The *external surface* is also broad and rough, for the attachment of ligaments. The *internal surface* presents two articular facets : the upper one, large and concave, articulates with the trapezoid ; the lower one, small and oval, with the base of the second metacarpal bone.

Hold the bone with the saddle-shaped surface downwards and the grooved surface away from you. The prominent, rough, non-articular surface points to the side to which the bone belongs.

Articulations.—With four bones : the scaphoid above, the trapezoid and second metacarpal bones internally, the first metacarpal below.

Attachment of Muscles.—Abductor pollicis, Flexor ossis metacarpi pollicis, and part of the Flexor brevis pollicis.

TRAPEZOID (fig. 111)

The **Trapezoid** is the smallest bone in the second row. It may be known by its wedge-shaped form, the broad end of the wedge forming the dorsal, the narrow end the palmar, surface ; and by its having four articular surfaces touching each other, and separated by sharp edges. The *superior surface*, quadrilateral in form, smooth, and slightly concave, articulates with the scaphoid. The *inferior surface* articulates with the upper end of the second metacarpal bone ; it is convex from side to side, concave from before backwards, and subdivided, by an elevated ridge, into two unequal lateral facets. The *posterior* or *dorsal* and *anterior* or *palmar surfaces* are rough, for the attachment of ligaments, the former being the larger of the two. The *external surface*, convex and smooth, articulates with the trapezium. The *internal surface* is concave and smooth in front, for articulation with the os magnum ; rough behind, for the attachment of an interosseous ligament.

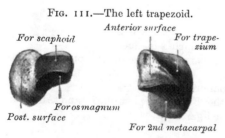

FIG. 111.—The left trapezoid.

Anterior surface

For scaphoid   For trapezium

For os magnum

Post. surface

For 2nd metacarpal

Hold the bone with the larger, non-articular surface towards you, and the smooth, quadrilateral articular surface upwards. The convex, articular surface will point to the side to which the bone belongs.*

* Occasionally in a badly marked bone there is some difficulty in ascertaining to which side the bone belongs ; the following method will sometimes be found useful. Hold the

Articulations.—With four bones : the scaphoid above, second metacarpal bone below, trapezium externally, os magnum internally.

Os Magnum (fig. 112)

The **Os Magnum** is the argest bone of the carpus, and occupies the centre of the wrist. It presents, above, a rounded portion or head, which is received into the concavity formed by the scaphoid and semilunar bones ; a constricted portion or neck ; and, below this, the body. The *superior surface* is rounded, smooth, and articulates with the semilunar. The *inferior surface* is divided by two ridges into three facets, for articulation with the second, third, and fourth metacarpal bones, that for the third (the middle facet) being the largest of the three. The *posterior* or *dorsal surface* is broad and rough ; the *anterior* or *palmar*, narrow,

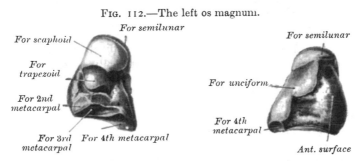

FIG. 112.—The left os magnum.

rounded, and also rough, for the attachment of ligaments and a part of the Adductor obliquus pollicis. The *external surface* articulates with the trapezoid by a small facet at its anterior inferior angle, behind which is a rough depression for the attachment of an interosseous ligament. Above this is a deep and rough groove, which forms part of the neck, and serves for the attachment of ligaments, bounded superiorly by a smooth, convex surface, for articulation with the scaphoid. The *internal surface* articulates with the unciform by a smooth, concave, oblong facet, which occupies its posterior and superior parts ; and is rough in front, for the attachment of an interosseous ligament.

Hold the bone with the broader, non-articular surface towards you, and the head upwards. The small, articular facet at the anterior inferior angle of the external surface will point to the side to which the bone belongs.

Articulations.—With seven bones : the scaphoid and semilunar above ; the second, third, and fourth metacarpal bones below ; the trapezoid on the radial side ; and the unciform on the ulnar side.

Attachment of Muscles.—Part of the Adductor obliquus pollicis.

Unciform (fig. 113)

The **Unciform** (uncus, *a hook* ; forma, *likeness*) may be readily distinguished by its wedge-shaped form, and the hook-like process which projects from its palmar surface. It is situated at the inner and lower angle of the carpus, with its base downwards, resting on the two inner metacarpal bones, and its apex directed upwards and outwards. The *superior surface*, the apex of the wedge, is narrow, convex, smooth, and articulates with the semilunar. The *inferior surface* articulates with the fourth and fifth metacarpal bones, the concave surface for each being separated by a ridge, which runs from before backwards. The *posterior* or *dorsal surface* is triangular and rough, for ligamentous attachment. The *anterior* or *palmar surface* presents, at its lower and inner side, a curved, hook-like process of bone, the *unciform process*, directed from the palmar surface forwards

bone with its broader, non-articular surface upwards, so that its sloping border is directed towards you. The border will slope to the side to which the bone belongs.

and outwards. It gives attachment, by its apex, to the annular ligament and the Flexor carpi ulnaris; by its inner surface to the Flexor brevis minimi digiti and the Flexor ossis metacarpi minimi digiti; and is grooved on its outer side, for the passage of the Flexor tendons into the palm of the hand. This is one of the four eminences on the front of the carpus to which the anterior annular ligament

FIG. 113.—The left unciform.

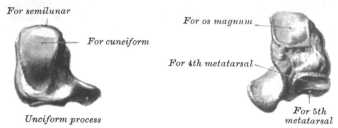

For semilunar

For cuneiform

For os magnum

For 4th metatarsal

Unciform process

For 5th metatarsal

is attached; the others being the pisiform internally, the oblique ridge of the trapezium and the tuberosity of the scaphoid externally. The *internal surface* articulates with the cuneiform by an oblong facet, cut obliquely from above, downwards and inwards. The *external surface* articulates with the os magnum by its upper and posterior part, the remaining portion being rough, for the attachment of ligaments.

Hold the bone with the hooked process away from you, and the articular surface, divided into two parts for the metacarpal bones, downwards. The concavity of the process will be on the side to which the bone belongs.

Articulations.—With five bones: the semilunar above, the fourth and fifth metacarpal below, the cuneiform internally, the os magnum externally.

Attachment of Muscles.—To three: the Flexor brevis minimi digiti, the Flexor ossis metacarpi minimi digiti, the Flexor carpi ulnaris.

THE METACARPUS

The **Metacarpal Bones** are five in number: they are long cylindrical bones, presenting for examination a shaft and two extremities.

Common Characters of the Metacarpal Bones

The **shaft** is prismoid in form, and curved longitudinally, so as to be convex in the longitudinal direction behind, concave in front. It presents three surfaces: two lateral and one posterior. The *lateral surfaces* are concave, for the attachment of the Interossei muscles, and separated from one another by a prominent anterior ridge. The *posterior* or *dorsal surface* presents in its distal half a smooth, triangular, flattened area which is covered, in the recent state, by the tendons of the Extensor muscles. This triangular surface is bounded by two lines, which commence in small tubercles situated on the dorsal aspect on either side of the digital extremity, and, running backwards, converge to meet together a little behind the centre of the bone and form a ridge which runs along the rest of the dorsal surface to the carpal extremity. This ridge separates two lateral, sloping surfaces for the attachment of the Dorsal interossei muscles.* To the tubercles on the digital extremities are attached the lateral ligaments of the metacarpophalangeal joints.

The **carpal extremity,** or **base,** is of a cuboidal form, and broader behind than in front: it articulates above with the carpus, and on each side with the adjoining metacarpal bones; its *dorsal* and *palmar surfaces* are rough, for the attachment of tendons and ligaments.

---

* By these sloping surfaces the metacarpal bones of the hand may be at once differentiated from the metatarsal bones of the foot.

The **digital extremity,** or **head,** presents an oblong surface markedly convex from before backwards, less so from side to side, and flattened laterally ; it articulates with the proximal phalanx ; it is broader, and extends farther forwards, on the palmar than on the dorsal aspect. It is longer in the antero-posterior than in the transverse diameter.   On either side of the head is a tubercle for the attachment of the lateral ligament of the metacarpo-phalangeal joint. The *posterior surface*, broad and flat, supports the Extensor tendons ; the *anterior surface* is grooved in the middle line for the passage of the Flexor tendons and marked on each side by an articular eminence continuous with the terminal articular surface.

PECULIAR CHARACTERS OF THE METACARPAL BONES

The **metacarpal bone of the thumb** (fig. 114) is shorter and wider than the rest, diverges to a greater degree from the carpus, and its *palmar surface* is directed inwards towards the palm.   The *shaft* is flattened and broad on its dorsal aspect, and does not present the ridge which is found on the other metacarpal bones ; it is concave from above downwards, on its palmar surface.   The *carpal extremity*, or *base*, presents a concavo-convex surface, for articulation with the trape-zium ; it has no lateral facets, but presents externally a tubercle for the insertion of the Extensor ossis metacarpi pollicis.   The *digital extremity* is less convex than that of the other metacarpal bones, broader from side to side than from before backwards.   It presents on its palmar aspect two distinct articular eminences for the two sesamoid bones in the tendons of the Flexor brevis pollicis ; the outer one being the larger of the two.

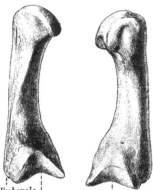

FIG. 114.—The first metacarpal. (Left.)

Tubercle
For trapezium   *For trapezium*

The side to which this bone belongs may be known by holding it in the position it occupies in the hand, with the carpal extremity upwards and the dorsal surface backwards ; the tubercle for the Extensor ossis metacarpi pollicis will point to the side to which it belongs.

Attachment of Muscles.—To four : the Flexor ossis metacarpi pollicis, the Extensor ossis metacarpi pollicis, the Flexor brevis pollicis, and the First dorsal interosseous.

The **metacarpal bone of the index finger** (fig. 115) is the longest, and its base the largest, of the other four.   Its *carpal extremity* is prolonged upwards and inwards, forming a prominent ridge.   The dorsal and palmar surfaces of this extremity are rough, for the attachment of tendons and ligaments.   It presents four articular facets : three on the upper aspect of the base ; the middle one of the three is the largest, concave from side to side, convex from before backwards, for articulation with the trapezoid ; the external one is a small, flat, oval facet, for articulation with the trapezium ; the internal one on the summit of the ridge is long and narrow, for articulation with the os magnum.   The fourth facet is on the inner or ulnar side of the extremity of the bone, and is for articulation with the third metacarpal bone.

The side to which this bone belongs is indicated by the absence of the lateral facet on the outer (radial) side of its base, so that if the bone is placed with its base towards the student, and the palmar surface upwards, the side on which there is no lateral facet will be that to which it belongs.

Attachment of Muscles.—To six : Flexor carpi radialis, Extensor carpi radialis longior, Adductor obliquus pollicis, First and Second dorsal interosseous, and First palmar interosseous.

The **metacarpal bone of the middle finger** (fig. 116) is a little smaller than the preceding; it presents a pyramidal eminence (the styloid process) on the radial side of its base (dorsal aspect), which extends upwards behind the os magnum; immediately below this, on the dorsal aspect, is a rough surface for the attachment of the Extensor carpi radialis brevior. The carpal articular facet is concave behind, flat in front, and articulates with the os magnum. On the radial side is a smooth, concave facet for articulation with the second metacarpal bone, and on the ulnar side two small, oval facets, for articulation with the fourth metacarpal.

The side to which this bone belongs is easily recognised by the styloid process on the radial side of its base. With the palmar surface uppermost, and the base towards the student, this process points towards the side to which the bone belongs.

Attachment of Muscles.—To six : Extensor carpi radialis brevior, Flexor carpi radialis, Adductor transversus pollicis, Adductor obliquus pollicis, and Second and Third dorsal interosseous.

Fig. 115.—The second metacarpal. (Left.)   Fig. 116.—The third metacarpal. (Left.)

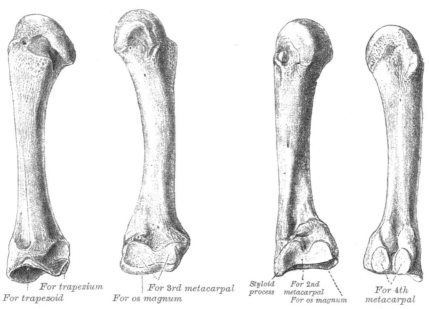

For trapezium   For 3rd metacarpal   Styloid   For 2nd   For 4th
For trapezoid   For os magnum   process   metacarpal   metacarpal
                                                       For os magnum

The **metacarpal bone of the ring-finger** (fig. 117) is shorter and smaller than the preceding, and its base small and quadrilateral; the carpal surface of the base presenting two facets, a large one externally for articulation with the unciform, and a small one internally for the os magnum. On the radial side are two oval facets, for articulation with the third metacarpal bone; and on the ulnar side, a single concave facet, for the fifth metacarpal.

If this bone is placed with the base towards the student, and the palmar surface upwards, the radial side of the base, which has two facets for articulation with the third metacarpal bone, will be on the side to which it belongs. If, as sometimes happens in badly marked bones, one of these facets is indistinguishable, the side may be known by selecting the surface on which the larger articular facet is present. This facet is for the fifth metacarpal bone, and would therefore be situated on the ulnar side—that is, the one to which the bone does *not* belong.

Attachment of Muscles.—To three : the Third and Fourth dorsal and Second palmar interosseous.

The **metacarpal bone of the little finger** (fig. 118) presents on its base one facet, which is concavo-convex, and which articulates with the unciform bone, and one lateral, articular facet, which articulates with the fourth metacarpal bone. On its ulnar side is a prominent tubercle, for the insertion of the tendon of the Extensor carpi ulnaris. The dorsal surface of the shaft is marked by an oblique ridge, which extends from near the ulnar side of the upper extremity to the radial side of the lower. The outer division of this surface serves for the attachment of the Fourth dorsal interosseous muscle; the inner division is smooth, and covered by the Extensor tendons of the little finger.

If this bone is placed with its base towards the student, and its palmar surface upwards, the side of the head which has a lateral facet will be that to which the bone belongs.

Attachment of Muscles.—To five: the Extensor carpi ulnaris, Flexor carpi ulnaris, Flexor ossis metacarpi minimi digiti, Fourth dorsal, and Third palmar interosseous.

FIG. 117.—The fourth metacarpal. (Left.)   FIG. 118.—The fifth metacarpal. (Left.)

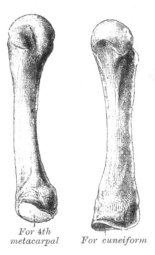

For 3rd metacarpal   For os magnum   For unciform   For 5th metacarpal   For 4th metacarpal   For cuneiform

Articulations.—Besides the phalangeal articulations, the first metacarpal bone articulates with the trapezium; the second with the trapezium, trapezoid, os magnum, and third metacarpal bone; the third with the os magnum, and second and fourth metacarpal bones; the fourth with the os magnum, unciform, and third and fifth metacarpal bones; and the fifth with the unciform and fourth metacarpal bone.

The *first* has no lateral facets on its carpal extremity; the *second* has no lateral facet on its radial side, but one on its ulnar side; the third has one on its radial and two on its ulnar side; the fourth has two on its radial and one on its ulnar side; and the fifth has only one on its radial side.

PHALANGES

The **Phalanges** (*internodia*) are the bones of the fingers; they are fourteen in number, three for each finger, and two for the thumb. They are long bones, and present for examination a shaft and two extremities. The *shaft* tapers from above downwards, is convex posteriorly, concave in front from above downwards, flat from side to side, and marked laterally by rough ridges, which give attachment to the fibrous sheaths of the Flexor tendons. The *metacarpal extremity* or *base*, in the first row presents an oval, concave, articular surface, broader from side to side than from before backwards; and the same extremity in the other two rows, a double concavity, separated by a longitudinal median ridge, extending from before backwards. The *digital extremities* are smaller than the bases, and terminate, in the first and second rows, in two small, lateral condyles, separated by a slight groove; the articular surface being prolonged farther forwards on the palmar than on the dorsal surface, especially in the first row.

The **Ungual Phalanges** are convex on their dorsal, flat on their palmar surfaces; they are recognised by their small size, and by a roughened, elevated surface of a horseshoe form on the palmar aspect of their ungual extremity, which serves to support the sensitive pulp of the finger.

Articulations.—The first row with the metacarpal bones and the second row of phalanges; the second row with the first and third; the third with the second row.

Attachment of Muscles.—To the base of the first phalanx of the thumb, five muscles: the Extensor brevis pollicis, Flexor brevis pollicis, Abductor pollicis, Adductor transversus and obliquus pollicis. To the second phalanx, two: the Flexor longus pollicis and the Extensor longus pollicis. To the base of the first phalanx of the index finger, the First dorsal and the First palmar interosseous; to that of the middle finger, the Second and Third dorsal interosseous; to that of the ring-finger, the Fourth dorsal and the Second palmar interosseous; and to that of the little finger, the Third palmar interosseous, the Flexor brevis minimi digiti, and Abductor minimi digiti. To the second phalanges, the Flexor sublimis digitorum, Extensor communis digitorum, and, in addition, the Extensor indicis to the index finger, the Extensor minimi digiti to the little finger. To the third phalanges, the Flexor profundus digitorum and Extensor communis digitorum.

Surface Form.—On the front of the wrist are two subcutaneous eminences, one on the radial side, the larger and flatter, produced by the tuberosity of the scaphoid and the ridge on the trapezium; the other on the ulnar side, caused by the pisiform bone. The tubercle of the scaphoid is to be felt just below and in front of the apex of the styloid process of the radius. It is best perceived by extending the hand on the forearm. Immediately below is to be felt another prominence, better marked than the tubercle; this is the ridge on the trapezium, which gives attachment to some of the short muscles of the thumb. On the inner side of the front of the wrist the pisiform bone is to be felt, forming a small but prominent projection in this situation. It is some distance below the styloid process of the ulna, and may be said to be just below the level of the styloid process of the radius. The rest of the front of the carpus is covered by tendons and the annular ligament, and entirely concealed, with the exception of the hooked process of the unciform, which can only be made out with difficulty. The back of the carpus is convex and covered by the Extensor tendons, so that none of the posterior surfaces of the bones are to be felt, with the exception of the cuneiform on the inner side. Below the carpus the dorsal surfaces of the metacarpal bones except the fifth are covered by tendons, and are scarcely visible except in very thin hands. The dorsal surface of the fifth is, however, subcutaneous throughout almost its whole length, and is plainly to be perceived and felt. In addition to this, slightly external to the middle line of the hand, is a prominence, frequently well marked, but occasionally indistinct, formed by the base of the metacarpal bone of the middle finger. The heads of the metacarpal bones are plainly to be felt and seen, rounded in contour and standing out in bold relief under the skin, when the fist is clenched. It should be borne in mind that when the fingers are flexed on the hand, the articular surfaces of the first phalanges glide off the heads of the metacarpal bones on to their anterior surfaces; so that the heads of these bones form the prominence of the knuckles and receive the force of any blow which may be given. The head of the third metacarpal bone is the most prominent, and receives the greater part of the shock of the blow. This bone articulates with the os magnum, so that the concussion is carried through this bone to the scaphoid and semilunar, with which the head of the os magnum articulates, and by these bones is transferred to the radius, along which it may be carried to the capitellum of the humerus. The enlarged extremities of the phalanges are to be plainly felt: they form the joints of the fingers. When the digits are bent, the proximal phalanges of the joints form prominences, which in the joint between the first and second phalanges are slightly hollowed, in accordance with the grooved shape of their articular surfaces, while in the last row the prominence is flattened and square-shaped. In the palm of the hand the four inner metacarpal bones are covered by muscles, tendons, and the palmar fascia, and no part of them but their heads is to be distinguished. With regard to the thumb, on the dorsal aspect, the base of the metacarpal bone forms a prominence, below the styloid process of the radius; the shaft is to be felt, covered by tendons, terminating at its head in a flattened prominence, in front of which can be felt the sesamoid bones.

Surgical Anatomy.—The carpal bones are little liable to fracture, except from extreme violence, when the parts are so comminuted as to necessitate amputation. Occasionally they are the seat of tuberculous disease. The metacarpal bones and the phalanges are not infrequently broken from direct violence. The first metacarpal bone is the one most commonly fractured, then the second, the fourth and the fifth, the third being the one least frequently broken. There are two diseases of the metacarpal bones and phalanges which require special mention on account of the frequency of their occurrence. One is tuberculous dactylitis, consisting in a deposit of tuberculous material in the medullary canal, expansion of the bone, with subsequent caseation and resulting necrosis. The other is chondroma, which is perhaps more frequently found in connection with the metacarpal bones and phalanges than with any other bones. The tumours are commonly multiple, and may spring either from the medullary canal, or from the periosteum.

Fig. 119.—Plan of the development of the hand.

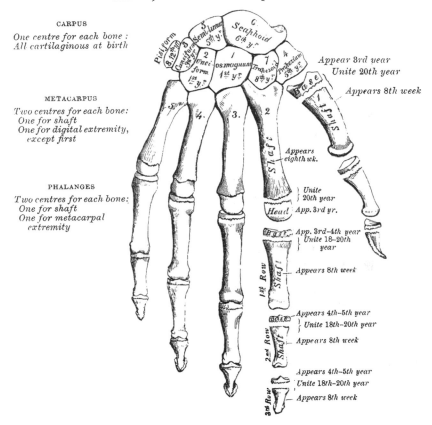

DEVELOPMENT OF THE BONES OF THE HAND

The **Carpal Bones** are each developed by a *single* centre. At birth, they are all cartilaginous. Ossification proceeds in the following order (fig. 119): In the os magnum and unciform an ossific point appears during the first year, the former preceding the latter; in the cuneiform, at the third year; in the trapezium and semilunar, at the fifth year, the former preceding the latter; in the scaphoid, at the sixth year; in the trapezoid, during the eighth year; and in the pisiform, about the twelfth year.

Occasionally an additional bone, the *os centrale*, is found in the carpus, lying between the scaphoid, trapezoid, and os magnum. During the second month of fœtal life it is represented by a small cartilaginous nodule, which, however, fuses with the cartilaginous scaphoid about the third month. Sometimes the styloid process of the third metacarpal is detached and forms an additional ossicle.

The **Metacarpal Bones** are each developed by *two* centres: one for the shaft, and one for the digital extremity, for the four inner metacarpal bones; one for the shaft, and one for the base, for the metacarpal bone of the thumb, which in this respect resembles the phalanges.* Ossification commences in the centre of the shaft about the eighth or ninth week, and gradually proceeds to either end of the bone; about the third year the digital extremities of the four inner meta-carpal bones, and the base of the first metacarpal, begin to ossify, and they unite about the twentieth year.

The **Phalanges** are each developed by *two* centres: one for the shaft, and one for the base. Ossification commences in the shaft, in all three rows, at about the eighth week, and gradually involves the whole of the bone excepting the upper extremity. Ossification of the base commences in the first row between the third and fourth years, and a year later in those of the second and third rows. The two centres become united, in each row, between the eighteenth and twentieth years.

In the ungual phalanges the centre for the shaft appears at the distal extremity of the phalanx, instead of at the middle of the shaft as is the case with the other phalanges.

THE LOWER EXTREMITY

The bones of the lower extremity consist of those of the pelvic girdle, of the thigh, of the leg, and of the foot.

THE PELVIC GIRDLE

The **Pelvic Girdle** consists of a single bone, the *os innominatum*, by which the thigh is connected to the trunk.

THE OS INNOMINATUM

The **Os Innominatum** (in, *not*; nomino, *I name*), or *nameless bone*, so called from bearing no resemblance to any known object, is a large, irregularly shaped, flat bone, constricted in the centre and expanded above and below. With its fellow of the opposite side, it forms the sides and anterior wall of the pelvic cavity. In young subjects it consists of three separate parts, which meet and form a large, cup-like cavity, the *acetabulum*, situated near the middle of the outer surface of the bone: and, although in the adult these have become united, it is usual to describe the bone as divisible into three portions—the ilium, the ischium, and the os pubis.

The **ilium,** so called from its supporting the flank (*ilia*), is the superior, broad and expanded portion which runs upwards from the acetabulum, and forms the prominence of the hip.

The **ischium** (ἰσχίον, *the hip*) is the inferior and strongest portion of the bone; it proceeds downwards from the acetabulum, expands into a large tuberosity, and then, curving forwards, forms, with the descending ramus of the os pubis, a large aperture, the obturator foramen.

The **os pubis** is that portion which extends inwards and downwards from the acetabulum to articulate in the middle line with the bone of the opposite side: it forms the front of the pelvis, supports the external organs of generation, and has received its name from the skin over it being covered with hair.

The **Ilium** presents for examination two surfaces—an external and an internal—a crest, and two borders—an anterior and a posterior.

* Allen Thomson has demonstrated the fact that the first metacarpal bone is often developed from three centres; that is to say, there is a separate nucleus for the distal end, forming a distinct epiphysis visible at the age of seven or eight years. He also states that there are traces of a proximal epiphysis in the second metacarpal bone. *Journal of Anatomy*, 1869.

External Surface or Dorsum of the Ilium (fig. 120).—The posterior part of this surface is directed backwards and outwards; its front part downwards and outwards. It is smooth, convex in front, deeply concave behind; bounded above by the crest, below by the upper border of the acetabulum; in front and behind, by the anterior and posterior borders. This surface is crossed in an arched direction by three semicircular lines—the superior, middle, and inferior curved lines. The superior curved line, the shortest of the three, commences at the crest, about

FIG. 120.—Right os innominatum. External surface.

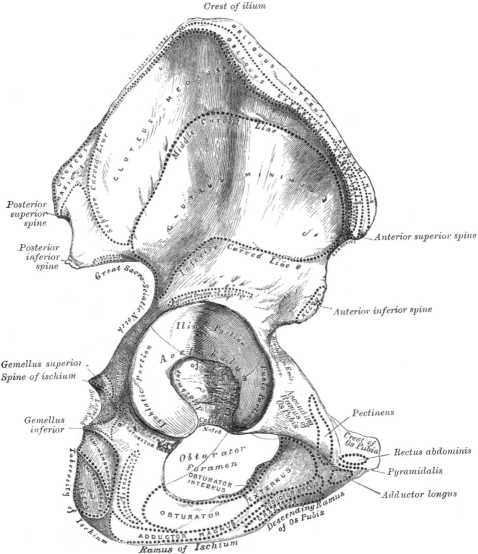

two inches in front of its posterior extremity; it is at first distinctly marked, but as it passes downwards and backwards to the upper part of the great sacro-sciatic notch, where it terminates, it becomes less marked, and is often altogether lost. Behind this line is a narrow semilunar surface, the upper part of which is rough and affords attachment to part of the Gluteus maximus; the lower part is smooth and has no muscular fibres attached to it. The middle curved line, the longest of the three, commences at the crest, about an inch behind its anterior extremity, and, taking a curved direction downwards and backwards, terminates at the upper part of the great sacro-sciatic notch. The space between the superior and middle curved lines and the crest is concave, and affords attachment to the Gluteus medius muscle. Near the central part of this line may often be observed the orifice of a nutrient foramen. The inferior curved line, the least distinct of the three, commences in front at the notch on the anterior border, and, taking a curved direction backwards and downwards, terminates at the middle of the great sacro-sciatic notch. The surface of bone included between the middle and inferior curved lines is concave from above downwards, convex from before backwards, and affords attachment to the Gluteus minimus muscle. Beneath the inferior curved line, and corresponding to the upper part of the acetabulum, is a roughened surface (sometimes a depression), to which is attached the reflected tendon of the Rectus femoris muscle.

The **Internal Surface** (fig. 121) of the ilium is bounded above by the crest; below it is continuous with the pelvic surfaces of the os pubis and ischium, a faint line only indicating the place of union; and before and behind it is bounded by the anterior and posterior borders. It presents a large, smooth, concave surface, called the *iliac fossa*, or *venter ilii*, which lodges the Iliacus muscle, and presents at its lower part the orifice of a nutrient canal; and below this a smooth, rounded border, the *linea ilio-pectinea*, which separates the iliac fossa from that portion of the internal surface which enters into the formation of the true pelvis, and which gives attachment to part of the Obturator internus muscle. Behind the iliac fossa is a rough surface, divided into two portions, an anterior and a posterior. The anterior or *auricular surface*, so called from its resemblance in shape to the ear, is coated with cartilage in the recent state, and articulates with a surface of similar shape on the side of the sacrum. The posterior portion is rough, for the attachment of the posterior sacro-iliac ligaments and for a part of the origin of the Erector and Multifidus spinæ.

The **crest** of the ilium is convex in its general outline and sinuously curved, being concave inwards in front, concave outwards behind. It is longer in the female than in the male, very thick behind, and thinner at the centre than at the extremities. It terminates at either end in a prominent eminence, the *anterior superior* and *posterior superior spinous process*. The surface of the crest is broad, and divided into an external lip, an internal lip, and an intermediate space About two inches behind the anterior superior spinous process there is a prominent tubercle on the outer lip. To the external lip are attached the Tensor fasciæ femoris, Obliquus externus abdominis, and Latissimus dorsi, and along its whole length the fascia lata; to the space between the lips, the Internal oblique; to the internal lip, the Transversalis, Quadratus lumborum, and Erector spinæ, the Iliacus, and the fascia iliaca.

The **anterior border** of the ilium is concave. It presents two projections, separated by a notch. Of these, the uppermost, situated at the junction of the crest and anterior border, is called the *anterior superior spinous process* of the ilium, the outer border of which gives attachment to the fascia lata, and the origin of the Tensor fasciæ femoris; its inner border, to the Iliacus; while its extremity affords attachment to Poupart's ligament and the origin of the Sartorius. Beneath this eminence is a notch which gives attachment to the Sartorius muscle, and across which passes the external cutaneous nerve. Below the notch is the *anterior inferior spinous process*, which terminates in the upper lip of the acetabulum; it gives attachment to the straight tendon of the Rectus femoris muscle and the ilio-femoral ligament. On the inner side of the anterior inferior spinous process is a broad, shallow groove, over which passes the Ilio-psoas muscle. This groove is bounded internally by an eminence, the *ilio-pectineal*, which marks the point of union of the ilium and os pubis.

The **posterior border** of the ilium, shorter than the anterior, also presents two projections separated by a notch, the *posterior superior* and the *posterior inferior spinous processes*. The former corresponds with that portion of the inner surface of the ilium which serves for the attachment of the oblique portion of the sacro-iliac ligaments and the Multifidus spinæ; the latter to the auricular portion which articulates with the sacrum. Below the posterior inferior spinous process is a deep notch, the *great sacro-sciatic*.

The **Ischium** forms the lower and back part of the os innominatum. It is divisible into a thick and solid portion, the *body*; a large, rough eminence, on

FIG. 121.—Right os innominatum. Internal surface.

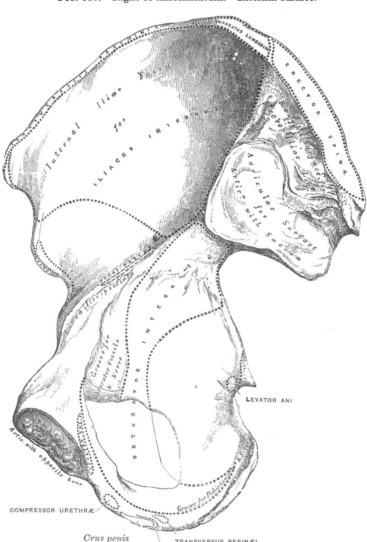

which the trunk rests in sitting, the *tuberosity*; and a thin part, which passes forwards and slightly upwards, the *ramus*.

The **body**, somewhat triangular in form, presents three surfaces, external, internal, and posterior; and three borders, external, internal, and posterior. The *external surface* corresponds to that portion of the acetabulum formed by the ischium; it is smooth and concave, and forms a little more than two-fifths of the acetabular cavity; its outer margin is bounded by a prominent rim or lip, the external border, to which the cotyloid fibro-cartilage is attached. Below the acetabulum, between it and the tuberosity, is a deep groove, along which the tendon of the Obturator externus muscle runs, as it passes outwards to be inserted into the digital fossa of the femur. The *internal surface* is smooth, concave, and enters into the formation of the lateral boundary of the true pelvic cavity. This surface is perforated by two or three large, vascular foramina, and affords attachment to part of the Obturator internus muscle. The *posterior surface* is quadrilateral in form, broad and smooth. Below, where it joins the tuberosity, it presents a groove continuous with that on the external surface for the tendon of the Obturator externus muscle. The lower edge of this groove is formed by the tuberosity of the ischium, and affords attachment to the Gemellus inferior muscle. This surface is limited, externally, by the margin of the acetabulum; behind, by the posterior border; it supports the Pyriformis, the two Gemelli, and the Obturator internus muscles, in their passage outwards to the great trochanter. The *external border* forms the prominent rim of the acetabulum and separates the posterior from the external surface. To it is attached the cotyloid fibro-cartilage. The *internal border* is thin, and forms the outer circumference of the obturator foramen. The *posterior border* of the body of the ischium presents, a little below the centre, a thin and pointed triangular eminence, the *spine of the ischium*, more or less elongated in different subjects; its external surface gives attachment to the Gemellus superior, its internal surface to the Coccygeus and Levator ani; while to the pointed extremity is connected the lesser sacro-sciatic ligament. Above the spine is a notch of large size, the *great sacro-sciatic*, converted into a foramen by the lesser sacro-sciatic ligament; it transmits the Pyriformis muscle, the gluteal vessels, the superior and inferior gluteal nerves, the sciatic vessels, the greater and lesser sciatic nerves, the internal pudic vessels and nerve, and the nerves to the Obturator internus and Quadratus femoris. Of these, the gluteal vessels and superior gluteal nerve pass out above the Pyriformis muscle, the other structures below it. Below the spine is a smaller notch, the *lesser sacro-sciatic*; it is smooth, coated in the recent state with cartilage, the surface of which presents two or three ridges corresponding to the subdivisions of the tendon of the Obturator internus, which winds over it. It is converted into a foramen by the sacro-sciatic ligaments, and transmits the tendon of the Obturator internus, the nerve which supplies that muscle, and the internal pudic vessels and nerve.

The **tuberosity** presents for examination three surfaces : external, internal, and posterior. The *external surface* is quadrilateral in shape and rough for the attachment of muscles. It is bounded above by the groove for the tendon of the Obturator externus; in front it is limited by the posterior margin of the obturator foramen, and below it is continuous with the ramus of the bone; behind, it is bounded by a prominent margin which separates it from the posterior surface. In front of this margin the surface gives attachment to the Quadratus femoris, and anterior to this to some of the fibres of origin of the Obturator externus. The lower part of the surface gives origin to part of the Adductor magnus. The *internal surface* forms part of the bony wall of the true pelvis. In front, it is limited by the posterior margin of the obturator foramen. Behind, it is bounded by a sharp ridge, for the attachment of a falciform prolongation of the great sacro-sciatic ligament; it sometimes presents a groove on the inner side of this ridge for the lodgment of the internal pudic vessels and nerve; and, more anteriorly, has attached the Transversus perinæi and Erector penis muscles. The *posterior surface* is divided into two portions: a lower, rough, somewhat triangular part, and an upper, smooth, quadrilateral portion. The anterior portion is subdivided by a prominent vertical ridge, passing from base to apex, into two parts; the outer one gives attachment to the Adductor magnus, the inner to the great sacro-sciatic ligament. The upper portion is subdivided into two facets by an oblique ridge, which runs downwards and outwards; from the upper and outer facet arises the Semi-membranosus; from the lower and inner, the Biceps and Semi-tendinosus.

The **ramus** is the thin, flattened part of the ischium, which ascends from the tuberosity upwards and inwards, and joins the descending ramus of the os pubis— their point of junction being indicated in the adult by a rough line. The outer surface of the ramus is rough, for the attachment of the Obturator externus muscle, and also some fibres of the Adductor magnus; its inner surface forms part of the anterior wall of the pelvis. Its inner border is thick, rough, slightly everted, forms part of the outlet of the pelvis, and presents two ridges and an intervening space. The ridges are continuous with similar ones on the descending ramus of the os pubis: to the outer one is attached the deep layer of the superficial perineal fascia, and to the inner the superficial layer of the triangular ligament of the urethra. If these two ridges are traced downwards, they will be found to join with each other just behind the point of origin of the Transversus perinæi muscle; here the two layers of fascia are continuous behind the posterior border of the muscle. To the intervening space, just in front of the point of junction of the ridges, is attached the Transversus perinæi muscle, and in front of this a portion of the crus penis vel clitoridis and the Erector penis vel clitoridis muscle. Its outer border is thin and sharp, and forms part of the inner margin of the obturator foramen.

The **Os Pubis** forms the anterior part of the os innominatum, and, with the bone of the opposite side, forms the front boundary of the true pelvic cavity. It is divisible into a body, an ascending and a descending ramus.

The **body** is somewhat quadrilateral in shape, and presents for examination two surfaces and three borders. The *anterior surface* is rough, directed downwards and outwards, and serves for the attachment of various muscles. To the upper and inner angle, immediately below the crest, is attached the Adductor longus; lower down, from without inwards, are attached the Obturator externus, the Adductor brevis, and the upper part of the Gracilis. The *posterior surface*, convex from above downwards, concave from side to side, is smooth, and forms part of the anterior wall of the pelvis. It gives attachment to the Levator ani, Obturator internus, a few muscular fibres prolonged from the bladder, and the pubo-prostatic ligaments. The *upper border* presents for examination a prominent tubercle, which projects forwards and is called the *spine*; to it is attached the outer pillar of the external abdominal ring and Poupart's ligament. Passing upwards and outwards from this is a prominent ridge, forming part of the *ilio-pectineal line* which marks the brim of the true pelvis: to it is attached a portion of the conjoined tendon of the Internal oblique and Transversalis muscles, Gimbernat's ligament and the triangular fascia of the abdomen. Internal to the spine of the os pubis is the *crest*, which extends from this process to the inner extremity of the bone. It affords attachment, anteriorly, to the conjoined tendon of the Internal oblique and Transversalis; and posteriorly, to the Rectus and Pyramidalis muscles. The point of junction of the crest with the inner border of the bone is called the *angle*; to it, as well as to the symphysis, is attached the internal pillar of the external abdominal ring. The *internal border* is articular; it is oval, covered by eight or nine transverse ridges, or a series of nipple-like processes arranged in rows, separated by grooves; they serve for the attachment of a thin layer of cartilage, placed between it and the central fibro-cartilage. The *outer border* presents a sharp margin, which forms part of the circumference of the obturator foramen and affords attachment to the obturator membrane.

The **ascending** or **superior ramus** extends from the body to the point of junction of the os pubis with the ilium, and forms the upper part of the circumference of the obturator foramen. It presents for examination a superior, inferior, and posterior surface, and an outer extremity. The *superior surface* presents a continuation of the ilio-pectineal line, already mentioned as commencing at the pubic spine. In front of this ridge, the surface of bone is triangular in form, wider externally than internally, smooth, and is covered by the Pectineus muscle. The surface is bounded externally by a rough eminence, the *ilio-pectineal*, which serves to indicate the point of junction of the ilium and os pubis, and gives attachment to the Psoas parvus, when this muscle is present. The triangular surface is bounded below by a prominent ridge, the *obturator crest*, which extends from the cotyloid notch to the spine of the os pubis. The *inferior surface* forms the upper boundary of the obturator foramen, and presents, externally, a broad and deep, oblique groove, for the passage of the obturator vessels and nerve; and internally, a sharp margin which forms part of the circumference of the obturator foramen, and to which the obturator membrane is attached. The *posterior surface* forms part of the anterior boundary of the true pelvis. It is smooth, convex from above downwards, and affords attachment to some fibres of the Obturator internus. The *outer extremity*, the thickest part of the ramus, forms one-fifth of the cavity of the acetabulum.

The **descending** or **inferior ramus** of the os pubis is thin and flattened. It passes outwards and downwards, becoming narrower as it descends and joins with the ramus of the ischium. Its *anterior surface* is rough, for the attachment of muscles—the Gracilis along its inner border; a portion of the Obturator externus where it enters into the formation of the foramen of that name; and between these two muscles, the Adductores brevis and magnus from within outwards. The *posterior surface* is smooth, and gives attachment to the Obturator internus, and, close to the inner margin, to the Compressor urethræ. The *inner border* is thick, rough, and everted, especially in females. It presents two ridges, separated by an intervening space. The ridges extend downwards, and are continuous with similar ridges on the ramus of the ischium; to the external one is attached the deep layer of the superficial perineal fascia, and to the internal one the superficial layer of the triangular ligament of the urethra. The *outer border* is thin and sharp, forms part of the circumference of the obturator foramen, and gives attachment to the obturator membrane.

The **cotyloid cavity,** or **acetabulum,** is a deep, cup-shaped, hemispherical depression, directed downwards, outwards, and forwards; formed internally by the os pubis, above by the ilium, behind and below by the ischium; a little less than two-fifths being formed by the ilium, a little more than two-fifths by the ischium, and the remaining fifth by the pubic bone. It is bounded by a prominent, uneven rim, which is thick and strong above, and serves for the attachment of the *cotyloid ligament*, which contracts its orifice, and deepens the surface for articulation. It presents below a deep notch, the *cotyloid notch*, which is continuous with a circular depression, the *fossa acetabuli*, at the bottom of the cavity: this depression is perforated by numerous apertures, lodges a mass of fat, and its margins, as well as those of the notch, serve for the attachment of the ligamentum teres. The notch is converted, in the natural state, into a foramen by a dense ligamentous band which passes across it. Through this foramen the nutrient vessels and nerves enter the joint.

The **obturator** or **thyroid foramen** is a large aperture, situated between the ischium and os pubis. In the male it is large, of an oval form, its longest diameter being obliquely from before backwards; in the female it is smaller, and more triangular. It is bounded by a thin, uneven margin, to which a strong membrane is attached; and presents, anteriorly, a deep groove, which runs from the pelvis obliquely inwards and downwards. This groove is converted into a foramen by the obturator membrane, and transmits the obturator vessels and nerve.

Structure.—This bone consists of much cancellous tissue, especially where it is thick, enclosed between two layers of dense, compact tissue. In the thinner parts of the bone, as at the bottom of the acetabulum and centre of the iliac fossa, it is usually semi-transparent, and composed entirely of compact tissue.

Development (fig. 122).—By *eight* centres: three primary—one for the ilium, one for the ischium, and one for the os pubis; and *five* secondary—one for the crest of the ilium, one for the anterior inferior spinous process (said to occur more frequently in the male than the female), one for the tuberosity of the ischium, one for the symphysis pubis (more frequent in the female than the male), and one or more for the Y-shaped piece at the bottom of the acetabulum. These various centres appear in the following order : First, in the ilium, at the lower part of the bone, immediately above the sciatic notch, at about the eighth or ninth week ; secondly, in the body of the ischium, at about the third month of fœtal life ; thirdly, in the body of the os pubis, between the fourth and fifth months. At birth, the three primary centres are quite separate, the crest, the bottom of the acetabulum, the ischial tuberosity, and the rami of the ischium and pubes being still cartilaginous. At about the seventh or eighth year, the rami of the os pubis and ischium are almost completely united by bone. About the thirteenth or fourteenth year, the three divisions of the bone have extended their growth into the bottom of the acetabulum, being separated from each other by a Y-shaped portion of cartilage, which now presents traces of ossification, often by two or more centres. One of these, the *os acetabuli*, appears about the age of

Fig. 122.—Plan of the development of the os innominatum.

By eight centres { *Three primary (Ilium, Ischium, and Os Pubis)* / *Five secondary* }

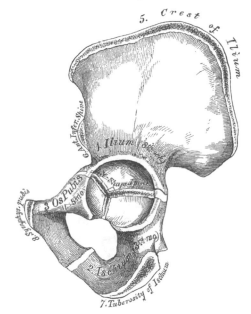

The three primary centres unite through Y-shaped piece about puberty.
Epiphyses appear about puberty, and unite about 25th year.

twelve, between the ilium and os pubis, and fuses with them about the age of eighteen. It forms the pubic part of the acetabulum. The ilium and ischium then become joined, and lastly the os pubis to the ischium, through the intervention of this Y-shaped portion. At about the age of puberty, ossification takes place in each of the remaining portions, and they become joined to the rest of the bone between the twentieth and twenty-fifth years. Separate centres are frequently found for the pubic and ischial spines.

Articulations.—With its fellow of the opposite side, the sacrum and femur.

Attachment of Muscles.—To the *ilium*, sixteen. To the outer lip of the crest, the Tensor vaginæ femoris, Obliquus externus abdominis, and Latissimus dorsi ; to the internal lip, the Iliacus, Transversalis, Quadratus lumborum, and Erector spinæ ; to the interspace between the lips, the Obliquus internus. To the outer surface of the ilium, the Gluteus maximus, Gluteus medius, Gluteus minimus, reflected tendon of the Rectus ; to the upper part of the great sacro-sciatic notch a portion of the Pyriformis ; to the internal surface, the Iliacus ; to that portion of the internal surface below the linea ilio-pectinea, the Obturator internus ; to the internal surface of the posterior superior spine, the Multifidus spinæ ; to the anterior border, the Sartorius and straight tendon of the Rectus.   To the *ischium*, thirteen.   To the outer surface of the ramus, the Obturator externus and Adductor magnus ; to the internal surface, the Obturator internus and Erector penis.   To the spine, the Gemellus superior, Levator ani, and Coccygeus. To the tuberosity, the Biceps, Semi-tendinosus, Semi-membranosus, Quadratus femoris, Adductor magnus, Gemellus inferior, Transversus perinæi, Erector penis. To the *os pubis*, sixteen: Obliquus externus, Obliquus internus, Transversalis, Rectus, Pyramidalis, Psoas parvus, Pectineus, Adductor magnus, Adductor longus, Adductor brevis, Gracilis, Obturator externus and internus, Levator ani, Compressor urethræ, and occasionally a few fibres of the Accelerator urinæ.

THE PELVIS (figs. 123, 124)

The **Pelvis**, so called from its resemblance to a basin (L. *pelvis*), is stronger and more massively constructed than either the cranial or thoracic cavity ; it is a bony ring, interposed between the lower end of the spine, which it supports, and the lower extremities, upon which it rests.   It is composed of four bones : the two ossa innominata, which bound it on either side and in front ; and the sacrum and coccyx, which complete it behind.

The pelvis is divided by an oblique plane passing through the prominence of the sacrum, the linea ilio-pectinea, and the upper margin of the symphysis pubis, into the false and true pelvis.

The **false pelvis** is the expanded portion of the cavity which is situated above this plane.   It is bounded on each side by the ossa ilii ; in front it is incomplete,

FIG. 123.—Male pelvis (adult).

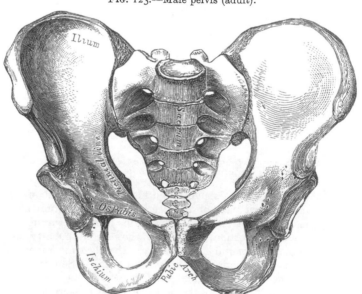

presenting a wide interval between the spinous processes of the ilia on either side, which is filled up in the recent state by the parietes of the abdomen ; behind, in the middle line, is a deep notch.   This broad, shallow cavity is fitted to support the intestines, and to transmit part of their weight to the anterior wall of the abdomen, and is, in fact, really a portion of the abdominal cavity.   The term false pelvis is incorrect, and this space ought more properly to be regarded as part of the hypogastric and iliac regions of the abdomen.

The **true pelvis** is that part of the pelvic cavity which is situated beneath the plane. It is smaller than the false pelvis, but its walls are more perfect. For

Fig. 124.— Female pelvis (adult).

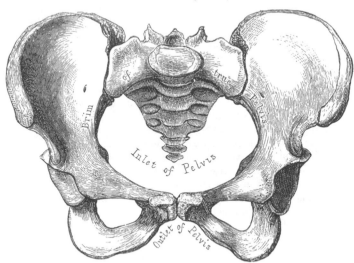

convenience of description, it is divided into a superior circumference or inlet, an inferior circumference or outlet, and a cavity.

The **superior circumference** forms the brim of the pelvis, the included space being called the *inlet*. It is formed by the linea ilio-pectinea, completed in front by the crests of the pubic bones, and behind by the anterior margin of the base of the sacrum and sacro-vertebral angle. The *inlet* of the pelvis is somewhat heart-shaped, obtusely pointed in front, diverging on either side, and encroached upon behind by the projection forwards of the promontory of the sacrum. It has three principal diameters : antero-posterior (sacro-pubic), transverse, and oblique. The antero-posterior extends from the sacro-vertebral angle to the symphysis pubis ; its average measurement is four inches in the male, four and three-quarters in the female. The transverse extends across the greatest width of the inlet, from the middle of the brim on one side to the same point on the opposite ; its average measurement is four and a half in the male, five and a quarter in the female. The oblique extends from the margin of the pelvis, corresponding to the ilio-pectineal eminence on one side, to the sacro-iliac articulation on the opposite side ; its average measurement is four and a quarter in the male, and five in the female.

The **cavity** of the true pelvis is bounded in front by the symphysis pubis ; behind, by the concavity of the sacrum and coccyx, which, curving forwards above and below, contracts the inlet and outlet of the canal ; and laterally it is bounded by a broad, smooth, quadrangular surface of bone, corresponding to the inner surface of the body of the ischium and that part of the ilium which is below the ilio-pectineal line. The cavity is shallow in front, measuring at the symphysis an inch and a half in depth, three inches and a half in the middle, and four inches and a half posteriorly. From this description, it will be seen that the cavity of the pelvis is a short, curved canal, considerably deeper on its posterior than on its anterior wall. This cavity contains, in the recent subject, the rectum, bladder, and part of the organs of generation. The rectum is placed at the back of the pelvis, and corresponds to the curve of the sacro-coccygeal column ; the bladder in front, behind the symphysis pubis. In the female, the uterus and vagina occupy the interval between these viscera.

The **lower circumference** of the pelvis is very irregular, and forms what is called the *outlet*. It is bounded by three prominent eminences : one posterior, formed by the point of the coccyx ; and one on each side, the tuberosities of the ischia. These eminences are separated by three notches : one in front, the *pubic arch*, formed by the convergence of the rami of the ischia and pubic bones on each side. The other notches, one on each side, are formed by the sacrum and coccyx behind, the ischium in front, and the ilium above : they are called the *sacro-sciatic notches* ; in the natural state they are converted into foramina by the lesser and greater sacro-sciatic ligaments. In the recent state, when the ligaments are *in situ*, the outlet of the pelvis is lozenge-shaped, bounded, in front, by the subpubic ligament and the rami of the ossa pubis and ischia ; on each side by the tuberosities of the ischia ; and behind, by the great sacro-sciatic ligaments and the tip of the coccyx.

The diameters of the outlet of the pelvis are two, antero-posterior and transverse. The *antero-posterior* extends from the tip of the coccyx to the lower part of the symphysis pubis ; its average measurement is three and a quarter inches in the male and five in· the female. The antero-posterior diameter varies with the length of the coccyx, and is capable of increase or diminution, on account of the mobility of that bone. The *transverse* extends from the posterior part of one ischiatic tuberosity to the same point on the opposite side ; the average measurement is three and a half inches in the male and four and three-quarters in· the female.*

Position of the Pelvis.—In the erect posture, the pelvis is placed obliquely with regard to the trunk of the body : the bony ring, which forms the brim of the true pelvis, is placed so as to form an angle of about 60° to 65° with the ground on which we stand. The pelvic surface of the symphysis pubis looks upwards and backwards, the concavity of the sacrum and coccyx downwards and forwards ; the base of the sacrum in well-formed female bodies being nearly four inches above the upper border of the symphysis pubis, and the apex of the coccyx a little more than half an inch above its lower border. In consequence of this obliquity of the pelvis, the line of gravity of the head, which passes through the middle of the odontoid process of the axis and through the points of junction of the curves of the vertebral column to the sacro-vertebral angle, descends towards the front of the cavity, so that it bisects a line drawn transversely through the middle of the heads of the thigh-bones. And thus the centre of gravity of the head is placed immediately over the heads of the thigh-bones on which the trunk is supported.

Axes of the Pelvis (fig. 125).—The plane of the inlet of the true pelvis will be represented by a line drawn from the base of the sacrum to the upper margin of the symphysis pubis. A line carried at right angles with this at its middle, would correspond at one extremity with the umbilicus and at the other with the middle of the coccyx : the axis of the inlet is therefore directed downwards and backwards. The axis of the outlet, produced upwards, would touch the base of the sacrum, and is therefore directed downwards and forwards. The axis of the cavity is curved like the cavity itself : this curve corresponds to the concavity of the sacrum and coccyx, the extremities being indicated by the central points of the inlet and outlet. A knowledge of the direction of these axes serves to explain the course of the fœtus in its passage through the pelvis during

---

* The measurements of the pelvis given above are, I believe, fairly accurate, but different measurements are given by various authors, no doubt due in a great measure to differences in the physique and stature of the population from whom the measurements have been taken. The following chart has been formulated to show the measurements of the pelvis, which are adopted by many obstetricians.—ED.

							A.P.	Obl.	Tr.
Inlet .	.	.	.	.	.	.	4	4½	5
Cavity	.	.	.	.	.	.	4½	4½	4½
Outlet	.	.	.	.	.	.	5	4½	4

parturition. It is also important to the surgeon, as indicating the direction of the force required in the removal of calculi from the bladder, by the sub-pubic operation, and as determining the direction in which instruments should be used in operations upon the pelvic viscera.

Differences between the Male and Female Pelvis.—The *female* pelvis, looked at as a whole, is distinguished from the *male* by the bones being more delicate, by its width being greater and its depth smaller. The whole pelvis is less massive, and its bones are lighter and more slender, and its muscular impressions are slightly marked. The iliac fossæ are shallow, and the anterior iliac spines widely separated; hence the greater prominence of the hips. The *inlet* in the female is larger than in the male; it is more nearly circular, and the sacro-vertebral angle projects less forwards. The *cavity* is shallower and wider; the sacrum is shorter, wider, and less curved; the obturator foramina are triangular and smaller in size than in the male. The *outlet* is larger and the coccyx more movable. The spines of the ischia project less inwards. The tuberosities of the ischia and the ace-

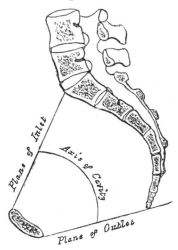

Fig. 125.—Vertical section of the pelvis, with lines indicating the axes of the pelvis.

tabula are wider apart. The *pubic arch* is wider and more rounded than in the male, where it is an angle rather than an arch. In consequence of this the width of the fore part of the pelvic outlet is much increased and the passage of the fœtal head facilitated.

The size of the pelvis varies, not only in the two sexes, but also in different members of the same sex. This does not appear to be influenced in any way by the height of the individual. Women of short stature, as a rule, have broad pelves. Occasionally the pelvis is equally contracted in all its dimensions, so much so that all its diameters measure an inch less than the average, and this even in well-formed women of average height. The principal divergences, however, are found at the inlet, and affect the relation of the antero-posterior to the transverse diameter. Thus we may have a pelvis the inlet of which is elliptical either in a transverse or antero-posterior direction; the transverse diameter in the former, and the antero-posterior in the latter, greatly exceeding the other diameters. Again, the inlet of the pelvis in some instances is seen to be almost circular.

The same differences are found in various races. European women are said to have the most roomy pelves. That of the negress is smaller, circular in shape, and with a narrow pubic arch. The Hottentots and Bushwomen possess the smallest pelves.

In the *fœtus*, and for several years after birth, the pelvis is small in proportion to that of the adult. The cavity is deep, and the projection of the sacro-vertebral angle less marked. The generally accepted opinion that the female pelvis does not acquire its sexual characters until after puberty has been shown by recent observations * to be erroneous, the characteristic differences between the male and female pelvis being distinctly indicated as early as the fourth month of fœtal life.

Surface Form.—The pelvic bones are so thickly covered with muscles that it is only at certain points that they approach the surface and can be felt through the skin. In front,

---

* Fehling, *Zeitschr. für Geburt. u. Gynäk.* Bd. ix. und x.; and Arthur Thomson, *Journal of Anatomy and Physiology*, vol. xxxiii.

the anterior superior spinous process is easily to be recognised; a portion of it is sub-cutaneous, and in thin subjects may be seen to stand out as a prominence at the outer extremity of the fold of the groin. In fat subjects its position is marked by an oblique depression among the surrounding fat, at the bottom of which the bony process may be felt. Proceeding upwards and outwards from this process, the crest of the ilium may be traced throughout its whole length, sinuously curved. It is represented, in muscular subjects, on the surface, by a groove or furrow, the *iliac furrow*, caused by the projection of fleshy fibres of the External oblique muscle of the abdomen; the iliac furrow lies slightly below the level of the crest. It terminates behind in the posterior superior spinous process, the position of which is indicated by a slight depres-sion on a level with the spinous process of the second sacral vertebra. Between the two posterior superior spinous processes, but at a lower level, is to be felt the spinous process of the third sacral vertebra (see page 22). Another part of the bony pelvis which is easily accessible to the touch is the tuberosity of the ischium, situated beneath the gluteal fold, and, when the hip is flexed, easily to be felt, as it is then to a great extent uncovered by muscle. Finally, the spine of the os pubis can always be readily felt, and constitutes an important surgical guide, especially in connection with the subject of hernia. It is nearly in the same horizontal line as the upper edge of the great trochanter. In thin subjects it is very apparent, but in the obese it is obscured by the pubic fat. It can, however, be detected by following up the tendon of origin of the Adductor longus muscle.

Surgical Anatomy.—There is arrest of development in the bones of the pelvis in cases of extroversion of the bladder; the anterior part of the pelvic girdle being deficient, the bodies of the pubic bones imperfectly developed, and the symphysis absent. 'The pubic bones are separated to the extent of from two to four inches, the superior rami shortened and directed forwards, and the obturator foramen diminished in size, narrowed, and turned outwards. The iliac bones are straightened out more than normal. The sacrum is very peculiar. The lateral curve, instead of being concave, is flattened out or even convex, with the ilio-sacral facets turned more outward than normal, while the vertical curve is straightened.' *

Fractures of the pelvis are divided into fractures of the false pelvis and of the true pelvis. Fractures of the false pelvis vary in extent; a small portion of the crest may be broken, or one of the spinous processes may be torn off, and this may be the result of muscular action; or the bone may be extensively comminuted. This latter accident is the result of some crushing violence, and may be complicated with fracture of the true pelvis. These cases may be accom-panied by injury to the intestine as it lies in the hollow of the bone, or to the iliac vessels as they course along the margin of the true pelvis. Fractures of the true pelvis generally occur through the ascending ramus of the os pubis and the ramus of the ischium, as this is the weakest part of the bony ring, and may be caused either by crushing violence applied in an antero-posterior direction, when the fracture occurs from direct force, or by compression laterally, when the acetabula are pressed together and the bone gives way in the same place from indirect violence. Occasionally the fracture may be double, occurring on both sides of the body. It is in these cases that injury to the contained viscera is liable to take place: the urethra, the bladder, the rectum, the vagina in the female, the small intestines, and even the uterus, have all been lacerated by a displaced fragment. Fractures of the acetabulum are occasionally met with: either a portion of the rim may be broken off, or a fracture may take place through the bottom of the cavity and the head of the femur be driven inwards and project into the pelvic cavity. Separation of the Y-shaped cartilage at the bottom of the acetabulum may also occur in the young subject, separating the bone into its three anatomical portions.

The sacrum is occasionally, but rarely, broken by direct violence—i.e. blows, kicks, or falls on the part. The lesion may be complicated with injury to the nerves of the sacral plexus, leading to paralysis and loss of sensation in the lower extremity, or to incontinence of fæces from paralysis of the sphincter ani.

The pelvic bones often undergo important deformity in rickets, the effects of which in the adult woman may interfere seriously with child-bearing. The deformity is due mainly to the weight of the spine and trunk, which presses on the sacro-vertebral angle and greatly increases it, so that the antero-posterior diameter of the pelvis is diminished. But, in addition to this, the weight of the viscera on the venter ilii causes them to expand and the tuberosities of the ischia to be incurved. In osteo-malacia also great deformity may occur. The weight of the trunk causes an increase in the sacro-vertebral angle, and a lessening of the antero-posterior diameter of the inlet, and at the same time the pressure of the acetabula on the heads of the thigh bones causes these cavities, with the adjacent bone, to be pushed upwards and backwards, so that the oblique diameters of the pelvis are also diminished, and the cavity of the pelvis assumes a triradiate shape, with the symphysis pubis pushed forwards.

---

* Wood. Heath's *Dictionary of Practical Surgery*, i. 426.

THE THIGH

The **Thigh** is that portion of the lower extremity which is situated between the pelvis and the knee. It consists in the skeleton of a single bone, the femur.

THE FEMUR

The **Femur** (femur, *the thigh*) is the longest,* largest, and strongest bone in the skeleton, and almost perfectly cylindrical in the greater part of its extent. In the erect posture it is not vertical, being separated from its fellow above by a considerable interval, which corresponds to the breadth of the pelvis, but inclining gradually downwards and inwards, so as to approach its fellow towards its lower part, for the purpose of bringing the knee-joint near the line of gravity of the body. The degree of this inclination varies in different persons, and is greater in the female than in the male, on account of the greater breadth of the pelvis. The femur, like other long bones, is divisible into a shaft and two extremities.

The **Upper Extremity** presents for examination a head, a neck, and the great and lesser trochanters.

The **head,** which is globular, and forms rather more than a hemisphere, is directed upwards, inwards, and a little forwards, the greater part of its convexity being above and in front. Its surface is smooth, coated with cartilage in the recent state, with the exception of an ovoid depression, which is situated a little below and behind its centre, and gives attachment to the ligamentum teres. The **neck** is a flattened pyramidal process of bone, which connects the head with the shaft. It varies in length and obliquity at various periods of life, and under different circumstances. The angle is widest in infancy, and becomes lessened during growth, so that at puberty it forms a gentle curve from the axis of the shaft. In the adult it forms an angle of about 130° with the shaft, but varies in inverse proportion to the development of the pelvis and the stature. In consequence of the prominence of the hips and widening of the pelvis in the female,

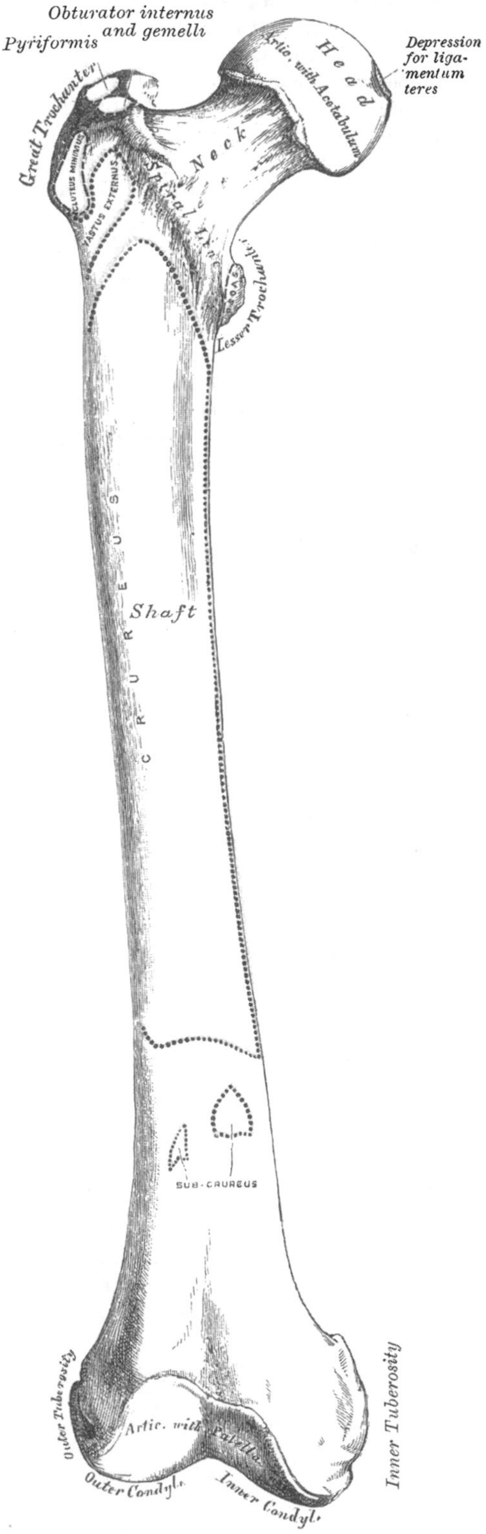

* In a man six feet high, it measures eighteen inches—one-fourth of the whole body.

FIG. 126.—Right femur. Anterior surface.

the neck of the thigh-bone forms more nearly a right angle with the shaft than it does in man. It has been stated that the angle diminishes in old age and the direction of the neck becomes horizontal, but this statement is founded on insufficient evidence. Sir George Humphry states that the angle decreases during the period of growth, but after full growth has been attained it does not usually undergo any change, even in old age. He further states that the angle varies considerably in different persons of the same age. It is smaller in short than in long bones, and when the pelvis is wide.* The neck is flattened from before backwards, contracted in the middle, and broader at its outer extremity, where it is connected with the shaft, than at its summit, where it is attached to the head. The vertical diameter of the outer half is increased by the thickening of the lower edge, which slopes downwards to join the shaft at the lesser trochanter, so that the outer half of the neck is flattened from before backwards, and its vertical diameter measures one-third more than the antero-posterior. The inner half is smaller, and of a more circular shape. The *anterior surface* of the neck is perforated by numerous vascular foramina. The *posterior surface* is smooth, and is broader and more concave than the anterior; it gives attachment to the posterior part of the capsular ligament of the hip-joint, about half an inch above the posterior intertrochanteric line. The *superior border* is short and thick, and terminates externally at the great trochanter; its surface is perforated by large foramina. The *inferior border*, long and narrow, curves a little backwards, to terminate at the lesser trochanter.

The **Trochanters** (τροχάω, *to run* or *roll*) are prominent processes of bone which afford leverage to the muscles which rotate the thigh on its axis. They are two in number, the great and the lesser.

The **Great Trochanter** is a large, irregular, quadrilateral eminence, situated at the outer side of the neck, at its junction with the upper part of the shaft. It is directed a little outwards and backwards, and, in the adult, is about three-quarters of an inch lower than the head. It presents for examination two surfaces and four borders. The *external surface*, quadrilateral in form, is broad, rough, convex, and marked by a prominent diagonal impression, which extends from the posterior superior to the anterior inferior angle, and serves for the attachment of the tendon of the Gluteus medius. Above the impression is a triangular surface, sometimes rough for part of the tendon of the same muscle, sometimes smooth for the interposition of a bursa between that tendon and the bone. Below and behind the diagonal line is a smooth, triangular surface, over which the tendon of the Gluteus maximus muscle plays, a bursa being interposed. The *internal surface* is of much less extent than the external, and presents at its base a deep depression, the *digital* or *trochanteric fossa*, for the attachment of the tendon of the Obturator externus muscle, and above and in front of this an impression for the attachment of the Obturator internus and Gemelli. The *superior border* is free; it is thick and irregular, and marked near the centre by an impression for the attachment of the Pyriformis. The *inferior border* corresponds to the point of junction of the base of the trochanter with the outer surface of the shaft; it is marked by a rough, prominent, slightly curved ridge, which gives attachment to the upper part of the Vastus externus muscle. The *anterior border* is prominent, somewhat irregular, as well as the surface of bone immediately below it; it affords attachment at its outer part to the Gluteus minimus. The *posterior border* is very prominent, and appears as a free, rounded edge, which forms the back part of the digital fossa.

The **Lesser Trochanter** is a conical eminence, which varies in size in different subjects; it projects from the lower and back part of the base of the neck. Its base is triangular, and connected with the adjacent parts of the bone by three well-marked borders: two of these are above—the *internal* continuous with the lower border of the neck, the *external* with the posterior intertrochanteric line—

* *Journal of Anatomy and Physiology.*

while the *inferior border* is continuous with the middle division of the linea aspera. Its summit, which is directed inwards and backwards, is rough, and gives insertion to the tendon of the Iliopsoas. The Iliacus is also inserted into the shaft below the lesser trochanter, between the Vastus internus in front and the Pectineus behind.

A well-marked prominence, of variable size, which projects from the upper and front part of the neck, at its junction with the great trochanter, is called the *tubercle of the femur*; it is the point of meeting of five muscles: the Gluteus minimus externally, the Vastus externus below, and the tendon of the Obturator internus and Gemelli above. Running obliquely downwards and inwards from the tubercle is the *spiral line* of the femur, or *anterior intertrochanteric line*; it winds round the inner side of the shaft, below the lesser trochanter, and terminates in the linea aspera, about two inches below this eminence. Its upper half is rough, and affords attachment to the ilio-femoral ligament of the hip-joint; its lower half is less prominent, and gives attachment to the upper part of the Vastus internus. Running obliquely downwards and inwards from the summit of the great trochanter on the posterior surface of the neck is a very prominent well-marked ridge, the *posterior intertrochanteric line*. Its upper half forms the posterior border of the great trochanter, and its lower half runs downwards and inwards to the upper and back part of the lesser trochanter. A slight ridge sometimes commences about the middle of the posterior intertrochanteric line, and passes vertically downwards for about two inches along the back part of the shaft: it is called the *linea quadrata*, and gives attachment to the Quadratus femoris and a few fibres of the Adductor magnus muscles.*

The **Shaft,** almost cylindrical in form, is a little broader above than in the centre, and somewhat flattened from before backwards below. It is slightly

---

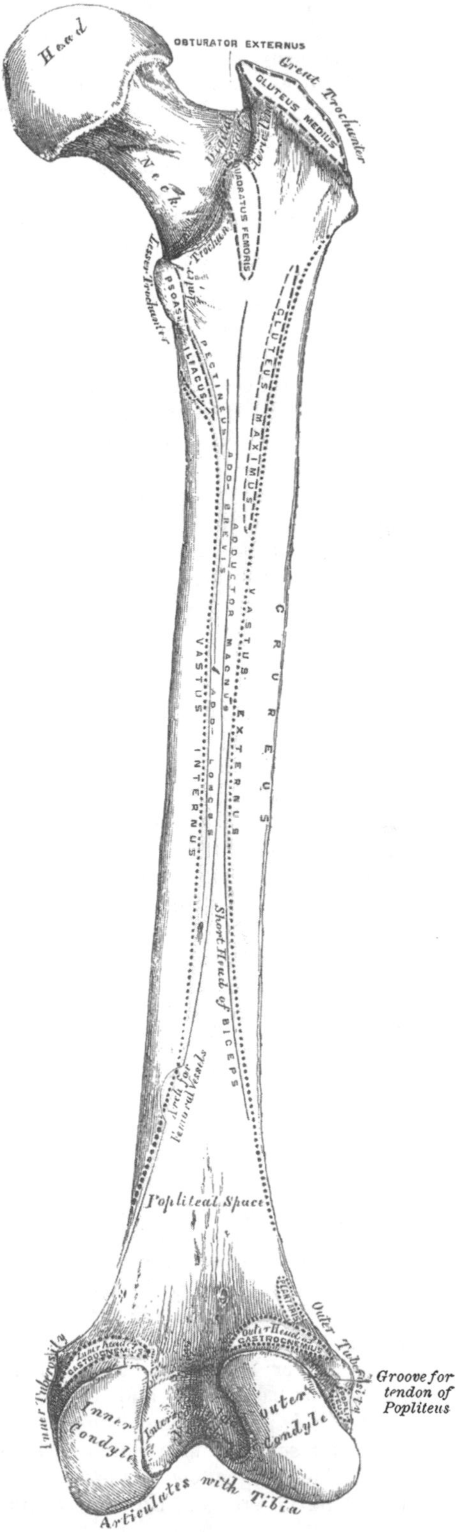

* Generally there is merely a slight thickening about the centre of the intertrochanteric line, marking the point of attachment of the Quadratus femoris. This is termed by some anatomists the *tubercle of the Quadratus.*

FIG. 127.—Right femur. Posterior surface.

arched, so as to be convex in front and concave behind, where it is strengthened by a prominent longitudinal ridge, the *linea aspera*. It presents for examination three borders, separating three surfaces. Of the three borders, one, the linea aspera, is posterior ; the other two are placed laterally.

The **linea aspera** (fig. 127) is a prominent longitudinal ridge or crest, on the middle third of the bone, presenting an external lip, an internal lip, and a rough intermediate space. Above, the crest is prolonged by three ridges. The most external one is very rough, and is continued almost vertically upwards to the base of the great trochanter. It is sometimes termed the *gluteal ridge*, and gives attachment to part of the Gluteus maximus muscle : its upper part is often elongated into a roughened crest, on which is a more or less well-marked, rounded tubercle, a rudimental third trochanter. The middle ridge, the least distinct, is continued to the base of the trochanter minor ; and the internal one is lost above in the spiral line of the femur. Below, the linea aspera is prolonged by two ridges, which enclose between them a triangular space, the *popliteal surface*, upon which rests the popliteal artery. Of these two ridges, the outer one is the more prominent, and descends to the summit of the outer condyle. The inner one is less marked, especially at its upper part, where it is crossed by the femoral artery. It terminates, below, at the summit of the internal condyle, in a small tubercle, the *Adductor tubercle*, which affords attachment to the tendon of the Adductor magnus.

To the inner lip of the linea aspera and its inner prolongation above and below is attached the Vastus internus ; and to the outer lip and its outer prolongation above is attached the Vastus externus. The Adductor magnus is attached to the linea aspera, to its outer prolongation above, and its inner prolongation below. Between the Vastus externus and the Adductor magnus are attached two muscles—viz. the Gluteus maximus above and the short head of the Biceps below. Between the Adductor magnus and the Vastus internus four muscles are attached : the Iliacus and Pectineus above ; the Adductor brevis and Adductor longus below. The linea aspera is perforated a little below its centre by the nutrient canal, which is directed obliquely upwards.

The two *lateral borders* of the femur are only slightly marked, the outer one extending from the anterior inferior angle of the great trochanter to the anterior extremity of the external condyle ; the inner one from the spiral line, at a point opposite the trochanter minor, to the anterior extremity of the internal condyle. The internal border marks the limit of attachment of the Crureus muscle internally.

The *anterior surface* includes that portion of the shaft which is situated between the two lateral borders. It is smooth, convex, broader above and below than in the centre, slightly twisted, so that its upper part is directed forwards and a little outwards, its lower part forwards and a little inwards. To the upper three-fourths of this surface the Crureus is attached ; the lower fourth is separated from the muscle by the intervention of the synovial membrane of the knee-joint and a bursa, and affords attachment to the Subcrureus to a small extent. The *external surface* includes the portion of bone between the external border and the outer lip of the linea aspera ; it is continuous above with the outer surface of the great trochanter, below with the outer surface of the external condyle : to its upper three-fourths is attached the outer portion of the Crureus muscle. The *internal surface* includes the portion of bone between the internal border and the inner lip of the linea aspera ; it is continuous above with the lower border of the neck, below with the inner side of the internal condyle : it is covered by the Vastus externus muscle.

The **Lower Extremity,** larger than the upper, is of a cuboid form, flattened from before backwards, and divided into two large eminences, the *condyles* (κόνδυλος, *a knuckle*), by an interval which presents a smooth depression in front called the *trochlea*, and a notch of considerable size behind—the *intercondyloid notch*. The *external condyle* is the more prominent anteriorly, and is the broader both in the antero-posterior and transverse diameters. The *internal condyle* is the narrower, longer, and more prominent inferiorly. The difference in the length of the two condyles is only observed when the bone is perpendicular, and depends upon the obliquity of the thigh-bones, in consequence of their separation above at the articulation with the pelvis. If the femur is held obliquely, the surfaces of the two condyles will be seen to be nearly horizontal. The two condyles are directly continuous in front, and form a smooth trochlear surface, which articulates with the patella. This surface presents a median groove, which extends downwards and backwards to the intercondyloid notch·; and two lateral convexities, of which the external is the broader, more prominent, and prolonged farther upwards upon the front of the outer condyle. The external border of this articular surface is also more prominent, and ascends higher than the internal one. The intercondyloid notch lodges the crucial ligaments; it is bounded laterally by the opposed surfaces of the two condyles, and in front by the lower end of the shaft.

Outer Condyle.—The *outer surface* of the external condyle presents, a little behind its centre, an eminence, the *outer tuberosity*; it is less prominent than the inner tuberosity, and gives attachment to the external lateral ligaments of the knee. Immediately beneath it is a groove which commences at a depression a little behind the centre of the lower border of this surface: the front part of this depression gives origin to the Popliteus muscle, the tendon of which is lodged in the groove during flexion of the knee. The groove is smooth, covered with cartilage in the recent state, and runs upwards and backwards to the posterior extremity of the condyle. The *inner surface* of the outer condyle forms one of the lateral boundaries of the intercondyloid notch, and gives attachment, by its posterior part, to the anterior crucial ligament. The *inferior surface* is convex, smooth, and broader than that of the internal condyle. The posterior extremity is convex and smooth : just above and to the outer side of the articular surface is a depression for the tendon of the outer head of the Gastrocnemius, above which is the origin of the Plantaris.

Inner Condyle.—The *inner surface* of the inner condyle presents a convex eminence, the *inner tuberosity*, rough, for the attachment of the internal lateral ligament. The *outer side* of the inner condyle forms one of the lateral boundaries of the intercondyloid notch, and gives attachment, by its anterior part, to the posterior crucial ligament. Its *inferior* or *articular surface* is convex, and presents a less extensive surface than the external condyle. Just above the articular surface of the condyle, behind, is a depression for the tendon of origin of the inner head of the Gastrocnemius.

Structure.—The shaft of the femur is a cylinder of compact tissue, hollowed by a large medullary canal. The cylinder is of great thickness and density in the middle third of the shaft, where the bone is narrowest and the medullary canal well formed ; but above and below this the cylinder gradually becomes thinner, owing to a separation of the layers of the bone into cancelli, which project into the medullary canal and finally obliterate it, so that the upper and lower ends of the shaft, and the articular extremities more especially, consist of cancellated tissue, invested by a thin compact layer.

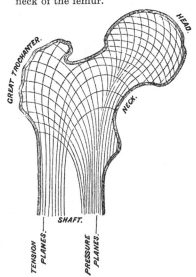

FIG. 128.—Diagram showing the arrangement of the cancelli of the neck of the femur.

The arrangement of the cancelli in the ends of the femur is remarkable. In the upper end they are arranged in two sets. One, starting from the top of the head, the upper surface of the neck, and the great trochanter, converge to the inner circumference of the shaft (fig. 128) ; these are placed in the direction of greatest pressure, and serve to support the vertical weight of the body.   The second set are planes of lamellæ intersecting the former nearly at right angles, and are situated in the line of the greatest tension—that is to say, along the lines in which the muscles and ligaments exert their traction.   In the head of the bone these planes are arranged in a curved form, in order to strengthen the bone when exposed to pressure in all directions.   In the midst of the cancellous tissue of the neck is a vertical plane of compact bone, the *femoral spur* (*calcar femorale*) which commences at the point where the neck joins the shaft midway between the lesser trochanter and the internal border of the shaft of the bone, and extends in the direction of the digital fossa (fig. 129).   This materially strengthens this portion of the bone.   Another point in connection with the structure of the neck

Fig. 129.—Calcar femorale.

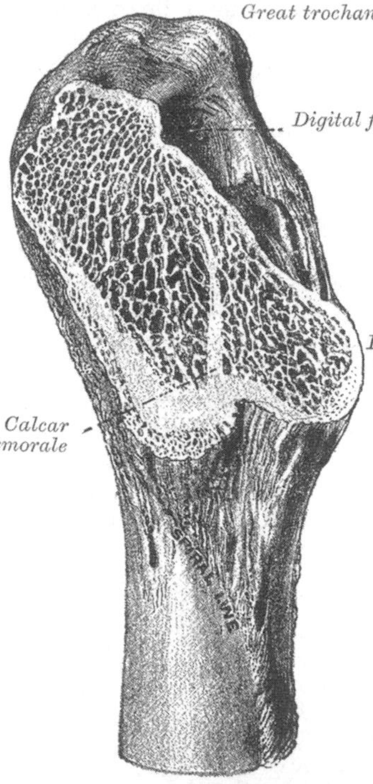

Great trochanter

Digital fossa

Lesser
trochanter

Calcar
femorale

Fig. 130.—Plan of the development of the
femur.   By five centres.

Appears at
4th year ;
joins shaft
about 18th yr.

Great Trochanter

Head

Appears at end
of 1st year ;
joins shaft about
18th year

Less. Troch.

Appears 13th–14th year ;
joins shaft about 18th
year

Shaft
5th w.

Appears at
9th month
(fœtal)

Joins shaft at 20th
year

Lower extremity

of the femur requires mention, especially on account of its influence on the production of fracture in this situation.   It will be noticed that a considerable portion of the great trochanter lies behind the level of the posterior surface of the neck, and if a section be made through the trochanter at this level, it will be seen that the posterior wall of the neck is prolonged into the trochanter. This prolongation is termed by Bigelow the ' true neck,'* and forms a thin, dense plate of bone, which passes beneath the posterior intertrochanteric ridge towards the outer surface of the bone.

In the lower end, the cancelli spring on all sides from the inner surface of the cylinder, and descend in a perpendicular direction to the articular surface, the cancelli being strongest and having a more accurately perpendicular course above the condyles.   In addition to this, however, horizontal planes of cancellous

* *Bigelow on the Hip*, p. 121.

tissue are to be seen, so that the spongy tissue in this situation presents an appearance of being mapped out into a series of rectangular areas.

Articulations.—With three bones : the os innominatum, tibia, and patella.

Development (fig. 130).—The femur is developed by *five* centres : one for the shaft, one for each extremity, and one for each trochanter. Of all the long bones, except the clavicle, it is the first to show traces of ossification ; this commences in the shaft, at about the seventh week of fœtal life ; the centres of ossification in the epiphyses appearing in the following order : First, in the lower end of the bone, at the ninth month of fœtal life * (from this the condyles and tuberosities are formed) ; in the head at the end of the first year after birth ; in the great trochanter, during the fourth year ; and in the lesser trochanter, between the thirteenth and fourteenth. The order in which the epiphyses are joined to the shaft is the reverse of that of their appearance : their junction does not commence until after puberty, the lesser trochanter being first joined, then the great, then the head, and, lastly, the inferior extremity (the first in which ossification commenced), which is not united until the twentieth year.

Attachment of Muscles.—To twenty-three. To the great trochanter : the Gluteus medius, Gluteus minimus, Pyriformis, Obturator internus, Obturator externus, Gemellus superior, Gemellus inferior, and Quadratus femoris. To the lesser trochanter : the Psoas magnus and the Iliacus below it. To the shaft : the Vastus externus, Gluteus maximus, short head of the Biceps, Adductor magnus, Pectineus, Adductor brevis, Adductor longus, Vastus internus, Crureus, and Subcrureus. To the condyles : the Gastrocnemius, Plantaris, and Popliteus.

Surface Form.—The femur is covered with muscles, so that in fairly muscular subjects the shaft is not to be detected through its fleshy covering, and the only parts accessible to the touch are the outer surface of the great trochanter and the lower expanded end of the bone. The external surface of the great trochanter is to be felt, especially in certain positions of the limb. Its position is generally indicated by a depression, owing to the thickness of the Gluteus medius and minimus, which project above it. When, however, the thigh is flexed, and especially if crossed over the opposite one, the trochanter produces a blunt eminence on the surface. The upper border is about on a level with the spine of the os pubis, and its exact level is indicated by a line drawn from the anterior superior spinous process of the ilium, over the outer side of the hip, to the most prominent point of the tuberosity of the ischium. This is known as Nélaton's line. The outer and inner condyles of the lower extremity are accessible to the touch. The outer one is more subcutaneous than the inner one, and readily felt. The tuberosity on it is comparatively little developed, but can be more or less easily recognised. The inner condyle is more thickly covered, and this gives a general convex outline to this part, especially when the knee is flexed. The tuberosity on it is easily felt, and at the upper part of the condyle the sharp tubercle for the insertion of the tendon of the Adductor magnus can be recognised without difficulty. When the knee is flexed, and the patella situated in the interval between the condyles and the upper end of the tibia, a part of the trochlear surface of the femur can be made out above the patella.

Surgical Anatomy.—There are one or two points about the ossification of the femur bearing on practice to which allusion must be made. It has been stated that the lower end of the femur is the only epiphysis in which ossification has commenced at the time of birth. The presence of this ossific centre is, therefore, a proof, in newly born children found dead, that the child has arrived at the full period of utero-gestation, and is always relied upon in medico-legal investigations. The position of the epiphysial line should be carefully noted. It is on a level with the Adductor tubercle, and the epiphysis does not, therefore, form the whole of the cartilage-clad portion of the lower end of the bone. It is essential to bear this point in mind in performing excision of the knee, since growth in length of the femur takes place chiefly from the lower epiphysis, and any interference with the epiphysial cartilage in a young child would involve such ultimate shortening of the limb, from want of growth, as to render the limb almost useless. Separation of the lower epiphysis may take place up to the age of twenty, at which time it becomes completely joined to the shaft of the bone ; but, as a matter of fact, few cases occur after the age of sixteen or seventeen. The epiphysis of the head of the femur is of interest principally on account of its being the seat of origin in a large number of cases of tuberculous disease of the hip-joint. The disease commences in the majority of cases in the highly vascular and growing tissue in the neighbourhood of the epiphysis, and from here extends into the joint,

* This is said to be the only epiphysis in which ossification begins before birth ; though according to some observers the centre for the upper epiphysis of the tibia also appears before birth.

Fractures of the femur are divided, like those of the other long bones, into fractures of the upper end; of the shaft; and of the lower end. The fractures of the upper end may be classified into (1) fracture of the neck; (2) fracture at the junction of the neck with the great trochanter; (3) fracture of the great trochanter; and (4) separation of the epiphysis, either of the head or of the great trochanter. The first of these, fracture of the neck, is usually termed intracapsular fracture, but this is scarcely a correct designation, as, owing to the attachment of the capsular ligament, the fracture may be partly within and partly without the capsule when the fracture occurs at the lower part of the neck. It generally occurs in old people, principally women, and usually from a very slight degree of indirect violence. Probably the main cause of the fracture taking place in old people is in consequence of the degenerative changes which the bone has undergone. Merkel believes that it is mainly due to the absorption of the calcar femorale. These fractures are occasionally impacted. As a rule they unite by fibrous tissue, but frequently no union takes place, and the surfaces of the fracture become smooth and eburnated.

Fractures at the junction of the neck with the great trochanter are usually termed extra-capsular, but this designation is also incorrect, as the fracture is partly within the capsule, owing to its attachment in front to the anterior intertrochanteric line, which is situated below the line of fracture. These fractures are produced by direct violence to the great trochanter, as from a blow or fall laterally on the hip. From the manner in which the accident is caused, the neck of the bone is driven into the trochanter, where it may remain impacted, or the trochanter may be split up into two or more fragments, and thus no fixation takes place.

Fractures of the great trochanter may be either ' oblique fracture through the trochanter major, without implicating the neck of the bone ' (Astley Cooper), or separation of the great trochanter. Most of the recorded cases of this latter injury occurred in young persons, and were probably cases of separation of the epiphysis of the great trochanter. Separation of the epiphysis of the head of the femur has been said to occur, but has probably never been verified by post-mortem examination.

Fractures of the shaft may occur at any part, but the most usual situation is at or near the centre of the bone. They may be caused by direct or indirect violence or by muscular action. Fractures of the upper third of the shaft are almost always the result of indirect violence, while those of the lower third are the result, for the most part, of direct violence. In the middle third fractures occur from both forms of injury in about equal proportions. Fractures of the shaft are generally oblique, but they may be transverse, longitudinal, or spiral. The transverse fracture occurs most frequently in children. The fractures of the lower end of the femur include transverse fracture above the condyles, the most common; and this may be complicated by a vertical fracture between the condyles, constituting the T-shaped fracture. In these cases the popliteal artery is in danger of being wounded. Oblique fracture separating either the internal or external condyle, and a longitudinal incomplete fracture between the condyles, may also take place.

The femur as well as the other bones of the leg are frequently the seat of acute necrosis in young children. This is no doubt due to their greater exposure to injury, which is often the exciting cause of this disease. Tumours are not infrequently found growing from the femur: the most common forms being sarcoma, which may grow either from the periosteum or from the medullary tissue within the interior of the bone; and exostosis, which is commonly found originating in the neighbourhood of the epiphysial cartilage of the lower end.

THE LEG

The skeleton of the Leg consists of three bones : the Patella, a large, sesamoid bone, placed in front of the knee; the Tibia ; and the Fibula.

THE PATELLA (figs. 131, 132)

The **Patella** (patella, *a small pan*) is a flat, triangular bone, situated at the anterior part of the knee-joint. It is usually regarded as a sesamoid bone, developed in the tendon of the Quadriceps extensor. It resembles these bones (1) in its being developed in a tendon ; (2) in its centre of ossification presenting a knotty or tuberculated outline ; (3) in its structure being composed mainly of dense cancellous tissue, as in the other sesamoid bones. It serves to protect the front of the joint, and increases the leverage of the Quadriceps extensor by making it act at a greater angle. It presents an anterior and posterior surface, three borders, and an apex.

The **anterior surface** is convex, perforated by small apertures, for the passage of nutrient vessels, and marked by numerous rough, longitudinal striæ. This surface is covered, in the recent state, by an expansion from the tendon of the Quadriceps extensor, which is continuous below with the superficial fibres of the ligamentum patellæ. It is separated from the integument by a bursa. The **posterior surface** presents a smooth, oval-shaped, articular surface, covered with cartilage in the recent state, and divided into two facets by a vertical ridge, which descends from the superior border towards the inferior angle of the bone. The ridge corresponds to the groove on the trochlear surface of the femur, and the two facets to the articular surfaces of the two condyles ; the outer facet, for articulation with the outer condyle, being the broader and deeper. This character serves to indicate the side to which the bone belongs.

FIG. 131.—Right patella. Anterior surface.

FIG. 132.—Right patella. Posterior surface.

Below the articular surface is a rough, convex, non-articular depression, the lower half of which gives attachment to the ligamentum patellæ ; the upper half being separated from the head of the tibia by adipose tissue.

The **superior border** is thick, and sloped from behind, downwards and forwards : it gives attachment to that portion of the Quadriceps extensor which is derived from the Rectus and Crureus muscles. The **lateral borders** are thinner, converging below. They give attachment to that portion of the Quadriceps extensor derived from the external and internal Vasti muscles.

The **apex** is pointed, and gives attachment to the ligamentum patellæ.

Structure.—It consists of a nearly uniform dense cancellous tissue, covered by a thin compact lamina. The cancelli immediately beneath the anterior surface are arranged parallel with it. In the rest of the bone they radiate from the posterior articular surface towards the other parts of the bone.

Development.—By a single centre, which makes its appearance, according to Béclard, about the third year. In two instances, I have seen this bone cartilaginous throughout, at a much later period (six years). More rarely, the bone is developed by two centres, placed side by side. Ossification is completed about the age of puberty.

Articulations.—With the two condyles of the femur.

Attachment of Muscles.—To four : the Rectus, Crureus, Vastus internus, and Vastus externus. These muscles, joined at their insertion, constitute the Quadriceps extensor cruris.

Surface Form.—The external surface of the patella can be seen and felt in front of the knee. In the extended position of the limb the internal border is a little more prominent than the outer, and if the Quadriceps extensor is relaxed, the bone can be moved from side to side and appears to be loosely fixed. If the joint is flexed, the patella recedes into the hollow between the condyles of the femur and the upper end of the tibia, and becomes firmly fixed against the femur.

Surgical Anatomy.—The main surgical interest about the patella is in connection with fractures, which are of common occurrence. They may be produced by muscular action, that is to say, by violent contraction of the Quadriceps extensor, while the limb is in a position of semi-flexion, so that the bone is snapped across the condyles ; or by direct violence, such as falls on the knee. In the former class of cases the fracture is transverse ; in the latter it may be oblique, longitudinal, stellate, or the bone variously comminuted. The principal interest in these cases attaches to their treatment. Owing to the wide separation of the fragments, and the difficulty there is in maintaining them in apposition, union takes place by fibrous tissue, and this may subsequently stretch, producing wide separation of the fragments and permanent lameness. Various plans, including opening the joint and suturing the fragments, have been advocated for overcoming this difficulty.

In the larger number of cases of fracture of the patella, the knee-joint is involved, the cartilage which covers its posterior surface being also torn. In some cases of fracture from direct violence, however, this need not necessarily happen, the lesion involving only the superficial part of the bone ; and, as Morris has pointed out, it is an anatomical possibility, in complete fracture, if the lesion involve only the lower and non-articular part of the bone, for it to take place without injury to the synovial membrane.

FIG. 133.—Bones of the right leg.
Anterior surface.

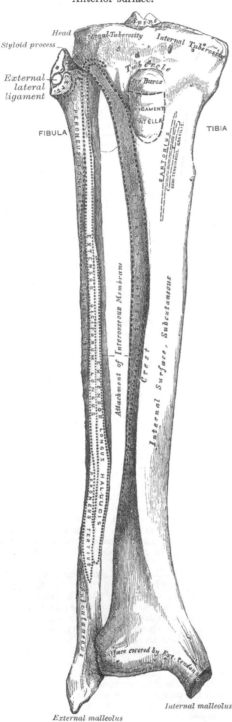

FIG. 133.—Bones of the right leg.
Anterior surface.

THE TIBIA (figs. 133, 134)

The **Tibia** (tibia, *a flute* or *pipe*) is situated at the front and inner side of the leg, and, excepting the femur, is the longest and largest bone in the skeleton. It is prismoid in form, expanded above, where it enters into the knee-joint, more slightly enlarged below. In the male, its direction is vertical, and parallel with the bone of the opposite side; but in the female it has a slightly oblique direction downwards and outwards, to compensate for the oblique direction of the femur inwards. It presents for examination a shaft and two extremities.

The **Upper Extremity,** or **Head,** is large, and expanded on each side into two lateral eminences, the *tuberosities. Superiorly,* the tuberosities present two smooth, concave surfaces, which articulate with the condyles of the femur; the internal articular surface is longer, deeper, and narrower than the external, oval from before backwards, to articulate with the internal condyle; the external one is broader and more circular, concave from side to side, but slightly convex from before backwards, especially at its posterior part, where it is prolonged on to the posterior surface for a short distance, to articulate with the external condyle. Between the two articular surfaces, and nearer the posterior than the anterior aspect of the bone, is an eminence, the *spinous process* of the tibia, surmounted on each side by a prominent tubercle, on to the lateral aspect of which the facets, just described, are prolonged; in front and behind the spinous process is a rough depression for the attachment of the anterior and posterior crucial ligaments and the semilunar fibro-cartilages. The *anterior surfaces* of the tuberosities are continuous with one another, forming a single large surface, which is somewhat flattened: it is triangular, broad above, and perforated by large vascular foramina: narrow below where it terminates in a prominent oblong elevation of large size, the *tubercle* of the tibia ; the lower half of this tubercle is rough, for the attachment of the ligamentum patellæ; the upper half presents a smooth facet supporting, in the recent state, a bursa which separates the ligament from the bone. *Posteriorly* the tuberosities are separated from each other by a shallow depression, the *popliteal notch*, which gives attachment to part of the posterior crucial ligament, and part of the posterior ligament of the knee-joint. The *inner tuberosity* presents posteriorly a deep transverse groove, for the insertion of one of the fasciculi of the tendon of the Semi-membranosus. Its *lateral surface* is convex, rough and prominent : it gives attachment to the internal lateral ligament. The *outer tuberosity* presents posteriorly a flat articular facet, nearly circular in form, directed downwards, backwards, and outwards, for articulation with the fibula. Its lateral surface is convex and rough, more prominent in front than the internal : it presents a prominent rough eminence, situated on a level with the upper border of the tubercle of the tibia at the junction of its anterior and outer surfaces, for the attachment of the ilio-tibial band. Just below this the Extensor longus digitorum and a slip from the Biceps are attached.

The **Shaft** of the tibia is of a triangular prismoid form, broad above, gradually decreasing in size to its most slender part, at the commencement of its lower fourth, where fracture most frequently occurs. It then enlarges again towards its lower extremity. It presents for examination three borders and three surfaces.

The **anterior border**, the most prominent of the three, is called the *crest of the tibia*, or, in popular language, the *shin* ; it commences above at the tubercle, and terminates below

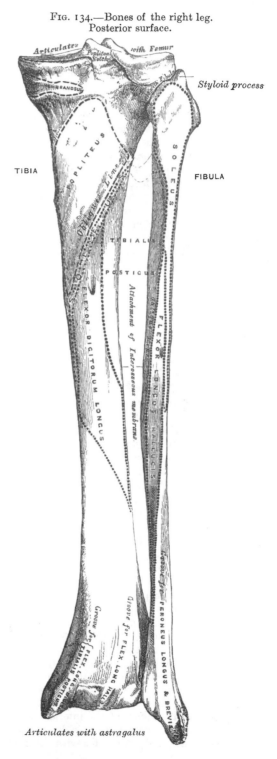

Fig. 134.—Bones of the right leg. Posterior surface.

TIBIA

FIBULA

Styloid *process*

Articulates with astragalus

at the anterior margin of the inner malleolus. This border is very prominent in the upper two-thirds of its extent, smooth and rounded below. It presents a very flexuous course, being usually curved outwards above, and inwards below ; it gives attachment to the deep fascia of the leg.

The **internal border** is smooth and rounded above and below, but more prominent in the centre ; it commences at the back part of the inner tuberosity, and terminates at the posterior border of the internal malleolus ; its upper part gives attachment to the internal lateral ligament of the knee to the extent of about two inches, and to some fibres of the Popliteus muscle ; its middle third, to some fibres of the Soleus and Flexor longus digitorum muscles.

The **external border, or interosseous ridge,** is thin and prominent, especially its central part, and gives attachment to the interosseous membrane ; it commences above in front of the fibular articular facet, and bifurcates below, to form the boundaries of a triangular rough surface, for the attachment of the interosseous ligament connecting the tibia and fibula.

The **internal surface** is smooth, convex, and broader above than below ; its upper third, directed forwards and inwards, is covered by the aponeurosis derived from the tendon of the Sartorius, and by the tendons of the Gracilis and Semitendinosus, all of which are inserted nearly as far forwards as the anterior border ; in the rest of its extent it is subcutaneous.

The **external surface** is narrower than the internal ; its upper two-thirds presents a shallow groove for the attachment of the Tibialis anticus muscle ; its lower third is smooth, convex, curves gradually forwards to the anterior aspect of the bone, and is covered from within outwards by the tendons of the following muscles : Tibialis anticus, Extensor proprius hallucis, Extensor longus digitorum.

The **posterior surface** (fig. 134) presents, at its upper part, a prominent ridge, the *oblique line* of the tibia, which extends from the back part of the articular facet for the fibula obliquely downwards, to the internal border, at the junction of its upper and middle thirds. It marks the lower limit of the insertion of the Popliteus muscle, and serves for the attachment of the popliteal fascia and part of the Soleus, Flexor longus digitorum, and Tibialis posticus muscles ; the triangular concave surface, above and to the inner side of this line, gives attachment to the Popliteus muscle. The middle third of the posterior surface is divided by a vertical ridge into two lateral halves : the ridge is well marked at its commencement at the oblique line, but becomes gradually indistinct below ; the inner and broader half gives attachment to the Flexor longus digitorum, the outer and narrower to part of the Tibialis posticus. The remaining part of the bone presents a smooth surface covered by the Tibialis posticus, Flexor longus digitorum, and Flexor longus hallucis muscles. Immediately below the oblique line is the medullary foramen, which is large and directed obliquely downwards.

The **Lower Extremity,** much smaller than the upper, presents five surfaces ; it is prolonged downwards, on its inner side into a strong process, the *internal malleolus*. The *inferior surface* of the bone is quadrilateral, and smooth for articulation with the astragalus. This surface is concave from before backwards, and broader in front than behind. It is traversed from before backwards by a slight elevation, separating two lateral depressions. It is narrow internally, where the articular surface becomes continuous with that on the inner malleolus. The *anterior surface* of the lower extremity is smooth and rounded above, and covered by the tendons of the Extensor muscles of the toes ; its lower margin presents a rough transverse depression, for the attachment of the anterior ligament of the ankle-joint ; the *posterior surface* presents a superficial groove directed obliquely downwards and inwards, continuous with a similar groove on the posterior surface of the astragalus, and serving for the passage of the tendon of the Flexor longus hallucis ; the *external surface* presents a triangular rough depression for the attachment of the inferior interosseous ligament connecting it with the fibula ; the lower part of this depression is smooth, covered with cartilage in the recent state, and articulates with the fibula. The surface is bounded by two prominent borders, continuous above with the interosseous ridge; they afford attachment to the anterior and posterior inferior tibio-fibular ligaments. The *internal surface* of the lower extremity is prolonged downwards to form a strong pyramidal process, flattened from without inwards—the *internal malleolus*. The *inner surface* of this process is convex and subcutaneous; its *outer surface* is smooth and slightly concave, and articulates with the astragalus; its *anterior border* is rough, for the attachment of the anterior fibres of the internal lateral or Deltoid ligament; its *posterior border* presents a broad and deep groove, directed obliquely downwards and inwards, which is occasionally double; this groove transmits the tendons of the Tibialis posticus and Flexor longus digitorum muscles. The *summit* of the internal malleolus is marked by a rough depression behind, for the attachment of the internal lateral ligament of the ankle-joint.

Structure.—Like that of the other long bones. At the junction of the middle and lower third, where the bone is smallest, the wall of the shaft is thicker than in other parts, in order to compensate for the smallness of the calibre of the bone.

Development.—By *three* centres (fig. 135): one for the shaft, and one for each extremity. Ossification commences in the centre of the shaft about the seventh week, and gradually extends towards either extremity. The centre for the upper epiphysis appears before or shortly after birth; it is flattened in form, and has a thin tongue-shaped process in front, which forms the tubercle. That for the lower epiphysis appears in the second year. The lower epiphysis joins the shaft at about the eighteenth, and the upper one about the twentieth year. Two additional centres occasionally exist, one for the tongue-shaped process of the upper epiphysis, which forms the tubercle, and one for the inner malleolus.

Articulations.—With three bones: the femur, fibula, and astragalus.

Attachment of Muscles.—To twelve: to the inner tuberosity, the Semi-membranosus; to the outer tuberosity, the Tibialis anticus and Extensor longus digitorum and Biceps; to the shaft, its internal surface, the Sartorius, Gracilis, and Semi-tendinosus; to its external surface, the Tibialis anticus; to its posterior surface, the Popliteus,

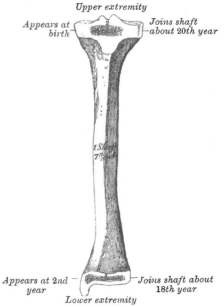

FIG. 135.—Plan of the development of the tibia. By three centres.

Upper extremity

Appears at birth — *Joins shaft about 20th year*

Appears at 2nd year — *Joins shaft about 18th year*

Lower extremity

Soleus, Flexor longus digitorum, and Tibialis posticus; to the tubercle, the ligamentum patellæ, by which the Quadriceps extensor muscle is inserted into the tibia. In addition to these muscles, the Tensor fasciæ femoris is inserted indirectly into the tibia, through the ilio-tibial band, and the Peroneus longus occasionally derives a few fibres of origin from the outer tuberosity.

Surface Form.—A considerable portion of the tibia is subcutaneous and easily to be felt. At the upper extremity the tuberosities are to be recognised just below the knee. The internal one is broad and smooth, and merges into the subcutaneous surface of the shaft below. The external one is narrower and more prominent, and on it, about midway between the apex of the patella and the head of the fibula, may be felt a prominent tubercle for the insertion of the ilio-tibial band. In front of the upper end of the bone, between the tuberosities, is the tubercle of the tibia, forming an oval eminence, which is continuous below with the anterior border or crest of the bone. This border can be felt, forming the prominence of the shin, in the upper two-thirds of its extent as a sharp and flexuous ridge, curved outwards above and inwards below. In the lower third of the leg the border disappears and the bone is concealed by the tendons of the muscles on the front of the leg. Internal to the anterior border is to be felt the broad internal surface of the tibia, slightly encroached upon by the muscles in front and behind. It commences above at the wide expanded inner tuberosity and terminates below at the internal malleolus. The internal malleolus is a broad prominence situated on a higher level and somewhat farther forwards than the external malleolus. It overhangs the inner border of the arch of the foot. Its anterior border is nearly straight, its posterior border presents a sharp edge, which forms the inner margin of the groove for the tendon of the Tibialis posticus muscle.

THE FIBULA (figs. 133, 134)

The **Fibula** (fibula, *a clasp*) is situated at the outer side of the leg. It is the smaller of the two bones, and, in proportion to its length, the most slender of all the long bones ; it is placed on the outer side of the tibia, with which it is connected above and below. Its upper extremity is small, placed towards the back of the head of the tibia, below the level of the knee-joint, and excluded from its formation ; the lower extremity inclines a little forwards, so as to be on a plane anterior to that of the upper end, projects below the tibia, and forms the outer ankle. It presents for examination a shaft and two extremities.

The **Upper Extremity, or Head,** is of an irregular quadrate form, presenting above a flattened articular facet, directed upwards, forwards and inwards, for articulation with a corresponding facet on the external tuberosity of the tibia. On the outer side is a thick and rough prominence, continued behind into a pointed eminence, the *styloid process*, which projects upwards from the posterior part of the head. The prominence gives attachment to the tendon of the Biceps muscle, and to the long external lateral ligament of the knee, the ligament dividing the tendon into two parts. The summit of the styloid process gives attachment to the short external lateral ligament. The remaining part of the circumference of the head is rough, for the attachment of muscles and ligaments. It presents in front a tubercle for the origin of the upper and anterior part of the Peroneus longus, and the adjacent surface gives attachment to the anterior superior tibio-fibular ligament ; and behind, another tubercle for the attachment of the posterior superior tibio-fibular ligament, and the upper fibres of the Soleus muscle.

The **shaft** presents four borders—the antero-external, the antero-internal, the postero-external, and the postero-internal ; and four surfaces—anterior, posterior, internal, and external.

The **antero-external border** commences above in front of the head, runs vertically downwards to a little below the middle of the bone, and then curving somewhat outwards, bifurcates so as to embrace the triangular subcutaneous surface immediately above the outer surface of the external malleolus. This border gives attachment to an intermuscular septum, which separates the extensor muscles on the anterior surface of the leg from the Peroneus longus and brevis muscles on the outer surface.

The **antero-internal border,** or **interosseous ridge,** is situated close to the inner side of the preceding, and runs nearly parallel with it in the upper third of its extent, but diverges from it so as to include a broader space in the lower two-thirds. It commences above just beneath the head of the bone (sometimes it is quite indistinct for about an inch below the head), and terminates below at the apex of a rough triangular surface immediately above the articular facet of the external malleolus. It serves for the attachment of the interosseous membrane, which separates the extensor muscles in front from the flexor muscles behind.

The **postero-external border** is prominent ; it commences above at the base of the styloid process, and terminates below in the posterior border of the outer malleolus. It is directed outwards above, backwards in the middle of its course, backwards and a little inwards below, and gives attachment to an aponeurosis which separates the Peronei muscles on the outer surface of the shaft from the flexor muscles on its posterior surface.

The **postero-internal border,** sometimes called the *oblique line,* commences above at the inner side of the head, and terminates by becoming continuous with the antero-internal border or interosseous ridge at the lower fourth of the bone. It is well marked and prominent at the upper and middle parts of the bone. It gives attachment to an aponeurosis which separates the Tibialis posticus from the Soleus above and the Flexor longus hallucis below.

The **anterior surface** is the interval between the antero-external and antero-internal borders. It is extremely narrow and flat in the upper third of its extent; broader and grooved longitudinally in its lower third; it serves for the attachment of three muscles, the Extensor longus digitorum, Peroneus tertius, and Extensor proprius hallucis.

The **external surface** is the space between the antero-external and postero-external borders. It is much broader than the preceding, and often deeply grooved, is directed outwards in the upper two-thirds of its course, backwards in the lower third, where it is continuous with the posterior border of the external malleolus. This surface is completely occupied by the Peroneus longus and brevis muscles.

The **internal surface** is the interval included between the antero-internal and the postero-internal borders. It is directed inwards, and is grooved for the attachment of the Tibialis posticus muscle.

The **posterior surface** is the space included between the postero-external and the postero-internal borders; it is continuous below with the rough triangular surface above the articular facet of the outer malleolus; it is directed backwards above, backwards and inwards at its middle, directly inwards below. Its upper third is rough, for the attachment of the Soleus muscle; its lower part presents a triangular rough surface, connected to the tibia by a strong interosseous ligament, and between these two points the entire surface is covered by the fibres of origin of the Flexor longus hallucis muscle. At about the middle of this surface is the nutrient foramen, which is directed downwards.

The **Lower Extremity,** or **external malleolus,** is of a pyramidal form, somewhat flattened from without inwards, and is longer, and descends lower, than the internal malleolus. Its *external surface* is convex, subcutaneous, and continuous with the triangular (also subcutaneous) surface on the outer side of the shaft. The *internal surface* presents in front a smooth triangular facet, broader above than below, and convex from above downwards, which articulates with a corresponding surface on the outer side of the astragalus. Behind and beneath the articular surface is a rough depression, which gives attachment to the posterior fasciculus of the external lateral ligament of the ankle. The *anterior border* is thick and rough, and marked below by a depression for the attachment of the anterior fasciculus of the external lateral ligament. The *posterior border* is broad and marked by a shallow groove, for the passage of the tendons of the Peroneus longus and brevis muscles. The *summit* is rounded, and gives attachment to the middle fasciculus of the external lateral ligament.

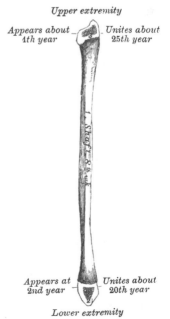

Fig. 136.—Plan of the development of the fibula. By three centres.

Upper extremity

Appears about 4th year — Unites about 25th year

Shaft & Head

Appears at 2nd year — Unites about 20th year

Lower extremity

In order to distinguish the side to which the bone belongs, hold it with the lower extremity downwards, and the broad groove for the Peronei tendons backwards, i.e. towards the holder : the triangular subcutaneous surface will then be directed to the side to which the bone belongs.

Articulations.—With two bones : the tibia and astragalus.

Development.—By *three* centres (fig. 136) : one for the shaft, and one for each extremity. Ossification commences in the shaft about the eighth week of fœtal life, a little later than in the tibia, and extends gradually towards the extremities. At birth both ends are cartilaginous. Ossification commences in the lower end in the second year, and in the upper one about the fourth year. The lower epiphysis, the first in which ossification commences, becomes united to the shaft about the twentieth year ; the upper epiphysis joins about the twenty-fifth year. Ossification appearing first in the lower epiphysis is contrary to the rule which prevails with regard to the commencement of ossification in epiphyses, viz. that that epiphysis towards which the nutrient artery is directed commences to ossify last ; but it follows the rule which prevails with regard to the union of epiphyses, by uniting first.

Attachment of Muscles.—To nine : to the head, the Biceps, Soleus, and Peroneus longus ; to the shaft, its anterior surface, the Extensor longus digitorum, Peroneus tertius, and Extensor proprius hallucis ; to the internal surface, the Tibialis posticus ; to the posterior surface, the Soleus and Flexor longus hallucis ; to the external surface, the Peroneus longus and brevis.

Surface Form.—The only parts of the fibula which are to be felt are the head, the lower part of the external surface of the shaft, and the external malleolus. The head is to be seen and felt behind and to the outer side of the outer tuberosity of the tibia. It presents a small, prominent, triangular eminence slightly above the level of the tubercle of the tibia. The external malleolus presents a narrow elongated prominence, situated on a plane posterior to the internal malleolus and reaching to a lower level. From it may be traced the lower third or half of the external surface of the shaft of the bone in the interval between the Peroneus tertius in front and the other two Peronei tendons behind.

Surgical Anatomy.—In fractures of the bones of the leg, both bones are usually fractured, but each bone may be broken separately, the fibula more frequently than the tibia. Fracture of both bones may be caused either by direct or indirect violence. When it occurs from indirect force, the fracture in the tibia is at the junction of the middle and lower third of the bone. Many causes conduce to render this the weakest part of the bone. The fracture of the fibula is usually at rather a higher level. These fractures present great variety, both as regards their direction and condition. They may be oblique, transverse, longitudinal, or spiral. When oblique, they are usually the result of indirect violence, and the direction of the fracture is from behind, downwards, forwards, and inwards in many cases, but may be downwards and outwards, or downwards and backwards. When transverse, the fracture is often at the upper part of the bone, and is the result of direct violence. The spiral fracture of the tibia usually commences as a vertical fissure, involving the ankle-joint, and is associated with fracture of the fibula higher up. It is the result of torsion, from twisting of the body while the foot is fixed.

Fractures of the tibia alone are almost always the result of direct violence, except where the malleolus is broken off by twists of the foot. Fractures of the fibula alone may arise from indirect or direct force, those of the lower end being usually the result of the former, and those higher up being caused by a direct blow on the part.

The tibia and fibula, like the femur, are frequently the seat of acute necrosis. Chronic abscess is more frequently met with in the cancellous tissue of the head and lower end of the tibia than in any other bone of the body. The abscess is of small size, very chronic, and probably the result of tuberculous osteitis in the highly vascular growing tissue at the end of the shaft near the epiphysial cartilage in the young subject.

The tibia is the bone which is most frequently and most extensively distorted in rickets. It gives way at the junction of the middle and lower third, its weakest part, and presents a curve forwards and outwards.

THE FOOT (figs. 137, 138)

The skeleton of the foot consists of three divisions : the Tarsus, Metatarsus, and Phalanges.

THE TARSUS

The bones of the **Tarsus** are seven in number : viz. the calcaneum or os calcis, astragalus, cuboid, navicular, internal, middle, and external cuneiform bones.

FIG. 137.—Bones of the right foot.   Dorsal surface.

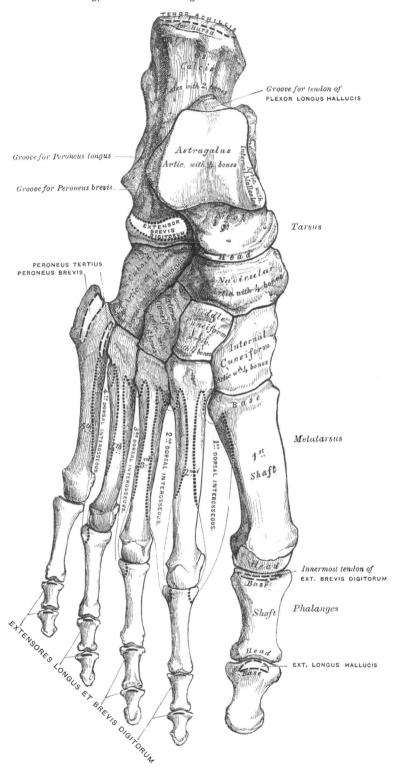

FIG. 138.—Bones of the right foot.   Plantar surface.

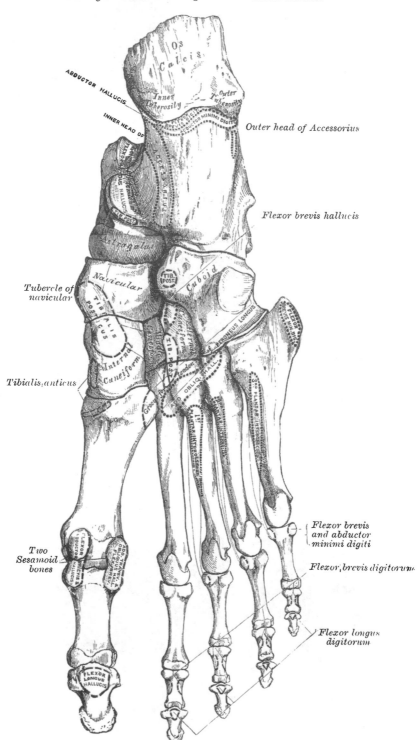

THE CALCANEUM (fig. 139)

The **Calcaneum,** or **Os Calcis** (calx, *the heel*), is the largest and strongest of the tarsal bones. It is irregularly cuboidal in form, having its long axis directed forwards and outwards. It is situated at the lower and back part of the foot, serving to transmit the weight of the body to the ground, and forming a strong lever for the muscles of the calf. It presents for examination six surfaces: superior, inferior, external, internal, anterior, and posterior.

The *superior surface* is formed behind by the upper aspect of that part of the os calcis which projects backwards to form the heel. It varies in length in

FIG. 139.—The left os calcis. A. Postero-external view. B. Antero-internal view.

A

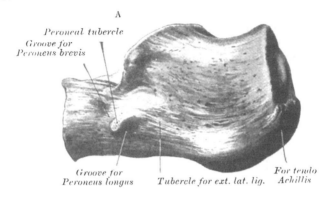

Peroneal tubercle
Groove for
Peroneus brevis

Groove for
Peroneus longus   *Tubercle for ext. lat. lig.*   *For tendo Achillis*

B

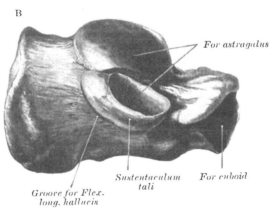

For astragalus

Groove for Flex.
long. hallucis   *Sustentaculum tali*   *For cuboid*

different individuals; is convex from side to side, concave from before backwards, and upon it rests a mass of adipose substance placed in front of the tendo Achillis. In the middle of the superior surface are two (sometimes three) articular facets, separated by a broad shallow groove, which is directed obliquely forwards and outwards, and is rough for the attachment of the interosseous ligament connecting the astragalus and os calcis. Of the two articular surfaces, the *external* is the larger, and situated on the body of the bone: it is of an oblong form, wider behind than in front and convex from before backwards. The *internal articular surface* is supported on a projecting process of bone, called the *lesser process* of the calcaneum (*sustentaculum tali*); it is also oblong, concave longitudinally, and sometimes subdivided into two parts, which differ in size and shape. More anteriorly is seen the upper surface of the *greater process,* marked by a rough depression for the attachment of numerous ligaments, and a tubercle for the origin of the Extensor brevis digitorum muscle.

The *inferior surface* is narrow, rough, uneven, wider behind than in front, and convex from side to side ; it is bounded posteriorly by two tubercles, separated by a rough depression ; the *external*, small, prominent, and rounded, gives attachment to part of the Abductor minimi digiti ; the *internal*, broader and larger, for the support of the heel, gives attachment, by its prominent inner margin, to the Abductor hallucis, and in front to the Flexor brevis digitorum muscles and plantar fascia ; the depression between the tubercles gives attachment to the Abductor minimi digiti. The rough surface in front of the tubercles gives attachment to the long plantar ligament, and to the outer head of the Flexor accessorius muscle ; while to a prominent tubercle nearer the anterior part of this surface, as well as to a transverse groove in front of it, is attached the short plantar ligament.

The *external surface* is broad, flat, and almost subcutaneous ; it presents near its centre a tubercle, for the attachment of the middle fasciculus of the external lateral ligament. At its upper and anterior part, this surface gives attachment to the external calcaneo-astragaloid ligament ; and in front of the tubercle it presents a narrow surface marked by two oblique grooves, separated by an elevated ridge which varies much in size in different bones ; it is named the *peroneal tubercle*, and gives attachment to a fibrous process from the external annular ligament. The *superior groove* transmits the tendon of the Peroneus brevis ; the *inferior*, the tendon of the Peroneus longus.

The *internal surface* is deeply concave ; it is directed obliquely downwards and forwards, and serves for the transmission of the plantar vessels and nerves into the sole of the foot ; it affords attachment to part of the Flexor accessorius muscle. At its upper and fore part it presents an eminence of bone, the *lesser process* or *sustentaculum tali*, which projects horizontally inwards, and to it a slip of the tendon of the Tibialis posticus is attached. This process is concave above, and supports the anterior articular surface of the astragalus ; below, it is grooved for the tendon of the Flexor longus hallucis. Its free margin is rough for the attachment of part of the internal lateral ligament of the ankle-joint.

The *anterior surface*, of a somewhat triangular form, articulates with the cuboid. It is concave from above, downwards and outwards, and convex in the opposite direction. Its inner border gives attachment to the inferior calcaneo-navicular ligament.

The *posterior surface* is rough, prominent, convex, and wider below than above. Its lower part is rough, for the attachment of the tendo Achillis and the tendon of the Plantaris muscle ; its upper part is smooth, and is covered by a bursa which separates the tendons from the bone.

Articulations.—With two bones : the astragalus and cuboid.

Attachment of Muscles.—To eight : part of the Tibialis posticus, the tendo Achillis, Plantaris, Abductor hallucis, Abductor minimi digiti, Flexor brevis digitorum, Flexor accessorius, and Extensor brevis digitorum.

THE ASTRAGALUS (fig. 140)

The **Astragalus** (ἀστράγαλος, *a die*) is the largest of the tarsal bones, next to the os calcis. It occupies the middle and upper part of the tarsus, supporting the tibia above, articulating with the malleoli on either side, resting below upon the os calcis, and joined in front to the navicular. This bone may easily be recognised by its large rounded head, by the broad articular facet on its upper convex surface, or by the two articular facets separated by a deep groove on its under concave surface. It presents six surfaces for examination.

The *superior surface* presents, behind, a broad smooth trochlear surface, for articulation with the tibia. The trochlea is broader in front than behind, convex from before backwards, slightly concave from side to side : in front of it is the upper surface of the neck of the astragalus ; rough for the attachment of ligaments. The *inferior surface* presents two articular facets separated by a deep groove. The groove runs obliquely forwards and outwards, becoming gradually broader and deeper in front : it corresponds with a similar groove upon the upper surface of the os calcis, and forms, when articulated with that bone, a canal, filled up in the recent state by the interosseous calcaneo-astragaloid ligament. Of the two articular facets, the posterior is the larger, of an oblong form, and deeply concave from side to side ; the anterior is shorter and narrower, of an elongated oval form, convex longitudinally, and often subdivided into two by an elevated ridge ; of these the posterior articulates with the lesser process of the os calcis ; the anterior, with the upper surface of the inferior calcaneo - navicular ligament. The *internal surface* presents at its upper part a pear-shaped articular facet for the inner malleolus, continuous above with the trochlear surface ; below the articular surface is a rough depression, for the attachment of the deep portion of the internal lateral ligament. The *external surface* presents a large triangular facet, concave from above downwards, for articulation with the external malleolus ; it is continuous above with the trochlear surface ; and in front of it is a rough depression for the attachment of the anterior fasciculus of the external lateral ligament of the ankle-joint. The *anterior surface*, convex and rounded, forms the *head* of the astragalus ; it is smooth, of an oval form, and directed obliquely inwards and downwards ; it articulates with the navicular. On its under and inner surface is a small facet, continuous in front with the articular surface of the head,

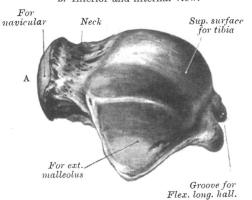

Fig. 140.—The left astragalus.
A. Superior and external view.
B. Inferior and internal view.

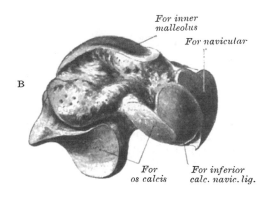

and behind with the smaller facet for the os calcis. This rests on the inferior calcaneo-navicular ligament, being separated from it by the synovial membrane, which is prolonged from the anterior calcaneo-astragaloid joint to the astragalo-navicular joint. The head is surrounded by a constricted portion, the *neck* of the astragalus. The *posterior surface* is narrow, and traversed by a groove, which runs obliquely downwards and inwards, and transmits the tendon of the Flexor longus hallucis, external to which is a prominent tubercle, to which the posterior fasciculus of the external lateral ligament is attached. This tubercle is sometimes separated from the rest of the astragalus, and is then known as the *os trigonum*. To the inner side of the groove is a second, but less marked tubercle.

To ascertain to which foot the bone belongs, hold it with the broad articular surface upwards, and the rounded head forwards ; the lateral triangular articular surface for the external malleolus will then point to the side to which the bone belongs.

Articulations.—With four bones : tibia, fibula, os calcis, and navicular.

THE CUBOID (fig. 141)

The **Cuboid** (κύβος, *a cube*; εἶδος, *like*) bone is placed on the outer side of the foot, in front of the os calcis, and behind the fourth and fifth metatarsal bones. It is of a pyramidal shape, its base being directed inwards, its apex outwards. It may be distinguished from the other tarsal bones ·by the existence of a deep groove on its under surface, for the tendon of the Peroneus longus muscle. It presents for examination six surfaces : three articular, and three non-articular.

The **non-articular surfaces** are the superior, inferior, and external. The *superior* or *dorsal surface*, directed upwards and outwards, is rough, for the attachment of numerous ligaments. The *inferior* or *plantar surface* presents in front a deep groove, which runs obliquely from without, forwards and inwards ; it lodges the tendon of the Peroneus longus, and is bounded behind by a prominent ridge, to which is attached the long calcaneo-cuboid ligament. The ridge terminates externally in an eminence, the *tuberosity of the cuboid*, the surface of which presents a convex facet, for articulation with the sesamoid bone of the tendon contained in the groove. The surface of bone behind the groove is rough, for the attachment of the short plantar ligament, a few fibres of the Flexor brevis hallucis and a fasciculus from the tendon of the Tibialis posticus. The *external surface*, the smallest and narrowest of the three, presents a deep notch formed by the commencement of the peroneal groove.

FIG. 141.—The left cuboid. A. Antero-internal view. B. Postero-external view.

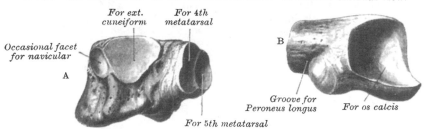

The **articular surfaces** are the posterior, anterior, and internal. The *posterior surface* is smooth, triangular, and concavo-convex, for articulation with the anterior surface of the os calcis. The *anterior*, of smaller size, but also irregularly triangular, is divided by a vertical ridge into two facets ; the inner one, quadrilateral in form, articulates with the fourth metatarsal bone ; the outer one, larger and more triangular, articulates with the fifth metatarsal. The *internal surface* is broad, rough, irregularly quadrilateral, presenting at its middle and upper part a smooth oval facet, for articulation with the external cuneiform bone ; and behind this (occasionally) a smaller facet, for articulation with the navicular ; it is rough in the rest of its extent, for the attachment of strong interosseous ligaments.

To ascertain to which foot the bone belongs, hold it so that its under surface, marked by the peroneal groove, looks downwards, and the large concavo-convex articular surface backwards, towards the holder : the narrow non-articular surface, marked by the commencement of the peroneal groove, will point to the side to which the bone belongs.

Articulations.—With four bones : the os calcis, external cuneiform, and the fourth and fifth metatarsal bones ; occasionally with the navicular.

Attachment of Muscles.—Part of the Flexor brevis hallucis and a slip from the tendon of the Tibialis posticus.

THE NAVICULAR (fig. 142)

The **Navicular** or **Scaphoid bone** is situated at the inner side of the tarsus, between the astragalus behind and the three cuneiform bones in front. It may be distinguished by its form, being concave behind, convex and subdivided into three facets in front.

The *anterior surface*, of an oblong form, is convex from side to side, and subdivided by two ridges into three facets, for articulation with the three cuneiform bones. The *posterior surface* is oval, concave, broader externally than internally, and articulates with the rounded head of the astragalus. The *superior surface* is convex from side to side, and rough for the attachment of ligaments. The *inferior* is irregular, and also rough for the attachment of ligaments. The *internal surface* presents a rounded tubercular eminence, the *tuberosity* of the navicular, the lower

FIG. 142.—The left navicular. A. Antero-external view. B. Postero-internal view.

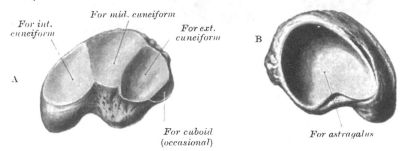

For int. cuneiform
For mid. cuneiform
For ext. cuneiform
A
B
For cuboid (occasional)
For astragalus

part of which projects, and gives attachment to part of the tendon of the Tibialis posticus. The *external surface* is rough, and irregular, for the attachment of ligamentous fibres, and occasionally presents a small facet for articulation with the cuboid bone.

To ascertain to which foot the bone belongs, hold it with the concave articular surface backwards, and the convex dorsal surface upwards; the external surface, i.e. the surface opposite the tubercle, will point to the side to which the bone belongs.

Articulations.—With four bones: astragalus and three cuneiform; occasionally also with the cuboid.

Attachment of Muscles.—Part of the Tibialis posticus.

THE CUNEIFORM BONES

The **Cuneiform Bones** have received their name from their wedge-like shape (cuneus, *a wedge*; forma, *likeness*). They form, with the cuboid, the anterior row of the tarsus, being placed between the navicular behind, the three innermost metatarsal bones in front, and the cuboid externally. They are called the *first*, *second*, and *third*, counting from the inner to the outer side of the foot, and, from their position, *internal*, *middle*, and *external*.

THE INTERNAL CUNEIFORM (fig. 143)

The **Internal Cuneiform** is the largest of the three. It is situated at the inner side of the foot, between the navicular behind and the base of the first metatarsal in front. It may be distinguished from the other two by its large size, and by its not presenting such a distinct wedge-like form. Without the others, it may be known by the large kidney-shaped anterior articulating surface, and by the prominence on the inferior or plantar surface for the attachment of the Tibialis posticus. It presents for examination six surfaces.

The *internal surface* is subcutaneous, and forms part of the inner border of the foot; it is broad, quadrilateral, and presents at its anterior inferior angle a smooth oval facet, into which the tendon of the Tibialis anticus is partially inserted; in the rest of its extent it is rough, for the attachment of ligaments. The

FIG. 143.—The left internal cuneiform.
A. Antero-internal view.   B. Postero-external view.

For 1st metatarsal   *For 2nd*   *For mid.*
                          metatarsal   *cuneiform*

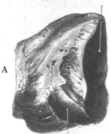
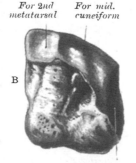

A   B

For tendon of
Tibialis ant.   *For navicular*

external surface is concave, presenting, along its superior and posterior borders, a narrow reversed L-shaped surface for articulation with the middle cuneiform behind, and second metatarsal bone in front: in the rest of its extent it is rough for the attachment of ligaments and part of the tendon of the Peroneus longus. The *anterior surface*, kidney-shaped, much larger than the posterior, articulates with the metatarsal bone of the great toe. The *posterior surface* is triangular, concave, and articulates with the innermost and largest of the three facets on the anterior surface of the navicular. The *inferior* or *plantar surface* is rough, and presents a prominent tuberosity at its back part for the attachment of part of the tendon of the Tibialis posticus. It also gives attachment in front to part of the tendon of the Tibialis anticus. The *superior surface* is the narrow pointed end of the wedge, which is directed upwards and outwards; it is rough for the attachment of ligaments.

To ascertain to which side the bone belongs, hold it so that its superior narrow edge looks upwards, and the long, kidney-shaped, articular surface forwards; the external surface, marked by its vertical and horizontal articular facets, will point to the side to which it belongs.

Articulations.—With four bones: navicular, middle cuneiform, first and second metatarsal bones.

Attachment of Muscles.—To three: the Tibialis anticus and posticus, and Peroneus longus.

THE MIDDLE CUNEIFORM (fig. 144)

The **Middle Cuneiform,** the smallest of the three, is of very regular wedge-like form, the broad extremity being placed upwards, the narrow end downwards. It is situated between the other two bones of the same name, and articulates with the navicular behind, and the second metatarsal in front. It is smaller than the external cuneiform bone, from which it may be further distinguished by the L-shaped articular facet, which runs round the upper and back part of its inner surface.

FIG. 144.—The left middle cuneiform.
    A. Antero-internal view.
    B. Postero-external view.

For int. cuneiform   *For navicular*

A   B

For 2nd metatarsal   *For ext. cuneiform*

The *anterior surface*, triangular in form, and narrower than the posterior, articulates with the base of the second metatarsal bone. The *posterior surface*, also triangular, articulates with the navicular. The *internal surface* presents a reversed L-shaped articular facet, running along the superior and posterior borders, for articulation with the internal cuneiform, and is rough in the rest of its extent for the attachment of ligaments. The *external surface* presents posteriorly a smooth facet for articulation with the external cuneiform bone. The *superior surface* forms the base of the wedge; it is quadrilateral, broader behind than in front, and rough for the attachment of ligaments. The *inferior surface*, pointed and tubercular, is also rough for ligamentous attachment, and for the insertion of a slip from the tendon of the Tibialis posticus.

To ascertain to which foot the bone belongs, hold its superior or dorsal surface upwards, the broadest edge being towards the holder: the smooth facet (limited to the posterior border) will then point to the side to which it belongs.

Articulations.—With four bones: navicular, internal and external cuneiform, and second metatarsal bone.

Attachment of Muscle.—A slip from the tendon of the Tibialis posticus is attached to this bone.

THE EXTERNAL CUNEIFORM (fig. 145)

The **External Cuneiform,** intermediate in size between the two preceding, is of a very regular wedge-like form, the broad extremity being placed upwards, the narrow end downwards. It occupies the centre of the front row of the tarsus between the middle cuneiform internally, the cuboid externally, the navicular behind, and the third metatarsal in front. It is distinguished from the internal cuneiform bone by its more regular wedge-like shape, and by the absence of the kidney-shaped articular surface: from the middle cuneiform, by the absence of the reversed L-shaped facet, and by the two articular facets which are present on both its inner and outer surfaces. It presents six surfaces for examination.

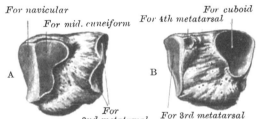

FIG. 145.--The left external cuneiform.
A. Postero-internal view.   B. Antero-external view.

For navicular
For mid. cuneiform   *For 4th metatarsal*
For cuboid
A   B
For 2nd metatarsal   *For 3rd metatarsal*

The *anterior surface*, triangular in form, articulates with the third metatarsal bone. The *posterior surface* articulates with the most external facet of the navicular, and is rough below for the attachment of ligamentous fibres. The *internal surface* presents two articular facets, separated by a rough depression: the anterior one, sometimes divided into two, articulates with the outer side of the base of the second metatarsal bone; the posterior one skirts the posterior border, and articulates with the middle cuneiform; the rough depression between the two gives attachment to an interosseous ligament. The *external surface* also presents two articular facets, separated by a rough non-articular surface; the anterior facet, situated at the superior angle of the bone, is small, and articulates with the inner side of the base of the fourth metatarsal bone; the posterior and larger one articulates with the cuboid; the rough, non-articular surface serves for the attachment of an interosseous ligament. The three facets for articulation with the three metatarsal bones are continuous with one another, and covered by a prolongation of the same cartilage; the facets for articulation with the middle cuneiform and navicular are also continuous, but that for articulation with the cuboid is usually separate. The *superior* or *dorsal surface* is of an oblong square form; its posterior external angle being prolonged backwards. The *inferior* or *plantar surface* is an obtuse rounded margin, and serves for the attachment of part of the tendon of the Tibialis posticus, part of the Flexor brevis hallucis, and ligaments.

To ascertain to which side the bone belongs, hold it with the broad dorsal surface upwards, the prolonged edge backwards; the separate articular facet for the cuboid will point to the proper side.

Articulations.—With six bones : the navicular, middle cuneiform, cuboid, and second, third, and fourth metatarsal bones.

Attachment of Muscles.—To two : part of the Tibialis posticus, and Flexor brevis hallucis.

The number of tarsal bones may be reduced owing to congenital ankylosis, which may occur between the os calcis and cuboid, the os calcis and navicular, the os calcis and astragalus, or the astragalus and navicular.

The Metatarsal Bones

The **Metatarsal Bones** are five in number, and are numbered one to five, in accordance with their position from within outwards ; they are long bones, and present for examination a shaft and two extremities.

Common Characters of the Metatarsal Bones

The **Shaft** is prismoid in form, tapers gradually from the tarsal to the phalangeal extremity, and is slightly curved longitudinally, so as to be concave below, slightly convex above.   The *posterior extremity*, or *base*, is wedge-shaped, articulating by its terminal surface with the tarsal bones, and by its lateral surfaces with the contiguous metatarsal bones : its dorsal and plantar surfaces being rough for the attachment of ligaments.   The *anterior extremity*, or *head*, presents a terminal rounded articular surface, oblong from above downwards, and extending farther backwards below than above.   Its sides are flattened, and present a depression, surmounted by a tubercle, for ligamentous attachment. Its under surface is grooved in the middle line for the passage of the Flexor tendon, and marked on each side by an articular eminence continuous with the terminal articular surface.

Peculiar Characters of the Metatarsal Bones

The **First** (fig. 146) is remarkable for its great thickness, but is the shortest of all the metatarsal bones.   The *shaft* is strong, and of well-marked prismoid form.   The *posterior extremity* presents, as a rule, no lateral articular facets, but occasionally on the outer side there is an oval facet, by which it articulates with the second metatarsal bones. Its terminal articular surface is of large size, kidney-shaped ; its circumference is grooved, for the tarso-metatarsal ligaments, and internally gives attachment to part of the tendon of the Tibialis anticus ; its inferior angle presents a rough oval prominence for the insertion of the tendon of the Peroneus longus.   The *head* is of large size ; on its plantar surface are two grooved facets, over which glide sesamoid bones ; the facets are separated by a smooth elevated ridge.

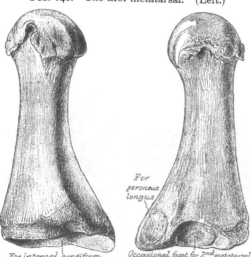

Fig. 146.—The first metatarsal.   (Left.)

For peroneus longus

For internal cuneiform   Occasional facet for 2nd metatarsal

This bone is known by the single kidney-shaped articular surface on its base ; the deeply grooved appearance of the plantar surface of its head ; and its great thickness, relatively to its length.   When it is placed in its natural position, the concave border of the kidney-shaped articular surface on its base points to the side to which the bone belongs.

Attachment of Muscles.—To three : part of the Tibialis anticus, the Peroneus longus, and the First dorsal interosseous.

The **Second** (fig. 147) is the longest and largest of the remaining metatarsal bones, being prolonged backwards into the recess formed between the three cuneiform bones. Its *tarsal extremity* is broad above, narrow and rough below. It presents four articular surfaces : one behind, of a triangular form, for articulation with the middle cuneiform ; one at the upper part of its internal lateral surface, for articulation with the internal cuneiform ; and two on its external lateral surface, an upper and lower, separated by a rough non-articular interval. Each of these articular surfaces is divided by a vertical ridge into two facets, thus making four facets ; the two anterior of these articulate with the third metatarsal ; the two posterior (sometimes continuous) with the external cuneiform.

FIG. 147.—The second metatarsal. (Left.)   FIG. 148.—The third metatarsal. (Left.)

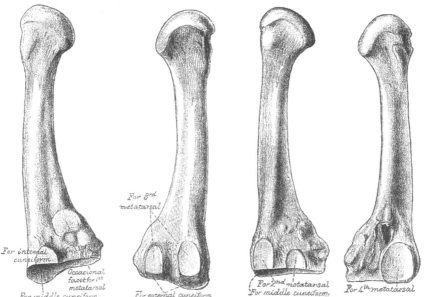

In addition to these articular surfaces, there is occasionally a fifth when this bone articulates with the first metatarsal bone. It is oval in shape, and is situated on the inner side of the shaft near the base.

The facets on the tarsal extremity of the second metatarsal bone serve at once to distinguish it from the rest, and to indicate the foot to which it belongs ; there being one facet at the upper angle of the internal surface, and two facets, each subdivided into two parts, on the external surface, pointing to the side to which the bone belongs. The fact that the two posterior subdivisions of these external facets sometimes run into one should not be forgotten.

Attachment of Muscles.—To four : the Adductor obliquus hallucis, First and Second dorsal interosseous, and a slip from the tendon of the Tibialis posticus, occasionally also a slip from the Peroneus longus.

The **Third** (fig. 148) articulates behind, by means of a triangular smooth surface, with the external cuneiform ; on its inner side, by two facets, with the second metatarsal ; and on its outer side, by a single facet, with the fourth metatarsal bone. The latter facet is of circular form, and situated at the upper angle of the base.

The third metatarsal bone is known by its having at its tarsal end two undivided facets on the inner side, and a single facet on the outer. This distinguishes it from the second metatarsal, in which the two facets, found on one side of its tarsal end, are each subdivided into two. The single facet (when the bone is put in its natural position) is on the side to which the bone belongs.

Attachment of Muscles.—To five : Adductor obliquus hallucis, Second and Third dorsal, and First plantar interosseous, and a slip from the tendon of the Tibialis posticus.

The **Fourth** (fig. 149) is smaller in size than the preceding ; its *tarsal extremity* presents an oblique quadrilateral surface for articulation with the cuboid ; a smooth facet on the inner side, divided by a ridge into an anterior portion for articulation with the third metatarsal, and a posterior portion for articulation with the external cuneiform ; on the outer side a single facet, for articulation with the fifth metatarsal.

The fourth metatarsal is known by its having a single facet on either side of the tarsal extremity, that on the inner side being divided into two parts. If this

FIG. 149.—The fourth metatarsal. (Left.) FIG. 150.—The fifth metatarsal. (Left.)

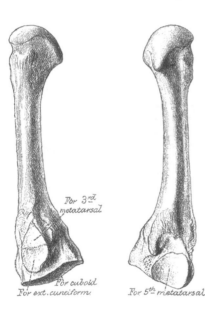

For 3rd metatarsal

For cuboid
For ext. cuneiform   For 5th metatarsal

For 4th metatarsal
For cuboid   Tuberosity

subdivision be not-recognisable, the fact that its tarsal end is bent somewhat outwards will indicate the side to which it belongs.

Attachment of Muscles.—To five : Adductor obliquus hallucis, Third and Fourth dorsal, and Second plantar interosseous, and a slip from the tendon of the Tibialis posticus.

The **Fifth** (fig. 150) is recognised by the tubercular eminence on the outer side of its base. It articulates behind, by a triangular surface cut obliquely from without inwards, with the cuboid ; and internally, with the fourth metatarsal.

The projection on the outer surface of this bone at its tarsal end at once distinguishes it from the others, and points to the side to which it belongs.

Attachment of Muscles.—To six : the Peroneus brevis, Peroneus tertius, Flexor brevis minimi digiti, Adductor transversus hallucis, Fourth dorsal, and Third plantar.

Articulations.—Each bone articulates with the tarsal bones by one extremity, and by the other with the first row of phalanges. The number of tarsal bones with which each metatarsal articulates is one for the first, three for the second, one for the third, two for the fourth, and one for the fifth.

PHALANGES

The **Phalanges** of the foot, both in number and general arrangement, resemble those in the hand ; there being two in the great toe, and three in each of the other toes.

The phalanges of the *first row* closely resemble those of the hand. The *shaft* is compressed from side to side, convex above, concave below. The *posterior extremity* is concave ; and the *anterior extremity* presents a trochlear surface, for articulation with the second phalanges.

The phalanges of the *second row* are remarkably small and short, but rather broader than those of the first row.

The *ungual* phalanges, in form, resemble those of the fingers ; but they are smaller, flattened from above downwards, presenting a broad base for articulation with the second row, and an expanded extremity for the support of the nail and end of the toe.

Articulations.—The first row, with the metatarsal bones behind and second phalanges in front ; the second row, of the four outer toes, with the first and third phalanges ; of the great toe, with the first phalanx ; the third row, of the four outer toes, with the second phalanges.

Attachment of Muscles.—To the first phalanges. Great toe, five muscles : innermost tendon of Extensor brevis digitorum, Abductor hallucis, Adductor obliquus hallucis, Flexor brevis hallucis, Adductor transversus hallucis. Second toe, three muscles : First and Second dorsal interosseous, and First lumbrical. Third toe, three muscles : Third dorsal and First plantar interosseous, and Second lumbrical. Fourth toe, three muscles : Fourth dorsal and Second plantar inter-osseous, and Third lumbrical. Fifth toe, four muscles : Flexor brevis minimi digiti, Abductor minimi digiti, and Third plantar interosseous, and Fourth lum-brical.—Second phalanges. Great toe : Extensor longus hallucis, Flexor longus hallucis. Other toes : Flexor brevis digitorum, one slip of the common tendon of the Extensor longus and brevis digitorum.*—Third phalanges : two slips from the common tendon of the Extensor longus and Extensor brevis digitorum, and the Flexor longus digitorum.

DEVELOPMENT OF THE FOOT (fig. 151)

The **Tarsal bones** are each developed by a single centre, excepting the os calcis, which has an epiphysis for its posterior extremity. The centres make their appearance in the following order : os calcis, at the sixth month of fœtal life ; astragalus, about the seventh month ; cuboid, at the ninth month ; external cuneiform, during the first year ; internal cuneiform, in the third year ; middle cuneiform and navicular in the fourth year. The epiphysis for the posterior tuberosity of the os calcis appears at the tenth year, and unites with the rest of the bone soon after puberty.

The **Metatarsal bones** are each developed by *two* centres : one for the shaft, and one for the digital extremity, in the four outer metatarsal ; one for the shaft, and one for the base, in the metatarsal bone of the great toe.† Ossification com-mences in the centre of the shaft about the ninth week, and extends towards either extremity. The centre in the proximal end of the first metatarsal bone appears about the third year ; the centre in the distal end of the other bones between the fifth and eighth years ; they become joined between the eighteenth and twentieth years.

The **Phalanges** are developed by *two* centres for each bone ; one for the shaft,

---

* Except the second phalanx of the fifth toe, which receives no slip from the Extensor brevis digitorum.

† As was noted in the first metacarpal bone, so in the first metatarsal, there is often to be observed a tendency to the formation of a second epiphysis in the distal extremity (see footnote, p. 142).

and one for the metatarsal extremity. The centre for the shaft appears about the tenth week, that for the epiphysis between the fourth and tenth years; they join the shaft about the eighteenth year.

CONSTRUCTION OF THE FOOT AS A WHOLE

The foot is constructed on the same principles as the hand, but modified to form a firm basis of support for the rest of the body when in the erect position. It is more solidly built up, and its component parts are less movable on each other than in the hand. This is especially the case with the great toe, which has to assist in supporting the body, and is therefore constructed with greater solidity;

FIG. 151.—Plan of the development of the foot.

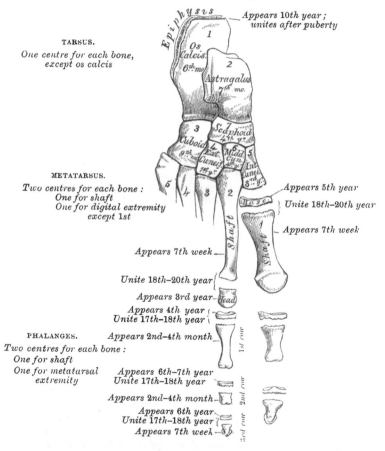

it lies parallel with the other toes, and has a very limited degree of mobility, whereas the thumb, which is occupied in numerous and varied movements, is constructed in such a manner as to permit of great mobility; its metacarpal bone is directed away from the others, so as to form an acute angle with the second, and it enjoys a considerable range of motion at its articulation with the carpus. The foot is placed at right angles to the leg—a position which is almost peculiar to man, and has relation to the erect position which he maintains. In order to allow of its supporting the weight of the whole body in this position, with the least expenditure of material, it is constructed in the form of an arch. This arch is not, however, made up of two equal limbs. The hinder one, which is made up of the os calcis and the posterior part of the astragalus, is about half the length of the anterior limb, and measures about three inches. The anterior limb consists of the rest of the tarsal and the metatarsal bones, and measures about seven inches. It may be said to consist of two parts, an inner segment made up of the head of the astragalus, the navicular, the three cuneiforms, and the three inner metatarsal bones ; and an outer segment composed of the os calcis, the cuboid and the two outer metatarsal bones. The summit of the arch is at the superior articular surface of the astragalus ; and its two extremities, that is to say the two piers on which the arch rests in standing, are the tubercles on the under surface of the os calcis posteriorly, and the heads of the metatarsal bones anteriorly. The weakest part of the arch is the joint between the astragalus and scaphoid, and here it is more liable to yield in those who are overweighted, and in those in whom the ligaments which complete and preserve the arch are relaxed. This weak point in the arch is braced on its concave surface by the inferior calcaneo-navicular ligament, which is more elastic than most other ligaments, and thus allows the arch to yield from jars or shocks applied to the anterior portion of the foot, and quickly restores it to its pristine condition. This ligament is supported internally by blending with the Deltoid ligament, and inferiorly by the tendon of the Tibialis posticus muscle, which is spread out into a fan-shaped insertion, and prevents undue tension of the ligament or such an amount of stretching as would permanently elongate it.

In addition to this longitudinal arch the foot presents a transverse arch, at the anterior part of the tarsus and hinder part of the metatarsus. This, however, can scarcely be described as a true arch, but presents more the character of a half-dome. The inner border of the central portion of the longitudinal arch is elevated from the ground, and from this point the bones arch over to the outer border, which is in contact with the ground, and, assisted by the longitudinal arch, produce a sort of rounded niche on the inner side of the foot, which gives the appearance of a transverse as well as a longitudinal arch.

The line of the foot, from the point of the heel to the toes, is not quite straight, but is directed a little outwards, so that the inner border is a little convex and the outer border concave. This disposition of the bones becomes more marked when the longitudinal arch of the foot is lost, as in the disease known under the name of ' flat-foot.'

Surface Form.—On the dorsum of the foot the individual bones are not to be distinguished, with the exception of the head of the astragalus, which forms a rounded projection in front of the ankle-joint when the foot is forcibly extended. The whole surface forms a smooth convex outline, the summit of which is the ridge formed by the head of the astragalus, the navicular, the middle cuneiform, and the second metatarsal bones ; from this it gradually inclines outwards and more rapidly inwards. On the inner side of the foot, the internal tuberosity of the os calcis and the ridge separating the inner from the posterior surface of the bone may be felt most posteriorly. In front of this, and below the internal malleolus, may be felt the projection of the sustentaculum tali. Passing forwards is the well-marked tuberosity of the navicular bone, situated about an inch or an inch and a quarter in front of the internal malleolus. Farther towards the front, the ridge formed by the base of the first metatarsal bone can be obscurely felt, and from this the shaft of the bone can be traced to the expanded head articulating with the base of the first phalanx of the great toe. Immediately beneath the base of this phalanx, the internal sesamoid bone is to be felt. Lastly, the expanded ends of the bones forming the last joint of the great toe are to be felt. On the outer side of the foot the most posterior bony point is the outer tuberosity of the os calcis, with the ridge separating the posterior from the outer surface of the bone. In front of this the greater part of the external surface of the os calcis is subcutaneous ; on it, below and in front of the external malleolus, may be felt the peroneal ridge, when this process is present. Farther forwards, the base of the fifth metatarsal bone forms a prominent and well-defined landmark, and in front of this the shaft of the bone, with its expanded head, and the base of the first phalanx, may be defined. The sole of the foot is almost entirely covered by soft parts, so that but few bony parts are to be made out, and these somewhat obscurely. The hinder part of the under surface of the os calcis and the heads of the metatarsal bones, with the exception of the first, which is concealed by the sesamoid bones, may be recognised.

Surgical Anatomy.—Considering the injuries to which the foot is subjected, it is surprising how seldom the tarsal bones are fractured. This is no doubt due to the fact that the tarsus is composed of a number of bones, articulated by a considerable extent of surface and joined together by very strong ligaments, which serve to break the force of violence applied to this part of the body. When fracture does occur, these bones being composed for the most part of a soft cancellous structure, covered only by a thin shell of compact tissue, are often extensively comminuted, especially as most of the fractures are produced by direct violence ; and, having only a very scanty amount of soft parts over them, the fractures are very often compound, and amputation is frequently necessary.

When fracture occurs in the anterior group of tarsal bones, it is almost invariably the result of direct violence ; but fractures of the posterior group, that is, of the calcaneum and astragalus, are most frequently produced by falls from a height on to the feet ; though fracture of the os calcis may be caused by direct violence or by muscular action. The posterior part of the bone, that is, the part behind the articular surfaces, is the most frequent seat of the fracture, though some few cases of fracture of the sustentaculum tali and of vertical fracture between the two articulating facets have been recorded. The neck of the astragalus, being the weakest part of the bone, is most frequently fractured, though fractures may occur in any part and almost in any direction. either associated or not with fracture of other bones.

In cases of club-foot, especially in congenital cases, the bones of the tarsus become altered in shape and size, and displaced from their proper positions. This is especially the case in congenital equino-varus, in which the astragalus, particularly about the head. becomes twisted and atrophied, and a similar condition may be present in the other bones. more especially the navicular. The tarsal bones are peculiarly liable to become the seat of tuberculous caries from comparatively trivial injuries. There are several reasons to account for this. They are composed of a delicate cancellated structure, surrounded by intricate synovial membranes. They are situated at the farthest point from the central organ of the circulation and exposed to vicissitudes of temperature ; and, moreover, on their dorsal surface are thinly clad with soft parts which have but a scanty blood-supply. And finally, after slight injuries, they are not maintained in a condition of rest to the same extent as some other parts of the body after similar injuries. Caries of the calcaneum or astragalus may remain limited to the one bone for a long period, but when one of the other bones is affected, the remainder frequently become involved, in consequence of the disease spreading through the large and complicated synovial membrane which is more or less common to these bones.

Amputation of the whole or a part of the foot is frequently required either for injury or disease. The principal amputations are as follows : (1) Syme's : amputation at the ankle-joint by a heel-flap, with removal of the malleoli and sometimes a thin slice from the lower end of the tibia. (2) Roux's : amputation at the ankle-joint by a large internal flap. (3) Pirogoff's amputation : removal of the whole of the tarsal bones, except the posterior part of the os calcis, and a thin slice from the tibia and fibula including the two malleoli. The sawn surface of the os calcis is then turned up and united to the similar surface of the tibia. (4) Subastragaloid amputation : removal of the foot below the astragalus through the joint between it and the os calcis. This operation has been modified by Hancock, who leaves the posterior third of the os calcis and turns it up against the denuded surface of the astragalus. This latter operation is of doubtful utility, and is rarely performed. (5) Chopart's or medio-tarsal : removal of the anterior part of the foot with all the tarsal bones except the os calcis and astragalus ; disarticulation being effected through the astragalo-scaphoid and calcaneo-cuboid joints. (6) Lisfranc's : amputation of the anterior part of the foot through the tarso-metatarsal joints. This has been modified by Hey, who disarticulated through the joints of the four outer metatarsal bones with the tarsus, and sawed off the projecting internal cuneiform ; and by Skey, who sawed off the base of the second metatarsal bone and disarticulated the others.

The bones of the tarsus occasionally require removal individually. This is especially the case with the astragalus and os calcis for disease limited to the one bone, or again the astragalus may require excision in cases of subastragaloid dislocation, or, as recommended by Mr. Lund, in cases of inveterate talipes. The cuboid has been removed for the same reason by Mr. Solly. But both these two latter operations have fallen very much into disuse, and have been superseded by resection of a wedge-shaped piece of bone from the outer side of the tarsus. Finally, Mickulicz and Watson have devised operations for the removal of more extensive portions of the tarsus. Mickulicz's operation consists in the removal of the os calcis and astragalus, along with the articular surfaces of the tibia and fibula, and also of the scaphoid and cuboid. The remaining portion of the tarsus is then brought into contact with the sawn surfaces of the tibia and fibula, and fixed there. The result is a position of the shortened foot resembling talipes equinus. Watson's operation is adapted to those cases where the disease is confined to the anterior tarsal bones. By two lateral incisions he saws through the bases of the metatarsal bones in front and opens up the joints between the scaphoid and astragalus, and the cuboid and os calcis, and removes the intervening bones.

The metatarsal bones and phalanges are nearly always broken by direct violence, and in the majority of cases the injury is the result of severe crushing accidents, necessitating amputation. The metatarsal bones, and especially that of the great toe, are frequently diseased, either in tuberculous subjects or in perforating ulcer of the foot.

SESAMOID BONES

These are small rounded masses, cartilaginous in early life, osseous in the adult, which are developed in those tendons which exert a great amount of pressure upon the parts over which they glide. It is said that they are more commonly found in the male than in the female, and in persons of an active muscular habit than in those who are weak and debilitated. They are invested throughout their whole surface by the fibrous tissue of the tendon in which they are found, excepting upon that side which lies in contact with the part over which they play, where they present a free articular facet. They may be divided into two kinds: those which glide over the articular surfaces of joints, and those which play over the cartilaginous facets found on the surfaces of certain bones.

The sesamoid bones of the joints in the upper extremity are two on the palmar surface of the metacarpo-phalangeal joint in the thumb, developed in the tendons of the Flexor brevis pollicis; occasionally one or two opposite the metacarpo-phalangeal articulations of the fore and little fingers; and, still more rarely, one opposite the same joints of the third and fourth fingers. In the lower extremity, the patella, which is developed in the tendon of the Quadriceps extensor; two small sesamoid bones, found in the tendons of the Flexor brevis hallucis, opposite the metatarso-phalangeal joint of the great toe; and occasionally one in the metatarso-phalangeal joint of the second toe, the little toe, and, still more rarely, the third and fourth toes.

Those found in the tendons which glide over certain bones occupy the following situations: one sometimes found in the tendon of the Biceps cubiti, opposite the tuberosity of the radius; one in the tendon of the Peroneus longus, where it glides through the groove in the cuboid bone; one, appearing late in life, in the tendon of the Tibialis anticus, opposite the smooth facet of the internal cuneiform bone; one is found in the tendon of the Tibialis posticus, opposite the inner side of the head of the astragalus; one in the outer head of the Gastrocnemius, behind the outer condyle of the femur; and one in the conjoined tendon of the Psoas and Iliacus, where it glides over the os pubis. Sesamoid bones are found occasionally in the tendon of the Gluteus maximus, as it passes over the great trochanter; and in the tendons which wind round the inner and outer malleoli.

THE ARTICULATIONS

THE various bones of which the Skeleton consists are connected together at different parts of their surfaces, and such a connection is designated by the name of *Joint* or *Articulation*. If the joint is *immovable*, as between the cranial and most of the facial bones, the adjacent margins of the bones are applied in almost close contact, a thin layer of fibrous membrane, the *sutural ligament*, and, at the base of the skull, in certain situations, a thin layer of cartilage being interposed. Where slight movement is required, combined with great strength, the osseous surfaces are united by tough and elastic fibro-cartilages, as in the joints between the vertebral 'bodies and in the interpubic articulation ; but in the *movable* joints, the bones forming the articulation are generally expanded for greater convenience of mutual connection, covered by *cartilage*, held together by strong bands or capsules of fibrous tissue, called *ligaments*, and partially lined by a membrane, the *synovial membrane*, which secretes a fluid to lubricate the various parts of which the joint is formed : so that the structures which enter into the formation of a movable joint are bone, cartilage, fibro-cartilage, ligament, and synovial membrane.

Bone constitutes the fundamental element of all the joints. In the long bones, the extremities are the parts which form the articulations ; they are generally somewhat enlarged, consisting of spongy cancellous tissue with a thin coating of compact substance. In the flat bones, the articulations usually take place at the edges ; and in the short bones at various parts of their surface. The layer of compact bone which forms the articular surface, and to which the cartilage is attached, is called the *articular lamella*. It is of white colour, extremely dense, and varies in thickness. Its structure differs from ordinary bone-tissue in this respect, that it contains no Haversian canals, and its lacunæ are much larger than in ordinary bone, and have no canaliculi. The vessels of the cancellous tissue, as they approach the articular lamella, turn back in loops, and do not perforate it ; this layer is consequently denser and firmer than ordinary bone, and is evidently designed to form an unyielding support for the articular cartilage.

The **cartilage,** which covers the articular surfaces of bone, and is called *articular*, will be found described, with the other varieties of cartilage, in the section on General Anatomy.

Ligaments consist of bands of various forms, serving to connect together the articular extremities of bones, and composed mainly of bundles of *white fibrous tissue* placed parallel with, or closely interlaced with one another, and presenting a white, shining silvery aspect. A ligament is pliant and flexible, so as to allow of the most perfect freedom of movement, but strong, tough, and inextensile, so as not to yield readily under the most severely applied force ; it is consequently well adapted to serve as the connecting medium between the bones. Some ligaments consist entirely of *yellow elastic tissue*, as the ligamenta subflava, which connect together the laminæ of adjacent vertebræ, and the ligamentum nuchæ in the lower animals. In these cases it will be observed that the elasticity of the ligament is intended to act as a substitute for muscular power.

Synovial membrane is a thin, delicate membrane of connective tissue, with branched connective-tissue corpuscles. Its secretion is thick, viscid, and glairy, like the white of an egg ; and is hence termed *synovia*. The synovial membranes found in the body admit of subdivision into three kinds—articular, bursal, and vaginal.

The *articular synovial membranes* are found in all the freely movable joints. In the fœtus this membrane is said, by Toynbee, to be continued over the surface of the cartilages ; but in the adult it is wanting, excepting at their circumference, upon which it encroaches for a short distance and to which it is firmly attached : it then invests the inner surface of the capsular or other ligaments enclosing the joint, and is reflected over the surface of any tendons passing through its cavity, as the tendon of the Popliteus in the knee, and the tendon of the Biceps in the shoulder. Hence the articular synovial membrane may be regarded as a short wide tube, attached by its open ends to the margins of the articular cartilages and covering the inner surface of the various ligaments which connect the articular surfaces, so that along with the cartilages it completely encloses the joint cavity. In some of the joints the synovial membrane is thrown into folds, which pass across the cavity. They are called *synovial ligaments*, and are especially distinct in the knee. In other joints there are flattened folds, subdivided at their margins into fringe-like processes, the vessels of which have a convoluted arrangement. These latter generally project from the synovial membrane near the margin of the cartilage, and lie flat upon its surface. They consist of connective tissue, covered with endothelium, and contain fat-cells in variable quantities, and, more rarely, isolated cartilage-cells. The larger folds often contain considerable quantities of fat. They were described, by Clopton Havers, as *mucilaginous glands*, and as the source of the synovial secretion. Under certain diseased conditions, similar processes are found covering the entire surface of the synovial membrane, forming a mass of pedunculated fibro-fatty growths, which projects into the joint. Similar structures are also found in some of the bursal and vaginal synovial membranes.

The *bursal synovial membranes* are found interposed between surfaces which move upon each other, producing friction, as in the gliding of a tendon, or of the integument over projecting bony surfaces. They admit of subdivision into two kinds, the *bursæ mucosæ* and the *bursæ synoviæ*. The *bursæ mucosæ* are large, simple, or irregular cavities in the subcutaneous areolar tissue, enclosing a clear viscid fluid. They are found in various situations, as between the integument and the front of the patella, over the olecranon, the malleoli, and other prominent parts. The *bursæ synoviæ* are found interposed between muscles or tendons as they play over projecting bony surfaces, as between the Glutei muscles and the surface of the great trochanter. They consist of a thin wall of connective tissue, partially covered by patches of cells, and contain a viscid fluid. Where one of these exists in the neighbourhood of a joint, it may communicate with its cavity, as in the case of the bursa between the tendon of the Psoas and Iliacus and the capsular ligament of the hip, or the one interposed between the under surface of the Subscapularis and the neck of the scapula.

The *vaginal synovial membranes* (*synovial sheaths*) serve to facilitate the gliding of tendons in the osseo-fibrous canals through which they pass. The membrane is here arranged in the form of a sheath, one layer of which adheres to the wall of the canal, and the other is reflected upon the surface of the contained tendon ; the space between the two free surfaces of the membrane containing synovia. These sheaths are chiefly found surrounding the tendons of the flexor and extensor muscles of the fingers and toes, as they pass through the osseo-fibrous canals in the hand or foot.

Synovia is a transparent, yellowish-white, or slightly reddish fluid, viscid like the white of egg, having an alkaline reaction and slightly saline taste. It consists, according to Frerichs, in the ox, of 94·85 water, 0·56 mucus and epithelium, 0·07 fat, 3·51 albumen and extractive matter, and 0·99 salts.

The articulations are divided into three classes : *synarthrosis*, or immovable ; *amphiarthrosis*, or mixed ; and *diarthrosis*, or movable joints.

1. Synarthrosis. Immovable Articulations

Synarthrosis includes all those articulations in which the surfaces of the bones are in almost direct contact, fastened together by an intervening mass of connective tissue, and in which there is no appreciable motion, as the joints between the bones of the cranium and face, excepting those of the lower jaw. The varieties of synarthrosis are four in number : Sutura, Schindylesis, Gomphosis, and Synchondrosis.

Sutura (*a seam*) is that form of articulation where the contiguous margins of flat bones are united by a thin layer of fibrous tissue. It is met with only in the skull. When the articulating surfaces are connected by a series of processes and indentations interlocked together, it is termed *sutura vera*; of which there are three varieties : sutura dentata, serrata, and limbosa. The surfaces of the bones are not in direct contact, being separated by a layer of membrane, continuous externally with the pericranium, internally with the dura mater. The *sutura dentata* (dens, *a tooth*) is so called from the tooth-like form of the projecting articular processes, as in the suture between the parietal bones. In the *sutura serrata* (serra, *a saw*) the edges of the two bones forming the articulation are serrated like the teeth of a fine saw, as between the two portions of the frontal bone. In the *sutura limbosa* (limbus, *a selvage*), besides the dentated processes, there is a certain degree of bevelling of the articular surfaces, so that the bones overlap one another, as in the suture between the parietal and frontal bones. When the articulation is formed by roughened surfaces placed in apposition with one another, it is termed the *false suture* (*sutura notha*), of which there are two kinds, the *sutura squamosa* (squama, *a scale*), formed by the overlapping of two contiguous bones by broad bevelled margins, as in the squamo-parietal (squamous) suture ; and the *sutura harmonia* (ἁρμονία, *a joining together*), where there is simple apposition of two contiguous rough bony surfaces, as in the articulation between the two superior maxillary bones, or of the horizontal plates of the palate bones.

Schindylesis (σχινδύλησις, *a fissure*) is that form of articulation in which a thin plate of bone is received into a cleft or fissure formed by the separation of two laminæ in another bone, as in the articulation of the rostrum of the sphenoid and perpendicular plate of the ethmoid with the vomer, or in the reception of the latter in the fissure between the superior maxillary and palate bones.

Gomphosis (γόμφος, *a nail*) is an articulation formed by the insertion of a conical process into a socket, as a nail is driven into a board ; this is not illustrated by any articulation between bones, properly so called, but is seen in the articulation of the teeth with the alveoli of the maxillary bones.

Synchondrosis.—Where the connecting medium is cartilage the joint is termed a synchondrosis. This is a temporary form of joint, for the cartilage becomes converted into bone before adult life. Such a joint is found between the epiphyses and shafts of long bones.

2. Amphiarthrosis. Mixed Articulations

In this form of articulation, the contiguous osseous surfaces are either connected together by broad flattened discs of fibro-cartilage, of a more or less complex structure, which adhere to the end of each bone, as in the articulation between the bodies of the vertebræ, and the pubic symphysis. This is termed **Symphysis**. Or, secondly, the bony surfaces are united by an interosseous ligament, as in the inferior tibio-fibular articulation. To this the term **Syndesmosis** is applied.

3. Diarthrosis. Movable Articulations

This form of articulation includes the greater number of the joints in the body, mobility being their distinguishing character. They are formed by the approximation of two contiguous bony surfaces, covered with cartilage, connected by ligaments, and lined by synovial membrane. The varieties of joints in this class have been determined by the kind of motion permitted in each. There are two varieties in which the movement is uniaxial, that is to say, all movements take place around one axis. In one form, the Ginglymus, this axis is practically speaking transverse ; in the other, the trochoid or pivot-joint, it is longitudinal. There are two varieties where the movement is biaxial, or around two horizontal axes at right angles to each other, or at any intervening axis between the two. These are the condyloid and saddle-joint. There is one form of joint where the movement is polyaxial, the enarthrosis or ball-and-socket joint. And finally there are the Arthrodia or Gliding joints.

Ginglymus or **Hinge-joint** (γίγγλυμος, *a hinge*).—In this form of joint the articular surfaces are moulded to each other in such a manner as to permit motion only in one plane, forwards and backwards ; the extent of motion at the same time being considerable. The direction which the distal bone takes in this motion is never in the same plane as that of the axis of the proximal bone, but there is always a certain amount of alteration from the straight line during flexion. The articular surfaces are connected together by strong lateral ligaments, which form their chief bond of union. The most perfect forms of ginglymus are the interphalangeal joints and the joint between the humerus and ulna ; the knee and ankle are less perfect, as they allow a slight degree of rotation or lateral movement in certain positions of the limb.

Trochoides (*pivot-joint*).—Where the movement is limited to rotation, the joint is formed by a pivot-like process turning within a ring, or the ring on the pivot, the ring being formed partially of bone, partly of ligament. In the superior radioulnar articulation, the ring is formed partly by the lesser sigmoid cavity of the ulna ; in the rest of its extent, by the orbicular ligament ; here, the head of the radius rotates within the ring. In the articulation of the odontoid process of the axis with the atlas, the ring is formed in front by the anterior arch of the atlas ; behind, by the transverse ligament ; here the ring rotates round the odontoid process.

Condyloid Articulations.—In this form of joint, an ovoid articular head, or condyle, is received into an elliptical cavity in such a manner as to permit of flexion and extension, adduction and abduction and circumduction, but no axial rotation. The articular surfaces are connected together by anterior, posterior, and lateral ligaments. An example of this form of joint is found in the wrist.

Articulations by Reciprocal Reception (*saddle-joint*).—In this variety the articular surfaces are concavo-convex ; that is to say, they are inversely convex in one direction and concave in the other. The movements are the same as in the preceding form ; that is to say, there is flexion, extension, adduction, abduction, and circumduction, but no axial rotation. The articular surfaces are connected by a capsular ligament. The best example of this form of joint is the carpo-metacarpal joint of the thumb.

Enarthrosis is that form of joint in which the distal bone is capable of motion around an indefinite number of axes, which have one common centre. It is formed by the reception of a globular head into a deep cup-like cavity (hence the name ' ball-and-socket '), the parts being kept in apposition by a capsular ligament strengthened by accessory ligamentous bands. Examples of this form of articulation are found in the hip and shoulder.

Arthrodia is that form of joint which admits of a gliding movement; it is formed by the approximation of plane surfaces, or one slightly concave, the other slightly convex ; the amount of motion between them being limited by the ligaments, or osseous processes, surrounding the articulation ; as in the articular processes of the vertebræ, the carpal joints, except that of the os magnum with the scaphoid and semilunar bones, and the tarsal joints with the exception of the joint between the astragalus and the navicular.

On the next page, in a tabular form, are the names, distinctive characters, and examples of the different kinds of articulations.

THE KINDS OF MOVEMENT ADMITTED IN JOINTS

The movements admissible in joints may be divided into four kinds : gliding, angular movement, circumduction, and rotation. These movements are often, however, more or less combined in the various joints, so as to produce an infinite variety, and it is seldom that we find only one kind of motion in any particular joint.

Synarthrosis, or Immovable Joint. Surfaces separated by fibrous membrane, without any intervening synovial cavity, and immovably connected with each other. As in joints of cranium and face (except the lower jaw).

Sutura. Articulation by processes and indentations interlocked together.

Sutura vera (true) articulate by indented borders.

Dentata, having tooth-like processes. As in interparietal suture.

Serrata, having serrated edges like the teeth of a saw. As in interfrontal suture.

Limbosa, having bevelled margins, and dentated processes. As in fronto-parietal suture.

Sutura notha (false) articulate by rough surfaces.

Squamosa, formed by thin bevelled margins, overlapping each other. As in squamo-parietal suture.

Harmonia, formed by the apposition of contiguous rough surfaces. As in intermaxillary suture.

Schindylesis.—Articulation formed by the reception of a thin plate of one bone into a fissure of another. As in articulation of rostrum of sphenoid with vomer.

Gomphosis.—Articulation formed by the insertion of a conical process into a socket. The teeth.

Amphiarthrosis, Mixed Articulation.

Symphysis.—Surfaces connected by fibro-cartilage, not separated by synovial membrane, and having limited motion. As in joints between bodies of vertebræ.

Syndesmosis.—Surfaces united by an interosseous ligament. As in the inferior tibio-fibular articulation.

Diarthrosis, Movable Joint.

Ginglymus.—Hinge-joint; motion limited to two directions, forwards and backwards. Articular surfaces fitted together so as to permit of movement in one plane. As in the inter-phalangeal joints and the joint between the humerus and the ulna.

Trochoides or Pivot-joint.—Articulation by a pivot process turning within a ring, or ring around a pivot. As in superior radio-ulnar articulation, and atlanto-axial joint.

Condyloid.—Ovoid head received into elliptical cavity. Movements in every direction except axial rotation. As the wrist-joint.

Reciprocal Reception (Saddle-joint).—Articular surfaces inversely convex in one direction and concave in the other. Movement in every direction except axial rotation. As in the carpo-metacarpal joint of the thumb.

Enarthrosis.—Ball-and-socket joint; capable of motion in all directions. Articulations by a globular head received into a cup-like cavity. As in hip- and shoulder-joints.

Arthrodia.—Gliding-joint; articulations by plane surfaces, which glide upon each other. As in carpal and tarsal articulations.

Gliding movement is the most simple kind of motion that can take place in a joint, one surface gliding or moving over another without any angular or rotatory movement. It is common to all movable joints ; but in some, as in most of the articulations of the carpus and tarsus, it is the only motion permitted. This movement is not confined to plane surfaces, but may exist between any two contiguous surfaces, of whatever form, limited by the ligaments which enclose the articulation.

Angular movement occurs only between the long bones, and by it the angle between the two bones is increased or diminished. It may take place in four directions : forwards and backwards, constituting flexion and extension, or inwards and outwards, from the mesial line of the body (or in the fingers and toes from the middle line of the hand or foot), constituting adduction and abduction. The strictly ginglymoid or hinge-joints admit of flexion and extension only. Abduction and adduction, combined with flexion and extension, are met with in the more movable joints ; as in the hip, shoulder, and metacarpal joint of the thumb, and partially in the wrist.

Circumduction is that limited degree of motion which takes place between the head of a bone and its articular cavity, while the extremity and sides of the limb are made to circumscribe a conical space, the base of which corresponds with the inferior extremity of the limb, the apex with the articular cavity ; this kind of motion is best seen in the shoulder- and hip-joints.

Rotation is the movement of a bone upon an axis, which is the axis of the pivot on which the bone turns, as in the articulation between the atlas and axis, where the odontoid process serves as a pivot around which the atlas turns ; or else is the axis of a pivot-like process which turns within a ring, as in the rotation of the radius upon the humerus.

Ligamentous Action of Muscles.—The movements of the different joints of a limb are combined by means of the long muscles which pass over more than one joint, and which, when relaxed and stretched to their greatest extent, act as elastic ligaments in restraining certain movements of one joint, except when combined with corresponding movements of the other—these latter movements being usually in the opposite direction. Thus the shortness of the hamstring muscles prevents complete flexion of the hip, unless the knee-joint is also flexed, so as to bring their attachments nearer together. The uses of this arrangement are threefold. 1. It co-ordinates the kinds of movement which are the most habitual and necessary, and enables them to be performed with the least expenditure of power. 'Thus in the usual gesture of the arms, whether in grasping or rejecting, the shoulder and the elbow are flexed simultaneously, and simultaneously extended,' in consequence of the passage of the Biceps and Triceps cubiti over both joints. 2. It enables the short muscles which pass over only one joint to act upon more than one. 'Thus, if the Rectus femoris remain tonically of such length that, when stretched over the extended hip, it compels extension of the knee, then the Gluteus maximus becomes, not only an extensor of the hip, but an extensor of the knee as well.' 3. It provides the joints with ligaments which, while they are of very great power in resisting movements to an extent incompatible with the mechanism of the joint, at the same time spontaneously yield when necessary. 'Taxed beyond its strength a ligament will be ruptured, whereas a contracted muscle is easily relaxed ; also, if neighbouring joints be united by ligaments, the amount of flexion or extension of each must remain in constant proportion to that of the other ; while, if the union be by muscles, the separation of the points of attachment of those muscles may vary considerably in different varieties of movement, the muscles adapting themselves tonically to the length required.' The quotations are from a very interesting paper by Dr. Cleland in the 'Journal of Anatomy and Physiology,' No. 1, 1866, p. 85 ; by whom I believe this important fact in the mechanism of joints was first clearly pointed out, though it has been independently observed afterwards by other anatomists. Dr. W. W. Keen points out how important it is 'that the surgeon should remember this ligamentous action of muscles in making passive motion—for instance, at the wrist after Colles' fracture. If the fingers be extended, the wrist can be flexed to a right angle. If, however, they be first flexed as in "making a fist," flexion at the wrist is quickly limited to from forty to fifty degrees in different persons, and is very painful beyond that point. Hence passive motion here should be made with the fingers extended. In the leg, when flexing the hip, the knee should be flexed.' Dr. Keen further points out that 'a beautiful illustration of this is seen in the perching of birds, whose toes are forced to clasp the perch by just such a passive ligamentous action so soon as they stoop. Hence they can go to sleep and not fall off the perch.'

The articulations may be arranged into those of the trunk, those of the upper extremity, and those of the lower extremity

ARTICULATIONS OF THE TRUNK

These may be divided into the following groups, viz. :—

I. Of the vertebral column.
II. Of the atlas with the axis.
III. Of the atlas with the occipital bone.
IV. Of the axis with the occipital bone.
V. Of the lower jaw.
VI. Of the ribs with the vertebræ.

VII. Of the cartilages of the ribs with the sternum, and with each other.
VIII. Of the sternum.
IX. Of the vertebral column with the pelvis.
X. Of the pelvis.

I. ARTICULATIONS OF THE VERTEBRAL COLUMN

The different segments of the spine are connected together by ligaments, which may be divided into five sets. 1. Those connecting the *bodies* of the vertebræ. 2. Those connecting the *laminæ*. 3. Those connecting the *articular processes*. 4. Those connecting the *spinous processes*. 5. Those connecting the *transverse processes*.

The articulations of the *bodies* of the vertebræ with each other form a series of amphiarthrodial joints ; those between the *articular processes* form a series of arthrodial joints.

1. THE LIGAMENTS OF THE BODIES

Anterior Common Ligament.   Posterior Common Ligament.
Intervertebral Substance.

The **Anterior Common Ligament** (figs. 152, 153, 160, 164) is a broad and strong band of fibres, which extends along the anterior surface of the bodies of the vertebræ, from the axis to the sacrum. It is broader below than above, thicker in the dorsal than in the cervical or lumbar regions, and somewhat thicker opposite the front of the body of each vertebra than opposite the intervertebral substance. It is attached, above, to the body of the axis by a pointed process, where it is continuous with the anterior atlanto-axial ligament, and is connected with the tendon of insertion of the Longus colli muscle, and extends down as far as the upper bone of the sacrum. It consists of dense longitudinal fibres, which are intimately adherent to the intervertebral substance, and the prominent margins of the vertebræ ; but less closely to the middle of the bodies. In the latter situation the fibres are exceedingly thick, and serve to fill up the concavities on their front surface, and to make the anterior surface of the spine more even. This ligament is composed of several layers of fibres, which vary in length, but are closely interlaced with each other. The most superficial or longest fibres extend between four or five vertebræ. A second, subjacent set extend between two or three vertebræ; while a third set, the shortest and deepest, extend from one vertebræ to the next. At the sides of the bodies the ligament consists of a few short fibres which pass from one vertebræ to the next, separated from the median portion by large oval apertures, for the passage of vessels.

The **Posterior Common Ligament** (figs. 152, 156) is situated within the spinal canal, and extends along the posterior surface of the bodies of the vertebræ, from the body of the axis above, where it is continuous with the occipito-axial ligament, to the sacrum below. It is broader above than below, and thicker in the dorsal than in the cervical or lumbar regions. In the situation of the intervertebral substance and contiguous margins of the vertebræ, where the ligament is more intimately adherent, it is broad, and presents a series of dentations with intervening concave margins; but it is narrow and thick over the centre of the bodies, from which it is separated by the *venæ basis vertebræ*. This ligament is composed of smooth, shining, longitudinal fibres, denser and more compact than those of the anterior ligament, and composed of a superficial layer occupying

FIG. 152.—Vertical section of two vertebræ and their ligaments, from the lumbar region.

Anterior common
ligament

Posterior
common
ligament

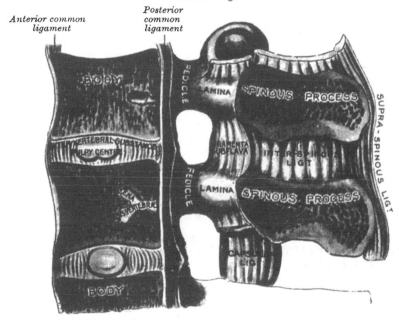

the interval between three or four vertebræ, and of a deeper layer which extends between one vertebra and the next adjacent to it. It is separated from the dura mater of the spinal cord by some loose connective tissue which is very liable to serous infiltration.

The **Intervertebral Substance** (figs. 152, 161) is a lenticular disc of composite structure interposed between the adjacent surfaces of the bodies of the vertebræ, from the axis to the sacrum, and forming the chief bond of connection between these bones. The discs vary in shape, size, and thickness, in different parts of the spine. In *shape* they accurately correspond with the surfaces of the bodies between which they are placed, being oval in the cervical and lumbar regions, and circular in the dorsal. Their *size* is greatest in the lumbar region. In *thickness* they vary not only in the different regions of the spine, but in different parts of the same disc: thus, they are thicker in front than behind in the cervical and lumbar regions, while they are uniformly thick in the dorsal region. The intervertebral discs form about one-fourth of the spinal column, exclusive of the first two vertebræ; they are not equally distributed, however, between the various bones; the dorsal portion of the spine having, in proportion to its length, a much smaller quantity than the cervical and lumbar regions, which necessarily gives to the latter parts greater pliancy and freedom of movement. The intervertebral discs are adherent, by their surfaces, to a thin layer of hyaline cartilage which covers the upper and under surfaces of the bodies of the vertebræ, and in which in early life the epiphysial plate develops; and by their circumference are closely connected in front to the anterior, and behind to the posterior common ligament; while in the dorsal region they are connected laterally, by means of the interarticular ligament, to the heads of those ribs which articulate with two vertebræ: they, consequently, form part of the articular cavities in which the heads of these bones are received.

Structure of the Intervertebral Substance.—The intervertebral substance is composed, at its circumference, of laminæ of fibrous tissue and fibro-cartilage; and, at its centre, of a soft, pulpy, highly elastic substance, of a yellowish colour, which rises up considerably above the surrounding level when the disc is divided horizontally. This pulpy substance, which is especially well developed in the lumbar region, is the remains of the chorda dorsalis, and, according to Luschka, contains a small synovial cavity in its centre. The laminæ are arranged concentrically one within the other, the outermost consisting of ordinary fibrous tissue, but the others and more numerous consisting of white fibro-cartilage. These plates are not quite vertical in their direction, those near the circumference being curved outwards and closely approximated; while those nearest the centre curve in the opposite direction, and are somewhat more widely separated. The fibres of which each plate is composed are directed, for the most part, obliquely from above downwards; the fibres of adjacent plates passing in opposite directions and varying in every layer: so that the fibres of one layer are directed across those of another, like the limbs of the letter X. This laminar arrangement belongs to about the outer half of each disc. The pulpy substance presents no concentric arrangement, and consists of a fine fibrous matrix, containing angular cells, united to form a reticular structure.

2. Ligaments connecting the Laminæ

Ligamenta Subflava.

The **Ligamenta Subflava** (fig. 152) are interposed between the laminæ of the vertebræ, from the axis to the sacrum. They are most distinct when seen from the interior of the spinal canal: when viewed from the outer surface they appear short, being overlapped by the laminæ. Each ligament consists of two lateral portions, which commence on each side at the root of either articular process, and pass backwards to the point where the laminæ converge to form the spinous process, where their margins are in contact and to a certain extent united; slight intervals being left for the passage of small vessels. These ligaments consist of yellow elastic tissue, the fibres of which, almost perpendicular in direction, are attached to the anterior surface of the lamina above, some distance from its inferior margin, and to the posterior surface, as well as to the margin of the lamina below. In the cervical region they are thin in texture, but very broad and long; they become thicker in the dorsal region; and in the lumbar acquire very considerable thickness. Their highly elastic property serves to preserve the upright posture, and to assist the spine in resuming it, after it has been flexed. These ligaments do not exist between the occiput and atlas, or between the atlas and axis.

3. Ligaments connecting the Articular Processes

Capsular.

The **Capsular Ligaments** (fig. 154) are thin and loose ligamentous sacs, attached to the contiguous margins of the articulating processes of each vertebra, through the greater part of their circumference, and completed internally by the ligamenta subflava. They are longer and looser in the cervical than in the dorsal or lumbar regions. The capsular ligaments are lined on their inner surface by synovial membrane.

4. LIGAMENTS CONNECTING THE SPINOUS PROCESSES

Supraspinous.   Interspinous.

The **Supraspinous Ligament** (fig. 152) is a strong fibrous cord, which connects together the apices of the spinous processes from the seventh cervical to the spinous processes of the sacrum. It is thicker and broader in the lumbar than in the dorsal region, and intimately blended, in both situations, with the neighbouring aponeurosis. The most superficial fibres of this ligament connect three or four vertebræ; those deeper-seated pass between two or three vertebræ; while the deepest connect the contiguous extremities of neighbouring vertebræ. It is continued upwards to the external occipital protuberance, as the ligamentum nuchæ, which, in the human subject, is thin, and forms merely an intermuscular septum.

The **Interspinous Ligaments** (fig. 152), thin and membranous, are interposed between the spinous processes. These ligaments extend from the root to the summit of each spinous process, connecting together their adjacent margins. They meet the ligamenta subflava in front and the supraspinous ligament behind. They are narrow and elongated in the dorsal region; broader, quadrilateral in form, and thicker in the lumbar region; and only slightly developed in the neck.

5. LIGAMENTS CONNECTING THE TRANSVERSE PROCESSES

Intertransverse.

The **Intertransverse Ligaments** consist of bundles of fibres, interposed between the transverse processes. In the cervical region they consist of a few irregular, scattered fibres; in the dorsal, they are rounded cords intimately connected with the deep muscles of the back; in the lumbar region they are thin and membranous.

Actions.—The movements permitted in the spinal column are, Flexion, Extension, Lateral Movement, Circumduction, and Rotation.

In *Flexion*, or movement of the spine forwards, the anterior common ligament is relaxed, and the intervertebral substances are compressed in front; while the posterior common ligament, the ligamenta subflava, and the inter- and supraspinous ligaments are stretched, as well as the posterior fibres of the intervertebral discs. The interspaces between the laminæ are widened, and the inferior articular processes of the vertebræ above glide upwards, upon the articular processes of the vertebræ below. Flexion is the most extensive of all the movements of the spine.

In *Extension*, or movement of the spine backwards, an exactly opposite disposition of the parts takes place. This movement is not extensive, being limited by the anterior common ligament, and by the approximation of the spinous processes.

Flexion and extension are free in the lower part of the lumbar region between the third and fourth and fourth and fifth lumbar vertebræ; above the third they are much diminished, and reach their minimum in the middle and upper part of the back. They increase again in the neck, the capability of motion backwards from the upright position being in this region greater than that of the motion forwards, whereas in the lumbar region the reverse is the case.

In *Lateral Movement*, the sides of the intervertebral discs are compressed, the extent of motion being limited by the resistance offered by the surrounding ligaments, and by the approximation of the transverse processes. This movement may take place in any part of the spine, but is most free in the neck and loins.

Circumduction is very limited, and is produced merely by a succession of the preceding movements.

Rotation is produced by the twisting of the intervertebral substances; this, although only slight between any two vertebræ, produces a considerable extent of movement, when it takes place in the whole length of the spine, the front of the upper part of the column being turned to one or the other side. This movement takes place only to a slight extent in the neck, but is freer in the upper part of the dorsal region, and is altogether absent in the lumbar region.

It is thus seen that the *cervical region* enjoys the greatest extent of each variety of movement, flexion and extension especially being very free. In the *dorsal region*, the three movements of flexion, extension, and circumduction are only permitted to a slight extent; while rotation is very free in the upper part and ceases below. In the lumbar region there is free flexion, extension, and lateral movement, but no rotation.

As Sir George Humphry has pointed out, the movements permitted are mainly due to the shape and position of the articular processes. In the loins the inferior articular processes are turned outwards and embraced by the superior; this renders rotation in this region of the spine impossible, while there is nothing to prevent a sliding upwards and downwards of the surfaces on each other so as to allow of flexion and extension. In the dorsal region, on the other hand, the articulating processes, by their direction and mutual adaptation, especially at the upper part of the series, permit of rotation, but prevent extension and flexion; while in the cervical region the greater obliquity and lateral slant of the articular processes allow not only flexion and extension, but also rotation.

The principal muscles which produce *flexion* are the Sterno-mastoid, Rectus capitis anticus major, and Longus colli; the Scaleni; the abdominal muscles and the Psoas magnus. *Extension* is produced by the fourth layer of the muscles of the back, assisted in the neck by the Splenius, Semispinalis dorsi et colli, and the Multifidus spinæ. *Lateral* motion is produced by the fourth layer of the muscles of the back, by the Splenius and the Scaleni, the muscles of one side only acting; and *rotation* by the action of the following muscles of one side only, viz. the Sterno-mastoid, the Rectus capitis anticus major, the Scaleni, the Multifidus spinæ, the Complexus, and the abdominal muscles.

II. Articulation of the Atlas with the Axis

The articulation of the Atlas with the Axis is of a complicated nature, comprising no fewer than four distinct joints. There is a pivot articulation between the odontoid process of the axis and the ring formed between the anterior arch of the atlas and the transverse ligament (see fig. 155). Here there are two joints: one in front between the posterior surface of the anterior arch of the atlas and the front of the odontoid process (the *atlanto-odontoid joint of Cruveilhier*); the other between the anterior surface of the transverse ligament and the back of the process (the *syndesmo-odontoid joint*). Between the articulating processes of the two bones there is a double arthrodia or gliding joint. The ligaments which connect these bones are the

Anterior Atlanto-axial.	Transverse.
Posterior Atlanto-axial.	Two Capsular.

The **Anterior Atlanto-axial Ligament** (fig. 153) is a strong membranous layer, attached, above, to the lower border of the anterior arch of the atlas; below, to the base of the odontoid process and front of the body of the axis. It is strengthened in the middle line by a rounded cord, which is attached, above, to the tubercle on the anterior arch of the atlas, and below to the body of the axis, being a continuation upwards of the anterior common ligament of the spine. These ligaments are in relation, in front, with the Recti antici majores.

The **Posterior Atlanto-axial Ligament** (fig. 154) is a broad and thin membranous layer, attached, above, to the lower border of the posterior arch of the atlas; below, to the upper edge of the laminæ of the axis. This ligament supplies the place of the ligamenta subflava, and is in relation, behind, with the Inferior oblique muscles.

The **Transverse Ligament** * (figs. 155, 156) is a thick, strong band, which arches across the ring of the atlas, and serves to retain the odontoid process in firm connection with its anterior arch. This ligament is flattened from before backwards, broader and thicker in the middle than at either extremity, and firmly attached on each side to a small tubercle on the inner surface of the lateral mass of the atlas. As it crosses the odontoid process, a small fasciculus is derived from its upper, and another from its lower border; the former passing upwards, to be inserted into the basilar process of the occipital bone; the latter, downwards, to be attached to the posterior surface of the body of the axis; hence, the whole ligament has received the name of *cruciform*. The transverse ligament divides the ring of the atlas into two unequal parts: of these, the posterior and larger

Fig. 153.—Occipito-atlantal and atlanto-axial ligaments. Front view.

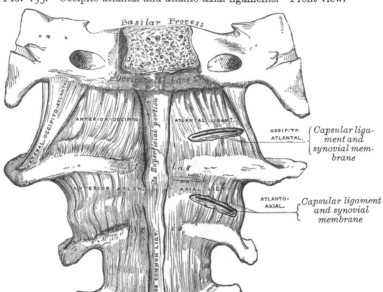

serves for the transmission of the cord and its membranes and the spinal accessory nerves; the anterior and smaller contains the odontoid process. Since the space between the anterior arch of the atlas and the transverse ligament is smaller at the lower part than the upper (because the transverse ligament embraces firmly the narrow neck of the odontoid process), this process is retained in firm connection with the atlas after all the other ligaments have been divided.

The **Capsular Ligaments** are two thin and loose capsules, connecting the lateral masses of the atlas with the superior articular surfaces of the axis, the fibres being strengthened at the posterior and inner part of the articulation by an *accessory ligament*, which is attached below to the body of the axis near the base of the odontoid process.

* It has been found necessary to describe the transverse ligament with those of the atlas and axis; but the student must remember that it is really a portion of the mechanism by which the movements of the head on the spine are regulated; so that the connections between the atlas and axis ought always to be studied together with those between the latter bones and the skull.

There are *four* **Synovial Membranes** in this articulation : one lining the inner surface of each of the capsular ligaments ; one between the anterior surface of the odontoid process and the anterior arch of the atlas, the *atlanto-odontoid joint* ; and one between the posterior surface of the odontoid process and the transverse ligament, the *syndesmo-odontoid joint*. The latter often communicates with those between the condyles of the occipital bone and the articular surfaces of the atlas.

Actions.—This joint allows the rotation of the atlas (and, with it, of the cranium) upon the axis, the extent of rotation being limited by the odontoid ligaments.

The principal muscles by which this action is produced are the Sterno-mastoid and Complexus of one side, acting with the Rectus capitis anticus major, Splenius, Trachelo-mastoid, Rectus capitis posticus major, and Inferior oblique of the other side.

Fig. 154.—Occipito-atlantal and atlanto-axial ligaments. Posterior view.

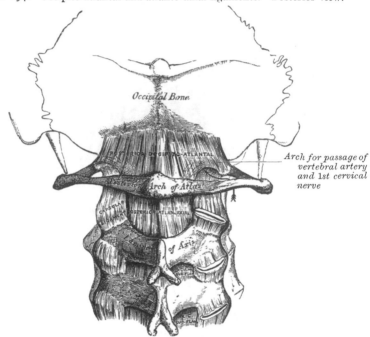

ARTICULATIONS OF THE SPINE WITH THE CRANIUM

The ligaments connecting the spine with the cranium may be divided into two sets, those connecting the occipital bone with the atlas, and those connecting the occipital bone with the axis.

III. ARTICULATION OF THE ATLAS WITH THE OCCIPITAL BONE

This articulation is a double condyloid joint.   Its ligaments are, the

    Anterior Occipito-atlantal.   Two Lateral Occipito-atlantal.
    Posterior Occipito-atlantal.   Two Capsular.

The **Anterior Occipito-atlantal Ligament** (fig. 153) is a broad membranous layer, composed of densely woven fibres, which passes between the anterior margin of the foramen magnum above, and the whole length of the upper border of the anterior arch of the atlas below. Laterally, it is continuous with the capsular ligaments. In the middle line in front it is strengthened by a strong, narrow, rounded cord, which is attached, above, to the basilar process of the occiput, and, below, to the tubercle on the anterior arch of the atlas. This ligament is in relation, in front, with the Recti antici minores; behind, with the odontoid ligaments.

The **Posterior Occipito-atlantal Ligament** (fig. 154) is a very broad but thin membranous lamina, intimately blended with the dura mater. It is connected, above, to the posterior margin of the foramen magnum; below, to the upper border of the posterior arch of the atlas. This ligament is incomplete at each side, and forms, with the superior intervertebral notch, an opening for the passage of the vertebral artery and suboccipital nerve. The fibrous band which arches over the artery and nerve sometimes becomes ossified. It is in relation, behind, with the Recti postici minores and Obliqui superiores; in front, with the dura mater of the spinal canal, to which it is intimately adherent.

The **Lateral Ligaments** are strong fibrous bands, directed obliquely upwards and inwards, attached above to the jugular process of the occipital bone; below, to the base of the transverse process of the atlas.

The **Capsular Ligaments** surround the condyles of the occipital bone, and connect them with the articular processes of the atlas; they consist of thin and loose capsules, which enclose the synovial membrane of the articulation.

Synovial Membranes.—There are two synovial membranes in this articulation: one lining the inner surface of each of the capsular ligaments. These

Fig. 155.—Articulation between odontoid process and atlas.

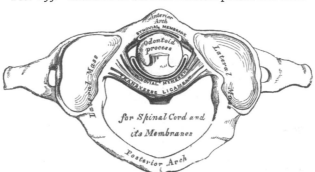

occasionally communicate with that between the posterior surface of the odontoid process and the transverse ligament.

Actions.—The movements permitted in this joint are flexion and extension, which give rise to the ordinary forward and backward nodding of the head, besides slight lateral motion to one or the other side. When either of these actions is carried beyond a slight extent, the whole of the cervical portion of the spine assists in its production. Flexion is mainly produced by the action of the Rectus capitis anticus major et minor and the Sterno-mastoid muscles; extension by the Rectus capitis posticus major et minor, the Superior oblique, the Complexus, Splenius, and upper fibres of the Trapezius. The Recti laterales are concerned in the lateral movement, assisted by the Trapezius, Splenius, Complexus, and the Sterno-mastoid of the same side, all acting together. According to Cruveilhier, there is a slight motion of rotation in this joint.

IV. Articulation of the Axis with the Occipital Bone

Occipito-axial.   Three Odontoid.

To expose these ligaments, the spinal canal should be laid open by removing the posterior arch of the atlas, the laminæ and spinous process of the axis, and the portion of the occipital bone behind the foramen magnum, as seen in fig. 156.

The **Occipito-axial Ligament** (*apparatus ligamentosus colli*) is situated within the spinal canal. It is a broad, strong band which covers the odontoid process and its ligaments, and appears to be a prolongation upwards of the posterior common ligament of the spine. It is attached, below, to the posterior surface of the body of the axis, and, becoming expanded as it ascends, is inserted into the basilar groove of the occipital bone, in front of the foramen magnum, where it becomes blended with the dura mater of the skull.

Relations.—By its anterior surface with the transverse ligament, by its posterior surface with the dura mater.

The **Odontoid** or **Check Ligaments** (*alar ligaments*) are strong, rounded, fibrous cords, which arise one on either side of the upper part of the odontoid process, and, passing obliquely upwards and outwards, are inserted into the rough depressions on the inner side of the condyles of the occipital bone. In the triangular interval left between these ligaments another strong fibrous cord (*ligamentum suspensorium* or *middle odontoid ligament*) may be seen, which

Fig. 156.—Occipito-axial and atlanto-axial ligaments. Posterior view, obtained by removing the arches of the vertebræ and the posterior part of the skull.

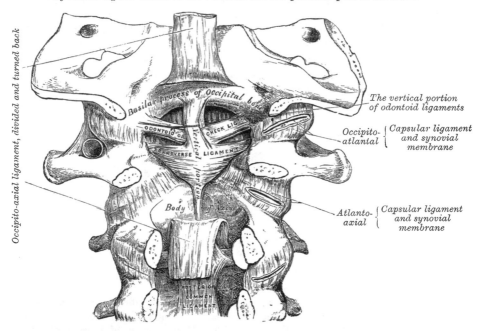

passes almost perpendicularly from the apex of the odontoid process to the anterior margin of the foramen magnum, being intimately blended with the deep portion of the anterior occipito-atlantal ligament, and upper fasciculus of the transverse ligament of the atlas.

Actions.—The odontoid ligaments serve to limit the extent to which rotation of the cranium may be carried; hence they have received the name of *check ligaments*.

In addition to these ligaments which connect the atlas and axis to the skull, the ligamentum nuchæ must be regarded as one of the ligaments by which the spine is connected with the cranium. It is described on a subsequent page.

Surgical Anatomy.—The ligaments which unite the component parts of the vertebræ together are so strong, and these bones are so interlocked by the arrangement of their articulating processes, that dislocation is very uncommon, and, indeed, unless accompanied by fracture, rarely occurs, except in the upper part of the neck. Dislocation of the occiput from the atlas has only been recorded in one or two cases; but dislocation of the atlas from the axis, with rupture of the transverse ligament, is much more common: it is the mode in which death is produced in many cases of execution by hanging. In the lower part of the neck—that is, below the third cervical vertebra—dislocation unattended by fracture occasionally takes place.

V. Articulation of the Lower Jaw (Temporo-Mandibular)

This is a ginglymo-arthrodial joint ; the parts entering into its formation on each side are, above, the anterior part of the glenoid cavity of the temporal bone and the eminentia articularis ; and, below, the condyle of the lower jaw. The ligaments are the following :—

External Lateral.	Stylo-mandibular.
Internal Lateral.	Capsular.

Interarticular Fibro-cartilage.

The **External Lateral Ligament** (fig. 157) is a short, thin, and narrow fasciculus attached, above, to the outer surface of the zygoma and to the tubercle on

Fig. 157.—Temporo-mandibular articulation. External view.

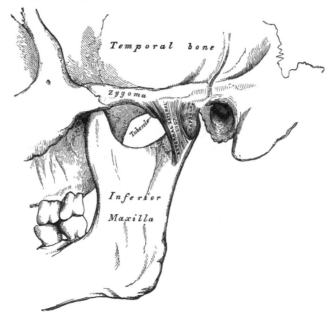

its lower border ; below, to the outer surface and posterior border of the neck of the lower jaw. It is broader above than below ; its fibres are placed parallel with one another, and directed obliquely downwards and backwards. Externally, it is covered by the parotid gland, and by the integument. Internally, it is in relation with the capsular ligament of which it is an accessory band, and not separable from it.

The **Internal Lateral Ligament** (*spheno-mandibular*) (fig. 158) is a flat thin band which is attached above to the spinous process of the sphenoid bone, and, becoming broader as it descends, is inserted into the lingula and margin of the dental foramen. Its outer surface is in relation, above, with the External pterygoid muscle ; lower down, it is separated from the neck of the condyle by the internal maxillary artery ; and still more inferiorly, the inferior dental vessels and nerve separate it from the ramus of the jaw. The inner surface is in relation with the Internal pterygoid. It is really the fibrous covering of a part of Meckel's cartilage.

The **Stylo-Mandibular Ligament** is a specialised band of the cervical fascia, which extends from near the apex of the styloid process of the temporal bone to the angle and posterior border of the ramus of the lower jaw, between the Masseter and Internal pterygoid muscles. This ligament separates the parotid from the submaxillary gland, and has attached to its inner side part of the fibres of origin of the Stylo-glossus muscle. Although usually classed among the ligaments of the jaw, it can only be considered as an accessory to the articulation.

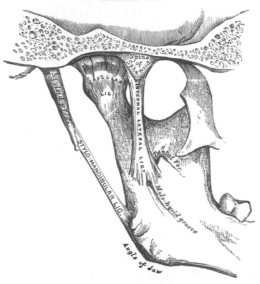

FIG. 158.—Temporo-mandibular articulation. Internal view.

The **Capsular Ligament** forms a thin and loose capsule, attached above to the circumference of the glenoid cavity and the articular surface immediately in front; below, to the neck of the condyle of the lower jaw. It consists of a few thin scattered fibres, and can hardly be considered as a distinct ligament; it is thickest at the back part of the articulation.*

The **Interarticular Fibro-cartilage** (fig. 159) is a thin plate of an oval form, placed horizontally between the condyle of the jaw and the glenoid cavity. Its upper surface is concavo-convex from before backwards, and a little convex transversely, to accommodate itself to the form of the glenoid cavity. Its under surface, where it is in contact with the condyle, is concave. Its circumference is connected to the capsular ligament; and in front to the tendon of the External pterygoid muscle. It is thicker at its circumference, especially behind, than at its centre. The fibres of which it is composed have a concentric arrangement, more apparent at the circumference than at the centre. Its surfaces are smooth. It divides the joint into two cavities, each of which is furnished with a separate synovial membrane.

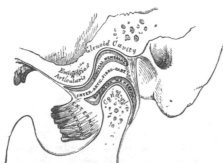

FIG. 159.—Vertical section of temporo-mandibular articulation.

The **Synovial Membranes**, two in number, are placed one above, and the other below, the fibro-cartilage. The upper one, the larger and looser of the two, is continued from the margin of the cartilage covering the glenoid cavity and eminentia articularis on to the upper surface of the fibro-cartilage. The lower one passes from the under surface of the fibro-cartilage to the neck of the condyle of the jaw, being prolonged downwards a little farther behind than in front. The interarticular cartilage is some-

---

* Sir G. Humphry describes the internal portion of the capsular ligament separately, as the short internal lateral ligament; and it certainly seems as deserving of a separate description as the external lateral ligament is.

times perforated in its centre, the two synovial sacs then communicate with each other.

The *nerves* of this joint are derived from the auriculo-temporal and masseteric branches of the inferior maxillary. The *arteries* are derived from the temporal branch of the external carotid.

Actions.—The movements permitted in this articulation are very extensive. Thus, the jaw may be depressed or elevated, or it may be carried forwards or backwards. It must be borne in mind that there are two distinct joints in this articulation—that is to say, one between the condyle of the jaw and the interarticular fibro-cartilage, and another between the fibro-cartilage and the glenoid fossa ; when the jaw is depressed, as in opening the mouth, the movements which take place in these two joints are not the same. In the lower compartment, that between the condyle and the fibro-cartilage, the movement is of a ginglymoid or hinge-like character, the condyle rotating on a transverse axis on the fibro-cartilage ; while in the upper compartment the movement is of a gliding character, the fibro-cartilage, together with the condyle, gliding forwards on to the eminentia articularis. These two movements take place simultaneously, the condyle and fibro-cartilage move forwards on the eminence, and at the same time the condyle revolves on the fibro-cartilage. In the opposite movement of shutting the mouth the reverse action takes place ; the fibro-cartilage glides back, carrying the condyle with it, and this at the same time revolves back to its former position. When the jaw is carried horizontally forwards, as in protruding the lower incisors in front of the upper, the movement takes place principally in the upper compartment of the joint, the fibro-cartilage, carrying with it the condyle, glides forwards on the glenoid fossa. This is because this movement is mainly effected by the External pterygoid muscles, which are inserted into both condyle and interarticular fibro-cartilage. The grinding or chewing movement is produced by the alternate movement of one condyle, with its fibro-cartilage, forwards and backwards, while the other condyle moves simultaneously in the opposite direction ; at the same time the condyle undergoes a vertical rotation on its own axis on the fibro-cartilage in the lower compartment. One condyle advances and rotates, the other condyle recedes and rotates, in alternate succession.

The lower jaw is *depressed* by its own weight, assisted by the Platysma, the Digastric, the Mylo-hyoid, and the Genio-hyoid. It is *elevated* by the anterior part of the Temporal, Masseter, and Internal pterygoid. It is drawn *forwards* by the simultaneous action of the External pterygoid, and the superficial fibres of the Masseter ; and it is drawn *backwards* by the deep fibres of the Masseter and the posterior fibres of the Temporal muscle. The grinding movement is caused by the alternate action of the two External pterygoids.

Surface Form.—The temporo-mandibular articulation is quite superficial, situated below the base of the zygoma, in front of the tragus and external auditory meatus, and behind the posterior border of the upper part of the Masseter muscle. Its exact position can be at once ascertained by feeling for the condyle of the jaw, the working of which can be distinctly felt in the movements of the lower jaw in opening and shutting the mouth. When the mouth is opened wide, the condyle advances out of the glenoid fossa on to the eminentia articularis and a depression is felt in the situation of the joint.

Surgical Anatomy.—The lower jaw is dislocated only in one direction—viz. forwards. The accident is caused by violence or muscular action. When the mouth is open, the condyle is situated on the eminentia articularis, and any sudden violence, or even a sudden muscular spasm as during a convulsive yawn, may displace the condyle forwards into the zygomatic fossa. The displacement may be unilateral or bilateral, according as one or both of the condyles are displaced. The latter of the two is the more common.

Sir Astley Cooper described a condition which he termed 'subluxation.' It occurs principally in delicate women, and is believed by some to be due to relaxation of the ligaments, permitting too free movement of the bone, and possibly some displacement of the fibro-cartilage. Others have believed that it is due to gouty or rheumatic changes in the joint. In close relation to the condyle of the jaw is the external auditory meatus and the tympanum ; any force, therefore, applied to the bone is liable to be attended with damage to these parts, or inflammation in the joint may extend to the ear, or on the other hand inflammation of the middle ear may involve the articulation and cause its destruction, thus leading to ankylosis of the joint. In children, arthritis of this joint may follow the exanthemata, and in adults occurs as the result of some constitutional conditions, as rheumatism or gout. The temporo-mandibular joint is also occasionally the seat of osteo-arthritis, leading to great suffering during efforts of mastication. A peculiar affection sometimes attacks the neck and condyle of the lower jaw, consisting in hypertrophy and elongation of these parts and consequent protrusion of the chin to the opposite side.

VI. Articulations of the Ribs with the Vertebræ

The articulations of the ribs with the vertebral column may be divided into two sets : 1. Those which connect the heads of the ribs with the bodies of the vertebræ (*costo-central*). 2. Those which connect the necks and tubercles of the ribs with the transverse processes (*costo-transverse*).

1. Articulations between the Heads of the Ribs and the Bodies of the Vertebræ (fig. 160)

These constitute a series of arthrodial joints, formed by the articulation of the heads of the ribs with the cavities on the contiguous margins of the bodies of the dorsal vertebræ and the intervertebral substance between them, except in the case of the first, tenth, eleventh, and twelfth ribs, where the cavity is formed by a single vertebra. The bones are connected by the following ligaments :—

Anterior Costo-vertebral or Stellate.
Capsular.   Interarticular.

The **Anterior Costo-vertebral** or **Stellate Ligament** connects the anterior part of the head of each rib with the sides of the bodies of two vertebræ, and the intervertebral disc between them. It consists of flat bundles of ligamentous fibres, which are attached to the anterior part of the head of the rib, just beyond the articular surface. The superior fibres pass upwards to be connected with the body of the vertebra above ; the inferior ones descend to the body of the vertebra below ; and the middle ones, the smallest and least distinct, pass horizontally inwards, to be attached to the intervertebral substance.

Relations. — In front, with the thoracic ganglia of the sympathetic, the pleura, and, on the right side, with the vena azygos major ; behind, with

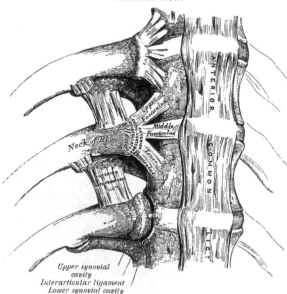

Fig. 160.—Costo-vertebral and costo-transverse articulations. Anterior view.

Upper synovial cavity
Interarticular ligament
Lower synovial cavity

the interarticular ligament and synovial membranes.

In the first rib, which articulates with a single vertebra, this ligament does not present a distinct division into three fasciculi ; its fibres, however, radiate, and are attached to the body of the last cervical vertebra, as well as to the body of the vertebra with which the rib articulates. In the tenth, eleventh, and twelfth ribs, which also articulate with a single vertebra, the division does not exist; but the fibres of the ligament in each case radiate and are connected with the vertebra above, as well as that with which the ribs articulate.

The **Capsular Ligament** is a thin and loose ligamentous bag, which surrounds the joint between the head of the rib and the articular cavity formed by the intervertebral disc and the adjacent vertebræ. It is very thin, firmly connected with the anterior ligament, and most distinct at the upper and lower parts of the articulation. Behind, some of its fibres pass through the intervertebral foramen to the back of the intervertebral disc. This is the analogue of the ligamentum conjugale of some mammals, which unites the heads of opposite ribs across the back of the intervertebral disc.

The **Interarticular Ligament** is situated in the interior of the joint. It consists of a short band of fibres, flattened from above downwards, attached by one extremity to the sharp crest which separates the two articular facets on the head

Fig. 161.—Costo-transverse articulation. Seen from above.

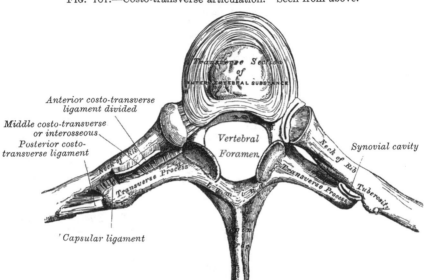

of the rib, and by the other to the intervertebral disc. It divides the joint into two cavities, which have no communication with each other. In the first, tenth, eleventh, and twelfth ribs, the interarticular ligament does not exist; consequently, there is but one synovial membrane.

The Synovial Membrane.—There are two synovial membranes in each of the articulations in which there is an interarticular ligament, one on each side of this structure.

2. ARTICULATIONS OF THE NECKS AND TUBERCLES OF THE RIBS WITH THE TRANSVERSE PROCESSES (fig. 161)

The articular portion of the tubercle of the rib and adjacent transverse process form an arthrodial joint.

In the *eleventh* and *twelfth ribs* this articulation is wanting.

The ligaments connecting these parts are the—

Anterior Costo-transverse.
Middle Costo-transverse (Interosseous).
Posterior Costo-transverse.
Capsular.

The **Anterior Costo-transverse Ligament** (*superior* or *long*) consists of two sets of fibres : the one (anterior) is attached below to the sharp crest on the upper border of the neck of each rib, and passes obliquely upwards and outwards, to the lower border of the transverse process immediately above ; the other (posterior) is attached below to the neck of the rib, and passes upwards and inwards to the base of the transverse process and outer border of the lower articular process of the vertebra above. This ligament is in relation, in front, with the intercostal vessels and nerves ; behind, with the Longissimus dorsi. Its *internal border* is thickened and free, and bounds an aperture through which pass the posterior branches of the intercostal vessels and nerves. Its *external border* is continuous with a thin aponeurosis, which covers the External intercostal muscle.

The *first rib* has no anterior costo-transverse ligament.

The **Middle Costo-transverse** or **Interosseous Ligament** consists of short but strong fibres, which pass between the rough surface on the posterior part of the neck of each rib and the anterior surface of the adjacent transverse process. In order fully to expose this ligament a horizontal section should be made across the transverse process and corresponding part of the rib ; or the rib may be forcibly separated from the transverse process, and its fibres put on the stretch.

In the *eleventh* and *twelfth ribs* this ligament is quite rudimentary or wanting.

The **Posterior Costo-transverse Ligament** is a short but thick and strong fasciculus, which passes obliquely from the summit of the transverse process to the rough non-articular portion of the tubercle of the rib. This ligament is shorter and more oblique in the upper than in the lower ribs. Those corresponding to the superior ribs ascend, while those of the inferior ribs descend slightly.

In the *eleventh* and *twelfth ribs* this ligament is wanting.

The **Capsular Ligament** is a thin, membranous sac attached to the circumference of the articular surfaces, and enclosing a small synovial membrane.

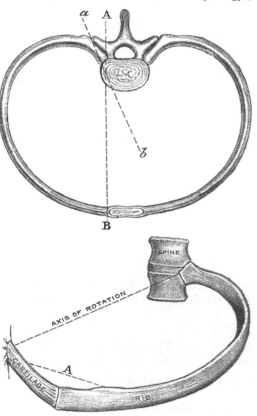

FIG. 162.—Diagrams showing the axis of rotation of the ribs in the movements of respiration. The one axis of rotation corresponds with a line drawn through the two articulations which the rib forms with the spine (*a*, *b*), and the other with a line drawn from the head of the rib to the sternum (A, B). (From Kirke's ' Handbook of Physiology.')

In the *eleventh* and *twelfth ribs* this ligament is absent.

Actions.—The heads of the ribs are so closely connected to the bodies of the vertebræ by the stellate and interarticular ligaments, and the necks and tubercles of the ribs to the transverse processes, that only a slight gliding movement of the articular surfaces on each other can take place in these articulations. The result of this gliding movement with respect to the six upper ribs, consists in an elevation of the front and middle portion of the rib, the hinder part being prevented from performing any upward movement by its close connection with the spine.   In this gliding movement the rib rotates on an axis corresponding with a line drawn through the two articulations, Costo-central and Costo-transverse, which the rib forms with the spine.   With respect to the seventh, eighth, ninth, and tenth ribs, each one, besides rotating in a similar manner to the upper six, also rotates on an axis corresponding with a line drawn from the head of the rib to the sternum. By the first movement—that of rotation of the rib on an axis corresponding with a line drawn through the two articulations which this bone forms with the spine— an elevation of the anterior part of the rib takes place, and a consequent enlargement of the antero-posterior diameter of the chest.   None of the ribs lie in a truly horizontal plane : they are all directed more or less obliquely, so that their anterior extremities lie on a lower level than their posterior, and this obliquity increases from the first to the seventh, and then again decreases.   If we examine any one rib—say, that in which there is the greatest obliquity—we shall see that it is obvious that as its sternal extremity is carried upwards, it must also be thrown forwards ; so that the rib may be regarded as a radius, moving on the vertebral joint as a centre, and causing the sternal attachment to describe an arc of a circle in the vertical plane of the body.   Since all the ribs are oblique and connected in front to the sternum by the elastic costal cartilages, they must have a tendency to thrust the sternum forwards, and so increase the antero-posterior diameter of the chest.   By the second movement—that of the rotation of the rib on an axis corresponding with a line drawn from the head of the rib to the sternum—an elevation of the middle portion of the rib takes place, and consequently an increase in the transverse diameter of the chest.   For the ribs not only slant downwards and forwards from their vertebral attachment, but they are also oblique in relation to their transverse plane—that is to say, their middle is on a lower level than either their vertebral or sternal extremities.   It results from this that when the ribs are raised, the centre portion is thrust outwards, somewhat after the fashion in which the handle of a bucket is thrust away from the side when raised to a horizontal position, and the lateral diameter of the chest is increased (see fig. 162).   The mobility of the different ribs varies very much.   The first rib is more fixed than the others, on account of the weight of the upper extremity and the strain of the ribs beneath ; but on the freshly dissected thorax it moves as freely as the rest.   From the same causes the movement of the second rib is also not very extensive.   In the other ribs, this mobility increases successively down to the last two, which are very movable.   The ribs are generally more movable in the female than in the male.

VII. Articulation of the Cartilages of the Ribs with the Sternum, etc. (fig. 163)

The articulations of the cartilages of the true ribs with the sternum are arthrodial joints, with the exception of the first, in which the cartilage is almost always directly united with the sternum, and which must, therefore, be regarded as a synarthrodial articulation.   The ligaments connecting them are—

Anterior Chondro-sternal.	Interarticular Chondro-sternal.
Posterior Chondro-sternal.	Anterior Chondro-xiphoid.
Capsular.	Posterior Chondro-xiphoid.

The **Anterior Chondro-sternal Ligament** is a broad and thin membranous band that radiates from the front of the inner extremity of the cartilages of the true ribs to the anterior surface of the sternum.   It is composed of fasciculi which pass in different directions.   The *superior fasciculi* ascend obliquely, the *inferior* pass obliquely downwards, and the *middle fasciculi* horizontally.   The superficial fibres of this ligament are the longest; they intermingle with the fibres of the

ligaments above and below them, with those of the opposite side, and with the
tendinous fibres of origin of the Pectoralis major, forming a thick fibrous membrane, which covers the surface of the sternum.   This is more distinct at the
lower than at the upper part.

The **Posterior Chondro-sternal Ligament,** less thick and distinct than the
anterior, is composed of fibres which radiate from the posterior surface of the
sternal end of the cartilages of the true ribs to the posterior surface of the
sternum, becoming blended with the periosteum.

The **Capsular Ligament** surrounds the joints formed between the cartilages
of the true ribs and the sternum.   It is very thin, intimately blended with the
anterior and posterior ligaments, and strengthened at the upper and lower part
of the articulation by a few fibres, which pass from the cartilage to the side of the
sternum.   These ligaments protect the synovial membranes.

The **Interarticular Chondro-sternal Ligaments.**—These are only found between
the second and third costal cartilages and the sternum.   The cartilage of the
second rib is connected with the sternum by means of an interarticular ligament,
attached by one extremity to the cartilage of the second rib, and by the other
extremity to the cartilage which unites the first and second pieces of the sternum.
This articulation is provided with two synovial membranes.   The cartilage of
the *third rib* is connected with the sternum by means of an interarticular
ligament which is attached by one extremity to the cartilage of the third rib,
and by the other extremity to the point of junction of the second and third
pieces of the sternum.   This articulation is provided with two synovial membranes.

The **Anterior Chondro-xiphoid.**—This is a band of ligamentous fibres, which
connects the anterior surface of the seventh costal cartilage, and occasionally also
that of the sixth, to the anterior surface of the ensiform appendix.   It varies in
length and breadth in different subjects.

The **Posterior Chondro-xiphoid** is a similar band of fibres on the internal or
posterior surface, though less thick and distinct.

Synovial Membranes.—There is no synovial membrane between the first
costal cartilage and the sternum, as this cartilage is directly continuous with the
sternum.   There are two synovial membranes both in the articulation of the
second and third costal cartilages to the sternum.   There is generally one
synovial membrane in each of the joints between the fourth, fifth, sixth, and
seventh costal cartilages to the sternum ; but it is sometimes absent in the sixth
and seventh chondro-sternal joints.   Thus there are *eight* synovial cavities on
each side in the articulations between the costal cartilages of the true ribs and
the sternum.   After middle life the articular surfaces lose their polish, become
roughened, and the synovial membranes appear to be wanting.   In old age, the
articulations do not exist, the cartilages of most of the ribs becoming continuous
with the sternum.

Actions.—The movements which are permitted in the chondro-sternal articulations are limited to elevation and depression, and these only to a slight extent.

ARTICULATIONS OF THE CARTILAGES OF THE RIBS WITH EACH OTHER
(INTERCHONDRAL) (fig. 163)

The contiguous borders of the sixth, seventh, and eighth, and sometimes the
ninth and tenth, costal cartilages articulate with each other by small, smooth,
oblong-shaped facets.   Each articulation is enclosed in a thin *capsular ligament*
lined by *synovial membrane,* and strengthened externally and internally by ligamentous fibres (interchondral ligaments), which pass from one cartilage to the
other.   Sometimes the fifth costal cartilage, more rarely that of the ninth, articulates, by its lower border, with the adjoining cartilage by a small oval facet ;
more frequently they are connected together by a few ligamentous fibres.
Occasionally, the articular surfaces above mentioned are wanting.

ARTICULATIONS OF THE RIBS WITH THEIR CARTILAGES (COSTO-CHONDRAL)
(fig. 163)

The outer extremity of each costal cartilage is received into a depression in the sternal end of the ribs, and the two are held together by the periosteum.

FIG. 163.—Chondro-sternal, chondro-xiphoid, and interchondral articulations. Anterior view.

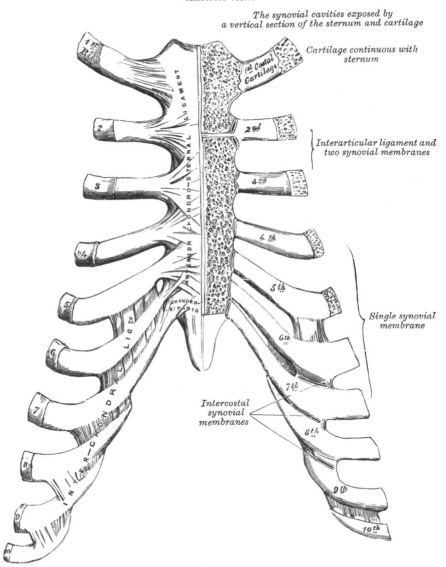

VIII. ARTICULATIONS OF THE STERNUM

The first piece of the sternum is united to the second either by an amphi-arthrodial joint—a single piece of true fibro-cartilage uniting the segments—or by a diarthrodial joint, in which each bone is clothed with a distinct lamina of cartilage, adherent on one side, free and lined with synovial membrane on the other. In the latter case, the cartilage covering the gladiolus is continued without interruption on to the cartilages of the second ribs. Mr. Rivington has found the diarthrodial form of joint in about one-third of the specimens examined by him, Mr. Maisonneuve more frequently. It appears to be rare in childhood, and is formed, in Mr. Rivington's opinion, from the amphiarthrodial form, by absorption. The diarthrodial joint seems to have no tendency to ossify at any age, while the amphiarthrodial is more liable to do so, and has been found ossified as early as thirty-four years of age. The two segments are further connected by an

Anterior Intersternal Ligament.
Posterior Intersternal Ligament.

The **Anterior Intersternal Ligament** consists of a layer of fibres, having a longitudinal direction ; it blends with the fibres of the anterior chondro-sternal

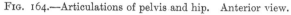

Fig. 164.—Articulations of pelvis and hip. Anterior view.

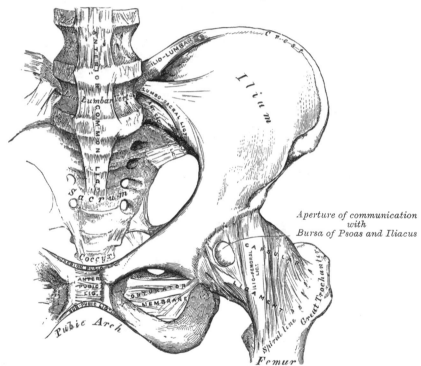

ligaments on both sides, and with the tendinous fibres of origin of the Pectoralis major. This ligament is rough, irregular, and much thicker below than above.

The **Posterior Intersternal Ligament** is disposed in a somewhat similar manner on the posterior surface of the articulation.

IX. Articulation of the Vertebral Column with the Pelvis

The ligaments connecting the last lumbar vertebra with the sacrum are similar to those which connect the segments of the spine with each other—viz.: 1. The continuation downwards of the anterior and posterior common ligaments. 2. The intervertebral substance connecting the flattened oval surfaces of the two bones and forming an amphiarthrodial joint. 3. Ligamenta subflava, connecting the arch of the last lumbar vertebra with the posterior border of the sacral canal. 4. Capsular ligaments connecting the articulating processes and forming a double arthrodia. 5. Inter- and supra-spinous ligaments.

The two proper ligaments connecting the pelvis with the spine are the lumbo-sacral and ilio-lumbar.

The **Lumbo-sacral Ligament** (fig. 164) is a short, thick, triangular fasciculus, which is connected above to the lower and front part of the transverse process of the last lumbar vertebra, passes obliquely outwards, and is attached below to the lateral surface of the base of the sacrum, becoming blended with the anterior sacro-iliac ligament. The ligament is in relation, in front, with the Psoas muscle.

The **Ilio-lumbar Ligament** (fig. 164) passes horizontally outwards from the apex of the transverse process of the last lumbar vertebra to the crest of the ilium immediately in front of the sacro-iliac articulation. It is of a triangular

FIG. 165.—Articulations of pelvis and hip. Posterior view.

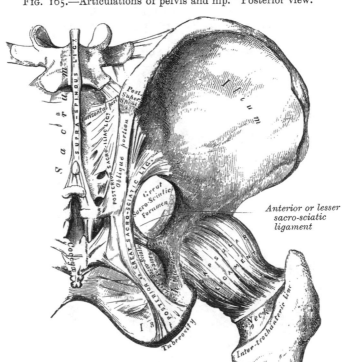

form, thick and narrow internally, broad and thinner externally. It is in relation, in front, with the Psoas muscle ; behind, with the muscles occupying the vertebral groove ; above, with the Quadratus lumborum.

X. ARTICULATIONS OF THE PELVIS

The ligaments connecting the bones of the pelvis with each other may be divided into four groups : 1. Those connecting the sacrum and ilium. 2. Those passing between the sacrum and ischium. 3. Those connecting the sacrum and coccyx. 4. Those between the two pubic bones.

1. ARTICULATIONS OF THE SACRUM AND ILIUM

The sacro-iliac articulation is an amphiarthrodial joint, formed between the lateral surfaces of the sacrum and the ilium. The anterior or auricular portion of each articular surface is covered with a thin plate of cartilage, thicker on the sacrum than on the ilium. These are in close contact with each other, and to a certain extent united together by irregular patches of softer fibro-cartilage, and at their upper and posterior part by fine fibres of interosseous fibrous tissue. In a considerable part of their extent, especially in advanced life, they are not connected together, but are separated by a space containing a synovial-like fluid, and hence the joint presents the characters of a diarthrosis.

The ligaments connecting these surfaces are the anterior and posterior sacro-iliac.

The **Anterior Sacro-iliac Ligament** (fig. 166) consists of numerous thin bands, which connect the anterior surfaces of the sacrum and ilium.

The **Posterior Sacro-iliac** (fig. 165) is a strong interosseous ligament, situated in a deep depression between the sacrum and ilium behind, and forming the chief bond of connection between those bones. It consists of numerous strong fasciculi, which pass between the bones in various directions. Three of these are of large size; the *two superior*, nearly horizontal in direction, arise from the first and second transverse tubercles on the posterior surface of the sacrum, and

FIG. 166.—Side view of pelvis, showing the great and lesser sacro-sciatic ligaments.

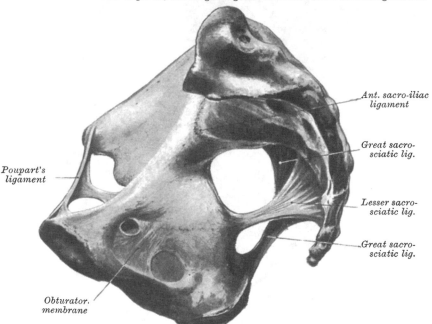

Ant. sacro-iliac ligament

Great sacro-sciatic lig.

Poupart's ligament

Lesser sacro-sciatic lig.

Great sacro-sciatic lig.

Obturator. membrane

are inserted into the rough, uneven surface at the posterior part of the inner surface of the ilium. The third fasciculus, oblique in direction, is attached by one extremity to the third transverse tubercle on the posterior surface of the sacrum, and by the other to the posterior superior spine of the ilium; it is sometimes called the *oblique sacro-iliac ligament.*

The position of the sacro-iliac joint is indicated by the posterior superior spine of the ilium. This process is immediately behind the centre of the articulation.

2. LIGAMENTS PASSING BETWEEN THE SACRUM AND ISCHIUM (fig. 166)

The Great Sacro-sciatic (Posterior).
The Lesser Sacro-sciatic (Anterior).

The **Great** or **Posterior Sacro-sciatic Ligament** is situated at the lower and back part of the pelvis. It is flat, and triangular in form; narrower in the middle than at the extremities; attached by its broad base to the posterior inferior spine of the ilium, to the fourth and fifth transverse tubercles of the sacrum, and to the lower part of the lateral margin of that bone and the coccyx. Passing obliquely downwards, outwards, and forwards, it becomes narrow and thick; and at its insertion into the inner margin of the tuberosity of the ischium, it increases in breadth, and is prolonged forwards along the inner margin of the ramus, forming what is known as the *falciform ligament.* The free concave edge of this prolongation has attached to it the obturator fascia, with which it forms a kind of groove, protecting the internal pudic vessels and nerve. One of its surfaces is turned towards the perinæum, the other towards the Obturator internus muscle.

The *posterior surface* of this ligament gives origin, by its whole extent, to fibres of the Gluteus maximus. Its *anterior surface* is united to the lesser sacro-sciatic ligament. Its *external border* forms, above, the posterior boundary of the great sacro-sciatic foramen, and, below, the posterior boundary of the lesser sacro-sciatic foramen. Its *lower border* forms part of the boundary of the perinæum. It is pierced by the coccygeal branch of the sciatic artery and coccygeal nerve.

The **Lesser** or **Anterior Sacro-sciatic Ligament,** much shorter and smaller than the preceding, is thin, triangular in form, attached by its apex to the spine of the ischium, and internally, by its broad base, to the lateral margin of the sacrum and coccyx, anterior to the attachment of the great sacro-sciatic ligament with which its fibres are intermingled.

It is in relation, *anteriorly,* with the Coccygeus muscle; *posteriorly,* it is covered by the great sacro-sciatic ligament, and crossed by the internal pudic vessels and nerve. Its *superior border* forms the lower boundary of the great sacro-sciatic foramen; its *inferior border,* part of the lesser sacro-sciatic foramen.

These two ligaments convert the sacro-sciatic notches into foramina. The *superior* or *great* sacro-sciatic foramen is bounded, in front and above, by the posterior border of the os innominatum; behind, by the great sacro-sciatic ligament; and below, by the lesser sacro-sciatic ligament. It is partially filled up, in the recent state, by the Pyriformis muscle which passes through it. Above this muscle, the gluteal vessels and superior gluteal nerve emerge from the pelvis; and below it, the sciatic vessels and nerves, the internal pudic vessels and nerve, the inferior gluteal nerve, and the nerves to the obturator internus and quadratus femoris. The *inferior* or *lesser* sacro-sciatic foramen is bounded, in front, by the tuber ischii; above, by the spine and lesser sacro-sciatic ligament; behind, by the greater sacro-sciatic ligament. It transmits the tendon of the Obturator internus muscle, its nerve, and the internal pudic vessels and nerve.

3. ARTICULATION OF THE SACRUM AND COCCYX

This articulation is an amphiarthrodial joint, formed between the oval surface at the apex of the sacrum, and the base of the coccyx. It is analogous to the joints between the bodies of the vertebræ, and is connected by similar ligaments. They are the

Anterior Sacro-coccygeal.   Lateral Sacro-coccygeal.
Posterior Sacro-coccygeal.   Interposed Fibro-cartilage.
Interarticular.

The **Anterior Sacro-coccygeal Ligament** consists of a few irregular fibres which descend from the anterior surface of the sacrum to the front of the coccyx, becoming blended with the periosteum.

The **Posterior Sacro-coccygeal Ligament** is a flat band, of a pearly tint, which arises from the margin of the lower orifice of the sacral canal, and descends to be inserted into the posterior surface of the coccyx. This ligament completes the lower and back part of the sacral canal. Its superficial fibres are much longer than the more deeply seated. This ligament is in relation, behind, with the Gluteus maximus.

The **Lateral Sacro-coccygeal Ligaments** connect the transverse processes of the coccyx to the lower lateral angles of the sacrum.

A **Fibro-cartilage** is interposed between the contiguous surfaces of the sacrum and coccyx ; it differs from that interposed between the bodies of the vertebræ in being thinner, and its central part firmer in texture. It is somewhat thicker in front and behind than at the sides. Occasionally, a synovial membrane is found when the coccyx is freely movable, which is more especially the case during pregnancy.

The **Interarticular Ligaments** are thin bands of ligamentous tissue, which connect the cornua of the two bones together.

The different segments of the coccyx are connected together by an extension downwards of the anterior and posterior sacro-coccygeal ligaments, a thin annular disc of fibro-cartilage being interposed between each of the bones. In the adult male, all the pieces become ossified ; but in the female, this does not commonly occur until a later period of life. The separate segments of the coccyx are first united, and at a more advanced age the joint between the sacrum and coccyx is obliterated.

Actions.—The movements which take place between the sacrum and coccyx, and between the different pieces of the latter bone, are forwards and backwards ; they are very limited. Their extent increases during pregnancy.

4. ARTICULATION OF THE OSSA PUBIS (SYMPHYSIS PUBIS) (fig. 167)

The articulation between the pubic bones is an amphiarthrodial joint, formed by the junction of the two oval articular surfaces of the ossa pubis. The ligaments of this articulation are the

Anterior Pubic.   Posterior Pubic.
Superior Pubic.   Subpubic.
                    Interpubic Disc.

The **Anterior Pubic Ligament** consists of several superimposed layers, which pass across the front of the articulation   The superficial fibres pass obliquely from one bone to the other, decussating and forming an interlacement with the fibres of the aponeurosis of the External oblique and the tendon of the Rectus muscles. The deep fibres pass transversely across the symphysis, and are blended with the fibro-cartilage.

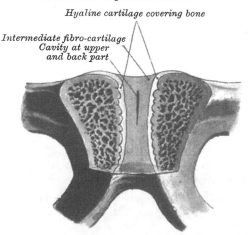

FIG. 167.—Vertical section of the symphysis pubis. Made near its posterior surface.

Hyaline cartilage covering bone

Intermediate fibro-cartilage
Cavity at upper
and back part

The **Posterior Pubic Ligament** consists of a few thin. scattered fibres, which unite the two pubic bones posteriorly.

The **Superior Pubic Ligament** is a band of fibres, which connects together the two pubic bones superiorly.

The **Subpubic Ligament** is a thick, triangular arch of ligamentous fibres, connecting together the two pubic bones below, and forming the upper boundary of the pubic arch. Above, it is blended with the interarticular fibro-cartilage ; laterally, it is united with the descending rami of the ossa pubis. Its fibres are closely connected, and have an arched direction.

The **Interpubic Disc** consists of a disc of cartilage and fibro-cartilage connecting the surfaces of the pubic bones in front. Each of the two surfaces is covered by a thin layer of hyaline cartilage, which is firmly connected to the bone by a series of nipple-like processes which accurately fit within corresponding depressions on the osseous surfaces. These opposed cartilaginous surfaces are connected together by an intermediate stratum of fibrous tissue and fibrocartilage which varies in thickness in different subjects. It often contains a cavity in its centre, probably formed by the softening and absorption of the fibro-cartilage, since it rarely appears before the tenth year of life, and is not lined by synovial membrane. It is larger in the female than in the male, but it is very questionable whether it enlarges, as was formerly supposed, during pregnancy. It is most frequently limited to the upper and back part of the joint; but it occasionally reaches to the front, and may extend the entire length of the cartilage. This cavity may be easily demonstrated by making a vertical section of the symphysis pubis near its posterior surface (fig. 167).

The **Obturator Membrane** is more properly regarded as analogous to the muscular fasciæ, with which it will be described.

ARTICULATIONS OF THE UPPER EXTREMITY

The articulations of the Upper Extremity may be arranged in the following groups: I. Sterno-clavicular articulation. II. Acromio-clavicular articulation. III. Ligaments of the Scapula. IV. Shoulder-joint. V. Elbow-joint. VI. Radio-ulnar articulations. VII. Wrist-joint. VIII. Articulations of the Carpal Bones. IX. Carpo-metacarpal articulations. X. Metacarpo-phalangeal articulations. XI. Articulations of the Phalanges.

I. Sterno-clavicular Articulation (fig. 168)

The **Sterno-clavicular** is regarded by most anatomists as an arthrodial joint; but Cruveilhier considers it to be an articulation by reciprocal reception. Probably the former opinion is the correct one; the varied movements, which the joint enjoys, being due to the interposition of an interarticular fibro-cartilage between

Fig. 168.—Sterno-clavicular articulation. Anterior view.

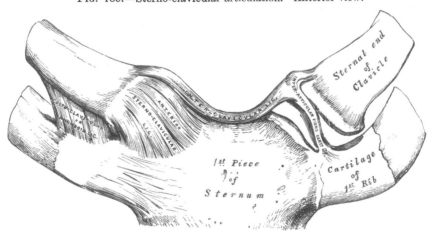

the joint surfaces. The parts entering into its formation are the sternal end of the clavicle, the upper and lateral part of the first piece of the sternum, and the cartilage of the first rib. The articular surface of the clavicle is much larger than that of the sternum, and invested with a layer of cartilage,* which is

* According to Bruch, the sternal end of the clavicle is covered by a tissue, which is rather fibrous than cartilaginous in structure.

considerably thicker than that on the latter bone. The ligaments of this joint are the

Capsular.   Interclavicular.
Anterior Sterno-clavicular.   Costo-clavicular (rhomboid).
Posterior Sterno-clavicular.   Interarticular fibro-cartilage.

The **Capsular Ligament** completely surrounds the articulation, consisting of fibres of varying degrees of thickness and strength. Those in front and behind are of considerable thickness, and form the anterior and posterior sterno-clavicular ligaments; but those above and below, especially in the latter situation, are thin and scanty, and partake more of the character of connective tissue than true fibrous tissue.

The **Anterior Sterno-clavicular Ligament** is a broad band of fibres, which covers the anterior surface of the articulation, being attached, above, to the upper and front part of the inner extremity of the clavicle; and, passing obliquely downwards and inwards, is attached, below, to the upper and front part of the first piece of the sternum. This ligament is covered, in front, by the sternal portion of the Sterno-cleido-mastoid and the integument; behind, it is in relation with the interarticular fibro-cartilage and the two synovial membranes.

The **Posterior Sterno-clavicular Ligament** is a similar band of fibres, which covers the posterior surface of the articulation, being attached, above, to the upper and back part of the inner extremity of the clavicle; and, passing obliquely downwards and inwards, is attached, below, to the upper and back part of the first piece of the sternum. It is in relation, in front, with the interarticular fibro-cartilage and synovial membranes; behind, with the Sterno-hyoid and Sterno-thyroid muscles.

The **Interclavicular Ligament** is a flattened band, which varies considerably in form and size in different individuals; it passes in a curved direction from the upper part of the inner extremity of one clavicle to the other, and is also attached to the upper margin of the sternum. It is in relation, in front, with the integument; behind, with the Sterno-thyroid muscles.

The **Costo-clavicular Ligament** (rhomboid) is short, flat, and strong: it is of a rhomboid form, attached, below, to the upper and inner part of the cartilage of the first rib, it ascends obliquely backwards and outwards, and is attached, above, to the rhomboid depression on the under surface of the clavicle. It is in relation, in front, with the tendon of origin of the Subclavius; behind, with the subclavian vein.

The **Interarticular fibro-cartilage** is a flat and nearly circular disc, interposed between the articulating surfaces of the sternum and clavicle. It is attached, above, to the upper and posterior border of the articular surface of the clavicle; below, to the cartilage of the first rib, at its junction with the sternum; and by its circumference to the anterior and posterior sterno-clavicular and interclavicular ligaments. It is thicker at the circumference, especially its upper and back part, than at its centre, or below. It divides the joint into two cavities, each of which is furnished with a separate synovial membrane.

Of the two **Synovial Membranes** found in this articulation, one is reflected from the sternal end of the clavicle, over the adjacent surface of the fibro-cartilage, and cartilage of the first rib; the other is placed between the articular surface of the sternum and adjacent surface of the fibro-cartilage; the latter is the larger of the two.

Actions.—This articulation is the centre of the movements of the shoulder, and admits of a limited amount of motion in nearly every direction—upwards, downwards, backwards, forwards, as well as circumduction. When these movements take place in the joint, the clavicle in its motion carries the scapula with it, this bone gliding on the outer surface of the chest. This joint therefore forms the centre from which all movements of the supporting arch of the shoulder originate, and is the only point of articulation of this part of the skeleton with the trunk. 'The movements attendant on elevation and depression of the shoulder take place between the clavicle and the interarticular fibro-cartilage, the bone rotating upon the ligament on an axis drawn from before backwards through its own articular facet. When the shoulder is moved forwards and backwards, the clavicle, with the interarticular fibro-cartilage, rolls to and fro on the articular surface of the sternum, revolving, with a sliding movement, round an axis drawn nearly vertically through the sternum. In the circumduction of the shoulder, which is compounded of these two movements, the clavicle revolves upon the interarticular fibro-cartilage, and the latter, with the clavicle, rolls upon the sternum.' * Elevation of the clavicle is principally limited by the costo-clavicular ligament ; depression, by the interclavicular. The muscles which *raise* the clavicle, as in shrugging the shoulders, are the upper fibres of the Trapezius, the Levator anguli scapulæ, the clavicular head of the Sterno-mastoid, assisted to a certain extent by the two Rhomboids, which pull the vertebral border of the Scapula backwards and upwards and so raise the clavicle. The *depression* of the clavicle is principally effected by gravity, assisted by the Subclavius, Pectoralis minor, and lower fibres of the Trapezius. It is drawn *backwards* by the Rhomboids and the middle and lower fibres of the Trapezius, and *forwards* by the Serratus magnus and Pectoralis minor.

Surface Form.—The position of the sterno-clavicular joint may be easily ascertained by feeling the enlarged sternal end of the collar-bone just external to the long, cord-like, sternal origin of the Sterno-mastoid muscle. If this muscle is relaxed by bending the head forwards, a depression just internal to the end of the clavicle, and between it and the sternum, can be felt, indicating the exact position of the joint, which is subcutaneous. When the arm hangs by the side, the cavity of the joint is V-shaped. If the arm is raised, the bones become more closely approximated, and the cavity becomes a mere slit.

Surgical Anatomy.—The strength of this joint mainly depends upon its ligaments, and it is owing to this, and to the fact that the force of the blow is generally transmitted along the long axis of the clavicle, that dislocation rarely occurs, and that the bone is generally broken rather than displaced. When dislocation does occur, the course which the displaced bone takes depends more upon the direction in which the violence is applied than upon the anatomical construction of the joint ; it may be either forwards, backwards, or upwards. The chief point worthy of note, as regards the construction of the joint, in regard to dislocations, is the fact that, owing to the shape of the articular surfaces being so little adapted to each other, and that the strength of the joint mainly depends upon the ligaments, the displacement when reduced is very liable to recur, and hence it is extremely difficult to keep the end of the bone in its proper place.

II. ACROMIO-CLAVICULAR ARTICULATION (fig. 169)

The **Acromio-clavicular** is an arthrodial joint, formed between the outer extremity of the clavicle and the inner margin of the acromion process of the scapula. Its ligaments are the

Superior Acromio-clavicular.
Inferior Acromio-clavicular.
Interarticular Fibro-cartilage.

Coraco-clavicular { Trapezoid and Conoid.

The **Superior Acromio-clavicular Ligament** is a quadrilateral band, which covers the superior part of the articulation, extending between the upper part of the outer end of the clavicle and the adjoining part of the upper surface of the acromion. It is composed of parallel fibres, which interlace with the aponeurosis of the Trapezius and Deltoid muscles ; below, it is in contact with the interarticular fibro-cartilage (when it exists) and the synovial membranes.

The **Inferior Acromio-clavicular Ligament,** somewhat thinner than the preceding, covers the under part of the articulation, and is attached to the adjoining surfaces of the two bones. It is in relation, above, with the synovial membranes, and in rare cases with the interarticular fibro-cartilage ; below, with the tendon of the Supraspinatus. These two ligaments are continuous with each other in front and behind, and form a complete capsule round the joint.

* Humphry, *On the Human Skeleton*, p. 402.

The **Interarticular Fibro-cartilage** is frequently absent in this articulation. When it exists, it generally only partially separates the articular surfaces, and occupies the upper part of the articulation. More rarely, it completely separates the joint into two cavities.

The Synovial Membrane.—There is usually only one synovial membrane in this articulation, but when a complete interarticular fibro-cartilage exists, there are two.

The **Coraco-clavicular Ligament** serves to connect the clavicle with the coracoid process of the scapula. It does not properly belong to this articulation, but as it forms a most efficient means in retaining the clavicle in contact with the acromial process, it is usually described with it. It consists of two fasciculi, called the *trapezoid* and *conoid ligaments*.

FIG. 169.—The left shoulder-joint, scapulo-clavicular articulation, and proper ligaments of scapula.

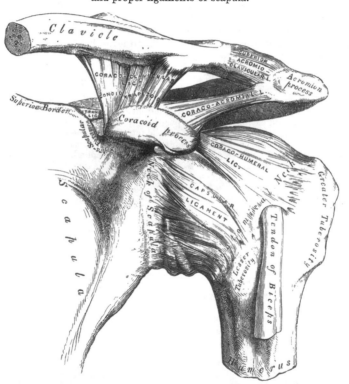

The *Trapezoid Ligament*, the anterior and external fasciculus, is broad, thin, and quadrilateral: it is placed obliquely between the coracoid process and the clavicle. It is attached, below, to the upper surface of the coracoid process; above, to the oblique line on the under surface of the clavicle. Its anterior border is free; its posterior border is joined with the conoid ligament: the two forming, by their junction, a projecting angle.

The *Conoid Ligament*, the posterior and internal fasciculus, is a dense band of fibres, conical in form, the base being directed upwards, the summit downwards. It is attached by its apex to a rough impression at the base of the coracoid process, internal to the preceding; above, by its expanded base, to the conoid tubercle on the under surface of the clavicle, and to a line proceeding internally from it for half an inch. These ligaments are in relation, in front, with the Subclavius and Deltoid; behind, with the Trapezius. They serve to limit rotation of the scapula; the Trapezoid limiting rotation forwards, and the Conoid backwards.

Actions.—The movements of this articulation are of two kinds. 1. A gliding motion of the articular end of the clavicle on the acromion. 2. Rotation of the scapula forwards and backwards upon the clavicle, the extent of this rotation being limited by the two portions of the coraco-clavicular ligament.

The acromio-clavicular joint has important functions in the movements of the upper extremity. It has been well pointed out by Sir George Humphry, that if there had been no joint between the clavicle and scapula, the circular movement of the scapula on the ribs (as in throwing the shoulders backwards or forwards) would have been attended with a greater alteration in the direction of the shoulder than is consistent with the free use of the arm in such positions, and it would have been impossible to give a blow straight forwards with the full force of the arm, that is to say, with the combined force of the scapula, arm, and fore-arm. 'This joint,' as he happily says, 'is so adjusted as to enable either bone to turn in a hinge-like manner upon a vertical axis drawn through the other, and it permits the surfaces of the scapula, like the baskets in a roundabout swing, to look the same way in every position, or nearly so.' Again, when the whole arch formed by the clavicle and scapula rises and falls (in elevation or depression of the shoulders), the joint between these two bones enables the scapula still to maintain its lower part in contact with the ribs.

Surface Form.— The position of the acromio-clavicular joint can generally be ascertained by the slightly enlarged extremity of the outer end of the clavicle, which causes it to project above the level of the acromion process of the scapula. Sometimes this enlargement is so considerable as to form a rounded eminence, which is easily to be felt. The joint lies in the plane of a vertical line passing up the middle of the front of the arm.

Surgical Anatomy.—Owing to the slanting shape of the articular surfaces of this joint, dislocation generally occurs downwards, that is to say, the acromion process of the scapula is dislocated under the outer end of the clavicle; but dislocation in the opposite direction has been described. The displacement is often incomplete, on account of the strong coraco-clavicular ligaments, which remain untorn. The same difficulty exists, as in the sterno-clavicular dislocation, in maintaining the ends of the bone in position after reduction.

III. Proper Ligaments of the Scapula (fig. 169)

The proper ligaments of the scapula are, the

Coraco-acromial.   Transverse.

The **Coraco-acromial Ligament** is a strong triangular band, extending between the coracoid and acromial processes. It is attached, by its apex, to the summit of the acromion just in front of the articular surface for the clavicle ; and by its broad base to the whole length of the outer border of the coracoid process. Its posterior fibres are directed inwards, its anterior fibres forwards and inwards. This ligament completes the vault formed by the coracoid and acromion processes for the protection of the head of the humerus. It is in relation, above, with the clavicle and under surface of the Deltoid ; below, with the tendon of the Supra-spinatus muscle, a bursa being interposed. Its outer border is continuous with a dense lamina that passes beneath the Deltoid upon the tendons of the Supra- and Infraspinatus muscles. This ligament is sometimes described as consisting of two marginal bands and a thinner intervening portion, the two bands being attached respectively to the apex and base of the coracoid process, and joining together at their attachment into the acromion process. When the Pectoralis minor is inserted, as sometimes is the case, into the capsule of the shoulder-joint instead of into the coracoid process, it passes between these two bands, and the intervening portion is then deficient.

The **Transverse** or **Coracoid** (*suprascapular*) **Ligament** converts the supra-scapular notch into a foramen. It is a thin and flat fasciculus, narrower at the middle than at the extremities, attached by one end to the base of the coracoid process, and by the other to the inner extremity of the scapular notch. The suprascapular nerve passes through the foramen ; the suprascapular vessels pass over the ligament.

An additional ligament (the *spino-glenoid*) is sometimes found on the scapula, stretching from the outer border of the spine to the margin of the glenoid cavity. When present, it forms an arch under which the suprascapular vessels and nerve pass as they enter the infraspinous fossa.

Movements of Scapula.—The scapula is capable of being moved upwards and downwards, forwards and backwards, or, by a combination of these movements, circumducted on the wall of the chest. The muscles which *raise* the scapula are the upper fibres of the Trapezius, the Levator anguli scapulæ, and the two Rhomboids ; those which *depress* it are the lower fibres of the Trapezius, the Pectoralis minor, and, through the clavicle, the Subclavius. The scapula is drawn *backwards* by the Rhomboids and the middle and lower fibres of the Trapezius, and *forwards* by the Serratus magnus and Pectoralis minor, assisted, when the arm is fixed, by the Pectoralis major. The mobility of the scapula is very considerable, and greatly assists the movements of the arm at the shoulder-joint. Thus, in raising the arm from the side, the Deltoid and Supraspinatus can only lift it to a right angle with the trunk, the further elevation of the limb being effected by the Trapezius and Serratus magnus moving the scapula on the wall of the chest. This mobility is of special importance in ankylosis of the shoulder-joint, the movements of this bone compensating to a very great extent for the immobility of the joint.

IV. Shoulder-Joint (fig. 169)

The **Shoulder** is an enarthrodial or ball-and-socket joint. The bones entering into its formation are the large globular head of the humerus, which is received into the shallow glenoid cavity of the scapula, an arrangement which permits of very considerable movement, while the joint itself is protected against displacement by the tendons which surround it and by atmospheric pressure. The ligaments do not maintain the joint surfaces in apposition, because when they alone remain the humerus can be separated to a considerable extent from the glenoid cavity ; their use therefore is to limit the amount of movement. Above, the joint is protected by an arched vault, formed by the under surface of the coracoid and acromion processes, and the coraco-acromial ligament. The articular surfaces are covered by a layer of cartilage : that on the head of the humerus is thicker at the centre than at the circumference, the reverse being the case in the glenoid cavity. The ligaments of the shoulder are, the

Capsular.	Transverse-humeral.
Coraco-humeral.	Glenoid.*

The **Capsular Ligament** completely encircles the articulation, being attached, above, to the circumference of the glenoid cavity beyond the glenoid ligament ; below, to the anatomical neck of the humerus, approaching nearer to the articular cartilage above than in the rest of its extent. It is thicker above and below than elsewhere, and is remarkably loose and lax, and much larger and longer than is necessary to keep the bones in contact, allowing them to be separated from each other more than an inch, an evident provision for that extreme freedom of movement which is peculiar to this articulation. It is strengthened, above, by the Supraspinatus ; below, by the long head of the Triceps ; behind, by the tendons of the Infraspinatus and Teres minor ; and in front, by the tendon of the Subscapularis. The capsular ligament usually presents three openings : one anteriorly, below the coracoid process, establishes a communication between the synovial membrane of the joint and a bursa beneath the tendon of the Sub-

---

* The long tendon of origin of the Biceps muscle also acts as one of the ligaments of this joint. See the observations on p. 193, on the function of the muscles passing over more than one joint.

scapularis muscle. The second, which is not constant, is at the posterior part, where a communication sometimes exists between the joint and a bursal sac belonging to the Infraspinatus muscle. The third is seen between the two tuberosities, for the passage of the long tendon of the Biceps muscle.

The **Coraco-humeral** is a broad band which strengthens the upper part of the capsular ligament. It arises from the outer border of the coracoid process, and passes obliquely downwards and outwards to the front of the great tuberosity of the humerus, being blended with the tendon of the Supraspinatus muscle. This ligament is intimately united to the capsular in the greater part of its extent.

FIG. 170.—Vertical sections through the shoulder-joint, the arm being vertical and horizontal. (After Henle.)

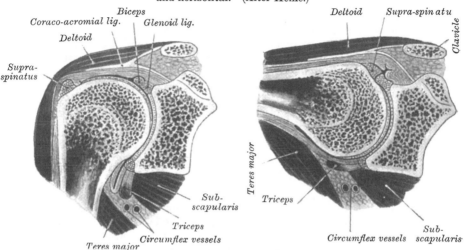

Supplemental Bands of the Capsular Ligament.—In addition to the coraco-humeral ligament, the capsular ligament is strengthened by supplemental bands in the interior of the joint. One of these bands is situated on the inner side of the joint, and passes from the inner edge of the glenoid cavity to the lower part of the lesser tuberosity of the humerus. This is sometimes known as *Flood's ligament*, and is supposed to correspond with the ligamentum teres of the hip-joint. A second of these bands is situated at the lower part of the joint, and passes from the under edge of the glenoid cavity to the under part of the neck of the humerus, and is known as *Schlemm's ligament*. A third, called the *glenohumeral ligament*, is situated at the upper part of the joint, and projects into its interior, so that it can only be seen when the capsule is opened. It is attached above to the apex of the glenoid cavity close to the root of the coracoid process, and passing downwards along the inner edge of the tendon of the Biceps, is attached below to the lesser tuberosity of the humerus, where it forms the inner boundary of the upper part of the bicipital groove. It is a thin, ribbon-like band, occasionally quite free from the capsule.

The Transverse Humeral Ligament.—This is a broad band of fibrous tissue passing from the lesser to the greater tuberosity of the humerus, and always limited to that portion of the bone which lies above the epiphysial line. It converts the bicipital groove into an osseo-aponeurotic canal, and is the analogue of the strong process of bone which connects the summits of the two tuberosities in the musk ox.

The Glenoid Ligament is a fibro-cartilaginous rim attached round the margin of the glenoid cavity. It is triangular on section, the thickest portion being fixed to the circumference of the cavity, the free edge being thin and sharp. It is continuous above with the long tendon of the Biceps muscle, which bifurcates at the upper part of the cavity into two fasciculi, and becomes continuous with the fibrous tissue of the glenoid ligament. This ligament deepens the cavity for articulation, and protects the edges of the bone. It is lined by the synovial membrane.

The **Synovial Membrane** is reflected from the margin of the glenoid cavity over the fibro-cartilaginous rim surrounding it ; it is then reflected over the internal surface of the capsular ligament, and covers the lower part and sides of the anatomical neck of the humerus as far as the cartilage covering the head of the bone. The long tendon of the Biceps muscle which passes through the capsular ligament is enclosed in a tubular sheath of synovial membrane, which is reflected upon it at the point where it perforates the capsule, and is continued around it as far as the summit of the glenoid cavity. The tendon of the Biceps is thus enabled to traverse the articulation, but it is not contained in the interior of the synovial cavity. The synovial membrane communicates with a large bursal sac beneath the tendon of the Subscapularis, by an opening at the inner side of the capsular ligament ; it also occasionally communicates with another bursal sac, beneath the tendon of the Infraspinatus, through an orifice at its posterior part. A third bursal sac, which does not communicate with the joint, is placed between the under surface of the Deltoid and the outer surface of the capsule.

The *Muscles* in relation with the joint are, above, the Supraspinatus ; below, the long head of the Triceps ; in front, the Subscapularis ; behind, the Infraspinatus and Teres minor ; within, the long tendon of the Biceps. The Deltoid is placed most externally, and covers the articulation on its outer side, as well as in front and behind.

The *Arteries* supplying the joint are articular branches of the anterior and posterior circumflex, and suprascapular.

The *Nerves* are derived from the circumflex and suprascapular.

Actions.—The shoulder-joint is capable of movement in every direction, forwards, backwards, abduction, adduction, circumduction, and rotation. The humerus is drawn *forwards* by the Pectoralis major, anterior fibres of the Deltoid, Coraco-brachialis, and by the Biceps, when the forearm is flexed ; *backwards* by the Latissimus dorsi, Teres major, posterior fibres of the Deltoid, and by the Triceps when the forearm is extended ; it is *abducted* (elevated) by the Deltoid and Supraspinatus ; it is *adducted* (depressed) by the Subscapularis, Pectoralis major, Latissimus dorsi, and Teres major ; it is *rotated outwards* by the Infraspinatus and Teres minor ; and it is *rotated inwards* by the Subscapularis, Latissimus dorsi, Teres major, and Pectoralis major.

The most striking peculiarities in this joint are : 1. The large size of the head of the humerus in comparison with the depth of the glenoid cavity, even when supplemented by the glenoid ligament. 2. The looseness of the capsule of the joint. 3. The intimate connection of the capsule with the muscles attached to the head of the humerus. 4. The peculiar relation of the Biceps tendon to the joint.

It is in consequence of the relative size of the two articular surfaces that the joint enjoys such free movement in every possible direction. When these movements of the arm are arrested in the shoulder-joint by the contact of the bony surfaces, and by the tension of the corresponding fibres of the capsule, together with that of the muscles acting as accessory ligaments, they can be carried considerably farther by the movements of the scapula, involving, of course, motion at the acromio- and sterno-clavicular joints. These joints are therefore to be regarded as accessory structures to the shoulder-joint.* The extent of these movements of the scapula is very considerable, especially in extreme elevation of the arm, which movement is best accomplished when the arm is thrown somewhat forwards and outwards, because the margin of the head of the humerus is by no means a true circle ; its greatest diameter is from the bicipital groove, downwards, inwards, and backwards, and the greatest elevation of the arm can

* See p. 221.

be obtained by rolling its articular surface in the direction of this measurement. The great width of the central portion of the humeral head also allows of very free horizontal movement when the arm is raised to a right angle, in which movement the arch formed by the acromion, the coracoid process, and the coraco-acromial ligament, constitutes a sort of supplemental articular cavity for the head of the bone.

The looseness of the capsule is so great that the arm will fall about an inch from the scapula when the muscles are dissected from the capsular ligament, and an opening made in it to remove the atmospheric pressure. The movements of the joint, therefore, are not regulated by the capsule so much as by the surrounding muscles and by the pressure of the atmosphere, an arrangement which ' renders the movements of the joint much more easy than they would otherwise have been, and permits a swinging, pendulum-like vibration of the limb when the muscles are at rest' (Humphry). The fact, also, that in all ordinary positions of the joint the capsule is not put on the stretch, enables the arm to move freely in all directions. Extreme movements are checked by the tension of appropriate portions of the capsule, as well as by the interlocking of the bones. Thus it is said that ' abduction is checked by the contact of the great tuberosity with the upper edge of the glenoid cavity, adduction by the tension of the coraco-humeral ligament' (Beaunis et Bouchard). Cleland* maintains that the limitations of movement at the shoulder-joint are due to the structure of the joint itself, the glenoid ligament fitting, in different positions of the elevated arm, into the anatomical neck of the humerus.

Cathcart* has pointed out that in abducting the arm and raising it above the head, the scapula rotates throughout the whole movement with the exception of a short space at the beginning and at the end ; that the humerus moves on the scapula not only from the hanging to the horizontal position but also in passing upwards as it approaches the vertical above ; that the clavicle moves not only during the second half of the movement but in the first as well, though to a less extent—i.e. the scapula and clavicle are concerned in the first stage as well as in the second ; and that the humerus is partly involved in the second as well as chiefly in the first.

The intimate union of the tendons of the four short muscles with the capsule converts these muscles into elastic and spontaneously acting ligaments of the joint, and it is regarded as being also intended to prevent the folds into which all portions of the capsule would alternately fall in the varying positions of the joint from being driven between the bones by the pressure of the atmosphere.

The peculiar relations of the Biceps tendon to the shoulder-joint appear to subserve various purposes. In the first place, by its connection with both the shoulder and elbow the muscle harmonises the action of the two joints, and acts as an elastic ligament in all positions, in the manner previously adverted to.† Next, it strengthens the upper part of the articular cavity, and prevents the head of the humerus from being pressed up against the acromion process, when the Deltoid contracts, instead of forming the centre of motion in the glenoid cavity. By its passage along the bicipital groove it assists in rendering the head of the humerus steady in the various movements of the arm. When the arm is raised from the side it assists the Supra- and Infraspinatus in rotating the head of the humerus in the glenoid cavity. It also holds the head of the bone firmly in contact with the glenoid cavity, and prevents its slipping over its lower edge, or being displaced by the action of the Latissimus dorsi and Pectoralis major, as in climbing and many other movements.

Surface Form.—The direction and position of the shoulder-joint may be indicated by a line drawn from the middle of the coraco-acromial ligament, in a curved direction, with its convexity inwards, to the innermost part of that portion of the head of the humerus which can be felt in the axilla when the arm is forcibly abducted from the side. When the arm hangs by the side, not more than one-third of the head of the bone is in contact with the glenoid cavity, and three-quarters of its circumference is in front of a vertical line drawn from the anterior border of the acromion process.

Surgical Anatomy.—Owing to the construction of the shoulder-joint and the freedom of movement which it enjoys, as well as in consequence of its exposed situation, it is more frequently dislocated than any other joint in the body. Dislocation occurs when the arm is abducted, and when, therefore, the head of the humerus presses against the lower and front part of the capsule, which is the thinnest and least supported part of the ligament. The rent in the capsule almost invariably takes place in this situation, and through it the head of the bone escapes, so that the dislocation in most instances is primarily subglenoid. The head of the bone does not usually remain in this situation, but generally assumes some other position, which varies according to the direction and amount of force producing the dislocation and the relative strength of the muscles in front and behind the joint. In consequence of the muscles at the back being stronger than those in front, and especially on account of the long head of the Triceps preventing the bone passing backwards, dislocation forwards is much more common than backwards. The most frequent position which the head of the humerus ultimately assumes is on the front of the neck of the scapula, beneath the coracoid process, and hence named subcoracoid dislocation. Occasionally, in consequence probably of a greater amount of force being brought to bear on the limb, the head is driven farther inwards, and rests on the upper part of the front of the chest, beneath the clavicle (subclavicular). Sometimes it remains in the position in which it was primarily displaced, resting on the axillary border of the scapula (subglenoid), and rarely it passes backwards and remains in the infraspinatus fossa, beneath the spine (subspinous).

The shoulder-joint is sometimes the seat of all those inflammatory affections, both acute and chronic, which attack joints, though perhaps less frequently than some other joints of equal size and importance. Acute synovitis may result from injury, rheumatism, or pyæmia, or may follow secondarily on acute epiphysitis of infants. It is attended with effusion into the joint, and when this occurs the capsule is evenly distended, and the contour of the joint rounded. Special projections may occur at the site of the openings in the capsular ligament. Thus a swelling may appear just in front of the joint, internal to the lesser tuberosity, from effusion into the bursa beneath the Subscapularis muscle ; or, again, a swelling which is sometimes bilobed may be seen in the interval between the Deltoid and Pectoralis major muscles, from effusion into the diverticulum, which runs down the bicipital groove with the tendon of the Biceps. The effusion into the synovial membrane can be best ascertained by examination from the axilla, where a soft, elastic, fluctuating swelling can usually be felt.

Tuberculous arthritis not infrequently attacks the shoulder-joint, and may lead to total destruction of the articulation, when ankylosis may result, or long-protracted suppuration may necessitate excision. This joint is also one of those which is most liable to be the seat of osteo-arthritis, and may also be affected in gout and rheumatism ; or in locomotor ataxy, when it becomes the seat of Charcot's disease.

Excision of the shoulder-joint may be required in cases of arthritis (especially the tuberculous form) which have gone on to destruction of the articulation ; in compound dislocations and fractures, particularly those arising from gunshot injuries, in which there has been extensive injury to the head of the bone ; in some cases of old unreduced dislocation, where there is much pain ; and possibly in some few cases of growth connected with the upper end of the bone. The operation is best performed by making an incision from the middle of the coraco-acromial ligament down the arm for about three inches : this will expose the bicipital groove and the tendon of the biceps, which may be either divided or hooked out of the way, according as to whether it is implicated in the disease or not. The capsule is freely opened, and the muscles attached to the greater and lesser tuberosities of the humerus divided. The head of the bone can then be thrust out of the wound and sawn off ; or divided with a narrow saw *in situ* and subsequently removed. The section should be made, if possible, just below the articular surface, so as to leave the bone as long as possible. The glenoid cavity must then be examined, and gouged if carious.

V. ELBOW-JOINT (figs. 171, 172)

The **Elbow** is a *ginglymus* or hinge-joint. The bones entering into its formation are the trochlea of the humerus, which is received into the greater sigmoid cavity of the ulna, and admits of the movements peculiar to this joint, viz. flexion and extension ; while the lesser, or radial, head of the humerus articulates with the cup-shaped depression on the head of the radius ; the circumference of the head of the radius articulates with the lesser sigmoid cavity of the ulna, allowing of the movement of rotation of the radius on the ulna, the chief action of the superior radio-ulnar articulation. The articular surfaces are covered with a thin layer of cartilage, and connected together by a capsular ligament of unequal thickness, being especially thickened on its two sides, and, to a less extent, in front and behind. These thickened portions are usually described as distinct ligaments under the following names:

Anterior.   Internal Lateral.
Posterior.   External Lateral.

The orbicular ligament of the upper radio-ulnar articulation must also be reckoned among the ligaments of the elbow.

The **Anterior Ligament** (fig. 171) is a broad and thin fibrous layer, which covers the anterior surface of the joint. It is attached to the front of the internal condyle and to the front of the humerus immediately above the coronoid and radial fossæ; below, to the anterior surface of the coronoid process of the ulna

FIG. 171.—Left elbow-joint, showing anterior and internal ligaments.

FIG. 172.—Left elbow-joint, showing posterior and external ligaments.

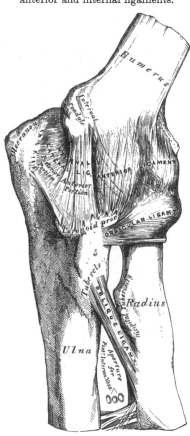
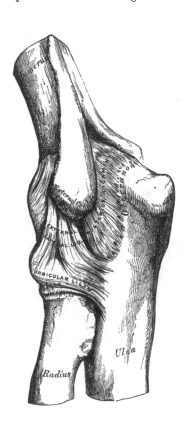

and orbicular ligament, being continuous on each side with the lateral ligaments. Its superficial fibres pass obliquely from the inner condyle of the humerus outwards to the orbicular ligament. The middle fibres, vertical in direction, pass from the upper part of the coronoid depression and become partly blended with the preceding, but mainly inserted into the anterior surface of the coronoid process. The deep or transverse set intersects these at right angles. This ligament is in relation, in front, with the Brachialis anticus, except at its outermost part; behind, with the synovial membrane.

The **Posterior Ligament** (fig. 172) is a thin and loose membranous fold, attached, above, to the lower end of the humerus, above and at the sides of the olecranon fossa; below, to the groove on the upper and outer surfaces of the olecranon. The superficial or transverse fibres pass between the adjacent margins of the olecranon fossa. The deeper portion consists of vertical fibres, some of which, thin and weak, pass from the upper part of the olecranon fossa to the margin of the olecranon ; others, thicker and stronger, pass from the back of the capitellum of the humerus to the posterior border of the lesser sigmoid cavity of the ulna. This ligament is in relation, behind, with the tendon of the Triceps and the Anconeus ; in front, with the synovial membrane.

The **Internal Lateral Ligament** (fig. 171) is a thick triangular band consisting of two portions, an anterior and posterior, united by a thinner intermediate portion. The *anterior portion*, directed obliquely forwards, is attached, above, by its apex, to the front part of the internal condyle of the humerus ; and, below, by its broad base, to the inner margin of the coronoid process. The *posterior portion*, also of triangular form, is attached, above, by its apex, to the lower and back part of the internal condyle ; below, to the inner margin of the olecranon. Between these two bands a few intermediate fibres descend from the internal condyle to blend with a transverse band of ligamentous tissue which bridges across the notch between the olecranon and coronoid processes. This ligament is in relation, internally, with the Triceps and Flexor carpi ulnaris muscles, and the ulnar nerve, and gives origin to part of the Flexor sublimis digitorum.

The **External Lateral Ligament** (fig. 172) is a short and narrow fibrous band, less distinct than the internal, attached, above, to a depression below the external condyle of the humerus ; below, to the orbicular ligament, some of its most posterior fibres passing over that ligament, to be inserted into the outer margin of the ulna. This ligament is intimately blended with the tendon of origin of the Supinator brevis muscle.

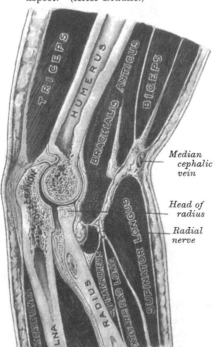

FIG. 173.—Sagittal section of the right elbow-joint, taken somewhat obliquely and seen from the radial aspect. (After Braune.)

Median cephalic vein

Head of radius

Radial nerve

The **Synovial Membrane** is very extensive. It covers the margin of the articular surface of the humerus, and lines the coronoid and olecranon fossæ on that bone : from these points, it is reflected over the anterior, posterior, and lateral ligaments ; and forms a pouch between the lesser sigmoid cavity, the internal surface of the orbicular ligament, and the circumference of the head of the radius. Projecting into the cavity is a crescentic fold of synovial membrane, between the radius and ulna, suggesting the division of the joint into two : one the humero-radial, the other the humero-ulnar.

Between the capsular ligament and the synovial membrane are three masses of fat : one, the largest, above the olecranon fossa, which is pressed into the fossa by the Triceps during flexion ; a second, over the coronoid fossa ; and a third over the radial fossa. These are pressed into their respective fossæ during extension.

The *Muscles* in relation with the joint are, in front, the Brachialis anticus ; behind, the Triceps and Anconeus ; externally, the Supinator brevis, and the common tendon of origin of the Extensor muscles ; internally, the common tendon of origin of the Flexor muscles, and the Flexor carpi ulnaris, with the ulnar nerve (fig. 173).

The *Arteries* supplying the joint are derived from the anastomosis between the superior profunda, inferior profunda, and anastomotica magna arteries, branches of the brachial, with the anterior, posterior, and interosseous recurrent branches of the ulna, and the recurrent branch of the radial. These vessels form a complete chain of inosculation around the joint.

The *Nerves* are derived from the ulnar, as it passes between the internal condyle and the olecranon ; a filament from the musculo-cutaneous (Rüdinger), and two from the median (Macalister).

Actions.—The elbow-joint comprises three different portions : viz. the joint between the ulna and humerus, that between the head of the radius and the humerus, and the superior radio-ulnar articulation, described below. All these articular surfaces are invested by a common synovial membrane, and the movements of the whole joint should be studied together. The combination of the movements of flexion and extension of the forearm with those of pronation and supination of the hand, which is ensured by the two being performed at the same joint, is essential to the accuracy of the various minute movements of the hand.

The portion of the joint between the ulna and humerus is a simple hinge-joint, and allows of movements of flexion and extension only. Owing to the obliquity of the trochlear surface of the humerus, this movement does not take place in a straight line ; so that when the forearm is extended and supinated, the axis of the arm and forearm is not in the same line, but the one portion of the limb forms an angle with the other, and the hand, with the forearm, is directed outwards. During flexion, on the other hand, the forearm and the hand tend to approach the middle line of the body, and thus enable the hand to be easily carried to the face. The shape of the articular surface of the humerus, with its prominences and depressions accurately adapted to the opposing surfaces of the olecranon, prevents any lateral movement. *Flexion* is produced by the action of the Biceps and Brachialis anticus, assisted by the muscles arising from the internal condyle of the humerus and the Supinator longus ; *extension*, by the Triceps and Anconeus, assisted by the extensors of the wrist and by the Extensor communis digitorum and Extensor minimi digiti.

The joint between the head of the radius and the capitellum or radial head of the humerus is an arthrodial joint. The bony surfaces would of themselves constitute an enarthrosis and allow of movement in all directions, were it not for the orbicular ligament by which the head of the radius is bound down firmly to the sigmoid cavity of the ulna, and which prevents any separation of the two bones laterally. It is to the same ligament that the head of the radius owes its security from dislocation, which would otherwise constantly occur, as a consequence of the shallowness of the cup-like surface on the head of the radius. In fact, but for this ligament, the tendon of the Biceps would be liable to pull the head of the radius out of the joint.* In complete extension, the head of the radius glides so far back on the outer condyle that its edge is plainly felt at the back of the articulation. Flexion and extension of the elbow-joint are limited by the tension of the structures on the front and back of the joint ; the limitation of flexion being also aided by the soft structures of the arm and forearm coming into contact.

In combination with any position of flexion or extension, the head of the radius can be rotated in the upper radio-ulnar joint, carrying the hand with it. The hand is directly articulated to the lower surface of the radius only, and the concave or sigmoid surface on the lower end of the radius travels round the lower end of the ulna. The latter bone is excluded from the wrist-joint (as will be seen in the sequel) by the interarticular fibro-cartilage. Thus, rotation of the head of the radius round an axis which passes through the centre of the radial head of the humerus imparts circular movement to the hand through a very considerable arc.

* Humphry, *op. cit.* p. 419.

Surface Form.—If the forearm be slightly flexed on the arm, a curved crease or fold with its convexity downwards may be seen running across the front of the elbow, extending from one condyle to the other. The centre of this fold is some slight distance above the line of the joint. The position of the radio-humeral portion of the joint can be at once ascertained by feeling for a slight groove or depression between the head of the radius and the capitellum of the humerus at the back of the articulation.

Surgical Anatomy.—From the great breadth of the joint, and the manner in which the articular surfaces are interlocked, and also on account of the strong lateral ligaments and the support which the joint derives from the mass of muscles attached to each condyle of the humerus, lateral displacement of the bones is very uncommon. Whereas antero-posterior dislocation, on account of the shortness of the antero-posterior diameter, the weakness of the anterior and posterior ligaments, and the want of support of muscles, occurs much more frequently. Dislocation backwards takes place when the forearm is in a position of extension, and forwards when in a position of flexion. For, in the former position, that of extension, the coronoid process is not interlocked into the coronoid fossa, and loses its grip to a certain extent, whereas the olecranon process is in the olecranon fossa, and entirely prevents displacement forwards. On the other hand, during flexion, the coronoid process is in the coronoid fossa, and prevents dislocation backwards, while the olecranon loses its grip and is not so efficient, as during extension, in preventing a forward displacement. When lateral dislocation does take place it is generally incomplete.

Dislocation of the elbow-joint is of common occurrence in children, far more common than dislocation of any other articulation, for, as a rule, fracture of a bone more frequently takes place, under the application of any severe violence, in young persons, than dislocation. In lesions of this joint there is often very great difficulty in ascertaining the exact nature of the injury.

The elbow-joint is occasionally the seat of acute synovitis. The synovial membrane then becomes distended with fluid, the bulging showing itself principally around the olecranon process, that is to say, on its inner and outer sides and above, in consequence of the laxness of the posterior ligament. Occasionally a well-marked triangular projection may be seen on the outer side of the olecranon, from bulging of the synovial membrane beneath the Anconeus muscle. Again, there is often some swelling just above the head of the radius, in the line of the radio-humeral joint. There is not generally much swelling at the front of the joint, though sometimes deep-seated fulness beneath the Brachialis anticus may be noted. When suppuration occurs the abscess usually points at one or other border of the Triceps muscle ; occasionally the pus discharges itself in front, near the insertion of the Brachialis anticus muscle. Chronic synovitis, usually of tuberculous origin, is of common occurrence in the elbow-joint : under these circumstances the forearm tends to assume the position of semi-flexion, which is that of greatest ease and relaxation of ligaments. It should be borne in mind, that should ankylosis occur in this or the extended position, the limb will not be nearly so useful as if ankylosed in a position of rather less than a right angle. Loose cartilages are sometimes met with in the elbow-joint, not so commonly, however, as in the knee ; nor do they, as a rule, give rise to such urgent symptoms, and rarely require operative interference. The elbow-joint is also sometimes affected with osteo-arthritis, but this affection is less common in this articulation than in some other of the larger joints.

Excision of the elbow is principally required for three conditions—viz. tuberculous arthritis, injury and its results, and faulty ankylosis—but may be necessary for some other rarer conditions, such as disorganising arthritis after pyæmia, unreduced dislocations, and osteo-arthritis. The results of the operation are, as a rule, more favourable than those of excision of any other joint, and it is one, therefore, that the surgeon should never hesitate to perform, especially in the first three of the conditions mentioned above. The operation is best performed by a single vertical incision down the back of the joint ; a transverse incision, over the outer condyle, being added if the parts are much thickened and fixed. A straight incision is made about four inches long, the mid-point of which is on a level with and a little to the inner side of the tip of the olecranon. This incision is made down to the bone, through the substance of the Triceps muscle. The operator with the point of his knife, and guarding the soft parts with his thumb-nail, separates them from the bone. In doing this there are two structures which he should carefully avoid : the ulnar nerve, which lies parallel to his incision, but a little internal, as it courses down between the internal condyle and the olecranon process ; and the prolongation of the Triceps into the deep fascia of the forearm over the Anconeus muscle. Having cleared the bones and divided the lateral and posterior ligaments, the forearm is strongly flexed and the ends of the bones turned out and sawn off. The section of the humerus should be through the base of the condyles, that of the ulna and radius should be just below the level of the lesser sigmoid cavity of the ulna and the neck of the radius. In this operation the object is to obtain such union as shall allow free motion of the bones of the forearm ; and, therefore, passive motion must be commenced early—that is to say, about the tenth day.

VI. Radio-ulnar Articulations

The articulation of the radius with the ulna is effected by ligaments, which connect together both extremities as well as the shafts of these bones. They may, consequently, be subdivided into three sets : 1. The superior radio-ulnar, which is a portion of the elbow-joint ; 2. the middle radio-ulnar ; and, 3. the inferior radio-ulnar articulations.

1. Superior Radio-ulnar Articulation

This articulation is a trochoid or pivot-joint. The bones entering into its formation are the inner side of the circumference of the head of the radius rotating within the lesser sigmoid cavity of the ulna. Its only ligament is the *annular* or *orbicular.*

The **Orbicular Ligament** (fig. 172) is a strong, flat band of ligamentous fibres, which surrounds the head of the radius, and retains it in firm connection with the lesser sigmoid cavity of the ulna. It forms about four-fifths of an osseo-fibrous ring, attached by each end to the extremities of the lesser sigmoid cavity, and is smaller at the lower part of its circumference than above, by which means the head of the radius is more securely held in its position. Its *outer surface* is strengthened by the external lateral ligament of the elbow, and affords origin to part of the Supinator brevis muscle. Its *inner surface* is smooth, and lined by synovial membrane. The synovial membrane is continuous with that which lines the elbow-joint.

Actions.—The movement which takes place in this articulation is limited to rotation of the head of the radius within the orbicular ligament, and upon the lesser sigmoid cavity of the ulna ; rotation forwards being called *pronation* ; rotation backwards, *supination.* Supination is performed by the Biceps and Supinator brevis, assisted to a slight extent by the Extensor muscles of the thumb and, in certain positions, by the Supinator longus. Pronation is performed by the Pronator radii teres and the Pronator quadratus, assisted, in some positions, by the Supinator longus.

Surface Form.—The position of the superior radio-ulnar joint is marked on the surface of the body by the little dimple on the back of the elbow which indicates the position of the head of the radius.

Surgical Anatomy.—Dislocation of the head of the radius alone is not an uncommon accident, and occurs most frequently in young persons from falls on the hand when the forearm is extended and supinated, the head of the bone being displaced forward. It is attended by rupture of the orbicular ligament. Occasionally a peculiar injury, which is supposed to be a subluxation, occurs in young children in lifting them from the ground by the hand or forearm. It is believed that the head of the radius is displaced downwards in the orbicular ligament, the upper border of which becomes folded over the head of the radius, between it and the capitellum of the humerus.

2. Middle Radio-ulnar Union

The interval between the shafts of the radius and ulna is occupied by two ligaments :

Oblique.   Interosseous.

The **Oblique** or **Round Ligament** (fig. 171) is a small, flattened, fibrous band, which extends obliquely downwards and outwards, from the tubercle of the ulna at the base of the coronoid process to the radius a little below the bicipital tuberosity. Its fibres run in the opposite direction to those of the interosseous ligament ; and it appears to be placed as a substitute for it in the upper part of the interosseous interval. This ligament is sometimes wanting.

The **Interosseous Membrane** is a broad and thin plane of fibrous tissue descending obliquely downwards and inwards, from the interosseous ridge on the radius to that on the ulna. It is deficient above, commencing about an inch beneath the tubercle of the radius; is broader in the middle than at either extremity; and presents an oval aperture just above its lower margin for the passage of the anterior interosseous vessels to the back of the forearm. This ligament serves to connect the bones, and to increase the extent of surface for the attachment of the deep muscles. Between its upper border and the oblique ligament an interval exists, through which the posterior interosseous vessels pass. Two or three fibrous bands are occasionally found on the posterior surface of this membrane, which descend obliquely from the ulna towards the radius, and which have consequently a direction contrary to that of the other fibres. It is in relation, *in front*, by its upper three-fourths, with the Flexor longus pollicis on the outer side, and with the Flexor profundus digitorum on the inner, lying upon the interval between which are the anterior interosseous vessels and nerve; by its lower fourth with the Pronator quadratus; *behind*, with the Supinator brevis, Extensor ossis metacarpi pollicis, Extensor brevis pollicis, Extensor longus pollicis, Extensor indicis; and, near the wrist, with the anterior interosseous artery and posterior interosseous nerve.

3. Inferior Radio-ulnar Articulation

This is a pivot-joint, formed by the head of the ulna received into the sigmoid cavity at the inner side of the lower end of the radius. The articular surfaces are covered by a thin layer of cartilage, and connected together by the following ligaments :

Anterior Radio-ulnar.   Posterior Radio-ulnar.
Interarticular Fibro-cartilage.

The **Anterior Radio-ulnar Ligament** (fig. 174) is a narrow band of fibres extending from the anterior margin of the sigmoid cavity of the radius to the anterior surface of the head of the ulna.

The **Posterior Radio-ulnar Ligament** (fig. 175) extends between similar points on the posterior surface of the articulation.

The **Interarticular Fibro-cartilage** (fig. 177) is triangular in shape, and is placed transversely beneath the head of the ulna, binding the lower end of this bone and the radius firmly together. Its periphery is thicker than its centre, which is thin and occasionally perforated. It is attached by its apex to a depression which separates the styloid process of the ulna from the head of that bone; and by its base, which is thin, to the prominent edge of the radius, which separates the sigmoid cavity from the carpal articulating surface. Its margins are united to the ligaments of the wrist-joint. Its *upper surface*, smooth and concave, articulates with the head of the ulna, forming an arthrodial joint; its *under*

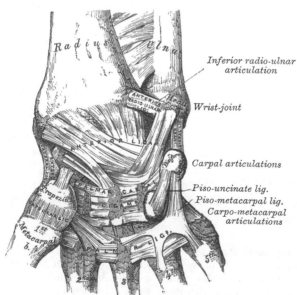

Fig. 174.—Ligaments of wrist and hand.   Anterior view.

Inferior radio-ulnar articulation

Wrist-joint

Carpal articulations

Piso-uncinate lig.
Piso-metacarpal lig.
Carpo-metacarpal articulations

surface, also concave and smooth, forms part of the wrist-joint and articulates with the cuneiform and inner part of the semilunar bones. Both surfaces are lined by a synovial membrane : the upper surface, by one peculiar to the radio-ulnar articulation ; the under surface, by the synovial membrane of the wrist.

The **Synovial Membrane** (fig. 177) of this articulation has been called, from its extreme looseness, the *membrana sacciformis* ; it extends horizontally inwards between the head of the ulna and the inter-articular fibro-carti-lage, and upwards between the radius and the ulna, forming here a very loose *cul-de-sac*. The quantity of synovia which it contains is usually considerable.

Actions. — The movement in the inferior radio-ulnar articulation is just the reverse of that in the superior radio-ulnar joint. It consists of a movement of rotation of the lower end of the radius round an axis which corresponds to the centre of the head of the ulna. When

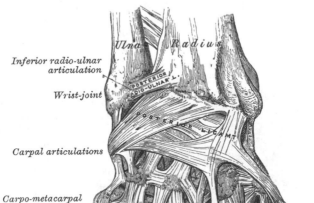

FIG. 175.—Ligaments of wrist and hand. Posterior view.

Inferior radio-ulnar articulation

Wrist-joint

Carpal articulations

Carpo-metacarpal articulations

the radius rotates forwards, *pronation* of the forearm and hand is the result ; and when backwards, *supination*. It will thus be seen that in pronation and supination of the forearm and hand the radius describes a segment of a cone, the axis of which extends from the centre of the head of the radius to the middle of the head of the ulna. In this movement, however, the ulna is not quite stationary, but rotates a little in the opposite direction. So that it also describes the segment of a cone, though of smaller size than that described by the radius. The movement which causes this alteration in the position of the head of the ulna takes place principally at the shoulder-joint by a rotation of the humerus, but possibly also to a slight extent at the elbow-joint.*

Surface Form.—The position of the inferior radio-ulnar joint may be ascertained by feeling for a slight groove at the back of the wrist, between the prominent head of the ulna and the lower end of the radius, when the forearm is in a state of almost complete pronation.

VII. RADIO-CARPAL OR WRIST-JOINT

The **Wrist** is a condyloid articulation. The parts entering into its formation are the lower end of the radius and under surface of the interarticular fibro-cartilage, which form together the receiving cavity ; and the scaphoid, semilunar, and cuneiform bones, which form the condyle. The articular surface of the radius and the under surface of the interarticular fibro-cartilage, the receiving cavity, form together a transversely elliptical concave surface. The articular surfaces of the scaphoid, semilunar, and cuneiform bones form together a smooth, convex surface, the *condyle*, which is received into the concavity above mentioned. All the bony surfaces of the articulation are covered with cartilage,

* See *Journ. of Anat. and Phys.* vol. xix. parts ii., iii. and iv.

and connected together by a capsule, which is divided into the following ligaments :—

| External Lateral. | Anterior. |
| Internal Lateral. | Posterior. |

The **External Lateral Ligament** (*radio-carpal*) (fig. 174) extends from the summit of the styloid process of the radius to the outer side of the scaphoid, some of its fibres being prolonged to the trapezium and annular ligament.

The **Internal Lateral Ligament** (*ulno-carpal*) is a rounded cord, attached, above, to the extremity of the styloid process of the ulna ; and dividing below into two fasciculi, which are attached, one to the inner side of the cuneiform bone, the other to the pisiform bone and annular ligament.

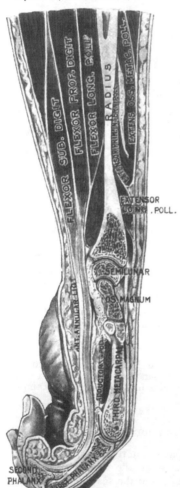

FIG. 176.—Longitudinal section of the right forearm, hand, and third finger, viewed from the ulnar aspect. (After Braune.)

The **Anterior Ligament** is a broad membranous band, attached, above, to the anterior margin of the lower end of the radius, its styloid process and the ulna ; its fibres pass downwards and inwards to be inserted into the palmar surface of the scaphoid, semilunar, and cuneiform bones, some of the fibres being continued to the os magnum. In addition to this broad membrane, there is a distinct rounded fasciculus, superficial to the rest, which passes from the base of the styloid process of the ulna to the semilunar and cuneiform bones. This ligament is perforated by numerous apertures for the passage of vessels, and is in relation, in front, with the tendons of the Flexor profundus digitorum and Flexor longus pollicis ; behind, with the synovial membrane of the wrist-joint.

The **Posterior Ligament** (fig. 175), less thick and strong than the anterior, is attached, above, to the posterior border of the lower end of the radius ; its fibres pass obliquely downwards and inwards, to be attached to the dorsal surface of the scaphoid, semilunar, and cuneiform bones, being continuous with those of the dorsal carpal ligaments. This ligament is in relation, behind, with the extensor tendons of the fingers ; in front, with the synovial membrane of the wrist.

The **Synovial Membrane** (fig. 177) lines the inner surface of the ligaments above described, extending from the lower end of the radius and interarticular fibro-cartilage above to the articular surfaces of the carpal bones below. It is loose and lax, and presents numerous folds, especially behind.

Relations.—The wrist-joint is covered in front by the flexor, and behind by the extensor tendons ; it is also in relation with the radial and ulnar arteries.

The *Arteries* supplying the joints are the anterior and posterior carpal branches of the radial and ulnar, the anterior and posterior interosseous, and some ascending branches from the deep palmar arch.

The *Nerves* are derived from the ulnar and posterior interosseous.

Action.—The movements permitted in this joint are flexion, extension, abduction, adduction, and circumduction. Its actions will be further studied with those of the carpus, with which they are combined.

Surface Form.—The line of the radio-carpal joint is on a level with the apex of the styloid process of the ulna.

Surgical Anatomy.—The wrist-joint is rarely dislocated, its strength depending mainly upon the numerous strong tendons which surround the articulation. Its security is further provided for by the number of small bones of which the carpus is made up, and which are united by very strong ligaments. The slight movement which takes place between the several bones serves to break the jars that result from falls or blows on the hand. Dislocation backwards, which is the more common, simulates to a considerable extent Colles's fracture of the radius, and is liable to be mistaken for it. The diagnosis can be easily made out by observing the relative position of the styloid processes of the radius and the ulna. In the natural condition the styloid process of the radius is on a lower level, i.e. nearer the ground, when the arm hangs by the side, than that of the ulna, and the same would be the case in dislocation. In Colles's fracture, on the other hand, the styloid process of the radius is on the same, or even a higher level than that of the ulna.

The wrist-joint is occasionally the seat of acute synovitis, the result of traumatism, or arising in the rheumatic or pyæmic state. When the synovial sac is distended with fluid, the swelling is greatest on the dorsal aspect of the wrist, showing a general fulness, with some bulging between the tendons. The inflammation is prone to extend to the intercarpal joints and to attack also the sheaths of the tendons in the neighbourhood. Chronic inflammation of the wrist is generally tuberculous, and often leads to similar disease in the synovial sheaths of adjacent tendons and of the intercarpal joints. The disease, therefore, when progressive, frequently leads to necrosis of the carpal bones, and the result is often unsatisfactory.

VIII. Articulations of the Carpus

These articulations may be subdivided into three sets :

1. The Articulations of the First Row of Carpal Bones.
2. The Articulations of the Second Row of Carpal Bones.
3. The Articulations of the Two Rows with each other.

1. Articulations of the First Row of Carpal Bones

These are arthrodial joints. The ligaments connecting the scaphoid, semilunar, and cuneiform bones are—

Dorsal. Palmar.
Two Interosseous.

The **Dorsal Ligaments** are placed transversely behind the bones of the first row ; they connect the scaphoid and semilunar, and the semilunar and cuneiform.

The **Palmar Ligaments** connect the scaphoid and semilunar, and the semilunar and cuneiform bones ; they are less strong than the dorsal, and placed very deeply below the anterior ligament of the wrist.

The **Interosseous Ligaments** (fig. 177) are two narrow bundles of fibrous tissue, connecting the semilunar bone, on one side with the scaphoid, and on the other with the cuneiform. They are on a level with the superior surfaces of these bones, and close the upper part of the spaces between them. Their upper surfaces are smooth, and form with the bones the convex articular surfaces of the wrist-joint.

The ligaments connecting the pisiform bone are—

Capsular. Two palmar ligaments.

The **Capsular Ligament** is a thin membrane which connects the pisiform bone to the cuneiform. It is lined with a separate synovial membrane.

The **two Palmar Ligaments** are two strong fibrous bands which connect the pisiform to the unciform, the *piso-uncinate*, and to the base of the fifth metacarpal bone, the *piso-metacarpal ligament* (fig. 174).

2. Articulations of the Second Row of Carpal Bones

These are also arthrodial joints. The articular surfaces are covered with cartilage, and connected by the following ligaments :

Dorsal.   Palmar.

Three Interosseous.

The **Dorsal Ligaments** extend transversely from one bone to another on the dorsal surface, connecting the trapezium with the trapezoid, the trapezoid with the os magnum, and the os magnum with the unciform.

The **Palmar Ligaments** have a similar arrangement on the palmar surface.

The **three Interosseous Ligaments,** much thicker than those of the first row, are placed one between the os magnum and the unciform, a second between the os magnum and the trapezoid, and a third between the trapezium and trapezoid. The first of these is much the strongest, and the third is sometimes wanting.

3. Articulations of the Two Rows of Carpal Bones with each other

The joint between the scaphoid, semilunar, and cuneiform, and the second row of the carpus, or the *mid-carpal joint*, is made up of three distinct portions ; in the centre the head of the os magnum and the superior surface of the unciform articulate with the deep cup-shaped cavity formed by the scaphoid and semilunar bones, and constitute a sort of ball-and-socket joint. On the outer side the trapezium and trapezoid articulate with the scaphoid, and on the inner side the unciform articulates with the cuneiform, forming gliding joints.

The ligaments are—

Anterior or Palmar.   External Lateral.

Posterior or Dorsal.   Internal Lateral.

The **Anterior** or **Palmar Ligament** consists of short fibres, which pass, for the most part, from the palmar surface of the bones of the first row to the front of the os magnum.

The **Posterior** or **Dorsal Ligament** consists of short, irregular bundles of fibres passing between the bones of the first and second row on the dorsal surface of the carpus.

The **Lateral Ligaments** are very short : they are placed, one on the radial, the other on the ulnar side of the carpus ; the former, the stronger and more distinct, connecting the scaphoid and trapezium, the latter the cuneiform and unciform ; they are continuous with the lateral ligaments of the wrist-joint. In addition to these ligaments, a slender interosseous band sometimes connects the os magnum and the scaphoid.

The **Synovial Membrane of the Carpus** is very extensive ; it passes from the under surface of the scaphoid, semilunar, and cuneiform bones to the upper surface of the bones of the second row, sending upwards two prolongations— between the scaphoid and semilunar, and the semilunar and cuneiform—sending downwards three prolongations between the four bones of the second row, which are further continued onwards into the carpo-metacarpal joints of the four inner metacarpal bones, and also for a short distance between the metacarpal bones. There is a separate synovial membrane between the pisiform and cuneiform bones.

Actions.—The articulation of the hand and wrist, considered as a whole, is divided into three parts : (1) the radius and the interarticular fibro-cartilage ; (2) the meniscus, formed by the scaphoid, semilunar, and cuneiform—the pisiform bone having no essential part in the movement of the hand ; (3) the hand proper, the metacarpal bones with the four carpal bones on which they are supported, viz. the trapezium, trapezoid, os magnum, and unciform. These three elements form two joints : (1) the superior (wrist-joint proper), between the meniscus and bones of the forearm ; (2) the inferior, between the hand and meniscus (transverse or mid-carpal joint).

(1) The articulation between the forearm and carpus is a true condyloid articulation, and therefore all movements but rotation are permitted. Flexion and extension are the most free, and of these a greater amount of extension than flexion is permitted on account of the articulating surfaces extending farther on the dorsal than on the palmar aspect of the carpal bones. In this movement the carpal bones rotate on a transverse axis drawn between the tips of the styloid processes of the radius and ulna. A certain amount of adduction (or ulnar flexion) and abduction (or radial flexion) is also permitted. Of these the former is considerably greater in extent than the latter. In this movement the carpus revolves upon an antero-posterior axis drawn through the centre of the wrist. Finally, circumduction is permitted by the consecutive movements of adduction, extension, abduction, and flexion, with intermediate movements between them. There is no rotation, but this is provided for by the supination and pronation of the radius on the ulna. The movement of *flexion* is performed by the Flexor carpi radialis, the Flexor carpi ulnaris, and the Palmaris longus ; *extension* by the Extensor carpi radialis longior et brevior and the Extensor carpi ulnaris ; *adduction* (ulnar flexion) by the Flexor carpi ulnaris and the Extensor carpi ulnaris ; and *abduction* (radial flexion) by the Extensors of the thumb, and the Extensor carpi radialis longior et brevior and the Flexor carpi radialis.

(2) The chief movements permitted in the transverse or mid-carpal joint are flexion and extension and a slight amount of rotation. In flexion and extension, which is the movement most freely enjoyed, the trapezium and trapezoid on the radial side and the unciform on the ulnar side glide forwards and backwards on the scaphoid and cuneiform respectively, while the head of the os magnum and the superior surface of the unciform rotate in the cup-shaped cavity of the scaphoid and semilunar. Flexion at this joint is freer than extension. A very trifling amount of rotation is also permitted, the head of the os magnum rotating round a vertical axis drawn through its own centre, while at the same time a slight gliding movement takes place in the lateral portions of the joint.

IX. CARPO-METACARPAL ARTICULATIONS

1. ARTICULATION OF THE METACARPAL BONE OF THE THUMB WITH THE TRAPEZIUM

This is a joint of reciprocal reception, and enjoys a great freedom of movement, on account of the configuration of its articular surfaces, which are saddle-shaped, so that, on section, each bone appears to be received into a cavity in the other, according to the direction in which they are cut. The joint is surrounded by a capsular ligament.

The **Capsular Ligament** is thick but loose, and passes from the circumference of the upper extremity of the metacarpal bone to the rough edge bounding the articular surface of the trapezium ; it is thickest externally and behind, and lined by a separate *synovial membrane*.

Movements.—In the articulation of the metacarpal bone of the thumb with the trapezium, the movements permitted are flexion, extension, adduction, abduction, and circumduction. When the joint is flexed, the metacarpal bone is brought in front of the palm, and the thumb is gradually turned to the fingers. It is by this peculiar movement that the tip of the thumb is opposed to the other digits ; for, by slightly flexing the fingers, the palmar surface of the thumb can be brought in contact with their palmar surfaces one after another.

2. Articulations of the Metacarpal Bones of the Four Inner Fingers with the Carpus

The joints formed between the carpus and four inner metacarpal bones are arthrodial joints. The ligaments are—

Dorsal. Palmar.
Interosseous.

The **Dorsal Ligaments,** the strongest and most distinct, connect the carpal and metacarpal bones on their dorsal surface. The second metacarpal bone receives two fasciculi, one from the trapezium, the other from the trapezoid; the third metacarpal receives two, one from the trapezoid, and one from the os magnum; the fourth two, one from the os magnum, and one from the unciform; the fifth receives a single fasciculus from the unciform bone, which is continuous with a similar ligament on the palmar surface, forming an incomplete capsule.

The **Palmar Ligaments** have a somewhat similar arrangement on the palmar surface, with the exception of the third metacarpal, which has three ligaments, an external one from the trapezium, situated above the sheath of the tendon of the Flexor carpi radialis; a middle one from the os magnum; and an internal one from the unciform.

The **Interosseous Ligaments** consist of short, thick fibres, which are limited to one part of the carpo-metacarpal articulation; they connect the contiguous inferior angles of the os magnum and unciform with the adjacent surfaces of the third and fourth metacarpal bones.

Fig. 177.—Vertical section through the articulations at the wrist, showing the five synovial membranes.

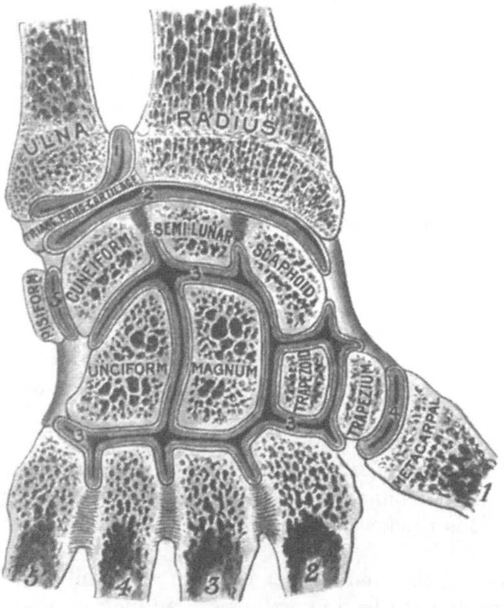

The **Synovial Membrane** is a continuation of that between the two rows of carpal bones. Occasionally, the articulation of the unciform with the fourth and fifth metacarpal bones has a separate synovial membrane.

The synovial membranes of the wrist and carpus (fig. 177) are thus seen to be five in number. The *first*, the *membrana sacciformis*, passes from the lower end of the ulna to the sigmoid cavity of the radius, and lines the upper surface of the interarticular fibro-cartilage. The *second* passes from the lower end of the radius and interarticular fibro-cartilage above, to the bones of the first row below. The *third*, the most extensive, passes between the contiguous margins of the two rows of carpal bones and between the bones of the second row to the carpal extremities of the four inner metacarpal bones. The *fourth*, from the margin of the trapezium to the metacarpal bone of the thumb. The *fifth*, between the adjacent margins of the cuneiform and pisiform bones.

Actions.—The movement permitted in the carpo-metacarpal articulations of the four inner fingers is limited to a slight gliding of the articular surfaces upon each other, the extent of which varies in the different joints. Thus the articulation of the metacarpal bone of the little finger is most movable, then that of the ring-finger. The metacarpal bones of the index and middle fingers are almost immovable.

3. ARTICULATIONS OF THE METACARPAL BONES WITH EACH OTHER

The carpal extremities of the four inner metacarpal bones articulate with one another at each side by small surfaces covered with cartilage, and connected together by dorsal, palmar, and interosseous ligaments.

The **Dorsal** and **Palmar Ligaments** pass transversely from one bone to another on the dorsal and palmar surfaces. The *Interosseous Ligaments* pass between their contiguous surfaces, just beneath their lateral articular facets.

The **Synovial Membrane** between the lateral facets is a reflection from that between the two rows of carpal bones.

The **Transverse Metacarpal Ligament** (fig. 178) is a narrow fibrous band, which passes transversely across the anterior surfaces of the digital extremities of the four inner metacarpal bones, connecting them together. It is blended anteriorly with the anterior (glenoid) ligament of the metacarpo-phalangeal articulations. To its posterior border is connected the fascia which covers the Interossei muscles. Its anterior surface is concave where the flexor tendons pass over it. Behind it the tendons of the Interossei muscles pass to their insertion.

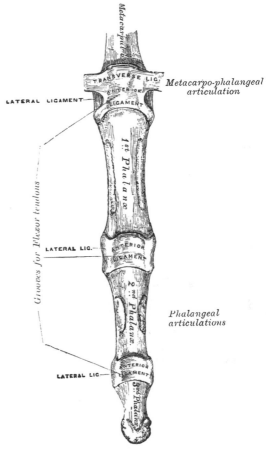

FIG. 178.—Articulations of the phalanges.

X. METACARPO-PHALANGEAL ARTICULATIONS (fig. 178)

These articulations are of the condyloid kind, formed by the reception of the rounded head of the metacarpal bone into a shallow cavity in the extremity of the first phalanx. The ligaments are—

Anterior.   Two Lateral.

The **Anterior Ligaments** (*Glenoid Ligaments* of Cruveilhier) are thick, dense, fibrous structures, placed upon the palmar surface of the joints in the intervals between the lateral ligaments, to which they are connected; they are loosely united to the metacarpal bone, but very firmly to the base of the first phalanges. Their palmar surface is intimately blended with the transverse metacarpal ligament, and presents a groove for the passage of the flexor tendons, the sheath surrounding which is connected to each side of the groove. By their deep surface, they form part of the articular surface for the head of the metacarpal bone, and are lined by a synovial membrane.

The **Lateral Ligaments** are strong, rounded cords, placed one on each side of the joint, each being attached by one extremity to the posterior tubercle on the side of the head of the metacarpal bone, and by the other to the contiguous extremity of the phalanx.

Actions.—The movements which occur in these joints are flexion, extension, adduction, abduction, and circumduction ; the lateral movements are very limited.

Surface Form.—The prominences of the knuckles do not correspond to the position of the joints either of the metacarpo-phalangeal or interphalangeal articulations. These prominences are invariably formed by the distal ends of the proximal bone of each joint, and the line indicating the position of the joint must be sought considerably in front of the middle of the knuckle. The usual rule for finding these joints is to flex the distal phalanx on the proximal one to a right angle ; the position of the joint is then indicated by an imaginary line drawn along the middle of the lateral aspect of the proximal phalanx.

XI. Articulations of the Phalanges

These are ginglymus joints. The ligaments are—

Anterior.   Two Lateral.

The arrangement of these ligaments is similar to those in the metacarpo-phalangeal articulations ; the extensor tendon supplies the place of a posterior ligament.

Actions.—The only movements permitted in the phalangeal joints are flexion and extension ; these movements are more extensive between the first and second phalanges than between the second and third. The movement of flexion is very considerable, but extension is limited by the anterior and lateral ligaments.

ARTICULATIONS OF THE LOWER EXTREMITY

The articulations of the Lower Extremity comprise the following groups : I. The hip-joint. II. The knee-joint. III. The articulations between the tibia and fibula. IV. The ankle-joint. V. The articulations of the tarsus. VI. The tarso-metatarsal articulations. VII. Articulations of the metatarsal bones with each other. VIII. The metatarso-phalangeal articulations. IX. The articulations of the phalanges.

I. Hip-Joint (fig. 179)

This articulation is an enarthrodial, or ball-and-socket joint, formed by the reception of the head of the femur into the cup-shaped cavity of the acetabulum. The articulating surfaces are covered with cartilage, that on the head of the femur being thicker at the centre than at the circumference, and covering the entire surface with the exception of a depression just below its centre for the ligamentum teres ; that covering the acetabulum is much thinner at the centre than at the circumference. It forms an incomplete cartilaginous ring, of a horse-shoe shape, being deficient below, where there is a circular depression, which is occupied in the recent state by a mass of fat, covered by synovial membrane. The ligaments of the joint are the

Capsular.   Teres.
Ilio-femoral.   Cotyloid.
           Transverse.

The **Capsular Ligament** is a strong, dense, ligamentous capsule, embracing the margin of the acetabulum above, and surrounding the neck of the femur below. Its *upper circumference* is attached to the acetabulum, two or three lines external to the cotyloid ligament, above and behind ; but in front, it is attached to the outer margin of the ligament, and opposite to the notch where the margin of the cavity is deficient, it is connected to the transverse ligament, and by a few fibres to the edge of the obturator foramen.   Its *lower circumference* surrounds the neck of the femur, being attached, in front, to the spiral or anterior intertrochanteric line ; above, to the base of the neck ; behind, to the neck of the bone, about half an inch above the posterior intertrochanteric line.   From this insertion the fibres are reflected upwards over the neck of the femur, forming a sort of tubular sheath (the *cervical reflection*), which blends with the periosteum, and can be traced as far as the articular cartilage.   On the surface of the neck of the femur some of these reflected fibres are raised into longitudinal folds, termed *retinacula*.   It is much thicker at the upper and fore part of the joint, where the greatest amount of resistance is required, than below and internally, where it is thin, loose, and longer than in any other part.   It consists of two sets of fibres, circular and longitudinal.   The circular fibres (*zona orbicularis*) are

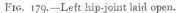

Fig. 179.—Left hip-joint laid open.

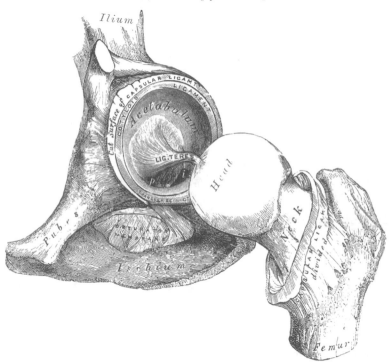

most abundant at the lower and back part of the capsule, and form a sling or collar around the neck of the femur.   Anteriorly they blend with the deep surface of the ilio-femoral ligament, and through its medium reach the anterior inferior spine of the ilium.   The longitudinal fibres are greatest in amount at the upper and front part of the capsule, where they form distinct bands, or accessory ligaments, of which the most important is the *ilio-femoral*.   The other accessory bands are known as the *pubo-femoral*, passing from the ilio-pectineal eminence to the front of the capsule ; and *ischio-capsular*, passing from the ischium, just below the acetabulum, to blend with the circular fibres at the lower part of the joint. The external surface (fig. 164, page 212) is rough, covered by numerous muscles, and separated in front from the Psoas and Iliacus by a synovial bursa, which not infrequently communicates, by a circular aperture, with the cavity of the joint. It differs from the capsular ligament of the shoulder in being much less loose and lax, and in not being perforated for the passage of a tendon.

The **Ilio-femoral Ligament** (figs. 164 and 180) is an accessory band of fibres extending obliquely across the front of the joint; it is intimately connected with the capsular ligament, and serves to strengthen it in this situation. It is attached, above, to the lower part of the anterior inferior spine of the ilium; and, diverging below, forms two bands, of which one passes downwards to be inserted into the lower part of the anterior intertrochanteric line; the other passes downwards and outwards to be inserted into the upper part of the same line and adjacent part of the neck of the femur. Between the two bands is a thinner part of the capsule. Sometimes there is no division, but the ligament spreads out into a flat triangular band which is attached below into the whole length of the anterior intertrochanteric line. This ligament is frequently called the Y-shaped ligament of Bigelow; and the outer or upper of the two bands is sometimes described as a separate ligament, under the name of the *ilio-trochanteric ligament*.

FIG. 180.—Hip-joint, showing the ilio-femoral ligament. (After Bigelow.)   FIG. 181.—Vertical section through hip-joint. (Henle.)

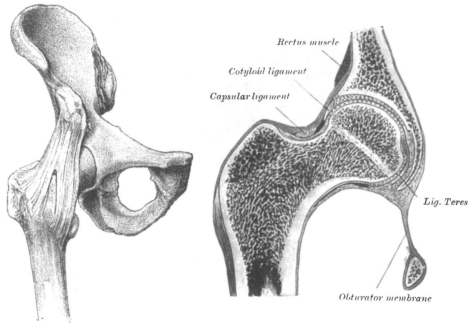

Rectus muscle

Cotyloid ligament

Capsular ligament

Lig. Teres

Obturator membrane

The **Ligamentum Teres** is a triangular band implanted by its apex into the depression a little behind and below the centre of the head of the femur, and by its broad base into the margins of the cotyloid notch, becoming blended with the transverse ligament. It is formed of connective tissue, surrounded by a tubular sheath of synovial membrane. Sometimes only the synovial fold exists, or the ligament may be altogether absent. The ligament is made tense when the hip is semiflexed, and the limb then adducted or rotated outwards; it is, on the other hand, relaxed when the limb is abducted. It has, however, but little influence as a ligament, though it may to a certain extent limit movement, and would appear to be merely a modification of the folds which in other joints fringe the margins of reflection of synovial membranes (see page 189).

The **Cotyloid Ligament** is a fibro-cartilaginous rim attached to the margin of the acetabulum, the cavity of which it deepens; at the same time it protects the edges of the bone, and fills up the inequalities on its surface. It bridges over the notch as the *transverse ligament*, and thus forms a complete circle, which closely surrounds the head of the femur, and assists in holding it in its place, acting as a sort of valve. It is prismoid on section, its base being attached to the margin of the acetabulum, and its opposite edge being free and sharp ; while its two surfaces are invested by synovial membrane, the external one being in contact with the capsular ligament, the internal one being inclined inwards so as to narrow the acetabulum, and embrace the cartilaginous surface of the head of the femur. It is much thicker above and behind than below and in front, and consists of close compact fibres, which arise from different points of the circumference of the acetabulum, and interlace with each other at very acute angles.

The **Transverse Ligament** is in reality a portion of the cotyloid ligament, though differing from it in having no cartilage cells among its fibres. It consists of strong, flattened fibres, which cross the notch at the lower part of the acetabulum, and convert it into a foramen. Thus an interval is left beneath the ligament for the passage of nutrient vessels to the joint.

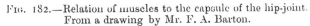

Fig. 182.—Relation of muscles to the capsule of the hip-joint.
From a drawing by Mr. F. A. Barton.

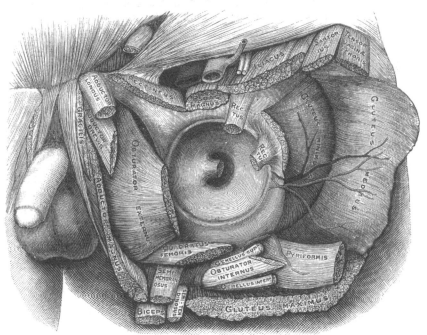

The **Synovial Membrane** is very extensive. Commencing at the margin of the cartilaginous surface of the head of the femur, it covers all that portion of the neck which is contained within the joint ; from the neck it is reflected on the internal surface of the capsular ligament, covers both surfaces of the cotyloid ligament and the mass of fat contained in the depression at the bottom of the acetabulum, and is prolonged in the form of a tubular sheath around the ligamentum teres, as far as the head of the femur. It sometimes communicates through a hole in the capsular ligament between the inner band of the Y-shaped ligament and the pubo-femoral ligament with a bursa situated on the under surface of the Ilio-psoas muscle.

The muscles in relation with the joint are, in front, the Psoas and Iliacus, separated from the capsular ligament by a synovial bursa ; above, the reflected head of the Rectus and Gluteus minimus, the latter being closely adherent to the capsule ; internally the Obturator externus and Pectineus ; behind, the Pyriformis, Gemellus superior, Obturator internus, Gemellus inferior, Obturator externus, and Quadratus femoris (fig. 182).

The arteries supplying the joint are derived from the obturator, sciatic, internal circumflex, and gluteal.

The nerves are articular branches from the sacral plexus, great sciatic, obturator, accessory obturator, and a filament from the branch of the anterior crural supplying the Rectus.

Actions.—The movements of the hip are very extensive, and consist of flexion, extension, adduction, abduction, circumduction, and rotation.

The hip-joint presents a very striking contrast to the shoulder-joint in the much more complete mechanical arrangements for its security and for the limitation of its movements. In the shoulder, as we have seen, the head of the humerus is not adapted at all in size to the glenoid cavity, and is hardly restrained in any of its ordinary movements by the capsular ligament. In the hip-joint, on the contrary, the head of the femur is closely fitted to the acetabulum for a distance extending over nearly half a sphere, and at the margin of the bony cup it is still more closely embraced by the cotyloid ligament, so that the head of the femur is held in its place by that ligament even when the fibres of the capsule have been quite divided (Humphry). The anterior portion of the capsule, described as the ilio-femoral ligament, is the strongest of all the ligaments in the body, and is put on the stretch by any attempt to extend the femur beyond a straight line with the trunk. That is to say, this ligament is the chief agent in maintaining the erect position without muscular fatigue ; for a vertical line passing through the centre of gravity of the trunk falls behind the centres of rotation in the hip-joints, and therefore the pelvis tends to fall backwards, but is prevented by the tension of the ilio-femoral and capsular ligaments. The security of the joint may be also provided for by the two bones being directly united through the ligamentum teres ; but it is doubtful whether this so-called ligament can have much influence upon the mechanism of the joint. Flexion of the hip-joint is arrested by the soft parts of the thigh and abdomen being brought into contact, when the leg is flexed on the thigh ; and by the action of the hamstring muscles when the leg is extended ; * extension by the tension of the ilio-femoral ligament and front of the capsule ; adduction by the thighs coming into contact ; adduction with flexion by the outer band of the ilio-femoral ligament, the outer part of the capsular ligament ; abduction by the inner band of the ilio-femoral ligament and the pubo-femoral band ; rotation outwards by the outer band of the ilio-femoral ligament ; and rotation inwards by the ischio-capsular ligament and the hinder part of the capsule. The muscles which *flex* the femur on the pelvis are, the Psoas, Iliacus, Rectus, Sartorius, Pectineus, Adductor longus and brevis, and the anterior fibres of the Gluteus medius and minimus. *Extension* is mainly performed by the Gluteus maximus, assisted by the hamstring muscles. The thigh is *adducted* by the Adductor magnus, longus and brevis, the Pectineus, the Gracilis, and lower part of the Gluteus maximus, and *abducted* by the Gluteus medius and minimus, and upper part of the Gluteus maximus. The muscles which *rotate* the thigh *inwards* are the anterior fibres of the Gluteus medius, the Gluteus minimus, and the Tensor fasciæ femoris ; while those which rotate it *outwards* are the posterior fibres of the Gluteus medius, the Pyriformis, Obturator externus and internus, Gemellus superior and inferior, Quadratus femoris, Iliacus, Gluteus maximus, the three Adductors, the Pectineus, and the Sartorius.

Surface Form.—A line drawn from the anterior superior spinous process of the ilium to the most prominent part of the tuberosity of the ischium (Nélaton's line) runs through the centre of the acetabulum, and would, therefore, indicate the level of the hip-joint ; or, in other words, the upper border of the great trochanter, which lies on Nélaton's line, is on a level with the centre of the hip-joint.

Surgical Anatomy.—In dislocation of the hip, 'the head of the thigh-bone may rest at any point around its socket' (Bryant) ; but whatever position the head ultimately assumes, the primary displacement is generally downwards and inwards, the capsule giving way at

---

* The hip-joint cannot be completely flexed, in most persons, without at the same time flexing the knee, on account of the shortness of the hamstring muscles.—Cleland, *Journ. of Anat. and Phys.* No. I. Old Series, p. 87.

its weakest—that is, its lower and inner—part. The situation that the head of the bone subsequently assumes is determined by the degree of flexion or extension, and of outward or inward rotation of the thigh at the moment of luxation, influenced, no doubt, by the ilio-femoral ligament, which is not easily ruptured. When, for instance, the head is forced backwards, this ligament forms a fixed axis, round which the head of the bone rotates, and is thus driven on to the dorsum of the ilium. The ilio-femoral ligament also influences the position of the thigh in the various dislocations: in the dislocations backwards it is tense, and produces inversion of the limb; in the dislocation on to the pubes, it is relaxed, and therefore allows the external rotators to evert the thigh; while in the thyroid dislocation it is tense, and produces flexion. The muscles inserted into the upper part of the femur, with the exception of the Obturator internus, have very little direct influence in determining the position of the bone. But Bigelow has endeavoured to show that the Obturator internus is the principal agent in determining whether, in the backward dislocations, the head of the bone shall be ultimately lodged on the dorsum of the ilium, or in or near the sciatic notch. In both dislocations the head passes, in the first instance, in the same direction; but, as Bigelow asserts, in the displacement on to the dorsum, the head of the bone travels up behind the acetabulum, in front of the muscle; while in the dislocation into the sciatic notch the head passes behind the muscle, and is therefore prevented from reaching the dorsum, in consequence of the tendon of the muscle arching over the neck of the bone, and it therefore remains in the neighbourhood of the sciatic notch. Bigelow distinguishes these two forms of dislocation by describing them as dislocations backwards, 'above and below' the Obturator internus.

The ilio-femoral ligament is rarely torn in dislocations of the hip, and this fact is taken advantage of by the surgeon in reducing these dislocations by manipulation. It is made to act as the fulcrum to a lever, of which the long arm is the shaft of the femur, and the short arm the neck of the bone.

The hip-joint is rarely the seat of acute synovitis from injury, on account of its deep position and its thick covering of soft parts. Acute inflammation may, and does, frequently occur as the result of constitutional conditions, as rheumatism, pyæmia, &c. When, in these cases, effusion takes place, and the joint becomes distended with fluid, the swelling is not very easy to detect on account of the thickness of the capsule and the depth of the articulation. It is principally to be found on the front of the joint, just internal to the ilio-femoral ligament; or behind, at the lower and back part. In these two places the capsule is thinner than elsewhere. Disease of the hip-joint is much more frequently of a chronic character, and is usually of a tuberculous origin. It begins either in the bones or in the synovial membrane. More frequently in the former, and probably, in most cases, in the growing, highly vascular tissue in the neighbourhood of the epiphysial cartilage. In this respect it differs very materially from tuberculous arthritis of the knee, where the disease usually commences in the synovial membrane. The reasons why the disease so frequently begins in this situation are two-fold: first, this part being the centre of rapid growth, its nutrition is unstable and apt to pass into inflammatory action; and, secondly, great strain is thrown upon it, from the frequency of falls and blows upon the hip, which causes crushing of the epiphysial cartilage or the cancellous tissue in its neighbourhood, with the results likely to follow such an injury. In addition to these, the depth of the joint protects it from the causes of synovitis.

In chronic hip-disease the affected limb assumes an altered position, the cause of which it is important to understand. In the early stage of a typical case, the limb is flexed, abducted, and rotated outwards. In this position all the ligaments of the joint are relaxed: the front of the capsule by flexion; the outer band of the ilio-femoral ligament by abduction; and the inner band of this ligament and the back of the capsule by rotation outwards. It is, therefore, the position of greatest ease. The condition is not quite obvious at first, upon examining a patient. If the patient is laid in the supine position, the affected limb will be found to be extended and parallel with the other. But it will be found that the pelvis is tilted downwards on the diseased side and the limb apparently longer than its fellow, and that the lumbar spine is arched forwards (lordosis). If now the thigh is abducted and flexed, the tilting downwards and the arching forwards of the pelvis disappear. The condition is thus explained. A limb which is flexed and abducted is obviously useless for progression, and, in order to overcome the difficulty, the patient depresses the affected side of his pelvis in order to produce parallelism of his limbs, and at the same time rotates his pelvis on its transverse horizontal axis, so as to direct the limb downwards, instead of forwards. In the later stages of the disease the limb becomes flexed and adducted and inverted. This position probably depends upon muscular action, at all events as regards the adduction. The Adductor muscles are supplied by the obturator nerve, which also largely supplies the joint. These muscles are therefore thrown into reflex action by the irritation of the peripheral terminations of this nerve in the inflamed articulation.

Osteo-arthritis is not uncommon in the hip-joint, and is said to be more common in the male than in the female, in whom the knee-joint is more frequently affected. It is a disease of middle age or advanced life.

Congenital dislocation is more commonly met with in the hip-joint than in any other articulation. The displacement usually takes place on to the dorsum ilii. It gives rise to extreme lordosis, and a waddling gait is noticed as soon as the child commences to walk.

Excision of the hip may be required for disease or for injury, especially gunshot. It may be performed either by an anterior incision or a posterior one. The former entails less interference with important structures, especially muscles, than the posterior one, but permits of less efficient drainage. In these days, however, when the surgeon aims at securing healing of his wound without suppuration, this second desideratum is not of so much importance. In the operation in front, the surgeon makes an incision three to four inches in length, starting immediately below and external to the anterior superior spinous process of the ilium, downwards and inwards between the Sartorius and Tensor fasciæ femoris, to the neck of the bone, dividing the capsule at its upper part. A narrow-bladed saw now divides the neck of the femur, and the head of the bone is extracted with sequestrum forceps. All diseased tissue is carefully removed with a sharp spoon or scissors, and the cavity thoroughly flushed out with a hot antiseptic fluid.

The posterior method consists in making an incision three or four inches long, commencing midway between the top of the great trochanter and the anterior superior spine, and ending over the shaft, just below the trochanter. The muscles are detached from the great trochanter, and the capsule opened freely. The head and neck are freed from the soft parts and the bone sawn through just below the top of the trochanter with a narrow saw. The head of the bone is then levered out of the acetabulum. In both operations, if the acetabulum is eroded, it must be freely gouged.

II. KNEE-JOINT

The knee-joint was formerly described as a ginglymus or hinge-joint, but is really of a much more complicated character. It must be regarded as consisting of three articulations in one : one between each condyle of the femur and the corresponding tuberosity of the tibia, which are condyloid joints, and one between the patella and the femur, which is partly arthrodial, but not completely so, since the articular surfaces are not mutually adapted to each other, so that the movement is not a simple gliding one. This view of the construction of the knee-joint receives confirmation from the study of the articulation in some of the lower mammals, where three synovial membranes are sometimes found, corresponding to these three subdivisions, either entirely distinct or only connected together by small communications. This view is further rendered probable by the existence of the two crucial ligaments within the joint, which must be regarded as the external and internal lateral ligaments of the inner and outer joints respectively. The existence of the ligamentum mucosum would further indicate a tendency to separation of the synovial cavity into two minor sacs, one corresponding to each joint.

The bones entering into the formation of the knee-joint are the condyles of the femur above, the head of the tibia below, and the patella in front. They are connected together by ligaments, some of which are placed on the exterior of the joint, while others occupy its interior.

External Ligaments	*Interior Ligaments*
Anterior, or Ligamentum Patellæ.	Anterior, or External Crucial.
Posterior, or Ligamentum Posticum Winslowii.	Posterior, or Internal Crucial.
	Two Semilunar Fibro-Cartilages.
Internal Lateral.	Transverse.
Two External Lateral.	Coronary.
Capsular.	Ligamentum mucosum } Processes of Synovial membrane.
	Ligamenta alaria }

The **Anterior Ligament** or **Ligamentum Patellæ** (fig. 183) is the central portion of the common tendon of the Extensor muscles of the thigh which is continued from the patella to the tubercle of the tibia, supplying the place of an anterior ligament. It is a strong, flat, ligamentous band, about three inches in length, attached, above, to the apex of the patella and the rough depression on its posterior surface ; below, to the lower part of the tubercle of the tibia ; its superficial fibres being continuous over the front of the patella with those of the tendon of the Quadriceps extensor. The lateral portions of the tendon of the Extensor muscles pass down on either side of the patella, attached to the borders of this bone and its ligament, to be inserted into the upper extremity of the tibia on each side of the tubercle ; externally, these portions merge into the capsular ligament.   They are termed *lateral patellar ligaments*.   The posterior surface of the ligamentum patellæ can usually be easily separated from the front of the capsular ligament.

The **Posterior Ligament (Ligamentum Posticum Winslowii)** (fig. 184) is a broad, flat, fibrous band, formed of fasciculi separated from one another by apertures for the passage of vessels and nerves.   It is attached above to the upper margin of the intercondyloid notch of the femur, and below to the posterior margin of the head of the tibia.   Superficial to the main part of the ligament is a strong fasciculus derived from the tendon of the Semi-membranosus, and passing from the back part of the inner tuberosity of the tibia obliquely upwards and

FIG. 183.—Right knee-joint.
Anterior view.

FIG. 184.—Right knee-joint.
Posterior view.

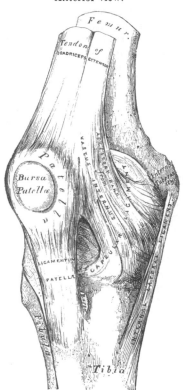

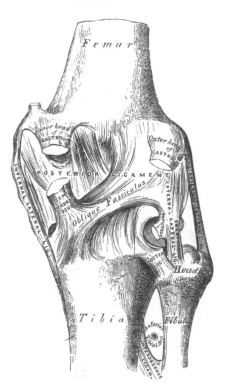

outwards to the back part of the outer condyle of the femur.   The posterior ligament forms part of the floor of the popliteal space, and the popliteal artery rests upon it.

The **Internal Lateral Ligament** is a broad, flat, membranous band, thicker behind than in front, and situated nearer to the back than the front of the joint. It is attached, above, to the inner tuberosity of the femur ; below, to the inner tuberosity and inner surface of the shaft of the tibia, to the extent of about two inches.   It is crossed, at its lower part, by the tendons of the Sartorius, Gracilis, and Semi-tendinosus muscles, a synovial bursa being interposed.   Its *deep surface* covers the anterior portion of the tendon of the Semi-membranosus, with which it is connected by a few fibres, the synovial membrane of the joint, and the inferior internal articular vessels and nerve ; it is intimately adherent to the internal semilunar fibro-cartilage.

The **Long External Lateral Ligament** is a strong, rounded, fibrous cord, situated nearer to the back than the front of the joint. It is attached, above, to the back part of the outer tuberosity of the femur; below, to the outer part of the head of the fibula. Its *outer surface* is covered by the tendon of the Biceps, which divides at its insertion into two parts, separated by the ligament. Passing beneath the ligament are the tendon of the Popliteus muscle, and the inferior external articular vessels and nerve.

The **Short External Lateral Ligament** is an accessory bundle of fibres placed behind and parallel with the preceding, attached, above, to the lower and back part of the outer tuberosity of the femur; below, to the summit of the styloid process of the fibula. This ligament is intimately connected with the capsular ligament, while passing beneath it are the tendon of the Popliteus muscle, and the inferior external articular vessels and nerve.

The **Capsular Ligament** consists of an exceedingly thin, but strong, fibrous membrane, which fills in the intervals left between the stronger bands above described, and is inseparably connected with them. In front it blends with and forms part of the lateral patellar ligaments, and fills in the interval between the anterior and lateral ligaments of the joint, with which latter structures it is closely connected. Behind, it is formed chiefly of vertical fibres, which arise above from the condyles and intercondyloid notch of the femur, and is connected below with the back part of the head of the tibia, being closely united with the origins of the Gastrocnemius, Plantaris, and Popliteus muscles. It passes in front of, but is inseparably connected with, the posterior ligament.

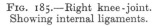

Fig. 185.—Right knee-joint. Showing internal ligaments.

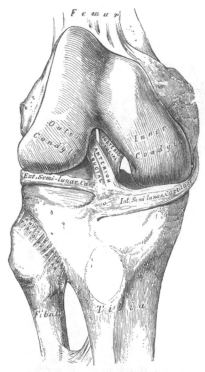

The **Crucial** are two interosseous ligaments of considerable strength, situated in the interior of the joint, nearer its posterior than its anterior part. They are called *crucial* because they cross each other somewhat like the lines of the letter X; and have received the names *anterior* and *posterior*, from the position of their attachment to the tibia.

The **Anterior** or **External Crucial Ligament** (fig. 185) is attached to the depression in front of the spine of the tibia, being blended with the anterior extremity of the external semilunar fibro-cartilage, and passing obliquely upwards, backwards, and outwards, is inserted into the inner and back part of the outer condyle of the femur.

The **Posterior** or **Internal Crucial Ligament** is stronger, but shorter and less oblique in its direction, than the anterior. It is attached to the back part of the depression behind the spine of the tibia, to the popliteal notch, and to the posterior extremity of the external semilunar fibro-cartilage; and passes upwards, forwards, and inwards, to be inserted into the outer and fore part of the inner condyle of the femur. It is in relation, in front, with the anterior crucial ligament; behind, with the capsular ligament.

The **Semilunar Fibro-cartilages** (fig. 186) are two crescentic lamellæ, which serve to deepen the surface of the head of the tibia, for articulation with the condyles of the femur. The circumference of each cartilage is thick, convex, and attached to the inside of the capsule of the knee ; the inner border is thin, concave, and free. Their upper surfaces are concave, and in relation with the condyles of the femur ; their lower surfaces are flat, and rest upon the head of the tibia. Each cartilage covers nearly the outer two-thirds of the corresponding articular surface of the tibia, leaving the inner third uncovered ; both surfaces are smooth, and invested by synovial membrane.

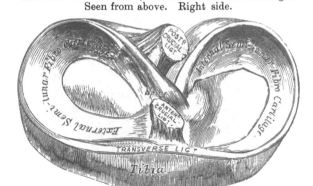

FIG. 186.—Head of tibia, with semilunar fibro-cartilages. Seen from above. Right side.

The **Internal Semilunar Fibro-cartilage** is nearly semicircular in form, a little elongated from before backwards, and broader behind than in front; its anterior extremity, thin and pointed, is attached to a depression on the anterior margin of the head of the tibia, in front of the anterior crucial ligament ; its posterior extremity is attached to the depression behind the spine, between the attachments of the external semilunar fibro-cartilage and the posterior crucial ligaments.

The **External Semilunar Fibro-cartilage** forms nearly an entire circle, covering a larger portion of the articular surface than the internal one. It is grooved on its outer side for the tendon of the Popliteus muscle. Its extremities, at their insertion, are interposed between the two extremities of the internal semilunar fibro-cartilage ; the anterior extremity being attached in front of the spine of the tibia to the outer side of, and behind, the anterior crucial ligament, with which it blends ; the posterior extremity being attached behind the spine of the tibia, in front of the posterior extremity of the internal semilunar fibro-cartilage. Just before its insertion posteriorly it gives off a strong fasciculus, the *ligament of Wrisberg*, which passes obliquely upwards and outwards, to be inserted into the inner condyle of the femur, close to the attachment of the posterior crucial ligament. Occasionally a small fasciculus is given off, which passes forwards to be inserted into the back part of the anterior crucial ligament. The external semilunar fibro-cartilage gives off from its anterior convex margin a fasciculus, which forms the transverse ligament.

The **Transverse Ligament** is a band of fibres which passes transversely from the anterior convex margin of the external semilunar fibro-cartilage to the anterior convex margin of the internal semilunar fibro-cartilage ; its thickness varies considerably in different subjects, and it is sometimes absent altogether.

The **Coronary Ligaments** are merely portions of the capsular ligament, which connect the circumference of each of the semilunar fibro-cartilages with the margin of the head of the tibia.

The **Synovial Membrane** of the knee-joint is the largest and most extensive in the body. Commencing at the upper border of the patella, it forms a short *cul-de-sac* beneath the Quadriceps extensor tendon of the thigh, on the lower part of the front of the shaft of the femur : this communicates with a synovial bursa interposed between the tendon and the front of the femur, by an orifice of variable size. On each side of the patella, the synovial membrane extends beneath the aponeurosis of the Vasti muscles, and more especially beneath that of the Vastus internus. Below the patella it is separated from the anterior ligament by the anterior part of the capsule and a considerable quantity of adipose tissue. In this situation it sends off a triangular prolongation, containing a few ligamentous fibres, which extends from the anterior part of the joint below the patella, to the front of the intercondyloid notch. This fold has been termed the **ligamentum mucosum**. It also sends off two fringe-like folds, called the **ligamenta alaria,** which extend from the sides of the ligamentum mucosum, upwards and laterally between the patella and femur. On either side of the joint, it passes downwards from the femur, lining the capsule to its point of attachment to the semilunar cartilages; it may then be traced over the upper surfaces of these cartilages to their free borders, and thence along their under surfaces to the tibia. At the back part of the external one it forms a *cul-de-sac* between the groove on its surface and the tendon of the Popliteus; it surrounds the crucial ligaments, and lines the inner surface of the ligaments which enclose the joint. The pouch of synovial membrane between the Extensor tendon and front of the femur is supported, during the movements of the knee, by a small muscle, the Subcrureus, which is inserted into the upper part of the capsular ligament.

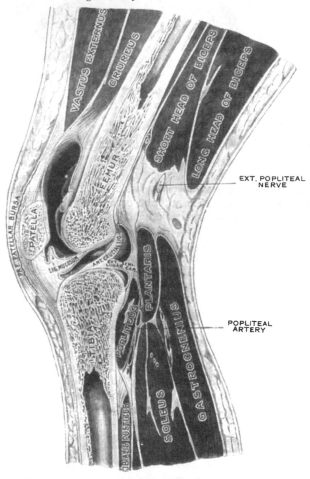

Fig. 187.—Longitudinal section through the middle of the right knee-joint. (After Braune.)

The folds of synovial membrane and the fatty processes contained in them act, as it seems, mainly as padding to fill up interspaces and obviate concussions. Sometimes the bursa beneath the Quadriceps extensor is completely shut off from the rest of the synovial cavity, thus forming a closed sac between the Quadriceps and the lower part of the front of the femur; or it may communicate with the synovial cavity by a minute aperture.

The bursæ about the knee-joint are the following:

In front there are three bursæ: one is interposed between the patella and the skin; another of small size between the upper part of the tuberosity of the tibia and the ligamentum patellæ; and a third between the lower part of the tuberosity of the tibia and the skin. On the outer side there are four bursæ: (1) one beneath the outer head of the Gastrocnemius (which sometimes communicates with the joint); (2) one above the external lateral ligament between it and the tendon of the Biceps; (3) one beneath the external lateral ligament between it and the tendon of the Popliteus (this is sometimes only an expansion from the next bursa) ; (4) one beneath the tendon of the Popliteus between it and the condyle of the femur, which is almost always an extension from the synovial membrane. On the inner side there are five bursæ : (1) one beneath the inner head of the Gastrocnemius, which sends a prolongation between the tendons of the Gastro-cnemius and Semi-membranosus. This bursa often communicates with the joint ; (2) one above the internal lateral ligament between it and the tendons of the Sartorius, Gracilis, and Semi-tendinosus ; (3) one beneath the internal lateral ligament between it and the tendon of the Semi-membranosus (this is sometimes only an expansion from the next bursa) ; (4) one beneath the tendon of the Semi-membranosus, between it and the head of the tibia ; (5) sometimes there is a bursa between the tendons of the Semi-membranosus and of the Semi-tendinosus.

Structures around the Joint.—In front, and at the sides, the Quadriceps extensor ; on the outer side, the tendons of the Biceps and the Popliteus and the external popliteal nerve ; on the inner side, the Sartorius, Gracilis, Semi-tendinosus and Semi-membranosus ; behind, an expansion from the tendon of the Semi-membranosus, the popliteal vessels, and the internal popliteal nerve, Popliteus, Plantaris, and inner and outer heads of the Gastrocnemius, some lymphatic glands, and fat.

The *Arteries* supplying the joint are derived from the anastomotica magna, a branch of the femoral, articular branches of the popliteal, anterior and posterior recurrent branches of the anterior tibial, and descending branch from the external circumflex of the profunda.

The *Nerves* are derived from the obturator, anterior crural, and external and internal popliteal.

Actions.—The knee-joint permits of movements of flexion and extension, and, in certain positions, of slight rotation inwards and outwards. The movement of flexion and extension does not, however, take place in a simple hinge-like manner, as in other joints, but is a complicated movement, consisting of a certain amount of gliding and rotation ; so that the same part of one articular surface is not always applied to the same part of the other articular surface, and the axis of motion is not a fixed one. If the joint is examined while in a condition of extreme flexion, the posterior part of the articular surfaces of the tibia will be found to be in contact with the posterior rounded extremities of the condyles of the femur ; and if a simple hinge-like movement were to take place, the axis, round which the revolving movement of the tibia occurs, would be in the back part of the condyle. If the leg is now brought forwards into a position of semi-flexion, the upper surface of the tibia will be seen to glide over the condyles of the femur, so that the middle part of the articular facets are in contact, and the axis of rotation must therefore have shifted forwards to nearer the centre of the condyles. If the leg is now brought into the extended position, a still further gliding takes place, and a further shifting forwards of the axis of rotation. This is not, however, a simple movement, but is accompanied by a certain amount of rotation outwards round a vertical axis drawn through the centre of the head of the tibia. This rotation is due to the greater length of the internal condyle and to the fact that the anterior portion of its articular surface is inclined obliquely outwards. In consequence of this, it will be seen that towards the close of the movement of extension, that is to say, just before complete extension is effected, the tibia glides obliquely upwards and outwards over this oblique surface on the inner condyle, and the leg is therefore necessarily rotated outwards. In flexion of the joint the converse of these movements takes place : the tibia glides back-wards round the end of the femur, and at the commencement of the movement the tibia is directed downwards and inwards along the oblique curve of the inner condyle, thus causing an inward rotation to the leg.

During flexion and extension, the patella moves on the lower end of the femur, but this movement is not a simple gliding one ; for if the articular surface of this bone is examined, it will be found to present on each side of the central vertical ridge two less marked transverse ridges, which divide the surface, except a small portion along the inner border, which is cut off by a slight vertical ridge, into six facets (see fig. 188), and therefore does not present a uniform curved surface, as would be the case if a simple gliding movement took place. These six facets—three on each side of the median vertical ridge—correspond to and denote the parts of the bone respectively in contact with the condyles of the femur during flexion, semi-flexion, and extension. In flexion, only the upper facets on the patella are in contact with the condyles of the femur; the lower two-thirds of the bone rests upon the mass of fat which occupies the space between the femur and tibia. In the semi-flexed position of the joint, the middle facets on the patella rest upon the most prominent portion of the condyles, and thus afford greater leverage to the Quadriceps by increasing its distance from the centre of motion. In complete extension the patella is drawn up so that only the lower facets are in contact with the articular surfaces of the condyles. The narrow strip along the inner border is in contact with the outer aspect of the internal condyle when the leg is fully flexed at the knee-joint. As in the elbow, so it is in the knee—the axis of rotation in flexion and extension is not precisely at right angles to the axis of the bone, but during flexion there is a certain amount of alteration of plane; so that, whereas in flexion the femur and tibia are in the same plane, in extension the one bone forms an angle of about ten degrees with the other. There is, however, this difference between the two extremities—that in the upper, during extension, the humeri are parallel, and the bones of the forearm diverge; in the lower, the femora converge below, and the tibiæ are parallel.

FIG. 188.—View of the posterior surface of the right patella. Showing diagrammatically the areas of contact with the femur in different positions of the knee.

In addition to the slight rotation during flexion and extension, the tibia enjoys an independent rotation on the condyles of the femur, in certain positions of the joint. The movement takes place between the interarticular fibro-cartilages and the tibia, whereas the movement of flexion and extension takes place between the interarticular fibro-cartilages and the femur. So that the knee may be said to consist of two joints, separated by the fibro-cartilages : an upper (menisco-femoral), in which flexion and extension take place ; and a lower (menisco-tibial), allowing of a certain amount of rotation. This latter movement can only take place in the semiflexed position of the limb, when all the ligaments are relaxed.

During *flexion* the ligamentum patellæ is put upon the stretch, as is also the posterior crucial ligament in extreme flexion. The other ligaments are all relaxed by flexion of the joint, though the relaxation of the anterior crucial ligament is very trifling. Flexion is only checked during life by the contact of the leg with the thigh. In the act of extending the leg upon the thigh the ligamentum patellæ is tightened by the Quadriceps extensor ; but when the leg is fully extended, as in the erect posture, the ligament becomes relaxed, so as to allow free lateral movement to the patella, which then rests on the front of the lower end of the femur. The other ligaments, with the exception of the posterior crucial, which is partly relaxed, are all on the stretch. When the limb has been brought into a straight line, extension is checked mainly by the tension of all the ligaments except the posterior crucial and ligamentum patellæ. The movements of *rotation*, of which the knee is capable, are permitted in the semi-flexed condition by the partial relaxation of both crucial ligaments, as well as the lateral ligaments. Rotation inwards appears to be limited by the tension of the anterior crucial ligament, and by the interlocking of the two ligaments ; but rotation outwards does not appear to be checked by either crucial ligament, since they uncross during the execution of this movement, but by the lateral ligaments, especially the internal. The main function of the crucial ligaments is to act as a direct bond of union between the tibia and femur, preventing the former bone from being carried too far backwards or forwards. Thus the anterior crucial ligament prevents the tibia being carried too far forwards by the extensor tendons, and the posterior crucial checks too great movement backwards by the flexors. They also assist the lateral ligaments in resisting any lateral bending of the joint. The interarticular cartilages are intended, as it seems, to adapt the surface of the tibia to the shape of the femur to a certain extent, so as to fill up the intervals which would otherwise be left in the varying positions of the joint, and to interrupt the jars which would be so frequently transmitted up the limb in jumping or falls on the feet; also to permit of the two varieties of motion, flexion and extension, and rotation, as explained above. The patella is a great defence to the knee-joint from any injury inflicted in front, and it distributes upon a large and tolerably even surface, during kneeling, the pressure which would otherwise fall upon the prominent ridges of the condyles; it also affords leverage to the Quadriceps extensor muscle to act upon the tibia, and Mr. Ward has pointed out * how this leverage varies in the various positions of the joint, so that the action of the muscle produces velocity at the expense of force in the commencement of extension, and, on the contrary, at the close of extension tends to diminish velocity, and therefore the shock to the ligaments at the moment tension of these structures takes place.

Extension of the leg on the thigh is performed by the Quadriceps extensor. *Flexion* by the hamstring muscles, assisted by the Gracilis and Sartorius, and, indirectly, by the Gastrocnemius, Popliteus, and Plantaris. *Rotation outwards* by the Biceps, and *rotation inwards* by the Popliteus, Semitendinosus, and, to a slight extent, the Semimembranosus, the Sartorius, and the Gracilis.

Surface Form.—The interval between the two bones entering into the formation of the knee-joint can always easily be felt. If the limb is extended, it is situated on a slightly higher level than the apex of the patella; but if the limb is slightly flexed, a knife carried horizontally backwards immediately below the apex of the patella would pass directly into the joint. When the knee-joint is distended with fluid, the outline of the synovial membrane at the front of the knee may be fairly well mapped out.

Surgical Anatomy.—From a consideration of the construction of the knee-joint, it would at first sight appear to be one of the least secure of any of the joints in the body. It is formed between the two longest bones, and therefore the amount of leverage which can be brought to bear upon it is very considerable; the articular surfaces are ill adapted to each other, and the range and variety of motion which it enjoys are great. All these circumstances tend to render the articulation very insecure; but, nevertheless, on account of the very powerful ligaments which bind the bones together, the joint is one of the strongest in the body, and dislocation from traumatism is of very rare occurrence. When, on the other hand, the ligaments have been softened or destroyed by disease, partial displacement is very liable to occur, and is frequently brought about by the mere action of the muscles displacing the articular surfaces from each other. The tibia may be dislocated in any direction from the femur—forwards, backwards, inwards, or outwards; or a combination of two of these dislocations may occur—that is, the tibia may be dislocated forwards and laterally, or backwards and laterally; and any of these dislocations may be complete or incomplete. As a rule, however, the antero-posterior dislocations are complete, the lateral ones incomplete.

One or other of the semilunar cartilages may become displaced and nipped between the femur and tibia. The accident is produced by a twist of the leg when the knee is flexed, and is accompanied by a sudden pain and fixation of the knee in a flexed position. The cartilage may be displaced either inwards or outwards: that is to say, either inwards towards the tibial spine, so that the cartilage becomes lodged in the intercondyloid notch; or outwards, so that the cartilage projects beyond the margin of the two articular surfaces.

Acute synovitis, the result of traumatism or exposure to cold, is very common in the knee, on account of its superficial position. When distended with fluid, the swelling shows itself above and at the sides of the patella, reaching about an inch or more above the trochlear surface of the femur, and extending a little higher under the Vastus internus than the Vastus externus. Occasionally the swelling may extend two inches or more. At the sides of the patella the swelling extends lower at the inner side than it does on the outer side. The lower level of the synovial membrane is just above the level of the upper part

---

* *Human Osteology*, p. 405.

of the head of the fibula. In the middle line it covers the upper third of the ligamentum patellæ, being separated from it, however, by the capsule and a pad of fat. Chronic synovitis principally shows itself in the form of pulpy degeneration of the synovial membrane, leading to tuberculous arthritis. The reasons why tuberculous disease of the knee usually commences in the synovial membrane appear to be the complex and extensive nature of this sac; the extensive vascular supply to it; and the fact that injuries are generally diffused and applied to the front of the joint rather than to the ends of the bones. Syphilitic disease not infrequently attacks the knee-joint. In the hereditary form of the disease it is usually symmetrical, attacking both joints, which become filled with synovial effusion, and is very intractable and difficult of cure. In the tertiary form of the disease, gummatous infiltration of the synovial membrane may take place. The knee is one of the joints most commonly affected with osteo-arthritis, and is said to be more frequently the seat of this disease in women than in men. The occurrence of the so-called loose cartilage is almost confined to the knee, though they are occasionally met with in the elbow, and, rarely, in some other joints. Many of them occur in cases of osteo-arthritis, in which calcareous or cartilaginous material is formed in one of the synovial fringes and constitutes the foreign body, and may or may not become detached, in the former case only meriting the usual term, 'loose' cartilage. In other cases they have their origin in the exudation of inflammatory lymph, and possibly, in some rare instances, a portion of the articular cartilage or one of the semilunar cartilage, becomes detached and constitutes the foreign body.

Genu valgum, or knock knee, is a common deformity of childhood, in which, owing to changes in and about the joint, the angle between the outer border of the tibia and femur is diminished, so that as the patient stands the two internal condyles of the femora are in contact, but the two internal malleoli of the tibiæ are more or less widely separated from each other. When, however, the knees are flexed to a right angle, the two legs are practically parallel with each other. At the commencement of the disease there is a yielding of the internal lateral ligament and other fibrous structures on the inner side of the joint; as a result of this there is a constant undue pressure of the outer tuberosity of the tibia against the outer condyle of the femur. This extra pressure causes arrest of growth and, possibly, wasting of the outer condyle, and a consequent tendency for the tibia to become separated from the internal condyle. To prevent this the internal condyle becomes depressed; probably, as was first pointed out by Mickulicz, by an increased growth of the lower end of the diaphysis on its inner side, so that the line of the epiphysis becomes oblique instead of transverse to the axis of the bone, with a direction downwards and inwards.

Excision of the knee-joint is most frequently required for tuberculous disease of this articulation, but is also practised in cases of disorganisation of the knee after rheumatic fever, pyæmia, &c., in osteo-arthritis, and in ankylosis. It is also occasionally called for in cases of injury, gunshot or otherwise. The operation is best performed either by a horseshoe incision, starting from one condyle, descending as low as the tubercle of the tibia, where it crosses the leg, and then carried upward to the other condyle; or by a transverse incision across the patella. In this latter incision the patella is either removed or sawn across, and the halves subsequently sutured together. The bone having been cleared, and in those cases where the operation is performed for tuberculous disease, all pulpy tissue having been carefully removed, the section of the femur is first made. This should never include, in children, more than, at the most, two-thirds of the articular surface, otherwise the epiphysial cartilage will be involved, with disastrous results as regards the growth of the limb. Afterwards a thin slice should be removed from the upper end of the tibia, not more than half an inch. If any diseased tissue still appears to be left in the bones, it should be removed with the gouge, rather than that a further section of the bones should be made.

III. Articulations between the Tibia and Fibula

The articulations between the tibia and fibula are effected by ligaments which connect both extremities, as well as the shafts of the bones. They may, consequently, be subdivided into three sets: 1. The Superior Tibio-fibular articulation. 2. The Middle Tibio-fibular ligament or interosseous membrane. 3. The Inferior Tibio-fibular articulation.

1. Superior Tibio-fibular Articulation

This articulation is an arthrodial joint. The contiguous surfaces of the bones present two flat, oval facets covered with cartilage, and connected together by the following ligaments:

Capsular.

Anterior Superior Tibio-fibular.   Posterior Superior Tibio-fibular.

The **Capsular Ligament** consists of a membranous bag, which surrounds the articulation, being attached round the margins of the articular facets on the tibia and fibula, and is much thicker in front than behind.

The **Anterior Superior Ligament** (fig. 185) consists of two or three broad and flat bands, which pass obliquely upwards and inwards from the front of the head of the fibula to the front of the outer tuberosity of the tibia.

The **Posterior Superior Ligament** (fig. 184) is a single thick and broad band, which passes upwards and inwards from the back part of the head of the fibula to the back part of the outer tuberosity of the tibia. It is covered by the tendon of the Popliteus muscle.

A **Synovial Membrane** lines this articulation, which at its upper and back part is occasionally continuous with that of the knee-joint.

2. MIDDLE TIBIO-FIBULAR LIGAMENT OR INTEROSSEOUS MEMBRANE

An interosseous membrane extends between the contiguous margins of the tibia and fibula, and separates the muscles on the front from those on the back of the leg. It consists of a thin, aponeurotic lamina composed of oblique fibres, which for the most part pass downwards and outwards between the interosseous ridges on the two bones ; some few fibres, however, pass in the opposite direction, downwards and inwards. It is broader above than below. Its upper margin does not quite reach the superior tibio-fibular joint, but presents a free concave border, above which is a large, oval aperture for the passage of the anterior tibial vessels forwards to the anterior aspect of the leg. At its lower part is an opening for the passage of the anterior peroneal vessels. It is continuous below with the inferior interosseous ligament; and is perforated in numerous parts for the passage of small vessels. It is in relation, in front, with the Tibialis anticus, Extensor longus digitorum, Extensor proprius hallucis, Peroneus tertius, and the anterior tibial vessels and nerve ; behind, with the Tibialis posticus and Flexor longus hallucis.

3. INFERIOR TIBIO-FIBULAR ARTICULATION

This articulation is formed by the rough, convex surface of the inner side of the lower end of the fibula, connected with a concave rough surface on the outer side of the tibia. Below, to the extent of about two lines, these surfaces are smooth, and covered with cartilage, which is continuous with that of the ankle-joint. The ligaments of this joint are—

Anterior Inferior Tibio-fibular.   Transverse or Inferior.
Posterior Inferior Tibio-fibular.   Inferior Interosseous.

The **Anterior Inferior Ligament** (fig. 190) is a flat, triangular band of fibres, broader below than above, which extends obliquely downwards and outwards between the adjacent margins of the tibia and fibula, on the front aspect of the articulation. It is in relation, in front, with the Peroneus tertius, the aponeurosis of the leg, and the integument ; behind, with the inferior interosseous ligament ; and lies in contact with the cartilage covering the astragalus.

The **Posterior Inferior Ligament,** smaller than the preceding, is disposed in a similar manner on the posterior surface of the articulation.

The **Transverse** or **Inferior Ligament** lies under cover of the posterior ligament, and is a strong, thick band of yellowish fibres which passes transversely across the back of the joint, from the external malleolus to the posterior border of the articular surface of the tibia, almost as far as its malleolar process. This ligament projects below the margin of the bones, and forms part of the articulating surface for the astragalus.

The **Inferior Interosseous Ligament** consists of numerous short, strong, fibrous bands, which pass between the contiguous rough surfaces of the tibia and fibula, and constitute the chief bond of union between the bones. This ligament is continuous, above, with the interosseous membrane.

The **Synovial Membrane** lining the articular surface is derived from that of the ankle-joint.

Actions.—The movement permitted in these articulations is limited to a very slight gliding of the articular surfaces one upon another.

IV. Ankle-Joint

The **Ankle** is a ginglymus, or hinge-joint. The bones entering into its formation are the lower extremity of the tibia and its malleolus, and the external malleolus of the fibula, which forms a mortise to receive the upper convex surface of the astragalus and its two lateral facets. The bony surfaces are covered with cartilage, and connected by a capsule, which in places forms thickened bands constituting the following ligaments :

Anterior.	Internal Lateral.
Posterior.	External Lateral.

The **Anterior Ligament** (fig. 189) is a broad, thin, membranous layer, attached, above, to the anterior margin of the lower extremity of the tibia ; below, to the

Fig. 189.—Ankle-joint : tarsal and tarso-metatarsal articulations.
Internal view. Right side.

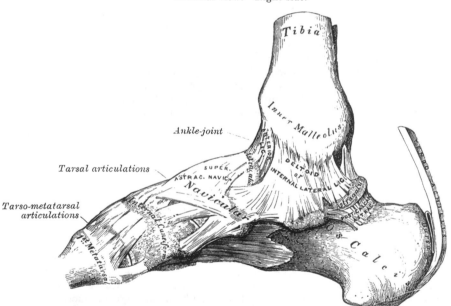

margin of the astragalus, in front of its articular surface. It is in relation, in front, with the Extensor tendons of the toes, with the tendons of the Tibialis anticus and Peroneus tertius, and the anterior tibial vessels and nerve ; behind, it lies in contact with the synovial membrane.

The **Posterior Ligament** is very thin, and consists principally of transverse fibres. It is attached, above, to the margin of the articular surface of the tibia, blending with the transverse tibio-fibular ligament ; below, to the astragalus, behind its superior articular facet. Externally, where a somewhat thickened band of transverse fibres is attached to the hollow on the inner surface of the external malleolus, it is thicker than internally.

The **Internal Lateral** or **Deltoid Ligament** is a strong, flat, triangular band, attached, above, to the apex and anterior and posterior borders of the inner malleolus. The most anterior fibres pass forwards to be inserted into the navicular bone and the inferior calcaneo-navicular ligament ; the middle descend almost perpendicularly to be inserted into the sustentaculum tali of the os calcis; and the posterior fibres pass backwards and outwards to be attached to the inner side of the astragalus. This ligament is covered by the tendons of the Tibialis posticus and Flexor longus digitorum muscles.

The **External Lateral Ligament** (fig. 190) consists of three distinctly specialised fasciculi of the capsule, taking different directions, and separated by distinct intervals, for which reason it is described by some anatomists as three distinct ligaments.*

The *anterior fasciculus* (anterior astragalo-fibular), the shortest of the three, passes from the anterior margin of the external malleolus, forwards and inwards, to the astragalus, in front of its external articular facet.

The *posterior fasciculus* (posterior astragalo-fibular), the most deeply seated, passes inwards from the depression at the inner and back part of the external malleolus to a prominent tubercle on the posterior surface of the astragalus. Its fibres are almost horizontal in direction.

Fig. 190.—Ankle-joint: tarsal and tarso-metatarsal articulations.
External view.   Right side.

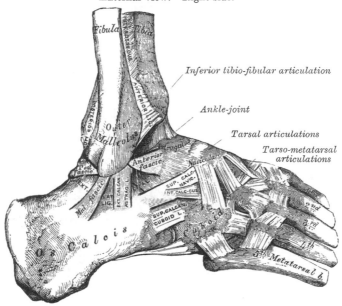

The *middle fasciculus* (calcaneo-fibular), the longest of the three, is a narrow, rounded cord, passing from the apex of the external malleolus downwards and slightly backwards to a tubercle on the outer surface of the os calcis.   It is covered by the tendons of the Peroneus longus and brevis.

The **Synovial Membrane** invests the inner surface of the ligaments, and sends a duplicature upwards between the lower extremities of the tibia and fibula for a short distance.

Relations.—The tendons, vessels, and nerves in connection with the joint are, in front, from within outwards, the Tibialis anticus, Extensor proprius hallucis, anterior tibial vessels, anterior tibial nerve, Extensor longus digitorum, and Peroneus tertius; behind, from within outwards, the Tibialis posticus, Flexor longus digitorum, posterior tibial vessels, posterior tibial nerve, Flexor longus hallucis; and, in the groove behind the external malleolus, the tendons of the Peroneus longus and brevis.

The *Arteries* supplying the joint are derived from the malleolar branches of the anterior tibial and the peroneal.

* Humphry *On the Skeleton*, p. 559.

The *Nerves* are derived from the anterior and posterior tibial.

Actions.—The movements of the joint are those of flexion and extension. Flexion consists in the approximation of the dorsum of the foot to the front of the leg, while in extension the heel is drawn up and the toes pointed downwards. The malleoli tightly embrace the astragalus in all positions of the joint, so that any slight degree of lateral movement which may exist is simply due to stretching of the inferior tibio-fibular ligaments, and slight bending of the shaft of the fibula. Of the ligaments, the internal, or deltoid, is of very great power— so much so, that it usually resists a force which fractures the process of bone to which it is attached. Its middle portion, together with the middle fasciculus of the external lateral ligament, binds the bones of the leg firmly to the foot, and resists displacement in every direction. Its anterior and posterior fibres limit extension and flexion of the foot respectively, and the anterior fibres also limit abduction. The posterior portion of the external lateral ligament assists the middle portion in resisting the displacement of the foot backwards, and deepens the cavity for the reception of the astragalus. The anterior fasciculus is a

FIG. 191.—Section of the right foot near its inner border, dividing the tibia, astragalus, navicular, internal cuneiform, and first metatarsal bone, and the first phalanx of the great toe.   (After Braune.)

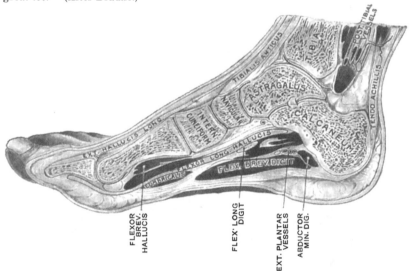

security against the displacement of the foot forwards, and limits extension of the joint. The movements of inversion and eversion of the foot, together with the minute changes in form by which it is applied to the ground or takes hold of an object in climbing, &c., are mainly effected in the tarsal joints; the one which enjoys the greatest amount of motion being that between the astragalus and os calcis behind, and the navicular and cuboid in front. This is often called the *transverse* or *medio-tarsal joint*, and it can, with the subordinate joints of the tarsus, replace the ankle-joint in a great measure when the latter has become ankylosed.

Extension of the tarsal bones upon the tibia and fibula is produced by the Gastrocnemius, Soleus, Plantaris, Tibialis posticus, Peroneus longus and brevis, Flexor longus digitorum, and Flexor longus hallucis. *Flexion*, by the Tibialis anticus, Peroneus tertius, Extensor longus digitorum, and Extensor proprius hallucis.* *Inversion*, in the extended position, is produced by the Tibialis anticus and posticus; and *eversion* by the Peronei.

* The student must bear in mind that the Extensor longus digitorum and Extensor proprius hallucis are *extensors* of the toes, but *flexors* of the ankle; and that the Flexor longus digitorum and Flexor longus hallucis are *flexors* of the toes, but *extensors* of the ankle.

Surface Form.—The line of the ankle-joint may be indicated by a transverse line drawn across the front of the lower part of the leg, about half an inch above the level of the tip of the internal malleolus.

Surgical Anatomy.—Displacement of the trochlear surface of the astragalus from the tibio-fibular mortise is not of common occurrence, as the ankle-joint is a very strong and powerful articulation, and great force is required to produce it. Nevertheless, dislocation does occasionally occur, both in an antero-posterior and a lateral direction. In the latter, which is the more common, fracture is a necessary accompaniment of the injury. The dislocation in these cases is somewhat peculiar, and is not a displacement in a horizontally lateral direction, such as usually occurs in lateral dislocations of ginglymoid joints, but the astragalus undergoes a partial rotation round an antero-posterior axis drawn through its own centre, so that the superior surface, instead of being directed upwards, is inclined more or less inwards or outwards according to the variety of the displacement.

The ankle-joint is more frequently sprained than any joint in the body, and this may lead to acute synovitis. In these cases, when the synovial sac is distended with fluid, the bulging appears principally in the front of the joint, beneath the anterior tendons, and on either side, between the Tibialis anticus and the internal lateral ligament on the inner side, and between the Peroneus tertius and the external lateral ligament on the outer side. In addition to this, bulging frequently occurs posteriorly, and a fluctuating swelling may be detected on either side of the tendo Achillis.

Chronic synovitis may result from frequent sprains, and when once this joint has been sprained it is more liable to a recurrence of the injury than it was before ; or it may be tuberculous in its origin, the disease usually commencing in the astragalus and extending to the joint, though it may commence as a synovitis the result probably of some slight strain in a tuberculous subject.

Excision of the ankle-joint is not often performed, for two reasons. In the first place, disease of the articulation for which this operation is indicated is frequently associated with disease of the tarsal bones, which prevents its performance ; and, secondly, the foot after excision is frequently of very little use ; far less, in fact, than after a Symes's amputation, which is often, therefore, a preferable operation in these cases. Excision may, however, be attempted in cases of tuberculous arthritis, in a young and otherwise healthy subject, where the disease is limited to the bones forming the joint. It may also be required after injury, where the vessels and nerves have not been damaged and the patient is young and free from visceral disease. The excision is best performed by two lateral incisions. One commencing two and a half inches above the external malleolus, carried down the posterior border of the fibula, round the end of the bone, and then forwards and downwards as far as the calcaneo-cuboid joint, midway between the tip of the external malleolus and the tuberosity on the fifth metatarsal bone. Through this incision the fibula is cleared, the external lateral ligament is divided, and the bone sawn through about half an inch above the level of the ankle-joint and removed. A similar curved incision is now made on the inner side of the foot, commencing two and a half inches above the lower end of the tibia, carried down the posterior border of the bone, round the internal malleolus and forwards and downwards to the tuberosity of the navicular bone. Through this incision the tibia is cleared in front and behind, the internal lateral, the anterior and posterior ligaments divided, and the end of the tibia protruded through the wound by displacing the foot outwards, and sawn off sufficiently high to secure a healthy section of bone. The articular surface of the astragalus is now to be sawn off or the whole bone removed. In cases where the operation is performed for tuberculous arthritis, the latter course is probably preferable, as the injury done by the saw is frequently the starting point of fresh caries ; and after removal of the whole bone the shortening is not appreciably increased, and the result as regards union appears to be as good as when two sawn surfaces of bone are brought into apposition.

V. ARTICULATIONS OF THE TARSUS

1. ARTICULATIONS OF THE OS CALCIS AND ASTRAGALUS

The articulations between the os calcis and astragalus are two in number—anterior and posterior. They are arthrodial joints. The bones are connected together by four ligaments :

External Calcaneo-astragaloid.	Posterior Calcaneo-astragaloid.
Internal Calcaneo-astragaloid.	Interosseous.

The **External Calcaneo-astragaloid Ligament** (fig. 190) is a short, strong fasciculus, passing from the outer surface of the astragalus, immediately beneath its external malleolar facet, to the outer surface of the os calcis. It is placed in front of the middle fasciculus of the external lateral ligament of the ankle-joint, with the fibres of which it is parallel.

The **Internal Calcaneo-astragaloid Ligament** is a band of fibres connecting the internal tubercle of the back of the astragalus with the back of the sustentaculum tali. Its fibres blend with those of the inferior calcaneo-navicular ligament.

The **Posterior Calcaneo-astragaloid Ligament** (fig. 189) connects the external tubercle of the astragalus with the upper and inner part of the os calcis; it is a short band, the fibres of which radiate from their narrow attachment to the astragalus.

The **Interosseous Ligament** forms the chief bond of union between the bones. It consists of numerous vertical and oblique fibres, attached by one extremity to the groove between the articulating facets on the under surface of the astragalus; by the other, to a corresponding depression on the upper surface of the os calcis. It is very thick and strong, being at least an inch in breadth from side to side, and serves to unite the os calcis and astragalus solidly together.

The **Synovial Membranes** (fig. 193) are two in number: one for the posterior calcaneo-astragaloid articulation; a second for the anterior calcaneo-astragaloid joint. The latter synovial membrane is continued forwards between the contiguous surfaces of the astragalus and navicular bones.

Actions.—The movements permitted between the astragalus and os calcis are limited to a gliding of the one bone on the other in a direction from before backwards, and from side to side.

2. ARTICULATIONS OF THE OS CALCIS WITH THE CUBOID

The ligaments connecting the os calcis with the cuboid are four in number:

Dorsal	{ Superior Calcaneo-cuboid. { Internal Calcaneo-cuboid (Interosseous).
Plantar	{ Long Calcaneo-cuboid. { Short Calcaneo-cuboid.

The **Superior Calcaneo-cuboid Ligament** (fig. 190) is a thin and narrow fasciculus, which passes between the contiguous surfaces of the os calcis and cuboid, on the dorsal surface of the joint.

The **Internal Calcaneo-cuboid (Interosseous) Ligament** (fig. 190) is a short, but thick and strong, band of fibres, arising from the os calcis, in the deep hollow which intervenes between it and the astragalus, and closely blended, at its origin, with the superior calcaneo-navicular ligament. It is inserted into the inner side of the cuboid bone. This ligament forms one of the chief bonds of union between the first and second rows of the tarsus.

The **Long Calcaneo-cuboid (Long Plantar) Ligament** (fig. 192), the more superficial of the two plantar ligaments, is the longest of all the ligaments of the tarsus: it is attached to the under surface of the os calcis, from near the tuberosities, as far forwards as the anterior tubercle; its fibres pass forwards to be attached to the ridge on the under surface of the cuboid bone, the more superficial fibres being continued onwards to the bases of the second, third, and fourth metatarsal bones. This ligament crosses the groove on the under surface of the cuboid bone, converting it into a canal for the passage of the tendon of the Peroneus longus.

The **Short Calcaneo-cuboid (Short Plantar) Ligament** lies nearer to the bones than the preceding, from which it is separated by a little areolar tissue. It is exceedingly broad, about an inch in length, and extends from the tubercle, and the depression in front of it, on the fore part of the under surface of the os calcis, to the inferior surface of the cuboid bone behind the peroneal groove.

Synovial Membrane.—The synovial membrane in this joint is distinct. It lines the inner surface of the ligaments.

Actions.—The movements permitted between the os calcis and cuboid are limited to a slight gliding upon each other.

3. The Ligaments connecting the Os Calcis and Navicular

Though these two bones do not directly articulate, they are connected by two ligaments:

Superior or External Calcaneo-navicular.
Inferior or Internal Calcaneo-navicular.

The **Superior** or **External Calcaneo-navicular** (fig. 190) arises, as already mentioned, with the internal calcaneo-cuboid in the deep hollow between the astragalus and os calcis; it passes forwards from the upper surface of the anterior extremity of the os calcis to the outer side of the navicular bone. These two ligaments resemble the letter Y, being blended together behind, but separated in front.

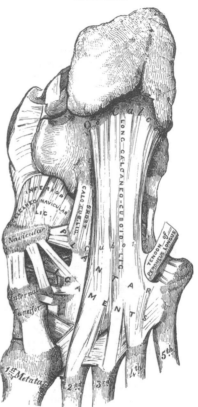

Fig. 192.—Ligaments of plantar surface of the foot.

The **Inferior** or **Internal Calcaneo-navicular** (fig. 192) is by far the larger and stronger of the two ligaments between these bones; it is a broad and thick band of fibres, which passes forwards and inwards from the anterior margin of the sustentaculum tali of the os calcis to the under surface of the navicular bone. This ligament not only serves to connect the os calcis and navicular, but supports the head of the astragalus, forming part of the articular cavity in which it is received. The *upper surface* presents a fibro-cartilaginous facet, lined by the synovial membrane continued from the anterior calcaneo-astragaloid articulation, upon which a portion of the head of the astragalus rests. Its *under surface* is in contact with the tendon of the Tibialis posticus muscle; * its inner border is blended with the fore part of the Deltoid ligament, thus completing the socket for the head of the astragalus.

Surgical Anatomy.—The inferior calcaneo-navicular ligament, by supporting the head of the astragalus, is principally concerned in maintaining the arch of the foot, and, when it yields, the head of the astragalus is pressed downwards, inwards, and forwards by the weight of the body, and the foot becomes flattened, expanded, and turned outwards, constituting the disease known as *flat-foot.* This ligament contains a considerable amount of elastic fibres, so as to give elasticity to the arch and spring to the foot; hence it is sometimes called the 'spring' ligament. It is supported, on its under surface, by the tendon of the Tibialis posticus, which spreads out at its insertion into a number of fasciculi, which are attached to most of the tarsal and metatarsal bones; this prevents undue stretching of the ligament, and is a protection against the occurrence of flat-foot.

4. Articulation of the Astragalus with the Navicular Bone

The articulation between the astragalus and navicular is an arthrodial joint: the rounded head of the astragalus being received into the concavity formed by the posterior surface of the navicular, the anterior articulating surface of the calcaneum, and the upper surface of the inferior calcaneo-navicular ligament,

* Mr. Hancock describes an extension of this ligament upwards on the inner side of the foot, which completes the socket of the joint in that direction. *Lancet*, 1866, vol. i. p. 618.

which fills up the triangular interval between these bones. The only ligament of this joint is the **superior astragalo-navicular.** It is a broad band, which passes obliquely forwards from the neck of the astragalus to the superior surface of the navicular bone. It is thin, and weak in texture, and covered by the Extensor tendons. The inferior calcaneo-navicular supplies the place of an inferior ligament.

The **Synovial Membrane** which lines the joint is continued forwards from the anterior calcaneo-astragaloid articulation.

Actions.—This articulation permits of considerable mobility; but its feebleness is such as to allow occasionally of dislocation of the other bones of the tarsus from the astragalus.

The *transverse tarsal* or *medio-tarsal joint* is formed by the articulation of the os calcis with the cuboid, and by the articulation of the astragalus with the navicular. The movement which takes place in this joint is more extensive than that in the other tarsal joints, and consists of a sort of rotation by means of which the sole of the foot may be slightly flexed and extended or carried inwards (inverted) and outwards (everted).

5. The Articulation of the Navicular with the Cuneiform Bones

The navicular is connected to the three cuneiform bones by

Dorsal and Plantar ligaments.

The **Dorsal Ligaments** are small, longitudinal bands of fibrous tissue, arranged as three bundles, one to each of the cuneiform bones. That bundle of fibres which connects the navicular with the internal cuneiform is continued round the inner side of the articulation to be continuous with the plantar ligament which connects these two bones.

The **Plantar Ligaments** have a similar arrangement to those on the dorsum. They are strengthened by processes given off by the tendon of the Tibialis posticus.

The **Synovial Membrane** of these joints is part of the great tarsal synovial membrane.

Actions.—The movements permitted between the navicular and cuneiform bones are limited to a slight gliding upon each other.

6. The Articulation of the Navicular with the Cuboid

The navicular bone is connected with the cuboid by

Dorsal, Plantar, and Interosseous ligaments.

The **Dorsal Ligament** consists of a band of fibrous tissue which passes obliquely forwards and outwards from the navicular to the cuboid bone.

The **Plantar Ligament** consists of a band of fibrous tissue which passes nearly transversely between these two bones.

The **Interosseous Ligament** consists of strong transverse fibres which pass between the rough non-articular portions of the lateral surfaces of these two bones.

The **Synovial Membrane** of this joint is part of the great tarsal synovial membrane.

Actions.—The movements permitted between the navicular and cuboid bones are limited to a slight gliding upon each other.

7. The Articulation of the Cuneiform Bones with each other

These bones are connected together by

Dorsal, Plantar, and Interosseous ligaments.

The **Dorsal Ligaments** consist of two bands of fibrous tissue which pass transversely: one connecting the internal with the middle cuneiform, and the other connecting the middle with the external cuneiform.

The **Plantar Ligaments** have a similar arrangement to those on the dorsum. They are strengthened by the processes given off from the tendon of the Tibialis posticus.

The **Interosseous Ligaments** consist of strong transverse fibres which pass between the rough non-articular portions of the lateral surfaces of the adjacent cuneiform bones.

The **Synovial Membrane** of these joints is part of the great tarsal synovial membrane.

Actions.—The movements permitted between the cuneiform bones are limited to a slight gliding upon each other.

8. The Articulation of the External Cuneiform Bone with the Cuboid

These bones are connected together by

Dorsal, Plantar, and Interosseous ligaments.

The **Dorsal Ligament** consists of a band of fibrous tissue which passes transversely between these two bones.

The **Plantar Ligament** has a similar arrangement. It is strengthened by a process given off from the tendon of the Tibialis posticus.

The **Interosseous Ligament** consists of strong transverse fibres which pass between the rough non-articular portions of the lateral surfaces of the adjacent sides of these two bones.

The **Synovial Membrane** of this joint is part of the great tarsal synovial membrane.

Actions.—The movements permitted between the external cuneiform and cuboid are limited to a slight gliding upon each other.

Nerve-supply.—All the joints of the tarsus are supplied by the anterior tibial nerve.

Surgical Anatomy.—In spite of the great strength of the ligaments which connect the tarsal bones together, dislocation at some of the tarsal joints does occasionally occur ; though, on account of the spongy character of the bones, they are more frequently broken than dislocated, as the result of violence. When dislocation does take place, it is most commonly in connection with the astragalus ; for not only may this bone be dislocated from the tibia and fibula at the ankle-joint, but the other bones may be dislocated from it, the trochlear surface of the bone remaining *in situ* in the tibio-fibular mortise. This constitutes what is known as the *sub-astragaloid* dislocation. Or, again, the astragalus may be dislocated from all its connections—from the tibia and fibula above, the os calcis below, and the navicular in front—and may even undergo a rotation, on either a vertical or a horizontal axis. In the former case the long axis of the bone becoming directed across the joint, so that the head faces the articular surface on one or other malleolus ; or, in the latter, the lateral surfaces becoming directed upwards and downwards, so that the trochlear surface faces to one or the other side. Finally, dislocation may occur at the medio-tarsal joint, the anterior tarsal bones being luxated from the astragalus and calcaneum. The other tarsal bones are also, occasionally, though rarely, dislocated from their connections.

VI. Tarso-metatarsal Articulations

These are arthrodial joints. The bones entering into their formation are four tarsal bones, viz. the internal, middle and external cuneiform, and the cuboid, which articulate with the metatarsal bones of the five toes. The metatarsal bone of the great toe articulates with the internal cuneiform ; that of the second is deeply wedged in between the internal and external cuneiform, resting against the middle cuneiform, and being the most strongly articulated of all the metatarsal bones ; the third metatarsal articulates with the extremity of the external cuneiform ; the fourth, with the cuboid and external cuneiform ; and the fifth, with the cuboid. The articular surfaces are covered with cartilage, lined by synovial membrane, and connected together by the following ligaments :

Dorsal. Plantar. Interosseous.

The **Dorsal Ligaments** consist of strong, flat, fibrous bands, which connect the tarsal with the metatarsal bones. The first metatarsal is connected to the internal cuneiform by a single broad, thin, fibrous band; the second has three dorsal ligaments, one from each cuneiform bone; the third has one from the external cuneiform; the fourth has two, one from the external cuneiform and one from the cuboid; and the fifth, one from the cuboid.

The **Plantar Ligaments** consist of longitudinal and oblique fibrous bands connecting the tarsal and metatarsal bones, but disposed with less regularity than on the dorsal surface. Those for the first and second metatarsal are the most strongly marked; the second and third metatarsal receive strong fibrous bands, which pass obliquely across from the internal cuneiform; the plantar ligaments of the fourth and fifth metatarsal consist of a few scanty fibres derived from the cuboid.

The **Interosseous Ligaments** are three in number : internal, middle, and external. The *internal* one is the strongest of the three, and passes from the outer extremity of the internal cuneiform to the adjacent angle of the second metatarsal. The *middle* one, less strong than the preceding, connects the external cuneiform with the adjacent angle of the second metatarsal. The *external* interosseous ligament connects the outer angle of the external cuneiform with the adjacent side of the third metatarsal.

The **Synovial Membrane** between the internal cuneiform bone and the first metatarsal bone is a distinct sac. The synovial membrane between the middle and external cuneiform behind, and the second and third metatarsal bones in front, is part of the great tarsal synovial membrane. Two prolongations are sent forwards from it, one between the adjacent sides of the second and third metatarsal bones, and one between the third and fourth metatarsal bones. The synovial membrane between the cuboid and the fourth and fifth metatarsal bones is a distinct sac. From it a prolongation is sent forwards between the fourth and fifth metatarsal bones.

Actions.—The movements permitted between the tarsal and metatarsal bones are limited to a slight gliding upon each other.

VII. Articulations of the Metatarsal Bones with each other

The base of the first metatarsal bone is not connected with the second metatarsal bone by any ligaments; in this respect it resembles the thumb.

The bases of the four outer metatarsal bones are connected by dorsal, plantar, and interosseous ligaments.

The **Dorsal Ligaments** consist of bands of fibrous tissue which pass transversely between the adjacent metatarsal bones.

The **Plantar Ligaments** have a similar arrangement to those on the dorsum.

The **Interosseous Ligaments** consist of strong transverse fibres which pass between the rough non-articular portions of the lateral surfaces.

The **Synovial Membrane** between the second and third, and the third and fourth, metatarsal bones is part of the great tarsal synovial membrane.

The synovial membrane between the fourth and fifth metatarsal bones is a prolongation of the synovial membrane of the cubo-metatarsal joint.

Actions.—The movement permitted in the tarsal ends of the metatarsal bones is limited to a slight gliding of the articular surfaces upon one another.

The Synovial Membranes in the Tarsal and Metatarsal Joints

The **Synovial Membranes** (fig. 193) found in the articulations of the tarsus and metatarsus are six in number : one for the posterior calcaneo-astragaloid articulation ; a second for the anterior calcaneo-astragaloid and astragalo-navicular articulations ; a third for the calcaneo-cuboid articulation ; and a fourth for the articulations of the navicular with the three cuneiform, the three cuneiform with each other, the external cuneiform with the cuboid, and the middle and external cuneiform with the bases of the second and third metatarsal bones, and the lateral surfaces of the second, third, and fourth metatarsal bones with each other ; a fifth for the internal cuneiform with the metatarsal bone of the great toe ; and a sixth for the articulation of the cuboid with the fourth and fifth metatarsal bones. A small synovial membrane is sometimes found between the contiguous surfaces of the navicular and cuboid bones.

Nerve-supply.—The nerves supplying the tarso-metatarsal joints are derived from the anterior tibial.

The *digital extremities* of all the metatarsal bones are connected together by the *transverse metatarsal ligament*.

The **Transverse Metatarsal Ligament** is a narrow fibrous band which passes transversely across the anterior extremities of all the metatarsal bones, connecting them together. It is blended anteriorly with the plantar (glenoid) ligament of the metatarso-phalangeal articulations. To its posterior border is connected the fascia covering the Interossei muscles. Its inferior surface is concave where

FIG. 193.—Oblique section of the articulations of the tarsus and metatarsus. Showing the six synovial membranes.

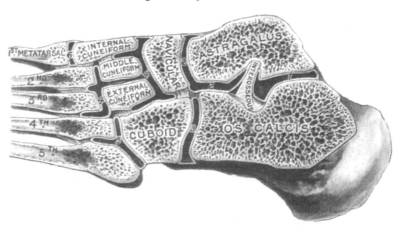

the flexor tendons pass over it. Above it the tendons of the Interossei muscles pass to their insertion. It differs from the transverse metacarpal ligament in that it connects the metatarsal bone of the great toe with the rest of the metatarsal bones.

VIII. METATARSO-PHALANGEAL ARTICULATIONS

The metatarso-phalangeal articulations are of the condyloid kind, formed by the reception of the rounded head of the metatarsal bone into a superficial cavity in the extremity of the first phalanx.

The ligaments are—

Plantar.   Two Lateral.

The **Plantar Ligaments** (Glenoid ligaments of Cruveilhier) are thick, dense, fibrous structures. Each is placed on the plantar surface of the joint in the interval between the lateral ligaments, to which they are connected ; they are loosely united to the metatarsal bone, but very firmly to the base of the first phalanges. Their plantar surface is intimately blended with the transverse metatarsal ligament, and presents a groove for the passage of the flexor tendons, the sheath surrounding which is connected to each side of the groove. By their deep surface, they form part of the articular surface for the head of the metatarsal bone, and are lined by a synovial membrane.

The **Lateral Ligaments** are strong, rounded cords, placed one on each side of the joint, each being attached, by one extremity, to the posterior tubercle on the side of the head of the metatarsal bone ; and, by the other, to the contiguous extremity of the phalanx.

The **Posterior Ligament** is supplied by the extensor tendon placed over the back of the joint.

Actions.—The movements permitted in the metatarso-phalangeal articulations are flexion, extension, abduction, and adduction.

IX. ARTICULATIONS OF THE PHALANGES

The articulations of the phalanges are ginglymus joints.
The ligaments are—

Plantar.   Two Lateral.

The arrangement of these ligaments is similar to those in the metatarso-phalangeal articulations : the extensor tendon supplies the place of a posterior ligament.

Actions.—The only movements permitted in the phalangeal joints are flexion and extension ; these movements are more extensive between the first and second phalanges than between the second and third. The movement of flexion is very considerable, but extension is limited by the plantar and lateral ligaments.

Surface Form.—The principal joints which it is necessary to distinguish, with regard to the surgery of the foot, are the medio-tarsal and the tarso-metatarsal joints. The joint between the astragalus and the navicular is best found by means of the tubercle of the navicular bone, for the line of the joint is immediately behind this process. If the foot is grasped and forcibly extended, a rounded prominence, the head of the astragalus, will appear on the inner side of the dorsum in front of the ankle-joint, and if a knife is carried downwards, just in front of this prominence and behind the line of the navicular tubercle, it will enter the astragalo-navicular joint. The calcaneo-cuboid joint is situated midway between the external malleolus and the prominent end of the fifth metatarsal bone. The plane of the joint is in the same line as that of the astragalo-navicular. The position of the joint between the fifth metatarsal bone and the cuboid is easily found by the projection of the fifth metatarsal bone, which is the guide to it. The direction of the line of the joint is very oblique. so that, if continued onwards, it would pass through the head of the first metatarsal bone. The joint between the fourth metatarsal bone and the cuboid and external cuneiform is the direct continuation inwards of the previous joint, but its plane is less oblique ; it would be represented by a line drawn from the outer side of the articulation to the middle of the first metatarsal bone. The plane of the joint between the third metatarsal bone and the external cuneiform is almost transverse. It would be represented by a line drawn from the outer side of the joint to the base of the first metatarsal bone. The tarso-metatarsal articulation of the great toe corresponds to a groove which can be felt by making firm pressure on the inner side of the foot one inch in front of the tubercle on the navicular bone ; and the joint between the second metatarsal bone and the middle cuneiform is to be found on the dorsum of the foot, half an inch behind the level of the tarso-metatarsal joint of the great toe. The line of the joints between the metatarsal bones and the first phalanges is about an inch behind the webs of the corresponding toes.

THE MUSCLES AND FASCIÆ*

THE Muscles are connected with the bones, cartilages, ligaments, and skin, either directly or through the intervention of fibrous structures, called tendons or aponeuroses. Where a muscle is attached to bone or cartilage, the fibres terminate in blunt extremities upon the periosteum or perichondrium, and do not come into direct relation with the osseous or cartilaginous tissue. Where muscles are connected with the skin, they either lie as a flattened layer beneath it, or are connected with its areolar tissue by larger or smaller bundles of fibres, as in the muscles of the face.

The muscles vary extremely in their form. In the limbs, they are of considerable length, especially the more superficial ones, the deep ones being generally broad; they surround the bones, and form an important protection to the various joints. In the trunk, they are broad, flattened, and expanded, forming the parietes of the cavities which they enclose; hence the reason of the terms, *long, broad, short,* &c., used in the description of a muscle.

There is a considerable variation in the arrangement of the fibres of certain muscles with reference to the tendons to which they are attached. In some, the fibres are parallel and run directly from their origin to their insertion; these are quadrilateral muscles, such as the Thyro-hyoid. A modification of these is found in the fusiform muscles, in which the fibres are not quite parallel, but slightly curved, so that the muscle tapers at each end; in their action, however, they resemble the quadrilateral muscles. Secondly, in other muscles the fibres are convergent; arising by a broad origin, they converge to a narrow or pointed insertion. This arrangement of fibres is found in the triangular muscles—e.g. the Temporal. In some muscles, which otherwise would belong to the quadrilateral or triangular type, the origin and insertion are not in the same plane, but the plane of the line of origin intersects that of their insertion: such is the case in the Pectineus muscle. Thirdly, in some muscles the fibres are oblique and converge, like the plumes of a pen, to one side of a tendon, which runs the entire length of the muscle. Such a muscle is rhomboidal or penniform, as the Peronei. A modification of these rhomboidal muscles is found in those cases where oblique fibres converge to both sides of a central tendon which runs down the middle of the muscle; these are called bipenniform, and an example is afforded in the Rectus femoris. Finally, we have muscles in which the fibres are arranged in curved bundles in one or more planes, as in the Sphincter muscles. The arrangement of the muscular fibres is of considerable importance in respect to their relative strength and range of movement. Those muscles where the fibres are long and few in number have great range, but diminished strength; where, on

---

* The Muscles and Fasciæ are described conjointly, in order that the student may consider the arrangement of the latter in his dissection of the former. It is rare for the student of anatomy in this country to have the opportunity of dissecting the fasciæ separately; and it is for this reason, as well as from the close connection that exists between the muscles and their investing sheaths, that they are considered together. Some general observations are first made on the anatomy of the muscles and fasciæ, the special description being given in connection with the different regions.

the other hand, the fibres are short and more numerous, there is great power, but lessened range.

Muscles differ much in size : the Gastrocnemius forms the chief bulk of the back of the leg, and the fibres of the Sartorius are nearly two feet in length, while the Stapedius, a small muscle of the internal ear, weighs about a grain, and its fibres are not more than two lines in length.

The names applied to the various muscles have been derived : 1, from their situation, as the Tibialis, Radialis, Ulnaris, Peroneus ; 2, from their direction, as the Rectus abdominis, Obliqui capitis, Transversalis ; 3, from their uses, as Flexors, Extensors, Abductors, &c. ; 4, from their shape, as the Deltoid, Trapezius, Rhomboideus ; 5, from the number of their divisions, as the Biceps, the Triceps ; 6, from their points of attachment, as the Sterno-cleido-mastoid, Sterno-hyoid, Sterno-thyroid.

In the description of a muscle, the term *origin* is meant to imply its more fixed or central attachment ; and the term *insertion* the movable point to which the force of the muscle is directed ; but the origin is absolutely fixed in only a very small number of muscles, such as those of the face, which are attached by one extremity to the bone, and by the other to the movable integument ; in the greater number, the muscle can be made to act from either extremity.

In the dissection of the muscles, the student should pay especial attention to the exact *origin, insertion,* and *actions* of each, and its more important *relations* with surrounding parts. An accurate knowledge of the points of attachment of the muscles is of great importance in the determination of their action. By a knowledge of the action of the muscles, the surgeon is able to explain the causes of displacement in various forms of fracture, and the causes which produce distortion in various deformities, and, consequently, to adopt appropriate treatment in each case. The relations, also, of some of the muscles, especially those in immediate apposition with the larger blood-vessels, and the surface-markings they produce, should be especially remembered, as they form useful guides in the application of a ligature to those vessels.

Tendons are white, glistening, fibrous cords, varying in length and thickness, sometimes round, sometimes flattened, of considerable strength, and devoid of elasticity. They consist almost entirely of white fibrous tissue, the fibrils of which have an undulating course parallel with each other, and are firmly united together. They are very sparingly supplied with blood-vessels, the smaller tendons presenting in their interior no trace of them. Nerves also are absent in the smaller tendons ; but the larger ones, as the tendo Achillis, receive nerves which accompany the nutrient vessels. The tendons consist principally of a substance which yields gelatin.

Aponeuroses are flattened or ribbon-shaped tendons, of a pearly-white colour, iridescent, glistening, and similar in structure to the tendons. They are destitute of nerves, and the thicker ones only sparingly supplied with blood-vessels.

The tendons and aponeuroses are connected, on the one hand, with the muscles ; and, on the other hand, with the movable structures, as the bones, cartilages, ligaments, and fibrous membranes (for instance, the sclerotic). Where the muscular fibres are in a direct line with those of the tendon or aponeurosis, the two are directly continuous, the muscular fibre being distinguishable from that of the tendon only by its striation. But where the muscular fibre joins the tendon or aponeurosis at an oblique angle, the former terminates, according to Kölliker, in rounded extremities, which are received into corresponding depressions on the surface of the latter, the connective tissue between the fibres being continuous with that of the tendon. The latter mode of attachment occurs in all the penniform and bipenniform muscles, and in those muscles the tendons of which commence in a membranous form, as the Gastrocnemius and Soleus.

The fasciæ (fascia, *a bandage*) are fibro-areolar or aponeurotic laminæ, of variable thickness and strength, found in all regions of the body, investing the softer and more delicate organs. The fasciæ have been subdivided, from the situation in which they are found, into two groups, superficial and deep.

The *superficial fascia* is found immediately beneath the integument over almost the entire surface of the body. It connects the skin with the deep or aponeurotic fascia, and consists of fibro-areolar tissue, containing in its meshes pellicles of fat in varying quantity. In the eyelids and scrotum, where adipose tissue is rarely deposited, this tissue is very liable to serous infiltration. The superficial fascia varies in thickness in different parts of the body: in the groin it is so thick as to be capable of being subdivided in several laminæ. Beneath the fatty layer of the superficial fascia, which is immediately subcutaneous, there is generally another layer of the same structure, comparatively devoid of adipose tissue, in which the trunks of the subcutaneous vessels and nerves are found, as the superficial epigastric vessels in the abdominal region, the radial and ulnar veins in the forearm, the saphenous veins in the leg and thigh, and the superficial lymphatic glands; certain cutaneous muscles also are situated in the superficial fascia, as the Platysma myoides in the neck, and the Orbicularis palpebrarum around the eyelids. This fascia is most distinct at the lower part of the abdomen, the scrotum, perinæum, and extremities; is very thin in those regions where muscular fibres are inserted into the integument, as on the side of the neck, the face, and around the margin of the anus. It is very dense in the scalp, in the palms of the hands, and soles of the feet, forming a fibro-fatty layer, which binds the integument firmly to the subjacent structure.

The superficial fascia connects the skin to the subjacent parts, facilitates the movement of the skin, serves as a soft nidus for the passage of vessels and nerves to the integument, and retains the warmth of the body, since the fat contained in its areolæ is a bad conductor of heat.

The *deep fascia* is a dense inelastic, unyielding fibrous membrane, forming sheaths for the muscles, and affording them broad surfaces for attachment. It consists of shining tendinous fibres, placed parallel with one another, and connected together by other fibres disposed in a rectilinear manner. It is usually exposed on the removal of the superficial fascia, forming a strong investment, which not only binds down collectively the muscles in each region, but gives a separate sheath to each, as well as to the vessels and nerves. The fasciæ are thick in unprotected situations, as on the outer side of a limb, and thinner on the inner side. The deep fasciæ assist the muscles in their action, by the degree of tension and pressure they make upon their surface: and, in certain situations, this is increased and regulated by muscular action, as, for instance, by the Tensor fasciæ femoris and Gluteus maximus in the thigh, by the Biceps in the upper and lower extremities, and Palmaris longus in the hand. In the limbs, the fasciæ not only invest the entire limb, but give off septa which separate the various muscles, and are attached beneath to the periosteum: these prolongations of fasciæ are usually spoken of as intermuscular septa.

The Muscles and Fasciæ may be arranged, according to the general division of the body, into those of the cranium, face, and neck; those of the trunk; those of the upper extremity; and those of the lower extremity.

MUSCLES AND FASCIÆ OF THE CRANIUM AND FACE

The muscles of the Cranium and Face consist of ten groups, arranged according to the region in which they are situated.

I. Cranial Region.	VI. Maxillary Region.
II. Auricular Region.	VII. Mandibular Region.
III. Palpebral Region.	VIII. Intermaxillary Region.
IV. Orbital Region.	IX. Temporo-mandibular Region.
V. Nasal Region.	X. Pterygo-mandibular Region.

The muscles contained in each of these groups are the following:

I. Cranial Region.
Occipito-frontalis.

II. Auricular Region.
Attrahens auriculam.
Attollens auriculam.
Retrahens auriculam.

III. Palpebral Region.
Orbicularis palpebrarum.
Corrugator supercilii.
Tensor tarsi.

IV. Orbital Region.
Levator palpebræ.
Rectus superior.
Rectus inferior.
Rectus internus.
Rectus externus.
Obliquus superior.
Obliquus inferior.

V. Nasal Region.
Pyramidalis nasi.
Levator labii superioris alæque nasi.
Dilatator naris posterior.
Dilatator naris anterior.

Compressor nasi.
Compressor narium minor.
Depressor alæ nasi.

VI. Maxillary Region.
Levator labii superioris.
Levator anguli oris.
Zygomaticus major.
Zygomaticus minor.

VII. Mandibular Region.
Levator labii inferioris.
Depressor labii inferioris.
Depressor anguli oris.

VIII. Intermaxillary Region.
Buccinator.
Risorius.
Orbicularis oris.

IX. Temporo-mandibular Region.
Masseter.
Temporal.

X. Pterygo-mandibular Region.
Pterygoideus externus.
Pterygoideus internus.

I. CRANIAL REGION—OCCIPITO-FRONTALIS

Dissection (fig. 194).—The head being shaved, and a block placed beneath the back of the neck, make a vertical incision through the skin from before backwards, commencing at the root of the nose in front, and terminating behind at the occipital protuberance;

FIG. 194.—Dissection of the head, face, and neck.

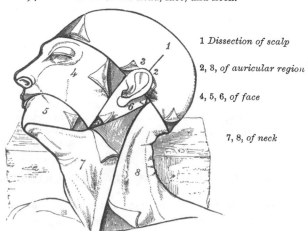

1 *Dissection of scalp*

2, 3, *of auricular region*

4, 5, 6, *of face*

7, 8, *of neck*

make a second incision in a horizontal direction along the forehead and round the side of the head, from the anterior to the posterior extremity of the preceding. Raise the skin in front, from the subjacent muscle, from below upwards; this must be done with great care, in order to avoid cutting through the vessels and nerves which lie immediately beneath the skin.

The Skin of the Scalp.—This is thicker than in any other part of the body. It is intimately adherent to the superficial fascia. The hair-follicles are very closely set together, and extend throughout the whole thickness of the skin. It also contains a number of sebaceous glands.

The **superficial fascia** in the cranial region is a firm, dense, fibro-fatty layer, intimately adherent to the integument, and to the Occipito-frontalis and its

FIG. 195.—Muscles of the head, face, and neck.

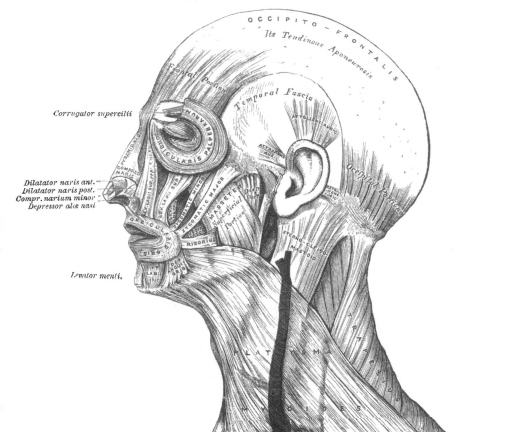

tendinous aponeurosis; it is continuous, behind, with the superficial fascia at the back part of the neck; and, laterally, is continued over the temporal fascia. It contains between its layers the superficial vessels and nerves and much granular fat.

The **Occipito-frontalis** (fig. 195) is a broad musculo-fibrous layer, which covers the whole of one side of the vertex of the skull, from the occiput to the eyebrow. It consists of two muscular slips, separated by an intervening tendinous aponeurosis. The *occipital portion* (sometimes called the *occipitalis* muscle) is thin, quadrilateral in form, and about an inch and a half in length ; it arises from the outer two-thirds of the superior curved line of the occipital bone, and from the mastoid portion of the temporal. Its fibres of origin are tendinous, but they soon become muscular, and ascend in a parallel direction to terminate in the tendinous aponeurosis. The *frontal portion* (sometimes called the *frontalis* muscle) is thin, of a quadrilateral form, and intimately adherent to the superficial fascia. It is broader, its fibres are longer, and their structure paler than the occipital portion. Its internal fibres are continuous with those of the Pyramidalis nasi. Its middle fibres become blended with the Corrugator supercilii and Orbicularis palpebrarum ; and the outer fibres are also blended with the latter muscle over the external angular process. According to Theile the innermost fibres are attached to the nasal bones, the outer to the external angular process of the frontal bone. From these attachments, the fibres are directed upwards and join the aponeurosis below the coronal suture. The inner margins of the frontal portions of the two muscles are joined together for some distance above the root of the nose ; but between the occipital portions there is a considerable, though variable, interval, which is occupied by the aponeurosis.

The **aponeurosis** covers the upper part of the vertex of the skull, being continuous across the middle line with the aponeurosis of the opposite muscle. Behind, it is attached, in the interval between the occipital origins, to the occipital protuberance and highest curved lines of the occipital bone ; in front, it forms a short and narrow prolongation between the frontal portions ; and on each side it has connected with it the Attollens and Attrahens auriculam muscles ; in this situation it loses its aponeurotic character, and is continued over the temporal fascia to the zygoma as a layer of laminated areolar tissue. This aponeurosis is closely connected to the integument by the firm, dense, fibro-fatty layer, which forms the superficial fascia : it is connected with the pericranium by loose cellular tissue, which allows of a considerable degree of movement of the integument.

Nerves.—The frontal portion of the Occipito-frontalis is supplied by the facial nerve ; its occipital portion by the posterior auricular branch of the facial.

Actions.—The frontal portion of the muscle raises the eyebrows and the skin over the root of the nose, and at the same time draws the scalp forwards, throwing the integument of the forehead into transverse wrinkles. The posterior portion draws the scalp backwards. By bringing alternately into action the frontal and occipital portions the entire scalp may be moved forwards and backwards. In the ordinary action of the muscles, the eyebrows are elevated, and at the same time the aponeurosis is fixed by the posterior portion, thus giving to the face the expression of surprise : if the action is more exaggerated, the eyebrows are still further raised, and the skin of the forehead thrown into transverse wrinkles, as in the expression of fright or horror.

II. Auricular Region (fig. 195)

Attrahens auriculam.   Attollens auriculam.
Retrahens auriculam.

These three small muscles are placed immediately beneath the skin around the external ear. In man, in whom the external ear is almost immovable, they are rudimentary. They are the analogues of large and important muscles in some of the mammalia.

Dissection.—This requires considerable care, and should be performed in the following manner : To expose the Attollens auriculam, draw the pinna or broad part of the ear downwards, when a tense band will be felt beneath the skin, passing from the side of the head to the upper part of the concha ; by dividing the skin over this band, in a direction from below upwards, and then reflecting it on each side, the muscle is exposed. To bring into view the Attrahens auriculam, draw the helix backwards by means of a hook, when the muscle will be made tense, and may be exposed in a similar manner to the preceding. To expose the Retrahens auriculam, draw the pinna forwards, when the muscle, being made tense, may be felt beneath the skin, at its insertion into the back part of the concha, and may be exposed in the same manner as the other muscles.

The **Attrahens auriculam** (*Auricularis anterior*), the smallest of the three, is thin, fan-shaped, and its fibres pale and indistinct ; they arise from the lateral edge of the aponeurosis of the Occipito-frontalis, and converge to be inserted into a projection on the front of the helix.

Relations.—*Superficially*, with the skin ; *deeply*, with the areolar tissue derived from the aponeurosis of the Occipito-frontalis, beneath which are the temporal artery and vein and the temporal fascia.

The **Attollens auriculam** (*Auricularis superior*), the largest of the three, is thin and fan-shaped : its fibres arise from the aponeurosis of the Occipito-frontalis, and converge to be inserted by a thin, flattened tendon into the upper part of the cranial surface of the pinna.

Relations.—*Superficially*, with the integument; *deeply*, with the areolar tissue derived from the aponeurosis of the Occipito-frontalis, beneath which is the temporal fascia.

The **Retrahens auriculam** (*Auricularis posterior*) consists of two or three fleshy fasciculi, which arise from the mastoid portion of the temporal bone by short aponeurotic fibres. They are inserted into the lower part of the cranial surface of the concha.

Relations.—*Superficially*, with the integument; *deeply*, with the mastoid portion of the temporal bone and the posterior auricular artery and nerve.

Nerves.—The Attrahens and Attollens auriculam are supplied by the temporal branch of the facial ; the Retrahens auriculam is supplied by the posterior auricular branch of the same nerve.

Actions.—In man, these muscles possess very little action : the Attrahens auriculam draws the ear forwards and upwards ; the Attollens auriculam slightly raises it ; and the Retrahens auriculam draws it backwards.

III. Palpebral Region (fig. 195)

Orbicularis palpebrarum.   Levator palpebræ.
Corrugator supercilii.   Tensor tarsi.

Dissection (fig. 194).—In order to expose the muscles of the face, continue the longitudinal incision, made in the dissection of the Occipito-frontalis, down the median line of the face to the tip of the nose, and from this point onwards to the upper lip; and carry another incision along the margin of the lip to the angle of the mouth, and transversely across the face to the angle of the jaw. Then make an incision in front of the external ear, from the angle of the jaw upwards, to join the transverse incision made in exposing the Occipito-frontalis. These incisions include a square-shaped flap, which should be removed in the direction marked in the figure, with care, as the muscles at some points are intimately adherent to the integument.

The **Orbicularis palpebrarum** is a sphincter muscle, which surrounds the circumference of the orbit and eyelids. It arises from the internal angular process of the frontal bone, from the nasal process of the superior maxillary bone in front of the lachrymal groove for the nasal duct, and from the anterior surface and borders of a short tendon, the *tendo oculi*, or *internal tarsal ligament*, placed at the inner angle of the orbit. From this origin, the fibres are directed outwards, forming a broad, thin, and flat layer, which covers the eyelids, surrounds the circumference of the orbit, and spreads out over the temple, and downwards on the cheek. The palpebral portion (ciliaris) of the Orbicularis is thin and pale ; it arises from the bifurcation of the tendo palpebrarum, and forms a series of concentric curves, which are inserted on the outer side of the eyelids into the external tarsal ligament. The orbital portion (orbicularis latus) is thicker and of a reddish colour: its fibres are well developed, and form complete ellipses. The upper fibres of this portion blend with the Occipito-frontalis and Corrugator supercilii.

Relations.—By its *superficial surface*, with the integument. By its *deep surface*, above, with the Occipito-frontalis and Corrugator supercilii, with which it is intimately blended, and with the supra-orbital vessels and nerve; below, it covers the lachrymal sac, and the origin of the Levator labii superioris alæque nasi, the Levator labii superioris, and the Zygomaticus minor muscles. *Internally*, it is occasionally blended with the Pyramidalis nasi. *Externally*, it lies on the temporal fascia. On the eyelids, it is separated from the conjunctiva by the Levator palpebræ, the tarsal ligaments, the tarsal plates, and the Meibomian glands.

The *tendo oculi* (internal tarsal ligament) is a short tendon, about two lines in length and one in breadth, attached to the nasal process of the superior maxillary bone in front of the lachrymal groove. Crossing the lachrymal sac, it divides into two parts, each division being attached to the inner extremity of the corresponding tarsal plate. As the tendon crosses the lachrymal sac, a strong aponeurotic lamina is given off from the posterior surface, which expands over the sac, and is attached to the ridge on the lachrymal bone. This is the reflected aponeurosis of the tendo oculi.

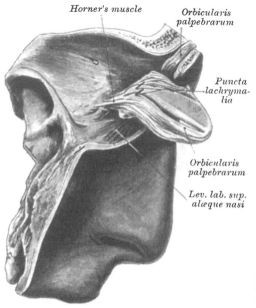

Fig. 196.—Horner's muscle. (From a preparation in the Museum of the Royal College of Surgeons of England.)

Horner's muscle   *Orbicularis palpebrarum*

Puncta lachrymalia

Orbicularis palpebrarum

Lev. lab. sup. alæque nasi

The *external tarsal ligament* is a much weaker structure than the tendo oculi. It is attached to the margin of the frontal process of the malar bone, and passes inwards to the outer commissure of the eyelids; it connects together the outer extremities of the two tarsal cartilages.

Use of Tendo oculi.—Besides giving attachment to part of the Orbicularis palpebrarum, and to the tarsal plates, it serves to suck the tears into the lachrymal sac, by its attachment to the sac. Thus, each time the eyelids are closed, the tendo oculi becomes tightened, through the action of the Orbicularis, and draws the wall of the lachrymal sac outwards and forwards, so that a vacuum is made in the sac, and the tears are sucked along the lachrymal canals into it.

The **Corrugator supercilii** is a small, narrow, pyramidal muscle, placed at the inner extremity of the eyebrow, beneath the Occipito-frontalis and Orbicularis palpebrarum muscles. It arises from the inner extremity of the superciliary ridge; whence its fibres pass upwards and outwards, and, passing between the palpebral and orbital portions of the Orbicularis palpebrarum, are inserted into the deep surface of the skin, opposite the middle of the orbital arch.

Relations.—By its *anterior surface*, with the Occipito-frontalis and Orbicularis palpebrarum muscles. By its *posterior surface*, with the frontal bone and supratrochlear nerve.

The **Levator palpebræ** will be described with the muscles of the orbital region.

The **Tensor tarsi** (Horner's muscle) (fig. 196) is a small thin muscle, about three lines in breadth and six in length, situated at the inner side of the orbit, behind the tendo oculi. It arises from the crest and adjacent part of the orbital surface of the lachrymal bone, and, passing across the lachrymal sac, divides into two slips, which cover the lachrymal canals, and are inserted into the tarsal plates internal to the puncta lachrymalia. Its fibres appear to be continuous with those of the palpebral portion of the Orbicularis palpebrarum; it is occasionally very indistinct.

Nerves.—The Orbicularis palpebrarum, Corrugator supercilii, and Tensor tarsi are supplied by the facial nerve. Recent investigations tend to show that the Orbicularis palpebrarum, Corrugator supercilii, and frontal part of the Occipito-frontalis, are in reality supplied by fibres of the third nerve, which descend through the pons Varolii to join the facial nerve.

Actions.—The Orbicularis palpebrarum is the sphincter muscle of the eyelids. The palpebral portion acts involuntarily, closing the lids gently, as in sleep or in blinking; the orbicular portion is subject to the will. When the entire muscle is brought into action, the skin of the forehead, temple, and cheek is drawn inwards towards the inner angle of the orbit, and the eyelids are firmly closed as in photophobia. When the skin of the forehead, temple, and cheek is thus drawn inwards by the action of the muscle it is thrown into folds, especially radiating from the outer angle of the eyelids, which give rise in old age to the so-called 'crow's feet.' The Levator palpebræ is the direct antagonist of this muscle; it raises the upper eyelid and exposes the globe. The Corrugator supercilii draws the eyebrow downwards and inwards, producing the vertical wrinkles of the forehead. It is the 'frowning' muscle, and may be regarded as the principal agent in the expression of suffering. The Tensor tarsi draws the eyelids and the extremities of the lachrymal canals inwards and compresses them against the surface of the globe of the eye; thus placing them in the most favourable situation for receiving the tears. It serves, also, to compress the lachrymal sac.

IV. Orbital Region (fig. 197)

Levator palpebræ superioris.   Rectus internus.
Rectus superior.   Rectus externus.
Rectus inferior.   Obliquus oculi superior.
                Obliquus oculi inferior.

Dissection.—To open the cavity of the orbit, remove the skull-cap and brain; then saw through the frontal bone at the inner extremity of the supra-orbital ridge, and externally at its junction with the malar. Break in pieces the thin roof of the orbit by a few slight blows of the hammer, and take it away; drive forward the superciliary portion of the frontal bone by a smart stroke, but do not remove it, as that would destroy the pulley of the Obliquus superior. When the fragments are cleared away, the periosteum of the orbit will be exposed; this being removed, together with the fat which fills the cavity of the orbit, the several muscles of this region can be examined. The dissection will be facilitated by distending the globe of the eye. In order to effect this, puncture the optic nerve near the eyeball with a curved needle, and push the needle onwards into the globe; insert the point of a blow-pipe through this aperture, and force a little air into the cavity of the eyeball; then apply a ligature round the nerve so as to prevent the air escaping. The globe being now drawn forwards, the muscles will be put upon the stretch.

The **Levator palpebræ superioris** is thin, flat, and triangular in shape. It arises from the under surface of the lesser wing of the sphenoid, above and in front of the optic foramen, from which it is separated by the origin of the Superior rectus. At its origin, it is narrow and tendinous, but soon becomes broad and fleshy, and finally terminates in a wide aponeurosis, which is inserted into the upper margin of the superior tarsal plate. From this aponeurosis a thin expansion is continued onwards, passing between the fibres of the Orbicularis to be inserted into the skin of the lid, and some deeper fibres blend with an expansion from the sheath of the Superior rectus muscle, and are with it prolonged into the conjunctiva.

Relations.—By its *upper surface*, with the frontal nerve, the supra-orbital artery, and the periosteum of the orbit and lachrymal gland; and, in the lid, with the inner surface of the tarsal ligament. By its *under surface*, with the Superior rectus; and, in the lid, with the conjunctiva. A small branch of the third nerve enters its under surface.

The **Superior rectus,** the thinnest and narrowest of the four Recti, arises from the upper margin of the optic foramen beneath the Levator palpebræ, and from the fibrous sheath of the optic nerve; and is inserted, by a tendinous expansion, into the sclerotic coat, about three or four lines from the margin of the cornea.

Relations.—By its *upper surface*, with the Levator palpebræ. By its *under surface*, with the optic nerve, the ophthalmic artery, the nasal nerve, and the branch of the third nerve which supplies it; and, in front, with the tendon of the Superior oblique, and the globe of the eye.

The **Inferior** and **Internal Recti** arise by a common tendon (the *ligament of Zinn*),* which is attached round the circumference of the optic foramen, except at its upper and outer part. The **External rectus** has two heads: the upper one arises from the outer margin of the optic foramen immediately beneath the Superior rectus; the lower head, partly from the ligament of Zinn, and partly from a small pointed process of bone on the lower margin of the sphenoidal

Fig. 197.—Muscles of the right orbit.

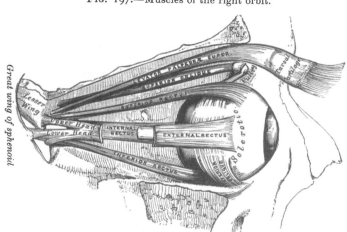

fissure. Each muscle passes forward in the position implied by its name, to be inserted by a tendinous expansion (the *tunica albuginea*) into the sclerotic coat, about three or four lines from the margin of the cornea. Between the two heads of the External rectus is a narrow interval, through which passes the third, the nasal branch of the ophthalmic division of the fifth and the sixth nerves, and the ophthalmic vein. Although nearly all of these muscles present a common origin and are inserted in a similar manner into the sclerotic coat, there are certain differences to be observed in them, as regards their length and breadth. The Internal rectus is the broadest; the External is the longest; and the Superior is the thinnest and narrowest.

The **Superior oblique** is a fusiform muscle, placed at the upper and inner side of the orbit, internal to the Levator palpebræ. It arises about a line above the inner margin of the optic foramen, and, passing forwards to the inner angle of the orbit, terminates in a rounded tendon, which plays in a ring or pulley,

* The ligament of Zinn ought, perhaps more appropriately, to be termed the *aponeurosis* or *tendon of Zinn*. Mr. C. B. Lockwood has described a somewhat similar structure on the under surface of the Superior rectus muscle, which is attached to the lesser wing of the sphenoid forming the upper and outer margin of the optic foramen. This *superior tendon* gives origin to the Superior rectus, the superior head of the External rectus, and the upper part of the Internal rectus.—*Journal of Anatomy and Physiology*, vol. xx. part i. p. 1.

formed by cartilaginous tissue attached to a depression beneath the internal angular process of the frontal bone, the contiguous surfaces of the tendon and ring being lined by a delicate synovial membrane, and enclosed in a thin fibrous investment. The tendon is reflected backwards, outwards, and downwards beneath the Superior rectus to the outer part of the globe of the eye, and is inserted into the sclerotic coat, behind the equator of the eyeball, the insertion of the muscle lying between the Superior and External recti.

Relations.—By its *upper surface*, with the periosteum covering the roof of the orbit, and the fourth nerve. The tendon, where it lies on the globe of the eye, is covered by the Superior rectus. By its *under surface*, with the nasal nerve, ethmoidal arteries, and the upper border of the Internal rectus.

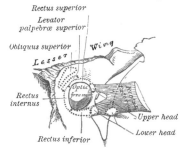

FIG. 198.—The relative position and attachment of the muscles of the left eyeball.

The **Inferior oblique** is a thin, narrow muscle, placed near the anterior margin of the orbit. It arises from a depression on the orbital plate of the superior maxillary bone, external to the lachrymal groove. Passing outwards, backwards, and upwards, between the Inferior rectus and the floor of the orbit, and then between the eyeball and the External rectus, it is inserted into the outer part of the sclerotic coat between the Superior and External recti, near to, but somewhat behind, the tendon of insertion of the Superior oblique.

Relations.—By its *ocular surface*, with the globe of the eye, and with the Inferior rectus. By its *orbital surface*, with the periosteum covering the floor of the orbit, and with the External rectus. Its borders look forwards and backwards ; the posterior one receives a branch of the third nerve.

Nerves.—The Levator palpebræ, Inferior oblique, and all the Recti excepting the External, are supplied by the third nerve; the Superior oblique, by the fourth ; the External rectus, by the sixth.

Actions.—The Levator palpebræ raises the upper eyelid, and is the direct antagonist of the Orbicularis palpebrarum. The four Recti muscles are attached in such a manner to the globe of the eye that, acting singly, they will turn it either upwards, downwards, inwards, or outwards, as expressed by their names. The movement produced by the Superior or Inferior rectus is not quite a simple one, for inasmuch as they pass obliquely outwards and forwards to the eyeball, the elevation or depression of the cornea must be accompanied by a certain deviation inwards, with a slight amount of rotation, which, however, is corrected by the Oblique muscles ; the Inferior oblique correcting the deviation inwards of the Superior rectus, and the Superior oblique that of the Inferior rectus. The contraction of the External and Internal recti, on the other hand, produces a purely horizontal movement. If any two contiguous recti of one eye act together they carry the globe of the eye in the diagonal of these directions, viz. upwards and inwards, upwards and outwards, downwards and inwards, or downwards and outwards. The movement of circumduction, as in looking round a room, is performed by the alternate action of the four Recti. The oblique muscles rotate the eyeball on its *antero-posterior axis*, this kind of movement being required for the correct viewing of an object when the head is moved laterally, as from shoulder to shoulder, in order that the picture may fall in all respects on the same part of the retina of each eye.*

---

* 'On the Oblique Muscles of the Eye in Man and Vertebrate Animals,' by John Struthers, M.D., in *Anatomical and Physiological Observations*. For a fuller account than our space allows of the various co-ordinated actions of the muscles of a single eye and of both eyes, the reader may be referred to Dr. M. Foster's *Text-book of Physiology*.

Fasciæ of the Orbit.—The connective tissue of the orbit is in various places condensed into thin membranous layers, which may be conveniently described as (1) the orbital fascia; (2) the sheath of the muscles; and (3) the covering of the eyeball.

(1) The *orbital fascia.* This forms the periosteum of the orbit. It is loosely connected to the bones, from which it can be readily separated. Behind, it is connected with the dura mater by processes which pass through the optic foramen and sphenoidal fissure, and with the sheath of the optic nerve. It front, it is connected with the periosteum at the margin of the orbit, and sends off a process which assists in forming the palpebral fascia. From its internal surface two processes are given off: one to enclose the lachrymal gland, the other to hold the pulley of the Superior oblique muscle in position.

(2) The sheaths of the muscles give off expansions to the margins of the orbit which limit the action of the muscles.

(3) The fascia covering the eyeball—Tenon's capsule—will be described in the sequel.

Surgical Anatomy.—The position and exact point of insertion of the tendons of the Internal and External recti muscles into the globe should be carefully examined from the front of the eyeball, as the surgeon is often required to divide the one or the other muscle for the cure of strabismus. In convergent strabismus, which is the more common form of the disease, the eye is turned inwards, requiring the division of the Internal rectus. In the divergent form, which is more rare, the eye is turned outwards, the External rectus being especially implicated. The deformity produced in either case is to be remedied by division of one or the other muscle. The operation is thus performed: the lids are to be well separated; the eyeball is rotated outwards or inwards, and the conjunctiva raised by a pair of forceps, and incised immediately beneath the lower border of the tendon of the muscle to be divided, a little behind its insertion into the sclerotic; the submucous areolar tissue is then divided, and into the small aperture thus made, a blunt hook is passed upwards between the muscle and the globe, and the tendon of the muscle and conjunctiva covering it, divided by a pair of blunt-pointed scissors. Or the tendon may be divided by a subconjunctival incision, one blade of the scissors being passed upwards between the tendon and the conjunctiva, and the other between the tendon and the sclerotic. The student, when dissecting these muscles, should remove on one side of the subject the conjunctiva from the front of the eye, in order to see more accurately the position of the tendons, while on the opposite side the operation may be performed.

V. Nasal Region (fig. 195)

Pyramidalis nasi.
Levator labii superioris alæque nasi.
Dilatator naris posterior.
Depressor alæ nasi.

Dilatator naris anterior.
Compressor nasi.
Compressor narium minor.

The **Pyramidalis nasi** is a small pyramidal slip placed over the nasal bone. Its origin is by tendinous fibres from the fascia covering the lower part of the nasal bone and upper part of the cartilage, where it blends with the Compressor nasi, and it is inserted into the skin over the lower part of the forehead between the two eyebrows, its fibres decussating with those of the Occipito-frontalis.

Relations.—By its *upper surface*, with the skin. By its *under surface*, with the frontal and nasal bones.

The **Levator labii superioris alæque nasi** is a thin triangular muscle, placed by the side of the nose, and extending between the inner margin of the orbit and upper lip. It arises by a pointed extremity from the upper part of the nasal process of the superior maxillary bone, and, passing obliquely downwards and outwards, divides into two slips, one of which is inserted into the cartilage of the ala of the nose; the other is prolonged into the upper lip, becoming blended with the Orbicularis oris and Levator labii superioris proprius.

Relations.—In front, with the integument; and with a small part of the Orbicularis palpebrarum above.

The **Dilatator naris posterior** is a small muscle, which is placed partly beneath the elevator of the nose and lip. It arises from the margin of the nasal notch of the superior maxilla, and from the sesamoid cartilages, and is inserted into the skin near the margin of the nostril.

The **Dilatator naris anterior** is a thin delicate fasciculus, passing from the cartilage of the ala of the nose to the integument near its margin. This muscle is situated in front of the preceding.

The **Compressor naris** is a small, thin, triangular muscle, arising by its apex from the superior maxillary bone, above and a little external to the incisive fossa : its fibres proceed upwards and inwards, expanding into a thin aponeurosis which is attached to the fibro-cartilage of the nose, and is continuous on the bridge of the nose with that of the muscle of the opposite side, and with the aponeurosis of the Pyramidalis nasi.

The **Compressor narium minor** is a small muscle, attached by one end to the alar cartilage, and by the other to the integument at the end of the nose.

The **Depressor alæ nasi** is a short, radiated muscle, arising from the incisive fossa of the superior maxilla ; its fibres ascend to be inserted into the septum, and back part of the ala of the nose. This muscle lies between the mucous membrane and muscular structure of the lip.

Nerves.—All the muscles of this group are supplied by the facial nerve.

Actions.—The Pyramidalis nasi draws down the inner angle of the eyebrows and produces transverse wrinkles over the bridge of the nose. The Levator labii superioris alæque nasi draws upwards the upper lip and ala of the nose ; its most important action is upon the nose, which it dilates to a considerable extent. The action of this muscle produces a marked influence over the countenance, and it is the principal agent in the expression of contempt and disdain. The two Dilatatores nasi enlarge the aperture of the nose. Their action in ordinary breathing is to resist the tendency of the nostrils to close from atmospheric pressure, but in difficult breathing they may be noticed to be in violent action, as well as in some emotions, as anger. The Depressor alæ nasi is a direct antagonist of the other muscles of the nose, drawing the ala of the nose downwards, and thereby constricting the aperture of the nares. The Compressor nasi depresses the cartilaginous part of the nose and draws the alæ together.

VI. Superior Maxillary Region (fig. 195)

Levator labii superioris.   Zygomaticus major.
Levator anguli oris.   Zygomaticus minor.

The **Levator labii superioris (proprius)** is a thin muscle of a quadrilateral form. It arises from the lower margin of the orbit immediately above the infra-orbital foramen, some of its fibres being attached to the superior maxilla, others to the malar bone ; its fibres converge to be inserted into the muscular substance of the upper lip.

Relations.—By its *superficial surface* above, with the lower segment of the Orbicularis palpebrarum ; below, it is subcutaneous. By its *deep surface*, it conceals the origin of the Compressor nasi and Levator anguli oris muscles, and the infra-orbital vessels and nerve, as they escape from the infra-orbital foramen.

The **Levator anguli oris** arises from the canine fossa, immediately below the infra-orbital foramen ; its fibres incline downwards and a little outwards, to be inserted into the angle of the mouth, intermingling with those of the Zygomaticus major, the Depressor anguli oris, and the Orbicularis.

Relations.—By its *superficial surface*, with the Levator labii superioris and the infra-orbital vessels and nerves. By its *deep surface*, with the superior maxilla, the Buccinator, and the mucous membrane.

The **Zygomaticus major** is a slender fasciculus, which arises from the malar bone, in front of the zygomatic suture, and descending obliquely downwards and inwards, is inserted into the angle of the mouth, where it blends with the fibres of the Levator anguli oris, the Orbicularis oris, and the Depressor anguli oris.

Relations.—By its *superficial surface*, with the subcutaneous adipose tissue. By its *deep surface*, with the Masseter and Buccinator muscles and the facial artery and vein.

The **Zygomaticus minor** arises from the malar bone, immediately behind the maxillary suture, and passing downwards and inwards, is continuous with the Orbicularis oris at the outer margin of the Levator labii superioris. It lies in front of the preceding.

Relations.—By its *superficial surface*, with the integument and the Orbicularis palpebrarum above. By its *deep surface*, with the Masseter, Buccinator, and Levator anguli oris, and the facial artery and vein.

Nerves.—This group of muscles is supplied by the facial nerve.

Actions.—The Levator labii superioris is the proper elevator of the upper lip, carrying it at the same time a little forwards. It assists in forming the naso-labial ridge, which passes from the side of the nose to the upper lip and gives to the face an expression of sadness. The Levator anguli oris raises the angle of the mouth and assists the Levator labii superioris in producing the naso-labial ridge. The Zygomaticus major draws the angle of the mouth backwards and upwards, as in laughing ; while the Zygomaticus minor, being inserted into the outer part of the upper lip and not into the angle of the mouth, draws it backwards, upwards, and outwards, and thus gives to the face an expression of sadness.

VII. Inferior Maxillary Region (fig. 195)

Levator labii inferioris (Levator menti).
Depressor labii inferioris (Quadratus menti).
Depressor anguli oris (Triangularis menti).

Dissection.—The Muscles in this region may be dissected by making a vertical incision through the integument from the margin of the lower lip to the chin ; a second incision should then be carried along the margin of the lower jaw as far as the angle, and the integument carefully removed in the direction shown in fig. 194.

The **Levator labii inferioris (Levator menti)** is to be dissected by everting the lower lip and raising the mucous membrane. It is a small conical fasciculus, placed on the side of the frænum of the lower lip. It arises from the incisive fossa, external to the symphysis of the lower jaw ; its fibres descend to be inserted into the integument of the chin.

Relations.—On its *inner surface*, with the mucous membrane ; in the *median line*, it is blended with the muscle of the opposite side ; and on its *outer side*, with the Depressor labii inferioris.

The **Depressor labii inferioris (Quadratus menti)** is a small quadrilateral muscle. It arises from the external oblique line of the lower jaw, between the symphysis and mental foramen, and passes obliquely upwards and inwards, to be inserted into the integument of the lower lip, its fibres blending with the Orbicularis oris, and with those of its fellow of the opposite side. It is continuous with the fibres of the Platysma at its origin. This muscle contains much yellow fat intermingled with its fibres.

Relations.—By its *superficial surface*, with part of the Depressor anguli oris, and with the integument, to which it is closely connected. By its *deep surface*, with the mental vessels and nerves, the mucous membrane of the lower lip, the labial glands, and the Levator menti, with which it is intimately united.

The **Depressor anguli oris (Triangularis menti)** is triangular in shape, arising, by its broad base, from the external oblique line of the lower jaw, whence its fibres pass upwards, to be inserted, by a narrow fasciculus, into the angle of the mouth. It is continuous with the Platysma at its origin, and with the Orbicularis oris and Risorius at its insertion, and some of its fibres are directly continuous with those of the Levator anguli oris.

Relations.—By its *superficial surface*, with the integument. By its *deep surface*, with the Depressor labii inferioris and Buccinator.

Nerves.—This group of muscles is supplied by the facial nerve.

Actions.—The Levator labii inferioris raises the lower lip, and protrudes it forwards, and at the same time wrinkles the integument of the chin, expressing doubt or disdain. The Depressor labii inferioris draws the lower lip directly downwards and a little outwards, as in the expression of irony. The Depressor anguli oris depresses the angle of the mouth, being the antagonist of the Levator anguli oris and Zygomaticus major; acting with these muscles, it will draw the angle of the mouth directly backwards.

VIII. INTERMAXILLARY REGION

Orbicularis oris.   Buccinator.   Risorius.

Dissection.—The dissection of these muscles may be considerably facilitated by filling the cavity of the mouth with tow, so as to distend the cheeks and lips; the mouth should then be closed by a few stitches, and the integument carefully removed from the surface.

The **Orbicularis oris** (fig. 195) is not a sphincter muscle like the Orbicularis palpebrarum, but consists of numerous strata of muscular fibres, having different directions, which surround the orifice of the mouth. These fibres are partially derived from the other facial muscles which are inserted into the lips, and are partly fibres proper to the lips themselves. Of the former, a considerable number are derived from the Buccinator and form the deeper stratum of the Orbicularis. Some of them, namely, those near the middle of the muscle, decussate at the angle of the mouth, those arising from the upper jaw passing to the lower lip, and those from the lower jaw to the upper lip. Other fibres of the muscle, situated at its upper and lower part, pass across the lips from side to side without decussation. Superficial to this stratum is a second, formed by the Levator and Depressor anguli oris, which cross each other at the angle of the mouth; those from the Depressor passing to the upper lip, and those from the Levator to the lower lip, along which they run to be inserted into the skin near the median line. In addition to these there are fibres from other muscles inserted into the lips, the Levator labii superioris, the Levator labii superioris alæque nasi, the Zygomatici, and the Depressor labii inferioris; these intermingle with the transverse fibres above described, and have principally an oblique direction. The proper fibres of the lips are oblique, and pass from the under surface of the skin to the mucous membrane, through the thickness of the lip. And in addition to these are fibres by which the muscle is connected directly with the maxillary bones and the septum of the nose. These consist, in the upper lip, of four bands, two of which (*Musculus incisivus superior*) arise from the alveolar border of the superior maxilla, opposite the lateral incisor tooth, and arching outwards on each side are continuous at the angles of the mouth with the other muscles inserted into this part. The two remaining muscular slips, called the *Naso-labialis*, connect the upper lip to the back of the septum of the nose: as they descend from the septum, an interval is left between them. It is this interval which forms the depression seen on the surface of the skin beneath the septum of the nose. The additional fibres for the lower segment (*Musculus incisivus inferior*) arise from the inferior maxilla, externally to the Levator labii inferioris, and arch outwards to the angles of the mouth, to join the Buccinator and the other muscles attached to this part.

Relations.—By its *superficial surface*, with the integument, to which it is closely connected. By its *deep surface*, with the buccal mucous membrane, the labial glands, and coronary vessels. By its *outer circumference*, it is blended with the numerous muscles which converge to the mouth from various parts of the face. Its *inner circumference* is free, and covered by mucous membrane.

The **Buccinator** (fig. 208) is a broad, thin muscle, quadrilateral in form, which occupies the interval between the jaws at the side of the face. It arises from the outer surface of the alveolar processes of the upper and lower jaws, corresponding to the three molar teeth ; and behind, from the anterior border of the pterygo-maxillary ligament. The fibres converge towards the angle of the mouth, where the central fibres intersect each other, those from below being continuous with the upper segment of the Orbicularis oris, and those from above with the inferior segment ; the highest and lowest fibres continue forward uninterruptedly into the corresponding segment of the lip without decussation.

Relations.—By its *superficial surface*, behind, with a large mass of fat, which separates it from the ramus of the lower jaw, the Masseter, and a small portion of the Temporal muscle ; anteriorly, with the Zygomatici, Risorius, Levator anguli oris, Depressor anguli oris, and Stenson's duct, which pierces it opposite the second molar tooth of the upper jaw ; the facial artery and vein cross it from below upwards ; it is also crossed by the branches of the facial and buccal nerves. By its *internal surface*, with the buccal glands and mucous membrane of the mouth.

The *pterygo-maxillary ligament* separates the Buccinator muscle from the Superior constrictor of the pharynx. It is a tendinous band, attached by one extremity to the apex of the internal pterygoid plate, and by the other to the posterior extremity of the internal oblique line of the lower jaw. Its *inner surface* corresponds to the cavity of the mouth, and is lined by mucous membrane. Its *outer surface* is separated from the ramus of the jaw by a quantity of adipose tissue. Its *posterior border* gives attachment to the Superior constrictor of the pharynx ; its *anterior border*, to the fibres of the Buccinator (see fig. 208).

The **Risorius (Santorini)** (fig. 195) consists of a narrow bundle of fibres, which arises in the fascia over the Masseter muscle and, passing horizontally forwards, is inserted into the skin at the angle of the mouth. It is placed superficial to the Platysma, and is broadest at its outer extremity. This muscle varies much in its size and form.

Nerves.—The muscles in this group are all supplied by the facial nerve. The buccal branch of the inferior maxillary nerve pierces the Buccinator muscle, and by some anatomists is regarded as partly supplying this muscle. Probably it merely pierces it on its way to the mucous membrane of the cheek.

Actions.—The Orbicularis oris in its ordinary action produces the direct closure of the lips ; by its deep fibres, assisted by the oblique ones, it closely applies the lips to the alveolar arch. The superficial part, consisting principally of the decussating fibres, brings the lips together and also protrudes them forwards. The Buccinators contract and compress the cheeks, so that, during the process of mastication, the food is kept under the immediate pressure of the teeth. When the cheeks have been previously distended with air, the Buccinator muscles expel it from between the lips, as in blowing a trumpet. Hence the name (*buccina*, a trumpet). The Risorius retracts the angles of the mouth, and produces the unpleasant expression which is sometimes seen in tetanus, and is known as 'risus sardonicus.'

IX. TEMPORO-MANDIBULAR REGION

Masseter.   Temporal.

Masseteric Fascia.—Covering the Masseter muscle, and firmly connected with it, is a strong layer of fascia, derived from the deep cervical fascia. Above, this fascia is attached to the lower border of the zygoma, and behind it covers the parotid gland, constituting the *parotid fascia*.

The **Masseter** is exposed by the removal of this fascia (fig. 195) ; it is a short, thick muscle, somewhat quadrilateral in form, consisting of two portions, superficial and deep. The *superficial portion*, the larger, arises by a thick, tendinous aponeurosis from the malar process of the superior maxilla, and from the anterior two-thirds of the lower border of the zygomatic arch: its fibres pass downwards and backwards, to be inserted into the angle and lower half of the outer surface of the ramus of the jaw. The *deep portion* is much smaller, and more muscular in texture; it arises from the posterior third of the lower border and the whole of the inner surface of the zygomatic arch; its fibres pass downwards and forwards, to be inserted into the upper half of the ramus and outer surface of the coronoid process of the jaw. The deep portion of the muscle is partly concealed, in front, by the superficial portion; behind, it is covered by the parotid gland. The fibres of the two portions are united at their insertion.

Relations.—By its *superficial surface*, with the Zygomatici, the parotid gland and Socia parotidis, and Stenson's duct; the branches of the facial nerve and the transverse facial vessels, which cross it; the masseteric fascia; the Risorius Santorini, Platysma myoides, and the integument. By its *deep surface*, with the Temporal muscle at its insertion, the ramus of the jaw, the Buccinator and the long

FIG. 199.—The Temporal muscle, the zygoma and Masseter having been removed.

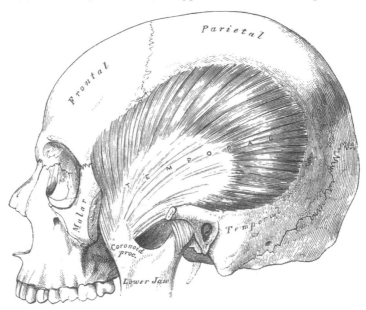

buccal nerve, from which it is separated by a mass of fat. The masseteric nerve and artery enter it on its under surface. Its *posterior margin* is overlapped by the parotid gland. Its *anterior margin* projects over the Buccinator muscle; and the facial vein lies on it below.

The *temporal fascia* is seen, at this stage of the dissection, covering in the Temporal muscle. It is a strong, fibrous investment, covered, on its outer surface, by the Attrahens and Attollens auriculam muscles, the aponeurosis of the Occipito-frontalis, and by part of the Orbicularis palpebrarum. The temporal vessels and the auriculo-temporal nerve cross it from below upwards. Above, it is a single layer, attached to the entire extent of the upper temporal ridge; but below, where it is attached to the zygoma, it consists of two layers, one of which is inserted into the outer, and the other into the inner border of the zygomatic arch. A small quantity of fat, the orbital branch of the temporal artery, and a filament from the orbital, or temporo-malar, branch of the superior maxillary nerve, are contained between these two layers. It affords attachment by its inner surface to the superficial fibres of the Temporal muscle.

Dissection.—In order to expose the Temporal muscle, remove the temporal fascia, which may be effected by separating it at its attachment along the upper border of the zygoma, and dissecting it upwards from the surface of the muscle. The zygomatic arch should then be divided, in front, at its junction with the malar bone; and, behind, near the external auditory meatus, and drawn downwards with the Masseter, which should be detached from its insertion into the ramus and angle of the jaw. The whole extent of the Temporal muscle is then exposed.

The **Temporal** (fig. 199) is a broad, radiating muscle, situated at the side of the head, and occupying the entire extent of the temporal fossa. It arises from the whole of the temporal fossa except that portion of it which is formed by the malar bone. Its attachment extends from the external angular process of the frontal in front, to the mastoid portion of the temporal behind; and from the curved line on the frontal and parietal bones above, to the pterygoid ridge on the great wing of the sphenoid below. It is also attached to the inner surface of the temporal fascia. Its fibres converge as they descend, and terminate in an aponeurosis, the fibres of which, radiated at its commencement, converge into a thick and flat tendon, which is inserted into the inner surface, apex, and anterior border of the coronoid process of the jaw, nearly as far forwards as the last molar tooth.

Relations.—By its *superficial surface*, with the integument, the Attrahens and Attollens auriculam muscles, the temporal vessels and nerves, the aponeurosis of the Occipito-frontalis, the temporal fascia, the zygoma, and Masseter. By its *deep surface*, with the temporal fossa, the External pterygoid and part of the Buccinator muscles, the internal maxillary artery, its deep temporal branches, and the deep temporal nerves. Behind the tendon are the masseteric vessels and nerve, and in front of it the buccal vessels and nerve. Its anterior border is separated from the malar bone by a mass of fat.

Nerves.—Both muscles are supplied by the inferior maxillary nerve.

X. PTERYGO-MANDIBULAR REGION (fig. 200).

External pterygoid.   Internal pterygoid.

Dissection.—The Temporal muscle having been examined, saw through the base of the coronoid process, and draw it upwards, together with the Temporal muscle, which should be detached from the surface of the temporal fossa. Divide the ramus of the jaw just below the condyle, and also, by a transverse incision extending across the middle, just above the dental foramen; remove the fragment of bone, and the Pterygoid muscles will be exposed.

The **External pterygoid** is a short, thick muscle, somewhat conical in form, which extends almost horizontally between the zygomatic fossa and the condyle of the jaw. It arises by two heads, separated by a slight interval: the *upper* arises from the inferior surface of the greater wing of the sphenoid and from the pterygoid ridge, which separates the zygomatic from the temporal fossa; the *lower* from the outer surface of the external pterygoid plate. Its fibres pass horizontally backwards and outwards, to be inserted into a depression in front of the neck of the condyle of the lower jaw, and into the corresponding part of the interarticular fibro-cartilage.

Relations.—By its *external surface*, with the ramus of the lower jaw, the internal maxillary artery, which crosses it,* the tendon of the Temporal muscle and the Masseter. By its *internal surface*, it rests against the upper part of the Internal pterygoid, the internal lateral ligament, the middle meningeal artery, and inferior maxillary nerve; by its *upper border* it is in relation with the temporal and masseteric branches of the inferior maxillary nerve; by its lower border it is in relation with the inferior dental and gustatory nerves. Through the interval between the two portions of the muscle, the buccal nerve emerges and the internal maxillary artery passes, when the trunk of this vessel lies on the muscle (see fig. 200).

* This is the usual relation, but in many cases the artery will be found below the muscle.

The **Internal pterygoid** is a thick, quadrilateral muscle, and resembles the Masseter in form. It arises from the pterygoid fossa, being attached to the inner surface of the external pterygoid plate, and to the grooved surface of the tuberosity of the palate bone, and by a second slip from the outer surface of the tuberosities of the palate and superior maxillary bones; its fibres pass downwards, outwards, and backwards, to be inserted, by a strong tendinous lamina, into the lower and back part of the inner side of the ramus and angle of the lower jaw, as high as the dental foramen.

Relations.—By its *external surface*, with the ramus of the lower jaw, from which it is separated, at its upper part, by the External pterygoid, the internal lateral ligament, the internal maxillary artery, the dental vessels and nerves, the lingual nerve, and a process of the parotid gland. By its *internal surface*, with the Tensor palati, being separated from the Superior constrictor of the pharynx by a cellular interval.

Nerves.—These muscles are supplied by the inferior maxillary nerve.

FIG. 200.—The Pterygoid muscles; the zygomatic arch and a portion of the ramus of the jaw having been removed.

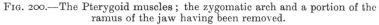
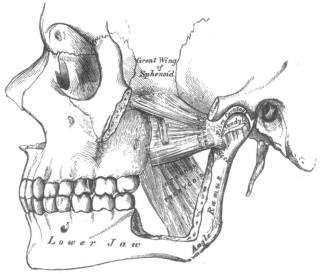

Actions.—The Temporal, Masseter, and Internal pterygoid raise the lower jaw against the upper with great force. The superficial portion of the Masseter assists the External pterygoid in drawing the lower jaw forwards upon the upper; the jaw being drawn back again by the deep fibres of the Masseter, and posterior fibres of the Temporal. The External pterygoid muscles are the direct agents in the trituration of the food, drawing the lower jaw directly forwards, so as to make the lower teeth project beyond the upper. If the muscle of one side acts, the corresponding side of the jaw is drawn forwards, and the other condyle remaining fixed, the symphysis deviates to the opposite side. The alternation of these movements on the two sides produces trituration.

Surface Form.—The outline of the muscles of the head and face cannot be traced on the surface of the body, except in the case of two of the masticatory muscles. Those of the head are thin, so that the outline of the bone is perceptible beneath them. Those of the face are small, covered by soft skin, and often by a considerable layer of fat, so that their outline is concealed; but they serve to round off and smooth prominent borders, and to fill up what would be otherwise unsightly angular depressions. Thus, the Orbicularis palpebrarum rounds off the prominent margin of the orbit, and the Pyramidalis nasi fills in the sharp depression beneath the glabella, and thus softens and tones down the abrupt depression which is seen on the unclothed bone. In like manner, the labial muscles, converging to the lips, and assisted by the superimposed fat, fill in the sunken hollow of the lower part of the face. Although the muscles of the face are usually described as arising from the bones, and inserted into the nose, lips, and corners of the mouth, they have fibres inserted into the skin of the face along their whole extent, so that almost every point of the skin of the face has its muscular fibre to move it; hence it is that when in action the facial muscles produce alterations in the skin-surface, giving rise to the formation of various folds or wrinkles, or otherwise altering the relative position of parts, so as to produce the varied expressions with which the face is endowed; hence these muscles are termed the 'Muscles of expression.' The only two muscles in this region which greatly influence surface-form are the Masseter and the Temporal. The Masseter is a quadrilateral muscle, which imparts fulness to the hinder part of the cheek. When the muscle is firmly contracted, as when the teeth are clenched, its outline is plainly visible; the anterior border forms a prominent vertical ridge, behind which is a considerable fulness, especially marked at the lower part of the muscle; this fulness is entirely lost when the mouth is opened, and the muscle no longer in a state of contraction. The Temporal muscle is fan-shaped, and fills the Temporal fossa, substituting for it a somewhat convex form, the anterior part of which, on account of the absence of hair over the temple, is more marked than the posterior, and stands out in strong relief when the muscle is in a state of contraction.

MUSCLES AND FASCIÆ OF THE NECK

The muscles of the Neck may be arranged into groups, corresponding with the region in which they are situated.

These groups are nine in number:

I. Superficial Region.
II. Depressors of the Os Hyoides and Larynx.
III. Elevators of the Os Hyoides and Larynx.
IV. Muscles of the Tongue.
V. Muscles of the Pharynx.
VI. Muscles of the Soft Palate.
VII. Muscles of the Anterior Vertebral Region.
VIII. Muscles of the Lateral Vertebral Region.
IX. Muscles of the Larynx.

The muscles contained in each of these groups are the following:

I. Superficial Region.
    Platysma myoides.
    Sterno-cleido-mastoid.

Infra-hyoid Region.
II. Depressors of the Os Hyoides and Larynx.
    Sterno-hyoid.
    Sterno-thyroid.
    Thyro-hyoid.
    Omo-hyoid.

Supra-hyoid Region.
III. Elevators of the Os Hyoides and Larynx.
    Digastric.
    Stylo-hyoid.
    Mylo-hyoid.
    Genio-hyoid.

Lingual Region.
IV. Muscles of the Tongue.
    Genio-hyo-glossus.
    Hyo-glossus.
    Chondro-glossus
    Stylo-glossus.
    Palato-glossus.

V. Muscles of the Pharynx.
    Inferior constrictor.
    Middle constrictor.
    Superior constrictor.
    Stylo-pharyngeus.
    Palato-pharyngeus.
    Salpingo-pharyngeus.

VI. Muscles of the Soft Palate.
    Levator palati.
    Tensor palati.
    Azygos uvulæ.
    Palato-glossus.
    Palato-pharyngeus.
    Salpingo-pharyngeus.

VII. Muscles of the Anterior Vertebral Region.
    Rectus capitis anticus major
    Rectus capitis anticus minor.
    Rectus capitis lateralis.
    Longus colli.

VIII. Muscles of the Lateral Vertebral Region.
    Scalenus anticus.
    Scalenus medius.
    Scalenus posticus.

IX. Muscles of the Larynx.
Included in the description of the Larynx.

I. SUPERFICIAL CERVICAL REGION

Platysma myoides.   Sterno-cleido-mastoid.

Dissection.—A block having been placed at the back of the neck, and the face turned to the side opposite that to be dissected, so as to place the parts upon the stretch, make two transverse incisions : one from the chin, along the margin of the lower jaw, to the mastoid process ; and the other along the upper border of the clavicle. Connect these by an oblique incision made in the course of the Sterno-mastoid muscle, from the mastoid process to the sternum ; the two flaps of integument having been removed in the direction shown in fig. 194, the superficial fascia will be exposed.

The **Superficial cervical fascia** is a thin, aponeurotic lamina, which is hardly demonstrable as a separate membrane.   Beneath it is found the Platysma myoides muscle.

The **Platysma myoides** (fig. 195) is a broad, thin plane of muscular fibres, placed immediately beneath the superficial fascia on each side of the neck.   It arises by thin, fibrous bands from the fascia covering the upper part of the Pectoral and Deltoid muscles ; its fibres pass over the clavicle, and proceed obliquely upwards and inwards along the side of the neck.   The anterior fibres interlace, below and behind the symphysis menti, with the fibres of the muscle of the opposite side ; the posterior fibres pass over the lower jaw, some of them being attached to the bone below the external oblique line, others passing on to be inserted into the skin and subcutaneous tissue of the lower part of the face, many of these fibres blending with the muscles about the angle and lower part of the mouth.   Sometimes fibres can be traced to the Zygomatic muscles, or to the margin of the Orbicularis oris.   Beneath the Platysma, the external jugular vein may be seen descending from the angle of the jaw to the clavicle.

Surgical Anatomy.—It is essential to remember the direction of the fibres of the Platysma, in connection with the operation of bleeding from the external jugular vein ; for if the point of the lancet is introduced in the direction of the muscular fibres, the orifice made will be filled up by the contraction of the muscle, and blood will not flow ; but if the incision is made across the course of the fibres, they will retract, and expose the orifice in the vein, and so allow the flow of blood.

Relations.—By its *external surface*, with the integument, to which it is united more closely below than above.   By its *internal surface*, with the Pectoralis major and Deltoid, and with the clavicle.   In the *neck*, with the external and anterior jugular veins, the deep cervical fascia, the superficial branches of the cervical plexus, the Sterno-mastoid, Sterno-hyoid, Omo-hyoid, and Digastric muscles.   Behind the Sterno-mastoid muscle, it covers in the posterior triangle of the neck.   On the *face*, it is in relation with the parotid gland, the facial artery and vein, and the Masseter and Buccinator muscles.

Actions.—The Platysma myoides produces a slight wrinkling of the surface of the skin of the neck, in an oblique direction, when the entire muscle is brought into action.   Its anterior portion, the thickest part of the muscle, depresses the lower jaw ; it also serves to draw down the lower lip and angle of the mouth on each side, being one of the chief agents in the expression of melancholy.

The **Deep cervical fascia** lies under cover of the Platysma myoides muscle and constitutes a complete investment for the neck.   It also forms a sheath for the carotid vessels, and, in addition, is prolonged deeply in the shape of certain processes or lamellæ, which come into close relation with the structures situated in front of the vertebral column.

The investing portion of the fascia is attached behind to the ligamentum nuchæ and to the spine of the seventh cervical vertebra.   Along this line it splits to enclose the Trapezius muscle, at the anterior border of which the two enclosing lamellæ unite and form a strong membrane, which extends forwards so as to roof in the posterior triangle of the neck.   Along the hinder edge of the Sterno-mastoid this membrane again divides to enclose this muscle, at the anterior edge of which it once more forms a single lamella, which roofs in the anterior triangle of the neck, and, reaching forwards to the middle line, is continuous with the corresponding part from the opposite side of the neck. In the middle line of the neck it is attached to the symphysis menti and body of the hyoid bone.

Above, the fascia is attached to the superior curved line of the occiput, to the mastoid process of the temporal, and to the whole length of the body of the jaw. Opposite the angle of the jaw the fascia is very strong, and binds the anterior

FIG. 201.—Section of the neck at about the level of the sixth cervical vertebra. Showing the arrangement of the deep cervical fascia.

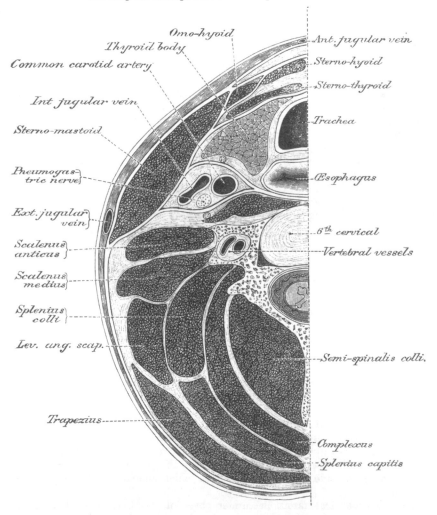

edge of the Sterno-mastoid firmly to that bone. Between the jaw and the mastoid process it ensheaths the parotid gland—the layer which covers the gland extending upwards under the name of the parotid fascia to be fixed to the zygomatic arch. From the layer which passes under the parotid a strong band, *the stylo-mandibular ligament*, reaches from the styloid process to the angle of the jaw.

Below, the fascia is attached to the acromion process, the clavicle, and manubrium sterni. Some little distance above the last, however, it splits into two layers, superficial and deep. The former is attached to the anterior border of the manubrium, the latter to its posterior border and to the interclavicular ligament. Between these two layers is a slit-like interval, the *suprasternal space*, or *space of Burns*. It contains a small quantity of areolar tissue, and sometimes a lymphatic gland ; the lower portions of the anterior jugular veins and their transverse connecting branch ; and also the sternal heads of the Sterno-mastoid muscles.

The fascia which lines the deep aspect of the Sterno-mastoid gives off certain important processes, viz.: (1) A process to envelop the tendon of the Omo-hyoid, and bind it down to the sternum and first costal cartilage. (2) A strong sheath, the carotid sheath, for the large vessels of the neck, enclosed within which are the carotid artery, internal jugular vein, the vagus, and descendens hypo-glossi nerves. (3) The prevertebral fascia, which extends inwards behind the carotid vessels, where it assists in forming their sheath, and passes in front of the prevertebral muscles. It thus forms the posterior limit of a fibrous compartment, which contains the larynx and trachea, the thyroid gland, and the pharynx and œsophagus. The prevertebral fascia is fixed above to the base of the skull, while below it is continued into the thorax in front of the Longus colli muscles. Parallel to the carotid vessels and along their inner aspect it gives off a thin lamina, the *bucco-pharyngeal fascia*, which closely invests the constrictor muscles of the pharynx, and is continued forward from the Superior constrictor on to the Buccinator. It is attached to the prevertebral layer by loose connective tissue only, and thus an easily distended space, the *retro-pharyngeal space*, is found between them. This space is limited above by the base of the skull, while below it extends behind the œsophagus into the thorax, where it is continued into the posterior mediastinum. The prevertebral fascia is prolonged downwards and out-wards behind the carotid vessels and in front of the Scaleni muscles, and forms a sheath for the brachial nerves and subclavian vessels in the posterior triangle of the neck, and, continued under the clavicle as the axillary sheath, is attached to the deep surface of the costo-coracoid membrane. Immediately above the clavicle an areolar space exists between the investing layer and the sheath of the subclavian vessels, and in it are found the lower part of the external jugular vein, the descending clavicular nerves, the suprascapular and transversalis colli vessels, and the posterior belly of the Omo-hyoid muscle. This space extends downwards behind the clavicle, and is limited below by the fusion of the costo-coracoid membrane with the anterior wall of the axillary sheath. (4) The pre-tracheal fascia, which extends inwards in front of the carotid vessels, and assists in forming the carotid sheath. It is further continued behind the depressor muscles of the hyoid bone, and, after enveloping the thyroid body, is prolonged in front of the trachea to meet the corresponding layer of the opposite side. Above, it is fixed to the hyoid bone, while below it is carried downwards in front of the trachea and large vessels at the root of the neck, and ultimately blends with the fibrous pericardium.

Surgical Anatomy.—The cervical fascia is of considerable importance from a surgical point of view. As will be seen from the foregoing description, it may be divided into three layers: (1) A superficial layer ; (2) a layer passing in front of the trachea, and forming with the superficial layer a sheath for the depressors of the hyoid bone ; (3) a prevertebral layer passing in front of the bodies of the cervical vertebræ, and forming with the second layer a space in which is contained the trachea, œsophagus, &c. The superficial layer forms a complete investment for the neck. It is attached behind to the ligamentum nuchæ and the spine of the seventh cervical vertebra: above it is attached to the external occipital protuberance, to the superior curved line of the occiput, to the mastoid process, to the zygoma and the lower jaw ; below it is attached to the manubrium sterni, the clavicle, the acromion process, and the spine of the scapula ; in front it blends with the fascia of the opposite side. This layer would oppose the extension of abscesses or new growths towards the surface, and pus forming beneath it would have a tendency to extend laterally. If it is in the posterior triangle, it might extend backwards under the Trapezius, forwards under the Sterno-mastoid, or downwards under the clavicle for some distance, until stopped by the junction of the cervical fascia to the Costo-coracoid mem-brane. If the pus is contained in the anterior triangle, it might find its way into the anterior mediastinum, being situated in front of the layer of fascia which passes down into the thorax to become continuous with the pericardium ; but owing to the less density and thickness of the fascia in this situation it more frequently finds its way through it and points above the sternum. The second layer of fascia is connected above with the hyoid bone. It passes down beneath the depressors and in front of the thyroid body and trachea to become continuous with the fibrous layer of the pericardium. Laterally it invests the great vessels of the neck and is connected with the superficial layer beneath the Sterno-mastoid. Pus forming beneath this layer would in all probability find its way into the posterior mediastinum. The third layer (the prevertebral fascia) is connected above to the base of the skull. Pus forming beneath this layer, in cases, for instance, of caries of the bodies of the cervical vertebræ, might extend towards the posterior and lateral part of the neck and point in this situation, or might perforate this layer of fascia and the pharyngeal fascia and point into the pharynx (retro-pharyngeal abscess).

In cases of cut throat the cervical fascia is of considerable importance. When the wound involves only the superficial layer the injury is usually trivial, the only special danger being injury to the external jugular vein, and the only special complication being diffuse cellulitis. But where the second of the two layers has been opened up, important structures may have been injured, which may lead to serious results.

It may be worth while mentioning that in Burns's space is contained the sternal head of origin of the Sterno-mastoid muscle, so that this space is opened in division of this tendon. The anterior jugular vein is also contained in the same space.

FIG. 202.—Muscles of the neck, and boundaries of the triangles.

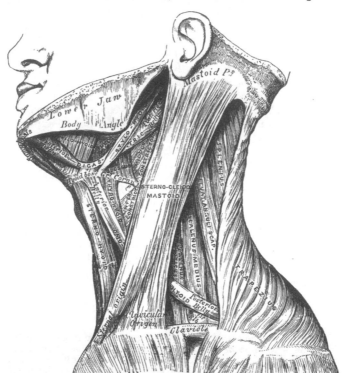

The **Sterno-mastoid** or **Sterno-cleido-mastoid** (fig. 202) is a large, thick muscle, which passes obliquely across the side of the neck, being enclosed between the two layers of the deep cervical fascia. It is thick and narrow at its central part, but is broader and thinner at each extremity. It arises, by two heads, from the sternum and clavicle. The *sternal portion* is a rounded fasciculus, tendinous in front, fleshy behind, which arises from the upper and anterior part of the first piece of the sternum, and is directed upwards, outwards, and backwards. The *clavicular portion* arises from the inner third of the superior border and anterior surface of the clavicle, being composed of fleshy and aponeurotic fibres; it is directed almost vertically upwards. These two portions are separated from one another, at their origin, by a triangular cellular interval; but become gradually blended, below the middle of the neck, into a thick, rounded muscle, which is inserted, by a strong tendon, into the outer surface of the mastoid process, from its apex to its superior border, and by a thin aponeurosis into the outer half of the superior curved line of the occipital bone. The Sterno-mastoid varies much in its extent of attachment to the clavicle : in one case the clavicular may be as narrow as the sternal portion ; in another, as much as three inches in breadth. When the clavicular origin is broad, it is occasionally subdivided into numerous slips, separated by narrow intervals. More rarely, the corresponding margins of the Sterno-mastoid and Trapezius have been found in contact. In the application of a ligature to the third part of the subclavian artery, it will be necessary, where the muscles come close together, to divide a portion of one or of both.

This muscle divides the quadrilateral space at the side of the neck into two triangles, an anterior and a posterior. The boundaries of the *anterior triangle* are, in front, the median line of the neck ; above, the lower border of the body of the jaw, and an imaginary line drawn from the angle of the jaw to the mastoid process ; behind, the anterior border of the Sterno-mastoid muscle. The apex of the triangle is at the upper border of the sternum. The boundaries of the *posterior triangle* are, in front, the posterior border of the Sterno-mastoid ; below, the middle third of the clavicle ; behind, the anterior margin of the Trapezius.* The apex corresponds with the meeting of the Sterno-mastoid and Trapezius on the occipital bone.

Relations.—By its *superficial surface*, with the integument and Platysma, from which it is separated by the external jugular vein, the superficial branches of the cervical plexus, and the anterior layer of the deep cervical fascia. By its *deep surface*, it is in relation with the Sterno-clavicular articulation ; a process of the deep cervical fascia ; the Sterno-hyoid, Sterno-thyroid, Omo-hyoid, posterior belly of the Digastric, Levator anguli scapulæ, Splenius and Scaleni muscles ; common carotid artery, internal and anterior jugular veins, commencement of the internal and external carotid arteries, the occipital, subclavian, transversalis colli, and suprascapular arteries and veins ; the phrenic, pneumogastric, hypoglossal, descendens and communicans hypoglossi nerves ; the spinal accessory nerve, which pierces its upper third ; the cervical plexus, parts of the thyroid and parotid glands and deep lymphatic glands.

Nerves.—The Platysma myoides is supplied by the facial nerve ; the Sterno-cleido-mastoid by the spinal accessory and deep branches of the cervical plexus.

Actions.—When only one Sterno-mastoid muscle acts, it draws the head towards the shoulder of the same side, assisted by the Splenius and the Obliquus capitis inferior of the opposite side. At the same time it rotates the head so as to carry the face towards the opposite side. If the head is fixed, the two muscles assist in elevating the thorax in forced inspiration.

Surface Form.—The anterior edge of the muscle forms a very prominent ridge beneath the skin, which it is important to notice, as it forms a guide to the surgeon in making the necessary incisions for ligature of the common carotid artery, and for œsophagotomy.

Surgical Anatomy.—The relations of the sternal and clavicular parts of the Sterno-mastoid should be carefully examined, as the surgeon is sometimes required to divide one or both portions of the muscle in *wry-neck*. One variety of this distortion is produced by spasmodic contraction or rigidity of the Sterno-mastoid ; the head being carried down towards the shoulder of the same side, and the face turned to the opposite side, and fixed in that position. When there is permanent shortening, subcutaneous division of the muscle is resorted to by some surgeons. This is performed by introducing a tenotomy knife beneath it, close to its origin, and dividing it from behind forwards while the muscle is put well upon the stretch. There is seldom any difficulty in dividing the sternal portion, by making a puncture on the inner side of the tendon, and then pushing a blunt tenotome behind it, and cutting forwards. In dividing the clavicular portion care must be taken to avoid wounding the external jugular vein, which runs parallel with the posterior border of the muscle in this situation, or the anterior jugular vein, which crosses beneath it. If the

* The anatomy of these triangles will be more exactly described with that of the vessels of the neck.

external jugular vein lies near the muscle, it is safer to make the first puncture at the outer side of the tendon, and introduce a blunt tenotome from without inwards. Many surgeons prefer dividing the muscle by the open method. An incision is made over either origin of the muscle, the tendon is exposed and a director is passed underneath it, and it is then divided. With care and attention to asepsis this plan of treatment is devoid of risk, and in this way the accidental division of the vessels can be avoided. Some of the fibres of the Sterno-mastoid muscle are occasionally torn during birth, especially in breech presentations; this is accompanied by hæmorrhage and formation of a swelling within the substance of the muscle. This is one of the causes of wry-neck; the scar tissue which is formed contracting and shortening the muscle.

II. Infra-hyoid Region (figs. 202, 203)

Depressors of the Os Hyoides and Larynx

Sterno-hyoid.	Thyro-hyoid.
Sterno-thyroid.	Omo-hyoid.

Dissection.—The muscles in this region may be exposed by removing the deep fascia from the front of the neck. In order to see the entire extent of the Omo-hyoid, it is necessary to divide the Sterno-mastoid at its centre, and turn its ends aside, and to detach the Trapezius from the clavicle and scapula. This, however, should not be done until the Trapezius has been dissected.

The **Sterno-hyoid** is a thin, narrow, riband-like muscle, which arises from the inner extremity of the clavicle, the posterior sterno-clavicular ligament, and the upper and posterior part of the first piece of the sternum; passing upwards and inwards, it is inserted, by short, tendinous fibres, into the lower border of the body of the os hyoides. This muscle is separated, below, from its fellow by a considerable interval; but the two muscles come into contact with one another in the middle of their course, and from this upwards, lie side by side. It sometimes presents, immediately above its origin, a transverse tendinous intersection, like those in the Rectus abdominis.

Relations.—By its *superficial surface*, below, with the sternum, the sternal end of the clavicle, and the Sterno-mastoid; and above, with the Platysma and deep cervical fascia. By its *deep surface*, with the Sterno-thyroid, Crico-thyroid, and Thyro-hyoid muscles, the thyroid gland, the superior thyroid vessels, the thyroid cartilage, the crico-thyroid and thyro-hyoid membranes.

The **Sterno-thyroid** is shorter and wider than the preceding muscle, beneath which it is situated. It arises from the posterior surface of the first bone of the sternum, below the origin of the Sterno-hyoid, and from the edge of the cartilage of the first rib, and occasionally of the second rib also; and is inserted into the oblique line on the side of the ala of the thyroid cartilage. This muscle is in close contact with its fellow at the lower part of the neck; and is occasionally traversed by a transverse or oblique tendinous intersection like those in the Rectus abdominis.

Relations.—By its *anterior surface*, with the Sterno-hyoid, Omo-hyoid, and Sterno-mastoid. By its *posterior surface*, from below upwards, with the trachea, vena innominata, common carotid (and on the right side the arteria innominata), the thyroid gland and its vessels, and the lower parts of the larynx and pharynx. The inferior thyroid vein lies along its inner border, a relation which it is important to remember in the operation of tracheotomy. On the left side the deep surface of the muscle is in relation to the œsophagus.

The **Thyro-hyoid** is a small, quadrilateral muscle appearing like a continuation of the Sterno-thyroid. It arises from the oblique line on the side of the thyroid cartilage, and passes vertically upwards to be inserted into the lower border of the body and greater cornu of the hyoid bone.

Relations.—By its *external surface*, with the Sterno-hyoid and Omo-hyoid muscles. By its *internal surface*, with the thyroid cartilage, the thyro-hyoid membrane, and the superior laryngeal vessels and nerve.

The **Omo-hyoid** passes across the side of the neck, from the scapula to the hyoid bone. It consists of two fleshy bellies, united by a central tendon. It arises from the upper border of the scapula, and occasionally from the transverse ligament which crosses the suprascapular notch ; its extent of attachment to the scapula varying from a few lines to an inch. From this origin, the posterior belly forms a flat, narrow fasciculus, which inclines forwards and slightly upwards across the lower part of the neck, behind the Sterno-mastoid muscle, where it becomes tendinous ; it then changes its direction, forming an obtuse angle, and terminates in the anterior belly, which passes almost vertically upwards, close to the outer border of the Sterno-hyoid, to be inserted into the lower border of the body of the os hyoides, just external to the insertion of the Sterno-hyoid. The central tendon of this muscle, which varies much in length and form, is held in position by a process of the deep cervical fascia, which includes it in a sheath. This process is prolonged down, to be attached to the clavicle and first rib. It is by this means that the angular form of the muscle is maintained.

FIG. 203.—Muscles of the neck. Anterior view.

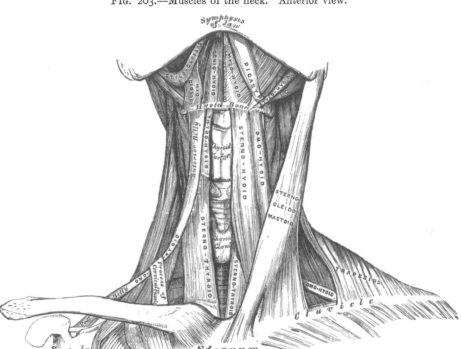

This muscle subdivides each of the two large triangles at the side of the neck into two smaller triangles ; the two posterior ones being the *posterior superior* or *occipital*, and the *posterior inferior* or *subclavian triangles* ; the two anterior, the *anterior superior* or *superior carotid*, and the *anterior inferior* or *inferior carotid triangles*.

Relations.—By its *superficial surface*, with the Trapezius, the Sterno-mastoid, deep cervical fascia, Platysma, and integument. By its *deep surface*, with the Scaleni muscles, phrenic nerve, lower cervical nerves, which go to form the brachial plexus, the suprascapular vessels and nerve, sheath of the common carotid artery and internal jugular vein, the Sterno-thyroid and Thyro-hyoid muscles.

Nerves.—The Thyro-hyoid is supplied by the hypoglossal ; the other muscles of this group by branches from the loop of communication between the descendens and communicans hypoglossi.

Actions.—These muscles depress the larynx and hyoid bone, after they have been drawn up with the pharynx in the act of deglutition. The Omo-hyoid muscles not only depress the hyoid bone, but carry it backwards, and to one or the other side. It is concerned especially in prolonged inspiratory efforts; for by tensing the lower part of the cervical fascia it lessens the inward suction of the soft parts, which would otherwise compress the great vessels and the apices of the lungs. The Thyro-hyoid may act as an elevator of the thyroid cartilage, when the hyoid bone ascends, drawing upwards the thyroid cartilage, behind the os hyoides. The Sterno-thyroid acts as a depressor of the thyroid cartilage.

III. Supra-hyoid Region (figs. 202, 203)

Elevators of the Os Hyoides—Depressors of the Lower Jaw

Digastric.	Mylo-hyoid.
Stylo-hyoid.	Genio-hyoid.

Dissection.—To dissect these muscles, a block should be placed beneath the back of the neck, and the head drawn backwards, and retained in that position. On the removal of the deep fascia, the muscles are at once exposed.

The **Digastric** consists of two fleshy bellies united by an intermediate, rounded tendon. It is a small muscle, situated below the side of the body of the lower jaw, and extending, in a curved form, from the side of the head to the symphysis of the jaw. The *posterior belly*, longer than the anterior, arises from the digastric groove on the inner side of the mastoid process of the temporal bone, and passes downwards, forwards, and inwards. The *anterior belly* arises from a depression on the inner side of the lower border of the jaw, close to the symphysis, and passes downwards and backwards. The two bellies terminate in the central tendon which perforates the Stylo-hyoid, and is held in connection with the side of the body and the greater cornu of the hyoid bone by a fibrous loop, lined by a synovial membrane. A broad aponeurotic layer is given off from the tendon of the Digastric on each side, which is attached to the body and great cornu of the hyoid bone; this is termed the *supra-hyoid aponeurosis*. It forms a strong layer of fascia between the anterior portion of the two muscles, and a firm investment for the other muscles of the supra-hyoid region which lie deeper.

The Digastric muscle divides the anterior superior triangle of the neck into two smaller triangles; the upper, or *submaxillary*, being bounded, above, by the lower border of the body of the jaw, and a line drawn from its angle to the mastoid process; below, by the posterior belly of the Digastric and the Stylo-hyoid muscles; in front, by the middle line of the neck, and anterior belly of the Digastric: the lower or *superior carotid triangle* being bounded above by the posterior belly of the Digastric, behind by the Sterno-mastoid, below by the Omo-hyoid.

Relations.—By its *superficial surface*, with the Mastoid process, the Platysma, Sterno-mastoid, part of the Splenius, Trachelo-mastoid, and Stylo-hyoid muscles, and the parotid gland. By its *deep surface*, the anterior belly lies on the Mylo-hyoid; the posterior belly on the Stylo-glossus, Stylo-pharyngeus, and Hyo-glossus muscles, the external carotid artery and its occipital, lingual, facial, and ascending pharyngeal branches, the internal carotid artery, internal jugular vein, and hypoglossal nerve.

The **Stylo-hyoid** is a small, slender muscle, lying in front of, and above, the posterior belly of the Digastric. It arises from the back and outer surface of the styloid process, near the base; and, passing downwards and forwards, is inserted into the body of the hyoid bone, just at its junction with the greater cornu, and immediately above the Omo-hyoid. This muscle is perforated, near its insertion, by the tendon of the Digastric.

Relations.—By its *superficial surface*, above with the parotid gland and deep cervical fascia; below it is superficial, being situated immediately beneath the deep cervical fascia. By its *deep surface*, with the posterior belly of the Digastric, the external carotid artery, with its lingual and facial branches, the Hyo-glossus muscle, and the hypoglossal nerve.

The Stylo-hyoid ligament.—In connection with the Stylo-hyoid muscle may be described a ligamentous band, the *Stylo-hyoid ligament*. It is a fibrous cord, often containing a little cartilage in its centre, which continues the styloid process down to the hyoid bone, being attached to the tip of the former and the small cornu of the latter. It is often more or less ossified, and in many animals forms a distinct bone, the *epihyal*.

The anterior belly of the Digastric should be removed, in order to expose the next muscle.

The **Mylo-hyoid** is a flat, triangular muscle, situated immediately beneath the anterior belly of the Digastric, and forming, with its fellow of the opposite side, a muscular floor for the cavity of the mouth. It arises from the whole length of the mylo-hyoid ridge of the lower jaw, extending from the symphysis in front to the last molar tooth behind. The posterior fibres pass inwards and slightly downwards, to be inserted into the body of the os hyoides. The middle and anterior fibres are inserted into a median fibrous raphé, extending from the symphysis of the lower jaw to the hyoid bone, where they join at an angle with the fibres of the opposite muscle. This median raphé is sometimes wanting; the muscular fibres of the two sides are then directly continuous with one another.

Relations.—By its *cutaneous* or *under surface*, with the Platysma, the anterior belly of the Digastric, the supra-hyoid aponeurosis, the submaxillary gland, submental vessels, and mylo-hyoid vessels and nerve. By its *deep* or *superior surface*, with the Genio-hyoid, part of the Hyo-glossus, and Stylo-glossus muscles, the hypoglossal and lingual nerves, the submaxillary ganglion, the sublingual gland, the deep portion of the submaxillary gland and Wharton's duct; the sublingual and ranine vessels, and the buccal mucous membrane.

Dissection.—The Mylo-hyoid should now be removed, in order to expose the muscles which lie beneath; this is effected by reflecting it from its attachments to the hyoid bone and jaw, and separating it by a vertical incision from its fellow of the opposite side.

The **Genio-hyoid** is a narrow, slender muscle, situated immediately beneath* the inner border of the preceding. It arises from the inferior genial tubercle on the inner side of the symphysis of the jaw, and passes downwards and backwards, to be inserted into the anterior surface of the body of the os hyoides. This muscle lies in close contact with its fellow of the opposite side, and increases slightly in breadth as it descends.

Relations.—It is covered by the Mylo-hyoid, and lies along the lower border of the Genio-hyo-glossus.

Nerves.—The Digastric is supplied: its anterior belly, by the mylo-hyoid branch of the inferior dental; its posterior belly, by the facial; the Stylo-hyoid, by the facial; the Mylo-hyoid, by the mylo-hyoid branch of the inferior dental; the Genio-hyoid, by the hypoglossal.

Actions.—This group of muscles performs two very important actions. They raise the hyoid bone, and with it the base of the tongue, during the act of deglutition; or, when the hyoid bone is fixed by its depressors and those of the larynx, they depress the lower jaw. During the first act of deglutition, when the mass is being driven from the mouth into the pharynx, the hyoid bone, and with it the

---

* This refers to the depth of the muscles from the skin in the order of dissection. In the erect position of the body each of these muscles lies above the preceding.

tongue, is carried upwards and forwards by the anterior belly of the Digastric, the Mylo-hyoid, and Genio-hyoid muscles. In the second act, when the mass is passing through the pharynx, the direct elevation of the hyoid bone takes place by the combined action of all the muscles; and after the food has passed, the hyoid bone is carried upwards and backwards by the posterior belly of the Digastric and Stylo-hyoid muscles, which assist in preventing the return of the morsel into the mouth.

IV. Lingual Region

Genio-hyo-glossus.   Chondro-glossus.
Hyo-glossus.   Stylo-glossus.
                  Palato-glossus.

Dissection.—After completing the dissection of the preceding muscles, saw through the lower jaw just external to the symphysis. Then draw the tongue forwards, and attach it, by a stitch, to the nose; when its muscles, which are thus put on the stretch, may be examined.

The **Genio-hyo-glossus** has received its name from its triple attachment to the jaw, hyoid bone, and tongue, but it would be better named the *Genio-glossus*, since its attachment to the hyoid bone is very slight or altogether absent. It is a flat, triangular muscle, placed vertically on either side of the middle line, its apex corresponding with its point of attachment to the lower jaw, its base with its insertion into the tongue and hyoid bone. It arises by a short tendon from the superior genial tubercle on the inner side of the symphysis of the jaw, immediately above the Genio-hyoid; from this point the muscle spreads out in a fan-like form, a few of the inferior fibres passing downwards, to be attached by a thin aponeurosis into the upper part of the body of the hyoid bone, a few fibres passing between the Hyo-glossus and Chondro-glossus to blend with the Constrictor muscles of the pharynx; the middle fibres passing backwards, and the superior ones upwards and forwards, to enter the whole length of the under surface of the tongue, from the base to the apex. The two muscles lie on either side of the median plane; behind, they are quite distinct from each other, and are separated at their insertion into the under surface of the tongue by a tendinous raphé, which extends through the middle of the organ; in front, the two muscles are more or less blended: distinct fasciculi are to be seen passing off from one muscle, crossing the middle line, and intersecting with bundles of fibres derived from the muscle on the other side (fig. 205).

Relations.—By its *internal surface* it is in contact with its fellow of the opposite side. By its *external surface*, with the Inferior lingualis, the Hyo-glossus, the lingual artery and hypoglossal nerve, the lingual nerve, and sublingual gland. By its *upper border*, with the mucous membrane of the floor of the mouth (frænum linguæ). By its *lower border*, with the Genio-hyoid.

The **Hyo-glossus** is a thin, flat, quadrilateral muscle, which arises from the side of the body, and whole length of the greater cornu of the hyoid bone, and passes almost vertically upwards to enter the side of the tongue, between the Stylo-glossus and Lingualis. Those fibres of this muscle which arise from the body are directed upwards and backwards, overlapping those arising from the greater cornu, which are directed upwards and forwards.

Relations.—By its *external surface*, with the Digastric, the Stylo-hyoid, Stylo-glossus, and Mylo-hyoid muscles, the submaxillary ganglion, the lingual and hypoglossal nerves, Wharton's duct, the ranine vein, the sublingual gland, and the deep portion of the submaxillary gland. By its *deep surface*, with the Stylo-hyoid ligament, the Genio-hyo-glossus, Lingualis, and Middle constrictor, the lingual vessels, and the glosso-pharyngeal nerve.

The **Chondro-glossus** is a distinct muscular slip, though it is sometimes described as a part of the Hyo-glossus, from which, however, it is separated by the fibres of the Genio-hyo-glossus, which pass to the side of the pharynx. It is about three-quarters to an inch in length, and arises from the inner side and base of the lesser cornu and contiguous portion of the body of the hyoid bone, and passes directly upwards to blend with the intrinsic muscular fibres of the tongue, between the Hyo-glossus and Genio-hyo-glossus. A small slip of muscular fibre is occasionally found, arising from the cartilago triticea in the thyro-hyoid ligament, and passing upwards and forwards to enter the tongue with the hindermost fibres of the Hyo-glossus.

The **Stylo-glossus,** the shortest and smallest of the three styloid muscles, arises from the anterior and outer side of the styloid process, near its apex, and from the stylo-mandibular ligament, to which its fibres, in most cases, are attached by a thin aponeurosis. Passing downwards and forwards between the internal and external carotid arteries, and becoming nearly horizontal in its direction, it divides upon

Fig. 204.—Muscles of the tongue. Left side.

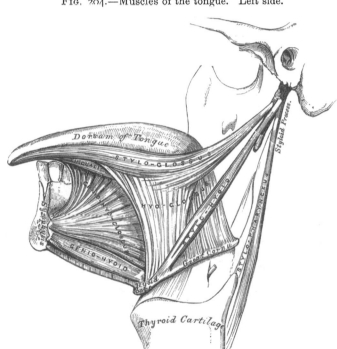

the side of the tongue into two portions : one longitudinal, which enters the side of the tongue near its dorsal surface, blending with the fibres of the Lingualis in front of the Hyo-glossus ; the other oblique, which overlaps the Hyo-glossus muscles and decussates with its fibres.

Relations.—By its *external surface,* from above downwards, with the parotid gland, the Internal pterygoid muscle, the lingual nerve, and the mucous membrane of the mouth. By its *internal surface,* with the tonsil, the Superior constrictor, and the Hyo-glossus muscle.

The **Palato-glossus,** or **Constrictor isthmi faucium,** although it is one of the muscles of the tongue, serving to draw its base upwards during the act of deglutition, is more nearly associated with the soft palate, both in its situation and function ; it will, consequently, be described with that group of muscles.

Nerves.—The Palato-glossus is probably innervated by the spinal accessory nerve, through the pharyngeal plexus ; the remaining muscles of this group by the hypoglossal.

Muscular substance of tongue.—The muscular fibres of the tongue run in various directions. These fibres are divided into two sets—Extrinsic and Intrinsic. The extrinsic muscles of the tongue are those which have their origin external to, and only their terminal fibres contained in, the substance of the organ. They are: the Stylo-glossus, the Hyo-glossus, the Palato-glossus, the Genio-hyo-glossus, and part of the Superior constrictor of the pharynx (Pharyngo-glossus). The intrinsic are those which are contained entirely within the tongue, and form the greater part of its muscular structure.

The tongue consists of symmetrical halves separated from each other in the middle line by a fibrous septum. Each half is composed of muscular fibres arranged in various directions, containing much interposed fat, and supplied by vessels and nerves.

Fig. 205.—Muscles of the tongue from below. (From a preparation in the Museum of the Royal College of Surgeons of England.)

Fig. 206.—Muscles on the dorsum of the tongue.

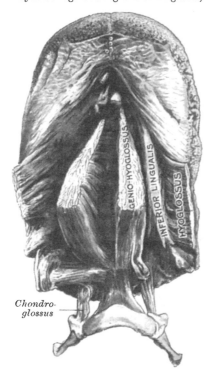

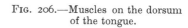

Chondro-glossus

Cut edge of Superior lingualis

To demonstrate the various fibres of the tongue, the organ should be subjected to prolonged boiling, in order to soften the connective tissue; the dissection may then be commenced from the dorsum (fig. 206). Immediately beneath the mucous membrane is a submucous, fibrous layer, into which the muscular fibres which terminate on the surface of the tongue are inserted. Upon removing this, with the mucous membrane, the first stratum of muscular fibres is exposed. This belongs to the group of intrinsic muscles, and has been named the *Superior lingualis* (*m. longitudinalis superior*). It consists of a thin layer of oblique and longitudinal fibres, which arise from the submucous fibrous layer, close to the Epiglottis, and from the fibrous septum, and pass forwards and outwards to the edges of the tongue. Between its fibres pass some vertical fibres derived from the Genio-hyo-glossus and from the vertical intrinsic muscle, which will be described later on. Beneath this layer is the second stratum of muscular fibres, derived principally from the extrinsic muscles. In front, it is formed by the fibres derived from the Stylo-glossus, running along the side of the tongue, and sending one set of fibres over the dorsum which runs obliquely forwards and inwards to the middle line, and another set of fibres, seen at a later period of the dissection, on to the under surface of the sides of the anterior part of the tongue, which runs forwards and inwards, between the fibres of the Hyo-glossus, to the middle line. Behind this layer of fibres, derived from the Stylo-glossus, are fibres derived from the Hyo-glossus, assisted by some few fibres of the Palato-glossus. The Hyo-glossus, entering the side of the under surface of the tongue, between the Stylo-glossus and Inferior lingualis, passes round its margin and spreads out into a layer on the dorsum, which occupies the middle third of the organ, and runs almost transversely inwards to the septum. It is reinforced by some fibres from the Palato-glossus; other fibres of this muscle pass more deeply and intermingle with the next layer. The posterior part of the second layer of the muscular fibres of the tongue is derived from those fibres of the Hyo-glossus which arise from the lesser cornu of the hyoid bone, and are here described as a separate muscle—the Chondro-glossus. The fibres of this muscle are arranged in a fan-shaped manner, and spread out over the posterior third of the tongue. Beneath this layer is the great mass of the intrinsic muscles of the tongue, intersected at right angles by the terminal fibres of one of the extrinsic muscles—the Genio-hyo-glossus. This portion of the tongue is paler in colour and softer in texture than that already described, and is some-times designated the medullary portion in contradistinction to the firmer superficial part, which is termed the cortical portion. It con-sists largely of transverse fibres, the *Transverse lingualis* (m. trans-versus linguæ), and of vertical fibres, the *Vertical lingualis* (m. verticalis linguæ). The Trans-verse lingualis forms the largest

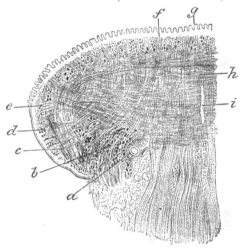

FIG. 207.—Coronal section of tongue. Showing intrinsic muscles. (Altered from Krause.)

a, lingual artery; b, Inferior lingualis, cut through; c, fibres of Hyo-glossus; d, oblique fibres of Stylo-glossus; e, in-sertion of Transverse lingualis; f, Superior lingualis; g, papillæ of tongue; h, vertical fibres of Genio-hyo-glossus intersecting Transverse lingualis; i, septum.

portion of the third layer of muscular fibres of the tongue. The fibres arise from the median septum, and pass outwards to be inserted into the submucous fibrous layer at the sides of the tongue. Intermingled with these transverse intrinsic fibres are transverse extrinsic fibres derived from the Palato-glossus and the Superior constrictor of the pharynx. These Transverse extrinsic fibres, however, run in the opposite direction, passing inwards towards the septum. Intersecting the transverse fibres are a large number of vertical fibres derived partly from the Genio-hyo-glossus and partly from intrinsic fibres, the *Vertical lingualis*. The fibres derived from the Genio-hyo-glossus enter the under surface of the tongue on each side of the median septum from base to apex. They ascend in a radiating manner to the dorsum, being inserted into the sub-mucous fibrous layer covering the tongue on each side of the middle line. The Vertical lingualis is found only at the borders of the fore part of the tongue, external to the fibres of the Genio-hyo-glossus. Its fibres extend from the upper to the under surface of the organ, decussating with the fibres of the other muscles, and especially with the Transverse lingualis. The fourth layer of muscular fibres of the tongue consists partly of extrinsic fibres derived from the Stylo-glossus, and partly of intrinsic fibres, the *Inferior lingualis* (m. longitudinalis inferior). At the sides of the under surface of the organ are some fibres derived from the Stylo-glossus, which, as it runs forwards at the side of the tongue, gives off fibres which, passing forwards and inwards between the fibres of the Hyo-glossus, form an inferior oblique stratum which joins in front with the anterior fibres of the Inferior lingualis. The Inferior lingualis is a longitudinal band, situated on the under surface of the tongue, and extending from the base to the apex of the organ. Behind, some of its fibres are connected with the body of the hyoid bone. It lies between the Hyo-glossus and the Genio-hyo-glossus, and in front of the Hyo-glossus it gets into relation with the Stylo-glossus, with the fibres of which it blends. It is in relation by its under surface with the ranine artery.

Surgical Anatomy.—The fibrous septum which exists between the two halves of the tongue is very complete, so that the anastomosis between the two lingual arteries is not very free, a fact often illustrated by injecting one-half of the tongue with coloured size, while the other half is left uninjected, or is injected with size of a different colour.

This is a point of considerable importance in connection with removal of one half of the tongue for cancer, an operation which is now frequently resorted to when the disease is strictly confined to one side of the tongue. If the mucous membrane is divided longi-tudinally exactly in the middle line, the tongue can be split into halves along the median raphé, without any appreciable hæmorrhage, and the diseased half can then be removed.

Actions.—The movements of the tongue, although numerous and complicated, may be understood by carefully considering the direction of the fibres of its muscles. The *Genio-hyo-glossi* muscles, by means of their posterior fibres, draw the base of the tongue forwards, so as to protrude the apex from the mouth. The anterior fibres draw the tongue back into the mouth. The whole length of these two muscles acting along the middle line of the tongue draw it downwards, so as to make it concave from side to side, forming a channel along which fluids may pass toward the pharynx, as in sucking. The *Hyo-glossi* muscles depress the tongue, and draw down its sides, so as to render it convex from side to side. The *Stylo-glossi* muscles draw the tongue upwards and backwards. The *Palato-glossi* muscles draw the base of the tongue upwards. With regard to the intrinsic muscles, both the Superior and Inferior linguales tend to shorten the tongue, but the former, in addition, turn the tip and sides upwards so as to render the dorsum concave, while the latter pull the tip downwards and cause the dorsum to become convex. The Transverse lingualis narrows and elongates the tongue, and the Vertical lingualis flattens and broadens it. The complex arrangement of the muscular fibres of the tongue, and the various directions in which they run, give to this organ the power of assuming the various forms necessary for the enuncia-tion of the different consonantal sounds; and Dr. Macalister states 'there is reason to believe that the musculature of the tongue varies in different races owing to the hereditary practice and habitual use of certain motions required for enunciating the several vernacular languages.'

V. PHARYNGEAL REGION

Inferior constrictor.	Superior constrictor.
Middle constrictor.	Stylo-pharyngeus.
Palato-pharyngeus.	(See next section.)
Salpingo-pharyngeus.	

Dissection (fig. 208).—In order to examine the muscles of the pharynx, cut through the trachea and œsophagus just above the sternum, and draw them upwards by dividing the loose areolar tissue connecting the pharynx with the front of the vertebral column. The parts being drawn well forwards, apply the edge of the saw immediately behind the styloid processes, and saw the base of the skull through from below upwards. The pharynx and mouth should then be stuffed with tow, in order to distend its cavity and render the muscles tense and easier of dissection.

The **Inferior constrictor,** the most superficial and thickest of the three constrictors, arises from the sides of the cricoid and thyroid cartilages. To the cricoid cartilage it is attached in the interval between the Crico-thyroid muscle in front, and the articular facet for the thyroid cartilage behind. To the thyroid cartilage it is attached to the oblique line on the side of the great ala, to the cartilaginous surface behind it, nearly as far as its posterior border, and to the inferior cornu. From these attachments the fibres spread backwards and inwards, to be inserted into the fibrous raphé in the posterior median line of the pharynx. The inferior fibres are horizontal, and continuous with the fibres of the œsophagus ; the rest ascend, increasing in obliquity, and overlap the Middle constrictor.

Relations.—It is covered by a thin membrane which surrounds the entire pharynx (bucco-pharyngeal fascia). *Behind*, it is in relation with the vertebral column and the prevertebral fascia and muscles ; *laterally*, with the thyroid gland, the common carotid artery, and the Sterno-thyroid muscle ; by its *internal surface*, with the Middle constrictor, the Stylo-pharyngeus, Palato-pharyngeus, the fibrous coat and mucous membrane of the pharynx. The internal laryngeal nerve and the laryngeal branch of the Superior Thyroid artery pass near the upper border, and the inferior, or recurrent laryngeal nerve, and the laryngeal branch of the Inferior Thyroid artery, beneath the lower border of this muscle, previous to their entering the larynx.

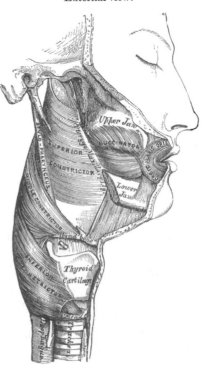

FIG. 208.—Muscles of the pharynx. External view.

The **Middle constrictor** is a flattened, fan-shaped muscle, smaller than the preceding. It arises from the whole length of the upper border of the greater cornu of the hyoid bone, from the lesser cornu, and from the stylo-hyoid ligament. The fibres diverge from their origin : the lower ones descending beneath the Inferior constrictor, the middle fibres passing transversely, and the upper fibres ascending and overlapping the Superior constrictor. The muscle is inserted into the posterior median fibrous raphé, blending in the middle line with the one of the opposite side.

Relations.—This muscle is separated from the Superior constrictor by the glosso-pharyngeal nerve and the Stylo-pharyngeus muscle and Stylo-hyoid ligament ; and from the Inferior constrictor by the superior laryngeal nerve. *Behind*, it lies on the vertebral column, the Longus colli, and the Rectus capitis anticus major. *On each side* it is in relation with the carotid vessels, the pharyngeal plexus, and some lymphatic glands. Near its origin it is covered by the Hyoglossus, from which it is separated by the lingual vessels. It lies upon the Superior constrictor, the Stylo-pharyngeus, the Palato-pharyngeus, the fibrous coat, and the mucous membrane of the pharynx.

The **Superior constrictor** is a quadrilateral muscle, thinner and paler than the other constrictors, and situated at the upper part of the pharynx. It arises from the lower third of the posterior margin of the internal pterygoid plate and its hamular process, from the contiguous portion of the palate bone and the reflected tendon of the Tensor palati muscle, from the pterygo-maxillary ligament, from the alveolar process above the posterior extremity of the mylo-hyoid ridge, and by a few fibres from the side of the tongue. From these points the fibres curve backwards, to be inserted into the median raphé, being also prolonged by means of a fibrous aponeurosis to the pharyngeal spine on the basilar process of the occipital bone. The superior fibres arch beneath the Levator palati and the Eustachian tube, the interval between the upper border of the muscle and the basilar process being deficient in muscular fibres, and closed by the pharyngeal aponeurosis. This interval is known as the *sinus of Morgagni*.

Relations.—By its *outer surface*, with the prevertebral fascia and muscles, the vertebral column, the internal carotid and ascending pharyngeal arteries, the internal jugular vein and pharyngeal venous plexus, the glosso-pharyngeal, pneumogastric, spinal accessory, hypoglossal, lingual, and sympathetic nerves, the Middle constrictor and internal pterygoid muscles, the Styloid process, the Stylo-hyoid ligament, and the Stylo-pharyngeus. By its *internal surface*, with the Palato-pharyngeus, the tonsil, the fibrous coat and mucous membrane of the pharynx.

The **Stylo-pharyngeus** is a long, slender muscle, round above, broad and thin below. It arises from the inner side of the base of the styloid process, passes downwards along the side of the pharynx between the Superior and Middle constrictors, and spreads out beneath the mucous membrane, where some of its fibres are lost in the Constrictor muscles; and others, joining with the Palato-pharyngeus, are inserted into the posterior border of the thyroid cartilage. The glosso-pharyngeal nerve runs on the outer side of this muscle, and crosses over it in passing forward to the tongue.

Relations.—*Externally*, with the Stylo-glossus muscle, the parotid gland, the external carotid artery, and the Middle constrictor. *Internally*, with the internal carotid, the internal jugular vein, the Superior constrictor, Palato-pharyngeus and mucous membrane.

Nerves.—The Constrictors are supplied by branches from the pharyngeal plexus, the Stylo-pharyngeus by the glosso-pharyngeal nerve, and the Inferior constrictor by additional branches from the external and recurrent laryngeal nerves.

Actions.—When deglutition is about to be performed, the pharynx is drawn upwards and dilated in different directions, to receive the food propelled into it from the mouth. The Stylo-pharyngei, which are much farther removed from one another at their origin than at their insertion, draw the sides of the pharynx upwards and outwards, and so increase its transverse diameter; its breadth in the antero-posterior direction being increased by the larynx and tongue being carried forwards in their ascent. As soon as the morsel is received in the pharynx, the Elevator muscles relax, the bag descends, and the Constrictors contract upon it, and convey it gradually downwards into the œsophagus. Besides its action in deglutition, the pharynx also exerts an important influence in the modulation of the voice, especially in the production of the higher tones.

VI. Palatal Region

Levator palati.	Palato-glossus.
Tensor palati.	Palato-pharyngeus.
Azygos uvulæ.	Salpingo-pharyngeus.

Dissection (fig. 209).—Lay open the pharynx from behind, by a vertical incision extending from its upper to its lower part, and partially divide the occipital attachment by a transverse incision on each side of the vertical one; the posterior surface of the soft palate is then exposed. Having fixed the uvula so as to make it tense, the mucous membrane and glands should be carefully removed from the posterior surface of the soft palate and the muscles of this part are at once exposed.

The **Levator palati** is a long, thick, rounded muscle, placed on the outer side of the posterior nares. It arises from the under surface of the apex of the petrous portion of the temporal bone, and from the inner surface of the cartilaginous portion of the Eustachian tube; after passing into the pharynx, above the upper concave margin of the Superior constrictor, it passes obliquely downwards and inwards, its fibres spreading out in the soft palate as far as the middle line, where they blend with those of the opposite side.

Relations.—*Externally*, with the Tensor palati, Superior constrictor, and Eustachian tube. *Internally*, with the mucous membrane of the pharynx. *Posteriorly*, with the posterior fasciculus of the Palato-pharyngeus, the Azygos uvulæ, and the mucous lining of the soft palate.

The **Circumflexus** or **Tensor palati** is a broad, thin, riband-like muscle, placed on the outer side of the Levator palati, and consisting of a vertical and a horizontal portion. The vertical portion arises by a flat lamella from the scaphoid fossa at the base of the internal pterygoid plate ; from the spine of the sphenoid and the

FIG. 209.—Muscles of the soft palate. The pharynx being laid open from behind.

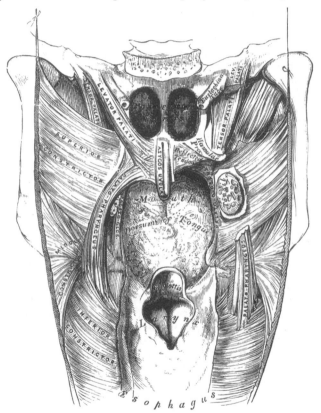

outer side of the cartilaginous portion of the Eustachian tube : descending vertically between the internal pterygoid plate and the inner surface of the Internal pterygoid muscle, it terminates in a tendon, which winds round the hamular process, being retained in this situation by some of the fibres of origin of the Internal pterygoid muscle. Between the hamular process and the tendon is a small bursa. The tendon or horizontal portion then passes horizontally inwards, and is inserted into a broad aponeurosis, the *palatine aponeurosis,* and into the transverse ridge on the horizontal portion of the palate bone.

Relations.—*Externally*, with the Internal pterygoid. *Internally*, with the Levator palati, from which it is separated by the Eustachian tube and Superior constrictor, and with the internal pterygoid plate. In the soft palate, its tendon and the palatine aponeurosis is anterior to that of the Levator palati, being covered by the Palato-glossus and the mucous membrane.

Palatine Aponeurosis.—Attached to the posterior border of the hard palate is a thin, firm, fibrous lamella which supports the muscles and gives strength to the soft palate. It is thicker above than below, where it becomes very thin and difficult to define. Laterally, it is continuous with the pharyngeal aponeurosis.

The **Azygos uvulæ** is not a single muscle, as would be inferred from its name, but a pair of narrow cylindrical fleshy fasciculi, placed on either side of the median line of the soft palate. Each muscle arises from the posterior nasal spine of the palate bone, and from the contiguous tendinous aponeurosis of the soft palate, and descends to be inserted into the uvula.

Relations.—*Anteriorly*, with the tendinous expansion of the Levatores palati; *behind*, with the posterior fasciculus of the Palato-pharyngeus and the mucous membrane.

The two next muscles are exposed by removing the mucous membrane from the pillars of the fauces throughout nearly their whole extent.

The **Palato-glossus (Constrictor isthmi faucium)** is a small fleshy fasciculus, narrower in the middle than at either extremity, forming, with the mucous membrane covering its surface, the anterior pillar of the soft palate. It arises from the anterior surface of the soft palate on each side of the uvula, and, passing downwards, forwards, and outwards in front of the tonsil, is inserted into the side of the tongue, some of its fibres spreading over the dorsum, and others passing deeply into the substance of the organ to intermingle with the Transversus linguæ. In the soft palate, the fibres of the muscle are continuous with those of the muscle of the opposite side.

The **Palato-pharyngeus** is a long, fleshy fasciculus, narrower in the middle than at either extremity, forming, with the mucous membrane covering its surface, the posterior pillar of the soft palate. It is separated from the Palato-glossus by an angular interval, in which the tonsil is lodged. It arises from the soft palate by an expanded fasciculus, which is divided into two parts by the Levator palati and Azygos uvulæ. The *posterior fasciculus* lies in contact with the mucous membrane, and joins with the corresponding muscle in the middle line; the *anterior fasciculus*, the thicker, lies in the soft palate between the Levator and Tensor, and joins in the middle line the corresponding part of the opposite muscle. Passing outwards and downwards behind the tonsil, the Palato-pharyngeus joins the Stylo-pharyngeus, and is inserted with that muscle into the posterior border of the thyroid cartilage, some of its fibres being lost on the side of the pharynx, and others passing across the middle line posteriorly, to decussate with the muscle of the opposite side.

Relations.—In the soft palate, its *posterior surface* is covered by mucous membrane, from which it is separated by a layer of palatine glands. By its *anterior surface* it is in relation with the Tensor palati. Where it forms the posterior pillars of the fauces, it is covered by mucous membrane, excepting on its outer surface. In the *pharynx* it lies between the mucous membrane and the Constrictor muscles.

The Salpingo-pharyngeus.—This muscle arises from the inferior part of the Eustachian tube near its orifice; it passes downwards and blends with the posterior fasciculus of the Palato-pharyngeus.

In a dissection of the soft palate from its posterior or nasal surface to its anterior or oral surface, the muscles would be exposed in the following order: viz. the posterior fasciculus of the Palato-pharyngeus, covered over by the mucous membrane reflected from the floor of the nasal fossæ; the Azygos uvulæ; the Levator palati; the anterior fasciculus of the Palato-pharyngeus; the aponeurosis of the Tensor palati, and the Palato-glossus covered over by a reflection from the oral mucous membrane.

Nerves.—The Tensor palati is supplied by a branch from the otic ganglion; the remaining muscles of this group are in all probability supplied by the internal branch of the spinal accessory, whose fibres are distributed along with certain branches of the pneumogastric through the pharyngeal plexus.* It is possible, however, that the Levator palati may be supplied by the facial through the Petrosal branch of the Vidian.

Actions.—During the *first stage* of deglutition, the morsel of food is driven back into the fauces by the pressure of the tongue against the hard palate : the base of the tongue being, at the same time, retracted, and the larynx raised with the pharynx, and carried forwards under it. During the *second stage* the entrance to the larynx is closed, not, as was formerly supposed, by the folding backwards of the epiglottis over it, but, as Anderson Stuart has shown, by the drawing forward of the arytenoid cartilages towards the cushion of the epiglottis—a movement produced by the contraction of the external thyro-arytenoid, the arytenoid and aryteno-epiglottidean muscles.

The morsel of food after leaving the tongue passes on to the posterior or laryngeal surface of the epiglottis, and glides along this for a certain distance ; † then the Palato-glossi muscles, the constrictors of the fauces, contract behind the food ; the soft palate is slightly raised by the Levator palati, and made tense by the Tensor palati ; and the Palato-pharyngei, by their contraction, pull the pharynx upwards over the morsel of food, and at the same time come nearly together, the uvula filling up the slight interval between them. By these means the food is prevented passing into the upper part of the pharynx or the posterior nares ; at the same time, the latter muscles form an inclined plane, directed obliquely downwards and backwards, along the under surface of which the morsel descends into the lower part of the pharynx. The Salpingo-pharyngeus raises the upper and lateral part of the pharynx ; i.e. that part which is above the point where the Stylo-pharyngeus is attached to the pharynx.

Surgical Anatomy.—The muscles of the soft palate should be carefully dissected, the relations they bear to the surrounding parts especially examined, and their action attentively studied upon the dead subject, as the surgeon is required to divide one or more of these muscles in the operation of staphyloraphy. Sir W. Fergusson was the first to show that in the congenital deficiency called *cleft palate*, the edges of the fissure are forcibly separated by the action of the Levatores palati and Palato-pharyngei muscles, producing very considerable impediment to the healing process after the performance of the operation for uniting their margins by adhesion ; he, consequently, recommended the division of these muscles as one of the most important steps in the operation. This he effected by an incision made with a curved knife introduced behind the soft palate. The incision is to be half-way between the hamular process and Eustachian tube, and perpendicular to a line drawn between them. This incision perfectly accomplishes the division of the Levator palati. The Palato-pharyngeus may be divided by cutting across the posterior pillar of the soft palate, just below the tonsil, with a pair of blunt-pointed curved scissors ; and the anterior pillar may be divided also. To divide the Levator palati, the plan recommended by Mr. Pollock is to be greatly preferred. The soft palate being put upon the stretch, a double-edged knife is passed through it just on the inner side of the hamular process, and above the line of the Levator palati. The handle being now alternately raised and depressed, a sweeping cut is made along the posterior surface of the soft palate, and the knife withdrawn, leaving only a small opening in the mucous membrane on the anterior surface. If this operation is performed on the dead body, and the parts afterwards dissected, the Levator palati will be found completely divided. In the present day, however, this division of the muscles, as part of the operation of staphyloraphy, is not so much insisted upon. All tension is prevented by making longitudinal incisions, on either side, parallel to the cleft and just internal to the hamular process, in such a position as to avoid the posterior palatine artery.

VII. Anterior Vertebral Region

Rectus capitis anticus major.	Rectus capitis lateralis.
Rectus capitis anticus minor.	Longus colli.

The **Rectus capitis anticus major** (fig. 210), broad and thick above, narrow below, appears like a continuation upwards of the Scalenus anticus. It arises by

---

* *Journal of Anatomy and Physiology*, vol. xxiii. p. 523.

† Walton (quoted by A. Stuart) maintains that the epiglottis is not essential to the deglutition even of liquids.

four tendinous slips from the anterior tubercles of the transverse processes of the third, fourth, fifth, and sixth cervical vertebræ, and ascends, converging towards its fellow of the opposite side, to be inserted into the basilar process of the occipital bone.

Relations.—By its *anterior surface*, with the pharynx, the sympathetic nerve, and the sheath enclosing the internal and common carotid artery, internal jugular vein, and pneumogastric nerve. By its *posterior surface*, with the Longus colli, the Rectus capitis anticus minor, and the upper cervical vertebræ.

The **Rectus capitis anticus minor** is a short, flat muscle, situated immediately behind the upper part of the preceding. It arises from the anterior surface of the lateral mass of the atlas, and from the root of its transverse process, and

FIG. 210.—The prevertebral muscles.

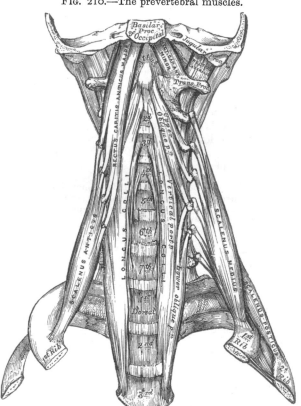

passing obliquely upwards and inwards, is inserted into the basilar process immediately behind the preceding muscle.

Relations.—By its *anterior surface*, with the Rectus capitis anticus major. By its *posterior surface*, with the front of the occipito-atlantal articulation.

The **Rectus capitis lateralis** is a short, flat muscle, which arises from the upper surface of the transverse process of the atlas, and is inserted into the under surface of the jugular process of the occipital bone.

Relations.—By its *anterior surface*, with the internal jugular vein. By its *posterior surface*, with the vertebral artery. On its *outer side* lies the occipital artery ; on its *inner side* the suboccipital nerve.

The **Longus colli** is a long, flat muscle, situated on the anterior surface of the spine, between the atlas and the third dorsal vertebra. It is broad in the middle, narrow and pointed at each extremity, and consists of three portions : a superior oblique, an inferior oblique, and a vertical portion. The *superior oblique portion* arises from the anterior tubercles of the transverse processes of the third, fourth, and fifth cervical vertebræ; and, ascending obliquely inwards, is inserted by a narrow tendon into the tubercle on the anterior arch of the atlas. The *inferior oblique portion*, the smallest part of the muscle, arises from the front of the bodies of the first two or three dorsal vertebræ; and, ascending obliquely outwards, is inserted into the anterior tubercles of the transverse processes of the fifth and sixth cervical vertebræ. The *vertical portion* lies directly on the front of the spine; it arises, below, from the front of the bodies of the upper three dorsal and lower three cervical vertebræ, and is inserted above into the front of the bodies of the second, third, and fourth cervical vertebræ above.

Relations.—By its *anterior surface*, with the prevertebral fascia, the pharynx, the œsophagus, the sympathetic nerve, the sheath of the great vessels of the neck, the inferior thyroid artery, and recurrent laryngeal nerve. By its *posterior surface*, with the cervical and dorsal portions of the spine. Its *inner border* is separated from the opposite muscle by a considerable interval below; but they approach each other above.

VIII. Lateral Vertebral Region

Scalenus anticus.   Scalenus medius.
Scalenus posticus.

The **Scalenus anticus** is a conical-shaped muscle, situated deeply at the side of the neck, behind the Sterno-mastoid. It arises from the anterior tubercles of the transverse processes of the third, fourth, fifth, and sixth cervical vertebræ, and descending, almost vertically, is inserted by a narrow, flat tendon into the Scalene tubercle on the inner border and upper surface of the first rib. The lower part of this muscle separates the subclavian artery and vein: the latter being in front, and the former, with the brachial plexus, behind.

Relations.—*In front*, with the clavicle, the Subclavius, Sterno-mastoid, and Omo-hyoid muscles, the transversalis colli, the suprascapular and ascending cervical arteries, the subclavian vein, and the phrenic nerve. By its *posterior surface*, with the Scalenus medius, pleura, the subclavian artery, and brachial plexus of nerves. It is separated from the Longus colli, on the inner side, by the vertebral artery. On the anterior tubercles of the transverse processes of the cervical vertebræ, between the attachments of the Scalenus anticus and Longus colli, lies the ascending cervical branch of the inferior thyroid artery.

The **Scalenus medius,** the largest and longest of the three Scaleni, arises from the posterior tubercles of the transverse processes of the lower six cervical vertebræ, and descending along the side of the vertebral column is inserted by a broad attachment into the upper surface of the first rib, behind the groove for the subclavian artery, as far back as the tubercle. It is separated from the Scalenus anticus by the subclavian artery below and the cervical nerves above. The posterior thoracic, or nerve of Bell, is formed in the substance of the Scalenus medius and emerges from it. The nerve to the Rhomboids also pierces it.

Relations.—By its *anterior surface*, with the Sterno-mastoid; it is crossed by the clavicle, the Omo-hyoid muscle, subclavian artery, and the cervical nerves. To its *outer side* is the Levator anguli scapulæ, and the Scalenus posticus muscle.

The **Scalenus posticus,** the smallest of the three Scaleni, arises, by two or three separate tendons, from the posterior tubercles of the transverse processes of the lower two or three cervical vertebræ, and, diminishing as it descends, is inserted by a thin tendon into the outer surface of the second rib, behind the attachment of the Serratus magnus. This is the most deeply placed of the three Scaleni, and is occasionally blended with the Scalenus medius.

Nerves.—The Rectus capitis anticus major and minor and the Rectus lateralis are supplied by the first cervical nerve, and from the loop formed between it and the second; the Longus colli and Scaleni, by branches from the anterior divisions of the lower cervical nerves (fifth, sixth, seventh, and eighth) before they form the brachial plexus. The Scalenus medius also receives a filament from the deep external branches of the cervical plexus.

Actions.—The Rectus anticus major and minor are the direct antagonists of the muscles at the back of the neck, serving to restore the head to its natural

FIG. 211.—Muscles of the neck. (From a preparation in the Museum of the Royal College of Surgeons of England.)

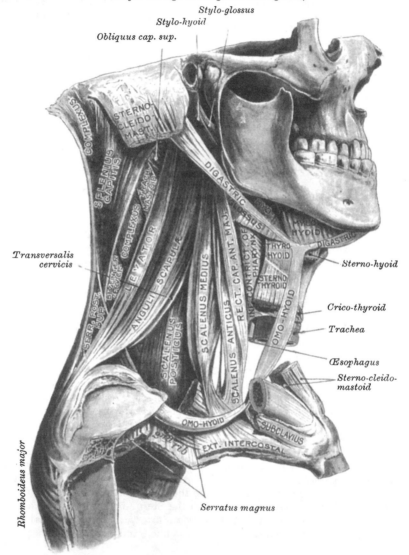

position after it has been drawn backwards. These muscles also serve to flex the head, and from their obliquity, rotate it, so as to turn the face to one or the other side. The Longus colli flexes and slightly rotates the cervical portion of the spine. The Scaleni muscles, when they take their fixed point from above, elevate the first and second ribs, and are, therefore, inspiratory muscles. When they take their fixed point from below, they bend the spinal column to one or the other side. If the muscles of both sides act, lateral movement is prevented, but the spine is slightly flexed. The Rectus lateralis, acting on one side, bends the head laterally.

Surface Form.—The muscles in the neck, with the exception of the Platysma myoides, are invested by the deep cervical fascia, which softens down their form, and is of considerable importance in connection with deep cervical abscesses and tumours, modifying the direction of their growth and causing them to extend laterally instead of towards the surface. The *Platysma myoides* does not influence surface form except it is in action, when it produces wrinkling of the skin of the neck, which is thrown into oblique ridges parallel with the fasciculi of the muscle. Sometimes this contraction takes place suddenly and repeatedly, as a sort of spasmodic twitching, the result of a nervous habit. The *Sterno-cleido-mastoid* is the most important muscle of the neck as regards its surface form. If the muscle is put into action by drawing the chin downwards and to the opposite shoulder, its surface form will be plainly outlined. The sternal origin will stand out as a sharply defined ridge, while the clavicular origin will present a flatter and less prominent outline. The fleshy middle portion will appear as an oblique roll or elevation, with a thick rounded anterior border gradually becoming less marked above. On the opposite side, i.e. on the side to which the head is turned, the outline is lost, its place being occupied by an oblique groove in the integument. When the muscle is at rest its anterior border is still visible, forming an oblique rounded ridge, terminating below in the sharp outline of the sternal head. The posterior border of the muscle does not show above the clavicular head. The anterior border is defined by drawing a line from the tip of the mastoid process to the sterno-clavicular joint. It is an important surface-marking in the operation of ligature of the common carotid artery and some other operations. Between the sternal and clavicular heads is a slight depression, most marked when the muscle is in action. This is bounded below by the prominent sternal extremity of the clavicle. Between the sternal origins of the two muscles is a V-shaped space, the *suprasternal notch*, more pronounced below, and becoming toned down above, where the Sterno-hyoid and Sterno-thyroid muscles, lying upon the trachea, become more prominent. Above the hyoid bone, in the middle line, the anterior belly of the *Digastric* to a certain extent influences surface form. It corresponds to a line drawn from the symphysis of the lower jaw to the side of the body of the hyoid bone, and renders this part of the hyo-mental region convex. In the posterior triangle of the neck, the posterior belly of the *Omo-hyoid*, when in action, forms a conspicuous object, especially in thin necks, presenting a cord-like form running across this region, almost parallel with, and a little above, the clavicle.

MUSCLES AND FASCIÆ OF THE TRUNK

The muscles of the Trunk may be arranged in four groups, corresponding with the region in which they are situated.

I. The Back.   III. The Abdomen.
II. The Thorax.   IV. The Perinæum.

I. MUSCLES OF THE BACK

The muscles of the back are very numerous, and may be subdivided into five layers.

FIRST LAYER
Trapezius.
Latissimus dorsi.

SECOND LAYER
Levator anguli scapulæ.
Rhomboideus minor.
Rhomboideus major.

THIRD LAYER
Serratus posticus superior.
Serratus posticus inferior.
Splenius capitis.
Splenius colli.

FOURTH LAYER
Sacral and Lumbar Regions
Erector spinæ.

Dorsal Region
Ilio-costalis.
Musculus accessorius ad ilio-costalem.
Longissimus dorsi.
Spinalis dorsi.

Cervical Region
Cervicalis ascendens.
Transversalis cervicis.
Trachelo-mastoid.
Complexus.
Biventer cervicis.
Spinalis colli.

FIFTH LAYER
Semispinalis dorsi.
Semispinalis colli.
Multifidus spinæ.

Rotatores spinæ.   Rectus capitis posticus major.
Supraspinales.   Rectus capitis posticus minor.
Interspinales.   Obliquus capitis inferior.
Extensor coccygis.   Obliquus capitis superior.
Intertransversales.

FIRST LAYER

Trapezius.   Latissimus dorsi.

Dissection (fig. 212).—Place the body in the prone position, with the arms extended over the sides of the table, and the chest and abdomen supported by several blocks, so as to render the muscles tense. Then make an incision along the middle line of the back from the occipital protuberance to the coccyx. Make a transverse incision from the upper end of this to the mastoid process; and a third incision from its lower end, along the crest of the ilium to about its middle. This large intervening space should, for convenience of dissection, be subdivided by a fourth incision, extending obliquely from the spinous process of the last dorsal vertebra, upwards and outwards, to the acromion process. This incision corresponds with the lower border of the Trapezius muscle. The flaps of integument are then to be removed in the direction shown in the figure.

The **superficial fascia** is exposed upon removing the skin from the back. It forms a layer of considerable thickness and strength, in which a quantity of granular pinkish fat is contained. It is continuous with the superficial fascia in other parts of the body. The **deep fascia** is a dense fibrous layer, attached to the occipital bone, the spines of the vertebræ, the crest of the ilium, and the spine of the scapula. It covers over the superficial muscles, forming sheaths for them, and in the neck forms the posterior part of the deep cervical fascia; in the thorax it is continuous with the deep fascia of the axilla and chest, and in the abdomen with that covering the abdominal muscles.

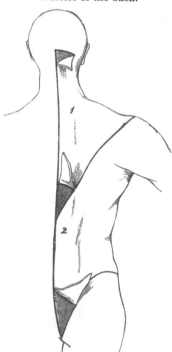

FIG. 212.—Dissection of the muscles of the back.

The **Trapezius** (fig. 213) is a broad, flat, triangular muscle, placed immediately beneath the skin and fascia, and covering the upper and back part of the neck and shoulders. It arises from the external occipital protuberance and the inner third of the superior curved line of the occipital bone; from the ligamentum nuchæ, the spinous process of the seventh cervical, and those of all the dorsal vertebræ; and from the corresponding portion of the supraspinous ligament. From this origin, the superior fibres proceed downwards and outwards, the inferior ones upwards and outwards; and the middle fibres, horizontally; and are inserted, the superior ones into the outer third of the posterior border of the clavicle; the middle fibres into the inner margin of the acromion process, and into the superior lip of the posterior border or crest of the spine of the scapula; the inferior fibres converge near the scapula, and terminate in a triangular aponeurosis, which glides over a smooth surface at the inner extremity of the spine, to be inserted into a tubercle at the outer part of this smooth surface. The Trapezius is fleshy in the greater part of its extent, but tendinous at its origin and insertion. At its occipital origin, it is connected to the bone by a thin fibrous lamina, firmly adherent to the skin, and wanting the lustrous, shining appearance of aponeuroses. At its origin from the spines of the vertebræ, it is

FIG. 213.—Muscles of the back. On the left side is exposed the first layer; on the right side, the second layer and part of the third.

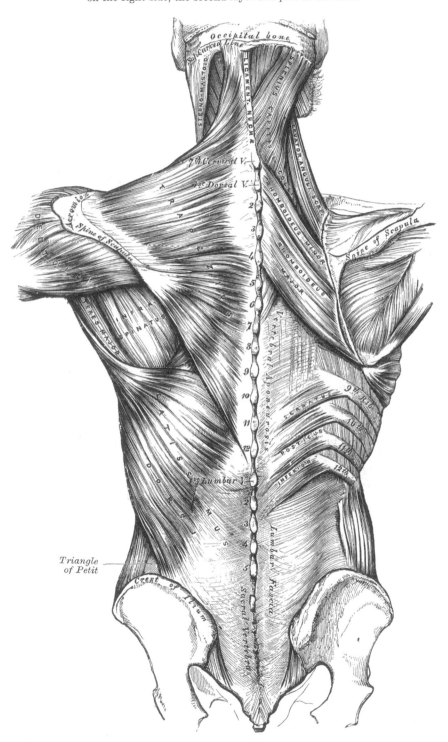

connected to the bones by means of a broad semi-elliptical aponeurosis, which occupies the space between the sixth cervical and the third dorsal vertebræ, and forms, with the aponeurosis of the opposite muscle, a tendinous ellipse. The rest of the muscle arises by numerous short tendinous fibres. If the Trapezius is dissected on both sides, the two muscles resemble a trapezium, or diamond-shaped quadrangle: two angles corresponding to the shoulders; a third to the occipital protuberance; and the fourth to the spinous process of the last dorsal vertebra.

The clavicular insertion of this muscle varies as to the extent of its attachment: it sometimes advances as far as the middle of the clavicle, and may occasionally become blended with the posterior edge of the Sterno-mastoid, or overlap it. This should be borne in mind in the operation for tying the third part of the subclavian artery.

Relations.—By its *superficial surface*, with the integument. By its *deep surface*, in the neck, with the Complexus, Splenius, Levator anguli scapulæ, and Rhomboideus minor; in the back, with the Rhomboideus major, Supraspinatus, Infraspinatus, the Vertebral aponeurosis (which separates it from the prolongations of the Erector spinæ), and the Latissimus dorsi. The superficial cervical artery, the spinal accessory nerve, and branches from the third and fourth cervical nerves, pass beneath the anterior border of this muscle. The anterior margin of its cervical portion forms the posterior boundary of the posterior triangle of the neck, the other boundaries being the Sterno-mastoid in front, and the clavicle below.

The **Ligamentum nuchæ** (fig. 213) is a fibrous membrane, which, in the neck, represents the supraspinous and interspinous ligaments of the lower vertebræ. It extends from the external occipital protuberance to the spinous process of the seventh cervical vertebra. From its anterior border a fibrous lamina is given off, which is attached to the external occipital crest, the posterior tubercle of the atlas, and the spinous process of each of the cervical vertebræ, so as to form a septum between the muscles on each side of the neck. In man it is merely the rudiment of an important elastic ligament, which, in some of the lower animals, serves to sustain the weight of the head.

The **Latissimus dorsi** is a broad flat muscle, which covers the lumbar and the lower half of the dorsal regions, and is gradually contracted into a narrow fasciculus at its insertion into the humerus. It arises by tendinous fibres from the spinous processes of the six inferior dorsal vertebræ and from the posterior layer of the lumbar fascia (see page 315), by which it is attached to the spines of the lumbar and sacral vertebræ, and to the supraspinous ligament. It also arises from the external lip of the crest of the ilium, behind the origin of the External oblique, and by fleshy digitations from the three or four lower ribs, which are interposed between similar processes of the External oblique muscle (fig. 218, page 331). From this extensive origin the fibres pass in different directions, the upper ones horizontally, the middle obliquely upwards, and the lower vertically upwards, so as to converge and form a thick fasciculus, which crosses the inferior angle of the scapula, and occasionally receives a few fibres from it. The muscle then curves around the lower border of the Teres major, and is twisted upon itself, so that the superior fibres become at first posterior and then inferior, and the vertical fibres at first anterior and then superior. It then terminates in a short quadrilateral tendon, about three inches in length, which, passing in front of the tendon of the Teres major, is inserted into the bottom of the bicipital groove of the humerus; its insertion extending higher on the humerus than that of the tendon of the Pectoralis major. The lower border of the tendon of this muscle is united with that of the Teres major, the surfaces of the two being separated by a bursa; another bursa is sometimes interposed between the muscle and the inferior angle of the scapula. This muscle at its insertion gives off an expansion to the deep fascia of the arm.

A muscular slip, the *axillary arch*, varying from 3 to 4 inches in length, and from ¼ to ¾ of an inch in breadth, occasionally arises from the upper edge of the Latissimus dorsi about the middle of the posterior fold of the axilla, and crosses the axilla in front of the axillary vessels and nerves, to join the under surface of the tendon of the Pectoralis major, the Coraco-brachialis, or the fascia over the Biceps. The position of this abnormal slip is a point of interest in its relation to the axillary artery, as it crosses the vessel just above the spot usually selected for the application of a ligature, and may mislead the surgeon during the operation. It may be easily recognised by the transverse direction of its fibres. Prof. Struthers found it, in 8 out of 105 subjects, occurring seven times on both sides.

There is usually a fibrous slip which passes from the lower border of the tendon of the Latissimus dorsi, near its insertion, to the long head of the Triceps. This is occasionally muscular, and is the representative of the *dorso-epitrochlearis* muscle of apes.

Relations.—Its *superficial surface* is subcutaneous, excepting at its upper part, where it is covered by the Trapezius, and at its insertion, where its tendon is crossed by the axillary vessels and the brachial plexus of nerves. By its *deep surface*, it is in relation with the Lumbar fascia, the Serratus posticus inferior, the lower external intercostal muscles and ribs, inferior angle of the scapula, Rhomboideus major, Infraspinatus, and Teres major. Its outer margin is separated below from the External oblique by a small triangular interval, the triangle of Petit; and another triangular interval exists between its upper border and the margin of the Trapezius, in which the Rhomboideus major muscle is exposed.

Nerves.—The Trapezius is supplied by the spinal accessory, and by branches from the anterior divisions of the third and fourth cervical nerves; the Latissimus dorsi by the middle or long subscapular nerve.

Second Layer

Levator anguli scapulæ. Rhomboideus minor.
Rhomboideus major.

Dissection.—The Trapezius must be removed, in order to expose the next layer; to effect this, detach the muscle from its attachment to the clavicle and spine of the scapula, and turn it back towards the spine.

The **Levator anguli scapulæ** is situated at the back part and side of the neck. It arises by tendinous slips from the transverse process of the atlas and from the posterior tubercles of the transverse processes of the second, third, and fourth cervical vertebræ; these, becoming fleshy, are united so as to form a flat muscle, which, passing downwards and backwards, is inserted into the posterior border of the scapula, between the superior angle and the triangular smooth surface at the root of the spine.

Relations.—By its *superficial surface*, with the integument, Trapezius, and Sterno-mastoid. By its *deep surface*, with the Splenius colli, Transversalis cervicis, Cervicalis ascendens, and Serratus posticus superior muscles, and with the posterior scapular artery and the nerve to the Rhomboids.

The **Rhomboideus minor** arises from the ligamentum nuchæ and spinous processes of the seventh cervical and first dorsal vertebræ. Passing downwards and outwards, it is inserted into the margin of the triangular smooth surface at the root of the spine of the scapula. This small muscle is usually separated from the Rhomboideus major by a slight cellular interval.

Relations.—By its *superficial (posterior) surface*, with the Trapezius. By its *deep surface*, with the same structures as the Rhomboideus major.

The **Rhomboideus major** is situated immediately below the preceding, the adjacent margins of the two being occasionally united. It arises by tendinous fibres from the spinous processes of the four or five upper dorsal vertebræ and the supraspinous ligament, and is inserted into a narrow tendinous arch, attached above to the lower part of the triangular surface at the root of the spine; below, to the inferior angle, the arch being connected to the border of the scapula by a thin membrane. When the arch extends, as it occasionally does, only a short distance, the muscular fibres are inserted into the scapula itself.

Relations.—By its *superficial (posterior) surface*, with the Latissimus dorsi. By its *deep (anterior) surface*, with the Serratus posticus superior, posterior scapular artery, the vertebral aponeurosis which separates it from the prolongations of the Erector spinæ, the Intercostal muscles, and ribs.

Nerves.—The Rhomboid muscles are supplied by branches from the anterior division of the fifth cervical nerve ; the Levator anguli scapulæ by the anterior division of the third and fourth cervical nerves, and frequently by a branch from the nerve to the Rhomboids.

Actions.—The movements effected by the preceding muscles are numerous, as may be conceived from their extensive attachment. The whole of the *Trapezius* when in action retracts the scapula and braces back the shoulder ; if the head is fixed, the upper part of the Trapezius will elevate the point of the shoulder, as in supporting weights ; when the lower fibres are brought into action they assist in depressing the bone. The middle and lower fibres of the muscle rotate the scapula, causing elevation of the acromion process. If the shoulders are fixed, both Trapezii, acting together, will draw the head directly backwards ; or if only one acts, the head is drawn to the corresponding side.

The *Latissimus dorsi,* when it acts upon the humerus, depresses it, draws it backwards, adducts, and at the same time rotates it inwards. It is the muscle which is principally employed in giving a downward blow, as in felling a tree or in sabre practice. If the arm is fixed, the muscle may act in various ways upon the trunk : thus, it may raise the lower ribs and assist in forcible inspiration ; or, if both arms are fixed, the two muscles may assist the Abdominal and great Pectoral muscles in suspending and drawing the whole trunk forwards, as in climbing or walking on crutches.

The *Levator anguli scapulæ* raises the superior angle of the scapula, assisting the Trapezius in bearing weights or in shrugging the shoulders. If the shoulder be fixed, the Levator anguli scapulæ inclines the neck to the corresponding side and rotates it in the same direction. The *Rhomboid* muscles carry the inferior angle backwards and upwards, thus producing a slight rotation of the scapula upon the side of the chest, the Rhomboideus major acting especially on the lower angle of the scapula, through the tendinous arch by which it is inserted. The Rhomboid muscles acting together with the middle and inferior fibres of the Trapezius will draw the scapula directly backwards towards the spine.

THIRD LAYER

Serratus posticus superior.   Serratus posticus inferior.

Splenius { Splenius capitis.
           Splenius colli.

Dissection.—To bring into view the third layer of muscles, remove the whole of the second, together with the Latissimus dorsi, by cutting through the Levator anguli scapulæ and Rhomboid muscles near their origin, and reflecting them downwards, and by dividing the Latissimus dorsi in the middle by a vertical incision carried from its upper to its lower part, and reflecting the two halves of the muscle.

The **Serratus posticus superior** is a thin, flat, quadrilateral muscle, situated at the upper and back part of the thorax. It arises by a thin and broad aponeurosis from the ligamentum nuchæ, and from the spinous processes of the last cervical and two or three upper dorsal vertebræ and from the supraspinous ligament. Inclining downwards and outwards, it becomes muscular, and is inserted, by four fleshy digitations, into the upper borders of the second, third, fourth, and fifth ribs, a little beyond their angles.

Relations.—By its *superficial surface*, with the Trapezius, Rhomboidei, and Levator anguli scapulæ. By its *deep surface*, with the Splenius, and the vertebral aponeurosis, which separates it from the prolongations of the Erector spinæ, and with the Intercostal muscles and ribs.

The **Serratus posticus inferior** is situated at the junction of the dorsal and lumbar regions : it is of an irregularly quadrilateral form, broader than the preceding, and separated from it by a considerable interval. It arises by a thin aponeurosis from the spinous processes of the last two dorsal and two or three upper lumbar vertebræ, and from the supraspinous ligaments. Passing obliquely upwards and outwards, it becomes fleshy, and divides into four flat digitations, which are inserted into the lower borders of the four lower ribs, a little beyond their angles. The thin aponeurosis of origin is intimately blended with the lumbar fascia.

Relations.—By its *superficial surface*, with the Latissimus dorsi. By its *deep surface*, with the Erector spinæ, ribs, and Intercostal muscles. Its upper margin is continuous with the vertebral aponeurosis.

The *Vertebral aponeurosis* is a thin, fibrous lamina, extending along the whole length of the back part of the thoracic region, serving to bind down the long Extensor muscles of the back which support the spine and head, and separate them from those muscles which connect the spine to the upper extremity. It consists of longitudinal and transverse fibres blended together, forming a thin lamella, which is attached in the median line to the spinous processes of the dorsal vertebræ ; externally, to the angles of the ribs ; and below, to the upper border of the Serratus posticus inferior and portion of the lumbar fascia, which gives origin to the Latissimus dorsi ; above, it passes beneath the Serratus posticus superior and the Splenius, and blends with the deep fascia of the neck.

The *Lumbar fascia* or *aponeurosis* (fig. 213), which may be regarded as the posterior aponeurosis of the Transversalis abdominis muscle, consists of three laminæ, which are attached as follows : the posterior layer, to the spines of the lumbar and sacral vertebræ and their supraspinous ligaments ; the middle, to the tips of the transverse processes of the lumbar vertebræ and their intertransverse ligaments ; the anterior, to the roots of the lumbar transverse processes. The posterior layer is continued above as the vertebral aponeurosis, while inferiorly it is fixed to the outer lip of the iliac crest. With this layer are blended the aponeurotic origin of the Serratus posticus inferior and part of that of the Latissimus dorsi. The middle layer is attached above to the last rib, and below to the iliac crest ; the anterior layer is fixed below to the ilio-lumbar ligament and iliac crest ; while above it is thickened to form the external arcuate ligament of the diaphragm, and stretches from the tip of the last rib to the transverse process of the first or second lumbar vertebra. These three layers, together with the vertebral column, enclose two spaces, the posterior of which is occupied by the Erector spinæ muscle, and the anterior by the Quadratus lumborum.

Now detach the Serratus posticus superior from its origin, and turn it outwards, when the Splenius muscle will be brought into view.

The **Splenius** is situated at the back of the neck and upper part of the dorsal region. At its origin, it is a single muscle, narrow, and pointed in form ; but it soon becomes broader, and divides into two portions, which have separate insertions. It arises, by tendinous fibres, from the lower half of the ligamentum nuchæ, from the spinous processes of the last cervical and of the six upper dorsal vertebræ, and from the supraspinous ligament. From this origin, the fleshy fibres proceed obliquely upwards and outwards, forming a broad flat muscle, which divides as it ascends into two portions, the Splenius capitis and Splenius colli.

The **Splenius capitis** is inserted into the mastoid process of the Temporal bone, and into the rough surface on the occipital bone just beneath the superior curved line.

The **Splenius colli** is inserted, by tendinous fasciculi, into the posterior tubercles of the transverse processes of the two or three upper cervical vertebræ.

The Splenius is separated from its fellow of the opposite side by a triangular interval, in which is seen the Complexus.

Relations.—By its *superficial surface*, with the Trapezius, from which it is separated below by the Rhomboidei and the Serratus posticus superior. It is covered at its insertion by the Sterno-mastoid, and at the lower and back part of the neck by the Levator anguli scapulæ. By its *deep surface*, with the Spinalis dorsi, Longissimus dorsi, Semispinalis colli, Complexus, Trachelo-mastoid, and Transversalis cervicis.

Nerves.—The Splenius is supplied from the external branches of the posterior divisions of the cervical nerves ; the Serratus posticus superior is supplied by the external branches of the posterior divisions of the upper dorsal nerves ; the Serratus posticus inferior by the external branches of the posterior divisions of the lower dorsal nerves.

Actions.—The Serrati are respiratory muscles. The Serratus posticus superior elevates the ribs ; it is therefore an inspiratory muscle ; while the Serratus inferior draws the lower ribs downwards and backwards, and thus elongates the thorax. It also fixes the lower ribs, thus aiding the downward action of the diaphragm and resisting the tendency which it has to draw the lower ribs upwards and forwards. It must therefore be regarded as a muscle of inspiration. This muscle is also probably a tensor of the vertebral aponeurosis. The Splenii muscles of the two sides, acting together, draw the head directly backwards, assisting the Trapezius and Complexus ; acting separately, they draw the head to one or the other side, and slightly rotate it, turning the face to the same side. They also assist in supporting the head in the erect position.

Fourth Layer

I. Erector spinæ.

a. Outer Column.	*b. Middle Column.*
Ilio-costalis.	Longissimus dorsi.
Musculus accessorius.	Transversalis cervicis.
Cervicalis ascendens.	Trachelo-mastoid.

c. Inner Column.
Spinalis dorsi.

II. Complexus.

Dissection.—To expose the muscles of the fourth layer, remove entirely the Serrati and the vertebral and lumbar fasciæ. Then detach the Splenius by separating its attachment to the spinous processes and reflecting it outwards.

The **Erector spinæ** (fig. 214), and its prolongations in the dorsal and cervical regions, fill up the vertebral groove on each side of the spine. It is covered in the lumbar region by the lumbar fascia ; in the dorsal region by the Serrati muscles and the vertebral aponeurosis ; and in the cervical region by a layer of cervical fascia continued beneath the Trapezius and the Splenius. This large muscular and tendinous mass varies in size and structure at different parts of the spine. In the sacral region, the Erector spinæ is narrow and pointed, and its origin chiefly tendinous in structure. In the lumbar region, the muscle becomes enlarged, and forms a large fleshy mass. In the dorsal region, it subdivides into three parts, which gradually diminish in size as they ascend to be inserted into the vertebræ and ribs.

The Erector spinæ arises from the anterior surface of a very broad and thick tendon, which is attached, internally, to the spines of the sacrum, to the spinous processes of the lumbar and the eleventh and twelfth dorsal vertebræ, and the supraspinous ligament ; externally, to the back part of the inner lip of the crest of the ilium, and to the series of eminences on the posterior part of the sacrum, which represent the transverse processes, where it blends with the great sacro-sciatic and

Fig. 214.—Muscles of the back. Deep layers.

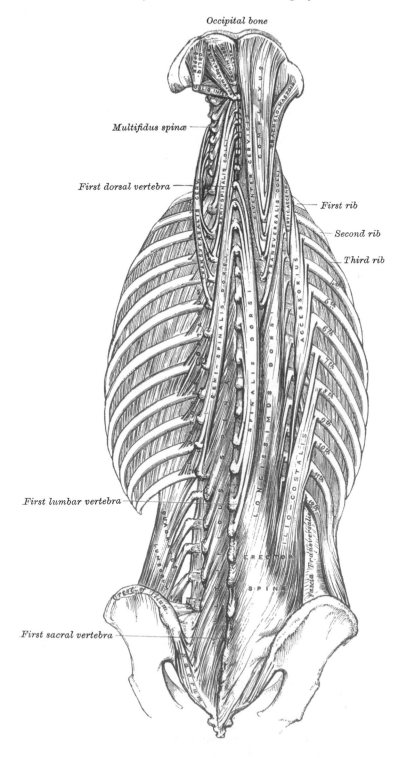

Occipital bone

Multifidus spinæ

First dorsal vertebra

First rib

Second rib

Third rib

First lumbar vertebra

First sacral vertebra

posterior sacro-iliac ligaments. Some of its fibres are continuous with the fibres of origin of the Gluteus maximus. The muscular fibres form a large fleshy mass, bounded in front by the transverse processes of the lumbar vertebræ, and by the middle lamella of the lumbar fascia. Opposite the last rib it divides into two parts, the Ilio-costalis and the Longissimus dorsi; the Spinalis dorsi is given off from the latter in the upper dorsal region.

The **Ilio-costalis** or **Sacro-lumbalis,** the external portion of the Erector spinæ, is inserted, generally, by six or seven flattened tendons, into the inferior borders of the angles of the six or seven lower ribs. The number of the tendons of this muscle is, however, very variable, and therefore the number of ribs into which it is inserted. Frequently it is found to possess nine or ten tendons, and sometimes as many tendons as there are ribs, and is then inserted into the angles of all the ribs. If this muscle is reflected outwards, it will be seen to be reinforced by a series of muscular slips, which arise from the angles of the ribs; by means of these the Ilio-costalis is continued upwards to the upper ribs, and the cervical portion of the spine. The accessory portions form two additional muscles, the Musculus accessorius and the Cervicalis ascendens.

The **Musculus accessorius ad ilio-costalem** arises, by separate flattened tendons, from the upper borders of the angles of the six lower ribs : these become muscular, and are finally inserted, by separate tendons, into the upper borders of the angles of the six upper ribs and into the back of the transverse process of the seventh cervical vertebra.

The **Cervicalis ascendens** * is the continuation of the Accessorius upwards into the neck; it is situated on the inner side of the tendons of the Accessorius, arising from the angles of the four or five upper ribs, and is inserted by a series of slender tendons into the posterior tubercles of the transverse processes of the fourth, fifth, and sixth cervical vertebræ.

The **Longissimus dorsi** is the middle and largest portion of the Erector spinæ. In the lumbar region, where it is as yet blended with the Ilio-costalis, some of the fibres are attached to the whole length of the posterior surface of the transverse processes and the accessory processes of the lumbar vertebræ, and to the middle layer of the lumbar fascia. In the dorsal region, the Longissimus dorsi is inserted, by long thin tendons, into the tips of the transverse processes of all the dorsal vertebræ, and into from seven to eleven of the lower ribs between their tubercles and angles. This muscle is continued upwards, to the cranium and cervical portion of the spine, by means of two additional muscles, the Transversalis cervicis and Trachelo-mastoid.

The **Transversalis cervicis** or **colli,** placed on the inner side of the Longissimus dorsi, arises by long thin tendons from the summits of the transverse processes of the six upper dorsal vertebræ, and is inserted by similar tendons into the posterior tubercles of the transverse processes of the cervical vertebræ from the second to the sixth inclusive.

The **Trachelo-mastoid** lies on the inner side of the preceding, between it and the Complexus muscle. It arises, by tendons, from the transverse processes of the five or six upper dorsal vertebræ, and the articular processes of the three or four lower cervical. The fibres form a small muscle, which ascends to be inserted into the posterior margin of the mastoid process, beneath the Splenius and Sterno-mastoid muscles. This small muscle is almost always crossed by a tendinous intersection near its insertion into the mastoid process.†

The **Spinalis dorsi** connects the spinous processes of the upper lumbar and the dorsal vertebræ together by a series of muscular and tendinous slips, which

---

* This muscle is sometimes called 'Cervicalis descendens.' The student should remember that these long muscles take their fixed point from above or from below, according to circumstances.

† These two muscles (Transversalis cervicis and Trachelo-mastoid) are sometimes described as one, having a common origin, but dividing above at their insertion. The Trachelo-mastoid is then termed the *Transversalis capitis.*

are intimately blended with the Longissimus dorsi. It is situated at the inner side of the Longissimus dorsi, arising, by three or four tendons, from the spinous processes of the first two lumbar and the last two dorsal vertebræ : these, uniting, form a small muscle, which is inserted, by separate tendons, into the spinous processes of the dorsal vertebræ, the number varying from four to eight. It is intimately united with the Semispinalis dorsi, which lies beneath it.

The *Spinalis colli* is a small muscle, connecting together the spinous processes of the cervical vertebræ, and analogous to the Spinalis dorsi in the dorsal region. It varies considerably in its size, and in its extent of attachment to the vertebræ, not only in different bodies, but on the two sides of the same body. It usually arises by fleshy or tendinous slips, varying from two to four in number, from the spinous processes of the fifth, sixth, and seventh cervical vertebræ, and occasionally from the first and second dorsal, and is inserted into the spinous process of the axis, and occasionally into the spinous processes of the two vertebræ below it. This muscle was found absent in five cases out of twenty-four.

Relations.—The Erector spinæ and its prolongations are bound down to the vertebræ and ribs in the lumbar and dorsal regions by the lumbar fascia and the vertebral aponeurosis. The inner part of these muscles covers the muscles of the fifth layer. In the neck, they are in relation, by their *superficial surface*, with the Trapezius and Splenius ; by their *deep surface*, with the Semispinalis dorsi et colli, and the Recti and Obliqui.

The **Complexus** is a broad thick muscle, situated at the upper and back part of the neck, beneath the Splenius, and internal to the Transversalis cervicis and Trachelo-mastoid. It arises, by a series of tendons, from the tips of the transverse processes of the upper six or seven dorsal and the last cervical vertebræ, and from the articular processes of the three cervical above this. The tendons, uniting, form a broad muscle, which passes obliquely upwards and inwards, and is inserted into the innermost depression between the two curved lines of the occipital bone. This muscle, about its middle, is traversed by a transverse tendinous intersection. The *Biventer cervicis* is a small fasciculus, situated on the inner side of the preceding, and in the majority of cases blended with it ; it has received its name from having a tendon intervening between two fleshy bellies. It is sometimes described as a part of the Complexus. It arises by from two to four tendinous slips, from the transverse processes of as many of the upper dorsal vertebræ, and is inserted, on the inner side of the Complexus, into the superior curved line of the occipital bone.

Relations.—The Complexus is covered by the Splenius and the Trapezius. It lies on the Rectus capitis posticus major and minor, the Obliquus capitis superior and inferior, and on the Semispinalis colli, from which it is separated by the profunda cervicis artery, the princeps cervicis artery, and branches of the posterior primary divisions of the cervical nerves. The Biventer cervicis is separated from its fellow of the opposite side by the ligamentum nuchæ.

Fifth Layer

Semispinalis dorsi.	Extensor coccygis.
Semispinalis colli.	Intertransversales.
Multifidus spinæ.	Rectus capitis posticus major.
Rotatores spinæ.	Rectus capitis posticus minor.
Supraspinales.	Obliquus capitis inferior.
Interspinales.	Obliquus capitis superior.

Dissection.—Remove the muscles of the preceding layer by dividing and turning aside the Complexus ; then detach the Spinalis and Longissimus dorsi from their attachments, divide the Erector spinæ at its connection below to the sacral and lumbar spines, and turn it outwards. The muscles filling up the interval between the spinous and transverse processes are then exposed.

The **Semispinalis dorsi** (fig. 214) consists of thin, narrow, fleshy fasciculi, interposed between tendons of considerable length. It arises by a series of small tendons from the transverse processes of the lower dorsal vertebræ, from the fifth or sixth to the tenth or eleventh·; and is inserted, by five or six tendons, into the spinous processes of the upper four dorsal and lower two cervical vertebræ.

The **Semispinalis colli,** thicker than the preceding, arises by a series of tendinous and fleshy fibres from the transverse processes of the upper five or six dorsal vertebræ, and is inserted into the spinous processes of four cervical vertebræ, from the axis to the fifth cervical. The fasciculus connected with the axis is the largest, and chiefly muscular in structure.

Relations.—By their *superficial surface*, from below upwards, with the Spinalis dorsi, Longissimus dorsi, Splenius, Complexus, the profunda cervicis artery, the princeps cervicis artery, and the internal branches of the posterior divisions of the first, second, and third cervical nerves. By their *deep surface*, with the Multifidus spinæ.

The **Multifidus spinæ** consists of a number of fleshy and tendinous fasciculi, which fill up the groove on either side of the spinous processes of the vertebræ, from the sacrum to the axis. In the sacral region, these fasciculi arise from the back of the sacrum, as low as the fourth sacral foramen, and from the aponeurosis of origin of the Erector spinæ ; from the inner surface of the posterior superior spine of the ilium, and posterior sacro-iliac ligaments ; in the lumbar regions, from the articular processes ; in the dorsal region, from the transverse processes ; and in the cervical region, from the articular processes of the three or four lower vertebræ. Each fasciculus, passing obliquely upwards and inwards, is inserted into the whole length of the spinous process of one of the vertebræ above. These fasciculi vary in length : the most superficial, the longest, pass from one vertebra to the third or fourth above ; those next in order pass from one vertebra to the second or third above ; while the deepest connect two contiguous vertebræ.

Relations.—By its *superficial surface*, with the Longissimus dorsi, Spinalis dorsi, Semispinalis dorsi, and Semispinalis colli. By its *deep surface*, with the laminæ and spinous processes of the vertebræ, and with the Rotatores spinæ in the dorsal region.

The **Rotatores spinæ** are found only in the dorsal region of the spine, beneath the Multifidus spinæ ; they are eleven in number on each side. Each muscle is small and somewhat quadrilateral in form ; it arises from the upper and back part of the transverse process, and is inserted into the lower border and outer surface of the lamina of the vertebra above, the fibres extending as far inwards as the root of the spinous process. The first is found between the first and second dorsal ; the last, between the eleventh and twelfth. Sometimes the number of these muscles is diminished by the absence of one or more from the upper or lower end.

The **Supraspinales** consist of a series of fleshy bands which lie on the spinous processes in the cervical region of the spine.

The **Interspinales** are short muscular fasciculi, placed in pairs between the spinous processes of the contiguous vertebræ, one on each side of the interspinous ligament. In the *cervical region*, they are most distinct, and consist of six pairs, the first being situated between the axis and third vertebra, and the last between the last cervical and the first dorsal. They are small narrow bundles, attached, above and below, to the apices of the spinous processes. In the *dorsal region*, they are found between the first and second vertebræ, and occasionally between the second and third ; and below, between the eleventh and twelfth. In the *lumbar region*, there are four pairs of these muscles in the intervals between the five lumbar vertebræ. There is also occasionally one in the interspinous space, between the last dorsal and first lumbar, and between the fifth lumbar and the sacrum.

The **Extensor coccygis** is a slender muscular fasciculus, occasionally present, which extends over the lower part of the posterior surface of the sacrum and coccyx. It arises by tendinous fibres from the last bone of the sacrum, or first piece of the coccyx, and passes downwards to be inserted into the lower part of the coccyx. It is a rudiment of the Extensor muscle of the caudal vertebræ of the lower animals.

The **Intertransversales** are small muscles placed between the transverse processes of the vertebræ. In the *cervical region* they are most developed, consisting of rounded muscular and tendinous fasciculi, which are placed in pairs, passing between the anterior and the posterior tubercles of the transverse processes of two contiguous vertebræ, separated from one another by the anterior division of the cervical nerve, which lies in the groove between them. In this region there are seven pairs of these muscles, the first pair being between the atlas and axis, and the last pair between the seventh cervical and first dorsal vertebræ. In the *dorsal region* they are least developed, consisting chiefly of rounded tendinous cords in the intertransverse spaces of the upper dorsal vertebræ ; but between the transverse processes of the lower three dorsal vertebræ, and between the transverse processes of the last dorsal and the first lumbar, they are muscular in structure. In the *lumbar region* they are arranged in pairs, on either side of the spine ; one set occupying the entire interspace between the transverse processes of the lumbar vertebræ, the *intertransversales laterales* ; the other set, *intertransversales mediales*, passing from the accessory process of one vertebra to the mammillary process of the next.

The **Rectus capitis posticus major** arises by a pointed tendinous origin from the spinous process of the axis, and, becoming broader as it ascends, is inserted into the inferior curved line of the occipital bone and the surface of bone immediately below it. As the muscles of the two sides pass upwards and outwards, they leave between them a triangular space, in which are seen the Recti capitis postici minores muscles.

Relations.—By its *superficial surface*, with the Complexus ; and, at its insertion, with the Superior oblique. By its *deep surface*, with part of the Rectus capitis posticus minor, the posterior arch of the atlas, the posterior occipito-atlantal ligament, and part of the occipital bone.

The **Rectus capitis posticus minor,** the smallest of the four muscles in this region, is of a triangular shape ; it arises by a narrow pointed tendon from the tubercle on the posterior arch of the atlas, and, becoming broader as it ascends, is inserted into the rough surface beneath the inferior curved line, nearly as far as the foramen magnum, nearer to the middle line than the preceding.

Relations.—By its *superficial surface*, with the Complexus and the Rectus capitis posticus major. By its *deep surface*, with the posterior occipito-atlantal ligament.

The **Obliquus capitis inferior,** the larger of the two Oblique muscles, arises from the apex of the spinous process of the axis, and passes outwards and slightly upwards, to be inserted into the lower and back part of the transverse process of the atlas.

Relations.—By its *superficial surface*, with the Complexus and with the posterior division of the second cervical nerve which crosses it. By its *deep surface*, with the vertebral artery, and posterior atlanto-axial ligament.

The **Obliquus capitis superior,** narrow below, wide and expanded above, arises by tendinous fibres from the upper surface of the transverse process of the atlas, joining with the insertion of the preceding, and, passing obliquely upwards and inwards, is inserted into the occipital bone, between the two curved lines, external to the Complexus.

Relations.—By its *superficial surface*, with the Complexus and Trachelo-mastoid and occipital artery. By its *deep surface*, with the posterior occipito-atlantal ligament.

The Suboccipital triangle.—Between the two oblique muscles and the Rectus capitis posticus major a triangular interval exists, the *suboccipital triangle*. This triangle is bounded, above and internally, by the Rectus capitis posticus major ;

above and externally, by the Obliquus capitis superior; below and externally, by the Obliquus capitis inferior. It is covered in by a layer of dense fibro-fatty tissue, situated beneath the Complexus muscle. The floor is formed by the posterior occipito-atlantal ligament and the posterior arch of the atlas. It contains the vertebral artery, as it runs in a deep groove on the upper surface of the posterior arch of the atlas; and the posterior division of the suboccipital nerve.

Nerves.—The third, fourth, and fifth layers of the muscles of the back are supplied by the posterior primary divisions of the spinal nerves.

Actions.—When both the Spinales dorsi contract, they extend the dorsal region of the spine; when only one muscle contracts, it helps to bend the dorsal portion of the spine to one side. The Erector spinæ, comprising the Ilio-costalis and the Longissimus dorsi with their accessory muscles, serves, as its name implies, to maintain the spine in the erect posture; it also serves to bend the trunk backwards when it is required to counterbalance the influence of any weight at the front of the body—as, for instance, when a heavy weight is suspended from the neck, or when there is any great abdominal distension, as in pregnancy or dropsy; the peculiar gait under such circumstances depends upon the spine being drawn backwards, by the counterbalancing action of the Erector spinæ muscles. The muscles which form the continuation of the Erector spinæ upwards steady the head and neck, and fix them in the upright position. If the Ilio-costalis and Longissimus dorsi of one side act, they serve to draw down the chest and spine to the corresponding side. The Cervicales ascendentes, taking their fixed points from the cervical vertebræ, elevate those ribs to which they are attached; taking their fixed points from the ribs, both muscles help to extend the neck; while one muscle bends the neck to its own side. The Transversales cervicis, when both muscles act, taking their fixed points from below, bend the neck backwards. The Trachelo-mastoid, when both muscles act, taking their fixed points from below, bend the head backwards; while, if only one muscle acts, the face is turned to the side on which the muscle is acting, and then the head is bent to the shoulder. The two Recti muscles draw the head backwards. The Rectus capitis posticus major, owing to its obliquity, rotates the cranium, with the atlas, round the odontoid process, turning the face to the same side. The Multifidus spinæ acts successively upon the different parts of the spine; thus, the sacrum furnishes a fixed point from which the fasciculi of this muscle act upon the lumbar region; these then become the fixed points for the fasciculi moving the dorsal region, and so on throughout the entire length of the spine; it is by the successive contraction and relaxation of the separate fasciculi of this and other muscles that the spine preserves the erect posture without the fatigue that would necessarily have been produced had this position been maintained by the action of a single muscle. The Multifidus spinæ, besides preserving the erect position of the spine, serves to rotate it, so that the front of the trunk is turned to the side opposite to that from which the muscle acts, this muscle being assisted in its action by the Obliquus externus abdominis. The Complexi draw the head directly backwards; if one muscle acts, it draws the head to one side, and rotates it so that the face is turned to the opposite side. The Superior oblique draws the head backwards; and, from the obliquity in the direction of its fibres, will slightly rotate the cranium, turning the face to the opposite side. The Obliquus capitis inferior rotates the atlas, and with it the cranium, round the odontoid process, turning the face to the same side. The Semispinales, when the muscles of the two sides act together, help to extend the spine; when the muscles of one side only act, they rotate the dorsal and cervical parts of the spine, turning the body to the opposite side. The Supraspinales and Interspinales by approximating the spinous process help to extend the spine. The Intertransversales approximate the transverse processes, and help to bend the spine to one side. The Rotatores spinæ assist the Multifidus spinæ to rotate the spine, so that the front of the trunk is turned to the side opposite to that from which the muscle acts.

Surface Form.—The surface forms produced by the muscles of the back are numerous and difficult to analyse unless they are considered in systematic order. The most superficial layer, consisting of large strata of muscular substance, influences to a certain extent the surface form, and at the same time reveals the forms of the layers beneath. The *Trapezius* at the upper part of the back, and in the neck, covers over and softens down the outline of the underlying muscles. Its anterior border forms the posterior boundary of the posterior triangle of the neck. It forms a slight undulating ridge which passes downwards and forwards from the occiput to the junction of the middle and outer third of the clavicle. The tendinous ellipse formed by a part of the origin of the two muscles at the back of the neck is always to be seen as an oval depression, more marked when the muscle is in action. A slight dimple on the skin opposite the interval between the spinous processes of the third and fourth dorsal vertebræ marks the triangular aponeurosis by which the inferior fibres are inserted into the root of the spine of the scapula. From this point the inferior border of the muscle may be traced as an undulating ridge to the spinous process of the twelfth dorsal vertebra. In like manner, the *Latissimus dorsi* softens down and modulates the underlying structures at the lower part of the back and lower part of the side of the chest. In this way it modulates the outline of the Erector spinæ; of the Serratus posticus inferior, which is sometimes to be discerned through it, and is sometimes entirely obscured by it; of part of the Serratus magnus and Superior oblique, which it covers, and of the convex oblique ridges formed by the ribs with the intervening intercostal spaces. The anterior border of the muscle is the only part which gives a distinct surface form. This border may be traced, when the muscle is in action, as a rounded edge, starting from the crest of the ilium, and passing obliquely forwards and upwards to the posterior border of the axilla, where it combines with the Teres major in forming a thick rounded fold, the posterior boundary of the axillary space. The muscles in the second layer influence to a very considerable extent the surface form of the back of the neck and upper part of the trunk. The *Levator anguli scapulæ* reveals itself as a prominent divergent line, running downwards and outwards, from the transverse processes of the upper cervical vertebræ to the angle of the scapula, covered over and toned down by the overlying Trapezius. The *Rhomboidei* produce, when in action, a vertical eminence between the vertebral border of the scapula and the spinal furrow, varying in intensity according to the condition of contraction or relaxation of the Trapezius muscle, by which they are for the most part covered. The lowermost part of the Rhomboideus major is uncovered by the Trapezius and forms on the surface an oblique ridge running upwards and inwards from the inferior angle of the scapula. Of the muscles of the third layer of the back, the *Serratus posticus superior* does not in any way influence surface form. The *Serratus posticus inferior*, when in strong action, may occasionally be revealed as an elevation beneath the Latissimus dorsi. The *Splenii* by their divergence serve to broaden out the upper part of the back of the neck and produce a local fulness in this situation, but do not otherwise influence surface form. Beneath all these muscles those of the fourth layer—the *Erector spinæ* and its continuations—influence the surface form in a decided manner. In the loins, the Erector spinæ, bound down by the lumbar fascia, forms a rounded vertical eminence, which determines the depth of the spinal furrow, and which below tapers to a point on the posterior surface of the sacrum and becomes lost there. In the back it forms a flattened plane which gradually becomes lost. In the neck the only part of this group of muscles which influences surface form is the *Trachelomastoid*, which produces a short convergent line across the upper part of the posterior triangle of the neck, appearing from under cover of the posterior border of the Sternomastoid and being lost below beneath the Trapezius.

II. Muscles and Fasciæ of the Thorax

The Muscles belonging exclusively to this region are few in number. They are, the

Intercostales externi.	Triangularis sterni.
Intercostales interni.	Levatores costarum.
Infracostales.	Diaphragm.

Intercostal fasciæ.—A thin but firm layer of fascia covers the outer surface of the External intercostal and the inner surface of the Internal intercostal muscles; and a third layer, the middle intercostal fascia, more delicate, is interposed between the two planes of muscular fibres. These are the intercostal fasciæ: they are best marked in those situations where the muscular fibres are deficient, as between the External intercostal muscles and sternum, in front; and between the Internal intercostals and spine, behind.

The **Intercostal muscles** (fig. 230) are two thin planes of muscular and tendinous fibres, placed one over the other, filling up the intercostal spaces, and being directed obliquely between the margins of the adjacent ribs. They have received the name 'external' and 'internal,' from the position they bear to one another. The tendinous fibres are longer and more numerous than the muscular : hence the walls of the intercostal spaces possess very considerable strength, to which the crossing of the muscular fibres materially contributes.

The **External intercostals** are eleven in number on each side. They extend from the tubercles of the ribs, behind, to the commencement of the cartilages of the ribs, in front, where they terminate in a thin membrane, the anterior intercostal membrane, which is continued forwards to the sternum. They arise from

FIG. 215.—Posterior surface of sternum and costal cartilages, showing Triangularis sterni muscle. (From a preparation in the Museum of the Royal College of Surgeons of England.)

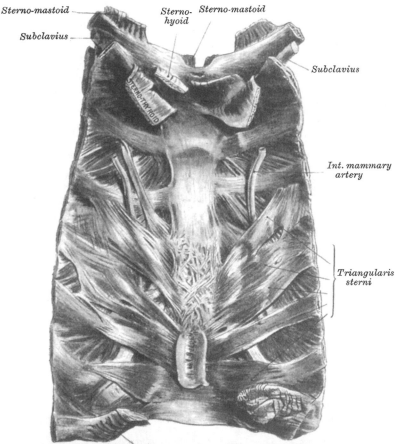

Sterno-mastoid

Sterno-hyoid   *Sterno-mastoid*

Subclavius

Sterno-mastoid

Subclavius

Int. mammary artery

Triangularis sterni

Transversalis abdominis

the lower border of each rib, and are inserted into the upper border of the rib below. In the two lowest spaces they extend to the end of the cartilages, and in the upper two or three spaces they do not quite extend to the ends of the ribs. Their fibres are directed obliquely downwards and forwards, in a similar direction to those of the External oblique muscle of the abdomen. They are thicker than the internal intercostals.

Relations.—By their *outer surface*, with the muscles which immediately invest the chest, viz. the Pectoralis major and minor, Serratus magnus, and Rhomboideus major, Serratus posticus superior and inferior, Scalenus posticus, Ilio-costalis, Longissimus dorsi, Cervicalis ascendens, Transversalis cervicis, Levatores costarum, Obliquus externus abdominis, and the Latissimus dorsi. By their *internal surface*, with the middle intercostal fascia, which separates them from the intercostal vessels and nerve, and the Internal intercostal muscles, and, behind, from the pleura.

The **Internal intercostals** are also eleven in number on each side. They commence anteriorly at the sternum, in the interspaces between the cartilages of the true ribs, and from the anterior extremities of the cartilages of the false ribs, and extend backwards as far as the angles of the ribs ; whence they are continued to the vertebral column by a thin aponeurosis, the posterior intercostal membrane. They arise from the ridge on the inner surface of each rib, as well as from the corresponding costal cartilage, and are inserted into the upper border of the rib below. Their fibres are directed obliquely downwards and backwards, passing in the opposite direction to the fibres of the External intercostal muscles.

Relations.—By their *external surface*, with the intercostal vessels and nerves, and the External intercostal muscles ; near the sternum, with the anterior intercostal membrane, and the Pectoralis major. By their *internal surface*, with the pleura costalis, Triangularis sterni, and Diaphragm.

The **Infracostales (subcostales)** consist of muscular and aponeurotic fasciculi, which vary in number and length : they are placed on the inner surface of the ribs, where the Internal intercostal muscles cease ; they arise from the inner surface of one rib, and are inserted into the inner surface of the first, second, or third rib below. Their direction is most usually oblique, like the Internal intercostals. They are most frequent between the lower ribs.

The **Triangularis sterni** (fig. 215) is a thin plane of muscular and tendinous fibres, situated upon the inner wall of the front of the chest. It arises from the lower third of the posterior surface of the sternum, from the posterior surface of the ensiform cartilage, and from the sternal ends of the costal cartilages of the three or four lower true ribs. Its fibres diverge upwards and outwards, to be inserted by digitations into the lower border and inner surfaces of the costal cartilages of the second, third, fourth, fifth, and sixth ribs. The lowest fibres of this muscle are horizontal in their direction, and are continuous with those of the Transversalis : those which succeed are oblique, while the superior fibres are almost vertical. This muscle varies much in its attachment, not only in different bodies, but on opposite sides of the same body.

Relations.—*In front*, with the sternum, ensiform cartilage, costal cartilages, Internal intercostal muscles, and internal mammary vessels. *Behind*, with the pleura, pericardium, and anterior mediastinum.

The **Levatores costarum** (fig. 214), twelve in number on each side, are small tendinous and fleshy bundles, which arise from the extremities of the transverse processes of the seventh cervical and eleven upper dorsal vertebræ, and, passing obliquely downwards and outwards, are inserted into the upper border of the rib below them, between the tubercle and the angle. The inferior Levatores divide into two fasciculi, one of which is inserted as above described ; the other fasciculus passes down to the second rib below its origin ; thus, each of the lower ribs receives fibres from the transverse processes of two vertebræ.

Nerves.—The muscles of this group are supplied by the intercostal nerves.

The **Diaphragm** (διάφραγμα, *a partition wall*) (fig. 216) is a thin musculo-fibrous septum, consisting of muscular fibres externally, which arise from the circumference of the thoracic cavity and pass upwards and inwards to converge to a central tendon. It is placed obliquely at the junction of the upper with the middle third of the trunk, and separates the thorax from the abdomen, forming the floor of the former cavity and the roof of the latter. It is elliptical, its longest diameter being from side to side, and somewhat fan-shaped ; the broad elliptical portion being horizontal, the narrow part, the *crura*, which represents the handle of the fan, vertical, and joined at right angles to the former. It is from this circumstance that some anatomists describe it as consisting of two portions, the upper or great muscle of the Diaphragm, and the lower or lesser muscle. It arises from the whole of the internal circumference of the thorax :

being attached, in front, by fleshy fibres to the ensiform cartilage; on either side, to the inner surface of the cartilages and bony portions of the six or seven inferior ribs, interdigitating with the Transversalis; and behind, to two aponeurotic arches, named the *ligamentum arcuatum externum et internum*; and by the crura, to the lumbar vertebræ. The fibres from these sources vary in length; those arising from the ensiform appendix are very short and occasionally aponeurotic; those from the ligamenta arcuata, and more especially those from the cartilages of the ribs at the side of the chest, are longer, describe well-marked curves as they ascend, and finally converge to be inserted into the circumference of the central tendon. Between the sides of the muscular slip from the ensiform appendix and the cartilages of the adjoining ribs, the fibres of the Diaphragm are deficient, the interval being filled by areolar tissue, covered on the thoracic

Fig. 216.—The Diaphragm. Under surface.

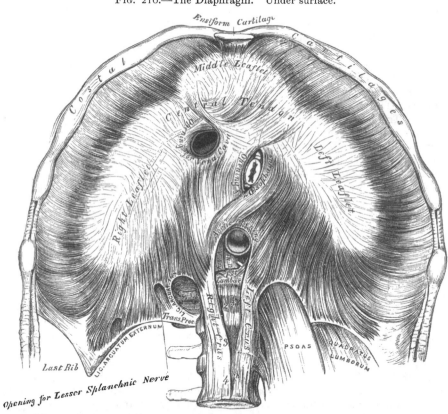

side by the pleuræ; on the abdominal, by the peritoneum.   This is, consequently, a weak point, and a portion of the contents of the abdomen may protrude into the chest, forming phrenic or diaphragmatic hernia, or a collection of pus in the mediastinum may descend through it, so as to point at the epigastrium.   A triangular gap is sometimes seen between the fibres springing from the internal and those arising from the external arcuate ligament.   When it exists the kidney is only separated from the pleura by fatty and areolar tissue.

The *ligamentum arcuatum internum* is a tendinous arch, thrown across the upper part of the Psoas magnus muscle, on each side of the spine.   It is connected, by one end, to the outer side of the body of the first or second lumbar vertebra, being continuous with the outer side of the tendon of the corresponding crus; and, by the other end, to the front of the transverse process of the first and sometimes also to that of the second, lumbar vertebra.

The *ligamentum arcuatum externum* is the thickened upper margin of the anterior lamella of the lumbar fascia ; it arches across the upper part of the Quadratus lumborum, being attached, by one extremity, to the front of the transverse process of the first lumbar vertebra ; and, by the other, to the apex and lower margin of the last rib.

The Crura.—The Diaphragm is connected to the spine by two *crura* or *pillars*, which are situated on the bodies of the lumbar vertebræ, on each side of the aorta. The crura, at their origin, are tendinous in structure : the right crus, larger and longer than the left, arising from the anterior ·surface of the bodies and intervertebral substances of the three or four upper lumbar vertebræ ; the left, from the two upper ; both blending with the anterior common ligament of the spine. These tendinous portions of the crura pass forwards and inwards and gradually converge to meet in the middle line, forming an arch, beneath which passes the aorta, vena azygos major, and thoracic duct. From this tendinous arch muscular fibres arise, which diverge, the outermost portion being directed upwards and outwards to the central tendon ; the innermost decussating in front of the aorta, and then diverging, so as to surround the œsophagus before ending in the central tendon. The fibres derived from the right crus are the most numerous, and pass in front of those derived from the left.

The *Central* or *Cordiform Tendon* of the Diaphragm is a thin but strong tendinous aponeurosis, situated at the centre of the vault formed by the muscle, immediately below the pericardium, with which it is partly blended. It is shaped somewhat like a trefoil leaf, consisting of three divisions, or leaflets, separated from one another by slight indentations. The right leaflet is the largest ; the middle one, directed towards the ensiform cartilage, the next in size ; and the left, the smallest. In structure, the tendon is composed of several planes of fibres, which intersect one another at various angles, and unite into straight or curved bundles—an arrangement which affords it additional strength.

The *Openings* connected with the Diaphragm are three large and several smaller apertures. The former are the aortic, the œsophageal, and the opening for the vena cava.

The *aortic opening* is the lowest and the most posterior of the three large apertures connected with this muscle, being at the level of the first lumbar vertebra. It is situated slightly to the left of the middle line, immediately in front of the bodies of the vertebræ ; and is, therefore, *behind* the Diaphragm, not in it. It is an osseo-aponeurotic aperture, formed by a tendinous arch thrown across the front of the bodies of the vertebræ, from the crus on one side to that on the other, and transmits the aorta, vena azygos major, and thoracic duct. Sometimes the vena azygos major is transmitted upwards through the right crus. Occasionally some tendinous fibres àre prolonged across the bodies of the vertebræ from the inner part of the lower end of the crura, passing behind the aorta, and thus converting the opening into a fibrous ring.

The *œsophageal opening* is situated at the level of the tenth dorsal vertebra ; it is elliptical in form, muscular in structure, and, formed by the decussating fibres of the two crura, is placed above, and, at the same time, anterior, and a little to the left of the preceding. It transmits the œsophagus and pneumogastric nerves and some small œsophageal arteries. The anterior margin of this aperture is occasionally tendinous, being formed by the margin of the central tendon.

The *opening for the vena cava* (*foramen quadratum*) is the highest, about on the level of the disc between the eighth and ninth dorsal vertebræ ; it is quadrilateral in form, tendinous in structure, and placed at the junction of the right and middle leaflets of the central tendon, its margins being adherent to the wall of the inferior vena cava.

The *right crus* transmits the greater and lesser splanchnic nerves of the right side ; the *left crus* transmits the greater and lesser splanchnic nerves of the left side, and the vena azygos minor. The gangliated cords of the sympathetic usually enter the abdominal cavity by passing behind the internal arcuate ligaments.

The *Serous Membranes* in relation with the Diaphragm are four in number : three lining its upper or thoracic surface ; one, its abdominal. The three serous membranes on its upper surface are the pleura on either side, and the serous layer of the pericardium, which covers the middle portion of the tendinous centre. The serous membrane covering its under surface is a portion of the general peritoneal membrane of the abdominal cavity.

The Diaphragm is arched, being convex towards the chest, and concave to the abdomen. The *right portion* forms a complete arch from before backwards, being accurately moulded over the convex surface of the liver, and having resting upon it the concave base of the right lung. The *left portion* is arched from before backwards in a similar manner ; but the arch is narrower in front, being encroached upon by the pericardium, and, at its summit, lower than the right by about three-quarters of an inch. It supports the base of the left lung, and covers the great end of the stomach, the spleen, and left kidney. At its circumference the Diaphragm is higher in the mesial line of the body than at either side ; but in the middle of the thorax, the central portion, which supports the heart, is on a lower level than the two lateral portions.

Nerves.—The Diaphragm is supplied by the phrenic and lower intercostal nerves and phrenic plexus of the sympathetic.

Actions.—The Intercostals are the chief agents in the movement of the ribs in ordinary respiration. When the first rib is elevated and fixed by the Scaleni, the External intercostals raise the other ribs, especially their fore part, and so increase the capacity of the chest from before backwards ; at the same time they evert their lower borders, and so enlarge the thoracic cavity transversely. The Internal intercostals, at the side of the thorax, depress the ribs, and invert their lower borders, and so diminish the thoracic cavity ; but at the fore part of the chest these muscles assist the External intercostals in raising the cartilages.* The Levatores costarum assist the External intercostals in raising the ribs. The Triangularis sterni draws down the costal cartilages ; it is therefore an expiratory muscle.

The Diaphragm is the principal muscle of inspiration. When in a condition of rest the muscle presents a domed surface, concave towards the abdomen ; and consists of a circumferential muscular and a central tendinous part. When the muscular fibres contract, they become less arched, or nearly straight, and thus cause the central tendon to descend, and in consequence the level of the chest wall is lowered, the vertical diameter of the chest being proportionately increased. In this descent the different parts of the tendon move unequally. The left leaflet descends to the greatest extent ; the right to a less extent, on account of the liver ; and the central leaflet the least, because of its connection to the pericardium. In descending the Diaphragm presses on the abdominal viscera, and so to a certain extent causes a projection of the abdominal wall ; but in consequence of these viscera not yielding completely, the central tendon becomes a

---

* The view of the action of the Intercostal muscles given in the text is that which is taught by Hutchinson (*Cycl. of Anat. and Phys.* art. 'Thorax'), and is usually adopted in our schools. It is, however, much disputed. Hamberger believed that the External intercostals act as elevators of the ribs, or muscles of inspiration, while the Internal act in expiration. Haller taught that both sets of muscles act in common—viz. as muscles of inspiration—and this view is adopted by many of the best anatomists of the Continent, and appears supported by many observations made on the human subject under various conditions of disease, and on living animals after the muscles have been exposed under chloroform. The reader may consult an interesting paper by Dr. Cleland in the *Journal of Anat. and Phys.* No. II., May 1867, p. 209, 'On the Hutchinsonian Theory of the Action of the Intercostal Muscles.' He refers to Henle, Luschka, Budge, and Bäumler, *Observations on the Action of the Intercostal Muscles*, Erlangen, 1860. (In *New Syd. Soc.'s Year-Book* for 1861, p. 69.) Dr. W. W. Keen has come to the conclusion, from experiments made upon a criminal executed by hanging, that the External intercostals are muscles of expiration, as they pulled the ribs down, while the Internal intercostals pulled the ribs up, and are muscles of inspiration. (*Trans. Coll. Phys. Philadelphia*, Third Series, vol. i., 1875, p. 97.)

fixed point, and enables the circumferential muscular fibres to act *from* it, and so elevate the lower ribs and expand the lower part of the thoracic cavity ; and Duchenne has shown that the Diaphragm has the power of elevating the ribs, to which it is attached, by its contraction, if the abdominal viscera are *in situ*, but that if these organs are removed, this power is lost. When at the end of inspiration, the Diaphragm relaxes, the thoracic walls return to their natural position in consequence of their elastic reaction and of the elasticity and weight of the displaced viscera.*

In all expulsive acts the Diaphragm is called into action, to give additional power to each expulsive effort. Thus, before sneezing, coughing, laughing, and crying ; before vomiting ; previous to the expulsion of the urine and fæces, or of the fœtus from the womb, a deep inspiration takes place.

The height of the Diaphragm is constantly varying during respiration, the muscles being carried upwards or downwards from the average level ; its height also varies according to the degree of distension of the stomach and intestines, and the size of the liver. After a forced expiration, the right arch is on a level, in front, with the fourth costal cartilage ; at the side, with the fifth, sixth, and seventh ribs ; and behind, with the eighth rib ; the left arch being usually from one to two ribs' breadth below the level of the right one. In a forced inspiration, it descends from one to two inches ; its slope would then be represented by a line drawn from the ensiform cartilage towards the tenth rib.

Muscles of Inspiration and Expiration.—The muscles which assist the action of the Diaphragm in ordinary tranquil inspiration are the Intercostals and the Levatores costarum, as above stated, and the Scaleni. When the need for more forcible action exists, the shoulders and the base of the scapula are fixed, and then the powerful muscles of forced inspiration come into play ; the chief of these are the Trapezius, the Pectoralis minor, the Serratus posticus superior and inferior, and the Rhomboidei. The lower fibres of the Serratus magnus may possibly assist slightly in dilating the chest by raising and everting the ribs. The Sterno-mastoid also, when the head is fixed, assists in forced inspiration, by drawing up the sternum, and by fixing the clavicle, and thus affording a fixed point for the action of the muscles of the chest. The Ilio-costalis and Quadratus lumborum assist in forced inspiration by fixing the last rib (see page 341).

The ordinary action of expiration is hardly effected by muscular force, but results from a return of the walls of the thorax to a condition of rest, owing to their own elasticity and to that of the lungs. Forced expiratory actions are performed mainly by the flat muscles (Obliqui and Transversalis) of the abdomen, assisted also by the Rectus. Other muscles of forced expiration are the Internal intercostals and the Triangularis sterni (as above mentioned).

III. Muscles of the Abdomen

The muscles of the abdomen may be divided into two groups : 1. The superficial muscles of the abdomen ; . 2. The deep muscles of the abdomen.

1. Superficial Muscles

The Muscles in this region are, the

Obliquus Externus.	Transversalis.
Obliquus Internus.	Rectus.
Pyramidalis.	

Dissection (fig. 217).—To dissect the abdominal muscles, make a vertical incision from the ensiform cartilage to the symphysis pubis ; a second incision from the umbilicus obliquely upwards and outwards to the outer surface of the chest, as high as the lower

---

* For a detailed description of the general relations of the Diaphragm, and its action, refer to Dr. Sibson's *Medical Anatomy*.

border of the fifth or sixth rib ; and a third, commencing midway between the umbilicus and pubes, transversely outwards to the anterior superior iliac spine, and along the crest of the ilium as far as its posterior third. Then reflect the three flaps included between these incisions from within outwards, in the lines of direction of the muscular fibres. If necessary, the abdominal muscle may be made tense by inflating the peritoneal cavity through the umbilicus.

The **Superficial fascia** of the abdomen consists over the greater part of the abdominal wall of a single layer of fascia, which contains a variable amount of fat ; but as this layer approaches the groin it is easily divisible into two, between which are found the superficial vessels and nerves and the superficial inguinal lymphatic glands. The superficial layer (*fascia of Camper*) is thick, areolar in texture, containing adipose tissue in its meshes, the quantity of which varies in different subjects. Below it passes over Poupart's ligament, and is continuous with the outer layer of the superficial fascia of the thigh. In the male this fascia is continued over the penis and outer surface of the cord to the scrotum, where it helps to form the dartos. As it passes to the scrotum it changes its character, becoming thin, destitute of adipose tissue, and of a pale reddish colour, and in the scrotum it acquires some involuntary muscular fibres. From the scrotum it may be traced backwards to be continuous with the superficial fascia of the perinæum. In the female, this fascia is continued into the labia majora. The deeper layer (*fascia of Scarpa*) is thinner and more membranous in character than the superficial layer. In the middle line, it is intimately adherent to the linea alba and to the symphysis pubis, and is prolonged on to the dorsum of the penis, forming the suspensory ligament ; above, it is continuous with the superficial fascia over the rest of the trunk ; below, it blends with the fascia lata of the thigh a little below Poupart's ligament ; and below and internally, it is continued over the penis and spermatic cord to the scrotum, where it helps to form the dartos. From the scrotum it may be traced backwards to be continuous with the deep layer of the superficial fascia of the perinæum. In the female, it is continued into the labia majora.

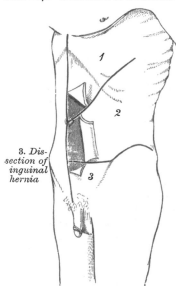

Fig. 217.—Dissection of abdomen.

3. *Dissection of inguinal hernia*

The **External** or **Descending oblique muscle** (fig. 218) is situated on the side and fore part of the abdomen ; being the largest and the most superficial of the three flat muscles in this region. It is broad, thin, and irregularly quadrilateral, its muscular portion occupying the side, its aponeurosis the anterior wall of the abdomen. It arises, by eight fleshy digitations, from the external surface and lower borders of the eight inferior ribs ; these digitations are arranged in an oblique line running downwards and backwards ; the upper ones being attached close to the cartilages of the corresponding ribs ; the lowest to the apex of the cartilage of the last rib ; the intermediate ones, to the ribs at some distance from their cartilages. The five superior serrations increase in size from above downwards, and are received between corresponding processes of the Serratus magnus ; the three lower ones diminish in size from above downwards, receiving between them corresponding processes from the Latissimus dorsi. From these attachments the fleshy fibres proceed in various directions. Those from the lowest ribs pass nearly vertically downwards, to be inserted into the anterior half of the outer lip of the crest of the ilium ; the middle and upper fibres, directed downwards and forwards, terminate in an aponeurosis, opposite a line drawn from the prominence of the ninth costal cartilage to the anterior superior spinous process of the ilium.

The **Aponeurosis of the External oblique** is a thin but strong membranous aponeurosis, the fibres of which are directed obliquely downwards and inwards. It is joined with that of the opposite muscle along the median line, and covers the whole of the front of the abdomen ; above, it is connected with the lower border of the Pectoralis major; below, its fibres are closely aggregated together, and

FIG. 218.—The External oblique muscle.

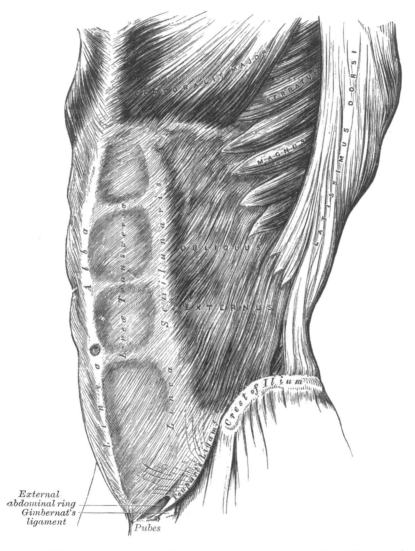

extend obliquely across from the anterior superior spine of the ilium to the spine of the os pubis and the linea ilio-pectinea.   In the median line, it interlaces with the aponeurosis of the opposite muscle, forming the linea alba, which extends from the ensiform cartilage to the symphysis pubis.

That portion of the aponeurosis which extends between the anterior superior spine of the ilium and the spine of the os pubis is a broad band, folded inwards, and continuous below with the fascia lata; it is called *Poupart's ligament*.   The portion which is reflected from Poupart's ligament at the spine of the os pubis

along the pectineal line is called *Gimbernat's ligament*. From the point of attachment of the latter to the pectineal line, a few fibres pass upwards and inwards,
behind the inner pillar of the ring, to the linea alba. They diverge as they
ascend, and form a thin triangular fibrous layer, which is called the *triangular
fascia of the abdomen*.

In the aponeurosis of the External oblique, immediately above the crest of
the os pubis, is a triangular opening, the *external abdominal ring*, formed by a
separation of the fibres of the aponeurosis in this situation.

Relations.—By its *external surface*, with the superficial fascia, superficial
epigastric and circumflex iliac vessels, and some cutaneous nerves. By its
internal surface, with the Internal oblique, the lower part of the eight inferior
ribs, and Intercostal muscles, the Cremaster, the spermatic cord in the male, and
round ligament in the female. Its *posterior border*, extending from the last rib
to the crest of the ilium, is fleshy throughout and free ; it is occasionally overlapped by the Latissimus dorsi, though generally a triangular interval exists
between the two muscles near the crest of the ilium, in which is seen a portion
of the Internal oblique. This triangle, *Petit's triangle*, is therefore bounded in
front by the External oblique, behind by the Latissimus dorsi, below by the
crest of the ilium, while its floor is formed by the Internal oblique (fig. 213).

The following parts of the aponeurosis of the External oblique muscle require
to be further described, viz. the external abdominal ring, the intercolumnar fibres
and fascia, Poupart's ligament, Gimbernat's ligament, and the triangular fascia
of the abdomen.

The **External abdominal ring.**—Just above, and to the outer side of the
crest of the os pubis, an interval is seen in the aponeurosis of the External
oblique, called the *External abdominal ring*. The aperture is oblique in direction,
somewhat triangular in form, and corresponds with the course of the fibres of
the aponeurosis. It usually measures from base to apex about an inch, and
transversely about half an inch. It is bounded below by the crest of the os pubis ;
above, by a series of curved fibres, the *intercolumnar*, which pass across the
upper angle of the ring, so as to increase its strength ; and on each side, by the
margins of the opening in the aponeurosis, which are called the columns or
pillars of the ring.

The external pillar, which is at the same time inferior from the obliquity of
its direction, is the stronger ; it is formed by that portion of Poupart's ligament
which is inserted into the spine of the os pubis ; it is curved so as to form a kind
of groove, upon which the spermatic cord rests. The internal or superior pillar
is a broad, thin, flat band which is attached to the front of the symphysis pubis,
interlacing with its fellow of the opposite side, that of the right side being
superficial.

The external abdominal ring gives passage to the spermatic cord in the male,
and round ligament in the female : it is much larger in men than in women, on
account of the large size of the spermatic cord, and hence the greater frequency
of inguinal hernia in men.

The **intercolumnar fibres** are a series of curved tendinous fibres, which arch
across the lower part of the aponeurosis of the External oblique. They have
received their name from stretching across between the two pillars of the external
ring, describing a curve with the convexity downwards. They are much thicker
and stronger at the outer margin of the external ring, where they are connected
to the outer third of Poupart's ligament, than internally, where they are inserted
into the linea alba. They are more strongly developed in the male than in the
female. The intercolumnar fibres increase the strength of the lower part of the
aponeurosis, and prevent the divergence of the pillars from one another.

These intercolumnar fibres, as they pass across the external abdominal ring,
are themselves connected together by delicate fibrous tissue, thus forming a fascia
which, as it is attached to the pillars of the ring, covers it in, and is called the
intercolumnar fascia. This **intercolumnar fascia** is continued down as a tubular

prolongation around the outer surface of the cord and testis, and encloses them in a distinct sheath, hence it is also called the *external spermatic fascia*.

The sac of an inguinal hernia, in passing through the external abdominal ring, receives an investment from the intercolumnar fascia.

If the finger is introduced a short distance into the external abdominal ring and the limb is then extended and rotated outwards, the aponeurosis of the External oblique, together with the iliac portion of the fascia lata, will be felt to become tense, and the external ring much contracted; if, on the contrary, the limb is flexed upon the pelvis and rotated inwards, this aponeurosis will become lax and the external abdominal ring sufficiently enlarged to admit the finger with comparative ease; hence the patient should always be put in the latter position when the taxis is applied for the reduction of an inguinal hernia, in order that the abdominal walls may be relaxed as much as possible.

Poupart's ligament, or the **crural arch,** is the lower border of the aponeurosis of the External oblique muscle, which extends from the anterior superior spine of the ilium to the spine of the os pubis. From this latter point it is reflected outwards to be attached to the pectineal line for about half an inch, forming Gimbernat's ligament. Its general direction is curved downwards towards the thigh, where it is continuous with the fascia lata. Its outer half is rounded and oblique in direction. Its inner half gradually widens at its attachment to the os pubis, is more horizontal in direction, and lies beneath the spermatic cord.

Nearly the whole of the space included between the crural arch and the innominate bone is filled in by the parts which descend from the abdomen into the thigh. These will be referred to again on a subsequent page.

Gimbernat's ligament is that part of the aponeurosis of the External oblique muscle which is reflected upwards and outwards from the spine of the os pubis to be inserted into the pectineal line. It is about half an inch in length, larger in the male than in the female, almost horizontal in direction in the erect posture, and of a triangular form with the base directed outwards. Its base, or outer margin, is concave, thin, and sharp, and lies in contact with the crural sheath; forming the inner boundary of the femoral ring. Its apex corresponds to the spine of the os pubis. Its posterior margin is attached to the pectineal line, and is continuous with the pubic portion of the fascia lata. Its anterior margin is continuous with Poupart's ligament. Its surfaces are directed upwards and downwards.

The **triangular fascia** of the abdomen is a layer of tendinous fibres of a triangular shape, which is attached by its apex to the pectineal line, where it is continuous with Gimbernat's ligament. It passes inwards beneath the spermatic cord, and expands into a somewhat fan-shaped fascia, lying behind the inner pillar of the external abdominal ring, and in front of the conjoined tendon, and interlaces with the ligament of the other side at the linea alba.

Ligament of Cooper.—This is a strong ligamentous band, which was first described by Sir Astley Cooper. It extends upwards and backwards from the base of Gimbernat's ligament along the ilio-pectineal line, to which it is attached. It is strengthened by the fascia transversalis, by the pectineal aponeurosis, and by a lateral expansion from the lower attachment of the linea alba (adminiculum lineae albae).

Dissection.—Detach the External oblique by dividing it across, just in front of its attachment to the ribs, as far as its posterior border, and separate it below from the crest of the ilium as far as the anterior superior spine; then separate the muscle carefully from the Internal oblique, which lies beneath, and turn it towards the opposite side.

The **Internal** or **Ascending oblique muscle** (fig. 219), thinner and smaller than the preceding, beneath which it lies, is of an irregularly quadrilateral form, and situated at the side and fore part of the abdomen. It arises, by fleshy fibres, from the outer half of Poupart's ligament, being attached to the groove on its upper surface; from the anterior two-thirds of the middle lip of the crest of the ilium, and from the posterior lamella of the lumbar fascia. From this origin the fibres diverge: those from Poupart's ligament, few in number, and paler in colour than the rest, arch downwards and inwards across the spermatic cord in the male and the round ligament in the female, and, becoming tendinous, are inserted, conjointly with those of the Transversalis, into the crest of the os pubis and pectineal line, to the extent of half an inch, forming what is known as the conjoined tendon of the Internal oblique and Transversalis; those from the anterior third of the iliac origin are horizontal in their direction, and, becoming tendinous along the lower fourth of the linea semilunaris, pass in front of the

FIG. 219.—The Internal oblique muscle.

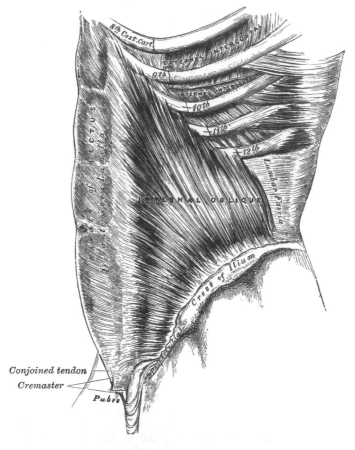

Rectus muscle to be inserted into the linea alba; those which arise from the middle third of the origin from the crest of the ilium pass obliquely upwards and inwards, and terminate in an aponeurosis, which divides at the outer border of the Rectus muscle into two lamellæ, which are continued forwards, in front and behind this muscle, to the linea alba: the posterior lamella being also connected to the cartilages of the seventh, eighth, and ninth ribs; the most posterior fibres pass almost vertically upwards, to be inserted into the lower borders of the cartilages of the three lower ribs, being continuous with the Internal intercostal muscles.

The conjoined tendon of the Internal oblique and Transversalis is inserted into the crest of the os pubis and pectineal line, immediately behind the external abdominal ring, serving to protect what would otherwise be a weak point in the abdominal wall. Sometimes this tendon is insufficient to resist the pressure from within, and is carried forward in front of the protrusion through the external ring, forming one of the coverings of direct inguinal hernia; or the hernia forces its way through the fibres of the conjoined tendon. The conjoined tendon is sometimes divided into an outer and an inner portion—the former being termed the *ligament of Hesselbach*; the latter, the *ligament of Henle*.

The aponeurosis of the Internal oblique is continued forward to the middle line of the abdomen, where it joins with the aponeurosis of the opposite muscle at the linea alba, and extends from the margin of the thorax to the os pubis. At the outer margin of the Rectus muscle, this aponeurosis, for the upper three-fourths of its extent, divides into two lamellæ, which pass, one in front and the other behind the muscle, enclosing it in a kind of sheath, and reuniting on its inner border at the linea alba; the anterior layer is blended with the aponeurosis of the External oblique muscle; the posterior layer with that of the Transversalis. Along the lower fourth, the aponeurosis passes altogether in front of the Rectus without any separation. Where the aponeurosis ceases to split, and passes altogether in front of the Rectus muscle, a deficiency is left in the sheath of the muscle behind; this is marked above by a sharp lunated margin, having its concavity downwards. This is known as the *semilunar fold of Douglas*.

Relations.—By its *external surface*, with the External oblique, Latissimus dorsi, spermatic cord, and external ring. By its *internal surface*, with the Transversalis muscle, the lower intercostal vessels and nerves, the ilio-hypogastric and the ilio-inguinal nerves. Near Poupart's ligament, it lies on the fascia transversalis, internal ring, and spermatic cord. Its lower border forms the upper boundary of the inguinal canal.

The **Cremaster muscle** is a thin muscular layer, composed of a number of fasciculi which arise from the inner part of Poupart's ligament where its fibres are continuous with those of the Internal oblique and also occasionally with the Transversalis. It passes along the outer side of the spermatic cord, descends with it through the external abdominal ring upon the front and sides of the cord, and forms a series of loops which differ in thickness and length in different subjects. Those at the upper part of the cord are exceedingly short, but they become in succession longer and longer, the longest reaching down as low as the testicle, where a few are inserted into the tunica vaginalis. These loops are united together by areolar tissue, and form a thin covering over the cord and testis, the *cremasteric fascia*. The fibres ascend along the inner side of the cord, and are inserted by a small pointed tendon into the crest of the os pubis and front of the sheath of the Rectus muscle.

It will be observed that the origin and insertion of the Cremaster is precisely similar to that of the lower fibres of the Internal oblique. This fact affords an easy explanation of the manner in which the testicle and cord are invested by this muscle. At an early period of fœtal life the testis is placed at the lower and back part of the abdominal cavity, but during its descent towards the scrotum, which takes place before birth, it passes beneath the arched fibres of the Internal oblique. In its passage beneath this muscle some fibres are derived from its lower part which accompany the testicle and cord into the scrotum. It occasionally happens that the loops of the Cremaster surround the cord, some lying behind as well as in front. It is probable that under these circumstances the testis, in its descent, passed through instead of beneath the fibres of the Internal oblique.

In the descent of an oblique inguinal hernia, which takes the same course as the spermatic cord, the Cremaster muscle forms one of its coverings. This muscle becomes largely developed in cases of hydrocele, and large old scrotal herniæ. The Cremaster muscle is only found in the male, but almost constantly in the female a few muscular fibres may be seen on the surface of the round ligament, which correspond to this muscle, and in cases of oblique inguinal hernia in the female a considerable amount of muscular fibre may be found covering the sac.

Dissection.—Detach the Internal oblique in order to expose the Transversalis beneath. This may be effected by dividing the muscle above, at its attachment to the ribs; below, at its connection with Poupart's ligament and the crest of the ilium; and behind, by a vertical incision extending from the last rib to the crest of the ilium. The muscle should previously be made tense by drawing upon it with the fingers of the left hand, and if its division is carefully effected, the cellular interval between it and the Transversalis, as well as the direction of the fibres of the latter muscle, will afford a clear guide to their separation; along the crest of the ilium the circumflex iliac vessels are interposed between them, and form an important guide in separating them. The muscle should then be thrown inwards towards the linea alba.

Fig. 220.—The Transversalis, Rectus, and Pyramidalis muscles.

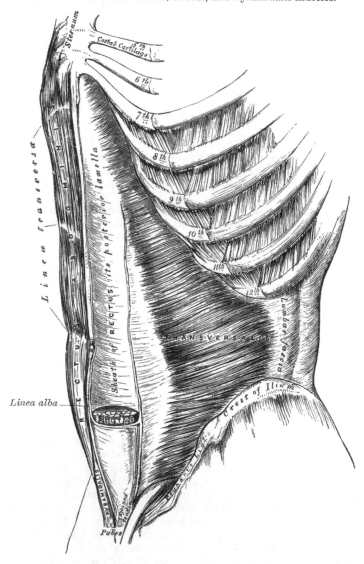

The **Transversalis muscle** (fig. 220), so called from the direction of its fibres, is the most internal flat muscle of the abdomen, being placed immediately beneath the Internal oblique. It arises by fleshy fibres from the outer third of Poupart's ligament; from the inner lip of the crest of the ilium, for its anterior three-fourths; from the inner surface of the cartilages of the six lower ribs, interdigitating with the Diaphragm; and from the lumbar fascia, which may be regarded as the posterior aponeurosis of the muscle (see p. 315). The muscle terminates in front in a broad aponeurosis, the lower fibres of which curve downwards and inwards, and are inserted, together with those of the Internal oblique, into the lower part of the linea alba, the crest of the os pubis and pectineal line, forming what is known as the conjoined tendon of the Internal oblique and Transversalis. Throughout the rest of its extent the aponeurosis passes horizontally inwards, and is inserted into the linea alba ; its upper three-fourths passing behind the Rectus muscle, blending with the posterior lamella of the Internal oblique ; its lower fourth passing in front of the Rectus.

Relations.—By its *external surface*, with the Internal oblique, the lower inter-costal nerves, and the inner surface of the cartilages of the lower ribs. By its *internal surface*, with the fascia transversalis, which separates it from the peritoneum. Its lower border forms the upper boundary of the inguinal canal.

Dissection.—To expose the Rectus muscle, open its sheath by a vertical incision extending from the margin of the thorax to the os pubis, and then reflect the two portions from the surface of the muscle, which is easily done, excepting at the lineæ transversæ, where so close an adhesion exists that the greatest care is requisite in separating them. Now raise the outer edge of the muscle, in order to examine the posterior layer of the sheath. By dividing the muscle in the centre, and turning its lower part downwards, the point where the posterior wall of the sheath terminates in a thin curved margin will be seen.

The **Rectus abdominis** is a long flat muscle, which extends along the whole length of the front of the abdomen, being separated from its fellow of the opposite side by the linea alba. It is much broader, but thinner, above than below, and arises by two tendons, the external or larger being attached to the crest of the os pubis ; the internal, smaller portion, interlacing with its fellow of the opposite side, and being connected with the ligaments covering the front of the symphysis pubis. The fibres ascend, and the muscle is inserted by three portions of unequal size into the cartilages of the fifth, sixth, and seventh ribs. The upper portion, attached principally to the cartilage of the fifth rib, usually has some fibres of insertion into the anterior extremity of the rib itself. Some fibres are occasionally connected with the costo-xiphoid ligaments, and side of the ensiform cartilage.

The Rectus muscle is traversed by tendinous intersections, three in number, which have received the name of *lineæ transversæ*. One of these is usually situated opposite the umbilicus, and two above that point ; of the latter, one corresponds to the extremity of the ensiform cartilage, and the other is about midway between the ensiform cartilage and the umbilicus. These intersections pass transversely or obliquely across the muscle in a zigzag course ; they rarely extend completely through its substance, sometimes pass only half-way across it, and are intimately adherent in front to the sheath in which the muscle is enclosed. Sometimes one or two additional lines may be seen, one usually below the umbilicus ; the position of the other, when it exists, is variable. These additional lines are for the most part incomplete.

The Rectus is enclosed in a sheath (fig. 221) formed by the aponeuroses of the Oblique and Transversalis muscles, which are arranged in the following manner. When the aponeurosis of the Internal oblique arrives at the outer margin of the Rectus, it divides into two lamellæ, one of which passes in front of the Rectus, blending with the aponeurosis of the External oblique ; the other, behind it, blending with the aponeurosis of the Transversalis ; and these, joining again at its inner border, are inserted into the linea alba. This arrangement of the aponeurosis exists along the upper three-fourths of the muscle : at the commencement of the lower fourth, the posterior wall of the sheath terminates in a thin curved margin, the *semilunar fold of Douglas*, the concavity of which looks downwards towards the pubes : the aponeuroses of all three muscles passing in front of the Rectus without any separation. The extremities of the fold of Douglas descend as pillars to the os pubis. The inner pillar is attached to the symphysis pubis ; the outer pillar, which is named by Braune the ligament of Hesselbach, passes downwards as a distinct band on the inner side of the internal abdominal ring, and there its fibres divide into two sets, internal and external ; the internal fibres are attached to the ascending ramus of the os pubis and the pectineal fascia ; the external ones pass to the Psoas fascia, to the deep surface of Poupart's ligament, and to the tendon of the Transversalis on the outer side of the ring. The Rectus muscle, in the situation where its sheath is deficient, is separated from the peritoneum by the transversalis fascia.

The **Pyramidalis** is a small muscle, triangular in shape, placed at the lower part of the abdomen, in front of the Rectus, and contained in the same sheath with that muscle. It arises by tendinous fibres from the front of the os pubis and the anterior pubic ligament ; the fleshy portion of the muscle passes upwards, diminishing in size as it ascends, and terminates by a pointed extremity, which is inserted into the linea alba, midway between the umbilicus and the os pubis. This muscle is sometimes found wanting on one or both sides ; the lower end of the Rectus then becomes proportionately increased in size. Occasionally it has been found double on one side, or the muscles of the two sides are of unequal size. Sometimes its length exceeds what is stated above.

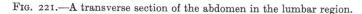

Fig. 221.—A transverse section of the abdomen in the lumbar region.

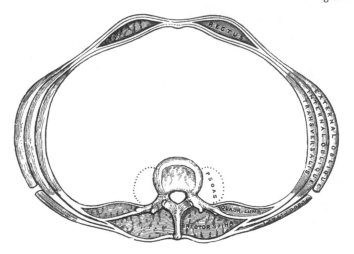

Besides the Rectus and Pyramidalis muscles, the sheath of the Rectus contains the superior and deep epigastric arteries, the terminations of the lumbar arteries and of the lower intercostal arteries and nerves.

Nerves.—The abdominal muscles are supplied by the lower intercostal nerves. The Transversalis and Internal oblique also receive filaments from the hypogastric branch of the ilio-hypogastric and sometimes from the ilio-inguinal. The Cremaster is supplied by the genital branch of the Genito-crural.

In the description of the abdominal muscles, mention has frequently been made of the linea alba, lineæ semilunares, and lineæ transversæ ; when the dissection of the muscles is completed, these structures should be examined.

The **linea alba** is a tendinous raphé seen along the middle line of the abdomen, extending from the ensiform cartilage to the symphysis pubis, to which it is attached. It is placed between the inner borders of the Recti muscles, and is formed by the blending of the aponeuroses of the Obliqui and Transversales muscles. It is narrow below, corresponding to the narrow interval existing between the Recti ; but broader above, as these muscles diverge from one another in their ascent, becoming of considerable breadth after great distension of the abdomen from pregnancy or ascites. It presents numerous apertures for the passage of vessels and nerves; the largest of these is the umbilicus, which in the fœtus transmits the umbilical vessels, but in the adult is obliterated, the cicatrix being stronger than the neighbouring parts; hence umbilical hernia occurs in the adult *near* the umbilicus, while in the fœtus it occurs *at* the umbilicus. The linea alba is in relation, in front, with the integument, to which it is adherent, especially at the umbilicus; behind, it is separated from the peritoneum by the transversalis fascia; and below, by the urachus, and the bladder when that organ is distended.

The **lineæ semilunares** are two curved tendinous lines placed one on each side of the linea alba. Each corresponds with the outer border of the Rectus muscle, extends from the cartilage of the ninth rib to the pubic spine, and is formed by the aponeurosis of the Internal oblique at its point of division to enclose the Rectus, where it is reinforced in front by the External oblique, and behind by the Transversalis.

The **lineæ transversæ** are narrow transverse lines which intersect the Recti muscles, as already mentioned; they connect the lineæ semilunares with the linea alba.

Actions.—The abdominal muscles perform a threefold action.

When the pelvis and thorax are fixed, they compress the abdominal viscera, by constricting the cavity of the abdomen, in which action they are materially assisted by the descent of the Diaphragm. By these means assistance is given in expelling the fœtus from the uterus, the fæces from the rectum, the urine from the bladder, and the contents of the stomach in vomiting.

If the pelvis and spine are fixed, these muscles compress the lower part of the thorax, materially assisting expiration. If the pelvis alone is fixed, the thorax is bent directly forward, when the muscles of both sides act, or to either side when those of the two sides act alternately, rotation of the trunk at the same time taking place to the opposite side.

If the thorax is fixed, these muscles, acting together, draw the pelvis upwards, as in climbing; or, acting singly, they draw the pelvis upwards, and bend the vertebral column to one side or the other. The Recti muscles, acting from below, depress the thorax, and consequently flex the vertebral column; when acting from above, they flex the pelvis upon the vertebral column. The Pyramidales are tensors of the linea alba.

The **fascia transversalis** is a thin aponeurotic membrane which lies between the inner surface of the Transversalis muscle and the extra-peritoneal fat. It forms part of the general layer of fascia which lines the interior of the abdominal and pelvic cavities, and is directly continuous with the iliac and pelvic fasciæ. In the inguinal region, the transversalis fascia is thick and dense in structure and joined by fibres from the aponeurosis of the Transversalis muscle, but it becomes thin and cellular as it ascends to the Diaphragm, and blends with the fascia covering this muscle. In front, it unites across the middle line with the fascia on the opposite side of the body, and behind it becomes lost in the fat which covers the posterior surfaces of the kidneys. Below, it has the following attachments: posteriorly, it is connected to the whole length of the crest of the ilium, between the attachments of the Transversalis and Iliacus muscles; between the anterior superior spine of the ilium and the femoral vessels it is connected to the posterior margin of Poupart's ligament, and is there continuous with the iliac fascia. Internal to the femoral vessels it is thin and attached to the os pubis and pectineal line, behind the conjoined tendon, with which it is united; and, corresponding to the point where the femoral vessels pass into the thigh, this fascia descends in front of them, forming the anterior wall of the crural sheath. Beneath Poupart's ligament it is strengthened by a band of fibrous tissue, which is only loosely connected to Poupart's ligament, and is specialised as the *deep crural arch*. The spermatic cord in the male and the round ligament in the female pass through this fascia; the point where they pass through is called the *internal abdominal ring*. This opening is not visible externally owing to a prolongation of the transversalis fascia on these structures, forming the *infundibuliform fascia*.

The **Internal** or **deep abdominal ring** is situated in the transversalis fascia, midway between the anterior superior spine of the ilium and the symphysis pubis, and about half an inch above Poupart's ligament. It is of an oval form, the extremities of the oval directed upwards and downwards, varies in size in different subjects, and is much larger in the male than in the female. It is bounded, above and externally, by the arched fibres of the Transversalis ; below and internally, by the deep epigastric vessels. It transmits the spermatic cord in the male and the round ligament in the female. From its circumference a thin funnel-shaped membrane, the *infundibuliform fascia*, is continued round the cord and testis, enclosing them in a distinct pouch.

When the sac of an oblique inguinal hernia passes through the internal or deep abdominal ring, the infundibuliform process of the transversalis fascia forms one of its coverings.

The **inguinal** or **spermatic canal** contains the spermatic cord in the male and the round ligament in the female. It is an oblique canal about an inch and a half in length, directed downwards and inwards, and placed parallel to and a little above Poupart's ligament. It commences above at the internal or deep abdominal ring, which is the point where the cord enters the spermatic canal, and terminates below at the external ring. It is bounded in front by the integument and superficial fascia, by the aponeurosis of the External oblique throughout its whole length, and by the Internal oblique for its outer third ; behind, by the triangular fascia, the conjoined tendon of the Internal oblique and Transversalis, transversalis fascia, and the subperitoneal fat and peritoneum ; above, by the arched fibres of the Internal oblique and Transversalis ; below, by Gimbernat's ligament, and by the union of the fascia transversalis with Poupart's ligament. The deep epigastric artery passes upwards and inwards behind the canal lying close to the inner side of the internal abdominal ring. The interval between this artery and the outer edge of the Rectus is named Hesselbach's triangle, the base of which is formed by Poupart's ligament.

That form of protrusion in which the intestine follows the course of the spermatic cord along the spermatic canal is called oblique inguinal hernia.

The **deep crural arch**.—Curving over the vessels, just at the point where they become femoral, on the abdominal side of Poupart's ligament and loosely connected with it, is a thickened band of fibres called the deep crural arch. It is apparently a thickening of the fascia transversalis, joining externally to the centre of Poupart's ligament, and arching across the front of the crural sheath to be inserted by a broad attachment into the spine of the os pubis and ilio-pectineal line, behind the conjoined tendon. In some subjects this structure is not very prominently marked, and not infrequently it is altogether wanting.

Surface Form.—The only two muscles of this group which have any considerable influence on surface form are the External oblique and Rectus muscles of the abdomen. With regard to the External oblique, the upper digitations of its origin from the ribs are well marked, intermingled with the serrations of the Serratus magnus ; the lower digitations are not visible, being covered by the thick border of the Latissimus dorsi. Its attachment to the crest of the ilium, in conjunction with the Internal oblique, forms a thick oblique roll, which determines the iliac furrow. Sometimes on the front of the lateral region of the abdomen an undulating outline marks the spot where the muscular fibres terminate and the aponeurosis commences. The outer border of the Rectus is defined by the *linea semilunaris*, which may be exactly determined by putting the muscle into action. It corresponds with a curved line, with its convexity outwards, drawn from the end of the cartilage of the ninth rib to the spine of the os pubis, so that the centre of the line, at or near the umbilicus, is three inches from the median line. The inner border of the Rectus corresponds to the *linea alba*, marked on the surface of the body by a groove, the *abdominal furrow*, which extends from the infrasternal fossa to, or to a little below, the umbilicus, where it gradually becomes lost. The surface of the Rectus presents three transverse furrows, the *lineæ transversæ*. The upper two of these, one opposite or a little below the tip of the ensiform cartilage, and another, midway between this point and the umbilicus, are usually well marked ; the third, opposite the umbilicus, is not so distinct. The umbilicus, situated in the linea alba, varies very much in position as regards its height. It is always situated above a zone drawn round the body opposite the highest point of the crest of the ilium, generally being about three-quarters of an inch to an inch above this line. It generally corresponds, therefore, to the fibro-cartilage between the third and fourth lumbar vertebræ.

2. DEEP MUSCLES OF THE ABDOMEN

Psoas magnus.   Iliacus.
Psoas parvus.   Quadratus lumborum.

The Psoas magnus, the Psoas parvus, and the Iliacus muscles, with the fascia covering them, will be described with the Muscles of the Lower Extremity (see page 391).

The fascia covering the Quadratus lumborum.—This is the most anterior of the three layers of the lumbar fascia. It is a thin layer of fascia which, passing over the anterior surface of the Quadratus lumborum, is attached, internally, to the bases of the transverse processes of the lumbar vertebræ ; below, to the ilio-lumbar ligament ; and above, to the apex and lower border of the last rib.

The portion of this fascia which extends from the transverse process of the first lumbar vertebra to the apex and lower border of the last rib, constitutes the ligamentum arcuatum externum.

The **Quadratus lumborum** (fig. 214, page 317) is situated in the lumbar region. It is irregularly quadrilateral in shape, and broader below than above. It arises by aponeurotic fibres from the ilio-lumbar ligament and the adjacent portion of the crest of the ilium for about two inches, and is inserted into the lower border of the last rib for about half its length and, by four small tendons, into the apices of the transverse processes of the four upper lumbar vertebræ. Occasionally a second portion of this muscle is found situated in front of the preceding. It arises from the upper borders of the transverse processes of three or four of the lower lumbar vertebræ, and is inserted into the lower margin of the last rib. The Quadratus lumborum is contained in a sheath formed by the anterior and middle lamellæ of the lumbar fasciæ.

Relations.—Its *anterior surface* (or rather the fascia which covers its anterior surface) is in relation with the colon, the kidney, the Psoas muscle, and the Diaphragm. Between the fascia and the muscle are the last dorsal, ilio-hypogastric and ilio-inguinal nerves. Its *posterior surface* is in relation with the middle lamella of the lumbar fascia, which separates it from the Erector spinæ. The Quadratus lumborum extends, however, beyond the outer border of the Erector spinæ.

Nerve-supply.—The anterior branches of the last dorsal and first lumbar nerves ; sometimes also a branch from the second lumbar nerve.

Actions.—The Quadratus lumborum draws down the last rib, and acts as a muscle of inspiration by helping to fix the origin of the Diaphragm. If the thorax and spine are fixed, it may act upon the pelvis, raising it towards its own side when only one muscle is put in action ; and when both muscles act together, either from below or above, they flex the trunk.

IV. MUSCLES OF THE PELVIC OUTLET

The muscles of this region are situated at the pelvic outlet in the ischio-rectal region and the perinæum. They include the following :

I. Muscles of the ischio-rectal region.
II. Muscles of the perinæum : A. In the Male ; B. In the Female.

I. Muscles of the Ischio-rectal Region

Corrugator cutis ani.　　　　　　　　Internal sphincter ani.
External sphincter ani.　　　　　　　Levator ani.
　　　　　　　　　　Coccygeus.

The Corrugator cutis ani.—Around the anus is a thin stratum of involuntary muscular fibre, which radiates from the orifice. Internally, the fibres fade off into the submucous tissue, while externally they blend with the true skin. By its contraction it raises the skin into ridges around the margin of the anus.

The **External sphincter ani** is a thin, flat plane of muscular fibres, elliptical in shape and intimately adherent to the integument surrounding the margin of the anus. It measures about three or four inches in length, from its anterior to its posterior extremity, being about an inch in breadth, opposite the anus. It arises from the tip and back of the coccyx, by a narrow tendinous band, and from the superficial fascia in front of that bone ; and is inserted into the central tendinous point of the perinæum, joining with the Transversus perinæi, the Levator ani, and the Accelerator urinæ. Like other sphincter muscles, it consists of two planes of muscular fibre, which surround the margin of the anus, and join in a commissure in front and behind, some fibres crossing from side to side in front and behind the anus.

Nerve-supply.—A branch from the anterior division of the fourth sacral and the inferior hæmorrhoidal branch of the internal pudic.

Actions.—The action of this muscle is peculiar : 1. It is, like other muscles, always in a state of tonic contraction, and having no antagonistic muscle it keeps the anal orifice closed. 2. It can be put into a condition of greater contraction under the influence of the will, so as to more firmly occlude the anal aperture in expiratory efforts, unconnected with defæcation. 3. Taking its fixed point at the coccyx, it helps to fix the central point of the perinæum, so that the Accelerator urinæ may act from this fixed point.

The **Internal sphincter ani** is a muscular ring which surrounds the lower extremity of the rectum for about an inch ; its inferior border being contiguous to, but quite separate from, the External sphincter. This muscle is about two lines in thickness, and is formed by an aggregation of the involuntary circular fibres of the intestine. It is paler in colour and less coarse in texture than the External sphincter.

Actions.—Its action is entirely involuntary. It helps the External sphincter to occlude the anal aperture.

The **Levator ani** (fig. 222) is a broad, thin muscle, situated on each side of the pelvis. It is attached to the inner surface of the sides of the true pelvis, and descending unites with its fellow of the opposite side to form the floor of the pelvic cavity. It supports the viscera in this cavity and surrounds the various structures which pass through it. It arises, in front, from the posterior surface of the body of the os pubis on the outer side of the symphysis ; posteriorly, from the inner surface of the spine of the ischium ; and between these two points, from the angle of division between the obturator and recto-vesical layers of the pelvic fascia at their under part. The fibres pass downwards to the middle line of the floor of the pelvis, and are inserted, the most posterior into the sides of the apex of the coccyx ; those placed more anteriorly unite with the muscle of the opposite side, in a median fibrous raphé, which extends between the coccyx and the margin of the anus. The middle fibres, which form the larger portion of the muscle, are inserted into the side of the rectum, blending with the fibres of the Sphincter muscles ; lastly, the anterior fibres, the longest, descend upon the side of the prostate gland to unite beneath it with the muscle of the opposite side, blending with the fibres of the External sphincter and Transversus perinæi muscles at the central tendinous point of the perinæum.

The anterior portion is occasionally separated from the rest of the muscle by connective tissue. From this circumstance, as well as from its peculiar relation with the prostate gland, descending by its side, and surrounding it as in a sling, it has been described by Santorini and others as a distinct muscle, under the name of Levator prostatæ. In the female, the anterior fibres of the Levator ani descend upon the side of the vagina.

Relations.—By its *inner* or *pelvic surface*, with the recto-vesical fascia, which separates it from the viscera of the pelvis and from the peritoneum. By its *outer* or *perineal surface*, it forms the inner boundary of the ischio-rectal fossa, and is covered by a thin layer of fascia, the *ischio-rectal* or *anal fascia*, given off from the obturator fascia. Its *posterior border* is free and separated from the Coccygeus muscle by a cellular interspace. Its *anterior border* is separated from the muscle of the opposite side by a triangular space, through which the urethra, and in the female the vagina, passes from the pelvis.

FIG. 222.—Side view of pelvis, showing Levator ani.
(From a preparation in the Museum of the Royal College of Surgeons of England.)

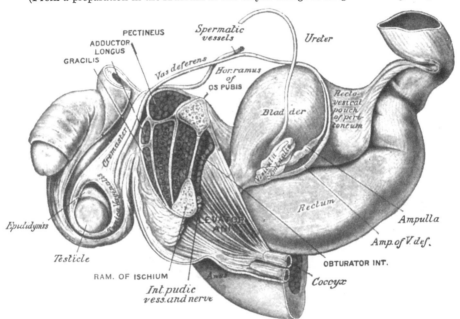

Nerve-supply.—A branch from the anterior division of the fourth sacral nerve and a branch from the pudic nerve, which is sometimes derived from the perineal, sometimes from the inferior hæmorrhoidal division.

Actions.—This muscle supports the lower end of the rectum and vagina, and also the bladder during the efforts of expulsion. It elevates and inverts the lower end of the rectum after it has been protruded and everted during the expulsion of the fæces. It is also a muscle of forced expiration.

The **Coccygeus** is situated behind and parallel with the preceding. It is a triangular plane of muscular and tendinous fibres, arising, by its apex, from the spine of the ischium and lesser sacro-sciatic ligament, and inserted, by its base, into the margin of the coccyx and into the side of the lower piece of the sacrum. It assists the Levator ani and Pyriformis in closing in the back part of the outlet of the pelvis.

Relations.—By its *inner* or *pelvic surface*, with the rectum. By its *external surface*, with the lesser sacro-sciatic ligament. The *lower border* is in relation with the posterior border of the Levator ani, but separated from it by a cellular interval : its *upper border* is in relation with the lower border of the Pyriformis, but separated from it by the sciatic and internal pudic vessels and nerve.

Nerve-supply.—A branch from the fourth and fifth sacral nerves.

Action.—The Coccygei muscles raise and support the coccyx, after it has been pressed backwards during defæcation or parturition.

II. a. Muscles and Fasciæ of the Perinæum in the Male

Transversus perinæi.	Erector penis.
Accelerator urinæ.	Compressor urethræ.

Superficial fascia.—The superficial fascia of the perinæum consists of two layers, superficial and deep, as in other regions of the body.

FIG. 223.—The perinæum.
The integument and superficial layer of superficial fascia reflected.

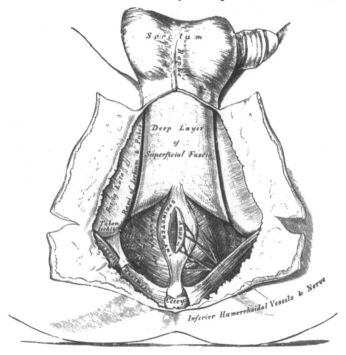

The *superficial layer* is thick, loose, areolar in texture, and contains much adipose tissue in its meshes, the amount of which varies in different subjects. In front, it is continuous with the dartos of the scrotum ; behind, it is continuous with the subcutaneous areolar tissue surrounding the anus ; and, on either side, with the same fascia on the inner side of the thighs. This layer should be carefully removed after it has been examined, when the deep layer will be exposed.

The *deep layer of superficial fascia* (Fascia of Colles) is thin, aponeurotic in structure, and of considerable strength, serving to bind down the muscles of the root of the penis. It is continuous, in front, with the dartos of the scrotum ; on either side it is firmly attached to the margins of the rami of the os pubis and ischium, external to the crus penis, and as far back as the tuberosity of the ischium ; posteriorly, it curves down behind the Transversus perinæi muscles to join the lower margin of the triangular ligament. This fascia not only covers the muscles in this region, but sends upwards a vertical septum from its deep surface, which separates the back part of the subjacent space into two, the septum being incomplete in front.

The central tendinous point of the Perinæum.—This is a fibrous point in the middle line of the perinæum, between the urethra and the rectum, and about half an inch in front of the anus. At this point four muscles converge and are attached : viz. the External sphincter ani, the Accelerator urinæ, and the two Transversi perinæi ; so that by the contraction of these muscles, which extend in opposite directions, it serves as a fixed point of support.

FIG. 224.—The muscles attached to the front of the pelvis.
(From a preparation in the Museum of the Royal College of Surgeons of England.)

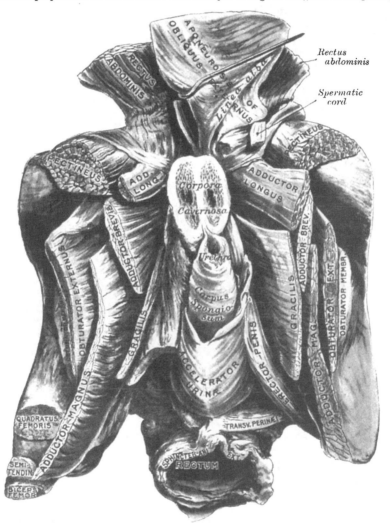

The **Transversus perinæi** is a narrow muscular slip, which passes more or less transversely across the back part of the perineal space. It arises by a small tendon from the inner and fore part of the tuberosity of the ischium, and, passing inwards, is inserted into the central tendinous point of the perinæum, joining in this situation with the muscle of the opposite side, the External sphincter ani behind, and the Accelerator urinæ in front.

Nerve-supply.—The perineal branch of the internal pudic.

Actions.—By their contraction they serve to fix the central tendinous point of the perinæum.

The **Accelerator urinæ** (*Ejaculator seminis*, or *Bulbo-cavernosus*) is placed in the middle line of the perinæum, immediately in front of the anus. It consists of two symmetrical halves, united along the median line by a tendinous raphé. It arises from the central tendon of the perinæum, and from the median raphé in front. From this point its fibres diverge like the plumes of a pen ; the most posterior form a thin layer, which is lost on the anterior surface of the triangular ligament ; the middle fibres encircle the bulb and adjacent parts of the corpus spongiosum, and join with the fibres of the opposite side, on the upper part of the corpus spongiosum, in a strong aponeurosis ; the anterior fibres, the longest and most distinct, spread out over the sides of the corpus cavernosum, to be inserted partly into that body, anterior to the Erector penis, occasionally extending to the os pubis ; partly terminating in a tendinous expansion, which covers the dorsal vessels of the penis. The latter fibres are best seen by dividing the muscle longitudinally, and dissecting it outwards from the surface of the urethra.

FIG. 225.—The superficial muscles and vessels of the perinæum.

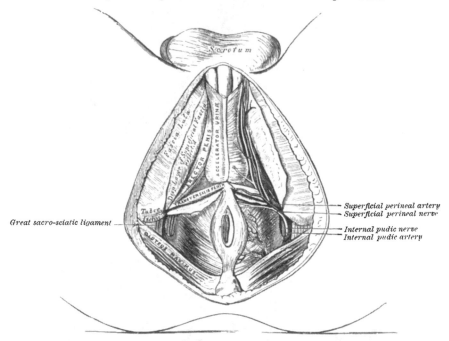

Action.—This muscle serves to empty the canal of the urethra, after the bladder has expelled its contents ; during the greater part of the act of micturition its fibres are relaxed, and it only comes into action at the end of the process. The middle fibres are supposed, by Krause, to assist in the erection of the corpus spongiosum, by compressing the erectile tissue of the bulb. The anterior fibres, according to Tyrrel, also contribute to the erection of the penis, as they are inserted into, and continuous with, the fascia of the penis, compressing the dorsal vein during the contraction of the muscle.

The **Erector penis** (*ischio-cavernosus*) covers part of the crus penis. It is an elongated muscle, broader in the middle than at either extremity, and situated on either side of the lateral boundary of the perinæum. It arises by tendinous and fleshy fibres from the inner surface of the tuberosity of the ischium, behind the crus penis ; from the surface of the crus ; and from the adjacent portion of the ramus of the ischium. From these points fleshy fibres succeed, which end in an aponeurosis which is inserted into the sides and under surface of the crus penis.

Nerve-supply.—The perineal branch of the internal pudic.

Actions.—It compresses the crus penis, and retards the return of the blood through the veins, and thus serves to maintain the organ erect.

Between the muscles just examined a triangular space exists, bounded internally by the Accelerator urinæ, externally by the Erector penis, and behind by the Transversus perinæi. The floor of this space is formed by the triangular ligament of the urethra (deep perineal fascia), and running from behind forwards in it are the superficial perineal vessels and nerves, the long pudendal nerve; the transverse perineal artery courses along the posterior boundary of the space on the Transversus perinæi muscle.

The **triangular ligament** (*Deep perineal fascia*) is stretched almost horizontally across the pubic arch, so as to close in the front part of the outlet of the pelvis. It consists of two dense membranous laminæ, which are united along their posterior borders, but are separated in front by intervening structures. The

FIG. 226.—Triangular ligament or deep perineal fascia.
On the left side the superficial layer has been removed.

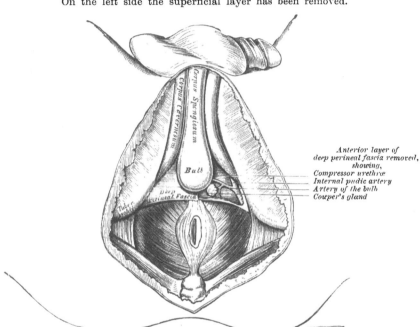

Anterior layer of deep perineal fascia removed, showing,
Compressor urethræ
Internal pudic artery
Artery of the bulb
Cowper's gland

superficial of these two layers, the *inferior layer of the triangular ligament,* is triangular in shape, about an inch and a half in depth. Its apex is directed forwards, and is separated from the subpubic ligament by an oval opening for the transmission of the dorsal vein of the penis. Its lateral margins are attached on each side to the rami of the ischium and os pubis, above the crura penis. Its base is directed towards the rectum, and connected to the central tendinous point of the perinæum. It is continuous with the deep layer of the superficial fascia behind the Transversus perinæi muscle, and with a thin fascia which covers the cutaneous surface of the Levator ani muscle (*anal* or *ischio-rectal* fascia).

This layer of the triangular ligament is perforated, about an inch below the symphysis pubis, by the urethra, the aperture for which is circular in form and about three or four lines in diameter; by the arteries to the bulb and the ducts of Cowper's glands close to the urethral orifice; by the arteries to the corpora cavernosa—one on each side close to the pubic arch and about halfway along the attached margin of the ligament; by the dorsal arteries and nerves of the penis near the apex of the ligament. Its base is also perforated by the superficial perineal vessels and nerves, while between its apex and the subpubic ligament the dorsal vein of the penis passes upwards into the pelvis.

If this superficial or inferior layer of the triangular ligament is detached on either side, the following structures will be seen between it and the deep layer : the dorsal vein of the penis ; the membranous portion of the urethra, and the Comp. essor urethræ muscle ; Cowper's glands and their ducts ; the pudic vessels and dorsal nerve of penis ; the artery and nerve of the bulb, and a plexus of veins.

The deep layer of the ligament (*superior layer of the triangular ligament*) is derived from the obturator fascia and stretches across the pubic arch. If the obturator fascia is traced inwards after covering the Obturator internus muscle, it will be found to be attached by some of its deeper or anterior fibres to the inner margin of the ischio-pubic ramus, while its superficial or posterior fibres pass over this attachment to become the superior layer of the triangular ligament. Behind, this layer of the fascia is continuous with the inferior layer and with the fascia of Colles, and in front it is separated from the apex of the prostate gland through the intervention of a prolongation of the recto-vesical fascia. It is pierced by the urethra, or rather consists of two halves which are separated in the middle line by the urethra passing between them.

The **Compressor urethræ** (*Constrictor urethræ*) surrounds the whole length of the membranous portion of the urethra, and is contained between the two layers of the triangular ligament. It arises, by aponeurotic fibres, from the junction of the rami of the os pubis and ischium, to the extent of half or three-quarters of an inch : each segment of the muscle passes inwards, and divides into two fasciculi, which surround the urethra from the prostate gland behind, to the bulbous portion of the urethra in front ; and unite, at the upper and lower surfaces of this tube, with the muscle of the opposite side, by means of a tendinous raphé.

Nerve-supply.—The perineal branch of the internal pudic.

Actions.—The muscles of both sides act together as a sphincter, compressing the membranous portion of the urethra. During the transmission of fluids they, like the Acceleratores urinæ, are relaxed, and only come into action at the end of the process to eject the last drops of the fluid.

II. B. MUSCLES OF THE PERINÆUM IN THE FEMALE (fig. 227)

Transversus perinæi.   Erector clitoridis.
Sphincter vaginæ.   Compressor urethræ.

The **Transversus perinæi** in the **female** is a narrow muscular slip, which passes more or less transversely across the back part of the perineal space. It arises by a small tendon from the inner and fore part of the tuberosity of the ischium, and, passing inwards, is inserted into the central point of the perinæum, joining in this situation with the muscle of the opposite side, the External sphincter ani behind, and the Sphincter vaginæ in front.

Nerve-supply.—The perineal branch of the internal pudic.

Action.—By their contraction they serve to fix the central tendinous point of the perinæum.

The **Sphincter vaginæ** surrounds the orifice of the vagina, and is analogous to the Accelerator urinæ in the male. It is attached posteriorly to the central tendinous point of the perinæum, where it blends with the External sphincter ani. Its fibres pass forwards on each side of the vagina, to be inserted into the corpora cavernosa of the clitoris, a fasciculus crossing over the body of the organ so as to compress the dorsal vein.

Nerve-supply.—The perineal branch of the internal pudic.

Actions.—It diminishes the orifice of the vagina. The anterior fibres contribute to the erection of the clitoris, as they are inserted into and are continuous with the fascia of the clitoris; compressing the dorsal vein during the contraction of the muscle.

The **Erector clitoridis** corresponds with the Erector penis in the male, but is smaller. It covers the unattached part of the crus clitoridis. It is an elongated muscle, broader at the middle than at either extremity, and situated on either side of the lateral boundary of the perinæum. It arises by tendinous and fleshy fibres from the inner surface of the tuberosity of the ischium, behind the crus clitoridis; from the surface of the crus; and from the adjacent portion of the ramus of the ischium. From these points fleshy fibres succeed, which end in an aponeurosis, which is inserted into the sides and under surface of the crus clitoridis.

Nerve-supply.—The perineal branch of the internal pudic.

Actions.—It compresses the crus clitoridis and retards the return of blood through the veins, and thus serves to maintain the organ erect.

FIG. 227.—Muscles of the female perinæum.

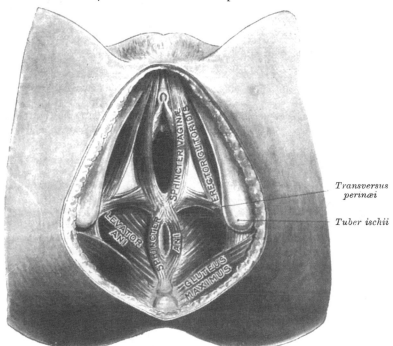

Transversus perinæi

Tuber ischii

The **triangular ligament** (*deep perineal fascia*) in the female is not so strong as in the male. It is attached to the pubic arch, its apex being-connected with the subpubic ligament. It is divided in the middle line by the aperture of the vagina, with the external coat of which it becomes blended, and in front of this is perforated by the urethra. Its posterior border is continuous, as in the male, with the deep layer of the superficial fascia around the Transversus perinæi muscle.

Like the triangular ligament in the male it consists of two layers, between which are to be found the following structures: the dorsal vein of the clitoris, a portion of the urethra and the Compressor urethræ muscle, the glands of Bartholin and their ducts; the pudic vessels and the dorsal nerve of the clitoris; the arteries of the bulbi vestibuli, and a plexus of veins.

The **Compressor urethræ** (*Constrictor urethræ*) arises on each side from the margin of the descending ramus of the os pubis. The fibres, passing inwards, divide into two sets: those of the fore part of the muscle are directed across the subpubic arch in front of the urethra to blend with the muscular fibres of the opposite side ; while those of the hinder and larger part pass inwards to blend with the wall of the vagina behind the urethra.

Nerve-supply.—The perineal branch of the internal pudic.

MUSCLES AND FASCIÆ OF THE UPPER EXTREMITY

The Muscles of the Upper Extremity are divisible into groups, corresponding with the different regions of the limb.

I. OF THE THORACIC REGION

1. *Anterior Thoracic Region*
Pectoralis major.   Pectoralis minor.
Subclavius.

2. *Lateral Thoracic Region*
Serratus magnus.

II. OF THE SHOULDER AND ARM

3. *Acromial Region*
Deltoid.

4. *Anterior Scapular Region*
Subscapularis.

5. *Posterior Scapular Region*
Supraspinatus.   Teres minor.
Infraspinatus.   Teres major.

6. *Anterior Humeral Region*
Coraco-brachialis.   Biceps.
Brachialis anticus.

7. *Posterior Humeral Region*
Triceps.   Subanconeus.

III. OF THE FOREARM

8. *Anterior Radio-Ulnar Region*
Superficial Layer.
Pronator radii teres.
Flexor carpi radialis.
Palmaris longus.
Flexor carpi ulnaris.
Flexor sublimis digitorum.

Deep Layer.
Flexor profundus digitorum.
Flexor longus pollicis.
Pronator quadratus.

9. *Radial Region*
Supinator longus.
Extensor carpi radialis longior.
Extensor carpi radialis brevior.

10. *Posterior Radio-Ulnar Region*
Superficial Layer.
Extensor communis digitorum.
Extensor minimi digiti.
Extensor carpi ulnaris.
Anconeus.

Deep Layer.
Supinator brevis.
Extensor ossis metacarpi pollicis.
Extensor brevis pollicis.
Extensor longus pollicis.
Extensor indicis.

IV. OF THE HAND

11. *Radial Region*
Abductor pollicis.
Flexor ossis metacarpi pollicis (Opponens pollicis).
Flexor brevis pollicis.
Adductor obliquus pollicis.
Adductor transversus pollicis.

12. *Ulnar Region*
Palmaris brevis.
Abductor minimi digiti.
Flexor brevis minimi digiti.
Flexor ossis metacarpi minimi digiti (Opponens minimi digiti).

13. *Middle Palmar Region*
Lumbricales.
Interossei palmares.
Interossei dorsales.

Dissection of Pectoral Region and Axilla (fig. 228).—The arm being drawn away from the side nearly at right angles with the trunk, and rotated outwards, make a vertical incision through the integument in the median line of the chest, from the upper to the lower part of the sternum; a second incision along the lower border of the Pectoral muscle, from the ensiform cartilage to the inner side of the axilla; a third, from the sternum along the clavicle, as far as its centre ; and a fourth, from the middle of the clavicle obliquely downwards, along the interspace between the Pectoral and Deltoid muscles, as low as the fold of the armpit.   The flap of integument is then to be dissected off in the direction indicated in the figure, but not entirely removed, as it should be replaced on completing the dissection.   If a transverse incision is now made from the lower end of the sternum to the side of the chest, as far as the posterior fold of the armpit, and the integument reflected outwards, the axillary space will be more completely exposed.

I. Muscles and Fasciæ of the Thoracic Region

1. *Anterior Thoracic Region*

Pectoralis major.   Pectoralis minor.
                    Subclavius.

The **superficial fascia** of the thoracic region is a loose cellulo-fibrous layer, enclosing masses of fat in its spaces. It is continuous with the superficial fascia of the neck and upper extremity above, and of the abdomen below. Opposite the mamma, it divides into two layers, one of which passes in front, the other behind that gland; and from both of these layers numerous septa pass into its substance, supporting its various lobes: from the anterior layer, fibrous processes pass forwards to the integument and nipple. These processes were called by Sir A. Cooper the *ligamenta suspensoria*, from the support they afford to the gland in this situation.

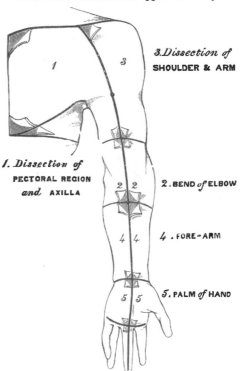

Fig. 228.—Dissection of upper extremity.

3. Dissection of **SHOULDER & ARM**

1. Dissection of **PECTORAL REGION** *and* **AXILLA**

2. **BEND** *of* **ELBOW**

4. **FORE-ARM**

5. **PALM** *of* **HAND**

The **deep fascia** of the thoracic region is a thin aponeurotic lamina, covering the surface of the great Pectoral muscle, and sending numerous prolongations between its fasciculi: it is attached, in the middle line, to the front of the sternum; and, above, to the clavicle; externally and below it becomes continuous with the fascia over the shoulder, axilla, and thorax. It is very thin over the upper part of the muscle, thicker in the interval between the Pectoralis major and Latissimus dorsi, where it closes in the axillary space, and divides at the outer margin of the latter muscle into two layers, one of which passes in front, and the other behind it; these proceed as far as the spinous processes of the dorsal vertebræ, to which they are attached. As the fascia leaves the lower edge of the, Pectoralis major to pass across the floor of the axilla it sends a layer upwards under cover of the muscle: this lamina splits to envelop the Pectoralis minor, at the upper edge of which it becomes continuous with the costo-coracoid membrane. The hollow of the armpit, seen when the arm is abducted, is mainly produced by the traction of this fascia on the axillary floor, and hence it is sometimes named the *suspensory ligament* of the axilla. At the lower part of the thoracic region this fascia is well developed, and is continuous with the fibrous sheath of the Recti muscles.

The **Pectoralis major** (fig. 229) is a broad, thick, triangular muscle, situated at the upper and fore part of the chest and in front of the axilla. It arises from the anterior surface of the sternal half of the clavicle; from half the breadth of the anterior surface of the sternum, as low down as the attachment of the cartilage of the sixth or seventh rib; this portion of its origin consists of aponeurotic

fibres, which intersect those of the opposite muscle; it also arises from the
cartilages of all the true ribs, with the exception, frequently, of the first, or of the
seventh, or both; and from the aponeurosis of the External oblique muscle of the
abdomen. The fibres from this extensive origin converge towards its insertion,
giving to the muscle a radiated appearance. Those fibres which arise from the
clavicle pass obliquely outwards and downwards, and are usually separated from
the rest by a cellular interval: those from the lower part of the sternum, and the
cartilages of the lower true ribs, pass upwards and outwards; while the middle

FIG. 229.—Muscles of the chest and front of the arm.   Superficial view.

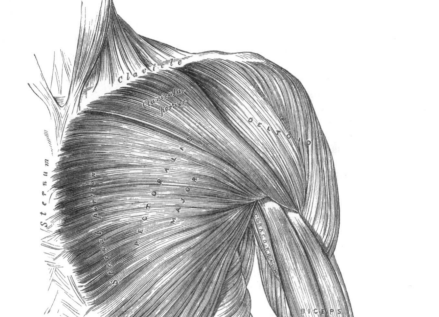

fibres pass horizontally. They all terminate in a flat tendon, about two inches
broad, which is inserted into the outer bicipital ridge of the humerus. This
tendon consists of two laminæ, placed one in front of the other, and usually
blended together below. The anterior, the thicker, receives the clavicular and
upper half of the sternal portion of the muscle; and its fibres are inserted in the
same order as that in which they arise; that is to say, the outermost fibres of
origin from the clavicle are inserted at the uppermost part of the tendon; the
upper fibres of origin from the sternum pass down to the lowermost part of this
anterior lamina of the tendon and extend as low as the tendon of the Deltoid and

join with it. The posterior lamina of the tendon receives the attachment of the lower half of the sternal portion and the deeper part of the muscle from the costal cartilages. These deep fibres, and particularly those from the lower costal cartilages, ascend the higher, turning backwards successively behind the superficial and upper ones, so that the tendon appears to be twisted. The posterior lamina reaches higher on the humerus than the anterior one, and from it an expansion is given off which covers the bicipital groove and blends with the capsule of the shoulder-joint. From the deepest fibres of this lamina at its insertion an expansion is given off which lines the bicipital groove of the humerus, while from the lower border of the tendon a third expansion passes downwards to the fascia of the arm.

Relations.—By its *anterior surface*, with the integument, the superficial fascia, the Platysma, some of the branches of the descending cervical nerves, the mammary gland, and the deep fascia. By its *posterior surface*; its *thoracic portion*, with the sternum, the ribs and costal cartilages, the costo-coracoid membrane, the Subclavius, Pectoralis minor, Serratus magnus, and the Intercostals; its *axillary portion* forms the anterior wall of the axillary space, and covers the axillary vessels and nerves, the Biceps and Coraco-brachialis muscles. Its *upper border* lies parallel with the Deltoid, from which it is separated by a slight interspace in which lie the cephalic vein and humeral branch of the acromial thoracic artery. Its *lower border* forms the anterior margin of the axilla, being at first separated from the Latissimus dorsi by a considerable interval; but both muscles gradually converge towards the outer part of the space.

Dissection.—Detach the Pectoralis major by dividing the muscle along its attachment to the clavicle, and by making a vertical incision through its substance a little external to its line of attachment to the sternum and costal cartilages. The muscle should then be reflected outwards, and its tendon carefully examined. The Pectoralis minor is now exposed, and immediately above it, in the interval between its upper border and the clavicle, a strong fascia, the *costo-coracoid membrane*.

The **costo-coracoid membrane** is a strong fascia, situated under cover of the clavicular portion of the Pectoralis major muscle. It occupies the interval between the Pectoralis minor and Subclavius muscles, and protects the axillary vessels and nerves. Traced upwards, it splits to enclose the Subclavius muscle, and its two layers are attached to the clavicle, one in front of and the other behind the muscle; the latter layer fuses with the deep cervical fascia and with the sheath of the axillary vessels. Internally, it blends with the fascia covering the first two intercostal spaces, and is attached also to the first rib internal to the origin of the Subclavius muscle. Externally, it is very thick and dense, and is attached to the coracoid process. The portion extending from its attachment to the first rib to the coracoid process is often whiter and denser than the rest; this is sometimes called the *costo-coracoid ligament*. Below this, it is thin, and at the upper border of the Pectoralis minor it splits into two layers to invest the muscle; from the lower border of the Pectoralis minor it is continued downwards to join the axillary fascia, and outwards to join the fascia over the short head of the Biceps. The costo-coracoid membrane is pierced by the cephalic vein, the acromial thoracic artery and vein, superior thoracic artery, and anterior thoracic nerves.

The **Pectoralis minor** (fig. 230) is a thin, flat, triangular muscle, situated at the upper part of the thorax, beneath the Pectoralis major. It arises by three tendinous digitations, from the upper margin and outer surface of the third, fourth, and fifth ribs, near their cartilages, and from the aponeurosis covering the Intercostal muscles; the fibres pass upwards and outwards, and converge to form a flat tendon, which is inserted into the inner border and upper surface of the coracoid process of the scapula.

Relations.—By its *anterior surface*, with the Pectoralis major, and the thoracic branches of the acromial thoracic artery. By its *posterior surface*, with the ribs, Intercostal muscles, Serratus magnus, the axillary space, and the axillary vessels and brachial plexus of nerves. Its upper border is separated from the clavicle by a triangular interval, broad internally, narrow externally, which is occupied by the costo-coracoid membrane. In this space are the first part of the axillary vessels and nerves. Running parallel to the lower border of the muscle is the long thoracic artery.

The costo-coracoid membrane should now be removed, when the Subclavius muscle will be seen.

The **Subclavius** is a small triangular muscle, placed in the interval between the clavicle and the first rib. It arises by a short, thick tendon from the first rib and its cartilage at their junction, in front of the rhomboid ligament; the fleshy fibres proceed obliquely upwards and outwards, to be inserted into a deep groove on the under surface of the clavicle.

FIG. 230.—Muscles of the chest and front of the arm, with the boundaries of the axilla.

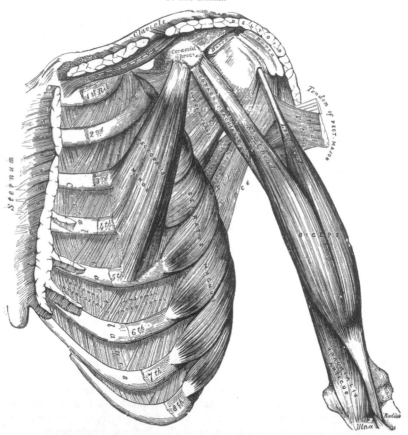

Relations.—By its *upper surface*, with the clavicle. By its *deep surface*, it is separated from the first rib by the subclavian vessels and brachial plexus of nerves. Its *anterior surface* is separated from the Pectoralis major by the costo-coracoid membrane, which, with the clavicle, forms an osseo-fibrous sheath in which the muscle is enclosed.

If the costal attachment of the Pectoralis minor is divided across, and the muscle reflected outwards, the axillary vessels and nerves are brought fully into view, and should be examined.

Nerves.—The Pectoral muscles are supplied by the anterior thoracic nerves; the Pectoralis major through these nerves receives filaments from all the spinal nerves entering into the formation of the brachial plexus; the Pectoralis minor receives its fibres from the eighth cervical and first dorsal nerves. The Sub-clavius is supplied by a filament from the fifth cervical nerve.

Actions.—If the arm has been raised by the Deltoid, the Pectoralis major will, conjointly with the Latissimus dorsi and Teres major, depress it to the side of the chest. If acting alone, it adducts and draws forwards the arm, bringing it across the front of the chest, and at the same time rotates it inwards. The Pectoralis minor depresses the point of the shoulder, drawing the scapula down-wards and inwards to the thorax, and throwing the inferior angle backwards. The Subclavius depresses the shoulder, drawing the clavicle downwards and forwards. When the arms are fixed, all three muscles act upon the ribs, drawing them upwards and expanding the chest, and thus becoming very important agents in forced inspiration. Asthmatic patients always assume an attitude which fixes the shoulders, so that all these muscles may be brought into action to assist in dilating the cavity of the chest.

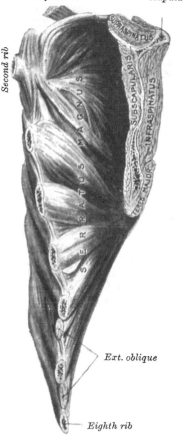

FIG. 231.—Serratus magnus. (From a preparation in the Museum of the Royal College of Surgeons of England.)

Slip of Serr. mag. to first rib   *Spine of scapula*

Ext. oblique

Eighth rib

2. *Lateral Thoracic Region*

Serratus magnus.

The **Serratus magnus** (fig. 231) is a thin, irregularly quadrilateral muscle, situated between the ribs and the scapula at the upper and lateral part of the chest. It arises by nine digitations or slips from the outer surface and upper border of the eight upper ribs (the second rib giving origin to two slips), and from the apo-neurosis covering the corresponding inter-costal muscles. From this extensive attachment the fibres pass backwards, closely applied to the chest-wall, and reach the vertebral border of the scapula, and are inserted into its ventral aspect in the following manner. The upper two digitations—i.e. the one from the first rib and the higher of the two from the second rib—converge to be inserted into a triangular area on the ventral aspect of the superior angle. The next two digitations spread out to form a thin triangular sheet, the base of which is directed backwards and is inserted into nearly the whole length of the ventral aspect of the vertebral border. The lower five digitations converge, as they pass backwards from the ribs, to form a fan-shaped structure, the apex of which is inserted, partly by muscular and partly by tendinous fibres, into a triangular impression on the ventral aspect of the inferior angle. The lower four slips interdigitate at their origin with the upper five slips of the External oblique muscle of the abdomen.

Relations.—This muscle is partly covered, in front, by the Pectoral muscles ; behind, by the Subscapularis. The axillary vessels and nerves lie upon its upper part, while its *deep surface* rests upon the ribs and Intercostal muscles.

Nerve.—The Serratus magnus is supplied by the posterior thoracic nerve, which is derived from the fifth, sixth, and generally the seventh cervical nerves.

Actions.—The Serratus magnus, as a whole, carries the scapula forwards, and at the same time raises the vertebral border of the bone. It is therefore concerned in the action of pushing. Its lower and stronger fibres move forwards the lower angle and assist the Trapezius in rotating the bone round an axis through its centre, and thus assist this muscle in raising the acromion and supporting weights upon the shoulder. It is also an assistant to the Deltoid in raising the arm, inasmuch as during the action of this latter muscle it fixes the scapula and so steadies the glenoid cavity on which the head of the humerus rotates. After the Deltoid has raised the arm to a right angle with the trunk, the Serratus magnus and the Trapezius, by rotating the scapula, raise the arm into an almost vertical position. It is possible that when the shoulders are fixed the lower fibres of the Serratus magnus may assist in raising and everting the ribs ; but it is not the important inspiratory muscle which it was formerly believed to be.

Surgical Anatomy.—When the muscle is paralysed, the vertebral border, and especially the lower angle of the scapula, leave the ribs and stand out prominently on the surface, giving a peculiar 'winged' appearance to the back. The patient is unable to raise the arm, and an attempt to do so is followed by a further projection of the lower angle of the scapula from the back of the thorax.

Dissection.—After completing the dissection of the axilla, if the muscles of the back have been dissected, the upper extremity should be separated from the trunk. Saw through the clavicle at its centre, and then cut through the muscles which connect the scapula and arm with the trunk, viz. the Pectoralis minor in front, Serratus magnus at the side, and the Levator anguli scapulæ, the Rhomboids, Trapezius, and Latissimus dorsi behind. These muscles should be cleaned and traced to their respective insertions. Then make an incision through the integument, commencing at the outer third of the clavicle, and extending along the margin of that bone, the acromion process, and spine of the scapula ; the integument should be dissected from above downwards and outwards, when the fascia covering the Deltoid is exposed (fig. 228, 3).

II. Muscles and Fasciæ of the Shoulder and Arm

The **superficial fascia** of the upper extremity is a thin cellulo-fibrous layer, containing the superficial veins and lymphatics, and the cutaneous nerves. It is most distinct in front of the elbow, where it contains very large superficial veins and nerves ; in the hand it is hardly demonstrable, the integument being closely adherent to the deep fascia by dense fibrous bands. Small subcutaneous bursæ are found in this fascia over the acromion, the olecranon, and the knuckles.

The deep fascia of the upper extremity comprises the aponeurosis of the shoulder, arm, and forearm, the anterior and posterior annular ligaments of the carpus, and the palmar fascia. These will be considered in the description of the muscles of the several regions.

3. *Acromial Region*
Deltoid.

The **deep fascia** covering the Deltoid (deltoid aponeurosis) is a fibrous layer, which covers the outer surface of the muscle, thick and strong behind, where it is continuous with the Infraspinatus fascia, thinner over the rest of its extent. It sends numerous prolongations between the fasciculi of the muscle. In front, it is continuous with the fascia covering the great Pectoral muscle ; behind, with that covering the Infraspinatus ; above, it is attached to the clavicle, the acromion, and spine of the scapula ; below, it is continuous with the deep fascia of the arm.

The **Deltoid** (fig. 229) is a large, thick, triangular muscle, which gives the rounded outline to the shoulder, and has received its name from its resemblance to the Greek letter Δ reversed. It surrounds the shoulder-joint in the greater part of its extent, covering it on its outer side, and in front and behind. It arises from the outer third of the anterior border and upper surface of the clavicle ;

from the outer margin and upper surface of the acromion process ; and from the lower lip of the posterior border of the spine of the scapula, as far back as the triangular surface at its inner end.   From this extensive origin the fibres converge towards their insertion, the middle passing vertically, the anterior obliquely backwards, the posterior obliquely forwards ; they unite to form a thick tendon, which is inserted into a rough triangular prominence on the middle of the outer side of the shaft of the humerus.   At its insertion the muscle gives off an expan-sion to the deep fascia of the arm.   This muscle is remarkably coarse in texture, and the arrangement of its muscular fibres is somewhat peculiar ; the central portion of the muscle—that is to say, the part arising from the acromion process —consists of oblique fibres, which arise in a bipenniform manner from the sides of tendinous intersections, generally four in number, which are attached above to the acromion process and pass downwards parallel to one another in the substance of the muscle.   The oblique muscular fibres thus formed are inserted into similar tendinous intersections, generally three in number, which pass upwards from the insertion of the muscle into the humerus and alternate with the descending septa. The portions of the muscle which arise from the clavicle and spine of the scapula are not arranged in this manner, but pass from their origin above, to be inserted into the margins of the inferior tendon.

Relations.—By its *superficial surface*, with the integument, the superficial and deep fasciæ, Platysma, and supra-acromial nerves.   Its *deep surface* is separated from the head of the humerus by a large sacculated synovial bursa, and covers the coracoid process, coraco-acromial ligament, Pectoralis minor, Coraco-brachialis, both heads of the Biceps, the tendon of the Pectoralis major, the insertions of the Supraspinatus, Infraspinatus, and Teres minor, the scapular and external heads of the Triceps, the circumflex vessels and nerve, and the humerus.   Its *anterior border* is separated at its upper part from the Pectoralis major by a cellular interspace, which lodges the cephalic vein and humeral branch of the acromial thoracic artery : lower down the two muscles are in close contact.   Its *posterior border* rests on the Infraspinatus and Triceps muscles.

Nerves.—The Deltoid is supplied by the fifth and sixth cervical through the circumflex nerve.

Actions.—The Deltoid raises the arm directly from the side, so as to bring it at right angles with the trunk.   Its anterior fibres, assisted by the Pectoralis major, draw the arm forwards ; and its posterior fibres, aided by the Teres major and Latissimus dorsi, draw it backwards.

Surgical Anatomy.—The Deltoid is very liable to atrophy, and when in this condition simulates dislocation of the shoulder-joint, as there is flattening of the shoulder and apparent prominence of the acromion process ; upon examination, however, it will be found that the relative position of the great tuberosity of the humerus to the acromion and coracoid process is unchanged.   Atrophy of the Deltoid may be due to disuse or loss of trophic influence, either from injury to the circumflex nerve or cord lesions, as in infantile paralysis.

Dissection.—Divide the Deltoid across, near its upper part, by an incision carried along the margin of the clavicle, the acromion process, and spine of the scapula, and reflect it downwards, when the structures under cover of it will be seen.

4. *Anterior Scapular Region*

Subscapularis.

The **subscapular fascia** is a thin membrane attached to the entire circum-ference of the subscapular fossa, and affording attachment by its inner surface to some of the fibres of the Subscapularis muscle : when this is removed, the Sub-scapularis muscle is exposed.

The **Subscapularis** (fig. 230) is a large triangular muscle, which fills up the subscapular fossa, arising from its internal two-thirds, with the exception of a narrow margin along the posterior border, and the surfaces at the superior and inferior angles which afford attachment to the Serratus magnus : it also arises from the lower two-thirds of the groove on the axillary border of the bone. Some fibres arise from tendinous laminæ, which intersect the muscle, and are attached to ridges on the bone ; and others from an aponeurosis, which separates the muscle from the Teres major and the long head of the Triceps.   The fibres pass outwards, and, gradually converging, terminate in a tendon, which is inserted into the lesser tuberosity of the humerus.   Those fibres which arise from the axillary border of the scapula are inserted into the neck of the humerus to the extent of an inch below the tuberosity.   The tendon of the muscle is in close contact with the anterior part of the capsular ligament of the shoulder-joint, and glides over a large bursa, which separates it from the base of the coracoid process. This bursa communicates with the cavity of the joint by an aperture in the capsular ligament.

Relations.—Its *anterior surface* forms a considerable part of the posterior wall of the axilla, and is in relation with the Serratus magnus, Coraco-brachialis, and Biceps, the axillary vessels and brachial plexus of nerves, and the subscapular vessels and nerves.   By its *posterior surface*, with the scapula and the capsular ligament of the shoulder-joint.   Its *lower border* is contiguous with the Teres major and Latissimus dorsi.

Nerves.—It is supplied by the fifth and sixth cervical nerves through the upper and lower subscapular nerves.

Actions.—The Subscapularis rotates the head of the humerus inwards ; when the arm is raised, it draws the humerus forwards and downwards.   It is a powerful defence to the front of the shoulder-joint, preventing displacement of the head of the bone.

5. *Posterior Scapular Region* (fig. 232)

Supraspinatus.	Teres minor.
Infraspinatus.	Teres major.

Dissection.—To expose these muscles, and to examine their mode of insertion into the humerus, detach the Deltoid and Trapezius from their attachment to the spine of the scapula and acromion process.   Remove the clavicle by dividing the ligaments connecting it with the coracoid process, and separate it at its articulation with the scapula : divide the acromion process near its root with a saw.   The fragments being removed, the tendons of the posterior Scapular muscles will be fully exposed, and can be examined.   A block should be placed beneath the shoulder-joint, so as to make the muscles tense.

The **supraspinous fascia** is a thick and dense membranous layer, which completes the osseo-fibrous case in which the Supraspinatus muscle is contained ; affording attachment, by its inner surface, to some of the fibres of the muscle. It is thick internally, but thinner externally under the coraco-acromial ligament. When this fascia is removed, the Supraspinatus muscle is exposed.

The **Supraspinatus muscle** occupies the whole of the supraspinous fossa, arising from its internal two-thirds, and from the strong fascia which covers its surface.   The muscular fibres converge to a tendon, which passes across the upper part of the capsular ligament of the shoulder-joint, to which it is intimately adherent, and is inserted into the highest of the three facets on the great tuberosity of the humerus.

Relations.—By its *upper surface*, with the Trapezius, the clavicle, the acromion, the coraco-acromial ligament, and the Deltoid.   By its *under surface*, with the scapula, the suprascapular vessels and nerve, and upper part of the shoulder-joint.

The **infraspinous fascia** is a dense fibrous membrane, covering in the Infraspinatus muscle and attached to the circumference of the infraspinous fossa ; it affords attachment, by its inner surface, to some fibres of that muscle.   At the point where the Infraspinatus commences to be covered by the Deltoid, this fascia divides into two layers : one layer passes over the Deltoid muscle, helping to form the Deltoid fascia already described ; the other passes beneath the Deltoid to the shoulder-joint.

The **Infraspinatus** is a thick triangular muscle, which occupies the chief part of the infraspinous fossa, arising by fleshy fibres from its internal two-thirds; and by tendinous fibres from the ridges on its surface : it also arises from a strong fascia which covers it externally, and separates it from the Teres major and minor. The fibres converge to a tendon, which glides over the external border of the spine of the scapula, and, passing across the posterior part of the capsular ligament of the shoulder-joint, is inserted into the middle facet on the great tuberosity of the humerus. The tendon of this muscle is occasionally separated from the spine of the scapula by a synovial bursa, which communicates with the synovial cavity of the shoulder-joint.

Relations.—By its *posterior surface*, with the Deltoid, the Trapezius, Latissimus dorsi, and the integument. By its *anterior surface*, with the

FIG. 232.—Muscles on the dorsum of the scapula and the Triceps.

scapula, from which it is separated by the suprascapular and dorsalis scapulæ vessels, and with the capsular ligament of the shoulder-joint. Its *lower border* is in contact with the Teres minor, and occasionally united with it, and with the Teres major.

The **Teres minor** is a narrow, elongated muscle, which arises from the dorsal surface of the axillary border of the scapula for the upper two-thirds of its extent, and from two aponeurotic laminæ, one of which separates this muscle from the Infraspinatus, the other from the Teres major ; its fibres pass obliquely upwards and outwards, and terminate in a tendon, which is inserted into the lowest of the three facets on the great tuberosity of the humerus, and, by fleshy fibres, into the humerus immediately below it. The tendon of this muscle passes across the posterior part of the capsular ligament of the shoulder-joint.

Relations.—By its *posterior surface*, with the Deltoid, and the integument. By its *anterior surface*, with the scapula, the dorsal branch of the subscapular artery, the long head of the Triceps, and the shoulder-joint. By its *upper border*, with the Infraspinatus. By its *lower border*, with the Teres major, from which it is separated anteriorly by the long head of the Triceps.

The **Teres major** is a thick but somewhat flattened muscle, which arises from the oval surface on the dorsal aspect of the inferior angle of the scapula, and from the fibrous septa interposed between it and the Teres minor and Infraspinatus; the fibres are directed upwards and outwards, and terminate in a flat tendon, about two inches in length, which is inserted into the inner bicipital ridge of the humerus. The tendon of this muscle, at its insertion into the humerus, lies behind that of the Latissimus dorsi, from which it is separated by a synovial bursa, the two tendons being, however, united along their lower borders for a short distance.

Relations.—By its *posterior surface*, with the Latissimus dorsi below, and the long head of the Triceps above. By its *anterior surface*, with the Subscapularis, Latissimus dorsi, Coraco-brachialis, short head of the Biceps, the axillary vessels, and brachial plexus of nerves. Its *upper border* is at first in relation with the Teres minor, from which it is afterwards separated by the long head of the Triceps. Its *lower border* forms, in conjunction with the Latissimus dorsi, part of the posterior boundary of the axilla. The Latissimus dorsi at first covers the origin of the Teres major, then wraps itself obliquely round its lower border, so that its tendon ultimately comes to lie in front of that of the Teres major.

Nerves.—The Supra- and Infraspinatus muscles are supplied by the fifth and sixth cervical nerves through the suprascapular nerve; the Teres minor, by the fifth cervical, through the circumflex; and the Teres major, by the fifth and sixth cervical, through the lower subscapular.

Actions.—The Supraspinatus assists the Deltoid in raising the arm from the side, and fixes the head of the humerus in the glenoid cavity. The Infraspinatus and Teres minor rotate the head of the humerus outwards: when the arm is raised, they assist in retaining it in that position and carrying it backwards. One of the most important uses of these three muscles is the great protection they afford to the shoulder-joint, the Supraspinatus supporting it above, and preventing displacement of the head of the humerus upwards, while the Infraspinatus and Teres minor protect it behind, and prevent dislocation backwards. The Teres major assists the Latissimus dorsi in drawing the humerus downwards and backwards, when previously raised, and rotating it inwards; when the arm is fixed it may assist the Pectoral and Latissimus dorsi muscles in drawing the trunk forwards.

6. *Anterior Humeral Region* (fig. 230)

Coraco-brachialis.   Biceps.   Brachialis anticus.

Dissection.—The arm being placed on the table, with the front surface uppermost, make a vertical incision through the integument along the middle line, from the outer extremity of the anterior fold of the axilla, to about two inches below the elbow-joint, where it should be joined by a transverse incision, extending from the inner to the outer side of the forearm; the two flaps being reflected on either side, the fascia should be examined (fig. 228).

The **deep fascia** of the arm is continuous with that covering the Deltoid and the great Pectoral muscles, by means of which it is attached, above, to the clavicle, acromion, and spine of the scapula; it forms a thin, loose, membranous sheath investing the muscles of the arm, and sends down septa between them; it is composed of fibres disposed in a circular or spiral direction, and connected together by vertical and oblique fibres. It differs in thickness at different parts, being thin over the Biceps, but thicker where it covers the Triceps, and over the condyles of the humerus: it is strengthened by fibrous aponeuroses, derived from the Pectoralis major and Latissimus dorsi on the inner side, and from the Deltoid externally. On either side it gives off a strong intermuscular septum, which is attached to the supra-condylar ridge and condyle of the humerus. These septa serve to separate the muscles of the anterior from those of the posterior brachial region. The *external intermuscular septum* extends from the lower part of the anterior bicipital ridge, along the external supra-condylar ridge, to the outer condyle; it is blended with the tendon of the Deltoid, gives attachment to the Triceps behind, to the Brachialis anticus, Supinator longus, and Extensor carpi radialis longior, in front; and is perforated by the musculo-spiral nerve and superior profunda artery. The *internal intermuscular septum*, thicker than the preceding, extends from the lower part of the posterior lip of the bicipital groove below the Teres major, along the internal supra-condylar ridge to the inner condyle; it is blended with the tendon of the Coraco-brachialis, and affords attachment to the Triceps behind and the Brachialis anticus in front. It is perforated by the ulnar nerve and the inferior profunda and anastomotic arteries. At the elbow, the deep fascia is attached to all the prominent points round the joint, viz. the condyles of the humerus and the olecranon process of the ulna, and is continuous with the deep fascia of the forearm. Just below the middle of the arm, on its inner side, in front of the internal intermuscular septum, is an oval opening in the deep fascia, which transmits the basilic vein and some lymphatic vessels. On the removal of this fascia, the muscles, vessels, and nerves of the anterior humeral region are exposed.

The **Coraco-brachialis,** the smallest of the three muscles in this region, is situated at the upper and inner part of the arm. It arises by fleshy fibres from the apex of the coracoid process, in common with the short head of the Biceps, and from the intermuscular septum between the two muscles; the fibres pass downwards, backwards, and a little outwards, to be inserted by means of a flat tendon into an impression at the middle of the inner surface and internal border of the shaft of the humerus between the origins of the Triceps and Brachialis anticus. It is perforated by the musculo-cutaneous nerve. The inner border of the muscle forms a guide to the position of the brachial artery in tying the vessel in the upper part of its course.

Relations.—By its *anterior surface*, with the Pectoralis major above, and at its insertion with the brachial vessels and median nerve which cross it. By its *posterior surface*, with the tendons of the Subscapularis, Latissimus dorsi, and Teres major, the inner head of the Triceps, the humerus, and the anterior circumflex vessels. By its *inner border*, with the brachial artery and the median and musculo-cutaneous nerves. By its *outer border*, with the short head of the Biceps and Brachialis anticus.

The **Biceps** (*Biceps flexor cubiti*) is a long fusiform muscle, occupying the whole of the anterior surface of the arm, and divided above into two portions or heads, from which circumstance it has received its name. The short head arises by a thick flattened tendon from the apex of the coracoid process, in common with the Coraco-brachialis. The long head arises from the upper margin of the glenoid cavity, and is continuous with the glenoid ligament. This tendon arches over the head of the humerus, being enclosed in a special sheath of the synovial membrane of the shoulder-joint; it then passes through an opening in the capsular ligament at its attachment to the humerus, and descends in the bicipital groove, in which it is retained by a fibrous prolongation from the tendon of the Pectoralis major. Each tendon is succeeded by an elongated muscular belly, and the two bellies, although closely applied to each other, can readily be separated until within about three inches of the elbow-joint. Here they end in a flattened tendon, which is inserted into the back part of the tuberosity of the radius, a synovial bursa being interposed between the tendon and the front of the tuberosity. As the tendon of the muscle approaches the radius it becomes twisted upon itself, so that its anterior surface becomes external and is applied to the tuberosity of the radius at its insertion: opposite the bend of the elbow the tendon gives off, from its inner side, a broad aponeurosis, the *bicipital fascia* (*semilunar fascia*), which passes obliquely downwards and inwards across the brachial artery, and is continuous with the deep fascia of the forearm (fig. 229). The inner border of this muscle forms a guide to the position of the vessel, in tying the brachial artery in the middle of the arm.*

Relations.—Its *anterior surface* is overlapped above by the Pectoralis major and Deltoid ; in the rest of its extent it is covered by the superficial and deep fasciæ and the integument. Its *posterior surface* rests above on the shoulder-joint and upper part of the humerus ; below it rests on the Brachialis anticus, with the musculo-cutaneous nerve intervening between the two, and on the Supinator brevis. Its *inner border* is in relation with the Coraco-brachialis, and overlaps the brachial vessels, and median nerve ; its *outer border*, with the Deltoid and Supinator longus.

The **Brachialis anticus** is a broad muscle, which covers the elbow-joint and the lower half of the front of the humerus. It is somewhat compressed from before backwards, and is broader in the middle than at either extremity. It arises from the lower half of the outer and inner surfaces of the shaft of the humerus ; and commences above at the insertion of the Deltoid, which it embraces by two angular processes. Its origin extends below, to within an inch of the margin of the articular surface, and is limited on each side by the external and internal borders of the shaft of the humerus. It also arises from the intermuscular septa on each side, but more extensively from the inner than the outer, from which it is separated below by the Supinator longus and Extensor carpi radialis longior. Its fibres converge to a thick tendon, which is inserted into a rough depression on the anterior surface of the coronoid process of the ulna, being received into an interval between two fleshy slips of the Flexor profundus digitorum.

Relations.—By its *anterior surface*, with the Biceps, the brachial vessels, musculo-cutaneous, and median nerves. By its *posterior surface*, with the humerus and front of the elbow-joint. By its *inner border*, with the Triceps, ulnar nerve, and Pronator radii teres, from which it is separated by the intermuscular septum. By its *outer border*, with the musculo-spiral nerve, radial recurrent artery, the Supinator longus, and Extensor carpi radialis longior.

Nerves.—The muscles of this group are supplied by the musculo-cutaneous nerve. The Brachialis anticus usually receives an additional filament from the musculo-spiral. The Coraco-brachialis receives its supply primarily from the seventh cervical, the Biceps and Brachialis anticus from the fifth and sixth cervical nerves.

Actions.—The Coraco-brachialis draws the humerus forwards and inwards, and at the same time assists in elevating it towards the scapula. The Biceps is a flexor of the forearm : it is also a powerful supinator, and serves to render tense the deep fascia of the forearm by means of the broad aponeurosis given off from its tendon. The Brachialis anticus is a flexor of the forearm, and forms an important defence to the elbow-joint. When the forearm is fixed, the Biceps and Brachialis anticus flex the arm upon the forearm, as is seen in efforts of climbing.

7. *Posterior Humeral Region*

Triceps.   Subanconeus.

The **Triceps** (*Triceps extensor cubiti*) (fig. 232) is situated on the back of the arm, extending the entire length of the posterior surface of the humerus. It is

* A third head to the Biceps is occasionally found (Theile says as often as once in eight or nine subjects), arising at the upper and inner part of the Brachialis anticus, with the fibres of which it is continuous, and inserted into the bicipital fascia and inner side of the tendon of the Biceps. In most cases this additional slip passes behind the brachial artery in its course down the arm. Occasionally the third head consists of two slips, which pass down, one in front, the other behind the artery, concealing the vessel in the lower half of the arm.

of large size, and divided above into three parts : hence its name. These three portions have been named (1) the middle, scapular, or long head ; (2) the external, or long humeral head ; and (3) the internal, or short humeral head.

The *middle* or *scapular head* arises, by a flattened tendon, from a rough triangular depression on the scapula, immediately below the glenoid cavity, being blended at its upper part with the capsular ligament ; the muscular fibres pass downwards between the two other portions of the muscle, and join with them in the common tendon of insertion.

The *external head* arises from the posterior surface of the shaft of the humerus, between the insertion of the Teres minor and the upper part of the musculo-spiral groove ; from the external border of the humerus and the external intermuscular septum : the fibres from this origin converge towards the common tendon of insertion.

The *internal head* arises from the posterior surface of the shaft of the humerus, below the groove for the musculo-spiral nerve : commencing above, narrow and pointed, below the insertion of the Teres major, and extending to within an inch of the trochlear surface : it also arises from the internal border of the humerus and from the back of the whole length of the internal and lower part of the external intermuscular septa. The fibres of this portion of the muscle are directed, some downwards to the olecranon, while others converge to the common tendon of insertion.

The *common tendon* of the Triceps commences about the middle of the back part of the muscle : it consists of two aponeurotic laminæ, one of which is subcutaneous and covers the posterior surface of the muscle for the lower half of its extent ; the other is more deeply seated in the substance of the muscle ; after receiving the attachment of the muscular fibres, they join together above the elbow, and are inserted, for the most part, into the back part of the upper surface of the olecranon process ; a band of fibres is, however, continued downwards, on the outer side, over the Anconeus, to blend with the deep fascia of the forearm. A small bursa, occasionally multilocular, is situated on the front part of this surface, beneath the tendon.

The long head of the Triceps descends between the Teres minor and Teres major, dividing the triangular space between these two muscles and the humerus into two smaller spaces, one triangular, the other quadrangular (fig. 232). The triangular space contains the dorsalis scapulæ vessels ; it is bounded by the Teres minor above, the Teres major below, and the scapular head of the Triceps externally : the quadrangular space transmits the posterior circumflex vessels and the circumflex nerve ; it is bounded by the Teres minor above, the Teres major below, the scapular head of the Triceps internally, and the humerus externally.

Relations.—By its *posterior surface*, with the Deltoid above : in the rest of its extent it is subcutaneous. By its *anterior surface*, with the humerus, musculo-spiral nerve, superior profunda vessels, and back part of the elbow-joint. Its *middle* or *long head* is in relation, behind, with the Deltoid and Teres minor ; in front, with the Subscapularis, Latissimus dorsi, and Teres major.

The **Subanconeus** is a name given to a few fibres from the under surface of the lower part of the Triceps muscle, which are inserted into the posterior ligament of the elbow-joint. By some authors it is regarded as the analogue of the Subcrureus in the lower limb, but it is not a separate muscle.

Nerves.—The Triceps is supplied by the seventh and eighth cervical nerves through the musculo-spiral nerve.

Actions.—The Triceps is the great extensor muscle of the forearm, serving, when the forearm is flexed, to extend the elbow-joint. It is the direct antagonist of the Biceps and Brachialis anticus. When the arm is extended, the long head of the muscle may assist the Teres major and Latissimus dorsi in drawing the humerus backwards and in adducting it to the thorax. The long head of the Triceps protects the under part of the shoulder-joint, and prevents displacement of the head of the humerus downwards and backwards. The Subanconeus draws up the posterior ligament during extension of the forearm.

Surgical Anatomy.—The existence of the band of fibres from the Triceps to the fascia of the forearm is of importance in excision of the elbow, and should always be carefully preserved from injury by the operator, as by means of these fibres the patient is enabled to extend the forearm, a movement which would otherwise mainly be accomplished by gravity, that is to say, by allowing the forearm to drop from its own weight.

III. MUSCLES AND FASCIÆ OF THE FOREARM

Dissection.—To dissect the forearm, place the limb in the position indicated in fig. 228, make a vertical incision along the middle line from the elbow to the wrist, and a transverse incision at the extremity of this; the superficial structures being removed, the deep fascia of the forearm is exposed.

The **deep fascia** of the forearm, continuous above with that enclosing the arm, is a dense, highly glistening aponeurotic investment, which forms a general sheath enclosing the muscles in this region; it is attached, behind, to the olecranon and posterior border of the ulna, and gives off from its inner surface numerous intermuscular septa, which enclose each muscle separately. Below, it is continuous in front with the anterior annular ligament, and forms a sheath for the tendon of the Palmaris longus muscle, which passes over the annular ligament to be inserted into the palmar fascia. Behind, near the wrist-joint, it becomes much thickened by the addition of many transverse fibres, and forms the posterior annular ligament. It consists of circular and oblique fibres, connected together by numerous vertical fibres. It is much thicker on the dorsal than on the palmar surface, and at the lower than at the upper part of the forearm, and is strengthened above by tendinous fibres derived from the Brachialis anticus and Biceps in front, and from the Triceps behind. Its deep surface gives origin to muscular fibres, especially at the upper part of the inner and outer sides of the forearm, and forms the boundaries of a series of conical-shaped cavities, in which the muscles are contained. Besides the vertical septa separating the individual muscles, transverse septa are given off both on the anterior and posterior surfaces of the forearm, separating the deep from the superficial layer of muscles. Numerous apertures exist in the fascia for the passage of vessels and nerves; one of these, of large size, situated at the front of the elbow, serves for the passage of a communicating branch between the superficial and deep veins.

The muscles of the forearm may be subdivided into groups corresponding to the region they occupy. One group occupies the inner and anterior aspect of the forearm, and comprises the Flexor and Pronator muscles. Another group occupies its outer side; and a third its posterior aspect. The two latter groups include all the Extensor and Supinator muscles.

8. *Anterior Radio-Ulnar Region*

The muscles in this region are divided for convenience of description into two groups or layers, superficial and deep.

Superficial layer

Pronator radii teres.	Palmaris longus.
Flexor carpi radialis.	Flexor carpi ulnaris.

Flexor sublimis digitorum.

These muscles take origin from the internal condyle of the humerus by a common tendon.

The **Pronator radii teres** arises by two heads. One, the larger and more superficial, arises from the humerus, immediately above the internal condyle, and from the tendon common to the origin of the other muscles; also from the fascia of the forearm, and intermuscular septum between it and the Flexor carpi radialis. The other head is a thin fasciculus, which arises from the inner side of the coronoid process of the ulna, joining the preceding at an acute angle. Between the two heads the median nerve enters the forearm. The muscle passes obliquely across the forearm from the inner to the outer side, and terminates in a flat tendon, which turns over the outer margin of the radius, and is inserted into a rough impression at the middle of the outer surface of the shaft of that bone.

Relations.—By its *anterior surface* throughout the greater part of its extent with the deep fascia ; at its insertion it is crossed by the radial vessels and nerve and covered by the Supinator longus. By its *posterior surface*, with the Brachialis anticus, Flexor sublimis digitorum, the median nerve, and ulnar artery : the small, or deep, head being interposed between the two latter structures. Its *outer border* forms the inner boundary of a triangular space, in which is placed the brachial artery, median nerve, and tendon of the Biceps muscle. Its *inner border* is in contact with the Flexor carpi radialis.

Surgical Anatomy.—This muscle, when suddenly brought into very active use, as in the game of lawn tennis, is apt to be strained, producing slight swelling, tenderness, and pain on putting the muscle into action. This is known as 'lawn-tennis arm.'

The **Flexor carpi radialis** lies on the inner side of the preceding muscle. It arises from the internal condyle by the common tendon, from the fascia of the forearm, and from the intermuscular septa between it and the Pronator radii teres, on the outside ; the Palmaris longus, internally ; and the Flexor sublimis digitorum, beneath. Slender and aponeurotic in structure at its commencement, it increases in size, and terminates in a tendon, which forms rather more than the lower half of its length. This tendon passes through a canal in the outer side of the annular ligament, runs through a groove in the os trapezium (which is converted into a canal by a fibrous sheath, and lined by a synovial membrane), and is inserted into the base of the metacarpal bone of the index finger, and by a slip into the base of the metacarpal bone of the middle finger. The radial artery lies between the tendon of this muscle and the Supinator longus, and may easily be tied in this situation.

Relations.—By its *superficial surface*, with the deep fascia and the integument. By its *deep surface*, with the Flexor sublimis digitorum, Flexor longus pollicis, and wrist-joint. By its *outer border*, with the Pronator radii teres and the radial vessels. By its *inner border*, with the Palmaris longus above, and the median nerve below.

The **Palmaris longus** is a slender, fusiform muscle, lying on the inner side of the preceding. It arises from the inner condyle of the humerus by the common tendon, from the deep fascia, and the intermuscular septa between it and the adjacent muscles. It terminates in a slender, flattened tendon, which passes over the upper part of the annular ligament, to end in the central part of the palmar fascia and lower part of the annular ligament, frequently sending a tendinous slip to the short muscles of the thumb. This muscle is often absent, and is subject to very considerable variations : it may be tendinous above and muscular below ; or it may be muscular in the centre, with a tendon above and below; or it may present two muscular bundles with a central tendon ; or finally it may consist simply of a mere tendinous band.

Relations.—By its *superficial surface*, with the deep fascia. By its *deep surface*, with the Flexor sublimis digitorum. *Internally*, with the Flexor carpi ulnaris. *Externally*, with the Flexor carpi radialis. The median nerve lies close to the tendon, just above the wrist, on its inner and posterior side.

The **Flexor carpi ulnaris** lies along the ulnar side of the forearm. It arises by two heads, connected by a tendinous arch, beneath which pass the ulnar nerve and posterior ulnar recurrent artery. One head arises from the inner condyle of the humerus by the common tendon ; the other from the inner margin of the olecranon and from the upper two-thirds of the posterior border of the ulna by an aponeurosis, common to it and the Extensor carpi ulnaris and Flexor profundus digitorum ; and from the intermuscular septum between it and the Flexor sublimis digitorum. The fibres terminate in a tendon, which occupies the anterior part of the lower half of the muscle, and is inserted into the pisiform bone, and is prolonged from this to the fifth metacarpal and unciform bones, by the piso-metacarpal and piso-uncinate ligaments : it is also attached by a few fibres to the annular ligament. The ulnar artery lies on the outer side of the tendon of

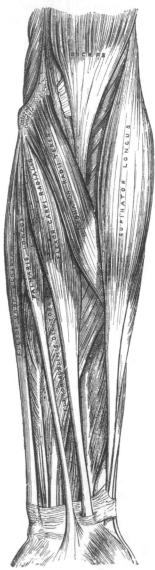

FIG. 233.—Front of the left forearm. Superficial muscles.

this muscle, in the lower two-thirds of the forearm ; the tendon forming a guide in tying the vessel in this situation.

Relations.—By its *superficial surface*, with the deep fascia, with which it is intimately connected for a considerable extent. By its *deep surface*, with the Flexor sublimis digitorum, the Flexor profundus digitorum, the Pronator quadratus, and the ulnar vessels and nerve. By its *outer* or *radial border*, with the Palmaris longus above, and the ulnar vessels and nerve below.

The **Flexor sublimis digitorum** (*perforatus*) is placed beneath the preceding muscles, which therefore must be removed in order to bring its attachment into view. It is the largest of the muscles of the superficial layer, and arises by three heads. One head arises from the internal condyle of the humerus by the common tendon, from the internal lateral ligament of the elbow-joint, and from the intermuscular septum common to it and the preceding muscles. The second head arises from the inner side of the coronoid process of the ulna, above the ulnar origin of the Pronator radii teres (fig. 100, p. 121). The third head arises from the oblique line of the radius, extending from the tubercle to the insertion of the Pronator radii teres. The fibres pass vertically downwards, forming a broad and thick muscle, which speedily divides into two planes of muscular fibres, superficial and deep : the superficial plane divides into two parts which end in tendons for the middle and ring fingers ; the deep plane also divides into two parts, which end in tendons for the index and little fingers, but previously to having done so, it gives off a muscular slip, which joins that part of the superficial plane which is intended for the ring finger. As the four tendons thus formed pass beneath the annular ligament into the palm of the hand, they are arranged in pairs, the superficial pair corresponding to the middle and ring fingers, the deep pair to the index and little fingers. The tendons diverge from one another as they pass onwards. Opposite the bases of the first phalanges each tendon divides into two slips, to allow of the passage of the corresponding tendon of the Flexor profundus digitorum ; the two portions of the tendon then unite and form a grooved channel for the reception of the accompanying deep flexor tendon. Finally they subdivide a second time, to be inserted into the sides of the second phalanges about their middle. After leaving the palm, these tendons accompanied by the deep flexor tendons lie in osseo-aponeurotic canals (fig. 234). These canals are formed by strong fibrous bands, which arch across the tendons, and are attached on each side to the margins of the phalanges. Opposite the middle of the proximal and second pha- langes the sheath is very strong, and the fibres pass transversely ; but opposite the joints it is much thinner, and the fibres pass obliquely. Each sheath is lined by a synovial membrane, which is reflected on the contained tendons.

Relations.—In the fore- arm, by its *superficial sur- face*, with the deep fascia and all the preceding superficial muscles ; by its *deep surface*, with the Flexor profundus digitorum, Flexor longus pol- licis, the ulnar vessels and nerve, and the median nerve. In the hand, its tendons are in relation, *in front*, with the palmar fascia, superficial palmar arch, and the branches of the median nerve ; *behind*, with the tendons of the deep Flexor and the Lumbricales.

Deep Layer

Flexor profundus digitorum.
Flexor longus pollicis.
Pronator quadratus.

Dissection.—Divide each of the superficial muscles at its centre, and turn either end aside ; the deep layer of muscles, to- gether with the median nerve and ulnar vessels, will then be exposed.

The **Flexor profundus digitorum** (*perforans*) (fig. 234) is situated on the ulnar side of the forearm, imme- diately beneath the superficial Flexors. It arises from the upper three-fourths of the anterior and inner surfaces of the shaft of the ulna, em- bracing the insertion of the Brachialis anticus above, and extending, below, to within a short distance of the Pronator quadratus. It also arises from a depression on the inner side of the coronoid process ;

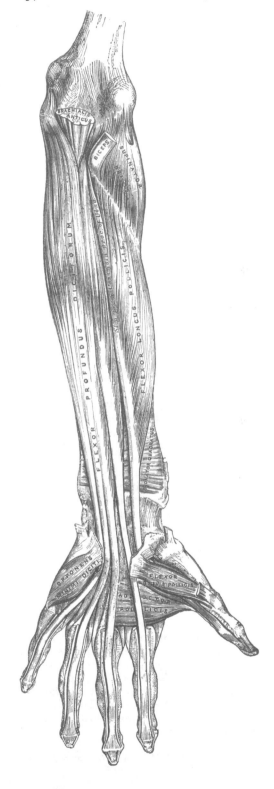

Fɪɢ. 234.—Front of the left forearm.   Deep muscles.

by an aponeurosis from the upper three-fourths of the posterior border of the ulna, in common with the Flexor and Extensor carpi ulnaris ; and from the ulnar half of the interosseous membrane. The fibres form a fleshy belly of considerable size, which divides into four tendons : these pass under the annular ligament beneath the tendons of the Flexor sublimis digitorum. Opposite the first phalanges, the tendons pass through the openings in the tendons of the Flexor sublimis digitorum, and are finally inserted into the bases of the last phalanges. The portion of the muscle for the index finger is usually distinct throughout, but the tendons for the three inner fingers are connected together by cellular tissue and tendinous slips, as far as the palm of the hand. The tendons of this muscle and those of the Flexor sublimis digitorum, while contained in the osseo-aponeurotic canals of the fingers, are invested in a synovial sheath, and are connected to each other, and to the phalanges, by slender, tendinous filaments, called *vincula accessoria tendinum*. One of these connects the deep tendon to the bone, before it passes through the superficial tendon ; a second connects the two tendons together, after the deep tendons have passed through ; and a third connects the deep tendon to the head of the second phalanx. This last consists largely of yellow elastic tissue, and may assist in drawing down the tendon after flexion of the finger.*

Four small muscles, the Lumbricales, are connected with the tendons of the Flexor profundus in the palm. They will be described with the muscles in that region.

Relations.—By its *superficial surface*, in the forearm, with the Flexor sublimis digitorum, the Flexor carpi ulnaris, the ulnar vessels and nerve, and the median nerve ; and in the hand, with the tendons of the superficial Flexor. By its *deep surface*, in the forearm, with the ulna, the interosseous membrane, the Pronator quadratus ; and in the hand, with the Interossei, Adductores pollicis, and deep palmar arch. By its *ulnar border*, with the Flexor carpi ulnaris. By its *radial border*, with the Flexor longus pollicis, the anterior interosseous vessels and nerve being interposed.

The **Flexor longus pollicis** is situated on the radial side of the forearm, lying on the same plane as the preceding. It arises from the grooved anterior surface of the shaft of the radius : commencing, above, immediately below the tuberosity and oblique line, and extending, below, to within a short distance of the Pronator quadratus. It also arises from the adjacent part of the interosseous membrane, and generally by a fleshy slip from the inner border of the coronoid process, or from the internal condyle of the humerus. The fibres pass downwards, and terminate in a flattened tendon, which passes beneath the annular ligament, is then lodged in the interspace between the outer head of the Flexor brevis pollicis and the Adductor obliquus pollicis, and, entering an osseo-aponeurotic canal similar to those for the other flexor tendons, is inserted into the base of the last phalanx of the thumb.

Relations.—By its *superficial surface*, with the Flexor sublimis digitorum, Flexor carpi radialis, Supinator longus, and radial vessels. By its *deep surface*, with the radius, interosseous membrane, and Pronator quadratus. By its *ulnar border*, with the Flexor profundus digitorum, from which it is separated by the anterior interosseous vessels and nerve.

The **Pronator quadratus** is a small, flat, quadrilateral muscle, extending transversely across the front of the radius and ulna, above their carpal extremities. It arises from the oblique or pronator ridge on the lower part of the anterior surface of the shaft of the ulna ; from the lower fourth of the anterior surface, and the anterior border of the ulna ; and from a strong aponeurosis which covers the inner third of the muscle. The fibres pass outwards and slightly downwards, to be inserted into the lower fourth of the anterior surface and anterior border of the shaft of the radius.

* Marshall, *Brit. and For. Med.-Chir. Rev.* 1853.

Relations.—By its *superficial surface,* with the Flexor profundus digitorum, the Flexor longus pollicis, Flexor carpi radialis, and the radial vessels. By its *deep surface,* with the radius, ulna, and interosseous membrane.

Nerves.—All the muscles of the superficial layer are supplied by the median nerve, excepting the Flexor carpi ulnaris, which is supplied by the ulnar. The Pronator radii teres and the Flexor carpi radialis derive their supply primarily from the sixth cervical; the Palmaris longus from the eighth cervical; the Flexor sublimis digitorum from the seventh and eighth cervical and first dorsal, and the Flexor carpi ulnaris from the eighth cervical and first dorsal nerves. Of the deep layer, the Flexor profundus digitorum is supplied by the eighth cervical and first dorsal through the ulnar and anterior interosseous branch of the median. The remaining two muscles, Flexor longus pollicis and Pronator quadratus, are also supplied by the eighth cervical and first dorsal through the anterior interosseous branch of the median.

Actions.—These muscles act upon the forearm, the wrist, and hand. The Pronator radii teres helps to rotate the radius upon the ulna, rendering the hand prone; when the radius is fixed, it assists the other muscles in flexing the forearm. The Flexor carpi radialis is one of the flexors of the wrist; when acting alone, it flexes the wrist, inclining it to the radial side. It can also assist in pronating the forearm and hand, and, by continuing its action, in bending the elbow. The Flexor carpi ulnaris is one of the flexors of the wrist; when acting alone, it flexes the wrist, inclining it to the ulnar side; and by continuing to contract, it bends the elbow. The Palmaris longus is a tensor of the palmar fascia. It also assists in flexing the wrist and elbow. The Flexor sublimis digitorum flexes first the middle, and then the proximal phalanx. It assists in flexing the wrist and elbow. The Flexor profundus digitorum is one of the flexors of the phalanges. After the Flexor sublimis has bent the second phalanx, the Flexor profundus flexes the terminal one; but it cannot do so until after the contraction of the superficial muscle. It also assists in flexing the wrist. The Flexor longus pollicis is a flexor of the phalanges of the thumb. When the thumb is fixed, it also assists in flexing the wrist. The Pronator quadratus helps to rotate the radius upon the ulna, rendering the hand prone.

9. *Radial Region* (fig. 235)

Supinator longus. Extensor carpi radialis longior.
Extensor carpi radialis brevior.

Dissection.—Divide the integument in the same manner as in the dissection of the anterior brachial region; and, after having examined the cutaneous vessels and nerves and deep fascia, remove all these structures. The muscles will then be exposed. The removal of the fascia will be considerably facilitated by detaching it from below upwards. Great care should be taken to avoid cutting across the tendons of the muscles of the thumb, which cross obliquely the larger tendons running down the back of the radius.

The **Supinator longus** (brachio-radialis) is the most superficial muscle on the radial side of the forearm: it is fleshy for the upper two-thirds of its extent, tendinous below. It arises from the upper two-thirds of the external supra-condylar ridge of the humerus, and from the external intermuscular septum, being limited above by the musculo-spiral groove. The fibres terminate above the middle of the forearm in a flat tendon, which is inserted into the outer side of the base of the styloid process of the radius.

Relations.—By its *superficial surface,* with the integument and fascia for the greater part of its extent; near its insertion it is crossed by the Extensor ossis metacarpi pollicis and the Extensor brevis pollicis. By its *deep surface,* with the humerus, the Extensor carpi radialis longior and brevior, the insertion of the Pronator radii teres, and the Supinator brevis. By its *inner border,* above the elbow, with the Brachialis anticus, the musculo-spiral nerve, and radial recurrent artery; and in the forearm with the radial vessels and nerve.

Fig. 235.—Posterior surface of the forearm. Superficial muscles.

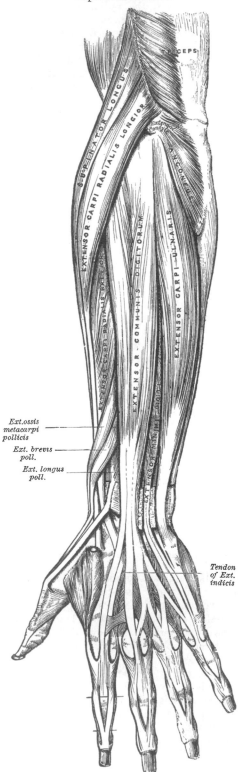

Ext. ossis metacarpi pollicis

Ext. brevis poll.

Ext. longus poll.

Tendon of Ext. indicis

The **Extensor carpi radialis longior** is placed partly beneath the preceding muscle. It arises from the lower third of the external supra-condylar ridge of the humerus, from the external intermuscular septum, and by a few fibres from the common tendon of origin of the Extensor muscles of the forearm. The fibres terminate at the upper third of the forearm in a flat tendon, which runs along the outer border of the radius, beneath the Extensor tendons of the thumb; it then passes through a groove common to it and the Extensor carpi radialis brevior, immediately behind the styloid process, and is inserted into the base of the metacarpal bone of the index finger, on its radial side.

Relations.—By its *superficial surface*, with the Supinator longus, and fascia of the forearm. Its *outer side* is crossed obliquely by the Extensor tendons of the thumb. By its *deep surface*, with the elbow-joint, the Extensor carpi radialis brevior, and back part of the wrist.

The **Extensor carpi radialis brevior** is shorter, as its name implies, and thicker than the preceding muscle, beneath which it is placed. It arises from the external condyle of the humerus, by a tendon common to it and the three following muscles: from the external lateral ligament of the elbow-joint; from a strong aponeurosis which covers its surface; and from the intermuscular septa between it and the adjacent muscles. The fibres terminate about the middle of the forearm in a flat tendon, which is closely connected with that of the preceding muscle, and accompanies it to the wrist, lying in the same groove on the posterior surface of the radius; it passes beneath the Extensor tendons of the thumb, then beneath the annular ligament, and, diverging somewhat from its fellow, is inserted into the base of the metacarpal bone of the middle finger, on its radial side.

The tendons of the two preceding muscles pass through the same compartment of the annular ligament, and are lubricated by a single synovial membrane, but are separated from each other by a small vertical ridge of bone as they lie in the groove at the back of the radius.

Relations.—By its *superficial surface*, with the Extensor carpi radialis longior, and with the Extensor muscles of the thumb which cross it. By its *deep surface*, with the Supinator brevis, tendon of the Pronator radii teres, radius, and wrist-joint. By its *ulnar border*, with the Extensor communis digitorum.

10. *Posterior Radio-Ulnar Region* (fig. 235)

The muscles in this region are divided for purposes of description into two groups or layers, superficial and deep.

Superficial Layer

Extensor communis digitorum.	Extensor carpi ulnaris.
Extensor minimi digiti.	Anconeus.

The **Extensor communis digitorum** is situated at the back part of the forearm. It arises from the external condyle of the humerus, by the common tendon ; from the deep fascia, and the intermuscular septa between it and the adjacent muscles. Just below the middle of the forearm it divides into three fleshy masses, from which tendons proceed ; these pass, together with the Extensor indicis, through a separate compartment of the annular ligament, lubricated by a synovial membrane. The tendons then diverge, the innermost one dividing into two ; and all, after passing across the back of the hand, are inserted into the second and third phalanges of the fingers in the following manner : the outermost tendon, accompanied by the Extensor indicis, goes to the index finger ; the second tendon is sometimes connected to the first by a thin transverse band, and receives a slip from the third tendon ; it goes to the middle finger ; the third tendon gives off the slip to the second, and receives a very considerable part of the fourth tendon ; the fourth, or innermost tendon, divides into two parts : one goes to join the third tendon, the other, reinforced by the Extensor minimi digiti, goes to the little finger. Each tendon opposite the metacarpo-phalangeal articulation becomes narrow and thickened, and gives off a thin fasciculus upon each side of the joint, which blends with the lateral ligaments and serves as the posterior ligament ; after having passed the joint, it spreads out into a broad aponeurosis, which covers the whole of the dorsal surface of the first phalanx ; and is reinforced, in this situation, by the tendons of the Interossei and Lumbricales. Opposite the first phalangeal joint this aponeurosis divides into three slips, a middle and two lateral : the former is inserted into the base of the second phalanx ; and the two lateral, which are continued onwards along the sides of the second phalanx, unite by their contiguous margins, and are inserted into the dorsal surface of the last phalanx. As the tendons cross the phalangeal joints, they furnish them with posterior ligaments.

Relations.—By its *superficial surface*, with the fascia of the forearm and hand, the posterior annular ligament, and integument. By its *deep surface*, with the Supinator brevis, the Extensor muscles of the thumb and index finger, the posterior interosseous vessels and nerve, the wrist-joint, carpus, metacarpus, and phalanges. By its *radial border*, with the Extensor carpi radialis brevior. By its *ulnar border*, with the Extensor minimi digiti and Extensor carpi ulnaris.

The **Extensor minimi digiti** is a slender muscle placed on the inner side of the Extensor communis, with which it is generally connected. It arises from the common tendon by a thin, tendinous slip ; and from the intermuscular septa between it and the adjacent muscles. Its tendon runs through a separate compartment in the annular ligament behind the inferior radio-ulnar joint, then divides into two as it crosses the hand, the outermost division being joined by the slip from the innermost tendon of the common extensor. The two slips thus formed spread into a broad aponeurosis, which after receiving a slip from the Abductor minimi digiti is inserted into the second and third phalanges. The tendon is situated on the ulnar side of, and somewhat more superficial than, the common extensor.

The **Extensor carpi ulnaris** is the most superficial muscle on the ulnar side of the forearm. It arises from the external condyle of the humerus, by the common tendon; by an aponeurosis from the posterior border of the ulna in common with the Flexor carpi ulnaris and the Flexor profundus digitorum; and from the deep fascia of the forearm. This muscle terminates in a tendon, which runs through a groove behind the styloid process of the ulna, passes through a separate compartment in the annular ligament, and is inserted into the prominent tubercle on the ulnar side of the base of the metacarpal bone of the little finger.

Relations.—By its *superficial surface*, with the deep fascia of the forearm. By its *deep surface*, with the ulna, and the muscles of the deep layer.

The **Anconeus** is a small, triangular muscle, placed behind and below the elbow-joint, and appears to be a continuation of the external portion of the Triceps. It arises by a separate tendon from the back part of the outer condyle of the humerus, and is inserted into the side of the olecranon, and upper fourth of the posterior surface of the shaft of the ulna; its fibres diverge from their origin, the upper ones being directed transversely, the lower obliquely inwards.

Relations.—By its *superficial surface*, with a strong fascia derived from the Triceps. By its *deep surface*, with the elbow-joint, the orbicular ligament, the ulna, and a small portion of the Supinator brevis.

Deep Layer (fig. 237)

Supinator brevis.	Extensor brevis pollicis.
Extensor ossis metacarpi pollicis.	Extensor longus pollicis.
Extensor indicis.	

The **Supinator brevis** (fig. 236) is a broad muscle, of a hollow cylindrical form, curved round the upper third of the radius. It consists of two distinct planes of muscular fibres, between which lies the posterior interosseous nerve. The two planes arise in common: the superficial one by tendinous, and the deeper by muscular fibres from the external condyle of the humerus; from the external lateral ligament of the elbow-joint, and the orbicular ligament of the radius; from the ridge on the ulna, which runs obliquely downwards from the posterior extremity of the lesser sigmoid cavity; from the triangular depression in front of it; and from a tendinous expansion which covers the surface of the muscle. The superficial fibres surround the upper part of the radius, and are inserted into the outer edge of the bicipital tuberosity and to the oblique line of the radius, as low down as the insertion of the Pronator radii teres. The upper fibres of the deeper plane form a sling-like fasciculus, which encircles the neck of the radius above the tuberosity and is attached to the back part of its inner surface: the greater part of this portion of the muscle is inserted into the posterior and external surface of the shaft, midway between the oblique line and the head of the bone. Between the insertion of the two planes, the posterior interosseous nerve lies on the shaft of the bone.

Relations.—By its *superficial surface*, with the superficial Extensor and Supinator muscles, and the radial vessels and nerve. By its *deep surface*, with the elbow-joint, the interosseous membrane, and the radius.

The **Extensor ossis metacarpi pollicis** is the most external and the largest of the deep Extensor muscles; it lies immediately below the Supinator brevis, with which it is sometimes united. It arises from the outer part of the posterior surface of the shaft of the ulna below the insertion of the Anconeus, from the interosseous membrane, and from the middle third of the posterior surface of the shaft of the radius. Passing obliquely downwards and outwards, it terminates in a tendon, which runs through a groove on the outer side of the styloid process of the radius, accompanied by the tendon of the Extensor brevis pollicis, and is

FIG. 236.—Supinator brevis. (From a preparation in the Museum of the Royal College of Surgeons of England.)

FIG. 237.—Posterior surface of the forearm. Deep muscles.

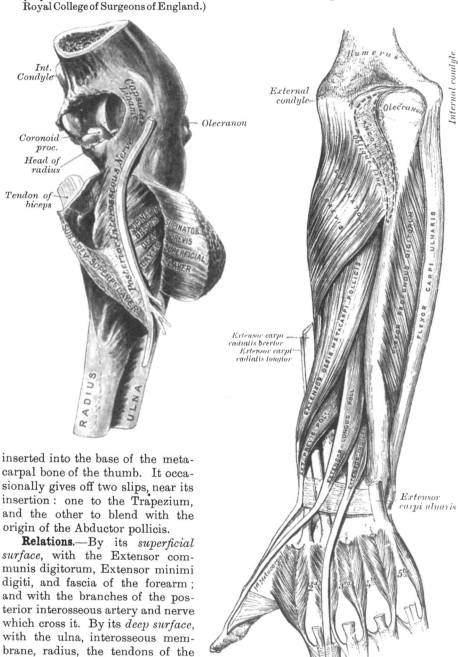

inserted into the base of the metacarpal bone of the thumb. It occasionally gives off two slips, near its insertion: one to the Trapezium, and the other to blend with the origin of the Abductor pollicis.

Relations.—By its *superficial surface*, with the Extensor communis digitorum, Extensor minimi digiti, and fascia of the forearm; and with the branches of the posterior interosseous artery and nerve which cross it. By its *deep surface*, with the ulna, interosseous membrane, radius, the tendons of the Extensor carpi radialis longior and brevior, which it crosses obliquely; and, at the outer side of the wrist, with the radial vessels. By its *upper border*, with the Supinator brevis. By its *lower border*, with the Extensor brevis pollicis.

The **Extensor brevis pollicis** (*Extensor primi internodii pollicis*), the smallest muscle of this group, lies on the inner side of the preceding. It arises from the posterior surface of the shaft of the radius, below the Extensor ossis metacarpi pollicis and from the interosseous membrane. Its direction is similar to that of the Extensor ossis metacarpi pollicis, its tendon passing through the same groove on the outer side of the styloid process, to be inserted into the base of the first phalanx of the thumb.

Relations.—The same as those of the Extensor ossis metacarpi pollicis.

The **Extensor longus pollicis** (*Extensor secundi internodii pollicis*) is much larger than the preceding muscle, the origin of which it partly covers in. It arises from the outer part of the posterior surface of the shaft of the ulna, below the origin of the Extensor ossis metacarpi pollicis, and from the interosseous membrane. It terminates in a tendon, which passes through a separate compartment in the annular ligament, lying in a narrow, oblique groove at the back part of the lower end of the radius. It then crosses obliquely the tendons of the Extensor carpi radialis longior and brevior, being separated from the other Extensor tendons of the thumb by a triangular interval, in which the radial artery is found ; and is finally inserted into the base of the last phalanx of the thumb.

Relations.—By its *superficial surface*, with the same parts as the Extensor ossis metacarpi pollicis. By its *deep surface*, with the ulna, interosseous membrane, the posterior interosseous nerve, radius, the wrist, the radial vessels, and metacarpal bone of the thumb.

The **Extensor indicis** is a narrow, elongated muscle, placed on the inner side of, and parallel with, the preceding. It arises from the posterior surface of the shaft of the ulna, below the origin of the Extensor longus pollicis and from the interosseous membrane. Its tendon passes with the Extensor communis digitorum through the same canal in the annular ligament, and subsequently joins the tendon of the Extensor communis which belongs to the index finger, opposite the lower end of the corresponding metacarpal bone, lying to the ulnar side of the tendon from the common extensor.

Relations.—The relations are similar to those of the preceding muscles.

Nerves.—The Supinator longus is supplied by the sixth, the Extensor carpi radialis longior by the sixth and seventh, and the Anconeus by the seventh and eighth cervical nerves, all through the musculo-spiral nerve ; the remaining muscles of the radial and posterior brachial region are supplied through the posterior interosseous nerve, the Supinator brevis being supplied by the sixth cervical, the Extensor carpi radialis brevior by the sixth and seventh cervical, and all the other muscles by the seventh cervical.

Actions.—The muscles of the radial and posterior brachial regions, which comprise all the Extensor and Supinator muscles, act upon the forearm, wrist, and hand ; they are the direct antagonists of the Pronator and Flexor muscles. The Anconeus assists the Triceps in extending the forearm. The chief action of the Supinator longus is that of a flexor of the elbow-joint, but in addition to this it may act both as a supinator and as a pronator : that is to say, if the forearm is forcibly pronated, it will act as a supinator, and bring the bones into a position midway between supination and pronation ; and, *vice versâ*, if the arm is forcibly supinated, it will act as a pronator, and bring the bones into the same position, midway between supination and pronation. The action of the muscle is therefore to throw the forearm and hand into the position they naturally occupy when placed across the chest. The Supinator brevis is a supinator : that is to say, when the radius has been carried across the ulna in pronation, and the back of the hand is directed forwards, this muscle carries the radius back again to its normal position on the outer side of the ulna, and the palm of the hand is again directed forwards. The Extensor carpi radialis longior extends the wrist and abducts the hand. It may also assist in bending the elbow-joint ; at all events it serves to fix or steady this articulation. The Extensor carpi radialis brevior assists the Extensor carpi radialis longior in extending the wrist, and may also act slightly as an abductor of the hand. The Extensor carpi ulnaris helps to extend the hand, but when acting alone inclines it towards the ulnar side : by its continued action it extends the elbow-joint. The Extensor communis digitorum extends the phalanges, then the wrist, and finally the elbow. It acts principally on the proximal phalanges, the middle and terminal phalanges being extended mainly by the Interossei and Lumbricales. It has also a tendency to separate the fingers as it extends them. The Extensor minimi digiti extends the little finger, and by its continued action assists in extending the wrist. It is owing to this muscle that the little finger can be extended or pointed while the others are flexed. The chief action of the Extensor ossis metacarpi pollicis is to carry the thumb outwards and backwards from the palm of the hand, and hence it has been called the *Abductor pollicis longus.* By its continued action it helps to extend and abduct the wrist. The Extensor brevis pollicis extends the proximal phalanx of the thumb. By its continued action it helps to extend and abduct the wrist. The Extensor longus pollicis extends the terminal phalanx of the thumb. By its continued action it helps to extend and abduct the wrist. The Extensor indicis extends the index finger, and by its continued action assists in extending the wrist. It is owing to this muscle that the index finger can be extended or pointed while the others are flexed.

Surgical Anatomy.—The tendons of the Extensor muscles of the thumb are liable to become strained, and their sheaths inflamed, after excessive exercise, producing a sausage-shaped swelling along the course of the tendons and giving a peculiar creaking sensation to the finger when the muscles act. In consequence of its often being caused by such movements as wringing clothes, it is known as ' washerwoman's sprain.'

IV. Muscles and Fasciæ of the Hand

The **Muscles of the Hand** are subdivided into three groups : 1. Those of the thumb, which occupy the radial side and produce the *thenar* eminence ; 2. Those of the little finger, which occupy the ulnar side and give rise to the *hypothenar* eminence ; 3. Those in the middle of the palm and within the interosseous spaces.

Dissection (fig. 228).—Make a transverse incision across the front of the wrist, and a second across the heads of the metacarpal bones : connect the two by a vertical incision in the middle line, and continue it through the centre of the middle finger. The anterior and posterior annular ligaments, and the palmar fascia, should then be dissected.

The **Anterior Annular Ligament** is a strong, fibrous band, which arches over the carpus, converting the deep groove on the front of the carpal bones into a canal, beneath which pass the flexor tendons of the fingers. It is attached, internally, to the pisiform bone and the hook of the unciform bone ; and externally, to the tuberosity of the scaphoid, and to the inner part of the anterior surface

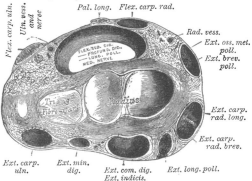

Fig. 238.—Transverse section through the wrist, showing the annular ligaments and the canals for the passage of the tendons.

and the ridge on the trapezium. It is continuous, above, with the deep fascia of the forearm, of which it may be regarded as a thickened portion ; and below, with the palmar fascia. It is crossed by the ulnar vessels and nerve, and the cutaneous branches of the median and ulnar nerves. At its outer extremity is the tendon of the Flexor carpi radialis, which lies in the groove on the trapezium between the attachments of the annular ligament to the bone.

It has inserted into its anterior surface a part of the tendon of the Palmaris longus and part of the tendon of the Flexor carpi ulnaris, and has arising from it, below, the small muscles of the thumb and little finger. Beneath it pass the tendons of the Flexor sublimis and profundus digitorum, the Flexor longus pollicis, and the median nerve.

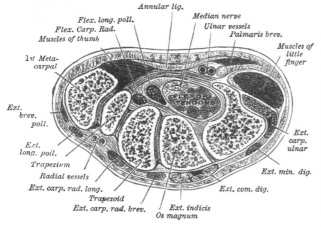

FIG. 239.—Transverse section through the carpus, showing the relative positions of the tendons, vessels, and nerves. (Henle.)

Annular lig.

Flex. long. poll.
Flex. Carp. Rad.
Muscles of thumb

Median nerve
Ulnar vessels
Palmaris brev.

1st Meta-carpal

Muscles of little finger

Ext. brev. poll.

E.xt. carp. ulnar

E.xt. long. poll.

Trapezium

Radial vessels

Ext. min. dig.

Ext. carp. rad. long.

Ext. com. dig.

Trapezoid
Ext. carp. rad. brev.
Os magnum

Ext. indicis

The **Synovial Membranes of the Flexor Tendons at the Wrist.**—There are two synovial membranes which enclose all the tendons as they pass beneath this ligament, one for the Flexor sublimis and profundus digitorum, the other for the Flexor longus pollicis. They extend up into the forearm for about an inch above the annular ligament, and downwards about halfway along the metacarpal bones, where they terminate in a blind diverticulum around each pair of tendons, with the exception of that of the thumb and those of the little finger; in these two digits the diverticulum is continued on and communicates with the synovial sheath of the tendons in the fingers. In the other three fingers the synovial sheath of the tendons begins as a blind pouch without communication with the large synovial sac (fig. 240).

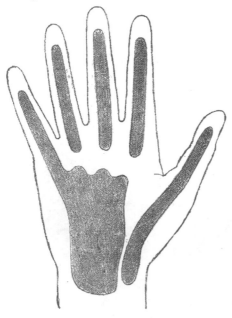

FIG. 240.—Diagram showing the arrangement of the synovial sheaths of the palm and fingers.

Surgical Anatomy.—This arrangement of the synovial sheaths explains the fact that thecal abscess in the thumb and little finger is liable to be followed by abscesses in the forearm, from extension of the inflammation along the continuous synovial sheaths. Ganglion is apt to occur in this situation, constituting 'compound palmar ganglion': it presents an hour-glass outline, with a swelling in front of the wrist and in the palm of the hand, and a constriction, corresponding to the annular ligament, between the two. The fluid can be forced from the one swelling to the other under the ligament.

The **Posterior Annular Ligament** is a strong, fibrous band, extending obliquely downwards and inwards across the back of the wrist, and consisting of the deep fascia of the back of the forearm, strengthened by the addition of some transverse fibres. It binds down the Extensor tendons in their passage to the fingers, being attached, internally, to the styloid process of the ulna, the cuneiform and pisiform bones; externally, to the margin of the radius; and, in its passage across the wrist, to the elevated ridges on the posterior surface of the radius. It presents six compartments for the passage of tendons, each of which is lined by a separate synovial membrane. These are, from without inwards : 1. On the outer side of the styloid process, for the tendons of the Extensor ossis metacarpi and Extensor brevis pollicis. 2. Behind the styloid process, for the tendons of the Extensor carpi radialis longior and brevior. 3. About the middle of the posterior surface of the radius, for the tendon of the Extensor longus pollicis. 4. To the inner side of the latter, for the tendons of the Extensor communis digitorum and Extensor indicis. 5. Opposite the interval between the radius

FIG. 241.—Palmar fascia.
(From a preparation in the Museum of the Royal College of Surgeons of England.)

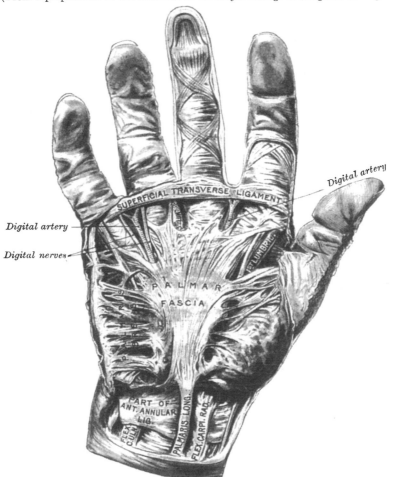

and ulna, for the Extensor minimi digiti. 6. Grooving the back of the ulna, for the tendon of the Extensor carpi ulnaris. The synovial membranes lining these sheaths are usually very extensive, reaching from above the annular ligament, down upon the tendons for a variable distance on the back of the hand.

The **deep palmar fascia** (fig. 241) forms a common sheath which invests the muscles of the hand. It consists of a central and two lateral portions.

The *central portion* occupies the middle of the palm, is triangular in shape, of great strength and thickness, and binds down the tendons and protects the vessels and nerves in this situation. It is narrow above, where it is attached to the lower margin of the annular ligament, and receives the expanded tendon of the Palmaris longus muscle. Below, it is broad and expanded, and divides into four slips, for the four fingers. Each slip gives off superficial fibres, which are inserted into the skin of the palm and finger, those to the palm joining the skin at the furrow corresponding to the metacarpo-phalangeal articulation, and those to the fingers passing into the skin at the transverse fold at the base of the fingers. The deeper part of each slip subdivides into two processes, which are inserted into the lateral margins of the anterior (glenoid) ligament of the metacarpo-phalangeal joint. From the sides of these processes, offsets are sent backwards, to be attached to the borders of the lateral surfaces of the metacarpal bones at their distal extremities. By this arrangement short channels are formed on the front of the lower ends of the metacarpal bones, through which the flexor tendons pass. Dr. W. W. Keen describes a fifth slip as frequently found passing to the thumb. The intervals left in the fascia, between the four fibrous slips, transmit the digital vessels and nerves, and the tendons of the Lumbricales. At the points of division of the palmar fascia into the slips above mentioned, numerous strong, transverse fibres bind the separate processes together. The palmar fascia is intimately adherent to the integument by dense fibro-areolar tissue forming the superficial palmar fascia, and gives origin by its inner margin to the Palmaris brevis; it covers the superficial palmar arch, the tendons of the flexor muscles, and the branches of the median and ulnar nerves; and on each side it gives off a vertical septum, which is continuous with the interosseous aponeurosis, and separates the lateral from the middle palmar group of muscles.

The *lateral portions* of the palmar fascia are thin, fibrous layers, which cover, on the radial side, the muscles of the ball of the thumb, and, on the ulnar side, the muscles of the little finger : they are continuous with the dorsal fascia, and in the palm with the central portion of the palmar fascia.

The **Superficial Transverse Ligament of the Fingers** is a thin, fibrous band, which stretches across the roots of the four fingers, and is closely attached to the skin of the clefts, and internally to the fifth metacarpal bone, forming a sort of rudimentary web. Beneath it the digital vessels and nerves pass onwards to their destination.

Surgical Anatomy.—The palmar fascia is liable to undergo contraction, producing a very inconvenient deformity, known as ' Dupuytren's contraction.' The ring and little fingers are most frequently implicated, but the middle, the index, and the thumb may be involved. The proximal phalanx is drawn down and cannot be straightened, and the two distal phalanges become similarly flexed as the disease advances.

11. *Radial Region* (figs. 242, 243)

Abductor pollicis.   Flexor brevis pollicis.
Opponens (Flexor ossis metacarpi) pollicis.   Adductor obliquus pollicis.
                    Adductor transversus pollicis.

The **Abductor pollicis** is a thin, flat muscle, placed immediately beneath the integument. It arises from the annular ligament, the tuberosity of the scaphoid, and the ridge of the trapezium, frequently by two distinct slips; and, passing outwards and downwards, is inserted by a thin, flat tendon into the radial side of the base of the first phalanx of the thumb, sending a slip to join the tendon of the Extensor longus pollicis.

Relations.—By its *superficial surface*, with the palmar fascia and superficialis volæ artery, which frequently perforates it. By its *deep surface*, with the Opponens pollicis, from which it is separated by a thin aponeurosis. Its *inner border* is separated from the Flexor brevis pollicis by a narrow cellular interval.

The **Opponens pollicis** (*Flexor ossis metacarpi pollicis*) is a small, triangular muscle, placed beneath the preceding. It arises from the palmar surface of the ridge on the trapezium and from the annular ligament, passes downwards and outwards, and is inserted into the whole length of the metacarpal bone of the thumb on its radial side.

Relations.—By its *superficial surface*, with the Abductor and Flexor brevis pollicis. By its *deep surface*, with the trapezio-metacarpal articulation. By its *inner border*, with the Adductor obliquus pollicis.

The **Flexor brevis pollicis** consists of two portions, outer and inner. The outer and more superficial portion arises from the outer two-thirds of the lower border of the annular ligament, and passes along the outer side of the tendon of the Flexor longus pollicis; and, becoming tendinous, has a sesamoid bone developed in its tendon, and is inserted into the outer side of the base of the first phalanx of the thumb. The inner and deeper portion of the muscle is very small, and arises from the ulnar side of the first metacarpal bone beneath the Adductor obliquus pollicis, and is inserted into the inner side of the base of the first phalanx with this muscle.

FIG. 242.—Muscles of thumb.
(From a preparation in the Museum of the Royal College of Surgeons of England.)

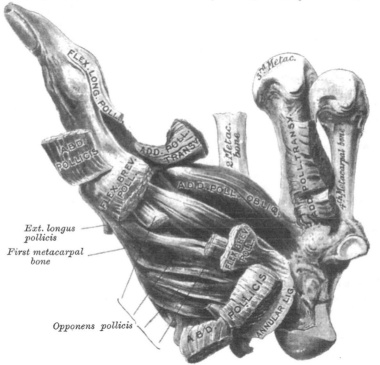

Ext. longus
pollicis

First metacarpal
bone

Opponens pollicis

Relations.—By its *superficial surface*, with the palmar fascia. By its *deep surface*, with the tendon of the Flexor longus pollicis. By its *external surface*, with the Opponens pollicis. *Behind*, with the Adductor obliquus pollicis.

The **Adductor obliquus pollicis** arises by several slips from the os magnum, the bases of the second and third metacarpal bones, the anterior carpal ligaments, and the sheath of the tendon of the Flexor carpi radialis. From this origin the greater number of fibres pass obliquely downwards and converge to a tendon, which, uniting with the tendons of the deeper portion of the Flexor brevis pollicis and the Adductor transversus, is inserted into the inner side of the base of the first phalanx of the thumb, a sesamoid bone being developed in the tendon of insertion. A considerable fasciculus, however, passes more obliquely outwards beneath the tendon of the long flexor to join the superficial portion of the short flexor and the Abductor pollicis.*

* This muscle was formerly described as the deep portion of the Flexor brevis pollicis.

Relations.—By its *superficial surface*, with the Flexor longus pollicis and the outer head of the Flexor brevis pollicis. Its *deep surface* is in relation with the deep palmar arch, which passes between the two adductors.

The **Adductor transversus pollicis** (fig. 242) is the most deeply seated of this group of muscles. It is of a triangular form, arising by its broad base from the lower two-thirds of the metacarpal bone of the middle finger on its palmar surface; the fibres, proceeding outwards, converge, to be inserted, with the inner part of the Flexor brevis pollicis, and the Adductor obliquus pollicis, into the ulnar side of the base of the first phalanx of the thumb. From the common tendon of insertion a slip is prolonged to the Extensor longus pollicis.

Relations.—By its *superficial surface*, with the Adductor obliquus pollicis, the tendons of the Flexor profundus, and the Lumbricales. Its *deep surface* covers the first two interosseous spaces, from which it is separated by a strong aponeurosis.

Three of these muscles of the thumb, the Abductor, the Adductor transversus, and the Flexor brevis pollicis, at their insertions, give off fibrous expansions which join the tendon of the Extensor longus pollicis. This permits of flexion of the proximal phalanx and extension of the terminal phalanx at the same time. These expansions, originally figured by Albinus, have been more recently described by M. Duchenne, 'Physiologie des Mouvements.'

Nerves.—The Abductor, Opponens, and outer head of the Flexor brevis pollicis are supplied by the sixth cervical through the median nerve; the inner head of the Flexor brevis, and the Adductors, by the eighth cervical through the ulnar nerve.

Actions.—The actions of the muscles of the thumb are almost sufficiently indicated by their names. This segment of the hand is provided with three extensors—an extensor of the metacarpal bone, an extensor of the first, and an extensor of the second phalanx; these occupy the dorsal surface of the forearm and hand. There are also three flexors on the palmar surface—a flexor of the metacarpal bone, a flexor of the proximal, and a flexor of the terminal phalanx; there is also an abductor and two adductors. The Abductor pollicis moves the metacarpal bone of the thumb outwards, that is, away from the index finger. The Flexor ossis metacarpi pollicis flexes the metacarpal bone, that is, draws it inwards over the palm, and at the same time rotates the bone, so as to turn the ball of the thumb towards the fingers, thus producing the movement of opposition. The Flexor brevis pollicis flexes and adducts the proximal phalanx of the thumb. The Adductores pollicis move the metacarpal bone of the thumb inwards, that is, towards the index finger. These muscles give to the thumb its extensive range of motion. It will be noticed, however, that in consequence of the position of the first metacarpal bone, these movements differ from the corresponding movements of the metacarpal bones of the other fingers. Thus extension of the thumb more nearly corresponds to the motion of abduction in the other fingers, and flexion to adduction.

12. *Ulnar Region* (fig. 243)

Palmaris brevis.
Abductor minimi digiti.   Flexor brevis minimi digiti.
Opponens (Flexor ossis metacarpi) minimi digiti.

The **Palmaris brevis** is a thin quadrilateral muscle, placed beneath the integument on the ulnar side of the hand. It arises, by tendinous fasciculi, from the annular ligament and palmar fascia; the fleshy fibres pass inwards, to be inserted into the skin on the inner border of the palm of the hand.

Relations.—By its *superficial surface*, with the integument, to which it is intimately adherent, especially by its inner extremity. By its *deep surface*, with the inner portion of the palmar fascia, which separates it from the ulnar vessels and nerve, and from the muscles of the ulnar side of the hand.

The **Abductor minimi digiti** is situated on the ulnar border of the palm of the hand. It arises from the pisiform bone and from the tendon of the Flexor carpi ulnaris, and terminates in a flat tendon, which divides into two slips; one is inserted into the ulnar side of the base of the first phalanx of the little finger.

FIG. 243.—Muscles of the left hand. Palmar surface.

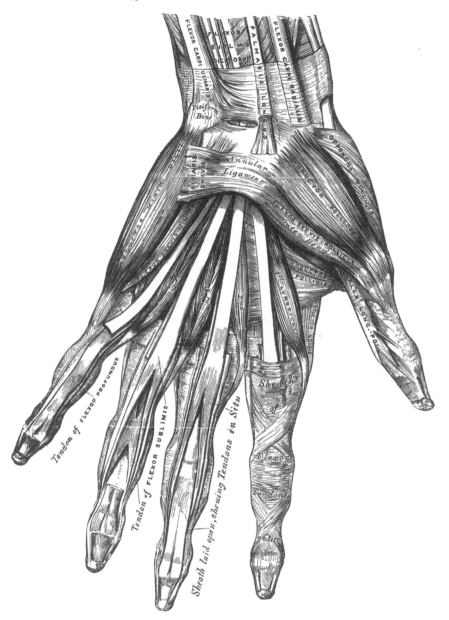

The other slip is inserted into the ulnar border of the aponeurosis of the Extensor minimi digiti.

Relations.—By its *superficial surface*, with the inner portion of the palmar fascia, and the Palmaris brevis. By its *deep surface*, with the Flexor ossis meta-carpi minimi digiti. By its *outer border*, with the Flexor brevis minimi digiti.

The **Flexor brevis minimi digiti** lies on the same plane as the preceding muscle, on its radial side. It arises from the convex aspect of the hook of the unciform bone, and anterior surface of the annular ligament, and is inserted into the inner side of the base of the first phalanx of the little finger. It is separated from the Abductor at its origin, by the deep branches of the ulnar artery and nerve. This muscle is sometimes wanting; the Abductor is then, usually, of large size.

Relations.—By its *superficial surface*, with the internal portion of the palmar fascia, and the Palmaris brevis. By its *deep surface*, with the Opponens. The deep branch of the ulnar artery and the corresponding branch of the ulnar nerve pass between the Abductor and Flexor brevis minimi digiti muscles.

The **Opponens (Flexor ossis metacarpi) minimi digiti** (fig. 234) is of a triangular form, and placed immediately beneath the preceding muscles. It arises from the convexity of the hook of the unciform bone, and contiguous portion of the annular ligament; its fibres pass downwards and inwards, to be inserted into the whole length of the metacarpal bone of the little finger, along its ulnar margin.

Relations.—By its *superficial surface*, with the Flexor brevis, and Abductor minimi digiti. By its *deep surface*, with the Interossei muscles in the fourth metacarpal space, the metacarpal bone, and the Flexor tendons of the little finger.

Nerves.—All the muscles of this group are supplied by the eighth cervical nerve through the ulnar nerve.

Actions.—The Abductor minimi digiti abducts the little finger from the middle line of the hand. It corresponds to a dorsal interosseous muscle. It also assists in flexing the proximal phalanx. The Flexor brevis minimi digiti abducts the little finger from the middle line of the hand. It also assists in flexing the proximal phalanx. The Opponens minimi digiti draws forwards the fifth metacarpal bone, so as to deepen the hollow of the palm. The Palmaris brevis corrugates the skin on the inner side of the palm of the hand.

13. *Middle Palmar Region*

<table>
<tr><td>Lumbricales.</td><td>Interossei dorsales.</td></tr>
<tr><td colspan="2" align="center">Interossei palmares.</td></tr>
</table>

The **Lumbricales** (fig. 243) are four small fleshy fasciculi, accessories to the deep Flexor muscle. They arise from the tendons of the deep Flexor: the first and second, from the radial side and palmar surface of the tendons of the index and middle fingers respectively; the third, from the contiguous sides of the tendons of the middle and ring fingers; and the fourth, from the contiguous sides of the tendons of the ring and little fingers. They pass to the radial side of the corresponding fingers, and opposite the metacarpo-phalangeal articulation each tendon is inserted into the tendinous expansion of the Extensor communis digitorum, covering the dorsal aspect of each finger.

The **Interossei muscles** (figs. 244, 245) are so named from occupying the intervals between the metacarpal bones, and are divided into two sets, a dorsal and palmar.

The **Dorsal interossei** are four in number, larger than the palmar, and occupy the intervals between the metacarpal bones. They are bipenniform muscles, arising by two heads from the adjacent sides of the metacarpal bones, but more extensively from the metacarpal bone of the finger into which the muscle is inserted. They are inserted into the bases of the first phalanges and into the aponeurosis of the common Extensor tendon. Between the double origin of each of these muscles is a narrow triangular interval, through the first of which passes the radial artery; through the other three passes a perforating branch from the deep palmar arch.

The *First dorsal interosseous muscle*, or *Abductor indicis*, is larger than the others. It is flat, triangular in form, and arises by two heads, separated by a fibrous arch, for the passage of the radial artery from the dorsum to the palm of the hand. The outer head arises from the upper half of the ulnar border of the first metacarpal bone; the inner head, from almost the entire length of the radial border of the second metacarpal bone; the tendon is inserted into the radial side of the index finger. The *second* and *third dorsal interossei* are inserted into the middle finger, the former into its radial, the latter into its ulnar side. The *fourth* is inserted into the ulnar side of the ring finger.

The **Palmar interossei,** three in number, are smaller than the Dorsal, and placed upon the palmar surface of the metacarpal bones, rather than between them. They arise from the entire length of the metacarpal bone of one finger, and are inserted into the side of the base of the first phalanx and aponeurotic expansion of the common Extensor tendon of the same finger.

The *first* arises from the ulnar side of the second metacarpal bone, and is inserted into the same side of the first phalanx of the index finger. The *second* arises from the radial side of the fourth metacarpal bone, and is inserted into the same side of the ring finger. The *third* arises from the radial side of the fifth

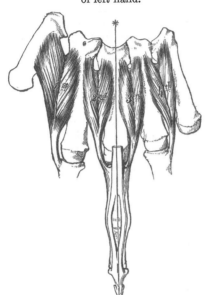

FIG. 244.—The Dorsal interossei of left hand.

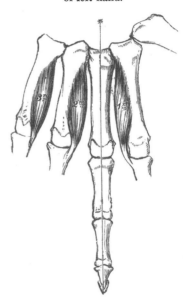

FIG. 245.—The Palmar interossei of left hand.

metacarpal bone, and is inserted into the same side of the little finger. From this account it may be seen that each finger is provided with two Interossei muscles, with the exception of the little finger, in which the Abductor muscle takes the place of one of the pair.

Nerves.—The two outer Lumbricales are supplied by the sixth cervical nerve, through the third and fourth digital branches of the median nerve: the two inner Lumbricales and all the Interossei are supplied by the eighth cervical nerve, through the deep palmar branch of the ulnar nerve. Brooks states that the third lumbrical received a twig from the median in twelve out of twenty-one cases.

Actions.—The Palmar interossei muscles adduct the fingers to an imaginary line drawn longitudinally through the centre of the middle finger; and the Dorsal interossei abduct the fingers from that line. In addition to this the Interossei, in conjunction with the Lumbricales, flex the first phalanges at the metacarpo-phalangeal joints, and extend the second and third phalanges in consequence of their insertion into the expansion of the Extensor tendons. The Extensor communis digitorum is believed to act almost entirely on the first phalanges.

Surface Form.—The *Pectoralis major* largely influences surface form and conceals a considerable part of the thoracic wall in front. Its sternal origin presents a festooned border, which bounds and determines the width of the sternal furrow. Its clavicular origin is somewhat depressed and flattened, and between the two portions of the muscle is often an oblique depression, which differentiates the one from the other. The outer margin of the muscle is generally well marked above, and bounds the infraclavicular fossa, a triangular interval which separates the Pectoralis major from the Deltoid. It gradually becomes less marked as it approaches the tendon of insertion, and is more closely blended with the Deltoid muscle. The lower border of the Pectoralis major forms the rounded anterior axillary fold, and corresponds with the direction of the fifth rib. The *Pectoralis minor* influences surface form. When the arm is raised, its lowest slip of origin produces a local fulness just below the border of the anterior fold of the axilla, and so serves to break the sharp line of the lower border of the Pectoralis major muscle which is produced when the arm is in this position. The origin of the *Serratus magnus* produces a very characteristic surface marking. When the arm is raised from the side, in a well-developed subject, the five or six lower serrations are plainly discernible, forming a zigzag line, caused by the series of digitations, which diminish in size from above downwards, and have their apices arranged in the form of a curve. When the arm is lying by the side, the first serration to appear, at the lower margin of the Pectoralis major, is the one attached to the fifth rib. The *Deltoid*, with the prominence of the upper extremity of the humerus, produces the rounded outline of the shoulder. It is rounder and fuller in front than behind, where it presents a somewhat flattened form. Its anterior border, above, presents a rounded, slightly curved eminence, which bounds externally the infraclavicular fossa ; below, it is closely united with the Pectoralis major. Its posterior border is thin, flattened, and scarcely marked above ; below, it is thicker and more prominent. When the muscle is in action, the middle portion becomes irregular, presenting alternate longitudinal elevations and depressions : the elevations corresponding to the fleshy portions ; the depressions, to the tendinous intersections of the muscle. The insertion of the Deltoid is marked by a depression on the outer side of the middle of the arm. Of the scapular muscles, the only one which materially influences surface form is the *Teres major*, which assists the Latissimus dorsi in forming the thick, rounded, posterior fold of the axilla. When the arm is raised, the Coraco-brachialis reveals itself as a long, narrow elevation, which emerges from under cover of the anterior fold of the axilla and runs downwards, internal to the shaft of the humerus. When the arm is hanging by the side, its front and inner part presents the prominence of the Biceps, bounded on either side by an intermuscular depression. This muscle determines the contour of the front of the arm, and extends from the anterior margin of the axilla to the bend of the elbow. Its upper tendons are concealed by the Pectoralis major and the Deltoid, and its lower tendon sinks into the space at the bend of the elbow. When the muscle is in a state of complete contraction—that is to say, when the forearm has been flexed and supinated—it presents a rounded convex form, bulged out laterally, and its length is diminished. On each side of the Biceps, at the lower part of the arm, the *Brachialis anticus* is discernible. On the outer side it forms a narrow eminence, which extends some distance up the arm along the border of the Biceps. On the inner side it shows itself only as a little fulness just above the elbow. On the back of the arm the long head of the *Triceps* may be seen as a longitudinal eminence emerging from under cover of the Deltoid, and gradually merging into the longitudinal flattened plane of the tendon of the muscle on the lower part of the back of the arm. The tendon of insertion of the muscle extends about halfway up the back of the arm, where it forms an elongated flattened plane when the muscle is in action. Under similar conditions the surface forms produced by the three heads of the muscle are well seen. On the anterior aspect of the elbow are to be seen two muscular elevations, one on each side, separated above, and converging below so as to form a triangular space. Of these, the inner elevation, consisting of the flexors and pronator, forms the prominence along the inner side and front of the forearm. It is a fusiform mass, pointed above at the internal condyle, and gradually tapering off below. The *Pronator radii teres*, the innermost muscle of the group, forms the boundary of the triangular space at the bend of the elbow. It is shorter, less prominent, and more oblique than the outer boundary. The most prominent part of the eminence is produced by the *Flexor carpi radialis*, the muscle next in order on the inner side of the preceding one. It forms a rounded prominence above, and may be traced downwards to its tendon, which can be felt lying on the front of the wrist, nearer to the radial than to the ulnar border and to the inner side of the radial artery. The *Palmaris longus* presents no surface marking above, but, below, is the most prominent tendon on the front of the wrist, standing out, when the muscle is in action, as a sharp tense cord beneath the skin. The *Flexor sublimis digitorum* does not directly influence surface form. The position of its four tendons on the front of the lower part of the forearm is indicated by an elongated depression between the tendons of the Palmaris longus and the Flexor carpi ulnaris. The *Flexor carpi ulnaris* occupies a small part of the posterior surface of the forearm, and is separated from the extensor and supinator group, which occupies the greater part of this surface, by the ulnar furrow, produced by the subcutaneous posterior border of the ulna. Its tendon can be perceived along the ulnar border of the front of the forearm, and is most marked when the hand is flexed and adducted. The deep muscles of the front of the forearm have no direct influence on surface form. The external group of muscles of the forearm, consisting of the extensors and supinators, occupy the outer, and a considerable portion of the posterior, surface of this region. It has a fusiform outline, which is altogether on a higher level than the pronato-flexor group. Its apex emerges from between the Triceps and Brachialis anticus muscles some distance above the elbow-joint, and acquires its greatest breadth opposite the external condyle, and thence gradually shades off into a flattened surface. About the middle of the forearm it divides into two longitudinal eminences which diverge from each other, leaving a triangular interval between them. The outer of these two groups of muscles consists of the Supinator longus and the Extensor carpi radialis longior et brevior, which forms a longitudinal eminence descending from the external condylar ridge in the direction of the styloid process of the radius. The other and more posterior group consists of the Extensor communis digitorum, the Extensor minimi digiti, and the Extensor carpi ulnaris. It commences above as a tapering form at the external condyle of the humerus, and is separated behind at its upper part from the Anconeus by a well-marked furrow; and below, from the pronato-flexor mass, by the ulnar furrow. In the triangular interval left between these two groups, the extensors of the thumb and index finger are seen. The only two muscles of this region which require special mention, as independently influencing surface form, are the Supinator longus and the Anconeus. The inner border of the *Supinator longus* forms the outer boundary of the triangular space at the bend of the elbow. It commences as a rounded border above the condyle, and is longer, less oblique, and more prominent than the inner boundary. Lower down, the muscle forms a full fleshy mass on the outer side of the upper part of the forearm, and below tapers into a tendon, which may be traced down to the styloid process of the radius. The *Anconeus* presents a distinct and characteristic surface form in the shape of a triangular, slightly elevated surface, immediately external to the subcutaneous posterior surface of the olecranon, and differentiated from the common extensor group by a well-marked oblique longitudinal depression. The upper angle of the triangle corresponds to the external condyle, and is marked by a depression or dimple in this situation. In the interval, caused by the divergence from each other of the two groups of muscles into which the extensor and supinator group is divided at the lower part of the forearm, an oblique elongated eminence is seen, caused by the emergence of two of the extensors of the thumb from their deep origin at the back of the forearm. This eminence, full above, and becoming flattened out and partially subdivided below, runs downwards and outwards over the back and outer surface of the radius to the outer side of the wrist-joint, where it forms a ridge, especially marked when the thumb is extended, which passes onwards to the posterior aspect of the thumb. The tendons of most of the Extensor muscles are to be seen and felt at the level of the wrist-joint. Most externally are the tendons of the Extensor ossis metacarpi pollicis and the Extensor brevis pollicis, forming a vertical ridge over the outer side of the joint from the styloid process of the radius to the thumb. Internal to this is the oblique ridge produced by the tendon of the Extensor longus pollicis, very noticeable when the muscle is in action. The Extensor carpi radialis longior is scarcely to be felt, but the Extensor carpi radialis brevior can be distinctly perceived, as a vertical ridge emerging from under the inner border of the tendon of the Extensor longus pollicis, when the hand is forcibly extended at the wrist. Internal to this, again, can be felt the tendons of the Extensor indicis, Extensor communis digitorum, and Extensor minimi digiti; the latter tendon being separated from those of the common extensor by a slight furrow. The muscles of the hand are principally concerned, as far as regards surface form, in producing the thenar and hypothenar eminences, and individually are not to be distinguished, on the surface, from each other. The *Adductor transversus pollicis* is, however, an exception to this; its anterior border gives rise to a ridge across the web of skin connecting the thumb to the rest of the hand. The thenar eminence is much larger and rounder than the hypothenar one, which presents a longer and narrower eminence along the ulnar side of the hand. When the *Palmaris brevis* is in action it produces a wrinkling of the skin over the hypothenar eminence, and a deep dimple on the ulnar border of the hand. The anterior extremities of the *Lumbrical* muscles help to produce the soft eminences just behind the clefts of the fingers, separated from each other by depressions corresponding to the flexor tendons in their sheaths. Between the thenar and hypothenar eminences, at the wrist-joint, is a slight groove or depression, widening out as it approaches the fingers; beneath this is the strong central part of the palmar fascia. Here are some furrows, which are pretty constant in their arrangement, and bear some resemblance to the letter **M**. One of these furrows passes obliquely outwards from the groove between the thenar and hypothenar regions near the wrist to the head of the metacarpal bone of the index finger. A second passes inwards, with a slight inclination upwards, from the termination of the first to the ulnar side of the hand. A third runs nearly parallel with the second and about three-quarters of an inch below it. Lastly, crossing these two latter furrows is an oblique furrow parallel with the first. The skin of the palm of the hand differs considerably from that of the forearm. At the wrist it suddenly becomes hard and dense, and covered with a thick layer of cuticle. The skin in the thenar region presents these characteristics less than elsewhere. In spite of this hardness and density the skin of the palm is exceedingly sensitive and very vascular.

It is destitute of hair, and no sebaceous follicles have been found in this region. Over the fingers the skin again becomes thinner, especially at the flexures of the joints, and over the terminal phalanges it is thrown into numerous ridges in consequence of the arrangement of the papillæ in it. These ridges form, in different individuals, distinctive and permanent patterns, which may be used for purposes of identification. The superficial fascia in the palm is made up of dense fibro-fatty tissue. This tissue binds down the skin so firmly to the deep palmar fascia that very little movement is permitted between the two. On the back of the hand the *Dorsal interossei* produce elongated swellings between the metacarpal bones. The first dorsal interosseous (Abductor indicis), when the thumb is closely adducted to the hand, forms a prominent fusiform bulging; the other interossei are not so marked.

Surgical Anatomy.—The student, having completed the dissection of the muscles of the upper extremity, should consider the effects likely to be produced by the action of the various muscles in fracture of the bones.

In considering the actions of the various muscles upon fractures of the upper extremity, the most common forms of injury have been selected both for illustration and description.

Fracture of the *middle of the clavicle* (fig. 246) is always attended with considerable displacement; the inner end of the outer fragment is displaced inwards and backwards, while the outer end of the same fragment is rotated forwards. The whole outer fragment is somewhat depressed.

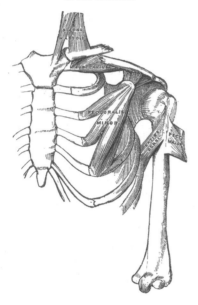

FIG. 246.—Fracture of the middle of the clavicle.

The displacement is produced as follows: *inwards*, by the muscles passing from the chest to the outer fragment of the clavicle, to the scapula, and to the humerus, viz. the Subclavius and the Pectoralis minor, and, to a less extent, the Pectoralis major and the Latissimus dorsi; *backwards*, in consequence of the rotation of the outer fragment. The Serratus magnus causes the scapula to rotate on the wall of the chest, this carries the acromion and outer end of the outer fragment of the clavicle forwards, and causes the piece of bone to rotate round a vertical axis through its centre, and so carries the inner end of the outer portion backwards. The depression of the whole outer fragment is produced by the weight of the arm and by the contraction of the Deltoid. The outer end of the inner fragment appears to be elevated, the skin being drawn tensely over it; this is owing to the depression of the outer fragment, as the inner fragment is usually kept fixed by the costo-clavicular ligament, and by the antagonism between the Sterno-mastoid and Pectoralis major muscles. But it may be raised by an unusually strong Sterno-mastoid, or by the inner end of the outer fragment getting below and behind it. The causes of displacement having been ascertained, it is easy to apply the appropriate treatment. The outer fragment is to be drawn outwards, and, together with the scapula, raised upwards to a level with the inner fragment, and retained in that position.

In fracture of the *acromial end of the clavicle*, between the conoid and trapezoid ligaments, only slight displacement occurs, as these ligaments, from their oblique insertion, serve to hold both portions of the bone in apposition. Fracture, also, of the *sternal end*, internal to the costo-clavicular ligament, is attended with only slight displacement, this ligament serving to retain the fragments in close apposition.

Fracture of the *acromion process* usually arises from violence applied to the upper and outer part of the shoulder; it is generally known by the rotundity of the shoulder being lost, from the Deltoid drawing the fractured portion downwards and forwards; and the displacement may easily be discovered by tracing the margin of the clavicle outwards, when the fragment will be found resting on the front and upper part of the head of the humerus. In order to relax the anterior and outer fibres of the Deltoid (the opposing muscle), the arm should be drawn forwards across the chest, and the elbow well raised, so that the head of the bone may press the acromion process upwards, and retain it in its position.

Fracture of the *coracoid process* is an extremely rare accident, and is usually caused by a sharp blow on the point of the shoulder. Displacement is here produced by the combined actions of the Pectoralis minor, short head of the Biceps, and Coraco-brachialis, the former muscle drawing the fragment inwards, and the latter directly downwards, the amount of displacement being limited by the connection of this process to the acromion

SURGICAL ANATOMY 387

by means of the coraco-acromial ligament. In many cases there appears to have been little or no displacement, from the fact that the coraco-clavicular ligament has remained intact, and has kept the separated fragment from displacement. In order to relax these muscles and replace the fragments in close apposition, the forearm should be flexed so as to relax the Biceps, and the arm drawn forwards and inwards across the chest so as to relax the Coraco-brachialis; the humerus should then be pushed upwards against the coraco-acromial ligament, and the arm retained in that position.

FIG. 247.—Fracture of the surgical neck of the humerus.

Fracture of the *surgical neck of the humerus* (fig. 247) is very common, is attended with considerable displacement, and its appearances correspond somewhat with those of dislocation of the head of the humerus into the axilla. The upper fragment is slightly elevated under the coraco-acromial ligament by the muscles attached to the greater and lesser tuberosities; the lower fragment is drawn inwards by the Pectoralis major, Latissimus dorsi, and Teres major; and the humerus is thrown obliquely outwards from the side by the Deltoid, and occasionally elevated so as to cause the upper end of the lower fragment to project beneath and in front of the coracoid process. The deformity is reduced by fixing the shoulder, and drawing the arm outwards and downwards. To counteract the opposing muscles, and to keep the fragments in position, a small conical-shaped pad should be placed in the axilla, and the arm bandaged to the side by a broad roller passed round the

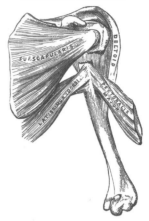

chest, in such a manner that the elbow is carried slightly forwards, so as to throw the upper end of the lower fragment backwards and outwards towards the head of the bone. The whole is then covered with a carefully moulded guttapercha or poroplastic shoulder cap.

In fracture of the *shaft of the humerus* below the insertion of the Pectoralis major, Latissimus dorsi, and Teres major, and above the insertion of the Deltoid, there is also considerable deformity, the upper fragment being drawn inwards by the first-mentioned muscles, and the lower fragment upwards and outwards by the Deltoid, producing shortening of the limb, and a considerable prominence at the seat of fracture, from the fractured ends of the bone riding over one another, especially if the fracture takes place in an oblique direction. The fragments may be brought into apposition by extension from the elbow, and retained in that position by adopting the same means as in the preceding injury.

In fractures of the *shaft of the humerus* immediately below the insertion of the Deltoid, the amount of deformity depends greatly upon the direction of the fracture. If it occurs in a transverse direction, only slight displacement takes place, the upper fragment being drawn a little forwards; but in oblique fracture, the combined actions of the Biceps and Brachialis anticus muscles in front, and the Triceps behind, draw upwards the lower fragment, causing it to glide over the upper fragment, either backwards or forwards, according to the direction of the fracture. Simple extension reduces the deformity, and the application of a shoulder cap and splints to the arm will retain the fragments in apposition. Care should be taken not to raise the elbow; but the forearm and hand may be supported in a sling.

FIG. 248.—Fracture of the humerus above the condyles.

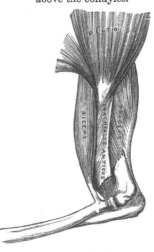

Fracture of the *humerus* (fig. 248) immediately above the condyles deserves very attentive consideration, as the general appearances correspond somewhat with those produced by separation of the epiphysis of the humerus, and with those of dislocation of the radius and ulna backwards. If the direction of the fracture is oblique, from above, downwards and forwards, the lower fragment is drawn upwards by the Brachialis anticus and Biceps in front, and the Triceps behind; and at the same time is drawn backwards behind the upper fragment by the Triceps. This injury may be diagnosed from dislocation by the increased mobility in fracture, the existence of crepitus, and the fact of the deformity being remedied by extension, on the discontinuance of which it is reproduced. The age of the patient is of importance in distinguishing this form of injury from separation of the epiphysis.

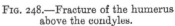

If fracture occurs in the opposite direction to that shown in fig. 248, the lower fragment is drawn upwards and forwards, causing a considerable prominence in front; and the upper fragment projects backwards beneath the tendon of the Triceps muscle.

Fracture of the *olecranon process* (fig. 249) is a frequent accident. The detached fragment is displaced upwards, by the action of the Triceps muscle, from half an inch to two inches; the prominence of the elbow is consequently lost, and a deep hollow is felt at the back part of the joint, which is much increased on flexing the limb. The patient at the same time loses, more or less, the power of extending the forearm. The treatment consists in relaxing the Triceps by extending the limb, and retaining it in the extended position by means of a long straight splint applied to the front of the arm; the fragments are thus brought into close apposition, and may be further approximated by drawing down the upper fragment. Union is generally ligamentous.

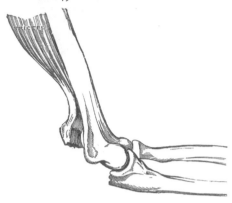

FIG. 249.—Fracture of the olecranon.

Fracture of the *neck of the radius* is an exceedingly rare accident, and is generally caused by direct violence. Its diagnosis is somewhat obscure, on account of the slight deformity visible, the injured part being surrounded by a large number of muscles; but the movements of pronation and supination are entirely lost. The upper fragment is drawn outwards by the Supinator brevis, its extent of displacement being limited by the attachment of the orbicular ligament. The lower fragment is drawn forwards and slightly upwards by the Biceps, and inwards by the Pronator radii teres, its displacement forwards and upwards being counteracted in some degree by the Supinator brevis. The treatment essentially consists in relaxing the Biceps, Supinator brevis, and Pronator radii teres muscles, by flexing the forearm, and placing it in a position midway between pronation and supination, extension having been previously made so as to bring the parts in apposition.

In fracture of the *radius* below the insertion of the Biceps, but above the insertion of the Pronator radii teres, the upper fragment is strongly supinated by the Biceps and Supinator brevis, and at the same time drawn forwards and flexed by the Biceps; the lower fragment is pronated and drawn inwards towards the ulna by the pronators. Thus there is extreme displacement with very little deformity. In treating such a fracture the arm must be put up in a position of supination, otherwise union will take place with great impairment of the movements of the hand. In fractures of the radius below the insertion of the Pronator radii teres (fig. 250), the upper fragment is drawn upwards by the Biceps, and inwards by the Pronator radii teres, holding a position midway between pronation and supination, and a degree of fulness in the upper half of the forearm is thus produced: the lower fragment is drawn downwards and inwards towards the ulna by the Pronator quadratus, and thrown into a state of pronation by the same muscle; at the same time, the Supinator longus, by elevating the styloid process, into which it is inserted, will serve to depress the upper end of the lower fragment still more towards the ulna. In order to relax the opposing muscles the forearm should be bent, and the limb placed in a position midway between pronation and supination; the fracture is then easily reduced by extension from the wrist and elbow: well-padded splints should be applied on both sides of the forearm from the elbow to the wrist; the hand being allowed to fall, will, by its own weight, counteract the action of the Pronator quadratus and Supinator longus, and elevate the lower fragment to the level of the upper one.

FIG. 250.—Fracture of the shaft of the radius.

In fracture of the *shaft of the ulna* the upper fragment retains its usual position, but the lower fragment is drawn outwards towards the radius by the Pronator quadratus, producing a well-marked depression at the seat of fracture, and some fulness on the dorsal and palmar surfaces of the forearm. The fracture is easily reduced by extension from the wrist and forearm. The forearm should be flexed, and placed in a position midway between pronation and supination, and well-padded splints applied from the elbow to the ends of the fingers.

In fracture of the *shafts of the radius and ulna together*, the lower fragments are drawn upwards, sometimes forwards, sometimes backwards, according to the direction of the fracture, by the combined actions of the Flexor and Extensor muscles, producing a degree of fulness on the dorsal or palmar surface of the forearm ; at the same time the two fragments are drawn into contact by the Pronator quadratus, the radius being in a state of pronation : the upper fragment of the radius is drawn upwards and inwards by the Biceps and Pronator radii teres to a higher level than the ulna ; the upper portion of the ulna is slightly elevated by the Brachialis anticus. The fracture may be reduced by extension from the wrist and elbow, and the forearm should be placed in the same position as in fracture of the ulna.

In fracture of the *lower end of the radius* (fig. 251) the displacement which is produced is very considerable, and bears some resemblance to dislocation of the carpus backwards,

FIG. 251.—Fracture of the lower end of the radius.

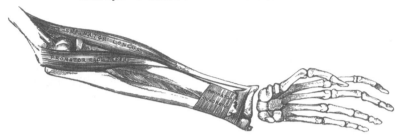

from which it should be carefully distinguished. The lower fragment is displaced backwards and upwards, but this displacement is probably due to the force of the blow driving the portion of the bone into this position and not to any muscular influence. The upper fragment projects forwards, often lacerating the substance of the Pronator quadratus, and is drawn by this muscle into close contact with the lower end of the ulna, causing a projection on the anterior surface of the forearm, immediately above the carpus, from the flexor tendons being thrust forwards. This fracture may be distinguished from dislocation by the deformity being removed on making sufficient extension, when crepitus may be occasionally detected ; at the same time, on extension being discontinued, the parts immediately resume their deformed appearance (see also page 130). The age of the patient will also assist in determining whether the injury is fracture or separation of the epiphysis. The treatment consists in flexing the forearm, and making powerful extension from the wrist and elbow, depressing at the same time the radial side of the hand, and retaining the parts in that position by well-padded *pistol-shaped* splints.

MUSCLES AND FASCIÆ OF THE LOWER EXTREMITY

The Muscles of the Lower Extremity are subdivided into groups, corresponding with the different regions of the limb

I. ILIAC REGION

Psoas magnus.
Psoas parvus.
Iliacus.

II. THIGH

1. *Anterior Femoral Region*

Tensor fasciæ femoris.
Sartorius.

Quadriceps extensor { Rectus.
Vastus externus.
Vastus internus.
Crureus.

Subcrureus.

2. *Internal Femoral Region*

Gracilis.
Pectineus.
Adductor longus.

Adductor brevis.
Adductor magnus.

3. *Gluteal Region*

Gluteus maximus.
Gluteus medius.
Gluteus minimus.
Pyriformis.
Obturator internus.
Gemellus superior.
Gemellus inferior.
Quadratus femoris.
Obturator externus.

4. *Posterior Femoral Region*

Biceps.
Semitendinosus.
Semimembranosus.

III. Leg

5. Anterior Tibio-fibular Region

Tibialis anticus.
Extensor proprius hallucis.
Extensor longus digitorum.
Peroneus tertius.

6. Posterior Tibio-fibular Region. Superficial Layer

Gastrocnemius.
Soleus.
Plantaris.

Deep Layer

Popliteus.
Flexor longus hallucis.
Flexor longus digitorum.
Tibialis posticus.

7. Fibular Region

Peroneus longus.
Peroneus brevis.

IV. Foot

8. Dorsal Region

Extensor brevis digitorum.

9. Plantar Region
First Layer

Abductor hallucis.
Flexor brevis digitorum.
Abductor minimi digiti.

Second Layer

Flexor accessorius.
Lumbricales.

Third Layer

Flexor brevis hallucis.
Adductor obliquus hallucis.
Flexor brevis minimi digiti.
Adductor transversus hallucis.

Fourth Layer

The Interossei.

I. Muscles and Fasciæ of the Iliac Region

Psoas magnus.   Psoas parvus.   Iliacus.

Dissection.—No detailed description is required for the dissection of these muscles. On the removal of the viscera from the abdomen, they are exposed, covered by the peritoneum and a thin layer of fascia, the iliac fascia.

The **iliac fascia** * is the aponeurotic layer which lines the back part of the abdominal cavity, and covers the Psoas and Iliacus muscles throughout their whole extent. It is thin above, and becomes gradually thicker below as it approaches the crural arch.

The *portion covering the Psoas* is attached, above, to the Ligamentum arcuatum internum ; internally, by a series of arched processes to the intervertebral substances, and prominent margins of the bodies of the vertebræ, and to the upper part of the sacrum ; the intervals so left, opposite the constricted portions of the bodies, transmitting the lumbar arteries and veins and filaments of the sympathetic cord. Externally, above the crest of the ilium, this portion of the iliac fascia is continuous with the anterior lamella of the lumbar fascia (see page 315), but below the crest of the ilium it is continuous with the fascia covering the Iliacus.

The *portion investing the Iliacus* is connected, externally, to the whole length of the inner border of the crest of the ilium ; and internally, to the brim of the true pelvis, where it is continuous with the periosteum ; and at the ilio-pectineal eminence it receives the tendon of insertion of the Psoas parvus, when that muscle exists. External to the femoral vessels, this fascia is intimately connected to the posterior margin of Poupart's ligament, and is continuous with the fascia transversalis. Immediately to the outer side of the femoral vessels the fascia iliaca is prolonged backwards and inwards from Poupart's ligament as a band, the *ilio-pectineal ligament*, which is attached to the ilio-pectineal eminence. This ligament divides the space between Poupart's ligament and the innominate bone into two parts, the inner of which transmits the femoral vessels, the outer the ilio-psoas and the anterior crural nerve (fig. 166). Internal to the vessels the iliac fascia is attached to the ilio-pectineal line behind the conjoined tendon, where it is

* The student must not confound this fascia with the *iliac portion of the fascia lata* (see p. 395).

again continuous with the transversalis fascia ; and corresponding to the point where the femoral vessels pass into the thigh, this fascia descends behind them forming the posterior wall of the femoral sheath. This portion of the iliac fascia which passes behind the femoral vessels is also attached to the ilio-pectineal line beyond the limits of the attachment of the conjoined tendon ; at this part it is continuous with the pubic portion of the fascia lata of the thigh. The external iliac vessels lie in front of the iliac fascia, but all the branches of the lumbar plexus behind it ; it is separated from the peritoneum by a quantity of loose areolar tissue.

The **Psoas magnus** (fig. 253) is a long fusiform muscle placed on the side of the lumbar region of the spine and margin of the pelvis. It arises from the front of the bases and lower borders of the transverse processes of the lumbar vertebræ by five fleshy slips ; also from the sides of the bodies, and the corresponding intervertebral substances of the last dorsal and all the lumbar vertebræ. The muscle is connected to the bodies of the vertebræ by five slips ; each slip is attached to the upper and lower margins of two vertebræ, and to the intervertebral substance between them ; the slips themselves being connected by tendinous arches which extend across the constricted part of the bodies, and beneath which pass the lumbar arteries and veins and filaments of the sympathetic cord. These tendinous arches also give origin to muscular fibres, and protect the blood-vessels and nerves from pressure during the action of the muscle. The first slip is attached to the contiguous margins of the last dorsal and first lumbar vertebræ ; the last to the contiguous margins of the fourth and fifth lumbar vertebræ, and to the intervertebral substance. From these points the muscle passes down across the brim of the pelvis, and, diminishing gradually in size, passes beneath Poupart's ligament, and terminates in a tendon, which, after receiving nearly the whole of the fibres of the Iliacus, is inserted into the lesser trochanter of the femur.

Relations.—In the lumbar region : By its *anterior surface*, which is placed behind the peritoneum, with the iliac fascia, the ligamentum arcuatum internum, the kidney, Psoas parvus, renal vessels, ureter, spermatic vessels, genito-crural nerve, and the colon. By its *posterior surface*, with the transverse processes of the lumbar vertebræ, and the Quadratus lumborum, from which it is separated by the anterior lamella of the lumbar fascia. The lumbar plexus is situated in the posterior part of the substance of the muscle. By its *inner side*, the muscle is in relation with the bodies of the lumbar vertebræ, the lumbar arteries, the ganglia of the sympathetic nerve, and their branches of communication with the spinal nerves ; the lumbar glands ; the vena cava inferior on the right, and the aorta on the left side, and along the brim of the pelvis with the external iliac artery. In the thigh it is in relation, in front, with the fascia lata ; behind, with the capsular ligament of the hip, from which it is separated by a synovial bursa, which frequently communicates with the cavity of the joint through an opening of variable size ; by its *inner border*, with the Pectineus and internal circumflex artery, and also with the femoral artery, which slightly overlaps it ; by its *outer border*, with the anterior crural nerve and Iliacus muscle.

The **Psoas parvus** is a long slender muscle, placed in front of the Psoas magnus. It arises from the sides of the bodies of the last dorsal and first lumbar vertebræ and from the intervertebral substance between them. It forms a small flat muscular bundle, which terminates in a long flat tendon, inserted into the ilio-pectineal eminence, and, by its outer border, into the iliac fascia. This muscle is often absent, and, according to Cruveilhier, sometimes double.

Relations.—It is covered by the peritoneum, and, at its origin, by the ligamentum arcuatum internum ; it rests on the Psoas magnus.

The **Iliacus** is a flat, triangular muscle, which fills up the whole of the iliac fossa. It arises from the upper two-thirds of this fossa, and from the inner margin of the crest of the ilium ; behind, from the ilio-lumbar ligament, and base of the sacrum ; in front, from the anterior superior and anterior inferior spinous processes of the ilium, from the notch between them, and by a few fibres from the capsule of the hip-joint. The fibres converge to be inserted into the outer side of the tendon of the Psoas, some of them being prolonged on to the shaft of the femur for about an inch below and in front of the lesser trochanter.*

Relations.—Within the abdomen : By its *anterior surface*, with the iliac fascia, which separates the muscle from the peritoneum, and with the external cutaneous nerve ; on the right side, with the cæcum ; on the left side, with the sigmoid flexure of the colon. By its *posterior surface*, with the iliac fossa. By its *inner border*, with the Psoas magnus, and anterior crural nerve. In the thigh, it is in relation, by its *anterior surface*, with the fascia lata, Rectus, Sartorius, and profunda femoris artery ; behind, with the capsule of the hip-joint, a synovial bursa common to it and the Psoas magnus being interposed.

Nerves.—The Psoas magnus is supplied by the anterior branches of the second and third lumbar nerves ; the Psoas parvus, when it exists, is supplied by the anterior branch of the first lumbar nerve ; and the Iliacus by the anterior branches of the second and third lumbar nerves through the anterior crural.

Actions.—The Psoas and Iliacus muscles, acting from above, flex the thigh upon the pelvis : acting from below, the femur being fixed, the muscles of both sides bend the lumbar portion of the spine and pelvis forwards. They also serve to maintain the erect position, by supporting the spine and pelvis upon the femur, and assist in raising the trunk when the body is in the recumbent posture.

The *Psoas parvus* is a tensor of the iliac fascia.

Surgical Anatomy.—In the iliac fascia there is no definite septum between the portions of fascia covering the Psoas and Iliacus respectively, and the fascia is only connected to the subjacent muscles by a quantity of loose connective tissue. When abscess forms beneath this fascia, as it is very apt to do, the matter is contained in an osseo-fibrous cavity which is closed on all sides within the abdomen, and is open only at its lower part, where the fascia is prolonged over the muscle into the thigh.

Abscess within the sheath of the Psoas muscle (*Psoas abscess*) is generally due to tuberculous caries of the bodies of the lower dorsal and lumbar vertebræ. When the disease is in the dorsal region, the matter tracks down the posterior mediastinum, in front of the bodies of the vertebræ, and, passing beneath the Ligamentum arcuatum internum, enters the sheath of the Psoas muscle, down which it passes as far as the pelvic brim ; it then gets beneath the iliac portion of the fascia, and fills up the iliac fossa. In consequence of the attachment of the fascia to the pelvic brim, it rarely finds its way into the pelvis, but passes by a narrow opening under Poupart's ligament into the thigh, to the outer side of the femoral vessels. It thus follows that a Psoas abscess may be described as consisting of four parts : (1) a somewhat narrow channel at its upper part, in the Psoas sheath ; (2) a dilated sac in the iliac fossa ; (3) a constricted neck under Poupart's ligament ; and (4) a dilated sac in the upper part of the thigh. When the lumbar vertebræ are the seat of the disease, the matter finds its way directly into the substance of the muscle. The muscular fibres are destroyed, and the nervous cords contained in the abscess are isolated and exposed in its interior ; the femoral vessels which lie in front of the fascia remain intact, and the peritoneum seldom becomes implicated. All Psoas abscesses do not, however, pursue this course : the matter may leave the muscle above the crest of the ilium, and tracking backwards may point in the loin (*lumbar abscess*) ; or it may point above Poupart's ligament in the inguinal region ; or it may follow the course of the iliac vessels into the pelvis, and, passing through the great sacro-sciatic notch, discharge itself on the back of the thigh ; or it may open into the bladder, or find its way into the perinæum.

II. MUSCLES AND FASCIÆ OF THE THIGH

1. *Anterior Femoral Region*

Tensor fasciæ femoris.	Quadriceps	Rectus.
Sartorius.	extensor	Vastus externus.
		Vastus internus.
		Crureus.

Subcrureus.

---

* The Psoas and Iliacus are sometimes regarded as a single muscle, the *Ilio-psoas,* having two heads of origin and a single insertion.

Dissection.—To expose the muscles and fasciæ in this region, make an incision along Poupart's ligament, from the anterior superior spine of the ilium to the spine of the os pubis; a vertical incision from the centre of this, along the middle of the thigh to below the knee-joint; and a transverse incision from the inner to the outer side of the leg, at the lower end of the vertical incision. The flaps of integument having been removed, the superficial and deep fasciæ should be examined. The more advanced student should commence the study of this region by an examination of the anatomy of femoral hernia, and Scarpa's triangle, the incisions for the dissection of which are marked out in fig. 252.

The **superficial fascia** forms a continuous layer over the whole of the thigh, consisting of areolar tissue, containing in its meshes much fat, and capable of being separated into two or more layers, between which are found the superficial vessels and nerves. It varies in thickness in different parts of the limb; in the groin it is thick, and the two layers are separated from one another by the superficial inguinal lymphatic glands, the internal saphenous vein, and several smaller vessels. One of these two layers, the superficial, is continuous above with the superficial fascia of the abdomen. The deep layer of the superficial fascia is a very thin, fibrous layer, best marked on the inner side of the long saphenous vein and below Poupart's ligament. It is placed beneath the subcutaneous vessels and nerves and upon the surface of the fascia lata. It is intimately adherent to the fascia lata a little below Poupart's ligament. It covers the saphenous opening in the fascia lata, being closely united to its circumference, and is connected to the sheath of the femoral vessels, corresponding to its under surface. The portion of fascia covering this aperture is perforated by the internal saphenous vein and by numerous blood and lymphatic vessels, hence it has been termed the *cribriform fascia*, the openings for these vessels having been likened to the holes in a sieve. The cribriform fascia adheres closely both to the superficial fascia and to the fascia lata, so that it is described by some anatomists as part of the fascia lata, but is usually considered (as in this work) as belonging to the superficial fascia. It is not until the cribriform fascia has been cleared away that the saphenous opening is seen, so that this opening does not in ordinary cases

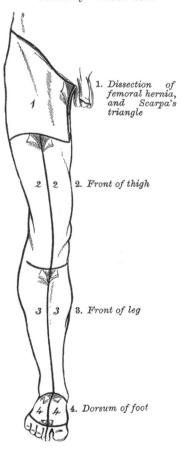

Fig. 252.—Dissection of lower extremity. Front view.

1. *Dissection of femoral hernia, and Scarpa's triangle*

2. *Front of thigh*

3. *Front of leg*

4. *Dorsum of foot*

exist naturally, but is the result of dissection. Mr. Callender, however, speaks of cases in which, probably as the result of pressure from enlarged inguinal lymphatic glands, the fascia has become atrophied, and a saphenous opening exists independent of dissection. A femoral hernia in passing through the saphenous opening receives the cribriform fascia as one of its coverings. A large subcutaneous bursa is found in the superficial fascia over the patella.

The **deep fascia** of the thigh is exposed on the removal of the superficial fascia, and is named, from its great extent, the *fascia lata*; it constitutes a uniform investment for the whole of this region of the limb, but varies in thickness in different parts; thus, it is thicker in the upper and outer part of the thigh, where it receives a fibrous expansion from the Gluteus maximus muscle, and where the

Fig. 253.—Muscles of the iliac and
anterior femoral regions.

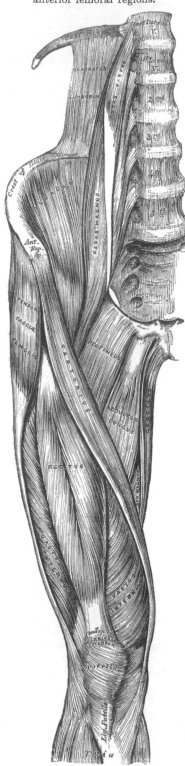

Tensor fasciæ femoris is inserted between its layers : it is very thin behind and at the upper and inner part, where it covers the Adductor muscles, and again becomes stronger around the knee, receiving fibrous expansions from the tendon of the Biceps externally, and from the Sartorius internally, and Quadriceps extensor cruris in front. The fascia lata is attached, above and behind, to the back of the sacrum and coccyx ; externally, to the crest of the ilium ; in front, to Poupart's ligament, and to the body of the os pubis ; and internally, to the descending ramus of the os pubis, to the ramus and tuberosity of the ischium, and to the lower border of the great sacro-sciatic ligament. From its attachment to the crest of the ilium it passes down over the Gluteus medius muscle to the upper border of the Gluteus maximus, where it splits into two layers, one passing superficial to and the other beneath this muscle. At the lower border of the muscle the two layers reunite. Externally, the fascia lata receives the greater part of the tendon of insertion of the Gluteus maximus, and becomes proportionately thickened. The portion of the fascia lata arising from the front part of the crest of the ilium, corresponding to the origin of the Tensor fasciæ femoris, passes down the outer side of the thigh as two layers, one superficial to and the other beneath this muscle ; these at its lower end become blended together into a thick and strong band, having first received the insertion of the muscle. This band is continued downwards, under the name of the *ilio-tibial band,* to be inserted into the external tuberosity of the tibia. Below, the fascia lata is attached to all the prominent points around the knee-joint, viz. the condyles of the femur, tuberosities of the tibia, and head of the fibula. On each side of the patella it is strengthened by transverse fibres given off from the lower part of the Vasti muscles, which are attached to and support this bone. Of these the outer are the stronger, and are continuous with the ilio-tibial band. From the inner surface of the fascia lata are given off two strong intermuscular septa, which are attached to the whole length of the linea aspera and its prolongations above and below : the external and stronger one, which extends from the insertion of the Gluteus maximus to the outer condyle, separates the Vastus externus in front from the short head of the Biceps behind, and gives partial origin to these muscles ; the inner one, the thinner of the two, separates the Vastus internus from the Adductor and Pectineus muscles. Besides these there are numerous smaller septa, separating the individual muscles, and enclosing each in a distinct sheath.

At the upper and inner part of the thigh, a little below Poupart's ligament, a large oval-shaped aperture is observed after the superficial fascia has been cleared off : it transmits the internal saphenous vein, and other smaller vessels, and is termed the *saphenous opening*. In order more correctly to consider the mode of formation of this aperture, the fascia lata in this part of the thigh is described as consisting of two portions, an iliac portion and a pubic portion.

The *iliac portion* is all that part of the fascia lata on the outer side of the saphenous opening. It is attached, externally, to the crest of the ilium and its anterior superior spine, to the whole length of Poupart's ligament as far internally as the spine of the os pubis, and to the pectineal line in conjunction with Gimbernat's ligament. From the spine of the os pubis it is reflected downwards and outwards, forming an arched margin, the *falciform process* or boundary (*superior cornu*) of the saphenous opening ; this margin overlies and is adherent to the anterior layer of the sheath of the femoral vessels : to its edge is attached the cribriform fascia ; and, below, it is continuous with the pubic portion of the fascia lata.

The *pubic portion* is situated at the inner side of the saphenous opening ; at the lower margin of this aperture it is continuous with the iliac portion ; traced upwards, it covers the surface of the Pectineus, Adductor longus, and Gracilis muscles, and, passing behind the sheath of the femoral vessels, to which it is closely united, is continuous with the sheath of the Psoas and Iliacus muscles, and is attached above to the ilio-pectineal line, where it becomes continuous with the iliac fascia. From this description it may be observed that the iliac portion of the fascia lata passes in front of the femoral vessels, and the pubic portion behind them, so that an apparent aperture exists between the two, through which the internal saphenous joins the femoral vein.*

The fascia should now be removed from the surface of the muscles. This may be effected by pinching it up between the forceps, dividing it, and separating it from each muscle in the course of its fibres.

The **Tensor fasciæ femoris** arises from the anterior part of the outer lip of the crest of the ilium, from the outer surface of the anterior superior spinous process, and part of the outer border of the notch below it, between the Gluteus medius and Sartorius, and from the surface of the fascia covering the Gluteus medius. It is inserted between two layers of the fascia lata about one-fourth down the outer side of the thigh. From the point of insertion the fascia is continued downwards to the external tuberosity of the tibia as a thickened band, the *ilio-tibial band*.

Relations.—By its *superficial surface*, with the fascia lata and the integument. By its *deep surface*, with the Gluteus medius, Rectus femoris, Vastus externus, and the ascending branches of the external circumflex artery. By its *anterior border*, with the Sartorius, from which it is separated below by a triangular space, in which is seen the Rectus femoris. By its *posterior border*, with the Gluteus medius.

The **Sartorius,** the longest muscle in the body, is flat, narrow, and ribbon-like ; it arises by tendinous fibres from the anterior superior spinous process of the ilium and the upper half of the notch below it, passes obliquely across the upper and anterior part of the thigh, from the outer to the inner side of the limb, then

* These parts will be again more particularly described with the anatomy of Hernia.

descends vertically, as far as the inner side of the knee, passing behind the inner condyle of the femur, and terminates in a tendon, which, curving obliquely forwards, expands into a broad aponeurosis, inserted, in front of the Gracilis and Semitendinosus, into the upper part of the inner surface of the shaft of the tibia, nearly as far forwards as the crest. The upper part of the tendon is curved backwards over the upper edge of the tendon of the Gracilis so as to be inserted behind it. An offset is derived from the upper margin of this aponeurosis, which blends with the fibrous capsule of the knee-joint, and another, given off from its lower border, blends with the fascia on the inner side of the leg.

The relations of this muscle to the femoral artery should be carefully examined, as it constitutes the chief guide in tying the vessel. In the upper third of the thigh it forms the outer side of a triangular space, *Scarpa's triangle*, the inner side of which is formed by the inner border of the Adductor longus, and the base, turned upwards, by Poupart's ligament; the femoral artery passes perpendicularly through the middle of this space from its base to its apex. In the middle third of the thigh, the femoral artery lies first along the inner border, and then behind the Sartorius.

Relations.—By its *superficial surface*, with the fascia lata and integument. By its *deep surface*, with the Rectus, Iliacus, Vastus internus, anterior crural nerve, sheath of the femoral vessels, Adductor longus, Adductor magnus, Gracilis, Semitendinosus, long saphenous nerve, and internal lateral ligament of the knee-joint.

The **Quadriceps extensor** includes the four remaining muscles on the front of the thigh. It is the great Extensor muscle of the leg, forming a large fleshy mass, which covers the front and sides of the femur, being united below into a tendon, attached to the patella, and above, subdivided into separate portions, which have received distinct names. Of these, one occupying the middle of the thigh, connected above with the ilium, is called the *Rectus femoris*, from its straight course. The other divisions lie in immediate connection with the shaft of the femur, which they cover from the trochanters to the condyles. The portion on the outer side of the femur is termed the *Vastus externus*; that covering the inner side, the *Vastus internus*; and that covering the front of the femur, the *Crureus*.

The **Rectus femoris** is situated in the middle of the anterior region of the thigh; it is fusiform in shape, and its superficial fibres are arranged in a bipenniform manner, the deep fibres running straight down to the deep aponeurosis. It arises by two tendons: one, the anterior or straight, from the anterior inferior spinous process of the ilium; the other, the posterior or reflected tendon, from a groove above the brim of the acetabulum; the two unite at an acute angle, and spread into an aponeurosis, which is prolonged downwards on the anterior surface of the muscle and from which the muscular fibres arise.* The muscle terminates in a broad and thick aponeurosis, which occupies the lower two-thirds of its posterior surface, and, gradually becoming narrowed into a flattened tendon, is inserted into the patella in common with the Vasti and Crureus.

Relations.—By its *superficial surface*, with the anterior fibres of the Gluteus minimus, the Tensor fasciæ femoris, the Sartorius, and the Iliacus; by its lower three-fourths, with the fascia lata. By its *posterior surface*, with the hip-joint, the external circumflex vessels, branches of the anterior crural nerve, and the Crureus and Vasti muscles.

The **Vastus externus** is the largest part of the Quadriceps extensor. It arises by a broad aponeurosis, which is attached to the upper half of the anterior intertrochanteric line, to the anterior and inferior borders of the root of the great

---

* Mr. W. R. Williams, in an interesting paper in the *Journ. of Anat. and Phys.* vol. xiii. p. 204, points out that the reflected tendon is the real origin of the muscle, and is alone present in early fœtal life. The direct tendon is merely an accessory band of condensed fascia. The paper will well repay perusal, though in some particulars I think the description in the text more generally accurate.—ED.

trochanter, to the outer lip of the gluteal ridge, and to the upper half of the outer lip of the linea aspera : this aponeurosis covers the upper three-fourths of the muscle, and from its inner surface many fibres take origin. A few additional fibres arise from the tendon of the Gluteus maximus, and from the external intermuscular septum between the Vastus externus and short head of the Biceps. The fibres form a large fleshy mass, which is attached to a strong aponeurosis, placed on the under surface of the muscle at its lower part : this becomes contracted and thickened into a flat tendon, which is inserted into the outer border of the patella, blending with the great Extensor tendon, and giving an expansion to the capsule of the knee-joint.

Relations.—By its *superficial surface*, with the Rectus, the Tensor fasciæ femoris, the fascia lata, and the tendon of the Gluteus maximus, from which it is separated by a synovial bursa. By its *deep surface*, with the Crureus, some large branches of the external circumflex artery and anterior crural nerve being interposed.

The **Vastus internus** and **Crureus** appear to be inseparably united, but when the Rectus femoris has been reflected a narrow interval will be observed extending upwards from the inner border of the patella between the two muscles. Here they can be separated, and the separation should be continued upwards as far as the lower part of the anterior intertrochanteric line, where, however, the two muscles are frequently continuous.

The **Vastus internus** arises from the lower half of the anterior intertrochanteric line, the spiral line, the inner lip of the linea aspera, the upper part of the internal supra-condylar line and the tendon of the Adductor magnus and internal intermuscular septum. Its fibres are directed downwards and forwards, and are chiefly attached to an aponeurosis which lies on the deep surface of the muscle and is inserted into the inner border of the patella and the Quadriceps extensor tendon, an expansion being sent to the capsule of the knee-joint.

The **Crureus** arises from the front and outer aspect of the shaft of the femur in its upper two-thirds and from the lower part of the external intermuscular septum. Its fibres end in a superficial aponeurosis, which forms the deep part of the Quadriceps extensor tendon.

Relations.—The inner edge of the Crureus is in contact with the anterior edge of the Vastus internus, but when separated from each other, as directed above, the latter muscle is seen merely to overlap the inner aspect of the femoral shaft without taking any fibres of origin from it. The Vastus internus is partly covered by the Rectus and Sartorius, but where these separate near the knee it becomes superficial, and produces a well-marked prominence above the inner aspect of the knee. In the middle third of the thigh it forms the outer wall of Hunter's canal, which contains the femoral vessels and the long saphenous nerve—the roof of the canal being formed by a strong fascia which extends from the Vastus internus to the Adductores longus and magnus. The Crureus is almost completely hidden by the Rectus femoris and Vastus externus. The deep surface of the two muscles is in relation to the femur and Subcrureus muscle. A synovial bursa is situated between the femur and the portion of the Quadriceps extensor tendon above the patella ; in the adult it communicates with the synovial cavity of the knee-joint.

The *tendons* of the different portions of the Quadriceps extensor unite at the lower part of the thigh, so as to form a single strong tendon, which is inserted into the upper part of the patella, some few fibres passing over it to blend with the Ligamentum patellæ. More properly, the patella may be regarded as a sesamoid bone, developed in the tendon of the Quadriceps ; and the Ligamentum patellæ, which is continued from the lower part of the patella to the tuberosity of the tibia, as the proper tendon of insertion of the muscle. A synovial bursa, the *post-patellar bursa*, is interposed between the tendon and the upper part of the tuberosity of the tibia ; and another, the *pre-patellar bursa*, is placed over the patella itself. This latter bursa often becomes enlarged, constituting ' housemaid's knee.'

The **Subcrureus** is a small muscle, usually distinct from the Crureus, but occasionally blended with it, which arises from the anterior surface of the lower part of the shaft of the femur, and is inserted into the upper part of the *cul-de-sac* of the capsular ligament which projects upwards beneath the Quadriceps for a variable distance. It sometimes consists of several separate muscular bundles.

Nerves.—The Tensor fasciæ femoris is supplied by the fourth and fifth lumbar and first sacral nerves through the superior gluteal nerve; the other muscles of this region, by the second, third, and fourth lumbar nerves, through branches of the anterior crural.

Actions.—The Tensor fasciæ femoris is a tensor of the fascia lata; continuing its action, the oblique direction of its fibres enables it to abduct and to rotate the thigh inwards. In the erect posture, acting from below, it will serve to steady the pelvis upon the head of the femur; and by means of the ilio-tibial band it steadies the condyles of the femur on the articular surfaces of the tibia, and assists the Gluteus maximus in supporting the knee in the extended position. The Sartorius flexes the leg upon the thigh, and, continuing to act, flexes the thigh upon the pelvis; it next rotates the thigh outwards. It was formerly supposed to adduct the thigh, so as to cross one leg over the other, and hence received its name of Sartorius, or tailor's muscle (*sartor*, a tailor), because it was supposed to assist in crossing the legs in the squatting position. When the knee is bent, the Sartorius assists the Semitendinosus, Semimembranosus, and Popliteus in rotating the tibia inwards. Taking its fixed point from the leg, it flexes the pelvis upon the thigh, and, if one muscle acts, assists in rotating the pelvis. The Quadriceps extensor extends the leg upon the thigh. The Rectus muscle assists the Psoas and Iliacus in supporting the pelvis and trunk upon the femur. It also assists in flexing the thigh on the pelvis, or if the thigh is fixed it will flex the pelvis. The Vastus internus draws the patella inwards as well as upwards.

Surgical Anatomy.—A few fibres of the Rectus muscle are liable to be ruptured from severe strain. This accident is especially liable to occur during the games of football and cricket, and is sometimes known as 'cricket thigh.' The patient experiences a sudden pain in the part, as if he had been struck, and the Rectus muscle stands out and is felt to be tense and rigid. The accident is often followed by considerable swelling from inflammatory effusion. Occasionally the Quadriceps extensor may be torn away from its insertion into the patella; or the tendon of the patella may be ruptured about an inch above the bone. This accident is caused in the same manner as fracture of the patella by muscular action is produced, viz. by a violent muscular effort to prevent falling while the knee is in a position of semiflexion. A distinct gap can be felt above the patella, and, owing to the retraction of the muscular fibres, union may fail to take place.

2. *Internal Femoral Region*

Gracilis.   Adductor longus.
Pectineus.   Adductor brevis.
                    Adductor magnus.

Dissection.—These muscles are at once exposed by removing the fascia from the fore part and inner side of the thigh. The limb should be abducted, so as to render the muscles tense and easier of dissection.

The **Gracilis** (figs. 253, 256) is the most superficial muscle on the inner side of the thigh. It is thin and flattened, broad above, narrow and tapering below. It arises by a thin aponeurosis from the lower half of the margin of the symphysis and the anterior half of the pubic arch. The fibres pass vertically downwards, and terminate in a rounded tendon, which passes behind the internal condyle of the femur, and, curving round the inner tuberosity of the tibia, becomes flattened, and is inserted into the upper part of the inner surface of the shaft of the tibia, below the tuberosity. A few of the fibres of the lower part of the tendon are prolonged into the deep fascia of the leg. The tendon of this muscle is situated immediately above that of the Semitendinosus, and its upper edge is overlapped by the tendon of the Sartorius with which it is in part blended. As it passes across the internal lateral ligament of the knee-joint, it is separated from it by a synovial bursa common to it and the Semitendinosus muscle.

FIG. 254.—Deep muscles of the internal femoral region.

Relations.—By its *superficial surface*, with the fascia lata and the Sartorius below; the internal saphenous vein crosses it obliquely near its lower part, lying superficial to the fascia lata. The internal saphenous nerve emerges between its tendon and that of the Sartorius. By its *deep surface*, with the Adductor brevis and the Adductor magnus, and the internal lateral ligament of the knee-joint.

The **Pectineus** (fig. 253) is a flat, quadrangular muscle, situated at the anterior part of the upper and inner aspect of the thigh. It arises from the linea ilio-pectinea, and to a slight extent from the surface of bone in front of it, between the pectineal eminence and spine of the os pubis, and from the fascia covering the anterior surface of the muscle; the fibres pass downwards, backwards, and outwards, to be inserted into a rough line leading from the lesser trochanter to the linea aspera.

Relations.—By its *anterior surface*, with the pubic portion of the fascia lata, which separates it from the femoral vessels and internal saphenous vein. By its *posterior surface*, with the capsular ligament of the hip-joint, the Adductor brevis and Obturator externus muscles, the obturator vessels and nerve being interposed. By its *outer border*, with the Psoas, a cellular interval separating them, through which pass the internal circumflex vessels. By its *inner border*, with the margin of the Adductor longus.

The **Adductor longus,** the most superficial of the three Adductors, is a flat triangular muscle, lying on the same plane as the Pectineus. It arises, by a flat narrow tendon, from the front of the os pubis, at the angle of junction of the crest with the symphysis; and soon expands into a broad fleshy belly, which, passing downwards, backwards, and outwards, is inserted, by an aponeurosis, into the linea aspera, between the Vastus internus and the Adductor magnus, with both of which it is usually blended.

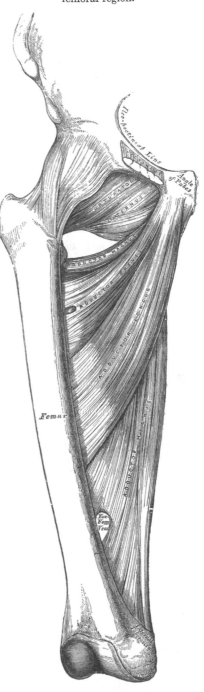

Relations.—By its *anterior surface*, with the fascia lata, the Sartorius, and, near its insertion, with the femoral artery and vein. By its *posterior surface*,

with the Adductor brevis and magnus, the anterior branches of the obturator
nerve, and with the profunda artery and vein near its insertion. By its *outer
border*, with the Pectineus. By its *inner border*, with the Gracilis.

The Pectineus and Adductor longus should now be divided near their origin, and
turned downwards, when the Adductor brevis and Obturator externus will be exposed.

The **Adductor brevis** is situated immediately behind the two preceding muscles.
It is somewhat triangular in form, and arises by a narrow origin from the outer
surface of the body and descending ramus of the os pubis, between the Gracilis
and Obturator externus. Its fibres, passing backwards, outwards, and downwards, are inserted, by an aponeurosis, into the lower part of the line leading
from the lesser trochanter to the linea aspera and the upper part of the same
line, immediately behind the Pectineus and upper part of the Adductor longus.

Relations.—By its *anterior surface*, with the Pectineus, Adductor longus,
profunda femoris artery, and anterior branches of the obturator nerve. By its
posterior surface, with the Adductor magnus, and posterior branch of the obturator
nerve. By its *outer border*, with the internal circumflex artery, the Obturator
externus, and conjoined tendon of the Psoas and Iliacus. By its *inner border*,
with the Gracilis and Adductor magnus. This muscle is pierced, near its
insertion, by the second or first and second perforating branches of the profunda
femoris artery.

The Adductor brevis should now be cut away near its origin, and turned outwards,
when the entire extent of the Adductor magnus will be exposed.

The **Adductor magnus** is a large triangular muscle, forming a septum between
the muscles on the inner and those on the back of the thigh. It arises from a
small part of the descending ramus of the os pubis, from the ramus of the
ischium, and from the outer margin of the inferior part of the tuberosity of the
ischium. Those fibres which arise from the ramus of the os pubis are very
short, horizontal in direction, and are inserted into the rough line leading from
the great trochanter to the linea aspera, internal to the Gluteus maximus ; those
from the ramus of the ischium are directed downwards and outwards with
different degrees of obliquity, to be inserted, by means of a broad aponeurosis,
into the linea aspera and the upper part of its internal prolongation below. The
internal portion of the muscle, consisting principally of those fibres which arise
from the tuberosity of the ischium, forms a thick fleshy mass consisting of coarse
bundles which descend almost vertically, and terminate about the lower third of
the thigh in a rounded tendon, which is inserted into the Adductor tubercle on
the inner condyle of the femur, being connected by a fibrous expansion to the
line leading upwards from the tubercle to the linea aspera. Between the two
portions of the muscle an interval is left, tendinous in front, fleshy behind, for
the passage of the femoral vessels into the popliteal space. The external portion
of the muscle at its attachment to the femur presents three or four osseo-
aponeurotic openings, formed by tendinous arches attached to the bone, from
which muscular fibres arise. The three superior of these apertures are for the
three perforating arteries, and the fourth, when it exists, for the terminal branch
of the profunda.

Relations.—By its *anterior surface*, with the Pectineus, Adductor brevis,
Adductor longus, and the femoral and profunda vessels and obturator nerve. By
its *posterior surface*, with the great sciatic nerve, the Gluteus maximus, Biceps,
Semitendinosus, and Semimembranosus. By its *superior* or *shortest border* it
lies parallel with the Quadratus femoris, the internal circumflex artery passing
between them. By its *internal* or *longest border*, with the Gracilis, Sartorius,
and fascia lata. By its *external* or *attached border*, it is inserted into the femur
behind the Adductor brevis and Adductor longus, which separate it from the
Vastus internus ; and in front of the Gluteus maximus and short head of the
Biceps, which separate it from the Vastus externus.

Nerves.—The three Adductor muscles and the Gracilis are supplied by the third and fourth lumbar nerves through the obturator nerve; the Adductor magnus receiving an additional branch from the sacral plexus through the great sciatic. The Pectineus is supplied by the second, third, and fourth lumbar nerves through the anterior crural, and by the accessory obturator, from the third lumbar, when it exists. Occasionally it receives a branch from the obturator nerve.*

Actions.—The Pectineus and three Adductors adduct the thigh powerfully; they are especially used in horse exercise, the flanks of the horse being grasped between the knees by the actions of these muscles. In consequence of the obliquity of their insertion into the linea aspera, they rotate the thigh outwards, assisting the external Rotators, and when the limb has been abducted, they draw it inwards, carrying the thigh across that of the opposite side. The Pectineus and Adductor brevis and longus assist the Psoas and Iliacus in flexing the thigh upon the pelvis. In progression, also, all these muscles assist in drawing forwards the hinder limb. The Gracilis assists the Sartorius in flexing the leg and rotating it inwards; it is also an adductor of the thigh. If the lower extremities are fixed, these muscles may take their fixed point from below and act upon the pelvis, serving to maintain the body in an erect posture; or, if their action is continued, to flex the pelvis forwards upon the femur.

Surgical Anatomy.—The Adductor longus is liable to be severely strained in those who ride much on horseback, or its tendon to be ruptured by suddenly gripping the saddle. And, occasionally, especially in cavalry soldiers, the tendon may become ossified, constituting the 'rider's bone.'

3. Gluteal Region

Gluteus maximus.	Obturator internus.
Gluteus medius.	Gemellus superior.
Gluteus minimus.	Gemellus inferior.
Pyriformis.	Quadratus femoris.

Obturator externus

Dissection (fig. 255).—The subject should be turned on its face, a block placed beneath the pelvis to make the buttocks tense, and the limbs allowed to hang over the end of the table, with the foot inverted, and the thigh abducted. Make an incision through the integument along the crest of the ilium to the middle of the sacrum and thence downwards to the tip of the coccyx, and carry a second incision from that point obliquely downwards and outwards to the outer side of the thigh, four inches below the great trochanter. The portion of integument included between these incisions is to be removed in the direction shown in the figure.

The **Gluteus maximus** (fig. 256), the most superficial muscle in the gluteal region, is a very broad and thick fleshy mass, of a quadrilateral shape, which forms the prominence of the nates. Its large size is one of the most characteristic points in the muscular system in man, connected as it is with the power he has of maintaining the trunk in the erect posture. In structure the muscle is remarkably coarse, being made up of muscular fasciculi lying parallel with one another, and collected together into large bundles, separated by deep cellular intervals. It arises from the superior curved line of the ilium, and the portion of bone, including the crest, immediately above and behind it; from the posterior surface of the lower part of the sacrum, the side of the coccyx, the aponeurosis of the Erector spinæ muscle, the great sacro-sciatic ligament, and the fascia covering the Gluteus medius. The fibres are directed obliquely downwards and outwards; those forming the upper and larger portion of the muscle, together with the superficial fibres of the lower portion, terminate in a thick tendinous lamina,

---

* Professor Paterson describes the Pectineus as consisting of two incompletely separated strata, of which the outer or dorsal stratum, which is constant, is supplied by the anterior crural nerve, or in its absence by the accessory obturator, with which it is intimately related; while the inner or ventral stratum, when present, is supplied by the obturator nerve. *Journal of Anatomy and Physiology*, vol. xxvi. p. 43.

which passes across the great trochanter, and is inserted in the fascia lata covering the outer side of the thigh; the deeper fibres of the lower portion of the muscle are inserted into the rough line leading from the great trochanter to the linea aspera between the Vastus externus and Adductor magnus.

Three *synovial bursæ* are usually found in relation with this muscle. One of these, of large size, and generally multilocular, separates it from the great trochanter. A second, often wanting, is situated on the tuberosity of the ischium. A third is found between the tendon of this muscle and the Vastus externus.

FIG. 255.—Dissection of lower extremity. Posterior view.

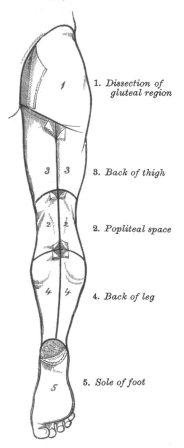

1. *Dissection of gluteal region*

3. *Back of thigh*

2. *Popliteal space*

4. *Back of leg*

5. *Sole of foot*

Relations.—By its *superficial surface*, with a thin fascia, which separates it from the subcutaneous tissue. By its *deep surface*, from above downwards, with the ilium, sacrum, coccyx, and great sacro-sciatic ligament, part of the Gluteus medius; Pyriformis, Gemelli, Obturator internus, Quadratus femoris, the tuberosity of the ischium, great trochanter, the origin of the Biceps, Semitendinosus, Semimembranosus, and Adductor magnus muscles. The superficial part of the gluteal artery reaches the deep surface of the muscle by passing between the Pyriformis and the Gluteus medius; the sciatic and internal pudic vessels and nerves, and muscular branches from the sacral plexus issue from the pelvis below the Pyriformis. The first perforating artery and the terminal branches of the internal circumflex artery are also found under cover of the muscle. Its *upper border* is thin, and connected with the Gluteus medius by the fascia lata. Its *lower border* is free and prominent.

Dissection.—Divide the Gluteus maximus near its origin by a vertical incision carried from its upper to its lower border; a cellular interval will be exposed, separating it from the Gluteus medius and External rotator muscles beneath. The upper portion of the muscle is to be altogether detached, and the lower portion turned outwards; the loose areolar tissue filling up the interspace between the trochanter major and tuberosity of the ischium being removed, the parts already enumerated as exposed by the removal of this muscle will be seen.

The **Gluteus medius** is a broad, thick, radiating muscle, situated on the outer surface of the pelvis. Its posterior third is covered by the Gluteus maximus; its anterior two-thirds by the fascia lata, which separates it from the integument. It arises from the outer surface of the ilium, between the superior and middle curved lines, and from the outer lip of that portion of the crest which is between them; it also arises from the dense fascia (gluteal aponeurosis) covering its outer surface. The fibres converge to a strong flattened tendon, which is inserted into the oblique line which traverses the outer surface of the great trochanter. A synovial bursa separates the tendon of the muscle from the surface of the trochanter in front of its insertion.

Relations.—By its *superficial surface*, with the Gluteus maximus behind, the Tensor fasciae femoris and deep fascia in front. By its *deep surface*, with the Gluteus minimus and the gluteal vessels and superior gluteal nerve. Its *anterior border* is blended with the Gluteus minimus. Its *posterior border* lies parallel with the Pyriformis, the gluteal vessels intervening.

This muscle should now be divided near its insertion and turned upwards, when the Gluteus minimus will be exposed.

The **Gluteus minimus,** the smallest of the three Glutei, is placed immediately beneath the preceding. It is fanshaped, arising from the outer surface of the ilium, between the middle and inferior curved lines, and behind, from the margin of the great sacrosciatic notch : the fibres converge to the deep surface of a radiated aponeurosis, which, terminating in a tendon, is inserted into an impression on the anterior border of the great trochanter. A synovial bursa is interposed between the tendon and the great trochanter.

Relations.—By its *superficial surface,* with the Gluteus medius, and the gluteal vessels and superior gluteal nerve. By its *deep surface,* with the ilium, the reflected tendon of the Rectus femoris, and capsular ligament of the hip-joint. Its *anterior margin* is blended with the Gluteus medius. Its *posterior margin* is in contact and sometimes joined with the tendon of the Pyriformis.

The **Pyriformis** is a flat muscle, pyramidal in shape, lying almost parallel with the posterior margin of the Gluteus medius. It is situated partly within the pelvis at its posterior part, and partly at the back of the hip-joint. It arises from the front of the sacrum by three fleshy digitations, attached to the portions of bone between the first, second, third, and fourth anterior sacral foramina, and also from the grooves leading from the foramina : a few fibres also arise from the

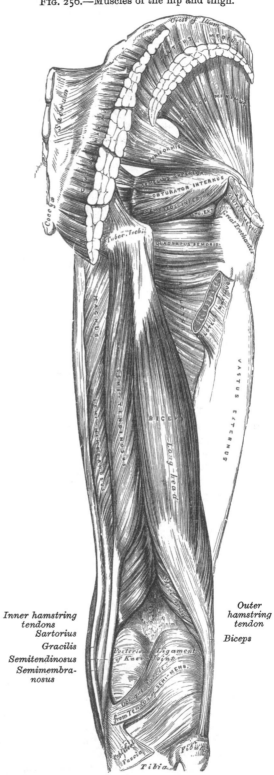

Fig. 256.—Muscles of the hip and thigh.

Inner hamstring tendons
Sartorius
Gracilis
Semitendinosus
Semimembranosus

Outer hamstring tendon
Biceps

margin of the great sacro-sciatic foramen, and from the anterior surface of the great sacro-sciatic ligament. The muscle passes out of the pelvis through the great sacro-sciatic foramen, the upper part of which it fills, and is inserted by a rounded tendon into the upper border of the great trochanter, behind, but often partly blended with, the tendon of the Obturator internus and Gemelli muscles.

Relations.—By its *anterior surface, within the pelvis*, with the Rectum (especially on the left side), the sacral plexus of nerves, and the branches of the internal iliac vessels; *external to the pelvis*, with the posterior surface of the ischium and capsular ligament of the hip-joint. By its *posterior surface, within the pelvis*, with the sacrum; and *external to it*, with the Gluteus maximus. By its *upper border*, with the Gluteus medius, from which it is separated by the gluteal vessels and superior gluteal nerve. By its *lower border*, with the Gemellus superior and Coccygeus; the sciatic vessels and nerves, the internal pudic vessels and nerve, and muscular branches from the sacral plexus, passing from the pelvis in the interval between the two muscles.

The **obturator membrane** (fig. 166) is a thin layer of interlacing fibres, which closes the obturator foramen. It is attached, externally, to the margin of the foramen; internally, to the posterior surface of the ischio-pubic ramus, below and internal to the margin of the foramen. It is occasionally incomplete, and presents at its upper and outer part a small canal, which is bounded below by a thickened band of fibres, for the passage of the obturator vessels and nerve. Both Obturator muscles are connected with this membrane.

Dissection.—The next muscle, as well as the origin of the Pyriformis, can only be seen when the pelvis is divided and the viscera removed.

The **Obturator internus,** like the preceding muscle, is situated partly within the cavity of the pelvis, and partly at the back of the hip-joint. It arises from the inner surface of the anterior and external wall of the pelvis, where it surrounds the greater part of the obturator foramen, being attached to the descending ramus of the os pubis and the ramus of the ischium, and at the side to the inner surface of the innominate bone below and behind the pelvic brim, reaching from the upper part of the great sacro-sciatic foramen above and behind to the thyroid foramen below and in front. It also arises from the inner surface of the obturator membrane except at its posterior part, from the tendinous arch which completes the canal for the passage of the obturator vessels and nerve and to a slight extent from the obturator layer of the pelvic fascia, which covers it. The fibres converge rapidly, and are directed backwards and downwards, and terminate in four or five tendinous bands, which are found on its deep surface; these bands are reflected at a right angle over the inner surface of the tuberosity of the ischium, which is grooved for their reception: the groove is covered with cartilage, and lined by a synovial bursa. The muscle leaves the pelvis by the lesser sacro-sciatic foramen; and the tendinous bands unite into a single flattened tendon, which passes horizontally outwards, and, after receiving the attachment of the Gemelli, is inserted into the fore part of the inner surface of the great trochanter in front of the Obturator externus. A synovial bursa, narrow and elongated in form, is usually found between the tendon of this muscle and the capsular ligament of the hip: it occasionally communicates with the bursa between the tendon and the tuberosity of the ischium, the two forming a single sac.

In order to display the peculiar appearances presented by the tendon of this muscle, it must be divided near its insertion and reflected inwards.

Relations.—*Within the pelvis*, this muscle is in relation, by its *anterior surface*, with the obturator membrane and inner surface of the anterior wall of the pelvis; by its *posterior surface*, with the pelvic and obturator fasciæ, which separate it from the Levator ani; and it is crossed by the internal pudic vessels and nerve. This surface forms the outer boundary of the ischio-rectal fossa. *External to the pelvis*, it is covered by the Gluteus maximus, crossed by the great sciatic nerve, and rests on the back part of the hip-joint. As the tendon of the Obturator internus emerges from the lesser sacro-sciatic foramen it is overlapped by the two Gemelli, while nearer its insertion the Gemelli pass in front of it and form a groove in which the tendon lies.

The **Gemelli** are two small muscular fasciculi, accessories to the tendon of the Obturator internus, which is received into a groove between them. They are called *superior* and *inferior*.

The **Gemellus superior,** the smaller of the two, arises from the outer surface of the spine of the ischium, and passing horizontally outwards becomes blended with the upper part of the tendon of the Obturator internus, and is inserted with it into the inner surface of the great trochanter. This muscle is sometimes wanting.

Relations.—By its *superficial surface,* with the Gluteus maximus and the sciatic vessels and nerves. By its *deep surface,* with the capsule of the hip-joint. By its *upper border,* with the lower margin of the Pyriformis. By its *lower border,* with the tendon of the Obturator internus.

The **Gemellus inferior** arises from the upper part of the tuberosity of the ischium, where it forms the lower edge of the groove for the Obturator internus tendon, and, passing horizontally outwards, is blended with the lower part of the tendon of the Obturator internus, and inserted with it into the inner surface of the great trochanter.

Relations.—By its *superficial surface,* with the Gluteus maximus and the sciatic vessels and nerves. By its *deep surface,* with the capsular ligament of the hip-joint. By its *upper border,* with the tendon of the Obturator internus. By its *lower border,* with the tendon of the Obturator externus and Quadratus femoris.

The **Quadratus femoris** is a short, flat muscle, quadrilateral in shape (hence its name), situated between the Gemellus inferior and the upper margin of the Adductor magnus. It arises from the upper part of the external lip of the tuberosity of the ischium, and, proceeding horizontally outwards, is inserted into the upper part of the linea quadrata, that is, the line which crosses the posterior intertrochanteric line. A synovial bursa is often found between the under surface of this muscle and the lesser trochanter, which it covers.

Relations.—By its *posterior surface,* with the Gluteus maximus and the sciatic vessels and nerves. By its *anterior surface,* with the tendon of the Obturator externus and trochanter minor, and with the capsule of the hip-joint. By its *upper border,* with the Gemellus inferior. Its *lower border* is separated from the Adductor magnus by the terminal branches of the internal circumflex vessels.

Dissection.—In order to expose the next muscle (the Obturator externus), it is necessary to remove the Psoas, Iliacus, Pectineus, and Adductores brevis and longus muscles from the front and inner side of the thigh; and the Gluteus maximus and Quadratus femoris from the back part. Its dissection should, consequently, be postponed until the muscles of the anterior and internal femoral regions have been examined.

The **Obturator externus** (fig. 257) is a flat, triangular muscle, which covers the outer surface of the anterior wall of the pelvis. It arises from the margin of bone immediately around the inner side of the obturator foramen, viz. from the body and ramus of the os pubis, and the ramus of the ischium ; it also arises from the inner two-thirds of the outer surface of the obturator membrane, and from the tendinous arch which completes the canal for the passage of the obturator vessels and nerves. The fibres from the pubic arch extend on to the inner surface of the bone, from which they obtain a narrow origin between the margin of the foramen and the attachment of the membrane. The fibres converging pass backwards, outwards, and upwards, and terminate in a tendon which runs across the back part of the hip-joint, and is inserted into the digital fossa of the femur.

Relations.—By its *anterior surface,* with the Psoas, Iliacus, Pectineus, Adductor magnus, and Adductor brevis ; and more externally, with the neck of the femur and capsule of the hip-joint. The obturator artery and vein lie between this muscle and the obturator membrane; the superficial part of the obturator nerve lies above the muscle, and the deep branch perforates it. By its *posterior surface*, with the obturator membrane and Quadratus femoris.

Nerves.—The Gluteus maximus is supplied by the fifth lumbar and first and second sacral nerves through the inferior gluteal nerve from the sacral plexus; the Gluteus medius and minimus, by the fourth and fifth lumbar and first sacral nerves through the superior gluteal; the Pyriformis is supplied by the first and second sacral nerves; the Gemellus inferior and Quadratus femoris by the last lumbar and first sacral nerve; the Gemellus superior and Obturator internus by the fifth lumbar and first and second sacral nerves, and the Obturator externus by the second, third, and fourth lumbar nerves through the obturator.

FIG. 257.—Obturator externus muscle.
(From a preparation in the Museum of the Royal College of Surgeons of England.)

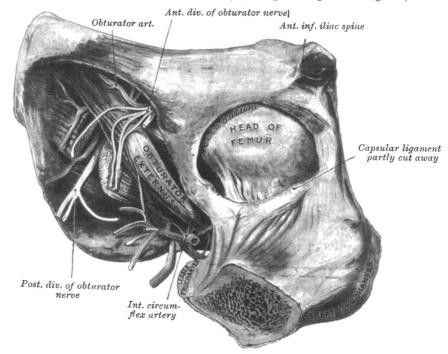

Actions.—The Gluteus maximus, when it takes its fixed point from the pelvis, extends the femur and brings the bent thigh into a line with the body. Taking its fixed point from below, it acts upon the pelvis, supporting it and the whole trunk upon the head of the femur, which is especially obvious in standing on one leg. Its most powerful action is to cause the body to regain the erect position after stooping, by drawing the pelvis backwards, being assisted in this action by the Biceps, Semitendinosus, and Semimembranosus. The Gluteus maximus is a tensor of the fascia lata, and by its connection with the ilio-tibial band it steadies the femur on the articular surface of the tibia during standing, when the Extensor muscles are relaxed. The lower part of the muscle also acts as an adductor and external rotator of the limb. The Gluteus medius and minimus abduct the thigh, when the limb is extended, and are principally called into action in supporting the body on one limb, in conjunction with the Tensor fasciæ femoris. Their anterior fibres, by drawing the great trochanter forwards, rotate the thigh inwards, in which action they are also assisted by the Tensor fasciæ femoris. The remaining muscles are powerful rotators of the thigh outwards. In the sitting posture, when the thigh is flexed upon the pelvis, their action as rotators ceases, and they become abductors, with the exception of the Obturator externus, which still rotates the femur outwards. When the femur is fixed, the Pyriformis and Obturator muscles serve to draw the pelvis forwards if it has been inclined backwards, and assist in steadying it upon the head of the femur.

4. *Posterior Femoral Region*

Biceps. Semitendinosus. Semimembranosus.
(*Hamstring muscles.*)

Dissection (fig. 255).—Make a vertical incision along the middle of the back of the thigh, from the lower fold of the nates to about three inches below the back of the knee-joint, and there connect it with a transverse incision, carried from the inner to the outer side of the leg. Make a third incision transversely at the junction of the middle with the lower third of the thigh. The integument having been removed from the back of the knee, and the boundaries of the popliteal space examined, the removal of the integument from the remaining part of the thigh should be continued, when the fascia and muscles of this region will be exposed.

The **Biceps** (*Biceps flexor cruris*) is a large muscle, of considerable length, situated on the posterior and outer aspect of the thigh (fig. 256). It arises by two heads. One, the long head, arises from the lower and inner impression on the back part of the tuberosity of the ischium, by a tendon common to it and the Semitendinosus, and from the lower part of the great sacro-sciatic ligament. The femoral, or short head, arises from the outer lip of the linea aspera, between the Adductor magnus and Vastus externus, extending up almost as high as the insertion of the Gluteus maximus ; from the outer prolongation of the linea aspera to within two inches of the outer condyle ; and from the external intermuscular septum. The fibres of the long head form a fusiform belly, which, passing obliquely downwards and a little outwards, terminates in an aponeurosis which covers the posterior surface of the muscle, and receives the fibres of the short head : this aponeurosis becomes gradually contracted into a tendon, which is inserted into the outer side of the head of the fibula, and by a small slip into the lateral surface of the external tuberosity of the tibia. At its insertion the tendon divides into two portions, which embrace the long external lateral ligament of the knee-joint. From the posterior border of the tendon a thin expansion is given off to the fascia of the leg. The tendon of this muscle forms the outer hamstring.

Relations.—By its *superficial surface*, with the Gluteus maximus, and the small sciatic nerve, the fascia lata, and integument. By its *deep surface*, with the Semimembranosus, Adductor magnus, and Vastus externus, the great sciatic nerve, and, near its insertion, with the external head of the Gastrocnemius, Plantaris, the superior external articular artery, and the external popliteal nerve.

The **Semitendinosus**, remarkable for the great length of its tendon, is situated at the posterior and inner aspect of the thigh. It arises from the lower and inner impression on the tuberosity of the ischium, by a tendon common to it and the long head of the Biceps ; it also arises from an aponeurosis which connects the adjacent surfaces of the two muscles to the extent of about three inches after their origin. It forms a fusiform muscle, which, passing downwards and inwards, terminates a little below the middle of the thigh in a long round tendon which lies along the inner side of the popliteal space, then curves around the inner tuberosity of the tibia, and is inserted into the upper part of the inner surface of the shaft of that bone, nearly as far forwards as its anterior border. At its insertion it gives off from its lower border a prolongation to the deep fascia of the leg. This tendon lies behind the tendon of the Sartorius, and below that of the Gracilis, to which it is united. A tendinous intersection is usually observed about the middle of the muscle.

Relations.—By its *superficial surface*, with the Gluteus maximus and fascia lata. By its *deep surface*, with the Semimembranosus, Adductor magnus, inner head of the Gastrocnemius, and internal lateral ligament of the knee-joint, the last being separated from the tendon by a bursa.

The **Semimembranosus,** so called from its membranous tendon of origin, is situated at the back-part and inner side of the thigh. It arises by a thick tendon from the upper and outer impression on the back part of the tuberosity of the ischium, above and to the outer side of the Biceps and Semitendinosus, and is inserted into the groove on the inner and back part of the inner tuberosity of the tibia, beneath the internal lateral ligament. The tendon of the muscle at its origin expands into an aponeurosis, which covers the upper part of its anterior surface : from this aponeurosis muscular fibres arise, and converge to another aponeurosis which covers the lower part of its posterior surface and contracts into the tendon of insertion. The tendon of the muscle at its insertion gives off certain fibrous expansions ; one of these, of considerable size, passes upwards and outwards to be inserted into the back part of the outer condyle of the femur, forming part of the posterior ligament of the knee-joint ; a second is continued downwards to the fascia which covers the Popliteus muscle ; while a few fibres join the internal lateral ligament of the joint.

The tendons of the two preceding muscles, with that of the Gracilis, form the inner hamstring.

Relations.—By its *superficial surface*, with the Gluteus maximus, Semitendinosus, Biceps, and fascia lata. By its *deep surface*, with the origin of the Quadratus femoris, popliteal vessels, Adductor magnus, and inner head of the Gastrocnemius, from which it is separated by a synovial bursa. By its *inner border*, with the Gracilis. By its *outer border*, with the great sciatic nerve, and its internal popliteal branch.

Nerves.—The muscles of this region are supplied by the first, second, and third sacral nerves through the great sciatic nerve.

Actions.—The hamstring muscles flex the leg upon the thigh. When the knee is semi-flexed, the Biceps, in consequence of its oblique direction downwards and outwards, rotates the leg slightly outwards ; and the Semitendinosus, and to a slight extent the Semimembranosus, rotate the leg inwards, assisting the Popliteus. Taking their fixed point from below, these muscles serve to support the pelvis upon the head of the femur, and to draw the trunk directly backwards, as in raising it from the stooping position or in feats of strength, when the body is thrown backwards in the form of an arch. When the leg is extended on the thigh they limit the amount of flexion of the trunk on the lower limbs.

Surgical Anatomy.—The tendons of these muscles occasionally require subcutaneous division in some forms of spurious ankylosis of the knee-joint dependent upon permanent contraction and rigidity of the flexor muscles, or from stiffening of the ligamentous and other tissues surrounding the joint, the result of disease. This is effected by putting the tendon upon the stretch, and inserting a narrow, sharp-pointed knife between it and the skin : the cutting edge being then turned towards the tendon, it should be divided, taking care that the wound in the skin is not at the same time enlarged. The relation of the external popliteal nerve to the tendon of the Biceps must always be borne in mind in dividing this tendon.

III. Muscles and Fasciæ of the Leg

These may be divided into three groups : those on the anterior, those on the posterior, and those on the outer side of the leg.

5. *Anterior Tibio-fibular Region*

Tibialis anticus.	Extensor longus digitorum.
Extensor proprius hallucis.*	Peroneus tertius.

* There is no such word as ' Hallux, -cis.' It is the result of some ignorant blunder, copied until it has become established by usage ; it has been thought better, therefore, to

Dissection (fig. 252).—The knee should be bent, a block placed beneath it, and the foot kept in an extended position; then make an incision through the integument in the middle line of the leg to the ankle, and continue it along the dorsum of the foot to the toes. Make a second incision transversely across the ankle, and a third in the same direction across the bases of the toes; remove the flaps of integument included between these incisions in order to examine the deep fascia of the leg.

The **Deep Fascia of the Leg** forms a complete investment to the muscles, but is not continued over the subcutaneous surfaces of the bones. It is continuous above with the fascia lata, receiving an expansion from the tendon of the Biceps on the outer side, and from the tendons of the Sartorius, Gracilis, and Semitendinosus on the inner side; in front, it blends with the periosteum covering the subcutaneous surface of the tibia, and with that covering the head and external malleolus of the fibula; below, it is continuous with the annular ligaments of the ankle. It is thick and dense in the upper and anterior part of the leg, and gives attachment, by its deep surface, to the Tibialis anticus and Extensor longus digitorum muscles; but thinner behind, where it covers the Gastrocnemius and Soleus muscles. Over the popliteal space it is much strengthened by transverse fibres, which stretch across from the inner to the outer hamstring muscles, and it is here perforated by the external saphenous vein. Its deep surface gives off, on the outer side of the leg, two strong intermuscular septa, which enclose the Peronei muscles, and separate them from the muscles on the anterior and posterior tibial regions and several smaller and more slender processes, which enclose the individual muscles in each region; at the same time a broad transverse intermuscular septum, called the *deep transverse fascia of the leg*, intervenes between the superficial and deep muscles in the posterior tibio-fibular region.

Remove the fascia by dividing it in the same direction as the integument, excepting opposite the ankle, where it should be left entire. Commence

retain it. According to Lewis and Short the word is ALLEX, masculine; genitive, ALLĬCIS, the great toe, and the correct rendering would be Extensor proprius allĭcis. It is a rare word, and is sometimes spelt, but not so correctly, ' Hallex.' It is used by Plautus, in the ' Pœnulus ' V. v. 31, of a little man, as we might say ' a hop-o'-my-thumb.' ' Tunc hic amator audes esse, *allex* viri ' (To think of you daring to make up to her, you hop-o'-my-thumb!). The word ' alex,' sometimes spelt ' allex,' a fish sauce, is probably a different word altogether. It is used by Horace and Pliny.

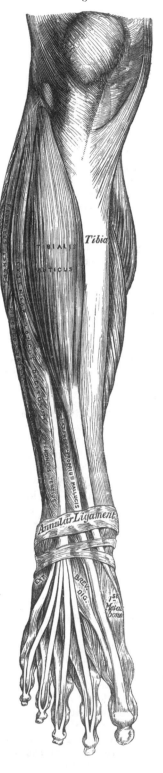

FIG. 258.—Muscles of the front of the leg.

the removal of the fascia from below, opposite the tendons, and detach it in the line of direction of the muscular fibres.

The **Tibialis anticus** is situated on the outer side of the tibia ; it is thick and fleshy at its upper part, tendinous below. It arises from the outer tuberosity and upper half or two-thirds of the external surface of the shaft of the tibia ; from the adjoining part of the interosseous membrane ; from the deep surface of the fascia ; and from the intermuscular septum between it and the Extensor longus digitorum : the fibres pass vertically downwards, and terminate in a tendon, which is apparent on the anterior surface of the muscle at the lower third of the leg. After passing through the innermost compartment of the anterior annular ligament, it is inserted into the inner and under surface of the internal cuneiform bone, and base of the metatarsal bone of the great toe.

Relations.—By its *anterior surface*, with the fascia, and with the annular ligament. By its *posterior surface*, with the interosseous membrane, tibia, ankle-joint, and inner side of the tarsus : this surface also overlaps the anterior tibial vessels and nerve in the upper part of the leg. By its *inner surface*, with the tibia. By its *outer surface*, with the Extensor longus digitorum, and Extensor proprius hallucis, and the anterior tibial vessels and nerve.

The **Extensor proprius hallucis** is a thin, elongated, and flattened muscle, situated between the Tibialis anticus and Extensor longus digitorum. It arises from the anterior surface of the fibula for about the middle two-fourths of its extent, its origin being internal to that of the Extensor longus digitorum ; it also arises from the interosseous membrane to a similar extent. The fibres pass downwards, and terminate in a tendon, which occupies the anterior border of the muscle, passes through a distinct compartment in the lower portion of the annular ligament, crosses the anterior tibial vessels near the bend of the ankle, and is inserted into the base of the last phalanx of the great toe. Opposite the metatarso-phalangeal articulation, the tendon gives off a thin prolongation on each side, which covers the surface of the joint. It usually sends an expansion, from the inner side of the tendon, to be inserted into the base of the first phalanx.

Relations.—By its *anterior surface*, with the fascia, and the anterior annular ligament. By its *posterior surface*, with the interosseous membrane, fibula, tibia, and ankle-joint. By its *outer side*, with the Extensor longus digitorum above, the dorsalis pedis vessels, anterior tibial nerve, and Extensor brevis digitorum below. By its *inner side*, with the Tibialis anticus and the anterior tibial vessels above. The muscle is external to the anterior tibial vessels in the upper part of the leg ; but in the lower third its tendon crosses over them, so that it lies internal to them on the dorsum of the foot.

The **Extensor longus digitorum** is an elongated, flattened, penniform muscle, situated the most externally of all the muscles on the fore part of the leg. It arises from the outer tuberosity of the tibia ; from the upper three-fourths of the anterior surface of the shaft of the fibula ; from the interosseous membrane ; from the deep surface of the fascia ; and from the intermuscular septa between it and the Tibialis anticus on the inner, and the Peronei on the outer side. The tendon enters a canal in the annular ligament, with the Peroneus tertius, and divides into four slips, which run across the dorsum of the foot, and are inserted into the second and third phalanges of the four lesser toes. The mode in which the tendons are inserted is the following : the three inner tendons opposite the metatarso-phalangeal articulation are joined, on their outer side, by a tendon of the Extensor brevis digitorum. They all receive a fibrous expansion from the Interossei and Lumbricales, and then spread out into a broad aponeurosis, which covers the dorsal surface of the first phalanx : this aponeurosis, at the articulation of the first with the second phalanx, divides into three slips, a middle one, which is inserted into the base of the second phalanx ; and two lateral slips, which, after uniting on the dorsal surface of the second phalanx, are continued onwards, to be inserted into the base of the third.

Relations.—By its *anterior surface*, with the fascia and the annular ligament. By its *posterior surface*, with the fibula, interosseous membrane, ankle-joint, and Extensor brevis digitorum. By its *inner side*, with the Tibialis anticus, Extensor proprius hallucis, and anterior tibial vessels and nerve. By its *outer side*, with the Peroneus longus and brevis.

The **Peroneus tertius** is a part of the Extensor longus digitorum, and might be described as its fifth tendon. The fibres belonging to this tendon arise from the lower fourth of the anterior surface of the fibula; from the lower part of the interosseous membrane; and from an intermuscular septum between it and the Peroneus brevis. The tendon, after passing through the same canal in the annular ligament as the Extensor longus digitorum, is inserted into the dorsal surface of the base of the metatarsal bone of the little toe. This muscle is sometimes wanting.

Nerves.—These muscles are supplied by the fourth and fifth lumbar and first sacral nerves through the anterior tibial nerve.

Actions.—The Tibialis anticus and Peroneus tertius are the direct flexors of the foot at the ankle-joint; the former muscle, when acting in conjunction with the Tibialis posticus, raises the inner border of the foot (i.e. inverts the foot); and the latter, acting with the Peroneus brevis and longus, draws the outer border of the foot upwards and the sole outwards (i.e. everts the foot). The Extensor longus digitorum and Extensor proprius hallucis extend the phalanges of the toes, and, continuing their action, flex the foot upon the leg. Taking their fixed point from below, in the erect posture, all these muscles serve to fix the bones of the leg in the perpendicular position, and give increased strength to the ankle-joint.

6. *Posterior Tibio-fibular Region*

Dissection (fig. 255).—Make a vertical incision along the middle line of the back of the leg, from the lower part of the popliteal space to the heel, connecting it below by a transverse incision extending between the two malleoli; the flaps of integument being removed, the fascia and muscles should be examined.

The muscles in this region of the leg are subdivided into two layers—superficial and deep. The superficial layer constitutes a powerful muscular mass, forming the calf of the leg. Their large size is one of the most characteristic features of the muscular apparatus in man, and bears a direct relation to his ordinary attitude and mode of progression.

Superficial Layer

Gastrocnemius.   Soleus.   Plantaris.

The **Gastrocnemius** is the most superficial muscle, and forms the greater part of the calf. It arises by two heads, which are connected to the condyles of the femur by two strong flat tendons. The inner and larger head arises from a depression at the upper and back part of the inner condyle and from the adjacent part of the femur. The outer head arises from an impression on the outer side of the external condyle and from the posterior surface of the femur immediately above the condyle. Both heads, also, arise by a few tendinous and fleshy fibres from the ridges which are continued upwards from the condyles to the linea aspera. Each tendon spreads out into an aponeurosis, which covers the posterior surface of that portion of the muscle to which it belongs; the muscular fibres of the inner head being thicker and extending lower than those of the outer. From the anterior surface of these tendinous expansions, muscular fibres are given off. The fibres in the median line, which correspond to the accessory portions of the muscle derived from the bifurcations of the linea aspera, unite at an angle upon a median tendinous raphé below: the remaining fibres converge to an aponeurosis which covers the anterior surface of the muscle, and this, gradually contracting, unites with the tendon of the Soleus, and forms with it the tendo Achillis.

Relations.—By its *superficial surface*, with the fascia of the leg, which separates it from the external saphenous vein and nerve. By its *deep surface*, with the posterior ligament of the knee-joint, the Popliteus, Soleus, Plantaris, popliteal vessels, and internal popliteal nerve. The tendon of the inner head corresponds with the back part of the inner condyle, from which it is separated by a synovial bursa, which, in some cases, communicates with the cavity of the knee-joint. The tendon of the outer head contains a sesamoid fibro-cartilage (rarely osseous), where it plays over the corresponding outer condyle; and one is occasionally found in the tendon of the inner head.

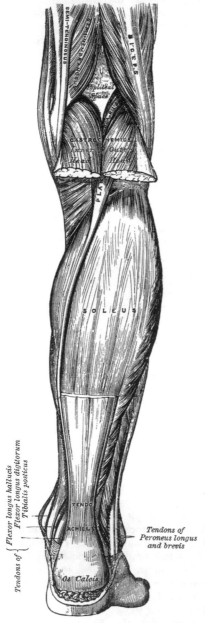

Fig. 259.—Muscles of the back of the leg. Superficial layer.

The Gastrocnemius should be divided across, just below its origin, and turned downwards, in order to expose the next two muscles.

The **Soleus** is a broad flat muscle situated immediately beneath the Gastrocnemius. It has received its name from its resemblance in shape to a sole-fish. It arises by tendinous fibres from the back part of the head of the fibula, and from the upper third of the posterior surface of its shaft; from the oblique line of the tibia, and from the middle third of its internal border; some fibres also arise from a tendinous arch placed between the tibial and fibular origins of the muscle, beneath which the popliteal vessels and internal popliteal nerve pass. The fibres pass backwards to an aponeurosis which covers the posterior surface of the muscle, and this gradually becoming thicker and narrower, joins with the tendon of the Gastrocnemius, and forms with it the tendo Achillis.

Relations.—By its *superficial surface*, with the Gastrocnemius and Plantaris. By its *deep surface*, with the Flexor longus digitorum, Flexor longus hallucis, Tibialis posticus, and posterior tibial vessels and nerve, from which it is separated by the transverse intermuscular septum or deep transverse fascia of the leg.

The **tendo Achillis,** the common tendon of the Gastrocnemius and Soleus,* is the thickest and strongest tendon in the body. It is about six inches in length, and commences about the middle of the leg, but receives fleshy fibres on its anterior surface, nearly to its lower end. Gradually becoming contracted below,

---

* These two muscles with a common tendon are by some anatomists classed together as one muscle, the *Triceps suræ,* the two heads of origin of the Gastrocnemius and the Soleus constituting the three heads of the Triceps, and the tendo Achillis the single tendon of insertion.

it is inserted into the lower part of the posterior surface of the os calcis, a synovial bursa being interposed between the tendon and the upper part of this surface. The tendon spreads out somewhat at its lower end, so that its narrowest part is usually about an inch and a half above its insertion. The tendon is covered by the fascia and the integument, and is separated from the deep muscles and vessels by a considerable interval filled up with areolar and adipose tissue. Along its outer side, but superficial to it, is the external saphenous vein.

The **Plantaris** is an extremely diminutive muscle, placed between the Gastrocnemius and Soleus, and remarkable for its long and delicate tendon. It arises from the lower part of the outer prolongation of the linea aspera, and from the posterior ligament of the knee-joint. It forms a small fusiform belly, about three or four inches in length, terminating in a long slender tendon which crosses obliquely between the two muscles of the calf, and, running along the inner border of the tendo Achillis, is inserted with it into the posterior part of the os calcis. This muscle is occasionally double, and is sometimes wanting. Occasionally, its tendon is lost in the internal annular ligament, or in the fascia of the leg.

Nerves.—The Gastrocnemius is supplied by the first and second sacral nerves, and the Plantaris by the fourth and fifth lumbar and first sacral nerves through the internal popliteal. The Soleus is supplied by the fifth lumbar and first and second sacral nerves through the internal popliteal and posterior tibial.

Actions.—The muscles of the calf are the chief extensors of the foot at the ankle-joint. They possess considerable power, and are constantly called into use in standing, walking, dancing, and leaping; hence the large size they usually present. In walking, these muscles draw powerfully upon the os calcis, raising the heel, and with it the entire body, from the ground; the body being thus supported on the raised foot, the opposite limb can be carried forwards. In standing, the Soleus, taking its fixed point from below, steadies the leg upon the foot and prevents the body from falling forwards, to which there is a constant tendency from the superincumbent weight. The Gastrocnemius, acting from below, serves to flex the femur upon the tibia, assisted by the Popliteus. The Plantaris is the rudiment of a large muscle which exists in some of the lower animals, and is continued over the os calcis to be inserted into the plantar fascia. In man it is an accessory to the Gastrocnemius, extending the ankle if the foot is free, or bending the knee if the foot is fixed. Possibly, acting from below, by its attachment to the posterior ligament of the knee-joint, it may pull that ligament backwards during flexion, and so protect it from being compressed between the two articular surfaces.

Deep Layer (fig. 260)

Popliteus.	Flexor longus digitorum.
Flexor longus hallucis.	Tibialis posticus.

Dissection.—Detach the Soleus from its attachment to the fibula and tibia, and turn it downwards, when the deep layer of muscles is exposed, covered by the deep transverse fascia of the leg.

The **Deep Transverse Fascia** of the leg is a transversely placed, intermuscular septum, between the superficial and deep muscles in the posterior tibio-fibular region. On either side it is connected to the margins of the tibia and fibula. Above, where it covers the Popliteus, it is thick and dense, and receives an expansion from the tendon of the Semimembranosus; it is thinner in the middle of the leg; but below, where it covers the tendons passing behind the malleoli, it is thickened and continuous with the internal annular ligament.

This fascia should now be removed, commencing from below opposite the tendons, and detaching it from the muscles in the direction of their fibres.

The **Popliteus** is a thin, flat, triangular muscle, which forms part of the floor of the popliteal space. It arises by a strong tendon about an inch in length, from a deep depression on the outer side of the external condyle of the femur and from the posterior ligament of the knee-joint; and is inserted into the inner two-thirds of the triangular surface above the oblique line on the posterior surface of the shaft of the tibia, and into the tendinous expansion covering the surface of the muscle. The tendon of the muscle is covered by that of the Biceps and by the external lateral ligament of the knee-joint; it grooves the posterior border of the external semilunar fibro-cartilage, and is invested by the synovial membrane of the knee-joint.

FIG. 260.—Muscles of the back of the leg. Deep layer.

Relations.—By its *superficial surface*, with the fascia covering it, which separates it from the Gastrocnemius, Plantaris, popliteal vessels, and internal popliteal nerve. By its *deep surface*, with the knee-joint and back of the tibia.

The **Flexor longus hallucis** is situated on the fibular side of the leg, and is the most superficial and largest of the three next muscles. It arises from the lower two-thirds of the posterior surface of the shaft of the fibula, with the exception of an inch at its lowest part; from the lower part of the interosseous membrane; from an intermuscular septum between it and the Peronei, externally; and from the fascia covering the Tibialis posticus internally. The fibres pass obliquely downwards and backwards, and terminate in a tendon which occupies nearly the whole length of the posterior surface of the muscle. This tendon occupies a groove on the posterior surface of the lower end of the tibia; it then lies in a second groove on the posterior surface of the astragalus, and finally in a third groove, beneath the sustentaculum tali of the os calcis, and passes into the sole of the foot, where it runs forwards between the two heads of the Flexor brevis hallucis, and is inserted into the base of the last phalanx of the great toe. The grooves in the astragalus and os calcis, which contain the tendon of the muscle, are converted by tendinous fibres into distinct canals, lined by synovial membrane; and as the tendon lies in the sole of the foot, it is connected to the common flexor by a tendinous slip.

Relations.—By its *superficial surface*, with the Soleus and tendo Achillis, from which it is separated by the deep transverse fascia. By its *deep surface*, with the fibula, Tibialis posticus, the peroneal vessels, the lower part of the interosseous membrane, and the ankle-joint. By its *outer border*, with the Peronei. By its *inner border*, with the Tibialis posticus, and posterior tibial vessels and nerve. In the sole of the foot it lies above the Abductor hallucis and Flexor longus digitorum.

The **Flexor longus digitorum** (*perforans*) is situated on the tibial side of the leg. At its origin it is thin and pointed, but gradually increases in size as it descends. It arises from the posterior surface of the shaft of the tibia, from immediately below the oblique line to within three inches of its extremity, internal to the tibial origin of the Tibialis posticus ; some fibres also arise from the fascia covering the Tibialis posticus. The fibres terminate in a tendon, which runs nearly the whole length of the posterior surface of the muscle. This tendon passes behind the internal malleolus, in a groove, common to it and the Tibialis posticus, but separated from the latter by a fibrous septum ; each tendon being contained in a special sheath lined by a separate synovial membrane. It then passes obliquely forwards and outwards, superficial to the internal lateral ligament into the sole of the foot (fig. 262), where, crossing superficially to the tendon of the Flexor longus hallucis,* to which it is connected by a strong tendinous slip, it becomes expanded, is joined by the Flexor accessorius, and finally divides into four tendons, which are inserted into the bases of the last phalanges of the four lesser toes, each tendon passing through a fissure in the tendon of the Flexor brevis digitorum opposite the bases of the first phalanges.

Relations.—In the leg : By its *superficial surface*, with the posterior tibial vessels and nerve, and the deep transverse fascia, which separates it from the Soleus muscle ; by its *deep surface*, with the tibia and Tibialis posticus. In the foot, it is covered by the Abductor hallucis and Flexor brevis digitorum, and crosses superficial to the Flexor longus hallucis.

The **Tibialis posticus** lies between the two preceding muscles, and is the most deeply seated of all the muscles in the leg. It commences above by two pointed processes, separated by an angular interval, through which the anterior tibial vessels pass forwards to the front of the leg. It arises from the whole of the posterior surface of the interosseous membrane, excepting its lowest part ; from the outer portion of the posterior surface of the shaft of the tibia, between the commencement of the oblique line above and the junction of the middle and lower third of the shaft below ; and from the upper two-thirds of the internal surface of the fibula ; some fibres also arise from the deep transverse fascia, and from the intermuscular septa separating it from the adjacent muscles on each side. This muscle, in the lower fourth of the leg, passes in front of the Flexor longus digitorum, and terminates in a tendon, which passes through a groove behind the inner malleolus, with the tendon of that muscle, but enclosed in a separate sheath ; it then passes through another sheath, over the internal lateral ligament into the foot, and then beneath the inferior calcaneo-navicular ligament, and is inserted into the tuberosity of the navicular and internal cuneiform bones. The tendon of this muscle contains a sesamoid fibro-cartilage, as it passes over the navicular bone, and gives off fibrous expansions, one of which passes backwards to the sustentaculum tali of the os calcis, others outwards to the middle and external cuneiform and cuboid, and some forwards to the bases of the second, third, and fourth metatarsal bones (fig. 263).

Relations.—By its *superficial surface*, with the Soleus, from which it is separated by the deep transverse fascia, the Flexor longus digitorum, the posterior tibial vessels and nerve, and the peroneal vessels. By its *deep surface*, with the interosseous ligament, the tibia, fibula, and ankle-joint.

Nerves.—The Popliteus is supplied by the fourth and fifth lumbar and first sacral nerves, through the internal popliteal ; the Flexor longus digitorum and Tibialis posticus by the fifth lumbar and first sacral ; and the Flexor longus hallucis by the fifth lumbar, and first and second sacral nerves through the posterior tibial.

Actions.—The Popliteus assists in flexing the leg upon the thigh ; when the leg is flexed, it will rotate the tibia inwards. It is especially called into action at the commencement of the act of bending the knee, inasmuch as it produces a

---

* That is, in the order of dissection of the sole of the foot.

slight inward rotation of the tibia, which is essential in the early stage of this movement. The Tibialis posticus is a direct extensor of the foot at the ankle-joint; acting in conjunction with the Tibialis anticus, it turns the sole of the foot inwards (i.e. inverts the foot), antagonising the Peronei, which turn it outwards (evert it). In the sole of the foot the tendon of the Tibialis posticus lies directly below the inferior calcaneo-scaphoid ligament, and is therefore an important factor in maintaining the arch of the foot. The Flexor longus digitorum and Flexor longus hallucis are the direct flexors of the phalanges, and, continuing their action, extend the foot upon the leg; they assist the Gastrocnemius and Soleus in extending the foot, as in the act of walking, or in standing on tiptoe. In consequence of the oblique direction of the tendon of the long flexor, the toes would be drawn inwards, were it not for the Flexor accessorius muscle, which is inserted into the outer side of its tendon, and draws it to the middle line of the foot during its action. Taking their fixed point from the foot, these muscles serve to maintain the upright posture by steadying the tibia and fibula, perpendicularly, upon the ankle-joint. They also serve to raise these bones from the oblique position they assume in the stooping posture.

7. Fibular Region

Peroneus longus.　　　　　　　　　　　　Peroneus brevis.

Dissection.—The muscles are readily exposed, by removing the fascia covering their surface, from below upwards, in the line of direction of their fibres.

The **Peroneus longus** is situated at the upper part of the outer side of the leg, and is the more superficial of the two muscles. It arises from the head and upper two-thirds of the outer surface of the shaft of the fibula, from the deep surface of the fascia, and from the intermuscular septa, between it and the muscles on the front, and those on the back of the leg: occasionally also by a few fibres from the outer tuberosity of the tibia. Between its attachment to the head and to the shaft of the fibula there is a small interval of bone from which no muscular fibres arise; through this gap the external popliteal nerve passes beneath the muscle. It terminates in a long tendon, which passes behind the outer malleolus, in a groove common to it and the tendon of the Peroneus brevis, behind which it lies, the groove being converted into a canal by a fibrous band, and the tendons invested by a common synovial membrane; it is then reflected obliquely forwards across the outer side of the os calcis, below its peroneal tubercle, being contained in a separate fibrous sheath, lined by a prolongation of the synovial membrane which lines the groove behind the malleolus. Having reached the outer side of the cuboid bone, it runs in a groove on the under surface of that bone, which is converted into a canal by the long calcaneo-cuboid ligament, and is lined by a synovial membrane: the tendon then crosses the sole of the foot obliquely, and is inserted into the outer side of the base of the metatarsal bone of the great toe and the internal cuneiform bone. Occasionally it sends a slip to the base of the second metatarsal bone. The tendon changes its direction at two points: first, behind the external malleolus; secondly, on the outer side of the cuboid bone; in both of these situations the tendon is thickened, and, in the latter, a sesamoid fibro-cartilage, or sometimes a bone, is usually developed in its substance.

Relations.—By its *superficial surface*, with the fascia and integument; by its *deep surface*, with the fibula, external popliteal nerve, the Peroneus brevis, os calcis, and cuboid bone; by its *anterior border*, with an intermuscular septum, which intervenes between it and the Extensor longus digitorum; by its *posterior border*, with an intermuscular septum, which separates it from the Soleus above and the Flexor longus hallucis below.

The **Peroneus brevis** lies beneath the Peroneus longus, and is shorter and smaller than it. It arises from the lower two-thirds of the external surface of the shaft of the fibula, internal to the Peroneus longus; and from the intermuscular septa, separating it from the adjacent muscles on the front and back part of the leg. The fibres pass vertically downwards, and terminate in a tendon, which runs in front of that of the preceding muscle, through the same groove behind the external malleolus, being contained in the same fibrous sheath, and lubricated by the same synovial membrane. It then passes through a separate sheath on the outer side of the os calcis, above that for the tendon of the Peroneus longus, the two tendons being here separated by the peroneal tubercle, and is finally inserted into the tuberosity at the base of the metatarsal bone of the little toe, on its outer side.

Relations.—By its *superficial surface*, with the Peroneus longus and the fascia of the leg and foot. By its *deep surface*, with the fibula and outer side of the os calcis.

Nerves.—The Peroneus longus and brevis are supplied by the fourth and fifth lumbar and first sacral nerves through the musculo-cutaneous branch of the external popliteal nerve.

Actions.—The Peroneus longus and brevis extend the foot upon the leg, in conjunction with the Tibialis posticus, antagonising the Tibialis anticus and Peroneus tertius, which are flexors of the foot. The Peroneus longus also everts the sole of the foot; hence the extreme eversion occasionally observed in fracture of the lower end of the fibula, where that bone offers no resistance to the action of this muscle. From the oblique direction of the Peroneus longus tendon across the sole of the foot it is an important agent in the maintenance of the transverse arch of the foot. Taking their fixed point below, the Peronei serve to steady the leg upon the foot. This is especially the case in standing upon one leg, when the tendency of the superincumbent weight is to throw the leg inwards : the Peroneus longus overcomes this tendency by drawing on the outer side of the leg, and thus maintains the perpendicular direction of the limb.

Surgical Anatomy.—The student should now consider the position of the tendons of the various muscles of the leg, their relation with the ankle-joint and surrounding blood-vessels, and especially their action upon the foot, as their rigidity and contraction give rise to one or other of the kinds of deformity known as *club foot*. The most simple and common deformity, and one that is rarely, if ever, congenital, is *talipes equinus*, the heel being raised by rigidity and contraction of the Gastrocnemius muscle, and the patient walking upon the ball of the foot. In *talipes varus*, the foot is forcibly adducted and the inner side of the sole raised, sometimes to a right angle with the ground, by the action of the Tibialis anticus and posticus. In *talipes valgus*, the outer edge of the foot is raised by the Peronei muscles, and the patient walks on the inner ankle. In *talipes calcaneus* the toes are raised by the Extensor muscles, the heel is depressed and the patient walks upon it. Other varieties of deformity are met with, as *talipes equino-varus, equino-valgus*, and *calcaneo-valgus*, whose names sufficiently indicate their nature. Of these, talipes equino-varus is the most common congenital form ; the heel is raised by the tendo Achillis, the inner border of the foot drawn upwards by the Tibialis anticus, the anterior two-thirds twisted inwards by the Tibialis posticus, and the arch increased by the contraction of the plantar fascia, so that the patient walks on the middle of the outer border of the foot. Each of these deformities may sometimes be successfully relieved by division of the opposing tendons and fascia: by this means the foot regains its proper position, and the tendons heal by the organisation of lymph thrown out between the divided ends. The operation is easily performed by putting the contracted tendon upon the stretch, and dividing it by means of a narrow, sharp-pointed knife inserted beneath it.

Rupture of a few of the fibres of the Gastrocnemius, or rupture of the Plantaris tendon, not uncommonly occurs, especially in men somewhat advanced in life, from some sudden exertion, and frequently occurs during the game of lawn tennis, and is hence known as ' lawn-tennis leg.' The accident is accompanied by a sudden pain, and produces a sensation as if the individual had been struck a violent blow on the part. The tendo Achillis is also sometimes ruptured. It is stated that John Hunter ruptured his tendo Achillis while dancing, at the age of forty.

IV. Muscles and Fasciæ of the Foot

The fibrous bands, or thickened portions of the fascia of the leg, which bind down the tendons in front of and behind the ankle in their passage to the foot, should now be examined ; they are termed the *annular ligaments*, and are three in number—anterior, internal, and external.

The **anterior annular ligament** consists of a superior or transverse portion, which binds down the Extensor tendons as they descend on the front of the tibia and fibula; and an inferior or Y-shaped portion, which retains them in connection with the tarsus, the two portions being connected by a thin intervening layer of fascia. The transverse portion is attached externally to the lower end of the fibula and internally to the tibia; above it is continuous with the fascia of the leg; it contains only one synovial sheath, for the tendon of the Tibialis anticus; the other tendons and the anterior tibial vessels and nerve passing beneath it, but without any distinct synovial sheath. The Y-shaped portion is placed in front of the ankle-joint, the stem of the Y being attached externally to the upper surface of the os calcis, in front of the depression for the interosseous ligament; it is directed inwards, as a double layer, one lamina passing in front, and the other behind, the tendons of the Peroneus tertius and Extensor longus digitorum. At the inner border of the latter tendon these two layers join together, forming a sort of loop or sheath in which the tendons are enclosed, surrounded by a synovial membrane. From the inner extremity of this loop the two limbs of the Y diverge: one passes upwards and inwards, to be attached to the internal malleolus, passing over the Extensor proprius hallucis and the vessels and nerves, but enclosing the Tibialis anticus and its synovial sheath by a splitting of its fibres. The other limb extends downwards and inwards to be attached to the inner border of the plantar fascia, and passes over the tendons of the Extensor proprius hallucis and Tibialis anticus and also the vessels and nerves. These two tendons are contained in separate synovial sheaths situated beneath the ligament.

The **internal annular ligament** is a strong fibrous band, which extends from the inner malleolus above to the internal margin of the os calcis below, converting a series of grooves in this situation into canals, for the passage of the tendons of the Flexor muscles and vessels into the sole of the foot. It is continuous by its upper border with the deep fascia of the leg, and by its lower border with the plantar fascia and the fibres of origin of the Abductor hallucis muscle. The four canals which it forms transmit, from within outwards, first, the tendon of the Tibialis posticus; second, the tendon of the Flexor longus digitorum; third, the posterior tibial vessels and nerve, which run through a broad space beneath the ligament; lastly, in a canal formed partly by the astragalus, the tendon of the Flexor longus hallucis. The canals for the tendons are lined by separate synovial membranes.

The **external annular ligament** extends from the extremity of the outer malleolus to the outer surface of the os calcis: it binds down the tendons of the Peroneus longus and brevis muscles in their passage beneath the outer ankle. The two tendons are enclosed in one synovial sheath.

Dissection of the Sole of the Foot.—The foot should be placed on a high block with the sole uppermost, and firmly secured in that position. Carry an incision round the heel and along the inner and outer borders of the foot to the great and little toes. This incision should divide the integument and thick layer of granular fat beneath, until the fascia is visible; the skin and fat should then be removed from the fascia in a direction from behind forwards, as seen in fig. 255.

The **plantar fascia,** the densest of all the fibrous membranes, is of great strength, and consists of pearly-white glistening fibres, disposed, for the most part, longitudinally: it is divided into a central and two lateral portions.

The *central portion,* the thickest, is narrow behind and attached to the inner tubercle of the os calcis, posterior to the origin of the Flexor brevis digitorum; and becoming broader and thinner in front, divides near the heads of the metatarsal bones into five processes, one for each of the toes. Each of these processes divides opposite the metatarso-phalangeal articulation into two strata, superficial and deep. The superficial stratum is inserted into the skin of the transverse sulcus which divides the toes from the sole. The deeper stratum divides into two slips which embrace the sides of the Flexor tendons of the toes, and blend with the sheaths of the tendons, and laterally with the transverse metatarsal ligament, thus forming a series of arches through which the tendons of the short and long flexors pass to the toes. The intervals left between the five processes allow the digital vessels and nerves, and the tendons of the Lumbricales muscles, to become superficial. At the point of division of the fascia into processes and slips, numerous transverse fibres are superadded, which serve to increase the strength of the fascia at this part by binding the processes together, and connecting them with the integument. The central portion of the plantar fascia is continuous with the lateral portions at each side, and sends upwards into the foot, at their point of junction, two strong vertical intermuscular septa, broader in front than behind, which separate the middle from the external and internal plantar group of muscles; from these again thinner transverse septa are derived, which separate the various layers of muscles in this region. The upper surface of this fascia gives attachment behind to the Flexor brevis digitorum muscle.

The *lateral portions* of the plantar fascia are thinner than the central piece, and cover the sides of the foot.

The *outer portion* covers the under surface of the Abductor minimi digiti; it is thick behind, thin in front, and extends from the outer tubercle of the os calcis, forwards, to the base of the fifth metatarsal bone, into the outer side of which it is attached; it is continuous internally with the middle portion of the plantar fascia, and externally with the dorsal fascia.

The *inner portion* is very thin, and covers the Abductor hallucis muscle; it is attached behind to the internal annular ligament, and is continuous around the side of the foot with the dorsal fascia, and externally with the middle portion of the plantar fascia.

8. *Dorsal Region*

Extensor brevis digitorum.

The **fascia** on the dorsum of the foot is a thin membranous layer, continuous above with the anterior margin of the annular ligament; it becomes gradually lost opposite the heads of the metatarsal bones, and on each side blends with the lateral portions of the plantar fascia; it forms a sheath for the tendons placed on the dorsum of the foot. On the removal of this fascia, the muscle and tendons of the dorsal region of the foot are exposed.

The **Extensor brevis digitorum** (fig. 258) is a broad thin muscle, which arises from the fore part of the upper and outer surfaces of the os calcis, in front of the groove for the Peroneus brevis; from the external calcaneo-astragaloid ligament; and from the common limb of the Y-shaped portion of the anterior annular ligament. It passes obliquely across the dorsum of the foot, and terminates in four tendons. The innermost, which is the largest, is inserted into the dorsal surface of the base of the first phalanx of the great toe, crossing the Dorsalis pedis artery; the other three, into the outer sides of the long Extensor tendons of the second, third, and fourth toes.

Relations.—By its *superficial surface*, with the fascia of the foot, the tendons of the Extensor longus digitorum and Peroneus tertius. By its *deep surface*, with the tarsal and metatarsal arteries and bones, and the Dorsal interossei muscles.

Nerves.—It is supplied by the anterior tibial nerve.

Actions.—The Extensor brevis digitorum is an accessory to the long Extensor, extending the phalanges of the four inner toes, but acting only on the first phalanx of the great toe. The obliquity of its direction counteracts the oblique movement given to the toes by the long Extensor, so that, both muscles acting together, the toes are evenly extended.

9. *Plantar Region*

The muscles in the plantar region of the foot may be divided into three groups, in a similar manner to those in the hand. Those of the internal plantar region are connected with the great toe, and correspond with those of the thumb; those of the external plantar region are connected with the little toe, and correspond with those of the little finger; and those of the middle plantar region are connected with the tendons intervening between the two former groups. But in order to facilitate the dissection of these muscles, it will be found more convenient to divide them into four layers, as they present themselves, in the order in which they are successively exposed.

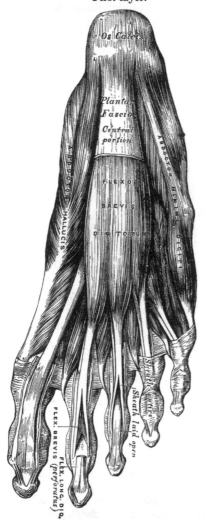

Fig. 261.—Muscles of the sole of the foot. First layer.

First Layer

Abductor hallucis.
Flexor brevis digitorum.
Abductor minimi digiti.

Dissection.—Remove the fascia on the inner and outer sides of the foot, commencing in front over the tendons, and proceeding backwards. The central portion should be divided transversely in the middle of the foot, and the two flaps dissected forwards and backwards.

The **Abductor hallucis** lies along the inner border of the foot. It arises from the inner tubercle on the under surface of the os calcis; from the internal annular ligament; from the plantar fascia; and from the intermuscular septum between it and the Flexor brevis digitorum. The fibres terminate in a tendon, which is inserted, together with the innermost tendon of the Flexor brevis hallucis, into the inner side of the base of the first phalanx of the great toe.

Relations.—By its *superficial surface*, with the plantar fascia. By its *deep surface*, with the Flexor brevis hallucis, the Flexor accessorius, and the tendons of the Flexor longus digitorum and Flexor longus hallucis, the Tibialis anticus and posticus, the plantar vessels and nerves. Its outer border is in relation to the Flexor brevis digitorum.

The **Flexor brevis digitorum** (*perforatus*) lies in the middle of the sole of the foot, immediately beneath * the plantar fascia, with which it is firmly united. It arises by a narrow tendinous process, from the inner tubercle of the os calcis, from the central part of the plantar fascia, and from the intermuscular septa between it and the adjacent muscles. It passes forwards, and divides into four tendons, one for each of the four outer toes. Opposite the bases of the first phalanges, each tendon divides into two slips, to allow of the passage of the corresponding tendon of the Flexor longus digitorum; the two portions of the tendon then unite and form a grooved channel for the reception of the accompanying long Flexor tendon. Finally, they divide a second time, to be inserted into the sides of the second phalanges about their middle. The mode of division

---

* That is, in the order of dissection of the sole of the foot.

of the tendons of the Flexor brevis digitorum, and their insertion into the phalanges, is analogous to the Flexor sublimis digitorum in the hand.

Relations.—By its *superficial surface*, with the plantar fascia. By its *deep surface*, with the Flexor accessorius, the Lumbricales, the tendons of the Flexor longus digitorum, and the external plantar vessels and nerve, from which it is separated by a thin layer of fascia. The *outer* and *inner borders* are separated from the adjacent muscles by means of vertical prolongations of the plantar fascia.

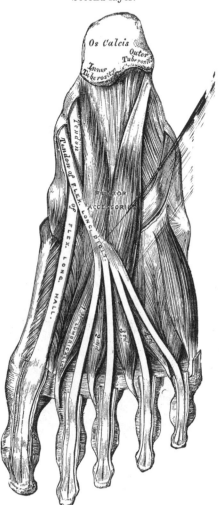

FIG. 262.—Muscles of the sole of the foot. Second layer.

Fibrous sheaths of the Flexor tendons.—These are not so well marked as in the fingers. The Flexor tendons of the toes as they run along the phalanges are retained against the bones by fibrous sheaths, forming osseo-aponeurotic canals. These sheaths are formed by strong fibrous bands, which arch across the tendons, and are attached on each side to the margins of the phalanges. Opposite the middle of the proximal and second phalanges the sheath is very strong, and the fibres pass transversely ; but opposite the joints it is much thinner, and the fibres pass obliquely. Each sheath is lined by a synovial membrane, which is reflected on the contained tendon.

The **Abductor minimi digiti** lies along the outer border of the foot. It arises, by a very broad origin, from the outer tubercle of the os calcis, from the under surface of the os calcis between the two tubercles, from the fore part of the inner tubercle, from the plantar fascia, and the intermuscular septum between it and the Flexor brevis digitorum. Its tendon, after gliding over a smooth facet on the under surface of the base of the fifth metatarsal bone, is inserted, with the short Flexor of the little toe, into the outer side of the base of the first phalanx of this toe.

Relations.—By its *superficial surface*, with the plantar fascia. By its *deep surface*, with the Flexor accessorius, the Flexor brevis minimi digiti, the long plantar ligament, and the tendon of the Peroneus longus. On its *inner side* are the external plantar vessels and nerve, and it is separated from the Flexor brevis digitorum by a vertical septum of fascia.

Dissection.—The muscles of the superficial layer should be divided at their origin, by inserting the knife beneath each, and cutting obliquely backwards, so as to detach them from the bone ; they should then be drawn forwards, in order to expose the second layer, but not cut away at their insertion. The two layers are separated by a thin membrane, the *deep plantar fascia*, on the removal of which is seen the tendon of the Flexor longus digitorum, the Flexor accessorius, the tendon of the Flexor longus hallucis, and the Lumbricales. The long Flexor tendons diverge from each other at an acute angle : the Flexor longus hallucis runs along the inner side of the foot, on a plane superior to that of the Flexor longus digitorum, the direction of which is obliquely outwards.

Second Layer

Flexor accessorius.   Lumbricales.

The **Flexor accessorius** arises by two heads, which are separated from each other by the long plantar ligament: the inner or larger, which is muscular, being attached to the inner concave surface of the os calcis, below the groove which lodges the tendon of the Flexor longus hallucis; the outer head, flat and tendinous, to the outer surface of the os calcis, in front of its lesser tubercle, and to the long plantar ligament: the two portions join at an acute angle, and are inserted into the outer margin and upper and under surfaces of the tendon of the Flexor longus digitorum, forming a kind of groove, in which the tendon is lodged.*

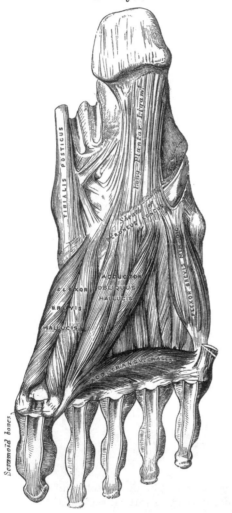

FIG. 263.—Muscles of the sole of the foot. Third layer.

Relations.—By its *superficial surface*, with the muscles of the superficial layer, from which it is separated by the external plantar vessels and nerves. By its *deep surface*, with the os calcis, and long calcaneo-cuboid ligament.

The **Lumbricales** are four small muscles, accessory to the tendons of the Flexor longus digitorum: they arise from the tendons of the long Flexor, as far back as their angles of division, each arising from two tendons, except the internal one. Each muscle terminates in a tendon, which passes forwards on the inner side of the four lesser toes, and is inserted into the expansion of the long Extensor tendon on the dorsum of the first phalanx of the corresponding toe

Dissection.—The Flexor tendons should be divided at the back part of the foot, and the Flexor accessorius at its origin, and drawn forwards, in order to expose the third layer.

Third Layer

Flexor brevis hallucis
Adductor obliquus hallucis.
Flexor brevis minimi digiti.
Adductor transversus hallucis.

The **Flexor brevis hallucis** arises, by a pointed tendinous process, from the inner part of the under surface of the cuboid bone, from the contiguous portion of the external cuneiform, and from the prolongation of the tendon of the Tibialis posticus, which is attached to that bone. The muscle divides, in front, into two portions, which are inserted into the inner and outer sides of the base of the first phalanx of the great toe, a sesamoid bone being developed in each tendon at its insertion. The inner portion of this muscle is blended with the Abductor hallucis previous to its insertion; the outer with the Adductor obliquus hallucis; and the tendon of the Flexor longus hallucis lies in a groove between them.

* According to Turner, the fibres of the Flexor accessorius end in aponeurotic bands' which contribute slips to the second, third, and fourth digits.

Relations.—By its *superficial surface*, with the Abductor hallucis and the tendon of the Flexor longus hallucis. By its *deep surface*, with the tendon of the Peroneus longus, and metatarsal bone of the great toe. By its *inner border*, with the Abductor hallucis. By its *outer border*, with the Adductor obliquus hallucis.

The **Adductor obliquus hallucis** is a large, thick, fleshy mass, passing obliquely across the foot, and occupying the hollow space between the four inner metatarsal bones. It arises from the tarsal extremities of the second, third, and fourth metatarsal bones, and from the sheath of the tendon of the Peroneus longus, and is inserted, together with the outer portion of the Flexor brevis hallucis, into the outer side of the base of the first phalanx of the great toe.

The small muscles of the great toe, the Abductor, Flexor brevis, Adductor obliquus, and Adductor transversus, like the similar muscles of the thumb, give off fibrous expansions, at their insertions, to blend with the long Extensor tendons.

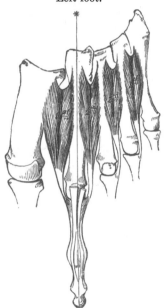

FIG. 264.—The Dorsal interossei.
Left foot.

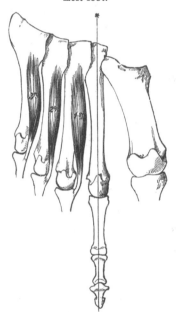

FIG. 265.—The Plantar interossei.
Left foot.

The **Flexor brevis minimi digiti** lies on the metatarsal bone of the little toe, and much resembles one of the Interossei. It arises from the base of the metatarsal bone of the little toe, and from the sheath of the Peroneus longus; its tendon is inserted into the base of the first phalanx of the little toe on its outer side. Occasionally some of the deeper fibres of the muscle are inserted into the outer part of the distal half of the fifth metatarsal bone; these are described by some as a distinct muscle, the Opponens minimi digiti.

Relations.—By its *superficial surface*, with the plantar fascia and tendon of the Abductor minimi digiti. By its *deep surface*, with the fifth metatarsal bone.

The **Adductor transversus hallucis** (*Transversus pedis*) is a narrow, flat, muscular fasciculus, stretched transversely across the heads of the metatarsal bones, between them and the Flexor tendons. It arises from the inferior metatarsophalangeal ligaments of the three outer toes, sometimes only from the third and fourth, and from the transverse ligament of the metatarsus; and is inserted into the outer side of the first phalanx of the great toe; its fibres being blended with the tendon of insertion of the Adductor obliquus hallucis.

Relations.—By its *superficial surface*, with the tendons of the long and short Flexors and Lumbricales. By its *deep surface*, with the Interossei.

Fourth Layer

The Interossei.

The **Interossei muscles** in the foot are similar to those in the hand, with this exception, that they are grouped around the middle line of the *second* toe, instead of the middle line of the *third* finger. They are seven in number, and consist of two groups, dorsal and plantar.

The **Dorsal interossei**, four in number, are situated between the metatarsal bones. They are bipenniform muscles, arising by two heads from the adjacent sides of the metatarsal bones between which they are placed; their tendons are inserted into the bases of the first phalanges, and into the aponeurosis of the common Extensor tendon. In the angular interval left between the heads of each muscle at its posterior extremity the perforating arteries pass to the dorsum of the foot; except in the First interosseous muscle, where the interval allows the passage of the communicating branch of the dorsalis pedis artery. The First dorsal interosseous muscle is inserted into the inner side of the second toe; the other three are inserted into the outer sides of the second, third, and fourth toes.

The **Plantar interossei**, three in number, lie beneath, rather than between, the metatarsal bones. They are single muscles, and are each connected with but one metatarsal bone. They arise from the bases and inner sides of the shaft of the third, fourth, and fifth metatarsal bones, and are inserted into the inner sides of the bases of the first phalanges of the same toes, and into the aponeurosis of the common Extensor tendon.

Nerves.—The Flexor brevis digitorum, the Flexor brevis and Abductor hallucis, and the innermost Lumbrical * are supplied by the internal plantar nerve. All the other muscles in the sole of the foot by the external plantar. The first dorsal interosseous muscle frequently receives an extra filament from the internal branch of the anterior tibial nerve on the dorsum of the foot, and the second dorsal interosseous a twig from the external branch of the same nerve.

Actions.—All the muscles of the foot act upon the toes, and in describing their action they may be grouped as Abductors, Adductors, Flexors, or Extensors. The *Abductors* are the Dorsal interossei, the Abductor hallucis, and the Abductor minimi digiti. The Dorsal interossei are abductors from an imaginary line passing through the axis of the second toe, so that the first muscle draws the second toe inwards, towards the great toe; the second muscle draws the same toe outwards; the third draws the third toe, and the fourth draws the fourth toe in the same direction. Like the interossei in the hand, each assists in flexing the proximal phalanx and extending the two terminal phalanges. The Abductor hallucis abducts the great toe from the others, and also flexes the proximal phalanx of this toe. And in the same way the action of the Abductor minimi digiti is twofold, as an abductor of this toe from the others, and also as a flexor of the proximal phalanx. The *Adductors* are the Plantar interossei, the Adductor obliquus hallucis, and the Adductor transversus hallucis. The plantar interosseous muscles adduct the third, fourth, and fifth toes, towards the imaginary line passing through the second toe, and by means of their insertion into the aponeurosis of the Extensor tendon they assist in flexing the proximal phalanx and extending the two terminal phalanges. The Adductor obliquus hallucis is chiefly concerned in adducting the great toe towards the second one, but also assists in flexing this toe. The Adductor transversus hallucis approximates all the toes and thus increases the curve of the transverse arch of the metatarsus. The *Flexors* are the Flexor brevis digitorum, the Flexor accessorius, the Flexor

---

* Formerly the two inner Lumbricales were described as being supplied by the internal plantar nerve. Brooks, however (*Journal of Anatomy*, vol. xxi. p. 575), in ten dissections found that in nine of them only the inner Lumbrical obtained its nerve supply from this source. In the tenth instance the first and second Lumbricales were supplied by both external and internal plantar.

brevis hallucis, the Flexor brevis minimi digiti, and the Lumbricales. The Flexor brevis digitorum flexes the second phalanges upon the first, and, continuing its action, flexes the first phalanges also, and brings the toes together. The Flexor accessorius assists the long Flexor of the toes and converts the oblique pull of the tendons of that muscle into a direct backward pull upon the toes. The Flexor brevis minimi digiti flexes the little toe and draws its metatarsal bone downwards and inwards. The Lumbricales, like the corresponding muscles in the hand, assist in flexing the proximal phalanx, and by their insertion into the long Extensor tendon aid that muscle in straightening the two terminal phalanges. The only muscle in the *Extensor* group is the Extensor brevis digitorum. It extends the first phalanx of the great toe and assists the long Extensor in extending the next three toes, and at the same time gives to the toes an outward direction when they are extended.

Surface Form.—Of the muscles of the thigh, those of the iliac region have no influence on surface form, while those of the anterior femoral region, being to a great extent superficial, largely contribute to the surface form of this part of the body. The *Tensor fasciæ femoris* produces a broad elevation immediately below the anterior portion of the crest of the ilium and behind the anterior superior spinous process. From its lower border, a longitudinal groove, corresponding to the ilio-tibial band, may be seen running down the outer side of the thigh to the outer side of the knee-joint. The *Sartorius* muscle, when it is brought into action by flexing the leg on the thigh, and the thigh on the pelvis, and rotating the thigh outwards, presents a well-marked surface form. At its upper part, where it constitutes the outer boundary of Scarpa's triangle, it forms a prominent oblique ridge, which becomes changed into a flattened plane below, and this gradually merges in a general fulness on the inner side of the knee-joint. When the Sartorius is not in action, a depression exists between the Extensor quadriceps and the Adductor muscles, running obliquely downwards and inwards from the apex of Scarpa's triangle to the inner side of the knee, which corresponds to this muscle. In the depressed angle formed by the divergence of the Sartorius and Tensor fasciæ femoris muscles, just below the anterior superior spinous process of the ilium, the *Rectus femoris* muscle appears, and, below this, determines to a great extent the convex form of the front of the thigh. In a well-developed subject, the borders of the muscle, when in action, are clearly to be defined. The *Vastus externus* forms a long flattened plane on the outer side of the thigh, traversed by the longitudinal groove formed by the ilio-tibial band. The *Vastus internus*, on the inner side of the lower half of the thigh, gives rise to a considerable prominence, which increases towards the knee and terminates somewhat abruptly in this situation with a full, curved outline. The *Crureus* and *Subcrureus* are completely hidden, and do not directly influence surface form. The *Adductor muscles*, constituting the internal femoral group, are not to be individually distinguished from each other, with the exception of the upper tendon of the Adductor longus and the lower tendon of the Adductor magnus. The upper tendon of the *Adductor longus*, when the muscle is in action, stands out as a prominent ridge, which runs obliquely downwards and outwards from the neighbourhood of the pubic spine, and forms the inner boundary of a flattened triangular space on the upper part of the front of the thigh, known as Scarpa's triangle. The lower tendon of the *Adductor magnus* can be distinctly felt as a short ridge extending down to the Adductor tubercle on the internal condyle, between the Sartorius and Vastus internus. The Adductor group of muscles fills in the triangular space at the upper part of the thigh, formed between the oblique femur and the pelvic wall, and to them is due the contour of the inner border of the thigh, the *Gracilis* largely contributing to the smoothness of the outline. These muscles are not marked off on the surface from those of the posterior femoral region by any intermuscular depression; but on the outer side of the thigh these latter muscles are defined from the Vastus externus by a distinct marking, corresponding to the external intermuscular septum. The *Gluteus maximus* and a part of the *Gluteus medius* are the only muscles of the buttock which influence surface form. The other part of the Gluteus medius, the Gluteus minimus, and the External rotators are completely hidden. The *Gluteus maximus* forms the full rounded outline of the buttock; it is more prominent behind, compressed in front, and terminates at its tendinous insertion in a depression immediately behind the great trochanter. Its lower border does not correspond to the gluteal fold, but is much more oblique, being marked by a line drawn from the side of the coccyx to the junction of the upper with the lower two-thirds of the thigh on the outer side. From beneath the lower margin of this muscle the *Hamstring muscles* appear, at first narrow and not well marked, but, as they descend, becoming more prominent and widened out, and eventually dividing into two well-marked ridges, which constitute the upper boundaries of the popliteal space, and are formed by the tendons of the inner and outer Hamstring muscles respectively. In the upper part of the thigh these muscles are not to be individually distinguished from each other; but lower down, the separation between the Semitendinosus and Semimembranosus is denoted by a slight intermuscular marking.

The external hamstring tendon formed by the *Biceps* is seen as a thick cord running down to the head of the fibula. The inner hamstring tendons comprise the Semitendinosus, the Semimembranosus, and the Gracilis. The *Semitendinosus* is the most internal of these, and can be felt, in certain positions of the limb, as a sharp cord; the *Semimembranosus* is thick, and the *Gracilis* is situated a little farther forwards than the other two.

All the muscles on the front of the leg appear to a certain extent somewhere on the surface, but the form of this region is mainly dependent upon the Tibialis anticus and the Extensor longus digitorum. The *Tibialis anticus* is well marked, and presents a fusiform enlargement at the outer side of the tibia, and projects beyond the crest of the shin bone. From the muscular mass, its tendon may be traced downward, standing out boldly, when the muscle is in action, on the front of the tibia and ankle-joint, and coursing down to its insertion along the inner border of the foot. A well-marked groove separates this muscle externally from the *Extensor longus digitorum*, which fills up the rest of the space between the upper part of the shaft of the tibia and fibula. This muscle does not present so bold an outline as the Tibialis anticus, and its tendon below, diverging from the tendon of the Tibialis anticus, forms with the latter a sort of plane, in which may be seen the tendon of the Extensor proprius hallucis. A groove on the outer side of the Extensor longus digitorum, seen most plainly when the muscle is in action, separates the tendon from a slight eminence corresponding to the *Peroneus tertius*. The fleshy fibres of the *Peroneus longus* are strongly marked at the upper part of the outer side of the leg, especially when the muscle is in action.

Fig. 266.—Fracture of the neck of the femur within the capsular ligament.

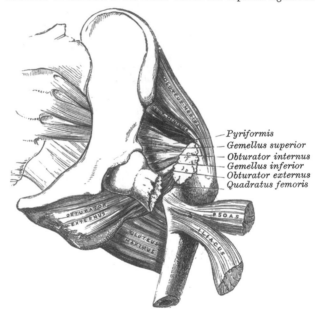

Pyriformis
Gemellus superior
Obturator internus
Gemellus inferior
Obturator externus
Quadratus femoris

It forms a bold swelling, separated by furrows from the Extensor longus digitorum in front and the Soleus behind. Below, the fleshy fibres terminate abruptly in a tendon which overlaps the more flattened form of the *Peroneus brevis*. At the External malleolus the tendon of the Peroneus brevis is more marked than that of the Peroneus longus. On the dorsum of the foot the tendons of the Extensor muscles, emerging from beneath the anterior annular ligament, spread out, and can be distinguished in the following order: the most internal and largest is the Tibialis anticus, then the Extensor proprius hallucis; next comes the Extensor longus digitorum, dividing into four tendons to the four outer toes; and lastly, most externally, is the Peroneus tertius. The flattened form of the dorsum of the foot is relieved by the rounded outline of the fleshy belly of the *Extensor brevis digitorum*, which forms a soft fulness on the outer side of the tarsus in front of the external malleolus, and by the Dorsal interossei, which bulge between the metatarsal bones. At the back of the knee is the popliteal space, bounded above by the tendons of the Hamstring muscles; below, by the two heads of the Gastrocnemius. Below this space is the prominent fleshy mass of the calf of the leg, produced by the *Gastrocnemius* and *Soleus*. When these muscles are in action, as in standing on tiptoe, the borders of the Gastrocnemius are well defined, presenting two curved lines, which converge to the tendon of insertion. Of these borders, the inner is more prominent than the outer. The fleshy mass of the calf terminates somewhat abruptly below in the tendo Achillis, which stands out prominently on the lower part of the back of the leg. It presents a somewhat tapering form in the upper three-fourths of its extent, but widens out slightly below. When the muscles of the calf are in action, the lateral portions of the *Soleus* may be seen, forming curved eminences, of which the outer is the longer, on either side of the Gastrocnemius. Behind the inner border of the lower part of the shaft of the tibia, a well-marked ridge, produced by the tendon of the Tibialis posticus, is visible when this muscle is in a state of contraction.

On the sole of the foot the superficial layer of muscles influences surface form; the *Abductor minimi digiti* most markedly. This muscle forms a narrow rounded elevation along the outer border of the foot, while the *Abductor hallucis* does the same, though to a less extent, on the inner side. The *Flexor brevis digitorum*, bound down by the plantar fascia, is not very apparent; it produces a flattened form, covered by the thickened skin of the sole, which is here thrown into numerous wrinkles.

Surgical Anatomy.—The student should now consider the effects produced by the action of the various muscles in fractures of the bones of the lower extremity. The more common forms of fracture are selected for illustration and description.

In fracture of the *neck of the femur internal to the capsular ligament* (fig. 266), the characteristic marks are slight shortening of the limb, and eversion of the foot, neither of which symptoms occurs, however, in some cases until some time after the injury. The eversion is caused by the weight of the limb rotating it outwards. The shortening is produced by the action of the Glutei, and by the Rectus femoris in front, and the Biceps, Semitendinosus and Semimembranosus behind.

FIG. 267.—Fracture of the femur below the trochanters.

FIG. 268.—Fracture of the femur above the condyles.

FIG. 269.—Fracture of the patella.

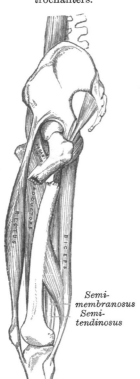

Semi-membranosus
Semi-tendinosus

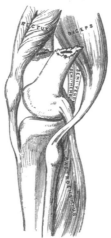

In fracture of the *femur just below the trochanters* (fig. 267), the upper fragment, the portion chiefly displaced, is tilted forwards almost at right angles with the pelvis, by the combined action of the Psoas and Iliacus; and, at the same time, everted and drawn outwards by the external rotator and Glutei muscles, causing a marked prominence at the upper and outer side of the thigh, and much pain from the bruising and laceration of the muscles. The limb is shortened, in consequence of the lower fragment being drawn upwards by the Rectus in front, and the Biceps, Semimembranosus and Semitendinosus behind; and is, at the same time, everted. This fracture may be reduced by two different methods; either by direct relaxation of all the opposing muscles, to effect which the limb should be put up in such a manner that the thigh is flexed on the pelvis and the leg on the thigh; or by overcoming the contraction of the muscles, by continued extension, which may be effected by means of the long splint.

Oblique fracture of the femur *immediately above the condyles* (fig. 268) is a formidable injury, and attended with considerable displacement. On examination of the limb, the lower fragment may be felt deep in the popliteal space, being drawn backwards by the Gastrocnemius and Plantaris muscles; and upwards by the Hamstring and Rectus muscles. The pointed end of the upper fragment is drawn inwards by the Pectineus and Adductor muscles, and tilted forwards by the Psoas and Iliacus, piercing the Rectus muscle, and occasionally the integument. Relaxation of these muscles, and direct approximation of the broken fragments, are effected by placing the limb on a double inclined plane. The greatest care is requisite in keeping the pointed extremity of the upper fragment in proper position ; otherwise, after union of the fracture, the power of extension of the limb is partially destroyed, from the Rectus muscle being held down by the fractured end of the bone, and from the patella, when elevated, being drawn upwards against the projecting fragment.

In fracture of the *patella* (fig. 269) the fragments are separated by the effusion which takes place into the joint, and possibly by the action of the Quadriceps extensor ; the extent of separation of the two fragments depending upon the degree of laceration of the ligamentous structures around the bone.

In oblique fracture of the *shaft of the tibia* (fig. 270), if the fracture has taken place obliquely from above, downwards and forwards, the fragments ride over one another, the

FIG. 270.—Oblique fracture of the shaft of the tibia.

FIG. 271.—Fracture of the fibula, with dislocation of the foot outwards—'Pott's Fracture.'

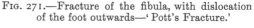

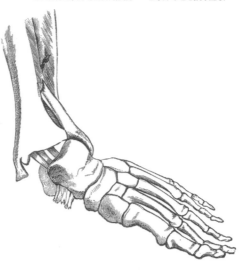

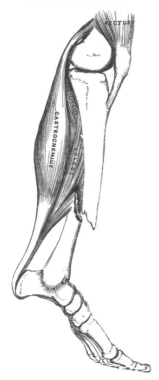

lower fragments being drawn backwards and upwards by the powerful action of the muscles of the calf ; the pointed extremity of the upper fragment projects forwards immediately beneath the integument, often protruding through it, and rendering the fracture a compound one. If the direction of the fracture is the reverse of that shown in the figure, the pointed extremity of the lower fragment projects forwards, riding upon the lower end of the upper one. By bending the knee, which relaxes the opposing muscles, and making extension from the ankle and counter-extension at the knee, the fragments may be brought into apposition. It is often necessary, however, in compound fracture, to remove a portion of the projecting bone with the saw before complete adaptation can be effected.

Fracture of the *fibula with dislocation of the foot outwards* (fig. 271), commonly known as 'Pott's Fracture,' is one of the most frequent injuries of the ankle-joint. The fibula is fractured about three inches above the ankle ; in addition to this the internal malleolus is broken off, or the deltoid ligament torn through, and the end of the tibia displaced from the corresponding surface of the astragalus. The foot is markedly everted, and the sharp edge of the upper end of the fractured malleolus presses strongly against the skin ; at the same time, the heel is drawn up by the muscles of the calf. This injury can generally be reduced by flexing the leg at right angles with the thigh, which relaxes all the opposing muscles, and by making extension from the ankle and counter-extension at the knee.

THE BLOOD-VASCULAR SYSTEM

THE blood-vascular system comprises the heart and blood-vessels with their contained fluid, the blood. The composition of the blood and the minute anatomy of the blood-vessels have already been considered in the section on Histology.

The Heart is the central organ of the entire system, and consists of a hollow muscle ; by its contraction the blood is pumped to all parts of the body through a complicated series of tubes, termed *arteries*. The arteries undergo enormous ramification in their course throughout the body, and end in very minute vessels, called *arterioles*, which in their turn open into a close-meshed network of microscopic vessels, termed *capillaries*. After the blood has passed through the capillaries it is collected into a series of larger vessels, called *veins*, by which it is again returned to the heart. The passage of the blood through the heart and blood-vessels constitutes what is termed the *circulation* of the blood, of which the following is an outline.

The human heart is divided by a septum into two halves, right and left, each half being further constricted into two cavities, the upper of the two being termed the *auricle* and the lower the *ventricle*. The heart therefore consists of four chambers or cavities, two forming the right half, the right auricle and right ventricle, and two the left half, the left auricle and left ventricle. The right half of the heart contains venous or impure blood ; the left, arterial or pure blood. From the cavity of the left ventricle the pure blood is carried into a large artery, the *aorta*, through the numerous branches of which it is distributed to all parts of the body, with the exception of the lungs. In its passage through the capillaries of the body the blood gives up to the tissues the materials necessary for their growth and nourishment, and at the same time receives from the tissues the waste products resulting from their metabolism, and in doing so becomes changed from arterial or pure blood into venous or impure blood, which is collected by the veins and through them returned to the right auricle of the heart. From this cavity the impure blood passes into the right ventricle, from which it is conveyed through the *pulmonary arteries* to the lungs. In the capillaries of the lungs it again becomes arterialised, and is then carried to the left auricle by the *pulmonary veins*. From this cavity it passes into that of the left ventricle, from which the cycle once more begins.

The course of the blood from the left ventricle through the body generally to the right side of the heart constitutes the greater or *systemic* circulation, while its passage from the right ventricle through the lungs to the left side of the heart is termed the lesser or *pulmonary* circulation.

It is necessary, however, to state that the blood which circulates through the spleen, pancreas, stomach, small intestine, and the greater part of the large intestine is not returned directly from these organs to the heart, but is collected into a large vein, termed the *portal vein*, by which it is carried to the liver. In the liver this vein divides, after the manner of an artery, and ultimately ends in capillary vessels, from which the rootlets of a series of veins, called the *hepatic veins*, arise ; these carry the blood into the inferior vena cava, which conveys it to the right auricle.

From this it will be seen that the blood contained in the portal vein passes through two sets of capillary vessels : (1) those in the spleen, pancreas, stomach, &c., and (2) those in the liver.

Speaking generally, the arteries may be said to contain pure, and the veins impure, blood. This is true of the systemic, but not of the pulmonary, vessels, since it has been seen that the impure blood is conveyed from the heart to the lungs by the pulmonary arteries, and the pure blood returned from the lungs to the heart by the pulmonary veins. Arteries, therefore, must be defined as vessels which convey blood *from* the heart, and veins as vessels which return blood *to* the heart.

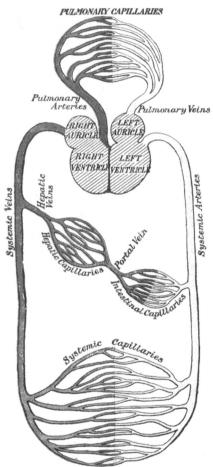

FIG. 272.—Diagram to show the course of the circulation of the blood.

The heart and lungs are contained within the cavity of the thorax, the walls of which afford them protection. The heart lies between the two lungs, and is there enclosed within a membranous bag, the *pericardium*, while each lung is invested by a serous membrane, the *pleura*. The skeleton of the thorax, and the shape and boundaries of the cavity, have already been described (page 93 *et seq.*).

The cavity of the thorax.—The capacity of the cavity of the thorax does not correspond with its apparent size externally, because (1) the space enclosed by the lower ribs is occupied by some of the abdominal viscera ; and (2) the cavity extends above the first rib into the neck. The size of the cavity of the thorax is constantly varying during life with the movements of the ribs and Diaphragm, and with the degree of distension of the abdominal viscera. From the collapsed state of the lungs, as seen when the thorax is opened, in the dead body, it would appear as if the viscera only partly filled the cavity of the thorax, but during life there is no vacant space, that which is seen after death being filled up by the expanded lungs.

The upper opening of the thorax.—The parts which pass through the upper opening of the thorax are, from before backwards in or near the middle line, the Sterno-hyoid and Sterno-thyroid muscles, the remains of the thymus gland, the trachea, œsophagus, thoracic duct, the inferior thyroid veins, and the Longus colli muscle of each side ; at the sides, the innominate artery, the left common carotid and left subclavian arteries, the internal mammary and superior intercostal arteries, the right and left innominate veins, the pneumogastric, cardiac, phrenic, and sympathetic nerves, the anterior branch of the first dorsal nerve, and the recurrent laryngeal nerve of the left side. The apex of each lung, covered by the pleura, also projects through this aperture, a little above the margin of the first rib.

The **lower opening of the thorax** is wider transversely than from before backwards. It slopes obliquely downwards and backwards, so that the cavity of the thorax is much deeper behind than in front. The Diaphragm (see page 325) closes in the opening, forming the floor of the thorax. The floor is flatter at the centre than at the sides, and is higher on the right side than on the left, corresponding in the dead body to the upper border of the fifth costal cartilage on the former, and to the corresponding part of the sixth costal cartilage on the latter. From the highest point on each side the floor slopes suddenly downwards

FIG. 273.—Pericardium, from in front. The sac has been distended with plaster. (From a preparation in the Museum of the Royal College of Surgeons of England.)

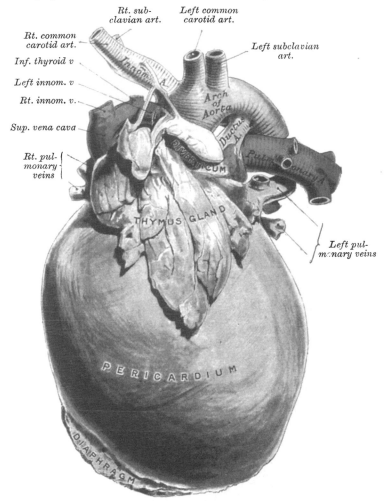

to the attachment of the Diaphragm to the ribs; this is more marked behind than in front, so that only a narrow space is left between it and the wall of the thorax.

THE PERICARDIUM

The **Pericardium** (figs. 273, 274) is a conical membranous sac, in which the heart and the commencement of the great vessels are contained. It is placed behind the sternum, and the cartilages of the third, fourth, fifth, sixth, and seventh ribs of the left side, in the interval between the pleuræ.

Its *apex* is directed upwards, and surrounds the great vessels about two inches above their origin from the base of the heart. Its *base* is attached to the central tendon and to the left part of the adjoining muscular structure of the

Diaphragm. *In front* it is separated from the sternum by the remains of the thymus gland above, and a little loose areolar tissue below ; and is covered by the margins of the lungs, especially the left. *Behind*, it rests upon the bronchi, the œsophagus, and the descending aorta. *Laterally*, it is covered by the pleuræ, and is in relation to the inner surface of the lungs ; the phrenic nerve, with its accompanying vessels, descends between the pericardium and pleura on either side.

Structure of the Pericardium.—The pericardium is a fibro-serous membrane, and consists, therefore, of two layers, an external fibrous and an internal serous.

FIG. 274.—Pericardium, from behind.
(From the same preparation as the preceding figure.)

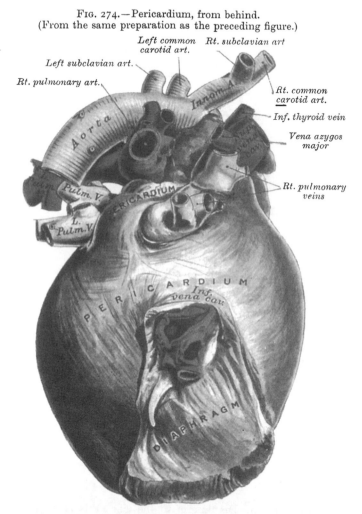

The *fibrous layer* is a strong, dense membrane. Above, it surrounds the great vessels arising from the base of the heart, on which it is continued in the form of tubular prolongations, which are gradually lost upon their external coats ; the strongest being that which encloses the aorta. It may be traced, over these vessels, to become continuous with the deep layer of the cervical fascia. In front the pericardium is connected to the posterior surface of the sternum by two fibrous bands, the *superior* and *inferior sterno-pericardiac ligaments* ; the upper passing to the manubrium, and the lower to the ensiform cartilage. On each side of the ascending aorta it sends upwards a diverticulum ; the one on the left side, somewhat conical in shape, passes upwards and outwards, between the arch of the aorta and the pulmonary artery, as far as the ductus arteriosus, where it terminates in a cæcal extremity which is attached by loose connective tissue to the obliterated duct (fig. 273). The one on the right side passes upwards and to the right, between the ascending aorta and vena cava superior, and also terminates in a cæcal extremity. Below, the fibrous layer is attached to the central tendon of the Diaphragm; and, on the left side, to its muscular fibres.

The vessels receiving fibrous prolongations from this membrane are, the aorta, the superior vena cava, the right and left pulmonary arteries, and the four pulmonary veins. As the inferior vena cava enters the pericardium through the central tendon of the Diaphragm, it receives no covering from the fibrous layer.

FIG. 275.—Front view of the thorax. The ribs and sternum are represented in relation to the lungs, heart, and other internal organs.

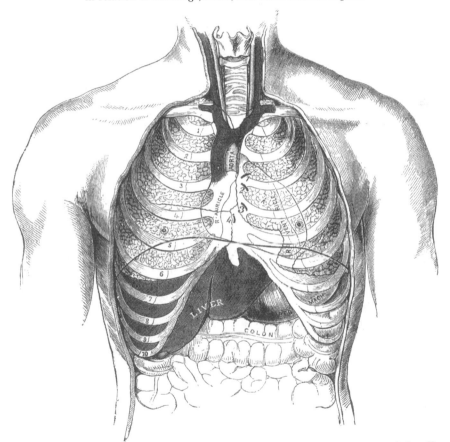

1. Pulmonary orifice.   2. Aortic orifice.   3. Left auriculo-ventricular orifice.   4. Right auriculo-ventricular orifice.

The *serous layer* invests the heart, and is then reflected on the inner surface of the pericardium. It consists, therefore, of a visceral and parietal portion. The former invests the surface of the heart, and the commencement of the great vessels, to the extent of an inch and an half from their origin; from these it is reflected upon the inner surface of the fibrous layer, lining, below, the upper surface of the central tendon of the Diaphragm. The serous membrane encloses the aorta and pulmonary artery in a single tube, so that a passage, termed the *transverse sinus* of the pericardium, exists between these vessels in front and the auricles behind. The membrane only partially covers the superior vena cava and the four pulmonary veins, and scarcely covers the inferior cava, as this vessel enters the heart almost directly after it has passed through the Diaphragm. Its inner surface is smooth and glistening and secretes a serous fluid which serves to facilitate the movements of the heart.

Arteries of the pericardium.—These are derived from the internal mammary, and its musculo-phrenic branch, and from the descending thoracic aorta.

Nerves of the pericardium.—These are branches from the vagus, the phrenic, and the sympathetic.

The vestigial fold of the pericardium.—Between the left pulmonary artery and subjacent pulmonary vein is a triangular fold of the serous pericardium ; it is known as the *vestigial fold of Marshall.* It is formed by the duplicature of the serous layer over the remnant of the lower part of the left superior cava (duct of Cuvier), which, after birth, becomes obliterated, and remains as a fibrous band stretching from the left superior intercostal vein to the left auricle, where it is continuous with a small vein, the oblique vein of Marshall, which opens into the coronary sinus.

Surgical Anatomy.—Paracentesis of the pericardium is sometimes required in cases of effusion into its cavity. The operation is best performed in the fifth intercostal space, one inch to the left of the sternum. The operation has been performed, however, in the fourth, sixth, and seventh spaces, and also on the right side of the sternum. Porter considers that by ' reason of the uncertain and varying relations of the pleura, and also of the anterior position of the heart, whenever the pericardial sac is distended with fluid, aspiration of the pericardium is a much more dangerous procedure than open incision when done by skilled hands.' He recommends that the operation should be done by resecting the fifth costal cartilage on the left side. By this means the surgeon avoids opening the pleural cavity, and secures continuous and free drainage, if the case is one of purulent pericarditis.

THE HEART

The **Heart** is a hollow muscular organ of a conical form, placed between the lungs, and enclosed in the cavity of the pericardium.

Position.—The heart is placed obliquely in the chest : the broad attached end, or base, is directed upwards, backwards, and to the right, and corresponds with the dorsal vertebræ, from the fifth to the eighth inclusive ; the apex is directed downwards, forwards, and to the left, and corresponds to the space between the cartilages of the fifth and sixth ribs, three-quarters of an inch to the inner side, and an inch and a half below the left nipple, or about three and a half inches from the middle line of the sternum The heart is placed behind the lower two-thirds of the sternum, and projects farther into the left than into the right half of the cavity of the chest, extending from the median line about three inches in the former direction, and only one and a half in the latter ; about one-third of the heart lies to the right and two-thirds to the left of the mesial plane. The anterior surface of the heart is round and convex, directed upwards and forwards, and formed chiefly by the right auricle and ventricle, together with a small part of the left ventricle. Its posterior surface, which looks downwards rather than backwards, is flattened, and rests upon the Diaphragm, and is formed chiefly by the left ventricle. The right or lower border is long, thin, and sharp ; the left or upper border short, but thick and round.

Size.—The heart, in the adult, measures five inches in length, three inches and a half in breadth in the broadest part, and two inches and a half in thickness. The prevalent weight, in the male, varies from ten to twelve ounces ; in the female, from eight to ten : its proportions to the body being as 1 to 169 in males ; 1 to 149 in females. The heart continues increasing in weight, and also in length, breadth, and thickness, up to an advanced period of life : this increase is more marked in men than in women.

Component Parts.—As has already been stated (page 429), the heart is subdivided by a muscular septum into two lateral halves, which are named respectively right and left ; and a transverse constriction subdivides each half of the organ into two cavities, the upper cavity on each side being called the *auricle*, the lower the *ventricle*. The course of the blood through the heart cavities and blood-vessels has already been described (page 429).

The division of the heart into four cavities is indicated by grooves upon its surface. The groove separating the auricles from the ventricles is called the *auriculo-ventricular groove*. It is deficient, in front, where it is crossed by the root of the pulmonary artery. It contains the trunks of the nutrient vessels of the heart. The auricular portion occupies the base of the heart, and is subdivided into two cavities by a median septum. The two ventricles are also separated into a right and left, by two furrows, the *interventricular grooves*, which are situated one on the anterior, the other on the posterior surface; these extend from the base of the ventricular portion to near the apex of the organ; the former being situated nearer to the left border of the heart, and the latter to the right. It follows, therefore, that the right ventricle forms the greater portion of the anterior surface of the heart, and the left ventricle more of its posterior surface.

Each of these cavities should now be separately examined.

Fig. 276.—The right auricle and ventricle laid open,
the anterior walls of both being removed.

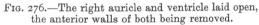

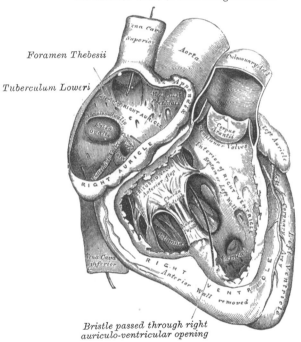

Bristle passed through right
auriculo-ventricular opening

The **Right Auricle** is a little larger than the left, its walls somewhat thinner, measuring about one line; and its cavity is capable of containing about two ounces. It consists of two parts : a principal cavity, the *sinus venosus*, or *atrium*, situated posteriorly, and an anterior, smaller portion, the *appendix auriculæ*.

The *sinus* is the large quadrangular cavity, placed between the two venæ cavæ; its walls are extremely thin; it is connected below with the right ventricle, and internally with the left auricle, being free in the rest of its extent.

The *appendix auriculæ*, so called from its fancied resemblance to a dog's ear, is a small conical muscular pouch, the margins of which present a dentated edge. It projects from the sinus forwards and to the left side, overlapping the root of the aorta.

To examine the interior of the right auricle, an incision should be made along its right border, from the entrance of the superior vena cava to that of the inferior. A second cut is to be made from the centre of this first incision to the tip of the auricular appendix, and the flaps raised.

The internal surface of the right auricle is smooth, except in the appendix and adjacent part of the anterior wall of the sinus venosus, where the muscular wall is thrown into parallel ridges, resembling the teeth of a comb and hence named the *musculi pectinati*. These end behind on a vertical smooth ridge, the *crista terminalis* of His, the position of which is indicated on the surface of the distended auricle by a furrow, the *sulcus terminalis* (His); this represents the line of fusion of the sinus venosus of the embryo with the primitive auricle proper.

It presents the following parts for examination :

Openings
- Superior cava.
- Inferior cava.
- Coronary sinus.
- Foramina Thebesii.
- Auriculo-ventricular.

Valves
- Eustachian.
- Coronary.

Fossa ovalis.
Annulus ovalis.
Tuberculum Loweri.
Musculi pectinati.

The *superior vena cava* returns the blood from the upper half of the body, and opens into the upper and back part of the auricle, the direction of its orifice being downwards and forwards.

The *inferior vena cava*, larger than the superior, returns the blood from the lower half of the body, and opens into the lowest part of the auricle, near the septum, the direction of its orifice being upwards and inwards. The direction of a current of blood through the superior vena cava would consequently be towards the auriculo-ventricular orifice ; while the direction of the blood through the inferior cava would be towards the auricular septum. This is the normal direction of the two currents in foetal life.

The *coronary sinus* opens into the auricle, between the inferior vena cava and the auriculo-ventricular opening. It returns the blood from the substance of the heart, and is protected by a semicircular fold of the lining membrane of the auricle, the *coronary valve*. The sinus, before entering the auricle, is considerably dilated—nearly to the size of the end of the little finger. Its wall is partly muscular, and, at its junction with the great coronary vein, is somewhat constricted, and furnished with a valve, consisting of two unequal segments.

The *foramina Thebesii* are numerous minute apertures, the mouths of small veins (*venæ cordis minimæ*), which open on various parts of the inner surface of the auricle. They return the blood directly from the muscular substance of the heart. Some of these foramina are minute depressions in the walls of the heart, presenting a closed extremity.

The *auriculo-ventricular opening* is the large oval aperture of communication between the auricle and the ventricle, to be presently described.

The *Eustachian valve* is situated between the anterior margin of the inferior vena cava and the auriculo-ventricular orifice. It is semilunar in form, its convex margin being attached to the wall of the vein ; its concave margin, which is free, terminating in two cornua, of which the left is attached to the anterior edge of the annulus ovalis ; the right being lost on the wall of the auricle. The valve is formed by a duplicature of the lining membrane of the auricle, containing a few muscular fibres.

In the foetus this valve is of large size, and serves to direct the blood from the inferior vena cava, through the foramen ovale, into the left auricle.

In the adult it is occasionally persistent, and may assist in preventing the reflux of blood into the inferior vena cava ; more commonly it is small, and its free margin presents a cribriform or filamentous appearance ; occasionally it is altogether wanting.

The *coronary valve* (valve of Thebesius) is a semicircular fold of the lining membrane of the auricle, protecting the orifice of the coronary sinus. It prevents the regurgitation of blood into the sinus during the contraction of the auricle. This valve is occasionally double.

The *fossa ovalis* is an oval depression, corresponding to the situation of the foramen ovale in the fœtus. It is situated at the lower part of the septum auricularum, above and to the left of the orifice of the inferior vena cava.

The *annulus ovalis* is the prominent oval margin of the foramen ovale. It is most distinct above, and at the sides; below, it is deficient. A small slit-like valvular opening is occasionally found, at the upper margin of the fossa ovalis, which leads upwards, beneath the annulus, into the left auricle, and is the remains of the aperture between the two auricles in the fœtus.

The *tuberculum Loweri* is a small projection on the right wall of the auricle, between the two venæ cavæ. It is most distinct in the hearts of quadrupeds; in man it is scarcely visible. It was supposed by Lower to direct the blood from the superior cava towards the auriculo-ventricular opening.

The **Right Ventricle** is triangular in form, and extends from the right auricle to near the apex of the heart. Its anterior or upper surface is rounded and convex, and forms the larger part of the front of the heart. Its under surface is flattened, rests upon the Diaphragm, and forms only a small part of the back of the heart. Its posterior wall is formed by the partition between the two ventricles, the *septum ventriculorum*, so that a transverse section of the cavity presents a semilunar outline. The surface of the septum is convex and bulges into the cavity of the right ventricle. Its upper and inner angle is prolonged into a conical pouch, the *infundibulum*, or *conus arteriosus*, from which the pulmonary artery arises. The walls of the right ventricle are thinner than those of the left, the proportion between them being as 1 to 3. The wall is thickest at the base, and gradually becomes thinner towards the apex. The cavity equals in size that of the left ventricle, and is capable of containing about three fluid ounces.*

To examine the interior of the right ventricle, its anterior wall should be turned downwards and to the right in the form of a triangular flap. This is accomplished by making two incisions: (1) from the pulmonary artery to the apex of the ventricle parallel to, but a little to the right of, the anterior interventricular furrow; (2) another, starting from the upper extremity of the first and carried outwards parallel to, but a little below, the auriculo-ventricular furrow; care being taken not to injure the auriculo-ventricular valve.

The following parts present themselves for examination:

Openings { Auriculo-ventricular.
{ Opening of the pulmonary artery.

Valves { Tricuspid.
{ Semilunar.

And a muscular and tendinous apparatus connected with the tricuspid valve:

Columnæ carneæ.   Chordæ tendineæ.

The *auriculo-ventricular orifice* is the large oval aperture of communication between the auricle and ventricle. It is situated at the base of the ventricle, near the right border of the heart. It is about an inch and a half in diameter,† oval from side to side, surrounded by a fibrous ring, covered by the lining

* Morrant Baker says that, 'taking the mean of various estimates, it may be inferred that each ventricle is able to contain four to six ounces of blood.'— Kirke's *Physiology*, 10th edition, p. 156.

† In the *Pathological Transactions*, vol. vi. p. 119, Dr. Peacock has given some careful researches upon the weight and dimensions of the heart in health and disease. He states, as the result of his investigations, that, in the healthy adult heart, the right auriculo-ventricular aperture has a mean circumference of 54·4 lines, or $4\frac{20}{24}$ inches; the left auriculo-ventricular aperture a mean circumference of 44·3 lines, or $3\frac{14}{24}$ inches; the pulmonic orifice of 40 lines, or $3\frac{13}{24}$ inches; and the aortic orifice of 35·5 lines, or $3\frac{4}{24}$ inches; but the dimensions of the orifices varied greatly in different cases, the right auriculo-ventricular aperture having a range of from 45 to 60 lines, and the others in the same proportion.

membrane of the heart; it is considerably larger than the corresponding aperture on the left side, being sufficient to admit the ends of four fingers. It is guarded by the tricuspid valve.

The *opening of the pulmonary artery* is circular in form, and situated at the summit of the conus arteriosus, close to the septum ventriculorum. It is placed above and on the left side of the auriculo-ventricular opening, upon the anterior aspect of the heart. Its orifice is guarded by the pulmonary semilunar valves.

The *tricuspid valve* consists of three segments of a triangular or trapezoidal shape, formed by a duplicature of the lining membrane of the heart, strengthened by a layer of fibrous tissue, which contains, according to Kürschner and Senac, muscular fibres. These segments are connected by their bases to the fibrous ring surrounding the auriculo-ventricular orifice, and by their sides with one another, so as to form a continuous annular membrane, which is attached round the margin of the auriculo-ventricular opening, their free margins and ventricular surfaces affording attachment to a number of delicate tendinous cords, the *chordæ tendineæ*. The largest and most movable segment is placed towards the left side of the auriculo-ventricular opening, interposed between that opening and the infundibulum ; hence it is called the *left* or *infundibular cusp*. Another segment corresponds to the right part of the front of the ventricle, the *right* or *marginal cusp* ; and a third to its posterior wall, the *posterior* or *septal cusp*. The central part of each segment is thick and strong : the lateral margins are thin and translucent. The chordæ tendineæ are connected with the adjacent margins of the principal segments of the valve, and are further attached to each segment in the following manner : 1. Three or four reach the attached margin of each segment, where they are continuous with the auriculo-ventricular tendinous ring. 2. Others, four to six in number, are attached to the central thickened part of each segment. 3. The most numerous and finest are connected with the marginal portion of each segment.

The *columnæ carneæ* are the rounded muscular columns which project from nearly the whole of the inner surface of the ventricle, excepting near the opening of the pulmonary artery, where the wall is smooth. They may be classified, according to their mode of connection with the ventricle, into three sets. The first set merely form prominent ridges on the inner surface of the ventricle, being attached by their entire length on one side, as well as by their extremities. The second set are attached by their two extremities, but are free in the rest of their extent ; while the third set (*musculi papillares*) are attached by one extremity to the wall of the heart, the opposite extremity giving attachment to the *chordæ tendineæ*. There are two papillary muscles, anterior and posterior : of these, the anterior is the larger ; its chordæ tendineæ are connected with the right and left segments of the valve. The posterior is not always single, but sometimes consists of two or three muscular columns ; its chordæ tendineæ are connected with the posterior and the right segments. In addition to these, some few chordæ may be seen springing directly from the ventricular septum, or from small eminences on it, and passing to the left and posterior segments. A fleshy band, well marked in the ox and some other animals, is frequently seen passing from the base of the anterior papillary muscle to the interventricular septum. From its attachments it may assist in preventing over-distension of the auricle, and so has been named the *moderator band*.

The right auriculo-ventricular orifice allows the blood to pass freely from the right auricle into the right ventricle, and it will be noted that the surface of the tricuspid valve next the blood-current is quite smooth. When the right ventricle contracts to force the blood into the pulmonary artery the segments of the tricuspid valve come together and close the auriculo-ventricular opening, and so prevent the blood from passing back into the auricle. The papillary muscles and chordæ tendineæ moor the segments of the valve, and prevent their being forced through into the auricle by the weight of blood behind them.

The *semilunar valves*, three in number,* guard the orifice of the pulmonary artery. They consist of three semicircular folds, two anterior (right and left) and one posterior, formed by a duplicature of the lining membrane, strengthened by fibrous tissue. They are attached, by their convex margins, to the wall of the artery, at its junction with the ventricle, the straight border being free, and directed upwards in the lumen of the vessel. The free margin of each is some-what thicker than the rest of the valve, is strengthened by a bundle of tendinous fibres, and presents, at its middle, a small projecting thickened nodule, called *corpus Arantii*, and consisting of bundles of interlacing connective-tissue fibres with branched connective-tissue cells and some few elastic fibres. From this nodule tendinous fibres radiate through the valve to its attached margin, and these fibres form a constituent part of its substance throughout its whole extent, excepting two narrow lunated portions, the *lunulæ*, placed one on each side of the nodule im-mediately adjoining the free margin ; here the valve is thin, and formed merely by the lining membrane. During the passage of the blood along the pulmonary artery these valves are opened, and the course of the blood along the tube is uninterrupted ; but during the ventricular diastole, when the current of blood along the pulmonary artery is checked, and partly thrown back by its elastic walls, these valves become immediately expanded, and effectually close the entrance of the tube. When the valves are closed, the lunated portions of each are brought into contact with one another by their opposed surfaces, the three corpora Arantii filling up the small triangular space that would be otherwise left by the approximation of the three semilunar valves.

Between the semilunar valves and the commencement of the pulmonary artery are three pouches or dilatations, one behind each valve. These are the pulmonary sinuses (*sinuses of Valsalva*). Similar sinuses exist between the semi-lunar valves of the aorta and the commencement of that vessel ; they are larger than the pulmonary sinuses. The blood, in its regurgitation towards the heart, finds its way into these sinuses, and so shuts down the valve-flaps.

In order to examine the interior of the left auricle, make an incision on the posterior surface of the auricle from the pulmonary veins on one side to those on the other, the incision being carried a little way into the vessels. Make another incision from the middle of the horizontal one to the appendix.

The **Left Auricle** is rather smaller than the right, but its walls are thicker, measuring about one line and a half ; it consists, like the right, of two parts, a principal cavity, or *sinus*, and an *appendix auriculæ*.

The *sinus* is cuboidal in form, and concealed, in front, by the pulmonary artery and aorta ; internally, it is separated from the right auricle by the septum auricularum ; behind, it receives on each side the two pulmonary veins, being free in the rest of its extent.

The *appendix auriculæ* is somewhat constricted at its junction with the auricle ; it is longer, narrower, and more curved than that of the right side, and its margins more deeply indented, presenting a kind of foliated appearance. Its direction is forwards and towards the right side, overlapping the root of the pulmonary artery.

The following parts present themselves for examination :

The openings of the four pulmonary veins.
Auriculo-ventricular opening.
Musculi pectinati.

The *pulmonary veins*, four in number, open, two into the right, and two into the left side of the auricle : they are not provided with valves. The two left veins frequently terminate by a common opening.

The *auriculo-ventricular opening* is the large oval aperture of communication

* The pulmonary semilunar valves have been found to be two in number instead of three (Dr. Hand, of St. Paul, Minn., in the *North-Western Med. and Surg. Journ.* July 1873), and the same variety is more frequently noticed in the aortic semilunar valves.

between the auricle and ventricle. It is rather smaller than the corresponding opening on the opposite side (see note, page 437).

The *musculi pectinati* are fewer in number and smaller than on the right side; they are confined to the inner surface of the appendix.

On the inner surface of the septum auricularum may be seen a lunated impression, bounded below by a crescentic ridge, the concavity of which is turned upwards. The depression is just above the fossa ovalis in the right auricle.

To examine the interior of the left ventricle, make an incision a little to the left of the anterior interventricular groove from the base to the apex of the heart, and carry it up thence a little to the left of the posterior interventricular groove, nearly as far as the auriculo-ventricular groove.

The **Left Ventricle** is longer and more conical in shape than the right ventricle, and on transverse section its cavity presents an oval or nearly

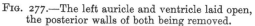

Fig. 277.—The left auricle and ventricle laid open, the posterior walls of both being removed.

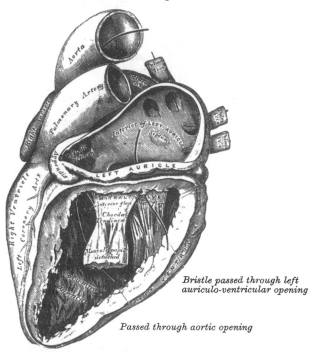

Bristle passed through left auriculo-ventricular opening

Passed through aortic opening

circular outline. It forms a small part of the anterior surface of the heart, and a considerable part of its posterior surface. It also forms the apex of the heart by its projection beyond the right ventricle. Its walls are much thicker than those of the right side, the proportion being as 3 to 1. They are thickest opposite the widest part of the ventricle, becoming gradually thinner towards the base, and also towards the apex, which is the thinnest part.

The following parts present themselves for examination :

Openings { Auriculo-ventricular. Aortic.   Valves { Mitral. Semilunar.

Chordæ tendineæ.   Columnæ carneæ.

The *auriculo-ventricular opening* is placed below and to the left of the aortic orifice. It is a little smaller than the corresponding aperture of the opposite side, admitting only two fingers ; but, like it, is broader in the transverse than in the antero-posterior diameter. It is surrounded by a dense fibrous ring, covered by the lining membrane of the heart, and guarded by the mitral valve.

The *aortic opening* is a circular aperture, in front and to the right side of the auriculo-ventricular, from which it is separated by one of the segments of the mitral valve. Its orifice is guarded by the semilunar valves. The portion of the ventricle immediately below the aortic orifice is often termed the *aortic vestibule* of Sibson. It possesses fibrous instead of muscular walls, and so does not collapse during the ventricular diastole; it thus gives space for the segments of the aortic valve during its closure.

The *mitral valve* is attached to the circumference of the auriculo-ventricular orifice in the same way that the tricuspid valve is on the opposite side. It is formed by a duplicature of the lining membrane, strengthened by fibrous tissue, and contains a few muscular fibres. It is larger in size, thicker, and altogether stronger than the tricuspid, and consists of two segments of unequal size. The larger segment is placed in front and to the right between the auriculo-ventricular and aortic orifices, the smaller to the left and behind the opening, close to the wall of the ventricle. Two smaller segments are usually found at the angles of junction of the larger. The mitral valve-flaps are furnished with chordæ tendineæ, the mode of attachment of which is precisely similar to those on the right side; but they are thicker, stronger, and less numerous.

The *semilunar valves* surround the orifice of the aorta; two are posterior (right and left) and one anterior: they are similar in structure, and in their mode of attachment, to those of the pulmonary artery. They are, however, larger, thicker, and stronger than those of the right side; the lunulæ are more distinct, and the corpora Arantii larger and more prominent. Opposite each segment the wall of the aorta presents a slight dilatation or bulging (sinus of Valsalva); they are larger than those at the commencement of the pulmonary artery.

The *columnæ carneæ* admit of a subdivision into three sets, like those upon the right side; but they are smaller, more numerous, and present a dense interlacement, especially at the apex, and upon the posterior wall. Those attached by one extremity only, the *musculi papillares*, are two in number, being connected one to the anterior, the other to the posterior wall; they are of large size, and terminate by free rounded extremities, from which the chordæ tendineæ arise.

The septum between the two ventricles is thick, especially below (fig. 278). At its upper part it suddenly tapers off and becomes destitute of muscular fibres, consisting only of fibrous tissue, covered by two layers of endocardium; and on the right side also covered, during diastole, by one of the flaps of the tricuspid valve. This upper portion is termed the *membranous part* of the septum, and is continued upwards and forms the septum between the aortic vestibule and the right auricle. It is derived from the lower part of the aortic septum of the fœtus, and an abnormal communication may exist at this part owing to defective development of this septum.

The *Endocardium* is a thin membrane which lines the internal surface of the heart; it assists in forming the valves by its reduplications, and is continuous with the lining membrane of the great blood-vessels. It is a smooth, transparent membrane, giving to the inner surface of the heart its glistening appearance. It is more opaque on the left than on the right side of the heart, thicker in the auricles than in the ventricles, and thickest in the left auricle. It is thin on the musculi pectinati, and on the columnæ carneæ; but thicker on the smooth part of the auricular and ventricular walls, and on the tips of the musculi papillares.

Structure.—The heart consists of muscular fibres, and of fibrous rings which serve for their attachment. It is closely covered by the visceral layer of the serous pericardium (*epicardium*), and its cavities are lined by the *endocardium*. Between these two membranes is the muscular wall of the heart, the *myocardium*.

The *fibrous rings* surround the auriculo-ventricular and arterial orifices: they are stronger upon the left than on the right side of the heart. The auriculo-ventricular rings serve for the attachment of the muscular fibres of the auricles and ventricles, and also for the mitral and tricuspid valves; the ring on the left side is closely connected, by its right margin, with the aortic arterial ring. Between these and the right auriculo-ventricular ring is a mass of fibrous tissue; and in some of the larger animals, as the ox and elephant, a nodule of bone, the *os cordis.*

The fibrous rings surrounding the arterial orifices serve for the attachment of the great vessels and semilunar valves. Each ring receives, by its ventricular margin, the attachment of the muscular fibres of the ventricles; its opposite margin presents three deep semicircular notches, within which the middle coat of the artery (which presents three convex semicircular segments) is firmly

FIG. 278.—Section of the heart, showing the interventricular septum.

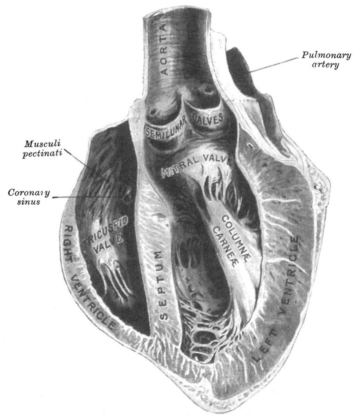

fixed; the attachment of the artery to its fibrous ring being strengthened by the thin cellular coat and serous membrane externally, and by the endocardium within. It is opposite the margins of these semicircular notches, in the arterial rings, that the endocardium, by its reduplication, forms the semilunar valves, the fibrous structure of the ring being continued into each of the segments of the valve at this part. The middle coat of the artery in this situation is thin, and the sides of the vessel are dilated to form the sinuses of Valsalva.

The *muscular structure of the heart* consists of bands of fibres, which present an exceedingly intricate interlacement. They are of a deep red colour, and marked with transverse striæ.

The muscular fibres of the heart admit of a subdivision into two groups, those of the auricles and those of the ventricles, which are quite independent of one another.

Fibres of the Auricles.—These are disposed in two layers—a superficial layer common to both cavities, and a deep layer proper to each. The *superficial fibres* are more distinct on the anterior surface of the auricles, across the bases of which they run in a transverse direction, forming a thin and incomplete layer. Some of these fibres pass into the septum auricularum. The *internal* or *deep fibres* proper to each auricle consist of two sets, looped and annular fibres. The *looped fibres* pass upwards over each auricle, being attached by their two extremities to the corresponding auriculo-ventricular rings, in front and behind. The *annular fibres* surround the whole extent of the appendices auricularum, and are continued upon the walls of the venæ cavæ and coronary sinus on the right side, and upon the pulmonary veins on the left side, at their connection with the heart. In the appendices they interlace with the longitudinal fibres.

Fibres of the Ventricles.—These are arranged in an exceedingly complex manner, and the accounts given by various anatomists differ considerably. This is probably due partly to the fact that the various layers of muscular fibre of which the heart is said to be composed are not independent, but their fibres are interlaced to a considerable extent, and therefore any separation into layers must be to a great extent artificial; and also partly to the fact, pointed out by Henle, that there are varieties in the arrangement owing to individual differences. If the epicardium and the subjacent fat are removed from a heart which has been subjected to prolonged boiling, so as to dissolve the connective tissues, the superficial fibres of the ventricles will be exposed. They will be seen to commence at the base of the heart, where they are attached to the tendinous rings around the orifices, and to pass obliquely downwards towards the apex, with a direction from right to left. At the apex the fibres turn suddenly inwards, into the interior of the ventricle, forming what is called the *vortex*. On the back of the heart it will be seen that the fibres pass continuously from one ventricle to the other over the interventricular groove; and the same thing will be noticed on the front of the heart at the upper and lower end of the anterior interventricular groove, but in the middle portion of this groove the fibres passing from one ventricle to the other are interrupted by fibres emerging from the septum along the groove; many of the superficial fibres pass in also at this groove to the septum. The vortex is produced, as stated above, by the sudden turning inwards of the superficial fibres in a peculiar spiral manner into the interior of the ventricle. Those fibres which descend on the posterior surface of the heart, enter the left ventricle at the vortex, and, ascending, form the posterior part of the inner layer of muscular fibres lining this cavity and the right (posterior) musculus papillaris; those fibres which descend on the front of the heart, to reach the apex, also pass, at the vortex, into the interior of the ventricle, where they form the remainder of the innermost layer of the ventricle and the left (anterior) musculus papillaris. The fibres forming the inner layer of the wall of the ventricle ascend to be attached to the fibrous rings around the orifices.

By dissection these superficial fibres may be removed as a thin stratum, and it will then be found that the ventricles are made up of oblique fibres, superimposed in layers one on the top of another, and assuming gradually a less oblique direction as they pass to the middle of the thickness of the ventricular wall, so that in the centre of the wall the fibres are transverse. Internal to this central transverse layer the fibres become oblique again, but in the opposite direction to the external ones. This division into distinct layers is, however, to a great extent artificial, as fibres cross from one layer to another, and have therefore to be divided in the dissection, and the change in the direction of the fibres is very gradual. These oblique fibres commence above at the fibrous rings at the base of the heart, and descending towards the apex they enter the septum near its lower end. In the septum the fibres which form the left ventricle may be traced in three directions. 1. Some pass upwards to be attached to the central mass of fibrous tissue. 2. Others pass through the septum to become continuous with the fibres of the right ventricle. 3. And the remainder pass through the septum to encircle the ventricle as annular fibres. Of the fibres of the right ventricle, some on entering the septum pass upwards to be attached to the central mass of fibrous tissue ; some entering the septum from behind pass forwards to become continuous with the fibres on the anterior surface of the left ventricle ; and others entering in front pass backwards to join the fibres on the posterior wall of the left ventricle. The septum therefore consists of three varieties of fibres : viz. annular fibres special to the left ventricle; ascending fibres, derived from both ventricles and ascending through the septum to the central fibro-cartilage ; and decussating fibres derived from the anterior wall of one ventricle and passing to the posterior wall of the other ventricle, or from the posterior wall of the right ventricle and passing to the anterior wall of the left. In addition to these fibres, there are a considerable number which appear to encircle both ventricles and which pass across the septum without turning into it.

Vessels and Nerves.—The *arteries* supplying the heart are the right and left coronary from the aorta.

The *veins* accompany the arteries, and terminate in the right auricle. They are, the anterior or great, posterior, left and anterior cardiac veins, the right or small, and the left or great, coronary sinuses and the venæ Thebesii (*venæ cordis minimæ*).

The *lymphatics* terminate in the thoracic and right lymphatic ducts.

The *nerves* are derived from the cardiac plexuses, which are formed partly from the cranial nerves, and partly from the sympathetic. They are freely distributed both on the surface and in the substance of the heart, the separate filaments being furnished with small ganglia.

Surface Form.—In order to show the extent of the heart in relation to the front of the chest, draw a line from the lower border of the second left costal cartilage, one inch from the sternum, to the upper border of the third right costal cartilage, half an inch from the sternum. This represents the base line, or upper limit of the organ. Take a point an inch and a half below, and three-quarters of an inch internal to the left nipple, that is, about three and a half inches to the left of the median line of the body. This represents the apex of the heart. Draw a line from this apex point, with a slight convexity downwards, to the junction of the seventh right costal cartilage to the sternum. This represents the lower limit of the heart. Join the right extremity of the first line—that is, the base line—with the right extremity of this line—that is, to the seventh right chrondro-sternal joint—with a slight curve outwards, so that it projects about an inch and a half from the middle line of the sternum. Lastly, join the left extremity of the base line and the apex point by a line curved slightly to the left.

The position of the various orifices is as follows : viz. the pulmonary orifice is situated in the upper angle formed by the articulation of the third left costal cartilage with the sternum ; the aortic orifice is a little below and internal to this, behind the left border of the sternum, close to the articulation of the third left costal cartilage to this bone. The left auriculo-ventricular opening is behind the sternum, rather to the left of the median line and opposite the fourth costal cartilages. The right auriculo-ventricular opening is a little lower, opposite the fourth interspace and in the middle line of the body (fig. 275).

A portion of the area of the heart thus mapped out is uncovered by lung, and therefore gives a dull note on percussion ; the remainder, being overlapped by the lung, gives a more or less resonant note. The former is known as the area of superficial cardiac dulness ; the latter, as the area of deep cardiac dulness. The area of superficial cardiac dulness is included between a line drawn from the centre of the sternum, on a level with the fourth costal cartilages, to the apex of the heart, and a line drawn from the same point down the lower third of the middle line of the sternum. Below, this area merges into the dulness which corresponds to the liver. Dr. Latham lays down the following rule as a sufficient practical guide for the definition of the portion of the heart which is uncovered by lung or pleura : ' Make a circle of two inches in diameter round a point midway between the nipple and the end of the sternum.'

Surgical Anatomy.—Wounds of the heart are often immediately fatal, but not necessarily so. They may be non-penetrating, when death may occur from hæmorrhage, if one of the coronary vessels has been wounded, or subsequently from pericarditis ; or, on the other hand, the patient may recover. Even a penetrating wound is not necessarily fatal, if the wound is a small one. A flap comprising the whole thickness of the thoracic wall may be made, the cavity of the pericardium opened, and the wound in the heart sutured. This has been done successfully

Peculiarities in the Vascular System of the Fœtus

The chief peculiarities in the heart of the fœtus are the direct communication between the two auricles through the foramen ovale, and the large size of the Eustachian valve. There are also several minor peculiarities. Thus, the position of the heart is vertical until the fourth month, when it commences to assume an oblique direction. Its size is also very considerable as compared with the body, the proportion at the second month being 1 to 50 ; at birth it is as 1 to 120 ; while in the adult the average is about 1 to 160. At an early period of fœtal life the auricular portion of the heart is larger than the ventricular, the right auricle being more capacious than the left ; but towards birth the ventricular portion becomes the larger. The thickness of both ventricles is, at first, about equal, but towards birth the left becomes the thicker of the two.

The *foramen ovale* is situated at the lower and back part of the septum auricularum, forming a communication between the auricles. It remains as a free oval opening until the middle period of fœtal life. About this period a fold grows up from the posterior wall of the auricle, to the left of the foramen ovale, and advances over the opening, so as to form a sort of valve, which allows the blood to pass only from the right to the left auricle, and not in the opposite direction.

The *Eustachian valve* is directed upwards on the left side of the opening of the inferior vena cava, and serves to direct the blood from this vessel through the foramen ovale into the left auricle.

The peculiarities in the arterial system of the fœtus are the communication between the pulmonary artery and the descending aorta by means of the *ductus arteriosus*, and the communication between the internal iliac arteries and the placenta by means of the *umbilical arteries*.

The *ductus arteriosus* is a short tube, about half an inch in length at birth, and of the diameter of a goose-quill. In the early condition it forms the continuation of the pulmonary artery, and opens into the descending aorta, just below the origin of the left subclavian artery ; and so conducts the chief part of the blood from the right ventricle into this vessel. When the branches of the pulmonary artery have become larger relatively to the ductus arteriosus, the latter is chiefly connected to the left pulmonary artery ; and the fibrous cord, which is all that remains of the ductus arteriosus in later life, will be found to be attached to the root of that vessel.

The *umbilical* or *hypogastric arteries* arise from the internal iliacs, in addition to the branches given off from those vessels in the adult. Ascending along the sides of the bladder to its apex, they pass out of the abdomen at the umbilicus and are continued along the umbilical cord to the placenta, coiling round the umbilical vein. They carry to the placenta the blood which has circulated in the system of the fœtus.

The peculiarity in the venous system of the fœtus is the communication established between the placenta and the liver and portal vein, through the umbilical vein ; and the inferior vena cava through the ductus venosus.

Fœtal Circulation

The blood destined for the nutrition of the fœtus is returned from the placenta to the fœtus by the umbilical vein. This vein enters the abdomen at the umbilicus, and passes upwards along the free margin of the suspensory ligament of the liver to the under surface of that organ, where it gives off two or three branches to the left lobe, one of which is of large size ; and others to the lobus quadratus and lobulus Spigelii. At the transverse fissure it divides into two branches : of these, the larger is joined by the portal vein, and enters the right lobe ; the smaller branch continues outwards, under the name of the ductus venosus, and joins the left hepatic vein at the point of junction of that vessel with the inferior vena cava.   The blood, therefore, which traverses the umbilical vein, reaches the inferior vena cava in three different ways.   The greater quantity circulates through the liver with the portal venous blood, before entering the vena cava by the hepatic veins; some enters the liver directly, and is also

FIG. 279.—Plan of the fœtal circulation.

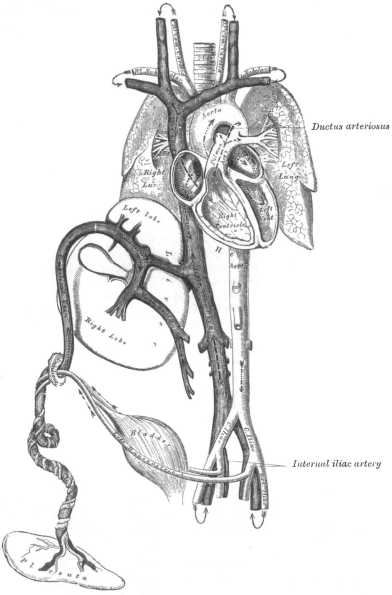

Ductus arteriosus

Internal iliac artery

In this plan the figured arrows represent the kind of blood, as well as the direction which it takes in the vessels. Thus—arterial blood is figured ≫·········➤ ; venous blood, ≫–––––➤ ; mixed (arterial and venous) blood, ≫·····—····➤

returned to the inferior cava by the hepatic veins: the smaller quantity passes directly into the vena cava, by the junction of the ductus venosus with the left hepatic vein.

In the inferior cava, the blood carried by the ductus venosus and hepatic veins becomes mixed with that returning from the lower extremities and wall of the abdomen. It enters the right auricle, and, guided by the Eustachian valve, passes through the foramen ovale into the left auricle, where it becomes mixed with a small quantity of blood returned from the lungs by the pulmonary veins. From the left auricle it passes into the left ventricle ; and from the left ventricle into the aorta, by means of which it is distributed almost entirely to the head and upper extremities, a small quantity being probably carried into the descending aorta. From the head and upper extremities the blood is returned by the tributaries of the superior vena cava to the right auricle, where it becomes mixed with a small portion of the blood from the inferior cava. From the right auricle it descends over the Eustachian valve into the right ventricle ; and from the right ventricle, passes into the pulmonary artery. The lungs of the fœtus being inactive, only a small quantity of the blood of the pulmonary artery is distributed to them, by the right and left pulmonary arteries, and is returned by the pulmonary veins to the left auricle : the greater part passes through the ductus arteriosus into the commencement of the descending aorta, where it becomes mixed with a small quantity of blood transmitted by the left ventricle into the aorta. Through this vessel it descends to supply the lower extremities and viscera of the abdomen and pelvis, the chief portion being, however, conveyed by the umbilical arteries to the placenta.

From the preceding account of the circulation of the blood in the fœtus, it will be seen :

1. That the placenta serves the purposes of nutrition and excretion, receiving the impure blood from the fœtus, and returning it charged with additional nutritive material.

2. That nearly the whole of the blood of the umbilical vein traverses the liver before entering the inferior cava ; hence the large size of this organ, especially at an early period of fœtal life.

3. That the right auricle is the point of meeting of a double current, the blood in the inferior cava being guided by the Eustachian valve into the left auricle, while that in the superior cava descends into the right ventricle. At an early period of fœtal life it is highly probable that the two streams are quite distinct ; for the inferior cava opens almost directly into the left auricle, and the Eustachian valve would exclude the current along the vein from entering the right ventricle. At a later period, as the separation between the two auricles becomes more distinct, it seems probable that some mixture of the two streams must take place.

4. The pure blood carried from the placenta to the fœtus by the umbilical vein, mixed with the blood from the portal vein and inferior cava, passes almost directly to the arch of the aorta, and is distributed by the branches of that vessel to the head and upper extremities : hence the large size and perfect development of those parts at birth.

5. The blood contained in the descending aorta, chiefly derived from that which has already circulated through the head and limbs, together with a small quantity from the left ventricle, is distributed to the lower extremities : hence the small size and imperfect development of these parts at birth.

CHANGES IN THE VASCULAR SYSTEM AT BIRTH

At birth, when respiration is established, an increased amount of blood from the pulmonary artery passes through the lungs, which now perform their office as respiratory organs, and, at the same time, the placental circulation is cut off. The foramen ovale becomes gradually closed by about the tenth day after birth : the valvular fold above mentioned becomes adherent to the margins of the foramen for the greater part of its circumference, but above a slit-like opening is left between the two auricles, and this sometimes persists.

The *ductus arteriosus* begins to contract immediately after respiration is established, becomes completely closed from the fourth to the tenth day, and ultimately degenerates into an impervious cord, which serves to connect the left pulmonary artery to the descending aorta.

Of the *umbilical* or *hypogastric arteries*, the portion continued on to the bladder from the trunk of the corresponding internal iliac remains pervious, as the superior vesical artery; and the part extending from the side of the bladder to the umbilicus becomes obliterated between the second and fifth days after birth, and projects as a fibrous cord towards the abdominal cavity, carrying on it a fold of peritoneum and separating two of the fossæ of the peritoneum, spoken of in the section on the surgical anatomy of direct inguinal hernia.

The *umbilical vein* and *ductus venosus* become completely obliterated between the second and fifth days after birth, and ultimately dwindle to fibrous cords; the former becoming the round ligament of the liver, the latter the fibrous cord which, in the adult, may be traced along the fissure of the ductus venosus.

THE ARTERIES

Arteries are cylindrical tubular vessels, which serve to convey blood from both ventricles of the heart to every part of the body. These vessels were named arteries (ἀήρ, *air*; τηρεῖν, *to contain*) from the belief entertained by the ancients that they contained air. To Galen is due the honour of refuting this opinion; he showed that these vessels, though for the most part empty after death, contain blood in the living body.

The distribution of the systemic arteries is like a highly ramified tree, the common trunk of which, formed by the aorta, commences at the left ventricle of the heart, the smallest ramifications corresponding to the circumference of the body and the contained organs. The arteries are found in nearly every part of the body, with the exception of the hair, nails, epidermis, cartilages, and cornea; and the larger trunks usually occupy the most protected situations, running, in the limbs, along the flexor side, where they are less exposed to injury.

There is considerable variation in the mode of division of the arteries : occasionally a short trunk subdivides into several branches at the same point, as may be observed in the cœliac and thyroid axes; or the vessel may give off several branches in succession, and still continue as the main trunk, as is seen in the arteries of the limbs; but frequently the division is dichotomous, as, for instance, the aorta dividing into the two common iliacs; and the common carotid into the external and internal.

The branches of arteries arise at very variable angles; some, as the superior intercostal arteries from the aorta, arise at an obtuse angle; others, as the lumbar arteries, at a right angle; or, as the spermatic, at an acute angle. An artery from which a branch is given off is smaller in size, but retains a uniform diameter until a second branch is derived from it. A branch of an artery is smaller than the trunk from which it arises; but if an artery divides into two branches, the combined area of the two vessels is, in nearly every instance, somewhat greater than that of the trunk; and the combined area of all the arterial branches greatly exceeds the area of the aorta; so that the arteries collectively may be regarded as a cone, the apex of which corresponds to the aorta, the base to the capillary system.

The arteries, in their distribution, communicate with one another, forming what is called an *anastomosis* (ἀνά, *between*; στόμα, *mouth*), or inosculation : and this communication is very free between the large as well as between the smaller branches. The anastomosis between trunks of equal size is found where great activity of the circulation is requisite, as in the brain; here the two vertebral arteries unite to form the basilar, and the two internal carotid arteries are connected by a short communicating trunk; it is also found in the abdomen, the intestinal arteries having very ample anastomoses between their larger branches. In the limbs, the anastomoses are most numerous and of largest size around the joints; the branches of an artery above inosculating with branches from the vessels below. These anastomoses are of considerable interest to the surgeon, as it is by their enlargement that a *collateral circulation* is established after the application of a ligature to an artery for the cure of aneurism. The smaller branches of arteries anastomose more frequently than the larger; and between the smallest twigs these inosculations become so numerous as to constitute a close network that pervades nearly every tissue of the body.

Throughout the body generally the larger arterial branches pursue a straight course; but in certain situations they are tortuous; thus he facial artery in its course over the face, and the arteries of the lips, are extremely tortuous in their course, to accommodate themselves to the movements of the parts. The uterine arteries are also tortuous, to accommodate themselves to the increase of size which the organ undergoes during pregnancy. Again, the internal carotid and vertebral arteries, previous to their entering the cavity of the skull, describe a series of curves, which are probably intended to diminish the velocity of the current of blood, by increasing the extent of surface over which it moves, and adding to the amount of impediment which is produced by friction.

The arteries are dense in structure, of considerable strength, highly elastic, and, when divided, they preserve, although empty, their cylindrical form.

In the description of the arteries, we shall first consider the efferent trunk of the pulmonic circulation, the pulmonary artery; and then the efferent trunk of the systemic circulation, the aorta, and its branches.

PULMONARY ARTERY (fig. 280)

The **pulmonary artery** conveys the venous blood from the right side of the heart to the lungs. It is a short, wide vessel, about two inches in length and 1¼ inch (30 mm.) in diameter, arising from the left side of the base (conus arteriosus) of the right ventricle, in front of the aorta. It extends obliquely upwards and backwards, passing at first in front of and then to the left of the ascending aorta, as far as the under surface of the arch, where it divides, about on a level with the intervertebral substance between the fifth and sixth dorsal vertebræ, into two branches of nearly equal size, the *right* and *left pulmonary arteries*.

Relations.—The whole of this vessel is contained, together with the ascending aorta, in the pericardium. It is enclosed with the aorta in a single tube of the serous pericardium, which is continued upwards upon them from the base of the heart and connects them together. The fibrous layer of the pericardium becomes gradually lost upon the external coat of its two branches. In front, the pulmonary artery is separated from the anterior extremity of the second left intercostal space by the pleura and left lung, in addition to the pericardium; it rests at first upon the ascending aorta, and higher up lies in front of the left auricle on a plane posterior to the ascending aorta. On each side of its origin is the appendix of the corresponding auricle and a coronary artery, the left coronary artery passing, in the first part of its course, behind the vessel.

The **right pulmonary artery,** longer and larger than the left, runs horizontally outwards, behind the ascending aorta and superior vena cava, to the root of the right lung, where it divides into two branches, of which the lower and larger supplies the middle and lower lobes; the upper and smaller is distributed to the upper lobe.

The **left pulmonary artery,** shorter and somewhat smaller than the right, passes horizontally in front of the descending aorta and left bronchus to the root of the left lung, where it divides into two branches for the two lobes.

The root of the left pulmonary artery is connected to the under surface of the arch of the aorta by a short fibrous cord, the *ligamentum arteriosum*; this is the remains of a vessel peculiar to fœtal life, the *ductus arteriosus*.

The terminal branches of the pulmonary artery will be described with the anatomy of the lung.

THE AORTA

The **aorta** (ἀορτή, *arteria magna*) is the main trunk of a series of vessels which convey the oxygenated blood to the tissues of the body for their nutrition. This vessel commences at the upper part of the left ventricle, where it is about 1⅓ inch in diameter, and after ascending for a short distance, arches backwards, and to the left side, over the root of the left lung, then descends within the thorax on the left side of the vertebral column, passes through the aortic opening

FIG. 280.—The arch of the aorta and its branches.

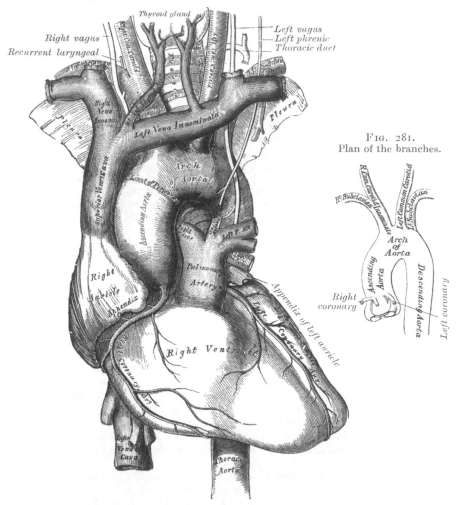

FIG. 281.
Plan of the branches.

in the Diaphragm, and entering the abdominal cavity, terminates, considerably diminished in size (about $\frac{7}{10}$ of an inch in diameter), opposite the lower border of the fourth lumbar vertebra, where it divides into the right and left common iliac arteries. Hence it is divided into the *ascending aorta*, the *arch* of the aorta, and the *descending aorta*, which last is again divided into *thoracic* and *abdominal aorta*, from the position of these parts.

ASCENDING AORTA

The **ascending aorta** is about two inches in length. It commences at the upper part of the base of the left ventricle, on a level with the lower border of the third costal cartilage behind the left half of the sternum; it passes obliquely upwards, forwards, and to the right, in the direction of the heart's axis, as high as the upper border of the second right costal cartilage, describing a slight curve in its course, and being situated, when distended, about a quarter of an inch behind the posterior surface of the sternum. A little above its commencement it is somewhat enlarged, and presents three small dilatations called the sinuses of Valsalva, opposite to which are attached the three semilunar valves, which serve the purpose of preventing any regurgitation of blood into the cavity of the ventricle. These valves are placed one in front and two behind. At the union of the ascending with the transverse part of the aorta the calibre of the vessel is increased, owing to a bulging outwards of its right wall. This dilatation is termed the *great sinus of the aorta*. A section of the aorta opposite this part has a somewhat oval figure; but below the attachment of the valves it is circular. This portion of the aorta is contained in the cavity of the pericardium, and, together with the pulmonary artery, is invested in a tube of serous membrane, continued on to them from the surface of the heart.

Relations.—The ascending aorta is covered at its commencement by the trunk of the pulmonary artery and the right auricular appendix, and, higher up, is separated from the sternum by the pericardium, the right pleura, and anterior margin of right lung, some loose areolar tissue, and the remains of the thymus gland; *behind*, it rests upon the right pulmonary artery and left auricle. On the *right side*, it is in relation with the superior vena cava and right auricle; on the *left side*, with the pulmonary artery.

PLAN OF THE RELATIONS OF THE ASCENDING AORTA

In front.

Pulmonary artery.
Right auricular appendix.
Pericardium.
Right pleura and lung.
Remains of thymus gland.

Right side.

Superior vena cava.
Right auricle.

Ascending Aorta.

Left side.

Pulmonary artery.

Behind.

Right pulmonary artery.
Left auricle.

Branches.—The only branches of the ascending aorta are the coronary arteries. They supply the heart, and are two in number, right and left, arising near the commencement of the aorta immediately above the free margin of the semilunar valves.

The **Right Coronary Artery,** about the size of a crow's quill, arises from the anterior sinus of Valsalva. It passes forwards between the pulmonary artery and the right auricular appendix, then runs obliquely to the right side, in the groove between the right auricle and ventricle, and curving around the right border of the heart, runs along its posterior surface as far as the posterior interventricular groove, where it divides into two branches, one of which (*transverse*) continues onwards in the groove between the left auricle and ventricle, and anastomoses with the left coronary ; the other (*descending*) courses along the posterior inter-ventricular furrow, supplying branches to both ventricles and to the septum, and anastomosing at the apex of the heart with the descending branches of the left coronary.

This vessel sends a large branch (*marginal*) along the thin margin of the right ventricle to the apex ; which in its course gives off numerous small branches to the anterior and posterior surfaces of the ventricle. It also gives off a branch (*infundibular*) which ramifies over the front part of the conus arteriosus of the right ventricle.

The **Left Coronary,** larger than the former, arises from the left posterior sinus of Valsalva ; it passes forwards between the pulmonary artery and the left auricular appendix, and divides into two branches. Of these, one (*transverse*) passes transversely outwards in the left auriculo-ventricular groove, and winds around the left border of the heart to its posterior surface, where it ana-stomoses with the transverse branch of the right coronary ; the other (*descending*) passes along the anterior interventricular groove to the apex of the heart, where it anastomoses with the descending branches of the right coronary. The left coronary supplies the left auricle and its appendix, gives branches to both ventricles, and numerous small branches to the pulmonary artery, and commencement of the aorta.*

Peculiarities.—These vessels occasionally arise by a common trunk, or their number may be increased to three ; the additional branch being of small size. More rarely, there are two additional branches.

ARCH OF THE AORTA

The **arch,** or **transverse aorta,** commences at the upper border of the second chondro-sternal articulation of the right side, and passes at first upwards and backwards and from right to left, and then from before backwards, to the left side of the lower border of the fourth dorsal vertebra behind. Its upper border is usually about an inch below the upper margin of the sternum.

Between the origin of the left subclavian artery and the attachment of the ductus arteriosus the lumen of the fœtal aorta is considerably narrowed, forming what is termed the *aortic isthmus,* while immediately beyond the ductus arte-riosus the vessel presents a fusiform dilatation which His has named the *aortic spindle*—the point of junction of the two parts being marked in the concavity of the arch by an indentation or angle. These conditions persist, to some extent, in the adult, where His found that the average diameter of the spindle exceeded that of the isthmus by 3 mm. (about one-eighth of an inch).

Relations.—Its *anterior surface* is covered by the pleuræ and lungs and the remains of the thymus gland, and crossed towards the left side by the left pneumogastric and phrenic nerves, and superficial cardiac branches of the left sympathetic and vagus, and by the left superior intercostal vein. Its *posterior surface* lies on the trachea, just above its bifurcation, on the great, or deep, cardiac plexus, the œsophagus, thoracic duct, and left recurrent laryngeal nerve. Its *upper border* is in relation with the left innominate vein ; and from its upper part are given off the innominate, left common carotid, and left subclavian arteries. Its *lower border* is in relation with the bifurcation of the pulmonary artery, the remains of the ductus arteriosus, which is connected with the left division of that vessel, and the superficial cardiac plexus ; the left recurrent laryngeal nerve winds round it from before backwards, while the left bronchus passes below it.

* According to Dr. Samuel West, there is a very free and complete anastomosis between the two coronary arteries (*Lancet,* June 2, 1883, p. 945). This, however, is not the view generally held by anatomists, for, with the exception of the anastomosis men-tioned above in the auriculo-ventricular and interventricular grooves, it is believed that the two arteries only communicate by very small vessels in the substance of the heart.

PLAN OF THE RELATIONS OF THE ARCH OF THE AORTA

Above.

Left innominate vein.
Innominate artery.
Left carotid.
Left subclavian.

In front.

Pleuræ and lungs.
Remains of thymus gland.
Left pneumogastric nerve.
Left phrenic nerve.
Superficial cardiac nerves.
Left superior intercostal vein.

Arch of
Aorta.

Behind.

Trachea.
Deep cardiac plexus.
Œsophagus.
Thoracic duct.
Left recurrent nerve.

Below.

Bifurcation of pulmonary artery.
Remains of ductus arteriosus.
Superficial cardiac plexus.
Left recurrent nerve.
Left bronchus.

Peculiarities.—The height to which the aorta rises in the chest is usually about an inch below the upper border of the sternum; but it may ascend nearly to the top of that bone. Occasionally it is found an inch and a half, more rarely two or even three inches below this point.

FIG. 282.—Relation of great vessels at base of heart, seen from above. (From a preparation in the Museum of the Royal College of Surgeons of England.)

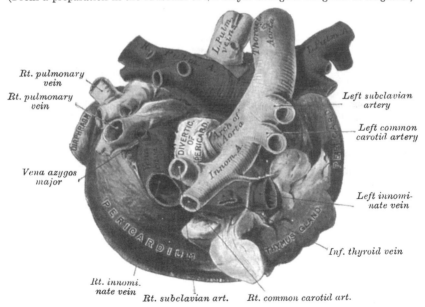

In Direction.—Sometimes the aorta arches over the root of the right instead of the left lung, as in birds, and passes down on the right side of the spine. In such cases all the viscera of the thoracic and abdominal cavities are transposed. Less frequently, the aorta, after arching over the root of the right lung, is directed to its usual position on the left side of the spine, this peculiarity not being accompanied by any transposition of the viscera.

In Conformation.—The aorta occasionally divides, as in some quadrupeds, into an ascending and a descending trunk, the former of which is directed vertically upwards, and subdivides into three branches, to supply the head and upper extremities. Sometimes the aorta subdivides soon after its origin into two branches, which soon reunite. In one of these cases the œsophagus and trachea were found to pass through the interval left by their division; this is the normal condition of the vessel in the reptilia.

Surgical Anatomy.—Of all the vessels of the arterial system, the aorta, and more especially its arch, is most frequently the seat of disease ; hence it is important to consider some of the consequences that may ensue from aneurism of this part.

It will be remembered that the ascending aorta is contained in the pericardium, just behind the sternum, being crossed at its commencement by the pulmonary artery and right auricular appendix, and having the right pulmonary artery behind, the vena cava on the right side, and the pulmonary artery and left auricle on the left side.

Aneurism of the ascending aorta, in the situation of the sinuses of Valsalva, in the great majority of cases, affects the anterior sinus ; this is mainly owing to the fact that the regurgitation of blood upon the sinuses takes place chiefly on the anterior aspect of the vessel. As the aneurismal sac enlarges, it may compress any or all of the structures in immediate proximity with it, but chiefly projects towards the right anterior side ; and, consequently, interferes mainly with those structures that have a corresponding relation with the vessel. In the majority of cases, it bursts into the cavity of the pericardium, the patient suddenly drops down dead, and, upon a post-mortem examination, the pericardial sac is found full of blood ; or it may compress the right auricle, or the pulmonary artery, and adjoining part of the right ventricle, and open into one or the other of these parts, or it may press upon the superior vena cava.

Aneurism of the ascending aorta, originating above the sinuses, most frequently implicates the right anterior wall of the vessel, where, as has been explained, there exists a normal dilatation, the great sinus of the aorta ; this is probably mainly owing to the blood being impelled against this part. The direction of the aneurism is also chiefly towards the right of the median line. If it attains a large size and projects forwards, it may absorb the sternum and the cartilages of the ribs, usually on the right side, and appear as a pulsating tumour on the front of the chest, just below the manubrium ; or it may burst into the pericardium, or may compress, or open into the right lung, the trachea, bronchi, or œsophagus.

Regarding the transverse aorta, the student is reminded that the vessel lies on the trachea, the œsophagus, and thoracic duct ; that the recurrent laryngeal nerve winds around it ; and that from its upper part are given off three large trunks, which supply the head, neck, and upper extremities. Now, an aneurismal tumour taking origin from the posterior part of the vessel, its most usual site, may press upon the trachea, impede the breathing, or produce cough, hæmoptysis, or stridulous breathing, or it may ultimately burst into that tube, producing fatal hæmorrhage. Again, its pressure on the laryngeal nerves may give rise to symptoms which so accurately resemble those of laryngitis, that the operation of tracheotomy has in some cases been resorted to, from the supposition that disease existed in the larynx ; or it may press upon the thoracic duct and destroy life by inanition ; or it may involve the œsophagus, producing dysphagia ; or may burst into the œsophagus, when fatal hæmorrhage will occur. Again, the innominate artery, or the subclavian, or left carotid, may be so obstructed by clots as to produce a weakness, or even a disappearance, of the pulse in one or the other wrist, or in the left temporal artery ; or the tumour may present itself at or above the manubrium, generally either in the median line, or to the right of the sternum, and may simulate an aneurism of one of the arteries of the neck.

Branches (figs. 280, 281).—The branches given off from the arch of the aorta are three in number : the innominate artery, the left common carotid, and the left subclavian.

Peculiarities. Position of the Branches.—The branches, instead of arising from the highest part of the arch (their usual position), may be moved more to the right, arising from the commencement of the transverse or upper part of the ascending portion ; or the distance from one another at their origin may be increased or diminished, the most frequent change in this respect being the approximation of the left carotid towards the innominate artery.

The Number of the primary branches may be reduced to a single vessel, or more commonly two : the left carotid arising from the innominate artery ; or (more rarely), the carotid and subclavian arteries of the left side arising from a left innominate artery. But the number may be increased to four, from the right carotid and subclavian arteries arising directly from the aorta, the innominate being absent. In most of these latter cases the right subclavian has been found to arise from the left end of the arch ; in other cases it was the second or third branch given off instead of the first. Another common form in which there are four primary branches is that in which the left vertebral artery arises from the arch of the aorta between the left carotid and subclavian arteries. Lastly, the number of trunks from the arch may be increased to five or six ; in these instances, the external and internal carotids arise separately from the arch, the common carotid being absent on one or both sides. In some cases, where six branches have been found, it has been due to a separate origin of the vertebral on both sides.

Number usual, Arrangement different.—When the aorta arches over to the right side, the three branches have an arrangement the reverse of what is usual, the innominate supplying the left side, and the carotid and subclavian (which arise separately) the right side. In other cases, where the aorta takes its usual course, the two carotids may be joined in a common trunk, and the subclavians arise separately from the arch, the right subclavian generally arising from the left end of the arch.*

In some instances other arteries are found to arise from the arch of the aorta. Of these the most common are the bronchial, one or both, and the thyroidea ima; but the internal mammary and the inferior thyroid have been seen to arise from this vessel.

INNOMINATE ARTERY

The **innominate artery (brachio-cephalic)** is the largest branch given off from the arch of the aorta. It arises, on a level with the upper border of the second right costal cartilage, from the commencement of the arch of the aorta in front of the left carotid, and, ascending obliquely to the upper border of the right sterno-clavicular articulation, divides into the right common carotid and right subclavian arteries. This vessel varies from an inch and a half to two inches in length.

Relations.—*In front*, it is separated from the first piece of the sternum by the Sterno-hyoid and Sterno-thyroid muscles, the remains of the thymus gland, the left innominate and right inferior thyroid veins which cross its root, and sometimes the inferior cervical cardiac branch of the right pneumogastric. *Behind*, it lies upon the trachea, which it crosses obliquely. On the *right side* is the right innominate vein, right pneumogastric nerve, and the pleura; and on the *left side*, the remains of the thymus gland, the origin of the left carotid artery, the left inferior thyroid vein, and the trachea.

Branches.—The innominate usually gives off no branches; but occasionally a small branch, the *thyroidea ima*, is given off from this vessel. It also sometimes gives off a *thymic* or *bronchial branch*. The **Thyroidea ima** ascends in front of the trachea to the lower part of the thyroid body, which it supplies. It varies greatly in size, and appears to compensate for deficiency or absence of one of the other thyroid vessels. It occasionally is found to arise from the right common carotid or from the aorta, the subclavian or internal mammary artery.

PLAN OF THE RELATIONS OF THE INNOMINATE ARTERY

In front.

Sternum.
Sterno-hyoid and Sterno-thyroid muscles.
Remains of thymus gland.
Left innominate and right inferior thyroid veins.
Inferior cervical cardiac branch from right pneumogastric nerve.

Right side.		*Left side.*
Right innominate vein.	Innominate Artery.	Remains of thymus.
Right pneumogastric nerve.		Left carotid.
Pleura.		Left inferior thyroid vein.
		Trachea.

Behind.
Trachea.

Peculiarities in point of Division.—When the bifurcation of the innominate artery varies from the point above mentioned, it sometimes ascends a considerable distance above the sternal end of the clavicle; less frequently it divides below it. In the former class of cases, its length may exceed two inches; and, in the latter, be reduced to an inch or less. These are points of considerable interest for the surgeon to remember in connection with the operation of tying this vessel.

Position.—When the aorta arches over to the right side, the innominate is directed to the left side of the neck instead of the right.

Collateral Circulation.—Allan Burns demonstrated, on the dead subject, the possibility of the establishment of the collateral circulation after ligature of the innominate artery, by tying and dividing that artery, after which, he says, ' Even coarse injection, impelled into the aorta, passed freely by the anastomosing branches into the arteries of the right arm, filling them and all the vessels of the head completely.' (*Surgical Anatomy of the*

* The anomalies of the aorta and its branches are minutely described by Krause in Henle's *Anatomy* (Brunswick, 1868), vol. iii. p. 203 *et seq.*

Head and Neck, p. 62.) The branches by which this circulation would be carried on are very numerous ; thus, all the communications across the middle line between the branches of the carotid arteries of opposite sides would be available for the supply of blood to the right side of the head and neck ; while the anastomosis between the superior intercostal of the subclavian and the first aortic intercostal (see *infra* on the collateral circulation after obliteration of the thoracic aorta) would bring the blood, by a free and direct course, into the right subclavian : the numerous connections, also, between the intercostal arteries and the branches of the axillary and internal mammary arteries would, doubtless, assist in the supply of blood to the right arm, while the deep epigastric, from the external iliac, would, by means of its anastomosis with the internal mammary, compensate for any deficiency in the vascularity of the wall of the chest.

Surgical Anatomy.—Although the operation of tying the innominate artery has been performed by several surgeons, for aneurism of the right subclavian extending inwards as far as the Scalenus, in only five instances, according to Mr. Jacobson, has the patient survived. Mott's patient, however, on whom the operation was first performed, lived nearly four weeks, and Graefe's more than two months. The chief danger of the operation appears to be the frequency of secondary hæmorrhage ; but in the present day, with the practice of aseptic surgery and our greater knowledge of the use of the ligature, more favourable results may be anticipated. Other causes of death after operation are pleurisy, pericarditis, and suppurative cellulitis. The main obstacles to the operation are, as the student will perceive from his dissection of this vessel, the deep situation of the artery behind and beneath the sternum, and the number of important structures which surround it in every part.

In order to apply a ligature to this vessel, the patient is to be placed upon his back with the thorax slightly raised, the head bent a little backwards, and the shoulder on the side of the aneurism strongly depressed, so as to draw out the artery from behind the sternum into the neck. An incision three or more inches long is then made along the anterior border of the Sterno-mastoid muscle, terminating at the sternal end of the clavicle. From this point, a second incision is carried about the same length along the upper border of the clavicle. The skin is then dissected back, and the Platysma divided on a director : the sternal end of the Sterno-mastoid is now brought into view, and a director being passed beneath it, and close to its under surface, so as to avoid any small vessels, it is to be divided ; in like manner the clavicular origin is to be divided throughout the whole or greater part of its attachment. By pressing aside any loose cellular tissue or vessels that may now appear, the Sterno-hyoid and Sterno-thyroid muscles will be exposed, and must be divided, a director being previously passed beneath them. The inferior thyroid veins may come into view, and must be carefully drawn either upwards or downwards, by means of a blunt hook, or tied with double ligatures and divided. After tearing through a strong fibro-cellular lamina, the right carotid is brought into view, and being traced downwards, the arteria innominata is arrived at. The left innominate vein should now be depressed ; the right innominate vein, the internal jugular vein, and the pneumogastric nerve drawn to the right side ; and a curved aneurism needle may then be passed around the vessel, close to its surface, and in a direction from below upwards and inwards ; care being taken to avoid the right pleural sac, the trachea, and cardiac nerves. The ligature should be applied to the artery as high as possible, in order to allow room between it and the aorta for the formation of the coagulum. The importance of avoiding the thyroid plexus of veins during the primary steps of the operation, and the pleural sac while including the vessel in the ligature, should be most carefully borne in mind. After the artery has been secured, the common carotid should be tied about half an inch above its origin, and also the thyroidea ima if the vessel is of any size. The severed muscles are united by buried sutures.

ARTERIES OF THE HEAD AND NECK

The artery which supplies the head and neck is the Common Carotid ; it ascends in the neck and divides into two branches : the External Carotid, supplying the superficial parts of the head and face and the greater part of the neck ; and the Internal Carotid, supplying to a great extent the parts within the cranial cavity.

COMMON CAROTID ARTERIES

The **common carotid arteries,** although occupying a nearly similar position in the neck, differ in position, and, consequently, in their relations at their origin. The right carotid arises from the innominate artery, behind the right sterno-clavicular articulation ; the left from the highest part of the arch of the aorta. The left carotid is, consequently, longer, and at its origin is contained within the thorax. The course and relation of that portion of the left carotid which intervenes between the arch of the aorta and the left sterno-clavicular articulation, will first be described (see fig. 280).

The *left carotid within the thorax* ascends obliquely outwards from the arch of the aorta to the root of the neck. *In front*, it is separated from the first piece of the sternum by the Sterno-hyoid and Sterno-thyroid muscles, the left innominate vein, and the remains of the thymus gland; *behind*, it lies on the trachea, œsophagus, and thoracic duct. *Internally*, it is in relation with the innominate artery, inferior thyroid veins and remains of Thymus gland; *externally*, with the left pneumogastric nerve, left pleura, and lung. The left subclavian artery is posterior and slightly external to it.

PLAN OF THE RELATIONS OF THE LEFT COMMON CAROTID.
THORACIC PORTION

In front.

Sternum.
Sterno-hyoid and Sterno-thyroid muscles.
Left innominate vein.
Remains of thymus gland.

Internally.	Left Common Carotid. Thoracic Portion.	*Externally.*
Innominate artery. Inferior thyroid veins. Remains of thymus gland.		Left pneumogastric nerve. Left pleura and lung. Left subclavian artery.

Behind.

Trachea.
Œsophagus.
Thoracic duct.
Left subclavian artery.

In the neck, the two common carotids resemble each other so closely, that one description will apply to both. Each vessel passes obliquely upwards, from behind the sterno-clavicular articulation, to a level with the upper border of the thyroid cartilage, opposite the fourth cervical vertebra, where it divides into the external and internal carotid; these names being derived from the distribution of the arteries to the external parts of the head and face, and to the internal parts of the cranium and orbit respectively.

At the lower part of the neck the two common carotid arteries are separated from each other by a very small interval, which contains the trachea; but at the upper part, the thyroid body, the larynx and pharynx project forwards between the two vessels, and give the appearance of their being placed farther back in this situation. The common carotid artery is contained in a sheath, derived from the deep cervical fascia, which also encloses the internal jugular vein and pneumogastric nerve, the vein lying on the outer side of the artery, and the nerve between the artery and vein, on a plane posterior to both. On opening the sheath, these three structures are seen to be separated from one another, each being enclosed in a separate fibrous investment.

Relations.—At the lower part of the neck the common carotid artery is very deeply seated, being covered by the integument, superficial fascia, Platysma, and deep cervical fascia, the Sterno-mastoid, Sterno-hyoid, and Sterno-thyroid muscles, and by the Omo-hyoid, opposite the cricoid cartilage; but in the upper part of its course, near its termination, it is more superficial, being covered merely by the integument, the superficial fascia, Platysma, deep cervical fascia, and inner margin of the Sterno-mastoid, and, when the latter is drawn backwards, it is seen to be contained in a triangular space, bounded behind by the Sterno-mastoid, above by the posterior belly of the Digastric, and below by the anterior belly of the Omo-hyoid. This part of the artery is crossed obliquely, from within outwards, by the sterno-mastoid artery; it is crossed also by the superior and middle thyroid veins, which terminate in the internal jugular ; and descending on its sheath in front is seen the descendens hypoglossi nerve, this filament being joined by one or two branches from the cervical nerves, which cross the vessel from without inwards. Sometimes the descendens hypoglossi is contained within the sheath. The middle thyroid vein crosses the artery about its middle, and the anterior jugular vein below ; the latter, however, is separated from the artery by the Sterno-hyoid and Sterno-thyroid muscles. *Behind*, the artery is separated from

Fig. 283.—Superficial dissection of the right side of the neck, showing the carotid and subclavian arteries.

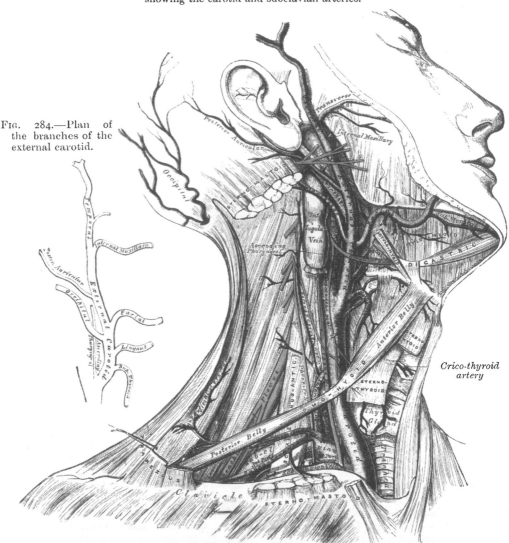

Fig. 284.—Plan of the branches of the external carotid.

Crico-thyroid artery

the transverse processes of the vertebræ by the Longus colli and Rectus capitis anticus major, the sympathetic nerve being interposed between it and the muscles. The recurrent laryngeal nerve and inferior thyroid artery cross behind the vessel at its lower part. *Internally*, it is in relation with the trachea and thyroid gland, the latter overlapping it, the inferior thyroid artery and recurrent laryngeal nerve being interposed : higher up, with the larynx and pharynx. On its *outer side* are placed the internal jugular vein and pneumogastric nerve.

At the lower part of the neck, the internal jugular vein on the right side diverges from the artery, but on the left side it approaches it, and often overlaps its lower part. This is an important fact to bear in mind during the performance of any operation on the lower part of the left common carotid artery.

PLAN OF THE RELATIONS OF THE COMMON CAROTID ARTERY

In front.

Integument, and superficial fascia. | Omo-hyoid.
Deep cervical fascia. | Descendens and Communicans hypoglossi
Platysma. | Sterno-mastoid artery. [nerves.
Sterno-mastoid. | Superior and middle thyroid veins.
Sterno-hyoid. | Anterior jugular vein.
Sterno-thyroid.

Externally.

Internal jugular vein.
Pneumogastric nerve.

Common Carotid.

Internally.

Trachea.
Thyroid gland.
Recurrent laryngeal nerve.
Inferior thyroid artery.
Larynx.
Pharynx.

Behind.

Longus colli | Sympathetic nerve.
Rectus capitis anticus major. | Inferior thyroid artery.
Recurrent laryngeal nerve.

Peculiarities as to Origin.—The *right common carotid* may arise above or below the upper border of the sterno-clavicular articulation. This variation occurs in one out of about eight cases and a half, and the origin is more frequently below than above: or the artery may arise as a separate branch from the arch of the aorta, or in conjunction with the left carotid. The *left common carotid* varies more frequently in its origin than the right. In the majority of abnormal cases it arises with the innominate artery, or, if the innominate artery is absent, the two carotids arise usually by a single trunk. It rarely joins with the left subclavian, except in cases of transposition of the arch.

Peculiarities as to point of Division.—The most important peculiarities of this vessel, from a surgical point of view, relate to its place of division in the neck. In the majority of abnormal cases, this occurs higher than usual, the artery dividing into two branches opposite the hyoid bone, or even higher; more rarely, it occurs below, opposite the middle of the larynx, or the lower border of the cricoid cartilage; and one case is related by Morgagni, where the common carotid, only an inch and a half in length, divided at the root of the neck. Very rarely, the common carotid ascends in the neck without any subdivision, the internal carotid being wanting; and in a few cases the common carotid has been found to be absent, the external and internal carotids arising directly from the arch of the aorta. This peculiarity existed on both sides in some instances, on one side in others.

Occasional Branches.—The common carotid usually gives off no branch previous to its bifurcation; but it occasionally gives origin to the superior thyroid, or its laryngeal branch, the ascending pharyngeal, the inferior thyroid, or, more rarely, the vertebral artery.

Surface Marking.—The carotid arteries are covered throughout their entire extent by the Sterno-mastoid muscle, but their course does not correspond to the anterior border of the muscle, which passes in a somewhat curved direction from the mastoid process to the sterno-clavicular joint. The course of the artery is indicated more exactly by a line drawn from the sternal end of the clavicle below, to a point midway between the angle of the jaw and the mastoid process above. The portion of this line below the level of the upper border of the thyroid cartilage would represent the course of the vessel.

Surgical Anatomy.—The operation of tying the common carotid artery may be necessary in a case of wound of that vessel or its branches, in aneurism, or in a case of pulsating tumour of the orbit or skull. If the wound involves the trunk of the common carotid, it will be necessary to tie the artery above and below the wounded part. But in cases of aneurism, or where one of the branches of the common carotid is wounded in an inaccessible situation, it may be judged necessary to tie the trunk. In such cases, the whole of the artery is accessible, and any part may be tied, except close to either end. When the case is such as to allow of a choice being made, the lower part of the carotid should never be selected as the spot upon which to place a ligature, for not only is the artery in this situation placed very deeply in the neck, but it is covered by three layers of muscles, and, on the left side, the internal jugular vein, in the great majority of cases, passes obliquely in front of it. Neither should the upper end be selected, for here the superior thyroid vein and its tributaries would give rise to very considerable difficulty in the application of a ligature. The point most favourable for the operation is that part of the vessel which is at the level of the cricoid cartilage. It occasionally happens that the carotid artery bifurcates below its usual position : if the artery be exposed at its point of bifurcation, both divisions of the vessel should be tied near their origin, in preference to tying the trunk of the artery near its termination ; and if, in consequence of the entire absence of the common carotid, or from its early division, two arteries, the external and internal carotids, are met with, the ligature should be placed on that vessel which is found on compression to be connected with the disease.

In this operation, the direction of the vessel and the inner margin of the Sterno-mastoid are the chief guides to its performance. The patient should be placed on his back with the head thrown back and turned slightly to the opposite side : an incision is to be made, three inches long, in the direction of the anterior border of the Sterno-mastoid, so that the centre corresponds to the level of the cricoid cartilage : after dividing the integument, superficial fascia, and Platysma, the deep fascia must be cut through on a director, so as to avoid wounding numerous small veins that are usually found beneath. The head may now be brought forwards so as to relax the parts somewhat, and the margins of the wound held asunder by retractors. The descendens hypoglossi nerve may now be exposed, and must be avoided, and the sheath of the vessel having been raised by forceps, is to be opened to a small extent over the artery at its inner side. The internal jugular vein may present itself alternately distended and relaxed ; this should be compressed both above and below, and drawn outwards, in order to facilitate the operation. The aneurism needle is passed from the outside, care being taken to keep the needle in close contact with the artery, and thus avoid the risk of injuring the internal jugular vein, or including the vagus nerve. Before the ligature is tied, it should be ascertained that nothing but the artery is included in it.

Ligature of the Common Carotid at the Lower Part of the Neck.—This operation is sometimes required in cases of aneurism of the upper part of the carotid, especially if the sac is of large size. It is best performed by dividing the sternal origin of the Sterno-mastoid muscle, but may be done in some cases, if the aneurism is not of very large size, by an incision along the anterior border of the Sterno-mastoid, extending down to the sterno-clavicular articulation, and by then retracting the muscle. The easiest and best plan, however, is to make an incision two or three inches long down the lower part of the anterior border of the Sterno-mastoid muscle to the sterno-clavicular joint, and a second incision, starting from the termination of the first, along the upper border of the clavicle for about two inches. This incision is made through the superficial and deep fascia, and the sternal origin of the muscle exposed. This is to be divided on a director and turned up, with the superficial structures, as a triangular flap. Some loose connective tissue is to be divided or torn through, and the outer border of the Sterno-hyoid muscle exposed. In doing this, care must be taken not to wound the anterior jugular vein, which crosses the muscle to reach the external jugular or subclavian vein. The Sterno-hyoid, and with it the Sterno-thyroid, are to be drawn inwards by means of a retractor, and the sheath of the vessel is exposed. This must be opened with great care on its inner or tracheal side, so as to avoid the internal jugular vein. This is especially necessary on the left side, where the artery is commonly overlapped by the vein. On the right side there is usually an interval between the artery and the vein, and not the same risk of wounding the latter.

The common carotid artery, being a long vessel without any branches, is particularly suitable for the performance of Brasdor's operation for the cure of an aneurism of the lower part of the vessel. Brasdor's procedure consists in ligaturing the artery on the distal side of the aneurism, and in the case of the common carotid there are no branches given off from the vessel between the aneurism and the site of the ligature, hence the flow of blood through the sac of the aneurism is diminished and cure takes place in the usual way, by the deposit of laminated fibrin.

Collateral Circulation.—After ligature of the common carotid, the collateral circulation can be perfectly established, by the free communication which exists between the carotid arteries of opposite sides, both without and within the cranium, and by enlargement of the branches of the subclavian artery on the side corresponding to that on which the vessel has been tied—the chief communication outside the skull taking place between the superior and inferior thyroid arteries, and the profunda cervicis and arteria princeps cervicis of the occipital ; the vertebral taking the place of the internal carotid within the cranium.

Sir A. Cooper had an opportunity of dissecting, thirteen years after the operation, the case in which he first successfully tied the common carotid (the second case in which he performed the operation).* The injection, however, does not seem to have been a successful one. It showed merely that the arteries at the base of the brain (circle of Willis) were much enlarged on the side of the tied artery, and that the anastomosis between the branches of the external carotid on the affected side and those of the same artery on the sound side was free, so that the external carotid was pervious throughout.

* *Guy's Hospital Reports*, i. 56.

External Carotid Artery

The **external carotid artery** (fig. 283) commences opposite the upper border of the thyroid cartilage, and, taking a slightly curved course, passes upwards and forwards, and then inclines backwards to the space between the neck of the condyle of the lower jaw and the external meatus, where it divides into the temporal and internal maxillary arteries. It rapidly diminishes in size in its course up the neck, owing to the number and large size of the branches given off from it. In the child, it is somewhat smaller than the internal carotid; but in the adult, the two vessels are of nearly equal size. At its commencement, this artery is more superficial, and placed nearer the middle line than the internal carotid, and is contained in the triangular space bounded by the Sterno-mastoid behind, the Omo-hyoid below, and the posterior belly of the Digastric and Stylo-hyoid above.

Relations.—It is covered by the skin, superficial fascia, Platysma, deep fascia, and anterior margin of the Sterno-mastoid, crossed by the hypoglossal nerve, and by the lingual and facial veins; it is afterwards crossed by the Digastric and Stylo-hyoid muscles, and higher up passes deeply into the substance of the parotid gland, where it lies beneath the facial nerve and the junction of the temporal and internal maxillary veins. *Internally* is the hyoid bone, wall of the pharynx, the superior laryngeal nerve, and the ramus of the jaw, from which it is separated by a portion of the parotid gland. *Externally*, in the lower part of its course, is the internal carotid artery. *Behind* it, near its origin, is the superior laryngeal nerve; and higher up, it is separated from the internal carotid by the Stylo-glossus and Stylo-pharyngeus muscles, the glosso-pharyngeal nerve, and part of the parotid gland.

PLAN OF THE RELATIONS OF THE EXTERNAL CAROTID

In front.

Skin, superficial fascia.
Platysma and deep fascia.
Anterior border of Sterno-mastoid.
Hypoglossal nerve.
Lingual and facial veins.
Digastric and Stylo-hyoid muscles.   [substance.
Parotid gland with facial nerve and·temporo-maxillary vein in its

Internally.

Hyoid bone.
Pharynx.
Superior laryngeal nerve.
Parotid gland.
Ramus of jaw.

External Carotid.

Externally.
Internal carotid artery.

Behind.

Superior laryngeal nerve.
Stylo-glossus.
Stylo-pharyngeus.
Glosso-pharyngeal nerve.
Parotid gland.

Surface Marking.—The position of the external carotid artery may be marked out with sufficient accuracy by a line drawn from the front of the meatus of the external ear to the side of the cricoid cartilage, slightly arching the line forwards.

Surgical Anatomy.—The application of a ligature to the external carotid may be required in cases of wound of this vessel, or of its branches when these cannot be tied, and in some cases of pulsating tumours of the scalp or face. The operation has not received the attention which it deserves, owing to the fear which surgeons have entertained of secondary hæmorrhage, on account of the number of branches given off from the vessel. Mr. Cripps, however, has shown that this fear is not well founded.* To tie this

* *Med.-Chir. Trans.* lxi. 229.

vessel near its origin, below the point where it is crossed by the Digastric, an incision about three inches in length should be made along the anterior margin of the Sterno-mastoid, from the angle of the jaw to the upper border of the thyroid cartilage. The ligature should be applied between the lingual and superior thyroid branches. To tie the vessel above the Digastric, between it and the parotid gland, an incision should be made, from the lobe of the ear to the great cornu of the os hyoides, dividing successively the skin, Platysma, and fascia. By drawing the Sterno-mastoid outwards, and the posterior belly of the Digastric and Stylo-hyoid muscles downwards, and separating them from the parotid gland, the vessel will be exposed, and a ligature may be applied to it. The circulation is at once re-established by the free communication between most of the large branches of the artery (facial, lingual, superior thyroid, occipital) and the corresponding arteries of the opposite side, and by the anastomosis of its branches with those of the internal carotid, and of the occipital with branches of the subclavian, &c.

Branches.—The external carotid artery gives off eight branches, which, for convenience of description, may be divided into four sets. (See fig. 284, Plan of the Branches.)

Anterior.	*Posterior.*	*Ascending.*	*Terminal.*
Superior Thyroid.	Occipital.	Ascending	Superficial Temporal.
Lingual.	Posterior Auricular.	Pharyngeal.	Internal Maxillary.
Facial.			

The student is here reminded that many variations are met with in the number, origin, and course of these branches in different subjects ; but the above arrangement is that which is found in the great majority of cases.

The **Superior Thyroid Artery** (figs. 283 and 288) is the first branch given off from the external carotid. being derived from that vessel just below the great cornu of the hyoid bone. At its commencement it is quite superficial, being covered by the integument, fascia, and Platysma, and is contained in the triangular space bounded by the Sterno-mastoid, Digastric, and Omo-hyoid muscles. After running upwards and inwards for a short distance, it curves downwards and forwards, in an arched and tortuous manner, to the upper part of the thyroid gland, passing beneath the Omo-hyoid, Sterno-hyoid, and Sterno-thyroid muscles and supplying them. It distributes numerous branches to the upper part of the gland, anastomosing with its fellow of the opposite side, and with the inferior thyroid arteries. The branches supplying the gland are generally three in number : one, the largest, supplies principally the anterior surface of the gland ; it courses along the inner border of the lobe as far as the upper border of the isthmus, and then passes in the substance of the isthmus to the middle line of the neck, where it anastomoses with the corresponding artery of the opposite side : a second branch courses along the external border of the lobe and supplies this portion of the gland, and the third passes to the posterior surface, the upper part of which it supplies. Besides the arteries distributed to the muscles by which it is covered and to the substance of the gland, the branches of the superior thyroid are the following :

Hyoid.	Superior laryngeal.
Superficial descending branch (Sterno-mastoid).	Crico-thyroid.

The **hyoid** (**infra-hyoid**) is a small branch which runs along the lower border of the os hyoides beneath the Thyro-hyoid muscle ; after supplying the muscles connected to that bone, it forms an arch, by anastomosing with the vessel of the opposite side.

The **superficial descending** or **Sterno-mastoid branch** runs downwards and outwards across the sheath of the common carotid artery, and supplies the Sterno-mastoid and neighbouring muscles and integument. There is frequently a separate branch from the external carotid distributed to the Sterno-mastoid muscle.

The **superior laryngeal,** larger than either of the preceding, accompanies the internal laryngeal nerve, beneath the Thyro-hyoid muscle ; it pierces the thyro-hyoid membrane, and supplies the muscles, mucous membrane, and glands of the larynx, anastomosing with the branch from the opposite side.

The **crico-thyroid** is a small branch which runs transversely across the crico-thyroid membrane, communicating with the artery of the opposite side.

Surgical Anatomy.—The superior thyroid, or one of its branches, is often divided in cases of cut throat, giving rise to considerable hæmorrhage. In such cases, the artery should be secured, the wound being enlarged for that purpose, if necessary. The operation may be easily performed, the position of the artery being very superficial, and the only structures of importance covering it being a few small veins. The operation of tying the superior thyroid artery, in bronchocele, has been performed in numerous instances with partial or temporary success. When, however, the collateral circulation between this vessel and the artery of the opposite side, and the inferior thyroid, is completely re-established, the tumour usually regains its former size, and hence the operation has been given up, especially as better results are obtained by other means. Both thyroid arteries on the same side, and indeed all the four thyroid arteries, have been tied in enlarged thyroid.

The position of the superficial descending branch is of importance in connection with the operation of ligature of the common carotid artery. It crosses and lies on the sheath of this vessel and may chance to be wounded in opening the sheath. The position of the crico-thyroid branch should be remembered, as it may prove the source of troublesome hæmorrhage during the operation of laryngotomy.

The **Lingual Artery** (fig. 288) arises from the external carotid between the superior thyroid and facial ; it first runs obliquely upwards and inwards to the great cornu of the hyoid bone ; it then curves downwards and forwards, forming a loop which is crossed by the hypoglossal nerve, and passing beneath the Digastric and Stylo-hyoid muscles, it runs horizontally forwards, beneath the Hyo-glossus, and finally, ascending almost perpendicularly to the tongue, turns forwards on its under surface as far as the tip, under the name of the *ranine artery.*

Relations.—Its first, or oblique, portion is superficial, being contained in the same triangular space as the superior thyroid artery, resting upon the Middle constrictor of the pharynx, and covered by the Platysma, and fascia of the neck. Its second, or curved, portion also lies upon the Middle constrictor, being covered at first by the tendon of the Digastric and the Stylo-hyoid muscle, and afterwards by the Hyo-glossus, the latter muscle separating it from the hypoglossal nerve. Its third, or horizontal, portion lies between the Hyo-glossus and Genio-hyo-glossus muscles. The fourth, or terminal, part, under the name of the *ranine*, runs along the under surface of the tongue to its tip : here it is very superficial, being covered only by the mucous membrane, and rests on the Lingualis, on the outer side of the Genio-hyo-glossus. The hypoglossal nerve crosses the lingual artery, and then becomes separated from it, in the second part of its course, by the Hyo-glossus muscle.

The branches of the lingual artery are : the

Hyoid.	Sublingual.
Dorsalis Linguæ.	Ranine.

The **hyoid branch** (**supra-hyoid**) runs along the upper border of the hyoid bone, supplying the muscles attached to it and anastomosing with its fellow of the opposite side.

The **dorsalis linguæ** (fig. 288) arises from the lingual artery beneath the Hyo-glossus muscle (which, in the figure, has been partly cut away, to show the vessel) ; it ascends to the dorsum of the tongue, and supplies the mucous membrane, the tonsil, soft palate, and epiglottis ; anastomosing with its fellow from the opposite side. This artery is frequently represented by two or three small branches.

The **sublingual**, which may be described as a branch of bifurcation of the lingual artery, arises at the anterior margin of the Hyo-glossus muscle, and runs forward between the Genio-hyo-glossus and the sublingual gland. It supplies the substance of the gland, giving branches to the Mylo-hyoid and neighbouring muscles, the mucous membrane of the mouth and gums. One branch runs behind the alveolar process of the lower jaw in the substance of the gum to anastomose with a similar artery from the other side.

The **ranine** may be regarded as the other branch of bifurcation, or, as is more usual, the continuation of the lingual artery; it runs along the under surface of the tongue, resting on the Inferior lingualis, and covered by the mucous membrane of the mouth; it lies on the outer side of the Genio-hyo-glossus, accompanied by the lingual nerve. On arriving at the tip of the tongue, it has been said to anastomose with the artery of the opposite side; but this is denied by Hyrtl. These vessels in the mouth are placed one on each side of the fraenum.

Surgical Anatomy.—The lingual artery may be divided near its origin in cases of cut throat, a complication that not infrequently happens in this class of wounds; or severe hæmorrhage, which cannot be restrained by ordinary means, may ensue from a wound, or deep ulcer, of the tongue. In the former case, the primary wound may be enlarged if necessary, and the bleeding vessel secured. In the latter case, it has been suggested that the lingual artery should be tied near its origin. Ligature of the lingual artery has been also occasionally practised, as a palliative measure, in cases of cancer of the tongue, in order to check the progress of the disease by starving the growth, and it is sometimes tied, as a preliminary measure to removal of the tongue. The operation is a difficult one, on account of the depth of the artery, the number of important parts by which it is surrounded, the loose and yielding nature of the parts upon which it is supported, and its occasional irregularity of origin. An incision is to be made in a curved direction from a finger's breadth external to the symphysis of the jaw downwards to the cornu of the hyoid bone, and then upwards to near the angle of the jaw. Care must be taken not to carry this incision too far backwards, for fear of endangering the facial vein. In the first incision the skin, superficial fascia, and Platysma will be divided, and the deep fascia exposed. This is then to be incised and the submaxillary gland exposed and pulled upwards by retractors. A triangular space is now exposed, bounded internally by the posterior border of the Mylo-hyoid muscle; below and externally, by the tendon of the Digastric; and above, by the hypoglossal nerve. The floor of the space is formed by the Hyo-glossus muscle, beneath which the artery lies. The fibres of this muscle are now to be cut through horizontally, and the vessel exposed, care being taken, while near the vessel, not to open the pharynx.

Troublesome hæmorrhage may occur in the division of the fraenum in children, if the ranine artery, which lies on each side of it, is wounded. The student should remember that the operation is always to be performed with a pair of blunt-pointed scissors, and the mucous membrane only is to be divided by a very superficial cut, which cannot endanger any vessel. The scissors, also, should be directed away from the tongue. Any further liberation of the tongue which may be necessary can be effected by tearing.

The **Facial Artery** (fig. 285) arises a little above the lingual, and passes obliquely upwards, beneath the Digastric and Stylo-hyoid muscles, and frequently beneath the hypoglossal nerve; it now runs forwards under cover of the body of the lower jaw, lodged in a groove on the posterior surface of the submaxillary gland; this may be called the cervical part of the artery. It then curves upwards over the body of the jaw at the anterior inferior angle of the Masseter muscle; passes forwards and upwards across the cheek to the angle of the mouth, then upwards along the side of the nose, and terminates at the inner canthus of the eye, under the name of the *angular artery*. This vessel, both in the neck and on the face, is remarkably tortuous: in the former situation, to accommodate itself to the movements of the pharynx in deglutition; and in the latter, to the movements of the jaw, and the lips and cheeks.

Relations.—*In the neck*, its origin is superficial, being covered by the integument, Platysma, and fascia; it then passes beneath the Digastric and Stylo-hyoid muscles, and part of the submaxillary gland. It lies upon the middle constrictor of the pharynx, and is separated from the Stylo-glossus and Hyo-glossus muscles by a portion of the submaxillary gland. *On the face*, where it passes over the body of the lower jaw, it is comparatively superficial, lying immediately beneath the Platysma. In this situation its pulsation may be distinctly felt, and compression of the vessel against the bone can be effectually made. In its course over the face, it is covered by the integument, the fat of the cheek, and, near the angle of the mouth, by the Platysma, Risorius, and Zygomatici muscles. It rests on the Buccinator, the Levator anguli oris, and the Levator labii superioris (sometimes piercing or else passing under this last muscle). The facial vein lies to the outer side of the artery, and takes a more direct course across the face, where it is separated from the artery by a considerable interval. In the neck it lies superficial to the artery. The branches of the facial nerve cross the artery, and the infra-orbital nerve lies beneath it.

The branches of this vessel may be divided into two sets : those given off below the jaw (cervical), and those on the face (facial).

Cervical Branches.	Facial Branches.
Inferior or Ascending palatine.	Muscular.
Tonsillar.	Inferior labial.
Submaxillary.	Inferior coronary.
Submental.	Superior coronary.
Muscular.	Lateral nasal.
	Angular.

The **inferior** or **ascending palatine** (fig. 289) passes up between the Styloglossus and Stylo-pharyngeus to the outer side of the pharynx, along which it is continued between the Superior constrictor and the Internal pterygoid to near the base of the skull. It supplies the neighbouring muscles, the tonsil, and Eustachian tube, and divides, near the Levator palati, into two branches : one follows the course of the Levator palati, and, winding over the upper border of the Superior constrictor, supplies the soft palate and the palatine glands, anastomosing with its fellow of the opposite side and with the posterior palatine branch of the internal maxillary artery ; the other pierces the Superior constrictor and supplies the tonsil, anastomosing with the tonsillar and ascending pharyngeal arteries.

The **tonsillar branch** (fig. 289) passes up between the Internal pterygoid and Stylo-glossus, and then ascends along the side of the pharynx, perforating the Superior constrictor, to ramify in the substance of the tonsil and root of the tongue.

The **submaxillary** or **glandular branches** consist of three or four large vessels, which supply the submaxillary gland, some being prolonged to the neighbouring muscles, lymphatic glands, and integument.

The **submental,** the largest of the cervical branches, is given off from the facial artery just as that vessel quits the submaxillary gland : it runs forwards upon the Mylo-hyoid muscle, just below the body of the jaw, and beneath the Digastric ; after supplying the surrounding muscles, and anastomosing with the sublingual artery by branches which perforate the Mylo-hyoid muscle, it arrives at the symphysis of the chin, where it turns over the border of the jaw and divides into a superficial and a deep branch ; the former passes between the integument and Depressor labii inferioris, supplies both, and anastomoses with the inferior labial. The deep branch passes between the latter muscle and the bone, supplies the lip, and anastomoses with the inferior labial and mental arteries.

The **muscular branches** are distributed to the Internal pterygoid and Stylohyoid in the neck, and to the Masseter and Buccinator on the face.

The **inferior labial** passes beneath the Depressor anguli oris, to supply the muscles and integument of the lower lip, anastomosing with the inferior coronary and submental branches of the facial, and with the mental branch of the inferior dental artery.

The **inferior coronary** is derived from the facial artery, near the angle of the mouth ; it passes upwards and inwards beneath the Depressor anguli oris, and, penetrating the Orbicularis oris muscle, runs in a tortuous course along the edge of the lower lip between this muscle and the mucous membrane, inosculating with the artery of the opposite side. This artery supplies the labial glands, the mucous membrane, and muscles of the lower lip ; and anastomoses with the inferior labial and the mental branch of the inferior dental artery.

The **superior coronary** is larger and more tortuous than the preceding. It follows the same course along the edge of the upper lip, lying between the mucous membrane and the Orbicularis oris, and anastomoses with the artery of the opposite side. It supplies the textures of the upper lip, and gives off in its course two or three vessels which ascend to the nose. One, named the *inferior artery of the septum,* ramifies on the septum of the nares as far as the point of the nose; another, the *artery of the ala,* supplies the ala of the nose.

The **lateral nasal** is derived from the facial, as that vessel is ascending along the side of the nose; it supplies the ala and dorsum of the nose, anastomosing

FIG. 285.—The arteries of the face and scalp.*

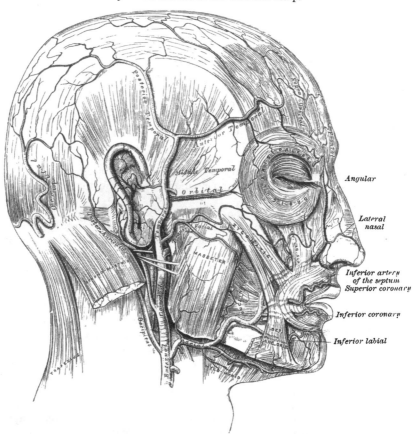

Angular

Lateral nasal

Inferior artery of the septum
Superior coronary

Inferior coronary

Inferior labial

with its fellow, the nasal branch of the ophthalmic, the inferior artery of the septum, the artery of the ala, and the infra-orbital.

The **angular artery** is the termination of the trunk of the facial; it ascends to the inner angle of the orbit, embedded in the fibres of the Levator labii superioris alæque nasi, and accompanied by a large vein, the *angular;* it distributes some branches on the cheek which anastomose with the infra-orbital, and, after supplying the lachrymal sac and Orbicularis palpebrarum muscle, terminates by anastomosing with the nasal branch of the ophthalmic artery.

The anastomoses of the facial artery are very numerous, not only with the vessel of the opposite side, but, in the neck, with the sublingual branch of the

* The muscular tissue of the lips must be supposed to have been cut away, in order to show the course of the coronary arteries.

lingual ; with the ascending pharyngeal ; with the posterior palatine, a branch of the internal maxillary, by its inferior or ascending palatine and tonsillar branches ; on the face, with the mental branch of the inferior dental as it emerges from the mental foramen ; with the transverse facial, a branch of the temporal ; with the infra-orbital, a branch of the internal maxillary ; and with the nasal branch of the ophthalmic.

Peculiarities.—The facial artery not infrequently arises by a common trunk with the lingual. This vessel is also subject to some variations in its size, and in the extent to which it supplies the face. It occasionally terminates as the submental, and not infrequently supplies the face only as high as the angle of the mouth or nose. The deficiency is then supplied by enlargement of one of the neighbouring arteries.

Surgical Anatomy.—The passage of the facial artery over the body of the jaw would appear to afford a favourable position for the application of pressure in cases of hæmorrhage from the lips, the result either of an accidental wound or during an operation ; but its application is useless, except for a very short time, on account of the free communication of this vessel with its fellow, and with numerous branches from different sources. In a wound involving the lip, it is better to seize the part between the fingers, and evert it, when the bleeding vessel may be at once secured with pressure-forceps. In order to prevent hæmorrhage in cases of removal of diseased growths from the part, the lip should be compressed on each side between the fingers and thumb, or by a pair of specially devised clamp-forceps, while the surgeon excises the diseased part. In order to stop hæmorrhage where the lip has been divided in an operation, it is necessary, in uniting the edges of the wound, to pass the sutures through the cut edges, almost as deep as its mucous surface ; by these means, not only are the cut surfaces more neatly and securely adapted to each other, but the possibility of hæmorrhage is prevented by including in the suture the divided artery. If the suture is, on the contrary, passed through merely the cutaneous portion of the wound, hæmorrhage occurs into the cavity of the mouth. The student should, lastly, observe the relation of the angular artery to the lachrymal sac, and it will be seen that, as the vessel passes up along the inner margin of the orbit, it ascends on its nasal side. In operating for fistula lachrymalis, the sac should always be opened on its outer side, in order that this vessel may be avoided.

The **Occipital Artery** (figs. 285, 286) arises from the posterior part of the external carotid, opposite the facial, near the lower margin of the Digastric muscle. At its origin, it is covered by the posterior belly of the Digastric and Stylo-hyoid muscles, and the hypoglossal nerve winds around it from behind forwards ; higher up, it passes across the internal carotid artery, the internal jugular vein, and the pneumogastric and spinal accessory nerves ; it then ascends to the interval between the transverse process of the atlas and the mastoid process of the temporal bone, and passes horizontally backwards, grooving the surface of the latter bone, being covered by the Sterno-mastoid, Splenius, Trachelo-mastoid, and Digastric muscles, and resting upon the Rectus lateralis, the Superior oblique, and Complexus muscles ; it then changes its course and passes vertically upwards, pierces the fascia which connects the cranial attachment of the Trapezius with the Sterno-mastoid, and ascends in a tortuous course over the occiput, as high as the vertex, where it divides into numerous branches. It is accompanied in the latter part of its course by the great occipital, and occasionally by a cutaneous filament from the suboccipital nerve

The branches given off from this vessel are :

Muscular.	Auricular.
Sterno-mastoid.	Meningeal.
Arteria princeps cervicis.	

The **muscular branches** supply the Digastric, Stylo-hyoid, Splenius, and Trachelo-mastoid muscles.

The **sterno-mastoid** is a large and constant branch, generally arising from the artery close to its commencement, but sometimes springing directly from the external carotid. It first passes downwards and backwards over the hypoglossal nerve, and enters the substance of the muscle, in company with the spinal accessory nerve.

The **auricular branch** supplies the back part of the concha. It frequently gives off a branch, which enters the skull through the mastoid foramen and supplies the dura mater, the diploë, and the mastoid cells.

The **meningeal branch** ascends with the internal jugular vein, and enters the skull through the foramen lacerum posterius, to supply the dura mater in the posterior fossa.

FIG. 286.—The occipital artery and its relations.
(From a dissection by Mr. Gerald S. Hughes.)

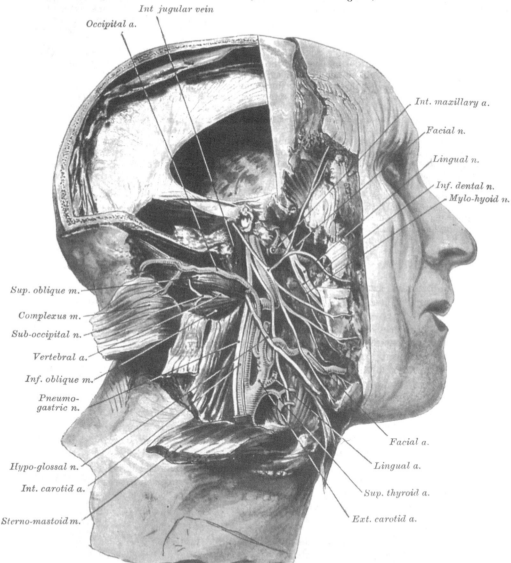

The **arteria princeps cervicis** (fig. 289), the largest branch of the occipital, descends along the back part of the neck, and divides into a superficial and deep portion. The former runs beneath the Splenius, giving off branches which perforate that muscle to supply the Trapezius, anastomosing with the superficial cervical artery, a branch of the Transversalis colli : the latter passes beneath the Complexus, between it and the Semispinalis colli, and anastomoses with branches from the vertebral and with the deep cervical artery, a branch of the superior intercostal. The anastomosis between these vessels serves mainly to establish the collateral circulation after ligature of the carotid or subclavian artery.

The cranial branches of the occipital artery are distributed upon the occiput : they are very tortuous, and lie between the integument and Occipito-frontalis, anastomosing with the artery of the opposite side, the posterior auricular and temporal arteries. They supply the back part of the Occipito-frontalis muscle, the integument, and pericranium.

The **Posterior auricular Artery** (fig. 285) is a small vessel which arises from the external carotid, above the Digastric and Stylo-hyoid muscles, opposite the apex of the styloid process. It ascends, under cover of the parotid gland, on the styloid process of the temporal bone, to the groove between the cartilage of the ear and the mastoid process, immediately above which it divides into its two terminal branches, the auricular and mastoid. Just before arriving at the mastoid process, this artery is crossed by the portio dura, and has beneath it the spinal accessory nerve.

Besides several small branches to the Digastric, Stylo-hyoid, and Sterno-mastoid muscles, and to the parotid gland, this vessel gives off three branches :

Stylo-mastoid. Auricular. Mastoid.

The **stylo-mastoid branch** enters the stylo-mastoid foramen, and supplies the tympanum, mastoid cells, and semicircular canals. In the young subject a branch from this vessel forms, with the tympanic branch from the internal maxillary, a vascular circle, which surrounds the membrana tympani, and from which delicate vessels ramify on that membrane. It anastomoses with the petrosal branch of the middle meningeal artery by a twig, which enters the hiatus Fallopii.

The **auricular branch,** one of the terminal branches, ascends behind the ear, beneath the Retrahens auriculam muscle, and is distributed to the back part of the cartilage of the ear, upon which it ramifies minutely, some branches curving round the margin of the fibro-cartilage, others perforating it, to supply its anterior surface. It anastomoses with the posterior branch of the superficial temporal and also with its anterior auricular branches.

The **mastoid branch** passes backwards, over the Sterno-mastoid muscle, to the scalp above and behind the ear. It supplies the posterior belly of the Occipito-frontalis muscle and the scalp in this situation. It anastomoses with the occipital artery.

The **Ascending pharyngeal Artery** (fig. 289), the smallest branch of the external carotid, is a long, slender vessel, deeply seated in the neck, beneath the other branches of the external carotid and the Stylo-pharyngeus muscle. It arises from the back part of the external carotid, near the commencement of that vessel, and ascends vertically between the internal carotid and the side of the pharynx, to the under surface of the base of the skull, lying on the Rectus capitis anticus major. Its branches may be subdivided into four sets :

Prevertebral. Pharyngeal. Tympanic. Meningeal.

The **prevertebral branches** are numerous small vessels, which supply the Recti capitis antici and Longus colli muscles, the sympathetic, hypoglossal, and pneumo-gastric nerves, and the lymphatic glands, anastomosing with the ascending cervical artery.

The **pharyngeal branches** are three or four in number. Two of these descend to supply the middle and inferior Constrictors and the Stylo-pharyngeus, ramifying in their substance and in the mucous membrane lining them. The largest of the pharyngeal branches passes inwards, running upon the Superior constrictor, and sends ramifications to the soft palate and tonsil, which take the place of the ascending palatine branch of the facial artery, when that vessel is of small size. A twig from this branch supplies the Eustachian tube.

The **tympanic branch** is a small artery which passes through a minute foramen in the petrous portion of the temporal bone, in company with the tympanic branch of the Glosso-pharyngeal nerve to supply the inner wall of the tympanum and anastomose with the other tympanic arteries.

The **meningeal branches** consist of several small vessels, which pass through foramina in the base of the skull, to supply the dura mater. One, the *posterior meningeal*, enters the cranium through the foramen lacerum posterius; a second passes through the foramen lacerum medium; and occasionally a third through the anterior condyloid foramen. They are all distributed to the dura mater.

Surgical Anatomy.—The ascending pharyngeal artery has been wounded from the throat; as in the case in which the stem of a tobacco-pipe was driven into the vessel, causing fatal hæmorrhage.

The **Superficial temporal artery** (fig. 285), the smaller of the two terminal branches of the external carotid, appears, from its direction, to be the continuation of that vessel. It commences in the substance of the parotid gland, in the interspace between the neck of the lower jaw and the external auditory meatus, crosses over the posterior root of the zygoma, passes beneath the Attrahens auriculam muscle, lying on the temporal fascia, and divides, about two inches above the zygomatic arch, into two branches, an anterior and a posterior.

The **anterior temporal** runs tortuously upwards and forwards to the forehead, supplying the muscles, integument, and pericranium in this region, and anastomoses with the supra-orbital and frontal arteries.

The **posterior temporal,** larger than the anterior, curves upwards and backwards along the side of the head, lying superficial to the temporal fascia, and inosculates with its fellow of the opposite side, and with the posterior auricular and occipital arteries.

The superficial temporal artery, as it crosses the zygoma, is covered by the Attrahens auriculam muscle, and by a dense fascia given off from the parotid gland: it is crossed by the temporo-facial division of the facial nerve and one or two veins, and is accompanied by the auriculo-temporal nerve, which lies behind it. Besides some twigs to the parotid gland, the articulation of the jaw, and the Masseter muscle, its branches are, the

Transverse Facial. Middle Temporal.
Anterior Auricular.

The **transverse facial** is given off from the temporal before that vessel quits the parotid gland; running forwards through its substance, it passes transversely across the face, between Stenson's duct and the lower border of the zygoma, and divides on the side of the face into numerous branches, which supply the parotid gland, the Masseter muscle, and the integument, anastomosing with the facial, masseteric, and infra-orbital arteries. This vessel rests on the Masseter, and is accompanied by one or two branches of the facial nerve. It is sometimes a branch of the external carotid.

The **middle temporal artery** arises immediately above the zygomatic arch, and, perforating the temporal fascia, gives branches to the Temporal muscle, anastomosing with the deep temporal branches of the internal maxillary. It occasionally gives off an **orbital** branch, which runs along the upper border of the zygoma, between the two layers of the temporal fascia, to the outer angle of the orbit. This branch, which may arise directly from the superficial temporal artery, supplies the Orbicularis palpebrarum, and anastomoses with the lachrymal and palpebral branches of the ophthalmic artery.

The **anterior auricular branches** are distributed to the anterior portion of the pinna, the lobule, and part of the external meatus, anastomosing with branches of the posterior auricular.

Surgical Anatomy.—Formerly the operation of arteriotomy was performed upon this vessel in cases of inflammation of the eye or brain, but now the operation is probably never performed. If the student will consider the relations of the trunk of the vessel, as it crosses the zygomatic arch, with the surrounding structures, he will observe that it is covered by a thick and dense fascia, crossed by one of the main divisions of the facial nerve and one or two veins, and accompanied by the auriculo-temporal nerve. Bleeding should not be performed in this situation, as much difficulty may arise from the dense fascia over the vessel preventing a free flow of blood, and considerable pressure is requisite afterwards to arrest the hæmorrhage. Again, a varicose aneurism may be formed by the accidental opening of one of the veins in front of the artery; or severe neuralgic pain may arise from the operation implicating one of the nervous filaments in the neighbourhood. The anterior branch, on the contrary, is subcutaneous, is a large vessel, and is readily compressed; it is consequently more suitable for the operation.

The **Internal maxillary** (fig. 287), the larger of the two terminal branches of the external carotid, arises from that vessel opposite the neck of the condyle of the lower jaw, and is at first embedded in the substance of the parotid gland; it passes

FIG. 287.—The internal maxillary artery, and its branches.

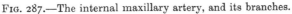

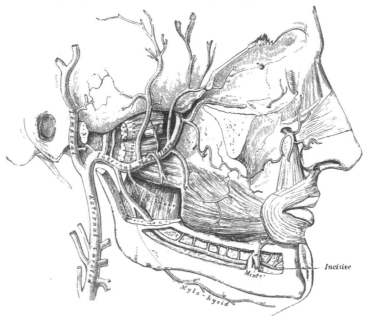

FIG. 288.—Plan of the branches.

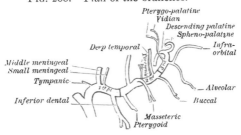

inwards between the ramus of the jaw and the internal lateral ligament, and then upon the outer surface of the External pterygoid muscle to the spheno-maxillary fossa to supply the deep structures of the face. For convenience of description, it is divided into three portions: a maxillary, a pterygoid, and a spheno-maxillary.

In the first part of its course (*maxillary portion*), the artery passes horizontally forwards and inwards, between the ramus of the jaw and the internal lateral ligament. The artery here lies parallel to and a little below the auriculo-temporal nerve; it crosses the inferior dental nerve, and lies along the lower border of the External pterygoid muscle.

In the second part of its course (*pterygoid portion*), it runs obliquely forwards and upwards upon the outer surface of the External pterygoid muscle, being covered by the ramus of the lower jaw, and lower part of the Temporal muscle; or it may pass on the inner surface of the External pterygoid muscle to reach the interval between its two heads, between which it passes to reach the spheno-maxillary fossa.

In the third part of its course (*spheno-maxillary portion*), it approaches the superior maxillary bone, and enters the spheno-maxillary fossa in the interval between the two heads of the External pterygoid, where it lies in relation with Meckel's ganglion, and gives off its terminal branches.

The branches of this vessel may be divided into three groups, corresponding with its three divisions.

Branches of the First or Maxillary Portion (fig. 288).

Tympanic (anterior).	Small meningeal.
Deep auricular.	Inferior dental.
Middle meningeal.	

The **tympanic branch** passes upwards behind the articulation of the lower jaw, enters the tympanum through the Glaserian fissure, and ramifies upon the membrana tympani, forming a vascular circle around the membrane with the stylo-mastoid artery, and anastomosing with the Vidian and the tympanic branch from the internal carotid.

The **deep auricular branch** often arises in common with the preceding. It passes upwards in the substance of the parotid gland, behind the temporo-maxillary articulation, pierces the cartilaginous or bony wall of the external auditory meatus, and supplies its cuticular lining and the outer surface of the membrana tympani.

The **middle meningeal** is the largest of the branches which supply the dura mater. It arises from the internal maxillary, between the internal lateral ligament and the neck of the jaw, and passes vertically upwards between the two roots of the auriculo-temporal nerve to the foramen spinosum of the sphenoid bone. On entering the cranium, it divides into two branches, anterior and posterior. The *anterior branch*, the larger, crosses the great ala of the sphenoid, and reaches the groove, or canal, in the anterior inferior angle of the parietal bone: it then divides into branches which spread out between the dura mater and internal surface of the cranium, some passing upwards over the parietal bone as far as the vertex, and others backwards to the occipital bone. The *posterior branch* crosses the squamous portion of the temporal, and on the inner surface of the parietal bone divides into branches which supply the posterior part of the dura mater and cranium. The branches of this vessel are distributed partly to the dura mater, but chiefly to the bones; they anastomose with the arteries of the opposite side, and with the anterior and posterior meningeal.

The middle meningeal on entering the cranium gives off the following collateral branches: 1. Numerous small vessels to the Gasserian ganglion, and to the dura mater in this situation. 2. A branch (*petrosal branch*), which enters the hiatus Fallopii, supplies the facial nerve, and anastomoses with the stylo-mastoid branch of the posterior auricular artery. 3. A minute *tympanic* branch, which runs in the canal for the Tensor tympani muscle, and supplies this muscle and the lining membrane of the canal. 4. Orbital branches, which pass through the sphenoidal fissure, or through separate canals in the great wing of the sphenoid to anastomose with the lachrymal or other branches of the ophthalmic artery. 5. Temporal or anastomotic branches, which pass through foramina in the great wing of the sphenoid, and anastomose in the temporal fossa with the deep temporal arteries.

Surgical Anatomy.—The middle meningeal is an artery of considerable surgical importance, as it may be injured in fractures of the temporal region of the skull, and the injury may be followed by considerable hæmorrhage between the bone and dura mater, which may cause compression of the brain, and require the operation of trephining for its relief. This artery crosses the anterior inferior angle of the parietal bone at a point 1½ inch behind the external angular process of the frontal bone, and 1¾ inch above the zygoma. From this point the anterior branch passes upwards and slightly backwards to the sagittal suture, lying about ½ inch to ¾ inch behind the coronal suture. The posterior branch passes upwards and backwards over the squamous portion of the temporal bone. In order to expose the artery as it lies in the canal in the parietal bone, a semilunar incision, with its convexity upwards, should be made, commencing an inch behind the external angular process, and carried backwards for 2 inches. The structures cut through are: (1) skin; (2) superficial fascia, with branches of the superficial temporal vessels and nerves; (3) the fascia continued down from the aponeurosis of the Occipito-frontalis; (4) the two layers of the temporal fascia; (5) the Temporal muscle; (6) the deep temporal vessels; (7) the pericranium; and (8) the bone.

The **small meningeal** is sometimes derived from the preceding. It enters the skull through the foramen ovale, and supplies the Gasserian ganglion and dura mater.

The **inferior dental** descends with the inferior dental nerve to the foramen on the inner side of the ramus of the jaw. It runs along the dental canal in the substance of the bone, accompanied by the nerve, and opposite the first bicuspid tooth divides into two branches, incisor and mental; the *former* is continued forwards beneath the incisor teeth as far as the symphysis, where it anastomoses with the artery of the opposite side; the *mental* branch escapes with the nerve at the mental foramen, supplies the structures composing the chin, and anastomoses with the submental, inferior labial, and inferior coronary arteries. Near its origin the inferior dental artery gives off a *lingual* branch, which descends with the lingual (gustatory) nerve and supplies the mucous membrane of the mouth. As the inferior dental artery enters the foramen, it gives off a *mylo-hyoid* branch, which runs in the mylo-hyoid groove, and ramifies on the under surface of the Mylo-hyoid muscle. The dental and incisor arteries during their course through the substance of the bone give off a few twigs which are lost in the cancellous tissue, and a series of branches which correspond in number to the roots of the teeth: these enter the minute apertures at the extremities of the fangs, and supply the pulp of the teeth.

BRANCHES OF THE SECOND OR PTERYGOID PORTION

Deep temporal.   Masseteric.
Pterygoid.   Buccal.

These branches are distributed, as their names imply, to the muscles in the maxillary region.

The **deep temporal branches,** two in number, anterior and posterior, each occupy that part of the temporal fossa indicated by its name. Ascending between the Temporal muscle and pericranium, they supply that muscle, and anastomose with the middle temporal artery; the anterior branch communicating with the lachrymal through small branches which perforate the malar bone and great wing of the sphenoid.

The **pterygoid branches,** irregular in their number and origin, supply the Pterygoid muscles.

The **masseteric** is a small branch which passes outwards, above the sigmoid notch of the lower jaw, to the deep surface of the Masseter. It supplies that muscle, and anastomoses with the masseteric branches of the facial and with the transverse facial artery.

The **buccal** is a small branch which runs obliquely forwards, between the Internal pterygoid and the ramus of the jaw, to the outer surface of the Buccinator, to which it is distributed, anastomosing with branches of the facial artery.

BRANCHES OF THE THIRD OR SPHENO-MAXILLARY PORTION

Alveolar.   Vidian.
Infra-orbital.   Pterygo-palatine.
Posterior or Descending palatine.   Naso- or Spheno-palatine.

The **alveolar** or **posterior dental branch** is given off from the internal maxillary, frequently by a common branch with the infra-orbital, just as the trunk of the vessel is passing into the spheno-maxillary fossa. Descending upon the tuberosity of the superior maxillary bone, it divides into numerous branches, some of which enter the posterior dental canals, to supply the molar and bicuspid teeth and the lining of the antrum, and others are continued forwards on the alveolar process to supply the gums.

The **infra-orbital** appears, from its direction, to be the continuation of the trunk of the internal maxillary. It often arises from that vessel by a common trunk with the preceding branch, and runs along the infra-orbital canal with the superior maxillary nerve, emerging upon the face at the infra-orbital foramen, beneath the Levator labii superioris. While contained in the canal, it gives off branches which ascend into the orbit, and assist in supplying the Inferior rectus and Inferior oblique muscles and the lachrymal gland. Other branches (*anterior dental*) descend through the anterior dental canals in the bone, to supply the mucous membrane of the antrum, and the front teeth of the upper jaw. On the face, some branches pass upwards to the inner angle of the orbit and the lachrymal sac, anastomosing with the angular branch of the facial artery ; other branches pass inwards towards the nose, anastomosing with the nasal branch of the ophthalmic ; and other branches descend beneath the Levator labii superioris, and anastomose with the transverse facial and buccal arteries.

The four remaining branches arise from that portion of the internal maxillary which is contained in the spheno-maxillary fossa.

The **descending palatine** descends through the posterior palatine canal with the anterior palatine branch of Meckel's ganglion, and, emerging from the posterior palatine foramen, runs forwards in a groove on the inner side of the alveolar border of the hard palate to the anterior palatine canal, where the terminal branch of the artery passes upwards through the foramen of Stenson to anastomose with the naso-palatine artery. Its branches are distributed to the gums, the mucous membrane of the hard palate, and the palatine glands. While it is contained in the palatine canal, it gives off branches, which descend in the accessory palatine canals to supply the soft palate and tonsil, anastomosing with the ascending palatine artery.

Surgical Anatomy.—The position of the descending palatine artery on the hard palate should be borne in mind in performing an operation for the closure of a cleft in the hard palate, as it is in danger of being wounded, and may give rise to formidable hæmorrhage. In one case, in which it was wounded, it was necessary to plug the posterior palatine canal in order to arrest the bleeding.

The **Vidian branch** passes backwards along the Vidian canal with the Vidian nerve. It is distributed to the upper part of the pharynx and Eustachian tube, sending a small branch into the tympanum, which anastomoses with the other tympanic arteries.

The **pterygo-palatine** is a very small branch, which passes backwards through the pterygo-palatine canal with the pharyngeal nerve, and is distributed to the upper part of the pharynx and Eustachian tube.

The **spheno-palatine** passes through the spheno-palatine foramen into the cavity of the nose, at the back part of the superior meatus, and divides into two branches : one internal, the *naso-palatine* or *artery of the septum*, passes obliquely downwards and forwards along the septum nasi, supplies the mucous membrane, and anastomoses in front with the terminal branch of the descending palatine. The external branches, two or three in number, supply the mucous membrane covering the lateral wall of the nose, the antrum, and the ethmoid and sphenoid cells.

The Triangles of the Neck

The student having considered the relative anatomy of the large arteries of the neck and their branches, and the relations they bear to the veins and nerves, should now examine these structures collectively, as they present themselves in certain regions of the neck, in each of which important operations are constantly being performed.

The side of the neck presents a somewhat quadrilateral outline, limited, above, by the lower border of the body of the jaw, and an imaginary line extending from the angle of the jaw to the mastoid process; below, by the prominent upper border of the clavicle; in front, by the median line of the neck; behind, by the anterior margin of the Trapezius muscle. This space is subdivided into two large triangles by the Sterno-mastoid muscle, which passes obliquely across the neck, from the sternum and clavicle below, to the mastoid process above. The triangular space in front of this muscle is called the *anterior triangle*; and that behind it, the *posterior triangle*.

Anterior Triangle of the Neck

The **anterior triangle** is bounded, in front, by a line extending from the chin to the sternum; behind, by the anterior margin of the Sterno-mastoid; its base, directed upwards, is formed by the lower border of the body of the jaw, and a line extending from the angle of the jaw to the mastoid process; its apex is below, at the sternum. This space is subdivided into three smaller triangles by the Digastric muscle above, and the anterior belly of the Omo-hyoid below. These smaller triangles are named, from below upwards, the inferior carotid, the superior carotid, and the submaxillary triangle.

The **Inferior Carotid Triangle** is bounded, in front, by the median line of the neck; behind, by the anterior margin of the Sterno-mastoid; above, by the anterior belly of the Omo-hyoid; and is covered by the integument, superficial fascia, Platysma, and deep fascia; ramifying between which are some of the descending branches of the superficial cervical plexus. Beneath these superficial structures are the Sterno-hyoid and Sterno-thyroid muscles, which, together with the anterior margin of the Sterno-mastoid, conceal the lower part of the common carotid artery.*

This vessel is enclosed within its sheath, together with the internal jugular vein and pneumogastric nerve; the vein lying on the outer side of the artery on the right side of the neck, but overlapping it below on the left side; the nerve lying between the artery and vein, on a plane posterior to both. In front of the sheath are a few filaments descending from the loop of communication between the descendens and communicantes hypoglossi; behind the sheath are seen the inferior thyroid artery, the recurrent laryngeal nerve, and the sympathetic nerve; and on its inner side, the trachea, the thyroid gland—much more prominent in the female than in the male—and the lower part of the larynx. By cutting into the upper part of this space, and slightly displacing the Sterno-mastoid muscle, the common carotid artery may be tied below the Omo-hyoid muscle.

---

* Therefore the common carotid artery and internal jugular vein are not, strictly speaking, contained in this triangle, since they are covered by the Sterno-mastoid muscle; that is to say, they lie behind the anterior border of that muscle, which forms the posterior border of the triangle. But as they lie very close to the structures which are really contained in the triangle, and whose position it is essential to remember in operating on this part of the artery, it has seemed expedient to study the relations of all these parts together.

The **Superior Carotid Triangle** is bounded, behind, by the Sterno-mastoid; below, by the anterior belly of the Omo-hyoid; and above, by the posterior belly of the Digastric muscle. It is covered by the integument, superficial fascia, Platysma, and deep fascia; ramifying between which are branches of the facial and superficial cervical nerves. Its floor is formed by parts of the Thyro-hyoid, Hyo-glossus, and the Inferior and Middle constrictor muscles of the pharynx. This space when dissected is seen to contain the upper part of the common carotid artery, which bifurcates opposite the upper border of the thyroid cartilage into the external and internal carotid. These vessels are occasionally somewhat concealed from view by the anterior margin of the Sterno-mastoid muscle, which overlaps them. The external and internal carotids lie side by side, the external being the more anterior of the two. The following branches of the external carotid are also met with in this space: the superior thyroid, running forwards and downwards; the lingual, directly forwards; the facial, forwards and upwards; the occipital, backwards; and the ascending pharyngeal directly upwards on the inner side of the internal carotid. The veins met with are: the internal jugular, which lies on the outer side of the common and internal carotid arteries; and veins corresponding to the above-mentioned branches of the external carotid— viz. the superior thyroid, the lingual, facial, ascending pharyngeal, and some- times the occipital—all of which accompany their corresponding arteries, and terminate in the internal jugular. The nerves in this space are the following: In front of the sheath of the common carotid is the descendens hypoglossi. The hypoglossal nerve crosses both the internal and external carotids above, curving round the occipital artery at its origin. Within the sheath, between the artery and vein, and behind both, is the pneumogastric nerve; behind the sheath, the sympathetic. On the outer side of the vessels, the spinal accessory nerve runs for a short distance before it pierces the Sterno-mastoid muscle; and on the inner side of the external carotid, just below the hyoid bone, may be seen the internal laryngeal nerve; and, still more inferiorly, the external laryngeal nerve. The upper part of the larynx and lower part of the pharynx are also found in the front part of this space.

The **Submaxillary Triangle** corresponds to the part of the neck immediately beneath the body of the jaw. It is bounded, above, by the lower border of the body of the jaw, and a line drawn from its angle to the mastoid process; below, by the posterior belly of the Digastric and Stylo-hyoid muscles; in front, by the anterior belly of the Digastric. It is covered by the integument, superficial fascia, Platysma, and deep fascia; ramifying between which are branches of the facial and ascending filaments of the superficial cervical nerves. Its floor is formed by the Mylo-hyoid and Hyo-glossus muscles. This space contains, in front, the submaxillary gland, superficial to which is the facial vein, while embedded in it is the facial artery and its glandular branches; beneath this gland, on the surface of the Mylo-hyoid muscle, are the submental artery and the mylo-hyoid artery and nerve. The posterior part of this triangle is separated from the anterior part by the stylo-maxillary ligament; it contains the external carotid artery, ascending deeply in the substance of the parotid gland; this vessel lies here in front of, and superficial to, the internal carotid, being crossed by the facial nerve, and gives off in its course the posterior auricular, temporal, and internal maxillary branches: more deeply are the internal carotid, the internal jugular vein, and the pneumo- gastric nerve, separated from the external carotid by the Stylo-glossus and Stylo- pharyngeus muscles, and the glosso-pharyngeal nerve.*

* The same remark will apply to this triangle as was made about the inferior carotid triangle. The structures enumerated as contained in the back part of the space lie, strictly speaking, beneath the muscles which form the posterior boundary of the triangle; but as it is very important to bear in mind their close relation to the parotid gland and its boundaries (on account of the frequency of surgical operations on this gland), all these parts are spoken of together.

Posterior Triangle of the Neck

The **posterior triangle** is bounded, in front, by the Sterno-mastoid muscle; behind, by the anterior margin of the Trapezius; its base corresponds to the middle third of the clavicle; its apex, to the occiput. The space is crossed, about an inch above the clavicle, by the posterior belly of the Omo-hyoid, which divides it unequally into two, an upper or occipital, and a lower or subclavian, triangle.

The **Occipital,** the larger division of the posterior triangle, is bounded, in front, by the Sterno-mastoid; behind, by the Trapezius; below, by the Omo-hyoid. Its floor is formed from above downwards by the Splenius, Levator anguli scapulæ, and the Middle and Posterior scaleni muscles. It is covered by the integument, the Platysma below, the superficial and deep fasciæ. The spinal accessory nerve is directed obliquely across the space from the Sterno-mastoid, which it pierces, to the under surface of the Trapezius; below, the descending branches of the cervical plexus and the transversalis colli artery and vein cross the space. A chain of lymphatic glands is also found running along the posterior border of the Sterno-mastoid, from the mastoid process to the root of the neck.

The **Subclavian,** the smaller division of the posterior triangle, is bounded, above, by the posterior belly of the Omo-hyoid; below, by the clavicle; its base, directed forwards, being formed by the Sterno-mastoid. The size of the subclavian triangle varies according to the extent of attachment of the clavicular portion of the Sterno-mastoid and Trapezius muscles, and also according to the height at which the Omo-hyoid crosses the neck above the clavicle. Its height also varies much, according to the position of the arm, being much diminished by raising the limb, on account of the ascent of the clavicle, and increased by drawing the arm downwards, when that bone is depressed. This space is covered by the integument, the Platysma, the superficial and deep fasciæ; and crossed by the descending branches of the cervical plexus. Just above the level of the clavicle, the third portion of the subclavian artery curves outwards and downwards from the outer margin of the Scalenus anticus, across the first rib, to the axilla. Sometimes this vessel rises as high as an inch and a half above the clavicle, or to any point intermediate between this and its usual level. Occasionally, it passes in front of the Scalenus anticus, or pierces the fibres of that muscle. The subclavian vein lies behind the clavicle, and is usually not seen in this space; but it occasionally rises as high up as the artery, and has even been seen to pass with that vessel behind the Scalenus anticus. The brachial plexus of nerves lies above the artery, and in close contact with it. Passing transversely behind the clavicle are the suprascapular vessels; and traversing its upper angle in the same direction, the transversalis colli artery and vein. The external jugular vein runs vertically downwards behind the posterior border of the Sterno-mastoid, to terminate in the subclavian vein; it receives the transverse cervical and suprascapular veins, which occasionally form a plexus in front of the artery, and a small vein which crosses the clavicle from the cephalic. The small nerve to the Subclavius muscle also crosses this triangle about its middle. A lymphatic gland is also found in the space. Its floor is formed by the first rib with the first digitation of the Serratus magnus.

Internal Carotid Artery

The **internal carotid artery** supplies the anterior part of the brain, the eye and its appendages, and sends branches to the forehead and nose. Its size, in the adult, is equal to that of the external carotid, though, in the child, it is larger than that vessel. It is remarkable for the number of curvatures that it presents in different parts of its course. It occasionally has one or two flexures near the base of the skull, while in its passage through the carotid canal and along the side of the body of the sphenoid bone it describes a double curvature which resembles the italic letter *s* placed horizontally. These curvatures most probably diminish the velocity of the current of blood, by increasing the extent of surface over which it moves, and adding to the amount of impediment produced from friction.

In considering the course and relations of this vessel, it may be conveniently divided into four portions : a cervical, petrous, cavernous, and cerebral.

Cervical Portion.—This portion of the internal carotid commences at the bifurcation of the common carotid, opposite the upper border of the thyroid cartilage, and runs perpendicularly upwards, in front of the transverse processes of the three upper cervical vertebræ, to the carotid canal in the petrous portion of the temporal bone. It is superficial at its commencement, being contained in the superior carotid triangle, and lying on the same level as the external carotid, but behind that artery, overlapped by the Sterno-mastoid, and covered by the deep fascia, Platysma, and integument : it then passes beneath the parotid gland, being crossed by the hypoglossal nerve, the Digastric and Stylo-hyoid muscles, and the occipital and posterior auricular arteries. Higher up, it is separated from the external carotid by the Stylo-glossus and Stylo-pharyngeus muscles, the glosso-pharyngeal nerve, and the pharyngeal branch of the pneumo-gastric. It is in relation, *behind*, with the Rectus capitis anticus major, the superior cervical ganglion of the sympathetic, and superior laryngeal nerve ; *externally*, with the internal jugular vein and pneumogastric nerve, the nerve lying on a plane posterior to the artery ; *internally*, with the pharynx, tonsil, the superior laryngeal nerve, and ascending pharyngeal artery. At the base of the skull the glosso-pharyngeal, vagus, spinal accessory, and hypoglossal nerves lie between the artery and the internal jugular vein.

PLAN OF THE RELATIONS OF THE INTERNAL CAROTID ARTERY IN THE NECK

In front.

Skin, superficial and deep fasciæ.
Platysma.
Sterno-mastoid.
Occipital and posterior auricular arteries.
Hypoglossal nerve.
Parotid gland.
Stylo-glossus and Stylo-pharyngeus muscles.
Glosso-pharyngeal nerve.
Pharyngeal branch of the pneumogastric.

Externally.

Internal jugular vein.
Pneumogastric nerve.

Internally.

Pharynx.
Superior laryngeal nerve.
Ascending pharyngeal artery.
Tonsil.

Behind.

Rectus capitis anticus major.
Sympathetic.
Superior laryngeal nerve.

Petrous Portion.—When the internal carotid artery enters the canal in the petrous portion of the temporal bone, it first ascends a short distance, then curves forwards and inwards, and again ascends as it leaves the canal to enter the cavity of the skull between the lingula and petrosal process. In this canal, the artery lies at first in front of the cochlea and tympanum ; from the latter cavity it is separated by a thin, bony lamella, which is cribriform in the young subject, and often absorbed in old age. Farther forwards it is separated from the Gasserian ganglion by a thin plate of bone, which forms the floor of the fossa for the ganglion and the roof of the horizontal portion of the canal. Frequently this bony plate is more or less deficient, and then the ganglion is separated from the artery by fibrous membrane. The artery is separated from the bony wall of the carotid canal by a prolongation of dura mater, and is surrounded by filaments of the carotid plexus, derived from the ascending branch of the superior cervical ganglion of the sympathetic, and a number of small veins.

FIG. 289.—The internal carotid and vertebral arteries. Right side.

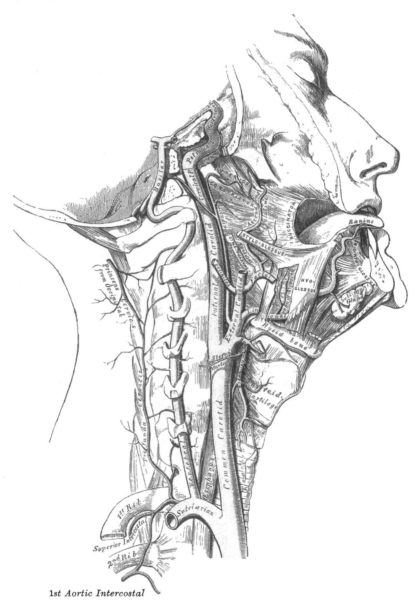

Cavernous Portion.—The internal carotid artery, in this part of its course, is situated between the layers of the dura mater forming the cavernous sinus, but covered by the lining membrane of the sinus. It at first ascends to the posterior clinoid process, then passes forwards by the side of the body of the sphenoid bone, and again curves upwards on the inner side of the anterior clinoid process, and perforates the dura mater forming the roof of the sinus. In this part of its course it is surrounded by filaments of the sympathetic nerve, and has in relation with it externally the sixth nerve.

Cerebral Portion.—Having perforated the dura mater, on the inner side of the anterior clinoid process, the internal carotid passes between the second and third cranial nerves to the anterior perforated spot at the inner extremity of the fissure of Sylvius, where it gives off its terminal or cerebral branches. This portion of the artery has the optic nerve on its inner side, and the third nerve externally.

Peculiarities.—The length of the internal carotid varies according to the length of the neck, and also according to the point of bifurcation of the common carotid. Its origin sometimes takes place from the arch of the aorta; in such rare instances, this vessel has been found to be placed nearer the middle line of the neck than the external carotid, as far upwards as the larynx, when the latter vessel crossed the internal carotid. The course of the vessel, instead of being straight, may be very tortuous. A few instances are recorded in which this vessel was altogether absent: in one of these the common carotid passed up the neck, and gave off the usual branches of the external carotid: the cranial portion of the internal carotid being replaced by two branches of the internal maxillary, which entered the skull through the foramen rotundum and ovale, and joined to form a single vessel.

Surgical Anatomy.—The cervical part of the internal carotid is very rarely wounded. Mr. Cripps, in an interesting paper in the ' Medico-Chirurgical Transactions,' compares the rareness of a wound of the internal carotid with one of the external carotid or its branches. It is, however, sometimes injured by a stab or gunshot wound in the neck, or even occasionally by a stab from within the mouth, as when a person receives a thrust from the end of a parasol, or falls down with a tobacco-pipe in his mouth. The relation of the internal carotid with the tonsil should be especially remembered, as instances have occurred in which the artery has been wounded during the operation of scarifying the tonsil, and fatal hæmorrhage has supervened. The indications for ligature are wounds, when the vessel should be exposed by a careful dissection and tied above and below the bleeding-point ; and aneurism, which if non-traumatic may be treated by ligature of the common carotid ; but if traumatic in origin, by exposing the sac and tying the vessel above and below. The incision for ligature of the cervical portion of the internal carotid should be made along the anterior border of the Sterno-mastoid, from the angle of the jaw to the upper border of the thyroid cartilage. The superficial structures being divided, and the Sterno-mastoid defined and drawn outwards, the cellular tissue must be carefully separated and the posterior belly of the Digastric and hypoglossal nerve sought for as guides to the vessel. When the artery is found, the external carotid should be drawn inwards and the Digastric muscle upwards, and the aneurism needle passed from without inwards.

Branches.—The branches given off from the internal carotid are :

From the Petrous portion	Tympanic (internal or deep).
From the Cavernous portion	Arteriæ receptaculi. Anterior meningeal. Ophthalmic.
From the Cerebral portion	Anterior cerebral. Middle cerebral. Posterior communicating. Anterior choroid.

The cervical portion of the internal carotid gives off no branches.

The **tympanic** is a small branch which enters the cavity of the tympanum, through a minute foramen in the carotid canal, and anastomoses with the tympanic branch of the internal maxillary, and with the stylo-mastoid artery.

The **arteriæ receptaculi** are numerous small vessels, derived from the internal carotid in the cavernous sinus ; they supply the pituitary body, the Gasserian ganglion, and the walls of the cavernous and inferior petrosal sinuses. Some of these branches anastomose with branches of the middle meningeal.

The **anterior meningeal** is a small branch which passes over the lesser wing of the sphenoid to supply the dura mater of the anterior fossa ; it anastomoses with the meningeal branch from the posterior ethmoidal artery.

The **ophthalmic artery** arises from the internal carotid, just as that vessel is emerging from the cavernous sinus, on the inner side of the anterior clinoid process, and enters the orbit through the optic foramen, below and on the outer side of the optic nerve. It then passes over the nerve to the inner wall of the orbit, and thence horizontally forwards, beneath the lower border of the Superior oblique muscle to a point behind the internal angular process of the frontal bone, where it divides into two terminal branches, the *frontal* and *nasal*. As the artery crosses the optic nerve it is accompanied by the nasal nerve, and is separated from the frontal nerve by the Rectus superior and Levator palpebræ superioris muscles

The branches of this vessel may be divided into an *orbital group*, which are distributed to the orbit and surrounding parts ; and an *ocular group*, which supply the muscles and globe of the eye.

Orbital Group.	*Ocular Group.*
Lachrymal.	Short ciliary.
Supra-orbital.	Long ciliary.
Posterior ethmoidal.	Anterior ciliary.
Anterior ethmoidal.	Arteria centralis retinæ.
Internal palpebral.	Muscular.
Frontal.	
Nasal.	

The *lachrymal* is one of the largest branches derived from the ophthalmic, arising close to the optic foramen : not infrequently it is given off from the artery

FIG. 290.—The ophthalmic artery and its branches, the roof of the orbit having been removed.

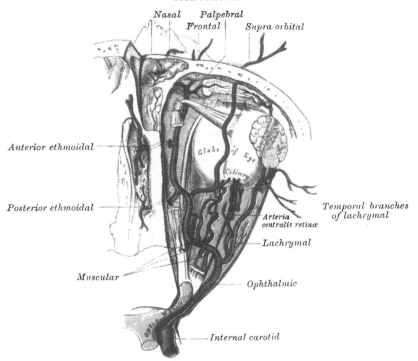

before it enters the orbit. It accompanies the lachrymal nerve along the upper border of the External rectus muscle, and is distributed to the lachrymal gland. Its terminal branches, escaping from the gland, are distributed to the eyelids and conjunctiva : of those supplying the eyelids, two are of considerable size and are named the *external palpebral* ; they run inwards in the upper and lower lids respectively and anastomose with the internal palpebral arteries, forming an arterial circle in this situation. The lachrymal artery gives off one or two *malar branches*, one of which passes through a foramen in the malar bone, to reach

the temporal fossa, and anastomoses with the deep temporal arteries ; the other
appears on the cheek through the malar foramen, and anastomoses with the

FIG. 291.—The arteries of the base of the brain.   The right half of the cerebellum
and pons have been removed.

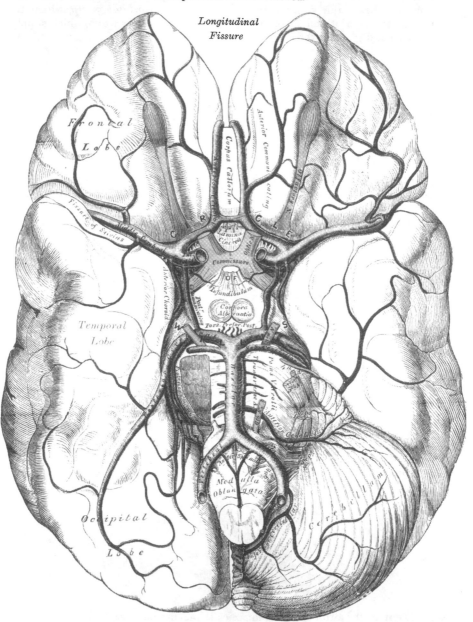

N.B.—It will be noticed that in the illustration the two anterior cerebral arteries
have been drawn at a considerable distance from each other; this makes the anterior
communicating artery appear very much longer than it really is.

transverse facial. A branch is also sent backwards, through the sphenoidal
fissure, to the dura mater, which anastomoses with a branch of the middle
meningeal artery.

Peculiarities.—The lachrymal artery is sometimes derived from one of the anterior branches of the middle meningeal artery.

The *supra-orbital artery* arises from the ophthalmic as that vessel is crossing over the optic nerve. Ascending so as to rise above all the muscles of the orbit, it runs forwards, with the supra-orbital nerve, between the periosteum and Levator palpebræ ; and, passing through the supra-orbital foramen, divides into a superficial and a deep branch, which supply the integument, the muscles, and the pericranium of the forehead, anastomosing with the frontal, the anterior branch of the temporal, and the artery of the opposite side. This artery in the orbit supplies the Superior rectus and the Levator palpebræ, and sends a branch inwards, across the pulley of the Superior oblique muscle, to supply the parts at the inner canthus. At the supra-orbital foramen it frequently transmits a branch to the diploë.

The *ethmoidal branches* are two in number : posterior and anterior. The former, which is the smaller, passes through the posterior ethmoidal foramen, supplies the posterior ethmoidal cells, and, entering the cranium, gives off a meningeal branch, which supplies the adjacent dura mater ; and nasal branches which descend into the nose through apertures in the cribriform plate, anastomosing with branches of the spheno-palatine. The anterior ethmoidal artery accompanies the nasal nerve through the anterior ethmoidal foramen, supplies the anterior ethmoidal cells and frontal sinuses, and, entering the cranium, gives off a meningeal branch, which supplies the adjacent dura mater ; and nasal branches, which descend into the nose, through the slit by the side of the crista galli, and running along the groove on the under surface of the nasal bone supply the skin of the nose.

The *palpebral arteries*, two in number, superior and inferior, arise from the ophthalmic, opposite the pulley of the Superior oblique muscle ; they leave the orbit to encircle the eyelids near their free margin, forming a superior and an inferior arch, which lie between the Orbicularis muscle and tarsal plates ; the superior palpebral inosculating, at the outer angle of the orbit, with the orbital branch of the temporal artery, and with the upper of the two external palpebral branches from the lachrymal artery : the inferior palpebral inosculating, at the outer angle of the orbit, with the lower of the two external palpebral branches from the lachrymal and with the transverse facial artery, and, at the inner side of the lid, with a branch from the angular artery. From this last anastomosis a branch passes to the nasal duct, ramifying in its mucous membrane, as far as the inferior meatus.

The *frontal artery*, one of the terminal branches of the ophthalmic, passes from the orbit at its inner angle, and, ascending on the forehead, supplies the integument, muscles, and pericranium, anastomosing with the supra-orbital artery, and with the artery of the opposite side.

The *nasal artery*, the other terminal branch of the ophthalmic, emerges from the orbit above the tendo oculi, and, after giving a branch to the upper part of the lachrymal sac, divides into two branches, one of which crosses the root of the nose, the *transverse nasal*, and anastomoses with the angular artery ; the other, the *dorsalis nasi*, runs along the dorsum of the nose, supplies its outer surface, and anastomoses with the artery of the opposite side, and with the lateral nasal branch of the facial.

The *ciliary arteries* are divisible into three groups, the short, long, and anterior. The *short ciliary arteries*, from six to twelve in number, arise from the ophthalmic, or some of its branches ; they surround the optic nerve as they pass forwards to the posterior part of the eyeball, pierce the sclerotic coat around the entrance of the nerve, and supply the choroid coat and ciliary processes. The *long ciliary arteries*, two in number, pierce the posterior part of the sclerotic at some little distance from the optic nerve, and run forwards, along each side of the eyeball, between the sclerotic and choroid, to the ciliary muscle, where they divide into two branches ; these form an arterial circle, the *circulus major*, around the circumference of the iris, from which numerous radiating branches pass forwards, in its substance, to its free margin, where they form a second arterial circle, the *circulus minor*, around its pupillary margin. The *anterior ciliary arteries* are derived from the muscular branches ; they pass to the front of the eyeball in company with the tendons of the Recti muscles, form a vascular zone beneath the conjunctiva, and then pierce the sclerotic a short distance from the cornea and terminate in the circulus major of the iris.

The *arteria centralis retinæ* is the first and one of the smallest branches of the ophthalmic artery. It runs for a short distance within the dural sheath of

FIG. 292.—Vascular area of the upper surface of the cerebrum. (After Duret.)

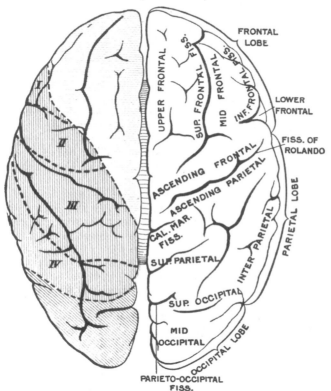

the nerve, but about half an inch behind the eyeball it pierces the optic nerve obliquely, and runs forward in the centre of its substance, and enters the globe of the eye through the porus opticus. Its mode of distribution will be described in the account of the anatomy of the eye.

The *muscular branches*, two in number, superior and inferior, frequently spring from a common trunk. The superior, the smaller, often wanting, supplies the Levator palpebræ, Superior rectus, and Superior oblique. The inferior, more constant in its existence, passes forwards, between the optic nerve and Inferior rectus, and is distributed to the External, Internal, and Inferior recti,

Blue indicates the distribution of the anterior cerebral ; yellow, the middle cerebral ; red, the posterior cerebral. I. The part supplied by the external and inferior frontal artery. II. The part supplied by the ascending frontal. III. The part supplied by the ascending parietal. IV. The part supplied by the parieto-sphenoidal artery.

and Inferior oblique. This vessel gives off most of the anterior ciliary arteries. Additional muscular branches are given off from the lachrymal and supra-orbital arteries or from the ophthalmic itself.

The **anterior cerebral** arises from the internal carotid, at the inner extremity of the fissure of Sylvius. It passes forwards and inwards across the anterior perforated space, above the optic nerve, to the commencement of the great longitudinal fissure. Here it comes into close relationship with the artery of the opposite side, and the two vessels are connected together by a short anastomosing trunk, about two lines in length, the *anterior communicating* artery. From this point, the two vessels run side by side in the longitudinal fissure, curve round the genu of the corpus callosum, and turning backwards continue along its upper surface to its posterior part, where they terminate by anastomosing with the posterior cerebral arteries.   In their course they give off the following branches :

Antero-median ganglionic.	Anterior internal frontal.
Inferior internal frontal.	Middle internal frontal.
	Posterior internal frontal.

The *antero-median ganglionic* is a group of small arteries which arise at the commencement of the anterior cerebral artery ; they pierce the anterior perforated space and lamina cinerea, and supply the head of the caudate nucleus.

The *inferior internal frontal*, two or three in number, are distributed to the orbital surface of the frontal lobe, where they supply the olfactory lobe, gyrus rectus, and internal orbital convolution.

FIG. 293.—Vascular area of the internal surface of the cerebrum.   (After Duret.)

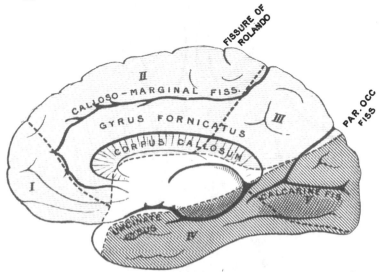

Blue indicates the distribution of the anterior cerebral artery ; red, the posterior cerebral.   I. The part supplied by the anterior and internal frontal.   II. The part supplied by the middle and internal frontal.   III. The part supplied by the posterior and internal frontal.   IV. The part supplied by the posterior temporal ; and V. the part supplied by the occipital, both terminal branches of the posterior cerebral.

The *anterior internal frontal branches* supply a part of the marginal convolution, and send branches over the edge of the hemisphere to the superior and middle frontal convolutions and upper part of the ascending frontal convolution. The *middle internal frontal branches* supply the corpus callosum, the convolution of the corpus callosum, the inner surface of the first frontal convolution, and the upper part of the ascending frontal convolution. The *posterior internal frontal branches* supply the lobus quadratus and adjacent outer surface of the hemisphere.

The *anterior communicating artery* is a short branch, about two lines in length, but of moderate size, connecting together the two anterior cerebral arteries across the longitudinal fissure.   Sometimes this vessel is wanting, the two arteries joining together to form a single trunk, which afterwards divides.   Or the vessel may be wholly, or partially, divided into two ; frequently it is longer and smaller than usual.   It gives off some of the antero-median ganglionic group of vessels, which are, however, principally derived from the anterior cerebral.

The **middle cerebral artery** (fig. 295), the largest branch of the internal carotid, passes obliquely outwards along the fissure of Sylvius, and, opposite the island of Reil, divides into its terminal branches.   The branches of the middle cerebral artery are :

Antero-lateral ganglionic.	Ascending frontal.
Inferior external frontal.	Ascending parietal.
	Parieto-temporal.

The *antero-lateral ganglionic branches* are a group of small arteries which arise at the commencement of the middle cerebral artery ; they pierce the anterior perforated space and supply the greater part of the caudate nucleus, the lenticular nucleus, the internal capsule, and a part of the optic thalamus. One artery of this group is of larger size than the rest, and is of special importance, as being the artery in the brain most frequently ruptured ; it has been termed by Charcot, the ' *artery of cerebral hæmorrhage.*' It passes up between the lenticular nucleus and the external capsule, and ultimately ends in the caudate nucleus. The *inferior external frontal* supplies the third or inferior frontal convolution (Broca's

FIG. 294.—Vascular area of the inferior surface of the cerebrum. (After Duret.)

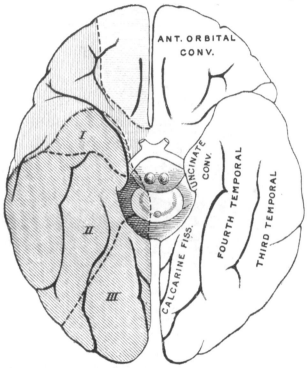

Blue indicates the distribution of the anterior cerebral ; yellow, the middle cerebral ; red, the posterior cerebral. I. The part supplied by the anterior temporal artery. II. The part supplied by the posterior temporal artery. III. The part supplied by the occipital artery.

convolution) and the outer part of the orbital surface of the frontal lobe. The *ascending frontal* supplies the ascending frontal convolution. The *ascending parietal* supplies the ascending parietal convolution and the lower part of the superior parietal convolution. The *parieto-temporal* supplies the supra-marginal, the superior, and part of the middle temporal convolutions, and the angular gyrus.

The **posterior communicating artery** arises from the back part of the internal carotid, runs directly backwards, and anastomoses with the posterior cerebral, a branch of the basilar. This artery varies considerably in size, being sometimes small, and occasionally so large that the posterior cerebral may be considered as arising from the internal carotid rather than from the basilar. It is frequently larger on one side than on the other side. From the posterior half of this vessel are given off a number of small branches, the *postero-median ganglionic branches*, which, with similar vessels from the posterior cerebral, pierce the posterior perforated space and supply the internal surfaces of the optic thalami and the walls of the third ventricle.

The **anterior choroid** is a small but constant branch, which arises from the back part of the internal carotid, near the posterior communicating artery.

Passing backwards and outwards between the temporal lobe and the crus cerebri, it enters the descending horn of the lateral ventricle through the transverse fissure and ends in the choroid plexus. It is distributed to the hippocampus major, corpus fimbriatum, velum interpositum, and choroid plexus.

THE BLOOD-VESSELS OF THE BRAIN

Recent investigations have tended to show that the mode of distribution of the vessels of the brain has an important bearing upon a considerable number of the anatomical lesions of which this part of the nervous system may be the seat; it therefore becomes important to consider a little more in detail the way in which the cerebral vessels are distributed.

FIG. 295.—The distribution of the middle cerebral artery. (After Charcot.)

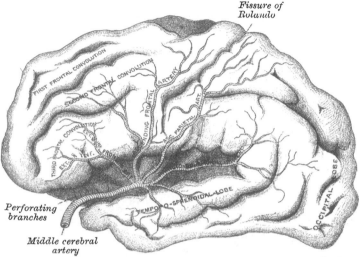

The cerebral arteries are derived from the internal carotid and the vertebral, which at the base of the brain form a remarkable anastomosis known as the *circle of Willis*. It is formed in front by the anterior cerebral arteries, branches of the internal carotid, which are connected together by the anterior communicating; behind by the two posterior cerebrals, branches of the basilar, which are connected on each side with the internal carotid by the posterior communicating (fig. 291, p. 482). The parts of the brain included within this arterial circle are the lamina cinerea, the commissure of the optic nerves, the infundibulum, the tuber cinereum, the corpora albicantia, and the posterior perforated space.

From the circle of Willis arise the three trunks which together supply each cerebral hemisphere. From its anterior part proceed the two anterior cerebrals; from its antero-lateral part the middle cerebrals, and from its posterior part the posterior cerebrals. Each of these principal arteries gives origin to two very different systems of secondary vessels. One of these systems has been named the *central ganglionic system*, and the vessels belonging to it supply the central ganglia of the brain; the other has been named the *cortical arterial system*, and its vessels ramify in the pia mater and supply the cortex and subjacent medullary matter. These two systems, though they have a common origin, do not communicate at any point of their peripheral distribution, and are entirely independent of each other. Though some of the arteries of the cortical system approach, at their terminations, the regions supplied by the central ganglionic system, no communication between the two sets of vessels takes place, and there is between the parts supplied by the two systems a borderland of diminished nutritive activity, where, it is said, softening is especially liable to occur in the brains of old people.

The central ganglionic system.—All the vessels belonging to this system are given off from the circle of Willis, or from the vessels immediately after their origin from it. So that if a circle is drawn at a distance of about an inch from the circle of Willis, it will include the origin of all the arteries belonging to this system (fig. 296). The vessels of this system form six principal groups : (I) the *antero-median group*, derived from the anterior cerebrals and anterior communicating ; (II) the *postero-median group*, from the posterior cerebrals and posterior communicating ; (III) the right and left *antero-lateral group*, from the middle cerebrals ; and (IV) the right and left *postero-lateral group*, from the posterior cerebrals, after they have wound round the crura cerebri. The vessels belonging to this system are larger than those of the cortical system, and are what Cohnheim has designated 'terminal' arteries—that is to say, vessels which

FIG. 296.—Diagram of the arterial circulation at the base of the brain. (After Charcot.)

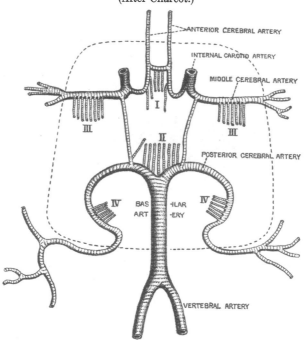

I. Antero-median group of ganglionic branches. II. Postero-median group. III. Right and left antero-lateral group. IV. Right and left postero-lateral group. The dotted line shows the limit of the ganglionic circle.

from their origin to their termination neither supply nor receive any anastomotic branch, so that, by one of the small vessels, only a limited area of the central ganglia can be injected, and the injection cannot be driven beyond the area of the part supplied by the particular vessel which is the subject of the experiment.

The cortical arterial system.—The vessels forming this system are the terminal branches of the anterior, middle, and posterior cerebral arteries, described above. These vessels divide and ramify in the substance of the pia mater, and give off nutrient arteries which penetrate the cortex perpendicularly. These nutrient vessels are divisible into two classes, the long and short. The *long*, or, as they are sometimes called, the *medullary arteries*, pass through the grey matter to penetrate the centrum ovale to the depth of about an inch and a half, without intercommunicating otherwise than by very fine capillaries, and thus constitute so many independent small systems. The *short vessels* are confined to the cortex, where they form with the long vessels a compact network in the middle zone of the grey matter, the outer and inner zones being sparingly supplied with blood (fig. 297). The vessels of the cortical arterial system are not so strictly 'terminal' as those of the central ganglionic system, but they approach this type very closely, so that injection of one area from the vessel of another area, though it may be possible, is frequently very difficult, and is only effected through vessels of small calibre. As a result of this, obstruction of one of the main branches, or its divisions, may have the effect of producing softening in a very limited area of the cortex.*

ARTERIES OF THE UPPER EXTREMITY

The artery which supplies the upper extremity continues as a single trunk from its commencement down to the elbow; but different portions of it have received different names, according to the region through which it passes. That

FIG. 297.—Distribution of the cortical arteries. (After Charcot.)

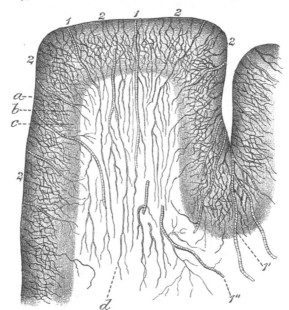

1. Medullary arteries. 1′. Group of medullary arteries in the sulcus between two adjacent convolutions. 1″. Arteries situated among the short association fibres. 2, 2. Cortical arteries. a. Capillary network with fairly wide meshes, situated beneath the pia mater. b. Network with more compact, polygonal meshes, situated in the cortex. c. Transitional network with wider meshes. d. Capillary network in the white matter.

part of the vessel which extends from its origin to the outer border of the first rib is termed the *subclavian* ; beyond this point to the lower border of the axilla, it is termed the *axillary* ; and from the lower margin of the axillary space to the bend of the elbow, it is termed *brachial*; here, the single trunk terminates by dividing into two branches, the *radial* and *ulnar*, an arrangement precisely similar to what occurs in the lower limb.

SUBCLAVIAN ARTERIES (fig. 298)

The **subclavian artery** on the right side arises from the innominate artery opposite the right sterno-clavicular articulation ; on the left side it arises from the

* The student who desires further information on this subject is referred to Charcot's *Localisation of Cerebral and Spinal Diseases*, p. 42 *et seq.*, whence the facts above given have been principally derived.

arch of the aorta. It follows, therefore, that these two vessels must, in the first part of their course, differ in their length, their direction, and their relation with neighbouring parts.

In order to facilitate the description of these vessels, more especially from a surgical point of view, each subclavian artery has been divided into three parts. The first portion, on the right side, passes upwards and outwards from the origin of the vessel to the inner border of the Scalenus anticus. On the left side it ascends nearly vertically, to gain the inner border of that muscle. The second part passes outwards, behind the Scalenus anticus ; and the third part passes from the outer margin of that muscle, beneath the clavicle, to the outer border of the first rib, where it becomes the axillary artery. The first portion of these two vessels differs so much in its course, and in its relation with neighbouring parts, that it will be described separately. The second and third parts are alike on the two sides.

First Part of the Right Subclavian Artery (figs. 280, 283, 298)

The **right subclavian artery** arises from the arteria innominata, opposite the upper part of the right sterno-clavicular articulation, and passes upwards and outwards to the inner margin of the Scalenus anticus muscle. In this part of its course it ascends a little above the clavicle, the extent to which it does so varying in different cases. It is covered, *in front*, by the integument, superficial fascia, Platysma, deep fascia, the clavicular origin of the Sterno-mastoid, the Sterno-hyoid, and Sterno-thyroid muscles, and another layer of the deep fascia. It is crossed by the internal jugular and vertebral veins, and by the pneumo-gastric and the cardiac branches of the sympathetic. A loop of the sympathetic nerve itself also crosses the artery, forming a ring around the vessels. The anterior jugular vein passes outwards in front of the artery but is not in contact with it, being separated from it by the Sterno-hyoid and Sterno-thyroid muscles. Below and behind the artery is the pleura, which separates it from the apex of the lung ; *behind* is the cord of the sympathetic nerve ; the recurrent laryngeal nerve winds round the lower and back part of the vessel.

Plan of Relations of First Portion of the Right Subclavian Artery

In front.

Skin, superficial fascia.
Platysma, deep fascia.
Clavicular origin of Sterno-mastoid.
Sterno-hyoid, and Sterno-thyroid.
Anterior jugular, internal jugular, and vertebral veins.
Pneumogastric and cardiac nerves.
Loop from the sympathetic.

Right Subclavian Artery. First Portion.

Beneath.

Pleura.
Recurrent laryngeal nerve.

Behind.

Recurrent laryngeal nerve.
Sympathetic.
Pleura and apex of lung.

First Part of the Left Subclavian Artery (fig. 280)

The **left subclavian artery** arises from the end of the arch of the aorta, opposite the fourth dorsal vertebra, and ascends nearly vertically to the inner margin of the Scalenus anticus muscle. This part of the vessel is, therefore, longer than the right, situated deeply in the cavity of the chest, and directed nearly vertically upwards instead of arching outwards like the vessel of the opposite side.

It is in relation, *in front*, with the pneumogastric, cardiac, and phrenic nerves, which lie parallel with it, the left carotid artery, left internal jugular and vertebral veins, and the commencement of the left innominate vein, and is covered by the Sterno-thyroid, Sterno-hyoid, and Sterno-mastoid muscles; *behind*, it is in relation with the œsophagus, thoracic duct, inferior cervical ganglion of the sympathetic, and Longus colli; higher up, however, the œsophagus and thoracic duct lie to its right side; the latter ultimately arching over the vessel to join the angle of union between the subclavian and internal jugular veins. To its *inner side* are the œsophagus, trachea, and thoracic duct; to its *outer side*, the left pleura and lung.

PLAN OF RELATIONS OF FIRST PORTION OF THE LEFT SUBCLAVIAN ARTERY

In front.

Pneumogastric, cardiac, and phrenic nerves.
Left carotid artery.   Thoracic duct.
Left internal jugular, vertebral, and innominate veins.
Sterno-thyroid, Sterno-hyoid, and Sterno-mastoid muscles.

Inner side.

Trachea.
Œsophagus.
Thoracic duct.

Left Subclavian Artery.

Outer side.

Pleura and left lung.

Behind.

Œsophagus and thoracic duct.
Inferior cervical ganglion of sympathetic.
Longus colli.

SECOND AND THIRD PARTS OF THE SUBCLAVIAN ARTERY (figs. 283, 298)

The **Second Portion of the Subclavian Artery** lies behind the Scalenus anticus muscle; it is very short, and forms the highest part of the arch described by that vessel.

Relations.—It is covered, *in front*, by the skin, superficial fascia, Platysma, deep cervical fascia, Sterno-mastoid, and the Scalenus anticus muscle. On the right side the phrenic nerve is separated from the second part of the artery by the Anterior scalene muscle, while on the left side the nerve crosses the first part of the artery immediately to the inner edge of the muscle. *Behind*, it is in relation with the pleura and the Middle scalene. *Above*, with the brachial plexus of nerves. *Below*, with the pleura. The subclavian vein lies below and in front of the artery, separated from it by the Scalenus anticus.

PLAN OF RELATIONS OF SECOND PORTION OF THE SUBCLAVIAN ARTERY

In front.

Skin and superficial fascia.
Platysma and deep cervical fascia.
Sterno-mastoid.
Phrenic nerve.
Scalenus anticus.
Subclavian vein.

Above.

Brachial plexus.

Subclavian Artery. Second Portion.

Below.

Pleura.

Behind.

Pleura and Middle Scalenus.

The **Third Portion of the Subclavian Artery** passes downwards and outwards from the outer margin of the Scalenus anticus to the outer border of the first rib, where it becomes the axillary artery. This portion of the vessel is the most superficial, and is contained in the Subclavian triangle (see p. 478).

Relations.—It is covered, *in front*, by the skin, the superficial fascia, the Platysma, the descending clavicular branches of the cervical plexus, and the deep

FIG. 298.—The subclavian artery, showing its relations.
(From a preparation in the Museum of the Royal College of Surgeons of England.)

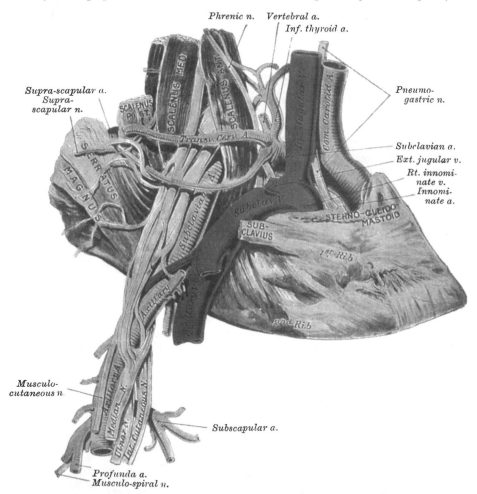

cervical fascia; by the clavicle, the Subclavius muscle, and the suprascapular artery and vein, and the transverse cervical vein; the nerve to the Subclavius muscle passes vertically downwards in front of the artery. The external jugular vein crosses it at its inner side, and receives the suprascapular and transverse cervical veins, which frequently form a plexus in front of it. The subclavian vein is below and in front of the artery, lying close behind the clavicle. *Behind*, it lies on the Middle scalene muscle and the lowest cord of the brachial plexus, formed by the union of the last cervical and first dorsal nerves. *Above* it, and to its outer side, is the brachial plexus, and Omo-hyoid muscle. *Below*, it rests on the upper surface of the first rib.

PLAN OF RELATIONS OF THIRD PORTION OF THE SUBCLAVIAN ARTERY

In front.

Skin and superficial fascia.
Platysma and deep cervical fascia.
Descending branches of cervical plexus.   Nerve to Subclavius muscle.
Subclavius muscle, suprascapular artery, and vein.
External jugular and transverse cervical veins.
Clavicle.

Above.

Brachial plexus.
Omo-hyoid.

Subclavian
Artery.
Third Portion.

Below.

First rib.

Behind.

Scalenus medius.
Lower cord of brachial plexus.

Peculiarities.—The subclavian arteries vary in their origin, their course, and the height to which they rise in the neck.

The origin of the right subclavian from the innominate takes place, in some cases, above the sterno-clavicular articulation; and occasionally, but less frequently, in the cavity of the thorax, below that joint.   Or the artery may arise as a separate trunk from the arch of the aorta.   In such cases it may be either the first, second, third, or even the last branch derived from that vessel; in the majority of cases, it is the first or last, rarely the second or third.   When it is the first branch, it occupies the ordinary position of the innominate artery; when the second or third, it gains its usual position by passing behind the right carotid; and when the last branch, it arises from the left extremity of the arch, at its upper or back part, and passes obliquely towards the right side, usually behind the trachea, œsophagus, and right carotid, sometimes between the œsophagus and trachea, to the upper border of the first rib, whence it follows its ordinary course.   In very rare instances, this vessel arises from the thoracic aorta, as low down as the fourth dorsal vertebra.   Occasionally, it perforates the Anterior scalene; more rarely it passes in front of that muscle.   Sometimes the subclavian vein passes with the artery behind the Anterior scalene.   The artery may ascend as high as an inch and a half above the clavicle, or any intermediate point between this and the upper border of the bone, the right subclavian usually ascending higher than the left.

The left subclavian is occasionally joined at its origin with the left carotid.

Surface Marking.—The course of the subclavian artery in the neck may be mapped out, by describing a curve, with its convexity upwards, at the base of the posterior triangle. The inner end of this curve corresponds to the sterno-clavicular joint, the outer end to the centre of the lower border of the clavicle.   The curve is to be drawn with such an amount of convexity that its mid-point reaches half an inch above the upper border of the clavicle.   The left subclavian artery is more deeply placed than the right in the first part of its course, and, as a rule, does not reach quite as high a level in the neck.   It should be borne in mind that the posterior border of the Sterno-mastoid muscle corresponds to the outer border of the Scalenus anticus, so that the third portion of the artery, that part most accessible for operation, lies immediately external to the posterior border of the Sterno-mastoid.

Surgical Anatomy.—The relations of the subclavian arteries of the two sides having been examined, the student should direct his attention to a consideration of the best position in which compression of the vessel may be effected, or in what situation a ligature may be best applied in cases of aneurism or wound.

Compression of the subclavian artery is required in cases of operations about the shoulder, in the axilla, or at the upper part of the arm; and the student will observe that there is only one situation in which it can be effectually applied, viz. where the artery passes across the upper surface of the first rib.   In order to compress the vessel in this situation, the shoulder should be depressed, and the surgeon grasping the side of the neck should press with his thumb in the angle formed by the posterior border of the Sterno-mastoid with the upper border of the clavicle, downwards, backwards, and inwards against the rib; if from any cause the shoulder cannot be sufficiently depressed, pressure may be made from before backwards, so as to compress the artery against the Middle scalenus and transverse process of the seventh cervical vertebra.   In appropriate cases, a preliminary incision may be made through the cervical fascia, and the finger may be pressed down directly upon the artery.

Ligature of the subclavian artery may be required in cases of wounds, or of aneurism in the axilla, or in cases of aneurism on the cardiac side of the point of ligature; and the third part of the artery is that which is most favourable for an operation, on account of its being comparatively superficial, and most remote from the origin of the large branches. In those cases where the clavicle is not displaced, this operation may be performed with comparative facility; but where the clavicle is pushed up by a large aneurismal tumour in the axilla, the artery is placed at a great depth from the surface, which materially increases the difficulty of the operation. Under these circumstances, it becomes a matter of importance to consider the height to which this vessel reaches above the bone. In ordinary cases, its arch is about half an inch above the clavicle, occasionally as high as an inch and a half, and sometimes so low as to be on a level with its upper border. If the clavicle is displaced, these variations will necessarily make the operation more or less difficult, according as the vessel is more or less accessible.

The chief points in the operation of tying the third portion of the subclavian artery are as follows: the patient being placed on a table in the supine position, with the head drawn over to the opposite side, and the shoulder depressed as much as possible, the integument should be drawn downwards over the clavicle, and an incision made through it, upon that bone, from the anterior border of the Trapezius to the posterior border of the Sterno-mastoid, to which may be added a short vertical incision meeting the inner end of the preceding. The object in drawing the skin downwards is to avoid any risk of wounding the external jugular vein, for as it perforates the deep fascia above the clavicle, it cannot be drawn downwards with the skin. The soft parts should now be allowed to glide up, and the cervical fascia divided upon a director, and if the interval between the Trapezius and Sterno-mastoid muscles be insufficient for the performance of the operation, a portion of one or both may be divided. The external jugular vein will now be seen towards the inner side of the wound: this and the suprascapular and transverse cervical veins which terminate in it should be held aside. If the external jugular vein is at all in the way and exposed to injury, it should be tied in two places and divided. The suprascapular artery should be avoided, and the Omo-hyoid muscle held aside if necessary. In the space beneath this muscle, careful search must be made for the vessel; a deep layer of fascia and some connective tissue having been divided carefully, the outer margin of the Scalenus anticus muscle must be felt for, and the finger being guided by it to the first rib, the pulsation of the subclavian artery will be felt as it passes over the rib. The sheath of the vessels having been opened, the aneurism needle may then be passed around the artery from above downwards and inwards so as to avoid including any of the branches of the brachial plexus. If the clavicle is so raised by the tumour that the application of the ligature cannot be effected in this situation, the artery may be tied above the first rib, or even behind the Scalenus anticus muscle; the difficulties of the operation in such a case will be materially increased, on account of the greater depth of the artery, and the alteration in position of the surrounding parts.

The second part of the subclavian artery, from being that portion which rises highest in the neck, has been considered favourable for the application of the ligature, when it is difficult to tie the artery in the third part of its course. There are, however, many objections to the operation in this situation. It is necessary to divide the Scalenus anticus muscle, upon which lies the phrenic nerve, and at the inner side of which is situated the internal jugular vein; and a wound of either of these structures might lead to the most dangerous consequences. Again, the artery is in contact, below, with the pleura, which must also be avoided; and, lastly, the proximity of so many of its large branches arising internal to this point must be a still further objection to the operation. In cases, however, where the sac of an axillary aneurism encroaches on the neck, it may be necessary to divide the outer half or two-thirds of the Scalenus anticus muscle, so as to place the ligature on the vessel at a greater distance from the sac. The operation is performed exactly in the same way as ligature of the third portion, until the Scalenus anticus is exposed, when it is to be divided on a director (never to a greater extent than its outer two-thirds), and it immediately retracts. The operation is therefore merely an extension of ligature of the third portion of the vessel.

In those cases of aneurism of the axillary or subclavian artery which encroach upon the outer portion of the Scalenus muscle to such an extent that a ligature cannot be applied in that situation, it may be deemed advisable, as a last resource, to tie the first portion of the subclavian artery. On the left side, this operation is almost impracticable; the great depth of the artery from the surface, its intimate relation with the pleura, and its close proximity to the thoracic duct and to so many important veins and nerves, present a series of difficulties which it is next to impossible to overcome.* On the right side, the operation is practicable, and has been performed, though never with success. The main objection to the operation in this situation is the smallness of the interval which usually exists between the commencement of the vessel and the origin of the nearest branch. The operation may be performed in the following manner: The patient being placed on the table in the supine position, with the neck extended, an incision should be made along the upper border of the inner part of the clavicle, and a second along the inner border of the Sterno-mastoid, meeting the former at an angle. The attachment of both heads of the

* The operation was, however, performed in New York by Dr. J. K. Rodgers, and the case is related in *A System of Surgery*, edited by T. Holmes, 2nd ed. vol. iii. pp. 620, &c.

Sterno-mastoid must be divided on a director, and turned outwards; a few small arteries and veins, and occasionally the anterior jugular, must be avoided, or, if necessary, ligatured in two places and divided, and the Sterno-hyoid and Sterno-thyroid muscles divided in the same manner as the preceding muscle. After tearing through the deep fascia with the finger-nail, the internal jugular vein will be seen crossing the subclavian artery; this should be pressed aside, and the artery secured by passing the needle from below upwards, by which the pleura is more effectually avoided. The exact position of the vagus, the recurrent laryngeal, the phrenic and sympathetic nerves should be remembered, and the ligature should be applied near the origin of the vertebral, in order to afford as much room as possible for the formation of a coagulum between the ligature and the origin of the vessel. It should be remembered, that the right subclavian artery is occasionally deeply placed in the first part of its course, when it arises from the left side of the aortic arch, and passes in such cases behind the œsophagus, or between it and the trachea.

Collateral Circulation.—After ligature of the third part of the subclavian artery, the collateral circulation is mainly established by three sets of vessels, thus described in a dissection :

' 1. A posterior set, consisting of the suprascapular and posterior scapular branches of the subclavian, anastomosing with the subscapular from the axillary.

' 2. An internal set, produced by the connection of the internal mammary on the one hand, with the superior and long thoracic arteries, and the branches from the subscapular on the other.

' 3. A middle or axillary set, consisting of a number of small vessels derived from branches of the subclavian, above; and, passing through the axilla, terminated either in the main trunk, or some of the branches of the axillary below. This last set presented most conspicuously the peculiar character of newly formed or, rather, dilated arteries, being excessively tortuous, and forming a complete plexus.

' The chief agent in the restoration of the axillary artery below the tumour was the subscapular artery, which communicated most freely with the internal mammary, suprascapular, and posterior scapular branches of the subclavian, from all of which it received so great an influx of blood as to dilate it to three times its natural size.' *

When a ligature is applied to the first part of the subclavian artery, the collateral circulation is carried on by : 1, the anastomosis between the superior and inferior thyroid ; 2, the anastomosis of the two vertebrals ; 3, the anastomosis of the internal mammary with the deep epigastric and the aortic intercostals ; 4, the superior intercostal anastomosing with the aortic intercostals ; 5, the profunda cervicis anastomosing with the princeps cervicis ; 6, the scapular branches of the thyroid axis anastomosing with the branches of the axillary ; and 7, the thoracic branches of the axillary anastomosing with the aortic intercostals.

Branches.—The branches given off from the subclavian artery are :

Vertebral.	Thyroid axis.
Internal mammary.	Superior intercostal.

On the left side all four branches generally arise from the first portion of the vessel ; but on the right side, the superior intercostal usually arises from the second portion of the vessel. On both sides of the body, the first three branches arise close together at the inner margin of the Scalenus anticus ; in the majority of cases, a free interval of from half an inch to an inch exists between the commencement of the artery and the origin of the nearest branch ; in a smaller number of cases, an interval of more than an inch exists, but it never exceeds an inch and three-quarters. In a very few instances, the interval has been found to be less than half an inch. The vertebral artery arises from the upper and posterior part of the artery, the internal mammary from the lower part of the

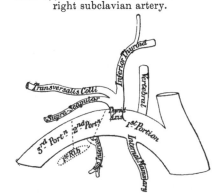

FIG. 299.—Plan of the branches of the right subclavian artery.

artery ; the thyroid axis from in front, and the superior intercostal from behind.

* *Guy's Hospital Reports,* vol. i. 1836. Case of axillary aneurism, in which Mr. Aston Key had tied the subclavian artery on the outer edge of the Scalenus muscle, twelve years previously.

The **Vertebral artery** (fig. 289) is generally the first and largest branch of the subclavian; it arises from the upper and back part of the first portion of the vessel, and, passing upwards, enters the foramen in the transverse process of the sixth cervical vertebra,* and ascends through the foramina in the transverse processes of all the vertebræ above this.   Above the upper border of the axis, it inclines outwards and upwards to the foramen in the transverse process of the atlas, through which it passes; it then winds backwards behind its articular process, runs in a deep groove on the upper surface of the posterior arch of this bone, and, passing beneath the posterior occipito-atlantal ligament, pierces the dura mater and arachnoid, and enters the skull through the foramen magnum. It then passes forwards and upwards, inclining from the lateral aspect to the front of the medulla oblongata.   It unites, in the middle line, with the vessel of the opposite side at the lower border of the pons Varolii to form the *basilar artery*.

Relations.—At its origin, it is situated behind the internal jugular and vertebral veins, and is crossed by the inferior thyroid artery: it lies between the Longus colli and Scalenus anticus muscles, having the thoracic duct in front of it on the left side.   It rests on the transverse process of the seventh cervical vertebra and the sympathetic nerve.   Within the foramina formed by the transverse processes of the vertebræ, it is accompanied by a plexus of nerves from the inferior cervical ganglion of the sympathetic, and is surrounded by a dense plexus of veins which unite to form the vertebral vein at the lower part of the neck.   It is situated in front of the cervical nerves, as they issue from the inter-vertebral foramina.   While winding round the articular process of the atlas, it is contained in a triangular space (*suboccipital triangle*) formed by the Rectus capitis posticus major, the Superior and the Inferior oblique muscles; and at this point is covered by the Complexus muscle.   The suboccipital nerve here lies between the artery and the bone.   Within the skull, as it winds round the medulla oblongata, it is placed between the hypoglossal nerve and the anterior root of the suboccipital nerve, beneath the first digitation of the ligamentum denticulatum, and finally ascends between the basilar process of the occipital bone and the anterior surface of the medulla oblongata.

The branches given off from this vessel may be divided into two sets—those given off in the neck, and those within the cranium.

Cervical Branches.	*Cranial Branches.*
Lateral spinal.	Posterior meningeal.
Muscular.	Anterior spinal.
	Posterior spinal.
	Posterior inferior cerebellar.
	Bulbar.

The **lateral spinal branches** enter the spinal canal through the intervertebral foramina, and divide into two branches.   Of these, one passes along the roots of the nerves to supply the spinal cord and its membranes, anastomosing with the other arteries of the spinal cord; the other divides into an ascending and a descending branch, which unite with similar branches from the artery above and below, so that two lateral anastomotic chains are formed on the posterior surface of the bodies of the vertebræ, near the attachment of the pedicles.   From these anastomotic chains branches are given off, to supply the periosteum and the bodies of the vertebræ, and to communicate with similar branches from the opposite side; from these latter small branches are given off which join similar branches above and below, so that a central anastomotic chain is formed on the posterior surface of the bodies of the vertebræ.

* The vertebral artery sometimes enters the foramen in the transverse process of the fifth vertebra.   Dr. Smyth, who tied this artery in the living subject, found it, in one of his dissections, passing into the foramen in the seventh vertebra.

Muscular branches are given off to the deep muscles of the neck, where the vertebral artery curves round the articular process of the atlas. They anastomose with the occipital, and with the ascending and deep cervical arteries.

The **posterior meningeal** are one or two small branches given off from the vertebral opposite the foramen magnum. They ramify between the bone and dura mater in the cerebellar fossæ, and supply the falx cerebelli.

The **anterior spinal** is a small branch, which arises near the termination of the vertebral, and, descending in front of the medulla oblongata, unites with its fellow of the opposite side at about the level of the foramen magnum. One of these vessels is usually larger than the other, but occasionally they are about equal in size. The single trunk, thus formed, descends on the front of the spinal cord, and is reinforced by a succession of small branches which enter the spinal canal through the intervertebral foramina; these branches are derived from the vertebral and ascending cervical of the inferior thyroid in the neck; from the intercostals in the dorsal region; and from the lumbar, ilio-lumbar, and lateral sacral arteries in the lower part of the spine. They unite, by means of ascending and descending branches, to form a single anterior median artery, which extends as far as the lower part of the spinal cord. This vessel is placed in the pia mater along the anterior median fissure; it supplies that membrane, and the substance of the cord, and sends off branches at its lower part to be distributed to the cauda equina, and ends on the central fibrous prolongation of the cord.

The **posterior spinal** arises from the vertebral, at the side of the medulla oblongata; passing backwards to the posterior aspect of the spinal cord, it descends on each side, lying behind the posterior roots of the spinal nerves; and is reinforced by a succession of small branches, which enter the spinal canal through the intervertebral foramina, and by which it is continued to the lower part of the cord, and to the cauda equina. Branches from these vessels form a free anastomosis round the posterior roots of the spinal nerves, and communicate, by means of very tortuous transverse branches, with the vessel of the opposite side. At its commencement it gives off an ascending branch, which terminates at the side of the fourth ventricle.

The **posterior inferior cerebellar artery** (fig. 291), the largest branch of the vertebral, winds backwards round the upper part of the medulla oblongata, passing between the origin of the pneumogastric and spinal accessory nerves, over the restiform body to the under surface of the cerebellum, where it divides into two branches: an internal one, which is continued backwards to the notch between the two hemispheres of the cerebellum; and an external one, which supplies the under surface of the cerebellum, as far as its outer border, where it anastomoses with the anterior inferior cerebellar and the superior cerebellar branches of the basilar artery. Branches from this artery supply the choroid plexus of the fourth ventricle.

The **bulbar arteries** comprise several minute vessels which spring from the vertebral and its branches and are distributed to the medulla oblongata.

Surgical Anatomy.—The vertebral artery has been tied in several instances: 1, for wounds or traumatic aneurism; 2, after ligature of the innominate, either at the same time to prevent hæmorrhage, or later on to arrest bleeding where it has occurred at the seat of ligature; and 3, in epilepsy. In these latter cases the treatment was recommended by Dr. Alexander, of Liverpool, in the hope that by diminishing the supply of blood to the posterior part of the brain and the spinal cord, a diminution or cessation of the epileptic fits would result. But, on account of the uncertainty as to what cases, if any, derived benefit from the operation, it has now been abandoned. The operation of ligature of the vertebral is performed by making an incision along the posterior border of the Sterno-mastoid muscle, just above the clavicle. The muscle is pulled to the inner side and the anterior tubercle of the transverse process of the sixth cervical vertebra sought for. A deep layer of fascia being now divided, the interval between the Scalenus anticus and the Longus colli, just below their attachment to the tubercle, is defined, and the artery and vein found in the interspace. The vein is to be drawn to the outer side, and the aneurism needle passed from without inwards. Drs. Ramskill and Bright have pointed out that severe pain at the back of the head may be symptomatic of disease of the vertebral artery just before it enters the skull. This is explained by the close connection of the artery with the suboccipital nerve in the groove on the posterior arch of the atlas. Disease of the same artery has been also said to affect speech, from pressure on the hypoglossal, where it is in relation with the vessel, leading to paralysis of the muscles of the tongue.

The **Basilar artery** (fig. 291), so named from its position at the base of the skull, is a single trunk formed by the junction of the two vertebral arteries; it extends from the posterior to the anterior border of the pons Varolii, lying in its median groove, under cover of the arachnoid. It ends by dividing into the two *posterior cerebral arteries*. Its branches are, on each side, the following:

Transverse.	Superior cerebellar.
Anterior inferior cerebellar.	Posterior cerebral.

The **transverse branches** supply the pons Varolii and adjacent parts of the brain; one branch, the *internal auditory*, accompanies the auditory nerve into the internal auditory meatus.

The **anterior inferior cerebellar artery** passes backwards across the crus cerebelli, to be distributed to the anterior border of the under surface of the cerebellum, anastomosing with the posterior inferior cerebellar branch of the vertebral.

The **superior cerebellar arteries** arise near the termination of the basilar. They pass outwards, immediately behind the third nerves, which separate them from the posterior cerebral, wind round the crura cerebri, close to the fourth nerve, and arriving at the upper surface of the cerebellum, divide into branches which ramify in the pia mater and, reaching the circumference of the cerebellum, anastomose with the branches of the inferior cerebellar arteries. Several branches are given to the pineal gland, the valve of Vieussens, and the velum interpositum.

The **posterior cerebral arteries,** the two terminal branches of the basilar, are larger than the preceding, from which they are separated near their origin by the third nerves. Passing outwards, parallel to the superior cerebellar artery, and receiving the posterior communicating from the internal carotid, they wind round the crura cerebri, and pass to the under surface of the occipital lobes of the cerebrum, and break up into branches for the supply of the temporal and occipital lobes. The branches of the posterior cerebral artery are:

Postero-median ganglionic.		⎧ Anterior temporal.
Posterior choroid.	Three terminal	⎨ Posterior temporal.
Postero-lateral ganglionic.		⎩ Occipital.

The *postero-median ganglionic branches* (fig. 296) are a group of small arteries which arise at the commencement of the posterior cerebral artery; these, with similar branches from the posterior communicating, pierce the posterior perforated space, and supply the internal surfaces of the optic thalami and the walls of the third ventricle. The *posterior choroid* enters the interior of the brain beneath the splenium of the corpus callosum, and supplies the velum interpositum and the choroid plexus. The *postero-lateral ganglionic branches* are a group of small arteries which arise from the posterior cerebral artery, after it has turned round the crus cerebri; they supply a considerable portion of the optic thalamus. The *terminal branches* are distributed as follows: the first (*anterior temporal*) to the uncinate gyrus; the second (*posterior temporal*) to the external occipital and the third temporal convolutions; and the third (*occipital*) to the inner and outer surfaces of the occipital lobe.

Circle of Willis.—The remarkable anastomosis which exists between the branches of the internal carotid and vertebral arteries at the base of the brain constitutes the *circle of Willis*. It is formed, in front, by the anterior cerebral arteries, branches of the internal carotid, which are connected together by the anterior communicating ; behind, by the two posterior cerebrals, branches of the basilar, which are connected on each side with the internal carotid by the posterior communicating arteries (fig. 291). It is by this anastomosis that the cerebral circulation is equalised, and provision made for effectually carrying it on if one or more of the branches are obliterated. The parts of the brain included within this arterial circle are : the lamina cinerea, the commissure of the optic nerves, the infundibulum, the tuber cinereum, the corpora albicantia, and the posterior perforated space.

The **Thyroid axis** (fig. 283) is a short thick trunk, which arises from the fore part of the first portion of the subclavian artery, close to the inner border of the Scalenus anticus muscle, and divides, almost immediately after its origin, into three branches, the *inferior thyroid, suprascapular,* and *transversalis colli.*

The **Inferior thyroid artery** passes upwards, in front of the vertebral artery and Longus colli muscle ; then turns inwards behind the sheath of the common carotid artery and internal jugular vein, and also behind the sympathetic nerve, the middle cervical ganglion resting upon the vessel, and reaching the lower border of the lateral lobe of the thyroid gland it divides into two branches, which supply the posterior and under part of the organ, and anastomose in its substance with the superior thyroid, and with the corresponding artery of the opposite side. The recurrent laryngeal nerve passes upwards, generally behind but occasionally in front of the artery. Its branches are : the

Inferior laryngeal.	Œsophageal.
Tracheal.	Ascending cervical.
Muscular.	

The **inferior laryngeal branch** ascends upon the trachea to the back part of the larynx, in company with the recurrent laryngeal nerve, and supplies the muscles and mucous membrane of this part, anastomosing with the branch from the opposite side, and with the laryngeal branch from the superior thyroid artery.

The **tracheal branches** are distributed upon the trachea, anastomosing below with the bronchial arteries.

The **œsophageal branches** are distributed to the œsophagus, and anastomose with the œsophageal branches of the aorta.

The **ascending cervical** is a small branch which arises from the inferior thyroid, just where that vessel is passing behind the common carotid artery, and runs up on the anterior tubercles of the transverse processes of the cervical vertebræ in the interval between the Scalenus anticus and Rectus capitis anticus major. It gives branches to the muscles of the neck, which anastomose with branches of the vertebral, and sends one or two branches into the spinal canal through the intervertebral foramina to be distributed to the spinal cord and its membranes, and to the bodies of the vertebræ in the same manner as the lateral spinal branches from the vertebral. It anastomoses with the ascending pharyngeal and occipital arteries.

The **muscular branches** supply the depressors of the hyoid bone, the Longus colli, the Scalenus anticus, and the Inferior constrictor of the pharynx.

Surgical Anatomy.—This artery has been tied, in conjunction with the superior thyroid, in cases of bronchocele. An incision is made along the anterior border of the Sterno-mastoid down to the clavicle. After the deep fascia has been divided, the Sterno-mastoid and carotid vessels are drawn outwards, and the carotid (Chassaignac's) tubercle sought for. The vessel will be found just below this tubercle, between the carotid sheath on the outer side and the trachea and œsophagus on the inner side. In passing the ligature, great care must be exercised to avoid including the recurrent laryngeal nerve, which is occasionally found crossing in *front* of the vessel.

The **Suprascapular artery (transversalis humeri)**, smaller than the transversalis colli, passes obliquely from within outwards, across the root of the neck.

It at first passes downwards and outwards across the Scalenus anticus and phrenic nerve, being covered by the Sterno-mastoid; it then crosses the subclavian artery and the cords of the brachial plexus, and runs outwards, behind and parallel with the clavicle and Subclavius muscle, and beneath the posterior belly of the Omo-hyoid, to the superior border of the scapula, where it passes over the transverse ligament of the scapula, which separates it from the suprascapular nerve, to the supraspinous fossa. In this situation it lies close to the bone, and ramifies between it and the Supraspinatus muscle, to which it supplies branches. It then passes downwards behind the neck of the scapula, to reach the infra-spinous fossa, where it anastomoses with the dor-salis scapulæ and posterior scapular arteries. Besides distributing branches to the Sterno-mastoid, Sub-clavius, and neighbouring muscles, it gives off a *supra-sternal branch*, which crosses over the sternal end of the clavicle to the

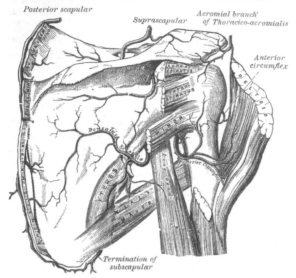

Fig. 300.—The scapular and circumflex arteries.

skin of the upper part of the chest; and a *supra-acromial branch*, which, piercing the Trapezius muscle, supplies the skin over the acromion, anastomosing with the acromial thoracic artery. As the artery passes over the transverse ligament of the scapula, a branch descends into the subscapular fossa, ramifies beneath that muscle, and anastomoses with the posterior and subscapular arteries. It also sends branches to the acromio-clavicular and shoulder joints, and a nutrient artery to the clavicle.

The **Transversalis colli** passes transversely outwards, across the upper part of the subclavian triangle, to the anterior margin of the Trapezius muscle, beneath which it divides into two branches, the *superficial cervical* and the *posterior scapular*. In its passage across the neck, it crosses in front of the phrenic nerve, Scaleni muscles, and the brachial plexus, between the divisions of which it sometimes passes, and is covered by the Platysma, Sterno-mastoid, Omo-hyoid, and Trapezius muscles.

The **superficial cervical** ascends beneath the anterior margin of the Trape-zius, distributing branches to it, and to the neighbouring muscles and glands in the neck, and anastomoses with the superficial branch of the arteria princeps cervicis.

The **posterior scapular** passes beneath the Levator anguli scapulæ to the superior angle of the scapula, and then descends along the posterior border of that bone as far as the inferior angle. In its course it is covered by the Rhomboid muscles, supplying them and the Latissimus dorsi and Trapezius, and ana-stomosing with the suprascapular and subscapular arteries, and with the posterior branches of some of the intercostal arteries.

Peculiarities.—The *superficial cervical* frequently arises as a separate branch from the thyroid axis; and the posterior scapular, from the third, more rarely from the second, part of the subclavian.

The **Internal mammary** (fig. 301) arises from the under surface of the first portion of the subclavian artery, opposite the thyroid axis. It passes downwards and inwards behind the costal cartilage of the first rib to the inner surface of the anterior wall of the chest, resting against the costal cartilages about half an inch

FIG. 301.—The internal mammary artery and its branches.

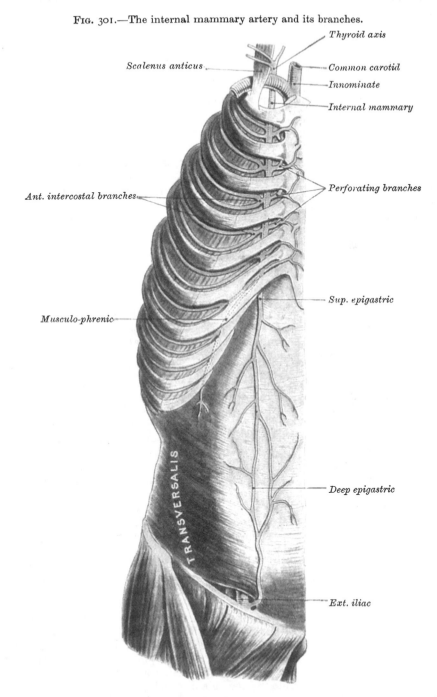

Thyroid axis

Scalenus anticus

Common carotid

Innominate

Internal mammary

Perforating branches

Ant. intercostal branches

Sup. epigastric

Musculo-phrenic

TRANSVERSALIS

Deep epigastric

Ext. iliac

from the margin of the sternum; and, at the interval between the sixth and seventh cartilages, divides into two branches, the *musculo-phrenic* and *superior epigastric*.

Relations.—At its origin it is covered by the internal jugular and subclavian veins, and as it enters the thorax is crossed from without inwards by the phrenic nerve, and then passes forwards close to the outer side of the innominate vein. In the upper part of the thorax, it lies behind the costal cartilages and Internal intercostal muscles, and is crossed by the terminations of the upper six intercostal nerves. At first it lies upon the pleura, but at the lower part of the thorax the Triangularis sterni separates the artery from this membrane. It has two venæ comites ; these unite into a single vein, which joins the innominate vein of its own side.

The branches of the internal mammary are :

Comes nervi phrenici (Superior phrenic).	Anterior intercostal.
Mediastinal.	Perforating.
Pericardiac.	Musculo-phrenic.
Sternal.	Superior epigastric.

The **comes nervi phrenici (superior phrenic)** is a long slender branch, which accompanies the phrenic nerve, between the pleura and pericardium, to the Diaphragm, to which it is distributed ; anastomosing with the other phrenic arteries from the internal mammary, and abdominal aorta.

The **mediastinal branches** are small vessels, which are distributed to the areolar tissue and lymphatic glands in the anterior mediastinum, and the remains of the thymus gland.

The **pericardiac branches** supply the upper part of the anterior surface of the pericardium, the lower part receiving branches from the musculo-phrenic artery.

The **sternal branches** are distributed to the Triangularis sterni, and to the posterior surface of the sternum.

The mediastinal, pericardiac, and sternal branches, together with some twigs from the comes nervi phrenici, anastomose with branches from the intercostal and bronchial arteries, and form a minute plexus beneath the pleura which has been named by Turner the *subpleural mediastinal plexus*.

The **anterior intercostal arteries** supply the five or six upper intercostal spaces. The branch corresponding to each space soon divides into two ; or the two branches may come off separately from the parent trunk. These small vessels pass outwards in the intercostal spaces, one lying near the lower margin of the rib above, and the other near the upper margin of the rib below, and anastomose with the intercostal arteries from the aorta. They are at first situated between the pleura and the Internal intercostal muscles, and then between the Internal and External intercostal muscles. They supply the Intercostal muscles and, by branches which perforate the External intercostal muscle, the Pectoral muscles and the mammary gland.

The **perforating arteries** correspond to the five or six upper intercostal spaces. They arise from the internal mammary, pass forwards through the intercostal spaces, and, curving outwards, supply the Pectoralis major and the integument. Those which correspond to the second, third, and fourth spaces are distributed to the mammary gland. In females, during lactation, these branches are of large size.

The **musculo-phrenic artery** is directed obliquely downwards and outwards, behind the cartilages of the false ribs, perforates the Diaphragm at the eighth or ninth rib, and terminates, considerably reduced in size, opposite the last intercostal space. It gives off anterior intercostal arteries to each of the intercostal spaces across which it passes ; these diminish in size as the spaces decrease in length, and are distributed in a manner precisely similar to the anterior intercostals from the internal mammary. The musculo-phrenic also gives branches to the lower part of the pericardium, and others which run backwards to the Diaphragm, and downwards to the abdominal muscles.

The **superior epigastric** continues in the original direction of the internal mammary ; it descends through the cellular interval between the costal and sternal attachments of the Diaphragm, and enters the sheath of the Rectus abdominis muscle, at first lying behind the muscle, and then perforating it and supplying it, and anastomosing with the deep epigastric artery from the external iliac. Some vessels perforate the sheath of the Rectus, and supply the muscles of the abdomen and the integument, and a small branch, which passes inwards upon the side of the ensiform appendix, anastomoses in front of that cartilage with the artery of the opposite side. It also gives some twigs to the Diaphragm, while from the artery of the right side small branches extend into the falciform ligament of the liver and anastomose with the hepatic artery.

Surgical Anatomy.—The course of the internal mammary artery may be defined by drawing a line across the six upper intercostal spaces, half an inch from and parallel with the sternum. The position of the vessel must be remembered, as it is liable to be wounded in stabs of the chest-wall. It is most easily reached by a transverse incision in the second intercostal space.

The **Superior intercostal** (fig. 289) arises from the upper and back part of the subclavian artery, behind the Anterior scalene muscle on the right side, and to the inner side of that muscle on the left side. Passing backwards, it gives off the *deep cervical branch*, and then descends behind the pleura in front of the necks of the first two ribs, and inosculates with the first aortic intercostal. As it crosses the neck of the first rib it lies to the inner side of the anterior division of the first dorsal nerve and to the outer side of the first thoracic ganglion of the sympathetic. In the first intercostal space, it gives off a branch which is distributed in a manner similar to the distribution of the aortic intercostals. The branch for the second intercostal space usually joins with one from the highest aortic intercostal. Each intercostal gives off a branch to the posterior spinal muscles, and a small one which passes through the corresponding intervertebral foramen to the spinal cord and its membranes.

The *deep cervical branch (profunda cervicis)* arises, in most cases, from the superior intercostal, and is analogous to the posterior branch of an aortic intercostal artery : occasionally it arises as a separate branch from the subclavian artery. Passing backwards, above the eighth cervical nerve and between the transverse process of the seventh cervical vertebra and the first rib, it runs up the back part of the neck, between the Complexus and Semispinalis colli muscles, as high as the axis, supplying these and adjacent muscles, and anastomosing with the deep branch of the arteria princeps cervicis of the occipital, and with branches which pass outwards from the vertebral. It gives off a special branch which enters the spinal canal through the intervertebral foramen between the seventh cervical and first dorsal vertebra.

ANATOMY OF THE AXILLA

The **Axilla** is a pyramidal space, situated between the upper and lateral part of the chest and the inner side of the arm.

Boundaries.—Its *apex*, which is directed upwards towards the root of the neck, corresponds to the interval between the first rib, the upper edge of the scapula, and the clavicle, through which the axillary vessels and nerves pass. The *base*, directed downwards, is formed by the integument, and a thick layer of fascia, the *axillary fascia*, extending between the lower border of the Pectoralis major in front, and the lower border of the Latissimus dorsi behind ; it is broad internally, at the chest, but narrow and pointed externally, at the arm. The *anterior boundary* is formed by the Pectoralis major and minor muscles, the former covering the whole of the anterior wall of the axilla, the latter covering only its central part. The space between the inner border of the Pectoralis minor and the clavicle is occupied by the costo-coracoid membrane. The *posterior boundary*, which extends somewhat lower than the anterior, is formed by the Subscapularis above, the Teres major and Latissimus dorsi below. On the *inner side* are the first four ribs with their corresponding Intercostal muscles, and part of the Serratus magnus. On the *outer side*, where the anterior and posterior boundaries converge, the space is narrow, and bounded by the humerus, the Coraco-brachialis and Biceps muscles.

Contents.—This space contains the axillary vessels, and brachial plexus of nerves, with their branches ; some branches of the intercostal nerves, and a large number of lymphatic glands, all connected together by a quantity of fat and loose areolar tissue.

Their position.—The axillary artery and vein, with the brachial plexus of nerves, extend obliquely along the outer boundary of the axillary space, from its apex to its base, and are placed much nearer the anterior than the posterior wall, the vein lying to the inner or thoracic side of the artery, and partially concealing it. At the fore part of the axillary space, in contact with the Pectoral muscles, are the thoracic branches of the axillary artery, and along the lower margin of the Pectoralis minor the long thoracic artery extends to the side of the chest. At the

Fig. 302.—The axillary artery and its branches.

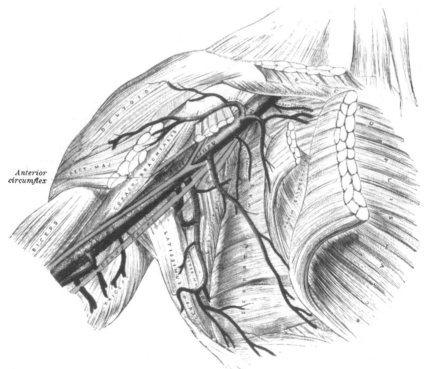

Anterior circumflex

back part, in contact with the lower margin of the Subscapularis muscle, are the subscapular vessels and nerves ; winding around the outer border of this muscle are the dorsalis scapulæ artery and veins ; and, close to the neck of the humerus, the posterior circumflex vessels and the circumflex nerve are seen curving backwards to the shoulder.

Along the inner or thoracic side no vessel of any importance exists, the upper part of the space being crossed merely by a few small branches from the superior thoracic artery. There are some important nerves, however, in this situation, viz. the posterior thoracic or external respiratory nerve, descending on the surface of the Serratus magnus, to which it is distributed ; and perforating the upper and anterior part of this wall, the intercosto-humeral nerve or nerves, passing across the axilla to the inner side of the arm.

The cavity of the axilla is filled by a quantity of loose areolar tissue, and a large number of small arteries and veins, all of which are, however, of inconsiderable size, and numerous lymphatic glands, the position and arrangement of which are described on a subsequent page.

Surgical Anatomy.—The axilla is a space of considerable surgical importance. It transmits the large vessels and nerves to the upper extremity, and these may be the seat of injury or disease : it contains numerous lymphatic glands which may require removal ; in it is a quantity of loose connective and adipose tissue which may be readily infiltrated with blood or inflammatory exudation, and it may be the seat of rapidly growing tumours. Moreover, it is covered at its base by thin skin, largely supplied with sebaceous and sweat glands, which is frequently the seat of small cutaneous abscesses and boils, and of eruptions due to irritation.

In suppuration in the axilla, the arrangement of the fasciæ plays a very important part in the direction which the pus takes. As described on page 353, the costo-coracoid membrane, after covering in the space between the clavicle and the upper border of the Pectoralis minor, splits to enclose this muscle, and, reblending at its lower border, becomes incorporated with the axillary fascia at the anterior fold of the axilla. This is known as the *clavi-pectoral fascia.* Suppuration may take place either superficial to or beneath this layer of fascia ; that is, either between the Pectorals or below the Pectoralis minor : in the former case, it would point either at the anterior border of the axillary fold, or in the groove between the Deltoid and the Pectoralis major ; in the latter, the pus would have a tendency to surround the vessels and nerves, and ascend into the neck, that being the direction in which there is least resistance. Its progress towards the skin is prevented by the axillary fascia ; its progress backwards, by the Serratus magnus ; forwards, by the clavi-pectoral fascia ; inwards, by the wall of the thorax ; and outwards, by the upper limb. The pus in these cases, after extending into the neck, has been known to spread through the superior opening of the thorax into the mediastinum.

In opening an axillary abscess, the knife should be entered in the floor of the axilla, midway between the anterior and posterior margins and near the thoracic side of the space. It is well to use a director and dressing forceps, after an incision has been made through the skin and fascia, in the manner directed by the late Mr. Hilton.

The student should attentively consider the relation of the vessels and nerves in the several parts of the axilla, for it is the almost universal plan, in the present day, to remove the glands from the axilla in operating for cancer of the breast. In performing such an operation, it will be necessary to proceed with much caution in the direction of the outer wall and apex of the space, as here the axillary vessels will be in danger of being wounded. Towards the posterior wall, it will be necessary to avoid the subscapular, dorsalis scapulæ, and posterior circumflex vessels. Along the anterior wall, it will be necessary to avoid the thoracic branches. In clearing out the axilla the axillary vein should be first defined and cleared by the fingers and an elevator up to the apex of the axilla, the Pectoralis major being pulled up by an assistant with a retractor. When the apex of the space is reached all fat and glands must be carefully removed and the whole axilla cleared by separating the tissues along the inner and posterior walls, so that when the proceeding is completed the axilla is cleared of all its contents except the main vessels and nerves.

AXILLARY ARTERY

The **Axillary artery,** the continuation of the subclavian, commences at the outer border of the first rib, and terminates at the lower border of the tendon of the Teres major muscle, where it takes the name of brachial. Its direction varies with the position of the limb : when the arm lies by the side of the chest, the vessel forms a gentle curve, the convexity being upwards and outwards ; when it is directed at right angles with the trunk, the vessel is nearly straight ; and when it is elevated still higher, the artery describes a curve, the concavity of which is directed upwards. At its commencement the artery is very deeply situated, but near its termination is superficial, being covered only by the skin and fascia. The description of the relations of this vessel is facilitated by its division into three portions : the first portion being that above the Pectoralis minor ; the second portion, behind ; and the third below, that muscle.

The **first portion of the axillary artery** is in relation, *in front*, with the clavicular portion of the Pectoralis major, the costo-coracoid membrane, the external anterior thoracic nerve, and the acromio-thoracic and cephalic veins ; *behind*, with the first intercostal space, the corresponding Intercostal muscle, the second and a portion of the third digitation of the Serratus magnus, and the posterior thoracic and internal anterior thoracic nerves ; on its *outer side*, with the brachial plexus, from which it is separated by a little cellular interval ; on its *inner*, or thoracic side, with the axillary vein which overlaps the artery.

RELATIONS OF THE FIRST PORTION OF THE AXILLARY ARTERY

In front.

Pectoralis major.
Costo-coracoid membrane.
External anterior thoracic nerve.
Acromio-thoracic and Cephalic veins.

Outer side.   *Inner side.*
Brachial plexus.   Axillary vein.

Behind.

First intercostal space, and Intercostal muscle.
Second and third digitations of Serratus magnus.
Posterior thoracic and internal anterior thoracic nerves.

The **second portion of the axillary artery** lies behind the Pectoralis minor. It is covered, *in front*, by the Pectoralis major and minor muscles ; *behind*, it is separated from the Subscapularis by a cellular interval ; on the *inner side* is the axillary vein, separated from the artery by the inner cord of the plexus and the internal anterior thoracic nerve. The brachial plexus of nerves surrounds the artery on three sides, and separates it from direct contact with the vein and adjacent muscles.

RELATIONS OF THE SECOND PORTION OF THE AXILLARY ARTERY

In front.

Pectoralis major and minor.

Outer side.   *Inner side.*
Outer cord of plexus.   Axillary vein.
                                 Inner cord of plexus.
                                 Internal anterior thoracic nerve.

Behind.

Subscapularis.
Posterior cord of plexus.

The **third portion of the axillary artery** lies below the Pectoralis minor. It is in relation, *in front*, with the lower part of the Pectoralis major above, but covered only by the integument and fascia below, where it is crossed by the inner head of the median nerve ; *behind*, with the lower part of the Subscapularis, and the tendons of the Latissimus dorsi and Teres major ; on its *outer side*, with the Coraco-brachialis ; on its *inner*, or thoracic side, with the axillary vein. The nerves of the brachial plexus bear the following relation to the artery in this part of its course : on the *outer side* is the median nerve, and the musculo-cutaneous for a short distance ; on the *inner side*, the ulnar (between the vein and artery) and lesser internal cutaneous nerves (to the inner side of the vein) ; in *front* is the internal cutaneous nerve, and *behind*, the musculo-spiral and circumflex, the latter extending only to the lower border of the Subscapularis muscle.

RELATIONS OF THE THIRD PORTION OF THE AXILLARY ARTERY

In front.

Integument and fascia.
Pectoralis major.
Inner head of median nerve.
Internal cutaneous nerve.

Outer side.

Coraco-brachialis.
Median nerve.
Musculo-cutaneous nerve.

Axillary
Artery.
Third Portion.

Inner side.

Ulnar nerve.
Axillary vein.
Lesser internal cutaneous nerve.

Behind.

Subscapularis.
Tendons of Latissimus dorsi and Teres major.
Musculo-spiral, and circumflex nerves.

Peculiarities.—The axillary artery, in about one case out of every ten, gives off a large branch, which forms either one of the arteries of the forearm, or a large muscular trunk. In the first set of cases, this artery is most frequently the radial (1 in 33), sometimes the ulnar (1 in 72), and, very rarely, the interosseous (1 in 506). In the second set of cases, the trunk has been found to give origin to the subscapular, circumflex, and profunda arteries of the arm. Sometimes, only one of the circumflex, or one of the profunda arteries, arose from the trunk. In these cases the brachial plexus surrounded the trunk of the branches, and not the main vessel.

Surface Marking.—The course of the axillary artery may be marked out by raising the arm to a right angle and drawing a line from the middle of the clavicle to the point where the tendon of the Pectoralis major crosses the prominence caused by the Coraco-brachialis as it emerges from under cover of the anterior fold of the axilla. The third portion of the artery can be felt pulsating beneath the skin and fascia, at the junction of the anterior with the middle third of the space between the anterior and posterior folds of the axilla, close to the inner border of the Coraco-brachialis.

Surgical Anatomy.—The student, having carefully examined the relations of the axillary artery in its various parts, should now consider in what situation compression of this vessel may be most easily effected, and the best position for the application of a ligature to it when necessary.

Compression of the vessel may be required in the removal of tumours, or in amputation of the upper part of the arm; and the only situation in which this can be effectually made is in the lower part of its course; by pressing on it in this situation from within outwards against the humerus, the circulation may be effectually arrested.

The axillary artery is perhaps more frequently lacerated than any other artery in the body, with the exception of the popliteal, by violent movements of the upper extremity, especially in those cases where its coats are diseased. It has occasionally been ruptured in attempts to reduce old dislocations of the shoulder-joint. This lesion is most likely to occur during the preliminary breaking down of adhesions, in consequence of the artery having become fixed to the capsule of the joint. Aneurism of the axillary artery is of frequent occurrence: a large percentage of the cases being traumatic in their origin, due to the violence to which it is exposed in the varied, extensive, and often violent movement of the limb.

The *application of a ligature to the axillary artery* may be required in cases of aneurism of the upper part of the brachial, or as a distal operation for aneurism of the subclavian; and there are only two situations in which it can be secured, viz. in the first and in the third parts of its course; for the axillary artery at its central part is so deeply seated, and, at the same time, so closely surrounded with large nervous trunks, that the application of a ligature to it in that situation would be almost impracticable.

In the *third part* of its course the operation is most simple, and may be performed in the following manner: The patient being placed on a bed, and the arm separated from the side, with the hand supinated, an incision is made through the integument forming the floor of the axilla, about two inches in length, a little nearer to the anterior than the posterior fold of the axilla. After carefully dissecting through the areolar tissue and fascia, the median nerve and axillary vein are exposed; the former having been displaced to the outer, and the latter to the inner side of the arm, the elbow being at the same time bent, so as to relax the structures and facilitate their separation, the ligature may be passed round the artery from the ulnar to the radial side. This portion of the artery is occasionally crossed by a muscular slip, the *axillary arch*, derived from the Latissimus dorsi, which may mislead the surgeon during an operation. The occasional existence of this muscular fasciculus was spoken of in the description of the muscles. It may easily be recognised by the transverse direction of its fibres.

The *first portion* of the axillary artery may be tied in cases of aneurism encroaching so far upwards that a ligature cannot be applied in the lower part of its course. Notwithstanding that this operation has been performed in some few cases, and with success, its performance is attended with much difficulty and danger. The student will remark that, in this situation, it would be necessary to divide a thick muscle, and, after incising the costo-coracoid membrane, the artery would be exposed at the bottom of a more or less deep space, with the cephalic and axillary veins in such relation with it as must render the application of a ligature to this part of the vessel particularly hazardous. Under such circumstances it is an easier, and, at the same time, more advisable operation, to tie the subclavian artery in the third part of its course.

The vessel can be best secured by a curved incision with the convexity downwards from a point half an inch external to the Sterno-clavicular joint to a point half an inch internal to the coracoid process. The limb is to be well abducted and the head inclined to the opposite side, and the incision carried through the superficial structures, care being taken of the cephalic vein at the outer angle of the incision. The clavicular origin of the Pectoralis major is then divided in the whole extent of the wound. The arm is now to be brought to the side, and the upper edge of the Pectoralis minor defined and drawn downwards. The costo-coracoid membrane is to be carefully divided on a director, close to the coracoid process, and the axillary sheath exposed ; this is to be opened with especial care on account of the vein overlapping the artery. The needle should be passed from below, so as to avoid wounding the vein.

In a case of wound of the vessel, the general practice of cutting down upon, and tying it above and below the wounded point, should be adopted in all cases.

Collateral Circulation after Ligature of the Axillary Artery.—If the artery be tied above the origin of the acromial thoracic, the collateral circulation will be carried on by the same branches as after the ligature of the subclavian ; if at a lower point, between the acromial thoracic and subscapular arteries, the latter vessel, by its free anastomoses with the other scapular arteries, branches of the subclavian, will become the chief agent in carrying on the circulation, to which the long thoracic, if it be below the ligature, will materially contribute, by its anastomoses with the intercostal and internal mammary arteries. If the point included in the ligature is below the origin of the subscapular artery, it will most probably also be below the origins of the two circumflex arteries. The chief agents in restoring the circulation will then be the subscapular and the two circumflex arteries anastomosing with the superior profunda from the brachial, which will be afterwards referred to as performing the same office after ligature of the brachial. The cases in which the operation has been performed are few in number, and no published account of dissections of the collateral circulation appears to exist.

The branches of the axillary artery are—

From first part { Superior thoracic. / Acromial thoracic.　　　*From second part* { Long thoracic. / Alar thoracic.

From third part { Subscapular. / Posterior circumflex. / Anterior circumflex.

The **superior thoracic** is a small artery, which arises from the axillary separately, or by a common trunk with the acromial thoracic. Running forwards and inwards along the upper border of the Pectoralis minor, it passes between it and the Pectoralis major to the side of the chest. It supplies these muscles, and the parietes of the thorax, anastomosing with the internal mammary and intercostal arteries.

The **acromial thoracic** is a short trunk, which arises from the fore part of the axillary artery, its origin being generally overlapped by the upper edge of the Pectoralis minor. Projecting forwards to the upper border of the Pectoralis minor, it divides into three sets of branches—*thoracic, acromial,* and *descending*. The *thoracic branches,* two or three in number, are distributed to the Serratus magnus and Pectoral muscles, anastomosing with the intercostal branches of the internal mammary. The *acromial branches* are directed outwards towards the acromion, supplying the Deltoid muscle, and anastomosing, on the surface of the acromion, with the suprascapular, and posterior circumflex arteries. The *descending* or *humeral branch* passes in the space between the Pectoralis major and Deltoid, in the same groove as the cephalic vein, and supplies both muscles. The artery also gives off a very small branch, the *clavicular*, which passes upwards to the Subclavius muscle.

The **long thoracic** passes downwards and inwards along the lower border of the Pectoralis minor to the side of the chest, supplying the Serratus magnus, the Pectoral muscles, and mammary gland, and sending branches across the axilla to the axillary glands and Subscapularis; it anastomoses with the internal mammary and intercostal arteries.

The **alar thoracic** is a small branch, which supplies the glands and areolar tissue of the axilla. Its place is frequently supplied by branches from some of the other thoracic arteries.

The **subscapular,** the largest branch of the axillary artery, arises opposite the lower border of the Subscapularis muscle, and passes downwards and backwards along its lower margin to the inferior angle of the scapula, where it anastomoses with the long thoracic and intercostal arteries and with the posterior scapular, a branch of the transversalis colli, from the thyroid axis of the subclavian. About an inch and a half from its origin it gives off a large branch, the *dorsalis scapulæ*, and terminates by supplying branches to the muscles in the neighbourhood.

The *dorsalis scapulæ* is given off from the subscapular about an inch and a half from its origin, and is generally larger than the continuation of the vessel. It curves round the axillary border of the scapula, leaving the axilla through the space between the Teres minor above, the Teres major below, and the long head of the Triceps externally (fig. 300), and enters the infraspinous fossa by passing under cover of the Teres minor, where it anastomoses with the posterior scapular and suprascapular arteries. In its course it gives off two sets of branches: one enters the subscapular fossa beneath the Subscapularis, which it supplies, anastomosing with the posterior scapular and suprascapular arteries; the other is continued along the axillary border of the scapula, between the Teres major and minor, and, at the dorsal surface of the inferior angle of the bone, anastomoses with the posterior scapular. In addition to these, small branches are distributed to the back part of the Deltoid muscle and the long head of the Triceps, anastomosing with an ascending branch of the superior profunda of the brachial.

The **circumflex arteries** wind round the neck of the humerus. The *posterior circumflex* (fig. 300), the larger of the two, arises from the back part of the axillary artery opposite the lower border of the Subscapularis muscle, and, passing backwards with the circumflex veins and nerve through the quadrangular space bounded by the Teres major and minor, the scapular head of the Triceps and the humerus, winds round the neck of that bone and is distributed to the Deltoid muscle and shoulder-joint, anastomosing with the anterior circumflex, and acromial thoracic arteries, and with the superior profunda branch of the brachial artery. The *anterior circumflex* (figs. 300, 302), considerably smaller than the preceding, arises nearly opposite that vessel, from the outer side of the axillary artery. It passes horizontally outwards, beneath the Coraco-brachialis and short head of the Biceps, lying upon the fore part of the neck of the humerus, and, on reaching the bicipital groove, gives off an ascending branch which passes upwards along the groove to supply the head of the bone and the shoulder-joint. The trunk of the vessel is then continued outwards beneath the Deltoid, which it supplies, and anastomoses with the posterior circumflex artery.

BRACHIAL ARTERY (fig. 303)

The **Brachial artery** commences at the lower margin of the tendon of the Teres major, and, passing down the inner and anterior aspect of the arm, terminates about half an inch below the bend of the elbow, where it divides into the *radial* and *ulnar* arteries. At first the brachial artery lies internal to the humerus; but as it passes down the arm it gradually gets in front of the bone, and at the bend of the elbow it lies midway between the two condyles.

Relations.—This artery is superficial throughout its entire extent, being covered, *in front*, by the integument, the superficial and deep fascia; the bicipital fascia separates it opposite the elbow from the median basilic vein; the median nerve crosses it at its middle. *Behind*, it is separated from the long head of the Triceps by the musculo-spiral nerve and superior profunda artery. It then lies upon the inner head of the Triceps, next upon the insertion of the Coraco-brachialis, and lastly on the Brachialis anticus. By its *outer side*, it is in relation with the commencement of the median nerve, and the Coraco-brachialis and Biceps muscles, which overlap the artery to a considerable extent. By its *inner side*, its upper half is in relation with the internal cutaneous and ulnar nerves, its lower half with the median nerve. The basilic vein lies on the inner side of the artery, but is separated from it in the lower part of the arm by the deep fascia. It is accompanied by two venæ comites, which lie in close contact with the artery, being connected together at intervals by short transverse communicating branches.

PLAN OF THE RELATIONS OF THE BRACHIAL ARTERY

In front.
Integument and fasciæ.
Bicipital fascia, median basilic vein.
Median nerve.
Overlapped by Coraco-brachialis and Biceps.

<table>
<tr><td>*Outer side.*
Median nerve (above).
Coraco-brachialis.
Biceps.</td><td></td><td>*Inner side.*
Internal cutaneous and ulnar nerve.
Median nerve (below).
Basilic vein.</td></tr>
</table>

Behind.
Triceps (long and inner heads).
Musculo-spiral nerve.
Superior profunda artery.
Coraco-brachialis.
Brachialis anticus.

ANATOMY OF THE BEND OF THE ELBOW

At the bend of the elbow the brachial artery sinks deeply into a triangular interval, the base of which is directed upwards, and may be represented by a line connecting the two condyles of the humerus; the sides are bounded, externally, by the inner edge of the Supinator longus; internally, by the outer margin of the Pronator radii teres; its floor is formed by the Brachialis anticus and Supinator brevis. This space contains the brachial artery, with its accompanying veins; the radial and ulnar arteries; the median and musculo-spiral nerves; and the tendon of the Biceps. The brachial artery occupies the middle line of this space, and divides opposite the neck of the radius into the radial and ulnar arteries; it is covered, *in front*, by the integument, the superficial fascia, and the median basilic vein, the vein being separated from direct contact with the artery by the bicipital fascia. *Behind*, it lies on the Brachialis anticus, which separates it from the elbow-joint. The median nerve lies on the inner side of the artery, close to it above; but separated from it below by the coronoid origin of the Pronator radii teres. The tendon of the Biceps lies to the outer side of the space, and the musculo-spiral nerve still more externally, situated upon the Supinator brevis, and partly concealed by the Supinator longus.

Peculiarities of the Brachial Artery as regards its Course.—The brachial artery, accompanied by the median nerve, may leave the inner border of the Biceps, and descend towards the inner condyle of the humerus, where it usually curves round a prominence of bone, the *supra-condylar process*, from which a fibrous arch is usually thrown over the artery; it then inclines outwards, beneath or through the substance of the Pronator radii teres muscle, to the bend of the elbow. The variation bears considerable analogy with the normal condition of the artery in some of the carnivora: it has been referred to in the description of the humerus (page 117).

As regards its Division.—Occasionally, the artery is divided for a short distance at its upper part into two trunks, which are united above and below. A similar peculiarity occurs in the main vessel of the lower limb.

The point of bifurcation may be above or below the usual point, the former condition being by far the more frequent. Out of 481 examinations recorded by Mr. Quain, some made on the right, and some on the left side of the body, in 386 the artery bifurcated in its normal position. In one case only was the place of division lower than usual, being two or three inches below the elbow-joint. 'In 94 cases out of 481, or about one in 5⅛, there were two arteries instead of one in some part, or in the whole of the arm.'

There appears, however, to be no correspondence between the arteries of the two arms, with respect to their irregular division; for in sixty-one bodies it occurred on one side only in forty-three; on both sides, in different positions, in thirteen; on both sides, in the same position, in five.

The point of bifurcation takes place at different parts of the arm, being most frequent in the upper part, less so in the lower part, and least so in the middle, the most usual point for the application of a ligature; under any of these circumstances, two large arteries would be found in the arm instead of one. The most frequent (in three out of four) of these peculiarities is the high division of the radial. That artery often arises from the inner side of the brachial, and runs parallel with the main trunk to the elbow, where it crosses it, lying beneath the fascia; or it may perforate the fascia, and pass over the artery immediately beneath the integument.

The ulnar sometimes arises from the brachial high up, and accompanies that vessel to the lower part of the arm, and then descends towards the inner condyle. In the forearm it generally lies beneath the deep fascia, superficial to the Flexor muscles; occasionally between the integument and deep fascia, and very rarely beneath the Flexor muscles.

The interosseous artery sometimes arises from the upper part of the brachial or axillary: as it passes down the arm, it lies behind the main trunk, and at the bend of the elbow regains its usual position.

In some cases of high division of the radial, the remaining trunk (ulnar-interosseous) occasionally passes, together with the median nerve, along the inner margin of the arm to the inner condyle, and then passing from within outwards, beneath or through the Pronator radii teres, regains its usual position at the bend of the elbow.

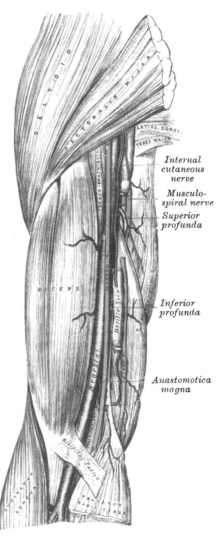

FIG. 303.—The brachial artery.

Occasionally, the two arteries representing the brachial are connected at the bend of the elbow by a short transverse branch, and are even sometimes reunited.

Sometimes, long slender vessels, *vasa aberrantia*, connect the brachial or axillary arteries with one of the arteries of the forearm, or a branch from them. These vessels usually join the radial.

*Varieties in Muscular Relations.**—The brachial artery is occasionally concealed, in some part of its course, by muscular or tendinous slips derived from the Coraco-brachialis, Biceps, Brachialis anticus, and Pronator radii teres muscles.

Surface Marking.—The direction of the brachial artery is marked by a line drawn

* See Struthers's *Anatomical and Physiological Observations.*

along the inner edge of the Biceps from the insertion of the Teres major muscle to the point midway between the condyles of the humerus.

Surgical Anatomy.—Compression of the brachial artery is required in cases of amputation and some other operations in the arm and forearm ; and it will be observed that it may be effected in almost any part of the course of the artery. If pressure is made in the upper part of the limb, it should be directed from within outwards ; and if in the lower part, from before backwards, as the artery lies on the inner side of the humerus above, and in front of it below. The most favourable situation is about the middle of the arm, where it lies on the tendon of the Coraco-brachialis on the inner flat side of the humerus.

The application of a ligature to the brachial artery may be required in cases of wound of the vessel, and in some cases of wound of the palmar arch. It is also sometimes necessary in cases of aneurism of the brachial, the radial, ulnar, or interosseous arteries. The artery may be secured in any part of its course. The chief guides in determining its position are the surface-markings produced by the inner margin of the Coraco-brachialis and Biceps, the known course of the vessel, and its pulsation, which should be carefully felt for before any operation is performed, as the vessel occasionally deviates from its usual position in the arm. In whatever situation the operation is performed, great care is necessary, on account of the extreme thinness of the parts covering the artery, and the intimate connection which the vessel has throughout its whole course with important nerves and veins. Sometimes a thin layer of muscular fibre is met with concealing the artery ; if such is the case, it must be cut across in order to expose the vessel.

In the upper third of the arm the artery may be exposed in the following manner : The patient being placed supine upon a table, the affected limb should be raised from the side, and the hand supinated. An incision about two inches in length should be made on the inner side of the Coraco-brachialis muscle, and the subjacent fascia cautiously divided, so as to avoid wounding the internal cutaneous nerve or basilic vein, which sometimes runs on the surface of the artery as high as the axilla. The fascia having been divided, it should be remembered that the ulnar and internal cutaneous nerves lie on the inner side of the artery, the median on the outer side, the latter nerve being occasionally superficial to the artery in this situation, and that the venæ comites are also in relation with the vessel, one on either side. These being carefully separated, the aneurism needle should be passed round the artery from the inner to the outer side.

If two arteries are present in the arm, in consequence of a high division, they are usually placed side by side ; and if they are exposed in an operation, the surgeon should endeavour to ascertain, by alternately pressing on each vessel, which of the two communicates with the wound or aneurism, when a ligature may be applied accordingly ; or if pulsation or hæmorrhage ceases only when both vessels are compressed, both vessels may be tied, as it may be concluded that the two communicate above the seat of disease, or are reunited.

It should also be remembered that two arteries may be present in the arm in a case of high division, and that one of these may be found along the inner intermuscular septum, in a line towards the inner condyle of the humerus ; or in the usual position of the brachial, but deeply placed beneath the common trunk ; a knowledge of these facts will suggest the precautions necessary in every case, and indicate 'the measures to be adopted when anomalies are met with.

In the middle of the arm the brachial artery may be exposed by making an incision along the inner margin of the Biceps muscle. The forearm being bent so as to relax the muscle, it should be drawn slightly aside, and the fascia carefully divided, when the median nerve will be exposed lying upon the artery (sometimes beneath) ; this being drawn inwards and the muscle outwards, the artery should be separated from its accompanying veins and secured. In this situation the inferior profunda may be mistaken for the main trunk, especially if enlarged, from the collateral circulation having become established ; this may be avoided by directing the incision externally towards the Biceps, rather than inwards or backwards towards the Triceps.

The lower part of the brachial artery is of interest in a surgical point of view, on account of the relation which it bears to the veins most commonly opened in venesection. Of these vessels, the median basilic is the largest and most prominent, and, consequently, the one usually selected for the operation. It should be remembered that this vein runs parallel with the brachial artery, from which it is separated by the bicipital fascia, and that care should be taken, in opening the vein, not to carry the incision too deeply, so as to endanger the artery.

Collateral Circulation.—After the application of a ligature to the brachial artery in the upper third of the arm, the circulation is carried on by branches from the circumflex and subscapular arteries, anastomosing with ascending branches from the superior profunda. If the brachial is tied *below* the origin of the profunda arteries, the circulation is maintained by the branches of the profundæ, anastomosing with the recurrent radial, ulnar, and interosseous arteries. Two cases are described by Mr. South,* in which the brachial

* Chelius's *Surgery*, vol. ii. p. 254. See also White's engraving, referred to by Mr. South, of the anastomosing branches after ligature of the brachial, in White's *Cases in Surgery.* Porta also gives a case (with drawings) of the circulation after ligature of both brachial and radial. (*Alterazioni Patologiche delle Arterie.*)

artery had been tied some time previously : in one 'a long portion of the artery had been obliterated, and sets of vessels are descending on either side from above the obliteration, to be received into others which ascend in a similar manner from below it. In the other, the obliteration is less extensive, and a single curved artery about as big as a crow-quill passes from the upper to the lower open part of the artery.'

The branches of the brachial artery are : the

Superior profunda.   Inferior profunda.
Nutrient.   Anastomotica magna.
Muscular.

The **superior profunda** arises from the inner and back part of the brachial, just below the lower border of the Teres major, and passes backwards to the interval between the outer and inner heads of the Triceps muscle, accompanied by the musculo-spiral nerve ; it winds round the back part of the shaft of the humerus in the spiral groove, between the outer head of the Triceps and the bone, to the outer side of the humerus, where it reaches the external intermuscular septum and divides into two terminal branches. One of these pierces the external intermuscular septum, and descends, in company with the musculo-spiral nerve, to the space between the Brachialis anticus and Supinator longus, where it anastomoses with the recurrent branch of the radial artery ; while the other, much the larger of the two, descends along the back of the external intermuscular septum to the back of the elbow-joint, where it anastomoses with the posterior interosseous recurrent, and across the back of the humerus with the posterior ulnar recurrent, the anastomotica magna, and inferior profunda (fig. 306). The superior profunda supplies the Triceps muscle and gives off a nutrient artery which enters the bone at the upper end of the musculo-spiral groove. Near its commencement it sends off a branch which passes upwards between the external and long heads of the Triceps muscle to anastomose with the posterior circumflex artery ; and, while in the groove, a small branch which accompanies a branch of the musculo-spiral nerve through the substance of the Triceps muscle and ends in the Anconeus below the outer condyle of the humerus.

The **nutrient artery** of the shaft of the humerus arises from the brachial, about the middle of the arm. Passing downwards, it enters the nutrient canal of that bone, near the insertion of the Coraco-brachialis muscle.

The **inferior profunda,** of small size, arises from the brachial, a little below the middle of the arm ; piercing the internal intermuscular septum, it descends on the surface of the inner head of the Triceps muscle, to the space between the inner condyle and olecranon, accompanied by the ulnar nerve, and terminates by anastomosing with the posterior ulnar recurrent and anastomotica magna. It sometimes supplies a branch to the front of the internal condyle, which anastomoses with the anterior ulnar recurrent.

The **anastomotica magna** arises from the brachial, about two inches above the elbow-joint. It passes transversely inwards upon the Brachialis anticus, and piercing the internal intermuscular septum, winds round the back part of the humerus between the Triceps and the bone, forming an arch above the olecranon fossa, by its junction with the posterior articular branch of the superior profunda. As this vessel lies on the Brachialis anticus, branches ascend to join the inferior profunda ; and others descend in front of the inner condyle, to anastomose with the anterior ulnar recurrent. Behind the internal condyle a branch is given off which anastomoses with the inferior profunda and posterior ulnar recurrent arteries and supplies the Triceps.

The **muscular** are three or four large branches, which are distributed to the muscles in the course of the artery. They supply the Coraco-brachialis, Biceps, and Brachialis anticus muscles.

The **Anastomosis around the Elbow-joint** (fig. 306).—The vessels engaged in this anastomosis may be conveniently divided into those situated *in front* of and *behind* the internal and external condyles. The branches anastomosing *in front* of the internal condyle are: The anastomotica magna, the anterior ulnar recurrent, and the anterior terminal branch of the inferior profunda. Those *behind* the internal condyle are: The anastomotica magna, the posterior ulnar recurrent, and the posterior terminal branch of the inferior profunda. The branches anastomosing *in front* of the external condyle are: The radial recurrent and the anterior terminal branch of the superior profunda. Those *behind* the external condyle (perhaps more properly described as being situated between the external condyle and the olecranon) are: The anastomotica magna, the interosseous recurrent, and the posterior terminal branch of the superior profunda. There is also a large arch of anastomosis above the olecranon, formed by the interosseous recurrent joining with the anastomotica magna and posterior ulnar recurrent (fig. 306).

From this description it will be observed that the anastomotica magna is the vessel most engaged, the only part of the anastomosis in which it is not employed being that *in front* of the external condyle.

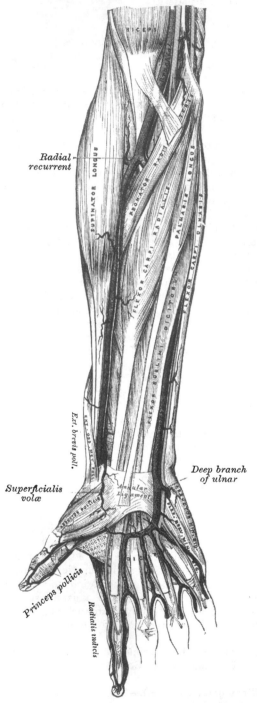

FIG. 304.—The radial and ulnar arteries.

Radial recurrent

Ext. brevis poll.

Superficialis volæ

Princeps pollicis

Radialis indicis

Deep branch of ulnar

RADIAL ARTERY (fig. 304)

The **Radial Artery** appears, from its direction, to be the continuation of the brachial, but, in size, it is smaller than the ulnar. It commences at the bifurcation of the brachial, just below the bend of the elbow, and passes along the radial side of the forearm to the wrist; it then winds backwards, round the outer side of the carpus, beneath the extensor tendons of the thumb to the upper end of the space between the metacarpal bones of the thumb and index finger, and, finally, passes forwards between the two heads of the First dorsal interosseous muscle, into the palm of the hand, where it crosses the metacarpal bones to the ulnar border of the hand, to form the deep palmar arch. At its termination, it inosculates with the deep branch of the ulnar artery.   The relations of this vessel may thus be conveniently divided into three parts, viz. in the forearm, at the back of the wrist, and in the hand.

Relations.—*In the forearm*, this vessel extends from opposite the neck of the radius to the fore part of the styloid process, being placed to the inner side of the shaft of the bone above, and in front of it below.   It is overlapped in the upper part of its course by the fleshy belly of the Supinator longus muscle; throughout the rest of its course it is superficial, being covered by the integument, the superficial and deep fasciæ.   In its course downwards, it lies upon the tendon of the Biceps, the Supinator brevis, the Pronator radii teres, the radial origin of the Flexor sublimis digitorum, the Flexor longus pollicis, the Pronator quadratus, and the lower extremity of the radius.   In the upper third of its course, it lies between the Supinator longus and the Pronator radii teres; in its lower two-thirds, between the tendons of the Supinator longus and the Flexor carpi radialis.   The radial nerve lies close to the outer side of the artery in the middle third of its course; and some filaments of the musculo-cutaneous nerve, after piercing the deep fascia, run along the lower part of the artery as it winds round the wrist.   The vessel is accompanied by venæ comites throughout its whole course.

PLAN OF THE RELATIONS OF THE RADIAL ARTERY IN THE FOREARM

In front.

Skin, superficial and deep fasciæ.
Supinator longus.

Inner side.   *Outer side.*
Pronator radii teres.   Radial Artery   Supinator longus.
Flexor carpi radialis.   in Forearm.   Radial nerve (middle third).

Behind.
Tendon of Biceps.
Supinator brevis.
Pronator radii teres.
Flexor sublimis digitorum.
Flexor longus pollicis.
Pronator quadratus.
Radius.

At the wrist, as it winds round the outer side of the carpus, from the styloid process to the first interosseous space, it lies upon the external lateral ligament, and then upon the scaphoid bone and trapezium, being covered by the extensor tendons of the thumb, subcutaneous veins, some filaments of the radial nerve, and the integument.   It is accompanied by two veins, and a filament of the musculo-cutaneous nerve.

In the hand, it passes from the upper end of the first interosseous space, between the heads of the Abductor indicis or First dorsal interosseous muscle, transversely across the palm, to the base of the metacarpal bone of the little finger, where it inosculates with the communicating branch from the ulnar artery, forming the *deep palmar arch*.   It lies upon the carpal extremities of the metacarpal bones and the Interossei muscles, being covered by the Adductor obliquus pollicis, the flexor tendons of the fingers, the Lumbricales, the Opponens and Flexor brevis minimi digiti.   Alongside of it is the deep branch of the ulnar nerve, but running in the opposite direction, that is to say, from within outwards.

Peculiarities.—The origin of the radial artery, according to Quain, is, in nearly one case in eight, higher than usual; more frequently arising from the axillary or upper part of the brachial, than from the lower part of this vessel.   The variations in the position of this vessel in the arm, and at the bend of the elbow, have been already mentioned.   In the forearm it deviates less frequently from its position than the ulnar. It has been found lying over the fascia instead of beneath it. It has also been observed on the surface of the Supinator longus, instead of under its inner border; and in turning round the wrist, it has been seen lying over, instead of beneath, the extensor tendons of the thumb.

Surface Marking.—The position of the radial artery in the forearm is represented by a line drawn from the outer border of the tendon of the Biceps in the centre of the hollow in front of the elbow-joint with a straight course to the inner side of the fore part of the styloid process of the radius.

Surgical Anatomy.—The radial artery is much exposed to injury in its lower third, and is frequently wounded by the hand being driven through a pane of glass, by the slipping of a knife or chisel held in the other hand, and such-like accidents. The injury is often followed by a traumatic aneurism, for which the old operation of laying open the sac and securing the vessel above and below is required.

The operation of tying the radial artery is required in cases of wounds either of its trunk, or of some of its branches, or for aneurism: and it will be observed, that the vessel may be exposed in any part of its course through the forearm without the division of any muscular fibres. The operation in the middle or inferior third of the forearm is easily performed; but in the upper third, near the elbow, it is attended with some difficulty, from the greater depth of the vessel, and from its being overlapped by the Supinator longus muscle.

To tie the artery in the upper third, an incision three inches in length should be made through the integument, in a line drawn from the centre of the bend of the elbow to the front of the styloid process of the radius, avoiding the branches of the median vein; the fascia of the arm being divided, and the Supinator longus drawn a little outwards, the artery will be exposed. The venæ comites should be carefully separated from the vessel, and the ligature passed from the radial to the ulnar side.

In the middle third of the forearm the artery may be exposed by making an incision of similar length on the inner margin of the Supinator longus. In this situation, the radial nerve lies in close relation with the outer side of the artery, and should, as well as the veins, be carefully avoided.

In the lower third, the artery is easily secured by dividing the integument and fascia in the interval between the tendons of the Supinator longus and Flexor carpi radialis muscles.

The branches of the radial artery may be divided into three groups, corresponding with the three regions in which the vessel is situated.

In the forearm.	*Wrist.*	*Hand.*
Radial recurrent.	Posterior carpal.	Princeps pollicis.
Muscular.	Metacarpal.	Radialis indicis.
Anterior carpal.	Dorsales pollicis.	Perforating.
Superficialis volæ.	Dorsalis indicis.	Palmar interosseous.
		Palmar recurrent.

The **radial recurrent** is given off immediately below the elbow. It ascends between the branches of the musculo-spiral nerve lying on the Supinator brevis, and then between the Supinator longus and Brachialis anticus, supplying these muscles and the elbow-joint, and anastomosing with the anterior terminal branch of the superior profunda.

The **muscular branches** are distributed to the muscles on the radial side of the forearm.

The **anterior carpal** is a small vessel which arises from the radial artery near the lower border of the Pronator quadratus, and, running inwards in front of the radius, anastomoses with the anterior carpal branch of the ulnar artery. In this way an arterial anastomosis, *anterior carpal arch,* is formed in front of the wrist: it is joined by branches from the anterior interosseous above, and by recurrent branches from the deep palmar arch below, and gives off branches which descend to supply the articulations of the wrist and carpus.

The **superficialis volæ** arises from the radial artery, just where this vessel is about to wind round the wrist. Running forwards, it passes between, occasionally over, the muscles of the thumb, which it supplies, and sometimes anastomoses with the palmar portion of the ulnar artery, completing the superficial palmar arch. This vessel varies considerably in size: usually it is very small, and terminates in the muscles of the thumb; sometimes it is as large as the continuation of the radial

The **posterior carpal** is a small vessel which arises from the radial artery beneath the extensor tendons of the thumb; crossing the carpus transversely to the inner border of the hand, it anastomoses with the posterior carpal branch of the ulnar, forming the *posterior carpal arch*, which is joined by the termination of the anterior interosseous artery. From this arch are given off descending branches, the *dorsal interosseous arteries* for the third and fourth interosseous spaces, which run forwards on the Third and Fourth dorsal interossei muscles and divide into dorsal digital branches which supply the adjacent sides of the middle, ring, and little fingers respectively, communicating with the digital arteries of the superficial palmar arch. At their origin they anastomose with the perforating branches from the deep palmar arch.

The **metacarpal (first dorsal interosseous branch)** arises beneath the extensor tendons of the thumb, sometimes with the posterior carpal artery; running forwards on the Second dorsal interosseous muscle, it communicates, behind, with the corresponding perforating branch of the deep palmar arch; and, in front, it divides into two *dorsal digital branches*, which supply the adjoining sides of the index and middle fingers, inosculating with the digital branch of the superficial palmar arch.

The **dorsales pollicis** are two small vessels which run along the sides of the dorsal aspect of the thumb. They arise separately, or occasionally by a common trunk, near the base of the first metacarpal bone.

The **dorsalis indicis**, also a small branch, runs along the radial side of the back of the index finger, sending a few branches to the Abductor indicis.

The **princeps pollicis** arises

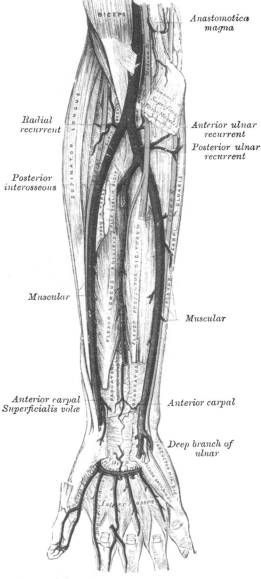

FIG. 305.—Ulnar and radial arteries. Deep view.

from the radial just as it turns inwards to the deep part of the hand; it descends between the Abductor indicis and Adductor obliquus pollicis, then between the Adductor transversus pollicis and Adductor obliquus pollicis, along the ulnar side of the metacarpal bone of the thumb, to the base of the first phalanx, where it divides into two branches, which run along the sides of the palmar aspect of the thumb, and form an arch on the palmar surface of the last phalanx from which branches are distributed to the integument and pulp of the thumb.

The **radialis indicis** arises close to the preceding, descends between the Abductor indicis and Adductor transversus pollicis, and runs along the radial side of the index finger to its extremity, where it anastomoses with the collateral digital artery from the superficial palmar arch. At the lower border of the Adductor transversus pollicis, this vessel anastomoses with the princeps pollicis, and gives a communicating branch to the superficial palmar arch.

The **perforating arteries,** three in number, pass backwards from the deep palmar arch between the heads of the last three Dorsal interossei muscles, to inosculate with the dorsal interosseous arteries.

The **palmar interosseous,** three or four in number, arise from the convexity of the deep palmar arch ; they run forwards upon the Interossei muscles, and anastomose at the clefts of the fingers with the digital branches of the superficial arch.

The **palmar recurrent branches** arise from the concavity of the deep palmar arch. They pass upwards in front of the wrist, supplying the carpal articulations and anastomosing with the anterior carpal arch.

ULNAR ARTERY (fig. 305)

The **Ulnar Artery,** the larger of the two terminal branches of the brachial, commences a little below the bend of the elbow, and crosses obliquely to the inner side of the forearm, at the commencement of its lower half ; it then runs along its ulnar border to the wrist, crosses the annular ligament on the radial side of the pisiform bone, and immediately beyond this bone divides into two branches, which enter into the formation of the superficial and deep palmar arches.

Relations in the Forearm.—In its *upper half*, it is deeply seated, being covered by all the superficial Flexor muscles, excepting the Flexor carpi ulnaris ; the median nerve is in relation with the inner side of the artery for about an inch and then crosses the vessel, being separated from it by the deep head of the Pronator radii teres ; it lies upon the Brachialis anticus and Flexor profundus digitorum muscles. In the *lower half* of the forearm, it lies upon the Flexor profundus, being covered by the integument, the superficial and deep fasciæ, and is placed between the Flexor carpi ulnaris and Flexor sublimis digitorum muscles. It is accompanied by two venæ comites ; the ulnar nerve lies on its inner side for the lower two-thirds of its extent, and a small branch from the nerve descends on the lower part of the vessel to the palm of the hand.

PLAN OF THE RELATIONS OF THE ULNAR ARTERY IN THE FOREARM

In front.

Superficial layer of flexor muscles. }
Median nerve.   } Upper half.
Superficial and deep fasciæ.   Lower half.

Inner side.   *Outer side.*

Flexor carpi ulnaris.   (Ulnar Artery)   Flexor sublimis digitorum.
Ulnar nerve (lower two-thirds).   (in Forearm.)

Behind.
Brachialis anticus.
Flexor profundus digitorum.

At the wrist (fig. 304) the ulnar artery is covered by the integument and fascia, and lies upon the anterior annular ligament. On its inner side is the pisiform bone. The ulnar nerve lies at the inner side, and somewhat behind the artery ; here the nerve and artery are crossed by a band of fibres, which extends from the pisiform bone to the anterior annular ligament.

Peculiarities.—The ulnar artery has been found to vary in its origin nearly in the proportion of one in thirteen cases, in one case arising lower than usual, about two or three inches below the elbow, and in all other cases much higher, the brachial being a more frequent source of origin than the axillary.

Variations in the position of this vessel are more frequent than in the radial. When its origin is normal, the course of the vessel is rarely changed. When it arises high up, it is almost invariably superficial to the flexor muscles in the forearm, lying commonly beneath the fascia, more rarely between the fascia and integument. In a few cases, its position was subcutaneous in the upper part of the forearm, and subaponeurotic in the lower part.

Surface Marking.—On account of the curved direction of the ulnar artery, the line on the surface of the body which indicates its course is somewhat complicated. First, draw a line from the front of the internal condyle of the humerus to the radial side of the pisiform bone ; the lower two-thirds of this line represents the course of the middle and lower third of the ulnar artery. Secondly, draw a line from the centre of the hollow in front of the elbow-joint to the junction of the upper and middle third of the first line ; this represents the course of the upper third of the artery.

Surgical Anatomy.—The application of a ligature to this vessel is required in cases of wound of the artery, or of its branches, or in consequence of aneurism. In the upper half of the forearm, the artery is deeply seated beneath the superficial flexor muscles, and the application of a ligature in this situation is attended with some difficulty. An incision is to be made in the course of a line drawn from the front of the internal condyle of the humerus to the outer side of the pisiform bone, so that the centre of the incision is three finger's breadths below the internal condyle. The skin and superficial fascia having been divided, and the deep fascia exposed, the white line which separates the Flexor carpi ulnaris from the other flexor muscles is to be sought for, and the fascia incised in this line. The Flexor carpi ulnaris is now to be carefully separated from the other muscles, when the ulnar nerve will be exposed and must be drawn aside. Some little distance below the nerve, the artery will be found accompanied by its venæ comites, and may be ligatured by passing the needle from within outwards. In the middle and lower third of the forearm, this vessel may be easily secured by making an incision on the radial side of the tendon of the Flexor carpi ulnaris : the deep fascia being divided, and the Flexor carpi ulnaris and its companion muscle, the Flexor sublimis, being separated from each other, the vessel will be exposed, accompanied by its venæ comites, the ulnar nerve lying on its inner side. The veins being separated from the artery, the ligature should be passed from the ulnar to the radial side, taking care to avoid the ulnar nerve.

The branches of the ulnar artery may be arranged in the following groups :

Forearm	Anterior ulnar recurrent.
	Posterior ulnar recurrent.
	Interosseous { Anterior interosseous. Posterior interosseous. }
	Muscular.
Wrist	Anterior carpal. Posterior carpal.
Hand	Deep palmar or communicating. Superficial palmar arch.

The **anterior ulnar recurrent** (fig. 305) arises immediately below the elbow-joint, passes upwards and inwards between the Brachialis anticus and Pronator radii teres, supplies twigs to those muscles, and, in front of the inner condyle, anastomoses with the anastomotica magna and inferior profunda.

The **posterior ulnar recurrent** is much larger, and arises somewhat lower than the preceding. It passes backwards and inwards, beneath the Flexor sublimis, and ascends behind the inner condyle of the humerus. In the interval between this process and the olecranon, it lies beneath the Flexor carpi ulnaris, and ascending between the heads of that muscle, in relation with the ulnar nerve, it supplies the neighbouring muscles and joint, and anastomoses with the inferior profunda, anastomotica magna, and interosseous recurrent arteries (fig. 306).

The **interosseous artery** (fig. 305) is a short trunk about half an inch in length, and of considerable size, which arises immediately below the tuberosity of the radius, and, passing backwards to the upper border of the interosseous membrane, divides into two branches, the *anterior* and *posterior interosseous*.

The **anterior interosseous** passes down the forearm on the anterior surface of the interosseous membrane, to which it is connected by a thin aponeurotic arch. It is accompanied by the interosseous branch of the median nerve, and over-lapped by the contiguous margins of the Flexor profundus digitorum and Flexor longus pollicis muscles, giving off in this situation muscular branches, and the nutrient arteries of the radius and ulna. At the upper border of the Pronator quadratus, a branch descends beneath the muscle, to anastomose in front of the carpus with the anterior carpal arch. The continuation of the artery passes behind the Pronator quadratus, and, piercing the interosseous membrane, reaches the back of the forearm, and anastomoses with the posterior interosseous artery (fig. 306). It then descends to the back of the wrist to join the posterior carpal arch. The anterior interosseous gives off a long, slender branch, the *median artery*, which accompanies the median nerve, and gives offsets to its substance. This artery is sometimes much enlarged, and accompanies the nerve into the palm of the hand.

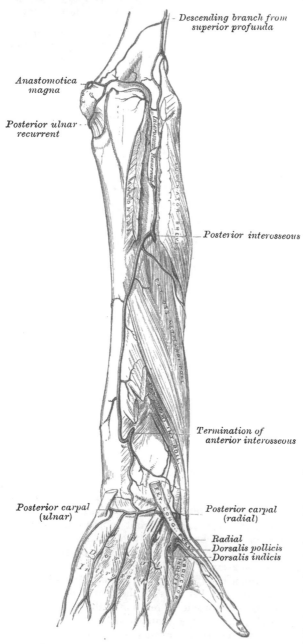

FIG. 306.—Arteries of the back of the forearm and hand.

Descending branch from superior profunda

Anastomotica magna

Posterior ulnar recurrent

Posterior interosseous

Termination of anterior interosseous

Posterior carpal (ulnar)

Posterior carpal (radial)

Radial
Dorsalis pollicis
Dorsalis indicis

The **posterior interosseous artery** passes backwards through the interval between the oblique ligament and the upper border of the interosseous membrane. It appears between the contiguous borders of the Supinator brevis and the Extensor ossis metacarpi pollicis, and runs down the back part of the forearm, between the superficial and deep layer of muscles, to both of which it distributes branches. At the lower part of the forearm it anastomoses with the termination of the anterior interosseous artery. Then, continuing its course over the head of the ulna, it joins the posterior carpal branch of the ulnar artery. This artery gives off, near its origin, the *interosseous recurrent branch*.

The **interosseous recurrent artery** is a large vessel which ascends to the interval between the external condyle and olecranon, on or through the fibres of the Supinator brevis, but beneath the Anconeus, anastomosing with a branch from the superior profunda, and with the posterior ulnar recurrent and ana-stomotica magna.

The **muscular branches** are distributed to the muscles along the ulnar side of the forearm.

The **anterior carpal** is a small vessel which crosses the front of the carpus beneath the tendons of the Flexor profundus, and inosculates with a corresponding branch of the radial artery.

The **posterior carpal** arises immediately above the pisiform bone, and winds backwards beneath the tendon of the Flexor carpi ulnaris; it passes across the dorsal surface of the carpus beneath the extensor tendons, anastomosing with a corresponding branch of the radial artery, and forming the *posterior carpal arch*. Immediately after its origin, it gives off a small branch, which runs along the ulnar side of the metacarpal bone of the little finger, forming one of the *metacarpal arteries*, and supplies the ulnar side of the dorsal surface of the little finger.

The **branch to the deep palmar arch** (**deep** or **communicating branch**) (fig. 305) passes deeply inwards between the Abductor minimi digiti and Flexor brevis minimi digiti, near their origins; it anastomoses with the termination of the radial artery, completing the deep palmar arch.

The continuation of the trunk of the ulnar artery in the hand forms the greater part of the superficial palmar arch.

The Superficial Palmar Arch (fig. 304)

The **Superficial Palmar Arch** is formed by the ulnar artery in the hand, and is completed on the outer side by this vessel anastomosing with a branch from the radialis indicis, though sometimes the arch is completed by the ulnar anastomosing with the superficialis volæ or princeps pollicis of the radial artery. The arch passes across the palm, describing a curve, with its convexity forwards, to the space between the ball of the thumb and the index finger, where the above mentioned anastomosis takes place.

Relations.—The superficial palmar arch is covered by the skin, the Palmaris brevis, and the palmar fascia. It lies upon the annular ligament, the Flexor brevis of the little finger, the tendons of the superficial flexor of the fingers, and the divisions of the median and ulnar nerves.

PLAN OF THE RELATIONS OF THE SUPERFICIAL PALMAR ARCH

In front.

Skin.
Palmaris brevis.
Palmar fascia.

Superficial
Palmar Arch.

Behind.

Annular ligament.
Flexor brevis of little finger.
Superficial flexor tendons.
Divisions of median and ulnar nerves.

The branches of the superficial palmar arch are the *digital.*

The **digital branches** (fig. 304), four in number, are given off from the convexity of the superficial palmar arch. They supply the ulnar side of the little finger, and the adjoining sides of the little, ring, middle, and index fingers ; the radial side of the index finger and thumb being supplied from the radial artery. The digital arteries at first lie superficial to the flexor tendons, but as they pass forwards with the digital nerves to the clefts between the fingers, they lie between them, and are there joined by the interosseous branches from the deep palmar arch. The digital arteries on the sides of the fingers lie beneath the digital nerves ; and, about the middle of the last phalanx, the two branches for each finger form an arch, from the convexity of which branches pass to supply the pulp of the finger.

Surface Marking.—The superficial palmar arch is represented by a curved line, starting from the outer side of the pisiform bone, and carried downwards as far as the middle third of the palm, and then curved outwards on a level with the upper end of the cleft between the thumb and index finger.

The deep palmar arch is situated about half an inch nearer to the carpus.

Surgical Anatomy.—Wounds of the palmar arches are of special interest, and are always difficult to deal with. When the superficial arch is wounded it is generally possible, by enlarging the wound if necessary, to secure the vessel and tie it ; or in cases where it is found impossible to encircle the vessel with a ligature, a pair of Wells's artery clips may be applied and left on for twenty-four or forty-eight hours. Wounds of the deep arch are not so easily dealt with. It may be possible to secure the vessel by forcipressure forceps, which may be left on ; or, failing this, the wound may be carefully plugged with gauze and an outside dressing carefully bandaged on. The plug should be allowed to remain untouched for three or four days. In wounds of the deep palmar arch, a ligature may be applied to the bleeding points, from the dorsum of the hand, by resection of the upper part of the third metacarpal bone. It is useless in these cases to ligature one of the arteries of the forearm alone, and indeed simultaneous ligature of both radial and ulnar arteries above the wrist is often unsuccessful, on account of the anastomosis carried on by the carpal arches. Therefore, upon the failure of pressure to arrest hæmorrhage, it is expedient to apply a ligature to the brachial artery.

ARTERIES OF THE TRUNK

The Descending Aorta

The **Descending Aorta** is [divided into two portions, the *thoracic* and *abdominal*, in correspondence with the two great cavities of the trunk in which it is situated.

The Thoracic Aorta

The **Thoracic Aorta** commences at the lower border of the fourth dorsal vertebra, on the left side, and terminates at the aortic opening in the Diaphragm, in front of the lower border of the last dorsal vertebra. At its commencement, it is situated on the left side of the spine ; it approaches the median line as it descends ; and, at its termination, lies directly in front of the column. The direction of this vessel being influenced by the spine, upon which it rests, it describes a curve which is concave forwards in the dorsal region. As the branches given off from it are small, the diminution in the size of the vessel is inconsiderable. It is contained in the back part of the posterior mediastinum.

Relations.—It is in relation, *in front*, from above downwards, with the root of the left lung, the pericardium, the œsophagus, and the Diaphragm : *behind*, with the vertebral column, and the vena azygos minor ; on the *right side*, with the vena azygos major, and thoracic duct ; on the *left side*, with the left pleura and lung. The œsophagus, with its accompanying nerves, lies on the right side of the aorta *above* ; but at the lower part of the thorax it gets in front of the aorta, and close to the Diaphragm is situated to its left side.

PLAN OF THE RELATIONS OF THE THORACIC AORTA

In front.

Root of left lung.
Pericardium.
Œsophagus.
Diaphragm.

Right side.		*Left side.*
Œsophagus (above). Vena azygos major. Thoracic duct.	Thoracic Aorta.	Pleura. Left lung. Œsophagus (below).

Behind.

Vertebral column.
Superior and inferior azygos minor veins.

The aorta is occasionally found to be obliterated at a particular spot, viz. at the junction of the arch with the thoracic aorta, just below the ductus arteriosus. Whether this is the result of disease, or of congenital malformation, is immaterial to our present purpose; it affords an interesting opportunity of observing the resources of the collateral circulation. The course of the anastomosing vessels, by which the blood is brought from the upper to the lower part of the artery, will be found well described in an account of two cases in the *Pathological Transactions*, vols. viii. and x. In the former (p. 162), Mr. Sydney Jones thus sums up the detailed description of the anastomosing vessels : ' The principal communications by which the circulation was carried on, were—Firstly, the internal mammary, anastomosing with the intercostal arteries, with the phrenic of the abdominal aorta by means of the musculo-phrenic and comes nervi phrenici, and largely with the deep epigastric. Secondly, the superior intercostal, anastomosing anteriorly by means of a large branch with the first aortic intercostal, and posteriorly with the posterior branch of the same artery. Thirdly, the inferior thyroid, by means of a branch about the size of an ordinary radial, forming a communication with the first aortic intercostal. Fourthly, the transversalis colli, by means of very large communications with the posterior branches of the intercostals. Fifthly, the branches (of the subclavian and axillary) going to the side of the chest were large, and anastomosed freely with the lateral branches of the intercostals.' In the second case also (vol. x. p. 97), Mr. Wood describes the anastomoses in a somewhat similar manner, adding the remark, that ' the blood which was brought into the aorta through the anastomoses of the intercostal arteries appeared to be expended principally in supplying the abdomen and pelvis; while the supply to the lower extremities had passed through the internal mammary and epigastrics.'

Surgical Anatomy.—The student should now consider the effects likely to be produced by aneurism of the thoracic aorta, a disease of common occurrence. When we consider the great depth of the vessel from the surface, and the number of important structures which surround it on every side, it may easily be conceived what a variety of obscure symptoms may arise from disease of this part of the arterial system, and how they may be liable to be mistaken for those of other affections. Aneurism of the thoracic aorta most usually extends backwards, along the left side of the spine, producing absorption of the bodies of the vertebræ, with curvature of the spine; while the irritation or pressure on the cord will give rise to pain, either in the chest, back, or loins, with radiating pain in the left upper intercostal spaces, from pressure on the intercostal nerves; at the same time the tumour may project backwards on each side of the spine, beneath the integument, as a pulsating swelling, simulating abscess connected with diseased bone; or it may displace the œsophagus, and compress the lung on one or the other side. If the tumour extend forward, it may press upon and displace the heart, giving rise to palpitation and other symptoms of disease of that organ; or it may displace, or even compress, the œsophagus, causing pain and difficulty of swallowing, as in stricture of that tube; and ultimately even open into it by ulceration, producing fatal hæmorrhage. If the disease extends to the right side, it may press upon the thoracic duct; or it may burst into the pleural cavity, or into the trachea or lung; and lastly, it may open into the posterior mediastinum.

BRANCHES OF THE THORACIC AORTA

Pericardiac.	Œsophageal.
Bronchial.	Posterior mediastinal.
Intercostal.	

The **pericardiac** are a few small vessels, irregular in their origin, distributed to the pericardium.

The **bronchial arteries** are the nutrient vessels of the lungs, and vary in number, size, and origin. That of the right side arises from the first aortic intercostal, or by a common trunk with the left bronchial, from the front of the thoracic aorta. Those of the left side, usually two in number, arise from the thoracic aorta, one a little lower than the other. Each vessel is directed to the back part of the corresponding bronchus along which it runs, dividing and subdividing along the bronchial tubes, supplying them, the cellular tissue of the lungs, the bronchial glands, and the œsophagus.

The **œsophageal arteries,** usually four or five in number, arise from the front of the aorta, and pass obliquely downwards to the œsophagus, forming a chain of anastomoses along that tube, anastomosing with the œsophageal branches of the inferior thyroid arteries above, and with ascending branches from the phrenic and gastric arteries below.

The **posterior mediastinal arteries** are numerous small vessels which supply the glands and loose areolar tissue in the mediastinum.

The **intercostal arteries** arise from the back of the aorta. They are usually nine in number on each side, the two superior intercostal spaces being supplied by the superior intercostal, a branch of the subclavian. The second space usually receives a considerable branch from the first aortic intercostal, which joins with the branch from the superior intercostal of the subclavian. The branch which runs along the lower border of the last rib is named the *subcostal artery*. The right intercostals are longer than the left, on account of the position of the aorta on the left side of the spine : they pass outwards, across the bodies of the vertebræ, to the intercostal spaces, being covered by the pleura, the œsophagus, thoracic duct, sympathetic nerve, and the vena azygos major ; the left passing outwards are crossed by the sympathetic ; the upper two are also crossed by the superior intercostal vein, the lower by the azygos minor veins. In each intercostal space the artery passes outwards, at first lying upon the External intercostal muscle, covered in front by the pleura and a thin fascia. It then passes between the two layers of Intercostal muscles, and having ascended obliquely to the lower border of the rib above it is continued forwards in the groove on its lower border and anastomoses with the anterior intercostal branches of the internal mammary. The first aortic intercostal anastomoses with the superior intercostal, and the last three pass between the abdominal muscles, inosculating with the epigastric in front, and with the phrenic and lumbar arteries. Each intercostal artery is accompanied by a vein and nerve, the former being above, and the latter below, except in the upper intercostal spaces, where the nerve is at first above the artery. The arteries are protected from pressure during the action of the Intercostal muscles by fibrous arches thrown across, and attached by each extremity to the bone. The lower intercostal arteries are continued anteriorly from the intercostal spaces into the abdominal wall, except the *subcostal*, which lies throughout its whole course in the abdominal wall, since it is placed below the last rib. They pass behind the costal cartilages between the Internal oblique and Transversalis muscle to the sheath of the Rectus, where they anastomose with the internal mammary and the deep epigastric arteries. Behind, the subcostal artery anastomoses with the first lumbar artery.

Each intercostal artery gives off the following branches :

Posterior or dorsal branch.   Spinal.
          Collateral intercostal.

The *posterior or dorsal branch* of each intercostal artery passes backwards to the inner side of the anterior costo-transverse ligament, and divides into an external and internal branch, which are distributed to the muscles and integument of the back.

The *spinal branch*, which enters the spinal canal through the intervertebral foramen, is distributed to the spinal cord and its membranes, and to the bodies of the vertebræ in the same manner as the lateral spinal branches from the vertebral.

The *collateral intercostal branch* comes off from the intercostal artery near the angle of the rib, and descends to the upper border of the rib below, along which it courses to anastomose with the anterior intercostal branch of the internal mammary.

Surgical Anatomy.—The position of the intercostal vessels should be borne in mind in performing the operation of paracentesis thoracis. The puncture should never be made nearer the middle line posteriorly than the angle of the rib, as the artery crosses the space

Fig. 307.—The abdominal aorta and its branches.

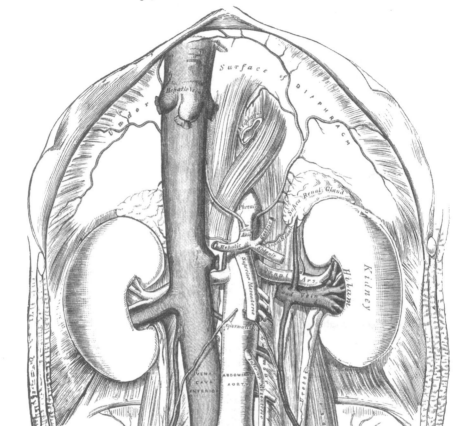

internal to this point. In the lateral portion of the chest, where the puncture is usually made, the artery lies at the upper part of the intercostal space, and therefore the puncture should be made just above the upper border of the rib forming the lower boundary of the space.

The Abdominal Aorta (fig. 307)

The **Abdominal Aorta** commences at the aortic opening of the Diaphragm, in front of the lower border of the body of the last dorsal vertebra, and, descending a little to the left side of the vertebral column, terminates on the body of the fourth lumbar vertebra, commonly a little to the left of the middle line,* where it divides into the two common iliac arteries. It diminishes rapidly in size, in consequence of the many large branches which it gives off. As it lies upon the bodies of the vertebræ, the curve which it describes is convex forwards, the greatest convexity corresponding to the third lumbar vertebra, which is a little above and to the left side of the umbilicus.

Relations.—It is covered, *in front*, by the lesser omentum and stomach, behind which are the branches of the cœliac axis, and the solar plexus : below these, by the splenic vein, the pancreas, the left renal vein, the transverse portion of the duodenum, the mesentery, and aortic plexus. *Behind*, it is separated from the lumbar vertebræ and intervening discs by the anterior common ligament and left lumbar veins. On the *right side* it is in relation with the inferior vena cava (the right crus of the Diaphragm being interposed above), the vena azygos major, thoracic duct, and right semilunar ganglion ; on the *left side*, with the sympathetic nerve, and left semilunar ganglion.

PLAN OF THE RELATIONS OF THE ABDOMINAL AORTA

In front.

Lesser omentum and stomach.
Branches of the cœliac axis and solar plexus.
Splenic vein.
Pancreas.
Left renal vein.
Transverse duodenum.
Mesentery.
Aortic plexus.

Right side.		*Left side.*
Right crus of Diaphragm	Abdominal Aorta.	Sympathetic nerve.
Inferior vena cava.		Left semilunar ganglion.
Vena azygos major.		
Thoracic duct.		
Right semilunar ganglion.		

Behind.

Left lumbar veins.
Vertebral column.

Surface Marking.—In order to map out the abdominal aorta on the surface of the abdomen, a line must be drawn from the middle line of the body, on a level with the distal extremity of the seventh costal cartilage, downwards and slightly to the left, so that it just skirts the umbilicus, to a zone drawn round the body opposite the highest point of the crest of the ilium. This point is generally half an inch below and to the left of the umbilicus, but as the position of this structure varies with the obesity of the individual, it is not a reliable landmark as to the situation of the bifurcation of the aorta.

Surgical Anatomy.—Aneurisms of the abdominal aorta near the cœliac axis communicate in nearly equal proportion with the anterior and posterior parts of the artery.

When an aneurismal sac is connected with the back part of the abdominal aorta, it usually produces absorption of the bodies of the vertebræ, and forms a pulsating tumour, that presents itself in the left hypochondriac or epigastric regions, and is accompanied by symptoms of disturbance in the alimentary canal. Pain is invariably present, and is usually of two kinds—a fixed and constant pain in the back, caused by the tumour pressing on or displacing the branches of the solar plexus and splanchnic nerves ; and a sharp lancinating pain, radiating along those branches of the lumbar nerves which are pressed on by the tumour ; hence the pain in the loins, the testes, the hypogastrium, and in the lower limb (usually of the left side). This form of aneurism usually bursts into the peritoneal cavity, or behind the peritoneum, in the left hypochondriac region ; or it may form a large aneurismal sac, extending down as low as Poupart's ligament ;

* Lord Lister, having accurately examined 30 bodies in order to ascertain the exact point of termination of this vessel, found it 'either absolutely, or almost absolutely, mesial in 15, while in 13 it deviated more or less to the left, and in 2 was slightly to the right.'—*System of Surgery*, edited by T. Holmes, 2nd ed. vol. v. p. 652.

hæmorrhage in these cases being generally very extensive, but slowly produced, and not rapidly fatal.

When an aneurismal sac is connected with the front of the aorta near the cœliac axis, it forms a pulsating tumour in the left hypochondriac or epigastric regions, usually attended with symptoms of disturbance of the alimentary canal, as sickness, dyspepsia, or constipation, and accompanied by pain, which is constant, but nearly always fixed, in the loins, epigastrium, or some part of the abdomen; the radiating pain being rare, as the lumbar nerves are seldom implicated. This form of aneurism may burst into the peritoneal cavity, or behind the peritoneum, between the layers of the mesentery, or, more rarely, into the duodenum; it rarely extends backwards so as to affect the spine.

The abdominal aorta has been tied several times, and although none of the patients permanently recovered, still, as one of them lived as long as ten days, the possibility of the re-establishment of the circulation may be considered to be proved. In the lower animals this artery has been often successfully tied. The vessel may be reached in several ways. In the original operation, performed by Sir A. Cooper, an incision was made in the linea alba, the peritoneum opened in front, the finger carried down among the intestines, towards the spine, the peritoneum again opened behind, by scratching through the mesentery, and the vessel thus reached. Or either of the operations, described below, for securing the common iliac artery, may, by extending the dissection a sufficient distance upwards, be made use of to expose the aorta. The chief difficulty in the dead subject consists in isolating the artery, in consequence of its great depth; but in the living subject, the embarrassment resulting from the proximity of the aneurismal tumour, and the great probability of disease in the vessel itself, add to the dangers and difficulties of this formidable operation so greatly, that it is very doubtful whether it ought ever to be performed.

The collateral circulation would be carried on by the anastomosis between the internal mammary and the deep epigastric; by the free communication between the superior and inferior mesenterics, if the ligature were placed above the latter vessel; or by the anastomosis between the inferior mesenteric and the internal pudic, when (as is more common) the point of ligature is below the origin of the inferior mesenteric; and possibly by the anastomoses of the lumbar arteries with the branches of the internal iliac.

The circulation through the abdominal aorta may be commanded, in thin persons, by firm pressure with the fingers. A tourniquet has been invented for this purpose, which is sometimes used in amputation at the hip-joint and some other operations.

BRANCHES OF THE ABDOMINAL AORTA

Phrenic.

Cœliac axis { Gastric. Hepatic. Splenic. }

Superior mesenteric.
Suprarenal.

Renal.
Spermatic in male.
Ovarian in female.
Inferior mesenteric.
Lumbar.
Sacra media.

The branches may be divided into two sets: 1. Those supplying the viscera. 2. Those distributed to the walls of the abdomen.

Visceral Branches.

Cœliac axis { Gastric. Hepatic. Splenic. }

Superior mesenteric.
Inferior mesenteric.
Suprarenal.

Renal.
Spermatic or Ovarian.

Parietal Branches.

Phrenic.
Lumbar.
Sacra media.

CŒLIAC AXIS (fig. 308)

To expose this artery, raise the liver, draw down the stomach, and then tear through the layers of the lesser omentum.

The **Cœliac axis** is a short thick trunk, about half an inch in length, which arises from the aorta, close to the margin of the opening in the Diaphragm, and passing nearly horizontally forwards (in the erect posture), divides into three large branches, the *gastric, hepatic,* and *splenic,* occasionally giving off one of the phrenic arteries.

Relations.—It is covered by the lesser omentum. On the *right side*, it is in relation with the right semilunar ganglion, and the lobus Spigelii; on the *left side*, with the left semilunar ganglion and cardiac end of the stomach. *Below*, it rests upon the upper border of the pancreas.

The **Gastric** or **Coronary artery,** the smallest of the three branches of the cœliac axis, passes upwards and to the left side, to the cardiac orifice of the stomach, distributing branches to the œsophagus, which anastomose with the aortic œsophageal arteries; others supply the cardiac end of the stomach, inosculating with branches of the splenic artery: it then passes from left to right, along the lesser curvature of the stomach to the pylorus, lying in its course between the layers of the lesser omentum, and giving branches to both surfaces of the organ: at its termination it anastomoses with the pyloric branch of the hepatic.

Fig. 308.—The cœliac axis and its branches, the liver having been raised, and the lesser omentum removed.

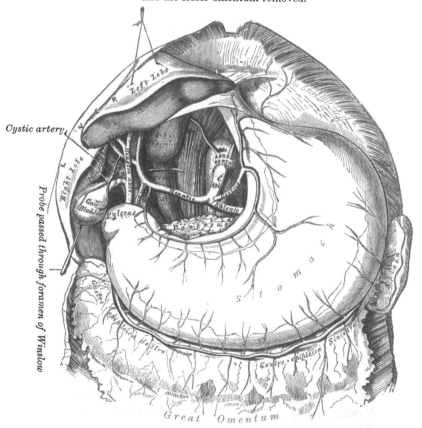

The **Hepatic artery** in the adult is intermediate in size between the gastric and splenic; in the fœtus, it is the largest of the three branches of the cœliac axis. It is first directed forwards and to the right, to the upper margin of the pyloric end of the stomach, forming the lower boundary of the foramen of Winslow. It then passes upwards between the layers of the lesser omentum, and in front of the foramen of Winslow, to the transverse fissure of the liver, where it divides into two branches, right and left, which supply the corresponding lobes of that organ, accompanying the ramifications of the vena portæ and hepatic ducts. The hepatic artery, in its course along the right border of the lesser omentum, is in relation with the ductus communis choledochus and portal vein, the duct lying to the right of the artery, and the vena portæ behind.

Its branches are : the

    Pyloric.

    Gastro-duodenalis { Gastro-epiploica dextra.
                        { Pancreatico-duodenalis superior.

    Cystic.

The **pyloric branch** arises from the hepatic, above the pylorus, descends to the pyloric end of the stomach, and passes from right to left along its lesser curvature, supplying it with branches, and inosculating with the gastric branches of the coronary artery.

Fig. 309.—The cœliac axis and its branches, the stomach having been raised and the transverse meso-colon removed.

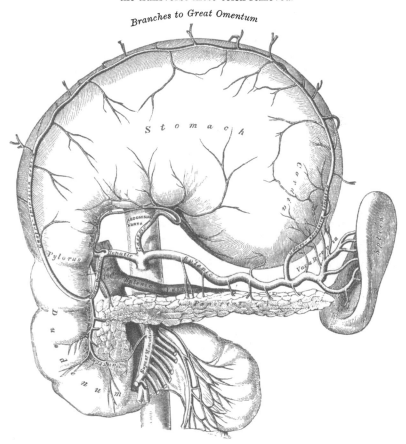

The **gastro-duodenalis** (fig. 309) is a short but large branch, which descends, near the pylorus, behind the first portion of the duodenum, and divides at the lower border of this viscus into two branches, the *gastro-epiploica dextra* and the *pancreatico-duodenalis superior*. Previous to its division, it gives off two or three small inferior pyloric branches to the pyloric end of the stomach and pancreas.

The **gastro-epiploica dextra** runs from right to left along the greater curvature of the stomach, between the layers of the great omentum, anastomosing about the middle of the lower border of the stomach with the gastro-epiploica sinistra from the splenic artery. This vessel gives off numerous branches, some of which ascend to supply both surfaces of the stomach, while others descend to supply the great omentum.

The **pancreatico-duodenalis superior** descends between the contiguous margins of the duodenum and pancreas. It supplies both these organs, and anastomoses with the inferior pancreatico-duodenal branch of the superior mesenteric artery, and with the pancreatic branches of the splenic.

The **cystic artery** (fig. 308), usually a branch of the right hepatic, passes upwards and forwards along the neck of the gall-bladder, and divides into two branches, one of which ramifies on its free surface, the other between it and the substance of the liver.

The **Splenic artery,** in the adult, is the largest of the three branches of the cœliac axis, and is remarkable for the extreme tortuosity of its course. It passes horizontally to the left side, behind the peritoneum and along the upper border of the pancreas, accompanied by the splenic vein, which lies below it; and, on arriving near the spleen, divides into branches, some of which enter the hilum of that organ to be distributed to its structure, while others are distributed to the pancreas and great end of the stomach. Its branches are : the

Pancreaticæ parvæ.   Gastric (Vasa brevia).
Pancreatica magna.   Gastro-epiploica sinistra.

The **pancreatic** are numerous small branches derived from the splenic as it runs behind the upper border of the pancreas, supplying its middle and left parts. One of these, larger than the rest, is given off from the splenic near the left extremity of the pancreas ; it runs from left to right near the posterior surface of the gland, following the course of the pancreatic duct, and is called the **pancreatica magna.** These vessels anastomose with the pancreatic branches of the pancreatico-duodenal arteries, derived from the hepatic on the one hand and the superior mesenteric on the other.

The **gastric (vasa brevia)** consist of from five to seven small branches, which arise either from the termination of the splenic artery, or from its terminal branches; and, passing from left to right, between the layers of the gastro-splenic omentum, are distributed to the great curvature of the stomach ; anastomosing with branches of the gastric and gastro-epiploica sinistra arteries.

The **gastro-epiploica sinistra,** the largest branch of the splenic, runs from left to right along the great curvature of the stomach, between the layers of the great omentum, and anastomoses with the gastro-epiploica dextra. In its course it distributes several branches to the stomach, which ascend upon both surfaces ; others descend to supply the omentum.

SUPERIOR MESENTERIC ARTERY (fig. 310)

In order to expose this vessel, raise the great omentum and transverse colon, draw down the small intestines, and cut through the peritoneum where the transverse meso-colon and mesentery join : the artery will then be exposed, just as it issues from beneath the lower border of the pancreas.

The **Superior Mesenteric artery** supplies the whole length of the small intestine, except the first part of the duodenum ; it also supplies the cæcum, ascending and transverse colon ; it is a vessel of large size, arising from the fore part of the aorta, about a quarter of an inch below the cœliac axis ; being covered at its origin by the splenic vein and pancreas. It passes forwards, between the pancreas and transverse portion of the duodenum, crosses in front of this portion of the intestine, and descends between the layers of the mesentery to the right iliac fossa, where, considerably diminished in size, it anastomoses with one of its own branches, viz. the ileo-colic. In its course it forms an arch, the convexity of which is directed forwards and downwards to the left side, the concavity backwards and upwards to the right. It is accompanied by the superior mesenteric vein, and is surrounded by the superior mesenteric plexus of nerves. Its branches are : the

Inferior pancreatico-duodenal.   Ileo-colic.
Vasa intestini tenuis.   Colica dextra.
Colica media.

The **inferior pancreatico-duodenal** is given off from the superior mesenteric, or from its first intestinal branch behind the pancreas. It courses to the right between the head of the pancreas and duodenum, and then ascends to anastomose with the superior pancreatico-duodenal artery. It distributes branches to the head of the pancreas and to the transverse and descending portions of the duodenum.

The **vasa intestini tenuis** arise from the convex side of the superior mesenteric artery. They are usually from twelve to fifteen in number, and are distributed to the jejunum and ileum. They run parallel with one another between the layers of the mesentery; each vessel dividing into two branches, which unite with a

FIG. 310.—The superior mesenteric artery and its branches.

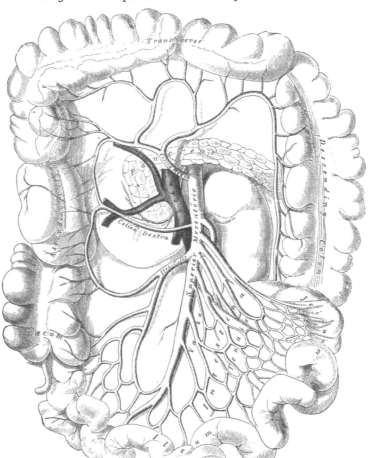

similar branch on each side, forming a series of arches, the convexities of which are directed towards the intestine. From this first set of arches branches arise, which again unite with similar branches from either side, and thus a second series of arches is formed; and from these latter, a third, and a fourth, or even fifth series of arches is constituted, diminishing in size the nearer they approach the intestine. From the terminal arches numerous small straight vessels arise which encircle the intestine, upon which they are distributed, ramifying between its coats. Throughout their course small branches are given off to the glands and other structures between the layers of the mesentery.

The **ileo-colic artery** is the lowest branch given off from the concavity of the superior mesenteric artery. It descends between the layers of the mesentery to the right iliac fossa, where it divides into two branches. Of these, the inferior division inosculates with the termination of the superior mesenteric artery, forming with it an arch, from the convexity of which branches proceed to supply the termination of the ileum, the cæcum and appendix cæci, and the ileo-cæcal valve. The superior division inosculates with the colica dextra, and supplies the commencement of the colon.

The **colica dextra** arises from about the middle of the concavity of the superior mesenteric artery, and, passing behind the peritoneum to the middle of the ascending colon, divides into two branches : a descending branch, which inosculates with the ileo-colic ; and an ascending branch, which anastomoses with the colica media. These branches form arches, from the convexity of which vessels are distributed to the ascending colon. The branches of this vessel are covered with peritoneum only on their anterior aspect.

The **colica media** arises from the upper part of the concavity of the superior mesenteric, and, passing forwards between the layers of the transverse meso-colon, divides into two branches : the one on the right side inosculating with the colica dextra ; that on the left side with the colica sinistra, a branch of the inferior mesenteric. From the arches formed by their inosculation, branches are distributed to the transverse colon. The branches of this vessel lie between two layers of the transverse meso-colon.

INFERIOR MESENTERIC ARTERY (fig. 311)

In order to expose this vessel, draw the small intestines and mesentery over to the right side of the abdomen, raise the transverse colon towards the thorax, and divide the peritoneum covering the front of the aorta.

The **Inferior Mesenteric artery** supplies the descending and sigmoid flexure of the colon, and the greater part of the rectum. It is smaller than the superior mesenteric, and arises from the left side of the aorta, between one and two inches above its division into the common iliacs. It passes downwards to the left iliac fossa, and then descends, between the layers of the meso-rectum, into the pelvis, under the name of the *superior hæmorrhoidal artery*. It lies at first in close relation with the left side of the aorta, and then passes as the superior hæmorrhoidal in front of the left common iliac artery. Its branches are : the

| Colica sinistra. | Sigmoid. | Superior hæmorrhoidal. |

The **colica sinistra** passes behind the peritoneum, in front of the left kidney, to reach the descending colon, and divides into two branches: an ascending branch, which inosculates with the colica media ; and a descending branch, which anastomoses with the sigmoid artery. From the arches formed by these inosculations, branches are distributed to the descending colon.

The **sigmoid artery** runs obliquely downwards across the Psoas muscle to the sigmoid flexure of the colon, and divides into branches which supply that part of the intestine ; anastomosing above, with the colica sinistra ; and below, with the superior hæmorrhoidal artery. This vessel is sometimes replaced by three or four small branches.

The **superior hæmorrhoidal artery,** the continuation of the inferior mesenteric, descends into the pelvis between the layers of the meso-rectum, crossing, in its course, the ureter and left common iliac vessels. It divides into two branches, which descend one on each side of the rectum, and about five inches from the anus break up into several small branches, which pierce the muscular coat of the bowel and run downwards, as straight vessels, placed at regular intervals from each other in the wall of the gut between its muscular and mucous coat, to the level of the internal sphincter ; here they form a series of loops around the lower end of the rectum, and communicate with the middle hæmorrhoidal arteries, branches of the internal iliac, and with the inferior hæmorrhoidal branches of the internal pudic.

The **Suprarenal arteries** are two small vessels which arise, one on each side of the aorta, opposite the superior mesenteric artery. They pass obliquely upwards and outwards, over the crura of the Diaphragm, to the under surface of the suprarenal capsules, to which they are distributed, anastomosing with capsular branches from the phrenic and renal arteries. In the adult these arteries are of small size; in the fœtus they are as large as the renal arteries.

RENAL ARTERIES (fig. 307)

The **Renal arteries** are two large trunks, which arise from the sides of the aorta, immediately below the superior mesenteric artery. Each is directed

FIG. 311.—The inferior mesenteric and its branches.

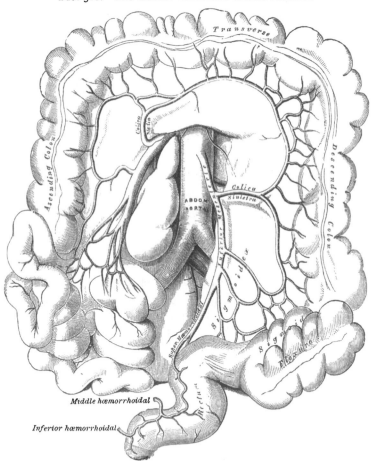

outwards, across the crus of the Diaphragm, so as to form nearly a right angle with the aorta. The right is longer than the left, on account of the position of the aorta; it passes behind the inferior vena cava. The left is somewhat higher than the right. Before reaching the hilum of the kidney, each artery divides into four or five branches; the greater number of which generally lie between the renal vein and ureter, the vein being in front, the ureter behind. Each vessel gives off some small branches to the suprarenal capsule, the ureter, and the surrounding cellular tissue and muscles. Frequently there is a second renal artery, which is given off from the abdominal aorta either above or below the renal artery proper, the former being the more common position. Instead of entering the kidney at the hilum, these accessory renal arteries usually pierce the upper or lower part of the gland.

SPERMATIC ARTERIES

The **Spermatic arteries** are distributed to the testes. They are two slender vessels, of considerable length, which arise from the front of the aorta, a little below the renal arteries. Each artery passes obliquely outwards and downwards, behind the peritoneum, resting on the Psoas muscle, the right spermatic lying in front of the inferior vena cava, the left behind the sigmoid flexure of the colon. It then crosses obliquely over the ureter and the lower part of the external iliac artery to reach the internal abdominal ring, through which it passes, and accompanies the other constituents of the spermatic cord along the inguinal canal to the scrotum, where it becomes tortuous, and divides into several branches, two or three of which accompany the vas deferens, and supply the epididymis, anastomosing with the artery of the vas deferens; others pierce the back part of the tunica albuginea, and supply the substance of the testis.

The **Ovarian arteries** (fig. 313) are the corresponding arteries in the female to the spermatic in the male. They supply the ovaries, are shorter than the spermatic, and do not pass out of the abdominal cavity. The origin and course of the first part of the artery are the same as the spermatic in the male, but on arriving at the margin of the pelvis the ovarian artery passes inwards, between the two layers of the broad ligament of the uterus, to be distributed to the ovary. One or two small branches supply the Fallopian tube; another passes on to the side of the uterus, and anastomoses with the uterine arteries. Other offsets are continued along the round ligament, through the inguinal canal, to the integument of the labium and groin.

At an early period of foetal life, when the testes or ovaries lie by the side of the spine, below the kidneys, the spermatic or ovarian arteries are short; but as these organs descend from the abdomen into the scrotum or pelvis, the arteries become gradually lengthened.

PHRENIC ARTERIES

The **Phrenic arteries** are two small vessels, which present much variety in their origin. They may arise separately from the front of the aorta, immediately above the cœliac axis, or by a common trunk, which may spring either from the aorta or from the cœliac axis. Sometimes one is derived from the aorta, and the other from one of the renal arteries. In only one out of thirty-six cases examined did these arteries arise as two separate vessels from the aorta. They diverge from one another across the crura of the Diaphragm, and then pass obliquely upwards and outwards upon its under surface. The left phrenic passes behind the œsophagus, and runs forwards on the left side of the œsophageal opening. The right phrenic passes behind the inferior vena cava, and ascends along the right side of the aperture for transmitting that vein. Near the back part of the central tendon each vessel divides into two branches. The internal branch runs forwards to the front of the thorax, supplying the Diaphragm, and anastomosing with its fellow of the opposite side, and with the musculo-phrenic and comes nervi phrenici, branches of the internal mammary. The external branch passes towards the side of the thorax, and inosculates with the intercostal arteries. The internal branch of the right phrenic gives off a few vessels to the inferior vena cava; and the left one some branches to the œsophagus. Each vessel also sends capsular branches to the suprarenal capsule of its own side. The spleen on the left side, and the liver on the right, also receive a few branches from these vessels.

LUMBAR ARTERIES

The **Lumbar arteries** are analogous to the intercostal. They are usually four in number on each side, and arise from the back part of the aorta, nearly at right angles with that vessel. They pass outwards and backwards, around the sides of the bodies of the lumbar vertebræ, behind the sympathetic nerve and the Psoas magnus muscle ; those on the right side being covered by the inferior vena cava, and the two upper ones on each side by the crura of the Diaphragm. In the interval between the transverse processes of the vertebræ each artery divides into a *dorsal* and an *abdominal branch.*

The *dorsal branch* gives off, immediately after its origin, a *spinal branch,* which enters the spinal canal ; it then continues its course backwards, between the transverse processes, and is distributed to the muscles and integument of the back, anastomosing with the similar branches of the adjacent lumbar arteries, and with the posterior branches of the intercostal arteries. The *spinal branch* enters the spinal canal through the intervertebral foramen, to be distributed to the spinal cord and its membranes and to the bodies of the vertebræ in the same manner as the lateral spinal branches from the vertebral (see page 496).

The *abdominal branches* pass outwards, having a variable relation to the Quadratus lumborum muscle. Most frequently the first branch passes in front of the muscle and the others behind it ; sometimes the order is reversed and the lowest branch passes in front of the muscle. At the outer border of the Quadratus they are continued between the abdominal muscles, and anastomose with branches of the epigastric and internal mammary *in front,* the intercostals *above,* and those of the ilio-lumbar and circumflex iliac *below.*

MIDDLE SACRAL ARTERY

The **Middle Sacral artery** is a small vessel, which arises from the back part of the aorta, just at its bifurcation. It descends upon the last lumbar vertebra, and along the middle line of the front of the sacrum, to the upper part of the coccyx, where it anastomoses with the lateral sacral arteries, and terminates in a minute branch, which runs down to the situation of the body immediately to be described as ' Luschka's gland.' From it, branches arise which run through the meso-rectum, to supply the posterior surface of the rectum. Other branches are given off on each side, which anastomose with the lateral sacral arteries, and send off small offsets, which enter the anterior sacral foramina.

The artery is the representative of the caudal prolongation of the aorta of animals, and its lateral branches correspond to the intercostal and lumbar arteries in the dorsal and lumbar regions.

Coccygeal gland, or Luschka's gland.—Lying near the tip of the coccyx in a small tendinous interval formed by the union of the Levator ani muscle of each side, and just above the coccygeal attachment of the Sphincter ani, is a small conglobate body, about as large as a lentil or a pea, first described by Luschka,* and named by him the *coccygeal gland.* Its most obvious connections are with the arteries of the part.

Structure.—It consists of a congeries of small arteries, with little aneurismal dilatations, derived from the middle sacral, and freely communicating with each other. These vessels are enclosed in one or more layers of polyhedral granular cells, and the whole structure is invested in a capsule of connective tissue which sends in trabeculæ, dividing [the interior into a number of spaces in which the vessels and cells are contained. Nerves pass into this little body from the sympathetic, but their mode of termination is unknown. Macalister believes the

---

* *Der Hirnanhang und die Steissdrüse des Menschen,* Berlin, 1860; *Anatomie des Menschen,* Tübingen, 1864, vol. ii. pt. 2, p. 187.

glomerulus of vessels 'consists of the condensed and convoluted metameric dorsal arteries of the caudal segments embedded in tissue which is possibly a small persisting fragment of the neurenteric canal.'

COMMON ILIAC ARTERIES

The abdominal aorta divides into the two **common iliac arteries**. The bifurcation usually takes place on the left side of the body of the fourth lumbar vertebra. The common iliac arteries are about two inches in length; diverging from the termination of the aorta, they pass downwards and outwards to the margin of the pelvis, and divide opposite the intervertebral substance, between the last lumbar vertebra and the sacrum, into two branches, the *external* and *internal iliac arteries*: the former supplying the lower extremity; the latter, the viscera and parietes of the pelvis.

The **right common iliac** is somewhat longer than the left, and passes more obliquely across the body of the last lumbar vertebra. In front of it are the peritoneum, the small intestines, branches of the sympathetic nerve, and, at its point of division, the ureter. *Behind*, it is separated from the fourth and fifth lumbar vertebræ, with the intervening intervertebral disc, by the two common iliac veins. On its *outer side*, it is in relation with the inferior vena cava, and the right common iliac vein, above; and the Psoas magnus muscle below.

The **left common iliac** is in relation, in front, with the peritoneum, branches of the sympathetic nerve, and the superior hæmorrhoidal artery; and is crossed at its point of bifurcation by the ureter. It rests on the bodies of the fourth and fifth lumbar vertebræ, with the intervening intervertebral disc. The left common iliac vein lies partly on the inner side, and partly beneath the artery; on its outer side, the artery is in relation with the Psoas magnus muscle.

PLAN OF THE RELATIONS OF THE COMMON ILIAC ARTERIES

In front.

Peritoneum.
Small intestines.
Sympathetic nerves.
Ureter.

In front.

Peritoneum.
Sympathetic nerves.
Superior hæmorrhoidal artery.
Ureter.

Right Common Iliac. (*Outer side.*) Vena cava. Right common iliac vein. Psoas muscle.	(*Inner side.*) Left common iliac vein.	**Left Common Iliac.** (*Outer side.*) Psoas magnus muscle.

Behind.

Fourth and fifth lumbar vertebræ.
Right and left common iliac veins.

Behind.

Fourth and fifth lumbar vertebræ.
Left common iliac vein.

Branches.—The common iliac arteries give off small branches to the peritoneum, Psoas magnus, ureters, and the surrounding cellular tissue, and occasionally give origin to the ilio-lumbar, or renal arteries.

Peculiarities.—The *point of origin* varies according to the bifurcation of the aorta. In three-fourths of a large number of cases, the aorta bifurcated either upon the fourth lumbar vertebra, or upon the intervertebral disc between it and the fifth; the bifurcation being, in one case out of nine below, and in one out of eleven above this point. In ten out of every thirteen cases, the vessel bifurcated within half an inch above or below the level of the crest of the ilium: more frequently below than above.

The *point of division* is subject to great variety. In two-thirds of a large number of cases it was between the last lumbar vertebra and the upper border of the sacrum ; being above that point in one case out of eight, and below it in one case out of six. The left common iliac artery divides lower down more frequently than the right.

The *relative length*, also, of the two common iliac arteries varies. The right common iliac was the longer in sixty-three cases ; the left in fifty-two ; while they were both equal in fifty-three. The length of the arteries varied in five-sevenths of the cases examined, from an inch and a half to three inches ; in about half of the remaining cases, the artery was longer, and in the other half, shorter : the minimum length being less than half an inch, the maximum four and a half inches. In two instances, the right common iliac has been found wanting, the external and internal iliacs arising directly from the aorta.

Surface Marking.—Draw a zone round the body opposite the highest part of the crest of the ilium ; in this line take a point half an inch to the left of the middle line. From this draw two lines to points midway between the anterior superior spines of the ilium and the symphysis pubis. These two diverging lines will represent the course of the common and external iliac arteries. Draw a second zone round the body corresponding to the level of the anterior superior spines of the ilium : the portion of the diverging lines between the two zones will represent the course of the common iliac artery ; the portion below the lower zone, that of the external iliac artery.

Surgical Anatomy.—The application of a ligature to the common iliac artery may be required on account of aneurism or hæmorrhage, implicating the external or internal iliacs. Now that the surgeon no longer dreads opening the peritoneal cavity, there can be no question that the easiest and best method of tying the artery is by a transperitoneal route. The abdomen is opened by an incision in either the semilunar line or the linea alba ; the intestines are drawn to one side and the peritoneum covering the artery divided. The sheath is then opened, and the needle passed from within outwards. On the right side great care must be exercised in passing the needle, since both the common iliac veins lie behind the artery. After the vessel has been tied the incision in the peritoneum over the artery should be sutured. Formerly there were two different methods by which the common iliac artery was tied, without opening the peritoneal cavity. 1. An anterior or iliac incision, by which the vessel is approached more directly from the front ; and 2. a posterior abdominal or lumbar incision, by which the vessel is reached from behind. If the surgeon select the iliac region, a curved incision, from five to eight inches in length according to the amount of fat, is made, commencing just outside the middle of Poupart's ligament and a finger's breadth above it, and carried outwards towards the anterior superior iliac spine, then upwards towards the ribs, and finally curving inwards towards the umbilicus. The abdominal muscles and transversalis fascia are divided, and the peritoneum raised upwards and inwards, until the Psoas is reached. The artery will be found on the inner side of this muscle, and is to be cleared with a director, especial care being taken on the right side, as here the common iliac veins lie behind the artery. The aneurism needle is to be passed from within outwards. But if the aneurismal tumour should extend high up in the abdomen, along the external iliac, it is better to select the posterior or lumbar operation, by making an incision partly in the abdomen, partly in the loin. The incision is commenced at the anterior extremity of the last rib, proceeding directly downwards to the ilium ; it is then curved forwards along the crest of the ilium, and a little above it, to the anterior superior spine of that bone. The abdominal muscles having been cautiously divided in succession, the transversalis fascia must be carefully cut through, and the peritoneum, together with the ureter, separated from the artery, and pushed aside ; the sacro-iliac articulation must then be felt for, and upon it the vessel will be felt pulsating, and may be fully exposed in close connection with its accompanying vein. On the right side, both common iliac veins, as well as the inferior vena cava, are in close connection with the artery, and must be carefully avoided. On the left side, the vein usually lies on the inner side, and behind the artery ; but it occasionally happens that the two common iliac veins are joined on the left instead of the right side, which would add much to the difficulty of an operation in such a case. The common iliac artery may be so short that danger may be apprehended from secondary hæmorrhage if a ligature is applied to it. It would be preferable, in such a case, to tie both the external and internal iliacs near their origin.

Collateral Circulation.—The principal agents in carrying on the collateral circulation after the application of a ligature to the common iliac are: the anastomoses of the hæmorrhoidal branches of the internal iliac with the superior hæmorrhoidal from the inferior mesenteric ; the anastomoses of the uterine and ovarian arteries, and of the vesical arteries of opposite sides ; of the lateral sacral with the middle sacral artery ; of the epigastric with the internal mammary, inferior intercostal, and lumbar arteries ; of the circumflex iliac with the lumbar arteries ; of the ilio-lumbar with the last lumbar artery ; of the obturator artery, by means of its pubic branch, with the vessel of the opposite side, and with the deep epigastric.

Compression of the Common Iliac Arteries.—The common iliac arteries are most efficiently compressed by Davy's lever. The instrument consists of a gum-elastic tube, about two feet long, in which fits a round wooden ' lever ' considerably longer than the tube. A small quantity of olive oil having been injected into the rectum, the gum-elastic tube, softened in hot water, is passed into the bowel sufficiently far to permit its pressing upon the common iliac artery as it lies in the groove between the last lumbar vertebra and the Psoas muscle. The wooden lever is then inserted into the tube, and the projecting end carried towards the opposite thigh and raised, when it acts as a lever of the first order, the anus being the fulcrum. In cases where the meso-rectum is abnormally short it may be impossible, without unjustifiable force, to compress the artery on the right side.

Internal Iliac Artery (fig. 312)

The **Internal Iliac artery** supplies the walls and viscera of the pelvis, the generative organs, and inner side of the thigh. It is a short, thick vessel, smaller in the adult than the external iliac, and about an inch and a half in length. It arises at the point of bifurcation of the common iliac, and, passing downwards to the upper margin of the great sacro-sciatic foramen, divides into two large trunks, an *anterior* and *posterior*; from its anterior division a partially obliterated cord, the *hypogastric artery*, extends forwards to the bladder.

Relations.—*In front* with the ureter, which separates it from the peritoneum. *Behind*, with the internal iliac vein, the lumbo-sacral cord, and Pyriformis muscle. By its *outer side*, near its origin, with the Psoas magnus muscle.

Plan of the Relations of the Internal Iliac Artery

In front.
Peritoneum.
Ureter.

Outer side.
Psoas magnus.

Internal
Iliac.

Behind.
Internal iliac vein.
Lumbo-sacral cord.
Pyriformis muscle.

In the fœtus, the internal iliac artery (*hypogastric*) is twice as large as the external iliac, and appears to be the continuation of the common iliac. Instead of dipping into the pelvis, it passes forwards to the bladder, and ascends along the sides of that viscus to its summit, to which it gives branches; it then passes upwards along the back part of the anterior wall of the abdomen to the umbilicus, converging towards its fellow of the opposite side. Having passed through the umbilical opening, the two arteries twine round the umbilical vein, forming with it the umbilical cord; and, ultimately, ramify in the placenta. The portion of the vessel within the abdomen is called the *hypogastric artery*; and that external to that cavity, the *umbilical artery.*

At birth, when the placental circulation ceases, the upper portion of the hypogastric artery, extending from the summit of the bladder to the umbilicus, contracts, and ultimately dwindles to a solid fibrous cord; but the lower portion, extending from its origin (in what is now the internal iliac artery) for about an inch and a half to the wall of the bladder, and thence to the summit of that organ, is not totally impervious, though it becomes considerably reduced in size, and serves to convey blood to the bladder, under the name of the *superior vesical artery.*

Peculiarities as regards Length.—In two-thirds of a large number of cases, the length of the internal iliac varied between an inch and an inch and a half; in the remaining third, it was more frequently longer than shorter, the maximum length being three inches, the minimum half an inch.

The lengths of the common and internal iliac arteries bear an inverse proportion to each other, the internal iliac artery being long when the common iliac is short, and *vice versâ.*

As regards its Place of Division.—The place of division of the internal iliac varies between the upper margin of the sacrum and the upper border of the sacro-sciatic foramen.

The arteries of the two sides in a series of cases often differed in length, but neither seemed constantly to exceed the other.

Surgical Anatomy.—The application of a ligature to the internal iliac artery may be required in cases of aneurism or hæmorrhage affecting one of its branches. The vessel may be secured by making an incision through the abdominal parietes in the iliac region, in a direction and to an extent similar to that for securing the common iliac ; the transversalis fascia having been cautiously divided, and the peritoneum pushed inwards from the iliac fossa towards the pelvis, the finger may feel the pulsation of the external iliac at

Fig. 312.—Arteries of the pelvis.

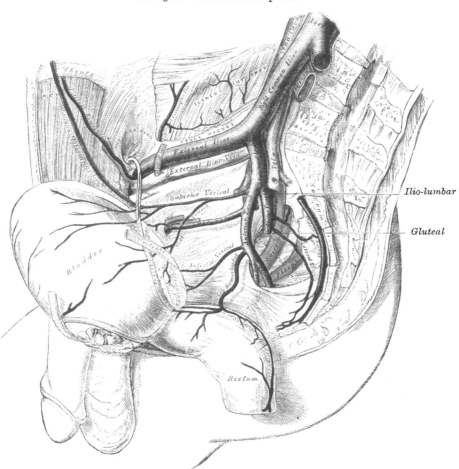

the bottom of the wound ; and, by tracing this vessel upwards, the internal iliac is arrived at, opposite the sacro-iliac articulation. It should be remembered that the vein lies behind, and on the right side, a little external to the artery, and in close contact with it ; the ureter and peritoneum, which lie in front, must also be avoided. The degree of facility in applying a ligature to this vessel will mainly depend upon its length. It has been seen that, in the great majority of the cases examined, the artery was short, varying from an inch to an inch and a half ; in these cases, the artery is deeply seated in the pelvis ; when, on the contrary, the vessel is longer, it is found partly above that cavity. If the artery is very short, as occasionally happens, it would be preferable to apply a ligature to the common iliac, or upon the external and internal iliacs at their origin.

Probably a better method of tying the internal iliac artery is by an abdominal section in the median line and reaching the vessel through the peritoneal cavity. This plan has been advocated by Dennis, of New York, on the following grounds : (1) It no way increases the danger of the operation ; (2) it prevents a series of accidents which have occurred during ligature of the artery by the older methods; (3) it enables the surgeon to ascertain the exact extent of disease in the main arterial trunk, and select his spot for the application of the ligature; and (4) it occupies much less time.

Collateral Circulation.—In Professor Owen's dissection of a case in which the internal iliac artery had been tied by Stevens ten years before death, for aneurism of the sciatic artery, the internal iliac was found impervious for about an inch above the point where the ligature had been applied; but the obliteration did not extend to the origin of the external iliac, as the ilio-lumbar artery arose just above this point. Below the point of obliteration, the artery resumed its natural diameter, and continued so for half an inch; the obturator, lateral sacral, and gluteal arising in succession from the latter portion. The obturator artery was entirely obliterated. The lateral sacral artery was as large as a crow's quill, and had a very free anastomosis with the artery of the opposite side, and with the middle sacral artery. The sciatic artery was entirely obliterated as far as its point of connection with the aneurismal tumour; but, on the distal side of the sac, it was continued down along the back of the thigh nearly as large in size as the femoral, being pervious about an inch below the sac by receiving an anastomosing vessel from the profunda.* The circulation was carried on by the anastomoses of the uterine and ovarian arteries; of the opposite vesical arteries; of the hæmorrhoidal branches of the internal iliac with those from the inferior mesenteric; of the obturator artery, by means of its pubic branch, with the vessel of the opposite side, and with the epigastric and internal circumflex; of the circumflex and perforating branches of the profunda femoris with the sciatic; of the gluteal with the posterior branches of the sacral arteries; of the ilio-lumbar with the last lumbar; of the lateral sacral with the middle sacral; and of the circumflex iliac with the ilio-lumbar and gluteal.

BRANCHES OF THE INTERNAL ILIAC

From the Anterior Trunk.

Superior vesical.
Middle vesical.
Inferior vesical.
Middle hæmorrhoidal.
Obturator.
Internal pudic.
Sciatic.

In female { Uterine.
{ Vaginal.

From the Posterior Trunk.

Ilio-lumbar.
Lateral sacral.
Gluteal.

The **superior vesical** is that part of the fœtal hypogastric artery which remains pervious after birth. It extends to the side of the bladder, distributing numerous branches to the apex and body of the organ. From one of these a slender vessel is derived, which accompanies the vas deferens in its course to the testis, where it anastomoses with the spermatic artery. This is the *artery of the vas deferens.* Other branches supply the ureter.

The **middle vesical,** usually a branch of the superior, is distributed to the base of the bladder and under surface of the vesiculæ seminales.

The **inferior vesical** arises from the anterior division of the internal iliac, frequently in common with the middle hæmorrhoidal, and is distributed to the base of the bladder, the prostate gland, and vesiculæ seminales. The branches distributed to the prostate communicate with the corresponding vessels of the opposite side.

The **middle hæmorrhoidal artery** usually arises together with the preceding vessel. It supplies the anus and parts outside the rectum, anastomosing with the other hæmorrhoidal arteries.

The **uterine artery** (fig. 313) passes inwards from the anterior trunk of the internal iliac to the neck of the uterus. Ascending, in a tortuous course on the side of this viscus, between the layers of the broad ligament, it distributes branches to its substance, anastomosing, near its termination, with a branch from the ovarian artery. It gives off branches to the cervex uteri (*cervical*), and branches which descend on the vagina, and, joining with branches from the vaginal arteries, form a median longitudinal vessel both in front and behind; these descend on the

* *Medico-Chirurgical Trans.* vol. xvi.

anterior and posterior surfaces of the vagina, and are named the *azygos arteries of the vagina*.

The **vaginal artery** is analogous to the inferior vesical in the male ; it descends upon the vagina, supplying its mucous membrane, and sending branches to the neck of the bladder and contiguous part of the rectum.   It assists in forming the azygos arteries of the vagina.

FIG. 313.—The arteries of the internal organs of generation of the female, seen from behind.   (After Hyrtl.)

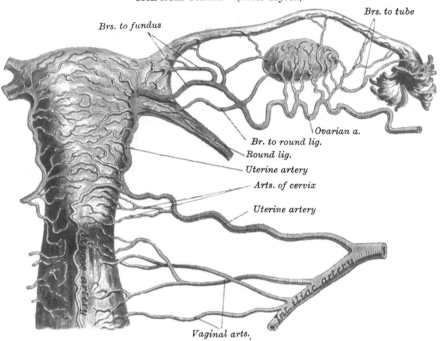

this vessel lies upon the pelvic fascia, beneath the peritoneum, and a little below the obturator nerve.

The **obturator artery** usually arises from the anterior trunk of the internal iliac, sometimes from the posterior.   It passes forwards, below the brim of the pelvis, to the upper part of the obturator foramen, and escaping from the pelvic cavity through a short canal, formed by a groove on the under surface of the ascending ramus of the os pubis, and the arched border of the obturator membrane, it divides into an internal and an external branch.   In the pelvic cavity

FIG. 314.—Variations in origin and course of obturator artery.

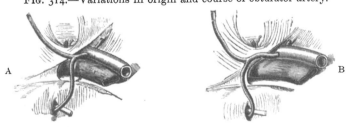

this vessel lies upon the pelvic fascia, beneath the peritoneum, and a little below the obturator nerve.

Branches.—*Within the pelvis*, the obturator artery gives off an *iliac branch* to the iliac fossa, which supplies the bone and the Iliacus muscle, and anastomoses with the ilio-lumbar artery ; a *vesical branch*, which runs backwards to supply the bladder ; and a *pubic branch*, which is given off from the vessel just before it leaves the pelvic cavity. This branch ascends upon the back of the os pubis, communicating with offsets from the deep epigastric artery, and with the corresponding vessel of the opposite side. It is placed on the inner side of the femoral ring. *External to the pelvis*, the obturator artery divides into an *internal* and an *external branch,* which are deeply situated beneath the Obturator externus muscle.

The *internal branch* curves downwards along the inner margin of the obturator foramen, lying beneath the Obturator externus muscle ; it distributes branches to the Obturator externus, Pectineus, Adductors, and Gracilis, and ana-stomoses with the external branch, and with the internal circumflex artery.

The *external branch* curves round the outer margin of the foramen, also lying beneath the Obturator externus muscle, to the space between the Gemellus inferior and Quadratus femoris, where it divides into two branches : one, the smaller, courses inwards around the lower margin of the foramen and anastomoses with the internal branch and with the internal circumflex ; the other inclines outwards in the groove below the acetabulum, and supplies the muscles attached to the tuberosity of the ischium and anastomoses with the sciatic artery. It sends a branch to the hip-joint through the cotyloid notch, which ramifies on the round ligament as far as the head of the femur.

Peculiarities.—In two out of every three cases the obturator arises from the internal iliac ; in one case in 3½ from the epigastric ; and in about one in seventy-two cases by two roots from both vessels. It arises in about the same proportion from the external iliac artery. The origin of the obturator from the epigastric is not commonly found on both sides of the same body.

When the obturator artery arises at the front of the pelvis from the epigastric, it descends almost vertically to the upper part of the obturator foramen. The artery in this course usually lies in contact with the external iliac vein, and on the outer side of the femoral ring (fig. 314, A) ; in such cases it would not be endangered in the operation for femoral hernia. Occasionally, however, it curves inwards along the free margin of Gimbernat's ligament (fig. 314, B), and under such circumstances would almost completely encircle the neck of a hernial sac (supposing a hernia to exist in such a case), and would be in great danger of being wounded if an operation were performed.

The **internal pudic** is the smaller of the two terminal branches of the anterior trunk of the internal iliac, and supplies the external organs of generation. Though the course of the artery is the same in the two sexes, the vessel is much smaller in the female than in the male, and the distribution of its branches somewhat different. The description of its arrangement in the male will first be given, and subsequently the differences which it presents in the female will be mentioned.

The **internal pudic artery in the male** passes downwards and outwards to the lower border of the great sacro-sciatic foramen, and emerges from the pelvis between the Pyriformis and Coccygeus muscles : it then crosses the spine of the ischium, and re-enters the pelvis through the lesser sacro-sciatic foramen. The artery now crosses the Obturator internus muscle, along the outer wall of the ischio-rectal fossa, being situated about an inch and a half above the lower margin of the ischial tuberosity. It is here contained in a sheath of the obturator fascia, and gradually approaches the margin of the ramus of the ischium, along which it passes forwards and upwards, pierces the base of the superficial layer of the triangular ligament of the urethra, and runs forwards along the inner margin of the ramus of the os pubis, and divides into its two terminal branches, the *dorsal artery of the penis* and the *artery of the corpus cavernosum.*

Relations.—In the first part of its course, within the pelvis, it lies in front of the Pyriformis muscle, the sacral plexus of nerves, and the sciatic artery, and on the outer side of the rectum (on the left side). As it crosses the spine of the ischium, it is covered by the Gluteus maximus and overlapped by the great sacro-sciatic ligament. Here the obturator nerve lies to the inner side and the nerve to the Obturator internus to the outer side of the vessel. In the pelvis, it lies on the outer side of the ischio-rectal fossa, upon the surface of the Obturator internus muscle, contained in a fibrous canal (canal of Alcock) formed by the splitting of the obturator fascia. It is accompanied by the pudic veins and the pudic nerve.

Peculiarities.—The internal pudic is sometimes smaller than usual, or fails to give off one or two of its usual branches; in such cases the deficiency is supplied by branches derived from an additional vessel, the *accessory pudic*, which generally arises from the internal pudic artery before its exit from the great sacro-sciatic foramen. It passes forwards along the lower part of the bladder and across the side of the prostate gland to the root of the penis, where it perforates the triangular ligament, and gives off the branches usually derived from the pudic artery. The deficiency most frequently met with is that in which the internal pudic ends as the artery of the bulb; the artery of the corpus cavernosum and arteria dorsalis penis being derived from the accessory pudic. Or the pudic may terminate as the superficial perineal, the artery of the bulb being derived, with the other two branches, from the accessory vessel. Occasionally the accessory pudic artery is derived from one of the other branches of the internal iliac, most frequently the inferior vesical or the obturator.

Fig. 315.—The internal pudic artery and its branches in the male.

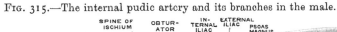

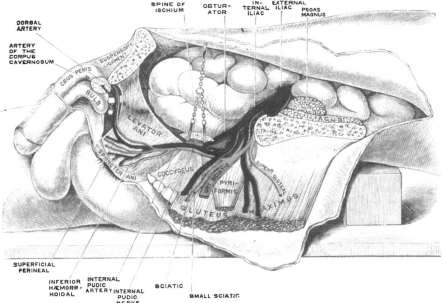

Surgical Anatomy.—The relation of the accessory pudic to the prostate gland and urethra is of the greatest interest in a surgical point of view, as this vessel is in danger of being wounded in the lateral operation of lithotomy. The student should also study the position of the internal pudic artery and its branches, when running a normal course, with regard to the same operation. The superficial and the transverse perineal arteries are, of necessity, divided in this operation, but the hæmorrhage from these vessels is seldom excessive; should a ligature be required, it can readily be applied on account of their superficial position. The artery of the bulb may be divided if the incision be carried too far forwards, and injury of this vessel may be attended with serious or even fatal consequences. The main trunk of the internal pudic artery may be wounded if the incision be carried too far outwards; but, being bound down by the strong obturator fascia, and under cover of the ramus of the ischium, the accident is not very likely to occur unless the vessel runs an anomalous course.

Branches.—The branches of the internal pudic artery are:

Muscular.	Transverse perineal.
Inferior hæmorrhoidal.	Artery of the bulb.
Superficial perineal.	Artery of the corpus cavernosum.

Dorsal artery of the penis.

The *muscular branches* consist of two sets: one given off in the pelvis; the other, as the vessel crosses the ischial spine. The former are several small offsets which supply the Levator ani, the Obturator internus, the Pyriformis, and the Coccygeus muscles. The branches given off outside the pelvis are distributed to the adjacent part of the Gluteus maximus and External rotator muscles. They anastomose with branches of the sciatic artery.

The *inferior hæmorrhoidal* are two or three small arteries which arise from the internal pudic as it passes above the tuberosity of the ischium. Crossing the ischio-rectal fossa, they are distributed to the muscles and integument of the anal region.

The *superficial perineal artery* supplies the scrotum, muscles and integument of the perinæum. It arises from the internal pudic, in front of the preceding branches, and turns upwards, crossing either over or under the Transversus perinæi muscle, and runs forwards, parallel to the pubic arch, in the interspace between the Accelerator urinæ and Erector penis muscles, both of which it supplies, and is finally distributed to the skin and dartos of the scrotum. In its passage through the perinæum, it lies beneath the superficial perineal fascia.

The *transverse perineal* is a small branch which arises either from the internal pudic, or from the superficial perineal artery as it crosses the Transversus perinæi muscle. It runs transversely inwards along the cutaneous surface of the Transversus perinæi muscle, which it supplies, as well as the structures between the anus and bulb of the urethra, and anastomoses with the one of the opposite side.

The *artery of the bulb* is a large but very short vessel, which arises from the internal pudic between the two layers of the triangular ligament, and, passing nearly transversely inwards between the fibres of the Compressor urethræ muscle, it pierces the bulb of the urethra, in which it ramifies. It gives off a small branch which descends to supply Cowper's gland.

Surgical Anatomy.—This artery is of considerable importance in a surgical point of view, as it is in danger of being wounded in the lateral operation of lithotomy, an accident usually attended in the adult with alarming hæmorrhage. The vessel is sometimes very small, occasionally wanting, or even double. It sometimes arises from the internal pudic earlier than usual, and crosses the perinæum to reach the back part of the bulb. In such a case the vessel could hardly fail to be wounded in the performance of the lateral operation of lithotomy. If, on the contrary, it should arise from an accessory pudic, it lies more forward than usual, and is out of danger in the operation.

The *artery of the corpus cavernosum*, one of the terminal branches of the internal pudic, arises from that vessel while it is situated between the two layers of the triangular ligament; it pierces the superficial layer, and entering the crus penis obliquely, it runs forwards in the centre of the corpus cavernosum, to which its branches are distributed.

The *dorsal artery of the penis* ascends between the crus and pubic symphysis, and, piercing the triangular ligament, passes between the two layers of the suspensory ligament of the penis, and runs forwards on the dorsum of the penis to the glans, where it divides into two branches, which supply the glans and prepuce. On the dorsum of the penis, it lies immediately beneath the integument, parallel with the dorsal vein and the corresponding artery of the opposite side. It supplies the integument and fibrous sheath of the corpus cavernosum, sending branches through the sheath to anastomose with the preceding vessel.

The **internal pudic artery in the female** is smaller than in the male. Its origin and course are similar, and there is considerable analogy in the distribution of its branches. The superficial perineal artery supplies the labia pudendi; the artery of the bulb supplies the bulbus vestibuli and the erectile tissue of the vagina; the artery of the corpus cavernosum supplies the cavernous body of the clitoris; and the arteria dorsalis clitoridis supplies the dorsum of that organ, and terminates in the glans and in the membranous fold corresponding to the prepuce of the male.

The **sciatic artery** (fig. 316), the larger of the two terminal branches of the anterior trunk of the internal iliac, is distributed to the muscles at the back of the pelvis. It passes down to the lower part of the great sacro-sciatic foramen, behind the internal pudic artery, resting on the sacral plexus of nerves and Pyriformis muscle, and escapes from the pelvis through this foramen between the Pyriformis and Coccygeus. It then descends in the interval between the trochanter major and tuberosity of the ischium, accompanied by the sciatic nerves, and covered by the Gluteus maximus, and is continued down the back of the thigh supplying the skin, and anastomosing with branches of the perforating arteries.

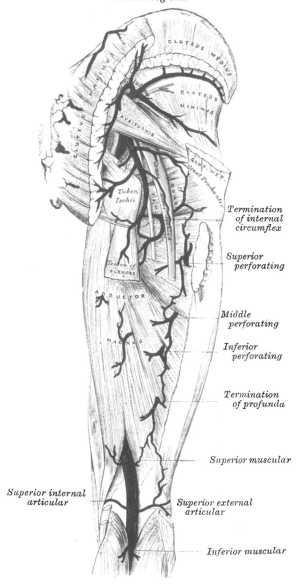

FIG. 316.— The arteries of the gluteal and posterior femoral regions.

Within the pelvis, it distributes branches [to the Pyriformis, Coccygeus, and Levator ani muscles ; some hæmorrhoidal branches, which supply the rectum, and occasionally take the place of the middle hæmorrhoidal artery ; and vesical branches to the base and neck of the bladder, vesiculæ seminales, and prostate gland. *External to the pelvis*, it gives off the following branches :

Coccygeal.
Inferior gluteal.
Comes nervi ischiadici.
Muscular.
Anastomotic.
Articular.

The *coccygeal branch* runs inwards, pierces the great sacro-sciatic ligament, and supplies the Gluteus maximus, the integument, and other structures on the back of the coccyx.

The *inferior gluteal branches*, three or four in number, supply the Gluteus maximus muscle, anastomosing with the gluteal artery in the substance of the muscle.

The *comes nervi ischiadici* is a long, slender vessel, which accompanies the great sciatic nerve for a short distance ; it then penetrates it, and runs in its substance to the lower part of the thigh.

The *muscular branches* supply the Gluteus maximus, anastomosing with the gluteal artery in the substance of the muscle : the external rotators, anastomosing with the internal pudic artery ; and the muscles attached to the tuberosity of the ischium, anastomosing with the external branch of the obturator and the internal circumflex arteries.

The *anastomotic artery* is directed downwards across the external rotators, and assists in forming the so-called *crucial anastomosis* by anastomosing with the superior perforating and the internal and external circumflex.

The *articular branch*, generally derived from the anastomotic, is distributed to the capsule of the hip-joint.

The **ilio-lumbar artery,** given off from the posterior trunk of the internal iliac, turns upwards and outwards between the obturator nerve and lumbo-sacral cord, to the inner margin of the Psoas muscle, behind which it divides into a lumbar and an iliac branch.

The *lumbar branch* supplies the Psoas and Quadratus lumborum muscles, anastomosing with the last lumbar artery, and sends a small spinal branch through the intervertebral foramen between the last lumbar vertebra and the sacrum, into the spinal canal, to supply the cauda equina.

The *iliac branch* descends to supply the Iliacus muscle ; some offsets, running between the muscle and the bone, anastomose with the iliac branch of the obturator ; one of these enters an oblique canal to supply the diploë, while others run along the crest of the ilium, distributing branches to the gluteal and abdominal muscles, and anastomose in their course with the gluteal, circumflex iliac, and external circumflex arteries.

The **lateral sacral arteries** (fig. 312) are usually two in number on each side, superior and inferior.

The *superior*, which is of large size, passes inwards, and, after anastomosing with branches from the middle sacral, enters the first or second anterior sacral foramen, is distributed to the contents of the sacral canal, and escaping by the corresponding posterior sacral foramen, supplies the skin and muscles on the dorsum of the sacrum, anastomosing with the gluteal.

The *inferior* passes obliquely across the front of the Pyriformis muscle and sacral nerves to the inner side of the anterior sacral foramina, descends on the front of the sacrum, and anastomoses over the coccyx with the sacra media and opposite lateral sacral artery. In its course it gives off branches, which enter the anterior sacral foramina ; these, after supplying the contents of the sacral canal, escape by the posterior sacral foramina, and are distributed to the muscles and skin on the dorsal surface of the sacrum, anastomosing with the gluteal.

The **gluteal artery** is the largest branch of the internal iliac, and appears to be the continuation of the posterior division of that vessel. It is a short, thick trunk, which passes out of the pelvis above the upper border of the Pyriformis muscle, and immediately divides into a *superficial* and *deep branch*. Within the pelvis, it gives off a few muscular branches to the Iliacus, Pyriformis, and Obturator internus, and just previous to quitting that cavity, a nutrient artery, which enters the ilium.

The *superficial branch* passes beneath the Gluteus maximus, and divides into numerous branches, some of which supply that muscle, while others perforate its tendinous origin, and supply the integument covering the posterior surface of the sacrum, anastomosing with the posterior branches of the sacral arteries.

The *deep branch* runs between the Gluteus medius and minimus, and subdivides into two. Of these, the *superior division*, continuing the original course of the vessel, passes along the upper border of the Gluteus minimus to the anterior superior spine of the ilium, anastomosing with the circumflex iliac and ascending branches of the external circumflex artery. The *inferior division* crosses the Gluteus minimus obliquely to the trochanter major, distributing branches to the Glutei muscles, and inosculates with the external circumflex artery. Some branches pierce the Gluteus minimus to supply the hip-joint.

Surface Marking.—The position of the three main branches of the internal iliac, the sciatic, internal pudic, and gluteal, which may occasionally be the object of surgical interference, is indicated on the surface in the following way: a line is to be drawn from the posterior superior iliac spine to the posterior superior angle of the great trochanter, with the limb slightly flexed and rotated inwards: the point of emergence of the *gluteal artery* from the upper part of the sciatic notch will correspond with the junction of the upper with the middle third of this line. A second line is to be drawn from the same point to the outer part of the tuberosity of the ischium; the junction of the lower with the middle third marks the point of emergence of the *sciatic* and *pudic arteries* from the great sciatic notch.

Surgical Anatomy.—Any of these three vessels may require ligaturing for a wound or for aneurism, which is generally traumatic. The *gluteal* artery is ligatured by turning the patient two-thirds over on to his face and making an incision from the posterior superior spine of the ilium to the upper and posterior angle of the great trochanter. This must expose the Gluteus maximus muscle, and its fibres are to be separated through the whole thickness of the muscle and pulled apart with retractors. The contiguous margins of the Gluteus medius and Pyriformis are now to be separated from each other, and the artery will be exposed emerging from the sciatic notch. In ligature of the *sciatic* artery, the incision should be made parallel with that for ligature of the gluteal, but one inch and a half lower down. After the fibres of the Gluteus maximus have been separated, the vessel is to be sought for at the lower border of the Pyriformis; the great sciatic nerve, which lies just above it, forming the chief guide to the artery.

EXTERNAL ILIAC ARTERY (fig. 312)

The **External Iliac artery** is larger in the adult than the internal iliac, and passes obliquely downwards and outwards along the inner border of the Psoas muscle, from the bifurcation of the common iliac to Poupart's ligament, where it enters the thigh and becomes the femoral artery.

Relations.—*In front*, with the peritoneum, subperitoneal areolar tissue, the termination of the ileum on the right side, and the sigmoid flexure on the left, and a thin layer of fascia, derived from the iliac fascia, which surrounds the artery and vein. At its origin it is occasionally crossed by the ureter. The spermatic vessels descend for some distance upon it near its termination, and it is crossed in this situation by the genital branch of the genito-crural nerve and the deep circumflex iliac vein; the vas deferens curves down along its inner side. *Behind*, it is in relation with the external iliac vein, which, at Poupart's ligament, lies at its inner side; on the left side the vein is altogether internal to the artery. *Externally*, it rests against the Psoas muscle, from which it is separated by the iliac fascia. The artery rests upon this muscle, near Poupart's ligament. Numerous lymphatic vessels and glands are found lying on the front and inner side of the vessel.

PLAN OF THE RELATIONS OF THE EXTERNAL ILIAC ARTERY

In front.

Peritoneum, intestines, and fascia.

Near Poupart's Ligament
- Lymphatic vessels and glands.
- Spermatic vessels.
- Genito-crural nerve (genital branch).
- Deep circumflex iliac vein.

Outer side.
Psoas magnus.
Iliac fascia.

External Iliac.

Inner side.
External iliac vein and vas deferens near Poupart's ligament.

Behind.
External iliac vein.
Psoas magnus.

Surface Marking.—The surface line indicating the course of the external iliac artery has been already given (see page 537).

Surgical Anatomy.—The application of a ligature to the external iliac may be required in cases of aneurism of the femoral artery, or for a wound of the artery. This vessel may be secured in any part of its course, excepting near its upper end, which is to be avoided on account of the proximity of the great stream of blood in the internal iliac, and near its lower end, which should also be avoided, on account of the proximity of the deep epigastric and circumflex iliac vessels. The patient having been placed in the supine position, an incision should be made, commencing below at a point about three-quarters of an inch above Poupart's ligament, and a little external to its middle, and running upwards and outwards, parallel to Poupart's ligament, to a point one inch internal and one inch above the anterior superior spine of the ilium. When the artery is deeply seated, more room will be required, and may be obtained by curving the incision from the point last named inwards towards the umbilicus for a short distance. Another mode of ligaturing the vessel is the plan advocated by Sir Astley Cooper, by making an incision close to Poupart's ligament from about half an inch outside the external abdominal ring to one inch internal to the anterior superior spine of the ilium. This incision, being made in the course of the fibres of the aponeurosis of the external oblique, is less likely to be followed by a ventral hernia, but there is danger of wounding the epigastric artery, and only the lower end of the vessel can be ligatured. Abernethy, who first tied this artery, made his incision in the course of the vessel. The abdominal muscles and transversalis fascia having been cautiously divided, the peritoneum should be separated from the iliac fossa and raised towards the pelvis ; and on introducing the finger to the bottom of the wound the artery may be felt pulsating along the inner border of the Psoas muscle. The external iliac vein is generally found on the inner side of the artery, and must be cautiously separated from it by the finger-nail, or handle of the knife, and the aneurism needle should be introduced on the inner side, between the artery and vein.

Ligature of the external iliac artery has recently been performed by a transperitoneal method. An incision four inches in length is made in the semilunar line, commencing about an inch below the umbilicus, and carried through the abdominal wall into the peritoneal cavity. The intestines are then pushed upwards and held out of the way by a broad abdominal retractor, and an incision made through the peritoneum at the margin of the pelvis in the course of the artery ; the vessel is secured in any part of its course which may seem desirable to the operator. The advantages of this operation appear to be, that if it is found necessary the common iliac artery can be ligatured instead of the external iliac without extension or modification of the incision ; and secondly, that the vessel can be ligatured without in any way interfering with the coverings of the sac. Possibly a disadvantage may exist in the greater risk of hernia after this method.

Collateral Circulation.—The principal anastomoses in carrying on the collateral circulation, after the application of a ligature to the external iliac, are—the ilio-lumbar with the circumflex iliac ; the gluteal with the external circumflex ; the obturator with the internal circumflex ; the sciatic with the superior perforating and circumflex branches of the profunda artery ; and the internal pudic with the external pudic. When the obturator arises from the epigastric, it is supplied with blood by branches, either from the internal iliac, the lateral sacral, or the internal pudic. The epigastric receives its supply from the internal mammary and inferior intercostal arteries, and from the internal iliac by the anastomoses of its branches with the obturator.

In the dissection of a limb, eighteen years after the successful ligature of the external iliac artery, by Sir A. Cooper, which is to be found in Guy's Hospital Reports, vol. i. p. 50, the anastomosing branches are described in three sets : *An anterior set.*—1. A very large branch from the ilio-lumbar artery to the circumflex iliac ; 2. another branch from the ilio-lumbar, joined by one from the obturator, and breaking up into numerous tortuous branches to anastomose with the external circumflex ; 3. two other branches from the obturator, which passed over the brim of the pelvis, communicated with the epigastric, and then broke up into a plexus to anastomose with the internal circumflex. *An internal set.*—Branches given off from the obturator, after quitting the pelvis, which ramified among the adductor muscles on the inner side of the hip-joint, and joined most freely with branches of the internal circumflex. *A posterior set.*—1. Three large branches from the gluteal to the external circumflex ; 2. several branches from the sciatic around the great sciatic notch to the internal and external circumflex, and the perforating branches of the profunda.

Branches.—Besides several small branches to the Psoas muscle and the neighbouring lymphatic glands, the external iliac gives off two branches of considerable size, the

Deep epigastric and Deep circumflex iliac.

The **deep epigastric artery** arises from the external iliac, a few lines above Poupart's ligament. It at first descends to reach this ligament, and then ascends obliquely along the inner margin of the internal abdominal ring, lying between the transversalis fascia and peritoneum, and, continuing its course upwards, it pierces the transversalis fascia, and, passing over the semilunar fold of Douglas, enters the sheath of the Rectus muscle. It then ascends on the posterior surface of the muscle, and finally divides into numerous branches, which anastomose, above the umbilicus, with the superior epigastric branch of the internal mammary and with the inferior intercostal arteries (fig. 301). The deep epigastric artery bears a very important relation to the internal abdominal ring as it passes obliquely upwards and inwards from its origin from the external iliac. In this part of its course it lies along the lower and inner margin of the ring, and beneath the commencement of the spermatic cord. As it passes to the inner side of the internal abdominal ring, it is crossed by the vas deferens in the male and the round ligament in the female.

The branches of this vessel are the following : the *cremasteric*, which accompanies the spermatic cord, and supplies the Cremaster muscle and other coverings of the cord, anastomosing with the spermatic artery ; a *pubic branch*, which runs along Poupart's ligament, and then descends behind the os pubis to the inner side of the femoral ring, and anastomoses with offsets from the obturator artery ; *muscular branches*, some of which are distributed to the abdominal muscles and peritoneum, anastomosing with the lumbar and circumflex iliac arteries : others perforate the tendon of the External oblique, and supply the integument, anastomosing with branches of the superficial epigastric.

Peculiarities.—The origin of the epigastric may take place from any part of the external iliac between Poupart's ligament and two inches and a half above it ; or it may arise below this ligament, from the common femoral, or from the deep femoral.

Union with branches.—It frequently arises from the external iliac, by a common trunk with the obturator. Sometimes the epigastric arises from the obturator, the latter vessel being furnished by the internal iliac, or the epigastric may be formed of two branches, one derived from the external iliac, the other from the internal iliac.

Surgical Anatomy.—The deep epigastric artery follows a line drawn from the middle of Poupart's ligament towards the umbilicus ; but shortly after this line crosses the linea semilunaris, the direction changes, and the course of the vessel is directly upwards in the line of junction of the inner third with the outer two-thirds of the Rectus muscle. It has important surgical relations, in addition to the fact that it is one of the principal means, through its anastomosis with the internal mammary, in establishing the collateral circulation after ligature of either the common or external iliac arteries. It lies close to the internal abdominal ring, and is therefore *internal* to an oblique inguinal hernia, but *external* to a direct inguinal hernia, as it emerges from the abdomen. It forms the outer boundary of Hesselbach's triangle. It is in close relationship with the spermatic cord, which lies in front of it in the inguinal canal, separated only by the transversalis fascia. The vas deferens hooks round its outer side.

The **deep circumflex iliac artery** arises from the outer side of the external iliac nearly opposite the epigastric artery. It ascends obliquely outwards behind Poupart's ligament, contained in a fibrous sheath formed by the junction of the transversalis and iliac fasciæ, to the anterior superior spinous process of the ilium. It then runs along the inner surface of the crest of the ilium to about its middle, where it pierces the Transversalis, and runs backwards between that muscle and the Internal oblique, to anastomose with the ilio-lumbar and gluteal arteries. Opposite the anterior superior spine of the ilium it gives off a large branch, which ascends between the Internal oblique and Transversalis muscles, supplying them, and anastomosing with the lumbar and epigastric arteries.

ARTERIES OF THE LOWER EXTREMITY

The artery which supplies the greater part of the lower extremity is the direct continuation of the external iliac. It continues as a single trunk from Poupart's ligament to the lower border of the Popliteus muscle, and here divides into two branches, the anterior and posterior tibial, an arrangement exactly similar to what occurs in the upper limb. For convenience of description, the upper part of the main trunk is named femoral, the lower part popliteal.

FEMORAL ARTERY (fig. 317)

The **Femoral artery** commences immediately behind Poupart's ligament, midway between the ·anterior superior spine of the ilium and the symphysis pubis, and passing down the fore part and inner side of the thigh, terminates at the opening in the Adductor magnus, at the junction of the middle with the lower third of the thigh, where it becomes the popliteal artery. The vessel, at the upper part of the thigh, lies in front of the hip-joint, just on a line with the innermost part of the head of the femur ; in the lower part of its course it is in close relation with the inner side of the shaft of the bone, and between these two parts, the vessel is some distance from the bone. In the upper third of the thigh it is contained in a triangular space, called *Scarpa's triangle*. In the middle third of the thigh it is contained in an aponeurotic canal, called *Hunter's canal*.

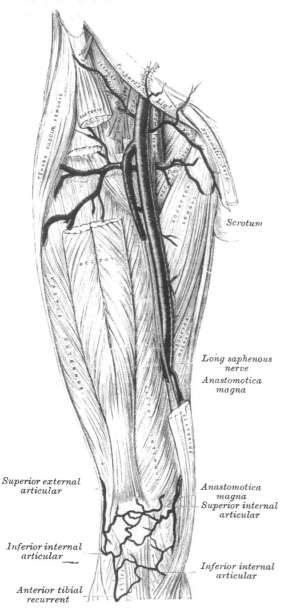

FIG. 317.—The femoral artery.

Scrotum

Long saphenous nerve

Anastomotica magna

Superior external articular

Anastomotica magna

Superior internal articular

Inferior internal articular

Inferior internal articular

Anterior tibial recurrent

Scarpa's Triangle.— Scarpa's triangle corresponds to the depression seen immediately below the fold of the groin. It is a triangular space, the apex of which is directed downwards, and the sides formed externally by the Sartorius, internally by the inner margin of the Adductor longus, and above by Poupart's ligament. The floor of the space is formed from without inwards by the Iliacus, Psoas, Pectineus (in some cases a small part of the Adductor brevis), and the Adductor longus muscles ; and it is divided into two nearly equal parts by the femoral vessels, which extend from the middle of its base to its apex : the artery giving off in this situation its cutaneous and profunda branches, the vein receiving the deep femoral and internal saphenous. On the outer side of the femoral artery is the anterior crural nerve dividing into its branches. Besides the vessels and nerves, this space contains some fat and lymphatics.

Hunter's Canal.—This is the aponeurotic space in the middle third of the thigh, extending from the apex of Scarpa's triangle to the femoral opening in the Adductor magnus muscle. It is bounded, externally, by the Vastus internus ; internally, by the Adductores longus and magnus muscles ; and covered in by a strong aponeurosis which extends transversely from the Vastus internus, across the femoral vessels to the Adductor longus and magnus ; lying on which aponeurosis is the Sartorius muscle. It contains the femoral artery and vein enclosed in their own sheath of areolar tissue, the vein being behind and on the outer side of the artery, and the internal or long saphenous nerve lying at first on the outer side and then in front of the vessels.

For convenience of description and also in reference to its surgical anatomy, the femoral artery is divided into a short trunk, about an inch and a half or two inches long, which is known as the *common femoral artery*, while the remainder of the vessel is termed the *superficial femoral*, to distinguish it from the *deep femoral* (profunda femoris), a large branch given off from the common femoral at its termination, and which by its derivation from the parent trunk marks the commencement of the superficial femoral artery.

The **common femoral artery** is very superficial, being covered by the skin and superficial fascia, superficial inguinal lymphatic glands, the iliac portion of the fascia lata, and the prolongation downwards of the Transversalis fascia, which forms the anterior part of the sheath of the vessels. It has in front of it filaments from the crural branch of the genito-crural nerve, the superficial circumflex iliac vein, and occasionally the superficial epigastric vein. It rests on the inner margin of the Psoas muscle, which separates it from the capsular ligament of the hip-joint, and a little lower on the Pectineus muscle ; and crossing behind it is the branch to the Pectineus from the anterior crural nerve. Separating the artery from the Pectineus muscle is the pubic portion of the fascia lata and the prolongation from the fascia covering the Iliacus muscle which forms the posterior layer of the sheath of the vessels. The anterior crural nerve lies about half an inch to the outer side of the common femoral artery, being separated from the artery by a small part of the Psoas muscle. To the inner side of the artery is the femoral vein, between the margins of the Pectineus and Psoas muscles. The two vessels are enclosed in a strong fibrous sheath, formed by the proper sheath of the vessels, strengthened by the fascia lata (see page 395) ; the artery and vein are separated, however, from one another by a thin fibrous partition.

PLAN OF THE RELATIONS OF THE COMMON FEMORAL ARTERY

In front.

Skin and superficial fascia.
Superficial inguinal glands.
Iliac portion of fascia lata.
Prolongation of transversalis fascia.
Crural branch of genito-crural nerve.
Superficial circumflex iliac vein.
Superficial epigastric vein.

Inner side.

Femoral vein.

Outer side.

Small part of Psoas muscle, separating the artery from the anterior crural nerve.

Behind.

Prolongation of fascia covering Iliacus muscle.
Pubic portion of fascia lata.
Nerve to Pectineus.
Psoas muscle.
Pectineus muscle.
Capsule of hip-joint.

The **superficial femoral artery** is only superficial where it lies in Scarpa's triangle. Here it is covered by the skin, superficial and deep fascia, and crossed by the internal cutaneous branch of the anterior crural nerve. In Hunter's canal it is more deeply seated, being covered by the integument, the superficial and deep fascia, the Sartorius and aponeurotic covering of Hunter's canal. The internal saphenous nerve crosses the artery from without inwards. Behind, the artery lies at its upper part on the femoral vein and the profunda artery and vein, which separate it from the Pectineus muscle, and lower down on the Adductor longus and Adductor magnus muscles. To the outer side is the long saphenous nerve and the nerve to the Vastus internus; the Vastus internus muscle and, at its lower part, the femoral vein. To the inner side is the Adductor longus above, and the Adductor magnus and Sartorius below.

PLAN OF THE RELATIONS OF THE SUPERFICIAL FEMORAL ARTERY

In front.

Skin, superficial and deep fasciæ.
Internal cutaneous nerve.
Sartorius.
Aponeurotic covering of Hunter's canal.
Internal saphenous nerve.

Inner side.

Adductor longus.
Adductor magnus.
Sartorius.

Superficial Femoral Artery.

Outer side.

Long saphenous nerve.
Nerve to Vastus internus.
Vastus internus.
Femoral vein (below).

Behind.

Femoral vein.
Profunda artery and vein.
Pectineus muscle.
Adductor longus.
Adductor magnus.

The *femoral vein* at Poupart's ligament lies close to the inner side of the artery, separated from it by a thin fibrous partition; but, lower down, it is behind it, and then to its outer side.

The *internal saphenous nerve* is situated on the outer side of the artery, in the middle third of the thigh, beneath the aponeurotic covering of Hunter's canal; but not usually within the sheath of the vessels. The internal cutaneous nerve passes obliquely across the upper part of the sheath of the femoral artery.

*Peculiarities.—Double femoral reunited.—*Several cases are recorded in which the femoral artery divided into two trunks below the origin of the profunda, and became reunited near the opening in the Adductor magnus, so as to form a single popliteal artery. One of them occurred in a patient operated upon for popliteal aneurism.

*Change of Position.—*A few cases have been recorded in which the femoral artery was situated at the back of the thigh, the vessel being continuous above with the internal iliac, escaping from the pelvis through the great sacro-sciatic foramen, and accompanying the great sciatic nerve to the popliteal space, where its division occurred in the usual manner. The external iliac in these cases was small, and terminated in the profunda.

*Position of the Vein.—*The femoral vein is occasionally placed along the inner side of the artery, throughout the entire extent of Scarpa's triangle; or it may be slit so that a large vein is placed on each side of the artery for a greater or less extent.

*Origin of the Profunda.—*This vessel occasionally arises from the inner side, and, more rarely, from the back of the common trunk; but the more important peculiarity, in a surgical point of view, is that which relates to the height at which the vessel arises from the femoral. In three-fourths of a large number of cases it arose between one or two inches below Poupart's ligament; in a few cases the distance was less than an inch; more rarely, opposite the ligament; and in one case above Poupart's ligament, from the external iliac. Occasionally, the distance between the origin of the vessel and

Poupart's ligament exceeds two inches, and in one case it was found to be as much as four inches.

Surface Marking.—The upper two-thirds of a line drawn from a point midway between the anterior superior spine of the ilium and the symphysis pubis to the prominent tuberosity on the inner condyle of the femur, with the thigh abducted and rotated outwards, will indicate the course of the femoral artery.

Surgical Anatomy.—*Compression* of the femoral artery, which is constantly requisite in amputations and other operations on the lower limb, and also for the cure of popliteal aneurism, is most effectually made immediately below Poupart's ligament. In this situation the artery is very superficial, and is merely separated from the ascending ramus of the os pubis by the Psoas muscle; so that the surgeon, by means of his thumb or a compressor, may effectually control the circulation through it. This vessel may also be compressed in the middle third of the thigh by placing a compress over the artery, beneath the tourniquet, and directing the pressure from within outwards, so as to compress the vessel against the inner side of the shaft of the femur.

The *application of a ligature* to the femoral artery may be required in cases of wound or aneurism of the arteries of the leg, the popliteal or femoral ; * and the vessel may be exposed and tied in any part of its course. The great depth of this vessel at its lower part, its close connection with important structures, and the density of its sheath, render the operation in this situation one of much greater difficulty than the application of a ligature at its upper part, where it is more superficial.

Ligature of the common femoral artery is not regarded with much favour, on account of the connection of large branches with it : viz. the deep epigastric and the deep circumflex iliac arising just above Poupart's ligament ; on account of the number of small branches which arise from it, in its short course, and on account of the uncertainty of the origin of the profunda femoris, which, if it arise high up, would be too close to the ligature for the formation of a firm coagulum. The profunda sometimes arises higher than the point above mentioned, and rarely between two or three inches (in one case four) below Poupart's ligament. It would appear, then, that the most favourable situation for the application of a ligature to the femoral is between four and five inches from its point of origin. In order to expose the artery in this situation, an incision, between three and four inches long, should be made in the course of the vessel, the patient lying in the recumbent position, with the limb slightly flexed and abducted, and rotated outwards. A large vein is frequently met with, passing in the course of the artery to join the internal saphenous vein ; this must be avoided, and the fascia lata having been cautiously divided, and the Sartorius exposed, that muscle must be drawn outwards, in order to fully expose the sheath of the vessels. The finger being introduced into the wound, and the pulsation of the artery felt, the sheath should be opened on the outer side of the vessel to a sufficient extent to allow of the introduction of the ligature, but no farther ; otherwise the nutrition of the coats of the vessel may be interfered with, or muscular branches which arise from the vessel at irregular intervals may be divided. In this part of the operation the long saphenous nerve and the nerve to the Vastus internus, which is in close relation with the sheath, should be avoided. The aneurism needle must be carefully introduced and kept close to the artery, to avoid the femoral vein, which lies behind the vessel in this part of its course.

To expose the artery, in Hunter's canal, an incision should be made through the integument, between three and four inches in length, a finger's breadth internal to the line of the artery, in the middle of the thigh—i.e. midway between the groin and the knee. The fascia lata having been divided, and the outer border of the Sartorius muscle exposed, it should be drawn inwards, when the strong fascia which is stretched across from the Adductors to the Vastus internus will be exposed, and must be freely divided ; the sheath of the vessels is now seen, and must be opened, and the artery secured by passing the aneurism needle between the vein and artery, in the direction from without inwards. The femoral vein in this situation lies on the outer side of the artery, the long saphenous nerve on its anterior and outer side.

It has been seen that the femoral artery occasionally divides into two trunks below the origin of the profunda. If, in the operation for tying the femoral, two vessels are met with, the surgeon should alternately compress each, in order to ascertain which vessel is connected with the aneurismal tumour, or with the bleeding from the wound, and that one only should be tied which controls the pulsation or hæmorrhage. If, however, it is necessary to compress both vessels before the circulation in the tumour is controlled, both should be tied, as it would be probable that they became reunited, as in the instances referred to above.

In wounds of the femoral artery the question of the mode of treatment is of considerable importance. If the wound in the superficial structures is a large one, the injured vessel must be exposed and tied ; but if the wound is a punctured one and the bleeding has ceased, the question will arise whether to cut down upon the artery or to trust to pressure. Mr. Cripps † advises, that if the wound is in the ' upper part of the thigh—that

---

* Ligature of the femoral artery has been also recommended and performed for elephantiasis of the leg and acute inflammation of the knee-joint.—Maunder, *Clin. Soc. Trans.* vol. ii. p. 37.   † Heath's *Dictionary of Practical Surgery*, vol. i. p. 525.

is to say, in a position where the femoral artery is comparatively superficial—the surgeon may enlarge the opening with a good prospect of finding the wounded vessel without an extensive or prolonged operation. If the wound be in the lower half of the thigh, owing to the greater depth of the artery, and the possibility of its being the popliteal that is wounded, the search is rendered a far more severe and hazardous operation, and it should not be undertaken until a thorough trial of pressure has proved ineffectual.'

Great care and attention are necessary for the successful application of pressure. The limb should be carefully bandaged from the foot upwards to the wound, which is not covered, and then onwards to the groin. The wound is then dusted with iodoform or boracic powder, and a conical pad applied over it. Rollers the thickness of the index finger are then placed along the course of the vessel above and below the wound, and the whole carefully bandaged to a back splint with a foot-piece.

Collateral Circulation.—When the common femoral is tied, the main channels for carrying on the circulation are the anastomoses of the gluteal and circumflex iliac arteries above with the external circumflex below ; of the obturator and sciatic above with the internal circumflex below ; and of the comes nervi ischiadici with the arteries in the ham.

The principal agents in carrying on the collateral circulation after ligature of the superficial femoral artery are, according to Sir A. Cooper, as follows :

'The arteria profunda formed the new channel for the blood.' 'The first artery sent off passed down close to the back of the thigh-bone, and entered the two superior articular branches of the popliteal artery.'

'The second new large vessel, arising from the profunda at the same part with the former, passed down by the inner side of the Biceps muscle, to a branch of the popliteal which was distributed to the Gastrocnemius muscle ; while a third artery, dividing into several branches, passed down with the sciatic nerve behind the knee-joint, and some of its branches united with the inferior articular arteries of the popliteal, with some recurrent branches of those arteries, with arteries passing to the Gastrocnemius, and, lastly, with the origin of the anterior and posterior tibial arteries.'

'It appears then that it is those branches of the profunda which accompany the sciatic nerve that are the principal supporters of the new circulation.' *

In Porta's work † (tab. xii. xiii.) is a good representation of the collateral circulation after ligature of the femoral artery. The patient had survived the operation three years. The lower part of the artery is at least as large as the upper ; about two inches of the vessel appear to have been obliterated. The external and internal circumflex arteries are seen anastomosing by a great number of branches with the lower branches of the femoral (muscular and anastomotica magna), and with the articular branches of the popliteal. The branches from the external circumflex are extremely large and numerous. One very distinct anastomosis can be traced between this artery on the outside and the anastomotica magna on the inside, through the intervention of the superior external articular artery, with which they both anastomose ; and blood reaches even the anterior tibial recurrent from the external circumflex by means of anastomosis with the same external articular artery. The perforating branches of the profunda are also seen bringing blood round the obliterated portion of the artery into long branches (muscular) which have been given off just below that portion. The termination of the profunda itself anastomoses most freely with the superior external articular. A long branch of anastomosis is also traced down from the internal iliac by means of the comes nervi ischiadici of the sciatic which anastomoses on the popliteal nerves with branches from the popliteal and posterior tibial arteries. In this case the anastomosis had been too free, since the pulsation and growth of the aneurism recurred, and the patient died after ligature of the external iliac.

There is an interesting preparation in the Museum of the Royal College of Surgeons of a limb on which John Hunter had tied the femoral artery fifty years before the patient's death. The whole of the superficial femoral and popliteal artery seems to have been obliterated. The anastomosis by means of the comes nervi ischiadici, which is shown in Porta's plate, is distinctly seen : the external circumflex, and the termination of the profunda artery, seem to have been the chief channels of anastomosis ; but the injection has not been a very successful one.

Branches.—The branches of the femoral artery are : the

Superficial epigastric.	⎧ External circumflex.
Superficial circumflex iliac.	Profunda ⎨ Internal circumflex.
Superficial external pudic.	⎩ Three perforating.
Deep external pudic.	Muscular.
Anastomotica magna.	

The **superficial epigastric** arises from the femoral, about half an inch below Poupart's ligament, and, passing through the saphenous opening in the fascia

* *Med.-Chir. Trans.* vol. ii. 1811.   † *Alterazioni patologiche delle Arterie.*

lata, ascends on to the abdomen, in the superficial fascia covering the External oblique muscle, nearly as high as the umbilicus. It distributes branches to the superficial inguinal glands, the superficial fascia, and the integument, anastomosing with branches of the deep epigastric.

The **superficial circumflex iliac,** the smallest of the cutaneous branches, arises close to the preceding, and, piercing the fascia lata, runs outwards, parallel with

FIG. 318.—Femoral artery and its branches.
(From a preparation in the Museum of the Royal College of Surgeons of England.)

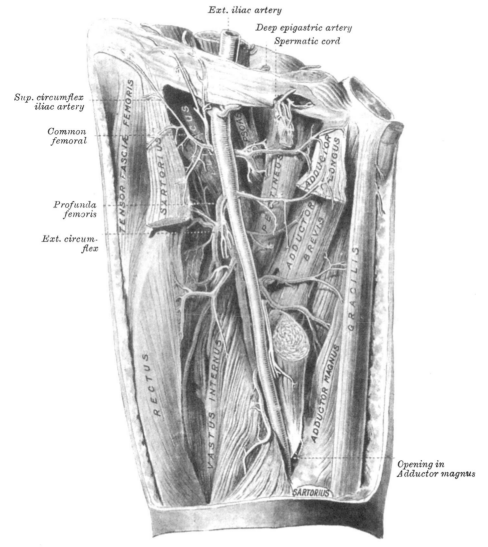

Poupart's ligament, as far as the crest of the ilium; it divides into branches which supply the integument of the groin, the superficial fascia, and the superficial inguinal lymphatic glands, anastomosing with the deep circumflex iliac, and with the gluteal and external circumflex arteries.

The **superficial external pudic** (superior) arises from the inner side of the femoral artery, close to the preceding vessels, and, after passing through the saphenous opening, courses inwards, across the spermatic cord or round ligament, to be distributed to the integument on the lower part of the abdomen, the penis and scrotum in the male, and the labium in the female, anastomosing with branches of the internal pudic.

The **deep external pudic** (inferior), more deeply seated than the preceding, passes inwards across the Pectineus and Adductor longus muscles, covered by the fascia lata, which it pierces at the inner border of the thigh, its branches being distributed, in the male, to the integument of the scrotum and perinæum ; and in the female to the labium, anastomosing with branches of the superficial perineal artery.

The **profunda femoris** (*deep femoral artery*) (fig. 318) nearly equals the size of the superficial femoral. It arises from the outer and back part of the femoral artery, from one to two inches below Poupart's ligament. It at first lies on the outer side of the superficial femoral, and then passes behind it and the femoral vein to the inner side of the femur, and, passing downwards beneath the Adductor longus, terminates at the lower third of the thigh in a small branch, which pierces the Adductor magnus (and from this circumstance is sometimes called the fourth perforating artery), and is distributed to the Flexor muscles on the back of the thigh, anastomosing with branches of the popliteal and inferior perforating arteries.

Relations.—*Behind*, it lies first upon the Iliacus, and then on the Pectineus, Adductor brevis, and Adductor magnus muscles. *In front*, it is separated from the superficial femoral artery, above by the femoral and profunda veins, and below by the Adductor longus. On its *outer side*, the origin of the Vastus internus separates it from the femur.

PLAN OF THE RELATIONS OF THE PROFUNDA ARTERY

In front.
Superficial femoral artery.
Femoral and profunda veins.
Adductor longus.

Outer side.
Vastus internus.

Profunda femoris.

Behind.
Iliacus.
Pectineus.
Adductor brevis.
Adductor magnus.

The profunda gives off the following named branches :

External circumflex.   Internal circumflex.   Four perforating.

The **external circumflex artery** supplies the muscles on the front of the thigh. It arises from the outer side of the profunda, passes horizontally outwards, between the divisions of the anterior crural nerve, and behind the Sartorius and Rectus muscles, and divides into three sets of branches, ascending, transverse, and descending.

The *ascending branch* passes upwards, beneath the Tensor fasciæ femoris muscle, to the outer side of the hip, anastomosing with the terminal branches of the gluteal and deep circumflex iliac arteries.

The *descending branches*, three or four in number, pass downwards, behind the Rectus, upon the Vasti muscles, to which they are distributed, one or two passing beneath the Vastus externus as far as the knee, anastomosing with the superior articular branches of the popliteal artery. These are accompanied by the branch of the anterior crural nerve to the Vastus externus.

The *transverse branch*, the smallest, passes outwards over the Crureus, pierces the Vastus externus, and winds round the femur to its back part, just below the great trochanter, anastomosing at the back of the thigh with the internal circumflex, sciatic, and superior perforating arteries.

The **internal circumflex artery,** smaller than the external, arises from the inner and back part of the profunda, and winds round the inner side of the femur, between the Pectineus and Psoas muscles. On reaching the upper border of the Adductor brevis, it gives off two branches, one of which passes inwards to be distributed to the Adductor muscles, the Gracilis, and Obturator externus, anastomosing with the obturator artery ; the other descends, and passes beneath the Adductor brevis, to supply it and the great Adductor ; while the continuation of the vessel passes backwards and divides into an ascending and a transverse branch (fig. 257). The *ascending branch* passes obliquely upwards upon the tendon of the Obturator externus and under cover of the Quadratus femoris towards the digital fossa, where it anastomoses with twigs from the gluteal and sciatic arteries. The *transverse branch*, larger than the ascending, appears between the Quadratus femoris and upper border of the Adductor magnus, anastomosing with the sciatic, external circumflex, and superior perforating arteries ('the *crucial anastomosis* '). Opposite the hip-joint, the artery gives off an articular vessel, which enters the joint beneath the transverse ligament ; and, after supplying the adipose tissue, passes along the round ligament to the head of the bone.

The **perforating arteries** (fig. 316), usually four in number, are so called from their perforating the tendon of the Adductor magnus muscle to reach the back of the thigh. They pass backwards close to the linea aspera of the femur under cover of small tendinous arches in the Adductor magnus. The first is given off above the Adductor brevis, the second in front of that muscle, and the third immediately below it.

The *first perforating artery* passes backwards between the Pectineus and Adductor brevis (sometimes perforates the latter) ; it then pierces the Adductor magnus close to the linea aspera. It gives off branches which supply the Adductor brevis, the Adductor magnus, the Biceps, and Gluteus maximus muscles, and anastomoses with the sciatic, internal and external circumflex, and middle perforating arteries.

The *second perforating artery*, larger than the first, pierces the tendons of the Adductor brevis and Adductor magnus muscles, and divides into ascending and descending branches, which supply the flexor muscles of the thigh, anastomosing with the first and third perforating. The second artery frequently arises in common with the first. The nutrient artery of the femur is usually given off from this branch.

The *third perforating artery* is given off below the Adductor brevis ; it pierces the Adductor magnus, and divides into branches which supply the flexor muscles of the thigh ; anastomosing above with the higher perforating arteries, and below with the terminal branches of the profunda and the muscular branches of the popliteal.

The *fourth perforating artery* is represented by the termination of the profunda femoris artery.

Muscular branches are given off from the superficial femoral throughout its entire course. They vary from two to seven in number, and supply chiefly the Sartorius and Vastus internus.

The **anastomotica magna** (fig. 319) arises from the femoral artery just before it passes through the tendinous opening in the Adductor magnus muscle, and immediately divides into a superficial and deep branch.

The *superficial branch* pierces the aponeurotic covering of Hunter's canal, and accompanies the long saphenous nerve to the inner side of the thigh. It passes between the Sartorius and Gracilis muscles, and, piercing the fascia lata, is distributed to the integument of the upper and inner part of the leg, anastomosing with the inferior internal articular.

The *deep branch* descends in the substance of the Vastus internus, lying in front of the tendon of the Adductor magnus, to the inner side of the knee, where it anastomoses with the superior internal articular artery and anterior recurrent

FIG. 319.—Side view of the popliteal artery.
(From a preparation in the Museum of the Royal College of Surgeons of England.)

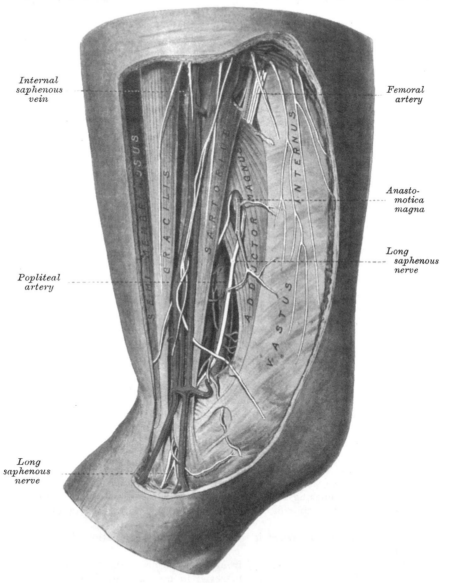

Internal
saphenous
vein

Femoral
artery

Anasto-
motica
magna

Long
saphenous
nerve

Popliteal
artery

Long
saphenous
nerve

branch of the anterior tibial.   A branch from this vessel crosses outwards above the articular surface of the femur, forming an anastomotic arch with the superior external articular artery, and supplies branches to the knee-joint.

POPLITEAL ARTERY

The **Popliteal artery** commences at the termination of the femoral at the opening in the Adductor magnus, and, passing obliquely downwards and outwards behind the knee-joint to the lower border of the Popliteus muscle, divides into the *anterior* and *posterior tibial arteries*. A portion of the artery lies in the popliteal space; but above and below, to a considerable extent, it is covered by the muscles which form the boundaries of the space, and is therefore beyond the confines of the hollow.

THE POPLITEAL SPACE (fig. 320)

Dissection.—A vertical incision about eight inches in length should be made along the back part of the knee-joint, connected above and below by a transverse incision from the inner to the outer side of the limb. The flaps of integument included between these incisions should be reflected in the direction shown in fig. 255, p. 402.

Boundaries.—The **popliteal space,** or the **ham,** is a lozenge-shaped space, widest at the back part of the knee-joint, and deepest above the articular end of the femur. It is bounded externally, above the joint, by the Biceps, and, below the joint, by the Plantaris and external head of the Gastrocnemius. Internally, above the joint, by the Semimembranosus, Semitendinosus, Gracilis, and Sartorius; below the joint, by the inner head of the Gastrocnemius.

Above, it is limited by the apposition of the inner and outer hamstring muscles; below, by the junction of the two heads of the Gastrocnemius. The floor is formed by the lower part of the posterior surface of the shaft of the femur, the posterior ligament of the knee-joint, the upper end of the tibia, and the fascia covering the Popliteus muscle, and the space is covered in by the fascia lata.

Contents.—It contains the popliteal vessels and their branches, together with the termination of the external saphenous vein, the internal and external popliteal nerves and some of their branches, the lower extremity of the small sciatic nerve, the articular branch from the obturator nerve, a few small lymphatic glands, and a considerable quantity of loose adipose tissue.

Position of contained parts.—The internal popliteal nerve descends in the middle line of the space, lying superficial and crossing the artery from without inwards. The external popliteal nerve descends on the outer side of the upper part of the space, lying close to the tendon of the Biceps muscle. More deeply at the bottom of the space are the popliteal vessels, the vein lying superficial to the artery, to which it is closely united by dense areolar tissue; it is a thick-walled vessel, and lies at first to the outer side of the artery, and then crosses it to gain the inner side below; sometimes the vein is double, the artery lying between two venæ comites, which are usually connected by short transverse branches. More deeply, and, at its upper part, close to the surface of the bone, is the popliteal artery, and passing off from it at right angles are its articular branches. The articular branch from the obturator nerve descends upon the popliteal artery to supply the knee; and occasionally there is found deep in the space an articular filament from the great sciatic nerve. The popliteal lymphatic glands, four or five in number, are found surrounding the artery: one usually lies superficial to the vessel; another is situated between it and the bone; and the rest are placed on either side of it.

The **popliteal artery,** in its course downwards from the aperture in the Adductor magnus to the lower border of the Popliteus muscle, rests first on the inner surface of the femur, and is then separated by a little fat from the hollowed popliteal surface of the bone; in the middle of its course, it rests on the posterior ligament of the knee-joint; and below, on the fascia covering the Popliteus muscle. *Superficially*, it is covered above by the Semimembranosus; in the middle of its course, by a quantity of fat, which separates it from the deep fascia and integument; and below, it is overlapped by the Gastrocnemius, Plantaris, and Soleus muscles, the popliteal vein, and the internal popliteal nerve. The popliteal vein, which is intimately attached to the artery, lies superficial and external to it above; it then crosses it and lies to its inner side. The internal popliteal nerve is still more superficial and external above, but below the joint it crosses the artery and lies on its inner side. *Laterally*, the artery is bounded by the muscles which are situated on either side of the popliteal space.

PLAN OF THE RELATIONS OF THE POPLITEAL ARTERY

In front.

Femur.
Ligamentum posticum.
Popliteus.

Inner side.		*Outer side.*
Semimembranosus.		Biceps.
Internal condyle.		Outer condyle.
Gastrocnemius (inner head).		Gastrocnemius (outer head).
		Plantaris.

Behind.

Semimembranosus.
Fascia.
Popliteal vein.
Internal popliteal nerve.
Gastrocnemius.
Plantaris.
Soleus.

Peculiarities in point of Division.—Occasionally the popliteal artery divides prematurely into its terminal branches; this unusual division occurs most frequently opposite the knee-joint. The anterior tibial under these circumstances may pass in front of the Popliteus muscle.

Unusual Branches.—The artery sometimes divides into the anterior tibial and peroneal, the posterior tibial being wanting, or very small. Occasionally the popliteal is found to divide into three branches, the anterior and posterior tibial, and peroneal.

Surface Marking.—The course of the upper part of the popliteal artery is indicated by a line drawn from the outer border of the Semimembranosus muscle at the junction of the middle and lower third of the thigh obliquely downwards to the middle of the popliteal space exactly behind the knee-joint. From this point it passes vertically downwards to the level of a line drawn through the lower part of the tubercle of the tibia.

Surgical Anatomy.—The popliteal artery is not infrequently the seat of injury. It may be torn by direct violence, as by the passage of a cart-wheel over the knee, or by hyper-extension of the knee; and in the dead body, at all events, the middle and internal coats may be ruptured by extreme flexion. It may also be lacerated by fracture of the lower part of the shaft of the femur, or by antero-posterior dislocation of the knee-joint. It has been torn in breaking down adhesions in cases of fibrous ankylosis of the knee, and is in danger of being wounded, and in fact has been wounded, in performing Macewen's operation of osteotomy of the lower end of the femur for genu valgum. In addition, Spencer records a case in which the popliteal artery was wounded from in front by a stab just below the knee, the knife passing through the interosseous space. The popliteal artery is more frequently the seat of aneurism than is any other artery in the body, with the exception of the thoracic aorta. This is due no doubt in a great measure to the amount of movement to which it is subjected, and to the fact that it is supported by loose and lax tissue only, and not by muscles as is the case with most arteries.

Ligature of the popliteal artery is required in cases of wound of that vessel, but for aneurism of the posterior tibial, it is preferable to tie the superficial femoral. The popliteal may be tied in the upper or lower part of its course; but in the middle of the ham the operation is attended with considerable difficulty, from the great depth of the artery, and from the extreme degree of tension of the lateral boundaries of the space.

In order to expose the vessel in the upper part of its course, the patient should be placed in the supine position, with the knee flexed and the thigh rotated outwards, so that it rests on its outer surface; an incision three inches in length, beginning at the junction of the middle and lower third of the thigh, is to be made parallel to and immediately behind the tendon of the Adductor magnus, and the skin, superficial and deep fascia divided. The tendon of the muscle is thus exposed, and is to be drawn forwards and the hamstring tendons backwards. A quantity of fatty tissue will now be opened up, in which the artery will be felt pulsating. This is to be separated with the point of a director until the artery is exposed. The vein and nerve will not be seen, as they lie to the outer side of the artery. The sheath is to be opened and the aneurism needle passed from before backwards, keeping its point close to the artery for fear of injuring the vein. The only structure to avoid is the long saphenous vein in the superficial incision. The upper part of the popliteal artery may also be tied by an incision on the back of the limb, along the outer margin of the Semimembranosus, but the operation is a more difficult one, as the internal popliteal nerve and the popliteal vein are first exposed, and great care has to be exercised in separating them from the artery.

To expose the vessel in the lower part of its course, where the artery lies between the two heads of the Gastrocnemius, the patient should be placed in the prone position with the limb extended. An incision should then be made through the integument in the middle line, commencing opposite the bend of the knee-joint, care being taken to avoid the external saphenous vein and nerve. After dividing the deep fascia, and separating some dense cellular membrane, the artery, vein, and nerve will be exposed, descending between the two heads of the Gastrocnemius. Some muscular branches of the popliteal should be avoided if possible, or, if divided, tied immediately. The leg being now flexed, in order the more effectually to separate the two heads of the Gastrocnemius, the nerve should be drawn inwards and the vein outwards, and the aneurism needle passed between the artery and vein from without inwards.

Branches.—The branches of the popliteal artery are : the

Muscular { Superior.		Superior external articular.
{ Inferior or Sural.		Azygos articular.
Cutaneous.		Inferior internal articular.
Superior internal articular.		Inferior external articular.

The **superior muscular branches,** two or three in number, arise from the upper part of the popliteal artery, and are distributed to the lower part of the Adductor magnus and flexor muscles of the thigh; anastomosing with the fourth perforating branch of the profunda.

The **inferior muscular (sural)** are two large branches, which are distributed to the two heads of the Gastrocnemius and to the Plantaris muscle. They arise from the popliteal artery opposite the knee-joint.

The **cutaneous branches** arise separately from the popliteal artery or from some of its branches; they descend between the two heads of the Gastrocnemius muscle, and, piercing the deep fascia, are distributed to the integument of the calf. One branch usually accompanies the short, or external, saphenous vein.

The **superior articular arteries,** two in number, arise one on each side of the popliteal, and wind round the femur immediately above its condyles to the front of the knee-joint. The *internal branch* winds inwards beneath the hamstring muscles, to which it supplies branches, above the inner head of the Gastrocnemius, and passing beneath the tendon of the Adductor magnus, divides into two branches, one of which supplies the Vastus internus, inosculating with the anastomotica magna and inferior internal articular; the other ramifies close to the surface of the femur, supplying it and the knee-joint, and anastomosing with the superior external articular artery. This branch is frequently of small size, a condition which is associated with an increase in the size of the anastomotica magna. The *external branch* passes above the outer condyle, beneath the tendon of the Biceps, and divides into a superficial and deep branch: the superficial branch supplies the Vastus externus, and anastomoses with the descending branch of the external circumflex, and the inferior external articular arteries; the deep branch supplies the lower part of the femur and knee-joint, and forms an anastomotic arch across the bone with the anastomotica magna and the inferior internal articular arteries.

The **azygos articular** is a small branch, arising from the popliteal artery opposite the bend of the knee-joint. It pierces the posterior ligament, and supplies the ligaments and synovial membrane in the interior of the articulation.

The **inferior articular arteries,** two in number, arise from the popliteal beneath the Gastrocnemius, and wind round the head of the tibia, below the joint. The *internal* one first descends along the upper margin of the Popliteus muscle, to which it gives branches; it then passes below the inner tuberosity, beneath the internal lateral ligament, at the anterior border of which it ascends to the front and inner side of the joint, to supply the head of the tibia and the articulation of the knee, anastomosing with the inferior external articular and superior internal articular arteries. The *external* one passes outwards above the head of the fibula, to the front of the knee-joint, passing in its course beneath the outer head of the Gastrocnemius, the external lateral ligament, and the tendon of the Biceps muscle, and divides into branches, which anastomose with the

FIG. 320.—The popliteal, posterior tibial, and peroneal arteries.

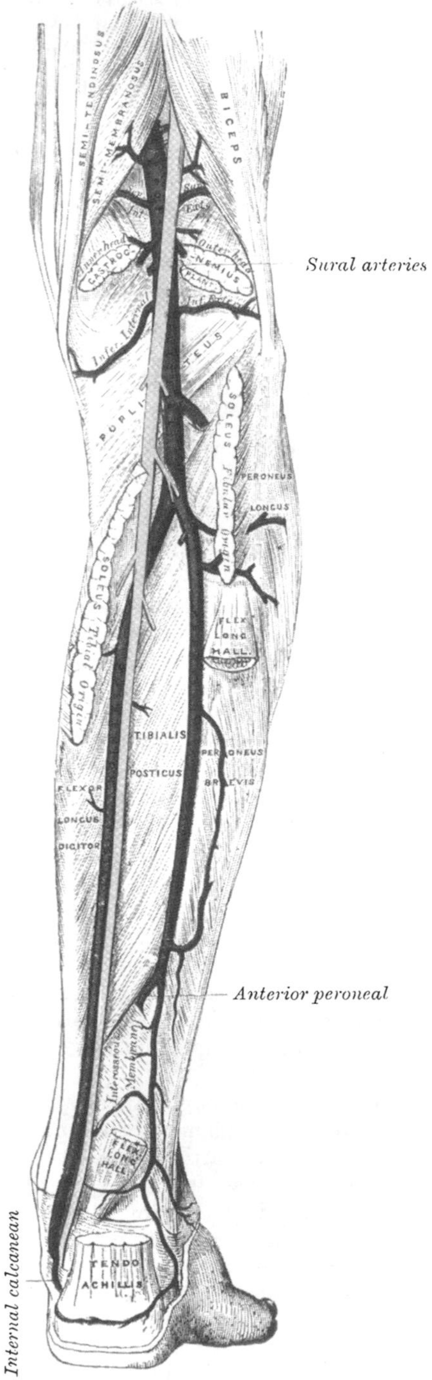

FIG. 321.—Anterior tibial and dorsalis pedis arteries.

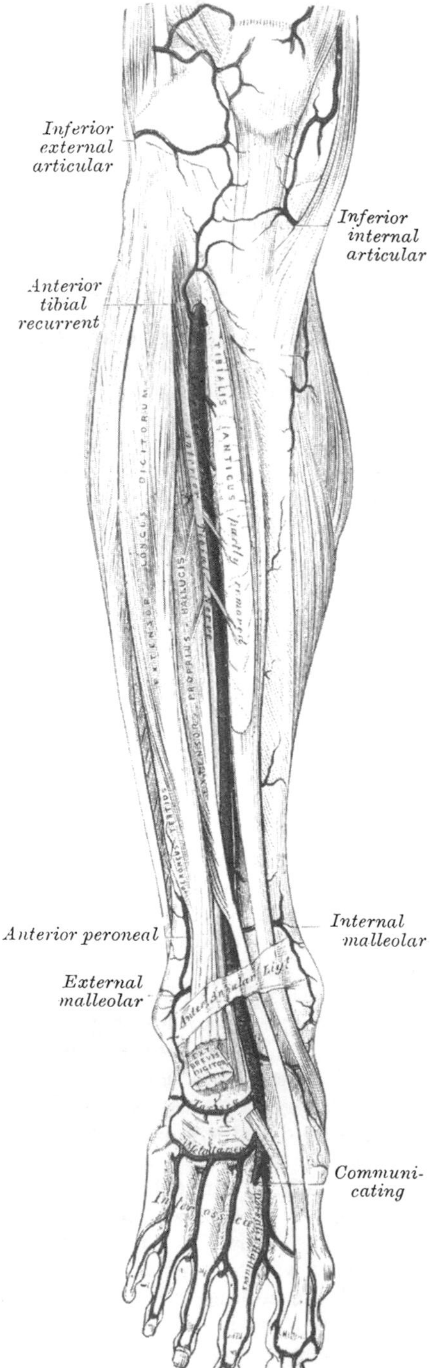

inferior internal articular artery, the superior external articular artery, and the anterior recurrent branch of the anterior tibial.

Circumpatellar anastomosis.—Around and above the patella, and on the contiguous ends of the femur and tibia, is a large network of vessels, forming a superficial and deep plexus. The superficial plexus is situated between the fascia and skin round about the patella; the deep plexus, which forms a close network of vessels, lies on the surface of the lower end of the femur and upper end of the tibia around their articular surfaces, and sends numerous offsets into the interior of the joint. The arteries from which this plexus is formed are the two internal and two external articular branches of the popliteal: the anastomotica magna; the terminal branch of the profunda; the descending branch from the external circumflex, and the anterior recurrent branch of the anterior tibial.

Anterior Tibial Artery (fig. 321)

The **Anterior Tibial artery** commences at the bifurcation of the popliteal, at the lower border of the Popliteus muscle, passes forwards between the two heads of the Tibialis posticus, and through the large oval aperture above the upper border of the interosseous membrane, to the deep part of the front of the leg: it here lies close to the inner side of the neck of the fibula; it then descends on the anterior surface of the interosseous membrane, gradually approaching the tibia; and, at the lower part of the leg, lies on this bone, and then on the anterior ligament of the ankle, to the bend of the ankle-joint, where it lies more superficially, and becomes the *dorsalis pedis*.

Relations.—In the upper two-thirds of its extent, it rests upon the interosseous membrane, to which it is connected by delicate fibrous arches thrown across it. In the lower third, upon the front of the tibia, and the anterior ligament of the ankle-joint. In the upper third of its course, it lies between the Tibialis anticus and Extensor longus digitorum; in the middle third, between the Tibialis anticus and Extensor proprius hallucis. At the bend of the ankle, it is crossed by the tendon of the Extensor proprius hallucis, and lies between it and the innermost tendon of the Extensor longus digitorum. It is covered, in the upper two-thirds of its course, by the muscles which lie on either side of it, and by the deep fascia; in the lower third, by the integument, anterior annular ligament, and fascia.

The anterior tibial artery is accompanied by two veins (venæ comites) which lie one on each side of the artery; the anterior tibial nerve, coursing round the outer side of the neck of the fibula, comes into relation with the outer side of the artery shortly after it has passed through the opening in the interosseous membrane; about the middle of the leg it is placed superficial to it; at the lower part of the artery the nerve is generally again on the outer side.

Plan of the Relations of the Anterior Tibial Artery

In front.

Integument, superficial and deep fasciæ.
Anterior tibial nerve.
Tibialis anticus (overlaps it in the upper part of the leg).
Extensor longus digitorum ⎱ (overlap it slightly).
Extensor proprius hallucis ⎰
Anterior annular ligament.

Inner side.		*Outer side.*
Tibialis anticus.		Anterior tibial nerve.
Extensor proprius hallucis (crosses it at its lower part).	 Anterior Tibial.	Extensor longus digitorum. Extensor proprius hallucis.

Behind.

Interosseous membrane.
Tibia.
Anterior ligament of ankle-joint.

Peculiarities in Size.—This vessel may be diminished in size, may be deficient to a greater or less extent, or may be entirely wanting, its place being supplied by perforating branches from the posterior tibial, or by the anterior division of the peroneal artery.

Course.—The artery occasionally deviates in its course towards the fibular side of the leg, regaining its usual position beneath the annular ligament at the front of the ankle. In two instances the vessel has been found to approach the surface in the middle of the leg, being covered merely by the integument and fascia below that point.

Surface Marking.—Draw a line from the inner side of the head of the fibula to midway between the two malleoli. In this line take a point one inch and a quarter below the head of the fibula, and the portion of the line below this point will mark the course of the artery.

Surgical Anatomy.—The anterior tibial artery may be tied in the upper or lower part of the leg. In the upper part the operation is attended with great difficulty, on account of the depth of the vessel from the surface. An incision, about four inches in length, should be made through the integument, midway between the spine of the tibia and the outer margin of the fibula, and the deep fascia exposed. The wound must now be carefully dried, its edges retracted, and the white line separating the Tibialis anticus from the Extensor longus digitorum sought for. When this has been clearly defined, the deep fascia is to be divided in this line, and the Tibialis anticus separated from adjacent muscles with the handle of the scalpel or a director until the interosseous membrane is reached. The foot is to be flexed in order to relax the muscles, and upon drawing them apart the artery will be found lying on the interosseous membrane with the nerve on its outer side or on the top of the artery. The nerve should be drawn outwards, and the venæ comites separated from the artery and the needle passed around it.

To tie the vessel in the lower third of the leg above the ankle-joint, an incision about three inches in length should be made through the integument between the tendons of the Tibialis anticus and Extensor proprius hallucis muscles, the deep fascia being divided to the same extent. The tendon on either side should be held aside, when the vessel will be seen lying upon the tibia, with the nerve on the outer side, and one of the venæ comites on either side.

Branches.—The branches of the anterior tibial artery are : the

Posterior recurrent tibial.	Muscular.
Superior fibular.	Internal malleolar.
Anterior recurrent tibial.	External malleolar.

The **posterior recurrent tibial** is not a constant branch, and is given off from the anterior tibial before that vessel passes through the interosseous space. It ascends beneath the Popliteus muscle, which it supplies, and anastomoses with the lower articular branches of the popliteal artery, giving off an offset to the superior tibio-fibular joint.

The **superior fibular** is sometimes given off from the anterior tibial, sometimes from the posterior tibial. It passes outwards, round the neck of the fibula, through the Soleus, which it supplies, and ends in the substance of the Peroneus longus muscle.

The **anterior recurrent tibial** arises from the anterior tibial, as soon as that vessel has passed through the interosseous space; it ascends in the Tibialis anticus muscle, and ramifies on the front and sides of the knee-joint, anastomosing with the articular branches of the popliteal, and with the anastomotica magna assisting in the formation of the circumpatellar plexus.

The **muscular branches** are numerous : they are distributed to the muscles which lie on each side of the vessel, some piercing the deep fascia to supply the integument, others passing through the interosseous membrane, and anastomosing with branches of the posterior tibial and peroneal arteries.

The **malleolar arteries** supply the ankle-joint. The *internal* arises about two inches above the articulation, and passes beneath the tendons of the Extensor proprius hallucis and Tibialis anticus, to the inner ankle, upon which it ramifies, anastomosing with branches of the posterior tibial and internal plantar arteries and with the internal calcanean from the posterior tibial. The *external* passes beneath the tendons of the Extensor longus digitorum and Peroneus tertius, and supplies the outer ankle, anastomosing with the anterior peroneal artery, and with ascending branches from the tarsal branch of the dorsalis pedis.

DORSALIS PEDIS ARTERY (fig. 321)

The **Dorsalis pedis,** the continuation of the anterior tibial, passes forwards from the bend of the ankle along the tibial side of the foot to the back part of the first intermetatarsal space, where it divides into two branches, the *dorsalis hallucis* and *communicating*.

Relations.—This vessel, in its course forwards, rests upon the astragalus, navicular, and middle cuneiform bones and the ligaments connecting them, being covered by the integument and fascia, anterior annular ligament, and crossed near its termination by the innermost tendon of the Extensor brevis digitorum. On its *tibial side* is the tendon of the Extensor proprius hallucis; on its *fibular side,* the innermost tendon of the Extensor longus digitorum, and the termination of the anterior tibial nerve. It is accompanied by two veins.

PLAN OF THE RELATIONS OF THE DORSALIS PEDIS ARTERY

In front.

Integument and fascia.
Anterior annular ligament.
Innermost tendon of Extensor brevis digitorum.

Tibial side.   Dorsalis   *Fibular side.*
Extensor proprius hallucis.   Pedis.   Extensor longus digitorum.
                                                  Anterior tibial nerve.

Behind.

Astragalus,
Navicular,   } and their ligaments.
Middle cuneiform,

Peculiarities in Size.—The dorsal artery of the foot may be larger than usual, to compensate for a deficient plantar artery; or it may be deficient in its terminal branches to the toes, which are then derived from the internal plantar; or its place may be supplied altogether by a large anterior peroneal artery.

Position.—This artery frequently curves outwards, lying external to the line between the middle of the ankle and the back part of the first interosseous space.

Surface Marking.—The dorsalis pedis artery is indicated on the surface of the dorsum of the foot by a line drawn from the centre of the space between the two malleoli to the back of the first intermetatarsal space.

Surgical Anatomy.—This artery may be tied, by making an incision through the integument, between two and three inches in length, on the fibular side of the tendon of the Extensor proprius hallucis, in the interval between it and the inner border of the short Extensor muscle. The incision should not extend farther forwards than the back part of the first intermetatarsal space, as the artery divides in that situation. The deep fascia being divided to the same extent, the artery will be exposed, the nerve lying upon its outer side.

Branches.—The branches of the dorsalis pedis are: the

Tarsal.   Dorsalis hallucis.
Metatarsal—Interosseous.   Communicating.

The **tarsal artery** arises from the dorsalis pedis, as that vessel crosses the navicular bone; it passes in an arched direction outwards, lying upon the tarsal bones, and covered by the Extensor brevis digitorum; it supplies that muscle and the articulations of the tarsus, and anastomoses with branches from the metatarsal, external malleolar, peroneal, and external plantar arteries.

The **metatarsal artery** arises a little anterior to the preceding; it passes outwards to the outer part of the foot, over the bases of the metatarsal bones, beneath the tendons of the short Extensor, its direction being influenced by its point of origin; and it anastomoses with the tarsal and external plantar arteries. This vessel gives off three branches, the *interosseous arteries,* which pass forwards upon the three outer Dorsal interossei muscles, and, in the clefts between the toes, divide into two dorsal collateral branches for the adjoining toes. At the back part of each interosseous space these vessels receive the posterior perforating branches from the plantar arch ; and at the fore part of each interosseous space, they are joined by the anterior perforating branches, from the digital arteries. The outermost interosseous artery gives off a branch which supplies the outer side of the little toe.

The **dorsalis hallucis, or first dorsal interosseous artery,** runs forwards along the outer border of the first metatarsal bone, and at the cleft between the first and second toes divides into two branches, one of which passes inwards, beneath the tendon of the Extensor proprius hallucis, and is distributed to the inner border of the great toe ; the outer branch bifurcates, to supply the adjoining sides of the great and second toes.

The **communicating artery** dips down into the sole of the foot, between the two heads of the First dorsal interosseous muscle, and inosculates with the termination of the external plantar artery, to complete the plantar arch. It here gives off its plantar digital branch, which is named the *arteria magna hallucis*. This artery passes forwards along the first interosseous space, and, after sending a branch along the inner side of the great toe, bifurcates for the supply of the adjacent sides of the great and second toes.

POSTERIOR TIBIAL ARTERY

The **posterior tibial** is an artery of large size, which extends obliquely downwards from the lower border of the Popliteus muscle, along the tibial side of the leg, to the fossa between the inner ankle and the heel, where it divides beneath the origin of the Abductor hallucis, on a level with a line drawn from the point of the internal malleolus to the centre of the convexity of the heel, into the *internal* and *external plantar arteries*. At its origin it lies opposite the interval between the tibia and fibula ; as it descends, it approaches the inner side of the leg, lying behind the tibia, and, in the lower part of its course, is situated midway between the inner malleolus and the tuberosity of the os calcis.

Relations.—It lies successively upon the Tibialis posticus, the Flexor longus digitorum the tibia, and the back part of the ankle-joint. It is *covered* by the deep transverse fascia, which separates it above from the Gastrocnemius and Soleus muscles : at its termination it is covered by the Abductor hallucis muscle. In the lower third, where it is more superficial, it is covered only by the integument and fascia, and runs parallel with the inner border of the tendo Achillis. It is accompanied by two veins, and by the posterior tibial nerve, which lies at first to the inner side of the artery, but soon crosses it, and is, in the greater part of its course, on its outer side.

PLAN OF THE RELATIONS OF THE POSTERIOR TIBIAL ARTERY

In front.
Tibialis posticus.
Flexor longus digitorum.
Tibia.
Ankle-joint.

Inner side.		*Outer side.*
Posterior tibial nerve, upper third.	Posterior Tibial.	Posterior tibial nerve, lower two-thirds.

Behind.
Integument and fascia.
Gastrocnemius.
Soleus.
Deep transverse fascia.
Posterior tibial nerve.
Abductor hallucis.

Behind the Inner Ankle, the tendons and blood-vessels are arranged, under cover of the internal annular ligament, in the following order, from within outwards : First, the tendons of the Tibialis posticus and Flexor longus digitorum, lying in the same groove, behind the inner malleolus, the former being the most internal. External to these is the posterior tibial artery, having a vein on either side ; and, still more externally, the posterior tibial nerve. About half an inch nearer the heel is the tendon of the Flexor longus hallucis.

Peculiarities in Size.—The posterior tibial is not infrequently smaller than usual, or absent, its place being supplied by a large peroneal artery, which passes inwards at the lower end of the tibia, and either joins the small tibial artery, or continues alone to the sole of the foot.

Surface Marking.—The course of the posterior tibial artery is indicated by a line drawn from a point one inch below the centre of the popliteal space to midway between the tip of the internal malleolus and the centre of the convexity of the heel.

Surgical Anatomy.—The *application of a ligature* to the posterior tibial may be required in cases of wound of the sole of the foot, attended with great hæmorrhage, when the vessel should be tied at the inner ankle. In cases of wound of the posterior tibial, it will be necessary to enlarge the opening so as to expose the vessel at the wounded point, excepting where the vessel is injured by a punctured wound from the front of the leg. In cases of aneurism from wound of the artery low down, the vessel should be tied in the middle of the leg. But in aneurism of the posterior tibial high up, it would be better to tie the femoral artery.

To tie the posterior tibial artery at the ankle, a semilunar incision, convex backwards, should be made through the integument, about two inches and a half in length, midway between the heel and inner ankle, or a little nearer the latter. The subcutaneous cellular tissue having been divided, a strong and dense fascia, the internal annular ligament, is exposed. This ligament is continuous above with the deep fascia of the leg, covers the vessels and nerves, and is intimately adherent to the sheaths of the tendons. This having been cautiously divided upon a director, the sheath of the vessels is exposed, and, being opened, the artery is seen with one of the venæ comites on each side. The aneurism needle should be passed round the vessel from the heel towards the ankle, in order to avoid the posterior tibial nerve, care being at the same time taken not to include the venæ comites.

The vessel may also be tied in the lower third of the leg by making an incision about three inches in length, parallel with the inner margin of the tendo Achillis. The internal saphenous vein being carefully avoided, the two layers of fascia must be divided upon a director, when the artery is exposed along the outer margin of the Flexor longus digitorum, with one of its venæ comites on either side, and the nerve lying external to it.

To tie the posterior tibial in the middle of the leg is a very difficult operation, on account of the great depth of the vessel from the surface. The patient being placed in the recumbent position, the injured limb should rest on its outer side, the knee being partially bent, and the foot extended, so as to relax the muscles of the calf. An incision about four inches in length should then be made through the integument, a finger's breadth behind the inner margin of the tibia, taking care to avoid the internal saphenous vein. The deep fascia having been divided, the margin of the Gastrocnemius is exposed, and must be drawn aside, and the tibial attachment of the Soleus divided, a director being previously passed beneath it. The artery may now be felt pulsating beneath the deep fascia, about an inch from the margin of the tibia. The fascia having been divided, and the limb placed in such a position as to relax the muscles of the calf as much as possible, the veins should be separated from the artery, and the aneurism needle passed round the vessel from without inwards, so as to avoid wounding the posterior tibial nerve.

Branches.—The branches of the posterior tibial artery are : the

Peroneal.	Muscular.
Nutrient.	Communicating.

Internal calcanean.

The **Peroneal artery** lies, deeply seated, along the back part of the fibular side of the leg. It arises from the posterior tibial, about an inch below the lower border of the Popliteus muscle, passes obliquely outwards to the fibula, and then descends along the inner border of that bone, contained in a fibrous canal between the Tibialis posticus and the Flexor longus hallucis, or in the substance of the latter muscle to the lower third of the leg, where it gives off the *anterior peroneal.* It then passes across the articulation between the tibia and fibula to the outer side of the os calcis, where it gives off its terminal branches, the *external calcanean.*

Relations.—This vessel rests at first upon the Tibialis posticus, and then, for the greater part of its course, in a fibrous canal between the origins of the Flexor longus hallucis and Tibialis posticus, covered or surrounded by the fibres of the Flexor longus hallucis. It is *covered*, in the upper part of its course, by the Soleus and deep transverse fascia ; *below*, by the Flexor longus hallucis.

PLAN OF THE RELATIONS OF THE PERONEAL ARTERY

In front.
Tibialis posticus.
Flexor longus hallucis.

Outer side.
Fibula.
Flexor longus hallucis.

Peroneal Artery.

Inner side.
Flexor longus hallucis.

Behind.
Soleus.
Deep transverse fascia.
Flexor longus hallucis.

Peculiarities in Origin.—The peroneal artery may arise three inches below the Popliteus, or from the posterior tibial high up, or even from the popliteal.

Its size is more frequently increased than diminished ; and then it either reinforces the posterior tibial by its junction with it, or altogether takes the place of the posterior tibial in the lower part of the leg and foot, the latter vessel only existing as a short muscular branch. In those rare cases where the peroneal artery is smaller than usual, a branch from the posterior tibial supplies its place ; and a branch from the anterior tibial compensates for the diminished anterior peroneal artery. In one case the peroneal artery has been found entirely wanting.

The anterior peroneal is sometimes enlarged, and takes the place of the dorsal artery of the foot.

The branches of the peroneal are : the

Muscular.
Nutrient.
Anterior peroneal.

Communicating.
Posterior peroneal.
External calcanean.

Muscular branches.—The peroneal artery, in its course, gives off branches to the Soleus, Tibialis posticus, Flexor longus hallucis, and Peronei muscles.

The *nutrient artery* supplies the fibula.

The *anterior peroneal* pierces the interosseous membrane, about two inches above the outer malleolus, to reach the fore part of the leg, and, passing down beneath the Peroneus tertius, to the outer ankle, ramifies on the front and outer side of the tarsus, anastomosing with the external malleolar and tarsal arteries.

The *communicating* is given off from the peroneal about an inch from its lower end, and, passing inwards, joins the communicating branch of the posterior tibial.

The *posterior peroneal* passes down behind the outer ankle to the back of the external malleolus, to terminate in branches which ramify on the outer surface and back of the os calcis.

The *external calcanean* are the terminal branches of the peroneal artery ; they pass to the outer side of the heel, and communicate with the external malleolar and, on the back of the heel, with the internal calcanean arteries.

The **nutrient artery** of the tibia arises from the posterior tibial, near its origin, and after supplying a few muscular branches, enters the nutrient canal of that bone, which it traverses obliquely from above downwards. This is the largest nutrient artery of bone in the body.

The **muscular branches** of the posterior tibial are distributed to the Soleus and deep muscles along the back of the leg.

The **communicating branch** runs transversely across the back of the tibia, about two inches above its lower end, beneath the Flexor longus hallucis, to join a similar branch of the peroneal.

The **internal calcanean** are several large arteries, which arise from the posterior tibial just before its division; they are distributed to the fat and integument behind the tendo Achillis and about the heel, and to the muscles on the inner side of the sole, anastomosing with the peroneal and internal malleolar and, on the back of the heel, with the external calcanean arteries.

The **Internal Plantar artery** (figs. 322, 323), much smaller than the external, passes forwards along the inner side of the foot. It is at first situated above * the Abductor hallucis, and then between it and the Flexor brevis digitorum,

FIG. 322.—The plantar arteries. Superficial view.   FIG. 323.—The plantar arteries. Deep view.

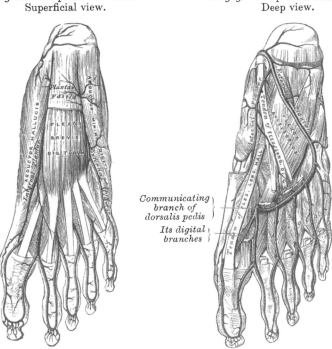

Communicating branch of dorsalis pedis
Its digital branches

both of which it supplies. At the base of the first metatarsal bone, where it has become much diminished in size, it passes along the inner border of the great toe, inosculating with its digital branch. Small superficial digital branches accompany the digital branches of the internal plantar nerve and join the plantar digital arteries of the three inner spaces.

The **External Plantar artery,** much larger than the internal, passes obliquely outwards and forwards to the base of the fifth metatarsal bone. It then turns obliquely inwards to the interval between the bases of the first and second metatarsal bones, where it anastomoses with the communicating branch from the dorsalis pedis artery, thus completing the *plantar arch.* As this artery passes outwards, it is first placed between the os calcis and Abductor hallucis, and then between the Flexor brevis digitorum and Flexor accessorius; and as it passes forwards to the base of the little toe, it lies more superficially between the Flexor

* This refers to the erect position of the body. In the ordinary position for dissection the artery is deeper than the muscle.

brevis digitorum and Abductor minimi digiti, covered by the deep fascia and integument. The remaining portion of the vessel is deeply situated; it extends from the base of the metatarsal bone of the little toe to the back part of the first interosseous space, and forms the plantar arch; it is convex forwards, lies upon the Interossei muscles, opposite the tarsal ends of the metatarsal bones, and is covered by the Adductor obliquus hallucis, the flexor tendons of the toes, and the Lumbricales.

Surface Marking.—The course of the internal plantar artery is represented by a line drawn from the mid-point between the tip of the internal malleolus and the centre of the convexity of the heel to the middle of the under surface of the great toe. The external plantar by a line from the same point to within a finger's breadth of the tuberosity of the fifth metatarsal bone. The plantar arch is indicated by a line drawn from this point : i.e. a finger's breadth internal to the tuberosity of the fifth metatarsal bone transversely across the foot to the back of the first interosseous space.

Surgical Anatomy.—Wounds of the plantar arch are always serious, on account of the depth of the vessel and the important structures which must be interfered with in an attempt to ligature it. They must be treated on similar lines to those of wounds of the palmar arches (see page 522). Delorme has shown that it may be ligatured from the dorsum of the foot in almost any part of its course by removing a portion of one of the three middle metatarsal bones.

Branches.—The plantar arch, besides distributing numerous branches to the muscles, integument, and fasciæ in the sole, gives off the following branches :

Posterior perforating. Digital—Anterior perforating.

The **posterior perforating** are three small branches, which ascend through the back part of the three outer interosseous spaces, between the heads of the Dorsal interossei muscles, and anastomose with the interosseous branches from the metatarsal artery.

The **digital branches** are four in number, and supply the three outer toes and half the second toe. The *first* passes outwards from the outer side of the plantar arch, and is distributed to the outer side of the little toe, passing in its course beneath the abductor and short flexor muscles. The *second, third,* and *fourth* run forwards along the interosseous spaces, and on arriving at the clefts between the toes divide into collateral branches, which supply the adjacent sides of the three outer toes and the outer side of the second. At the bifurcation of the toes, each digital artery sends upwards, through the fore part of the corresponding interosseous space, a small branch, which inosculates with the interosseous branches of the metatarsal artery. These are the *anterior perforating branches.*

From the arrangement already described of the distribution of the vessels to the toes, it will be seen that both sides of the three outer toes, and the outer side of the second toe, are supplied by branches from the plantar arch; both sides of the great toe, and the inner side of the second, are supplied by the communicating branch of the dorsalis pedis.

THE VEINS

The veins are the vessels which serve to return the blood from the capillaries of the different parts of the body to the heart. They consist of two distinct sets of vessels, the *pulmonary* and *systemic*.

The **Pulmonary Veins** are concerned in the circulation in the lungs. Unlike other vessels of this kind, they contain arterial blood, which they return from the lungs to the left auricle of the heart.

The **Systemic Veins** are concerned in the general circulation; they return the venous blood from the body generally to the right auricle of the heart.

The **Portal Vein,** an appendage to the systemic venous system, is confined to the abdominal cavity, returning the venous blood from the viscera of digestion, and carrying it to the liver by a single trunk of large size, the *vena portæ*. This vessel ramifies in the substance of the liver and breaks up into a minute network of capillaries. These capillaries then re-collect to form the hepatic veins, by which the blood is conveyed to the inferior vena cava.

The veins, like the arteries, are found in nearly every tissue of the body. They commence by minute plexuses which receive the blood from the capillaries. The branches which have their commencement in these plexuses unite together into trunks, and these, in their passage towards the heart, constantly increase in size as they receive tributaries, or join other veins. The veins are larger and altogether more numerous than the arteries; hence, the entire capacity of the venous system is much greater than that of the arterial; the pulmonary veins excepted, which only slightly exceed in capacity the pulmonary arteries. From the combined area of the smaller venous branches being greater than the main trunks, it results that the venous system represents a cone, the summit of which corresponds to the heart, its base to the circumference of the body. In form, the veins are perfectly cylindrical like the arteries, their walls being collapsed when empty, and the uniformity of their surface being interrupted at intervals by slight constrictions, which indicate the existence of valves in their interior. They usually retain, however, the same calibre as long as they receive no branches.

The veins communicate very freely with one another, especially in certain regions of the body; and this communication exists between the larger trunks as well as between the smaller branches. Thus, in the cavity of the cranium, and between the veins of the neck, where obstruction would be attended with imminent danger to the cerebral venous system, we find that the sinuses and larger veins have large and very frequent anastomoses. The same free communication exists between the veins throughout the whole extent of the spinal canal, and between the veins composing the various venous plexuses in the abdomen and pelvis, as the spermatic, uterine, vesical, and prostatic.

Veins have thinner walls than arteries, the difference in thickness being due to the small amount of elastic and muscular tissues which the veins contain. The superficial veins usually have thicker coats than the deep veins, and the veins of the lower limb are thicker than those of the upper.

The minute structure of these vessels has been described in the section on General Anatomy.

The systemic veins are subdivided into three sets: superficial, deep, and sinuses.

The **Superficial** or **Cutaneous Veins** are found between the layers of the superficial fascia, immediately beneath the integument; they return the blood from these structures, and communicate with the deep veins by perforating the deep fascia.

The **Deep Veins** accompany the arteries, and are usually enclosed in the same sheath with those vessels. With the smaller arteries—as the radial, ulnar, brachial, tibial, peroneal—they exist generally in pairs, one lying on each side of the vessel, and are called *venæ comites*. The larger arteries—as the axillary, subclavian, popliteal, and femoral—have usually only one accompanying vein. In certain organs of the body, however, the deep veins do not accompany the arteries; for instance, the veins in the skull and spinal canal, the hepatic veins in the liver, and the larger veins returning blood from the osseous tissue.

Sinuses are venous channels, which, in their structure and mode of distribution, differ altogether from the veins. They are found only in the interior of the skull, and consist of channels formed by a separation of the two layers of the dura mater; their outer coat consisting of fibrous tissue, their inner of an endothelial layer continuous with the lining membrane of the veins.

THE PULMONARY VEINS

The *Pulmonary Veins* return the arterial blood from the lungs to the left auricle of the heart. They are four in number, two for each lung. The pulmonary differ from other veins in several respects: 1. They carry arterial instead of venous blood. 2. They are destitute of valves. 3. They are only slightly larger than the arteries they accompany. 4. They accompany those vessels singly. They commence in a capillary network, upon the walls of the air-cells, where they are continuous with the capillary ramifications of the pulmonary artery, and, uniting together, form one vessel for each lobule. These vessels, uniting successively, form a single trunk for each lobe, three for the right, and two for the left lung. The vein from the middle lobe of the right lung generally unites with that from the upper lobe, forming two trunks on each side, which open separately into the left auricle. Occasionally, they remain separate; there are then three veins on the right side. Not infrequently, the two left pulmonary veins terminate by a common opening.

Within the lung, the branches of the pulmonary artery are *in front*, the veins *behind*, and the bronchi *between* the two.

At the root of the lung, the veins are *in front*, the artery *in the middle*, and the bronchus *behind*.

Within the pericardium, their anterior surface is invested by the serous layer of this membrane. The right pulmonary veins pass behind the right auricle and ascending aorta and superior vena cava; the left pass in front of the thoracic aorta, with the left pulmonary artery.

THE SYSTEMIC VEINS

The systemic veins may be arranged into three groups: 1. Those of the head and neck, upper extremity and thorax, which terminate in the superior vena cava. 2. Those of the lower extremity, abdomen, and pelvis, which terminate in the inferior vena cava. 3. The cardiac veins, which open directly into the right auricle of the heart.

VEINS OF THE HEAD AND NECK

The veins of the head and neck may be subdivided into three groups: 1. The veins of the exterior of the head and face. 2. The veins of the neck. 3. The veins of the diploë and the interior of the cranium.

VEINS OF THE EXTERIOR OF THE HEAD AND FACE

The veins of the exterior of the head and face are : the

Frontal.	Temporal.
Supra-orbital.	Internal maxillary.
Angular.	Temporo-maxillary.
Facial.	Posterior auricular.

Occipital.

The **Frontal vein** commences on the anterior part of the skull by a venous plexus which communicates with the anterior tributaries of the temporal vein. The veins converge to form a single trunk, which runs downwards near the middle line of the forehead parallel with the vein of the opposite side, and unites with it at the root of the nose, by a transverse branch, called the *nasal arch*. Occasionally the frontal veins join to form a single trunk, which bifurcates at the root of the nose into the two angular veins. At the root of the nose the veins diverge, and join the *supra-orbital vein*, at the inner angle of the orbit, to form the *angular vein*.

The **Supra-orbital vein** commences on the forehead, communicating with the anterior temporal vein, and runs downwards and inwards, superficial to the Occipito-frontalis muscle, receiving tributaries from the neighbouring structures, and joins the frontal vein at the inner angle of the orbit to form the *angular vein*.

The **Angular vein** formed by the junction of the frontal and supra-orbital veins runs obliquely downwards and outwards, on the side of the root of the

FIG. 324.—Veins of the head and neck.

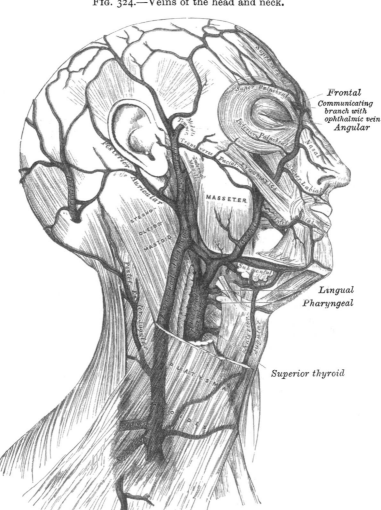

nose, and receives the veins of the ala nasi on its inner side, and the superior palpebral veins on its outer side ; it moreover communicates with the ophthalmic vein, thus establishing an important anastomosis between this vessel and the cavernous sinus. Some small veins from the dorsum of the nose terminate in the nasal arch.

The **Facial vein** commences at the side of the root of the nose, being a direct continuation of the angular vein. It lies behind and follows a less tortuous course than the facial artery. It passes obliquely downwards and outwards, beneath the Zygomaticus major and minor muscles, descends along the anterior border of the Masseter, crosses over the body of the lower jaw, with the facial artery, and, passing obliquely outwards and backwards, beneath the Platysma and cervical fascia, unites with the anterior division of the temporo-maxillary vein to form a trunk of large size (*common facial vein*) which enters the internal jugular. From near its termination a communicating branch often runs down the anterior border of the Sterno-mastoid to join the lower part of the anterior jugular.

Tributaries.—The facial vein receives, near the angle of the mouth, communicating tributaries of considerable size (the *deep facial* or *anterior internal maxillary vein*) from the pterygoid plexus. It is also joined by the inferior palpebral, the superior and inferior labial veins, the buccal veins from the cheek, and the masseteric veins. Below the jaw it receives the submental ; the inferior palatine, which returns the blood from the plexus around the tonsil and soft palate ; the submaxillary vein, which commences in the submaxillary gland ; and, generally, the ranine vein.

Surgical Anatomy.—There are some points about the facial vein which render it of great importance in surgery. It is not so flaccid as are most superficial veins, and, in consequence of this, remains more patent when divided. It has, moreover, no valves. It communicates freely with the intracranial circulation, not only at its commencement by its tributaries, the angular and supra-orbital veins, communicating with the ophthalmic vein, a tributary of the cavernous sinus, but also by its deep branch, which communicates through the pterygoid plexus with the cavernous sinus by branches which pass through the foramen ovale and foramen lacerum medium (see page 583). These facts have an important bearing upon the surgery of some diseases of the face. For on account of its patency the facial vein favours septic absorption, and therefore any phlegmonous inflammation of the face following a poisoned wound is liable to set up thrombosis in the facial vein, and detached portions of the clot may give rise to purulent foci in other parts of the body. And on account of its communications with the cerebral sinuses, these thrombi are apt to extend upwards into them, and so induce a fatal issue.

The **Temporal vein** commences by a minute plexus on the side and vertex of the skull, which communicates with the frontal and supra-orbital veins in front, the corresponding vein of the opposite side, and the posterior auricular and occipital veins behind. From this network, anterior and posterior branches are formed which unite above the zygoma, forming the trunk of the vein. This trunk is joined in this situation by a large vein, the *middle temporal*, which receives the blood from the substance of the Temporal muscle and pierces the fascia at the upper border of the zygoma. The temporal vein then descends between the external auditory meatus and the condyle of the jaw, enters the substance of the parotid gland, and unites with the internal maxillary vein to form the temporo-maxillary vein.

Tributaries.—The temporal vein receives in its course some parotid veins, an articular branch from the articulation of the jaw, anterior auricular veins from the external ear, and a vein of large size, the *transverse facial*, from the side of the face. The middle temporal vein, previous to its junction with the temporal vein, receives a branch, the *orbital vein*, which is formed by some external palpebral branches, and passes backwards between the layers of the temporal fascia.

The **Internal maxillary vein** is a vessel of considerable size, receiving branches which correspond with those of the internal maxillary artery. Thus it receives the middle meningeal veins, the deep temporal, the pterygoid, masseteric, buccal, alveolar, some palatine veins, and the inferior dental. These branches form a large plexus, the *pterygoid*, which is placed between the Temporal and External pterygoid, and partly between the Pterygoid muscles. This plexus communicates very freely with the facial vein, and with the cavernous sinus, by branches through the foramen Vesalii, foramen ovale, and foramen lacerum medium, at the base of the skull. The trunk of the vein then passes backwards, behind the neck of the lower jaw, and unites with the temporal vein, forming the temporo-maxillary vein.

The **Temporo-maxillary vein,** formed by the union of the temporal and internal maxillary veins, descends in the substance of the parotid gland on the outer surface of the external carotid artery, between the ramus of the jaw and the Sterno-mastoid muscle, and divides into two branches, an anterior, which passes inwards to join the facial vein, and a posterior, which is joined by the posterior auricular vein and becomes the external jugular.

The **Posterior auricular vein** commences upon the side of the head, by a plexus which communicates with the tributaries of the temporal and occipital veins. The vein descends behind the external ear and joins the posterior division of the temporo-maxillary vein, forming the external jugular. This vessel receives the stylo-mastoid vein, and some tributaries from the back part of the external ear.

The **Occipital veins** commence at the back part of the vertex of the skull, by a plexus, in a similar manner to the other veins. These unite and form one or two veins, which follow the course of the occipital artery, passing deeply beneath the muscles of the back part of the neck, and terminate in the internal jugular, occasionally in the external jugular vein. As these veins pass across the mastoid portion of the temporal bone, one of them receives the mastoid vein, and thus establishes a communication with the lateral sinus.

VEINS OF THE NECK

The veins of the neck, which return the blood from the head and face, are the

External jugular.	Anterior jugular.
Posterior external jugular.	Internal jugular.
Vertebral.	

The **External jugular vein** receives the greater part of the blood from the exterior of the cranium and deep parts of the face, being formed by the junction of the posterior division of the temporo-maxillary with the posterior auricular vein. It commences in the substance of the parotid gland, on a level with the angle of the lower jaw, and runs perpendicularly down the neck, in the direction of a line drawn from the angle of the jaw to the middle of the clavicle. In its course it crosses the Sterno-mastoid muscle, and runs parallel with its posterior border as far as its attachment to the clavicle, where it perforates the deep fascia, and terminates in the subclavian vein, on the outer side of or in front of the Scalenus anticus muscle. In the neck it is separated from the Sterno-mastoid by the investing layer of the deep cervical fascia, and is covered by the Platysma, the superficial fascia, and the integument. This vein is crossed about its middle by the superficialis colli nerve, and its upper half is accompanied by the auricularis magnus nerve. The external jugular vein varies in size, bearing an inverse proportion to that of the other veins of the neck ; it is occasionally double. It is provided with two pairs of valves, the lower pair being placed at its entrance into the subclavian vein, the upper pair in most cases about an inch and a half above the clavicle. The portion of vein between the two sets of valves is often dilated, and is termed the *sinus*. These valves do not prevent the regurgitation of the blood, or the passage of injection from below upwards.*

Surgical Anatomy.—Venesection used formerly to be performed on the external jugular vein, but is now probably never resorted to. The anatomical point to be remembered in performing this operation is to cut across the fibres of the Platysma myoides in opening the vein, so that by their contraction they will expose the orifice in the vein and so allow the flow of blood.

Tributaries.—This vein receives the occipital occasionally, the posterior external jugular, and, near its termination, the suprascapular and transverse

---

* The student may refer to an interesting paper by Dr. Struthers, 'On Jugular Venesection in Asphyxia, anatomically and experimentally considered, including the Demonstration of Valves in the Veins of the Neck,' in the *Edinburgh Medical Journal* for November 1856.

cervical veins. It communicates with the anterior jugular, and, in the substance of the parotid, receives a large branch of communication from the internal jugular.

The **Posterior external jugular vein** commences in the occipital region and returns the blood from the integument and superficial muscles in the upper and back part of the neck, lying between the Splenius and Trapezius muscles. It runs down the back part of the neck, and opens into the external jugular just below the middle of its course.

The **Anterior jugular vein** commences near the hyoid bone from the convergence of several superficial veins from the submaxillary region. It passes down between the median line and the anterior border of the Sterno-mastoid, and, at the lower part of the neck, passes beneath that muscle to open into the termination of the external jugular, or into the subclavian vein (fig. 331). This vein varies considerably in size, bearing almost always an inverse proportion to the external jugular. Most frequently there are two anterior jugulars, a right and left ; but occasionally only one. This vein receives some laryngeal veins, and occasionally a small thyroid vein. Just above the sternum, the two anterior jugular veins communicate by a transverse trunk, which receives tributaries from the inferior thyroid veins. It also communicates with the internal jugular. There are no valves in this vein.

The **Internal jugular vein** collects the blood from the interior of the cranium, from the superficial parts of the face, and from the neck. It commences just external to the jugular foramen, at the base of the skull, being formed by the coalescence of the lateral and inferior petrosal sinuses (fig. 329). At its origin it is somewhat dilated, and this dilatation is called the *sinus*, or *bulb*, of the internal jugular vein. It runs down the side of the neck in a vertical direction, lying at first on the outer side of the internal carotid, and then on the outer side of the common carotid, and at the root of the neck unites with the subclavian vein to form the innominate vein. The internal jugular vein, at its commencement, lies upon the Rectus capitis lateralis, and behind the internal carotid artery and the nerves passing through the jugular foramen ; lower down, the vein and artery lie upon the same plane, the glosso-pharyngeal and hypoglossal nerves passing forwards between them ; the pneumogastric descends between and behind them in the same sheath, and the spinal accessory passes obliquely outwards, behind or in front of the vein. At the root of the neck the vein of the right side is placed at a little distance from the artery ; on the left side, it usually lies over the artery at its lower part. The right internal jugular vein crosses the first part of the subclavian artery. The vein is of considerable size, but varies in different individuals, the left one being usually the smaller. It is provided with a pair of valves, which are placed at its point of termination, or from half to three-quarters of an inch above it.

Tributaries.—The vein receives in its course the facial, lingual, pharyngeal, superior and middle thyroid veins, and sometimes the occipital. At its point of junction with the common facial vein it becomes greatly increased in size.

The **lingual veins** commence on the dorsum, sides, and under surface of the tongue, and, passing backwards, following the course of the lingual artery and its branches, terminate in the internal jugular. Sometimes the ranine vein, which is a branch of considerable size, commencing below the tip of the tongue, joins the lingual. Generally, however, it passes backwards, crosses the Hyo-glossus muscle in company with the hypoglossal nerve, and joins the facial.

The **pharyngeal vein** commences in a minute plexus, the *pharyngeal*, at the back part and sides of the pharynx, and, after receiving meningeal tributaries, and the Vidian and spheno-palatine veins, terminates in the internal jugular. It occasionally opens into the facial, lingual, or superior thyroid vein.

The **superior thyroid vein** commences in the substance and on the surface of the thyroid gland, by tributaries corresponding with the branches of the superior thyroid artery, and terminates in the upper part of the internal jugular vein. It receives the superior laryngeal and crico-thyroid veins.

The **middle thyroid vein** collects the blood from the lower part of the lateral lobe of the thyroid gland, and, being joined by some veins from the larynx and trachea, terminates in the lower part of the internal jugular vein.

The **facial** and **occipital veins** have been described above.

Surgical Anatomy.—The internal jugular vein occasionally requires ligature in cases of septic thrombosis of the lateral sinus from suppuration in the middle ear, in order to prevent septic emboli being carried into the general circulation. This operation has been performed recently in several cases, with the most satisfactory results. The cases are generally those of chronic disease of the middle ear, with discharge of pus which perhaps has existed for many years. The patient is seized with acute septic inflammation, spreading to the mastoid cells, and consequent on this, septic thrombosis of the lateral sinus extending to the internal jugular vein. Such cases are always extremely grave, for there is a danger of a portion of the septic clot being detached and causing septic embolism in the thoracic viscera. If the condition is suspected, the sinus should be at once explored by trephining at a point an inch behind the centre of the external auditory meatus and a quarter of an inch above Reid's base line. The condition of the sinus is then investigated, and if it is found to be thrombosed, the surgeon should at once proceed to ligature the internal jugular vein, by an incision along the anterior border of the sterno-mastoid, the centre of which is on a level with the greater cornu of the hyoid bone. The vein should be ligatured in two places and divided between. After the vessel has been secured and divided, the lateral sinus is to be thoroughly cleared out, and by removing the ligature from the upper end of the divided vein, all septic clots removed by syringing from the sinus through the vein. If hæmorrhage occurs from the distal end of the sinus, it can be arrested by careful plugging with antiseptic gauze.

The **Vertebral vein** commences in the occipital region, by numerous small tributaries, from the deep muscles at the upper and back part of the neck ; these pass outwards and enter the foramen in the transverse process of the atlas, and descend, forming a dense plexus around the vertebral artery, in the canal formed by the foramina in the transverse processes of the cervical vertebræ. This plexus unites at the lower part of the neck into two main trunks, one of which emerges from the foramen in the transverse process of the sixth cervical vertebra, and the other through that of the seventh, and, uniting, form a single vessel, which terminates at the root of the neck in the back part of the innominate vein near its origin, its mouth being guarded by a pair of valves. On the right side, it crosses the first part of the subclavian artery.

Tributaries.—The vertebral vein receives in its course a vein from the inside of the skull through the posterior condyloid foramen ; muscular veins, from the muscles in the prevertebral region ; dorsi-spinal veins, from the back part of the cervical portion of the spine ; meningo-rachidian veins, from the interior of the spinal canal ; the anterior and posterior vertebral veins ; and close to its termination it is joined by a small vein from the first intercostal space which accompanies the superior intercostal artery.

The **anterior vertebral vein** commences in a plexus around the transverse processes of the upper cervical vertebræ, descends in company with the ascending cervical artery between the Scalenus anticus and Rectus capitis anticus major muscles, and opens into the vertebral vein just before its termination.

The **posterior vertebral vein** (the **deep cervical**) accompanies the profunda cervicis artery, lying between the Complexus and Semispinalis colli. It commences in the suboccipital region by communicating branches from the occipital vein and tributaries from the deep muscles at the back of the neck. It receives tributaries from the plexuses around the spinous processes of the cervical vertebræ, and terminates in the lower end of the vertebral vein.

Veins of the Diploë

The diploë of the cranial bones is channelled in the adult by a number of tortuous canals, which are lined by a more or less complete layer of compact tissue.

The veins they contain are large and capacious, their walls being thin, and formed only of endothelium resting upon a layer of elastic tissue, and they present, at irregular intervals, pouch-like dilatations, or *culs-de-sac*, which serve as reservoirs for the blood. These are the veins of the diploë; they can only be displayed by removing the outer table of the skull.

In adult life, as long as the cranial bones are distinct and separable, these veins are confined to the particular bones ; but in old age, when the sutures are united, they communicate with each other, and increase in size. These vessels communicate, in the interior of the cranium, with the meningeal veins, and with

Fig. 325.—Veins of the diploë as displayed by the removal of the outer table of the skull.

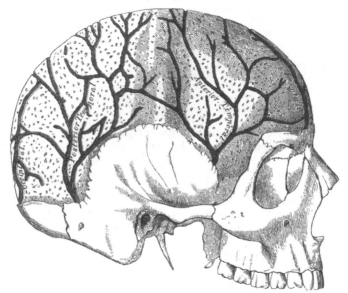

the sinuses of the dura mater ; and, on the exterior of the skull, with the veins of the pericranium. They are divided into the *frontal*, which opens into the supra-orbital vein by an aperture in the supra-orbital notch ; the *anterior temporal*, which is confined chiefly to the frontal bone, and opens into one of the deep temporal veins, after escaping by an aperture in the great wing of the sphenoid ; the *posterior temporal*, which is confined to the parietal bone, and terminates in the lateral sinus by an aperture at the posterior inferior angle of the parietal bone ; and the *occipital*, the largest of the four, which is confined to the occipital bone, and opens either into the occipital vein, or internally into the lateral sinus or torcular Herophili.

Cerebral Veins

The **Cerebral veins** are remarkable for the extreme thinness of their coats in consequence of the muscular tissue in them being wanting, and for the absence of valves. They may be divided into two sets : the superficial which are placed on the surface, and the deep veins which occupy the interior of the organ.

The **Superficial cerebral veins** ramify upon the surface of the brain, being lodged in the sulci, between the convolutions, a few running across the convolutions. They receive branches from the substance of the brain, and terminate in the sinuses. They are named, from the position they occupy, superior, median, and inferior cerebral veins.

The **superior cerebral veins,** eight to twelve in number on each side, return the blood from the convolutions on the superior surface of the hemisphere; they pass forwards and inwards towards the great longitudinal fissure, where they receive the *median cerebral veins*; near their termination, they become invested with a tubular sheath of the arachnoid membrane, and open into the superior longitudinal sinus, in the opposite direction to the course of the current of the blood.

The **median cerebral veins** return the blood from the convolutions of the mesial surface of the corresponding hemisphere; they open into the superior cerebral veins; or occasionally into the inferior longitudinal sinus.

The **inferior cerebral veins** ramify on the lower part of the outer and on the under surface of the cerebral hemisphere. Some, collecting tributaries from the under surface of the anterior lobes of the brain, terminate in the cavernous sinus. One vein of large size, the *middle cerebral vein*, commences on the under surface of the temporal lobe, and, running along the fissure of Sylvius, opens into the cavernous sinus. Another large vein, the *great anastomotic vein of Trolard*, commences on the parietal lobe, runs along the horizontal limb of the fissure of Sylvius, and opens into the anterior part of the cavernous sinus under the lesser wing of the sphenoid. Others commence on the under surface of the base of the brain, and unite to form from three to five veins, which open into the superior petrosal and lateral sinuses from before backwards.

The **Deep cerebral,** or **ventricular veins** (*venæ Galeni*), are two in number. They are formed by the union of two veins, the *vena corporis striati*, and the *choroid vein*, on either side. They run backwards, parallel with one another, between the layers of the velum interpositum, and pass out of the brain at the great transverse fissure, between the posterior extremity, or *splenium*, of the corpus callosum and the tubercula quadrigemina, to enter the straight sinus. The two veins usually unite to form one, the *vena magna Galeni*, before opening into the straight sinus. Just before their union they receive the basilar vein.

The **vena corporis striati** commences in the groove between the corpus striatum and thalamus opticus, receives numerous veins from both of these parts, and unites behind the anterior pillar of the fornix with the choroid vein, to form one of the venæ Galeni.

The **choroid vein** runs along the whole length of the outer border of the choroid plexus, receiving veins from the hippocampus major, the fornix and corpus callosum, and unites, at the anterior extremity of the choroid plexus, with the vein of the corpus striatum.

The **Basilar vein** commences at the anterior perforated space at the base of the brain by the union of a small anterior cerebral vein, which courses backwards between the anterior lobes of the cerebrum, with the deep Sylvian vein, which descends through the lower part of the Sylvian fissure. It passes backwards round the crus cerebri, receiving the inferior striate vein from the corpus striatum, interpeduncular veins from the interpeduncular space, ventricular veins from the middle cornu of the lateral ventricles, and tributaries from the uncinate convolution, and enters the vein of Galen just before its junction with the vein of the opposite side.

The **Cerebellar veins** occupy the surface of the cerebellum, and are disposed in three sets, superior, inferior, and lateral. The *superior* pass partly forwards and inwards, across the superior vermiform process, to terminate in the straight sinus and the venæ Galeni, partly outwards to the lateral and superior petrosal sinuses. The *inferior cerebellar veins*, of large size, terminate in the lateral, superior petrosal, and occipital sinuses.

The perivascular lymphatics alluded to in the section on General Anatomy are especially found in connection with the vessels of the brain. These vessels are enclosed in a sheath. which acts as a lymphatic channel, through which the lymph is carried to the subarachnoid and subdural spaces, from which it is returned into the general circulation.

Sinuses of the Dura Mater

The sinuses of the dura mater are venous channels, analogous to the veins, their outer coat being formed by the dura mater; their inner, by a continuation of the lining membrane of the veins. They are fourteen in number, and are divided into two sets: 1. Those situated at the upper and back part of the skull; 2. Those at the base of the skull. The former are: the

Superior longitudinal sinus.   Straight sinus.
Inferior longitudinal sinus.   Lateral sinuses.
                    Occipital sinus.

The **Superior longitudinal sinus** occupies the attached margin of the falx cerebri. Commencing at the foramen cæcum, through which, in the child, it constantly communicates by a small branch with the veins of the nasal fossæ, it runs from before backwards, grooving the inner surface of the frontal, the adjacent margins of the two parietal, and the superior division of the crucial ridge of the occipital bone, and terminates by opening into the torcular Herophili. The sinus is triangular in form, narrow in front, and gradually increases in size as it passes backwards. On examining its inner surface, it presents the internal openings of

Fig. 326.—Vertical section of the skull, showing the sinuses of the dura mater.

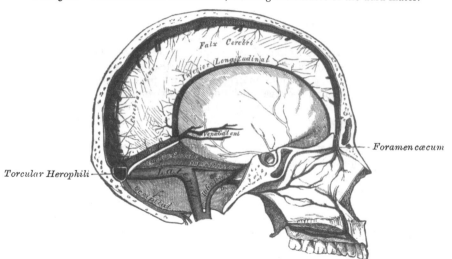

the superior cerebral veins, which run, for the most part, from behind forwards, and open chiefly at the back part of the sinus, their orifices being concealed by fibrous folds; numerous fibrous bands (*chordæ Willisii*) are also seen, extending transversely across the inferior angle of the sinus; and, lastly, some small, white, projecting bodies, the *glandulæ Pacchioni*. This sinus receives the superior cerebral veins, numerous veins from the diploë and dura mater, and, at the posterior extremity of the sagittal suture, veins from the pericranium, which pass through the parietal foramina.

The *torcular Herophili*, or *confluence of the sinuses*, is the dilated extremity of the superior longitudinal sinus. It is of irregular form, and is lodged on one side (generally the right) of the internal occipital protuberance. From it the lateral sinus of the side to which it is deflected is derived. It receives also the blood from the occipital sinus.

The **Inferior longitudinal sinus,** more correctly described as the *inferior longitudinal vein*, is contained in the posterior part of the free margin of the falx cerebri. It is of a cylindrical form, increases in size as it passes backwards, and terminates in the straight sinus. It receives several veins from the falx cerebri, and occasionally a few from the mesial surface of the hemispheres.

The **Straight sinus** is situated at the line of junction of the falx cerebri with the tentorium. It is triangular in form, increases in size as it proceeds backwards, and runs obliquely downwards and backwards from the termination of the inferior longitudinal sinus to the lateral sinus of the opposite side to that into which the superior longitudinal sinus is prolonged. It communicates by a cross branch with the torcular Herophili. Besides the inferior longitudinal sinus, it receives the venæ Galeni and the superior cerebellar veins. A few transverse bands cross its interior.

The **Lateral sinuses** are of large size, and are situated in the attached margin of the tentorium cerebelli. They commence at the internal occipital protuberance; one, generally the right, being the direct continuation of the superior longitudinal sinus, the other of the straight sinus. They pass outwards and forwards, describing a slight curve with its convexity upwards, to the base of the petrous portion of the temporal bone, then curve downwards and inwards on each side to reach the jugular foramen, where they terminate in the internal jugular vein. Each sinus rests, in its course, upon the inner surface of the occipital, the posterior inferior angle of the parietal, the mastoid portion of the temporal, and on the occipital, again, just before its termination. These sinuses are frequently of unequal size, that formed by the superior longitudinal sinus being the larger, and they increase in size as they proceed from behind forwards. The horizontal portion is of a triangular form, the curved portion semicylindrical. Their inner surface is smooth, and not crossed by the fibrous bands found in the other sinuses. These sinuses receive the blood from the superior petrosal sinuses at the base of the petrous portion of the temporal bone, and they unite with the inferior petrosal sinus, just external to the jugular foramen, to form the internal jugular vein (fig. 329). They communicate with the veins of the pericranium by means of the mastoid and posterior condyloid veins, and they receive some of the inferior cerebral and inferior cerebellar veins, and some veins from the diploë. The *petro-squamous sinus*, when present, runs backwards along the junction of the petrous and squamous-temporal, and opens into the lateral sinus.

The **Occipital** is the smallest of the cranial sinuses. It is generally single, but occasionally there are two. It is situated in the attached margin of the falx cerebelli. It commences by several small veins around the margin of the foramen magnum, one of which joins the termination of the lateral sinus; it communicates with the posterior spinal veins, and terminates in the torcular Herophili.

The sinuses at the base of the skull are: the

Cavernous sinuses.	Superior petrosal sinuses.
Circular sinus.	Inferior petrosal sinuses.
Transverse sinus.	

The **Cavernous sinuses** are named from their presenting a reticulated structure, due to their being traversed by numerous interlacing filaments. They are two in number, of irregular form, larger behind than in front, and are placed one on each side of the sella turcica, extending from the sphenoidal fissure to the apex of the petrous portion of the temporal bone; they receive anteriorly the ophthalmic veins through the sphenoidal fissure, and open behind into the petrosal sinuses. On the inner wall of each sinus is found the internal carotid artery, accompanied by filaments of the carotid plexus and by the sixth nerve; and on its outer wall, the third, fourth, and ophthalmic division of the fifth nerve. These parts are separated from the blood flowing along the sinus by the lining membrane, which is continuous with the inner coat of the veins. The cavernous sinuses receive some of the cerebral veins, and also a small sinus, the *spheno-parietal*, which extends inwards on the under aspect of the lesser wing of the sphenoid; they communicate with the lateral sinuses by means of the superior and inferior petrosal sinuses, and with the facial vein through the ophthalmic vein. They also communicate with each other by means of the circular sinus.

Surgical Anatomy.— An arterio-venous communication may be established between the cavernous sinus and the carotid artery, as it lies in it, giving rise to a pulsating tumour in the orbit.   These communications may be the result of injury, such as a bullet wound, a stab, or a blow or fall sufficiently severe to cause a fracture of the base of the skull in this situation ; or they may occur idiopathically, from the rupture of an aneurism or a diseased condition of the internal carotid artery.   The disease begins with sudden noise and pain in the head, followed by exophthalmos, swelling and congestion of the lids and conjunctivæ, and development of a pulsating tumour at the margin of the orbit, with thrill and the characteristic bruit; accompanying these symptoms there may be impairment of sight, paralysis of the iris and orbital muscles, and pain of varying intensity. In some cases the opposite orbit becomes affected by the passage of the arterial blood into the opposite sinus by means of the circular sinus.   Or the arterial blood may find its way through the emissary veins (see page 583) into the pterygoid plexus, and thence into the veins of the face.   Pulsating tumours of the orbit may also be due to traumatic aneurism of one of the orbital arteries, and symptoms resembling those of pulsating tumour may be produced by pressure on the ophthalmic vein, as it enters the sinus, by an aneurism of the internal carotid artery.

The **Ophthalmic Veins** are two in number, superior and inferior.

The *superior ophthalmic vein* connects the angular vein at the inner angle of the orbit with the cavernous sinus ; it pursues the same course as the ophthalmic artery, and receives tributaries corresponding to the branches derived from that vessel.   Forming a short single trunk, it passes through the inner extremity of the sphenoidal fissure, and terminates in the cavernous sinus.

Fɪɢ. 327.—Plan showing the relative position of the structures in the right cavernous sinus, viewed from behind.

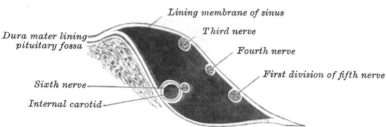

Lining membrane of sinus

Dura mater lining pituitary fossa

Third nerve

Fourth nerve

First division of fifth nerve

Sixth nerve

Internal carotid

The *inferior ophthalmic vein* receives the veins from the floor of the orbit, and either passes out of the orbit through the spheno-maxillary fissure to join the pterygoid plexus of veins ; or else, passing backwards through the sphenoidal fissure, it enters the cavernous sinus, either by a separate opening, or, more frequently, in common with the superior ophthalmic vein.

The **Circular sinus** is formed by two transverse vessels, the *anterior* and *posterior intercavernous sinuses*, which connect together the two cavernous sinuses ; the one passing in front and the other behind the pituitary body, and thus forming with the cavernous sinuses a venous circle around that body.   The anterior one is usually the larger of the two, and one or other is occasionally found to be absent.

The **Superior petrosal sinus** is situated along the superior border of the petrous portion of the temporal bone, in the front part of the attached margin of the tentorium.   It is small and narrow, and connects together the cavernous and lateral sinuses at each side.   It receives some cerebellar and inferior cerebral veins, and veins from the tympanic cavity.

The **Inferior petrosal sinus** is situated in the groove formed by the junction of the posterior border of the petrous portion of the temporal with the basilar process of the occipital.   It commences in front at the termination of the cavernous sinus, and behind joins the lateral sinus after it has passed through the jugular foramen ; the junction of these two sinuses forming the commencement of the internal jugular vein.   The inferior petrosal sinus receives the veins from the internal ear and also veins from the medulla, pons, and under surface of the cerebellum.

The junction of the two sinuses takes place at the lower border of, or just external to, the jugular foramen. The exact relation of the parts to one another in the foramen is as follows: the inferior petrosal sinus is in front, with the meningeal branch of the ascending pharyngeal artery, and is directed obliquely downwards and backwards; the lateral sinus is situated at the back part of the foramen with a meningeal branch of the occipital artery, and between the two are the glosso-pharyngeal, pneumogastric, and spinal accessory nerves. These three sets of structures are divided from each other by two processes of fibrous tissue. The junction of the sinuses takes place superficial to the nerves, so that these latter lie a little internal to the venous channels in the foramen (see fig. 329). These sinuses are semicylindrical in form.

The **Transverse sinus**, or **basilar sinus**, consists of several interlacing veins between the layers of the dura mater over the basilar process of the occipital bone, which serve to connect the two inferior petrosal sinuses. With them the anterior spinal veins communicate.

Fig. 328.—The sinuses at the base of the skull.

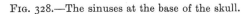
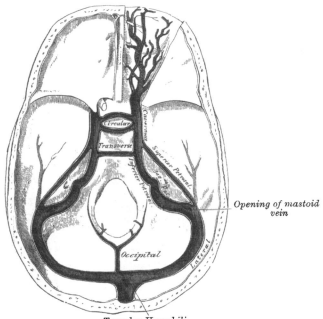

Torcular Herophili

Emissary veins.—The emissary veins are vessels which pass through apertures in the cranial wall and establish communications between the sinuses inside the skull and the veins external to it. Some of these are always present, others only occasionally so. They vary much in size in different individuals. The principal emissary veins are the following: 1. A vein, almost always present, which passes through the mastoid foramen and connects the lateral sinus with the posterior auricular or with an occipital vein. 2. A vein, which passes through the parietal foramen and connects the superior longitudinal sinus with the veins of the scalp. 3. A plexus of minute veins, which passes through the anterior condyloid foramen and connects the occipital sinus with the vertebral vein and deep veins of the neck. 4. An inconstant vein, which passes through the posterior condyloid foramen and connects the lateral sinus with the deep veins of the neck. 5. One or two veins of considerable size, which pass through the foramen ovale and connect the cavernous sinus with the pterygoid and pharyngeal plexuses. 6. Two or three small veins, which pass through the foramen lacerum medium and connect the cavernous sinus with the pterygoid and pharyngeal plexuses. 7. There is sometimes a small vein passing through the foramen of Vesalius connecting the same parts. 8. A plexus of veins passing through the carotid canal and connecting the cavernous sinus with the internal jugular vein.

Surgical Anatomy.—These emissary veins are of great importance in surgery. In addition there are, however, other communications between the intra- and extra-cranial circulation. As, for instance, the communication of the angular and supra-orbital veins with the ophthalmic vein, at the inner angle of the orbit (page 573) and the communication of the veins of the scalp with the diploic veins (page 578). Through these communications inflammatory processes commencing on the outside of the skull may travel inwards, leading to osteo-phlebitis of the diploë and inflammation of the membranes of the brain. To this in former days was to be attributed one of the principal dangers of scalp wounds and other injuries of the scalp.

FIG. 329.—Relation of structures in jugular foramen.

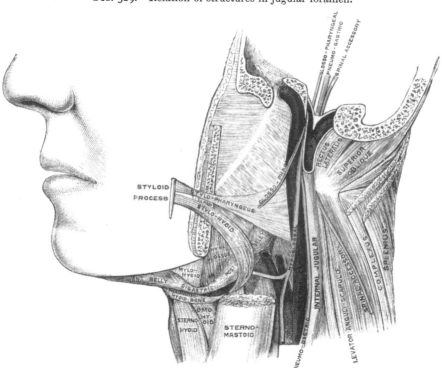

By means of these emissary veins blood may be abstracted almost directly from the intra-cranial circulation. For instance, leeches applied behind the ear abstract blood almost directly from the lateral sinus, by means of the vein passing through the mastoid foramen. Again, epistaxis in children will frequently relieve severe headache, the blood which flows from the nose being derived from the longitudinal sinus by means of the vein passing through the foramen cæcum, which is another communication between the intra-cranial and extra-cranial circulation, and is constantly found in children.

VEINS OF THE UPPER EXTREMITY AND THORAX

The Veins of the Upper Extremity are divided into two sets, *superficial* and *deep.*

The **Superficial veins** are placed immediately beneath the integument between the two layers of superficial fascia.

The **Deep veins** accompany the arteries, and constitute the venæ comites of those vessels.

Both sets of vessels are provided with valves, which are more numerous in the deep than in the superficial.

The superficial veins of the upper extremity are : the

> Superficial veins of the hand.
> Anterior ulnar.
> Posterior ulnar.
> Common ulnar.
> Radial.
> Median.
> Median cephalic.
> Median basilic.
> Basilic.
> Cephalic.

The **Superficial veins of the hand and fingers** are principally situated on the dorsal surface, and form two plexuses, an inner and outer, on the back of the hand. The inner plexus is formed by the veins from the little finger (*vena salvatella*), the ring finger, and the ulnar side of the middle finger ; from it the anterior and posterior ulnar veins are derived. The outer plexus is formed by veins from the thumb, the index finger, and radial side of the middle finger ; from it the radial vein is derived. These two plexuses communicate on the back of the hand, forming the superficial arch of veins in this situation. The superficial veins from the palm of the hand form a plexus in front of the wrist, from which the median vein is derived.

The **Anterior ulnar vein** commences on the anterior surface of the ulnar side of the hand and wrist, and ascends along the anterior surface of the ulnar side of the forearm to the bend of the elbow, where it joins with the posterior ulnar vein to form the common ulnar. Occasionally it opens separately into the median basilic vein. It communicates with branches of the median vein in front, and with the posterior ulnar behind.

The **Posterior ulnar vein** commences on the posterior surface of the ulnar side of the wrist. It runs on the posterior surface of the ulnar side of the forearm, and just below the elbow unites with the anterior ulnar vein to form the common ulnar, or else joins the median basilic to form the basilic. It communicates with the deep veins of the palm by a branch which emerges from beneath the Abductor minimi digiti muscle.

The **Common ulnar** is a short trunk which is not constant. When it exists

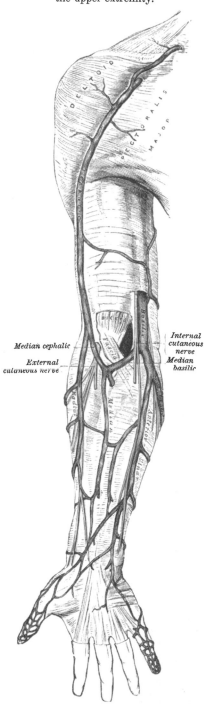

FIG. 330.—The superficial veins of the upper extremity.

Median cephalic

External cutaneous nerve

Internal cutaneous nerve

Median basilic

it is formed by the junction of the two preceding veins, and, passing upwards and outwards, joins the median basilic to form the basilic vein. When it does not exist, the anterior and posterior ulnar veins open separately into the median basilic vein.

The **Radial vein** commences from the dorsal surface of the wrist, communicating with the deep veins of the palm by a branch which passes through the first interosseous space. It forms a large vessel, which ascends along the radial side of the forearm, and receives numerous veins from both its surfaces. At the bend of the elbow it unites with the median cephalic to form the cephalic vein.

The **Median vein** ascends on the front of the forearm, and communicates with the anterior ulnar and radial veins. At the bend of the elbow it receives a branch of communication from the deep veins, and divides into two branches, the median cephalic and median basilic, which diverge from each other as they ascend.

The **Median cephalic,** usually the smaller of the two, passes outwards in the groove between the Supinator longus and Biceps muscles, and joins with the radial to form the cephalic vein. The branches of the external cutaneous nerve pass beneath this vessel.

The **Median basilic vein** passes obliquely inwards, in the groove between the Biceps and Pronator radii teres, and joins the common ulnar to form the basilic vein. This vein passes in front of the brachial artery, from which it is separated by a fibrous expansion (the *bicipital fascia*) which is given off from the tendon of the Biceps to the fascia covering the Flexor muscles of the forearm. Filaments of the internal cutaneous nerve pass in front as well as behind this vessel.*

Venesection is usually performed at the bend of the elbow, and as a matter of practice the largest vein in this situation is commonly selected. This is usually the median basilic, and there are anatomical advantages and disadvantages in selecting this vein. The advantages are, that in addition to its being the largest, and therefore yielding a greater supply of blood, it is the least movable and can be easily steadied on the bicipital fascia on which it rests. The disadvantages are, that it is in close relationship with the brachial artery, separated only by the bicipital fascia; and formerly, when venesection was frequently practised, arterio-venous aneurism was no uncommon result of this practice. Another disadvantage is, that the median basilic is crossed by some of the branches of the internal cutaneous nerve, and these may be divided in the operation, giving rise to 'traumatic neuralgia of extreme intensity' (Tillaux).

The **Basilic vein** is of considerable size, formed by the coalescence of the common ulnar vein with the median basilic. It passes upwards along the inner side of the Biceps muscle, pierces the deep fascia a little below the middle of the arm, and, ascending in the course of the brachial artery to the lower border of the tendons of the Latissimus dorsi and Teres major muscles, it is continued onwards as the axillary vein.

The **Cephalic vein** courses along the outer border of the Biceps muscle, lying in the same groove with the upper external cutaneous branch of the musculo-spiral nerve, to the upper third of the arm; it then passes in the interval between the Pectoralis major and Deltoid muscles, lying in the same groove with the descending or humeral branch of the acromial-thoracic artery. It pierces the costo-coracoid membrane, and crossing the axillary artery, it terminates in the axillary vein just below the clavicle. This vein is occasionally connected with the external jugular or subclavian, by a branch which passes from it upwards in front of the clavicle.

The **Deep veins of the upper extremity** follow the course of the arteries, forming their venæ comites. They are generally two in number, one lying on

---

* Cruveilhier says: 'Numerous varieties are observed in the disposition of the veins of the elbow; sometimes the common median vein is wanting; but in those cases, its two branches are furnished by the radial vein, and the cephalic is almost always in a rudimentary condition. In other cases, only two veins are found at the bend of the elbow, the radial and ulnar, which are continuous, without any demarcation, with the cephalic and basilic.'

each side of the corresponding artery, and they are connected at intervals by short transverse branches.

There are two digital veins, accompanying each artery along the sides of the fingers : these, uniting at their base, pass along the interosseous spaces in the palm, and terminate in the two venæ comites which accompany the superficial palmar arch. Branches from these vessels on the radial side of the hand accompany the superficialis volæ, and on the ulnar side terminate in the deep ulnar veins. The deep ulnar veins, as they pass in front of the wrist, communicate with the interosseous and superficial veins, and, at the elbow, unite with the deep radial veins to form the venæ comites of the brachial artery.

The **Interosseous veins** accompany the anterior and posterior interosseous arteries. The anterior interosseous veins commence in front of the wrist, where they communicate with the deep radial and ulnar veins ; at the upper part of the forearm they receive the posterior interosseous veins, and terminate in the venæ comites of the ulnar artery.

The **Deep palmar veins** accompany the deep palmar arch, being formed by tributaries which accompany the ramifications of that vessel. They communicate with the deep ulnar veins at the inner side of the hand, and on the outer side terminate in the venæ comites of the radial artery. At the wrist, they receive a dorsal and a palmar tributary from the thumb, and unite with the deep radial veins. Accompanying the radial artery, these vessels terminate in the venæ comites of the brachial artery.

The **Brachial veins** are placed one on each side of the brachial artery, receiving tributaries corresponding with the branches given off from that vessel ; at the lower margin of the Subscapularis, they join the axillary vein.

These deep veins have numerous anastomoses, not only with each other, but also with the superficial veins.

The **Axillary vein** is of large size, and is the continuation upwards of the basilic vein. It commences at the lower border of the tendons of the Teres major and Latissimus dorsi, increases in size as it ascends, by receiving tributaries corresponding with the branches of the axillary artery, and terminates immediately beneath the clavicle at the outer border of the first rib, where it becomes the subclavian vein. This vessel is covered in front by the Pectoral muscles and costo-coracoid membrane, and lies on the thoracic side of the axillary artery, which it partially overlaps. Near its termination it receives the cephalic vein. This vein is provided with a pair of valves, opposite the lower border of the Subscapularis muscle ; valves are also found at the termination of the cephalic and subscapular veins.

Surgical Anatomy.—There are several points of surgical interest in connection with the axillary vein. Being more superficial, larger, and slightly overlapping the axillary artery, it is more liable to be wounded in the operation of extirpation of the axillary glands, especially as these glands, when diseased, are apt to become adherent to the vessel. When wounded, there is always a danger of air being drawn into its interior, and death resulting. This is due not only to the fact that it is near the thorax and therefore liable to be influenced by the respiratory movements, but also because it is adherent by its anterior surface to the costo-coracoid membrane, and therefore if wounded is likely to remain patulous and favour the chance of air being sucked in. This adhesion of the vein to the fascia prevents its collapsing, and therefore favours the furious bleeding which takes place in these cases.

To avoid wounding the axillary vein in the extirpation of glands from the axilla, no sharp cutting instruments should be used after the axillary cavity has been freely exposed, and care should be taken to use no undue force in isolating the glands. Should the vein be so embedded in the malignant deposit that the latter cannot be removed without taking away a part of the vein, this must be done, the vessel having been first ligatured above and below.

The **Subclavian vein,** the continuation of the axillary, extends from the outer border of the first rib to the inner end of the clavicle, where it unites with the internal jugular to form the innominate vein. It is in relation, in front, with the clavicle and Subclavius muscle ; behind and above, with the subclavian artery, from which it is separated internally by the Scalenus anticus muscle and phrenic nerve. Below, it rests in a depression on the first rib and upon the pleura. Above, it is covered by the cervical fascia and integument.

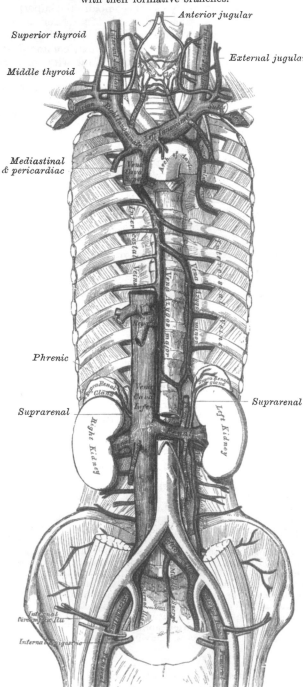

FIG. 331.—The venæ cavæ and azygos veins, with their formative branches.

Anterior jugular

Superior thyroid

External jugular

Middle thyroid

Mediastinal & pericardiac

Phrenic

Suprarenal

Suprarenal

Right Kidney

Left Kidney

The subclavian vein occasionally rises in the neck to a level with the third part of the sub-clavian artery, and in two instances has been seen passing with this vessel behind the Sca-lenus anticus. This vessel is usually pro-vided with valves about an inch from its termina-tion in the innominate, just external to the entrance of the external jugular vein.

Tributaries.—It re-ceives the external and anterior jugular veins and a small branch from the cephalic, outside the Scalenus; and on the inner side of that muscle, the internal jugular vein. At the angle of junction with the internal jugular, the left subclavian vein receives the thoracic duct; while the right subclavian vein receives the right lymphatic duct.

The **Innominate** or **Brachio-cephalic veins** (fig. 331) are two large trunks, placed one on each side of the root of the neck, and formed by the union of the internal jugular and subclavian veins of the corresponding side.

The **Right innomi-nate vein** is a short vessel, an inch in length, which commences at the inner end of the clavicle, and, passing almost vertically down-wards, joins with the left innominate vein just below the cartilage of the first rib, close to the right border of the sternum, to form the superior vena cava. It lies superficial and external to the innominate artery; on its right side is the phrenic nerve, and the pleura is here interposed between it and the apex of the lung. This vein, at the angle of junction of the internal jugular with the subclavian, receives the right vertebral vein; and, lower down, the right internal mammary, right inferior thyroid, and sometimes the right superior intercostal veins.

The **Left innominate vein,** about two and a half inches in length, and larger than the right, passes from left to right across the upper and front part of the chest, at the same time inclining downwards, to unite with its fellow of the opposite side, forming the *superior vena cava.* It is in relation, in front, with the first piece of the sternum, from which it is separated by the Sterno-hyoid and Sterno-thyroid muscles, the thymus gland or its remains, and some loose areolar tissue. Behind, it lies across the roots of the three large arteries arising from the arch of the aorta. This vessel is joined by the left vertebral, left internal mammary, left inferior thyroid, and the left superior intercostal veins, and occasionally some thymic and pericardiac veins. There are no valves in the innominate veins.

Peculiarities.—Sometimes the innominate veins open separately into the right auricle; in such cases the right vein takes the ordinary course of the superior vena cava; but the left vein—*left superior vena cava,* as it is termed—after communicating by a small branch with the right one, passes in front of the root of the left lung and, turning to the back of the heart, receives the cardiac veins, and terminates in the back of the right auricle. This occasional condition in the adult is due to the persistence of the early fœtal condition, and is the normal state of things in birds and some mammalia.

The **internal mammary veins,** two in number to each artery, follow the course of that vessel, and receive branches corresponding with those derived from it. The two veins of each side unite into a single trunk, which terminates in the innominate vein.

The **inferior thyroid veins,** two, frequently three or four, in number, arise in the venous plexus on the thyroid body, communicating with the middle and superior thyroid veins. They form a plexus in front of the trachea, behind the Sterno-thyroid muscles. From this plexus, a left vein descends and joins the left innominate trunk, and a right vein passes obliquely downwards and outwards across the innominate artery to open into the right innominate vein, just at its junction with the superior vena cava. These veins receive œsophageal, tracheal, and inferior laryngeal veins, and are provided with valves at their termination in the innominate veins.

The **Superior intercostal veins** (right and left) drain the blood from two or three intercostal spaces below the first. The *right* vein passes downwards and inwards and opens into the vena azygos major; the *left* runs across the transverse aorta and opens into the left innominate vein. It usually receives the left bronchial and left superior phrenic vein, and communicates below with the vena azygos minor superior. The *highest intercostal vein,* i.e. from the first space, opens directly into the corresponding vertebral or innominate vein.

The **Superior vena cava** receives the blood which is conveyed to the heart from the whole of the upper half of the body. It is a short trunk, varying from two inches and a half to three inches in length, formed by the junction of the two innominate veins. It commences immediately below the cartilage of the first rib close to the sternum on the right side, and, descending vertically, enters the pericardium about an inch and a half above the heart, and terminates in the upper part of the right auricle opposite the upper border of the third right costal cartilage. In its course it describes a slight curve, the convexity of which is turned to the right side.

Relations.—*In front,* with the pericardium and process of cervical fascia which is continuous with it; this separates it from the thymus gland, and from the sternum; *behind,* with the root of the right lung. On its *right side,* with the phrenic nerve and right pleura; on its *left side,* with the commencement of the innominate artery and the ascending part of the aorta. The portion contained within the pericardium is covered by the serous layer of that membrane, in its anterior three-fourths. It receives the vena azygos major just before it enters the pericardium, and several small veins from the pericardium and parts in the mediastinum. The superior vena cava has no valves.

The **Azygos veins** connect together the superior and inferior venæ cavæ, supplying the place of those vessels in the part of the chest which is occupied by the heart.

The larger, or *right azygos vein* (vena azygos major), commences opposite the first or second lumbar vertebra, by a branch from the right lumbar veins (the *ascending lumbar*); sometimes by a branch from the right renal vein, or from the inferior vena cava. It enters the thorax through the aortic opening in the Diaphragm, and passes along the right side of the vertebral column to the fourth dorsal vertebra, where it arches forward over the root of the right lung, and terminates in the superior vena cava, just before that vessel enters the pericardium. While passing through the aortic opening of the Diaphragm, it lies with the thoracic duct on the right side of the aorta; and in the thorax, it lies upon the intercostal arteries, on the right side of the aorta and thoracic duct, and is partly covered by pleura.

Tributaries.—It receives the lower ten intercostal veins of the right side, the upper two or three of these opening first of all into the right superior intercostal vein. It receives the azygos minor veins, several œsophageal, mediastinal, and pericardial veins; near its termination, the right bronchial vein; and generally the right superior intercostal vein. A few imperfect valves are found in this vein; but its tributaries are provided with complete valves.

The intercostal veins on the left side, below the three upper intercostal spaces, usually form two trunks, named the left lower, and the left upper, azygos veins.

The *left lower*, or *smaller azygos vein* (vena azygos minor), commences in the lumbar region, by a branch from one of the lumbar veins (*ascending lumbar*), or from the left renal. It passes into the thorax, through the left crus of the Diaphragm, and, ascending on the left side of the spine, as high as the ninth dorsal vertebra, passes across the column, behind the aorta and thoracic duct, to terminate in the right azygos vein. It receives the four or five lower intercostal veins of the left side, and some œsophageal and mediastinal veins.

The *left upper azygos vein* varies inversely with the size of the left superior intercostal. It receives veins from the intercostal spaces between the left superior intercostal vein and highest tributary of the left lower azygos. They are usually three or four in number, and join to form a trunk which ends in the right azygos vein, or in the left lower azygos. It sometimes receives the left bronchial vein. When this vein is small, or altogether wanting, the left superior intercostal vein will extend as low as the fifth or sixth intercostal space.

Surgical Anatomy.—In obstruction of the inferior vena cava, the azygos veins are one of the principal means by which the venous circulation is carried on, connecting as they do the superior and inferior venæ cavæ and communicating with the common iliac veins by the ascending lumbar veins and with many of the tributaries of the inferior vena cava.

The *bronchial veins* return the blood from the substance of the lungs; that of the right side opens into the vena azygos major, near its termination; that of the left side, into the left superior intercostal vein or left upper azygos vein.

THE SPINAL VEINS

The numerous venous plexuses placed upon and within the spine may be arranged into four sets:

1. Those placed on the exterior of the spinal column (*the dorsi-spinal veins*).

2. Those situated in the interior of the spinal canal, between the vertebræ and the theca vertebralis (*meningo-rachidian veins*).

3. The veins of the bodies of the vertebræ (*venæ basis vertebrarum*).

4. The veins of the spinal cord (*medulli-spinal*).

1. The **Dorsi-spinal veins** commence by small branches, which receive their blood from the integument of the back of the spine, and from the muscles in the vertebral grooves. They form a complicated network, which surrounds the spinous processes, the laminæ, and the transverse and articular processes of all the vertebræ. At the bases of the transverse processes, they communicate, by means of ascending and descending branches, with the veins surrounding the contiguous vertebræ, and they join with the veins in the spinal canal by branches which perforate the ligamenta subflava. Other branches pass obliquely forwards between the transverse processes, and communicate with the intraspinal veins through the intervertebral foramina. They terminate by joining the vertebral veins in the neck, the intercostal veins in the thorax, and the lumbar and sacral veins in the loins and pelvis.

2. The **Meningo-rachidian veins.**—The principal veins contained in the spinal canal are situated between the theca vertebralis and the vertebræ. They consist

FIG. 332.—Transverse section of a dorsal vertebra, showing the spinal veins.

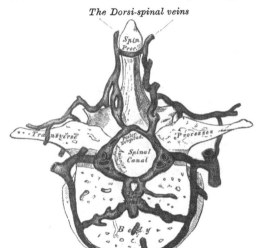

of two longitudinal plexuses, one of which runs along the posterior surface of the bodies of the vertebræ (*anterior longitudinal spinal veins*). The other plexus (*posterior longitudinal spinal veins*) is placed on the inner or anterior surface of the laminæ of the vertebræ.

The *Anterior longitudinal spinal veins* consist of two large, tortuous veins, which extend along the whole length of the vertebral column, from the foramen magnum, where they communicate by a venous ring around that opening, to the base of the coccyx, being placed one on each side of the posterior surface of the bodies of the vertebræ, along the margin of the posterior common ligament. These veins communicate together opposite each vertebra by transverse trunks, which pass beneath the ligament, and receive the large *venæ basis vertebrarum* from the interior of the body of each vertebra. The anterior longitudinal spinal veins are least developed in the cervical and sacral regions. They are not of uniform size throughout, being alternately enlarged and constricted. At the intervertebral foramina they communicate with the dorsi-spinal veins, and with the vertebral veins in the neck, with the intercostal veins in the dorsal region, and with the lumbar and sacral veins in the corresponding regions.

The *Posterior longitudinal spinal veins*, smaller than the anterior, are situated one on each side, between the inner surface of the laminæ and the theca vertebralis. They communicate (like the anterior), opposite each vertebra, by transverse trunks ; and with the anterior longitudinal veins, by lateral transverse branches, which pass from behind forwards. These veins, by branches which perforate the ligamenta subflava, join with the dorsi-spinal veins. From them branches are given off, which pass through the intervertebral foramina and join the vertebral, intercostal, lumbar, and sacral veins.

3. The **Veins of the bodies of the vertebræ** (*venæ basis vertebrarum*) emerge from the foramina on their posterior surface, and join the transverse trunk connecting the anterior longitudinal spinal veins. They are contained in large, tortuous channels in the substance of the bones, similar in every respect to those found in the diploë of the cranial bones. These canals lie parallel to the upper and lower surfaces of the bones. They commence by small openings on the front and sides of the bodies of the vertebræ, through which communicating branches from the veins external to the bone pass into its substance, and converge to the principal canal, which is sometimes double towards its posterior part,

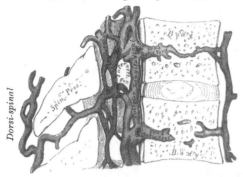

FIG. 333.—Vertical section of two dorsal vertebræ, showing the spinal veins.

and open into the corresponding transverse branch uniting the anterior longitudinal veins. They become greatly developed in advanced age.

4. The **Veins of the spinal cord** (*medulli-spinal*) consist of a minute, tortuous, venous plexus which covers the entire surface of the cord, being situated between the pia mater and arachnoid. These vessels emerge chiefly from the median furrows, and are largest in the lumbar region. Near the base of the skull they unite, and form two or three small trunks, which communicate with the vertebral veins, and then terminate in the inferior cerebellar veins, or in the inferior petrosal sinuses. Each of the spinal nerves is accompanied by a branch as far as the intervertebral foramina, where it joins the other veins from the spinal canal.

There are no valves in the spinal veins.

VEINS OF THE LOWER EXTREMITY, ABDOMEN, AND PELVIS

The Veins of the Lower Extremity are subdivided, like those of the upper, into two sets, superficial and deep : the superficial veins being placed beneath the integument, between the two layers of superficial fascia ; the deep veins accompanying the arteries, and forming the venæ comites of those vessels. Both sets of veins are provided with valves, which are more numerous in the deep than in the superficial set. These valves are also more numerous in the lower than in the upper limb.

The **Superficial veins of the lower extremity** are the *internal* or *long saphenous*, and the *external* or *short saphenous*.

On the dorsum of the foot is a venous arch, situated in the superficial structures over the anterior extremities of the metatarsal bones. It has its convexity directed forwards, and receives digital tributaries from the upper surface of the toes ; at its concavity it is joined by numerous small veins, which form a plexus on the dorsum of the foot. The arch terminates internally in the long saphenous, externally in a short saphenous vein.

The **internal** or **long saphenous vein** (fig. 334) commences at the inner side of the arch on the dorsum of the foot ; it ascends in front of the inner malleolus, and along the inner side of the leg, behind the inner margin of the tibia, accompanied by the internal saphenous nerve. At the knee, it passes backwards behind the inner condyle of the femur, ascends along the inside of the thigh, and, passing through the saphenous opening in the fascia lata, terminates in the femoral vein about an inch and a half below Poupart's ligament. This vein receives

FIG. 334.—The internal or long saphenous vein and its branches.

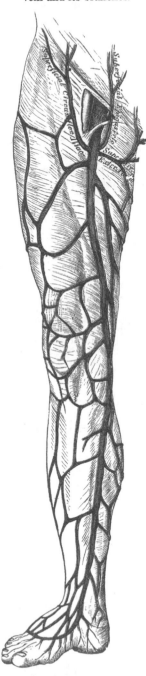

FIG. 335.
External or short saphenous vein.

in its course cutaneous tributaries from the leg and thigh, and at the saphenous opening the superficial epigastric, superficial circumflex iliac, and external pudic veins. The veins from the inner and back part of the thigh frequently unite to form a large vessel, which enters the main trunk near the saphenous opening; and sometimes those on the outer side of the thigh join to form another large vessel; so that occasionally three large veins are seen converging from different parts of the thigh towards the saphenous opening. The internal saphenous vein communicates in the foot with the internal plantar vein ; in the leg, with the posterior tibial veins, by branches which perforate the tibial origin of the Soleus muscle, and also with the anterior tibial veins ; at the knee, with the articular veins ; in the thigh, with the femoral vein by one or more branches.   The valves in this vein vary from two to six in number ; they are more numerous in the thigh than in the leg.

The **external** or **short saphenous vein** (fig. 335) commences at the outer side of the arch on the dorsum of the foot; it ascends behind the outer malleolus, and along the outer border of the tendo Achillis, across which it passes at an acute angle to reach the middle line of the posterior aspect of the leg.   Passing directly upwards, it perforates the deep fascia in the lower part of the popliteal space, and terminates in the popliteal vein, between the heads of the Gastrocnemius muscle.*   It receives numerous large tributaries from the back part of the leg, and communicates with the deep veins on the dorsum of the foot, and behind the outer malleolus.   Before it perforates the deep fascia, it gives off a communicating branch, which passes upwards and inwards to join the internal saphenous vein.   This vein has a variable number of valves, from three to nine (Gay), one of which is always found near its termination in the popliteal vein. The external saphenous nerve lies close beside this vein.

Surgical Anatomy.—The saphena veins are of considerable surgical importance, since a varicose condition of these vessels is more frequently met with than those in other parts of the body, except perhaps the spermatic and hæmorrhoidal veins.   The course of the internal saphenous is in front of the tip of the malleolus, over the subcutaneous surface of the lower end of the tibia, and then along the internal border of this bone, to the back part of the internal condyle of the femur, whence it follows the course of the Sartorius muscle and is represented on the surface by a line drawn from the posterior border of the Sartorius, on a level with the internal condyle, to the saphenous opening.   The short saphenous lies behind the external malleolus, and from this follows the middle line of the calf to just below the ham.   It is not generally so apparent beneath the skin as the internal saphenous.   Both these veins in the leg are accompanied by nerves, the internal saphenous being joined by its companion nerve just below the level of the knee-joint.   No doubt much of the pain of varicose veins in the leg is due to this fact.   On the Continent the internal saphenous vein as it rests on the tibia just above the malleolus is sometimes selected for venesection.

The **Deep veins of the lower extremity** accompany the arteries and their branches, and are called the *venæ comites* of those vessels.

The **external** and **internal plantar veins** unite to form the **posterior tibial veins**.   These latter accompany the posterior tibial artery, and are joined by the *peroneal veins*.

The **anterior tibial veins** are formed by a continuation upwards of the venæ comites of the dorsalis pedis artery.   They pass between the tibia and fibula, through the large oval aperture above the interosseous membrane, and form, by their junction with the posterior tibial, the *popliteal* vein.

The valves in the deep veins are very numerous.

The **popliteal vein** is formed by the junction of the venæ comites of the anterior and posterior tibial vessels ; it ascends through the popliteal space to the tendinous aperture in the Adductor magnus, where it becomes the femoral vein.   In the lower part of its course, it is placed internal to the artery ; between the heads of the Gastrocnemius, it is superficial to that vessel ; but above the knee-joint, it is close to its outer side.   It receives the sural veins from the Gastrocnemius muscle, the articular veins, and the external saphenous.   The valves in this vein are usually four in number.

---

* Mr. Gay calls attention to the fact that the external saphenous vein often (he says nvariably) penetrates the fascia at or about the point where the tendon of the Gastrocnemius commences, and runs below the fascia in the rest of its course, or sometimes among the muscular fibres, to join the popliteal vein.   See Gay on *Varicose Disease of the Lower Extremities*, p. 24, where there is also a careful and elaborate description of the branches of the saphena veins.

The **Femoral vein** accompanies the femoral artery through the upper two-thirds of the thigh. In the lower part of its course it lies external to the artery; higher up, it is behind it; and at Poupart's ligament, it lies to its inner side, and on the same plane. It receives numerous muscular tributaries, and about an inch and a half below Poupart's ligament it is joined by the profunda femoris: near its termination it is joined by the internal saphenous vein. The valves in this vein are four or five in number.

The **External iliac vein** commences at the termination of the femoral, beneath the crural arch, and passing upwards along the brim of the pelvis, terminates opposite the sacro-iliac synchondrosis, by uniting with the internal iliac to form the common iliac vein. On the right side, it lies at first along the inner side of the external iliac artery; but as it passes upwards, gradually inclines behind it. On the left side, it lies altogether on the inner side of the artery. It receives, immediately above Poupart's ligament, the deep epigastric and deep circumflex iliac veins and a small pubic vein, corresponding to the pubic branch of the obturator artery. According to Friedreich, it frequently contains one and sometimes two valves.

The deep epigastric veins.—Two veins accompany the deep epigastric artery; they usually unite into a single trunk before their termination in the external iliac vein.

The deep circumflex iliac veins.—Two veins accompany the deep circumflex iliac artery. These unite into a single trunk which crosses the external iliac artery just above Poupart's ligament, and terminates in the external iliac vein.

The **Internal iliac vein** is formed by the venæ comites of the branches of the internal iliac artery, the umbilical arteries of the fœtus excepted. It receives the blood from the exterior of the pelvis by the gluteal, sciatic, internal pudic, and obturator veins; and from the organs in the cavity of the pelvis by the hæmorrhoidal and vesico-prostatic plexuses in the male, and the uterine and vaginal plexuses in the female. The vessels forming these plexuses are remarkable for their large size, their frequent anastomoses, and the number of valves which they contain. The internal iliac vein lies at first on the inner side, and then behind the internal iliac artery, and terminates opposite the sacro-iliac articulation, by uniting with the external iliac to form the common iliac vein. This vessel has no valves.

The *internal pudic veins* (*venæ comites*) have the same course as the internal pudic artery. They receive tributaries corresponding to the branches of the artery except the tributary corresponding to the dorsal artery of the penis, that is, the dorsal vein of the penis, which opens into the prostatic plexus.

The *hæmorrhoidal plexus* surrounds the lower end of the rectum, being formed by the superior hæmorrhoidal veins, tributaries of the inferior mesenteric. It commences as a series of dilated pouches, about twelve in number, which are arranged circularly at the verge of the anus and are connected by transverse branches. From these pouches veins, about six in number, pass upwards in a straight direction in the submucous tissue for about three inches; they then pierce the muscular coat and become arranged in a circular manner at right angles to the long axis of the gut, and eventually unite to form the superior hæmorrhoidal vein.

Surgical Anatomy.—The veins of this plexus are apt to become dilated and varicose, and form piles. This is due to several anatomical reasons: the vessels are contained in very loose, lax connective tissue, so that they get less support from surrounding structures than most other veins, and are less capable of resisting increased blood pressure; the condition is favoured by gravitation, being influenced by the erect posture, either sitting or standing, and by the fact that the superior hæmorrhoidal and portal veins have no valves; the veins pass through muscular tissue and are liable to be compressed by its contraction, especially during the act of defæcation; they are affected by every form of portal obstruction.

The *vesico-prostatic plexus* surrounds the neck and base of the bladder and prostate gland. It communicates with the hæmorrhoidal plexus behind, and receives the dorsal vein of the penis, which enters the pelvis beneath the sub-pubic ligament. This plexus is supported upon the sides of the bladder by a reflection of the pelvic fascia. The veins composing it are very liable to become varicose, and often contain hard, earthy concretions, called *phleboliths.*

Surgical Anatomy.—This plexus, is wounded in the lateral operation of lithotomy, and it is through it that septic matter finds its way into the general circulation after this operation.

The *dorsal vein of the penis* is a vessel of large size, which returns the blood from the body of that organ. At first it consists of two branches, which are con-tained in the groove on the dorsum of the penis, and it receives veins from the glans penis, the corpus spongiosum, and numerous superficial veins ; these unite into a single trunk, which passes between the two parts of the suspensory liga-ment of the penis, and through an aperture below the subpubic ligament, and divides into two branches, which enter the prostatic plexus.

The *vaginal plexus* surrounds the vagina, being especially developed at the orifice of the canal ; it communicates with the vesical plexus in front, and with the hæmorrhoidal plexus behind.

The *uterine plexus* is situated along the sides and superior angles of the uterus, between the layers of the broad ligament, receiving, during pregnancy, large venous canals (the *uterine sinuses*) from the substance of the placenta. The veins composing this plexus anastomose frequently with each other and with the ovarian veins. They are not tortuous like the arteries.

The **Common iliac veins** are formed by the union of the external and internal iliac veins in front of the sacro-iliac articulation ; passing obliquely upwards towards the right side, they terminate upon the intervertebral substance between the fourth and fifth lumbar vertebræ, where the veins of the two sides unite at an acute angle to form the inferior vena cava. The *right common iliac* is shorter than the left, nearly vertical in its direction, and ascends behind and then to the outer side of its corresponding artery. The *left common iliac*, longer and more oblique in its course, is at first situated on the inner side of the corresponding artery, and then behind the right common iliac. Each common iliac receives the ilio-lumbar, and sometimes the lateral sacral veins. The left receives, in addition, the middle sacral vein. No valves are found in these veins.

The **middle sacral veins** accompany the corresponding artery along the front of the sacrum, and join to form a single vein which terminates in the left common iliac vein ; occasionally in the angle of junction of the two iliac veins.

Peculiarities.—The left common iliac vein, instead of joining with the right in its usual position, occasionally ascends on the left side of the aorta as high as the kidney, where, after receiving the left renal vein, it crosses over the aorta, and then joins with the right vein to form the vena cava. In these cases, the two common iliacs are connected by a small communicating branch at the spot where they are usually united.*

The **Inferior vena cava** returns to the heart the blood from all the parts below the Diaphragm. It is formed by the junction of the two common iliac veins, on the right side of the fifth lumbar vertebra. It passes upwards along the front of the spine, on the right side of the aorta, and, having reached the under surface of the liver, is contained in a groove on its posterior surface. It then perforates the central tendon of the Diaphragm, enters the pericardium, where it is covered for a very short distance by its serous layer, and terminates in the lower and back part of the right auricle. At its termination in the auricle it is provided with a valve, the *Eustachian*, which is of large size during fœtal life.

Relations.—*In front*, from below upwards, with the mesentery, right spermatic artery, transverse portion of the duodenum, the pancreas, portal vein, and the posterior surface of the liver, which partly and occasionally completely surrounds it ; *behind*, with the vertebral column, the right crus of the Diaphragm, the right

* See two cases which have been described by Mr. Walsham in the *St. Bartholomew's Hospital Reports*, vols. xvi. and xvii.

renal and lumbar arteries, right semilunar ganglion ; on the *left side*, with the aorta.

Peculiarities.—In Position.—This vessel is sometimes placed on the left side of the aorta, as high as the left renal vein, after receiving which it crosses over to its usual position on the right side ; or it may be placed altogether on the left side of the aorta, as far upwards as its termination in the heart: in such cases, the abdominal and thoracic viscera, together with the great vessels, are all transposed.

Point of Termination.—Occasionally the inferior vena cava joins the right azygos vein, which is then of large size. In such cases, the superior cava receives the whole of the blood from the body before transmitting it to the right auricle, except the blood from the hepatic veins, which passes directly into the right auricle.

Tributaries.—It receives in its course the following veins :

| Lumbar. | Renal. | Phrenic. |
| Right spermatic. | Suprarenal. | Hepatic. |

The **lumbar veins,** four in number on each side, collect the blood by dorsal tributaries from the muscles and integument of the loins, and by abdominal tributaries from the walls of the abdomen, where they communicate with the epigastric veins. At the spine, they receive veins from the spinal plexuses, and then pass forwards, round the sides of the bodies of the vertebræ beneath the Psoas magnus, and terminate at the back part of the inferior cava. The left lumbar veins are longer than the right, and pass behind the aorta. The lumbar veins are connected together by a longitudinal vein which passes in front of the transverse processes of the lumbar vertebræ, and is called the *ascending lumbar.* It forms the most frequent origin of the corresponding vena azygos, and serves to connect the common iliac, ilio-lumbar, lumbar, and azygos veins of the corresponding side of the body.

The **spermatic veins** emerge from the back of the testis, and receive tributaries from the epididymis ; they unite and form a convoluted plexus, called the *spermatic plexus (plexus pampiniformis),* which forms the chief mass of the cord ; the vessels composing this plexus are very numerous, and ascend along the cord in front of the vas deferens ; below the external abdominal ring they unite to form three or four veins, which pass along the inguinal canal, and, entering the abdomen through the internal abdominal ring, coalesce to form two veins, which ascend on the Psoas muscle, behind the peritoneum, lying one on each side of the spermatic artery ; these unite to form a single vein, which opens on the right side into the inferior vena cava, at an acute angle ; on the left side into the left renal vein, at a right angle. The spermatic veins are provided with valves.* The left spermatic vein passes behind the sigmoid flexure of the colon, and is thus exposed to pressure from the contents of that bowel.

Surgical Anatomy.—The spermatic veins are very frequently varicose, constituting the disease known as *varicocele.* Though it is quite possible that the originating cause of this affection may be a congenital abnormality either in the size or number of the veins of the pampiniform plexus, still it must be admitted that there are many anatomical reasons why these veins should become varicose: viz. the imperfect support afforded to them by the loose tissue of the scrotum ; their great length ; their vertical course ; their dependent position ; their plexiform arrangement in the scrotum, with their termination in one small vein in the abdomen ; their few and imperfect valves ; and the fact that they may be subjected to pressure in their passage through the abdominal wall.

The **ovarian veins** are analogous to the spermatic in the male; they form a plexus near the ovary and in the broad ligament and Fallopian tube, communicating with the uterine plexus. They terminate in the same way as the spermatic veins in the male. Valves are occasionally found in these veins. These vessels, like the uterine veins, become much enlarged during pregnancy.

* Rivington has pointed out that a valve is usually found at the orifices of both the right and left spermatic veins. When no valves exist at the opening of the left spermatic vein into the left renal vein, valves are generally present in the left renal vein within a quarter of an inch from the orifice of the spermatic vein. (*Journal of Anatomy and Physiology*, vol. vii. p. 163.)

The **renal veins** are of large size, and placed in front of the renal arteries.*
The left is longer than the right, and passes in front of the aorta, just below the
origin of the superior mesenteric artery. It receives the left spermatic, the left
inferior phrenic, and, generally, the left suprarenal veins. It opens into the vena
cava, a little higher than the right.

The **suprarenal veins** are two in number : that on the right side terminates
in the vena cava ; that on the left side, in the left renal or phrenic vein.

The **phrenic veins** follow the course of the phrenic arteries. The *two superior*,
of small size, accompany the phrenic nerve and comes nervi phrenici artery, and
join the internal mammary. The *two inferior phrenic veins* follow the course of
the phrenic arteries, and terminate, the right in the inferior vena cava, the left
in the left renal vein.

The **hepatic veins** commence in the substance of the liver, in the capillary
terminations of the portal vein and hepatic artery : these tributaries, gradually
uniting, usually form three large veins, which converge towards the posterior
surface of the liver, and open into the inferior vena cava, while that vessel is
situated in the groove at the back part of this organ. Of these three veins, one
from the right, and another from the left lobe, open obliquely into the inferior
vena cava ; that from the middle of the organ and lobulus Spigelii having a
straight course. The hepatic veins run singly, and are in direct contact with the
hepatic tissue. They are destitute of valves.

Portal System of Veins

The portal venous system is composed of four large veins which collect the
venous blood from the viscera of digestion (stomach, intestine, and pancreas) and
from the spleen. The trunk formed by their union (*vena portæ*) enters the liver
and ramifies throughout its substance after the manner of an artery and ends
in capillaries, from which the blood is collected into the hepatic veins, which
terminate in the inferior vena cava. The branches in this vein are in all cases
single, and destitute of valves.

The veins forming the portal system are : the

Superior mesenteric.	Inferior mesenteric.
Splenic.	Gastric.

Cystic.

The **superior mesenteric vein** returns the blood from the small intestines, and
from the cæcum and ascending and transverse portions of the colon, corresponding with the distribution of the branches of the superior mesenteric artery. The
large trunk formed by the union of these branches ascends along the right side
and in front of the corresponding artery, passes in front of the transverse portion
of the duodenum, and unites, behind the upper border of the pancreas, with the
splenic vein to form the vena portæ. It receives the right gastro-epiploic vein.

The **splenic vein** commences by five or six large branches, which return the
blood from the substance of the spleen. These unite to form a single vessel,
which passes from left to right, grooving the upper and back part of the pancreas,
below the artery, and terminates at its greater end by uniting at a right angle
with the superior mesenteric to form the vena portæ. The splenic vein is of
large size, and not tortuous like the artery. It receives the vasa brevia from the
left extremity of the stomach, the left gastro-epiploic vein, pancreatic branches
from the pancreas, the pancreatico-duodenal vein, and the inferior mesenteric vein.

The **inferior mesenteric vein** returns the blood from the rectum, sigmoid
flexure, and descending colon, corresponding with the ramifications of the

* The student may observe that all veins above the Diaphragm, which do not lie on
the same plane as the arteries which they accompany, lie in front of them : and that all
veins below the Diaphragm, which do not lie on the same plane as the arteries which they
accompany, lie behind them, except the renal and profunda femoris vein.

branches of the inferior mesenteric artery. It lies to the left of the artery, and ascends beneath the peritoneum in the lumbar region ; it passes behind the transverse portion of the duodenum and pancreas, and terminates in the splenic vein. Its hæmorrhoidal branches inosculate with those of the internal iliac, and thus establish a communication between the portal and the general venous system.*

FIG. 336.—Portal vein and its branches.

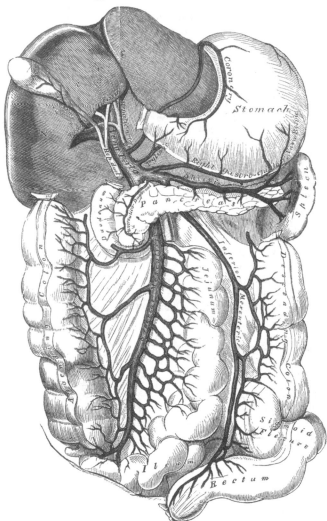

NOTE.—In this diagram the right gastro-epiploic vein opens into the splenic vein ; generally it empties itself into the superior mesenteric, close to its termination.

The **gastric veins** are two in number : one, a small vein, corresponds to the pyloric branch of the hepatic artery ; the other, considerably larger, corresponds

* Besides this anastomosis between the portal vein and the branches of the vena cava other anastomoses between the portal and systemic veins are formed by the communication between the gastric veins and the œsophageal veins which empty themselves into the vena azygos minor ; between the left renal vein and the veins of the intestines, especially of the colon and duodenum ; between the veins of the round ligament of the liver and the portal veins ; and between the superficial branches of the portal veins of the liver and the phrenic veins, as pointed out by Mr. Kiernan. (See *Physiological Anatomy*, by Todd and Bowman, 1859, vol. ii. p. 348.)

to the gastric artery. The former (*pyloric*, Walsham) runs along the lesser curvature of the stomach towards the pyloric end, receives branches from the pylorus and duodenum, and ends in the vena portæ. The latter (*coronary*, Walsham) begins near the pylorus, runs along the lesser curvature of the stomach, towards the œsophageal opening, and then passes across the front of the spine from left to right to end in the vena portæ, at a point a little above the junction of the pyloric vein.

The **Portal Vein** is formed by the junction of the superior mesenteric and splenic veins, their union taking place in front of the vena cava, and behind the upper border of the head of the pancreas. Passing upwards through the right border of the lesser omentum to the under surface of the liver, it enters the transverse fissure, where it is somewhat enlarged, forming the *sinus* of the portal vein, and divides into two branches, which accompany the ramifications of the hepatic artery and hepatic duct throughout the substance of the liver. Of these two branches the right is the larger, but the shorter, of the two. The portal vein is about three or four inches in length, and, while contained in the lesser omentum, lies behind and between the common bile duct and the hepatic artery, the former being to the right, the latter to the left. These structures are accompanied by filaments of the hepatic plexus of nerves, and numerous lymphatics, surrounded by a quantity of loose areolar tissue (*capsule of Glisson*), and placed between the layers of the lesser omentum.

The **Cystic Vein.**—The vena portæ generally receives the cystic vein, although it sometimes terminates in the right branch of the vena portæ.

The portal vein divides, in the substance of the liver, like an artery, and its minute ramifications end in capillaries, from which the blood is carried to the inferior vena cava by the hepatic veins ; these veins also collect the blood which has been brought to the liver by the hepatic artery. It will therefore be seen that the blood which is carried to the liver by the portal vein passes through two sets of capillary vessels, viz. : (1) the capillaries in the stomach, intestine, pancreas, and spleen, and (2) the capillaries of the portal vein in the liver.

CARDIAC VEINS

The veins which return the blood from the substance of the heart are : the

Great cardiac vein.   Anterior cardiac veins.
Posterior cardiac vein.   Right or small coronary vein.
Left cardiac veins.   Coronary sinus.
Venæ Thebesii.

The **Great cardiac vein** (sometimes called the *Coronary vein*) is a vessel of considerable size, which commences at the apex of the heart, and ascends along the anterior interventricular groove to the base of the ventricles. It then curves to the left side, around the auriculo-ventricular groove, between the left auricle and ventricle, to the back part of the heart, and opens into the left extremity of the coronary sinus, its aperture being guarded by two valves. It receives, in its course, tributaries from both ventricles, but especially the left, and also from the left auricle ; one of these, ascending along the thick margin of the left ventricle, is of considerable size. The vessels joining it are provided with valves.

The **Posterior cardiac vein** (sometimes called the *Middle cardiac vein*) commences by small tributaries, at the apex of the heart, communicating with those of the preceding. It ascends along the posterior interventricular groove to the base of the heart, and terminates in the coronary sinus, its orifice being guarded by a valve. It receives the veins from the posterior surface of both ventricles.

The **Left cardiac veins** are three or four small vessels, which collect the blood from the posterior surface of the left ventricle, and open into the lower border of the coronary sinus.

The **Anterior cardiac veins** are three or four small vessels, which collect the blood from the anterior surface of the right ventricle. One of these (the *vein of Galen*), larger than the rest, runs along the right border of the heart. They open separately into the lower part of the right auricle.

The **Right** or **small coronary vein** runs along the groove between the right auricle and ventricle, to open into the right extremity of the coronary sinus. It receives blood from the back part of the right auricle and ventricle.

The **Coronary sinus** is that portion of the anterior or great cardiac vein which is situated in the posterior part of the left auriculo-ventricular groove. It is about an inch in length, presents a considerable dilatation, and is covered by the muscular fibres of the left auricle. It receives the veins enumerated above, and an *oblique vein* from the back part of the left auricle, the remnant of the obliterated left Cuvierian duct of the fœtus, described by Mr. Marshall. The great coronary sinus terminates in the right auricle, between the inferior vena cava and the auriculo-ventricular aperture, its orifice being guarded by a semilunar fold of the lining membrane of the heart, the *Thebesian valve*. All the veins joining this vessel, excepting the oblique vein above mentioned, are provided with valves.

The **Venæ Thebesii** (*venæ cordis minimæ*) are numerous minute veins, which return the blood directly from the muscular substance, without entering the venous current. They open by minute orifices (*foramina Thebesii*) on the inner surface of the right auricle.

THE LYMPHATIC SYSTEM

THE Lymphatic system includes not only the lymphatic vessels and the glands through which they pass, but also the *lacteal* or *chyliferous vessels*. The lacteals are the lymphatic vessels of the small intestine, and differ in no respect from the lymphatics generally, excepting that they contain a milk-white fluid, the *chyle*, during the process of digestion, and convey it into the blood through the thoracic duct.

The **lymphatics** have derived their name from the appearance of the fluid contained in their interior (*lympha*, water). They are also called *absorbents*, from the property they possess of absorbing certain materials from the tissues and conveying them into the circulation.

The lymphatics are exceedingly delicate vessels, the coats of which are so transparent that the fluid they contain is readily seen through them. They retain a nearly uniform size, being interrupted at intervals by constrictions, which give them a knotted or beaded appearance. These constrictions are due to the presence of valves in their interior. Lymphatics have been found in nearly every texture and organ of the body which contains blood-vessels. Such non-vascular structures as cartilage, the nails, cuticle, and hair have none, but with these exceptions it is probable that eventually all parts will be found to be permeated by these vessels.

The lymphatics are arranged into a superficial and a deep set. The *superficial* lymphatics, on the surface of the body, are placed immediately beneath the integument, accompanying the superficial veins; they join the deep lymphatics in certain situations by perforating the deep fascia. In the interior of the body they lie in the submucous areolar tissue, throughout the whole length of the gastro-pulmonary and genito-urinary tracts; and in the subserous tissue in the cranial, thoracic, and abdominal cavities. The method of their origin has been described along with the other details of their minute anatomy. Here it will be sufficient to say that a plexiform network of minute lymphatics may be found interspersed among the proper elements and blood-vessels of the several tissues; the vessels composing which, as well as the meshes between them, are much larger than those of the capillary plexus. From these networks small vessels emerge, which pass, either to a neighbouring gland, or to join some larger lymphatic trunk. The *deep* lymphatics, fewer in number, and larger than the superficial, accompany the deep blood-vessels. Their mode of origin is probably similar to that of the superficial vessels. The lymphatics of any part or organ exceed the veins in number, but in size they are much smaller. Their anastomoses also, especially those of the large trunks, are more frequent, and are effected by vessels equal in diameter to those which they connect, the continuous trunks retaining the same diameter.

The **lymphatic** or **absorbent glands,** named also *conglobate glands*, are small, solid, glandular bodies, situated in the course of the lymphatic and lacteal vessels. In size they vary from a hemp-seed to an almond, and their colour, on section, is of a pinkish-grey tint, excepting the bronchial glands, which in the adult are mottled with black. Each gland has a layer or capsule of cellular tissue investing it, from which prolongations dip into its substance, forming partitions. The lymphatic and lacteal vessels pass through these bodies in their passage to the thoracic and lymphatic ducts. A lymphatic or lacteal vessel, previous to entering a gland, divides into several small branches, which are named *afferent vessels*. As they enter, their external coat becomes continuous with the capsule of the gland, and the vessels, much thinned, and consisting only of their internal or endothelial coat, pass into the gland, and branch out upon and in the tissue of the capsule; these branches opening into the lymph sinuses of the gland. From these sinuses fine branches proceed to form a plexus, the vessels of which unite to form a single *efferent vessel*, which, on emerging from the gland, is again invested with an external coat. Further details on the minute anatomy of the lymphatic vessels and glands will be found in the section on General Anatomy.

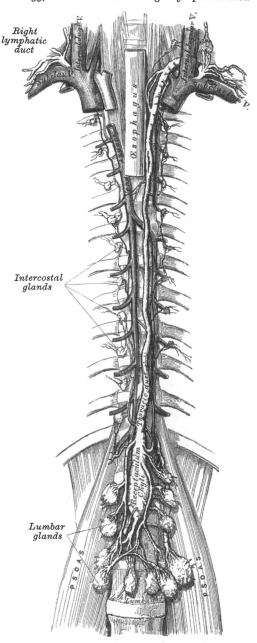

Fig. 337.—The thoracic and right lymphatic duct.

THORACIC DUCT

The **thoracic duct** (fig. 337) conveys the great mass of the lymph and chyle into the blood. It is the common trunk of all the lymphatic vessels of the body, excepting those of the right side of the head, neck, and thorax, and right upper extremity, the right lung, right side of the heart, and the convex surface of the liver. It varies in length from fifteen to eighteen inches in the adult, and extends from the second lumbar vertebra to the root of the neck. It commences in the abdomen by a triangular dilatation, the *receptaculum chyli* (*reservoir* or *cistern of Pecquet*), which is situated upon the front of the body of the second lumbar vertebra, to the right side and behind the aorta, by the side of the right crus of the Diaphragm. It ascends into the thorax through the aortic opening in the Diaphragm, lying to the right of the aorta, and is placed in the posterior mediastinum in front of the vertebral column, lying between the aorta and vena azygos major. Opposite the fourth dorsal vertebra, it inclines towards the left side and ascends behind the arch of the aorta, on the left side of the œsophagus, and behind the first portion of the left subclavian artery to the upper orifice of the thorax. Opposite the seventh cervical vertebra, it turns outwards, and then curves downwards over the subclavian artery, and in front of the Scalenus anticus muscle, so as to form an arch; and terminates in the left subclavian vein, at its angle of junction with the left internal jugular vein. The thoracic duct, at its commencement, is about equal in size to the diameter of a goose-quill, diminishes considerably in its calibre in the middle of the thorax, and is again dilated just before its termination. It is generally flexuous in its course, and constricted at intervals so as to present a varicose appearance. The thoracic duct not infrequently divides in the middle of its course into two branches of unequal size, which soon reunite, or into several branches which form a plexiform interlacement. It occasionally divides, at its upper part, into two branches, of which the one on the left side terminates in the usual manner, while that on the right opens into the right subclavian vein, in connection with the right lymphatic duct. The thoracic duct has numerous valves throughout its whole course, but they are more numerous in the upper than in the lower part : at its termination it is provided with a pair of valves, the free borders of which are turned towards the vein, so as to prevent the passage of venous blood into the duct.

Tributaries.—The thoracic duct, at its commencement, receives four or five large trunks from the abdominal lymphatic glands, and also the trunk of the lacteal vessels. Within the thorax, it is joined by the lymphatic vessels from the left half of the wall of the thoracic cavity, the lymphatics from the sternal and intercostal glands, those of the left lung, left side of the heart, trachea, and œsophagus; and, just before its termination, it receives the lymphatics of the left side of the head and neck, and left upper extremity.

Structure.—The thoracic duct is composed of three coats, which differ in some respects from those of the lymphatic vessels. The *internal coat* consists of a single layer of flattened lanceolate-shaped endothelial cells, with serrated borders ; of a subendothelial layer similar to that found in the arteries ; and an elastic fibrous coat, the fibres of which run in a longitudinal direction. The *middle coat* consists of a longitudinal layer of white connective tissue with elastic fibres, external to which are several laminæ of muscular tissue, the fibres of which are for the most part disposed transversely, but some are oblique or longitudinal, and intermixed with elastic fibres. The *external coat* is composed of areolar tissue, with elastic fibres and isolated fasciculi of muscular fibres.

The **Right Lymphatic Duct** is a short trunk, about half an inch in length, and a line or a line and a half in diameter. It terminates in the right subclavian vein, at its angle of junction with the right internal jugular vein. Its orifice is guarded by two semilunar valves, which prevent the passage of venous blood into the duct.

Tributaries.—It receives the lymph from the right side of the head and neck, the right upper extremity, the right side of the thorax, the right lung and right side of the heart, and from part of the convex surface of the liver.

LYMPHATICS OF THE HEAD, FACE, AND NECK

The **lymphatic glands of the head** (fig. 338) are arranged in the following groups : (1) The *occipital*, one or two in number, placed at the back of the head close to the occipital artery. (2) The *posterior auricular* or *mastoid*, usually two in number, situated on the insertion of the Sterno-mastoid to the mastoid process. Both these sets of glands are affected in cutaneous eruptions and other diseases of the scalp. (3) The *parotid* or *pre-auricular*, some of which are superficial to, and others are embedded in, the substance of the parotid gland. (4) The *buccal*, one or more, placed on the surface of the Buccinator muscle. (5) The *internal maxillary*, beneath the ramus of the jaw. (6) The *lingual*, two or three in number, lying on the Hyo-glossus and Genio-hyo-glossus.   (7) The *retro-pharyngeal*, lying one on each side of the middle line in front of the Rectus capitis anticus major.

The **lymphatic vessels of the scalp** are divided into an *anterior* and a *posterior set*, which follow the course of the temporal and occipital vessels.   The *temporal* accompany the temporal artery in front of the ear, to the parotid lymphatic glands, from which they proceed to the lymphatic glands of the neck.   The *occipital* follow the course of the occipital artery, descend to the occipital and posterior auricular lymphatic glands, and finally join the cervical glands.

The **lymphatic vessels of the face** are divided into two sets, *superficial* and *deep*.

FIG. 338.—The superficial lymphatics and glands of the head, face, and neck.

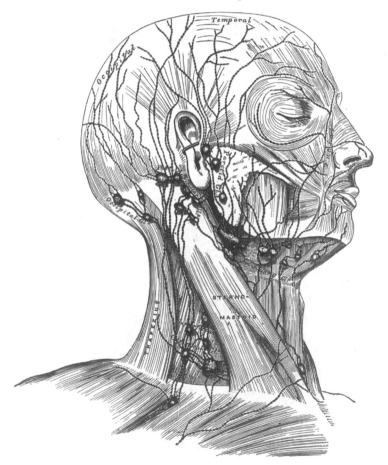

The **superficial lymphatic vessels of the face** are more numerous than those of the head, and commence over its entire surface.   Those from the frontal region accompany the frontal vessels ; they then pass obliquely across the face, running with the facial vein, pass through the buccal glands on the surface of the Bucci-nator muscle, and join the submaxillary lymphatic glands.   The latter receive the lymphatic vessels from the lips, and are often found enlarged in cases of malignant disease of those parts.

The **lymphatic vessels of the cranium** consist of two sets, the *meningeal* and *cerebral*.   The *meningeal lymphatics* accompany the meningeal vessels, escape through foramina at the base of the skull, and join the deep cervical lymphatic glands.   The *cerebral lymphatics* are described by Eshmann as being situated between the arachnoid and pia mater, as well as in the choroid plexuses of the lateral ventricles; they accompany the trunks of the carotid and vertebral arteries, and probably pass through foramina at the base of the skull, to terminate in the deep cervical glands. They have not at present been demonstrated in the dura mater, or in the substance of the brain.

The lymphatics of the orbit and of the temporal and zygomatic fossæ run with the branches of the internal maxillary artery to the maxillary glands, and afterwards to the deep cervical.

FIG. 339.—The deep lymphatics and glands of the neck and thorax.

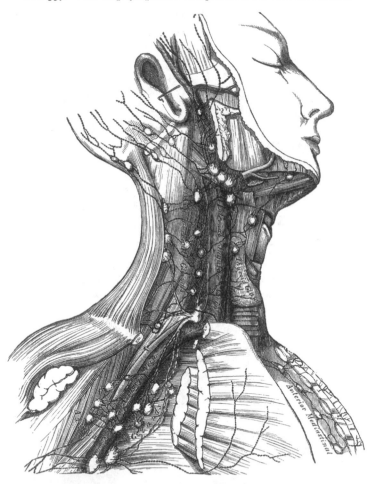

The lymphatics of the nose can be injected from the subdural and subarachnoid spaces. They terminate in the retro-pharyngeal and supra-hyoid glands. The lymphatics of the tongue chiefly accompany the ranine vein first to the lingual glands and from these to the deep cervical. Those from the anterior part of the tongue and floor of the mouth pierce the Mylo-hyoid muscles and so reach the submaxillary glands. From the upper part of the pharynx the lymphatics pass to the retro-pharyngeal glands, from the lower part to the deep cervical glands. From the larynx two sets of vessels arise: an upper, piercing the thyro-hyoid membrane and joining the superior set of deep glands; and a lower, perforating the crico-thyroid membrane to join the lower set of deep cervical glands. The lymphatics of the thyroid body accompany the superior and inferior thyroid arteries, and open partly into the upper and partly into the lower set of deep cervical glands.

The **Lymphatic glands of the neck** are divided into two sets, *superficial* and *deep*.

The **superficial cervical glands** may be arranged in three sets : (1) the *submaxillary*, eight to ten in number, situated beneath the body of the lower jaw in the submaxillary triangle ; (2) *supra-hyoid*, one or two in number, situated in the middle line of the neck, between the anterior bellies of the two Digastric muscles ; and (3) *cervical*, placed in the course of the external jugular vein between the Platysma and deep fascia. They are most numerous at the root of the neck, in the triangular interval between the clavicle, the Sterno-mastoid, and the Trapezius, where they are continuous with the axillary glands. A few small glands are also found on the front and sides of the larynx.

The **deep cervical glands** (fig. 339) are numerous and of large size ; they form a chain along the sheath of the carotid artery and internal jugular vein, lying by the side of the pharynx, œsophagus, and trachea, and extending from the base of the skull to the thorax, where they communicate with the lymphatic glands in that cavity. They are subdivided into two sets : an *upper*, ten to twenty in number, situated about the bifurcation of the common carotid and along the upper part of the internal jugular vein ; and a *lower*, ten to fifteen in number, clustered around the lower part of the internal jugular vein, and extending outwards into the supra-clavicular fossa, where they are continuous with the axillary glands. Internally, this set is continuous with the mediastinal glands.

The **superficial** and **deep cervical lymphatic vessels** are a continuation of those already described on the cranium and face. After traversing the glands in those regions, they pass through the chain of glands which lie along the sheath of the carotid vessels, being joined by the lymphatics from the pharynx, œsophagus, larynx, trachea, and thyroid gland. At the lower part of the neck, after receiving some lymphatics from the thorax, they unite into a single trunk, which terminates, on the left side, in the thoracic duct ; on the right side, in the right lymphatic duct.

Surgical Anatomy.—The cervical glands are very frequently the seat of tuberculous disease. This condition is most usually set up by some lesion in those parts from which they receive their lymph. This excites inflammation, which subsequently takes on a tuberculous character. It is very desirable therefore for the surgeon, in dealing with these cases, to possess a knowledge of the relation of the respective groups of glands to the periphery. The following table is extracted from Mr. Treves's work on ' Scrofula and its Gland Diseases.'

Scalp.—Posterior part = suboccipital and mastoid glands. Frontal and parietal portions = parotid glands.

Lymphatic vessels from the scalp also enter the superficial cervical set of glands.

Skin of face and neck = submaxillary, parotid, and superficial cervical glands.

External ear = superficial cervical glands.

Lower lip = submaxillary and supra-hyoid glands.

Buccal cavity = submaxillary and upper set of deep cervical glands.

Gums of lower jaw = submaxillary glands.

Tongue.—Anterior portion = supra-hyoid and submaxillary glands. Posterior portion = upper set of deep cervical glands.

Tonsils and palate = upper set of deep cervical glands.

Pharynx.—Upper part = parotid and retro-pharyngeal glands. Lower part = upper set of deep cervical glands.

Larynx, orbit, and roof of mouth = upper set of deep cervical glands.

Nasal fossæ = retro-pharyngeal glands, upper set of deep cervical glands. Some lymphatic vessels from posterior part of the fossæ enter the parotid glands.

LYMPHATICS OF THE UPPER EXTREMITY

The **Lymphatic Glands of the Upper Extremity** (fig. 340) are divided into two sets, *superficial* and *deep*.

The **superficial lymphatic glands** are few and of small size. There are occasionally two or three in front of the elbow, and one or two above the internal condyle of the humerus, near the basilic vein, while one or two may be found lying beside the cephalic vein between the Pectoralis major and Deltoid muscles.

The **deep lymphatic glands** are few in number and are subdivided into those in the forearm, the arm, and the axilla. In the *forearm*, a few small ones are occasionally found in the course of the radial and ulnar vessels. In the *arm*, there is a chain of small glands along the inner side of the brachial artery. One, sometimes two, fairly constant glands are situated a little above and in front of the inner condyle of the humerus. In the *axilla* they are of large size, and

FIG. 340.—The superficial lymphatics and glands of the upper extremity.

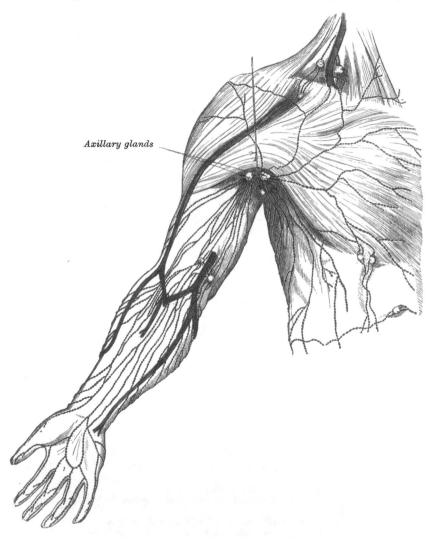

Axillary glands

usually ten or twelve in number. A chain of these glands surrounds the axillary vessels, embedded in a quantity of loose areolar tissue; they receive the lymphatic vessels from the arm; others are dispersed in the areolar tissue of the axilla: the remainder are arranged in two series, a small chain running along the lower border of the Pectoralis major, receiving the lymphatics from the front of the chest and mamma; and others are placed along the lower margin of the posterior wall of the axilla, which receive the lymphatics from the integument of the back. Two or three subclavian or infra-clavicular lymphatic glands are placed immediately beneath the clavicle; it is through these that the axillary and deep cervical glands communicate with each other. The efferent vessels from the axillary glands may be from one to three or four in number. They accompany the subclavian vein into the neck, and end, on the right side, by joining the right lymphatic duct, on the left side by opening into the thoracic duct.

Surgical Anatomy.—In malignant diseases, tumours, or other affections implicating the upper part of the back and shoulder, the front of the chest and mamma, the upper part of the front and side of the abdomen, or the hand, forearm, and arm, the axillary glands are liable to be found enlarged.

The **Lymphatic vessels of the upper extremity** are divided into two sets, *superficial* and *deep*.

The **superficial lymphatic vessels of the upper extremity** commence on the fingers; two vessels running along either side of each finger, one on the palmar and the other on the dorsal surface. Those on the palmar surface form an arch in the palm of the hand, from which are derived two sets of vessels, which pass up the forearm, taking the course of the subcutaneous veins. The lymphatics from the dorsal surface of the fingers form a plexus on the back of the hand, and, winding around the inner and outer borders of the forearm, unite with those in front. Those from the inner border of the hand accompany the ulnar veins along the inner side of the forearm to the bend of the elbow, where they are joined by some lymphatics from the outer side of the forearm; they then follow the course of the basilic vein, communicate with the glands immediately above the elbow, and terminate in the axillary glands, joining with the deep lymphatics. The superficial lymphatics from the outer and back part of the hand accompany the radial veins to the bend of the elbow. They are less numerous than the preceding. At the bend of the elbow, the greater number join the basilic group; the rest ascend with the cephalic vein on the outer side of the arm, some crossing the upper part of the Biceps obliquely, to terminate in the axillary glands, while one or two accompany the cephalic vein in the cellular interval between the Pectoralis major and Deltoid, and enter the subclavian lymphatic glands.

The **deep lymphatic vessels of the upper extremity** accompany the deep blood-vessels. In the forearm they consist of four sets, corresponding with the radial, ulnar, and interosseous arteries; they pass through the glands occasionally found in the course of those vessels, and communicate at intervals with the superficial lymphatics. In their course upwards, some of them pass through the glands which lie upon the brachial artery; they then enter the axillary and subclavian glands, and at the root of the neck terminate, on the left side, in the thoracic duct, and on the right side in the right lymphatic duct.

LYMPHATICS OF THE LOWER EXTREMITY

The **Lymphatic glands of the lower extremity** are divided into two sets, *superficial* and *deep*. The superficial are confined to the inguinal region, forming the *superficial inguinal lymphatic glands*.

The **superficial inguinal lymphatic glands,** placed immediately beneath the integument, are of large size, and vary from eight to ten in number. They are divisible into two groups: an upper *oblique* set, disposed irregularly along Poupart's ligament, which receive the lymphatic vessels from the integument of the scrotum, penis, parietes of the abdomen, perineal and gluteal regions, and the mucous membrane of the urethra; and an inferior *vertical* set, two to five in number, which surround the saphenous opening in the fascia lata; a few being sometimes continued along the saphenous vein to a variable extent. This latter group receive the superficial lymphatic vessels from the lower extremity. Leaf * figures some of the efferent vessels from these glands as terminating directly in the veins of this region.

* *The Surgical Anatomy of the Lymphatic Glands,* 1898.

FIG. 341.—The superficial lymphatics and glands of the lower extremity.

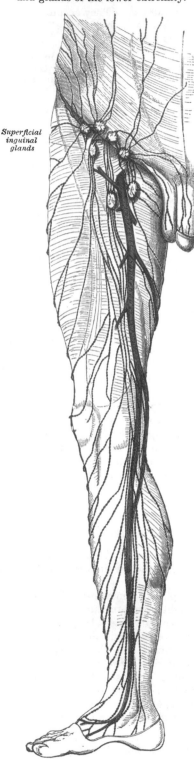

Superficial inguinal glands

Surgical Anatomy.—These glands frequently become enlarged in diseases implicating the parts from which their lymphatics originate. Thus in malignant or syphilitic affections of the prepuce and penis, or of the labia majora in the female, in cancer scroti, in abscess in the perinæum, or in any other diseases affecting the integument and superficial structures in those parts, or the sub-umbilical part of the abdominal wall, or the gluteal region, the upper chain of glands is almost invariably enlarged, the lower chain being implicated in diseases affecting the lower limb.

The **deep lymphatic glands** are: the anterior tibial, popliteal, deep inguinal, gluteal, and ischiatic.

The **anterior tibial gland** is not constant in its existence. It is generally found by the side of the anterior tibial artery, upon the interosseous membrane at the upper part of the leg. Occasionally, two glands are found in this situation.

The **popliteal glands,** four or five in number, are of small size; they surround the popliteal vessels, embedded in the cellular tissue and fat of the popliteal space.

The **deep inguinal glands** are placed beneath the deep fascia around the femoral artery and vein. They are of small size, and communicate with the superficial inguinal glands through the saphenous opening.

The **gluteal** and **ischiatic glands** are placed, the former above, the latter below, the Pyriformis muscle, resting on their corresponding vessels as they pass through the great sacro-sciatic foramen.

The **Lymphatic vessels of the lower extremity,** like the veins, may be divided into two sets, *superficial* and *deep*.

The **superficial lymphatic vessels** are placed beneath the integument in the superficial fascia, and are divisible into two groups: an internal group, which follow the course of the internal saphenous vein; and an external group, which accompany the external saphenous. The *internal group*, the larger, commence on the inner side and dorsum of the foot; they pass, some in front, and some behind the inner ankle, run up the leg with the internal saphenous vein, pass with it behind the inner condyle of the femur, and accompany it to the groin, where they terminate in the group of superficial inguinal lymphatic glands which surround the saphenous opening.   Some of the efferent vessels from these glands pierce the cribriform fascia and sheath of the femoral vessels, and terminate in a lymphatic gland contained in the femoral canal, thus establishing a communication between the lymphatics of the lower extremity and those of the trunk ; others pierce the fascia lata, and join the deep inguinal glands.   The *external group* arise from the outer side of the foot, ascend in front of the leg, and, just below the knee, cross the tibia from without inwards, to join the lymphatics on the inner side of the thigh.   Others commence on the outer side of the foot, pass behind the outer malleolus, and accompany the external saphenous vein along the back of the leg, where they enter the popliteal glands.

The **deep lymphatic vessels of the lower extremity** are few in number, and accompany the deep blood-vessels.   In the leg, they consist of three sets, the anterior tibial, peroneal, and posterior tibial, which accompany the corresponding blood-vessels, two or three to each artery ; they ascend with the blood-vessels, and enter the lymphatic glands in the popliteal space ; the efferent vessels from these glands accompany the femoral vein, and join the deep inguinal glands ; from these, the vessels pass beneath Poupart's ligament, and communicate with the chain of glands surrounding the external iliac vessels.

The deep lymphatic vessels of the gluteal and ischiatic regions follow the course of the blood-vessels, and join the gluteal and ischiatic glands at the great sacro-sciatic foramen.

Lymphatics of the Pelvis and Abdomen

The **Lymphatic glands in the pelvis** are : the external iliac, the internal iliac, and the sacral.   Those of the abdomen are the lumbar and cœliac glands.

The **external iliac glands** form an uninterrupted chain round the external iliac vessels, three being placed round the commencement of the vessels just behind the crural arch.   They communicate below with the deep inguinal lymphatic glands, and above with the lumbar glands.

The **internal iliac glands** surround the internal iliac vessels ; they receive the lymphatic vessels corresponding to the branches of the internal iliac artery, and communicate with the lumbar glands.

The **sacral glands** occupy the sides of the anterior surface of the sacrum, some being situated in the meso-rectal fold.   These and the internal iliac glands are affected in malignant disease of the bladder, rectum, or uterus.

The **lumbar glands** are very numerous ; they are situated on the front of the lumbar vertebræ, surrounding the common iliac vessels, the aorta, and vena cava ; they receive the lymphatic vessels from the lower extremities and pelvis, as well as from the testes and some of the abdominal viscera ; the efferent vessels from these glands unite into a few large trunks, which, with the lacteals, form the commencement of the thoracic duct.   In addition to these there are a few small *lateral lumbar glands*, which lie between the transverse processes of the vertebræ, behind the Psoas muscle, and receive lymphatics from the back.   In some cases of malignant disease these glands become enormously enlarged, completely surrounding the aorta and vena cava, and occasionally greatly contracting the calibre of those vessels.   In all cases of malignant disease of the testis, and in malignant disease of the lower limb, before any operation is attempted, careful examination of the abdomen should be made, in order to ascertain if any enlargement exists ; and if any should be detected, all operative measures should be avoided as fruitless.

The **cœliac glands,** nearly twenty in number, surround the cœliac axis and lie in front of the aorta near the origin of that vessel.   They receive the lymphatic vessels from a large part of the liver, from the spleen, pancreas, and stomach.   Their efferent vessels join the lacteals from the intestine and open into the receptaculum chyli.

The **Lymphatic vessels of the abdomen and pelvis** may be divided into two sets, *superficial* and *deep*.

The **superficial lymphatic vessels of the walls of the abdomen and pelvis** follow the course of the superficial blood-vessels. Those derived from the integument of the lower part of the abdomen below the umbilicus follow the course of the superficial epigastric vessels and converge to the superior group of

FIG. 342.—The deep lymphatic vessels and glands of the abdomen and pelvis.

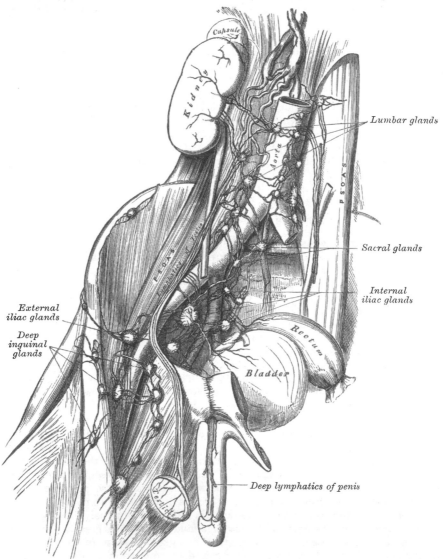

the superficial inguinal glands; a deeper set accompany the deep epigastric vessels, and communicate with the external iliac glands. The superficial lymphatics from the sides of the lumbar part of the abdominal wall wind round the crest of the ilium, accompanying the superficial circumflex iliac vessels, to join the superior group of the superficial inguinal glands; the greater number, however, run backwards along with the ilio-lumbar and lumbar vessels, to join the lateral lumbar glands.

The **superficial lymphatic vessels of the gluteal region** turn horizontally round the outer side of the nates, and join the superficial inguinal glands.

The **superficial lymphatic vessels of the scrotum and perinæum** follow the course of the external pudic vessels, and terminate in the superficial inguinal glands.

The **superficial lymphatic vessels of the penis** occupy the sides and dorsum of the organ, the latter receiving the lymphatics from the skin covering the glans penis; they all converge to the upper chain of the superficial inguinal glands. The **deep lymphatic vessels of the penis** follow the course of the internal pudic vessels, and join the internal iliac glands.

In the female, the lymphatic vessels of the mucous membrane of the labia, nymphæ, and clitoris terminate in the upper chain of the inguinal glands.

The **deep lymphatic vessels of the abdomen and pelvis** take the course of the principal blood-vessels. Those of the parietes of the pelvis, which accompany the gluteal, ischiatic, and obturator vessels, follow the course of the internal iliac artery, and ultimately join the lumbar lymphatics.

The efferent vessels from the inguinal glands enter the pelvis beneath Poupart's ligament, where they lie in close relation with the femoral vein; they then pass through the chain of glands surrounding the external iliac vessels, and finally terminate in the lumbar glands. They receive the deep epigastric and circumflex iliac lymphatics.

The **lymphatic vessels of the bladder** arise from the entire surface of the organ; * the greater number run beneath the peritoneum on its posterior surface, and, after passing through the lymphatic glands in that situation, join with the lymphatics from the prostate and vesiculæ seminales, and enter the internal iliac glands.

The **lymphatic vessels of the rectum** are of large size; after passing through some small glands that lie upon its outer wall and in the meso-rectum, they pass to the sacral glands.

The **lymphatic vessels of the uterus** consist of two sets, *superficial* and *deep*; the former being placed beneath the peritoneum, the latter in the substance of the organ. The lymphatics of the cervix uteri, together with those from the greater part of the vagina, enter the internal iliac and sacral glands; those from the body and fundus of the uterus pass outwards in the broad ligaments, and, being joined by the lymphatics from the ovaries, broad ligaments, and Fallopian tubes, ascend with the ovarian vessels to open into the lumbar glands; the lymphatics from the lower part of the vagina join those of the external genitals and pass to the superficial inguinal glands. In the unimpregnated uterus they are small; but during gestation they become very greatly enlarged.

The **lymphatic vessels of the testicle** consist of two sets, *superficial* and *deep*; the former commence on the surface of the tunica vaginalis, the latter in the epididymis and body of the testis. They form several large trunks, which ascend with the spermatic cord, and, accompanying the spermatic vessels into the abdomen, terminate in the lumbar glands; hence |the enlargement of these glands in malignant disease of the testis.

The **lymphatic vessels of the kidney** arise on the surface, and also in the interior of the organ; they join at the hilum, and, after receiving the lymphatic vessels from the ureter and suprarenal capsules, open into the lumbar glands.

The **lymphatic vessels of the liver** are divisible into two sets, *superficial* and *deep*. The former arise in the subperitoneal areolar tissue over the entire surface of the organ. Those on the convex surface may be divided into four groups: 1. Those which pass from behind forwards, consisting of three or four branches, which ascend in the falciform ligament, and unite to form a single trunk, which passes up between the fibres of the Diaphragm, behind the ensiform cartilage, to enter the anterior mediastinal glands, and finally ascends to the root of

---

* Curnow states that they are confined to the base of the organ.

the neck, to terminate in the right lymphatic duct.   2. Another group, which also incline from behind forwards, are reflected over the anterior margin of the liver, to its under surface, and thence pass along the longitudinal fissure to the glands in the gastro-hepatic omentum.   3. A third group incline outwards to the right lateral ligament, and, uniting into one or two large trunks, pierce the Diaphragm, and run along its upper surface to enter the anterior mediastinal glands ; or, instead of entering the thorax, turn inwards across the crus of the Diaphragm, and open into the commencement of the thoracic duct.   4. The fourth group incline outwards from the surface of the left lobe of the liver to the left lateral ligament, pierce the Diaphragm, and, passing forwards, terminate in the glands in the anterior mediastinum.

The *superficial lymphatics on the under surface of the liver* are divided into three sets : 1. Those on the right side of the gall-bladder enter the lumbar glands. 2. Those surrounding the gall-bladder form a remarkable plexus ; they accompany the hepatic vessels, and open into the glands in the gastro-hepatic omentum. 3. Those on the left of the gall-bladder pass to the œsophageal glands, and to the glands which are situated along the lesser curvature of the stomach.

The *deep lymphatics* accompany the branches of the portal vein and the hepatic artery and duct through the substance of the liver ; passing out at the transverse fissure, they enter the lymphatic glands along the lesser curvature of the stomach and behind the pancreas, or join with one of the lacteal vessels previous to its termination in the thoracic duct.

The **lymphatic glands of the stomach** are of small size ; they are placed along the upper part of the lesser and towards the pyloric end of the greater curvature.

The **lymphatic vessels of the stomach** consist of two sets, *superficial* and *deep* ; the former originating in the subserous, and the latter in the submucous coat. They follow the course of the blood-vessels, and may consequently be arranged into three groups.   The *first group* accompany the gastric vessels along the lesser curvature to the cardiac orifice, receiving branches from both surfaces of the organ, and pass to the cœliac glands.   The *second group* pass from the great end of the stomach, accompanying the vasa brevia, and enter the splenic lymphatic glands.   The *third group* run along the greater curvature with the right gastro-epiploic vessels towards the pylorus, and, receiving the lymphatics from the upper part of the duodenum, terminate in the cœliac glands.

The **lymphatic glands of the spleen** occupy the hilum.   Its *lymphatic vessels* consist of two sets, superficial and deep ; the former are placed beneath its peritoneal covering, the latter in the substance of the organ ; they accompany the blood-vessels, passing through a series of small glands, and, after receiving the lymphatics from the pancreas, ultimately pass into the cœliac glands.

The *lymphatics of the pancreas* also enter the cœliac glands.

The Lymphatic System of the Intestines

The **lymphatic glands of the small intestine** are placed between the layers of the mesentery, occupying the meshes formed by the superior mesenteric vessels, and hence called *mesenteric glands*.   They vary in number from a hundred to a hundred and fifty ; and in size, from that of a pea to that of a small almond.* These glands are most numerous, and largest above, the glands of the jejunum being more numerous than those of the ileum.   This latter group becomes enlarged and infiltrated with deposit in cases of fever accompanied with ulceration of the intestines.

The **lymphatic glands of the large intestine** are much less numerous than the mesenteric glands ; they are situated along the vascular arches formed by the arteries previous to their distribution, and even sometimes upon the intestine itself.   They are fewest in number along the transverse colon, where they form an uninterrupted chain with the mesenteric glands.

* Leaf (*op. cit.*) says it is very common to find not more than forty or fifty.

The **lymphatic vessels of the small intestine** are called **lacteals** from the milk-white fluid they usually contain ; they consist of two sets, superficial and deep : the former lie between the layers of the muscular coat and between the muscular and peritoneal coats, taking a longitudinal course along the outer side of the intestine ; the latter occupy the submucous tissue, and course transversely round the intestine, accompanied by the branches of the mesenteric vessels ; they pass between the layers of the mesentery, enter the mesenteric glands, and finally unite to form two or three large trunks which terminate separately in the receptaculum chyli ; frequently, however, they first unite to form a single large trunk, termed the *intestinal lymphatic trunk*.

The **lymphatic vessels of the large intestine** consist of two sets : those of the cæcum, ascending and transverse colon, which after passing through their proper glands, enter the mesenteric glands ; and those of the descending colon, sigmoid flexure, and rectum, which pass to the lumbar glands.

The Lymphatics of the Thorax

The **Lymphatic glands of the thoracic wall** are : the intercostal, internal mammary, anterior mediastinal, and posterior mediastinal.

The **intercostal glands** are small, and situated on each side of the spine, near the costo-vertebral articulations ; they vary from one to three in each space.

The **sternal** or **internal mammary glands** are placed at the anterior extremity of each intercostal space, by the side of the internal mammary vessels.

The **anterior mediastinal glands** are placed in the loose areolar tissue of the anterior mediastinum, some lying upon the Diaphragm in front of the pericardium, and others round the great vessels at the base of the heart.

The **posterior mediastinal glands** are situated in the areolar tissue in the posterior mediastinum, forming a continuous chain by the side of the aorta and œsophagus ; they communicate on each side with the intercostal, below with the lumbar, and above with the deep cervical glands.

The **Superficial lymphatic vessels of the front of the thorax** run across the great Pectoral muscle, and those on the back part of this cavity lie upon the Trapezius and Latissimus dorsi ; they all converge to the axillary glands. The lymphatics from the greater part of the mammary gland pass outwards to the lower border of the Pectoralis major muscle, where they enter a chain of small glands, situated in the axillary space along the lower border of its anterior boundary. Some few lymphatics from the inner side of the mammary gland pass through the intercostal spaces to reach the anterior mediastinal glands.

The **Deep lymphatic vessels of the thoracic wall** are : the intercostal, internal mammary, and diaphragmatic.

The **intercostal lymphatic vessels** follow the course of the intercostal vessels, receiving lymphatics from the intercostal muscles and pleura ; they pass backwards to the spine, and unite with lymphatics from the back part of the thorax and spinal canal. After traversing the intercostal glands, they pass down the spine, and terminate in the thoracic duct.

The **internal mammary lymphatic vessels** follow the course of the internal mammary vessels ; they commence in the muscles of the abdomen above the umbilicus, communicating with the epigastric lymphatics, ascend between the fibres of the Diaphragm at its attachment to the ensiform appendix, and in their course behind the costal cartilages are joined by the intercostal lymphatics ; they terminate on the right side in the right lymphatic duct, on the left side in the thoracic duct.

The **lymphatic vessels of the Diaphragm** follow the course of their corresponding vessels, and terminate, some in front in the anterior mediastinal and internal mammary glands, some behind in the intercostal and posterior mediastinal lymphatics.

The **Lymphatic glands of the viscera of the thorax** are the bronchial glands.

The **bronchial glands** are situated round the bifurcation of the trachea and roots of the lungs. They are ten or twelve in number, the largest being placed opposite the bifurcation of the trachea, the smallest round the bronchi and their primary divisions for some little distance within the substance of the lungs. In infancy they present the same appearance as lymphatic glands in other situations; in the adult they assume a brownish tinge, and in old age a deep black colour. Occasionally, they become sufficiently enlarged to compress and narrow the canals of the bronchi; and they are often the seat of tuberculous deposits.

The **superior mediastinal** or **cardiac glands** lie in front of the transverse aorta and left innominate vein; this group consists of numerous large glands. They receive the lymph from the pericardium, heart, and thymus gland.

The **lymphatic vessels of the lung** consist of two sets, *superficial* and *deep*: the former are placed beneath the pleura, forming a minute plexus, which covers the outer surface of the lung; the latter accompany the blood-vessels, and run along the bronchi: they both terminate at the root of the lungs in the bronchial glands. The efferent vessels from these glands, two or three in number, ascend upon the trachea to the root of the neck, traverse the tracheal and œsophageal glands, and terminate on the left side in the thoracic duct, and on the right side in the right lymphatic duct.

The **cardiac lymphatic vessels** consist of two sets, *superficial* and *deep*: the former arise in the subserous areolar tissue of the surface, and the latter in the deeper tissues of the heart. They follow the course of the coronary vessels: those of the right side unite into a trunk at the root of the aorta, which, ascending across the arch of that vessel, communicates with one or more of the cardiac glands, and passes backwards to the trachea, upon which it ascends, to terminate at the root of the neck in the right lymphatic duct. Those of the left side unite into a single vessel at the base of the heart, which, passing along the pulmonary artery, and traversing some glands at the root of the aorta, ascends on the trachea to terminate in the thoracic duct.

The **thymic lymphatic vessels** arise from the under surface of the thymus gland, and enter the superior mediastinal glands, from which they emerge as two vessels: these terminate, one on each side, in the corresponding internal jugular vein.

The **lymphatic vessels of the œsophagus** form a plexus round that tube, traverse the glands in the posterior mediastinum, and after communicating with the pulmonary lymphatic vessels near the roots of the lungs, terminate in the thoracic duct.

THE NERVOUS SYSTEM

THE Nervous System is composed: 1. Of a series of large centres of nerve-matter, called, collectively, the *cerebro-spinal centres* or *cerebro-spinal axis*. 2. Of smaller centres, termed *ganglia*. 3. Of nerves connected either with the cerebro-spinal axis or the ganglia. And 4. Of certain modifications of the peripheral terminations of the nerves forming the organs of the external senses.

The **Cerebro-spinal axis** consists of the brain or encephalon and the spinal cord, which are contained within the skull and spinal canal. The brain and its membranes will be first considered, and then the spinal cord and its coverings.

THE MEMBRANES OF THE BRAIN

Dissection.—To examine the brain with its membranes, the skull-cap must be removed. In order to effect this, saw through the external table, the section commencing, in front, about an inch above the margin of the orbit, and extending, behind, to a little above the level with the occipital protuberance. Then break the internal table with the chisel and hammer, to avoid injuring the investing membranes or brain; loosen, and forcibly detach the skull-cap, when the dura mater will be exposed. The adhesion between the bone and the dura mater is very intimate, and much more so in the young subject than in the adult.

The membranes of the brain are: the dura mater, arachnoid membrane, and pia mater.

DURA MATER

The **Dura Mater** is a thick and dense, inelastic fibrous membrane, which lines the interior of the skull. Its outer surface is rough and fibrillated, and adheres closely to the inner surface of the bones, forming their internal periosteum, this adhesion being most marked opposite the sutures and at the base of the skull. Its inner surface is smooth, and lined by a layer of endothelium. It sends four processes inwards, into the cavity of the skull, for the support and protection of the different parts of the brain : and is prolonged to the outer surface of the skull, through the various foramina which exist at the base, and thus becomes continuous with the pericranium ; its fibrous layer forms sheaths for the nerves which pass through these apertures. At the base of the skull, it sends a fibrous prolongation into the foramen cæcum ; it sends a series of tubular prolongations round the filaments of the olfactory nerves as they pass through the cribriform plate, and also round the nasal nerve as it passes through the nasal slit ; a prolongation is also continued through the sphenoidal fissure into the orbit, and another is continued into the same cavity through the optic foramen forming a sheath for the optic nerve, which is continued as far as the eyeball. In the posterior fossa it sends a process into the internal auditory meatus, ensheathing the facial and auditory nerves ; another through the jugular foramen, forming a sheath for the structures which pass through this opening, and a third through the anterior condyloid foramen. Around the margin of the foramen magnum it is closely adherent to the bone, and is continuous with the dura mater lining the spinal canal. In certain situations, as already mentioned (page 571), the fibrous layers of this membrane separate, to form sinuses for the passage of venous blood. Upon the outer surface of the dura mater, in the situation of the longitudinal sinus, may be seen numerous small whitish bodies, the *glandulæ Pacchioni.*

Structure.—The dura mater consists of white fibrous tissue with connective-tissue cells and elastic fibres arranged in flattened laminæ which are imperfectly separated by lacunar spaces and blood-vessels into two layers, *endosteal* and *meningeal.* The endosteal layer is the internal periosteum for the cranial bones, and contains the blood-vessels for their supply. At the margin of the foramen magnum it becomes continuous with the periosteum lining the spinal canal. The meningeal or supporting layer is lined on its inner surface by a layer of nucleated endothelium, similar to that found on serous membranes: these cells were formerly regarded as belonging to the arachnoid membrane. By its reduplication the meningeal layer forms the falx cerebri, the tentorium and falx cerebelli, and the diaphragma sellæ. The two layers are connected by fibres which intersect each other obliquely.

Its *arteries* are very numerous, but are chiefly distributed to the bones. Those found in the anterior fossa are the anterior meningeal branches of the anterior and posterior ethmoidal and internal carotid, and a branch from the middle meningeal. In the middle fossa are the middle and small meningeal branches of the internal maxillary, a branch from the ascending pharyngeal, which enters the skull through the foramen lacerum medium basis cranii; branches from the internal carotid, and a recurrent branch from the lachrymal. In the posterior fossa are meningeal branches from the occipital: one of which enters the skull through the jugular foramen, and the other through the mastoid foramen; the posterior meningeal, from the vertebral; occasionally meningeal branches from the ascending pharyngeal, which enter the skull, one at the jugular foramen, the other at the anterior condyloid foramen; and a branch from the middle meningeal.

The *veins*, which return the blood from the dura mater, and partly from the bones, anastomose with the diploic veins. These vessels terminate in the various sinuses, with the exception of two which accompany the middle meningeal artery, and pass out of the skull at the foramen spinosum to join the internal maxillary vein; above they communicate with the superior longitudinal sinus. Many of the meningeal veins do not open directly into the sinuses, but indirectly through a series of ampullæ termed *venous lacunæ.* These are found on each side of the superior longitudinal sinus, especially near its middle portion, and are often invaginated by Pacchionian bodies; they also exist near the lateral and straight sinuses. They communicate with the underlying cerebral veins, and also with the diploic and emissary veins.

The *nerves* of the dura mater are filaments from the Gasserian ganglion, from the ophthalmic, superior maxillary, inferior maxillary, vagus, and hypoglossal nerves, and from the sympathetic.

Processes of the Dura Mater.—The processes of the dura mater, sent inwards into the cavity of the skull, are four in number: the falx cerebri, the tentorium cerebelli, the falx cerebelli, and the diaphragma sellæ.

The *falx cerebri*, so named from its sickle-like form, is a strong arched process of the dura mater, which descends vertically in the longitudinal fissure between the two hemispheres of the brain. It is narrow in front, where it is attached to the crista galli of the ethmoid bone; and broad behind, where it is connected with the upper surface of the tentorium. Its upper margin is convex, and attached to the inner surface of the skull, in the middle line, as far back as the internal occipital protuberance; it contains the superior longitudinal sinus. Its lower margin is free, concave, and presents a sharp curved edge, which contains the inferior longitudinal sinus.

The *tentorium cerebelli* is an arched lamina of dura mater, elevated in the middle, and inclining downwards towards the circumference. It covers the upper surface of the cerebellum, and supports the occipital lobes of the brain, and prevents them pressing upon the cerebellum. It is attached, behind, by its convex border to the transverse ridges upon the inner surface of the occipital bone, and there encloses the lateral sinuses; in front, to the superior margin of the petrous portion of the temporal bone on either side, enclosing the superior petrosal sinuses; and at the apex of this bone, the free or internal border and the attached or external border meet, and crossing one another, are continued forwards, to be attached to the anterior and posterior clinoid processes respectively. Along the middle line of its upper surface the posterior border of the falx cerebri is attached, the straight sinus being placed at their point of junction. Its anterior border is free and concave, and bounds a large oval opening for the transmission of the crura cerebri.

The *falx cerebelli* is a small triangular process of dura mater, received into the indentation between the two lateral lobes of the cerebellum behind. Its base is attached, above, to the under and back part of the tentorium; its posterior margin to the lower division of the vertical crest on the inner surface of the occipital bone. As it descends, it sometimes divides into two smaller folds, which are lost on the sides of the foramen magnum.

The *diaphragma sellæ* is a horizontal process formed by a reduplication of the meningeal layer of the dura mater. It forms a small circular fold, which constitutes a roof for the sella turcica. This almost completely covers the pituitary body, presenting merely a small central opening for the infundibulum to pass through.

Arachnoid Membrane

The **arachnoid** (ἀράχνη, εἶδος, *like a spider's web*), so named from its extreme thinness, is a delicate membrane which envelops the brain, lying between the pia mater internally and the dura mater externally; from this latter membrane it is separated by a space, the *subdural space*.

It invests the brain loosely, being separated from direct contact with the cerebral substance by the pia mater, and a quantity of loose areolar tissue, the *subarachnoidean*. On the upper surface of the cerebrum the arachnoid is thin and transparent, and may be easily demonstrated by injecting a stream of air beneath it by means of a blowpipe; it passes over the convolutions without dipping down into the sulci between them. At the base of the brain the arachnoid is thicker, and slightly opaque towards the central part; it covers the anterior lobes, and extends across between the two temporal lobes so as to leave a considerable interval between it and the brain, the *anterior subarachnoidean space* (Cisterna pontis); it is in contact with the pons and under surface of the cerebellum; but between the hemispheres of the cerebellum and the medulla oblongata another considerable interval is left between it and the brain, called the *posterior subarachnoidean space* (Cisterna magna). These two spaces communicate together across the crura cerebelli. Other smaller cisternæ are found in various positions; and all communicate freely with one another. The arachnoid membrane surrounds the nerves which arise from the brain, and encloses them in loose sheaths as far as their point of exit from the skull.

The *subarachnoid space* is the interval between the arachnoid and pia mater. It is not, properly speaking, a *space*, for it is occupied everywhere by a spongy tissue consisting of trabeculæ of delicate connective tissue, which pass from the pia mater to the arachnoid, and in the meshes of which the subarachnoid fluid is contained. This so-called space is small on the surface of the hemispheres; but at the base of the brain the subarachnoid tissue is less abundant and its meshes larger, where it forms the Cisternæ pontis et magna mentioned above. In addition to these two large spaces, a third is formed on the upper surface of the corpus callosum, for the arachnoid stretches across from one cerebral hemisphere to the other immediately beneath the free border of the falx cerebri, and thus leaves a space in which the anterior cerebral arteries are contained. Another space is found in the fissure of Sylvius, for the arachnoid stretches across from the anterior to the middle lobe of the brain, without dipping down to the bottom of the fissure, and in this space the middle cerebral artery ramifies. The subarachnoid space communicates with the general ventricular cavity of the brain by means of three openings : one of these is in the middle line at the inferior boundary of the fourth ventricle, where an opening in the pia-matral covering of this cavity, the *foramen of Majendie*, exists and permits the passage of fluid from the one space to the other. The other two communications are at the extremities of the lateral recesses of the fourth ventricle, behind the upper roots of the glosso-pharyngeal nerves ; they are named the *foramina of Key and Retzius*. It is stated by Merkel that the lateral ventricles also communicate with the subarachnoid space at the apices of their descending horns.

The subdural space also contains fluid ; this is, however, small in quantity compared with the cerebro-spinal fluid and is probably of the nature of lymph.

The *cerebro-spinal fluid* fills up the subarachnoid space. It is a clear, limpid fluid, having a saltish taste, and a slightly alkaline reaction. According to Lassaigne, it consists of 98·5 parts of water, the remaining 1·5 per cent. being solid matters, animal and saline. It varies in quantity, being most abundant in old persons, and is quickly reproduced. Its chief use is probably to afford mechanical protection to the nervous centres, and to prevent the effects of concussions communicated from without.

Structure.—The arachnoid consists of bundles of white fibrous and elastic tissue intimately blended together. Its outer surface is covered with a layer of endothelium. Vessels of considerable size, but few in number, and, according to Bochdalek, a rich plexus of nerves derived from the motor division of the fifth, the facial, and the spinal accessory nerves, are found in the arachnoid.

GLANDULÆ PACCHIONI OR ARACHNOID VILLI

The **glandulæ Pacchioni** are numerous small whitish granulations usually collected into clusters of variable size, which are found in the following situations: 1. Upon the outer surface of the dura mater, in the vicinity of the superior longitudinal sinus, being received into little depressions on the inner surface of the calvarium. 2. On the inner surface of the dura mater. 3. In the superior longitudinal sinus. 4. On the pia mater, near the margin of the hemispheres.

These bodies are not glandular in structure, but simply enlarged normal villi of the arachnoid. In their growth they appear to perforate the dura mater, and when of large size they cause absorption of the bone, and come to be lodged in pits or depressions on the inner table of the skull.| Their manner of growth is as follows : at an early period they project through minute holes in the inner layer of the dura mater, which open into large venous spaces situated in the tissues of the membrane, on either side of the longitudinal sinus and communicating with it. In their onward growth the villi push the outer layer of the dura mater before them, and this forms over them a delicate membranous sheath. In structure they consist of spongy trabecular tissue, covered over by a membrane, which is continuous with the arachnoid. The space between these two coverings, derived from the dura mater and arachnoid respectively, corresponds to and is continuous with the subdural space. The spongy tissue of which they are composed is continuous with the trabecular tissue of the subarachnoid space ; so that fluid injected into the subarachnoid space finds its way into the Pacchionian bodies ; and through their coverings filters into the superior longitudinal sinus. They are supposed to be the means by which excess of cerebro-spinal fluid is got rid of, when its quantity is increased above normal.

These bodies are not found in infancy, and very rarely until the third year. They are usually found after the seventh year; and from this period they increase in number as age advances. Occasionally they are wanting.

PIA MATER

The **pia mater** is a vascular membrane, and derives its blood from the internal carotid and vertebral arteries. It consists of a minute plexus of blood-vessels, held together by an extremely fine areolar tissue. It invests the entire surface of the brain, dipping down between the convolutions and laminæ, and is prolonged into the interior, forming the velum interpositum and the choroid plexuses of the lateral and fourth ventricles. Upon the surfaces of the hemispheres, where it covers the grey matter of the convolutions, it is very vascular, and gives off from its inner surface a multitude of minute vessels, which extend perpendicularly for some distance into the cerebral substance. At the base of the brain, in the situation of the anterior and posterior perforated spaces, a number of long straight vessels are given off, which pass through the white matter to reach the grey substance in the interior. On the cerebellum the membrane is more delicate, and the vessels from its inner surface are shorter. The pia mater of the spinal cord is thicker, firmer, and less vascular than that of the brain, and as it is traced upwards over the medulla it is seen to preserve these characters. At the upper border of the medulla it is prolonged over the lower half of the fourth ventricle, forming a covering for it (tela choroidea inferior) before it is reflected on to the under surface of the cerebellum.

According to Frohmann and Arnold, this membrane contains numerous lymphatic vessels. Its nerves are derived from the sympathetic, and also from the third, fifth, sixth, facial, glosso-pharyngeal, pneumogastric, and spinal accessory. They accompany the branches of the arteries.

THE BRAIN

GENERAL CONSIDERATIONS AND DIVISIONS

The **encephalon** or **brain** is that portion of the cerebro-spinal axis which is contained in the cavity of the cranium. For purposes of description it may be divided into five parts, as follows: (1) the two cerebral hemispheres; (2) the inter-brain; (3) the mid-brain; (4) the pons Varolii and cerebellum; and (5) the medulla oblongata. If the student will refer to the section on the Development of the Brain he will find that these five portions correspond fairly accurately to the five secondary cerebral vesicles, of which the brain at an early period of embryonal life consisted: the prosencephalon, or *first* vesicle, by means of a protrusion from its front part on either side forms the cerebral hemispheres and the lateral ventricles; the remainder of the prosencephalon, together with the *second* vesicle, the thalamencephalon, form the inter-brain and third ventricle; the *third* vesicle, the mesencephalon, forms the mid-brain, or that portion which connects the inter-brain and hemispheres above with the pons Varolii below, and the cavity of the vesicle forms the aqueduct of Sylvius, or iter a tertio ad quartum ventriculum; the *fourth* vesicle, the epencephalon, becomes the future pons Varolii and cerebellum, and its cavity forms the upper half of the fourth ventricle; and finally, the *fifth* vesicle, the metencephalon, develops into the medulla oblongata, and its cavity forms the lower half of the fourth ventricle. It will thus be seen that the five divisions of the encephalon mentioned above correspond to the five secondary cerebral vesicles, with the exception of the first two, which together form the cerebral hemispheres and the inter-brain. In consequence of this these two portions of the brain are sometimes grouped together as the *cerebrum*.

I. The Hemispheres of the Brain

General Considerations.—The two hemispheres constitute the largest portion of the encephalon, and, together with the parts derived from the thalamencephalon, form what is called by some writers the *fore-brain*. They occupy the whole of the vault of the skull, and consist of a central cavity, in either hemisphere, surrounded by exceedingly thick and convoluted walls of nervous tissue. The under surface or base of the cerebrum is of an irregular form, resting in front on the anterior and middle fossæ of the skull and behind upon the tentorium cerebelli. The upper surface is of an ovoid form, broader behind than in front, convex in general outline, and divided into two lateral halves or hemispheres, right and left, by the *great longitudinal fissure*, which extends throughout the entire length of the cerebrum in the middle line, reaching down to the base of the brain in front and behind, but interrupted in the middle by a broad transverse commissure of white matter, the *corpus callosum*, which connects the two hemispheres together.

The Surface of the Cerebrum.—Each hemisphere presents an outer convex surface, filling the concavity of the corresponding half of the vault of the cranium ; an inner, flattened surface, which is vertical and directed towards the corresponding surface of the opposite hemisphere (the two forming the sides of the longitudinal fissure) ; and an under surface or base, of an irregular form, which rests in front on the anterior and middle fossæ of the base of the skull, and behind upon the tentorium cerebelli. The hemispheres are composed of an outer stratum of grey matter, called the *cortical substance*. It is thrown into a number of creases or infoldings, which are termed *fissures* and *sulci*, and these separate the surface into a number of irregular eminences, named *convolutions* or *gyri*.

The infoldings or creases are of two kinds, *fissures* and *sulci*. The fissures are of large size, and appear early in fœtal life ; they are few in number, nearly constant in their arrangement, and are produced by infoldings of the entire thickness of the wall of the prosencephalon, producing corresponding elevations in the interior of the ventricle, and hence are termed *complete fissures*. They comprise (*a*) the hippocampal, or dentate fissure ; (*b*) the anterior part of the calcarine fissure; (*c*) the collateral fissure. The *sulci* are more numerous ; they are superficial depressions of the grey matter, which is folded inwards and only indents the central white-substance. They produce no corresponding elevations in the interior of the ventricle, and are therefore spoken of as *incomplete fissures*. They are fairly constant in their arrangement, and have received names indicative of their position and direction, but at the same time vary, within certain limits, in different individuals. They are similar, without being absolutely identical, on the two sides of the brain. It therefore follows that the gyri or convolutions which lie between these sulci are fairly constant in their general arrangement.

The number and extent of the convolutions, as well as the depth of the intervening sulci, appear to bear a close relation to the intellectual power of the individual, as is shown in their increasing complexity of arrangement as one ascends from the lowest mammalia up to man, where they present a most complex arrangement. Again, in the child, at birth, before the intellectual faculties are exercised, the convolutions are simpler, and the sulci between them shallower, than in the adult. In old age, when the mental faculties have diminished in activity, they become less prominently marked. By their arrangement the convolutions are adapted to increase the amount of grey matter without occupying much additional space, while they also afford a greater extent of surface for the termination of white fibres in grey matter.

It will be convenient, in the first instance, to describe the fissure which separates the two hemispheres from each other, and those which divide each hemisphere into its larger divisions.

The Longitudinal Fissure (fig. 343).—This great fissure separates the cerebrum into two hemispheres, and reaches from the front to the back of the organ ; it contains a vertical process of the dura mater, the falx cerebri (page 618). In front and behind, it extends from the top to the bottom of the cerebrum, and completely separates the two hemispheres, but its middle portion only separates the hemispheres for about half their vertical extent, the floor of this part of the fissure being formed by the great central white commissure, the *corpus callosum*, which connects the two hemispheres together.

The remaining fissures are situated in one or other of the two hemispheres, with the exception of the transverse fissure, one half of which is contained in each hemisphere.

Sylvian Fissure (fig. 344).—This fissure is a well-marked cleft on the base and side of the hemisphere. Starting at the base of the brain in a depression, the *vallecula Sylvii*, in which is situated the anterior perforated space, it passes outwards to the external surface of the hemisphere. It here gives off a short *anterior limb*, which

Fig. 343.—Upper surface of the brain, the arachnoid having been removed.

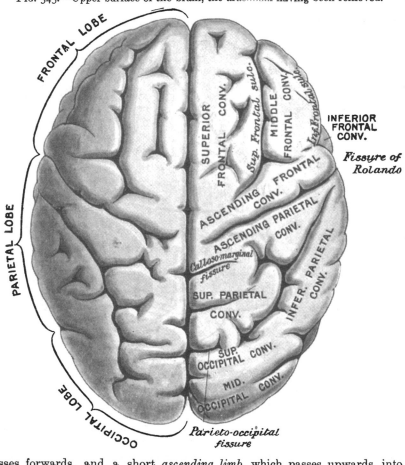

passes forwards, and a short *ascending limb*, which passes upwards into the inferior frontal convolution. It is then continued backwards as the *horizontal limb*, which terminates by an upward inflexion in the parietal lobe. It occupies the middle third of the lateral surface of the hemisphere.

The **Fissure of Rolando** is situated about the middle of the outer surface of the hemisphere, and, coursing obliquely downwards and forwards, divides the surface of the hemisphere into approximately equal parts. It commences at or near the longitudinal fissure, a little behind its mid-point, and runs sinuously downwards and forwards, to terminate a little above the horizontal limb of the fissure of

Sylvius, and about half an inch behind the ascending limb of the same fissure. It forms two chief curves : the upper or *superior genu* is concave forwards and upwards, while the lower or *inferior genu* has its concavity directed backwards.

The **parieto-occipital fissure** is only seen to a slight extent on the outer surface of the hemisphere, being situated for the most part on its mesial aspect. The portion on the outer surface is called the *external parieto-occipital fissure*, to distinguish it from the part continued on to the internal surface, which is termed the *internal parieto-occipital fissure*. The external parieto-occipital fissure commences about midway between the posterior extremity or occipital pole of the brain and the fissure of Rolando, and runs downwards and outwards for about an inch.

These three fissures divide the external surface of the hemisphere into four lobes : the *frontal*, the *parietal*, the *occipital*, and the *temporal*. To these must be added (1) the central lobe, or island of Reil, which is situated deeply in the Sylvian fissure, and (2) the olfactory lobe, which is found at the base of the brain and was formerly described under the name of the olfactory nerve.

Fɪɢ. 344.—Fissures and lobes on the external surface of the cerebral hemispheres.

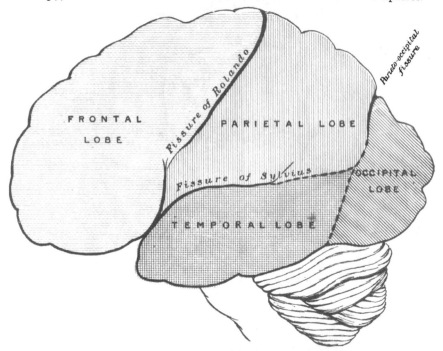

The Lobes on the External Surface.—The lobes on the external surface have received their names from the bones of the skull with which they are most nearly in relation, but it must be borne in mind that they do not correspond in shape or limit with the bone after which they are named. The division is, moreover, to a certain extent artificial, as will be seen from the following description. If a line is drawn in continuation of the external parieto-occipital fissure downwards and outwards to the lower border of the hemisphere, it will impinge on a slight notch, the *pre-occipital notch*, and if a second line is prolonged backwards from the horizontal part of the fissure of Sylvius to join the first, the division of the outer surface of the hemisphere into four lobes will be accomplished (fig. 344). The portion in front of the fissure of Rolando is the *frontal* lobe ; that behind the fissure of Rolando and above the fissure of Sylvius is the *parietal* lobe ; the portion behind the parieto-occipital fissure and its continuation is the *occipital* lobe ; and the part below the fissure of Sylvius and in front of the occipital lobe is the *temporal* lobe.

The **fissures and lobes of the Mesial and Tentorial surfaces.**—The mesial surface of the cerebrum can only be fully viewed by dividing the corpus callosum and the structures beneath it longitudinally in the middle line; in order to expose the tentorial surface, the pons Varolii, cerebellum, and medulla must be removed, by division of the crus cerebri on either side. When this has been done, a section such as is represented in fig. 345 will be shown. The parts in the centre, below the corpus callosum, belong to the interior of the brain, and will be disregarded for the present, while the lobes and fissures of the remaining portion of the hemisphere are considered. The fissures are five in number, in addition to a small part of the fissure of Sylvius, the commencement of which is seen, separating the frontal and temporal lobes. These fissures are named the *calloso-marginal,* the *internal parieto-occipital,* the *calcarine,* the *collateral,* and the *dentate* or *hippocampal.*

The **calloso-marginal fissure** commences below the anterior extremity of the corpus callosum; it at first runs forwards and upwards, parallel with the rostrum of the corpus callosum, and, winding round in front of the genu of that body, it

FIG. 345.—Fissures and lobes on the internal surface of the cerebral hemispheres.

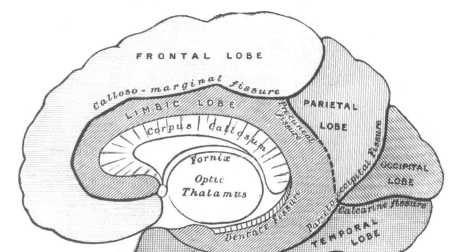

continues from before backwards, between the upper margin of the hemisphere and the convolution of the corpus callosum, to about midway between the anterior and posterior extremities of the brain, where it ascends to reach the upper margin of the hemisphere, a short distance behind the superior extremity of the fissure of Rolando.

The **internal parieto-occipital** extends in an oblique direction downwards and forwards to join the calcarine fissure, on a level with the hinder end of the corpus callosum.

The **calcarine fissure** commences, usually by two branches, close to the posterior extremity of the hemisphere. These soon unite, and the fissure runs nearly horizontally forwards, and is joined by the parieto-occipital fissure, and continues as far as the posterior extremity of the corpus callosum, a little below the level of which it terminates. Its anterior part causes the prominence, in the interior of the brain, known as the hippocampus minor or calcar avis.

The **collateral fissure** is situated on the tentorial surface, below and external to the preceding, being separated from it by the sub-collateral or uncinate gyrus.

It runs forwards, from the posterior extremity of the brain, nearly as far as the tip of the temporal lobe. It lies below the posterior and descending horns of the lateral ventricle, and its middle part causes the prominence, in the interior of the brain, known as the eminentia collateralis.

The **dentate** or **hippocampal fissure** commences immediately behind the posterior extremity of the corpus callosum, and runs forward to terminate at the recurved part of the hippocampal gyrus. It causes the prominence of the hippocampus major in the descending horn of the lateral ventricle. In addition to these fissures, which are constant, there is frequently an irregular broken fissure, which appears to be a continuation backwards of the posterior part of the callosomarginal fissure, before it ascends to reach the upper edge of the hemisphere. This has been termed the *post-limbic fissure*. These fissures map off portions of

FIG. 346.—Convolutions and sulci on the external surface of the cerebral hemisphere.

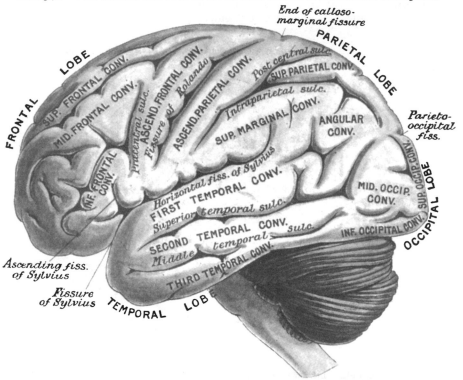

the internal and tentorial surfaces of the hemispheres, which form parts of the lobes found on the external surface. That portion which lies in front and above the calloso-marginal fissure belongs almost entirely to the frontal lobe; its posterior extremity, which extends for a short distance behind the upper end of the fissure of Rolando, forms a small part of the parietal lobe; that portion which lies above the internal parieto-occipital fissure and behind the callosomarginal fissure forms a part of the parietal lobe; that between the parieto-occipital fissure above and the calcarine fissure below is a portion of the occipital lobe; and all the region below the calcarine fissure behind and the collateral fissure in front belongs to the temporal lobe. The remainder of the mesial and tentorial surfaces of the hemisphere constitute what Broca termed the *limbic lobe*, which is subsequently referred to (page 631).

The surface of the hemisphere has thus been divided into its different parts, viz. : the frontal, the parietal, the occipital, the temporal, the limbic, the central lobe

or island of Reil, and the olfactory lobes.   Each of these lobes is further subdivided
into convolutions or gyri by smaller fissures, which, though less constant in their
arrangement than the fissures already described, have a fairly definite course.

1. The **frontal lobe**.—On its *external* surface the frontal lobe presents three
sulci, which divide it into four convolutions (fig. 346).   The *precentral sulcus*
runs upwards through this lobe, parallel to the lower half of the fissure of
Rolando.   It is frequently broken or interrupted by annectant gyri.   It limits
a convolution, which lies between it and the fissure of Rolando, and which is
called the *ascending frontal convolution*.   From it two sulci, the *superior* and
inferior frontal, run forwards and downwards, and divide the remainder of
the outer surface of the lobe into three parallel principal convolutions, named
respectively the *superior, middle,* and *inferior frontal convolutions*.

FIG. 347.—Convolutions and sulci on the under surface of the anterior lobe.

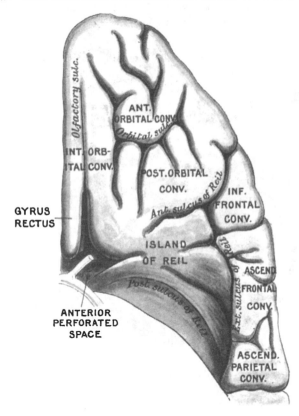

The *ascending frontal convolution* is a simple convolution, bounded in front
by the precentral sulcus, behind by the fissure of Rolando, and extending from
the upper margin of the hemisphere above to a little behind the bifurcation of the
fissure of Sylvius below.

The *superior frontal convolution* is situated between the margin of the longitudinal fissure and the superior frontal sulcus.   It extends above on to the inner
aspect of the hemisphere, forming the greater part of the marginal convolution,
and in front on to the orbital surface, forming the internal orbital convolution.
It is usually more or less completely subdivided into two by an antero-
posterior sulcus, the *sulcus frontalis mesialis* of Cunningham, which, however, is
frequently interrupted and broken into several parts by bridging convolutions.

The *middle frontal convolution* is situated between the superior and inferior frontal sulci, and extends from the precentral sulcus on to the orbital surface of the lobe, where it forms the anterior orbital convolution. The middle frontal convolution is frequently subdivided into two by a sagitally directed sulcus, the *sulcus frontalis medius* of Eberstaller.

The *inferior frontal convolution* is situated below the inferior frontal sulcus, and extends forwards from the lower part of the precentral sulcus, on to the under surface of the lobe, where it forms the posterior orbital convolution. The inferior frontal convolution is subdivided by the anterior and ascending limbs of the fissure of Sylvius into three parts, viz.: (1) *anterior* or *pars orbitalis*, below the anterior limb of the fissure; (2) *middle* or *pars triangularis* ('cap' of Broca), between the two limbs; and (3) *posterior* or *pars basalis*, behind the ascending limb.

The left inferior frontal convolution is, as a rule, more highly developed than the right, and is named the *convolution of Broca*, from the fact that in 1861 Broca discovered that it was the centre for language.

Fig. 348.—Convolutions and sulci on the internal surface of cerebral hemispheres.

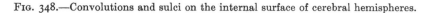
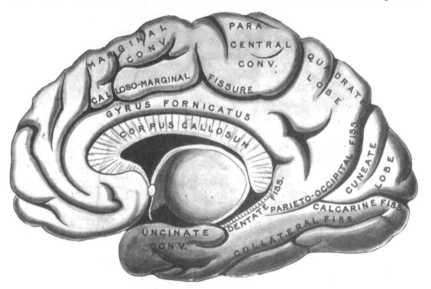

The *under surface* of the frontal lobe rests on the orbital plate of the frontal bone, and is sometimes named the *orbital lobe* (fig. 347). It is divided into three convolutions by a well-marked sulcus, the *orbital* or *tri-radiate sulcus*. These are named, from their position, the *internal, anterior,* and *posterior orbital convolutions,* and are the continuations respectively of the superior, middle, and inferior frontal convolutions of the external surface. The internal orbital convolution presents a well-marked antero-posterior groove or sulcus, the *olfactory sulcus,* for the olfactory tract; and the portion internal to this is named the *gyrus rectus,* and is continuous with the marginal gyrus, presently to be described. The *mesial* or *internal surface* of the frontal lobe is occupied by a single curved convolution, which from its situation is termed the *marginal gyrus* (fig. 348). It commences in front of the anterior perforated space, runs along the margin of the longitudinal fissure on the mesial surface of the orbital lobe, where it is continuous with the internal orbital convolution; it then ascends, and runs backwards to the point where the calloso-marginal fissure turns upwards to reach the superior border of the hemisphere. An oval portion at the posterior part of this convolution is sometimes marked off by a vertical fissure, and is distinguished as the *paracentral gyrus*, because it is continuous with the convolutions in front and behind the central fissure or fissure of Rolando.

2. The **parietal lobe.**—On its external surface the parietal lobe presents for examination two sulci and three convolutions.

The *intra-parietal sulcus* commences close to the horizontal limb of the fissure of Sylvius, about midway between the fissure of Rolando and the upturned extremity of the fissure of Sylvius. It first runs upwards parallel to and behind the lower half of the fissure of Rolando, and then turns backwards, extending nearly to the termination of the external parieto-occipital fissure, where it sometimes becomes continuous with the superior occipital sulcus. The ascending portion of this sulcus separates off a convolution, the *ascending parietal*, which lies between it and the fissure of Rolando, while the horizontal portion divides the remainder of the external surface of the parietal lobe into two other convolutions, the *superior* and *inferior parietal*.

The *post-central sulcus* is a slightly marked groove, which is sometimes a branch of the intra-parietal sulcus, being given off where the ascending portion of this sulcus turns backwards. It lies parallel to and behind the upper part of the fissure of Rolando, and separates the ascending from the superior parietal convolution.*

The *ascending parietal convolution* is bounded in front by the fissure of Rolando, behind by the ascending portion of the intra-parietal sulcus and by the post-central sulcus. It extends from the great longitudinal fissure above to the horizontal limb of the fissure of Sylvius below. It lies parallel with the ascending frontal convolution, with which it is connected below, and also, sometimes, above the termination of the fissure of Rolando.

The *superior parietal convolution* is bounded in front by the post-central sulcus, which lies between it and the previous convolution, but with which it is usually connected above the upper extremity of the sulcus; behind, it is bounded by the external parieto-occipital fissure, below the termination of which it is joined to the occipital lobe by a narrow convolution, the *first annectant gyrus*; below, it is separated from the inferior parietal convolution by the horizontal portion of the intra-parietal sulcus; and above, it is continuous on the inner surface of the hemisphere with the quadrate lobe.

The *inferior parietal convolution* is that portion of the parietal lobe which is situated between the ascending portion of the intra-parietal sulcus in front, the horizontal portion of the same sulcus above, the horizontal limb of the fissure of Sylvius below, and the posterior boundary of the parietal lobe behind. It is divided into two convolutions by an indistinct groove. One, the *supra-marginal*, lies behind the ascending part of the intra-parietal sulcus and above the horizontal limb of the fissure of Sylvius, over the extremity of which it arches. It is connected in front with the ascending parietal convolution below the intra-parietal sulcus, and behind with the superior temporal convolution round the

* Professor Cunningham describes these two sulci, the intra-parietal and post-central, somewhat differently. He regards them as both belonging to the intra-parietal sulcus, which he divides into three parts: the ascending portion of the intra-parietal, as described above, he terms the *ramus verticalis inferior*; the horizontal portion as the *ramus horizontalis*; while the post-central sulcus he denominates the *ramus verticalis superior*. He states that considerable variability is exhibited in the relation to each other of these different parts of the intra-parietal sulcus, but that the one in which the three parts of the sulcus are confluent is by far the most constant condition. Sometimes, however, the three parts of the sulcus may be separate, or the ramus horizontalis confluent with the ramus verticalis inferior, the ramus verticalis superior remaining separate; or, again, the vertical limbs may be confluent and the horizontal limb separate; or, finally, the ramus horizontalis may be joined to the lower end of the ramus verticalis superior, while the lower vertical ramus is separate. The prolongation of the intra-parietal sulcus into the occipital lobe, which sometimes exists, he calls the *ramus occipitalis*. In the majority of cases, however, the occipital ramus is separated from the main portion of the intra-parietal sulcus by a superficial or deep bridging convolution. (*Journal of Anatomy and Physiology*, vol. xxiv. part ii. p. 135.)

posterior extremity of the fissure of Sylvius. The other, the *angular*, is united anteriorly with the foregoing, while posteriorly it is continuous with the middle temporal convolution by a process which curves round the superior temporal or parallel sulcus. It is connected with the occipital lobe by the *second annectant gyrus.*

The *internal* or *mesial* surface of the parietal lobe is continuous with the external surface over the upper edge of the hemisphere. It is of small size, and forms one square-shaped convolution, which from its shape is termed the *quadrate lobe*. From its situation above the cuneate lobe it is sometimes named *precuneus.*

3. The **occipital lobe.**—The occipital lobe is divided on its *external* surface into three convolutions by two indistinct sulci, the *superior* and *middle occipital sulci*. They are directed backwards across the lobe, being frequently small and ill marked ; the superior is sometimes continuous with the horizontal portion of the intra-parietal sulcus.

The *superior occipital convolution* is situated above the superior sulcus, and is connected to the superior parietal convolution by the *first annectant gyrus.*

The *middle occipital convolution* is situated between the superior and middle occipital sulci, and is connected to the angular convolution by the *second annectant gyrus*, and to the middle temporal convolution by the *third annectant gyrus.*

The *inferior occipital convolution* is situated below the middle occipital sulcus, and is sometimes separated from the external occipito-temporal convolution on the under surface of the hemisphere by an inconstant sulcus, the *inferior occipital sulcus*. It is connected to the inferior temporal convolution by the *fourth annectant gyrus.*

The *internal* or *mesial* surface of the occipital lobe presents a triangular convolution, which is known as the *cuneus* or *cuneate lobule*. It is situated between the internal parieto-occipital and calcarine fissures, which, as already mentioned, meet some distance behind the posterior extremity of the corpus callosum.

4. The **temporal lobe,** sometimes called the temporo-sphenoidal lobe, presents an outer and an inferior surface. The *outer surface* is subdivided by two fissures, named respectively the first and second temporal sulci. The *first temporal sulcus* is well marked, and runs from before backwards through the temporal lobe parallel with, but some little distance below, the horizontal limb of the fissure of Sylvius, and hence is often termed the *parallel sulcus*. The *second temporal sulcus* takes the same direction as the first, but is situated at a lower level, and is often interrupted by one or more bridging convolutions. These two sulci subdivide this surface of the temporal lobe into three convolutions. The *first* or *superior temporal convolution* is situated between the horizontal limb of the fissure of Sylvius and the first temporal sulcus, and is continuous behind with the supramarginal convolution. The *second* or *middle temporal convolution* lies between the first and second temporal sulci, and is continued behind into the angular and middle occipital convolutions. The *third* or *inferior temporal convolution* is placed below the second temporal sulcus : it is connected posteriorly with the inferior occipital convolution, and is also prolonged on to the under or tentorial surface of the temporal lobe, where it is limited internally by the third temporal sulcus, about to be described.

The inferior or tentorial surface presents two fissures, viz. : the third temporal sulcus and the collateral fissure—the latter of which has already been described (page 625). The *third temporal sulcus* extends from near the occipital pole behind, to near the anterior extremity of the temporal lobe in front, but is, however, frequently subdivided by bridging gyri. The convolutions on the inferior surface are (1) the *fourth temporal* or *subcollateral convolution* (sometimes called the *external occipito-temporal*), situated between the third temporal sulcus and the collateral fissure ; and (2) the *subcalcarine convolution* or *lingual lobule*, lying between the calcarine fissure above and the posterior part of the collateral fissure

below, and continuous in front with the hippocampal convolution, the latter forming part of the limbic lobe.*

5. The **central lobe** or **island of Reil** (fig. 349) lies deeply in the Sylvian fissure,
and can only be seen when the lips of that fissure are widely separated, since it is
overlapped and hidden by the convolutions which bound the fissure. These
convolutions are termed the *opercula of the insula* ; they are separated from each
other by the three limbs of the Sylvian fissure, and named the orbital, frontal,
fronto-parietal, and temporal opercula. It is almost surrounded by a deep
limiting sulcus, which separates it from the frontal, parietal, and temporal lobes.
When the opercula have been removed the insula presents the form of a triangular
eminence ; its apex is directed downwards and inwards towards the anterior
perforated space, and is continuous in front with the posterior orbital convolution
and behind with the hippocampal convolution. It is divided into a *pre-central*
and a *post-central* lobe by the *sulcus centralis*, which runs backwards and upwards
from the apex of the insula. The pre-central lobe is further subdivided by
shallow sulci into three or four short convolutions, the *gyri breves*, while the
post-central lobe is named the *gyrus longus* and is often bifurcated at its upper

FIG. 349.—The Island of Reil. Left side. The overlapping parts of the hemisphere
have been removed. (Dalton.)

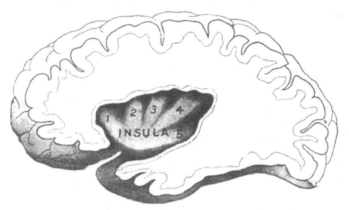

1, 2, 3, gyri breves ; 4, 5, gyrus longus, bifurcated at its upper extremity.

extremity. The grey matter of the insula is continuous with that of the different
opercula, while its mesial surface corresponds with the lenticular nucleus of the
corpus striatum.

6. The **Limbic lobe.**—The term limbic lobe (*grande lobe limbique*) was introduced by Broca in 1878, and under it he included two convolutions, viz. the callosal
and hippocampal, which together arch round the corpus callosum and the
hippocampal fissure. These he separated on the morphological ground that
they are well developed in animals possessing a keen sense of smell (osmatic
animals), such as the dog and fox. To the lobe thus defined the following parts
must be added, viz.: the laminæ of the septum lucidum, together with the
fornix and its fimbriæ, which may be regarded as forming an inner or deep arch ;
the peduncles and longitudinal striæ of the corpus callosum, together with the
gyrus dentatus, which form a middle arch, while the outer arch is constituted by
the callosal and hippocampal convolutions : the first two arches are separated
from each other by the corpus callosum.

* It will be seen from this description that the tentorial surface of the occipital lobe is
regarded as forming part of the same surface as the temporal lobe. The boundary
between the occipital and temporal lobes on the tentorial surface is purely artificial, and if
represented by a line drawn upwards and inwards from the pre-occipital notch would cut
both the subcollateral and subcalcarine gyri.

Convolutions of the limbic lobe.—(1) The *callosal convolution, gyrus forni-catus*, or *gyrus cinguli* is an arch-shaped convolution, lying in close relation to the superficial surface of the corpus callosum, from which it is separated by a slit-like fissure, the *callosal fissure*. It commences below the rostrum of the corpus callosum, curves round in front of the genu, extends along the upper surface of the body, and finally turns downwards behind the splenium, where it is connected by a narrow isthmus with the gyrus hippocampi. It is separated from the marginal convolution by the calloso-marginal sulcus, from the quadrate lobe by the post-limbic sulcus, and from the subcalcarine convolution by the calcarine fissure.

(2) The *hippocampal convolution* (gyrus hippocampi) is bounded above by the hippocampal or dentate fissure, and below by the anterior part of the collateral fissure. Behind, it is continuous superiorly, through the isthmus, with the callosal convolution, and inferiorly with the subcalcarine or lingual convolution. Its anterior extremity is recurved in the form of a hook, and is named the *uncus*. Running in the substance of the callosal and hippocampal convolutions, and connecting them together, is a tract of arched fibres, named the *cingulum*. The outer root of the olfactory tract passes into the anterior extremity of the hippo-campal convolution, and the inner root into the commencement of the callosal convolution, so that these two convolutions, with the addition of the olfactory tract, present a racquet-like appearance—the olfactory tract constituting the handle and the two convolutions the circumference of the blade.

(3) The *dentate convolution* (formerly named the *dentate fascia*) is situated above the gyrus hippocampi, from which it is separated by the hippocampal or dentate fissure. It is covered by the fimbria, and is a narrow, elongated convolu-tion, the free surface of which presents a notched or toothed appearance, hence its name. Posteriorly it is prolonged as a delicate lamina, the *fasciola cinerea*, around the splenium of the corpus callosum, and becomes continuous on the upper surface of that body with its mesial and lateral longitudinal striæ. Anteriorly it is prolonged into the notch produced by the recurving of the uncus, where it forms a sharp curve; from here it can be traced as a delicate band (*band of Giacomini*) over the uncus, on the outer surface of which it is lost.

The remaining structures which contribute to the formation of the limbic lobe will be subsequently described.

7. The **olfactory lobe** is situated on the under surface of the frontal lobe. It is rudimentary in man and some other mammals, but in vertebrates generally it is well developed, and consists of a distinct extension of the cerebral hemisphere, enclosing a portion of the anterior horn of the lateral ventricle. In man it is long and slender, and may be described as consisting of two parts, the *anterior* and *posterior olfactory lobules*.

The *anterior olfactory lobule* is made up of : (1) the olfactory bulb; (2) the olfactory tract; (3) the trigonum olfactorium; and (4) the area of Broca.

(1) The *olfactory bulb* is an oval mass of a reddish-grey colour, which rests on the cribriform plate of the ethmoid bone, and forms the anterior expanded extremity of the olfactory tract. Its under surface receives the olfactory nerves, which pass upwards through the cribriform plate from the olfactory region of the nose. Its minute structure will be subsequently described.

(2) The *olfactory tract* is a band of white matter, triangular on section, the apex being directed upwards. It lies in the olfactory sulcus on the under surface of the frontal lobe. Traced backwards, it is seen to divide into two roots, an outer and an inner. The *outer root* passes across the outer part of the anterior perforated space to the nucleus amygdalæ and the anterior part of the gyrus hippocampi. The *inner root* turns sharply inwards, and ends partly in Broca's area and partly in the callosal convolution ; in other words, the inner root is con-tinuous with one extremity and the outer root with the other extremity of the limbic lobe.

(3) The *trigonum olfactorium* is situated between the diverging roots of the

olfactory tract, and is sometimes described as the *middle* or *grey root* of the tract. It is part of an area of grey matter, which forms the base of the anterior olfactory lobule ; another portion of it is termed (4) the *area of Broca* ; and a third portion, of no special significance, is situated external to the outer root of the olfactory tract. This area of grey matter is bounded internally and posteriorly by a fissure (*fissura prima*) which separates it from the peduncle of the corpus callosum and from the posterior olfactory lobule.   The area of Broca is continuous with the gyrus fornicatus.

The *posterior olfactory lobule* or *anterior perforated space* is marked off from the anterior lobule by the fissura prima, and is situated at the commencement of

FIG. 350.—Base of the brain.

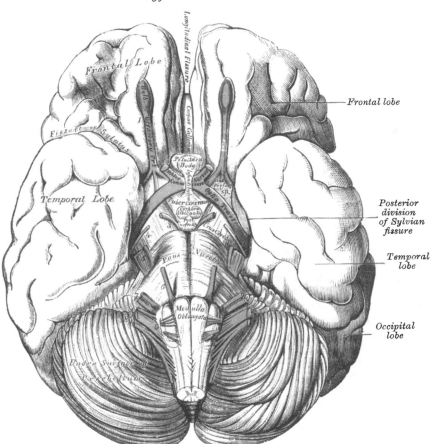

the fissure of Sylvius.   Internally, it is bounded by the peduncle of the corpus callosum, and is continuous with the lamina cinerea.   Posteriorly it is bounded by the optic tract, and it is partially concealed by the temporal lobe which overlaps it.   It has received the name of anterior perforated space from its being perforated by numerous openings, which transmit blood-vessels to the interior of the brain, and it corresponds to the under surface of the lenticular nucleus and part of the claustrum.

Under surface or base of the encephalon (fig. 350).—Having considered the surface of the hemispheres, the student should direct his attention to the base of the brain, before commencing the study of the component parts which make up the two hemispheres.

The base of the brain presents for examination the under surfaces of the frontal

and temporal lobes ; the structures contained in the interpeduncular space, with
the crura cerebri or cerebral peduncles ; the under surfaces of the pons Varolii,
cerebellum, and medulla oblongata ; and the superficial origins of the cranial
nerves.

The various objects exposed to view (with the exception of the origins of the
cranial nerves, which will be considered in another section) in the middle line and
on either. side of the middle line, are here arranged in the order they are met
with from before backwards.

In the Middle Line	*On each side of the Middle Line*
Longitudinal fissure.	Frontal lobe.
Rostrum and peduncles of	Olfactory lobe.
corpus callosum.	Fissure of Sylvius.
Lamina cinerea.	Optic tracts.
Optic commissure.	Crus cerebri.
Tuber cinereum.	Temporal lobe.
Infundibulum.	Hemisphere of cerebellum.
Pituitary body.	
Corpora albicantia.	
Posterior perforated space.	
Pons Varolii.	
Medulla oblongata.	

The **longitudinal fissure** partially separates the two hemispheres from each
other.   It divides completely the anterior portions of the two frontal lobes ; and
on raising the cerebellum and pons, it will be seen to completely separate the
two occipital lobes : of these two portions of the longitudinal fissure, that which
separates the occipital lobes is the longer.   The intermediate part of the fissure
is filled up by the great transverse band of white matter, the *corpus callosum*.
In the fissure between the two frontal lobes the anterior cerebral arteries ascend
on the corpus callosum.

The **corpus callosum** terminates at the base of the brain by a concave margin,
which is connected with the tuber cinereum through the intervention of a thin
layer of grey substance, the *lamina cinerea*.   This may be exposed by gently
raising and drawing back the optic commissure.   A white band may be observed
on each side, passing backwards from the under surface of the corpus callosum,
across the posterior margin of the anterior perforated space to the hippocampal
gyrus, where each meets the corresponding outer root of the olfactory tract : these
bands are called the *peduncles of the corpus callosum*.   They may be traced
upwards around the genu to become continuous with the *striæ longitudinales*
on its upper surface.   Laterally, this portion of the corpus callosum extends into
the frontal lobe.

The **lamina cinerea** is a thin layer of grey substance, extending backwards
from the termination of the corpus callosum above the optic commissure to the
tuber cinereum ; it is continuous on each side with the grey matter of the anterior
perforated space, and forms the anterior part of the inferior boundary of the third
ventricle.

The **optic commissure** is situated in the middle line, immediately in front of
the tuber cinereum and below the lamina cinerea ; that is to say, the commissure
is superficial to the lamina in the order of dissection when the base is uppermost.
It is the point of junction between the two optic tracts, and will be described
with the cranial nerves.   Immediately behind the diverging optic tracts, and
between them and the peduncles of the cerebrum (*crura cerebri*), is a lozenge-
shaped interval, the *interpeduncular space*, which is bounded behind by the pons
Varolii, and in which are found the following parts : the tuber cinereum, infundibulum, pituitary body, corpora albicantia, and the posterior perforated space.

The **tuber cinereum** is an eminence of grey matter, situated between the optic tracts, and extending from the corpora albicantia to the optic commissure, to which it is attached ; it is connected with the surrounding parts of the cerebrum, forms part of the floor of the third ventricle, and is continuous with the grey substance in that cavity.   From the middle of its under surface a conical tubular process of grey matter, about two lines in length, is continued downwards and forwards to be attached to the posterior lobe of the pituitary body. This is the **infundibulum,** and its canal, which is funnel-shaped, communicates with the third ventricle.

The **pituitary body** (*hypophysis cerebri*) is a small, reddish-grey, vascular mass, weighing from five to ten grains, and of an oval form, situated in the sella turcica, where it is retained by a process of dura mater, named the diaphragma sellæ.   This process covers in the sella turcica, and has a small hole in its centre through which the infundibulum passes.

Structure.– The pituitary body is very vascular, and consists of two lobes, separated from one another by a fibrous lamina. Of these, the anterior is the larger, of an oblong form, and somewhat concave behind, where it receives the posterior lobe, which is round. The two lobes differ both in development and structure. The *anterior lobe*, of a dark, reddish-brown colour, is developed from the epiblast of the buccal cavity, and resembles to a considerable extent, in microscopic structure, the thyroid body. It consists of a number of isolated vesicles and slightly convoluted tubules, lined by epithelium and united together by a very vascular connective tissue. The epithelium is columnar and occasionally ciliated. The alveoli sometimes contain a colloid material, similar to that found in the thyroid body, and their walls are surrounded by a close network of lymphatic and capillary blood-vessels. The *posterior lobe* is developed as an outgrowth from the embryonic brain, and during fœtal life contains a cavity which communicates through the infundibulum with the cavity of the third ventricle. In the adult it becomes firmer and more solid, and consists of a sponge-like connective tissue arranged in the form of reticulating bundles, between which are branched cells, some of them containing pigment. In the lower animals the two lobes are quite distinct, and it is only in the mammalia that they become fused together.

The **corpora albicantia** or **mammillaria** are two small, round, white masses, each about the size of a pea, placed side by side immediately behind the tuber cinereum, and connected with each other across the mesial plane. They are mainly formed by the anterior crura of the fornix, which, after descending to the base of the brain, are twisted upon themselves to form loops, and constitute the white covering of the corpora albicantia. A second fasciculus, the *bundle of Vicq d'Azyr,* converges from the optic thalamus, and enters the anterior part of each body on its dorso-mesial surface. They are composed externally of white substance, and internally of grey matter ; the nerve-cells of the grey matter are arranged in two sets, inner and outer, the cells of the former set being the smaller. They are also connected to the tegmentum by a small bundle of fibres, the peduncle of the mammillary body. At an early period of fœtal life they are blended together into one large mass, but become separated about the seventh month. In most vertebrates there is only one median corpus albicans.

The **posterior perforated space** (*pons Tarini*) corresponds to a whitish-grey fossa placed between the corpora albicantia in front, the pons Varolii behind, and the crus cerebri on either side. It forms the posterior part of the floor of the third ventricle, and is perforated by numerous small orifices for the passage of the postero-median ganglionic branches of the posterior cerebral and posterior communicating arteries.

The **pons Varolii** is situated immediately behind the two crura of the cerebrum. It consists of a broad band of white fibres, which passes transversely from one cerebellar hemisphere to the other ; the band becoming narrower as it enters the cerebellum.   In the middle line on its under surface a narrow groove runs from before backwards and accommodates the basilar artery.

The **medulla oblongata** emerges from the posterior border of the pons Varolii; it is pyramidal in form, and is continuous below with the cervical portion of the spinal cord.   It is marked on its ventral surface by a median fissure, continuous below with the anterior median fissure of the cord, and on either side by secondary fissures and columns, which will be described in the sequel.

The **frontal lobe.**—The under surface of the frontal lobe, sometimes named the orbital lobe, is seen on the anterior part of the base of the brain on either side of the median line.   It has already been described (page 628).

The **fissure of Sylvius** at the base of the brain separates the frontal from the temporal lobe, and lodges the middle cerebral artery.   It has also been described (page 623).

The **optic tracts** are well-marked flattened bands of fibres, which run obliquely across the crus cerebri on either side, and unite anteriorly to form the optic commissure.   They will be described in connection with the cranial nerves.

Fig. 351.—Transverse vertical section of the brain, through the fore part of the foramen magnum, looked at from the front.   (After Hirschfeld and Leveillé.)

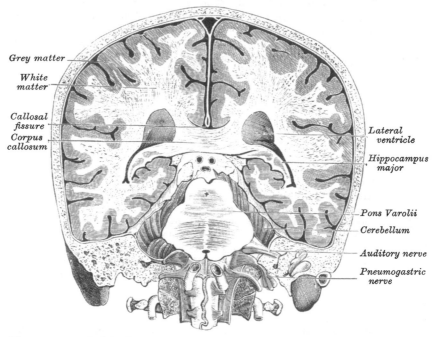

The **crura cerebri** (*peduncles of the cerebrum*) are two thick cylindrical bundles of white matter, which appear in front of the anterior border of the pons, and diverge as they pass forwards and outwards to enter the under surface of each hemisphere.   Each crus is about three-quarters of an inch in length, and is about the same in breadth anteriorly, but somewhat less posteriorly.   They are marked upon their surface with longitudinal striæ, and each is crossed, just before entering the hemisphere, by the fourth nerve and the optic tract, the latter of which is adherent by its upper surface to the peduncle.

The **temporal lobe.**—The under surface of the temporal lobe is visible at the base of the brain, on either side of the crura and the structures contained in the interpeduncular space.   It is separated anteriorly from the frontal lobe by the fissure of Sylvius, and behind is limited by the anterior border of the lateral hemispheres of the cerebellum.   The fissures and lobes on its surface have already been described (page 630).

The **hemispheres of the cerebellum** are situated on either side of the middle line, and cover the occipital lobes of the cerebrum, when viewed from the base.

The cerebellum differs much in appearance from the rest of the encephalon, being of a darker colour, while its convolutions are smaller and narrower, and arranged like the leaves of a book, and hence called *folia*.

General arrangement of the parts composing the cerebrum.—Each hemisphere, as already stated, consists of a central cavity, the *lateral ventricle*, surrounded by thick and convoluted walls of nervous tissue.

Interior of the cerebrum.—If the upper part of either hemisphere is removed with a knife, about half an inch above the level of the corpus callosum, its internal white matter will be exposed. It is an oval-shaped centre, of white substance, surrounded on all sides by a narrow convoluted margin of grey matter, which presents an equal thickness in nearly every part. This white central mass has

FIG. 352.—Section of the brain. Made on a level with the corpus callosum.

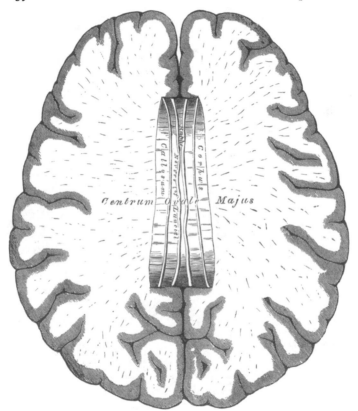

been called the *centrum ovale minus*. Its surface is studded with numerous minute red dots (*puncta vasculosa*), produced by the escape of blood from divided blood-vessels. In inflammation or great congestion of the brain these are very numerous, and of a dark colour. If the remaining portion of one hemisphere is slightly separated from the other, a broad band of white substance will be observed, connecting them at the bottom of the longitudinal fissure; this is the *corpus callosum*. The margins of the hemispheres which overlap this portion of the brain are called the *labia cerebri*. Each labium is part of the callosal convolution already described; and the space between it and the upper surface of the corpus callosum is termed the *callosal fissure* (fig. 348). The hemispheres should now be sliced off to a level with the upper surface of the corpus callosum, when the white substance of that structure will be seen connecting the two hemispheres. The large expanse of medullary matter now exposed, surrounded by the convoluted margin of grey substance, is called the *centrum ovale majus of Vieussens* (fig. 352).

The **corpus callosum.**—The corpus callosum is a thick stratum of transversely directed nerve-fibres, by which probably almost every part of one hemisphere is connected with the corresponding part of the other hemisphere. The fibres of this body, when they pass from it into the hemispheres, radiate in various directions, to terminate in the grey matter of the periphery. It thus connects the two hemispheres of the brain, forming their great transverse commissure, and at the same time roofs in the lateral ventricles. The best conception of its size and form is obtained by making an anterior posterior vertical section through the

FIG. 353.—Vertical median section of the encephalon, showing the parts in the middle line.

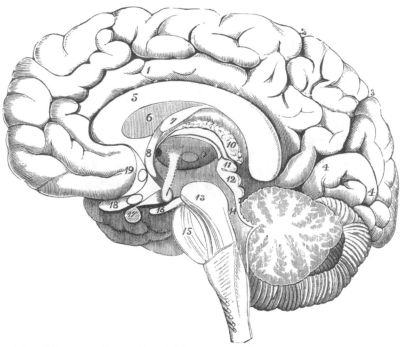

1. Convolution of the corpus callosum. Above it is the calloso-marginal fissure.
2. Fissure of Rolando.
3. The parieto-occipital fissure.
4. 4 point to the calcarine fissure, which is just above the numbers. Between 2 and 3 are the convolutions of the quadrate lobe. Between 3 and 4 is the cuneate lobe.
5. The corpus callosum.
6. The septum lucidum.
7. The fornix.
8. Anterior crus of the fornix, descending to the base of the brain, and turning on itself to form the corpus albicans. The bundle of Vicq d'Azyr is indicated by a dotted line.
9. The optic thalamus. Behind the anterior crus of the fornix, a shaded part indicates the foramen of Monro ; in front of the number an oval mark shows the position of the grey or middle commissure.
10. The velum interpositum.
11. The pineal gland.
12. The corpora quadrigemina.
13. The crus cerebri.
14. The valve of Vieussens (to the right of the number).
15. The pons Varolii.
16. The third nerve.
17. The pituitary body.
18. The optic nerve.
19 points to the anterior commissure, indicated by the oval outline behind the number.

centre of the brain (fig. 353). It is then seen to be a long, thick, irregularly flattened arch ; in front taking a sharp bend, the *genu*, and dipping downwards and backwards to the base of the brain by a reflected portion, the *rostrum*, which is connected with the lamina cinerea ; behind it terminates by a rounded end, which is folded over and is named the *splenium*. It is about four inches in length, and extends to within an inch and a half of the anterior, and two inches and a half of the posterior extremity of the cerebrum. It is somewhat broader behind than in front, and is thicker at either end than in its central part, being thickest behind. The reflected anterior portion of the corpus callosum is called the *beak* or *rostrum*; it becomes gradually thinner as it descends, and is attached by its lateral margins to the frontal lobes. At its termination, in addition to joining the lamina cinerea, the corpus callosum gives off two bands of white substance, the peduncles of the corpus callosum, already described (page 634).

Posteriorly, the corpus callosum forms a thick rounded fold, called the *splenium* or *pad*, which is free for a little distance as it curves forwards, and is then continuous by its under surface with the fornix. The splenium overlaps the mesencephalon, but is separated from it by the pia mater, which is prolonged forwards to form the velum interpositum. On its upper surface, the structure of the corpus callosum is very apparent, being collected into coarse transverse bundles. Along the middle line is a longitudinal depression, the so-called *raphé*, bounded laterally by two or more slightly elevated longitudinal bands, called the *striæ longitudinales* or *nerves of Lancisi*; and, still more externally, other longitudinal striæ are seen, beneath the callosal convolutions. These are the *striæ longitudinales laterales*, or *tænia tectæ*. On each side of the middle line the under surface of the corpus callosum forms the roof of the lateral ventricles, while in the mesial plane it is continuous behind with the fornix, being separated from it in front by the septum lucidum, which forms a vertical partition between the two ventricles. On each side the fibres of the corpus callosum extend into the substance of the hemispheres, connecting them together. The greater thickness of the two extremities of this commissure is explained by the fact that the fibres from the anterior and posterior parts of each hemisphere cannot pass directly across, but have to take a curved direction. The part of the corpus callosum which curves forwards on each side from the genu into the frontal lobe and covers the front part of the anterior cornu of the lateral ventricle is called the *forceps anterior* or *minor*. The part which curves backwards from each side of the splenium into the occipital lobe is known as the *forceps posterior* or *major*. Between these two parts on each side is the main body of the fibres, which extend laterally into the temporal lobe and cover in the body of the lateral ventricle. These are known as the *tapetum* or *mat*.

An incision should now be made through the corpus callosum, on either side of the raphé, when two large irregular cavities will be exposed, which extend through a great part of the length of each hemisphere. These are the lateral ventricles.

The **lateral ventricles** (fig. 354).—The lateral ventricles, two in number, right and left, are irregular cavities situated in the lower and inner parts of the cerebral hemisphere, one on either side of the middle line. They are separated from each other by a mesial vertical partition, the *septum lucidum*, but communicate with the third ventricle and indirectly with each other through the *foramen of Monro*. They are lined by a thin, diaphanous membrane, the *ependyma*, which is covered by ciliated epithelium, and are moistened by a serous fluid, which, even in health, may be secreted in considerable amount. Each lateral ventricle consists of a central cavity or *body*, and three prolongations from it, termed *cornua*. The *anterior* cornu curves forwards and outwards into the frontal lobe; the *posterior* backwards and inwards into the occipital lobe; and the *middle* descends into the temporal lobe.

The **central cavity** or **body** of the lateral ventricle is situated in the lower part of the parietal lobe. It is an irregularly curved cavity, triangular in shape on transverse section, and presents a roof, a floor, and an inner wall. Its *roof* is formed by the under surface of the corpus callosum; its *inner wall* is the septum lucidum, which separates it from the opposite ventricle, and connects the under surface of the corpus callosum with the fornix; its *floor* is formed by the following parts, enumerated in their order of position, from before backwards: the *caudate nucleus* of the *corpus striatum, tænia semicircularis, optic thalamus, choroid plexus,* one half of the *fornix* and its *posterior pillar.*

The **anterior cornu** passes forwards and outwards, with a slight inclination downwards, from the foramen of Monro into the frontal lobe, curving round the anterior extremity of the caudate nucleus. It is bounded above by the corpus callosum, and below by the upper surface of its reflected portion, the rostrum. It is bounded internally by the anterior portion of the septum lucidum, and externally by the head of the caudate nucleus of the corpus striatum. Its apex reaches the posterior surface of the genu of the corpus callosum.

The **posterior cornu** curves backwards into the substance of the occipital lobe, its direction being backwards and outwards, and then inwards; its concavity is therefore directed inwards. Its roof is formed by the fibres of the corpus callosum passing to the temporal and occipital lobes. On its inner wall is seen a longitudinal eminence, which is an involution of the ventricular wall produced by the calcarine sulcus; this is called the *hippocampus minor*, or *calcar avis*. Just above

FIG. 354.—The lateral ventricles of the brain.

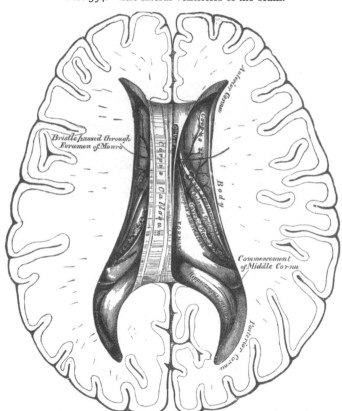

this the forceps major of the corpus callosum, sweeping round to enter the occipital lobe, causes another projection, which is known as the *bulb of the posterior horn*. The hippocampus minor and bulb of the posterior horn are extremely variable in their degree of development, being in some cases ill defined, while in others they are unusually prominent.

Between the middle and posterior cornu is a triangular area, called the *trigonum ventriculi* (see Descending Horn).

The **middle** or **descending cornu,** the largest of the three, traverses the temporal lobe of the brain, forming in its course a remarkable curve round the back of the optic thalamus. It passes at first backwards, outwards, and downwards, and then curves round the crus cerebri, forwards and inwards, to within an inch of the apex of the temporal lobe, its direction being fairly well indicated on the surface of the brain by that of the parallel sulcus. Its upper boundary, or roof, is formed chiefly by the under surface of the tapetum of the corpus callosum, but the tail of the nucleus caudatus of t˙e corpus striatum and the tænia semicircularis are also prolonged into it, and extend forwards in the roof of the descending horn to its extremity, where they end in a mass of grey matter, the *amygdaloid nucleus* ; this nucleus is merely a localised thickening of the adjacent grey cortex. Its lower boundary, or floor, presents for examination the following parts : the *hippocampus major, pes hippocampi, eminentia collateralis* or *pes accessorius, corpus fimbriatum*, prolonged from the posterior pillar of the fornix, and the *choroid plexus*. Along the mesial aspect of the descending cornu there is a cleft-like opening, which is the lower part of the *transverse fissure*, through which the choroid plexus of the pia mater is invaginated into the ventricle, but covered by the ependyma, which is pushed in before it.

FIG. 355.—Middle part of a horizontal section through the cerebrum at the level of the dotted line in the small figure of one hemisphere. (From Ellis, after Dalton.)

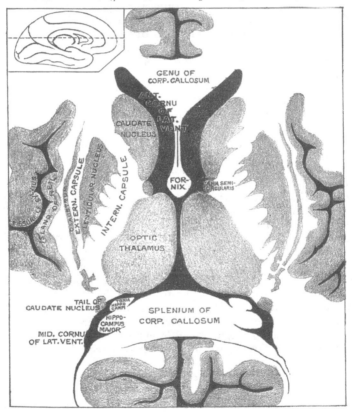

The **corpus striatum** has received its name from the striped appearance which its section presents, in consequence of diverging white fibres being mixed with the grey matter which forms the greater part of its substance. The larger portion of this body is embedded in the white substance of the hemisphere, and is therefore external to the ventricle. It is termed the *extra-ventricular portion*, or the *nucleus lenticularis* ; a part, however, is visible in the ventricle and its anterior cornu : this is the *intra-ventricular portion*, or the *nucleus caudatus*.

The **nucleus caudatus** (fig. 355) is a pear-shaped, highly arched mass of grey matter; its broad extremity is directed forwards into the fore part of the body and anterior cornu of the lateral ventricle ; its narrow end is directed outwards and backwards on the outer side of the optic thalamus ; it is continued downwards into the roof of the descending cornu, where it terminates in the *nucleus amygdalæ*, a collection of grey matter in the apex of the temporal lobe. It is covered by the lining of the ventricle, and crossed by some veins of considerable size. It is separated from the extra-ventricular portion, in the greater part of its extent, by a lamina of white matter, which is called the *internal capsule*, but the two portions of the corpus striatum are united in front.

The **nucleus lenticularis,** or extra-ventricular portion of the corpus striatum, is only seen in sections of the hemisphere. When divided horizontally, it presents, to some extent, the appearance of a biconvex lens, while a vertical transverse section of it gives a somewhat triangular outline. It does not extend as far forwards or backwards as the nucleus caudatus. It is bounded externally by a lamina of white matter called the *external capsule*, on the outer surface of which is a thin layer of grey matter termed the *claustrum*. The claustrum presents ridges and furrows on its outer surface, corresponding to the convolutions and sulci of the island of Reil, from which it is separated by a thin white lamina.

Upon making a transverse vertical section through the middle of the nucleus lenticularis it is seen to present two white lines, parallel with its lateral border, which divide it up into three zones, of which the outer and largest is of a reddish colour and is known as the *putamen*, while the two inner are paler and of a yellowish tint and are termed the *globus pallidus*. All three zones are marked by fine radiating white fibres, which are most distinct in the putamen. The grey matter of the corpus striatum is traversed by nerve-fibres, some of which are believed to originate in it. The cells are multipolar, both large and small; those of the lenticular nucleus containing yellow pigment.

The **internal capsule** is formed by fibres of the crusta of the crus cerebri, supplemented by fibres derived from the corpus striatum and optic thalamus on each side. In horizontal section (fig. 355) it is seen to be somewhat abruptly curved, with its convexity inwards; the prominence of the curve is called the *genu*, and projects between the caudate nucleus and the optic thalamus. The portion in front of the genu is termed the *anterior limb*, and separates the lenticular from the caudate nucleus; the portion behind the genu is the *posterior limb*, and separates the lenticular nucleus from the optic thalamus. The internal capsule is composed largely of fibres, which, derived from the crusta of the crus cerebri, are continued through it to the cortex of the cerebral hemispheres, the fibres of the anterior limb passing to the frontal region; those from the genu and the anterior two-thirds of the posterior limb pass to the Rolandic area of the cortex, while those in the hindermost third of the same limb pass to the temporo-occipital region. In addition to these, there are fibres which terminate in the corpus striatum and the optic thalamus; and other fibres, derived from the grey matter of these two bodies, from the subthalamic region,* and from the hemisphere of the opposite side through the corpus callosum, which pass through the internal capsule to the cerebral cortex.

The **external capsule** is a lamina of white matter, situated on the outer side of the lenticular nucleus, between it and the claustrum, and is continuous with the internal capsule below and behind the lenticular nucleus. It is made up of fibres derived partly from the anterior white commissure and partly from the sub-thalamic region.

The **claustrum** is a thin layer of grey matter, situated on the outer surface of the external capsule. On transverse section it is seen to be triangular, with its apex directed upwards and its base downwards. Its inner surface, which is contiguous to the outer capsule, is smooth, but its outer surface presents ridges and furrows which correspond with the convolutions and sulci of the island of Reil, with which it is in close relationship. The claustrum is regarded as a detached portion of the grey matter of the island of Reil, from which it is separated by a layer of white fibres, the *capsula extrema* or *band of Baillarger*. Its cells are small

---

* The *subthalamic region* is the mass upon which the optic thalamus rests, and is an extension forwards of the tegmentum of the mesencephalon (see page 653).

and spindle-shaped, and contain yellow pigment ; they are similar to those found in the deepest layer of the cortex.

The **tænia semicircularis** is a narrow, whitish band of medullary substance, situated in the depression between the caudate nucleus and the optic thalamus. Anteriorly its fibres are partly continued into the anterior pillar of the fornix ; some, however, pass over the anterior commissure to the grey matter between the caudate nucleus and septum lucidum, while others penetrate the caudate nucleus. Posteriorly it is continued into the roof of the middle or descending horn of the lateral ventricle, at the extremity of which it enters the *nucleus amygdalæ*, an oval mass of grey matter, situated in the roof of the lower extremity of the descending horn. Like the corpus striatum, it is formed by a localised thickening of the grey matter of the cortex cerebri. Superficial to it is a large vein, *vena corporis striati*, which receives numerous small veins from the surface of the corpus striatum and optic thalamus ; it runs forwards and passes through the foramen of Monro to join the corresponding vena Galeni. On the surface of the vein of the corpus striatum is a narrow band of white fibres, named the *lamina cornea*.

The remains of the corpus callosum should now be removed in order to expose the fornix.

The **fornix** (figs. 353, 354) is a longitudinal, arch-shaped lamella of white matter, situated beneath the corpus callosum, with which it is continuous behind, but separated in front by the septum lucidum. It may be described as consisting of two symmetrical halves, one for either hemisphere. The two portions are not united to each other in front and behind, but their central parts are joined together in the middle line. The two anterior, separated parts are called the *anterior pillars (columnæ fornicis)* ; the intermediate united portions constitute the *body of the fornix* ; and the posterior parts, which are also separated from each other, are called the *posterior pillars (crura fornicis)*.

The *body of the fornix* is triangular, narrow in front, broad behind. Its upper surface is connected, in the median line, to the septum lucidum in front, and the corpus callosum behind ; laterally this surface forms part of the floor of each lateral ventricle. Its under surface rests upon the velum interpositum, which separates it from the third ventricle and the inner portion of the upper surface of the optic thalami. Its outer edge, on each side, is free, and is connected with the choroid plexuses.

The *anterior pillars* arch downwards towards the base of the brain, separated from each other by a narrow interval. They are composed of white fibres, which descend through the grey matter in the lateral wall of the third ventricle, and are placed immediately behind the anterior commissure. At the base of the brain, each pillar becomes twisted upon itself to form a loop, somewhat resembling the figure of 8. The lowest part of the loop constitutes the white matter of the corresponding *corpus albicans*, from which the fibres can apparently be traced upwards and backwards, as the *bundle of Vicq d'Azyr*, into the substance of the corresponding optic thalamus (fig. 353). It must be stated, however, that there is probably no direct continuity between this bundle and the anterior pillar of the fornix—the latter possibly ending in the grey matter of the corpus albicans. The anterior crura of the fornix are joined in their course by the peduncles of the pineal gland and the superficial fibres of the tænia semicircularis, and receive fibres from the septum lucidum. Zuckerkandl describes an *olfactory fasciculus*, which becomes detached from the main portion of the anterior pillar of the fornix, and passes downwards, in front of the anterior commissure, to the base of the brain, where it divides into two bundles, one joining the inner root of the olfactory tract ; the other, the peduncle of the corpus callosum, and through it reaching the hippocampal convolution.

Between the anterior pillars of the fornix and the anterior extremity of the optic thalamus, an oval aperture is seen on each side ; this is the foramen of Monro (fig. 359). The two openings descend towards the middle line, and lead into the upper part of the third ventricle. Through this foramen the lateral ventricles communicate with the third ventricle, and consequently with each other ; through it also the two choroid plexuses become joined with each other across the middle line. The boundaries of the opening are, above and in front, the anterior pillars of the fornix ; behind, the anterior extremity of the optic thalamus.

The *posterior pillars* are the backward prolongations of the two halves of the body of the fornix. They are flattened bands, and, at their commencement, are intimately connected by their upper surfaces with the corpus callosum. Diverging from one another, each curves round the posterior extremity of the optic thalamus, and then passes downwards and forwards into the descending horn of the lateral ventricle. Here it lies along the concavity of the hippocampus major, on the surface of which some of its fibres are spread out, while the remainder are continued, as the *corpus fimbriatum* or *tænia hippocampi*, into the hook or uncus of the hippocampal convolution. Upon examining the under surface of the fornix, between its diverging posterior pillars, a triangular portion of the under surface of the corpus callosum may be seen. On it are a number of curved or oblique lines passing between the two pillars of the fornix. This portion has been termed the *lyra*, from the fancied resemblance it bears to a harp.

The **anterior commissure** is a bundle of white fibres, placed in front of the anterior pillars of the fornix, and appears to connect together the corpora striata. On transverse section it is seen to be oval in shape, its long diameter being vertical in direction and measuring about one-fifth of an inch. Its fibres can be traced backwards and downwards through the globus pallidus and below the putamen on each side into the substance of the temporal lobe. It serves in this way to connect the two temporal lobes, but it also contains fibres from the olfactory tract of the opposite side, the decussation of which in the anterior commissure may serve to explain the condition of crossed anosmia, e.g. where there is a lesion in one temporal lobe with a loss of smell in the olfactory area of the opposite side of the nose.

The **septum lucidum** is a thin, double, vertically placed partition, which forms the internal boundary of the body and anterior horn of the lateral ventricle. It consists of two distinct laminæ, separated in part of their extent by a narrow chink or interval, called the *fifth ventricle*. It is a thin, semitransparent septum, attached, above, to the under surface of the corpus callosum ; below, to the anterior part of the fornix behind, and the reflected portion of the corpus callosum in front. It is triangular in form, broad in front and narrow behind ; its inferior angle corresponds with the upper part of the anterior commissure. The outer surface of each lamina is directed towards the lateral ventricle, and is covered by the ependyma of that cavity, while its mesial surface bounds the cavity of the fifth ventricle.

Fifth ventricle.—The fifth ventricle was originally a part of the great longitudinal fissure, which has become shut off by the union of the hemispheres in the formation of the corpus callosum above and the fornix below. Each half of the septum is therefore formed by the median wall of the hemisphere, and consists of an internal layer of grey matter, derived from the grey matter of the cortex, and an external layer of white substance continuous with the white matter of the cerebral hemispheres. The fifth ventricle differs from the other ventricles of the brain, inasmuch as it is not developed from the cavity of the cerebral vesicles, it is not lined by ciliated epithelium but by altered pia mater, and it does not communicate with the general ventricular cavity ; further, the fluid it contains is of the nature of lymph.

The structures on the floor of the descending horn of the lateral ventricle will now be described.

The **hippocampus major,** or **cornu Ammonis** (fig. 356), is a white eminence, about two inches in length, of a curved elongated form, extending throughout the entire length of the floor of the descending horn of the lateral ventricle. At its lower extremity it becomes enlarged, and presenting two or three rounded elevations with intervening depressions, it resembles the paw of an animal, and is called the *pes hippocampi*. If a transverse section is made through the hippocampus major, it will be seen that this eminence is produced by the folding of the cortex of the brain to form the dentate (hippocampal) sulcus. To the outer side and parallel with the hippocampus major an elongated eminence, the *eminentia collateralis*, is frequently recognised. It corresponds with the middle part of the collateral fissure, and its size depends on the direction and depth of this fissure. The main mass of the hippocampus major consists of grey matter, but on its ventricular surface is a thin layer of white matter, known as the *alveus*, which is continuous with the corpus fimbriatum of the fornix and is covered by the ependyma of the ventricle. Dr. J. G. Macarthy, of McGill University, Montreal, has shown * that, if the alveus and superficial strata of grey matter be reflected from the surface of the hippocampus by an incision carried along its convexity, the 'core' of the hippocampus, as he terms it, presents in many cases a corrugated or crimped appearance.

The **corpus fimbriatum** or **fimbria**(*tænia hippocampi*) has already been mentioned

FIG. 356.—Transverse section of the middle horn of the lateral ventricle.
(From a drawing by Mr. F. A. Barton.)

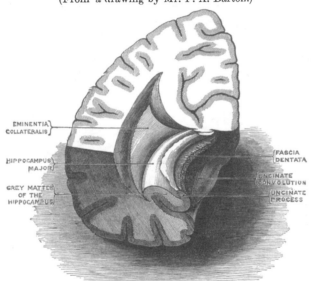

as a part of the posterior pillar of the fornix. It consists of a narrow white band, which is placed immediately below the choroid plexus, and is attached by its deep surface to the white matter (alveus) of the hippocampus major, as it courses through the descending cornu of the lateral ventricle. It can be traced as far as the uncus or hook of the hippocampal gyrus. Its inner margin is free, and rests upon the dentate convolution, from which it is separated by a slit-like fissure, the *fimbrio-dentate fissure*. Its outer margin is attenuated and irregular, and forms the line along which the ependyma is reflected over the choroid plexus as the latter is invaginated through the inferior part of the transverse fissure. When the choroid plexus is pulled away it carries the ependyma with it, and the descending horn opens on to the surface of the brain through the transverse fissure. If now the inner border of the corpus fimbriatum be raised, a notched band of grey matter, the gyrus dentatus, will be exposed; this has already been described as forming part of the limbic lobe (page 632).

The **choroid plexus** is a highly vascular, fringe-like structure, which is situated partly in the body and partly in the descending cornu of the lateral ventricle.

* *Journal of Anat. and Phys.* vol. xxxiii. 1899.

It will be desirable to consider these two portions separately, in order to get a just conception of how they are formed.

The portion in the body of the ventricle is the vascular, fringed border of a triangular fold of pia mater, the *velum interpositum*, which lies on the under surface of the fornix and forms the roof of the third ventricle. It will be remembered that the developing brain vesicles are covered by pia mater. As the prolongation from the first vesicle, which is to form the cerebral hemispheres, increases in size, it grows backwards and downwards and covers the other vesicles, with the result that the pia mater covering the hemisphere comes in contact with that covering the upper surface of the second vesicle (fig. 357).* A portion of the two layers which are in contact forms the velum interpositum. Immediately above is the body of the fornix, which is formed by the fusion of the cerebral hemispheres in the middle line, and below is the cavity of the second vesicle (the third ventricle), with the optic thalamus on either side (fig. 359). Just beyond the free lateral border of the fornix, between it and the tænia semicircularis, is a portion of the first cerebral vesicle, which is not developed into

FIG. 357.—Diagram showing the mode of formation of the velum interpositum.

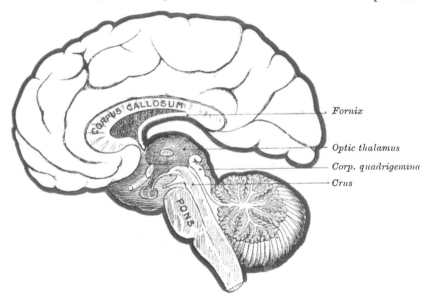

Fornix

Optic thalamus

Corp. quadrigemina

Crus

nervous matter but is made up only of ependyma covered by pia mater. The vessels of this portion of the highly vascular pia mater become dilated and prolonged, and grow into the ventricle, pushing the ependyma before them, and forming an irregular congeries of vessels, apparently encroaching on the cavity of the lateral ventricle, but in reality being external to it, because they are separated from it by the lining membrane of the cavity, the ependyma. This vascular structure is the choroid plexus of the body of the ventricle.

The part of the choroid plexus seen in the descending cornu is formed in exactly the same way, viz. by an ingrowth of the vessels of the pia mater into the cavity, pushing the ependyma before it, at a part of the wall of the horn where there is a similar absence of nervous tissue and where it consists simply of pia mater and ependyma in close contact. This portion lies between the corpus fimbriatum in the floor and the tænia semicircularis in the roof of the descending horn. This area, destitute of nervous matter, is continuous with the area in the body of the

* In the diagram the two layers are represented as being separated from each other, for the sake of clearness.

ventricle, from which the choroid plexus of this region originated, and in it the vessels of its pia mater increase, and, invaginating the ependyma, appear in the descending horn as its choroid plexus. In the body of the ventricle the choroid plexus is really the vascular fringed margin of the velum; beyond the posterior margin of the velum the plexus of the descending horn is continuous with the pia mater on the surface of the gyrus hippocampi; the two portions of the plexus are, however, directly continuous with each other. The gap or cleft through which the invagination of the pia mater takes place is called the *transverse fissure*.

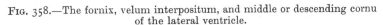

FIG. 358.—The fornix, velum interpositum, and middle or descending cornu of the lateral ventricle.

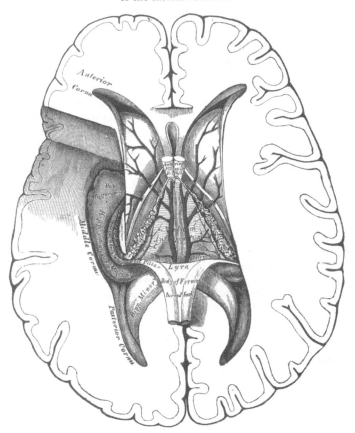

In front, the choroid plexus of the lateral ventricle is small and tapering, and communicates with that of the opposite side through the *foramen of Monro*. In structure, it consists of minute and highly vascular villous processes, containing an afferent and efferent vessel, and covered by a single layer of flattened epithelium, the cells of which often contain a yellowish fat molecule. The anterior choroidal artery is derived from the internal carotid, and enters the ventricle at the extremity of the descending cornu, and after ramifying in the plexus, sends branches into the adjacent parts of the brain. The posterior choroidal arteries, one or two in number, are derived from the posterior cerebral artery, and reach the plexus by passing forwards under the splenium of the corpus callosum. The veins of the choroid plexus unite to form a prominent vein, which courses from behind forwards to the foramen of Monro, and joins with the vein of the corpus striatum to form the corresponding vein of Galen.

The **transverse fissure** is not a real fissure or cleft, because it is filled by the invagination of the pia mater, forming the velum interpositum and the choroid plexuses, covered by the lining of the ventricular cavities. If this involution of pia mater is pulled out, the ventricular lining will necessarily be torn away with it, and a cleft-like space will be left on either side, extending from the foramen of Monro to the bottom of the descending horn of the lateral ventricle. The upper part of this cleft, that is to say, the part nearest the foramen of Monro, is between the lateral border of the body of the fornix and the optic thalamus ; below this, at the commencement of the middle horn, it is between the commencing corpus fimbriatum of the fornix and the pulvinar of the optic thalamus ; and lower still, in the descending horn, between the corpus fimbriatum on the floor and the tænia semicircularis in the roof of the cornu. Posteriorly the transverse fissure opens between the splenium of the corpus callosum above, and the corpora quadrigemina and pineal gland below. Through the fissure the venæ Galeni emerge to join the straight sinus.

The **velum interpositum** or **tela choroidea superior** (fig. 358) is a vascular membrane, and is a prolongation of the pia mater into the interior of the brain through the middle part of the transverse fissure. It is of a triangular form, and separates the under surface of the body and posterior pillars of the fornix from the cavity of the third ventricle. Laterally it covers the inner part of the upper surface of the optic thalamus. Its posterior border or base lies beneath the splenium of the corpus callosum above, and the optic thalamus, the corpora quadrigemina, and pineal body below. Its anterior extremity, or apex, ends just behind the anterior pillars of the fornix, where it is connected with the anterior extremities of the choroid plexuses, which are here united through the foramen of Monro, and are then prolonged backwards on the under surface of the velum as the *choroid plexuses of the third ventricle* ; in front, these plexuses of the third ventricle lie close to the middle line, but diverge from each other behind. The lateral margins of the velum interpositum form the choroid plexuses of the lateral ventricles. It is supplied by the anterior and posterior choroidal arteries, already described. The veins of the velum interpositum, the *venæ Galeni*, two in number, run between its layers, each being formed by the union of the vein of the corpus striatum with the choroid vein. The venæ Galeni unite posteriorly into a single trunk, the *vena magna Galeni*, which terminates in the straight sinus (fig. 326).

II. The Inter-brain

The **inter-brain** (thalamencephalon) is the region of the third ventricle, and comprises the parts developed from the second cerebral vesicle, together with that portion of the first vesicle which is not concerned in the formation of the cerebral hemispheres.

The inter-brain is connected above and in front with the cerebral hemispheres ; behind, with the mid-brain or mesencephalon. On its upper surface it is entirely concealed from view, as it is covered by those portions of the internal surfaces of the cerebral hemispheres which have fused together to form the corpus callosum and the fornix, and is separated from the latter by the two layers of pia mater which form the velum interpositum. Inferiorly it reaches the base of the brain, forming the structures contained in the interpeduncular space.

The third ventricle is the cavity of the inter-brain (fig. 359). It is a narrow median crevice between the two optic thalami, which constitute the side walls of the inter-brain. Its roof is formed by the velum interpositum, from which are suspended the choroid plexuses of the third ventricle. Its floor, somewhat oblique in its direction, is formed, from before backwards, by the tuber cinereum, with its infundibulum and pituitary body ; the corpora albicantia ; the posterior perforated space ; and the tegmenta of the crura cerebri. Its sides are formed by the optic thalami, and are limited above by a delicate band of white fibres, the *stria pinealis*, which runs along the junction of the mesial and upper surfaces of the optic thalamus to join the anterior pillars of the fornix. Its sides are somewhat convex, so that in the middle of the ventricle the two lateral walls are almost in contact, and are here united across the middle line by a band of grey nervous matter, the *middle, grey,* or *soft* commissure. The ventricle is bounded in front by the anterior pillars of the fornix and the lamina cinerea; behind by the pineal gland, the posterior commissure, and the upper end of the iter a tertio ad quartum ventriculum. The cavity is much deeper in front than behind, and presents a recess at its anterior part, which lies over the optic commissure and is therefore termed the *optic recess*. Behind and below this is the conical depression of the infundibulum, passing downwards and forwards to the pituitary body. At its posterior extremity the cavity forms another and smaller recess, which extends into

FIG. 359.—The third and fourth ventricles.

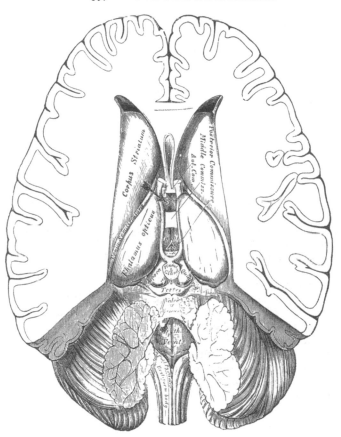

An arrow has been placed in the position of the foramen of Monro.

the stalk of the pineal gland, and is termed the *pineal recess*. At its upper and anterior part, immediately behind the anterior pillars of the fornix and in front of the optic thalamus, is an opening, the *foramen of Monro*, by which this ventricle communicates with the lateral ventricle on either side. The roof of the cavity is limited in front and behind by transverse bands of white matter, known respectively as the *anterior* and *posterior commissures*. The former has already been described in connection with the corpus striatum (page 644).

The *middle* or *soft commissure* consists almost entirely of grey matter. It connects the two optic thalami, and is continuous with the grey matter lining the anterior part of the third ventricle. It is frequently broken in examining the brain, and might then be supposed to be wanting ; it is sometimes double.

The *posterior commissure* is a rounded band of white fibres, which stretches across from one optic thalamus to the other, overlying the upper end of the aqueduct of Sylvius, or iter a tertio ad quartum ventriculum. It is usually described as belonging to the inter-brain, but would appear to belong in part to the mid-brain, since some of its fibres are commissural and connect the anterior corpora quadrigemina to the fillet of the opposite side (see below). In addition there are other decussating fibres, which come from the tegmentum of the crus cerebri on one side, and decussate with those of the opposite side in the posterior commissure, and passing through the optic thalamus reach the cerebral hemispheres. Fibres have also been described as taking their origin in the pineal body and ganglion habenulæ, and passing across to the posterior longitudinal bundle and oculo-motor nucleus of the opposite side ; these fibres occupy the ventral part of the commissure, and receive their myelin sheath before those in its dorsal part. But to a certain extent the posterior commissure belongs to the inter-brain, since it contains fibres which serve as commissural fibres between the two optic thalami.

The **optic thalami** are two large oblong masses, situated on either side of the third ventricle, and lying between the diverging portions of the corpora striata. They are composed mainly of grey matter, but their free surfaces are coated with a thin layer of white nervous tissue. They present outer and under surfaces, which are not free, but are blended with contiguous parts of the brain, and upper, inner, and posterior surfaces, which are free. The anterior extremity is narrow, and forms the posterior boundary of the foramen of Monro. The outer surface is in contact with the posterior limb of the internal capsule, which separates it from the lenticular nucleus. The *inferior* surface rests upon and is continuous with the tegmentum of the crus cerebri. Its upper surface is free, and is separated from the caudate nucleus by a furrow which lodges the lamina cornea, the vein of the corpus striatum, and the tænia semicircularis. It is divided into an outer and an inner part by a groove which runs from behind, forwards and inwards. The outer part forms a portion of the floor of the lateral ventricle, and is covered by the ependyma of that cavity ; it terminates in front in a tubercle, the *anterior tubercle* of the optic thalamus. The inner part is covered by the velum interpositum, which separates it from the fornix, and is excluded from both the lateral and third ventricles by the reflection of the lining of these cavities, and is therefore destitute of an ependymal covering.

The *internal* surface forms the lateral wall of the third ventricle, and running along its upper border is the peduncle of the pineal gland, from which the ependyma of the third ventricle is reflected on to the under surface of the velum interpositum. The *posterior* surface projects beyond the level of the corpora quadrigemina, and forms a well-marked rounded prominence, the *posterior tubercle* or *pulvinar*. The pulvinar is continued externally into a second eminence, the *external geniculate body*, which is placed above and to the outer side of the *internal geniculate body*, and from which it is separated by the superior brachium, one of the roots of the optic tract.

The optic thalamus is formed chiefly of grey matter, which is arranged in two masses, the *outer* and *inner nuclei*, and these are partially separated from each other by an **S**-shaped vertical lamina of white matter, called the *internal medullary lamina*. This is named internal in contradistinction to a second or *external medullary lamina* of white matter, which coats the outer surface of the optic thalamus and connects it with the internal capsule. The inner nucleus is connected with the corresponding nucleus of the opposite side through the middle commissure of the third ventricle. The external nucleus, which is the larger of the two, extends backwards into the pulvinar. The grey matter of the optic thalamus contains large multipolar and fusiform cells, and is traversed in every direction by numerous nerve-fibres.

The optic thalamus is intimately connected with the following structures :

1. It constitutes a relay for the greater number of the fibres of the tegmentum of the crus cerebri.

2. The pulvinar receives many of the fibres of the optic tract.

3. It is connected with the cerebral cortex, (a) through the *anterior stalk of the optic thalamus*, which passes from the anterior extremity of the thalamus through the anterior limb of the internal capsule to the frontal lobe ; (b) through the *posterior stalk* or *optic radiations*, consisting of fibres which take their origin in the pulvinar and are transmitted through the extreme posterior part of the internal capsule to the occipital lobe ; (c) through the *inferior stalk* or *ansa peduncularis*, made up of fibres which leave the inferior surface of the thalamus and end in the temporal lobe ; (d) through fibres which pass from the external surface of the thalamus to the parietal lobe.

4. With the corpus striatum. The fibres destined for the caudate nucleus leave the external surface ; those for the lenticular nucleus, the inferior aspect of the thalamus.

5. With the corpus albicans through the bundle of Vicq d'Azyr.

In connection with the optic thalamus two small nuclei of grey matter require consideration : (1) One of these, the *anterior nucleus*, is situated in the *anterior tubercle* of the optic thalamus. This nucleus receives the fibres (bundle of Vicq d'Azyr) which take origin in the cells of the corpus albicans (see page 635). Though this bundle of fibres appears to be the direct continuation of the anterior pillar of the fornix through the corpus albicans to the optic thalamus, it is believed to have no histological continuity with it. The fibres of the anterior pillar of the fornix form synapses in the corpus albicans around the cells which give origin to the bundle of Vicq d'Azyr, and thus an indirect communication only is established between the fornix and the optic thalamus. (2) The second grey nucleus lies in a depressed space, the *trigonum habenulæ*, situated between the pulvinar and the posterior part of the peduncle of the pineal gland. It is termed the *ganglion of the habenula*. It receives fibres from the peduncles of the pineal body, and sends off others, which pass to a small collection of grey matter, situated between the diverging crura cerebri, and named the *ganglion interpedunculare*.

The **pineal gland** (*epiphysis cerebri*), so named from its peculiar shape (pinus, a *fir-cone*), is a small reddish-grey body, conical in shape (hence its synonym, *conarium*), placed immediately above and behind the posterior commissure and between the anterior corpora quadrigemina, on which it rests. It is covered by the velum interpositum, which intervenes between it and the splenium of the corpus callosum. It is an upgrowth from the second cerebral vesicle (hence the name epiphysis), and is at first hollow, but soon becomes solid and loses its connection with the ventricular cavity. It is retained in its position by a duplicature of pia mater, derived from the under surface of the velum interpositum, which almost completely invests it. The pineal gland is about four lines in length and from two to three in width at its base, and is said to be larger in the child than in the adult, and in the female than in the male. It is attached on either side by a flattened stalk of white matter, the *pedunculus conarii*. This stalk consists of two laminæ, upper and lower, separated by a little recess, the *pineal recess* (see page 649). The lower lamina is prolonged into the posterior commissure. The upper divides into two strands, the *peduncles of the pineal gland*, or *striæ pinealis* ; these extend on either side along the optic thalamus at the junction of its mesial and upper surfaces (see page 650) to the anterior pillars of the fornix, with which they blend. The two stalks join together at their posterior extremity, in front of the pineal gland, forming a sort of festoon, and the base of the gland is connected to their posterior margin at the point of junction.

Structure.—The pineal gland consists of a number of follicles, lined by epithelium, and connected together by ingrowths of connective tissue. The follicles contain a transparent viscid fluid and a quantity of sabulous matter, named *acervulus cerebri*, composed of phosphate and carbonate of lime, phosphate of magnesia and ammonia, with a little animal matter. These concretions are almost constant in their existence, and are present at all periods of life. They are found upon the surface of the pineal body and occasionally upon its peduncles.

Morphologically the pineal gland is regarded as the homologue of the structure termed the *pineal eye* of the lizards. In these reptiles the epiphysis cerebri is attached by an elongated stalk and projects through the parietal foramen. Its extremity lies immediately under the epidermis, and on microscopic examination presents, in a rudimentary fashion, structures similar to those found in the eyeball.

III. THE MID-BRAIN

The **mid-brain**, or **mesencephalon,** is the constricted portion of the brain which connects the pons Varolii with the inter-brain and hemispheres, and hence it is frequently called the *isthmus cerebri*. It is developed from the third cerebral vesicle, the cavity of which becomes the aqueduct of Sylvius. It comprises the crura cerebri, the corpora quadrigemina, the geniculate bodies, and the Sylvian aqueduct. Its direction is from before backwards and downwards. In front and above it is continuous with the inter-brain; below with the pons. Its two surfaces are ventral and dorsal. They are free, but concealed; the ventral surface by the apices of the temporal lobes, which overlap it; the dorsal, by the overhanging cerebral hemispheres. The *ventral* surface, when exposed by drawing aside the temporal lobes, is seen to consist of two cylindrical bundles of white matter, which emerge from the pons and diverge as they pass forwards and outwards to enter the inner and under part of either hemisphere. They are the *crura cerebri* or *cerebral peduncles*, and between them is a triangular area, already described as part of the interpeduncular space (see page 634); near the point of divergence of the crura the roots of the third nerve are seen to emerge in several bundles from a groove, the *sulcus oculo-motorius*. The *dorsal* surface is not visible until a considerable portion of the cerebral hemispheres and other overlying structures have been removed. It then presents four rounded eminences placed in pairs, two in front and two behind, and separated from one another by a crucial depression. These are termed the *corpora* or *tubercula quadrigemina*. The ventral and dorsal surfaces meet on the side of the mid-brain, and are separated from each other by a furrow, the *lateral groove*, which runs from below upwards and forwards (fig. 359).

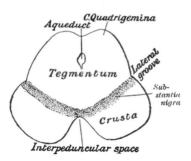

FIG. 360.—Transverse section of the mid-brain.

Aqueduct *C.Quadrigemina*

Tegmentum *Lateral groove*

Substantia nigra

Crusta

Interpeduncular space

If a cross section be made through the mesencephalon (fig. 360) it will be seen that each lateral half is divided into two unequal portions by a lamina of deeply pigmented grey matter, named the *substantia nigra*; of these the postero-superior portion is named the *tegmentum*, and the antero-inferior the *crusta* or *pes*. The substantia nigra is curved on section with its concavity upwards, and extends from the lateral groove externally to the oculo-motor sulcus internally. The two crustæ are quite separate from one another, but the two halves of the tegmentum are joined to each other in the mesial plane by a prolongation forwards of the raphe or median septum of the pons. Laterally the tegmenta are free, but dorsally they blend with the corpora quadrigemina.

Crustæ.—The crustæ, which are two in number, separated by the interpeduncular space, are semilunar in section, and consist of longitudinal bundles of white fibres, which may be divided into three principal sets: (1) Those occupying the outer third of the crusta are believed to arise from the cells of the nuclei pontis, grey nuclei in the pons Varolii, and pass through the posterior part of the internal capsule to the cerebral cortex of the occipital and temporal lobes. (2) The fibres occupying the middle third of the crusta take their origin in the cells of the Rolandic area of the cortex, and, converging to the internal capsule, pass down through its genu and through the anterior two-thirds of its posterior limb to the crusta, from which they are prolonged through the pons into the anterior pyramid of the medulla oblongata. (3) The origin, from below, of the fibres occupying the inner third of the crusta is uncertain, though by some they are believed to arise in the crusta itself, from the cells of the locus niger. Above, they pass through the anterior part of the internal capsule to the cerebral cortex of the frontal lobe. In addition to these three sets of longitudinal fibres, a well-marked bundle, defined by having an oblique direction, must be noted. This is named the *mesial fillet*. It arises from the fillet (see below), and at the lower part of the crusta is situated at its mesial border ; as it ascends it courses obliquely outwards to reach the lateral border of the middle group of fibres (pyramidal tract) and becomes lost in the subthalamic region.

The **tegmentum,** or that portion of the mid-brain which is superior to the substantia nigra, consists of longitudinally directed strands of white fibres, which are separated from each other by transversely arched fibres. There is also a considerable quantity of grey matter. It thus forms a peculiar reticulated structure, which has been named *formatio reticularis*, and is similar to a like structure in the pons and medulla, with which it is continuous. In some parts of the tegmentum the longitudinal fibres are arranged in fairly well defined tracts, which are as follows : 1. The *posterior longitudinal bundle*, which is composed of large nerve-fibres, and lies on either side of the median line, just below the aqueduct. These fibres are continued upwards from the anterior column of the cord, in which they probably form short longitudinal commissures between its different segments. They pass through the pyramid of the medulla, then form the posterior longitudinal bundle of the pons and enter the tegmentum ; here they give off fibres to the nuclei of the third and fourth cranial nerves: At the front of the mid-brain some of the fibres of the posterior longitudinal bundle enter the posterior commissure and there decussate (see page 650); others pass upwards to the subthalamic region. 2. Fibres from the superior cerebellar peduncle. These lie on either side of the middle line of the tegmentum, and, as they pass through it, decussate with each other, so that the fibres of one half of the cerebellum pass to the opposite half of the cerebrum. Having crossed to the opposite side, the bundle of fibres passes upwards and forwards, enclosing a mass of grey matter, the *red nucleus*, or *nucleus of the tegmentum*, from which it probably receives fibres, and eventually passes into the optic thalamus. 3. The *fillet*. This takes its chief origin in the medulla, and passes through the pons to the mid-brain, as will be described in the sequel. It forms a considerable bundle of longitudinal fibres in the ventral part of the tegmentum, and divides into three parts—the *upper, mesial,* and *lower fillet*. The *upper* fillet passes to the upper pair of corpora quadrigemina and the occipital region of the cerebral hemisphere. The *mesial* fillet has already been alluded to in the description of the crusta. After separating from the rest of the fillet its fibres assume an oblique direction, and are eventually lost in the subthalamic region. The *lower* fillet, also called *lemniscus*, is situated in the ventral part of the tegmentum, through which it passes obliquely and emerges at its side, and after crossing the superior peduncle of the cerebellum, passes to the inferior quadrigeminal bodies. It is reinforced by some fibres from the superior medullary velum. 4. Fibres from the olivary nucleus, which pass in a longitudinal direction through the reticular formation of the tegmentum and are continued onwards into the internal capsule.

The *red nucleus*, or *nucleus of the tegmentum*, is a tract of grey matter situated on either side of the middle line, and is composed of numerous large cells, which are deeply pigmented. It is pierced by the fibres of the third nerve, and prolonged above into the posterior part of the subthalamic region.

The **substantia nigra**.—This, as already stated, is a layer of deeply pigmented grey matter, which separates the crusta from the tegmentum. It is thicker internally than externally, where it is partially divided up by the mesial fillet passing from the tegmentum to the crusta. It is traversed at its inner part by some of the fibres of origin of the third cranial nerve. The cells are small and multipolar, and are characterised by containing a large amount of dark pigment granules.

The **corpora** or **tubercula quadrigemina** are four rounded eminences placed in pairs, two in front and two behind, and separated from one another by a crucial depression. They are situated on the dorsal surface of the mid-brain, immediately behind the third ventricle and posterior commissure, and beneath the splenium of the corpus callosum. The *anterior* or *upper* pair, sometimes called the *nates*, are the larger. They are oval, their long diameter being directed forwards and outwards, and are of a grey colour. The *posterior* or *lower* pair, called the *testes*, are hemispherical in form, and lighter in colour than the preceding. From the outer side of each of these eminences, a prominent white band, termed *brachium*, is continued forwards and outwards. Those from the nates (*brachia anteriora*) pass obliquely outwards between the pulvinar and the inner geniculate bodies into the external geniculate bodies. Those from the testes (*brachia posteriora*) lose themselves beneath an oval prominence on either side of the corpora quadrigemina, called the *internal geniculate body*. The corpora quadrigemina are larger in the lower animals than in man. In fishes, reptiles, and birds, they are hollow, and only two in number (*corpora bigemina*) ; they represent the anterior quadrigeminals of mammals. In these lower animals the corpora bigemina are frequently termed the *optic lobes*, because of their connection with the optic tracts. In the mammalia they are four in number, and solid. In the human fœtus all four bodies are differentiated by the fifth month, and form at this time a considerable proportion of the brain.

The corpora quadrigemina are composed of white matter externally, and grey matter within. The *posterior* pair consist almost entirely of grey matter, covered over by a very thin stratum of white substance. Beneath the grey matter is a thin layer of white fibres, forming a part of the lower fillet. This separates the grey matter of the posterior corpora quadrigemina from the central grey matter of the aqueduct. The *anterior* pair are covered superficially by a thin stratum of white matter, the *stratum zonale*, the fibres of which are fine and arranged transversely. Beneath this is the *stratum cinereum*, a layer of grey matter which resembles a cup, semilunar in shape, thicker in the centre, and thinning off towards the margins, and consisting of numerous multipolar cells, for the most part of small size, embedded in a fine network of nervefibres. Below this again is the *stratum opticum*, or *upper grey-white layer*, characterised by the large amount of fine nerve-fibres which intersect the grey matter. These fibres vary in size in different parts of the layer, but have for the most part a longitudinal direction. The nerve-cells between the fibres are larger, and send their axis-cylinder processes into the next stratum. Finally there is the *stratum lemnisci*, or *deep grey-white layer*, which separates the rest of the body from the grey matter around the aqueduct. It consists of fibres partly derived from the upper fillet and partly from the cells of the preceding layer. Interspersed among these fibres are nerve-cells of large size.

In close relationship with the corpora quadrigemina are the **superior peduncles of the cerebellum**. They emerge from the upper and mesial part of the hemispheres of the cerebellum, and run upwards and forwards to the corpora quadrigemina, with which they come in close contact. They then pass under these bodies, through the tegmentum (vide supra), and enter the optic thalamus.

The **corpora geniculata** are two small, oblong masses on each side, situated behind and beneath the posterior end of the optic thalamus, and named, from their position, *corpus geniculatum externum* and *internum*. These two bodies are separated from each other by the brachium anterius of the anterior quadrigeminal body. It is convenient and customary to describe these two bodies together, but the student should bear in mind that the corpus geniculatum externum belongs in reality to the optic thalamus ; the corpus geniculatum internum alone being a part of the mid-brain. The *external* geniculate body is of a dark colour, and presents a laminated arrangement, consisting of alternate layers of grey and white matter. Its cells are large, multipolar, and pigmented ; their processes are intimately related with the visual area in the cerebral cortex of the occipital region. It is believed that the intercellular grey matter of these bodies is composed, to a considerable extent, of the terminations of the optic nerve, which form synapses around the cells. The *internal* geniculate body is smaller in size, lighter in colour, and does not present a laminated arrangement. It receives the posterior brachium from the inferior quadrigeminal body, and some of the fibres of the optic tract appear to enter it. The internal geniculate bodies are connected with each other through the optic commissure by a band of fibres named *Gudden's commissure* (see page 701). The anterior quadrigeminal body, the pulvinar, and the external geniculate body are intimately concerned with vision. They constitute the lower cerebral centre for the optic nerve-fibres which end in them. Extirpation of the eyes in newly born animals entails an arrest of their development, but has no effect on the posterior quadrigeminal body or the internal geniculate body. These latter also are well developed in the mole, where the superior quadrigeminal body is rudimentary.

The **aqueduct of Sylvius, or iter a tertio ad quartum ventriculum.**—This is a narrow canal, about half an inch in length, situated between the corpora quadrigemina and the tegmentum, and connecting the third with the fourth ventricle. Its shape on transverse section varies, being T-shaped below, triangular above, and oval about the middle of its course. It is lined by columnar ciliated epithelium, and surrounded by a layer of grey matter, called the *central grey matter of the aqueduct*, which is continuous with the grey matter of the third and fourth ventricles. This grey matter is separated above from that of the corpora quadrigemina by the stratum lemnisci ; below it, is the posterior longitudinal bundle and the formatio reticularis of the tegmentum. The central grey matter is more abundant below the canal than above it. Here are certain defined groups of cells, which are connected with the roots of the third, fourth, and fifth cranial nerves.

Subthalamic region.—One other structure, to which allusion has already been made, requires mention in this connection ; it is the *subthalamic region*. It is a prolongation forwards of the tegmentum of the crus cerebri, which becomes continuous with the lower surface of the optic thalamus. Towards the anterior part of the crus cerebri the tegmentum becomes thinned out, and is blended with the superjacent portion of the optic thalamus. To this region, the name *subthalamic tegmental region* has been given. In front it is lost at the base of the brain in the grey matter of the anterior perforated space, and is continuous with the grey matter of the floor of the third ventricle. The subthalamic tegmental region contains a forward prolongation of the red nucleus, and consists from above downwards of three layers : (1) *stratum dorsale*, which is directly applied to the under surface of the optic thalamus, and consists of fine longitudinal fibres ; (2) *zona incerta*, a continuation forwards of the formatio reticularis of the tegmentum ; and (3) the *corpus subthalamicum*, a mass of grey matter, which on section presents a lenticular shape, and lies immediately above the substantia nigra.

STRUCTURE OF THE CEREBRUM

The **cerebrum**, like the other parts of the great nerve centre, is composed of grey and white matter. In order to give some general idea of its construction, at all events in part, it may be compared, for the sake of illustration, to a tree, the trunk of which divides into two main divisions, and these break up into smaller branches, which finally end in twigs, to which are attached the leaves, forming an investment to the branches and covering the whole tree. The trunk is represented by the medulla oblongata as it passes through the foramen magnum; the two main divisions by the crura cerebri, which break up into smaller branches; these diverge from each other, dividing and subdividing, until they reach the surface of the hemispheres, where they terminate in single nerve-fibres, which are continuous with the basal axial cylinder processes of the nerve-cells, the representatives of the leaves. These cells are arranged on the surface, resembling a cap, covering the hemispheres, and constitute the cerebral cortex. But here the analogy ends, for in the cerebrum there are, in addition to this cortex, other masses of grey matter situated in the middle of the brain; and other white fibres besides the diverging ones that have been mentioned, and which serve either to connect the two cerebral hemispheres, or to unite different structures in the same hemisphere.

The **white matter of the cerebrum** consists of medullated fibres, varying in size and arranged in bundles, separated by neuroglia. They may be divided into three distinct systems, according to the course they take. 1. Projection or peduncular fibres, which connect the hemisphere with the medulla oblongata and cord. 2. Transverse or commissural fibres, which unite together the two hemispheres. 3. Association fibres, which connect different structures in the same hemisphere. These are, in many instances, collateral branches of the projection fibres, but others are the axons of independent cells.

1. The **projection** or **peduncular fibres** consist of fibres which pass either to or from the cord. They form the longitudinal fibres of the pons, and at its upper border divide into two main groups, which, diverging from each other, constitute the *crura cerebri* or *cerebral peduncles*. In the crura cerebri, as has been before described, the diverging fibres are arranged in two strata, which are separated by the substantia nigra; the ventral or superficial stratum forming the *crusta* of these bodies, and the dorsal or deeper stratum, the *tegmentum*. The fibres derived from these two sources take a different course, and will have to be separately considered.

The fibres of the *crusta* are derived from the pyramid of the medulla, and are continued upwards through the pons; they are reinforced in their passage through the crus by accessory fibres, derived from the central grey nucleus around the Sylvian aqueduct and from the substantia nigra. When they emerge from the crus, most of the fibres pass through the internal capsule, and when they leave it, spread out forwards, upwards, and backwards, forming a series of radiating fibres, the *corona radiata*, which proceed to the cortex. As the fibres pass through the internal capsule they give off branches to the optic thalamus and to the caudate and lenticular nuclei of the striate body, and other fibres, derived especially from the first of these ganglia, form a part of the corona radiata, and pass to the cortex of the cerebral hemispheres. The fibres of the *tegmentum* are continuous with those longitudinal fibres of the pons which are derived from the nucleus gracilis and nucleus cuneatus, and from the formatio reticularis of the medulla. They are reinforced by fibres from the corpora quadrigemina and the corpora geniculata, and from the superior peduncle of the cerebellum. Some of the fibres are continued directly to the cerebral cortex, but the majority pass to the subthalamic region, and either end there or in the substance of the optic thalamus—the connection with the cortex being effected by means of fibres which arise in the optic thalamus. They spread out to form part of the corona radiata, and are distributed especially to the cortex of the temporal and occipital lobes.

2. The **transverse** or **commissural fibres** connect the two hemispheres. They include: (a) the transverse fibres of the corpus callosum; (b) the anterior commissure; (c) the posterior commissure, and have already been described.

3. **Association fibres.**—These connect different structures in the same hemisphere, and are of two kinds: (1) those which unite adjacent convolutions, *short association fibres*; (2) those which pass between more distant parts in the same hemisphere, *long association fibres*.

The *short association fibres* are situated immediately beneath the grey substance of the cortex of the hemispheres, and connect together adjacent convolutions.

The *long association fibres* include the following : (*a*) the uncinate fasciculus ; (*b*) the cingulum ; (*c*) the superior longitudinal fasciculus ; (*d*) the inferior longitudinal fasciculus ; (*e*) the perpendicular fasciculus ; and (*f*) the fornix.

(*a*) The *uncinate fasciculus* passes across the bottom of the Sylvian fissure, and connects the convolutions of the frontal lobe with the anterior end of the temporal lobe.

(*b*) The *cingulum* is a band of white matter which encircles the hemisphere in an antero-posterior direction, lying in the substance of the convolution of the corpus callosum. Commencing in front at the anterior perforated space, it passes forwards and upwards parallel with the rostrum, winds round the genu, runs in the convolution from before backwards, immediately above the corpus callosum, turns round its posterior extremity, and passes into the hippocampus major, through which it courses to its anterior extremity.

(*c*) The *superior longitudinal fasciculus* runs along the convex surface of the hemisphere, and connects the frontal and occipital and the frontal and temporal lobes.

(*d*) The *inferior longitudinal fasciculus* is a collection of fibres which connects the temporal and occipital lobes, running along the outer wall of the descending and posterior cornua of the lateral ventricle.

(*e*) The *perpendicular fasciculus* runs vertically through the front part of the occipital lobe, and connects the inferior parietal lobule with the fourth temporal convolution.

(*f*) The *fornix* connects the hippocampal convolution with the corpus albicans, and, by means of the bundle of Vicq d'Azyr, with the optic thalamus (see page 651). Through the fibres of the lyra it probably also unites the opposite hippocampal convolutions.

The **grey matter of the cerebrum** is disposed in two great groups : (1) The grey matter of the cerebral cortex. (2) The grey matter of the basal ganglia, the nucleus caudatus and the nucleus lenticularis of the corpus striatum ; the claustrum and the amygdaloid nucleus. They are, with the exception of the amygdaloid nucleus, situated to the inner side of the island of Reil, and form with this convolution the oldest part of the hemisphere, for they are the first parts of the encephalon to be differentiated in the development of the individual. They are simply semi-detached local thickenings of the grey cortex, and having already been described require no further mention in this place. The optic thalamus is not reckoned as a basal ganglion, but as belonging to the thalamencephalon.

GREY MATTER OF THE CORTEX

On examining a section through one of the convolutions of the Rolandic area with a lens, it is seen to consist of alternating white and grey layers thus disposed from the surface inwards : (1) a thin layer of white substance ; (2) a layer of grey substance ; (3) a second layer of white substance (outer band of Baillarger or band of Gennari) ; (4) a second grey layer ; (5) a third white layer (inner band of Baillarger) ; (6) a third grey layer, which rests on the medullary substance of the convolution.

The cortex is made up of nerve-cells which vary in size and shape, and of nerve-fibres, which are either medullated or naked axis-cylinders, embedded in a matrix of neuroglia.

Nerve-cells.—According to Cajal, whose description is now generally accepted, the nerve-cells are arranged in four layers, named from the surface inwards as follows : (1) the molecular layer, (2) the layer of small pyramidal cells, (3) the layer of large pyramidal cells, (4) the layer of polymorphous cells.

The **molecular layer.**—In this layer the cells are polygonal, triangular, or fusiform in shape. Each polygonal cell gives off some four or five dendrites, while its axon may arise directly from the cell or from one of its dendrites. The axons and dendrites of these cells ramify in the molecular layer. Each triangular cell gives off two or three dendrites, from one of which the axon arises, the dendrites and the axon ramifying in the molecular layer. The fusiform cells are placed with their long axes parallel to the surface and are mostly bipolar, each pole being prolonged into a dendrite, which runs horizontally for some distance and furnishes ascending branches. Their axons, two or three in number, arise from the dendrites, and, like them, take a horizontal course, giving off numerous ascending collaterals. The distribution of the axons and dendrites of all three sets of cells is limited to the molecular layer.

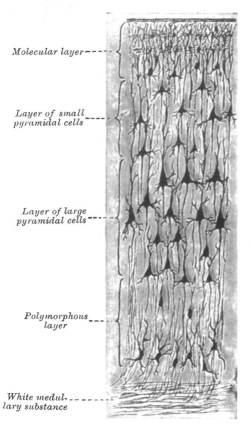

FIG. 361.—The four layers of cells in the cerebral cortex. (After Cajal.) Modified from Testut.

Molecular layer

Layer of small pyramidal cells

Layer of large pyramidal cells

Polymorphous layer

White medullary substance

The **layer of small** and the **layer of large pyramidal cells.** — The cells in these two layers may be studied together, since, with the exception of the difference in size and the more superficial position of the smaller cells, they resemble each other. The body of each cell is pyramidal in shape, its base being directed to the deeper parts and its apex towards the surface. It contains granular pigment, and stains deeply with ordinary reagents. The nucleus is nucleolated, of large size, and round or oval in shape. The base of the cell gives off the axis-cylinder, and this passes into the central white substance, giving off collaterals in its course, and is distributed as a projection, commissural, or association fibre. Both the apical and basal parts of the cell give off dendrites. The apical dendrite is directed towards the surface, and ends in the molecular layer by dividing into numerous branches, all of which may be seen, when prepared by the silver or methylene-blue method, to be studded with projecting bristle-like processes. The larger pyramidal cells, especially in the Rolandic area, may exceed $50\,\mu$ in length and $40\,\mu$ in breadth, and are termed *giant cells*.

Layer of polymorphous cells.—The cells in this layer, as their name implies, are very irregular in contour, the commonest varieties being of a spindle, star, oval, or triangular shape. Their dendrites are directed outwards, towards, but do not reach, the molecular layer; their axons pass into the subjacent white matter.

There are two other kinds of cells in the cerebral cortex, but their axons pass in a direction opposite to that of the pyramidal and polymorphous cells, among which they lie. They are: (*a*) the cells of Golgi, the axons of which do not become medullated, but divide immediately after their origin into a large number of branches, which are directed towards the surface of the cortex ; (*b*) the cells of

Martinotti, which are chiefly found in the polymorphous layer. Their dendrites are short, and may have an ascending or descending course, while their axons pass out into the molecular layer and form an extensive horizontal arborisation.

Nerve-fibres.—These fill up a large part of the intervals between the cells, and may be medullated or non-medullated—the latter comprising the axons of the smallest pyramidal cells and the cells of Golgi. In their direction the fibres may be either transverse (tangential or horizontal) or vertical (radial). The *transverse fibres* run parallel to the surface of the hemisphere, intersecting the vertical fibres at a right angle. They consist of several strata, of which the following are the most important: (1) a stratum of white fibres covering the superficial aspect of the molecular layer; (2) the external band of Baillarger, or band of Gennari, which runs through the layer of large pyramidal cells; (3) the internal band of Baillarger, which intervenes between the layer of large pyramidal cells and the polymorphous layer. According to Cajal, the transverse fibres consist of (a) the collaterals of the pyramidal and polymorphous cells and of the cells of Martinotti; (b) the arborisations of the axons of Golgi's cells; (c) the collaterals and terminal arborisations of the projection, commissural, or association fibres. The *vertical fibres*.—Some of these, viz. the axons of the pyramidal and polymorphous cells, are directed towards the central white matter, while others, the terminations of the commissural, projection, or association fibres, pass outwards to end in the cortex. The axons of the cells of Martinotti are also ascending fibres.

In certain parts of the cortex this typical structure is departed from. The chief of these regions are: (1) the occipital lobe, (2) the hippocampus major, (3) the dentate convolution, and (4) the olfactory bulb.

Special Types of Grey Matter

1. In the cuneus and the calcarine fissure of the occipital lobe Cajal has recently described as many as nine layers. Here the inner band of Baillarger is absent; the outer band of Baillarger (band of Gennari) is, on the other hand, of considerable thickness. If a section be examined microscopically, an additional layer is seen to be interpolated between the molecular layer and the layer of small pyramidal cells. This extra layer consists of two or three strata of fusiform cells, the long axes of which are at right angles to the surface. Each cell gives off two dendrites, external and internal, from the latter of which the axon arises and passes into the white central substance. In the layer of small pyramidal cells, fusiform cells, identical with the above, are seen, as well as ovoid or star-like cells with ascending axons (cells of Martinotti). This area of the cortex forms the visual centre, and it has been shown by Dr. J. S. Bolton * that in old standing cases of optic atrophy the thickness of Gennari's band is reduced by nearly 50 per cent.

2. In the hippocampus major the molecular layer is very thick and contains a large number of Golgi cells. It has been divided into three strata: (a) S. convolutum or S. granulosum, containing many tangential fibres; (b) S. lacunosum, presenting numerous lymphatic or vascular spaces; (c) S. radiatum, exhibiting a rich plexus of fibrils. The two layers of pyramidal cells are condensed into one, and the cells are mostly of large size. The axons of the cells in the polymorphous layer may run in an ascending, descending, or horizontal direction. Between the polymorphous layer and the ventricular ependyma is the white substance of the alveus.

3. In the rudimentary dentate convolution the molecular layer contains some pyramidal cells, while the layer of pyramidal cells is almost entirely represented by small ovoid cells.

4. The **olfactory bulb.**—In many of the lower animals this contains a cavity which communicates through the hollow olfactory stalk with the cavity of the lateral ventricle. In man the original cavity is filled up by neuroglia and its wall becomes thickened, but much more so on its ventral than on its dorsal aspect.

* *Phil. Trans. of Royal Society*, Series B, vol. 193, p. 165.

Its dorsal part contains a small amount of grey and white matter, but it is scanty and ill defined. A section through the ventral part shows it to consist of the following layers from without inwards. (1) A layer of olfactory nerve-fibres, which are the non-medullated axons prolonged from the olfactory cells of the nose, and which reach the bulb by passing through the cribriform plate of the ethmoid bone. At first they cover the bulb, and then penetrate it to end by forming synapses with the dendrons of the mitral cells, presently to be described. (2) *Glomerular layer.*—This contains numerous spheroidal reticulated enlargements, termed *glomeruli,* which are produced by the branching and arborisation of the processes of the olfactory nerve-fibres with the descending dendrite of the mitral cells. (3) *Molecular layer.*—This is formed of a matrix of neuroglia, embedded in which are the *mitral cells.* These cells are pyramidal in shape, and the basal part of each gives off a thick dendron, which descends into the glomerular layer, where it arborises as indicated above, and others which interlace with similar dendrites of neighbouring mitral cells. The axons pass through the next layer into the white matter of the bulb, from which, after becoming bent on themselves at a right angle, they are continued into the olfactory tract. (4) *Nerve-fibre layer.*— This lies next the central core of neuroglia, and its fibres consist of the axons or afferent processes of the mitral cells which are passing to the brain ; some efferent fibres are, however, also present, and terminate in the molecular layer, but nothing is known as to their exact origin.

IV. The Hind-brain

The **hind-brain,** or **epencephalon,** comprises those parts which are developed from the fourth cerebral vesicle ; namely, the pons, the cerebellum, and the upper half of the fourth ventricle.

Pons Varolii

The **pons Varolii** (*tuber annulare*) is the bond of union of the various segments of the encephalon, connecting the cerebrum above, the medulla oblongata below, and the cerebellum behind. It is situated above the medulla oblongata, below the crura cerebri, and between the hemispheres of the cerebellum. It is about an inch in length and in thickness, and about an inch and a half in width. It presents four surfaces : *superior,* which is attached, by direct continuation of fibres, to the mid-brain ; *inferior,* which is continuous with the medulla oblongata ; while the *anterior* or *ventral* and the *posterior or dorsal* surfaces are free.

The *anterior* or *ventral* surface is very prominent, markedly convex from side to side, and less so from before backwards. It consists of transverse white fibres, which arch like a bridge across the middle line, and on either side are gathered together into a compact mass, forming the *middle peduncle of the cerebellum.* Above and below it presents a well-defined border ; below, its transverse fibres slightly overlap the pyramidal bodies of the medulla, which disappear into its substance ; above, the transverse fibres slightly overlap the crura cerebri which emerge from it. This surface rests upon the clivus of the sphenoid bone, and presents in the middle line a longitudinal groove, wider in front than behind, in which rests the basilar artery.

The *posterior* or *dorsal* surface of the pons is free, but is concealed from view by the cerebellum. It forms the upper part of the floor of the fourth ventricle, and will be described with this cavity.

Structure.—Transverse sections of the pons Varolii show that it consists of two parts, which differ in appearance and structure from each other : the anterior or *ventral* portion consists, for the most part, of fibres arranged in transverse and longitudinal bundles with a small amount of grey matter ; the posterior or dorsal portion is a continuation of the reticular formation of the medulla, and is called the *tegmental* portion, as most of its constituents are continued into the tegmentum of the crus cerebri.

The anterior or ventral part consists of three layers of fibres: 1. superficial transverse fibres; 2. longitudinal fibres; 3. deep transverse fibres. These three layers are not, however, completely differentiated from each other, for some transverse fibres may be seen between the bundles of the longitudinal fibres (fig. 362).

1. The *superficial transverse fibres*, consisting of a rather thick layer on the ventral surface of the pons, cross the middle line, and proceeding laterally are collected into a large rounded bundle of fibres on each side. This bundle, with the addition of some transverse fibres from the deeper part of the pons, forms the *middle peduncle* of the cerebellum of the corresponding side.

2. The *longitudinal fibres* enter the pons below as a single mass, which forms the continuation upwards of the fibres of the pyramids of the medulla; as they ascend they become broken up into bundles by some of the transverse fibres, and

FIG. 362.—Superficial dissection of the medulla oblongata and pons. (Ellis.)

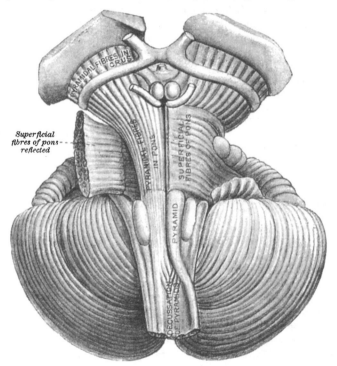

are continued into the crusta of the mid-brain. They lie on either side of the middle line, and cause a bulging of the superficial transverse fibres on the ventral surface of the pons, with a longitudinal mesial groove between them. This is the groove, mentioned above, in which the basilar artery is received. As the fibres ascend they are increased in number, being reinforced by others derived from the nerve-cells in the deep transverse strata.

3. The *deep transverse fibres* form a thicker layer than the superficial set, and there is much grey matter between them. The fibres pass from the middle line, where they interlace with those from the opposite side, and coursing to the lateral borders of the pons, they, for the most part, curve dorsally, and assist the superficial transverse fibres in forming the middle peduncle of the cerebellum. Some of the fibres join the nerve-cells which are situated in the grey matter of this layer, and in addition nerve-fibres derived from others of these cells pass off to join the longitudinal fibres (see above).

The *tegmental* or *dorsal portion* of the pons is chiefly constituted by a continuation upwards of the reticular formation and grey matter of the medulla. It is subdivided into lateral halves by a median raphé continuous with that of the medulla, but this does not extend into the ventral half of the pons, being here obliterated by the transverse fibres.

The dorsal portion of the pons, like the ventral, contains both transverse and longitudinal fibres. The transverse fibres are collected into a distinct bundle, which, from its shape, is sometimes termed the *trapezium* or *corpus trapezoides*. It consists of fibres which proceed laterally to become connected with the cells of the accessory auditory nucleus. The longitudinal fibres, which are continuous with those of the medulla, are mostly collected into two bundles on either side. One of these lies between the corpus trapezoides and the formatio reticularis of the pons, and is a continuation upwards of the sensory tracts; it is termed the *fillet*. The other bundle is situated more dorsally, near the floor of the fourth ventricle; it is the *posterior longitudinal bundle*, and contains both ascending and descending fibres. Other longitudinal fibres, which are more diffusely distributed, arise from the cells of the grey matter of the pons itself. The greater part of the dorsal portion of the pons is, as stated above, a continuation upwards of the formatio reticularis of the medulla, and like it presents, on transverse section, viewed under a moderate magnifying power, a reticular appearance. In addition to the grey matter, which presents a number of small reticularly arranged masses, with nerve-cells, there are some important collections of nerve-cells, which require mention.

1. The **superior olivary nucleus** is a small isolated collection of grey matter, situated on the dorsal surface of the outer part of the trapezium. In structure it resembles the inferior olivary nucleus of the medulla, presently to be described, and is situated immediately above it. The nerve-fibres derived from its cells pass into the trapezium, and, as stated above, cross the middle line and enter the accessory auditory nucleus of the other side. The other collections of nerve-cells in the formatio reticularis of the pons are nuclei from which some of the cranial nerves arise.

2. **Nuclei of the fifth nerve.**—The nuclei of the fifth nerve in the pons are two in number: one for the motor root and the other for the sensory. The *motor* nucleus is situated in the higher portion of the pons, close under the dorsal surface and along the line of the lateral margin of the fourth ventricle. The *sensory* nucleus lies external to the motor one, beneath the superior peduncle of the cerebellum, which forms the lateral boundary of the upper half of the fourth ventricle. Some of the fibres from these nuclei pass to the raphé of the pons, and thence probably to the higher parts of the brain; the rest form the nerve-roots of the motor and sensory parts of the fifth nerve respectively. They pass through the pons to emerge on its ventral surface at its lateral and constricted portion, nearer its superior than its inferior margin. It must be mentioned that the whole of the roots of the fifth nerve are not formed from these nuclei. The sensory root is partly formed by a long tract of fibres, known as the *ascending* root, which can be traced through the pons and medulla to the upper part of the spinal cord. The motor root, in like manner, is partly formed by a long tract of fibres, which passes downwards from the grey matter in the floor of the Sylvian aqueduct and which is termed the *descending* root.

3. The **nucleus of the sixth nerve** is situated beneath the floor of the fourth ventricle, on either side of the middle line. It lies close to the root of the facial nerve, immediately to be described, being a little external to and beneath it, and corresponds to the upper half of the fasciculus teres of the floor of the fourth ventricle (fig. 371). The fibres pass through the substance of the pons, and emerge at the lower margin of this structure, between it and the upper end of the medulla.

4. The **nucleus of the facial nerve** is of elongated form, and is situated deeply in the reticular formation below the floor of the fourth ventricle and dorsal to

the superior olivary nucleus. The roots of the nerve derived from it pursue a remarkably tortuous course in the substance of the pons. At first they pass backwards and inwards till they reach the floor of the fourth ventricle, close to the median groove, where they are collected into a rounded bundle. This passes upwards and forwards, producing an elevation (*fasciculus teres*) in the floor of the ventricle, and then takes a sharp bend and arches outwards through the substance of the pons to emerge at its lower border in the interval between the olivary and restiform bodies of the medulla.

5. The **nuclei of the auditory nerve** are two in number, dorsal and ventral. The *dorsal* nucleus is principally situated in the medulla, but is prolonged upwards into the pons, where it lies beneath the upper half of the floor of the fourth ventricle. The *ventral* or accessory nucleus is also partly contained in the medulla and partly in the pons. In the medulla it is situated on the antero-external surface of the restiform body, lying between the vestibular and cochlear divisions of the auditory nerve, the latter being to its outer side. In the pons it is seen to lie beyond the boundary of the fourth ventricle on the outer and ventral aspect of the restiform body. A third nucleus (*nucleus of Deiters*) is sometimes termed the outer nucleus of the auditory nerve. It is situated below the outer angle of the fourth ventricle, and contains multipolar nerve-cells of large size. The root of the auditory nerve consists of two portions, lateral and mesial, which pass, one to the outer and the other to the inner side of the restiform body, those from the lateral part arising mainly from the ventral nucleus, those from the mesial part arising from the dorsal auditory nucleus. They emerge at the lower border of the pons, in the groove between the olivary and restiform bodies.

The **Nuclei Pontis.**—In addition to these nuclei of grey matter, which have been described as being situated in the tegmental or dorsal portion of the pons, there are small masses of grey matter, as mentioned above, in the anterior or ventral portion. These are known as the *nuclei pontis*, and consist of small multipolar nerve-cells, scattered between the bundles of transverse fibres.

THE CEREBELLUM

The **Cerebellum** is contained in the inferior occipital fossæ, and is situated beneath the occipital lobes of the cerebrum, from which it is separated by the tentorium cerebelli. In form, it is oblong, and flattened from above downwards, its great diameter being from side to side. It measures from three and a half to four inches transversely, two to two and a half inches from before backwards, and is about two inches thick in the centre, and about six lines at the circumference. It consists of grey and white matter : the former, darker than that of the cerebrum, occupies the surface ; the latter, the interior. The surface of the cerebellum is not convoluted like that of the cerebrum, but is traversed by numerous curved furrows or sulci, which vary in depth at different parts, and separate the laminæ of which it is composed.

Lobes of the Cerebellum.—The cerebellum consists of three parts or lobes, a median and two lateral. They are all continuous with each other, and are substantially the same in structure. The median portion is called the *worm* or *vermiform process*, from the annulated appearance which it presents, owing to transverse ridges and furrows upon it. On the upper surface of the cerebellum, the worm is only slightly elevated above the level of the lateral portions, but on the under surface it is sunk almost out of sight in a deep depression, which is called the *vallecula*. The lateral parts are called *hemispheres* ; they attain a considerable size, overlapping and obscuring the inferior part of the worm. Below and behind they are separated by a deep notch (*posterior cerebellar notch, incisura marsupialis*), and in front by a broader, shallower notch (*anterior cerebellar notch, incisura semilunaris*). The *anterior* notch lies close to the pons and upper part of the medulla, and its upper edge encircles the posterior corpora quadrigemina. The *posterior* notch is free, and contains, in the recent state, the upper part of the falx cerebelli. The sides of the notches are formed by the margins of the hemispheres, while the bottom of the notches is formed by the anterior and posterior extremities of the worm respectively. The cerebellum is characterised by its laminated or foliated appearance; it is everywhere marked by deep, transverse, somewhat curved fissures, which lie close together, and extend for a considerable depth into the substance of the cerebellum, dividing it into a series of layers or leaves. Upon making sections across the laminæ it will be seen that the folia, though differing in appearance from the convolutions of the cerebrum, are homologous with them, inasmuch as they consist of a central white substance with a covering or cortex of grey matter.

The largest and deepest fissure is the *great horizontal fissure*. It commences in front at the pons, and passes horizontally round the free margin of the hemisphere to the middle line behind, and divides the cerebellum into an upper and lower portion. Several secondary but deep fissures separate the cerebellum

FIG. 363.—Upper surface of the cerebellum. (Schäfer.)

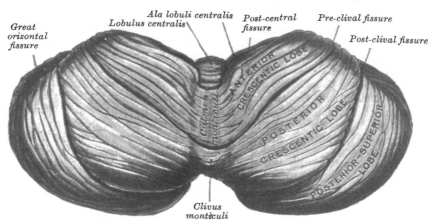

into lobes, and these are further subdivided by shallower sulci, which separate the individual folia or laminæ from each other.

The cerebellum is connected to the cerebrum, pons, and medulla by three pairs of peduncles, which will be described in the sequel: a *superior* pair, connect it with the cerebrum; a *middle* pair, with the pons; and an *inferior* pair, with the medulla.

Upper Surface of Cerebellum (fig. 363).—The superior surface of the cerebellum is somewhat elevated in the middle line and sloped towards its circumference, its hemispheres being connected together by an elevated median portion or lobe, the *superior worm* or *superior vermiform process*. The surface is traversed by four curved fissures, which are named, from their situation in front or behind two prominent lobes of the worm, the central lobe and the clivus, (1) the pre-central fissure, (2) the post-central fissure, (3) the pre-clival fissure, and (4) the post-clival fissure. These four fissures divide the entire upper surface of the cerebellum into five lobes, but the portion of the lobe in the worm has received a different name from that in the hemisphere, though the two are continuous with each other. The five lobes in the worm are named from before backwards: (1) the *lingula*, (2) the *lobulus centralis*, (3) the *culmen monticuli*, (4) the *clivus monticuli*, and (5) the *folium cacuminis*. The five lobes in the hemispheres are named from before backwards: (1) the *frænulum*, (2) the *ala lobuli centralis*, (3) *anterior crescentic*, (4) *posterior crescentic*, and (5) *posterior superior*. The arrangement of these fissures and lobules will be understood by reference to the

accompanying schematic arrangement, in which the lobules are named in order
from before backwards, with the fissures which separate them :

UPPER SURFACE OF THE CEREBELLUM

Worm	*Hemisphere*
Lingula.	Frænulum.

Pre-central fissure

Lobulus centralis.	Ala lobuli centralis.

Post-central fissure

Culmen monticuli.	Anterior crescentic lobe.

Pre-clival fissure

Clivus monticuli.	Posterior crescentic lobe.

Post-clival fissure

Folium cacuminis.	Posterior superior lobe.

The **lingula** is a tongue-shaped process of the cerebellum, which lies in front
of the lobulus centralis and is partially or completely concealed by it. It is in
relation, in front, with the valve of Vieussens, on the dorsal surface of which it
rests and with which it is connected ; its white matter being continuous with that
of the valve. At either side the lingula gradually shades off, and is prolonged
only for a short distance into the hemispheres, where it forms the **frænulum.**
This does not stretch beyond the superior peduncle of the cerebellum over which
it lies.

The **lobulus centralis.**—The lobulus centralis is a small square lobe, situated
in the anterior notch. It overlaps the lingula and is in turn partially concealed
by the culmen monticuli. Laterally the lobulus centralis extends along the upper
and anterior part of each hemisphere, where it forms a wing-like prolongation, the
ala lobuli centralis.

The **culmen monticuli** is much larger than the two lobes just described, and
constitutes, with the succeeding lobe, the clivus, the bulk of the upper worm. In
front it partially overlaps and obscures the lobulus centralis, and behind it is
separated from the clivus by the *pre-clival fissure*. It forms the most prominent
part of the upper worm, and is marked on its surface by three or four secondary
fissures, dividing it up into smaller lobules. Laterally it is continuous with the
anterior crescentic lobe of the hemispheres, which is distinctly differentiated
from the posterior crescentic lobe by the pre-clival fissure ; though the two were
formerly classed together as the quadrate lobe of the lateral hemisphere.

The **clivus monticuli** is of considerable size, and, as stated above, forms with
the culmen the major part of the superior worm. It consists of a group of
laminæ, which in front are separated from the culmen by the pre-clival fissure
and behind appear to be almost continuous with the folium cacuminis, especially
in the median line ; but it will be found, on careful examination, to be separated
from it by a well-defined fissure, the *post-clival* fissure. Laterally this lobe is
continued into the hemispheres as the **posterior crescentic lobe,** which is somewhat
semilunar in shape, and, with the anterior crescentic lobe, constitutes the greater
part of the upper surface of the hemispheres.

The **folium cacuminis** is a short and narrow, concealed band at the posterior
extremity of the worm, consisting apparently of a single folium, but in reality
marked on its upper and under surfaces by secondary fissures. Laterally it
expands in either hemisphere into a considerable lobe, which is semilunar in
shape, and is situated at the postero-superior part of the hemisphere and bounded

below by the great horizontal fissure. It is named the **posterior superior lobe** and occupies the posterior third of the upper surface of the hemisphere, forming its rounded postero-lateral border.

The Under Surface of the Cerebellum (fig. 364) presents in the middle line the

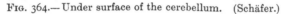

FIG. 364.— Under surface of the cerebellum. (Schäfer.)

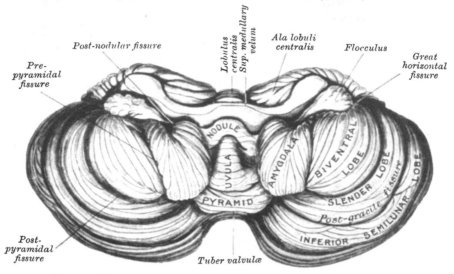

inferior worm, buried in the vallecula, and separated from the hemispheres by lateral grooves. Here, as on the upper surface, there are deep fissures, dividing it into separate segments or lobes, but the arrangement is more complicated, and the relation of the segments of the worm to those of the hemisphere is less clearly marked. The fissures are three in number, but are not so regularly

FIG. 365.—Diagram showing fissures on under surface of the cerebellum.

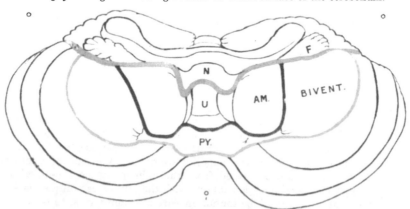

F. Flocculus; N. Nodule; U. Uvula; Py. Pyramid; Am. Amygdala; Bivent. Biventral lobe.

disposed as those on the upper surface (fig. 365). They are named, from their relation to the pyramid and nodule, two of the lobes on the under surface of the worm, (1) *post-nodular*, (2) *pre-pyramidal*, and (3) *post-pyramidal fissures*. The part of the worm in front of the post-nodular fissure is termed the *nodule*, and the lobule in the hemisphere corresponding with this is the *flocculus*. The next lobe is situated between the post-nodular and pre-pyramidal fissures. In the vermiform process it is known as the *uvula*, and its lateral expansion in the hemisphere is named the *amygdala* or *tonsil*. The lobule of the worm between the pre- and post-pyramidal fissures is the *pyramid*, and its corresponding part in the hemisphere is the *biventral* or *digastric* lobe. Finally, behind the post-pyramidal fissure in the worm is a small lobe, the *tuber valvulæ* or *tuber posticum*; this, in the hemispheres, expands into a large lobe, which occupies at least two-thirds of the inferior surface of the cerebellum, and is subdivided into two by a secondary fissure, named the *post-gracile* fissure. The anterior of the two subdivisions is named the *slender* lobe; and the posterior, the *inferior semilunar* or *postero-inferior lobe*. These fissures and lobes are here arranged, from before backwards, in a schematic form:

UNDER SURFACE OF THE CEREBELLUM

Worm   *Hemisphere*

Nodule.   Flocculus.

Post-nodular fissure

Uvula.   Amygdala.

Pre-pyramidal fissure

Pyramid.   Biventral lobe.

Post-pyramidal fissure

Tuber valvulæ.   { Slender lobe.
                                  { *Post-gracile fissure.*
                                  { Inferior semilunar lobe.

The **chief fissures of the under surface,** as stated above, are three in number, and are not so regularly disposed as on the upper surface. (1) The *post-nodular fissure* in the worm courses transversely across it, separating the nodule in front from the uvula behind. When it reaches the hemispheres it passes in front of the amygdala, and then crosses between the flocculus in front and the biventral lobe behind, and joins the anterior end of the great horizontal fissure. (2) The *pre-pyramidal fissure* crosses the worm between the uvula in front and the pyramid behind, then curves laterally behind the amygdala, and passes forwards along the outer border of this lobe, between it and the biventral lobe, to join the post-nodular sulcus. (3) The *post-pyramidal fissure* passes across the worm behind the pyramid and in front of the tuber valvulæ, and in the hemispheres courses behind the amygdala and biventral lobes, and then along the outer border of the biventral lobe to the post-nodular sulcus. It cuts off at least two-thirds of the inferior surface of the hemisphere. From it a secondary sulcus springs, and coursing forwards and outwards divides this surface into two parts and falls into the great horizontal fissure. This sulcus is termed the *post-gracile fissure*.

THE LOBES OF THE INFERIOR SURFACE OF THE CEREBELLUM

The **Nodule** and **Flocculus.**—The nodule is a distinct prominence, forming the anterior extremity of the inferior worm. It projects into the roof of the fourth ventricle, and can only be distinctly seen after the cerebellum has been separated from the medulla and pons. On each side of the nodule is a thin layer of white substance, named the *inferior medullary velum*. It is semilunar in form, its convex border being continuous with the white substance of the cerebellum; it extends on either side as far as the flocculus, which it connects with the nodule. The flocculus is a prominent, irregular lobule, situated just in front of the biventral lobe, between it and the middle peduncle of the cerebellum. It is subdivided into a few small laminæ, and is connected to the inferior medullary velum by its central white core.

The **Uvula** and **Amygdalæ.**—The uvula occupies a considerable portion of the inferior worm; it is separated on either side from the amygdala by a deep groove, the *sulcus valleculæ*, at the bottom of which it is connected to the amygdala by a commissure of grey matter, indented on its surface, and called the *furrowed band*. It is marked on its surface by three or four transverse fissures. The amygdalæ or tonsils are rounded masses, situated in the lateral hemispheres. Each lies in a deep fossa between the uvula and the biventral lobe ; this fossa is known by the name of the *bird's nest (nidus avis)*.

The **Pyramid** and **Biventral Lobes.**—The pyramid is a conical projection, forming the largest prominence of the lower worm. It is separated from the hemispheres by the sulcus valleculæ, across which it is connected to the biventral lobe by an indistinct band of grey matter, analogous to the furrowed band already described. The biventral lobe is triangular in shape, with the apex pointing inwards and backwards to become joined by the connecting band to the pyramid. The external border is separated from the slender lobe by the post-pyramidal fissure. The base is directed forwards, and is on a line with the anterior border of the amygdala, and is separated from the flocculus by the post-nodular fissure.

The **Tuber Valvulæ**, or **Tuber Posticum**, and **Posterior Inferior Lobes.**—The tuber valvulæ is the posterior division of the inferior worm. It is of small size, and laterally spreads out into the large posterior inferior lobes of the hemispheres. These lobes, which, as stated above, comprise at least two-thirds of the inferior surface of the hemisphere, are divided into two by the post-gracile fissure. The anterior lobe is named the *slender lobe*, and the posterior, the *inferior semilunar lobe*. Both these lobes show a tendency to subdivision into two ; that of the slender lobe is well marked, and its subdivisions are sometimes described as distinct lobes and named the *anterior* and *posterior slender lobes*, the fissure between them being termed the *intra-gracile fissure*.

Internal Structure of the Cerebellum

The cerebellum consists of white and grey matter.

The **White Matter.**—If a sagittal section (fig. 366) is made through either hemisphere of the cerebellum, the interior will be found to consist of a central stem of white matter, which contains in its interior a grey mass, the *corpus dentatum*. From the surface of this central stem a series of plates of medullary matter are detached, which, covered with grey matter, form the laminæ. In consequence of the main branches from the central stem dividing and subdividing, the section presents a characteristic appearance, which is named the *arbor vitæ*. If a vertical section is made in the median plane of the cerebellum it will be found that the central stem divides into two main branches, which, from their direction, may be named respectively the vertical and the horizontal branch. The *vertical* branch passes upwards to the culmen, where it subdivides freely, some of its ramifications passing forwards and upwards to the central lobe. The *horizontal* branch passes backwards to the folium cacuminis, considerably diminished in size, in consequence of having given off large secondary branches : one, from its upper surface, ascends to the clivus ; the others descend, and enter the lobes in the inferior vermiform process, the tuber valvulæ, the pyramid, the uvula, and the nodule. It is not necessary to describe in detail the various divisions of the white matter, as they correspond to the lobes on the surface.

The white matter of the cerebellum includes two varieties of nerve-matter : (1) the *peduncular fibres*, which are directly continuous with those of the peduncles of the cerebellum ; (2) the fibres proper (*fibræ propriæ*) of the cerebellum itself.

The **Peduncles of the Cerebellum.**—From the anterior part of each hemisphere arise three large processes or peduncles—superior, middle, and inferior—by which the cerebellum is connected with the rest of the encephalon.

The **superior peduncles** form the upper lateral boundaries of the floor of the fourth ventricle. As they extend forwards and upwards they converge on the dorsal aspect of the ventricle, and thus assist to roof it in. They may be traced as far as the corpora quadrigemina, under which they pass. They enter the upper and mesial part of the medullary substance of the hemispheres, beneath the ala lobuli centralis and the fraenulum, and pass to a great extent into the interior of the corpus dentatum, though some of their fibres wind round it and reach the grey cortical matter, especially of the inferior surface.

The fibres of the superior peduncles mainly emerge from the hilum of the corpus dentatum ; others come from the cortex and probably also from the smaller nuclei in the central white substance. The majority of the fibres decussate with those of the opposite peduncle below the corpora quadrigemina, and pass to the red nucleus of the tegmentum, from which a relay is prolonged through the optic thalamus to the cerebral cortex. Fibres also connect the spinal cord with the cerebellum through its superior peduncles ; these are chiefly derived from the antero-lateral ascending cerebellar tract of Gowers.

The **Valve of Vieussens,** or **Superior Medullary Velum.**—Stretched across from one superior peduncle to the other is a thin, transparent lamina of white matter, the *valve of Vieussens* ; on to the dorsal surface of its lower half the folia

Fig. 366.—Sagittal section of the cerebellum, near the point of junction of the worm with the hemisphere. (Schäfer.)

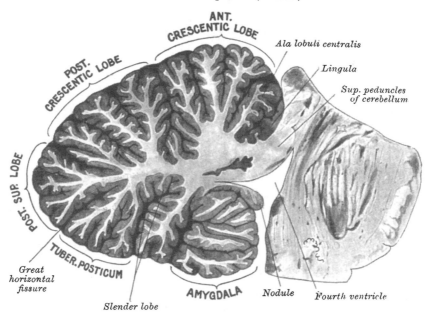

of the lingula are prolonged. It forms with the superior peduncles the roof of the upper part of the fourth ventricle, and is continuous with the central white stem of the cerebellum. It is narrow above, where it passes beneath the corpora quadrigemina, and broader below, at its connection with the white substance of the superior worm of the cerebellum. A slight elevated ridge descends upon the upper part of the valve from between the lower corpora quadrigemina, and on either side of this may be seen the fourth nerve.

The **middle peduncles** are the largest of the three pairs. They consist of a mass of curved fibres, which are, as already described, comprises most of the transverse fibres of the pons. They enter the cerebellum between the margins of the great horizontal fissure at the anterior notch, and the fibres spread out in all directions : some passing to the upper part, and some to the lower part of the hemisphere, while others pass to its middle region. Of the fibres contained in the middle peduncles many are commissural between the two hemispheres of the cerebellum ; others apparently end in the grey matter ; others have been described as giving fibres to the posterior longitudinal bundle, and through it to the nuclei of the third, fourth, and sixth nerves. Cajal describes still another set, which have their origin in the grey reticular formation of the pons, and which pass partly into the peduncle of the same side and partly into that of the opposite side.

The **inferior peduncles** connect the cerebellum with the medulla oblongata. As the restiform bodies of the latter, they will be described in the sequel. They pass upwards and outwards, forming part of the lateral wall of the fourth ventricle, and enter the cerebellum beneath the middle peduncle; passing upwards, they end in the grey cortex of the upper surface of the hemisphere, some being prolonged into the white matter of the superior vermiform process. The following are the chief sets of fibres in the inferior peduncles: (1) from the direct cerebellar tract of the spinal cord; (2) from the gracile and cuneate nuclei (crossed and uncrossed fibres); (3) from the opposite olivary body of the medulla; (4) fibres to the nuclei of the fifth, eighth, ninth, and tenth nerves; (5) descending cere-

FIG. 367.—Diagrammatic representation of the cells of the cerebellum. (Modified from Foster's ' Physiology.')

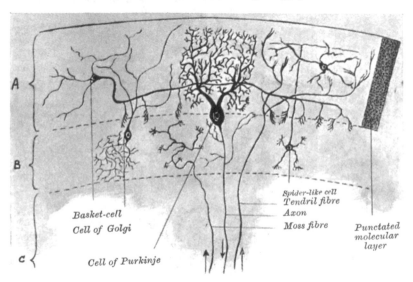

Basket-cell
Cell of Golgi
Cell of Purkinje

Spider-like cell
Tendril fibre
Axon
Moss fibre

Punctated molecular layer

A. molecular layer ; B. nuclear layer ; C. white matter.

bellar fibres which pass down the restiform body and antero-lateral column of the cord to terminate around the cells in the anterior horn of the cord.

The **fibræ propriæ of the cerebellum** are of two kinds : (1) *commissural fibres*, which cross the middle line to connect the opposite halves of the cerebellum, some at the anterior part, and others at the posterior part of the vermiform process ; (2) *arcuate* or *association fibres*, which connect adjacent laminæ with each other.

The **grey matter** of the cerebellum is found in two situations : (1) on the surface, forming the cortex ; (2) as independent masses in the interior.

1. The **grey matter of the cortex** presents a characteristic foliated appearance, due to the series of laminæ which are given off from the central white matter ; these in their turn give off secondary laminæ, which are covered with grey matter. This arrangement gives to the cut surface of the organ a foliated appearance (fig. 366). Externally, the cortex is covered by pia mater; internally, is the medullary centre, consisting mainly of nerve-fibres.

Microscopic appearance of the cortex.—The cortex presents a remarkable structure, consisting of two distinct layers, viz. an external grey molecular layer, and an internal rust-coloured, granular layer. Between the two layers is an incomplete stratum of the characteristic cells of the cerebellum, the *corpuscles of Purkinje.*

The *external grey* or *molecular layer* (figs. 367, 368) consists of fibres and cells. The nerve-fibres are delicate fibrillæ, and are derived from the following sources : (a) the dendrites and axon collaterals of Purkinje's cells ; (b) fibres from cells in the granular layer ; (c) fibres from the central white substance of the cerebellum ; (d) fibres derived from cells in the molecular layer itself. In addition to these are other fibres, which have a vertical direction. These are the processes of large glia-cells, situated in the granular layer. They pass outwards to the periphery of the grey matter, where they expand into little conical enlargements, which form a sort of limiting membrane beneath the pia mater, analogous to the membrana limitans interna in the retina, formed by the fibres of Müller.

FIG. 368.—Vertical section through the grey matter of the human cerebellum. Magnified about 100 diameters. (Klein and Noble Smith.)

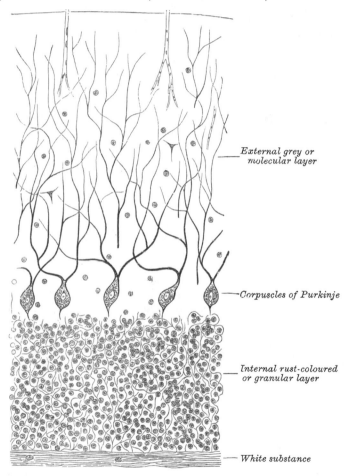

External grey or molecular layer

Corpuscles of Purkinje

Internal rust-coloured or granular layer

White substance

The *cells* of the molecular layer are small, and are arranged in two strata, an outer and an inner. They all possess branching axis-cylinder processes ; those of the inner layer run for some distance horizontally, i.e. parallel with the surface of the folia, giving off at intervals collaterals, which pass in a vertical direction towards the cell-bodies of Purkinje's corpuscles, around which they become enlarged, and ramify like a basket. Hence these cells of the inner layer are named *basket-cells*.

The *corpuscles of Purkinje* (fig. 368) are flask-shaped cells, situated at the junction of the molecular and granular layers, their bases resting against the latter. From the bottom of the flask the axis-cylinder process arises ; this passes through the granular layer, and, becoming medullated, is continued as a nerve-fibre in the medullary substance beneath. This axon gives off fine collaterals as it passes through the granular layer, some of which run back into the molecular layer. From the neck of the flask numerous dendrites are given off, which branch in an antler-like manner in the molecular layer and terminate in free extremities.

The *internal rust-coloured* or *granular layer* (fig. 368) is characterised by containing numerous small nerve-cells or granules of a reddish-brown colour, together with many nerve-fibrils. Most of the cells are nearly spherical and provided with short dendrites, which spread out in a spider-like manner in the granular layer. Their axons pass outwards into the molecular layer, and, bifurcating at right angles, run horizontally for some distance. In the outer part of the granular layer are also to be observed some larger cells, of the type termed *Golgi cells* (fig. 367). Their axons undergo frequent division as soon as they leave the nerve-cells, and pass into the granular layer, while their dendrites ramify chiefly in the molecular layer.

Finally, in the grey matter of the cerebellar cortex fibres are to be seen which come from the white centre and penetrate the cortex. The cell origin of these fibres is unknown, though it is believed that it is probably in the grey matter of the spinal cord. Some of these fibres end in the granular layer, by dividing into numerous branches, on which are to be seen peculiar moss-like appendages ; hence they have been termed by Ramón y Cajal the 'moss fibres : ' they form an arborescence around the cells of the granular layer. Other fibres derived from the medullary centre can be traced into the molecular layer, where their branches cling around the dendrites of Purkinje's cells, and hence they have been named the *clinging* or *tendril fibres*.

2. The **independent centres of grey matter in the cerebellum** are four in number on each side : one is of large size, and is known as the corpus dentatum ; the other three, much smaller, are situated near the middle of the cerebellum, and are known as the nucleus emboliformis, nucleus globosus, and nucleus fastigii.

The *corpus dentatum* or *ganglion of the cerebellum* is situated a little to the inner side of the centre of the stem of the white matter of the hemisphere. It consists of an irregularly folded lamina of a greyish-yellow colour, containing white fibres, and presenting on its antero-internal aspect an opening, the hilum, from which most of the fibres of the superior cerebellar peduncle emerge.

The *nucleus emboliformis* is a mass of grey matter placed immediately to the inner side of the corpus dentatum, and partly covering its hilum. The *nucleus globosus* is an elongated mass of grey matter, directed antero-posteriorly, and placed to the inner side of the preceding. The *nucleus fastigii* is somewhat larger than the other two, and is situated close to the middle line at the anterior end of the superior vermiform process, and immediately over the roof of the fourth ventricle, from which it is separated by a thin layer of white matter. It is known as the *roof nucleus of Stilling*.

Weight of the Cerebellum.—Its average weight in the male is about 5 oz. 4 drs. It attains its maximum weight between the twenty-fifth and fortieth years, its increase in weight after the fourteenth year being relatively greater in the female than in the male. The proportion between the cerebellum and cerebrum is, in the male, as 1 to 8½, and in the female as 1 to 8. In the infant the cerebellum is proportionately much smaller than in the adult, the relation between it and the cerebrum being, according to Chaussier, between 1 to 13, and 1 to 26 ; by Cruveilhier the proportion was found to be 1 to 20.

V. The Medulla Oblongata (fig. 370)

The **medulla oblongata** or **metencephalon**, known also as the **spinal bulb,** is the lowest division of the encephalon, and is continuous with the spinal cord.

It is developed from the fifth cerebral vesicle, the cavity of which forms the lower half of the fourth ventricle. It extends from the lower margin of the pons Varolii to a plane passing transversely just below the decussation of the pyramids, at which level the spinal cord commences. This plane corresponds to the lower margin of the foramen magnum. The upper limit of the medulla is marked off from the pons Varolii on its ventral aspect by the abrupt lower margin of the latter.

The medulla oblongata is directed from above obliquely downwards and backwards; its ventral surface rests on the basilar groove of the occipital bone, while its dorsal surface is received into the fossa between the hemispheres of the cerebellum, and forms the lower part of the floor of the fourth ventricle. It is pyramidal in shape, its broad extremity directed upwards, its lower end being narrow at its point of connection with the cord. It measures an inch in length, three-quarters of an inch in breadth at its widest part, and half an inch in thickness. Its surface is marked, in the median line, in front and behind, by an *anterior* and a *posterior median fissure*, which are continuous with similar fissures on the anterior and posterior surfaces of the cord. The anterior fissure contains a fold of pia mater, and terminates just below the pons in a *cul-de-sac*, the *foramen cæcum* of Vicq d'Azyr. It is interrupted at its lower part by some bundles of fibres, which cross obliquely from one side to the other, forming the *decussation of the pyramids*. The posterior is a deep but narrow fissure, continued upwards to about the middle of the medulla, where it expands into the fourth ventricle.

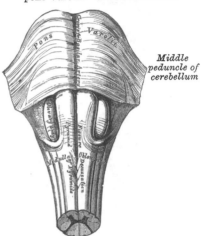

Fig. 369.—Medulla oblongata and pons Varolii. Anterior surface.

Middle peduncle of cerebellum

These two fissures divide the medulla into two symmetrical halves, each half presenting elongated eminences, which are continuous with the columns of the cord. By taking the lines along which some of the cranial nerves emerge from the medulla, as landmarks, the surface of this portion of the nervous system may be divided into three columns, in the same way as the spinal cord is divided into three columns by the lines corresponding to the points of exit of the anterior and posterior roots of the spinal nerves. The anterior column comprises that portion which is situated between the anterior median fissure and the fibres of origin of the hypoglossal nerve: this column is called the *pyramid*. The lateral column comprises that portion which is situated between the fibres of origin of the hypoglossal nerve and the fibres of origin of the glosso-pharyngeal, pneumogastric, and spinal accessory nerves. In the lower part of the medulla this column is single, and is called the *lateral tract*; but in the upper part an oval-shaped body comes forward between it and the pyramid, and pushes aside the lateral tract. This is called the *olivary body*. The posterior column comprises that portion which is situated between the fibres of origin of the glosso-pharyngeal, pneumogastric, and spinal accessory nerves and the posterior median fissure. It is marked by slight furrows dividing it into smaller columns, and these in the lower part of the medulla are named, from without inwards, the *funiculus of Rolando*, the *funiculus cuneatus*, and the *funiculus gracilis*; in the upper part of the medulla, the funiculus of Rolando and the funiculus cuneatus appear to become fused together, forming a single body, called the *restiform body* (fig. 370).

The **pyramids** are two pyramidal bundles of white matter, placed one on either side of the anterior median fissure, and separated from the olivary body by a slight depression, from which the roots of the hypoglossal nerve emerge. At the lower border of the pons these bodies are somewhat constricted, and are here crossed by a band of arched fibres, the *ponticulus* of Arnold; below this they become enlarged, and then taper as they descend to their lower extremity. The fibres of which these pyramids are composed may be arranged in two bundles : an outer, continuous below with the direct pyramidal tract of the anterior column of the same side of the spinal cord, and an inner, continuous with the crossed pyramidal tract of the lateral column of the opposite side of the cord. As will be subsequently mentioned, the direct pyramidal tract in the cord lies next to the anterior median fissure, but as the crossed pyramidal tract of the cord ascends to the medulla it decussates with its fellow of the opposite side across the anterior median fissure, and so dis-

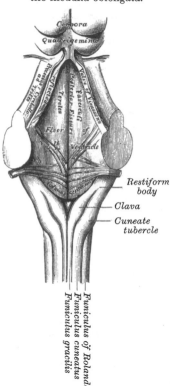

FIG. 370.—Posterior surface of the medulla oblongata.

places laterally the direct pyramidal tract, and ascends, after decussation, through the medulla to its inner or mesial side. This decussation is usually spoken of as the *decussation of the pyramids*, but it must be borne in mind that it is only a portion of the fibres of the pyramid which decussate; namely, those derived from the crossed pyramidal tract of the cord ; the outermost fibres, derived from the anterior column of the cord, do not decussate. Each pyramid enters the substance of the pons in one bundle, and may be traced through it, after breaking up into several smaller fasciculi, into the corresponding crus cerebri.

The **lateral column,** in the lower part of the medulla, is of the same width as the lateral column of the cord, and appears on the surface to be a direct continuation of it. As a matter of fact it is only a part of the lateral column of the spinal cord which is continued upwards into this column ; for the crossed pyramidal tract passes into the pyramid of the opposite side, and the direct cerebellar tract of the lateral column of the cord passes into the restiform body. The rest of the lateral column of the cord, that is to say, the antero-lateral ground bundle and the antero-lateral cerebellar tract, can be traced upwards into this area. In the upper part of the medulla, the lateral tract, on account of the interpolation of the olivary body, becomes almost concealed by this body.

The **olivary body** is a prominent oval mass, situated on the outer side of the pyramid, from which it is separated by a slight groove, along which the fibres of the hypoglossal nerve emerge. It is separated externally from the restiform body by a longitudinal, narrow band of fibres, prolonged upwards from the lateral tract, and by a groove, from which the glosso-pharyngeal, pneumogastric, and spinal accessory nerves arise. It is equal in breadth to the pyramid ; it is broader above than below, and is about half an inch in length, being separated above from the pons Varolii by a slight depression, in which a band of arched fibres is sometimes to be seen. Numerous white fibres (*superficial arciform fibres*) are seen winding across the lower half of the pyramid and the olivary body to enter the restiform body.

The **funiculus of Rolando** is a longitudinal prominence on the outer side of the lateral tract. It begins at the lower end of the medulla by a tapering extremity, and has, apparently, no corresponding column in the cord. It gradually enlarges as it ascends, and forms, at a level with the lower border of the olivary body, a considerable prominence, known as the *tubercle of Rolando*. This is caused by the substantia gelatinosa of Rolando of the cord gradually finding its way to the surface, so as to form a prominence there. About half an inch below the pons the funiculus of Rolando appears to blend with the funiculus cuneatus. In front, it is separated from the lateral tract by a distinct groove, the continuation upwards of the postero-lateral groove of the cord ; behind, the separation from the funiculus cuneatus is much less distinct.

The **funiculus cuneatus** is the direct continuation upwards of the postero-lateral column (tract of Burdach) of the cord. It is situated between the funiculus of Rolando and the funiculus gracilis. It enlarges as it ascends, and forms, opposite the lower extremity of the fourth ventricle, a slight eminence or enlargement, the *cuneate tubercle*, which is best marked in children. Above this point it disappears from the surface.

The **funiculus gracilis** is the direct continuation upwards of the postero-median column of the cord (tract of Goll). It is a narrow white band, placed parallel to and along the side of the posterior median fissure. It is separated from the funiculus cuneatus by a slight groove, continuous with that on the surface of the cord, which marks off the postero-median column. At first the funiculi of the two sides lie in close contact on either side of the posterior median fissure. Opposite the apex of the fourth ventricle each presents an enlargement, the *clava* ; they then diverge and form the lateral boundaries of the lower part of the fourth ventricle, and gradually tapering off become no longer traceable.

The **restiform body.**—The upper part of the posterior area of the medulla is occupied by the *restiform body*. It appears, at first sight, as if this body were the direct continuation upwards of the funiculus cuneatus and the funiculus of Rolando, and it was formerly described as such. This, however, is not so, for the restiform body is largely formed by a set of fibres, the *external arcuate fibres*, which issue from the anterior median fissure and will presently be described. They pass laterally over the pyramid and olive, and assist in forming the restiform body. There is also a narrow strand of fibres, derived from the lateral column of the cord, the *direct cerebellar tract*, which joins the above-mentioned arcuate fibres. These two sets of fibres, reinforced by the *internal arcuate fibres* from the opposite side of the medulla, form the restiform body.

The restiform bodies are the largest prominences of the medulla, and are placed between the lateral tracts in front and the funiculus cuneatus behind, from both of which they are separated by slight grooves. As they ascend they diverge from each other, assist in forming the lower part of the lateral boundaries of the fourth ventricle, and then enter the corresponding hemisphere of the cerebellum, forming its inferior peduncles.

The **posterior surface of the medulla oblongata** forms part of the floor of the fourth ventricle. This portion is of a triangular form, bounded on each side by the diverging funiculi graciles and cuneati and restiform bodies. The divergence of these two funiculi and of the restiform bodies, together with the opening out of the posterior fissure and central canal of the spinal cord, displays in the floor of the ventricle the grey matter of the medulla, which is continuous below with the grey matter of the cord. In the middle line is seen a longitudinal furrow, which divides this part of the ventricle into right and left halves, and is continuous below with the central canal of the cord.

The **arciform** or **arcuate** fibres, which have been mentioned as forming part of the restiform body, are found in the upper half of the medulla, crossing its surface and also traversing its substance. They are divided for purposes of description into two sets—external and internal. The *external* or *superficial arciform fibres* have already been alluded to as crossing the pyramid and olivary body on each side. They emerge from the anterior median fissure, and if traced into it are found to enter the raphé and cross to the opposite side, after which their further course is a matter of some doubt. After emerging from the anterior median fissure they cross the pyramid and olivary body, often concealing from view the upper part of the cuneate and Rolandic funiculi, and enter the restiform body. As they cross the olivary body they are reinforced by some of the internal arciform fibres, which come to the surface on the inner side of, or through, this structure. The *internal arciform fibres* are described with the microscopic anatomy of the medulla.

It is advisable, at this stage, to take up the consideration of the cavity of the fourth ventricle, an acquaintance with which will render the description of the internal structure of the medulla oblongata more intelligible.

THE FOURTH VENTRICLE (fig. 371)

The fourth ventricle is lozenge- or diamond-shaped ; that is to say, it is composed of two triangles, with their bases in contact. The sides of the lower triangle are formed by the divergence of the funiculi graciles, funiculi cuneati, and restiform bodies of the medulla on either side. As these columns pass upwards in the medulla they turn outwards from the median line, and, diverging from each other, form the lateral boundaries of the lower half of the fourth ventricle. In like manner the sides of the upper triangle are formed by the convergence of the superior peduncles of the cerebellum. These peduncles are separated below by a somewhat wide interval, but as they pass upwards and forwards towards the corpora quadrigemina they gradually converge and ultimately come into contact with each other. This cavity is therefore bounded laterally by the superior peduncles of the cerebellum in its upper half, and by the funiculi graciles, the funiculi cuneati, and the restiform bodies in its lower half. It presents four angles. The upper angle reaches as high as the upper border of the pons, and corresponds with the lower opening of the aqueduct of Sylvius, by which this ventricle communicates with the third ventricle. The lower angle is on a level with the lower border of the olivary body, and is continuous with the central canal of the spinal cord. From the resemblance that it bears to the point of a writing pen it has been named the *calamus scriptorius*. Its lateral angles extend for some distance between the medulla and the cerebellum, each forming a pointed *lateral recess*.

The *roof* of the fourth ventricle is formed from above downwards by the following structures : a part of the superior peduncles of the cerebellum, the superior medullary velum, the inferior medullary velum, the tela choroidea inferior, the obex, and the ligula.

The **superior peduncles of the cerebellum,** when they emerge from the medullary substance of its hemispheres, pass upwards and forwards, forming the lateral boundaries of the upper half of the fourth ventricle, but, converging as they approach the corpora quadrigemina, the mesial portions of the peduncles form a part of the roof of the cavity, in consequence of the ventricle extending to a slight extent underneath the peduncles.

The **superior medullary velum** (**valve of Vieussens**).—In the angular interval left between the two superior peduncles is a thin lamina of white matter, continuous with the white centre of the cerebellum, which bridges across from one peduncle to the other, and so completes the roof of the superior part of the ventricle. This is the *superior medullary velum*, or *valve of Vieussens*. Its dorsal surface is covered by the folia of the lingula, already described (page 665).

The **inferior medullary velum** is a thin layer of white substance, prolonged from the white centre of the medulla on either side of the nodule, which assists in forming a part of the roof of the fourth ventricle, stretching over it towards its lateral angles. It is continuous with the white substance of the cerebellum by its convex edge, while its thin concave margin is apparently free. In reality, however, it is continuous with the epithelium of the ventricle, which is prolonged downwards from the velum to the edge of the ligula.

The **tela choroidea inferior** is a layer of pia mater, which covers in the lower part of the fourth ventricle below the inferior medullary velum. Superiorly it is reflected on to the under surface of the cerebellum, while inferiorly it is continued on to the restiform bodies and lower part of the medulla. This part of the roof of the ventricle contains no nervous matter, but consists merely of the ventricular epithelium covered by pia mater. The tela choroidea inferior, like the superior, really consists of two layers, which become more or less adherent, viz. that covering the under surface of the cerebellum and that covering the epithelium. It also possesses a pair of choroid plexuses, which project into the ventricular cavity, invaginating before them the epithelial lining. Each plexus consists of a *vertical* portion, which extends forwards, near the middle line, from the foramen of Majendie, and of a *transverse* part, which passes outwards into the lateral recess of the ventricle as far as the foramina of Key and Retzius. The two plexuses present the form of a T, the vertical limb of which is, however, double, ⅂Γ. The tela does not form a complete membrane, for in it there are three openings, one in the middle line at the inferior angle of the ventricle, just above the position of the opening of the central canal of the cord; this is the *foramen of Majendie*: the other two are at the extremities of the lateral recesses of the ventricle, and are named the *foramina of Key and Retzius* (see page 620). Through these foramina the ventricles of the brain communicate with the subarachnoid space.

FIG. 371.—Floor of the fourth ventricle. Diagrammatic.

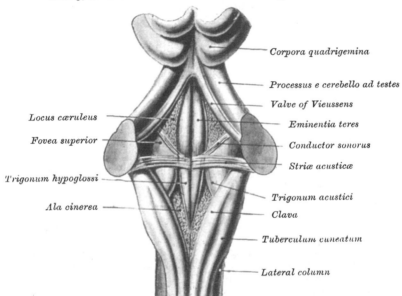

Corpora quadrigemina

Processus e cerebello ad testes

Valve of Vieussens

Locus cæruleus

Eminentia teres

Fovea superior

Conductor sonorus

Striæ acusticæ

Trigonum hypoglossi

Trigonum acustici

Ala cinerea

Clava

Tuberculum cuneatum

Lateral column

The **obex** is a thin, triangular lamina of grey matter, continuous below with the anterior grey commissure of the cord, which fills in the angle between the two diverging funiculi graciles for a short distance.

The **ligula** (*tæniæ*) are narrow bands of white matter, which project from the internal border of the funiculi graciles. They at first run upwards and forwards, and then turn outwards over the restiform bodies, as far as the lateral recesses of the ventricle. Their inner borders are continuous with the epithelial roof of the ventricle.

The **floor of the fourth ventricle** (fig. 371) is rhomboidal in shape, and is traversed by a vertical median fissure, the *sulcus longitudinalis medianus*. At its widest part, opposite the level of the lateral recesses, it is marked by some transverse white lines, the *striæ medullaris* or *striæ acusticæ*. These consist of white fibres, which emerge from the longitudinal sulcus, and pass outwards across the floor of the ventricle.

These striæ divide the floor of the ventricle into two triangles, inferior and superior. The *inferior triangle*, or lower half of the floor, presents above an angular groove, the *fovea inferior*, the apex of which is at the striæ, while the two limbs diverge below, and form the sides of a triangular, dark area, termed the *ala cinerea*, which becomes elevated into a prominence below (*eminentia cinerea*). This area corresponds with the nuclei of the vagus and glosso-pharyngeal nerves, and is therefore termed the *trigonum vagi*. A second triangular area lies between the inner limb of the fovea and the median sulcus; its base is directed upwards, and limited by the striæ medullaris. It is termed the *trigonum hypoglossi*, because it corresponds in position to the tract of nerve-cells from which the hypoglossal nerve takes origin. A third triangular area, to the outer side of the fovea inferior, is named the *trigonum acustici*. It lies between the groove forming the outer boundary of the fovea inferior and the lateral wall of the ventricle, and, like the trigonum hypoglossi, has its base directed upwards. Here it is continuous with a prominence, the *tuberculum acusticum*, which extends into the anterior part of the floor of the ventricle.

The *superior triangle*, or upper half of the floor of the fourth ventricle, i.e. the part above the striæ medullaris, presents in the middle line the continuation of the median longitudinal sulcus. On either side of this is a spindle-shaped longitudinal eminence, prominent in its centre, but less so above and below. This is the *eminentia teres*, and is produced by an underlying bundle of white fibres, the *funiculus teres*, formed, in part at all events, by the fibres of the facial nerve. Immediately above and to the outer side of the eminentia teres is an angular depression, the *fovea superior*; this is sometimes crossed by a whitish band of fibres, the *conductor sonorus*, which is connected below with the striæ medullaris. Above the fovea is a bluish depressed area, the *locus cæruleus*. Its colour is due to some pigmented nerve-cells, showing through the white covering of the floor. These pigmented cells are named the *substantia ferruginosa*, and in them one of the roots of the fifth nerve terminates.

The **lining membrane** of the fourth ventricle is continuous above with that of the third, through the aqueduct of Sylvius, and below with that of the central canal of the spinal cord. The cavity of the ventricle communicates below with the subarachnoidean space by means of the foramen of Majendie and the foramina of Key and Retzius, already described.

Internal Structure of the Medulla Oblongata (fig. 372)

If the cranial nerves emerging from the medulla are traced into its substance, it will be seen that they divide each half into three wedge-shaped areas, which are named the anterior, lateral, and posterior areas of the medulla, and each of which corresponds to one of the subdivisions already described on the surface of this portion of the encephalon.

The **anterior area** comprises that portion which is situated between the anterior median fissure and the fibres of origin of the hypoglossal nerve. On the surface of the medulla this area corresponds to the pyramid.

The **lateral area** is situated between the fibres of origin of the hypoglossal nerve on the one hand, and the fibres of the glosso-pharyngeal, pneumogastric, and spinal accessory nerves on the other. On the surface of the medulla, in its lower part, this area is single, and is called the *lateral tract*; but in the upper part an oval-shaped body, the *olivary body*, comes forward between it and the pyramid, pushing aside the lateral tract.

The **posterior area** comprises that portion which is situated between the fibres of origin of the glosso-pharyngeal, pneumogastric, and spinal accessory nerves, and the posterior median fissure. On the surface of the medulla this area is marked by slight furrows, splitting it up into smaller columns; those in the lower part of the medulla are named, from without inwards, the funiculus of Rolando, the funiculus cuneatus, and the funiculus gracilis; in the upper part of the medulla they are replaced by the restiform body. Finally, the halves of the medulla are separated from each other by a median septum or raphé.

Each of these three areas is made up of grey and white matter, the former being derived for the most part from that of the cord. In like manner the white matter is partly made up of longitudinal fibres continuous with those of the cord, and partly of transverse fibres which intersect them.

In order to understand the internal structure of the medulla, it is necessary to describe the appearances as they are seen in the upper and lower portions of the medulla, since they differ considerably in these two parts.

The **lower part of the medulla.**—The first change in the internal structure is caused by the passage of the fibres of the crossed pyramidal tract obliquely through the grey matter of the anterior horn. As stated above, the pyramid is composed of fibres derived from the direct pyramidal tract of the anterior column of the cord of the same side, and from the crossed pyramidal tract of the lateral column of the opposite half of the cord. Those fibres which are derived from the

FIG. 372.—Section of the medulla oblongata at about the middle of the olivary body. (Schwalbe.)

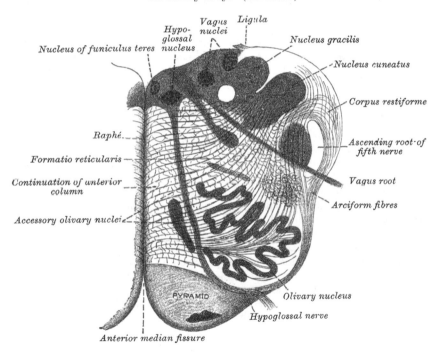

direct pyramidal tract and which in the cord lie close to the median fissure are in the medulla placed to the outer side of the pyramid, being pushed aside, as it were, by the interpolation of the fibres derived from the crossed pyramidal tract, which are much more numerous. The crossed pyramidal fibres ascend from the lateral column of the spinal cord, and, passing through the anterior grey cornu and across the middle line, form the inner part of the pyramid. In consequence of this passage of white fibres through its substance the anterior grey cornu is broken up into a coarse network, while one portion of it, the caput cornu, is entirely separated from the rest; only the base of the cornu remains intact, close to the ventro-lateral aspect of the central canal. The caput cornu, thus separated, is displaced laterally, and comes to lie close to the caput cornu posterioris, which has also shifted its position. In consequence of this breaking up of the greater part of the anterior grey cornu by white fibres a coarse network is formed in the anterior and lateral areas of the medulla, which is named the *formatio reticularis*.

The posterior cornu also undergoes somewhat similar changes. It becomes subdivided by the passage through it of the sensory fibres of the columns of Goll and Burdach. These pass across to the opposite anterior area of the medulla, where they are seen to lie immediately on the dorsal aspect of the pyramids. In their passage through the posterior horns of grey matter the latter become subdivided, in a manner somewhat similar to what has been seen to occur in the anterior horns. This crossing of the sensory fibres is termed the *superior pyramidal* or *sensory decussation*. The caput cornu is displaced outwards, so as almost to reach the surface, where it forms a projection, the *funiculus Rolando*, which enlarges above into a distinct prominence, the *tubercle of Rolando*. Above the level of the tubercle of Rolando the caput cornu is separated from the surface by a band of fibres, termed the *ascending root of the fifth nerve*, and by the external arcuate fibres. The neck of the cornu becomes broken up into a reticular formation by the decussation of the columns of Goll and Burdach, and by this means the caput is separated from the rest of the grey matter. The base of the cornu increases in size, and, as the central canal expands into the fourth ventricle, becomes pushed outwards, and portions of it extend into the funiculi graciles and cuneati, and produce externally the eminences of the clava and cuneate tubercle. A third portion of the base becomes separated from the rest, and is placed outside the nucleus of the funiculus cuneatus. This is called the *accessory cuneate nucleus*, and is supposed to be a continuation upwards of Clarke's vesicular column of the cord.

The **upper part of the medulla.**—The upper part of the medulla comprises the portion which enters into the formation of the floor of the fourth ventricle, where, in fact, the upper end of the central canal has opened out into this cavity. In this region the formatio reticularis is confined chiefly to the anterior and lateral areas. In the ventral portion of the posterior area there is only a small amount of reticular formation, but in addition to this there are individual masses of cells scattered among the longitudinal fibres.

The **formatio reticularis** is situated in the medulla, behind the pyramid and olivary body, extending laterally as far as the restiform bodies, and dorsally to within a short distance of the floor of the fourth ventricle. It presents a peculiar reticulated appearance, from which it derives its name, and which is due to the intersection of bundles of fibres running at right angles to each other, some being longitudinal, others transverse. The formatio reticularis presents a different appearance in the anterior area from what it does in the lateral area. In the former there is almost an entire absence of nerve-cells in the reticulated network, and hence it is known as the *formatio reticularis alba*; whereas, in the lateral area, the nerve-cells are numerous, and, as a consequence, this part is known as the *formatio reticularis grisea*. In the substance of the formatio reticularis is a small nucleus of grey matter. It is situated near the dorsal aspect of the hilum of the olivary nucleus, and has been named the *inferior central nucleus*. The fibres of the formatio reticularis are longitudinal and transverse. In the anterior area the longitudinal fibres may be arranged in two well-defined sets: (1) one set lies immediately behind the pyramid and is named the *fillet* or *lemniscus*. The fibres of the fillet are chiefly derived from the cells of the gracile and cuneate nuclei, and may therefore be regarded as relay fibres of the columns of Goll and Burdach of the spinal cord, which terminate in synapses around the cells of the gracile and cuneate nuclei. They are prolonged inwards and forwards across the middle line forming the superior pyramidal or sensory decussation (decussation of the fillet); (2) the other set is continued from the antero-lateral ground bundle of the cord, and a portion of these fibres forms the *posterior longitudinal bundle* already referred to (page 674). Both these sets of fibres are continued upwards into the pons and mid-brain. The longitudinal fibres of the reticular formation in the lateral area are not arranged in distinct bundles.

They are derived from the lateral column of the cord, after the crossed pyramidal tract has passed over to the opposite side. The longitudinal fibres of the posterior area are merely indeterminate fibres of the formatio reticularis, with the exception of two distinct bundles, which may be regarded as ascending roots of the fifth and vago-glosso-pharyngeal nerves ; the latter is termed the *funiculus solitarius.*

The *transverse* fibres of the reticular formation are the arched or arcuate fibres. The *external arciform fibres* have already been described. The *internal arciform fibres* are more numerous than the superficial set; they traverse nearly the whole area of the upper half of the medulla, except the pyramid. They pass from the raphé ; some become superficial and join the external arciform fibres, while others remain deep and pass to the olivary body, the restiform body, and to the nuclei of the funiculus cuneatus and funiculus gracilis.

Independent Nuclei.—In the upper part of the medulla are several independent nuclei of grey matter, which may be divided into two sets : (1) those which are traceable from the grey matter of the spinal cord ; and (2) those which are not represented in the cord. The former are the hypoglossal nucleus, the nucleus of the funiculus teres, and those of the auditory, glosso-pharyngeal, vagus, and spinal accessory nerves. The latter are the nucleus of the olivary body and the accessory olivary nuclei. In addition to these, small collections of grey matter are to be found in the median septum or raphe.

The hypoglossal nucleus.—The base of the anterior horn, which in the lower part of the medulla was situated on the ventro-lateral aspect of the central canal, now approaches the floor of the ventricle, where it lies close to the median sulcus under the funiculus teres. In it is a column of large nerve-cells, from which the roots of the hypoglossal nerve are derived. It is accordingly designated the *hypoglossal nucleus.*

The auditory nuclei.—Towards the upper part of the medulla, a considerable tract of grey matter may be found lying immediately beneath that portion of the floor of the fourth ventricle which is known as the trigonum acustici. This is the *dorsal* or *inner auditory nucleus,* and it lies just external to the vago-glosso-pharyngeal nucleus. In addition to this, there is a small collection of nerve-cells on the ventral surface of the restiform body, between the two roots of the auditory nerve, which is known as the *accessory* or *ventral auditory nucleus.* On the outer side of the restiform body is a mass of cells associated with the cochlear root of the auditory nerve. This mass is termed the *lateral acoustic tubercle* or *ganglion radicis cochlearis.*

Nuclei of the glosso-pharyngeal and vagus nerves.—These are two in number, *principal* and *accessory.* The *principal* nucleus of the two nerves lies beneath that portion of the floor of the fourth ventricle which is known as the ala cinerea and fovea inferior. They form an oblong mass of grey matter, above the nucleus of the spinal accessory and lateral to the hypoglossal nucleus. The *accessory* nuclei are situated in the reticular formation of the posterior area, some distance from the floor of the fourth ventricle. They consist of a pear-shaped mass of cells, which is connected with the rest of the grey matter by a sort of stalk or peduncle, and was formerly known as the *nucleus ambiguus.*

Nucleus of the spinal accessory nerve.—This nucleus consists of a group of cells, which is situated partly in the lower part of the medulla at the base of the posterior horn and close to the central canal. It extends upwards, lying beneath the lower part of the floor of the fourth ventricle and on the outer side of the hypoglossal nucleus.

The nucleus of the olivary body.—When the olivary body is cut across, it is seen to be covered externally by white fibres, and internally to consist of a grey layer. This grey layer is the *nucleus of the olivary body,* or, as it is sometimes called, the *corpus dentatum of the olive.* It is composed of a thin, wavy lamina, which is arranged in the form of a hollow capsule, open at its upper and inner part, and presenting a zigzag or dentated outline. Microscopically examined, the olivary nucleus is seen to consist of small rounded yellowish nerve-cells embedded in a matrix of neuroglia and fine nerve-fibres. White fibres, which can be traced from the raphé, and are probably derived from the opposite olive, enter the interior of the capsule by the aperture at its upper or inner part, constituting the olivary peduncle.

The fibres of the olivary peduncle as they enter the body diverge, and some are lost in the grey matter of the olivary nucleus; others pass through it, and of these some turn backwards to join the restiform body, and pass to the cerebellum as internal arcuate fibres; while others pierce the white matter of the olivary body and, reaching the surface, are continued to the restiform body as external arcuate fibres. The fibres of the olivary peduncle connect the olivary nucleus with the cerebral hemisphere of the same side. The nucleus is also connected to the anterior horn of the same side of the cord; and with the opposite cerebellar hemisphere through the internal arcuate fibres. Removal of one cerebellar hemisphere is followed by atrophy of the opposite olivary nucleus.

Accessory olivary nuclei.—Two small isolated masses of grey matter are to be found, one on the mesial and the other on the dorsal aspect of the corpus denta-tum. These are the *mesial* and *lateral accessory olivary nuclei*. They are con-nected with the restiform body by some of the internal arcuate fibres. The fibres of the hypoglossal nerve, as they traverse the bulb, pass between the mesial accessory nucleus and the chief olivary nucleus.

The raphé.—The raphé is situated in the middle line of the medulla, above the decussation of the pyramids. It consists of nerve-fibres intermingled with nerve-cells. The fibres have different directions, which can only be seen in suitable microscopic sections, thus: 1. Some are antero-posterior; these in front are continuous with the superficial arciform fibres. 2. Some are longitudinal; these are derived from the arciform fibres, which on entering the raphé change their direction and become longitudinal. 3. Some are oblique; these are continuous with the deep arciform fibres which pass from the raphé.

The nerve-cells of the raphé are multipolar; some are connected with the antero-posterior fibres, others with the superficial arcuate fibres.

Weight of the Encephalon.—The average weight of the brain, in the adult male, is 49½ oz., or a little more than 3 lb. avoirdupois; that of the female, 44 oz.; the average difference between the two being from 5 to 6 oz. The pre-vailing weight of the brain, in the male, ranges between 46 oz. and 53 oz.; and, in the female, between 41 oz. and 47 oz. In the male, the maximum weight out of 278 cases was 65 oz. and the minimum weight 34 oz. The maximum weight of the adult female brain, out of 191 cases, was 56 oz., and the minimum weight 31 oz. According to Luschka, the average weight of a man's brain is 1,424 grammes (about 45 oz.), of a woman's 1,272 grammes (about 41 oz.); and according to Krause, 1,570 grammes (about 48½ oz.) for the male, and 1,350 grammes (about 43 oz.) for the female. It appears that the weight of the brain increases rapidly up to the seventh year, more slowly to between sixteen and twenty, and still more slowly to between thirty and forty, when it reaches its maximum. As age advances and the mental faculties decline, the brain diminishes slowly in weight, to the extent of about an ounce for each subsequent decennial period. These results apply alike to both sexes.

The size of the brain was formerly said to bear a general relation to the intellectual capacity of the individual. Cuvier's brain weighed rather more than 64 oz., that of the late Dr. Abercrombie 63 oz., and that of Dupuytren 62½ oz. On the other hand, the brain of an idiot seldom weighs more than 23 oz. But these facts are by no means conclusive, and it is well known that these weights have been equalled by the brains of persons who never displayed any remarkable intellect. Dr. Haldennan, of Cincinnati, has recorded the case of a mulatto, aged 45, whose brain weighed 68⅜ oz.; he had been a slave, and was never regarded as particularly intelligent; he was illiterate, but is said to have been reserved, meditative, and economical. Dr. Ensor, district medical officer at Port Eliza-

beth, reports that the brain of Carey, the Irish informer, weighed 61 oz.
M. Nikiforoff has published an article on the subject of the weight of brains in
the *Novosti*. According to him, the weight of the brain has no influence whatever on the mental faculties. It ought to be remembered that the significance
of the weight of the brain should depend upon the proportion it bears to the
dimensions of the whole body and to the age of the individual. It is equally
important to know what was the cause of death, for long illness or old age
exhausts the brain. To define the real degree of development of the brain, it is
therefore necessary to have a knowledge of the condition of the whole body ;

FIG. 373.—Side view of the brain of man, showing the localisation of
various functions. (After Ferrier.)

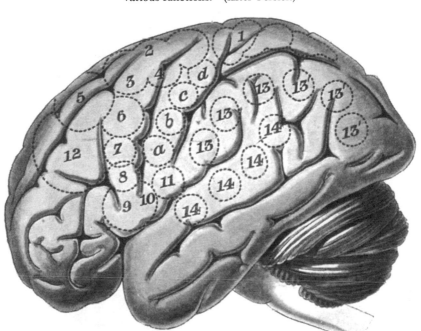

1. Centre for movements of opposite leg and foot. 2, 3, 4. Centres for complex movements of the arms and legs, as in
swimming. 5. Extension forwards of the arm and hand. 6. Supination of the hand and flexion of the forearm.
7, 8. Elevators and depressors of the angle of the mouth. 9, 10. Movements of the lips and tongue. 11. Retraction
of the angle of the mouth. 12. Movements of the eyes. 13, 13′. Vision. 14. Hearing. *a, b, c, d*. Movements of the
wrists and fingers.

and, as this is usually lacking, the mere record of weight possesses little
significance.

The human brain is heavier than that of all the lower animals, excepting the
elephant and whale. The brain of the former weighs from eight to ten pounds ;
and that of a whale, in a specimen seventy-five feet long, weighed rather more
than five pounds.

Cerebral Localisation and Topography.—Physiological and pathological research
have now gone far to prove that the surface of the brain may be mapped out into a series
of definite areas, each one of which is intimately connected with some well-defined
function. And this is especially true with regard to the convolutions on either side of the
fissure of Rolando, which are believed by most physiologists of the present day to be concerned in motion, those grouped around the fissure being associated with movements of
the extremities of the opposite side of the body, and those around the lower end of the
fissure being related to movements of the mouth and tongue.

This is not the place, nor can space be given here, to describe these localities. But the
two accompanying woodcuts from Ferrier (figs. 373, 374) have been introduced, and will
serve to indicate the position of some of the more important areas.

The relation of the principal fissures and convolutions of the cerebrum to the outer surface of the scalp has been the subject of much investigation, and many systems have been devised by which one may localise these parts from an examination of the external surface of the head.

These plans can only be regarded as approximately correct for several reasons: in the first place, because the relations of the convolutions and sulci to the surface are found to be very variable in different individuals; secondly, because the surface area of the scalp is greater than the surface area of the brain, so that lines drawn on the one cannot correspond exactly to sulci or convolutions on the other; and thirdly, because the sulci and convolutions in two individuals are never precisely alike. Nevertheless, the principal fissures and convolutions can be mapped out with sufficient accuracy for all practical

FIG. 374.—Top view of the brain of man, showing the localisation of various functions. (After Ferrier.).

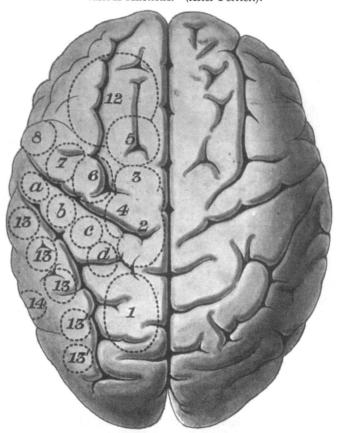

1. Centre for movements of opposite leg and foot. 2, 3, 4. Centres for complex movements of the arms and legs, as in swimming. 5. Extension forwards of the arm and hand. 6. Supination of the hand and flexion of the forearm. 7, 8. Elevators and depressors of the angle of the mouth. 9, 10. Movements of the lips and tongue. 11. Retraction of the angle of the mouth. 12. Movements of the eyes. 13, 13'. Vision. 14. Hearing. a, b, c, d. Movements of the wrists and fingers.

purposes, so that any particular convolution can be generally exposed by removing with the trephine a certain portion of the skull's area.

The various landmarks on the outside of the skull, which can be easily felt, and which serve as indications of the position of the parts beneath, have been already referred to (see p. 87), and the relation of the fissures and convolutions to these landmarks is as follows:

Longitudinal Fissure.—This corresponds to a line drawn from the glabella at the root of the nose to the external occipital protuberance.

The Fissure of Sylvius.—The position of the fissure of Sylvius and its horizontal limb is marked by a line starting from a point one inch and a quarter behind the external angular process of the frontal bone to a point three-quarters of an inch below the most prominent point of the parietal eminence. The first three-quarters of an inch will represent the main fissure, the remainder the horizontal limb. The bifurcation of the fissure is, therefore, two inches behind and about a quarter of an inch above the level of the external angular process. The ascending limb of the fissure passes upwards from this point parallel to, and immediately behind, the coronal suture.

Fissure of Rolando.—To find the upper end of the fissure of Rolando, a measurement should be taken from the glabella to the external occipital protuberance. The position of the top of the sulcus will be, measuring from in front, 55·6 per cent. of the whole distance from the glabella to the external occipital protuberance. Professor Thane adopts a somewhat simpler method. He divides the distance from the glabella to the external occipital protuberance over the top of the head into two equal parts, and having thus defined the middle point of the vertex, he takes half an inch behind it as the top of the sulcus. This is not quite so accurate as the former method, but it is sufficiently so for all practical purposes, and, on account of its simplicity, is very generally adopted. From this point the fissure runs downwards and forwards for 3¾ inches, its axis making an angle of 67° with the middle line. Cunningham states that this angle more nearly averages 71·5°.

FIG. 375.—Drawing to illustrate cranio-cerebral topography. (Taken from a cast in the Museum of the Royal College of Surgeons of England, prepared by Professor Cunningham.)

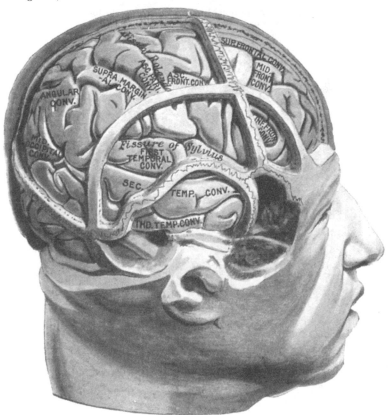

In order to mark this groove, two strips of metal may be employed: one, the shorter, being fixed to the middle of the other at the angle mentioned. If the longer strip is now placed along the sagittal suture so that the junction of the two strips is over the point corresponding to the top of the furrow, the shorter, oblique strip will indicate the direction, and 3¾ inches will mark the length of the furrow. Dr. Wilson has devised an instrument, called a cystometer, which combines the scale of measurements for localising the fissure with data for representing its length and direction.* Professor Thane gives the lower end of the furrow as 'close to the posterior limb, and about half an inch behind the bifurcation of the fissure of Sylvius.' So that, according to this anatomist, a line drawn from a point half an inch behind the mid-point between the glabella and external occipital protuberance to this spot would mark out the fissure of Rolando. Dr. Reid adopts a different method

---

* *Lancet*, vol. i. 1888, p. 408.

(fig. 376). He first indicates, on the surface, the longitudinal fissure and the horizontal limb of the fissure of Sylvius (p. 684). He then draws two perpendicular lines from his 'base-line' (that is, a line from the lowest part of the infra-orbital margin through the middle of the external auditory meatus to the back of the head) to the top of the cranium, one (D E, fig. 376) from the depression in front of the external auditory meatus, and the other (F G, fig. 376) from the posterior border of the mastoid process at its root. He has thus described on the surface of the head a four-sided figure (F D G E, fig. 376), and a diagonal line from the posterior superior angle to the anterior perpendicular line where it is crossed by the fissure of Sylvius will represent the furrow.

The *parieto-occipital fissure* on the upper surface of the cerebrum runs outwards at right angles to the great longitudinal fissure for about an inch, from a point ⅓th of an inch in front of the lambda (posterior fontanelle). Reid states that if the horizontal limb of the fissure of Sylvius be continued onwards to the sagittal suture, the last inch of this line will indicate the position of the sulcus.

The *precentral sulcus* begins four-fifths of an inch in front of the middle of the fissure of Rolando, and extends nearly, but not quite, to the horizontal limb of the fissure of Sylvius.

FIG. 376.—Relation of the principal fissures and convolutions of the cerebrum to the outer surface of the scalp. (Reid.)

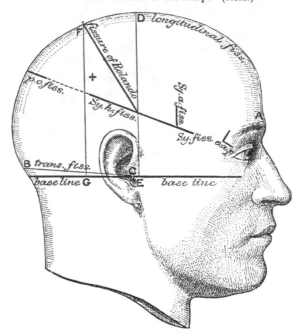

The *superior frontal fissure* runs backwards from the supra-orbital notch, parallel with the line of the longitudinal fissure to two-fifths of an inch in front of the line indicating the position of the fissure of Rolando.

The *inferior frontal fissure* follows the course of the superior temporal ridge on the frontal bone.

The *intraparietal fissure* begins on a level with the junction of the middle and lower third of the fissure of Rolando, on a line carried across the head from the back of the root of one auricle to that of the other. After passing upwards, it curves backwards, lying parallel to the longitudinal fissure, midway between it and the parietal eminence; it then curves downwards to end midway between the posterior fontanelle and the parietal eminence.

THE SPINAL CORD AND ITS MEMBRANES

Dissection.—To dissect the cord and its membranes, it will be necessary to lay open the whole length of the spinal canal. For this purpose, the muscles must be separated from the vertebral grooves, so as to expose the spinous processes and laminæ of the vertebræ; and the latter must be sawn through on each side, close to the roots of the transverse processes, from the third or fourth cervical vertebra above, to the sacrum below. The vertebral arches having been displaced by means of a chisel, and the separate fragments removed, the dura mater will be exposed, covered by a plexus of veins and a quantity of loose areolar tissue, often infiltrated with serous fluid. The arches of the upper vertebræ are best divided by means of a strong pair of cutting bone-forceps.

MEMBRANES OF THE CORD

The membranes which envelop the spinal cord are three in number. The most external is the *dura mater*, a strong fibrous membrane, which forms a loose sheath around the cord. The most internal is the *pia mater*, a cellulo-vascular membrane which closely invests the entire surface of the cord. Between the two is the *arachnoid membrane*, a non-vascular membrane, which envelops the cord, and is connected to the pia mater by slender filaments of connective tissue.

The **Dura Mater** of the cord represents only the meningeal or supporting layer of the cranial dura mater. The endocranial or endosteal layer ceases at the foramen magnum posteriorly, but reaches as low as the third cervical vertebra in front; below these levels its place is taken by the periosteum. It forms a loose sheath which surrounds the cord, and is separated from the bony walls of the spinal canal by a quantity of loose areolar tissue, and a plexus of veins. The situation of the veins between the dura mater of the cord and the periosteum of the vertebræ corresponds therefore to that of the cranial sinuses between the endocranial and supporting layers. It is attached to the circumference of the foramen magnum, and to the axis and third cervical vertebra; it is also fixed to the posterior common ligament, especially near the lower end of the spinal canal, by fibrous slips; it extends below as far as the second or third piece of the sacrum; here it becomes impervious, and, ensheathing the filum terminale, descends to the back of the coccyx, where it blends with the periosteum. The dura mater is much larger than is necessary for its contents, and its size is greater in the cervical and lumbar regions than in the dorsal. Its inner surface is smooth. On each side may be seen the double openings which transmit the two roots of the corresponding spinal nerve, the fibrous layer of the dura mater being continued in the form of a tubular prolongation on them as they pass through these apertures. These prolongations of the dura mater are short in the upper part of the spine, but become gradually longer below, forming a number of tubes of fibrous membrane, which enclose the sacral nerves, and are contained in the spinal canal.

The chief peculiarities of the dura mater of the cord, as compared with that investing the brain, are the following:

The dura mater of the cord is not adherent to the bones of the spinal canal, which have an independent periosteum.

It does not send partitions into the fissures of the cord, as in the brain.

Its fibrous laminæ do not separate, to form venous sinuses, as in the brain.

Structure.—The dura mater consists of white fibrous and elastic tissue, arranged in bands or lamellæ, which, for the most part, are parallel with one another and have a longitudinal arrangement. Its internal surface is covered by a layer of endothelial cells, which gives this surface its smooth appearance. It is sparingly supplied with vessels; and some few nerves have been traced into it.

The **Arachnoid** is exposed by slitting up the dura mater, and reflecting that membrane to either side (fig. 377). It is a thin, delicate, tubular membrane, which invests the surface of the cord, and is connected to the pia mater by slender filaments of connective tissue. Above, it is continuous with the cerebral arachnoid; on each side it is continued on the various nerves, so as to form a sheath for them as they pass outwards to the intervertebral foramina. The outer surface of the arachnoid is in contact with the inner surface of the dura mater, and the two are, here and there, joined together by isolated connective-tissue trabeculæ, especially on the posterior surface of the cord. For the most part, however, the membranes are not connected together, and the interval between them is named the *subdural space*. The inner surface of the arachnoid is separated from the pia mater by a considerable interval, which is called the *subarachnoidean space*. The space is largest at the lower part of the spinal canal, and encloses the mass of nerves which form the cauda equina. Superiorly it is continuous with the cranial subarachnoid space, and communicates with the general ventricular cavity of the brain, by means of the openings in the pia mater, in the roof of the fourth ventricle (*foramen of Majendie* and *foramina of Key and Retzius*). It contains an abundant serous secretion, the *cerebro-spinal fluid*. This secretion is sufficient in amount to expand the arachnoid membrane, so as to completely fill up the whole of the space included in the dura mater. The subarachnoidean space is occupied by trabeculæ of delicate connective tissue, connecting the pia mater on the one hand with the arachnoid membrane on the other. This is named *subarachnoid tissue*. In addition to this it is partially subdivided by a longitudinal membranous partition, the *septum posticum*, which serves to connect the arachnoid with the pia mater, opposite the posterior median fissure of the spinal cord, a partition which is incomplete, and cribriform in structure, consisting of bundles of white fibrous tissue, interlacing with each other. This space is to be regarded as, in reality, a great lymph-space, from which the lymph carried to it by the perivascular lymphatics is conveyed back into the circulation.

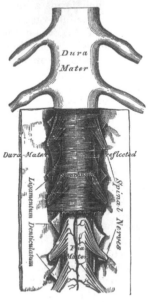

FIG. 377.—The spinal cord and its membranes.

Structure.—The arachnoid is a delicate membrane, made up of closely arranged interlacing bundles of connective tissue in several layers.

The **Pia Mater** of the cord is exposed on the removal of the arachnoid (fig. 377). It covers the entire surface of the cord, to which it is very intimately adherent, forming its neurilemma, and sending a process downwards into its anterior fissure. It also forms a sheath for each of the filaments of the spinal nerves, and invests the nerves themselves. A longitudinal fibrous band extends along the middle line on its anterior surface, called by Haller the *linea splendens*; and a somewhat similar band, the *ligamentum denticulatum*, is situated on each side. At the point where the cord terminates, the pia mater becomes contracted, and is continued down as a long, slender filament (*filum terminale*), which descends through the centre of the mass of nerves forming the cauda equina. It perforates the dura about the level of the second or third lumbar vertebræ, receiving a sheath from it, and extends downwards as far as the base of the coccyx, where it blends with the periosteum. It assists in maintaining the cord in its position during the movements of the trunk, and is from this circumstance called the *central ligament* of the spinal cord. It contains a little grey nervous substance, which may be traced for some distance into its upper part, and is accompanied by a small artery and vein. At the upper part of the cord, the pia mater presents a greyish, mottled tint, which is owing to yellow or brown pigment-cells scattered among the elastic fibres.

FIG. 378.—Transverse section of the spinal cord and its membranes.

Structure.—The pia mater of the cord is less vascular in structure, but thicker and denser, than the pia mater of the brain, with which it is continuous.

It consists of two layers: an outer composed of bundles of connective-tissue fibres, arranged for the most part longitudinally; and an inner consisting of stiff bundles of the same tissue, which present peculiar angular bends, and is covered on both surfaces by a layer of endothelium. Between the two layers are a number of cleft-like lymphatic spaces, which communicate with the subarachnoid cavity, and a number of blood-vessels, which are enclosed in a perivascular sheath, derived from the inner layer of the pia mater, into which the lymphatic spaces open. It is also supplied with nerves, which are derived from the sympathetic.

The **Ligamentum Denticulatum** (fig. 377) is a narrow fibrous band, situated on each side of the spinal cord, throughout its entire length, and separating the anterior from the posterior roots of the spinal nerves. It has received its name from the serrated appearance which it presents. Its inner border is continuous with the pia mater, at the side of the cord. Its outer border presents a series of triangular, dentated serrations, the points of which are fixed at intervals to the dura mater. These serrations are twenty-one in number, on each side, the first being attached to the dura mater, opposite the margin of the foramen magnum between the vertebral artery and the hypoglossal nerve; and the last near the lower end of the cord. Its use is to support the cord in the fluid by which it is surrounded.

Surgical Anatomy.—Evidence of great value in the diagnosis of meningitis may be obtained by puncturing the theca of the cord and withdrawing some of the cerebro-spinal fluid, and the operation is regarded by some as curative, under the supposition that the draining away of the cerebro-spinal fluid relieves the patient by diminishing the intra-cranial pressure. The operation is performed by inserting a trocar, of the smallest size, between the lamina of the third and fourth or the fourth and fifth lumbar vertebræ through the ligamenta subflava. The spinal cord even of a child at birth does not reach below the third lumbar vertebra, and therefore the canal may be punctured between the third and fourth vertebra without any risk of injuring its contents. The point of puncture is indicated by laying the child on its side and dropping a perpendicular line from the highest point of the crest of the ilium; this will cross the upper border of the spine of the fourth lumbar vertebra, and will indicate the level at which the trocar should be inserted a little to one side of the median line.

THE SPINAL CORD (fig. 379)

The **Spinal Cord** (*medulla spinalis*) is the cylindrical elongated part of the cerebro-spinal axis which is contained in the vertebral canal. Its length is usually about seventeen or eighteen inches, and its weight, when divested of its membranes and nerves, about one ounce and a half; its proportion to the encephalon being about 1 to 33. It does not nearly fill the canal in which it is contained, its investing membranes being separated from the surrounding walls by areolar tissue and a plexus of veins. It occupies, in the adult, the upper two-thirds of the vertebral canal, extending from the upper border of the atlas to the lower border of the body of the first lumbar vertebra, where it terminates in a slender filament of grey substance, which is continued for some distance into the *filum terminale*. In the fœtus, before the third month it extends to the bottom of the sacral canal; but after this period it gradually recedes from below, as the growth of the bones composing the canal is more rapid in proportion than that of the cord; so that in the child at birth the cord extends as far as the third lumbar vertebra. Its position varies also according to the degree of curvature of the spinal column, being raised somewhat in flexion of the spine. On examining its surface, it presents a difference in its diameter in different parts, being marked by two enlargements, an upper or cervical, and a lower or lumbar. The *cervical enlargement* extends from about the third cervical to the first or second dorsal vertebra: its greatest diameter is in the transverse direction (13 mm.), and it corresponds with the origin of the nerves which supply the upper extremities. The *lumbar enlargement* is situated opposite the last two or three dorsal vertebræ, and corresponds with the origin of the nerves

FIG. 379.—Posterior view of the spinal cord *in situ*.

which supply the lower extremities. Below the lumbar enlargement the cord gradually tapers to form a cone, the *conus medullaris*, the apex of which is continuous with the filum terminale. In form, the spinal cord is a cylinder, flattened before and behind.

Fissures.—It presents on its anterior surface, along the middle line, a longitudinal fissure, the *anterior median fissure*; and, on its posterior surface, another fissure, which also extends along the entire length of the cord, the *posterior median fissure*. These fissures penetrate through the greater part of the thickness of the cord, and incompletely divide it into symmetrical halves, united in the middle line by a transverse band of nervous substance, the *commissure*.

The **Anterior Median Fissure** is wider, but of less depth than the posterior, ex-

FIG. 380.—Spinal cord. Side view. Plan of the fissures and columns.

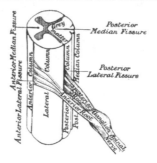

tending into the cord for about one-third of its thickness, and is deepest at the lower part of the cord. It contains a prolongation from the pia mater; and its floor is formed by the *anterior white commissure*, which is perforated by numerous blood-vessels, passing to the centre of the cord.

The **Posterior Median Fissure** is not an actual fissure, as the space between the lateral halves of the posterior part of the cord is crossed by connective tissue and numerous blood-vessels, so that no actual hiatus exists, and there is consequently no prolongation of the pia mater into it. It extends into the cord to about one half its depth, and its floor is formed by the *posterior grey commissure*.

Lateral Fissures.—On each side of the posterior median fissure, along the line of attachment of the posterior roots of the nerves, a delicate fissure may be seen, leading down to the grey matter which approaches the surface in this situation; this is called the *postero-lateral fissure* of the spinal cord. On the posterior surface of the spinal cord, between the posterior median fissure and the postero-lateral fissure on each side, is a slight longitudinal furrow (*posterior intermediate furrow*), marking off two slender tracts, the *postero-median and postero-lateral columns*. These are most distinct in the cervical region, but are stated by Foville to exist throughout the whole length of the cord. On each side of the anterior median fissure the anterior roots of the spinal nerves emerge from the cord, not in one vertical line, but by separate bundles which occupy an area of some width. This is called, by some anatomists, the *antero-lateral fissure* of the cord, although no actual fissure exists in this situation.

Columns of the Cord.—Each half of the spinal cord is thus divided into four columns: an anterior column, a lateral column, a posterior column, and a postero-median column. This division, however, is very imperfect, since the limit between the so-called anterior and lateral columns cannot be defined on account of the bundles of the anterior roots being spread over a considerable area. It is therefore customary to divide each half of the spinal cord into two

FIG. 381.—Transverse sections of the cord.

Opposite middle of cervical region.

Opposite middle of dorsal region.

Opposite lumbar region.

FIG. 382.—From a transverse section through the spinal cord of a calf. Magnified about 180 diameters, showing part of the central canal and the tissue immediately around it, viz. the central grey matter. (Klein and Noble Smith.)

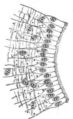

The canal is lined with epithelium, composed of ciliated, more or less conical, cells; in most instances a filamentous process passes from the cell into the tissue underneath. This tissue contains, in a hyaline matrix, a network of fibrils; most of these run horizontally; others have a longitudinal course, and appear therefore here cut transversely, i.e. as small dots. The nuclei correspond to the cells of the neuroglia, the cell-substance not being shown. Both the nuclei of the neuroglia cells and those of the epithelium contain three or more large disc-shaped particles.

columns, separated by the postero-lateral groove: (1) a small *posterior column*, which is bounded internally by the posterior median fissure, and externally by the postero-lateral fissure, and (2) a large *antero-lateral column*, which comprises the rest of the cord. The posterior column is further divided, at all events at its upper part, by the posterior intermediate septum, into a postero-median column and a postero-lateral column.

Structure of the Cord.—If a transverse section of the spinal cord be made, it will be seen to consist of white and grey nervous substance. The white matter is situated externally, and constitutes the greater part. The grey substance occupies the centre, and is so arranged as to present on the surface of the section two crescentic masses, placed one in each lateral half of the cord, united together by a transverse band of grey matter, the *grey commissure*. Each crescentic mass has an anterior (ventral) and posterior (dorsal) horn. The posterior horn is long and narrow, and approaches the surface at the postero-lateral fissure, near which it presents a slight enlargement, the *caput cornu*: from this it tapers, to form the *apex cornu*, which at the surface of the cord becomes continuous with some of the fibres of the posterior roots of the spinal nerves. The anterior horn is short and thick, and does not quite reach the surface, but extends towards the point of

attachment of the anterior roots of the nerves.   Its margin presents a dentate or
stellate appearance.   Owing to the projections towards the surface of the
anterior and posterior horns of the grey matter, each half of the cord is divided,
more or less completely, into three columns, anterior, middle, and posterior : the
anterior and middle being joined to form the antero-lateral column, as the
anterior horn does not quite reach the surface.

The **commissure of the spinal cord** is composed of white and grey matter, and
is therefore divided into the white and grey commissures.   The *white commissure*
is situated at the bottom of the anterior median fissure, and is formed of
medullated nerve-fibres, which pass between the grey matter of the anterior
horn and the anterior white column of the one side into similar parts on the
other.   The fibres are oblique in direction ; many which enter at the posterior
part of the commissure on the one side leave it at the anterior part of the commissure on the other, and *vice versâ*, a decussation taking place in the middle line.

The *grey commissure*, which connects the two crescentic masses of grey
matter, is separated from the bottom of the anterior median fissure by the
anterior white commissure.   It consists of transverse medullated nerve-fibres,
with a considerable quantity of neuroglia between them.   The fibres when they
reach the lateral crescents diverge : some pass backwards to the posterior roots ;
others spread out, at various angles, into the crescent.

Running through the grey commissure of the whole length of the cord is a
minute canal, which is barely visible to the naked eye in the human cord, but is
proportionally larger in some of the lower vertebrata.   It is called the *central
canal* ; it opens above into the fourth ventricle, and terminates below in a
somewhat dilated extremity.   It is surrounded by an area of neuroglia, which,
in the recent state, has a gelatinous appearance, and in which there are no nerve-
fibres.   This is sometimes called the *substantia gelatinosa centralis*.   When
hardened in alcohol or chromic salts it has a finely reticulated appearance.   The
canal is lined in the fœtus by columnar ciliated epithelium, but in the adult
the cilia have disappeared, and the canal is filled with their remains.

The mode of arrangement of the grey matter, and its amount in proportion to
the white, vary in different parts of the cord.   Thus, the posterior horns are long
and narrow, in the cervical region ; short and narrower, in the dorsal ; short, but
wider, in the lumbar region.   In the cervical region, the crescentic portions are
small, and the white matter more abundant than in any other region of the cord.
In the dorsal region, the grey matter is least developed, the white matter being
also small in quantity.   In the lumbar region, the grey matter is more abundant
than in any other region of the cord.   Towards the lower end of the cord, the
white matter gradually ceases.   The crescentic portions of the grey matter soon
blend into a single mass, which forms the only constituent of the extreme point
of the cord.

Minute Anatomy of the Cord.—The cord consists of an outer part, composed
of medullated nerve-fibres, which is the *white substance* ; and of a central part,
the *grey matter* : both supported by a peculiar kind of connective tissue, called
neuroglia.

The **neuroglia** consists of a homogeneous transparent matrix, of a network of
very delicate fibrillæ, and of stellate or branched cells, the *neuroglia cells*.

In addition to forming a ground substance, in which the nerve-fibres, nerve-
cells, and blood-vessels are embedded, a considerable accumulation of neuroglia
takes place in three situations—(1) on the surface of the cord, beneath the pia
mater ; (2) around the central canal, the *substantia gelatinosa centralis* ; and
(3) as a cap over the extremity of the posterior horn, forming the *substantia
cinerea gelatinosa*.

The **white substance of the cord** consists of medullated nerve-fibres, mostly
disposed longitudinally, with blood-vessels and neuroglia.   When stained with
carmine it presents a very striking appearance on transverse section.   It is seen
to be studded all over with minute dots, surrounded by a white area (fig. 387).

This is due to the longitudinal medullated fibres seen on section. The dot is the axis cylinder, the white area the substance of Schwann. Externally, the neuroglia forms a sheath closely investing the outer surface of the cord immediately beneath the pia mater; from it numerous septa pass inwards and separate the respective bundles of fibres and extend between the individual nerve-fibres, acting as a supporting medium, in which they are embedded.

There are, however, also oblique and transverse fibres in the white substance. These principally consist of (1) the fibres of the white commissure; (2) horizontal or oblique fibres passing from the roots of the nerves into the grey matter; and (3) fibres leaving the grey matter and pursuing a longer or shorter horizontal course.

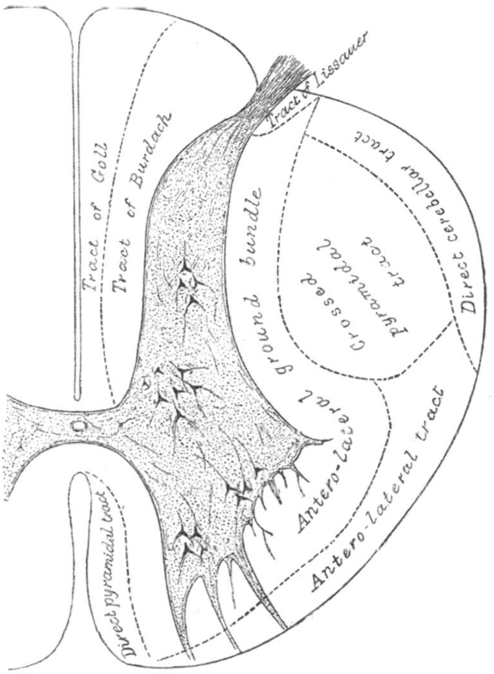

Conducting tracts.—It is impossible to trace the course of the nerve-fibres in their passage through the cord; but the investigation of pathological lesions has shown that the white matter of the cord consists of certain columns or tracts of fibres; for it has been found that separate lesions are strictly limited to certain well-determined parts of the cord without involving neighbouring regions. That these parts or fasciculi correspond to so many distinct anatomical systems, each endowed with special functions, seems abundantly proved by the researches of Flechsig and others on the development of the spinal cord during the later periods of utero-gestation and in the newly born infant. By these researches several tracts can be traced along the greater part of the cord and into or from the encephalon. Thus (1) in the antero-lateral column of the cord, on either side of the anterior median fissure, a portion of the column may be divided off as the *direct*

FIG. 383.—Columns of the cord.

pyramidal tract (*Fasciculus of Türck*). This tract is only found in the upper part of the cord, it gradually diminishes as it is traced downwards, and disappears about the middle of the dorsal region. It consists of centrifugal or descending fibres, which can be traced downwards from the pyramid of the medulla of the same side, and are derived from the motor area of the cerebral cortex. The fibres of this tract decussate in their course down the cord, passing across the middle line through the anterior white commissure; this explains the gradual diminution and eventual disappearance of the tract. (2) In the hinder part of the antero-lateral column is a somewhat triangular area, larger than the preceding, which is named the *crossed pyramidal tract*. This also consists of descending fibres, which are derived from the pyramid of the medulla of the opposite side, and which have crossed in the decussation of the pyramids. The fibres are derived from the motor area of the cerebral cortex of the opposite side. Thus it will be seen that all the fibres from the motor area, which descend through the internal capsule, the crus cerebri and the pons Varolii to the pyramidal body of the medulla, decussate; some at the upper part of the cord, and these descend through it as the crossed pyramidal tract; and others, which descend as the direct pyramidal tract and cross through the anterior commissure of the cord to reach the crossed pyramidal tract of the opposite side. Although this is the usual method of describing the crossing of the direct pyramidal tract in the cord, it seems probable that its fibres cross in the anterior commissure and pass directly to the anterior horn of grey matter, to end by forming synapses around its cells. (3) The *antero-lateral ascending tract* (Gower's tract) is an extensive crescent-shaped strand which skirts the circumference of the anterior three-quarters of the antero-lateral column of the cord. Behind, where it is thickest, it lies in the angle formed by the direct cerebellar and crossed pyramidal tracts, becoming narrower as it passes forwards towards the direct pyramidal tract. It consists of centripetal or ascending fibres, which arise from cells situated at the base of the posterior horn and which cross to the opposite side of the cord in the anterior grey commissure. They can be traced upwards through the medulla and pons to the cerebellum, reaching the latter through its superior peduncles. If the spinal cord is divided in the cervical region some scattered fibres in this column degenerate in a downward direction. This would seem to prove therefore that it contains some descending fibres, which are believed to be derived from the same side of the cerebellum. (4) The *direct cerebellar tract* is situated at the circumference of the cord behind the preceding and external to the crossed pyramidal tract, occupying a narrow area which extends backwards as far as the postero-lateral fissure, or nearly so. It commences at the level of the upper lumbar

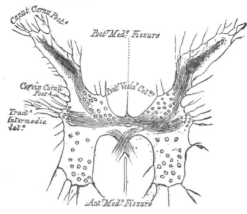

Fig. 384.—Transverse section of the grey substance of the spinal cord, near the middle of the dorsal region. Magnified 13 diameters.

region, and increases in size as it ascends and passes through the restiform body of the medulla to the cerebellum. Its fibres are derived from the cells of the posterior vesicular column of Clarke in the grey matter of the cord. (5) Close to the point where the posterior roots enter the cord, in the antero-lateral column, is a small collection of fibres, which is known as the *tract of Lissauer*; it is formed by some of the fibres of the posterior roots which run upwards in the tract for a short distance, and then enter the posterior horn of the grey matter. (6) The rest of the antero-lateral column of the spinal cord is occupied by the *antero-lateral ground bundle*. It surrounds the anterior cornu and separates the antero-lateral tract and the crossed pyramidal tract from the grey matter of the cord. It consists of (*a*) longitudinal commissural fibres, which unite the groups of cells in the grey matter with one another; (*b*) of fibres which pass across the anterior commissure from the grey matter of the opposite side; and (*c*) horizontal fibres belonging to the anterior roots of the nerves, which pass through it before leaving the cord.

In the posterior column of the cord there are two tracts. They are marked off from each other by the posterior intermediate furrow on the surface of the cord. The part which has been described previously as the posterior median column pretty nearly corresponds to the one tract, the *tract of Goll*, and the remainder of the posterior column corresponds to the other, the *tract of Burdach*. (7) The *tract of Goll* increases as it ascends, and consists of long, but small, fibres derived from the posterior roots of the spinal nerves, which ascend to the medulla oblongata, where they end in the nucleus gracilis. (8) The *tract of Burdach* consists of shorter, but larger, fibres than the preceding; they are however, derived from the same source, the posterior roots; some ascend only for a short distance in the tract, and then enter the grey matter and come into close relationship with the cells of the posterior vesicular column of Clarke; others incline towards the mesial plane, and, entering Goll's column, can be traced as far as the medulla. In the cervical and upper dorsal regions there is contained in the substance of Burdach's column a small strand of fibres, called the *descending comma tract*. It presents, on transverse section, the appearance of a comma, the blunt extremity of which is directed forward. The fibres forming it probably represent in part descending portions of the dorsal nerve-roots, together with descending commissural fibres within the cord itself. A small strand of similar descending fibres is seen, in the lower part of the cord, lying in the inner part of Goll's column.

Fig. 385.—Transverse section of the grey substance of the spinal cord through the middle of the lumbar enlargement. On the left side of the figure groups of large cells are seen; on the right side, the course of the fibres is shown without the cells. Magnified 13 diameters.

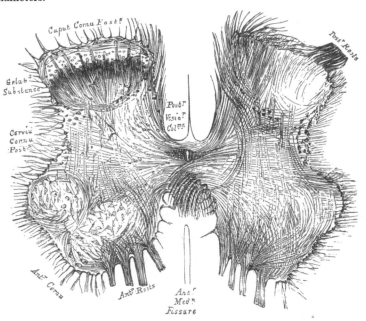

The **grey substance of the cord** occupies its central part in the shape of two crescentic horns, joined together by the grey commissure. Each of these crescents has an anterior or ventral and a posterior or dorsal cornu.

The *posterior horn* consists of a slightly narrowed portion, at its base, where it is connected with the rest of the grey substance—this is the *cervix cornu*; from this it gradually expands into the main part of the horn, the *caput cornu*, around its extremity is a lamina or layer of gelatinous material, which covers the head like a cap, and from this it tapers almost to a point, which approaches the surface of the cord at the postero-lateral groove.

The *gelatinous substance* is a peculiar accumulation of neuroglia (Klein) similar to that found around the central canal (page 692), and has been named by Rolando the *substantia cinerea gelatinosa*. It probably takes its origin from the columnar cells which line the posterior part of the embryonic spinal canal.

The *anterior horn of the grey substance* in the cervical and lumbar swellings, where it gives origin to the motor nerves of the extremities, is much larger than in any other region, and contains several distinct groups of large and variously shaped cells.

In addition to this, a *lateral horn* is found projecting outwards from the lateral region of the grey matter on a level with the grey commissure in the upper part of the dorsal region of the cord; in the cervical and lumbar regions this lateral horn blends with the anterior horn, which thus becomes broad and expanded. From the concavity of the crescent, between the anterior and posterior horns, processes of grey matter extend into the white substance, where they divide and anastomose to form a network, termed the *formatio reticularis*.

The grey commissure contains the central canal, and is situated behind the white commissure, which separates it from the bottom of the anterior median fissure.

The **grey substance of the cord** consists

FIG. 386.—Longitudinal section of the white and grey substance of the spinal cord, through the middle of the lumbar enlargement. Magnified 14 diameters.

Posterior roots

Post. column

Gelat. substance

Grey substance

Anterior column

Anterior roots

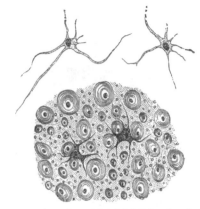

FIG. 387.—Transverse section through the white matter of the spinal cord of a calf. Magnified about 300 diameters. (Klein and Noble Smith.)

In the upper part are shown two isolated flattened nucleated cells of the neuroglia, under a somewhat higher power than the rest. In the bulk of the figure the nerve-fibres are seen in transverse section. They are of different sizes, and possess a laminated medullary sheath surrounding the axis cylinder, which was deeply stained in the preparation, and is here represented by a black dot. The nerve-fibres are embedded in the neuroglia. Among the neuroglia are also seen two branched connective-tissue cells— neuroglia cells.

of—(1) nerve-fibres of variable but smaller average diameter than those of the white columns; (2) nerve-cells of various shapes and sizes, with from two to eight processes; (3) blood-vessels and connective tissue.

The *nerve-fibres* of the grey matter of the posterior horn are for the most part composed of a dense interlacement of minute fibrils, intermingled with nerves of a larger size. This interlacement is formed partly by the axons and dendrites of the cells of the grey matter, and partly by fibres which enter the grey matter and which come from various sources.

The *nerve-cells* of the grey matter are collected into groups as seen on transverse section, but they really form columns of cells placed longitudinally; or else they are found scattered throughout the whole of the grey matter.

In the anterior horn the cells consist of two chief groups: one mesial, the more constant, near the anterior column; the other lateral, near the lateral column. A second lateral group is present in the cervical and lumbar enlargements At the base of the posterior horn on its inner side, adjoining the grey commissure, is a group of nerve-cells, called *Clarke's posterior vesicular column*, which extends from the eighth cervical to the second lumbar nerve.

At the junction of the anterior and posterior cornu, in the outer portion of the grey matter, is a third group of cells, the *lateral cell column*; this is best seen in the dorsal region. In certain regions of the cord these cells extend in among the fibres of the white matter of the lateral column, and give rise to the lateral horn. In addition to these groups a few large scattered cells are found in the posterior horn, and in the substantia gelatinosa of Rolando.

Origin of the Spinal Nerves.—The roots of the spinal nerves are attached to the surface of the cord, opposite the horns of grey matter.

The *posterior nerve-root* enters the cord in two bundles, *mesial* and *lateral*. The mesial strand consists of coarse fibres which enter the outer part of the column of Burdach. The lateral strand is sometimes divided into a middle and an external bundle. The former contains large fibres, and passes through the gelatinous substance of Rolando into the posterior horn. The external bundle consists of fine fibres which assume a longitudinal direction in Lissauer's tract. All the posterior root-fibres divide into ascending and descending branches on entering the cord, and these in their turn give off collaterals. The fibres and their collaterals terminate by forming arborescences, some around the cells in the posterior horn, and others around the cells of Clarke's column, while the long ascending branches pass up in the columns of Goll and Burdach, and end by arborising around the cells in the gracile and cuneate nuclei. Some of the fibres, however, pass to the grey matter of the opposite horn, and others to the anterior horn of the same side of the cord.

Anterior nerve-roots.—The majority of the fibres of the anterior nerve-roots are the continuations outwards of the axons of the large or small multipolar cells in the anterior horn of grey matter. Some, however, appear to pass across in the anterior white commissure to the cells in the anterior horn of the opposite side, while others extend backwards to the posterior horn and outwards to the lateral column of the same side.

NERVE-TRACTS

The anatomy of the various parts of the central nervous system having been described, a short account will now be given of the course taken by its more important nerve-tracts, and of the direction in which impulses pass along them. Before doing so, however, it is necessary to refer to the methods employed in elucidating this complex subject. All nerve-fibres may be regarded as outgrowths from nerve-cells, and it is found that if a nerve-fibre be cut, the portion of it which is severed from the cell undergoes degeneration and becomes atrophied. Until recent years it was believed that the cell itself showed no change under such circumstances. This, however, is not the case, for if a nerve, the sciatic for instance, be divided in an animal, and after an interval of some weeks the animal be injected with methylene-blue and killed, it will be seen, on examining sections of the lumbar region of the spinal cord, that the cells are stained imperfectly or not at all, owing to a diminution, or, it may be, an entire disappearance of the chromatin, a substance which, in a normal cell, shows marked affinity for staining reagents. Further, the body of the cell is swollen, the nucleus displaced towards the periphery, and the part of the axon still attached to the altered cell is diminished in size and somewhat atrophied. Under favourable conditions the cell is capable of reassuming its normal appearance, and the axon may commence to grow. This method of injecting with methylene-blue is of great value in determining the origin of nerve-fibres from their cells. Again, stimulation of certain localised areas of the brain or of the tracts arising from them is followed by the contraction of the muscles of the body. These cortical centres of the motor tracts are situated in the convolutions adjacent to the fissure of Rolando. When the stimulus is applied to one part the muscles of the hind limb contract, while other portions control the movements of the fore limb, &c. Destruction of these parts entails loss of function, paralysis of muscles, and degeneration of the tracts below the seat of injury. During life injury and disease may give rise to symptoms resembling either the effects of stimulation or those of destruction; and after death the tracts, or the centres and the tracts, are seen to be degenerated or otherwise altered. Further, by observing the development of the nervous system during the growth of the embryo, the fact is disclosed that all axis-cylinders do not acquire a medullary sheath at one and the same time. Speaking generally, it may be said that afferent fibres become medullated before efferent, and that in the case of the latter myelination occurs earlier in the brain than in the cord. By watching the effects of these different processes the functions of a considerable part of the brain and of the nerves leading from or to it have been determined.

MOTOR, EFFERENT, OR DESCENDING TRACT

The constituent fibres of this tract are the axis-cylinder processes of cells situated in the cortex of the convolutions around the fissure of Rolando. At first they are somewhat widely diffused, but as they descend through the

FIG. 388.—Dorsal roots entering cord and dividing into ascending and descending branches. (Van Gehuchten.)

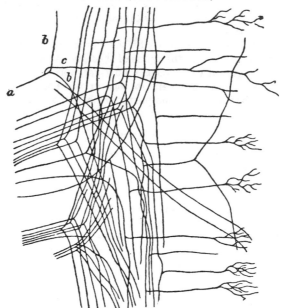

a, Stem-fibre ; *b, b*, ascending and descending limbs of bifurcation ; *c*, collateral arising from stem-fibre.

corona radiata they gradually approach each other and pass between the lenticular nucleus and optic thalamus in the genu and anterior two-thirds of the posterior limb of the internal capsule. Proceeding downwards they next occupy the middle of the pes or crusta of the crus cerebri, and enter the pons Varolii, where the transverse fibres of this body not only conceal them but divide them up into irregular bundles. Eventually they reach the medulla, and here the motor tracts form the anterior pyramids which lie one on each side of the median fissure. The transit of the fibres from the medulla is effected by two paths The fibres nearest to the anterior median fissure cross the middle line, forming the *decussation of the pyramids*, and descend in the opposite side of the cord as the indirect or crossed pyramidal tract. Throughout the length of the spinal cord fibres from this column pass into the grey matter, to terminate by ramifying around the cells of the anterior horn. The more laterally placed portion of the motor tract does not decussate in the medulla, but descends as the direct or uncrossed pyramidal tract ; these fibres, however, end in the anterior grey horn of the opposite side of the spinal cord by passing across in the anterior white commissure. Further, it must be remembered that many fibres which descend in and constitute part of the motor tract decussate before reaching the medulla, and terminate by forming synapses with the nuclei of the cranial nerves situated near the aqueduct of Sylvius, in the pons or in the medulla itself. There is considerable variation in the extent to which decussation takes place in the medulla, the commonest condition being that in which about three-fourths of the fibres decussate in the medulla and the remainder in the cord.

OTHER DESCENDING TRACTS

1. From the cortex of the frontal lobe, anterior to the Rolandic area, fibres arise which descend through the anterior limb of the internal capsule and enter the crusta, where they lie to the inner side of the pyramidal tract; finally they enter, and end in, the pons.

2. Descending fibres also take origin in the temporo-occipital cortex and pass through the posterior limb of the internal capsule behind the fibres from the Rolandic area. They pass through the crusta, where they lie to the outer side of the same tract, and end in the pons.

3. A small tract arises from the cells of the caudate nucleus and descends to end in the substantia nigra or pons. In the crus cerebri it lies immediately above the motor tract, which is on its ventral aspect.

SENSORY, AFFERENT, ASCENDING TRACT

The course taken by those fibres of the posterior nerve-roots which ascend has been arrived at by dividing the nerve-roots between their ganglia and their entrance into the spinal cord and subsequently examining the degenerated areas. It has been found that the fibres pursue an oblique course, being situated at first in the outer part of Burdach's column; higher up they occupy the middle of this column, being displaced inwards by the accession of other entering fibres, while still higher they enter and are continued upwards in the column of Goll. The upper cervical fibres do not reach the column of Goll, but are entirely confined to that of Burdach. The degeneration method proves that the localisation of these fibres is very precise : the sacral nerves lying to the inner side of Goll's column and near its periphery; the lumbar nerves to their outer side ; the dorsal nerves still more laterally; while the cervical nerves are confined to the outer part of Burdach's column.

On reaching the medulla these ascending fibres end by arborising around the cells in the gracile and cuneate nuclei, and the further upward course of the tract is effected by the axis-cylinder processes of these cells. These new fibres decussate in the medulla, dorsal to the crossing of the motor tract, in what is termed the *superior pyramidal decussation*, the *sensory decussation*, or *decussation of the fillet*; terms which are synonymous. Having crossed the middle line they ascend through the pons and tegmentum of the crus cerebri, and reaching the ventral surface of the optic thalamus, the majority end either in the subthalamic region or in the optic thalamus, but a small proportion is continued directly into the brain cortex. From the grey matter of the optic thalamus the fibres of the third link in the chain arise. They pass through the internal capsule and end in the cerebral cortex : those which go to the fronto-parietal cortex being situated in the extreme front part of the] anterior limb of the internal capsule, while in the hinder extremity of the posterior limb other fibres pass to their distribution in the temporal and occipital cortex.

OTHER ASCENDING TRACTS

The direct cerebellar tract begins about the level of the second lumbar vertebra, and is the continuation upwards of the axis cylinders of Clarke's column. At the upper end of the cord it passes into the restiform body and through this reaches the cerebellum. This tract seems to lose some of its fibres in the cord, since the area of its degeneration resulting from a section of the lower part of the cord diminishes from below upwards; only some of its fibres therefore pass directly to the cerebellum. On the other hand the tract is reinforced by an accession of fibres from the cord itself, so that its transverse area is greater above than below.

The antero-lateral ascending tract of Gower arises in the cord, probably as the axis cylinders of cells situated in the posterior horn. Passing across the middle line through the anterior grey commissure the fibres ascend in the antero-lateral column of the cord, and ultimately reach the cerebellum through its superior peduncles.*

CRANIAL NERVES

The cranial nerves arise from some part of the cerebro-spinal centre, and are transmitted through foramina in the base of the cranium. They have been named numerically, according to the order in which they pass through the dura mater lining the base of the skull. Other names are also given to them derived from the parts to which they are distributed, or from their functions. Taken in their order, from before backwards, they are as follows :

1st. Olfactory.	7th. Facial (Portio dura).
2nd. Optic.	8th. Auditory (Portio mollis).
3rd. Motor oculi.	9th. Glosso-pharyngeal.
4th. Trochlear (Pathetic).	10th. Pneumogastric or Vagus.
5th. Trifacial (Trigeminus).	11th. Spinal accessory.
6th. Abducent.	12th. Hypoglossal.

All the cranial nerves are connected to some part of the surface of the brain. This is termed their *superficial* or *apparent origin*. But their fibres may, in all cases, be traced deeply into the substance of the brain to some special centre of grey matter, termed a *nucleus*. This is called their *deep* or *real origin*. The nerves, after emerging from the brain at their apparent origin, pass through foramina or tubular prolongations in the dura mater, leave the skull through foramina in its base, and pass to their final distribution.

FIRST NERVE (fig. 350, page 633)

The **First cranial** or the **Olfactory nerves** (*nn. olfactorii*), the special nerves of the sense of smell, are about twenty in number. They are given off from the under surface of the olfactory bulb, an oval mass of a greyish colour, which rests on the cribriform plate of the ethmoid bone, and forms the anterior expanded extremity of a slender process of brain-substance, named the *olfactory tract*. The olfactory tract and bulb have already been described (page 632).

Each nerve is surrounded by a tubular prolongation from the dura mater and pia mater : the former being lost on the periosteum lining the nose; the latter, in the neurilemma of the nerve. The nerves, as they enter the nares, are divisible into two groups : an inner group, larger than those on the outer wall, spread out over the upper third of the septum ; and an outer set, which is distributed over the superior turbinated bone, and the surface of the ethmoid in front of it. As the filaments descend, they unite in a plexiform network, and are believed by most observers to terminate by becoming continuous with the deep extremities of the olfactory cells.

The olfactory differs in structure from other nerves, in being composed exclusively of non-medullated fibres. They are deficient in the white substance of Schwann, and consist of axis cylinders, with a distinct nucleated sheath, in

---

* Testut describes the ascending column of Gower as joining with the fillet, and through it being carried to the cerebral cortex.

which there are, however, fewer nuclei than in ordinary non-medullated fibres. The olfactory centre in the cortex is not definitely known. It is generally associated with the temporal lobe, where it probably includes the gyrus hippocampi, uncus, and hippocampus major. It is further described as comprising the part of the callosal convolution which lies below the genu and rostrum of the corpus callosum, and also the posterior part of the orbital surface of the frontal lobe.

Surgical Anatomy.—In severe injuries to the head, the olfactory bulb may become separated from the olfactory nerves, thus producing loss of smell (*anosmia*), and with this a considerable loss in the sense of taste, as much of the perfection of the sense of taste is due to the sapid substances being also odorous, and simultaneously exciting the sense of smell.

SECOND NERVE (fig. 389)

The **Second** or **Optic nerve** (*n. opticus*), the special nerve of the sense of sight, is distributed exclusively to the eyeball. The nerves of opposite sides are connected together at the commissure, and from the back of the commissure they may be traced to the brain, under the name of the *optic tracts*.

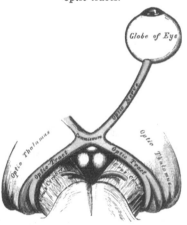

FIG. 389.—The left optic nerve and optic tracts.

The *optic tract*, at its connection with the brain, is divided into two bands, external and internal. The *external* band is the larger; it arises from the external geniculate body and from the under part of the pulvinar of the optic thalamus, and is partly continuous with the brachium of the anterior or upper quadrigeminal body. The *internal* band curves round the crusta, and passes beneath the internal geniculate body, with which it is connected, and then appears to lose itself in the brachium of the posterior or inferior quadrigeminal body. The fibres by which it is connected to the internal geniculate body are merely commissural, forming part of Gudden's commissure. From this origin the tract winds obliquely across the under surface of the crus cerebri, in the form of a flattened band, and is attached to the crus by its anterior margin. It then assumes a cylindrical form, and, as it passes forwards, is connected with the tuber cinereum and lamina cinerea. It finally joins with the tract of the opposite side to form the *optic commissure*.

The *commissure* or *chiasma*, somewhat quadrilateral in form, rests upon the olivary eminence and on the anterior part of the diaphragma sellæ, being bounded, above, by the lamina cinerea; behind, by the tuber cinereum; on either side, by the anterior perforated space. Within the commissure, the optic nerves of the two sides undergo a partial decussation. The fibres which form the inner margin of each tract and posterior part of the commissure have no connection with the optic nerves. They simply pass across the commissure from one hemisphere of the brain to the other, and connect the internal geniculate bodies of the two sides. They are known as the *commissure of Gudden*. The remainder and principal part of the commissure consists of two sets of fibres, crossed and uncrossed.

FIG. 390.—Course of the fibres in the optic commissure.

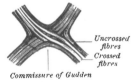

Uncrossed fibres
Crossed fibres

Commissure of Gudden

The *crossed*, which are the more numerous, occupy the central part of the chiasma, and pass from the optic tract of one side to the optic nerve of the other, decussating in the commissure with similar fibres of the opposite tract.   The uncrossed fibres occupy the outer part of the chiasma, and pass from the tract of one side to the nerve of the same side.*

The great majority of the fibres of the optic nerve consist of the afferent axons of nerve-cells in the retina.   Some few, however, are efferent fibres, and grow out from the brain.   The afferent fibres end in arborisations around the cells in the external geniculate body, pulvinar, and upper quadrigeminal body, which are sometimes termed the *lower visual centres*.   From these nuclei other fibres are prolonged to the *cortical visual centre*, which, according to most observers, is situated in the cuneus, and probably also in the lingual lobule of the occipital lobe.

It should be stated that some fibres are detached from the optic tract, and pass through the crus cerebri to the nucleus of the third nerve.   These fibres are small, and may be regarded as afferent branches for the sphincter pupillæ and ciliary muscles.   Other fibres pass to the cerebellum through its superior peduncles, while others, again, are lost in the pons.

The *optic nerves* arise from the fore part of the commissure, and, diverging from one another, become rounded in form and firm in texture, and are enclosed in a sheath derived from the pia mater and arachnoid.   As each nerve passes through the corresponding optic foramen, it receives a sheath from the dura mater ; and as it enters the orbit this sheath divides into two layers, one of which becomes continuous with the periosteum of the orbit; the other forms the proper sheath of the nerve, and surrounds it as far as the sclerotic.   The nerve passes forwards and outwards through the cavity of the orbit, pierces the sclerotic and choroid coats at the back part of the eyeball, about one-eighth of an inch to the nasal side of its centre, and expands into the retina.   A small artery, the *arteria centralis retinæ*, perforates the optic nerve a little behind the globe, and runs along its interior in a tubular canal of fibrous tissue.   It supplies the inner surface of the retina, and is accompanied by corresponding veins.

Surgical Anatomy.—The optic nerve is peculiarly liable to become the seat of neuritis or undergo atrophy in affections of the central nervous system, and as a rule the pathological relationship between the two affections is exceedingly difficult to trace.   There are, however, certain points in connection with the anatomy of this nerve which tend to throw light upon the frequent association of these affections with intracranial disease : (1) From its mode of development, and from its structure, the optic nerve must be regarded as a prolongation of the brain-substance, rather than as an ordinary cerebro-spinal nerve.   (2) As it passes from the brain, it receives sheaths from the three cerebral membranes, a perineural sheath from the pia mater ; an intermediate sheath from the arachnoid ; and an outer sheath from the dura mater, which is also connected with the periosteum, as it passes through the optic foramen.   These sheaths are separated from each other by spaces, which communicate with the subdural and subarachnoid spaces respectively.   The innermost or perineural sheath sends a process around the arteria centralis retinæ into the interior of the nerve, and enters intimately into its structure.   Thus inflammatory affection of the meninges or of the brain may readily extend themselves along these spaces, or along the interstitial connective tissue in the nerve.

The course of the fibres in the optic commissure has an important pathological bearing, and has been the subject of much controversy.   Microscopic examination, experiments, and pathology all seem to point to the fact that there is a partial decussation of the fibres, each optic tract supplying the corresponding half of each eye, so that the right tract supplies the right half of each eye, and the left tract the left half of each eye.   At the same time Charcot believes, and his view has met with general acceptation, that the fibres which do not decussate at the optic commissure have already decussated in the corpora quadrigemina, so that lesion of the cerebral centre of one side causes complete blindness of the opposite eye, because both sets of decussating fibres are destroyed. Whereas should one tract, say the right, be destroyed by disease, there will be blindness of the right half of both retinæ.

An antero-posterior section through the commissure would divide the decussating fibres, and would therefore produce blindness of the inner half of each eye ; while a section at the margin of the side of the optic commissure would produce blindness of the external half of the retina of the same side.

* A specimen of congenital absence of the optic commissure is to be found in the Museum of the Westminster Hospital.   See also Henle, *Nervenlehre*, p. 393, ed. 2.

The optic nerve may also be affected in injuries or diseases involving the orbit; in fractures of the anterior fossa of the base of the skull; in tumours of the orbit itself, or those invading this cavity from neighbouring parts.

THIRD NERVE (figs. 391, 392, 393)

The **Third** or **Motor oculi nerve** (*n. oculo-motorius*) supplies all the muscles of the orbit, except the Superior oblique and External rectus; it also supplies, through its connection with the ciliary ganglion, the Sphincter muscle of the iris and the Ciliary muscle. It is rather a large nerve, of rounded form and firm texture.

Its apparent origin is from the inner surface of the crus cerebri, immediately in front of the pons Varolii. The *deep origin* may be traced through the substantia nigra and tegmentum of the crus to a nucleus situated on either side of the median line beneath the floor of the aqueduct of Sylvius. The nucleus of the third nerve also receives fibres from the sixth nerve of the opposite side. These will be again referred to in the description of the latter nerve. The nucleus of the third nerve, considered from a physiological standpoint, can be sub-divided into several smaller groups of cells, each group controlling a particular muscle. The nerves to the different muscles appear to take their origin from before backwards, as follows: Inferior oblique, Inferior rectus, Superior rectus and Levator palpebræ, Internal rectus; while from the anterior end of the nucleus the fibres for accommodation and for the Sphincter pupillæ take their origin.

On emerging from the brain, the nerve is invested with a sheath of pia mater, and enclosed in a prolongation from the arachnoid. It passes between the superior cerebellar and posterior cerebral arteries, and then pierces the dura mater in front of and external to the posterior clinoid process, passing between the two processes from the free and attached borders of the tentorium,

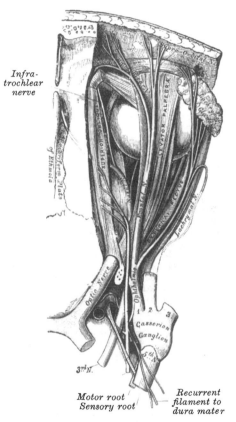

FIG. 391.—Nerves of the orbit. Seen from above.

Infra-trochlear nerve

Motor root
Sensory root
Recurrent filament to dura mater

which are prolonged forwards to be connected with the anterior and posterior clinoid processes of the sphenoid bone. It passes along the outer wall of the cavernous sinus, above the other orbital nerves, receiving in its course one or two filaments from the cavernous plexus of the sympathetic, and a communicating branch from the first division of the fifth. It then divides into two branches, which enter the orbit through the sphenoidal fissure, between the two heads of the External rectus muscle. On passing through the fissure, the nerve is placed below the fourth and the frontal and lachrymal branches of the ophthalmic nerve, and has passing between its two divisions the nasal nerve.

The *superior division*, the smaller, passes inwards over the optic nerve, and supplies the Superior rectus and Levator palpebræ. The *inferior division*, the larger, divides into three branches. One passes beneath the optic nerve to the Internal rectus; another, to the Inferior rectus; and the third, the longest of the three, passes forwards between the Inferior and External recti to the Inferior oblique. From this latter a short, thick branch is given off to the lower part of

FIG. 392.—Plan of the motor oculi nerve. (After Flower.)

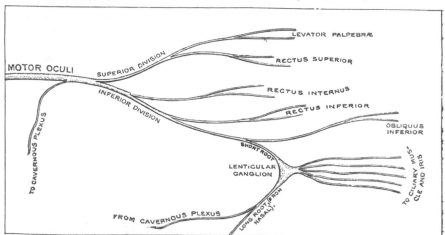

the lenticular ganglion, and forms its inferior root. It also gives off one or two filaments to the Inferior rectus. All these branches enter the muscles on their ocular surface, with the exception of the nerve to the Inferior oblique, which enters the muscle at its posterior border.

Surgical Anatomy.—Paralysis of the third nerve may be the result of many causes: as cerebral disease; conditions causing pressure on the cavernous sinus; periostitis of the bones entering into the formation of the sphenoidal fissure. It results, when complete, in (1) ptosis, or drooping of the upper eyelid, in consequence of the Levator palpebræ being paralysed; (2) external strabismus, on account of the unopposed action of the External rectus muscle, which is not supplied by the third nerve, and is not therefore paralysed; (3) dilatation of the pupil, because the sphincter fibres of the iris are paralysed; (4) loss of power of accommodation, as the Sphincter pupillæ, the Ciliary muscle, and the Internal rectus are paralysed; (5) slight prominence of the eyeball, owing to most of its muscles being relaxed. Occasionally paralysis may affect only a part of the nerve, that is to say, there may be, for example, a dilated and fixed pupil, with ptosis, but no other signs. Irritation of the nerve causes spasm of one or other of the muscles supplied by it; thus, there may be internal strabismus from spasm of the Internal rectus; accommodation for near objects only from spasm of the Ciliary muscle; or myosis, contraction of the pupil, from irritation of the Sphincter of the pupil.

FOURTH NERVE (fig. 391)

The **Fourth** or **Trochlear nerve** (*n. trochlearis*), the smallest of the cranial nerves, supplies the Superior oblique muscle.

Its *apparent origin*, at the base of the brain, is on the outer side of the crus cerebri, just in front of the pons Varolii, but the fibres can be traced backwards behind the corpora quadrigemina to the valve of Vieussens, on the upper surface of which the two nerves decussate. Its *deep origin* may be traced to a nucleus in the floor of the aqueduct of Sylvius immediately below that of the third nerve, with which it is continuous.

Emerging from the upper end of the valve of Vieussens, the nerve is directed outwards across the superior peduncle of the cerebellum, and then winds forwards round the outer side of the crus cerebri, immediately above the pons Varolii, pierces the dura mater in the free border of the tentorium cerebelli, just behind, and external to, the posterior clinoid process, and passes forwards in the outer wall of the cavernous sinus, between the third nerve and the ophthalmic division of the fifth. It crosses the third nerve and enters the orbit, through the sphenoidal fissure. It now becomes the highest of all the nerves, lying at the inner extremity of the fissure internal to the frontal nerve. In the orbit it passes inwards, above the origin of the Levator palpebræ, and finally enters the orbital surface of the Superior oblique muscle. In the outer wall of the cavernous sinus this nerve is not infrequently blended with the ophthalmic division of the fifth.

Branches of communication.—In the outer wall of the cavernous sinus it receives some filaments from the cavernous plexus of the sympathetic. In the sphenoidal fissure it occasionally gives off a branch to assist in the formation of the lachrymal nerve. *Branches of distribution.*—It gives off a recurrent branch, which passes backwards between the layers of the tentorium, dividing into two or three filaments which may be traced as far back as the wall of the lateral sinus.

Surgical Anatomy.—The fourth nerve when paralysed causes loss of function in the Superior oblique, so that the patient is unable to turn his eye downwards and outwards. Should the patient attempt to do this, the eye of the affected side is twisted inwards, producing diplopia or double vision. Accordingly, it is said that the first symptom of loss of function in this muscle is giddiness when going down hill or in descending stairs, owing to the double vision induced by the patient looking at his steps while descending.

FIFTH NERVE

The **Fifth** or **Trifacial nerve** (*n. trigeminus*) is the largest cranial nerve. It resembles a spinal nerve : (1) in arising by two roots ; (2) in having a ganglion developed on its posterior root ; and (3) in its function, since it is a compound nerve. It is the great sensory nerve of the head and face, and the motor nerve of the muscles of mastication. Its upper two divisions are entirely sensory, the third division is partly sensory and partly motor.

It arises by two roots : of these the anterior is the smaller, and is the motor root ; the posterior, the larger and sensory. Its *superficial origin* is from the side of the pons Varolii, nearer to the upper than the lower border. The smaller root consists of three or four bundles ; the larger root consists of numerous bundles of fibres, varying in number from seventy to a hundred. The two roots are separated from one another by a few of the transverse fibres of the pons. The *deep origin* of the larger or sensory root is chiefly from a long tract in the medulla, the *lower sensory nucleus*, which is continuous below with the substantia gelatinosa of Rolando. The fibres from this nucleus form the so-called *ascending root of the fifth* ; they pass upwards through the pons and join with fibres from the locus cæruleus or *upper sensory nucleus*, which is situated to the outer side of the nucleus, from which the lower part of the motor root takes origin. The *deep origin* of the smaller or motor root is derived partly from a nucleus embedded in the grey matter of the upper part of the floor of the fourth ventricle and partly from a collection of nerve-cells situated at the side of the aqueduct of Sylvius, from which the fibres pass downwards under the name of the *descending root of the fifth*. The real origin of the sensory root is from the Gasserian ganglion, which corresponds with the ganglion on a spinal nerve (see Development of Spinal Nerves in section on Embryology).

The two roots of the nerve pass forwards below the tentorium cerebelli as it bridges over the notch on the inner part of the superior border of the petrous portion of the temporal bone : they then run between the bone and the dura mater to the apex of the petrous portion of the temporal bone, where the fibres of the sensory root form a large semilunar ganglion (*Gasserian*), while the motor root passes beneath the ganglion without having any connection with it, and joins, outside the cranium, with one of the trunks derived from it.

The **Gasserian** or **semilunar ganglion** * is lodged in an osteo-fibrous space, the *cavum Meckelii*, near the apex of the petrous portion of the temporal bone. It is of somewhat crescentic form, with its convexity turned forwards. Its upper surface is intimately adherent to the dura mater. Besides the small or motor root, the large superficial petrosal nerve lies underneath the ganglion.

Branches of communication.—This ganglion receives, on its *inner side*, filaments from the carotid plexus of the sympathetic. *Branches of distribution.*—It gives off minute branches to the tentorium cerebelli and the dura mater, in the middle fossa of the cranium. From its *anterior border*, which is directed forwards and outwards, three large branches proceed: the *ophthalmic, superior maxillary*, and *inferior maxillary*. The ophthalmic and superior maxillary consist exclusively of fibres derived from the larger root and ganglion, and are solely nerves of common sensation. The third division, or inferior maxillary, is joined outside the cranium by the motor root. This, therefore, strictly speaking, is the only portion of the fifth nerve which can be said to resemble a spinal nerve.

OPHTHALMIC NERVE (figs. 391, 393, 394)

The **Ophthalmic** (*n. ophthalmicus*), or first division of the fifth, is a sensory nerve. It supplies the eyeball, the lachrymal gland, the mucous lining of the eye and nasal fossæ, and the integument of the eyebrow, forehead, and nose. It is the smallest of the three divisions of the fifth, rising from the upper part of the Gasserian ganglion. It is a short, flattened band, about an inch in length, which passes forwards along the outer wall of the cavernous sinus, below the other nerves, and just before entering the orbit, through the sphenoidal fissure, divides into three branches, *lachrymal, frontal*, and *nasal*.

Branches of communication.—The ophthalmic nerve is joined by filaments from the cavernous plexus of the sympathetic, communicates with the third and sixth nerves, and is not infrequently joined with the fourth.

Branches of distribution.—It gives off recurrent filaments which pass between the layers of the tentorium, and then divides into

Lachrymal.   Frontal.   Nasal.

The **lachrymal** is the smallest of the three branches of the ophthalmic. It sometimes receives a filament from the fourth nerve, but this is possibly derived from the branch of communication which passes from the ophthalmic to the fourth. It passes forwards in a separate tube of dura mater, and enters the orbit through the narrowest part of the sphenoidal fissure. In the orbit it runs along the upper border of the External rectus muscle, with the lachrymal artery, and communicates with the temporo-malar branch of the superior maxillary. It enters the lachrymal gland and gives off several filaments, which supply the gland and the conjunctiva. Finally, it pierces the superior palpebral ligament, and terminates in the integument of the upper eyelid, joining with filaments of the facial nerve. The lachrymal nerve is occasionally absent, when its place is taken by the temporal branch of the superior maxillary. Sometimes the latter branch is absent, and a continuation of the lachrymal is substituted for it.

The **frontal** is the largest division of the ophthalmic, and may be regarded both from its size and direction, as the continuation of the nerve. It enters the orbit above the muscles, through the sphenoidal fissure, and runs forwards along the middle line, between the Levator palpebræ and the periosteum. Midway

* A Viennese anatomist, Raimund Balthasar Hirsch (1765), was the first who recognised the ganglionic nature of the swelling on the sensory root of the fifth nerve, and called it, in honour of his otherwise unknown teacher, Jon. Laur. Gasser, the 'Ganglion Gasseri.' Julius Casserius, whose name is given to the musculo-cutaneous nerve of the arm, was Professor at Padua, 1545–1605. See Hyrtl, *Lehrbuch der Anatomie*, p. 895 and p. 55.

between the apex and base of the orbit it divides into two branches, supratrochlear and supraorbital.

The *supratrochlear branch*, the smaller of the two, passes inwards, above the pulley of the Superior oblique muscle, and gives off a descending filament, which joins with the infratrochlear branch of the nasal nerve. It then escapes from the orbit between the pulley of the Superior oblique and the supra-orbital foramen, curves up on to the forehead close to the bone, ascends beneath the Corrugator supercilii and Occipito-frontalis muscles, and dividing into branches, which pierce these muscles, it supplies the integument of the lower part of the forehead on either side of the middle line and sends filaments to the conjunctiva and skin of the upper lid.

The *supraorbital branch* passes forwards through the supraorbital foramen, and gives off, in this situation, palpebral filaments to the upper eyelid. It then ascends upon the forehead, and terminates in cutaneous and pericranial branches. The *cutaneous branches*, two in number, an inner and an outer, supply the integument of the cranium as far back as the occiput. They are at first situated

FIG. 393.—Nerves of the orbit and ophthalmic ganglion. Side view.

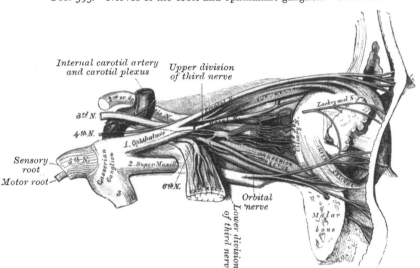

beneath the Occipito-frontalis, the inner branch perforating the frontal portion of the muscle, the outer branch its tendinous aponeurosis. The *pericranial branches* are distributed to the pericranium over the frontal and parietal bones.

The **nasal nerve** is intermediate in size between the frontal and lachrymal, and more deeply placed than the other branches of the ophthalmic. It enters the orbit between the two heads of the External rectus, and passes obliquely inwards across the optic nerve, beneath the Superior rectus and Superior oblique muscles, to the inner wall of the orbit, where it passes through the anterior ethmoidal foramen, and entering the cavity of the cranium, traverses a shallow groove on the front of the cribriform plate of the ethmoid bone, and passes down, through the slit by the side of the crista galli, into the nose, where it divides into two branches, an internal and an external. The *internal branch* supplies the mucous membrane near the fore part of the septum of the nose. The *external branch* descends in a groove on the inner surface of the nasal bone, and supplies a few filaments to the mucous membrane covering the fore part of the outer wall of the nares as far as the inferior spongy bone; it then leaves the cavity of the nose, between the lower border of the nasal bone and the upper lateral cartilage of the nose, and, passing down beneath the Compressor nasi, supplies the integument of the ala and the tip of the nose, joining with the facial nerve.

The branches of the nasal nerve are, the *ganglionic, ciliary,* and *infra-trochlear.*

The *ganglionic* is a slender branch, about half an inch in length, which usually arises from the nasal, between the two heads of the External rectus. It passes forwards on the outer side of the optic nerve, and enters the postero-superior angle of the ciliary ganglion, forming its superior or long root. It is sometimes joined by a filament from the cavernous plexus of the sympathetic, or from the superior division of the third nerve.

The *long ciliary nerves,* two or three in number, are given off from the nasal as it crosses the optic nerve. They join the short ciliary nerves from the ciliary ganglion, pierce the posterior part of the sclerotic, and, running forwards between it and the choroid, are distributed to the Ciliary muscle, iris, and cornea.

The *infratrochlear branch* is given off just before the nasal nerve passes through the anterior ethmoidal foramen. It runs forwards along the upper border of the Internal rectus, and is joined, beneath the pulley of the Superior oblique, by a filament from the supratrochlear nerve. It then passes to the inner angle of the eye, and supplies the integument of the eyelids and side of the nose, the conjunctiva, lachrymal sac, and caruncula lachrymalis.

Ophthalmic Ganglion (figs. 393, 394)

Connected with the three divisions of the fifth nerve are four small ganglia. With the first division is connected the *ophthalmic ganglion*; with the second division, the *spheno-palatine* or *Meckel's ganglion*; and with the third, the *otic* and *submaxillary ganglia.* All the four receive sensory filaments from the fifth, and motor and sympathetic filaments from various sources; these filaments are called the *roots of the ganglia.*

The **Ophthalmic, Lenticular,** or **Ciliary ganglion** is a small, quadrangular, flattened ganglion, of a reddish-grey colour, and about the size of a pin's head, situated at the back part of the orbit between the optic nerve and the External rectus muscle, lying generally on the outer side of the ophthalmic artery. It is enclosed in a quantity of loose fat, which makes its dissection somewhat difficult.

Its *branches of communication,* or *roots,* are three, all of which enter its posterior border. One, the long or sensory root, is derived from the nasal branch of the ophthalmic, and joins its superior angle. The second, the short or motor root, is a short, thick nerve, occasionally divided into two parts, which is derived from the branch of the third nerve to the Inferior oblique muscle, and is connected with the inferior angle of the ganglion. The third, the sympathetic root, is a slender filament from the cavernous plexus of the sympathetic. This is frequently blended with the long root, though it sometimes passes to the ganglion separately. According to Tiedemann, this ganglion receives a filament of communication from the spheno-palatine ganglion.

Its *branches of distribution* are the short ciliary nerves. These are delicate filaments, from six to ten in number, which arise from the fore part of the ganglion in two bundles, connected with its superior and inferior angles; the lower bundle is the larger. They run forwards with the ciliary arteries in a wavy course, one set above and the other below the optic nerve, and are accompanied by the long ciliary nerves from the nasal. They pierce the sclerotic at the back part of the globe, pass forwards in delicate grooves on its inner surface, and are distributed to the Ciliary muscle, iris, and cornea. Tiedemann has described one small branch as penetrating the optic nerve with the arteria centralis retinæ.

Superior Maxillary Nerve (fig. 395)

The **Superior maxillary** (*n. maxillaris*), or second division of the fifth, is a sensory nerve. It is intermediate, both in position and size, between the

FIG. 394.—Plan of the fifth cranial nerve. (After Flower.)

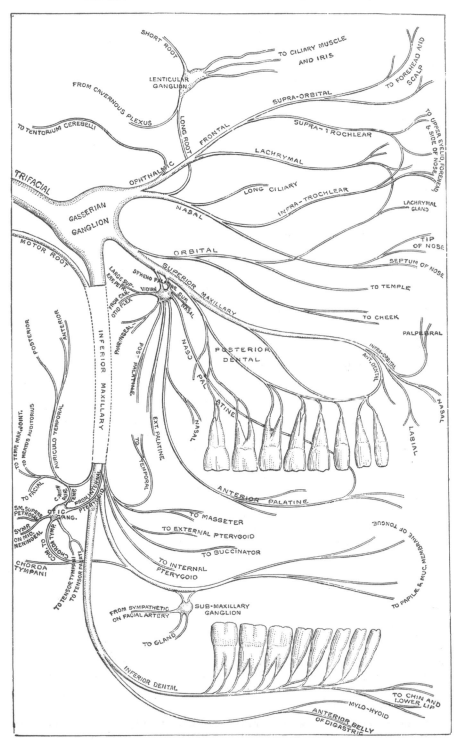

ophthalmic and inferior maxillary. It commences at the middle of the Gasserian ganglion as a flattened plexiform band, and passing horizontally forwards, it leaves the skull through the foramen rotundum, where it becomes more cylindrical in form, and firmer in texture. It then crosses the spheno-maxillary fossa, enters the orbit through the spheno-maxillary fissure, traverses the infra-orbital canal in the floor of the orbit, and appears upon the face at the infra-orbital foramen.* At its termination, the nerve lies beneath the Levator labii superioris muscle, and divides into a leash of branches, which spread out upon the side of the nose, the lower eyelid, and upper lip, joining with filaments of the facial nerve.

Branches of distribution.—The branches of this nerve may be divided into four groups : 1. Those given off in the cranium. 2. Those given off in the spheno-maxillary fossa. 3. Those in the infra-orbital canal. 4. Those on the face.

In the cranium . .	Meningeal
Spheno-maxillary fossa	⎰ Orbital or temporo-malar. ⎱ Spheno-palatine. Posterior superior dental.
Infra-orbital canal .	⎰ Middle superior dental. ⎱ Anterior superior dental.
On the face . .	⎰ Palpebral. ⎱ Nasal. Labial.

The **meningeal branch** is given off directly after its origin from the Gasserian ganglion ; it accompanies the middle meningeal artery and supplies the dura mater.

The **orbital** or **temporo-malar branch** arises in the spheno-maxillary fossa, enters the orbit by the spheno-maxillary fissure, and divides at the back of that cavity into two branches, temporal and malar.

The *temporal branch* runs in a groove along the outer wall of the orbit (in the malar bone), receives a branch of communication from the lachrymal, and, passing through a foramen in the malar bone, enters the temporal fossa. It ascends between the bone and substance of the Temporal muscle, pierces this muscle and the temporal fascia about an inch above the zygoma, and is distributed to the integument covering the temple and side of the forehead, communicating with the facial and auriculo-temporal branch of the inferior maxillary nerve. As it pierces the temporal fascia, it gives off a slender twig, which runs between the two layers of the fascia to the outer angle of the orbit.

The *malar branch* passes along the external inferior angle of the orbit, emerges upon the face through a foramen in the malar bone, and, perforating the Orbicularis palpebrarum muscle, supplies the skin on the prominence of the cheek, and is named *subcutaneus malæ*. It joins with the facial nerve and with the palpebral branches of the superior maxillary.

The **spheno-palatine branches,** two in number, descend to the spheno-palatine ganglion.

The **posterior superior dental branches** arise from the trunk of the nerve just as it is about to enter the infra-orbital canal ; they are generally two in number, but sometimes arise by a single trunk, and immediately divide and pass downwards on the tuberosity of the superior maxillary bone. They give off several twigs to the gums and neighbouring parts of the mucous membrane of the cheek (*superior gingival branches*). They then enter the posterior dental canals on the zygomatic surface of the superior maxillary bone, and, passing from behind forwards in the substance of the bone, communicate with the middle dental nerve, and give off branches to the lining membrane of the antrum and three twigs to each of the molar teeth. These twigs enter the foramina at the apices of the fangs, and supply the pulp.

---

* After it enters the infra-orbital canal, the nerve is frequently called the *infra-orbital.*

The **middle superior dental branch** is given off from the superior maxillary nerve in the back part of the infra-orbital canal, and runs downwards and forwards in a special canal in the outer wall of the antrum to supply the two bicuspid teeth. It communicates with the posterior and anterior dental branches. At its point of communication with the posterior branch is a slight thickening which has received the name of the *ganglion of Valentin*; and at its point of communication with the anterior branch is a second enlargement, which is called the *ganglion of Bochdalek*. It is probable that neither of these is a true ganglion.

The **anterior superior dental branch,** of large size, is given off from the superior maxillary nerve just before its exit from the infra-orbital foramen; it enters a

FIG. 395.—Distribution of the second and third divisions of the fifth nerve and submaxillary ganglion.

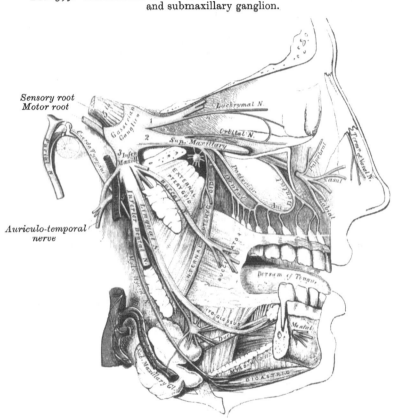

special canal in the anterior wall of the antrum, and divides into a series of branches which supply the incisor and canine teeth. It communicates with the middle dental nerve, and gives off a *nasal branch*, which passes through a minute canal into the nasal fossa, and supplies the mucous membrane of the fore part of the inferior meatus and the floor of this cavity, communicating with the nasal branches from Meckel's ganglion.

The **palpebral branches** pass upwards beneath the Orbicularis palpebrarum. They supply the integument and conjunctiva of the lower eyelid with sensation, joining at the outer angle of the orbit with the facial nerve and malar branch of the orbital.

The **nasal branches** pass inwards; they supply the integument of the side of the nose, and join with the nasal branch of the ophthalmic.

The **labial branches,** the largest and most numerous, descend beneath the

Levator labii superioris, and are distributed to the integument of the upper lip, the mucous membrane of the mouth, and labial glands.

All these branches are joined, immediately beneath the orbit, by filaments from the facial nerve, forming an intricate plexus, the *infra-orbital*.

SPHENO-PALATINE GANGLION (fig. 396)

The **spheno-palatine ganglion** (*Meckel's*), the largest of the cranial ganglia, is deeply placed in the spheno-maxillary fossa, close to the spheno-palatine foramen. It is triangular or heart-shaped, of a reddish-grey colour, and is situated just below the superior maxillary nerve as it crosses the fossa.

Its branches of communication.—Like the other ganglia of the fifth nerve, it possesses a motor, a sensory, and a sympathetic root. Its *sensory root* is derived from the superior maxillary nerve through its two spheno-palatine branches. These branches of the nerve, given off in the spheno-maxillary fossa, descend to the ganglion. Their fibres, for the most part, pass in front of the ganglion, as they proceed to their destination, in the palate and nasal fossa, and are not incorporated in the ganglionic mass ; some few of the fibres, however, enter the ganglion, constituting its sensory root. Its *motor root* is derived from the facial nerve through the large superficial petrosal nerve ; and its *sympathetic root*, from the carotid plexus, through the large deep petrosal nerve. These two nerves join together to form a single nerve, the *Vidian*, before their entrance into the ganglion.

The *large superficial petrosal branch* (*nervus petrosus superficialis major*) is given off from the geniculate ganglion of the facial nerve in the aqueductus Fallopii ; it passes through the hiatus Fallopii ; enters the cranial cavity, and runs forwards contained in a groove on the anterior surface of the petrous portion of the temporal bone, lying beneath the dura mater. It then enters the cartilaginous substance which fills in the foramen lacerum medium basis cranii, and joining with the large deep petrosal branch forms the Vidian nerve.

The *large deep petrosal branch* (*nervus petrosus profundus*) is given off from the carotid plexus, and runs through the carotid canal on the outer side of the internal carotid artery. It then enters the cartilaginous substance which fills in the foramen lacerum medium, and joins with the large superficial petrosal nerve to form the Vidian.

The *Vidian nerve*, formed in the cartilaginous substance which fills in the middle lacerated foramen by the junction of the two preceding nerves, passes forwards, through the Vidian canal, with the artery of the same name, and is joined by a small ascending branch, the *sphenoidal branch*, from the otic ganglion. Finally, it enters the spheno-maxillary fossa, and joins the posterior angle of Meckel's ganglion.

Its *branches of distribution* are divisible into four groups : *ascending*, which pass to the orbit ; *descending*, to the palate ; *internal*, to the nose ; and *posterior branches*, to the pharynx and nasal fossæ.

The *ascending branches* are two or three delicate filaments, which enter the orbit by the spheno-maxillary fissure, and supply the periosteum. According to Luschka, some filaments pass through foramina in the suture between the os planum of the ethmoid and frontal bones to supply the mucous membrane of the posterior ethmoidal and sphenoidal sinuses.

The *descending* or *palatine branches* are distributed to the roof of the mouth, the soft palate, tonsil, and lining membrane of the nose. They are almost a direct continuation of the spheno-palatine branches of the superior maxillary nerve, and are three in number : anterior, middle, and posterior.

The *anterior* or *large palatine nerve* descends through the posterior palatine canal, emerges upon the hard palate at the posterior palatine foramen, and passes forwards through a groove in the hard palate, nearly as far as the incisor teeth. It supplies the gums, the mucous membrane and glands of the hard palate, and communicates in front with the termination of the naso-palatine nerve. While in the posterior palatine canal, it gives off *inferior nasal branches*, which enter the nose through openings in the palate bone, and ramify over the inferior turbinated bone and middle and inferior meatuses; and, at its exit from the canal, a palatine branch is distributed to both surfaces of the soft palate.

The *middle* or *external palatine nerve* descends, through one of the accessory palatine canals, distributing branches to the uvula, tonsil, and soft palate. It is occasionally wanting,

The *posterior* or *small palatine nerve* descends with a minute artery through the small posterior palatine canal, emerging by a separate opening behind the posterior palatine foramen. It supplies the Levator palati and Azygos uvulæ muscles,* the soft palate, tonsil, and uvula. The middle and posterior palatine join with the tonsillar branches of the glosso-pharyngeal to form the plexus around the tonsil (*circulus tonsillaris*).

FIG. 396.—The spheno-palatine ganglion and its branches.

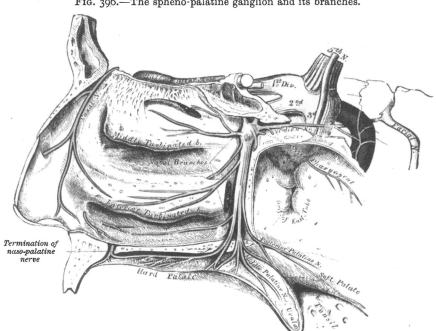

The *internal branches* are distributed to the septum and outer wall of the nasal fossæ. They are the superior nasal (anterior) and the naso-palatine.

The *superior nasal branches* (*anterior*), four or five in number, enter the back part of the nasal fossa by the spheno-palatine foramen. They supply the mucous membrane covering the superior and middle spongy bones, and the lining of the posterior ethmoidal cells, a few being prolonged to the upper and back part of the septum.

The *naso-palatine nerve* (*Cotunnius*) also enters the nasal fossa through the spheno-palatine foramen, and passes inwards across the roof of the nose, below the orifice of the sphenoidal sinus, to reach the septum; it then runs obliquely downwards and forwards along the lower part of the septum, to the anterior palatine foramen, lying between the periosteum and mucous membrane. It descends to the roof of the mouth through the anterior palatine canal. The two nerves are here contained in separate and distinct canals, situated in the inter-

* It is probable that this is not the true motor supply to these muscles, but that they are supplied by the spinal accessory through the pharyngeal plexus.

maxillary suture, and termed the *foramina of Scarpa*, the left nerve being usually anterior to the right one. In the mouth, they become united, supply the mucous membrane behind the incisor teeth, and join with the anterior palatine nerve. The naso-palatine nerve furnishes a few small filaments to the mucous membrane of the septum.

The *posterior branches* are, the pharyngeal (pterygo-palatine), and the upper posterior nasal branches.

The *pharyngeal nerve* (*pterygo-palatine*) is a small branch arising from the back part of the ganglion, being generally blended with the Vidian nerve. It passes through the pterygo-palatine canal with the pterygo-palatine artery, and is distributed to the mucous membrane of the upper part of the pharynx, behind the Eustachian tube.

The *upper posterior nasal branches* are a few twigs given off from the posterior part of the ganglion, which runs backwards in the sheath of the Vidian nerve to the mucous membrane at the back part of the roof, septum, and superior meatus of the nose and that covering the end of the Eustachian tube.

INFERIOR MAXILLARY NERVE (fig. 395)

The **Inferior maxillary nerve** (*n. mandibularis*) distributes branches to the teeth and gums of the lower jaw, the integument of the temple and external ear, the lower part of the face and lower lip, and the muscles of mastication; it also supplies the tongue with a large branch. It is the largest of the three divisions of the fifth, and is made up of two roots : a large or sensory root proceeding from the inferior angle of the Gasserian ganglion ; and a small or motor root, which passes beneath the ganglion, and unites with the sensory root, just after its exit from the skull through the foramen ovale. Immediately beneath the base of the skull, this nerve divides into two trunks, anterior and posterior. Previous to its division, the primary trunk gives off from its inner side a recurrent (meningeal) branch, and the nerve to the Internal pterygoid muscle.

The **recurrent branch** is given off directly after its exit from the foramen ovale. It passes backwards into the skull through the foramen spinosum with the middle meningeal artery. It divides into two branches, anterior and posterior, which accompany the main divisions of the artery and supply the dura mater. The posterior branch also supplies the mucous lining of the mastoid cells. The anterior branch communicates with the meningeal branch of the superior maxillary nerve.

The **internal pterygoid nerve,** given off from the inferior maxillary, previous to its division, is intimately connected at its origin with the otic ganglion. It is a long and slender branch, which passes inwards to enter the deep surface of the Internal pterygoid muscle.

The *anterior* and smaller division, which receives nearly the whole of the motor root, divides into branches, which supply the muscles of mastication. They are, the masseteric, deep temporal, buccal, and external pterygoid.

The **masseteric branch** passes outwards, above the External pterygoid muscle, in front of the temporo-mandibular articulation and behind the tendon of the temporal muscle ; it crosses the sigmoid notch with the masseteric artery, to the deep surface of the Masseter muscle, in which it ramifies nearly as far as its anterior border. It occasionally gives a branch to the Temporal muscle, and a filament to the articulation of the jaw.

The **deep temporal branches,** two in number, anterior and posterior, supply the deep surface of the Temporal muscle. The *posterior branch,* of small size, is placed at the back of the temporal fossa. It is sometimes joined with the masseteric branch. The *anterior branch* is frequently given off from the buccal nerve ; it is reflected upwards, at the pterygoid ridge of the sphenoid, to the front of the temporal fossa. Sometimes there are three deep temporal branches ; the third branch (*middle deep temporal*) passes outwards above the External pterygoid muscle, and runs upwards on the bone to enter the deep surface of the Temporal muscle.

The **buccal branch** passes forwards between the two heads of the External pterygoid, and downwards beneath the inner surface of the coronoid process of the lower jaw, or through the fibres of the Temporal muscle, to reach the surface of the Buccinator, upon which it divides into a superior and an inferior branch. It gives a branch to the External pterygoid during its passage through that muscle, and a few ascending filaments to the Temporal muscle, one of which occasionally joins with the anterior branch of the deep temporal nerve. The *upper branch* supplies the integument and upper part of the Buccinator muscle, joining with the facial nerve round the facial vein. The *lower branch* passes forwards to the angle of the mouth; it supplies the integument and Buccinator muscle, as well as the mucous membrane lining the inner surface of that muscle, and joins the facial nerve.*

The **external pterygoid nerve** is most frequently derived from the buccal, but it may be given off separately from the anterior trunk of the nerve. It enters the muscle on its inner surface.

The *posterior* and larger division of the inferior maxillary nerve is for the most part sensory, but receives a few filaments from the motor root. It divides into three branches: auriculo-temporal, lingual (gustatory), and inferior dental.

The **auriculo-temporal nerve** generally arises by two roots, between which the middle meningeal artery passes. It runs backwards beneath the External pterygoid muscle to the inner side of the neck of the lower jaw. It then turns upwards with the temporal artery, between the external ear and condyle of the jaw, under cover of the parotid gland, and, escaping from beneath this structure, ascends over the zygoma, and divides into two temporal branches.

The *branches of communication* are with the facial and with the otic ganglion. The branches of communication with the facial, usually two in number, pass forwards, from behind the neck of the condyle of the jaw, to join this nerve at the posterior border of the Masseter muscle. They form one of the principal branches of communication between the facial and the fifth nerve. The filaments of communication with the otic ganglion are derived from the commencement of the auriculo-temporal nerve.

The *branches of distribution* are:

Anterior auricular.	Articular.
Branches to the meatus auditorius.	Parotid.
Superficial temporal.	

The *anterior auricular branches* are usually two in number. They supply the front of the upper part of the pinna, being distributed principally to the skin covering the front of the helix and tragus.

Branches to the meatus auditorius, two in number, enter the canal between the bony and cartilaginous portion of the meatus. They supply the skin lining the meatus; the upper one sending a filament to the membrana tympani.

A *branch to the temporo-mandibular articulation* is usually derived from the auriculo-temporal nerve.

The *parotid branches* supply the parotid gland.

The *superficial temporal* accompanies the temporal artery to the vertex of the skull, and supplies the integument of the temporal region, communicating with the facial nerve, and the temporal branch of the temporo-malar, from the superior maxillary.

The **lingual nerve (gustatory)** supplies the papillæ and mucous membrane of the anterior two-thirds of the tongue. It is deeply placed throughout the whole

---

* There seems to be no reason to doubt that the branch supplying the Buccinator muscle is entirely a nerve of ordinary sensation, and that the true motor supply of this muscle is from the facial.

of its course. It lies at first beneath the External pterygoid muscle, together with the inferior dental nerve, being placed to the inner side of this nerve, and is occasionally joined to it by a branch which may cross the internal maxillary artery. The chorda tympani also joins it at an acute angle in this situation. The nerve then passes between the Internal pterygoid muscle and the inner side of the ramus of the jaw, and crosses obliquely to the side of the tongue over the Superior constrictor and Stylo-glossus muscles, and then between the Hyo-glossus muscle and deep part of the submaxillary gland ; the nerve finally runs across Wharton's duct, and along the side of the tongue to its apex, lying immediately beneath the mucous membrane.

The *branches of communication* are with the facial through the chorda tympani, the inferior dental and hypoglossal nerves, and the submaxillary ganglion. The branches to the submaxillary ganglion are two or three in number ; those connected with the hypoglossal nerve form a plexus at the anterior margin of the Hyo-glossus muscle.

The *branches of distribution* supply the mucous membrane of the mouth, the gums, the sublingual gland, the filiform and fungiform papillæ and mucous membrane of the tongue ; the terminal filaments communicate, at the tip of the tongue, with the hypoglossal nerve.

The **inferior dental** is the largest of the three branches of the inferior maxillary nerve. It passes downwards with the inferior dental artery, at first beneath the External pterygoid muscle, and then between the internal lateral ligament and the ramus of the jaw to the dental foramen. It then passes forwards in the dental canal of the inferior maxillary bone, lying beneath the teeth, as far as the mental foramen, where it divides into two terminal branches, incisor and mental.

The branches of the inferior dental are, the mylo-hyoid, dental, incisive and mental.

The *mylo-hyoid* is derived from the inferior dental just as that nerve is about to enter the dental foramen. It descends in a groove on the inner surface of the ramus of the jaw, in which it is retained by a process of fibrous membrane. It reaches the under surface of the Mylo-hyoid muscle, which it supplies together with the anterior belly of the Digastric.

The *dental branches* supply the molar and bicuspid teeth. They correspond in number to the fangs of those teeth : each nerve entering the orifice at the point of the fang, and supplying the pulp of the tooth.

The *incisive branch* is continued onwards within the bone to the middle line, and supplies the canine and incisor teeth.

The *mental branch* emerges from the bone at the mental foramen, and divides beneath the Depressor anguli oris into two or three branches ; one descends to supply the skin of the chin, and another (sometimes two) ascends to supply the skin and mucous membrane of the lower lip. These branches communicate freely with the facial nerve.

Two small ganglia are connected with the inferior maxillary nerve : the otic with the trunk of the nerve ; and the submaxillary with its lingual branch.

OTIC GANGLION (fig. 397)

The **Otic ganglion** (*Arnold's*) is a small, oval-shaped, flattened ganglion of a reddish-grey colour, situated immediately below the foramen ovale, on the inner surface of the inferior maxillary nerve, surrounding the origin of the internal pterygoid nerve. It is in relation, *externally*, with the trunk of the inferior maxillary nerve, at the point where the motor root joins the sensory portion ; *internally*, with the cartilaginous part of the Eustachian tube, and the origin of the Tensor palati muscle ; *behind* it is the middle meningeal artery.

Branches of communication.—This ganglion is connected with the internal pterygoid branch of the inferior maxillary nerve by two or three short, delicate filaments. From this it may obtain a motor root, and possibly also a sensory root, as these filaments from the nerve to the Internal pterygoid may contain sensory fibres. It communicates with the glosso-pharyngeal and facial nerves, through the small superficial petrosal nerve continued from the tympanic plexus, and through this communication it probably receives its sensory root from the glosso-pharyngeal and its motor root from the facial; its communication with the sympathetic is effected by a filament from the plexus surrounding the middle meningeal artery. The ganglion also communicates with the auriculo-temporal nerve. This is probably a branch from the glosso-pharyngeal which passes to the ganglion, and through it and the auriculo-temporal nerve to the parotid gland. A slender filament (*sphenoidal*) ascends from it to the Vidian nerve.

Fig. 397.—The otic ganglion and its branches.

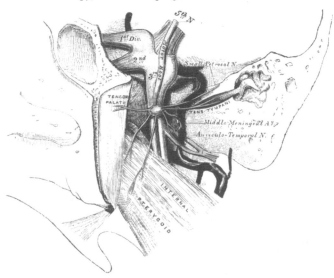

Its *branches of distribution* are, a filament to the Tensor tympani, and one to the Tensor palati. The former passes backwards, on the outer side of the Eustachian tube; the latter arises from the ganglion, near the origin of the internal pterygoid nerve, and passes forwards. The fibres of these nerves are, however, mainly derived from the nerve to the Internal pterygoid muscle. It also gives off a small communicating branch to the chorda tympani.

SUBMAXILLARY GANGLION (fig. 395)

The **submaxillary ganglion** is of small size, fusiform in shape, and situated above the deep portion of the submaxillary gland, near the posterior border of the Mylo-hyoid muscle, being connected by filaments with the lower border of the lingual (gustatory) nerve.

Branches of communication.—This ganglion is connected with the lingual (gustatory) nerve by a few filaments which join it separately, at its fore and back part. It also receives a branch from the chorda tympani, by which it communicates with the facial; and communicates with the sympathetic by filaments from the sympathetic plexus around the facial artery.

Branches of distribution.—These are five or six in number; they arise from the lower part of the ganglion, and supply the mucous membrane of the mouth and Wharton's duct, some being lost in the submaxillary gland. The branch of communication from the lingual to the fore part of the ganglion is by some regarded as a branch of distribution, by which filaments of the chorda tympani pass from the ganglion to the nerve, and by it are conveyed to the sublingual gland and the tongue.

Surface Marking.—It will be seen from the above description that the terminal branches of the three divisions of the fifth nerve emerge from foramina in the bones of the skull and face on to the face: the terminal branch of the first division emerging through the supra-orbital foramen; that of the second through the infra-orbital foramen; and the third through the mental foramen. The supra-orbital foramen is situated at the junction of the internal and middle thirds of the supra-orbital arch. If a straight line is drawn from this point to the lower border of the inferior maxillary bone, so that it passes between the two bicuspid teeth of the lower jaw, it will pass over the infra-orbital and mental foramina; the former being situated about one centimetre ($\frac{2}{5}$ of an inch) below the margin of the orbit, and the latter varying in position according to the age of the individual. In the adult it is midway between the upper and lower borders of the inferior maxillary bone; in the child it is nearer the lower border, and in the edentulous jaw of old age it is close to the upper margin.

Surgical Anatomy.—The fifth nerve may be affected in its entirety; or its sensory or motor root may be affected; or one of its primary main divisions. In injury to the sensory root there is anæsthesia of the half of the face on the side of the lesion, with the exception of the skin over the parotid gland; insensibility of the conjunctiva, followed by destructive inflammation of the cornea, partly from loss of trophic influence, and partly from the irritation produced by the presence of foreign bodies on it, which are not perceived by the patient, and therefore not expelled by the act of winking; dryness of the nose, loss to a considerable extent of the sense of taste, and diminished secretion of the lachrymal and salivary glands. In injury to the motor root, there is impaired action of the lower jaw, from paralysis of the muscles of mastication on the affected side.

The fifth nerve is often the seat of neuralgia, and each of the three divisions has been divided, or a portion of the nerve excised, for this affection. The supra-orbital nerve may be exposed by making an incision an inch and a half in length along the supra-orbital margin below the eyebrow, which is to be drawn upwards, the centre of the incision corresponding to the supra-orbital notch. The skin and Orbicularis palpebrarum having been divided, the nerve can be easily found emerging from the notch, and lying in some loose cellular tissue. It should be drawn up by a blunt hook and divided, or, what is better, a portion of it removed. The infra-orbital nerve has been divided at its exit by an incision on the cheek; or the floor of the orbit has been exposed, the infra-orbital canal opened up, and the anterior part of the nerve resected; or the whole nerve, together with Meckel's ganglion as far back as the foramen rotundum, has been removed. This latter operation, though undoubtedly a severe proceeding, appears to have been followed by the best results. The operation is performed as follows: the superior maxillary bone is first exposed by a T-shaped incision, one limb passing along the lower margin of the orbit, the other from the centre of this vertically down the cheek to the angle of the mouth. The nerve is to be found, divided, and a piece of silk tied to it as a guide. A small trephine ($\frac{1}{2}$-inch) is applied to the bone, below, but including the infra-orbital foramen, and the antrum opened. The trephine is now applied to the posterior wall of the antrum, and the spheno-maxillary fossa exposed. The infra-orbital canal is opened up from below by fine cutting-pliers or a chisel, and the nerve drawn down into the trephine hole, it being held on the stretch by means of the piece of silk; it is severed with fine curved scissors as near the foramen rotundum as possible, any branches coming off from the ganglion being also divided.[*] The mental branch of the inferior dental nerve has been divided at its exit from the foramen by an incision made through the mucous membrane where it is reflected from the alveolar process on to the lower lip; or a portion of the trunk of the inferior dental nerve has been resected by an incision on the cheek through the Masseter muscle, exposing the outer surface of the ramus of the jaw. A trephine was then applied over the position of the inferior dental foramen and the outer table removed, so as to expose the inferior dental canal. The nerve was dissected out of the portion of the canal exposed, and, having been divided after its exit from the mental foramen, it was, by traction on the end exposed in the trephine hole, drawn out entire, and cut off as high up as possible.[†] The inferior dental nerve has also been divided by an incision within the mouth, the bony point guarding the inferior dental foramen forming the guide to the nerve. The buccal nerve may be divided by an incision through the mucous membrane of the mouth and the Buccinator just in front of the anterior border of the ramus of the lower jaw (Stimson).

The lingual (gustatory) nerve is occasionally divided with the view of relieving the pain in cancerous disease of the tongue. This may be done in that part of its course where it lies below and behind the last molar tooth. If a line is drawn from the middle of the crown of the last molar tooth to the angle of the jaw it will cross the nerve, which lies about half an inch behind the tooth, parallel to the bulging alveolar ridge on the inner side of the body of the bone. If the knife is entered three-quarters of an inch behind and below

---

* Camochan, *Amer. Journ. Med. Science*, 1858, p. 136.
† Mears, *Trans. Amer. Surg. Assoc.* vol. ii. p. 469.

the last molar tooth, and carried down to the bone, the nerve will be divided. Hilton divided it opposite the second molar tooth, where it is covered only by the mucous membrane, and Lucas pulls the tongue forwards and over to the opposite side, when the nerve can be seen standing out as a firm cord under the mucous membrane by the side of the tongue, and can be easily seized with a sharp hook and divided, or a portion excised. This is a simple enough operation on the cadaver, but when the disease is extensive and has extended to the floor of the mouth, as is generally the case when division of the nerve is required, the operation is not practicable.

Sixth Nerve (fig. 393)

The **Sixth** or **Abducent Nerve** supplies the External rectus muscle.

Its *superficial origin* is by several filaments from the constricted part of the pyramid, close to the pons, or from the lower border of the pons itself, in the groove between this body and the medulla. Its *deep origin* is from the upper part of the floor of the fourth ventricle, close to the median line, beneath the eminentia teres.

From the nucleus of the sixth nerve fibres pass through the posterior longitudinal bundle to the oculo-motor nucleus of the opposite side and into the third nerve, along which they are carried to the Internal rectus muscle. The External rectus of one eye and the Internal rectus of the other may therefore be said to

Fig. 398.—Relations of structures passing through the sphenoidal fissure.

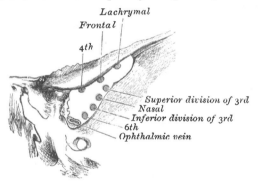

receive their nerves from the same nucleus—a factor of great importance in connection with the conjugate movements of the eyeball, and one that may explain certain paralytic phenomena of the Recti muscles, which are often associated with lesions in the pons.

The nerve pierces the dura mater on the basilar surface of the sphenoid bone, runs through a notch immediately below the posterior clinoid process, and enters the cavernous sinus. It passes forwards through the sinus, lying on the outer side of the internal carotid artery. It enters the orbit through the sphenoidal fissure, and lies above the ophthalmic vein, from which it is separated by a lamina of dura mater. It then passes between the two heads of the External rectus, and is distributed to that muscle on its ocular surface.

Branches of communication.—It is joined by several filaments from the carotid and cavernous plexuses, and by one from the ophthalmic nerve.

The sixth nerve, together with the third, fourth, and the ophthalmic division of the fifth, as they pass to the orbit, bear a certain relation to each other in the cavernous sinus, at the sphenoidal fissure, and in the cavity of the orbit, which will now be described.

In the *cavernous sinus* (fig. 327), the third, fourth, and ophthalmic division of the fifth are placed on the outer wall of the sinus, in their numerical order, both from above downwards, and from within outwards. The sixth nerve lies at the outer side of the internal carotid artery. As these nerves pass forwards to the sphenoidal fissure, the third and fifth nerves become divided into branches, and the sixth approaches the rest; so that their relative position becomes considerably changed.

In the *sphenoidal fissure* (fig. 398), the fourth, and the frontal and lachrymal divisions of the ophthalmic lie upon the same plane, the former being most internal, the latter external; and they enter the cavity of the orbit above the muscles. The remaining nerves enter the orbit between the two heads of the External rectus. The superior division of the third is the highest of these; beneath this lies the nasal branch of the ophthalmic; then the inferior division of the third; and the sixth lowest of all.

In the *orbit*, the fourth, and the frontal and lachrymal divisions of the ophthalmic, lie on the same plane immediately beneath the periosteum, the fourth nerve being internal and resting on the Superior oblique, the frontal resting on the Levator palpebræ, and the lachrymal on the External rectus. Next in order comes the superior division of the third nerve lying immediately beneath the Superior rectus, and then the nasal branch of the ophthalmic, crossing the optic nerve from the outer to the inner side of the orbit. Beneath these is found the optic nerve, surrounded in front by the ciliary nerves, and having the lenticular ganglion on its outer side, between it and the External rectus. Below the optic is the inferior division of the third, and the sixth, which lies on the outer side of the orbit.

Surgical Anatomy.—The sixth nerve is more frequently involved in fractures of the base of the skull than any other of the cranial nerves. The result of paralysis of this nerve is internal or convergent squint. When injured so that its function is destroyed, there is, in addition to the paralysis of the External rectus muscle, often a certain amount of contraction of the pupil, because some of the sympathetic fibres to the radiating muscle of the iris pass along with this nerve.

SEVENTH NERVE (figs. 399, 400, 401)

The **Seventh** or **Facial Nerve** (*portio dura*) is the motor nerve of all the muscles of expression in the face, and of the Platysma and Buccinator; the muscles of the External ear; the posterior belly of the Digastric, and the Stylo-hyoid. Its chorda tympani branch is the nerve of taste for the anterior two-thirds of the tongue and the vaso-dilator nerve of the submaxillary and sublingual glands; its tympanic branch supplies the Stapedius.

Its *superficial origin* is from the upper end of the medulla oblongata, in the groove between the olivary and restiform bodies. Its *deep origin* is from a nucleus

FIG. 399.—The course and connections of the facial nerve in the temporal bone.

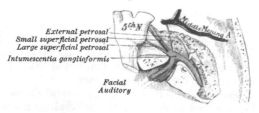

situated in the reticular formation of the lower part of the pons, a little external and ventral to the nucleus of the sixth nerve. From this origin the fibres pursue a curved course in the substance of the pons. They first pass backwards and inwards, and then turn upwards and forwards forming the funiculus teres, which produces an eminence (eminentia teres) on the floor of the fourth ventricle, and finally bend sharply downwards and outwards round the upper end of the nucleus of origin of the sixth nerve, to reach their superficial origin between the olivary and restiform bodies. From the nucleus of the third nerve some fibres arise which descend in the posterior longitudinal bundle and join the facial just before it leaves the pons ; these fibres are said to supply the anterior belly of the Occipito-frontalis, Orbicularis palpebrarum, and the Corrugator supercilii, as these muscles have been observed to escape paralysis in lesions of the nucleus of the facial nerve.

The auditory nerve (*portio mollis*) lies to its outer side; and between the two is a small fasciculus, the *pars intermedia* of Wrisberg, which arises from the medulla and joins the facial nerve in the internal auditory meatus. The deep origin of the pars intermedia is from the upper end of the nucleus of the glosso-pharyngeal nerve, and at its emergence it is frequently connected with both nerves.

The pars intermedia may be regarded as the sensory root of the facial nerve, analogous to the sensory root of the fifth, and its real nucleus of origin would then consist of the geniculate ganglion.

The facial nerve, firmer, rounder, and smaller than the auditory, passes forwards and outwards upon the middle peduncle of the cerebellum, and enters the internal auditory meatus with the auditory nerve. Within the meatus, the facial nerve lies in a groove along the upper and anterior part of the auditory nerve, and the pars intermedia is placed between the two, and joins the inner angle of the geniculate ganglion. Occasionally a few of its fibres pass into the auditory nerve. Beyond the ganglion its fibres are generally regarded as forming the chorda tympani.

At the bottom of the meatus, the facial nerve enters the aqueductus Fallopii, and follows the course of that canal through the petrous portion of the temporal bone, from its commencement at the internal meatus, to its termination at the stylo-mastoid foramen. It is at first directed outwards between the cochlea and vestibule towards the inner wall of the tympanum; it then bends suddenly backwards and arches downwards behind the tympanum to the stylo-mastoid foramen. At the point where it changes its direction, it presents a reddish gangliform swelling (*intumescentia ganglioformis*, or *geniculate ganglion*). On emerging from the stylo-mastoid foramen, it runs forwards in the substance of the parotid gland, crosses the external carotid artery, and divides behind the ramus of the lower jaw into two primary branches, *temporo-facial* and *cervico-facial*, from which numerous offsets are distributed over the side of the head, face, and upper part of the neck, supplying the superficial muscles in these regions. As the primary branches and their offsets diverge from each other, they present somewhat the appearance of a bird's claw ; hence the name of *pes anserinus* is given to the divisions of the facial nerve in and near the parotid gland.

The communications of the facial nerve may be thus arranged :

In the internal auditory meatus .	With the auditory nerve.
From the geniculate ganglion .	With Meckel's ganglion by the large superficial petrosal nerve. With the otic ganglion by the small superficial petrosal nerve. With the sympathetic on the middle meningeal by the external superficial petrosal nerve.
In the Fallopian aqueduct .	With the auricular branch of the pneumo-gastric.
At its exit from the stylo-mastoid foramen . . .	With the glosso-pharyngeal. „ pneumogastric. „ auricularis magnus. „ auriculo-temporal.
Behind the ear	With the small occipital.
On the face	With the three divisions of the fifth.
In the neck	With the superficial cervical.

In the internal auditory meatus some minute filaments pass between the facial and auditory nerves.

Opposite the hiatus Fallopii, the gangliform enlargement on the facial nerve communicates with Meckel's ganglion by means of the large superficial petrosal nerve, which forms its motor root ; with the otic ganglion, by the small superficial petrosal nerve: and with the sympathetic filaments accompanying the middle meningeal artery, by the external petrosal (Bidder).   From the gangliform enlargement, according to Arnold, a twig is sent back to the auditory nerve. Just before the facial nerve emerges from the stylo-mastoid foramen, it generally receives a twig of communication from the auricular branch of the pneumogastric.

After its exit from the stylo-mastoid foramen, it sends a twig to the glosso-pharyngeal, another to the pneumogastric nerve, and communicates with the great auricular branch of the cervical plexus, with the auriculo-temporal branch of the inferior maxillary nerve in the parotid gland, with the small occipital behind the ear, on the face with the terminal branches of the three divisions of the fifth, and in the neck with the transverse cervical.

BRANCHES OF DISTRIBUTION

Within the aqueductus Fallopii .	{ Tympanic, to the Stapedius muscle. { Chorda Tympani.	
At its exit from the stylo-mastoid foramen	{ Posterior Auricular. { Digastric. { Stylo-hyoid.	
On the face	{ Temporo-facial	{ Temporal. { Malar. { Infra-orbital.
	{ Cervico-facial.	{ Buccal. { Supramaxillary. { Inframaxillary.

The **tympanic branch** arises from the nerve opposite the pyramid ; it passes through a small canal in the pyramid, and supplies the Stapedius muscle.

The **chorda tympani** is given off from the facial as it passes vertically down-wards at the back of the tympanum, about a quarter of an inch before its exit from the stylo-mastoid foramen.   It passes from below upwards and forwards in a distinct canal, and enters the cavity of the tympanum, through an aperture (*iter chordæ posterius*) on its posterior wall between the opening of the mastoid cells and the attachment of the membrana tympani, and becomes invested with mucous membrane.   It passes forwards through the cavity of the tympanum, be-tween the fibrous and mucous layers of the membrana tympani, and over the handle of the malleus, emerging from that cavity through a foramen at the inner end of the Glaserian fissure, which is called the *iter chordæ anterius*, or *canal of Huguier*. It then descends between the two Pterygoid muscles, meets the lingual nerve at an acute angle, and accompanies it to the submaxillary gland ; part of it then joins the submaxillary ganglion ; the rest is continued onwards through the muscular substance of the tongue to the mucous membrane covering its anterior two-thirds. A few of its fibres probably pass through the submaxillary ganglion to the sub-lingual gland.   Before joining the lingual nerve it receives a small communicating branch from the otic ganglion.   As already stated, the chorda tympani nerve is by many regarded as the continuation of the pars intermedia of Wrisberg.

The **posterior auricular nerve** arises close to the stylo-mastoid foramen, and passes upwards in front of the mastoid process, where it is joined by a filament from the auricular branch of the pneumogastric, and communicates with the mastoid branch of the great auricular and with the small occipital.   As it ascends between the meatus and mastoid process it divides into two branches. The *auricular branch* supplies the Retrahens auriculam and the small muscles on the cranial surface of the pinna.   The *occipital branch*, the larger, passes back-wards along the superior curved line of the occipital bone, and supplies the occipital portion of the Occipito-frontalis.

The **digastric branch** usually arises by a common trunk with the Stylo-hyoid branch; it divides into several filaments, which supply the posterior belly of the Digastric; one of these perforates that muscle to join the glosso-pharyngeal nerve.

The **stylo-hyoid** is a long slender branch, which passes inwards, entering the Stylo-hyoid muscle about its middle.

The **temporo-facial**, the larger of the two terminal branches, passes upwards and forwards through the parotid gland, crosses the external carotid artery and temporo-maxillary vein, and passes over the neck of the condyle of the jaw, being connected in this situation with the auriculo-temporal branch of the inferior maxillary nerve, and divides into branches, which are distributed over the temple and upper part of the face; these are divided into three sets: temporal, malar, and infra-orbital.

FIG. 400.—Plan of the facial nerve.

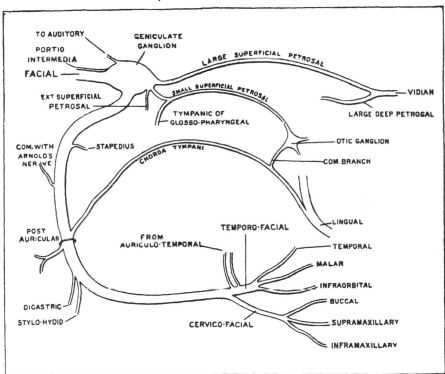

The *temporal branches* cross the zygoma to the temporal region, supplying the Attrahens and Attollens auriculam muscles, and join with the temporal branch of the temporo-malar, a branch of the superior maxillary, and with the auriculo-temporal branch of the inferior maxillary. The more anterior branches supply the frontal portion of the Occipito-frontalis, the Orbicularis palpebrarum, and Corrugator supercilii muscles, joining with the supra-orbital and lachrymal branches of the ophthalmic.

The *malar branches* pass across the malar bone to the outer angle of the orbit, where they supply the Orbicularis palpebrarum muscle, joining with filaments from the lachrymal nerve; others supply the lower eyelid, joining with filaments of the malar branch (*subcutaneous malæ*) of the superior maxillary nerve.

The *infra-orbital*, of larger size than the rest, pass horizontally forwards to be distributed between the lower margin of the orbit and the mouth. The *superficial branches* run beneath the skin and above the superficial muscles of the face which they supply: some branches are distributed to the Pyramidalis nasi, joining at the inner angle of the orbit with the infratrochlear and nasal branches of the ophthalmic. The *deep branches* pass beneath the Zygomatici and the Levator labii superioris, supplying them and the Levator anguli oris, and form a plexus (*infra-orbital*) by joining with the infra-orbital branch of the superior maxillary nerve and the buccal branches of the cervico-facial. This branch also supplies the Levator labii superioris alæque nasi and the small muscles of the nose.

The **cervico-facial division of the facial nerve** passes obliquely downwards and forwards through the parotid gland, crossing the external carotid artery. In this situation it is joined by branches from the great auricular nerve. Opposite the

FIG. 401.—The nerves of the scalp, face, and side of the neck.

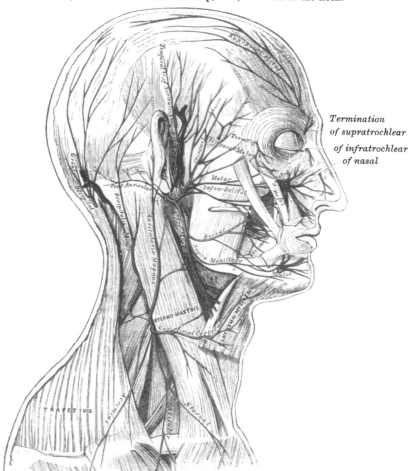

angle of the lower jaw it divides into branches which are distributed on the lower half of the face and upper part of the neck. These may be divided into three sets : buccal, supramaxillary, and inframaxillary.

The *buccal branches* cross the Masseter muscle. They supply the Buccinator and Orbicularis oris, and join with the infra-orbital branches of the temporo-facial division of the nerve, and with filaments of the buccal branch of the inferior maxillary nerve.

The *supramaxillary (mandibular) branches* pass forwards beneath the Platysma and Depressor anguli oris, supplying the muscles of the lower lip and chin, and communicating with the mental branch of the inferior dental nerve.

The *inframaxillary* (*cervical*) *branches* run forward beneath the Platysma, and form a series of arches across the side of the neck over the suprahyoid region. One of these branches descends vertically to join with the superficial cervical nerve from the cervical plexus ; others supply the Platysma.

Surgical Anatomy.—The facial nerve is more frequently paralysed than any of the other cranial nerves. The paralysis may depend either upon (1) central causes, i.e. blood clots or intracranial tumours pressing on the nerve before its entrance into the internal auditory meatus. It is also one of the nerves involved in 'bulbar paralysis.' Or (2) it may be paralysed in its passage through the petrous bone, by damage due to middle ear disease, or by fractures of the base. Or (3) it may be affected at or after its exit from the stylo-mastoid foramen. This is commonly known as 'Bell's paralysis.' It may be due to exposure to cold or to injury of the nerve, either from accidental wounds of the face, or during some surgical operation, as removal of parotid tumours, opening of abscesses, or operations on the lower jaw.

When the cause is central, the sixth nerve is usually paralysed as well, and there is also hemiplegia on the opposite side. In these cases the electrical reactions are the same as in health ; whereas, when the paralysis is in the course of the nerve, the reaction is usually lost. When the nerve is paralysed in the petrous bone, in addition to the paralysis of the muscles of expression, there is loss of taste in the anterior part of the tongue, and the patient is unable to recognise the difference between bitters and sweets, acids and salines, from involvement of the chorda tympani. The mouth is dry, because the salivary glands are not secreting ; and the sense of hearing is affected from paralysis of the Stapedius. When the cause of the paralysis is from fracture of the base of the skull, the auditory nerve and the petrosal nerves, which are connected with the intumescentia ganglioformis, are also involved. When the injury to the nerve is after its exit from the stylo-mastoid foramen, all the muscles of expression, except the Levator palpebræ, together with the posterior belly of the Digastric and Stylo-hyoid, are paralysed. There is smoothness of the forehead, and the patient is unable to frown ; the eyelids cannot be closed and the lower lid droops, so that the punctum is no longer in contact with the globe, and the tears run down the cheek ; there is smoothness of the cheek and loss of the naso-labial furrow ; the nostril cannot be dilated ; the mouth is drawn to the sound side, and there is inability to whistle ; food collects between the cheek and gum from paralysis of the Buccinator. The facial nerve is at fault in cases of so-called 'histrionic spasm,' which consists in an almost constant and uncontrollable twitching of the muscles of the face. This twitching is sometimes so severe as to cause great discomfort and annoyance to the patient, and to interfere with sleep, and for its relief the facial nerve has been stretched. The operation is performed by making an incision behind the ear, from the root of the mastoid process to the angle of the jaw. The parotid is turned forwards and the dissection carried along the anterior border of the Sterno-mastoid muscle and mastoid process, until the upper border of the posterior belly of the Digastric is found. The nerve is parallel to this on about a level of the middle of the mastoid process. When found, the nerve must be stretched by passing a blunt hook beneath it and pulling it forwards and outwards. Too great force must not be used, for fear of permanent injury to the nerve.

EIGHTH NERVE

The **Eighth** or **Auditory Nerve** (*portio mollis*) is the special nerve of the sense of hearing, being distributed exclusively to the internal ear.

Origin of the Eighth Nerve.—The eighth nerve consists of two sets of fibres, which, although differing in their central connections, are both concerned in the transmission of afferent impulses from the internal ear to the medulla and pons, and from there, by means of new fibres which arise from collections of grey matter in these structures, to the cerebrum and cerebellum. One set of fibres forms the vestibular root of the nerve, and arises from the cells in the ganglion of Scarpa ; the other set constitutes the cochlear root, and takes origin from the cells in the ganglion spirale or ganglion of Corti. At its connection with the brain the eighth nerve occupies the groove between the pons and medulla, where it is situated between the restiform body, which is behind, and the seventh nerve, which is in front.

Vestibular or ventral root.—The fibres of this root enter the medulla to the inner side of those of the cochlear root, and pass between the restiform body, which is to their outer side, and the inferior root of the fifth, which lies to their inner side. They then divide into an ascending and a descending set. The fibres of the latter end by arborising round the cells of the internal nucleus which is situated in the *trigonum acustici* in the floor of the fourth ventricle.

The ascending fibres either end in the same manner or in the *external nucleus*, which is situated to the outer side of the trigonum acustici and farther from the ventricular floor. It is described as consisting of two parts, an inner, the *nucleus of Deiters*, and an outer, the *nucleus of Bechteren*. Some of the axons of the cells of the external nucleus, and possibly also of the internal nucleus, are continued upwards through the restiform body to the roof nuclei of the opposite side of the cerebellum, to which also are prolonged other fibres of the vestibular root without undergoing a relay in the nuclei of the medulla. A second set of fibres from the internal and external nuclei end partly in the tegmentum, while the remainder ascend in the posterior longitudinal bundle to arborise around the nuclei of the oculo-motor nerve.

Cochlear or dorsal root.—This part of the nerve is placed externally to the vestibular root. Its fibres end in two nuclei, one of which, the *accessory nucleus*, lies immediately in front of the restiform body; the other, the *tuberculum acusticum*, somewhat to its outer side.

The striæ acusticæ or medullary striæ are the axons of the cells of the tuberculum acusticum. They pass backwards and inwards over the restiform body, and across the floor of the fourth ventricle towards the middle line. Here they dip into the substance of the pons, to end around the cells of the *superior olive* of the same or opposite side. There are, however, other fibres, and these are both direct and crossed, which do not arborise around the tegmental nuclei but pass into the lateral fillet. The cells of the accessory nucleus give origin to fibres which pass transversely in the pons and constitute the trapezium. The description given as to the mode of ending of the striæ acusticæ is applicable to that of the trapezoid fibres, viz. around the cells of the superior olive or of the *trapezoid nucleus* (which lies ventral to the olive) of the same or opposite side, while others, crossed or uncrossed, pass directly into the lateral fillet.

If the further connections of the cochlear nerve of one side, say the left, are considered, it is found that they lie to the outer side of the main sensory tract, the fillet, and are therefore termed the *lateral fillet*. The fibres comprising the left lateral fillet arise in the superior olive or trapezoid nucleus of the same or opposite side, while others are the uninterrupted fibres already alluded to, and these are either crossed or uncrossed, the former being the axons of the cells of the right accessory nucleus or of the cells of the right tuberculum acusticum, while the latter are derived from the same cells of the left side. In the upper part of the fillet there is a collection of nerve-cells, the *nucleus of the fillet*, around the cells of which some of the fibres arborise and from the cells of which axons originate to continue upwards the tract of the lateral fillet. The ultimate ending of the left lateral fillet is partly in the quadrigeminal bodies of the same or opposite side, while the remainder of the fibres ascend in the posterior limb of the internal capsule to reach the first and perhaps the second left temporal convolution.

The auditory nerve contains a few efferent fibres which arise in the quadrigeminal bodies, the nucleus of the lateral fillet, trapezoid nucleus, and superior olive.

The auditory nerve after leaving the medulla passes forwards across the posterior border of the middle peduncle of the cerebellum, in company with the facial nerve, from which it is partially separated by a small artery (auditory). It then enters the internal auditory meatus with the facial nerve. At the bottom of the meatus it receives one or two filaments from the facial nerve, and then divides into its two branches, *cochlear* and *vestibular*. The auditory nerve is soft in texture (hence the name *portio mollis*), and is destitute of neurilemma. The distribution of the auditory nerve in the internal ear will be found described along with the anatomy of that organ in a subsequent page.

Surgical Anatomy.—The auditory nerve is frequently injured, together with the facial nerve, in fractures of the middle fossa of the base of the skull implicating the internal auditory meatus. The nerve may be either torn across, producing permanent deafness, or it may be bruised or pressed upon by extravasated blood or inflammatory exudation, when the deafness will in all probability be temporary. The nerve may also be injured by violent blows on the head without fracture, and deafness may arise from loud explosions from dynamite, &c., probably from some lesion of this nerve, which is more liable to be injured than the other cranial nerves on account of its structure. The test that the nerve is destroyed, and that the deafness is not due to some lesion of the auditory apparatus, is obtained by placing a vibrating tuning-fork on the head. The vibrations will be heard in cases where the auditory apparatus is at fault, but not in cases of destruction of the auditory nerve.

NINTH PAIR (figs. 402, 403, 404)

The **Ninth** or **Glosso-pharyngeal Nerve** is distributed, as its name implies, to the tongue and pharynx, being the nerve of ordinary sensation to the mucous membrane of the pharynx, fauces, and tonsil; and the nerve of taste to all parts of the tongue to which it is distributed.

Its *superficial origin* is by three or four filaments, closely connected together, from the upper part of the medulla oblongata, in the groove between the olivary and the restiform body.

Its *deep origin* may be traced through the fasciculi of the lateral tract, to three different sources : (1) some of the fibres may be traced to a nucleus of grey matter at the lower part of the floor of the fourth ventricle beneath the inferior fovea; (2) others may be traced downwards into the *funiculus solitarius*, a rounded bundle of fibres in the lower part of the medulla, commencing immediately above the decussation of the pyramids. These fibres have not yet been distinctly traced to cells; (3) a third set of fibres take origin from the cells of the *nucleus ambiguus*. This nucleus is situated some distance from the floor of the fourth ventricle and lies slightly internal to the inferior fovea. It gives origin to the motor branches of the glosso-pharyngeal and vagus, and to the bulbar part of the spinal accessory.

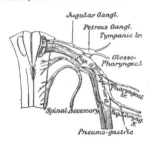

FIG. 402.—Origin, ganglia, and communications of the ninth, tenth, and eleventh cranial nerves.

The real origin of the sensory fibres of the glosso-pharyngeal must be looked for in the jugular and petrosal ganglia which are developed from the neural crest.

From its superficial origin, it passes outwards across the flocculus, and leaves the skull at the central part of the jugular foramen, in a separate sheath of the dura mater, external to and in front of the pneumogastric and spinal accessory nerves (fig. 329). In its passage through the jugular foramen, it grooves the lower border of the petrous portion of the temporal bone; and, at its exit from the skull, passes forwards between the jugular vein and internal carotid artery, and descends in front of the latter vessel, and beneath the styloid process and the muscles connected with it, to the lower border of the Stylo-pharyngeus. The nerve now curves inwards, forming an arch on the side of the neck and lying upon the Stylo-pharyngeus and Middle constrictor of the pharynx. It then passes beneath the Hyo-glossus, and is finally distributed to the mucous membrane of the fauces and base of the tongue, and the mucous glands of the mouth and tonsil.

In passing through the jugular foramen, the nerve presents, in succession, two gangliform enlargements. The superior, the smaller, is called the *jugular ganglion*; the inferior and larger, the *petrous ganglion*, or the *ganglion of Andersch*.

The **superior**, or **jugular**, **ganglion** is situated in the upper part of the groove in which the nerve is lodged during its passage through the jugular foramen. It is of very small size, and involves only the lower part of the trunk of the nerve. It is usually regarded as a segmentation from the lower ganglion.

The **inferior, or petrous, ganglion** is situated in a depression in the lower border of the petrous portion of the temporal bone; it is larger than the former, and involves the whole of the fibres of the nerve. From this ganglion arise those filaments which connect the glosso-pharyngeal with the pneumogastric and sympathetic nerves.

The *branches of communication* are with the pneumogastric, sympathetic, and facial.

The branches to the pneumogastric are two filaments, arising from the petrous ganglion, one to its auricular branch, and one to the upper ganglion of the pneumogastric.

The branch to the sympathetic, also arising from the petrous ganglion, is connected with the superior cervical ganglion.

The branch of communication with the facial perforates the posterior belly of the Digastric. It arises from the trunk of the nerve below the petrous ganglion, and joins the facial just after its exit from the stylo-mastoid foramen.

The *branches of distribution* are, the tympanic, carotid, pharyngeal, muscular, tonsillar, and lingual.

The **tympanic branch** (*Jacobson's nerve*) arises from the petrous ganglion, and enters a small bony canal in the lower surface of the petrous portion of the temporal bone; the lower opening of which is situated on the bony ridge which separates the carotid canal from the jugular fossa. It ascends to the tympanum, enters that cavity by an aperture in its floor close to the inner wall, and divides into branches, which are contained in grooves upon the surface of the promontory, forming the tympanic plexus. This plexus gives off (1) the greater part of the small superficial petrosal nerve; (2) a branch to join the great superficial petrosal nerve; and (3) branches to the tympanic cavity, all which will be described in connection with the anatomy of the ear.

The **carotid branches** descend along the trunk of the internal carotid artery as far as its commencement, communicating with the pharyngeal branch of the pneumogastric, and with branches of the sympathetic.

The **pharyngeal branches** are three or four filaments which unite, opposite the Middle constrictor of the pharynx, with the pharyngeal branches of the pneumo-gastric and sympathetic nerves, to form the pharyngeal plexus, branches from which perforate the muscular coat of the pharynx to supply the muscles and mucous membrane.

The **muscular branch** is distributed to the Stylo-pharyngeus.

The **tonsillar branches** supply the tonsil, forming a plexus (*circulus tonsillaris*) around this body, from which branches are distributed to the soft palate and fauces, where they communicate with the palatine nerves.

The **lingual branches** are two in number: one supplies the circumvallate papillæ and the mucous membrane covering the surface of the base of the tongue; the other perforates its substance, and supplies the mucous membrane and follicular glands of the posterior half of the tongue and communicates with the lingual nerve.

Tenth Pair (figs. 403, 404)

The **Tenth or Pneumogastric Nerve** (*nervus vagus* or *par vagum*) has a more extensive distribution than any of the other cranial nerves, passing through the neck and thorax to the upper part of the abdomen. It is composed of both motor and sensory fibres. It supplies the organs of voice and respiration with motor and sensory fibres; and the pharynx, œsophagus, stomach, and heart with motor fibres.

Its *superficial origin* is by eight or ten filaments from the groove between the olivary and the restiform bodies below the glosso-pharyngeal; its *deep origin* may be traced through the fasciculi of the medulla, to terminate in a nucleus of grey matter, the *nucleus vagi*, at the lower part of the floor of the fourth ventricle,

FIG. 403.—Plan of the glosso-pharyngeal, pneumogastric, and spinal accessory nerves. (After Flower.)

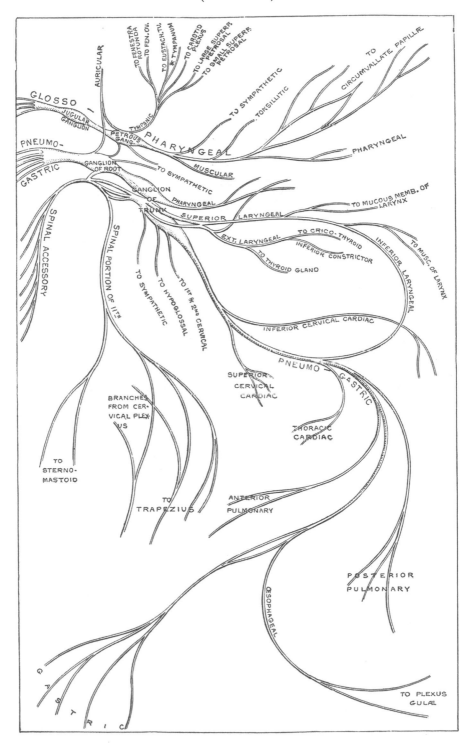

beneath the ala cinerea, below and continuous with the nucleus of origin of the glosso-pharyngeal. In addition to this a few fibres pass into the funiculus solitarius, and others into the nucleus ambiguus or accessory vagal nucleus.

FIG. 404.— Course and distribution of the ninth, tenth, and eleventh cranial nerves.

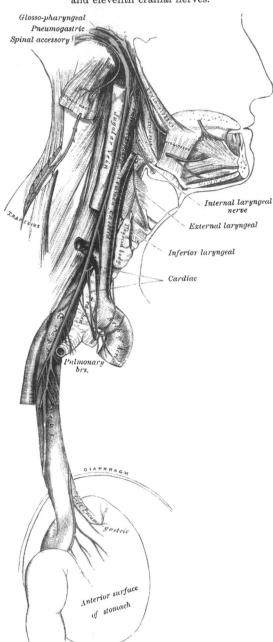

Glosso-pharyngeal
Pneumogastric
Spinal accessory

Internal laryngeal nerve

External laryngeal

Inferior laryngeal

Cardiac

Pulmonary brs.

DIAPHRAGM

Anterior surface of stomach

The real origin of the sensory fibres of the vagus is to be found in the cells of the ganglia on the nerve, viz. the ganglion of the root and the ganglion of the trunk.

The filaments become united, and form a flat cord, which passes outwards beneath the flocculus to the jugular foramen, through which it emerges from the cranium. In passing through this opening, the pneumo-gastric accompanies the spinal accessory, being contained in the same sheath of dura mater with it, a membranous septum separating them from the glosso-pharyngeal, which lies in front (fig. 329). The nerve in this situation presents a well-marked gan-glionic enlargement, which is called *jugular ganglion*, or the *ganglion of the root of the pneumogastric*: to it the accessory part of the spinal accessory nerve is connected by one or two filaments. After the exit of the nerve from the jugular foramen the nerve is joined by the ac-cessory portion of the spinal accessory, and enlarges into a second gangliform swelling, called the *ganglion inferius*, or the *ganglion of the trunk of the nerve*; through which the fibres of the spinal acces-sory pass unchanged, being principally distributed to the pharyngeal and superior laryngeal branches of the vagus, but some of the fila-ments from it are continued into the trunk of the vagus below the ganglion, to be distributed with the recurrent laryngeal nerve and probably also with the cardiac nerves.

The nerve passes vertically down the neck within the sheath of the carotid vessels lying between the internal carotid artery and internal jugular vein as far as the thyroid cartilage, and then between the same vein and the common carotid to the root of the neck. Here the course of the nerve becomes different on the two sides of the body.

On the *right side*, the nerve passes across the subclavian artery between it and the right innominate vein, and descends by the side of the trachea to the back part of the root of the lung, where it spreads out in a plexiform network (*posterior pulmonary plexus*), from the lower part of which two cords descend upon the œsophagus, upon which they divide, forming, with branches from the opposite nerve, the œsophageal plexus (*plexus gulæ*); below, these branches are collected into a single cord, which runs along the back part of the œsophagus, enters the abdomen, and is distributed to the posterior surface of the stomach, joining the left side of the solar plexus, and sending filaments to the splenic plexus and a considerable branch to the cœliac plexus.

On the *left side*, the pneumogastric nerve enters the chest between the left carotid and subclavian arteries, behind the left innominate vein. It crosses the arch of the aorta, and descends behind the root of the left lung, forming the *posterior pulmonary plexus*, and along the anterior surface of the œsophagus, where it unites with the nerve of the right side in forming the *plexus gulæ*, to the stomach, distributing branches over its anterior surface, some extending over the great *cul-de-sac*, and others along the lesser curvature. Filaments from these branches enter the gastro-hepatic omentum, and join the hepatic plexus.

The *ganglion of the root* is of a greyish colour, circular in form, about two lines in diameter, and resembles the ganglion on the large root of the fifth nerve.

Connecting branches.—To this ganglion the accessory portion of the spinal accessory nerve is connected by several delicate filaments; it also has a communicating twig with the petrous ganglion of the glosso-pharyngeal, with the facial nerve by means of its auricular branch, and with the sympathetic by means of an ascending filament from the superior cervical ganglion.

The *ganglion of the trunk* (inferior) is a plexiform cord, cylindrical in form, of a reddish colour, and about an inch in length; it involves the whole of the fibres of the nerve, and passing through it is the accessory portion of the spinal accessory nerve, which blends with the pneumogastric below the ganglion, and is then principally continued into its pharyngeal and superior laryngeal branches.

Connecting branches.—This ganglion is connected with the hypoglossal, the superior cervical ganglion of the sympathetic, and the loop between the first and second cervical nerves.

The *branches of the pneumogastric* are:

In the jugular fossa	Meningeal. Auricular.
In the neck	Pharyngeal. Superior laryngeal. Recurrent laryngeal. Cervical cardiac.
In the thorax	Thoracic cardiac. Anterior pulmonary. Posterior pulmonary. Œsophageal.
In the abdomen	Gastric.

The **meningeal branch** is a recurrent filament given off from the ganglion of the root in the jugular foramen. It passes backwards, and is distributed to the dura mater covering the posterior fossa of the base of the skull.

The **auricular branch** (*Arnold's*) arises from the ganglion of the root, and is joined soon after its origin by a filament from the petrous ganglion of the glosso-pharyngeal; it passes outwards behind the jugular vein, and enters a small canal on the outer wall of the jugular fossa. Traversing the substance of the temporal bone, it crosses the aqueductus Fallopii about two lines above its termination at the stylo-mastoid foramen ; here it gives off an ascending branch, which joins the facial : the continuation of the nerve reaches the surface by passing through the auricular fissure between the mastoid process and the external auditory meatus, and divides into two branches, one of which communicates with the posterior auricular nerve, while the other supplies the integument at the back part of the pinna and the posterior part of the external auditory meatus.

The **pharyngeal branch,** the principal motor nerve of the pharynx, arises from the upper part of the inferior ganglion of the pneumogastric. It consists principally of filaments from the accessory portion of the spinal accessory ; it passes across the internal carotid artery to the upper border of the Middle constrictor, where it divides into numerous filaments, which join with those from the glosso-pharyngeal, superior laryngeal (its external branch), and sympathetic, to form the pharyngeal plexus, from which branches are distributed to the muscles and mucous membrane of the pharynx and the muscles of the soft palate. From the pharyngeal plexus a minute filament is given off, which descends and joins the hypoglossal nerve as it winds round the occipital artery.

The **superior laryngeal** is the nerve of sensation to the larynx. It is larger than the preceding, and arises from the middle of the inferior ganglion of the pneumogastric. It consists principally of filaments from the accessory portion of the spinal accessory. In its course it receives a branch from the superior cervical ganglion of the sympathetic. It descends, by the side of the pharynx, behind the internal carotid, where it divides into two branches, the external and internal laryngeal.

The *external laryngeal branch*, the smaller, descends by the side of the larynx, beneath the Sterno-thyroid, to supply the Crico-thyroid muscle. It gives branches to the pharyngeal plexus and the Inferior constrictor, and communicates with the superior cardiac nerve, behind the common carotid.

The *internal laryngeal branch* descends to the opening in the thyro-hyoid membrane, through which it passes with the superior laryngeal artery, and is distributed to the mucous membrane of the larynx. A small branch communicates with the recurrent laryngeal nerve. The branches to the mucous membrane are distributed, some in front to the epiglottis, the base of the tongue, and the epiglottidean glands ; while others pass backwards, in the aryteno-epiglottidean fold, to supply the mucous membrane surrounding the superior orifice of the larynx, as well as the membrane which lines the cavity of the larynx as low down as the vocal cord. The filament which joins with the recurrent laryngeal descends beneath the mucous membrane on the inner surface of the thyroid cartilage, where the two nerves become united.

The **inferior** or **recurrent laryngeal,** so called from its reflected course, is the motor nerve of the larynx. It arises on the right side, in front of the subclavian artery ; winds from before backwards round that vessel, and ascends obliquely to the side of the trachea, behind the common carotid, and either in front of or behind the inferior thyroid artery. On the left side, it arises in front of the arch of the aorta, and winds from before backwards round the aorta at the point where the remains of the ductus arteriosus are connected with it, and then ascends to the side of the trachea. The nerves on both sides ascend in the groove between the trachea and œsophagus, and, passing under the lower border of the Inferior constrictor muscle, enter the larynx behind the articulation of the inferior cornu of the thyroid cartilage with the cricoid, being distributed to all the muscles of the larynx, excepting the Crico-thyroid. It communicates with the superior laryngeal nerve, and gives off a few filaments to the mucous membrane of the lower part of the larynx.

The recurrent laryngeal, as it winds round the subclavian artery and aorta, gives off several cardiac filaments, which unite with the cardiac branches from the pneumogastric and sympathetic. As it ascends in the neck it gives off œsophageal branches, more numerous on the left than on the right side, which supply the mucous membrane and muscular coat of the œsophagus ; tracheal branches to the mucous membrane and muscular fibres of the trachea ; and some pharyngeal filaments to the Inferior constrictor of the pharynx.

The **cervical cardiac branches,** two or three in number, arise from the pneumo-gastric, at the upper and lower part of the neck.

The *superior branches* are small, and communicate with the cardiac branches of the sympathetic. They can be traced to the great or deep cardiac plexus.

The *inferior branches*, one on each side, arise at the lower part of the neck, just above the first rib. On the right side, this branch passes in front or by the side of the arteria innominata, and communicates with one of the cardiac nerves proceeding to the great or deep cardiac plexus. On the left side, it passes in front of the arch of the aorta, and joins the superficial cardiac plexus.

The **thoracic cardiac branches,** on the right side, arise from the trunk of the pneumogastric, as it lies by the side of the trachea, and from its recurrent laryngeal branch ; but on the left side from the recurrent nerve only ; passing inwards, they terminate in the deep cardiac plexus.

The **anterior pulmonary branches,** two or three in number, and of small size, are distributed on the anterior aspect of the root of the lungs. They join with filaments from the sympathetic, and form the *anterior pulmonary plexus.*

The **posterior pulmonary branches,** more numerous and larger than the anterior, are distributed on the posterior aspect of the root of the lung : they are joined by filaments from the third and fourth (sometimes also first and second) thoracic ganglia of the sympathetic, and form the *posterior pulmonary plexus.* Branches from both plexuses accompany the ramifications of the air-tubes through the substance of the lungs.

The **œsophageal branches** are given off from the pneumogastric both above and below the pulmonary branches. The lower are more numerous and larger than the upper. They form, together with branches from the opposite nerve, the *œsophageal plexus* or *plexus gulæ.* From this plexus branches are distributed to the back of the pericardium.

The **gastric branches** are the terminal filaments of the pneumogastric nerve. The nerve on the right side is distributed to the posterior surface of the stomach, and joins the left side of the cœliac plexus and the splenic plexus. The nerve on the left side is distributed over the anterior surface of the stomach, some filaments passing across the great *cul-de-sac*, and others along the lesser curvature. They unite with branches of the right nerve and with the sympathetic, some filaments passing through the lesser omentum to the hepatic plexus.

Surgical Anatomy.—The laryngeal nerves are of considerable importance in consider-ing some of the morbid conditions of the larynx. When the peripheral terminations of the superior laryngeal nerve are irritated by some foreign body passing over them, reflex spasm of the glottis is the result. When the trunk of this same nerve is pressed upon, by, for instance, a goitre or an aneurism of the upper part of the carotid, there is a peculiar dry, brassy cough. When the nerve is paralysed, there is anæsthesia of the mucous membrane of the larynx, so that foreign bodies can readily enter the cavity, and, in consequence of its supplying the Crico-thyroid muscle, the vocal cords cannot be made tense, and the voice is deep and hoarse. Paralysis of the superior laryngeal nerves may be the result of bulbar paralysis ; may be a sequel to diphtheria when both nerves are usually involved ; or it may, though less commonly, be caused by the pressure of tumours or aneurisms, when the paralysis is generally unilateral. Irritation of the inferior laryngeal nerves produces spasm of the muscles of the larynx. When both these recurrent nerves are paralysed, the vocal cords are motionless, in the so-called 'cadaveric position' —that is to say, in the position in which they are found in ordinary tranquil respiration ; neither closed as in phonation, nor open as in deep inspiratory efforts. When one recurrent nerve is paralysed, the cord of the same side is motionless, while the opposite one crosses the middle line to accommodate itself to the affected one ; hence phonation is present, but the voice is altered and weak in timbre. The recurrent laryngeal nerves may be paralysed in bulbar paralysis or after diphtheria, when the paralysis usually affects both sides ; or they may be affected by the pressure of aneurisms of the aorta, innominate or subclavian arteries ; by mediastinal tumours ; by bronchocele ; or by cancer of the upper part of the œsophagus, when the paralysis is often unilateral.

Eleventh Pair (figs. 403, 404)

The **Eleventh** or **Spinal Accessory Nerve** consists of two parts : one, the accessory part to the vagus, and the other the spinal portion.

The **bulbar** or **accessory part** is the smaller of the two. Its *superficial origin* is by four or five delicate filaments from the side of the medulla, below the roots of the vagus. Its *deep origin* may be traced to a nucleus of grey matter at the back of the medulla, dorso-lateral to the hypoglossal nucleus, and extending as far down as the intermedio-lateral tract of the spinal cord. It passes outwards to the jugular foramen, where it interchanges fibres with the spinal portion or becomes united to it for a short distance; it is also connected, in the foramen, with the upper ganglion of the vagus by one or two filaments. It then passes through the foramen, and becoming again separated from the spinal portion it is continued over the surface of the ganglion of the trunk of the vagus, being adherent to its surface, and is distributed principally to the pharyngeal and superior laryngeal branches of the pneumogastric. Through the pharyngeal branch it probably supplies the muscles of the soft palate (see page 304). Some few filaments from it are continued into the trunk of the vagus below the ganglion, to be distributed with the recurrent laryngeal nerve and probably also with the cardiac nerves.

The **spinal portion** is firm in texture. Its *superficial origin* is by several filaments from the lateral tract of the cord, as low down as the sixth cervical nerve. Its *deep origin* may be traced to the intermedio-lateral tract of the grey matter of the cord. This portion of the nerve ascends between the ligamentum denticulatum and the posterior roots of the spinal nerves, enters the skull through the foramen magnum, and is then directed outwards to the jugular foramen, through which it passes, lying in the same sheath as the pneumogastric, but separated from it by a fold of the arachnoid. In the jugular foramen, it receives one or two filaments from the accessory portion. At its exit from the jugular foramen, it passes backwards, either in front of or behind the internal jugular vein, and descends obliquely behind the Digastric and Stylo-hyoid muscles to the upper part of the Sterno-mastoid. It pierces that muscle, and passes obliquely across the posterior triangle, to terminate in the deep surface of the Trapezius. This nerve gives several branches to the Sterno-mastoid during its passage through it, and joins in its substance with branches from the second cervical, which supply the muscle. In the posterior triangle it joins with the second and third cervical nerves, while beneath the Trapezius it forms a sort of plexus with the third and fourth cervical nerves, and from this plexus fibres are distributed to the muscle.

Surgical Anatomy.—In cases of spasmodic torticollis in which all previous palliative treatment has failed, and the spasms are so severe as to undermine the patient's health, division or excision of a portion of the spinal accessory nerve has been resorted to. This may be done along either the anterior or posterior border of the Sterno-mastoid muscle. The former operation consists in making an incision from the apex of the mastoid process, three inches in length, along the anterior border of the Sterno-mastoid muscle. The anterior border of the muscle is defined and pulled backwards, so as to stretch the nerve, which is then to be sought for, beneath the Digastric muscle, about two inches below the apex of the mastoid process. The other operation is performed by making an incision along the posterior border of the muscle, so that the centre of the incision corresponds to the middle of this border of the muscle. The superficial structures having been divided and the border of the muscle defined, the nerve is to be sought for as it emerges from the muscle to cross the occipital triangle. When found, it is to be traced upwards through the muscle, and a portion of it excised above the point where it gives off its branches to the Sterno-mastoid. In this operation one of the descending branches of the superficial cervical plexus is liable to be mistaken for the nerve.

Twelfth Pair (figs. 405, 406)

The **Twelfth** or **Hypoglossal Nerve** is the motor nerve of the tongue.

Its *superficial origin* is by several filaments from ten to fifteen in number, from the groove between the pyramidal and olivary bodies of the medulla, in a continuous line with the anterior roots of the spinal nerves. Its *deep origin* can be traced to a nucleus of grey matter (*trigonum hypoglossi*) on the floor of the fourth ventricle.

The filaments of this nerve are collected into two bundles, which perforate the dura mater separately, opposite the anterior condyloid foramen, and unite together after their passage through it. In those cases in which the anterior condyloid foramen in the occipital bone is double, these two portions of the nerve are separated by the small piece of bone which divides the foramen. The nerve descends almost vertically to a point corresponding with the angle of the jaw. It is at first deeply seated beneath the internal carotid artery and internal jugular vein, and intimately connected with the pneumogastric nerve; it then passes

FIG. 405.—Plan of the hypoglossal nerve.

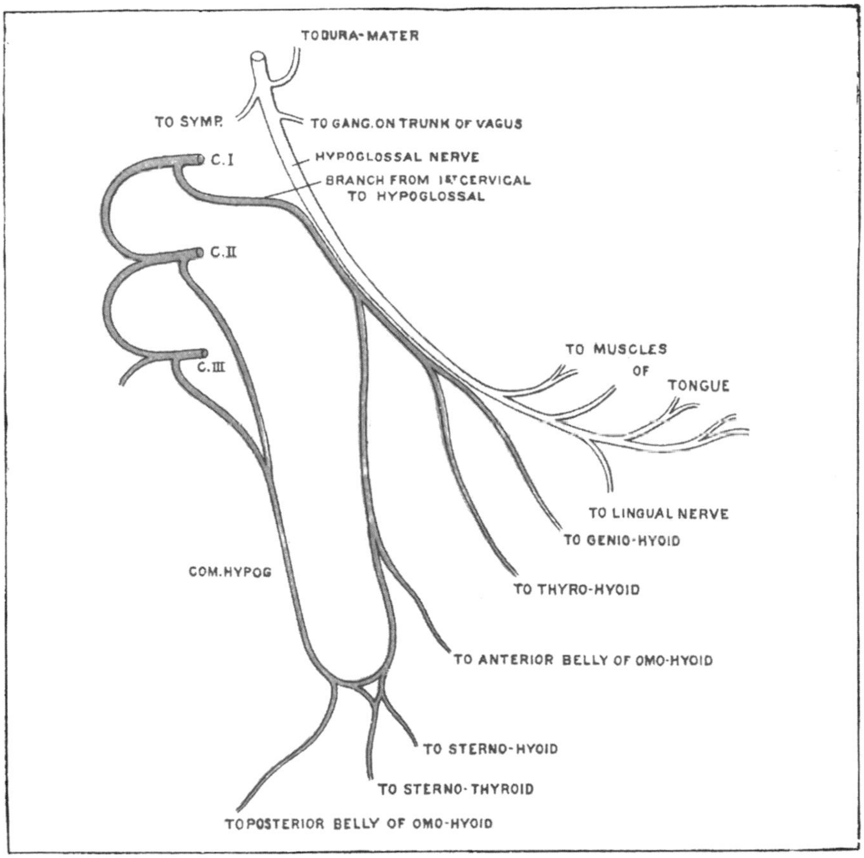

forwards between the vein and artery, and lower down in the neck becomes superficial below the Digastric muscle. The nerve then loops round the occipital artery, and crosses the external carotid and its lingual branch below the tendon of the Digastric muscle. It passes beneath the tendon of the Digastric, the Stylo-hyoid, and the Mylo-hyoid muscles, lying between the last-named muscle and the Hyo-glossus, and communicates at the anterior border of the Hyo-glossus with the lingual (gustatory) nerve; it is then continued forwards in the fibres of the Genio-hyo-glossus muscle as far as the tip of the tongue, distributing branches to its muscular substance.

The *branches of communication* are : with the

Pneumogastric.   First and second cervical nerves.
Sympathetic.   Lingual (gustatory).

The communication with the pneumogastric takes place close to the exit of the nerve from the skull, numerous filaments passing between the hypoglossal and lower ganglion of the pneumogastric through the mass of connective tissue which unites the two nerves. It also communicates with the pharyngeal plexus by a minute filament as it winds round the occipital artery.

The communication with the sympathetic takes place opposite the atlas by branches derived from the superior cervical ganglion, and in the same situation the nerve is joined by a filament derived from the loop connecting the first two cervical nerves.

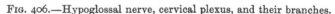

FIG. 406.—Hypoglossal nerve, cervical plexus, and their branches.

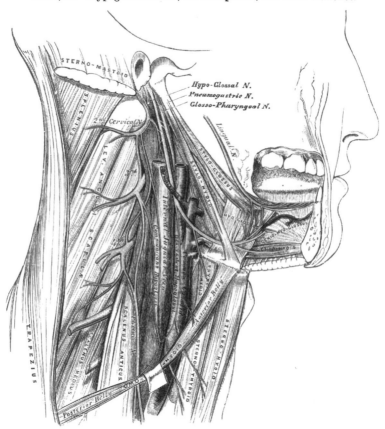

The communication with the lingual (gustatory) takes place near the anterior border of the Hyo-glossus muscle by numerous filaments which ascend upon it.

The *branches of distribution* are: the

Meningeal.   Thyro-hyoid.
Descendens hypoglossi.   Muscular.

Meningeal branches.—As the hypoglossal nerve passes through the anterior condyloid foramen it gives off, according to Luschka, several filaments to the dura mater in the posterior fossa of the base of the skull; these filaments are probably derived from a branch which passes from the first cervical nerve to the hypoglossal nerve.

The **descendens hypoglossi** is a long slender branch, which quits the hypo-glossal where it turns round the occipital artery. It consists mainly of fibres which pass to the hypoglossal from the first and second cervical nerves in the above-mentioned communication. It descends in front of or within the sheath of the carotid vessels, giving off a branch to the anterior belly of the Omo-hyoid, and then joins the communicating branches from the second and third cervical nerves, just below the middle of the neck, to form a loop, the *ansa hypoglossi*. From the convexity of this loop branches pass to supply the Sterno-hyoid, Sterno-thyroid, and the posterior belly of the Omo-hyoid. According to Arnold, another filament descends in front of the vessels into the chest, and joins the cardiac and phrenic nerves.

The **thyro-hyoid** is a small branch, arising from the hypoglossal near the posterior border of the Hyo-glossus ; it passes obliquely across the great cornu of the hyoid bone, and supplies the Thyro-hyoid muscle.

The **muscular branches** are distributed to the Stylo-glossus, Hyo-glossus, Genio-hyoid, and Genio-hyo-glossus muscles. At the under surface of the tongue numerous slender branches pass upwards into the substance of the organ to supply its intrinsic muscles.

Surgical Anatomy.—The hypoglossal nerve is an important guide in the operation of ligature of the lingual artery (see page 464). It runs forwards on the Hyo-glossus just above the great cornu of the hyoid bone, and forms the upper boundary of the triangular space in which the artery is to be sought for by cutting through the fibres of the Hyo-glossus.

THE SPINAL NERVES

The **spinal nerves** are so called because they take their origin from the spinal cord, and are transmitted through the intervertebral foramina on either side of the spinal column. There are thirty-one pairs of spinal nerves, which are arranged into the following groups, corresponding to the region of the spine through which they pass :

Cervical	8 pairs
Dorsal	12 ,,
Lumbar	5 ,,
Sacral	5 ,,
Coccygeal	1 pair

It will be observed that each group of nerves corresponds in number with the vertebræ in that region, except the cervical and coccygeal.

Each spinal nerve arises by two roots, an anterior or motor root, and a posterior or sensory root, the latter being distinguished by a ganglion, termed the *spinal* ganglion.

ROOTS OF THE SPINAL NERVES

The Anterior Roots.—The *superficial origin* is from the antero-lateral columns of the cord, corresponding to the situation of the anterior cornu of grey matter. Each root is composed of from four to eight filaments.

The *deep origin* can be traced through the antero-lateral column ; the roots, after penetrating horizontally through the longitudinal fibres of this tract, enter the grey substance, where their fibrils diverge in several directions : some, passing inwards, are continued across the anterior commissure in front of the central canal, to become continuous with the axis-cylinder processes of the large cells of the anterior cornu of the opposite side ; others terminate in the mesial group of cells of the anterior column of the same side ; other fibrils pass outwards, to become continuous with the axis-cylinder processes of the group of cells in the lateral part of the anterior column.

The Posterior Roots.—The *superficial origin* is from the postero-lateral fissure of the cord. The *real origin* of these fibres is from the nerve-cells in the posterior root ganglion, from which they can be traced into the cord in two main bundles, the course of which has already been studied (page 697).

The *anterior roots* are smaller than the posterior, devoid of ganglionic enlargement, and their component fibrils are collected into two bundles, near the intervertebral foramina.

The *posterior roots* of the nerves are larger, but the individual filaments are finer and more delicate than those of the anterior. As their component fibrils pass outwards, towards the aperture in the dura mater, they coalesce into two bundles, receive a tubular sheath from that membrane, and enter the ganglion which is developed upon each root.

The posterior root of the first cervical nerve forms an exception to these characters. It is smaller than the anterior, has occasionally no ganglion developed upon it, and when the ganglion exists it is often situated within the dura mater.

Ganglia of the Spinal Nerves

A **ganglion** is developed upon the posterior root of each of the spinal nerves. These ganglia are of an oval form, and of a reddish colour ; they bear a proportion in size to the nerves upon which they are formed, and are placed in the intervertebral foramina, external to the point where the nerves perforate the dura mater. Each ganglion is bifid internally, where it is joined by the two bundles

FIG. 407.—Diagram to show the composition of a peripheral nerve-trunk. (Böhn and Davidoff.)

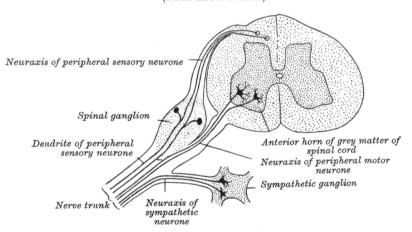

Neuraxis of peripheral sensory neurone

Spinal ganglion

Dendrite of peripheral sensory neurone

Anterior horn of grey matter of spinal cord
Neuraxis of peripheral motor neurone
Sympathetic ganglion

Nerve trunk *Neuraxis of sympathetic neurone*

of the posterior root, the two portions being united into a single mass externally· The ganglion upon the first and second cervical nerves forms an exception to these characters, being placed on the arches of the vertebræ over which the nerves pass. The ganglia of the sacral nerves are placed within the spinal canal ; and that on the coccygeal nerve, also in the canal, is situated at some distance from the origin of the posterior root.

Distribution of the Spinal Nerves

Immediately beyond the ganglion the two roots coalesce, their fibres intermingle, and the trunk thus formed constitutes the *spinal nerve*; it passes out of the intervertebral foramen, and divides into a posterior division for the supply of the posterior part of the body, and an anterior division for the supply of the anterior part of the body ; each containing fibres from both roots.

Before dividing, each spinal nerve gives off a small *recurrent* or *meningeal* branch, which is joined by a filament from the communicating branch of the sympathetic, which connects the ganglion with the anterior division. It passes inwards through the intervertebral foramen and supplies the dura mater, sending branches to the bones and ligaments.

The **posterior divisions of the spinal nerves** are generally smaller than the anterior ; they arise from the trunk resulting from the union of the roots, in the intervertebral foramina ; and, passing backwards, divide into internal and external branches, which are distributed to the muscles and integument behind the spine. The first cervical, the fourth and fifth sacral, and the coccygeal, do not divide into external and internal branches.

The **anterior divisions of the spinal nerves** supply the parts of the body in front of the spine, including the limbs. They are for the most part larger than the posterior divisions. Each division is connected by a slender filament with the sympathetic. In the dorsal region the anterior divisions of the spinal nerves are quite separate from each other, and are uniform in their distribution ; but in the cervical, lumbar, and sacral regions they form intricate plexuses previous to their distribution.

POINTS OF EMERGENCE OF THE SPINAL NERVES

The roots of the spinal nerves from their origin in the cord run obliquely downwards to their point of exit from the intervertebral foramina : the amount of obliquity varying in different regions of the spine, and being greater in the lower than the upper part. The level of their emergence from the cord is within certain limits variable, and of course does not correspond to the point of emergence of the nerve from the intervertebral foramina. The accompanying table, from Macalister, shows as accurately as can be shown the relation of these points of origin from the spinal cord to the bodies and spinous processes of the vertebræ.

Level of Body of	No of Nerve	Level of tip of Spine of	Level of Body of	No. of Nerve	Level of tip of Spine of
C. 1	C. 1	—	D. 8	D. 9	7 d.
2	2	—	9	10	8 d.
	3	1 c.	10	11	9 d.
3	4	2 c.	—	12	10 d.
4	5	3 c.	11	L. 1	11 d.
5	6	4 c.	—	2	—
6	7	5 c.	12	3	12 d.
—	8	6 c.	—	4	
7	D. 1	7 c.		5	
D. 1	2	1 d.		S. 1	—
2	3	—	L. 1	2	
3	4	2 d.		3	
4	5	3 d.		4	1 L.
5	6	4 d.	—	5	
6	7	5 d.	—	C. 1	
7	8	6 d.	L. 2	—	—

CERVICAL NERVES

The **roots of the cervical nerves** increase in size from the first to the fifth, and then remain the same size to the eighth. The posterior roots bear a proportion to the anterior of 3 to 1, which is much greater than in any other region, the individual filaments being also much larger than those of the anterior roots. The posterior root of the first cervical is an exception to this rule : it is smaller than the anterior root. In direction, the roots of the cervical are less oblique than those of the other spinal nerves. The first cervical nerve is directed a little upwards and outwards; the second is horizontal; the others are directed obliquely downwards and outwards, the lowest being the most oblique, and consequently longer than the upper, the distance between their place of origin and their point of exit from the spinal canal never exceeding the depth of one vertebra.

The *trunk of the first cervical nerve (suboccipital)* leaves the spinal canal, between the occipital bone and the posterior arch of the atlas; the second, between the posterior arch of the atlas and the lamina of the axis; and the eighth (the last), between the last cervical and first dorsal vertebræ.

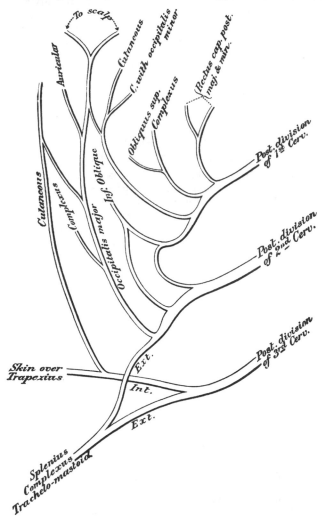

FIG. 408.—Posterior divisions of the upper cervical nerves.

Each nerve, at its exit from the intervertebral foramen, divides into a posterior and an anterior division. The anterior divisions of the four upper cervical nerves form the cervical plexus. The anterior divisions of the four lower cervical nerves, together with the first dorsal, form the brachial plexus.

POSTERIOR DIVISIONS OF THE CERVICAL NERVES (fig. 408)

The **posterior division of the first cervical,** or **sub-occipital, nerve** differs from the posterior divisions of the other cervical nerves in not dividing into an internal and external branch. It is larger than the anterior division, and escapes from the spinal canal between the occipital bone and the posterior arch of the atlas, lying beneath the vertebral artery. It enters the suboccipital triangle formed by the Rectus capitis posticus major, the Obliquus superior, and Obliquus inferior, and supplies the Recti and Obliqui muscles, and the Complexus. From the branch which supplies the Inferior oblique a filament is given off which joins the second cervical nerve. This nerve also occasionally gives off a cutaneous filament, which accompanies the occipital artery, and communicates with the occipitalis major and minor nerves.

The **posterior division of the second cervical nerve** is three or four times greater than the anterior division, and the largest of all the posterior cervical divisions. It emerges from the spinal canal between the posterior arch of the atlas and lamina of the axis, below the Inferior oblique. It supplies a twig to this muscle, and receives a communicating filament from the first cervical. It then divides into an internal and an external branch.

The *internal branch*, called, from its size and distribution, the *occipitalis major*, ascends obliquely inwards between the Obliquus inferior and Complexus, and pierces the latter muscle and the Trapezius near their attachments to the cranium. It is now joined by a filament from the posterior division of the third cervical nerve, and, ascending on the back part of the head with the occipital artery, divides into two branches, which supply the integument of the scalp as far forwards as the vertex, communicating with the occipitalis minor. It gives off an auricular branch to the back part of the ear, and muscular branches to the Complexus. The *external branch* is often joined by the external branch of the posterior division of the third, and supplies the Splenius, Trachelo-mastoid, and Complexus.

The **posterior division of the third cervical** is smaller than the preceding, but larger than the fourth; it differs from the posterior divisions of the remaining cervical nerves in supplying an additional filament, the third occipital nerve, to the integument of the occiput. The posterior division of the third nerve, like the others, divides into an internal and external branch. The *internal branch* passes between the Complexus and Semispinalis, and, piercing the Splenius and Trapezius, supplies the skin over the latter muscle; the *external branch* joins with that of the posterior division of the second to supply the Splenius, Trachelo-mastoid, and Complexus.

The *third occipital nerve* arises from the internal or cutaneous branch beneath the Trapezius; it then pierces that muscle, and supplies the skin on the lower and back part of the head. It lies to the inner side of the occipitalis major, with which it is connected.

The posterior division of the suboccipital nerve and the internal branches of the posterior divisions of the second and third cervical nerves are occasionally joined beneath the Complexus by communicating branches. This communication is described by Cruveilhier as the *posterior cervical plexus*.

The **posterior divisions of the fourth, fifth, sixth, seventh, and eighth cervical nerves** (fig. 415) pass backwards, and divide, behind the Intertransversales muscles, into internal and external branches. The *internal branches*, the larger, are distributed differently in the upper and lower part of the neck. Those derived from the fourth and fifth nerves pass between the Complexus and Semispinalis muscles, and, having reached the spinous processes, perforate the aponeurosis of the Splenius and Trapezius, and are continued outwards to the integument over the Trapezius, while those derived from the three lowest cervical nerves are the smallest, and are placed beneath the Semispinalis colli, which they supply, and then pass into the Interspinales, Multifidus spinæ and Complexus, and send twigs through this latter muscle to supply the integument near the spinous processes (Hirschfeld). The *external branches* supply the muscles at the side of the neck, viz. the Cervicalis ascendens, Transversalis colli, and Trachelo-mastoid.

ANTERIOR DIVISIONS OF THE CERVICAL NERVES

The **anterior division of the first** or **suboccipital nerve** is of small size. It escapes from the spinal canal through a groove upon the posterior arch of the atlas. In this groove it lies beneath the vertebral artery, to the inner side of the Rectus capitis lateralis. As it crosses the foramen in the transverse process of the atlas, it receives a filament from the sympathetic. It then descends, in front of this process, to communicate with an ascending branch from the second cervical nerve.

Communicating filaments from the loop between this nerve and the second join the pneumogastric, the hypoglossal, and sympathetic, and some branches are distributed to the Rectus lateralis and the two Anterior recti. The fibres which communicate with the hypoglossal, simply pass through the latter nerve to become for the most part the descendens|hypoglossi. According to Valentin, the anterior division of the suboccipital distributes filaments to the occipito-atlantal articulation, and mastoid process of the temporal bone.

The **anterior division of the second cervical nerve** escapes from the spinal canal, between the posterior arch of the atlas and the lamina of the axis, and, passing forwards on the outer side of the vertebral artery, divides, in front of the Intertransverse muscle, into an ascending branch, which joins the first cervical; and one or two descending branches which join the third. It gives off the small occipital; a branch to assist in forming the great auricular; another to assist in forming the superficial cervical; one of the communicantes hypoglossi, and a filament to the Sterno-mastoid, which communicates in the substance of the muscle with the spinal accessory.

The **anterior division of the third cervical nerve** is double the size of the preceding. At its exit from the intervertebral foramen, it passes downwards and outwards beneath the Sterno-mastoid, and divides into two branches. The ascending branch joins the anterior division of the second cervical; the descending branch passes down in front of the Scalenus anticus, and communicates with the fourth. It gives off the larger part of the great auricular and superficial cervical nerves; one of the communicantes hypoglossi; a branch to the supraclavicular nerves; a filament to assist in forming the phrenic; and muscular branches to the Levator anguli scapulæ and Trapezius; this latter nerve communicates beneath the muscle with the spinal accessory. Sometimes the nerve to the Scalenus medius is derived from this source.

The **anterior division of the fourth cervical** is of the same size as the preceding. It receives a branch from the third, sends a communicating branch to the fifth cervical, and, passing downwards and outwards, divides into numerous filaments, which cross the posterior triangle of the neck, forming the supraclavicular nerves. It gives a branch to the phrenic nerve, while it is contained in the intertransverse space, and sometimes a branch to the Scalenus medius muscle. It also gives a branch to the Levator anguli scapulæ and to the Trapezius, which unites with the branch given off from the third nerve, and communicates beneath the muscle with the spinal accessory.

The **anterior divisions of the fifth, sixth, seventh, and eighth cervical nerves** are remarkable for their size. They are much larger than the preceding nerves, and are all of equal dimensions. They assist in the formation of the brachial plexus.

CERVICAL PLEXUS

The cervical plexus (fig. 409) is formed by the anterior divisions of the four upper cervical nerves. It is situated opposite the four upper cervical vertebræ, resting upon the Levator anguli scapulæ and Scalenus medius muscles, and covered in by the Sterno-mastoid.

Its branches may be divided into two groups, *superficial* and *deep*, which may be thus arranged :

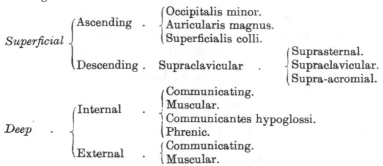

SUPERFICIAL BRANCHES OF THE CERVICAL PLEXUS

The **Occipitalis minor** (fig. 415) arises from the second cervical nerve, sometimes also from the third; it curves round the posterior border of the Sterno-mastoid, and ascends, running parallel to the posterior border of the muscle, to the back part of the side of the head. Near the cranium it perforates the deep fascia, and is continued upwards along the side of the head behind the ear, supplying the integument, and communicating with the occipitalis major, the auricularis magnus, and with the posterior auricular branch of the facial.

This nerve gives off an *auricular branch*, which supplies the integument of the upper and back part of the auricle, communicating with the mastoid branch of the auricularis magnus.

This branch is occasionally derived from the great occipital nerve. The occipitalis minor varies in size; it is occasionally double.

The **Auricularis Magnus** is the largest of the ascending branches. It arises from the second and third cervical nerves, winds round the posterior border of the Sterno-mastoid, and, after perforating the deep fascia, ascends upon that muscle beneath the Platysma to the parotid gland, where it divides into facial, auricular, and mastoid branches.

The *facial branches* pass across the parotid, and are distributed to the integument of the face over the parotid gland; others penetrate the substance of the gland, and communicate with the facial nerve.

FIG. 409.—Plan of the cervical plexus.

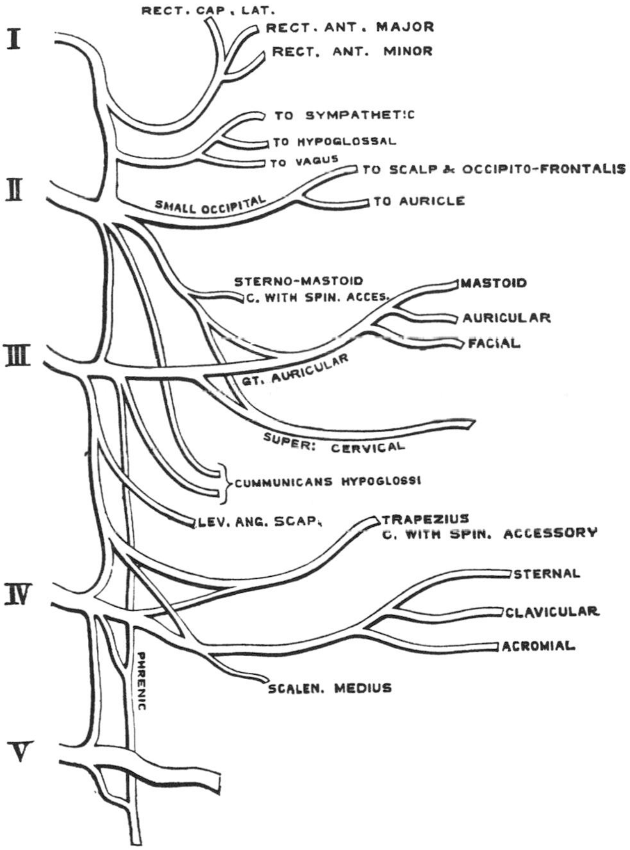

The *auricular branches* ascend to supply the integument of the back of the pinna, except at its upper part, communicating with the auricular branches of the facial and pneumogastric nerves. A filament pierces the pinna to reach its outer surface, where it is distributed to the lobule and lower part of the concha.

The *mastoid branch* communicates with the occipitalis minor and the posterior auricular branch of the facial, and is distributed to the integument behind the ear.

The **Superficialis Colli** arises from the second and third cervical nerves, turns round the posterior border of the Sterno-mastoid about its middle, and, passing obliquely forwards beneath the external jugular vein to the anterior border of the muscle, perforates the deep cervical fascia, and divides beneath the Platysma into two branches, which are distributed to the antero-lateral parts of the neck.

The *ascending branch* gives a filament which accompanies the external jugular vein; it then passes upwards to the submaxillary region, and divides into branches, some of which form a plexus with the cervical branches of the facial nerve beneath the Platysma; others pierce that muscle, and are distributed to the integument of the upper half of the neck, at its fore part, as high as the chin.

The *descending branch* (occasionally represented by two or more filaments) pierces the Platysma, and is distributed to the integument of the side and front of the neck, as low as the sternum.

The **Descending** or **supraclavicular** branches arise from the third and fourth cervical nerves: emerging beneath the posterior border of the Sterno-mastoid, they descend in the posterior triangle of the neck beneath the Platysma and deep cervical fascia. Near the clavicle they perforate the fascia and Platysma to become cutaneous, and are arranged, according to their position, into three groups.

The *inner* or *suprasternal branches* cross obliquely over the external jugular vein and the clavicular and sternal attachments of the Sterno-mastoid, and supply the integument as far as the median line. They furnish one or two filaments to the sterno-clavicular joint.

The *middle* or *supraclavicular branches* cross the clavicle, and supply the integument over the Pectoral and Deltoid muscles, communicating with the cutaneous branches of the upper intercostal nerves.

The *external* or *supra-acromial branches* pass obliquely across the outer surface of the Trapezius and the acromion, and supply the integument of the upper and back part of the shoulder.

Deep Branches of the Cervical Plexus. Internal Series

The **communicating branches** consist of several filaments, which pass from the loop between the first and second cervical nerves in front of the atlas to the pneumogastric, hypoglossal, and sympathetic; of branches from all four cervical nerves to the superior cervical ganglion of the sympathetic, together with a branch from the fourth to the fifth cervical.

Muscular branches supply the Anterior recti and Rectus lateralis muscles; they proceed from the first cervical nerve, and from the loop formed between it and the second.

The **Communicantes Hypoglossi** (fig. 406) consist usually of two filaments, one being derived from the second, and the other from the third cervical. These filaments pass downwards on the outer side of the internal jugular vein, cross in front of the vein a little below the middle of the neck, and form a loop with the descendens hypoglossi in front of the sheath of the carotid vessels (see page 737). Occasionally, the junction of these nerves takes place within the sheath.

The **Phrenic Nerve** (*internal respiratory of Bell*) arises chiefly from the fourth cervical nerve with a few filaments from the third and a communicating branch from the fifth. It descends to the root of the neck, running obliquely across the front of the Scalenus anticus and beneath the Sterno-mastoid, the posterior belly of the Omo-hyoid, and the Transversalis colli and suprascapular vessels. It next passes over the first part of the subclavian artery, between it and the subclavian vein, and, as it enters the chest, crosses the internal mammary artery near its origin. Within the chest, it descends nearly vertically in front of the root of the lung, and by the side of the pericardium, between it and the mediastinal portion of the pleura, to the Diaphragm, where it divides into branches, which separately pierce that muscle, and are distributed to its under surface.

The two phrenic nerves differ in their length, and also in their relations at the upper part of the thorax.

The *right nerve* is situated more deeply, and is shorter and more vertical in direction than the left; it lies on the outer side of the right vena innominata and superior vena cava.

The *left nerve* is rather longer than the right, from the inclination of the heart to the left side, and from the Diaphragm being lower on this than on the opposite side. It enters the thorax behind the left innominate vein, and crosses in front of the vagus and the arch of the aorta and the root of the lung. In the thorax each phrenic nerve is accompanied by a branch of the internal mammary artery, the comes nervi phrenici.

Each nerve supplies filaments to the pericardium and pleura, and near the chest is joined by a filament from the sympathetic, and, occasionally, by one from the union of the descendens hypoglossi with the spinal nerves: this filament is found, according to Swan, only on the left side. It frequently receives a filament from the nerve to the Subclavius muscle. Branches have been described as passing to the peritoneum.

From the *right nerve*, one or two filaments pass to join in a small ganglion with phrenic branches of the solar plexus: and branches from this ganglion are distributed to the hepatic plexus, the suprarenal capsule, and inferior vena cava. From the *left nerve*, filaments pass to join the phrenic plexus of the sympathetic, but without any ganglionic enlargement.

Deep Branches of the Cervical Plexus. External Series

Communicating branches.—The deep branches of the external series of the cervical plexus communicate with the spinal accessory nerve, in the substance of the Sterno-mastoid muscle, in the posterior triangle, and beneath the Trapezius.

Muscular branches are distributed to the Sterno-mastoid, Trapezius, Levator anguli scapulæ, and Scalenus medius.

The branch for the Sterno-mastoid is derived from the second cervical, the Trapezius and Levator anguli scapulæ receive branches from the third and fourth. The Scalenus medius is derived sometimes from the third, sometimes from the fourth, and occasionally from both nerves.

The Brachial Plexus (fig. 410)

The **Brachial Plexus** is formed by the union of the anterior divisions of the four lower cervical and the greater part of the first dorsal nerves, receiving usually a fasciculus from the fourth cervical nerve, and frequently one from the second dorsal nerve. It extends from the lower part of the side of the neck to the axilla. It is very broad and presents little of a plexiform arrangement at its commencement. It is narrow opposite the clavicle, becomes broad, and forms a more dense interlacement in the axilla, and divides opposite the coracoid process into numerous branches for the supply of the upper limb. The nerves which form the plexus are all similar in size, and their mode of communication is subject to considerable variation, so that no one plan can be given as applying to every case. The following appears, however, to be the most constant arrangement: The fifth and sixth cervical unite together soon after their exit from the intervertebral foramina to form a common trunk. The eighth cervical and first dorsal also unite to form one trunk. So that the nerves forming the plexus, as they lie on the Scalenus medius, external to the outer border of the Scalenus anticus, are blended into three trunks: an upper one formed by the junction of the fifth and sixth cervical nerves; a middle one, consisting of the seventh cervical nerve; and a lower one, formed by the junction of the eighth cervical and first dorsal nerves. As they pass beneath the clavicle, each of these three trunks divides into two branches, an *anterior* and a *posterior*.* The anterior divisions of the upper and middle

---

* The posterior division of the lower trunk is very much smaller than the others, and is frequently derived entirely from the eighth cervical nerve.

trunks then unite to form a common cord, which is situated on the outer side of the middle part of the axillary artery, and is called the *outer cord* of the brachial plexus. The anterior division of the lower trunk passes down on the inner side of the axillary artery in the middle of the axilla, and forms the *inner cord* of the brachial plexus. The posterior divisions of all three trunks unite to form the *posterior cord* of the brachial plexus, which is situated behind the second portion of the axillary artery. From this posterior cord are given off the two lower subscapular nerves; the upper subscapular nerve being given off from the posterior division of the upper trunk prior to its junction with the posterior divisions of the lower and middle trunks. The posterior cord divides into the circumflex and musculo-spiral nerves.

FIG. 410.—Plan of the brachial plexus.

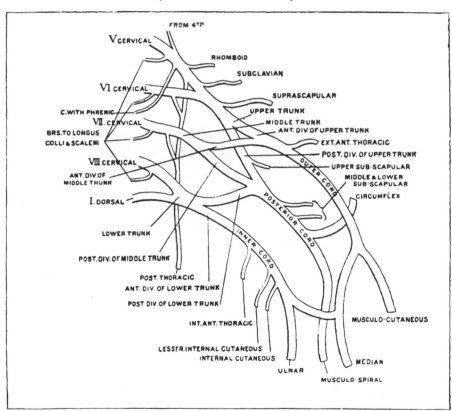

The brachial plexus communicates with the cervical plexus by a branch from the fourth to the fifth nerve, and with the phrenic nerve by a branch from the fifth cervical, which joins that nerve on the Anterior scalenus muscle: the fifth and sixth cervical nerves are joined by filaments to the middle cervical ganglion of the sympathetic, the seventh and eighth cervical to its inferior ganglion, and the first dorsal nerve to its first thoracic ganglion, close to their exit from the intervertebral foramina.

Relations.—*In the neck,* the brachial plexus lies in the posterior triangle, being covered by the skin, Platysma, and deep fascia: it is crossed by the posterior belly of the Omo-hyoid and by the Transversalis colli artery. When the posterior scapular artery arises from the third part of the subclavian, it usually passes between the roots of the plexus. It lies at first between the Anterior and Middle scaleni muscles, and then above and to the outer side of the subclavian artery: it next passes behind the clavicle and Subclavius muscle, lying upon the first serration of the Serratus magnus, and the Subscapularis muscles. *In the axilla* it is placed on the outer side of the first portion of the axillary artery; it surrounds the artery in the second part of its course, one cord lying upon the outer side of that vessel, one on the inner side, and one behind it; and at the lower part of the axillary space gives off its terminal branches to the upper extremity.

Branches.—The branches of the brachial plexus are arranged into two groups, viz. those given off above the clavicle, and those below that bone.

BRANCHES ABOVE THE CLAVICLE

Communicating.	Posterior thoracic.
Muscular.	Suprascapular.

The **communicating branch** with the phrenic is derived from the fifth cervical nerve or from the loop between the fifth and sixth; it joins the phrenic on the Anterior scalenus muscle. The communications with the sympathetic have already been referred to.

The **muscular branches** supply the Longus colli, Scaleni, Rhomboidei, and Subclavius muscles. Those for the Longus colli and Scaleni arise from the four lower cervical nerves at their exit from the intervertebral foramina. The rhomboid branch arises from the fifth cervical, pierces the Scalenus medius, and passes beneath the Levator anguli scapulæ, which it occasionally supplies, to the Rhomboid muscles. The nerve to the Subclavius is a small filament, which arises from the fifth cervical at its point of junction with the sixth nerve; it descends in front of the third part of the subclavian artery to the Subclavius muscle, and is usually connected by a filament with the phrenic nerve.

The **posterior thoracic nerve** (*long thoracic, external respiratory of Bell*) (fig. 413) supplies the Serratus magnus, and is remarkable for the length of its course. It sometimes arises by two roots, from the fifth and sixth cervical nerves immediately after their exit from the intervertebral foramina, but generally by three roots from the fifth, sixth, and seventh nerves. These unite in the substance of the Middle scalenus muscle, and, after emerging from it, the nerve passes down behind the brachial plexus and the axillary vessels, resting on the outer surface of the Serratus magnus. It extends along the side of the chest to the lower border of that muscle, supplying filaments to each of its digitations.

The **suprascapular nerve** (fig. 414) arises from the cord formed by the fifth and sixth cervical nerves; passing obliquely outwards beneath the Trapezius and the Omo-hyoid, it enters the supraspinous fossa below the transverse or· supra-scapular ligament, and, passing beneath the Supraspinatus muscle, curves round the external border of the spine of the scapula to the infraspinous fossa. In the supraspinous fossa it gives off two branches to the Supraspinatus muscle, and an articular filament to the shoulder-joint; and in the infraspinous fossa it gives off two branches to the Infraspinatus muscle, besides some filaments to the shoulder-joint and scapula.

BRANCHES BELOW THE CLAVICLE

The branches given off below the clavicle are derived from the three cords of the brachial plexus, in the following manner:

From the outer cord arise the external anterior thoracic nerve, the musculo-cutaneous, and the outer head of the median.

From the inner cord arise the internal anterior thoracic nerve, the internal cutaneous, the lesser internal cutaneous (nerve of Wrisberg), the ulnar and inner head of the median.

From the posterior cord arise two of the three subscapular nerves, the third taking origin from the posterior division of the trunk formed by the fifth and sixth cervical nerves; the cord then divides into the musculo-spiral and circumflex nerves.

These may be arranged according to the parts they supply:

To the chest . . . . . Anterior thoracic.

To the shoulder . . . .
{ Subscapular.
{ Circumflex.

To the arm, forearm, and hand
⎧ Musculo-cutaneous.
⎪ Internal cutaneous.
⎪ Lesser internal cutaneous.
⎨ Median.
⎪ Ulnar.
⎩ Musculo-spiral.

The fasciculi of which these nerves are composed may be traced through the plexus to the spinal nerves from which they originate. They are as follows:

External anterior thoracic from 5th, 6th, and 7th cervical.
Internal anterior thoracic ,,   8th cervical and 1st dorsal.
Subscapular   ,,   5th, 6th, 7th, and 8th cervical.
Circumflex   ,,   5th and 6th cervical.
Musculo-cutaneous   ,,   5th and 6th cervical.
Internal cutaneous   ,,   8th cervical and 1st dorsal.
Lesser internal cutaneous   ,,   1st dorsal.
Median   ,,   6th, 7th, and 8th cervical, and 1st dorsal.
Ulnar   ,,   8th cervical and 1st dorsal.
Musculo-spiral   ,,   6th, 7th, and 8th cervical; sometimes also
                                 from the 5th.

The **Anterior Thoracic Nerves** (fig. 413), two in number, supply the Pectoral muscles.

The *external* or superficial nerve, the larger of the two, arises from the outer cord of the brachial plexus, through which its fibres may be traced to the fifth, sixth, and seventh cervical nerves. It passes inwards, across the axillary artery and vein, pierces the costo-coracoid membrane, and is distributed to the under surface of the Pectoralis major. It sends down a communicating filament to join the internal nerve, which forms a loop round the inner side of the axillary artery.

The *internal* or deep nerve arises from the inner cord, and through it from the eighth cervical and first dorsal. It passes behind the first part of the axillary artery, then curves forwards between the axillary artery and vein, and joins with the filament from the anterior nerve. It then passes to the under surface of the Pectoralis minor muscle, where it divides into a number of branches, which supply the muscle on its under surface. Some two or three branches pass through the muscle to supply the Pectoralis major.

The **Subscapular Nerves,** three· in number, supply the Subscapularis, Teres major, and Latissimus dorsi muscles. The fasciculi of which they are composed may be traced to the fifth, sixth, seventh, and eighth cervical nerves.

The *upper subscapular nerve*, the smallest, enters the upper part of the Subscapularis muscle; this nerve is frequently represented by two branches.

The *lower subscapular nerve* enters the axillary border of the Subscapularis, and terminates in the Teres major. The latter muscle is sometimes supplied by a separate branch.

The *middle* or *long subscapular*, the largest of the three, follows the course of the subscapular artery, along the posterior wall of the axilla to the Latissimus dorsi, through which it may be traced as far as its lower border.

The **Circumflex Nerve** (fig. 414) supplies some of the muscles, and the integument of the shoulder, and the shoulder-joint. It arises from the posterior cord of the brachial plexus, in common with the musculo-spiral nerve, and its fibres may be traced through the posterior cord to the fifth and sixth cervical nerves.

It is at first placed behind the axillary artery, between it and the Subscapularis muscle, and passes downwards and outwards to the lower border of that muscle. It then winds backwards, in company with the posterior circumflex artery, through a quadrilateral space, bounded above by the Teres minor, below by the Teres

FIG. 411.—Cutaneous nerves of right upper extremity.   Anterior view.

FIG. 412.—Cutaneous nerves of right upper extremity.   Posterior view.

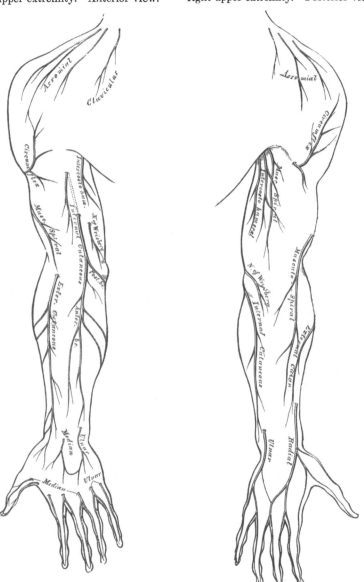

major, internally by the long head of the Triceps, and externally by the neck of the humerus, and divides into two branches.

The *upper branch* winds round the surgical neck of the humerus, beneath the Deltoid, with the posterior circumflex vessels, as far as the anterior border of that muscle, supplying it, and giving off cutaneous branches, which pierce the muscle and ramify in the integument covering its lower part.

The *lower branch*, at its origin, distributes filaments to the Teres minor and back part of the Deltoid muscles. Upon the filament to the former muscle an oval enlargement usually exists. The nerve then pierces the deep fascia, and supplies the integument over the lower two-thirds of the posterior surface of the Deltoid, as well as that covering the long head of the Triceps.

The circumflex nerve, before its division, gives off an articular filament, which enters the shoulder-joint below the Subscapularis.

The **Musculo-cutaneous Nerve** (fig. 413) (*external cutaneous* or *perforans Casserii*)* supplies some of the muscles of the arm, and the integument of the forearm. It arises from the outer cord of the brachial plexus, opposite the lower border of the Pectoralis minor, receiving filaments from the fifth and sixth cervical nerves. It perforates the Coraco-brachialis muscle, passes obliquely between the Biceps and Brachialis anticus, to the outer side of the arm, and, a little above the elbow, winds round the outer border of the tendon of the Biceps, and, perforating the deep fascia, becomes cutaneous. This nerve, in its course through the arm, supplies the Coraco-brachialis, Biceps, and the greater part of the Brachialis anticus muscles. The branch to the Coraco-brachialis is given off from the nerve close to its origin, and in some instances, especially in early life, as a separate filament from the outer cord of the plexus. The branches to the Biceps and Brachialis anticus are given off after the nerve has pierced the Coraco-brachialis. The nerve also sends a small branch to the bone, which enters the nutrient foramen with the accompanying artery, and a filament, from the branch supplying the Brachialis anticus, to the elbow-joint.

The cutaneous portion of the nerve passes behind the median cephalic vein, and divides, opposite the elbow-joint, into an anterior and a posterior branch.

The *anterior branch* descends along the radial border of the forearm to the wrist, and supplies the integument over the outer half of its anterior surface. At the wrist-joint it is placed in front of the radial artery, and some filaments, piercing the deep fascia, accompany that vessel to the back of the wrist, supplying the carpus. The nerve then passes downwards to the ball of the thumb, where it terminates in cutaneous filaments. It communicates with a branch from the radial nerve, and with the palmar cutaneous branch of the median.

The *posterior branch* passes downwards, along the back part of the radial side of the forearm, to the wrist. It supplies the integument of the lower third of the forearm, communicating with the radial nerve and the external cutaneous branch of the musculo-spiral.

The musculo-cutaneous nerve presents frequent irregularities. It may adhere for some distance to the median and then pass outwards, beneath the Biceps, instead of through the Coraco-brachialis. Frequently some of the fibres of the median run for some distance in the musculo-cutaneous and then leave it to join their proper trunk. Less frequently the reverse is the case, and the median sends a branch to join the musculo-cutaneous. Instead of piercing the Coraco-brachialis the nerve may pass under it or through the Biceps. Occasionally it gives a filament to the Pronator teres, and it has been seen to supply the back of the thumb when the radial nerve was absent.

The **Internal Cutaneous Nerve** (fig. 413) is one of the smallest branches of the brachial plexus. It arises from the inner cord in common with the ulnar and internal head of the median, and, at its commencement, is placed on the inner side of the axillary, and afterwards of the brachial artery. It derives its fibres from the eighth cervical and first dorsal nerves. It passes down the inner side of the arm, pierces the deep fascia with the basilic vein, about the middle of the limb, and, becoming cutaneous, divides into two branches, anterior and posterior.

* See footnote, p. 706.

FIG. 413.—Nerves of the left upper extremity.

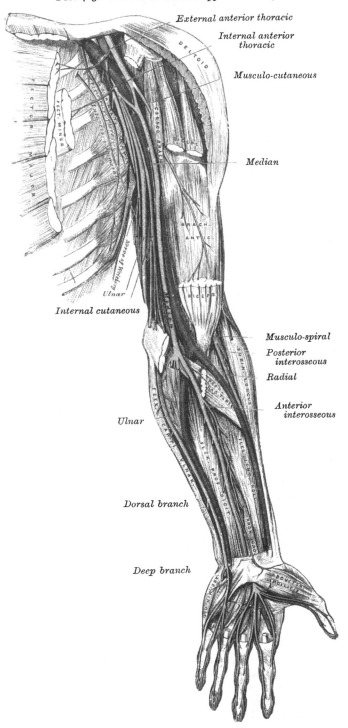

External anterior thoracic

Internal anterior thoracic

Musculo-cutaneous

Median

Musculo-spiral

Posterior interosseous

Radial

Anterior interosseous

Internal cutaneous

Ulnar

Ulnar

Dorsal branch

Deep branch

This nerve gives off, near the axilla, a cutaneous filament, which pierces the fascia and supplies the integument covering the Biceps muscle, nearly as far as the elbow. This filament lies a little external to the common trunk, from which it arises.

The *anterior branch*, the larger of the two, passes usually in front of, but occasionally behind, the median basilic vein. It then descends on the anterior surface of the ulnar side of the forearm, distributing filaments to the integument as far as the wrist, and communicating with a cutaneous branch of the ulnar nerve.

The *posterior branch* passes obliquely downwards on the inner side of the basilic vein, in front of, or over, the internal condyle of the humerus to the back of the forearm, and descends on the posterior surface of its ulnar side as far as the wrist, distributing filaments to the integument. It communicates, above the elbow, with the lesser internal cutaneous, and above the wrist with the dorsal cutaneous branch of the ulnar nerve (Swan).

The **Lesser Internal Cutaneous Nerve** (*nerve of Wrisberg*) is distributed to the integument on the inner side of the arm (fig. 413). It is the smallest of the branches of the brachial plexus, and arising from the inner cord, with the internal cutaneous and ulnar nerves, receives its fibres from the first dorsal nerve. It passes through the axillary space, at first lying behind, and then on the inner side of the axillary vein, and communicates with the intercosto-humeral nerve. It descends along the inner side of the brachial artery to the middle of the arm, where it pierces the deep fascia, and is distributed to the integument of the back part of the lower third of the arm, extending as far as the elbow, where some filaments are lost in the integument in front of the inner condyle, and others over the olecranon. It communicates with the posterior branch of the internal cutaneous nerve.

In some cases the nerve of Wrisberg and intercosto-humeral are connected by two or three filaments, which form a plexus at the back part of the axilla. In other cases, the intercosto-humeral is of large size, and takes the place of the nerve of Wrisberg, receiving merely a filament of communication from the brachial plexus, which represents the latter nerve. In other cases, this filament is wanting, the place of the nerve of Wrisberg being supplied entirely from the intercosto-humeral.

The **Median Nerve** (fig. 413) has received its name from the course it takes along the middle of the arm and forearm to the hand, lying between the ulnar and the musculo-spiral and radial nerves. It arises by two roots, one from the outer, and one from the inner cord of the brachial plexus ; these embrace the lower part of the axillary artery, uniting either in front or on the outer side of that vessel. It receives filaments from the sixth, seventh, and eighth cervical and the first dorsal nerves. As it descends through the arm, it lies at first on the outer side of the brachial artery, crosses that vessel in the middle of its course, usually in front, but occasionally behind it, and lies on its inner side to the bend of the elbow, where it is placed beneath the bicipital fascia, and is separated from the elbow-joint by the Brachialis anticus. *In the forearm* it passes between the two heads of the Pronator radii teres, and descends beneath the Flexor sublimis, lying on the Flexor profundus, to within two inches above the annular ligament, where it becomes more superficial, lying between the tendons of the Flexor sublimis and Flexor carpi radialis, beneath, and rather to the radial side of, the tendon of the Palmaris longus, covered by the integument and fascia. It then passes beneath the annular ligament into the hand. In its course through the forearm it is accompanied by a branch of the anterior interosseous artery.

Branches.—With the exception of the nerve to the Pronator teres, which some-times arises above the elbow-joint, the median nerve gives off no branches in the arm. *In the forearm* its branches are, muscular, anterior interosseous, and palmar cutaneous, and, according to Rüdinger and Macalister, two articular twigs to the elbow-joint.

The *muscular branches* supply all the superficial muscles on the front of the forearm, except the Flexor carpi ulnaris. These branches are derived from the nerve near the elbow.

The *anterior interosseous* supplies the deep muscles on the front of the forearm, except the inner half of the Flexor profundus digitorum. It accompanies the anterior interosseous artery along the interosseous membrane, in the interval between the Flexor longus pollicis and Flexor profundus digitorum muscles, both of which it supplies, and terminates below in the Pronator quadratus and wrist-joint.

The *palmar cutaneous branch* arises from the median nerve at the lower part of the forearm. It pierces the fascia above the annular ligament, and descending over that ligament, divides into two branches: of which the *outer* supplies the skin over the ball of the thumb, and communicates with the anterior cutaneous branch of the musculo-cutaneous nerve; and the *inner* supplies the integument of the palm of the hand, communicating with the cutaneous branch of the ulnar.

In the palm of the hand, the median nerve is covered by the integument and palmar fascia, and crossed by the superficial palmar arch. It rests upon the tendons of the flexor muscles. In this situation it becomes enlarged, somewhat flattened, of a reddish colour, and divides into two branches. Of these, the *external* supplies a muscular branch to some of the muscles of the thumb, and digital branches to the thumb and index finger; the *internal* supplies digital branches to the contiguous sides of the index and middle, and of the middle and ring fingers.

The *branch to the muscles of the thumb* is a short nerve, which divides to supply the Abductor, Opponens, and the superficial head of the Flexor brevis pollicis muscles; the remaining muscles of this group being supplied by the ulnar nerve.

The *digital branches* are five in number. The *first* and *second* pass along the borders of the thumb, the external branch communicating with branches of the radial nerve. The *third* passes along the radial side of the index finger, and supplies the First lumbrical muscle. The *fourth* subdivides to supply the adjacent sides of the index and middle fingers, and sends a branch to the Second lumbrical muscle. The *fifth* supplies the adjacent sides of the middle and ring fingers, and communicates with a branch from the ulnar nerve.

Each digital nerve, opposite the base of the first phalanx, gives off a dorsal branch, which joins the dorsal digital nerve from the radial, and runs along the side of the dorsum of the finger, to end in the integument over the last phalanx. At the end of the finger, the digital nerve divides into a palmar and a dorsal branch: the former of which supplies the extremity of the finger, and the latter ramifies round and beneath the nail. The digital nerves, as they run along the fingers, are placed superficially to the digital arteries.

The **Ulnar nerve** (fig. 413) is placed along the inner or ulnar side of the upper limb, and is distributed to the muscles and integument of the forearm and hand. It is smaller than the median, behind which it is placed, diverging from it in its course down the arm. It arises from the inner cord of the brachial plexus, in common with the inner head of the median and the internal cutaneous nerve, and derives its fibres from the eighth cervical and first dorsal nerves. At its commencement it lies to the inner side of the axillary artery, and holds the same relation with the brachial artery to the middle of the arm. From this point it runs obliquely across the internal head of the Triceps, pierces the internal inter-muscular septum, and descends to the groove between the internal condyle and the olecranon, accompanied by the inferior profunda artery. *At the elbow*, it rests upon the back of the inner condyle, and passes into the forearm between the two heads of the Flexor carpi ulnaris. *In the forearm*, it descends in a perfectly straight course along its ulnar side, lying upon the Flexor profundus digitorum, its upper half being covered by the Flexor carpi ulnaris, its lower half lying on the outer side of the muscle, covered by the integument and fascia.

The ulnar artery, in the upper third of its course, is separated from the ulnar nerve by a considerable interval ; but in the rest of its extent the nerve lies to its inner side. *At the wrist* the ulnar nerve crosses the annular ligament on the outer side of the pisiform bone, to the inner side of, and a little behind the ulnar artery, and immediately beyond this bone divides into two branches, superficial and deep palmar.

The branches of the ulnar nerve are :

In the forearm
- Articular (elbow).
- Muscular.
- Cutaneous.
- Dorsal cutaneous.
- Articular (wrist).

In the hand
- Superficial palmar.
- Deep palmar.

The *articular branches* to the elbow-joint consist of several small filaments. They arise from the nerve as it lies in the groove between the inner condyle and olecranon.

The *muscular branches* are two in number : one supplying the Flexor carpi ulnaris ; the other, the inner half of the Flexor profundus digitorum. They arise from the trunk of the nerve near the elbow.

The *cutaneous branch* arises from the ulnar nerve about the middle of the forearm, and divides into two branches.

One branch (frequently absent) pierces the deep fascia near the wrist, and is distributed to the integument, communicating with a branch of the internal cutaneous nerve.

The second branch (*palmar cutaneous*) lies on the ulnar artery, which it accompanies to the hand, some filaments entwining round the vessel ; it ends in the integument of the palm, communicating with branches of the median nerve.

The *dorsal cutaneous branch* arises about two inches above the wrist ; it passes backwards beneath the Flexor carpi ulnaris, perforates the deep fascia, and, running along the ulnar side of the back of the wrist and hand, divides into branches ; one of these supplies the inner side of the little finger ; a second supplies the adjacent sides of the little and ring fingers ; a third joins the branch of the radial nerve which supplies the adjoining sides of the middle and ring fingers, and assists in supplying them ; a fourth is distributed to the metacarpal region of the hand, communicating with a branch of the radial nerve.

On the little finger the dorsal digital branches only extend as far as the base of the terminal phalanx, and on the ring finger as far as the base of the second phalanx ; the more distal parts of these digits are supplied by dorsal branches derived from the palmar digital branches of the ulnar.

The *superficial palmar branch* supplies the Palmaris brevis, and the integument on the inner side of the hand, and terminates in two digital branches, which are distributed, one to the ulnar side of the little finger, the other to the adjoining sides of the little and ring fingers, the latter communicating with a branch from the median. The digital branches are distributed to the fingers in the same manner as the digital branches of the median.

The *deep palmar branch*, accompanied by the deep branch of the ulnar artery, passes between the Abductor and Flexor brevis minimi digiti muscles ; it then perforates the Opponens minimi digiti and follows the course of the deep palmar arch beneath the flexor tendons. At its origin it supplies the muscles of the little finger. As it crosses the deep part of the hand, it sends two branches to each interosseous space, one for the Dorsal and one for the Palmar interosseous muscle, the branches to the Second and Third palmar interossei supplying filaments to the two inner Lumbrical muscles. At its termination between the thumb and index finger, it supplies the Adductores transversus et obliquus pollicis and the inner head of the Flexor brevis pollicis. It also sends articular filaments to the wrist-joint.

It will be remembered that the inner part of the Flexor profundus digitorum is supplied by the ulnar nerve; the two inner Lumbricales, which are connected with the tendons of this part of the muscle, are therefore supplied by the same nerve. The outer part of the Flexor profundus is supplied by the median nerve: the two outer Lumbricales, which are connected with the tendons of this part of the muscle, are therefore supplied by the same nerve. Brooks states that in twelve instances out of twenty-one he found that the third lumbrical received a twig from the median nerve, in addition to its branch from the ulnar.

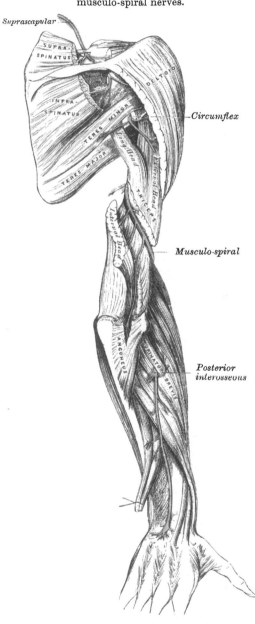

FIG. 414.—The suprascapular, circumflex, and musculo-spiral nerves.

The **Musculo-spiral nerve** (fig. 414), the largest branch of the brachial plexus, supplies the muscles of the back part of the arm and forearm, and the integument of the same parts, as well as that of the back of the hand. It arises from the posterior cord of the brachial plexus, of which it may be regarded as the continuation. It receives filaments from the sixth, seventh, and eighth, and sometimes also from the fifth cervical nerves. At its commencement it is placed behind the axillary and upper part of the brachial arteries, passing down in front of the tendons of the Latissimus dorsi and Teres major. It winds round the humerus in the musculo-spiral groove with the superior profunda artery, passing from the inner to the outer side of the bone, between the internal and external heads of the Triceps muscle. It pierces the external intermuscular septum, and descends between the Brachialis anticus and Supinator longus to the front of the external condyle, where it divides into the radial and posterior interosseous nerves.

The branches of the musculo-spiral nerve are:

Muscular.
Cutaneous.
Radial.
Posterior interosseous.

The *muscular branches* are divided into internal, posterior, and external; they supply the Triceps, Anconeus, Supinator longus, Extensor carpi radialis longior, and Brachialis anticus. These branches are derived from the nerve, at the inner side, back part, and outer side of the arm.

The internal muscular branches supply the inner and middle heads of the Triceps muscle. That to the inner head of the Triceps is a long, slender filament, which lies close to the ulnar nerve, as far as the lower third of the arm, and is therefore frequently spoken of as the *ulnar collateral.*

The posterior muscular branch, of large size, arises from the nerve in the groove between the Triceps and the humerus. It divides into branches, which supply the inner and outer heads of the Triceps and the Anconeus muscles. The branch for the latter muscle is a long, slender filament, which descends in the substance of the Triceps to the Anconeus.

The external muscular branches supply the Supinator longus, Extensor carpi radialis longior, and (usually) the outer part of the Brachialis anticus.

The *cutaneous branches* are three in number, one internal and two external.

The internal cutaneous branch arises in the axillary space, with the inner muscular branch. It is of small size, and passes through the axilla to the inner side of the arm, supplying the integument on its posterior aspect nearly as far as the olecranon. In its course it crosses beneath the intercosto-humeral, with which it communicates.

The two external cutaneous branches perforate the outer head of the Triceps, at its attachment to the humerus. The upper and smaller one passes to the front of the elbow, lying close to the cephalic vein, and supplies the integument of the lower half of the arm on its anterior aspect. The lower branch pierces the deep fascia below the insertion of the Deltoid, and passes down along the outer side of the arm and elbow, and then along the back part of the radial side of the forearm to the wrist, supplying the integument in its course, and joining, near its termination, with the posterior cutaneous branch of the musculo-cutaneous nerve.

The **radial nerve** passes along the front of the radial side of the forearm to the commencement of its lower third. It lies at first a little to the outer side of the radial artery, concealed beneath the Supinator longus. In the middle third of the forearm, it lies beneath the same muscle, in close relation with the outer side of the artery. It quits the artery about three inches above the wrist, passes beneath the tendon of the Supinator longus, and, piercing the deep fascia at the outer border of the forearm, divides into two branches.

The external branch, the smaller of the two, supplies the integument of the radial side and ball of the thumb, joining with the anterior branch of the musculo-cutaneous nerve.

The internal branch communicates, above the wrist, with the posterior cutaneous branch from the musculo-cutaneous, and, on the back of the hand, forms an arch with the dorsal cutaneous branch of the ulnar nerve. It then divides into four digital nerves, which are distributed as follows : the first supplies the ulnar side of the thumb ; the second, the radial side of the index finger ; the third, the adjoining sides of the index and middle fingers ; and the fourth, the adjacent borders of the middle and ring fingers.* The latter nerve communicates with a filament from the dorsal branch of the ulnar nerve.

The **posterior interosseous nerve** winds to the back of the forearm round the outer side of the radius, passes between the two planes of fibres of the Supinator brevis, and is prolonged downwards between the superficial and deep layer of muscles, to the middle of the forearm. Considerably diminished in size, it descends on the interosseous membrane, beneath the Extensor longus pollicis, to the back of the carpus, where it presents a gangliform enlargement from which filaments are distributed to the ligaments and articulations of the carpus. It supplies all the muscles of the radial and posterior brachial regions, excepting the Anconeus, Supinator longus, and Extensor carpi radialis longior.

---

* According to Hutchinson, the digital nerve to the thumb reaches only as high as the root of the nail : the one to the forefinger as high as the middle of the second phalanx : and the one to the middle and ring fingers not higher than the first phalangeal joint. (*London Hos. Gaz.* vol. iii. p. 319.)

Surgical Anatomy.—The brachial plexus may be ruptured by traction on the limb, leading to complete paralysis. In these cases the lesion would appear to be rather a tearing away of the nerves from the spinal cord, than a solution of continuity of the nerve-fibres themselves. In the axilla, any of the nerves forming the brachial plexus may be injured in a wound of this part, the median being the one which is most frequently damaged from its exposed position; and the musculo-spiral, on account of its sheltered and deep position, being the least often wounded. The brachial plexus in the axilla is often damaged from the pressure of a crutch, producing the condition known as ' crutch paralysis.' In these cases the musculo-spiral appears most frequently to be the nerve which is chiefly implicated; the ulnar nerve being the one that appears to suffer next in frequency.

The *circumflex nerve* is of particular surgical interest. On account of its course round the surgical neck of the humerus, it is liable to be torn in fractures of this part of the bone, and in dislocations of the shoulder-joint, leading to paralysis of the deltoid, and, according to Erb, inflammation of the shoulder-joint is liable to be followed by a neuritis of this nerve from extension of the inflammation to it.

Mr. Hilton takes the circumflex nerve as an illustration of a law which he lays down, that ' the same trunks of nerves whose branches supply the groups of muscles moving a joint, furnish also a distribution of nerves to the skin over the insertions of the same muscles, and the interior of the joint receives its nerves from the same source.' In this way he explains the fact that an inflamed joint becomes rigid, because the same nerves which supply the interior of the joint supply the muscles also which move that joint.

The *median nerve* is liable to injury in wounds of the forearm. When paralysed, there is loss of flexion of the second phalanges of all the fingers, and of the terminal phalanges of the index and middle fingers. Flexion of the terminal phalanges of the ring and little fingers is effected by that portion of the Flexor profundus digitorum which is supplied by the ulnar nerve. There is power to flex the proximal phalanges through the Interossei. The thumb cannot be flexed or opposed, and is maintained in a position of extension and adduction. All power of pronation is lost. The wrist can be flexed, if the hand is first adducted, by the action of the Flexor carpi ulnaris. There is loss or impairment of sensation on the palmar surface of the thumb, index, middle, and outer half of ring fingers, and on the dorsal surface of the same fingers over the last two phalanges; except in the thumb, where the loss of sensation would be limited to the back of the last phalanx. In order to expose the median nerve, for the purpose of stretching, an incision should be made along the radial side of the tendon of the Palmaris longus, which serves as a guide to the nerve.

The *ulnar nerve* is also liable to be injured in wounds of the forearm. When paralysed, there is loss of power of flexion in the ring and little fingers; there is impaired power of ulnar flexion and adduction; there is inability to spread out the fingers from paralysis of the Interossei; and there is inability to adduct the thumb. Sensation is lost, or impaired, in the skin supplied by the nerve. In order to expose the nerve in the lower part of the forearm, an incision should be made along the outer border of the tendon of the Flexor carpi ulnaris, and the nerve will be found lying on the ulnar side of the ulnar artery.

The *musculo-spiral nerve* is probably more frequently injured than any other nerve of the upper extremity. In consequence of its close relationship to the humerus, as it lies in the musculo-spiral groove, it is frequently torn or injured in fractures of this bone, or subsequently involved in the callus that may be thrown out around a fracture and thus pressed upon and its functions interfered with. It is also liable to be contused against the bone by kicks or blows, or to be divided by wounds of the arm. When paralysed, the hand is flexed at the wrist and lies flaccid. This is known as ' drop wrist.' The fingers are also flexed, and on an attempt being made to extend them, the last two phalanges only will be extended, through the action of the Interossei; the first phalanges remaining flexed. There is no power of extending the wrist. Supination is completely lost when the forearm is extended on the arm, but is possible to a certain extent if the forearm is flexed so as to allow of the action of the Biceps. The power of extension of the forearm is lost on account of paralysis of the Triceps. The best position in which to expose the nerve, for the purpose of stretching, is to make an incision along the inner border of the Supinator longus, just above the level of the elbow-joint. The skin and superficial structures are to be divided and the deep fascia exposed. The white line in this structure indicating the border of the muscle is to be defined, and the deep fascia divided in this line. By now raising the Supinator longus, the nerve will be found lying beneath it, on the Brachialis anticus.

DORSAL NERVES (fig. 415)

The **dorsal nerves** are twelve in number on each side. The first appears between the first and second dorsal vertebræ, and the twelfth between the last dorsal and first lumbar.

The *roots of the dorsal nerves* are of small size, and vary but slightly from the second to the last. Both roots are very slender; the posterior roots only slightly exceeding the anterior in thickness. They gradually increase in length from above downwards, and, in the lower part of the dorsal region, pass down in contact with the spinal cord for a distance equal to the height of, at least, two vertebræ, before they emerge from the spinal canal. They then join in the inter-vertebral foramen, and, at their exit, divide into two primary divisions, a posterior (dorsal) and an anterior (intercostal).

The first, the second, and the last dorsal nerves are peculiar in some respects.

POSTERIOR DIVISIONS OF THE DORSAL NERVES

The **posterior divisions of the dorsal nerves,** which are smaller than the anterior, pass backwards between the transverse processes, and divide into internal and external branches.

The *internal branches of the six upper nerves* pass inwards between the Semi-spinalis dorsi and Multifidus spinæ muscles, which they supply ; and then, piercing the origins of the Rhomboidei and Trapezius muscles, become cutaneous by the side of the spinous processes and ramify in the integument. The internal branches of *the six lower nerves* are distributed to the Multifidus spinæ, without giving off any cutaneous filaments.

The *external branches* increase in size from above downwards. They pass through the Longissimus dorsi, to the cellular interval between it and the Ilio-costalis, and supply those muscles, as well as their continuations upwards to the head, and the Levatores costarum ; the five or six lower nerves also give off cutaneous filaments, which pierce the Serratus posticus inferior and Latissimus dorsi, in a line with the angles of the ribs, and then ramify in the integument.

The *cutaneous branches of the posterior primary divisions of the dorsal nerves* are twelve in number. The six upper cutaneous nerves are derived from the internal branches of the posterior divisions of the dorsal nerves. They pierce the origins of the Rhomboidei and Trapezius muscles, and become cutaneous by the side of the spinous processes, and then ramify in the integument. They are frequently furnished with gangliform enlargements. The six lower cutaneous nerves are derived from the external branches of the posterior divisions of the dorsal nerves. They pierce the Serratus posticus inferior and Latissimus dorsi, in a line with the angles of the ribs, and then ramify in the integument.

ANTERIOR DIVISIONS OF THE DORSAL NERVES

The **anterior divisions of the dorsal nerves** (*intercostal nerves*) are twelve in number on each side. They are, for the most part, distributed to the parietes of the thorax and abdomen, separately from each other, without being joined in a plexus ; in which respect they differ from the other spinal nerves. Each nerve is connected with the adjoining ganglia of the sympathetic by one or two fila-ments. The intercostal nerves may be divided into two sets, from the difference they present in their distribution. The six upper, with the exception of the first and the intercosto-humeral branch of the second, are limited in their distribution to the parietes of the chest. The six lower supply the parietes of the chest and abdomen, the last one sending a cutaneous filament to the buttock.

The first dorsal nerve.—The anterior division of the first dorsal nerve divides into two branches : one, the larger, leaves the thorax in front of the neck of the first rib, and enters into the formation of the brachial plexus ; the other and smaller branch runs along the first intercostal space, forming the *first intercostal nerve,* and terminates on the front of the chest, by forming the first anterior cutaneous nerve of the thorax. Occasionally this anterior cutaneous branch is wanting. The first intercostal nerve as a rule gives off no lateral cutaneous branch ; but sometimes a small branch is given off, which communicates with the intercosto-humeral. It frequently receives a connecting twig from the second dorsal nerve, which passes upwards over the neck of the second rib.

FIG. 415.— Superficial and deep distribution of the posterior divisions of the spinal nerves (after Hirschfeld and Leveillé). On the left side the cutaneous branches are represented lying on the superficial layer of muscles. On the right side the superficial muscles have been removed, the Splenius capitis and Complexus divided in the neck, and the Erector spinæ divided and partly removed in the back, so as to expose the posterior divisions of the spinal nerves near their origin.

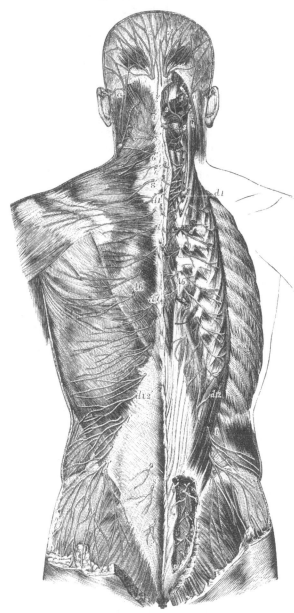

a a. Lesser occipital nerve from the cervical plexus. 1. External muscular branches of the first cervical nerve, and union by a loop with the second. 2. Placed on the Rectus capitis posticus major muscle, marks the great occipital nerve, passing round the short muscles and piercing the Complexus : the external branch is seen to the outside. 3. External branch from the posterior division of the third nerve. 3'. Its internal branch, sometimes called the third occipital. 4' to 8'. The internal branches of the several corresponding nerves on the left side. The external branches of these nerves, proceeding to muscles, are displayed on the right side. *d* 1 to *d* 6 and thence to *d* 12. External muscular branches of the posterior divisions of the 12 dorsal nerves on the right side. *d* 1' to *d* 6'. The internal cutaneous branches of the six upper dorsal nerves on the left side. *d* 7' to *d* 12'. Cutaneous twigs from the external branches of the six lower dorsal nerves. *l l.* External branches from the posterior divisions of several lumbar nerves on the right side, piercing the muscles, the lower descending over the gluteal region. *l' l'.* The same, more superficially, on the left side. *s s.* The issue and union by loops of the posterior divisions of four sacral nerves on the right side. *s' s'.* Some of those distributed to the skin on the left side.

The upper dorsal nerves.—The anterior divisions of the second, third, fourth, fifth, and sixth dorsal nerves and the small branch from the first dorsal are confined to the parietes of the thorax, and are named *upper* or *pectoral intercostal nerves*. They pass forwards in the intercostal spaces with the intercostal vessels, being situated below them. At the back of the chest they lie between the pleura and the External intercostal muscles, but are soon placed between the two planes of Intercostal muscles as far as the middle of the rib. They then enter the substance of the Internal intercostal muscles, and, running amidst their fibres as far as the costal cartilages, they gain the inner surface of the muscles and lie between them and the pleura. Near the sternum, they cross in front of the internal mammary artery and Triangularis sterni muscle, pierce the Internal intercostal muscles, the anterior intercostal membrane, and Pectoralis major muscle, and supply the integument of the front of the chest and over the mammary gland, forming the anterior cutaneous nerves of the thorax ; the branch from the second nerve is joined with the supraclavicular nerves of the cervical plexus.

Branches.—Numerous slender muscular filaments supply the Intercostals, the Infracostales, the Levatores costarum, Serratus posticus superior, and Triangularis sterni muscles. Some of these branches, at the front of the chest, cross the costal cartilages from one to another intercostal space.

Lateral cutaneous nerves.—These are derived from the intercostal nerves, midway between the vertebræ and sternum ; they pierce the External intercostal and Serratus magnus muscles, and divide into two branches, anterior and posterior.

The *anterior branches* are reflected forwards to the side and the fore part of the chest, supplying the integument of the chest and mamma ; those of the fifth and sixth nerves supply the upper digitations of the External oblique.

The *posterior branches* are reflected backwards, to supply the integument over the scapula and Latissimus dorsi.

The lateral cutaneous branch of the second intercostal nerve is of large size, and does not divide, like the other nerves, into an anterior and a posterior branch. It is named, from its origin and distribution, the *intercosto-humeral nerve* (fig. 413). It pierces the External intercostal muscle, crosses the axilla to the inner side of the arm, and joins with a filament from the nerve of Wrisberg. It then pierces the fascia, and supplies the skin of the upper half of the inner and back part of the arm, communicating with the internal cutaneous branch of the musculo-spiral nerve. The size of this nerve is in inverse proportion to the size of the other cutaneous nerves, especially the nerve of Wrisberg. A second intercosto-humeral nerve is frequently given off from the third intercostal. It supplies filaments to the armpit and inner side of the arm.

The lower dorsal nerves.—The anterior divisions of the seventh, eighth, ninth, tenth, and eleventh dorsal nerves are continued anteriorly from the intercostal spaces into the abdominal wall, and the twelfth dorsal is continued throughout its whole course in the abdominal wall, since it is placed below the last rib ; hence these nerves are named *lower* or *abdominal intercostal nerves*. They have (with the exception of the last) the same arrangement as the upper ones as far as the anterior extremities of the intercostal spaces, where they pass behind the costal cartilages, and between the Internal oblique and Transversalis muscles, to the sheath of the Rectus, which they perforate. They supply the Rectus muscle, and terminate in branches which become subcutaneous near the linea alba. These branches are named the anterior cutaneous nerves of the abdomen. They are directed outwards as far as the lateral cutaneous nerves, supplying the integument of the front of the belly. The lower intercostal nerves supply the Intercostals, Serratus posticus inferior, and Abdominal muscles, and, about the middle of their course, give off lateral cutaneous branches, which pierce the External intercostal and External oblique muscles, in the same line as the lateral cutaneous nerves of the thorax, and divide into anterior and posterior branches, which are distributed to the integument of the abdomen and back ; the anterior branches supply the digitations of the External oblique muscle, and extend downwards and forwards nearly as far as the margin of the Rectus : the posterior branches pass backwards to supply the skin over the Latissimus dorsi.

The **last dorsal** is larger than the other dorsal nerves. Its anterior division runs along the lower border of the last rib, and passes under the external arcuate ligament of the Diaphragm. It then runs in front of the Quadratus lumborum, perforates the Transversalis, and passes forwards between it and the Internal oblique, to be distributed in the same manner as the lower intercostal nerves. It communicates with the ilio-hypogastric branch of the lumbar plexus, and is frequently connected with the first lumbar nerve by a slender branch, the *dorsi-lumbar nerve*, which descends in the substance of the Quadratus lumborum. It gives a branch to the Pyramidalis muscle.

The *lateral cutaneous branch of the last dorsal* is remarkable for its large size ; it perforates the Internal and External oblique muscles, passes downwards over the crest of the ilium in front of the iliac branch of the ilio-hypogastric (fig. 422), and is distributed to the integument of the front part of the gluteal region, some of its filaments extending as low down as the trochanter major. It does not divide into an anterior and posterior branch like the other lateral cutaneous branches of the intercostal nerves.

Surgical Anatomy.—The lower seven intercostal nerves and the ilio-hypogastric from the first lumbar nerve, supply the skin of the abdominal wall. They run downwards and inwards fairly equidistant from each other. The sixth and seventh supply the skin over the ' pit of the stomach ' ; the eighth corresponds to about the position of the middle linea transversa; the tenth to the umbilicus ; and the ilio-hypogastric supplies the skin over the pubes and external abdominal ring. There are several points of surgical import- ance about the distribution of these nerves, and it is important to remember their origin and course, for in many diseases affecting the nerve-trunks at or near the origin, the pain is referred to their peripheral terminations. Thus, in Pott's disease of the spine, children will often be brought to the surgeon suffering from pain in the belly. This is due to the fact that the nerves are irritated at the seat of disease as they issue from the spinal canal. When the irritation is confined to a single pair of nerves, the sensation complained of is often a feeling of constriction, as if a cord were tied round the abdomen, and in these cases the situation of the sense of constriction may serve to localise the disease in the spinal column. In other cases where the bone disease is more extensive, and two or more nerves are involved, a more general, diffused pain in the abdomen is complained of. A similar condition is sometimes present in affections of the cord itself, as in tabes dorsalis.

Again, it must be borne in mind that the same nerves which supply the skin of the abdomen supply also the planes of muscle, which constitute the greater part of the abdominal wall. Hence it follows that any irritation applied to the peripheral termina- tions of the cutaneous branches in the skin of the abdomen is immediately followed by reflex contraction of the abdominal muscles. A good practical illustration of this may sometimes be seen in watching two surgeons examine the abdomen of the same patient. One, whose hand is cold, causes an immediate reflex contraction of the abdominal muscles, so that the belly wall becomes rigid and not nearly so suitable for examination; the other, who has taken the precaution to warm his hand, examines the abdomen without exciting any reflex contraction. The supply of both muscles and skin from the same source is of importance in protecting the abdominal viscera from injury. A blow on the abdomen, even of a severe character, will do no injury to the viscera if the muscles are in a condition of firm contraction ; whereas in cases where the muscles have been taken unawares, and the blow has been struck while they were in a state of rest, an injury insufficient to produce any lesion of the abdominal wall has been attended with rupture of some of the abdominal contents. The importance therefore of immediate reflex contraction upon the receipt of an injury cannot be over-estimated, and the intimate association of the cutaneous and muscular fibres in the same nerve produces a much more immediate response on the part of the muscles to any peripheral stimulation of the cutaneous filaments than would be the case if the two sets of fibres were derived from independent sources.

Again, the nerves supplying the abdominal muscles and skin derived from the lower intercostal nerves are intimately connected with the sympathetic supplying the abdominal viscera, through the lower thoracic ganglia from which the splanchnic nerves are derived. In consequence of this, in laceration of the abdominal viscera and in acute peritonitis, the muscles of the belly wall become firmly contracted, and thus as far as possible preserve the abdominal contents in a condition of rest.

LUMBAR NERVES

The **lumbar nerves** are five in number on each side. The first appears between the first and second lumbar vertebræ, and the last between the last lumbar and the base of the sacrum.

The **roots of the lumbar nerves** are the largest, and their filaments the most numerous, of all the spinal nerves, and they are closely aggregated together upon the lower end of the cord. The anterior roots are the smaller; but there is not the same disproportion between them and the posterior roots as in the cervical nerves. The roots of these nerves have a vertical direction, and are of considerable length, more especially the lower ones, since the spinal cord does not extend beyond the first lumbar vertebra. The roots become joined in the intervertebral foramina; and the nerves, so formed, divide at their exit into two divisions, posterior and anterior.

Posterior Divisions of the Lumbar Nerves

The **posterior divisions of the lumbar nerves** (fig. 415) diminish in size from above downwards; they pass backwards between the transverse processes, and divide into internal and external branches.

The *internal branches*, the smaller, pass inwards close to the articular processes of the vertebræ, and supply the Multifidus spinæ and Interspinales muscles.

The *external branches* supply the Erector spinæ and Intertransverse muscles. From the three upper branches, cutaneous nerves are derived which pierce the aponeurosis of the Latissimus dorsi muscle, and descend over the back part of the crest of the ilium, to be distributed to the integument of the gluteal region; some of the filaments passing as far as the trochanter major.

Anterior Divisions of the Lumbar Nerves

The **anterior divisions of the lumbar nerves** increase in size from above downwards. At their origin, they communicate with the lumbar ganglia of the sympathetic by long, slender filaments, which accompany the lumbar arteries round the sides of the bodies of the vertebræ, beneath the Psoas muscle. The nerves pass obliquely outwards behind the Psoas magnus, or between its fasciculi, distributing filaments to it and the Quadratus lumborum. The anterior divisions of the four upper nerves are connected together in this situation by anastomotic loops, and form the *lumbar plexus*. The anterior division of the fifth lumbar, joined with a branch from the fourth, descends across the base of the sacrum to join the anterior division of the first sacral nerve, and assist in the formation of the sacral plexus. The cord resulting from the union of the fifth lumbar and the branch from the fourth is called the *lumbo-sacral cord*.

Lumbar Plexus (fig. 416)

The **Lumbar plexus** is formed by the loops of communication between the anterior divisions of the four upper lumbar nerves. The plexus is narrow above, and often connected with the last dorsal by a slender branch, the *dorsi-lumbar nerve*; it is broad below, where it is joined to the sacral plexus by the lumbosacral cord. It is situated in the substance of the Psoas muscle near its posterior part, in front of the transverse processes of the lumbar vertebræ.

The mode in which the plexus is arranged varies in different subjects. It differs from the brachial plexus in not forming an intricate interlacement, but the several nerves of distribution arise from one or more of the spinal nerves, somewhat in the following manner: The first lumbar nerve receives a branch from the last dorsal, and gives off a larger branch, which subdivides into the iliohypogastric and ilio-inguinal; a communicating branch which passes down to the second lumbar nerve ; and a third branch which unites with a branch of the second lumbar to form the genito-crural nerve. The second, third, and fourth lumbar nerves divide into an anterior and posterior division. The anterior division of the second divides into two branches, one of which joins with the above-mentioned branch of the first nerve to form the genito-crural ; the other unites with the anterior division of the third nerve, and a part of the anterior division of the fourth nerve to form the obturator nerve. The remainder of the anterior division of the fourth nerve passes down to communicate with the fifth lumbar nerve. The posterior divisions of the second and third nerve divide into two branches, a smaller branch from each uniting to form the external cutaneous nerve, and a larger branch from each, which join with the whole of the posterior division of the fourth lumbar nerve to form the anterior crural. The accessory obturator, when it exists, is formed by the union of two small branches given off from the third and fourth nerves.

From this arrangement it follows that the ilio-hypogastric and ilio-inguinal are derived entirely from the first lumbar nerve ; the genito-crural from the first and second nerves ; the external cutaneous from the second and third ; the anterior crural and obturator by fibres derived from the second, third, and fourth ; and the accessory obturator, when present, from the third and fourth.

The branches of the lumbar plexus are : the

> Ilio-hypogastric.
> Ilio-inguinal.
> Genito-crural.
> External cutaneous.
> Obturator.
> Accessory obturator.
> Anterior crural.

The **Ilio-hypogastric nerve** arises from the first lumbar nerve. It emerges from the outer border of the Psoas muscle at its upper part, and crosses obliquely in front of the Quadratus lumborum to

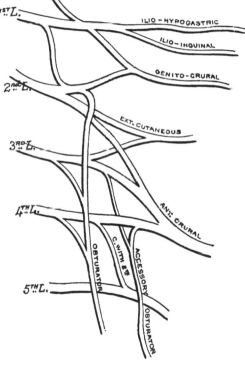

FIG. 416.—Plan of the lumbar plexus.

the crest of the ilium. It then perforates the Transversalis muscle at its posterior part, near the crest of the ilium, and divides between it and the Internal oblique into two branches, iliac and hypogastric.

The *iliac branch* pierces the Internal and External oblique muscles immediately above the crest of the ilium, and is distributed to the integument of the gluteal region, behind the lateral cutaneous branch of the last dorsal nerve (fig. 422). The size of this nerve bears an inverse proportion to that of the cutaneous branch of the last dorsal nerve.

The *hypogastric branch* (fig. 418) continues onwards between the Internal oblique and Transversalis muscles. It then pierces the Internal oblique, and becomes cutaneous by perforating the aponeurosis of the External oblique, about an inch above, and a little to the outer side of the external abdominal ring, and is distributed to the integument of the hypogastric region.

The ilio-hypogastric nerve communicates with the last dorsal and ilio-inguinal nerves.

The **Ilio-inguinal nerve,** smaller than the preceding, arises with it from the first lumbar nerve. It emerges from the outer border of the Psoas just below the ilio-hypogastric, and, passing obliquely across the Quadratus lumborum and Iliacus muscles, perforates the Transversalis, near the fore part of the crest of the ilium, and communicates with the ilio-hypogastric nerve between that muscle and the Internal oblique. The nerve then pierces the Internal oblique, distributing filaments to it, and accompanying the spermatic cord through the external abdominal ring, and is distributed to the integument of the upper and inner part of the thigh; and to the scrotum in the male, and to the labium majus in the female. The size of this nerve is in inverse proportion to

FIG. 417.—The lumbar plexus and its branches.

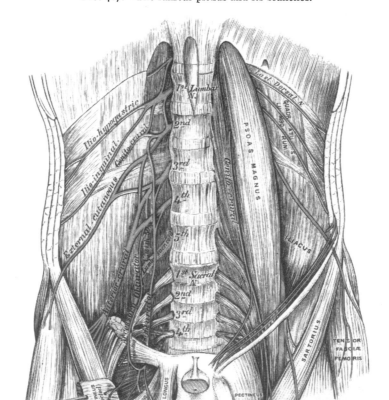

that of the ilio-hypogastric. Occasionally it is very small, and ends by joining the ilio-hypogastric; in such cases, a branch from the ilio-hypogastric takes the place of the ilio-inguinal, or the latter nerve may be altogether absent.

The **Genito-crural nerve** arises from the first and second lumbar nerves. It passes obliquely through the substance of the Psoas, and emerges from its inner border, close to the vertebral column, opposite the disc between the third and fourth lumbar vertebræ; it then descends on the surface of the Psoas muscle, under cover of the peritoneum, and divides into a genital and crural branch.

The *genital branch* passes outwards on the Psoas magnus, and pierces the fascia transversalis, or passes through the internal abdominal ring; it then descends along the back part of the spermatic cord to the scrotum, and supplies, in the male, the Cremaster muscle. In the female, it accompanies the round ligament, and is lost upon it.

The *crural branch* descends on the external iliac artery, sending a few filaments round it, and, passing beneath Poupart's ligament to the thigh, enters the sheath of the femoral vessels, lying superficial and a little external to the femoral artery. It pierces the anterior layer of the sheath of the vessels, and, becoming superficial by passing through the fascia lata, it supplies the skin of the anterior aspect of the thigh as far as midway between the pelvis and knee. On the front of the thigh it communicates with the outer branch of the middle cutaneous nerve, derived from the anterior crural.

A few filaments from this nerve may be traced on to the femoral artery; they are derived from the nerve as it passes beneath Poupart's ligament.

The **External cutaneous nerve** arises from the second and third lumbar nerves. It emerges from the outer border of the Psoas muscle about its middle, and crosses the Iliacus muscle obliquely, towards the anterior superior spine of the ilium. It then passes under Poupart's ligament and over the Sartorius muscle into the thigh, where it divides into two branches, anterior and posterior.

The *anterior branch* descends in an aponeurotic canal formed in the fascia lata, becomes superficial about four inches below Poupart's ligament, and divides into branches which are distributed to the integument along the anterior and outer part of the thigh, as far down as the knee. This nerve occasionally communicates with a branch of the long saphenous nerve in front of the knee-joint.

The *posterior branch* pierces the fascia lata, and subdivides into branches which pass backwards across the outer and posterior surface of the thigh, supplying the integument from the crest of the ilium as far as the middle of the thigh.

The **Obturator nerve** supplies the Obturator externus and Adductor muscles of the thigh, the articulations of the hip and knee, and occasionally the integument of the thigh and leg. It arises by three branches: from the second, the third, and the fourth lumbar nerves. Of these, the branch from the third is the largest, while that from the second is often very small. It descends through the inner fibres of the Psoas muscle, and emerges from its inner border near the brim of the pelvis; it then runs along the lateral wall of the pelvis, above the obturator vessels, to the upper part of the obturator foramen, where it enters the thigh, and divides into an anterior and a posterior branch, separated by some of the fibres of the Obturator externus (fig. 257), and lower down by the Adductor brevis muscle.

The *anterior branch* (fig. 419) passes down in front of the Adductor brevis, being covered by the Pectineus and Adductor longus; and at the lower border of the latter muscle communicates with the internal cutaneous and internal saphenous nerves, forming a kind of plexus. It then descends upon the femoral artery, upon which it is finally distributed. The nerve, near the obturator foramen, gives off an articular branch to the hip-joint. Behind the Pectineus, it distributes muscular branches to the Adductor longus and Gracilis, and usually to the Adductor brevis, and in rare cases to the Pectineus, and receives a communicating branch from the accessory obturator nerve.

Occasionally the communicating branch to the internal cutaneous and internal saphenous nerves is continued down, as a cutaneous branch, to the thigh and leg. When this is so, this occasional cutaneous branch emerges from beneath the lower border of the Adductor longus, descends along the posterior margin of the Sartorius to the inner side of the knee, where it pierces the deep fascia, communicates with the long saphenous nerve, and is distributed to the integument of the inner side of the leg as low down as its middle. When this communicating branch is small, its place is supplied by the internal cutaneous nerve.

The *posterior branch of the obturator nerve* pierces the Obturator externus, sending branches to supply it, and passes behind the Adductor brevis on the front of the Adductor magnus, where it divides into numerous muscular branches,

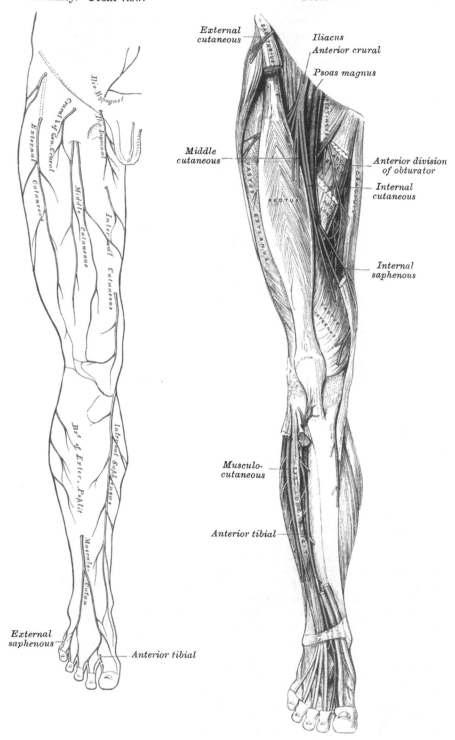

FIG. 418.—Cutaneous nerves of lower extremity.   Front view.

FIG. 419.—Nerves of the lower extremity. Front view.

which supply the Adductor magnus, and the Adductor brevis when the latter does not receive a branch from the anterior division of the nerve. One of the branches gives off a filament to the knee-joint.

The *articular branch for the knee-joint* is sometimes absent; it perforates the lower part of the Adductor magnus, and enters the popliteal space; it then descends upon the popliteal artery, as far as the back part of the knee-joint, where it perforates the posterior ligament, and is distributed to the synovial membrane. It gives filaments to the artery in its course.

Accessory obturator nerve (fig. 417) is not constantly present. It is of small size, and arises by separate filaments from the third and fourth lumbar nerves. It descends along the inner border of the Psoas muscle, crosses the ascending ramus of the os pubis, and passes under the outer border of the Pectineus muscle, where it divides into numerous branches. One of these supplies the Pectineus, penetrating its under surface; another is distributed to the hip-joint; while a third communicates with the anterior branch of the obturator nerve. When this nerve is absent, the hip-joint receives two branches from the obturator nerve. Occasionally it is very small and becomes lost in the capsule of the hip-joint.

The **Anterior crural nerve** (figs. 417, 419) is the largest branch of the lumbar plexus. It supplies muscular branches to the Iliacus, Pectineus, and all the muscles on the front of the thigh, excepting the Tensor fasciæ femoris; cutaneous filaments to the front and inner side of the thigh, and to the leg and foot; and articular branches to the hip and knee. It arises from the second, third, and fourth lumbar nerves. It descends through the fibres of the Psoas muscle, emerging from it at the lower part of its outer border; and passes down between it and the Iliacus, and beneath Poupart's ligament, into the thigh, where it becomes somewhat flattened, and divides into an anterior and a posterior part. Under Poupart's ligament, it is separated from the femoral artery by a portion of the Psoas muscle, and lies beneath the iliac fascia.

Within the abdomen the anterior crural nerve gives off from its outer side some small branches to the Iliacus, and a branch to the femoral artery, which is distributed upon the upper part of that vessel. The origin of this branch varies: it occasionally arises higher than usual, or it may arise lower down in the thigh.

External to the pelvis the following branches are given off:

From the Anterior Division	*From the Posterior Division*
Middle cutaneous.	Long saphenous.
Internal cutaneous.	Muscular.
Muscular.	Articular.

The *middle cutaneous nerve* (fig. 418) pierces the fascia lata (generally the Sartorius also) about three inches below Poupart's ligament, and divides into two branches, which descend in immediate proximity along the fore part of the thigh, to supply the integument as low as the front of the knee, where it communicates with the internal cutaneous and the patellar branch of the internal saphenous nerve, to form the patellar plexus. In the upper part of the thigh the outer division of the middle cutaneous communicates with the crural branch of the genito-crural nerve.

The *internal cutaneous nerve* passes obliquely across the upper part of the sheath of the femoral artery, and divides in front, or at the inner side of that vessel, into two branches, anterior and posterior or internal.

The *anterior branch* runs downwards on the Sartorius, perforates the fascia lata at the lower third of the thigh, and divides into two branches: one of which supplies the integument as low down as the inner side of the knee; the other crosses to the outer side of the patella, communicating in its course with the nervus cutaneus patellæ, a branch of the internal saphenous nerve.

The *posterior* or *internal branch* descends along the inner border of the

Sartorius muscle to the knee, where it pierces the fascia lata, communicates with
the long saphenous nerve, and gives off several cutaneous branches. The nerve
then passes down the inner side of the leg, to the integument of which it is
distributed. This nerve, beneath the fascia lata, at the lower border of the
Adductor longus, joins in a plexiform network by uniting with branches of the
long saphenous and obturator nerves (fig. 419). When the communicating
branch from the obturator nerve is large and continued to the integument of the
leg, the inner branch of the internal cutaneous is small, and terminates at the
plexus, occasionally giving off a few cutaneous filaments.

The internal cutaneous nerve, before dividing, gives off a few filaments, which
pierce the fascia lata, to supply the integument of the inner side of the thigh,
accompanying the long saphenous vein. One of these filaments passes through
the saphenous opening ; a second becomes subcutaneous about the middle of the
thigh ; and a third pierces the fascia at its lower third.

Muscular branches of the anterior division.—The *nerve to the Pectineus* is often
duplicated ; it arises from the anterior crural immediately below Poupart's
ligament, and passes inwards behind the femoral sheath to enter the anterior
surface of the muscle. The *nerve to the Sartorius* arises in common with the
middle cutaneous.

The *long* or *internal saphenous nerve* is the largest of the cutaneous branches
of the anterior crural. It approaches the femoral artery where this vessel passes
beneath the Sartorius, and lies in front of it, beneath the aponeurotic covering of
Hunter's canal, as far as the opening in the lower part of the Adductor magnus.
It then quits the artery, and descends vertically along the inner side of the knee,
beneath the Sartorius, pierces the fascia lata, opposite the interval between the
tendons of the Sartorius and Gracilis, and becomes subcutaneous. The nerve
then passes along the inner side of the leg, accompanied by the internal saphenous
vein, descends behind the internal border of the tibia, and, at the lower third
of the leg, divides into two branches : one continues its course along the margin
of the tibia, terminating at the inner ankle ; the other passes in front of the ankle,
and is distributed to the integument along the inner side of the foot, as far
as the great toe, communicating with the internal branch of the musculo-
cutaneous nerve.

Branches.—The long saphenous nerve, about the middle of the thigh, gives
off a communicating branch which joins the plexus formed by the obturator and
internal cutaneous nerves.

At the inner side of the knee it gives off a large patellar branch (*nervus
cutaneus patellæ*) which pierces the Sartorius and fascia lata, and is distributed
to the integument in front of the patella. This nerve communicates above the
knee with the anterior branch of the internal cutaneous and with the middle
cutaneous ; below the knee, with other branches of the long saphenous ; and, on
the outer side of the joint, with branches of the external cutaneous nerve, forming
a plexiform network, the *plexus patellæ*. The cutaneous nerve of the patella is
occasionally small, and terminates by joining the internal cutaneous, which
supplies its place in front of the knee.

Below the knee, the branches of the long saphenous nerve are distributed
to the integument of the front and inner side of the leg, communicating with
the cutaneous branches from the internal cutaneous, or from the obturator nerve.

The *muscular branches of the posterior division* supply the four parts of the
Quadriceps extensor muscle.

The *branch to the Rectus muscle* enters its under surface high up, sending off
a small filament to the hip-joint.

The *branch to the Vastus externus*, of large size, follows the course of the
descending branch of the external circumflex artery to the lower part of the
muscle. It gives off an articular filament to the knee-joint.

The *branch to the Vastus internus* is a long branch which runs down on the
outer side of the femoral vessels in company with the internal saphenous nerve for

its upper part. It enters the muscle about its middle, and gives off a filament, which can usually be traced downwards on the surface of the muscle to the knee-joint.

The *branch to the Crureus* enters the muscle on its anterior surface about the middle of the thigh, and sends a filament through the muscle to the Subcrureus and the knee-joint.

The *articular branch to the hip-joint* is derived from the nerve to the Rectus.

The *articular branches to the knee-joint* are three in number. One, a long, slender filament, is derived from the nerve to the Vastus externus ; it penetrates the capsular ligament of the joint on its anterior aspect. Another is derived from the nerve to the Vastus internus. It can usually be traced downwards on the surface of this muscle to near the joint ; it then penetrates the muscular fibres, and accompanies the deep branch of the anastomotica magna artery, pierces the capsular ligament of the joint on its inner side, and supplies the synovial membrane. The third branch is derived from the nerve to the Crureus.

THE SACRAL AND COCCYGEAL NERVES

The *sacral nerves* are five in number on each side. The four upper ones pass from the sacral canal, through the sacral foramina ; the fifth through the foramen between the sacrum and coccyx.

The *roots of the upper sacral nerves* are the largest of all the spinal nerves ; while those of the lowest sacral and coccygeal nerve are the smallest. They are longer than those of any of the other spinal nerves, on account of the spinal cord not extending beyond the first lumbar vertebra. From their great length, and the appearance they present in connection with their attachment to the spinal cord, the roots of origin of these nerves are called collectively the *cauda equina*.

Each sacral and coccygeal nerve separates into two divisions, posterior and anterior.

The **posterior divisions of the sacral nerves** (fig. 420) are small, diminish in size from above downwards, and emerge, except the last, from the sacral canal by the posterior sacral foramina.

The *three upper ones* are covered, at their exit from the sacral canal, by the Multifidus spinæ, and divide into internal and external branches.

The *internal branches* are small, and supply the Multifidus spinæ.

The *external branches* join with one another, and with the last lumbar and fourth sacral nerves, in the form of loops on the posterior surface of the sacrum. From these loops branches pass to the outer surface of the great sacro-sciatic ligament, where they form a second series of loops beneath the Gluteus maximus. Cutaneous branches from this second series of loops, usually two or three in number, pierce the Gluteus maximus along a line drawn from the posterior superior spine of the ilium to the tip of the coccyx. They supply the integument over the posterior part of the gluteal region.

The *posterior divisions of the two lower sacral nerves* are situated below the Multifidus spinæ. They are of small size, and do not divide into internal and external branches, but join with each other, and with the coccygeal nerve, so as to form loops on the back of the sacrum, filaments from which supply the Extensor coccygis and the integument over the coccyx.

The **coccygeal nerve** divides into its anterior and posterior division in the spinal canal. The *posterior division* is the smaller. It does not divide, but receives, as already mentioned, a communicating branch from the last sacral, and is lost in the integument over the back of the coccyx.

The **anterior divisions of the sacral nerves** diminish in size from above downwards. The four upper ones emerge from the anterior sacral foramina : the anterior division of the fifth, after emerging from the spinal canal through its terminal opening, curves forwards between the sacrum and the coccyx. All the anterior sacral nerves communicate with the sacral ganglia of the sympathetic, at their exit from the sacral foramina. The *first nerve*, of large size, unites with the lumbo-sacral cord, formed by the fifth lumbar and a branch from the fourth. The *second*, equal in size to the preceding, and a *third*, about one-fourth the size of the second, unite with this trunk, and form, with a small fasciculus from the fourth, the *sacral plexus*, a visceral branch being given off from the third nerve to the bladder.

The **fourth anterior sacral nerve** sends a branch to join the sacral plexus. The remaining portion of the nerve divides into visceral and muscular branches ; and a communicating filament descends to join the fifth sacral nerve. The *visceral branches* are distributed to the viscera of the pelvis, communicating with the sympathetic nerve. These branches ascend upon the rectum and bladder, and in the female upon the vagina, communicating with branches of the sym-

Fig. 420.—The posterior sacral nerves.

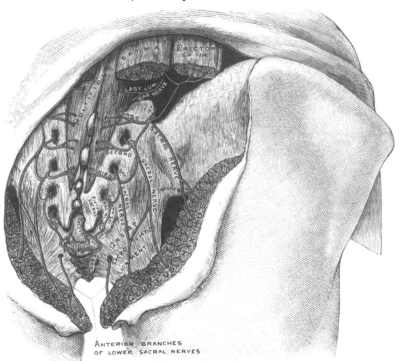

pathetic from the pelvic plexus. The *muscular branches* are distributed to the Levator ani, Coccygeus, and Sphincter ani. The branch to the Sphincter ani pierces the Levator ani, so as to reach the ischio-rectal fossa, where it is found lying in front of the coccyx. Cutaneous filaments arise from the latter branch, which supply the integument between the anus and coccyx. Another cutaneous branch is frequently given off from this nerve, though sometimes from the pudic (Schwalbe). It perforates the great sacro-sciatic ligament, and, winding round the lower border of the Gluteus maximus, supplies the skin over the lower and inner part of this muscle.

The **fifth anterior sacral nerve,** after passing from the lower end of the sacral canal, curves forwards through the fifth sacral foramen, formed between the lower part of the sacrum and the transverse process of the first piece of the coccyx. It pierces the Coccygeus muscle, and descends upon its anterior surface to near the tip of the coccyx, where it again perforates the muscle, to be distributed to the integument over the back part and side of the coccyx. This nerve communicates above with the fourth sacral, and below with the coccygeal nerve, and supplies the Coccygeus muscle.

The **anterior division of the coccygeal nerve** is a delicate filament which escapes at the termination of the sacral canal; it passes downwards behind the rudimentary transverse process of the first piece of the coccyx, and curves forwards, through the notch between the first and second pieces, piercing the Coccygeus muscle and descending on its anterior surface to near the tip of the coccyx, where it again pierces the muscle, to be distributed to the integument over the back part and side of the coccyx. It is joined by a branch from the fifth anterior sacral as it descends on the surface of the Coccygeus muscle.

SACRAL PLEXUS (fig. 421)

The **Sacral plexus** is formed by the lumbo-sacral cord, the anterior divisions of the three upper sacral nerves, and part of that of the fourth. These nerves

FIG. 421.—Side view of pelvis, showing sacral nerves.

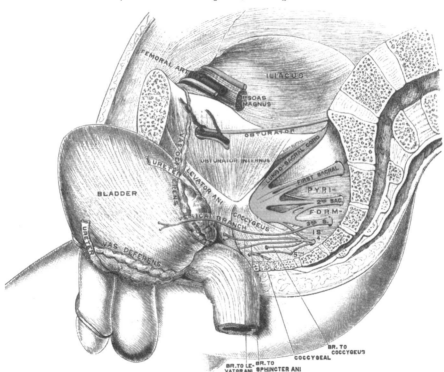

proceed in different directions: the upper ones obliquely downwards and outwards, the lower one nearly horizontally, and they all unite into two cords: an *upper* and larger which is formed by the lumbo-sacral cord with the first, second, and the greater part of the third sacral nerves; and a *lower* and smaller formed by the remainder of the third, with a portion of the fourth sacral nerve. The upper cord is prolonged into the great sciatic nerve and the lower into the pudic. Frequently a small filament is given off from the second sacral nerve to join the lower cord.

The sacral plexus is triangular in form, its base corresponding with the exit of the nerves from the sacrum, its apex with the lower part of the great sacro-sciatic foramen. It rests upon the anterior surface of the Pyriformis, and is covered in front by the pelvic fascia, which separates it from the sciatic and pudic branches of the internal iliac artery, and from the viscera of the pelvis.

The branches of the sacral plexus are :

Collateral branches
{ Muscular.
Superior gluteal.
Inferior gluteal.
Small sciatic.
Perforating cutaneous.

Terminal branches
{ Pudic.
Great sciatic.

The **Muscular branches** supply the Pyriformis, Obturator internus, the two Gemelli, and the Quadratus femoris. The branch to the Pyriformis arises from the upper two sacral nerves before they enter the plexus ; the branch to the Obturator internus arises at the junction of the lumbo-sacral and first sacral nerves : it passes out of the pelvis through the great sacro-sciatic foramen below the Pyriformis, crosses the spine of the ischium, and re-enters the pelvis through the lesser sacro-sciatic foramen to enter the inner surface of the Obturator internus ; the branch to the Gemellus superior arises in common with the nerve to the Obturator internus : it enters the muscle at the upper part of its posterior surface ; the small branch to the Gemellus inferior and Quadratus femoris also arises from the upper part of the plexus : it passes through the great sacro-sciatic foramen below the Pyriformis, and courses down beneath the great sciatic nerve, the Gemelli and tendon of the Obturator internus, and supplies the muscles on their deep or anterior surface. It gives off an articular branch to the hip-joint. A second articular branch is occasionally derived from the upper part of the sacral plexus.

The **Superior gluteal nerve** (fig. 423) arises from the back part of the lumbo-sacral cord, with some filaments from the first sacral nerve ; it passes from the pelvis through the great sacro-sciatic foramen above the Pyriformis muscle, accompanied by the gluteal vessels, and divides into a superior and an inferior branch.

The *superior branch* follows the line of origin of the Gluteus minimus, and supplies the Gluteus medius.

The *inferior branch* crosses obliquely between the Gluteus minimus and medius, distributing filaments to both these muscles, and terminates in the Tensor fasciæ femoris, extending nearly to its lower end.

The **Inferior gluteal** arises from the lumbo-sacral cord, and first and second sacral nerves, and is intimately connected with the small sciatic at its origin. It passes out of the pelvis through the great sciatic notch, beneath the Pyriformis muscle, and dividing into a number of branches enters the Gluteus maximus muscle, on its under surface.

The **Small sciatic nerve** (fig. 423) supplies the integument of the perinæum and back part of the thigh and leg. It is usually formed by the union of two branches, which arise from the second and third nerves of the sacral plexus. It issues from the pelvis through the great sacro-sciatic foramen below the Pyriformis muscle, descends beneath the Gluteus maximus with the sciatic artery, and at the lower border of that muscle passes along the back part of the thigh, beneath the fascia lata and over the long head of the Biceps, to the lower part of the popliteal region, where it pierces the fascia and becomes cutaneous. It then accompanies the external saphenous vein to about the middle of the leg, its terminal filaments communicating with the external saphenous nerve.

The branches of the small sciatic nerve are all cutaneous, and are grouped as follows : gluteal, perineal, and femoral.

The *gluteal cutaneous branches* (*ascending*) consist of two or three filaments, which turn upwards round the lower border of the Gluteus maximus to supply the integument covering the lower and outer part of that muscle.

The *perineal cutaneous branches* are distributed to the skin at the upper and inner side of the thigh, on its posterior aspect. One branch, longer than the rest, the *inferior pudendal*, curves forwards below the tuber ischii, pierces the fascia

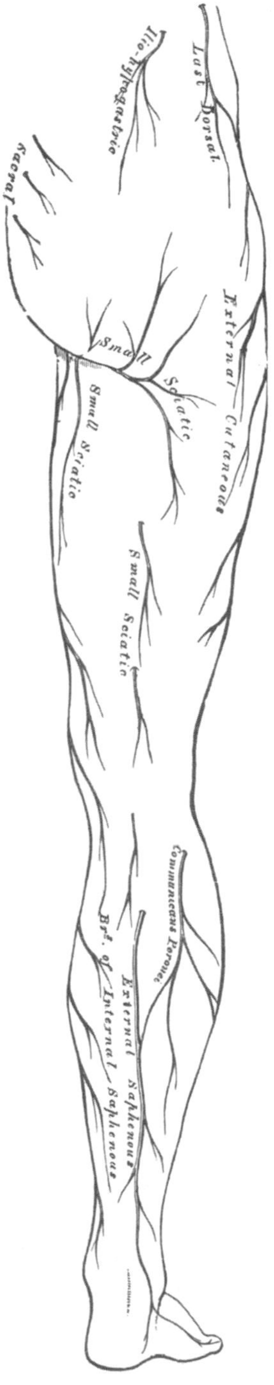

Fig. 422.—Cutaneous nerves of lower extremity. Posterior view.

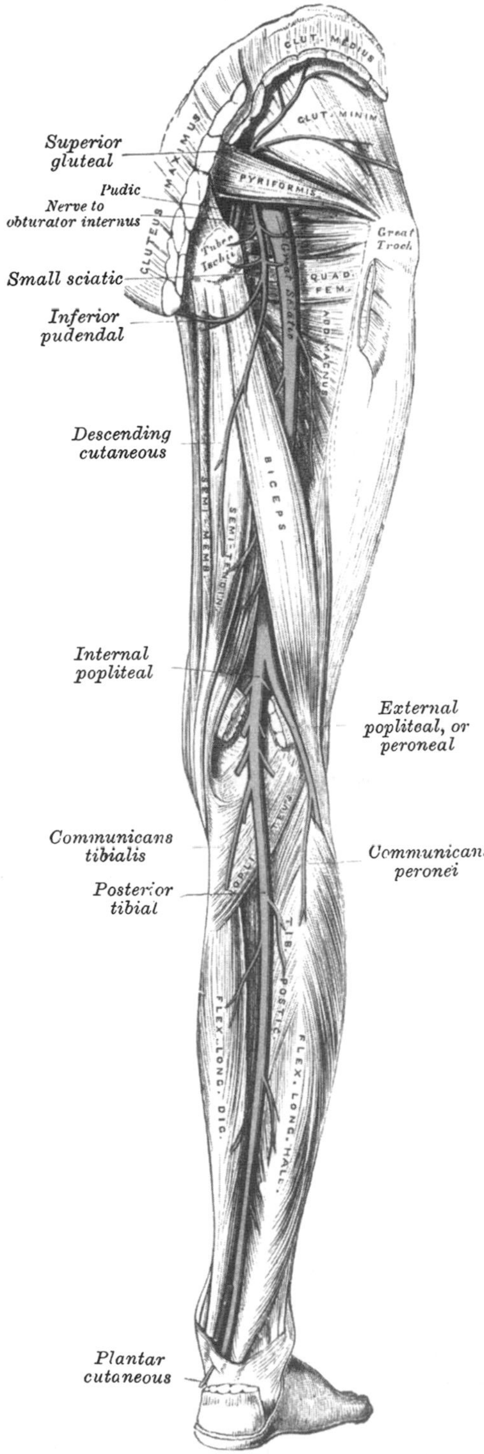

Fig. 423.—Nerves of the lower extremity.* Posterior view.

* N.B.—In this diagram the communicans tibialis and communicans peronei are|not in their normal position. They have been displaced by the removal of the superficial muscles.

lata, and passes forwards beneath the superficial fascia of the perinæum to be distributed to the integument of the scrotum in the male and the labium in the female, communicating with the superficial perineal and inferior hæmorrhoidal nerves.

The *femoral cutaneous branches* (*descending*) are numerous filaments, derived from both sides of the nerves, which are distributed to the back, inner and outer sides of the thigh, to the skin covering the popliteal space, and to the upper part of the leg.

The **Perforating cutaneous nerve** usually arises from the second and third sacral nerves, and is of small size. It is continued backwards through the great sacro-sciatic ligament, and, winding round the lower border of the Gluteus maximus, supplies the integument covering the inner and lower part of that muscle.

The **Pudic nerve** is the direct continuation of the lower cord of the sacral plexus, and derives its fibres from the third and fourth sacral nerves and frequently from the second also. It leaves the pelvis, through the great sacro-sciatic foramen, below the Pyriformis. It then crosses the spine of the ischium, and re-enters the pelvis through the lesser sacro-sciatic foramen. It accompanies the pudic vessels upwards and forwards along the outer wall of the ischio-rectal fossa, being contained in a sheath of the obturator fascia, termed *Alcock's canal*, and divides into two terminal branches, the perineal nerve, and the dorsal nerve of the penis or clitoris. Before its division, it gives off the inferior hæmorrhoidal nerve.

The *inferior hæmorrhoidal nerve* is occasionally derived separately from the sacral plexus. It passes across the ischio-rectal fossa, with its accompanying vessels, towards the lower end of the rectum, and is distributed to the Sphincter ani externus and to the integument round the anus. Branches of this nerve communicate with the inferior pudendal and superficial perineal nerves at the fore part of the perinæum.

The *perineal nerve*, the inferior and larger of the two terminal branches of the pudic, is situated below the pudic artery. It accompanies the superficial perineal artery in the perinæum, dividing into cutaneous and muscular branches.

The cutaneous branches (superficial perineal) are two in number, posterior and anterior. The *posterior* or *external branch* pierces the base of the triangular ligament of the urethra, and passes forwards along the outer side of the urethral triangle in company with the superficial perineal artery ; it is distributed to the skin of the scrotum. It communicates with the inferior hæmorrhoidal, the inferior pudendal, and the other superficial perineal nerve. The *anterior* or *internal branch* also pierces the base of the triangular ligament, and passes forwards nearer to the middle line, to be distributed to the inner and back part of the scrotum. Both these nerves supply the labia majora in the female.

The muscular branches are distributed to the Transversus perinæi, Accelerator urinæ, Erector penis, and Compressor urethræ. A distinct branch is given off from the nerve to the Accelerator urinæ, which pierces this muscle, and supplies the corpus spongiosum, ending in the mucous membrane of the urethra. This is the nerve to the bulb.

The *dorsal nerve of the penis* is the deepest division of the pudic nerve ; it accompanies the pudic artery along the ramus of the ischium ; it then runs forwards along the inner margin of the ramus of the os pubis, between the superficial and deep layers of the triangular ligament. Piercing the superficial layer it gives a branch to the corpus cavernosum, and passes forwards, in company with the dorsal artery of the penis, between the layers of the suspensory ligament on to the dorsum of the penis, along which it is carried as far as the glans to which it is distributed.

In the female the dorsal nerve is very small, and supplies the clitoris.

The **Great sciatic nerve** (fig. 423) supplies nearly the whole of the integument of the leg, the muscles of the back of the thigh, and those of the leg and foot.

It is the largest nervous cord in the body, measuring three-quarters of an inch in breadth, and is the continuation of the upper division of the sacral plexus. It passes out of the pelvis through the great sacro-sciatic foramen, below the Pyriformis muscle. It descends between the trochanter major and tuberosity of the ischium, along the back part of the thigh to about its lower third, where it divides into two large branches, the *internal* and *external popliteal nerves*.

This division may take place at any point between the sacral plexus and the lower third of the thigh. When the division occurs at the plexus, the two nerves descend together, side by side; or they may be separated, at their commencement, by the interposition of part or the whole of the Pyriformis muscle. As the nerve descends along the back of the thigh, it rests upon the posterior surface of the ischium, the nerve to the Quadratus femoris, and the External rotator muscles, in company with the small sciatic nerve and artery, being covered by the Gluteus maximus; lower down, it lies upon the Adductor magnus, and is covered by the long head of the Biceps.

The *branches* of the nerve, before its division, are articular and muscular.

The *articular branches* arise from the upper part of the nerve; they supply the hip-joint, perforating the posterior part of its fibrous capsule posteriorly. These branches are sometimes derived from the sacral plexus.

The *muscular branches* are distributed to the flexors of the leg: viz. the Biceps, Semitendinosus, and Semimembranosus, and a branch to the Adductor magnus. These branches are given off beneath the Biceps muscle.

The **Internal popliteal nerve,** the larger of the two terminal branches of the great sciatic, descends along the back part of the thigh, through the middle of the popliteal space to the lower part of the Popliteus muscle, where it passes with the artery beneath the arch of the Soleus, and becomes the posterior tibial. It is overlapped by the hamstring muscles above, and then becomes more superficial, and lies to the outer side of, and some distance from, the popliteal vessels; opposite the knee-joint, it is in close relation with the vessels, and crosses to the inner side of the artery. Below, it is overlapped by the Gastrocnemius.

The *branches* of this nerve are, articular, muscular, and a cutaneous branch, the *communicans tibialis nerve.*

The *articular branches,* usually three in number, supply the knee-joint; two of these branches accompany the superior and inferior internal articular arteries; and a third, the azygos articular artery.

The *muscular branches,* four or five in number, arise from the nerve as it lies between the two heads of the Gastrocnemius muscle; they supply that muscle, the Plantaris, Soleus, and Popliteus. The filaments which supply the Popliteus turn round its lower border and are distributed to its deep surface.

The *communicans tibialis* descends between the two heads of the Gastrocnemius muscle, and, about the middle of the back of the leg, pierces the deep fascia, and joins a communicating branch (*communicans peronei*) from the external popliteal nerve to form the external, or short, saphenous (fig. 422).

The *external saphenous nerve,* formed by the communicating branches of the internal and external popliteal nerves, passes downwards and outwards near the outer margin of the tendo Achillis, lying close to the external saphenous vein, to the interval between the external malleolus and the os calcis. It winds round the outer malleolus, and is distributed to the integument along the outer side of the foot and little toe, communicating on the dorsum of the foot with the musculo-cutaneous nerve. In the leg, its branches communicate with those of the small sciatic.

The **Posterior tibial nerve** (fig. 423) commences at the lower border of the Popliteus muscle, and passes along the back part of the leg with the posterior tibial vessels to the interval between the inner malleolus and the heel, where it divides into the *external* and *internal plantar nerves.* It lies upon the deep muscles of the leg, and is covered in the upper part by the muscles of the calf, lower down by the skin and fascia. In the upper part of its course, it lies to the inner side of the posterior tibial artery; but it soon crosses that vessel, and lies to its outer side as far as the ankle. In the lower third of the leg, it is placed parallel with the inner margin of the tendo Achillis.

The *branches of the posterior tibial nerve* are muscular, calcaneo-plantar, and articular.

The *muscular branches* arise either separately or by a common trunk from the upper part of the nerve. They supply the Soleus, Tibialis posticus, Flexor longus digitorum, and Flexor longus hallucis muscles; the branch to the latter muscle accompanying the peroneal artery. The branch to the Soleus enters its deep surface, while the branch which this muscle receives from the internal popliteal enters its superficial aspect.

The *calcaneo-plantar (internal calcanean) branch* perforates the internal annular ligament, and supplies the integument of the heel and inner side of the sole of the foot.

The *articular branch* is given off just above the bifurcation of the nerve, and supplies the ankle-joint.

The **internal plantar nerve** (fig. 424), the larger of the two terminal branches of the posterior tibial, accompanies the internal plantar artery along the inner side of the foot. From its origin at the inner ankle it passes beneath the Abductor hallucis, and then forwards between this muscle and the Flexor brevis digitorum; it divides opposite the bases of the metatarsal bones into four digital branches, and communicates with the external plantar nerve.

FIG. 424.—The plantar nerves.

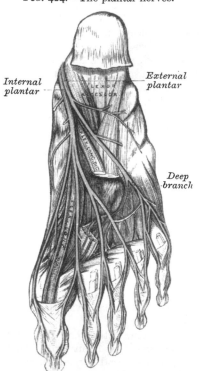

Internal plantar — — *External plantar*

Deep branch

Branches.—In its course, the internal plantar nerve gives off *cutaneous branches*, which pierce the plantar fascia, and supply the integument of the sole of the foot; *muscular branches*, which supply the Abductor hallucis and Flexor brevis digitorum; *articular branches* to the articulations of the tarsus and metatarsus; and *four digital branches*. The three outer branches pass between the divisions of the plantar fascia in the clefts between the toes: the first (innermost) branch becomes cutaneous farther back between the Abductor hallucis and Flexor brevis digitorum. They are distributed in the following manner: The *first* supplies the inner border of the great toe, and sends a filament to the Flexor brevis hallucis muscle; the *second* bifurcates, to supply the adjacent sides of the great and second toes, sending a filament to the First lumbrical muscle;* the *third digital branch* supplies the adjacent sides of the second and third toes; the *fourth* supplies the corresponding sides of the third and fourth toes, and receives a communicating branch from the external plantar nerve. Each digital nerve gives off cutaneous and articular filaments; and opposite the last phalanx sends a dorsal branch, which supplies the structures around the nail, the continuation of the nerve being distributed to the ball of the toe. It will be observed, that the distribution of these branches is precisely similar to that of the median nerve in the hand.

* See footnote, p. 424.

The **external plantar nerve,** the smaller of the two, completes the nervous supply to the structures of the sole of the foot, being distributed to the little toe and one-half of the fourth, as well as to most of the deep muscles, its distribution being similar to that of the ulnar in the hand. It passes obliquely forwards with the external plantar artery to the outer side of the foot, lying between the Flexor brevis digitorum and Flexor accessorius; and, in the interval between the former muscle and Abductor minimi digiti, divides into a superficial and a deep branch. Before its division, it supplies the Flexor accessorius and Abductor minimi digiti.

The *superficial branch* separates into two digital nerves: one, the smaller of the two, supplies the outer side of the little toe, the Flexor brevis minimi digiti, and the two Interosseous muscles of the fourth metatarsal space; the other and larger digital branch supplies the adjoining sides of the fourth and fifth toes, and communicates with the internal plantar nerve.

The *deep* or *muscular branch* accompanies the external plantar artery into the deep part of the sole of the foot, beneath the tendons of the Flexor muscles and Adductor obliquus hallucis, and supplies all the Interossei (except those in the fourth metatarsal space), the three outer Lumbricales, the Adductor obliquus hallucis, and the Adductor transversus hallucis.

The **External popliteal** or **peroneal nerve** (fig. 423), about one-half the size of the internal popliteal, descends obliquely along the outer side of the popliteal space to the head of the fibula, close to the inner margin of the Biceps muscle. It is easily felt beneath the skin behind the head of the fibula, at the inner side of the tendon of the Biceps. It passes between the tendon of the Biceps and outer head of the Gastrocnemius muscle, winds round the neck of the fibula, between the Peroneus longus and the bone, and divides beneath the muscle into the anterior tibial and musculo-cutaneous nerves.

The *branches of the peroneal nerve,* previous to its division, are articular and cutaneous.

The *articular branches* are three in number; two of these accompany the superior and inferior external articular arteries to the outer side of the knee. The upper one occasionally arises from the great sciatic nerve before its bifurcation. The third (*recurrent*) articular nerve is given off at the point of division of the peroneal nerve; it ascends with the anterior recurrent tibial artery through the Tibialis anticus muscle to the front of the knee, which it supplies.

The *cutaneous branches,* two or three in number, supply the integument along the back part and outer side of the leg; one of these, larger than the rest, the *communicans peronei,* arises near the head of the fibula, crosses the external head of the Gastrocnemius to the middle of the leg, and joins with the communicans tibialis to form the external saphenous. This nerve occasionally exists as a separate branch, which is continued down as far as the heel.

The **Anterior tibial nerve** (fig. 419) commences at the bifurcation of the peroneal nerve, between the fibula and upper part of the Peroneus longus, passes obliquely forwards beneath the Extensor longus digitorum to the fore part of the interosseous membrane, and gets into relation with the anterior tibial artery above the middle of the leg; it then descends with the artery to the front of the ankle-joint, where it divides into an external and an internal branch. This nerve lies at first on the outer side of the anterior tibial artery, then in front of it, and again at its outer side at the ankle-joint.

The *branches of the anterior tibial nerve,* in its course through the leg, are the muscular branches to the Tibialis anticus, Extensor longus digitorum, Peroneus tertius, and Extensor proprius hallucis muscles, and an *articular branch* to the ankle-joint.

The *external* or *tarsal branch of the anterior tibial* passes outwards across the tarsus, beneath the Extensor brevis digitorum, and, having become enlarged like the posterior interosseous nerve at the wrist, supplies the Extensor brevis digitorum. From the enlargement three minute *interosseous branches* are given off, which supply the tarsal joints and the metatarso-phalangeal joints of the second, third. and fourth toes. The first of these sends a filament to the second dorsal interosseous muscle.

The *internal branch*, the continuation of the nerve, accompanies the dorsalis pedis artery along the inner side of the dorsum of the foot, and, at the first interosseous space, divides into two branches, which supply the adjacent sides of the great and second toes, communicating with the internal branch of the musculocutaneous nerve. Before it divides it gives off an *interosseous branch* to the first space, which supplies the metatarso-phalangeal joint of the great toe and sends a filament to the First dorsal interosseous muscle.

The **Musculo-cutaneous nerve** (fig. 419) supplies the muscles on the fibular side of the leg, and the integument of the dorsum of the foot. It passes forwards between the Peronei muscles and the Extensor longus digitorum, pierces the deep fascia at the lower third of the leg, on its front and outer side, and divides into two branches. This nerve, in its course between the muscles, gives off muscular branches to the Peroneus longus and brevis, and cutaneous filaments to the integument of the lower part of the leg.

The *internal branch of the musculo-cutaneous nerve* passes in front of the ankle-joint, and divides into two branches, one of which supplies the inner side of the great toe, the other, the adjacent sides of the second and third toes. It also supplies the integument of the inner ankle and inner side of the foot, communicating with the internal saphenous nerve, and joining with the anterior tibial nerve, between the great and second toes.

The *external branch*, the smaller, passes along the outer side of the dorsum of the foot, and divides into two branches, the inner being distributed to the contiguous sides of the third and fourth toes, the outer to the opposed sides of the fourth and fifth toes. It also supplies the integument of the outer ankle and outer side of the foot, communicating with the short saphenous nerve.

The branches of the musculo-cutaneous nerve supply all the toes excepting the outer side of the little toe, and the adjoining sides of the great and second toes, the former being supplied by the external saphenous, and the latter by the internal branch of the anterior tibial. It frequently happens, however, that some of the outer branches of the musculo-cutaneous are absent, their place being then taken by branches of the external saphenous nerve.

Surgical Anatomy.—The lumbar plexus passes through the Psoas muscle, and therefore in psoas abscess, any or all of its branches may be irritated, causing severe pain in the part to which the irritated nerves are distributed. The genito-crural nerve is the one which is most frequently implicated. This nerve is also of importance as it is concerned in one of the principal reflexes employed in the investigation of diseases of the spine. If the skin over the inner side of the thigh just below Poupart's ligament, the part supplied by the crural branch of the genito-crural nerve, be gently tickled in a male child, the testicle will be noticed to be drawn upwards, through the action of the Cremaster muscle, supplied by the genital branch of the same nerve. The same result may sometimes be noticed in adults, and can almost always be produced by severe stimulation. This reflex, when present, shows that the portion of the cord from which the first and second lumbar nerves are derived is in a normal condition.

The anterior crural nerve is in danger of being injured in fractures of the true pelvis, since the fracture most commonly takes place through the ascending ramus of the os pubis, at or near the point where this nerve crosses the bone. It is also liable to be injured in fractures and dislocations of the femur, and is likely to be pressed upon, and its functions impaired, in some tumours growing in the pelvis. Moreover, on account of its superficial position, it is exposed to injury in wounds and stabs in the groin. When this nerve is paralysed, the patient is unable to flex his hip completely, on account of the loss of motion in the Iliacus; or to extend the knee on the thigh, on account of paralysis of the Quadriceps extensor cruris; there is complete paralysis of the Sartorius, and partial paralysis of the Pectineus. There is loss of sensation down the front and inner side of the thigh, except in that part supplied by the crural branch of the genito-crural, and by the ilio-inguinal. There is also loss of sensation down the inner side of the leg and foot as far as the ball of the great toe.

The obturator nerve is of special surgical interest. It is rarely paralysed alone, but occasionally in association with the anterior crural. The principal interest attached to it is in connection with its supply to the knee; pain in the knee being symptomatic of many diseases in which the trunk of this nerve, or one of its branches, is irritated. Thus it is well known that in the earlier stages of hip-joint disease the patient does not complain of pain in that articulation, but on the inner side of the knee, or in the knee-joint itself, both these articulations being supplied by the obturator nerve, the final distribution of the nerve being to the knee-joint. Again, the same thing occurs in sacro-iliac disease : pain is complained of in the knee-joint, or on its inner side. The obturator nerve is in close relationship with the sacro-iliac articulation, passing over it, and, according to some anatomists, distributing filaments to it. Again, in cancer of the sigmoid flexure, and even in cases where masses of hardened fæces are impacted in this portion of the gut, pain is complained of in the knee. The left obturator nerve lies beneath the sigmoid flexure, and is readily pressed upon and irritated when disease exists in this part of the intestine. Finally, pain in the knee forms an important diagnostic sign in obturator hernia. The hernial protrusion as it passes out through the opening in the obturator membrane presses upon the nerve and causes pain in the parts supplied by its peripheral filaments. When the obturator nerve is paralysed, the patient is unable to press his knees together or to cross one leg over the other, on account of paralysis of the Adductor muscles. Rotation outwards of the thigh is impaired from paralysis of the Obturator externus. Sometimes there is loss of sensation in the upper half of the inner side of the thigh.

The great sciatic nerve is liable to be pressed upon by various forms of pelvic tumours, giving rise to pain along its trunk, to which the term sciatica is applied. Tumours growing from the pelvic viscera, or bones, aneurisms of some of the branches of the internal iliac artery, calculus in the bladder when of large size, accumulation of fæces in the rectum, may all cause pressure on the nerve inside the pelvis, and give rise to sciatica. Outside the pelvis exposure to cold, violent movements of the hip-joint, exostoses or other tumours growing from the margin of the sacro-sciatic foramen, may also give rise to the same condition. When paralysed there is loss of motion in all the muscles below the knee, and loss of sensation in the same situation, except the upper half of the back of the leg, supplied by the small sciatic and the upper half of the inner side of the leg, when the communicating branch of the obturator is large (see page 765).

The sciatic nerve has been frequently cut down upon and stretched, or has been acu-punctured for the relief of sciatica. The nerve has also been stretched in cases of locomotor ataxy, the anæsthesia of leprosy, &c. In order to define it on the surface, a point is taken at the junction of the middle and lower third of a line stretching from the posterior superior spine of the ilium to the outer part of the tuber ischii, and a line drawn from this to the middle of the upper part of the popliteal space. The line must be slightly curved with its convexity outwards, and as it passes downwards to the lower border of the Gluteus maximus is slightly nearer the tuber ischii than the great trochanter, as it crosses a line drawn between these two points. The operation of stretching the sciatic nerve is performed by making an incision over the course of the nerve about the centre of the thigh. The skin, superficial structures, and deep fascia having been divided, the interval between the inner and outer hamstrings is to be defined, and these muscles pulled inwards and outwards with retractors. The nerve will be found a little to the inner side of the Biceps. It is to be separated from the surrounding structures, hooked up with the finger, and stretched by steady and continuous traction for two or three minutes. The sciatic nerve may also be stretched by what is known as the 'dry' plan. The patient is laid on his back, the foot is extended, the leg flexed on the thigh, and the thigh strongly flexed on the abdomen. While the thigh is maintained in this position, the leg is forcibly extended to its full extent, and the foot as fully flexed on the leg.

The position of the external popliteal, close behind the tendon of the Biceps on the outer side of the ham, should be remembered in subcutaneous division of the tendon. After it is divided, a cord often rises up close beside it, which might be mistaken for a small undivided portion of the tendon, and the surgeon might be tempted to reintroduce his knife and divide it. This must never be done, as the cord is the external popliteal nerve, which becomes prominent as soon as the tendon is divided.

THE SYMPATHETIC NERVOUS SYSTEM

The **Sympathetic Nervous System** consists of (1) a series of ganglia, connected together by intervening cords, extending from the base of the skull to the coccyx, one on each side of the middle line of the body, partly in front and partly on each side of the vertebral column ; (2) of three great gangliated plexuses, or aggregations of nerves and ganglia, situated in front of the spine in the thoracic, abdominal, and pelvic cavities respectively ; (3) of smaller ganglia situated in relation with the abdominal viscera ; and (4) of numerous nerve-fibres. These latter are of two kinds : *communicating*, by which the ganglia communicate with each other and with the cerebro-spinal nerves; and *distributory*, supplying the internal viscera and the coats of the blood-vessels.

Each **gangliated cord** may be traced upwards from the base of the skull into its cavity by an ascending branch, which passes through the carotid canal, forms a plexus on the internal carotid artery, and communicates with the ganglia on the first and second divisions of the fifth nerve. According to some anatomists, the two cords are joined, at their cephalic extremities, by these ascending branches communicating in a small ganglion (the *ganglion of Ribes*), situated upon the anterior communicating artery. The ganglia of these cords are distinguished as cervical, dorsal, lumbar, and sacral, and except in the neck they correspond pretty nearly in number to the vertebræ against which they lie. They may be thus arranged :

Cervical portion	. .	3 pairs of ganglia.
Dorsal　　,,	. .	12　　,,　　　,,
Lumbar　,,	. .	4　　,,　　　,,
Sacral　　,,	. .	4 or 5　,,　　　,,

In the neck they are situated in front of the transverse processes of the vertebræ; in the dorsal region, in front of the heads of the ribs; in the lumbar region, on the sides of the bodies of the vertebræ; and in the sacral region, in front of the sacrum. As the two cords pass into the pelvis, they converge and unite together in a single ganglion (*ganglion impar*), placed in front of the coccyx. Each ganglion may be regarded as a distinct centre, and, in addition to its branches of distribution, possesses also branches of communication, which communicate with other ganglia and with the cerebro-spinal nerves.

The branches of communication between the ganglia are composed of grey and white nerve-fibres, the latter being continuous with those fibres of the spinal nerves which pass to the ganglia.

The branches of communication between the ganglia and the cerebro-spinal nerves also consist of white and grey nerve-fibres, which may be contained in separate filaments or united in a single branch; the former proceeding from the spinal nerve *to* the ganglion, the latter passing *from* the ganglion to the spinal nerve, so that a double interchange takes place between the two systems. While grey communicating fibres pass from all the sympathetic ganglia to all the spinal nerves, it would appear that the white communicating fibres from the spinal nerves to the sympathetic only exist in the dorsal and upper lumbar regions.

The **three great gangliated plexuses** are situated in front of the spine in the thoracic, abdominal, and pelvic regions, and are named respectively, the *cardiac*, the *solar* or *epigastric*, and the *hypogastric plexus*. They consist of collections of nerves and ganglia; the nerves being derived from the gangliated cords, and from the cerebro-spinal nerves. They distribute branches to the viscera.

Smaller ganglia are also found lying amidst the nerves, some of them of microscopic size, in certain viscera—as, for instance, in the heart, the stomach, and the uterus. They serve as additional centres for the origin of nerve-fibres.

The **branches of distribution** derived from the gangliated cords, from the prevertebral plexuses, and also from the smaller ganglia, are principally destined for the blood-vessels and thoracic and abdominal viscera, supplying the involuntary muscular fibres of the coats of the vessels and the hollow viscera, and the secreting cells, as well as the muscular coats of the vessels in the glandular viscera.

CERVICAL PORTION OF THE GANGLIATED CORD

The cervical portion of the gangliated cord consists of three ganglia on each side, which are distinguished, according to their position, as the superior, middle, and inferior cervical.

The **Superior cervical ganglion,** the largest of the three, is placed opposite the second and third cervical vertebræ. It is of a reddish-grey colour, and usually

FIG. 425.—The sympathetic nerve.

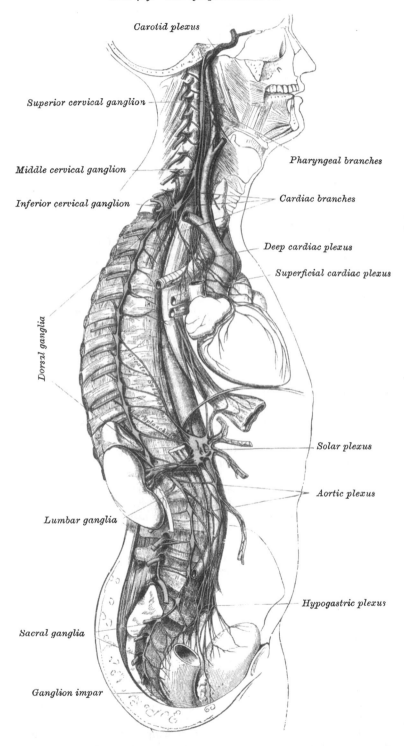

Carotid plexus

Superior cervical ganglion

Pharyngeal branches

Middle cervical ganglion

Inferior cervical ganglion

Cardiac branches

Deep cardiac plexus

Superficial cardiac plexus

Dorsal ganglia

Solar plexus

Aortic plexus

Lumbar ganglia

Hypogastric plexus

Sacral ganglia

Ganglion impar

fusiform in shape ; sometimes broad and flattened, and occasionally constricted at intervals, so as to give rise to the opinion that it consists of the coalescence of several smaller ganglia ; and it is usually believed that it is formed by the coalescence of the four ganglia, corresponding to the four upper cervical nerves. It is in relation, in front, with the sheath of the internal carotid artery, and internal jugular vein ; behind, it lies on the Rectus capitis anticus major muscle.

Its branches may be divided into superior, inferior, external, internal, and anterior.

The *superior branch* appears to be a direct prolongation of the ganglion. It is soft in texture, and of a reddish colour. It ascends by the side of the internal carotid artery, and, entering the carotid canal in the temporal bone, divides into two branches, which lie, one on the outer, and the other on the inner side of that vessel.

The *outer branch*, the larger of the two, distributes filaments to the internal carotid artery, and forms the *carotid plexus.*

The *inner branch* also distributes filaments to the internal carotid, and, continuing onwards, forms the *cavernous plexus.*

The **carotid plexus** is situated on the outer side of the internal carotid. Filaments from this plexus occasionally form a small gangliform swelling on the under surface of the artery, which is called the carotid ganglion. The carotid plexus communicates with the Gasserian ganglion, with the sixth nerve, and the spheno-palatine ganglion, and distributes filaments to the wall of the carotid artery, and to the dura mater (Valentin) ; while in the carotid canal it communicates with Jacobson's nerve (tympanic branch of the glosso-pharyngeal).

The *communicating branches with the sixth nerve* consist of one or two filaments which join that nerve as it lies upon the outer side of the internal carotid. Other filaments are also connected with the Gasserian ganglion. The communication with the spheno-palatine ganglion is effected by a branch, the *large deep petrosal*, which is given off from the plexus on the outer side of the artery, and which passes through the cartilage filling up the foramen lacerum medium, and joins the great superficial petrosal to form the Vidian nerve. The Vidian nerve then proceeds along the pterygoid or Vidian canal to the spheno-palatine ganglion. The communication with Jacobson's nerve is effected by two branches, one of which is called the *small deep petrosal nerve*, and the other the *carotico-tympanic* ; the latter may consist of two or three delicate filaments.

The **cavernous plexus** is situated below, and internal to that part of the internal carotid which is placed by the side of the sella Turcica, in the cavernous sinus, and is formed chiefly by the internal division of the ascending branch from the superior cervical ganglion. It communicates with the third, the fourth, the ophthalmic division of the fifth, and the sixth nerves, and with the ophthalmic ganglion, and distributes filaments to the wall of the internal carotid. The branch of communication with the third nerve joins it at its point of division ; the branch to the fourth nerve joins it as it lies on the outer wall of the cavernous sinus ; other filaments are connected with the under surface of the trunk of the ophthalmic nerve ; and a second filament of communication joins the sixth nerve.

The filament of connection with the ophthalmic ganglion arises from the anterior part of the cavernous plexus ; it accompanies the nasal nerve, or continues forwards as a separate branch.

The terminal filaments from the carotid and cavernous plexuses are prolonged along the internal carotid, forming plexuses which entwine round the cerebral and ophthalmic arteries ; along the former vessel they may be traced on to the pia mater ; along the latter, into the orbit, where they accompany each of the subdivisions of the vessel, a separate plexus passing, with the arteria centralis retinæ, into the interior of the eyeball. The filaments prolonged on to the

Fig. 426.—Plan of the cervical portion of the sympathetic. (After Flower.)

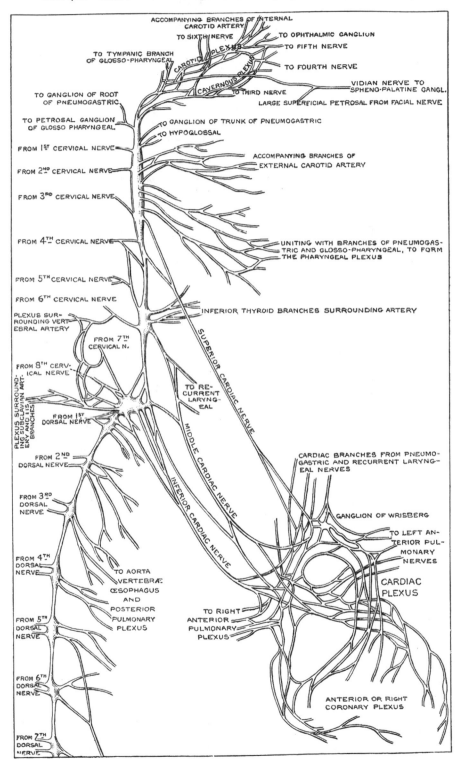

anterior communicating artery form a small ganglion, the *ganglion of Ribes*,* which serves, as mentioned above, to connect the sympathetic nerves of the right and left sides.

The **inferior** or **descending branch of the superior cervical ganglion** communicates with the middle cervical ganglion.

The **external branches** are numerous, and communicate with the cranial nerves and with the four upper spinal nerves. Sometimes the branch to the fourth spinal nerve may come from the cord connecting the upper and middle cervical ganglia. The branches of communication with the cranial nerves consist of delicate filaments, which pass from the superior cervical ganglion to the ganglion of the trunk of the pneumogastric, and to the hypoglossal nerve. A separate filament from the cervical ganglion subdivides and joins the petrosal ganglion of the glosso-pharyngeal, and the ganglion of the root of the pneumogastric in the jugular foramen.

The **internal branches** are three in number : the *pharyngeal, laryngeal,* and *superior cardiac nerve.* The *pharyngeal branches* pass inwards to the side of the pharynx, where they join with branches from the glosso-pharyngeal, pneumogastric, and external laryngeal nerves to form the *pharyngeal plexus.* The *laryngeal branches* unite with the superior laryngeal nerve and its branches.

The *superior cardiac nerve (nervus superficialis cordis)* arises by two or more branches from the superior cervical ganglion, and occasionally receives a filament from the cord of communication between the first and second cervical ganglia. It runs down the neck behind the common carotid artery, lying upon the Longus colli muscle ; and crosses in front of the inferior thyroid artery, and recurrent laryngeal nerve.

The *right superior cardiac nerve*, at the root of the neck, passes either in front of or behind the subclavian artery, and along the arteria innominata, to the back part of the arch of the aorta, where it joins the deep cardiac plexus. This nerve, in its course, is connected with other branches of the sympathetic ; about the middle of the neck it receives filaments from the external laryngeal nerve ; lower down, one or two twigs from the pneumogastric ; and as it enters the thorax it is joined by a filament from the recurrent laryngeal. Filaments from this nerve communicate with the thyroid branches from the middle cervical ganglion.

The *left superior cardiac nerve*, in the chest, runs by the side of the left common carotid artery, and in front of the arch of the aorta, to the superficial cardiac plexus ; but occasionally it passes behind the aorta, and terminates in the deep cardiac plexus.

The **anterior branches** ramify upon the external carotid artery and its branches, forming round each a delicate plexus, on the nerves composing which small ganglia are occasionally found. The plexuses accompanying some of these arteries have important communications with other nerves. That surrounding the external carotid is connected with the branch of the facial nerve to the Stylohyoid muscle ; that surrounding the facial communicates with the submaxillary ganglion by one or two filaments ; and that accompanying the middle meningeal artery sends offsets which pass to the otic ganglion and to the geniculate ganglion of the facial nerve (external petrosal).

The **Middle cervical ganglion** (*thyroid ganglion*) is the smallest of the three cervical ganglia, and is occasionally altogether wanting. It is placed opposite the sixth cervical vertebra, usually upon, or close to, the inferior thyroid artery ; hence the name ' thyroid ganglion,' assigned to it by Haller. It is probably formed by the coalescence of two ganglia corresponding to the fifth and sixth cervical nerves.

Its *superior branches* ascend to communicate with the superior cervical ganglion.

Its *inferior branches* descend to communicate with the inferior cervical ganglion.

* The existence of this ganglion is doubted by some observers.

Its *external branches* pass outwards to join the fifth and sixth spinal nerves. These branches are not constantly found.

Its *internal branches* are the thyroid and the middle cardiac nerve.

The *thyroid branches* are small filaments, which accompany the inferior thyroid artery to the thyroid gland; they communicate, on the artery, with the superior cardiac nerve, and, in the gland, with branches from the recurrent and external laryngeal nerves.

The *middle cardiac nerve* (*nervus cardiacus magnus*), the largest of the three cardiac nerves, arises from the middle cervical ganglion, or from the cord between the middle and inferior ganglia. On the right side it descends behind the common carotid artery; and at the root of the neck passes either in front of or behind the subclavian artery; it then descends on the trachea, receives a few filaments from the recurrent laryngeal nerve, and joins the right side of the deep cardiac plexus. In the neck, it communicates with the superior cardiac and recurrent laryngeal nerves. On the left side, the middle cardiac nerve enters the chest between the left carotid and subclavian arteries, and joins the left side of the deep cardiac plexus.

The **Inferior Cervical Ganglion** is situated between the base of the transverse process of the last cervical vertebra and the neck of the first rib, on the inner side of the superior intercostal artery. Its form is irregular; it is larger in size than the preceding, and frequently joined with the first thoracic ganglion. It is probably formed by the coalescence of two ganglia which correspond to the two last cervical nerves.

Its *superior branches* communicate with the middle cervical ganglion.

Its *inferior branches* descend, some in front of, others behind the subclavian artery, to join the first thoracic ganglion.

Its *internal branch* is the inferior cardiac nerve.

The *inferior cardiac nerve* (*nervus cardiacus minor*) arises from the inferior cervical or first thoracic ganglion. It passes down behind the subclavian artery and along the front of the trachea, to join the deep cardiac plexus. It communicates freely behind the subclavian artery with the recurrent laryngeal and middle cardiac nerves.

The *external branches* consist of several filaments, some of which communicate with the seventh and eighth spinal nerves; others accompany the vertebral artery along the vertebral canal, forming a plexus round the vessel, supplying it with filaments, which are continued up the vertebral and basilar to the cerebral arteries. The branches communicate with the cervical spinal nerves.

Thoracic Portion of the Gangliated Cord

The thoracic portion of the gangliated cord consists of a series of ganglia, which usually correspond in number to that of the vertebræ; but, from the occasional coalescence of two, their number is uncertain. These ganglia are placed on each side of the spine, resting against the heads of the ribs, and covered by the pleura costalis; the last two are, however, anterior to the rest, being placed on the side of the bodies of the eleventh and twelfth dorsal vertebræ. The ganglia are small in size, and of a greyish colour. The first, larger than the rest, is of an elongated form, and frequently blended with the last cervical. They are connected together by cord-like prolongations from their substance.

The *external branches* from each ganglion, usually two in number, communicate with each of the dorsal spinal nerves.

The *internal branches from the five or six upper ganglia* are very small; they supply filaments to the thoracic aorta and its branches, besides small branches to the bodies of the vertebræ and their ligaments. Branches from the third and fourth, and sometimes, also, from the first and second ganglia form part of the posterior pulmonary plexus.

The *internal branches from the six or seven lower ganglia* are large and white in colour; they distribute filaments to the aorta, and unite to form the three splanchnic nerves. These are named the *great*, the *lesser*, and the *smallest* or *renal splanchnic*.

The *great splanchnic nerve* is of a white colour, firm in texture, and bears a marked contrast to the ganglionic nerves. It is formed by branches from the thoracic ganglia between the fifth or sixth and the ninth or tenth, but the fibres in the higher roots may be traced upwards in the sympathetic cord as far as the first or second thoracic ganglia. These roots unite to form a large round cord of considerable size. It descends obliquely inwards in front of the bodies of the vertebræ along the posterior mediastinum, perforates the crus of the Diaphragm, and terminates in the semilunar ganglion of the solar plexus, distributing filaments to the renal and suprarenal plexus.

The *lesser splanchnic nerve* is formed by filaments from the tenth and eleventh ganglia, and from the cord between them. It pierces the Diaphragm with the preceding nerve, and joins the solar plexus. It communicates in the chest with the great splanchnic nerve, and occasionally sends filaments to the renal plexus.

The *smallest*, or *renal splanchnic nerve* arises from the last ganglion, and, piercing the Diaphragm, terminates in the renal plexus and lower part of the solar plexus. It occasionally communicates with the preceding nerve.

A striking analogy appears to exist between the splanchnic and the cardiac nerves. The cardiac nerves are three in number; they arise from the three cervical ganglia, and are distributed to a large and important organ in the thoracic cavity. The splanchnic nerves, also three in number, are connected probably with all the dorsal ganglia, and are distributed to important organs in the abdominal cavity.

LUMBAR PORTION OF THE GANGLIATED CORD

The lumbar portion of the gangliated cord is situated in front of the vertebral column, along the inner margin of the Psoas muscle. It consists usually of four ganglia, connected together by interganglionic cords. The ganglia are of small size, of a greyish colour, shaped like a barleycorn, and placed much nearer the median line than the thoracic ganglia.

The *superior* and *inferior branches of the lumbar ganglia* serve as communicating branches between the chain of ganglia in this region. They are usually single, and of a white colour.

The *external branches* communicate with the lumbar spinal nerves. From the situation of the lumbar ganglia, these branches are longer than in the other regions. They are usually two in number from each ganglion, but their connection with the spinal nerves is not so uniform as in other regions. They accompany the lumbar arteries around the sides of the bodies of the vertebræ, passing beneath the fibrous arches from which some of the fibres of the Psoas muscle arise.

Of the *internal branches*, some pass inwards, in front of the aorta, and help to form the aortic plexus. Other branches descend in front of the common iliac arteries, and join over the promontory of the sacrum, helping to form the hypogastric plexus. Numerous delicate filaments are also distributed to the bodies of the vertebræ, and the ligaments connecting them.

PELVIC PORTION OF THE GANGLIATED CORD

The pelvic portion of the gangliated cord is situated in front of the sacrum, along the inner side of the anterior sacral foramina. It consists of four or five small ganglia on each side, connected together by interganglionic cords. Below, these cords converge and unite on the front of the coccyx, by means of a small ganglion (the *coccygeal ganglion*, or *ganglion impar*).

The *superior* and *inferior branches* are the cords of communication between the ganglia above and below.

The *external branches*, exceedingly short, communicate with the sacral nerves. They are two in number from each ganglion. The coccygeal nerve communicates with either the last sacral, or the coccygeal ganglion.

The *internal branches* communicate, on the front of the sacrum, with the corresponding branches from the opposite side ; some, from the first two ganglia, pass to join the pelvic plexus, and others form a plexus, which accompanies the middle sacral artery and sends filaments to the coccygeal gland.

THE GREAT PLEXUSES OF THE SYMPATHETIC

The great plexuses of the sympathetic are the large aggregations of nerves and ganglia, above alluded to, situated in the thoracic, abdominal, and pelvic cavities respectively. From them are derived the branches which supply the viscera.

Cardiac Plexus

The cardiac plexus is situated at the base of the heart, and is divided into a *superficial part*, which lies in the concavity of the arch of the aorta, and a *deep part*, which lies between the trachea and aorta. The two plexuses are, however, closely connected.

The **great** or **deep cardiac plexus** (*plexus magnus profundus*, Scarpa) is situated in front of the trachea at its bifurcation, above the point of division of the pulmonary artery, and behind the arch of the aorta. It is formed by the cardiac nerves derived from the cervical ganglia of the sympathetic, and the cardiac branches of the recurrent laryngeal and pneumogastric. The only cardiac nerves which do not enter into the formation of this plexus are the left superior cardiac nerve, and the inferior cervical cardiac branch from the left pneumogastric.

The branches from the *right side* of this plexus pass, some in front of, and others behind, the right pulmonary artery ; the former, the more numerous, transmit a few filaments to the anterior pulmonary plexus, and are then continued onwards to form part of the anterior coronary plexus ; those behind the pulmonary artery distribute a few filaments to the right auricle, and are then continued onwards to form part of the posterior coronary plexus.

The branches from the *left side* of the deep cardiac plexus distribute a few filaments to the superficial cardiac plexus, to the left auricle of the heart, and to the anterior pulmonary plexus, and then pass on to form the greater part of the posterior coronary plexus.

The **superficial (anterior) cardiac plexus** lies beneath the arch of the aorta, in front of the right pulmonary artery. It is formed by the left superior cardiac nerve, the inferior cervical cardiac branches of the left (and occasionally the right) pneumogastric, and filaments from the deep cardiac plexus. A small ganglion (*cardiac ganglion of Wrisberg*) is occasionally found connected with these nerves at their point of junction. This ganglion, when present, is situated immediately beneath the arch of the aorta, on the right side of the ductus arteriosus. The superficial cardiac plexus forms the chief part of the anterior coronary plexus, and several filaments pass along the pulmonary artery to the left anterior pulmonary plexus.

The **posterior** or **right coronary plexus** is chiefly formed by filaments prolonged from the left side of the deep cardiac plexus, and by a few from the right side. It surrounds the branches of the coronary artery at the back of the heart, and its filaments are distributed with those vessels to the muscular substance of the ventricles.

The **anterior** or **left coronary plexus** is formed chiefly from the superficial cardiac plexus, but receives filaments from the deep cardiac plexus. Passing forwards between the aorta and pulmonary artery, it accompanies the left coronary artery on the anterior surface of the heart.

Valentin has described nervous filaments ramifying under the endocardium ; and Remak has found, in several mammalia, numerous small ganglia on the cardiac nerves, both on the surface of the heart and in its muscular substance.

EPIGASTRIC OR SOLAR PLEXUS (figs. 425, 427)

The **Epigastric** or **Solar plexus** supplies all the viscera in the abdominal cavity. It consists of a great network of nerves and ganglia situated behind the stomach and in front of the aorta and crura of the Diaphragm. It surrounds the cœliac axis and root of the superior mesenteric artery, extending downwards as low as the pancreas, and outwards to the suprarenal capsules. This plexus, and the ganglia connected with it, receive the great and small splanchnic nerves of both sides, and some filaments from the right pneumogastric. It distributes filaments, which accompany, under the name of plexuses, all the branches from the front of the abdominal aorta.

Of the ganglia of which the solar plexus is partly composed the principal are the two **semilunar ganglia,** which are situated one on each side of the plexus, and are the largest ganglia in the body. They are large irregular gangliform masses, formed by the aggregation of smaller ganglia, having interspaces between them. They are situated in front of the crura of the Diaphragm, close to the suprarenal capsules : the one on the right side lies beneath the inferior vena cava ; the upper part of each ganglion is joined by the greater splanchnic nerve, and to the inner side of each the branches of the solar plexus are connected.

From the epigastric or solar plexus are derived the following :

Phrenic or Diaphragmatic plexus.
Suprarenal plexus.
Renal plexus.
Spermatic plexus.
　　　　　　　　Aortic plexus.

Cœliac plexus ⎰ Gastric plexus.
　　　　　　　 ⎨ Hepatic plexus.
　　　　　　　 ⎱ Splenic plexus.
Superior mesenteric plexus.

The **phrenic plexus** accompanies the phrenic artery to the Diaphragm, which it supplies ; some filaments passing to the suprarenal capsule. It arises from the upper part of the semilunar ganglion, and is larger on the right than on the left side. It receives one or two branches from the phrenic nerve. In connection with this plexus, on the right side, at its point of junction with the phrenic nerve, is a small ganglion (*ganglion diaphragmaticum*). This ganglion is placed on the under surface of the Diaphragm, near the suprarenal capsule. Its branches are distributed to the inferior vena cava, suprarenal capsule, and the hepatic plexus. There is no ganglion on the left side.

The **suprarenal plexus** is formed by branches from the solar plexus, from the semilunar ganglion, and from the phrenic and great splanchnic nerves, a ganglion being formed at the point of junction of the latter nerve. It supplies the suprarenal capsule. The branches of this plexus are remarkable for their large size, in comparison with the size of the organ they supply.

The **renal plexus** is formed by filaments from the solar plexus, the outer part of the semilunar ganglion, and the aortic plexus. It is also joined by filaments from the lesser and smallest splanchnic nerves. The nerves from these sources, fifteen or twenty in number, have numerous ganglia developed upon them. They accompany the branches of the renal artery into the kidney ; some filaments on the right side being distributed to the inferior vena cava, and others to the spermatic plexus, on both sides.

The **spermatic plexus** is derived from the renal plexus, receiving branches from the aortic plexus.   It accompanies the spermatic vessels to the testes.

Fig. 427.—Lumbar portion of the gangliated cord, with the solar and hypogastric plexuses.   (After Henle.)

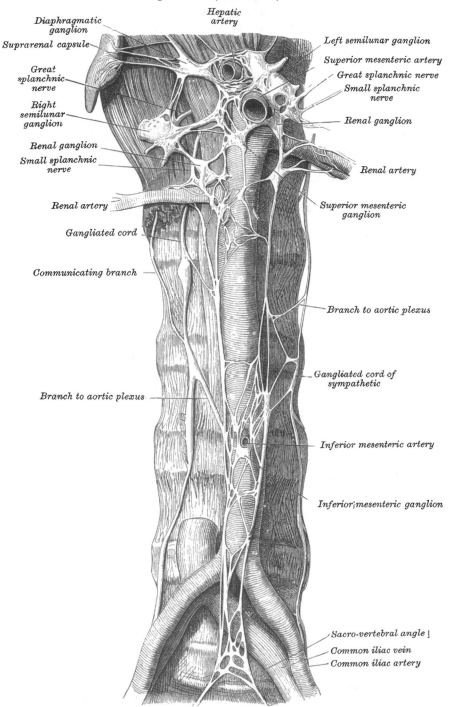

Diaphragmatic ganglion

Suprarenal capsule

Hepatic artery

Left semilunar ganglion

Superior mesenteric artery

Great splanchnic nerve

Great splanchnic nerve

Small splanchnic nerve

Right semilunar ganglion

Renal ganglion

Renal ganglion

Small splanchnic nerve

Renal artery

Renal artery

Superior mesenteric ganglion

Gangliated cord

Communicating branch

Branch to aortic plexus

Gangliated cord of sympathetic

Branch to aortic plexus

Inferior mesenteric artery

Inferior mesenteric ganglion

Sacro-vertebral angle |

Common iliac vein

Common iliac artery

In the female, the **ovarian plexus** is distributed to the ovaries and fundus of the uterus.

The **cœliac plexus,** of large size, is a direct continuation from the solar plexus : it surrounds the cœliac axis, and subdivides into the gastric, hepatic, and splenic plexuses. It receives branches from the lesser splanchnic nerves, and, on the left side, a filament from the right pneumogastric.

The *gastric* or *coronary plexus* accompanies the gastric artery along the lesser curvature of the stomach, and joins with branches from the left pneumogastric nerve. It is distributed to the stomach.

The *hepatic plexus*, the largest offset from the cœliac plexus, receives filaments from the left pneumogastric and right phrenic nerves. It accompanies the hepatic artery, ramifying in the substance of the liver upon its branches, and upon those of the vena portæ.

Branches from this plexus accompany all the divisions of the hepatic artery. Thus there is a *pyloric plexus* accompanying the pyloric branch of the hepatic, which joins with the gastric plexus and pneumogastric nerves. There is also a *gastro-duodenal plexus*, which subdivides into the *pancreatico-duodenal plexus*, which accompanies the pancreatico-duodenal artery, to supply the pancreas and duodenum, joining with branches from the mesenteric plexus ; and a *gastro-epiploic plexus*, which accompanies the right gastro-epiploic artery along the greater curvature of the stomach, and anastomoses with branches from the splenic plexus. A *cystic plexus*, which supplies the gall-bladder, also arises from the hepatic plexus, near the liver.

The *splenic plexus* is formed by branches from the cœliac plexus, the left semilunar ganglia, and from the right pneumogastric nerve. It accompanies the splenic artery and its branches to the substance of the spleen, giving off, in its course, filaments to the pancreas (*pancreatic plexus*), and the *left gastro-epiploic plexus*, which accompanies the gastro-epiploica sinistra artery along the convex border of the stomach.

The **superior mesenteric plexus** is a continuation of the lower part of the great solar plexus, receiving a branch from the junction of the right pneumo-gastric nerve with the cœliac plexus. It surrounds the superior mesenteric artery, which it accompanies into the mesentery, and divides into a number of secondary plexuses, which are distributed to all the parts supplied by the artery, viz. pancreatic branches to the pancreas ; intestinal branches, which supply the whole of the small intestine ; and ileo-colic, right colic, and middle colic branches, which supply the corresponding parts of the great intestine. The nerves composing this plexus are white in colour and firm in texture, and have numerous ganglia developed upon them near their origin.

The **aortic plexus** is formed by branches derived, on each side, from the solar plexus and the semilunar ganglia, receiving filaments from some of the lumbar ganglia. It is situated upon the sides and front of the aorta, between the origins of the superior and inferior mesenteric arteries. From this plexus arise part of the spermatic, the inferior mesenteric, and the hypogastric plexuses ; and it distributes filaments to the inferior vena cava.

The *inferior mesenteric plexus* is derived chiefly from the left side of the aortic plexus. It surrounds the inferior mesenteric artery, and divides into a number of secondary plexuses, which are distributed to all the parts supplied by the artery, viz. the left colic and sigmoid plexuses, which supply the descending and sigmoid flexure of the colon : and the superior hæmorrhoidal plexus, which supplies the upper part of the rectum, and joins in the pelvis with branches from the pelvic plexus.

HYPOGASTRIC PLEXUS

The **Hypogastric plexus** supplies the viscera of the pelvic cavity. It is situated in front of the promontory of the sacrum, between the two common iliac arteries, and is formed by the union of numerous filaments, which descend on each side from the aortic plexus, and from the lumbar ganglia. This plexus contains no evident ganglia ; it bifurcates, below, into two lateral portions, which form the *pelvic plexus.*

PELVIC PLEXUS

The **Pelvic Plexus** (sometimes called *inferior hypogastric*), which supplies the viscera of the pelvic cavity, is situated at the sides of the rectum in the male, and at the sides of the rectum and vagina in the female. It is formed by a continuation of the hypogastric plexus, by branches from the second, third, and fourth sacral nerves, and by a few filaments from the first two sacral ganglia. At the point of junction of these nerves, small ganglia are found. From this plexus numerous branches are distributed to all the viscera of the pelvis. They accompany the branches of the internal iliac artery.

The **inferior hæmorrhoidal plexus** arises from the back part of the pelvic plexus. It supplies the rectum, joining with branches of the superior hæmorrhoidal plexus.

The **vesical plexus** arises from the fore part of the pelvic plexus. The nerves composing it are numerous, and contain a large proportion of spinal nerve-fibres. They accompany the vesical arteries, and are distributed to the side and base of the bladder. Numerous filaments also pass to the vesiculæ seminales and vas deferens ; those accompanying the vas deferens join, on the spermatic cord, with branches from the spermatic plexus.

The **prostatic plexus** is continued from the lower part of the pelvic plexus. The nerves composing it are of large size. They are distributed to the prostate gland, vesiculæ seminales, and erectile structure of the penis. The nerves supplying the erectile structure of the penis consist of two sets, the small and large cavernous nerves. They are slender filaments, which arise from the fore part of the prostatic plexus, and, after joining with branches from the internal pudic nerve, pass forwards beneath the pubic arch.

The *small cavernous nerves* perforate the fibrous covering of the penis, near its root.

The *large cavernous nerve* passes forwards along the dorsum of the penis, joins with the dorsal branch of the pudic nerve, and is distributed to the corpus cavernosum and spongiosum.

The **vaginal plexus** arises from the lower part of the pelvic plexus. It is lost on the walls of the vagina, being distributed to the erectile tissue at its anterior part and to the mucous membrane. The nerves composing this plexus contain, like the vesical, a large proportion of spinal nerve-fibres.

The **uterine plexus** arises from the upper part of the pelvic plexus, above the point where the branches from the sacral nerves join the plexus. Its branches accompany the uterine arteries to the side of the organ between the layers of the broad ligament, and are distributed to the cervix and lower part of the body of the uterus, penetrating its substance.

Other filaments pass separately to the body of the uterus and Fallopian tube.

Branches from the plexus accompany the uterine arteries into the substance of the uterus. Upon these filaments ganglionic enlargements are found.

ORGANS OF SPECIAL SENSE

THE Organs of the Senses are five in number : viz. those of Touch, of Taste, of Smell, of Hearing, and of Sight. The skin, which is the principal seat of the sense of touch, has been described in the section on General Anatomy. The remaining four are the Organs of Special Sense.

THE TONGUE

The tongue is the organ of the special sense of taste. It is situated in the floor of the mouth, in the interval between the two lateral portions of the body of the lower jaw.

Its base, or root, is directed backwards, and connected with the os hyoides by the Hyo-glossi and Genio-hyo-glossi muscles and the hyo-glossal membrane ; with the epiglottis by three folds (*glosso-epiglottic*) of mucous membrane ; with the soft palate by means of the anterior pillars of the fauces ; and with the pharynx by the Superior constrictors and the mucous membrane. Its apex, or tip, thin and narrow, is directed forwards against the inner surface of the lower incisor teeth. The under surface of the tongue is connected with the lower jaw by the Genio-hyo-glossi muscles ; from its sides, the mucous membrane is reflected to the inner surface of the gums ; and from its under surface on to the floor of the mouth, where, in the middle line, it is elevated into a distinct vertical fold, the *frænum linguæ*. To the outer side of the frænum is a slight fold of the mucous membrane, the *plica fimbriata*, the free edge of which exhibits a series of fringe-like processes.

The tip of the tongue, part of the under surface, its sides, and dorsum are free.

The dorsum of the tongue is convex, marked along the middle line by a raphé, which divides it into symmetrical halves ; this raphé terminates behind, about an inch from the base of the organ, in a depression, the *foramen cæcum*, from which a shallow groove, the *sulcus terminalis* of His, runs outwards and forwards on each side to the lateral margin of the tongue. The part of the dorsum of the tongue in front of this groove, forming about two-thirds of its upper surface, is rough and covered with papillæ ; the posterior third is smoother, and contains numerous muciparous glands and lymphoid follicles.

Structure of the tongue.—The tongue is partly invested by mucous membrane and a submucous fibrous layer. It consists of symmetrical halves, separated from each other, in the middle line, by a fibrous septum. Each half is composed of muscular fibres arranged in various directions (page 298), containing much interposed fat, and supplied by vessels and nerves.

The **mucous membrane** invests the entire extent of the free surface of the tongue. On the dorsum it is thicker behind than in front, and is continuous with the sheaths of the muscles attached to it, through the submucous fibrous layer. On the under surface of the organ, where it is thin and smooth, it can be traced on each side of the frænum, through the ducts of the submaxillary and the sub-lingual glands. As it passes over the borders of the organ, it gradually assumes its papillary character.

The **structure of the mucous membrane of the tongue** differs in different parts. That covering the under surface of the organ is thin, smooth, and identical in structure with that lining the rest of the oral cavity. The mucous membrane covering the tongue behind the foramen cæcum and sulcus terminalis is thick and freely movable over the subjacent parts. It contains a large number of lymphoid follicles, which together constitute what is sometimes termed the *lingual tonsil*. Each follicle forms a rounded eminence, the centre of which is perforated by a minute orifice leading into a funnel-shaped cavity or recess; around this recess are grouped numerous oval or rounded nodules of lymphoid tissue, each enveloped by a capsule derived from the submucosa, while opening into the

FIG. 428.—Upper surface of the tongue.

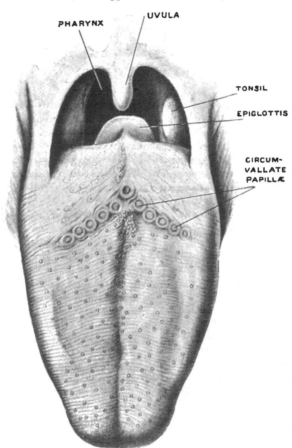

bottom of the recesses are also seen the ducts of mucous glands. The mucous membrane on the anterior part of the dorsum of the tongue is thin and intimately adherent to the muscular tissue, and covered with minute eminences, *the papillæ* of the tongue. It consists of a layer of connective tissue, the *corium* or *mucosa*, supporting numerous papillæ, and covered, as well as the papillæ, with epithelium.

The epithelium is of the scaly variety, like that of the epidermis. It covers the free surface of the tongue, as may be demonstrated by maceration or boiling, when it can be easily detached entire: it is much thinner than on the skin: the intervals between the large papillæ are not filled up by it, but each papilla has a separate investment from root to summit. The deepest cells may sometimes be detached as a separate layer, corresponding to the rete mucosum, but they never contain colouring matter.

The *corium* consists of a dense felt-work of fibrous connective tissue, with numerous elastic fibres, firmly connected with the fibrous tissue forming the septa

Filiform

Artery   *Vein*

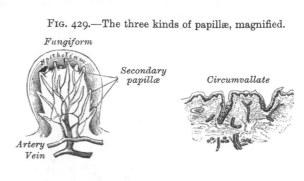

FIG. 429.—The three kinds of papillæ, magnified.

Fungiform

Epithelium

Secondary papillæ

Circumvallate

Artery Vein

between the muscular bundles of the tongue. It contains the ramifications of the numerous vessels and nerves from which the papillæ are supplied, large plexuses of lymphatic vessels, and the glands of the tongue.

The papillæ of the tongue.—These are papillary projections of the corium. They are thickly distributed over the anterior two-thirds of its upper surface, giving to it its characteristic roughness. The varieties of papillæ met with are the papillæ maximæ (*circumvallatæ*), papillæ mediæ (*fungiformes*), papillæ minimæ (*conicæ* or *filiformes*), and papillæ simplices.

The *papillæ maximæ* (circumvallatæ) are of large size, and vary from eight to twelve in number. They are situated at the back part of the dorsum of the tongue, near its base, forming a row on each side, which, running backwards and inwards, meet in the middle line, like the two lines of the letter V inverted. Each papilla consists of a projection of mucous membrane from $\frac{1}{20}$ to $\frac{1}{12}$ of an inch wide, attached to the bottom of a cup-shaped depression of the mucous membrane; the papilla is in shape like a truncated cone; the smaller end being directed downwards and attached to the tongue, the broader part or base projecting on the surface and being studded with numerous small secondary papillæ, which, however, are covered by a smooth layer of the epithelium. The cup-shaped depression forms a kind of fossa round the papilla, having a circular margin of about the same elevation, covered with smaller papillæ. Immediately behind the apex of the V is the foramen cæcum, mentioned above. This, according to His, represents the remains of the invagination

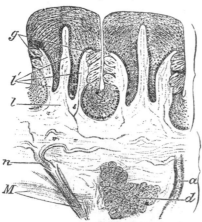

FIG. 430.—Circumvallate papillæ of tongue of rabbit, showing position of taste-buds. (Stöhr.)

a. Duct of gland.   *d.* Serous gland.   *g.* Taste-buds.
l. Primary septa, and *l'*, secondary septa, of papillæ.
n. Medullated nerve.   *M.* Muscular fibres.

which forms the median rudiment of the thyroid body, and for a time opens by a duct, the *thyro-glossal duct*, on to the dorsum of the tongue. It may extend downwards towards the hyoid bone. Kanthack, however, disputes this view.*

* *Journal of Anat. and Physiol.* 1891.

The *papillæ mediæ* (fungiformes), more numerous than the preceding, are scattered irregularly and sparingly over the dorsum of the tongue; but are found chiefly at its sides and apex. They are easily recognised, among the other papillæ, by their large size, rounded eminences, and deep red colour. They are narrow at their attachment to the tongue, but broad and rounded at their free extremities, and covered with secondary papillæ. Their epithelial investment is very thin.

The *papillæ minimæ* (conicæ or filiformes) cover the anterior two-thirds of the dorsum of the tongue. They are very minute, more or less conical or filiform in shape, and arranged in lines corresponding in direction with the two rows of the papillæ circumvallatæ; excepting at the apex of the organ, where their direction is transverse. Projecting from their apices are numerous filiform processes, or secondary papillæ; these are of a whitish tint, owing to the thickness and density of the epithelium of which they are composed, and which has here undergone a peculiar modification, the cells having become cornified and elongated into dense, imbricated, brush-like processes. They contain also a number of elastic fibres, which render them firmer and more elastic than the papillæ of mucous membrane generally.

Simple papillæ, similar to those of the skin, cover the whole of the mucous membrane of the tongue, as well as the larger papillæ. They consist of closely set, microscopic elevations of the corium, containing a papillary loop, covered by a layer of epithelium.

Structure of the papillæ.—The papillæ apparently resemble in structure those of the cutis, consisting of a cone-shaped projection of connective tissue, covered with a thick layer of squamous epithelium, and contain one or more capillary loops, among which nerves are distributed in great abundance. If the epithelium is removed, it will be found that they are not simple elevations like the papillæ of the skin, for the surface of each is studded with minute conical processes of the mucous membrane, which form secondary papillæ (Todd and Bowman). In the papillæ circumvallatæ, the nerves are numerous and of large size; in the papillæ fungiformes, they are also numerous, and terminate in a plexiform network, from which brush-like branches proceed; in the papillæ filiformes, their mode of termination is uncertain. Buried in the epidermis of the papillæ circumvallatæ, and in some of the fungiformes, are certain peculiar bodies called *taste-buds.** They are flask-like in shape, their broad base resting on the corium, and their neck opening by an orifice, the *gustatory pore,* between the cells of the epithelium. They are formed by two kinds of cells: supporting cells and gustatory cells. The *supporting cells* are mostly arranged like the staves of a cask, and form an outer envelope for the bud. Some, however, are found in the interior of the bud between the gustatory cells. The *gustatory cells* occupy the central portion of the bud; they are spindle-shaped, and each possesses a large spherical nucleus near the middle of the cell. The peripheral end of the cell terminates at the gustatory pore in a fine hair-like filament, the *gustatory hair.* The central process passes towards the deep extremity of the bud, and there ends in a single or bifurcated varicose filament, which was formerly supposed to be continuous with the terminal fibril of a nerve; the investigations of Lenhossék and others would seem to prove, however, that this is not so, but that the nerve-fibrils after losing their medullary sheaths enter the taste-bud, and terminate in fine

FIG. 431.—Taste-buds.

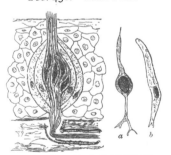

a. Supporting cell.   *b.* Gustatory cell.

* These bodies are also found in considerable numbers at the side of the base of the tongue, just in front of the anterior pillars of the fauces, and also on the posterior surface of the epiglottis and anterior surface of the soft palate.

extremities between the gustatory cells. Other nerve-fibrils may be seen ramifying between the cortical cells and terminating in fine extremities; these, however, are believed to be nerves of ordinary sensation and not gustatory.

Glands of the tongue.—The tongue is provided with mucous and serous glands.

The *mucous glands* are similar in structure to the labial and buccal glands. They are found especially at the back part behind the circumvallate papillæ, but are also present at the apex and marginal parts. In connection with these glands, a special one has been described by Blandin or Nuhn. It is situated near the apex of the tongue on either side of the frænum, and is covered over by a fasciculus of muscular fibres derived from the Stylo-glossus and Inferior lingualis. It is from half an inch to nearly an inch long, and about the third of an inch broad. It has from four to six ducts, which open on the under surface of the apex.

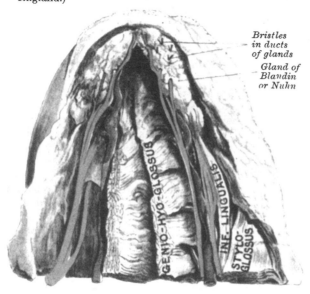

FIG. 432.—Under surface of tongue, showing position and relations of gland of Blandin or Nuhn. (From a preparation in the Museum of the Royal College of Surgeons of England.)

Bristles in ducts of glands

Gland of Blandin or Nuhn

GENIO-HYO-GLOSSUS

INF. LINGUALIS

STYLO-GLOSSUS

Lingual nerve   Ranine artery

The *serous glands* occur only at the back of the tongue in the neighbourhood of the taste-buds, their ducts opening for the most part into the fossæ of the circumvallate papillæ. These glands are racemose, the duct branching into several minute ducts, which terminate in alveoli, lined by a single layer of more or less columnar epithelium. Their secretion is of a watery nature, and probably assists in the distribution of the substance to be tasted over the taste area. (Ebner.)

The *fibrous septum* consists of a vertical layer of fibrous tissue, extending throughout the entire length of the middle line of the tongue, from the base to the apex, though not quite reaching the dorsum. It is thicker behind than in front, and occasionally contains a small fibro-cartilage, about a quarter of an inch in length. It is well displayed by making a vertical section across the organ.

The **hyo-glossal membrane** is a strong fibrous lamina, which connects the under surface of the base of the tongue to the body of the hyoid bone. This membrane receives, in front, some of the fibres of the Genio-hyo-glossi.

Vessels of the tongue.—The *arteries of the tongue* are derived from the lingual, the facial, and ascending pharyngeal. The *veins* open into the internal jugular.

Muscles of the tongue.—The muscular fibres of the tongue run in various directions. These fibres are divided into two sets, Extrinsic and Intrinsic, which have already been described (page 298).

The *lymphatic vessels from the tongue* pass to one or two small glands situated on the Hyo-glossus muscle in the submaxillary region, and thence to the deep glands of the neck.

The *nerves of the tongue* are five in number in each half: the lingual branch of the fifth, which is distributed to the papillæ at the fore part and sides of the tongue, and forms the nerve of ordinary sensibility for its anterior two-thirds ; the chorda tympani, which runs in the sheath of the lingual, is generally regarded as the nerve of taste for the same area ; the lingual branch of the glosso-pharyngeal, which is distributed to the mucous membrane at the base and sides of the tongue, and to the papillæ circumvallatæ, and which supplies both sensory and gustatory filaments to this region ; the hypoglossal nerve, which is the motor nerve to the muscular substance of the tongue ; and the superior laryngeal, which sends some fine branches to the root near to the epiglottis.   Sympathetic filaments also pass to the tongue from the nervi molles on the lingual and other arteries supplying it.

FIG. 433.—Under surface of tongue, showing the distribution of nerves to this organ (From a preparation in the Museum of the Royal College of Surgeons of England.)

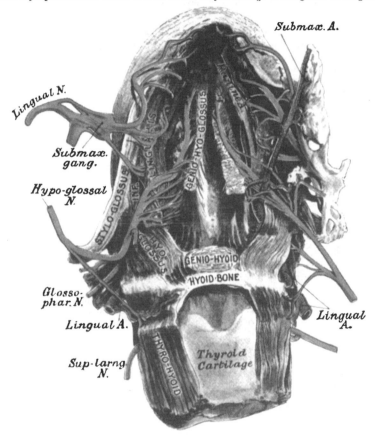

Surgical Anatomy.—The diseases to which the tongue is liable are numerous, and its surgical anatomy of importance, since any or all the structures of which it is composed—muscles, connective tissue, mucous membrane, glands, vessels, nerves, and lymphatics—may be the seat of morbid changes.   It is not often the seat of congenital defects, though a few cases of vertical cleft have been recorded, and it is occasionally, though much more rarely than is commonly supposed, the seat of ' tongue tie,' from shortness of the frænum (see page 464).

There is, however, one not uncommon condition, which must be regarded as congenital, though it does not sometimes evidence itself until a year or two after birth.   This is an enlargement of the tongue which is due primarily to a dilatation of the lymph-channels and a greatly increased development of the lymphatic tissue throughout the organ.   This is often aggravated by inflammatory changes induced by injury or exposure, and the tongue may assume enormous dimensions and hang out of the mouth, giving the child an imbecile expression.   The treatment consists in excising a ∨-shaped portion and bringing the cut surfaces together with deeply placed silver sutures.   Compression has been resorted to in some cases and with success, but it is difficult to apply. Acute inflammation of the tongue, which may be caused by injury and the introduction of some septic or irritating matter, is attended by great swelling from infiltration of its connective tissue, which is in considerable quantity. This renders the patient incapable of swallowing or speaking, and may seriously impede respiration. It may run on to suppuration, and the formation of an acute abscess. Chronic abscess, which has been mistaken for cancer, may also occur in the substance of the tongue.

The mucous membrane of the tongue may become chronically inflamed, and presents different appearances in different stages of the disease, to which the terms leucoplakia, psoriasis, and ichthyosis have been given.

The tongue, being very vascular, is often the seat of nævoid growths, and these have a tendency to increase rapidly.

The tongue is frequently the seat of ulceration, which may arise from many causes, as from the irritation of jagged teeth, dyspepsia, tubercle, syphilis, and cancer. Of these the cancerous ulcer is the most important and probably also the most common. The variety is the squamous epithelioma, which soon develops into an ulcer with an indurated base. It produces great pain, which speedily extends to all parts supplied with sensation by the fifth nerve, especially to the region of the ear. The pain in these cases is conducted to the ear and temporal region by the lingual nerve, and from it to the other branches of the inferior maxillary nerve, especially the auriculo-temporal. Possibly pain in the ear itself may be due to implication of the fibres of the glosso-pharyngeal nerve, which by its tympanic branch is conducted to the tympanic plexus.

Cancer of the tongue may necessitate removal of a part or the whole of the organ, and many different methods have been adopted for its excision. It may be removed from the mouth by the écraseur or the scissors. Probably the better method is by the scissors, usually known as Whitehead's method. The mouth is widely opened with a gag; the tongue transfixed with a stout silk ligature, by which to hold and make traction on it, and the reflection of mucous membrane from the tongue to the jaw, and the insertion of the Genio-hyo-glossus first divided with a pair of curved, blunt scissors. The Palato-glossus is also divided. The tongue can now be pulled well out of the mouth. The base of the tongue is cut through by a series of short snips, each bleeding vessel being dealt with as soon as divided, until the situation of the main artery is reached. The remaining undivided portion of tissue is to be seized with a pair of Wells' forceps; the tongue removed, and the vessel secured. In the event of the artery being accidentally injured, hæmorrhage can be at once controlled, by passing two fingers over the dorsum of the tongue as far as the epiglottis, and dragging the root of the tongue forcibly forwards.

In cases where the disease is confined to one side of the tongue, this operation may be modified by splitting the tongue down the centre and removing only the affected half. In cases where the submaxillary glands are involved, Kocher's operation should be performed. He removes the tongue from the neck, having performed a preliminary tracheotomy, by an incision from near the lobule of the ear, down the anterior border of the Sterno-mastoid to the level of the great cornu of the hyoid bone, then forwards to the body of the hyoid bone, and upwards to near the symphysis of the jaw. The lingual artery is now secured, and by a careful dissection the submaxillary lymphatic glands and the tongue removed. Regnoli advocated the removal of the tongue by a semilunar incision in the submaxillary triangle, along the line of the lower jaw, and a vertical incision from the centre of the semilunar one backwards to the hyoid bone. Care must be taken not to carry the first incision too far backwards, so as to wound the facial arteries. The tongue is thus reached through the floor of the mouth, pulled out through the external incision, and removed with the écraseur or knife. The great objection to this operation is that all the muscles which raise the hyoid bone and larynx are divided, and that therefore the movements of deglutition and respiration are interfered with.

Finally, where both sides of the floor of the mouth are involved in the disease, or where very free access is required on account of the extension backwards of the disease to the pillars of the fauces and the tonsil, or where the lower jaw is involved, the operation recommended by Syme must be performed. This is done by an incision through the central line of the lip, across the chin, and down as far as the hyoid bone. The lower jaw is sawn through at the symphysis, and the two halves of the bone forcibly separated from each other. The mucous membrane is separated from the bone, and the Genio-hyo-glossi detached from the bone, and the Hyo-glossi divided. The tongue is then drawn forwards, and removed close to its attachment to the hyoid bone. Any glands which are enlarged can be removed, and if the bone is implicated in the disease, it can also be removed by freeing it from the soft parts externally and internally, and making a second section with the saw beyond the diseased part.

Formerly many surgeons before removing the tongue performed a preliminary tracheotomy: (1) to prevent blood entering the air passages; and (2) to allow the patient to breathe through the tube and not inspire air which had passed over a sloughy wound, and which was loaded with septic organisms and likely to induce septic pneumonia. By the judicious use of iodoform, this secondary evil may be obviated, and the preliminary tracheotomy is now usually dispensed with.

THE NOSE

The nose is the peripheral organ of the sense of smell : by means of the peculiar properties of its nerves, it protects the lungs from the inhalation of deleterious gases, and assists the organ of taste in discriminating the properties of food.

The organ of smell consists of two parts : one external, the *outer nose* ; the other internal, the *nasal fossæ*.

The *outer nose* (nasus externus) is the more anterior and prominent part of the organ of smell. Of a triangular form, it is directed downwards, and projects from the centre of the face, immediately above the upper lip. Its summit, or *root*, is connected directly with the forehead. Its inferior part, or *base*, presents two elliptical orifices, the nostrils or anterior nares, separated from each other by an antero-posterior septum, the *columna*. The margins of these orifices are provided with a number of stiff hairs, or *vibrissæ*, which arrest the passage of foreign substances carried with the current of air intended for respiration. The

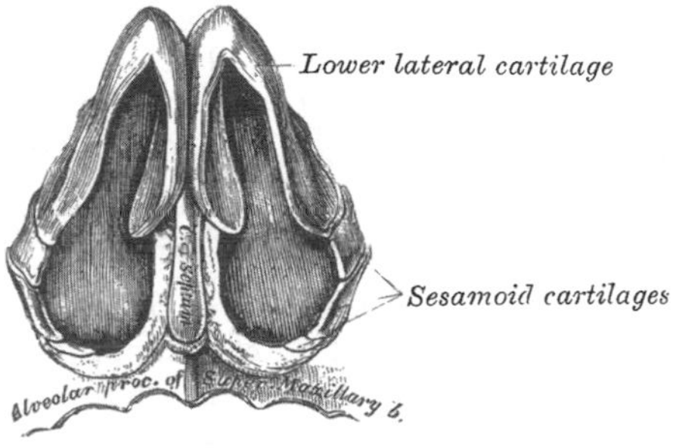

Fig. 434.—Cartilages of the nose, seen from below.

Fig. 435.—Cartilages of the nose. Side view.

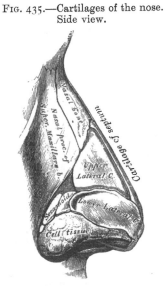

Lower lateral cartilage

Sesamoid cartilages

lateral surfaces of the nose form, by their union in the middle line, the *dorsum*, the direction of which varies considerably in different individuals. The lateral surface terminates below in a rounded eminence, the *ala nasi*.

The nose is composed of a framework of bones and cartilages, the latter being slightly acted upon by certain muscles. It is covered externally by the integument, internally by mucous membrane, and supplied with vessels and nerves.

The *bony framework* occupies the upper part of the organ : it consists of the nasal bones, and the nasal processes of the superior maxillary.

The *cartilaginous framework* consists of five pieces, the two upper and the two lower lateral cartilages, and the cartilage of the septum.

The *upper lateral cartilages* are situated below the free margin of the nasal bones : each cartilage is flattened, and triangular in shape. Its anterior margin is thicker than the posterior, and continuous with the cartilage of the septum. Its posterior margin is attached to the nasal process of the superior maxillary and nasal bones. Its inferior margin is connected by fibrous tissue with the lower lateral cartilage ; one surface is turned outwards, the other inwards towards the nasal cavity.

The *lower lateral cartilages* are two thin, flexible plates, situated immediately below the preceding, and bent upon themselves in such a manner as to form the inner and outer walls of each orifice of the nostril. The portion which forms the inner wall, thicker than the rest, is loosely connected with the same part of the opposite cartilage, and forms a small part of the columna. Its inferior border, free, rounded, and projecting, forms, with the thickened integument and subjacent tissue, and the corresponding parts of the opposite side, the septum mobile nasi. The part which forms the outer wall is curved to correspond with the ala of the nose; it is oval and flattened, narrow behind, where it is connected with the nasal process of the superior maxilla by a tough fibrous membrane, in which are found three or four small cartilaginous plates (sesamoid cartilages), *cartilagines minores.* Above, it is connected by fibrous tissue to the upper lateral cartilage and front part of the cartilage of the septum; below, it falls short of the margin of the nostril, the ala being formed by dense cellular tissue covered by skin. In front the lower lateral cartilages are separated by a notch which corresponds with the point of the nose.

The *cartilage of the septum* is somewhat quadrilateral in form, thicker at its margins than at its centre, and completes the separation between the nasal fossæ in front. Its anterior margin, thickest above, is connected with the nasal bones, and is continuous with the anterior margins of the two upper lateral cartilages. Below, it is connected to the inner portions of the lower lateral cartilages by fibrous tissue. Its posterior margin is connected with the perpendicular lamella of the ethmoid; its inferior margin with the vomer and the palate processes of the superior maxillary bones.

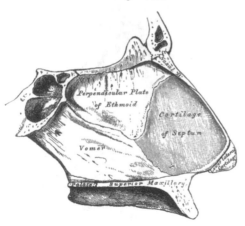

Fig. 436.—Bones and cartilages of septum of nose. Right side.

It may be prolonged backwards (especially in children) for some distance between the vomer and perpendicular plate of the ethmoid, forming what is termed the *processus sphenoidalis.* The septal cartilage does not reach as far as the lowest part of the nasal septum. This is formed by the inner portions of the lower lateral cartilages and by the skin; it is freely movable, and hence is termed the *septum mobile nasi.*

These various cartilages are connected to each other, and to the bones by a tough fibrous membrane, which allows the utmost facility of movement between them.

The *muscles of the nose* are situated beneath the integument: they are (on each side) the Pyramidalis nasi, the Levator labii superioris alæque nasi, the Dilatator naris, anterior and posterior, the Compressor nasi, the Compressor narium minor, and the Depressor alæ nasi. They have been described above (page 278).

The *integument* covering the dorsum and sides of the nose is thin, and loosely connected with the subjacent parts; but where it forms the tip and the alæ of the nose it is thicker and more firmly adherent, and is furnished with a large number of sebaceous follicles, the orifices of which are usually very distinct.

The *mucous membrane*, lining the interior of the nose, is continuous with the skin externally, and with that which lines the nasal fossæ within.

The *arteries of the nose* are the lateralis nasi from the facial, and the inferior artery of the septum from the superior coronary, which supply the alæ and septum;

the sides and dorsum being supplied from the nasal branch of the ophthalmic and the infra-orbital.

The *veins of the nose* terminate in the facial and ophthalmic.

The *nerves* for the muscles of the nose are derived from the facial, while the skin receives its branches from the infra-orbital, infratrochlear, and nasal branches of the ophthalmic.

NASAL FOSSÆ

The nasal fossæ are two irregular cavities situated in the middle of the face, and extending from before backwards. They open in front by the two anterior nares, and terminate, behind, by the posterior nares in the naso-pharynx. The *anterior nares* are somewhat pear-shaped apertures, each measuring about one inch antero-posteriorly and half an inch transversely at their widest part. The *posterior nares* are two oval openings, which are smaller in the living or recent subject than in the skeleton, because they are narrowed by the mucous membrane. Each measures an inch in the vertical, and half an inch in the transverse direction in a well-developed adult skull.

For the description of the bony boundaries of the nasal fossæ, see section on Osteology (page 84).

Inside the aperture of the nostril is a slight dilatation, the *vestibule*, which extends as a small pouch, the *ventricle*, towards the point of the nose. The fossa, above and behind the vestibule, has been divided into two parts: an *olfactory* portion, consisting of the upper and central part of the septum and probably the superior turbinated bone, and a *respiratory* portion, which comprises the rest of the fossa.

Outer wall.—The sphenoidal air sinus opens into the *spheno-ethmoidal recess*, a narrow recess above the superior turbinated bone. The posterior ethmoidal cells open into the front and upper part of the superior meatus. On raising or cutting away the middle turbinated bone the outer wall of the middle meatus is fully exposed, and presents (1) a rounded elevation, termed the *bulla ethmoidalis*, opening on or immediately above which are the orifices of the middle ethmoidal cells; (2) a deep, narrow, curved groove, in front of the bulla ethmoidalis, termed the *hiatus semilunaris*, into which the anterior ethmoidal cells and the antrum of Highmore open, the orifice of the latter being placed near the level of its roof. The middle meatus is prolonged, above and in front, into the *infundibulum*, which leads into the frontal sinus. The anterior extremity of the meatus is continued into a depressed area, which lies above the vestibule and is named the *atrium*. The *nasal duct* opens into the anterior part of the inferior meatus, the opening being frequently overlapped by a fold of mucous membrane.

The inner wall or septum is frequently more or less deflected from the mesial plane, thus limiting the size of one fossa and increasing that of the other. Ridges or spurs of bone growing outwards from the septum are also sometimes present. Immediately over the incisive foramen at the lower edge of the cartilage of the septum a depression, the *naso-palatine recess*, may be seen. In the septum close to this recess a minute orifice may be discerned: it leads into a blind pouch, the rudimentary *organ of Jacobson*, which is well developed in some of the lower animals, and is supported by a plate of cartilage, the *cartilage of Jacobson*.

The *mucous membrane* lining the nasal fossæ is called *pituitary*, from the nature of its secretion; or *Schneiderian*, from Schneider, the anatomist who first showed that the secretion proceeded from the mucous membrane, and not, as was formerly imagined, from the brain. It is intimately adherent to the periosteum or perichondrium, over which it lies. It is continuous externally with the skin through the anterior nares, and with the mucous membrane of the naso-pharynx through the posterior nares. From the nasal fossæ its continuity with the conjunctiva may be traced, through the nasal duct and lachrymal canals; with the lining membrane of the tympanum and mastoid cells, through the Eustachian tube; and with the frontal, ethmoidal, and sphenoidal sinuses, and the antrum of Highmore, through the several openings in the meatuses. The mucous membrane is thickest, and most vascular, over the turbinated bones. It is also thick over the septum; but, in the intervals between the spongy bones, and on the floor of the nasal fossæ, it is very thin. Where it lines the various sinuses and the antrum of Highmore, it is thin and pale.

Owing to the great thickness of this membrane, the nasal fossæ are much narrower, and the turbinated bones, especially the lower ones, appear larger and more prominent than in the skeleton. From the same circumstance, also, the various apertures communicating with the meatuses are considerably narrowed.

The vestibule is lined by modified skin, and contains hairs or vibrissæ which guard the entrance of the nostril.

Structure of the mucous membrane.—The epithelium covering the mucous

Fig. 437.—Transverse vertical section of the nasal fossæ.

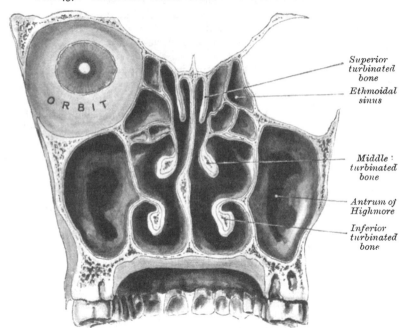

Superior
turbinated
bone

Ethmoidal
sinus

Middle
turbinated
bone

Antrum of
Highmore

Inferior
turbinated
bone

membrane differs in its character according to the functions of the part of the nose in which it is found. In the respiratory portion of the nasal cavity the epithelium is columnar and ciliated. Interspersed among the columnar ciliated cells are goblet or mucin cells, while between their bases are found smaller pyramidal cells. In this region, beneath the epithelium and its basement membrane, is a fibrous layer infiltrated with lymph-corpuscles, so as to form in many parts a diffuse adenoid tissue, and beneath this a nearly continuous layer of smaller and larger glands, some mucous and some serous, the ducts of which open upon the surface. In the olfactory region the mucous membrane is yellowish in colour and the epithelial cells are columnar and non-ciliated; they are of two kinds, supporting cells and olfactory cells. The *supporting cells* contain oval nuclei, situated in the deeper parts of the cells: the free surface of each cell presents a sharp outline, and its deep extremity is prolonged into a process which runs inwards, branching to communicate with similar processes from neighbouring cells, so as to form a network in the deep part of the mucous membrane. Lying between these central processes of the supporting cells are a large number of spindle-shaped cells, the *olfactory cells*, which consist of a large spherical nucleus surrounded by a small amount of granular protoplasm, and possessing two processes, of which one runs outwards between the columnar epithelial cells, and

FIG. 438.—Section of the olfactory mucous membrane.   (Cadiat.)

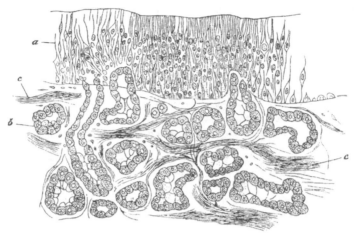

a. Epithelium. *b*. Glands of Bowman. *c*. Nerve bundles.

projects on the surface of the mucous membrane as a fine, hair-like process, the *olfactory hair*; the other or deep process runs inwards, is frequently beaded like a nerve-fibre, and is believed by most observers to be in connection with one of the terminal filaments of the olfactory nerve. Beneath the epithelium, extending through the thickness of the mucous membrane, is a layer of tubular, often branched, glands, the *glands of Bowman*, identical in structure with serous glands.

The *arteries of the nasal fossæ* are the anterior and posterior ethmoidal, from the ophthalmic, which supply the ethmoidal cells, frontal sinuses, and roof of the nose; the spheno-palatine, from the internal maxillary, which supplies the mucous membrane covering the spongy bones, the meatuses and septum; the inferior artery of the septum, from the superior coronary of the facial; and the infraorbital and alveolar branches of the internal maxillary, which supply the lining membrane of the antrum. The ramifications of these vessels form a close, plexiform network, beneath and in the substance of the mucous membrane.

FIG. 439.—Nerves of septum of nose.
Right side.

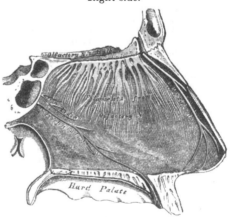

The *veins of the nasal fossæ* form a close cavernous-like network beneath the mucous membrane. This cavernous appearance is especially well marked over the lower part of the septum and over the middle and inferior turbinated bones. Some of the veins pass, with those accompanying the spheno-palatine artery, through the spheno-palatine foramen; and others, through the alveolar branch, to join the facial vein; some accompany the ethmoidal arteries, and terminate in the ophthalmic vein; and, lastly, a few communicate with the veins in the interior of the skull, through the foramina in the cribriform plate of the ethmoid bone, and the foramen cæcum.

The *lymphatics* can be injected from the subdural and subarachnoid spaces, and form a plexus in the superficial portion of the mucous membrane. The lymph is drained partly into one or two glands which lie near the great cornu of the hyoid bone and partly into a gland situated in front of the axis.

The *nerves* are : the olfactory, the nasal branch of the ophthalmic, filaments from the anterior dental branch of the superior maxillary, the Vidian, the naso-palatine, descending anterior palatine, and nasal branches of Meckel's ganglion.

The *olfactory*, the special nerve of the sense of smell, is distributed to the olfactory region, already referred to (p. 801).

The *nasal branch of the ophthalmic* distributes filaments to the fore part of the septum, and outer wall of the nasal fossæ.

Filaments from the anterior dental branch of the superior maxillary supply the inferior meatus and inferior turbinated bone.

The *Vidian nerve* supplies the upper and back part of the septum, and superior spongy bone ; and the *upper nasal branches* from the spheno-palatine ganglion have a similar distribution.

The *naso-palatine nerve* supplies the middle of the septum.

The *larger*, or *anterior palatine nerve*, supplies the *lower nasal branches* to the middle and lower spongy bones.

Surgical Anatomy.—Instances of congenital deformity of the nose are occasionally met with, such as complete absence of the nose, an aperture only being present ; or perfect development on one side. and suppression or malformation on the other ; or there may be imperfect apposition of the nasal bones, so that the nose presents a median cleft or furrow. Deformities which have been acquired are much more common, such as flattening of the nose, the result of syphilitic necrosis ; or imperfect development of the nasal bones in cases of congenital syphilis ; or a lateral deviation of the nose may result from fracture. The skin over the alæ and tip of the nose is thick and closely adherent to subjacent parts. Inflammation of this part is therefore very painful, on account of the tension. It is largely supplied with blood, and, the circulation here being terminal, vascular engorgement is liable to occur, especially in women at the menopause, and in both sexes from disorders of digestion, exposure to cold, &c. The skin of the nose also contains a lárge number of sebaceous follicles, and these, as the result of intemperance, are apt to become affected and the nose reddened, congested, and irregularly swollen. To this the term 'grog-blossom' is popularly applied. In some of these cases there is enormous hypertrophy of the skin and subcutaneous tissues, producing pendulous masses, termed lipomata nasi. Epithelioma and rodent ulcer may attack the nose, the latter being the more common of the two. Lupus and syphilitic ulceration frequently attack the nose, and may destroy the whole of the cartilaginous portion. In fact, lupus vulgaris begins more frequently on the ala of the nose than in any other situation.

Cases of congenital occlusion of one or both nostrils, or adhesion between the ala and septum, may occur and may require immediate operation, since the obstruction much interferes with sucking. Bony closure of the posterior nares may also occur.

To examine the nasal cavities, the head should be thrown back and the nose drawn upwards, the parts being dilated by some form of speculum. It can also be examined with the little finger or a probe, and in this way foreign bodies detected. A still more extensive examination can be made by Rouge's operation, which was introduced for the cure of ozæna, by the removal of any dead bone which may be present in this disease. The whole framework of the nose is lifted up by an incision made inside the mouth, through the junction of the upper lip with the bone ; the septum nasi and the lateral cartilages are divided with strong scissors till the anterior nares are completely exposed. The posterior nares can be explored by reflected light from the mouth, by which the posterior nares can be illuminated. The examination is very difficult to carry out, and, as a rule, sufficient information regarding the presence of foreign bodies or tumours in the naso-pharynx can be obtained by the introduction of the finger behind the soft palate through the mouth. The septum of the nose may be displaced or may deviate from the middle line : this may be the result of an injury or of some congenital defect in its development ; in the latter case the deviation usually occurs along the line of union of the vomer and mesethmoid and rarely occurs before the seventh year. Sometimes the deviation may be so great that the septum may come in contact with the outer wall of the nasal fossæ, and may even become adherent to it, thus producing complete obstruction Perforation of the septum is not an uncommon affection, and may arise from severa causes : syphilitic or tuberculous ulceration, blood tumour or abscess of the sepum and especially in workmen exposed to the vapour of bichromate of potash, from the irritating and corrosive action of the fumes. When small, the perforation may cause a peculiar whistling sound during respiration. When large, it may lead to the falling in of the bridge of the nose.

Epistaxis is a very common affection in children. It is rarely of much consequence, and will almost always subside, but in the more violent hæmorrhages of later life it may be necessary to plug the posterior nares. In performing this operation it is desirable to remember the size of the posterior nares. A ready method of regulating the size of the plug to fit the opening is to make it of the same size as the terminal phalanx of the thumb of the patient to be operated on.

Nasal polypus is a very common disease, and presents itself in three forms: the gelatinous, the fibrous, and the malignant. The first is by far the most common. It grows from the mucous membrane of the outer wall of the nasal fossa, where there is an abundant layer of highly vascular submucous tissue; rarely from the septum, where the mucous membrane is closely adherent to the cartilage and bone, without the intervention of much, if any, submucous tissue. Their most common seat is probably the middle turbinated bone. The fibrous polypus generally grows from the base of the skull behind the posterior nares, or from the roof of the nasal fossæ. The malignant polypi, both sarcomatous or carcinomatous, may arise in the nasal cavities and the nasopharynx; or they may originate in the antrum, and protrude through its inner wall into the nasal fossa.

Rhinoliths, or nose-stones, may sometimes be found in the nasal cavities, from the formation of phosphate of lime, either upon a foreign body or a piece of inspissated secretion.

THE EYE

The eyeball is contained in the cavity of the orbit. In this situation it is securely protected from injury, while its position is such as to ensure the most extensive range of sight. It is acted upon by numerous muscles, by which it is capable of being directed to different parts; it is supplied by vessels and nerves, and is additionally protected in front by several appendages, such as the eyebrow, eyelids, &c.

The eyeball is embedded in the fat of the orbit, but is surrounded by a thin membranous sac, the *capsule of Tenon*, which isolates it, so as to allow of free movement.

The **capsule of Tenon** consists of a thin membrane which envelops the eyeball from the optic nerve to the ciliary region, separating it from the orbital fat and forming a socket in which it plays. Its inner surface is smooth, and is in contact with the outer surface of the sclerotic, the *perisclerotic lymph-space* only intervening. This lymph-space is continuous with the subdural and subarachnoid spaces, and is traversed by delicate bands of connective tissue which extend between the capsule and the sclerotic. The capsule is perforated behind by the ciliary vessels and nerves and by the optic nerve, being continuous with the sheath of the latter. In front it blends with the ocular conjunctiva, and with it is attached to the ciliary region of the eyeball. It is perforated by the muscles which move the eyeball and on each it sends a tubular sheath. The sheath of the Superior oblique is carried as far as the fibrous pulley of that muscle; that on the Inferior oblique reaches as far as the floor of the orbit, to which it gives off a slip. The sheaths on the recti are gradually lost in the perimysium, but they give off important expansions. The expansion from the Superior rectus blends with the tendon of the Levator palpebræ; that of the Inferior rectus is attached to the inferior tarsal plate. These two recti, therefore, will exercise some influence on the movements of the eyelids. The expansions from the sheaths of the Internal and External recti are strong, especially the one from the latter muscle, and are attached to the lachrymal and malar bones respectively. As they probably check the action of these two recti they have been named the *internal and external check ligaments*.

Lockwood has also described a thickening of the lower part of the capsule of Tenon, which he has named the *suspensory ligament of the eye*. It is slung like a hammock below the eyeball, being expanded in the centre and narrow at its extremities, which are attached to the malar and lachrymal bones respectively.*

The eyeball is composed of segments of two spheres of different sizes. The anterior segment is one of a small sphere, and forms about one-sixth of the eyeball. It is more prominent than the posterior segment, which is one of a much larger sphere, and forms about five-sixths of the globe. The segment of the larger sphere is opaque, and formed by the sclerotic, the tunic of protection to the eyeball; the smaller sphere is transparent, and formed by the cornea. The term *anterior pole* is applied to the central point of the anterior curvature of the eyeball, and that of *posterior pole* to the central point of its posterior curvature; a line joining the two poles forms its *sagittal axis*. The axes of the eyeballs are nearly parallel, and therefore do not correspond to the axes of the orbits, which are directed outwards. The optic nerves follow the direction of the axes of the orbits, and are therefore not parallel; each enters its eyeball about 1 mm. below and 3 mm. to the inner or nasal side of the posterior pole. The eyeball measures rather more in its transverse and antero-posterior diameters than in its vertical diameter, the former amounting to nearly an inch, the latter to about nine-tenths of an inch.

The eyeball is composed of three investing tunics and of three refracting media.

TUNICS OF THE EYE

From without inwards the three tunics are:

1. Sclerotic and Cornea.
2. Choroid, Ciliary Body, and Iris.
3. Retina.

I. THE SCLEROTIC AND CORNEA

The sclerotic and cornea (fig. 440) form the external tunic of the eyeball; they are essentially fibrous in structure, the sclerotic being opaque, and forming the posterior five-sixths of the globe; the cornea, which forms the remaining sixth, being transparent.

The **Sclerotic** (σκληρός, *hard*) has received its name from its extreme density and hardness; it is a firm, unyielding, fibrous membrane, serving to maintain the form of the globe. It is much thicker behind than in front. Its *external surface* is of a white colour, and is in contact with the inner surface of the capsule of Tenon; it is quite smooth, except at the points where the Recti and Obliqui muscles are inserted into it, and its anterior part is covered by the conjunctival membrane: hence the whiteness and brilliancy of the front of the eyeball. Its *inner surface* is stained of a brown colour, marked by grooves, in which are lodged the ciliary nerves and vessels; it is loosely connected by an exceedingly fine cellular tissue (*lamina fusca*) with the outer surface of the choroid, an extensive lymph-space (*perichoroidal*) intervening between the sclerotic and choroid. Behind, it is pierced by the optic nerve, and is continuous with its fibrous sheath, which is derived from the dura mater. At the point where the optic nerve passes through the sclerotic, this tunic forms a thin cribriform lamina (the *lamina cribrosa*); the minute orifices in this region serve for the transmission of the nervous filaments, and the fibrous septa dividing them from one another are continuous with the membranous processes which separate the bundles of nerve-fibres. One of these openings, larger than the rest, occupies the centre of the lamella; it transmits the arteria centralis retinæ to the interior of the eyeball. Around the cribriform lamella are numerous small apertures for the transmission

* See a paper by C. B. Lockwood, *Journal of Anatomy and Physiology*, vol. xx. part i. p. 1.

of the ciliary vessels and nerves, and about midway between the margin of the cornea and the entrance of the optic nerve are four or five large apertures, for the transmission of veins (*venæ vorticosæ*). In front, the fibrous tissue of the sclerotic is directly continuous with that of the cornea by direct continuity of tissue, but the opaque sclerotic slightly overlaps the outer surface of the transparent cornea.

Structure.—The sclerotic is formed of white fibrous tissue intermixed with fine elastic fibres, and of flattened connective-tissue corpuscles, some of which are pigmented, contained in cell-spaces between the fibres. These fibres are aggregated into bundles, which are arranged chiefly in a longitudinal direction. It yields gelatin on boiling. Its vessels are not numerous, the capillaries being of small size, uniting at long and wide intervals. Its nerves are derived from the ciliary nerves, but their exact mode of ending is not known.

FIG. 440.—A horizontal section of the eyeball. (Allen.)

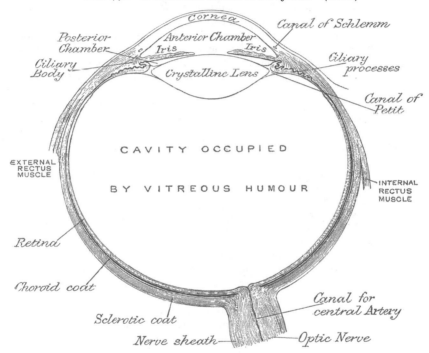

The **Cornea** is the projecting transparent part of the external tunic of the eyeball, and forms the anterior sixth of the globe. It is almost circular in shape, occasionally a little broader in the transverse than in the vertical direction. It is convex anteriorly, and projects forwards from the sclerotic in the same manner that a watch-glass does from its case. Its degree of curvature varies in different individuals, and in the same individual at different periods of life; it is more prominent in youth than in advanced life, when it becomes flattened. The cornea is dense and of uniform thickness throughout; its posterior surface is perfectly circular in outline, and exceeds the anterior surface slightly in extent, from the latter being overlapped by the sclerotic.

Structure.—The cornea consists of four layers: namely (1) several strata of epithelial cells, continuous with those of the conjunctiva; (2) a thick central fibrous structure, the *substantia propria*; (3) a homogeneous elastic lamina; and (4) a single layer of endothelial cells, forming part of the lining membrane of the anterior chamber of the eyeball.

The *conjunctival epithelium*, which covers the front of the cornea proper, consists of several strata of epithelial cells. The deepest layer is columnar:

then follow two or three layers of polyhedral cells, the majority of which present finger-like processes (i.e. prickle-cells) similar to those found in the cuticle. Lastly, there are three or four layers of scaly epithelium, with flattened nuclei.

The *proper substance of the cornea* is fibrous, tough, unyielding; perfectly transparent, and continuous with the sclerotic. It is composed of about sixty flattened lamellæ, superimposed one on another. These lamellæ are made up of bundles of modified connective tissue, the fibres of which are directly continuous with the fibres of the sclerotic. The fibres of each lamella are for the most part parallel with each other; those of alternating lamellæ at right angles to each other. Fibres, however, frequently pass from one lamella to the next.

The lamellæ are connected with each other by an interstitial cement-substance, in which are spaces, the *corneal spaces*. These are stellate in shape and have numerous offsets, by which they communicate with each other. Each contains a cell, the *corneal corpuscle*, which resembles in form the space in which it is lodged; it does not, however, entirely fill it.

Immediately beneath the conjunctival epithelium, the cornea proper presents certain characteristics which have led some anatomists to regard it as a distinct membrane, and it has been named by Bowman the *anterior elastic lamina*. It differs, however, from the true elastic lamina or membrane of Descemet in many essential particulars, presenting evidence of fibrillar structure, and not having the same tendency to curl inwards, or to undergo fracture, when detached from the other layers of the cornea. It consists of extremely closely interwoven fibrils, similar to those found in the rest of the cornea proper, but contains no corneal corpuscles. It ought, therefore, to be regarded as a part of the proper tissue of the cornea.*

The *posterior elastic lamina* (*membrane of Descemet or Demours*), which covers the proper structure of the cornea behind, presents no structure recognisable under the microscope. It consists of an elastic, and perfectly transparent homogeneous membrane, of extreme thinness, which is not rendered opaque by either water, alcohol, or acids. It is very brittle, but its most remarkable property is its extreme elasticity, and the tendency which it presents to curl up, or roll upon itself, with the attached surface innermost, when separated from the proper substance of the cornea. Its use appears to be (as suggested by Dr. Jacob) 'to preserve the requisite permanent correct curvature of the flaccid cornea proper.'

At the margin of the cornea this posterior elastic membrane breaks up into fibres to form a reticular structure at the outer angle of the anterior chamber, the intervals between the fibres forming small cavernous spaces, the *spaces of Fontana*. These little recesses communicate with a circular canal in the substance of the sclerotic close to its junction with the cornea. This is the *canal of Schlemm*, or *sinus venosus scleræ*; it communicates internally with the anterior chamber through the spaces of Fontana, and externally with the scleral veins. Some of the fibres of this reticulated structure are continued into the front of the iris, forming the *ligamentum pectinatum iridis*; while others are connected with the fore part of the sclerotic and choroid.

The *endothelial lining of the aqueous chamber* covers the posterior surface of the elastic lamina, is reflected on to the front of the iris, and also lines the spaces of Fontana. It consists of a single layer of polygonal flattened transparent nucleated cells, similar to those lining other serous cavities.

Arteries and Nerves.—The cornea is a non-vascular structure, the capillary vessels terminating in loops at its circumference. Lymphatic vessels have not as yet been demonstrated in it, but are represented by the channels in which the bundles of nerves run; these are lined by an endothelium, and are continuous with the cell-spaces. The nerves are numerous, twenty-four to thirty-six in number (Kölliker); forty to forty-five (Waldeyer and Sümisch): they are derived

---

* This layer has been called by Reichert the 'anterior limiting layer,' a name which appears more applicable to it than that of 'anterior elastic lamina.'

from the ciliary nerves, and enter the laminated tissue of the cornea. They ramify throughout its substance in a delicate network, and their terminal filaments form a firm and closer plexus on the surface of the cornea proper beneath the epithelium. This is termed the *subepithelial plexus*, and from it fibrils are given off which ramify between the epithelial cells, forming a network which is termed the *intra-epithelial plexus*.

Dissection.—In order to separate the sclerotic and cornea, so as to expose the second tunic, the eyeball should be immersed in a small vessel of water, and held between the finger and thumb. The sclerotic is then carefully incised, in the equator of the globe, till the choroid is exposed. One blade of a pair of probe-pointed scissors is now introduced through the opening thus made, and the sclerotic divided around its entire circumference, and removed in separate portions. The front segment being then drawn forwards, the handle of the scalpel should be pressed gently against it at its connection with the iris, and these being separated, a quantity of perfectly transparent fluid will escape; this is the aqueous humour. In the course of the dissection, the ciliary nerves may be seen lying in the loose cellular tissue between the choroid and sclerotic, or contained in delicate grooves on the inner surface of the latter membrane.

II. The Choroid, Ciliary Body, and Iris

The Second Tunic of the Eye (*tunica vasculosa oculi*) is formed from behind forwards by the choroid, the ciliary body, and the iris.

The choroid is the vascular and pigmentary tunic of the eyeball, investing the posterior five-sixths of the globe, and extending as far forwards as the ora serrata of the retina; the ciliary body connects the choroid to the circumference of the iris. The iris is the circular muscular septum, which hangs vertically behind the cornea, presenting in its centre a large rounded aperture, the *pupil*.

The **Choroid** is a thin, highly vascular membrane, of a dark brown or chocolate colour, which invests the posterior five-sixths of the globe, and is pierced behind by the optic nerve, and in this situation is firmly adherent to the sclerotic. It is thicker behind than in front. Externally, it is loosely connected by the lamina fusca with the inner surface of the sclerotic. Its inner surface is attached to the retina.

Structure.—The choroid consists mainly of a dense capillary plexus and of small arteries and veins, carrying the blood to and returning it from this plexus. On its external surface,

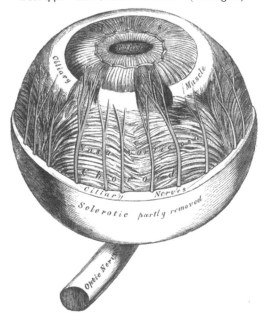

FIG. 441.—The choroid and iris. (Enlarged.)

i.e. the surface next the sclerotic, is a thin membrane, the *lamina suprachoroidea*, composed of delicate non-vascular lamellæ—each lamella consisting of a network of fine elastic fibres among which are branched pigment-cells. The spaces between the lamellæ are lined by endothelium, and open freely into the perichoroidal lymph-space, which, in its turn, communicates with the perisclerotic space by the perforations in the sclerotic through which the vessels and nerves are transmitted.

Internal to this is the *choroid proper*, and in consequence of the small arteries and veins being arranged on the outer surface of the capillary network, it is customary to describe this as consisting of two layers : the outermost, composed of small arteries and veins, with pigment-cells interspersed between them ; and the inner, consisting of a capillary plexus. The *external layer* or *lamina vasculosa* consists, in part, of the larger branches of the short ciliary arteries which run forwards between the veins, before they bend inwards to terminate in the capillaries ; but is formed principally of veins, which are named, from their arrangement, *venæ vorticosæ.* They converge to four or five equidistant trunks, which pierce the sclerotic midway between the margin of the cornea and the entrance of the optic nerve. Interspersed between the vessels are dark star-shaped pigment-cells, the offsets from which, communicating with similar branchings from neighbouring cells, form a delicate network or stroma, which towards the inner surface of the choroid loses its pigmentary character. The *internal layer* consists of an exceedingly fine capillary plexus, formed by the short

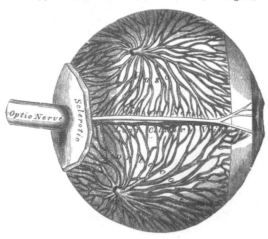

Fig. 442.—The veins of the choroid. (Enlarged.)

ciliary vessels, and is known as the *lamina chorio-capillaris* or *tunica Ruyschiana.* The network is close, and finer at the hinder part of the choroid than in front. About half an inch behind the cornea its meshes become larger, and are continuous with those of the ciliary processes. These two laminæ are connected by an *intermediate stratum*, which is destitute of pigment-cells and consists of fine elastic fibres. On the inner surface of the lamina chorio-capillaris is a very thin, structureless, or, according to Kölliker, faintly fibrous membrane, called the *lamina basalis* or membrane of Bruch ; it is closely connected with the stroma of the choroid, and separates it from the pigmentary layer of the retina.

Tapetum.—This name is applied to the iridescent appearance which is seen in the outer and posterior part of the choroid of many animals.

The ciliary body should now be examined. It may be exposed, either by detaching the iris from its connection with the Ciliary muscle, or by making a transverse section of the globe, and examining it from behind.

The **ciliary body** comprises the orbiculus ciliaris, the ciliary processes, and the Ciliary muscle.

The *orbiculus ciliaris* is a zone of about one-sixth of an inch in width, directly continuous with the anterior part of the choroid ; it presents numerous ridges arranged in a radial manner.

The *ciliary processes* are formed by the plaiting and folding inwards of the various layers of the choroid (i.e. the choroid proper and the lamina basalis) at its anterior margin, and are received between corresponding foldings of the suspensory ligament of the lens, thus establishing a connection between the choroid and inner tunic of the eye. They are arranged in a circle, and form a sort of plaited frill behind the iris, round the margin of the lens. They vary between sixty and eighty in number, lie side by side, and may be divided into large and small ; the latter, consisting of about one-third of the entire number, are situated in the spaces between the former, but without regular alternation. The larger processes are each about one-tenth of an inch in length, and are attached by their periphery to three or four of the ridges of the orbiculus ciliaris, and are continuous with the layers of the choroid : the opposite margin is free, and rests upon the circumference of the lens. Their anterior surface is turned towards the back of the iris, with the circumference of which they are continuous. The posterior surface is connected with the suspensory ligament of the lens.

Structure.—The ciliary processes are similar in structure to the choroid, but the vessels are larger, and have chiefly a longitudinal direction. They are covered on their inner surface by two strata of black pigment-cells, which are continued forwards from the retina, and are named the *pars ciliaris retinæ*. In the stroma of the ciliary processes there are also stellate pigment-cells, which, however, are not so numerous as in the choroid itself.

The *Ciliary muscle* (Bowman) consists of unstriped fibres : it forms a greyish, semitransparent, circular band, about one-eighth of an inch broad, on the outer surface of the fore part of the choroid. It is thickest in front, and gradually becomes thinner behind. It consists of two sets of fibres, radiating and circular. The former, much the more numerous, arise at the point of junction of the cornea and sclerotic, and partly also from the ligamentum pectinatum iridis, and, passing backwards, are attached to the choroid opposite to the ciliary processes. One bundle, according to Waldeyer, is continued backwards to be inserted into the sclerotic. The circular fibres are internal to the radiating ones and to some

Fig. 443.—The arteries of the choroid and iris.
The sclerotic has been mostly removed. (Enlarged.)

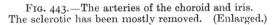

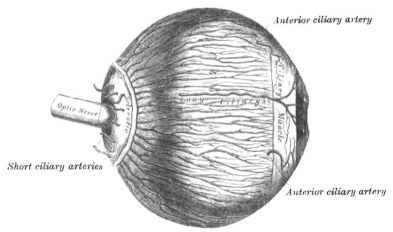

extent unconnected with them, and have a circular course around the attachment of the iris. They are sometimes called the 'ring muscle' of Müller, and were formerly described as the ciliary ligament. They are well developed in hypermetropic, but are rudimentary or absent in myopic eyes. The Ciliary muscle is admitted to be the chief agent in accommodation, i.e. in adjusting the eye to the vision of near objects. Bowman believed that this was effected by its compressing the vitreous body, and so causing the lens to advance ; but the view which now prevails is that the contraction of the muscle, by drawing on the ciliary processes, relaxes the suspensory ligament of the lens, thus allowing the anterior surface of the lens to become more convex. The pupil is at the same time slightly contracted.*

The **Iris** (*iris*, a rainbow) has received its name from its various colours in different individuals. It is a thin, circular-shaped, contractile curtain, suspended in the aqueous humour behind the cornea, and in front of the lens, being perforated a little to the nasal side of its centre by a circular aperture, the *pupil*, for the transmission of light. By its circumference it is continuous with the ciliary body, and is also connected with the posterior elastic lamina of the cornea by means of

* See explanation and diagram in Power's *Illustrations of some of the Principal Diseases of the Eye*, p. 590.

the pectinate ligament; its inner or free edge forms the margin of the pupil; its surfaces are flattened, and look forwards and backwards, the anterior towards the cornea, the posterior towards the ciliary processes and lens. The anterior surface of the iris is variously coloured in different individuals, and marked by lines which converge towards the pupil. The posterior surface is of a deep purple tint, from being covered by two layers of pigmented, columnar epithelium, which are continuous posteriorly with the pars ciliaris retinæ. This pigmented epithelium is termed the *pars iridica retinæ*, though it is sometimes named *uvea*, from its resemblance in colour to a ripe grape.

Structure.—The iris is composed of the following structures :

1. In front is a layer of flattened endothelial cells placed on a delicate hyaline basement-membrane. This layer is continuous with the epithelial layer covering the membrane of Descemet, and in men with dark coloured irides the cells contain pigment-granules.

FIG. 444.—Section of the eye, showing the relations of the cornea, sclerotic, and iris, together with the Ciliary muscle and the spaces of Fontana near the angle of the anterior chamber. (Waldeyer.)

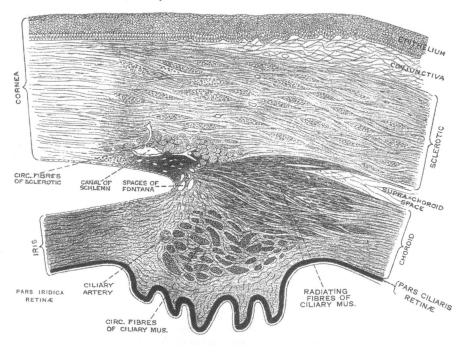

2. *Stroma.*—The stroma consists of fibres and cells. The former are made up of delicate bundles of fibrous tissue, of which some few fibres have a circular direction at the circumference of the iris ; but the chief mass consists of fibres radiating towards the pupil. They form, by their interlacement, a delicate mesh, in which the vessels and nerves are contained. Interspersed between the bundles of are connective tissue numerous branched cells with fine processes. Many of them in dark eyes contain pigment-granules, but in blue eyes and the pink eyes of albinos they are unpigmented.

3. The *muscular fibre* is involuntary, and consists of circular and radiating fibres. The *circular fibres* (sphincter pupillæ) surround the margin of the pupil on the posterior surface of the iris, like a sphincter, forming a narrow band about one-thirtieth of an inch in width ; those near the free margin being closely aggregated ; those more external somewhat separated, and forming less complete circles. The *radiating fibres* (dilator pupillæ) converge from the circumference

towards the centre, and blend with the circular fibres near the margin of the pupil. These fibres are regarded by some as elastic, not muscular.

4. *Pigment.*—The situation of the pigment-cells differs in different irides. In the various shades of blue eyes, the only pigment-cells are several layers of small round or polyhedral cells, filled with dark pigment, situated on the posterior surface of the iris, and continuous with the pigmentary lining of the ciliary processes. The colour of the eye in these individuals is due to this colouring matter showing more or less through the texture of the iris. In the albino, even this pigment is absent. In the grey, brown, and black eye there are, as mentioned above, pigment-granules to be found in the cells of the stroma and in the epithelial layer on the front of the iris; to these the dark colour of the eye is due.

The *arteries of the iris* are derived from the long and anterior ciliary, and from the vessels of the ciliary processes (see page 483). The long ciliary arteries, two in number, having reached the attached margin of the iris, divide into an upper and lower branch, and encircling the iris, anastomose with corresponding branches from the opposite side; into this vascular zone (circulus major) the anterior ciliary pour their blood. From this zone vessels converge to the free margin of the iris, and these communicate by branches from one to another and thus form a second zone (circulus minor) in this situation.

The *nerves of the choroid and iris* are derived from the ciliary branches of the lenticular ganglion, and the long ciliary from the nasal branch of the ophthalmic division of the fifth. They pierce the sclerotic around the entrance of the optic nerve, and run forwards in the perichoroidal space, and supply the blood-vessels of the choroid. After reaching the iris they form a plexus around its attached margin; from this are derived non-medullated fibres which terminate in the circular and radiating muscular fibres. Their exact mode of termination has not been ascertained. Other fibres from the plexus terminate in a network on the anterior surface of the iris. The fibres derived from the motor root of the lenticular ganglion (third nerve) supply the circular fibres, while those derived from the sympathetic supply the radiating fibres.

Membrana pupillaris.—In the fœtus, the pupil is closed by a delicate transparent vascular membrane, the *membrana pupillaris*, which divides the space in which the iris is suspended into two distinct chambers. This membrane contains numerous minute vessels, continued from the margin of the iris to those on the front part of the capsule of the lens. These vessels have a looped arrangement, and converge towards each other without anastomosing. Between the seventh and eighth month the membrane begins to disappear by its gradual absorption, from the centre towards the circumference, and at birth only a few fragments remain. It is said sometimes to remain permanent and produce blindness.

III. The Retina

The **Retina** is a delicate nervous membrane, upon the surface of which the images of external objects are received. Its outer surface is in contact with the choroid; its inner with the vitreous body. Behind, it is continuous with the optic nerve; it gradually diminishes in thickness from behind forwards; and, in front, extends nearly as far as the ciliary body, where it appears to terminate in a jagged margin, the *ora serrata.* Here the nervous tissues of the retina end, but a thin prolongation of the membrane extends forwards over the back of the ciliary processes and iris, forming the *pars ciliaris retinæ* and *pars iridica retinæ* already referred to. This forward prolongation consists of the pigmentary layer of the retina together with a stratum of columnar epithelium. The retina is soft, semitransparent, and of a purple tint in the fresh state, owing to the presence of a colouring material named *rhodopsin* or *visual purple*; but it soon becomes clouded, opaque, and bleached when exposed to sunlight. Exactly in the centre of the posterior part of the retina, corresponding to the axis of the eye, and at a point in which the sense of vision is most perfect, is an oval yellowish spot, called, after its discoverer, the *yellow spot* or *macula lutea* of Sömmerring; having a central depression, the *fovea centralis*. The retina in the situation of the fovea centralis is exceedingly thin, and the dark colour of the choroid is distinctly seen through it; so that it presents more the appearance of a foramen, and hence the name 'foramen of Sömmerring' at first given to it. It exists only in man, the quadrumana, and some saurian reptiles. About one-eighth of an inch (3 mm.) to the inner side of the yellow spot is the point of entrance of the optic nerve (*porus opticus*); here the nervous substance is slightly raised so as to form an eminence (*colliculus nervi optici*); the arteria centralis retinæ pierces its centre. This is the only part of the surface of the retina from which the power of vision is absent, and is termed the 'blind spot.'

Structure.—The retina is an exceedingly complex structure, and, when examined microscopically by means of sections made perpendicularly to its surface, is found to consist of ten layers, which are named from within outwards, as follows:

1. Membrana limitans interna.
2. Layer of nerve-fibres (stratum opticum).
3. Ganglionic layer, consisting of nerve-cells.
4. Inner molecular, or plexiform, layer.
5. Inner nuclear layer, or layer of inner granules.
6. Outer molecular, or plexiform, layer.
7. Outer nuclear layer, or layer of outer granules.
8. Membrana limitans externa.
9. Jacob's membrane (layer of rods and cones).
10. Pigmentary layer (tapetum nigrum).

1. The *membrana limitans interna* is the most internal layer of the retina, and is in contact with the hyaloid membrane of the vitreous humour. It is derived from the supporting framework of the retina, with which tissue it will be described.

2. The *layer of nerve-fibres* is formed by the expansion of the optic nerve. This nerve passes through all the other layers of the retina, except the membrana limitans interna, to reach its destination. As the nerve passes through the lamina cribrosa of the sclerotic coat, the fibres of which it is composed lose their medullary sheaths and are continued onwards, through the choroid and retina, as simple axis cylinders. When these non-medullated fibres reach the internal surface of the retina, they radiate from their point of entrance over the surface of the retina, grouped in bundles, and in many places, according to Michel, arranged in plexuses. Most of the fibres in this layer are centripetal, and are the direct continuations of the axis-cylinder processes of the cells of the next layer, but a few of them (centrifugal fibres) pass through it and the next succeeding layer to ramify in the inner molecular and inner nuclear layers, where they terminate in enlarged extremities (fig. 448 1, *m*). The layer is thickest at the optic nerve entrance, and gradually diminishes in thickness towards the ora serrata.

3. The *ganglionic layer* consists of a single layer of large ganglion-cells; except in the macula lutea, where there are several strata. The cells are somewhat flask-shaped; their rounded internal margin resting on the preceding layer and sending off an axon which is prolonged as a nerve-fibre into the fibrous layer. From the opposite extremity numerous thicker processes (dendrites) extend into the inner molecular layer, where they branch out into flattened arborisations at different levels (fig. 448 VII). The ganglion-cells vary much in size, and the dendrites of the smaller ones as a rule arborise in the inner molecular layer as soon as they enter it; while the processes of the larger cells ramify close to the inner nuclear layer.

4. The *inner molecular layer* is made up of a dense reticulum of minute fibrils, formed by the interlacement of the dendrites of the ganglion-cells with those of the cells contained in the next layer, immediately to be described. Within the reticulum formed by these fibrils a few branched spongioblasts are sometimes embedded.

5. The *inner nuclear layer* is made up of a number of closely packed cells, of which there are three different kinds. (1) A large number of oval cells, which are commonly regarded as bipolar nerve-cells, and are much more numerous

FIG. 445.—The arteria centralis retinæ, yellow spot, &c.,
the anterior half of the eyeball being removed. (Enlarged.)

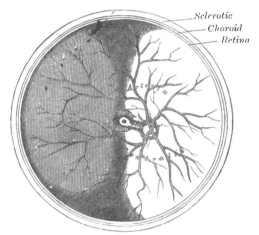

Sclerotic
Choroid
Retina

than either of the other kind. They each consist of a large oval body placed vertically to the surface, and containing a distinct nucleus : they are surrounded by a small amount of protoplasm, which is prolonged into two processes ; one of these passes inwards into the inner molecular layer, is varicose in appearance, and ends in a terminal ramification, which is often in close proximity to the

FIG. 446.   FIG. 447.

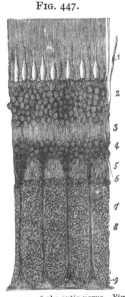

Vertical sections of the human retina. Fig. 446, half an inch from the entrance of the optic nerve. Fig. 447, close to the latter. 1. Layer of rods and cones, Jacob's membrane, bounded underneath by the *membrana limitans externa*. 2. Outer nuclear layer. 3. Outer molecular layer. 4. Inner nuclear layer. 5. Inner molecular layer. 6. Ganglionic layer. 7. Layer of nerve-fibres. 8. Sustentacular fibres of Müller. 9. Their attachment to the *membrana limitans interna*.

ganglion-cells (fig. 448-1, c). The other process passes outwards, into the outer molecular layer, and there breaks up into a number of branches. According to Cajal there are two varieties of these bipolar cells : one in which the outer process arborises around the knobbed ends of the rod-fibres, and the inner around the cells of the ganglionic layer; these he calls *rod-bipolars* (fig. 448 1, *c, d*); the others are those in which the outer process breaks up in a horizontal ramification, in contact with the end of a cone-fibre; these are the *cone-bipolars*, and their inner process breaks up into its terminal ramifications in the inner molecular layer (fig. 448 1, *e*). (2) At the innermost part of this inner nuclear layer is a stratum of cells, which are named by Cajal *amacrine* cells, from the fact that they have no axis-cylinder process, but they give a number of short protoplasmic processes, which extend into the inner molecular layer and there ramify (fig. 448 1, *h*). There are also at the outermost part of this layer some cells, the processes of which extend into and ramify in the outer molecular layer. These are the *horizontal* cells of Cajal. (3) Some few cells are also found in this layer, connected with the fibres of Müller, and will be described with those structures.

6. The *outer molecular layer* is much thinner than the inner molecular layer; but, like it, consists of a dense network of minute fibrils, derived from the processes of the horizontal cells of the preceding layer and the outer processes of the bipolar cells, which ramify in it, forming arborisations around the ends of the rod-fibres and with the branched foot-plates of the cone-fibres.

7. *The outer nuclear layer.*—Like the inner nuclear layer, this layer contains several strata of clear oval nuclear bodies; they are of two kinds, and on account of their being respectively connected with the rods and cones of Jacob's membrane, are named rod-granules and cone-granules. The *rod-granules* are much the more numerous, and are placed at different levels throughout the layer. Their nuclei present a peculiar cross-striped appearance, and prolonged from either extremity of the granule is a fine process: the outer process is continuous with a single rod of Jacob's membrane; the inner passes inwards towards the outer molecular layer and terminates in an enlarged extremity, and is embedded in the tuft into which the outer process of the rod-bipolars break up. In its course it presents numerous varicosities. The *cone-granules*, fewer in number than the rod-granules, are placed close to the membrana limitans externa, through which they are continuous with the cones of Jacob's membrane. They do not present any cross-striping, but contain a pyriform nucleus, which almost completely fills the cell. From their inner extremity a thick process passes inwards to the outer molecular layer upon which it rests by a somewhat pyramidal enlargement, from which are given off numerous fine fibrils, which enter the outer molecular layer, where they come in contact with the outer processes of the cone-bipolars.

DESCRIPTION OF FIGURE 448.

I. Section of the dog's retina. *a.* Cone-fibre. *b.* Rod-fibre and nucleus. *c d.* Bipolar cells (inner granules) with vertical ramification of outer processes destined to receive the enlarged ends of rod-fibres. *e.* Bipolars with flattened ramification for ends of cone-fibres. *f.* Giant bipolar with flattened ramification. *g.* Cell sending a neuron or nerve-fibre process to the outer molecular layer. *h.* Amacrine cell with diffuse arborisation in inner molecular layer. *i.* Nerve-fibrils passing to outer molecular layer. *j.* Centrifugal fibres passing from nerve-fibre layer to inner molecular layer. *m.* Nerve-fibril passing into inner molecular layer. *n.* Ganglionic cells.
II. Horizontal or basal cells of the outer molecular layer of the dog's retina. A. Small cell with dense arborisation. B. Large cell, lying in inner nuclear layer but with its processes branching in the outer molecular. *a.* Its horizontal neuron. C. Medium-sized cell of the same character.
III. Cells from the retina of the ox. *a.* Rod-bipolars with vertical arborisations. *b c d e.* Cone-bipolars with horizontal ramification of outer process. *h.* Cells lying on the outer surface of the outer molecular layer, and ramifying within it. *i j m.* Amacrine cells within the substance of the inner molecular layer.
IV. Neurons or axis-cylinder processes belonging to horizontal cells of the outer molecular layer, one of them, *b*, ending in a close ramification at *a*.
V. Nervous elements connected with the inner molecular layer of the ox's retina. A. Amacrine cell, with long processes ramifying in the outermost stratum. B. Large amacrine with thick processes ramifying in second stratum. C. Flattened amacrine with long and fine processes ramifying mainly in the first and fifth strata. D. Amacrine with radiating tuft of fibrils destined for third stratum. E. Large amacrine, with processes ramifying in fifth stratum. F. Small amacrine, branching in second stratum. G H. Other amacrines destined for fourth stratum. *a.* Small ganglion-cell sending its processes to fourth stratum. *b.* A small ganglion-cell with ramifications in three strata. *c.* A small cell ramifying ultimately in first stratum. *d.* A medium-sized ganglion-cell ramifying in fourth stratum. *e.* Giant-cell, branching in third stratum. *f.* A bi-stratified cell ramifying in second and fourth strata.
VI. Amacrines and ganglion-cells from the dog. A. Amacrine with radiating tuft. B. Large amacrine passing to third stratum. C and G. Small amacrines with radiations in second stratum. F. Small amacrine passing to third stratum. D. Amacrine with diffuse arborisation. E. Amacrine belonging to fourth stratum. *a d e g.* Small ganglion-cells, ramifying in various strata. *b f.* Large ganglion-cells showing two different characters of arborisation. *i.* Bi-stratified cell.
VII. Amacrines and ganglion-cells from the dog. A B C. Small amacrines ramifying in middle of molecular layer. *b d g h i.* Small ganglion-cells showing various kinds of arborisation. *f.* A larger cell, similar in character to *g*, but with longer branch. *a c e.* Giant-cells with thick branches ramifying in the first, second, and third layers L L. Ends of bipolars branching over ganglion-cells.

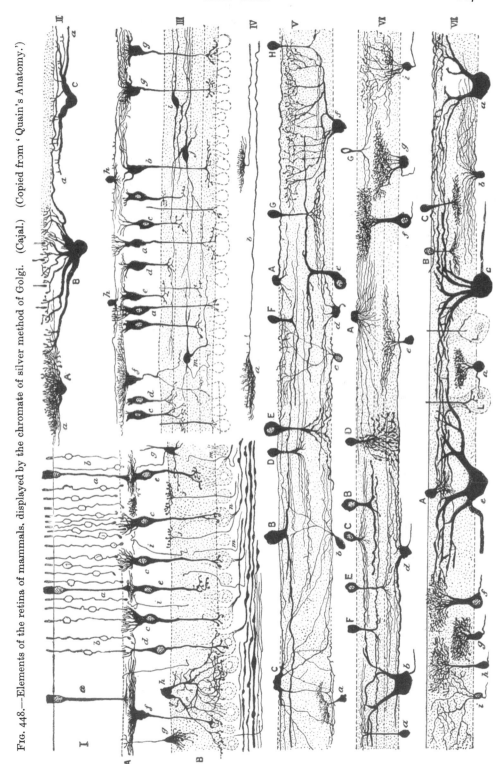

Fig. 448.—Elements of the retina of mammals, displayed by the chromate of silver method of Golgi. (Cajal.) (Copied from 'Quain's Anatomy.')

8. *The membrana limitans externa.*—This layer, like the membrana limitans interna, is derived from the fibres of Müller, with which structures it will be described.

9. *Jacob's membrane (layer of rods and cones).*—The elements which compose this layer are of two kinds, *rods* and *cones,* the former being much more numerous than the latter. The *rods* are of nearly uniform size, and arranged perpendicularly to the surface. Each rod consists of two portions, an outer and inner, which are of about equal length. The segments differ from each other as regards refraction and in their behaviour with colouring reagents, the inner portion becoming stained by carmine, iodine, &c., the outer portion remaining unstained with these reagents, but staining yellowish-brown with osmic acid. The outer portion of each rod is marked by transverse striæ, and is made up of a number of thin discs superimposed on one another. It also exhibits faint longitudinal markings. The inner portion of each rod, at its deeper part where it is joined to the outer process of the rod-granule, is indistinctly granular ; its more superficial part presents a longitudinal striation, being composed of fine, bright, highly refracting fibrils. The visual purple or rhodopsin is only found in the outer segments of the rods.

The *cones* are conical or flask-shaped, their broad ends resting upon the membrana limitans externa, the narrow pointed extremity being turned to the choroid. Like the rods, they are made up of two portions, outer and inner ; the outer portion is a short conical process, which, like the outer segment of the rods, presents transverse striæ. The inner portion resembles the inner portion of the rods in structure, presenting a superficial striated and deeper granular part ; but differs from it in size, being bulged out laterally and presenting a flask shape. The chemical and optical characters of the two portions are identical with those of the rods.

10. *The pigmentary layer, or tapetum nigrum.*—The most external layer of the retina, formerly regarded as a part of the choroid, consists of a single layer of hexagonal epithelial cells, loaded with pigment-granules. They are smooth externally, where they are in contact with the choroid, but internally they are prolonged into fine, straight processes, which extend between the rods, this being especially the case when the eye is exposed to light. In the eyes of albinos, the cells of the pigmentary layer are present, but they contain no colouring matter. In many of the mammals also, as in the horse, and many of the carnivora, there is no pigment in the cells of this layer, and the choroid possesses a beautiful iridescent lustre, which is termed the *tapetum lucidum.*

Supporting framework of the retina.—Almost all these layers of the retina are connected together by a supporting framework, formed by the *fibres of Müller,* or *radiating fibres,* from which the membrana limitans interna et externa are derived. These fibres are found stretched between the two limiting layers, 'as columns between a floor and a ceiling,' and passing through all the nervous layers, except Jacob's membrane. Each commences on the inner surface of the retina by a conical hollow base, which sometimes contains a spheroidal body, staining deeply with hæmatoxylin, the edges of the bases of adjoining fibres being united and thus forming a boundary line, which is the membrana limitans interna. As they pass through the nerve-fibre and ganglionic layers they give off few lateral branches ; in the inner nuclear layer they give off numerous lateral processes for the support of the inner granules, while in the outer nuclear layer they form a network around the rod and cone-fibrils, and unite to form the external limiting membrane at the bases of the rods and cones. In the inner nuclear layer each fibre of Müller presents a clear oval nucleus, which is sometimes situated at the side of, sometimes altogether within, the fibre.

Macula lutea and fovea centralis.—The structure of the retina at the yellow spot presents some modifications. In the macula lutea (1) the nerve-fibres are wanting as a continuous layer ; (2) the ganglionic layer consists of several strata of cells, instead of a single layer; (3) in Jacob's membrane there are no rods, but only cones, and these are longer and narrower than in other parts ; and (4) in the outer nuclear layer there are only cone-fibres, which are very long and arranged in curved lines.   At the fovea centralis the only parts which exist are (1) the cones of Jacob's membrane ; (2) the outer nuclear layer, the cone-fibres of which are almost horizontal in direction ; (3) an exceedingly thin inner granular layer ; (4) the pigmentary layer, which is thicker and its pigment more pronounced than elsewhere.   The colour of the macula seems to imbue all the layers except Jacob's membrane ; it is of a rich yellow, deepest towards the centre, and does not appear to consist of pigment-cells, but simply a staining of the constituent parts.

At the ora serrata the nervous layers of the retina terminate abruptly, and the retina is continued onwards as a single layer of elongated columnar cells covered by

Fig. 449.—The layers of the retina (diagrammatic).   (After Schultze.)

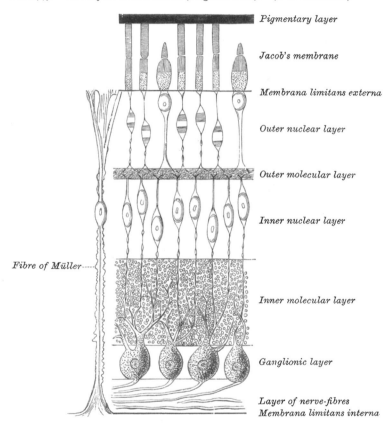

the pigmentary layer.   This prolongation is known as the *pars ciliaris retinæ,* and can be traced forwards from the ciliary processes on to the back of the iris, where it is termed the *pars iridica retinæ* or *uvea.*

From the description given of the nervous elements of the retina it will be seen that there is no direct continuity between the structures which form its different layers except between the ganglionic and nerve-fibre layers, the majority of the nerve-fibres being formed of the axons of the ganglionic cells.   In the inner molecular layer the dendrites of the ganglionic layer interlace with those of the cells of the inner nuclear layer, while in the outer molecular layer a like synapsis occurs between the processes of the inner granules and the rod and cone elements.

The *arteria centralis retinæ* and its accompanying vein pierce the optic nerve, and enter the globe of the eye through the porus opticus. It immediately bifurcates into an upper and a lower branch, and each of these again divides into an inner or nasal and an outer or temporal branch, which at first run between the hyaloid membrane and the nervous layer; but they soon enter the latter, and pass forwards, dividing dichotomously. From these branches a minute capillary plexus is given off, which does not extend beyond the inner nuclear layer. The macula receives small twigs from the temporal branches and others directly from the central artery; these do not, however, reach as far as the fovea centralis, which has no blood-vessels. The branches of the arteria centralis retinæ do not anastomose with each other—in other words, they are 'terminal arteries.' In the fœtus, a small vessel passes forwards, through the vitreous humour, to the posterior surface of the capsule of the lens.

REFRACTING MEDIA

The Refracting media are three, viz. :

Aqueous humour.   Vitreous body.   Crystalline lens.

I. AQUEOUS HUMOUR

The **aqueous humour** completely fills the anterior and posterior chambers of the eyeball. It is small in quantity (scarcely exceeding, according to Petit, four or five grains in weight), has an alkaline reaction, in composition is little more than water, less than one-fiftieth of its weight being solid matter, chiefly chloride of sodium.

The *anterior chamber* is the space bounded in front by the cornea; behind, by the front of the iris. The *posterior chamber* is a narrow chink between the peripheral part of the iris, the suspensory ligament of the lens, and the ciliary processes.

In the adult, these two chambers communicate through the pupil; but in the fœtus of the seventh month, when the pupil is closed by the membrana pupillaris, the two chambers are quite separate.

II. VITREOUS BODY

The **vitreous body** forms about four-fifths of the entire globe. It fills the concavity of the retina, and is hollowed in front, forming a deep concavity, the *fossa patellaris*, for the reception of the lens. It is perfectly transparent, of the consistence of thin jelly, and is composed of an albuminous fluid enclosed in a delicate transparent membrane, the *membrana hyaloidea*. It has been supposed, by Hannover, that from its inner surface numerous thin lamellæ are prolonged inwards in a radiating manner, forming spaces in which the fluid is contained. In the adult, these lamellæ cannot be detected even after careful microscopic examination in the fresh state, but in preparations hardened in weak chromic acid, it is possible to make out a distinct lamellation at the periphery of the body; and in the fœtus a peculiar fibrous texture pervades the mass, the fibres joining at numerous points, and presenting minute nuclear granules at their point of junction. In the centre of the vitreous humour, running from the entrance of the optic nerve to the posterior surface of the lens, is a canal, filled with fluid and lined by a prolongation of the hyaloid membrane. This is the *canal of Stilling*, which in the embryonic vitreous humour conveyed the minute vessel from the central artery of the retina to the back of the lens. The fluid from the vitreous body resembles nearly pure water; it contains, however, some salts, and a little albumen.

The hyaloid membrane encloses the whole of the vitreous humour. In front of the ora serrata it is thickened by the accession of radial fibres and is termed the *zonule of Zinn* or *zonula ciliaris*. It presents a series of radially arranged furrows, in which the ciliary processes are accommodated and to which they are adherent, as evidenced by the fact that when removed some of their pigment remains attached to the zonule. The zonule of Zinn splits into two layers, one of which is thin and lines the fossa patellaris, the other is named the *suspensory ligament of the lens*; it is thicker, and passes over the ciliary body to be attached to the capsule of the lens a short distance in front of its equator. Scattered and delicate fibres are also attached to the region of the equator itself. This ligament retains the lens in position, and is relaxed by the contraction of the radial fibres of the Ciliary muscle, so that the lens is allowed to become more convex. Behind the suspensory ligament there is a sacculated canal, the *canal of Petit*, which encircles the equator of the lens and which can be easily inflated through a fine blow-pipe inserted through the suspensory ligament.

In the fœtus, the centre of the vitreous humour presents the canal of Stilling, already referred to, which transmits a minute artery to the capsule of the lens. *In the adult*, no vessels penetrate its substance; so that its nutrition must be carried on by the vessels of the retina and ciliary processes, situated upon its exterior.

III. CRYSTALLINE LENS

The **crystalline lens,** enclosed in its capsule, is situated immediately behind the pupil, in front of the vitreous body, and encircled by the ciliary processes, which slightly overlap its margin.

The *capsule of the lens* is a transparent, highly elastic, and brittle membrane, which closely surrounds the lens. It rests, behind, in the fossa patellaris in the fore part of the vitreous body; in front, it is in contact with the free border of the iris, this latter receding from it at the circumference, thus forming the posterior chamber of the eye; and it is retained in its position chiefly by the suspensory ligament of the lens, already described. The capsule is much thicker in front than behind, and structureless in texture; when ruptured, the edges roll up with the outer surface innermost, like the elastic lamina of the cornea.

FIG. 450.—The crystalline lens, hardened and divided. (Enlarged.)

The anterior surface of the lens is covered by a single layer of transparent, polygonal, nucleated cells. At the circumference of the lens, these cells undergo a change in form : they become elongated, and Babucin states that he can trace the gradual transition of the cells into proper lens-fibres, with which they are directly continuous. There is no epithelium on the posterior surface.

In the fœtus, a small branch from the arteria centralis retinæ runs forwards, as already mentioned, through the vitreous humour to the posterior part of the capsule of the lens, where its branches radiate and form a plexiform network, which covers its surface, and they are continuous round the margin of the capsule with the vessels of the pupillary membrane, and with those of the iris. In the adult no vessels enter its substance.

The *lens* is a transparent, bi-convex body, the convexity being greater on the posterior than on the anterior surface. The central points of its anterior and posterior surfaces are known as its *anterior* and *posterior poles*. It measures from 9 to 10 mm. in the transverse diameter, and about 4 mm. in the anteroposterior. It consists of concentric layers, of which the external in the fresh state are soft and easily detached (*substantia corticalis*) ; those beneath are firmer, the central ones forming a hardened nucleus (*nucleus lentis*). These laminæ are best demonstrated by boiling, or immersion in alcohol, and consist of minute parallel fibres, which are hexagonal prisms, the edges being dentated, and the dentations fitting accurately into each other; their breadth is about $\frac{1}{3000}$th of an inch. Faint lines radiate from the anterior and posterior poles to the circumference of the lens. In the adult there may be six or more of these, but in the fœtus they are only three in number and diverge from each other at angles of 120° (fig. 451). On the anterior surface one line ascends vertically and the other two diverge downwards and outwards. On the posterior surface one ray descends vertically

FIG. 451.—Diagram to show the direction and arrangement of the radiating lines on the front and back of the fœtal lens. (A) from the front. (B) from the back.

 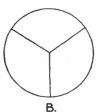

A.   B.

and the other two diverge upwards. They correspond with the free edges of an equal number of septa in the lens, along which the ends of the lens fibres come into apposition and are joined by transparent amorphous substance. The fibres run in a curved manner from the septa on the anterior surface to those on the posterior surface. No fibres pass from pole to pole, but they are arranged in such a way that fibres which commence near the pole on the one aspect of the lens terminate near the peripheral extremity of the plane on the other, and *vice versâ*. The fibres of the outer layers of the lens each contain a nucleus, which together form a layer (nuclear layer) on the surface of the lens, most distinct towards its circumference.

The *changes produced in the lens by age* are the following:

In the fœtus, its form is nearly spherical, its colour of a slightly reddish tint; it is not perfectly transparent, and is so soft as to break down readily on the slightest pressure.

In the adult, the posterior surface is more convex than the anterior; it is colourless, transparent, and firm in texture.

In old age it becomes flattened on both surfaces, slightly opaque, of an amber tint, and it also increases in density.

The *arteries of the globe of the eye* are the short, long, and anterior ciliary arteries, and the arteria centralis retinæ. They have been already described (see page 484).

The *ciliary veins* are seen on the outer surface of the choroid, and are named, from their arrangement, the *venæ vorticosæ*. They converge to four or five equidistant trunks which pierce the sclerotic midway between the margin of the cornea and the entrance of the optic nerve. Another set of veins accompany the anterior ciliary arteries and open into the ophthalmic vein.

The *ciliary nerves* are derived from the nasal branch of the ophthalmic nerve and from the ciliary or ophthalmic ganglion.

Surgical Anatomy.—From a surgical point of view the cornea may be regarded as consisting of three layers: (1) of an external epithelial layer, developed from the epiblast, and continuous with the external epithelial covering of the rest of the body, and therefore in its lesions resembling those of the epidermis; (2) of the cornea proper, derived from the mesoblast, and associated in its diseases with the fibro-vascular structures of the body; and (3) the posterior elastic layer with its endothelium, also derived from the mesoblast and having the characters of a serous membrane, so that inflammation of it resembles inflammation of the other serous and synovial membranes of the body.

The cornea contains no blood-vessels except at its periphery, where numerous delicate loops, derived from the anterior ciliary arteries, may be demonstrated on the anterior surface of the cornea. The rest of the cornea is nourished by lymph, which gains access to the proper substance of the cornea and the posterior layer through the spaces of Fontana. This lack of a direct blood-supply renders the cornea very apt to inflame in the cachectic and ill-nourished. In cases of granular lids, there is a peculiar affection of the cornea, called *pannus*, in which the anterior layers of the cornea become vascularised, and a rich network of blood-vessels may be seen upon it ; and in interstitial keratitis new vessels extend into the cornea, giving it a pinkish hue, to which the term ' salmon patch ' is applied. The cornea is richly supplied with nerves, derived from the ciliary, which enter the cornea through the fore part of the sclerotic and form plexuses in the stroma, terminating between the epithelial cells by free ends or in corpuscles. In cases of glaucoma the ciliary nerves may be pressed upon as they course between the choroid and sclerotic, and the cornea becomes anæsthetic. The sclerotic has very few blood-vessels and nerves. The blood-vessels are derived from the anterior ciliary, and form an open plexus in its substance. As they approach the corneal margin this arrangement is peculiar. Some branches pass through the sclerotic to the ciliary body ; others become superficial and lie in the episcleral tissue, and form arches, by anastomosing with each other, some little distance behind the corneal margin. From these arches numerous straight vessels are given off, which run forwards to the cornea,· forming its marginal plexus. In inflammation of the sclerotic and episcleral tissue these vessels become conspicuous, and form a pinkish zone of straight vessels radiating from the corneal margin, commonly known as the *zone of ciliary injection*. In inflammation of the iris and ciliary body this zone is present, since the sclerotic speedily becomes involved when these structures are inflamed. But in inflammation of the cornea the sclerotic is seldom much affected, though the cornea and sclerotic are structurally continuous. This would appear to be due to the fact, that the nutrition of the cornea is derived from a different source from that of the sclerotic. The sclerotic may be ruptured subcutaneously without any laceration of the conjunctiva, and the rupture usually occurs near the corneal margin, where the tunic is thinnest. It may be complicated with lesions of adjacent parts—laceration of the choroid, retina, iris, or suspensory ligament of the lens—and is then often attended with hæmorrhage into the anterior chamber, which masks the nature of the injury. In some cases the lens has escaped through the rent in the sclerotic and has been found under the conjunctiva. Wounds of the sclerotic are always dangerous, and are often followed by inflammation, suppuration, and by sympathetic ophthalmia.

One of the functions of the choroid is to provide nutrition for the retina, and to convey vessels and nerves to the ciliary body and iris. Inflammation of the choroid is therefore followed by grave disturbance in the nutrition of the retina, and is attended with early interference with vision. In its diseases it bears a considerable analogy to those which affect the skin, and, like it, is one of the places from which melanotic sarcomata may grow. These tumours contain a large amount of pigment in their cells, and grow only from those parts where pigment is naturally present. The choroid may be ruptured, without injury to the other tunics, as well as participating in general injuries of the eyeball. In cases of uncomplicated rupture, the injury is usually at its posterior part, and is the result of a blow on the front of the eye. It is attended by considerable hæmorrhage, which for a time may obscure vision, but, in most cases, this is restored, as soon as the blood is absorbed.

The iris is the seat of a malformation, termed *coloboma*, which consists in a deficiency or cleft, which in a great number of cases is clearly due to an arrest in development. In these cases it is found at the lower aspect, extending directly downwards from the pupil, and the gap frequently extends through the choroid to the entrance of the optic nerve. In some rarer cases the gap is found in other parts of the iris, and is then not associated with any deficiency of the choroid. The iris is abundantly supplied with blood-vessels and nerves, and is therefore very prone to become inflamed. And when inflamed, in consequence of the intimate relationship which exists between the vessels of the iris and choroid, this latter tunic is very apt to participate in the inflammation. And, in addition, inflammation of adjacent structures, the cornea and sclerotic, is apt to spread into the iris. The iris is covered with epithelium, and partakes of the character of a serous membrane, and, like these structures, is liable to pour out a plastic exudation, when inflamed, and contract adhesions, either to the cornea in front (*synechia anterior*), or to the capsule of the lens behind (*synechia posterior*). In iritis the lens may become involved, and the condition known as secondary cataract may be set up. Tumours occasionally commence in the iris ; of these, cysts, which are usually congenital, and sarcomatous tumours, are the most common and require removal. Gummata are not infrequently found in this situation. In some forms of injury of the eyeball, as the impact of a spent shot, the rebound of a twig, or a blow with a whip, the iris may be detached from the ciliary muscle, the amount of detachment varying from the slightest degree to separation of the whole iris from its ciliary connection.

The retina, with the exception of its pigment layer and its vessels, is perfectly transparent, so as to be invisible when examined by the ophthalmoscope, so that its diseased conditions are recognised by its loss of transparency. In retinitis, for instance, there is more or less dense and extensive opacity of its structure, and not infrequently extravasations of blood into its substance.   Hæmorrhages may also take place into the retina, from rupture of a blood-vessel without inflammation.

The retina may become displaced from effusion of serum between it and the choroid, or by blows on the eyeball, or may occur without apparent cause in progressive myopia, and in this case the ophthalmoscope shows an opaque, tremulous cloud. Glioma, a form of sarcoma, and essentially a disease of early life, is occasionally met with in connection with the retina.

The lens has no blood-vessels, nerves, or connective tissue in its structure, and therefore is not subject to those morbid changes to which tissues containing these structures are liable.   It does however present certain morbid or abnormal conditions of various kinds. Thus, variations in shape, absence of the whole or a part of the lens, and displacements are among its congenital defects. Opacities may occur from injury, senile changes, malnutrition, or errors in growth or development.   Senile changes may take place in the lens, impairing its elasticity and rendering it harder than in youth, so that its curvature can only be altered to a limited extent by the Ciliary muscle.   And, finally, the lens may be dislocated or displaced by blows upon the eyeball ; and its relations to surrounding structures altered by adhesions or the pressure of new growths.

There are two particular regions of the eye which require special notice : one of these is known as the 'filtration area,' and the other as the 'dangerous area.'   The *filtration area* is the circumcorneal zone immediately in front of the iris.   Here are situated the cavernous spaces of Fontana, which communicate with the canal of Schlemm, through which the chief transudation of fluid from the eye is now believed to take place.   The *dangerous area of the eye* is the region in the neighbourhood of the ciliary body, and wounds or injuries in this situation are peculiarly dangerous.   For inflammation of the ciliary body is liable to spread to many of the other structures of the eye, especially to the iris and choroid, which are intimately connected by nervous and vascular supplies.   Moreover, wounds which involve the ciliary region are especially liable to be followed by sympathetic ophthalmia, in which destructive inflammation of one eye is excited by some irritation in the other.

APPENDAGES OF THE EYE

The appendages of the eye (*tutamina oculi*) include the eyebrows, the eyelids, the conjunctiva, and the lachrymal apparatus, viz. the lachrymal gland, the lachrymal sac, and the nasal duct.

The **eyebrows** (*supercilia*) are two arched eminences of integument, which surmount the upper circumference of the orbit on each side, and support numerous short, thick hairs, directed obliquely on the surface.   In structure, the eyebrows consist of thickened integument, connected beneath with the Orbicularis palpebrarum, Corrugator supercilii, and Occipito-frontalis muscles.   These muscles serve, by their action on this part, to control to a certain extent the amount of light admitted into the eye.

The **eyelids** (*palpebræ*) are two thin, movable folds, placed in front of the eye, protecting it from injury by their closure.   The upper lid is the larger, and the more movable of the two, and is furnished with a separate elevator muscle, the *Levator palpebræ superioris*.   When the eyelids are open, an elliptical space (*fissura palpebrarum*) is left between their margins, the angles of which correspond to the junction of the upper and lower lids, and are called *canthi*.

The *outer canthus* is more acute than the inner, and the lids here lie in close contact with the globe : but the *inner canthus* is prolonged for a short distance inwards towards the nose, and the two lids are separated by a triangular space, the *lacus lacrimalis*.   At the commencement of the lacus lacrimalis, on the margin of each eyelid, is a small conical elevation, the *lachrymal papilla*, the apex of which is pierced by a small orifice, the *punctum lacrimale*, the commencement of the lachrymal canal.

The *eyelashes* (*cilia*) are attached to the free edges of the eyelids ; they are short, thick, curved hairs, arranged in a double or triple row at the margin of the lids : those of the upper lid, more numerous and longer than the lower, curve upwards ; those of the lower lid curve downwards, so that they do not interlace in closing the lids.   Near the attachment of the eyelashes are the openings of a number of glands, *glands of Moll*, arranged in several rows close to the free margin of the lid.   They are regarded as enlarged and modified sweat-glands.

Structure of the eyelids.—The eyelids are composed of the following structures taken in their order from without inwards :

Integument, areolar tissue, fibres of the Orbicularis muscle, tarsal plate and its ligament, Meibomian glands and conjunctiva. The upper lid has, in addition, the aponeurosis of the Levator palpebræ.

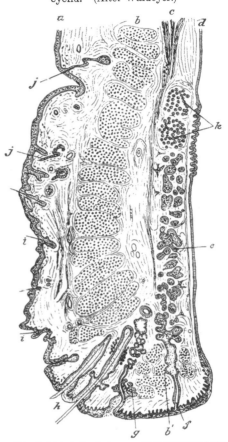

FIG. 452.—Vertical section through the upper eyelid. (After Waldeyer.)

The *integument* is extremely thin, and continuous at the margin of the lids with the conjunctiva.

The *subcutaneous areolar tissue* is very lax and delicate, seldom contains any fat, and is extremely liable to serous infiltration.

The *fibres of the Orbicularis muscle*, where they cover the palpebræ, are thin, pale in colour, and possess an involuntary action.

The *tarsal plates* are two thin elongated plates of dense connective tissue, about an inch in length. They are placed one in each lid, contributing to their form and support.

The *superior*, the larger, is of a semilunar form, about one-third of an inch in breadth at the centre, and becoming gradually narrowed at each extremity. To the anterior surface of this plate the aponeurosis of the Levator palpebræ is attached.

The *inferior tarsal* plate, the smaller, is thinner, and of an elliptical form.

The *free* or *ciliary margin* of these plates is thick, and presents a perfectly straight edge. The *attached orbital margin* is connected to the circumference of the orbit by the fibrous membrane of the lids with which it is continuous.

a. Skin. *b.* Orbicularis palpebrarum. *b'.* Marginal fasciculus of orbicularis (ciliary bundle). *c.* Levator palpebræ. *d.* Conjunctiva. *e.* Tarsal plate. *f.* Meibomian gland. *g.* Sebaceous gland. *h.* Eyelashes. *i.* Small hairs of skin. *j.* Sweat-glands. *k.* Posterior tarsal glands.

The outer angle of each plate is attached to the malar bone by the external tarsal ligament. The inner angles of the two plates terminate at the commencement of the lacus lacrimalis ; they are attached to the nasal process of the superior maxilla by the internal tarsal ligament or tendo oculi.

The *palpebral* ligaments are membranous expansions situated one in each lid, and are attached marginally to the edge of the orbit, where they are continuous with the periosteum. The superior ligament blends with the tendon of the Levator palpebræ, the inferior with the inferior tarsal plate. Externally the two ligaments fuse to form the external tarsal ligament, just referred to ; internally they are much thinner and, becoming separated from the internal tarsal ligament, are fixed to the lachrymal bone immediately behind the lachrymal sac. Together, the ligaments form an incomplete septum, the *septum orbitale*, which is perforated by the vessels and nerves which pass from the orbital cavity to the face and scalp.

The *Meibomian glands* (fig. 453) are situated upon the inner surface of the eyelids, between the tarsal plates and conjunctiva, and may be distinctly seen through the mucous membrane on everting the eyelids, presenting the appearance of parallel strings of pearls. They are about thirty in number in the upper eyelid, and somewhat fewer in the lower. They are embedded in grooves in the inner surface of the tarsal plates, and correspond in length with the breadth of each plate; they are, consequently, longer in the upper than in the lower eyelid. Their ducts open on the free margin of the lids by minute foramina, which correspond in number to the follicles. The use of their secretion is to prevent adhesions of the lids.

FIG. 453.—The Meibomian glands &c. seen from the inner surface of the eyelids.

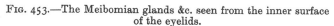

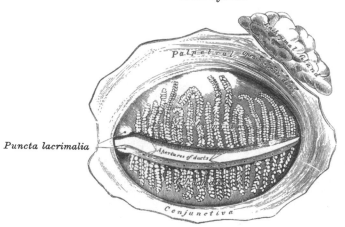

Structure of the Meibomian glands.—These glands are a variety of the cutaneous sebaceous glands, each consisting of a single straight tube or follicle, having a cæcal termination, and with numerous small secondary follicles opening into it. The tubes consist of basement-membrane, lined at the mouths of the tubes by stratified epithelium; the deeper parts of the tubes and the secondary follicles are lined by a layer of polyhedral cells. They are thus identical in structure with the sebaceous glands.

The **conjunctiva** is the mucous membrane of the eye. It lines the inner surface of the eyelids, and is reflected over the fore part of the sclerotic and cornea. In each of these situations, its structure presents some peculiarities.

The *palpebral portion of the conjunctiva* is thick, opaque, highly vascular, and covered with numerous papillæ, its deeper parts presenting a considerable amount of lymphoid tissue. At the margin of the lids it becomes continuous with the lining membrane of the ducts of the Meibomian glands, and, through the lachrymal canals, with the lining membrane of the lachrymal sac and nasal duct. At the outer angle of the upper lid the lachrymal ducts open on its free surface; and at the inner angle of the eye it forms a semilunar fold, the *plica semilunaris*. The folds formed by the reflection of the conjunctiva from the lids on to the eye are called the *superior* and *inferior palpebral folds*, the former being the deeper of the two. Upon the *sclerotic*, the conjunctiva is loosely connected to the globe; it becomes thinner, loses its papillary structure, is transparent, and only slightly vascular in health. Upon the *cornea*, the conjunctiva consists only of epithelium, constituting the anterior layer of the cornea (conjunctival epithelium) already described (see page 807). *Lymphatics* arise in the conjunctiva in a delicate zone around the cornea, from which the vessels run to the ocular conjunctiva.

At the point of reflection of the conjunctiva from the lid on to the globe of the eye, termed the *fornix conjunctivæ*, are a number of mucous glands, which are much convoluted. They are chiefly found in the upper lid. Other glands, analogous to lymphoid follicles, and called by Henle *trachoma glands*, are found in the conjunctiva, and, according to Strohmeyer, are chiefly situated near the inner canthus of the eye. They were first described by Brush, in his description of Peyer's patches of the small intestine, as 'identical structures existing in the under eyelid of the ox.'

The nerves in the conjunctiva are numerous and form rich plexuses. According to Krause, they terminate in a peculiar form of tactile corpuscle, which he terms 'terminal bulb.'

The *caruncula lacrimalis* is a small, reddish, conical-shaped body, situated at the inner canthus of the eye, and filling up the small triangular space in this situation, the *lacus lacrimalis*. It consists of a small island of skin containing sebaceous and sweat glands, and is the source of the whitish secretion which constantly collects at the inner angle of the eye. A few slender hairs are attached to its surface. On the outer side of the caruncula is a slight semilunar fold of mucous membrane, the concavity of which is directed towards the cornea; it is called the *plica semilunaris*. Müller found smooth muscular fibres in this fold, and in some of the domesticated animals a thin plate of cartilage has been discovered. This structure is considered to be the rudiment of the third eyelid in birds, the *membrana nictitans*.

LACHRYMAL APPARATUS (fig. 454)

The lachrymal apparatus consists of the lachrymal gland, which secretes the tears, and its excretory ducts, which convey the fluid to the surface of the eye. This fluid is carried away by the lachrymal canals into the lachrymal sac, and along the nasal duct into the cavity of the nose.

The **lachrymal gland** is lodged in a depression at the outer angle of the orbit, on the inner side of the external angular process of the frontal bone. It is of an oval form, about the size and shape of an almond. Its upper convex surface is in contact with the periosteum of the orbit, to which it is connected by a few fibrous bands. Its under concave surface rests upon the convexity of the eyeball, and upon the Superior and External recti muscles. Its vessels and nerves enter its posterior border, while its anterior margin is closely adherent to the back part of the upper eyelid, where it is covered to a slight extent by the reflection of the conjunctiva. The fore part of the gland is separated from the rest by a fibrous septum; hence it is sometimes described as a separate lobe, called the *palpebral portion of the gland* (*accessory gland of Rosenmüller*). Its ducts, from six to twelve in number, run obliquely beneath the mucous membrane for a short distance, and, separating from each other, open by a series of minute orifices on the upper and outer half of the con-

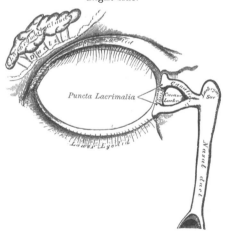

FIG. 454.—The lachrymal apparatus. Right side.

junctiva, near its reflection on to the globe. These orifices are arranged in a row, so as to disperse the secretion over the surface of the membrane.

Structure of the lachrymal gland.—In structure and general appearance the lachrymal resembles the serous salivary glands (page 869). In the recent state the cells are so crowded with granules that their limits can hardly be defined. They contain an oval nucleus, and the cell protoplasm is finely fibrillated.

The **lachrymal canals** commence at the minute orifices, *puncta lacrimalia*, on the summit of a small conical elevation, the *lachrymal papilla*, seen on the margin of the lids, at the outer extremity of the lacus lacrimalis. The *superior canal*, the smaller and shorter of the two, at first ascends, and then bends at an acute angle, and passes inwards and downwards to the lachrymal sac. The *inferior canal* at first descends, and then, abruptly changing its course, passes almost horizontally inwards to the lachrymal sac. They are dense and elastic in structure and somewhat dilated at their angle. The mucous membrane is covered with scaly epithelium.

The **lachrymal sac** is the upper dilated extremity of the nasal duct, and is lodged in a deep groove formed by the lachrymal bone and nasal process of the superior maxillary. It is oval in form, its upper extremity being closed in and rounded, while below it is continued into the nasal duct. It is covered by a fibrous expansion derived from the tendo oculi, and on its deep surface it is crossed by the Tensor tarsi muscle (Horner's muscle, page 274), which is attached to the ridge on the lachrymal bone.

Structure.—It consists of a fibrous elastic coat, lined internally by mucous membrane : the latter being continuous, through the lachrymal canals, with the mucous lining of the conjunctiva, and through the nasal duct with the pituitary membrane of the nose.

The **nasal duct** is a membranous canal, about three-quarters of an inch in length, which extends from the lower part of the lachrymal sac to the inferior meatus of the nose, where it terminates by a somewhat expanded orifice, provided with an imperfect valve, the *valve of Hasner*, formed by the mucous membrane. It is contained in an osseous canal, formed by the superior maxillary, the lachrymal, and the inferior turbinated bones, is narrower in the middle than at either extremity, and takes a direction downwards, backwards, and a little outwards. It is lined by mucous membrane, which is continuous below with the pituitary lining of the nose. This membrane in the lachrymal sac and nasal duct is covered with columnar epithelium as in the nose. This epithelium is in places ciliated.

Surface Form.—The palpebral fissure, or opening between the eyelids, is elliptical in shape, and differs in size in different individuals and in different races of mankind. In the Mongolian races, for instance, the opening is very small, merely a narrow fissure, and this makes the eye appear small in these races, whereas the size of the eyeball is relatively very constant. The normal direction of the fissure is slightly oblique, in a direction upwards and outwards, so that the outer angle is on a slightly higher level than the inner. This is especially noticeable in the Mongolian races, in whom, owing to the upward projection of the malar bone and the shortness of the external angular process of the frontal bone, the tarsal plate of the upper lid is raised at its outer part, and gives an oblique direction to the palpebral fissure.

When the eyes are directed forwards, as in ordinary vision, the upper part of the cornea is covered by the upper lid, and the lower margin of the cornea corresponds to the level of the lower lid, so that about the lower three-fourths of the cornea is exposed under ordinary circumstances. On the margins of the lids, about a quarter of an inch from the inner canthus, are two small openings, the *puncta lacrimalia*, the commencement of the lachrymal canals. They are best seen by everting the eyelids. In the natural condition they are in contact with the conjunctiva of the eyeball, and are maintained in this position by the Tensor tarsi muscle, so that the tears running over the surface of the globe easily find their way into the lachrymal canals. The position of the lachrymal sac into which the canals open is indicated by a little tubercle (page 88), which is plainly to be felt on the lower margin of the orbit. The lachrymal sac lies immediately above and to the inner side of this tubercle, and a knife passed through the skin in this situation would open the cavity. The position of the lachrymal sac may also be indicated by the tendo oculi, or internal tarsal ligament. If both lids be drawn outwards so as to tighten the skin at the inner angle, a prominent cord will be seen beneath the tightened skin. This is the *tendo oculi*, which lies immediately over the lachrymal sac, bisecting it, and thus forming a useful guide to its situation. A knife entered immediately beneath the tense cord would open the lower part of the sac. A probe introduced through this opening can be readily passed downwards through the duct into the inferior meatus of the nose. The direction of the duct is downwards, outwards, and backwards, and this course should be borne in mind in passing the probe, otherwise the point may be driven through the thin bony walls of the canal. A convenient plan is to direct the probe in such a manner, that if it were pushed onwards it would strike the first molar tooth of the lower jaw on the same side of the body. In other words, the surgeon standing in front of his patient should carry in his mind the position of the first molar tooth, and should push his probe onwards in such a way as if he desired to reach this structure.

Beneath the internal angular process of the frontal bone, the pulley of the Superior oblique muscle of the eye can be plainly felt by pushing the finger backwards between the upper and inner angle of the eye and the roof of the orbit ; passing backwards and outwards from this pulley the tendon can be felt for a short distance.

Surgical Anatomy.—The eyelids are composed of various tissues, and consequently are liable to a variety of diseases. The skin which covers them is exceedingly thin and delicate, and is supported on a quantity of loose and lax subcutaneous tissue, which contains no fat. In consequence of this it is very freely movable, and is liable to be drawn down by the contraction of neighbouring cicatrices, and thus produce an eversion of the lid, known as *ectropion*. Inversion of the lids (*entropion*) from spasm of the Orbicularis palpebrarum or from chronic inflammation of the palpebral conjunctiva may also occur. The eyelids are richly supplied with blood, and are often the seat of vascular growths, such as nævi. Rodent ulcer also frequently commences in this situation. The loose cellular tissue beneath the skin is liable to become extensively infiltrated either with blood or inflammatory products, producing very great swelling. Even from very slight injuries to this tissue, the extravasation of blood may be so great as to produce considerable swelling of the lids and complete closure of the eye, and the same is the case when inflammatory products are poured out. The follicles of the eyelashes or the sebaceous glands associated with these follicles may be the seat of inflammation, constituting the ordinary 'sty.' The Meibomian glands are affected in the so-called 'tarsal tumour :' the tumour, according to some, being caused by the retained secretion of these glands ; by others it is believed to be a neoplasm connected with the gland. The ciliary follicles are liable to become inflamed, constituting the disease known as *blepharitis ciliaris* or ' blear eye.' Irregular or disorderly growth of the eyelashes not infrequently occurs ; some of them being turned towards the eyeball and producing inflammation and ulceration of the cornea, and possibly eventually complete destruction of the eye. The Orbicularis palpebrarum may be the seat of spasm, either in the form of slight quivering of the lids ; or repeated twitchings, most commonly due to errors of refraction in children ; or more continuous spasm, due to some irritation of the fifth or seventh cranial nerve. The Orbicularis may be paralysed, generally associated with paralysis of the other facial muscles. Under these circumstances, the patient is unable to close the lids, and, if he attempts to do so, rolls the eyeball upwards under the upper lid. The tears overflow from displacement of the lower lid, and the conjunctiva and cornea, being constantly exposed and the patient being unable to wink, become irritated from dust and foreign bodies. In paralysis of the Levator palpebræ superioris there is drooping of the upper eyelid and other symptoms of implication of the third nerve. The eyelids may be the seat of bruises, wounds, or burns. In these latter injuries, adhesions of the margins of the lids to each other, or adhesion of the lids to the globe may take place. The eyelids are sometimes the seat of emphysema, after fracture of some of the thin bones forming the inner wall of the orbit. If shortly after such an injury the patient blows his nose, air is forced from the nostril through the lacerated structures into the connective tissue of the eyelids, which suddenly swell up and present the peculiar crackling characteristic of this affection.

Foreign bodies frequently get into the conjunctival sac and cause great pain, especially if they come in contact with the corneal surface, during the movements of the lid and the eye on each other. The conjunctiva is frequently involved in severe injuries of the eyeball, but is seldom ruptured alone ; the most common form of injury to the conjunctiva alone is from a burn, either from fire, strong acids, or lime. In these cases union is liable to take place between the eyelid and the eyeball. The conjunctiva is often the seat of inflammation arising from many different causes, and the arrangement of the conjunctival vessels should be remembered as affording a means of diagnosis between this condition and injection of the sclerotic, which is present in inflammation of the deeper structures of the globe. The inflamed conjunctiva is bright red ; the vessels are large and tortuous, and greatest at the circumference, shading off towards the corneal margin ; they anastomose freely and form a dense network, and they can be emptied or displaced by gentle pressure.

The lachrymal gland is occasionally, though rarely, the seat of inflammation, either acute or chronic ; it is also sometimes the seat of tumours, benign or malignant, and for these may require removal. This may be done by an incision through the skin, just below the eyebrow ; and the gland, being invested with a special capsule of its own, may be isolated and removed, without opening the general cavity of the orbit. The canaliculi may be obstructed, either as a congenital defect, or by some foreign body, as an eyelash or a dacryolith, causing the tears to run over the cheek. The canaliculi may also become occluded as a result of burns or injury ; overflow of the tears may in addition result from deviation of the puncta, or from chronic inflammation of the lachrymal sac. This latter condition is set up by some obstruction to the nasal duct, frequently occurring in tuberculous subjects. In consequence of this the tears and mucus accumulate in the lachrymal sac, distending it. Suppuration in the lachrymal sac is sometimes met with ;

this may be the sequel of a chronic inflammation ; or may occur after some of the eruptive fevers, in cases where the lachrymal passages were previously quite healthy. It may lead to lachrymal fistula.

THE EAR

The organ of hearing is divisible into three parts : the external ear, the middle ear or tympanum, and the internal ear or labyrinth.

THE EXTERNAL EAR

The **external ear** consists of an expanded portion named *pinna* or *auricle*, and the auditory canal, or *meatus*. The former serves to collect the vibrations of the air by which sound is produced ; the latter conducts those vibrations to the tympanum.

The **pinna** or **auricle** (fig. 455) is of an ovoid form, with its larger end directed upwards. Its outer surface is irregularly concave, directed slightly forwards,

Fig. 455.—The pinna, or auricle. Outer surface.

and presents numerous eminences and depressions which result from the foldings of its fibro-cartilaginous element. To each of these, names have been assigned. Thus the external prominent rim of the auricle is called the *helix*. Another curved prominence, parallel with and in front of the helix, is called the *antihelix* ; this bifurcates above, so as to inclose a triangular depression, the *fossa of the antihelix* (*fossa triangularis*). The narrow curved depression between the helix and antihelix is called the *fossa of the helix* (*Scapha*) ; the antihelix describes a curve round a deep, capacious cavity, the *concha*, which is partially divided into two parts by *crus helicis* or the commencement of the helix ; the upper part is termed the *cymba conchæ*, the lower part the *caivum conchæ*. In front of the concha, and projecting backwards over the meatus, is a small pointed eminence, the *tragus* : so called from its being generally covered on its under surface with a tuft of hair, resembling a goat's beard.

Opposite the tragus, and separated from it by a deep notch (*incisura intertragica*), is a small tubercle, the *antitragus*. Below this is the *lobule*, composed of tough areolar and adipose tissue, wanting the firmness and elasticity of the rest of the pinna. Where the helix turns downwards a small tubercle, the *tubercle of Darwin*, is frequently seen. This tubercle is very evident about the sixth month of fœtal life ; at this stage the human pinna has a close resemblance to that of some of the adult monkeys.

The cranial surface of the pinna presents elevations which correspond to the depressions on its outer surface and after which they are named, e.g. eminentia conchæ, eminentia triangularis, &c.

Structure of the pinna.—The pinna is composed of a thin plate of yellow fibro-cartilage, covered with integument, and connected to the surrounding parts by the extrinsic ligaments and muscles ; and to the commencement of the external auditory canal by fibrous tissue.

The *integument* is thin, closely adherent to the cartilage, and covered with hairs furnished with sebaceous glands, which are most numerous in the concha and scaphoid fossa. The hairs are most numerous and largest on the tragus and antitragus.

The *cartilage of the pinna* consists of one single piece : it gives form to this part of the ear, and upon its surface are found all the eminences and depressions above described. It does not enter into the construction of all parts of the auricle; thus it does not form a constituent part of the lobule; it is deficient, also, between the tragus and beginning of the helix, the notch between them being filled up by dense fibrous tissue. At the front part of the pinna, where the helix bends upwards, is a small projection of cartilage, called the *spina helicis*, while the lower part of the helix is prolonged downwards as a tail-like process, the *cauda helicis*; this is separated from the antihelix by a fissure, the *fissura antitragohelicina*. The cartilage of the pinna presents several intervals or fissures in its substance, which partially separate the different parts. The fissure of the helix is a short vertical slit, situated at the fore part of the pinna. Another fissure, the fissure of the tragus, is seen upon the anterior surface of the tragus. The cartilage of the pinna is very pliable, elastic, of a yellowish colour, and belongs to that form of cartilage which is known under the name of yellow fibro-cartilage.

FIG. 456.—The muscles of the pinna.

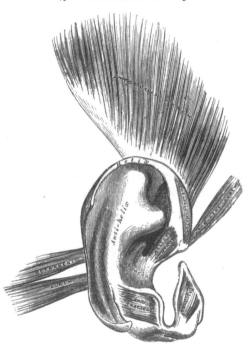

The *ligaments of the pinna* consist of two sets: 1. The extrinsic set, or those connecting it to the side of the head. 2. The intrinsic set, or those connecting various parts of its cartilage together.

The *extrinsic ligaments*, the most important, are two in number, anterior and posterior. The *anterior ligament* extends from the spina helicis and tragus to the root of the zygoma. The *posterior ligament* passes from the posterior surface of the concha to the outer surface of the mastoid process of the temporal bone.

The chief *intrinsic ligaments* are: (1) a strong fibrous band, stretching across from the tragus to the commencement of the helix, completing the meatus in front, and partly encircling the boundary of the concha; and (2) a band which extends between the antihelix and the cauda helicis. Other less important bands are found on the cranial surface of the pinna.

The *muscles of the pinna* (fig. 456) consist of two sets: 1. The *extrinsic*, which connect it with the side of the head, moving the pinna as a whole, viz. the Attollens, Attrahens, and Retrahens auriculam (page 273); and 2. the *intrinsic*, which extend from one part of the auricle to another, viz.:

Helicis major.	Antitragicus.
Helicis minor.	Transversus auriculæ.
Tragicus.	Obliquus auriculæ.

The *Musculus helicis major* is a narrow vertical band of muscular fibres, situated upon the anterior margin of the helix. It arises, below, from the cauda helicis, and is inserted into the anterior border of the helix, just where it is about to curve backwards. It is pretty constant in its existence.

The *Musculus helicis minor* is an oblique fasciculus, which covers the crus helicis.

The *Tragicus* is a short, flattened band of muscular fibres situated upon the outer surface of the tragus, the direction of its fibres being vertical.

The *Antitragicus* arises from the outer part of the antitragus : its fibres are inserted into the cauda helicis and antihelix. This muscle is usually very distinct.

The *Transversus auriculæ* is placed on the cranial surface of the pinna. It consists of scattered fibres, partly tendinous and partly muscular, extending from the convexity of the concha to the prominence corresponding with the groove of the helix.

The *Obliquus auriculæ* (Tod) consists of a few fibres extending from the upper and back part of the concha to the convexity immediately above it.

The *arteries of the pinna* are the posterior auricular from the external carotid, the anterior auricular from the temporal, and an auricular branch from the occipital artery.

The *veins* accompany the corresponding arteries.

The *nerves* are : the auricularis magnus, from the cervical plexus ; the auricular branch of the pneumogastric ; the auriculo-temporal branch of the inferior

FIG. 457.—A front view of the organ of hearing.   Right side.

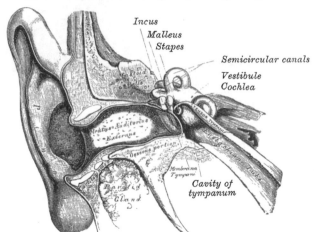

maxillary nerve ; the occipitalis minor from the cervical plexus, and the occipitalis major or internal branch of the posterior division of the second cervical nerve. The muscles of the pinna are supplied by the facial nerve.

The **Auditory Canal** (*meatus auditorius externus*) extends from the bottom of the concha to the membrana tympani (fig. 457). It is about an inch and a half in length if measured from the tragus ; from the bottom of the concha its length is about an inch. It forms a sort of **S**-shaped curve, and is directed at first inwards, forwards, and slightly upwards (*pars externa*) ; it then passes inwards and backwards (*pars media*), and lastly is carried inwards, forwards, and slightly downwards (*pars interna*). It forms an oval cylindrical canal, the greatest diameter being in the vertical direction at the external orifice, but in the transverse direction at the tympanic end. It presents two constrictions, one near the inner end of the cartilaginous portion, and another, the *isthmus*, in the osseous portion, about three-quarters of an inch from the bottom of the concha. The membrana tympani, which occupies the termination of the meatus, is obliquely directed, in consequence of which the floor of the canal is longer than the roof, and the anterior wall longer than the posterior. The auditory canal is formed partly by cartilage and membrane, and partly by bone, and is lined by skin.

The *cartilaginous portion* is about one-third of an inch (8 mm.) in length, it is formed by the cartilage of the pinna, prolonged inwards, and firmly attached to the circumference of the auditory process of the temporal bone.   The cartilage is deficient at its upper and back part, its place being supplied by fibrous membrane.   This part of the canal is rendered extremely movable by two or three deep fissures (*incisuræ Santorini*), which extend through the cartilage in a vertical direction.

The *osseous portion* is about two-thirds of an inch (16 mm.) in length, and narrower than the cartilaginous portion.   It is directed inwards and a little forwards, forming a slight curve in its course, the convexity of which is upwards and backwards.   Its inner end, which communicates, in the dry bone, with the cavity of the tympanum, is smaller than the outer and sloped, the anterior wall projecting beyond the posterior about two lines ; it is marked, except at its upper part, by a narrow groove, the *sulcus tympanicus*, for the insertion of the membrana tympani.   Its outer end is dilated and rough in the greater part of its circumference, for the attachment of the cartilage of the pinna.   Its vertical transverse section is oval, the greatest diameter being from above downwards. The front and lower parts of this canal are formed by a curved plate of bone, which, in the fœtus, exists as a separate ring (*annulus tympanicus*), incomplete at its upper part.   See section on Osteology (page 40).

The *skin* lining the meatus is very thin, adheres closely to the cartilaginous and osseous portions of the tube, and covers the surface of the membrana tympani, forming its outer layer.   After maceration, the thin pouch of epidermis, when withdrawn, preserves the form of the meatus.   In the thick subcutaneous tissue of the cartilaginous part of the meatus are numerous ceruminous glands, which secrete the ear-wax.   They resemble in structure sweat-glands, and their ducts open on the surface of the skin.

Relations of the meatus.—In front of the osseous part is the condyle of the mandible, which, however, is separated from the cartilaginous part by the retromandibular part of the parotid gland.   The movements of the jaw influence to some extent the lumen of this latter portion.   Behind the osseous part are the mastoid air-cells, separated from it by a thin layer of bone.

The *arteries* supplying the meatus are branches from the posterior auricular, internal maxillary, and temporal.

The *nerves* are chiefly derived from the auriculo-temporal branch of the inferior maxillary nerve and the auricular branch of the pneumogastric.

Surface Form.—At the point of junction of the osseous and cartilaginous portions, the tube forms an obtuse angle, which projects into the tube at its antero-inferior wall.   This produces a sort of constriction in this situation, and renders it a narrow portion of the canal—an important point to be borne in mind in connection with the presence of foreign bodies in the ears.   The cartilaginous is connected to the bony part by fibrous tissue, which renders the outer part of the tube very movable, and therefore by drawing the pinna upwards and backwards the canal is rendered almost straight.   At the external orifice are a few short, crisp hairs, which serve to prevent the entrance of small particles of dust, or flies or other insects.   In the external auditory meatus the secretion of the ceruminous glands serves to catch any small particles which may find their way into the canal, and prevent their reaching the membrana tympani, where their presence might excite irritation. In young children the meatus is very short, the osseous part being very deficient, and consisting merely of a bony ring (the *annulus tympanicus*), which supports the membrana tympani.   In the fœtus, the osseous part is entirely absent.   The shortness of the canal in children should be borne in mind in introducing the aural speculum, so that it be not pushed in too far, at the risk of injuring the membrana tympani ; indeed, even in the adult the speculum should never be introduced beyond the constriction which marks the junction of the osseous and cartilaginous portions. · In using this instrument it is advisable that the pinna should be drawn upwards, backwards, and a little outwards, so as to render the canal as straight as possible, and thus assist the operator in obtaining, by the aid of reflected light, a good view of the membrana tympani.   Just in front of the membrane is a well-marked depression, situated on the floor of the canal, and bounded by a somewhat prominent ridge ; in this foreign bodies may become lodged.   By aid of the speculum, combined with traction of the auricle upwards and backwards, the whole of the membrana tympani is rendered visible.   It is a pearly-grey membrane, slightly glistening in the adult, placed obliquely, so as to form with the floor of the meatus a very acute angle (about 55°), while with the roof it forms an obtuse angle.   At birth it is more horizontal. situated in almost the same plane as the base of the skull.   About midway between the anterior and posterior margins of the membrane, and extending from the centre obliquely upwards, is a reddish-yellow streak ; this is the handle of the malleus, which is inserted into the membrane.   At the upper part of this streak, close to the roof of the meatus, a little white, rounded prominence is plainly to be seen; this is the processus brevis of the malleus, projecting against the membrane.   The membrana tympani does not present a plane surface ; on the contrary, its centre is drawn inwards, on account of its connection with the handle of the malleus, and thus the external surface is rendered concave.

MIDDLE EAR, OR TYMPANUM

The **middle ear,** or **tympanum,** is an irregular cavity, compressed from without inwards, and situated within the petrous bone.   It is placed above the jugular fossa ; the carotid canal lying in front, the mastoid cells behind, the meatus auditorius externally, and the labyrinth internally.   It is filled with air, and communicates with the naso-pharynx by the Eustachian tube.   The tympanum is traversed by a chain of movable bones, which connect the membrana tympani with the labyrinth, and serve to convey the vibrations communicated to the membrana tympani across the cavity of the tympanum to the internal ear.

The tympanic cavity consists of two parts : the *atrium* or *tympanic cavity* proper, opposite the tympanic membrane, and the *attic* or *epitympanic* recess, above the level of the upper part of the membrane ; the latter contains the upper half of the malleus and the greater part of the incus.   Its diameter, including the attic, measures about 15 mm. vertically and transversely.   From without inwards it measures about 6 mm. above and 4 mm. below ; opposite the centre of the tympanic membrane it is only about 2 mm.   It is bounded externally by the membrana tympani and meatus ; internally, by the outer surface of the internal ear ; and communicates, behind, with the mastoid antrum and through it with the mastoid cells ; and in front with the Eustachian tube and canal for the Tensor tympani.   Its roof and floor are formed by thin osseous laminæ, the one forming the roof being a thin plate situated on the anterior surface of the petrous portion of the temporal bone, close to its angle of junction with the squamous portion of the same bone.

The **roof** (*paries tegmentalis*) is broad, flattened, and formed of a thin plate of bone (*tegmen tympani*), which separates the cranial and tympanic cavities.   It is prolonged backwards so as to roof in the mastoid antrum ; it is also carried forwards to cover in the canal for the Tensor tympani muscle.

The **floor** (*paries jugularis*) is narrow, and is separated by a thin plate of bone (*fundus tympani*) from the jugular fossa.   It presents, near the inner wall, a small aperture for the passage of Jacobson's nerve.

The **outer wall** is formed mainly by the membrana tympani, partly by the ring of bone into which this membrane is inserted.   This ring of bone is incomplete at its upper part, forming a notch (*incisura Rivini*).   Close to it are three small apertures : the iter chordæ posterius, the Glaserian fissure, and the iter chordæ anterius.

The *iter chordæ posterius* is in the angle of junction between the posterior and external walls of the tympanum, immediately behind the membrana tympani and on a level with the upper end of the handle of the malleus ; it leads into a minute canal, which descends in front of the aqueductus Fallopii, and terminates in that canal near the stylo-mastoid foramen.   Through it the chorda tympani nerve enters the tympanum.

The *Glaserian fissure* opens just above and in front of the ring of bone into which the membrana tympani is inserted ; in this situation it is a mere slit about a line in length.   It lodges the long process and anterior ligament of the malleus, and gives passage to the tympanic branch of the internal maxillary artery.

The *iter chordæ anterius* is seen at the inner end of the preceding fissure ; it leads into a canal (*canal of Huguier*), which runs parallel with the Glaserian fissure.   Through it the chorda tympani nerve leaves the tympanum.

The **internal wall of the tympanum** (*paries labyrinthica*) (fig. 458) is vertical

in direction, and looks directly outwards.   It presents for examination the following parts :

Fenestra ovalis.   Promontory.
Fenestra rotunda.   Ridge of the aqueductus Fallopii.

The *fenestra ovalis* is a reniform opening leading from the tympanum into
the vestibule ; its long diameter is directed horizontally, and its convex border is
upwards.   The opening in the recent state is occupied by the base of the stapes,
which is connected to the margin of the foramen by its annular ligament.

The *fenestra rotunda* is an aperture placed at the bottom of a funnel-shaped
depression, leading into the cochlea.   It is situated below and rather behind the
fenestra ovalis, from which it is separated by a rounded elevation, the *promontory* ;
it is closed in the recent state by a membrane (*membrana tympani secundaria*,
Scarpa).   This membrane is concave towards the tympanum, convex towards the
cochlea.   It consists of three layers : the external, or mucous, derived from the
mucous lining of the tympanum ; the internal from the lining membrane of the
cochlea ; and an intermediate, or fibrous layer.

FIG. 458.—View of inner wall of tympanum.   (Enlarged.)

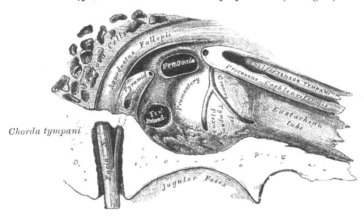

The *promontory* is a rounded hollow prominence, formed by the projection
outwards of the first turn of the cochlea ; it is placed between the fenestræ, and
is furrowed on its surface by three small grooves, which lodge branches of the
tympanic plexus.   A minute spicule of bone frequently connects the promontory
to the pyramid.

The *rounded eminence of the aqueductus Fallopii*, the prominence of the bony
canal in which the facial nerve is contained, traverses the inner wall of the tympanum above the fenestra ovalis, and behind that opening curves nearly vertically
downwards along the posterior wall.

The **posterior wall of the tympanum** (*paries mastoidea*) is wider above than
below, and presents for examination the

Opening of the antrum.   Pyramid.

The *opening of the antrum* is a large irregular aperture, which extends
backwards from the epitympanic recess and leads into a considerable air space,
the *antrum mastoideum* (see page 36).   The antrum communicates with large
irregular cavities contained in the interior of the mastoid process, the *mastoid
air-cells*.   These cavities vary considerably in number, size, and form ; they
are lined by mucous membrane, continuous with that lining the cavity of the
tympanum.

The *pyramid* is a conical eminence, situated immediately behind the fenestra
ovalis, and in front of the vertical portion of the eminence above described ; it is
hollow in the interior, and contains the Stapedius muscle ; its summit projects

forwards towards the fenestra ovalis, and presents a small aperture, which transmits the tendon of the muscle. The cavity in the pyramid is prolonged into a minute canal, which communicates with the aqueductus Fallopii, and transmits the nerve which supplies the Stapedius.

The **anterior wall of the tympanum** (*paries carotica*) is wider above than below; it corresponds with the carotid canal, from which it is separated by a thin plate of bone, perforated by the tympanic branch of the internal carotid artery. It presents for examination the

Canal for the Tensor tympani.   Orifice of the Eustachian tube.

The processus cochleariformis.

The orifice of the canal for the Tensor tympani and the orifice of the Eustachian tube are situated at the upper part of the anterior wall, being separated from each other by a thin, delicate, horizontal plate of bone, the *processus cochleariformis*. These canals run from the tympanum forwards, inwards, and a little downwards, to the retiring angle between the squamous and petrous portions of the temporal bone.

The *canal for the Tensor tympani* is the superior and the smaller of the two ; it is rounded and lies beneath the forward prolongation of the tegmen tympani. It extends on to the inner wall of the tympanum and ends immediately above the fenestra ovalis. The processus cochleariformis passes backwards below this part of the canal, forming its outer wall and floor; it expands above the anterior extremity of the fenestra ovalis and terminates by curving outwards so as to form a pulley over which the tendon passes.

The *Eustachian tube* is the channel through which the tympanum communicates with the pharynx. Its length is an inch and a half (36 mm.), and its direction downwards, forwards, and inwards, forming an angle of about 45° with the sagittal plane and one of from 30° to 40° with the horizontal plane. It is formed partly of bone, partly of cartilage and fibrous tissue.

The *osseous portion* is about half an inch in length. It commences in the anterior wall of the tympanum, below the processus cochleariformis, and, gradually narrowing, terminates at the angle of junction of the petrous and squamous portions, its extremity presenting a jagged margin which serves for the attachment of the cartilaginous portion.

The *cartilaginous portion*, about an inch in length, is formed of a triangular plate of elastic fibro-cartilage, the apex of which is attached to the margin of the inner extremity of the osseous canal, while its base lies directly under the mucous membrane of the naso-pharynx, where it forms an elevation or cushion behind the pharyngeal orifice of the tube. The upper edge of the cartilage is curled upon itself, being bent outwards so as to present on transverse section the appearance of a hook ; a groove or furrow is thus produced, which opens below and externally, and this part of the canal is completed by fibrous membrane. The cartilage is fixed to the base of the skull, and lies in a groove between the petrous-temporal and the greater wing of the sphenoid ; this groove ends opposite the middle of the internal pterygoid plate. The cartilaginous and bony portions of the tube are not in the same plane, the former inclining downwards a little more than the latter. The diameter of the canal is not uniform throughout, being greatest at the pharyngeal orifice and least at the junction of the bony and cartilaginous portions, where it is named the *isthmus* ; it again expands somewhat as it approaches the tympanic cavity. The position and relations of the pharyngeal orifice are described with the anatomy of the naso-pharynx. Through this canal the mucous membrane of the pharynx is continuous with that which lines the tympanum. The mucous membrane is covered with ciliated epithelium and is thin in the osseous portion, while in the cartilaginous portion it contains many mucous glands and near the pharyngeal orifice a considerable amount of adenoid tissue, which has been named by Gerlach the *tube-tonsil*. The tube is opened during deglutition by the Salpingo-pharyngeus and Dilator tubæ muscles.

The **membrana tympani** separates the cavity of the tympanum from the bottom of the external meatus. It is a thin, semi-transparent membrane, nearly oval in form, somewhat broader above than below, and directed very obliquely downwards and inwards so as to form an angle of about 55° with the floor of the meatus. The greater part of its circumference is thickened to form an annular ring which is fixed in a groove, the *sulcus tympanicus*, at the inner extremity of the meatus. This sulcus is deficient superiorly at the incisure or notch of Rivinus. From the extremities of this notch two bands, the *anterior* and *posterior malleolar folds*, are prolonged to the short process of the malleus. The small, somewhat triangular part of the membrane situated above these folds is lax and thin, and is named the *membrana flaccida* of Shrapnell; in it a small orifice is sometimes seen. The handle of the malleus is firmly attached to the inner aspect of the membrane as far as its centre, which it draws inwards towards the cavity of the tympanum. The outer surface of the membrane is thus concave, and the most depressed part of this concavity is named the *umbo* or *navel*.

Structure.—This membrane is composed of three layers, an external (cuticular), a middle (fibrous), and an internal (mucous). The *cuticular lining* is derived from the integument lining the meatus. The fibrous layer consists of two strata, an external of *radial fibres*, which diverge from the handle of the malleus, and an internal of *circular fibres*, which are plentiful around the circumference but sparse and scattered near the centre of the membrane. Branched or *dendritic* fibres, as pointed out by Grüber, are also present, especially in the posterior half of the membrane.

The arteries are derived from the deep auricular branch of the internal maxillary, which ramifies beneath the cuticular layer and from the stylo-mastoid branch of the posterior auricular and tympanic branch of the internal maxillary, which are distributed on the mucous surface. The superficial veins open into the external jugular ; those on the mucous surface drain themselves partly into the lateral sinus and veins of the dura mater and partly into a plexus on the Eustachian tube. The membrane receives its nervous supply from the auriculo-temporal branch of the inferior maxillary, the auricular branch of the vagus, and the tympanic branch of the glosso-pharyngeal.

OSSICLES OF THE TYMPANUM (fig. 459).

The tympanum is traversed by a chain of movable bones, three in number, the *malleus, incus*, and *stapes*. The first is attached to the membrana tympani, the last to the fenestra ovalis, the incus being placed between the two, and connected to both by delicate articulations.

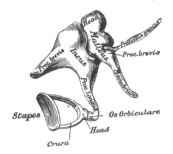

FIG. 459.—The small bones of the ear seen from the outside. (Enlarged.)

The **Malleus,** so named from its fancied resemblance to a hammer, consists of a head, neck, and three processes : the handle or manubrium, the processus gracilis, and the processus brevis.

The head is the large upper extremity of the bone ; it is oval in shape, and articulates posteriorly with the incus, being free in the rest of its extent. The facet for articulation with the incus is constricted near the middle, and is divided by a ridge into an upper, larger, and lower, lesser part, which form nearly a right angle with each other. Opposite the constriction the lower margin of the facet projects in the form of a process, the *cog-tooth* or *spur* of the malleus.

The *neck* is the narrow contracted part just beneath the head ; and below this is a prominence, to which the various processes are attached.

The *manubrium* is a vertical process of bone, which is connected by its outer margin with the membrana tympani. It is directed downwards, inwards, and backwards ; it decreases in size towards its extremity, where it is curved slightly forwards, and flattened from within outwards. On the inner side, near its upper end, is a slight projection, into which the tendon of the Tensor tympani is inserted.

The *processus gracilis* is a long and very delicate process, which passes from the eminence below the neck forwards and outwards to the Glaserian fissure, to which it is connected by ligamentous fibres. In the fœtus this is the longest process of the malleus, and is in direct continuity with the cartilage of Meckel.

The *processus brevis* is a slight conical projection, which springs from the root of the manubrium ; it is directed outwards, and is attached to the upper part of the tympanic membrane.

The **Incus** has received its name from its supposed resemblance to an anvil, but it is more like a bicuspid tooth, with two roots, which differ in length, and are widely separated from each other. It consists of a body and two processes.

The *body* is somewhat quadrilateral but compressed laterally. On its anterior surface is a deeply concavo-convex facet, which articulates with the head of the malleus ; in the fresh state it is covered with cartilage and the joint lined with synovial membrane.

The two processes diverge from one another nearly at right angles.

The *short process*, somewhat conical in shape, projects nearly horizontally backwards, and articulates with a depression, the *fossa incudis*, in the lower and back part of the epitympanic recess.

The *long process*, longer and more slender than the preceding, descends nearly vertically behind and parallel to the handle of the malleus, and, bending inwards, terminates in a rounded globular projection, the *os orbiculare* or *lenticular process*, which is tipped with cartilage, and articulates with the head of the stapes. In the fœtus the os orbiculare exists as a separate bone.

The **Stapes,** so called from its close resemblance to a stirrup, consists of a head, neck, two crura, and a base.

The *head* presents a depression, tipped with cartilage, which articulates with the os orbiculare.

The *neck*, the constricted part of the bone succeeding the head, receives the insertion of the Stapedius muscle.

The *two crura* diverge from the neck and are connected at their extremities by a flattened oval-shaped plate (the *base*), which forms the foot-plate of the stirrup and is fixed to the margin of the fenestra ovalis by ligamentous fibres. Of the two crura the anterior is shorter and less curved than the posterior.

Ligaments of the Ossicula.—These small bones are connected with each other, and with the walls of the tympanum, by ligaments, and moved by small muscles. The articular surfaces of the malleus and incus, and the orbicular process of the incus and head of the stapes, are covered with cartilage, connected together by delicate capsular ligaments, and lined by synovial membrane. The ligaments connecting the ossicula with the walls of the tympanum are five in number : three for the malleus, one for the incus, and one for the stapes.

The *anterior ligament of the malleus* was formerly described by Sömmering as a muscle (*Laxator tympani*). It is now, however, believed by most observers to consist of ligamentous fibres only. It is attached by one extremity to the neck of the malleus, just above the processus gracilis, and by the other to the anterior wall of the tympanum, close to the Glaserian fissure, some of its fibres being prolonged through the fissure to reach the spine of the sphenoid.

The *superior ligament of the malleus* is a delicate, round bundle of fibres which descends perpendicularly from the roof of the epitympanic recess to the head of the malleus.

The *external ligament of the malleus* is a triangular plane of fibres passing from the posterior part of the notch in the tympanic ring (*incisura Rivini*) to the short process of the malleus.

The *posterior ligament of the incus* is a short, thick, ligamentous band which connects the extremity of the short process of the incus to the posterior and lower part of the epitympanic recess, near the margin of the opening of the mastoid cells.

The inner surface and the circumference of the base of the stapes are covered with hyaline cartilage, and the *annular ligament of the stapes* connects the circumference of the base to the margin of the fenestra ovalis.

A *superior ligament of the incus* has been described by Arnold, but it is little more than a fold of mucous membrane.

The **muscles of the tympanum** are two :

Tensor tympani.   Stapedius.

The *Tensor tympani*, the larger, is contained in the bony canal, above the osseous portion of the Eustachian tube, from which it is separated by the processus cochleariformis. It arises from the under surface of the petrous bone, from the cartilaginous portion of the Eustachian tube, and from the osseous canal in which it is contained. Passing backwards through the canal, it terminates in a slender tendon which enters the tympanum and makes a sharp bend outward round the extremity of the processus cochleariformis, and is inserted into the handle of the malleus, near its root. It is supplied by a branch from the otic ganglion.

The *Stapedius* arises from the side of a conical cavity, hollowed out of the interior of the pyramid ; its tendon emerges from the orifice at the apex of the pyramid, and, passing forwards, is inserted into the neck of the stapes. Its surface is aponeurotic, its interior fleshy ; and its tendon occasionally contains a slender bony spine, which is constant in some mammalia. It is supplied by the tympanic branch of the facial nerve.

Actions.—The Tensor tympani draws the membrana tympani inwards, and thus heightens its tension. The Stapedius draws the head of the stapes backwards, and thus causes the base of the bone to rotate on a vertical axis drawn through its own centre : in doing this the back part of the base would be pressed inwards towards the vestibule, while the fore part would be drawn from it. It probably compresses the contents of the vestibule.

The **mucous membrane of the tympanum** is continuous with the mucous membrane of the pharynx, through the Eustachian tube. It invests the ossicula, and the muscles and nerves contained in the tympanic cavity ; forms the internal layer of the membrana tympani, and the outer layer of the membrana tympani secundaria, and is reflected into the mastoid antrum and cells, which it lines throughout. It forms several vascular folds, which extend from the walls of the tympanum to the ossicles ; of these one descends from the roof of the tympanum to the head of the malleus and upper margin of the body of the incus, a second invests the Stapedius muscle : other folds invest the chorda tympani nerve and the Tensor tympani muscle. These folds separate off pouch-like cavities, and give the interior of the tympanum a somewhat honey-comb appearance. One of these pouches is well marked, viz. the *pouch of Prussak*, which lies between the neck of the malleus and the membrana flaccida. In the tympanum this membrane is pale, thin, slightly vascular and covered for the most part with columnar ciliated epithelium, but that covering the pyramid, ossicula, and membrana tympani possesses a flattened non-ciliated epithelium. In the antrum and mastoid cells its epithelium is also non-ciliated. In the osseous portion of the Eustachian tube the membrane is thin ; but in the cartilaginous portion it is very thick, highly vascular, covered with ciliated epithelium, and provided with numerous mucous glands.

The **arteries supplying the tympanum** are six in number. Two of them are larger than the rest, viz. the tympanic branch of the internal maxillary, which supplies the membrana tympani ; and the stylo-mastoid branch of the posterior auricular, which supplies the back part of the tympanum and mastoid cells. The smaller branches are—the petrosal branch of the middle meningeal, which enters through the hiatus Fallopii ; a branch from the ascending pharyngeal and another from the Vidian, which accompany the Eustachian tube; and the tympanic branch from the internal carotid, given off in the carotid canal and perforating the thin anterior wall of the tympanum.

The **veins of the tympanum** terminate in the pterygoid plexus and in the superior petrosal sinus.

The **nerves of the tympanum** constitute the tympanic plexus, which ramifies upon the surface of the promontory. The plexus is formed by (1) the tympanic branch of the glosso-pharyngeal ; (2) the small deep petrosal nerve ; (3) the small superficial petrosal nerve ; and (4) a branch which joins the great superficial petrosal.

The *tympanic branch of the glosso-pharyngeal* (Jacobson's nerve) enters the tympanum by an aperture in its floor close to the inner wall and divides into branches, which ramify on the promontory and enter into the formation of the plexus. The *small deep petrosal nerve* from the carotid plexus of the sympathetic passes through the wall of the carotid canal, and joins the branches of Jacobson's nerve. The branch to the great superficial petrosal passes through an opening on the inner wall of the tympanum in front of the fenestra ovalis. The *small superficial petrosal nerve*, derived from the otic ganglion, passes through a foramen in the middle fossa of the base of the skull (sometimes the foramen ovale), passes backwards and enters the petrous bone through a small aperture, situated external to the hiatus Fallopii on the anterior surface of this bone ; it then courses downwards through the bone, and, passing by the gangliform enlargement of the facial nerve, receives a connecting filament from it and enters the tympanic cavity, where it communicates with Jacobson's nerve, and assists in forming the tympanic plexus.

The *branches of distribution* of the tympanic plexus are distributed to the mucous membrane of the tympanum ; one special branch passing to the fenestra ovalis, another to the fenestra rotunda, and a third to the Eustachian tube. The small superficial petrosal may be looked upon as a branch from the plexus to the otic ganglion.

In addition to the tympanic plexus there are the nerves supplying the muscles. The Tensor tympani is supplied by a branch from the third division of the fifth through the otic ganglion, and the Stapedius by the tympanic branch of the facial.

The *chorda tympani* nerve crosses the tympanic cavity. It is given off from the facial, as it passes vertically downwards at the back of the tympanum, about a quarter of an inch before its exit from the stylo-mastoid foramen. It passes from below upwards and forwards in a distinct canal, and enters the cavity of the tympanum through an aperture, *iter chordæ posterius*, already described (page 834), and becomes invested with mucous membrane. It passes forwards, through the cavity of the tympanum, crossing internal to the membrana tympani and over the handle of the malleus to the anterior inferior angle of the tympanum, and emerges from that cavity through the *iter chordæ anterius*, or *canal of Huguier*. It is invested by the fold of mucous membrane already mentioned, and therefore lies between the mucous and fibrous layers of the membrana tympani.

INTERNAL EAR, OR LABYRINTH

The **internal ear** is the essential part of the organ of hearing, receiving the ultimate distribution of the auditory nerve. It is called the **labyrinth,** from the complexity of its shape, and consists of two parts : the *osseous labyrinth*, a series of cavities channelled out of the substance of the petrous bone, and the *membranous labyrinth*, the latter being contained within the former.

The Osseous Labyrinth

The **osseous labyrinth** consists of three parts: the *vestibule, semicircular canals,* and *cochlea.* These are cavities hollowed out of the substance of the bone, and lined by periosteum; they contain a clear fluid, perilymph, or liquor Cotunnii, in which the membranous labyrinth is situated.

The **Vestibule** (fig. 460) is the common central cavity of communication between the parts of the internal ear. It is situated on the inner side of the tympanum, behind the cochlea, and in front of the semicircular canals. It is somewhat ovoidal in shape from before backwards, flattened from within outwards, and measures about one-fifth of an inch from before backwards, as well as from above downwards, and about one-eighth of an inch from without inwards. On its *outer* or *tympanic wall* is the fenestra ovalis, closed, in the recent state, by the base of the stapes, and its annular ligament. On its *inner wall,* at the fore part, is a small circular depression, *fovea hemispherica* or *recessus sphæricus,* which is perforated, at its anterior and inferior part, by several minute holes (*macula cribrosa media*) for the passage of filaments of the auditory nerve to the saccule;

Fig. 460.—The osseous labyrinth laid open. (Enlarged.)

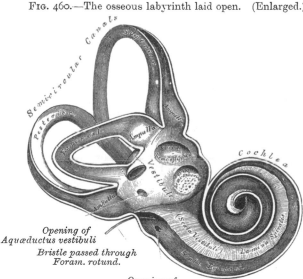

Opening of
Aquæductus vestibuli
Bristle passed through
Foram. rotund.

Opening of
Aquæductus cochleæ

and behind this depression is a vertical ridge, the *crista vestibuli.* This ridge bifurcates below to enclose a small depression, the *fossa cochlearis,* which is perforated by a number of holes for the passage of filaments of the auditory nerve which supply the posterior end of the ductus cochlearis. At the hinder part of the inner wall is the orifice of the *aquæductus vestibuli,* which extends to the posterior surface of the petrous portion of the temporal bone. It transmits a small vein, and contains a tubular prolongation of the lining membrane of the vestibule, the *ductus endo-lymphaticus,* which ends in a *cul-de-sac* between the layers of the dura mater within the cranial cavity. On the *upper wall* or *roof* is a transversely oval depression, *fovea semi-elliptica,* separated from the fovea hemispherica by the crista vestibuli already mentioned. *Behind,* the semicircular canals open into the vestibule by five orifices. In *front* is an elliptical opening, which communicates with the scala vestibuli of the cochlea by an orifice, *apertura scalæ vestibuli cochleæ.*

The **semicircular canals** are three bony canals, situated above and behind the vestibule. They are of unequal length, compressed from side to side, and describe the greater part of a circle. They measure about one-twentieth of an inch in diameter, and each presents a dilatation at one end, called the *ampulla*, which measures more than twice the diameter of the tube. These canals open into the vestibule by five orifices, one of the apertures being common to two of the canals.

The *superior semicircular canal* is vertical in direction, and is placed transversely to the long axis of the petrous portion of the temporal bone ; on the anterior surface of which its arch forms a round projection. It describes about two-thirds of a circle. Its outer extremity, which is ampullated, communicates by a distinct orifice with the upper part of the vestibule ; the opposite end of the canal, which is not dilated, joins with the corresponding part of the posterior canal to form the *crus commune*, which opens into the upper and inner part of the vestibule.

The *posterior semicircular canal*, also vertical in direction, is directed backwards, nearly parallel to the posterior surface of the petrous bone ; it is the longest of the three, its ampullated end commences at the lower and back part of the vestibule, its opposite end joining to form the common canal already mentioned.

The *external* or *horizontal canal* is the shortest of the three, its arch being directed outwards and backwards ; thus each semicircular canal stands at right angles to the other two. Its ampullated end corresponds to the upper and outer angle of the vestibule, just above the fenestra ovalis, where it opens close to the ampullary end of the superior canal ; its opposite end opens by a distinct orifice at the upper and back part of the vestibule.

The **Cochlea** bears some resemblance to a common snail-shell : it forms the anterior part of the labyrinth, is conical in form, and placed almost horizontally in front of the vestibule ; its apex is directed forwards and outwards, with a slight inclination downwards, towards the upper and front part of the inner wall of the tympanum ; its base corresponds with the anterior depression at the bottom of the internal auditory meatus, and is perforated by numerous apertures for the passage of the cochlear division of the auditory nerve. It measures nearly a quarter of an inch (5 mm.) from base to apex, and its breadth across the base is somewhat greater (about 9 mm.). It consists of a conical-shaped central axis, the *modiolus* or *columella* ; of a canal, the inner wall of which is formed by the central axis, wound spirally around it for two turns and three-quarters, from the base to the apex, and of a delicate lamina (the *lamina spiralis ossea*) which projects from the modiolus, and, following the windings of the canal, partially subdivides into two. In the recent state certain membranous layers are attached to the free border of this lamina, which project into the canal and completely separate it into two passages, which, however, communicate with each other at the apex of the modiolus by a small opening, named the *helicotrema*.

The *modiolus* or *columella* is the central axis or pillar of the cochlea. It is conical in form, and extends from the base to the apex of the cochlea. Its base is broad, and appears at the bottom of the internal auditory meatus, where it corresponds with the area cochleæ ; it is perforated by numerous orifices, which transmit filaments of the cochlear division of the auditory nerve, the nerves for the first turn and a half being transmitted through the foramina of the tractus spiralis foraminosus ; the fibres for the apical turn passing up through the foramen centrale. The foramina of the tractus spiralis foraminosus pass up through the modiolus and successively bend outwards to reach the attached margin of the lamina spiralis ossea. Here they become enlarged, and by their apposition form a spiral canal (*canalis spiralis modioli*), which follows the course of the attached margin of the lamina spiralis ossea and lodges the ganglion spirale. The foramen centrale is continued as a canal up the middle of the modiolus to its apex. The axis diminishes rapidly in size in the second and succeeding coil.

The bony canal of the cochlea (fig. 461) takes two turns and three-quarters round the modiolus. It is a little over an inch in length, (about 30 mm.) and diminishes gradually in size from the base to the summit, where it terminates in a *cul-de-sac*, the *cupola*, which forms the apex of the cochlea. The commencement of this canal is about the tenth of an inch in diameter ; it diverges from the modiolus towards the tympanum and vestibule, and presents three openings. One, the *fenestra rotunda*, communicates with the tympanum ; in the recent state this aperture is closed by a membrane, the *membrana tympani secundaria*. Another aperture, of an elliptical form, enters the vestibule. The third is the aperture of the aquæductus cochleæ, leading to a minute funnel-shaped canal, which opens on the basilar surface of the petrous bone and transmits a small vein, and also forms a communication between the subarachnoidean space of the skull and the perilymph contained in the scala tympani.

The *lamina spiralis ossea* is a bony shelf or ledge which projects outwards from the modiolus into the interior of the spiral canal, and, like the canal,

FIG. 461.—The cochlea laid open. (Enlarged.)

takes two and three-quarter turns round the modiolus. It reaches about halfway towards the outer wall of the spiral tube, and partially divides its cavity into two passages or scalæ, of which the upper is named the *scala vestibuli* while the lower is termed the *scala tympani*. Near the summit of the cochlea the lamina terminates in a hook-shaped process, the *hamulus*, which assists to form the boundary of a small opening, the *helicotrema*, by which the two scalæ communicate with each other. From the canalis spiralis modioli numerous foramina pass outwards through the osseous spiral lamina as far as its outer or free edge. In the lower part of the first turn a second bony lamina (*lamina spiralis secundaria*) projects inwards from the outer wall of the bony tube ; it does not, however, reach the primary osseous spiral lamina, so that if viewed from the vestibule a narrow fissure, the *fissura vestibuli*, is seen between them.

The Membranous Labyrinth (fig. 462)

The **membranous labyrinth** is contained within the bony cavities just described, having the same general form as the cavities in which it is contained, though considerably smaller, being separated from the bony walls by a quantity of fluid, the *perilymph*. It does not, however, float loosely in this fluid, but in places is fixed to the walls of the cavity. The membranous sac contains fluid, the *endolymph*, and on it the ramifications of the auditory nerve are distributed.

Within the osseous vestibule the membranous labyrinth does not quite preserve the form of the bony cavity, but presents two membranous sacs, the *utricle* and the *saccule*. The *utricle* is the larger of the two, of an oblong form, compressed laterally, and occupies the upper and back part of the vestibule, lying in contact with the fovea semi-elliptica and the part below it. That portion which is lodged in the fovea forms a sort of pouch or *cul-de-sac*, the floor and anterior wall of which are much thicker than elsewhere, and form the *macula acustica utricularis*, which receives the utricular filaments of the auditory nerve and has attached to its internal surface a layer of calcareous particles (otoliths).　The cavity of the utricle communicates behind with the membranous semicircular canals by five orifices.　From its anterior wall is given off a small canal, which joins with a canal from the saccule to form the *ductus endólymphaticus*.

The *saccule* is the smaller of the two véstibular sacs ; it is globular in form, lies in the fovea hemispherica near the opening of the scala vestibuli of the cochlea.　Its anterior part exhibits an oval thickening, the *macula saccularis*, to which are distributed the saccular filaments of the auditory nerve.　Its cavity does not directly communicate with that of the utricle.　From the posterior wall is given off a canal, which joins with a similar canal given off from the utricle to form the *ductus endo-lymphaticus*.　This duct passes along the aqueductus vestibuli and ends in a blind pouch on the posterior surface of the petrous portion of the temporal bone, where it is in contact with the dura mater.　From the lower part of the saccule a short tube, the *canalis reuniens* of Hensen, passes downwards and outwards to open into the ductus cochlearis (fig. 462).

The *membranous semicircular canals* are about one-third the diameter of the osseous canals, but in number, shape, and general form they are precisely similar,

FIG. 462.—The membranous labyrinth.　(Enlarged.)

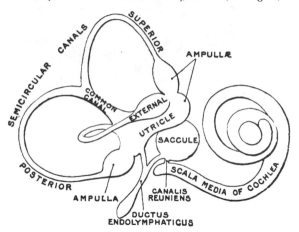

and present at one end an ampullary enlargement.　They open by five orifices into the utricle, one opening being common to two canals.　In the ampullæ the wall is thickened, and projects into the cavity as a fiddle-shaped, transversely placed elevation, the *septum transversum*, in which the nerves end.

The utricle, saccule, and membranous canals are held in position by numerous fibrous bands which stretch across the space between them and the bony walls.

Structure.—The walls of the utricle, saccule, and semicircular canals consist of three layers.　The *outer layer* is a loose and flocculent structure, apparently composed of ordinary fibrous tissue, containing blood-vessels and pigment-cells analogous to those in the pigment coat of the retina.　The *middle layer*, thicker and more transparent, bears some resemblance to the hyaloid membrane, but it presents on its internal surface, especially in the semicircular canals, numerous papilliform projections, and, on the addition of acetic acid, presents an appearance of longitudinal fibrillation and elongated nuclei.　The *inner layer* is formed of polygonal nucleated epithelial cells.　In the maculæ of the utricle and saccule, and in the transverse septa of the ampullæ of the canals, the middle coat is thickened and the epithelium is columnar, and consists of *supporting cells* and *hair-cells*, the free ends of the latter being surmounted by a long, tapering filament (auditory hair) which projects into the cavity.　The filaments of the auditory nerve enter these parts, and having pierced the outer and thickened middle layer, they lose their medullary sheath, and their axis cylinders ramify between the hair-cells.

Two small rounded bodies termed *otoliths*, and consisting of a mass of minute crystalline grains of carbonate of lime, held together in a mesh of delicate fibrous tissue, are contained in the walls of the utricle and saccule opposite the distribution of the nerves. A calcareous material is also, according to Bowman, sparingly scattered in the cells lining the ampullæ of the semicircular canals.

The **membranous cochlea, ductus cochlearis,** or **scala media** consists of a spirally arranged tube enclosed in the bony canal of the cochlea and lying along its outer wall. The manner in which it is formed will now be described.

The osseous spiral lamina, as above stated, extends only part of the distance between the modiolus and the outer bony wall of the cochlea. A membrane, the *membrana basilaris*, stretches from its free edge to the outer wall of the cochlea, and completes the roof of the scala tympani. A second and more delicate membrane, the *membrane of Reissner*, extends from the thickened periosteum covering the lamina spiralis ossea to the outer wall of the cochlea, to which it is attached at some little distance above the membrana basilaris. A canal is thus shut off between the scala tympani below and the scala vestibuli above; this is the *membranous canal of the cochlea, ductus cochlearis,* or *scala media*. It is triangular on transverse section, its roof being formed by the membrane of

FIG. 463.— Floor of scala media, showing the organ of Corti, &c.

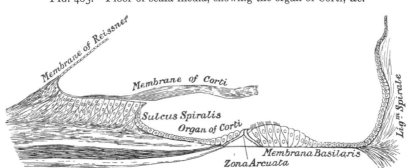

Reissner, its outer wall by the periosteum which lines the bony canal, and its floor by the membrana basilaris, and the outer part of the lamina spiralis ossea, on the former of which is placed the organ of Corti. Reissner's membrane is thin and homogeneous, and is covered on its upper and under surfaces by a layer of epithelium. The periosteum, which forms the outer wall of the ductus cochlearis, is greatly thickened and altered in character, forming what is called the *ligamentum spirale*. It projects inwards below as a triangular prominence, the *crista basilaris*, which gives attachment to the outer edge of the membrana basilaris, and immediately above which is a concavity, the *sulcus spiralis externus*. The upper portion of the ligamentum spirale contains numerous capillary loops and small blood-vessels, and forms what is termed the *stria vascularis*.

The lamina spiralis ossea consists of two plates of bone extending outwards; between these are the canals for the transmission of the filaments of the auditory nerve. On the upper plate of that part of the osseous spiral lamina which is outside Reissner's membrane the periosteum is thickened to form the *limbus laminæ spiralis*, and this terminates externally in a concavity, the *sulcus spiralis internus*, which presents, on section, the form of the letter **C**; the upper part of the letter, formed by the overhanging extremity of the limbus, is named the *labium vestibulare*; the lower part, prolonged and tapering, is called the *labium tympanicum*, and is perforated by numerous foramina (*foramina nervosa*) for the passage of the cochlear nerves. Externally, the labium tympanicum is continuous with the membrana basilaris. The upper surface of the labium vestibulare is intersected at right angles by a number of furrows, between which are numerous elevations ; these present the appearance of teeth along the free margin of the labium, and have been named by Huschke the *auditory teeth*. The basilar membrane may be divided into two areas, inner and outer. The inner is thin, and is named the *zona arcuata* : it supports the organ of Corti. The outer is thicker and striated, and is termed the *zona pectinata*. The under surface of the membrane is covered by a layer of vascular connective tissue. One of these vessels is somewhat larger than the rest, and is named the *vas spirale* ; it lies below Corti's tunnel.

Organ of Corti.*—This organ (fig. 464) is situated upon the inner part of the membrana basilaris, and appears at first sight as a papilla, winding spirally throughout the whole length of the ductus cochlearis, from which circumstance it has been designated the *papilla spiralis*. More accurately viewed, it is seen to be composed of a remarkable arrangement of cells, which may be likened to the keyboard of a pianoforte. Of these cells, the two central ones are rodlike bodies, and are called the inner and outer *rods of Corti*. They are placed on the basilar

FIG. 464.—Section through the organ of Corti. Magnified. (G. Retzius.)

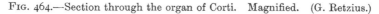
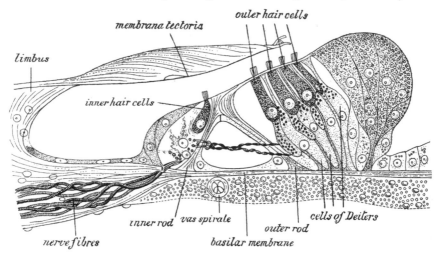

membrane, at some little distance from each other, but are inclined towards each other, so as to meet at their opposite extremities, and form a series of arches roofing over a minute tunnel, the *tunnel of Corti*, between them and the basilar membrane, which ascends spirally through the whole length of the cochlea.

The *inner rods*, some 6,000 in number, are more numerous than the outer ones, and rest on the basilar membrane, close to the labium tympanicum ; they project obliquely upwards and outwards, and terminate above in expanded extremities, which resemble in shape the upper end of the ulna, with its sigmoid cavity, coronoid and olecranon processes. On the outer side of the rod, in the angle formed between it and the basilar membrane, is a nucleated mass of protoplasm ; while on the inner side is a row of epithelial cells (*inner hair-cells*), surmounted by a brush of fine, stiff, hairlike processes. On the inner side of these cells are two or three rows of columnar supporting cells, which are continuous with the cubical cells lining the sulcus spiralis internus.

The *outer rods*, numbering about 4,000, also rest by a broad foot on the basilar membrane ; they incline upwards and inwards, and their upper extremity resembles the head and bill of a swan : the back of the head fitting into the con

* Corti's original paper is in the *Zeitschrift f. Wissen. Zool.* iii. 109.

cavity—the analogue of the sigmoid cavity—of one or more of the internal rods, and the bill projecting outwards as a phalangeal process of the membrana reticularis, presently to be described.

In the head of these outer rods is an oval portion, where the fibres of which the rod appears to be composed are deficient, and which stains more deeply with carmine than the rest of the rod. At the base of the rod, on its internal side— that is to say, in the angle formed by the rod with the basilar membrane— is a similar protoplasmic mass to that found on the outer side of the base of the inner rod ; these masses of protoplasm are probably the undifferentiated portions of the cells from which the rods are developed. External to the outer rod are three or four successive rows of epithelial cells, more elongated than those found on the internal side of the inner rod, but, like them, furnished with minute hairs or cilia. These are termed the *outer hair-cells*, in contradistinction to the *inner hair-cells* above referred to. There are about 12,000 outer hair-cells, and about 3,500 inner hair-cells.

FIG. 465.—Longitudinal section of the cochlea, showing the relations of the scalæ, the ganglion spirale, &c.

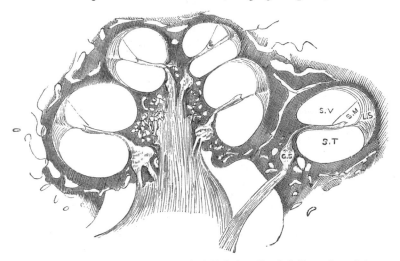

S. V. Scala vestibuli.   S. T. Scala tympani.   S. M. Scala media.   L. S. Ligamentum spirale.
G. S. Ganglion spirale.

The hair-cells are somewhat oval in shape ; their free extremities are on a level with the heads of Corti's rods, and from each some twenty fine hairlets project and are arranged in the form of a crescent, the concavity of which opens inwards. The deep ends of the cells are rounded and contain large nuclei : they only reach as far as the middle of Corti's rods, and are in contact with the ramifications of the nervous filaments. Between the rows of the outer hair-cells are rows of supporting cells, called the *cells of Deiters* ; their expanded bases are planted on the basilar membrane, while their opposite ends present a clubbed extremity or *phalangeal* process. Immediately to the outer side of Deiters's cells are some five or six rows of columnar cells, the *supporting cells of Hensen*. Their bases are narrow, while their upper parts are expanded and form a rounded elevation on the floor of the ductus cochlearis. The columnar cells lying outside Hensen's cells are termed the *cells of Claudius*. A space is seen between the outer rods of Corti and the adjacent hair-cells ; this is called the *space of Nuel*.

The *lamina reticularis* or *membrane of Kölliker* is a delicate framework perforated by rounded holes. It extends from the inner rods of Corti to the external row of the outer hair-cells, and is formed by several rows of 'minute fiddle-shaped cuticular structures,' called *phalanges*, between which are circular apertures containing the free ends of the hair-cells. The innermost row of phalanges consists of the phalangeal processes of the outer rods of Corti ; the outer rows are formed by the modified free ends of Deiters's cells.

Covering over these structures, but not touching them, is the *membrana tectoria*, or membrane of Corti, which is attached to the vestibular surface of the lamina spiralis close to the attachment of the membrane of Reissner. It is thin near its inner margin, and overlies the auditory teeth of Huschke. Its outer half is thick, and along its lower edge, opposite the inner hair-cells, is a clear band, named *Hensen's stripe*. Externally, the membrane becomes much thinner, and is attached to the outer row of Deiters's cells (Retzius).

The *inner surface of the osseous labyrinth* is lined by an exceedingly thin fibro-serous membrane, analogous to a periosteum, from its close adhesion to the inner surfaces of these cavities, and performing the office of a serous membrane by its free surface. It lines the vestibule, and from this cavity is continued into the semicircular canals and the scala vestibuli of the cochlea, and through the helicotrema into the scala tympani. A delicate tubular process is prolonged along the aqueduct of the vestibule to the inner surface of the dura mater. This membrane is continued across the fenestra ovalis and rotunda, and consequently has no communication with the lining membrane of the tympanum. Its attached surface is rough and fibrous, and closely adherent to the bone ; its free surface is smooth and pale, covered with a layer of epithelium, and secretes a thin, limpid fluid, the *aqua labyrinthi, liquor Cotunnii,* or *perilymph* (Blainville).

The scala media is closed above and below. The upper blind extremity is termed the *lagena*, and is attached to the cupola at the upper part of the helicotrema ; the lower end is lodged in the recessus cochlearis of the vestibule. Near this blind extremity, the scala media receives the *canalis reuniens of Hensen* (fig. 462), a very delicate canal, by which the ductus cochlearis is brought into continuity with the saccule.

The **arteries of the labyrinth** are the internal auditory, from the basilar, and the stylo-mastoid, from the posterior auricular. The internal auditory divides at the bottom of the internal meatus into two branches : cochlear and vestibular.

The cochlear branch subdivides into from twelve to fourteen twigs, which traverse the canals in the modiolus, and are distributed, in the form of a capillary network, in the substance of the lamina spiralis.

The vestibular branches accompany the nerves, and are distributed, in the form of a minute capillary network, in the substance of the membranous labyrinth.

The **veins** (auditory) of the vestibule and semicircular canals accompany the arteries, and, receiving those of the cochlea at the base of the modiolus, terminate in the posterior part of the superior petrosal sinus or in the lateral sinus.

The **auditory nerve,** the special nerve of the sense of hearing, divides, at the bottom of the internal auditory meatus, into two branches, the cochlear and vestibular.

The *vestibular nerve*, the posterior of the two, presents, as it lies in the internal auditory meatus, a ganglion, the *ganglion of Scarpa* ; it divides into three branches, which pass through minute openings at the upper and back part of the bottom of the meatus (*area vestibularis superior*), and, entering the vestibule, are distributed to the utricle and to the ampulla of the external and superior semicircular canals.

The nervous filaments enter the ampullary enlargements opposite the septum transversum, and arborise around the hair-cells. In the utricle and saccule the nerve-fibres pierce the membrana propria of the maculæ, and end in arborisations round the hair-cells.

The *cochlear nerve* gives off the branch to the saccule, the filaments of which are transmitted from the internal auditory meatus through the foramina of the *area vestibularis inferior*, which lies at the lower and back part of the floor of the meatus. It also gives off the branch for the ampulla of the posterior semicircular canal, which leaves the meatus through the *foramen singulare*.

The rest of the cochlear nerve divides into numerous filaments at the base of the modiolus ; those for the basal and middle coils pass through the foramina in the tractus foraminosus, those for the apical coil through the canalis centralis, and the nerves bend outwards to pass between the lamellæ of the osseous spiral lamina. Occupying the spiral canal of the modiolus is the *ganglion spirale*, consisting of bipolar nerve-cells, which really constitute the true cells of origin of this nerve, one pole being prolonged centrally to the brain and the other peripherally to the hair-cells of Corti's organ. Reaching the outer edge of the osseous spiral lamina, they pass through the foramina in the labium tympanicum, and end, some by arborising around the bases of the inner hair-cells, while others pass between Corti's rods and through the tunnel, to terminate in a similar manner in relation to the outer hair-cells.

Surgical Anatomy.—Malformations, such as imperfect development of the external parts, absence of the meatus, or supernumerary auricles, are occasionally met with. Or the pinna may present a congenital fistula, which is due to defective closure of the first visceral cleft, or rather of that portion of it which is not concerned in the formation of the Eustachian tube, tympanum, and meatus. The skin of the auricle is thin and richly supplied with blood, but in spite of this it is frequently the seat of frost-bite, due to the fact that it is much exposed to cold, and lacks the usual underlying subcutaneous fat found in most other parts of the body. A collection of blood is sometimes found between the cartilage and perichondrium (*hæmatoma auris*), usually the result of traumatism, but not necessarily due to this cause. It is said to occur most frequently in the ears of the insane. Keloid sometimes grows in the auricle around the puncture made for earrings, and epithelioma occasionally affects this part. Deposits of urate of soda are often met with in the pinna in gouty subjects.

The external auditory meatus can be most satisfactorily examined by light reflected down a funnel-shaped speculum ; by gently moving the latter in different directions the whole of the canal and membrana tympani can be brought into view. The points to be noted are, the presence of wax or foreign bodies ; the size of the canal ; and the condition of the membrana tympani. Accumulation of wax is often the cause of deafness, and may give rise to very serious consequences, causing ulceration of the membrane and even absorption of the bony wall of the canal. Foreign bodies are not infrequently introduced into the ear by children, and, when situated in the first portion of the canal, may be removed with tolerable facility by means of a minute hook or loop of fine wire, aided by reflected light ; but when they have slipped beyond the narrow middle part of the meatus, their removal is in no wise easy, and attempts to effect it, in inexperienced hands, may be followed by destruction of the membrana tympani and possibly the contents of the tympanum. The calibre of the external auditory canal may be narrowed by inflammation of its lining membrane, running on to suppuration ; by periostitis ; by polypi, sebaceous tumours, and exostoses. The membrana tympani, when seen in a healthy ear, ' reflects light strongly, and, owing to its peculiar curvature, presents a bright area of triangular shape at its lower and anterior portion.' From the apex of this, proceeding upwards and slightly forwards, is a white streak formed by the handle of the malleus, while near the upper part of the membrane may be seen a slight projection, caused by the short process of the malleus. In disease, alterations in colour, lustre, curvature or inclination, and perforation must be noted. Such perforations may be caused by a blow, a loud report, a wound, or as the result of suppuration in the middle ear.

The upper wall of the meatus is separated from the cranial cavity by a thin plate of bone ; the anterior wall is separated from the temporo-mandibular joint and parotid gland by the bone forming the glenoid fossa ; and the posterior wall is in relation with the mastoid cells, hence inflammation of the external auditory meatus may readily extend to the membranes of the brain, to the temporo-mandibular joint, or to the mastoid cells ; and, in addition to this, blows on the chin may cause fracture of the wall of the meatus.

The nerves supplying the meatus are, the auricular branch of the pneumogastric, the auriculo-temporal, and the auricularis magnus. The connections of these nerves explain the fact of the occurrence, in cases of any irritation of the meatus, of constant coughing and sneezing, from implication of the pneumogastric, or of yawning from implication of the auriculo-temporal. No doubt also the association of earache with toothache in cancer of the tongue is due to implication of the same nerve, a branch of the fifth, which supplies also the teeth and the tongue. The vessels of the meatus and membrana tympani are derived from the posterior auricular, temporal, and internal maxillary arteries. The upper half of the membrana tympani is much more richly supplied with blood than the lower half. For this reason, and also to avoid the chorda tympani nerve and ossicles, incisions through the membrane should be made at the lower and posterior part.

The principal point in connection with the surgical anatomy of the tympanum is its relations to other parts. Its roof is formed by a thin plate of bone, which, with the dura mater, is all that separates it from the temporal lobe of the brain. Its floor is immediately above the jugular fossa behind and the carotid canal in front. Its posterior wall presents the openings of the mastoid cells. On its anterior wall is the opening of the Eustachian tube. Thus it follows that in disease of the middle ear we may get subdural abscess, septic meningitis, or abscess of the cerebrum or cerebellum, from extension of the inflammation through the bony roof; thrombosis of the lateral sinus, with or without pyæmia, by extension through the floor; or mastoid abscess, by extension backwards. In addition to this we may get fatal hæmorrhage from the internal carotid in destructive changes of the middle ear. And in throat disease we may get the inflammation extending up the Eustachian tube to the middle ear. The Eustachian tube is accessible from the nose. If the nose and mouth be closed, and an attempt made to expire air, a sense of pressure with dulness of hearing is produced in both ears, from the air finding its way up the Eustachian tube and bulging out the membrana tympani. During the act of swallowing, the pharyngeal orifice of the tube, which is normally closed, is opened, probably by the action of the Dilator tubæ muscle. This fact was employed by Politzer in devising an easy method of inflating the tube. The nozzle of an indiarubber syringe is inserted into the nostril; the patient takes a mouthful of water and holds it in his mouth; both nostrils are closed with the finger and thumb to prevent the escape of air, and the patient is then requested to swallow; as he does so, the air is forced out of the syringe into his nose, and is driven into the Eustachian tube, which is now open. The impact of the air against the membrana tympani can be heard, if the membrane is sound, by means of a piece of indiarubber tubing, one end of which is inserted into the meatus of the patient's ear, the other into that of the surgeon. The direct examination of the Eustachian tube is made by the Eustachian catheter. This is passed along the floor of the nostril, close to the septum, with the point touching the floor, to the posterior wall of the pharynx. When this is felt, the catheter is to be withdrawn about half an inch, and the point rotated outwards through a quarter of a circle and pushed again slightly backwards, when it will enter the orifice of the tube, and will be found to be caught, and air forced into the catheter will be heard impinging on the tympanic membrane, if the ears of the patient and surgeon are connected by an indiarubber tube.

ORGANS OF DIGESTION

THE Apparatus for the Digestion of the Food consists of the alimentary canal and of certain accessory organs.

The **alimentary canal** is a musculo-membranous tube, about thirty feet in length, extending from the mouth to the anus, and lined throughout its entire extent by mucous membrane. It has received different names in the various parts of its course : at its commencement is the *mouth*, where provision is made for the mechanical division of the food (*mastication*), and for its admixture with a fluid secreted by the salivary glands (*insalivation*); beyond this are the organs of deglutition, the *pharynx* and the *œsophagus*, which convey the food into that part of the alimentary canal (the *stomach*) in which the principal chemical changes occur, and in which the reduction and solution of the food take place ; in the small intestines, the nutritive principles of the food (the *chyle*) are separated, by its admixture with the bile, pancreatic and intestinal fluids, from that portion which passes into the large intestine, most of which is expelled from the system.

Alimentary Canal

Mouth.	Small intestine	Duodenum. Jejunum. Ileum.
Pharynx.		
Œsophagus.		
Stomach.	Large intestine	Cæcum. Colon. Rectum.

Accessory Organs

Teeth.

Salivary glands	Parotid.	Liver.
	Submaxillary.	Pancreas.
	Sublingual.	Spleen.

The **mouth** (*oral* or *buccal cavity*) is placed at the commencement of the alimentary canal; it is a nearly oval-shaped cavity, in which the mastication of the food takes place (fig. 466). It consists of two parts : an outer, smaller portion, the vestibule (vestibulum oris), and an inner, larger part, the cavity proper of the mouth (cavum oris proprium).

The *vestibulum oris* is a slit-like aperture, bounded in front and laterally by the lips and cheeks ; behind and internally by the gums and teeth. Above and below it is limited by the reflection of the mucous membrane from the lips and cheeks to the gum covering the upper and lower alveolar arch respectively. It receives the secretion from the parotid glands, and communicates, when the jaws are closed, with the cavum oris by an aperture on each side behind the wisdom teeth.

The *cavum oris proprium* is bounded laterally and in front by the alveolar arches with their contained teeth ; behind it communicates with the pharynx by a constricted aperture termed the *isthmus faucium*. It is roofed in by the hard and soft palate, while the greater part of the floor is formed by the tongue, the remainder being completed by the reflection of the mucous membrane from the sides and under surface of the tongue to the gum lining the inner aspect of the mandible. It receives the secretion from the submaxillary and sublingual glands.

The *mucous membrane* lining the mouth is continuous with the integument at the free margin of the lips, and with the mucous lining of the pharynx behind; it is of a rose-pink tinge during life, and very thick where it covers the hard parts bounding the cavity. It is covered by stratified epithelium.

The **lips** are two fleshy folds which surround the orifice of the mouth, formed externally of integument and internally of mucous membrane, between which are found the Orbicularis oris muscle, the coronary vessels, some nerves, areolar tissue, and fat, and numerous small labial glands. The inner surface of each lip

FIG. 466.—Sectional view of the nose, mouth, pharynx, &c.

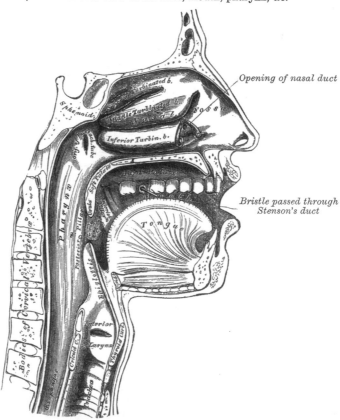

Opening of nasal duct

Bristle passed through
Stenson's duct

is connected in the middle line to the gum of the corresponding jaw by a fold of mucous membrane, the *frænum labii superioris* and *inferioris*—the former being the larger of the two.

The *labial glands* are situated between the mucous membrane and the Orbicularis oris, round the orifice of the mouth. They are rounded in form, about the size of small peas, their ducts opening by small orifices upon the mucous membrane. In structure they resemble the salivary glands.

The **cheeks** form the sides of the face, and are continuous in front with the lips. They are composed externally of integument; internally of mucous membrane; and between the two of a muscular stratum, besides a large quantity of fat, areolar tissue, vessels, nerves, and buccal glands.

The *mucous membrane* lining the cheek is reflected above and below upon the gums, and is continuous behind with the lining membrane of the soft palate.

Opposite the second molar tooth of the upper jaw is a papilla, the summit of which presents the aperture of the duct of the parotid gland. The principal muscle of the cheek is the Buccinator; but numerous other muscles enter into its formation: viz. the Zygomatici, Risorius Santorini, and Platysma myoides.

The *buccal glands* are placed between the mucous membrane and Buccinator muscle: they are similar in structure to the labial glands, but smaller. Two or three of larger size than the rest are placed between the Masseter and Buccinator muscles; their ducts open in the mouth opposite the last molar tooth. They are called *molar glands*.

The **gums** are composed of a dense fibrous tissue, closely connected to the periosteum of the alveolar processes, and surrounding the necks of the teeth. They are covered by smooth and vascular mucous membrane, which is remarkable for its limited sensibility. Around the necks of the teeth this membrane presents numerous fine papillæ; and from this point it is reflected into the alveolus, where it is continuous with the periosteal membrane lining that cavity.

THE TEETH

The human subject is provided with two sets of teeth, which make their appearance at different periods of life. The first set appear in childhood, and are called the *temporary, deciduous,* or *milk teeth.* The second set, which also appear at an early period, continue until old age, and are named *permanent.*

The *temporary teeth* are twenty in number: four incisors, two canines, and four molars, in each jaw.

The *permanent teeth* are thirty-two in number: four incisors (two central and two lateral), two canines, four bicuspids, and six molars, in each jaw.

The dental formulæ may be represented as follows:

Temporary Teeth

	mol.	can.	in.	in.	can.	mol.	
Upper jaw . .	2	I	2	2	I	2	} Total 20
Lower jaw . .	2	I	2	2	I	2	

Permanent Teeth

	mol.	bic.	can.	in.	in.	can.	bic.	mol.	
Upper jaw .	3	2	I	2	2	I	2	3	} Total 32
Lower jaw .	3	2	I	2	2	I	2	3	

General Characters.—Each tooth consists of three portions: the *crown,* or *body,* projecting above the gum; the *root,* or *fang,* entirely concealed within the alveolus; and the *neck,* the constricted portion, between the other two.

The *roots of the teeth* are firmly implanted within the alveoli; these depressions are lined with periosteum, which is reflected on to the tooth at the point of the fang, and covers it as far as the neck. At the margin of the alveolus, the periosteum becomes continuous with the fibrous structure of the gums.

In consequence of the curve of the dental arch, such terms as anterior, posterior, internal and external, as applied to the teeth, are misleading and confusing. Special terms are therefore applied to the different surfaces of a tooth: that surface which is directed towards the lips or cheek is known as the *labial* surface; that which is directed towards the tongue is described as the *lingual* surface; that surface which is directed towards the mesial line, supposing the teeth were arranged in a straight line outwards from the central incisor, is known as the *proximal* surface; while that which is directed away from the mesial line is called the *distal* surface.

The teeth in the upper jaw form a larger arch than those in the lower jaw, so that they slightly overlap those of the inferior maxilla both in front and at the sides in the normal condition. In consequence of the greater width of the upper central incisors over those of the lower, the other teeth in the upper jaw are thrown somewhat distally and the two sets do not quite correspond to each other when the mouth is closed : thus the canine tooth of the upper jaw rests partly on the canine of the lower jaw and partly on the first premolar, and in the molar teeth the cusps of the teeth of the upper jaw lie behind the corresponding teeth of the lower jaw. The two series, however, terminate pretty nearly at the same point behind ; this is mainly due to the smaller size of the molars in the upper jaw.

PERMANENT TEETH

The **incisors,** or cutting teeth, are so named from their presenting a sharp cutting edge, adapted for biting the food. They are eight in number, and form the four front teeth in each jaw.

The *crown* is directed vertically, and is chisel-shaped, being bevelled at the expense of its lingual surface, so as to present a sharp horizontal cutting edge, which, before being subject to attrition, presents three small prominent points separated by two slight notches. It is convex, smooth, and highly polished on its labial surface ; concave on its lingual surface, where it is frequently

FIG. 467.—Permanent teeth. Right side. (Burchard.)

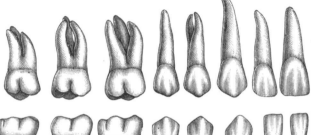

marked by slight longitudinal furrows. The one in the centre, being the most prominent, is known as the *basal ridge* or *cingulum.*

The *neck* is constricted.

The *fang* is long, single, conical, transversely flattened, thicker before than behind, and slightly grooved on each side in the longitudinal direction.

The *incisors of the upper jaw* are altogether larger and stronger than those of the lower jaw. They are directed obliquely downwards and forwards. The two central ones are larger than the two lateral, and the root is more rounded.

The *incisors of the lower jaw* are smaller than the upper : the two central ones are smaller than the two lateral, and are the smallest of all the incisor teeth. They are placed vertically in the jaw. These are somewhat bevelled in front, where they have been worn down by contact with the overlapping edge of the upper teeth. The cingulum is absent.

The **canine teeth** are four in number : two in the upper, and two in the lower jaw ; one being placed distally to each lateral incisor. They are larger and stronger than the incisors, especially the root, which sinks deeply into the jaw, and causes a well-marked prominence upon its surface.

The crown is large and conical, very convex on its labial surface, a little hollowed and uneven on its lingual surface, and tapering to a blunted point or cusp, which rises beyond the level of the other teeth.

The *root* is single, but longer and thicker than that of the incisors, conical in form, compressed laterally, and marked by a slight groove on each side.

The *upper canine teeth* (popularly called *eye-teeth*) are larger and longer than the two lower, and situated a little distally to them.

The *lower canine teeth* are placed mesially to the upper, so that their summits correspond to the interval between the upper canine tooth and the neighbouring incisors on each side.

The **bicuspid teeth** (premolars, small, or false molars) are eight in number : four in each jaw, two being placed distally to each of the canine teeth. They are smaller and shorter than the canine.

FIG. 468.—Deciduous teeth. Left side.

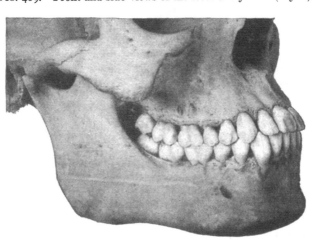

The *crown* is compressed proximo-distally, and surmounted by two pyramidal eminences, or cusps, separated by a groove ; hence their name, *bicuspid*. Of the two cusps the labial is larger and more prominent than the lingual.

The *neck* is oval.

The *root* is generally single, compressed, and presents a deep groove on each side, which indicates a tendency in the root to become double. The apex is generally bifid.

The *upper bicuspids* are larger, and present a greater tendency to the division of their roots than the lower ; this is especially marked in the second upper bicuspid.

The **molar teeth** (*multicuspidati*, true or large molars) are the largest of the permanent set and are adapted from the great breadth of their crowns for grinding and pounding the food. They are twelve in number : six in each jaw, three being placed distally to the second bicuspids.

The *crown* is nearly cubical in form, convex on its labial and lingual surfaces, flattened on its proximal and distal aspects ; the upper surface being surmounted

FIG. 469.—Front and side views of the teeth and jaws. (Cryer.)

by four or five tubercles, or cusps (four in the upper, five in the lower molars), separated from each other by a crucial depression ; hence their name, *multicuspid*.

The *neck* is distinct, large, and rounded.

The *root* is subdivided into from two to five fangs, each of which presents an aperture at its summit.

The crown of the *first molar tooth* in the upper jaw has usually four, but occasionally five cusps; the root consists of three fangs, widely separated from one another, two being labial, the other lingual.

The crown of the first molar tooth in the lower jaw is larger than that of the upper; it has five cusps, and its root consists of two fangs, one being placed proximally, the other distally : they are both compressed from before backwards, and grooved on their contiguous faces, indicating a tendency to division.

The *second molar* is a little smaller than the first.

The crown has three or four cusps in the upper, and usually five in the lower jaw.

The root has three fangs in the upper jaw, and two in the lower, the characters of which are similar to the preceding tooth.

The *third molar tooth* is called the *wisdom-tooth* (*dens sapientiæ*), from its late appearance through the gum.

Its crown is nearly as large as, sometimes even larger than, the second molar, but is smaller than the first. In the upper jaw it is usually furnished with three cusps, the two lingual ones being blended ; in the lower jaw there are five cusps as in the other molars.

The root is generally single, short, conical, slightly curved, and grooved so as to present traces of a subdivision into three fangs in the upper, and two in the lower jaw.

Temporary Teeth

The **temporary** or **milk teeth** are smaller, but, generally speaking, resemble in form those of the permanent set. The hinder of the two temporary molars is the largest of all the milk teeth, and is succeeded by the second permanent bicuspid. The first upper molar has only three cusps—two labial, one lingual ; the second upper molar has four cusps. The first lower molar has four cusps ; the second lower molar has five. The fangs of the temporary molar teeth are smaller and more divergent than those of the permanent set, but in other respects bear a strong resemblance to them.

Structure of the Teeth

On making a vertical section of a tooth (fig. 470), a cavity will be found in the interior. This cavity is situated in the interior of the crown and the centre of each fang, and opens by a minute orifice at the extremity of the latter. The shape of the cavity corresponds somewhat with that of the tooth ; it forms what is called the *pulp cavity*, and contains a soft, highly vascular, and sensitive substance, the *dental pulp*. The pulp consists of a loose connective tissue consisting of fine fibres and cells ; it is richly supplied with vessels and nerves, which enter the cavity through the small aperture at the point of each fang. The cells of the pulp are partly found permeating the matrix, and partly arranged as a layer on the wall of the pulp cavity. These latter cells are named the *odontoblasts of Waldeyer*. These cells, during the development of the tooth, are columnar in shape, but later on, after the dentine is fully formed, they become flattened and resemble the osteoblasts found in the osteogenetic layer of the periosteum of bone. They have two fine processes, the outer or distal one passing into a dental tubule, the inner being continuous with the processes of the connective-tissue cells of the pulp matrix.

The solid portion of the tooth consists of three distinct structures, viz. the proper dental substance, which forms the larger portion of the tooth, the *ivory* or *dentine* ; a layer which covers the exposed part of the crown, the *enamel* ; and a thin layer, which is disposed on the surface of the fang, the *cement* or *crusta petrosa*.

The **ivory,** or **dentine** (fig. 472), forms the principal mass of a tooth ; in its central part is the cavity enclosing the pulp. It is a modification of osseous tissue, from which it differs, however, in structure. On microscopic examination it is seen to consist of a number of minute wavy and branching tubes, having distinct parietes. They are called the *dentinal tubules*, and are embedded in a dense homogeneous substance, the *intertubular tissue*.

The *dentinal tubules* (fig. 470) are placed parallel with one another, and open at their inner ends into the pulp cavity. In their course to the periphery they present two or three curves, and are twisted on themselves in a spiral direction. The direction of these tubes varies : they are vertical in the upper portion of the crown, becoming oblique and then horizontal in the neck and upper part of the root, while towards the lower part of the root they are inclined downwards. The

FIG. 470.—Vertical section of a tooth *in situ* (15 diameters).

FIG. 471.—Vertical section of a molar tooth.

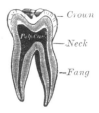

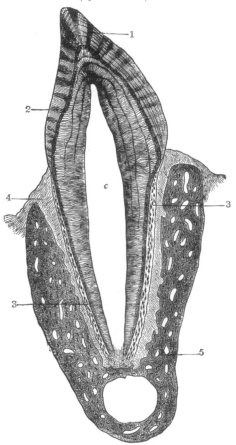

FIG. 472.—Vertical section of a bicuspid tooth. (Magnified.)

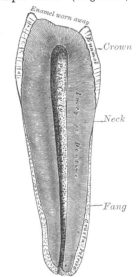

c. is placed in the pulp cavity, opposite the cervix or neck of the tooth ; the part above it is the crown, that below is the root (fang). 1. Enamel with radial and concentric markings. 2. Dentine with tubules and incremental lines. 3. Cement or crusta petrosa, with bone corpuscles. 4. Dental periosteum. 5. Bone of lower jaw.

tubules, at their commencement, are about $\frac{1}{4500}$ of an inch in diameter ; in their course they divide and subdivide dichotomously, so as to give to the cut surface of the dentine a striated appearance. From the sides of the tubes, especially in the fang, ramifications of extreme minuteness are given off, which join together in loops in the intertubular substance, or terminate in small dilatations, from which branches are given off. Near the periphery of the dentine, the finer ramifications of the tubules terminate imperceptibly by free ends. The dentinal tubules have comparatively thick walls, consisting of an elastic homogeneous membrane, the *dentinal sheath* of Neumann, which resists the action of acids ; they contain slender cylindrical prolongations, first described by Tomes, and named *Tomes's fibres* or *dentinal fibres*. These dentinal fibres are analogous to the soft contents of the canaliculi of bone.

The *intertubular substance* is translucent, and contains the chief part of the earthy matter of the dentine. In it are a number of fine fibrils, which are continuous with the fibrils of the dental pulp (Mummery). After the earthy matter has been removed by steeping a tooth in weak acid, the animal basis remaining may be torn into laminæ which run parallel with the pulp cavity, across the direction of the tubes. A section of dry-dentine often displays a series of somewhat parallel lines—the *incremental lines of Salter*. These lines are composed of imperfectly calcified dentine arranged in layers. In consequence of the imperfection in the calcifying process, little irregular cavities are left, termed *interglobular spaces*. These spaces are found especially towards the outer surface of the dentine, where they form a layer, which is sometimes known as the *granular layer* (fig. 473). They have received their name from the fact that they are surrounded by minute nodules or globules of dentine. Other curved lines may be seen parallel to the surface. These are the *lines of Schreger*, and are due to the optical effect of simultaneous curvature of the dentinal fibres.

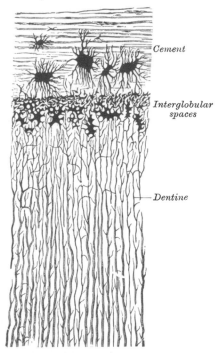

Fig. 473.—Transverse section of a portion of the root of a canine tooth. (Magnified 300 diameters.)

Cement

Interglobular spaces

Dentine

Chemical Composition.—According to Berzelius and Bibra, dentine consists of 28 parts of animal and 72 of earthy matter. The animal matter is resolvable by boiling into gelatin. The earthy matter consists of phosphate of lime, carbonate of lime, a trace of fluoride of calcium, phosphate of magnesia, and other salts.

The **enamel** is the hardest and most compact part of a tooth, and forms a thin crust over the exposed part of the crown, as far as the commencement of the fang. It is thickest on the grinding surface of the crown, until worn away by attrition, and becomes thinner towards the neck. It consists of a congeries of minute hexagonal rods or columns. They lie parallel with one another, resting by one extremity upon the dentine, which presents a number of minute depressions for their reception; and forming the free surface of the crown by the other extremity. The columns are directed vertically on the summit of the crown, horizontally at the sides; they are about $\frac{1}{5500}$ of an inch in diameter, and pursue a more or less wavy course. Each column is a six-sided prism and presents numerous dark transverse shadings; these shadings are probably due to the manner in which the columns are developed in successive stages, producing shallow constrictions, as will be explained in the sequel. Another series of lines, having a brown appearance, and denominated the *parallel striæ* or *coloured lines of Retzius*, are seen on a section of the enamel. According to Ebner, they are produced by air in the interprismatic spaces.

Numerous minute interstices intervene between the enamel fibres near their dentinal surface, a provision calculated to allow of the permeation of fluids from the dentinal tubule into the substance of the enamel. It is a disputed point whether the dentinal fibres penetrate a certain distance between the rods of the enamel or not. No nutritive canals exist in the enamel.

Chemical Composition.—According to Bibra, enamel consists of 96·5 per cent. of earthy matter, and 3·5 per cent. of animal matter. The earthy matter consists of phosphate of lime, with traces of fluoride of calcium, carbonate of lime, phosphate of magnesia, and other salts.

The **crusta petrosa** or **cement** is disposed as a thin layer on the roots of the teeth, from the termination of the enamel, as far as the apex of the fang, where it is usually very thick. In structure and chemical composition it resembles bone. It contains, sparingly, the lacunæ and canaliculi which characterise true bone : the lacunæ placed near the surface have the canaliculi radiating from the side of the lacunæ towards the periodontal membrane ; and those more deeply placed join with the adjacent dental tubules. In the thicker portions of the crusta petrosa, the lamellæ and Haversian canals peculiar to bone are also found.

As age advances, the cement increases in thickness, and gives rise to those bony growths, or exostoses, so common in the teeth of the aged ; the pulp cavity also becomes partially filled up by a hard substance, intermediate in structure between dentine and bone (*osteo-dentine*, Owen ; *secondary dentine*, Tomes). It appears to be formed by a slow conversion of the dental pulp, which shrinks, or even disappears.

DEVELOPMENT OF THE TEETH

In describing the development of the teeth, the mode of formation of the temporary or milk teeth must first be considered, and then that of the permanent series.

Development of the temporary teeth.—The development of these teeth begins at a very early period of fœtal life—about the sixth week. It commences as a thickening of the epithelium along the line of the future jaw ; the thickening being due to a rapid multiplication of the more deeply situated epithelial cells. As these cells multiply they extend into the subjacent mesoblast, and thus form a semicircular ridge or strand of cells, enclosed by mesoblast. About the seventh week a longitudinal splitting or cleavage of this strand of cells takes place, and it becomes divided into two strands ; the separation beginning in front and extending laterally : the process occupying four or five weeks. Of the two strands thus formed, the *outer* or *labial* forms the future labio-dental furrow, and is therefore termed the *labio-dental strand* : while the other, the *inner* or *lingual*, is the ridge of cells in connection with which the teeth, both temporary and permanent, are developed. Hence it is known as the *dental lamina* or *common dental germ*. It forms a flat band of cells, which grows into the substance of the embryonic jaw, at first horizontally inwards, and then, as the teeth develop, vertically, i.e. upwards in the upper jaw, and downwards in the lower jaw. While still maintaining a horizontal direction, it has two edges : one, the *attached edge*, which is continuous with the epithelium lining the mouth ; the other, the *free edge*, projecting inwards, and embedded in the mesoblastic tissue of the embryonic jaw. Along its line of attachment to the buccal epithelium is a shallow groove, the *dental furrow*.

About the ninth week this dental lamina begins to develop enlargements along its free border. These are ten in number in each jaw, and each corresponds to a future milk tooth. They consist of masses of epithelial cells ; and the cells of the deeper part—that is, the part farthest from the margin of the jaw—increase rapidly and spread out in all directions. Each mass thus comes to assume a flask shape, connected by a narrow neck, embraced by mesoblast, with the general epithelial lining of the mouth. They are now known as *special dental germs*. After a time the lower expanded portion, or body of the flask, inclines outwards, so as to form an angle with the superficial constricted portion, which is sometimes known as the *neck* of the special dental germ. About the tenth week the mesoblastic tissue beneath these special dental germs becomes differentiated into papillæ; these grow upwards, and come in contact with the epithelial cells of the special dental germs, which become folded over them like a hood or cap. There is, then, at this stage a papilla or papillæ, which have already begun to assume somewhat the shape of the crown of the future tooth, and from which the dentine and pulp of the milk teeth are formed, surmounted by a dome or cap of epithelial cells, from which the enamel is derived.

In the meantime, while these changes have been going on, the dental lamina has been extending backwards behind the special dental germ corresponding to the second molar tooth of the temporary set, and at about the seventeenth week it presents an enlargement, the special dental germ for the first permanent molar, soon followed by the formation of a papilla in the mesoblastic tissue for the same tooth. This is followed by a further extension backwards of the dental lamina; the formation of another enlargement and its corresponding papilla about the sixth month after birth for the second molar. And finally the process is repeated for the third molar, its papilla appearing about the fifth year of life.

FIG. 474.—Vertical section of the inferior maxilla of an early human fœtus.
(Magnified 25 diameters.)

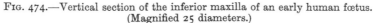

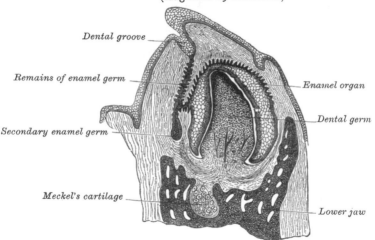

Dental groove   *Enamel organ*

Remains of enamel germ

Dental germ

Secondary enamel germ

Meckel's cartilage   *Lower jaw*

After the formation of the special dental germs, the dental lamina undergoes atrophic changes and becomes cribriform, except on the lingual and lateral aspects of each of the special germs of the temporary teeth, where it undergoes a local thickening, forming the special dental germ of each of the successional permanent teeth—i.e. the ten anterior ones in each jaw. Here the same process goes on as was described in connection with those of the milk teeth: that is, they recede into the substance of the gum behind the germs of the temporary teeth. As they recede they become flask-shaped, form an expansion at their distal extremity, and finally meet a papilla, which has been formed in the mesoblast, just in the same manner as was the case in the temporary teeth. The apex of the papilla indentates the dental germ, which encloses it, and forming a cap for it, becomes converted into the enamel, while the papilla forms the dentine and pulp of the permanent tooth.

The special dental germs consist at first of rounded or polyhedral epithelial cells; after the formation of the papillæ, these cells undergo a differentiation into three classes. Those which are in immediate contact with the papilla become elongated, and form a layer of well-marked columnar epithelium coating the papilla. They are the cells which form the enamel fibres, and are therefore termed *enamel cells* or *adamantoblasts*. The outer layer of cells of the special dental germ, which are in contact with the inner surface of the dentinal sac, presently to be described, are much shorter, cubical in form, and are named the *external enamel epithelium*. All the intermediate round cells of the dental germ between these two layers undergo a peculiar change. They become stellate in shape and develop processes, which unite to form a network into which fluid is secreted, which has the appearance of a jelly, and to which the name of enamel pulp is given. This transformed special dental germ is now known under the name of *enamel organ*.

While these changes are going on, a sac is formed around each enamel organ from the surrounding mesoblastic tissue. This is known as the *dental sac*, and is a vascular membrane of connective tissue. It grows up from below, and thus encloses the whole tooth germ; as it grows it causes the neck of the enamel organ to atrophy and disappear; so that all communication between the enamel organ and the superficial epithelium is cut off. At this stage there are vascular papillæ surmounted by inverted caps of epithelial cells, the whole being surrounded by membranous sacs. The cap consists of an internal layer of cells—the enamel cells or adamantoblasts—in contact with the papilla; of an external layer of cells—the external enamel epithelium—lining the interior of the dentinal sac; and of an intermediate mass of stellate cells, with anastomosing process—the enamel pulp (fig. 475).

Formation of the enamel.—The enamel is formed exclusively from the enamel cells or adamantoblasts of the original special dental germ, either by direct calcification of the columnar cell, which becomes elongated into the hexagonal rods of the enamel; or, as is believed by some, as a secretion from the adamantoblasts, within which calcareous matter is subsequently deposited.

The process begins at the apex of each cusp, at the end of the enamel cells, in contact with the dental papilla. Here a fine globular deposit takes place, being apparently shed from the end of the adamantoblasts. It is known by the name of *enamel droplet*, and resembles keratin in its resistance to the action of mineral acids. This droplet then calcifies and forms the first layer of the

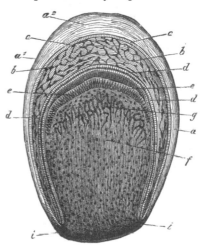

Fig. 475.—Dental sac of a human embryo at an advanced stage of development. Partly diagrammatic.

a. Wall of the sac, formed of connective tissue, with its outer stratum *a¹* and its inner *a²*. *b*. Enamel organ with its papillary and parietal layer of cells. *c, d*. The enamel-membrane and enamel-prisms. *e*. Dentine cells. *f*. Dental germ and capillaries. *g, i*. Transition of the wall of the follicle into the tissue of the dental germ.

enamel; a second droplet now appears and calcifies, and so on; successive droplets of keratin-like material are shed from the adamantoblasts and form successive layers of enamel, the adamantoblast gradually receding as each layer is produced, until at the termination of the process they have almost disappeared. The intermediate cells of the enamel pulp atrophy and disappear, so that the newly formed calcified material and the external enamel epithelium come into apposition. This latter layer, however, soon disappears on the emergence of the tooth beyond the gum. After its disappearance the crown of the tooth is still covered by a distinct membrane, which remains persistent for some time. This is known as the *cuticula dentis*, or *Nasmyth's membrane*, and is believed to be the last-formed layer of enamel derived from the *adamantoblast*, which has not become calcified. It forms a horny layer, which may be separated from the subjacent calcified mass by the action of strong acids. It is marked by the hexagonal impressions of the enamel prisms, and when stained by nitrate of silver, shows the characteristic appearance of epithelium.

Formation of the dentine.—While these changes are taking place in the epithelium to form the enamel, contemporaneous changes are occurring in the differentiated mesoblast of the dental papillæ which result in the formation of the dentine. As before stated, the first germ of the dentine consists in the formation of papillæ, corresponding in number to the teeth, from the soft mesoblastic tissue which bounds the depressions containing the special enamel germ. The papillæ grow upwards into the enamel germ and become covered by it, both being enclosed in a vascular connective tissue, the dentinal sac, in the manner above described. Each papilla then constitutes the formative pulp from which the dentine and permanent pulp are developed; it consists of rounded cells, and is very vascular, and soon begins to assume the shape of the tooth which is to be developed from it. The next step is the development of the odontoblasts, which have a relation to the development of the teeth similar to that of the osteoblasts in the formation of bone. They are formed first from the cells of the periphery of the papilla—that is to say, from the cells in immediate contact with the adamantoblasts of the special dental germ. These cells become elongated; one end of the elongated cell resting against the epithelium of the special dental germs, the other being tapered and often branched. By the direct transformation of the peripheral end of these cells, or by a secretion from it, a layer of uncalcified matrix is formed which caps the cusp or cusps, if there are more than one, of the papillæ. In this matrix islets of calcification make their appearance, and coalescing form a continuous layer of calcified material, which, capping each cusp, forms the first layer of dentine. The odontoblasts having thus formed the first layer, retire towards the centre of the papilla, and as they do so form successive layers of dentine from their peripheral extremities—that is to say, they form the dentinal matrix in which calcification subsequently takes place. As they thus retire from the periphery of the papilla, they leave behind them filamentous processes of cell protoplasm, provided with finer side processes; these are surrounded on all sides by calcified material, and thus form the dentinal tubules, and, by their side branches, the anastomosing tubules, whereby the dentinal tubules communicate: the processes of protoplasm contained within them, forming the dentinal fibres (Tomes' fibres) which, as mentioned above, are found within the tubules. In this way the entire thickness of the dentine is formed, each tubule being completed throughout its whole length by a single odontoblast. The central part of the papilla does not undergo calcification; its cells proliferate, nerve-fibres are developed in it, and it remains persistent as the pulp of the tooth. In this process of formation of dentine it has been shown that an uncalcified matrix is first formed, and that in this islets of calcification appear, which subsequently blend together to form a cap to each cusp: in like manner a succession of layers become formed, which become ultimately blended with each other. In certain places this blending is not complete, portions of the matrix remaining uncalcified between the successive layers; this gives rise, in the macerated tooth, to little spaces, which are the interglobular spaces alluded to above.

Formation of the cement.—The root of the tooth begins to be formed shortly before the crown emerges through the gum, but is not completed until some time afterwards. It is formed by a downward growth of the epithelium of the dental germ, which extends almost as far as the situation of the apex of the future fang, and determines the form of this portion of the tooth. This fold of epithelium is known as the *epithelial sheath,* and on its papillary surface odontoblasts are formed, which in turn form dentine, so that the dentine formation is identical in the crown and root of the tooth. After the dentine of the root has been formed, the vascular tissues of the dental sac begin to break through the epithelial sheath, and spread over the surface of the fang as a layer of bone-forming material. In this osteoblasts form, and the process of ossification goes on in identically the same manner as in the ordinary intra-membranous ossification of bone. In this way the cement is formed, and is simply ordinary bone, containing canaliculi and lacunæ.

Formation of the alveoli.—About the fourteenth week of embryonic life the dental lamina becomes enclosed in a trough or groove of mesoblastic tissue, which at first is common to all the dental germs, but subsequently becomes divided, by bony septa, into loculi, each loculus containing the special dental germ of a temporary tooth and its corresponding permanent tooth. After birth each cavity becomes subdivided, so as to form separate loculi (the future alveoli) for the milk tooth and its corresponding permanent tooth. Although at one time the whole of the growing tooth is contained in the cavity of the alveolus it never completely encloses it, since there is always an aperture over the top of the crown filled by soft tissue, by which the dental sac is connected with the surface of the gum, and which in the permanent teeth is called the *gubernaculum dentis*.

Development of the permanent teeth.—The permanent teeth as regards their development may be divided into two sets : (1) those which replace the temporary teeth, and which, like them, are ten in number ; these are the *successional permanent teeth* ; and (2) those which have no temporary predecessors, but are superadded at the back of the temporary dental series. These are three in number on either side in each jaw, and are termed *superadded permanent teeth*. They are the three molars of the permanent set, the molars of the temporary set being replaced by the premolars or bicuspids of the permanent set. The development of the successional permanent teeth—the ten anterior ones in either jaw—will first be considered. As already stated, the original dental lamina, after the formation of the special dental germs of the temporary teeth, undergoes atrophic changes, except at one spot behind and lateral to each of the special germs of the milk teeth ; here a local thickening takes place, and forms the special dental germ of each successional permanent tooth. In this, identically the same changes go on as took place in the special germs of the temporary teeth : they elongate and recede into the gum behind the germs of the milk tooth ; a papilla springs up from the mesoblastic tissue to meet them, and the two become enclosed in a dental sac ; the formation of the dentine and enamel takes place in the same way as in the temporary teeth. During their development the permanent teeth, enclosed in their sacs, come to be placed on the lingual side and at a lower level than the temporary teeth, that is to say, at a farther distance from the margin of the future gum, and, as already stated, are separated from them by bony partitions. As the crown of the permanent tooth grows, absorption of these bony partitions and of the fang of the temporary tooth takes place, through the agency of osteoclasts, which appear at this time, and finally nothing but the crown of the temporary tooth remains. This is shed or removed, and the permanent tooth takes its place.

The superadded permanent teeth are developed in the manner already described, by extensions backward of the posterior part of the dental lamina of the immediately preceding tooth. About the seventeenth week, that portion of the common dental germ of the last temporary tooth which lies behind the tooth, and which has remained unaltered, is prolonged backwards, and forms the special dental germ of the first permanent molar, and into this a papilla projects. In a similar manner about the fourth month after birth, a further extension backwards having taken place, a second enlargement occurs, into which a papilla projects about the sixth month, and thus the rudiment of the second molar is formed. Finally a third enlargement takes place, posterior to the other two, for the third molar, and its papilla becomes visible about the fifth year.

Eruption.—When the calcification of the different tissues of the tooth is sufficiently advanced to enable it to bear the pressure to which it will be afterwards subjected, its eruption takes place ; the tooth making its way through the gum. The gum is absorbed by the pressure of the crown of the tooth against it, which is itself pressed up by the increasing size of the fang. At the same time the septa between the dental sacs, at first fibrous in structure, ossify, and constitute the alveoli ; these firmly embrace the necks of the teeth, and afford them a solid basis of support.

The eruption of the temporary teeth commences at the seventh month, and is complete about the end of the second year, those of the lower jaw preceding those of the upper.

The following, according to C. S. Tomes, is the most usual time of eruption :

Lower central incisors	6 to 9 months.
Upper incisors	8 to 10 months.
Lower lateral incisors and first molars . . .	15 to 21 months.
Canines	16 to 20 months.
Second molars	20 to 24 months.

Calcification of the permanent teeth proceeds in the following order in the lower jaw ; in the upper jaw it takes place a little later : First molar, soon after birth ; the central incisor, lateral incisor, and canine, about six months after birth ; the bicuspids, at the second year, or a little later ; second molar, end of second year ; third molar, about the twelfth year.

The eruption of the permanent teeth takes place at the following periods, the teeth of the lower jaw preceding those of the upper by a short interval :

First molars	6th year.
Two middle incisors	7th year.
Two lateral incisors	8th year.
First bicuspids	9th year.
Second bicuspids	10th year.
Canines	11th to 12th year.
Second molars	12th to 13th year.
'Wisdom' teeth	17th to 25th year.

Towards the sixth year, before the shedding of the temporary teeth begins, there are twenty-four teeth in each jaw, viz. the ten temporary teeth and the crowns of all the permanent teeth except those of the third molars.

The Palate

The **palate** forms the roof of the mouth : it consists of two portions, the hard palate in front, the soft palate behind.

The **hard palate** is bounded in front and at the sides by the alveolar arches and gums ; behind, it is continuous with the soft palate. It is covered by a dense structure formed by the periosteum and mucous membrane of the mouth, which are intimately adherent. Along the middle line is a linear ridge or raphé, which terminates anteriorly in a small papilla, corresponding with the inferior opening of the anterior palatine fossa. On either side and in front of the raphé the mucous membrane is thick, pale in colour, and corrugated ; behind, it is thin, smooth, and of a deeper colour : it is covered with squamous epithelium, and furnished with numerous glands (palatal glands), which lie between the mucous membrane and the surface of the bone.

The **soft palate** (*velum pendulum palati*) is a movable fold, suspended from the posterior border of the hard palate, and forming an incomplete septum between the mouth and pharynx. It consists of a fold of mucous membrane, enclosing muscular fibres, an aponeurosis, vessels, nerves, adenoid tissue, and mucous glands. When occupying its usual position (i.e. relaxed and pendent), its anterior surface is concave, continuous with the roof of the mouth, and marked by a medium ridge or raphé, which indicates its original separation into two lateral halves. Its posterior surface is convex, and continuous with the mucous membrane covering the floor of the posterior nares. Its upper border is attached to the posterior margin of the hard palate, and its sides are blended with the pharynx. Its lower border is free.

Hanging from the middle of its lower border is a small, conical-shaped, pendulous process, the *uvula* ; and arching outwards and downwards from the base of the uvula on each side are two curved folds of mucous membrane, containing muscular fibres, called the *arches* or *pillars of the soft palate* or *pillars of the fauces.*

The *anterior pillars* run downwards, outwards, and forwards to the sides of the base of the tongue, and are formed by the projection of the Palato-glossi muscles, covered by mucous membrane.

The *posterior pillars* are nearer to each other, and larger than the anterior ; they run downwards, outwards, and backwards to the sides of the pharynx, and are formed by the projection of the Palato-pharyngei muscles, covered by mucous membrane. The anterior and posterior pillars are separated below by a triangular interval, in which the tonsil is lodged.

The space left between the arches of the palate on the two sides is called the *isthmus of the fauces.* It is bounded, above, by the free margin of the soft palate ; below, by the back of the tongue ; and on each side, by the pillars of the fauces and the tonsil.

The *mucous membrane of the soft palate* is thin, and covered with squamous epithelium on both surfaces, excepting near the orifice of the Eustachian tube, where it is columnar and ciliated.* Beneath the mucous membrane on the oral surface of the soft palate is a considerable amount of adenoid tissue. The palatine glands form a continuous layer on its posterior surface and round the uvula.

The *aponeurosis of the soft palate* is a thin but firm fibrous layer attached above to the posterior border of the hard palate, and becoming thinner towards the free margin of the velum. Laterally, it is continuous with the pharyngeal aponeurosis. It forms the framework of the soft palate, and is joined by the tendon of the Tensor palati muscle.

The *muscles of the soft palate* are five on each side : the Levator palati, Tensor palati, Azygos uvulæ, Palato-glossus, and Palato-pharyngeus (see page 302). The following is the relative position of these structures in a dissection of the soft palate from the posterior or nasal to the anterior or oral surface. Immediately beneath the nasal mucous membrane is a thin stratum of muscular fibres, the posterior fasciculus of the Palato-pharyngeus muscle, joining with its fellow of the opposite side in the middle line. Beneath this is the Azygos uvulæ, consisting of two rounded fleshy fasciculi, placed side by side in the median line of the soft palate. Next comes the aponeurosis of the Levator palati joining with the muscle of the opposite side in the middle line. Fourthly, the anterior fasciculus of the Palato-pharyngeus, thicker than the posterior, and separating the Levator palati from the next muscle, the Tensor palati. This muscle terminates in a tendon which, after winding round the hamular process, expands into a broad aponeurosis in the soft palate, anterior to the other muscles which have been enumerated. Finally we have a thin muscular stratum, the Palato-glossus muscle, placed in front of the aponeurosis of the Tensor palati, and separated from the oral mucous membrane by adenoid tissue.

The **Tonsils** (*amygdalæ*) are two prominent bodies situated one on each side of the fauces, between the anterior and posterior pillars of the soft palate. They are of a rounded form, and vary considerably in size in different individuals. A recess, the *fossa supra-tonsillaris*, may be seen, directed upwards and backwards, above the tonsil. His regards this as the remains of the lower part of the second visceral cleft. It is covered by a fold of mucous membrane termed the *plica triangularis.* Externally the tonsil is in relation with the inner surface of the Superior constrictor, to the outer side of which is the Internal pterygoid muscle. The internal carotid artery lies behind and to the outer side of the tonsil, and nearly an inch (20 to 25 mm.) distant from it. It corresponds to the angle of the lower jaw. Its *inner surface* presents from twelve to fifteen orifices,

* According to Klein, the mucous membrane on the nasal surface of the soft palate in the fœtus is covered throughout by columnar ciliated epithelium, which subsequently becomes squamous ; and some anatomists state that it is covered with columnar ciliated epithelium, except at its free margin, throughout life.

leading into small recesses, from which numerous follicles branch out into the substance of the gland. These follicles are lined by a continuation of the mucous membrane of the pharynx, covered with epithelium ; around each follicle is a layer of closed capsules embedded in the submucous tissue. These capsules are analogous to those of Peyer's glands, consisting of adenoid tissue. No openings from the capsules into the follicles can be recognised. They contain a thick greyish secretion. Surrounding each follicle is a close plexus of lymphatic vessels. From these plexuses the lymphatic vessels pass to the deep cervical glands in the upper part of the neck, which frequently become enlarged in affections of these organs.

The *arteries* supplying the tonsil are the dorsalis linguæ from the lingual, the ascending palatine and tonsillar from the facial, the ascending pharyngeal from the external carotid, the descending palatine branch of the internal maxillary, and a twig from the small meningeal.

The *veins* terminate in the tonsillar plexus, on the outer side of the tonsil.

The *nerves* are derived from Meckel's ganglion, and from the glosso-pharyngeal.

THE SALIVARY GLANDS (fig. 476)

The principal salivary glands communicating with the mouth, and pouring their secretion into its cavity, are the parotid, submaxillary, and sublingual.

The **parotid gland,** so called from being placed near the ear (παρά, near ; οὖς, ὠτός, the ear), is the largest of the three salivary glands, varying in weight from half an ounce to an ounce. It lies upon the side of the face, immediately below and in front of the external ear. It is limited above by the zygoma ; below, by the angle of the jaw, and by a line drawn between it and the mastoid process ; anteriorly, it extends to a variable extent over the Masseter muscle ; posteriorly, it is bounded by the external meatus, the mastoid process, and the Sterno-mastoid and Digastric muscles, slightly overlapping the two muscles.

Its *anterior surface* is grooved to embrace the posterior margin of the ramus of the lower jaw, and advances forwards beneath the ramus, between the two Pterygoid muscles and superficial to the ramus over the Masseter muscle. Its *outer surface*, slightly lobulated, is covered by the integument and parotid fascia, and has one or two lymphatic glands resting on it. Its *inner surface* extends deeply into the neck, by means of two large processes, one of which dips behind the styloid process, and projects beneath the mastoid process and the Sterno-mastoid muscle ; the other is situated in front of the styloid process, and passes into the back part of the glenoid fossa, behind the articulation of the lower jaw. The structures passing through the parotid gland are, the external carotid artery, giving off its three terminal branches : the posterior auricular artery emerges from the gland behind ; the temporal artery above ; the transverse facial, a branch of the temporal, in front ; and the internal maxillary winds through it as it passes inwards, behind the neck of the jaw. Superficial to the external carotid is the trunk formed by the union of the temporal and internal maxillary veins : a branch, connecting this trunk with the internal jugular, also passes through the gland. It is also traversed by the facial nerve and its branches, which emerge at its anterior border ; branches of the great auricular nerve pierce the gland to join the facial, and the auriculo-temporal branch of the inferior maxillary nerve emerges from the upper part of the gland. The internal carotid artery and internal jugular vein lie close to its deep surface.

The **duct of the parotid gland** (*Stenson's*) is about two inches and a half in length. It commences by numerous branches from the anterior part of the gland, crosses the Masseter muscle, and at its anterior border dips down into the substance of the Buccinator muscle, which it pierces ; it then runs for a short distance obliquely forwards between the Buccinator and mucous membrane of the mouth, and opens upon the inner surface of the cheek by a small orifice, opposite the second molar tooth of the upper jaw. While crossing the Masseter it receives the duct of a small detached portion of the gland, *socia parotidis*, which occasionally exists as a separate lobe, just beneath the zygomatic arch. In this position it has the transverse facial artery above it and some branches of the facial nerve below it.

Structure.—The parotid duct is dense, of considerable thickness, and its canal about the size of a crow-quill, but at its orifice on the inner aspect of the cheek its lumen is greatly reduced in size ; it consists of an external or fibrous coat, of considerably density, containing contractile fibres, and of an internal or mucous coat lined with short columnar epithelium.

Surface Form.—The direction of the duct corresponds to a line drawn across the face about a finger's breadth below the zygoma, that is, from the lower margin of the concha to midway between the free margin of the upper lip and the ala of the nose.

FIG. 476.—The salivary glands.

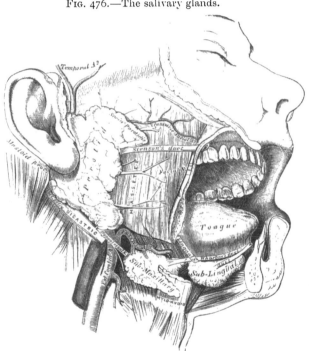

Vessels and Nerves.—The *arteries* supplying the parotid gland are derived from the external carotid, and from the branches given off by that vessel in or near its substance. The *veins* empty themselves into the external jugular, through some of its tributaries. The *lymphatics* terminate in the superficial and deep cervical glands, passing in their course through two or three lymphatic glands, placed on the surface and in the substance of the parotid. The *nerves* are derived from the plexus of the sympathetic on the external carotid artery, the facial, the auriculo-temporal, and great auricular nerves.

It is probable that the branch from the auriculo-temporal nerve is derived from the glosso-pharyngeal through the otic ganglion (see page 717). At all events in some of the lower animals this has been proved experimentally to be the case.

The **submaxillary gland** is situated below the jaw, in the anterior part of the submaxillary triangle of the neck. It is irregular in form, and weighs about two drachms. It is covered by the integument, Platysma, deep cervical fascia, and the body of the lower jaw, corresponding to a depression on the inner surface of that bone; and lies upon the Mylo-hyoid, Hyo-glossus, and Stylo-glossus muscles, a portion of the gland passing beneath the posterior border of the Mylo-hyoid. In front of it is the anterior belly of the Digastric; behind it is separated from the parotid gland by the stylo-maxillary ligament, and from the sublingual gland in front by the Mylo-hyoid muscle. The facial artery lies embedded in a groove in its posterior and upper border.

The **duct of the submaxillary gland** (*Wharton's*) is about two inches in length, and its walls are much thinner than those of the parotid duct. It commences by numerous branches from the deep portion of the gland which lies on the upper surface of the Mylo-hyoid muscle, and passes forwards and inwards between the Mylo-hyoid and the Hyo-glossus and Genio-hyo-glossus muscles, then between the sublingual gland and the Genio-hyo-glossus, and opens by a narrow orifice on the summit of a small papilla, at the side of the frænum linguæ. On the Hyo-glossus muscle it lies between the lingual and hypoglossal nerves, but at the anterior border of the muscle it crosses under the lingual nerve, and is then placed above it.

Vessels and Nerves.—The *arteries* supplying the submaxillary gland are branches of the facial and lingual. Its *veins* follow the course of the arteries. The *nerves* are derived from the submaxillary ganglion, through which it receives filaments from the chorda tympani of the facial and lingual branch of the inferior maxillary, sometimes from the mylo-hyoid branch of the inferior dental, and from the sympathetic.

The **sublingual gland** is the smallest of the salivary glands. It is situated beneath the mucous membrane of the floor of the mouth, at the side of the frænum linguæ, in contact with the inner surface of the lower jaw, close to the symphysis. It is narrow, flattened, in shape somewhat like an almond, and weighs about a drachm. It is in relation, *above*, with the mucous membrane; *below*, with the Mylo-hyoid muscle; *in front*, with the depression on the side of the symphysis of the lower jaw, and with its fellow of the opposite side; *behind*, with the deep part of the submaxillary gland; and *internally*, with the Genio-hyo-glossus, from which it is separated by the lingual nerve and Wharton's duct. Its excretory ducts (*ducts of Rivini*) are from eight to twenty in number; some join Wharton's duct; others open separately into the mouth, on the elevated crest of mucous membrane, caused by the projection of the gland, on either side of the frænum linguæ. One or more join to form a tube, which opens into the Whartonian duct: this is called the *duct of Bartholin*.

Vessels and Nerves.—The sublingual gland is supplied with blood from the sublingual and submental arteries. Its nerves are derived from the lingual.

Structure of Salivary Glands.—The salivary are compound racemose glands, consisting of numerous lobes, which are made up of smaller lobules, connected together by dense areolar tissue, vessels, and ducts. Each lobule consists of the ramifications of a single duct, 'branching frequently in a tree-like manner,' the branches terminating in dilated ends or alveoli on which the capillaries are distributed. These alveoli, however, as Pflüger points out, are not necessarily spherical, though sometimes they assume that form; sometimes they are perfectly cylindrical, and very often they are mutually compressed. The alveoli are enclosed by a basement-membrane, which is continuous with the membrana propria of the duct. It presents a peculiar reticulated structure, having the appearance of a basket with open meshes, and consisting of a network of branched and flattened nucleated cells.

The alveoli of the salivary glands are of two kinds, which differ in the appearance of their secreting cells, in their size, and in the nature of their secretion. The one variety secretes a ropy fluid, which contains mucin, and has therefore been named the *mucous*; while the other secretes a thinner and more watery fluid, which contains serum-albumin, and has been named *serous* or *albuminous*. The sublingual gland may be regarded as an example of the former variety: the parotid of the latter. The submaxillary is of the mixed variety, containing both mucous and serous alveoli, the latter, however, preponderating.

Both alveoli are lined by cells, and it is by the character of these cells that the nature of the gland is chiefly to be determined. In addition, however, the alveoli of the serous glands are smaller than those of the mucous ones.

The **cells in the mucous alveoli** are spheroidal in shape, glassy, transparent, and dimly striated in appearance. The nucleus is usually situated in the part of the cell which is next the basement-membrane, against which it is sometimes flattened. The most remarkable peculiarity presented by these cells is, that they give off an extremely fine process, which is curved in a direction parallel to the surface of the alveolus, lies in contact with the membrana propria, and overlaps the processes of neighbouring cells. The cells contain a quantity of mucin, to which their clear, transparent appearance is due.

Here and there in the alveoli are seen peculiar half-moon-shaped bodies, lying between the cells and the membrana propria of the alveolus. They are termed the *crescents of Gianuzzi*, or the *demilunes of Heidenhain* (fig. 477), and are composed of polyhedral granular cells, which Heidenhain regards as young epithelial cells destined to supply the place of those salivary cells which have undergone disintegration. This view, however, is not accepted by Klein.

FIG. 477.—A highly magnified section of the submaxillary gland of the dog, stained with carmine. (Kölliker.)

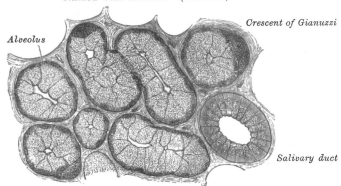

Serous alveoli.—In the serous alveoli the cells almost completely fill the cavity, so that there is hardly any lumen perceptible. Instead of presenting the clear transparent appearance of the cells of the mucous alveoli, they present a granular appearance, due to distinct granules, of an albuminous nature, embedded in a closely reticulated protoplasm. The ducts which originate out of the alveoli are lined at their commencement by epithelium which differs little from the pavement type. As the ducts enlarge, the epithelial cells change to the columnar type, and the part of the cell next the basement-membrane is finely striated. The lobules of the salivary glands are richly supplied with blood-vessels which form a dense network in the interalveolar spaces. Fine plexuses of nerves are also found in the interlobular tissue. The nerve-fibrils pierce the basement-membrane of the alveoli, and end in branched varicose filaments between the secreting cells. There is no doubt that ganglia are to be found in some salivary glands in connection with the nerve-plexuses in the interlobular tissue; they are to be found in the submaxillary, but not in the parotid.

In the submaxillary and sublingual glands the lobes are larger and more loosely united than in the parotid.

Mucous Glands.—Besides the salivary glands proper, numerous other glands are found in the mouth. They appear to secrete mucus only, which serves to keep the mouth moist during the intervals of the salivary secretion, and which is mixed with that secretion in swallowing. Many of these glands are found at the posterior part of the dorsum of the tongue, behind the circumvallate papillæ, and also along its margins as far forward as the apex.* Others lie around and in the tonsil between its crypts, and a large number are present in the soft palate. These glands are of the ordinary compound racemose type.

Surface Form.—The orifice of the mouth is bounded by the lips : two thick, fleshy folds covered externally by integument and internally by mucous membrane, and consisting of muscles, vessels, nerves, areolar tissue, and numerous small glands. The size of the orifice of the mouth varies considerably in different individuals, but seems to bear a close relation to the size and prominence of the teeth. Its corners correspond pretty accurately to the outer border of the canine teeth. In the Mongolian tribes, where the front teeth are large and inclined forward, the mouth is large ; and this, combined with the thick and everted lips, which appear to be associated with prominent teeth, gives to the Negro's face much of the peculiarity by which it is characterised. The smaller teeth, and the slighter prominence of the alveolar arch of the more highly civilised races, render the orifice of the mouth much smaller, and thus a small mouth is an indication of intelligence, and is regarded as an evidence of the higher civilisation of the individual.

Upon looking into the mouth, the first thing to be noted is the tongue, the upper surface of which will be seen occupying the floor of the cavity. This surface is convex, and is marked along the middle line by a raphé, which divides it into two symmetrical portions. The anterior two thirds are rough and studded with papillæ ; the posterior third, smooth and tuberculated, is covered by numerous glands which project from the surface. Upon raising the tongue, the mucous membrane which invests the upper surface may be traced covering the sides of the under surface and then reflected over the floor of the mouth on to the inner surface of the lower jaw, a part of which it covers. As it passes over the borders of the tongue it changes its character, becoming thin and smooth, and losing the papillæ which are to be seen on the upper surface. In the middle line the mucous membrane on the under surface of the tip of the tongue forms a distinct fold, the *frænum linguæ*, by which this organ is connected to the symphysis of the jaw. Occasionally it is found that this frænum is rather shorter than natural, and, acting as a bridle, prevents the complete protrusion of the tongue. When this condition exists and an attempt is made to protrude the organ, the tip will be seen to remain buried in the floor of the mouth, and the dorsum of the tongue is rendered very convex, and more or less extruded from the mouth : at the same time a deep furrow will be noticed to appear in the middle line of the anterior part of the dorsum. Sometimes, a little external to the frænum, the ranine vein may be seen immediately beneath the mucous membrane. The corresponding artery, being more deeply placed, does not come into view, nor can its pulsation be felt with the finger. On either side of the frænum, in the floor of the mouth, is a longitudinal elevation or ridge, produced by the projection of the sublingual gland, which lies immediately beneath the mucous membrane. Close to the attachment of the frænum to the tip of the tongue may be seen on either side the slit-like orifices of Wharton's ducts, into which a fine probe may be passed without much difficulty. By everting the lips the smooth mucous membrane lining them may be examined, and may be traced from them on to the outer surface of the alveolar arch. In the middle line, both of the upper and lower lip, a small fold of mucous membrane passes from the lip to the bone, constituting the *fræna* ; these are not so large as the frænum linguæ. By pulling outwards the angle of the mouth the mucous membrane lining the cheeks can be seen, and on it may be perceived a little papilla which marks the position of the orifice of Stenson's duct—the duct of the parotid gland. The exact position of the orifice of the duct will be found to be opposite the second molar tooth of the upper jaw. The introduction of a probe into this duct is attended with considerable difficulty. The teeth are the next objects which claim our attention upon looking into the mouth. There are, as stated above, ten in either jaw in the temporary set, and sixteen in the permanent set. The gums, in which they are implanted, are dense, firm, and vascular.

At the back of the mouth is seen the *isthmus of the fauces*, or, as it is popularly called, ' the throat :' this is the space between the pillars of the fauces on either side, and is the means by which the mouth communicates with the pharynx. Above, it is bounded by the soft palate, the anterior surface of which is concave and covered with mucous membrane, which is continuous with that lining the roof of the mouth. Projecting downwards from the middle of its lower border is a conical-shaped projection, the *uvula*. On either side of the isthmus of the fauces are the anterior and posterior pillars, formed by the Palato-glossus and Palato-pharyngeus muscles respectively, covered over by mucous membrane. Between

* It has been shown by Ebner that many of these glands open into the trenches around the circumvallate papillæ, and that their secretion is more watery than that of ordinary mucous glands. He supposes that they assist in the more rapid distribution of the substance to be tasted over the region where the special apparatus of the sense of taste is situated.

the two pillars on either side is situated the tonsil. The extirpation of this body is not unattended with danger of hæmorrhage. Dr. Weir has stated that he believes that when hæmorrhage occurs after their removal, it arises from one of the palatine arteries having been wounded. These vessels are large: they lie in the muscular tissue of the palate, and when wounded are constantly exposed to disturbance from the contraction of the palatine muscles. The vessels of the tonsil, Dr. Weir states, are small, and lie in the soft tissue, and readily contract when wounded.

When the mouth is wide open a prominent tense fold of mucous membrane may be seen and felt extending upwards and backwards from the position of the fang of the last molar tooth to the posterior part of the hard palate. This is caused by the Pterygo-maxillary ligament which is attached by one extremity to the apex of the internal pterygoid plate, and by the other to the posterior extremity of the mylo-hyoid ridge of the lower jaw. It connects the Buccinator with the Superior constrictor of the pharynx. The fang of the last molar tooth indicates the position of the lingual (gustatory) nerve, where it is easily accessible, and can with readiness be divided in cases of cancer of the tongue (see page 718). On the inner side of the last molar tooth we can feel the hamular process of the internal pterygoid plate of the sphenoid bone, around which the tendon of the Tensor palati plays. The exact position of this process is of importance in performing the operation of staphy-loraphy. About one-third of an inch in front of the hamular process and the same distance directly inwards from the last molar tooth is the situation of the opening of the posterior palatine canal, through which emerges the posterior or descending palatine branch of the internal maxillary artery, and one of the descending palatine nerves from Meckel's ganglion. The exact position of the opening on the subject may be ascertained by driving a needle through the tissues of the palate in this situation, when it will be at once felt to enter the canal. The artery emerging from the opening runs forwards in a groove in the bone, just internal to the alveolar border of the hard palate, and may be wounded in the operation for the cure of cleft palate. Under these circumstances the palatine canal may require plugging. By introducing the finger into the mouth the anterior border of the coronoid process of the jaw can be felt, and is especially prominent when the jaw is dislocated. By throwing the head well back a considerable portion of the posterior wall of the pharynx may be seen through the isthmus faucium, and on introducing the finger the anterior surfaces of the bodies of the upper cervical vertebræ may be felt immediately beneath the thin muscular stratum forming the wall of the pharynx. The finger can be hooked round the posterior border of the soft palate, and by turning it forwards, the posterior nares, separated by the septum, can be felt, or the presence of any adenoid or other growths in the naso-pharynx ascertained.

THE PHARYNX

The **pharynx** is that part of the alimentary canal which is placed behind the nose, mouth, and larynx. It is a musculo-membranous tube, somewhat conical in form, with the base upwards, and the apex downwards, extending from the under surface of the skull to the level of the cricoid cartilage in front, and that of the intervertebral disc between the fifth and sixth cervical vertebræ behind.

The pharynx is about four inches and a half in length, and broader in the transverse than in the antero-posterior diameter. Its greatest breadth is opposite the cornua of the hyoid bone; its narrowest point at its termination in the œsophagus. It is limited, *above*, by the body of the sphenoid and basilar process of the occipital bone; *below*, it is continuous with the œsophagus; *posteriorly*, it is connected by loose areolar tissue with the cervical portion of the vertebral column, and the Longi colli and Recti capitis antici muscles; *anteriorly*, it is incomplete, and is attached in succession to the internal pterygoid plate, the pterygo-maxillary ligament, the lower jaw, the tongue, hyoid bone, and thyroid and cricoid cartilages; *laterally*, it is connected to the styloid processes and their muscles, and is in contact with the common and internal carotid arteries, the internal jugular veins, and the glosso-pharyngeal, pneumogastric, hypoglossal, and sympathetic nerves, and above with a small part of the Internal pterygoid muscles.

It has seven openings communicating with it: the two posterior nares, the two Eustachian tubes, the mouth, larynx, and œsophagus.

The pharynx may be subdivided from above downwards into three parts, nasal, oral and laryngeal. The nasal part of the pharynx (*pars nasalis*) or naso-pharynx lies behind the nose and above the level of the soft palate: it differs from the two lower parts of the tube in that its cavity always remains patent. In front it communicates through the choanæ with the nasal fossæ. On its lateral wall is the pharyngeal orifice of the Eustachian tube, which presents the appearance of a vertical cleft bounded behind by a firm prominence, the *cushion*, caused by the inner extremity of the cartilage of the tube impinging on the deep surface of the mucous membrane. A vertical fold of mucous membrane, the *plica salpingo-pharyngea*, stretches from the lower part of the cushion to the pharynx ; it contains the Salpingo-pharyngeus muscle. A second and smaller mucous fold may be seen stretching from the upper part of the cushion to the palate, the *plica salpingo-palatina*. Behind the orifice of the Eustachian tube is a deep recess, the *fossa of Rosenmüller*, which represents the remains of the upper part of the second branchial cleft.

The oral part of the pharynx (*pars oralis*) reaches from the soft palate to the level of the hyoid bone. It opens anteriorly, through the isthmus faucium, into the mouth, while in its lateral wall, between the two pillars of the fauces, is the tonsil.

The laryngeal part of the pharynx (*pars laryngea*) reaches from the hyoid bone to the lower border of the cricoid cartilage, where it is continuous with the œsophagus. In front it presents the triangular aperture of the larynx, the base of which is directed forwards and is formed by the epiglottis, while its lateral boundaries are constituted by the aryteno-epiglottidean folds. On either side of the laryngeal orifice is a recess, termed the *sinus pyriformis* ; it is bounded internally by the aryteno-epiglottidean fold, externally by the thyroid cartilage and thyro-hyoid membrane.

Structure.—The pharynx is composed of three coats : mucous, fibrous, and muscular.

The *pharyngeal aponeurosis*, or *fibrous coat*, is situated between the mucous and muscular layers. It is thick above where the muscular fibres are wanting, and is firmly connected to the basilar process of the occipital and petrous portion of the temporal bones. As it descends it diminishes in thickness, and is gradually lost. It is strengthened posteriorly by a strong fibrous band, which is attached above to the pharyngeal spine on the under surface of the basilar portion of the occipital bone, and passes downwards, forming a median raphé, which gives attachment to the Constrictor muscles of the pharynx.

The *mucous coat* is continuous with that lining the Eustachian tubes, the nares, the mouth, and the larynx. In the naso-pharynx it is covered by columnar ciliated epithelium ; in the buccal and laryngeal portions the epithelium is of the squamous variety. Beneath the mucous membrane are found racemose mucous glands ; they are especially numerous at the upper part of the pharynx around the orifices of the Eustachian tubes. Throughout the pharynx are also numerous crypts or recesses, the walls of which are surrounded by lymphoid tissue, similar to what is found in the tonsils. Across the back part of the pharyngeal cavity, between the two Eustachian tubes, a considerable mass of this tissue exists, and has been named the *pharyngeal tonsil*. Above this in the middle line is an irregular, flask-shaped depression of the mucous membrane, extending up as far as the basilar process of the occipital bone. It is known as the *bursa pharyngea*, and was regarded by Luschka as the remains of the diverticulum, which is concerned in the development of the anterior lobe of the pituitary body. Other anatomists believe that it is connected with the formation of the pharyngeal tonsils.

The *muscular coat* has been already described (page 300).

Surgical Anatomy.—The internal carotid artery is in close relation with the pharynx, so that its pulsations can be felt through the mouth. It has been occasionally wounded by sharp-pointed instruments. introduced into the mouth and thrust through the wall of the pharynx. In aneurism of this vessel in the neck, the tumour necessarily bulges into the pharynx, as this is the direction in which it meets with the least resistance, nothing lying between the vessel and the mucous membrane except the thin Constrictor muscle, whereas on the outer side there is the dense cervical fascia, the muscles descending from the styloid process and the margin of the Sterno-mastoid.

The mucous membrane of the pharynx is very vascular, and is often the seat of inflammation, frequently of a septic character, and dangerous on account of its tendency to spread to the larynx. On account of the tissue which surrounds the pharyngeal wall being loose and lax, the inflammation is liable to spread through it far and wide, extending downwards into the posterior mediastinum along the œsophagus. Abscess may form in the connective tissue behind the pharynx, between it and the vertebral column, constituting what is known as retro-pharyngeal abscess. This is most commonly due to caries of the cervical vertebræ ; but may also be caused by suppuration of a lymphatic gland, which is situated in this position opposite the axis, and which receives lymphatics from the nares ; by a gumma ; or by acute pharyngitis. In these cases the pus may be easily evacuated by an incision, with a guarded bistoury, through the mouth, but, for aseptic reasons, it is desirable that the abscess should be opened from the neck. In some instances this is perfectly easy ; the abscess can be felt bulging at the side of the neck and merely requires an incision for its relief, but this is not always so, and then an incision should be made along the posterior border of the Sterno-mastoid and the deep fascia divided. A director is now to be inserted into the wound, the forefinger of the left hand being introduced into the mouth and pressure made upon the swelling. This acts as a guide, and the director is to be pushed onwards until pus appears in the groove. A pair of sinus forceps are now inserted along the director and the opening into the cavity dilated.

Foreign bodies not infrequently become lodged in the pharynx, and most usually at its termination at about the level of the cricoid cartilage, just beyond the reach of the finger, as the distance from the arch of the teeth to the commencement of the œsophagus is about six inches.

THE ŒSOPHAGUS

The **œsophagus**, or **gullet**, is a muscular canal, about nine inches in length, extending from the pharynx to the stomach. It commences at the upper border of the cricoid cartilage, opposite the intervertebral disc between the fifth and sixth cervical vertebræ, descends along the front of the spine, through the posterior mediastinum, passes through the Diaphragm, and, entering the abdomen, terminates at the cardiac orifice of the stomach, opposite the tenth dorsal vertebra or the intervertebral disc between the tenth and eleventh dorsal vertebræ. The general direction of the œsophagus is vertical ; but it presents two or three slight curves in its course. At its commencement it is placed in the median line ; but it inclines to the left side as far as the root of the neck, gradually passes to the middle line again, and finally again deviates to the left as it passes forwards to the œsophageal opening of the Diaphragm. The œsophagus also presents an antero-posterior flexure, corresponding to the curvature of the cervical and thoracic portions of the spine. It is the narrowest part of the alimentary canal, being most contracted at its commencement, and at the point where it passes through the Diaphragm.

Relations.—*In the neck*, the œsophagus is in relation, *in front*, with the trachea ; and at the lower part of the neck, where it projects to the left side, with the thyroid gland and thoracic duct ; *behind*, it rests upon the vertebral column and Longi colli muscles ; *on each side* it is in relation with the common carotid artery (especially the left, as it inclines to that side), and part of the lateral lobes of the thyroid gland ; the recurrent laryngeal nerves ascend between it and the trachea.

In the thorax, it is at first situated a little to the left of the median line ; it then passes behind the aortic arch, separated from it by the trachea, and descends in the posterior mediastinum, along the right side of the aorta, nearly to the Diaphragm, where it passes in front and a little to the left of the artery, previous to entering the abdomen. It is in relation, *in front*, with the trachea, the arch of the aorta, the left carotid and left subclavian arteries, which incline towards its left side, the left bronchus, the pericardium, and the Diaphragm ; *behind*, it rests upon the vertebral column, the Longi colli muscles, the right intercostal arteries, and the vena azygos minor ; and below, near the Diaphragm, upon the front of the aorta ; *laterally*, it comes in contact with both pleuræ, especially with the left pleura above and the right pleura below : it overlaps the vena azygos major, which lies on its right side, while the descending aorta is placed on its left side. The pneumogastric nerves descend in close contact with it, the right nerve passing down behind, and the left nerve in front of it ; the two nerves uniting to form a plexus (the *plexus gulæ*) around the tube.

In the lower part of the posterior mediastinum the thoracic duct lies to the right side of the œsophagus ; higher up, it is placed behind it, and, crossing about the level of the fourth dorsal vertebra, is continued upwards on its left side.

Structure.—The œsophagus has three coats : an external or muscular ; a middle or areolar ; and an internal or mucous coat.

The *muscular coat* is composed of two planes of fibres of considerable thickness : an external longitudinal, and an internal circular.

The *longitudinal fibres* are arranged, at the commencement of the tube, in three fasciculi : one in front, which is attached to the vertical ridge on the posterior surface of the cricoid cartilage ; and one at each side, which is continuous with the fibres of the Inferior constrictor : as they descend they blend together, and form a uniform layer, which covers the outer surface of the tube.

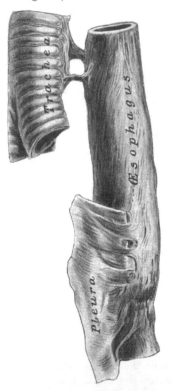

FIG. 478.—Accessory muscular fibres between the œsophagus and pleura, and œsophagus and trachea. (From a preparation in the Museum of the Royal College of Surgeons of England.)

Accessory slips of muscular fibres are described by Cunningham as passing between the œsophagus and the left pleura, where it covers the thoracic aorta (almost always), or the root of the left bronchus (usually), or the back of the pericardium, as well as other still more rare accessory fibres. In fig. 478, taken from a dissection in the Museum of the Royal College of Surgeons of England, several of these accessory slips may be seen passing from the œsophagus to the pleura, and two slips to the back of the trachea just above its bifurcation.

The *circular fibres* are continuous above with the Inferior constrictor ; their direction is transverse at the upper and lower parts of the tube, but oblique in the central part.

The muscular fibres in the upper part of the œsophagus are of a red colour, and consist chiefly of the striped variety ; but below, they consist for the most part of involuntary muscular fibre.

The *areolar coat* connects loosely the mucous and muscular coats.

The *mucous coat* is thick, of a reddish colour above, and pale below. It is disposed in longitudinal folds, which disappear on distension of the tube. Its surface is studded with minute papillæ, and it is covered throughout with a thick layer of stratified pavement epithelium. Beneath the mucous membrane, between it and the areolar coat, is a layer of longitudinally arranged non-striped muscular fibres. This is the *muscularis mucosæ*. At the commencement it is absent, or only represented by a few scattered bundles ; lower down it forms a considerable stratum.

The *œsophageal glands* are numerous small compound racemose glands scattered throughout the tube : they are lodged in the submucous tissue, and open upon the surface by a long excretory duct. They are most numerous at the lower part of the tube, where they form a ring round the cardiac orifice.

Vessels of the œsophagus.—The arteries supplying the œsophagus are derived from the inferior thyroid branch of the thyroid axis of the subclavian ; from the descending thoracic aorta, from the gastric branch of the cœliac axis, and from the left inferior phrenic of the abdominal aorta. They have for the most part a longitudinal direction.

Nerves of the œsophagus.—The nerves are derived from the pneumogastric and from the sympathetic ; they form a plexus, in which are groups of ganglion-cells, between the two layers of the muscular coats, and also a second plexus in the submucous tissue.

Surgical Anatomy.—The relations of the œsophagus are of considerable practical interest to the surgeon, as he is frequently required, in cases of stricture of this tube, to dilate the canal by a bougie, when it is of importance that the direction of the œsophagus, and its relations to surrounding parts, should be remembered. In cases of malignant disease of the œsophagus, where its tissues have become softened from infiltration of the morbid deposit, the greatest care is requisite in directing the bougie through the strictured part, as a false passage may easily be made, and the instrument may pass into the mediastinum, or into one or the other pleural cavity, or even into the pericardium.

The student should also remember that obstruction of the œsophagus, and consequent symptoms of stricture, are occasionally produced by an aneurism of some part of the aorta pressing upon this tube. In such a case, the passage of a bougie could only hasten the fatal issue.

In passing a bougie the left forefinger should be introduced into the mouth, and the epiglottis felt for, care being taken not to throw the head too far backwards. The bougie is then to be passed beyond the finger until it touches the posterior wall of the pharynx. The patient is now asked to swallow, and at the moment of swallowing the bougie is passed gently onwards, all violence being carefully avoided.

It occasionally happens that a foreign body becomes impacted in the œsophagus, which can neither be brought upwards nor moved downwards. When all ordinary means for its removal have failed, excision is the only resource. This, of course, can only be performed when it is not very low down. If the foreign body is allowed to remain, extensive inflammation and ulceration of the œsophagus may ensue. In one case the foreign body ultimately penetrated the intervertebral substance, and destroyed life by inflammation of the membranes and substance of the cord.

The operation of œsophagotomy is thus performed. The patient being placed upon his back, with the head and shoulders slightly elevated, an incision, about four inches in length, should be made on the left side of the trachea, from the thyroid cartilage downwards, dividing the skin, Platysma, and deep fascia. The edges of the wound being separated, the Omo-hyoid muscle should, if necessary, be divided, and the fibres of the Sterno-hyoid and Sterno-thyroid muscles drawn inwards ; the sheath of the carotid vessels, being exposed, must be drawn outwards, and retained in that position by retractors : the œsophagus will now be exposed, and should be divided over the foreign body, which can then be removed. Great care is necessary to avoid wounding the thyroid vessels, the thyroid gland, and the laryngeal nerves.

The œsophagus may be obstructed not only by foreign bodies, but also by changes in its coats, producing stricture, or by pressure on it from without of new growths or aneurism, &c.

The different forms of stricture are : (1) the spasmodic, usually occurring in nervous women, and intermittent in character, so that the dysphagia· is not constant; (2) fibrous, due to cicatrisation after injuries, such as swallowing corrosive fluids or boiling water ; and (3) malignant, usually epitheliomatous in its nature. This is situated generally either at the upper end of the tube, opposite the cricoid cartilage, or at its lower end at the cardiac orifice, but is also occasionally found at that part of the tube where it is crossed by the left bronchus.

The operation of œsophagostomy has occasionally been performed in cases where the stricture in the œsophagus is at the upper part, with a view to making a permanent opening below the stricture through which to feed the patient, but the operation has been far from successful, and the risk of setting up diffuse inflammation in the loose planes of connective tissue deep in the neck is so great that it would appear to be better, if any operative interference is undertaken, to perform gastrostomy. The operation is performed in the same manner as œsophagotomy, but the edges of the opening in the œsophagus are stitched to the skin incision.

THE ABDOMEN

The **Abdomen** is the largest cavity in the body. It is of an oval form, the extremities of the oval being directed upwards and downwards : the upper one being formed by the under surface of the Diaphragm, the lower by the upper concave surface of the Levatores ani. In order to facilitate description, it is artificially divided into two parts : an upper and larger part, the *abdomen proper* ; and a lower and smaller part, the *pelvis*. These two cavities are not separated from each other, but the limit between them is marked by the brim of the true

FIG. 479.—Topography of thoracic and abdominal viscera.

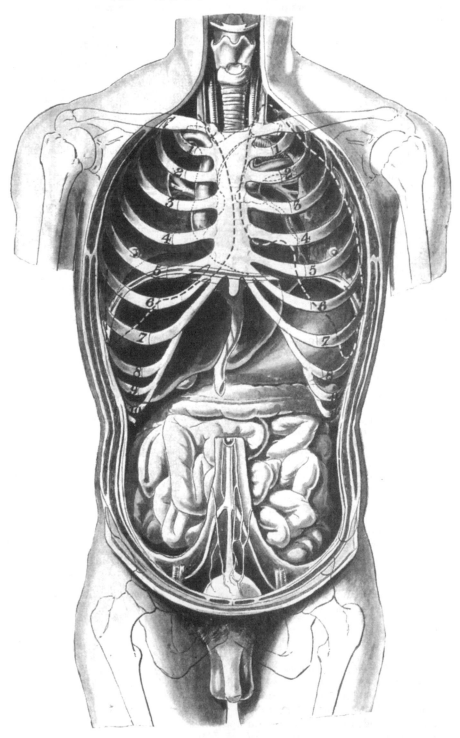

pelvis. The space is wider above than below, and measures more in the vertical than in the transverse diameter.

The abdomen proper differs from the other great cavities of the body in being bounded for the most part by muscles and fasciæ, so that it can vary in capacity and shape according to the condition of the viscera which it contains; but, in addition to this, the abdomen varies in form and extent with age and sex. In the adult male, with moderate distension of the viscera, it is oval or barrel-shaped, but at the same time flattened from before backwards. In the adult female, with a fully developed pelvis, it is conical with the apex above, and in young children it is conical with the apex below.

Boundaries.—The boundary between the thorax and abdomen is the Diaphragm. This muscle forms a dome over the abdomen, and the cavity extends high into the bony thorax, reaching to the level of the junction of the fourth costal cartilage with the sternum. The lower end of the abdomen is limited by the structures which clothe the inner surface of the bony pelvis, principally the Levatores ani and Coccygei muscles on either side. These muscles are sometimes termed the *Diaphragm of the pelvis.* The **abdomen proper** is bounded *in front*, and *at the sides*, by the lower ribs, the abdominal muscles, and the venter ilii; *behind*, by the vertebral column and the Psoas and Quadratus lumborum muscles; *above*, by the Diaphragm; *below*, by the brim of the pelvis. The muscles forming the boundaries of the cavity are lined upon their inner surface by a layer of fascia, differently named according to the part which it covers.

The abdomen contains the greater part of the alimentary canal; some of the accessory organs to digestion, viz. the liver and pancreas; the spleen, the kidneys, and suprarenal capsules. Most of these structures, as well as the wall of the cavity in which they are contained, are covered by an extensive and complicated serous membrane, the *peritoneum.*

The *apertures* found in the walls of the abdomen, for the transmission of structures to or from it, are, the *umbilicus*, for the transmission (in the fœtus) of the umbilical vessels; the *caval opening* in the Diaphragm, for the transmission of the inferior vena cava; the *aortic opening*, for the passage of the aorta, vena azygos major, and thoracic duct; and the *œsophageal opening*, for the œsophagus and pneumogastric nerves. *Below*, there are two apertures on each side: one for the passage of the femoral vessels, and the other for the transmission of the spermatic cord in the male, and the round ligament in the female.

Regions.—For convenience of description of the viscera, as well as of reference to the morbid conditions of the contained parts, the abdomen is artificially divided into nine regions. Thus, if two circular lines are drawn round the body, the one through the extremities of the ninth ribs, where they join their costal cartilages, and the other through the highest point of the crests of the ilia, the abdominal cavity is divided into three zones—an upper, a middle, and a lower. If two parallel lines are drawn perpendicularly upwards from the centre of Poupart's ligament, each of these zones is subdivided into three parts— a middle and two lateral.*

The middle region of the upper zone is called the *epigastric* (ἐπί, over; γαστήρ, *the stomach*); and the two lateral regions, the *right* and *left hypochondriac* (ὑπό, *under*; χόνδροι, *the cartilages*). The central region of the middle zone is the

---

* Anatomists are far from agreed as to the best method of subdividing the abdominal cavity. Cunningham suggests that the lower line should encircle the body on a level with the highest point of the iliac crest as seen from the front—a point corresponding with a prominent tubercle on the outer lip of the iliac crest about two inches behind the anterior superior spine. Addison,* in a careful analysis of the abdominal viscera in forty subjects, adopts the following lines : (1) a median. from symphysis pubis to ensiform cartilage ; (2) two lateral lines drawn vertically through a point midway between the anterior superior iliac spine and the symphysis pubis ; (3) an upper transverse line halfway between the symphysis pubis and the supra-sternal notch ; and (4) a lower transverse line midway between the last and the upper border of the symphysis pubis.

umbilical; and the two lateral regions, the *right* and *left lumbar*. The middle region of the lower zone is the *hypogastric* or *pubic region*; and the lateral regions are the *right* and *left inguinal* or *iliac*. The viscera contained in these different regions are the following (fig. 480):

Right Hypochondriac

The greater part of right lobe of the liver, the hepatic flexure of the colon, and part of the right kidney.

Epigastric Region

The greater part of the stomach, including both cardiac and pyloric orifices, the left lobe and part of the right lobe of the liver and the gall-bladder, the pancreas, the duodenum, the suprarenal capsules, and parts of the kidneys.

Left Hypochondriac

The fundus of the stomach, the spleen and extremity of the pancreas, the splenic flexure of the colon, and part of the left kidney.

Right Lumbar

Ascending colon, part of the right kidney, and some convolutions of the small intestines.

Umbilical Region

The transverse colon, part of the great omentum and mesentery, transverse part of the duodenum, and some convolutions of the jejunum and ileum, and part of both kidneys.

Left Lumbar

Descending colon, part of the omentum, part of the left kidney, and some convolutions of the small intestines.

Right Inguinal (Iliac)

The cæcum and vermiform appendix.

Hypogastric Region

Convolutions of the small intestines, the bladder in children, and in adults if distended, and the uterus during pregnancy.

Left Inguinal (Iliac)

Sigmoid flexure of the colon.

If the anterior abdominal wall is reflected in the form of four triangular flaps by means of vertical and transverse incisions—the former from the ensiform cartilage to the symphysis pubis, the latter from flank to flank at the level of the umbilicus—the abdominal or peritoneal cavity is freely opened into and the contained viscera are in part exposed.*

Above and to the right side is the liver, situated chiefly under the shelter of the right ribs and their cartilages, but extending across the middle line and reaching for some distance below the level of the ensiform cartilage. Below and to the left of the liver is the stomach, from the lower border of which an apron-like fold of peritoneum, the *great omentum*, descends for a varying distance, and obscures, to a greater or lesser extent, the other viscera. Below it, however, some of the coils of the small intestine can generally be seen, while in the right and left iliac regions respectively the *cæcum* and the *sigmoid flexure* of the colon are exposed. The bladder occupies the anterior part of the pelvis, and, if distended, will project above the symphysis pubis; the rectum lies in the concavity of the sacrum, but is usually obscured by the coils of the small intestine.

If the stomach is followed from left to right it will be found to be continuous with the first part of the small intestine, or *duodenum*, the point of continuity being marked by a thickened ring which indicates the position of the pyloric valve. The duodenum passes towards the under surface of the liver, and then curving

* It must be borne in mind that, although the term abdominal cavity is used, there is, under normal conditions, only a potential cavity or lymph-space, since the viscera are everywhere in contact with the parietes.

downwards, is lost to sight. If, however, the great omentum be thrown upwards over the chest, the terminal part of the duodenum will be observed passing across the spine towards the left side, where it becomes continuous with the *coils of the small intestine*. These measure some twenty feet in length, and if followed downwards will be seen to end in the right iliac fossa by opening into the *cæcum* or commencement of the *large intestine*. From the cæcum the large intestine takes an arched course, passing at first upwards on the right side, then across the middle line and downwards on the left side, and forming respectively the *ascending, transverse, and descending parts of the colon*. In the left iliac region it makes still another bend, the *sigmoid flexure*, and then follows the curve of the sacrum as the *rectum*.

FIG. 480.—The regions of the abdomen and their contents.
(Edge of costal cartilages in dotted outline.)

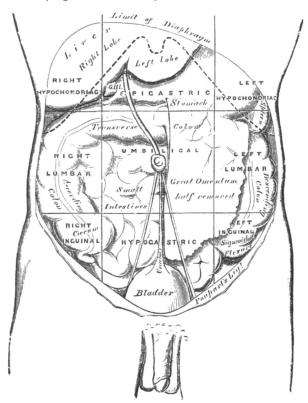

The *spleen* lies behind the stomach in the left hypochondriac region, and may be in part exposed by pulling the stomach over towards the right side.

The glistening appearance of the deep surface of the abdominal wall and of the exposed viscera is due to the fact that the former is lined and the latter more or less completely covered by a serous membrane, the *peritoneum*.

THE PERITONEUM

The peritoneum is the largest serous membrane in the body, and consists, in the male, of a closed sac, a part of which is applied against the abdominal parietes, while the remainder is reflected over the contained viscera. In the female the peritoneum is not a closed sac, since the free extremities of the Fallopian tubes open directly into the peritoneal cavity. The former constitutes the *parietal*, the latter the *visceral* part of the peritoneum. The *free surface* of the membrane is smooth, covered by a layer of flattened endothelium, and lubricated by a small quantity of serous fluid. Hence the viscera can glide freely against the wall of the cavity or upon one another with the least possible amount of friction. Its *attached surface* is rough, being connected to the viscera and inner surface of the parietes by means of *areolar tissue*, termed the *subserous areolar tissue*. The parietal portion is loosely connected with the fascia lining the abdomen and pelvis, but more closely to the under surface of the Diaphragm and also in the middle line of the abdomen.

The peritoneum differs from the other serous membranes of the body in presenting a much more complex arrangement—an arrangement which can only be clearly understood by following the changes which take place in the alimentary canal during its development; and therefore the student is advised to preface his study of the peritoneum by reviewing the chapter dealing with this subject in the section on Embryology.

To trace the continuity of the membrane from one viscus to another, and from the viscera to the parietes, it is necessary to follow its reflections in the vertical and horizontal directions, and in doing so it matters little where a start is made.

If the stomach is drawn downwards a fold of peritoneum will be seen stretching from its lesser curvature to the transverse fissure of the liver (fig. 481). This is the *gastro-hepatic*, or *small omentum*, and consists of two layers ; these, on being traced downwards, split to envelop the stomach, covering respectively its anterior and posterior surfaces. At the greater curvature of the stomach they again come into contact and are continued downwards in front of the transverse colon, forming the anterior two layers of the *great* or *gastro-colic* omentum. Reaching the free edge of this fold they are reflected upwards as its two posterior layers, and thus the great omentum consists of four layers of peritoneum. Followed upwards the two posterior layers separate so as to enclose the transverse colon, above which they once more come into contact and pass backwards to the abdominal wall as the *transverse mesocolon*. Reaching the abdominal wall about the level of the transverse part of the duodenum the two layers of the transverse mesocolon become separated from each other and take different directions ; the upper or anterior layer ascends (ascending layer of transverse mesocolon) in front of the pancreas, and its further course will be followed presently. The lower or posterior layer is carried downwards, as the anterior layer of the *mesentery*, by the superior mesenteric vessels to the small intestine around which it may be followed and subsequently traced upwards as the posterior layer of the mesentery to the abdominal wall. From the posterior abdominal wall it sweeps downwards over the aorta into the pelvis, where it invests the first part of the rectum and attaches it to the front of the sacrum by a fold termed the *mesorectum*. Leaving first the sides and then the front of the second part of the rectum it is reflected on to the back of the bladder, and, after covering the posterior and upper aspects of this viscus, is carried by the urachus and obliterated hypogastric arteries on to the posterior surface of the anterior abdominal wall. Between the rectum and bladder it forms a pouch, the *recto-vesical pouch*, bounded on each side by a crescentic or *semilunar fold* ; the bottom of this pouch is about on a level with the middle of the vesiculæ seminales, i.e. three inches or so from the orifice of the anus. When the bladder is distended the peritoneum is carried up with the expanded viscus, so that a considerable part of the anterior surface of the latter lies directly against the abdominal wall without the intervention of the peritoneal membrane.

In the female the peritoneum is reflected from the rectum on to the upper part of the posterior vaginal wall, forming the *recto-vaginal pouch* or *pouch of Douglas*. It is then carried over the posterior aspect and fundus of the uterus on to its anterior surface, which it covers as far as the junction of the body and cervix uteri, forming here a second, but shallower depression, the *utero-vesical pouch*. It is alro reflected from the sides of the uterus to the lateral walls of the pelvis as two expanded folds, the *broad ligaments of the uterus*, in the free margin of each of which can be felt a thickened cord-like structure, the *Fallopian tube*.

On following the parietal peritoneum upwards on the back of the anterior abdominal wall it is seen to be reflected around a fibrous band, the *ligamentum teres* or *obliterated umbilical vein*, which reaches from the umbilicus to the under surface of the liver.   Here the membrane forms a somewhat triangular fold, the *falciform* or *suspensory ligament* of the liver, which attaches the upper and anterior surfaces of that organ to the Diaphragm and abdominal wall.   With the exception of the line of attachment of this ligament the peritoneum covers the under surface of the anterior part of the Diaphragm and is reflected from it on to the upper surface of the liver as the *anterior* or *superior layer of the coronary ligament*. Covering the upper and anterior surfaces of the liver it is reflected round its sharp margin on to its under surface as far as the transverse fissure, where it is continuous with the anterior layer of the small omentum from which a start was made.   The posterior layer of this omentum is carried backwards from the transverse fissure over the under surface and Spigelian lobe of the liver and is then reflected, as the *posterior* or *inferior layer of the coronary ligament*, on to the Diaphragm and is prolonged downwards over the pancreas to become continuous with the ascending layer of the transverse mesocolon.   Between the two layers of the coronary ligament there is a triangular surface of the liver which is devoid of peritoneum; it is named the *bare area* of the liver, and is attached to the Diaphragm by connective tissue.   If, however, the two layers of the coronary ligaments are traced towards the right and left margins of the liver they approach each other, and, ultimately fusing, they form the right and left lateral ligaments of the liver and attach its right and left lobes respectively to the Diaphragm.

If the small omentum is followed towards the right side it is seen to form a distinct free edge around which its anterior and posterior layers are continuous with each other and between which are situated the portal vein, hepatic artery, and bile duct.   If the finger is introduced behind this free edge it passes through a somewhat constricted ring, the *foramen of Winslow*.   This is the communication between what are termed the greater and lesser sacs of the peritoneum and has the following boundaries : in front, the free edge of the gastro-hepatic omentum with the portal vein, hepatic artery, and bile duct between its two layers ; behind, the vena cava inferior ; above, the Spigelian and caudate lobes of the liver ; below, the duodenum and the hepatic artery, as the latter passes forwards and upwards from the cœliac axis.

The *lesser sac of the peritoneum* therefore lies behind the small omentum and has the following dimensions : above, it is limited by the portion of the liver which lies behind the transverse fissure ; below, it extends downwards into the great omentum, reaching, in the fœtus, as far as its free edge ; in the adult, however, its vertical extent is limited by adhesions between the layers of the omentum.   In front, it is bounded by the small omentum, stomach, and anterior two layers of the great omentum ; behind, by the two posterior layers of the great omentum, the transverse colon, and ascending layer of the transverse mesocolon which passes upwards in front of the pancreas as far as the posterior surface of the liver.   Laterally the lesser sac reaches from the foramen of Winslow on the right side as far as the spleen on the left, where it is limited by the lieno-renal ligament.   The extent of the lesser sac and its relations to surrounding parts can be definitely made out by tearing through the small omentum and inserting the hand through the opening thus made.

It should be stated that during a considerable part of fœtal life the transverse colon is suspended from the posterior abdominal wall by a mesentery of its own— the two posterior layers of the great omentum, passing, at this stage, in front of the colon.   This condition sometimes persists throughout adult life, but as a rule adhesion occurs between the mesentery of the transverse colon and the posterior layer of the great omentum, with the result that the colon appears to receive its peritoneal covering by the splitting of the two posterior layers of the latter fold.

In addition to tracing the peritoneum vertically, it is necessary to trace it horizontally. If this is done below the transverse colon, the circle is extremely simple, as it includes only the greater sac of the peritoneum (fig. 481). Above the level of the transverse colon the arrangement is more complicated, on account of the existence of the two sacs.

Starting from the linea alba, below the level of the transverse colon, and tracing the continuity in a horizontal direction to the right, the peritoneum covers the internal surface of the abdominal wall almost as far as the anterior border of the Quadratus lumborum muscle; it encloses the cæcum, and is reflected over the sides and anterior surface of the ascending colon, fixing it to the abdominal wall, from which it can be traced over the kidney to the front of the bodies of the vertebræ. It then passes along the mesenteric vessels to invest the small

FIG. 481.—The reflections of the peritoneum, as seen in a vertical section of the abdomen.

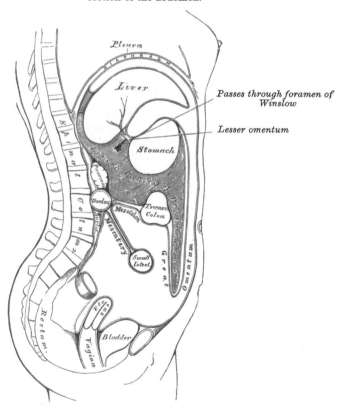

intestine, and back again to the spine, forming the mesentery, between the layers of which are contained the mesenteric blood-vessels, nerves, lacteals, and glands. Lastly, it passes over the left kidney to the sides and anterior surface of the descending colon, and, reaching the abdominal wall, is continued along it to the middle line of the abdomen.

Above the transverse colon (fig. 482) the peritoneum can be traced, forming the greater and lesser cavities, and their communication through the foramen of Winslow can be demonstrated. Commencing in the middle line of the abdomen, the membrane may be traced lining its anterior wall, and sending a process backwards to encircle the obliterated umbilical vein (the round ligament of the liver), forming the falciform or longitudinal ligament of the liver. Continuing its course to the right, it is reflected over the front of the upper part of the right kidney, across the vena cava inferior and aorta, and over the left kidney to the hilum of the spleen, forming the anterior layer ·of the *lieno-renal ligament*, the posterior layer being formed by the termination of the *cul-de-sac* of the greater cavity between the kidney and spleen. From the hilum of the spleen it is reflected to the stomach, forming the posterior layer of the *gastro-splenic omentum*. It covers the posterior surface of the stomach, and from its lesser curvature it passes around the portal vein, hepatic artery, and bile duct, and back again to the stomach, as the lesser omentum, and thus it forms the anterior boundary of the foramen of Winslow. It now covers the front of the stomach, and upon reaching the cardiac extremity it passes to the hilum of the spleen, forming the anterior layer of the gastro-splenic omentum. From the hilum of the spleen it can be traced over the surface of this organ to which it gives a serous covering; it is then reflected from the posterior border of the hilum on to the left kidney, forming the posterior layer of the lieno-renal ligament.

FIG. 482.—Transverse section of peritoneum.

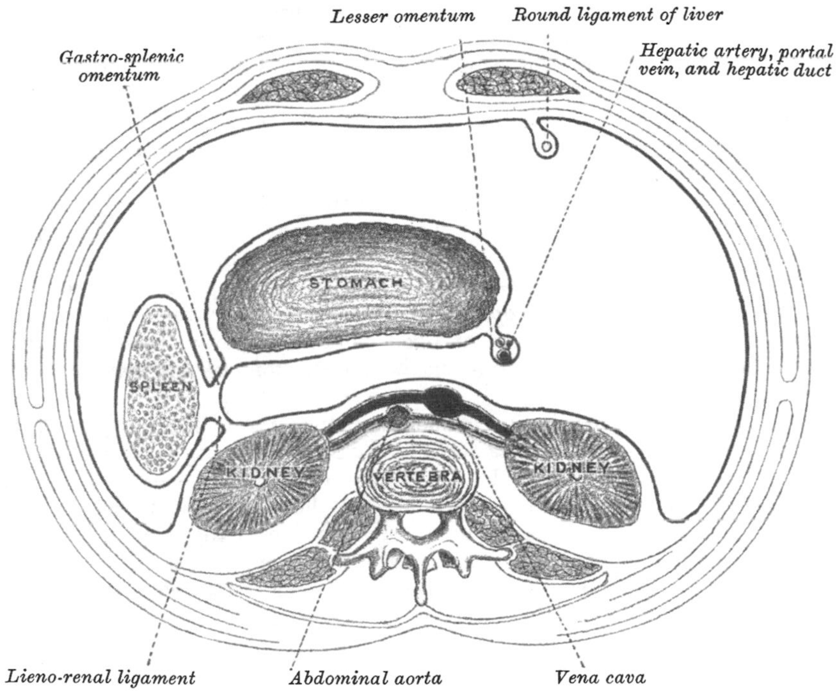

Numerous folds, formed by the peritoneum, extend between the various organs or connect them to the parietes. These serve to hold them in position, and, at the same time, enclose the vessels and nerves proceeding to each part. Some of these folds are called *ligaments*, such as the ligaments of the liver and the false ligaments of the bladder. Others, which connect certain parts of the intestine with the abdominal wall, constitute the *mesenteries*; and lastly, those which proceed from the stomach to certain viscera in its neighbourhood are called *omenta*.

The **Ligaments**, formed by folds of the peritoneum, include those of the liver, spleen, bladder, and uterus. They will be found described with their respective organs.

The **Omenta** are: the lesser omentum, the great omentum, and the gastro-splenic omentum.

The *lesser omentum (gastro-hepatic)* is the duplicature which extends between the transverse fissure of the liver and the lesser curvature of the stomach. It is extremely thin, and consists of two layers of peritoneum : that is, the two layers covering respectively the anterior and posterior surfaces of the stomach. When these two layers reach the lesser curvature of the stomach, they join together and ascend as the double fold to the transverse fissure of the liver ; to the left of this fissure the double fold is attached to the fissure of the ductus venosus as far as the Diaphragm, where the two layers separate to embrace the end of the œsophagus. At the right border the lesser omentum is free, and the two layers of which it is composed are continuous. The anterior layer, which belongs to the greater sac, turns round the hepatic vessels to become continuous with the posterior layer belonging to the lesser one. They here form a free, rounded margin, which contains between its layers the hepatic artery, the common bile-duct, the portal vein, lymphatics, and the hepatic plexus of nerves—all these structures being enclosed in loose areolar tissue, called *Glisson's capsule* Between the layers where they are attached to the stomach lie the gastric artery and the pyloric branch of the hepatic, anastomosing with it.

The *great omentum (gastro-colic)* is the largest peritoneal fold. It consists of four layers of peritoneum, two of which descend from the stomach, one from its anterior, the other from its posterior surface, and, uniting at its lower border, descend in front of the small intestines, sometimes as low down as the pelvis ; they then turn upon themselves, and ascend again as far as the transverse colon, where they separate and enclose that part of the intestine. These separate layers may be easily demonstrated in the young subject, but in the adult they are more or less inseparably blended. The left border of the great omentum is continuous with the gastro-splenic omentum ; its right border extends as far only as the duodenum. The great omentum is usually thin, presents a cribriform appearance, and always contains some adipose tissue, which in fat subjects accumulates in considerable quantity. Its use appears to be to protect the intestines from the cold, and to facilitate their movement upon each other during their vermicular action. Between its two anterior layers is the anastomosis between the right and left gastro-epiploic arteries.

The *gastro-splenic omentum* is the fold which connects the margins of the hilum of the spleen to the *cul-de-sac* of the stomach, being continuous, by its lower border with the great omentum. It contains the vasa brevia vessels.

The **Mesenteries** are : the mesentery proper, the transverse mesocolon, the sigmoid mesocolon, and the mesorectum. In addition to these there are some-times present an ascending and a descending mesocolon.

The *mesentery* (μέσον ἔντερον), so called from being connected to the middle of the cylinder of the small intestine, is the broad fold of peritoneum which connects the convolutions of the jejunum and ileum with the posterior wall of the abdomen. Its *root*, the part connected with the vertebral column, is narrow, about six inches in length, and directed obliquely from the left side of the second lumbar vertebra to the right sacro-iliac symphysis (fig. 483). Its intestinal border is much longer ; and here its two layers separate so as to enclose the intestine, and form its peritoneal coat. Its breadth, between its vertebral and intestinal border, is about eight inches. Its *upper border* is continuous with the under surface of the transverse mesocolon : its *lower border*, with the peritoneum covering the cæcum and ascending colon. It serves to retain the small intestines in their position, and contains between its layers the mesenteric vessels and nerves, the lacteal vessels, and mesenteric glands.

In most cases the peritoneum covers only the front and sides of the ascending and descending parts of the colon. Sometimes, however, these are surrounded by the serous membrane and attached to the posterior abdominal wall by an ascending and a descending mesocolon respectively. At the place where the transverse colon turns downwards to form the descending colon, a fold of perito-neum is continued to the under surface of the Diaphragm opposite the tenth and eleventh ribs. This is the *phreno-colic ligament* ; it passes below the spleen, and serves to support this organ, and therefore it has received the second name of *sustentaculum lienis*.

The *transverse mesocolon* is a broad fold, which connects the transverse colon to the posterior wall of the abdomen. It is formed by the two ascending or posterior layers of the great omentum, which, after separating to surround the transverse colon, join behind it, and are continued backwards to the spine, where they diverge in front of the duodenum. This fold contains between its layers the vessels which supply the transverse colon.

FIG. 483.—Diagram devised by Dr. Delépine to show the lines along which the peritoneum leaves the wall of the abdomen to invest the viscera.

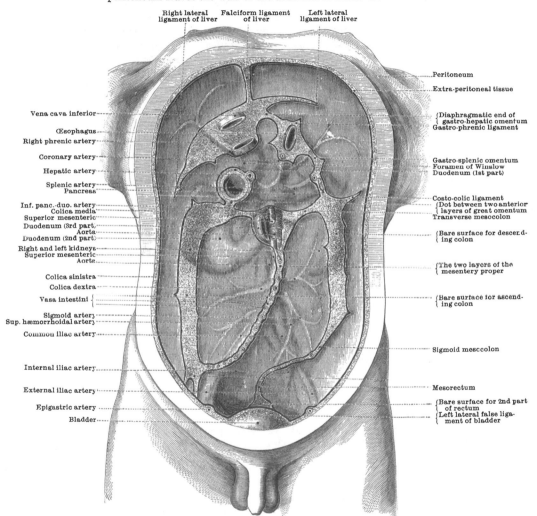

The *sigmoid mesocolon* is the fold of peritoneum which retains the sigmoid flexure in connection with the left iliac fossa.

The *mesorectum* is the narrow fold which connects the upper part of the rectum with the front of the sacrum. It contains the superior hæmorrhoidal vessels.

The *appendices epiploicæ* are small pouches of the peritoneum filled with fat and situated along the colon and upper part of the rectum. They are chiefly appended to the transverse colon.

Retro-peritoneal fossæ.—In certain parts of the abdominal cavity there are recesses of peritoneum forming *culs-de-sac* or pouches, which are of surgical interest in connection with the possibility of the occurrence of retro-peritoneal hernia. One of these is the lesser sac of the peritoneum, which may be regarded as a recess of peritoneum through the foramen of Winslow, in which a hernia may take place, but there are several others, of smaller size, which require mention.

These recesses or fossæ may be divided into three groups, viz.: (1) the duodenal fossæ; (2) pericæcal fossæ; and (3) the intersigmoid fossa.

1 *Duodenal fossæ.*—Moynihan has described no less than nine fossæ as occurring in the neighbourhood of the duodenum. Three of these are fairly constant, and are the only ones which require mention. (*a*) The *inferior duodenal fossa* is the most constant of all the peritoneal fossæ in this region, being present in from 70 to 75 per cent. of cases. It is situated opposite the third lumbar vertebra on the left side of the ascending portion of the duodenum. The opening into the fossa is directed upwards, and is bounded by a thin sharp fold of peritoneum with a concave margin, called the *inferior duodenal fold*. The tip of the index finger introduced into the fossa under the fold passes some little distance up behind the ascending or fourth portion of the duodenum. (*b*) The *superior duodenal fossa* is the next most constant pouch or recess, being present in from 40 to 50 per cent. of cases. It often co-exists with the inferior one, and its orifice looks downwards, in the opposite direction to the preceding fossa. It lies to the left of the ascending portion of the duodenum. It is bounded by the free edge of the *superior duodenal fold*, which presents a semilunar margin; to the right it is blended with the peritoneum covering the ascending duodenum, and to the left with the peritoneum covering the perirenal tissues. The fossa is bounded in front by the superior duodenal fold; behind by the second lumbar vertebra; to the right by the duodenum. Its depth is 2 c.m., and it terminates in the angle formed by the left renal vein crossing the aorta. This fossa is of importance, as it is in relation with the inferior mesenteric vein : that is to say, the vein almost always corresponds to the line of union of the superior duodenal fold with the posterior parietal peritoneum. (*c*) The *duodeno-jejunal fossa* can be seen by pulling the jejunum downwards and to the right, after the transverse colon has been pulled upwards. It will appear as an almost circular opening, looking downwards and to the right, and bounded by two free borders or folds of peritoneum, the *duodeno-mesocolic ligaments*. The opening admits the little finger into the fossa, to the depth of from 2 to 3 c.m. The fossa is bounded above by the pancreas, to the right by the aorta, and to the left by the kidney ; beneath is the left renal vein. The fossa exists in from 15 to 20 per cent. of cases, and has never yet been found in conjunction with any other form of duodenal fossa.

2. *Pericæcal fossæ.*—There are at least three pouches or recesses to be found in the neighbourhood of the cæcum, which are termed *pericæcal fossæ*. (1) The *ileo-colic fossa* (superior ileo-cæcal) is formed by a fold of peritoneum, the ileo-colic fold, arching over a branch of the ileo-colic artery, which supplies the ileo-colic junction, and appears to be the direct continuation of the artery. The fossa is a narrow chink situated between the ileo-colic fold in front, and the mesentery of the small intestine, the ileum, and a small portion of the cæcum behind. (2) The *ileo-cæcal fossa* (inferior ileo-cæcal) is situated behind the angle of junction of the ileum and cæcum. It is formed by a fold of peritoneum (the ileo-cæcal fold or bloodless fold of Treves), the upper border of which is attached to the ileum, opposite its mesenteric attachment, and the lower border, passing over the ileo-cæcal junction, joins the mesentery of the appendix, and sometimes the appendix itself ; hence this fold is sometimes called the ileo-appendicular. Between this fold and the mesentery of the vermiform appendix is the ileo-cæcal fossa. It is bounded above by the posterior surface of the ileum and the mesentery ; in front and below by the ileo-cæcal fold, and behind by the upper part of the

mesentery of the appendix. (3) The *subcæcal fossa* (retro-cæcal) is situated immediately behind the cæcum, which has to be raised to bring it into view. It varies
much in size and extent. In some cases it is sufficiently large to admit the
index finger, and extends upwards behind the ascending colon in the direction of
the kidney : in others it is merely a shallow depression. It is bounded and formed
by two folds : one, the *parieto-colic*, which is attached by one edge to the abdominal
wall from the lower border of the kidney to the iliac fossa and by the other to the
postero-external aspect of the colon ; and the other, *mesenterico-parietal*, which is
in reality the insertion of the mesentery into the iliac fossa. In some instances
the subcæcal fossa is double.

3. The *Intersigmoid fossa* is constant in the fœtus and during infancy, but
disappears in a certain percentage of cases as age advances. Upon drawing the
sigmoid flexure upwards, the left surface of the sigmoid mesocolon is exposed,
and on it will be seen a funnel-shaped recess of the peritoneum, lying on the
external iliac vessels, in the interspace between the Psoas and Iliacus muscles.
This is the orifice leading to the fossa intersigmoidea, which lies behind the
sigmoid mesocolon, and in front of the parietal peritoneum. The fossa varies in
size ; in some instances it is a mere dimple, whereas in others it will admit the
whole of the index finger.

Any of these fossæ may be the site of a retro-peritoneal hernia. The pericæcal fossæ are of especial interest, because hernia of the vermiform appendix
frequently takes place into one of them, and may there become strangulated. The
presence of these pouches also explains the course which pus has been known to
take in cases of perforation of the appendix, where it travels upwards behind the
ascending colon as far as the Diaphragm.*

THE STOMACH

The **Stomach** is the principal organ of digestion. It is the most dilated part
of the alimentary canal, and is situated between the termination of the œsophagus
and the commencement of the small intestine. Its form is somewhat pyriform
with the large end (*fundus*) directed upwards and the small end bent to the
right. It is situated in the left hypochondriac and epigastric regions, and is
placed, in part, immediately behind the anterior wall of the abdomen and
beneath the Diaphragm. Viewing the stomach from in front it appears that the
right margin of the œsophagus is continued downwards as the upper two-thirds
of the lesser curvature of the stomach, the remaining third of this border bending
sharply backwards and to the right, to complete the smaller curvature (fig. 484).

The greater curvature begins at the left
border of the termination of the œsophagus
in a somewhat acute angle ; it then passes
upwards and to the left to the under surface of the Diaphragm, with which it lies in
contact for some distance, and then sweeps
downwards with a convexity to the left, and,
continued across the middle line of the body,
finally turns upwards and backwards, to
terminate at the commencement of the small
intestine. It will thus be seen that the
stomach may be divided into a main or
cardiac portion, the long axis of which is
directed downwards, with a little inclination
forwards and to the right, and a smaller or
pyloric portion, the long axis of which is horizontal with an inclination backwards.
Of the two openings, the *cardiac orifice*, by which it communicates with the

FIG. 484.—Diagrammatic outline
of the stomach.

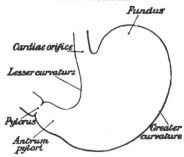

* On the anatomy of these fossæ, see the *Arris and Gale Lectures* by Moynihan, 1899.

œsophagus, is situated slightly to the left of the middle line of the body to the right of the *fundus*, or dilated upper extremity of the stomach, and is directed downwards ; the other, the *pyloric orifice*, by which it communicates with the small intestine, is on a lower plane, close to the right of the mid-line, and looks directly backwards.

The stomach has two surfaces, called anterior and posterior, and two borders, termed the greater and lesser curvatures.

Surfaces.—With regard to the so-called anterior and posterior surfaces of the stomach, it must be borne in mind that these names are not strictly correct, as the anterior surface has a certain amount of inclination upwards and the posterior downwards.

The *anterior* surface has a somewhat flattened appearance when the stomach is empty, but when it is full the surface becomes convex. It is in relation with the Diaphragm ; the thoracic wall formed by the anterior parts of the seventh, eighth, and ninth ribs of the left side ; the left lobe of the liver ; and the anterior abdominal wall. Between the part covered by the liver and that covered by the left ribs there is a triangular segment of the anterior wall of the stomach, which is in contact with the abdominal wall and is the only part of the stomach which is visible when the abdominal wall is removed and the viscera allowed to remain *in situ*. It is of about 40 sq. c.m. and is of great importance to the surgeon, as the stomach can readily be reached in this situation. Occasionally the transverse colon may be found lying in front of the lower part of the anterior surface of the stomach. The whole of this surface of the stomach is covered by peritoneum.

The *posterior* surface of the stomach is in relation with the Diaphragm, the gastric surface of the spleen, the left supra-renal capsule, the upper part of the left kidney, the anterior surface of the pancreas, the splenic flexure of the colon, and the ascending layer of the transverse mesocolon. These structures form a shallow concavity or *bed* on which this surface of the stomach rests. The transverse mesocolon intervenes between the stomach and the duodeno-jejunal junction and commencement of the ileum. Its greater curvature is in relation with the transverse colon and has attached to it the anterior two layers of the great omentum. Almost the whole of this surface is covered with peritoneum, but behind the cardiac orifice there is a small portion of the stomach which is uncovered by peritoneum and is in contact with the Diaphragm and frequently with the upper portion of the left supra-renal capsule.

The *lesser curvature* of the stomach extends between the cardiac and pyloric orifices along the right border of the organ. It descends in front of the left crus of the Diaphragm, along the left side of the eleventh and twelfth dorsal vertebræ, and then turning to the right it crosses the first lumbar vertebræ and ascends to the pylorus. It gives attachment to the two layers of the gastro-hepatic omentum, between which blood-vessels and lymphatics pass to reach the organ.

The *greater curvature* is directed to the left, and is four or five times as long as the lesser curvature. Starting from the cardiac orifice it forms an arch to the left with its convexity upwards, the highest point of which is on a level with the costal cartilage of the sixth rib of the left side. It then passes nearly straight downwards, with a slight convexity to the left, as low as the costal cartilage of the ninth rib and then turns to the right to end at the pylorus. As it crosses the median line the lowest edge of the greater curvature is about two fingers' breadth above the umbilicus. The lower part of the greater curvature gives attachment to the two anterior layers of the great omentum, between which layers, vessels and lymphatics pass to the organ.

The *cardiac orifice* is the opening by which the œsophagus communicates with the stomach. It is therefore sometimes termed the *œsophageal opening*. It is the most fixed part of the stomach, and is situated about two inches below the highest part of the fundus on a level with the body of the tenth or eleventh dorsal vertebra to the left and a little in front of the aorta. This would correspond on the anterior surface of the body to the articulation of the seventh left costal cartilage to the sternum.

The *pyloric orifice* communicates with the duodenum, the aperture being guarded by a valve. Its position varies with the movements of the stomach. When the stomach is empty the pylorus is situated just to the right of the median line of the body, on a level with the upper border of the first lumbar vertebra. On the anterior surface of the body its position would be indicated by a point one inch below the tip of the ensiform cartilage and a little to the right. As the stomach becomes distended the pylorus moves to the right, and in a fully distended stomach may be situated two or three inches to the right of the median line. Near the pylorus the stomach frequently exhibits a slight dilatation, which is named the *antrum pylori.*

The size of the stomach varies considerably in different subjects. When moderately distended its greatest length, from the top of the fundus to the lowest part of the greater curvature, is from ten to twelve inches ; and its diameter at the widest part from four to five inches. The distance between the two orifices is three to six inches, and the measurement from the anterior to the posterior wall three and a half inches. Its weight, according to Clendinning, is about four ounces and a half, and its capacity in the adult male is five to eight pints.

Alterations in Position.—There is no organ in the body the position and connections of which present such frequent alterations as the stomach. *When empty*, it lies at the back part of the abdomen, some distance from the surface. Its pyloric end is situated close to or very slightly to the right of the middle line, covered in front by the left lobe of the liver, and being on a level with the first lumbar vertebra. When empty, the stomach assumes a more or less cylindrical form, especially noticeable at its pyloric end. *When the stomach is distended*, its surfaces, which are flattened when the organ is empty, become convex. The greater curvature is elevated and carried forwards, so that the anterior surface is turned more or less upwards and the posterior surface downwards, and the stomach brought well against the anterior wall of the abdomen. Its fundus expands and rises considerably above the level of the cardiac orifice ; in doing this the Diaphragm is forced upwards, contracting the cavity of the chest; hence the dyspnœa complained of, from inspiration being impeded. The apex of the heart is also tilted upwards ; hence the oppression in this region and the palpitation experienced in extreme distension of the stomach. The left lobe of the liver is pushed to the right side. When the stomach becomes distended the change in the position of the pylorus is very considerable ; it is shifted to the right, some two or three inches from the median line, and lies under cover of the liver, near the neck of the gall-bladder. In consequence of the distension of the stomach the lesser *cul-de-sac* bulges over the pylorus, concealing it from view, and causing it to undergo a rotation, so that its orifice is directed backwards. *During inspiration* the stomach is displaced downwards by the descent of the Diaphragm, and elevated by the pressure of the abdominal muscles during expiration. *Pressure from without*, as from tight lacing, pushes the stomach down towards the pelvis. In disease, also, the position and connection of the organ may be greatly changed, from the accumulation of fluid in the chest or abdomen, or from alteration in size of any of the surrounding viscera. *Variations according to age.*—In an early period of development the stomach is vertical, and in the new-born child it is more vertical than later on in life, as owing to the large size of the liver it is more pushed over to the left side of the abdomen, and the whole of the anterior surface is covered by the left lobe of this organ.

On looking into the pyloric end of the stomach, the mucous membrane is found projecting inwards in the form of a circular fold, the *pyloric valve,* leaving a narrow circular aperture, about half an inch in diameter, by which the stomach communicates with the duodenum.

The *pyloric valve* is formed by a reduplication of the mucous membrane of the stomach, containing numerous circular muscular fibres, which are aggregated into a thick circular ring; the longitudinal fibres and serous membrane being continued over the fold without assisting in its formation.

Structure.—The wall of the stomach consists of four coats : serous, muscular, areolar, and mucous, together with vessels and nerves.

The *serous coat* is derived from the peritoneum, and covers the entire surface of the organ, excepting along the greater and lesser curvatures, at the points of attachment of the greater and lesser omenta ; here the two layers of peritoneum leave a small triangular space, along which the nutrient vessels and nerves pass. On the posterior surface of the stomach, close to the cardiac orifice, there is also a small area uncovered by peritoneum, where the organ is in contact with the under surfaces of the Diaphragm.

The *muscular coat* (fig. 486) is situated immediately beneath the serous covering, to which it is closely connected. It consists of three sets of fibres—longitudinal, circular, and oblique.

The *longitudinal fibres* are most superficial; they are continuous with the longitudinal fibres of the œsophagus, radiating in a stellate manner from the cardiac orifice. They are most distinct along the curvatures, especially the lesser, but are very thinly distributed over the surfaces. At the pyloric end they are more thickly distributed, and continuous with the longitudinal fibres of the small intestine.

FIG. 485.—The mucous membrane of the stomach and duodenum with the bile ducts.

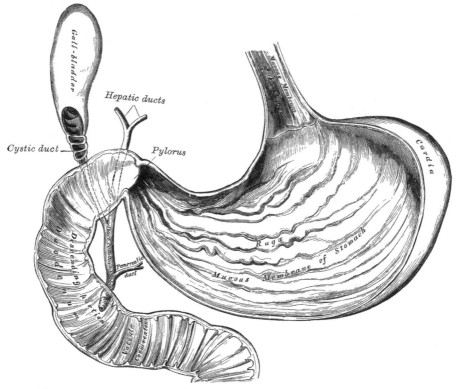

The *circular fibres* form a uniform layer over the whole extent of the stomach beneath the longitudinal fibres. At the pylorus they are most abundant, and are aggregrated into a circular ring, which projects into the cavity, and forms, with the fold of mucous membrane covering its surface, the *pyloric valve*. They are continuous with the circular fibres of the œsophagus.

The *oblique fibres* are limited chiefly to the cardiac end of the stomach, where they are disposed as a thick uniform layer, covering both surfaces, some passing obliquely from left to right, others from right to left, round the cardiac end.

The *areolar* or *submucous coat* consists of a loose, filamentous, areolar tissue, connecting the mucous and muscular layers. It supports the blood-vessels previous to their distribution to the mucous membrane : hence it is sometimes called the *váscular coat*.

The *mucous membrane* is thick; its surface smooth, soft, and velvety. In the fresh state it is of a pinkish tinge at the pyloric end, and of a red or reddish-brown colour over the rest of its surface. In infancy it is of a brighter hue, the vascular redness being more marked. It is thin at the cardiac extremity, but thicker towards the pylorus. During the contracted state of the organ it is thrown into numerous plaits or rugæ, which, for the most part, have a longitudinal direction, and are most marked towards the lesser end of the stomach, and along the greater curvature (fig. 485). These folds are entirely obliterated when the organ becomes distended.

Structure of the Mucous Membrane.—When examined with a lens, the inner surface of the mucous membrane presents a peculiar honeycomb appearance from being covered with small shallow depressions or alveoli, of a polygonal or hexagonal form, which vary from $\frac{1}{100}$ to $\frac{1}{200}$ of an inch in diameter, and are separated by slightly elevated ridges. In the bottom of the alveoli are seen the orifices of minute tubes, the *gastric glands*, which are situated perpendicularly side by side throughout the entire substance of the mucous membrane. The surface of the mucous membrane of the stomach is covered by a single layer of columnar epithelium; it lines the alveoli, and also for a certain

FIG. 486.—The muscular coat of the stomach.

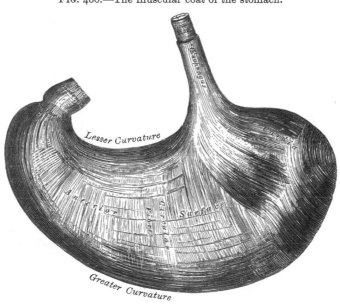

distance the mouths of the gastric glands. This epithelium commences very abruptly at the cardiac orifice, where the cells suddenly change in character from the stratified epithelium of the œsophagus. The cells are elongated, and consist of two parts, the inner or attached portions being granular, and the outer or free parts being clear and occupied by a muco-albuminous substance.

The gastric glands are of two kinds, which differ from each other in structure, and it is believed also in the nature of their secretion. They are named respectively *pyloric* and *cardiac* or *oxyntic glands.* They are both tubular in character, and are formed of a delicate basement-membrane, lined by epithelium. The basement-membrane consists of flattened transparent endothelial cells, with processes which extend between and support the epithelium. The *pyloric glands* (fig. 487) are most numerous at the pyloric end of the stomach, and from this fact have received their name. They consist of two or three short, closed tubes opening into a common duct, the external orifice of which is situated at the bottom of an alveolus. The cæcal tubes are wavy, and are of about equal length with the duct. The tubes and duct are lined throughout with epithelium, the duct being lined by columnar cells, continuous with the epithelium lining the surface of the mucous membrane of the stomach, the tubes with shorter and more cubical cells which are finely granular. The *cardiac glands* (fig. 488) are found all over the surface of the stomach, but occur most numerously at the cardiac end. Like the pyloric glands they consist of a duct, into which open two or more cæcal tubes. The duct, however, in these glands is shorter than in the other variety, sometimes not amounting to more than one-sixth of the whole length of the gland; it is lined throughout by columnar epithelium. At the point where the terminal tubes open into the duct, and which is termed the neck, the epithelium alters, and consists of short columnar or polyhedral, granular cells, which almost fill the tube, so that the lumen becomes suddenly constricted, and is continued down as a very fine channel. They are known as the *chief* cells or *central* cells of the glands. Between these cells and the basement-membrane are found other darker granular-looking cells, studded throughout the tube at intervals, and giving it a beaded or varicose appearance. These are known

Fig. 487.—Pyloric gland.   Fig. 488.—Cardiac gland.

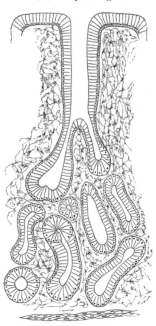 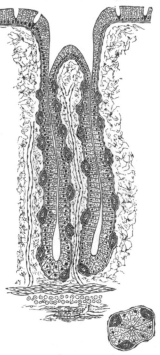

as the *parietal* or *oxyntic* cells. Between the glands the mucous membrane consists of a connective-tissue framework, with lymphoid tissue. In places, this latter tissue, especially in early life, is collected into little masses, which to a certain extent resemble the solitary glands of the intestine, and are by some termed the *lenticular* glands of the stomach. They are not, however, so distinctly circumscribed as the solitary glands. Beneath the mucous membrane, and between it and the submucous coat, is a thin stratum of involuntary muscular fibre (*muscularis mucosæ*), which in some parts consists only of a single longitudinal layer; in others of two layers, an inner, circular, and an outer, longitudinal.

Vessels and Nerves.—The arteries supplying the stomach are : the gastric, the pyloric and right gastro-epiploic branches of the hepatic, the left gastro-epiploic and vasa brevia from the splenic. They supply the muscular coat, ramify in the submucous coat, and are finally distributed to the mucous membrane. The arrangement of the vessels in the mucous membrane is somewhat peculiar. The arteries break up at the base of the gastric tubules into a plexus of fine capillaries which run upwards between the tubules, anastomosing with each other, and ending in a plexus of larger capillaries, which surround the mouths of the tubes, and also form hexagonal meshes around the alveoli. From these latter the *veins* arise, and pursue a straight course downwards, between the tubules, to the submucous tissue ; they terminate either in the splenic and superior, mesenteric veins, or directly in the portal vein. The *lymphatics* are numerous ; they consist of a superficial and deep set, which pass through the lymphatic glands found along the two curvatures of the organ. The *nerves* are the terminal branches of the right and left pneumogastric, the former being distributed upon the back, and the latter upon the front part of the organ. A great number of branches from the sympathetic also supply the organ.

Surface Form.—The stomach lies for the most part in the left hypochondriac region, but also slightly in the epigastric region, and is partly in contact with the abdominal wall, partly under cover of the lower ribs on the left side, and partly under the left lobe of the liver. Its cardiac orifice corresponds to the articulation of the seventh left costal cartilage with the sternum. The pyloric orifice is in a vertical line drawn from the right border of the sternum, two and a half or three inches below the level of the sterno-xiphoid articulation. According to Braune, when the stomach is distended, the pylorus moves considerably to the right, as much sometimes as three inches. The fundus of the stomach reaches, on the left side, as high as the level of the sixth costal cartilage of the left side, being a little below and behind the apex of the heart. The portion of the stomach which is in contact with the abdominal walls, and is therefore accessible for opening in the operations of gastrotomy and gastrostomy, is represented by a triangular space, the base of which is formed by a line drawn from the tip of the tenth costal cartilage on the left side to the tip of the ninth costal cartilage on the right, and the sides by two lines drawn from the extremity of the eighth costal cartilage on the left side to the ends of the base line.

Surgical Anatomy.—Operations on the stomach are frequently performed. By 'gastrotomy' is meant an incision into the stomach for the removal of a foreign body, the opening being immediately afterwards closed—in contradistinction to 'gastrostomy,' the making of a more or less permanent fistulous opening. *Gastrotomy* is probably best performed by an incision in the linea alba, especially if the foreign body is large, by a cut from the ensiform cartilage to the umbilicus ; but may be performed by an incision over the body itself, where this can be felt, or by one of the incisions for gastrostomy, to be mentioned immediately. The peritoneal cavity is opened, and the point at which the stomach is to be incised decided upon. This portion is then brought out of the abdominal wound and sponges carefully packed around. The stomach is now opened by a transverse incision and the foreign body extracted. The wound in the stomach is then closed by Lembert's sutures, i.e. by sutures passed through the peritoneal and muscular coats in such a way that the peritoneal surfaces on each side of the wound are brought into apposition, and in this way the wound is closed. *Gastrostomy* was formerly done in two stages by the *direct* method. The first stage consisted in opening the abdomen, drawing up the stomach into the external wound, and fixing it there ; and the second stage, performed from two to four days afterwards, consisted in opening the stomach. The operation is now done by a *valvular* method. An incision is commenced opposite the eighth intercostal space, two inches from the median line, and carried downwards for three inches. By this incision the fibres of the Rectus muscle are exposed and these are separated from each other in the same line with a steel director. The posterior layer of the sheath, the transversalis fascia and the peritoneum, are then divided, and the peritoneal cavity opened. The anterior wall of the stomach is now seized and drawn out of the wound and a silk suture passed through its muscular and serous coats at the point selected for opening the viscus. This is held by an assistant so that a long conical diverticulum of the stomach protrudes from the external wound, and the parietal peritoneum and the posterior layer of the sheath of the rectus are sutured to it. A second incision is made through the skin, over the margin of the costal cartilage, above and a little to the outer side of the first incision. With a pair of dressing forceps a track is made under the skin through the subcutaneous tissue from the one opening to the other and the diverticulum of the stomach is drawn along this track by means of the suture inserted into it, so that its apex appears at the second opening. A small perforation is now made into the stomach through this protruding apex and its margins carefully and accurately sutured to the margin of the external wound. The remainder of this incision and the whole of the first incision are then closed in the ordinary way and the wound dressed.

In cases of gastric ulcer perforation sometimes takes place, and this was formerly regarded as an almost fatal complication. In the present day, by opening the abdomen and closing the perforation, which is generally situated on the anterior surface of the stomach, a considerable percentage of cases are cured, provided the operation is undertaken within twelve or fifteen hours after the perforation has taken place. The opening is best closed by bringing the peritoneal surfaces on either side into apposition by means of Lembert's sutures.

Excision of the pylorus has occasionally been performed, but the results of this operation are by no means favourable, and, in cases of cancer of the pylorus, before operative proceedings are undertaken, the tumour has become so fixed and has so far implicated surrounding parts that removal of the pylorus is impossible and gastro-enterostomy has to be substituted. The object of this operation is to make a fistulous communication between the stomach, on the cardiac side of the disease, and the small intestine, as high up as is possible.

Digital dilatation of the pylorus for simple stricture was first performed by Loreta. He exposed the stomach and opened it by a transverse incision near the pylorus. He then inserted the forefingers of both hands and passed these through the pylorus and stretched it with some degree of force. The operation has now, however, dropped out of use and been replaced by pyloro-plasty. Thisconsists in making a longitudinal incision from the stomach through the pylorus into the duodenum, and converting this longitudinal incision into a transverse one by traction at the centre of the incision, and retaining it permanently in this position by sutures.

THE SMALL INTESTINE

The small intestine is a convoluted tube, extending from the pylorus to the ileo-cæcal valve, where it terminates in the large intestine. It is about twenty feet in length,* and gradually diminishes in size from its commencement to its termination. It is contained in the central and lower part of the abdominal cavity, and is surrounded above and at the sides by the large intestine; a portion of it extends below the brim of the pelvis and lies in front of the rectum; it is in relation, in front, with the great omentum and abdominal parietes; and connected to the spine by a fold of peritoneum, the mesentery. The small intestine is divisible into three portions—the duodenum, the jejunum, and ileum.

The **duodenum** has received its name from being about equal in length to the breadth of twelve fingers (ten inches). It is the shortest, the widest, and the most fixed part of the small intestine. Its course presents a remarkable curve, which in the adult, as regards the greater part of its extent, is U-shaped; though sometimes, in consequence of the transverse portion being very short or altogether wanting, it partakes more of the character of the letter V. In children, up to the age of about seven, the duodenum is annular; its two extremities are on about the same level; and between them it describes a regular curve embracing the head of the pancreas, the neck of which lies between the two extremities of the ring.

In the adult the course of the duodenum is as follows : commencing at the pylorus the direction of the first portion depends upon the amount of distension of the stomach and therefore upon the position of the pylorus. When the stomach is empty and the pylorus situated at the right of the upper border of the first lumbar vertebra, it is nearly horizontal and transverse; but where the stomach is distended, in consequence of the alteration of the position of the pylorus to the right the proximal end of the duodenum also becomes altered in position, while the distal end remains fixed and the direction of this portion of the bowel is now antero-posterior. Whether directed transversely or antero-posteriorly, it reaches the under surface of the liver, where it takes a sharp curve and descends along the right side of the vertebral column, for a variable distance, generally to the body of the fourth lumbar vertebra. It now takes a second bend, and passes across the front of the vertebral column from right to left and finally ascends on the left side of the vertebral column and aorta to the level of the upper border of the second lumbar vertebra and there terminates in the jejunum. As it unites with the jejunum it often turns abruptly forwards, forming the *duodeno-jejunal angle*

---

* Treves states that, in one hundred cases, the average length of the small intestine in the adult male was 22 feet 6 inches, and in the adult female 23 feet 4 inches : but that it varies very much, the extremes in the male being 31 feet 10 inches in one case, and 15 feet 6 inches in another, a difference of over 15 feet. He states that he has convinced himself that the length of the bowel is independent, in the adult, of age, height, and weight.

From the above description it will be seen that the duodenum may be divided for purposes of description into four portions—superior, descending, transverse, and ascending.

FIG. 489.—Relations of duodenum, pancreas and spleen.
(From a cast by Professor Birmingham.*)

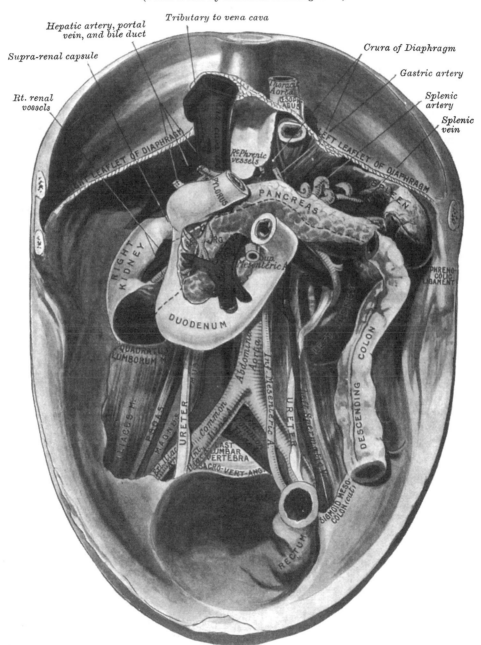

The dotted line represents the line of attachment of the transverse mesocolon.

The first or *superior portion* (fig. 489) is very variable in length, but is usually estimated as being about two inches. Beginning at the pylorus it ends at the

* In the subject from which the cast was taken the left kidney was lower than normal.

neck of the gall-bladder. It is the most movable of the four portions. It is almost completely covered by peritoneum derived from the two layers of the lesser omentum, but a small part of its posterior surface near the neck of the gall-bladder and the inferior vena cava is uncovered. It is in such close relation with the gall-bladder that it is usually found to be stained by bile after death, especially on its anterior surface. It is in relation above and in front with the quadrate lobe of the liver and the gall-bladder; behind with the gastro-duodenal artery, the common bile duct, and the vena porta; and below with the head of the pancreas.

The second or *descending portion* is between three and four inches in length, and extends from the neck of the gall-bladder on a level with the first lumbar vertebra along the right side of the vertebral column as low as the body of the fourth lumbar vertebra. It is crossed in its middle third by the transverse colon, the posterior surface of which is uncovered by peritoneum and is connected to the duodenum by a small quantity of connective tissue (fig. 483). The portions of the descending part of the duodenum above and below this interspace are named the supra- and infra-colic portions, and are covered in front by peritoneum. The right side of the supra-colic portion is covered by peritoneum derived from the anterior surface of the right kidney, the left side of the same portion being covered by the peritoneum forming the lesser sac. The infra-colic part is covered by the right leaf of the mesentery. Posteriorly the descending portion of the duodenum is uncovered by peritoneum. It is in relation, in front, with the transverse colon, and above this with the liver; behind with the front of the right kidney, to which it is connected by loose areolar tissue, the renal vessels and the vena cava inferior; at its inner side is the head of the pancreas, and the ductus communis choledochus; to its outer side is the hepatic flexure of the colon. The common bile-duct and the pancreatic duct perforate the inner side of this portion of the intestine obliquely, some three or four inches below the pylorus. The relations of the second part of the duodenum to the right kidney present considerable variations.

The third or *transverse portion* (pre-aortic portion) varies much in length; when the duodenum assumes the ordinary U-shaped form, it measures from two to three inches; but when it presents the rarer V-shaped form it is practically wanting or very much reduced in length. It commences at the right side of the fourth lumbar vertebra and passes from right to left, with a slight inclination upwards, in front of the great vessels and crura of the Diaphragm, and ends in the fourth portion just to the left of the abdominal aorta. It is crossed by the superior mesenteric vessels and mesentery. Its front surface is covered by the anterior layer of the mesentery, but near the middle line it is separated from this layer of the mesentery by the superior mesenteric vessels as they cross this portion of the duodenum. Its posterior surface is uncovered by peritoneum, except towards its left extremity, where the posterior layer of the mesentery may sometimes be found covering it to a variable extent. This surface rests upon the aorta, the vena cava inferior, and the crura of the Diaphragm. By its upper surface this portion of the duodenum is in relation with the head of the pancreas.

The fourth or *ascending portion* of the duodenum is about two inches long. It ascends on the left side of the vertebral column and aorta, as far as the level of the upper border of the second lumbar vertebra, where it turns abruptly forwards to become the jejunum, forming the *duodeno-jejunal flexure*. It is covered entirely in front and partly at the sides by peritoneum, derived from the left portion of the mesentery. It touches the left kidney, slightly overlapping its inner margin, and rests upon the left crus of the Diaphragm.

The first part of the duodenum, as stated above, is somewhat movable, but the rest is practically fixed and is bound down to neighbouring viscera and the posterior abdominal wall by the peritoneum. In addition to this, the fourth part of the duodenum and the duodeno-jejunal flexure is further bound down and fixed by a structure to which the name of *musculus suspensorius duodeni* has been given. This structure commences in the connective tissue around the cœliac axis and left crus of the Diaphragm, and passes downwards to be inserted into the superior border of the duodeno-jejunal curve and a part of the ascending duodenum, and from this it is continued into the mesentery. It possesses, according to Treitz, plain muscular fibres mixed with the fibrous tissue, of which it is principally made up. It is of little importance as a muscle, but acts as a suspensory ligament.

Vessels and Nerves.—The *arteries* supplying the duodenum are the pyloric and pancreatico-duodenal branches of the hepatic, and the inferior pancreatico-duodenal branch of the superior mesenteric. The *veins* terminate in the splenic and superior mesenteric. The *nerves* are derived from the solar plexus.

Jejunum and Ileum.—The remainder of the small intestine from the termination of the duodenum is named *jejunum* and *ileum*; the former term being given to the upper two-fifths and the latter to the remaining three-fifths. There is no morphological line of distinction between the two, and the division is arbitrary; but at the same time it must be noted that the character of the intestine gradually undergoes a change from the commencement of the jejunum to the termination of the ileum, so that a portion of the bowel taken from these two situations would present characteristic and marked differences. These are briefly as follows :

The *jejunum*, which derives its name from the Latin word *jejunus* (empty), because it was formerly supposed to be empty after death, is wider, its diameter being about one inch and a half, and is thicker, more vascular, and of a deeper colour than the ileum, so that a given length weighs more. Its valvulæ conniventes are large and thickly set and its villi are larger than in the ileum. The glands of Peyer are almost absent in the upper part of the jejunum, and in the lower part are less frequently found than in the ileum, and are smaller and tend to assume a circular form. Brunner's glands are only found in the upper part of the jejunum. By grasping the jejunum between the finger and thumb the valvulæ conniventes can be felt through the walls of the gut ; these being absent in the lower part of the ileum, it is possible in this way to distinguish the upper from the lower part of the small intestine.

The *ileum*, so called from the Greek word εἰλεῖν (to twist), on account of its numerous coils and convolutions, is narrow, its diameter being one inch and a quarter, and its coats thinner and less vascular than those of the jejunum. It possesses but few valvulæ conniventes, and they are small and disappear entirely towards its lower end, but Peyer's patches are larger and more numerous. The jejunum for the most part occupies the umbilical and left iliac regions, while the ileum occupies chiefly the umbilical, hypogastric, right iliac, and pelvic regions, and terminates in the right iliac fossa by opening into the inner side of the commencement of the large intestine. The jejunum and ileum are attached to the posterior abdominal wall by an extensive fold of peritoneum, the *mesentery*, which allows the freest motion, so that each coil can accommodate itself to changes in form and position. The mesentery is fan-shaped ; its posterior border, about six inches in length, is attached to the abdominal wall from the left side of the second lumbar vertebræ to the right iliac fossa (fig. 483). Its length is about eight inches from its commencement to its termination at the intestine, and it is rather longer about its centre than at either end of the bowel. According to Lockwood it tends to increase in length as age advances. Between the two layers of which it is composed are contained blood-vessels, nerves, lacteals, and lymphatic glands, together with a variable amount of fat.

Meckel's diverticulum.—Occasionally there may be found connected with the lower part of the ileum, on an average of about three and a half feet from its termination, a blind diverticulum or tube, varying in length. It is attached to and communicates with the lumen of the bowel by one extremity, and by the other is unattached or may be connected with the abdominal wall or some other portion of the intestine by a fibrous band. This is Meckel's diverticulum, and represents the remains of the vitelline or omphalo-mesenteric duct, the duct of communication between the umbilical vesicle and the alimentary canal in early fœtal life.

Structure.—The wall of the small intestine is composed of four coats—serous, muscular, areolar, and mucous.

The *serous coat* is derived from the peritoneum. The first or ascending portion of the duodenum is almost completely surrounded by this membrane near its pyloric end, but only in front at the other extremity; the second or descending portion is covered by it in front, except where it is carried off by the transverse colon; and the third or transverse portion lies behind the peritoneum, which passes over it, without being closely incorporated with the other coats of this part of the intestine, and is separated from it in the middle line by the superior mesenteric artery. The remaining portion of the small intestine is surrounded by the peritoneum, excepting along its attached or mesenteric border; here a space is left for the vessels and nerves to pass to the gut.

The *muscular coat* consists of two layers of fibres, an external or longitudinal, and an internal or circular layer. The *longitudinal fibres* are thinly scattered over the surface of the intestine, and are more distinct along its free border. The *circular fibres* form a thick, uniform layer; they surround the cylinder of the intestine in the greater part of its circumference, and are composed of plain muscle-cells of considerable length. The muscular coat is thicker at the upper than at the lower part of the small intestine.

The *areolar* or *submucous coat* connects together the mucous and muscular layers. It consists of loose, filamentous areolar tissue, which forms a nidus for the subdivision of the nutrient vessels, previous to their distribution to the mucous surface.

The *mucous membrane* is thick and highly vascular at the upper part of the small intestine, but somewhat paler and thinner below. It consists of the following structures: next the areolar or submucous coat is a layer of unstriped muscular fibres, the *muscularis mucosæ*; internal to this is a quantity of retiform tissue, enclosing in its meshes lymph-corpuscles, and in which the blood-vessels and nerves ramify. Lastly, a basement-membrane, supporting a single layer of epithelial cells, which throughout the intestines are columnar in character. They are granular in appearance, and possess a clear oval nucleus. At their superficial or unattached end they present a distinct layer of highly refracting material, marked by vertical striæ, which were formerly believed to be minute channels, by which the chyle was taken up into the interior of the cell, and by them transferred to the lacteal vessels of the mucous membrane.

The mucous membrane presents for examination the following structures, contained within it or belonging to it:

Valvulæ conniventes.	Duodenal glands.
Villi.	Glands { Solitary glands.
Simple follicles.	Peyer's or Agminate glands.

The **valvulæ conniventes** (valves of Kerkring) are large folds or valvular flaps projecting into the lumen of the bowel. They are composed of reduplications or folds of the mucous membrane, the two layers of the fold being bound together by submucous tissue; they contain no muscular fibres, and, unlike the folds in the stomach, they are permanent, and are not obliterated when the intestine is distended. The majority extend transversely across the cylinder of the intestine for about one-half or two-thirds of its circumference, but some form complete circles, and others have a spiral direction; the latter usually extend a little more than once round the bowel, but occasionally two or three times. The spiral arrangement is the characteristic one of the shark family of fishes. The larger folds are about one-third of an inch in depth at their broadest part; but the greater number are of smaller size. The larger and smaller folds alternate with each other. They are not found at the commencement of the duodenum, but begin to appear about one or two inches beyond the pylorus. In the lower part of the descending portion, below the point where the bile and pancreatic ducts enter the intestine, they are very large and closely approximated. In the transverse portion of the duodenum and upper half of the jejunum they are large and numerous; and from this point, down to the middle of the ileum, they diminish considerably in size. In the lower part of the ileum they almost entirely disappear; hence the comparative thinness of this portion of the intestine, as compared with the duodenum and jejunum. The valvulæ conniventes retard the passage of the food along the intestines, and afford a more extensive surface for absorption.

The **villi** are minute, highly vascular processes, projecting from the mucous membrane of the small intestine throughout its whole extent, and giving to its

FIG. 490.—Diagrammatic section of a villus. (Watney.)

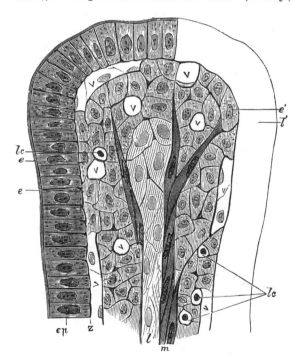

ep. Epithelium only partially shaded in. *l.* Central chyle-vessel; the cells forming the vessel have been less shaded to distinguish them from the cells of the parenchyma of the villus. *m.* Muscle-fibres running up by the side of the chyle-vessel. It will be noticed that each muscle-fibre is surrounded by the reticulum, and by this reticulum the muscles are attached to the cells forming the membrana propria, as at *e'*, or to the reticulum of the villus. *lc.* Lymph-corpuscles, marked by a spherical nucleus and a clear zone of protoplasm. *l'.* Upper limit of the chyle-vessel. *e, e, e'.* Cells forming the membrana propria. It will be seen that there is hardly any difference between the cells of the parenchyma, the endothelium of the *upper* part of the chyle-vessel, and the cells of the membrana propria. V. Blood-vessels. *z.* Dark line at the base of the epithelium formed by the reticulum. It will be seen that the reticulum penetrates between all the other elements of the villus. The reticulum contains thickenings or 'nodal points.' The diagram shows that the cells of the upper part of the villus are larger and contain a larger zone of protoplasm than those of the lower part. The cells of the upper part of the chyle-vessel differ somewhat from those of the lower part, in that they more nearly resemble the cells of the parenchyma.

surface a velvety appearance. In shape, according to Rauber, they are short and leaf-shaped in the duodenum, tongue-shaped in the jejunum, and filiform in the ileum. They are largest and most numerous in the duodenum and jejunum, and become fewer and smaller in the ileum. Krause estimates their number in the upper part of the small intestine at from fifty to ninety in a square line; and in the lower part from forty to seventy; the total number for the whole length of the intestine being about four millions.

Structure of the villi (fig. 490).—The structure of the villi has been studied by many eminent anatomists. The description here followed is that of Watney,[*] whose researches have an important bearing on the physiology of the absorption of fat, which is the peculiar function of this part of the intestine.

The essential parts of a villus are: the lacteal vessel, the blood-vessels, the epithelium, the basement-membrane, and muscular tissue of the mucosa, these structures being supported and held together by retiform lymphoid tissue.

These structures are arranged in the following manner. Situated in the centre of the villus is the lacteal, terminating near the summit in a blind extremity; running along this vessel are unstriped muscular fibres; surrounding it is a plexus of capillary vessels, the whole being enclosed by a basement-membrane, and covered by columnar epithelium. Those structures which are contained within the basement-membrane—namely, the lacteal, the muscular tissue, and the blood-vessels—are surrounded and enclosed by a delicate reticulum which forms the matrix of the villus, and in the meshes of which are found large flattened cells, with an oval nucleus, and, in smaller numbers, lymph-corpuscles. These

Fig. 491.—Villi of small intestine. (Cadiat.)

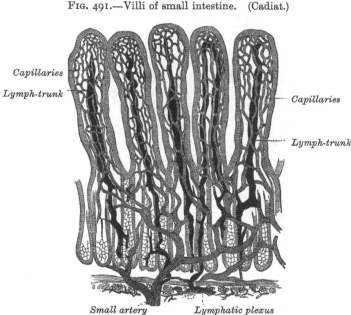

Capillaries

Lymph-trunk

Capillaries

Lymph-trunk

Small artery　　　　*Lymphatic plexus*

latter are to be distinguished from the larger cells of the villus by their behaviour with reagents, by their size, and by the shape of their nucleus, which is spherical. Transitional forms, however, of all kinds are met with between the lymph-corpuscle and the proper cells of the villus. Nerve-fibres are contained within the villi: they form ramifications throughout the reticulum.

The *lacteals* are in some cases double, and in some animals multiple. Situated in the axis of the villi, they commence by dilated cæcal extremities near to, but not quite at, the summit of the villus. The walls are composed of a single layer of endothelial cells, the interstitial substance between the cells being continuous with the reticulum of the matrix.

The *muscular fibres* are derived from the muscularis mucosæ, and are arranged in longitudinal bundles around the lacteal vessel, extending from the base to the summit of the villus, and giving off, laterally, individual muscle-cells,

* *Phil. Trans.* vol. clxvi. pt. ii.

which are enclosed by the reticulum, and by it are attached to the basement-membrane.

The *blood-vessels* form a plexus between the lacteal and the basement-membrane, and are enclosed in the reticular tissue. In the interstices of the capillary plexus, which they form, are contained the cells of the villus.

These structures are surrounded by the basement-membrane, which is made up of a stratum of endothelial cells, and upon which is placed a layer of columnar epithelium. The reticulum of the matrix is continuous through the basement-membrane (that is, through the interstitial substance between the individual endothelial cells) with the interstitial cement-substance of the columnar cells on the surface of the villus. Thus we are enabled to trace a direct continuity between the interior of the lacteal and the surface of the villus by means of the reticular tissue, and it is along this path that the chyle passes in the process of absorption by the villi. That is to say, it passes first of all into the columnar epithelial cells, and, escaping from them, is carried into the reticulum of the villus, and thence into the central lacteal.

FIG. 492.—Longitudinal section of crypts of Lieberkühn. Goblet-cells seen among the columnar epithelial cells. (Klein and Noble Smith.)

The **simple follicles,** or *crypts of Lieberkühn* (figs. 492, 493), are found in considerable numbers over every part of the mucous membrane of the small intestine. They consist of minute tubular depressions of the mucous membrane, arranged perpendicularly to the surface, upon which they open by small circular apertures. They may be seen with the aid of a lens, their orifices appearing as minute dots, scattered between the villi. Their walls are thin, consisting of a basement-membrane lined by columnar epithelium, and covered on their exterior by capillary vessels.

The **duodenal** or **Brunner's glands** are limited to the duodenum and commencement of the jejunum. They are small, flattened, granular bodies embedded in the submucous areolar tissue, and open upon the surface of the mucous membrane by minute excretory ducts. They are most numerous and largest near the pylorus. They are small compound acino-tubular glands, and much resemble the small glands

FIG. 493.—Transverse section of crypts of Lieberkühn. (Klein and Noble Smith.)

which are found in the mucous membrane of the mouth. They are believed by Watney to be direct continuations of the pyloric glands of the stomach. They consist of a number of tubular alveoli, lined by epithelium, and opening by a single duct on the inner surface of the intestine.

The **solitary glands** (*glandulæ solitariæ*) are found scattered throughout the mucous membrane of the small intestine, but are most numerous in the lower part of the ileum. They are small, round, whitish bodies, from half a line to a line in diameter. Their free surface is covered with villi, and each gland is surrounded by the openings of the follicles of Lieberkühn. They are now recognised as lymph-follicles, and consist of a dense interlacing retiform tissue closely packed with lymph-corpuscles, and permeated with an abundant capillary network (fig. 494). The interspaces of the retiform tissue are continuous with larger lymph-spaces at the base of the gland, through which they communicate with the lacteal system. They are situated partly in the submucous tissue, partly in the mucous membrane, where they form slight projections of its epithelial layer, after having penetrated the muscularis mucosæ. The villi which are situated on them are generally absent from the very summit (or 'cupola,' as Frey calls it) of the gland.

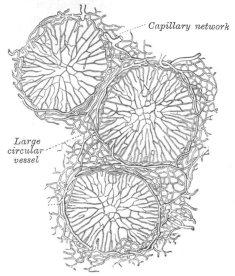

FIG. 494.—Transverse section through the equatorial plane of three of Peyer's follicles from the rabbit.

Capillary network

Large circular vessel

Peyer's glands (agminated glands) (figs. 494 to 497) may be regarded as aggregations of solitary glands, forming circular or oval patches from twenty to thirty in number, and varying in length from half an inch to four inches. They are largest and most numerous in the ileum. In the lower part of the jejunum they are small, of a circular form, and few in number. They are occasionally seen in the duodenum. They are placed lengthwise in the intestine, and are situated in the portion of the tube most distant from the attachment of the mesentery. Each patch is formed of a group of the above-described solitary glands covered with mucous membrane, and in almost every respect are similar in structure to them. They do not, however, as a rule, possess villi on their free surface. Each patch is surrounded by a circle of the crypts of Lieberkühn. They are best marked in the young subject, becoming indistinct in middle age, and sometimes altogether disappearing in advanced life. They are largely supplied with blood-vessels,

FIG. 495.—Patch of Peyer's glands. From the lower part of the ileum.

FIG. 496.—A portion of the above magnified.

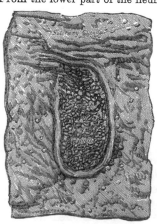

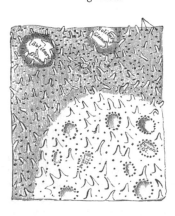

which form an abundant plexus around each follicle and give off fine branches which permeate the lymphoid tissue in the interior of the follicle. The lacteal plexuses which are found throughout the small intestine are especially abundant around these patches ; here they form rich plexuses with sinuses around the glands (fig. 497).

Vessels and Nerves.—The jejunum and ileum are supplied by the superior mesenteric artery, the branches of which, having reached the attached border

of the bowel, run between the serous and muscular coats, with frequent inosculations to the free border, where they also anastomose with other branches

FIG. 497.—Vertical section of one of Peyer's patches from man,
injected through its lymphatic canals.

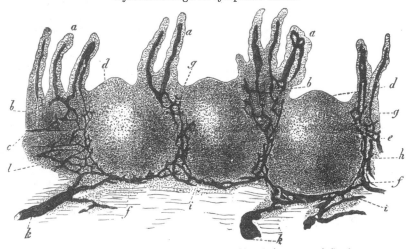

a. Villi with their chyle-passages. *b.* Follicles of Lieberkühn. *c.* Muscularis mucosæ. *d.* Cupola or apex of solitary
glands. *e.* Mesial zone of glands. *f.* Base of glands. *g.* Points of exit of the chyle-passages from the villi, and
entrance into the true mucous membrane. *h.* Retiform arrangement of the lymphatics in the mesial zone.
i. Course of the latter at the base of the glands. *k.* Confluence of the lymphatics opening into the vessels of the
submucous tissue. *l.* Follicular tissue of the latter.

FIG. 498.—Meissner's plexus.   (Klein and Noble Smith.)

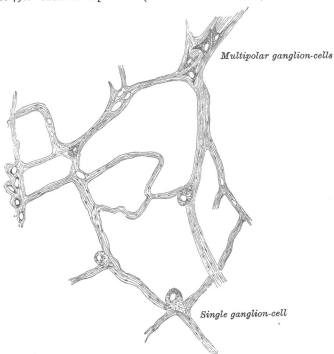

Multipolar ganglion-cells

Single ganglion-cell

running round the opposite surface of the gut. From these vessels numerous
branches are given off, which pierce the muscular coat, supplying it and forming
an intricate plexus in the submucous tissue. From this plexus minute vessels

pass to the glands and villi of the mucous membrane. The veins have a similar course and arrangement to the arteries. The *lymphatics of the small intestines* (lacteals) are arranged in two sets, those of the mucous membrane, and those of the muscular coat. The lymphatics of the villi commence in these structures in the manner described above, and form an intricate plexus in the mucous and sub-mucous tissue, being joined by the lymphatics from the lymph-spaces at the bases of the solitary glands, and from this pass to larger vessels at the mesenteric border of the gut. The lymphatics of the muscular coat are situated to a great extent between the two layers of muscular fibres, where they form a close plexus, and throughout their course communicate freely with the lymphatics from the mucous membrane, and empty themselves in the same manner into the commencement of the lacteal vessels at the attached border of the gut.

The *nerves of the small intestines* are derived from the plexuses of sympathetic nerves around the superior mesenteric artery. From this source they run to a plexus of nerves and ganglia situated between the circular and longitudinal muscular fibres (*Auerbach's plexus*), from which the nervous branches are distributed to the muscular coats of the intestine. From this plexus a secondary plexus is derived (*Meissner's plexus*), and is formed by branches which have perforated the circular muscular fibres (fig. 498). This plexus lies between the muscular and mucous coats of the intestine. It is also gangliated, and from it the ultimate fibres pass to the muscularis mucosæ and to the villi and mucous membrane.

THE LARGE INTESTINE

The large intestine extends from the termination of the ileum to the anus. It is about five feet in length, being one-fifth of the whole extent of the intestinal canal. It is largest at its commencement at the cæcum, and gradually diminishes

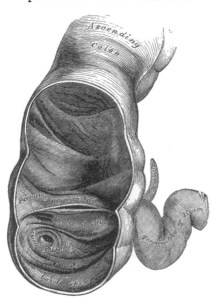

FIG. 499.—The cæcum and colon laid open to show the ileo-cæcal valve.

as far as the rectum, where there is a dilatation of considerable size just above the anus. It differs from the small intestine in its greater size, its more fixed position, its sacculated form, and in possessing certain appendages to its external coat, the *appendices epiploicæ*. Further, its longitudinal muscular fibres do not form a continuous layer around the gut but are arranged in three longitudinal bands or *tæniæ*. The large intestine, in its course, describes an arch, which surrounds the convolutions of the small intestine. It commences in the right inguinal region, in a dilated part, the *cæcum*. It ascends through the right lumbar and hypochondriac regions to the under surface of the liver; it here takes a bend (the *hepatic flexure*) to the left, and passes transversely across the abdomen on the confines of the epigastric and umbilical regions, to the left hypochondriac region; it then bends again (the *splenic flexure*), and descends through the left lumbar region to the left iliac fossa, where it becomes convoluted, and forms the sigmoid flexure; finally it enters the pelvis, and descends along its posterior wall to the anus. The large intestine is divided into the cæcum, colon, and rectum.

The **cæcum** (cæcus, *blind*) is the large blind pouch, or *cul-de-sac*, situated below the ileo-cæcal valve, in which the large intestine commences (fig. 499). Its blind end is directed downwards, and its open end upwards, communicating directly with the colon, of which this blind pouch appears to be the beginning or head, and hence the old name *caput cæcum coli* was applied to it. Its size is variously estimated by different authors, but on an average it may be said to be two and a half inches in length and three in breadth. It is situated in the right iliac fossa, above the outer half of Poupart's ligament: it rests on the Ilio-psoas muscle and lies immediately behind the abdominal wall. As a rule, it is entirely enveloped on all sides by peritoneum, but in a certain number of cases (6 per cent., Berry) the peritoneal covering is not complete, so that a small portion of the upper end of the posterior surface is uncovered and connected to the iliac fascia by connective tissue. The cæcum lies quite free in the abdominal cavity and enjoys a considerable amount of movement, so that it often becomes herniated down the right inguinal canal, and has occasionally been found in an inguinal hernia on the left side. The cæcum varies in shape, but, according to Treves, in man it may be classified under one of four types. In early fœtal life it is short, conical, and broad at the base, with its apex turned upwards and inwards towards the ileo-cæcal junction. It then resembles the cæcum of some of the monkey tribe, e.g. Mangabey monkey. As the fœtus grows the cæcum increases in length more than in breadth, so that it forms a longer tube than in the primitive form and without the broad base, but with the same inclination inwards of the apex towards the ileo-cæcal junction. This form is seen in others of the monkey tribe : e.g. the spider monkey. As development goes on, the lower part of the tube ceases to grow and the upper part becomes greatly increased, so that at birth there is a narrow tube, the vermiform appendix, hanging from a conical projection, the cæcum. This is the infantile form, and as it may persist throughout life, in about 2 per cent. of cases, it is regarded by Treves as the *first* of his four types of human cæca. The cæcum is conical and the appendix rises from its apex. The three longitudinal bands start from the appendix and are equidistant from each other. In the *second* type, the conical cæcum has become quadrate by the growing out of a saccule on either side of the anterior longitudinal band. These saccules are of equal size, and the appendix arises from between them, instead of from the apex of a cone. This type is found in about 3 per cent. of cases. The *third* type is the normal type of man. Here the two saccules, which in the second type were uniform, have grown at unequal rates : the right with greater rapidity than the left. In consequence of this an apparently new apex has been formed by the growing downwards of the right saccule, and the original apex, with the appendix attached, is pushed over to the left towards the ileo-cæcal junction. The three longitudinal bands still start from the base of the appendix, but they are now no longer equidistant from each other, because the right saccule has grown between the anterior and postero-external bands, pushing them over to the left. This type occurs in about 90 per cent. of cases. The *fourth* type is merely an exaggerated condition of the third ; the right saccule is still larger, and at the same time the left saccule has become atrophied, so that the original apex of the cæcum, with the appendix, is close to the ileo-cæcal junction, and the anterior band courses inwards to the same situation. This type is present in about 4 per cent. of cases.

The **vermiform appendix** is a long, narrow, worm-shaped tube, which starts from what was originally the apex of the cæcum, and may pass in several directions : upwards behind the cæcum ; to the left behind the ileum and mesentery ; or downwards and inwards into the true pelvis. It varies from one to nine inches in length, its average being about three inches. It is retained in position by a fold of peritoneum derived from the left leaf of the mesentery, which forms a mesentery for it. This is triangular in shape, but does not extend the whole length of the tube, but leaves the distal third free and completely covered by peritoneum. Between its two layers lies a considerable branch of the ileo-colic artery, the artery of the appendix. Its canal is small, extends throughout the whole length of the tube, and communicates with the cæcum by an orifice which is placed below and behind the ileo-cæcal opening. It is sometimes guarded, according to Gerlach, by a semilunar valve formed by a fold of mucous membrane, but this is by no means constant. Its coats are the same as those of the intestine: serous, muscular, submucous and mucous, the latter containing an abundant supply of retiform tissue, especially in young subjects.

It is stated that the vermiform appendix tends to undergo obliteration as an involution change of a functionless organ.

The **ileo-cæcal valve** (*valvula Bauhini*).—The lower end of the ileum terminates by opening into the inner and back part of the large intestine, at the point of junction of the cæcum with the colon. The opening is guarded by a valve, consisting of two semilunar segments, an upper or colic and lower or cæcal, which project into the lumen of the large intestine. The upper one, nearly horizontal in direction, is attached by its convex border to the point of junction of the ileum with the colon ; the lower segment, which is more concave and longer, is attached to the point of junction of the ileum with the cæcum. At each end of the aperture the two segments of the valve coalesce, and are continued as a narrow membranous ridge around the canal for a short distance, forming the *fræna* or *retinacula* of the valve. The left or anterior end of the aperture is rounded ; the right or posterior is narrow and pointed.

Each segment of the valve is formed by a reduplication of the mucous membrane and of the circular muscular fibres of the intestine, the longitudinal fibres and peritoneum being continued uninterruptedly across from one portion of the intestine to the other. When these are divided or removed, the ileum may be drawn outwards, and all traces of the valve will be lost, the ileum appearing to open into the large intestine by a funnel-shaped orifice of large size.

The surface of each segment of the valve directed towards the ileum is covered with villi, and presents the characteristic structure of the mucous membrane of the small intestine ; while that turned towards the large intestine is destitute of villi, and marked with the orifices of the numerous tubular glands peculiar to the mucous membrane of the large intestine. These differences in structure continue as far as the free margin of the valve.

When the cæcum is distended, the margins of the opening are approximated so as to prevent any reflux into the ileum.

The **colon** is divided into four parts—the ascending, transverse, descending, and the sigmoid flexure.

The **ascending colon** is smaller than the cæcum, with which it is continuous. It passes upwards, from its commencement at the cæcum, opposite the ileo-cæcal valve, to the under surface of the right lobe of the liver, on the right of the gall-bladder, where it is lodged in a shallow depression, the impressio colica ; he e it bends abruptly inwards to the left, forming the *hepatic flexure*. It is retained in contact with the posterior wall of the abdomen by the peritoneum, which covers its anterior surface and sides, its posterior surface being connected by loose areolar tissue with the Quadratus lumborum and Transversalis muscles, and with the front of the lower and outer part of the right kidney (figs. 500 and 501). Sometimes the peritoneum almost completely invests it, and forms a distinct but narrow mesocolon.* It is in relation, in front, with the convolutions of the ileum and the abdominal parietes.

* Treves states that, after a careful examination of one hundred subjects, he found that in fifty-two there was neither an ascending nor a descending mesocolon. In twenty-two there was a descending mesocolon, but no trace of a corresponding fold on the other side. In fourteen subjects there was a mesocolon to both the ascending and the descending segments of the bowel ; while in the remaining twelve there was an ascending mesocolon, but no corresponding fold on the left side. It follows, therefore, that in performing lumbar colotomy a mesocolon may be expected upon the left side in 36 per cent. of all cases, and on the right in 26 per cent. (*The Anatomy of the Intestinal Canal and Peritoneum in Man*, 1885, p. 55.)

The **transverse colon,** the longest part of the large intestine, passes transversely from right to left across the abdomen, opposite the confines of the

FIG. 500.—Diagram of the relations of the large intestine and kidney, from behind.

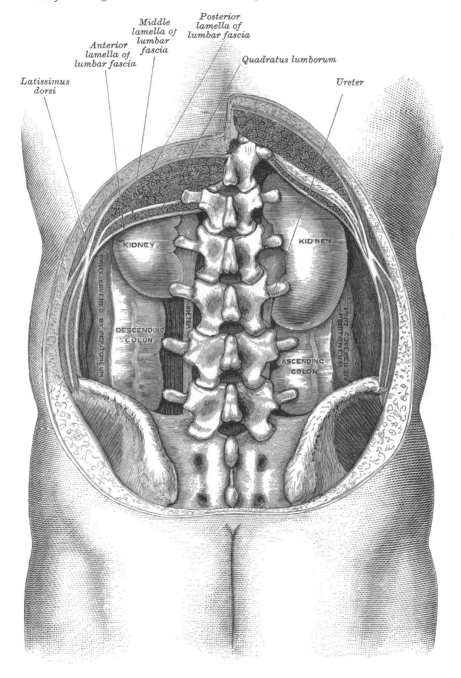

epigastric and umbilical zones, into the left hypochondriac region, where it curves downwards beneath the lower end of the spleen, forming the *splenic flexure*. In its course it describes an arch, the concavity of which is directed backwards towards the vertebral column and a little upwards; hence the name *transverse arch of the colon*. This is the most movable part of the colon, being almost completely invested by peritoneum, and connected to the spine behind by a large and wide duplicature of that membrane, the *transverse mesocolon*. It is in relation, by its upper surface, with the liver and gall-bladder, the great curvature of the stomach, and the lower end of the spleen; by its under surface, with the small intestines; by its anterior surface, with the anterior layers of the great omentum and the abdominal parietes; its posterior surface on the right side is in relation with the second portion of the duodenum, and on the left is in contact with some of the convolutions of the jejunum and ileum.

The **descending colon** passes downwards through the left hypochondriac and lumbar regions along the outer border of the left kidney. At the lower end of the kidney it turns inwards towards the outer border of the Psoas muscle, along which it descends to the crest of the ileum, where it terminates in the sigmoid flexure. At its commencement it is connected with the Diaphragm by a fold of peritoneum, the *phreno-colic* ligament (see page 884). It is retained in position by the peritoneum, which covers its anterior surface and sides, its posterior surface being connected by areolar tissue with the outer border of the left kidney, and the Quadratus lumborum and Transversalis muscles (figs. 500, 501). It is smaller in calibre and more deeply placed than the ascending colon, and is more frequently covered with peritoneum on its posterior surface than the ascending colon (Treves).

The **sigmoid flexure** is the narrowest part of the colon : it is situated in the left iliac fossa, commencing from the termination of the descending colon, at the margin of the crest of the ileum, and ending in the rectum at the brim of the true pelvis opposite the left sacro-iliac symphysis. It curves in the first place forwards, downwards, and inwards for about two inches and then forms a loop, which varies in length and position and which terminates in the rectum.* The first portion is in close relation with the iliac fascia, and is covered by peritoneum on its sides and anterior surface only. The loop is entirely surrounded by peritoneum, and is retained in its place by a loose fold of peritoneum, the *sigmoid mesocolon*, which connects it to the Psoas muscle. This loop, which normally hangs downwards, sometimes into the true pelvis, is very movable, and may be displaced upwards in cases of distension of the pelvic viscera. The sigmoid flexure is in relation in front with the small intestines and abdominal parietes. The sigmoid mesocolon is attached to a line running downwards and inwards from the crest of the ileum, across the Psoas muscle, to become continuous with the *mesorectum* near the bifurcation of the common iliac artery (fig. 483). In its left layer is the intersigmoid fossa (see page 887).

The **rectum** is the terminal part of the large intestine, and extends from the sigmoid flexure to the anal orifice. The superior limit cannot be determined precisely, since there is no point of demarcation between the sigmoid flexure and the first part of the rectum; but the brim of the true pelvis, opposite the left sacro-iliac joint, is arbitrarily given as its point of commencement. From this point it passes downwards, backwards, and to the right to the level of the third sacral vertebra, where it lies in the middle line. This is the *first part* of the rectum. The *second part* curves forwards and is continued downwards as far as the apex of the prostate gland, about an inch in front of the tip of the coccyx. From this point the bowel is directed backwards, and, passing downwards, terminates at the anal orifice. This is the third portion of the rectum, or, as described by Symington, the *anal canal*. It will be seen, therefore, that the rectum presents two antero-posterior curves : the first, with its convexity back-

---

* Treves describes the sigmoid flexure somewhat differently. He includes in his description of this portion of the bowel the upper part of the rectum, and makes it terminate opposite the third portion of the sacrum. Instead of forming a sigmoid curve, he describes it as a large loop or bend, more like the Greek letter Ω (omega).

wards, is due to the conformation of the sacro-coccygeal column, and represents the arc of a circle, the centre of which is opposite the third sacral vertebra.   The

FIG. 501.—The relations of the viscera and large vessels of the abdomen. (Seen from behind, the last dorsal vertebra being well raised.)

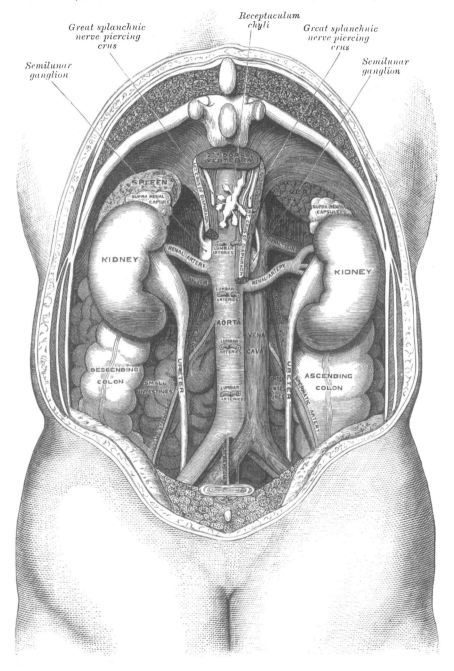

lower one has its convexity forwards, and is angular.   Its centre corresponds to a line drawn between the anterior parts of the ischial tuberosities.   Two lateral curves are also described: the one to the right, opposite the junction of the third and fourth sacral vertebræ; the other to the left, opposite the sacro-coccygeal articulation. They are of little importance.

The length of the rectum is about eight inches. The first part is four inches, the second three, and the third one to one and a half, being rather longer in the male than in the female. The rectum is narrower in its upper part than the sigmoid flexure, but is capable of considerable distension. In the lower part of the second portion it becomes a transverse slit, its anterior and posterior walls lying close together when the tube is empty, on account of the organs in the front part of the pelvis pushing the rectum backwards on the sacrum and coccyx. The third part of the rectum, the anal canal, is also a slit, with, however, an antero-posterior direction, so that its lateral walls are in apposition (fig. 502).

The first portion of the rectum is almost completely surrounded by peritoneum, and is connected to the anterior surface of the sacrum by a double fold, called the *mesorectum*, which is continuous above with the sigmoid mesocolon.

FIG. 502.—Coronal section through the anal canal. (Symington.)

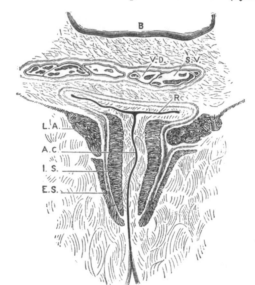

B. Cavity of bladder.   VD. Vas deferens.   SV. Seminal vesicle.   R. Second part of rectum.   AC. Anal canal.
LA. Levator ani.   IS. Internal sphincter.   ES. External sphincter.

The mesorectum is triangular in shape, the apex of which ends below at the third sacral vertebræ; between its two layers is the superior hæmorrhoidal artery. The second portion has no mesorectum, but is covered in front and laterally by peritoneum at its upper part; gradually the peritoneum leaves its sides, and about an inch above the prostate is reflected from the anterior surface of the bowel on to the posterior wall of the bladder in the male, and the upper fifth of the posterior wall of the vagina in the female, forming the recto-vesical and recto-vaginal pouches respectively. The third portion of the rectum has no peritoneal covering. The level at which the peritoneum leaves the anterior wall of the rectum to be reflected on to the viscus in front of it is of considerable importance from a surgical point of view, in connection with removal of the lower part of the rectum. It is higher in the male than in the female. In the former the height of the recto-vesical pouch is about three inches, that is to say, the height to which an ordinary index finger can reach from the anus. In the female the height of the recto-vaginal pouch is about $2\frac{1}{4}$ inches from the anal orifice.

The *first portion* of the rectum is in relation, behind, with the mesorectum and the superior hæmorrhoidal artery, the left Pyriformis muscle, and left sacral plexus of nerves, which separate it from the anterior surface of the upper sacral vertebræ ; to its left side are the branches of the left internal iliac artery and the left ureter ; in front it is separated, in the male, from the posterior surface of the bladder ; in the female, from the posterior surface of the uterus and its appendages, by some convolutions of the small intestine, and frequently by the sigmoid flexure of the colon. The *second portion* of the rectum is in relation, in front, in the male, with the recto-vesical pouch, the triangular portion of the base of the bladder, the vesiculæ seminales, and vasa differentia, and more anteriorly with the under surface of the prostate. In the female, with the posterior wall of the vagina below, and the recto-vaginal pouch above, in which are some convolutions of small intestine. The *third portion* or *anal canal* is invested by the Internal sphincter, supported by the Levatores ani muscles, and surrounded at its termination by the External sphincter ; in the empty condition it presents the appearance of a longitudinal slit. In the male it is separated from the membranous portion and bulb of the urethra by a triangular space ; and in the female it is separated from the lower end of the vagina by the perineal body. Laterally is the fat in the ischio-rectal fossæ.

Structure.—The large intestine has four coats—serous, muscular, areolar, and mucous.

The *serous coat* is derived from the peritoneum, and invests the different portions of the large intestine to a variable extent. The cæcum is completely covered by the serous membrane, except in a small percentage of cases (5 or 6 per cent.), where a small portion of the upper end of the posterior surface is uncovered. The ascending and descending colon are usually covered only in front and at the sides ; a variable amount of the posterior surface is uncovered.* The transverse colon is almost completely invested, the parts corresponding to the attachment of the great omentum and transverse mesocolon being alone excepted. The sigmoid flexure is completely surrounded, except along the line to which the sigmoid mesocolon is attached. The upper part of the rectum is completely invested by the peritoneum, except along the attachment of the mesorectum ; the middle portion is covered only on its anterior surface, and part of its sides in the upper portion ; and the lower portion is entirely devoid of any serous covering. In the course of the colon and upper part of the rectum, the peritoneal coat is thrown into a number of small pouches filled with fat, called *appendices epiploicæ.* They are chiefly appended to the transverse colon.

The *muscular coat* consists of an external longitudinal and an internal circular layer of muscular fibres.

The *longitudinal fibres*, although found to a certain extent all round the intestine, do not form a uniform layer over the whole surface of the large intestine. In the cæcum and colon they are especially collected into three flat longitudinal bands or tæniæ, each being about half an inch in width. These bands commence at the attachment of the vermiform appendix, which is surrounded by a uniform layer of longitudinal muscular fibres, to the cæcum : one, the posterior, is placed along the attached border of the intestine ; the anterior, the largest, corresponds along the arch of the colon to the attachment of the great omentum, but is in front in the ascending and descending colon and sigmoid flexure ; the third, or lateral band, is found on the inner side of the ascending and descending colon, and on the under aspect of the transverse colon. These bands are nearly one-half shorter than the other coats of the intestine, and serve to produce the sacculi which are characteristic of the cæcum and colon ; accordingly, when they are dissected off, the tube can be lengthened, and its sacculated character becomes lost. In the sigmoid flexure the longitudinal fibres become more scattered ; but upon its lower part, and round the rectum, they spread out and form a layer,

* See footnote, p. 906.

which completely encircles this portion of the gut, but is thicker on the anterior and posterior surfaces, where it forms two bands, than on the lateral surfaces. In addition to the muscular fibres of the bowels, two bands of plain muscular tissue arise from the second and third coccygeal vertebræ, and pass downwards and forwards to blend with the longitudinal muscular fibres on the posterior wall of the anal canal. These are known as the *recto-coccygeal* muscles.

The *circular fibres* form a thin layer over the cæcum and colon, being especially accumulated in the intervals between the sacculi; in the rectum they form a thick layer, especially at its lower end, where they become numerous, and constitute the Internal sphincter.

The *areolar coat* connects the muscular and mucous layers closely together.

The *mucous membrane*, in the cæcum and colon, is pale, smooth, destitute of villi, and raised into numerous crescentic folds which correspond to the intervals between the sacculi. In the rectum it is thicker, of a darker colour, more vascular, and connected loosely to the muscular coat, as in the œsophagus. When the lower part of the rectum is contracted, its mucous membrane is thrown into a number of folds, some of which, near the anus, are longitudinal in direction, and are effaced by the distension of the gut. Besides these there are certain permanent folds, of a semilunar shape, known as Houston's valves.* They are usually three in number; sometimes a fourth is found, and occasionally only two are present. One is situated near the commencement of the rectum, on the right side; another extends inwards from the left side of the tube, opposite the middle of the sacrum; the largest and most constant one projects backwards from the fore part of the rectum, opposite the base of the bladder. When a fourth is present, it is situated about an inch above the anus on the back of the rectum. These folds are about half an inch in width, and contain some of the circular fibres of the gut. In the empty state of the intestine they overlap each other, as Houston remarks, so effectually as to require considerable manœuvring to conduct a bougie or the finger along the canal of the intestine. Their use seems to be, 'to support the weight of fæcal matter, and prevent its urging towards the anus, where its presence always excites a sensation demanding its discharge.'

As in the small intestine, the mucous membrane consists of a muscular layer, the muscularis mucosæ; of a quantity of retiform tissue in which the vessels ramify; of a basement-membrane and epithelium, which is of the columnar variety, and exactly resembles the epithelium found in the small intestine. The mucous membrane of this portion of the bowel presents for examination simple follicles and solitary glands.

The *simple follicles* are minute tubular prolongations of the mucous membrane arranged perpendicularly, side by side, over its entire surface; they are longer, more numerous, and placed in much closer apposition than those of the small intestine; and they open by minute rounded orifices upon the surface, giving it a cribriform appearance.

The *solitary glands* (fig. 503) in the large intestine are most abundant in the *cæcum* and *vermiform appendix*, but are irregularly scattered also over the rest of the intestine. They are similar to those of the small intestine.

Vessels and Nerves.—The arteries supplying the large intestine give off large branches, which ramify between the muscular coats, supplying them, and, after dividing into small vessels in the submucous tissue, pass to the mucous membrane. The rectum is supplied mainly by the superior hæmorrhoidal branch of the inferior mesenteric, but also at its lower end by the middle hæmorrhoidal from the internal iliac, and the inferior hæmorrhoidal from the pudic artery. The superior hæmorrhoidal, the continuation of the superior mesenteric, divides into two branches, which run down either side of the rectum to within about five inches of the anus; they here split up into about six branches, which pierce the

* *Dublin Hosp. Reports*, vol. v. p. 163.

muscular coat and descend between it and the mucous membrane in a longitudinal direction, parallel with each other as far as the Internal sphincter, where they anastomose with the other hæmorrhoidal arteries and form a series of loops around the anus. The veins of the rectum commence in a plexus of vessels which surrounds the lower extremity of the intestinal canal. In the vessels forming this plexus are small saccular dilatations just within the margin of the anus ; from it about six vessels of considerable size are given off. These ascend between the muscular and mucous coats for about five inches, running parallel to each other ; they then pierce the muscular coat, and, by their union, form a single trunk, the superior hæmorrhoidal vein. This arrangement is termed the *hæmorrhoidal plexus* ; it communicates with the tributaries of the middle and inferior hæmorrhoidal veins at its commencement, and thus a communication is established between the systemic and portal circulations. The nerves are derived from the plexuses of the sympathetic nerve around the branches of the superior and inferior mesenteric arteries that are distributed to the large intestine. They are distributed in a similar way to those in the small intestine. The lymphatic vessels of the large intestine are found in the submucosa, where they form a wide meshed network, and also, more deeply seated, beneath the simple follicles. Those

FIG. 503.—Minute structure of large intestine.

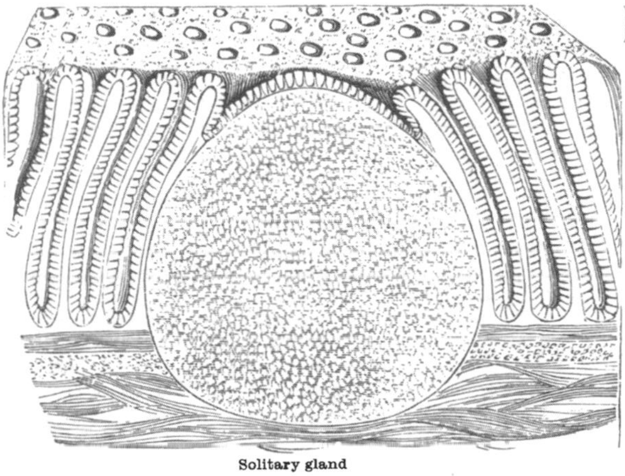

Surface of mucous membrane with opening of Lieberkühn's follicles

Lieberkühn's follicles

Muscularis mucosæ (two layers

Submucous connective tissue

Solitary gland

from the colon open into the mesenteric glands ; those from the sigmoid flexure into the lumbar glands ; those from the rectum enter the glands which are situated in the hollow of the sacrum ; and those around the anus open into the glands in the groin.

Surface Form.—The coils of the small intestine occupy the front of the abdomen, below the transverse colon, and are covered more or less completely by the great omentum. For the most part the coils of the jejunum occupy the left side of the abdominal cavity, i.e. the left lumbar and inguinal regions and the left half of the umbilical region ; while the coils of the ileum are situated to the right, in the right lumbar and inguinal regions, in the right half of the umbilical region, and also the hypogastric. The cæcum is situated in the right inguinal region. Its position varies slightly, but the mid-point of a line drawn from the anterior superior spinous process of the ilium to the symphysis pubis will about mark the middle of its lower border. It is comparatively superficial. From it the ascending colon passes upwards through the right lumbar and hypochondriac regions, and becomes more deeply situated as it ascends to the hepatic flexure, which is deeply placed, under cover of the liver. The transverse colon crosses the belly transversely on the confines of the umbilical and epigastric regions ; its lower border being on a level slightly above the umbilicus, its upper border just below the greater curvature of the stomach. The splenic flexure of the colon is situated behind the stomach in the left hypochondrium, and is on a higher level than the hepatic flexure. The descending colon is deeply seated, passing down through the left hypochondriac and lumbar regions to the sigmoid flexure, which is situated in the left inguinal region and can be felt in thin persons, with relaxed abdominal walls, rolling under the fingers when empty, and when distended forming a distinct tumour. The position of the base of the vermiform appendix is indicated by a point two inches from the anterior superior spinous process of the ilium, in a line drawn from this process to the umbilicus. This is known as *McBurney's spot*. Another mode of defining the position of the base of the appendix is to draw a line between the anterior superior spines of the ilia and marking the point where this line intersects the right semilunar line.

Upon introducing the finger into the rectum, the membranous portion of the urethra can be felt, if an instrument has been introduced into the bladder, exactly in the middle line ; behind this the prostate gland can be recognised by its shape and hardness and any enlargement detected ; behind the prostate the fluctuating wall of the bladder when full can be felt, and if thought desirable it can be tapped in this situation ; on either side and behind the prostate the vesiculæ seminales can be readily felt, especially if enlarged by tuberculous disease. Behind, the coccyx is to be felt, and on the mucous membrane one or two of Houston's folds. The ischio-rectal fossæ can be explored on either side, with a view to ascertaining the presence of deep-seated collections of pus. Finally, it will be noted that the finger is firmly gripped by the sphincter for about an inch up the bowel. By gradual dilatation of the sphincter, the whole hand can be introduced into the rectum so as to reach the descending colon. This method of exploration is rarely, however, required for diagnostic purposes.

Surgical Anatomy.—The small intestines are much exposed to injury, but, in consequence of their elasticity and the ease with which one fold glides over another, they are not so frequently ruptured as would otherwise be the case. Any part of the small intestine may be ruptured, but probably the most common situation is the transverse duodenum, on account of its being more fixed than other portions of the bowel, and because it is situated in front of the bodies of the vertebræ, so that if this portion of the intestine is struck by a sharp blow, as from the kick of a horse, it is unable to glide out of the way, but is compressed against the bone and so lacerated. Wounds of the intestine sometimes occur. If the wound is a small puncture, under, it is said, three lines in length, no extravasation of the contents of the bowel takes place. The mucous membrane becomes everted and plugs the little opening. The bowels, therefore, may be safely punctured with a fine capillary trocar, in cases of excessive distension of the intestine with gas, without fear of extravasation. A longitudinal wound gapes more than a transverse, owing to the greater amount of circular muscular fibres. The small intestine, and most frequently the ileum, may become strangulated by internal bands, or through apertures, normal or abnormal. The bands may be formed in several different ways : they may be old peritoneal adhesions from previous attacks of peritonitis ; or an adherent omentum from the same cause ; or the band may be formed by Meckel's diverticulum, which has contracted adhesions at its distal extremity ; or it may be the result of the abnormal attachment of some normal structure, as the adhesion of two appendices epiploicæ, or an adherent vermiform appendix or Fallopian tube. Intussusception or invagination of the small intestine may take place in any part of the jejunum and ileum, but the most frequent situation is at the ileo-cæcal valve, the valve forming the apex of the entering tube. This form may attain great size, and it is not uncommon in these cases to find the valve projecting from the anus. Stricture, the impaction of foreign bodies, and twisting of the gut (*volvulus*) may lead to intestinal obstruction.

Resection of a portion of the intestine may be required in cases of gangrenous gut ; in cases of intussusception ; for the removal of new growth in the bowel ; in dealing with artificial anus ; and in cases of rupture. The operation is termed *enterectomy*, and is performed as follows : the abdomen having been opened and the amount of bowel requiring removal having been determined upon, the gut must be clamped on either side of this portion in order to prevent the escape of any of the contents of the bowel during the operation. The portion of bowel is then separated above and below by means of scissors. If the portion removed is small, it may be simply removed from the mesentery at its attachment and the bleeding vessels tied ; but if it is large it will be necessary to remove also a triangular piece of the mesentery, and, having secured the vessels, suture the cut edges of this structure together. The surgeon then proceeds to unite the cut ends of the bowel together by the operation of what is termed end-to-end anastomosis. There are many ways of doing this, which may be divided into two classes : one where the anastomosis is made by means of some mechanical appliance, such as Murphy's button, or one of the forms of decalcified bone bobbins ; and the other, where the operation is performed by suturing the ends of the bowel in such a manner that the peritoneum covering the free divided ends of the bowel is brought into contact, so that speedy union may ensue.

The vermiform appendix is very liable to become inflamed. This condition may be set up by the appendix becoming twisted, owing to the shortness of its mesentery, in consequence of distension of the cæcum. As the result of this its blood supply, which is mainly through one large artery running in the mesentery, becomes interfered with. Again, in rarer cases, the inflammation is set up by the impaction of a solid mass of fæces or a foreign body in it. The inflammation may result in ulceration and perforation, or if the torsion is very acute in gangrene of the appendix. These conditions may require operative interference, and in cases of recurrent attacks of appendicitis it is generally advisable to remove this diverticulum of the bowel. In external hernia the ileum is the portion of bowel most frequently herniated. When a part of the large intestine is involved it is usually the cæcum, and this may occur even on the left side. In some few cases the vermiform appendix has been the part implicated in cases of strangulated hernia, and has given rise to serious symptoms of obstruction. Occasionally ulceration of the duodenal glands may occur in cases of burns, but is not a very common complication. The ulcer may perforate one of the large duodenal vessels and may cause death from hæmorrhage, or it may perforate the coats of the intestine and produce fatal septic peritonitis. The diameter of the large intestine gradually diminishes from the cæcum, which has the greatest diameter of any part of the bowel, to the point of junction of the sigmoid flexure with the rectum, at or a little below which point stricture most commonly occurs, and diminishes in frequency as one proceeds upwards to the cæcum. When distended by some obstruction low down, the outline of the large intestine can be defined throughout nearly the whole of its course—all, in fact, except the hepatic and splenic flexures, which are more deeply placed ; the distension is most obvious in the two flanks and on the front of the abdomen just above the umbilicus. The cæcum, however, is that portion of the bowel which is, of all, most distended. It sometimes assumes enormous dimensions, and has been known to give way from the distension, causing fatal peritonitis. The hepatic flexure and the right extremity of the transverse colon are in close relationship with the liver, and abscess of this viscus sometimes bursts into the gut in this situation. The gall-bladder may become adherent to the colon, and gall-stones may find their way through into the gut, where they may become impacted or may be discharged per anum. The mobility of the sigmoid flexure renders it more liable to become the seat of a volvulus or twist than any other part of the intestine. It generally occurs in patients who have been the subjects of habitual constipation, and in whom therefore the mesosigmoid flexure is elongated. The gut at this part being loaded with fæces, from its weight falls over the gut below, and so gives rise to the twist.

The surgical anatomy of the rectum is of considerable importance. There may be congenital malformations due to arrest or imperfect development. Thus, there may be no inflection of the epiblast (see page [135]), and consequently a complete absence of the anus ; or the hind-gut may be imperfectly developed, and there may be an absence of the rectum, though the anus is developed ; or the inflection of the epiblast may not communicate with the termination of the hind-gut from want of solution of continuity in the septum which in early foetal life exists between the two. The mucous membrane is thick and but loosely connected to the muscular coat beneath, and thus favours prolapse, especially in children. The vessels of the rectum are arranged, as mentioned above, longitudinally, and are contained in the loose cellular tissue between the mucous and muscular coats, and receive no support from surrounding tissues, and this favours varicosity. Moreover, the veins, after running upwards in a longitudinal direction for about five inches in the submucous tissue, pierce the muscular coats, and are liable to become constricted at this spot by the contraction of the muscular wall of the gut. In addition to this there are no valves in the superior hæmorrhoidal veins, and the vessels of the rectum are placed in a dependent position, and are liable to be pressed upon and obstructed by hardened fæces. The anatomical arrangement, therefore, of the hæmorrhoidal vessels explains the great tendency to the occurrence of piles. The presence of the Sphincter ani is of surgical importance, since it is the constant contraction of this muscle which prevents an ischio-rectal abscess from healing, and causes it to become a fistula. Also the reflex contraction of this muscle is the cause of the severe pain complained of in fissure of the anus. The relations of the peritoneum to the rectum are of importance in connection with the operation of removal of the lower end of the rectum for malignant disease. This membrane gradually leaves the rectum as it descends into the pelvis ; first leaving its posterior surface, then the sides, and then the anterior surface, to become reflected, in the male on to the posterior wall of the bladder, forming the recto-vesical pouch, and in the female on to the posterior wall of the vagina, forming Douglas's pouch. The recto-vesical pouch of peritoneum extends to within three inches from the anus, so that it is not desirable to remove more than two and a half inches of the entire circumference of the bowel for fear of the risk of opening the peritoneum. When, however, the disease is confined to the posterior surface of the rectum, or extends farther in this direction, a greater amount of the posterior wall of the gut may be removed, as the peritoneum does not extend on this surface to a lower level than five inches from the margin of the anus. The recto-vaginal or Douglas's pouch in the female extends somewhat lower than the recto-vesical pouch of the male, and therefore it is necessary to remove a less length of the tube in this sex. Of recent years, however, much more extensive operations have been done for the removal of cancer of the rectum, and in these the peritoneal cavity has necessarily been opened. If, in these cases, the opening is plugged with antiseptic wool until the operation is completed and then the edges of the wound in the peritoneum accurately brought together with sutures, no evil result appears to follow. For cases of cancer of the rectum which are too low to be reached by abdominal section, and too high to be removed by the ordinary operation from below, Kraske has devised an operation which goes by his name. The patient is placed on his right side and an incision is made from the second sacral spine to the anus. The soft parts are now separated from the back of the left side of the sacrum as far as its left margin, and the greater and lesser sacro-sciatic ligaments are divided. A portion of the lateral mass of the sacrum, commencing on the left border at the level of the third posterior sacral foramen, and running downwards and inwards through the fourth foramen to the cornu, is now cut away with a chisel. The left side of the wound being now forcibly drawn outwards, the whole of the rectum is brought into view, and the diseased portion can be removed, leaving the anal portion of the gut, if healthy. The two divided ends of the gut can then be approximated and sutured together in front, the posterior part being left open for drainage.

The colon frequently requires opening in cases of intestinal obstruction, and by some surgeons this operation is performed in cases of cancer of the rectum, as soon as the disease is recognised, in the hope that the rate of growth may be retarded by removing the irritation produced by the passage of fæcal matter over the diseased surface. The operation of colotomy may be performed either in the inguinal or lumbar region; but inguinal colotomy has in the present day almost superseded the lumbar operation. The main reason for preferring this operation is that a spur-shaped process of the mesocolon can be formed which prevents any fæcal matter finding its way past the artificial anus, and becoming lodged on the diseased structures below. The sigmoid flexure being almost entirely surrounded by peritoneum, a coil can be drawn out of the wound and the greater part of its calibre removed, leaving the remainder attached to the mesocolon, which forms a spur, much the same as in an artificial anus caused by sloughing of the gut after a strangulated hernia, and this prevents any fæcal matter finding its way from the gut above the opening into that below. The operation is performed by making an incision two or three inches in length from a point one inch internal to the anterior superior spinous process of the ilium, parallel to Poupart's ligament. The various layers of abdominal muscles are cut through, and the peritoneum opened and sewn to the external skin. The sigmoid flexure is now sought for, and pulled out of the wound and fixed by passing a needle threaded with carbolised silk through the mesocolon close to the gut, and then through the abdominal wall. The intestine is now sewn to the skin all round, the sutures passing only through the serous and muscular coats. The wound is dressed, and on the second to the fourth day, according to the requirements of the case, the protruded coil of intestine is opened and removed with scissors.

Lumbar colotomy is performed by placing the patient on the side opposite to the one to be operated on, with a firm pillow under the loin. A line is then drawn from the anterior superior to the posterior superior spine of the ilium, and the mid-point of this line (Heath) or half an inch behind the mid-point (Allingham) is taken, and a line drawn vertically upwards from it to the last rib. This line represents, with sufficient correctness, the position of the normal colon. An oblique incision four inches in length is now made midway between the last rib and the crest of the ilium, so that its centre bisects the vertical line, and the following parts successively divided: (1) The skin, superficial fascia, with cutaneous vessels and nerves and deep fascia   2) The posterior fibres of the External oblique and anterior fibres of the Latissimus dorsi.   3) The Internal oblique.   (4) The lumbar fascia and the external border of the Quadratus lumborum. The edges of the wound are now to be held apart with retractors, and the transversalis fascia will be exposed. This is to be opened with care, commencing at the posterior angle of the incision. If the bowel is distended, it will bulge into the wound, and no difficulty will be found in dealing with it. If, however, the gut is empty, this bulging will not take place, and the colon will have to be sought for. The guides to it are the lower end of the kidney, which will be plainly felt, and the outer edge of the Quadratus lumborum. The bowel having been found, is to be drawn well up into the wound, and it may be opened at once and the margins of the opening stitched to the skin at the edge of the wound; or, if the case is not an urgent one, it may be retained in this position by two harelip pins passed through the muscular coat, the rest of the wound closed, and the bowel opened in three or four days, when adhesion of the bowel to the edges of the wound has taken place.

THE LIVER

The **Liver** is the largest gland in the body, and is situated in the upper and right part of the abdominal cavity, occupying almost the whole of the right hypochondrium, the greater part of the epigastrium, and extending into the left hypochondrium as far as the mammary line. In the male it weighs from fifty to sixty ounces, in the female from forty to fifty. It is relatively much larger in the fœtus than in the adult, constituting, in the former, about one-eighteenth, and in the latter, about one thirty-sixth of the entire body weight. Its greatest transverse measurement is from eight to nine inches. Vertically, near its lateral or right surface, it measures about six or seven inches, while its greatest antero-posterior diameter is on a level with the upper end of the right kidney and is from four to five inches. Opposite the vertebral column its measurement from before backwards is reduced to about three inches. Its consistence is that of a soft solid ; it is, however, friable and easily lacerated ; its colour is a dark reddish-brown, and its specific gravity is 1·05.

To obtain a correct idea of its shape it must be hardened *in situ*, and it will then be seen to present the appearance of a wedge, the base of which is directed to the right and the thin edge towards the left. Symington describes its shape as that ' of a right-angled triangular prism with the right angles rounded off.' It possesses five surfaces, viz. : superior, inferior, anterior, posterior, and lateral.

The superior and anterior surfaces are separated from each other by a thick rounded border, and are attached to the Diaphragm and anterior abdominal wall by a triangular or falciform fold of peritoneum, the *suspensory* or *falciform ligament*, which divides the liver into two unequal parts, termed the right and left lobes. Except along the line of attachment of this ligament to the liver, the superior and anterior surfaces are covered by peritoneum.

The *superior surface* (fig. 504) comprises a part of both lobes, and, as a whole, is convex, and fits under the vault of the Diaphragm ; its central part, however, presents a shallow depression, which corresponds with the position of the heart on the upper surface of the Diaphragm. It is separated from the anterior, posterior,

FIG. 504.—The liver. Upper surface. (Drawn from His's models.)

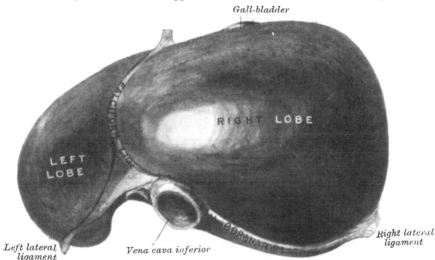

Gall-bladder

RIGHT LOBE

LEFT LOBE

Left lateral ligament   Vena cava inferior   Right lateral ligament

and lateral surfaces by thick, rounded borders. Its left extremity is separated from the under surface by a prominent sharp margin.

The *anterior surface* is large and triangular in shape, comprising also a part of both lobes. It is directed forwards, and the greater part of it is in contact with the Diaphragm, which separates it from the right lower ribs and their cartilages. In the middle line it lies behind the ensiform cartilage, to the left of which it is protected by the seventh and eighth left costal cartilages. In the angle between the diverging rib cartilages of opposite sides the anterior surface is in contact with the abdominal wall. It is separated from the inferior surface by a sharp margin, and from the superior and lateral surfaces by thick rounded borders.

The *lateral* or *right surface* is convex from before backwards and slightly so from above downwards. It is directed towards the right side, forming the base of the wedge, and lies against the lateral portion of the Diaphragm, which separates it from the lower part of the left pleura and lung, outside which are the right costal arches from the seventh to the eleventh inclusive.

Its *under* or *visceral surface* (figs. 505, 506) is uneven, concave, directed downwards and backwards and to the left, and is in relation with the stomach and duodenum, the hepatic flexure of the colon, and the right kidney and suprarenal capsule. The surface is divided by a longitudinal fissure into a right and left lobe, and is almost completely invested by peritoneum; the only parts where this covering is absent are where the gall-bladder is attached to the liver and at the transverse fissure, where the two layers of the lesser omentum are separated from each other by the blood-vessels and duct of the viscus. The under surface of the left lobe presents behind and to the left a depression where it is moulded over the cardiac part of the stomach, and to the right and near the centre a rounded eminence, the *tuber omentale*, which fits into the concavity of the lesser curvature, lying in front of the anterior layer of the lesser omentum. The under surface of the right lobe is divided into two unequal portions by a fossa, which lodges the gall-bladder, the *fossa vesicalis*; the portion to the left, the smaller of the two, is somewhat oblong in shape, its antero-posterior diameter being greater than its transverse. It is known as the *quadrate lobe*, and is in relation with the pyloric end of the stomach and the first portion of the duodenum. The portion of the under surface of the right lobe to the right of the fossa vesicalis presents two shallow concave impressions, one situated behind the other, the two being separated by a ridge. The anterior of these two impressions, the *impressio colica*, is produced by

FIG. 505.—The liver. Posterior and inferior surfaces. (Drawn from His's models.)

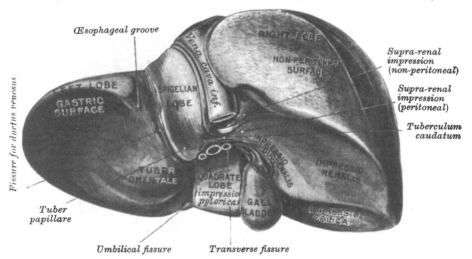

the hepatic flexure of the colon ; the posterior, the *impressio renalis*, is occupied by the upper end of the right kidney. To the inner side of the latter impression is a third and slightly marked impression, lying between it and the neck of the gall-bladder. This is caused by the second portion of the duodenum, and is known as the *impressio duodenalis*. Just in front of the vena cava is a narrow strip of liver tissue, the caudate lobe, which connects the right inferior angle of the Spigelian lobe to the under surface of the right lobe. Immediately below it is the foramen of Winslow.

The *posterior surface* is rounded and broad behind the right lobe, but narrow on the left. Over a large part of its extent it is not covered by peritoneum ; this uncovered portion is about three inches broad, and is in direct contact with the Diaphragm. It is marked off from the upper surface by the line of reflection of the upper or anterior layer of the coronary ligament. It is in the same way marked off from the under surface of the liver by the line of reflection of the lower layer of the coronary ligament. In its centre this posterior surface is deeply notched for the vertebral column and crura of the Diaphragm, and to the right of this it is indented for the inferior vena cava, which is often partly embedded in its substance. Close to the right of this indentation and immediately above the renal impression is a small triangular depressed area (*impressio suprarenalis*), the greater part of which is devoid of peritoneum ; it lodges the right suprarenal capsule. To the left of the inferior vena cava is the *Spigelian lobe*, which lies between the fissure for the vena cava and the fissure for the ductus venosus. Below and in front it projects and forms part of the posterior boundary of the transverse fissure. Here, to the right, it is connected with the under surface of the right lobe of the liver by the caudate lobe, and to the left it presents a tubercle, the *tuber papillare*. It is opposite the tenth and eleventh dorsal vertebræ, and rests upon the aorta and crura of the Diaphragm, being covered by the peritoneum of the lesser sac. This lobe is nearly vertical in position, and is directed backwards : it is longer from above downwards than from side to side, and is somewhat concave in the transverse direction. On the posterior surface to the left of the Spigelian lobe is a groove indicating the position of the œsophageal orifice of the stomach.

The *inferior border* is thin and sharp, and marked opposite the attachment of the falciform ligament by a deep notch, the *umbilical notch*, and opposite the cartilage of the ninth rib by a second notch for the fundus of the gall-bladder. In adult males this border usually corresponds with the margin of the ribs in the right nipple line ; but in women and children it usually projects below the ribs.

The *left extremity of the liver* is thin and flattened from above downwards.

Fissures (fig. 505).—Five fissures are seen upon the under and posterior surfaces of the liver, which serve to divide it into five lobes. They are, the umbilical fissure, the fissure of the ductus venosus, the transverse fissure, the fissure for the gall-bladder, and the fissure for the inferior vena cava. They are arranged in the form of the letter **H**. The left limb of the **H** is known as the *longitudinal fissure*. The right limb is formed in front by the *fissure for the gall-bladder*, and behind by the *fissure for the inferior vena cava* ; these two fissures are separated from each other by the caudate lobe. The connecting bar of the **H** is the *transverse* or *portal fissure*. It separates the quadrate lobe in front from the caudate and Spigelian lobes behind.

The *longitudinal fissure* is a deep groove, which extends from the notch on the anterior margin of the liver to the upper border of the posterior surface of the organ. It separates the right and left lobes ; the transverse fissure joins it, at right angles, and divides it into two parts. The anterior part is called the *umbilical fissure* ; it is deeper than the posterior, and lodges the umbilical vein in the fœtus, and its remains (the round ligament) in the adult ; the posterior part contains the ductus venosus, and is known as the *fissure of the ductus venosus*. This fissure lies between the quadrate lobe and the left lobe of the liver, and is often partially bridged over by a prolongation of the hepatic substance, the *pons hepatis*.

The *fissure of the ductus venosus* is the back part of the longitudinal fissure, and is situated mainly on the posterior surface of the liver. It lies between the left lobe and the lobe of Spigelius. It lodges in the fœtus the ductus venosus, and in the adult a slender fibrous cord, the obliterated remains of that vessel.

The *transverse* or *portal fissure* is a short but deep fissure, about two inches in length, extending transversely across the under surface of the left portion of the right lobe, nearer to its posterior surface than its anterior border. It joins, nearly at right angles, with the longitudinal fissure, and separates the quadrate lobe in front from the caudate and Spigelian lobes behind. By the older anatomists this fissure was considered the gateway (*porta*) of the liver ; hence the large vein which enters at this fissure was called the *portal vein*. Besides this vein, the fissure transmits the hepatic artery and nerves, and the hepatic duct and lymphatics. At their entrance into the fissure, the hepatic duct lies in front and to the right, the hepatic artery to the left, and the portal vein behind and between the duct and artery.

The *fissure for the gall-bladder* (*fossa vesicalis*) is a shallow, oblong fossa, placed on the under surface of the right lobe, parallel with the longitudinal fissure. It extends from the anterior free margin of the liver, which is notched for its reception, to the right extremity of the transverse fissure.

The *fissure for the inferior vena cava* is a short deep fissure, occasionally a complete canal, in consequence of the substance of the liver surrounding the vena cava. It extends obliquely upwards from the lobus caudatus, which separates it from the transverse fissure, on the posterior surface of the liver, and separates the Spigelian from the right lobe. On slitting open the inferior vena cava the orifices of the hepatic veins will be seen opening into this vessel at its upper part, after perforating the floor of this fissure.

Lobes.—The lobes of the liver, like the ligaments and fissures, are five in number—the right lobe, the left lobe, the lobus quadratus, the lobus Spigelii, and the lobus caudatus, the last three being merely parts of the right lobe.

The *right lobe* is much larger than the left; the proportion between them being as six to one. It occupies the right hypochondrium, and is separated from the left lobe, on its upper and anterior surfaces by the falciform ligament; on its under and posterior surfaces by the longitudinal fissure; and in front by the umbilical notch. It is of a somewhat quadrilateral form, its under and posterior surfaces being marked by three fissures—the transverse fissure, the fissure for the gall-bladder, and the fissure for the inferior vena cava, which

FIG. 506.—Posterior and under surfaces of the liver. (From Ellis.)

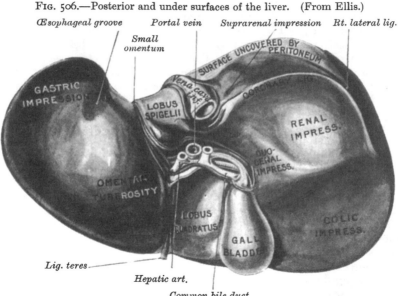

separate its left part into three smaller lobes—lobus Spigelii, lobus quadratus, and lobus caudatus. On it are seen four shallow impressions, one in front (*impressio colica*), for the hepatic flexure of the colon; a second behind (*impressio renalis*), for the right kidney; a third internal, between the last-named and the gall-bladder (*impressio duodenalis*), for the second part of the duodenum; and a fourth on its posterior surface, for the suprarenal capsule (*impressio suprarenalis*).

The *lobus quadratus*, or square lobe, is situated on the under surface of the right lobe, bounded in front by the inferior margin of the liver; behind by the transverse fissure; on the right, by the fissure of the gall-bladder; and on the left, by the umbilical fissure.

The *lobus Spigelii* is situated upon the posterior surface of the right lobe of the liver. It looks directly backwards, and is nearly vertical in direction. It is bounded, above, by the upper layer of the coronary ligament; below, by the transverse fissure; on the right, by the fissure for the vena cava; and, on the left, by the fissure for the ductus venosus. Its left upper angle forms part of the groove for the oesophagus.

The *lobus caudatus*, or tailed lobe, is a small elevation of the hepatic substance extending obliquely outwards, from the lower extremity of the lobus Spigelii to the under surface of the right lobe. It is situated behind the transverse fissure, and separates the fissure for the gall-bladder from the commencement of the fissure for the inferior vena cava.

The *left lobe* is smaller and more flattened than the right. It is situated in the epigastric and left hypochondriac regions. Its upper surface is slightly convex ; its under surface is concave, and presents a shallow depression for the stomach (*gastric impression*). This is situated in front of the groove for the œsophagus, and is separated from the longitudinal fissure by the *omental tuberosity*, which lies against the small omentum and lesser curvature of the stomach.

Ligaments.—The liver is connected to the under surface of the Diaphragm and the anterior wall of the abdomen by five ligaments, four of which are peritoneal folds ; the fifth is a round, fibrous cord, resulting from the obliteration of the umbilical vein. These ligaments are the falciform, two lateral, coronary, and round. It is also attached to the lesser curvature of the stomach by the gastro-hepatic or small omentum (see page 884).

The *falciform ligament* (broad or suspensory ligament) is a broad and thin antero-posterior peritoneal fold, falciform in shape, its base being directed downwards and backwards, its apex upwards and backwards. It is attached by one margin to the under surface of the Diaphragm, and the posterior surface of the sheath of the right Rectus muscle as low down as the umbilicus ; by its hepatic margin it extends from the notch on the anterior margin of the liver, as far back as its posterior surface. It consists of two layers of peritoneum closely united together. Its base or free edge contains the round ligament between its layers.

The *lateral ligaments* (fig. 504), two in number, right and left, are triangular in shape. They are formed by the apposition of the upper and lower layers of the coronary ligament, and extend from the Diaphragm to the liver—the right being attached to the border between its lateral and inferior surfaces, the left, the longer of the two, to the upper surface of the left lobe, where it lies in front of the œsophageal opening in the Diaphragm.

The *coronary ligament* connects the posterior surface of the liver to the Diaphragm. It is formed by the reflection of the peritoneum from the Diaphragm on to the upper and lower margins of the posterior surface of the organ. The coronary ligament consists of two layers, which are continuous on each side with the lateral ligaments ; and, in front, with the falciform ligament. Between the layers, a large triangular area is left uncovered by peritoneum, and is connected to the Diaphragm by firm areolar tissue.

The *round ligament* (ligamentum teres) is a fibrous cord resulting from the obliteration of the umbilical vein. It ascends from the umbilicus, in the free margin of the falciform ligament, to the notch in the anterior border of the liver, from which it may be traced along the longitudinal fissure on the under surface of the liver ; on the posterior surface it is continued as the obliterated ductus venosus as far back as the inferior vena cava.

Vessels.—The vessels connected with the liver are also five in number : they are, the hepatic artery, the portal veins, the hepatic vein, the hepatic duct, and the lymphatics.

The *hepatic artery* and *portal vein*, accompanied by numerous lymphatics and nerves, ascend to the transverse fissure, between the layers of the gastro-hepatic omentum. The *hepatic duct*, lying in company with them, descends from the transverse fissure between the layers of the same omentum. The relative position of the three structures is as follows : The hepatic duct lies to the right, the hepatic artery to the left, and the portal vein behind and between the other two. These are enveloped in a loose areolar tissue, the *capsule of Glisson* which accompanies the vessels in their course through the *portal canals*, in the interior of the organ.

The *hepatic veins* convey the blood from the liver. They commence in the substance of the liver, in the capillary terminations of the portal vein and hepatic artery ; these tributaries, gradually uniting, usually form three veins, which converge towards the posterior surface of the liver and open into the inferior vena cava, while that vessel is situated in the groove at the back part of this organ.   Of these three veins, one from the right and another from the left lobe open obliquely into the vena cava : that from the middle of the organ and lobus Spigelii having a straight course.

The hepatic veins have very little cellular investment ; what there is binds their parietes closely to the walls of the canals through which they run ; so that, on section of the organ, these veins remain widely open and solitary, and may be easily distinguished from the branches of the portal vein, which are more or less collapsed, and always accompanied by an artery and duct.   The hepatic veins are destitute of valves.

Structure.— The substance of the liver is composed of lobules, held together by an extremely fine areolar tissue, and of the ramifications of the portal vein, hepatic duct, hepatic artery, hepatic veins, lymphatics, and nerves ; the whole being invested by a serous and a fibrous coat.

The *serous coat* is derived from the peritoneum, and invests the greater part of the surface of the organ.   It is intimately adherent to the fibrous coat.

The *fibrous coat* lies beneath the serous investment, and covers the entire surface of the organ.   It is difficult of demonstration, excepting where the serous coat is deficient.   At the transverse fissure it is continuous with the capsule of Glisson, and, on the surface of the organ, with the areolar tissue separating the lobules.

The *lobules* form the chief mass of the hepatic substance ; they may be seen either on the surface of the organ, or by making a section through the gland. They are small granular bodies, about the size of a millet seed, measuring from one-twentieth to one-tenth of an inch in diameter.   In the human subject their outline is very irregular ; but in some of the lower animals (for example, the pig) they are well-defined, and, when divided transversely, have a polygonal outline. If divided longitudinally they are more or less foliated or oblong.   The bases of the lobules are clustered round the smallest radicles (*sublobular*) of the hepatic veins, to which each is connected by means of a small branch which issues from the centre of the lobule (*intralobular*).   The remaining part of the surface of each lobule is imperfectly isolated from the surrounding lobules by a thin stratum of areolar tissue, in which is contained a plexus of vessels (the *interlobular plexus*) and ducts.   In some animals, as the pig, the lobules are completely isolated one from another by the interlobular areolar tissue.

If one of the sublobular veins be laid open, the bases of the lobules may be seen through the thin wall of the vein on which they rest, arranged in the form of a tesselated pavement, the centre of each polygonal space presenting a minute aperture, the mouth of an intralobular vein (fig. 507).

Microscopic appearance.—Each lobule is composed of a mass of cells (*hepatic cells*), surrounded by a dense capillary plexus, composed of vessels which penetrate from the circumference to the centre of the lobule, and terminate in a single straight vein, which runs through its centre, to open at its base into one of the radicles of the hepatic vein.   Between the cells are also the minute commencements of the bile-ducts.   Therefore, in the lobule we have all the essentials of a secreting gland ; that is to say : (1) *cells*, by which the secretion is formed ; (2) *blood-vessels*, in close relation with the cells, containing the blood from which the secretion is derived ; (3) *ducts*, by which the secretion, when formed, is carried away.   Each of these structures will have to be further considered.

(1) The *hepatic cells* are of more or less spheroidal form ; but may be rounded, flattened, or many-sided from mutual compression.   They vary in size from $\frac{1}{1000}$ to $\frac{1}{2000}$ of an inch in diameter.   They consist of a honeycomb network without any cell-wall (Klein), and contain one or sometimes two distinct nuclei. In the nucleus is a highly refracted nucleolus with granules.   Embedded in the

honeycomb network are numerous yellow particles, the colouring-matter of the bile, and oil-globules. The cells adhere together by their surfaces so as to form rows, which radiate from the centre to the circumference of the lobules.* As stated above, they are the chief agents in the secretion of the bile.

(2) *The blood-vessels.*—The blood in the capillary plexus, around the l. er-cells, is brought to the liver principally by the portal vein, but also to a certain extent by the hepatic artery. For the sake of clearness, the distribution of the blood derived from the hepatic artery may be considered first.

The *hepatic artery*, entering the liver at the transverse fissure with the portal vein and hepatic duct, ramifies with these vessels through the portal canals. It gives off *vaginal branches*, which ramify in the capsule of Glisson, and appear to be destined chiefly for the nutrition of the coats of the large vessels, the ducts, and the investing membranes of the liver. It also gives off *capsular branches*, which reach the surface of the organ, terminating in its fibrous coat in stellate plexuses. Finally it gives off *interlobular branches*, which form a plexus on the outer side of each lobule, to supply its wall and the accompanying bile-ducts. From this, lobular branches enter the lobule and end in the capillary network

FIG. 507.—Longitudinal section of an hepatic vein. (After Kiernan.)   FIG. 508.—Longitudinal section of a small portal vein and canal. (After Kiernan.)

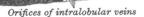
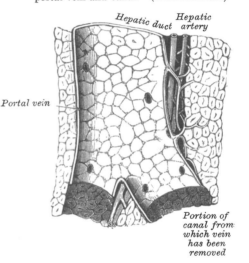

Portal vein

Hepatic duct   *Hepatic artery*

Portion of canal from which vein has been removed

Orifices of intralobular veins

between the cells. Some anatomists, however, doubt whether it transmits any blood directly to the capillary network.

The *portal vein* also enters at the transverse fissure and runs through the portal canals, enclosed in Glisson's capsule, dividing into branches in its course, which finally break up into a plexus (the *interlobular plexus*) in the interlobular spaces. In their course these branches receive the vaginal and capsular veins, corresponding to the vaginal and capsular branches of the hepatic artery (fig. 508). Thus it will be seen that all the blood carried to the liver by the portal vein and hepatic artery, except perhaps that derived from the interlobular branches of the hepatic artery, directly or indirectly, finds its way into the interlobular plexus. From this plexus the blood is carried into the lobule by fine branches which pierce its wall, and then converge from the circumference to the centre of the lobule, forming a number of converging

* Delépine states that there are evidences of the arrangement of these cells in the form of columns, which form tubes with narrow lumina branching from terminal bile ducts. This branching is evidenced by a divergence of the columns from lines extending between adjacent portal vessels. The columns of cells group round terminal bile ducts and not round the so-called intralobular veins. (*Lancet*, vol. i. 1895, p. 1254.)

vessels, which are connected by transverse branches (fig. 509). In the interstices of the network of vessels thus formed are situated, as before said, the liver-cells ; and here it is that, the blood being brought into intimate connection with the liver-cells, the bile is secreted. Arrived at the centre of the lobule, all these minute vessels empty themselves into one vein, of considerable size, which runs down the centre of the lobule from apex to base, and is called the *intralobular vein*. At the base of the lobule this vein opens directly into the *sublobular vein*, with which the lobule is connected, and which, as before mentioned, is a radicle of the hepatic vein. The sublobular veins, uniting into larger and larger trunks, end at last in the hepatic veins, which do not receive any intralobular veins. Finally, the hepatic veins, as mentioned on page 598, converge to form three large trunks which open into the inferior vena cava, while that vessel is situated in the fissure appropriated to it at the back of the liver.

(3) *The ducts.*—Having shown how the blood is brought into intimate relation with the hepatic cells in order that the bile may be secreted, it remains now only to consider the way in which the secretion, having been formed, is carried away.

FIG. 509.—Horizontal section of liver (dog).

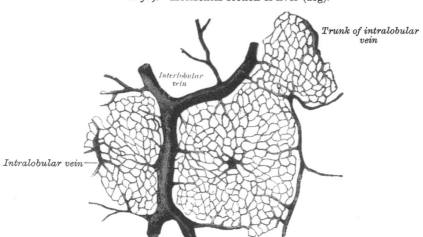

Trunk of intralobular vein

Interlobular vein

Intralobular vein

Several views have prevailed as to the mode of origin of the hepatic ducts ; it seems, however, to be clear that they commence by little passages which are formed between the cells, and which have been termed *intercellular biliary passages* or *bile capillaries*. These passages are merely little channels or spaces left between the contiguous surfaces of two cells, or in the angle where three or more liver-cells meet (fig. 510), and it seems doubtful whether there is any delicate membrane forming the wall of the channel. The channels thus formed radiate to the circumference of the lobule, and, piercing its wall, form a plexus (*interlobular*) between the lobules. From this plexus ducts are derived which pass into the portal canals, become enclosed in Glisson's capsule, and, accompanying the portal vein and hepatic artery (fig. 511), join with other ducts to form two main trunks, which leave the liver at the transverse fissure, and by their union form the hepatic duct.

Structure.—The coats of the smallest biliary ducts, which lie in the interlobular spaces, are a connective tissue coat, in which are muscle-cells, arranged both circularly and longitudinally, and an epithelial layer, consisting of short columnar cells. In the larger ducts, which lie in the portal canals, there are a number of orifices disposed in two longitudinal rows, which were formerly regarded as the openings of mucous glands, but which are merely the orifices of tubular recesses. They occasionally anastomose, and from the sides of them saccular dilatations are given off.

Lymphatics of the liver.—The lymphatics in the substance of the liver commence in lymphatic spaces around the capillaries of the lobules ; they accompany the vessels of the interlobular plexus, often enclosing and surrounding them. These unite and form larger vessels, which run in the portal canals, enclosed in Glisson's capsule, and emerge at the portal fissure to be distributed in the manner described. Other superficial lymphatics form a close plexus, under the peritoneum, where this membrane covers the liver, and pass in various directions through the ligaments of the liver (page 613).

Nerves of the liver.—The nerves of the liver derived from the pneumogastric and sympathetic enter the liver at the transverse fissure and accompany the vessels and ducts to the interlobular spaces. Here, according to Korolkow, the medullated fibres are distributed almost exclusively to the coats of the blood-vessels ; while the non-medullated enter the lobules and ramify between the cells.

FIG. 510.—Section of liver.

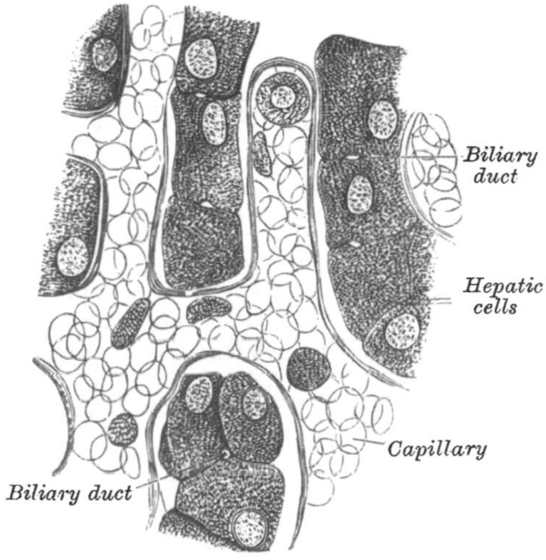

Biliary duct

Hepatic cells

Capillary

Biliary duct

FIG. 511.—A transverse section of a small portal canal and its vessels. (After Kiernan.)

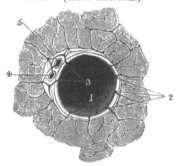

1. Portal vein. 2. Interlobular branches.
3. Vaginal branches. 4. Hepatic duct.
5. Hepatic artery.

EXCRETORY APPARATUS OF THE LIVER

The excretory apparatus of the liver consists of (1) the *hepatic duct*, which, as we have seen, is formed by the junction of the two main ducts, which pass out of the liver at the transverse fissure, and are formed by the union of the bile capillaries ; (2) the *gall-bladder*, which serves as a reservoir for the bile ; (3) the *cystic duct*, which is the duct of the gall-bladder ; and (4) the *common bile duct*, formed by the junction of the hepatic and cystic ducts.

The hepatic duct.—Two main trunks of nearly equal size issue from the liver at the transverse fissure, one from the right, the other from the left lobe ; these unite to form the hepatic duct, which then passes downwards and to the right for about an inch and a half, between the layers of the lesser omentum, where it is joined at an acute angle by the cystic duct, and so forms the ductus communis choledochus. The hepatic duct, as it descends from the transverse fissure of the liver, between the two layers of the lesser omentum, lies in company with the hepatic artery and portal vein.

The **gall-bladder** is the reservoir for the bile ; it is a conical or pear-shaped musculo-membranous sac, lodged in a fossa on the under surface of the right lobe of the liver, and extending from near the right extremity of the transverse fissure to the anterior border of the organ. It is about four inches in length, one inch in breadth at its widest part, and holds from eight to ten drachms. It is divided into a fundus, body, and neck. The *fundus*, or broad extremity, is directed downwards, forwards, and to the right, and projects beyond the anterior border of the liver; the *body* and *neck* are directed upwards and backwards to the left. The upper surface of the gall-bladder is attached to the liver by connective tissue and vessels. The under surface is covered by peritoneum, which is reflected on to it from the surface of the liver. Occasionally the whole of the organ is invested by the serous membrane, and is then connected to the liver by a kind of mesentery.

Relations.—The *body of the gall-bladder* is in relation, by its upper surface, with the liver, to which it is connected by areolar tissue and vessels; by its under surface, with the commencement of the transverse colon; and farther back, with the upper end of the descending portion of the duodenum or sometimes with the pyloric end of the stomach, or first portion of the duodenum. The *fundus* is completely invested by peritoneum; it is in relation, in front, with the abdominal parietes, immediately below the ninth costal cartilage; behind with the transverse arch of the colon. The *neck* is narrow, and curves upon itself like the letter S; at its point of connection with the cystic duct it presents a well-marked constriction.

When the gall-bladder is distended with bile or calculi, the fundus may be felt through the abdominal parietes, especially in an emaciated subject: the relations of this sac will also serve to explain the occasional occurrence of abdominal biliary fistulæ, through which biliary calculi may pass out, and of the passage of calculi from the gall-bladder into the stomach, duodenum, or colon, which occasionally happens.

Structure.—The gall-bladder consists of three coats—serous, fibrous and muscular, and mucous.

The *external* or *serous coat* is derived from the peritoneum; it completely invests the fundus, but covers the body and neck only on their under surface.

The *fibro-muscular coat* is a thin but strong layer, which forms the framework of the sac, consisting of dense fibrous tissue, which interlaces in all directions, and is mixed with plain muscular fibres, which are disposed chiefly in a longitudinal direction, a few running transversely.

The *internal* or *mucous coat* is loosely connected with the fibrous layer. It is generally tinged with a yellowish-brown colour, and is everywhere elevated into minute rugæ, by the union of which numerous meshes are formed; the depressed intervening spaces having a polygonal outline. The meshes are smaller at the fundus and neck, being most developed about the centre of the sac. Opposite the neck of the gall-bladder the mucous membrane projects inwards in the form of oblique ridges or folds, forming a sort of screw-like valve.

The mucous membrane is covered with columnar epithelium, and secretes an abundance of thick viscid mucus; it is continuous through the hepatic duct with the mucous membrane lining the ducts of the liver, and through the ductus communis choledochus with the mucous membrane of the alimentary canal.

The **cystic duct,** the smallest of the three biliary ducts, is about an inch and a half in length. It passes obliquely downwards and to the left from the neck of the gall-bladder, and joins the hepatic duct to form the common bile duct. It lies in the gastro-hepatic omentum in front of the vena portæ, the hepatic artery lying to its left side. The mucous membrane lining its interior is thrown into a series of crescentic folds, from five to twelve in number, similar to those found in the neck of the gall-bladder. They project into the duct in regular succession, and are directed obliquely round the tube, presenting much the appearance of a continuous spiral valve. When the duct is distended, the spaces between the folds are dilated, so as to give to its exterior a sacculated appearance.

The **ductus communis choledochus,** or *common bile duct,* the largest of the three, is the common excretory duct of the liver and gall-bladder. It is about three inches in length, of the diameter of a goose-quill, and formed by the junction of the cystic and hepatic ducts.

It descends along the right border of the lesser omentum behind the first portion of the duodenum, in front of the vena portæ, and to the right of the hepatic artery; it then passes between the pancreas and descending portion of the duodenum, and, running for a short distance along the right side of the pancreatic duct, near its termination, passes, with it, obliquely between the mucous and muscular coats. The two ducts open by a common orifice upon the summit of a papilla, situated at the inner side of the descending portion of the duodenum, a little below its middle and about three or four inches below the pylorus.

Structure.—The coats of the large biliary ducts are, an external or fibrous, and an internal or mucous. The fibrous coat is composed of strong fibro-areolar tissue, with a certain amount of muscular tissue, arranged, for the most part, in a circular manner around the duct. The mucous coat is continuous with the lining membrane of the hepatic ducts and gall-bladder, and also with that of the duodenum; and, like the mucous membrane of these structures, its epithelium is of the columnar variety. It is provided with numerous mucous glands, which are lobulated and open by minute orifices, scattered irregularly in the larger ducts. The coats of the smallest biliary ducts, which lie in the interlobular spaces, are a connective-tissue coat, in which, according to Heidenhain, are muscle-cells, arranged both circularly and longitudinally, and an epithelial layer, consisting of short columnar cells.

Surface Relations.—The liver is situated in the right hypochondriac and the epigastric regions, and is moulded to the arch of the Diaphragm. In the greater part of its extent it lies under cover of the lower ribs and their cartilages, but in the epigastric region it comes in contact with the abdominal wall, in the subcostal angle. The *upper limit of the right lobe of the liver* may be defined in the middle line by the junction of the mesosternum with the ensiform cartilage; on the right side the line must be carried upwards as far as the fifth rib cartilage in the line of the nipple and then downwards to reach the seventh rib at the side of the chest. The *upper limit of the left lobe* may be defined by continuing this line to the left with an inclination downwards to a point about two inches to the left of the sternum on a level with the sixth left costal cartilage. The *lower limit of the liver* may be indicated by a line drawn half an inch below the lower border of the thorax on the right side as far as the ninth right costal cartilage, and thence obliquely upwards across the subcostal angle to the eighth left costal cartilage. A slight curved line with its convexity to the left from this point, i.e. the eighth left costal cartilage, to the termination of the line indicating the upper limit, will denote the left margin of the liver. The fundus of the gall-bladder approaches the surface behind the anterior extremity of the ninth costal cartilage, close to the outer margin of the right Rectus muscle.

It must be remembered that the liver is subject to considerable alterations in position, and the student should make himself acquainted with the different circumstances under which this occurs, as they are of importance in determining the existence of enlargement or other diseases of the organ.

Its position varies according to the posture of the body. In the erect position in the adult male, the edge of the liver projects about half an inch below the lower edge of the right costal cartilages, and its anterior border can be often felt in this situation if the abdominal wall is thin. In the supine position the liver gravitates backwards, and recedes above the lower margin of the ribs, and cannot then be detected by the finger. In the prone position it falls forward, and can then generally be felt in a patient with loose and lax abdominal walls. Its position varies also with the ascent or descent of the Diaphragm. In a deep inspiration the liver descends below the ribs; in expiration it is raised behind them. Again, in emphysema, where the lungs are distended, and the Diaphragm descends very low, the liver is pushed down: in some other diseases, as phthisis, where the Diaphragm is much arched, the liver rises very high up. Pressure from without, as in tight-lacing, by compressing the lower part of the chest, displaces the liver considerably: its anterior edge often extending as low as the crest of the ilium; and its convex surface is often at the same time deeply indented from the pressure of the ribs. Again, its position varies greatly according to the greater or less distension of the stomach and intestines. When the intestines are empty, the liver descends in the abdomen; but when they are distended, it is pushed upwards. Its relations to surrounding organs may also be changed by the growth of tumours, or by collections of fluid in the thoracic or abdominal cavities.

Surgical Anatomy.—On account of its large size, its fixed position, and its friability, the liver is more frequently ruptured than any of the abdominal viscera. The rupture may vary considerably in extent, from a slight scratch to an extensive laceration completely through its substance, dividing it into two parts. Sometimes an internal rupture, without laceration of the peritoneal covering, takes place, and such injuries are most susceptible of repair; but small tears of the surface may also heal; when, however, the laceration is extensive, death usually takes place from hæmorrhage, on account of the fact that the hepatic veins are contained in rigid canals in the liver-substance and are unable to contract, and are moreover unprovided with valves. The liver may also be torn by the end of a broken rib perforating the Diaphragm. The liver may be injured by stabs or other punctured wounds, and when these are inflicted through the chest wall both pleural and peritoneal cavities may be opened up, and both lung and liver be wounded. In cases of wound of the liver from the front, hernia of a part of this viscus may take place, but can generally easily be replaced. In cases of laceration of the liver, when there is evidence that bleeding is going on, the abdomen must be opened, the laceration sought for, and the bleeding arrested. This may be done temporarily by introducing the forefinger into the foramen of Winslow and placing the thumb on the gastro-hepatic omentum and compressing the hepatic artery and portal vein between the two. Any bleeding points can then be seen and tied and the margins of the laceration, if small, brought together and sutured by means of a blunt curved needle passed from one side of the wound to the other. All sutures must be passed before any are tied, and this must be done with the greatest gentleness as the liver substance is very friable. When the laceration is extensive it must be packed with iodoform gauze, the end of which is allowed to hang out of the external wound. Abscess of the liver is of not infrequent occurrence, and may open in many different ways on account of the relations of this viscus to other organs. Thus it has been known to burst into the lungs and the pus coughed up, or into the stomach and the pus vomited; it may burst into the colon, or into the duodenum; or, by perforating the Diaphragm, it may empty itself into the pleural cavity. Frequently it makes its way forwards, and points on the anterior abdominal wall, and finally it may burst into the peritoneal or pericardiac cavities. Abscesses of the liver frequently require opening, and this must be done by an incision in the abdominal wall, in the thoracic wall, or in the lumbar region, according to the direction in which the abscess is tracking. The incision through the abdominal wall is to be preferred when possible. The abdominal wall is incised over the swelling, and unless the peritoneum is adherent, sponges are packed all round the exposed liver surface and the abscess opened, if deeply seated preferably by the thermo-cautery. Hydatid cysts are more often found in the liver than in any other of the viscera. The reason of this is not far to seek. The embryo of the egg of the tænia echinococcus, being liberated in the stomach by the disintegration of its shell, bores its way through the gastric walls and usually enters a blood-vessel, and is carried by the blood-stream to the hepatic capillaries, where its onward course is arrested, and where it undergoes development into the fully formed hydatid. Tumours of the liver have recently been subjected to surgical treatment by removal of a portion of the organ. The abdomen is opened and the diseased portion of liver exposed; the circulation is controlled by compressing the portal vein and the hepatic artery in the gastro-hepatic omentum and a wedge-shaped portion of liver containing the tumour removed; the divided vessels are ligatured and the cut surfaces brought together and sutured in the manner directed above.

When the *gall-bladder* or one of its main ducts is ruptured, which may occur independently of laceration of the liver, death usually occurs from peritonitis. If the symptoms have led to the performance of a laparotomy and a rent is found, it should be sutured if small, or the gall-bladder removed if it is extensive. If the cystic duct is torn, its intestinal end must be closed and the gall-bladder removed. In rupture of either of the other ducts, the only thing which can be done is to provide for free drainage, in the hope that a biliary fistula may form.

The gall-bladder may become distended in cases of obstruction of its duct or the common bile duct, or from a collection of gall-stones within its anterior, thus forming a large tumour. The swelling is pear-shaped, and projects downwards and forwards to the umbilicus. It moves with respiration, since it is attached to the liver. To relieve this condition, the gall-bladder must be opened and the gall-stones removed. The operation is performed by an incision, two or three inches long in the right semilunar line, commencing at the costal margin. The peritoneal cavity is opened, and the tumour having been found, sponges are packed round it to protect the peritoneal cavity, and it is aspirated. When the contained fluid has been evacuated the flaccid bladder is drawn out of the abdominal wound and its wall incised to the extent of an inch; any gall-stones in the bladder are now removed and the interior of the sac sponged dry. If the case is one of obstruction of the duct, an attempt must be made to dislodge the stone by manipulation through the wall of the duct; or it may be crushed from without by the fingers or carefully padded forceps. If this does not succeed, the safest plan is to incise the duct, extract the stone, and close the incision by fine sutures in two layers. After all obstruction has been removed, four courses are open to the surgeon: 1. The wound in the gall-bladder may be at once sewn up, the organ returned into the abdominal cavity, and the external incision closed. 2. The edges of the incision in the gall-bladder may be sutured to the external wound, and a fistulous communication established between the gall-bladder and the exterior; this fistulous opening usually closes in the course of a few weeks. 3. The gall-bladder may be connected with the intestinal canal, preferably the duodenum, by means of a lateral anastomosis; this is known as cholecystenterostomy. 4. The gall-bladder may be completely removed.

THE PANCREAS

Dissection.—The pancreas may be exposed for dissection in three different ways:
1. By raising the liver, drawing down the stomach, and tearing through the gastro-hepatic omentum and the ascending layer of the transverse mesocolon. 2. By raising the stomach, the arch of the colon, and great omentum, and then dividing the inferior layer of the transverse mesocolon and raising its ascending layer. 3. By dividing the two layers of peritoneum, which descend from the great curvature of the stomach to form the great omentum; turning the stomach upwards, and then cutting through the ascending layer of the transverse mesocolon (see fig. 481, page 882).

The **Pancreas** (παν-κρέας, *all flesh*) is a compound racemose gland, analogous in its structure to the salivary glands, though softer and less compactly arranged than those organs. It is long and irregularly prismatic in shape, and has been compared to a human or a dog's tongue: its right extremity being broad, is called the *head*—this is connected to the main portion of the organ, the *body*, by a slight constriction, the *neck*; while its left extremity gradually tapers to form

FIG. 512.—The pancreas and its relations.

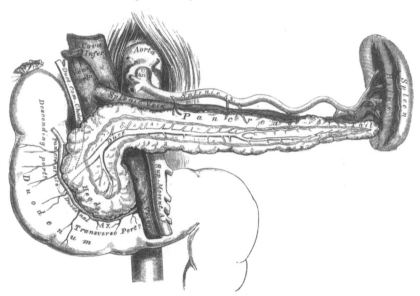

the *tail.* It is situated transversely across the posterior wall of the abdomen, at the back of the epigastric and left hypochondriac regions. Its length varies from five to six inches, its breadth is an inch and a half, and its thickness from half an inch to an inch, being greater at its right extremity and along its upper border. Its weight varies from two to three and a half ounces, but it may reach six ounces.

The *right extremity* or *head of the pancreas* (fig. 512) is shaped like the head of a hammer, being elongated both above and below; it is flattened from before backwards, and conforms to the whole concavity of the duodenum, which is slightly overlapped by it. The anterior surface near its left border is crossed by the superior mesenteric vessels, and at its lower end it is crossed by the transverse colon and its mesocolon. Behind, the head of the pancreas is in relation with the inferior vena cava, the left renal vein, the right crus of the Diaphragm, and the aorta. The common bile duct descends behind, between the duodenum and pancreas; and the pancreatico-duodenal artery descends in front between the same parts.

The *neck of the pancreas* is about an inch long, and passes upwards and forwards to the left, having the first part of the duodenum above it, and the termination of the fourth portion below. It lies in front of the commencement of the vena portæ, and is grooved on the right by the gastro-duodenal and superior pancreatico-duodenal arteries. The pylorus lies just above it.

The *body and tail of the pancreas* are somewhat prismatic in shape, and have three surfaces: anterior, posterior, and inferior.

The *anterior surface* is somewhat concave, and is covered by the posterior surface of the stomach which rests upon it, the two organs being separated by the lesser sac of the peritoneum. At its right extremity there is a well-marked prominence, called by His the *omental tuberosity*.

Fig. 513.—The duodenum and pancreas.
The liver has been lifted up and the greater part of the stomach removed. (Testut.)

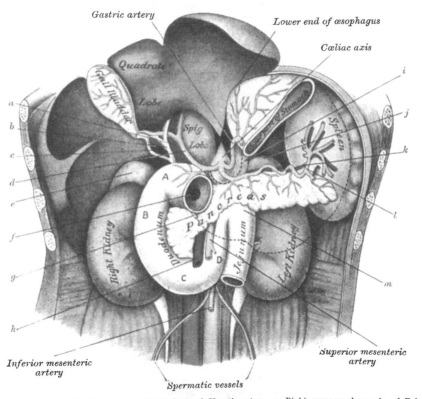

a. Portal vein. b. Hepatic duct. c. Cystic duct. d. Hepatic artery. e. Right supra-renal capsule. f. Pyloric orifice. g. Right gastro-epiploic artery. h. Superior mesenteric vein. i. Left crus of diaphragm. j. Left supra-renal capsule. k. Splenic vein. l. Splenic artery. m. Duodeno-jejunal junction. A, B, C, D. The four portions of the duodenum.

The *posterior surface* is separated from the vertebral column by the aorta, the splenic vein, the left kidney and its vessels, the left suprarenal capsule, the pillars of the Diaphragm, and the origin of the superior mesenteric artery.

The *inferior surface* is narrow, and lies upon the duodeno-jejunal flexure and on some coils of the jejunum; its left extremity rests on the splenic flexure of the colon.

The *superior border* of the body is blunt and flat to the right; narrow and sharp to the left, near the tail. It commences to the right in the omental tuberosity, and is in relation with the cœliac axis, from which the hepatic artery courses to the right just above the gland, while the splenic branch runs in a groove along this border to the left.

The *anterior border* is the position where the two layers of the transverse mesocolon separate: the one passing upwards in front of the anterior surface, the other backwards below the inferior surface.

The *lesser end* or *tail of the pancreas* is narrow; it extends to the left as far as the lower part of the inner aspect of the spleen.

Birmingham describes the body of the pancreas as projecting forwards as a prominent ridge into the abdominal cavity and forming a sort of shelf on which the stomach lies. He says: 'The portion of the pancreas to the left of the middle line has a very considerable antero-posterior thickness; as a result the anterior surface is of considerable extent, it looks strongly upwards and forms a large and important part of the shelf. As the pancreas extends to the left towards the spleen it crosses the upper part of the kidney, and is so moulded on to it that the top of the kidney forms an extension inwards and backwards of the upper surface of the pancreas and extends the bed in this direction. On the other hand,

Fig. 514.—Transverse section through the middle of the first lumbar vertebra, showing the relations of the pancreas. (Braune.)

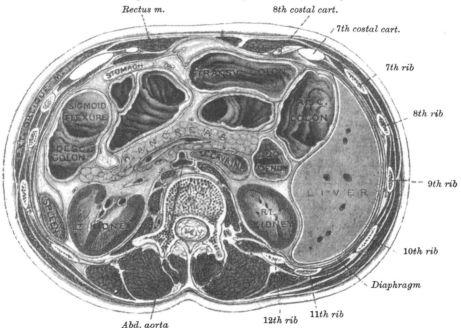

the extremity of the pancreas comes in contact with the spleen in such a way that the plane of its upper surface runs with little interruption upwards and backwards into the concave gastric surface of the spleen, which completes the bed behind and to the left, and running upwards, forms a partial cap for the wide end of the stomach'* (see fig. 489).

The principal excretory duct of the pancreas, called the **pancreatic duct** or **canal of Wirsung,** from its discoverer, extends transversely from left to right through the substance of the pancreas. In order to expose it, the superficial portion of the gland must be removed. It commences by the junction of the small ducts of the lobules situated in the tail of the pancreas, and, running from right to left through the body, it constantly receives the ducts of the various lobules composing the gland. Considerably augmented in size, it reaches the neck, and turning obliquely downwards, backwards and to the right, it comes into relation with the common bile duct, lying to its left side; leaving the head of

* *Journal of Anatomy and Physiology*, vol. xxxi. pt. i, p. 102.

the gland it passes very obliquely through the mucous and muscular coats of the duodenum, and terminates by an orifice common to it and the ductus communis choledochus upon the summit of an elevated papilla, situated at the inner side of the descending portion of the duodenum, three or four inches below the pylorus.

Sometimes the pancreatic duct and ductus communis choledochus open separately into the duodenum. Occasionally there is an accessory duct, which is given off from the canal of Wirsung in the neck of the pancreas and passes horizontally to the right to open into the duodenum about an inch above the orifice of the main duct. This is known as the *ductus pancreaticus accessorius* or *ductus Santorini*.

The pancreatic duct, near the duodenum, is about the size of an ordinary quill : its walls are thin, consisting of two coats, an external fibrous and an internal mucous ; the latter is smooth, and furnished near its termination with a few scattered follicles.

In **structure,** the pancreas resembles the salivary glands. It differs from them, however, in certain particulars, and is looser and softer in its texture. It is not enclosed in a distinct capsule, but is surrounded by areolar tissue, which dips into its interior, and connects together the various lobules of which it is composed. Each lobule, like the lobules of the salivary glands, consists of one of the ultimate ramifications of the main duct, terminating in a number of cæcal pouches or alveoli, which are tubular and somewhat convoluted. The minute ducts connected with the alveoli are narrow and lined with flattened cells. The alveoli are almost completely filled with secreting cells, so that scarcely any lumen is visible. In some animals those cells which occupy the centre of the alveolus are spindle-shaped, and are known as the *centro-acinar cells of Langerhans.* The true secreting cells which line the wall of the alveolus are very characteristic. They are columnar in shape and present two zones : an outer one clear and finely striated next the basement-membrane, and an inner granular one next the lumen. During activity the granular zone occupies the greater part of the cell : before the cells are called into action, while in a condition of rest, the outer or clear zone is the larger. In some secreting cells of the pancreas is a spherical mass, staining more easily than the rest of the cells ; this is termed the *paranucleus,* and is believed to be an extension from the nucleus. The connective tissue between the alveoli presents in certain parts collections of cells, which are termed *inter-alveolar cell-islets.*

Vessels and Nerves.—The *arteries of the pancreas* are derived from the splenic and the pancreatico-duodenal branches of the hepatic and the superior mesenteric. Its *veins* open into the splenic and superior mesenteric veins. Its *lymphatics* terminate in the lumbar glands. Its *nerves* are filaments from the splenic plexus.

Surface Form.—The pancreas lies in front of the second lumbar vertebra, and can sometimes be felt, in emaciated subjects, when the stomach and colon are empty, by making deep pressure in the middle line about three inches above the umbilicus.

Surgical Anatomy.—The pancreas presents but little of surgical importance. It is occasionally the seat of cancer, which usually affects the head or duodenal end, and therefore often speedily involves the common bile duct, leading to persistent jaundice. Cysts are also occasionally found in it, which may present in the epigastric region, above and to the right of the umbilicus, and may require opening and draining. The fluid in them contains some of the elements of the pancreatic secretion, and is very irritating, so that, if allowed to come in contact with the skin of the abdominal wall, it is likely to produce intractable eczema. It has been said that the pancreas is the only abdominal viscus which has never been found in a hernial protrusion but even this organ has been found, in company with other viscera, in rare cases of diaphragmatic hernia. The pancreas has been known to become invaginated into the intestine, and portions of the organ have sloughed off. In cases of excision of the pylorus great care must be exercised to avoid wounding the pancreas, as the escape of the pancreatic fluid may be attended with serious results. According to Billroth, it is likely, in consequence of its peptonising qualities, to dissolve the cicatrix of the stomach.

The Spleen

The **Spleen** belongs to that class of bodies which are known as *ductless glands*. It is probably related to the blood-vascular system, but in consequence of its anatomical relationship to the stomach and its physiological relationship to the liver it is convenient to describe it in this place. It is situated principally in the left hypochondriac region, its upper and inner extremity extending into the epigastric region ; lying between the fundus of the stomach and the Diaphragm. It is the largest of the ductless glands, and measures some five or six inches in length. It is of an oblong, flattened form, soft, of very brittle consistence, highly vascular, and of a dark purplish colour.

Surfaces.—The *external* or *phrenic surface* is convex, smooth, and is directed upwards, backwards, and to the left, except at its upper end, where it is directed slightly inwards. It is in relation with the under surface of the Diaphragm, which separates it from the eighth, ninth, tenth, and eleventh ribs of the left side, and in part from the lower border of the left lung and pleura.

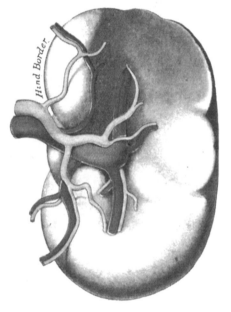

FIG. 515.—The spleen, showing its gastric and renal surfaces. (Testut.)

The *internal surface* is concave, and divided by a ridge into an anterior or larger, and a posterior or smaller portion.

The *anterior portion* of the internal surface or *gastric surface*, which is directed forwards and inwards, is broad and concave, and is in contact with the posterior wall of the great end of the stomach ; and below this with the tail of the pancreas. It presents near its inner border a long fissure, termed the *hilum*. This is pierced by several irregular apertures, for the entrance and exit of vessels and nerves.

The *posterior portion* of the internal surface or *renal surface* is directed inwards and downwards. It is somewhat flattened, does not reach as high as the gastric surface, is considerably narrower than the latter, and is in relation with the upper part of the outer surface of the left kidney and occasionally with the left suprarenal capsule.

The *upper end* is directed inwards, towards the vertebral column, where it lies on a level with the eleventh dorsal vertebra. The *lower end*, sometimes termed the *basal surface*, is flat, triangular in shape, and rests upon the splenic flexure of the colon and the phreno-colic ligament, and is generally in contact with the tail of the pancreas. The *anterior* border is free, sharp, and thin, and is often notched, especially below. It separates the phrenic from the gastric surface. The *posterior* border is more rounded and blunter than the anterior. It separates the renal portion of the internal surface from the phrenic surface. It corresponds to the lower border of the eleventh rib and lies between the Diaphragm and left kidney. The *internal* border is the name sometimes given to the ridge which separates the renal and gastric portions of the internal surface.

The spleen is almost entirely surrounded by peritoneum, which is firmly adherent to its capsule, and is held in position by two folds of this membrane : one, the *lieno-renal ligament*, is derived from the layers of peritoneum forming the greater and lesser sacs, where they come into contact between the left kidney and the spleen. Between its two layers the splenic vessels pass (fig. 482); the second, the *gastro-splenic omentum*, also formed of two layers, derived from the greater and lesser sacs respectively, where they meet between the spleen and stomach (fig. 482). Between these two layers run the vasa brevia of the splenic artery and vein. It is also supported by the phreno-colic ligament, upon which its lower end rests (see page 884).

The size and weight of the spleen are liable to very extreme variations at different periods of life, in different individuals, and in the same individual under different conditions. *In the adult*, in whom it attains its greatest size, it is usually about five inches in length, three inches in breadth, and an inch or an inch and a half in thickness, and weighs about seven ounces. *At birth*, its weight, in proportion to the entire body, is almost equal to what is observed in the adult, being as 1 to 350 : while in the adult it varies from 1 to 320 and 400. *In old age*, the organ not only decreases in weight, but decreases considerably in proportion to the entire body, being as 1 to 700. The size of the spleen is increased during and after digestion, and varies considerably according to the

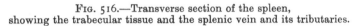

FIG. 516.—Transverse section of the spleen,
showing the trabecular tissue and the splenic vein and its tributaries.

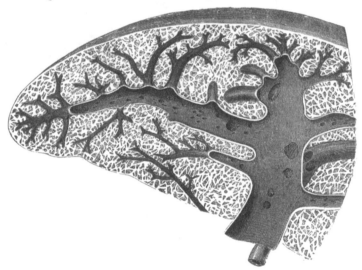

state of nutrition of the body, being large in highly fed, and small in starved animals. In intermittent and other fevers it becomes much enlarged, weighing occasionally from 18 to 20 pounds.

Frequently in the neighbourhood of the spleen, and especially in the gastro-splenic and great omenta, small nodules of splenic tissue may be found, either isolated or connected to the spleen by thin bands of splenic tissue. They are known as *supernumerary* or *accessory spleens*. They vary in size from a pea to a plum.

Structure.—The spleen is invested by two coats—an external serous, and an internal fibro-elastic coat.

The *external* or *serous coat* is derived from the peritoneum ; it is thin, smooth and in the human subject intimately adherent to the fibro-elastic coat. It invests the entire organ, except at the places of its reflection on to the stomach and Diaphragm and at the hilum.

The *fibro-elastic coat* forms the framework of the spleen. It invests the organ, and at the hilum is reflected inwards upon the vessels in the form of sheaths. From these sheaths, as well as from the inner surface of the fibro-elastic coat, numerous small fibrous bands, *trabeculæ* (fig. 517), are given off in all directions; these uniting, constitute the framework of the spleen. This resembles a sponge-like material, consisting of a number of small spaces or *areolæ*, formed by the trabeculæ, which are given off from the inner surface of the capsule, or from the sheaths prolonged inwards on the blood-vessels. In these spaces or areolæ is contained the *splenic pulp*.

The proper coat, the sheaths of the vessels and the trabeculæ, consist of a dense mesh of white and yellow elastic fibrous tissues, the latter considerably predominating. It is owing to the presence of this tissue that the spleen possesses a considerable amount of elasticity, which allows of the very great variations in size that it presents under certain circumstances. In addition to these constituents of this tunic, there is found in man a small amount of non-striped muscular fibre; and in some mammalia (e.g. dog, pig, and cat) a very considerable amount, so that the trabeculæ appear to consist chiefly of muscular tissue. It is probably owing to this structure that the spleen exhibits, when acted upon by the galvanic current, faint traces of contractility.

The *proper substance of the spleen* or *spleen pulp* is a soft mass of a dark reddish-brown colour, resembling grumous blood. When examined, by means of a thin section, under the microscope, it is found to consist of a number of branching cells, and of an intercellular substance. The cells are connective-tissue corpuscles, and have been named the *sustentacular* or *supporting cells of the pulp*. The processes of these branching cells communicate with each other, thus forming a delicate reticulated tissue in the interior of the areolæ formed by the trabeculæ of the capsule, so that each primary space may be considered to be divided into a number of smaller spaces by the junction of the processes of the branching corpuscles. These secondary spaces contain blood, in which, however, the white corpuscles are found to be in larger proportion than they are in ordinary blood. The sustentacular cells are either small, uni-nucleated, or larger, multi-nucleated cells; they do not become deeply stained with carmine, like the cells of the Malpighian bodies, presently to be described (W. Müller), but like them they possess amœboid movements (Cohnheim). In many of them may be seen deep red or reddish-yellow granules of various sizes, which present the characters of the hæmatin of the blood. Sometimes, also, unchanged blood-discs are seen included in these cells, but more frequently blood-discs are found which are altered both in form and colour. In fact, blood-corpuscles in all stages of dis-integration may be noticed to occur within them. Klein has pointed out that sometimes these cells, in the young spleen, contain a proliferating nucleus; that is to say, the nucleus is of large size, and presents a number of knob-like projections, as if small nuclei were budding from it by a process of gemmation. This observation is of importance, as it may explain one possible source of the colourless blood-corpuscles.

The interspaces or areolæ formed by the framework of the spleen are thus filled by a delicate reticulum of branched connective-tissue corpuscles, the interstices of which are occupied by blood, and in which the blood-vessels terminate in the manner now to be described.

Blood-vessels of the Spleen.—The splenic artery is remarkable for its large size in proportion to the size of the organ, and also for its tortuous course. It divides into six or more branches, which enter the hilum of the spleen and ramify throughout its substance (fig. 517), receiving sheaths from an involution of the external fibrous tissue. Similar sheaths also invest the nerves and veins.

Each branch runs in the transverse axis of the organ, from within outwards, diminishing in size during its transit, and giving off in its passage smaller branches, some of which pass to the anterior, others to the posterior part. These ultimately leave the trabecular sheaths, and terminate in the proper substance of the spleen in small tufts or pencils of minute arterioles, which open into the interstices of the reticulum formed by the branched sustentacular cells. Each of the larger branches of the artery supplies chiefly that region of the organ in which the branch ramifies, having no anastomosis with the majority of the other branches.

The *arterioles*, supported by the minute trabeculæ, traverse the pulp in all directions in bundles or pencilli of straight vessels. Their external coat, on leaving the trabecular sheaths, consists of ordinary connective tissue, but it gradually undergoes a transformation, becomes much thickened, and converted into a lymphoid material.* This change is effected by the conversion of the connective tissue into a lymphoid tissue; the bundles of connective tissue becoming looser and laxer, their fibrils more delicate, and containing in their interstices an abundance of lymph-corpuscles (W. Müller). This lymphoid material is supplied with blood by minute vessels derived from the artery with which it is in contact, and which terminates by breaking up into a network of capillary vessels.

The altered coat of the arterioles, consisting of lymphoid tissue, presents here and there thickenings of a spheroidal shape, the *Malpighian bodies of the spleen*. These bodies vary in size from about $\frac{1}{100}$ of an inch to $\frac{1}{25}$ of an inch in diameter. They are merely local expansions or hyperplasiæ of the lymphoid tissue, of which the external coat of the smaller arteries of the spleen is formed.

FIG. 517.—Transverse section of the human spleen, showing the distribution of the splenic artery and its branches.

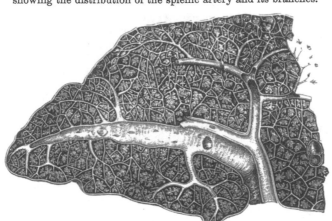

They are most frequently found surrounding the arteriole, which thus seems to tunnel them, but occasionally they grow from one side of the vessel only, and present the appearance of a sessile bud growing from the arterial wall. Klein, however, denies this, and says it is incorrect to describe the Malpighian bodies as isolated masses of adenoid tissue, but that they are always formed around an artery, though there is generally a greater amount on one side than on the other, and that, therefore, in transverse sections, the artery, in the majority of cases, is found in an eccentric position. These bodies are visible to the naked eye on the surface of a fresh section of the organ, appearing as minute dots of a semi-opaque whitish colour in the dark substance of the pulp. In minute structure they resemble the adenoid tissue of lymphatic glands, consisting of a delicate reticulum, in the meshes of which lie ordinary lymphoid cells.

The reticulum of the tissue is made up of extremely delicate fibrils, and is comparatively open in the centre of the corpuscle, becoming closer at the periphery of the body. The cells which it encloses, like the supporting cells of the pulp, are possessed of amœboid movements, but when treated with

* According to Klein, it is the sheath of the small vessel which undergoes this transformation, and forms a 'solid mass of adenoid tissue which surrounds the vessel like a cylindrical sheath.' (*Atlas of Histology*, p. 424.)

carmine become deeply stained, and can be easily recognised from those of the pulp.

The arterioles terminate in capillaries, which traverse the pulp in all directions; their walls become much attenuated, lose their tubular character, and the cells of the lymphoid tissue of which they are composed become altered, presenting a branched appearance, and acquiring processes which are directly connected with the processes of the sustentacular cells of the pulp (fig. 519). In this manner the capillary vessels terminate, and the blood flowing through them finds its way into the interstices of the reticulated tissue formed by the branched connective-tissue corpuscles of the splenic pulp. Thus the blood passing through the spleen is brought into intimate relation with the elements of the pulp, and no doubt undergoes important changes.

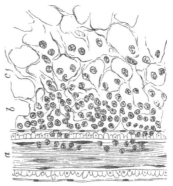

FIG. 518.—Part of a Malpighian capsule of the spleen of man. (Klein and Noble Smith.)

a. Arterial branch in longitudinal section.
b. Adenoid tissue, still containing the lymph-corpuscles; only their nuclei are shown.
c. Adenoid reticulum, the lymph-corpuscles accidentally removed.

After these changes have taken place the blood is collected from the interstices of the tissue by the rootlets of the veins, which commence much in the same way as the arteries terminate. Where a vein is about to commence the connective-tissue corpuscles of the pulp arrange themselves in rows, in such a way as to form an elongated space or sinus. They become changed in shape, being elongated and spindle-shaped, and overlap each other at their extremities. They thus form a sort of endothelial lining of the path or sinus, which is the radicle of a vein. On the outer surface of these cells are seen delicate transverse lines or markings, which are due to minute elastic fibrillæ arranged in a circular manner around the sinus. Thus the channel obtains a continuous external investment, and gradually

FIG. 519.—Section of spleen, showing the termination of the small blood-vessels.

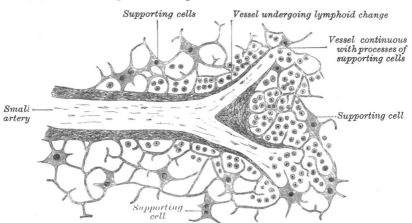

becomes converted into a small vein, which after a time presents a coat of ordinary connective-tissue, lined by a layer of fusiform epithelial cells, which are continuous with the supporting cells of the pulp. The smaller veins unite to form larger ones, which do not accompany the arteries, but soon enter the trabecular sheaths of the capsule, and by their junction form six or more branches, which emerge from the hilum, and, uniting, constitute the splenic vein, the largest radicle of the vena portæ.

The veins are remarkable for their numerous anastomoses, while the arteries hardly anastomose at all.

The lymphatics originate in two ways, i.e. from the sheaths of the arteries and in the trabeculæ. The former accompany the blood-vessels, the latter pass to the superficial lymphatic plexus which may be seen on the surface of the organ. The two sets communicate in the interior of the organ. They pass through the lymphatic glands at the hilum, and terminate in the thoracic duct.

The *nerves* are derived from branches of the right and left semilunar ganglia, and from the right pneumogastric nerve.

Surface Form.—The spleen is situated under cover of the ribs of the left side, being separated from them by the Diaphragm, and above by a small portion of the lower margin of the left lung and pleura. Its position corresponds to the eighth, ninth, tenth, and eleventh ribs. It is placed very obliquely. ' It is oblique in two directions, viz. from above downwards and outwards, and also from above downwards and forwards ' (Cunningham). ' Its highest and lowest points are on a level respectively with the ninth dorsal and first lumbar spines ; its inner end is distant about an inch and a half from the median plane of the body, and its outer end about reaches the mid-axillary line ' (Quain).

Surgical Anatomy.—Injury of the spleen is less common than that of the liver, on account of its protected situation and connections. It may be ruptured by direct or indirect violence ; torn by a broken rib ; or injured by a punctured or gunshot wound. When the organ is enlarged, the chance of rupture is increased. The great risk is hæmorrhage, owing to the vascularity of the organ, and the absence of a proper system of capillaries. The injury is not, however, necessarily fatal, and this would appear to be due, in a great measure, to the contractile power of its capsule, which narrows the wound and prevents the escape of blood. In cases where the diagnosis is clear and the symptoms indicate danger to life, laparotomy must be performed, and if the hæmorrhage cannot be stayed by ordinary surgical methods, the spleen must be removed. The spleen may become displaced, producing great pain from stretching of the vessels and nerves, and this may require removal of the organ. The spleen may become enormously enlarged in certain diseased conditions, such as ague, leukæmia, syphilis, valvular disease of the heart, or without any obtainable history of previous disease. It may also become enlarged in lymphadenoma, as a part of a general blood-disease. In these cases the tumour may fill a considerable part of the abdomen and extend into the pelvis, and may be mistaken for ovarian or uterine disease.

The spleen is sometimes the seat of cystic tumours, especially hydatids, and of abscess. These cases require treatment by incision and drainage ; and in abscess great care must be taken, if there are no adhesions between the spleen and abdominal cavity, to prevent the escape of any of the pus into the peritoneal cavity. If possible the operation should be performed in two stages. Sarcoma and carcinoma are occasionally found in the spleen, but very rarely as a primary disease.

Extirpation of the spleen has been performed for wounds or injuries, in floating spleen, in simple hypertrophy, and in leukæmic enlargement; but in these latter cases the operation is now regarded as unjustifiable, as every case in which it has been performed has terminated fatally. The incision is best made in the left semilunar line ; the spleen is isolated from its surroundings, and the pedicle transfixed and ligatured in two portions, before the tumour is turned out of the abdominal cavity, if this is possible, so as to avoid any traction on the pedicle, which may cause tearing of the splenic vein. In applying the ligature care must be taken not to include the tail of the pancreas, and in lifting out the organ to avoid rupturing the capsule.

ORGANS OF VOICE AND RESPIRATION

THE LARYNX

THE Larynx is the organ of voice, and is placed at the upper part of the air-passage. It is situated between the trachea and base of the tongue, at the upper and fore part of the neck, where it forms a considerable projection in the middle line. On either side of it lie the great vessels of the neck ; behind, it forms part of the boundary of the pharynx, and is covered by the mucous membrane lining that cavity. Its vertical extent corresponds to the fourth, fifth, and sixth cervical vertebræ, but it is placed somewhat higher in the female and also during childhood. In infants between six and twelve months of age Symington found that the tip of the epiglottis was a little above the level of the cartilage, between the odontoid process and body of the axis, and that between infancy and adult life the larynx descends for a distance equal to two vertebral bodies and two intervertebral discs. According to Sappey the average measurements of the adult larynx are as follows :

	In males	In females
Vertical diameter	44 mm.	36 mm.
Transverse diameter	43 ,,	41 ,,
Antero-posterior diameter	36 ,,	26 ,,
Circumference	136 ,,	112 ,,

Until puberty there is no marked difference between the larynx of the male and that of the female. In the latter its further increase in size is only slight, whereas in the former it is great ; all the cartilages are enlarged and the thyroid becomes prominent as the *pomum Adami* in the middle line of the neck, while the length of the glottis is nearly doubled.

The larynx is broad above, where it presents the form of a triangular box, flattened behind and at the sides, and bounded in front by a prominent vertical ridge. Below, it is narrow and cylindrical. It is composed of cartilages, which are connected together by ligaments and moved by numerous muscles. It is lined by mucous membrane, which is continuous above with that lining the pharynx and below with that of the trachea.

The **Cartilages of the Larynx** are nine in number, three single, and three pairs :

Thyroid.	Two Arytenoid.
Cricoid.	Two Cornicula Laryngis.
Epiglottis.	Two Cuneiform.

The **thyroid** (θυρεός, *a shield*) is the largest cartilage of the larynx. It consists of two lateral lamellæ or alæ, united at an acute angle in front, forming a vertical projection in the middle line which is prominent above, and called the *pomum Adami*. This projection is subcutaneous, more distinct in the male than in the female, and occasionally separated from the integument by a bursa mucosa.

Each lamella is quadrilateral in form. Its *outer surface* presents an oblique ridge, which passes downwards and forwards from a tubercle, situated near the root of the superior cornu, to a small tubercle near the anterior part of the lower border. This ridge gives attachment to the Sterno-thyroid and Thyro-hyoid muscles, and the portion of cartilage included between it and the posterior border to part of the Inferior constrictor muscle.

The *inner surface* of each ala is smooth, slightly concave, and covered by mucous membrane above and behind ; but in front, in the receding angle formed by their junction, are attached the epiglottis, the true and false vocal cords, the Thyro-arytenoid and Thyro-epiglottidean muscles, and the thyro-epiglottidean ligament.

The *upper border of the thyroid cartilage* is sinuously curved, being concave at its posterior part, just in front of the superior cornu, then rising into a convex outline, which dips in front to form the sides of a notch, the *thyroid notch*, in the middle line, immediately above the pomum Adami. This border gives attachment throughout its whole extent to the thyro-hyoid membrane.

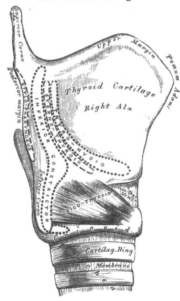

FIG. 520.—Side view of the thyroid and cricoid cartilages.

The *lower border* is nearly straight in front, but behind, close to the cornu, is concave. It is connected to the cricoid cartilage, in and near the median line, by the middle portion of the crico-thyroid membrane ; and, on each side, by the Crico-thyroid muscle.

The *posterior borders*, thick and rounded, terminate, above, in the *superior cornua*, and below, in the *inferior cornua*. The two superior cornua are long and narrow, directed upwards, backwards, and inwards, and terminate in conical extremities, which give attachment to the lateral thyro-hyoid ligament. The two inferior cornua are short and thick ; they pass downwards, with a slight inclination forwards and inwards, and present, on their inner surfaces, a small oval articular facet for articulation with the side of the cricoid cartilage. The posterior border receives the insertion of the Stylo-pharyngeus and Palato-pharyngeus muscles on each side.

During infancy the alæ of the thyroid cartilage are joined to each other by a narrow, lozenge-shaped strip, named the *intra-thyroid cartilage*. This strip extends from the upper to the lower border of the cartilage in the middle line, and is distinguished from the alæ by being more transparent and more flexible.

The **cricoid cartilage** is so called from its resemblance to a signet ring (κρίκος, *a ring*). It is smaller, but thicker and stronger than the thyroid cartilage, and forms the lower and back part of the cavity of the larynx. It consists of two parts : a quadrate portion, situated behind, and a narrow ring or arch, one-fourth or one-fifth the depth of the posterior part, situated in front. The posterior square portion rapidly narrows at the sides of the cartilage, at the expense of the upper border, into the anterior portion.

Its *posterior portion* is very deep and broad, and measures from above downwards about an inch (2–3 c.m.) ; it presents, on its posterior surface, in the middle line, a vertical ridge for the attachment of the longitudinal fibres of the œsophagus ; and on either side a broad depression for the Crico-arytenoideus posticus muscle.

Its *anterior portion* is narrow and convex, and measures vertically about one-fourth or one-fifth of an inch (·5–·7 c.m.) ; it affords attachment externally in front and at the sides to the Crico-thyroid muscles, and behind, to part of the Inferior constrictor.

At the point of junction of the posterior quadrate portion with the rest of the cartilage is a small round elevation, for articulation with the inferior cornu of the thyroid cartilage.

The *lower border* of the cricoid cartilage is horizontal, and connected to the upper ring of the trachea by fibrous membrane.

Its *upper border* is directed obliquely upwards and backwards, owing to the great depth of the posterior surface. It gives attachment, in front, to the middle portion of the crico-thyroid membrane; at the sides, to the lateral portion of the same membrane and to the lateral Crico-arytenoid muscle; behind, it presents, in the middle, a shallow notch, and on each side of this is a smooth, oval surface, directed upwards and outwards, for articulation with the arytenoid cartilage.

The *inner surface* of the cricoid cartilage is smooth, and lined by mucous membrane.

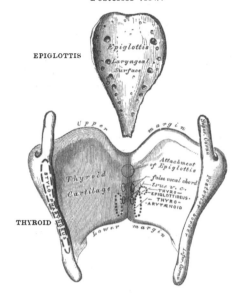

FIG. 521.—The cartilages of the larynx. Posterior view.

The **arytenoid cartilages** are so called from the resemblance they bear, when approximated, to the mouth of a pitcher (ἀρύταινα, *a pitcher*). They are two in number, and situated at the upper border of the cricoid cartilage, at the back of the larynx. Each cartilage is pyramidal in form, and presents for examination three surfaces, a base, and an apex.

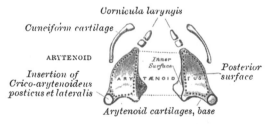

The *posterior surface* is triangular, smooth, concave, and gives attachment to the Arytenoid muscle.

The *anterior* or *external surface* is somewhat convex and rough. It presents rather below its centre a transverse ridge, to the inner extremity of

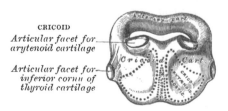

which is attached the false vocal cord, and to the outer part, as well as the surfaces above and below, is attached the Thyro-arytenoid muscle.

The *internal surface* is narrow, smooth, and flattened, covered by mucous membrane, and forms the lateral boundary of the respiratory part of the glottis.

The *base* of each cartilage is broad, and presents a concave smooth surface, for articulation with the cricoid cartilage. Two of its angles require special mention: the *external*, which is short, rounded, and prominent, projects backwards and outwards, and is termed the *muscular process*, from receiving the insertion of the Posterior and Lateral crico-arytenoid muscles. The *anterior angle*, also prominent, but more pointed, projects horizontally forwards, and gives attachment to the true vocal cord. This angle is called the *vocal process*.

The *apex* of each cartilage is pointed, curved backwards and inwards, and surmounted by a small conical, cartilaginous nodule, the *corniculum laryngis*.

The **cornicula laryngis** (*cartilages of Santorini*) are two small conical nodules, consisting of white fibro-cartilage, which articulate with the summit of the arytenoid cartilages and serve to prolong them backwards and inwards. To them are attached the aryteno-epiglottidean folds. They are sometimes united to the arytenoid cartilages.

The **cuneiform cartilages** (*cartilages of Wrisberg*) are two small, elongated, cartilaginous bodies, placed one on each side, in the fold of mucous membrane which extends from the apex of the arytenoid cartilage to the side of the epiglottis (*aryteno-epiglottidean fold*); they give rise to small whitish elevations on the inner surface of the mucous membrane, just in front of the arytenoid cartilages.

The **epiglottis** is a thin lamella of fibro-cartilage, of a yellowish colour, shaped like a leaf, and placed behind the tongue in front of the superior opening of the larynx. Its free extremity is broad and rounded; its attached part is long, narrow, and connected to the receding angle between the two alæ of the thyroid cartilage, just below the median notch, by a long, narrow, ligamentous band, the *thyro-epiglottic ligament*. It is also connected to the posterior surface of the body of the hyoid bone by an elastic ligamentous band, the *hyo-epiglottic ligament*.

Its *anterior* or *lingual surface* is curved forwards towards the tongue, and covered at its upper, free part by mucous membrane, which is reflected on to the sides and base of the organ, forming a median and two lateral folds, the *glosso-epiglottidean folds*. The depressions between the epiglottis and the base of the tongue, on either side of the median fold, are named the *valleculæ*. The lower part of its anterior surface lies behind the hyoid bone, the thyro-hyoid membrane, and upper part of the thyroid cartilage, but is separated from these structures by a mass of fatty tissue.

Its *posterior* or *laryngeal surface* is smooth, concave from side to side, concavo-convex from above downwards; its lower part projects backwards as an elevation, the *tubercle* or *cushion*; when the mucous membrane is removed, the surface of the cartilage is seen to be studded with a number of small mucous glands, which are lodged in little pits upon its surface. To its sides the aryteno-epiglottidean folds are attached.

Structure.—The cuneiform cartilages, the epiglottis, and the apices of the arytenoids are composed of yellow fibro-cartilage, which shows little tendency to calcification; on the other hand the thyroid, cricoid, and the greater part of the arytenoids consist of hyaline cartilage, and become more or less ossified as age advances. Ossification commences about the twenty-fifth year in the thyroid cartilage, somewhat later in the cricoid and arytenoids; by the sixty-fifth year these cartilages may be completely converted into bone. The cornicula laryngis consist of white fibro-cartilage, which becomes osseous about the seventieth year.

Ligaments.—The ligaments of the larynx are *extrinsic*, i.e. those connecting the thyroid cartilage and epiglottis with the hyoid bone, and the cricoid cartilage with the trachea; and *intrinsic*, those which connect the several cartilages of the larynx to each other.

The ligaments connecting the thyroid cartilage with the hyoid bone are three in number—the thyro-hyoid membrane, and the two lateral thyro-hyoid ligaments.

The *thyro-hyoid membrane*, or *middle thyro-hyoid ligament*, is a broad, fibro-elastic, membranous layer, attached below to the upper border of the thyroid cartilage, and above to the posterior border of the body and greater cornua of the hyoid bone, thus passing behind the postero-inferior surface of the hyoid, and being separated from it by a synovial bursa, which facilitates the upward movement of the larynx during deglutition. It is thicker in the middle line than at either side, and is pierced, in the latter situation, by the superior laryngeal vessels and the internal branch of the superior laryngeal nerve. Its anterior surface is in relation with the Thyro-hyoid, Sterno-hyoid, and Omo-hyoid muscles, and with the body of the hyoid bone.

The *two lateral thyro-hyoid ligaments* are rounded, elastic cords, which pass between the superior cornua of the thyroid cartilage and the extremities of the greater cornua of the hyoid bone. A small cartilaginous nodule (*cartilago triticea*), sometimes bony, is frequently found in each.

The ligament connecting the epiglottis with the hyoid bone is the *hyo-epiglottic*. In addition to this extrinsic ligament, the epiglottis is connected to the tongue by the three glosso-epiglottidean folds of mucous membrane, which may also be considered as extrinsic ligaments of the epiglottis.

The *hyo-epiglottic ligament* is an elastic band, which extends from the anterior surface of the epiglottis, near its apex, to the upper border of the body of the hyoid bone.

The ligaments connecting the thyroid cartilage to the cricoid are also three in number—the crico-thyroid membrane, and the capsular ligaments.

The *crico-thyroid membrane* is composed mainly of yellow elastic tissue. It consists of three parts, a central, triangular portion and two lateral portions. The central part is thick and strong, narrow above and broadening out below. It connects together the contiguous margins of the thyroid and cricoid cartilages. It is convex, concealed on each side by the Crico-thyroid muscle, but subcutaneous in the middle line ; it is crossed horizontally by a small anastomotic arterial arch, formed by the junction of the two crico-thyroid arteries. The lateral portions are thinner and lie close under the mucous membrane of the larynx. They extend from the superior border of the cricoid cartilage to the inferior margin of the true vocal cords, with which they are continuous.

The lateral portions are lined internally by mucous membrane, and covered by the lateral Crico-arytenoid and Thyro-arytenoid muscles.

A *capsular ligament* encloses the articulation of the inferior cornu of the thyroid with the cricoid cartilage on each side. The articulation is lined by synovial membrane.

The ligaments connecting the arytenoid cartilages to the cricoid are two *capsular ligaments*, and two *posterior crico-arytenoid ligaments*. The *capsular ligaments* are thin and loose capsules attached to the margin of the articular surfaces ; they are lined internally by synovial membrane. The *posterior crico-arytenoid ligaments* extend from the cricoid to the inner and back part of the base of the arytenoid cartilage.

The ligament connecting the epiglottis with the thyroid cartilage is the thyro-epiglottic.

The *thyro-epiglottic ligament* is a long, slender, elastic cord which connects the apex of the epiglottis with the receding angle of the thyroid cartilage, immediately beneath the median notch, above the attachment of the vocal cords.

The *crico-tracheal ligament* connects the cricoid cartilage with the first ring of the trachea. It resembles the fibrous membrane which connects the cartilaginous rings of the trachea to each other.

Interior of the Larynx.—The *cavity of the larynx* extends from its superior aperture to the lower border of the cricoid cartilage. It is divided into two parts by the projection inwards of the true vocal cords, between which is a narrow triangular fissure or chink, the *rima glottidis*. The portion of the cavity of the larynx above the true vocal cords, sometimes called the *vestibule*, is broad and triangular in shape, and corresponds to the interval between the alæ of the thyroid cartilage ; it contains the false vocal cords, and between these and the true vocal cords are the ventricles of the larynx. The portion below the true vocal cords widens out, and is at first of an elliptical and lower down of a circular form, and is continuous with the tube of the trachea.

The *superior aperture of the larynx* (fig. 522) is a triangular or cordiform opening, wide in front, narrow behind, and sloping obliquely downwards and backwards. It is bounded, in front, by the epiglottis ; behind, by the apices of he arytenoid cartilages and the cornicula laryngis ; and laterally, by a fold of mucous membrane, enclosing ligamentous and muscular fibres, stretched between the sides of the epiglottis and the apices of the arytenoid cartilages : these are the *aryteno-epiglottidean folds,* on the margins of which the cuneiform cartilages form more or less distinct whitish prominences.

The *rima glottidis* is the elongated fissure or chink between the inferior or true vocal cords in front, and between the bases and vocal processes of the arytenoid carti-lages behind. It is therefore frequently subdivided into an anterior interligamentous or *vocal* portion (*glottis vocalis*) and a posterior intercartilaginous or *respiratory* part (*glottis respiratoria*). Posteriorly it is limited by the mucous membrane passing between the arytenoid cartilages. The vocal portion averages about three-fifths of the length of the entire aperture. It is the narrowest part of the cavity of the

FIG. 522.—Larynx, viewed from above.   (Testut.)

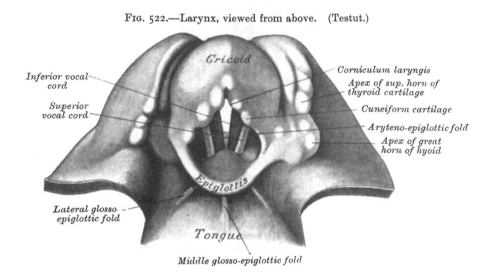

Inferior vocal cord

Superior vocal cord

Lateral glosso-epiglottic fold

Corniculum laryngis

Apex of sup. horn of thyroid cartilage

Cuneiform cartilage

Aryteno-epiglottic fold

Apex of great horn of hyoid

Middle glosso-epiglottic fold

larynx, and its level corresponds with the bases of the arytenoid cartilages. Its length, in the male, measures rather less than an inch (20–25 mm.) ; in the female it is shorter by 5 or 6 mm., or three lines. The width and shape of the rima glottidis vary with the movements of the vocal cords and arytenoid cartilages during respiration and phonation. In the condition of rest, i.e. when these structures are uninfluenced by muscular action, as in quiet respiration, the glottis vocalis is triangular, with its apex in front and its base behind—the latter being repre-sented by a line, about 8 mm. long, connecting the anterior extremities of the vocal processes, while the inner surfaces of the arytenoids are parallel to each other, and hence the glottis respiratoria is rectangular. During extreme adduc-tion of the cords, as in the emission of a high note, the glottis vocalis is reduced to a linear slit by the apposition of the cords, while the glottis respiratoria is triangular, its apex corresponding to the anterior extremities of the vocal processes of the arytenoids, which are approximated by the inward rotation of the cartilages. Conversely in extreme abduction of the cords, as in forced inspiration, the arytenoids and their vocal processes are rotated outwards, and the glottis respiratoria is triangular in shape but with its apex directed backwards.

In this condition the entire glottis is somewhat lozenge-shaped, the sides of the glottis vocalis diverging from before backwards, those of the glottis respiratoria diverging from behind forwards—the widest part of the aperture corresponding with the attachment of the cords to the vocal processes.*

The *superior* or *false vocal cords*, so called because they are not directly concerned in the production of the voice, are two thick folds of mucous membrane, enclosing a narrow band of fibrous tissue, the *superior thyro-arytenoid ligament*, which is attached in front to the angle of the thyroid cartilage immediately below the attachment of the epiglottis, and behind to the anterior surface of the arytenoid cartilage. The lower border of this ligament, enclosed in mucous membrane, forms a free crescentic margin, which constitutes the upper boundary of the ventricle of the larynx.

The *inferior* or *true vocal cords*, so called from their being concerned in the production of sound, are two strong bands (*inferior thyro-arytenoid ligaments*), covered on their surface by a thin layer of mucous membrane. Each ligament consists of a band of yellow elastic tissue, attached in front to the depression between the two alæ of the thyroid cartilage, and behind to the anterior angle (vocal process) of the base of the arytenoid. Its lower border is continuous with the thin lateral part of the crico-thyroid membrane. Its upper border forms the lower boundary of the ventricle of the larynx. Externally, the Thyro-arytenoideus muscle lies parallel with it. It is covered internally by mucous membrane, which is extremely thin, and closely adherent to its surface.

The *ventricle of the larynx* is an oblong fossa, situated between the superior and inferior vocal cords on each side, and extending nearly their entire length. This fossa is bounded, above, by the free crescentic edge of the superior vocal cord; below, by the straight margin of the true vocal cord; externally, by the mucous membrane covering the corresponding Thyro-arytenoideus muscle. The anterior part of the ventricle leads up by a narrow opening into a cæcal pouch of mucous membrane of variable size, called the *laryngeal pouch*.

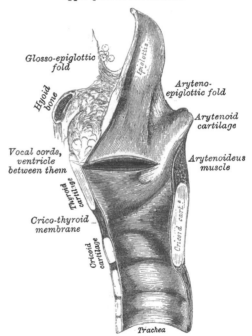

FIG. 523.—Vertical section of the larynx and upper part of the trachea.

Glosso-epiglottic fold

Epiglottis

Hyoid bone

Aryteno-epiglottic fold

Arytenoid cartilage

Vocal cords, ventricle between them

Arytenoideus muscle

Thyroid cartilage

Crico-thyroid membrane

Cricoid cart.

Cricoid cartilage

Trachea

The *sacculus laryngis*, or laryngeal pouch, is a membranous sac, placed between the superior vocal cord and the inner surface of the thyroid cartilage, occasionally extending as far as its upper border or even higher: it is conical in form, and curved slightly backwards. On the surface of its mucous membrane are the openings of sixty or seventy mucous glands, which are lodged in the submucous areolar tissue. This sac is enclosed in a fibrous capsule, continuous below with the superior thyro-arytenoid ligament: its laryngeal surface is covered by the Aryteno-epiglottideus inferior muscle (*Compressor sacculi laryngis*,

* On the shape of the rima glottidis, in the various conditions of breathing and speaking, see Czermak, *On the Laryngoscope*, translated for the *New Sydenham Society*.

Hilton) ; while its exterior is covered by the Thyro-arytenoideus and Thyro-epiglottideus muscles. These muscles compress the sacculus laryngis, and discharge the secretion it contains upon the chordæ vocales, the surfaces of which it is intended to lubricate.

Muscles.—The intrinsic muscles of the larynx are eight in number, five of which are the muscles of the vocal cords and rima glottidis, and three are connected with the epiglottis.

The five muscles of the vocal cords and rima glottidis are : the

Crico-thyroid.   Crico-arytenoideus lateralis.
Crico-arytenoideus posticus.   Arytenoideus.
                                         Thyro-arytenoideus.

The *Crico-thyroid* is triangular in form, and situated at the fore part and side of the cricoid cartilage. It arises from the front and lateral part of the cricoid cartilage ; its fibres diverge, passing obliquely upwards and outwards, to be inserted into the lower border of the thyroid cartilage, and into the anterior border of the lower cornua.

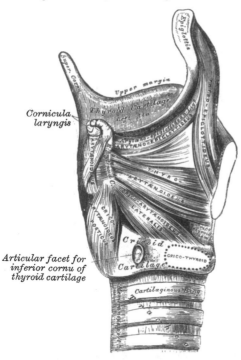

FIG. 524.—Muscles of larynx. Side view. Right ala of thyroid cartilage removed.

Cornicula laryngis

Articular facet for inferior cornu of thyroid cartilage

The inner borders of these two muscles are separated in the middle line by a triangular interval, occupied by the central part of the crico-thyroid membrane.

The *Crico-arytenoideus posticus* arises from the broad depression occupying each lateral half of the posterior surface of the cricoid cartilage ; its fibres pass upwards and outwards, converging to be inserted into the outer angle (muscular process) of the base of the arytenoid cartilage. The upper fibres are nearly horizontal, the middle oblique, and the lower almost vertical.*

Crico - arytenoideus lateralis is smaller than the preceding, and of an oblong form. It arises from the upper border of the side of the cricoid cartilage, and, passing obliquely upwards and backwards, is inserted into the muscular process of the arytenoid cartilage, in front of the preceding muscle.

The *Arytenoideus* is a single muscle, filling up the posterior concave surface of the arytenoid cartilages. It arises from the posterior surface and outer border of one arytenoid cartilage, and is inserted into the corresponding parts of the opposite cartilage. It consists of three planes of fibres, two oblique and one transverse. The *oblique fibres*, the more superficial, form two fasciculi, which pass from the base of one cartilage to the apex of the opposite one. The *trans-*

* Merkel, of Leipsic, has described a muscular slip which occasionally extends between the outer border of the posterior surface of the cricoid cartilage and the posterior margin of the inferior cornu of the thyroid ; this he calls the ' Musculus kerato-cricoideus.' It is not found in every larynx, and when present exists usually only on one side, but is occasionally found on both sides. Sir Wm. Turner (*Edinburgh Medical Journal*, Feb. 1860) states that it is found in about one case in five. Its action is to fix the lower horn of the thyroid cartilage backwards and downwards, opposing in some measure the part of the Crico-thyroid muscle which is connected to the anterior margin of the horn.

verse fibres, the deeper and more numerous, pass transversely across between the two cartilages; hence the Arytenoideus was formerly considered as three muscles, the transverse and the two oblique. A few of the oblique fibres are occasionally continued round the outer margin of the cartilage, and blend with the Thyro-arytenoid or the Aryteno-epiglottideus muscle.

The *Thryo-arytenoideus* is a broad, flat muscle, which lies parallel with the outer side of the true vocal cord. It arises in front from the lower half of the receding angle of the thyroid cartilage, and from the crico-thyroid membrane. Its fibres pass backwards and outwards, to be inserted into the base and anterior surface of the arytenoid cartilage. This muscle consists of two fasciculi.* The *inner* or *inferior portion*, the thicker, is inserted into the vocal process of the arytenoid cartilage, and into the adjacent portion of its anterior surface; it lies parallel with the true vocal cord, to which it is adherent. The *outer* or *superior fasciculus*, the thinner, is inserted into the anterior surface and outer border of the arytenoid cartilage above the preceding fibres; it lies on the outer side of the sacculus laryngis, immediately beneath the mucous membrane.†

The muscles of the epiglottis are: the

Thyro-epiglottideus.
Aryteno-epiglottideus superior.
Aryteno-epiglottideus inferior.

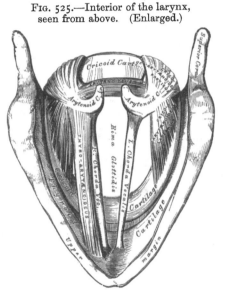

FIG. 525.—Interior of the larynx, seen from above. (Enlarged.)

The *Thyro-epiglottideus* is a delicate fasciculus, which arises from the inner surface of the thyroid cartilage, just external to the origin of the Thyro-arytenoid muscle, of which it is sometimes described as a part, and spreads over the outer surface of the sacculus laryngis: some of its fibres are lost in the aryteno-epiglottidean fold, while others are continued forwards to the margin of the epiglottis (*Depressor epiglottidis*).

The *Aryteno-epiglottideus superior* consists of a few delicate muscular fasciculi, which arise from the apex of the arytenoid cartilages, and become lost in the fold of mucous membrane extending between the arytenoid cartilage and the side of the epiglottis (*aryteno-epiglottidean fold*).

The *Aryteno-epiglottideus inferior* (*Compressor sacculi laryngis*, Hilton) arises from the arytenoid cartilage, just above the attachment of the superior vocal cord; passing forwards and upwards, it spreads out upon the anterior surface of the epiglottis. This muscle is separated from the preceding by an indistinct areolar interval.‡

* Henle describes these two portions as separate muscles, under the names of External and Internal thyro-arytenoid.

† Luschka has described a small but fairly constant muscle as the *Arytenoideus rectus*. It is attached below to the posterior, concave surface of the arytenoid cartilage, beneath the Arytenoideus muscle, and passing upwards, emerges at the upper border of this muscle, and is inserted into the posterior surface of the cartilage of Santorini. (*Anatomy*, by Hyrtl, page 718.)

‡ MUSCULUS TRITICEO-GLOSSUS. Bochdalek jun. (*Prager Vierteljahreschrift*, 2nd part, 1866) describes a muscle hitherto entirely overlooked, except for a brief statement in Henle's *Anatomy*, which arises from the nodule of cartilage (*corpus triticeum*) in the posterior thyro-hyoid ligament, and passes forwards and upwards to enter the tongue along with the Hyo-glossus muscle. He met with this muscle eight times in twenty-two subjects. It occurred in both sexes, sometimes on both sides, sometimes on one only.

Actions.—In considering the action of the muscles of the larynx, they may be conveniently divided into two groups, viz.: 1. Those which open and close the glottis. 2. Those which regulate the degree of tension of the vocal cords.

1. The muscles which open the glottis are the Crico-arytenoidei postici; and those which close it are the Arytenoideus and the Crico-arytenoidei laterales. 2. The muscles which regulate the tension of the vocal cords are the Crico-thyroidei, which tense and elongate them; and the Thyro-arytenoidei, which relax and shorten them. The Thyro-epiglottideus is a depressor of the epiglottis, and the Aryteno-epiglottidei constrict the superior aperture of the larynx, compress the sacculi laryngis, and empty them of their contents.

The *Crico-arytenoidei postici* separate the chordæ vocales, and, consequently, open the glottis, by rotating the arytenoid cartilages outwards around a vertical axis passing through the crico-arytenoid joints; so that their vocal processes and the vocal cords attached to them become widely separated.

The *Crico-arytenoidei laterales* close the glottis, by rotating the arytenoid cartilages inwards, so as to approximate their vocal processes.

The *Arytenoideus muscles* approximate the arytenoid cartilages, and thus close the opening of the glottis, especially at its back part.

The *Crico-thyroid muscles* produce tension and elongation of the vocal cords. This is effected as follows: The thyroid cartilage is fixed by its Extrinsic muscles, then the Crico-thyroid muscles, when they act, draw upwards the front of the cricoid cartilage and so depress the posterior portion, which carries with it the arytenoid cartilages, and thus elongate the vocal cords.

The *Thyro-arytenoidei muscles*, consisting of two parts having different attachments and different directions, are rather complicated as regards their action. Their main use is to draw the arytenoid cartilages forwards towards the thyroid, and thus shorten and relax the vocal cords. But, owing to the connection of the inner portion with the vocal cord, this part, if acting separately, is supposed to modify its elasticity and tension, and the outer portion, being inserted into the outer part of the anterior surface of the arytenoid cartilage, may rotate it inwards, and thus narrow the rima glottidis by bringing the two cords together.

The *Thyro-epiglottidei* may depress the epiglottis; they assist in compressing the sacculi laryngis. The *Aryteno-epiglottideus superior* constricts the superior aperture of the larynx, when it is drawn upwards, during deglutition. The *Aryteno-epiglottideus inferior*, together with some fibres of the Thyro-arytenoidei, compress the sacculus laryngis.

The **Mucous Membrane of the Larynx** is continuous above with that lining the mouth and pharynx, and is prolonged through the trachea and bronchi into the lungs. It lines the posterior surface and the anterior part of the upper surface of the epiglottis, to which it is closely adherent, and forms the aryteno-epiglottidean folds which form the lateral boundaries of the superior aperture of the larynx. It lines the whole of the cavity of the larynx; forms, by its reduplication, the chief part of the superior, or false, vocal cord; and, from the ventricle, is continued into the sacculus laryngis. It is then reflected over the true vocal cords, where it is thin, and very intimately adherent; covers the inner surface of the crico-thyroid membrane and cricoid cartilage; and is ultimately continuous with the lining membrane of the trachea. The fore part of the anterior surface and the upper half of the posterior surface of the epiglottis, the upper part of the aryteno-epiglottidean folds, and the true vocal cords are covered by stratified squamous epithelium; all the rest of the laryngeal mucous membrane is covered by columnar ciliated cells.

Glands.—The mucous membrane of the larynx is furnished with numerous muciparous glands, the orifices of which are found in nearly every part; they are very numerous upon the epiglottis, being lodged in little pits in its substance; they are also found in large numbers along the posterior margin of the aryteno-epiglottidean fold, in front of the arytenoid cartilages, where they are termed the *arytenoid glands*. They exist also in large numbers upon the inner surface of the sacculus laryngis. None are found on the free edges of the vocal cords.

Vessels and Nerves.—The *arteries of the larynx* are the laryngeal branches derived from the superior and inferior thyroid. The *veins* accompany the arteries: those accompanying the superior laryngeal artery join the superior

thyroid vein, which opens into the internal jugular vein ; while those accompanying the inferior laryngeal artery join the inferior thyroid vein, which opens into
the innominate vein.   The lymphatics consist of two sets, superior and inferior.
The former accompany the superior laryngeal artery and pierce the thyro-hyoid
membrane, to terminate in the glands situated near the bifurcation of the common
carotid artery.   The latter pass through the crico-thyroid membrane, and open
into one or two glands lying either in front of that membrane or to the side of
the cricoid cartilage.   The nerves are derived from the internal and external
laryngeal branches of the superior laryngeal nerve, from the inferior or recurrent
laryngeal, and from the sympathetic.   The internal laryngeal nerve is almost
entirely sensory, but some motor filaments are said to be carried by it to the
Arytenoideus muscle.   It divides into a branch which is distributed to both
surfaces of the epiglottis, a second to the aryteno-epiglottidean folds, and a third,
the largest, which supplies the mucous membrane over the back of the larynx
and communicates with the recurrent laryngeal.   The external laryngeal nerve
supplies the Crico-thyroid muscle.   The recurrent laryngeal passes upwards under
the lower border of the Inferior constrictor, and enters the larynx between the
cricoid and thyroid cartilages.   It supplies all the muscles of the larynx
except the Crico-thyroid and part of the Arytenoideus.   The sensory branches of
the laryngeal nerves form subepithelial plexuses, from which fibres ascend to end
between the cells covering the mucous membrane.

Over the posterior surface of the epiglottis, in the aryteno-epiglottidean folds,
and less regularly in some other parts, taste-buds, similar to those in the tongue,
are found.

The Trachea (fig. 526)

The **trachea,** or **windpipe,** is a cartilaginous and membranous cylindrical tube,
flattened posteriorly, which extends from the lower part of the larynx, on a level
with the sixth cervical vertebra, to opposite the fourth, or sometimes the fifth, dorsal
vertebra, where it divides into the two bronchi, one for each lung.   The trachea
measures about four inches and a half in length ; its diameter, from side to side,
is from three-quarters of an inch to an inch, being always greater in the male
than in the female.

Relations.—The anterior surface of the trachea is convex, and covered, *in the
neck,* from above downwards, by the isthmus of the thyroid gland, the inferior
thyroid veins, the arteria thyroidea ima (when that vessel exists), the Sterno-
hyoid and Sterno-thyroid muscles, the cervical fascia, and, more superficially, by
the anastomosing branches between the anterior jugular veins ; *in the thorax,* it
is covered from before backwards by the first piece of the sternum, the remains
of the thymus gland, the left innominate vein, the arch of the aorta, the innominate and left carotid arteries, and the deep cardiac plexus.   Posteriorly it is in
relation with the œsophagus ; laterally, *in the neck,* it is in relation with the
common carotid arteries, the lateral lobes of the thyroid gland, the inferior
thyroid arteries, and recurrent laryngeal nerves ; and, *in the thorax,* it lies in the
upper part of the interpleural space (superior mediastinum), and is in relation on
the right to the pleura and right vagus, and near the root of the neck to the
innominate artery ; on its left side are the recurrent laryngeal nerve, the aortic
arch, the left common carotid and subclavian arteries.

The **Right Bronchus,** wider, shorter, and more vertical in direction than the
left, is about an inch in length, and enters the right lung opposite the fifth dorsal
vertebra.   The vena azygos major arches over it from behind ; and the right
pulmonary artery lies below and then in front of it.   About three-quarters of an
inch from its commencement it gives off a branch to the upper lobe of the right
lung.   This is termed the *eparterial* branch, because it is given off above the right
pulmonary artery.   The bronchus now passes below the artery, and is known as
the *hyparterial* branch. It divides into two branches for the middle and lower lobes.

The **Left Bronchus** is smaller and longer than the right, being nearly

two inches in length.   It enters the root of the left lung, opposite the sixth
dorsal vertebra, about an inch lower than the right bronchus.   It passes beneath
the arch of the aorta, crosses in front of the œsophagus, the thoracic duct, and the
descending aorta, and has the left pulmonary artery lying at first above, and then
in front of it.   The left bronchus has no branch corresponding to the eparterial
branch of the right bronchus, and therefore it has been supposed by some that

Fig. 526.—Front view of cartilages of larynx ; the trachea and bronchi.

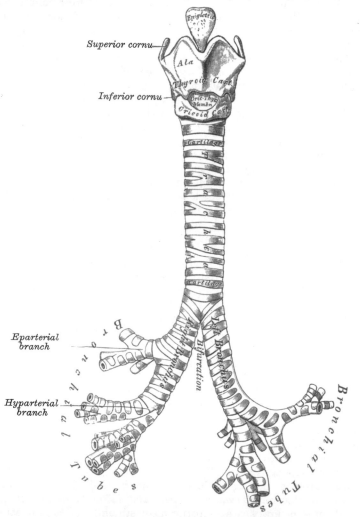

there is no upper lobe to the left lung, but that the so-called upper lobe. corresponds to the middle lobe of the right lung.

When the bronchi enter the lung they appear to divide into nearly equal
branches at the root of the lung, but a somewhat similar arrangement to what is
found in many animals may be made out where each bronchus passes downwards
and backwards towards the extremity of the lower lobe, giving off four branches
at intervals in two directions, dorsally and ventrally, and, in addition, accessory
branches, which arise from the front of the bronchus and pass mesially and
dorsally into the inferior lobe.   In the right bronchus the *first ventral* branch
supplies the middle lobe, the other three and all the dorsal going to the lower
lobe ; in the left bronchus, the *first ventral* supplies the superior lobe, and all the
others, both ventral and dorsal, go to the lower lobe.

If a transverse section is made across the trachea, a short distance above its point of bifurcation, and a bird's-eye view taken of its interior (fig. 527), the septum placed at the bottom of the trachea and separating the two bronchi will be seen to occupy the left of the median line, and the right bronchus appears to be a more direct continuation than the left, so that any solid body dropping into the trachea would naturally be directed towards the right bronchus. This tendency is aided by the larger size of the right tube as compared with its fellow. This fact serves to explain why a foreign body in the trachea more frequently falls into the right bronchus.*

Structure.—The trachea is composed of imperfect cartilaginous rings, fibrous membrane, muscular fibres, mucous membrane, and glands.

The **cartilages** vary from sixteen to twenty in number : each forms an imperfect ring, which surrounds about two-thirds of the cylinder of the trachea, being imperfect behind, where the tube is completed by fibrous membrane. The cartilages are placed horizontally above each other, separated by narrow membranous intervals. They measure about two lines in depth, and half a line in thickness. Their outer surfaces are flattened, but internally they are convex, from being thicker in the middle than at the margins. Two or more of the cartilages often unite, partially or completely, and are sometimes bifurcated at their extremities. They are highly elastic, but sometimes become calcified in advanced life. In the right bronchus the cartilages vary in number from six to eight ; in the left, from nine to twelve. They are shorter and narrower than those of the trachea. The peculiar cartilages are the first and the last.

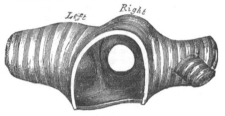

Fig. 527.—Transverse section of the trachea, just above its bifurcation, with a bird's-eye view of the interior.

The *first cartilage* is broader than the rest, and sometimes divided at one end ; it is connected by fibrous membrane with the lower border of the cricoid cartilage, with which, or with the succeeding cartilage, it is sometimes blended.

The *last cartilage* is thick and broad in the middle, in consequence of its lower border being prolonged into a triangular hook-shaped process, which curves downwards and backwards between the two bronchi. It terminates on each side in an imperfect ring, which encloses the commencement of the bronchi. The cartilage above the last is somewhat broader than the rest at its centre.

The fibrous membrane.—The cartilages are enclosed in an elastic fibrous membrane, which forms a double layer ; one layer, the thicker of the two, passing over the outer surface of the ring, the other over the inner surface : at the upper and lower margins of the cartilages these two layers blend together to form a single membrane, which connects the rings one with another. They are thus, as it were, embedded in the membrane. In the space behind, between the extremities of the rings, the membrane forms a single distinct layer.

The **muscular fibres** are disposed in two layers, longitudinal and transverse. The *longitudinal fibres* are the most external, and consist merely of a few scattered longitudinal bundles of fibres.

The *transverse fibres* (Trachealis muscle, Todd and Bowman), the most internal, form a thin layer, which extends transversely between the ends of the cartilages and the intervals between them at the posterior part of the trachea. The muscular fibres are of the unstriped variety.

The **Mucous membrane** is continuous above with that of the larynx, and below with that of the bronchi. Microscopically, it consists of areolar and lymphoid

---

* Reigel asserts that the entry of a foreign body into the *left* bronchus is by no means so infrequent as is generally supposed. See also *Med.-Chir. Trans.* vol. lxxi. p. 121.

tissue, and presents a well-marked basement-membrane, supporting a layer of columnar, ciliated epithelium, between the deeper ends of which are smaller triangular cells, the bases of which, often branched, are attached to the basement-membrane. These triangular cells are mucus-secreting, and may be seen as goblet or chalice cells when their contents have been discharged. In the deepest part of the mucous membrane, and especially between the mucous and submucous layers, longitudinally arranged fibres are very abundant and form a distinct layer.

The **Tracheal glands** are found in great abundance at the posterior part of the trachea. They are racemose glands, and consist of a basement-membrane lined by columnar mucus-secreting cells. They are situated at the back of the trachea, outside the layer of muscular tissue, between it and the outer fibrous layer. Their excretory ducts pierce the muscular and inner fibrous layers, and pass through the submucous and mucous layers to open on the surface of the mucous membrane. Some glands of smaller size are also found at the sides of the trachea, between the layers of fibrous tissue connecting the rings, and others immediately beneath the mucous coat. The secretion from these glands serves to lubricate the inner surface of the trachea.

Vessels and Nerves.—The trachea is supplied with blood by the inferior thyroid arteries. The *veins* terminate in the thyroid venous plexus. The *nerves* are derived from the pneumogastric and its recurrent branches, and from the sympathetic.

Surface Form.—In the middle line of the neck, some of the cartilages of the larynx can be readily distinguished. In the receding angle below the chin, the hyoid bone can easily be made out (see page 93), and a finger's breadth below it is the pomum Adami, the prominence between the upper borders of the two alæ of the thyroid cartilage. About an inch below this, in the middle line, is a depression corresponding to the crico-thyroid space, in which the operation of laryngotomy is performed. This depression is bounded below by a prominent arch, the anterior ring of the cricoid cartilage, below which the trachea can be felt, though it is only in the emaciated that the separate rings can be distinguished. The lower part of the trachea is not easily made out, for as it descends in the neck it takes a deeper position, and is farther removed from the surface. The level of the vocal cords corresponds to the middle of the anterior margin of the thyroid cartilage.

With the laryngoscope the following structures can be seen : The base of the tongue, and the upper surface of the epiglottis, with the glosso-epiglottic ligaments ; the superior aperture of the larynx ; bounded on either side by the aryteno-epiglottidean folds, in which may be seen two rounded eminences, corresponding to the cornicula and cuneiform cartilages. Beneath these the true and false vocal cords, with the ventricle between them. Still deeper, the cricoid cartilage, and some of the anterior parts of the rings of the trachea, and sometimes, in deep inspiration, the bifurcation of the trachea.

Surgical Anatomy.—*Foreign bodies* often find their way into the air-passages. These may be either large soft substances, as a piece of meat, which may become lodged in the upper aperture of the larynx, or in the rima glottidis, and cause speedy suffocation unless rapidly got rid of, or unless an opening is made into the air-passages below, so as to enable the patient to breathe. Smaller bodies, frequently of a hard nature, such as cherry or plum stones, small pieces of bone, buttons, &c., may find their way through the rima glottidis into the trachea or bronchus, or may become lodged in the ventricle of the larynx. The dangers then depend not so much upon the mechanical obstruction as upon the spasm of the glottis which they excite from reflex irritation. When lodged in the ventricle of the larynx, they may produce very few symptoms beyond sudden loss of voice or alteration in the voice sounds, immediately following the inhalation of the foreign body. When, however, they are situated in the trachea, they are constantly striking against the vocal cords during expiratory efforts, and produce attacks of dyspnœa from spasm of the glottis. When lodged in the bronchus, they usually become fixed there, and, occluding the lumen of the tube, cause a loss of the respiratory murmur on the affected side, which is, as stated above, more often the right.

Beneath the mucous membrane of the upper part of the air-passages, there is a considerable amount of submucous tissue, which is liable to become much swollen from effusion in inflammatory affections, constituting the disease known as 'œdema of the glottis.' This effusion does not extend below the level of the vocal cords, on account of the fact that the mucous membrane is closely adherent to these structures, without the intervention of any submucous tissue. So that, in cases of this disease, in which it is necessary to open the air-passages to prevent suffocation, the operation of laryngotomy is sufficient.

Chronic laryngitis is an inflammation of the mucous glands of the larynx, which occurs in those who speak much in public, and is known as 'Clergyman's sore throat.' It is due to the dryness induced by the large amount of cold air drawn into the air-passages during prolonged speaking, which excites increased activity of the mucous glands to keep the parts moist, and this eventually terminates in inflammation of these structures.

Ulceration of the larynx may occur from syphilis, either as a superficial ulceration, or from the softening of a gumma ; from tuberculous disease (laryngeal phthisis), or from malignant disease (epithelioma).

The air-passages may be opened in two different situations : through the crico-thyroid membrane (*laryngotomy*), or in some part of the trachea (*tracheotomy*) ; and to these some surgeons have added a third method, by opening the crico-thyroid membrane and dividing the cricoid cartilage with the upper ring of the trachea (*laryngo-tracheotomy*).

Laryngotomy is anatomically the more simple operation : it can readily be performed, and should be employed in those cases where the air-passages require opening in an emergency for the relief of some sudden obstruction to respiration. The crico-thyroid membrane is very superficial, being covered only in the middle line by the skin, superficial fascia, and the deep fascia. On each side of the middle line it is also covered by the Sterno-hyoid and Sterno-thyroid muscles, which diverge from each other at their upper parts, leaving a slight interval between them. On these muscles rest the anterior jugular veins. The only vessel of any importance in connection with this operation is the crico-thyroid artery, which crosses the crico-thyroid membrane, and which may be wounded, but rarely gives rise to any trouble. The operation is performed thus : the head being thrown back and steadied by an assistant, the finger is passed over the front of the neck, and the crico-thyroid depression felt for. A vertical incision is then made through the skin, in the middle line over this spot, and carried down through the fascia until the crico-thyroid membrane is exposed. A cross cut is then made through the membrane, close to the upper border of the cricoid cartilage, so as to avoid, if possible, the crico-thyroid artery, and a tracheotomy tube introduced. It has been recommended, as a more rapid way of performing the operation, to make a transverse instead of a longitudinal cut through the superficial structures, and thus to open at once the air-passages. It will be seen, however, that in operating in this way the anterior jugular veins would be in danger of being wounded.

Tracheotomy may be performed either above or below the isthmus of the thyroid body : or this structure may be divided and the trachea opened behind it.

The isthmus of the thyroid gland usually crosses the second and third rings of the trachea ; along its upper border is frequently to be found a large transverse communicating branch between the superior thyroid veins ; and the isthmus itself is covered by a venous plexus, formed between the thyroid veins of the opposite sides. Theoretically, therefore, it is advisable to avoid dividing this structure in opening the trachea.

Above the isthmus the trachea is comparatively superficial, being covered by the skin, superficial fascia, deep fascia, Sterno-hyoid and Sterno-thyroid muscles, and a second layer of the deep fascia, which, attached above to the lower border of the hyoid bone, descends beneath the muscles to the thyroid body, where it divides into two layers and encloses the isthmus.

Below the isthmus the trachea lies much more deeply, and is covered by the Sterno-hyoid and the Sterno-thyroid muscles, and a quantity of loose areolar tissue, in which is a plexus of veins, some of them of large size ; they converge to two trunks, the inferior thyroid veins, which descend on either side of the median line on the front of the trachea, and open into the innominate vein. In the infant the thymus gland ascends a variable distance along the front of the trachea ; and opposite the episternal notch the windpipe is crossed by the left innominate vein. Occasionally also in young subjects the innominate artery crosses the tube obliquely above the level of the sternum. The thyroidea ima artery, when that vessel exists, passes from below upwards, along the front of the trachea.

From these observations it must be evident that the trachea can be more readily opened above than below the isthmus of the thyroid body.

Tracheotomy above the isthmus is performed thus : the patient should, if possible, be laid on his back on a table in a good light. A pillow is to be placed under the shoulders and the head thrown back and steadied by an assistant. The surgeon standing on the right side of his patient makes an incision from an inch and a half to two inches in length in the median line of the neck from the top of the cricoid cartilage. The incision must be made exactly in the middle line so as to avoid the anterior jugular veins, and after the superficial structures have been divided, the interval between the Sterno-hyoid muscles must be found, the raphé divided, and the muscles drawn apart. The lower border of the cricoid cartilage must now be felt for, and the upper part of the trachea exposed from this point downwards in the middle line. Bose has recommended that the layer of fascia in front of the trachea should be divided transversely at the level of the lower border of the cricoid cartilage, and, having been seized with a pair of forceps, pressed downwards with the handle of the scalpel. By this means the isthmus of the thyroid gland is depressed and is saved from all danger of being wounded, and the trachea cleanly exposed. The trachea is now transfixed with a sharp hook and drawn forwards in order to steady it, and is then opened by inserting the knife into it and dividing the two or three upper rings from below upwards. If the trachea is to be opened below the isthmus, the incision must be made from a little below the cricoid cartilage to the top of the sternum.

In the child the trachea is smaller, more deeply placed, and more movable than in the adult. In fat or short-necked people, or in those in whom the muscles of the neck are prominently developed, the trachea is more deeply placed than in the opposite conditions.

A portion of the larynx or the whole of it has been removed for malignant disease. The results which have been obtained from the removal of the whole of it have not been very satisfactory, and the cases in which the operation is justifiable are very few. It may be removed by a median incision through the soft parts; freeing the cartilages from the muscles and other structures in front; separating the larynx from the trachea below, and dissecting off the deeper structure from below upwards.

THE PLEURÆ

Each lung is invested, upon its external surface, by an exceedingly delicate serous membrane, the **pleura,** which encloses the organ as far as its root, and is then reflected upon the inner surface of the thorax. The portion of the serous membrane investing the surface of the lung and dipping into the fissures between its lobes, is called the *pleura pulmonalis* (visceral layer of pleura), while that which lines the inner surface of the chest is called the *pleura costalis* (parietal layer of pleura). The space between these two layers is called the *cavity of the pleura,* but it must be borne in mind that in the healthy condition the two layers are in contact and there is no real cavity, until the lung becomes collapsed and a separation of it from the wall of the chest takes place. Each pleura is therefore a shut sac, one occupying the right, the other the left half of the thorax ; and they are perfectly separate from each other. The two pleuræ do not meet in the middle line of the chest, excepting anteriorly opposite the second and third pieces of the sternum ; a space being left between them, which contains all the viscera of the thorax excepting the lungs ; this is the *mediastinum.*

Reflections of the Pleura (fig. 528).—Commencing at the sternum, the pleura passes outwards, lines the costal cartilages, the inner surface of the ribs and Intercostal muscles, and at the back part of the thorax passes over the thoracic ganglia and their branches, and is reflected upon the sides of the bodies of the vertebræ, where it is separated by a narrow interval, the *posterior mediastinum,* from the opposite pleura. From the vertebral column the pleura passes to the side of the pericardium, which it covers to a slight extent ; it then covers the back part of the root of the lung, from the lower border of which a triangular fold descends vertically by the side of the posterior mediastinum to the Diaphragm. This fold is the broad ligament of the lung, the *ligamentum latum pulmonis,* and serves to retain the lower part of that organ in position. *From the root,* the pleura may be traced over the convex surface of the lung, the summit and base, and also over the sides of the fissures between the lobes on to its inner surface and the front part of its root ; from this it is reflected on to the pericardium, and from it to the back of the sternum. *Below,* it covers the upper surface of the Diaphragm, and extends, in front, as low as the costal cartilage of the seventh rib ; at the side of the chest, to the lower border of the tenth rib on the left side and to the upper border of the same rib on the right side ; and behind, it reaches as low as the twelfth rib, and sometimes even as low as the transverse process of the first lumbar vertebra. *Above,* its apex projects, in the form of a *cul-de-sac,* through the superior opening of the thorax into the neck, extending from one to two inches above the margin of the first rib, and receives the summit of the corresponding lung ; this sac is strengthened, according to Sibson, by a dome-like expansion of fascia, attached in front to the posterior border of the first rib, and behind to the anterior border of the transverse process of the seventh cervical vertebra. This is covered and strengthened by a few spreading muscular fibres derived from the Scaleni muscles.

In the front of the chest, where the parietal layer of the pleura is reflected backwards to the pericardium, the two pleural sacs are in contact for a considerable extent. At the upper part of the chest, behind the manubrium, they are not in contact; the point of reflection being represented by a line drawn from the sterno-clavicular articulation to the mid-point of the junction of the manubrium to the body of the sternum. From this point the two pleuræ descend in close contact to the level of the fourth costal cartilage. Here the line of reflection on the right side is continued onwards in nearly a straight line to the lower end of the gladiolus and then turns outwards, while on the left side the line of reflection diverges outwards, so that opposite the seventh cartilage it is about three-quarters of an inch from the left border of the sternum. It, however, always extends considerably farther over the pericardium than the corresponding lung. The lower limit of the pleura is on a considerably lower level than the lower limit of the lung, but does not extend to the attachment of the Diaphragm, so that below the line of reflection of the pleura from the chest wall on to the Diaphragm the latter is in direct contact with the rib cartilages and the Internal intercostal

FIG. 528.—A transverse section of the thorax, showing the relative position of the viscera, and the reflections of the pleuræ.

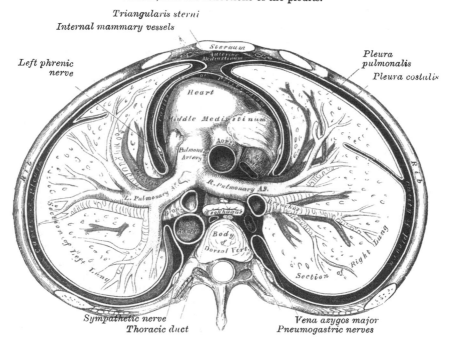

muscles. Moreover, in ordinary inspiration the thin margin of the base of the lung does not extend as low as the line of pleural reflection, with the result that the costal and diaphragmatic pleura are here in contact, the narrow slit between the two being termed the *phrenico-costal sinus*. A similar condition exists behind the sternum and rib cartilages, where the anterior thin margin of the lung falls short of the line of pleural reflection, and where the slit-like cavity between the two layers of pleura forms what is sometimes called the *costo-mediastinal sinus*.

The inner surface of the pleura is smooth, polished, and moistened by a serous fluid ; its outer surface is intimately adherent to the surface of the lung, and to the pulmonary vessels as they emerge from the pericardium ; it is also adherent to the upper surface of the Diaphragm : throughout the rest of its extent it is somewhat thicker, and may be separated from the adjacent parts with extreme facility.

The right pleural sac is shorter, wider, and reaches higher in the neck than the left.

Vessels and Nerves.—The *arteries of the pleura* are derived from the inter-costal, the internal mammary, the musculo-phrenic, thymic, pericardiac, and bronchial vessels. The *veins* correspond to the arteries. The *lymphatics* are very numerous. The *nerves* are derived from the phrenic and sympathetic (Luschka). Kölliker states that nerves accompany the ramification of the bronchial arteries in the pleura pulmonalis.

Surgical Anatomy.—In operations upon the kidney, it must be borne in mind that the pleura may sometimes extend below the level of the last rib, and may therefore be opened in these operations, especially when the last rib is removed in order to give more room.

THE MEDIASTINUM

The **Mediastinum** is the space left in and near the median line of the chest by the non-approximation of the two pleuræ. It extends from the sternum in front to the spine behind, and contains all the viscera in the thorax excepting the lungs. The mediastinum may be divided for purposes of description into two parts : an upper portion, above the upper level of the pericardium, which is named the *superior mediastinum* (Struthers) ; and a lower portion, below the upper level of the pericardium. This lower portion is again subdivided into three parts : that which contains the pericardium and its contents, the *middle mediastinum* ; that which is in front of the pericardium, the *anterior mediastinum* ; and that which is behind the pericardium, the *posterior mediastinum.*

The **superior mediastinum** is that portion of the interpleural space which lies above the upper level of the pericardium between the manubrium sterni in front, and the upper dorsal vertebræ behind. It is bounded below by a plane passing backwards from the junction of the manubrium and gladiolus sterni to the lower part of the body of the fourth dorsal vertebra, and laterally by the lungs and pleuræ. It contains the origins of the Sterno-hyoid and Sterno-thyroid muscles and the lower ends of the Longi colli muscles ; the arch of the aorta ; the inno-minate, the thoracic portion of the left carotid and subclavian arteries ; the upper half of the superior vena cava and the innominate veins, and the left superior intercostal vein ; the pneumogastric, cardiac, phrenic, and left recurrent laryngeal nerves ; the trachea, œsophagus, and thoracic duct ; the remains of the thymus gland and some lymphatic glands.

The **anterior mediastinum** is bounded in front by the sternum, on each side by the pleura, and behind by the pericardium. It is narrow above, but widens out a little below, and, owing to the oblique course taken by the left pleura, it is directed from above obliquely downwards and to the left. Its anterior wall is formed by the left Triangularis sterni muscle and the fifth, sixth, and seventh left costal cartilages. It contains a quantity of loose areolar tissue, some lymphatic vessels which ascend from the convex surface of the liver, two or three lymphatic glands (anterior mediastinal glands), and the small mediastinal branches of the internal mammary artery.

The **middle mediastinum** is the broadest part of the interpleural space. It contains the heart, enclosed in the pericardium, the ascending aorta, the lower half of the superior vena cava, with the vena azygos major opening into it, the bifurcation of the trachea and the two bronchi, the pulmonary artery dividing into its two branches and the right and left pulmonary veins, the phrenic nerves, and some bronchial lymphatic glands.

The **posterior mediastinum** (fig. 529) is an irregular triangular space running parallel with the vertebral column ; it is bounded in front by the pericardium and roots of the lungs, behind by the vertebral column from the lower border of the fourth dorsal vertebra, and on either side by the pleura. It contains the descending thoracic aorta, the greater and lesser azygos veins, the pneumo-gastric and splanchnic nerves, the œsophagus, thoracic duct, and some lymphatic glands.

The Lungs

The **Lungs** are the essential organs of respiration ; they are two in number, placed one on each side of the chest, separated from each other by the heart and other contents of the mediastinum. Each lung is conical in shape, and presents for examination an apex, a base, two borders, and two surfaces (see fig. 531).

FIG. 529.—The posterior mediastinum.

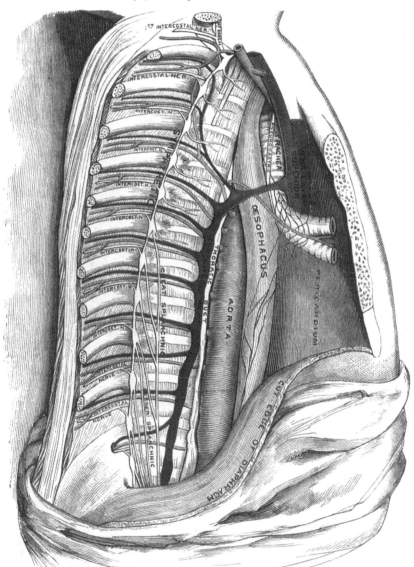

The *apex* forms a tapering cone, which extends into the root of the neck about an inch to an inch and a half above the level of the first rib.

The *base* is broad, concave, and rests upon the convex surface of the Diaphragm, which separates the right lung from the upper surface of the right lobe of the liver and the left lung from the upper surface of the left lobe of the liver, the stomach, and spleen ; its circumference is thin, and projects for some distance into the phrenico-costal sinus of the pleura, between the lower ribs and the costal attachment of the Diaphragm, extending lower down externally and behind than in front.

The *external* or *thoracic surface* is smooth, convex, of considerable extent, and corresponds to the form of the cavity of the chest, being deeper behind than in front.

The *inner surface* is concave. It presents, in front, a depression corresponding to the convex surface of the pericardium ; and behind, a deep fissure (the hilum pulmonis), which gives attachment to the root of the lung.

The *posterior border* is rounded and broad, and is received into the deep concavity on either side of the spinal column. It is much longer than the anterior border, and projects, below, into the phrenico-costal sinus.

The *anterior border* is thin and sharp, overlaps the front of the pericardium, and is projected into the costo-mediastinal sinus of the pleura. The anterior

FIG. 530.—Transverse section through the upper margin of the third dorsal vertebra. (Braune.)

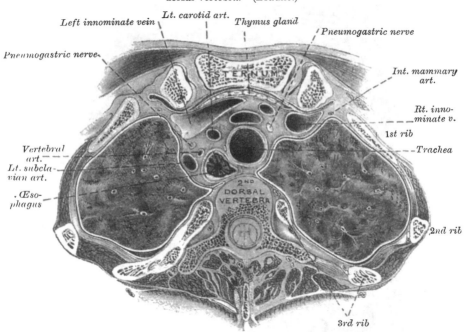

border of the right lung is almost vertical ; that of the left presents, below, an angular notch, the *incisura cardiaca*, into which the heart and pericardium are received.

Each lung is divided into two lobes, an upper and lower, by a long and deep fissure, which extends from the upper part of the posterior border of the organ, about three inches from its apex, downwards and forwards to the lower part of its anterior border. This fissure penetrates nearly to the root. In the right lung the upper lobe is partially subdivided by a second and shorter fissure, which extends almost horizontally forwards from the middle of the preceding to the anterior margin of the organ, marking off a small triangular portion, the middle lobe.

The *right lung* is the larger and heavier ; it is broader than the left, owing to the inclination of the heart to the left side ; it is also shorter by an inch, in consequence of the Diaphragm rising higher on the right side to accommodate the liver.

The roots of the lungs.—A little above the middle of the inner surface of each lung, and nearer its posterior than its anterior border, is its root, by which the lung is connected to the heart and the trachea. The root is formed by the bronchial tube, the pulmonary artery, the pulmonary veins, the bronchial arteries and veins, the pulmonary plexuses of nerves, lymphatics, bronchial glands, and areolar tissue, all of which are enclosed by a reflection of the pleura. The root of the right lung lies behind the superior vena cava and ascending portion of the aorta, and below the vena azygos major. That of the left lung passes beneath the arch of the aorta and in front of the descending aorta; the phrenic nerve and the anterior pulmonary plexus lie in front of each, and the pneumogastric and posterior pulmonary plexus behind each.

The chief structures composing the root of each lung are arranged in a similar manner from before backwards on both sides, viz.: the two pulmonary veins in

FIG. 531.—Front view of the heart and lungs.

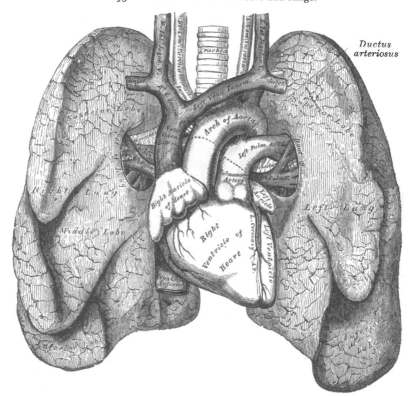

front; the pulmonary artery in the middle; and the bronchus, together with the bronchial vessels, behind. From above downwards, on the two sides, their arrangement differs, thus:

On the right side their position is—bronchus, pulmonary artery, pulmonary veins; but on the left side their position is—pulmonary artery, bronchus, pulmonary veins. It should be noted that the entire right bronchus does not lie above the right pulmonary artery, but only its eparterial branch (see page 949), which passes to the upper lobe of the right lung; the divisions of the bronchus for the middle and lower lobes lie below the artery.

The *weight* of both lungs together is about forty-two ounces, the right lung being two ounces heavier than the left; but much variation is met with according to the amount of blood or serous fluid they may contain. The lungs are heavier in the male than in the female, their proportion to the body being, in the former, as 1 to 37, in the latter as 1 to 43. The specific gravity of the lung-tissue varies from 0·345 to 0·746, water being 1,000.

The *colour* of the lungs at birth is a pinkish-white; in adult life, a dark slate-colour, mottled in patches; and as age advances, this mottling assumes a black colour. The colouring matter consists of granules of a carbonaceous substance deposited in the areolar tissue near the surface of the organ. It increases in quantity as age advances, and is more abundant in males than in females. The posterior border of the lung is usually darker than the anterior.

The *surface* of the lung is smooth, shining, and marked out into numerous polyhedral spaces, indicating the lobules of the organ: the area of each of these spaces is crossed by numerous lighter lines.

The *substance* of the lung is of a light, porous, spongy texture; it floats in water, and crepitates when handled, owing to the presence of air in the tissue: it is also highly elastic; hence the collapsed state of these organs when they are removed from the closed cavity of the thorax.

Structure.—The lungs are composed of an external serous coat, a subserous areolar tissue, and the pulmonary substance or parenchyma.

The *serous coat* is derived from the pleura; it is thin, transparent, and invests the entire organ as far as the root.

The *subserous areolar tissue* contains a large proportion of elastic fibres; it invests the entire surface of the lung, and extends inwards between the lobules.

The *parenchyma* is composed of lobules, which, although closely connected together by an interlobular areolar tissue, are quite distinct from one another, and may be teased asunder without much difficulty in the fœtus. The lobules vary in size: those on the surface are large, of pyramidal form, the base turned towards the surface; those in the interior smaller, and of various forms. Each lobule is composed of one of the ramifications of a bronchial tube and its terminal air-cells, and of the ramifications of the pulmonary and bronchial vessels, lymphatics, and nerves; all of these structures being connected together by areolar tissue.

The *bronchus*, upon entering the substance of the lung, divides and subdivides bipinnately throughout the entire organ. Sometimes three branches arise together, and occasionally small lateral branches are given off from the sides of a main trunk. Each of the smaller subdivisions of the bronchi enters a pulmonary lobule, and is termed a *lobular bronchial tube* or *bronchiole*. Its wall now begins to present irregular dilatations, *air-cells* or *alveoli*, at first sparingly and on one side of the tube only, but as it proceeds onwards these dilatations become more numerous and surround the tube on all sides, so that it loses its cylindrical character. The bronchiole now becomes enlarged, and is termed the *atrium* or *alveolar passage*; from it are given off, on all sides, ramifications, called *infundibula*, which are closely beset in all directions by *alveoli* or *air-cells*. Within the lungs the bronchial tubes are circular, not flattened, and present certain peculiarities of structure.

Changes in the structure of the bronchi in the lungs.—1. *In the lobes of the lungs.*—In the lobes of the lungs the following changes take place. The *cartilages* are not imperfect rings, but consist of thin laminæ, of varied form and size, scattered irregularly along the sides of the tube, being most distinct at the points of division of the bronchi. They may be traced into tubes, the diameter of which is only one-fourth of a line. Beyond this point the tubes are wholly membranous. The fibrous coat is continued into the smallest ramifications of the bronchi. The muscular coat is disposed in the form of a continuous layer of annular fibres, which may be traced upon the smallest bronchial tubes, and consists of the unstriped variety of muscular tissue. The mucous membrane lines the bronchi and its ramifications throughout, and is covered with columnar ciliated epithelium.

2. *In the lobules of the lung.*—In the lobular bronchial tubes and in the infundibula the following changes take place. The muscular tissue begins to

FIG. 532.—The roots of the lungs and posterior pulmonary plexus, seen from behind.

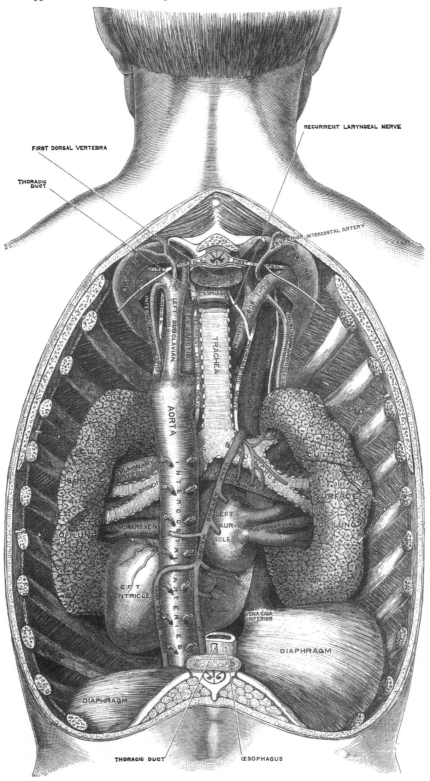

disappear, so that in the infundibula there is scarcely a trace of it. The fibrous coat becomes thinner and degenerates into areolar tissue. The epithelium becomes non-ciliated and flattened. This occurs gradually; thus, in the lobular bronchial tubes, patches of non-ciliated, flattened epithelium may be found scattered among the columnar ciliated epithelium; then these patches of non-ciliated flattened epithelium become more and more numerous, until in the infundibula and air-cells all the epithelium is of the non-ciliated pavement variety. In addition to these flattened cells, there are small polygonal granular cells in the air-sacs in clusters of two or three, between the others.

The air-cells are small polyhedral recesses, composed of a fibrillated connective tissue, and surrounded by a few involuntary muscular and elastic fibres. Free in their cavity are to be seen under the microscope granular, rounded, amœboid cells (eosinophile leucocytes), often containing carbonaceous particles. The air-cells are well seen on the surface of the lung, and vary from $\frac{1}{200}$th to $\frac{1}{70}$th of an inch in diameter; being largest on the surface at the thin borders and at the apex, and smallest in the interior.

The *pulmonary artery* conveys the venous blood to the lungs; it divides into branches which accompany the bronchial tubes, and terminates in a dense capillary network, upon the walls of the intercellular passages and air-cells. In the lung, the branches of the pulmonary artery are usually above and in front of a bronchial tube, the vein below.

The *pulmonary capillaries* form plexuses which lie immediately beneath the mucous membrane, in the walls and septa of the air-cells, and of the infundibula. In the septa between the air-cells the capillary network forms a single layer. The capillaries form a very minute network, the meshes of which are smaller than the vessels themselves; * their walls are also exceedingly thin. The arteries of neighbouring lobules are independent of each other, but the veins freely anastomose together.

The *pulmonary veins* commence in the pulmonary capillaries, the radicles coalescing into larger branches, which run along through the substance of the lung, independently from the minute arteries and bronchi. After freely communicating with other branches they form large vessels, which ultimately come into relation with the arteries and bronchial tubes, and accompany them to the hilum of the organ. Finally they open into the left auricle of the heart, conveying oxygenated blood to be eventually distributed to all parts of the body by the aorta.

The *bronchial arteries* supply blood for the nutrition of the lung; they are derived from the thoracic aorta or from the upper aortic intercostal arteries, and, accompanying the bronchial tubes, are distributed to the bronchial glands, and upon the walls of the larger bronchial tubes and pulmonary vessels. Those supplying the bronchial tubes form a capillary plexus in the muscular coat, from which branches are given off to form a second plexus in the mucous coat. This plexus communicates with branches of the pulmonary artery, and empties itself into the pulmonary vein. Others are distributed in the interlobular areolar tissue, and terminate partly in the deep, partly in the superficial, bronchial veins. Lastly, some ramify upon the surface of the lung, beneath the pleura, where they form a capillary network.

The *bronchial vein* is formed at the root of the lung, receiving superficial and deep veins corresponding to branches of the bronchial artery. It does not, however, receive all the blood supplied by the artery, as some of it passes into the pulmonary vein. It terminates on the right side in the vena azygos major, and on the left side in the superior intercostal or left upper azygos vein.

The **lymphatics** consist of a superficial and deep set; they terminate at the root of the lung, in the bronchial glands.

---

* The meshes are only 0·002''' to 0·008''' in width, while the vessels are 0·003''' to 0·005'''. (Kölliker, *Human Microscopic Anatomy*.)

Nerves.—The lungs are supplied from the anterior and posterior pulmonary plexuses, formed chiefly by branches from the sympathetic and pneumogastric. The filaments from these plexuses accompany the bronchial tubes, upon which they are lost. Small ganglia are found upon these nerves.

Surface Form.—The apex of the lung is situated in the neck, behind the interval between the two heads of origin of the Sterno-mastoid. The height to which it rises above the clavicle varies very considerably, but is generally about one inch. It may, however, extend as much as an inch and a half or an inch and three-quarters, or, on the other hand, it may scarcely project above the level of this bone. In order to mark out the anterior margin of the lung, a line is to be drawn from the apex point, one inch above the level of the clavicle, and rather nearer the posterior than the anterior border of the Sterno-mastoid muscle, downwards and inwards across the sterno-clavicular articulation and first piece of the sternum until it meets, or almost meets, its fellow of the other side at the level of the articulation of the manubrium and gladiolus. From this point the two lines are to be drawn downwards, one on either side of the mesial line and close to it, as far as the level of the articulation of the fourth costal cartilages to the sternum. From here the two lines diverge, the left at first passing outwards with a slight inclination downwards, and then taking a bend downwards with a slight inclination outwards to the apex of the heart, and thence to the sixth costo-chondral articulation. The direction of the anterior border of this part of the left lung is denoted with sufficient accuracy by a curved line, with its convexity directed upwards and outwards from the articulation of the fourth right costal cartilage of the sternum to the fifth intercostal space, an inch and a half below, and three-quarters of an inch internal to the left nipple. The continuation of the anterior border of the right lung is marked by a prolongation of its line from the level of the fourth costal cartilages vertically downwards as far as the sixth, when it slopes off along the line of the sixth costal cartilage to its articulation with the rib.

The lower border of the lung is marked out by a slightly curved line, with its convexity downwards, from the articulation of the sixth costal cartilage to its rib to the spinous process of the tenth dorsal vertebra. If vertical lines are drawn downwards from the nipple, the mid-axillary line, and the apex of the scapula, while the arms are raised from the sides, they should intersect this convex line, the first at the sixth, the second at the eighth, and the third at the tenth rib. It will thus be seen that the pleura (see page 955) extends farther down than the lung, so that it may be wounded, and a wound pass through its cavity into the Diaphragm, and even injure the abdominal viscera, without the lung being involved.

The posterior border of the lung is indicated by a line drawn from the level of the spinous process of the seventh cervical vertebra, down either side of the spine, corresponding to the costo-vertebral joints as low as the spinous process of the tenth dorsal vertebra. The trachea bifurcates opposite the spinous process of the fourth dorsal vertebra, and from this point the two bronchi are directed outwards.

The position of the great fissure in the right lung may be indicated by a line drawn from the fourth dorsal vertebra round the side of the chest to the anterior margin of the lung opposite the seventh rib, and the smaller or secondary fissure by a line drawn from the preceding, where it bisects the mid-axillary line, to the junction of the fourth costal cartilage to the sternum. The great fissure in the left lung is a little higher, extending from the third dorsal vertebra round the side of the chest to reach the anterior margin of the lung opposite the sixth costal cartilage.

Surgical Anatomy.—The lungs may be wounded or torn in three ways. (1) By compression of the chest, without any injury to the ribs. (2) By a fractured rib penetrating the lung. (3) By stabs, gunshot wounds, &c.

The first form, where the lung is ruptured by external compression without any fracture of the ribs, is very rare, and usually occurs in young children, and affects the root of the lung, i.e. the most fixed part, and thus, implicating the great vessels, is frequently fatal. It would seem *a priori* a most unusual injury, and its exact mode of causation is difficult to interpret. The probable explanation is that immediately before the compression is applied a deep inspiration is taken and the lungs are fully inflated; owing then to spasm of the glottis at the moment of compression, the air is unable to escape from the lung, which is not able to recede, and consequently gives way.

In the second variety, when the wound in the lung is produced by the penetration of a broken rib, both the pleura costalis and pulmonalis must necessarily be injured, and consequently the air taken into the wounded air-cells may find its way through these wounds into the cellular tissue of the parietes of the chest. This it may do without collecting in the pleural cavity; the two layers of the pleura are so intimately in contact that the air passes straight through from the wounded lung into the subcutaneous tissue. Emphysema constitutes therefore the most important sign of injury to the lung in cases of fracture of the ribs. Pneumothorax, or air in the pleural cavity, is much more likely to occur in injuries to the lung of the third variety, that is to say, from external wounds, from stabs, gunshot injuries, and such like, in which case air passes either from the wound of the lung or from the external wound into the cavity of the pleura during the respiratory movements. In these cases there is generally no emphysema of the sub-cutaneous tissue unless the external wound is small and valvular, so that the air is drawn into the wound during inspiration, and then forced into the cellular tissue around during expiration because it cannot escape from the external wound. Occasionally in wounds of the parietes of the chest no air finds its way into the cavity of the pleura, because the lung at the time of the accident protrudes through the wound and blocks the opening. This occurs where the wound is large, and constitutes one form of *hernia* of the lung. Another form of hernia of the lung occurs, though very rarely, after wounds of the chest wall, when the wound has healed and the cicatrix subsequently yields from the pressure of the viscus behind. It forms a globular, elastic, crepitating swelling, which enlarges during expiratory efforts, falls in during inspiration, and disappears on holding the breath.

THE THYROID GLAND

The **thyroid gland** is classified with the thymus, suprarenal capsules, and spleen, under the head of *ductless glands*, i.e. glands which do not possess an excretory duct. From its situation in connection with the trachea and larynx, the thyroid body is usually described with those organs, although it takes no part in the function of respiration. It is situated at the front and sides of the neck, and consists of two lateral lobes connected across the middle line by a narrow transverse portion, the *isthmus*.

The weight of the gland is somewhat variable but is usually about one ounce. It is somewhat heavier in the female, in whom it becomes enlarged during menstruation and pregnancy.

The lobes are conical in shape, the apex of each being directed upwards and outwards as far as the junction of the middle with the lower third of the thyroid cartilage ; the base looks downwards, and is on a level with the fifth or sixth tracheal ring.

The *external* or *superficial surface* is convex, and covered by the skin, the superficial and deep fascia, the Sterno-mastoid, the anterior belly of the Omo-hyoid, the Sterno-hyoid and Sterno-thyroid muscles, and beneath the last muscle by the pre-tracheal layer of the deep fascia, which forms a capsule for the gland.

The *deep* or *internal surface* is moulded over the underlying structures, viz. the thyroid and cricoid cartilages, the trachea, the inferior constrictor and posterior part of the Crico-thyroid muscles, the œsophagus, (particularly on the left side of the neck), the superior and inferior thyroid arteries, and the recurrent laryngeal nerves.

Its *anterior border* is thin, and inclines obliquely from above downwards and inwards towards the middle line of the neck, while the *posterior border* is thick and overlaps the common carotid artery. Each lobe is about two inches in length, it greatest width is about one inch and a quarter, and its thickness about three quarters of an inch.

The *isthmus* connects the lower third of the two lateral lobes ; it measures about half an inch in breadth, and the same in depth, and usually covers the second and third rings of the trachea. Its situation presents, however, many variations, a point of importance in the operation of tracheotomy. In the middle line of the neck it is covered by the skin and fascia, and close to the middle line, on either side, by the Sterno-hyoid. Across its upper border runs a branch of the superior thyroid artery ; at its lower border are the inferior thyroid veins. Sometimes the isthmus is altogether wanting.

A third lobe, of conical shape, called the *pyramid*, occasionally arises from the upper part of the isthmus, or from the adjacent portion of either lobe, but most commonly the left, and ascends as high as the hyoid bone. It is occasionally quite detached, or divided into two or more parts, or altogether wanting.

A few muscular bands are occasionally found attached, above, to the body of the hyoid bone, and below to the isthmus of the gland, or its pyramidal process These form a muscle, which was named by Sömmerring the *Levator glandulæ thyroideæ*.

Small detached portions of thyroid tissue (*accessory thyroid*) are sometimes found above the isthmus, and their presence is readily explained by a reference to the manner in which the gland is developed. They represent isolated portions of the median thyroid rudiment. (See section on Embryology.)

Structure.—The thyroid body is invested by a thin capsule of connective tissue, which projects into its substance and imperfectly divides it into masses of irregular form and size. When the organ is cut into, it is of a brownish-red colour, and is seen to be made up of a number of closed vesicles, containing a yellow glairy fluid, and separated from each other by intermediate connective tissue.

According to Baber, the vesicles of the thyroid of the adult animal are generally closed cavities; but in some young animals (e.g. young dogs) the vesicles are more or less tubular and branched. This appearance he supposes to be due to the mode of growth of the gland, and merely indicating that an increase in the number of vesicles is taking place. Each vesicle is lined by a single layer of epithelium, the cells of which, though differing somewhat in shape in different animals, have always a tendency to assume a columnar form. Between

FIG. 533.—Minute structure of thyroid. From a transverse section of the thyroid of a dog. (Semi-diagrammatic.) (Baber.)

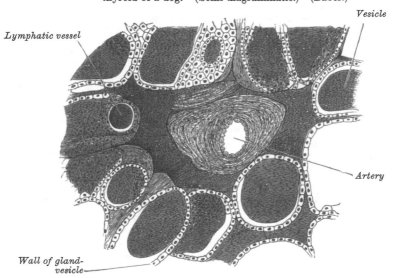

Vesicle

Lymphatic vessel

Artery

Wall of gland-vesicle

the epithelial cells exists a delicate reticulum. The vesicles are of various sizes and shapes, and contain as a normal product a viscid, homogeneous, semi-fluid, slightly yellowish material, which frequently contains blood; red corpuscles are found in it in various stages of disintegration and decolorisation, the yellow tinge being probably due to the hæmoglobin, which is thus set free from the coloured corpuscles. Baber has also described in the thyroid gland of the dog large round cells ('parenchymatous cells'), each provided with a single oval-shaped nucleus, which migrate into the interior of the gland-vesicles.

The capillary blood-vessels form a dense plexus in the connective tissue around the vesicles, between the epithelium of the vesicles and the endothelium of the lymph-spaces, which latter surround a greater or smaller part of the circumference of the vesicle. These lymph-spaces empty themselves into lymphatic vessels which run in the interlobular connective tissue, not uncommonly surrounding the arteries which they accompany, and communicate with a network in the capsule of the gland. Baber has found in the lymphatics of the thyroid a viscid material which is morphologically identical with the normal constituent of the vesicle.

Vessels and Nerves.—The *arteries* supplying the thyroid are the superior and inferior thyroid, and sometimes an additional branch (thyroidea media, or ima) from the innominate artery, or the arch of the aorta, which ascends upon the front of the trachea. The arteries are remarkable for their large size and frequent anastomoses. The *veins* form a plexus on the surface of the gland, and on the front of the trachea, from which arise the superior, middle, and inferior thyroid veins ; the two former terminating in the internal jugular, the latter in the innominate vein. The *lymphatics* are numerous, of large size, and terminate in the thoracic and right lymphatic ducts. The *nerves* are derived from the middle and inferior cervical ganglia of the sympathetic.

Surgical Anatomy.—The thyroid gland is subject to enlargement, which is called goitre. This may be due to hypertrophy of any of the constituents of the gland. The simplest (parenchymatous goitre) is due to an enlargement of the follicles. The *fibroid* is due to increase of the interstitial connective tissue. The *cystic* is that form in which one or more large cysts are formed from dilatation and possibly coalescence of adjacent follicles. The *pulsating goitre* is where the vascular changes predominate over the parenchymatous, and the vessels of the gland are especially enlarged. Finally, there is *exophthalmic goitre* (Graves's disease), where there is great vascularity and often pulsation accompanied by exophthalmos, palpitation, and rapid pulse.

For the relief of these growths various operations have been resorted to : such as injection of tincture of iodine or perchloride of iron, especially applicable to the cystic form of the disease ; ligature of the thyroid arteries ; excision of the isthmus ; and extirpation of a part or the whole of the gland. This latter operation is one of difficulty, and when the entire gland has been removed the operation has been followed by a condition resembling myxœdema. In removing the organ great care must be taken to avoid tearing the capsule, as if this happens the gland-tissue bleeds profusely. The thyroid arteries should be ligatured before an attempt is made to remove the mass, and in ligaturing the inferior thyroid the position of the recurrent laryngeal nerve must be borne in mind, so as not to include it in the ligature. A large number of cases of what were formerly supposed to be goitre are now known to be cases of adenomatous enlargement, where an adenoma, starting in one part of the gland, gradually spreads and involves the whole organ.

Parathyroids.—These are small rounded, brownish-red bodies, with an average diameter of about a quarter of an inch, situated in or near the thyroid gland, from which, however, they differ in structure, being composed of masses of cells arranged in a more or less columnar fashion with numerous intervening capillaries. They are divided from their situation into *external* and *internal*. The former, usually two in number, are situated, one on each side, in relation to the postero-internal surface of the lateral lobe ; sometimes they are duplicated. The latter, also usually two in number, are placed one in each lateral lobe, generally near its mesial surface.

FIG. 534.—Minute structure of thymus gland.

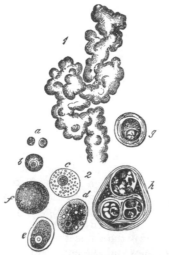

1 Upper portion of the thymus of a fœtal pig of 2″ in length, showing the bud-like lobuli and glandular elements. 2. Cells of the thymus, mostly from a man. *a.* Free nuclei. *b.* Small cells. *c.* Larger. *d.* Larger, with oil-globules, from the ox. *e, f.* Cells completely filled with fat, at *f* without a nucleus. *g, h.* Concentric bodies. *g.* An encapsulated nucleated cell. *h.* A composite structure of a similar nature.

THE THYMUS GLAND

The **thymus gland** is a temporary organ, attaining its full size at the end of the second year, when it ceases to grow, and gradually dwindles, until at puberty it has almost disappeared. If examined when its growth is most active, it will be found to consist of two lateral lobes placed in close contact along the middle line, situated partly in the superior mediastinum, partly in the neck, and extending from the fourth costal cartilage upwards, as high as the lower border of the thyroid gland. It is covered by the sternum, and by the origins of the Sterno-hyoid and Sterno-thyroid muscles. Below, it rests upon the pericardium, being separated from the arch of the aorta and great vessels by a layer of fascia. In the neck it lies on the front and sides of the trachea, behind the Sterno-hyoid and Sterno-thyroid muscles. The two lobes generally differ in size; they are occasionally united, so as to form a single mass; and sometimes separated by an intermediate lobe. The thymus is of a pinkish-grey colour, soft and lobulated on its surfaces. It is about two inches in length, one and a half in breadth below, and about three or four lines in thickness. At birth it weighs about half an ounce.

FIG. 535.—Minute structure of thymus gland. Follicle of injected thymus from calf, four days old, slightly diagrammatic, magnified about 50 diameters. The large vessels are disposed in two rings, one of which surrounds the follicle, the other lies just within the margin of the medulla. (Watney.)

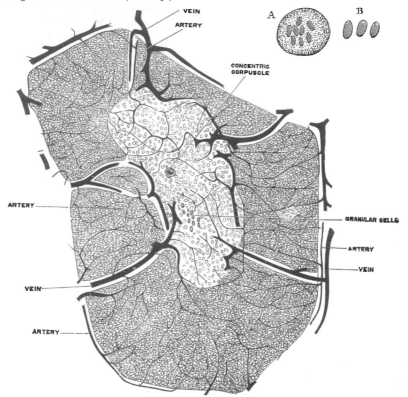

A and B. From thymus of camel, examined without addition of any reagent. Magnified about 400 diameters. A. Large colourless cell, containing small oval masses of hæmoglobin. Similar cells are found in the lymph-glands, spleen, and medulla of bone. B. Coloured blood-corpuscles.

Structure.—Each lateral lobe is composed of numerous lobules, held together by delicate areolar tissue; the entire gland being enclosed in an investing capsule of a similar but denser structure. The primary lobules vary in size from a pin's head to a small pea, and are made up of a number of small nodules or follicles, which are irregular in shape and are more or less fused together, especially towards the interior of the gland. Each follicle consists of a medullary and cortical portion, which differ in many essential particulars from each other. The *cortical portion* is mainly composed of lymphoid cells, supported by a delicate reticulum. In addition to this reticulum, of which traces only are found in the medullary portion, there is also a network of finely branched cells, which is continuous with a similar network in the medullary portion. This network forms an adventitia to the blood-vessels. In the *medullary portion* there are but few lymphoid cells, but there are, especially towards the centre, granular cells and concentric corpuscles. The granular cells are rounded or flask-shaped masses, attached (often by fibrillated extremities) to blood-vessels and to newly formed connective tissue. The concentric corpuscles are composed of a central mass, consisting of one or more granular cells, and of a capsule which is formed of epithelioid cells, which are continuous with the branched cells forming the network mentioned above.

Each follicle is surrounded by a capillary plexus, from which vessels pass into the interior, and radiate from the periphery towards the centre, and form a second zone just within the margin of the medullary portion. In the centre of the medulla there are very few vessels, and they are of minute size.

Watney has made the important observation that hæmoglobin is found in the thymus, either in cysts or in cells situated near to, or forming part of, the concentric corpuscles. This hæmoglobin varies from granules to masses exactly resembling coloured blood-corpuscles, oval in the bird, reptile, and fish ; circular in all mammals, except in the camel. He has also discovered, in the lymph issuing from the thymus, similar cells to those found in the gland, and, like them, containing hæmoglobin, in the form of either granules or masses. From these facts he arrives at the physiological conclusion that the thymus is one source of the coloured blood-corpuscles.

Vessels and Nerves.—The *arteries* supplying the thymus are derived from the internal mammary, and from the superior and inferior thyroid. The *veins* terminate in the left innominate vein, and in the thyroid veins. The *lymphatics* are of large size, arise in the substance of the gland, and are said to terminate in the internal jugular vein. The *nerves* are exceedingly minute ; they are derived from the pneumogastric and sympathetic. Branches from the descendens hypoglossi and phrenic reach the investing capsule, but do not penetrate into the substance of the gland.

THE URINARY ORGANS

THE KIDNEYS

THE Kidneys, two in number, are situated in the back part of the abdomen, and are for the purpose of separating from the blood certain materials which, when dissolved in a quantity of water, also separated from the blood by the kidneys, constitute the urine.

They are placed in the loins, one on each side of the vertebral column, behind the peritoneum, and surrounded by a mass of fat and loose areolar tissue. Their upper extremity is on a level with the upper border of the twelfth dorsal vertebra, their lower extremity on a level with the third lumbar. The right kidney is usually on a slightly lower level than the left, probably on account of the vicinity of the liver.

Each kidney is about four inches in length, two to two and a half in breadth, and rather more than one inch in thickness. The left is somewhat longer, though narrower, than the right. The weight of the kidney in the adult male varies from 4½ ounces to 6 ounces, in the adult female from 4 to 5½ ounces. The combined weight of the two kidneys in proportion to the body is about 1 in 240.

The kidney has a characteristic form. It is flattened on its sides, and presents at one part of its circumference a hollow. It is larger at its upper than its lower extremity. It presents for examination two surfaces, two borders, and an upper and lower extremity.

Its *anterior surface* is convex, looks forwards and outwards, and is partially covered by peritoneum. The right kidney, in its upper three-fourths, is in contact with the posterior part of the under surface of the right lobe of the liver, on which it produces a concave impression, the *impressio renalis* (page 918). Towards its inner border it is covered by the second part of the duodenum, while its lower and outer part is in relation with the hepatic flexure of the colon. The relation of the second part of the duodenum to the front of the right kidney is a varying one. The left kidney is covered above by the posterior surface of the stomach, below the stomach by the pancreas, behind which are the splenic vessels. Its lower half is in contact with some of the coils of the small intestine and sometimes with the third part of the duodenum. Near its outer border the anterior surface lies behind the spleen and the splenic flexure of the colon.

The kidneys are partly covered in front by peritoneum and partly uncovered. On the right kidney, the *hepatic area*, that is to say, that portion of the kidney which produces the renal impression on the liver, is covered by peritoneum, which therefore separates the kidney from the liver : the *duodenal* and *colic areas* are not peritoneal, and these structures are connected to the kidney by loose connective tissue ; at the lower and inner extremity is a small area, the *mesocolic area*, which is covered by a layer of peritoneum of the greater sac and by the colic vessels. On the left kidney, the *gastric area* is covered by the peritoneum of the lesser sac ; the *pancreatic* and *colic areas* are non-peritoneal : while as on the right side, at the lower and inner extremity, is an area, *mesocolic*, which is covered by the peritoneum of the greater sac and by the colic vessels.

The posterior surface of the kidney is flatter than the anterior and is directed backwards and inwards. It is entirely devoid of peritoneal covering, being embedded in areolar and fatty tissue. It lies upon the Diaphragm, the anterior layer of the lumbar aponeurosis, the external and internal arcuate ligaments, the Psoas and Transversalis muscles, one or two of the upper lumbar arteries, the last dorsal, ilio-hypogastric, and ilio-inguinal nerves. The right kidney rests upon the twelfth rib, the left usually on the eleventh and twelfth. The Diaphragm separates the kidney from the pleura as it dips down to form the phrenico-costal sinus, but frequently the muscular fibres of the Diaphragm are defective or absent over a triangular area immediately above the external arcuate ligament, and when this is the case the perirenal areolar tissue is in immediate apposition with the diaphragmatic pleura.

The *external border* is convex, and is directed outwards and backwards, towards the postero-lateral wall of the abdomen. On the left side it is in contact, at its upper part, with the spleen.

The *internal border* is concave, and is directed forwards and a little downwards. It presents a deep longitudinal fissure, bounded by a prominent overhanging anterior and posterior lip. This fissure is named the *hilum*, and allows of the passage of the vessels, nerves, and ureter into and out of the kidney.

The *superior extremity*, directed slightly inwards as well as upwards, is thick and rounded, and is surmounted by the suprarenal capsule, which covers also a small portion of the anterior surface.

The *inferior extremity*, directed a little outwards as well as downwards, is smaller and thinner than the superior. It extends to within two inches of the crest of the ilium.

At the hilum of the kidney the relative position of the main structures passing into and out of the kidney is as follows : the vein is in front, the artery in the middle, and the duct or ureter behind and towards the lower part. By a knowledge of these relations, the student may distinguish between the right and left kidney. The kidney is to be laid on the table, before the student, on its posterior surface, with its lower extremity towards the observer—that is to say, with the ureter *behind* and *below* the other vessels ; the hilum will then be directed to the side to which the kidney belongs.

General Structure of the Kidney.—The kidney is surrounded by a distinct investment of fibrous tissue, which forms a firm, smooth covering to the organ. It closely invests it, but can be easily stripped off, in doing which, however, numerous fine processes of connective tissue and small blood-vessels are torn through. Beneath this coat, a thin wide-meshed network of unstriped muscular fibre forms an incomplete covering to the organ. When the fibrous coat is stripped off, the surface of the kidney is found to be smooth and even, and of a deep red colour.

In infants, fissures extending for some depth may be seen on the surface of the organ, a remnant of the lobular construction of the gland. The kidney is dense in texture, but is easily lacerable by mechanical force. In order to obtain a knowledge of the structure of the gland, a vertical section must be made from its convex to its concave border, and the loose tissue and fat removed from around the vessels and the excretory duct (fig. 536). It will be then seen that the kidney consists of a central cavity, surrounded at all parts but one by the proper kidney-substance. This central cavity is called the *sinus*, and is lined by a prolongation of the fibrous coat of the kidney, which enters through a longitudinal fissure, the *hilum* (before mentioned), which is situated at that part of the cavity which is not surrounded by kidney structure. Through this fissure the blood-vessels of the kidney and its excretory duct pass, and therefore these structures, upon entering the kidney, are contained within the sinus. The excretory duct or *ureter*, after entering, dilates into a wide, funnel-shaped sac, named the *pelvis*. This divides into two or three tubular divisions, which subdivide into several short, truncated branches, named *calices* or *infundibula*, all of which are contained in the central cavity of the kidney. The blood-vessels of the kidney, after passing through the hilum, are contained in the sinus or central cavity, lying between its lining membrane and the excretory apparatus before entering the kidney-substance.

This central cavity, as before mentioned, is surrounded on all sides, except at the hilum, by the substance of the kidney, which is at once seen to consist of two parts, viz. of an external, granular, investing part, which is called the *cortical portion*, and of an internal part, the *medullary portion*, made up of a number of dark-coloured pyramidal masses, with their bases resting on the cortical part, and their apices converging towards the centre, where they form prominent papillæ, which project into the interior of the calices.

The cortical substance is of a bright reddish-brown colour, soft, granular, and easily lacerable. It is found everywhere immediately beneath the capsule, and is seen to extend in an arched form over the base of each medullary pyramid. The

FIG. 536.—Vertical section of kidney.

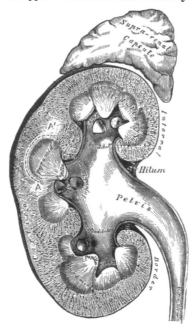

FIG. 537.—Plan of uriniferous tubes.

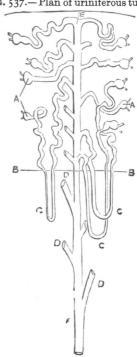

A A. Malpighian bodies. B B. Margin of medullary structure. C C C. Loops of Henle. D D D. Straight tubes cut off. E. Commencing straight tubes. F. Termination of straight tube.

part separating the sides of any two pyramids through which the arteries and nerves enter, and the veins and lymphatics emerge from the kidney, is called a *cortical column* or *columnæ Bertini* (A A′, fig. 536); while that portion which stretches from one cortical column to the next, and intervenes between the base of the pyramid and the capsule (which is marked by the dotted line, extending from A to A′ in fig. 536), is called a *cortical arch*, the depth of which varies from a third to half an inch.

The *medullary substance*, as before said, is seen to consist of red-coloured, striated, conical masses, the *pyramids of Malpighi*; the number of which, varying from eight to eighteen, corresponds to the number of lobes of which the organ in the fœtal state is composed. The base of each pyramid is surrounded by a cortical arch and directed towards the circumference of the kidney; the sides are contiguous with the cortical columns; while the apex, known as the *papilla* or *mammilla* of the kidney, projects into one of the calices of the ureter, each calyx receiving two or three papillæ.

These two parts, *cortical* and *medullary*, so dissimilar in appearance, are very similar in structure, being made up of urinary tubes and blood-vessels, united and bound together by a connecting matrix or stroma.

Minute Anatomy.—The *tubuli uriniferi*, of which the kidney is for the most part made up, commence in the cortical portion of the kidney, and, after pursuing a very circuitous course through the cortical and medullary parts of the kidney, finally terminate at the apices of the Malpighian pyramids by open mouths (fig. 537), so that the fluid which they contain is emptied into the dilated extremity of the ureter contained in the sinus of the kidney. If the surface of one of the papillæ is examined with a lens, it will be seen to be studded over with a number of small depressions from sixteen to twenty in number, and in a fresh kidney, upon pressure being made, fluid will be seen to exude from these depressions. They are the orifices of the tubuli uriniferi, which terminate in this situation. They commence in the cortical portion of the kidney as the *Malpighian bodies*, which are small rounded masses, varying in size, but of an average of about $\frac{1}{120}$ of an inch in diameter. They are of a deep red colour, and are found only in the cortical portion of the kidney. Each of these little bodies is composed of two parts : a central glomerulus of vessels, called a *Malpighian tuft* ; and a mem-

Fig. 538.—Minute structure of kidney.   Fig. 539.—Malpighian body.

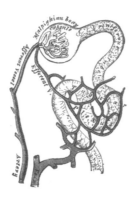

branous envelope, the *Malpighian capsule*, or *capsule of Bowman*, which latter is a small pouch-like commencement of a uriniferous tubule.

The *Malpighian tuft*, or vascular glomerulus, is a network of convoluted capillary blood-vessels, held together by scanty connective tissue and grouped into from two to five lobules. This capillary network is derived from a small arterial twig, the *afferent vessel*, which pierces the wall of the capsule, generally at a point opposite that at which the latter is connected with the tube ; and the resulting vein, the *efferent vessel*, emerges from the capsule at the same point. The afferent vessel is usually the larger of the two (fig. 538). The *Malpighian*, or *Bowman's capsule*, which surrounds the glomerulus, is formed of a hyaline membrane, supported by a small amount of connective tissue, which is continuous with the connective tissue of the tube. It is lined on its inner surface by a layer of squamous epithelial cells, which are reflected from the lining membrane on to the glomerulus, at the point of entrance or exit of the afferent and efferent vessels. The whole surface of the glomerulus is covered with a continuous layer of the same cells, on a delicate supporting membrane, which with the cells dips in between the lobules of the glomerulus, closely surrounding them (fig. 539). Thus between the glomerulus and the capsule a space is left, forming a cavity lined by a continuous layer of cells, which varies in size according to the state of secretion and the amount of fluid present in it. The cells, as above stated, are squamous in the adult, but in the fœtus and young subject they are polyhedral or even columnar.

The *tubuli uriniferi*, commencing in the Malpighian bodies, in their course present many changes in shape and direction, and are contained partly in the medullary and partly in the cortical portions of the organ. At their junction with the Malpighian capsule they present a somewhat constricted portion, which is termed the *neck*. Beyond this the tube becomes convoluted, and pursues a considerable course in the cortical structure, constituting the *proximal convoluted tube*. After a time the convolutions disappear, and the tube approaches the medullary portion of the kidney in a more or less spiral manner. This section of the tube has been called the *spiral tube of Schachowa*. Throughout this portion of their course the tubuli uriniferi have been contained entirely in the cortical structure, and have presented a pretty uniform calibre. They now enter the medullary portion, suddenly become much smaller, quite straight in direction, and dip down for a variable depth into the pyramids, constituting the *descending limb of Henle's loop*. Bending on themselves, they form a kind of loop, the *loop of Henle*, and re-ascending, they become suddenly enlarged and again spiral in direction, forming the *ascending limb of Henle's loop*, and re-enter the cortical structure. This portion of the tube does not present a uniform calibre, but becomes narrower as it ascends, and irregular in outline. As a narrow tube it enters the cortex and ascends for a short distance, when it again becomes dilated, irregular, and angular. This section is termed the *irregular tubule*; it terminates in a convoluted tube, which exactly resembles the proximal convoluted tubule, and is called the *distal convoluted tubule*. This again terminates in a narrow *curved tube*, which enters the straight or collecting tube.

Each *straight*, otherwise called a *collecting* or *receiving tube*, commences by a small orifice on the summit of one of the papillæ, thus opening and discharging its contents into the interior of one of the calices. Traced into the substance of the pyramid, these tubes are found to run from apex to base, dividing dichotomously in their course and slightly diverging from each other. Thus dividing and subdividing, they reach the base of the pyramid, and enter the cortical structure greatly increased in number. Upon entering the cortical portion they continue a straight course for a variable distance, and are arranged in groups, called *medullary rays*, several of these groups corresponding to a single pyramid. The tubes in the centre of the group are the longest, and reach almost to the surface of the kidney, while the external ones are shorter, and advance only a short distance into the cortex. In consequence of this arrangement the cortical portion presents a number of conical masses, the apices of which reach the periphery of the organ, and the bases are applied to the medullary portion. These are termed the *pyramids of Ferrein*. As they run through the cortical portion, the straight tubes receive on either side the curved extremity of the convoluted tubes, which, as stated above, commence at the Malpighian bodies.

It will be seen from the above description that there is a continuous series of tubes, from their commencement in the Malpighian bodies to their termination at the orifices on the apices of the pyramids of Malpighi; and that the urine, the secretion of which commences in the capsule, finds its way through these tubes into the calices of the kidney, and so into the ureter. Commencing at the capsule, the tube first presents a narrow constricted portion (1) the *neck*. 2. It forms a wide convoluted tube, the *proximal convoluted tube*. 3. It becomes spiral, the *spiral tubule of Schachowa*. 4. It enters the medullary structure as a narrow, straight tube, the *descending limb of Henle's loop*. 5. Forming a loop and becoming dilated, it ascends somewhat spirally, and, gradually diminishing in calibre, again enters the cortical structure, the *ascending limb of Henle's loop*. 6. It now becomes irregular and angular in outline, the *irregular tubule*. 7. It then becomes convoluted, the *distal convoluted tubule*. 8. Diminishing in size, it forms a curve, the *curved tubule*. 9. Finally it joins a straight tube, the *straight collecting tube,* which is continued downwards through the medullary substance to open at the apex of a pyramid.

The Tubuli Uriniferi : their Structure.—The tubuli uriniferi consist of basement-membrane lined with epithelium. The epithelium varies considerably in different sections of the uriniferous tubes. In the neck the epithelium is continuous with that lining the Malpighian capsule, and like it consists of flattened cells with an oval nucleus (fig. 540, A). The cells are, however, very indistinct and difficult to trace, and the tube has here the appearance of a simple basement-membrane unlined by epithelium. In the proximal convoluted tubule and the spiral tubule of Schachowa the epithelium is polyhedral in shape, the sides of the cells not being straight, but fitting into each other, and in some animals so fused together that it is impossible to make out the lines of junction. In the human

FIG. 540.—Uriniferous tube.

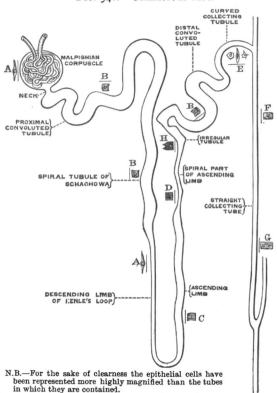

N.B.—For the sake of clearness the epithelial cells have been represented more highly magnified than the tubes in which they are contained.

kidney the cells often present an angular projection of the surface next the basement-membrane. These cells are made up of more or less rod-like fibres, which rest by one extremity on the basement-membrane, while the other projects towards the lumen of the tube. This gives to the cells the appearance of distinct striation (Heidenhain) (fig. 540, B). In the descending limb of Henle's loop the epithelium resembles that found in the Malpighian capsule and the commencement of the tube, consisting of flat transparent epithelial plates with an oval nucleus (figs. 540, A; 541). In the ascending limb, on the other hand, the cells partake more of the character of those described as existing in the proximal convoluted tubule, being polyhedral in shape, and presenting the same appearance of striation. The nucleus, however, is not situated in the centre of the cell, but near the lumen (fig. 540, C). After the ascending limb of Henle's loop becomes narrower upon entering the cortical structure, the striation appears to be confined to the outer part of the cell; at all events it is much more distinct in this situation; the nucleus, which appears flattened and angular, being still situated near the lumen (fig. 540, D). In the irregular tubule, the cells undergo a still further change becoming very angular, and presenting thick bright rods or markings, which render the striation much more distinct than in any other section of the urinary tubules (fig. 540, H). In the distal convoluted tubule the epithelium appears to be somewhat similar to that which has been described as existing in the proximal convoluted tubule, but presents a peculiar refractive appearance (fig. 540, B). In the curved tubule, just before its entrance into the straight collecting tube, the epithelium varies greatly as regards the shape of the cells, some being angular with short processes, others spindle-shaped, others polyhedral (fig. 540, E).

In the straight tubes the epithelium is more or less columnar: in its papillary portion the cells are distinctly columnar and transparent (figs. 542, 543); but as the tube approaches the cortex the cells are less uniform in shape; some are polyhedral, and others angular with short processes (fig. 540, F and G).

The Renal Blood-vessels.—The kidney is plentifully supplied with blood by the renal artery, a large offset of the abdominal aorta. Previously to entering the kidney, each artery divides into four or five branches, which are distributed

FIG. 541.*—Longitudinal section of Henle's descending limb.

FIG. 542.—Longitudinal section of straight tube.

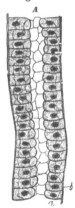

a. Membrana propria.
b. Epithelium.

a. Cylindrical or cubical epithelium.
b. Membrana propria.

to its substance. At the hilum these branches lie between the renal vein and ureter, the vein being in front, the ureter behind. Each vessel gives off some small branches to the suprarenal capsules, the ureter, and the surrounding cellular tissue and muscles. Frequently there is a second renal artery, which is given off from the abdominal aorta at a lower level, and supplies the lower portion of the kidney. It is termed the *inferior renal artery*. The branches of the renal artery, while in the sinus, give off a few twigs for the nutrition of the surrounding tissues, and terminate in the *arteriæ propriæ renales*, which enter the kidney proper in the columns of Bertini. Two of these pass to each pyramid of Malpighi, and run along its sides for its entire length, giving off as they advance the afferent vessels of the Malpighian bodies in the columns. Having arrived at the bases of the pyramids, they make a bend in their course, so as to lie between the bases of the pyramids and the cortical arches, where they break up into two distinct sets of branches devoted to the supply of the remaining portions of the kidney.

The *first set*, the *interlobular arteries* (figs. 544, 545, B), are given off at right angles from the side of the arteriæ propriæ renales looking towards the cortical

* From the *Handbook for the Physiological Laboratory*.

substance, and passing directly outwards between the pyramids of Ferrein they reach the capsule, where they terminate in the capillary network of this part. In their outward course they give off lateral branches; these are the *afferent vessels* for the Malpighian bodies (see page 972), and, having pierced the capsule,

FIG. 543.—Transverse section of pyramidal substance of kidney of pig, the blood-vessels of which are injected.

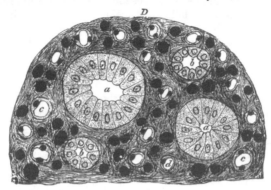

a. Large collecting tube, cut across, lined with cylindrical epithelium. *b.* Branch of collecting tube, cut across, lined with epithelium with shorter cylinders. *c* and *d.* Henle's loops cut across. *e.* Blood-vessels cut across. D. Connective-tissue ground-substance.

end in the Malpighian tufts. From each tuft the corresponding renal efferent arises, and, having made its egress from the capsule near to the point where the afferent vessel entered, breaks up into a number of branches, which form a dense *venous plexus* around the adjacent urinary tubes (fig. 546).

FIG. 544.—Diagrammatical sketch of the blood-vessels of the kidney.

FIG. 545.—A portion of fig. 544 enlarged. (The references are the same.)

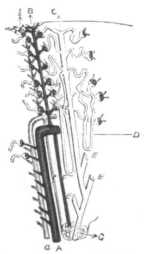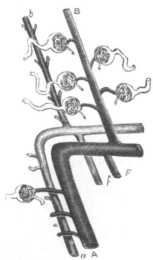

A *a.* Proper renal artery and vein, the former giving off the renal afferents, the latter receiving the renal efferents. B *b.* Interlobular artery and vein, the latter commencing from the stellate veins, and receiving branches from the plexus around the tubuli contorti, the former giving off renal afferents. c. Straight tube, surrounded by tubuli contorti, with which it communicates, as more fully shown in fig. 537. D. Margin of medullary substance. E E E. Receiving tubes, cut off. F *f.* Arteriolæ et venæ rectæ, the latter arising from (G) the plexus at the medullary apex.

The *second set of branches* from the arteriæ propriæ renales are for the supply of the medullary pyramids, which they enter at their bases; and, passing straight through their substance to their apices, terminate in the venous plexuses found in that situation. They are called the *arteriolæ rectæ* (figs. 544, 545, F).

The *renal veins* arise from *three sources*—the veins beneath the capsule, the plexuses around the convoluted tubules in the cortical arches, and the plexuses situated at the apices of the pyramids of Malpighi. The veins beneath the capsule are stellate in arrangement, and are derived from the capillary network of the capsule, into which the terminal branches of the interlobular arteries break up. These join to form the *venæ interlobulares*, which pass inwards between the pyramids of Ferrein, receive branches from the plexuses around the convoluted tubules, and, having arrived at the bases of the Malpighian pyramids, join with the venæ rectæ, next to be described (figs. 544, 545, *b*).

The *venæ rectæ* are branches from the plexuses at the apices of the medullary pyramids, formed by the terminations of the arteriolæ rectæ. They pass outwards in a straight course between the tubes of the medullary structure, and joining, as above stated, the venæ interlobulares, form the proper renal veins (figs. 544, 545, *f*).

These vessels, *venæ propriæ renales*, accompany the arteries of the same name, running along the entire length of the sides of the pyramids; and, having received in their course the efferent vessels from the Malpighian bodies in the adjacent cortical structure, quit the kidney substance to enter the sinus. In this cavity they inosculate with the corresponding veins from the other pyramids to form the *renal vein*, which emerges from the kidney at the hilum and opens into the inferior vena cava; the left being longer than the right, from having to cross in front of the abdominal aorta.

Nerves of the Kidney.—The nerves of the kidney, although small, are about fifteen in number. They have small ganglia developed upon them, and are derived from the renal plexus, which is formed by branches from the solar plexus, the lower and outer part of the semilunar ganglion and aortic plexus, and from the lesser and smallest splanchnic nerves. They communicate with the spermatic plexus, a circumstance which may explain the occurrence of pain in the testicle in affections of the kidney. So far as they have been traced, they seem to accompany the renal artery and its branches, but their exact mode of termination is not known.

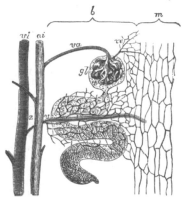

FIG. 546.—Diagrammatic representation of the blood-vessels in the substance of the cortex of the kidney.

m. Region of the medullary ray. *b.* Region of the tortuous portion of the tubules. *ai.* Arteria interlobularis. *vi.* Vena interlobularis. *va.* Vas afferens. *gl.* Glomerulus. *ve.* Vas efferens. *vz.* Venous twig of the interlobularis. (From Ludwig, in Stricker's ' Handbook.')

The *lymphatics* consist of a superficial and deep set, which terminate in the lumbar glands.

Connective tissue, or *intertubular stroma.*—Although the tubules and vessels are closely packed, a certain small amount of connective tissue, continuous with the capsule, binds them firmly together. This tissue was first described by Goodsir, and subsequently by Bowman. Ludwig and Zawarykin have observed distinct fibres passing around the Malpighian bodies; and Henle has seen them between the straight tubes composing the medullary structure.

Surface Form.—The kidneys, being situated at the back part of the abdominal cavity and deeply placed, cannot be felt unless enlarged or misplaced. They are situated on the confines of the epigastric and umbilical regions internally, with the hypochondriac and lumbar regions externally. The left is somewhat higher than the right. According to Morris, the position of the kidney may be thus defined: *Anteriorly.* ' 1. A horizontal line through the umbilicus is below the lower edge of each kidney. 2. A vertical line carried upwards to the costal arch from the middle of Poupart's ligament has one third of the kidney to its outer side, and two thirds to its inner side, i.e. between this line and the median line of the body.' In adopting these lines it must be borne in mind that the axes of the kidneys are not vertical, but oblique, and if continued upwards would meet about the ninth dorsal vertebra. *Posteriorly.* The upper end of the left kidney would be defined by a line drawn horizontally outwards from the spinous process of the eleventh dorsal vertebra, and its lower end by a point two inches above the iliac crest. The right kidney would be half to three-quarters of an inch lower. Morris lays down the following rules for indicating the position of the kidney on the posterior surface of the body : ' 1. A line parallel with, and one inch from, the spine between the lower edge of the tip of the spinous process of the eleventh dorsal vertebra, and the lower edge of the spinous process of the third lumbar vertebra.   2. A line from the top of this first line outwards at right angles to it for 2¾ inches.   3. A line from the lower end of the first transversely outwards for 2¾ inches. 4. A line parallel to the first, and connecting the outer extremities of the second and third lines just described.'

The hilum of the kidney lies about two inches from the middle line of the back at the level of the spinous process of the first lumbar vertebra.

Surgical Anatomy.—Malformations of the kidney are not uncommon.   There may be an entire absence of one kidney, though, according to Morris, the number of these cases is ' excessively small : ' or there may be congenital atrophy of one kidney, when the kidney is very small, but usually healthy in structure.   These cases are of great importance, and must be duly taken into account, when nephrectomy is contemplated.   A more common malformation is where the two kidneys are fused together.   They may be only joined together at their lower ends by means of a thick mass of renal tissue, so as to form a horse-shoe-shaped body or they may be completely united, forming a disc-like kidney, from which two ureters descend into the bladder.   These fused kidneys are generally situated in the middle line of the abdomen, but may be misplaced as well.

One or both kidneys may be misplaced as a congenital condition, and remain fixed in this abnormal position.   They are then very often misshapen.   They may be situated higher or lower than normal or removed farther from the spine than usual or they may be displaced into the iliac fossa, over the sacro-iliac joint, on to the promontory of the sacrum, or into the pelvis between the rectum and bladder or by the side of the uterus.   In these latter cases they may give rise to very serious trouble.   The kidney may also be misplaced as a congenital condition, but may not be fixed.   It is then known as a *floating kidney.* It is believed to be due to the fact that the kidney is completely enveloped by peritoneum which then passes backwards to the spine as a double layer, forming a mesonephron, which permits of movement taking place.   The kidney may also be misplaced as an acquired condition ; in these cases the kidney is mobile in the tissues by which it is surrounded, either moving in its capsule or else moving with the capsule in the perinephritic tissues. This condition is known as *movable kidney*, and is more common in the female than the male.   Other malformations are the persistence of the fœtal lobulation ; the presence of two pelves or two ureters to the one kidney.   In some rare instances a third kidney may be present.

The kidney is embedded in a large quantity of loose fatty tissue, and is only partially covered by peritoneum ; hence rupture of this organ is not nearly so serious an accident as rupture of the liver or spleen, since the extravasation of blood and urine which follows is, in the majority of cases, outside the peritoneal cavity.   Occasionally the kidney may be bruised by blows in the loin, or by being compressed between the lower ribs and the ilium when the body is violently bent forwards.   This is followed by a little transient hæmaturia, which, however, speedily passes off.   Occasionally, when rupture involves the pelvis of the kidney or the commencement of the ureter, this duct may become blocked, and hydro-nephrosis follow.

The loose cellular tissue around the kidney may be the seat of suppuration, constituting *perinephritic abscess.*   This may be due to injury, to disease of the kidney itself, or to extension of inflammation from neighbouring parts.   The abscess may burst into the pleura, constituting empyema ; into the colon or bladder ; or may point externally in the groin or loin.   Tumours of the kidney, of which perhaps sarcoma, in children, is the most common, may be recognised by their position and fixity ; by the resonant colon lying in front of it ; by their not moving with respiration ; and by their rounded outline not presenting a notched anterior margin like the spleen, with which they are most likely to be confounded.   The examination of the kidney should be bimanual ; that is to say, one hand should be placed in the flank and firm pressure made forwards, while the other hand is buried in the abdominal wall, over the situation of the organ.   Manipulation of the kidney frequently produces a peculiar sickening sensation, with sometimes faintness.

The kidney is mainly held in position by the mass of fatty matter in which it is embedded.   If this fatty matter is loose or lax or is absorbed, the kidney may become movable and may give rise to great pain.   This condition occurs, therefore, in badly nourished people, or in those who have become emaciated from any cause, and is more common in women than in men.   It must not be confounded with the *floating kidney* ; this is a congenital condition due to the development of a mesonephron, which permits the organ to move more or less freely.   The two conditions cannot, however, be distinguished until the abdomen is opened or the kidney explored from the loin.

The kidney has, of late years, been frequently the seat of surgical interference.   It may be exposed for exploration or the evacuation of pus (nephrotomy) ; it may be incised for the removal of stone (nephro-lithotomy); it may be sutured when movable or floating (nephroraphy); or it may be removed (nephrectomy).

The kidney may be exposed either by a lumbar or abdominal incision. The operation is best performed by a lumbar incision, except in cases of very large tumours, or of wandering kidneys, with a loose mesonephron, on account of the advantages which it possesses of not opening the peritoneum, and affording admirable drainage. It may be performed either by an oblique, a vertical, or a transverse incision. Perhaps the preferable, as affording the best means for exploring the whole surface of the kidney, is an incision from the tip of the last rib, backwards to the edge of the Erector spinae. This incision must not be quite parallel to the rib, but its posterior end must be at least three-quarters of an inch below it, lest the pleura be wounded. This cut is quite sufficient for an exploration of the organ. Should it require removal, a vertical incision can be made downwards to the crest of the ilium, along the outer border of the Quadratus lumborum. The structures divided are, the skin, the superficial fascia with the cutaneous nerves, the deep fascia, the posterior border of the External oblique muscle of the abdomen, and the outer border of the Latissimus dorsi; the Internal oblique and the posterior aponeurosis of the Transversalis muscle; the outer border of the Quadratus lumborum; the deep layer of the lumbar fascia, and the transversalis fascia. The fatty tissue around the kidney is now exposed to view and must be separated by the fingers, or a director, in order to reach the kidney.

The abdominal operation is best performed by an incision in the linea semilunaris on the side of the kidney to be removed, as recommended by Langenbuch; the kidney is then reached from the outer side of the colon, ascending or descending, as the case may be, and the vessels of the colon are not interfered with. If the incision is made in the linea alba, the kidney is reached from the inner side of the colon, and the vessels running to supply it must necessarily be interfered with. The incision is made of varying length according to the size of the kidney, commencing just below the costal arch. The abdominal cavity is opened. The intestines are held aside, and the outer layer of the mesocolon incised, so that the fingers can be introduced behind the peritoneum and the renal vessels sought for. These are then to be ligatured: if tied separately, care must be taken to ligature the artery first. The kidney must now be enucleated, and the vessels and ureter divided, and the latter tied, or if thought necessary stitched to the edge of the wound.

THE URETERS

The **Ureters** are the two tubes which conduct the urine from the kidneys into the bladder. They commence within the sinus of the kidneys, by a number of short truncated branches, the *calices* or *infundibula*, which unite, either directly or indirectly, to form a dilated pouch, the *pelvis*, from which the ureter, after passing through the hilum of the kidney, descends to the bladder. The *calices* are cup-like tubes encircling the apices of the Malpighian pyramids; but inasmuch as one calyx may include two or even more papillæ, their number is generally less than the pyramids themselves, the former being from seven to thirteen, while the latter vary from eight to eighteen. These calices converge into two or three tubular divisions, which by their junction form the *pelvis*, or dilated portion of the ureter. The portion last mentioned, where the pelvis merges into the ureter proper, is found opposite the spinous process of the first lumbar vertebra, in which situation it is accessible behind the peritoneum (see fig. 500, page 907).

The *ureter proper* is a cylindrical, membranous tube, about sixteen inches in length, and of the diameter of a goose-quill, extending from the pelvis of the kidney to the bladder. Its course is obliquely downwards and inwards through the lumbar region into the cavity of the pelvis, where it passes downwards, forwards, and inwards across that cavity to the base of the bladder, into which it then opens by a constricted orifice, after having passed obliquely for nearly an inch between its muscular and mucous coats.

Relations.—In its course it rests upon the Psoas muscle, being covered by the peritoneum, and crossed obliquely, from within outwards, by the spermatic vessels; the right is crossed by the branches of the mesenteric arteries which are distributed to the ascending, and the left by those for the descending colon; the right ureter lying close to the outer side of the inferior vena cava. Opposite the first piece of the sacrum it crosses either the common or external iliac artery, lying behind the ileum on the *right* side and the sigmoid flexure of the colon on the *left*. In the pelvis it enters the posterior false ligament of the bladder, below the obliterated hypogastric artery, the vas deferens in the male passing between it and the bladder. In the female the ureter passes along the side of the neck of the uterus and upper part of the vagina. At the base of the bladder it is situated about two inches from its fellow : lying, in the male, about an inch and a half from the vesical orifice of the urethra, at one of the posterior angles of the trigone.

Structure.—The *ureter* is composed of three coats—a fibrous, muscular, and mucous.

The *fibrous coat* is the same throughout the entire length of the duct, being continuous at one end with the capsule of the kidney at the floor of the sinus ; while at the other it is lost in the fibrous structure of the bladder.

In the pelvis of the kidney the *muscular coat* consists of two layers, longitudinal and circular ; the longitudinal fibres become lost upon the sides of the papillæ at the extremities of the calices ; the circular fibres may be traced, surrounding the medullary structure in the same situation. In the ureter proper the muscular fibres are very distinct, and are arranged in three layers—an external longitudinal, a middle circular, and an internal layer, less distinct than the other two, but having a general longitudinal direction. According to Kölliker this internal layer is only found in the neighbourhood of the bladder.

The *mucous coat* is smooth, and presents a few longitudinal folds which become effaced by distension. It is continuous with the mucous membrane of the bladder below, while it is prolonged over the papillæ of the kidney above. Its epithelium is of a peculiar character, and resembles that found in the bladder. It is known by the name of 'transitional' epithelium (see fig. [14], page [14]). It consists of several layers of cells, of which the innermost—that is to say, the cells in contact with the urine—are quadrilateral in shape, with a concave margin on their outer surface, into which fits the rounded end of the cells of the second layer. These, the intermediate cells, more or less resemble columnar epithelium, and are pear-shaped, with a rounded internal extremity which fits into the concavity of the cells of the first layer, and a narrow external extremity which is wedged in between the cells of the third layer. The external or third layer consists of conical or oval cells varying in number in different parts, and presenting processes which extend down into the basement-membrane.

The *arteries* supplying the ureter are branches from the renal, spermatic, internal iliac, and inferior vesical.

The *nerves* are derived from the inferior mesenteric, spermatic, and pelvic plexuses.

Surgical Anatomy.—Subcutaneous rupture of the ureter is not a common accident, but occasionally occurs from a sharp, direct blow on the abdomen, as from the kick of a horse. It may be either torn completely across or only partially divided, and, as a rule, the peritoneum escapes injury. If torn completely across, the urine collects in the retro-peritoneal tissues ; if it is not completely divided, the lumen of the tube may become obstructed and hydro-nephrosis or pyo-nephrosis result. The ureter may be accidentally wounded in some abdominal operations ; if this should happen the divided ends must be sutured together, or. failing to accomplish this, the upper end must be implanted into the bladder or the intestine.

SUPRARENAL CAPSULES

The **Suprarenal Capsules** belong to the class of ductless glands. They are two small flattened bodies, of a yellowish colour, situated at the back part of the abdomen, behind the peritoneum, and immediately above and in front of the upper end of each kidney ; hence their name. The right one is somewhat triangular in shape, bearing a resemblance to a cocked hat ; the left is more semilunar, usually larger and placed at a higher level than the right. They vary in size in different individuals, being sometimes so small as to be scarcely detected : their usual size is from an inch and a quarter to nearly two inches in length, rather less in width, and from two to three lines in thickness. Their average weight is from one to one and a half drachms each.

Relations.—The relations of the suprarenal capsules differ on the two sides of the body. The *right suprarenal* is roughly triangular in shape, its angles pointing upwards, downwards, and outwards. It presents two surfaces for examination, an anterior and posterior. The *anterior surface* presents two areas, separated by a furrow, the *hilum*: one area, occupying about one third of the whole surface, is situated above and internally; it is depressed, uncovered by peritoneum, and is in contact in front with the posterior surface of the right lobe of the liver, and along its inner border with the inferior vena cava; the remaining area is elevated, and is divided into a non-peritoneal portion, in contact with

FIG. 548.—Minute structure of suprarenal capsule.

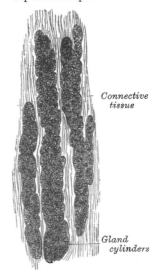

FIG. 547.—Vertical section of the suprarenal capsule. (From Elberth, in Stricker's 'Manual.')

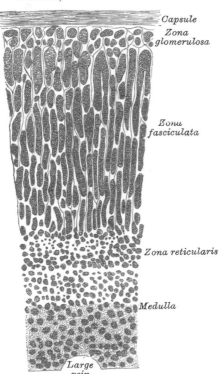

FIG. 549.—Minute structure of suprarenal capsule.

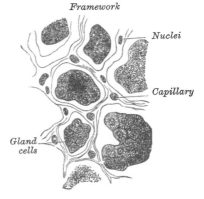

the hepatic flexure of the duodenum, and a portion covered by the peritoneum forming the hepato-renal fold. The *posterior surface* is slightly convex, and rests upon the Diaphragm. The *base* is concave, and is in contact with the upper end and the adjacent part of the anterior surface of the kidney. The *left suprarenal* is crescentic in shape, its concavity being adapted to the upper end of the left kidney. It presents an inner border which is convex, and an outer which is concave; its upper border is narrow, and its lower rounded. Its *anterior surface* presents two areas: an upper one, covered by the peritoneum forming the lesser sac, which separates it from the cardiac end of the stomach and to a small extent from the superior extremity of the spleen; and a lower one, which is in contact with the pancreas and splenic artery, and is therefore not covered by the peritoneum. Its *posterior surface* presents a vertical ridge, which divides it into two areas. The ridge lies in the sulcus between the kidney and crus of the Diaphragm, while the area on either side of it lies on these parts respectively; the outer area, which is thin, resting on the kidney, and the inner and smaller area resting on the left crus of the Diaphragm. The surface of the suprarenal gland is surrounded by areolar tissue containing much fat, and closely invested by a thin fibrous coat, which is difficult to remove, on account of the numerous fibrous processes and vessels which enter the organ through the furrows on its anterior surface and base.

Small accessory suprarenals are often to be found in the connective tissue around the suprarenals. The smaller of these, on section, show a uniform surface, but in some of the larger a distinct medulla can be made out.

Structure.—On making a perpendicular section, the gland is seen to consist of two substances—external or cortical, and internal or medullary. The former, which constitutes the chief part of the organ, is of a deep yellow colour. The medullary substance is soft, pulpy, and of a dark brown or black colour, whence the name *atrabiliary capsules* formerly given to these organs. In the centre is often seen a space, not natural, but formed by the breaking down after death of the medullary substance.

The **cortical portion** consists chiefly of narrow columnar masses, placed perpendicularly to the surface. This arrangement is due to the disposition of the capsule, which sends processes into the interior of the gland; these are connected with each other by transverse bands, so as to form a series of inter-communicating spaces. These spaces are of slight depth near the surface of the organ, so that there the section somewhat resembles a net; this is termed the *zona glomerulosa*; but they become much deeper or longer farther in, so as to resemble pipes or tubes placed endwise, the *zona fasciculata*. Still deeper down, near the medullary part, the spaces become again of small extent; this is named the *zona reticularis*. These processes or trabeculæ, derived from the capsule and forming the framework of the spaces, are composed of fibrous connective tissue, with longitudinal bundles of unstriped muscular fibres. Within the interior of the spaces are contained groups of polyhedral cells, which are finely granular in appearance, and contain a spherical nucleus, and not in-frequently fat globules. These groups of cells do not entirely fill the spaces in which they are contained, but between them and the trabeculæ of the framework is a channel, which is believed to be a lymph path or sinus, and which communicates with certain passages between the cells composing the group. The lymph path is supposed to open into a plexus of efferent lymphatic vessels which are con-tained in the capsule.

In the **medullary portion,** the fibrous stroma seems to be collected together into a much closer arrangement, and forms bundles of connective tissue which are loosely applied to the large plexus of veins of which this part of the organ mainly consists. In the interstices lie a number of cells compared by Frey to those of columnar epithelium. They are coarsely granular, do not contain any fat, and some of them are branched. Luschka has affirmed that these branches are connected with the nerve-fibres of a very intricate plexus which is found in the medulla; this statement has not been verified by other observers, for the tissue of the medullary substance is less easy to make out than that of the cortical, owing to its rapid decomposition.

The numerous arteries which enter the suprarenal bodies, from the sources mentioned below, penetrate the cortical part of the gland, where they break up into capillaries in the fibrous septa, and these converge to the very numerous veins of the medullary portion, which are collected together into the suprarenal vein, which usually emerges as a single vessel from the centre of the gland.

The *arteries* supplying the suprarenal capsules are numerous and of large size; they are derived from the aorta, the phrenic, and the renal; they subdivide into numerous minute branches previous to entering the substance of the gland.

The *suprarenal vein* returns the blood from the medullary venous plexus and receives several branches from the cortical substance ; it opens on the right side into the inferior vena cava, on the left side into the renal vein.

The *lymphatics* terminate in the lumbar glands.

The *nerves* are exceedingly numerous, and are derived from the solar and renal plexuses, and, according to Bergmann, form the phrenic and pneumogastric nerves. They enter the lower and inner part of the capsule, traverse the cortex, and terminate around the cells of the medulla. They have numerous small ganglia developed upon them, from which circumstance the organ has been conjectured to have some function in connection with the sympathetic nervous system.

THE PELVIS

The **cavity of the pelvis** is that part of the general abdominal cavity which is below the level of the linea ilio-pectinea and the promontory of the sacrum.

Boundaries.—It is bounded, behind, by the sacrum, the coccyx, the Pyriformis muscles, and the great sacro-sciatic ligaments ; in front and at the sides, by the ossa pubis and ischia, covered by the Obturator muscles ; above, it communicates with the cavity of the abdomen ; and below, the outlet is closed by the triangular ligament, the Levatores ani and Coccygei muscles, and the visceral layer of the pelvic fascia, which is reflected from the wall of the pelvis on to the viscera.

Contents.—The viscera contained in this cavity are the urinary bladder, the rectum, some of the generative organs peculiar to each sex, and some convolutions of the small intestines : they are partially covered by the peritoneum, and supplied with blood-vessels, lymphatics, and nerves.

THE BLADDER

The **bladder** is the reservoir for the urine. It is a musculo-membranous sac, situated in the pelvis, behind the pubes, and in front of the rectum in the male ; the cervix uteri and vagina intervening between it and that intestine in the female. The shape, position, and relations of the bladder are greatly influenced by age, sex, and the degree of distension of the organ. *During infancy* it is conical in shape, and projects above the upper border of the ossa pubis into the hypogastric region. *In the adult*, when quite empty and contracted, it is cup-shaped, and on vertical median section its cavity, with the adjacent portion of the urethra, presents a Y-shaped cleft, the stem of the Y corresponding to the urethra. It is placed deeply in the pelvis, flattened from before backwards, and reaches as high as the upper border of the symphysis pubis. When slightly distended, it has a rounded form, and is still contained within the pelvic cavity ; and when greatly distended it is ovoid in shape, rising into the abdominal cavity, and often extending nearly as high as the umbilicus. It is larger in its vertical diameter than from side to side, and its long axis is directed from above obliquely downwards and backwards, in a line directed from some point between the symphysis pubis and umbilicus (according to its distension) to the end of the coccyx. The bladder, when distended, is slightly curved forwards towards the anterior wall of the abdomen, so as to be more convex behind than in front. In the female it is larger in the transverse than in the vertical diameter, and its capacity is said to be greater than in the male.* When moderately distended, it measures about five inches in length, and three inches across, and the ordinary amount which it contains is about a pint.

* According to Henle, the bladder is considerably smaller in the female than in the male.

The bladder is divided for purposes of description into a superior, an antero-inferior, and two lateral surfaces, a base or fundus and a summit or apex.

The **superior** or **abdominal surface** is entirely free, and is covered throughout by peritoneum. It looks almost directly upwards into the abdominal cavity, and extends in an antero-posterior direction from the apex to the base of the bladder. It is in relation with the small intestine and sometimes with the sigmoid flexure, and in the female, with the uterus. On each side, in the male, a portion of the vas deferens is in contact with the hinder part of this surface, lying beneath the peritoneum.

The **antero-inferior** or **pubic surface** looks downwards and forwards. In the undistended condition it is uncovered by peritoneum, and is in relation with the

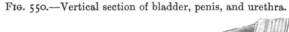

FIG. 550.—Vertical section of bladder, penis, and urethra.

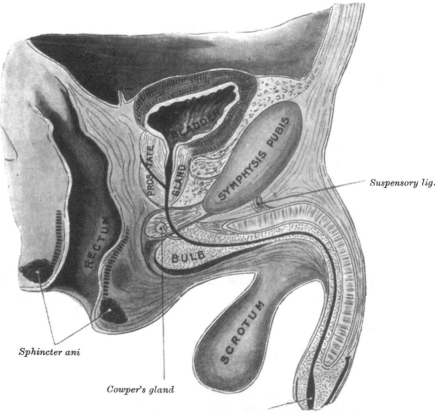

Suspensory lig.

Sphincter ani

Cowper's gland

Fossa navicularis

Obturator internus muscle on each side, with the recto-vesical fascia, and anterior true ligaments of the bladder. It is separated from the body of the pubis by a triangular interval, the *space of Retzius*, occupied by fatty tissue. As the bladder ascends into the abdominal cavity during distension the distance between its apex and the umbilicus is necessarily diminished, and the urachus is thus relaxed; so that, instead of passing directly upwards to the umbilicus, it descends first on the upper part of the anterior surface of the bladder, and then, curving upwards, ascends on the back of the abdominal wall. The peritoneum, which follows the urachus, thus comes to form a pouch of varying depth between the anterior surface of the viscus and the abdominal wall. Thus, when the bladder is distended, the upper part of its anterior surface is in relation to the urachus and is covered by peritoneum. The lower part of its anterior surface, for a distance of about two inches above the symphysis pubis, is devoid of peritoneum, and is in contact with the abdominal wall.

The **lateral surfaces** are covered behind and above by peritoneum, which extends as low as the level of the obliterated hypogastric artery ; below and in front of this, these surfaces are uncovered by peritoneum, and are separated from the Levatores ani muscles and walls of the pelvis by a quantity of loose areolar tissue containing fat. In front this surface is connected to the recto-vesical fascia by a broad expansion on either side, the *lateral true ligaments*. The vas deferens crosses the hinder part of the lateral surface obliquely, and passes between the ureter and the bladder.

FIG. 551.—Vertical median section of the male pelvis. (Henle.)

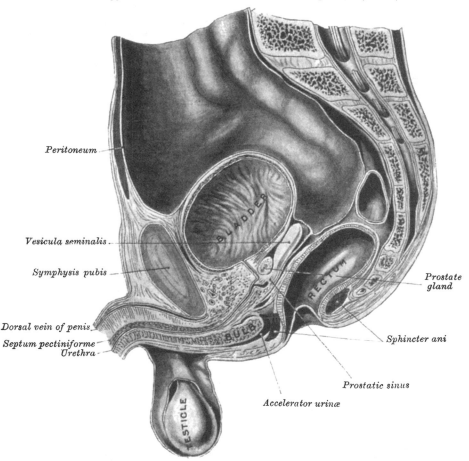

The **fundus** or **base** is directed downwards and backwards, and is partly covered by peritoneum and partly uncovered. In the male the upper portion, to within about an inch and a half of the prostate, is covered by the recto-vesical pouch of peritoneum. The lower part is in direct contact with the anterior wall of the second part of the rectum and the vesiculæ seminales and vasa deferentia. The ureters enter the bladder at the upper part of its base, about two inches above the prostate gland.

The portion of the bladder in relation with the rectum corresponds to a triangular space, bounded, below, by the prostate gland ; above, by the recto-vesical fold of the peritoneum ; and on each side, by the vesicula seminalis and vas deferens. It is separated from direct contact with the rectum by the recto-

vesical fascia.   When the bladder is very full, the peritoneal fold is raised with
it, and the distance between its reflection and the anus is about four inches ; but
this distance is much diminished when the bladder is empty and contracted.   *In
the female*, the base of the bladder is connected to the anterior aspect of the cervix
uteri by areolar tissue, and is adherent to the anterior wall of the vagina.   Its
upper surface is separated from the anterior surface of the body of the uterus by
the utero-vesical pouch of peritoneum.

The so-called *neck (cervix) of the bladder* is the point of commencement of
the urethra ; there is, however, no tapering part, which would constitute a true
neck, but the bladder suddenly contracts to the opening of the urethra.   In the

FIG. 552.—Frontal section of the lower part of the abdomen.
Viewed from the front.   (Braune.)

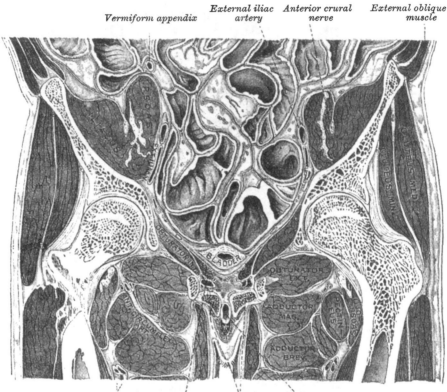

Vermiform appendix   *External iliac*   *Anterior crural*   *External oblique*
                              artery   *nerve*   *muscle*

Profunda vessels   Levator ani   Corpora   Urethra
                                    cavernosa

male its direction is oblique in the erect posture, and it is surrounded by the
prostate gland.   In the female its direction is obliquely downwards and
forwards.

The **urachus** is the obliterated remains of the tubular canal of the *allantois*,
which exists in the embryo, and a portion of which becomes expanded to form
the bladder (see section on Embryology).   It passes upwards, from the apex of
the bladder, between the transversalis fascia and peritoneum, to the umbilicus,
becoming thinner as it ascends.   It is composed of fibrous tissue, mixed with
plain muscular fibres.   On each side of it is placed a fibrous cord, the obliterated
portion of the hypogastric artery, which, passing upwards from the side of the
bladder, approaches the urachus above its summit.   In the infant, at birth, it is
occasionally found pervious, so that the urine escapes at the umbilicus, and
calculi have been found in its canal.

Ligaments.—The bladder is retained in its place by ligaments, which are divided into true and false. The true ligaments are five in number : two anterior, two lateral, and the urachus. The false ligaments, also five in number, are formed by folds of the peritoneum.

The *anterior true ligaments* (*pubo-prostatic*) extend from the back of the ossa pubis, one on each side of the symphysis, to the front of the neck of the bladder, over the anterior surface of the prostate gland. These ligaments are formed by the recto-vesical fascia, and contain a few muscular fibres prolonged from the bladder.

The *lateral true ligaments*, also formed by the recto-vesical fascia, are broader and thinner than the preceding. They are attached to the lateral parts of the prostate, and to the sides of the base of the bladder.

The *urachus* is the fibro-muscular cord already mentioned, extending between the summit of the bladder and the umbilicus. It is broad below, at its attachment to the bladder, and becomes narrower as it ascends.

The *false ligaments of the bladder* are two posterior, two lateral, and one superior.

The *two posterior* pass forwards, in the male, from the sides of the rectum ; in the female, from the sides of the uterus, to the posterior and lateral aspect of the

FIG. 553.—Superficial layer of the epithelium of the bladder. Composed of polyhedral cells of various sizes, each with one, two, or three nuclei. (Klein and Noble Smith.)

FIG. 554.—Deep layers of epithelium of bladder, showing large club-shaped cells above, and smaller, more spindle-shaped cells below—each with an oval nucleus. (Klein and Noble Smith.)

bladder : they form the lateral boundaries of the recto-vesical fold of the peritoneum, and contain the obliterated hypogastric arteries, and the ureters, together with vessels and nerves.

The *two lateral ligaments* are reflections of the peritoneum, from the iliac fossæ and lateral walls of the pelvis to the sides of the bladder.

The *superior ligament* (*ligamentum suspensorium*) is the prominent fold of peritoneum extending from the summit of the bladder to the umbilicus. It is carried off from the bladder by the urachus and the obliterated hypogastric arteries.

Structure.—The bladder is composed of four coats—serous, muscular, submucous, and mucous.

The *serous coat* is partial, and derived from the peritoneum. It invests the superior surface and the upper part of the lateral surfaces and base, and is reflected from these parts on to the abdominal and pelvic walls.

The *muscular coat* consists of three layers of unstriped muscular fibre : an external layer, composed of fibres having for the most part a longitudinal arrangement ; a middle layer, in which the fibres are arranged, more or less, in a circular manner ; and an internal layer, in which the fibres have a general longitudinal arrangement.

The *fibres of the external longitudinal layer* arise from the posterior surface of the body of the os pubis in both sexes (*musculi pubo-vesicalis*), and in the male from the adjacent part of the prostate gland and its capsule. They pass, in a more or less longitudinal manner, up the anterior surface of the bladder, over its apex, and then descend along its posterior surface to its base, where they become attached to the prostate in the male, and to the front of the vagina in the female. At the sides of the bladder the fibres are arranged obliquely and intersect one another. This layer has been named the *detrusor urinæ muscle*.

The fibres of the *middle circular layer* are very thinly and irregularly scattered on the body of the organ, and though to some extent placed transversely to the long axis of the bladder, are for the most part arranged obliquely. Towards the lower part of the bladder, round the cervix and commencement of the urethra, they are disposed in a thick circular layer, forming the *sphincter vesicæ*, which is continuous with the muscular fibres of the prostate gland.

The *internal longitudinal layer* is thin, and its fasciculi have a reticular arrangement, but with a tendency to assume for the most part a longitudinal direction.

Two bands of oblique fibres, originating behind the orifices of the ureters, converge to the back part of the prostate gland, and are inserted, by means of a fibrous process, into the middle lobe of that organ. They are the *muscles of the ureters*, described by Sir C. Bell, who supposed that during the contraction of the bladder they serve to retain the oblique direction of the ureters, and so prevent the reflux of the urine into them.

The *submucous coat* consists of a layer of areolar tissue, connecting together the muscular and mucous coats, and intimately united to the latter.

The *mucous coat* is thin, smooth, and of a pale rose colour. It is continuous above through the ureters with the lining membrane of the uriniferous tubes, and below with that of the urethra. It is connected loosely to the muscular coat by a layer of areolar tissue, and is therefore thrown into folds or *rugæ* when the bladder is empty. The epithelium covering it is of the transitional variety, consisting of a superficial layer of polyhedral flattened cells, each with one, two, or three nuclei ; beneath these, a stratum of large club-shaped cells, with the narrow extremity directed downwards and wedged in between smaller spindle-shaped cells, containing an oval nucleus (figs. 553, 554). There are no true glands in the mucous membrane of the bladder, though certain mucous follicles which exist, especially near the neck of the bladder, have been regarded as such.

Objects seen on the inner surface.—Upon the inner surface of the bladder are seen the orifices of the ureters, the trigone, and the commencement of the urethra.

The orifices of the ureters.—These are situated at the base of the trigone, being distant from each other about two inches ; they are about an inch and a half from the base of the prostate and the commencement of the urethra.

The *trigonum vesicæ*, or *trigone vesical*, is a triangular smooth surface, with the apex directed forwards, situated at the base of the bladder, immediately behind the urethral orifice. It is paler in colour than the rest of the interior, and never presents any rugæ, even in the collapsed condition of the organ, owing to the intimate adhesion of its mucous membrane to the subjacent tissue. It is bounded at each posterior angle by the orifice of the ureter, and in front by the orifice of the urethra. Projecting from the lower and anterior part of the bladder, and reaching to the orifice of the urethra, is a slight elevation of mucous membrane, called the *uvula vesicæ*. It is formed by a thickening of the submucous tissue.

The *arteries* supplying the bladder are the superior, middle, and inferior vesical in the male, with additional branches from the uterine and vaginal in the female. They are all derived from the anterior trunk of the internal iliac. The obturator and sciatic arteries also supply small visceral branches to the bladder.

The *veins* form a complicated plexus round the neck, sides, and base of the bladder, and terminate in the internal iliac vein.

The *lymphatics* form two plexuses, one in the muscular and another in the submucous coat ; they are most numerous in the neighbourhood of the trigone.

They accompany the blood-vessels, and ultimately terminate in the internal iliac glands.

The *nerves* are derived from the pelvic plexus of the sympathetic and from the third and the fourth sacral nerves; the former supplying the upper part of the organ, the latter its base and neck. According to F. Darwin the sympathetic fibres have ganglia connected with them, which send branches to the vessels and muscular coat.

Surface Form.—The surface form of the bladder varies with its degree of distension and under other circumstances. In the young child it is represented by a conical figure, the apex of which, even when the viscus is empty, is situated in the hypogastric region, about an inch above the level of the symphysis pubis. In the adult, when the bladder is empty, its apex does not reach above the level of the upper border of the symphysis pubis, and the whole organ is situated in the pelvis; the neck, in the male, corresponding to a line drawn horizontally backwards, through the symphysis a little below its middle. As the bladder becomes distended it gradually rises out of the pelvis into the abdomen and forms a swelling in the hypogastric region, which is perceptible to the hand, as well as to percussion. In extreme distension it reaches into the umbilical region. Under these circumstances the lower part of its anterior surface, for a distance of about two inches above the symphysis pubis, is closely applied to the abdominal wall, without the intervention of peritoneum, so that it can be tapped by an opening in the middle line just above the symphysis pubis, without any fear of wounding the serous membrane. When the rectum is distended, the prostatic portion of the urethra is elongated and the bladder lifted out of the pelvis and the peritoneum pushed upwards. Advantage is taken of this by some surgeons in performing the operation of suprapubic cystotomy. The rectum is distended by an indiarubber bag, which is introduced into this cavity empty, and then filled with ten or twelve ounces of water. If now the bladder is injected with about half a pint of some antiseptic fluid, it will appear above the pubes plainly perceptible to the sight and touch. The peritoneum will be pushed out of the way, and an incision three inches long may be made in the linea alba, from the symphysis pubis upwards, without any great · risk of wounding the peritoneum. Other surgeons object to the employment of this bag, as its use is not unattended with risk, and because it causes pressure on the prostatic sinuses and produces congestion of the vessels over the bladder and a good deal of venous hæmorrhage.

When distended the bladder can be felt in the male, from the rectum, behind the prostate, and fluctuation can be perceived by a bimanual examination, one finger being introduced into the rectum and the distended bladder tapped on the front of the abdomen with the finger of the other hand. This portion of the bladder, that is, the portion felt in the rectum by the finger, is also uncovered by peritoneum, and the bladder may here be punctured from the rectum, in the middle line, without risk of wounding the serous membrane.

Surgical Anatomy.—A defect of development, in which the bladder is implicated, is known under the name of *extroversion of the bladder*. In this condition the lower part of the abdominal wall and the anterior wall of the bladder are wanting, so that the posterior surface of the bladder presents on the abdominal surface, and is pushed forwards by the pressure of the viscera within the abdomen, forming a red vascular tumour, on which the openings of the ureters are visible. The penis, except the glans, is rudimentary and is cleft on its dorsal surface, exposing the floor of the urethra, a condition known as *epispadias*. The pelvic bones are also arrested in development (see page 154).

The bladder may be ruptured by violence applied to the abdominal wall, when the viscus is distended, without any injury to the bony pelvis, or it may be torn in cases of fracture of the pelvis. The rupture may be either intraperitoneal or extraperitoneal, that is, may implicate the superior surface of the bladder in the former case, or one of the other surfaces in the latter. Rupture of the antero-inferior surface alone is, however, very rare. Until recently intraperitoneal rupture was uniformly fatal, but now abdominal section and suturing the rent with Lembert's suture is resorted to, with a very considerable amount of success. The sutures are inserted only through the peritoneal and muscular coats in such a way as to bring the serous surfaces at the margins of the wound into apposition, and one is inserted just beyond each end of the wound. The bladder should be tested as to whether it is water-tight before closing the external incision.

The muscular coat of the bladder undergoes hypertrophy in cases in which there is any obstruction to the flow of urine. Under these circumstances the bundles of which the muscular coat consists become much increased in size, and, interlacing in all directions, give rise to what is known as the *fasciculated bladder*. Between these bundles of muscular fibres the mucous membrane may bulge out, forming sacculi, constituting the *sacculated bladder*, and in these little pouches phosphatic concretions may collect, forming *encysted calculi*. The mucous membrane is very loose and lax, except over the trigone, to allow of the distension of the viscus.

Various forms of tumour have been found springing from the wall of the bladder. The innocent tumours are the papilloma and the mucous polypus arising from the mucous membrane; the fibrous, from the submucous tissue; and the myoma, originating in the muscular tissue; and, very rarely, dermoid tumours, the exact origin of which it is difficult to explain. Of the malignant tumours, epithelioma is the most common, but sarcomata are occasionally found in the bladder of children.

Puncture of the bladder may be performed either above the symphysis pubis or through the rectum, in both cases without wounding the peritoneum. The former plan is generally to be preferred, since in puncture by the rectum a permanent fistula may be left from abscess forming between the rectum and the bladder; or pelvic cellulitis may be set up; moreover, it is exceedingly inconvenient to keep a cannula in the rectum. In some cases, in performing this operation the recto-vesical pouch of peritoneum has been wounded, inducing fatal peritonitis. The operation, therefore, has been almost completely abandoned.

THE MALE URETHRA

The **urethra in the male** extends from the neck of the bladder to the meatus urinarius at the end of the penis. It presents a double curve in the flaccid state of the penis (fig. 550), but in the erect state of this organ it forms only a single curve, the concavity of which is directed upwards. Its length varies from eight to nine inches; and it is divided into three portions, the *prostatic, membranous,* and *spongy,* the structure and relations of which are essentially different. Except during the passage of the urine or semen, the urethra is a mere transverse cleft or slit, with its upper and under surfaces in contact. At the meatus urinarius the slit is vertical, and in the prostatic portion somewhat arched.

The **prostatic portion** is the widest and most dilatable part of the canal. It passes through the prostate gland, from its base to the apex, lying nearer its anterior than its posterior surface. It is about an inch and a quarter in length; the form of the canal is spindle-shaped, being wider in the middle than at either extremity, and narrowest below, where it joins the membranous portion. A transverse section of the canal as it lies in the prostate is horse-shoe in shape, the convexity being directed forwards (fig. 556), since the direction of the canal is nearly vertical.

Upon the floor of the canal is a narrow longitudinal ridge, the *verumontanum,* or *caput gallinaginis,* formed by an elevation of the mucous membrane and its subjacent tissue. It is eight or nine lines in length, and a line and a half in height; and contains, according to Kobelt, muscular and erectile tissues. When distended, it may serve to prevent the passage of the semen backwards into the bladder. On each side of the verumontanum is a slightly depressed fossa, the *prostatic sinus,* the floor of which is perforated by numerous apertures, the *orifices of the prostatic ducts* from the lateral lobes of the gland; the ducts of the middle lobe open behind the verumontanum. At the fore part of the verumontanum, in the middle line, is a depression, the *sinus pocularis (vesicula prostatica)*; and upon or within its margins are the slit-like openings of the ejaculatory ducts. The sinus pocularis forms a *cul-de-sac* about a quarter of an inch in length, which runs upwards and backwards in the substance of the prostate behind the middle lobe; its prominent anterior wall partly forms the verumontanum. Its walls are composed of fibrous tissue, muscular fibres, and mucous membrane, and numerous small glands open on its inner surface. It has been called by Weber, who discovered it, the *uterus masculinus,* from its being developed from the united lower ends of the atrophied Müllerian ducts, and therefore homologous with the uterus and vagina in the female.

The **membranous portion of the urethra** extends between the apex of the prostate and the bulb of the corpus spongiosum. It is the narrowest part of the canal (excepting the meatus), and measures three-quarters of an inch along its upper, and half an inch along its lower surface, in consequence of the bulb projecting backwards beneath it. Its anterior concave surface is placed about an inch below and behind the pubic arch, from which it is separated by the dorsal vessels and nerves of the penis, and some muscular fibres. Its posterior convex surface is separated from the rectum by a triangular space, which constitutes the perinæum. The membranous portion of the urethra lies between the inferior and superior layers of the triangular ligament. As it pierces the inferior layer, the fibres around the opening are prolonged over the tube. It is also surrounded by the Compressor urethræ muscle.

The **spongy portion** is the longest part of the urethra, and is contained in the corpus spongiosum. It is about six inches in length, and extends from the termination of the membranous portion to the meatus urinarius. Commencing just below the triangular ligament, it inclines downwards for a short distance; it next ascends for about half its length, and then, in the flaccid condition of the penis, it bends suddenly downwards. It is narrow, and of uniform size in the body of the penis, measuring about a quarter of an inch in diameter; being dilated behind, within the bulb; and again anteriorly within the glans penis, where it forms the *fossa navicularis*.

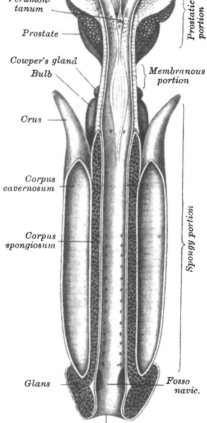

FIG. 555.—The male urethra, laid open on its anterior (upper) surface. (Testut.)

The *bulbous portion* is a name given, in some descriptions of the urethra, to the posterior part of the spongy portion contained within the bulb.

The *meatus urinarius* is the most contracted part of the urethra; it is a vertical slit, about three lines in length, bounded on each side by two small labia.

The inner surface of the lining membrane of the urethra, especially on the floor of the spongy portion, presents the orifices of numerous mucous glands and follicles situated in the submucous tissue, and named the *glands of Littré*. They vary in size, and their orifices are directed forwards, so that they may easily intercept the point of a catheter in its passage along the canal. One of these lacunæ, larger than the rest, is situated on the upper surface of the fossa navicularis, about an inch and a half from the orifice; it is called the *lacuna magna*. Into the bulbous portion are found opening the ducts of Cowper's glands.

Structure.—The urethra is composed of a continuous mucous membrane, supported by a submucous tissue which connects it with the various structures through which it passes.

The *mucous coat* forms part of the genito-urinary mucous membrane. It is continuous with the mucous membrane of the bladder, ureters, and kidneys; externally, with the integument covering the glans penis; and is prolonged into the ducts of the glands which open into the urethra, viz. Cowper's glands and the prostate gland; and into the vasa deferentia and vesiculæ seminales, through the ejaculatory ducts. In the spongy and membranous portions the mucous membrane is arranged in longitudinal folds when the tube is empty. Small papillæ are found upon it, near the orifice; and its epithelial lining is of the columnar variety, excepting near the meatus, where it is squamous.

The *submucous tissue* consists of a vascular erectile layer; outside which is a layer of unstriped muscular fibres, arranged in a circular direction, which separates the mucous membrane and submucous tissue from the tissue of the corpus spongiosum.

Surgical Anatomy.—The urethra may be ruptured by the patient falling astride of any hard substance and striking his perinæum, so that the urethra is crushed against the pubic arch. Bleeding will at once take place from the urethra, and this, together with the bruising in the perinæum and the history of the accident, will at once point to the nature of the injury.

The surgical anatomy of the urethra is of considerable importance in connection with the passage of instruments into the bladder. Otis was the first to point out that the urethra is capable of great dilatability, so that, excepting through the external meatus, an instrument corresponding to 18 English gauge (29 French) can usually be passed without damage. The orifice of the urethra is not so dilatable, and therefore frequently requires slitting. A recognition of this dilatability caused Bigelow to very considerably modify the operation of lithotrity and introduce that of litholapaxy. In passing catheters, especially fine ones, the point of the instrument should be kept as far as possible along the upper wall of the canal, as it is otherwise very liable to enter one of the lacunæ. Stricture of the urethra is a disease of very common occurrence, and is generally situated in the spongy part of the urethra, most commonly in the bulbous portion, just in front of the membranous urethra, but in a very considerable number of cases in the penile or ante-scrotal part of the canal.

FEMALE BLADDER AND URETHRA

The **bladder** is situated at the anterior part of the pelvis. It is in relation, *in front*, with the symphysis pubis ; *behind*, with the utero-vesical pouch of peritoneum which separates it from the body of the uterus ; its *base* lies in contact with the connective tissue in front of the cervix uteri and upper part of the vagina. *Laterally*, is the recto-vesical fascia. The bladder is said by some anatomists to be larger in the female than in the male. At any rate it does not rise above the symphysis pubis till more distended than in the male, but this is perhaps owing to the more capacious pelvis rather than to its being of actually larger size.

THE URETHRA

The **urethra** is a narrow membranous canal, about an inch and a half in length, extending from the neck of the bladder to the meatus urinarius. It is placed beneath the symphysis pubis, embedded in the anterior wall of the vagina and its direction is obliquely downwards and forwards, its course being slightly curved, the concavity directed forwards and upwards. Its diameter when undilated is about a quarter of an inch. The urethra perforates the triangular ligament, and its external orifice is situated directly in front of the vaginal opening and about an inch behind the glans clitoridis.

Structure.—The urethra consists of three coats : muscular, erectile, and mucous.

The *muscular coat* is continuous with that of the bladder ; it extends the whole length of the tube, and consists of a circular stratum of muscular fibres. In addition to this, between the two layers of the triangular ligament, the female urethra is surrounded by the Compressor urethræ, as in the male.

A *thin layer of spongy erectile tissue*, containing a plexus of large veins, inter-mixed with bundles of unstriped muscular fibre, lies immediately beneath the mucous coat.

The *mucous coat* is pale, continuous externally with that of the vulva, and internally with that of the bladder. It is thrown into longitudinal folds, one of which, placed along the floor of the canal, resembles the verumontanum in the male urethra. It is lined by laminated epithelium, which becomes transitional near the bladder. Its external orifice is surrounded by a few mucous follicles.

The urethra, from not being surrounded by dense resisting structures, as in the male, admits of considerable dilatation, which enables the surgeon to remove, with considerable facility, calculi, or other foreign bodies, from the cavity of the bladder.

MALE ORGANS OF GENERATION

THE PROSTATE GLAND

THE **Prostate Gland** (προΐστημι, *to stand before*) is a pale, firm, partly glandular
and partly muscular body, which is placed immediately below the neck of the
bladder and around the commencement of the urethra. It is placed in the pelvic
cavity, behind the lower part of the symphysis pubis, and above the deep layer of
the triangular ligament, and rests upon the rectum, through which it may be

FIG. 556.—Transverse section of the prostate gland; showing the urethra,
with the eminence of the caput gallinaginis; beneath it the sinus pocularis
and ejaculatory ducts. (Enlarged.)

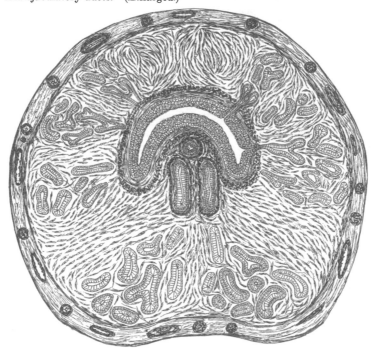

distinctly felt, especially when enlarged. In shape and size it resembles a
chestnut.

Its *base* is directed upwards, and is situated immediately below the neck of
the bladder.

Its *apex* is directed downwards to the deep layer of the triangular ligament,
which it touches.

Its *posterior surface* is flattened, marked by a slight longitudinal furrow, and
rests on the second part of the rectum, and is distant about one inch and a half
from the anus.

Its *anterior surface* is convex, and placed about three-quarters of an inch behind the pubic symphysis, from which it is separated by a plexus of veins and a quantity of loose fat. It is connected to the pubic bone on either side by the pubo-prostatic ligaments.

The *lateral surfaces* are prominent, and are covered by the anterior portions of the Levatores ani muscles, which are, however, separated from the gland by a plexus of veins.

The prostate measures about an inch and a half transversely at the base, an inch in its antero-posterior diameter, and an inch and a quarter in its vertical diameter. Its weight is about four and a half drachms. It is held in its position by the anterior ligaments of the bladder (*pubo-prostatic*); by the deep layer of the triangular ligament, which invests the commencement of the membranous portion of the urethra and prostate gland; and by the anterior portions of the Levatores ani muscles, which pass backwards from the os pubis and embrace the sides of the prostate. These portions of the Levatores ani, from the support they afford to the prostate, are named the *Levator prostatæ.*

The prostate consists of two lateral lobes and a middle lobe.

The *two lateral lobes* are of equal size, separated by a deep notch behind, and by a slight furrow upon their anterior and posterior surfaces, which indicates the bilobed condition of the organ in some animals.

The third, or *middle lobe*, is a small transverse band, occasionally a rounded or triangular prominence, placed between the two lateral lobes, at the posterior part of the organ. It lies immediately beneath the neck of the bladder, behind the commencement of the urethra, and above the ejaculatory ducts, which pass through the gland between its middle and lateral lobes. Its existence is not constant, but it is occasionally found at an early period of life, as well as in adults, and in old age.

The prostate gland is perforated by the urethra and the ejaculatory ducts. The urethra usually lies along the junction of its anterior with its middle third. The ejaculatory ducts pass obliquely downwards and forwards through the posterior part of the prostate, and open into the prostatic portion of the urethra.

Structure.—The prostate is immediately enveloped by a thin but firm fibrous capsule, distinct from that derived from the recto-vesical fascia, and separated from it by a plexus of veins. Its substance is of a pale reddish-grey colour, of great density, and not easily torn. It consists of glandular substance and muscular tissue.

The muscular tissue, according to Kölliker, constitutes the proper stroma of the prostate; the connective tissue being very scanty, and simply forming thin trabeculæ between the muscular fibres, in which the vessels and nerves of the gland ramify. The muscular tissue is arranged as follows: immediately beneath the fibrous capsule is a dense layer, which forms an investing sheath for the gland; secondly, around the urethra, as it lies in the prostate, is another dense layer of circular fibres, continuous above with the internal layer of the muscular coat of the bladder, and below blending with the fibres surrounding the membranous portion of the urethra. Between these two layers, strong bands of muscular tissue, which decussate freely, form meshes in which the glandular structure of the organ is embedded. In that part of the gland which is situated in front of the urethra the muscular tissue is especially dense, and there is here little or no gland tissue; while in that part which is behind the urethra the muscular tissue presents a wide-meshed structure, which is densest at the base of the gland—that is, near the bladder—becoming looser and more sponge-like towards the apex of the organ.

The *glandular substance* is composed of numerous follicular pouches, opening into elongated canals, which join to form from twelve to twenty small excretory ducts. The follicles are connected together by areolar tissue, supported by prolongations from the fibrous capsule and muscular stroma, and enclosed in a delicate capillary plexus. The epithelium lining of both the canals and the terminal vesicles is of the columnar variety. The prostatic ducts open into the floor of the prostatic portion of the urethra.

Vessels and Nerves.—The *arteries* supplying the prostate are derived from the internal pudic, vesical, and hæmorrhoidal. Its *veins* form a plexus around the sides and base of the gland ; they receive in front the dorsal vein of the penis, and terminate in the internal iliac vein. The *nerves* are derived from the pelvic plexus.

Surgical Anatomy.—The relation of the prostate to the rectum should be noted : by means of the finger introduced into the gut, the surgeon detects enlargement or other disease of this organ ; he can feel the apex of the gland, which is the guide to Cock's operation for stricture ; he is enabled also by the same means to direct the point of a catheter, when its introduction is attended with difficulty either from injury or disease of the membranous or prostatic portions of the urethra. When the finger is introduced into the bowel, the surgeon may, in some cases, especially in boys, learn the position, as well as the size, of a calculus in the bladder; and in the operation for its removal, if, as is not infrequently the case, it should be lodged behind an enlarged prostate, it may be displaced from its position by pressing upwards the base of the bladder from the rectum. The prostate gland is occasionally the seat of suppuration, due either to injury, gonorrhœa, or tuberculous disease. The gland is enveloped in a dense unyielding capsule, which determines the course of the abscess, and also explains the great pain which is present in the acute form of the disease. The abscess most frequently bursts into the urethra, the direction in which there is least resistance, but may occasionally burst into the rectum, or more rarely in the perinæum. In advanced life the prostate sometimes becomes considerably enlarged and projects into the bladder so as to impede the passage of the urine. According to Dr. Messer's researches, conducted at Greenwich Hospital, it would seem that such obstruction exists in 20 per cent. of all men over sixty years of age. In some cases the enlargement affects principally the lateral lobes, which may undergo considerable enlargement without causing much inconvenience. In other cases it would seem that the middle lobe enlarges most, and even a small enlargement of this lobe may act injuriously, by forming a sort of valve over the urethral orifice, preventing the passage of the urine, and the more the patient strains, the more completely will it block the opening into the urethra. In consequence of the enlargement of the prostate, a pouch is formed at the base of the bladder behind the projection, in which water collects, and cannot be entirely expelled. It becomes decomposed and ammoniacal, and leads to cystitis. For this condition 'prostatectomy' is sometimes done. The bladder is opened by an incision above the symphysis pubis, the mucous membrane incised, and the enlarged and projecting middle lobe enucleated.

COWPER'S GLANDS

Cowper's Glands are two small, rounded, and somewhat lobulated bodies, of a yellow colour, about the size of peas, placed behind the fore part of the membranous portion of the urethra, between the two layers of the triangular ligament. They lie close above the bulb, and are enclosed by the transverse fibres of the Compressor urethræ muscle. Their existence is said to be constant: they gradually diminish in size as age advances.

Structure.—Each gland consists of several lobules, held together by a fibrous investment. Each lobule consists of a number of acini, lined by columnar epithelial cells, opening into one duct, which, joining with the ducts of other lobules outside the gland, form a single excretory duct. The excretory duct of each gland, nearly an inch in length, passes obliquely forwards beneath the mucous membrane, and opens by a minute orifice on the floor of the bulbous portion of the urethra.

THE PENIS

The **Penis** consists of a root, body, and extremity or *glans penis.*

The *root* is firmly connected to the rami of the os pubis and ischium by two strong tapering fibrous processes, the *crura*; and to the front of the symphysis pubis by the *suspensory ligament*, a strong band of fibrous tissue which passes downwards from the front of the symphysis pubis to the upper surface of the root of the penis, where it splits into two portions and blends with the fascial sheath of the organ.

The *extremity*, or *glans penis*, presents the form of an obtuse cone, flattened from above downwards. At its summit is a vertical fissure, the orifice of the urethra (*meatus urinarius*). The base of the glans forms a rounded projecting border, the *corona glandis*; and behind the corona is a deep constriction, the *cervix*. Upon both of these parts, numerous small sebaceous glands are found, the *glandulæ Tysonii odoriferæ.** They secrete a sebaceous matter of very peculiar odour, which probably contains caseine, and becomes easily decomposed.

The *body of the penis* is the part between the root and extremity. In the flaccid condition of the organ it is cylindrical, but when erect has a triangular prismatic form with rounded angles, the broadest side being turned upwards, and called the *dorsum*. The body is covered by integument, and contains in its interior a large portion of the urethra. The integument covering the penis is remarkable for its thinness, its dark colour, its looseness of connection with the deeper parts of the organ, and its containing no adipose tissue. At the root of the penis, the integument is continuous with that upon the pubes and scrotum : and at the neck of the glans it leaves the surface, and becomes folded upon itself to form the *prepuce*.

The internal layer of the prepuce is attached behind to the cervix, and approaches in character to a mucous membrane ; from the cervix it is reflected over the glans penis, and at the meatus urinarius is continuous with the mucous lining of the urethra.

The integument covering the glans penis contains no sebaceous glands ; but projecting from its free surface are a number of small, highly sensitive papillæ. At the back part of the meatus urinarius a fold of mucous membrane passes backwards to the bottom of a depressed raphé, where it is continuous with the prepuce ; this fold is termed the *frænum præputii*.

Structure of the Penis.—The penis is composed of a mass of erectile tissue, enclosed in three cylindrical fibrous compartments. Of these, two, the *corpora cavernosa*, are placed side by side along the upper part of the organ ; the third, or *corpus spongiosum*, encloses the urethra, and is placed below.

The **Corpora Cavernosa** form the chief part of the body of the penis. They consist of two fibrous cylindrical tubes, placed side by side, and intimately connected along the median line for their anterior three-fourths, while at their back part they separate from each other to form the *crura*, which are two strong tapering fibrous processes firmly connected to the rami of the os pubis and ischium. Each crus commences by a blunt-pointed process in front of the tuberosity of the ischium ; and, before its junction with its fellow to form the body of the penis, it presents a slight enlargement, named by Kobelt the *bulb of the corpus cavernosum*. Just beyond this point they become constricted, and retain an equal diameter to their anterior extremity, where they form a single rounded end, which is received into a fossa in the base of the glans penis. A median groove on the upper surface lodges the dorsal vein of the penis, and the groove on the under surface receives the corpus spongiosum. The root of the penis is connected to the symphysis pubis by the suspensory ligament.

Structure.— The corpora cavernosa are surrounded by a strong fibrous envelope, consisting of two sets of fibres : the one, longitudinal in direction, being common to the two corpora cavernosa, and investing them in a common covering ; the other, internal, circular in direction, being proper to each corpus cavernosum. The internal circular fibres of the two corpora cavernosa form, by their junction in the mesial plane, an incomplete partition or septum between the two bodies.

The *septum* between the two corpora cavernosa is thick and complete behind, but in front it is incomplete, and consists of a number of vertical bands, which are arranged like the teeth of a comb, whence the name which it has received,

---

* Stieda (*Comptes-rendus du XII Congrès International de Médecine*, Moscow, 1897) asserts that Tyson's glands are never found on the corona glandis, and that what have hitherto been mistaken for glands are really large papillæ.

septum pectiniforme. These bands extend between the dorsal and the urethral surface of the corpora cavernosa. This fibrous investment is extremely dense, of considerable thickness, and consists of bundles of shining white fibres, with an admixture of well-developed elastic fibres, so that it is possessed of great elasticity.

From the internal surface of the fibrous envelope, as well as from the sides of the septum, are given off a number of bands or cords, which cross the interior of the corpora cavernosa in all directions, subdividing them into a number of separate compartments, and giving the entire structure a spongy appearance. These bands and cords are called *trabeculæ*, and consist of white fibrous tissue, elastic fibres, and plain muscular fibres. In them are contained numerous arteries and nerves.

The component fibres of which the trabeculæ are composed are larger and stronger round the circumference than at the centre of the corpora cavernosa; they are also thicker behind than in front. The interspaces, on the contrary, are larger at the centre than at the circumference, their long diameter being directed transversely; they are largest anteriorly. They are occupied by venous blood, and are lined by a layer of flattened cells similar to the endothelial lining of veins.

FIG. 557.—From the peripheral portion of the corpus cavernosum penis under a low magnifying power. (Copied from Langer.)

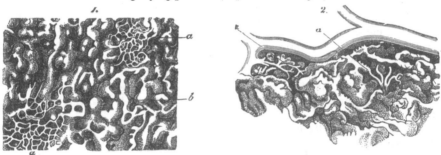

1. *a.* Capillary network. *b.* Cavernous spaces. 2. Connection of the arterial twigs (*a*) with the cavernous spaces.

The whole of the structure of the corpora cavernosa, contained within the fibrous sheath, consists therefore of a sponge-like tissue of areolar spaces, freely communicating with each other and filled with venous blood. The spaces may therefore be regarded as large cavernous veins.

The arteries bringing the blood to these spaces are the arteries of the corpora cavernosa and branches from the dorsal artery of the penis, which perforate the fibrous capsule, along the upper surface, especially near the fore part of the organ.

These arteries on entering the cavernous structure divide into branches, which are supported and enclosed by the trabeculæ. Some of these terminate in a capillary network, the branches of which open directly into the cavernous spaces; others assume a tendril-like appearance, and form convoluted and somewhat dilated vessels, which were named by Müller *helicine arteries.* They project into the spaces, and from them are given off small capillary branches to supply the trabecular structure. They are bound down in the spaces by fine fibrous processes, and are more abundant in the back part of the corpora cavernosa (fig. 557).

The blood from the cavernous spaces is returned by a series of vessels, some of which emerge in considerable numbers from the base of the glans penis and converge on the dorsum of the organ to form the dorsal vein; others pass out on the upper surface of the corpora cavernosa and join the dorsal vein; some emerge from the under surface of the corpora cavernosa and, receiving branches from the corpus spongiosum, wind round the sides of the penis to terminate in the dorsal vein ; but the greater number pass out at the root of the penis and join the prostatic plexus.

The **Corpus Spongiosum** encloses the urethra, and is situated in the groove on the under surface of the corpora cavernosa. It commences posteriorly below the superficial layer of the triangular ligament of the urethra, between the diverging crura of the corpora cavernosa, where it forms a rounded enlargement, the *bulb* ; and terminates, anteriorly, in another expansion, the *glans penis*, which overlaps the anterior rounded extremity of the corpora cavernosa. The central portion, or body of the corpus spongiosum, is cylindrical, and tapers slightly from behind forwards.

The *bulb* varies in size in different subjects ; it receives a fibrous investment from the superficial layer of the triangular ligament, and is surrounded by the Accelerator urinæ muscle. The urethra enters the bulb nearer its upper than its lower surface, being surrounded by a layer of erectile tissue, a thin prolongation of which is continued backwards round the membranous and prostatic portions of the canal to the neck of the bladder, lying between the two layers of muscular tissue. The portion of the bulb below the urethra presents a partial division into two lobes, being marked externally by a linear raphé, while internally there projects, for a short distance, a thin fibrous septum, which is more distinct in early life.

Structure.—The corpus spongiosum consists of a strong fibrous envelope, enclosing a trabecular structure, which contains in its meshes erectile tissue. The fibrous envelope is thinner, whiter in colour, and more elastic than that of the corpora cavernosa. The trabeculæ are more delicate, nearly uniform in size, and the meshes between them smaller than in the corpora cavernosa : their long diameter, for the most part, corresponding with that of the penis. The external envelope or outer coat of the corpus spongiosum is formed partly of unstriped muscular fibre, and a layer of the same tissue immediately surrounds the canal of the urethra.

The *lymphatics of the penis* consist of a superficial and a deep set ; the former are derived from a dense network on the skin of the glans and prepuce and from the mucous membrane of the urethra, and terminate in the superficial inguinal glands ; the latter emerge from the corpora cavernosa and corpus spongiosum, and, passing beneath the pubic arch, join the deep lymphatics of the pelvis.

The *nerves* are derived from the internal pudic nerve and the pelvic plexus. On the glans and bulb some filaments of the cutaneous nerves have Pacinian bodies connected with them, and, according to Krause, many of them terminate in a peculiar form of end-bulb (see page [50]).

Surgical Anatomy.—The penis occasionally requires removal for malignant disease. Usually, removal of the ante-scrotal portion is all that is necessary, but sometimes it is requisite to remove the whole organ from its attachment to the rami of the ossa pubis and ischia. The former operation is performed either by cutting off the whole of the anterior part of the penis with one sweep of the knife ; or, what is better, cutting through the corpora cavernosa from the dorsum, and then separating the corpus spongiosum from them, dividing it at a level nearer the glans penis. The mucous membrane of the urethra is then slit up, and the edges of the flap attached to the external skin, in order to prevent contraction of the orifice, which would otherwise take place. The vessels which require ligature are the two dorsal arteries of the penis, the arteries of the corpora cavernosa, and the artery of the septum. When the entire organ requires removal, the patient is placed in the lithotomy position, and an incision is made through the skin and subcutaneous tissue round the root of the penis, and carried down the median line of the scrotum as far as the perinæum. The two halves of the scrotum are then separated from each other, and a catheter having been introduced into the bladder as a guide, the spongy portion of the urethra below the triangular ligament is separated from the corpora cavernosa and divided, the catheter having been withdrawn just behind the bulb. The suspensory ligament is now severed, and the crura separated from the bone with a periosteum scraper, and the whole penis removed. The membranous portion of the urethra, which has not been removed, is now to be attached to the skin at the posterior extremity of the incision in the perinæum. The remainder of the wound is to be brought together, free drainage being provided for.

THE TESTES AND THEIR COVERINGS (fig. 558)

The **Testes** are two glandular organs, which secrete the semen ; they are situated in the scrotum, being suspended by the spermatic cords.   At an early period of fœtal life the testes are contained in the abdominal cavity, behind the peritoneum.   Before birth they descend to the inguinal canal, along which they pass with the spermatic cord, and, emerging at the external abdominal ring they descend into the scrotum, becoming invested in their course by numerous coverings derived from the serous, muscular, and fibrous layers of the abdominal parietes, as well as by the scrotum.   The coverings of the testes are : the

Skin \
Dartos / Scrotum.

Intercolumnar, or External spermatic fascia.

Cremasteric fascia.

Infundibuliform, or Fascia propria (Internal spermatic fascia).

Tunica vaginalis.

FIG. 558.—Transverse section through the left side of the scrotum and the left testicle. The sac of the tunica vaginalis represented in a distended condition.   (Delépine.)

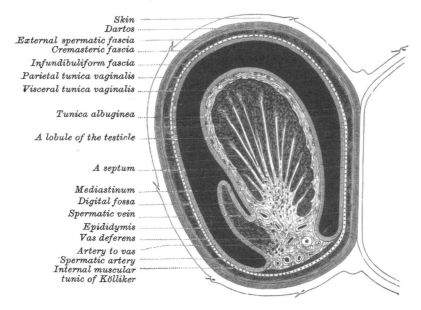

The **scrotum** is a cutaneous pouch which contains the testes and part of the spermatic cords.   It is divided on its surface into two lateral portions by a median line, or *raphé*, which is continued forwards to the under surface of the penis, and backwards along the middle line of the perinæum to the anus.   Of these two lateral portions the left is longer than the right, and corresponds with the greater length of the spermatic cord on the left side.   Its external aspect varies under different circumstances : thus, under the influence of warmth, and in old and debilitated persons, it becomes elongated and flaccid ; but, under the influence of cold, and in the young and robust, it is short, corrugated, and closely applied to the testes.

The scrotum consists of two layers, the integument and the dartos.

The **integument** is very thin, of a brownish colour, and generally thrown into folds or rugæ.   It is provided with sebaceous follicles, the secretion of which has a peculiar odour, and is beset with thinly scattered, crisp hairs, the roots of which are seen through the skin.

The **dartos** is a thin layer of loose reddish tissue, endowed with contractility : it forms the proper tunic of the scrotum, is continuous, around the base of the scrotum, with the two layers of the superficial fascia of the groin and perinæum, and sends inwards a distinct septum, *septum scroti*, which divides it into two cavities for the testes, the septum extending between the raphé and the under surface of the penis, as far as its root.

The dartos is closely united to the skin externally, but connected with the subjacent parts by delicate areolar tissue, upon which it glides with the greatest facility. The dartos is very vascular, and consists of a loose areolar tissue, containing unstriped muscular fibre, but no fat. Its contractility is slow, and excited by cold and mechanical stimuli, but not by electricity.

The **intercolumnar fascia** is a thin membrane, which is derived from the pillars of the external abdominal ring during the descent of the testis in the fœtus, and is prolonged downwards around the surface of the cord and testis. It is separated from the dartos by loose areolar tissue, which allows of considerable movement of the latter upon it, but is intimately connected with the succeeding layers.

The **cremasteric fascia** consists of scattered bundles of muscular fibres (*Cremaster muscle*), connected together into a continuous covering by intermediate areolar tissue. The muscular fibres are continuous with the lower border of the Internal oblique muscle (see page 335).

The **infundibuliform fascia** is a thin membranous layer, which loosely invests the surface of the cord. It is a continuation downwards of the fascia transversalis. Beneath it is a quantity of loose connective tissue which connects this layer of fascia with the spermatic cord and posterior part of the testicle. This connective tissue is continuous above with the subserous areolar tissue of the abdomen. These two layers, the infundibuliform fascia and the tissue beneath it, are known collectively as the *fascia propria*.

The **tunica vaginalis** is described with the testis.

Vessels and Nerves.—The *arteries* supplying the coverings of the testis are : the superficial and deep external pudic, from the femoral ; the superficial perineal branch of the internal pudic ; and the cremasteric branch from the epigastric. The *veins* follow the course of the corresponding arteries. The *lymphatics* terminate in the inguinal glands. The *nerves* are the ilio-inguinal branch of the lumbar plexus, the two superficial perineal branches of the internal pudic nerve, the inferior pudendal branch of the small sciatic nerve, and the genital branch of the genito-crural nerve.

The **spermatic cord** extends from the internal or deep abdominal ring, where the structures of which it is composed converge, to the back part of the testicle. In the abdominal wall the cord passes obliquely along the inguinal canal, lying at first beneath the Internal oblique, and upon the fascia transversalis ; but nearer the pubes, it rests upon Poupart's ligament, having the aponeurosis of the External oblique in front of it, and the conjoined tendon behind it. It then escapes at the external ring, and descends nearly vertically into the scrotum. The left cord is rather longer than the right, consequently the left testis hangs somewhat lower than its fellow.

Structure of the spermatic cord.—The spermatic cord is composed of arteries, veins, lymphatics, nerves, and the excretory duct of the testicle. These structures are connected together by areolar tissue, and invested by the layers brought down by the testicle in its descent.

The *arteries of the cord* are : the spermatic, from the aorta ; the artery of the vas deferens, from the superior vesical ; and the cremasteric, from the deep epigastric.

The *spermatic artery*, a branch of the abdominal aorta, escapes from the abdomen at the internal or deep abdominal ring, and accompanies the other constituents of the spermatic cord along the inguinal canal and through the external abdominal ring into the scrotum. It then descends to the testicle, and, becoming tortuous, divides into several branches, two or three of which accompany the vas deferens and supply the epididymis, anastomosing with the artery of the vas deferens : others pierce the back of the tunica albuginea and supply the substance of the testis.

The *cremasteric artery* is a branch of the deep epigastric artery. It accompanies the spermatic cord and supplies the Cremaster muscle and other coverings of the cord, anastomosing with the spermatic artery.

The *artery of the vas deferens*, a branch of the superior vesical, is a long slender vessel, which accompanies the vas deferens, ramifying upon the coats of that duct, and anastomosing with the spermatic artery near the testis.

The *spermatic veins* emerge from the back of the testis, and receive tributaries from the epididymis : they unite and form a convoluted plexus (*plexus pampiniformis*), which forms the chief mass of the cord ; the vessels composing this plexus are very numerous, and ascend along the cord in front of the vas deferens ; below the external or superficial abdominal ring they unite to form three or four veins, which pass along the inguinal canal, and, entering the abdomen through the internal or deep abdominal ring, coalesce to form two veins. These again unite to form a single vein, which opens on the right side into the inferior vena cava, at an acute angle, and on the left side into the renal vein at a right angle.

The *lymphatic vessels* terminate in the lumbar glands.

The *nerves* are the spermatic plexus from the sympathetic, joined by filaments from the pelvic plexus which accompany the artery of the vas deferens.

Surgical Anatomy.—The scrotum forms an admirable covering for the protection of the testicle. This body, lying suspended and loose in the cavity of the scrotum and surrounded by a serous membrane, is capable of great mobility, and can therefore easily slip about within the scrotum, and thus avoid injuries from blows or squeezes. The skin of the scrotum is very elastic and capable of great distension, and on account of the looseness and amount of subcutaneous tissue, the scrotum becomes greatly enlarged in cases of œdema, to which this part is especially liable on account of its dependent position. The scrotum is frequently the seat of epithelioma ; this is no doubt due to the rugæ on its surface, which favour the lodgment of dirt, and this, causing irritation, is the exciting cause of the disease, which is especially common in chimney-sweeps from the lodgment of soot. The scrotum is also the part most frequently affected by elephantiasis.

On account of the looseness of the subcutaneous tissue, considerable extravasations of blood may take place from very slight injuries. It is therefore generally recommended never to apply leeches to the scrotum, since they may lead to considerable ecchymosis, but rather to puncture one or more of the superficial veins of the scrotum in cases where local blood-letting from this part is judged to be desirable. The muscular fibre in the dartos causes contraction and considerable diminution in the size of a wound of the scrotum, as after the operation of castration, and is of assistance in keeping the edges together, and covering the exposed parts.

THE TESTES

The **testes** are suspended in the scrotum by the spermatic cords. As the left spermatic cord is rather longer than the right one, the left testicle hangs somewhat lower than its fellow. Each gland is of an oval form, compressed laterally, and having an oblique position in the scrotum ; the upper extremity being directed forwards and a little outwards ; the lower, backwards and a little inwards ; the anterior convex border looks forwards and downwards ; the posterior or straight border, to which the cord is attached, backwards and upwards.

The anterior border and lateral surfaces, as well as both extremities of the organ, are convex, free, smooth, and invested by the visceral layer of the tunica vaginalis. The posterior border, to which the cord is attached, receives only a partial investment from that membrane. Lying upon the outer edge of this posterior border is a long, narrow, flattened body, named, from its relation to the testis, the *epididymis* (δίδυμος, testis). It consists of a central portion, or *body*, an upper enlarged extremity, the head or *globus major* ; and a lower pointed extremity, the tail, or *globus minor*. The globus major is intimately connected with the upper end of the testicle by means of its efferent ducts ; and the globus minor is connected with its lower end by cellular tissue, and a reflection of the tunica vaginalis.   The outer surface and upper and lower ends of the epididymis are free and covered by serous membrane ; the body is also completely invested by it, excepting along its posterior border, and between the body and the testicle is a pouch or *cul-de-sac*, named the *digital fossa*.   The epididymis is connected to the back of the testis by a fold of the serous membrane.   Attached to the upper end of the testis, close to the globus major, are two small pedunculated bodies. One of them is pear-shaped, and attached by its narrow stalk, the other is small and sessile ; they are believed to be the remains of the upper extremity of the Müllerian duct (page [143]), and are termed the *hydatids of Morgagni* ; some observers, however, regard the stalked hydatid as being a rudiment of the pronephros.   When the testicle is removed from the body, the position of the vas deferens, on the posterior surface of the testicle and inner side of the epididymis, marks the side to which the gland has belonged.

Size and weight.—The average dimensions of this gland are from one and a half to two inches in length, an inch in breadth, and an inch and a quarter in the antero-posterior diameter ; and the weight varies from six to eight drachms, the left testicle being a little the larger.

The testis is invested by three tunics, the tunica vaginalis, tunica albuginea, and tunica vasculosa.

The **tunica vaginalis** is the serous covering of the testis.   It is a pouch of serous membrane, derived from the peritoneum during the descent of the testis in the foetus from the abdomen into the scrotum.   After its descent, that portion of the pouch which extends from the internal ring to near the upper part of the gland becomes obliterated, the lower portion remains as a shut sac, which invests the outer surface of the testis, and is reflected on to the internal surface of the scrotum ; hence it may be described as consisting of a visceral and parietal portion.

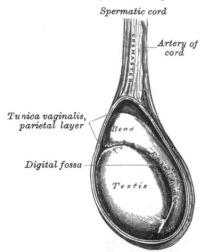

FIG. 559.—The testis *in situ*, the tunica vaginalis having been laid open.

Spermatic cord

Artery of cord

Tunica vaginalis, parietal layer

Digital fossa

The *visceral portion of the tunica vaginalis* covers the outer surface of the testis, as well as the epididymis, connecting the latter to the testis by means of a distinct fold.   From the posterior border of the gland it is reflected on to the internal surface of the scrotum.

The *parietal portion of the tunica vaginalis* is far more extensive than the visceral portion, extending upwards for some distance in front, and on the inner side of the cord, and reaching below the testis.   The inner surface of the tunica vaginalis is free, smooth, and covered by a layer of endothelial cells.   The interval between the visceral and parietal layers of this membrane constitutes the cavity of the tunica vaginalis.

The obliterated portion of the pouch may generally be seen as a fibro-cellular thread lying in the loose areolar tissue around the spermatic cord ; sometimes this may be traced as a distinct band from the upper end of the inguinal canal, where it is connected with the peritoneum, down to the tunica vaginalis ; sometimes it gradually becomes lost on the spermatic cord.   Occasionally no trace of it can be detected.   In some cases it happens that the pouch of peritoneum does not become obliterated, but the sac of the peritoneum communicates with the tunica vaginalis.   This may give rise to one of the varieties of oblique inguinal hernia (page 1035). Or in other cases the pouch may contract, but not become entirely obliterated ; it then forms a minute canal leading from the peritoneum to the tunica vaginalis.*

The **tunica albuginea** is the fibrous covering of the testis. It is a dense fibrous membrane, of a bluish-white colour, composed of bundles of white fibrous tissue, which interlace in every direction. Its outer surface is covered by the tunica vaginalis, except at the points of attachment of the epididymis to the testicle, and along its posterior border, where the spermatic vessels enter the gland. This membrane surrounds the glandular structure of the testicle, and, at its posterior border, is reflected into the interior of the gland, forming an incomplete vertical septum, called the *mediastinum testis* (*corpus Highmorianum*).

The *mediastinum testis* extends from the upper, nearly to the lower extremity of the gland, and is wider above than below. From the front and sides of this septum numerous slender fibrous cords and imperfect septa (*trabeculæ*) are given off, which radiate towards the surface of the organ, and are attached to the inner surface of the tunica albuginea. They therefore divide the interior of the organ into a number of incomplete spaces, which are somewhat cone shaped, being broad at their bases at the surface of the gland, and becoming narrower as they converge to the mediastinum. The mediastinum supports the vessels and ducts of the testis in their passage to and from the substance of the gland.

The **tunica vasculosa** is the vascular layer of the testis, consisting of a plexus of blood-vessels, held together by delicate areolar tissue. It covers the inner surface of the tunica albuginea and the different septa in the interior of the gland, and therefore forms an internal investment to all the spaces of which the gland is composed.

Structure.—The glandular structure of the testis consists of numerous lobules (*lobuli testis*). Their number, in a single testis, is estimated by Berres at 250, and by Krause at 400. They differ in size according to their position, those in the middle of the gland being larger and longer. The lobules are conical in shape, the base being directed towards the circumference of the organ, the apex towards the mediastinum. Each lobule is contained in one of the intervals between the fibrous cords and vascular processes which extend between the mediastinum testis and the tunica albuginea, and consists of from one to three, or more, minute convoluted tubes, the *tubuli seminiferi*. The tubes may be separately unravelled, by careful dissection under water, and may be seen to commence either by free cæcal ends or by anastomotic loops. The total number of tubes is considered by Monro to be about 300, and the length of each about sixteen feet; by Lauth, their number is estimated at 840, and their average length two feet and a quarter. Their diameter varies from $\frac{1}{200}$ to $\frac{1}{150}$ of an inch. The tubuli are pale in colour in early life, but in old age they acquire a deep yellow tinge, from containing much fatty matter. Each tube consists of a basement layer, formed of epithelioid cells united edge to edge, outside which are other layers of flattened cells arranged in interrupted laminæ, which give to the tube an appearance of striation in cross section. The cells of the outer layers gradually pass into the interstitial tissue. Within the basement-membrane are epithelial cells arranged in several irregular layers, which are not always clearly separated, but which may be arranged in three different groups. Among these cells may be seen the *spermatozoa* in different stages of development. 1. Lining the basement-membrane and forming the outer zone is a layer of cubical cells, with small nuclei; these are known as the *lining cells* or *spermatogonia*. The nucleus of some of them may be seen to be in the process of indirect division (*karyokinesis*, page [3]), and in consequence of this daughter cells are formed,

---

* It is recorded that in the post-mortem examination of Sir Astley Cooper, this minute canal was found on both sides of the body. Sir Astley Cooper states that when a student he suffered from inguinal hernia ; probably this was of the congenital variety, and the canal found after death was the remains of the one down which the hernia travelled. (*Lancet*, vol. ii. 1824, p. 116.)

which constitute the second zone. 2. Within this first layer is to be seen a number of larger cells with clear nuclei, arranged in two or three layers; these are the *intermediate cells* or *spermatocytes*. Most of these cells are in a condition of karyokinetic division, and the cells which result from this division form those of the next layer, the *spermatoblasts* or *spermatids*. 3. The third layer of cells therefore consists of the spermatoblasts or spermatids, and each of these, without further subdivision, becomes a *spermatozoon*. They are ill-defined granular masses of protoplasm, of an elongated form, with a nucleus, which becomes the head of the future spermatozoon. In addition to these three layers of cells others are seen, which are termed the *supporting cells*, or *cells of Sertoli*. They are elongated and columnar, and project inwards from the basement-membrane towards the lumen of the tube. They give off numerous lateral branches, which form a reticulum for the support of the three groups of cells just described. As development of the spermatozoa proceeds the latter group themselves around the inner extremities of the supporting cells. The nuclear part of the spermatozoon, which is partly embedded in the supporting cell, is differentiated to form the head of the spermatozoon, while the cell protoplasm becomes lengthened out to form the middle piece and tail, the latter projecting into the lumen of the tube. Ultimately the heads are separated and the spermatozoa are set free.

Spermatogenesis.—The stages in the development of the spermatozoa are as follows : The spermatogonia become enlarged to form the spermatocytes, and each spermatocyte subdivides into two cells, and each of these again divides into two spermatids or young spermatozoa, so that the spermatocyte gives origin to four spermatozoa.

The process of spermatogenesis bears a close relation to that of maturation of the ovum. The spermatocyte is equivalent to the immature ovum. It undergoes subdivision and ultimately gives origin to four spermatozoa, each of which contains, therefore, only one-fourth of the chromatin elements of the nucleus of the spermatocyte. In the process of maturation of the ovum its nucleus divides, one half being extended as the first polar body. The remaining half of the nucleus again subdivides, one half being extended as the second polar body. The portion of the nucleus which is retained to form the female pronucleus of the now matured ovum contains, therefore, only one-fourth of the chromatin elements of the original nucleus, and thus the spermatozoon and the matured ovum, so far as their nuclear elements are concerned, may be regarded as of the same morphological value.

The tubules are enclosed in a delicate plexus of capillary vessels, and are held together by an intertubular connective tissue, which presents large interstitial spaces lined by endothelium, which are believed to be the rootlets of the lymphatic vessels of the testis.

In the apices of the lobules, the tubuli become less convoluted, assume a nearly straight course, and unite together to form from twenty to thirty larger ducts, of about $\frac{1}{50}$ of an inch in diameter, and these, from their straight course, are called *vasa recta*.

The *vasa recta* enter the fibrous tissue of the mediastinum, and pass upwards and backwards, forming, in their ascent, a close network of anastomosing tubes which are merely channels in the fibrous stroma, lined by flattened epithelium and having no proper walls ; this constitutes the *rete testis*. At the upper end of the mediastinum, the vessels of the rete testis terminate in from twelve to fifteen or twenty ducts, the *vasa efferentia* : they perforate the tunica albuginea, and carry the seminal fluid from the testis to the epididymis. Their course is at first straight ; they then become enlarged, and exceedingly convoluted, and form a series of conical masses, the *coni vasculosi*, which together constitute the globus major of the epididymis. Each cone consists of a single convoluted duct, from six to eight inches in length, the diameter of which gradually decreases from the testis to the epididymis. Opposite the bases of the cones the efferent vessels open at narrow intervals into a single duct, which constitutes, by its complex convolutions, the body and globus minor of the epididymis. When the convolutions of this tube are unravelled, it measures upwards of twenty feet in length ; it increases in diameter and thickness as it approaches the vas deferens. The convolutions are held together by fine areolar tissue, and by bands of fibrous tissue.

The vasa recta are of smaller diame er than the seminal tubes, and have very thin parietes. They, like the channels of the rete testis, are lined by a single layer of flattened epithelium. The vasa efferentia and the tube of the epididymis have walls of considerable thickness, on account of the presence in them of muscular tissue, which is principally arranged in a circular manner. These tubes are lined by columnar ciliated epithelium.

The **vas deferens,** the excretory duct of the testis, is the continuation of the epididymis. Commencing at the lower part of the globus minor, it ascends along the posterior border of the testis and inner side of the epididymis, and along the back part of the spermatic cord, through the inguinal canal to the internal or deep abdominal ring. From the ring it curves round the outer side of the epigastric artery, crosses the external iliac vessels, and descends into the pelvis at the side of the bladder ; it arches backwards and downwards to its base, crossing over the obliterated hypogastric artery, and to the inner side of the ureter. At the base of the bladder it lies between that viscus and the rectum, running along the inner border of the vesicula seminalis. In this situation it becomes enlarged and sacculated, forming the *ampulla* ; and then, becoming narrowed at the base of the prostate, unites with the duct of the vesicula seminalis to form the ejaculatory duct. The vas deferens presents a hard and cord-like sensation to the fingers ; it is about two feet in length, of cylindrical form, and about a line and a quarter in diameter. Its walls are dense, measuring one-third of a line ; and its canal is extremely small, measuring about half a line.

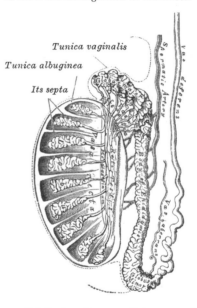

FIG. 560.—Vertical section of the testicle, to show the arrangement of the ducts.

Tunica vaginalis
Tunica albuginea
Its septa

Structure.—The vas deferens consists of three coats : 1. An external or areolar coat. 2. A muscular coat, which in the greater part of the tube consists of two layers of unstriped muscular fibre : an outer, longitudinal in direction, and an inner, circular ; but in addition to these, at the commencement of the vas deferens, there is a third layer, consisting of longitudinal fibres, placed internal to the circular stratum, between it and the mucous membrane. 3. An internal, or mucous coat, which is pale, and arranged in longitudinal folds ; its epithelial covering is of the columnar variety.

A long narrow tube, the *vas aberrans of Haller*, is occasionally found connected with the lower part of the canal of the epididymis, or with the commencement of the vas deferens. It extends up into the cord for about two or three inches, where it terminates by a blind extremity, which is occasionally bifurcated. Its length varies from an inch and a half to fourteen inches, and sometimes it becomes dilated towards its extremity ; more commonly it retains the same diameter throughout. Its structure is similar to that of the vas deferens. Occasionally it is found unconnected with the epididymis.

Organ of Giraldes.—This term is applied to a small collection of convoluted tubules, situated in front of the lower part of the cord above the globus major of the epididymis. These tubes are lined with columnar ciliated epithelium, and probably represent the remains of a part of the Wolffian body.

Surgical Anatomy.—The testicle frequently requires removal for malignant disease; in tuberculous disease; in cystic disease; in cases of large hernia testis, and in some instances of incompletely descended or misplaced testicles. The operation of castration has also been, during the last few years, performed for enlargement of the prostate; for it has been found that removal of the testicles is followed by very rapid and often considerable diminution in the size of the prostate. The operation is, however, one of severity, and is frequently followed by death in these cases, performed, as it necessarily is, in old men. Reginald Harrison has proposed to substitute for it excision of a portion of the vasa deferentia. The operation of castration is a comparatively simple one. An incision is made into the tunica vaginalis from the external ring to the bottom of the scrotum. The coverings are shelled off the organ, and the mesorchium, stretching between the back of the testicle and the scrotum, divided. The cord is then isolated, and an aneurism needle, armed with a double ligature, passed under it, as high as is thought necessary, and the cord tied in two places, and divided between the ligatures. Sometimes, in cases of malignant disease, it is desirable to open the inguinal canal and tie the cord as near the internal abdominal ring as possible.

Spermatozoa.—The spermatozoa are minute thread-like bodies, which constitute the essential elements of the semen. Each consists of a *head*, a middle piece or *body*, and an elongated filament or *tail*. The head, on surface view, appears oval in shape, but if seen in profile it is narrow and pointed at its free end. It represents the modified nucleus of the spermatid, and consists chiefly of chromatin, and so stains readily with nuclear reagents; it is covered by a thin cap of protoplasm. The body is a short cylindrical or conical piece, intervening between the head and tail, and is therefore sometimes spoken of as the intermediate segment. The tail is about four times the combined lengths of the head and body; its terminal part is extremely fine, and is named the *end-piece*. Contained within the body and tail is an axial filament, surrounded, except in the end-piece, by a thin layer of protoplasm; this axial filament terminates just below the head in a rounded knob or button. In virtue of their tails, which act as propellers, the spermatozoa, in the fresh condition, are capable of free movement, and if placed in favourable surroundings (e.g. in the female passages) may retain their vitality for some days or even weeks.

Vesiculæ Seminales

The **seminal vesicles** are two lobulated membranous pouches, placed between the base of the bladder and the rectum, serving as reservoirs for the semen, and secreting a fluid to be added to the secretion of the testicles. Each sac is somewhat pyramidal in form, the broad end being directed backwards, and the narrow end forwards towards the prostate. They measure about two and a half inches in length, about five lines in breadth, and two or three lines in thickness. They vary, however, in size, not only in different individuals, but also in the same individual on the two sides. Their *upper surface* is in contact with the base of the bladder, extending from near the termination of the ureters to the base of the prostate gland. Their *under surface* rests upon the rectum, from which they are separated by the recto-vesical fascia. Their *posterior extremities* diverge from each other. Their *anterior extremities* are pointed, and converge towards the base of the prostate gland, where each joins with the corresponding vas deferens to form the ejaculatory duct. Along the inner margins of each vesicula runs the enlarged and convoluted vas deferens. The inner borders of the vesiculæ, and the corresponding vas deferens, form the lateral boundaries of a triangular space, limited behind by the recto-vesical peritoneal fold; the portion of the bladder included in this space rests on the rectum.

Each vesicula consists of a single tube, coiled upon itself, and giving off several irregular cæcal diverticula; the separate coils, as well as the diverticula, being connected together by fibrous tissue. When uncoiled, this tube is about the diameter of a quill, and varies in length from four to six inches; it terminates posteriorly in a *cul-de-sac*; its anterior extremity becomes constricted into a narrow straight duct, which joins with the corresponding vas deferens, and forms the ejaculatory duct.

The **ejaculatory ducts,** two in number, one on each side, are formed by the junction of the ducts of the vesiculæ seminalis with the vasa deferentia. Each duct is about three-quarters of an inch in length; it commences at the base of the prostate, and runs forwards and downwards between its middle and lateral lobes, and, along the side of the sinus pocularis, to terminate by a separate slit-like orifice close to or just within the margins of the sinus. The ducts diminish in size, and also converge towards their termination.

Structure.—The vesiculæ seminalis are composed of three coats : an *external* or *areolar*; a *middle* or *muscular coat*, which is thinner than in the vas deferens,

FIG. 561.—Base of the bladder, with the vasa deferentia and vesiculæ seminales.

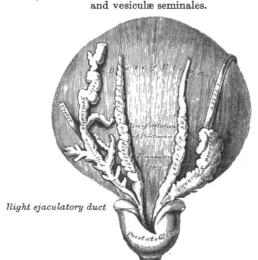

Right ejaculatory duct

arranged in two layers, an outer, longitudinal, and inner, circular; an *internal* or *mucous coat*, which is pale, of a whitish-brown colour, and presents a delicate reticular structure, like that seen in the gall-bladder, but the meshes are finer. The epithelium is columnar.

The coats of the ejaculatory ducts are extremely thin. They are : an *outer fibrous layer*, which is almost entirely lost after their entrance into the prostate ; a *layer of muscular fibres*, consisting of an outer thin circular and an inner longitudinal layer, and *mucous membrane*.

Vessels and Nerves.—The *arteries* supplying the vesiculæ seminales are derived from the middle and inferior vesical and middle hæmorrhoidal. The veins and lymphatics accompany the arteries. The nerves are derived from the pelvic plexus.

Surgical Anatomy.—The vesiculæ seminales are often the seat of an extension of the disease in cases of tuberculous disease of the testicle, and should always be examined from the rectum before coming to a decision with regard to castration in this affection.

FEMALE ORGANS OF GENERATION

EXTERNAL ORGANS

THE **external organs of generation in the female** are : the mons Veneris, the labia majora and minora, the clitoris, the meatus urinarius, and the orifice of the vagina. The term 'vulva' or ' pudendum,' as generally applied, includes all these parts.

FIG. 562.--The vulva. External female organs of generation.

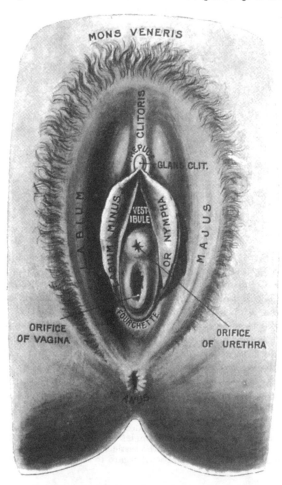

The **mons Veneris** is the rounded eminence in front of the pubic symphysis formed by a collection of fatty tissue beneath the integument. It becomes covered with hair at the time of puberty.

The **labia majora** are two prominent longitudinal cutaneous folds, extending downwards from the mons Veneris to the anterior boundary of the perinæum, and enclosing the common urino-sexual opening. Each labium has two surfaces, an outer, which is pigmented and covered with strong, crisp hairs ; and an inner, which is smooth and is beset with large sebaceous follicles and is continuous with the genito-urinary mucous tract ; between the two there is a considerable quantity of areolar tissue, fat, and a tissue resembling the dartos of the scrotum, besides vessels, nerves, and glands. The labia are thicker in front, where they form by their meeting the *anterior commissure*. Posteriorly they are not really joined, but appear to become lost in the neighbouring integument, terminating close to, and nearly parallel with each other. Together with the connecting skin between them, they form the posterior commissure, or posterior boundary of the vulval orifice.

FIG. 563.—Vertical median section of the female pelvis.

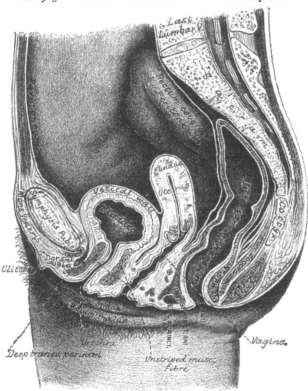

The interval between the posterior commissure and the anus, about an inch to an inch and a quarter in length, constitutes the perinæum. The *fourchette* is the anterior edge of the perinæum, and between it and the hymen is a depression, the *fossa navicularis*. The labia correspond to the scrotum in the male.

The **labia minora,** or **nymphæ,** are two small cutaneous folds, situated within the labia majora, and extending from the clitoris obliquely downwards, outwards, and backwards for about an inch and a half on each side of the orifice of the vagina, between which and the labia majora they are lost. Anteriorly, the two labia minora meet, and form the *frænum of the clitoris*. The prepuce of the clitoris, passing backwards on each side, is inserted, as it were, into the labia minora, but is not actually a part of them. The nymphæ are really modified skin. Their internal surfaces have numerous sebaceous follicles.

The **clitoris** is an erectile structure, analogous to the corpora cavernosa of the penis. It is situated beneath the anterior commissure, partially hidden between the anterior extremities of the labia minora. It is connected to the rami of the os pubis and ischium on each side by a crus; the body is short and concealed beneath the labia; the free extremity, or *glans clitoridis*, is a small rounded tubercle, consisting of spongy erectile tissue, and highly sensitive. It is provided, like the penis, with a suspensory ligament, and with two small muscles, the Erectores clitoridis, which are inserted into the crura of the clitoris. The clitoris consists of two corpora cavernosa, composed of erectile tissue enclosed in a dense layer of fibrous membrane, united together along their inner surfaces by an incomplete fibrous pectiniform septum.

Between the clitoris and the entrance of the vagina is a triangular smooth surface, bounded on each side by the nymphæ; this is the *vestibule*.

The orifice of the urethra (**meatus urinarius**) is situated at the back part of the vestibule, about an inch below the clitoris, and near the margin of the vagina, surrounded by a prominent elevation of the mucous membrane. Below the meatus urinarius is the orifice of the vagina, more or less closed in the virgin by a membranous fold, the *hymen*.

The **hymen** varies much in shape. Its commonest form is that of a ring, generally broadest posteriorly; sometimes it is represented by a semilunar fold, with its concave margin turned towards the pubes. A complete septum stretched across the lower part of the vaginal orifice is called 'imperforate hymen.' Occasionally it is cribriform, or its free margin forms a membranous fringe, or it may be entirely absent. It may persist after copulation, so that it cannot be considered as a test of virginity. After parturition, the small rounded elevations known as the *carunculæ myrtiformes* are found as the remains of the hymen.

Glands of Bartholin.—On each side of the commencement of the vagina, and behind the hymen, is a round or oblong body, of a reddish-yellow colour, and of the size of a horse-bean, analogous to Cowper's gland in the male. It is called the *gland of Bartholin*. Each gland opens by means of a long single duct, immediately external to the hymen, in the angle or groove between it and the nympha.

Bulbi vestibuli.—Extending from the clitoris, along either side of the vestibule, and lying a little above the nymphæ, are two large oblong masses, about an inch in length, consisting of a plexus of veins, enclosed in a thin layer of fibrous membrane. These bodies are narrow in front, rounded below, and are connected with the crura of the clitoris and rami of the pubes: they are termed by Kobelt the *bulbi vestibuli*; and he considers them analogous to the bulb of the corpus spongiosum in the male. Immediately in front of these bodies is a smaller venous plexus, continuous with the bulbi vestibuli behind and the glans clitoridis in front: it is called by Kobelt the *pars intermedia,* and is considered by him as analogous to that part of the body of the corpus spongiosum which immediately succeeds the bulb.

INTERNAL ORGANS

The **internal organs of generation** are: the vagina, the uterus and its appendages, the Fallopian tubes, the ovaries and their ligaments.

The **vagina** extends from the vulva to the uterus. It is situated in the cavity of the pelvis, behind the bladder, and in front of the rectum. Its direction is curved upwards and backwards, at first in the line of the outlet, and afterwards in that of the axis of the cavity of the pelvis. Its walls are ordinarily in contact, and its usual shape on transverse section is that of an **H**, the transverse limb being slightly curved forwards or backwards, while the lateral limbs are somewhat convex towards the median line. Its length is about two and a half inches along its anterior wall, and from three to three and a half inches along its posterior wall. It is constricted at its commencement, dilated in the middle, and narrowed near its uterine extremity; it surrounds the vaginal portion of the cervix uteri, a short distance from the os, its attachment extending higher up on the posterior than on the anterior wall of the uterus.

Relations.—Its *anterior surface* is in relation with the base of the bladder, and with the urethra. Its *posterior surface* is connected for the lower three-fourths of its extent to the anterior wall of the rectum, the upper fourth being separated from that tube by the recto-vaginal pouch of peritoneum, or pouch of Douglas, between the vagina and rectum. Its sides are enclosed between the Levatores ani muscles.

Structure.—The vagina consists of an internal mucous lining, of a muscular coat, and between the two of a layer of erectile tissue.

The *mucous membrane* is continuous above with that lining the uterus. Its inner surface presents, along the anterior and posterior walls, a longitudinal ridge or raphé, called the *columns of the vagina*, and numerous transverse ridges or rugæ, extending outwards from the raphé on either side. These rugæ are divided by furrows of variable depth, giving to the mucous membrane the appearance of being studded over with conical projections or papillæ ; they are most numerous near the orifice of the vagina, especially in females before parturition. The epithelium covering the mucous membrane is of the squamous variety. The submucous tissue is very loose, and contains numerous large veins, which by their anastomoses form a plexus, together with smooth muscular fibres derived from the muscular coat ; it is regarded by Gussenbauer as an erectile tissue. It contains a number of mucous crypts, but no true glands.

The *muscular coat* consists of two layers : an external longitudinal, which is far the stronger, and an internal circular layer. The longitudinal fibres are continuous with the superficial muscular fibres of the uterus. The strongest fasciculi are those attached to the recto-vesical fascia on each side. The two layers are not distinctly separable from each other, but are connected by oblique decussating fasciculi, which pass from the one layer to the other. In addition to this, the vagina at its lower end is surrounded by a band of striped muscular fibres, the *sphincter vaginæ* (see page 348).

External to the muscular coat is a layer of connective tissue, containing a large plexus of blood-vessels.

The *erectile tissue* consists of a layer of loose connective tissue, situated between the mucous membrane and the muscular coat ; embedded in it is a plexus of large veins, and numerous bundles of unstriped muscular fibres, derived from the circular muscular layer. The arrangement of the veins is similar to that found in other erectile tissues.

THE UTERUS

The **uterus** is the organ of gestation, receiving the fecundated ovum in its cavity, retaining and supporting it during the development of the fœtus, and becoming the principal agent in its expulsion at the time of parturition.

In the virgin state it is pear-shaped, flattened from before backwards, and situated in the cavity of the pelvis, between the bladder and the rectum ; it is retained in its position by the round and broad ligaments on each side, and projects into the upper end of the vagina below. Its upper end, or base, is directed upwards and forwards ; its lower end, or apex, downwards and backwards, in the line of the axis of the inlet of the pelvis. It therefore forms an angle with the vagina, since the direction of the vagina corresponds to the axis of the cavity and outlet of the pelvis. The uterus measures about three inches in length, two in breadth at its upper part, and nearly an inch in thickness, and it weighs from an ounce to an ounce and a half.

It consists of two parts : (1) the *body*, with its upper broad extremity, the *fundus* ; and (2) the *cervix*, or *neck*, which is situated partly above and partly in the vagina. The fundus is placed below the level of the brim of the pelvis, and its direction varies with the condition of the bladder.

The division between the body and cervix is indicated externally by a slight constriction, and by the reflection of the peritoneum from the anterior surface of the uterus on to the bladder, and internally by a narrowing of the canal, called the *internal os*.

The *body* gradually narrows from the fundus to the neck. Its anterior surface is flattened, covered by peritoneum, which becomes separated from it at its union with the cervix, in order to form the utero-vesical pouch, which lies between the uterus and bladder. Its *posterior surface* is convex transversely, covered by peritoneum throughout, and separated from the rectum by some convolutions of the intestine. Its *lateral margins* are concave, and give attachment to the Fallopian tube above, the round ligament below and in front of this, and the ligament of the ovary behind both of these structures.

FIG. 564.—Douglas's pouch.
(From a preparation in the Museum of the Royal College of Surgeons of England.)

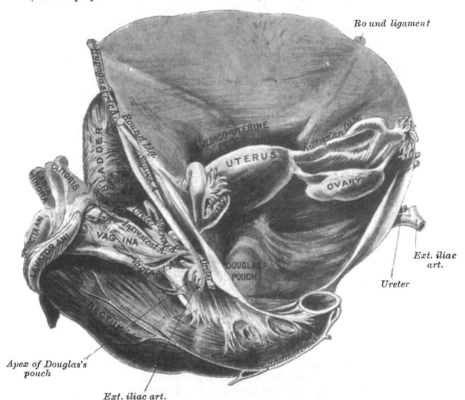

The *cervix* is the lower constricted segment of the uterus ; around its circumference is attached the upper end of the vagina, which extends upwards a greater distance behind than in front.

The supravaginal portion is not covered by peritoneum in front ; a pad of cellular tissue is interposed between it and the bladder. Behind, the peritoneum is extended over it. The vaginal portion is the rounded lower end projecting into the vagina. On its surface is a small aperture, the *os uteri*, generally circular in shape, but sometimes oval or almost linear. The margin of the opening is, in the absence of past parturition or disease, quite smooth.

Ligaments.—The ligaments of the uterus are eight in number : one anterior ; one posterior ; two lateral or broad ; two sacro-uterine—all these being formed of peritoneum : and, lastly, two round ligaments.

The *anterior ligament* (vesico-uterine) is reflected on to the bladder from the front of the uterus, at the junction of the cervix and body.

The *posterior ligament* (recto-uterine) passes from the posterior wall of the uterus over the upper fourth of the vagina, and thence on to the rectum and sacrum. It thus forms a pouch called *Douglas's pouch* (fig. 564), the boundaries of which are, in front, the posterior wall of the uterus, the supravaginal cervix, and the upper fourth of the vagina; behind, the rectum and sacrum; above, the small intestine; and laterally, the sacro-uterine ligaments.

The *two lateral* or *broad ligaments* pass from the sides of the uterus to the lateral walls of the pelvis, forming a septum across the pelvis, which divides

Fig. 565.—Side view of the female pelvic organs.
(From a preparation in the Museum of the Royal College of Surgeons of England.)

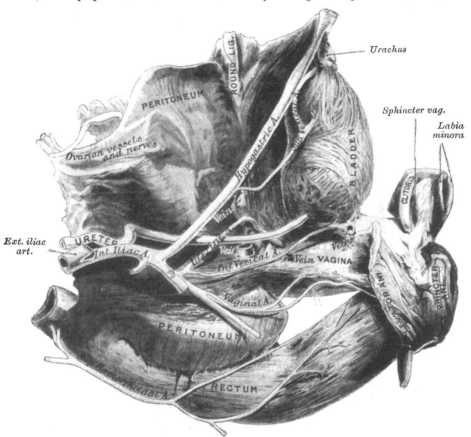

that cavity into two portions. In the anterior part are contained the bladder, urethra, and vagina; in the posterior part, the rectum. Between the two layers of each broad ligament are contained: (1) the Fallopian tube superiorly; (2) the round ligament; (3) the ovary and its ligament; (4) the parovarium, or organ of Rosenmüller; (5) connective tissue; (6) unstriped muscular fibre; and (7) blood-vessels and nerves. The Fallopian tube is contained in a special fold of the broad ligament, which is attached to the part of the ligament near the ovary, and is known by the name of the *mesosalpinx*. Between the fimbriated extremity of the tube and the lower attachment of the broad ligament is a concave rounded margin, called the *infundibulo-pelvic ligament* (fig. 567).

The *sacro-uterine ligaments* pass from the second and third bones of the sacrum, downwards and forwards on the lateral aspects of the rectum to be attached one on each side of the uterus at the junction of the supravaginal cervix and the body, this point corresponding internally to the position of the os internum.

The *round ligament* will be described in the sequel.

The **cavity of the uterus** is small in comparison with the size of the organ : that portion of the cavity which corresponds to the body is triangular, flattened from before backwards, so that its walls are closely approximated, and having its base directed upwards towards the fundus. At each superior angle is a funnel-shaped cavity, which constitutes the remains of the division of the body of the uterus into two cornua ; and at the bottom of each cavity is the minute orifice of the Fallopian tube. At the inferior angle of the uterine cavity is a small constricted opening, the internal orifice (*ostium internum*), which leads into the cavity of the cervix.

The **cavity of the cervix** is somewhat fusiform, flattened from before backwards, broader at the middle than at either extremity, and communicates, below, with the vagina. The wall of the canal presents, anteriorly and posteriorly, a longitudinal column, from which proceed a number of small oblique columns, giving the appearance of branches from the stem of a tree ; and hence the name *arbor vitæ uterina* applied to it. These folds usually become very indistinct after the first labour.

Structure.—The uterus is composed of three coats : an external serous coat, a middle or muscular, and an internal mucous coat.

The *serous coat* is derived from the peritoneum ; it invests the fundus and the whole of the posterior surface of the uterus ; but covers the anterior surface only as far as the junction of the body and cervix. In the lower fourth of the posterior surface the peritoneum, though covering the uterus, is not closely connected with it, being separated from it by a layer of loose cellular tissue and some large veins.

The *muscular coat* forms the chief bulk of the substance of the uterus. In the unimpregnated state it is dense, firm, of a greyish colour, and cuts almost like cartilage. It is thick opposite the middle of the body and fundus, and thin at the orifices of the Fallopian tubes. It consists of bundles of unstriped muscular fibres, disposed in layers, intermixed with areolar tissue, blood-vessels, lymphatic vessels, and nerves. In the impregnated state the muscular tissue becomes more prominently developed, and is disposed in three layers—external, middle, and internal.

The external layer is placed beneath the peritoneum, disposed as a thin plane on the anterior and posterior surfaces It consists of fibres, which pass transversely across the fundus, and, converging at each superior angle of the uterus, are continued on to the Fallopian tube, the round ligament, and the ligament of the ovary : some passing at each side into the broad ligament, and others running backwards from the cervix into the sacro-uterine ligaments.

The middle layer of fibres, which is thickest, presents no regularity in its arrangement, being disposed longitudinally, obliquely, and transversely. It contains most blood-vessels.

The internal or deep layer consists of circular fibres arranged in the form of two hollow cones, the apices of which surround the orifices of the Fallopian tubes, their bases intermingling with one another on the middle of the body of the uterus. At the internal os these circular fibres form a distinct sphincter.

The *mucous membrane* is thin, smooth, and closely adherent to the subjacent tissue. It is continuous, through the fimbriated extremity of the Fallopian tubes, with the peritoneum ; and, through the os uteri, with the lining of the vagina.

In the body of the uterus it is smooth, soft, of a pale red colour, lined by columnar ciliated epithelium, and presents, when viewed with a lens, the orifices of numerous tubular follicles, arranged perpendicularly to the surface. It is unprovided with any submucosa, but is intimately connected with the innermost layer of the muscular coat, which by some anatomists is regarded as the muscularis mucosæ. In structure the corium differs from ordinary mucous membrane, consisting of an embryonic nucleated and highly cellular form of connective tissue in which run numerous large lymphatics. In it are the tube-like *uterine glands*, which are of small size in the unimpregnated uterus, but shortly after impregnation become enlarged and elongated, presenting a contorted or waved appearance towards their closed extremities, which reach into the muscularis, and may be single or bifid. They consist of a delicate membrane, lined by an epithelium, which becomes ciliated towards the orifices. The changes which take place in the mucous membrane of the impregnated uterus are more fully dealt with in the section on Embryology.

In the cervix the mucous membrane is sharply differentiated from that of the uterine cavity. It is thrown into numerous oblique ridges, which diverge from an anterior and posterior longitudinal raphé, presenting an appearance which

FIG. 566.—The arteries of the internal organs of generation of the female, seen from behind. (After Hyrtl.)

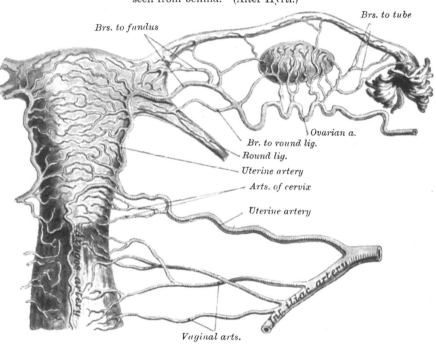

has received the name of *arbor vitæ*. In the upper two-thirds of the canal, the mucous membrane is provided with numerous deep glandular follicles, which secrete a clear viscid alkaline mucus; and, in addition, extending through the whole length of the canal, are a variable number of little cysts, presumably follicles, which have become occluded and distended with retained secretion. They are called the *ovula Nabothi*. The mucous membrane covering the lower half of the cervical canal presents numerous papillæ. The epithelium of the upper two-thirds is cylindrical and ciliated, but below this it loses its cilia, and gradually changes to squamous epithelium close to the external os.

Vessels and Nerves.—The *arteries of the uterus* are the uterine, from the internal iliac; and the ovarian, from the aorta. They are remarkable for their tortuous course in the substance of the organ, and for their frequent anastomoses. The termination of the ovarian artery meets the termination of the uterine artery, and forms an anastomotic trunk from which branches are given off to supply the uterus, their disposition being, as shown by Sir John Williams, circular.   The *veins* are of large size, and correspond with the arteries.   In the impregnated uterus these vessels form the *uterine sinuses*, consisting of the lining membrane of the veins adhering to the walls of canals channelled through the substance of the uterus.   They terminate in the uterine plexuses. The *lymphatics* of the body terminate in the lumbar glands, those of the cervix in the pelvic glands.   The *nerves* are derived from the inferior hypogastric and ovarian plexuses, and from the third and fourth sacral nerves.

The form, size, and situation of the uterus vary at different periods of life and under different circumstances.

In the fœtus the uterus is contained in the abdominal cavity, projecting beyond the brim of the pelvis.   The cervix is considerably larger than the body.

At puberty the uterus is pyriform in shape, and weighs from eight to ten drachms.   It has descended into the pelvis, the fundus being just below the level of the brim of this cavity.   The arbor vitæ is distinct, and extends to the upper part of the cavity of the organ.

The position of the uterus in the adult is liable to considerable variation, depending chiefly on the condition of the bladder and rectum.   When the bladder is empty the entire uterus is directed forwards, and is at the same time bent on itself at the junction of the body and cervix, so that the body lies upon the bladder.   As the latter fills the uterus gradually becomes more and more erect, until with a fully distended bladder the fundus may be directed backwards towards the sacrum.

During menstruation the organ is enlarged, and more vascular, its surfaces rounder ; the os externum is rounded, its labia swollen, and the lining membrane of the body thickened, softer, and of a darker colour.   According to Sir J. Williams, at each recurrence of menstruation a molecular disintegration of the mucous membrane takes place, which leads to its complete removal, only the bases of the glands embedded in the muscle being left.   At the cessation of menstruation, by a proliferation of the remaining structures a fresh mucous membrane is formed.

During pregnancy the uterus becomes enormously enlarged, and in the ninth month reaches the epigastric region.   The increase in size is partly due to growth of pre-existing muscle, and partly to development of new fibres.

After parturition the uterus nearly regains its usual size, weighing about an ounce and a half ; but its cavity is larger than in the virgin state ; the external orifice is more marked ; its edges present a fissured surface ; its vessels are tortuous ; and its muscular layers are more defined.

In old age the uterus becomes atrophied, and paler and denser in texture ; a more distinct constriction separates the body and cervix.   The ostium internum, and, occasionally, the vaginal orifice often become obliterated, and its labia almost entirely disappear.

Surgical Anatomy.—The uterus may require removal in cases of malignant disease or for fibroid tumours.   Carcinoma is the most common form of malignant disease of the uterus, though cases of sarcoma do occur.   It may show itself either as a columnar carcinoma or as a squamous carcinoma ; the former commencing either in the cervix or body of the uterus, the latter always commencing in the epithelial cells of the mucous covering of the vaginal surface of the cervix.   The columnar form may be treated in the early stage, before fixation has taken place, by removal of the uterus, either through the vagina or by means of abdominal section.   The former operation is the better of the two, and is attended by a much smaller death rate.   Vaginal hysterectomy is performed by placing the patient in the lithotomy position and introducing a large duckbill speculum.   The cervix is then seized with a volsellum and pulled down as far as possible and the mucous membrane of the vagina incised around the cervix and as near to it as the disease will allow, especially in front, where the ureters are in danger of being wounded.   A pair of dressing forceps are then pushed through into Douglas's pouch and opened sufficiently to allow of the introduction of the two fore-fingers, by means of which the opening is dilated laterally as far as the sacro-uterine ligaments.   A somewhat similar proceeding is adopted in front, but here the bladder has to be separated from the anterior wall of the uterus for about an inch before the vesico-uterine fold of peritoneum can be reached.   This is done by carefully burrowing upwards with a director and stripping the tissues off the anterior uterine wall.   When the vesico-uterine pouch has been opened and the opening dilated laterally, the uterus remains attached only by the broad ligaments, in which are contained the vessels that supply the uterus.   Before division of the ligaments, these vessels have to be dealt with.   The fore-finger of the left hand is introduced into Douglas's pouch, and an aneurism needle, armed with a long silk ligature, is inserted into the vesico-uterine pouch, and is pushed through the broad ligament about an inch above its lower level and at some distance from the uterus.   One end of the ligature is now pulled through the anterior opening, and in this way we have the lowest inch of the broad ligament, in which is contained the uterine artery (fig. 566), enclosed in a ligature. This is tied tightly, and the operation is repeated on the other side.   The broad ligament is then divided on either side, between the ligature and the uterus, to the extent to which it has been constricted. By traction on the volsellum which grasps the cervix, the uterus can be pulled considerably further down in the vagina, and a second inch of the broad ligament is treated in a similar way. This second ligature will embrace the pampiniform plexus of veins, and, when the broad ligament has been divided on either side, it will be found that a third ligature can be made to pass over the Fallopian tube and top of the broad ligament, after the uterus has been dragged down as far as possible. After the third ligature has been tied and the structures between it and the uterus divided, this organ will be freed from all its connections and can be removed from the vagina. This canal is then sponged out and lightly dressed with gauze; no sutures being used. The gauze may be removed at the end of the second day. In squamous epithelioma, amputation of the cervix is all that is necessary in those cases where the disease is recognised before it has invaded the walls of the vagina or the neighbouring broad ligaments. The operation consists in removing a wedge-shaped piece of the uterus, including the cervix, through the vagina and attaching the cut surfaces of the stump to the anterior and posterior vaginal walls, so as to prevent retraction. In the treatment of uterine fibroids which require operative interference, removal of the whole of the uterus together with the tumours through an abdominal incision gives the most satisfactory results; for, if the tumour is large, its size acts as a barrier to its safe delivery through the pelvis and genital passages. After the abdomen has been opened the uterine vessels are secured and the broad ligaments divided in a similar manner to that employed in vaginal hysterectomy, except that the proceeding is commenced from above. When the two first ligatures have been tied, and the broad ligament divided, it will be found that the uterus can be raised out of the pelvis. A transverse incision is now made through the peritoneum, where it is reflected from the anterior surface of the uterus on to the back of the bladder and the serous membrane peeled from the surface of the uterus until the vagina is reached. The anterior wall of this canal is then cut across. The uterus is now turned forwards and the peritoneum at the bottom of Douglas's pouch incised transversely, and the posterior wall of the vagina cut across, until it meets the incision on the anterior wall. The uterus is now almost free, and is held only by the lower part of the broad ligament on either side, containing the uterine artery. A third ligature is made to encircle this, and, after having been tied, the structures are divided between the ligature and the uterus. The organ can now be removed. The vagina is plugged with gauze, and the external wound closed in the usual way. The vagina acts as a drain, and therefore the opening into it is not sutured.

APPENDAGES OF THE UTERUS

The **appendages of the uterus** are the Fallopian tubes, the ovaries and their ligaments, and the round ligaments. They are placed in the following order: in front is the round ligament; the Fallopian tube occupies the upper margin of the broad ligament; the ovary and its ligament are behind both.

FIG. 567.—Uterine appendages, seen from behind. (Henle.)

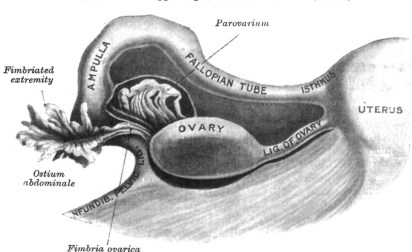

The **Fallopian tubes,** or **oviducts,** convey the ova from the ovaries to the cavity of the uterus. They are two in number, one on each side, situated in the upper margin of the broad ligament, extending from each superior angle of the uterus to the side of the pelvis. Each tube is about four inches in length; and is described as consisting of three portions : (1) the *isthmus,* or inner constricted third ; (2) the *ampulla,* or outer dilated portion, which curves over the ovary ; and (3) the *infundibulum,* with its *ostium abdominale,* surrounded by fimbriæ,

FIG. 568.—The uterus and its appendages. Posterior view. The parts have been somewhat displaced from their proper position in the preparation of the specimen ; thus the right ovary has been raised above the Fallopian tube, and the fimbriated extremities of the tubes have been turned upwards and outwards. (From a preparation in the Museum of the Royal College of Surgeons of England.)

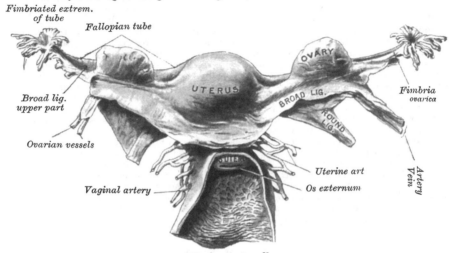

one of which is attached to the ovary, the *fimbria ovarica.* The general direction of the Fallopian tube is outwards, backwards, and downwards. The uterine opening is minute, and will only admit a fine bristle ; the abdominal opening is somewhat larger. In connection with the fimbriæ of the Fallopian tube, or with the broad ligament close to them, there is frequently one or more small vesicles floating on a long stalk of peritoneum. These are termed the *hydatids of Morgagni.*

Structure.—The Fallopian tube consists of three coats : serous, muscular, and mucous.

The *external* or *serous coat* is peritoneal.

The *middle* or *muscular coat* consists of an external longitudinal and an internal circular layer of muscular fibres continuous with those of the uterus.

The *internal* or *mucous coat* is continuous with the mucous lining of the uterus, and, at the free extremity of the tube, with the peritoneum. It is thrown into longitudinal folds, which in the outer, larger part of the tube, or ampulla, are much more extensive than in the narrow canal of the isthmus. The lining epithelium is columnar ciliated. This form of epithelium is also found on the inner surface of the fimbriæ ; while on the outer or serous surfaces of these processes the epithelium gradually merges into the endothelium of the peritoneum.

The **ovaries** (*testes muliebres,* Galen) are analogous to the testes in the male. They are oval-shaped bodies, of an elongated form, flattened from above downwards, situated one on each side of the uterus, in the posterior part of the broad ligament behind and below the Fallopian tubes. Each ovary is connected, by its anterior straight margin, to the broad ligament ; by its lower extremity, to the uterus by a proper ligament, the *ligament of the ovary* ; and by its upper end to the fimbriated extremity of the Fallopian tube by the ovarian fimbria ; its mesial and lateral surfaces and posterior convex border are free. The ovaries are of a greyish-pink colour, and present either a smooth or puckered, uneven surface. They are each about an inch and a half in length, three-quarters of an inch in width, and about a third of an inch thick, and weigh from one to two drachms.

The exact position of the ovary has been the subject of considerable difference of opinion, and writers differ much as to what is to be regarded as the normal position. The fact appears to be that it is differently placed in different individuals. Hasse has described it as being situated with its long axis transverse, or almost transverse, to the pelvic cavity. Schultze, on the other hand, believes that its long axis is antero-posterior. Kölliker asserts that the truth lies between these two views, and that the ovary is placed obliquely in the pelvis, its long axis lying parallel to the external iliac vessels, with its surfaces directed inwards and outwards, and its convex free border upwards. His has made some important observations on this subject, and his views are largely accepted. He teaches that the uterus rarely lies symmetrically in the middle of the pelvic cavity, but is generally inclined to one or other side, most frequently to the left, in the proportion of three to two. The position of the two ovaries varies according to the inclination of the uterus. When the uterus is inclined to the left, the ovary of this side lies with its long axis vertical and with one side closely applied to the outer wall of the pelvis ; while the ovary of the opposite side, being dragged upon by the inclination of the uterus, lies obliquely, its outer extremity being retained in close apposition to the side of the pelvis by the infundibulo-pelvic ligament (page 1014). When, on the other hand, the uterus is inclined to the right, the position of the two ovaries is exactly reversed, the right being vertical and the left oblique. In whichever position the ovary is placed, the Fallopian tube forms a loop around it, the uterine half ascending obliquely over it, and the outer half, including the dilated extremity, descending and bulging freely behind it. From this extremity the fimbriæ pass upwards on to the ovary and closely embrace it.

Waldeyer * states, as the result of the examination of fifty female subjects, ranging from early childhood to advanced age, that the ovary 'lies on the lateral pelvic wall and vertically when the woman takes the erect posture.' Its *tubal* extremity is near the external iliac vein ; its uterine end is directed downwards, while the Fallopian tube overlies it so as to cover it on its medial face entirely or nearly so. Its convex margin looks downwards and backwards towards the pelvic cavity and rectum, while its straight margin or hilum lies laterally on the pelvic wall attached to the mesosalpinx. He also finds that it lies in a distinct but shallow groove (*fossa ovarii*) limited above by the hypogastric artery and below by the ureter, in such a manner that the ureter lies along the convex margin of the ovary, and the hypogastric artery passes near the hilum or straight margin.

Structure.—The ovary consists of a number of Graafian vesicles, embedded in the meshes of a stroma or framework, and invested by a serous covering derived from the peritoneum.

Serous Covering.—Though the investing membrane of the ovary is derived from the peritoneum, it differs essentially from that structure, inasmuch as its epithelium consists of a single layer of columnar cells, instead of the flattened endothelial cells on other parts of the membrane ; this has been termed the *germinal epithelium of Waldeyer*, and gives to the surface of the ovary a dull grey aspect instead of the shining smoothness of serous membranes generally.

Stroma.—The stroma is a peculiar soft tissue, abundantly supplied with bloodvessels, consisting for the most part of spindle-shaped cells with a small amount of ordinary connective tissue. These cells have been regarded by some anatomists as unstriped muscle-cells, which, indeed, they most resemble (His) ; by others as connective-tissue cells (Waldeyer, Henle, and Kölliker). On the surface of the organ this tissue is much condensed, and forms a layer composed of short

* *Journal of Anat. and Physiol.* vol. xxxii.

connective-tissue fibres, with fusiform cells between them. This was formerly regarded as a distinct fibrous covering, and was termed the *tunica albuginea*, but is nothing more than a condensed layer of the stroma of the ovary.

Graafian vesicles.—Upon making a section of an ovary, numerous round transparent vesicles of various sizes are to be seen ; they are the *Graafian vesicles*, or ovisacs containing the ova. Immediately beneath the superficial covering is a layer of stroma, in which are a large number of minute vesicles, of uniform size, about $\frac{1}{100}$ of an inch in diameter. These are the Graafian vesicles in their earliest condition, and the layer where they are found has been termed the *cortical layer*. They are especially numerous in the ovary of the young child. After puberty, and during the whole of the child-bearing period, large and mature, or almost mature, Graafian vesicles are also found in the cortical layer in small numbers, and also 'corpora lutea,' the remains of vesicles which have burst and are undergoing atrophy and absorption. Beneath this superficial stratum, other large and more mature Graafian vesicles are found embedded in the ovarian stroma. These increase in size as they recede from the surface towards a highly vascular stroma in the centre of the organ, termed the *medullary substance* (*zona vasculosa*, Waldeyer). This stroma forms the tissue of the hilum by which the ovary is attached, and through which the blood-vessels enter: it does not contain any Graafian vesicles.

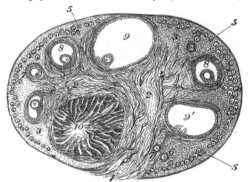

FIG. 569.—Section of the ovary. (After Schrön.)

Outer covering. 1'. Attached border. 2. Central, stroma. 3. Peripheral stroma. 4. Blood-vessels. 5. Graafian follicles in their earliest stage. 6, 7, 8. More advanced follicles. 9. An almost mature follicle. 9'. Follicle from which the ovum has escaped. 10. Corpus luteum.

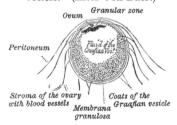

FIG. 570.—Section of the Graafian vesicle. (After Von Baer.)

The larger *Graafian vesicles* consist of an external fibro-vascular coat, connected with the surrounding stroma of the ovary by a network of blood-vessels ; and an internal coat, named *ovicapsule*, which is lined by a layer of nucleated cells, called the *membrana granulosa*. The fluid contained in the interior of the vesicles is transparent and albuminous, and in it is suspended the ovum. In that part of the mature Graafian vesicle which is nearest the surface of the ovary, the cells of the membrana granulosa are connected into a mass which projects into the cavity of the vesicle. This is termed the *discus proligerus*, and in this the ovum is embedded.*

The ova are formed from the germinal epithelium on the surface of the ovary. This becomes thickened, and in it are seen some cells which are larger and more rounded than the rest; these are termed the *primordial ova*. The germinal epithelium grows downwards in the form of tubes or columns, termed the *egg tubes* of Pflüger, into the ovarian stroma, which grows outwards between the tubes, and ultimately cuts them off from the germinal epithelium. These tubes are further subdivided into rounded *nests* or groups, each containing a primordial ovum which undergoes further development and growth while the surrounding cells of the nest form the epithelium of the Graafian follicle.

The development and maturation of the Graafian vesicles and ova continue

* For a description of the ovum, see page [78].

uninterruptedly from puberty to the end of the fruitful period of woman's life, while their formation commences before birth. Before puberty the ovaries are small, the Graafian vesicles contained in them are disposed in a comparatively thick layer in the cortical substance; here they present the appearance of a large number of minute closed vesicles, constituting the early condition of the Graafian vesicles; many, however, never attain full development, but shrink and disappear. At puberty the ovaries enlarge, and become more vascular, the Graafian vesicles are developed in greater abundance, and their ova are capable of fecundation.

Discharge of the ovum.—The Graafian vesicles, after gradually approaching the surface of the ovary, burst: the ovum and fluid contents of the vesicles are liberated, and escape on the exterior of the ovary, passing thence into the Fallopian tube.*

In the fœtus, the ovaries are situated, like the testes, in the lumbar region, near the kidneys. They may be distinguished from those bodies at an early period by their elongated and flattened form, and by their position, which is at first oblique and then nearly transverse. They gradually descend into the pelvis.

Lying above the ovary in the broad ligament between it and the Fallopian tube is the *organ of Rosenmüller,* called also the *parovarium* or *epoöphoron.* This is the remnant of a fœtal structure, the development of which is described on page [142]. In the adult it consists of a few closed convoluted tubes, lined with epithelium, which converge towards the ovary at one end and at the other are united by a longitudinal tube, which is the homologue of the *duct of Gärtner* in the cow. This duct terminates in a bulbous enlargement (see fig. [182]). The parovarium is connected at its uterine extremity with the remains of the Wolffian duct. A few scattered rudimentary tubules, best seen in the child, are situated in the broad ligament between the parovarium and the uterus. These constitute the *paroöphoron of Waldeyer.*

The **ligament of the ovary** is a rounded cord, which extends from each superior angle of the uterus to the inner extremity of the ovary; it consists of fibrous tissue and a few muscular fibres derived from the uterus.

The **round ligaments** are two rounded cords between four and five inches in length, situated between the layers of the broad ligament in front of and below the Fallopian tube. Commencing on each side at the superior angle of the uterus, this ligament passes forwards, upwards, and outwards through the internal abdominal ring, along the inguinal canal to the labium majus, in which it becomes lost. The round ligament consists principally of muscular tissue, prolonged from the uterus; also of some fibrous and areolar tissue, besides blood-vessels and nerves, enclosed in a duplicature of peritoneum, which, in the fœtus, is prolonged in the form of a tubular process for a short distance into the inguinal canal. This process is called the *canal of Nuck.* It is generally obliterated in the adult, but sometimes remains pervious even in advanced life. It is analogous to the peritoneal pouch which precedes the descent of the testis.

Vessels and Nerves.—The *arteries of the ovaries and Fallopian tubes* are the ovarian from the aorta. They enter the attached border, or hilum, of the ovary. The *veins* follow the course of the arteries; they form a plexus near the ovary, the *pampiniform plexus.* The *nerves* are derived from the inferior hypogastric or pelvic plexus, and from the ovarian plexus, the Fallopian tube receiving a branch from one of the uterine nerves.

MAMMARY GLANDS

The **mammæ,** or breasts, secrete the milk, and are accessory glands of the generative system. They exist in the male as well as in the female; but in the former only in the rudimentary state, unless their growth is excited by peculiar

---

* This is effected either by application of the tube to the ovary, or by a curling upwards of the fimbriated extremity, so that the ovum is caught as it falls.

circumstances. In the female they are two large hemispherical eminences situated towards the lateral aspect of the pectoral region, corresponding to the intervals between the third and sixth or seventh ribs, and extending from the side of the sternum to the axilla. Their weight and dimensions differ at different periods of life, and in different individuals. Before puberty they are of small size, but enlarge as the generative organs become more completely developed. They increase during pregnancy, and especially after delivery, and become atrophied in old age. The left mamma is generally a little larger than the right. Their bases are nearly circular, flattened or slightly concave, and have their long diameter directed upwards and outwards towards the axilla ; they are separated from the Pectoral muscles by a layer of fascia. The outer surface of the mamma is convex, and presents, just below the centre, a small conical prominence, the nipple (*mammilla*). The surface of the nipple is dark-coloured, and surrounded by an *areola* having a coloured tint. In the virgin the areola is of a delicate rosy hue ;

FIG. 571.—Dissection of the lower half of the female breast during the period of lactation. (From Luschka.)

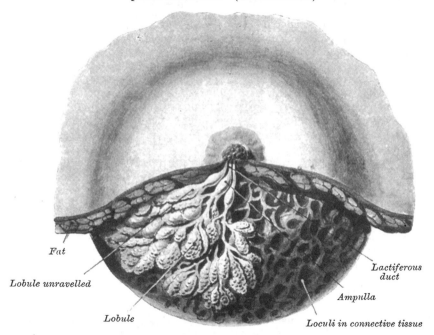

Fat

Lobule unravelled

Lobule

Lactiferous duct

Ampulla

Loculi in connective tissue

about the second month after impregnation it enlarges and acquires a darker tinge, which increases as pregnancy advances, becoming in some cases of a dark brown, or even black colour. This colour diminishes as soon as lactation is over, but is never entirely lost throughout life. These changes in the colour of the areola are of importance in forming a conclusion in a case of suspected first pregnancy.

The **nipple** is a cylindrical or conical eminence, capable of undergoing a sort of erection from mechanical excitement, a change mainly due to the contraction of its muscular fibres. It is of a pink or brownish hue, its surface wrinkled and provided with papillæ; and it is perforated by numerous orifices, the apertures of the lactiferous ducts. Near the base of the nipple, and upon the surface of the areola, are numerous sebaceous glands, which become much enlarged during lactation, and present the appearance of small tubercles beneath the skin. These glands secrete a peculiar fatty substance, which serves as a protection to the integument of the nipple during the act of sucking. The nipple consists of numerous vessels, intermixed with plain muscular fibres, which are principally arranged in a circular manner around the base : some few fibres radiating from base to apex.

Structure.—The mamma consists of gland-tissue ; of fibrous tissue, connecting its lobes ; and of fatty tissue in the intervals between the lobes. The gland-tissue, when freed from fibrous tissue and fat, is of a pale reddish colour, firm in texture, circular in form, flattened from before backwards, thicker in the centre than at the circumference, and presenting several inequalities on its surface, especially in front. It consists of numerous lobes, and these are composed of lobules, connected together by areolar tissue, blood-vessels, and ducts. The smallest lobules consist of a cluster of rounded alveoli, which open into the smallest branches of the lactiferous ducts ; these ducts uniting form larger ducts, which terminate in a single canal, corresponding with one of the chief subdivisions of the gland. The number of excretory ducts varies from fifteen to twenty ; they are termed the *tubuli lactiferi*, or *galactophori*. They converge towards the areola, beneath which they form dilatations, or *ampullæ*, which serve as reservoirs for the milk, and, at the base of the nipple, become contracted, and pursue a straight course to its summit, perforating it by separate orifices considerably narrower than the ducts themselves. The ducts are composed of areolar tissue, with longitudinal and transverse elastic fibres ; muscular fibres are entirely absent ; their mucous lining is continuous, at the point of the nipple, with the integument. The epithelium of the mammary gland differs according to the state of activity of the organ. In the gland of a woman who is not pregnant or suckling, the alveoli are very small and solid, being filled with a mass of granular polyhedral cells. During pregnancy the alveoli enlarge, and the cells undergo rapid multiplication. At the commencement of lactation, the cells in the centre of the alveolus undergo fatty degeneration, and are eliminated in the first milk, as *colostrum corpuscles*. The peripheral cells of the alveolus remain, and form a single layer of granular, short columnar cells, with a spherical nucleus, lining the limiting membrana propria. These cells during the state of activity of the gland, are capable of forming, in their interior, oil-globules, which are then ejected into the lumen of the alveolus, and constitute the milk-globules.

The *fibrous tissue* invests the entire surface of the breast, and sends down septa between its lobes, connecting them together.

The *fatty tissue* surrounds the surface of the gland, and occupies the interval between its lobes. It usually exists in considerable abundance, and determines the form and size of the gland. There is no fat immediately beneath the areola and nipple.

Vessels and Nerves.—The *arteries* supplying the mammæ are derived from the thoracic branches of the axillary, the intercostals, and internal mammary. The *veins* describe an anastomotic circle round the base of the nipple, called by Haller the *circulus venosus*. From this, large branches transmit the blood to the circumference of the gland, and end in the axillary and internal mammary veins. The *lymphatics*, for the most part, run along the lower border of the Pectoralis major to the axillary glands ; some few, from the inner side of the breast, perforate the intercostal spaces and empty themselves into the anterior mediastinal glands. The *nerves* are derived from the anterior and lateral cutaneous nerves of the thorax.

THE SURGICAL ANATOMY OF
INGUINAL HERNIA

Dissection (fig. 217).—For dissection of the parts concerned in inguinal hernia, a male subject, free from fat, should always be selected. The body should be placed in the supine position, the abdomen and pelvis raised by means of blocks placed beneath them, and the lower extremities rotated outwards, so as to make the parts as tense as possible. If the abdominal walls are flaccid, the cavity of the abdomen should be inflated through an aperture made at the umbilicus. An incision should be made along the middle line, from a little below the umbilicus to the symphysis pubis, and continued along the front of the scrotum ; and a second incision, from the anterior superior spine of the ilium to just below the umbilicus. These incisions should divide the integument ; and the triangular-shaped flap included between them should be reflected downwards and outwards, when the superficial fascia will be exposed.

The superficial fascia of the abdomen.—This, over the greater part of the abdominal wall, consists of a single layer of fascia, which contains a variable amount of fat ; but as it approaches the groin it is easily divisible into two layers, between which are found the superficial vessels and nerves and the superficial inguinal lymphatic glands.

The *superficial layer* (*fascia of Camper*) is thick, areolar in texture, containing adipose tissue in its meshes, the quantity of which varies in different subjects. Below, it passes over Poupart's ligament and is continuous with the outer layer of the superficial fascia of the thigh. In the male, this fascia is continued over the penis and over the outer surface of the cord to the scrotum, where it helps to form the dartos. As it passes to the penis and over the cord to the scrotum it changes its character, becoming thin, destitute of adipose tissue, and of a pale, reddish colour; and in the scrotum it acquires some involuntary muscular fibres. From the scrotum it may be traced backwards, to be continuous with the superficial fascia of the perinæum. In the female this fascia is continued into the labia majora.

The *hypogastric branch of the ilio-hypogastric nerve* perforates the aponeurosis of the External oblique muscle about an inch above and a little to the outer side of the external abdominal ring, and is distributed to the integument of the hypogastric region.

The *ilio-inguinal nerve* escapes at the external abdominal ring, and is distributed to the integument of the upper and inner part of the thigh ; to the scrotum in the male and to the labium in the female.

The *superficial epigastric artery* arises from the femoral about half an inch below Poupart's ligament, and, passing through the saphenous opening in the fascia lata, ascends on to the abdomen, in the superficial fascia covering the External oblique muscle, nearly as high as the umbilicus. It distributes branches to the superficial inguinal lymphatic glands, the superficial fascia, and the integument, anastomosing with branches of the deep epigastric and internal mammary arteries.

The *superficial circumflex iliac artery*, the smallest of the cutaneous branches, arises close to the preceding, and, piercing the fascia lata, runs outwards, parallel with Poupart's ligament, as far as the crest of the ilium, dividing into branches which supply the superficial inguinal lymphatic glands, the superficial fascia and the integument, anastomosing with the deep circumflex iliac and with the gluteal and external circumflex arteries.

The *superficial external pudic (superior) artery* arises from the inner side of the femoral artery, close to the preceding vessels, and, after passing through the saphenous opening, courses inwards, across the spermatic cord, to be distributed to the integument on the lower part of the abdomen, the penis and scrotum in the male, and the labium in the female, anastomosing with branches of the internal pudic.

The superficial veins.—The veins accompanying these superficial vessels are

FIG. 572.—Inguinal hernia.   Superficial dissection.

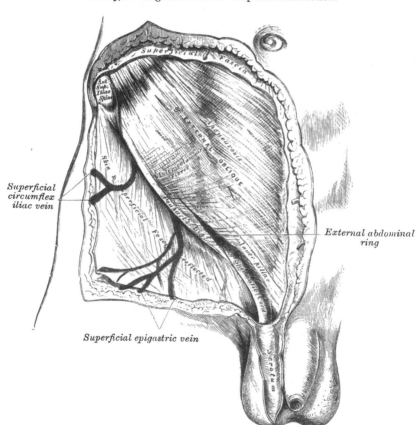

usually much larger than the arteries : they terminate in the internal saphenous vein.

The *superficial inguinal lymphatic glands* are placed immediately beneath the integument, are of large size, and vary from eight to ten in number.   They are divisible into two groups : an upper, disposed irregularly along Poupart's ligament, which receives the lymphatic vessels from the integument of the scrotum, penis, parietes of the abdomen, perineal and gluteal regions, and the mucous membrane of the urethra ; and an inferior group, which surround the saphenous opening in the fascia lata, a few being sometimes continued along the saphenous vein to a variable extent.   This latter group receives the superficial lymphatic vessels from the lower extremity.

The *deep layer of the superficial fascia (fascia of Scarpa)* is thinner and more membranous in character than the superficial layer. In the middle line it is intimately adherent to the linea alba; above, it is continuous with the superficial fascia over the rest of the trunk; below, it blends with the fascia lata of the thigh, a little below Poupart's ligament; below and internally, in the male, it is continued over the penis and over the outer surface of the cord to the scrotum, where it helps to form the dartos. From the scrotum it may be traced backwards to be continuous with the base of the triangular ligament of the urethra. In the female it is continuous with the labia majora.

The scrotum is a cutaneous pouch, which contains the testes and part of the spermatic cords, and into which an inguinal hernia frequently descends (see page 999).

The **aponeurosis of the External oblique muscle.**—This is a thin, but strong membranous aponeurosis, the fibres of which are directed obliquely downwards and inwards. That portion of the aponeurosis which extends between the anterior superior spine of the ilium and the spine of the os pubis is a broad band, folded inwards and continuous below with the fascia lata; it is called *Poupart's ligament*. The portion which is reflected from Poupart's ligament at the spine of the os pubis, along the pectineal line, is called *Gimbernat's ligament*. From the point of attachment of the latter to the pectineal line, a few fibres pass upwards and inwards, behind the inner pillar of the ring to the linea alba. They diverge as they ascend, and form a thin, triangular, fibrous band, which is called the *triangular fascia of the abdomen*.

The **external or superficial abdominal ring.**—Just above and to the outer side of the crest of the os pubis, an interval is seen in the aponeurosis of the External oblique, called the *external abdominal ring*. This aperture is oblique in direction, somewhat triangular in form, and corresponds with the course of the fibres of the aponeurosis. It usually measures from base to apex about an inch, and transversely about half an inch. It is bounded below by the crest of the os pubis; above by a series of curved fibres, the *intercolumnar*, which pass across the upper angle of the ring, so as to increase its strength; and on either side, by the margins of the opening in the aponeurosis, which are called the *columns* or *pillars of the ring*.

The *external pillar*, which, at the same time, is *inferior* from the obliquity of its direction, is the stronger; it is formed by that portion of Poupart's ligament which is inserted into the spine of the os pubis; it is curved, so as to form a kind of groove, upon which the spermatic cord rests.

The *internal* or *superior pillar* is a broad, thin, flat band, which is attached to the front of the body of the os pubis, interlacing with its fellow of the opposite side, in front of the symphysis pubis, that of the right side being superficial.

The external abdominal ring gives passage to the spermatic cord in the male, and round ligament in the female; it is much larger in men than in women, on account of the large size of the spermatic cord, and hence the great frequency of inguinal hernia in men.

The *intercolumnar fibres* are a series of curved tendinous fibres, which arch across the lower part of the aponeurosis of the External oblique. They have received their name from stretching across between the two pillars of the external ring; they increase the strength of the lower part of the aponeurosis, and prevent the divergence of the pillars from one another. They are thickest below, where they are connected to the outer third of Poupart's ligament; and are inserted into the linea alba, describing a curve, with the convexity downwards. They are much thicker and stronger at the outer angle of the external ring than internally, and are more strongly developed in the male than in the female. These intercolumnar fibres, as they pass across the external abdominal ring, are themselves connected together by delicate fibrous tissue, thus forming a fascia which, as it is attached to the pillars of the ring, covers it in, and is called the *intercolumnar fascia*. This intercolumnar fascia is continued downwards as a tubular prolongation around the outer surface of the cord and testis, and encloses them in a distinct sheath; hence it is also called the *external spermatic* or *intercolumnar fascia*. The sac of an inguinal hernia, in passing through the external abdominal ring, receives an investment from the intercolumnar fascia.

If the finger is introduced a short distance into the external ring, and then if the limb is extended and rotated outwards, the aponeurosis of the External oblique, together with the iliac portion of the fascia lata, will be felt to become tense, and the external ring much contracted; if the limb is, on the contrary, flexed upon the pelvis and rotated inwards, this aponeurosis will become lax, and the external ring sufficiently enlarged to admit the finger with comparative ease; hence the patient should always be put in the latter position when the taxis is applied for the reduction of an inguinal hernia, in order that the abdominal walls may be relaxed as much as possible.

Fig. 573.—Inguinal hernia.
Dissection showing the Internal oblique and Cremaster.

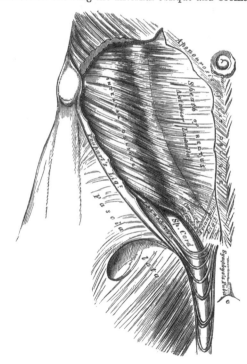

The aponeurosis of the External oblique should be removed by dividing it across in the same direction as the external incisions, and reflecting it downwards and outwards; great care is requisite in separating it from the aponeurosis of the muscle beneath. The lower part of the Internal oblique and the Cremaster are then exposed, together with the inguinal canal, which contains the spermatic cord (fig. 573). The mode of insertion of Poupart's and Gimbernat's ligaments into the os.pubis should also be examined.

Poupart's ligament, or the crural arch, is the lower border of the aponeurosis of the External oblique muscle, which extends from the anterior superior spine of the ilium to the spine of the os pubis. From this latter point it is reflected outwards to be attached to the pectineal line for about half an inch, forming Gimbernat's ligament. Its general direction is curved downwards towards the thigh, where it is continuous with the fascia lata. Its outer half is rounded and oblique in direction: its inner half gradually widens at its attachment to the os pubis, is more horizontal in direction, and lies beneath the spermatic cord.

Gimbernat's ligament (fig. 581) is that portion of the aponeurosis of the External oblique muscle which is reflected upwards and outwards from the spine of the os pubis to be inserted into the pectineal line. It is about half an inch in length, larger in the male than in the female, almost horizontal in direction in the erect posture, and of a triangular form, with the base directed outwards. Its base or outer margin is concave, thin, and sharp, and lies in contact with the femoral sheath, forming the inner boundary of the crural ring (see fig. 581). Its apex corresponds to the spine of the os pubis. Its posterior margin is attached to the pectineal line, and is continuous with the pubic portion of the fascia lata. Its anterior margin is continuous with Poupart's ligament.

The **triangular fascia of the abdomen** is a band of tendinous fibres, of a triangular shape, which is attached by its apex to the pectineal line, where it is continuous with Gimbernat's ligament. It passes inwards beneath the spermatic cord, and expands into a somewhat fan-shaped fascia, lying behind the inner pillar of the external abdominal ring, and in front of the conjoined tendon, and interlaces with the ligament of the other side at the linea alba.

The **Internal oblique muscle** has been previously described (page 333). The part which is now exposed is partly muscular and partly tendinous in structure. Those fibres which arise from Poupart's ligament, few in number, and paler in colour than the rest, arch downwards and inwards across the spermatic cord, and, becoming tendinous, are inserted conjointly with those of the Transversalis into the crest of the os pubis and pectineal line, forming what is known as the conjoined tendon of the Internal oblique and Transversalis. This tendon is inserted immediately behind the inguinal canal and external abdominal ring, serving to protect what would otherwise be a weak point in the abdominal wall. Sometimes this tendon is insufficient to resist the pressure from within, and is carried forwards in front of the protrusion through the external ring, forming one of the coverings of direct inguinal hernia; or the hernia forces its way through the fibres of the conjoined tendon.

The **Cremaster** is a thin muscular layer, composed of a number of fasciculi which arise from the middle of Poupart's ligament at the inner side of the Internal oblique, being connected with that muscle, and also occasionally with the Transversalis. It passes along the outer side of the spermatic cord, descends with it through the external ring upon the front and sides of the cord, and forms a series of loops, which differ in thickness and length in different subjects. Those at the upper part of the cord are exceedingly short, but they become in succession longer and longer, the longest reaching down as low as the testicle, where a few are inserted into the tunica vaginalis. These loops are united together by areolar tissue, and form a thin covering over the cord and testis, the *cremasteric fascia*. The fibres ascend along the inner side of the cord, and are inserted by a small pointed tendon into the crest of the os pubis, and the front of the sheath of the Rectus muscle.

It will be observed that the origin and insertion of the Cremaster is precisely similar to that of the lower fibres of the Internal oblique. This fact affords an easy explanation of the manner in which the testicle and cord are invested by this muscle. At an early period of foetal life the testis is placed at the lower and back part of the abdominal cavity, but during its descent towards the scrotum, which takes place before birth, it passes beneath the arched border of the Internal oblique. In its passage beneath this muscle some fibres are derived from its lower part, which accompany the testicle and cord into the scrotum.

It occasionally happens that the loops of the Cremaster surround the cord, some lying behind as well as in front. It is probable that, under these circumstances, the testis, in its descent, passes through, instead of beneath, the fibres of the Internal oblique.

In the descent of an oblique inguinal hernia, which takes the same course as the spermatic cord, the Cremaster muscle forms one of its coverings. This muscle becomes largely developed in cases of hydrocele and large, old scrotal herniæ. No such muscle exists in the female, but an analogous structure is developed in those cases where an oblique inguinal hernia descends beneath the margin of the Internal oblique.

The Internal oblique should be detached from Poupart's ligament, separated from the Transversalis to the same extent as in the previous incisions, and reflected inwards on to the sheath of the Rectus (fig. 574). The deep circumflex iliac vessels, which lie between these two muscles, form a valuable guide to their separation.

The **Transversalis muscle** has been previously described (page 336). The portion which is now exposed is partly muscular and partly tendinous in structure; it arises from the outer third of Poupart's ligament, its fibres curve downwards and inwards, and are inserted together with those of the Internal oblique into the lower part of the linea alba, into the crest of the os pubis and the pectineal line, forming what is known as the conjoined tendon of the Internal

FIG. 574.—Inguinal hernia. Dissection showing the Transversalis muscle, the transversalis fascia, and the internal abdominal ring.

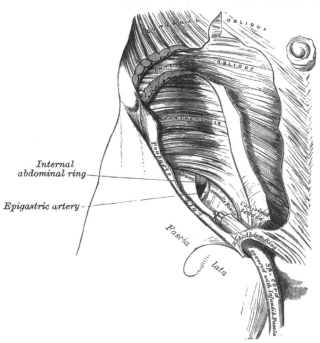

oblique and Transversalis. Between the lower border of this muscle and Poupart's ligament, a space is left, in which is seen the transversalis fascia.

The **inguinal** or **spermatic canal** contains the spermatic cord in the male, and the round ligament in the female. It is an oblique canal, about an inch and a half in length, directed downwards and inwards, and placed parallel with, and a little above, Poupart's ligament. It commences, above, at the internal or deep abdominal ring, which is the point where the cord enters the inguinal canal; and terminates, below, at the external or superficial ring. It is bounded, *in front*, by the integument and superficial fascia, by the aponeurosis of the External oblique throughout its whole length, and by the Internal oblique for its outer third; *behind*, by the triangular fascia, the conjoined tendon of the Internal oblique and Transversalis, transversalis fascia, and the subperitoneal fat and peritoneum; *above*, by the arched fibres of the Internal oblique and Transversalis; *below*, by the union of the transversalis fascia with Poupart's ligament. That form of hernia in which the intestine follows the course of the spermatic cord along the inguinal canal is called *oblique inguinal hernia*.

The **transversalis fascia** is a thin aponeurotic membrane, which lies between the inner surface of the Transversalis muscle and the peritoneum. It forms part of the general layer of fascia which lines the interior of the abdominal and pelvic cavities, and is directly continuous with the iliac and pelvic fasciæ.

In the inguinal region, the transversalis fascia is thick and dense in structure and joined by fibres from the aponeurosis of the Transversalis muscle; but it becomes thin and cellular as it ascends to the Diaphragm. Below, it has the following attachments : external to the femoral vessels it is connected to the posterior margin of Poupart's ligament, and is there continuous with the iliac fascia. Internal to the vessels it is thin, and attached to the os pubis and pectineal line behind the conjoined tendon, with which it is united; and, corresponding to the points where the femoral vessels pass into the thigh, this fascia descends in front of them, forming the anterior wall of the femoral sheath. The spermatic cord in the male, and the round ligament in the female, pass through this fascia ; the point where they pass through is called the internal or deep abdominal ring. This opening is not visible externally owing to a prolongation of the transversalis fascia on these structures, forming the infundibuliform fascia.

The **internal** or **deep abdominal ring** is situated in the transversalis fascia, midway between the anterior superior spine of the ilium and the symphysis pubis, and about half an inch above Poupart's ligament. It is of an oval form, its long diameter being directed upwards and downwards ; it varies in size in different subjects, and is much larger in the male than in the female. It is bounded, above and externally, by the arched fibres of the Transversalis muscle, below and internally by the deep epigastric vessels. It transmits the spermatic cord in the male, and the round ligament in the female. From its circumference a thin, funnel-shaped membrane, the *infundibuliform fascia*, is continued round the cord and testis, enclosing them in a distinct pouch. When the sac of an oblique inguinal hernia passes through the internal or deep abdominal ring, the infundibuliform fascia constitutes one of its coverings.

The subperitoneal areolar tissue.—Between the transversalis fascia and the peritoneum is a quantity of loose areolar tissue. In some subjects it is of considerable thickness, and loaded with adipose tissue. Opposite the internal ring it is continued round the surface of the cord, forming a loose sheath for it.

The *deep epigastric artery* arises from the external iliac artery a few lines above Poupart's ligament. It at first descends to reach this ligament, and then ascends obliquely along the inner margin of the internal or deep abdominal ring, lying between the transversalis fascia and the peritoneum ; and, passing upwards, pierces the transversalis fascia and enters the sheath of the Rectus muscle by passing over the semilunar fold of Douglas. Consequently the deep epigastric artery bears a very important relation to the internal abdominal ring, as it passes obliquely upwards and inwards, from its origin from the external iliac. In this part of its course it lies along the lower and inner margin of the internal ring, and beneath the commencement of the spermatic cord. At its commencement it is crossed by the vas deferens in the male, and by the round ligament in the female.

The **peritoneum,** corresponding to the inner surface of the internal ring, presents a well-marked depression, the depth of which varies in different subjects. A thin fibrous band is continued from it along the front of the cord for a variable distance, and becomes ultimately lost. This is the remains of the pouch of peritoneum, which in the fœtus precedes the cord and testis into the scrotum and the obliteration of which commences soon after birth. In some cases the fibrous band can only be traced a short distance ; but occasionally it may be followed, as a fine cord, as far as the upper end of the tunica vaginalis. Sometimes the tube of peritoneum is only closed at intervals, and presents a sacculated appearance ; or a single pouch may extend along the whole length of the cord, which may be closed above ; or the pouch may be directly continuous with the peritoneum by an opening at its upper part.

In the female fœtus the peritoneum is also prolonged in the form of a tubular process for a short distance into the inguinal canal. This process is called the *canal of Nuck*. It is generally obliterated in the adult, but sometimes it remains pervious even in advanced life.

In order to understand the relation of the peritoneum to inguinal hernia, it is necessary to view the anterior abdominal wall from its internal aspect, when it will be seen as shown in fig. 575. Between the upper margin of the front of the pelvis and the umbilicus, the peritoneum, when viewed from behind, will be seen to be raised into five vertical folds, with intervening depressions,

FIG. 575.—Posterior view of the anterior abdominal wall in its lower half. The peritoneum is in place, and the various cords are shining through. (After Joessel.)

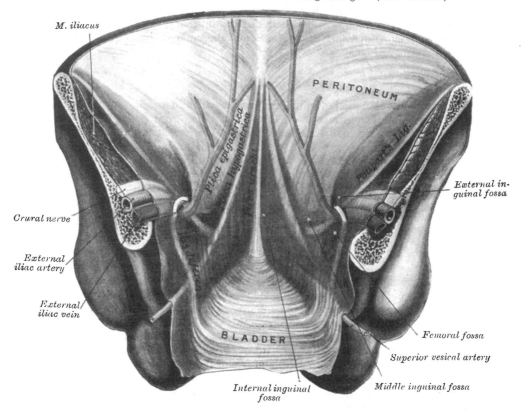

by more or less prominent bands which converge to the umbilicus. One of these is situated in the median line, and is caused by the urachus, the remnant of the allantois ; it extends from the summit of the bladder to the umbilicus. The fold of peritoneum covering it is known as the *plica urachi*. On either side of this is a prominent band, caused by the obliterated hypogastric artery, which extends from the side of the bladder obliquely upwards and inwards to the umbilicus. This is covered by a fold of peritoneum, which is known as the *plica hypogastrica*. To either side of these three cords is the deep epigastric artery, which ascends obliquely upwards and inwards from a point midway between the symphysis pubis and the anterior superior spine of the ilium to the semilunar fold of Douglas, in front of which it disappears. It is covered by a fold of peritoneum, which is known as the *plica epigastrica*. Between these raised folds are depressions of the peritoneum, constituting so-called fossæ.

The most internal, between the plica urachi and the plica hypogastrica, is known as the *internal inguinal fossa* (fovea supravesicalis). The middle one is situated between the plica hypogastrica and the plica epigastrica, and is termed the *middle inguinal fossa* (fovea inguinalis mesialis). The external one is external to the plica epigastrica, and is known as the *external inguinal fossa* (fovea inguinalis lateralis). Occasionally the deep epigastric artery corresponds in position to the obliterated hypogastric artery, and then there is but one fold on each side of the middle line, and the two external fossæ are merged into one. In the usual condition of the parts the floor of the external inguinal fossa corresponds to the internal abdominal ring, and into this fossa an oblique inguinal hernia descends. To the inner side of the plica epigastrica are the two internal fossæ, and through either of these a direct hernia may descend, as will be explained in the sequel (page 1036). The whole of this space, that is to say, the space between the deep epigastric artery, the margin of the Rectus and Poupart's ligament, is commonly known as *Hesselbach's triangle*. These three depressions or fossæ are situated above the level of Poupart's ligament, and in addition to them is another below the ligament, corresponding to the position of the femoral ring, and into which a femoral hernia descends.

INGUINAL HERNIA

Inguinal hernia is that form of protrusion which makes its way through the abdomen in the inguinal region.

There are two principal varieties of inguinal hernia—external or oblique, and internal or direct.

External or *oblique inguinal hernia*, the more frequent of the two, takes the same course as the spermatic cord. It is called *external*, from the neck of the sac being on the outer or iliac side of the deep epigastric artery.

Internal or *direct inguinal hernia* does not follow the same course as the cord, but protrudes through the abdominal wall, on the inner or pubic side of the deep epigastric artery.

OBLIQUE INGUINAL HERNIA

In **oblique inguinal hernia** the intestine escapes from the abdominal cavity at the internal ring, pushing before it a pouch of peritoneum, which forms the hernial sac (fig. 577, A). As it enters the inguinal canal it receives an investment from the subserous areolar tissue, and is enclosed in the infundibuliform process of the transversalis fascia. In passing along the inguinal canal it displaces upwards the arched fibres of the Transversalis and Internal oblique muscles, and is surrounded by the fibres of the Cremaster. It then passes along the front of the cord, and escapes from the inguinal canal at the external ring, receiving an investment from the intercolumnar fascia. Lastly, it descends into the scrotum, receiving coverings from the superficial fascia and the integument.

The coverings of this form of hernia, after it has passed through the external ring, are, from without inwards, the integument, superficial fascia, intercolumnar fascia, Cremaster muscle, infundibuliform fascia, subserous areolar tissue, and peritoneum.

This form of hernia lies in front of the vessels of the spermatic cord, and seldom extends below the testis, on account of the intimate adhesion of the coverings of the cord to the tunica vaginalis.

The *seat of stricture* in oblique inguinal hernia is either at the external ring; in the inguinal canal, caused by the fibres of the Internal oblique or Transversalis; or at the internal ring; most frequently in the latter situation. If it is situated at the external ring, the division of a few fibres at one point of its circumference is all that is necessary for the replacement of the hernia. If in the inguinal canal, or at the internal ring, it may be necessary to divide the aponeurosis of the

External oblique, so as to lay open the inguinal canal. In dividing the stricture, the direction of the incision should be upwards.

When the intestine passes along the inguinal canal, and escapes from the external ring into the scrotum, it is called *complete oblique inguinal*, or *scrotal hernia*. If the intestine does not escape from the external ring, but is retained in the inguinal canal, it is called *incomplete inguinal hernia*, or *bubonocele*. In each of these cases, the coverings which invest it will depend upon the extent to which it descends in the inguinal canal.

There are some other varieties of oblique inguinal hernia depending upon congenital defects in the processus vaginalis. The testicle in its descent from the

FIG. 576.—Oblique inguinal hernia, showing its various coverings. (From a preparation in the Museum of the Royal College of Surgeons of England.)

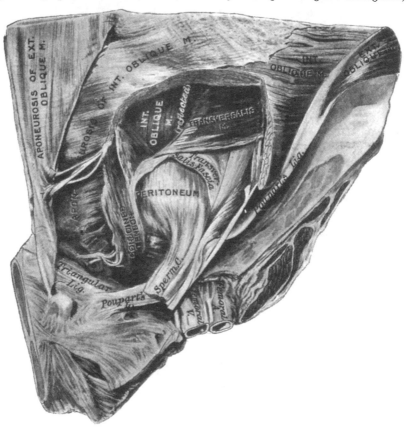

abdomen into the scrotum is preceded by a pouch of peritoneum, which about the period of birth becomes shut off from the general peritoneal cavity by a closure of that portion of the pouch which extends from the internal abdominal ring to near the upper part of the testicle; the lower portion of the pouch remaining persistent as the tunica vaginalis. It would appear that this closure commences at two points, viz. at the internal abdominal ring, and at the top of the epididymis, and gradually extends until, in the normal condition, the whole of the intervening portion is converted into a fibrous cord. From failure in the completion of this process, variations in the relation of the hernial protrusion to the testicle and tunica vaginalis are produced, which constitute distinct varieties of inguinal hernia, and which have received separate names, and are of surgical importance. These are congenital, infantile, encysted, and hernia of the funicular process.

Congenital hernia (fig. 577, B).—Where the pouch of peritoneum which precedes the cord and testis in its descent remains patent throughout, and is

FIG. 577.—Varieties of oblique inguinal hernia.

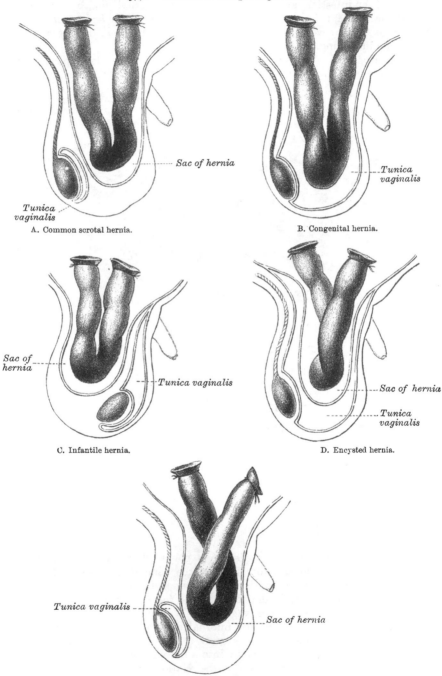

A. Common scrotal hernia.

B. Congenital hernia.

C. Infantile hernia.

D. Encysted hernia.

E. Hernia into the funicular process.

unclosed at any point, the cavity of the tunica vaginalis communicates directly with the peritoneum. The intestine descends along this pouch into the cavity of the tunica vaginalis, which constitutes the sac of the hernia, and the gut lies in contact with the testicle.

Infantile and encysted hernia.—Where the pouch of peritoneum is occluded at the internal ring only, and remains patent throughout the rest of its extent, two varieties of oblique inguinal hernia may be produced, which have received the names of infantile and encysted hernia. In the *infantile* form (fig. 577, c) the bowel pressing upon the septum and the peritoneum in its immediate neighbourhood causes it to yield and form a sac, which descends behind the tunica vaginalis; so that, in front of the bowel, there are three layers of peritoneum, the two layers of the tunica vaginalis and its own sac. In the *encysted* form (fig. 577, D) pressure in the same position—namely, at the occluded spot in the pouch—causes the septum to yield and form a sac which projects *into* and not *behind* the tunica vaginalis, as in the infantile form, and thus it constitutes a sac within a sac, so that in front of the bowel there are two layers of peritoneum, one layer of the tunica vaginalis and its own sac.

Hernia into the funicular process (fig. 577, E).—Where the pouch of peritoneum is occluded at the lower point only, that is, just above the testicle, the intestine descends into the pouch of peritoneum as far as the testicle, but is prevented from entering the sac of the tunica vaginalis by the septum which has formed between it and the pouch, so that it resembles the congenital form in all respects, except that, instead of enveloping the testicle, that body can be felt below the rupture.

DIRECT INGUINAL HERNIA

In **direct inguinal hernia** the protrusion makes its way through some part of the abdominal wall internal to the epigastric artery.

At the lower part of the abdominal wall is a triangular space (*Hesselbach's triangle*), bounded, externally, by the deep epigastric artery; internally by the margin of the Rectus muscle; below, by Poupart's ligament (fig. 575). The conjoined tendon is stretched across the inner two-thirds of this space, the remaining portion of the space having only the subperitoneal areolar tissue, and the transversalis fascia between the peritoneum and the aponeurosis of the External oblique muscle.

In some cases the hernial protrusion escapes from the abdomen on the outer side of the conjoined tendon, pushing before it the peritoneum, the subserous areolar tissue, and the transversalis fascia. It then enters the inguinal canal, passing along nearly its whole length, and finally emerges from the external ring, receiving an investment from the intercolumnar fascia. The coverings of this form of hernia are precisely similar to those investing the oblique form, with the insignificant difference that the infundibuliform fascia is replaced by a portion derived from the general layer of the transversalis fascia.

In other cases, and this is the more frequent variety, the hernia is either forced through the fibres of the conjoined tendon, or the tendon is gradually distended in front of it, so as to form a complete investment for it. The intestine then enters the lower end of the inguinal canal, escapes at the external ring, lying on the inner side of the cord, and receives additional coverings from the superficial fascia and the integument. This form of hernia has the same coverings as the oblique variety, excepting that the conjoined tendon is substituted for the Cremaster and the infundibuliform fascia is replaced by a portion derived from the general layer of the transversalis fascia.

The difference between the position of the neck of the sac in these two forms of direct inguinal hernia has been referred, with some probability, to a difference in the relative positions of the obliterated hypogastric artery and the deep epigastric artery. When the course of the obliterated hypogastric artery corresponds pretty nearly with that of the deep epigastric, the projection of these arteries towards the cavity of the abdomen produces two fossæ in the peritoneum.

The bottom of the external fossa of the peritoneum corresponds to the position of the internal abdominal ring, and a hernia which distends and pushes out the peritoneum lining this fossa is an oblique hernia. When, on the other hand, the obliterated hypogastric artery lies considerably to the inner side of the deep epigastric artery, corresponding to the outer margin of the conjoined tendon, it divides the triangle of Hesselbach into two parts, so that three depressions will be seen on the inner surface of the lower part of the abdominal wall: viz. an external one, on the outer side of the deep epigastric artery; a middle one, between the deep epigastric and the obliterated hypogastric arteries; and an internal one, on the inner side of the obliterated hypogastric artery (see page 1033). In such a case a hernia may distend and push out the peritoneum forming the bottom of either fossa. When the hernia distends and pushes out the peritoneum forming the bottom of the external fossa, it is an oblique or external inguinal hernia. When the hernia distends and pushes out the peritoneum forming the bottom of either the middle or the internal fossa, it is a direct or internal hernia. The anatomical difference between these two forms of direct or internal inguinal hernia is that, when the hernia protrudes through the middle fossa—that is, the fossa between the deep epigastric and the obliterated hypogastric arteries—it will enter the upper part of the inguinal canal; consequently its covering will be the same as those of an oblique hernia, with the significant difference that the infundibuliform fascia is replaced by a portion derived from the general layer of the transversalis fascia, whereas when the hernia protrudes through the internal fossa it is either forced through the fibres of the conjoined tendon, or the tendon is gradually distended in front of it, so as to form a complete investment for it. The intestine then enters the lower part of the inguinal canal, and escapes from the external abdominal ring lying on the inner side of the cord.

This form of hernia has the same coverings as the oblique variety, excepting that the conjoined tendon is substituted for the Cremaster, and the infundibuliform fascia is replaced by a portion derived from the general layer of the transversalis fascia.

The *seat of stricture* in both varieties of direct hernia is most frequently at the neck of the sac, or at the external ring. In that form of hernia which perforates the conjoined tendon, it not infrequently occurs at the edges of the fissure through which the gut passes. In dividing the stricture, the incision should in all cases be directed upwards.*

If the hernial protrusion passes into the inguinal canal, but does not escape from the external abdominal ring, it forms what is called *incomplete direct hernia*. This form of hernia is usually of small size, and in corpulent persons very difficult of detection.

Direct inguinal hernia is of much less frequent occurrence than the oblique, their comparative frequency being, according to Cloquet, as one to five. It occurs far more frequently in men than in women, on account of the larger size of the external ring in the former sex. It differs from the oblique in its smaller size and globular form, dependent most probably on the resistance offered to its progress by the transversalis fascia and conjoined tendon. It differs also in its position, being placed over the os pubis, and not in the course of the inguinal canal. The deep epigastric artery runs on the outer or iliac side of the neck of the sac, and the spermatic cord along its external and posterior side, not directly behind it, as in oblique inguinal hernia.

* In all cases of inguinal hernia, whether oblique or direct, it is proper to divide the stricture directly upwards; the reason of this is obvious, for by cutting in this direction the incision is made parallel to the deep epigastric artery—either external to it, in the oblique variety; or internal to it, in the direct form of hernia; and thus all chance of wounding the vessel is avoided. If the incision was made outwards, the artery might be divided if the hernia was direct; and if made inwards it would stand an equal chance of injury if the case was one of oblique inguinal hernia.

SURGICAL ANATOMY OF FEMORAL HERNIA

The dissection of the parts comprised in the anatomy of femoral hernia should be performed, if possible, upon a female subject free from fat. The subject should lie upon its back; a block is first placed under the pelvis, the thigh everted, and the knee slightly bent, and retained in this position. An incision should then be made from the anterior superior spinous process of the ilium along Poupart's ligament to the symphysis pubis; a second incision should be carried transversely across the thigh about six inches beneath the preceding; and these are to be connected together by a vertical one carried along the inner side of the thigh. These several incisions should divide merely the integument; this is to be reflected outwards, when the superficial fascia will be exposed.

The **superficial fascia** forms a continuous layer over the whole of the thigh, consisting of areolar tissue, containing in its meshes much fat, and capable of being separated into two or more layers, between which are found the superficial vessels and nerves. It varies in thickness in different parts of the limb. In the groin it is thick, and the two layers are separated from one another by the superficial inguinal lymphatic glands, the internal saphenous vein, and several smaller vessels. One of these layers, the superficial, is continuous with the superficial fascia of the abdomen.

The superficial layer should be detached by dividing it across in the same direction as the external incisions; its removal will be facilitated by commencing at the lower and inner angle of the space, detaching it at first from the front of the internal saphenous vein, and dissecting it off from the anterior surface of that vessel and its tributaries; it should then be reflected outwards, in the same manner as the integument. The cutaneous vessels and nerves, and superficial inguinal glands, are then exposed, lying upon the deep layer of the superficial fascia. These are the internal saphenous vein, and the superficial epigastric, superficial circumflex iliac, and superficial external pudic vessels, as well as numerous lymphatics ascending with the saphenous vein to the inguinal glands.

The *internal* or *long saphenous vein* ascends along the inner side of the thigh, and, passing through the saphenous opening in the fascia lata, terminates in the femoral vein about an inch and a half below Poupart's ligament. This vein receives, at the saphenous opening, the superficial epigastric, the superficial circumflex iliac, and the superficial external pudic veins.

The *superficial external pudic artery* (superior) arises from the inner side of the femoral artery, and, after passing through the saphenous opening, courses inwards, across the spermatic cord, to be distributed to the integument on the lower part of the abdomen, the penis and scrotum in the male, and the labium in the female, anastomosing with branches of the internal pudic.

The *superficial epigastric artery* arises from the femoral, about half an inch below Poupart's ligament, and, passing through the saphenous opening in the fascia lata, ascends on to the abdomen, in the superficial fascia covering the External oblique muscle, nearly as high as the umbilicus. It distributes branches to the superficial inguinal lymphatic glands, the superficial fascia, and the integument, anastomosing with branches of the deep epigastric and internal mammary arteries.

The *superficial circumflex iliac artery*, the smallest of the cutaneous branches, arises close to the preceding, and, piercing the fascia lata, runs outwards parallel with Poupart's ligament, as far as the crest of the ilium, dividing into branches which supply the superficial inguinal lymphatic glands, the superficial fascia and the integument of the groin, anastomosing with the deep circumflex iliac, and with the gluteal and external circumflex arteries.

The superficial veins.—The veins accompanying these superficial arteries are usually much larger than the arteries: they terminate in the internal or long saphenous vein at the saphenous opening.

The *superficial inguinal lymphatic glands*, placed immediately beneath the integument, are of large size, and vary from eight to ten in number. They are divisible into two groups: an upper, disposed irregularly along Poupart's ligament, which receives the lymphatic vessels from the integument of the scrotum, penis, parietes of the abdomen, perineal and gluteal regions, and the mucous membrane of the urethra; and an inferior group, which surrounds the saphenous opening in

FIG. 578.—Femoral hernia.   Superficial dissection.

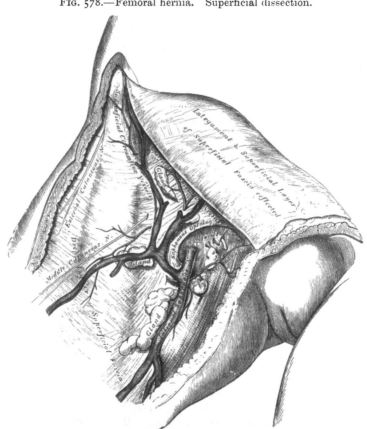

the fascia lata, a few being sometimes continued along the saphenous vein to a variable extent.   This latter group receives the superficial lymphatic vessels from the lower extremity.

The *ilio-inguinal nerve* arises from the first lumbar nerve. It escapes at the external abdominal ring, and is distributed to the integument of the upper and inner part of the thigh: to the scrotum in the male, and to the labium in the female.   The size of this nerve is in inverse proportion to that of the ilio-hypogastric.   Occasionally it is very small, and ends by joining the ilio-hypogastric: in such cases a branch of the ilio-hypogastric takes the place of the ilio-inguinal, or the latter nerve may be altogether absent.   The *crural branch of the genito-crural nerve* passes along the inner margin of the Psoas muscle, beneath Poupart's ligament, into the thigh, entering the sheath of the femoral vessels, and lying superficial and a little external to the femoral artery.   It pierces the anterior layer of the sheath of the vessels, and, becoming superficial by passing through the fascia lata, it supplies the skin of the anterior aspect of the thigh as far as midway between the pelvis and knee. On the front of the thigh it communicates with the outer branch of the middle cutaneous nerve, derived from the anterior crural.

The *deep layer of the superficial fascia* is a very thin fibrous layer, best marked on the inner side of the long saphenous vein and below Poupart's ligament. It is placed beneath the subcutaneous vessels and nerves, and upon the surface of the fascia lata, to which it is intimately adherent at the lower margin of Poupart's ligament. It covers the saphenous opening in the fascia lata, is closely united to its circumference, and is connected to the sheath of the femoral vessels corresponding to its under surface. The portion of fascia covering this aperture is perforated by the internal saphenous vein, and by numerous blood and lymphatic

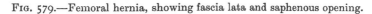

FIG. 579.—Femoral hernia, showing fascia lata and saphenous opening.

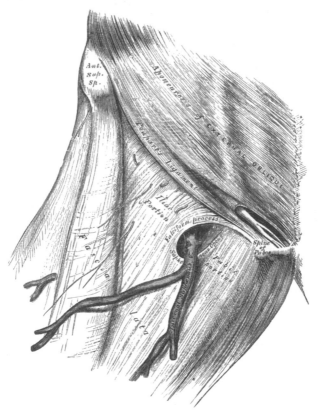

vessels; hence it is termed the *cribriform fascia*, the openings for these vessels having been likened to the holes in a sieve. The cribriform fascia adheres closely both to the superficial fascia and to the fascia lata, so that it is described by some anatomists as a part of the fascia lata, but it is usually considered (as in this work) as belonging to the superficial fascia. It is not till the cribriform fascia has been cleared away that the saphenous opening is seen, so that this opening does not in ordinary cases exist naturally, but is the result of dissection. A femoral hernia, in passing through the saphenous opening, receives the cribriform fascia as one of its coverings.

The deep layer of superficial fascia, together with the cribriform fascia, having been removed, the fascia lata is exposed.

The **fascia lata** has been already described with the muscles of the front of the thigh (page 393). At the upper and inner part of the thigh, a little below

Poupart's ligament, a large oval-shaped aperture is observed after the superficial fascia has been cleared away; it transmits the internal saphenous vein and other smaller vessels, and is called the *saphenous opening*. In order the more correctly to consider the mode of formation of this aperture, the fascia lata in this part of the thigh is described as consisting of two portions, an iliac portion and a pubic portion.

The *iliac portion* is all that part of the fascia lata on the outer side of the saphenous opening. It is attached externally to the crest of the ilium and its anterior superior spine; to the whole length of Poupart's ligament; and to the pectineal line in conjunction with Gimbernat's ligament. From the spine of the os pubis it is reflected downwards and outwards, forming an arched margin, the outer boundary, or falciform process or superior cornu of the saphenous opening. This margin overlies and is adherent to the anterior layer of the sheath of the femoral vessels; to its edge is attached the cribriform fascia, and, below, it is continuous with the pubic portion of the fascia lata.

The *pubic portion of the fascia lata* is situated at the inner side of the saphenous opening; at the lower margin of this aperture it is continuous with the iliac portion; traced upwards, it covers the surface of the Pectineus, Adductor longus, and Gracilis muscles; and, passing behind the sheath of the femoral vessels, to which it is closely united, is continuous with the sheath of the Psoas and Iliacus muscles, and is attached above to the ilio-pectineal line, where it becomes continuous with the fascia covering the Iliacus muscle. From this description it may be observed that the iliac portion of the fascia lata passes in front of the femoral vessels and the pubic portion behind them, so that an apparent aperture consequently exists between the two, through which the internal saphenous joins the femoral vein.

The **saphenous opening** is an oval-shaped aperture, measuring about an inch and a half in length, and half an inch in width. It is situated at the upper and inner part of the front of the thigh, below Poupart's ligament, and is directed obliquely downwards and outwards.

Its *outer margin* is of a semilunar form, thin, strong, sharply defined, and lies on a plane considerably anterior to the inner margin. If this edge is traced upwards, it will be seen to form a curved elongated process, the *falciform process*, or *superior cornu*, which ascends in front of the femoral vessels, and, curving inwards, is attached to Poupart's ligament and to the spine of the os pubis and pectineal line, where it is continuous with the pubic portion. If traced downwards, it is found continuous with another curved margin, the concavity of which is directed upwards and inwards: this is the inferior cornu of the saphenous opening, and is blended with the pubic portion of the fascia lata covering the Pectineus muscle.

The *inner boundary of the opening* is on a plane posterior to the outer margin and behind the level of the femoral vessels; it is much less prominent and defined than the outer, from being stretched over the subjacent Pectineus muscle. It is through the saphenous opening that a femoral hernia passes after descending along the crural canal.

If the finger is introduced into the saphenous opening while the limb is moved in different directions, the aperture will be found to be greatly constricted on extending the limb or rotating it outwards, and to be relaxed on flexing the limb and inverting it: hence the necessity for placing the limb in the latter position in employing the taxis for the reduction of a femoral hernia.

The iliac portion of the fascia lata, but not its falciform process, should now be removed by detaching it from the lower margin of Poupart's ligament, carefully dissecting it from the subjacent structures, and turning it inwards, when the sheath of the femoral vessel is exposed descending beneath Poupart's ligament (fig. 580).

Poupart's ligament, or the **crural arch,** is the lower border of the aponeurosis of the External oblique muscle, which extends from the anterior superior spine of the ilium to the spine of the os pubis. From this latter point it is reflected outwards, to be attached to the pectineal line for about half an inch, forming Gimbernat's ligament. Its general direction is curved downwards towards the thigh, where it is continuous with the fascia lata. Its outer half is rounded and oblique in direction. Its inner half gradually widens at its attachment to the os pubis, is more horizontal in direction, and lies beneath the spermatic cord. Nearly the whole of the space included between the crural arch and the innominate bone is filled in by the parts which descend from the abdomen into the thigh (fig. 581). The outer half of the space is occupied by the Iliacus and Psoas muscles, together with the external cutaneous and anterior crural nerves. The pubic half

FIG. 580.—Femoral hernia. Iliac portion of fascia lata removed, and sheath of femoral vessels and femoral canal exposed.

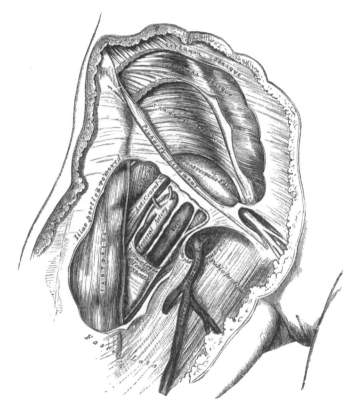

of the space is occupied by the femoral vessels included in their sheath; a small oval-shaped interval existing between the femoral vein and the inner wall of the sheath, which is occupied merely by a little loose areolar tissue, a few lymphatic vessels, and occasionally by a small lymphatic gland; this is the femoral ring, through which the gut descends in femoral hernia.

Gimbernat's ligament (figs. 581, 582) is that part of the aponeurosis of the External oblique muscle which is reflected backwards and outwards, from the spine of the os pubis, to be inserted into the pectineal line. It is about half an inch in length, larger in the male than in the female, almost horizontal in direction in the erect posture, and of a triangular form, with the base directed outwards. Its *base*, or outer margin, is concave, thin, and sharp, and lies in contact with the femoral sheath. Its *apex* corresponds to the spine of the os pubis. Its *posterior margin* is attached to the pectineal line, and is continuous with the pubic portion of the fascia lata. Its *anterior margin* is continuous with Poupart's ligament.

Femoral sheath.—The femoral or crural sheath is a continuation downwards of the fasciæ that line the abdomen, the transversalis fascia passing down in front of the femoral vessels, and the iliac fascia descending behind them; these fasciæ are directly continuous on the iliac side of the femoral artery, but a small space exists between the femoral vein and the point where they are continuous on the pubic side of that vessel, which constitutes the femoral or crural canal. The femoral sheath is closely adherent to the contained vessels about an inch below the saphenous opening, being blended with the areolar sheath of the vessels, but opposite Poupart's ligament it is much larger than is required to contain them; hence the funnel-shaped form which it presents. The outer border of the sheath is perforated by the genito-crural nerve. Its inner border

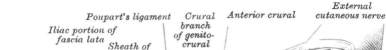

FIG. 581.—Structures which pass beneath the crural arch.

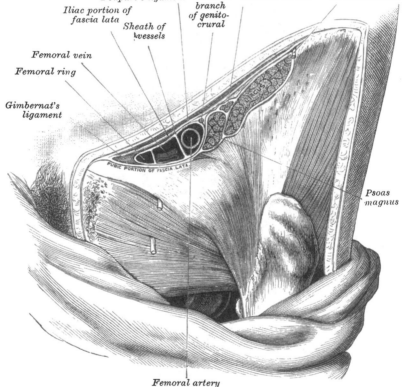

is pierced by the internal saphenous vein and numerous lymphatic vessels. In front, it is covered by the iliac portion of the fascia lata; and behind it is the pubic portion of the same fascia.

If the anterior wall of the sheath is removed, the femoral artery and vein are seen lying side by side, a thin septum separating the two vessels, while another septum may be seen lying just internal to the vein, and cutting off a small space between the vein and the inner wall of the sheath. The septa are stretched between the anterior and posterior walls of the sheath, so that each vessel is enclosed in a separate compartment. The interval left between the vein and the inner wall of the sheath is not filled up by any structure, excepting a little loose areolar tissue, a few lymphatic vessels, and occasionally by a small lymphatic gland; this is the *femoral* or *crural canal*, through which the intestine descends in femoral hernia.

Deep crural arch.—Passing across the front of the femoral sheath on the abdominal side of Poupart's ligament, and closely connected with it, is a thickened band of fibres, called the *deep crural arch*. It is apparently a thickening of the transversalis fascia, joining externally to the centre of Poupart's ligament, and arching across the front of the crural sheath, to be inserted by a broad attachment into the pectineal line, behind the conjoined tendon. In some subjects this structure is not very prominently marked, and not infrequently it is altogether wanting.

The **crural canal** is the narrow interval between the femoral vein and the inner wall of the femoral sheath. It exists as a distinct canal only when the sheath has been separated from the vein by dissection, or by the pressure of a hernia or tumour. Its length is from a quarter to half an inch, and it extends from Gimbernat's ligament to the upper part of the saphenous opening.

Its *anterior wall* is very narrow, and formed by a continuation downwards of the transversalis fascia, under Poupart's ligament, covered by the falciform process of the fascia lata.

Its *posterior wall* is formed by a continuation downwards of the iliac fascia covering the pubic portion of the fascia lata.

Its *outer wall* is formed by the fibrous septum separating it from the inner side of the femoral vein.

Its *inner wall* is formed by the junction of the processes of the transversalis and iliac fasciæ which form the inner side of the femoral sheath, and lies in contact, at its commencement, with the outer edge of Gimbernat's ligament.

This canal has two orifices—an upper one, the *femoral* or *crural ring*, closed by the septum crurale; and a lower one, the *saphenous opening*, closed by the cribriform fascia.

The **femoral** or **crural ring** (fig. 582) is the upper opening of the femoral canal, and leads into the cavity of the abdomen. It is bounded, in front, by Poupart's ligament and the deep crural arch; behind, by the os pubis, covered by the Pectineus muscle, and the pubic portion of the fascia lata; internally, by the base of Gimbernat's ligament, the conjoined tendon, the transversalis fascia, and the deep crural arch; externally, by the fibrous septum lying on the inner side of the femoral vein. The femoral ring is of an oval form; its long diameter, directed transversely, measures about half an inch, and is larger in the female than in the male, which is one of the reasons of the greater frequency of femoral hernia in the former sex.

Position of parts around the ring.—The spermatic cord in the male, and round ligament in the female, lie immediately above the anterior margin of the femoral ring, and may be divided in an operation for femoral hernia if the incision for the relief of the stricture is not of limited extent. In the female this is of little importance, but in the male the spermatic artery and vas deferens may be divided.

The *femoral vein* lies on the outer side of the ring.

The *deep epigastric artery*, in its passage upwards and inwards from the external iliac artery, passes across the upper and outer angle of the crural ring, and is consequently in danger of being wounded if the stricture is divided in a direction upwards and outwards.

The *communicating branch* between the deep epigastric and obturator arteries lies in front of the ring.

The circumference of the ring is thus seen to be bounded by vessels in every part, excepting internally and behind. It is in the former position that the stricture is divided in cases of strangulated femoral hernia.

The *obturator artery*, when it arises by a common trunk with the deep epigastric, which occurs once in every three subjects and a half, bears a very important relation to the crural ring. In most cases it descends on the inner side of the external iliac vein to the obturator foramen, and will consequently lie on the outer side of the crural ring, where there is no danger of its being wounded in the operation for dividing the stricture in femoral hernia (see fig. 314, page

541, A). Occasionally, however, the obturator artery curves along the free margin of Gimbernat's ligament in its passage to the obturator foramen; it would, consequently, skirt along the greater part of the circumference of the crural ring, and could hardly avoid being wounded in the operation (see fig. 314, page 541, B).

Septum crurale.—The femoral ring is closed by a layer of condensed areolar tissue, called, by J. Cloquet, the *septum crurale*. This serves as a barrier to the protrusion of a hernia through this part. Its upper surface is slightly concave, and supports a small lymphatic gland, by which it is separated from the subserous areolar tissue and peritoneum. Its under surface is turned towards the femoral canal. The septum crurale is perforated by numerous apertures for the passage of lymphatic vessels, connecting the deep inguinal lymphatic glands with those surrounding the external iliac artery.

The size of the femoral canal, the degree of tension of its orifices, and, consequently, the degree of constriction of a hernia, vary according to the position

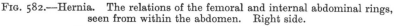

FIG. 582.—Hernia. The relations of the femoral and internal abdominal rings, seen from within the abdomen. Right side.

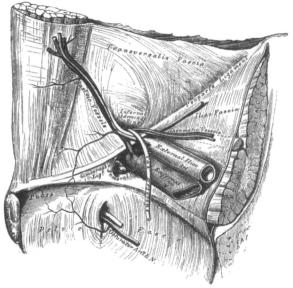

of the limb. If the leg and thigh are extended, abducted, or everted, the femoral canal and its orifices are rendered tense, from the traction on these parts by Poupart's ligament and the fascia lata, as may be ascertained by passing the finger along the canal. If, on the contrary, the thigh is flexed upon the pelvis, and at the same time adducted and rotated inwards, the femoral canal and its orifices become considerably relaxed; for this reason the limb should always be placed in the latter position when the application of the taxis is made in attempting the reduction of a femoral hernia.

The *subperitoneal areolar tissue* is continuous with the subserous areolar tissue of surrounding parts. It is usually thickest and most fibrous where the iliac vessels leave the abdominal cavity. It covers over the small interval (crural ring) on the inner side of the femoral vein. In some subjects it contains a considerable amount of adipose tissue. In such cases, where it is protruded forwards in front of the sac of a femoral hernia, it may be mistaken for a portion of omentum. The peritoneum lining the portion of the abdominal wall between Poupart's ligament and the brim of the pelvis is similar to that lining any other portion of the abdominal wall, being very thin. It has here no natural aperture for the escape of intestine.

Descent of the hernia.—From the preceding description it follows that the femoral ring must be a weak point in the abdominal wall : hence it is that, when violent or long-continued pressure is made upon the abdominal viscera, a portion of intestine may be forced into it, constituting a femoral hernia ; and the changes in the tissues of the abdomen which are produced by pregnancy, together with the larger size of this aperture in the female, serve to explain the frequency of this form of hernia in women.

When a portion of the intestine is forced through the femoral ring, it carries before it a pouch of peritoneum, which forms what is called the *hernial sac* ; it receives an investment from the subserous areolar tissue, and from the septum crurale, and descends vertically along the crural canal in the inner compartment of the sheath of the femoral vessels as far as the saphenous opening ; at this point it changes its course, being prevented from extending farther down the sheath, on account of its narrowing and close contact with the vessels, and also from the close attachment of the superficial fascia and crural sheath to the lower part of the circumference of the saphenous opening ; the tumour is, consequently, directed forwards, pushing before it the cribriform fascia, and then curves upwards on to the falciform process of the fascia lata and lower part of the tendon of the External oblique, being covered by the superficial fascia and integument. While the hernia is contained in the femoral canal, it is usually of small size, owing to the resisting nature of the surrounding parts ; but when it has escaped from the saphenous opening into the loose areolar tissue of the groin it becomes considerably enlarged. The direction taken by a femoral hernia in its descent is at first downwards, then forwards and upwards ; this should be borne in mind, as in the application of the taxis for the reduction of a femoral hernia pressure should be directed in the reverse order.

Coverings of the hernia.—The coverings of a femoral hernia, from within outwards, are : peritoneum, subserous areolar tissue, the septum crurale, crural sheath, cribriform fascia, superficial fascia, and integument.*

Varieties of femoral hernia.—If the intestine descends along the femoral canal only as far as the saphenous opening, and does not escape from this aperture, it is called *incomplete femoral hernia*. The small size of the protrusion in this form of hernia, on account of the firm and resisting nature of the canal in which it is contained, renders it an exceedingly dangerous variety of the disease, from the extreme difficulty of detecting the existence of the swelling, especially in corpulent subjects. The coverings of an incomplete femoral hernia would be, from without inwards : integument, superficial fascia, falciform process of fascia lata, crural sheath, septum crurale, subserous areolar tissue, and peritoneum. When, however, the hernial tumour protrudes through the saphenous opening, and directs itself forwards and upwards, it forms a *complete femoral hernia*. Occasionally the hernial sac descends on the iliac side of the femoral vessels, or in front, or even sometimes behind them.

The *seat of stricture of a femoral hernia* varies : it may be in the peritoneum at the neck of the hernial sac ; in the greater number of cases it would appear to be at the point of junction of the falciform process of the fascia lata with the lunated edge of Gimbernat's ligament ; or at the margin of the saphenous opening in the thigh. The stricture should in every case be divided in a direction upwards and inwards ; and the extent necessary in the majority of cases is about two or three lines. By these means, all vessels or other structures of importance, in relation with the neck of the hernial sac, will be avoided.

* Sir Astley Cooper has described an investment for femoral hernia, under the name of 'fascia propria,' lying immediately external to the peritoneal sac, but frequently separated from it by more or less adipose tissue. Surgically it is important to remember the existence (at any rate, the occasional existence) of this layer, on account of the ease with which an inexperienced operator may mistake the fascia for the peritoneal sac, and the contained fat for omentum. Anatomically, this fascia appears identical with what is called in the text 'subserous areolar tissue,' the areolar tissue being thickened and caused to assume a membranous appearance by the pressure of the hernia.

SURGICAL ANATOMY OF THE PERINÆUM

Dissection.—The student should select a well-developed muscular subject free from fat, and the dissection should be commenced early, in order that the parts may be examined in as recent a state as possible. A staff having been introduced into the bladder, and the subject placed in the position shown in fig. 583, the scrotum should be raised upwards, and retained in that position, and the rectum moderately distended with tow.

The **perinæum** corresponds to the inferior aperture or outlet of the pelvis. Its deep boundaries are, in front, the pubic arch and subpubic ligament; behind, the tip of the coccyx; and on each side, the rami of the os pubis and ischium, the tuberosities of the ischium, and great sacro-sciatic ligaments. The space included by these boundaries is somewhat lozenge-shaped, and is limited on the surface of the body by the scrotum in front, by the buttocks behind, and on each side by the inner side of the thighs. A line drawn transversely between the anterior part of the tuberosity of the ischium, on each side, in front of the anus, divides this space into two portions. The anterior portion contains the penis and urethra, and is called the *perinæum proper* or *genito-urinary region.* The posterior portion contains the termination of the rectum, and is called the *ischio-rectal* or *anal region.*

ISCHIO-RECTAL REGION

The **ischio-rectal region** contains the termination of the rectum and a deep fossa, filled with fat, on each side of the intestine, between it and the tuberosity of the ischium : this is called the *ischio-rectal fossa.*

The ischio-rectal region presents in the middle line the *aperture of the anus* ; around this orifice the integument is thrown into numerous folds, which are obliterated on distension of the intestine. The integument is of a dark colour, continuous with the mucous membrane of the rectum, and provided with numerous follicles, which occasionally inflame and suppurate, and may be mistaken for fistulæ. The veins round the margin of the anus are occasionally much dilated, forming a number of hard pendent masses, of a dark bluish colour, covered partly by mucous membrane and partly by the integument. These tumours constitute the disease called *external piles.*

Dissection (fig. 583).—Make an incision through the integument, along the median line, from the base of the scrotum to the anterior extremity of the anus : carry it round the margins of this aperture to its posterior extremity, and continue it backwards to about an inch behind the tip of the coccyx. A transverse incision should now be carried across the base of the scrotum, joining the anterior extremity of the preceding ; a second, carried in the same direction, should be made in front of the anus ; and a third at the posterior extremity of the first incision. These incisions should be sufficiently extensive to enable the dissector to raise the integument from the inner side of the thighs. The flaps of skin corresponding to the ischio-rectal region should now be removed. In dissecting the integument from this region great care is required, otherwise the Corrugator cutis ani and External sphincter will be removed, as they are intimately adherent to the skin.

The **superficial fascia** is exposed on the removal of the skin : it is very thick, areolar in texture, and contains much fat in its meshes. In it are found ramifying two or three branches of the perforating cutaneous nerve ; these turn round the inferior border of the Gluteus maximus, and are distributed to the integument around the anus.

In this region, and connected with the lower end of the rectum, are four muscles : the Corrugator cutis ani ; the two Sphincters, External and Internal ; and the Levator ani.

These muscles have been already described (see page 342).

The **ischio-rectal fossa** is situated between the lower end of the rectum and the tuberosity of the ischium. It is triangular in shape ; its base, directed to the surface of the body, is formed by the integument of the ischio-rectal region ; its *apex*, directed upwards, corresponds to the point of division of the obturator fascia, and the thin membrane given off from it, which covers the outer surface of the Levator ani (ischio-rectal or anal fascia). Its dimensions are about an inch in breadth at the base, and about two inches in depth, being deeper behind than in front. It is bounded, *internally*, by the Sphincter ani, Levator ani, and

Fig. 583.—Dissection of perinæum and ischio-rectal region.

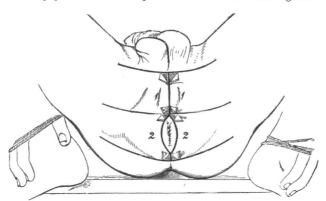

Coccygeus muscles ; *externally*, by the tuberosity of the ischium and the obturator fascia, which covers the inner surface of the Obturator internus muscle ; *in front*, it is limited by the line of junction of the superficial fascia with the base of the triangular ligament ; and *behind*, by the margin of the Gluteus maximus, and the great sacro-sciatic ligament. This space is filled with a large mass of adipose tissue which explains the frequency with which abscesses in the neighbourhood of the rectum burrow to a considerable depth.

If the subject has been injected, on placing the finger on the outer wall of this fossa, the internal pudic artery, with its accompanying veins and the two divisions of the nerve, will be felt about an inch and a half above the margin of the ischiatic tuberosity, but approaching nearer the surface as they pass forwards along the inner margin of the pubic arch. These structures are enclosed in a sheath (canal of Alcock) formed by the obturator fascia, the perineal nerve lying below the artery and the dorsal nerve of the penis above it (fig. 315). Crossing the space transversely about its centre, are the inferior hæmorrhoidal vessels and nerves, which are distributed to the integument of the anus, and to the muscles of the lower end of the rectum. These vessels are occasionally of large size, and may give rise to troublesome hæmorrhage when divided in the operation of lithotomy, or in that for fistula in ano. At the back part of this space, near the coccyx, may be seen a branch of the fourth sacral nerve ; and at the fore part of the space, the superficial perineal vessels and nerves can be seen for a short distance.

THE PERINÆUM PROPER IN THE MALE

The perineal space is of a triangular form ; its deep boundaries are limited, laterally, by the rami of the pubic bones and ischia, meeting, in front, at the pubic arch : behind, by an imaginary transverse line, extending between the anterior parts of the tuberosities of the ischia. The lateral boundaries are, in the adult, from three inches to three inches and a half in length ; and the base from two inches to three inches and a half in breadth ; the average extent of the space being two inches and three-quarters.

The variations in the diameter of this space are of extreme interest in connection with the operation of lithotomy, and the extraction of a stone from the cavity of the bladder. In those cases where the tuberosities of the ischia are near together it would be necessary to make the incisions in the lateral operation of lithotomy less oblique than if the tuberosities were widely separated, and the perineal space, consequently, wider. The perinæum is subdivided by the median raphé into two equal parts. Of these, the left is the one in which the operation of lithotomy is performed.

In the middle line the perinæum is convex, and corresponds to the bulb of the urethra. The skin covering it is of a dark colour, thin, freely movable upon the subjacent parts, and covered with sharp crisp hairs, which should be removed before the dissection of the part is commenced. In front of the anus a prominent line commences, the *raphé*, continuous in front with the raphé of the scrotum.

Upon removing the skin and superficial structures from this region, in the manner shown in fig. 583, a plane of fascia will be exposed, covering in the triangular space and stretching across from one ischio-pubic ramus to the other. This is the *deep layer of the superficial fascia*, or *fascia of Colles*. It has already been described (page 344). It is a layer of considerable strength, and encloses and covers a space in which are contained muscles, vessels, and nerves. It is continuous, in front, with the dartos of the scrotum ; on each side, it is firmly attached to the margin of the ischio-pubic ramus and to the tuberosity of the ischium ; and posteriorly, it curves behind the Transversi perinæi muscles to join the base of the triangular ligament.

It is between this layer of fascia and the triangular ligament of the urethra that extravasation of urine most frequently takes place in cases of rupture of the urethra. The triangular ligament of the urethra (see page 347) is attached to the ischio-pubic rami, and in front to the subpubic ligament. It is clear, therefore, that when extravasation of fluid takes place between these two layers, it cannot pass backwards, because the two layers are continuous with each other around the Transversi perinæi muscles ; it cannot extend laterally, on account of the connection of both these layers to the rami of the os pubis and ischium ; it cannot find its way into the pelvis, because the opening into this cavity is closed by the triangular ligament, and, therefore, so long as these two layers remain intact, the only direction in which the fluid can make its way is forwards into the areolar tissue of the scrotum and penis, and thence on to the anterior wall of the abdomen.

When the deep layer of the superficial fascia is removed, a space is exposed, between this fascia and the triangular ligament, in which are contained the superficial perineal vessels and nerves and some of the muscles connected with the penis and urethra, viz. : in the middle line, the Accelerator urinæ ; on each side, the Erector penis ; and behind, the Transversus perinæi ; together with the crura of the corpora cavernosa and the bulb of the corpus spongiosum. Here also is seen the *central tendinous point of the perinæum*. This is a fibrous point in the middle line of the perinæum between the urethra and the rectum, being about half an inch in front of the anus. At this point four muscles converge and are attached : viz. the External sphincter ani, the Accelerator urinæ, and the two Transversi perinæi muscles ; so that by the contraction of these muscles, which extend in opposite directions, it serves as a fixed point of support.

The Accelerator urinæ, the Erector penis, and the Transversus perinæi muscles have been already described (page 344). They form a triangular space, bounded, internally, by the Accelerator urinæ ; externally, by the Erector penis ; and behind, by the Transversus perinæi. The floor of this space is formed by the triangular ligament of the urethra ; and running from behind forwards in it are the superficial perineal vessels and nerves, and the transverse perineal artery coursing along the posterior boundary of the space, on the Transversus perinæi muscle.

The Accelerator urinæ and Erector penis should now be removed, when the triangular ligament of the urethra will be exposed, stretching across the front of the outlet of the pelvis. The urethra is seen perforating its centre, just behind the bulb ; and on each side is the crus penis, connecting the corpus cavernosum with the rami of the ischium and os pubis.

The **triangular ligament,** which has already been described (see page 347), consists of two layers, the inferior superficial layer of which is now exposed. It is united to the superior or deep layer behind, but is separated in front by a subfascial space in which are contained certain structures.

FIG. 584.—The superficial muscles and vessels of the perinæum.

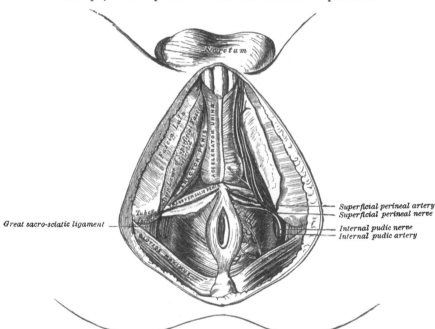

Great sacro-sciatic ligament

Superficial perineal artery
Superficial perineal nerve

Internal pudic nerve
Internal pudic artery

The *inferior layer of the triangular ligament* consists of a strong fibrous membrane, the fibres of which are disposed transversely, which stretches across from one ischio-pubic ramus to the other, and completely fills in the pubic arch ; it is attached in front to the subpubic ligament, except just in the centre, where a small interspace is left for the dorsal vein of the penis. In the erect position of the body it is almost horizontal. It is perforated by the urethra in the middle line, and on each side of the urethral opening by the ducts of Cowper's glands and by the arteries of the bulb ; in front, and external to this, by the artery of the corpus cavernosum, immediately before this vessel enters the crus penis. Near its apex the ligament is perforated by the termination of the pudic artery and by the dorsal nerve of the penis. The *crura penis* are exposed, lying superficial to this ligament. They will be seen to be attached by blunt-pointed processes to the rami of the os pubis and ischium, in front of the tuberosities, and passing forwards and inwards, joining to form the body of the penis. In the middle line the bulb and corpus spongiosum are exposed by the removal of the Accelerator urinæ muscle.

If the superficial layer of the triangular ligament is detached on either side, the deep perineal interspace will be exposed and the following parts will be seen, between it and the deep layer of the ligament : the subpubic ligament in front, close to the symphysis pubis ; the dorsal vein of the penis ; the membranous portion of the urethra and the Compressor urethræ muscle ; Cowper's glands and their ducts ; the pudic vessels and the dorsal nerve of the penis ; the artery and nerve of the bulb, and a plexus of veins.

The *superior layer of the triangular ligament*, or *deep perineal fascia*, is derived from the obturator fascia, and is continuous with it along the pubic arch.   Behind, it joins with the superficial layer of the triangular ligament, and is continuous with the anal fascia.   Above it is the prostate gland, supported by the anterior fibres of the Levator ani, which act as a sling for the gland and form the Levator prostatæ muscle.   The superior layer of the triangular ligament is continuous round

Fig. 585.—Deep perineal fascia.   On the left side the inferior layer
has been removed.

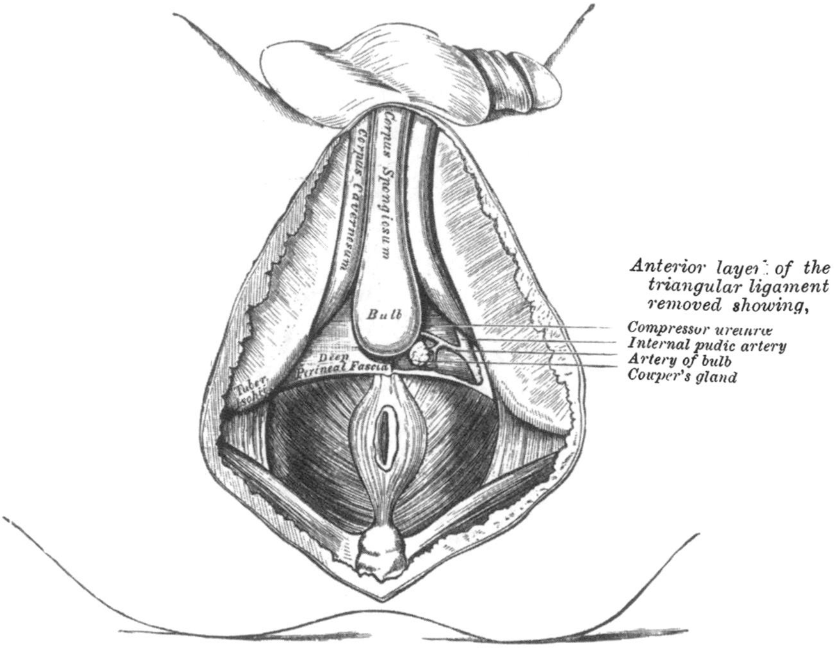

Anterior layer of the
triangular ligament
removed showing,

Compressor urethræ
Internal pudic artery
Artery of bulb
Cowper's gland

the anterior free edge of this muscle with the recto-vesical layer covering the prostate gland.   The superior layer of the triangular ligament is perforated by the urethra.   Between the two layers of the triangular ligament are situated the membranous part of the urethra, enveloped by the Compressor urethræ muscle ; the ducts of Cowper's glands ; the arteries to the bulb ; the pudic vessels and the dorsal nerve of the penis.   The membranous part of the urethra is about three-quarters of an inch in length, and passes downwards and forwards behind the symphysis pubis, from which it is distant about an inch.   It is the narrowest part of the tube, and is enveloped, as has already been stated, by the Compressor urethræ muscle.

The **Compressor urethræ** has already been described (page 348).   In addition to this muscle and immediately beneath it, *circular muscular fibres* surround the membranous portion of the urethra from the bulb in front to the prostate behind, and are continuous with the muscular fibres of the bladder.   These fibres are involuntary.

Cowper's glands are situated immediately below the membranous portion of the urethra, close behind the bulb and below the artery of the bulb.

The **pudic vessels** and **dorsal nerve of the penis** are placed along the inner margin of the pudic arch (pages 542 and 774).

The **artery of the bulb** passes transversely inwards, from the internal pudic along the base of the triangular ligament, between its two layers, accompanied by a branch of the pudic nerve (page 544). If the deep layer of the triangular ligament is removed and the crus penis of one side detached from the bone, the under or perineal surface of the Levator ani is brought fully into view. This muscle, with the triangular ligament in front and the Coccygeus and Pyriformis behind, closes the outlet of the pelvis.

The Levator ani and Coccygeus muscles have already been described (page 342).

FIG. 586.—A view of the position of the viscera at the outlet of th pelvis.

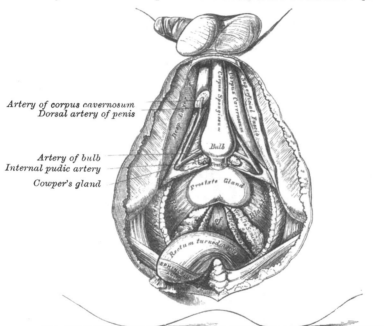

Position of the Viscera at the Outlet of the Pelvis.—Divide the central tendinous point of the perinæum, separate the rectum from its connections by dividing the fibres of the Levator ani, which descend upon the sides of the prostate gland, and draw the gut backwards towards the coccyx, when the under surface of the prostate gland, the neck and base of the bladder, the vesiculæ seminales, and the vasa deferentia will be exposed.

The **prostate gland** is a pale, firm, glandular body which is placed immediately below the neck of the bladder, around the commencement of the urethra. It is placed in the pelvic cavity, behind the lower part of the symphysis pubis, above the deep layer of the triangular ligament, and rests upon the rectum, through which it may be distinctly felt, especially when enlarged. In shape and size it resembles a chestnut. Its base is directed upwards towards the neck of the bladder. Its apex is directed downwards to the deep layer of the triangular ligament, which it touches.

Its posterior surface is smooth, marked by a slight longitudinal furrow, and rests on the second part of the rectum, to which it is connected by areolar tissue. Its anterior surface is flattened, marked by a slight longitudinal furrow, and placed about three-quarters of an inch behind the pubic symphysis. It measures about an inch and a half in its transverse diameter at the base, an inch in its antero-posterior diameter, and three-quarters of an inch in depth. Hence the greatest extent of incision that can be made in it without dividing its substance completely across, is obliquely backwards and outwards. This is the direction in which the incision is made in the lateral operation of lithotomy.

Above the prostate a small triangular portion of the bladder is seen, bounded, in front and below, by the prostate gland ; above, by the recto-vesical fold of the peritoneum ; on each side, by the vesicula seminalis and the vas deferens. It is separated from direct contact with the rectum by the recto-vesical fascia. The relation of this portion of the bladder to the rectum is of extreme interest to the surgeon. In cases of retention of urine this portion of the organ is found projecting into the rectum, between three and four inches from the margin of the anus, and may be easily perforated, without injury to any important parts ; this portion of the bladder is, consequently, occasionally selected for the performance of the operation of tapping the bladder.

FIG. 587.—A transverse section of the pelvis, showing the pelvic fascia from behind.

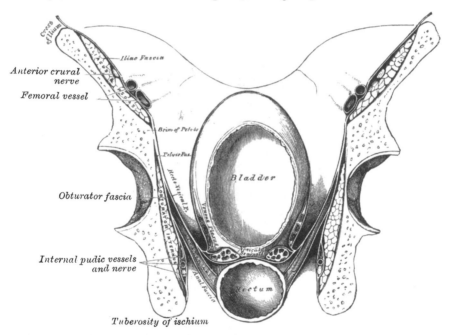

Surgical Anatomy.—The student should consider the position of the various parts in reference to the lateral operation of lithotomy. This operation is performed on the left side of the perinæum, as it is most convenient for the right hand of the operator. A grooved staff having been introduced into the bladder, the first incision is commenced mid-way between the anus and the back of the scrotum (i.e. in an ordinary adult perinæum, about an inch and a half in front of the anus), a little on the left side of the raphé, and carried obliquely backwards and outwards to midway between the anus and tuberosity of the ischium. The incision divides the integument and superficial fascia, the inferior hæmor-rhoidal vessels and nerves, and the superficial and transverse perineal vessels. If the forefinger of the left hand is thrust upwards and forwards into the wound, pressing at the same time the rectum inwards and backwards, the staff may be felt in the membranous portion of the urethra. The finger is fixed upon the staff, and the structures covering it are divided with the point of the knife, which must be directed along the groove towards the bladder, the edge of the knife being turned outwards and backwards, dividing in its course the membranous portion of the urethra, and part of the left lobe of the prostate gland, to the extent of about an inch. The knife is then withdrawn, and the forefinger of the left hand passed along the staff into the bladder. The position of the stone having been ascertained, the staff is to be withdrawn, and the forceps introduced over the finger into the bladder. If the stone is very large, the opposite side of the prostate may be notched before the forceps is introduced: the finger is now withdrawn, and the blades of the forceps opened and made to grasp the stone, which must be extracted by slow and cautious undulating movements.

Parts divided in the operation.—The various structures divided in this operation are as follows: the integument, superficial fascia, inferior hæmorrhoidal vessels and nerves and probably the superficial perineal vessels and nerves, the posterior fibres of the Accelerator urinæ, the Transversus perinæi muscle and artery, the triangular ligament, the anterior fibres of the Levator ani, part of the Compressor urethræ, the membranous and prostatic portions of the urethra, and part of the prostate gland.

Parts to be avoided in the operation.—In making the necessary incisions in the perinæum for the extraction of a calculus, the following parts should be avoided. The primary incision should not be made too near the middle line, for fear of wounding the bulb of the corpus spongiosum or the rectum; nor too far externally, otherwise the pudic artery may be implicated as it ascends along the inner border of the pubic arch. If the incisions are carried too far forwards, the artery of the bulb may be divided; if carried too far backwards, the entire breadth of the prostate and neck of the bladder may be cut through, which allows the urine to become infiltrated behind the pelvic fascia into the loose areolar tissue between the bladder and rectum, instead of escaping externally: diffuse inflammation is consequently set up, and peritonitis, from the close proximity of the recto-vesical peritoneal fold, is the result. If, on the contrary, only the anterior part of the prostate is divided, the urine makes its way externally, and there is less danger of infiltration taking place.

During the operation it is of great importance that the finger should be passed into the bladder *before* the staff is removed; if this is neglected, and if the incision made in the prostate and neck of the bladder is too small, great difficulty may be experienced in introducing the finger afterwards; and in the child, where the connections of the bladder to the surrounding parts are very loose, the force made in the attempt is sufficient to displace the bladder upwards into the abdomen, out of the reach of the operator. Such a proceeding has not infrequently occurred, producing the most embarrassing results, and total failure of the operation.

It is necessary to bear in mind that the arteries in the perinæum occasionally take an abnormal course. Thus the artery of the bulb, when it arises, as sometimes happens, from the pudic opposite the tuber ischii, is liable to be wounded in the operation for lithotomy, in its passage forwards to the bulb. The accessory pudic may be divided near the posterior border of the prostate gland, if this is completely cut across; and the prostatic veins, especially in people advanced in life, are of large size, and give rise, when divided, to troublesome hæmorrhage.

The Female Perinæum

The **female perinæum** presents certain differences from that of the male, in consequence of the whole of the structures which constitute it being perforated in the middle line by the vulvo-vaginal passage.

The **superficial fascia,** as in the male, consists of two layers: of which the superficial one is continuous with the superficial fascia over the rest of the body; and the deep layer, corresponding to the fascia of Colles in the male, is like it attached to the ischio-pubic ramus, and in front is continued forwards through the labia majora to the inguinal region. It is of less extent than the male, in consequence of being perforated by the aperture of the vulva.

On removing this fascia the muscles of the female perinæum, which have already been described (page 348), are exposed. The Sphincter vaginæ, corresponding to the Accelerator urinæ in the male, consists of an attenuated plane of fibres, forming an orbicular muscle around the orifice of the vagina, instead of being united in a median raphé, as in the male. The Erector clitoridis is proportionately reduced in size, but differs in no other respect; and the Transversus perinæi is similar to the muscle of the same name in the male.

The triangular ligament of the urethra is not so strongly marked as in the male. It transmits the urethra and the tube of the vagina.

The **Compressor urethræ** (*Transversus perinæi profundus*) corresponds with the Compressor urethræ in the male. It arises from the ischio-pubic ramus, and, passing inwards, its anterior fibres blend with the muscle of the opposite side in front of the urethra; its middle fibres, the most numerous, are inserted into the side of the vagina, and the posterior fibres join the central point of the perinæum.

The distribution of the internal pudic artery is the same as in the male (see page 544), and the pudic nerve has also a similar arrangement, the dorsal nerve being, however, very small and supplying the clitoris.

The corpus spongiosum is divided into two lateral halves, which are represented by the *bulbi vestibuli* and *partes intermediales* (see page 1010).

The **perineal body** fills up the interval between the lower part of the vagina and the rectum. Its base is covered by the skin lying between the anus and vagina on what is called the 'perinæum.' Its anterior surface lies behind the posterior vaginal wall, and its posterior surface lies in front of the anterior rectal wall and the anus. It measures about an inch and a quarter from before backwards, and laterally extends from one tuberosity of the ischium to the other. In it are situated the muscles belonging to the external organs of generation. Through its centre runs the transverse perineal septum, which is of great strength

Fig. 588.—Side view of the pelvic viscera of the male subject, showing the pelvic and perineal fascia.

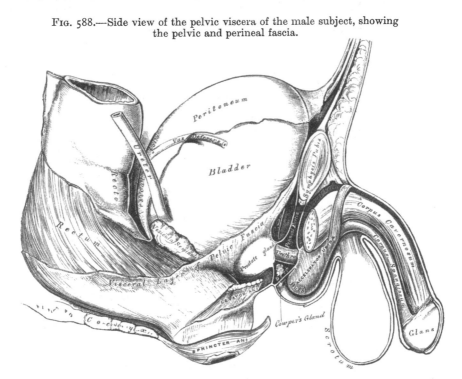

in women, and forms on either side, behind the posterior commissure, a hard, ill-defined body, consisting of connective tissue, with much yellow elastic tissue and interlacing bundles of involuntary muscular fibres, in which the voluntary muscles of the perinæum are inserted.

PELVIC FASCIA

The **pelvic fascia** (fig. 588) is a thin membrane which lines the whole of the cavity of the pelvis, and is continuous over the back part of the ilio-pectineal line with the iliac fascia. It is attached to the brim of the pelvis, for a short distance, at the side of the cavity, and to the inner surface of the bone round the attachment of the Obturator internus. At the posterior border of this muscle, it is continued backwards as a very thin membrane in front of the Pyriformis muscle and sacral nerves to the front of the sacrum. In front it follows the attachment of the Obturator internus to the bone, arches beneath the obturator vessels, completing the orifice of the obturator canal, and at the front of the pelvis is attached to the lower part of the symphysis pubis. At the level of a line extending from the lower part of the symphysis pubis to the spine of the ischium, is a thickened whitish band, termed the *white line* ; this marks the attachment of the Levator ani muscle to the pelvic fascia, and corresponds to its point of division into two layers, the *obturator* and *recto-vesical*.

The **obturator fascia** descends and covers the Obturator internus muscle. It is a direct continuation of the parietal pelvic fascia below the white line above mentioned, and is attached to the pubic arch, the ischial tuberosities, and to the margin of the great sacro-sciatic ligaments. This fascia forms a canal for the pudic vessels and nerve in their passage forwards to the perinæum, and gives off a thin membrane which covers the perineal aspect of the Levator ani muscle, called the *ischio-rectal (anal) fascia*. From its attachment to the rami of the os pubis and ischium a process is given off which is continuous with a similar process from the opposite side, so as to close the front part of the outlet of the pelvis, forming the deep layer of the triangular ligament.

The **recto-vesical fascia** (*visceral layer of the pelvic fascia*) descends into the pelvis upon the upper surface of the Levator ani muscle, and invests the prostate, bladder, and rectum. From the inner surface of the symphysis pubis a short rounded band is continued, on each side of the middle line, to the upper surface of the prostate and neck of the bladder, forming the pubo-prostatic or anterior true ligaments of the bladder. At the side, this fascia is connected to the prostate, enclosing this gland and the vesico-prostatic plexus of veins, and is continued on to the side of the bladder, forming the lateral true ligaments of the organ. Another prolongation invests the vesiculæ seminales, and passes across between the bladder and rectum, being continuous with the same fascia of the opposite side. Another thin prolongation is reflected round the surface of the lower end of the rectum. The Levator ani muscle arises from the point of division of the pelvic fascia ; the visceral layer of the fascia descending upon and being intimately adherent to the upper surface of the muscle, while the under surface of the muscle is covered by a thin layer derived from the obturator fascia, called the *ischio-rectal* or *anal fascia*. In the female, the vagina perforates the recto-vesical fascia, and receives a prolongation from it.

INDEX

ABDOMEN, 875; apertures found in, 877; boundaries of, 877; lymphatics of, 611; muscles of, 329; regions of, 877; viscera of, 878

Abdominal aorta, 525; branches of, 527; surgical anatomy of, 526; surface marking of, 526; muscles, 329; ring, external, 332, 1026; internal, 340, 1030; stalk, (footnote) [92]; viscera, position of, 878

Abdomino-thoracic arch, 102

Abducent nerve, 719; surgical anatomy of, 720

Abductor hallucis, 420

Abductor indicis muscle, 382

Abductor minimi digiti muscle, (hand) 381, (foot) 421

Abductor pollicis muscle, 378

Aberrant duct of testis, 1005

Absorbent glands, 602

Absorbents, 602

Accelerator urinæ muscle, 346

Accessorius ad ilio-costalem muscle, 318

Accessorius pedis, 422

Accessory gland of Rosenmüller, 827; obturator nerve, 767; olivary nuclei, 682; descending palatine canals, 61; pudic artery, 543; processes, 11; thyroid glands, 965

Acervulus cerebri, 651

Acetabulum, 148

Achromatic spindle, [3]

Achromatin, [2]

Acromial end of clavicle, fracture of, 386

Acromial region, muscles of, 356; thoracic artery, 508

Acromio-clavicular joint, 219; surface form of, 221; surgical anatomy of, 221

Acromion process, 110; fracture of, 386

Actions of Muscles. See each Group of Muscles

Adamantoblasts, 860

Adductor brevis muscle, 400; longus muscle, 399; magnus muscle, 400; obliquus pollicis, 379; transversus pollicis, 380; obliquus hallucis, 423; transversus hallucis, 423

Adductor tubercle, 158

Adenoid connective tissue, [19]

Adipose tissue, [19]

Adminicum lineæ albæ, 333

Afferent nerves, [49]

Afferent vessels of kidney, 976

Air-cells, 962

Ala cinerea, 678, 681; lobuli centralis, 665; nasi, 799; artery of, 466

Alæ of vomer, 64; of sacrum, 17

Alar lamina, [107]; ligaments, 202; of knee, 250; thoracic artery, 509

Alcock, canal of, 543, 1047

Alimentary canal, 851; development of, [129]; subdivisions of, 851

Allantoic vessels, [92]; vesicle, [92]

Allantois, [92]

Alveolar artery, 474; process, 55, 67

Alveoli of lower jaw, 67; of upper jaw, 55; formation of, 863; of stomach, 891

Alveus, 645

Amacrine cells of retina, 816

Amnion, [90]; true, [91]; false, [91]

Amniotic cavity, [91]; primitive, [83]

Amphiarthrosis, 190

Ampullæ of semicircular canals, 842; of tubuli lactiferi, 1023

Amygdalæ, 865; of cerebellum, 667

Amygdaloid nucleus, 641

Anal fascia, 343, 1055; canal, 908

Anaphase of karyokinesis, [3]

Anastomosis of arteries, 448

Anastomotica magna of brachial, 513; of femoral, 557

Anatomical neck of humerus, 114; fracture of, 120

Anconeus muscle, 372

Andersch, ganglion of, 727

Aneurisms of abdominal aorta, 526; of arch of aorta, 453; of thoracic aorta, 523

Angle of jaw, 67; of Ludovic, 102; of os pubis, 147; of rib, 99; sacro-vertebral, 14

Angular artery, 466; convolution, 630; movement, 193; process, external, 31; internal, 31; vein, 573

Animal cell, [1]; constituent of bone, [30]

Ankle-joint, 256; relations of tendons and vessels to, 257; surface form of, 258; surgical anatomy of, 258

Annectant gyrus, 629, 630

Annular ligament of radius and ulna, 231; of wrist, anterior, 375; posterior, 376; of ankle, anterior, 418; external, 418; internal, 418; of stapes, 839

Annulus ovalis, 437; tympanicus, 833

Ansa hypoglossi, 737; peduncularis, 651

Anterior annular ligament, (wrist) 375, (ankle) 418; chamber of eye, 820; crescentic lobe, 665; crural nerve, 767; surgical anatomy of, 778; dental canal, 53; ethmoidal cells, 48; fontanelle, 49; fossa of skull, 72; nasal spine, 56; nerve-roots, 697; palatine canal, 56; palatine fossa, 56, 77; perforated space, 633; region of skull, 81; triangle of neck, 475

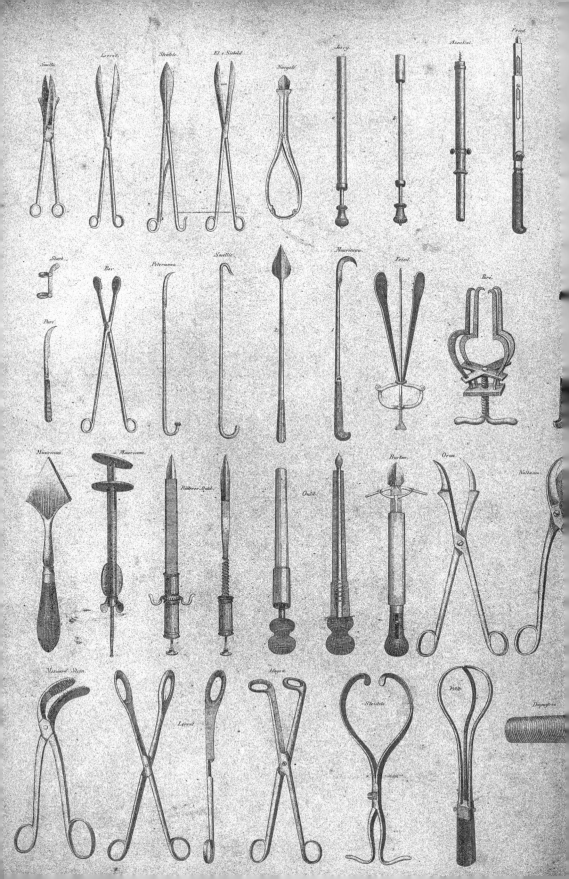